Contents

ASM Handbook®

Volume 20
Materials Selection and Design

Prepared under the direction of the
ASM International Handbook Committee

George E. Dieter, Volume Chair

Scott D. Henry, Assistant Director of Reference Publications
Steven R. Lampman, Technical Editor
Grace M. Davidson, Manager of Handbook Production
Bonnie R. Sanders, Chief Copy Editor
Randall L. Boring and Kathleen S. Dragolich, Production Coordinators
Amy E. Hammel, Editorial Assistant
William W. Scott, Jr., Director of Technical Publications

Editorial Assistance
Nikki DiMatteo
Kelly Ferjutz
Heather Lampman
Mary Jane Riddlebaugh

The Materials Information Society

First printing, December 1997

This book is a collective effort involving hundreds of technical specialists. It brings together a wealth of information from world-wide sources to help scientists, engineers, and technicians solve current and long-range problems.

Great care is taken in the compilation and production of this Volume, but it should be made clear that NO WARRANTIES, EXPRESS OR IMPLIED, INCLUDING, WITHOUT LIMITATION, WARRANTIES OF MERCHANTABILITY OR FITNESS FOR A PARTICULAR PURPOSE, ARE GIVEN IN CONNECTION WITH THIS PUBLICATION. Although this information is believed to be accurate by ASM, ASM cannot guarantee that favorable results will be obtained from the use of this publication alone. This publication is intended for use by persons having technical skill, at their sole discretion and risk. Since the conditions of product or material use are outside of ASM's control, ASM assumes no liability or obligation in connection with any use of this information. No claim of any kind, whether as to products or information in this publication, and whether or not based on negligence, shall be greater in amount than the purchase price of this product or publication in respect of which damages are claimed. THE REMEDY HEREBY PROVIDED SHALL BE THE EXCLUSIVE AND SOLE REMEDY OF BUYER, AND IN NO EVENT SHALL EITHER PARTY BE LIABLE FOR SPECIAL, INDIRECT OR CONSEQUENTIAL DAMAGES WHETHER OR NOT CAUSED BY OR RESULTING FROM THE NEGLIGENCE OF SUCH PARTY. As with any material, evaluation of the material under enduse conditions prior to specification is essential. Therefore, specific testing under actual conditions is recommended.

Nothing contained in this book shall be construed as a grant of any right of manufacture, sale, use, or reproduction, in connection with any method, process, apparatus, product, composition, or system, whether or not covered by letters patent, copyright, or trademark, and nothing contained in this book shall be construed as a defense against any alleged infringement of letters patent, copyright, or trademark, or as a defense against liability for such infringement.

Comments, criticisms, and suggestions are invited, and should be forwarded to ASM International.

Library of Congress Cataloging-in-Publication Data

ASM International

ASM handbook.
Vols. 1–2 have title: Metals handbook.
Includes bibliographical references and indexes.
Contents: v. 1. Properties and selection—irons, steels, and high-performance alloys—v. 2. Properties and selection—nonferrous alloys and special-purpose materials—[etc.]—v. 20. Materials selection and design.

1. Metals—Handbooks, manuals, etc. 2. Metal-work—Handbooks, manuals, etc. I. ASM International. Handbook Committee. II. Metals Handbook.
TA459.M43 1990 620.1'6 90-115
ISBN 0-87170-377-7 (v.1)
SAN 204-7586

ISBN 0-87170-386-6

ASM International®
Materials Park, OH 44073-0002

Printed in the United States of America

Foreword

Handbooks published by ASM International have long been the premier reference sources on the properties, processing, and applications of metals and nonmetallic engineering materials. The fundamental purpose of these handbooks is to provide authoritative information and data necessary for the appropriate selection of materials to meet critical design and performance criteria. *ASM Handbook,* Volume 20 takes the next step by focusing in detail on the *processes* of materials selection and engineering design and by providing tools, techniques, and resources to help optimize these processes. Information of this type has been provided in other handbook volumes—most notably in Volume 3 of the 9th Edition *Metals Handbook*—but never to the impressive scope and depth of this handbook.

Volume 20 reflects the increasingly interrelated nature of engineering product development, encompassing design, materials selection and processing, and manufacturing and assembly. Many of the articles in this volume describe methods for coordinating or integrating activities that traditionally have been viewed as isolated, self-contained steps in a linear process. Other articles focus on specific design and materials considerations that must be addressed to achieve particular design and performance objectives. As in all *ASM Handbook* volumes, the emphasis is on providing practical information that will help engineers and technical personnel perform their jobs.

The creation of this multidisciplinary volume has been a complex and demanding task. It would not have been possible without the leadership of Volume Chair George E. Dieter. We are grateful to Dr. Dieter for his efforts in developing the concept for this volume, organizing an outstanding group of contributors, and guiding the project through to completion. Special thanks are also due to the Section Chairs, to the members of the ASM Handbook Committee, and to the ASM editorial and production staff. We are especially grateful to the more than two hundred authors and reviewers who contributed their time and expertise to create this extraordinary information resource.

George Krauss
President
ASM International

Michael J. DeHaemer
Managing Director
ASM International

Policy on Units of Measure

By a resolution of its Board of Trustees, ASM International has adopted the practice of publishing data in both metric and customary U.S. units of measure. In preparing this Handbook, the editors have attempted to present data in metric units based primarily on Système International d'Unités (SI), with secondary mention of the corresponding values in customary U.S. units. The decision to use SI as the primary system of units was based on the aforementioned resolution of the Board of Trustees and the widespread use of metric units throughout the world.

For the most part, numerical engineering data in the text and in tables are presented in SI-based units with the customary U.S. equivalents in parentheses (text) or adjoining columns (tables). For example, pressure, stress, and strength are shown both in SI units, which are pascals (Pa) with a suitable prefix, and in customary U.S. units, which are pounds per square inch (psi). To save space, large values of psi have been converted to kips per square inch (ksi), where 1 ksi = 1000 psi. The metric tonne ($kg \times 10^3$) has sometimes been shown in megagrams (Mg). Some strictly scientific data are presented in SI units only.

To clarify some illustrations, only one set of units is presented on artwork. References in the accompanying text to data in the illustrations are presented in both SI-based and customary U.S. units. On graphs and charts, grids corresponding to SI-based units usually appear along the left and bottom edges. Where appropriate, corresponding customary U.S. units appear along the top and right edges.

Data pertaining to a specification published by a specification-writing group may be given in only the units used in that specification or in dual units, depending on the nature of the data. For example, the typical yield strength of steel sheet made to a specification written in customary U.S. units would be presented in dual units, by the sheet thickness specified in that specification might be presented only in inches.

Data obtained according to standardized test methods for which the standard recommends a particular system of units are presented in the units of that system. Wherever feasible, equivalent units are also presented. Some statistical data may also be presented in only the original units used in the analysis.

Conversions and rounding have been done in accordance with ASTM Standard E 380, with attention given to the number of significant digits in the original data. For example, an annealing temperature of 1570 °F contains three significant digits. In this case, the equivalent temperature would be given as 855 °C; the exact conversion to 854.44 °C would not be appropriate. For an invariant physical phenomenon that occurs at a precise temperature (such as the melting of pure silver), it would be appropriate to report the temperature as 961.93 °C or 1763.5 °F. In some instances (especially in tables and data compilations), temperature values in °C and °F are alternatives rather than conversions.

The policy of units of measure in this Handbook contains several exceptions to strict conformance to ASTM E 380; in each instance, the exception has been made in an effort to improve the clarity of the Handbook. The most notable exception is the use of g/cm^3 rather than kg/m^3 as the unit of measure for density (mass per unit volume).

SI practice requires that only one virgule (diagonal) appear in units formed by combination of several basic units. Therefore, all of the units preceding the virgule are in the numerator and all units following the virgule are in the denominator of the expression; no parentheses are required to prevent ambiguity.

Preface

All engineers who are concerned with the development of products or the design of machines and structures must be knowledgeable about the materials from which they are made. After all, the selection of the correct material for a design is a key step in the design process because it is the crucial decision that links the computer calculations and the lines on an engineering drawing with a working design. At the same time, the rapid progress in materials science and engineering has made a large number of materials—metals, polymers, ceramics, and composites—of potential interest to the designer. Thus, the range of materials available to the engineer is much larger than ever before. This presents the opportunity for innovation in design by utilizing these materials in products that provide greater performance at lower cost. To achieve this requires a more rational process for materials selection than is normally used.

Materials engineers have traditionally been involved in helping to select materials. In most cases, this is done more or less in isolation from the actual design process. Sometimes the materials expert becomes involved only when the design fails. In the past ten years, mostly in response to the pressures of international competitiveness, new approaches to product design and development have arisen to improve quality, drive down cost, and reduce product cycle time. Generally called concurrent engineering, it uses product development teams of experts from all functions—design, manufacturing, marketing, and so forth—to work together from the start of the product design project. This opens new opportunities for better material selection. It also has resulted in the development of new computer-based design tools. If materials engineers are to play an important future role in product development, they need to be more familiar with the design process and these design tools.

Thus, Volume 20 of *ASM Handbook* is aimed at two important groups: materials professionals and design professionals. As a handbook on materials selection and design, it is unique. No other handbook deals with this subject area in this way, bridging the gaps between two vital but often distant areas of expertise. The Handbook is divided into seven sections:

- The Design Process
- Criteria and Concepts in Design
- Design Tools
- The Materials Selection Process
- Effects of Composition, Processing, and Structure on Materials Properties
- Properties versus Performance of Materials
- Manufacturing Aspects of Design

Emphasis throughout is on concepts and principles, amply supported by examples and case histories. This is not a handbook of material property data, nor is it a place to find detailed discussion of specific material selection problems. Other volumes in the *ASM Handbook* series often provide this type of information.

Section 1, "The Design Process," sets the stage for the materials engineer to better understand and participate in the product design process. The context of design within a manufacturing firm is described, and the role of the materials engineer in design is discussed. Emphasis is placed on methods for conceptual and configuration design, including the development of a product specification. Methods for creative generation of conceptual designs and for evaluation of conceptual and configuration alternatives are introduced. Learning to work effectively in cross-functional teams is discussed.

Section 2, "Criteria and Concepts in Design" deals with design concepts and methods that are important for a complete understanding of engineering design. The list is long: concurrent engineering, including QFD; codes and standards; statistical aspects of design; reliability in design; life-cycle engineering; design for quality; robust design (the Taguchi approach); risk and hazard analysis; human factors in design; design for the environment (green design); safety; and product liability and design.

Section 3 considers "Design Tools." This section provides an overview of the computer-aided engineering tools that are finding wide usage in product design. This includes the fundamentals of computer-aided design, and the use of computer-based methods in mechanism dynamics, stress analysis (finite element analysis), fluid and heat transfer analysis, and electronic design. Also considered are computer methods for design optimization and tolerance analysis. Finally, the section ends with discussions of the document packages necessary for design and of methods for rapid prototyping.

Section 4, "The Materials Selection Process," lays out the complexity of the materials selection problem and describes various methodologies for the selection of materials. Included are Ashby's material property charts and performance indices, the use of decision matrices, and computer-aided methods. Also discussed are the use that can be made of value analysis and failure analysis in solving a materials selection problem. The close interrelationship of materials selection and economic issues and processing are reinforced in separate articles.

Section 5, "Effects of Composition, Processing, and Structure on Materials Properties," is aimed chiefly at the design engineer who is not a materials specialist. It is a "mini-textbook" on materials science and engineering, with a strong engineering flavor and oriented chiefly at explaining mechanical properties and behavior in terms of structure. The role that processing plays in influencing structure is given emphasis. The articles in this Section cover metallic alloys, ceramics, engineering plastics, and composite materials. The Section concludes with an article on places to find materials information and properties.

Section 6, "Properties versus Performance of Materials," features articles that attempt to cross the materials/design gap in a way that the designer will understand how the material controls properties and the materials engineer will become more familiar with real-world operating conditions. Again, emphasis is mostly on mechanical behavior and includes articles on design for static structures, fatigue, fracture toughness, and high temperature. Other articles consider design for corrosion resistance, oxidation, wear, and electronic and magnetic applications. Separate articles consider the special concerns when designing with brittle materials, plastics, and composite materials.

Section 7, "Manufacturing Aspects of Design," focuses on the effects of manufacturing processes on the properties and the costs of product designs. The section contains articles on design for manufacture and assembly (DFM and DFA), general guidelines for selecting processes, modeling of processes, and cost estimation in manufacturing. Individual articles deal with design for casting, deformation processes, powder processing, machining, joining, heat treatment, residual stresses, and surface finishing. Articles also deal with design for ceramic processing, plastics processing, and composite manufacture.

This Handbook would not have been possible without the dedicated hard work of the chairmen of the sections: John R. Dixon, University of Massachusetts (retired); Bruce Boardman, Deere & Company; Kenneth H. Huebner, Ford Motor Company; Richard W. Heckel, Michigan Technological University (retired); David A. Woodford, Materials Performance analysis Inc.; and Howard A. Kuhn, Concurrent Technologies Corporation. Special thanks goes to several individuals who did work well beyond the normal call of duty in reviewing manuscripts: Serope Kalpakjian, John A. Schey, and Charles O. Smith. I wish to thank all of the busy people who agreed to author articles for the Handbook. The high rate of acceptance, from both the design community and the materials community, is a strong indicator of the importance of the need that *ASM Handbook,* Volume 20, fills.

George E. Dieter
University of Maryland

Authors and Contributors

Peter Andresen
General Electric Corporate Research and Development Center

Michael F. Ashby
Cambridge University

Anne-Marie M. Baker
University of Massachusetts

Charles A. Barrett
NASA Lewis Research Center

Carol M.F. Barry
University of Massachusetts

Raymond Bayer
Tribology Consultant

Michael Blinn
Materials Characterization Laboratory

Bruce E. Boardman
Deere & Company Technical Center

Geoffrey Boothroyd
Boothroyd Dewhurst Inc.

David L. Bourell
The University of Texas at Austin

James G. Bralla
Manufacturing Consultant

Bruce L. Bramfitt
Bethlehem Steel Corporation

Peter R. Bridenbaugh
Alcoa Technical Center

Eric W. Brooman
Concurrent Technologies Corporation

Ronald N. Caron
Olin Corporation

Umesh Chandra
Concurrent Technologies Corporation

Joel P. Clark
Massachusetts Institute of Technology

Don P. Clausing
Massachusetts Institute of Technology

Thomas H. Courtney
Michigan Technological University

Mark Craig
Variation Systems Analysis, Inc.

James E. Crosheck
CADSI

Shaun Devlin
Ford Motor Company

Donald L. Dewhirst
Ford Motor Company

R. Judd Diefendorf
Clemson University

George E. Dieter
University of Maryland

John R. Dixon
University of Massachusetts

William E. Dowling, Jr.
Ford Motor Company

Stephen F. Duffy
Cleveland State University

Lance A. Ealey
McKinsey & Company

Peter Elliot
Corrosion and Materials Cosultancy Inc.

Mahmoud M. Farag
American University in Cairo

Frank R. Field III
Massachusetts Institute of Technology

B. Lynn Ferguson
Deformation Control Technology

Shirley Fleischmann
Grand Valley State University

F. Peter Ford
General Electric Corporate Research and Development Center

Theodore C. Fowler
Fowler & Whitestone

Victor A. Greenhut
Rutgers—The State University of New Jersey

Daniel C. Haworth
General Motors Research and Development Center

Richard W. Heckel
Michigan Technological University

David P. Hoult
Massachusetts Institute of Technology

Kenneth H. Huebner
Ford Motor Company

Thomas A. Hunter
Forensic Engineering Consultants, Inc.

Lesley A. Janosik
NASA Lewis Research Center

Geza Kardos
Carleton University

Erhard Krempl
Rensselaer Polytechnic Institute

Howard A. Kuhn
Concurrent Technologies Corporation

Richard C. Laramee
Intermountain Design, Inc.

John MacKrell
CIMdata

Arnold R. Marder
Lehigh University

C. Lawrence Meador
Massachusetts Institute of Technology

Edward Muccio
Ferris State University

Peter O'Rourke
Los Alamos National Laboratory

Kevin N. Otto
Massachusetts Institute of Technology

Nagendra Palle
Ford Motor Company

Anand J. Paul
Concurrent Technologies Corporation

Thomas S. Piwonka
The University of Alabama

Hans H. Portisch
Krupp VDM Austria GmbH

Raj Ranganathan
General Motors Corporation

Richard C. Rice
Battelle Columbus

Mark L. Robinson
Hamilton Precision Metals

Richard Roth
Massachusetts Institute of Technology

Eugene Rymaszewski
Rensselaer Polytechnic Institute

K. Sampath
Concurrent Technologies Corporation

Howard Sanderow
Management and Engineering Technologies

Jon Schaeffer
General Electric Aircraft Engines

John A. Schey
University of Waterloo

James Smialek
NASA Lewis Research Center

Charles O. Smith
Engineering Consultant

Douglas E. Smith
Ford Motor Company

Preston G. Smith
New Product Dynamics

James T. Staley
Alcoa Technical Center

David A. Stephenson
General Motors Corporation

Henry Stoll
Northwestern University

Charles L. Thomas
University of Utah

Gerald Trantina
General Electric Corporate Research and Development Center

B. Lee Tuttle
GMI Engineering and Management Institute

George F. Vander Voort
Buehler Ltd.

Anthony J. Vizzini
University of Maryland

Gary A. Vrsek
Ford Motor Company

Volker Weiss
Syracuse University

Jack H. Westbrook
Brookline Technologies

James C. Williams
General Electric Aircraft Engines

Roy Williams
Materials Characterization Laboratory

Kristin L. Wood
University of Texas

David A. Woodford
Materials Performance analysis, Inc.

Reviewers

John Abraham
Purdue University

Robert M. Aiken, Jr.
Case Western Reserve University

David J. Albert
Albert Consulting Group

C. Wesley Allen
CWA Engineering

William Anderson
Automated Analysis Corporation

Harry W. Antes
SPS Technologies (retired)

William R. Apblett
Amet Engineering

Michael F. Ashby
Cambridge University

Carl Baker
Pacific Northwest National Laboratory

H. Barry Bebb
Barry Bebb & Associates

James Birchmeier
General Motors Corporation

Neil Birks
University of Pittsburgh

Peter J. Blau
Oak Ridge National Laboratory

Omer W. Blodgett
Lincoln Electric Company

Geoffrey Boothroyd
Boothroyd Dewhurst Inc.

David L. Bourell
University of Texas at Austin

Rodney R. Boyer
Boeing Company

Bruce L. Bramfitt
Bethlehem Steel Corporation

Charlie R. Brooks
The University of Tennessee

Eric W. Brooman
Concurrent Technologies Corporation

William L. Brown
Caterpillar Inc.

Myron E. Browning
Matrix Technologies

George C. Campbell
Ford Motor Company

Barry H. Carden
Charter Oak Consulting Group, Inc.

Ronald N. Caron
Olin Corporation

Craig D. Clauser
Consulting Engineers Inc.

Don P. Clausing
Massachusetts Institute of Technology

Lou Cohen
Independent Consultant

Arthur Cohen
Copper Development Association Inc.

Thomas H. Courtney
Michigan Technological University

Eugene E. Covert
Massachusetts Institute of Technology

Margaret D. Cramer
IMO Pumps, IMO Industries Inc.

Richard Crawford
University of Texas

Robert C. Creese
West Virginia University

Frank W. Crossman
Lockheed Martin Advanced Technology Center

Charles J. Crout
Forging Developments International, Inc.

David Cutherell
Design Edge

Fran Cverna
ASM International

Edward J. Daniels
Argonne National Laboratory

Craig V. Darragh
The Timken Company

Randall W. Davis
McDonnell Douglas Helicopter Systems

Rudolph Deanin
University of Massachusetts-Lowell

John J. deBarbadillo
Inco Alloys International

Donald L. Dewhirst
Ford Motor Company

George E. Dieter
University of Maryland

John R. Dixon
University of Massachusetts

Keith A. Ellison
Wilson & Daleo Inc.

William J. Endres
University of Michigan

Steven Eppinger
Massachusetts Institute of Technology

Georges Fadel
Clemson University

Abdel Aziz Fahmy
North Carolina State University

Mahmoud M. Farag
The American University in Cairo

Mattison K. Ferber
Oak Ridge National Laboratory

Stephen Freiman
National Institute of Standards and Technology

Peter A. Gallerani
Integrated Technologies, Inc.

Murray W. Garbrick
Lockheed Martin Corporation

Michelle M. Gauthier
Raytheon Electronic Systems

T.B. Gibbons
ABB-CE Power Plant Laboratories

Brian Gleeson
The University of New South Wales

Raphael Haftka
University of Florida

Larry D. Hanke
Materials Evaluation and Engineering, Inc.

Richard W. Heckel
Michigan Technological University

Alfredo Herrera
McDonnell Douglas Helicopter Systems

Barry S. Hindin
Battelle Columbus Division

David Hoeppner
University of Utah

Maurice Howes
IIT Research Institute

Kenneth H. Huebner
Ford Motor Company

M.W. Hyer
Virginia Polytechnic Institute and State University

Serope Kalpakjian
Illinois Institute of Technology

Geza Kardos
Carleton University

Theodoulos Z. Kattamis
University of Connecticut

J. Gilbert Kaufman
Aluminum Association

Michael Kemen
Attwood Corporation

Robert D. Kissinger
GE Aircraft Engines

William D. Kline
GE Aircraft Engines

Lawrence J. Korb
Metallurgical Consultant

Paul J. Kovach
Stress Engineering Services, Inc.

Jesa Kreiner
California State University, Fullerton

Howard A. Kuhn
Concurrent Technologies Corporation

Joseph V. Lambert
Lockheed Martin

Richard C. Laramee
Intermountain Design Inc.

David E. Laughlin
Carnegie Mellon University

Alan Lawley
Drexel University

Peter W. Lee
The Timken Company

Keith Legg
Rowan Catalyst Inc.

Richard L. Lehman
Rutgers—The State University of New Jersey

Iain LeMay
Metallurgical Consulting Services Ltd.

James H. Lindsay
General Motors R&D Center

Carl R. Loper, Jr.
The University of Wisconsin-Madison

Kenneth Ludema
University of Michigan

John MacKrell
CIMdata, Inc.

Arnold R. Marder
Lehigh University

Lee S. Mayer
Cessna Aircraft Company

Anna E. McHale
Consultant

Gerald H. Meier
University of Pittsburgh

A. Mikulec
Ford Motor Company

M.R. Mitchell
Rockwell International Science Center

James G. Morris
University of Kentucky

Edward Muccio
Ferris State University

Mary C. Murdock
Buffalo State College

James A. Murray
Independent Consultant

John S. Nelson
Pennsylvania Steel Technologies, Inc.

Glenn B. Nordmark
Consultant

David LeRoy Olson
Colorado School of Mines

Joel Orr
Orr Associates International

Kevin N. Otto
Massachusetts Institute of Technology

William G. Ovens
Rose-Hulman Institute of Technology

Charles Overby
Ohio University

Leander F. Pease III
Powder-Tech Associates, Inc.

Thomas S. Piwonka
The University of Alabama

Michael Poccia
Eastman Kodak Company

Hans H. Portisch
Krupp VDM Austria GmbH

Tom Priestley
Analogy Inc.

Louis J. Pulgrano
DuPont Company

Chandra Putcha
California State University, Fullerton

Donald W. Radford
Colorado State University

James A. Rains, Jr.
General Motors Corporation

Harold S. Reemsnyder
Bethlehem Steel Corporation

Michael Rigdon
Institute for Defense Analyses

David A. Rigney
The Ohio State University

Ana Rivas
Case Western Reserve University

J. Barry Roach
Welch Allyn, Inc.

Mark L. Robinson
Hamilton Precision Metals, Inc.

Gerald J. Roe
Bethlehem Steel Corporation

Edwin Ruh
Ruh International Inc.

John Rumble
National Institute of Standards and Technology

Jerry Russmann
Deere & Company

C.O. Ruud
The Pennsylvania State University

Edmund F. Rybicki
The University of Tulsa

K. Sampath
Concurrent Technologies Corporation

John A. Schey
University of Waterloo

Julie M. Schoenung
California State Polytechnic University, Ponoma

Marlene Schwarz
Polaroid Corporation

S.L. Semiatin
Air Force Materials Directorate, Wright Laboratory

Donald P. Seraphim
Rainbow Displays & Company

Sheri D. Sheppard
Stanford University

John A. Shields, Jr.
CSM Industries, Inc.

Allen W. Sindel
Sindel & Associates

M. Singh
NYMA, Inc., NASA Lewis Research Center

James L. Smialek
NASA Lewis Research Center

Charles O. Smith
Engineering Consultant

Robert S. Sproule
Consulting Engineer

James T. Staley
Alcoa Technical Center

Edgar A. Starke, Jr.
University of Virginia

Henry Stoll
Northwestern University

Brent Strong
Brigham Young University

Gary S. Strumolo
Ford Motor Company

John Sullivan
Ford Motor Company

Thomas F. Talbot
Consulting Engineer

Raj B. Thakkar
A.O. Smith Automotive Products Company

Thomas Thurman
Rockwell Avionics and Communications

Tracy S. Tillman
Eastern Michigan University

Peter Timmins
Risk Based Inspection Inc.

George E. Totten
Union Carbide Corporation

Marc Tricard
Norton Company

R.C. Tucker, Jr.
Praxair Surface Technologies, Inc.

Floyd R. Tuler
Alcan Aluminum Corporation

George F. Vander Voort
Buehler Ltd.

Garret N. Vanderplaats
Vanderplaats Research & Development, Inc.

Jack R. Vinson
University of Delaware

Anthony M. Waas
University of Michigan

John Walters
Scientific Forming Technologies Corporation

Harry W. Walton
The Torrington Company

Paul T. Wang
Alcoa Technical Center

Colin Wearring
Variation Systems Analysis, Inc.

David C. Weckman
University of Waterloo

David W. Weiss
University of Maryland

Volker Weiss
Syracuse University

Jack H. Westbrook
Brookline Technologies

Bruce A. Wilson
McDonnell Douglas Corporation

Ronald Wolosewicz
Rockwell Graphic Systems

Kristin L. Wood
University of Texas

David A. Woodford
Materials Performance analysis, Inc.

Michael G. Wyzgoski
General Motors R&D Center

Ren-Jye Yang
Ford Motor Company

Steven B. Young
Trent University

David C. Zenger
Worcester Polytechnic Institute

Contents

Section 7: Manufacturing Aspects of Design 667
Chairperson: Howard A. Kuhn, Concurrent Technologies Corporation

Section 1: The Design Process

The Role of the Materials Engineer in Design

Bruce Boardman, Deere and Company Technical Center
James C. Williams, General Electric Aircraft Engines
Peter R. Bridenbaugh, Aluminum Company of America Technical Center

THE ROLE of the materials engineer in the design and manufacture of today's highly sophisticated products is varied, complex, exciting, and always changing. Because it is not always the metallurgical or materials engineer who specifies the material, this *ASM Handbook* on materials selection and design is prepared to benefit all engineers who are involved with selecting materials with their related processes that lead to a ready-to-assemble manufactured component. This article discusses the various roles and responsibilities of materials engineers in a product realization organization and suggests new and different ways in which materials engineers may benefit their organization. Insights into use of the remainder of this Volume are also offered.

Materials selection specialists have been practicing their art since the beginning of recorded time. The first caveman, searching for food, required an implement that would not break during use. Although wood, stone, and bone were the only structural materials available, there were still choices: hard wood versus soft wood, and hard stones and flint, which would sharpen when broken, versus soft stones. While prehistoric man learned only from experience, learning nevertheless took place, and the art of materials selection became a valued skill within the community. As other materials, such as copper and iron, became available, the skill became almost mystical, with knowledge passed down from father to son, until the middle to late 19th century. By then the blacksmith had replaced the alchemist. At this point, the blacksmith had become the local expert in materials selection and shaping and was recognized as a valuable and enabling member of the community.

The role of the materials selection expert has evolved. Today when we think of materials selection specialists, we think of those who have been formally trained as metallurgical or materials engineers. But as discussed below, there are many more engineers involved in materials selection than those with the title metallurgist, materials engineer, or materials scientist. Modern engineered materials are now available that have attractive but complex properties. Therefore, it is becoming essential to develop a much closer working relationship between those who design a component and those who advise the designer on materials selection. In fact, the most efficient structural designs are now generated by incorporating, from the beginning, the complex properties of modern engineered materials into the design synthesis step (matching form to function).

The actual selection of a material to satisfy a design need is effectively performed every day in literally dozens of different ways by people of many different backgrounds. The selection process can range from simply re-specifying a previously used material (or one used by a competitor) through finite element analyses or modeling routines to precisely identify property requirements. Additionally, the selection may be done by someone formally trained in metallurgy and materials science or by designers themselves. There is no unique individual role when it comes to materials selection.

Today, the selection of the material and its processing, product design, cost, availability, recycleability, and performance in final product form have become inseparable. As a result, more and more companies are forming integrated product development (IPD) teams to ensure that all needed input is obtained concurrently. Whether it is used in a small company (which frequently, from lack of resources, is forced to work in the IPD mode) or a large company (who may have to create a "skunk works"), the IPD approach has been shown to lead to a better result and to achieve this result faster. The integration of material, process, and product design relies on individuals who are trained in materials selection and can work in a team environment. Often, it is the materials specialist, familiar with the frequent, conflicting needs of design, production, and marketing, who can assume the role of mediator to focus on the final product. We hope that this point will be made clearly in this Volume.

Attempting to define a single role for the individual who actually selects a material for a design is not possible. That individual frequently assumes roles that cross many engineering and manufacturing disciplines. Starting with the initial design and material choice, through prototype manufacture and testing, and continuing to final production, the materials selection specialist is an essential team member. As more companies shrink their in-house, captive manufacturing and assembly operations, the role of the materials selection function may increasingly be outsourced, along with the actual manufacturing activity. This possibility can create opportunities for the materials selection specialist, but it can also create risk for the "virtual manufacturers."

Worldwide, the vast majority of manufacturing firms are small and cannot afford the luxury of a formally trained materials scientist or materials selection specialist. Rather, they have individuals trained in many areas, one of which is materials. In a smaller enterprise, these individuals actually select materials as a part of their daily design activity. Whether that training was gained as a part of another degree program, as part of a community college associates program, on the job, or as the result of a series of ASM International's Materials Engineering Institute courses, the result is the development of an individual trained in the many and varied facets of materials selection. For most products and materials applications this practice works quite well. However, for high-performance products, where understanding the subtleties of materials performance can be the defining difference, this practice can lead to a less than optimal result. The emergence of agile manufacturing and rapid response scenarios, coupled with ongoing developments in new and tailored materials, further specializes the critical function of materials selection.

Before proceeding into detail about the many roles of the materials engineer, it is appropriate to summarize the content of the remainder of this Volume to help guide readers to the portions most important to their specific interests. The volume is divided into seven instructional sections, which are summarized in Table 1 and discussed further in the following paragraphs.

Table 1 Overview of the Sections in *ASM Handbook*, Vol 20, *Materials Selection and Design*

Section title	Summary
1. The Design Process	This section offers insights into the several roles that must be played by the materials selection expert. It also reviews the process and methods that may be applied to enhance and improve the effectiveness of the design process.
2. Criteria and Concepts in Design	This section goes into detail on many of the "soft" issues related to design, process, safety, manufacturability, and quality. These issues are not historically a part of the design and material selection process, because they do not relate to the quantifiable properties (e.g., strength or toughness) or attributes (e.g., wear or corrosion resistance) that determine the ability of a material to perform the desired function. Nevertheless, they are of critical importance, because parts and assemblies must be made with well-understood variance, consistent processing, and the expectation that the part will perform safely and reliably in the ultimate customer's application.
3. Design Tools	This section details the tools associated with a state-of-the-art design process. Included are discussions on paper and paperless drawings, adding tolerances, computer-aided drafting and computer-aided design, rapid prototyping, modeling, finite element methods, optimization methods, and documenting and communicating the design to others.
4. The Materials Selection Process	This section begins the details of what steps and methods are actually required to properly select a material and its corresponding manufacturing process. Topics included are an overview of the process, technical and economic issues, the Ashby materials selection charts, use of decision matrices, computer-aided materials selection, the relationship between materials properties and processing, and the use of value analysis and failure analysis.
5. Effects of Composition, Processing, and Structure on Materials Properties	The science of materials selection is introduced in this section as the relationships between different families of materials (e.g., metals, ceramics, plastics) are discussed. Additionally the effects of thermal and mechanical processing on performance properties of materials are discussed. Sources of materials data are also listed in this section.
6. Properties versus Performance of Materials	This section details and discusses the actual properties needed for specific general types of design (e.g., structural, optical, magnetic, electronic) as well as accepted design processes and methodology for prevention of several common performance needs (e.g., corrosion, fatigue, fracture toughness, high temperature, wear, oxidation). Additionally there is discussion relating to design with brittle materials, plastics, and composite materials, and for surface treatments.
7. Manufacturing Aspects of Design	This section discusses what may be the most important aspects of a successful design: how the conceptual ideas are cost effectively converted into hardware. The majority of commonly used manufacturing processes are discussed in detail in a series of separate chapters, but ultimately, the designer and materials selection expert must merge these thermal and mechanical processes into a description of the properties and attributes of the final part. Techniques for computer-based modeling and costing are also discussed. Additionally, there is discussion about the effect of processing on several of the common nonmetallic materials and the control of residual stresses resulting from manufacturing. Finally, this section includes a discussion on designing for ease of assembly of the many parts that may be involved in a final product, ready for delivery to the ultimate customer.

The Design Process

Section 1 of this Volume shows that the process of materials selection during design can take many paths. As already suggested, the task may simply be to design a "new" part that is nearly identical to an existing part and is expected to be used in similar ways. In this case, it may be possible to use the same material and processing as were used for the existing part. Alternatively, the task may be to design and select material for a new part for which there is no prior history. Obviously, this is a much more complex task and requires knowledge of loads, load distributions, environmental conditions, and a host of other performance factors (including customer expectations) and manufacturing-related factors.

In addition to a knowledge of the required performance characteristics, the materials selector must be able to define and account for manufacturing-induced changes in material properties. Different production methods, as well as controlled and uncontrolled thermal and mechanical treatments, will have varying effects on the performance properties and the cost of the final part or assembly. Hence, the materials specialist must also work with the value engineering function to achieve the lowest cost consistent with customer value. Often, it is by relating the varying effects of manufacturing processes to customer needs that one manufacturer develops a product that has an advantage over another, using essentially the same material and process combinations.

While the effects of manufacturing-induced changes to performance properties are covered in a later section (as well as in other *ASM Handbook* Volumes), it is critical to understand and accept that the choice of manufacturing processes is frequently not under the direct control of the materials selection expert. In fact, by the time the concept and initial configuration of a design is committed to paper, or to a computer-aided design (CAD) system, the manufacturing processes and sequence of processes required to produce a product cost effectively are normally fixed. They are no longer variables that can be controlled without redesign.

The above approach generally follows the path that George Dieter refers to as a "process first approach" in his article "Overview of the Materials Selection Process" in this Volume. Unfortunately, it has been common for designers, inadvertently, to create parts with geometric features that place severe restrictions on the selection of manufacturing processes, with even less freedom remaining for material selection (Table 4 in Dieter's article demonstrates this point). The use of "design for manufacturability" concepts and IPD teams is beginning to eliminate this undesirable practice. Until the IPD approach is in common use, an alternative, referred to as a "materials first approach," may be useful. The materials first approach depends on a thorough understanding of the service environment and advocates choices based on properties that satisfy those performance needs (Table 3 in Dieter's article provides a useful starting point). Similarly, overly restrictive selection of the material independently limits the manufacturing processes available. This is all the more reason to use IPD methods.

As suggested above, the use of a cross-functional IPD team to translate the desired performance requirements into a design concept usually yields the best result most quickly. Such a team contains the expertise to decide between the use of steel sheet, machined forgings, nonferrous castings, or reinforced polymers as well as to select the processing and joining methods. Table 2 summarizes many common specialties required to define materials, processes, and manufacturing methods for making cost effective parts and assemblies that meet the customer's expectations. These decisions are not, by themselves, sufficient to ensure a successful design, but the use of cross-functional teams to concurrently consider design, materials, manufacturing processes, and final cost provides superior customer value. Obviously, no individual design exercise will contain one member from each specialty; in many practical cases, each member can represent multiple specialties.

Criteria and Concepts in Design

Material selection involves more than meeting minimum property requirements for strength, fatigue, toughness, corrosion resistance, or wear resistance. There are numerous options for product design and materials selection, and frequently they cannot be quantified. This precludes the use of mathematical optimization routines and shifts the emphasis to experience. Experience is essential in dealing with these "soft issues" related to qualitative non-property considerations.

The design must be producible. This means robust processes must be selected that have known statistical variation and will yield features or complete parts that lie well within the specification limits. This design for manufacturability approach is becoming popular, is an integral part of an IPD team's tool box, and has been demonstrated to be effective in improving quality and reducing cost.

Table 2 Typical specialties involved during an "ideal" materials selection process

General area	Specialty
Materials science	Metals
	Plastics
	Ceramics
	Coatings
	Chemistry
	Electrochemistry
Processing	Forging
	Casting
	Welding
	Hot forming
	Cold forming
	Molding
	Machining
	Sintering
	Heat treatment
Cost analysis	Purchasing (supply management)
	Process engineering
	Industrial engineering
	Life cycle costing
Design	(Specific to application)
Quality assurance	Inspection
	Statistics
	Reliability
	Field test
	Customer
Other	Marketing
	Legal
	Environmental

Designing to minimize the total costs to the consumer during the expected product life (the life cycle cost) is yet another challenge. These costs include raw material, production, use, maintenance (scheduled or otherwise), and disposal or recycling costs. Some of these cost elements are unknown. This is where the combination of the art and skill of engineering faces its most severe test.

Similar issues arise when the safety, product liability, and warranty cost exposure aspects of product design and material selection are concerned. In many cases, alternate designs or materials could be chosen with no measurable difference. However, there are also many cases where a particular design and/or material choice could prevent an undesirable product failure mode. An understanding of how a part, assembly, or entire structure can fail and the ramifications of that failure is essential in providing a safe and reliable design. A well-known example is the failure of one material in a ductile mode while another fails in a brittle mode. The former could provide that extra margin of safety by giving a warning that there is an impending failure while the latter fails catastrophically without warning. Knowing the ways a product can fail and the safety ramifications of each failure mode will go a long way to minimizing the consequences of failure if the product is used in a manner that exceeds the design intent. Failure mode and effects analysis (FMEA) can help in this regard.

Product success requires that the appearance and function of the product must meet the customer's approval. Normally these are design factors, but material selection and surface finish can be equally important. Consumers' tastes often change with time; for instance, current camera customers prefer a dull or matte black finish instead of brightly finished ones. Numerous materials-related solutions to accommodate this change in buying patterns were proposed, including anodizing, painting, and changing the substrate material from metal to plastic.

The camera example leads into a discussion of designing for the environment. The growing environmental and regulatory demand to consider the entire life cycle of a product could require the manufacturer to recover and recycle the product and process waste materials. This places renewed emphasis on considering all options. Changing the materials or the manufacture of the camera mentioned above involves designing an environmentally friendly product. Changing from chromium plating appears to be environmentally friendly, but today's chrome plating units are being constructed to operate in a zero discharge mode, so there is no obvious gain from eliminating the chrome. The anodizing process can be just as clean. Paint, on the other hand, is suffering severe scrutiny over both emissions during the painting process as well as subsequent mishandling by the consumer. And, changing the camera body to plastic is not necessarily a good solution because the recycling infrastructure is not yet adequate on a global level to effectively reclaim the material.

Another design factor is the repairability of a product. Automobiles are not intended to have accidents, but they do. Design and material selection only for initial cost and performance factors has led to the widespread use of one-piece plastic parts that are not repairable in many cases. Any product that costs more to repair than the owner finds acceptable will eventually suffer in the marketplace.

The second Section of this Handbook, "Criteria and Concepts in Design," provides significant additional detail about factors that must be considered during the conceptual stage of design. While many of these factors are not quantifiable, they affect the ultimate cost and ability of the design to satisfy customer expectations. Often, it is the materials engineer who is best equipped to integrate and account for these soft issues, which can be one of the deciding factors in the marketplace. Unfortunately, the pressure of design schedules can squeeze the time allotted for a thorough selection of material and process. The materials engineer must guard against this.

Design Tools

Once the concept and geometry of a part or assembly have been determined, the designer proceeds to the detailed manufacturing design phase. The output from this phase is a physical blueprint or electronic CAD file from which the part will be manufactured. This output contains input for the materials engineer in the form of material selection and processing notes that will guide the manufacturing activity and ultimately may evolve into formal material and processing specifications.

Section 3, "Design Tools," contains numerous articles relating to the functions required to pass from the conceptual stage to a detailed and optimized design. These articles introduce concepts for CAD, tolerancing, optimizing, documenting, and prototyping. A common thread between all of these aspects is that the designer requires sets of validated material and processing properties. Again, the materials engineer is an important resource. While there are numerous sources of basic materials data, few sources take into consideration the inherent differences between manufacturing facilities. It is the materials engineer, familiar with the required manufacturing processes and how they individually and collectively affect the ultimate properties of the material, who leads the process of translating handbook data into anticipated product performance.

The need to produce a prototype part that accurately represents the future parts, including manufacturing process capability, is another factor that complicates the design process. While a prototype can be machined from a block of wrought metal, the properties of this first part will not be the same as those of the production parts if casting, forming, or powder consolidation processes are ultimately used to produce the required shape. The machined prototype will be useful for testing, fit, design functionality, and the determination of service loads, but it will provide little information about ultimate fatigue life, fracture toughness, or other environmental needs. Driven by this need, new methods of rapid prototyping continue to be developed. In a very few cases, techniques are available to quickly produce accurate prototypes that equal final production parts. Continuing with the example of machined versus cast parts, a replica of the part can be machined from expanded polystyrene and the lost foam casting method can be used to produce a "real" casting. This casting possesses all of the significant characteristics of the yet-to-be manufactured production parts. More details of these technologies can be found in the article "Rapid Prototyping" in this Volume. The materials engineer will often be asked to evaluate the degree to which the prototype can be expected to represent the production parts. Failure to include this comparison step can result in retro design under duress, schedule delays, and increased cost.

The Materials Selection Process

Ultimately, the design reaches the stage where final material selection is required. At that time, knowledge of both mechanical and environmental requirements is essential. During the conceptual design stage, only general data were required about materials properties, materials processing effects, and performance parameters. These broad descriptions need to be refined into specific performance requirements, including the processing steps that will ensure this performance. The materials engineer provides guidance based on knowledge of the properties of the base materials and knowledge of the relationships between the material processing and the final properties.

The materials engineer's knowledge of the processes available within the manufacturing facility and the property changes due to the mechanical or thermomechanical processes can simplify the choices between cost, manufacture, environment, and many other issues. Section 4, "The Materials Selection Process," provides details on many of the issues and steps required to finally arrive at the optimal material selection.

Effects of Composition, Processing, and Structure on Materials Properties

Few product lines require a thorough knowledge of all the different materials, compositions, structure, and processing relationships contained in Section 5. However, materials engineers must know which of these to apply to their operations and have general knowledge about the others. In many cases, the materials and process content of a product can be used to differentiate it in the marketplace. Therefore, it is important for the materials engineer to possess the education and background to become expert in new materials and material processes as they emerge so that the company's new products will be competitive.

Properties versus Performance of Materials

Up to this point, the subject of performance has been referred to only in passing or as something that is known and will be satisfied by the material and processing combination chosen. Obviously, that is a gross oversimplification.

Section 6 addresses the more significant relationships between properties and performance. For simplicity, these subjects are presented individually. In reality there are usually several limiting, and often competing, property-related performance criteria. Bridges and boilers, for example, require strength, modulus, fatigue, fracture, corrosion, thermal expansion, and so on. It is the role of the materials engineer to integrate these many factors into a successful product.

Detailed discussion of the methods used to determine the minimum materials properties required to meet desired product characteristics is not included here. In general, the methods for determining the minimum required performance properties are well beyond the scope of this Volume, or perhaps any single handbook. Fortunately, the vast majority of products designed are derived from existing products, so the materials engineer has a good idea of service conditions and product requirements. An accurate and complete understanding of a customer's intended use of a product is essential to the design and manufacture of a successful product. This information is the heart of the product design specification discussed in the article "Conceptual and Configuration Design of Products and Assemblies" in Section 1.

Also missing from Section 6 is any reference to methods for testing new or prototype parts, assemblies, or products in service-based conditions. Since Wohler's pioneering explanation of fatigue in railroad axles over one hundred years ago, there has been continuous advancement in the understanding of service environments, recording of service conditions (loads, strains, strain rates, corrosion, temperature, etc.), and accelerated laboratory testing methods to understand the effect of these conditions. From Wohler's simple axle test unit, to laboratory-sized material property test coupons, to full-scale automobile or airplane test beds, there has been a competitive need for something other than placing a product in the hands of the consumer and waiting (possibly years) to learn if it was underdesigned (premature failure and safety or liability issues), overdesigned (too heavy or expensive), or appropriately designed. Adding to the complexity is the fact that many consumers do not have similar or well-defined operating envelopes, resulting in large variations in service loads and lifetimes. Dealing with this uncertainty is one of the major challenges for a designer.

Manufacturing Aspects of Design

Section 7, "Manufacturing Aspects of Design," introduces articles on manufacturing-related factors besides properties, including cost. As previously stated, the manufacturing processes capable of producing a specific part design are restricted, if not fixed, at the time of conceptual design. Combine this with the fact that, for most parts, costs are related to manufacturing and assembly, and it becomes apparent that choosing the "best" design is highly dependent on choosing the "best" manufacturing method. Since this decision is made early in the process, it becomes important for designers to avail themselves of the manufacturing expertise provided by the materials engineer.

Once the manufacturing process has been identified, there is still the need to optimize the process, determine its capability, and understand the effect(s) that the process will have on a material and its properties. Computer modeling is making significant contributions to our understanding of the effects of processing on properties, as well as which steps in the processing sequence are most important to control in order to consistently produce high-quality parts that meet the design intent. The articles "Design for Quality" and "Robust Design" in Section 2 provide additional detail on the needs and methods used for process control. It is worth noting that, in almost every example, quality improvements also lead to cost reductions by reducing rejections, downstream rework, inventory requirements, warranty costs, and disappointed customers.

Section 7 provides detail on methods for optimizing the majority of manufacturing processes for several specific material classes. Probably the most challenging, as well as the most needed, are modeling methods for predicting what will happen on a microstructural basis during manufacturing operations such as heat treatment, forging, and casting. Only through an understanding of the time-temperature profile, and its relationship to non-isothermal cooling and/or solidification of a material, can the materials engineer predict final microstructures, including any transformation and/or thermally induced stresses.

Overview of the Design Process

John R. Dixon, University of Massachusetts (Professor Emeritus)

THE ROLE OF ENGINEERING DESIGN in a manufacturing firm is to transform relatively vague marketing goals into the specific information needed to manufacture a product or machine that will make the firm a profit. This information is in the form of drawings, computer-aided design (CAD) data, notes, instructions, and so forth.

Figure 1 shows that engineering design takes place approximately between marketing and manufacturing within the total product realization process of a firm. Engineering design, however, is not an isolated activity. It influences, and is influenced by, all the other parts of a manufacturing business.

In the past, the interrelatedness of design with other product realization functions was not sufficiently recognized. New design processes and methods involve the use of cross-functional teams and constant, effective two-way communications with all those who contribute to product realization in a firm.

A discussion of engineering design benefits from distinguishing between parts and assemblies. Though a few products consist of only one part—a straight wrench or paper clip, for example—most products are assemblies of parts. The process of designing assemblies is described in the article "Conceptual and Configuration Design of Products and Assemblies" in this Volume.

Distinguishing between special-purpose assemblies and standard components is also helpful. A standard component is an assembly that is manufactured in quantity for use in many other products. Examples are motors, switches, gear boxes, and so forth.

As assemblies are designed, a repeated (or recursive) process takes place in which the product is decomposed into subassemblies and finally into individual parts or standard components. (See the section "Engineering Conceptual Design" in this article.) Then to complete the design, the individual parts must be designed, manufactured, and assembled. The process of designing parts is described in the article "Conceptual and Configuration Design of Parts" in this Volume.

The design of a part involves selection of a material and a complementary manufacturing process. The majority of parts used in products today are either injection molded plastics, stamped ferrous metals, or die-cast nonferrous metals. Of course, many other material-process combinations are also in use. Some parts are made by a sequence of processes, such as casting followed by selective machining. Materials and process selection are described in the Sections "The Materials Selection Process" and "Manufacturing Aspects of Design" in this Volume.

The above paragraphs point out several important and unique requirements imposed on the engineering design process. An obvious one is that parts must be designed for manufacturing as well as for functionality, a requirement that has generated a body of knowledge called design for manufacturing (DFM). Another obvious requirement is that to obtain a final product, parts must be assembled. This has fostered the special field of design for assembly (DFA). Though it is not so obvious, a consideration overriding both DFM and DFA is that assemblies and parts should be designed in a way that results in the minimum total number of parts possible (Ref 1). A smaller part count almost always will result in lower total product cost when all costs are considered, including costs of materials, tooling, processing, assembly, inventory, overhead, and so forth.

Of course, engineering designers must design products that not only can be economically manufactured and assembled, but they also must function as intended. This requires selecting and understanding the physical principles by which the product will operate. Moreover, proper function requires special attention to tolerances. These two considerations are called designing for function and fit. However, designers must consider a myriad of other issues as well: installation, maintenance, service, environment, disposal, product life, reliability, safety, and others. The phrase design for X (DFX) refers to all these other issues (Ref 2).

Designing for DFM, DFA, minimum parts, function, fit, and DFX is still not all that is required of the engineering designer. Products also must be designed for marketing and profit, that is, for the customer and for the nature of the market-

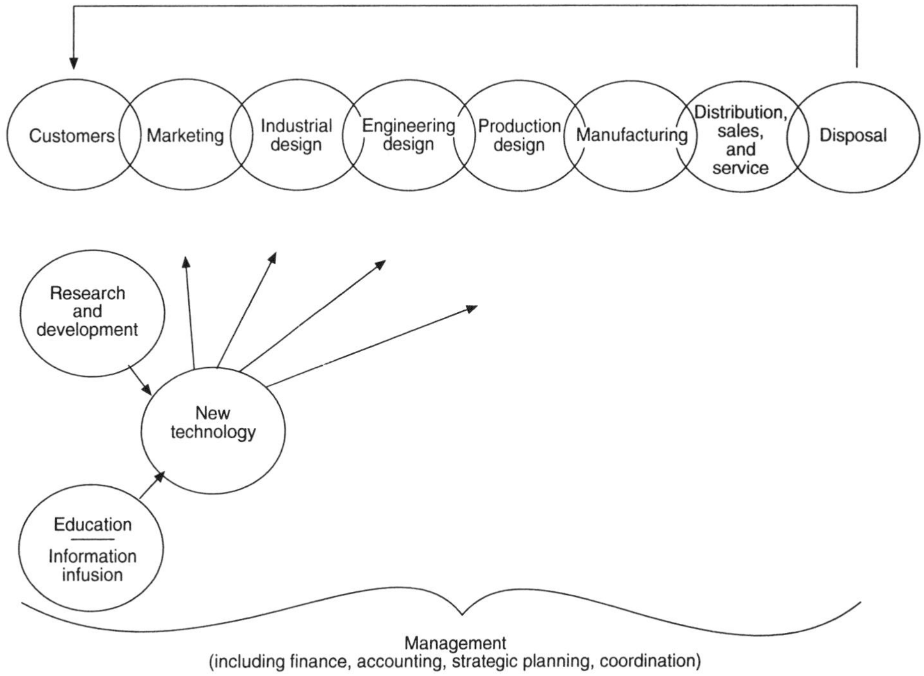

Fig. 1 Engineering design as a part of the product realization process

place. Designers, therefore, must be aware of what features customers want, and what customers consider to be quality in a product. In addition, marketing considerations must include cost, quality, and, increasingly important, time—that is, when the product will reach the marketplace.

Designers also should recognize that the processes by which parts and products are made, and the conditions under which they are used, are variable. Designing so that products are robust under these variabilities is another design requirement.

Designing a complex product or even a relatively simple one with all these requirements and considerations in mind is a tough and complex task. Therefore, finding creative, effective solutions to the many problems that are encountered throughout the process is essential to competitive success. Creative problem solving is especially important early in the design process when conceptual alternatives are generated, and choices are made that essentially fix the nature and character of the product. Creative problem solving in a design context is discussed in the article "Creative Concept Development" in this Volume.

A great deal of varied knowledge is needed to perform design competently and quickly. Thus design is usually a team effort involving people from marketing, several branches of engineering, and manufacturing. The formulation, organization, and operation of such design teams are discussed in the article "Cross-Functional Design Teams" in this Volume.

The remainder of this article presents an overview of the engineering design process. Though the process is extremely complex, distinct stages of design activities can be identified and described (Ref 3). The first stage is how marketing goals, often vague or subjective, are translated into quantitative, objective engineering requirements to guide the rest of the engineering design process.

From Marketing Goals to Engineering Requirements

The goal of this first stage in the engineering design process is to translate a marketing idea into specific engineering terms. Accomplishing this translation involves an understanding and communication among marketing people, industrial designers, engineering designers, and customers.

Industrial Design. The industrial design process creates the first broadly functional description of a product together with its essential visual conception. Artistic renderings of proposed new products are made, and almost always physical models are developed. Models at this stage are usually very rough, nonfunctional ones showing only external form, color, and texture, though some also may have a few moving parts.

Though practices vary, it is strongly advised that industrial design be a cooperative effort of the industrial designers and engineers, as well as materials, manufacturing, and marketing people. Industrial designers consider marketing, aesthetics, company image, and style when creating a proposed size and shape for a product. Engineering designers, on the other hand, are concerned with how to get all the required functional parts into the limited size and shape proposed. Another issue requiring cooperation may be choosing materials for those parts that consumers can see or handle. Both design engineers and manufacturing engineers, of course, are concerned with how the product is to be made within the required cost and time constraints.

The phrase *product marketing concept* describes fairly well the results of industrial design. The product marketing concept includes all information about the product essential to its marketing. On the other hand, the design at this stage should contain as little information as possible about engineering design and manufacturing in order to allow as much freedom as possible to the engineering design phases that follow. Such a policy is called *least commitment,* and it is a good policy at all stages of product realization. The idea is to allow as much freedom as possible for downstream decisions so that engineers are free to develop the best possible solutions unconstrained by unnecessary commitments made at previous stages.

A least commitment policy, for example, means that materials should not be specified this early in the design process unless the material choice has clear marketing implications. This often happens for those parts of the product that customers see and handle.

The Engineering Design Specification. The engineering design specification, also called the product design specification (PDS) (Ref 4), is described in detail in the article "Conceptual and Configuration Design of Products and Assemblies" in this Volume. Though different products require different kinds of information in their specification, essential categories are common to all. Regardless of how it is organized, an engineering specification in one way or another must contain information in two major categories:

- *In-use purposes* are related to the anticipated users and misusers (i.e., the customers) of the product including the primary intended purpose(s) to which users will put the product, any unintended purposes to which the product may be put (given that human beings behave the way they do), and any special features or secondary functions required or desired.
- *Functional requirements* are qualitative or quantitative goals and limits placed on product performance, the environmental and other conditions under which the product is to perform, physical attributes, process technologies, aesthetics, and business issues like time and cost.

Though the initial engineering design specification should be as complete and accurate as possible, it must also be recognized that a specification is never fully completed. Indeed, a specification is normally subjected to a certain amount of change throughout the design process. However, if changes cause significant redesign, they often can be very expensive and time consuming and affect the final product quality. Moreover maintaining the connection between engineering characteristics and customer requirements is crucial.

Engineering Stages

A design is information. As a product is designed, the information known and recorded about it increases and becomes more detailed. Though no formal theoretical foundation exists for identifying specific stages of design information content, some stages are intuitively obvious (Ref 3) and include:

- *Stage 1:* the product marketing concept
- *Stage 2:* the engineering (or physical) concept
- *Stage 3:* for parts, the configuration design
- *Stage 4:* the parametric design

The information contained in a product marketing concept is described in the section "From Marketing Goals to Engineering Requirements" in this article. The other stages are discussed in sections that follow.

Some references (e.g., Ref 5) expand the conceptual stage into two separate stages called conceptual and embodiment design and then include the configuration design of parts as a part of detail design.

Guided Iteration

For all of the stages of engineering design, that is, stages 2, 3, and 4 listed above, the problem solving methodology employed is called guided iteration (Ref 3). The steps in the guided iteration process, illustrated in Fig. 2, are formulation of the problem; generation of alternative solutions; evaluation of the alternatives; and if none is acceptable, redesign guided by the results of the evaluations. This methodology is fundamental to design processes. It is repeated hundreds or thousands of times during a product design. It is used again and again in recursive fashion for the conceptual stage to select materials and processes, to configure parts, and to assign numerical values to dimensions and tolerances (i.e., parametric design). See Fig. 3.

The action of a designer that adds to the information content of the design is a decision based on evaluation results. Some say, therefore, that design *is* decision making. This statement is true to some extent, but it does not illuminate how design decisions are made. They are made by guided iteration. That is, the additional information needed to advance the design is made explicit in a problem formulation. Alternative ways of providing that information are generated, and the alternatives are evaluated. Finally a decision is made about the acceptability of the alternatives. Thus decision making in design, repeated over and over again in all stages, is to accept, revise, or reject a proposed alternative.

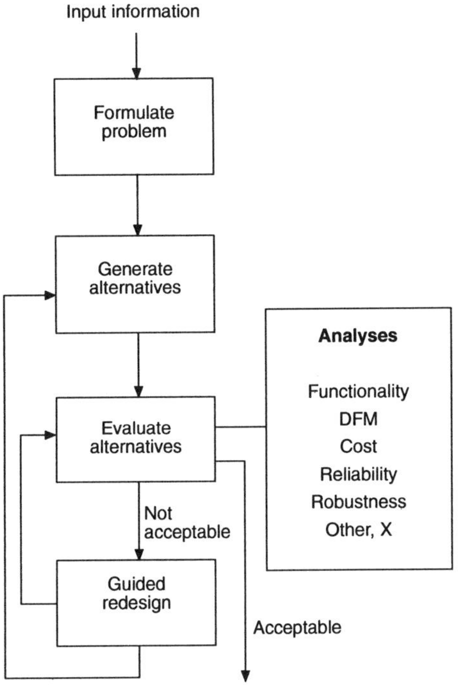

Fig. 2 Guided iteration methodology

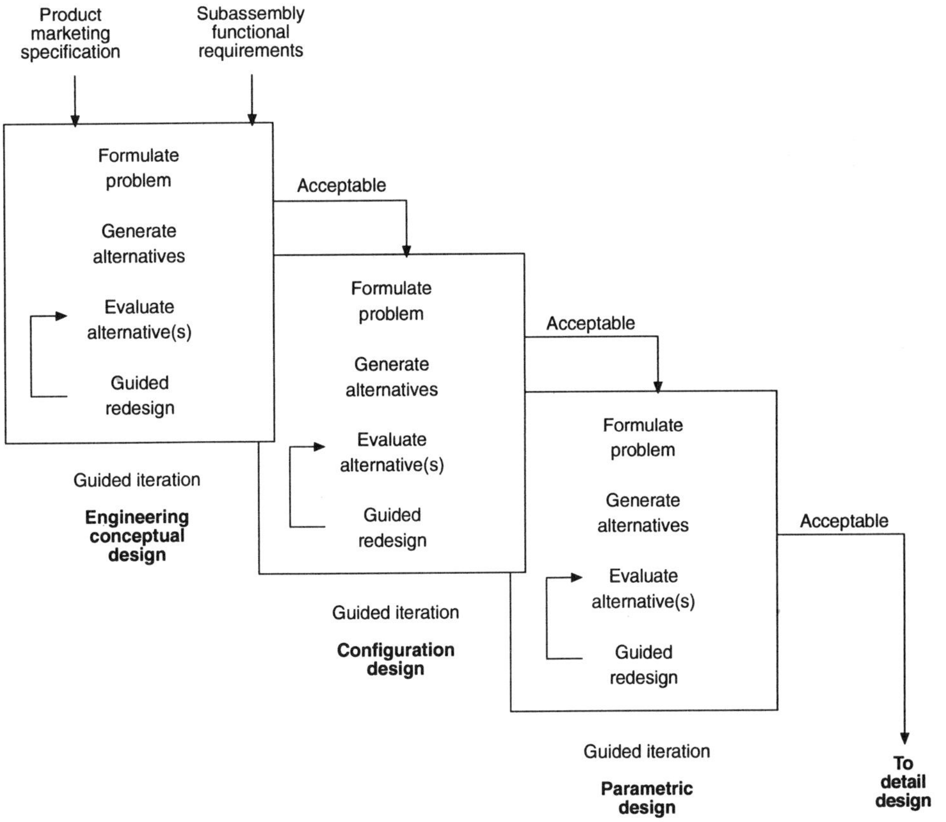

Fig. 3 Guided iteration used for conceptual, configuration, and parametric design

It is important to note how critical the generation of alternatives and their evaluation is to the decision about acceptability. If an alternative has not been considered, it cannot be evaluated and accepted. If the evaluations performed are incorrect, careless, or have failed to consider all the issues, then a poor decision may be made. All the steps in guided iteration must be well done every time they are done in order to obtain the best possible design result.

Engineering Conceptual Design

With an engineering design specification prepared to the extent that is feasible, the next stage of design is to determine the physical concept by which the product will function. (Hopefully, the concept has not been dictated by the specification in violation of the least commitment policy.) The physical concept includes the physical principles by which the product will work and an abstract physical embodiment that will employ the principles to accomplish the desired functionality.

As a very simple example of the meaning of these terms, suppose the required function is simply to support a load over an open space. One physical principle derived from beam theory is that longitudinal tensile and compressive stresses within a bending member can support a transverse load. The physical embodiment that uses this effect is a long, slender member of uniform cross section; here it is called a beam. Note in this example how the physical principle is integral to the embodiment. If only purely tension or compression stresses were used to support the load, an embodiment called a truss, which employs only tension and compression, might have resulted. In other words, though there is not usually a unique embodiment for implementing a physical concept, a concept and its embodiment are inextricably linked.

When a product is more complex, it consists of an assembly of subassemblies and parts. Then the physical concept is not so simple as in the above examples, and the embodiment must identify a set of principal functional subassemblies. For example, for an automobile, the subassemblies identified might be the engine, drivetrain, frame, body, suspension system, and steering system. The physical principles by which a product will work are specified by including sufficient information in its embodiment about how each of these functional subassemblies will interact with all the others to accomplish the required product functions.

The term *decomposition* is generally used to describe the part of the design process that identifies the subassemblies comprising a product or larger assembly. That is, in the conceptual design of an automobile, it could be decomposed into the engine, drivetrain, frame, and so forth.

Decomposition can be performed in two ways. (a) It can be done first purely in terms of functions. Physical embodiments are selected to fulfill the functions. (b) Alternatively, it can be done directly in terms of physical embodiments with the functions remaining more or less implicit. Most design is done as in (b). However, there are very good reasons for proceeding as in (a), that is, in function-first fashion (Ref 5, 6). In the automobile, for example, the function of the engine is to convert a source of on-board energy to rotational mechanical power. This function need not be provided by the usual internal combustion engine; instead it could be provided by an electric motor, a turbine powered by compressed gas, human-powered pedals, and many other alternatives. In the case of an automobile, the available alternative sources of power are very familiar. In a new, less-familiar product, however, the advantage of function-first decomposition is that it stimulates designers to consider many ways of fulfilling a given function instead of choosing the most common embodiment that comes to mind.

For an initial embodiment, it is usually sufficient to perform only one level of functional or physical decomposition, but all subassemblies thus created will ultimately, as a part of their own conceptual design, be decomposed again and again. For example, a lawn mower engine may be decomposed into, among other things, an engine block and a carburetor. Then in turn, the carburetor may be decomposed into, among other things, a float and a cover. Thus the process of conceptual decomposition repeats (or recurs) until no new subassemblies are created, that is, until only parts or standard components are obtained. See Fig. 4.

Generating Conceptual Design Alternatives. A large number of alternative physical concepts should be generated for evaluation in terms of the requirements because the selection of the best possible conceptual alternative is a crucial

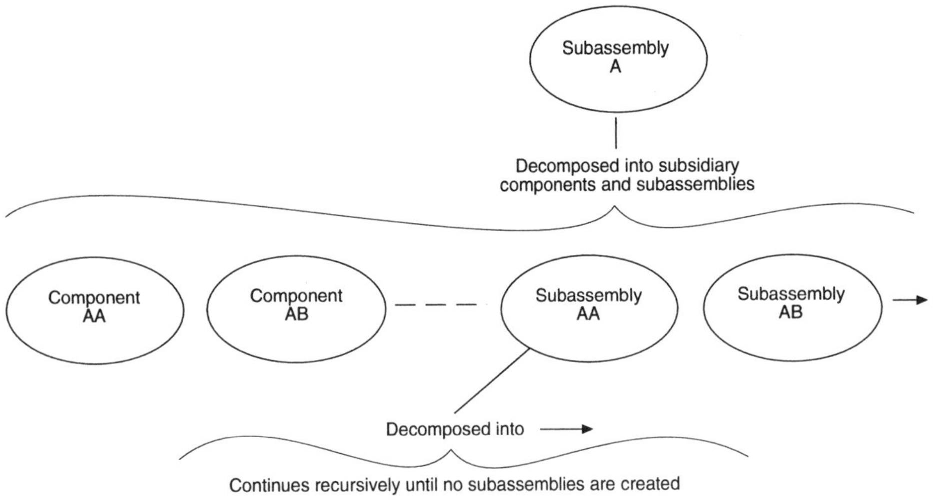

Fig. 4 Model of recursive decomposition

step in obtaining the best possible solution. Nonoptimal choices at this stage are extremely costly in time and money if they have to be corrected later. Unfortunately, there is a human tendency, strong in some designers and design organizations, to pass quickly through the engineering conceptual stage by considering only the one or two possible conceptual solutions that are most familiar to the people involved. This procedure very often ignores other possible solutions that may be superior; that is, ones that may be found by business competitors who are more thorough.

Evaluating Conceptual Design Alternatives. Evaluation of proposed conceptual designs is a crucial step. There is a significant difference between having a design and having the best competitive design (Ref 7). This distinction is often missed by people in marketing, management, and manufacturing.

Evaluation must be incisively and knowledgeably done, and all the issues must be considered as thoroughly as possible. Here again, the tendency in some firms is to perform only quick, subjective evaluations. Unfortunately, a common evaluation process used is, "I like this one best!" However, the design decision can be only as good as the evaluations performed, and good evaluation methods are available. See, for example, Ref 4.

Guided Redesign of Conceptual Alternatives. All of the methods available for comparison and evaluation of physical concepts indicate in general, qualitative terms which alternatives are best. In addition, and at least as important, the methods also illuminate the specific characteristics of proposed alternatives that are weak or strong. Thus evaluation directs the attention of designers to the changes or refinements that are needed to improve the alternatives. After such improvements are made, the alternatives can be reevaluated and then redesigned.

After evaluation and redesign, if none of the generated alternatives is acceptable, the search for new alternatives must be resumed. This search, too, can now be guided by the evaluation results. Thorough evaluation develops a great deal of useful information about the design. In particular, after evaluation, designers know why the alternatives generated so far are unacceptable, and thus they know why different principles, technologies, materials, or manufacturing processes are needed. Such knowledge is important in guiding the renewed search for concepts that will have a better chance of fulfilling the requirements of the engineering design specification.

It is important to appreciate that the engineering conceptual design process, from development of an engineering specification through generation of alternatives, evaluation of alternatives, to guided redesign, essentially must be repeated for each subassembly that is created as the product is decomposed and through as many levels of decomposition as needed to get to individual parts or standard components. Each subassembly has its own special functionality and engineering requirements, which are not the same as those of the product as a whole. For large products, the complexity that results from all these design processes inside design processes, and so forth can be astounding. Keeping track of all the interactions is a monumental task, especially as changes are made that may propagate throughout the design. Thus clear, written documentation is essential throughout the process, and this documentation is particularly critical for effective and efficient teamwork.

Design of Assemblies Compared to Design for Assembly. Discussion so far has been about design of assemblies. Design for assembly (DFA) involves mainly the design of parts so that they can be handled easily and inserted properly into place during the assembly process; these concepts are addressed in the article "Design for Manufacture and Assembly" in this Volume. Design for assembly does involve some design of assembly issues, such as the paramount issue of designing for the minimum number of parts. Also, if the assembly is to be done automatically, assemblies should be designed so that all parts are insertable from a single direction.

There are, however, issues in design of assemblies that have little to do with design for assembly. One of these is called stack-up, meaning the way tolerances can add up in an assembly. (See the article "Dimensional Management and Tolerance Analysis" in this Volume.) Designers obviously must be aware of such issues: establishing tolerances requires attention to both functionality and manufacturability.

The Configuration Design of Special-Purpose Parts

As described in the preceding section, the engineering conceptual design process decomposes a product into layers of nested subassemblies and ultimately into standard components and special-purpose parts. Often an enormous number of special parts have to be designed, manufactured, and assembled into a subassembly for final assembly into the product. This section contains a discussion of the first stage in the design of these parts: designing their configuration.

Is this Part Necessary? The starting place for designing a part is to try to eliminate the part. Readers are referred to Ref 8 by Boothroyd and Dewhurst for a relatively easy method for determining whether a proposed part is actually needed as a separate part. As pointed out above, one complex part is almost always less expensive overall than two or more simpler parts. However, this general rule may have exceptions and must be examined. The added complexity, for example, may delay production while the more complex tooling is being made (Ref 9).

If a part is necessary and a standard part can be used, then it is usually more economical to specify the standard part instead of designing and manufacturing a special-purpose part.

What is a Configuration? Ultimately, designers must determine exact numerical values for the dimensions and tolerances of parts, that is, perform parametric design. However, before this step can be done, parts are configured. A part configuration specifies the features of a part (see the bulleted list below) and their arrangement and connectivity, but the part configuration does not specify exact dimensions. The configuration can and should be evaluated as a configuration before its final dimensions and tolerances are established.

Features of parts include:

- *Walls* of various kinds (flat, curved, tapered, and so forth)
- *Add-ons* to walls, such as holes, bosses, notches, grooves, and ribs
- *Solid elements,* such as rods, cubes, tubes or spheres
- *Intersections* among the walls, add-ons, and solid elements

As with engineering conceptual design, designing a part configuration is done by guided iteration.

Formulating the Problem. Designing a part requires an engineering design specification for the part. The functions and other requirements for a part are not, in general, the same as those of the subassembly or product into which the part will be assembled. However, the engineering specification of a part will contain the same types of information as listed in the section "From Marketing Goals to Engineering Requirements" in this article.

Each part in a product is important to the whole, but each part also has a life (e.g., a functionality) of its own. Products and to some extent their subassemblies have a wide variety of unique functions, but there are only a relatively limited number of (mostly technical) functions for parts to perform. These include, for example, supporting forces, providing a barrier, providing a passage, providing a location, aiding manufacture, and adding strength or stiffness. Since reducing the number of parts is always an important goal, it is always helpful to combine as many such technical functions as possible into a single part.

The Configuration Requirements Sketch. Designing a part can be done only by sketching, whether on paper or in a CAD system. To begin, a designer must know the interactions that the part has with other parts and subassemblies. These interactions include forces (loads and available support areas), energy or material flows, and physical matings or other spatial requirements (e.g., certain spaces may be unavailable to the part). A sketch that shows these interactions to approximate scale is a very helpful starting place and is called a configuration requirement sketch.

Generating Alternative Configuration Solutions. There may be dozens or even hundreds of possible part configurations. Often too many exist to consider generating all the possible ones for evaluation. Thus the generation of alternatives must be limited by qualitative physical reasoning and by reasoning about manufacturability.

Qualitative essentially means reasoning without numbers though orders of magnitude of numbers are certainly involved. Thus qualitative reasoning fits configuration design evaluation well because configurations are themselves largely without numbers. Nevertheless, even without numbers, the basis of qualitative reasoning is still rooted in fundamental physical principles. Qualitative reasoning is far more objective and useful than guesses or feelings. It can be used to generate configurations that, once dimensions are added, will make efficient use of materials, avoid common failure modes, promote or restrict heat transfer, and so forth.

It should be remembered that, though designers are ultimately responsible for decisions made during design, others are available for input all along the way. This is one advantage of cross-functional teams, and experts and consultants from outside the team also can be called in.

Materials at the Configuration Stage. At this point in the part design process, it is necessary to decide upon a manufacturing process and at least a class of materials (e.g., aluminum, thermoplastic, steel). However, unless the information is needed for evaluation of the configurations, selection of the exact material (e.g., the particular aluminum alloy or thermoplastic) should be postponed consistent with least commitment until the parametric stage. Consultation with materials and manufacturing experts is, of course, strongly advised. It should also be remembered that some material choices have marketing implications as well. Other factors, such as recycling concerns and existing business relationships, also may be relevant.

Evaluating Design for Manufacturability at the Part Configuration Stage. In addition to qualitative physical reasoning about functionality, effective part configurations are strongly influenced by manufacturing issues. In stamping, injection molding, and die casting, for example, the part configuration is strongly related to die costs (Ref 3).

Design for manufacturability guidelines is determined by the physical nature of the manufacturing process involved. Descriptions of a number of manufacturing processes are presented in the Section "Manufacturing Aspects of Design" in this Volume. For assembly, also see especially Ref 1 and 8.

In considering DFM guidelines, designers should remember that reducing part count is an overriding concern. Thus complications that reduce part count are generally preferable to simplified designs with more parts. Of course, when part count is minimum, then making parts easy to manufacture is desirable.

Redesigning. The evaluations for functionality (including material use) and for DFM will guide the redesign of prospective configurations.

Tolerances at the Configuration Stage. Determining tolerances of part designs so that the parts will both function well and be manufacturable also has important implications at the configuration stage. Increasing the number and tightness of specified tolerances causes a corresponding increase in the cost and difficulty of manufacturing.

Methods for Parametric Design

Evaluations of concepts and configurations are based primarily on qualitative reasoning about physical principles and manufacturing processes. In parametric design, however, numerical computations become much more important. The attributes of parts identified at the configuration stage become the design variables for parametric design, and their values must now be determined. These values are mostly, though not exclusively, numerical. Relative processing costs (as distinguished from tooling costs) are sensitive to the exact values assigned so that relative processing costs must now be considered along with functionality as a part of parametric design (Ref 3).

Most parametric design methods can be applied to special-purpose parts and to standard parts and standard assemblies. A number of powerful methods are available for the parametric design of components and small assemblies, including guided iteration, optimization (see the article "Design Optimization" in this Volume), and statistical methods (see the articles "Statistical Aspects of Design" and "Robust Design" in this Volume).

Tolerances at the Parametric Stage of Design. At the configuration stage, the concern is with reducing the relative tightness and number of tolerances that must be assigned to obtain the required functionality. At the parametric stage, actual tolerance values are assigned. Not only do the values strongly influence functionality, they also have a strong influence on processing costs.

Why Methods for Parametric Design Are Needed. At the parametric design stage, the tendency in practice is to avoid the use of formal methods. Experienced designers tend to rely on experience and what has worked previously. If experience has resulted in general knowledge and understanding that can be applied in new situations and it works, then that is fine. However, the slow adaptation of Taguchi's approach (Ref 10, 11) (or some method) of robust design by United States industry was an important factor enabling foreign competitors to design and produce reliable products that captured a number of important markets (see the article "Robust Design" in this Volume). The lesson is that experience that merely works does not necessarily work well enough to beat competitors who are continually learning new and better methods.

Of course, some dimensions are determined exactly by manufacturing considerations, space or weight concerns, and other such limits.

Guided Iteration for Parametric Design of Components. As with engineering conceptual and configuration design, parametric design problems for components can be solved by the general method of guided iteration. The specific methods used to implement these steps in parametric design are different from the methods used in conceptual and configuration design.

Problem formulation in parametric design requires identification of the design variables, including their range of allowed values; identification of the performance parameters whose values will be computed or measured to evaluate the performance of trial designs; the performance criteria to be included in the evaluation; and the analysis methods that will be used to compute values for the evaluation parameters.

Generation of alternatives in parametric design requires selecting an initial design, that is, an initial set of values for the design variables. Consideration of DFM issues may guide or limit these values.

Evaluation in parametric design requires computation of the values for the performance parameters as well as selection and implementation of a method for evaluating the overall quality of the trial design. Since multiple evaluation criteria usually exist, this step requires considering how to obtain an overall evaluation from the separate evaluations of each criterion.

Redesign in parametric requires that new values be selected for the design variables so that a new trial design can be evaluated. Obtaining the new values is guided by the evaluations of the preceding trial or trials.

Reference 3 discusses this subject in detail for interested readers.

Optimization Methods for Parametric Design. Optimization is a well-developed field of study that is the subject of entire courses and books. Excellent texts and reference books on optimization are available for use by designers; an example is Ref 12. The more technically advanced manufacturing firms will likely have optimization experts with which designers and design teams can consult. Computer programs are also available, for example, the optimization software programs Optdes and OptdesX (trademarks of Design Synthesis, Inc., East Provo, UT).

Though there are exceptions, in general, optimization methods are useful when the following conditions are met. The design variables are all numeric and continuous. In this case, optimization methods are likely to be effective and efficient. If not, optimization can still possibly be used, but some adaptations will be required. A single function (called the "criterion function" or "objective function") can be written in terms of the design variables that express the overall quality or goodness of a trial design. Often this single function is cost, though in some cases it can be weight, efficiency, robustness, or some other performance factor.

Suboptimization. In complex, realistic parametric design problems, an appropriate criterion function often cannot be readily formulated to meet the conditions required by optimization techniques. Nevertheless, sometimes certain subparts of problems can be solved by optimization. This is called suboptimization.

Suboptimization can be effective and helpful with a stipulation. One cannot in general optimize a whole problem solution by dividing it up into subproblems, each of which is suboptimized separately. Suboptimization of all the subparts of a system does not in general lead to optimization of the whole system. The degree to which the subsystems are coupled is the degree to which suboptimization is suboptimal. Still, suboptimization can be advantageous in situations where any adverse effect from a suboptimized section on the whole system is negligible or acceptable.

Statistically Based and Taguchi Approaches for Parametric Design. Methods from the field of statistics and design of experiments (Ref 13, 14) also can be used to assist performing parametric design in some cases. Only the so-called Taguchi approach (Ref 10, 11) is introduced here because it is fairly easily applied and because its use is now fairly common. Moreover, it has a good record of successful application.

Robustness. The overall evaluation criterion in Taguchi's techniques is called robustness. Robustness refers to how consistently a component or product performs under variable conditions in its environment and as it wears during its lifetime. The variable conditions under which a product must function may include, for example, a range of temperatures, humidity, or input conditions (e.g., voltages, flow rates). Robustness also refers to the degree that the performance of a product is immune to normal variations in manufacture, that is, to variations in materials and processing.

Noise Factors. The terms noise or noise factors are commonly used for the uncontrollable variable conditions of environment, wear, and manufacture. Thus another way to describe robustness is to say that it is the degree to which the performance of a product is insensitive to noise factors.

Control Factors. Noise factors, which the designer cannot control, are not to be confused with the design variables, whose values the designer can control. Design variables are called control factors in the Taguchi approaches. Though designers have no control over the noise factors, the ranges over which noise factors vary are usually reasonably predictable.

Strategies to Achieve Robustness. To achieve robustness in the face of the environmental and other noise factors, two different strategies may be followed. One strategy is to design the product so that the performance of sensitive parts is insulated from the noise conditions (e.g., provide thermal or vibration insulation). Alternatively, steps might be taken to remove or reduce the source of a noise (e.g., eliminate the cause of temperature variations or the source of vibrations). Both insulating the part or product from the noise and eliminating the source of the noise are called "reduce the noise" strategies.

A second design strategy is to accept the noise but reduce its consequences. In this approach, the product is designed so that its lifetime performance is as insensitive to the noises as possible. For example, instead of thermally insulating the part or parts whose performance is sensitive to temperature, those parts can be designed so their performance is not significantly impaired by the expected temperature variations.

Often, of course, both "reduce the noise" and "reduce the consequences" strategies may be used simultaneously, but reducing the noise is usually a considerably more expensive solution.

Taguchi techniques have a built-in trade-off methodology for selecting the set of control factors that results in the best combination of performance and robustness given the conditions of noise in manufacturing and use. The methodology maximizes the signal-to-noise ratio. Though it is a reasonable criterion, designers using the Taguchi methods have no control over it. Nevertheless, the Taguchi techniques have a very good track record for producing excellent overall results.

A disadvantage of the Taguchi method is that only a few values of the design variables over a limited range can be considered. Another disadvantage is that in many cases, experimentation is required to obtain the performance results. When the cost and time required for experimentation are large, the disadvantage is obvious. Where analysis and/or simulation can be used instead of experimentation, they usually will be both quicker and less expensive. Moreover, the use of statistical methods and proper design of experiments can, in most cases, make experimentation more efficient, and these methods can be applied to analytical models and numerical simulations as well as to hardware.

Using the Taguchi approach is not the only statistical approach to achieving robust designs that also perform well (Ref 13, 14). Robustness (that is, variability of performance) often can be included when using guided iteration and optimization for parametric design.

Additional information about the Taguchi methods is provided in the article "Robust Design" in this Volume.

Best Practices of Product Realization

This section very briefly describes a number of the practices, called best practices, used by successful firms to achieve the goals of quality, cost, time-to-market, and marketing flexibility (Ref 15).

Traditionally, cost was considered of paramount importance, and no one would argue that cost is unimportant. However, in the 1980s, quality, defined very broadly, became equally or possibly more important. Unfortunately, in many firms, quality was rather narrowly viewed as exclusively related to manufacturing or production instead of to design and the rest of the product realization process. This limited view results in a serious error because it ignores the fact that many factors determining quality are largely decided in the design stages, especially the early design stages. For a good discussion of quality issues, see Ref 16.

What Business Are You Really In? In some (especially older) manufacturing firms, a strong cultural bent exists toward the belief that the business of the firm is to manufacture things, specifically the things that constitute their current product or product line. However, the business of such a firm is not really a particular product or product line. Rather, their business is the service that these products perform for customers. For example, the business of a company manufacturing pencils is not making pencils; it is providing the service that pencils provide their users. That is, the business of such a firm is to provide the service that enables people to record their thoughts and other information onto hard copy.

A firm manufacturing pencils may never decide to manufacture ballpoint pens, word processors, or speech recognition systems, but at least viewing their business as a service provider reveals who their competitors are (not just other pencil manufacturers) and who they may be in the future. It also gives the firm an incentive for inventing the next popular thought-recording product, even if it is only a better pencil. Thus manufacturing firms should determine and become conscious of what service their products perform for customers, and what the customer values about that service and the way it is provided.

Sources of New Ideas. There are four primary sources of ideas for new or revised products in firms: customers, employees, benchmarking, and new technology.

Competitive manufacturing businesses require constant feedback from the customers who buy, sell, repair, or use the products of the company. If a design engineer is looking for positive new ideas as well as for shortcomings of current products, then he or she must get out personally and talk to the customers throughout the design process, and after. Design engineer communication with customers through field trials, field observations, focus groups, and interviews is important to excellent design results.

Employees in the factory, shops, and offices are also an extremely valuable source of new ideas for products and product improvements. Good practice requires that there must be a believable, financially rewarding, well understood, and low threshold (easy to use) mechanism for employees to get their new product, product improvement, and process improvement ideas heard and seriously considered.

Benchmarking (Ref 17, 18) also stimulates engineers and others in a firm to see and discover new product ideas and new ways of viewing both design and manufacturing. Benchmarking can emcompass studies of both competitor and non-competitor products and processes.

Keeping abreast of new technologies and methodologies in materials, manufacturing, design, engineering, and management is another important source of ideas for new and improved products. Coupling new technological information with the search for new or improved product ideas is an essential part of the product development process that is not, strictly speaking, engineering design defined here, but it is important if engineering designers are to produce the best possible products for a company.

Cross-Functional Teams. It would be difficult to overstate the importance of cross-functional teams and teamwork to the implementation of effective, modern design practices and methods. The close cooperation of different disciplines is especially important to realizing the benefits of DFA and DFM and to ensuring that designs are consistent with marketing and business considerations. In this Volume, the topic of cross-functional teams is discussed in detail in the article "Cross-Functional Design Teams." It is also covered in Ref 19.

Focus on Quality. The most competitive companies recognize that quality is crucial to competitiveness and that quality cannot be built into or inspected into a product unless it is first designed into a product. Time-to-market is also recognized as a critical factor in profitability, and development times can be significantly shortened through appropriate management and engineering design approaches (e.g., concurrent design and design for manufacturing). See the article "Concurrent Engineering" in this Volume. Finally, competitive firms know that quality, time-to-market, and cost are all interrelated. None should be sacrificed for the other.

The most competitive firms tend to have established metrics (i.e., measurements) that indicate their performance regarding quality, cost, and time-to-market. One way to help establish such metrics is through competitive benchmarking. See Ref 17 and 18.

Competitive benchmarking involves a detailed look at the products and processes (both design and manufacturing processes) of the very best competitors of a company. Competing products can be purchased, taken apart, and analyzed for cost, performance, and manufacturability. Out of this process, metrics can be established for the products and processes of a company, and performance can be measured against these metrics.

Concurrent Engineering, Design for X (DFX), and Design for Manufacturing (DFM). Concurrent design attempts to organize the product realization process so as to have as much information and knowledge available about all the issues in the life of a product at all stages of the design process. This is also referred to as design for X, where X stands for the customer, robustness, manufacturing (including tooling, assembly, processing), environment, safety, reliability, inspectability, maintenance and service, shipping, disposability, and all the other issues in the life cycle of the designed object and its production (Ref 2).

Design for the Customer. Quality function deployment (QFD) (Ref 20) is a method for deriving the desired engineering characteristics of a product from customer input or of transforming customer inputs into engineering requirements. A technique for implementing QFD, called the house of quality (Ref 20), is generally used to perform the product or design evaluation and to guide the redesign for improved customer satisfaction.

DFA and DFM. The importance of DFA and DFM to product realization has already been indicated. However, lip service to DFA and DFM is not sufficient. There is much to know about both of them, and the firm has to acquire and apply that knowledge to their processes.

Design for Robustness. As with DFA and DFM, design for robustness requires more than lip service. The knowledge of what it is and how to do it actually must be brought into a firm and used if its benefits on product quality are to be realized.

Physical Prototyping Policies. Reducing the number of planned prototypes (e.g., from three to two) will save a great deal of time (Ref 21) because design engineers, who know ideas will get tested in prototypes, are prone to take risks in their initial designs. But the product realization process is not the time to take risks. Risky ideas should be developed and tested in the laboratory before they are incorporated into product development programs.

Strategic Use of Computational Prototyping and Simulations. Modern computational methods employing computers make it possible to reduce or even eliminate more expensive and more time-consuming physical prototyping. Computer-aided design, solid modeling, finite element methods, and many kinds of simulation programs are used by best practice firms to improve quality and reduce design and development time. See the articles "Computer-Aided Design" and "Rapid Prototyping" in this Volume.

Exacting Control of Processes. The previous idea of quality assurance was to inspect parts and assemblies after they had been produced. The new best practice is to control processes so rigorously that inspection is unnecessary. Methods of statistical process control (SPC) have been developed for this purpose and are in widespread use (Ref 22).

Intimate Involvement of Vendors. Dozens, hundreds, or even thousands of vendors may be involved in the manufacture of certain products and machines. Previous practice was to prepare specifications that vendors must meet with their products and that were used to obtain competitive bids from a number of competing vendors. The present practice is to employ only one or two vendors and to involve them in the product design, especially as it relates to the parts and subassemblies to be supplied by the vendor.

REFERENCES

1. G. Boothroyd, *Assembly Automation and Product Design,* Marcel Dekker, 1992
2. D.A. Gatenby, Design for "X"(DFX) and CAD/CAE, *Proceedings of the 3rd International Conference on Design for Manufacturability and Assembly,* 6–8 June 1988, (Newport, RI)
3. J.R. Dixon and C. Poli, *Engineering Design and Design for Manufacturing,* Field Stone Publishers, 1995
4. S. Pugh, *Total Design: Integrating Methods for Successful Product Engineering,* Addison-Wesley, 1991
5. G. Pahl and W. Beitz, *Engineering Design,* K. Wallace, Ed., The Design Council, 1984
6. E. Crossley, A Shorthand Route to Design Creativity, *Mach. Des.,* April 10, 1980
7. C.W. Allen, personal communication, 1993
8. G. Boothroyd and P. Dewhurst, *Product Design for Assembly,* Boothroyd Dewhurst, Inc., 1989
9. K.T. Ulrich et al., "Including the Value of Time in Design for Manufacturing," MIT Sloan School of Management Working Paper No. 3243-91-MSA, Dec, 1991
10. G. Taguchi, *The Development of Quality Engineering,* The American Supplier Institute, Vol 1 (No. 1), Fall, 1988
11. G. Taguchi and D. Clausing, Robust Quality, *Harvard Business Review,* Jan-Feb, 1990
12. P.Y. Papalambros and D.J. Wilde, *Principles of Optimal Design,* Cambridge University Press, John Wiley & Sons, 1989
13. G.E.P. Box, S. Bisgaard, and C. Fung, An Explanation and Critique of Taguchi's Contributions to Quality Engineering, *Qual. Reliab. Int.,* Vol 4, 1988, p 121–131
14. G. Box and S. Bisgaard, Statistical Tools for Improving Designs, *Mech. Eng.,* Jan, 1988
15. National Research Council, *Improving Engineering Design: Designing for Competitive Advantage,* National Academy Press, 1991

16. D.A. Garvin, Competing on the Eight Dimensions of Quality, *Harvard Business Review,* Nov/Dec 1987, p 101–109
17. R.C. Camp, *Benchmarking,* ASCQ Quality Press, 1989
18. F.G. Tucker, How to Measure Yourself Against the Best, *Harvard Business Review,* Jan/Feb 1987, p 8-10
19. P.G. Smith and D.G. Reinertsen, *Developing Products in Half the Time,* Van Nostrand Reinhold, 1991
20. D.R. Hauser and D. Clausing, The House of Quality, *Harvard Business Review,* May–June, 1988
21. M.B. Wall, K. Ulrich, and W.C. Flowers, Making Sense of Prototyping Technologies for Product Design, *Proceedings, Design Theory and Methodology Conference,* DE Vol 31, ASME, April, 1991
22. R. Galezian, *Process Control: Statistical Principles and Tools,* Quality Alert Institute, 1991

Conceptual and Configuration Design of Products and Assemblies

Kevin N. Otto, Massachusetts Institute of Technology
Kristin L. Wood, The University of Texas

COMPETITIVE DESIGN of new products is the key capability that companies must master to remain in business. It requires more than good engineering, it is fraught with risks and opportunities, and it requires effective judgment about technology, the market, and time. Several recent business decisions give insight to these claims:

- To avoid losing market share, all U.S. commercial airplane manufacturers have offered contracts to deliver aircraft at prices that are below current cost (Ref 1). The companies are betting that they can remain profitable through improvement of their products and processes.
- In the early 1980s, Sony offered an improved magnetic videotape recording technology, the Betamax system. Although it offered better magnetic media performance, it did not satisfy customers, who rather were more concerned with low cost, large selection of entertainment, and standardization.
- In 1996, both Ford and Toyota launched new family sedans. Three years earlier, each had torn apart and thoroughly analyzed each other's cars. Ford decided to increase the options in its Taurus, matching Toyota's earlier Camry, while Toyota decided to decrease the options in its Camry, matching Ford's earlier Taurus.

There is clearly a need to apply statistically sound methods to evaluating the intended customer population for a product. It is equally important to design into the product what is required to meet customer demands, applying rigorous methods for incorporating the best technologies.

This article describes an integrated set of structured methods (Fig. 1) that were developed to address these needs. The methods start with identifying the customer population for the product and developing a representation of the feature demands of this group. Based on this representation, a functional architecture is established for the new product, defining what it must do. The next step is to identify competitive products and analyze how they perform as they do. This competitive benchmarking is then used to create a customer-driven specification for the product, through a process known as quality function deployment. From this specification, different technologies and components can be systematically explored and selected through functional models. With a preliminary concept selected, the functional model can be refined into a physically based parametric model that can be optimized to establish geometric and physical targets. This model may then be detailed and established as the alpha prototype of a new product.

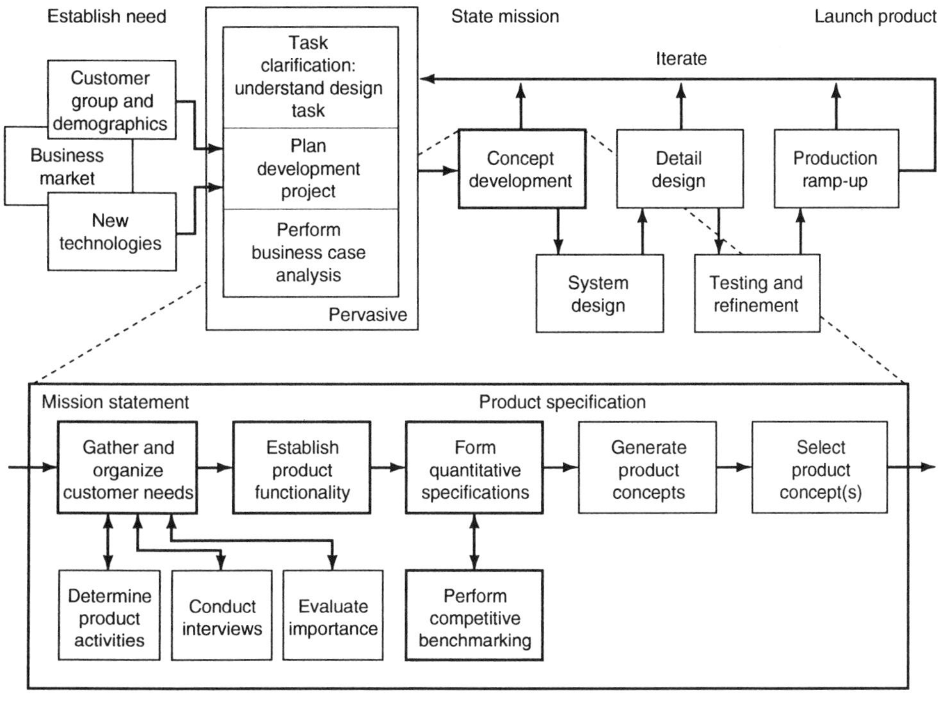

Fig. 1 The concept and configuration development process. "Pervasive" activities occur throughout product development.

Task Clarification

Conceptual and configuration design of products, as depicted in Fig. 1, begins and ends with customers, emphasizing quality processes and artifacts throughout. Intertwined with the focus on customers and quality are a number of technical and business concerns. We thus initiate the conceptual design process with task clarification: understanding the design task and mission, questioning the design efforts and organization, and investigating the business and technological market. Task clarification sets the foundation for solving a design task, where the foundation is continually revisited to find weak points and to seek structural integrity of a design team approach. It occurs not only at the beginning of the process, but throughout.

Table 1 Summary of the Harvard business case method

Process step	Description
1. Problem statement	What market problem are you addressing, fixing, improving, making more efficient, etc.? This should be limited to one sentence, two at the most. Only one problem can be addressed. If the problem is complex, with many interrelated subproblems, the problem should be clarified and refined to the basic (atomic) or kernel problems.
2. Assumptions	Discuss any limiting assumptions made in preparing the business case proposal, such as product costs, direction of the industry/department, etc. This step provides a clear statement of the scope of work.
3. Major factors	List, briefly, major factors of the environment that affect the decision. This may be the state of the business (capital constraint), critical business needs or directions (strategies), etc.
4. Minor factors	List, briefly, factors that should be considered, but that do not seem to have a significant effect on the problem
5. Alternatives	List concrete or hypothesized alternatives (minimum of three) to address the problem or opportunity defined by the problem statement, assumptions, and major factors. Two or three sentences should be sufficient. Under each alternative list the advantages and disadvantages of each.
6. Discussion of alternatives	Review each of the alternatives with respect to the stated problem, assumptions, major and minor factors. Compare alternatives and discuss the relative merits of each (in terms of cost savings/avoidance, cycle time reduction, increase in quality, and head count reduction). From this discussion, a clear leader among the alternatives (i.e., the most feasible alternative) should be identified.
7. Recommendation	State your recommendation. There should be no need to defend it; this should have been covered in the last section. If needed, elaborate on the recommendation to add clarity.
8. Implementation	Describe the implementation plan. Include resource requirements: financial, human, space, etc. Describe the time frame requirements, controls, and measurements that will be needed to ensure that goals are met. Measurements should be tied directly to solving the problem, and adequate tracking mechanisms should be used to quantify the success of the project. Contingency plans should be developed that address any high-risk aspects of the solution.

Mission Statement and Technical Questioning

A mission statement and technical clarification of the task are important first steps in the conceptual design process. They are intended to:

- Focus design efforts
- Define goals
- Define timelines for task completion
- Provide guidelines for the design process, to prevent conflicts within the design team and concurrent engineering organization

The first step in task clarification is usually to gather additional information. The following questions need to be answered, not once but continually through the life cycle of the design process (Ref 2):

- What is the problem really about?
- What implicit expectations and desires are involved?
- Are the stated customer needs, functional requirements, and constraints truly appropriate?
- What avenues are open for creative design?
- What avenues are limited or not open for creative design? Are there limitations on scope?
- What characteristics/properties must the product have?
- What characteristics/properties must the product not have?
- What aspects of the design task can and should be quantified?
- Do any biases exist in the chosen task statement or terminology? Has the design task been posed at the appropriate level of abstraction?
- What are the technical and technological conflicts inherent in the design task?

It is surprising how often a great deal of time (and money) are wasted because no one took time at the front end of a project to really understand the problem. To obtain this understanding, the design of any product or service must begin with a complete understanding of the customers' needs, as discussed in the section "Understanding and Satisfying the Customer" in this article.

The tangible result of technical questioning is a clear statement of the design team's mission. Following is a sample template for a mission statement (Ref 3):

Product description	One concise and focused sentence
Key business or humanitarian goals	Schedule Gross margin/profit or break-even point Market share Advancement of human needs
Primary market	Brief phrase of market sector/group
Secondary market	List of secondary markets, currently or perceived
Assumptions	Key assumptions or uncontrolled factors, to be confirmed by customer(s)
Stakeholders	One- to five-word statements of customer sets
Avenues for creative design	Identify key areas for innovation
Scope limitations	List of limitations that will reign back the design team from "solving the world"

The mission statement should not be used as a mere statement of "parenthood." Instead it should be used as a "passport," "calling card," and "banner," stating the design team's intentions. When interviewing customers, meeting with potential suppliers, or carrying out design reviews, members of the design team should make the mission statement the lead item of discussion.

Business Case Analysis: Understanding the Financial Market

Technical questioning is only one side of the proverbial design coin. Understanding the business market represents the other side, especially to complete the mission statement. During any conceptual and configuration design effort, a product's market must be clarified through the development of a business case analysis. A number of financial assessment techniques exist at varying levels of detail. Two notable generic techniques are the "Economics of Product Development Projects" (Ref 3) and the Harvard business case method (Ref 4–6). This section explains how the Harvard business case method can be used to understand the potential impact of product development. A summary is shown in Table 1. Application of the methodology is described below for a simple mechanical product: a fingernail clipper.

Fingernail clippers are widely used by several markets: consumers (the primary market), professional salons, and domestic pet manicurists. Assume that a company seeks to improve its current product offering for the consumer market, in order to increase its market share in a complementary product line, fingernail polish. The mission is to design a fingernail clipper for comfortable use by either the left or right hand. It is assumed that comfort, low cost, reliability (consistently remove nails with a simple finger force throughout the product's life), and storage compactness are the driving market needs (to be confirmed or revised through customer interviews). The corporation also seeks only a 30% gross margin, since the goal is to increase the market share of fingernail polish.

A number of solutions exist for addressing both the technical and process issues associated with a fingernail clipper product development. A business case may be derived for each of the possible solutions. However, the intent during the early stages of conceptual and configuration design is not to study all possible alternatives in detail, but rather to determine whether a minimal benefit to the business will be realized by improving the clipper (i.e., comfort, cost, reliability, and compactness). As such, this example concentrates on steps 5 and 6 of the Harvard business case method (Table 1), where only one generic alternative is considered. A device solution (i.e., a new, generic, hypothetical clipper design) is the alternative considered, emphasizing the possibilities of reduced cost and higher reliability through compactness and fewer components.

These possible benefits call for a "break-even" financial analysis for the clipper problem. This analysis answers the question: "Is the hypothetical clipper with less material (compactness) and fewer components feasible as a business venture?" It begins with a summary of the current costs for fingernail clipper development (Table 2), as projected from the current product. Because these costs continually change with new technology and market forces, the actual costs have been multiplied by a random factor. The important issue is the cost of the current clipper operation relative to the cost of the hypothesized solution. The cost projections are based on 750,000 clippers, with a product distribution of 80% small clippers, for fingernails, and 20% large clippers, for toenails. The average cost for

Table 2 Current costs scenario (fingernail clipper example)

Category	Projected cost, $	Cost per product, $/clipper
Labor costs		
Small clipper		
Assembly	60,000	0.10
Handling	36,000	0.06
Large clipper		
Assembly	16,500	0.11
Handling	10,500	0.07
Total	**123,000**	**0.17**
Fabrication costs		
Small clipper		
Materials	96,000	0.16
Piece-parts	72,000	0.12
Tooling	6,000	0.01
Large clipper		
Materials	30,000	0.20
Piece-parts	21,000	0.14
Tooling	4,500	0.03
Total	**229,500**	**0.31**
Subtotal (ongoing costs)	**352,500**	...
Engineering costs		
Total	**173,600**	**0.23**
Total costs	**526,100**	**0.71**

Table 3 Proposed costs scenario (fingernail clipper example)

Category	Projected cost, $	Cost per product, $/clipper
Labor costs		
Small clipper		
Assembly	30,000	0.05
Handling	30,000	0.05
Large clipper		
Assembly	9,000	0.06
Handling	7,500	0.05
Total	**76,500**	...
Fabrication costs		
Small clipper		
Materials	66,000	0.11
Piece-parts	24,000	0.04
Tooling	24,000	0.04
Large clipper		
Materials	21,000	0.14
Piece-parts	9,000	0.06
Tooling	10,500	0.07
Total	**154,500**	...
Subtotal (ongoing costs)	**231,000**	...
Engineering costs		
Total	**187,000**	**0.25**
Total costs	**418,000**	...

Table 4 Break-even cost analysis (fingernail clipper example)

Issue	Analysis result
Estimated payback period for development costs	6 months
Projected savings for first 100,000 products	$16,200
Projected cost savings for next 650,000 products	$105,300
Expected average cycle time savings for each 100,000 product lot	38% of current work days

this distribution is $0.31 per product for fabrication, $0.17 for labor, and $0.23 for engineering time.

For the purpose of comparison, the adopted concept for this analysis is a "generic," hypothesized clipper with reduced parts. It is assumed that suitable component and fabrication technology exists for this concept. Such a device would require fewer materials and components and less assembly and labor, but tool costs would potentially increase due to higher precision in the cutter alignment. Based on this new concept, Table 3 lists the expected costs for 750,000 products (same distribution of small and large clippers), again multiplied by a random factor). One-time development (engineering) costs account for $187,000 (increase in tooling design), and projected ongoing fabrication and engineering costs account for $231,000 ($154,500 fabrication plus $76,500 labor), compared with current product ongoing costs of $352,500.

It is then necessary to compare the current and proposed costs, to determine the payback period and cost savings. Table 4 shows the results of this break-even analysis. The payback period is 6 months, with a potential savings of $121,500 for 750,000 products. These results are extremely encouraging. Significant cycle time and cost savings may be achieved for the business if a suitable fingernail clipper concept can be developed. Because of these potential savings, the project should be carried to the next stage: conceptual design and prototype.

Implications. While only a subset of the Harvard business case method is illustrated above, the potential impact is impressive. A "go/no-go" decision may be made early in the product development process, provided that financial information exists for the current market and projected costs can be readily assumed for hypothesized concepts. Such decisions should be made in parallel with technical and industrial design clarifications. Also, they should continually be reviewed and updated as new information becomes available, especially as concrete product configurations are derived.

A critical aspect of the Harvard business case method is that cost data are needed in order to predict a product's potential return on investment of a product. If an entirely new product or family of products is under development, cost data may not exist. The Harvard business case method can still be applied if data can be obtained from a similar or analogous product, or from very rough estimates of preliminary product layouts.

Understanding and Satisfying the Customer

Having clarified the business opportunity, a firm should determine whether there is actual demand for a new or revised product. Too many technology development initiatives are undertaken with no basis for market acceptance other than management belief. The "technologist's problem" (the belief that if the developer thinks the technology is valuable, everyone else should also) is unfortunately very common in the engineering community. Akia Morita, founder of Sony Corporation, once boasted that, "Our plan is to lead the public to new products rather than ask them what they want. The public does not know what is possible, we do" (Ref 7). One result is products such as the Betamax. While the technology-push approach can and does sometimes work, it is also clear that considering the customer's desires will pull product development into better directions and amplify success.

It is important to recognize that "the customer" is only a statistical concept. There are countless potential buyers. Several tasks, discussed below, must be completed in order to develop a statistically valid customer needs list.

Gathering Customer Need Data

Reference 8 is an excellent management science reference on customer requirements. Reference 9 provides a total quality management perspective.

Some of the techniques for constructing a list of customer needs include using the product, circulating questionnaires, holding focus group discussions, and conducting interviews.

Using the Product. The design teams goes to the locations where their or their competitor's product is used, and they use the product as the customer would. If customer tasks can be easily understood and undertaken by the design team, and the design team is small, then this approach is effective. It is costly, though, for projects with either large design teams or highly skilled customer tasks that require training. Further, it does not address the need to document customer needs.

Questionnaires. The design team makes a list of criteria that they consider relevant to customers' concerns, and customers rank the product on these criteria. Alternatively, the design team forms a list of questions for customers to answer. The problem is that what the design team considers most important is not necessarily what customers consider most important.

Focus Groups. In a focus group discussion, a moderator facilitates a session with a group of customers who examine, use, and discuss the product. Usually this is done in the design team's environment, typically in a room with a two-way mirror so that the design team can observe the customers during the session. This session can be video- or audiotaped for later examination.

Interviews. Sometimes a design team asks an interviewer to discuss the product with a single customer, typically in the environment where the product is used by the customer. Again, the interview can be video- or audiotaped for later examination.

Griffin and Hauser (Ref 10) found that interviews are the most effective technique for uncovering information per amount of effort. They also report that for consumer product design projects,

Clipper project				
Customer data				
Customer:		Interviewer:	KNO	
Customer ID:	KNO5	Date:	9/3/95	
Willing to follow up?	No	Location:	Cambridge, MA	
Type of user:	Middle claass, white male, traveling			
Question/prompt	Customer statement	**Interpreted need**	**Weight**	**Activity**
When usually use?	In the evening in hotel			
	Keep in my shaving bag	Reasonably compact	Must	Store
How big is that?	About 3" 2" 6",and I have a lot of things in it, it is always full			
Size of things is important?	Very important, I look for the smallest size of everything			
	So I dig it out of my bag, and carry it to the bed, where I usually clip my nails.	Striking appearance	Nice	Prepare for filing
		Lightweight	Nice	
	Spin handle and rotate simultaneously	Easy to open file	Should	
Do you file?	Yes, I file at an angle with a vertical and angular motion	File at an angle	Must	File nails
	With file between thumb and index finger, and clipper body in fist			
		Easy to close file	Should	Unprepare for filing
	Rotate file back in place			
		Easy to open clipper	Should	Prepare for clipping
	Rotate handle into position	Easy to hold clipper	Should	
	Grab in hand using thumb and index finger, with tail up against middle finger edge			
	Position nail to be cut on bottom blade	Easy to align clipper	Nice	Clipping
		Low clippering squeeze force	Nice	
	Squeeze finger and thumb to cut	Blade shape curved	Nice	
	Reposition blades and make a second cut	Clip nails	Must	
	Reposition blades for final cut	Nail falls predictably	Nice	
	Catch cut nail			
	Toss in garbage	Dispose of nails		
	Rotate handle back	Easy to close clipper	Should	Return from clipping
	Spin handle to closed position			
	Put back in bag			

Fig. 2 Customer need collection form (fingernail clipper example)

properly interviewing nine customers for one hour each uncovers over 90% of customer needs. Assuming a homogeneous market, interviews beyond the ninth subject tend to uncover very few new customer needs.

Using an interview sheet with canned questions does not work well for eliciting customer needs. It is much better to state nothing other than the single request, "Walk me through a typical session using the product." Typically the interview starts with the customer approaching the product in storage, before even using it. Where is it stored? What must the customer do to obtain it from storage and prepare it for use? How is it unpacked from the box and assembled? Ideally, when the customer does any motion or thought processing at all, the customer should describe it to the interviewer. This should be continued through the product use, followed by cleanup and re-storage. Here are some useful questions for prompting conversation during silent moments (Ref 3):

- What do you like or dislike about this product?
- What issues do you consider when purchasing this product?
- What improvements would you make to this product?

If a company is contemplating the development of a new product (e.g., a new technology with no existing products on the market), the above questions work well as a starting point. Because the customer cannot walk through the use of an actual product, an analogous device should be used, even a blob of clay, so that customers can manipulate a substance when describing their desires.

Here are some general hints for effective customer interviews (Ref 3):

- *Go with the flow.* Do not try to stick closely to any interview guide, including this one.
- *Use visual stimuli and props.* Bring any tangentially related product, and ask about it.
- *Suppress preconceived notions about the product technology.* The customers will make assumptions about the technology, but the interviewer should avoid biasing the discussion with any assumptions about how the product will be designed or used. That leads to speculation, not facts.
- *Have the customer demonstrate.* It usually unveils new information.
- *Be alert for surprises and latent needs.* The interviewer should pursue any surprise answer with follow-up questions. This usually uncovers latent needs, needs that the customer is not consciously aware of and that are otherwise hard to uncover.
- *Watch for nonverbal information.* Words cannot communicate all product sensations.

Figure 2 shows a form for collecting customer interview data, completed for the fingernail clipper example. The first two columns are completed during the actual interviews. The first column documents any interviewer's prompts to the customer, what might have been said to get the response, if anything. The second column documents the raw data, what the customer said *in his or her own words.* No interpretation should be made by the interviewer when completing these columns.

The other columns are completed as soon as possible after the interview. In the third column, customer statements from the second column are interpreted in a structured noun-verb format (though not rigidly so). When making these interpretations, it is important to express what the product must do, not how the product might do it. Positive rather than negative phrasing should be used, to keep the interpretations focused on actual needs and not on whether a product is satisfying them. Finally, the words "must" and "should" should not be used in the statements. Rather, these qualifications should be incorporated into subsequent importance ratings, which constitute the fourth column.

In the fourth column, the importance of the customer's stated needs is interpreted using five ordered ratings: *must, good, should, nice,* and *not* (Ref 11). A *must* rating is used when a customer absolutely must have this feature, generally when it is the determining criterion in purchasing the product. *Must* ratings will act as constraints. The *not* rating is for features that the customer never uses and does not care about.

Compiling Customer Needs

From interviews or other customer data, multiple lists of customer needs must be compiled into one. The design team should transcribe each need onto an index card, then group the cards into categories to make an affinity diagram (Fig. 3).

An alternative approach is to have a few customers sort the cards as explained above. This prevents the customer data from being biased by the design team. Next, a matrix is created in which the needs are listed down the rows and are repeated across the columns. The matrix is filled in with entries (i,j), the number of times that need i appears with need j. From this matrix, a statistical hierarchical cluster analysis can be performed, converting the matrix into a tree structure where each need is arranged next to the need statements or clusters that are "closest." The design team then parses the tree into a two- or three-level structure with exemplar labels for the branches. Reference 8 provides details. This approach is believed to be a more complete way to parse the need statements, although more costly.

Ranking Customer Needs

Once the customer needs have been compiled, numerical importance rankings must be established. A design team should take care, however, because the typical customer population will be multi-modal, with segments that have different importance weightings. Multi-modal populations present systems-level choices about which product options to offer. Methods for designing a product family to meet such demands is a subject of active research (Ref 12, 13).

A traditional approach to rating customer needs is to compare the number of subjects who mention a need to the total number of subjects. For example:

$$\omega_{CR_i} = \frac{\text{Number of times mentioned}}{\text{Number of subjects}} \qquad \text{(Eq 1)}$$

where ω_{CR_i} is the ith interpreted customer need importance rank. This ranking is flawed, as it includes a measure of need obviousness, as opposed to need importance. A need may not be important but may be very obvious, and so every subject mentions it. Because of this concern, the design team typically reviews the different statements in column 2 of the customer response sheets (Fig. 2) to raise and lower the result from Eq 1. This approach is less than quantitative, takes excessive time, and is hard to justify.

A good approach to forming an importance ranking for a population is to send the list of customer needs to a random sample of customers (generally at least 100), asking them to rank the importance of each need. This approach can provide a sound statistical sample. However, any determination of statistical importance must incorporate two phases. First, a decision must be made as to whether the customer need is a *hard constraint* that must be satisfied, or an *objective* that can be traded off versus other customer needs. The former must be separated from the latter and accounted for differently by the design team.

To separate out customer needs that are hard constraints, each need is examined one at a time, and the number of *must* responses is compared to the total number of subjects. Clearly, if every subject flags the need as a *must*, then that need must be satisfied. But if only a fraction of the subjects indicate that the product must satisfy a need, a decision must be made about what fraction should be used before interpreting the customer need as a constraint.

To answer this question, statistical outlier analysis can be applied to determine when a "few" *musts* are outlier responses that are not worth flagging. A *must-confidence percentage level*, C_{must}, can be defined as the desired customer response percentage about the median needed to switch the customer need from an objective to a constraint. C_{must} is bounded between zero and one. Note that although C_{must} is a confidence percentage level, the approach here does not presume normally distributed data. No confidence intervals have been mentioned, only confidence percentages. Often engineers feel comfortable with $C_{must} = 0.999$, corresponding to three confidence intervals, when operating with normally distributed data. Such a value of C_{must} is excessive here, as it can create excessive constraints for the design team. It will force many customer needs to carry infinite importance as *musts*.

To establish whether a customer need is a constraint, tabulate the importance responses into the five categories described above. One can calculate how many subjects need to provide *must* responses for a customer need to be considered a *must:*

$$(1 - C_{must})(N - 1) < N_{must_i} \qquad \text{(Eq 2)}$$

where N is the total number of subjects and N_{must_i} is the number of subjects who provided *must* responses on the ith need. If N_{must_i} is less than the left-hand side of Eq 2, then the need becomes a constraint with a *must* importance rating. This test is very simple to implement and use.

Having separated the customer needs into *must* and *non-must* categories, relative importance ratings can now be placed on the *non-must* needs in the traditional way. As one approach, first convert the subjective importance ratings into numerical equivalents. A typical transformation used is:

Importance rating	Numerical equivalent	
	Option 1	Option 2
Must	9	1.0
Good	7	0.7
Should	5	0.5
Nice	3	0.3
Not mentioned	0	0

This mapping is always a subjective interpretation-conversion that a design team must agree upon in order to convert linguistic customer expressions into numbers. This subjectivity is a customer modeling issue that will arise with any approach. Once the mapping is established, the importance assigned to each customer need can be calculated by the average

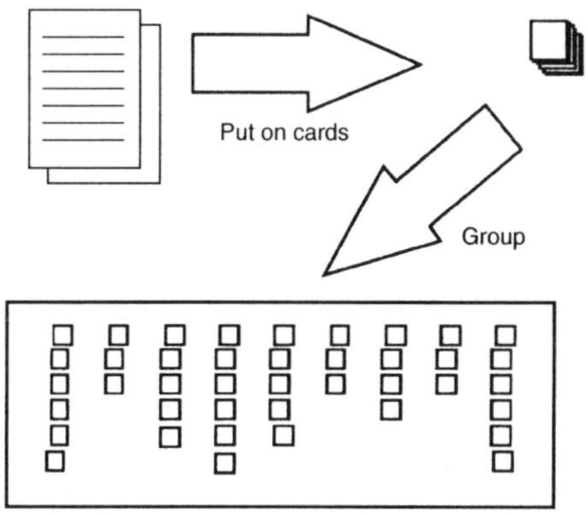

Fig. 3 Converting the set of customer needs into an affinity diagram

$$\omega_{CR_i} = \frac{\sum_j \omega_{CR_{i,j}}}{\text{Number of subjects}} \qquad \text{(Eq 3)}$$

where $\omega_{CR_{ij}}$ is the numerical importance rating for the ith need assigned by the jth customer. The result of Eq 3 can be linearly scaled to any other numerical range desired, and the information contained will remain unchanged. The variance of Eq 3 across the subject pool can also profitably serve as an uncertainty indicator to establish significant figures.

Other methods for determining importance are detailed in Ref 8. One is an anchored measure approach, whereby the customer respondent first determines the most important need, which is assigned a "10." Then each lower importance need is ranked relative to that need. Another approach is to require that the importance ranks add up to a constant figure, such as 100. This method can work in small sets of needs, but otherwise it can be confusing and cause the customer to rank low-priority needs inaccurately.

All of these direct approaches may suffer from scaling problems: it may be that customers think they know what they want, but they do not purchase accordingly. Therefore, some experts advocate a revealed preference approach, where the design team describes different product features as numbers, then asks for purchasing preferences on entire products (not features), and then uses regression analysis to fit importance coefficients from the products to the features. However, Hauser and Griffin (Ref 10) report poor results with the revealed preference approach.

Gathering Customer Use Data

A customer engages in many distinct activities when using any non-trivial consumer product. A product is purchased, transported, assembled, stored, removed for use, initialized, used in different ways in different environments, perhaps modified by the user, periodically cleaned or maintained, and disposed. Different customer use patterns should be captured and communicated to the design team, as they can give rise to different product forms. Further, capturing the customer use patterns helps to ensure that customer needs have been gathered for each different activity.

To form the possible use patterns, it is important to capture the activity sequence for each customer by completing the column 5 of the customer data sheet (Fig. 2). Typically, a sequence of customer statements will have one associated activity. *Activities* differs from *customer needs* in that activities label what the customer is doing (not the product) when a group of needs are expressed about the product. After column 5 is completed, the activities are transcribed onto index cards and sorted into another affinity diagram (Fig. 3). The number of cards for each activity can indicate how typical each is. The initial and final activities should be highlighted, so that the team can help system-engineer the environment within which the product will be used.

The affinity diagram communicates to the design team what the customer does with the product. It helps ensure that a design team is aware of all customer life cycle product needs. The activity diagram can be expanded "upstream" to capture sales, distribution, and manufacturing activities, or "downstream" to capture activities such as disposal. Incorporating such life cycle and manufacturing concerns remains an active research area.

In addition, the affinity list is useful in categorizing customer needs. Typically, customer needs are grouped into abstract categories, based on arbitrary interpretations by the design team or customers. Experience has shown that the activities list provides a more meaningful way to group customer needs.

Clipper project		
Customer requirements		
Interviewer(s): KNO		
Date 9-7-95		
5	Number of customers	
0–5	Weighting scale	
75%	Must confidence	
Average customer		
	Male/Female, age 20-60 middle class	
	Not in the hair or nail business	
	Customer requirement	Weight
Purchase		
	Cost	4
Transport in package		
Unpackage		
Chain keys		
	Act as keychain	0
Store		
	Compact storage	4
	Non-snag storage	1
	Lightweight	2
	Striking appearance	3
Prepare to file/pick		
	Easy to open	2
Filing/picking		
	Files nails	2
	Picks nails	Must
	File rough	1
	File holds filed dust	0
	File has a picking tip	Must
Return from filing/picking		
	Easy to close file	2
Prepare for clipping		
	Easy to open clipper	4
Clip nails		
	Easy to align clipper	2
	Easy to hold clipper	3
	Body contoured to hold	3
	Curved blade shape	3
	Wide handle	2
	Low squeeze force	1
	Blade can act as pusher	1
	Blade can act as a file	1
	Clips nails	Must
	Clips toe nails	1
	Clips hangnails	1
	Sharp blade	1
	Nails fall predictably	1
	Stores cut nails	1
	Easy to clean	0
Return from clipping		
	Easy to close clipper	4
Throw away		

Fig. 4 Partial customer needs list (fingernail clipper example)

Summarizing Customer Needs

Customer needs must be summarized into a document that the design team can reference. A sample is shown in Fig. 4. Among other items, the header information contains a brief description of a typical customer. This description should have been given to the design team from management originally or through the mission statement, but it should be refined, based on what is now known about customer needs. The customer description for the fingernail clipper is "male or female, age 20 to 60, carries clipper on the run, not employed in the beauty industry." Basically, the customer description includes the core demographic description applied when customers were selected for interviews, and any relevant product-specific information. This information is particularly important for segmented customer populations.

In the sample form, customer needs are listed by activity. In turn, the activities are arranged in serial order, with parallel activities ordered by importance. Within each activity group, customer needs should be similarly arranged. The numerical importance rating for each need should be included in the second column. Customer needs that are *musts* should have this indicated in the second column. The scale (0 to 1, 1 to 10, etc.) and normalization (relative to most important goal, relative to a sum, etc.) should be indicated on the top of the form, if there is no company standard.

In addition to satisfying customer needs, some products must satisfy other requirements, typically legislative or manufacturing. These can be represented as additional requirements in the customer needs list. Typically all have *must* ratings, since they must be met in order for the product to be legally sold or physically produced.

Non-customer requirements can be incorporated in the customer needs list as deemed appropriate. Alternatively, a specification sheet may be added for non-customer requirements, organizing the requirements according to topic. References 14 and 15 provide detailed examples of how to create a specification sheet.

Functional Decomposition: Modeling, Analysis, and Structure

Design teams need a representation of what the customer wants from the product, that is, a model of how the product should function in order to satisfy the customer. Functionally, all products *do something*. Products therefore accept *inputs* and operate to produce *outputs* (i.e., the desired performance). We can model any product, assembly, subassembly, or component as a *system,* with inputs and outputs that traverse a system boundary. The essence of such a model is the need-function-form definition of engineering design. In the sections below, we construct the necessary machinery for understanding and representing design function, according to a system perspective.

Why Functional Decomposition?

While methods to carry out market studies or to gather customer needs are widely accepted, methods to generate concepts are typically left to the whims of the design team. The transition from customer needs to concrete solutions is seen as more of an art than a science (Ref 16). In fact, for many consumer products, there is a tendency to seek form solutions directly based on the previous experience of the design team members. With ever-shrinking cycle times and budgets, and with ever-expanding demands for quality, this approach has a number of limitations. Most notably, the links between customer needs and design concepts are, at best, indirect or implicit. They exist only in the minds of the designers. As such, customer needs are relegated to the criteria for evaluating concepts, not the direct catalysts for generating concepts.

Over the last 20 years, new methods for engineering design have emerged that focus on mapping customer needs to functional descriptions, or mapping these descriptions to sets of technologies that satisfy the underlying functional requirements (Ref 3, 14, 15, 17, 18). When combined, these methods have a number of intrinsic advantages:

- The emphasis is on *what* has to be achieved by a new concept or redesign, not on *how* it is to be achieved. By so doing, a component- and form-independent expression of the design task may be achieved so that the design team can comprehensively search for solutions.
- Creativity is enhanced by the ability to decompose problems and manipulate partial solutions (Ref 17). When a design task is decomposed into its functional elements, solutions to each element are more apparent due to the reduction of complexity and extraneous information.
- Functions or sets of functions may be derived or generated directly from customer needs. These functions define clear boundaries to associate assemblies or subassemblies of the final design solutions. In turn, these boundaries provide a basis for allocating resources to concurrent engineering efforts and for seeking modular concepts.

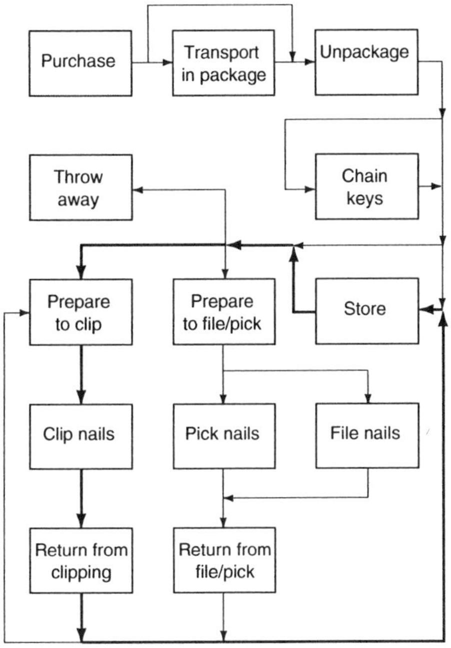

Fig. 5 Completed activity diagram (fingernail clipper example)

- Functional modeling provides a natural forum for abstracting a design task. Many levels of functional abstractions may be created, from a very high-level single-function statement to detailed functional statements for the subsystems of a design.
- By mapping customer needs first to function and then to form, more solutions may be systematically generated to solve the design problem. "If one generates one idea it will probably be a poor idea; if one generates twenty ideas, one good idea might exist for further development" (Ref 17).
- Needs mapped to function and then to form promote set-based concurrent engineering processes (Ref 19). Feasible regions of technology may be explicitly defined based on functional requirements. Tradeoffs may also be explored in parallel among a wide array of radical and known solutions, since a common functional description is driving the design effort directly from the voice of the customer.

The next section describes a systematic approach for establishing the functionality of a new design or redesign. The fingernail clipper provides a running example to clarify the approach.

Establishing Functionality and Product Architecture

The procedures described above yielded a clear statement of customer needs, organized to establish priorities for the design effort. Functional modeling begins the systematic process of transforming these needs into a clear specification of the design task. It also initiates the conceptual design phase, wherein an array of solutions are sought.

Phase 1: Develop Process Descriptions as Activity Diagrams. To start the functional modeling process, an important tool is to specify the *process* by which the product being designed will be functionally implemented. A process or process description, in this sense, includes three phases: *preparation, execution,* and *conclusion* (Ref 20–22). Within each phase, we network high-level user activities to show the full life cycle of a product, from purchase to recycling or disposal. As high-level activities are listed in each phase, a number of product characteristics are chosen.

Here, the process description is represented as an activity diagram (Fig. 5). This diagram clearly shows a number of process choices that will influence the final design. For example, the activities of "picking" and "filing" are process choices for improving the customer's fingernail appearance through mechanical contact. Chemical "soaking" process choices or others might be chosen as alternatives. They would lead to different activities, functional descriptions, and, ultimately, product architectures and components. Figure 5 does not include manufacturing activities such as packaging and transport, sales functions, or disposal, but depending on the scope of the design task, it could have included these activities.

Phase 2: Formulate Subfunctions through Task Listing and Black Box Modeling. Using the activity diagram and the list of customer needs, a *function structure* for the product can be formulated (Fig. 6). A function structure is defined as an input-output model that maps energy, material, and signal flows to a transformed and desired state. Function structure modeling (Ref 3, 14, 15, 17, 20, 23, 24) has historically been used to create a form-independent product expression. We extend common function structure modeling to include a mapping of customer needs to subfunction sequences (called *task listing*), a method for aggregating subfunctions, and a comparison of a functional decomposition with customer needs.

The first step is to identify primary flows associated with the product activities. A *flow* is a physical phenomenon (e.g., material, energy, or signal) that is intrinsic to a product operation or subfunction. In the context of input-output modeling, a flow enters an operation or subfunction, is manipulated by the subfunction, and exits in a new state. For example, an operation may be to pressurize a fluid. Two critical flows for this operation are an energy to execute pressure change and the fluid material being operated upon. A list of some common energy flows is given in Table 5 (Ref 25, 26).

Considering the fingernail clipper example, a subset of the customer needs are:

Need	Effect	Importance
Cost	Not inexpensive	4
Compact	Not compact	4
Files well	Does not file well	2
Cuts well	Does not cut well	Must
Easy to open/close	Not easy to open/close	4
Easy to hold	Not easy to hold	3
Comfortable	Not comfortable	3
Sharp cutting surface	Not a sharp surface	3

We now translate the customer needs to energy, material, or signal flows of the product when effects are exhibited or are expected to be exhibited during product use. Cost is not an aspect of using the product, so we do not model "not inexpensive." Primary flows associated with "not compact" are

Table 5 Examples of common energy flows

displacement	magnetism
power	pneumatic power
deflection/strain	oscillating force
electricity	sound (acoustic)
human force	linear force
friction force	stress
solar energy	torque
light	heat
rotation	noise
time	hydraulic power
vibration	pressure

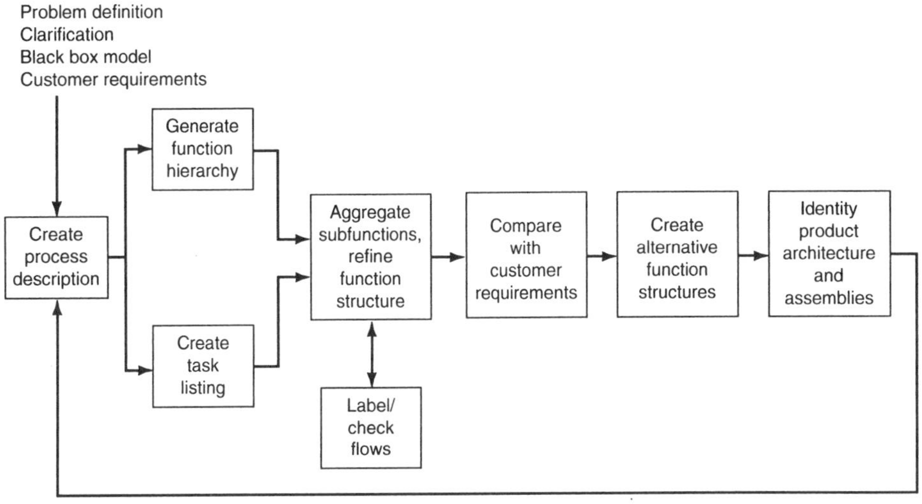

Fig. 6 Schematic representation of the function modeling and analysis process

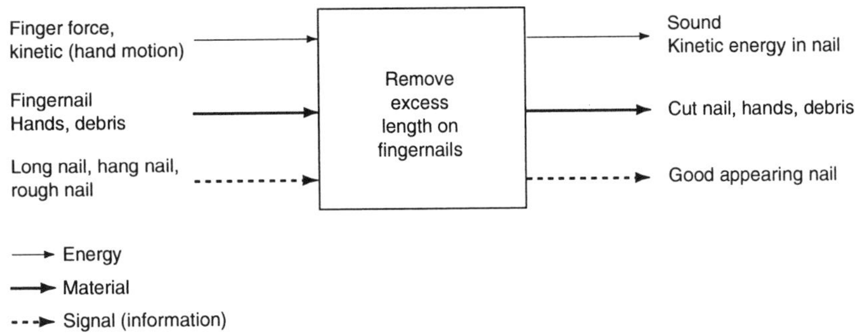

Fig. 7 Black box model (fingernail clipper example)

the user's hands, the fingernail dimensions, and storage compartments (e.g., pants pockets, wallets, or purses). These flows are material in nature and capture capacity in terms of "volume." Primary flows for the remaining customer needs include:

- "Does not file well"—hand motion (energy), fingernail (material), fingernail roughness (signal)
- "Does not cut well"—generated cutting force (energy), finger force (energy), and fingernail (material)
- "Not easy to open/close"—hand movements (energy) and hand (material)
- "Not easy to hold"—finger force (energy) and hand (material)
- "Not comfortable"—finger force (energy) and hand (material)
- "Not a sharp cutting surface"—generated cutting force (energy) and fingernail (material)

To document the mapping of customer needs to flows, a "black box" model of the product is developed. A black box model lists all input and output flows for the primary, high-level function of the design task, stated in active verb-noun phrases. Figure 7 illustrates a black box model for the fingernail clipper task. This model must now be refined and decomposed to identify the basic product or device functions that will satisfy the overall function and needs.

For each of the flows, the next step is to identify a sequence of subfunctions and specific user operations. A subfunction, in this case, is an active verb paired with a noun that represents the causal reason behind a product behavior. An operation is a specific action by the user that is needed to complete the function structure. References 17, 23, 25, and 26 provide lists of appropriate verbs and nouns to use in functional analysis, and Table 6 summarizes typical classes of engineering functions. A useful approach for generating subfunctions is to trace the flow as it is transformed from its initial creation state to its final expected state when it leaves the product's system boundary. This approach may be executed by *playacting* the flow (becoming the flow) or brainstorming a hierarchy of functions that must process the flow.

For example, a customer need, expressed in the customer's voice, may exist for "cuts nail well." A suitable flow for addressing this need is a force flow that ultimately acts on the nail material flow. Through playacting these flows, a subfunction sequence may be of the form: capture force, apply force, transform to larger force, transmit force as motion, guide motion, cut material, stop motion, release force, dampen reaction to force, etc. Figure 8 illustrates task lists for three sets of customer needs, and the corresponding flows.

Phase 3: Aggregate Subfunctions into a Refined Function Structure. The subfunctions for the full set of customer needs are aggregated (combined) to represent the functions of the entire product. This step is accomplished by appropriately connecting flows between each sequence and adding subfunctions that interact or provide control states.

Aggregation and refinement of the function structure ends when two criteria are met:

- The subfunctions are "atomic" (i.e., they can be fulfilled by a *single, basic* solution principle that satisfies the function).
- The level of detail is sufficient to address the customer needs.

The first criterion provides a basis for choosing the depth of functional analysis (e.g., a subfunction of "control motion in three dimensions" should be refined to "control motion in three rotations and translations"—six total functions—since a single, basic form solution in fingernail clipper technology does not provide 3D control). The second criterion ensures that time is not wasted refining a function structure to the level of miscellaneous and secondary product components, such as fastening.

For the fingernail clipper design effort, an aggregated function structure is shown in Fig. 9. Notice that subfunctions and flows are combined for overlapping or redundant functionality from Fig. 8. User functions are listed outside of the system boundary for clarity.

Phase 4: Validate the Functional Decomposition. Once the design team has completed the subfunction aggregation, functional modeling

Table 6 Basic device/product function (active verb) classes

Function class	Definition	Basic functions	Alternatives (synonyms)
Channel	To cause a direction of travel or path for a material or energy	Import	Input, receive, allow, form entrance
		Export	Discharge, eject, dispose, remove
		Transport	Channel, lift, move
		Transmit	Conduct, convey, transfer
		Stop	Insulate, resist, protect, shield
		Guide	Direct, straighten, steer
Store/supply	To accumulate or provide material or energy	Store	Collect, contain, reserve
		Supply	Fill, provide, replenish
Connect	To join two or more flows (materials or energies)	Couple	Connect, assemble, join
		Mix	Add, pack, blend, combine, coalesce
Branch	To cause a material or energy to no longer be joined or mixed	Branch	Divide, diverge, switch, valve
		Filter	Strain, filtrate, purify, percolate, clear
		Separate	Release, detach, disconnect, disassemble, release, subtract
		Remove	Polish, cut, sand, drill, lathe
		Distribute	Scatter, disperse, diffuse, empty
		Dissipate	Absorb, dampen, dispel, diffuse
Control magnitude	To alter or maintain the size or amplitude of material or energy	Actuate	Start, initiate
		Regulate	Control, allow, prevent, enable, disable, limit, interrupt
		Change	Increase, decrease, amplify, reduce, normalize, multiply, scale, rectify
		Form	Compact, crush, shape
Convert	To change from one form of flow type to another (e.g., one energy type to another energy type)	Convert	Transform, gyrate, liquefy, solidify, evaporate, condense, integrate, differentiate, process
Support	To secure firmly a material or energy path into a defined location	Stabilize	Steady
		Secure	Attach, mount, hold, fasten, lock
		Position	Orient
		Translate	...
		Rotate	Turn, spin
		Allow degrees of freedom	Constrain, unsecure, unlock
Signal	To provide information to, within, or out of the system boundary	Sense	Perceive, recognize, discern, check, locate, verify
		Indicate	Mark
		Display	...
		Measure	Calculate, compare, count
		Clear	...

and analysis comes to a closure through two verification steps:

1. All major flows between the subfunctions are labeled and checked according to their state of transformation. This ensures validity and continuity, perhaps leading to the addition of further functional representations. Table 7 lists the pertinent questions, checks, guidelines, and actions implemented at this stage.
2. The customer needs list is reviewed, and the subfunction or sequence of subfunctions is identified that satisfies each customer need. Needs not covered by the function structure require further analysis, and subfunctions not satisfying a need require confirmation of their incorporation. This verification typically adds more subfunctions to the network, while simplifying or removing others that really do not apply.

Consider the following validation for the fingernail clipper task:

Need	Effect	Validation (applicable subfunction(s))
Cost	Not inexpensive	Not applicable (criterion for evaluating design concepts)
Compact	Not compact	Apply finger force, form filing surface, form grasping surface, cut nail
Files well	Does not file well	Form filing surface, form grasping surface, orient surface, slide over nail
Cuts well	Does not cut well	Convert to large force, cut nail
Easy to open/close	Not easy to open/close	Guide to nail, move to cut nail
Easy to hold	Not easy to hold	Apply finger force, release force, release motion
Comfortable	Not comfortable	Apply finger force
Sharp cutting surface	Not a sharp surface	Cut nail

The subfunctions listed for the validation combine to represent the customer need. For example, "compact" will ultimately be governed by the solution principles chosen for the "apply finger force" subfunction. The size of these solution principles determines the overall compactness of the final fingernail clipper. This customer need will need to be balanced with "cuts well" and the "cut nail" subfunction, since a minimum size will be needed to cut a nail.

Phase 5: Establish and Identify Product Architecture and Assemblies. With the functional decomposition verified, it becomes important to identify product assemblies that can be addressed by individual designers or cross-functional design teams. The key elements of this phase are to define collection of functions (chunks) that will form assemblies in the product and to clarify the interactions and interfaces between these chunks. By so doing, a product team will have a basis for choosing between *modular* and *integral* architectures (Ref 3, 27). They will also have a basis to choose parallel design tasks for the product development, where only the interaction and interface information need to be shared continuously among the subteams.

A simple process for establishing the product architecture and assemblies includes the following steps:

1. Using the functional decomposition of a product, cluster the subfunctions or elements in the function structure. Dashed boxes around the clusters of subfunctions and a title will serve as an appropriate representation. These clusters are chosen by identifying parallel subfunction chains (each parallel track is a candidate cluster), subfunction chains that have common energy types as flows, and subfunction chains that have only simple interactions with each other.
2. Create a rough spatial layout (block diagram with a reference frame) for the product. This layout is meant to show the relative position of each cluster in order to clarify spatial interactions and interfaces (the boundaries between clusters).
3. Define interactions, interfaces, and performance characteristics between each cluster.

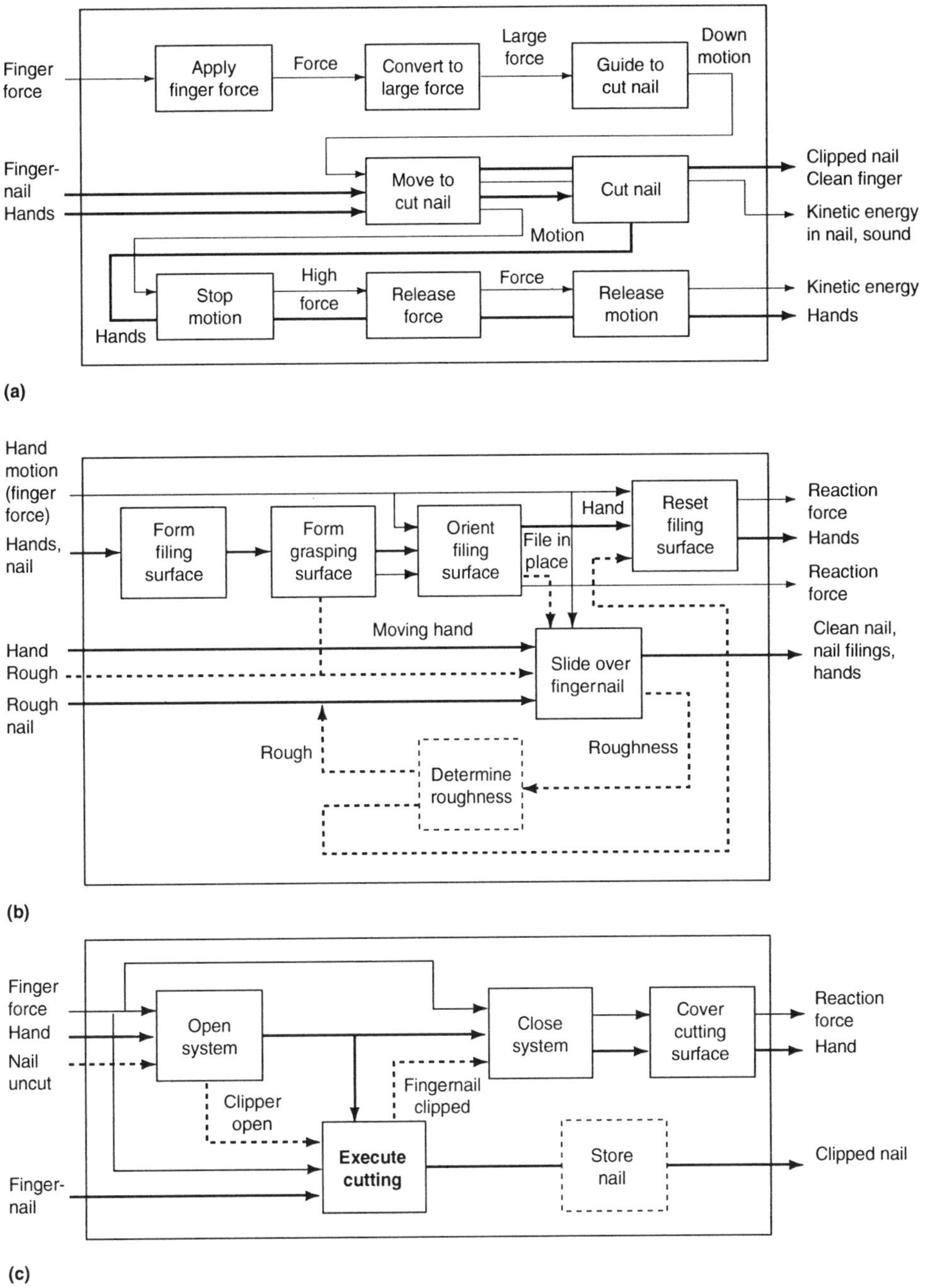

Fig. 8 Task lists for three sets of customer needs (fingernail clipper example). (a) Customer needs: cuts well, easy to hold, comfortable. Flows: finger force, nail, hands. (b) Customer need: file well. Flows: nail, hand motion, roughness signal. (c) Customer need: opens and closes easily. Flows: nail, hand

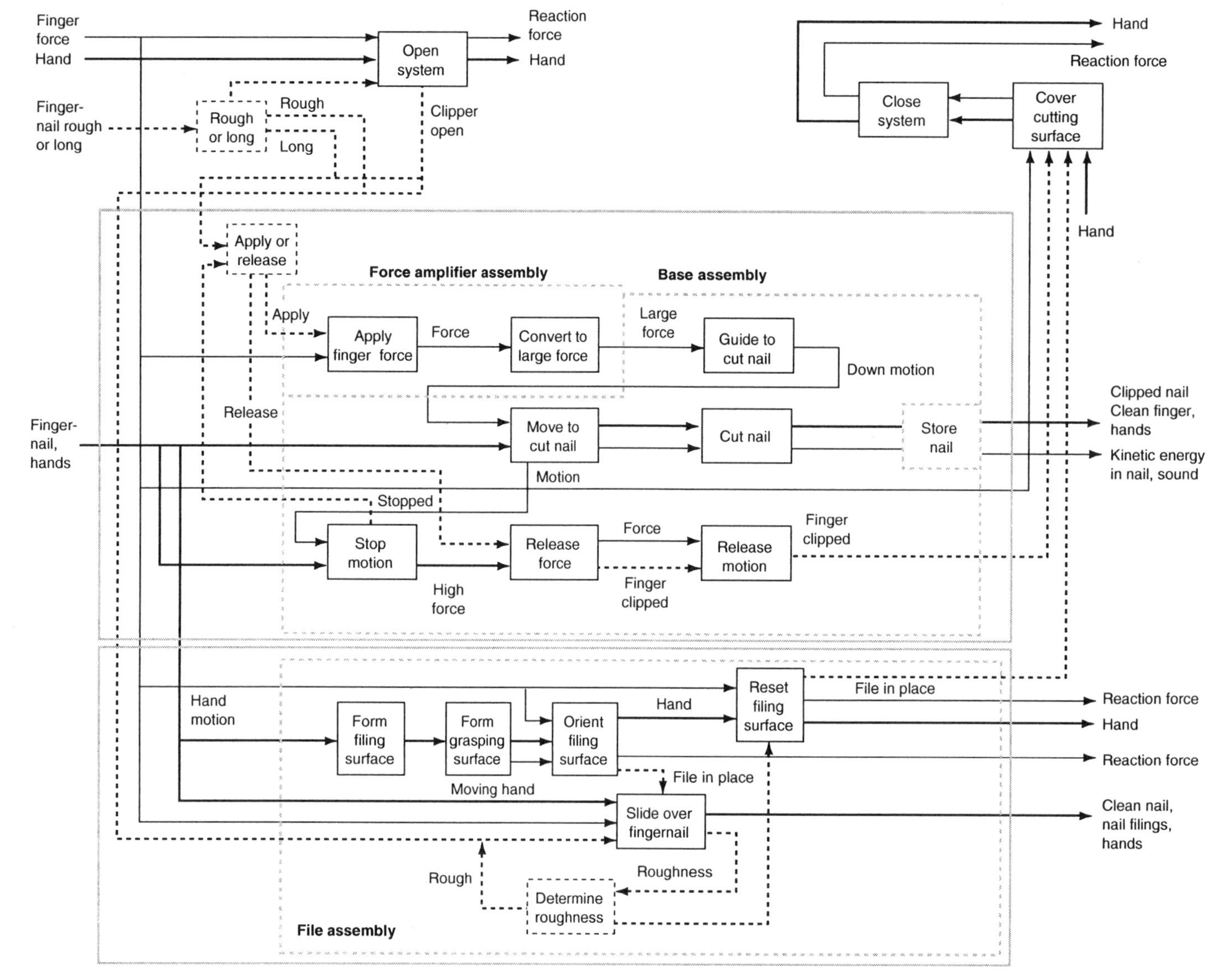

Fig. 9 Refined function structure (fingernail clipper example)

Four types of interactions exist as flows across interfaces: spatial (geometry), energy, information, and material. These interactions represent what must be shared across interfaces within the performance requirements.

Figure 9 shows the application of the first of these steps to the fingernail clipper. Three primary assemblies are identified: the base, force amplifier, and file. Important interactions include the hand flow of the user and attachments to allow relative motion for opening and closing activities. These assemblies may now be designed relatively independently, if a modular architecture is chosen. Reference 27 presents further examples, including spatial layouts.

Competitive Benchmarking

To remain competitive, a design team must compare any proposed concept with the competition. Typically a corporate group exists to tear down the competitor's product, estimate costs, plot trends, make predictions on requirements, and work with the design teams. These efforts uncover the clever things that the competition has spent effort on, uncover the principles behind how they work, and predict costs.

It may be appropriate to benchmark products before customer needs analysis and functional decomposition. Alternatively, if a team does not wish to bias these initial analyses, it can update the results with benchmarking data.

Reference 28 presents benchmarking methods and an overview of a reverse engineering meth-

Part #		Part name	Quantity	Finish	Function	Physical parameters
A1	Arm assembly					
	001	Actuator arm	1	Chrome	Transmit finger force	2 in., shaped
					Input from finger	0.25 in. pivot
	002	Pin	1	Chrome	Pivot	0.13 in. round
A2	Cutter assembly					
	003	Blade arms	2	Chrome	Cut nails	2 × 0.44 × 0.13 in.
					Spring action	0.13 in. gap
A3	File assembly					
	004	File	1	Chrome	File nail	Scored, 1.50 × 0.25 × 0.06 in.
	005	Pivot rivet	1	Chrome	Attach file	0.19 in. rivet

Fig. 10 Reverse engineering bill of materials (fingernail clipper example)

odology at a large company. Reference 29 presents an approach to benchmarking and a case study from the computer industry. Reference 30 presents a ten-step approach to benchmarking company operational practices. In this section we present a method that is focused on product design, using the customer demands and the product functions. The steps in this method are:

1. Identify design issues.
2. Identify competitive and related products.
3. Conduct an information search.
4. Prepare for product teardowns.
5. Evaluate distribution and installation.
6. Take apart and measure the assemblies.
7. Write a bill of materials.
8. Plot industry trends.

Step 1: Identify Design Issues. First, it must be clear what problems the design team is facing on the current project. If this is a new project, the technical form issues may be unknown, so it will be worthwhile to investigate the market, competitors, and competitive products. If this is a redesign project, an investigation can ask of the previous design team:

- What was difficult for you?
- What design problem did you solve that you are proud of?
- What related technologies are you interested in?

The deliverable from this step is a list of keywords with explanations about what information to gather.

Step 2: Identify Competitive and Related Products. Once the product function is known, one must examine the sales outlets for products that address those functions. For consumer products, sales outlets are typically retail stores. One must list all competitors and their different product models, and all related products in their portfolio. If a competitor has a family of products under a common product architecture (they use identical components for some aspects of each product, but different components for niche demands), one should detail this information, as it can indicate the competitor's preferred market segments.

This step should only identify the competitors, by company names and product names. The list should be screened by highlighting the particular names that appear most crucial for the design team to fully understand.

Step 3: Conduct an Information Search. The importance of an information search cannot be overstated. The wealth of information available about business operations across the globe is amazing. A design team must understand the market demand for product features and what the competition is doing to meet it. The team should gather information on the products and related products, the functions they perform, and the targeted market segments. All keywords associated with these three categories should be formed from step 2 and used in informational searches.

Most business persons are happy to discuss the market and non-competitive business units. Though most will not provide strategic information about their own companies, many people are willing to tell all about their competitors. Suppliers will usually discuss their customers, if it appears that you might provide an additional sale. Once people understand you are designing or redesigning a new product, they naturally want to get involved with new orders, and they will help you as much as they legally can. Pursuit of information beyond that point is unethical, and not necessary. Most people are happy to share information, so simple honesty and a friendly attitude can get you a long way.

Public sources of information include the library and electronic wire services. University libraries are filled with technical engineering modeling references. Also, librarians have expertise in uncovering obtuse references with limited initial information. Particular references of interest include:

- *Thomas Register of Companies.* This is a "yellow pages" for manufacturing related businesses. The Thomas Register lists vendors by product.
- *Market Share Reporter.* Published every year by International Thomson Publishers, this book summarizes the market research of Gale Research Inc. It comprises market research reports from the periodicals literature. It includes corporate market shares, institutional shares (not-for-profits), and brand market shares.
- U.S. Department of Commerce, National Institute of Standards and Technology. This U.S. government branch provides, among other things, national labor rates for all major countries. This proves very useful for determining competitors' manufacturing costs.

On the World Wide Web, databases can be searched using keywords for a small fee. In the United States, these include:

- *Lexis/Nexis.* These two databases are the most comprehensive full-text online sources of news and business information. *Lexis* provides legal information, and *Nexis* focuses on business. They both provide abstracts and full text from public news sources, including newspapers, magazines, wire services, newsletters, journals, company and industry analyst reports, and broadcast transcripts.
- *Dialog Information Services Databases.* Dialog is the world's largest online information research service, containing millions of documents. These include databases with simple abstracts as well as the full text. For the most part, the articles come from business journals and include information on company histories, competitive intelligence, new product development efforts, sales and earnings forecasts, market share projections, research and development expenditures, financial activities, demographics, socioeconomic activities, government regulations, and events that affect the business environment.

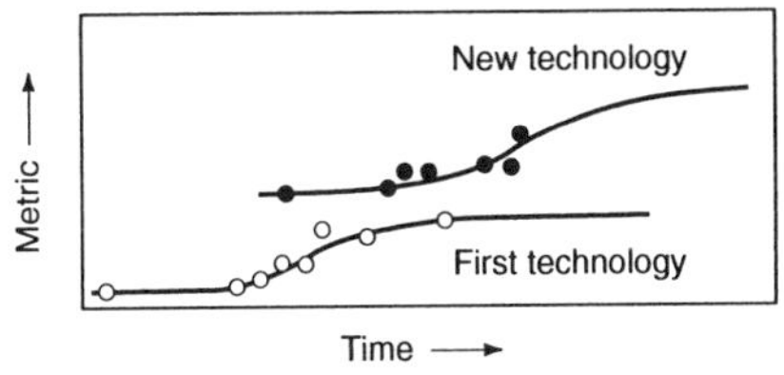

Fig. 11 Technological innovations plotted as S-curves

Table 7 Function structure organization and flow checks

Question	Check/guideline/action
Are physical laws maintained?	For example, validate conservation of mass and energy.
What is the new state of flow for each subfunction? Is this state valid and correct?	Label input/output states.
Are system and subsystem (assembly) boundaries clearly shown?	Define system boundary from process description or activity diagram. Subsystems (assemblies) starting point—define as a group of functions where the energy flow does not change (e.g., no relative motion).
Are all functions (noun) form-independent?	If not, the function should be generalized.
Are the subfunctions atomic? Can each subfunction be replaced by a device or structure that only performs that function?	Refine subfunction into a set of atomic subfunctions, without adding unnecessary detail.
Does an alternative ordering or placement or number of subfunctions exist?	If so, develop alternative function structures or subfunction structures.
Are all subfunctions product or device functions?	If not, the subfunction(s) is a user function (not performed by the product). Convert the subfunction to the device functions that must support the user function. If leaving the user function in the function structure adds clarity, double box it to show a distinction.
Are certain functions included as "wishes"?	If so, place a dashed box around the function to show that it is auxiliary or secondary.
Are certain functions "prolific," meaning that they affect or are affected by flows of many other subfunctions (e.g., "support impact loads")?	If so, place a ground symbol on the subfunction and include only primary flows through it.
Is the entire function structure sequential (i.e., the subfunctions are connected one after another in a dependent sequence or chain)?	If so, product parallelism (assemblies/subsystems) have not been identified. All of the customer needs are dependent. Parallelism of functions that do not directly depend on one another should be separated.
Do redundant functions exist?	Remove or combine redundant functions.
Do functions exist outside the system boundary?	If so, they should be clearly distinguished and maintained for clarity only, such as a user function.

Subfunction	Specifications
Open system	Opening force
	Opening gap
	Opening grip area
	Opening surfact friction
Rough or long	...
Apply or release	...
Apply finger force	Fingerpoint surface friction
	Fingerpoint cupness
Convert to large force	Force gain
	Motion reduction
Guide to cut nail	Blade visibility
	Blade curvature
Move to cut nail	Open blade opening width
Cut nail	Sharpness
	Hardness
	Flatness
Store nail	Storage volume
Stop motion	Click sound
	Stop compliance
Release force	Expansion force
Release motion	...
Form filing surface	Opening friction torque
	Finger opening surface area
	Finger opening friction
	Finger pushing area
	Open alignment force
Form grasping surface	Grip area
	Grip surface friction
Orient filing surface	Filed nail visibility
Slide over fingernail	Filing surface area
	File roughness, left to right
	File roughness, right to left
Reset filing surface	Closed alignment force
Determine roughness	...
Cover cutting surface	Closed blade opening width
Close system	Closed arm force

Fig. 12 Generating metrics for subfunctions (fingernail clipper example)

Most major manufacturers have a presence on the World Wide Web. Much information can be gathered simply and at no cost, particularly technical information. Material properties of industrial brand plastics and metals, for example, are easily found. Topical search engines are readily available.

After examining trade journals and uncovering which competitors have bragged about new innovations, it is useful to gather the patents on these new innovations. However, patent searches based on company names are difficult, since companies typically "bury" their patents by filing under the individual names of designers. Uncovering individual patents is usually done by refined topical searches. As much information as possible should be supplied to the person searching the patents.

Vendors of original equipment manufacture (OEM) components also are a valuable source of information. Cost quotes using the competition's ordering quantities are effective at uncovering OEM part costs. Usually one can also obtain unsolicited information about the competition. Persons in the industry can provide good pointers on non-obvious cost, technical, and market drivers. Overseas labor rates, shipping rates, and reliability of supply from particular geographic areas are examples of "inside knowledge" that might be uncovered from public sources when one knows what to look for. University faculty specializing in a relevant technical area can provide invaluable assistance in design or serve as design reviewers. All of these and other sources must be leveraged to determine as much as possible about what customers want, what competitors are supplying, and the dynamics of the market.

Step 4: Prepare for Product Teardowns. In the steps described above, the team establishes a list of products that are worth tearing apart and analyzing. This list should include products that can provide technical solutions to design needs. Typically, it includes the least expensive model on the market, the most expensive model, the most popular model, and models that have particular technical features.

Next, one must clarify what data are required, such as:

- Primary function
- Quantity of parts per product unit
- Dimensions
- Maximum, minimum, and average material thicknesses
- Weight
- Material
- Color/finish
- Manufacturing process, including sufficient information for a design for manufacturing analysis

Next, one should identify all tools that will be required to complete teardown. In addition, it is necessary to identify all sensors and test equipment required (e.g., camera, a videotape of the product operating, multimeter, hardness tester, optical sensor, flow meter). Typically, one docu-

		Open system	– Opening force	– Opening gap	– Opening grip area	– Opening surface friction	Rough or long	Apply or release	Apply finger force	– Fingerpoint surface friction	– Fingerpoint cupness	Convert to large force	– Force gain	– Motion reduction	Guide to cut nail	– Blade visbility	– Blade curvature	Move to cut nail	– Open blade opening width	Cut nail	– Sharpness	– Hardness	– Finger squeeze area	– Flatness	Store nail	– Storage volume	Stop motion	– Click sound	– Stop compliance	Release force	– Expansion force	Release motion	Form filing surface	– Opening friction torque	– Finger opening surface area	– Finger opening friction	– Finger pushing area	– Open alignment force	Form grasping surface	– Grip area	– Grip surface area	Orient filing surface	– Field nail visibility	Slide over finger nail	– Filing surface area	– File roughness, left to right	– File roughness. right to left	Reset filing surface	– Closed alignment force	Determine roughness	Cover cutting surface	– Closed blade opening width	Close system	– Closed arm force	Material finish	Keychain	Weight	Material cost
Purchase																																																										
– Cost	4																				△	○		△																																		⊙
Transport in package																																																										
Unpackage																																																										
Chain keys																																																										
– Act as keychain	0																																																						⊙		
Store																																																										
– Compact storage	4				△									△												⊙									△		△			△					△							△						
– Non-snag storage	1																																																○			⊙		⊙				
– Lightweight	2												△									△							△																												⊙	
– Striking appearance	3										△						△																																						⊙			
Prepare to file/pick																																																										
– Easy to open file	2																																	⊙	○	○	○	○											○									
Filing																																																										
– Files nails	2																																											○	⊙	○											
– File rough	1																																												⊙	○											
Return from filing/picking																																																										
– Easy to close file	2																																	⊙	○	○	○	○											○									
Prepare for clipping																																																										
– Easy to open clipper	4		⊙	○	○	○																																																				
Clip nails																																																										
– Easy to align clipper	2															○	○		○																																							
– Easy to hold clipper	3									○																																																
– Body contoured to hold	3										⊙																																															
– Curved blade shape	3																																																									
– Wide handle	2																						⊙																																			
– Low squeeze force	1												⊙																																													
– Blade can act as pusher	1															⊙																																										
– Blade can act as a file	1																																										○		○	⊙	○											
– Clips nails	Must												○								⊙	○																																				
– Clips toe nails	1												○								⊙	○																																				
– Clips hang nails	1																																																									
– Sharp blade	1																				⊙																																					
– Nails fall predictably	1															⊙																																										
– Stores cut nails	1																								⊙																																	
– Easy to clean	0																○		⊙																																							
Return from clipping																																																										
– Easy to close clipper	4		⊙	○	○	○																																																△				
Throw away																																																										

Fig. 13 House of quality matrix (fingernail clipper example)

ments this information in a written or electronic report, similar to professional laboratory experiments.

Step 5: Evaluate Distribution and Installation. Important factors in product development are the means used to acquire parts, contain parts and ship, distribute, and market the product. These must also be examined as a part of the benchmarking process. The packaging of the product should be examined and reported to the design team, because often it is quite expensive. Installation instructions and procedures should be examined for costs and effectiveness.

Step 6: Take Apart and Measure the Assemblies. Disassembly is the most obvious step in reverse engineering. However, to be effective, it must be coordinated with measurements. The first step is to take pictures and measurements on the whole assembly before disassembly. Then the assembly is torn down, pictures are taken in an exploded view, and the parts and assemblies are measured.

Step 7: Write a Bill of Materials. A written form should be completed that details the product. A good format is shown in Fig. 10, where the column headings are the data identified in step 4. The bill of materials should include photographs, in sequence, and an exploded-view computer-aided design drawing.

Step 8: Plot Industry Trends. The last step is to arrange and transform the data for clear understanding of implications for the design or redesign task. This process should include:

- Categorization of the market
- Categorization of technical solutions
- Benchmarking of technical solutions
- Benchmarking of competitors

Market categorization is typically done by socioeconomic status and percentage of the market.

Technical solutions should be categorized by function structure modeling. This often poses difficulty, since competitors may include functionality that the design team chooses not to include. Nevertheless, function structure modeling can provide a means to list all known technical solutions to particular subfunctions.

Benchmarking of technical solutions and companies is more readily completed, but it requires a time history of product measurements. All technological innovations manifest themselves into the market along "S-curve" timelines, as shown in Fig. 11. As an example, consider clock speed for microprocessors. For all the different microprocessors on the market, one can plot the clock speed as a function of the time when the product was introduced. The metric values will naturally fall on an S-curve. First, the values are low and widely spaced: not much innovation is occurring in the market. Next, rapid innovation occurs, and many products are launched. The lower leg of the "S" is forming. Eventually, however, engineers cannot extract more performance. The slope of the "S" tops out, and the curve becomes flatter.

Industry trends are critical for a competitive company to understand. If the market is becoming more competitive, the company must understand that it needs to invest in product and process quality, or lose. If the market technology is topping out (the top of the "S"), the company needs to begin to invest in a new technology, to "jump" to the next S-curve, higher on the scale of the metric. Plotting trends provides all of this information. Clearly, trend charting of competitive data is a necessary and culminating business activity as part of product benchmarking.

Subfunction	Current	Solution principles		
Opening				
Open clipper	Spin and flip			
Determine rough or long				
Clip nail				
Determine apply or release				
Apply finger force	Shaped top, bent bottom	Shaped top and bottom	Hand grip	Finger holes
Convert to large force	Pivot	Linkage	Concentrate it	Pliers
Stop motion	Teeth hit	Peg		
Release force	Spring of bent body	Linear spring	Coil spring	Leaf spring
File nail				
Move file into place	Pivot out file	File on arm	Slide arm out	Flip out file
Slide over nail				
Determine roughness				
Return file in place	Pivot back file	(File on top arm)		
Closing				
Close clipper	Flip and spin			
Hold shut	Spring of bent body	Spring of bent body		

Fig. 14 Partial morphological matrix (fingernail clipper example)

Forming Quantitative Specifications

Having established the function structure/architecture and benchmarked competitive product performance, the design team must associate each subfunction with at least one line item in a product functional *specification sheet.* This sheet should give functional specifications about what the product must do, not necessarily specifications for purchasing components. The specification sheet should also rank each specification according to its importance.

The most common approaches to forming specifications are total quality management (TQM) methods (Ref 9, 31, 32), in particular quality function deployment (QFD) (Ref 33) and the house of quality (Ref 34). For the most part, these methods provide a means to agree on a proposed list of specifications and set target values on the variables. What these and other tools do not provide is a means to identify what vari-

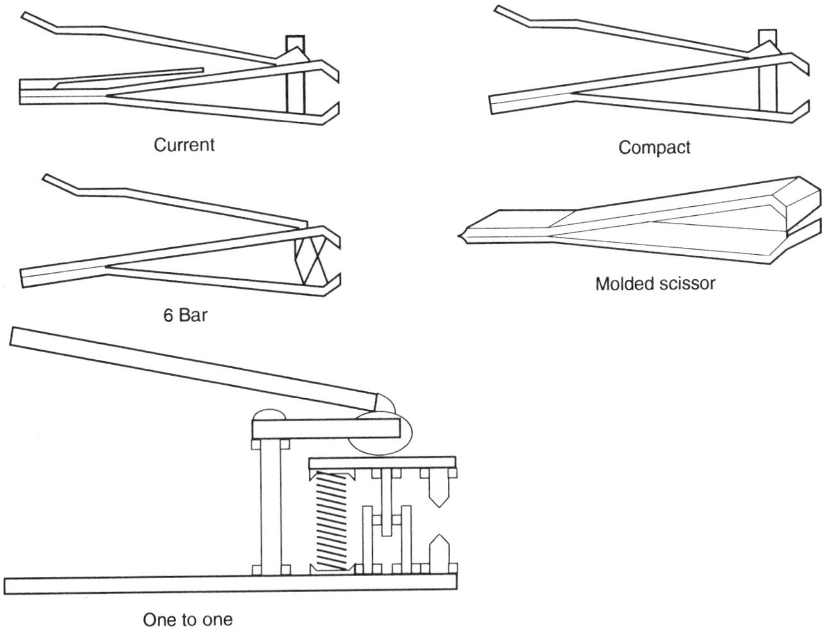

Fig. 15 Sample concept variants (fingernail clipper example)

ables should be used as specifications. How does one identify the important measurable variables?

Many researchers and practicing engineers have developed methods for forming specifications. Reference 3 contains an excellent chapter about forming product specifications and using the house of quality. Reference 34 is a classic article on the house of quality. Reference 33 is a text on QFD, complete with case studies. Reference 8 presents several methods for selecting among different variables. In all of these methods, the approach for generating variables is simply to be creative. In this section we review methods for generating and tracking specification variables.

Definition of Specifications

Typical definitions state that an engineering specification includes both a *metric* and a *value* (Ref 3). A metric is the quantity of discussion. The value is a target for the metric (either a particular number, a range, or an inequality).

These statements are easy to understand when operating with numerically real-valued metrics. It is not the case that all specifications must be real-valued, however. A specification can be generated using a collection of colors, smells, or other sets of elements that are hard to quantify.

In general, design team members subjectively interpret metrics and target values differently, based on their past experience. To be valid, a specification must have three characteristics that all team members implicitly or explicitly agree on:

- It must come from a set of different possible elements (or values).
- These elements must collectively be *comprehensive.*
- These elements must be *measurable.*

A set is comprehensive if, by knowing the value, the members of the design team can interpret the extent of achievement of the associated informal reason for setting the specification. For example, if a numerical specification of "20" for machinability is not agreed on or understood by the design team, it is not a good specification of machinability. The specification must be grounded with understood points (Ref 11), such as associating "50" with the machinability of 1020 steel.

Formal definitions of measurability exist (Ref 11), but essentially the design team must be able to order the set of values, at least partially. Otherwise, the set is no more formal than a person's subjective interpretations. It must be understood that "30" is better than or equal to "20," for example. This is not an issue with quantitative numbers, but it is with non-numeric datasets. (See the article "Computer-Aided Materials Selection" in this Volume.)

Identifying Specification Metrics

To establish an initial set of engineering specifications, a design team should begin by listing each subfunction. For each subfunction, a relevant product subsystem (either architectural assembly or physical subsystem of a redesign) is examined. From these sources, a means to "instrument" the product subsystem to measure the functional flows in and out of the sub-function should be considered. Depending on the product, this instrumentation can rely on engineering data acquisition, or it can be as simple as touch, feel, or look. A partial list for the fingernail clipper is shown in Fig. 12, with the process assumption of a mechanical cutting surface (a blade).

This approach generally produces better results than other approaches, in that the sub-functions are more quantified than customer needs. The approach clearly relies on the creativity of the design team, but the creativity is composed of two stages: conceiving of the function structure itself and conceiving of how the flows of each subfunction can be "instrumented" for measurement on the product subsystem. It is less of a conceptual leap to generate measurable metrics for an independently generated set of subfunctions, each associated with a subsystem, than to generate measurable metrics for each customer need directly.

Next, the metrics generated must have target values assigned to them. This assignment is completed by examining the relevant customer needs associated with each metric. In general, establishing a target may require some calibration of the metric. For example, once it is understood that handle temperature is a good metric to represent comfort of the customer, it may be necessary to test different handle temperatures with customers to determine the highest acceptable temperature value.

After these steps, a relevant hierarchical set of functional specifications is completed. It can be collected into a house of quality matrix, for example, to verify and communicate how customer needs are being met. The house of quality can be used to document the product design targets of the different team members working in concert.

Ranking Specification Importance

Finally, the importance of each specification can be calculated through the house of quality.

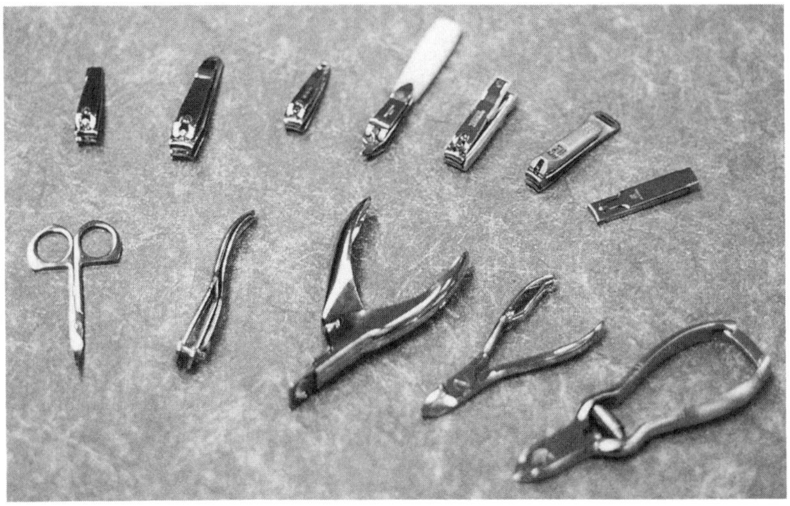

Fig. 16 Fingernail clippers currently on the market

Batter Up Project
Customer Needs

Interviewer(s): Seventh Inning Stretch Team
Date: 9/95
10 Number of customers
0–5 Weighting scale
75% Must confidence

Average customer:
Male/Female, age 12–21 yrs. old, student
Person with severe mental/physical disabilities

Customer need	Weight
Manufacturing	2
Easy to build/duplicate	4
Affordable	3
Set up	
Can be used in several positions	4
Useful for different grip abilities	1
Adjustable for different skills	Must
Various social settings	3
Operation	
Hits ball consistently	Must
Operates easily	3
Ball flies differently	3
User is independent	5
Resets itself	2
Stimulates sensory modes	
Visual	1
Tactile	3
Auditory	4
Store/maintain	
Compact transport/storage	4
Simple maintenance	3

Fig. 17 Partial customer needs list (Batter Up! example)

This calculation is carried out in the usual manner by inserting relationship values in the house of quality matrix to relate customer needs to engineering specifications. Then a suitable algebra is used to combine relationship ratings with customer need importance ratings, resulting in importance ratings for individual specifications.

The first step is to map customer needs to the specifications. In typical house of quality manner, the relationships are captured by a set of symbols, described below. There is always a question concerning how to make these assignments, and in this application, the problem is even more complex because of the additional *must* constraints. When some of the customer needs have *must* ratings and others do not, how can symbols be used to make importance ratings?

We propose the following system. One considers each specification one at a time and assigns symbols based on the following subjective rules:

Symbol	Meaning
⊙	Attaining the target on this specification will totally satisfy the customer need
○	Attaining the target on this specification will satisfy at least half of the customer need
Δ	Attaining the target on this specification will satisfy less than half of the customer need
(Blank)	Attaining the target on this specification will do nothing toward satisfying the customer need

The linguistic values of "half" and "nothing" define both the subjective weighting criteria and the level of resolution that the design team must take care to agree upon.

The next step is to map the *must* ratings of customer needs into *must* assignments for the specifications. The specifications with *must* ratings then act as functional constraints for the design team—they must be met.

It is not immediately clear what overall algebra is appropriate for mapping customer importance ratings (Eq 2 and 3) into importance ratings for the specifications. In particular, the means to combine *must* and non-*must* ratings is not obvious. On the other hand, it is clear that if a *must* customer need has a ⊙ relationship with a specification, then that specification deserves a *must* rating. For the remaining specifications that derive from that *must* customer need, the *must* ratings need not be treated as such, because they are already "covered" with a specification. That *must* customer need can be assigned the highest numerical importance rating in Eq 3 for use in ranking the remaining non-*must* specifications.

Given these insights, the best approach is to insist that every *must* customer need should have at least one ⊙ entry in its mappings to the specifications. That specification can serve as a direct proxy measurement for the *must* customer need.

If no specification exists that can be directly mapped with an ⊙ entry, the specification list should be re-examined. If the condition persists, the team must agree on the set of specifications that predominantly map to the *must* customer need. All of these specifications should then be assigned *must* ratings, keeping in mind that an excessive number of *must* ratings places excessive demands on a design team.

Table 8 Partial morphological matrix (Batter Up! example)

Functions	Mechanical	Hydraulic	Electrical	Miscellaneous
Adapt to handedness	Gears, pulleys, wheels, pegs, pins	Water/air pressure	Solenoid, servo motor	...
Accept person	Walking, rolling to product	Riding a water wave, flying	...	...
Accept bat	Slot, groove, hole, guide rail, ridge	Fluid suction	...	Magnetic attraction
Position bat	Pegs, axis movement	Piston cylinder	Solenoid, servo motor	Magnets
Secure bat	Clamp, vice, belt, brace, latch	Fluid suction	Electrical charge (static cling)	Velcro, magnets, adhesion
Initiate bat swing	Conveyor, lift, mechanical impulse, catapult	Waterjet, piston cylinder	Electrical impulse	Magnetic repulsion
Guide bat	Guide rails, parallel plates, friction, loop on bat, string, hole in bat	Jet stream	...	Magnetic restriction/attraction
Channel bat	Human input, gravity, string	Air flow, water pressure	...	Magnets, explosion
Accept ball	Mount, cartridge, hoop	Tube	...	...
Adjust ball height	Jack, pegs, pulley system, lever	Air pressure	Electric current	Magnets, explosion
Maintain ball height	Clamp, mount	Air pressure, fluid suction	...	Magnets, adhesion
Propel ball	Loop, impact	Air suction, jet stream, water pressure	...	Magnetic attraction, explosion
Accept energy	Lever, four bar, crankshaft, rope	Tube, pipe, fan, windmill	Electrical inlets	Metal surfaces, panels, fuses
Store energy	Translational/rotational spring, material deformation, rubber band, pendulum, bat movement	Fluid column, compressed air (balloon, bladder)	Batteries, capacitor	Magnetic field, solar panels, chemical
Transform energy	Gears, belt/sprocket, lever, four bar, cam, rack and pinion, universal joint	Piston cylinder	Motor, generator	...
Transmit energy	Linkages bearings	Pipe, volume deformation	Wires, volt potential energy	Magnetic field
Initiate vibration	Button, lever, string, switch, key		Laser	
Control vibration	Shaker, unbalanced motor, virotube, centrifuge	Pneumatic, wave machine, water pulse, bubbles	Sound waves (Bass)	Magnets
Terminate vibration	Button, lever, string, switch, key		Laser	...
Make sound	Chimes, bell, drums, siren, klaxon	Whistle, horn, water waves	Speaker	Chemical reaction, explosion
Stop bat	Block, slow with spring, sponge, pillow, rubber stopper, friction	Water/air resistance	...	Magnets
Reset function	Reapply functions: position bat, secure bat, accept ball, adjust ball height, accept energy, store energy			
Product stability	Weight, clamps, stakes, balance, strength of materials	Suction cups	...	Gyros, magnets, adhesion of part joints
Universal design			Global function	
Inclusion			Global function	

Once the specifications that act as constraints are identified, the subsequent specifications can be assigned numerical importance ratings. This process is usually a simple exercise of multiplication and addition. The causality values are given numerical equivalents, typically:

Symbol	Numerical equivalent: Option 1	Numerical equivalent: Option 2
⊙	9	1.0
○	5	0.5
Δ	3	0.3
(Blank)	0	0

Any *must* customer needs are assigned the maximum importance value, typically 9 out of 9 or 1.0 out of 1.0. Then the numerical importance of each specification is determined through a weighted sum across the needs:

$$\omega_i = \sum_j T_{ij}\, \omega_{CR_j} \qquad \text{(Eq 4)}$$

where T_{ij} is a numerical equivalent taken from the table above. The results can then be linearly normalized to any selected scale.

The results for the fingernail clipper are shown and summarized in Fig. 13, the house of quality

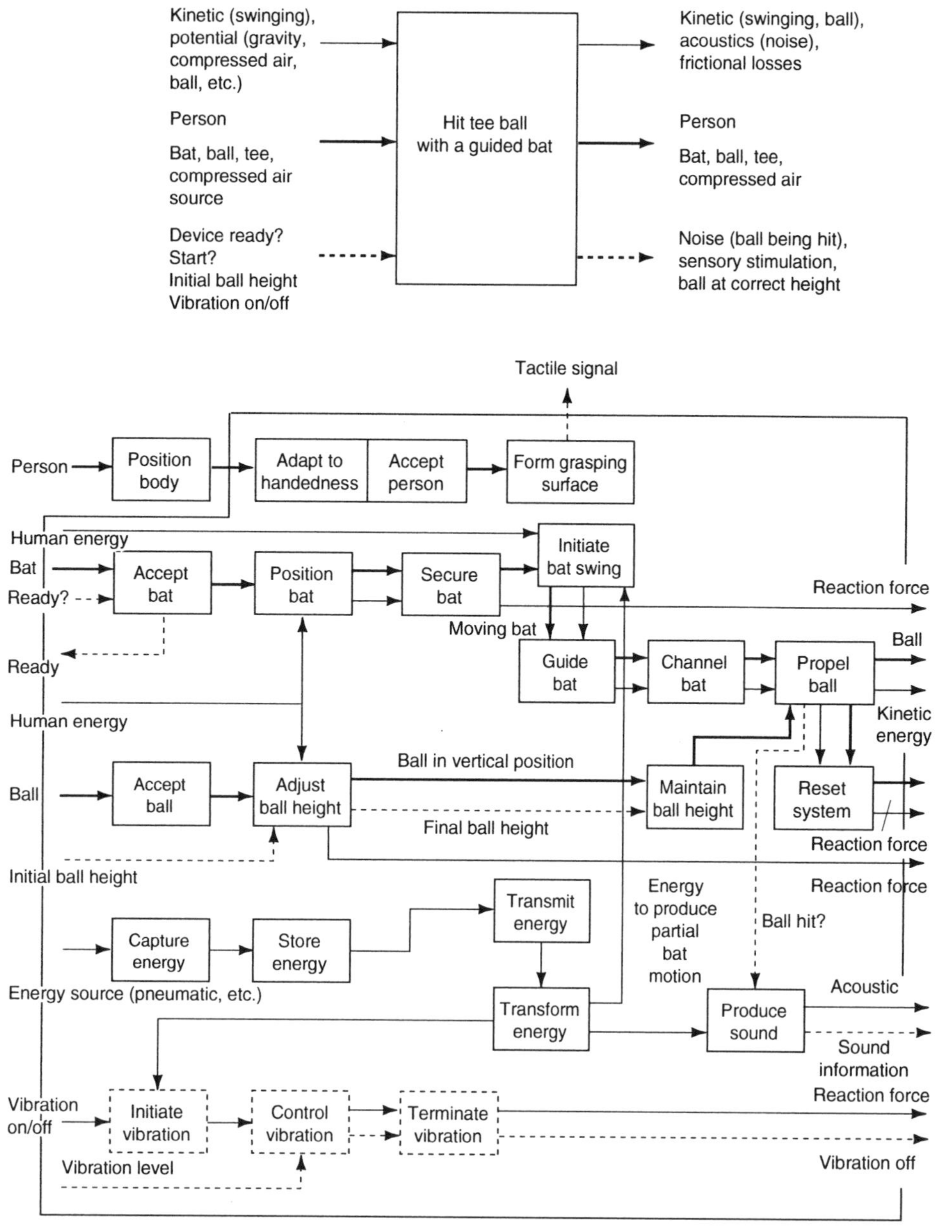

Fig. 18 Black box model and refined function structure (Batter Up! example)

matrix for the clipper. Note the propagation of the *must* ratings onto the engineering specifications. These can now be readily modeled into hard constraints for the design team. The other numerically rated specifications become objectives that can be traded off in design decision making.

Generating Design Configurations and Form Concepts

Figure 13 illustrates a crisp, quantitative specification for product design. Its roots are grounded in customer needs, and its structure takes into account the established product functionality. However, this specification does not complete the conceptual design process. Form solutions must be generated for the decomposed subfunctions, combined into a finite set of concept variants, and then selected using the quantified specifications exemplified by Fig. 13. Other articles in this handbook document detailed methods for executing these tasks. As a preview, however, let's consider the fingernail clipper example and generate solutions that may be derived as a direct consequence of our systematic approach.

Various methods for concept generation are documented in the literature (Ref 2, 3, 14–18, 35). For the purpose of the fingernail clipper design, we seek to develop concepts according to a four-step process:

1. Generate as many solution principles to each of the subfunctions (Fig. 9) as possible.
2. Prune the number of solution principles using the criteria of Fig. 13.
3. Combine the remaining solution principles into a number of feasible concept variants that satisfy the entire set of product functions.
4. Identify avenues for function sharing in each concept variant to reduce part count and complexity.

Figures 14 and 15 illustrate a subset of the results from this process. The morphological matrix contained in Fig. 14 shows a breadth of form solutions to the decomposed functions. In turn, Fig. 15 shows some possible combinations into concept variants. Because the concepts are generated directly from the functional specification, the likelihood of satisfying customer needs is high. Figure 16 shows the range of fingernail clippers that are actually on the market.

Product Application Example

This section presents a brief example of successful product design. The product, known as Batter Up!, was an original design (Ref 36) that carved out a new market, initiating the first data point on an S-curve.

Children with disabilities need recreational opportunities and skills as part of their educational development (Ref 37). This general need led to the mission statement "to design a device and/or process that will enhance the recreational activities of students with severe mental and physical disabilities." After customer interviews at Rosedale School in the Austin Independent School District (Ref 36), the mission statement was refined to "the design of a product that would enhance tee ball for the students, while maintaining their integration in the community." The primary market for the device was thought to be educational institutions. Initial analysis showed that institutions should be able to fabricate the device themselves so that the cost would be under \$100.

Customer needs analysis (10 customers) for this application gave the results shown in Fig. 17. Customers wanted a product that would guide a child's bat to hit a softball consistently and that would be adaptable to changing skill levels. Beyond these "musts," the critical needs identified were that the product should be adjustable in height for children in wheelchairs and children of various sizes, be usable without adult assistance, provide auditory stimulation for cause-effect learning, be easily storaged and transported, and be easy to build/duplicate.

A black box model and refined function structure for this product are shown in Fig. 18. The system boundary for this functional analysis included only the operational phase, tee ball. In the function structure, notice the convergence of an external energy source and the human energy from the user. These energy flows, in addition to guidance functions, provide the primary means of satisfying the operational needs of students with disabilities.

A house of quality matrix for the product included measurements of the volume, mass, and number of piece parts. In addition, the length of

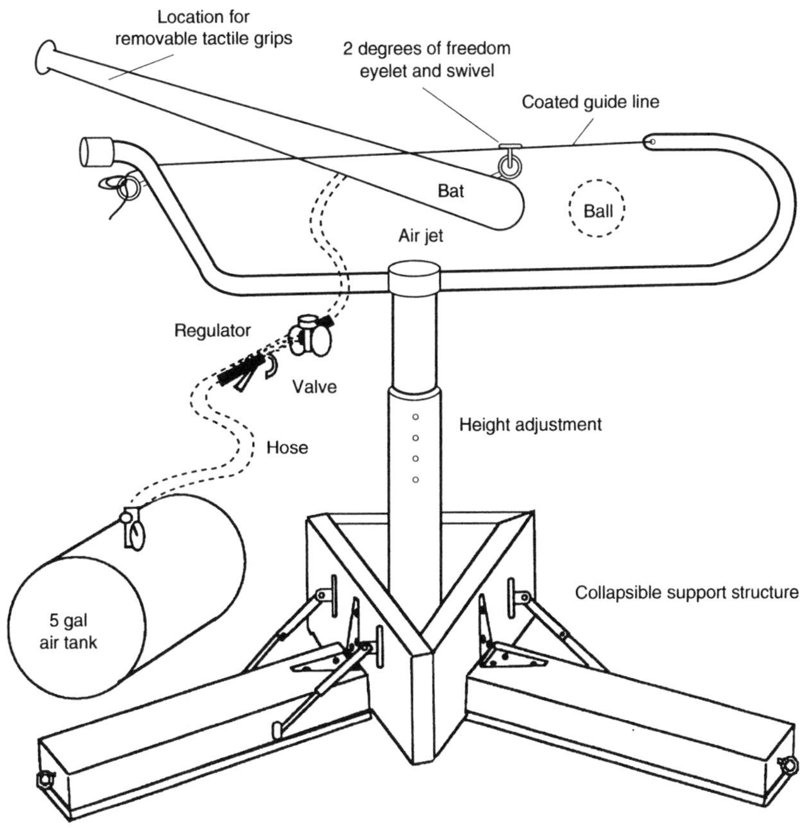

Fig. 19 Design concept for beta functional prototype (Batter Up! example)

bat grips, operating noise, range of bat height, range of handle movement, and range of bat speeds provided a quantified specification of the functions in Fig. 18.

Table 8 shows a partial morphological analysis of the main functions in the function structure. Solution principles are listed discursively according to energy domain. A wide variety of mechanical and electromechanical solutions are shown for the application, especially for the external energy source. These solution principles led to a number of concept variants.

Figure 19 illustrates the selected concept. This concept, based on a pneumatic energy source and a guide line, was designed to make use of basic hardware in consideration of the fabrication limitations of wood shops in secondary schools. After prototype testing, this assistive device was successfully installed in the Austin Independent School District.

Conclusion

Product concepts begin and end with the customer. A concise and quantified specification is essential to developing suitable concepts, and this article presents a systematic method for creating such a specification. We begin with the technical and business market, probing the design task for legitimacy. Customer needs analysis, functional decomposition, and competitive benchmarking are then presented as methods for directly mapping customer statements to functional requirements, in the context of competitors' products. Results of these methods lead to a house of quality matrix, forming the product specification and setting the playing field for concept generation. Further information is provided in the articles "Conceptual and Configuration Design of Parts" and "Creative Concept Development" in this Volume.

A common criticism of design methodology is that it is explanatory and thought-provoking, but not relevant to actual practice. This criticism is becoming an antiquated view. Leading industrial companies are constantly seeking more structured approaches to their product development processes, especially in the era of concurrent development, intensive computing, and downsizing of workforces. An effective structured method allows not just one expert to understand and complete a task, but many others as well, and it provides for all decisions to be documented in case of subsequent redesign.

REFERENCES

1. Wall Street Journal, 24 April 1995, p 1
2. G. Altshuller, *Creativity as an Exact Science,* Gordon & Breach Publishers, 1984
3. K. Ulrich and S. Eppinger, *Product Design and Development,* McGraw-Hill, 1995
4. M.P. McNair, *The Case Method at the Harvard Business School,* McGraw-Hill, 1954
5. R. Ratliff, K.L. Wood, C. Sumrell, and R.H. Crawford, "A Conceptual Design Study of the EVT Process of In-Circuit Tests," IBM Technical Report, IBM ECAT, Austin, TX, May 1993
6. R. Ronstadt, *The Art of Case Analysis: A Guide to the Diagnosis of Business Situations,* Lord Publishing, Natick, MA, 1988
7. V. Barabba and G. Zaltman, *Hearing the Voice of the Market: Competitive Advantage through the Creative Use of Market Information,* Harvard Business School Press, 1991
8. G. Urban and J. Hauser, *Design and Marketing of New Products,* 2nd ed., Prentice-Hall, 1993
9. S. Shiba, A. Graham, and D. Walden, *A New American TQM,* Productivity Press, Cambridge, MA, 1995
10. A. Griffin and J. Hauser, The Voice of the Customer, *Marketing Sci.,* Vol 12 (No. 1), Winter 1993
11. K. Otto, Measurement Methods for Product Evaluation, *Res. Eng. Des.,* Vol 7, 1995, p 86–101
12. K. Ulrich, The Role of Product Architecture in the Manufacturing Firm, *Res. Policy,* Vol 24, 1995
13. M. Fisher, A. Jain, and J. MacDuffie, Strategies for Product Variety: Lessons from the Auto Industry, *Designing the Firm,* E. Bowen and B. Kogut, Ed., Oxford Press, New York, 1995
14. G. Pahl and W. Beitz, *Engineering Design: A Systematic Approach,* Springer Verlag, 1984
15. N. Cross, *Engineering Design Methods—Strategies for Product Design,* 2nd ed., John Wiley & Sons, Chichester, 1994
16. J. Dixon and C. Poli, *Engineering Design and Design for Manufacturing—A Structured Approach,* Field Stone Publishers, Conway, MA, 1995
17. D. Ullman, *The Mechnical Design Process,* McGraw-Hill, 1992
18. V. Hubka and W. Ernst Eder, *Theory of Technical Systems,* Springer-Verlag, Berlin, 1984
19. D. Sobek and A. Ward, Principles from Toyota's Set-Based Concurrent Engineering Process, *Proc. Design Theory and Methodology Conf.,* ASME, Aug 1996
20. V. Hubka, M. Andreason, and W. Eder, *Practical Studies in Systematic Design,* Butterworths, London, 1988
21. K. Otto, Forming Product Design Specifications, *Proc. Design Theory and Methodology Conf.,* ASME, in press
22. K. Otto and K. Wood, A Reverse Engineering and Redesign Methodology for Product Evolution, *Proc. Design Theory and Methodology Conf.,* ASME, Aug 1996
23. L. Miles, *Techniques of Value Analysis and Engineering,* McGraw-Hill, 1972
24. Y. Shimomura, H. Takeda, and T. Tomiyama, Representation of Design Object Based on the Functional Evolution Process Model, *Proc. Design Theory and Methodology Conf.,* ASME, Sept 1995
25. M. Hundal, A Systematic Design Method for Developing Function Structure, Solutions, and

Concept Variants, *Mechanism and Machine Theory,* Vol 25 (No. 3), 1990, p 243–256

26. A. Little, "A Reverse Engineering Toolbox for Functional Product Disassembly and Measurement," Master's thesis, The University of Texas, Austin, TX, 1997
27. D.A. Cutherell, Product Architecture. *The PDMA Handbook of New Product Development,* M.D. Rosenau, Ed., John Wiley and Sons, 1996
28. K. Ingle, *Reverse Engineering,* McGraw-Hill, 1994
29. A. Thornton and D. Meeker, Benchmarking within the Product Development Process, *Proc. 1995 Design Technical Conferences,* Vol DE-83, ASME, 1995, p 639–648
30. R. Camp, *Benchmarking: The Search for Industries' Best Practices That Lead to Superior Performance,* Quality Press, 1989
31. K. Ishikawa, *Guide to Quality Control,* 10th ed., Asian Productivity Org., Hong Kong, 1992
32. D. Clausing, *Total Quality Development,* ASME Press, New York, 1994
33. Y. Akao, *Quality Function Deployment: Integrating Customer Requirements into Product Design,* Productivity Press, Cambridge, MA, 1990
34. J. Hauser and D. Clausing, The House of Quality, *Harvard Bus. Rev.,* May–June 1988, p 63–73
35. M. French, *Conceptual Design for Engineers,* Springer-Verlag, London, 1985
36. G. Faulkner, A. Yaunitell, and K. Jafar, "Designing Fun and Independence into Tee Ball," ME 392M Project Report, The University of Texas, Dept. of Mechanical Engineering, Austin, TX, 1995
37. D. Auxter, J. Pyfer, and C. Huettig, *Principles and Methods of Adapted Physical Education and Recreation,* 7th ed., Mosby-Year Book, 1993

SELECTED REFERENCES

- M. Asimow, *Introduction to Design,* Prentice Hall, 1962
- S. Ashley, If at First You Don't Succeed, Redesign, *Mech. Eng.,* Vol 116 (No. 3), 1994
- P. Asthana, Jumping the Technology S-Curve, *IEEE Spectrum,* June 1995, p 49–54
- D. Clausing, *Total Quality Development,* ASME Press, 1994
- G. Dieter, *Engineering Design: A Materials and Processing Approach,* 2nd ed., McGraw-Hill, 1991
- V. Hubka, *Principles of Engineering Design,* Butterworth & Co., London, 1982
- D. Lefever, "Integration of Design for Assembly Techniques into Reverse Engineering," Master's thesis, The University of Texas, 1995
- D. Lefever and K. Wood, Design for Assembly Techniques in Reverse Engineering and Redesign, *Proc. 1996 Design Theory and Methodology Conf.,* ASME, Aug 1996
- S. Love, *Planning and Creating Successful Engineered Designs,* Advanced Professional Development, Los Angeles, 1986
- J. Miller, "New Shades of Green," Master's thesis, The University of Texas, Austin, TX, 1995
- J. Nevins and D. Whitney, *Concurrent Design of Products and Processes,* McGraw-Hill, 1989
- S. Pugh, *Total Design,* Addison-Wesley, 1991
- D. Ullman, "A New View of Function Modeling," paper presented at International Conference on Engineering Design (The Hague), 17–19 Aug 1993
- J. Warfield, *Societal Systems,* Intersystems Publications, Salinas, CA, 1989

Conceptual and Configuration Design of Parts

John R. Dixon, Professor Emeritus, University of Massachusetts Amherst

PRODUCTS ARE DESIGNED by a recursive process that decomposes the initial abstract physical and functional concept of a whole product into subassemblies, and then decomposes the subassemblies further into more subassemblies, and so on until all the elements that describe the product are either individual parts or standard modules. As recursive decomposition proceeds, the description of the product—that is, the design—becomes less and less abstract and more and more detailed. Ultimately, the specific standard modules and standard parts to be used must be selected by the designer, and any nonstandard (i.e., special-purpose) parts must be designed. (A *standard module* is a product manufactured in quantity for use in a number of other products. Examples are motors, clutches, switches, pumps, and the like. A *standard part* is a part manufactured in quantity for use in a number of products. Examples are screws, bolts, brackets, beams, shafts, and the like. Most standard parts and standard modules are offered for sale through the catalogs of their manufacturers.)

This article concerns the conceptual and configuration design of *special-purpose parts*—parts that are designed and manufactured especially for use in a particular product (Ref 1). In the design of special-purpose parts, the following issues must be considered, and the discussion here is organized in terms of these issues:

- Functionality
- The relationship of the part to the whole assembly or subassembly, including their design for assembly
- The material from which the part will be made and the process by which it will be manufactured
- Developing the configuration
- Space, fit, and tolerances
- Design for manufacture (DFM)

Although these issues can be discussed separately and sequentially, designers must consider them simultaneously (or at least repeatedly and iteratively) as parts are designed. Beginning a part design, however, requires a clear understanding of the function of the part to be designed.

Part Functions

At their most abstract level, parts are initially conceived as the result of the functional or physical decomposition of an assembly or subassembly. Thus, parts contribute to the overall functionality of their parent assembly. However, unless the product consists of only one part, the function of a part is never identical to the function of the product as a whole. Parts have their own special, usually limited and highly technical, functions. Only in synergistic concert do the functions of parts produce the function of their assembly or product. Readers are referred to the article "Conceptual and Configuration Design of Products and Assemblies" in this Volume for a complete discussion of product and assembly functionality.

Table 1 is a list—comprehensive, but not necessarily complete—of functions served by parts. There are also a few examples of the types of parts and features of parts that perform the listed functions. Some parts, of course, are designed to serve more than one function.

In addition to such overall part functionality, special features are almost always designed into parts to aid in their manufacture or to reduce material cost. Examples are listed in Table 2.

Relationship of Parts to Assemblies

By performing their own limited functions, parts contribute to the functionality of their parent assembly or subassembly. However, parts are expensive, sometimes very expensive. Not only are they expensive to design—a typical part in a complex assembly can require three to four designer-months—but then the parts must be manufactured. This requires the design and manufacture of tooling as well as the cost of materials and processing. Then the parts must be assembled, either manually or with the aid of machines.

Design and manufacturing costs for parts are substantial, but they are not all the costs—and sometimes not even the largest. Parts must be stored, kept track of, and often distributed for regional or local inventories. The overhead costs

Table 1 Functions served by parts

Function	Examples of part types or features
Transmit or support force(s) or torque(s)	Brackets, beams, struts, columns, bolts, springs, bosses, knobs Levers, wheels, rollers, handles Parts that fasten, hold, or clamp, such as bolts, screws, nails
Transmit or convert energy	
Heat	Heat fins, electric resistance heating elements
Mechanical power	Shafts, connecting rods, gears
Electricity	Wires, lightbulb elements, resistors
Provide a barrier (for example: reflect, cover, enclose, or protect)	
Light	Walls, plugs, caps
Heat	Thermal insulators, thermal reflecting surfaces
Electricity	Electrical insulators, magnetic shields
Sound	Walls, sound-absorbing wall surfaces
Control motion	Cams, grooves, slots, gears
Allow passage (of light, rods, shafts, wires, pipes, etc.)	Holes, windows, grooves
Control or regulate the passage of	
Fluids	Nozzles, orifices, pipes, ducts
Light	Shutters, wheels
Indicate	Clock hands, instrument needles, colors, embossing
Locate or guide	Grooves, holes, bosses, tabs, slots

Table 2 Special features designed into parts to aid manufacturing or to reduce material cost

Function	Examples of part features
Aid manufacturing	Fillets, gussets, ribs, slots, holes
Add strength or rigidity (e.g., stiffen)	Ribs, fillets, gussets, rods
Reduce material use	Windows or holes through walls, ribs that allow thinner walls, slots
Provide a connection or contiguity (so the part can be a single part)	Walls, rods, ribs, gussets, tubes

associated with a given part varies depending on the nature of the part and the business, but they can be large.

The design rule resulting from these considerations is that the number of parts in a product should be minimized. Thus, after the function of a part has been articulated, the next step in the design process is to determine whether the part is *really* necessary. Can the function be eliminated, or can the function be accomplished by combining parts? For specific guidance in addressing this issue, readers are referred to the article "Design for Manufacture and Assembly" in this Volume and to Ref 2.

There is a tendency for designers to underestimate the importance of reducing part count in a product, especially when combining part functions results in parts that are very complex. Though precise calculations are not possible, all indications are—everything considered—that a single complex part is less expensive than two simpler parts. Moreover, this conclusion is generally true even when the combined part becomes extremely complex. As long as a single complex part can be manufactured and assembled, it is almost always the preferred design.

There is a proviso. If the single, more complex part is going to cause a costly time-to-market delay during tooling and production ramp-up, then the savings may not be realized. See Ref 3.

To illustrate, one firm designed a new single part that would replace eleven injection-molded parts. Not surprisingly, the new part was incredibly complex. When it came time to produce it, the extreme complexity made rapid, high-yield production impossible. Therefore, the design was modified resulting in two parts, and new tooling made. Subsequent production went very well. Fortunately in this case, the time delay required to redesign the new part, and to get new tooling, was not critical to marketing the product. Thus, in this case the net savings in tooling, manufacturing, assembly costs, and overhead created by reducing eleven parts to two were tremendous.

Design for manufacturing requires designers to know the limitations of manufacturing processes, so they will not design parts that are unnecessarily difficult or expensive to produce. As pointed out in the article "Design for Manufacture and Assembly," however, reducing part count requires that designers know not only the limitations of processes but also their full capabilities, so that savings derived from employing fewer parts, even if they are more complex, can be realized.

Material and Process Selection

Generally the design of a part cannot begin, and can never proceed very far, without a decision being made about the material and manufacturing process to be employed. This subject is covered thoroughly in the articles "Relationship between Materials Selection and Processing," and "Manufacturing Processes and Their Selection," in this Volume.

Because this subject is discussed so completely elsewhere in this Volume, it will not be discussed here. Readers should note, however, that materials and process selection is a critical and early step in the design of every part.

Determining the Configuration

The preliminary steps in part design have been described above: identifying the required functions of the part, determining whether the part is really necessary, and selecting the material and manufacturing process by which the part will be made. Next, designers must make decisions about how the part will be configured.

To configure means to select the elements of which an object will be composed and to arrange the relative positions and connectivity of those elements. Exact dimensions, precise shapes, and tolerances are not explicitly included in the configuration design.

The "elements" that comprise a part are called features. The types of features that can make up a part include:

- *Walls* of various kinds such as flat, curved, and so forth
- *Add-ons* to walls such as holes, bosses, notches, grooves, ribs, and so forth
- *Solid* elements such as rods, cubes, tubes, spheres, and so forth
- *Intersections* among the walls, add-ons, and solid elements

There are usually several—and sometimes many—ways a part can be configured. As always in design, it is best to proceed by first generating a number of alternatives and then evaluating, modifying, and again evaluating the alternatives before accepting one of them. In part design, the only feasible way to get started with this process is by creating sketches of alternative configurations. Because parts are usually three dimensional, this requires designers to possess good three-dimensional visualization and drawing abilities. The sketching can be done with a computer if the available modeling software supports quick generation of the alternative possibilities. At this stage in the design, sketches need not be formal and dimensions need be only approximate. The goal here is only to determine what features and arrangement of features will result in the best overall design once dimensions and tolerances are assigned.

Making sketches of a part can begin only after the space allowed for the part and the physical contacts required for it are defined, at least approximately. Thus the process begins with a sketch of the surroundings of the part. This sketch should be roughly to scale, but only roughly because the design of the surrounding parts may not be complete, or will be subject to revision. This sketch will show not only the material surroundings but also the approximate locations of external forces, flows, energy transfers, and the like that the part must accommodate. Thus sketching begins by showing the essential surroundings of the part and locates loads, possible support points or areas, heat or other energy flows, adjacent parts, forbidden spaces, and so on.

With the surroundings defined as much as is practical, the sketching of alternative part configurations can begin. Even in relatively simple cases, however, there may be far too many possible configurations for all to be considered. The generation of alternatives must be guided and limited by qualitative physical reasoning about the functionality of the part and about manufacturing issues. These two subjects are discussed in the next two sections of this article.

One way to think about a configuration is to note that it establishes the attributes of the part that will require subsequent dimensioning.

Qualitative Physical Reasoning to Guide Generation of Part Configuration Alternatives

Considering geometric issues only has the potential for generating a very large number of alternative configurations. The generation process must be guided and limited in a way that provides a smaller set of those alternative configurations that are the most likely to be successful. Then this smaller set can be subjected to more detailed evaluation and redesign. This guidance is provided by qualitative physical reasoning about the functionality and manufacturability of the configuration. Even though at the configuration stage of part design actual dimensions have not been determined—the sizes and spatial relationships of the features are still only approximate—knowledge of physical principles and manufacturing processes can still be applied to help create the most effective alternative designs for further evaluation.

Qualitative in this context means, essentially, without numbers—though orders of magnitude of numbers are usually involved. The basis of qualitative reasoning is rooted in fundamental physical principles. It is much, much more objective than guesses or feelings. To illustrate, the following paragraphs discuss some of the configuration issues that can be guided by qualitative physical reasoning.

Material Use. Good configurations will enable the use of as little material as possible; that is, they can use material efficiently to achieve their functionality. The I-beam configuration is a good example of material efficiency when the function is to support a load across an open space. By placing the most material away from the neutral axis of the beam, it makes maximum use of the material to carry the load.

In the case of a part whose function is to transmit a force, a very common part function, stresses in the part should be as uniform as possible throughout, and as high as is safe. Having relatively lightly stressed material in some sections of such a part is a waste of material, at least from a functional point of view in supporting forces. Thus, configurations are best that have the potential for uniformly high stresses throughout. This generally means that tension and compression are better than bending for supporting and transmitting forces.

There are exceptions to this general guideline, however, because manufacturing must also be considered. In a forging, for example, it may be that reducing the amount of material used results in the need for more dies and higher loads (to move more material greater distances and to produce the thin sections). The added dies and processing costs may be greater than the cost of material saved.

Columns can fail in compression, but mostly they fail by either lateral or torsional buckling. The latter is a special concern because it is often neglected, allowing columns to be designed that are perfectly safe in compression and even lateral buckling, but which fail (sometimes surprisingly) by twisting.

It is known from strength of materials principles that the cross-sectional second moment of area is a key to buckling resistance. For lateral buckling of a column, the minimum moment, considering a 360° realm of possibilities, is the only one that matters. Thus a configuration that has a more-or-less constant moment of inertia regardless of orientation will be the most efficient. Clearly, a round (or at least symmetrical), configuration is the most efficient.

For torsional buckling, it is the torsional second moment of area that matters. Intersections of walls increase this moment somewhat, but the most effective configuration against torsion is to create sections that are hollow. This has the effect of moving material away from the axis of twist, somewhat analogous to the logic of an I-beam. Thus if torsional buckling is a concern, a hollow symmetrical section may be most efficient.

Brackets and Trusses. A bracket is a part whose function is to support a load or loads, transferring the force to an adjacent supporting structure, usually a wall of some kind. Brackets often must be designed with spatial constraints in mind; that is, there are regions of space that may not be invaded by the bracket material. But whatever the constraints, qualitative rules can help guide the generation of effective alternatives. A few examples of such rules follow.

Tension and compression are more efficient ways to transmit forces from one place to another than is bending. When bending is required, it is best to make the "beam" as short as possible and make the cross section as much like an I-beam or box as possible.

In most trusses, loads are carried by compression and tension; thus trusses tend naturally to be efficient configurations from the standpoint of material use.

Sometimes, a trusslike structure must be designed to connect loads at oddly distributed points to supports at other locations. Configurations that have the potential for using as much tension and compression as possible (as distinguished from bending) are desirable. Thus placing members in the same directions as the loads can produce a more effective configuration.

Inside Corners. To reduce local stress concentrations, configurations that allow for rounded inside corners or that can add on fillets are best. As an added benefit, rounded inside corners almost always support ease of manufacturability.

Heat Transfer. The configuration of parts whose function involves the transfer of heat can also be evaluated, in part, by qualitative physical reasoning. Heat is transferred by conduction, convection, and/or radiation. To the extent that the heat transfer mechanism employed is conduction, configurations that result in the lowest possible temperature gradients are the best. This is accomplished by providing the potential for using a high thermal conductivity material and by providing for a suitably large, direct path. Configurations that provide opportunities for reducing contact resistances to adjacent parts involved in the heat flow are also helpful.

When convection is the primary mechanism, increasing the heat flow requires a configuration that provides large surface areas adjacent to the surrounding medium. (This is the reason fin configurations are often used.) If the configuration can help induce turbulence in the surrounding medium, that is also helpful. In the case of finlike parts, the surface area is naturally large, but designers must also consider the need for conduction of heat to the extremities of the fin. Keeping a uniformly high temperature would require a tapered fin, but this may raise manufacturing difficulties. Fins that are too long and thin cannot be manufactured easily, but this is an issue that will get sorted out during the parametric design stage.

Radiation is an important heat-transfer mechanism requiring that the heat exchanging surfaces be as large as possible, and be "visible" to each other. The materials should have the potential for being made with surfaces of high emissivity and absorptivity.

Whenever possible, of course, configurations that can make use of multiple heat-transfer mechanisms may be advantageous, but trade-offs here are not always obvious until computations can be made at the parametric design stage.

Insulating against Heat Transfer. To insulate against heat transfer, designers must try to prevent or reduce all three of the heat-transfer mechanisms: conduction, convection, and radiation. To reduce conduction, configurations that can reduce or eliminate direct paths for heat conduction, especially when the material has a high thermal conductivity, must be used.

To block convection, the circulation of the surrounding gas, usually air, must be inhibited. Most insulating materials thus create a large number of very small pockets for gas, which (if it cannot circulate to produce convection heat transfer) is a poor conductor of heat.

An interesting configuration to discuss here in the context of qualitative reasoning about configurations is the evacuated wall bottle for storing hot or cold liquids. This configuration virtually eliminates both conduction and convection through the sides and bottom by employing an evacuated chamber in which there is no material to conduct heat or air to circulate in convection. Moreover, by using very low emissivity surfaces, it reduces radiation greatly. Almost all of the heat loss from such a bottle is through the corked opening, which should therefore be kept as small as possible.

Residential windows make another interesting example of configurations that must resist heat transfer. Configurations of metal-frame windows that allow a direct metal (or even wood) path from outside to inside are poor, of course. The concept of providing an air space between two panes in such windows is based on qualitative reasoning about heat transfer: the air space is an insulator, but it does not interfere with light flow.

Qualitative Reasoning about Manufacturing to Guide Generation of Part Configuration Alternatives

Configurations of parts are also strongly influenced by manufacturing issues, and it is not necessary or desirable to wait until dimensions and tolerances have been established to consider manufacturing issues in part design. This is one important reason the manufacturing process must be selected early in part design, and it is also the reason designers must be familiar with manufacturing processes.

This Volume contains a number of articles describing various manufacturing processes in considerable detail, so they will not be discussed in detail again here. See also Chapter 3 in Ref 1.

The following paragraphs describe typical DFM guidelines for several of the most widely used processes. The list is by no means complete, and is intended only to indicate the kinds of issues that can be considered in guiding the generation and evaluation of part configurations.

Qualitative DFM Guidelines for Injection-Molded, Compression-Molded, Transfer-Molded, and Die-Cast Parts. Injection molding, die casting, compression molding and transfer molding are all internal flow processes, followed first by cooling and solidification, and then followed by ejection from the mold. Thus, part configurations for these processes should ideally be designed so that: (a) the internal flow can be smooth and fill the cavity evenly, (b) cooling, and hence solidification, can be rapid to shorten cycle time and uniform to reduce warpage, and (c) ejection can be accomplished with as little tooling complexity as possible.

In designing parts to be made by injection molding, die casting, compression molding, and transfer molding, designers must decide—as a part of their design—the direction of mold closure and the location of the parting surface. Though these decisions are tentative, and advice should be sought from a manufacturing expert, it is really impossible to do much DFM in these processes until they are made.

An easy-to-manufacture part must be easily ejected from the die, and dies will be less expensive if they do not require special moving parts (such as moving side cores) that must be activated in order to allow parts to be ejected. Because undercuts require side cores, part configurations without undercuts are less costly to mold and cast. With knowledge of the mold closure direction and parting surface, designers can make tentative decisions about location(s) of features (holes, projections, etc.) in order to avoid undercuts wherever possible.

Because of the need for resin or metal to flow through the die cavity, part configurations that allow for relatively smooth and easy internal flow paths with low flow resistance are desirable. For example, sharp corners and sudden changes or large differences in wall thickness should be avoided because they create flow problems. Such features also make uniform cooling difficult.

Thick walls or heavy sections will slow the cooling process. This is especially true with plastic-molding processes because plastic is a poor thermal conductor. Thus, part configurations that will not result in thick walls or other thick sections are less costly to produce.

In addition, every effort should be made to design part configurations that will enable walls of uniform, or nearly uniform, thickness to be designed. If there are both thick and thin sections in a part, solidification may proceed unevenly, causing difficult-to-control internal stresses and warping. Remember, too, that solidification time, and hence total cycle time, is largely determined by the thickest section.

Large or complex parts may require two or more gates through which resin or metal will flow into the mold. There will then be fusion lines in the part where the streams meet inside the mold. The line of fusion may be a weak region, and it may also be visible. Therefore, designers who suspect that multiple gates may be needed for a part should discuss these issues with manufacturing experts as early as possible in the design process. With proper design and planning, the location of the fusion lines can usually be controlled as needed for appearance and functionality.

DFM Guidelines for Stamped Parts. Because of the short cycle times per part, tooling and material are generally the dominating cost factors for stampings. For example, at production volumes less than 100,000, the proportion of part cost due to tooling is about 75%, while the proportion of part cost due to processing is less than 2%. Even at production volumes of about 2,000,000, the proportion of part cost due to processing still remains low (less than 5%). At such large production volumes, the greatest proportion of part cost (about 75%) is due to material cost.

For large production volumes, designing part configurations to reduce the amount of scrap can be important. However, in most cases, it is the manufacturing process designer who designs the stamping process (often referred to as developing the "strip layout" for the part) to achieve minimum scrap. Therefore, the discussion below is primarily about the relationship between part design and tooling cost.

The number of distinct features in a stamped part configuration should be kept to a minimum. The reason is that as the number of distinct features increases, the number of die stations increases and both die construction costs and die material costs increase.

The necessity for closely spaced features (holes, slots, ribs, etc.) should be avoided when possible. If features must be spaced closely together (less than three sheet thicknesses), then there may be insufficient clearance for the punches. Even if space permits, however, the die sections become thin, making them susceptible to breakage, and punch breakage can occur due to metal deformation around any closely spaced features during piercing. As a result, two stations are required for each type of closely spaced feature; each station creates alternating features.

Configurations that require the use of narrow cutouts and narrow projections should be avoided. If a narrow projection is present, then to separate the link from the strip at one station requires a blanking punch with a narrow cutout along with a die containing a narrow projection. In this situation, the narrow section of the die would be easily susceptible to damage.

The number of bend stages in a part should be kept to a minimum. A U-shaped part, with both bends in the same direction, will have both bends created at the same time at one die station. An equivalent Z-shaped part, with bends in opposite directions, will require that each bend be separately created at two different die stations. Thus, bends in opposite directions create increased tool construction and die material costs.

Bend angles greater than 90° should be avoided whenever possible. To create a bend angle greater than 90° requires two die stations. At the first station the part is bent through 90°, while at the next station the part is bent past 90°. Clearly, then, parts with a bend angle greater than 90° require more costly tooling. Thus, configurations that require side-action features should also be avoided, if feasible. Features such as holes, whose shape and/or location from a bend line must be accurately located, must be created after bending. This requires that one or more additional die stations be added, thereby increasing tooling cost. Also, because the bends themselves cannot be closely controlled, they make poor reference points for important or close tolerances. In general, to keep side-action features to a minimum, the tolerances of feature dimensions that must be referred to bend lines should be generous.

Design for Manufacturing Guidelines for Forged Parts. The nature of the forging process—solid metal is squeezed and moved within a die set to form a part—leads to the following broad DFM guidelines for consideration at the configuration stage.

Materials should be used that are relatively easy to deform because they will require fewer dies, shorten the processing cycle, and require a smaller hammer or press.

Part shapes that provide smooth and easy external flow paths are desirable. Thus, corners with generous radii are desirable. In addition, tall thin projections should be avoided because such projections require large forces (hence large presses and/or hammers) and more preforming stages (hence more dies), cause rapid die wear, and result in increased processing cycle time.

Ribs, if any, should be able to be widely spaced. Spacing between longitudinal ribs should be greater than the rib height; spacing between radial ribs should be greater than 30 °C. Closely spaced ribs can result in greater die wear and an increase in the number of dies required to produce the part.

Internal undercuts, and external undercuts caused by projections, must be avoided because they are impossible to form by the movement of solid metal. External undercuts that are the result of holes should be avoided because they increase both die costs and processing costs.

Qualitative Reasoning on Design for Manufacturing Aluminum Extrusions. Most DFM considerations for aluminum extrusions evolve from the difficulty of forcing the metal to flow uniformly from a large round billet through small complex die openings. This leads to the following "rules" that can help guide the design of more easily extrudable aluminum parts.

Sections with both thick and thin sections are to be avoided. Metal tends to flow faster where thicker sections occur, giving rise to distortions in the extruded shape. Designs should be used that will function with all the walls as uniform in thickness as possible. Designs requiring slugs of material should be avoided. Long, thin wall sections should also be avoided, because such shapes are difficult to keep straight and flat. If such sections are absolutely necessary, then the addition of ribs to the walls will help distribute the flow evenly.

Hollow sections are quite feasible, though they cost about 10% more per pound produced. The added cost is often compensated for by the added torsional stiffness that the hollow shape provides. It is best if hollow sections can have a longitudinal plane of symmetry.

"Semihollow" features should be avoided. A semihollow feature is one that requires the die to contain a very thin—and hence relatively weak—neck.

Design for Assembly Guidelines for Part Design at the Configuration Stage. Readers are referred to the article "Design for Manufacture and Assembly" in this Volume and to the references included there.

Tolerances

The issues of tolerances are at the intersection between the requirements of functionality and capabilities of manufacturing processes. Proper tolerances are crucial to the proper functioning, reliability, and long life of products. However, at the same time a common cause of excessive manufacturing cost is the specification by designers of too many tolerances and/or tighter-than-necessary tolerances.

Mass-produced parts cannot all be produced to any exact dimensions specified by designers. In a production run, regardless of the process used, there will always be variations in dimensions from nominal. Some of the reasons are: tools and dies wear, processing conditions change slightly during production, and raw materials vary in composition and purity. Modern methods for controlling processes are achieving increasingly consistent and accurate dimensions, but variations of some frequency and magnitude are inevitable.

Designers can, in order to get the functionality required, specify the range of dimensional variations in parts that reach the assembly line. However, the stricter these limitations, and the more dimensions subject to special tolerance limitations, the more expensive the part will be to produce.

Fortunately, the functionality of parts seldom, if ever, requires that all or even most dimensions of parts be controlled beyond that obtained by normal manufacturing practices. To achieve the desired functionality (and other requirements) of a part, generally only a few dimensions and other geometric characteristics (e.g., straightness or flatness) may require especially accurate control, while others can be allowed to vary to a greater degree. Thoughtful configuration design can result in parts and assemblies in which the number of characteristics requiring greater than standard dimensional control is minimal. Also, those dimensions that do require critical control may be of a type or at a location where they are more easily controlled during manufacturing.

Every manufacturing process has standard or commercial tolerances. These are the tolerance levels that can be produced with the normal attention paid to process control and inspection. Though standard tolerance values are not always completely and accurately defined—and available in print for designers—they are generally well known to manufacturing process engineers. See also Fig. 5 in the article "Overview of the Materials Selection Process" in this Volume.

Parts designed so that there are no requirements tighter than standard will be the least expensive to produce. Moreover, and this is an important point for designers, the *number* of tolerances that must be critically controlled, whether standard or tighter, is also crucial to ease and cost of manufacturing. Controlling one or two critical dimensions, unless they are of an especially difficult type or extremely tight, is often relatively easy to do if the other dimensions of the part do not need special control.

At the configuration stage of design for parts, even before exact dimensions are assigned, designers can and should think ahead about the dimensions or other characteristics (e.g., flatness) that will need tight tolerancing to achieve the desired functionality. Configurations should be developed, if possible, that reduce the number of such dimensions or characteristics, and configurations avoided that will make tight tolerances more difficult to control. The following paragraphs provide some guidance for several of the most used manufacturing processes.

Injection Molding and Die Casting. In every manufacturing process, certain features, when present and involved in a tolerance specification, have special influence. For example, in injection molding and die casting, the base angle of an unsupported projection or wall will be difficult to control precisely. Thus tolerances applied to dimensions that relate to the end of such projections will also be difficult to control. The unsupported projection or wall is called a tolerance influence feature. Other tolerance influence features in injection molding and die casting include:

- Undercuts
- Unsupported walls and unsupported significant projections
- Supported walls and supported significant projections
- Primitive connections or intersections
- Parting plane

Some of these features (e.g., undercuts) are clearly identifiable at the configuration stage. The existence of others can be verified only after dimensions are added. However, even in these cases, it is often apparent at the configuration stage, even without exact dimensions, that the potential for adverse influence on tolerancing exists.

Trade associations—for example, the Society of the Plastics Industry (SPI)—also describe the features of parts that affect the ability of processes to hold tolerances easily. The Society of the Plastics Industry considers two types of tolerances: dimensional tolerances such as size and location, and geometric tolerances that refer to orientation.

By listing the above tolerance influence features, it is not implied that the presence of a feature in a part necessarily leads to the tolerancing difficulties or added costs. The variable capability of molders and the quality of equipment make exact prediction impossible. It is implied, however, that the presence of the listed features in a part configuration is a signal that designers should investigate their possible impact on tolerancing. Moreover, the number of such features involved in tolerancing should be kept as small as possible.

Stamped Parts. The tolerance influence features in stamping are:

- Side-action features (undercuts)
- Closely spaced features
- Multiple bends

Tolerances involving multiple bends (e.g., on dimensions across one or more bends) are difficult to control in stamped parts because the bends themselves are difficult to control exactly, in part due to springback of the sheet after bending.

Aluminum Extrusions. Configuration guidelines for ease of tolerance control in aluminum extrusions include:

- Avoid tight straightness tolerances on long, thin wall sections. Add small ribs to help achieve straightness.
- Angles of unsupported wall sections are difficult to control precisely. Hence, tolerances on size dimensions involving long legs are difficult to control. If necessary for tolerance control, support long legs with webs connecting to other parts of the extrusion. (This may create a more costly hollow shape, but if tolerances are critical, it may be necessary.)
- Using uniform wall thickness throughout the cross section makes all dimensions easier to control.
- Rounding inside corners makes angles easier to control than when corners are sharp.
- The dimensions of symmetrical shapes are generally easier to control than nonsymmetrical shapes.

Note how some of the issues here are similar to those in injection molding; the reason is that both are internal flow processes.

Forging. Shrinkage and warpage contribute largely to the difficulties of controlling tolerances in forging. By designing forged parts so that they have simple shapes and common proportions, designers can help minimize these effects. It is also important to provide for well-rounded corners, fillets, and edges. Thin sections, such as webs, should be kept to a minimum in number and should be made as thick as possible.

As in injection-molded and die-cast parts, dimensions that extend across the forging die closure parting plane are more difficult to control than those on one side only of the parting plane.

Design for Tolerances: General Guidelines for Configuration Stage. In addition to the process-specific guidelines outlined above, there are some design guidelines relevant to tolerances that apply to all processes:

- Designers should consider tolerances with not only the function in mind, but also with knowledge of the standard tolerance capabilities of the proposed manufacturing process.
- Designers should minimize the need for and use of tolerances; not every dimension requires a special tolerance. The fewer tolerances that are specified, the easier the part will be to manufacture.
- Where tolerances are required, designers should attempt to avoid the need for tolerances that are tighter than standard for the proposed process.
- If nonstandard tolerances cannot be avoided, then the number of them should be minimized.

Table 3 Questions for revealing part configuration design risks

Factor	Questions
What are the most likely ways the part might fail in service?	
Excessive stress	Can the part be dimensioned to keep stresses below yield levels? Add ribs? Use stronger material?
Fatigue	If there will be cyclic loads, can the configuration be dimensioned so as to keep the internal stresses below the fatigue limit?
Stress concentrations	Can the part be dimensioned to keep local stress concentrations low?
Buckling	If buckling is a possibility, can the configuration be dimensioned to prevent it?
Unexpected shocks or loads	What unexpected dynamic loads might be encountered in service or in assembly? Can these be handled by the configuration?
What are the most likely ways the part might not meet its expected functionality?	
Tolerances	Is the configuration such that functionality will be especially sensitive to the actual tolerances that can be expected in a production situation? Are too many special (tight) tolerances required to make the part work well?
Creep	If creep is a possibility, will it result in loss of functionality?
Strain and deformation	If functional performance is sensitive to retention of size and shape, can the configuration be dimensioned to preserve the required integrity?
Thermal deformations	Might thermal expansion or contraction cause the configuration to deform so that function will be impaired?
Handling and assembly	Might there be unforeseen difficulties with handling and assembly?
Dimensions	Might the part end up being dimensioned so that assumptions about assembleability become invalid?
Tangling	Might the parts tangle if dimensioned in some way?
Will the available production machines be able to make the part?	
Production runs	Are the desired production runs consistent with the machines and expected costs?
Tooling wear	Is tooling wear or maintenance a possible problem that will impact part cost or performance?
Weld lines(a)	If the process is a flow process, can weld lines be located appropriately?
Other design and materials factors	
Geometric compatibility	Is the part geometrically compatible with its adjoining parts? What could go wrong is this regard? If there is a small change in this part, or in an adjoining part, can the configuration accommodate the change without major redesign? What about the effects of tolerances of the *adjoining* parts? Or on the assembly as a whole?
Materials	Is the material selected compatible with the configuration and the manufacturing process? Is surface finish properly accounted for? Will standard raw material supplies be of adequate quality? Has the material been thoroughly investigated for its use in *this* particular application? Are there previous uses in similar applications? Have experts on the properties and processing of the material been consulted? Is the material compatible with the rest of the product?
Designer and design team knowledge	Has every possible, unfortunate, unlikely, unlucky, even stupid "What if . . . " situation been considered? Are there aspects of the part design where the designer or design team is working without adequate knowledge? Where is the design based on insufficient knowledge of materials, forces, flows, temperatures, environment, etc.? Where are there guesses, hopes, fears, and assumptions instead of knowledge: Materials? Stresses? Fastening methods? Manufacturing process? Tolerances? Costs? Adjoining parts? Environmental conditions?

(a) A weld line is formed when a material flow must divide—say around a hole—and then rejoin. The weld lines tend to be weaker and more subject to fatigue failures.

- Avoid redundant dimensions and consider stack-up effects in evaluating tolerances.

Evaluating Trial Part Configurations

Once the set of the most practical part configurations has been generated using the guidelines presented above, a more formal evaluation should be performed. The evaluation can be done by Pugh's method (Ref 4) or by other methods presented in Ref 1. In any method, the criteria on which the alternative configurations are compared will include the following:

- *Functionality:* Can exact dimensions and tolerances be imposed that will enable the part to perform its function properly and reliably?
- *Use of materials:* When dimensions are imposed, will the configuration provide for efficient use of all the required material?
- *Mechanical failure:* When dimensions are imposed, can the risks of failure from mechanical causes such as fatigue, excessive stress, buckling, and so forth, be made suitably low?
- *Analyzability:* Does the configuration enable analyses to be performed for stresses, vibrations, heat flow, and so forth?
- *Manufacturability:* Can the selected manufacturing process hold the tolerances that will be needed for the configuration to meet the required functionality? Does the configuration allow for ease of handling and insertion for assembly? Are there special issues that will influence the time required for tooling and production?

It should also be noted that the tooling costs for parts can be compared quite well when the configuration has been established. Exact dimensions are not required. (See Ref 1, Chapters 11 and 12.)

Looking for Risks

All evaluation methods in some sense balance expected benefits against potential risks. Therefore, the notion of *explicitly* looking for technical and manufacturing risks in a design is useful. The idea is to look for any previously unquestioned assumptions that have been made while generating the configuration and while doing the evaluations—that is, to look for issues that may so far have been overlooked.

Risk cannot be avoided completely. No matter how conservative a configuration design is, there will be some risk, some places where absolutely complete information does not exist. Indeed, good designs will have some reasonable risk or they will be too conservative, too costly, too heavy, and so on. However, designers cannot count on good luck: there are many more ways for a design to fail than there are for it to succeed. Thus, risks must be sought out, faced, and evaluated. Table 3 lists questions intended to reveal risks related to part configurations. Additional information is provided in the article "Risk and Hazard Analysis in Design" in this Volume.

Redesign Guided by Evaluations

Whatever method of formal part design evaluation and risk analysis is used, the results should point clearly to those aspects of prospective designs that need redesign, or to the need for generating more alternatives. Thus redesign is guided directly by the evaluations. This is the reason that, to get excellent part designs, careful, complete formal evaluation is essential.

REFERENCES

1. J. R. Dixon and C. Poli, *Engineering Design and Design for Manufacturing*, Field Stone Publishers, 1995
2. G. Boothroyd, P. Dewhurst, and W. Knight, *Product Design for Manufacturing and Assembly*, Marcel Dekker, 1994
3. K. Ulrich, et al., "Including the Value of Time in Design for Manufacturing Decision Making," Working Paper #3243-91-MSA, Massachusetts Institute of Technology, Sloan School of Management, 1991
4. S. Pugh, *Total Design: Integrating Methods for Successful Product Engineering*, Addison-Wesley, 1991

Creative Concept Development

B. Lee Tuttle, GMI Engineering and Management Institute

MOST ENGINEERS are effectively skilled in utilizing their critical thinking (analytical thinking) skills in the "solution" of product engineering problems. However, the creative thinking (divergent thinking) skills inherent in every human being are often not implemented by the engineer in solving new product designs. The process of creativity in any domain, product design, or artistic creation, involves an iterative application of both creative thinking skills and critical thinking skills to eventually generate, develop, and refine an option that satisfies a human need in a new manner. Without the application of creativity to product design, engineers will continue to improve and refine current forms of products until they become perfect for manufacture, assembly, service, and usage. If an engineer applied only critical thinking skills to the phonograph (record) player, he would design a great phonograph player. If he employed both the creative thinking skills and the critical thinking skills to the product problem of storing and playing sound, he might create the tape player or the CD player! The evolution in product design will occur only with the application of creativity to the product design process.

Definitions of Creativity

Many people persist in perpetuating the myth that creativity is a mysterious human ability that is associated with the blinding discovery of a unique product. Creativity is not an inherited talent unique to special people. Creativity is an operation that occurs in the human mind of all individuals, not just certain so-called "gifted" individuals. Creative engineers, chemists, doctors, and accountants are all normal people who use specific abilities and skills to a greater degree than most people do.

Creativity is a very difficult concept to define in terms that can be understood by all individuals. Some of this confusion arises from the approach that researchers have taken in defining what they mean by creativity. Creativity is an extremely complex phenomenon that encompasses the person, the process, the product, and the place (Ref 1). Many definitions of creativity focus on only one dimension of the multifaceted concept of creativity.

To illustrate the complexity of the creativity, some definitions are given here:

- "Creativity is a process that results in novelty which is accepted as useful, tenable, or satisfying by a significant group of others at some point in time." (Ref 2)
- "Creativity, a word that should be reserved to name a complex, multifaceted phenomenon is misused to name only one part of a phenomenon. The multifaceted conception of creativity should consider four factors: person (personality characteristics or traits of creative people); process (elements of motivation, perception, learning, thinking, and communicating); product (ideas translated into tangible forms); and press (the relationship between human beings and their environment)." (Ref 1)
- "[Creativity] is that mental process by which man combines and recombines his past experience, possibly with some distortion, in such a way that he arrives at new patterns, new configurations, and arrangements that better solve some need of mankind." (Ref 3)
- "Creativity is a process of becoming sensitive to problems, deficiencies, gaps in knowledge, missing elements, disharmonies, and so on; identifying the difficulty; searching for solutions, making guesses, or formulating hypotheses about the deficiencies; testing and retesting these hypotheses and possibly modifying and retesting them; finally communicating the results." (Ref 4)

Despite the lack of a formal approved engineering definition of creativity, there appears to be some common threads to the definitions:

- The processing of information in the human mind
- The critical analysis of the problem (need)
- A finite time for the incubation of the problem
- The generation of unusual combinations of disparate concepts drawn from the manipulation of past knowledge

Although each of these elements may be critical to the creative process, the generation of new ideas is a skill that can be taught or retaught to adults to develop their ability to generate new novel products. This style of divergent thinking that shows ways to accomplish a problem solution is termed creative thinking or lateral thinking (Ref 5).

Creative Thinking

The creativity process is a total problem resolution process that yields a truly novel solution to a problem. In order to effect such a novel solution a person must utilize two thinking styles: analytical thinking (vertical thinking, convergent thinking, judgmental thinking) and creative thinking (lateral thinking, divergent thinking, generative thinking). While individuals are born with both

Table 1 Characteristics of vertical thinking and lateral thinking

Vertical thinking (logical thinking)	Lateral thinking (creative thinking)
1. Only one correct solution (selective)	1. Many possible solutions (generative)
2. Is analytical (judgmental)	2. Is provocative (nonjudgmental)
3. Movement is made in a sequential, rule based manner	3. Movement is made in a chaotic random pattern
4. If a positive decision cannot be made at a step, thinking stops cold	4. If a positive decision cannot be made at a step, thinking jumps
5. Follows only the most likely solution path	5. Follows all paths
6. Only relevant facts are considered	6. All facts are considered
7. Deals only with reality as science knows it today (reality)	7. Can create its own reality (fantasy)
8. Moves only if a clear direction is exposed	8. Moves *specifically* to generate a direction
9. Classifications and labels are sacred	9. Reclassifies objects to generate ideas

Source: Ref 5, 6

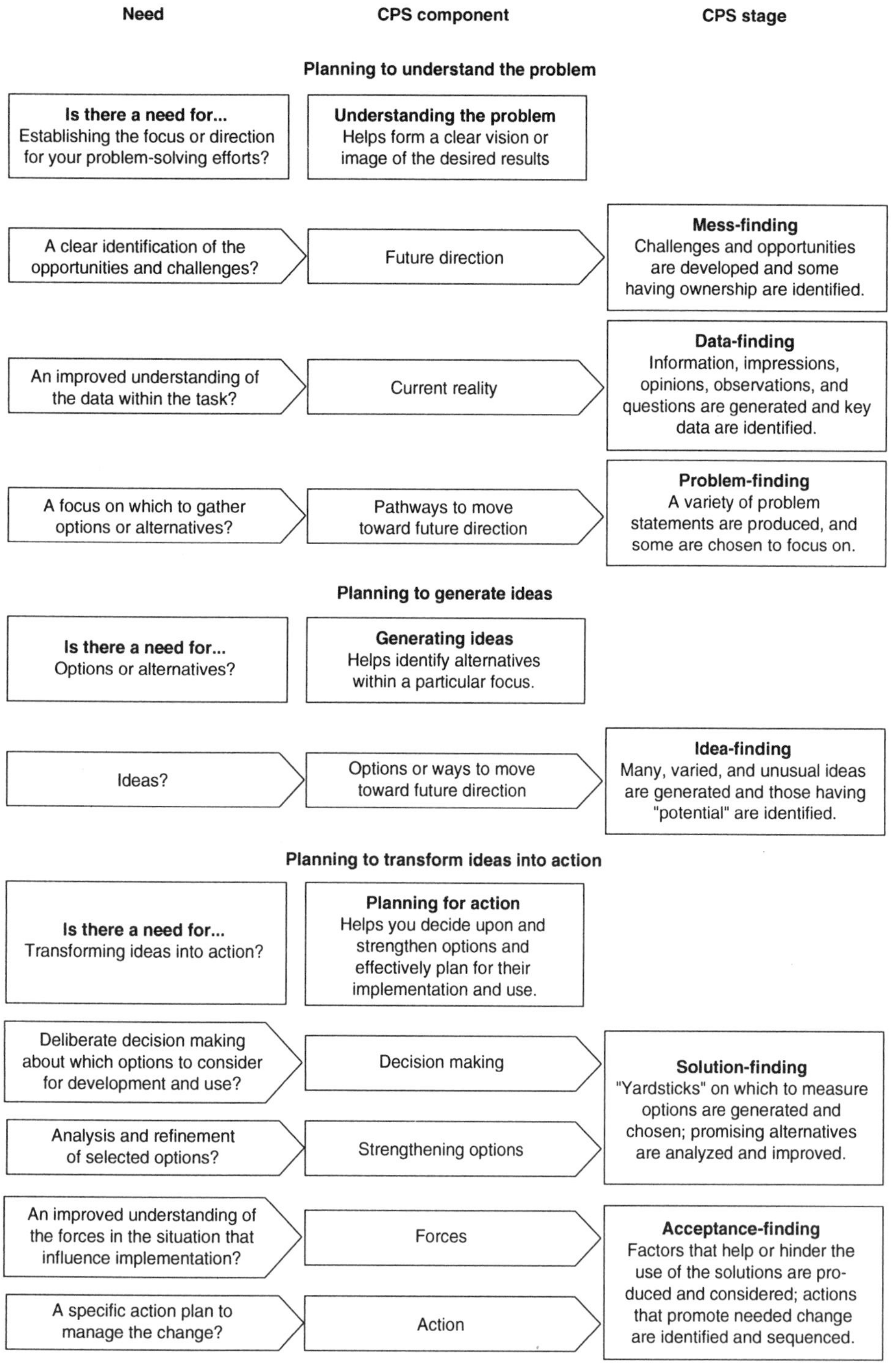

Fig. 1 Generalized flow chart for the Creative Problem Solving process. Source: Ref 9

styles of thinking equally imbued, the educational system to which they are subjected to from kindergarten to college forces them to use and develop analytical thinking. Throughout their educational experience, the creative thinking skills are suppressed until many people feel uncomfortable using those skills.

In vertical thinking one moves forward with an analysis of an idea in sequential steps after a positive decision has been made about the idea until a final decision is reached about the concept. Once a negative decision is reached about the idea, the process stops cold in its tracks. The mind must then fast reverse along the analysis trail until the original concept statement is reached. All thinking stops now.

In lateral thinking, one's mind moves in many different directions, combining different bits of information into new patterns until several different solution concepts become apparent. Once an idea is exposed, the mind rockets off to another idea.

The difference between vertical thinking and lateral thinking are summarized in Table 1, which is a compilation of work done by DeBono (Ref 5) and Hajcak (Ref 6). From this summary it is evident that vertical thinking evaluates the suitability of an idea for application to a problem in the context of the criteria at that moment. On the other hand, lateral thinking generates many potential solutions to a problem that may or may not be effective when applied to the problem today.

Since the development of the lateral thinking abilities of most people has been suppressed, tools must be used to rekindle these thinking skills. These tools, which are called creative thinking tools, concept generators, and idea triggers, will be discussed later.

Creative Problem Solving

If creative thinking is thinking in a creative fashion, then what is creative problem solving and what makes it different from creative thinking? Creative thinking produces many ordinary and new means of potentially resolving a need. However, an idea developed from creative thinking is not ready for the marketplace until it has been refined by the application of critical thinking to refine the idea. Creative problem solving is the systematic combination of both creative thinking and critical thinking in an iterative manner that generates potential solutions, refines some solutions, and eventually implements one novel solution to a human need. Not all problem solving needs to be creative problem solving. When a problem is clearly defined and known analytical tools are available to facilitate a solution, then creative problem solving is not needed or warranted.

An example of a process style for utilizing both creative thinking skills and critical thinking skills in problem solving would be the Creative Problem Solving (CPS) process. (Note that when the words Creative Problem Solving are capitalized, the name refers to a specific process of problem development and solution.) The CPS process refers to a specific problem-solving methodology originally developed by Osborn (Ref 7) and evolved by colleagues, Isaksen, Dorval, and Treffinger at the Center for Studies in Creativity at Buffalo State College through the last 20 years (Ref 8). They describe CPS as a broadly flexible process that provides an organizing framework for specific creative thinking tools and critical thinking tools to help design and develop new and useful outcomes. Through this system, productive thinking tools can be applied to understanding problems and opportunities; generating many, varied, and unusual ideas; and evaluating, developing, and implementing potential solutions. A generalized flow chart of the CPS process is shown in Fig. 1. CPS has three components: understanding the problem, generating ideas, and planning for action. Within each component there are one to three distinct stages that utilize both creative thinking and convergent thinking. The CPS process employs some specific structured divergent thinking and convergent thinking tools to move the thought process toward a final solution.

Creative Problem Solving is a total problem-solving process. The creative thinking tools that are contained in this article are divergent thinking tools that can be used in the idea generation, criteria generation, and acceptance generation phases of the CPS process. Because the focus of this article is on the concept generation tools in product design, the convergent thinking tools that can be used in the idea generation, criteria generation, and acceptance generation stages of the CPS process will not be described.

Design for Manufacture and Assembly

Design for manufacture and assembly (DFMA) is a total problem-solving process applied to the solution of product design problems beginning with the product function specification and ending with the implementation of the product by the customer. As such, the DFMA process utilizes the creative thinking tools in the generation of ideas for the functions of the product, ideas for the processing of individual parts, ideas for the joining of parts, and ideas for the finish of the parts. Like any process of creative problem solving, DFMA utilizes an iterative process of both divergent thinking and convergent thinking to move the concept toward an acceptable solution. Further information about DFMA and the complete product development process is given in the Section "Manufacturing Aspects of Design" in this Volume.

Understanding the Product Problem

Often, an engineer will make a product proposal by specifying the solution to the problem. That is, the engineer will present a sketch of the product, such as the jar opener shown in Fig. 2(a), and say, "Design this jar opener for me." Unknowingly, the engineer has trapped the designer with a visual image of the intended product. In order to facilitate the exploration of the multiple solutions to the product, the physical manifestation of the product must be transformed into something that has no visual image. Sketching is the language of the engineers who traditionally do not express themselves very well verbally. However, it is the verbal explanation of the product function that is needed to move one level backward in abstraction from the physical manifestation of the product, which really represents the solution to the product problem, to the verbal image, which represents the problem statement or need of the customer.

Transforming the Product

The physical manifestation of the product, the mechanical mechanism, represents *how* to accomplish the need of the customer. What we must do is to move the product upward in abstraction and transform it to the verbal side of our brain by specifying *what* the function of the product is. Pahl and Beitz in Germany have been the leaders in the development of this approach to product design (Ref 10). The product functional specification is developed by asking *what* function the product performs and *who* receives the output of the product.

The operation of the product at the product level where it interfaces with the operator and the operatee can be expressed using a simple terminology of two or three words: a verb/noun combination or a verb/noun/noun combination. The verb indicates what is being done by the product, and the noun indicates who is receiving the action. It is critically important at this stage in product specification to clearly distinguish between *what* is being done by the product and *how* it is being done by the specific product variant that is conceived in the initial concept.

The product level functional specification for the product shown in Fig. 2(a) would be "open/jar (lids)." The use of divergent thinking at this stage would reveal that the customer desires the product to be adjustable for different jars and to be portable. This problem would best be broken down into smaller subproblems called subfunctional groups.

Breaking the Product into Subfunctional Groups

In the solution of most product designs, a product function can effectively be broken down into simpler subfunctional groups (SFG) (Ref 11, 12). The SFG is a group of parts that operate together to perform one of the subfunctions of the product. A subfunction of the product is a lower-level function that must be accomplished together with other subfunctions of the product in order to accomplish the total function of the product. Two or more SFGs operate together in series or in parallel to accomplish the overall product function. The definition of the SFG represents the statement of a smaller design problem that within itself still provides the breadth for multiple creative solutions to that problem. The eventual search for solutions to the product problem will normally occur sequentially in each of the SFGs. Depending on the final solutions for each of the SFGs, the SFGs will operate in series or in parallel with each other to eventually accomplish the overall function of the product. At the SFG level, the specification of the function is often accomplished more readily by expressing it in the terminology of a verb/noun/noun. The verb indicates what is being done, the first noun identifies what is doing the action, and the second noun identifies what is receiving the action. The object of the functional specification of the SFG is to specify the interrelationship among several groups of parts that operate to perform the product function. By subdividing the product design into smaller "subproblems," the designer achieves a level of problem abstraction that can be readily understood, but is still abstract enough to permit multiple different solutions to the problem.

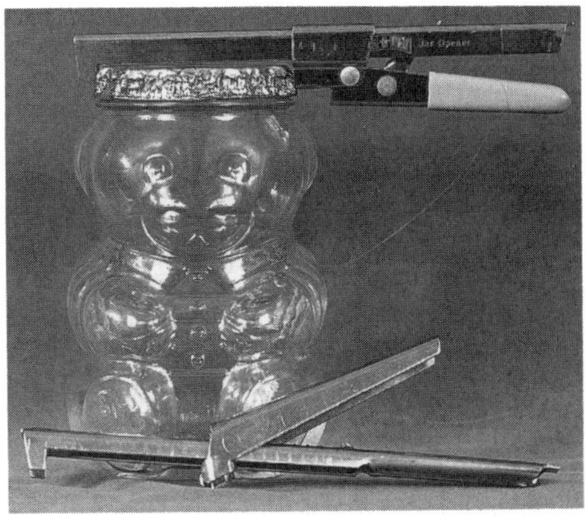

(a)

(b)

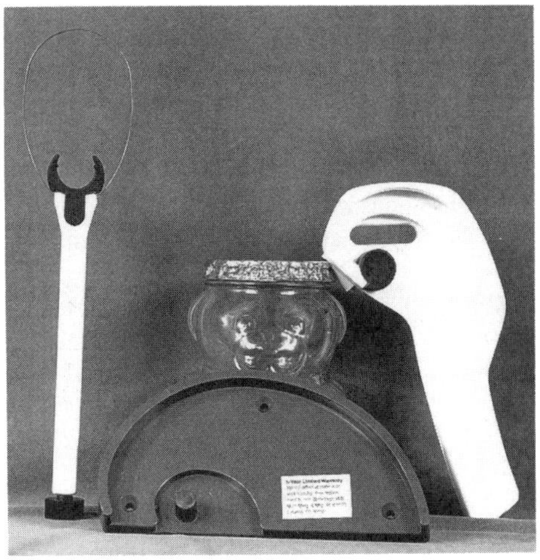

(c)

Fig. 2 Jar opener design concepts. (a) Two variants of a jar opener. (b) Alternative mechanisms. (c) Additional concepts

At other times it is more appropriate to indicate the function of a product with three terms where there is an interrelationship among three objects. The verb indicates what is being performed, the first noun identifies the object initiating the action, and the second noun indicates the object receiving the action.

The jar opener can be effectively broken down into three subfunctional groups:

1. Adjust/Jaw 1/Jaw 2
2. Grip/Lid (jar)
3. Rotate/Lid/Jar

These three SFGs do not represent the only product decomposition or specification that could be developed. Indeed, one of the critical elements in the design process is to define and then redefine the SFGs until an open-enough definition is achieved to create a wide spectrum of design concepts. The product functional design specification (PFDS) involves first the application of divergent thinking skills to generate different verbal visions of the product function. Second, the convergent thinking skills are employed to refine and reform the verbal functional definition of the product into the proper function terminology of verb/noun to finalize the product functional design specification (PFDS). Third, the various functions are refined into the minimum number of product functions and subfunctional groups that express the operation of the product.

When divergent thinking skills are applied to each subfunctional group to stimulate new, novel methods of performing the function and then these subfunctional concepts are melded together to form complete products, the physical appearance of the products change drastically. Figure 2(b) shows several totally different forms of the jar-opener product. If the designer's mind had been focused on the original physical concept of the product would he have developed the "strap wrench" opener shown in the right corner? Certainly, if he had focused on the physical manifestation of the product, he would not have developed the continuously variable gripping device with no moving parts shown in the left corner. Figure 2(c) illustrates some other versions of a jar opener. The opener shown in the foreground of the jar has transformed the "rotate/jar" function to the operator's hand and the "grip/jar" function to the kitchen cabinet base. The opener on the right has accomplished the "adjust/jaw 1/jaw 2" function to a curvilinear surface that has no parts and accomplished the "adjust/jaw 1/jaw 2" function in one smooth operation with the twist of the hand. On the left side of the picture is a modern strap wrench-style opener in which both ends of the strap are retracted into the handle in a smooth continuous action to accomplish "adjust/jaw 1/jaw 2" and then "grip/lid." If the only tool in the engineer's toolbox was the design for assembly (DFA) tool, then these new, novel product concepts would have never seen the light of discovery. Divergent thinking gives birth to new product methods. Design for assembly refines the concepts into manufacturable and assemblable products.

Creative Concept Generation Tools

The concept generating tools are often arbitrarily described as individual or group tools. The categories individual and group refer to the context in which the tools were originally developed rather than the specific applications of the tools today.

Individual Tools

Individual tools rely on free association or forced relationships to stimulate the germination of an idea in an individual's mind. Group tools are those that were specifically developed to focus the interaction of the individual persons in the group toward greater concept generation. Because many group techniques rely on the interaction of the minds of diverse persons to stimulate the flow of ideas, those techniques do not lend themselves to idea generation by individuals. However, many individual techniques, particularly forced random stimulation techniques, can be successfully adapted for use inside of a group ideation technique.

Checklists are not, as their name may imply, lists that permit an individual (or a group) to check off novel, creative concepts for a new product. In the process of evolving his brainstorming tool from an advertising conference room tool to a general-usage problem-solving ideation tool, Osborn found that many individuals needed some stimulus to get them started generating ideas. He developed a "checklist" of questions that could be used, or discarded, by an individual desiring to generate ideas (Ref 7). The purpose of a checklist is to give the designer a "check" of the present concept (or already developed concepts) relative to some metrics by addressing some open-ended questions to the problem, such as "What if we ... to the product?" "Can we ... the product? "What can we ... in the product to make it better?" The blank can be filled in with one of the words from the checklist. Some typical checklist words include *enlarge, reduce, add, eliminate, move, rotate, substitute* (to make product more powerful, lighter, quieter, etc.), and *invert.*

Perhaps the simplest form of checklist questions is given by writing instructors for use in developing a composition. Namely, Who?, What?, When?, Where?, How?, and Why?

Because idea-spurring checklists are very general, many individuals find it difficult to modify and adapt the questions to their problem statement. If a product design group will be generating many products that are similar but different, they should generate their own set of product development idea-spurring questions directed toward their brand of product that will be more meaningful to them.

Metaphors. The word "metaphor" comes from the Greek word *metaphora,* derived from the word meta meaning "over," and therein, "to carry" (Ref 13). It refers to a particular style of figurative speech in which the attributes of one object are transferred over to an object that does not normally possess those attributes itself. The objective of figurative speech in language is to create more vivid images of the transferred object in the mind of the reader. The objective of using a metaphorical transformation in creative concept generation is to drive the idea into a new realm. Psychologists who have studied some great inventors discovered that they came up with their discovery by transforming their perspective of the problem through the use of a metaphor (Ref 14).

The use of metaphors is illustrated by the sentence, "Edison forged new products in his New Jersey laboratory from the raw materials of his mind's image." Edison certainly was not a giant metallic forging hammer pounding out ideas from a new material called image. The visual image created by the metaphors in the Edison sentence will only exist if the reader has a previous knowledge of manufacturing processes. The development of the metaphor as an idea trigger should follow some simple guidelines given below:

- The metaphor should make a transference to the problem at hand from a subject area with which the problem solvers have some familiarity.
- The metaphor should be directed toward the function of the product to be designed.
- Metaphors that develop a unique image using one of the five sensory images (visual, aural, aromatic, tactic, or taste) have been shown to be most effective.
- Once a good metaphor is generated, the engineer should look with this new set of eyes at the original product to imagine new ideas.

The success of the metaphorical transformation of the problem statement lies in the language ability of the user. Many people have difficulty making comparisons between dissimilar objects and thus cannot benefit from analogies. Other people can see all the colors of the rainbow in their mind's eye and are stimulated by metaphors. As with any tool, the generation of metaphors may become a useful idea trigger if the designer practices the generation of metaphors separate from the sessions in which the designer wishes to generate ideas.

Forced random stimulation is an idea-generating technique in which an object is chosen at random, the attributes or functions of that object are force-fit one at a time into the problem statement in an attempt to generate new ideas (Ref 15).

The first step is to select a random object. A very effective method is to select an object, without looking, from a box of miniature objects or tokens of real life objects. The tokens that have been placed in the box are all carefully screened to see that they have positive connotations to most people. Such a box has been called an idea box or an idea factory by various creative practitioners. In the absence of an idea box, the object can be chosen as follows:

- Select one of the primary colors and choose the first object of that color seen in the room.
- Open a catalog at random and select the object printed in the upper (or lower) right (or left) corner of the page.
- Open the dictionary to a page given by the last three digits of a telephone number and select the word from the top whose position is the

same as the fifth and seventh numbers of the telephone number.

- Telephone a child and ask the child to describe his/her favorite toy.
- Open a magazine to a page at random and look at the object in the advertisement on that page.

The DFMA team now focuses on the object itself. They play with it. They imagine using it around the home. They envision other people using it. They now write five to ten attributes (descriptors, images, impressions, sounds, or feelings) that come to mind about this object. (Originally when this tool was developed, the team was asked to list the functions of the random object because the tool was being used in the functionally driven design approach. However, the listing of the functions of the random object proved to be too restrictive and not free-flowing. Thus this approach was dropped.) At this time the object is discarded from view and mind, and the attributes are used.

The first attribute is now forced at the problem statement in an effort to generate some new ideas. The problem statement should be approached by stating: What would happen to the problem if we put a (describe the attribute) on (into, near, over, under, in front of, or behind) the problem? Once ideas stop flowing from the first attribute, then the team moves on to the next attribute until all the attributes have been used individually. Then, attributes can be combined together in pairs to force further ideas. The condensed Forced Random Stimulation Worksheet is shown in Fig. 3 (Ref 15).

A team should not be discouraged if the random object does not stimulate any new concepts. When a reasonable effort has been expended on a set of attributes for one random object, then a second random object can be selected and the process repeated. If the technique does not work for a specific team, then other concept-generation techniques should be used.

Product Improvement Checklist. The Product Improvement Checklist, or PICL (pronounced "pickle"), was developed by VanGundy as an unrelated, forced relationship tool for product improvement (Ref 16). It has been successfully applied to real physical products such as toasters, lawn mowers, and electric drills. The PICL tool consists of two series of idea-spurring questions that are displayed on a 22 × 27 in. worksheet or recorded in an electronic file.

Basic PICL uses 91 idea-stimulating questions to prompt new concepts. These questions are organized into categories such as Who?, What?, Where?, When?, and Why? Typical questions are: Who uses the product? Who buys the product? What need does it fulfill? Where is it used?

The questions posed by the Basic PICL chart are employed as follows:

1. Read the product problem statement.
2. Read a question and write down any ideas that come to mind.
3. When ideas stop flowing, move onto another question.

FORCED RANDOM STIMULATION
WORKSHEET

RANDOM METHOD (Select):

Idea Box   Toy Box   Dictionary   Color Object

RANDOM OBJECT: ____________________

LIST ATTRIBUTES:

1. ____________	2. ____________
3. ____________	4. ____________
5. ____________	6. ____________
7. ____________	8. ____________
9. ____________	10. ____________

ATTRIBUTE #1 IDEAS:

ATTRIBUTE #2 IDEAS:

ATTRIBUTE #3 IDEAS:

ATTRIBUTE #4 IDEAS:

ATTRIBUTE #5 IDEAS:

ATTRIBUTE #6 IDEAS:

ATTRIBUTE #7 IDEAS:

ATTRIBUTE #8 IDEAS:

ATTRIBUTE #9 IDEAS:

ATTRIBUTE #10 IDEAS:

COMBINED ATTRIBUTES # ___ & # ___ IDEAS:

COMBINED ATTRIBUTES # ___ & # ___ IDEAS:

© B. Lee Tuttle 1994

Fig. 3 Condensed force random stimulation worksheet

4. When all questions have been answered, evaluate the responses for application to the problem.

PICL Juice is the name given to the 576 idea-stimulating words distributed at random among four categories on the PICL chart. The four major categories of word stimulators are: Try To, Make It, Think Of, and Take Away or Add. A word in each category is selected at random and combined with the title of the category to stimulate ideas. Examples of idea-spurring phrases are:

- *Try to:* do it backwards, separate it, spread it, classify it, roll it, stack it
- *Make it:* porous, hard, flexible, cylindrical, cold, swivel, diverge, opaque
- *Think of:* pogo sticks, scissors, drills, closures, brief cases, plows, Frisbees
- *Take away or add:* rollers, handles, Velcro, sand, movement, springs, music, reflectors

PICL Juice can be applied in the following manner to a product problem:

1. Read the product problem statement.
2. Select a category. Combine the category words one at a time with the random words in that category.
3. Write down any ideas that come to mind. Move on to a new word when ideas stop flowing.

Additional idea-spurring phrases can be generated by selecting at random a word from each category and then applying that phrase of four words to the product statement. For example, you might combine: twist, cylindrical, dental floss, and wheels. Perhaps, the product should have cylindrical wheels made of dental floss or dental floss wrapped around the wheels for traction.

The use of idea stimuli unrelated to the focus of the problem increases the ability of the tool to

Table 2 The SCAMPER questions

S	What can be **SUBSTITUTED?** Who else instead? What else instead? What other material? What other process? What other place?
C	What can be **COMBINED** with it? What about a blend, an alloy, an assortment, an ensemble? Combine units? Combine purposes? Combine appeals? Combine ideas? Combine concepts?
A	**ADAPT?** What else like this? What other ideas does it suggest? Does the past offer a parallel?
M	**MODIFY?** New twist? Change meaning, color, motion, sound, odor, form, shape? Other changes? **MAGNIFY?** What to add? More time? Greater frequency? Stronger? Higher? Longer? Thicker? Extra value? Duplicate? **MULTIPLY?** Exaggerate?
P	**PUT TO OTHER USES?** New ways to use as is? Other uses if modified? Other places to use? Other people to reach?
E	**ELIMINATE?** Minimize? What to eliminate? What to subtract? Smaller? Condensed? Miniature? Lower? Shorter? Lighter? Split up? Streamline? Understate?
R	**REARRANGE?** Interchange components? Other pattern? Other layout? Other sequence? Transpose cause and effect?

Source: Ref 17

develop unique ideas. Because there are an infinite number of word combinations, the ability to generate a large number of novel ideas is high. VanGundy states that the major disadvantage of PICL is that many people have trouble using forced relationships when the stimuli are not related to the problem. This type of idea trigger is best utilized by people who have the ability to tolerate ambiguity and see connections between seemingly unrelated objects.

SCAMPER. Eberle developed some techniques to assist children in rekindling their four phases of divergent thinking (fluent thinking, flexible thinking, original thinking, and elaborative thinking) (Ref 17). He recognized that people continually draw from their personal experiences in their mind and adapt, combine, rearrange, and modify these bits of information to form new ideas. Although this process appears to happen randomly in the subconscious mind, he wanted a technique to bring this to the conscious mind and thus increase the probability of novel ideas.

He took some of the idea-spurring questions posed by Osborn in his book *Applied Imagination* (Ref 7) and modified these questions to form an easily remembered list called SCAMPER. The final form of the SCAMPER questions are shown in Table 2.

The SCAMPER questions are applied to the problem as follows:

1. Read aloud the first SCAMPER question.
2. Write down ideas or sketch ideas that are spurred by the question.
3. Rephrase the question and apply it to the other features of the product problem.
4. Continue applying the questions until ideas cease to flow.

Because the idea-spurring questions are generalized questions, they sometimes will not apply to a specific product problem. If a question fails to evoke ideas, the thinker should move enthusiastically along toward the next question without feeling uncomfortable. The generalized nature of the questions requires that the concept developer using it be of a mind to make very strange connections between objects.

In new product development the SCAMPER questions may best be applied to the individual subfunctions of the product rather than to the product as a whole.

Group Tools

Those divergent thinking tools that are classified as group tools usually rely on the interaction between the ideas of diverse people to stimulate the flow of ideas from each individual participant. These tools may use free association, or forced relationship association, to initiate the flow of ideas; however, it is the sparks emitted when the verbal (visual) flint strikes the iron mind that illuminate new ideas in the listener's mind! The effective application of a group tool relies on the ability of a facilitator to keep the ideas flowing from all participants and to prevent a single individual from dominating the ideation process.

Brainstorming is perhaps the best known and yet most poorly understood group creative thinking technique. In today's society the term "brainstorm" has come to mean any effort to think out a problem. In the 1950s, Osborn developed group ideation sessions at his advertising agency that they named "Brainstorm Sessions," literally a tempest in the mind. Osborn had discovered that most people felt meetings were for evaluation of concepts and not the generation of options. To facilitate the rapid-fire exchange of ideas among the participants, Osborn developed the following rules of a brainstorming session (Ref 7):

- *Criticism is ruled out.* Adverse judgment of ideas must be withheld until later.
- *"Free-wheeling" is welcomed.* The wilder the idea, the better. It is easier to tame down than to think up.
- *Quantity is wanted.* The greater the number of ideas, the more the likelihood of useful ideas.
- *Combination and improvement are sought.* In addition to contributing ideas of their own, participants should suggest how ideas of others can be turned into better ideas or how two or more ideas can be joined into still another idea.

Although many people claim to hold "brainstorming" sessions, few group sessions are truly brainstorming sessions. Unless the rules of conduct espoused by Osborn are strictly followed, these sessions are merely meetings. The objective of a brainstorming session is to use the disconnected ideas of individuals to trigger new ideas in each participant. The technique relies heavily on the group interaction to create a rapid-fire exchange of ideas.

Preparing and Conducting a Brainstorming Session. In preparation for a brainstorming session Osborn would select a diverse group of 8 to 12 individuals. The group would include people with a good technical knowledge of the problem and people with little technical knowledge about the problem. Although brainstorming does not require an extensive training process, the procedures and rules of the process would be explained to each member of the group. In preparation for the ideation session, Osborn would send out an expanded problem statement providing group members with a full description of the problem and inviting them to jot down any ideas that occurred before the meeting.

At the beginning of the actual brainstorming session, the facilitator explains the four rules of brainstorming again to the participants. A brief informational discussion is then held to clarify the problem statement. The group is coached by the facilitator in the implementation of the brainstorming process on several warmup exercises on critical problem topics. When the group is ready to begin work on the formal problem, the facilitator writes the problem statement on the flip chart. Ideas are shouted out as rapidly as they come to mind by each participant. They are recorded by a scribe on the flip charts or on a tape recorder for transcription later. After 15 to 30 minutes of intense ideation, most people lose interest. To prevent boredom with the concept of brainstorming, the facilitator leader should close the meeting when he or she sees that the group's energy is waning. At the end of the session, the facilitator should encourage participants to continue to generate ideas on the problem. Additional ideas can be sent to the facilitator for inclusion in the pool of ideas.

Evolution in Brainstorming Processes. The brainstorming tool has been developed into many variations by different practitioners of creativity through the years. The operation of the traditional brainstorming itself has undergone evolutionary changes through the years since Osborn developed it in the 1950s. During their development of the Creative Problem Solving process, Isaksen, Dorval, and Treffinger have implemented several changes in the brainstorming tool (Ref 8). More recent CPS research has shown that ideas flourish better in a smaller group of five to seven participants rather than the eight to twelve participants suggested by Osborn. The gathering and sharing of ideas can be sped up by having the participants write their ideas on self-adhesive note paper and read them aloud to the group. The notes are then passed to the facilitator for display on the wall. Other concept-generating tools, particularly those employing unrelated stimuli, have been introduced within the framework of the brainstorming session.

A significant development in the original brainstorming process occurred when the group was employed for culling, sorting, screening, and selecting among the ideas that were generated in the ideation session. Isaksen, Dorval, and Treffinger have shown the value of using critical thinking tools to bring some degree of closure to each brainstorming session in order to move the problem-solving process forward.

The traditional brainstorming tool can be applied to many types of problems. For example, it

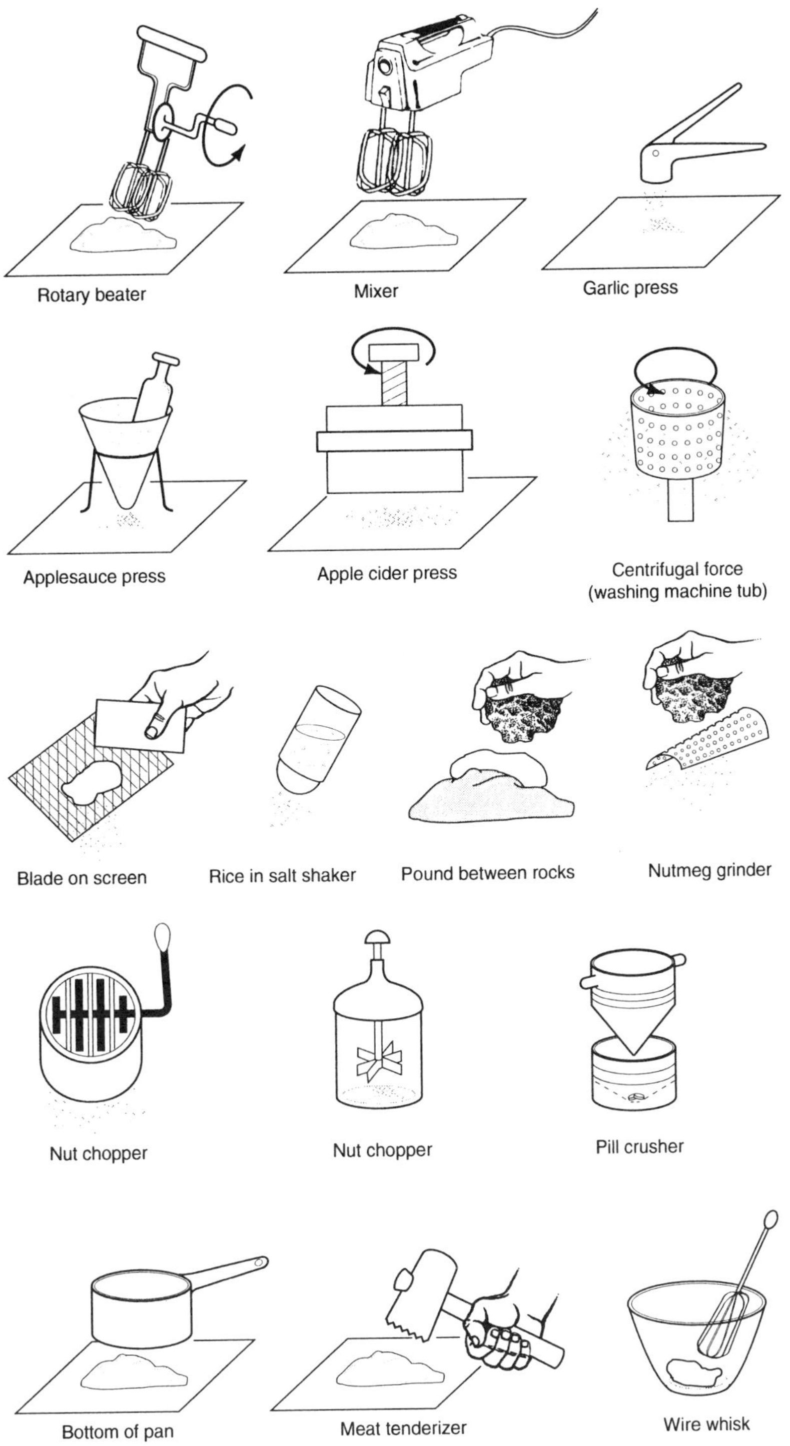

Fig. 4 Concepts generated for the attribute "break lumps." Courtesy of Christina Earnhardt, GMI Engineering and Management Institute

can be used to generate methods to accomplish the functions of a new product, develop criteria for the final selection of the product concepts, and evolve multiple methods of packaging the product for shipment to the customer.

Attribute Listing. Crawford discovered in his study of the creative process in new product development that the creation of a new product consisted generally of shifting an attribute from an unrelated product to the product to be created (Ref 18). The attributes which Crawford referred to could be the physical features of the product, the operational functions of the product, and the aesthetic visions of the product. If a product has been defined in a functional specification that utilizes subfunctional groups, then the attributes should be the functions of the subfunctional group. For those people who are visual thinkers, then the physical features of the product that achieve these functions might be the attributes. Once the attributes are listed, then each attribute is explored for ideas separately from the other attributes.

A prescription for the implementation of this technique might be :

1. Define the product in terms of its functional attribute or physical features.
2. List each function on a piece of paper. Generate as many alternate means of creating the function (feature) as possible.
3. Move on to each successive function (feature) when the ideas for each one stop flowing.
4. Combine different new methods of achieving each attribute to generate a totally new product.

One of the problems that is often encountered in this method is that the ideas do not flow for a specific attribute. If this is encountered, then the combination of an idea-spurring tool such as SCAMPER is appropriate to get the ideas to flow.

A student examined a new design for a flour sifter using the product functional specification that divided the product into four subfunctional groups: contain (release)/flour, break/lumps, sift/flour, and grip/hand. The various alternative means for accomplishing the function "break lumps" are shown in Fig. 4 expressed both in the verbal thinking style and in the visual thinking style. The actual idea implemented by the student employed a combination of the nut chopper and the pill crusher.

The attribute listing technique is best applied to product design when the functional or physical features of the product are defined first. The search for alternate means of achieving each attribute is best made using an idea-spurring tool in addition to the usually employed free association.

Morphological analysis (matrix analysis) was developed by Zwicky as a structured tool to force comparisons among the various attributes (parameters, functions) of a problem (Ref 19). The tool raises the problem solver one level above the basic attribute listing approach by rigorously forcing the problem solver to consider all potential solution combinations no matter how unlikely.

In applying the tool, the following steps should be followed:

1. The problem should be stated in a few (three to six) attributes. In product development these would represent functions or physical features.
2. Arrange the major attributes as the sides of a matrix (two-dimensional comparison) or the sides of a cube (three-dimensional comparison).
3. Using free-association divergent thinking, generate a list of methods of accomplishing the first attribute. Repeat the free-association divergent thinking process for each attribute until multiple concepts are available for each attribute.
4. Review the lists for each attribute. Discard any that are inappropriate for the problem until the lists have the same number.

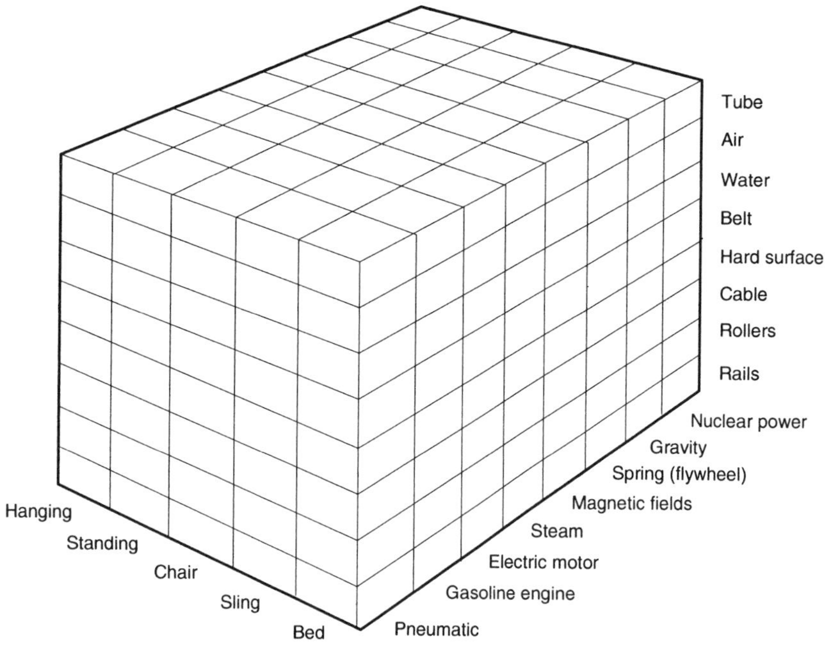

Fig. 5 Morphological analysis for a personal transportation problem. Source: Ref 18

5. Write one method of achieving the function in each row (column) of the matrix.
6. Systematically combine one solution concept from each attribute to identify a new product.
7. Record the identity of all the methods of achieving the functions of the product for later evaluation.

An excellent example of this technique applied to a product development problem is the personal transportation system described by Parnes and Haring in Crawford's book, *The Techniques of Creative Thinking* (Ref 18). The three parameters of the transportation system to be investigated are: provide power, support passenger, support carriage. Figure 5 shows a matrix on which are listed some of the alternate methods for achieving each subfunction of the total product. The product configurations are developed by combining the various methods for each subfunctional group represented by each of the cells of the cube. (That is 320 product ideas in all.)

A few of the ideas can be considered to demonstrate the direction of the tool. One product concept might involve: electric motor, chair, and rails. That defines a very old electric trolley car. Another product concept might be: electric motor, standing, and cable. That might define a cable car in San Francisco. In the Alps, that might define a cable car. Another product variation might be: gasoline engine, belt, and lying. What might this transporter be?

Morphological Matrix. The morphological analysis tool as developed by Zwicky (Ref 19) had the major disadvantage that when all possible combinations are considered a very large number of ideas had to be either hard or soft selected to achieve convergence. The underlying concept of the tool was modified by Parnes and later further refined by Treffinger, Isaksen, and Dorval to a morphological matrix format (Ref 8, 9).

The morphological matrix can be employed as follows:

1. Construct a matrix with the 4 to 6 functions of the product at the head of each column. Each column should have about ten rows beneath it.
2. Beginning with the first function, generate ten methods of performing that function. Record each method, one per line, down the column.
3. Continue generating and recording methods of accomplishing each of the functions until all the functions have been developed.
4. Randomly select one method of performing each function from each column to develop a new product level concept. The random selection can be facilitated by the use of a four- to six-digit random number to sequence the selection of the functional methods from each column. When the individual or group feels that they have enough new combined ideas to consider, they should end the random combination of ideas.

The morphological analysis (matrix) tool is an analytical idea generation tool that appeals to the adaptive style of many engineers. It generates a new product level concept by combining existing methods of accomplishing each of the subfunctional groups (parameters) of the product in new, novel ways. Because the ideas in each attribute (function) category are achieved by free association, they are not as wild as might be achieved by a forced-relationship tool. However, the act of combining subfunctional group concepts in a random manner yields some unusual if not improbable products.

Syntectics. William J.J. Gordon and George M. Prince worked in the Invention Design Group of the Arthur D. Little Company in the early 1950s. Rather than inventing the new product themselves, they helped the engineers at the client company develop the product concepts themselves. Gordon and Prince became more interested in the process of generating ideas for new products than in actually inventing new products. Their interest in creative concept generation led them to form their own company, Synectics, Inc. (Cambridge, MA). Their name for the creative concept development process that they developed was Synectics, a Greek word meaning the joining together of seemingly unrelated and different concepts to create a solution (Ref 20). Gordon and Prince continued to develop their process of concept generation by conducting continuing studies of how the idea-generation teams thought during the product invention sessions. Eventually the Synectics process developed from a concept generation technique into a full creative problem-solving process similarly to the manner in which brainstorming developed into the Creative Problem Solving process. The emphasis here is on the original Synectics tool for concept generation.

The essence of the Synectics process has been described as follows:

1. Make the strange familiar.
2. Make the familiar strange.
3. Transform the strange into familiar again.

The original Synectics research identified three characteristics of good creative concept generators:

- Creative efficiency in people can be increased if they understand the psychological process by which their mind operates.
- In the creative process, the emotional component is more important than the intellectual, the irrational more important than the rational.
- It is the emotional, irrational elements that can and must be understood in order to increase the probability of success in creative concept generation (Ref 20).

In any concept generation situation the first step is to formulate the problem statement or the product functional specification. One of the early discoveries of the Synectics research was that too many engineers took the problem statement at face value. The first step requires that the team develop alternative product specifications and select the one that eventually describes the product as it really is. Gordon and Prince found that when engineers reformulated the problem statement into terms that were familiar to them they were better able to find solutions to the problem.

The second step was one of the startling discoveries of the Synectics research. Namely that efficient generative problem solvers were those who purposely distorted the image of the problem statement in their mind until it became far removed from the original frame of reference. These engineers focused their energy on solving this distorted problem without the constraints of the original frame of reference for the problem.

Freed from the constraints of normal places, they generated numerous ideas for the distorted problem.

The last step was to take the solutions to the distorted problem and force fit them back into the original problem situation. Synectics research has shown that many of the problem solutions generated by the distant formulated problem could be force fit to the original problem (Ref 21, 22).

The heart of the original Synectics research and the tool itself was the study and development of the ways in which people distorted the original problem. The work of Gordon and Prince showed that the product designers transformed their original problem into a new problem through the use of four basic types of analogies: personal analogy, direct analogy, symbolic analogy, and fantasy analogy.

Personal Analogy. In this type of analogy, the individual places himself/herself in the position of the machine and performs the function of the product. A student design team was once observed in the design studio to be pivoting in a circle, raising up their mouths open at one point in the rotation and falling back to the ground with their mouth closed at another point in the rotation. They reported that once they had assumed the identity of the parts of the toy fishing game, they had a different perspective of how the product was operating and what it was doing.

Direct analogy is the one that engineers used most often in the original Synectics research because it involves a comparison between different technologies or looking for a solution in a totally different scientific field. Many inventions have been created by drawing an analogy between the mechanical device and a phenomenon in the field of botany, biology, or anatomy. George de Mestral was working on a special button for pompier's coats that would permit them to fasten and unfasten them rapidly with gloved hands. When he returned to his home one evening after a stroll though the woods, he noticed some cockleburs clinging to his pants. As he removed these from the pants, he observed the strong fastening that they made with the pants. He pressed them back to the pants and observed that they again clung to the pants. He went to his workshop and studied the cockleburs and the fabric of his pants under the microscope. Eventually he developed the Velcro-type fastener in which one side has the stiff, barbed hooks like those on the burrs and the other side has soft loops like the fabric of his pants.

Symbolic analogy actually tries to transform the problem statement from a verbal image to a visual image using a generalized image to represent a particular function of the product. The objective is to indicate a functional relationship among the various parts of a problem without defining the physical embodiment of the interfaces between each element in the product design. A mechanical engineer developing a hydraulic system for a die casting machine would represent the hydraulic system by a block (function) diagram without physically specifying the details of each function block. Typically, several physical mechanisms could perform the function in each element block of the block diagram. Pahl and Beitz encourage engineers to develop a function structure diagram of the product functions using symbols or "pictures" representing the characteristic operation performed by that portion of the product (Ref 10, 11).

Fantasy Analogy. The least-used analogy (at least in the view of Gordon and Prince) was the fantasy analogy. The fantasy analogy moves the mind into the realm of time and space where physical laws as they are known do not exist. The creators of science fiction novels exist in this realm all the time. Individuals who are successful in employing this analogy will develop the most novel ideas for a product.

Analogy Development. The development of the analogies employed in the Synectics method is guided by a trained facilitator. Because the key element to the idea-generating system is the analogical relationship between the original problem and the distorted problem, the abilities of the facilitator to generate meaningful analogies for the team members is critical for success.

Applying Creative Concept Generation in New Product Design

Most product design engineers would say that the "creative" portion of the product development process occurs during the early concept discovery of the process. Indeed, this may be the norm for some engineers. However, engineers must realize that the purpose of their analysis (critical thinking) is to identify opportunities for improvement in the product form, manufacturing process, assembly process, and product function. Once these opportunities for improvement have been identified by a specific engineering analysis tool, only the creative thinking skills can generate potential methods to accomplish these opportunities. The improvement of a product can only be accomplished with the balanced application of the analytical skills and the creative skills of the engineer.

Creative Thinking in Design for Assembly

Once the functionality of the new product concept has been evaluated, the embodiment of the product concept needs to be refined for ease of assembly and manufacture. The engineering team should perform a design for assembly analysis concentrating at the subassembly level to identify which parts must be separate physical parts in the given mechanism and which parts can be physically fused together into new mega parts. (Further details about DFA methodology are provided in the article "Design for Manufacture and Assembly" in this Volume.)

Once the critical thinking questions are answered for the Theoretical Minimum Part count, then the engineering team needs to use their divergent thinking skills to develop different ways to combine the nonseparate parts into the separate parts. Figure 6 illustrates a simple pneumatic piston actuator from a copy machine that has been often used to illustrate the methodology of the DFA analysis technique. Figure 7 illustrates a DFA linear redesign concept matrix for that product. The original product contained seven individual parts. The theoretical minimum part count for the product indicated the product could be made from as few as two or three parts. Each of the design concept columns indicates various combinations of the original parts that could be achieved for products with 4, 3, or 2 parts. Concept 3 would fuse the physical features of the cover and the screws into one piece (C); the piston, the piston stop, and the spring into a second piece (B), and the main block into a third piece (A). The combination of the zeros and ones in the theoretical minimum part count column may or may not involve divergent thinking skills. However, generating a physical shape for the newly merged parts in this product will call upon the divergent thinking skills of any engineer.

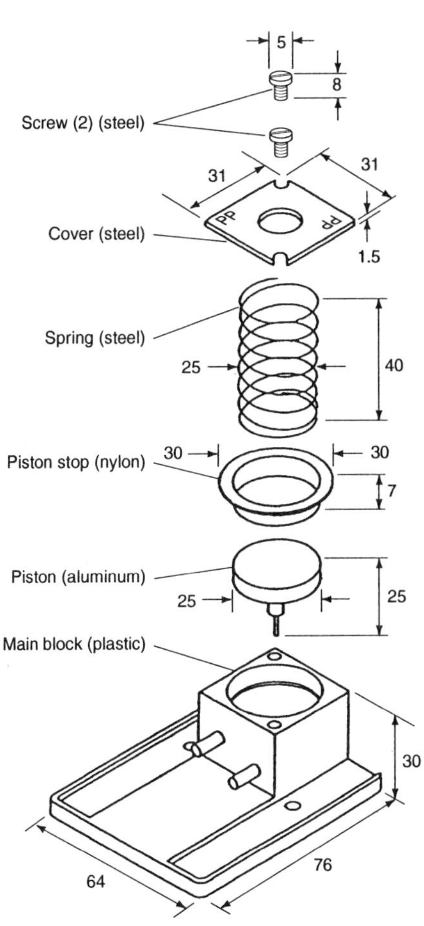

Fig. 6 Pneumatic piston subassembly. Dimensions given in millimeters. Source: Ref 23

Divergent Thinking in Manufacturing

Many practitioners equate the design for manufacture (DFM) analysis with only the manufacturing process for individual parts. The DFM analysis is concerned with the primary manufac-

Theoretical minimum part count	Name of assembly: pneumatic piston		Design concepts				
	Part name		1	2	3	4	5
1	Main block		A	A	A	A	
1	Piston		B	B	B	B	
0	Piston stop		D	C	B	B	
1/0	Spring		C	C	B	B	
0	Cover		D	C	C	B	
0	Screws		D	C	C	B	
		Part count	4	3	3	2	

Fig. 7 Design for assembly redesign concept matrix for the part shown in Fig. 6

turing process, the secondary manufacturing processes, the finishing processes, and the assembly process for the parts within each subassembly. After the DFA analysis is performed and number of parts in the product concept is minimized, the engineering team is faced with developing a manufacturing processing system that can physically produce that shape. Without the divergent thinking skills of the manufacturing engineer, many of the products on the market today could not be made from the concepts developed by the DFA engineers.

Conclusions

The process of innovation and evolution in product design involves the integrated application of both convergent thinking (critical thinking) and divergent thinking (generative thinking) in an iterative continuous process flowing both from customer and process back to the customer and processor. Some tools to stimulate divergent thinking are described in this article. Other tools and methods of creative thinking are contained in the books described in the list of References. Each product design team should seek those divergent thinking tools that are functional for them and apply those tools at appropriate junctures in the product development process.

REFERENCES

1. M. Rhodes, An Analysis of Creativity, *Phi Delta Kappan*, Vol 42, 1961, p 305–310
2. M.I. Stein and S.J. Heinze, *Creativity and the Individual*, Free Press, 1960
3. J.E. Arnold, The Creative Engineer, *Creative Engineering*, American Society of Mechanical Engineers, 1956
4. E.P. Torrance, *Norms and Technical Manual for the Tolerance Tests of Creative Thinking*, Scholastic Testing Service, Bensonville, IL, 1974
5. E. De Bono, *Lateral Thinking: Creativity Step by Step*, Harper and Row, 1970
6. E. Hajcak and T. Garwood, *Expanding Creative Imagination*, Institute for the Study and Development of Human Potential, West Chester, PA, 1981
7. A.F. Osborn, *Applied Imagination*, Charles Scribner's Sons, 1979
8. S.G. Isaksen, K.B. Dorval, and D.J. Treffinger, *Creative Approaches to Problem Solving*, Kendall/Hunt Publishing, 1995
9. S.G. Isaksen, *Creative Problem Solving*, The Creative Problem Solving Group—Buffalo, Williamsville, NY
10. G. Pahl and W. Beitz, *Engineering Design: A Systematic Approach*, The Design Council, London, 1988
11. B.L. Tuttle, "Design for Function: A Cornerstone for DFMA," International Forum on DFMA, Newport, RI, June 1991
12. N. Cross, *Engineering Design Methods*, 2nd ed., John Wiley & Sons, Sept 1994
13. T. Hawkes, *Metaphor*, Routledge, 1989
14. L.V. Williams, *Teaching for the Two Sided Mind*, Simon & Schuster, 1983
15. B.L. Tuttle, *DFMA/A Practicum Manual*, GMI Engineering & Management Institute, Flint, MI, 1995
16. A.B. VanGundy, Jr., *Techniques of Structured Problem Solving*, Van Nostrand Reinhold, 1988
17. R. Eberle, *SCAMPER: Games for Imagination Development*, D.O.K. Press, Buffalo, NY, 1990
18. R.P. Crawford, *The Techniques of Creative Thinking*, Fraser Publishing, 1954
19. F. Zwicky, *Discovery, Invention and Research through the Morphological Approach*, Macmillan, 1969
20. W.J.J. Gordon, *Synectics*, Harper & Brothers, 1961
21. W.J.J. Gordon, Some Source Material in Discovery by Analogy, *J. Creative Behavior*, Vol 8 (No. 4), 1975, p 239–257
22. T. Poze, Analogical Connections—The Essence of Creativity, *J. Creative Behavior*, Vol 17 (No. 4), 1984, p 240–258
23. G. Boothroyd and P. Dewhurst, *Product Design for Assembly*, Boothroyd Dewhurst, Inc., Wakefield, RI, 1987

Cross-Functional Design Teams

Preston G. Smith, New Product Dynamics, Portland, Oregon

THE TERM *TEAMS* is used heavily in industry today, often with little more than a hope behind it. However, as companies strive for greater productivity and responsiveness to market changes, effective teams often play a central role in initiating organizational change. Such real teams may occur in any part of the business, but this article focuses on the particular issues arising in using teams in the product design process.

The most effective design teams generally involve a clearly delineated group of individuals who work full time on the specified project from its beginning until market introduction. The team comprises not only research and development professionals but also manufacturing and marketing members, and often members from quality, finance, or field service. These teams cut across traditional organizational boundaries, thus changing traditional reporting and decision-making relationships. Team members often report to the team leader for the duration of the project and are physically located together (co-located). Although these characteristics can increase productivity and responsiveness greatly, each also represents a major challenge in organizational change for most companies.

Specifically, such team characteristics encourage the use of generalists as team members, thus creating challenges in incorporating specialists, such as materials engineers or scientists. This article provides special coverage on alternative roles for such specialists whose expertise is essential to the success of the project but whose involvement with the team may violate some of the above characteristics.

Background: The Changing Role of Product Design and Development in Industry

Most manufacturing companies today are under heavy pressure to succeed, even to survive. Service industries have taken a dominant role in commerce, much manufacturing has moved offshore, and many manufactured goods, especially materials, have become commodities. In addition, environmental and product liability issues complicate manufacturing operations. All of this is occurring with a rising tempo, as evidenced by market shifts and other external demands that occur ever more frequently.

The Growing Importance of New Products. Senior managers often see new products as the key to coping with this chaotic environment. New products promise higher profit margins, opportunities to avoid commodity product status by creating market niches and added value, and an avenue for revitalizing the corporate image. New products are no longer just something done in research and development but have become central to the plans of the corporation. Many business leaders go beyond this by deciding to use new product development as the keystone in a broader plan of fundamental improvements in how their companies operate.

An Emphasis on Productivity and Responsiveness. Two thrusts come from these management desires:

- A requirement for consistently successful new products in a less predictable environment
- A requirement to obtain these products ever more quickly while using fewer financial and human resources

Design, or more broadly, development, teams have an effect on the product success requirement, but increasingly they are being considered essential to achieving productivity and time-to-market goals. This optimism regarding teams is well founded: many stories have appeared in trade and business magazines and research journals describing how cross-functional teams have brought new products to market far more quickly and inexpensively than more traditional organizational approaches to product development.

As discussed in a later section, a team is not the answer to every development project, but teams have demonstrated their power to improve development effectiveness dramatically. This article covers the characteristics of such teams, how to staff and organize them, and the critical role of specialists, such as materials specialists, in working with such teams.

Types of Teams

Team is a heavily used and abused term in the workplace today. Any identifiable group of workers is generally labeled a team, and teams form in the sales, accounting, and research departments and from the factory floor to the executive suite. Seldom does calling a group a team change the way in which work gets done.

Effective teams can exist anywhere in the organization, but teams that deliver superior performance exhibit certain characteristics (Ref 1):

- A small (fewer than ten), well-defined group with complementary skills
- A meaningful purpose, specific goals, and agreement on concrete operating principles for reaching the goals
- Mutual accountability for results and joint ownership of work products

Teams and Meetings. Katzenbach and Smith (Ref 1) distinguish teams that make or do things from those that recommend things or ones that run or manage things. Product development teams are of the type that do things, and it is essential to recognize that the doing gets done mostly between team meetings. Development team meetings are to assess what got done, solve problems, and set plans for doing the next work. Although meetings are an essential tool of teams, if the team equates itself with meetings and depends on meetings to get work done, progress will be slow. In effective teams, meetings tend to be highly spontaneous and largely transparent. These teams demand far more of their members than just participating in scheduled meetings.

Special Characteristics of Cross-Functional Development Teams. Three traits of product development make development teams particularly challenging ones to set up and manage: (a) most of those involved are professional knowledge workers; (b) a broad range of professional skills is needed, including engineering, science, marketing, manufacturing, and finance; and (c) innovation is an uncertain activity. Although some exceptions exist (Ref 1, 2), most of the team literature treats simpler situations, such as assembly plant operations or mortgage application processing. Consequently, the literature is of limited use here; this article relies more on tools that the author and his colleagues have seen work well in other product development settings.

One insight from this experience in helping clients set up development teams is that the organizations doing best at it are those that have already tried other kinds of teams. They simply have a greater appreciation for the difficulties involved and the training required.

Staffing a Development Team

Much like a cooking recipe, this "recipe" first lists the ingredients (the staffing issues) and then moves on to directions for combining them (the organizational issues).

The Team Leader. Choosing a team leader is the most important decision management will make in setting up a development team. Two criteria should guide the choice. One is that, because product development amounts to an obstacle course, the leader must be strong enough to figure out how to overcome the obstacles and work the existing system. The second is that the leader must operate from a business perspective, not a particular functional perspective, such as engineering or marketing.

If the team leader cannot deal effectively with the obstacles, then management must step in, which destroys the team's value and morale. Similarly, if the leader operates from a particular functional perspective, other functional managers will step in to ensure the participation of their function, again undermining the team's integrity. Neither of these situations provides the high-quality problem-solving and decision-making infrastructure desired.

In addition, a leader should have a strong, customer-centered vision of the product and sense of project direction. This is crucial in providing the leader with a touchstone for making the countless daily decisions that can deflect the team from its course. Leadership, then, is the ability to transform this vision into action.

Clearly, another essential requirement is a leader with excellent people skills, including communication (listening and providing ongoing performance feedback), conflict management, and the ability to influence others throughout the organization. A key part of people skills is giving credit and exposure to team members, rather than the leader accepting it.

From Which Department? For highly technical products, it is natural to choose a technical person as team leader. It seems that only a technical person will understand the design adequately. Others, with a longer view, might argue that only a marketer could provide the customer-focused guidance needed for marketplace success. Similarly, manufacturing might make a case for a manufacturing person as leader because a manufacturable product is essential.

Unfortunately, all of this discussion misses the point. No company has enough candidates for the demanding team leader job, so no company can afford to restrict its search to one function. Besides, the qualified person is someone who thinks and operates as a general manager, not a functional specialist.

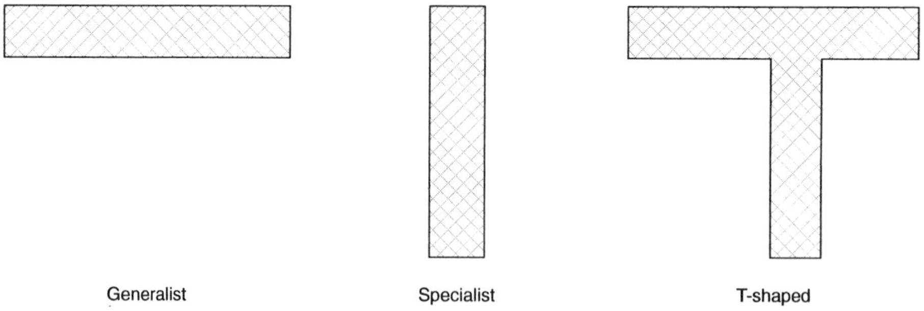

Fig. 1 T-shaped individual. The horizontal direction portrays breadth of experience, and vertical indicates depth of specialization.

Team Members. While much has been written about leaders and leadership, little guidance is available on selecting team members. Kelley (Ref 3) makes the point that the criteria for selecting team members are remarkably similar to those for team leaders. In particular, a development team needs self-starters able to work without supervision and individuals who will present their thoughts independently. Groupthink is dangerous on a development team, and the best defense is team members with the strength of conviction to present contrary views.

Another key criterion is a willingness to share information and credit. A member who tries to build his or her own self-worth by withholding information or credit is disastrous on a development team.

Generalists Versus Specialists. In the development of sophisticated products, the tendency is to think of using highly specialized people who can contribute that something extra that will yield a competitive success in the marketplace. Usually, the recognition, compensation, and promotion systems of a company reinforces this bias toward specialists.

Unfortunately, specialists create several difficulties on a team, including scheduling problems, lack of commitment to the project, and lack of a solid understanding of project objectives and customer desires. Therefore, the bias in selecting team members should swing toward generalists who have a firm grasp of the job to be done and can be engaged for the duration of the project. The ideal member is the so-called T-shaped individual, one who has depth in a crucial area but is also able and willing to handle many other jobs, often under the direction of others, when their specialty is not needed (see Fig. 1).

Figure 2 is a staffing chart for a simple product developed by a company preferring specialists. Each bar represents one individual on the team, and the height of the bar indicates this individual's degree of dedication to the project, that is, the number of hours he or she spent on it compared against the total number of hours possible for the duration of the project. Specifically, five people on the tail end of the chart are purchasing specialists, each permitted to purchase only a specific commodity.

The company represented in Fig. 2 has moved toward generalists. It uses fewer members on a team, but each is involved at a high level of dedication. Communication, coordination, and commitment have improved accordingly.

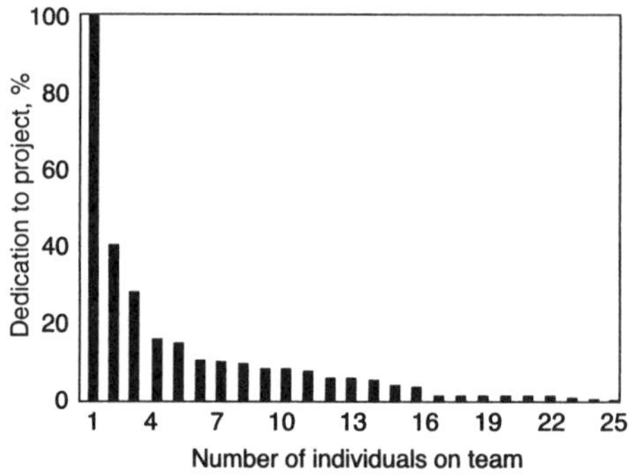

Fig. 2 Staffing diagram for a project that depended on many specialists, most of whom contributed less than 10 percent of their time to the project. Source: Ref 4

Clearly, the specialist-generalist issue applies to a materials specialist whose expertise may be needed for a small portion of the project.

Team Selection Process. To enhance commitment to the project, team members should have a say in whether or not they want to be on a team; in essence, they should volunteer (Ref 4, p 127–128).

Normally, the team leader recruits team members after management recruits the leader. Recruiting team members is a negotiating process between the team leader and management because management will be unable to release certain members requested by the leader.

Suppliers on the Team. To leverage their resources, manufacturers are turning increasingly to suppliers to provide larger portions of their products. Also, there is a trend toward forming strong alliances with a few key suppliers rather than working with many at arms length to avoid being held hostage by a single supplier.

Product development is not as far along as production in making these transitions, but the changes are definitely occurring in product development as well. What this means for product development is that supplier personnel are joining their customers' development teams just as if they were employees of the customer. This practice has become routine for automobile manufacturers where suppliers are involved at many different levels (Ref 5).

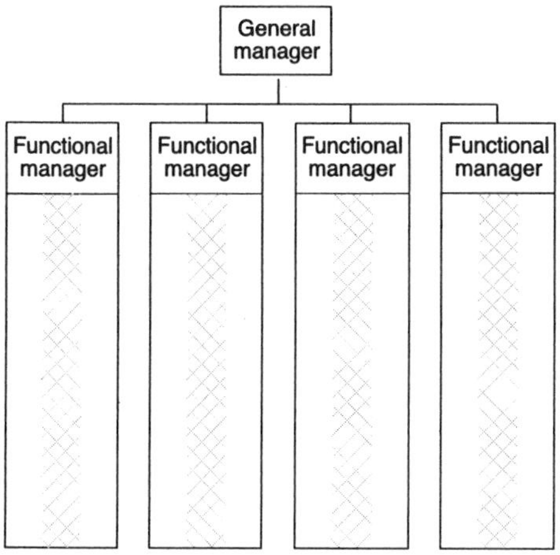

Fig. 3 A functional organization, in which authority rests with the functional managers. Source: Ref 4

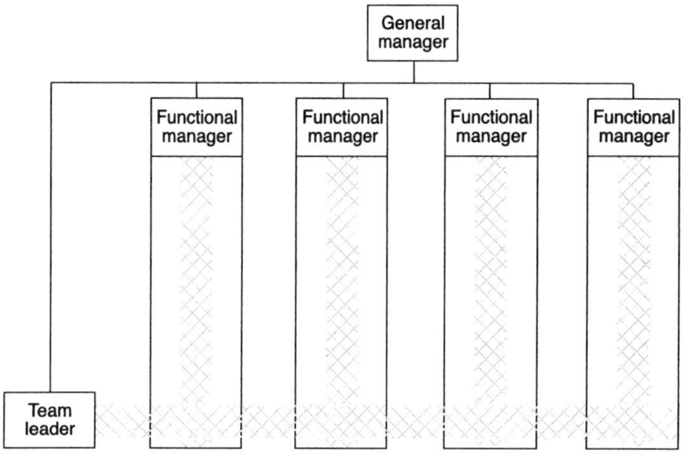

Fig. 4 A balanced matrix, where the team leader and functional managers have equal authority over team members. Source: Ref 4

Suppliers should be considered as team members when they have essential technical expertise to contribute, when their parts are critical to the cost or schedule of the product, or when the customer's design of a part will affect the supplier's ability to produce it reliably.

Clearly, many different levels of supplier involvement are possible. It is important to be flexible in molding each circumstance to fit the requirements. When supplier involvement is planned, the previously covered concerns about specialists should be kept in mind. A few key suppliers involved heavily are better than many involved superficially.

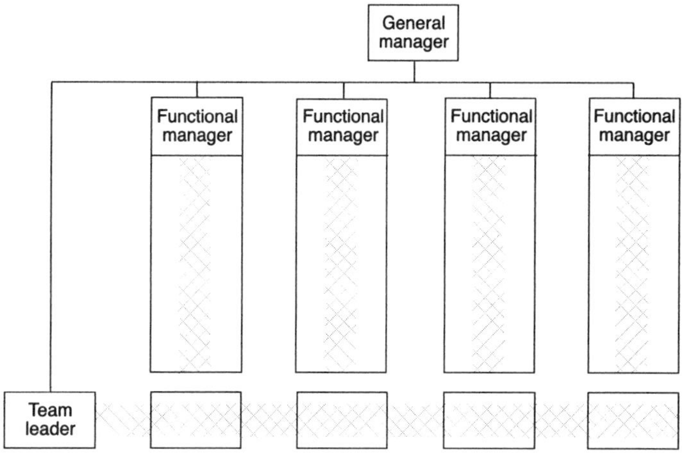

Fig. 5 A separate project organization, in which members report solely to the team leader. Source: Ref 4

Organizing a Development Team

Every organization has its formal organization depicted on the organization chart. Each also has an informal organization, the linkages by which things actually get done, decisions get made, and information flows. These systems have evolved over time to serve the primary needs of the firm. Due to need and tradition, most of these organizational structures are vertically (functionally) oriented. Although this vertical structure may be best for many corporate activities, it does not work well for developing innovative new products, which require heavy horizontal information flow.

Fortunately, corporate organizational structures are becoming more horizontal as firms delayer, decentralize, empower workers, and move toward team-based activity. The increasing emphasis on new products encourages this shift. However, the growing need for new products is outpacing changes in inertia-bound organizational structures. Usually, this suggests a bias toward structures for product development that are more horizontal and team based than the familiar ones. The change will require some organizational inventing and pioneering. Such organizational innovation is far more likely to take root if it is planned and set up before initiating a project.

Products of today are often complex, which means a development team must incorporate several types of technical expertise. Consider something as commonplace as a telephone set. Developing a new one requires electrical, mechanical, and software engineers, acoustics and materials experts, industrial design and ergonomics, and manufacturing process expertise. In addition, marketing, purchasing, and finance will be key participants. Thousands of decisions lie ahead, and thousands of problems await solutions. For the set to be a commercial success, the developers must reach delicate cross-functional balances repeatedly.

The present task is to provide an environment, that is, a team, to address such cross-functional problems and decisions quickly and effectively. Without such a team, the more vertical communication infrastructure in a company is likely to degrade the quality of the new product, add to its cost, and delay it.

Candidate Organizational Forms. It is helpful to think of the possible organizational forms as spanning a spectrum, from the functional one (strongly vertical) in Fig. 3, through the balanced matrix (Fig. 4), to the separate project shown in Fig. 5. The critical parameter that varies in these charts is the degree of control and influence the team leader has over individuals on the team compared with that held by the functional managers. In Fig. 3, there is no team leader, so all decisions flow through functional managers. In the balanced matrix, the team leader and functional managers hold equal power over team members. In Fig. 5, the team leader has unquestioned authority over those assigned to the project.

Important points on this spectrum occur between the illustrated ones. For example, between the charts displayed in Fig. 3 and 4 is a so-called lightweight team leader form, in which a team leader exists but has less clout than the functional managers. This is a popular and often dangerous form because organizations have moved to it from the functional form, thinking they have arrived at teams but not realizing that they really need to take more steps. Lightweight teams are often impotent, as the label suggests, and the leader often becomes frustrated. Between Fig. 4 and 5 is the heavyweight team leader form, a powerful one used by Honda, among others.

Figures 4 and 5 illustrate another key point. The team leader reports to a general manager, not to a functional manager, such as the vice president of engineering. Recall the earlier discussion about the team leader functioning as a general manager so that he or she would integrate the viewpoints of all functional managers. If the team leader reports to a functional manager, the project will take on the orientation of that function. The other functional managers will get involved to inject their opinions, bringing back the shortcomings of the functional form.

Selecting the Best Form for a Project. Every organizational form has its pros and cons. For example, the functional form is superior for maintaining consistency between products in a company's product line. But it is poor at facilitating communication across the functions involved in developing an innovative new product. Conversely, the separate project form excels at such cross-functional communication but is weak in cross-project coordination. The balanced matrix provides some of both but introduces potential conflicts because individuals on the team essentially have two equal bosses tugging at them.

The solution to this dilemma is to choose the form with strengths that most closely match the primary objectives of a particular project, then recognize the shortcomings of the chosen form, and put compensating mechanisms in place to handle them. For example, many firms introduce cross-functional project communication into the functional form by having weekly team meetings. (The earlier warning about trying to run a team through meetings should be noted.)

A consequence of this approach to organizational design is that each project will have its own structural form based on the specific objectives of that project. This makes the organization chart more complex but enables each project to use the most effective organizational tools available.

In general, a form closer to the separate project should be used for innovative, new-to-the-world products, and more functionally oriented forms should be used for more routine product upgrades (Ref 6).

There is nothing magical about the terminology used here, for instance the *heavyweight team leader* form. Other jargon is used, such as *core teams*. What really matters is how members are involved day-to-day, which is the next topic.

Full-Time, End-to-End Involvement. Another important characteristic of effective development teams is that, to the greatest extent possible, each member serves from the beginning of the project to its end and is involved full time for that period. Handoffs from person to person or from department to department mean breaks in the continuity of vital information. Engineers, according to a stereotype that is partially true, often want to redesign whatever they receive from someone else.

Full-time involvement (also called dedication) translates into higher commitment and accountability and into greater focus on key objectives of the project, such as the desires of key customers. By using full-time people, fewer people can handle the project, with the benefit that communication becomes far simpler. If a full-time member cannot be justified, their role should be defined carefully (Ref 4, p 142).

Full-time, end-to-end involvement is much easier to accomplish with generalists. This is one benefit of using generalists on a team, as discussed earlier.

The first person to be dedicated full time for the duration of the project should be the team leader. Part-time involvement in this key position is particularly ineffective.

The Power and Difficulties of Co-Location. Once a leader is selected, team members are recruited, an organizational form is chosen, and the degree of dedication expected from each member is established, then the last decision to be made is where to locate this crew. The basic choices are to leave members in the place where they were before the team formed or to physically locate them close together; this latter choice is called co-location.

The argument for co-location is that product development, especially for highly innovative products, requires a great deal of cross-functional communicating, problem solving, and decision making. Placing the participants close together simplifies these activities greatly. Project focus and easy access to project-related materials, such as products of the competitors, are additional advantages.

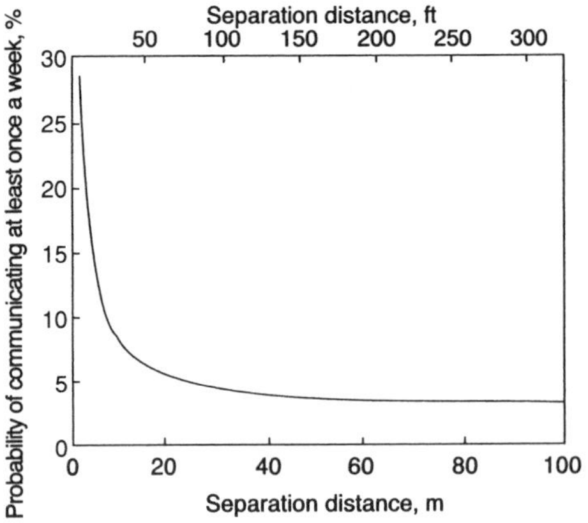

Fig. 6 Effect of separation distance on communication between team members. Communication is much more likely to occur if team members are located within about 10 m (30 ft) of one another. Source: Ref 7

Figure 6 illustrates the basic case for co-location. These data from several research and development sites show how likely individuals are to communicate about technical matters, depending on their separation. Note that the "knee" of the curve is at about 10 m (30 ft), which suggests that there is great value in having team members close enough to overhear conversations of one another.

Thus, true co-location means that team members are within conversational distance, not just in the same building or on the same floor. As discussed earlier, this team includes members from marketing and manufacturing, not just the research and development portion of the team. In the author's experience in working with over a hundred product development teams, this type of co-location is a powerful tool to shorten development cycle time dramatically.

Although the benefits of co-location are great, the resistance can be equally great in many organizations. Those who have tried it appreciate its benefits and would always use it again if effective project communication were critical. Many who have not tried it are skeptical, often due to personal reasons, such as lack of privacy; see Ref 4 (p 145–150, 271–272).

Co-Location Versus Electronic Team Linkages. The data in Fig. 6 are from Ref 7, which is over 20 years old. Many engineers in high-tech industries discount Fig. 6, asserting that modern electronic means of communication, for instance, faxes, e-mail, and videoconferencing, have superseded the need for physical co-location. Figure 6 suggests that the threshold (10 m, 30 ft) is so low that people are not willing to work very hard to communicate. If they have to take the effort to dial the phone, compose a message on their computer, or arrange a videoconference, they will instead just make this mini decision themselves. After a while, poor mini decisions pile up.

Electronic communications have two other shortcomings. One is that they are not very fast; the inherent delays in phone tag and its e-mail equivalent are commonplace. The more fundamental weakness is a lack of communication quality. The words themselves account for less than half of what a message communicates, most of the communication being attributed to intonation, body language, and timing. To various extents, all of the electronic media filter out this vital information. Even the current resolution of videoconferencing fails to pick up many clues.

Electronic media certainly have their value, but their limitations diminish their ability to facilitate rapid, effective team progress. Being aware of the limitations will help the team to compensate for them.

The Role of Rewards and Other Motivators. Many researchers and authors have addressed the effectiveness of motivators, such as compensation, recognition, and promotion in improving corporate productivity. This is a difficult subject about which to be definitive, and much of the available material is contradictory. However, two general observations apply to cross-functional development teams.

One is that these systems ultimately have to come into alignment with the behavior desired of the team, or the team will revert to traditional ways of operating. For example, if the culture punishes mistakes, then the behavior change sought, learning from mistakes and getting beyond mistakes quickly, will not occur. The new products developed by the team will not likely be very innovative in such a risk-averse environment. Similarly, if team cooperation is the desired outcome, individuals should not be rewarded.

Second, substantial dependence on rewards to achieve results is likely to backfire. In the author's experience, clients who focus on rewards usually have other, more fundamental difficulties, such as overbearing top management, and superficial fixes with rewards will not overcome the fundamental issue. In the end, team members must be motivated intrinsically by an interest in the work itself, and extrinsic motivators will have limited effect. For a sobering view of this subject, see Ref 8.

The Specialist's Role on a Development Team

An assumption underlying this article is that the reader is probably a materials specialist or manager who is reading it concerning their involvement on a cross-functional development team. Thus, the specialist's role needs specific attention here.

Balancing Team Needs with the Specialist's Needs. The dilemma of the specialist was covered earlier: the specialist's expertise is often needed to provide the technical product innovativeness essential to marketplace success, but the specialist introduces several complications in managing a high-performance development team. Thus, the specialist's role is one of those organizational design factors that should be resolved by first satisfying the major project objectives, then identifying known weaknesses in the specialist's role, and compensating for these. This means that the best solution is likely to differ every time.

The Specialist on a Weak Team. A weak team, for example, a functional organization or a lightweight team leader form, is really just a variety of specialists being guided by functional managers. Consequently, technical specialists fit into these forms quite naturally, but they also contribute to all of the shortcomings of these forms.

Whatever the organizational form, a chronic weakness of highly specialized technical people on development projects is that they often have little contact with the customer for which they are designing. They need to get into the field rather than rely on filtered information from others. For example, a plastics specialist working on a new type of plastic body panel resin for automobiles should spend time in body shops, car wash establishments, and shopping mall parking lots to see firsthand just how cars get used and abused.

The Specialist on a Strong Team. The specialist's role dilemma is most evident in the stronger team forms. Fortunately, there are options for how the specialist can contribute to the team.

Joining the Team Option. If the specialist's expertise constitutes a major contribution to the project, this person should be a regular, dedicated, co-located member of the team for at least most of the development and testing. The specialist should be a T-shaped individual, as discussed earlier, to justify end-to-end, full-time involvement. Limited involvement would mean that this person will be gone when problems associated with his or her design choices begin to appear later.

Expert Contributor Option. This is a popular middle ground, but it must be treated with care to get a quality, responsive contribution from the specialist. This individual is not a member of the team (trying to include such associates to help them feel more involved will simply dilute the significance of the team).

Therefore, a regular member of the team acts as a liaison to the specialist, and clear objectives, deliverables, and due dates are established for each task. The liaison should monitor progress closely, watching for slippage due to the specialist's other activities or lack of understanding of project goals. The specialist must spend enough time with the team that he or she can experience firsthand what the team is about. Team meetings may not be the place for specialists to get this direct exposure.

The expert contributor option simply provides a contracted deliverable, much like a supplier's, and should be managed accordingly.

Expert Advisor Option. An expert advisor acts as a consultant to the project and is expected to deliver competent professional advice, based on one's field of expertise. It is the team's responsibility, not the specialist's, to be sure this advice fits with team objectives and to identify contextual shortcomings in it. For example, if an automotive plastics specialist suggests a certain resin, it is the team's responsibility to ascertain that this resin is suitable for Siberia and Saudi Arabia, where they may intend to sell their cars.

If the specialist's advice is critical to the success or schedule of the project, then the specialist's participation should be arranged in advance.

Conclusions

Unlike much of the other material covered in the *ASM Handbook*, this article covers subjects without a strong scientific basis. There are few firm rules, and the best solution will depend greatly on the specific circumstances involved. Much of the supporting evidence is anecdotal, as in the case of co-location, for example.

However, this does not mean that there are no preferred solutions. Some solutions are far more powerful and effective than others, so it is definitely worth struggling with the issues to find the solution that works best in a specific situation. Individuals forming a design team should form their objectives, analyze the existing data to select an approach, and then *do something*. In making progress, action is preferable to inaction.

Initial team "experiments" should be operated on a manageable scale where the risk is reasonable, and they should involve the most enthusiastic people to initiate change. See Ref 4, Ch 15, for further information on making such changes. Results should be monitored, and adjustments should be made on an ongoing basis. See Ref 9.

For more detailed coverage of the material in this article, see Ref 4, especially Ch 7 and 8.

REFERENCES

1. J.R. Katzenbach and D.L. Smith, *The Wisdom of Teams*, Harper Business, 1993
2. G.M. Parker, *Cross-Functional Teams*, Jossey-Bass, 1994
3. R.E. Kelley, In Praise of Followers, *Harvard Business Review*, Vol 66 (No. 6), Nov-Dec 1988, p 142–148
4. P.G. Smith and D.G. Reinertsen, *Developing Products in Half the Time*, Van Nostrand Reinhold, 1995
5. R.R. Kamath and J.K. Liker, A Second Look at Japanese Product Development, *Harvard Business Review*, Vol 72 (No. 6), Nov-Dec 1994, p 154–170
6. E.M. Olson, O.C. Walker, and R.W. Ruekert, Organizing for Effective Product Development: The Moderating Effect of Product Innovativeness, *Journal of Marketing*, Vol 59 (No. 1), Jan 1995, p 48–62
7. T.J. Allen, *Managing the Flow of Technology*, The MIT Press, 1977, p 239
8. A. Kohn, *Punished by Rewards: The Trouble with Gold Stars, Incentive Plans, A's, Praise, and Other Bribes*, Houghton Mifflin, 1993
9. P.G. Smith, Your Product Development Process Demands Ongoing Improvement, *Research-Technology Management*, Vol 39 (No. 2), March-April 1996, p 37–44

Section 2: Criteria and Concepts in Design

Concurrent Engineering

Don Clausing, Massachusetts Institute of Technology

CONCURRENT ENGINEERING is product development that is done by concurrently utilizing all of the relevant information in making each decision. It replaces an approach to product development in which one type of information was predominant in making each sequential decision.

Sequential engineering, the product development style that was dominant in the United States during the 1960s and 1970s, emphasizes expertise. Decisions are taken by specialists. Each decision-making activity considers primarily only one specialty. Then each decision is passed on to another group of specialists for further decisions. This style does provide much expertise in the decisions. However, the integrated result leaves much to be desired. It has become caricatured as work that is thrown over the wall from one group of specialists to the next group.

Concurrent engineering came into practice in many companies in the United States during the 1980s. By 1990 it was widely recognized as the desired form of product development in most industries. Concurrent engineering emphasizes holistic decision making, with each decision taking into account the knowledge of all of the relevant specialties. The viewpoints of all of the specialties are integrated by being made responsive to the voice of the customer. By focusing expert knowledge on the needs of the customers, concurrent engineering avoids suboptimization and introspective specialized objectives that are irrelevant to the customers. Thus concurrent engineering provides products that customers want and will purchase.

Concurrent Process

The concurrent process utilizes all relevant expertise in making each decision.

Holistic Decisions. All product development decisions must take into account the following three aspects:

- *Product functionality:* the ability of the product to perform
- *Production capability:* the ability to produce the new product
- *Field-support capability:* the ability to support the product after it is shipped

If a product functions well but is difficult to produce or difficult to support in the field, it is a weak product. In the concurrent process, all three viewpoints are taken into account in making each decision. This is suggested in Fig. 1. The three functional activities are shown occurring concurrently. In the older style they occurred sequentially.

In addition to the specializations that are displayed in Fig. 1, many other specializations separate development people. Examples are mechanical engineers, electrical engineers, and computer scientists. Concurrent engineering is intended to integrate all forms of specialization.

Of course, all of the several aspects of product development cannot be kept in mind at the same time. Concurrent decision making is a matter of degree or of the time scale of iteration. In sequential engineering, the tendency is to make all of the functional decisions first. Then it is thrown over the wall for production decisions. Subsequently this is repeated for field-support decisions. This leads to massive iteration loops, with a time scale of months or years.

Concurrent engineering still requires iteration. However, the time scale for the iteration is typically minutes or hours, not months. Also, all of the relevant people are involved throughout. Therefore, the knowledge from other specialties does not come as a surprise after thinking has become solidified around a viewpoint based on personal expertise. This closing of minds tends to happen in sequential engineering. The interleaving of knowledge to form a holistic fabric is the key feature of concurrent engineering.

For example, a team of three people are working together to make the critical decisions about a key part for a machine. It is essentially a beam to support small mechanisms. First they consider the deflection of the beam and the need to keep its deflection less than 0.05 mm—the functional viewpoint. After half an hour they have an initial design for a stiffening rib. Then they work for 45 min on the production capability. Should the beam be an aluminum die casting or fiber-reinforced polymer? They decide on an aluminum die casting and adjust the processing to minimize distortion due to residual stresses. The details of the rib geometry are refined for aluminum die casting. Then for half an hour the team of three considers the field-support need to be able to readily remove the part in the field for servicing. Thus one could say that they are working sequentially. However, the time scale of each sequence is very small. Also, all three people are involved in all decisions so that they develop commitment to the decisions. This is the concurrent process.

Example 1: Problems Encountered in Sequential Engineering of a Materials Innovation. In the 1960s, dual-hardness armor was developed at the United States Steel Research Center in Monroeville, Pennsylvania. This was a marvelous accomplishment. However, in the sequential style that was dominant in that era, the functionality of the armor plate was emphasized for a long period of time before production capability was seriously addressed.

Traditional armor plate has two conflicting requirements: It must be very hard, to shatter the projectile, but also tough, so that the armor plate is not shattered. Traditional armor tries to satisfy both requirements with one parameter, the composition/microstructure of the armor. However, one parameter is insufficient to satisfy two requirements.

At the United States Steel Research Center, an invention was made that introduced two parameters to satisfy the two requirements: dual-hardness armor. This combined a hard front plate with a tougher rear plate. The two were diffusion bonded together to form one solid plate.

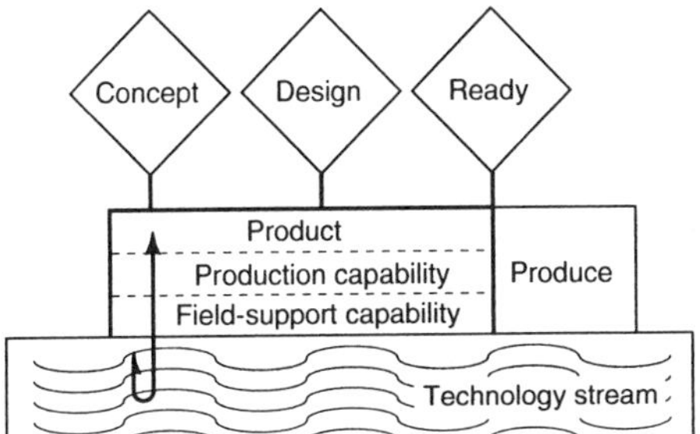

Fig. 1 Concurrent engineering process

The functional development was done with plates that were 1 ft in diameter—easy to make in laboratory equipment. The first production trials with 4 ft by 6 ft plates revealed a spherical bow (curvature) that was five times the specified limit.

The source of the bow was obvious. During the heat treatment, the dual-hardness plate behaved in a way similar to a bimetallic strip. The problem had not attracted serious attention during the functional development because the deviation from flatness varies as the square of the span of the plate. In 1 ft spans, the deviation did not look bad. However, the larger plates had 25 times as much deviation from flatness, an unacceptable bow.

After much time had been spent concentrating on the functional development, the production capability became the prime focus. The first attempt was to build a curved quenching die. This bent the hot plate in the opposite direction from the quenching curvature. The intent was that the two curvatures would cancel each other. This worked in the 1 ft plates. However in the larger plates, another problem arose.

After they were quenched, the larger plates appeared to attempt to reach a flat condition. However, they were elastically unstable. They snapped through the flat position. There were two stable positions, one on each side of flat. This was the familiar oil-can phenomenon.

A review of the applied mechanics literature revealed a paper that provided the needed analysis. Spherically curved plates with a sufficiently large ratio of span to thickness are unstable. They snap through the flat position. However during this analytical work, it was recognized that the differential strains that caused the objectionable bow were very small, less than 0.1%. This led to another invention.

The invention was simply to compress the softer, concave side of the plate. This expanded it laterally until it was the same dimension as the hard side, which eliminated the bow. A rolling-mill roll was prepared with bumps on it, something like a sheepsfoot roller. When the bumps pressed against the concave side of the dual-hardness plate, they made very slight impressions (Brinelling), which were sufficient to flatten the plates. The localized deformation provided the needed flattening without requiring high values of rolling force. This was very satisfactory on the laboratory plates. As no potential instability was involved, this process would undoubtedly have worked on production plates also.

However, another difficulty intervened. When the two inventors (the dual-hardness-armor inventor and the flattening-process inventor) went to a production rolling mill (factory) to plan production trials, they were rebuffed. "Run armor plate through our rolling mill — no way!" The mill operators were of course accustomed to rolling soft steel. Armor plate seemed out of bounds to them. The production-operations people had been involved too late to build up confidence in the approach.

This is the way sequential development work is done. It usually has problems of the type that are revealed in this example. Concurrent engineering does much better. If this project were done with concurrent engineering, a multifunctional team would be formed early in the project. The team would consist of:

- Metallurgist, the inventor of the functional capability (dual-hardness armor)
- Materials scientist/applied mechanician, the inventor of the flattening process
- Production mill engineer

In concurrent engineering, the same work would be done, but it would be done concurrently. This would address the technical and production aspects of the bowing problem much sooner. It would also gain the confidence and commitment of all three members of the team.

Note that the production process for flattening the bowed plates was eventually addressed in two stages:

- Analytical, which revealed the fundamental nature of the problem and led to a solution
- Production operations

Often future production problems can be addressed analytically at very early stages of development. This is an important practice in concurrent engineering. To make it fully effective, the production-operations people should be included on the team during the analysis so that their concerns can be adequately addressed and their commitment developed.

Multifunctional Teams

The concurrent process is carried out by a multifunctional team that integrates the specialties (functions). There are five types of organizational configuration for product development:

- Functional structure
- Lightweight product manager
- Heavyweight product manager
- Project execution team
- The independent product development team

The first two organizational types emphasize functional expertise and are of less interest in today's complex competitive world. The last three of these configurations are all referred to as multifunctional teams although they differ substantially from one another.

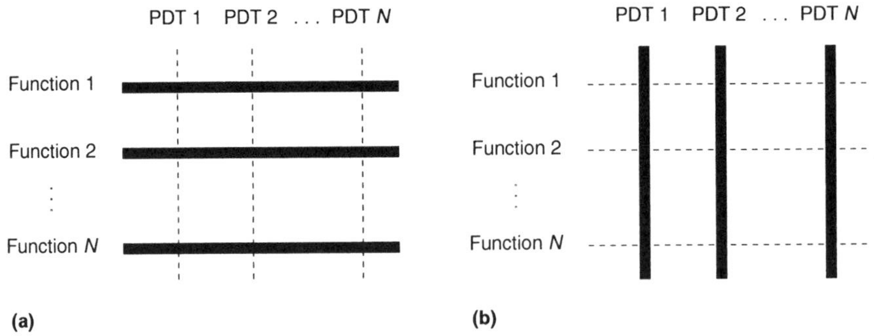

Fig. 2 Product design teams versus functional roles. (a) Organizational configuration with strong functional roles. (b) Organizational configuration with strong product design teams

The heavyweight product manager form of organization still has the players in functional homes. The multifunctional team aspect is achieved by the heavyweight product manager, who has enough power in decision making to focus the people on the product and its customers.

In the project execution team, many of the members have functional homes, but are seconded into the team organization for the duration, or at least a significant fraction of the duration, of a product development program. This achieves even more focus on the product and its customers than the heavyweight product manager form of organization. It also enables development people to return periodically to functional homes to maintain their functional competence.

In the independent product development team (PDT), the product development people do not have functional homes. Their complete organizational role is as a member of a PDT. This generates even greater focus on the product and its customers than the project execution team (although it tends to cause some career development problems for the team members).

In reality the different forms of organization lie on a continuum that relates the relative strength of the functions and the PDTs (Fig. 2). In the functional structure organizational configuration, the functions have the power, as suggested in Fig. 2(a). Progressing through the five types of organizational configuration, the functions (rows) become weaker, and the PDTs (columns) become stronger until in the independent PDT configuration the PDTs have all of the power (Fig. 2b). In the heavyweight product manager configuration—the first one that is typically referred to as a multifunctional team—the PDTs and the functions usually have roughly equal power.

If the products are fairly simple, the market is changing only slowly, and technical expertise is very important, then the functional organization works well. However, for complex products in a rapidly changing market with only normal technical-expertise requirements, the functional organization has proven unsatisfactory; it is too slow and weak in coping with complexity.

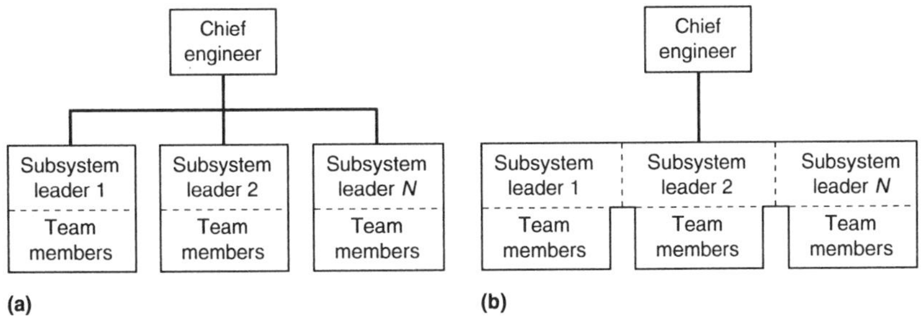

Fig. 3 Product design team configurations. (a) Subsystem teams. (b) Team of subsystem leaders

The independent PDT configuration has the great advantage of strong focus on each product and its customers. It has three potential dangers:

- Functional obsolescence
- Weak learning across the PDTs
- Weak technology development

One approach to overcoming these three problems is to have a technology development organization that is responsible for avoiding the three dangers. These three dangers must be guarded against even when the concept of the independent PDT is working very well.

The trend since 1980 has been toward the project execution team and the independent PDT configurations. They provide the needed focus on the product and its customers.

Subsystem Teams. For relatively simple products, the multifunctional team is sufficient. However, for more complex products, the organizational configuration will have a team for each subsystem, as shown in Fig. 3(a). For still more complex products, the organization can have added features, for example, another layer, modules, can lie between the Chief Engineer and the subsystem teams.

Higher Level Teams. An extension of the subsystem teams is indicated in Fig. 3(b). In this organizational configuration, the subsystem leaders also work as a team to make decisions. For example, resources are shifted from one subsystem to another as required. The practice of this higher level of multifunctional teamwork is still lagging behind the success of the multifunctional subsystem teams.

It is now increasingly important to extend the multifunctional team along the complete value chain. The early form of this has been called supplier involvement or something similar. However, the trend is to more equal partnerships, that is, the extension of multifunctional teamwork along the value chain. A critical decision is the design of the value-chain enterprise. Enterprise here usually means more than one corporation. This is a further challenge to the concept of the multifunctional team. Additional information about team-based activities is provided in the article "Cross-Functional Design Teams" in this Volume.

In summary, the central concept of concurrent engineering is the multifunctional team that concurrently brings all of their specialized knowledge to bear in making each decision. Next the decision-making style of the multifunctional team is considered. As they carry out the concurrent process, their decision-making style is critical to success.

Requirements, Concepts, and Improvement

The multifunctional product development teams make many types of decisions. All of them follow a three-step decision-making process: requirements, concepts, and improvement.

1. *Requirements:* What the product needs to be and do is assessed.
2. *Concepts:* The concept with the best potential to satisfy the requirements is determined.
3. *Improvement:* The details of the concept are rapidly improved to approach its full potential.

This three-step decision process is applied repeatedly, starting with a very broad scope and converging to the most detailed decisions. Early in the total development process, the decisions will be about product families and platforms. Near the end of the development, the decisions will be very detailed, for example, how a hole should be drilled in a part. The earlier decisions become requirements for the later, more detailed decisions. For example, after the team has decided that a part will be made of aluminum, the decision about the hole must be for an aluminum part. The fact that it is faster to make a hole in a polymer is then no longer relevant. Aluminum has become a requirement. The concept of the hole (cast, drilled, reamed, broached, etc.) must be responsive to the requirement of aluminum. (Of course, the machinability of the material was taken into account in the earlier decision to select aluminum.) The style that teams use to make the decisions about requirements, concepts, and improvement can range from ad hoc to structured decision making.

Structured Decision Making

Simply having a multifunctional team improves decision making in three ways:

- All relevant information is brought to bear.
- Iteration loops are short.
- The commitment of all parties makes success possible.

However, experience has shown that problems still exist with decision making by multifunctional teams.

The default decision-making style of teams is slightly caricatured as follows. Everyone sits around a table, talks, and waves their hands. At the end of the meeting, they write down on a flip chart some words that point to the decisions that were made. This ad hoc style of decision making leads to non-vigilant information processing.

Teams in their decision making need to achieve common understanding, consensus, and commitment. When teams engage in the ad hoc decision-making process that they tend to fall into by default, they seldom do well in achieving common understanding, consensus, and commitment. This is especially true for complex systems.

By definition, multifunctional teams lack common understanding as they start their work. The diversity of knowledge is what makes a multifunctional team necessary. Only after they develop a reasonable set of common understandings are they able to engage together in sound decision making.

Consensus and commitment are needed for the decisions to be applied in the future. A great weakness of the ad hoc approach to decision making is that the words on the flip chart at the end of the day are often ignored in subsequent decision making.

Experience and research have shown that structured decision making leads to vigilant information processing, which achieves common understanding, consensus, and commitment. There is still some concern that structured methods will inhibit creativity or have other negative effects. As Aristotle pointed out, the golden mean is usually desirable. It is possible to go too far with structured methods. However, in the opposite direction lies non-vigilant information processing and chaos and confusion, especially in the face of complexity. The judicious application of structured methods enables sound decisions and makes it possible to reduce complexity to workable tasks.

The structured decision-making processes described below enable the multifunctional team to make high-quality decisions in a reasonably short time. These high-quality decisions save much time by avoiding the later rework that plagues the projects that try to make do with ad hoc decisions.

Requirements Development

There are two types of requirements development by multifunctional teams. One, quality function deployment (QFD), is market driven. The second, functional analysis, is an engineering analysis that is driven internally.

Quality Function Deployment (QFD). QFD starts with the needs of the customers as expressed in the voice of the customer. The PDT then deploys these requirements from the voice of the customer to the factory floor. This deploy-

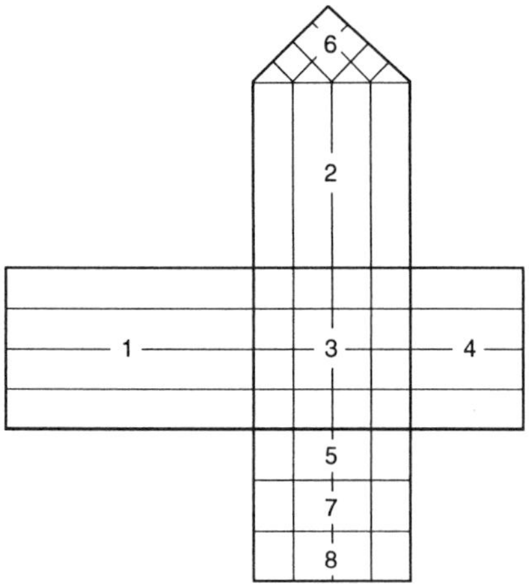

Fig. 4 Organization of a quality function deployment (QFD) "house of quality" product planning table. 1, voice of customer input; 2, specifications for new product; 3, translation between voice of customer and product specifications; 4, market research information; 5, technical benchmarking; 6, intersection of specifications (to examine conflicts or synergy); 7, weighting of importance of each characteristic; 8, quantitative goals for new product

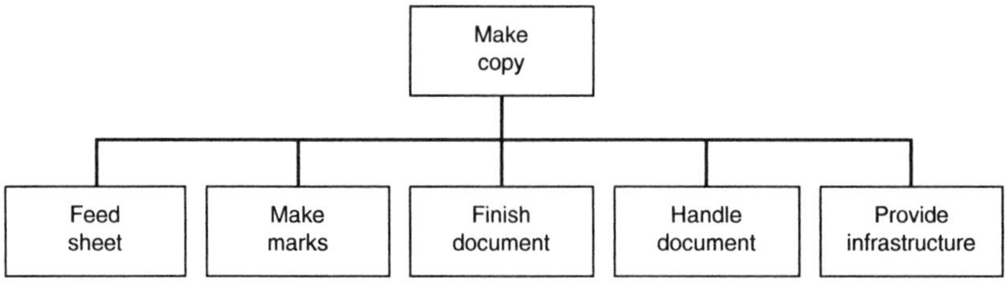

Fig. 5 Example of a functional tree for a photocopier

ment is done in small logical steps that are feasible to do.

QFD uses large visible displays to help the team in its decision making. These large visible displays are matrices (input/output tables) that are usually on large sheets of paper, approximately two meters square, that are fastened to a wall. This becomes the focus of the team decision making. It enables the team to refresh their memory about the relevant information, see the relationships among the information, and to record the decisions as they are made.

The first matrix in the deployment chain is for product planning and is called the "house of quality." Its input is the voice of the customer, and its output is the quantified specifications for the new product.

The product planning matrix is referred to as a house simply because it has the shape of a house (Fig. 4). Then each field within the matrix is called a room. Room 1 is the voice of the customer, the input. Rooms 2 and 8 contain the output. Room 2 contains the names of the specifications for the new product. Room 8 contains the quantified goals for the new product. For example, for a car, one column would specify the horsepower, while another column would specify the required life.

The other rooms in the house of quality are to help the PDT to make the decisions that deploy the objectives from room 1 to rooms 2 and 8. Room 3 is used to understand, model, and verify the translation from the voice of the customer (room 1) to the corporate requirements (room 2).

Rooms 4 and 5 are market research and benchmarking results, which serve as the basis for deciding the quantitative goals in room 8. Room 4 is benchmarking by customers (market research), as well as strategic goal setting. Room 5 is benchmarking by technical tests. Then the quantitative goals in room 8 are selected to be sufficient improvements over the benchmarks.

Rooms 6 and 7 are used to help plan the development project. Each cell in room 6 is the intersection of two columns, and the question becomes whether the requirements conflict or are synergistic. For example, hardness and toughness conflict. By recognizing all conflicts early, the PDT can plan the development project to apply the resources to achieve success in both characteristics, not simply settling for the existing tradeoff. Room 7 contains an estimate of the importance of each characteristic and an estimate of the difficulty in achieving the quantified specification. Other project-management information can be included in room 7. The combination of rooms 6 and 7 helps the PDT to plan the development project, that is, which objectives should have the most resources. Thus the project is made responsive to the needs.

Once the product has been planned, the PDT has a clear statement of what the product must be and do. Furthermore, the PDT is committed to the goals because they worked together to develop them.

Next the goals are deployed to the next stage of decision making. The exact path that is followed depends on the complexity of the product. Also, it is necessary to select the product concept before the requirements can be completely deployed into more detail.

Before the deployment path is considered, the other form of requirements development is described. Additional information about QFD is provided in the article "Conceptual and Configuration Design of Products and Assemblies" in this Volume.

Functional analysis is a traditional engineering practice, often used in the past in value engineering. It is even more useful when it is used from the beginning of the development project in conjunction with QFD.

The basic approach is the construction of a functional tree; an example is shown in Fig. 5. The top of the functional tree contains the primary function to be performed by the whole product. The requirement for a copier is to make copy. Then the PDT members ask the question *how*? This leads to the next level of functions. Feed sheet is one of the "hows" to make copy. Each function is stated as a verb and a noun. Sometimes adjectives are added to clarify the meaning of the noun. By simply answering how, the PDT deploys the functional requirements into more detail. (The complete functional tree for a copier has several hundred boxes. Although this might seem to imply thousands of requirements in the accompanying QFD matrices, in actual practice the requirements in each QFD matrix are prioritized so that a typical matrix has roughly 30 requirements.)

As with QFD, it is not possible to develop the complete set of requirements until some concept selection has been completed. For example, the requirements at the next more detailed level will depend on the selection of xerography or ink jet as the marking concept.

Integration of QFD and Functional Analysis. Functional analysis is a sparse, engineering statement of requirements. QFD is much more detailed and starts with the voice of the customer. The basic concept of integrating the two is simple. The statement that the requirement for a copier is to make copy includes many other caveats. It should cost less than a penny per copy. The copier should not take too much space. It should be easy to use, etc. These amplifications of the functional requirement are captured in the columns of the QFD matrix. The amplifications that are most important to the development project are captured in the QFD matrix. Other amplifications of the functional requirement are more mundane and fall into the category of knowledge-based engineering or standards. (Knowledge-based engineering emphasizes engineering practices, art, and solutions that have a record of success. These techniques often are implemented with the help of computers.) This relationship is indicated in Fig. 6.

As indicated in Fig. 6, the column headings in the QFD matrix can be thought of as having two sources:

- The rows of the matrix
- The associated functional requirement

There is not at this time a detailed methodology for executing this integrated view of requirements. Usually it is sufficient for the PDT to prepare the QFD matrices and the functional trees, and have the relevant knowledge-based engineering and standards available. Then the PDT considers all three types of requirements:

- Functional tree
- QFD matrix
- Knowledge-based engineering and standards

The relationships among the three are suggested by

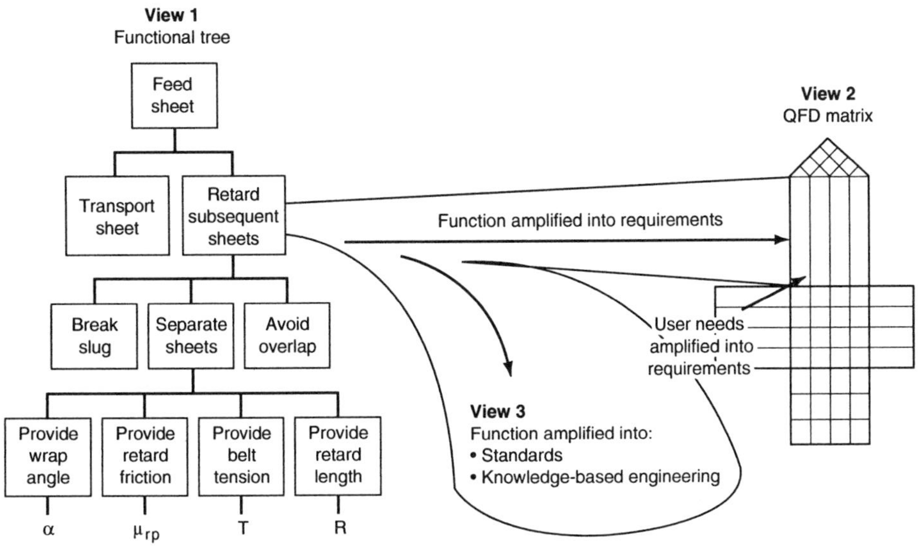

Fig. 6 Three views of product requirements

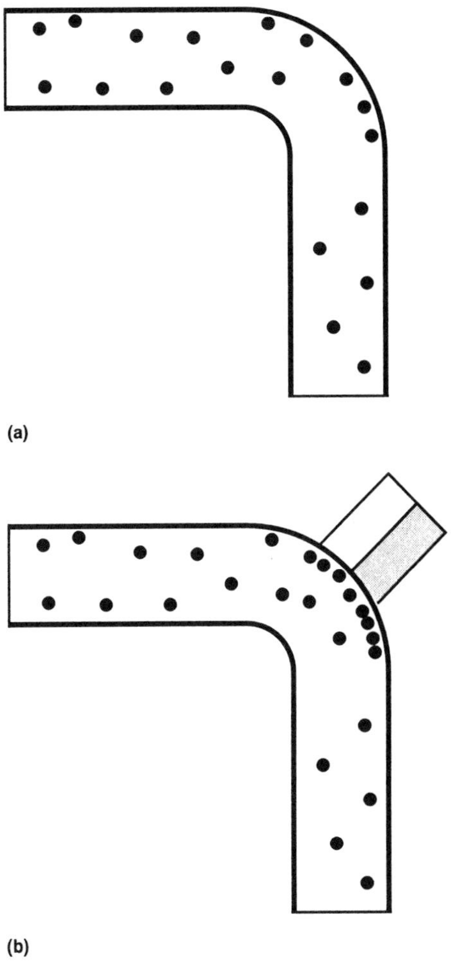

Fig. 7 Design problem and solution. (a) Problem: Shot wears pipe at turn. The contradiction is that a coating appears to be needed, but is not a good solution because of added cost and short life. (b) Solution: Magnets are used to form a continuously replenishable protective layer of shot.

Fig. 6. The important operational step is for the team to come to a consensus on a consistent message while considering all three types of requirements.

The requirements lead to the generation and selection of the concept.

Concept Development and Selection

The PDT selects the concept in response to the requirements. First, concepts must be found or generated. Then starting with the initial set of concepts, the PDT progressively develops and selects concepts until they converge on the dominant concept.

The generation of new concepts, invention, has long been said to be the one area to which structured methods cannot be applied. The best approach has been to immerse the creative people in the definition of the requirements. (Necessity is the mother of invention.) In response to the emerging statement of the requirements, the creative people will spontaneously generate new concepts. This is still a good approach. However, in the mid-1990s, a structured approach is also starting to be used in the United States. Additional information is provided in the article "Creative Concept Development" in this Volume.

TRIZ. A structured method for invention that was developed in the former USSR is starting to be used in the United States and western Europe. It is called theory of inventive problem solving (TIPS), or TRIZ (the Russian equivalent of TIPS), or related names. Although the experience with this approach is brief, it appears to be very promising and is attracting many early applications.

TRIZ was developed by Genrikh Altshuller in the USSR, who studied a huge number of patents and observed patterns in invention. Some of his key observations are:

- Conflicts are the mother of invention.
- Standard solutions occur in diverse fields.
- Evolution of technological systems follow certain patterns.
- Systematic application of scientific effects aids invention.

An example will help to communicate the basic approach of TRIZ. In Fig. 7(a) shot is flowing through a pipe. A basic contradiction is displayed. The TRIZ solution is shown in Fig. 7(b). The use of one of the interfacing materials to form the interface is one of the standard approaches that has emerged from the study of the huge number of patents. Mastery of TRIZ involves learning these standard approaches and the other patterns that have emerged from the long study of invention.

However the concepts are formed, it is important to build a set of competing concepts that are each responsive to the requirements. These are used as the starting concepts in the Pugh concept selection process.

Pugh Concept Selection. The Pugh concept selection process starts with the initial set of concepts and the requirements, and helps the PDT to develop and select the best concept. Although it is called selection, the concept that is selected is seldom exactly one of the initial concepts. "Evolution" describes the successful practice better than does "selection." The team further develops the concepts as the result of the insights that this process helps them to recognize.

The Pugh concept selection process is carried out with a large matrix, Fig. 8. The rows are the criteria, and the columns are the concepts. The criteria are based on the QFD matrix and the functional tree. The PDT typically develops 15 to 20 criteria to be used in the Pugh matrix.

One concept is selected as the datum (reference concept), and the other concepts are evaluated with respect to the datum. This is done one row at a time. This pair-wise comparison is essential to the success of this process. Pair-wise comparison is much superior to abstract evaluation. Each concept is simply judged to be better than (+), the same as (S), or inferior to (–) the datum concept.

Although the scores are shown added up in Fig. 8, the main value of the Pugh concept selection process is that it helps the team to develop insight into both the criteria (row headings) and the concepts. These insights lead to clarification of the criteria and to improved concepts. The new concepts that are generated during this process are often hybrids of the initial concepts.

This is an iterative process. Typically the matrix is run one or two times during a session. The datum is changed between runs. There are explicit efforts to eliminate negatives and reinforce positives. After another week or two of work, the team comes together again to run the matrix. The process is complete when the team converges on the dominant concept in which they have confidence and to which they are committed.

The Pugh concept selection process leads to the selection of good concepts without the application of excessive resources. The good concept and the confidence of the team are essential to the successful development of the new product. An example of the Pugh method applied to materials selection is provided in the article "Use of Decision Matrices in Materials Selection" in this Volume.

Integration of Requirements Development and Concept Development and Selection

Figure 9(a) displays the integration of requirements and concept. In the spirit of concurrent engineering, the requirements for the product and the requirements for the production process are considered together in selecting the concept. The concept includes both the product itself and the production capability to produce the product. (Field-support capability is also included, but for simplicity it is omitted from this description.)

The arrows with loops in Fig. 9(a) indicate that the QFD form of the requirements and the functional tree form of the requirements are developed together and made to be consistent. This is done for both the product and for the production capability. The production process capability QFD matrix has for its input (rows) the product requirements (columns) of the product QFD matrix. This integrates the product and its production capability, and makes both responsive to the customers.

In reality, Fig. 9(a) is highly idealized with respect to actual practice. The typical concurrent practice goes through the steps of Fig. 9(a) iteratively. Usually the product requirements and concept are done first; then the production capability is considered. If the production appears to be straightforward, it is often not considered until later. Functional trees are often postponed until later. Each multifunctional team needs to find a path through Fig. 9(a) that is effective for them. However, the best practice is to integrate the elements of Fig. 9(a) as much as possible; that is to do all of the steps with iteration loops that are very short in duration.

After the steps of Fig. 9(a) are completed, the production decisions are deployed to the factory floor in the QFD operations matrix, as implied in Fig. 9(b). The process QFD is process engineering, while the operations QFD generates the nitty-gritty information that is needed for factory operation.

For a simple product (e.g., a knife with one or two parts), Fig. 9(b) is all that is needed. However for complex products, the requirements and concepts must be developed in several levels, such as total system, subsystem, and piece part or component. The decisions at each level must be relatively complete before proceeding very far at the next level. For example, the total system architecture (TSA) must be selected before the requirements are deployed from the total system level to the subsystem level.

The basic concept is displayed in Fig. 9(c). First the requirements are developed, and the concept is selected at one level. Then in the context of the selected concept, the requirements are deployed down to the next level, and the process is repeated. This is done for as many levels as is necessary, which will depend on the complexity of the system that is being developed. A typical set of levels for one complex system is: total system, module, subsystem, subassembly, piece part, and piece-part feature. This hierarchy also extends upward to include platforms and families of products.

Concepts that are responsive to the requirements are selected for each system level by the process that is displayed in Fig. 9(c). These concepts have high potential. However, some of them will be innovative but embryonic. They need improvement to let their full potential bloom. The improvement is needed rapidly to

	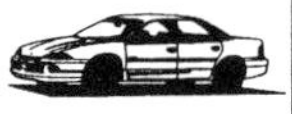	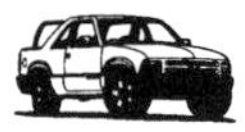		
Responsiveness	Datum	+	+	+
Braking	Datum	–	–	–
Ease of manufacturing	Datum	S	–	–
Safety	Datum	+	+	–
Risk	Datum	–	–	–
Complexity	Datum	S	–	–
Total		+ 2	+ 2	+ 1
		S 2	S 0	S 0
		– 2	– 4	–5

Fig. 8 Example of a Pugh concept selection matrix. +, better than; S, same as; and –, inferior to the datum

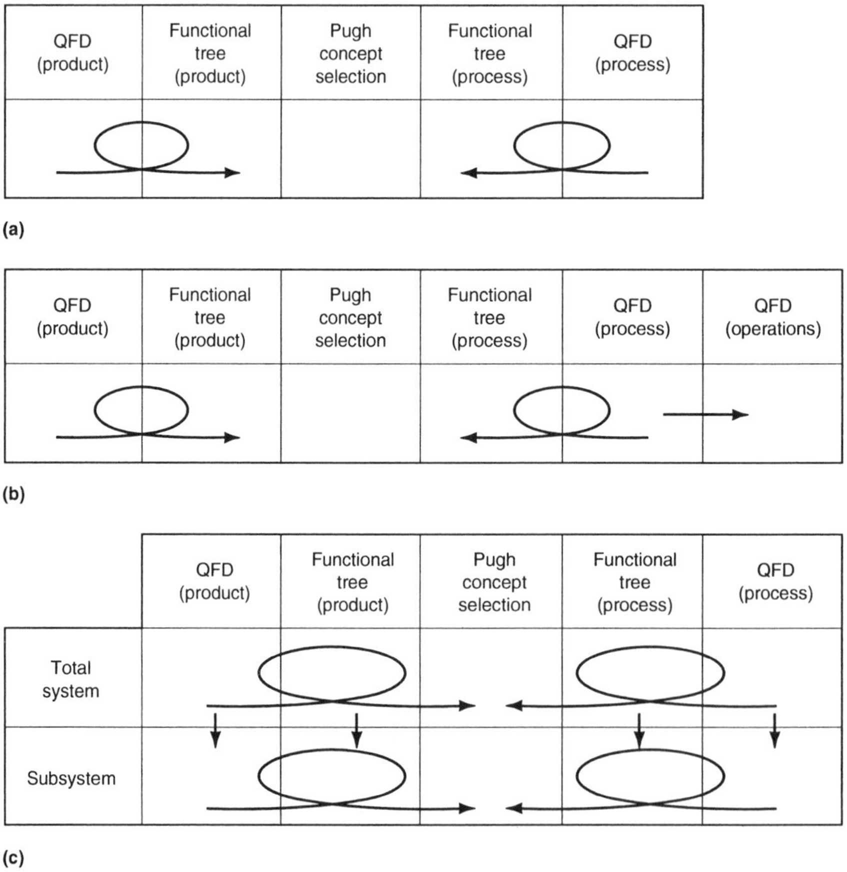

Fig. 9 Integration of product requirements and concept development. (a) Idealized representation of steps for developing product requirements, concept, and production capability. (b) Steps including deployment to the factory floor (operations QFD). (c) Deployment through the levels of the system

satisfy the ever shortening time-to-market requirements.

Improvement of Concepts

After a concept is selected, it must be improved to achieve something close to the full potential of the concept. Of course, if the concept was already improved close to its potential limit for a previous product, then it does not have to be improved again. This is a big advantage of reusability. Nevertheless, some elements of a product are usually new concepts that have not previously been systematically improved.

Improvement to achieve high quality and reliability has two aspects:

- Improving performance to make it close to ideal
- Minimizing mistakes

Very different approaches are needed for these two aspects.

Robust Design. Robustness means that the performance of the system is always acceptably close to the ideal function of the system. Even the best of concepts does not do this in its initial configuration and trials. Systematic optimization rapidly improves the performance from the initial promising but inadequate state to a level that captures most of the potential of the concept.

Metallurgists have long practiced a simple form of this systematic improvement. The concept of a new alloy is developed on the basis of prior knowledge until the composition and heat treatment are approximately as desired. Then designed experiments are used to improve the alloy to its final design. The initial composition can be thought of as the concept. Then rapid improvement is done. The emphasis has been on the design of the experiments. The improvement of an alloy is relatively simple.

The rapid improvement of a complex system is much more difficult than fine tuning an alloy. First of all, the emphasis must shift. The design of the experiments is challenging, but is relatively the simpler part of the entire process. The greater challenge is to define the objective in a way that will be effective. A second challenge is to conduct the improvement activity in a way that is time efficient. People with experience in the simple improvement of alloys usually have had some difficulty in making the switch to the improvement of complex systems. It is a big mistake to think of robust design as just a small extension of the design of experiments (DOE).

Although there were pockets of robust design activity in the United States before 1980, it became a well-recognized industrial practice when Dr. G. Taguchi came to the United States and introduced his comprehensive and well-developed approach. Taguchi developed his approach as a practicing engineer in industry. It is uniquely well suited to competitive product development.

Taguchi's quality engineering has two major elements:

- *Parameter design:* optimization of the nominal values of the critical parameters of the system to make the performance robust
- *Tolerance design*: the selection of the best precision levels around the nominal values

It is critically important to do parameter design first. Tolerance design selects the best precision levels around the optimized target values of the critical parameters. Tolerance design alone, no matter how well done, is never a substitute for parameter design.

Parameter Design. A critical parameter is simply a parameter that strongly affects the performance of the system and for which the PDT can select the nominal value. In Taguchi's nomenclature, these are control parameters because the PDT has control of the nominal value. For a steel alloy, the amount of nickel is a critical control parameter. Fault trees (the negative of functional trees) have been found to be very useful in helping PDTs to identify the critical parameters.

Complex systems typically have hundreds of critical parameters: dimensions, forces, friction coefficients, electrical characteristics, etc. One big designed experiment with hundreds of parameters is not feasible. Therefore, most of the optimization of robustness is done at the subsystem or subassembly level. The number of control parameters in one designed experiment is commonly in the range of 4 to 15.

Time-efficient improvement is achieved by an appropriate tradeoff between time and certainty. One could run many repetitions of the same system setup in order to be more certain (higher confidence level) of the result. However, this is a poor tradeoff, which will take too much time. Usually there is a fixed amount of time to complete the improvement of the robustness of the system, typically a few months. The available time can be used to optimize a few parameters with great certainty. Alternatively, many more parameters can be optimized with somewhat less certainty. The latter is the best approach. Therefore, unless repetitions are very easy and quick to do, they are not used. Putting it in another way, between any two trials of the system the values of the critical control parameters are systematically changed. (The changes are defined by an orthogonal array, a table of designed experiments.) This provides much more information than repetitions.

In addition to the parameters that the PDT can control, there is another type of parameter that affects the performance of the system. For a product that operates out of doors, the ambient temperature is an example. The PDT cannot control such a parameter during operation by customers. Still they must be taken into account during the optimization. Taguchi calls such parameters noise parameters. There are three types of noise parameters:

- *Environmental:* Ambient temperature is an example.
- *Manufacturing:* No two units of production are exactly alike.
- *Deterioration:* This will cause further changes in the parts of the system.

Note that the last two types of noises, manufacturing and deterioration, are essentially the same, that is, deviations of the values of the critical control parameters away from the nominal value. The manufacturing type includes the initial deviations within the product as it is shipped. The deterioration type includes further changes as the result of use and time.

It is very important to introduce noises when doing the optimization of robustness. One of the primary objectives is to make the system robust against the noises. Thus the noises are intentionally introduced into the optimization. This is problem prevention, not waiting for the problem to appear much later. In the absence of an explicit approach to robustness, the noises are usually ignored until late in the development project; that is, the system is initially pampered. When noises are introduced during production-readiness testing, the bad effects are observed when it is too late to effectively cope with them. Parameter design introduces the noises very early so that the optimization can mitigate the effects of the noises. This avoids many problems and much rework later in the development project.

The introduction of noises often requires some ingenuity and development. A paper feeder is an example. One failure mode is the feeding of two sheets of paper. Many good types of papers have little tendency to have this failure mode. Testing with such papers has little value for the improvement of the robustness of the performance. Eventually it was recognized that alternating sheets from two different types of paper created a special paper stack with a useful amount of noise; that is, it challenged the feeder with a tendency to multifeed. Developing such noise conditions is often essential to the efficient optimization of robustness.

The most important element in the successful practice of robust design is the characterization of the performance. As an example, consider an electrical resistor. Its performance is usually characterized by the resistance, R. However, the value of R is not important to quality and reliability. Any nominal value of R that is desired is easily achieved. The characterization R already assumes a linear relationship between voltage and current, with R being the slope of the assumed straight-line relationship between voltage and current. The specific value of R is not of primary interest in optimizing the robustness of the resistor. Rather the quality of the straight-line relationship between voltage and current is important. Therefore, voltage is plotted versus current with two noise conditions: noise values that cause small current and noise values that cause large current. The most robust resistor is the one that has the least deviation from a straight line, which is the ideal performance of the resistor. The smallest value of the ratio of the deviation from the straight line divided by the slope of the straight line is needed. After further analysis, the square of the ratio is taken. Therefore, the ratio of the average of the square of the deviations (averaged over all data points) is divided by the square of the slope of the best-fit straight line through the

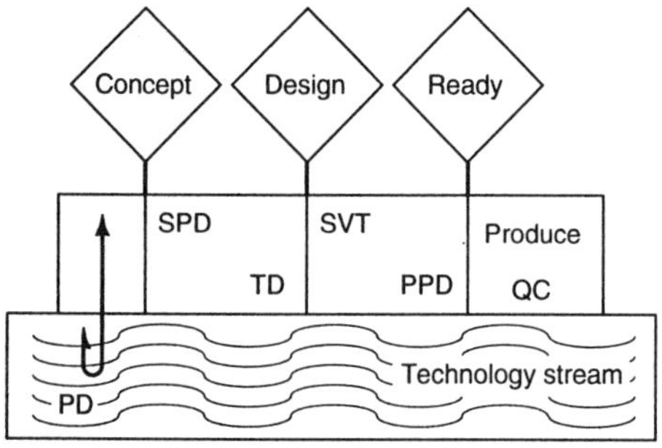

Fig. 10 Timing of Taguchi robust design steps. PD, parameter design (new product and process technologies); SPD, system (product) parameter design; TD, tolerance design; SVT, system verification test; PPD, process parameter design; QC, on line quality control (factory floor)

data. This is the measure of robustness. Taguchi developed a set of such metrics to which he gave the name signal-to-noise (SN) ratios. Larger values of any SN ratio represent more robustness.

The most important steps in robust design are:

1. Define the ideal performance – often not simple to do
2. Select the best SN definition to characterize the deviations from ideal performance
3. Develop the sets of noises that will cause the performance to deviate from the ideal

After some experience, the use of the design of experiments to rapidly increase the SN value is relatively straightforward.

As an example, a subsystem is considered for which the PDT has identified the 13 most critical control parameters. Initial judgments are made of the best nominal values for each of the 13 control factors. Seeking improvement, a larger value and a smaller value are selected as representing feasible but significant changes to the initial design. This gives three candidate values for each of the 13 critical control factors. The total number of candidate sets of values is 3^{13}, which is 1,594,323. Even a relatively simple subsystem gives a large number of candidates from which the PDT must select the best one, or better yet, quickly pick one of the best candidates. This requires systematic trials, that is, designed experiments. A standard orthogonal array is found that has 13 columns and 27 rows. The 13 critical control factors are assigned to the 13 columns. Each row defines one candidate for the critical values of the 13 parameters. The 27 rows define a balanced set of 27 candidates from the total of 1,594,323 candidates. For each of the selected 27 candidates, the appropriate sets of noises are applied and the performance is determined, either analytically or experimentally. The SN ratio (robustness) of each of the 27 candidates is calculated. A simple interpolation among the 27 values of the SN ratio predicts the candidate from the total of 1,594,323 that is probably the best. Typically the PDT iterates two or three times. The control factors that had little effect are dropped, and some others are introduced. The ranges of the values for the control factors are reduced to fine tune the optimization. Then a confirmation trial is conducted to verify the magnitude of the improvement. It is very important to do this parameter design early and quickly.

The results of the parameter design are best captured in a critical parameter drawing. This drawing shows the system (usually a subsystem) with only as much detail as is needed to make the critical parameters clear, and it shows the values of the critical functional parameters that have been optimized. These then become specifications for the detailed design. By constraining the detailed design to the optimized values of the critical functional parameters, the robust performance is ensured.

Tolerance Design. The optimization of robustness (SN value) often brings very large improvements. After the nominal values of the critical control factors are optimized, tolerance design is done. Of course, most of tolerance design is guided by standards and knowledge-based engineering. However, some decisions require more in-depth analysis. The primary step in tolerance design is to select the production process (or the precision of a purchased component) that provides the best tradeoff between initial manufacturing cost and quality loss in the field. Taguchi developed methods for this analysis. After the production process is selected, tolerances are calculated to be put on the drawings and other specifications. However, selecting the production process is the most important step in tolerance design, as it controls the inherent precision.

The timing of robust design is critical for success. The optimization of robustness must be done early to achieve the benefits of problem prevention. As shown in Fig. 10, most of the optimization of robustness (parameter design) should be done to new technologies before they are pulled out of the stream of new technologies and integrated into any specific product. Any remaining product parameter design (SPD) is done early in the product program, before detailed design has progressed very far. The final verification of the robustness is done in the system verification test (SVT). The SVT is usually performed on the first total-system prototypes, which are made after the detailed design has been completed. (In the concept phase, the decisions of the first row of Fig. 9(c), total system decisions, are made. In the design phase, the decisions of the second row of Fig. 9(c), subsystem decisions, and the decisions at the other more detailed levels are made. In the readiness phase, the decisions are deployed to the factory floor, as indicated at the right of Fig. 9(b). Also in the readiness phase, mistakes are eliminated.)

Robust design is very important. Robust systems provide customer satisfaction, because they work well in the hands of the customer. They have lower costs because they are less sensitive to variations. Robust subsystems and components can be readily integrated into new systems because they are robust against the noises that are introduced by new interfaces. Most important of all, the early optimization of robustness reduces time to market by eliminating much of the rework that has traditionally plagued the latter stages of product development. This section is a brief introduction to the subject that has emphasized the primary features. Additional information is provided in the article "Robust Design" in this Volume.

Mistake minimization is completely different from the optimization of robustness. Robustness optimization is done for concepts that are new, for which the best values of the critical functional parameters are unknown. Mistake minimization applies to system elements for which there is experience and a satisfactory design approach is known, but was not applied. Examples range from a simple dimensional error to a gear that is mounted on a cantilever shaft that is too long. The excessive deflection of the shaft causes too much gear noise and wear. The design of gears and shafts is well understood, so one that has a problem is a mistake. The mistake could be a simple numerical error. It could be that the person (or computer program) with the necessary knowledge was not readily available.

The first approach is to avoid making mistakes by using a combination of:

- Knowledge-based engineering (and standards)
- Concurrent engineering (multifunctional teams)
- Reusability

Knowledge-based engineering helps to design standard elements, such as gears and shafts, using design rules and computers to implement proven approaches. Multifunctional teams help to avoid mistakes by having the needed expertise available. (A common source of mistakes is that the knowledgeable person was not involved in the design.) Reuse of proven subsystems, which have demonstrated that they are not plagued with mistakes, will also reduce mistakes.

Despite all of the best efforts to avoid the occurrence of mistakes, some mistakes will still occur. Then they must be rooted out of prototypes of the system by the problem solving process. This process is basically:

- Identify problems.
- Determine the root causes of the problems.
- Eliminate the root causes while ensuring that no new problems are being introduced.

Failure-modes-and-effects analysis (FMEA) is very useful in doing this (see the article "Risk and Hazard Analysis in Design" in this Volume).

The combination of robust design and mistake minimization will achieve excellent system quality and reliability. It is important to recognize that reliability is not a separate subject above and beyond robust design and mistake minimization. The traditional field that is called reliability is primarily devoted to keeping score of reliability and projecting it into the future based on certain assumptions. Robust design and mistake minimization achieve early and rapid improvement of reliability. This will rapidly develop new concepts to capture their full potential.

Conclusions

The fundamental core of concurrent engineering is the multifunctional team that carries out the concurrent process to make holistic decisions. These decisions integrate the many diverse specialties to develop a product that provides customer satisfaction.

Simply having a multifunctional team improves the decision making by bringing all of the relevant information to bear on each decision. The concurrent process also gains the commitment of all of the participants to the decisions, which leads to effective implementation. However, multifunctional teams can improve their decision making relative to the ad hoc approach into which it is all too natural to fall. The judicious application of structured methods described in this article and elsewhere in this Volume enables sound decisions and makes it possible to reduce complexity to workable tasks.

A multifunctional team that makes holistic decisions by using the best structured decision-making processes while concentrating on both customer satisfaction and business goals provides the greatest leverage for the abilities of the product development people. The products that they develop will:

- Be quick to market
- Satisfy customers
- Have constrained costs
- Be flexible in responding to changes in the marketplace

The ultimate purpose of concurrent engineering is to provide products that customers want and will purchase.

SELECTED REFERENCES

- K. Clark and T. Fujimoto, *Product Development Performance*, Harvard Business School Press, 1991
- D. Clausing, *Total Quality Development*, ASME Press, 1994
- D. Clausing, EQFD and Taguchi, Effective Systems Engineering, *First Pacific Rim Symposium on Quality Deployment*, Macquarie University Graduate School of Management (Sydney, Australia) 15–17 Feb 1995
- L. Cohen, *Quality Function Deployment*, Addison Wesley, 1995
- M. Phadke, *Quality Engineering Using Robust Design*, Prentice Hall, 1989
- S. Pugh, *Total Design*, Addison Wesley, 1990
- S. Pugh, *Creating Innovative Products Using Total Design*, Addison Wesley, 1996

Designing to Codes and Standards

Thomas A. Hunter, Forensic Engineering Consultants Inc.

REGARDLESS OF THE MATERIAL to be used, most design projects are exercises in creative problem solving. If the project is a very advanced one, pushing the boundaries of available technical knowledge, there are few guidelines available for the designer. In such instances, basic science, intuition, and discussions with peers are common approaches that combine to produce an approach to solving the problem. With the application of skill, daring, a little bit of luck, money, and patience, a workable solution usually emerges.

However, most design projects just are not that challenging or different from what has been done in the past. In mechanical and structural design, for example, a tremendous amount of solid experience has been accumulated into what has been called *good practices*. Historically, such information was carefully guarded and was often kept secret. With the passage of time, however, these privately developed methods of solving design problems became common knowledge, ever more firmly established. Eventually they evolved into published standards of practice. Some government entities, acting under their general duty to preserve general welfare and to protect life and property from harm, added the standards to their legal bases. This gave the added weight of authority to the standards development movement.

In some cases, use of a standard may be optional to the designer. In others, adherence to standard requirements may be mandatory, with the full backing of the legal system to enforce it. In any case, as soon as the problem has been defined, a competent designer should make a survey of any existing standards that may apply to the given problem. There are two obvious advantages to this effort. First, the standards may give valuable guidance to the problem solution. Second, conformance to standards can avoid later legal complications with product liability lawyers.

Historical Background

Anyone who has taken a course in elementary physics has been taught about the "fundamental" quantities of mass, length, and time. When the metric system of measurements was established in 1790, a standard was set for the unit of length: one ten-millionth of the distance from one of the earth's poles to the equator. It was, by definition, the meter. However, there were a couple of problems with it. Because there was no way to make such an actual measurement at that time, there was a certain degree of error, and the standard suffered from a lack of portability. Some improvement was made in 1889, when an international convention on weights and measures agreed that the standard meter would be defined by the distance between two marks on a metal bar. This improved both accuracy and portability, and this standard was used until 1960. Then the standard changed to the wavelength of an orange-red line in the spectrum of Krypton 86. In 1983 the standard of length changed again, this time to a measurement based on the speed of light in a vacuum. The point here is that even the most basic standard units are subject to change as methods of measurement become more and more refined.

While basic standards change only infrequently, technical standards and codes are all subject to more frequent modification. The thousands of published standards and codes are reviewed and updated periodically, many of them on an annual basis. Therefore, when making the recommended survey of applicable standards, the designer should check to make certain they are the most current ones. In addition, because of the periodic review process, it is advisable to query the publisher of the standard to find out if a revised version is being worked on, and if it may be released before the design is scheduled for completion. Obviously, to avoid instant obsolescence, any oncoming changes should be factored into the decisions made by the designer.

The Need for Codes and Standards

The information contained in codes and standards is of major importance to designers in all disciplines. As soon as a design problem has been defined, a key component in the formulation of a solution to the problem should be the collection of available reference materials; codes and standards are an indispensable part of that effort. Use of codes and standards can provide guidance to the designer as to what constitutes good practice in that field and ensure that the product conforms to applicable legal requirements.

The fundamental need for codes and standards in design is based on two concepts, interchangeability and compatibility. When manufactured articles were made by artisans working individually, each item was unique and the craftsman made the parts to fit each other. When a replacement part was required, it had to be made specially to fit. However, as the economy grew and large numbers of an item were required, the handcrafted method was grossly inefficient. Economies of scale dictated that parts should be as nearly identical as possible, and that a usable replacement part would be available in case it was needed. The key consideration was that the replacement part had to be interchangeable with the original one.

Large-scale production was not possible until Eli Whitney invented the jig. Although he is best remembered for his invention in 1793 of a machine for combing the seeds out of cotton, the gin (which any good mechanic could copy it and many did), Whitney made his most valuable contribution with the jig. Its use enabled workers to replicate parts to the same dimensions over and over, thus ensuring that the parts produced were interchangeable.

Before the Civil War, the Union Army issued a purchase order for 100 rifles, but included a unique requirement that all the rifles had to be assembled, fired, taken apart, the parts commingled, and then reassembled into 100 working rifles. Interchangeability was the key problem. Whitney saw that the jig was the solution. By using jigs, Whitney was the only bidder able to meet the requirement. With that, the industrial age of large-scale production was on its way.

Standardization of parts within a particular manufacturing company to ensure interchangeability is only one part of the industrial production problem. The other part is compatibility. What happens when parts from one company, working to their standards, have to be combined with parts from another company, working to *their* standards? Will parts from company A fit with parts from company B? Yes, but only if the parts are compatible. In other words, the standards of the two companies must be the same.

Examples of problems resulting from lack of compatibility are common. For years, railroads each had their own way of determining local times. A particular method may have been useful for the one railroad that used it, but wrecks and confusion demanded that standard times be developed. There used to be several different threads used on fire hose couplings and hydrants. All of them worked, but emergency equipment from one town could not be used to assist an adjoining town in case of need. So a national standard was agreed upon.

Any international traveler knows that the frequency and voltage of electric power supplies vary from one country to another. Some are 110 V, others 220. Some are 50 Hz, others 60. In addition, all the connecting plugs are different. Even the side of a road on which one drives presents compatibility problems. Approximately 50 countries, notably the United Kingdom and Japan, use the left side; other countries use the right lane. With the global market for automobiles, manufacturers must produce two different versions to meet the incompatible local market requirements. Perhaps someday there will be a global standard, but the costs of any changeover will be enormous. This situation points out the near-irreversibility of somewhat arbitrary standardization decisions. Because of the relative permanence of their decisions, standards writers bear a particularly heavy burden of responsibility.

Purposes and Objectives of Codes and Standards

The protection of general welfare is one of the common reasons for the establishment of a government agency. The purpose of codes is to assist that government agency in meeting its obligation to protect the general welfare of the population it serves. The objectives of codes are to prevent damage to property and injury to or loss of life by persons. These objectives are accomplished by applying accumulated knowledge to the avoidance, reduction, or elimination of definable hazards.

Before going any further, the reader needs to understand the differences between "codes" and "standards." Which items are codes and which are standards? One of the several dictionary definitions for "code" is "any set of standards set forth and enforced by a local government for the protection of public safety, health, etc., as in the structural safety of buildings (building code), health requirements for plumbing, ventilation, etc. (sanitary or health code), and the specifications for fire escapes or exits (fire code)." "Standard" is defined as "something considered by an authority or by general consent as a basis of comparison; an approved model."

As a practical matter, codes tell the user *what* to do and *when* and under what circumstances to do it. Codes are often legal requirements that are adopted by local jurisdictions that then enforce their provisions. Standards tell the user *how* to do it and are usually regarded only as recommendations that do not have the force of law. As noted in the definition for code, standards are frequently collected as reference information when codes are being prepared. It is common for sections of a local code to refer to nationally recognized standards. In many instances, entire sections of the standards are adopted into the code by reference, and then become legally enforceable. A list of such standards is usually given in an appendix to the code.

How Standards Develop

Whenever a new field of economic activity emerges, inventors and entrepreneurs scramble to get into the market, using a wide variety of approaches. After a while the chaos decreases, and a consensus begins to form as to what constitutes "good practice" for that economic activity.

By that time, the various companies in the field have worked out their own methods of design and production and have prepared "in-house" standards that are used by engineering, purchasing, and manufacturing to ensure uniformity and quality of their product. In time, members of the industry may form an association to work together to expand the scope of their proprietary standards to cover the entire industry. A "trade" or "industry" standard may be prepared, one of its purposes being to promote compatibility among various components. This is usually done on a consensus basis. However, this must be done very carefully because compatibility within an industry may be regarded as collusion by the justice department, resulting in an antitrust action being filed. A major example of this entire process is the recent growth of the Internet, where compatibility plays a primary function in the formulation of networks, but so far regulators have used a light hand.

As an industry matures, more and more companies get involved as suppliers, subcontractors, assemblers, and so forth. Establishing national trade practices is the next step in the standards development process. This is usually done through the American National Standards Institute (ANSI), which provides the necessary forum. A sponsoring trade association will request that ANSI review its standard. A review group is then formed that includes members of many groups other than the industry, itself. This expands the area of consensus and is an essential feature of the ANSI process.

ANSI circulates copies of the proposed standard to all interested parties, seeking comments. A time frame is set up for receipt of comments, after which a Board of Standards Review considers the comments and makes what it considers necessary changes. After more reviews, the standard is finally issued and published by ANSI, listed in their catalog, and available to anyone who wishes to purchase a copy. A similar process is used by the International Standards Organization (ISO), which began to prepare an extensive set of worldwide standards in 1996.

One of the key features of the ANSI system is the unrestricted availability of its standards. Company, trade, or other proprietary standards may not be available to anyone outside that company or trade, but ANSI standards are available to everyone. With the wide consensus format and easy accessibility, there is no reason for designers to avoid the step of searching for and collecting any and all standards applicable to their particular projects.

Types of Codes

There are two broad types of codes: performance codes and specification or prescriptive codes. Performance codes state their regulations in the form of what the specific requirement is supposed to achieve, not what method is to be used to achieve it. The emphasis is on the result, not on how the result is obtained. Specification or prescriptive codes state their requirements in terms of specific details and leave no discretion to the designer. There are many of each type in use.

Trade codes relate to several public welfare concerns. For example, the plumbing, ventilation, and sanitation codes relate to health. The electrical codes relate to property damage and personal injury. Building codes treat structural requirements that ensure adequate resistance to applied loads. Mechanical codes are involved with both proper component strength and avoidance of personal injury hazards. All of these codes, and several others, provide detailed guidance to designers of buildings and equipment that will be constructed, installed, operated, or maintained by persons skilled in those particular trades.

Safety codes, on the other hand, treat only the safety aspects of a particular entity. The Life Safety Code, published by the National Fire Protection Association (NFPA) as their Standard No. 101, sets forth detailed requirements for safety as it relates to buildings. Architects and anyone else concerned with the design of buildings and structures must be familiar with the many No. 101 requirements. In addition to the Life Safety Code, NFPA publishes hundreds of other standards, which are collected in a 12-volume set of paperbound volumes known as the National Fire Codes. These are revised annually, and a set of loose-leaf binders are available under a subscription service that provides replacement pages for obsolete material. Three additional loose-leaf binders are available for recommended practices, manuals, and guides to good engineering practice.

The National Safety Council publishes many codes that contain recommended practices for reducing the frequency and severity of industrial accidents. Underwriters' Laboratories (UL) prepares hundreds of detailed product safety standards and testing procedures that are used to certify that the product meets their requirements. In contrast to the ANSI standards, UL standards are written in-house and are not based on consensus. However, UL standards are available to anyone who orders them, but some are very expensive.

Professional society codes have been developed, and several have wide acceptance. The American Society of Mechanical Engineers (ASME) publishes the Boiler and Pressure Vessel Code, which has been used as a design standard for many decades. The Institute of Electrical and Electronic Engineers (IEEE) publishes a series of books that codify recommended good practices in various areas of their discipline. The Society of Automotive Engineers (SAE) publishes hundreds of standards relating to the design and safety requirements for vehicles and their appurtenances. The American Society for Testing and Materials (ASTM) publishes thousands of standards relating to materials and the methods of testing to ensure compliance with the requirements of the standards.

Statutory codes are those prepared and adopted by some governmental agency, either local, state, or federal. They have the force of law and contain enforcement provisions, complete with license requirements and penalties for violations. There are literally thousands of these, each applicable within its geographical area of jurisdiction.

Fortunately for designers, most of the statutory codes are very similar in their requirements, but there can be substantial local or state variations. For example, California has far more severe restrictions on automotive engine emissions than other states. Local building codes often have detailed requirements for wind or snow loads. Awareness of these local peculiarities by designers is mandatory.

Regulations. Laws passed by legislatures are written in general and often vague language. To implement the collective wisdom of the lawmakers, the agency staff then comes in to write the regulations that spell out the details. A prime example of this process is the Occupational Safety and Health Act (OSHA), which was passed by the U.S. Congress, then sent to the Department of Labor for administration. The regulations were prepared under title 29 of the U.S. Code, published for review and comment in the Federal Register, and issued as legal minimum requirements for design of any products intended for use in any U.S. workplace. Several states have their own departments of labor and issue supplements or amendments to the federal regulations that augment and sometimes exceed the minimums set by OSHA. Again, recognition of the local regulatory design requirements is a must for all design professionals in that field.

Types of Standards

Proprietary (in-house) standards are prepared by individual companies for their own use. They usually establish tolerances for various physical factors such as dimensions, fits, forms, and finishes for in-house production. When outsourcing is used, the purchasing department will usually use the in-house standards in the terms and conditions of the order. Quality assurance provisions are often in-house standards, but currently many are being based on the requirements of ISO 9000. Operating procedures for material review boards are commonly based on in-house standards. It is assumed that designers, as a function of their jobs, are intimately familiar with their own employer's standards.

Industry consensus standards, such as those prepared by ANSI and the many organizations that work with ANSI, have already been discussed. A slightly abridged list of ANSI-sponsoring industry groups and their areas of concern will be given under the following section on Codes and Standards Preparation Organizations.

Government specification standards for federal, state, and local entities involve literally thousands of documents. Because government purchases involve such a huge portion of the national economy, it is important that designers become familiar with standards applicable to this enormous market segment. To make certain that the purchasing agency gets precisely the product it wants, the specifications are drawn up in elaborate detail. Failure to comply with the specifications is cause for rejection of the seller's offer, and there are often stringent inspection, certification, and documentation requirements included.

It is important for designers to note that government specifications, particularly Federal specifications, contain a section that sets forth other documents that are incorporated by reference into the body of the primary document. These other documents are usually federal specifications, federal and military standards (which are different from specifications), and applicable industrial or commercial standards. They are all part of the package, and a competent designer must be familiar with all branches of what is called the specification tree. The MIL standards and Handbooks for a particular product line should be a basic part of the library of any designers working in the government supply area. General Services Administration (GSA) procurement specifications have a format similar to the military specifications and cover all nonmilitary items.

Product definition standards are published by the National Institute of Standards and Technology under procedures of the Department of Commerce. An example of a widely used Product Standard (PS) is the American Softwood Lumber Standard, PS 20. It establishes the grading rules, names of specific varieties of soft wood, and sets the uniform lumber sizes for this very commonly used material. The Voluntary Standards Program uses a consensus format similar to that used by ANSI. The resulting standard is a public document. Because it is a voluntary standard, compliance with its provisions is optional unless the Product Standard document is made a part of some legal agreement.

Commercial standards (denoted by the letters CS) are published by the Commerce Department for articles considered to be commodities. Commingling of such items is commonplace, and products of several suppliers may be mixed together by vendors. The result can be substantial variations in quality. To provide a uniform basis for fair competition, the Commercial Standards set forth test methods, ratings, certifications, and labeling requirements. When the designer intends to use commodity items as raw materials in the proposed product, a familiarity with the CS documents is mandatory.

Testing and certification standards are developed for use by designers, quality assurance agencies, industries, and testing laboratories. The leading domestic publisher of such standards is the American Society for Testing and Materials (ASTM). Its standards number several thousand and are published in a set of 70 volumes divided into 15 separate sections. The standards are developed on a consensus basis with several steps in the review process. Initial publication of a standard is on a tentative basis; such standards are marked with a T until finally accepted. Periodic reviews keep the requirements and methods current. Because designers frequently call out ASTM testing requirements in their materials specifications, the designer should routinely check ASTM listings to make certain the applicable version is being called for.

International standards have been proliferating rapidly for the past decade. This has been in response to the demands of an increasingly global economy for uniformity, compatibility, and interchangeability—demands for which standards are ideally suited. Beginning in 1987, the International Standards Organization (ISO) attacked one of the most serious international standardization problems, that of quality assurance and control. These efforts resulted in the publication of the ISO 9000 Standard for Quality Management. This has been followed by ISO 14000 for Environmental Management Standards, which is directed at international environmental problems. The ISO has several Technical Committees (TC) that publish handbooks and standards in their particular fields. Examples are the ISO Standards Handbooks on Mechanical Vibration and Shock, Statistical Methods for Quality Control, and Acoustics. All of these provide valuable information for designers of products intended for the international market.

Design standards are available for many fields of activity, some esoteric, many broad based. Take marinas for example. Because it has so many recreational boaters, the state of California has prepared comprehensive and detailed design standards for marinas. These standards have been widely adopted by other states. Playgrounds and their equipment have several design standards that relate to the safety of their users. Of course one of the biggest applications of design standards is to the layout, marking, and signage of public highways. Any serious design practitioner in those and many other fields must be cognizant of the prevailing design standards.

Physical reference standards, such as those for mass, length, time, temperature, and so forth, are of importance to designers of instruments and precision equipment of all sorts. Testing, calibration, and certification of such products often call for reference to national standards that are maintained by the National Institute for Standards and Technology (NIST) in Gaithersburg, MD, or to local standards that have had their accuracy certified by NIST. Designers of high precision products should be aware of the procedures to be followed to ensure traceability of local physical standards back to the NIST.

Codes and Standards Preparation Organizations

U.S. Government Documents. For Federal government procurement items, other than for the Department of Defense, the Office of Federal Supply Services of the General Services Administration issues the Index of Federal Specifications, Standards and Commercial Item Descriptions every April. It is available from the Superintendent of Documents, U.S. Government Printing Office. Washington, D.C. 20402.

General Services Administration item specifications are available from GSA Specifications Unit (WFSIS), 7th and D Streets SW, Washington, D.C. 20407.

Specifications and standards of the Department of Defense are obtainable from the Naval Publications and Forms Center, 5801 Tabor Avenue, Philadelphia, PA 19120.

To order documents issued by the National Institute of Standards and Technology it is first necessary to obtain the ordering number of the desired document. You get this from NIST Publication and Program Inquiries, E128 Administration Bldg., NIST, Gaithersburg, MD 20899. With the ordering number, the documents are available from the Government Printing Office, Washington, D.C. 20402, or the National Technical Information Service, Springfield, VA 22161.

Underwriters' Laboratories documents can be obtained from Underwriters' Laboratories, Inc., Publications Stock, 333 Pfingsten Road, Northbrook, IL 60062.

ASTM Standards. Publications of the American Society for Testing and Materials can be ordered from ASTM, 100 Barr Harbor Drive, West Conshohocken, PA 19428.

National Fire Codes and other NFPA publications can be ordered from the National Fire Protection Association, 1 Batterymarch Park, Quincy, MA 02269-9101.

Building codes are issued by three organizations. The southern states use the Standard Building Code published by the Southern Building Code Congress International, Inc. (SBCCI), 900 Montclair Road, Birmingham, AL 35213-1206. The western states use the Uniform Building Code published by the International Conference of Building Officials (ICBO), 5360 Workman Mill Road, Whittier CA 90601-2298. The central and eastern states use the BOCA National Building Code obtainable from Building Officials and Code Administrators International, Inc. (BOCA), 4051 West Flossmoor Road, Country Club Hills, IL 60478-5795. A separate building code, applicable only to one and two family dwellings, is published by the Council of American Building Officials (CABO), 5203 Leesburg Pike, Falls Church, VA 22041, as a joint effort of SBCCI, BOCA, and ICBO and is obtainable from any of them.

The International Mechanical Code is published by the International Code Council, Inc., as a joint effort of the BOCA membership. It is intended to be compatible with the requirements of the Standard, Uniform, and National Building Codes and can be obtained from any CABO organization.

The International Plumbing Code is also published by the International Code Council as a CABO joint effort and is obtainable from any member organization.

The Model Energy Code is published under the auspices of CABO as a joint effort of BOCA, SBCCI, and ICBO with heavy input from the American Society of Heating, Refrigerating, and Air Conditioning Engineers, Inc. (ASHRAE) and the Illuminating Engineering Society of North America (IESNA). Copies are obtainable from any CABO member.

ANSI Documents. The American National Standards Institute (ANSI), 11 West 42nd Street, New York, NY 10036, publishes and supplies all American National Standards. The American National Standards Institute also publishes a catalog of all their publications and distributes catalogs of standards published by 38 other ISO member organizations. They also distribute ASTM and ISO standards and English language editions of Japanese Standards, Handbooks, and Materials Data Books. ANSI does not handle publications of the British Standards Institute or the standards organizations in Germany and France.

As mentioned previously, there are many organizations that act as sponsors for the standards that ANSI prepares under their consensus format. The sponsors are good sources for information on forthcoming changes in standards and should be consulted by designers wishing to avoid last-minute surprises. Listings in the ANSI catalog will have the acronym for the sponsor given after the ANSI/ symbol. For example, the standard for Letter Designations for Radar Frequency Bands, sponsored by the IEEE as their standard 521, issued in 1984, and revised in 1990, is listed as ANSI/IEEE 521-1984(R1990). All of one sponsor's listings are grouped under one heading in alphabetical order by organization. The field of interest of each sponsor is usually obvious from the name of the organization. Table 1 is slightly abridged from the full acronym tabulation in the ANSI catalog. Addresses and phone numbers have been obtained from listings in association directories. ANSI does not give that data.

Standards Information Services. Copies of standards and information about documents published by the more than 350 code- and standard-generating organizations in the United States and several other countries can be obtained from resellers such as Global Engineering Documents, Englewood, CO. They provide information on CD-ROM, magnetic tape, microfilm, or microfiche formats. Similar services exist in many countries throughout the world.

Designer's Responsibility

As soon as a designer has been able to establish a solid definition of the problem at hand, and to formulate a promising solution to it, the next logical step is to begin the collection of available reference materials such as codes and standards. This is a key part of the background phase of the design effort. Awareness of the existence and applicability of codes and standards is a major responsibility of the designer.

A primary component of the reference materials collection will be the codes and standards of which the designer is aware and that are known to be applicable to the design problem. As pointed out previously, there are several readily accessible sources for the myriad reference documents that the designer may review and examine to decide which ones are applicable.

One of the designer's responsibilities in the background phase is to make certain that the collection of reference codes and standards is both complete and comprehensive. Considering the enormous amount of information available, and the ease of access to it, this can be a formidable task. However, a designer's failure to acquire a complete and comprehensive collection of applicable standards is ill-advised in today's litigious environment. In addition, failure of the designer to meet the requirements set forth in the standards can be considered professional malpractice.

If the designer's product goes into production and enters the marketplace, the maker of the product hopes that it will be accepted by purchasers and will be an economic success. The purchasers, on the other hand, hope that the product will meet their expectations. Among their expectations are: first, that the product will perform its intended function and, second, that it will do no harm to them personally or to their property. In other words, the purchaser expects the product to be safe to use in its ordinarily intended manner of use. This expectation of safety extends even to some uses never intended or even conceived of by the designer (misuse) and to instances of deliberate overloading of the product (abuse). Thus, one of the designer's responsibilities is to eliminate the possibility that the product will do harm. If any of these customer expectations of the product are violated and harm occurs the result may be a legal action based on the laws of torts.

Torts, in the legal sense, are simply acts of wrongdoing. The failure of a product to perform as intended is not a tort, just a bad product. However, if the product does harm to any person or property, that may be considered a wrongful act. Recovery for damages caused by the wrongful act can be obtained through the courts by filing a lawsuit.

If the suit results in a finding that the product was defective in some way, and that the defect was related to the causation of the personal injury or property damage, then monetary damages may be assessed against the maker of that product. That is, the maker is liable for the resulting harm caused by that product. This is the part of the tort

Table 1 Sponsoring organizations for standards published by the American National Standards Institute

Acronym	Organization
AAMA	American Apparel Manufacturers Association. 2500 Wilson Blvd., Arlington, VA 22201 (703) 524-1864
AAMA	American Architectural Manufacturers Association. 1540 E. Dundee Rd., Palatine, IL 60067 (708) 202-1350
AAMI	Association for the Advancement of Medical Instrumentation. 3330 Washington Blvd., Arlington, VA 22201 (703) 525-4890
AASHTO	American Association of State Highway and Transportation Officials. 444 N. Capitol St., N.W., Washington, D.C. 20001 (202) 624-5800
AATCC	American Association of Textile Chemists and Colorists. P.O. Box 12215, Research Triangle Park, NC 22709-2215 (919) 549-8141
ABMA	American Bearing Manufacturers Association and Anti-Friction Bearing Manufacturers Association (AFBMA). 1900 Arch St., Philadelphia, PA 19103 (215) 564-3484
...	American Boat and Yacht Council. 3069 Solomon's Island Rd., Edgewater, MD 21037-1416 (410) 956-1050
ACI	American Concrete Institute. P.O. Box 19150, Detroit, MI 48219 (313) 532-2600
ADA	American Dental Association. 211 E. Chicago Ave., Chicago, IL 60611 (312) 440-2500
AGA	American Gas Association. 1515 Wilson Blvd., Arlington, VA 22209 (703) 841-8400
AGMA	American Gear Manufacturers Association. 1500 King St., Alexandria, VA 22314 (703) 684-0211
AHAM	Association of Home Appliance Manufacturers. 20 W. Wacker Dr., Chicago, IL 60606 (312) 984-5800
AIA	Automated Imaging Association. 900 Victor's Way, Ann Arbor, MI 48106 (313) 994-6088
AIAA	American Institute of Aeronautics and Astronautics. 370 L'Enfant Promenade, S.W., Washington, D.C. 20024 (202) 646-7400
AIIM	Association for Information and Image Management. 1100 Wayne Ave., Silver Spring, MD 20910 (301) 587-8202
AISC	American Institute of Steel Construction, Inc. 1 E. Wacker Dr., Chicago, IL 60601-2001 (312) 670-2400
ANS	American Nuclear Society. 555 N. Kensington Ave., La Grange Park, IL 60525 (708) 352-6611
API	American Petroleum Institute. 1220 L St., N.W., Washington, D.C. 20005 (202) 682-8000
ARI	Air-Conditioning and Refrigeration Institute. 4301 N. Fairfax Dr., Arlington, VA 22203 (703) 524-8800
ASAE	American Society of Agricultural Engineers. 2950 Niles Rd., St. Joseph, MI 49085-9659 (616) 429-0300
ASCE	American Society of Civil Engineers. 1015 15th St., N.W., Washington, D.C. 20005 (202) 789-2200
ASHRAE	American Society of Heating, Refrigerating and Air-Conditioning Engineers. 1791 Tullie Circle, N.E., Atlanta, GA 30329 (404) 636-8400
ASME	American Society of Mechanical Engineers. 345 E. 47th St., New York, NY 10017 (212) 705-7722
ASQC	American Society for Quality Control. 611 E. Wisconsin Ave., Milwaukee, WI 53201 (414) 272-8575
ASSE	American Society of Sanitary Engineering. P.O. Box 40362, Bay Village, OH 44140 (216) 835-3040
AWS	American Welding Society. 550 LeJeune Rd., N.W., Miami, FL 33126 (305) 443-9353
AWWA	American Water Works Association. 6666 W. Quincy Ave., Denver, CO 80235 (303) 794-7711
BHMA	Builders Hardware Manufacturers Association. 355 Lexington Ave., New York, NY 10017 (212) 661-4261
CEMA	Conveyor Equipment Manufacturers Association. 9384-D Forestwood Ln., Manassas, VA 22110 (703) 330-7079
CGA	Compressed Gas Association. 1725 Jefferson Davis Highway, Arlington, VA 22202-4100 (703) 412-0900
CRSI	Concrete Reinforcing Steel Institute. 933 Plum Grove Rd., Schaumburg, IL 60173 (708) 517-1200
DHI	Door and Hardware Institute. 14170 Newbrook Dr., Chantilly, VA 22021-2223 (703) 222-2010
EIA	Electronic Industries Association. 2500 Wilson Blvd., Arlington, VA 22201 (703) 907-7550
FCI	Fluid Controls Institute. P.O. Box 9036, Morristown, NJ 07960 (201) 829-0990
HI	Hydraulic Institute. 9 Sylvan Way, Parsippany, NJ 07054-3802 (201) 267-9700
HTI	Hand Tools Institute. 25 North Broadway, Tarrytown, NY 10591 (914) 332-0040
ICEA	Insulated Cable Engineers Association. P.O. Box 440, South Yarmouth, MA 02664 (508) 394-4424
IEC	International Electrotechnical Commission. Geneva, Switzerland. Communications: c/o ANSI 11 W. 42nd St., New York, NY 10036 (212) 642-4900
IEEE	Institute of Electrical and Electronics Engineers. 345 E. 47th St., New York, NY 10017 (212) 705-7900
IESNA	Illuminating Engineering Society of North America. 120 Wall St., New York, NY 10005-4001 (212) 248-5000
IPC	Institute for Interconnecting and Packaging Electronic Circuits. 2215 Sanders Rd., Northbrook, IL 60062-6135 (708) 509-9700
ISA	Instrument Society of America. P.O. Box 12277 Research Triangle Park, NC 27709 (919) 549-8411
ISDI	Insulated Steel Door Institute. 30200 Detroit Rd., Cleveland, OH 44145-1967 (216) 899-0010
ISO	International Organization for Standardization. Geneva, Switzerland. Communications: c/o ANSI, 11 W. 42nd St., New York, NY 10036 (212) 642-4900
NAAMM	National Association of Architectural Metal Manufacturers. 11 S. La Salle St., Chicago, IL 60603 (312) 201-0101
NAPM	National Association of Photographic Manufacturers. 550 Mamaroneck Ave., Harrison, NY 10528 (914) 698-7603
NEMA	National Electrical Manufacturers Association. 1300 N. 17th St., Rosslyn, VA 22209 (703) 841-3200
NFoPA	National Forest Products Association. 1111 19th St., N.W., Washington, D.C. 20036 (202) 463-2700
NFiPA	National Fire Protection Association. 1 Batterymarch Park, P.O. Box 9101, Quincy, MA 02269-9101 (617) 770-3000
NFlPA	National Fluid Power Association. 3333 N. Mayfair Rd., Milwaukee, WI 53222-3219 (414) 778-3344
NISO	National Information Standards Organization. 4733 Bethesda Ave., Bethesda, MD 20814 (301) 654-2512
NSF	National Sanitation Foundation, International. 4201 Wilson Blvd., Arlington, VA 22230 (703) 306-1070
NSPI	National Spa and Pool Institute. 2111 Eisenhower Ave., Alexandria, VA 22314 (703) 838-0083
OPEI	Outdoor Power Equipment Institute, Inc. 341 S. Patrick St., Alexandria, VA 22314 (703) 549-7600
RESNA	Rehabilitation Engineering and Assistive Technology Society of North America. 1700 N. Moore St., Arlington, VA 22209-1903 (703) 524-6686
RIA	Robotic Industries Association. 900 Victors Way, Ann Arbor, MI 48106 (313) 994-6088
RMA	Rubber Manufacturers Association. 1400 K St., N.W., Washington, D.C. 20005 (202) 682-4800
SAAMI	Sporting Arms and Ammunition Manufacturers Institute. Flintlock Ridge Office Center, 11 Mile Hill Rd., Newtown, CT 06470 (203) 426-4358
SAE	Society of Automotive Engineers. 400 Commonwealth Dr., Warrendale, PA 15096 (412) 776-4841
SIA	Scaffold Industries Association. 14039 Sherman Way, Van Nuys, CA 91405-2599 (818) 782-2012
SMA	Screen Manufacturers Association. 2545 S. Ocean Blvd., Palm Beach, FL 33480-5453 (407) 533-0991
SMPTE	Society of Motion Picture and Television Engineers. 595 W. Hartsdale Ave., White Plains, NY 10607 (914) 761-1100
SPI	The Society of the Plastics Industry, Inc. 1275 K St., N.W., Washington, D.C. 20005 (202) 371-5200
TIA	Telecommunications Industries Association. 2001 Pennsylvania Ave., N.W., Washington, D.C. 20006-4912 (202) 457-4912

arena treated by the product liability laws. Further discussion of this subject is given in the article "Products Liability and Design" in this Volume.

One of the most commonly used allegations of product defect, and one of the easiest ones to prove, is that the designer failed to recognize and observe requirements set forth in applicable standards. Such a situation is extremely hard to defend and frequently results in the court making what is called a summary judgment for the plaintiff.

How does a designer avoid such situations? The best method is through frequent and thorough design reviews. Part of being thorough is being aware of applicable codes and standards and taking their requirements into consideration at the first design review session. Of course, design reviews should cover several areas: material selection, the processing of the material during the manufacturing cycle, quality assurance, costs, and several other factors must all be considered and trade-offs made to secure the optimal solution to the design problem. A tedious and often contentious process to be sure, but design reviews help to define the problems very clearly.

The designer's responsibility to avoid doing harm requires that during the review process a special effort is made to discover and define all potential sources of harm inherent in the proposed design. That is the hazard recognition phase of the design effort. Some hazards may be open and obvious. The challenge is to ferret out the hidden or unusual hazards that can cause problems later.

Once the designer has recognized the hazards, the next, best, and most obvious step is to design the hazards out of the product. Sometimes that is not entirely possible, so the second-best approach must be used. This is to figure out some way of mitigating the hazard by adding a guard to protect the user from the recognized hazard. If a suitable guard cannot be designed, then the third and least effective approach to hazard mitigation is used: placing warnings on the product. There are even standards for doing that. They are given in ANSI Z535.4-1991, "Product Safety Signs and Labels."

Related information is provided in the articles "Safety in Design" and "Products Liability and Design" in this Volume.

Statistical Aspects of Design

Richard C. Rice, Battelle Columbus

FOR MANY YEARS engineers have designed components and structures using best available estimates of material properties, operating loads, and other design parameters. Once established, most of these estimated values have commonly been treated as fixed quantities or constants. This approach is called deterministic, in that each set of input parameters allows the determination of one or more output parameters, where those output parameters might include a prediction of factors such as operating stress, strain, deflection, deformation, wear rate, fatigue strength, creep strength, or service life. In reality, virtually all material properties and design parameters exhibit some statistical variability and uncertainty that influence the adequacy of a design.

It is common practice that almost all engineered components or structures are designed with the expectation that only a small percentage of the units that are produced will fail within the warranty period. In the case of structures that are sold with warranty provisions, warranty costs are directly tied to failure rates within that period. Invariably, greater-than-anticipated failure rates lead to extraordinary warranty costs. In addition, high recall rates, even on noncritical structural components, often lead to buyer perceptions that the product, as a whole, is unreliable and perhaps unsafe.

Assume, for example, that the average service life of structure A is 20% greater than that of structure B (the good news). However, assume also that the variability in service lives for structure A is twice that of structure B (the bad news). For simplicity, consider that this variability has been shown to conform to a normal distribution. In spite of the inferiority of structure B in average performance, the first failures out of a sample (or fleet) of 1000 units for structure A would be expected to occur at service lives approximately 17% less often than for structure B, making structure B more desirable in terms of predicted service life to initial "fleet" failures. This simple case will be illustrated in detail later in this article.

The example just described requires that a significant data base exist, one that allows accurate estimates of average properties and realistic estimates of the variability in those properties. In a typical design scenario, the data base available to define material, design, and operating parameters is limited. Handbook data on similar materials and operating loads collected on a similar design may be all that is available in some instances. In such cases, lower- or upper-bound parameter estimates (whichever are seen as most critical) are often used in combination to produce what are believed to be conservative performance estimates. With regard to safe operating loads or stresses, industry standard design factors or "safety factors" can also be applied. With limited data, the actual conservatism of individual parameter estimates can vary widely, and the ultimate degree of conservatism in the performance estimates is unknown. This situation can lead to either an unconservative design with unacceptably high failure rates, or a very conservative design that provides the required performance with unnecessarily high product costs.

Fundamentally, designing to prevent service failures is a statistical problem. In simplistic terms, an engineered component fails when the resistance to failure is less than the imposed service condition. Depending on the structure and its performance requirements, the definition of failure varies; it could be buckling, permanent deformation, tensile failure, fatigue cracking, loss of cross section due to wear, corrosion, or erosion, or fracture due to unstable crack growth. In any of these cases, the failure resistance of a large number of components of a particular design is a random variable, and the nature of this random variable often changes with time. The imposed service condition for these components is also a random variable; it too can change with time. The intersection of these two random variables at any point in time represents the expected failure percentage and provides a measure of component reliability, as shown in Fig. 1.

There are numerous texts on statistics, as well as many references on engineering design. A number of sources have combined these two disciplines in addressing the statistical aspects of design; some of these sources are cited in the Selected References at the end of this article.

This article presents some of the statistical aspects of design from an engineer's perspective. Some statistical terms are clarified first because many engineers have not worked in the field of statistics enough to put these terms into day-to-day engineering practice. Commonly used statis-

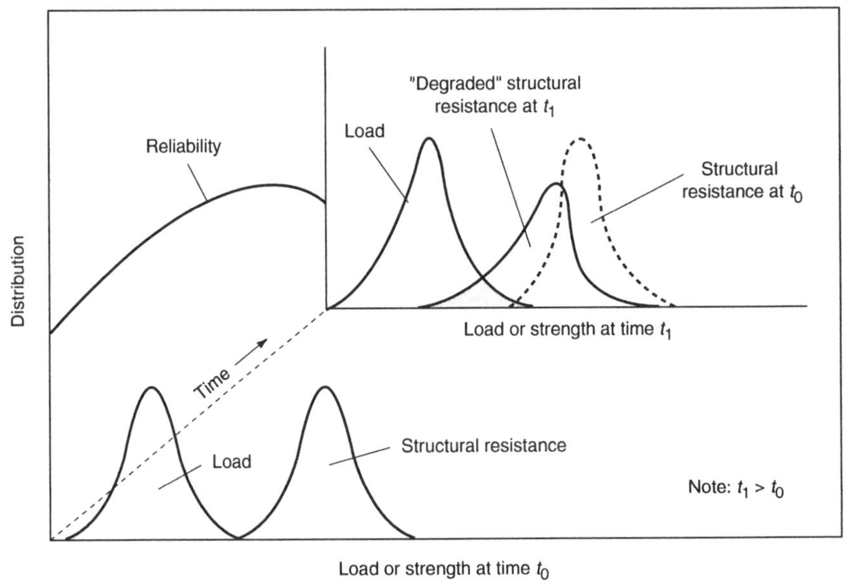

Fig. 1 Reliability as it relates to statistical distributions of structural integrity and applied loads

tical distributions are reviewed next in the section "Statistical Distributions Applied to Design," with the primary goal of providing some guidance on practical engineering applications for these distributions. The section that follows, "Statistical Procedures," describes some basic statistical procedures that can be used to address questions of variability and uncertainty in an engineering analysis; some example problems are included. The final section, "Related ASTM Engineering Statistics Standards," is provided as an easy reference guide; it has a table listing relevant statistics standards published by the American Society of Testing and Materials.

Clarification of Statistical Terms

Random Variables. Any collection of test coupons, parts, components, or structures designed to the same set of specifications or standards will exhibit some variability in performance from one unit to the other. Performance can be measured by a wide variety of parameters, or some combination of those parameters, as discussed earlier. In any case, these measures of performance are not controlled, although there is often an attempt to optimize them to maximize performance, within prescribed cost constraints. Because these measures of performance are not controlled and subject to inherent random variability, they are commonly called random variables.

The tensile strength of a structural material is a practical example of a random variable. Given a single heat and lot of a material manufactured to a public specification, such as an ASTM, SAE/AMS, or DoD specification, repeated tests to determine the tensile strength of that material will produce varied results. This will be true even if the individual tests are performed identically, within the limits of engineering accuracy.

Repeated "identical" tests to determine any engineering property of a material will produce results showing some degree of variability, which means they are all random variables. Properties such as hardness, elastic modulus, and coefficient of thermal expansion tend to show relatively low variability when repeat precision measurements are made. Other material properties show more variability, such as crack initiation, fatigue life, fracture toughness, post-heat-treatment residual stress, and creep rupture strength.

The apparent randomness or variability of some material properties is also related to the complexity of the test needed to develop the property estimates. These properties can also vary significantly from one supplier to another, or even from one heat to another for a single supplier. An important goal in an engineering/statistical analysis is to develop an accurate estimate of the material/component variability that will be represented in production.

Beyond material variability, there are many other random variables that should be considered in an engineering/statistical analysis. Actual service loads often show a great deal of variability, and this variability must be accounted for in a comprehensive assessment of structural performance. Unfortunately, reliable estimates of average service loads, let alone statistically characterized service loads, are often not readily available. End-user processing of a supplier's material or component (e.g., heat treatment, coating, shot peening), and manufacturing (e.g., machining, riveting, spot welding, forming) all add variability in the performance of the final product, and are, in themselves, random variables.

Table 1 Representative fatigue data showing variability in cycles to failure

Life interval, 10^6 cycles	Number of failures	Cycles to failure, 10^6 cycles
0.0–0.5	1	0.425
0.5–1.0	5	0.583, 0.645, 0.77, 0.815, 0.94
1.0–1.5	7	1.01, 1.09, 1.11, 1.21, 1.30, 1.41, 1.49
1.5–2.0	4	1.61, 1.70, 1.85, 1.97
2.0–2.5	2	2.19, 2.32
2.5–3.0	2	2.65, 2.99
3.0–3.5	1	3.42
3.5–4.0	0	...
4.0–4.5	0	...
4.5–5.0	1	4.66
Total observations	**23**	

In some cases, with random variables such as service loads, the engineer has little control over their variability and must simply characterize these random variables as accurately as possible and account for this variability when making service-life estimates. For example, in the case of wheel/rail loads for different kinds of rail service, there is not only significant variability in loads within a given railroad, but significant differences in the distribution of different severity loads for different kinds of rail service as shown in Fig. 2. If one is to realistically assess the structural integrity of a rail system or rail vehicle, it is necessary to characterize accurately the statistical variations in loading that apply to that system. The same is certainly true for any other transportation system or any operating system subjected to variable loads.

In other cases, with random variables such as processing and manufacturing procedures, the manufacturing or process engineer has considerable control over their variability, and a realistic goal is to minimize these "nuisance" variables to the point where they are insignificant or at least controllable from an engineering perspective.

Many engineering quantities are continuous random variables, although some others are discrete, such as the possible failure modes of a component. There are a finite number of ways that a component can fail (e.g., fatigue, corrosion, overload, brittle fracture), and those failure modes are not defined over a continuum (although service failures of engineering components often do occur due to a combination of causes). Initial designs are validated against the failure mode(s) considered to be most likely. Accumulated fleet service records for critical components eventually allow the statistical quantification of the probability of occurrence of different failure modes.

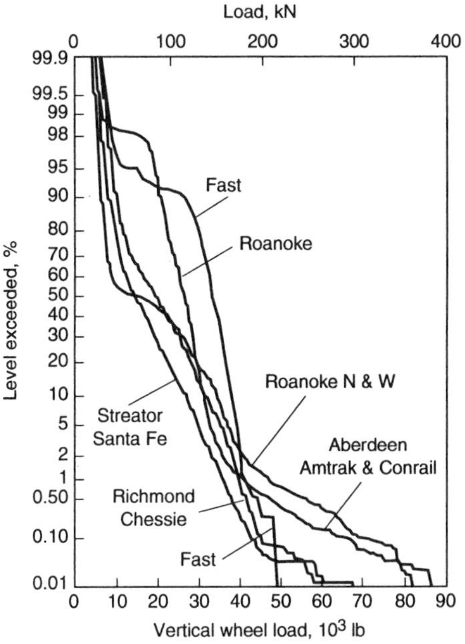

Fig. 2 Statistical characterization of wheel/rail loads on concrete ties

Density Functions. In statistical terms, a density function is simply a function that shows the probability that a random variable will have any one of its possible values. Consider, for example, the distribution of fatigue lives for a material as shown in Table 1. Assume that these 23 observations were generated from a series of replicate tests, that is, repeated tests under the same simulated service conditions. A substantial range in fatigue lives resulted from these tests, with the greatest fatigue life being more than ten times the lowest observation.

The resulting approximate density function for these data is shown in Fig. 3. This figure shows the number of fatigue life observations within uniform cycles-to-failure intervals. Each interval of the histogram shows the frequency of occurrence of fatigue failures within the interval. It is evident that the probability of occurrence of a fatigue failure for this material and test condition is not constant over the range of possible fatigue lives. If 300 observations were available, instead of 23, the shape of the histogram would tend to

Table 2 Cumulative distribution of fatigue failures from Table 1

	Cycle interval × 10^6									
	0.0–0.5	0.5–1.0	1.0–1.5	1.5–2.0	2.0–2.5	2.5–3.0	3.0–3.5	3.5–4.0	4.0–4.5	4.5–5.0
No. of failures	1	5	7	4	2	2	1	0	0	1
Cumulative failures	1	6	13	17	19	21	22	22	22	23

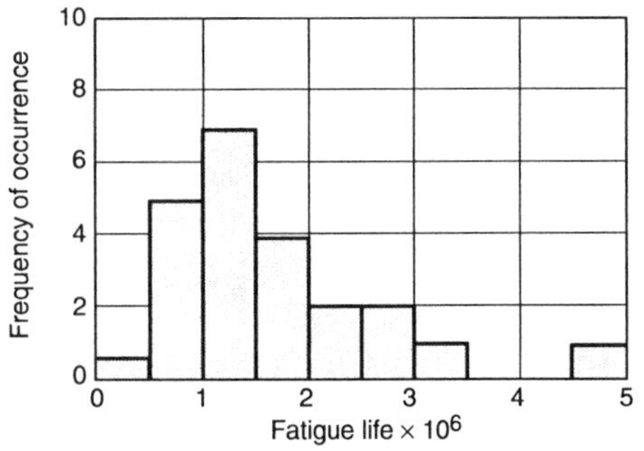

Fig. 3 Histogram of fatigue data from Table 1 showing approximate density function

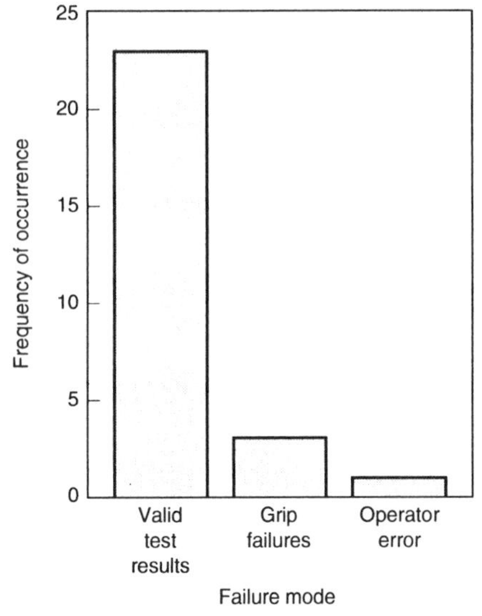

Fig. 4 Histogram of failure modes with their approximate discrete density function

stabilize. As the number of observations increase, "bumps" in the frequency diagram (as in Fig. 3) caused by random variations in fatigue life would tend to disappear, and the shape will begin to resemble that of a continuous function. A mathematical representation of such a distribution is called a density function.

In the case just described, the possible values of the random variable (fatigue life) are continuous and the resulting density function is, therefore, continuous. With discrete random variables the density function is discontinuous. For example, suppose that it was necessary to generate 27 test results before achieving 23 valid fatigue failures—one specimen might have been lost due to operator error, and three specimens could have failed in the grips. The approximate density function for the discrete random variable, failure mode, is shown for this case in Fig. 4.

Cumulative Distribution Functions. Plots of experimental data as density functions, as shown in Fig. 2 and 3, provide some useful statistical information. Inferences can be made regarding the central tendencies of the data and the overall variability in the data. However, additional information can be obtained from a data sample like the one summarized in Table 1, by representing the data cumulatively, as in Table 2, and plotting these data on probability paper, as shown in Fig. 5. This is done by ranking the observations from lowest to highest and assigning a probability of failure to each ranked value. These so-called median ranks can be obtained from tables of these values from a statistical text, such as Ref 1, or can be computed for small samples from the following approximate formula:

$$\text{Median or 50\% rank} = \frac{j - 0.3}{n + 0.4} \quad \text{(Eq 1)}$$

where j is the failure order number and n is the sample size.

For example, the first rank value in this case can be approximated as (1 – 0.3)/(23 + 0.4) = 0.030. Each of the fatigue lives, from lowest to highest, has been plotted in this manner in Fig. 5, which shows the range of fatigue lives on normal probability paper.

The best-fit representation of the cumulative data is known as a cumulative distribution function (CDF). As the name implies, a CDF provides an estimate of the cumulative percentage of total observations that can be expected at a particular value of the random variable (in this case fatigue life). In this case, where normal probability paper has been used, the data should fall in a straight line if the underlying population of fatigue lives is normally distributed. Because the data clearly do not follow a straight line, it is a fairly safe assumption that the underlying distribution is not normal. This result is not too surprising considering the lack of symmetry of the histogram of

Table 3 Random sample statistics drawn from a normal distribution with a mean value of 100 and a standard deviation of 5

Observation No.	Value of observation	Estimated average	Estimated standard deviation	Estimate of error: Average, %	Estimate of error: Standard deviation, %
1	97.99	97.99	...	...	...
2	100.81	99.40	1.41	–0.60	–71.81
3	100.47	99.75	1.26	–0.25	–74.86
4	106.08	101.34	2.95	1.34	–41.05
5	111.21	103.31	4.75	3.31	–5.05
6	98.10	102.44	4.75	2.44	–5.03
7	100.84	102.21	4.43	2.21	–11.37
8	106.16	102.71	4.35	2.71	–13.08
9	96.33	102.00	4.56	2.00	–8.77
10	105.94	102.39	4.49	2.39	–10.28
11	97.91	101.98	4.47	1.98	–10.65
12	98.41	101.69	4.39	1.69	–12.20
13	92.55	100.98	4.87	0.98	–2.59
14	102.00	101.06	4.70	1.06	–5.99
15	100.95	101.05	4.54	1.05	–9.17
16	91.06	100.42	5.02	0.42	0.37
17	98.40	100.31	4.89	0.31	–2.17
18	98.76	100.22	4.77	0.22	–4.66
19	109.37	100.70	5.07	0.70	1.40
20	94.80	100.41	5.11	0.41	2.13
21	106.41	100.69	5.15	0.69	2.90
22	103.53	100.82	5.06	0.82	1.22
23	96.28	100.62	5.04	0.62	0.72
24	103.70	100.75	4.97	0.75	–0.64
25	101.99	100.80	4.87	0.80	–2.52
26	103.34	100.90	4.80	0.90	–3.92
27	102.49	100.96	4.72	0.96	–5.52
28	96.60	100.80	4.71	0.80	–5.83
29	105.19	100.95	4.70	0.95	–6.09
30	100.58	100.94	4.62	0.94	–7.66
31	100.95	100.94	4.54	0.94	–9.16
32	99.06	100.88	4.48	0.88	–10.35
33	97.86	100.79	4.44	0.79	–11.11
34	97.42	100.69	4.42	0.69	–11.69
35	94.88	100.53	4.46	0.53	–10.84
36	103.26	100.60	4.42	0.60	–11.62
37	100.51	100.60	4.36	0.60	–12.83
38	110.72	100.87	4.60	0.87	–8.08
39	100.18	100.85	4.54	0.85	–9.24
40	91.76	100.62	4.70	0.62	–6.00
41	110.71	100.87	4.90	0.87	–2.07
42	95.90	100.75	4.90	0.75	–2.07
43	95.69	100.63	4.90	0.63	–2.02
44	105.49	100.74	4.90	0.74	–2.06
45	100.28	100.73	4.84	0.73	–3.15
46	94.05	100.59	4.89	0.59	–2.24
47	101.35	100.60	4.84	0.60	–3.26
48	102.16	100.63	4.79	0.63	–4.17
49	95.81	100.54	4.79	0.54	–4.18
50	104.25	100.61	4.77	0.61	–4.57

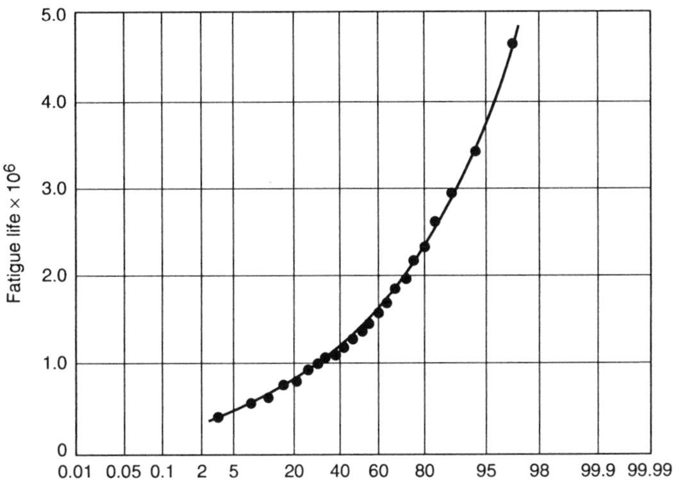

Fig. 5 Cumulative distribution function for fatigue data from Table 1 based on an assumed normal distribution

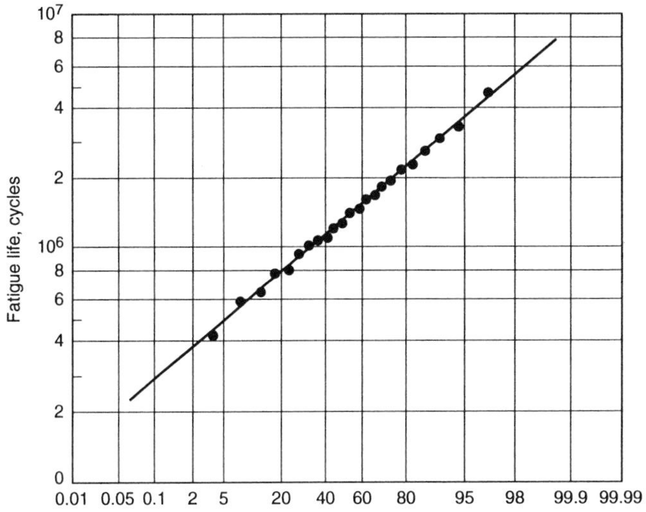

Fig. 6 Cumulative distribution function for fatigue data from Table 1 based on an assumed log-normal distribution

fatigue failures, shown earlier in Fig. 3, and most engineer's familiarity with the "bell-shaped" symmetry of a normal distribution.

An alternative is the effect of creating a CDF based on the logarithms of fatigue lives, as shown in Fig. 6. Using exactly the same data set, and making a simple transformation of the random variable, it is possible to see that the underlying distribution representing fatigue life (in this case) could very well be log normal.

Two statistical data points of significance can be drawn from Fig. 6. First, an estimate of the geometric average value of the sample can be obtained by examining the intersection of the line at a failure percentage of 50%, which in this case is a fatigue life of approximately 1.4 million cycles. Second, an estimate of the standard deviation of the sample can be obtained by examining the difference (in log life) between the 50th percentile and the 16th percentile (15.87 percentile to be exact). The 16th percentile is significant for a normal distribution because it corresponds to 1 standard deviation below the mean (of course the 84th percentile has the same connotation, i.e., 1 standard deviation above the mean), as shown in Table 3. Because the fatigue life at the 16th percentile is approximately 750,000 cycles, the standard deviation in $\log_{10}$(life) is approximately 0.271 [$\log_{10}(1.4 \times 10^6) - \log^{10}(7.5 \times 10^5) = 6.146 - 5.875 = 0.271$].

The same procedure could be followed with other distributional assumptions and other probability paper, such as Weibull probability paper. Although not statistically rigorous, plotting of CDFs on different types of probability paper can be very helpful in evaluating whether or not it is reasonable to assume a particular statistical distribution.

Sample versus Population Parameters. When performing a deterministic analysis, the input parameters are defined to represent the material in some specific way. For example, the input parameter may be the average or typical yield strength of a material or a specification minimum value. The implicit assumption is that, given an infinite number of observations of the strength of this material, the assumed average or minimum values would match the "real" values for this infinite population. Of course, the best that can be done is to generate a finite number of observations to characterize the performance of this material (in this case, yield strength); these observations are considered a statistical sample of this never-attainable, infinite population. Intuitively, an increase in the number of observations (the sample size), should increase the accuracy of the sample estimate of the real, or population material properties.

The mean, or arithmetic average, of a sample is defined as:

$$\bar{x} = \frac{\sum_{i=1}^{n} x_i}{n} \qquad \text{(Eq 2)}$$

where x_i represents the ith value of n total observations, and the Greek symbol, Σ, indicates a summation of all of the values from the first value to the nth value. For example, the average elongation value for an aluminum casting might be found to be 4.85, as shown below, based on the following sample of six experimental observations: 3.6, 5.0, 6.1, 4.7, 5.5, and 4.2:

$$e = \frac{3.6 + 5.0 + 6.1 + 4.7 + 5.5 + 4.2}{6}$$

A commonly accepted measure of the scatter, or variability in a sample, is the sample standard deviation, which is a nonnegative value calculated as follows:

$$s = \sqrt{\frac{\sum_{i=1}^{n} (x_i - \bar{x})^2}{n-1}} \qquad \text{(Eq 3)}$$

It is clear from Eq 3 that a sample showing low variability, with observations closely clustered around the sample average, will produce a small standard deviation, while a sample showing high variability, with broadly dispersed observations, will produce a large standard deviation. For example, for the small sample of elongation values just noted, the standard deviation would be approximately 0.896. This value of nearly 1 for the sample standard deviation, compared to an average value of less than 5, shows a rather high level of relative variability in this sample.

Another measure of variability commonly used by statisticians is the sample variance. The sample variance is simply the sample standard deviation squared. The standard deviation is probably used most often by engineers because its magnitude relative to the mean provides a useful measure of the relative intrinsic variability in material properties, as discussed in the next section of this article "Statistical Distributions Applied to Design."

Because increased sample sizes invariably mean increased cost, there is an obvious trade-off to be made between accuracy of sample statistics and cost. The required precision of these sample statistics, and in turn the required sample sizes, are generally tied to the desired level of reliability of the final component or structure. For example, it is common practice in the design of single-load-path (nonredundant) aircraft structures to design them using 99% exceedance, 95% confidence (which translates to less than 1 failure in 100, 19 times out of 20), lower tolerance limits on yield and ultimate strength, based on at least 10 heats and lots of the material (Ref 2). Minimum sample sizes for these calculations have been set at 100 observations when the underlying distribution can be defined with high confidence and at 300 observations when the underlying distribution cannot be defined with 95% confidence. The reasons for some of these sample size requirements will become apparent in subsequent discussions.

The issue of sample size also becomes relevant when making comparisons in performance among similar materials, processes, or components. The ability to distinguish relatively subtle differences in performance at high levels of statistical confidence depends on large sample sizes. An example of this is shown in Fig. 7, where the required sample size to detect a drop in performance, with 80% confidence, is shown. For generality, the drop in performance is shown in normalized terms, that is, in tenths of a standard deviation. To put this chart into practical terms, consider a material that displays scatter in accordance with a normal distribution, with a historical mean tensile strength of 100 ksi and a standard deviation of 5 ksi. This chart shows that, in order to say (at an 80% confidence level) that a 2.5 ksi drop in the mean tensile strength (perhaps due to a processing change) had occurred, a sample size of at least 25 observations would be required.

Figure 7 also shows that even larger sample sizes are required to detect similar differences in the lower "tail" of the distribution. The first and tenth percentiles noted in Fig. 7 correspond to the 1% and 10% failure percentages for the sample. For example, the expected first percentile in fatigue life for the material characterized in Fig. 6 would be approximately 330,000 cycles. This lower portion of the CDF is known as the lower tail of the distribution. Large sample sizes are required to detect statistically significant differences in the lower tails of two samples because these lower tail values depend not only on uncertainty in the mean value, but also on uncertainty in the standard deviation.

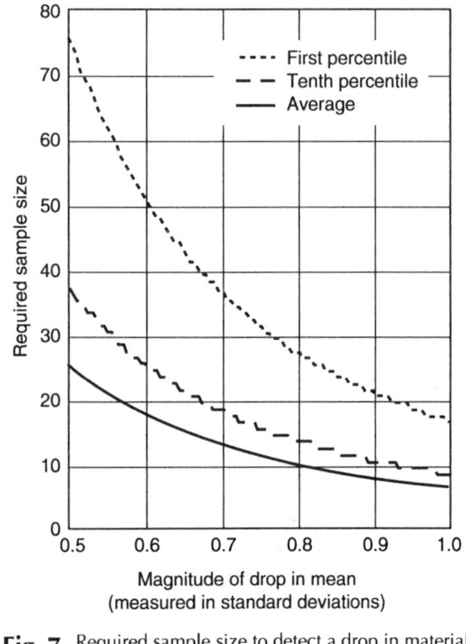

Fig. 7 Required sample size to detect a drop in material performance (with 80% confidence)

Because the standard deviation is basically the slope of the CDF, which "rotates" about the 50% failure point (which is itself uncertain), it is clear that the uncertainty in the first percentile is greater than that of the tenth percentile. This is shown schematically in Fig. 8, where the combined uncertainties in the mean (50% failure point) and the standard deviation (slope) of the CDF are added together to represent the overall uncertainty in the CDF for different failure percentages.

Figure 8 demonstrates the difference between a population and a sample. If the population were perfectly known (which it never is), the mean and standard deviation would be precisely defined and the CDF for the population would be known, as represented by the dark line. In the real world, only samples of the true population are available, which in turn says that the population mean and standard deviation are never known with certainty, meaning that the true CDF can only be bounded with statistical limits as shown in Fig. 8.

Variability versus Uncertainty. Variability and uncertainty are terms sometimes used interchangeably. However, it is useful when talking about statistical aspects of design to draw a distinction between them. Uncertainty is defined in most dictionaries as "the condition of being in doubt" or something similar to this. In the context of this discussion, uncertainty can be considered the bounds within which the "true" engineering result, such as the average fatigue life of a component, can be expected to fall. To establish this uncertainty in quantitative terms, it is necessary to identify and quantify all of the factors that contribute to that uncertainty. If the bounds are wide, it suggests a high level of uncertainty in the actual outcome. Generally speaking, the uncertainty in an engineering result is broad in the early stages of a design, and this uncertainty decreases significantly as more experimentation and analysis work is done. This uncertainty is commonly described in terms of confidence limits, tolerance limits, or prediction limits. Each of these limits are quantified in statistical terms later in this discussion.

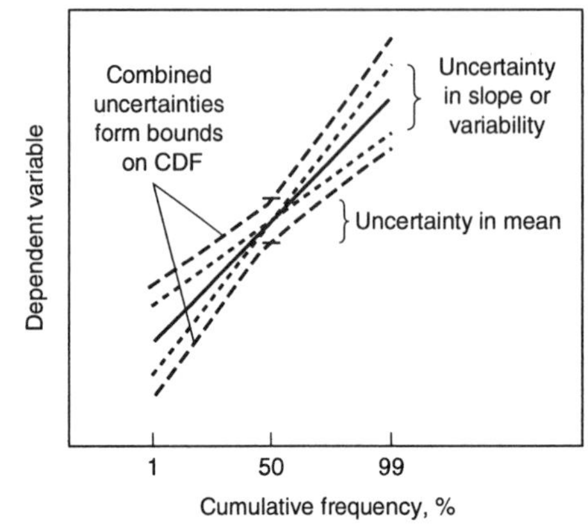

Fig. 8 Approximate confidence limits on a cumulative distribution function showing uncertainty in the mean and standard deviation

Variability is an important element of an uncertainty calculation. The meaning of the term can be drawn directly from its root word, variable. Virtually all elements of a design process can be described as random variables, with some measure of central tendency or average performance and some measure of scatter or variability. For continuous variables, this variability is generally quantified in terms of a sample standard deviation, or sample variance, which is simply the standard deviation squared.

Another commonly used measure of variability is the coefficient of variation, which is defined as the sample mean divided by the sample standard deviation and generally is expressed as a percentage. This normalized parameter allows direct comparisons of relative variability of materials or products that display significantly different mean properties. For example, two different aluminum alloys might have tensile strengths that are described statistically as follows:

Aluminum alloy	Average tensile strength, ksi	Standard deviation, ksi
No. 1	70	2.5
No. 2	75	4.0

Alloy No. 1 displays a coefficient of variation (COV) of approximately 3.6%, while alloy No. 2 displays a COV of approximately 5.3%. The higher variability of alloy No. 2, as represented by a COV nearly 50% higher than alloy No. 1, could lead to a higher level of uncertainty in the performance of a structure made from the second alloy in comparison to the first.

Precision versus Bias. Precision and bias are statistical terms that are important when considering the suitability of an engineering test method or making a comparison of the relative merits between two or more procedures or processes (Ref 3). A statement concerning the precision of a test method provides a measure of the variability that can be expected between test results when the method is followed appropriately by a competent laboratory. Precision can be defined as the reciprocal of the sample standard deviation, which means that a decrease in the scatter of a test method as represented by a smaller standard deviation of the test results leads directly to an increase in the precision. Conversely, the greater the variability or scatter of the test results, the lower the precision.

The test results of two different processes (e.g., heat treatments and their effect on tensile strength) can also be statistically compared, and a statement made regarding the precision of one process compared to another. Such a statement of relative precision is only valid if no other potentially significant variables, such as different test machines, laboratories, or machining practices are involved in the comparison.

When evaluating the precision of a test method, it is generally best to select a material that is relatively uniform and to evaluate the method at several different levels of severity, as discussed in

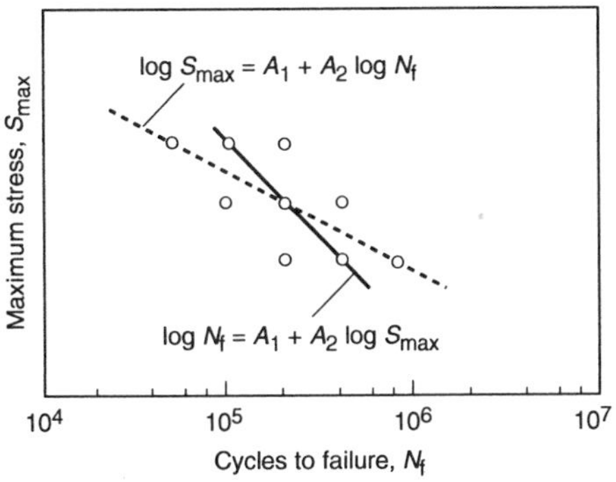

Fig. 9 Illustration of the effect of choice of dependent and independent variables in representing *S-N* fatigue data

ASTM E 691 (Ref 4). In any case, there is never a single value of precision that can be associated with a particular test method; typically, precision statements apply only to the limited condition they represent, such as a single laboratory, a single machine, a single operator, or even a single location of samples taken from a large sheet, plate, or lay-up of material.

The bias of a set of measurements quantifies the difference in those test results from an accepted reference value or standard. The concept of a bias can also be used to describe the consistent difference between two operators, test machines, testing periods, or laboratories. The term accuracy is sometimes used as a synonym for bias, but it is possible to identify a bias between two sets of measurements or procedures without necessarily having any awareness of which set is most accurate. In order to make quantitative statements about absolute bias or accuracy, it is necessary to have a known baseline or reference point; such fixed points are seldom available in the real world of an engineer.

Independent versus Dependent Variables. Many engineering analyses involve the prediction of some outcome based on a set of predefined input conditions. A very simple example is the prediction of stress in the elastic range for a metal, based on a measured strain and an estimated or measured value for elastic modulus. Another more complex example is the prediction of the fatigue resistance of a part based on estimated local stress or strain amplitudes and experimentally determined fatigue parameters. In general, relationships such as these are developed through a regression analysis, in which the predicted outcome, y, is estimated from a series of terms written as a function of one or more input quantities, x_i, and regression parameters, A_i, as follows:

$$y = A_0 + A_1 f_1(x) + A_2 f_2(x) \ldots \qquad \text{(Eq 4)}$$

In this expression, the predicted result, y, is the dependent variable, while the defined quantity, x, is the independent variable. In some cases, multiple independent variables are needed in combination to realistically estimate the independent variable(s). (Generally, the dependent variable is taken to be the variable with largest measurement error.)

Independent and dependent variables sometimes get reversed in engineering analyses, with significant potential ramifications. A simple illustration of the erroneous mean trends that can be predicted if the independent and dependent variables are reversed follows. Assume that triplicate fatigue test results have been generated at three evenly spaced stress levels, as shown in Fig. 9. Because the stress amplitude for each test was controlled, and therefore is known with little error, it represents the independent variable. The resulting fatigue lives were not controlled, and significant scatter in fatigue lives was observed at each stress level. Therefore, fatigue life represents the predicted or dependent variable in the analysis. If a linear regression is performed with stress as the independent variable, the continuous curve results; at each stress level the scatter in fatigue lives about the mean curve is minimized. If a linear regression is performed with fatigue life as the independent variable, the dashed curve results; at each fatigue life the "scatter" in stresses about the mean curve is minimized. Clearly, the dashed line underestimates the high-stress mean fatigue lives and overestimates the low-stress mean fatigue lives. This effect is certainly less when the scatter is lower or the overall range of the true independent variable is increased, but additional replication would have a negligible effect on the misrepresented mean curve. Therefore, an analytical expression involving fatigue life as the dependent variable is generally the most realistic from a statistical perspective.

When performing any type of correlation or regression analysis in addressing an engineering problem, the choice of dependent and independent variables should be made very carefully. Perhaps the simplest way to think about it is to consider that a regression analysis is meant to predict values of the dependent variable for specific, defined values of the independent variable(s). It only makes sense to predict values that are unknown at the outset of an experiment or analysis; these values are dependent on the selected or controlled (independent) variables.

Why Not Just Assume Normality and Forget the Complications? The first assumption that most engineers will make when applying statistics to an engineering problem is that the property in question is normally distributed. The normal distribution is relatively simple mathematically, is well understood, and is well characterized. The engineering statistics applications where nonnormal statistical distributions and nonparametric (distribution-free) approaches are used probably do not total the number of applications based on normal statistics.

The assumption of normality is a reasonable one in many cases where only mean trends or average properties are of interest. The importance of verifying this assumption increases when properties removed from the mean or average are being addressed, such as an estimated first percentile value of a design factor or a value 2 or 3 standard deviations below the mean.

The significance of the assumed distribution on estimated lower-bound properties can be demonstrated by a simple example. Case 1: the fatigue data presented in Table 1 are assumed to be normally distributed. Case 2: the fatigue data presented in Table 1 are assumed to be log-normally distributed. Calculate a lower-bound estimate of the fatigue behavior for this material for both cases based on two standard deviations below the sample mean.

Case 1:	$\bar{x}$ =		$\bar{N}$ =	1.659×10^6
	s =			9.98×10^5
	$\bar{x} - 2s$ =		N_{lower} =	-3.37×10^5
Case 2:	$\bar{x}$ =	6.146,	$\bar{N} = 10^{6.146}$ =	1.40×10^6
	s =			0.271
	$\bar{x} - 2s$ =	5.604,	$N_{lower} = 10^{5.604}$ =	4.02×10^5

In Case 1 a nonsensical answer results, because the lower-bound fatigue life estimate is less than 0. In case 2 a workable answer results, based on a realistic representation of the underlying distribution of the random variable. (When dealing with transformed variables, such as in case 2, it is important to remember to untransform the variable after performing the statistical calculations.) Clearly, the assumptions that are made regarding underlying statistical distributions can have a major impact on the meaningfulness of the answers that result.

Statistical Distributions Applied to Design

Statistical distributions as applied to engineering design can be discussed in terms of two categories: continuous distributions and discrete distributions.

Continuous Distributions

Normal Distribution. Lacking more detailed information regarding the nature of an engineering random variable, it is often assumed that its variability can be represented by a normal distribution. The normal distribution has been used to model a wide variety of physical, mechanical, and chemical properties such as tensile and yield strength of some metallic alloys, temperature variations over a period of time, and the reaction rate of chemical reactions.

Two of the main reasons for the popularity of the normal distribution are obvious—the normal distribution is the best known of all statistical distributions, and sample estimates of its parameters are easily computed. Also, many sample statistics have a normal distribution for large sample sizes. Another pragmatic reason is that it works pretty well in many cases, especially if an accurate representation of very low probability events is not required.

The density function for a normal distribution is defined as follows:

$$f(x) = \frac{1}{\sigma\sqrt{2\pi}} \exp\left[-\frac{(x-\mu)^2}{2\sigma^2}\right] \quad -\infty \le x \le \infty \qquad \text{(Eq 5)}$$

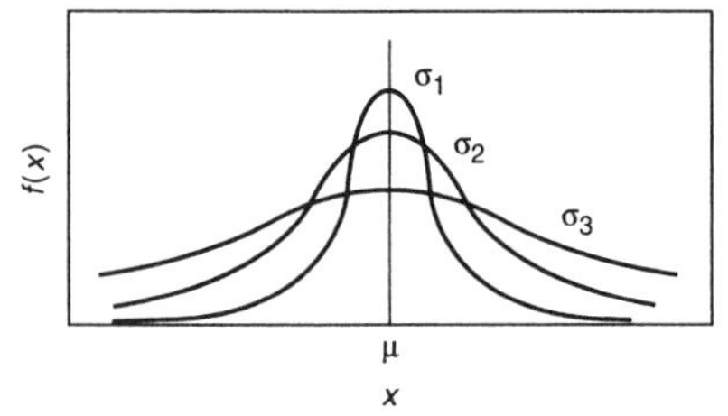

Fig. 10 Approximate effect of changing the variability in the normal distribution with no change in the mean value

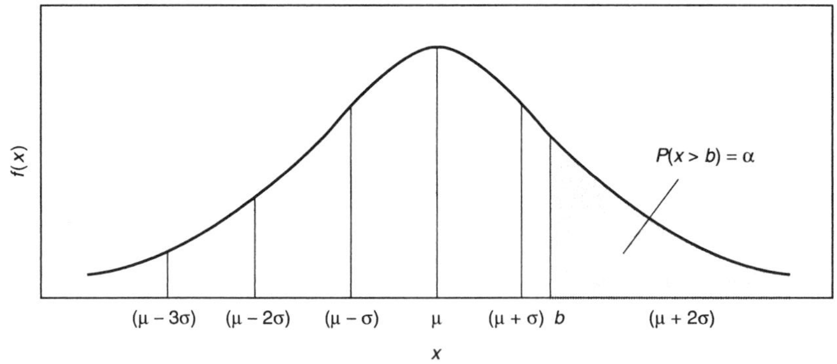

Fig. 12 Illustration of the standardized normal distribution

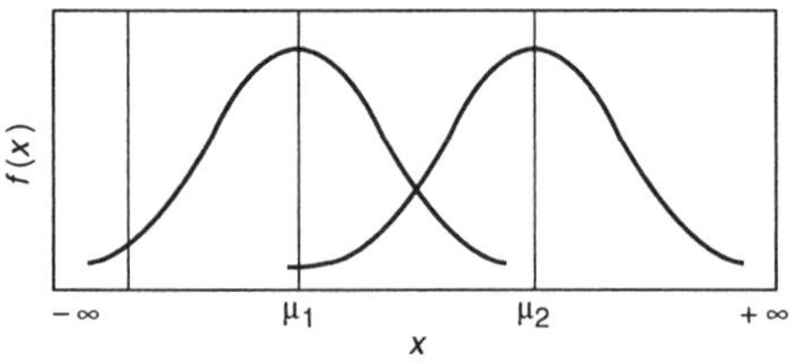

Fig. 11 Approximate effect of changing the mean value in the normal distribution with no change in the variability

where x is the random variable, $f(x)$ is the probability density function of x, μ is the population mean, σ is the population standard deviation, and exp stands for e (2.718) raised to the power in brackets. Normal distributions having the same mean but different standard deviations are shown in Fig. 10, while normal distributions having different means but the same standard deviations are shown in Fig. 11.

The only parameters required to define a normal distribution are the mean and the standard deviation. If a material property is normally distributed, relatively accurate estimates of the mean value of the material can be obtained with small sample sizes. Larger sample sizes are required to obtain good estimates of the standard deviation of a normally distributed property. This point is illustrated in Table 3, where sample statistics are tabulated for a random variable that is normally distributed, with a true mean of 100 and a true standard deviation of 5. With as few as five observations, a fairly stable estimate of the population mean is obtained, but rather large fluctuations in the estimated standard deviation of the population are still evident. Larger sample sizes tend to improve the accuracy of the estimated values for the population mean and standard deviation, although it is evident that even sample sizes as large as 50 do not provide truly precise estimates of either statistical parameter.

Clearly, the normal distribution is symmetric about the mean, in that the upper tail is a mirror image of the lower tail when reflected about the mean. Notice in Eq 5 that the normal distribution is not bounded by 0 on the lower extreme. This means that the frequency of occurrence of low probability events of a material property represented by a normal distribution may not be well represented, especially if that property displays a large coefficient of variation.

Equation 5 applies to the probability of occurrence of a single value of x. If the probability of occurrence of any value of x greater than b is desired, Eq 5 can be integrated as follows:

$$P(x > b) = \int_{b}^{\infty} \frac{1}{\sigma\sqrt{2\pi}} \exp\left[-\frac{(x-\mu)^2}{2\sigma^2}\right] dx \qquad \text{(Eq 6)}$$

Because the integral of this equation is not available in a simple form, it is usually solved through the use of tables of "normal deviates." This process is simplified if a simple substitution is made, wherein a new variable, z, is defined:

$$\frac{b-\mu}{\sigma} = z_b \qquad \text{(Eq 7)}$$

and Eq 6 is rewritten in normalized form as follows:

$$P(z > z_b) = \int_{z_b}^{\infty} \frac{1}{\sqrt{2\pi}} \exp\left[-\frac{z^2}{2}\right] dz \qquad \text{(Eq 8)}$$

The region from z_b to ∞ is the shaded area shown in Fig. 12. Clearly, a z_b of 0 is equivalent to the mean, which gives an α value of 0.50, representing half of the area under the symmetrical normal curve. The shaded area in Fig. 12 is tabulated in many statistics textbooks and is repeated here as Table 4.

Log-Normal Distribution. Probably the second-most-often-assumed distribution for engineering random variables is the log-normal distribution. Of course, the log-normal distribution is really just a special case of the normal distribution, where the quantities that are assumed to be normally distributed are the logarithms of the original observations.

The log-normal distribution does offer some interesting attributes that are useful in representing some engineering random variables. First, the log-normal distribution is bounded at 0, which makes it useful in some cases to represent random variables that cannot take on negative values (such as fatigue cycles to failure). Second, the log-normal distribution is inherently positively skewed (in the untransformed variable space), with an elongated upper tail compared to the lower tail, as shown in Fig. 13. The extent of the skewness depends on the coefficient of variation in the untransformed data, with the least skewness being evident in data with low COVs.

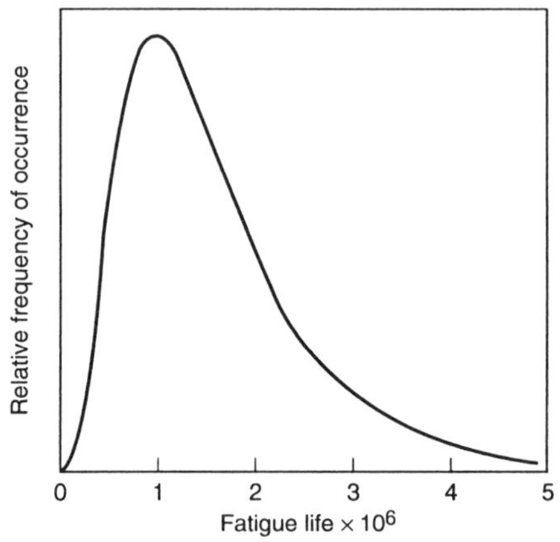

Fig. 13 Density function for a log-normal distribution

Because of the nature of the log-normal distribution, it is used most often to model life phenomena where the cycles or time to failure for a given condition tend to be skewed when viewed in normal space (Ref 5, 6). When using the log-normal distribution to represent variability in fatigue data it is important to recognize that the variability tends to increase with increasing mean fatigue lives (Ref 7, 8).

Weibull distribution is used frequently in representing engineering random variables because it is very flexible. It was originally proposed (Ref 9) to represent fatigue data, but it is now used to represent many other types of engineering data. For the modeling of fatigue strength at a given life, it has been argued that only the three-parameter Weibull distribution is appropriate (Ref 10). One of the reasons that the Weibull distribution is popular is that data can be plotted on Weibull paper and the conformance of these data to the Weibull distribution can be evaluated by the linearity of the CDF in the same way as a normal distribution.

General Three-Parameter Weibull. The general, three-parameter form of the Weibull distribution density function is:

$$f(x) = \frac{b}{\theta}\left[\frac{x-x_0}{\theta}\right]^{b-1} \exp\left[-\left(\frac{x-x_0}{\theta}\right)^{b}\right] \qquad \text{(Eq 9)}$$

Table 4 Area under the normal curve from −∞ to the mean plus *z* standard deviations

For negative values of *z*, subtract the tabular value from unity.

z_b	0.00	0.01	0.02	0.03	0.04	0.05	0.06	0.07	0.08	0.09
0.0	0.5000	0.5040	0.5080	0.5120	0.5160	0.5199	0.5239	0.5279	0.5319	0.5359
0.1	0.5398	0.5438	0.5478	0.5517	0.5557	0.5596	0.5636	0.5675	0.5714	0.5753
0.2	0.5793	0.5832	0.5871	0.5910	0.5948	0.5987	0.6026	0.6064	0.6103	0.6141
0.3	0.6179	0.6217	0.6255	0.6293	0.6331	0.6368	0.6406	0.6443	0.6480	0.6517
0.4	0.6554	0.6591	0.6628	0.6664	0.6700	0.6736	0.6772	0.6808	0.6844	0.6879
0.5	0.6915	0.6950	0.6985	0.7019	0.7054	0.7088	0.7123	0.7157	0,7190	0.7224
0.6	0.7257	0.7291	0.7324	0.7357	0.7389	0.7422	0.7454	0.7486	0.7517	0.7549
0.7	0.7580	0.7611	0.7642	0.7673	0.7704	0.7734	0.7764	0.7794	0.7823	0.7852
0.8	0.7881	0.7910	0.7939	0.7967	0.7995	0.8023	0.8051	0.8078	0.8106	0.8133
0.9	0.8159	0.8186	0.8212	0.8238	0.8264	0.8289	0.8315	0.8340	0.8365	0.8389
1.0	0.8413	0.8438	0.8461	0.8485	0.8508	0.8531	0.8554	0.8557	0.8599	0.8621
1.1	0.8643	0.8665	0.8686	0.8708	0.8729	0.8749	0.8770	0.8790	0.8810	0.8820
1.2	0.8849	0.8869	0.8888	0.8907	0.8925	0.8944	0.8962	0.8980	0.8997	0.9015
1.3	0.9032	0.9049	0.9066	0.9082	0.9099	0.9115	0.9131	0.9147	0.9162	0.9177
1.4	0.9192	0.9207	0.9222	0.9236	0.9251	0.9265	0.9279	0.9292	0.9306	0.9319
1.5	0.9332	0.9345	0.9357	0.9370	0.9382	0.9394	0.9406	0.9418	0.9429	0.9441
1.6	0.9452	0.6463	0.9474	0.9484	0.9495	0.9505	0.9515	0.9525	0.9535	0.9545
1.7	0.9554	0.9564	0.9573	0.9582	0.9591	0.9599	0.9608	0.9616	0.9625	0.9633
1.8	0.9641	0.9649	0.9656	0.9664	0.9671	0.9678	0.9686	0.9693	0.9699	0.9706
1.9	0.9713	0.9719	0.9726	0.9732	0.9738	0.9744	0.9750	0.9756	0.9761	0.9767
2.0	0.9772	0.9778	0.9783	0.9788	0.9793	0.9798	0.9803	0.9808	0.9812	0.9817
2.1	0.9821	0.9826	0.9830	0.9834	0.9838	0.9842	0.9846	0.9850	0.9854	0.9857
2.2	0.9861	0.9864	0.9868	0.9871	0.9875	0.9878	0.9881	0.9884	0.9887	0.9890
2.3	0.9893	0.9896	0.9898	0.9901	0.9904	0.9906	0.9909	0.9911	0.9913	0.9916
2.4	0.9918	0.9920	0.9922	0.9925	0.9927	0.9929	0.9931	0.9932	0.9934	0.9936
2.5	0.9938	0.9940	0.9941	0.9943	0.9945	0.9946	0.9948	0.9949	0.9951	0.9952
2.6	0.9953	0.9955	0.9956	0.9957	0.9959	0.9960	0.9961	0.9962	0.9963	0.9964
2.7	0.9965	0.9966	0.9967	0.9968	0.9969	0.9970	0.9971	0.9972	0.9973	0.9974
2.8	0.9974	0.9975	0.9976	0.9977	0.9977	0.9978	0.9979	0.9979	0.9980	0.9981
2.9	0.9981	0.9982	0.9982	0.9983	0.9984	0.9984	0.9985	0.9985	0.9986	0.9986
3.0	0.9987	0.9987	0.9987	0.9988	0.9988	0.9989	0.9989	0.9989	0.9990	0.9990
3.1	0.9990	0.9991	0.9991	0.9991	0.9992	0.9992	0.9992	0.9992	0.9993	0.9993
3.2	0.9993	0.9993	0.9994	0.9994	0.9994	0.9994	0.9994	0.9995	0.9995	0.9995
3.3	0.9995	0.9995	0.9995	0.9996	0.9996	0.9996	0.9996	0.9996	0.9996	0.9997
3.4	0.9997	0.9997	0.9997	0.9997	0.9997	0.9997	0.9997	0.9997	0.9997	0.9998

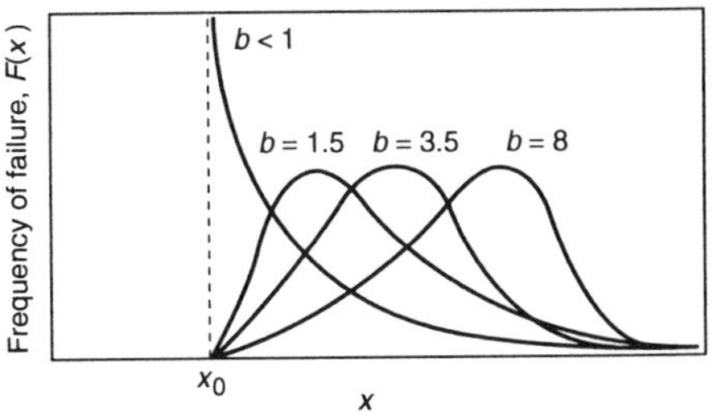

Fig. 14 Density function for the Weibull distribution with different shape factors

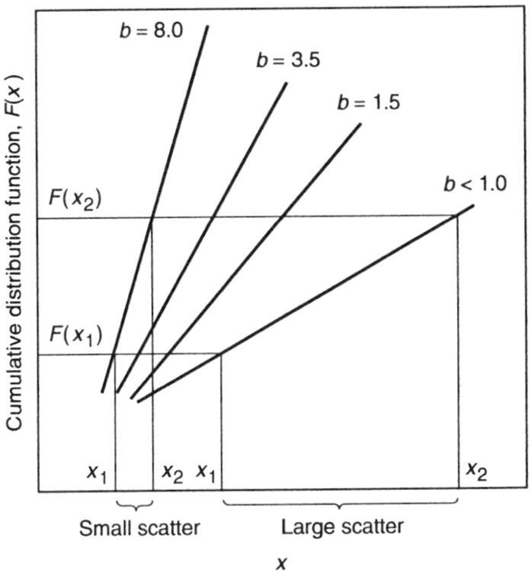

Fig. 15 CDF plots for various shape factors on Weibull probability paper

where the parameters are x_0, b, and θ, and they are defined as follows: x_0 is the expected minimum value of x, and is often referred to as the threshold parameter, b is the so-called Weibull slope or shape parameter, and θ is the so-called characteristic or scale parameter.

The Weibull density function can take on dramatically different shapes, depending on the chosen or optimized value of the shape parameter, b, as shown in Fig. 14. Values of b less than 1 produce an exponentially decaying density function, while values of b greater than 1 provide a peaked density function, with a maximum value removed from x_0. A shape factor of approximately 3.5 provides a nearly symmetrical distribution that looks somewhat like a normal distribution density function with a "pinned" lower tail. Shape factors between 1 and 3.5 provide positively skewed density functions with long upper tails, while shape factors above 3.5 provide negatively skewed density functions with long lower tails.

The relatively complex form of the density function given in Eq 9 simplifies considerably when converted to a cumulative distribution function:

$$F(x) = 1 - \exp\left[-\left(\frac{x - x_0}{\theta}\right)^b\right] \quad \text{(Eq 10)}$$

Two-Parameter Weibull. Equations 9 and 10 are simplified considerably when the threshold parameter is set to 0. Only the cumulative distribution function for the two-parameter Weibull distribution will be represented here.

$$F(x) = 1 - \exp\left[-\left(\frac{x}{\theta}\right)^b\right] \quad \text{(Eq 11)}$$

This form of the Weibull distribution is known, quite logically, as the two-parameter Weibull. There are certain kinds of engineering properties that can be well represented (or at least conservatively represented) by a two-parameter Weibull density function. For example, it is common practice to assume that the strength of polymer matrix composite materials can be represented by a two-parameter Weibull distribution (Ref 11). One significant thing to remember regarding the two-parameter Weibull distribution is that the lower tail is "pinned" at 0, meaning that very low probability, lower-tail, events predicted from this distribution may be very (possibly unrealistically) low in magnitude. The two-parameter Weibull does offer the advantage that the threshold parameter does not have to be estimated, allowing a closed-form determination of the remaining two parameters.

By taking the double natural logarithm of Eq 11 it is possible to show that:

$$\ln \ln \frac{1}{1 - F(x)} = b(\ln x) - b \ln \theta \quad \text{(Eq 12)}$$

which is the equation for a straight line in ln x, with a slope of b and intercept $-b$ ln θ. Example Weibull CDF plots for a range of different b values are given in Fig. 15.

One way to evaluate whether a random variable follows a Weibull distribution and, if so whether the location parameter is nonzero is to plot the available data on Weibull probability paper, as shown in Fig. 16. This can be done, as shown earlier, by ranking the data, and assigning median ranks to the observations in ascending order. These values can then be plotted, assuming different values for the location parameter, x_0. If the CDF is linear for some nonzero value of x_0, there is evidence that a three-parameter Weibull distribution with a location parameter of x_0 would provide a reasonable representation of this random variable.

Exponential distribution can be considered a special case of the Weibull distribution (although the latter was derived from the former). In this case, the Weibull distribution has a threshold parameter, x_0, of 0, and a slope parameter, b, of unity. In this sense, the exponential distribution is a one-parameter Weibull distribution, with the only optimized or measured parameter being θ, the scale parameter. This distribution has been

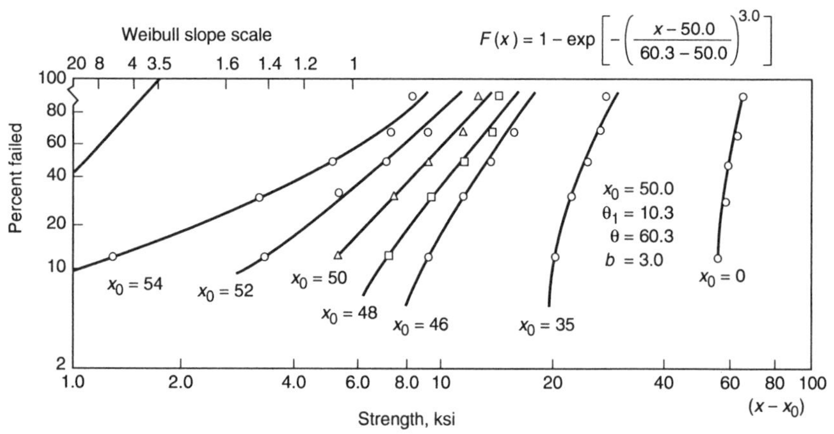

Fig. 16 Modified Weibull plot for determination of Weibull parameters

found to be useful in the analysis of failure rates of complete systems or assemblies, such as light bulbs, water heaters, and automobile transmissions, or where their failures result from chance occurrence alone, such as a wheel hitting a large pothole.

The density function for an exponential distribution can be written as:

$$f(t) = \frac{1}{\theta} e^{-t/\theta} \quad \text{(Eq 13)}$$

where $f(t)$ is the probability of failure, t is the operating time (the independent variable), θ is the characteristic life or scale parameter, and e is the base of natural logarithms (2.718).

The CDF for an exponential distribution:

$$F(t) = 1 - e^{-t/\theta} \quad \text{(Eq 14)}$$

can be plotted on Weibull probability paper, with a slope of unity.

As an example of the use of the exponential distribution, assume that the failure rate of a heat pump is known to follow an exponential distribution, and it has been found that 20% of these heat pumps fail in ten years. How long will it take for 50% of the heat pumps to fail?

In this case the variable is time, t. At the end of ten years $F(t) = 1 - e^{-t/\theta} = 0.20$. Solving for θ gives 44.84 years. Knowing θ and calculating t, based on an $F(t) = 0.50$, which corresponds to the B_{50} life, a value of 31.1 years is obtained. This calculation is straightforward. In the real world, however, the assumption of an underlying exponential distribution would have to be verified or tested in some manner, and the failure rate of 20% in ten years would have to be treated as an estimate, based on some finite sample, rather than a fixed quantity.

Discrete Distributions

Binomial distribution may be applicable where the random variable is discrete and takes on nonnegative integer values. It is used commonly in developing sampling plans for periodic inspections of manufacturing or material quality. It is also used in Bernouilli trials to evaluate the probability of occurrence of particular events, such as the likelihood that a defect of a certain size will be present in x components, with a known distribution of defects in the overall population.

For example, assume that it is known that defects sufficient to cause a downgrade in the usage of a component are present in 5% of the overall population. If these substandard parts are produced randomly, the likelihood of picking a part off of the line and finding it defective would be 0.05. The likelihood of randomly picking two successive parts off of the line that were defective would be $0.05 \times 0.05 = 0.0025$. This is based on the fact that the probability of the simultaneous occurrence of independent events equals the product of their individual probabilities of occurrence.

In large sample sizes, there are many different combinations of events that can occur, each with their respective probabilities of occurrence. Given an experiment consisting of n independent trials, with an overall probability of successful occurrence being p, and unsuccessful occurrence being $q = 1 - p$, the probability that the number of successful occurrences, r, during n trials is equal to r_1, and is given by:

$$p(r = r_1) = \binom{n}{r_1} p^{r_1} q^{n-r_1} \quad \text{for } r_1 = 0, 1, 2, \ldots, n \quad \text{(Eq 15)}$$

where:

$$\binom{n}{r_1} = \frac{n!}{r_1!(n - r_1)!} \quad \text{(Eq 16)}$$

and $n!$, $r_1!$, and $(n - r_1)!$ are factorials of the respective quantities. The mean value of this binomial distribution is np and the standard deviation is $\sqrt{npq}$.

For example, given a population containing 10% defective, what is the probability of finding three defective components in a sample of four components? In this case, the number of trials, $n = 4$, the value of $p = 0.10$, and $q = 0.90$. The probability that three defective components will be found is given by Eq 15:

$$p(r = 3) = \binom{4}{3}(0.10^3)(0.90^{4-3})$$

$$= \frac{4!}{3!\,1!}(0.10^3)(0.90^1) = 4(0.001)(0.9) = 0.0036$$

When evaluating the probability of all defectives or no defectives in a sample, the factorial of 0 will be required. The factorial of 0 is 1.

One very practical engineering application of the binomial distribution is in the calculation of nonparametric lower tolerance limit values on mechanical properties, such as are presented in MIL-HDBK-5 (Ref 2) for metallic materials and in MIL-HDBK-17 (Ref 6) for composite materials. These lower limit values are computed based on probabilities derived from the binomial distribution for specific rank values. For example, a so-called B value, which corresponds to a 90% exceedance level at 95% confidence, is associated with the lowest rank value in a sample of 29 observations. Similarly, the fifth rank value from the lowest corresponds to a B value in a sample of 89 observations.

Other Discrete Distributions. Two other common variations of the binomial distribution are the hypergeometric and the Poisson distributions. The hypergeometric distribution is used for the same kinds of applications as the binomial distribution, except that for the binomial the percentage of defective items is assumed to be constant throughout the experiment. This implies that the sample from which the items are drawn is essentially infinite compared to the number of items taken. Refer to one of the statistical textbooks in the References and Selected References at the end of this article for a detailed development of the hypergeometric distribution.

The Poisson distribution differs from the binomial and hypergeometric distributions in that the number of times that an event occurs is not known, but the probability of occurrence of the event is known. For example, one could determine how often a driver hit a curb in a given period of time, but it is not possible to quantify how many times the driver did not hit the curb in that same period of time. In this case, all one needs to know is the average number of occurrences of the event (hitting the curb) and the probabilities of observing a certain number of these events in a given period of time can be calculated. Again, refer to one of the statistical textbooks in the References and Selected References for a detailed development of the Poisson distribution.

In order to design a meaningful experiment or properly analyze a set of data, it is important to make a realistic choice regarding the statistical distribution that will be used to represent the data. There is no foolproof test or method available that an experimenter or analyst can apply to determine *a priori* the best distribution to use. In the absence of information that may aid in determining the suitability of a particular distribution, a normal distribution is often assumed. However, in the case of life data, a log-normal or Weibull

Table 5 0.975 fractiles of the *F*-distribution associated with n_1 and n_2 degrees of freedom, $F_{0.975}(n_1,n_2)$

	n_1																		
n_2	1	2	3	4	5	6	7	8	9	10	12	15	20	24	30	40	60	120	∞
1	647.8	799.5	864.2	899.6	921.8	937.1	948.2	956.7	963.3	968.6	976.7	984.9	993.1	997.2	1001	1006	1010	1014	1018
2	38.51	39.00	39.17	39.25	39.30	39.33	39.36	39.37	39.39	39.40	39.41	39.43	39.45	39.46	39.45	39.47	39.48	39.49	39.50
3	17.44	16.04	15.44	15.10	14.88	14.73	14.62	14.54	14.47	14.42	14.34	14.25	14.17	14.12	14.08	14.04	13.99	13.95	13.90
4	12.22	10.65	9.98	9.60	9.36	9.20	9.07	8.98	8.90	8.84	8.75	8.66	8.56	8.51	8.46	8.41	8.36	8.31	8.26
5	10.01	8.43	7.76	7.39	7.15	6.98	6.85	6.76	6.68	6.62	6.52	6.43	6.33	6.28	6.23	6.18	6.12	6.07	6.02
6	8.81	7.26	6.60	6.23	5.99	5.82	5.70	5.60	5.52	5.46	5.37	5.27	5.17	5.12	5.07	5.01	4.96	4.90	4.95
7	8.07	6.54	5.89	5.52	5.29	5.12	4.99	4.90	4.82	4.76	4.67	4.57	4.47	4.42	4.36	4.31	4.25	4.20	4.14
8	7.57	6.06	5.42	5.05	4.82	4.65	4.53	4.43	4.36	4.30	4.20	4.10	4.00	3.95	3.89	3.84	3.78	3.73	3.67
9	7.21	5.71	5.08	4.72	4.48	4.32	4.20	4.10	4.03	3.96	3.87	3.77	3.67	3.61	3.56	3.51	3.45	3.39	3.33
10	6.94	5.46	4.83	4.47	4.24	4.07	3.95	3.95	3.78	3.72	3.62	3.52	3.42	3.37	3.31	3.26	3.20	3.14	3.08
11	6.72	5.26	4.63	4.28	4.04	3.88	3.76	3.66	3.59	3.53	3.43	3.33	3.23	3.17	3.12	3.06	3.00	2.94	2.88
12	6.55	5.10	4.47	4.12	3.89	3.73	3.61	3.51	3.44	3.37	3.28	3.18	3.07	3.02	2.96	2.91	2.85	2.79	2.72
13	6.41	4.97	4.35	4.00	3.77	3.60	3.48	3.39	3.31	3.25	3.15	3.05	2.95	2.89	2.84	2.78	2.72	2.66	2.60
14	6.30	4.86	4.24	3.89	3.66	3.50	3.38	3.29	3.21	3.15	3.05	2.95	2.84	2.79	2.73	2.67	2.61	2.55	2.49
15	6.20	4.77	4.15	3.80	3.58	3.41	3.29	3.20	3.12	3.06	2.96	2.86	2.76	2.70	2.64	2.59	2.52	2.46	2.40
16	6.12	4.69	4.08	3.73	3.50	3.34	3.22	3.12	3.05	2.99	2.89	2.79	2.69	2.63	2.57	2.51	2.45	2.38	2.32
17	6.04	4.62	4.01	3.66	3.44	3.28	3.16	3.06	2.98	2.92	2.82	2.72	2.62	2.56	2.50	2.44	2.38	2.32	2.25
18	5.98	4.56	3.95	3.61	3.38	3.22	3.10	3.01	2.93	2.87	2.77	2.67	2.56	2.50	2.44	2.38	2.32	2.26	2.19
19	5.92	4.51	3.90	3.56	3.33	3.17	3.05	2.96	2.88	2.82	2.72	2.62	2.51	2.45	2.39	2.33	2.27	2.20	2.13
20	5.87	4.46	3.86	3.51	3.29	3.13	3.01	2.91	2.84	2.77	2.68	2.57	2.46	2.41	2.35	2.29	2.22	2.16	2.09
21	5.83	4.42	3.82	3.48	3.25	3.09	2.97	2.87	2.80	2.73	2.64	2.53	2.42	2.37	2.31	2.25	2.18	2.11	2.04
22	5.79	4.38	3.78	3.44	3.22	3.05	2.93	2.84	2.76	2.70	2.60	2.50	2.39	2.33	2.27	2.21	2.14	2.08	2.00
23	5.75	4.25	3.75	3.41	3.18	3.02	2.90	2.81	2.73	2.67	2.57	2.47	2.36	2.30	2.24	2.18	2.11	2.04	1.97
24	5.72	4.32	3.72	3.38	3.15	2.99	2.87	2.78	2.70	2.64	2.54	2.44	2.33	2.27	2.21	2.15	2.08	2.01	1.94
25	5.69	4.29	3.69	3.35	3.13	2.97	2.85	2.75	2.68	2.61	2.51	2.41	2.30	2.24	2.19	2.12	2.05	1.98	1.91
26	5.66	4.27	3.67	3.33	3.10	2.94	2.82	2.73	2.65	2.59	2.49	2.39	2.28	2.22	2.16	2.09	2.03	1.95	1.88
27	5.63	4.24	3.65	3.31	3.08	2.92	2.80	2.71	2.63	2.57	2.47	2.36	2.25	2.19	2.13	2.07	2.00	1.93	1.85
28	5.61	4.22	3.63	3.29	3.06	2.90	2.78	2.69	2.61	2.55	2.45	2.34	2.23	2.17	2.11	2.05	1.98	1.91	1.83
29	5.59	4.20	3.61	3.27	3.04	2.89	2.76	2.67	2.59	2.53	2.43	2.32	2.21	2.15	2.09	2.03	1.96	1.99	1.81
30	5.57	4.18	3.59	3.25	3.03	2.87	2.75	2.65	2.57	2.51	2.41	2.31	2.20	2.14	2.07	2.01	1.94	1.87	1.79
40	5.42	4.05	3.46	3.13	2.90	2.74	2.62	2.53	2.45	2.39	2.29	2.18	2.07	2.01	1.94	1.88	1.80	1.72	1.64
60	5.29	3.93	3.34	3.01	2.79	2.63	2.51	2.41	2.33	2.27	2.17	2.06	1.94	1.88	1.82	1.74	1.67	1.58	1.49
120	5.15	3.80	3.23	2.89	2.67	2.52	2.39	2.30	2.22	2.16	2.05	1.94	1.82	1.76	1.69	1.61	1.53	1.43	1.31
∞	5.02	3.69	3.12	2.79	2.57	2.41	2.29	2.19	2.11	2.05	1.94	1.83	1.71	1.64	1.57	1.48	1.39	1.27	1.00

n_1, degrees of freedom for numerator; n_2, degrees of freedom for denominator

distribution is usually assumed, lacking evidence to the contrary. For more extensive guidance, the reader should refer to Ref 1.

Statistical Procedures

Many different statistical techniques may be useful in analysis of mechanical-property data. This section presents brief descriptions of procedures that are used frequently (Ref 2). More detailed descriptions of these and other statistical techniques and tables in their various forms can be found in a number of workbooks and texts (for example, Ref 12).

Goodness-of-Fit Tests

This section contains a discussion and illustration of a method that can be used to establish whether or not a sample can realistically be assumed to follow a normal distribution. Similar, but more complex procedures are detailed elsewhere (Ref 2) for the three-parameter Weibull distribution.

The method that is presented is based on the "Anderson-Darling" goodness-of-fit test statistic for testing normality. This test is objective and indicates (at a 5% risk of error in the cases considered here) whether the sample is drawn from a normal distribution. Unfortunately, this test may reject the hypothesis of normality even though the distribution may provide a reasonable approximation within the lower tail. This is especially true for large data sets, where small deviations from the assumed distribution will be "detected." Consequently, some engineering judgment should be employed after using this, or any other goodness-of-fit test.

After a goodness-of-fit test has been performed (especially if the distributional assumption has been rejected), it is often useful to form a cumulative probability plot of the data to illustrate graphically the degree to which the assumed distribution fits the data. A method that can be used to develop normal probability plots is presented following the next section. In what follows, sample size is denoted by n, sample observations by $x_1, \ldots, x_n$, and sample observations ordered from least to greatest by $x_{(1)}, \ldots, x_{(n)}$.

The Anderson-Darling test for normality can be used to determine whether the density function that fits a given set of data can be approximated by a normal curve. The test involves a numerical comparison of the cumulative distribution function for observed data with that for the fitted normal curve over the entire range of the property being measured. The normalized deviation of individual data points from the sample average can be computed as follows:

$$z_{(i)} = (x_{(i)} - \bar{x})/s \qquad i = 1, \ldots, n \qquad \text{(Eq 17)}$$

where $x_{(i)}$ is the ith smallest sample observation, $\bar{x}$ is the sample average, and s is the sample standard deviation. Equations for computing sample statistics were presented earlier in the section "Sample versus Population Parameters."

The Anderson-Darling test statistic is:

$$AD = \left[\sum_{i=1}^{n} \frac{1-2i}{n} \left[\ln (F_o(z_{(i)})) + \ln (1 - F_o(z_{(n+1-i)})) \right] \right] - n \qquad \text{(Eq 18)}$$

where F_o is the standard normal distribution function (Ref 2). If:

$$AD > 0.752/(1 + 0.75/n + 2.25/n^2) \qquad \text{(Eq 19)}$$

it can be concluded (at a 5% risk of error) that the population from which the sample was drawn is not normally distributed. Otherwise, the hypothesis that the population is normally distributed is not rejected. For further information on this test procedure, see Ref 13 and 14.

The same procedure can be used to test the normality of the residuals:

$$e_i = y_i - (a + bx_i) \qquad i = 1, \ldots, n \tag{Eq 20}$$

from a regression assuming uniformity of variance of the residuals over the range of the independent variable. When calculating the test statistic *AD*, define:

$$z_{(i)} = e_{(i)}/s_y \qquad i = 1, \ldots, n \tag{Eq 21}$$

where $e_{(i)}$, $i = 1, \ldots, n$ are the ordered residuals from smallest to largest and s_y is the root mean square error of the regression defined in the section "Data-Regression Techniques." The justification for this procedure can be found in Ref 15.

Normal Probability Plot. Arithmetic normal probability paper is recommended for graphic illustration of the degree to which a normal distribution fits a set of data. Logarithmic normal probability paper can be used to determine whether the distribution of data could be normalized by a logarithmic transformation. One axis is scaled in units of the property measured, and the other is a nonlinear scale of probability.

If there are relatively few data points, all of them can be plotted individually on the chart. With larger samples, it is only necessary to plot enough points to define the curve.

The rank of each point selected for plotting is equal to the number of lower test points plus the plotted point plus one-half the number of other points equal to the plotted point (if there are points within the sample at this rank that are equal within engineering accuracy). Cumulative probability (*P*), in percent, is equal to the rank times 100, divided by one more than the total number of test points:

$$P \text{ (in percent)} = \frac{(\text{rank}) \cdot (100)}{n + 1} \tag{Eq 22}$$

The measured value of each test point is plotted versus its cumulative probability and a straight line is drawn to represent the normal distribution. This line can be established by plotting any two points from the normal distribution curve for $\bar{x}$ and s (for example, $\bar{x} - 3s$ at 0.13% probability, and $\bar{x} + 3s$ at 99.87% probability) and connecting these two points.

The plotted points should then be compared with the line to determine whether there is a reasonably good fit. With sample sizes on the order of 100 test points, only those points lying between about 10 and 90% probability should be considered in making this evaluation. When sample sizes of less than 50 are examined, the greatest focus should be placed on the observations between 20 and 80% probability.

If normal probability paper is unavailable, a normal probability plot can be formed by plotting the measured value of each test point versus $\bar{x} + s\,F_o^{-1}\,(P/100)$ where F_o^{-1} is the inverse standard normal cumulative distribution function. The line representing the fitted normal distribution is the line passing through the points with equal horizontal and vertical coordinates. If the horizontal axis is labeled with cumulative probabilities (*P* values) rather than $F_o^{-1}\,(P/100)$ values, the plot will be a normal probability plot.

Tests of Significance

A test of significance is used to make a decision on a statistical basis. In this section, the *F*- and *t*-tests are described for use in determining whether the populations from which two samples are drawn are identical. Other, more comprehensive procedures, such as the *k*-sample Anderson-Darling procedure are described elsewhere (Ref 2).

The *F*-test is normally used first to determine whether two sample variances differ significantly or not (with a 5% risk of error in the case examined here). If the two sample variances do not differ significantly, the *t*-test is used next to determine whether the sample means differ significantly. If either the sample variances or the sample means differ significantly, it can be concluded that the populations from which the two samples were drawn are not statistically identical. Otherwise, the hypothesis that the two populations are identical should not be rejected. These tests are appropriate to use when (1) the observations within each sample are taken randomly from a single population of possible observations, and (2) the characteristic measured is normally distributed within each population. To carry out a similar procedure without requiring the assumption of an underlying normal distribution, or if three or more samples are to be compared, the *k*-sample Anderson-Darling test can be employed. This test is a nonparametric procedure and simply tests the hypothesis that populations from which the samples are drawn are identical.

The *F*-Test: Evaluation of Differences in Variability between Samples. The *F*-test is used to determine whether two products differ with regard to their variability. Consider products 1 and 2; they might represent two different processes, thickness ranges, or test directions. The statistics for the samples drawn from these products are:

Attribute	Product 1	Product 2
Sample size	n_1	n_2
Sample standard deviation	s_1	s_2
Sample mean	$\bar{x}_1$	$\bar{x}_2$

F is the ratio of the two sample variances:

$$F = s_1^2/s_2^2$$

If the true variances of products A and B are identical at a significance level of $\alpha = 0.05$, *F* should lie within the interval defined by:

$$F_{0.975} \text{ (for } n_1 - 1 \text{ and } n_2 - 1 \text{ degrees of freedom)}$$

and

$$1/F_{0.975} \text{ (for } n_2 - 1 \text{ and } n_1 - 1 \text{ degrees for freedom)}$$

If *F* does not lie within this interval, it can be concluded at a 95% confidence level (2.5% in each tail of the two-sided test) that the two products differ with regard to their variability. Values of $F_{0.975}$ can be obtained from Table 5. Values of $F_{0.975}$ can also be computed from the following analytical expression:

$$F_{0.975} = \exp\left[2\delta\left(1 + \frac{z^2 - 1}{3} - 4\sigma^2\right)^{1/2} + 2\sigma z\left(1 + \frac{\sigma^2\,(z^2 - 3)}{6}\right)\right] \tag{Eq 23}$$

where $z = 1.96$, $\delta = 0.5\,[1/(n_2 - 1) - 1/(n_1 - 1)]$, $\sigma^2 = 0.5\,[1/(n_2 - 1) + 1/(n_1 - 1)]$, n_1 = degrees of freedom for the numerator, and n_2 = degrees of freedom for the denominator. This expression is accurate to within 0.4% for n_1 degrees of freedom greater than or equal to 10 and n_2 degrees of freedom greater than or equal to 16.

Example 1: *F*-Test Computation. Assume that the following sample statistics were reported:

Attribute	Product 1	Product 2
Sample size	20	30
Sample standard deviation, ksi	4.0	5.0
Sample mean, ksi	100.0	102.0

Perform an *F*-test as follows:

$$F = s_1^2/s_2^2 = 4^2/5^2 = 0.64$$

$$df_1 = n_1 - 1 = 19$$

$$df_2 = n_2 - 1 = 29$$

where *df* is degrees of freedom. From interpolation within Table 5 (or from Eq 23) it can be found that:

$$F_{0.975\,(12,29)} = 2.23$$

and

$$1/F_{0.975\,(29,19)} = 1/2.40 = 0.42$$

Because 0.64 lies within the interval of 0.42 to 2.23, there is insufficient statistical evidence to conclude that products 1 and 2 differ with regard to their variability.

The *t*-Test: Evaluation of Differences in Means Between Samples. The *t*-test is used to determine whether two products differ with regard to their average properties. If they do, it can be concluded that the two products do not belong to the same population.

In making the *t*-test, it is assumed that the variances of two products are nearly equal, as first determined from the *F*-test. If the *F*-test shows that the variances are significantly different, there is no need to conduct the *t*-test.

Consider products 1 and 2 as before. $D_{\bar{x}}$ is the absolute difference between the two sample means:

$$D_{\bar{x}} = |\,\bar{x}_1 - \bar{x}_2\,| \tag{Eq 24}$$

If the true means of products 1 and 2 are identical, $D_{\bar{x}}$ should not exceed u, which is determined as

Table 6 0.950 and 0.975 fractiles of the *t*-distribution associated with degrees of freedom (*df*)

df	$t_{0.95}$	$t_{0.975}$
1	6.314	12.706
2	2.920	4.303
3	2.353	3.182
4	2.132	2.776
5	2.015	2.571
6	1.943	2.447
7	1.895	2.365
8	1.860	2.306
9	1.833	2.262
10	1.812	2.228
11	1.796	2.201
12	1.782	2.179
13	1.771	2.160
14	1.761	2.145
15	1.753	2.131
16	1.746	2.120
17	1.740	2.110
18	1.734	2.101
19	1.729	2.093
20	1.725	2.086
21	1.721	2.080
22	1.717	2.074
23	1.714	2.069
24	1.711	2.064
25	1.708	2.060
26	1.706	2.056
27	1.703	2.052
28	1.701	2.048
29	1.699	2.045
30	1.697	2.042
40	1.684	2.021
50	1.676	2.009
60	1.671	2.000
80	1.664	1.990
100	1.660	1.984
120	1.658	1.980
200	1.653	1.972
500	1.648	1.965
∞	1.645	1.960

Note: The following equations may be used to compute 0.95 and 0.975 fractiles of the *t*-distribution in lieu of using table values:
$t_{0.95} = 1.645 + \exp\,[0.377 - 0.990\ln(df) + 1.15/df]$
$t_{0.975} = 1.960 + \exp\,[0.779 - 0.980\ln(df) + 1.57/df]$
These approximations are accurate to within 0.5% for $df \geq 4$.

indicated by the following equation for a significance level of $\alpha = 0.05$.

$$u = t_{0.975}\, s_p \sqrt{\frac{n_1 + n_2}{n_1 \cdot n_2}} \qquad \text{(Eq 25)}$$

where $t_{0.975}$ has $n_1 + n_2 - 2$ degrees of freedom, and :

$$s_p = \sqrt{\frac{(n_1 - 1)\cdot s_1^2 + (n_2 - 1)\cdot s_2^2}{n_1 + n_2 - 2}} \qquad \text{(Eq 26)}$$

Values of $t_{0.975}$ can be found through interpolation within Table 6. Values of $t_{0.975}$ can also be calculated from the approximate formula given in the footnote in Table 6. This formula is accurate to within 0.5% for degrees of freedom greater than or equal to 4.

Example 2: *t*-Test Computation. The following sample statistics are the same as those presented earlier:

Attribute	Product 1	Product 2
Sample size	20	30
Sample standard deviation, ksi	4.0	5.0
Sample mean, ksi	100.0	102.0

Using the sample statistics presented earlier, it has already been determined that the variances of products 1 and 2 do not differ significantly. The *t*-test computations to test the sample means are as follows:

$$df = n_1 + n_2 - 2 = 48$$

$$t_{0.975}\ (\text{for 48 } df) = 2.011\ (\text{from Table 6})$$

$$s_p = \sqrt{\frac{(n_1 - 1)s_1^2 + (n_2 - 1)s_2^2}{n_1 + n_2 - 2}}$$

$$= \sqrt{\frac{(19)\,(4)^2 + (29)\,(5)^2}{48}} = 4.63 \text{ ksi}$$

$$\sqrt{\frac{n_1 + n_2}{n_1 \cdot n_2}} = \sqrt{\frac{20 + 30}{(20)\cdot(30)}} = 0.2887$$

$$D_{\bar{x}} = |\,\bar{x}_1 - \bar{x}_2\,| = 2.0 \text{ ksi}$$

$$u = 2.011\,(4.63)\,(0.2887) = 2.7$$

Because $D_{\bar{x}}$ (2.0) is not greater than u (2.7), it can be concluded that there is no reason to believe that products 1 and 2 differ with regard to their average strength. On the basis of both tests in this example, the conclusion would be that the two products can be assumed to be drawn from the same population.

Data-Regression Techniques

When it is suspected that the average of one measured value varies linearly or curvilinearly with some other measured value, a regression analysis is often used to investigate and describe the relationship between the two quantities. Examples are effect of product thickness on tensile strength, effect of temperature on yield strength, and effect of stress on cycles to failure or time to rupture. Mathematical techniques for performing a simple linear-regression analysis are contained in the following section. Statistical tests to determine whether or not a straight line adequately describes the data are described in the section "Testing the Adequacy of a Regression." An example is then presented using hypothetical data to illustrate the regression calculations.

Although the relationship between the dependent and independent variables may not be linear, it may be possible to linearize the relationship through a simple transformation of variables. In this case, it is important to perform all calculations within the transformed space and then untransform the results to the original engineering units for final presentation.

Least-Squares Linear Regression. Linear regression is appropriate when there is an approximate linear relationship between two measurable characteristics. Such a relationship is expressed algebraically by an equation that, in the case of two measurable characteristics x and y, has the form:

$$y = \alpha + \beta x + \varepsilon \qquad \text{(Eq 27)}$$

where x is the independent variable, y is the dependent variable, α is the true intercept of the regression equation, β is the true slope of the regression equation, and ε is the measurement or experimental error by which y differs from the ideal linear relationship.

Aside from the error term, ε, this is the equation of a straight line. The parameter α determines the point where this line intersects the y-axis, and the β represents its slope. The variables x and y can represent either direct measurements or some transformation measurements of the characteristics under consideration. Knowing or assuming such an approximate linear relationship, the problem becomes one of estimating the parameters α and β of the regression equations. It is necessary to have a random sample consisting of n pairs of observations, which is denoted by $(x_1, y_1), (x_2, y_2), \ldots, (x_n, y_n)$. Such a sample can be represented graphically by n points plotted on a coordinate system, in which x is plotted horizontally and y vertically. A subjective solution can be obtained by drawing a line that, by visual inspection, appears to fit the points satisfactorily. An objective solution is given by the method of least squares.

The method of least squares is a numerical procedure for obtaining a line having the property that the sum of squares of vertical deviations of the sample points from this line is less than that for any other line. In this analysis, the least-squares line is represented by the equation:

$$\hat{y} = a + bx \qquad \text{(Eq 28)}$$

in which $\hat{y}$ is the predicted value of y for any value of x, and a and b are estimates of the parameters α and β in the true regression equation obtained by the least-squares method presented below.

It can be shown with the aid of calculus that the values of a and b that minimize the sum of squares of the vertical deviations are given by the formulas:

$$a = \frac{\Sigma y - b\Sigma x}{n} \qquad \text{(Eq 29)}$$

$$b = \frac{s_{xy}}{s_{xx}} \qquad \text{(Eq 30)}$$

where

$$s_{xy} = \Sigma xy - \frac{\Sigma x\,\Sigma y}{n} \qquad \text{(Eq 31)}$$

and

$$s_{xx} = \Sigma x^2 - \frac{(\Sigma x)^2}{n} \qquad \text{(Eq 32)}$$

The root mean square error of y is expressed as:

$$s_y = \sqrt{\frac{\Sigma(y - \hat{y})^2}{\bar{n} - 2}} \quad \text{(Eq 33)}$$

where $\hat{y}$ is the predicted value of y defined above. This quantity is an estimate of the standard deviation of the distribution of y about the regression line. A convenient computational formula for s_y is

$$s_y = \sqrt{\frac{s_{yy} - b^2 s_{xx}}{n - 2}} \quad \text{(Eq 34)}$$

where

$$s_{yy} = \Sigma y^2 - \frac{(\Sigma y)^2}{n} \quad \text{(Eq 35)}$$

The quantity $R^2 = (b^2 s_{xx})/s_{yy}$ measures the proportion of total variation in the y data, about its average, that is explained by the regression. An R^2 equal to 1 indicates that the regression model describes data perfectly, which is rare in practice. R^2 provides a rough idea of how well data is described by a linear regression. A more precise determination of the adequacy of a linear regression is discussed in the next section.

Testing the Adequacy of a Regression. It is possible that the relationship between dependent variable y and independent variable x may not be approximately linear. In that case, a straight line would not "fit" the data very well. It is also possible that the relationships between x and y, although approximately linear, may not be very strong. In such a case, estimated slope parameter b would not be significantly different from 0. Both the lack of fit and the significance of a linear-regression equation can be evaluated through an analysis of variance as described in this section.

The analysis of variance for testing lack of fit and significance of regression is based on the assumption that measurement errors, ε_i, in the approximate linear relationship between y_i and x_i, that is,

$$\varepsilon_i = y_i - (\alpha + \beta x_i) \qquad i = 1, 2, \ldots, n \quad \text{(Eq 36)}$$

are independent and normally distributed with an overall mean of 0 and a constant variance of σ^2. Assuming uniformity of measurement error variance over the range of the independent variable, the normality assumption concerning unobservable ε_i can be checked by performing the Anderson-Darling test for normality on the observable residuals:

$$e_i = y_i - (a + bx_i) \qquad i = 1, 2, \ldots, n \quad \text{(Eq 37)}$$

where a and b are the least-squares estimates of α and β. See the section "Anderson Darling Test for Normality" in this article for more details. By plotting the residuals e_i against the respective x_i, an informal check on the assumption of constant variance σ^2 is possible as well. In such a plot residuals should vary approximately equally over the range of x_i values.

Table 7 Analysis of variance for testing lack of fit and significance of linear regression

Source of variation	Degrees of freedom (linear)	Sum of square, SS	Mean squares, MS	F_{calc}
Regression	1	SSR	MSR	F_1
Error	$n-2$	SSE	MSE	
Lack of fit	$k-2$	$SSLF$	$MSLF$	F_2
Pure error	$n-k$	$SSPE$	$MSPE$	
Total	$n-1$	SST		

$SSR = b^2 s_{xx}$
$SST = s_{yy}$
$SSE = SST - SSR$

The analysis of variance table for testing the lack of fit and significance of a linear regression is shown in Table 7. The sums of squares for the three primary lines of the analysis of variance table (Regression, Error, and Total) are calculated using quantities defined in the section "Least-Squares Linear Regression" in this article.

If multiple observations at one or more values of the independent variable x are available, and observations are made at three or more distinct x values, then it is possible to evaluate the adequacy of a line in describing the relationship between x and y. The notation in testing for this lack of fit is summarized in the analysis of variance table above. Let Y_{ij} denote the jth data value at the ith x level, k represent the number of distinct x levels for which there is data, and n_i represent the number of data values at the ith x level. Denote the total of the observations at the ith x level by:

$$T_i = \sum_{j=1}^{n_i} Y_{ij} \qquad \text{for } i = 1, \ldots, k \quad \text{(Eq 38)}$$

and note that:

$$n = \sum_{i=1}^{k} n_i \quad \text{(Eq 39)}$$

is used to denote the total number of observations. The lack of fit and pure error sums of squares are computed as:

$$SSLF = \sum_{i=1}^{k} (T_i^2/n_i) - (\Sigma y)^2/n - SSR \quad \text{(Eq 40)}$$

and

$$SSPE = SSE - SSLF$$

The sums of squares are divided by the corresponding degrees of freedom to compute mean squares as shown below:

$$MSR = SSR$$

$$MSE = SSE/(n-2)$$

$$MSLF = SSLF/(k-2)$$

$$MSPE = SSPE/(n-k)$$

These mean squares are used to compute two F statistics that test for lack of fit and significance of regression. (Note: If there is only one data value for each x level, i.e., $n_i = 1$ for $i = 1, 2, \ldots, k$, or if $k = 2$, then it is not possible to test for lack of fit.)

The two F statistics, F_1 and F_2, are defined as ratios of the mean squares as specified below:

$$F_1 = MSR/MSE$$

$$F_2 = MSLF/MSPE$$

F_2 and Table 8 are used to test for lack of fit. If F_2 is greater than the 95th percentile of F distribution with $k - 2$ numerator degrees of freedom and $n - k$ denominator degrees of freedom (from Table 8), then there is significant lack of fit. In this case it can be concluded (with a 5% risk of error) that linear regression does not adequately describe the relationship between x and y. Otherwise, lack of fit can be considered insignificant and a linear-regression model can be assumed.

If lack of fit is not significant, the significance of regression can be tested using F_1 and Table 8. If F_1 is greater than the 95th percentile of F distribution with 1 numerator degree of freedom and $n - 2$ denominator degrees of freedom (from Table 8), then regression is significant. Otherwise, the regression is not significant and x is considered to have little or no predictive value for y.

Example 3: Linear Regression Computations. In this example, x represents thickness and y, the tensile yield strength values determined from a group of tensile tests. Values of x and y are:

x	y
0.100	121
0.100	119
0.200	114
0.200	108
0.300	112
0.300	108
0.400	112
0.400	106
0.500	101
0.500	99

Table 8 0.950 fractiles of the *F*-distribution associated with n_1 and n_2 degrees of freedom, $F_{0.950}(n_1,n_2)$

	n_1																		
n_2	1	2	3	4	5	6	7	8	9	10	12	15	20	24	30	40	60	120	∞
1	161.4	199.5	215.7	224.6	230.2	234.0	236.8	238.9	240.5	241.9	243.9	245.9	248.0	249.0	250.1	251.1	252.2	253.2	254.3
2	18.51	19.00	19.16	19.25	19.30	19.33	19.35	19.37	19.38	19.40	19.41	19.43	19.45	19.45	19.46	19.47	19.48	19.49	19.51
3	10.13	9.55	9.28	9.12	9.01	8.94	8.89	8.85	8.81	8.79	8.74	8.70	8.66	8.64	8.62	8.59	8.57	8.55	8.53
4	7.71	6.94	6.59	6.39	6.26	6.16	6.09	6.04	6.00	5.96	5.91	5.86	5.80	5.77	5.75	5.72	5.69	5.66	5.63
5	6.61	5.79	5.41	5.19	5.05	4.95	4.88	4.82	4.77	4.74	4.68	4.62	4.56	4.53	4.50	4.46	4.43	4.40	4.37
6	5.99	5.14	4.76	4.53	4.39	4.28	4.21	4.15	4.10	4.06	4.00	3.94	3.87	3.84	3.81	3.77	3.74	3.70	3.67
7	5.59	4.74	4.35	4.12	3.97	3.87	3.79	3.73	3.68	3.64	3.57	3.51	3.44	3.41	3.38	3.34	3.30	3.27	3.23
8	5.32	4.46	4.07	3.84	3.69	3.58	3.50	3.44	3.39	3.35	3.28	3.22	2.15	3.12	3.08	3.04	3.01	2.97	2.93
9	5.12	4.26	3.86	3.63	3.48	3.37	3.29	3.23	3.18	3.14	3.07	3.01	2.94	2.90	2.86	2.83	2.79	2.75	2.71
10	4.96	4.10	3.71	3.48	3.33	3.22	3.14	3.07	3.02	2.98	2.91	2.85	2.77	2.74	2.70	2.66	2.62	2.58	2.54
11	4.84	3.98	3.59	3.36	3.20	3.09	3.01	2.95	2.90	2.85	2.79	2.72	2.65	2.61	2.57	2.53	2.49	2.45	2.40
12	4.75	3.89	3.49	3.26	3.11	3.00	2.91	2.85	2.80	2.75	2.69	2.62	2.54	2.51	2.47	2.43	2.38	2.34	2.30
13	4.67	3.81	3.41	3.18	3.03	2.92	2.83	2.77	2.71	2.67	2.60	2.53	2.46	2.42	2.38	2.34	2.30	2.25	2.21
14	4.60	3.74	3.34	3.11	2.96	2.85	2.76	2.70	2.65	2.60	2.53	2.46	2.39	2.35	2.31	2.27	2.22	2.18	2.13
15	4.54	3.68	3.29	3.06	2.90	2.79	2.71	2.64	2.59	2.54	2.48	2.40	2.33	2.29	2.25	2.20	2.16	2.11	2.07
16	4.49	3.63	3.24	3.01	2.85	2.74	2.66	2.59	2.54	2.49	2.42	2.35	2.28	2.24	2.19	2.15	2.11	2.06	2.01
17	4.45	3.59	3.20	2.96	2.81	2.70	2.61	2.55	2.49	2.45	2.38	2.31	2.23	2.19	2.15	2.10	2.06	2.01	1.96
18	4.41	3.55	3.16	2.93	2.77	2.66	2.58	2.51	2.46	2.41	2.34	2.27	2.19	2.15	2.11	2.06	2.02	1.97	1.92
19	4.38	3.52	3.13	2.90	2.74	2.63	2.54	2.48	2.42	2.38	2.31	2.23	2.16	2.11	2.07	2.03	1.98	1.93	1.88
20	4.35	3.49	3.10	2.87	2.71	2.60	2.51	2.45	2.39	2.35	2.28	2.20	2.12	2.08	2.04	1.99	1.95	1.90	1.84
21	4.32	3.47	3.07	2.84	2.68	2.57	2.49	2.42	2.37	2.32	2.25	2.18	2.10	2.05	2.01	1.96	1.92	1.87	1.81
22	4.30	3.44	3.05	2.82	2.66	2.55	2.46	2.40	2.34	2.30	2.23	2.15	2.07	2.03	1.98	1.94	1.89	1.84	1.78
23	4.28	3.42	3.03	2.80	2.64	2.53	2.44	2.37	2.32	2.27	2.20	2.13	2.05	2.01	1.96	1.91	1.86	1.81	1.76
24	4.26	3.40	3.01	2.78	2.62	2.51	2.42	2.36	2.30	2.25	2.18	2.11	2.03	1.98	1.94	1.89	1.84	1.79	1.73
25	4.24	3.39	2.99	2.76	2.60	2.49	2.40	2.34	2.28	2.24	2.16	2.09	2.01	1.96	1.92	1.87	1.82	1.77	1.71
26	4.23	3.37	2.98	2.74	2.59	2.47	2.39	2.32	2.27	2.22	2.15	2.07	1.99	1.95	1.90	1.85	1.80	1.75	1.69
27	4.21	3.35	2.96	2.73	2.57	2.46	2.37	2.31	2.25	2.20	2.13	2.06	1.97	1.93	1.88	1.84	1.79	1.73	1.67
28	4.20	3.34	2.95	2.71	2.56	2.45	2.36	2.29	2.24	2.19	2.12	2.04	1.96	1.91	1.87	1.82	1.77	1.71	1.65
29	4.18	3.33	2.93	2.70	2.55	2.43	2.35	2.28	2.22	2.18	2.10	2.03	1.94	1.90	1.85	1.81	1.75	1.70	1.64
30	4.17	3.32	2.92	2.69	2.53	2.42	2.33	2.27	2.21	2.16	2.09	2.01	1.93	1.89	1.84	1.79	1.74	1.68	1.62
40	4.08	3.23	2.84	2.61	2.45	2.34	2.25	2.18	2.12	2.08	2.00	1.92	1.84	1.79	1.74	1.69	1.64	1.58	1.51
60	4.00	3.15	2.76	2.53	2.37	2.25	2.17	2.10	2.04	1.99	1.92	1.84	1.75	1.70	1.65	1.59	1.53	1.47	1.39
120	3.92	3.07	2.68	2.45	2.29	2.18	2.09	2.02	1.96	1.91	1.83	1.75	1.66	1.61	1.55	1.50	1.43	1.35	1.25
∞	3.84	3.00	2.61	2.37	2.21	2.10	2.01	1.94	1.88	1.83	1.75	1.67	1.57	1.52	1.46	1.39	1.32	1.22	1.00

n_1, degrees of freedom for numerator; n_2, degrees of freedom for denominator

From these data, the following quantities can be calculated:

$$n = 10$$
$$\Sigma x = 3$$
$$\Sigma y = 1100.0$$
$$\Sigma x^2 = 1.1$$
$$\Sigma y^2 = 121452$$
$$\Sigma xy = 321.6$$
$$(\Sigma x)^2 = 9$$
$$(\Sigma y)^2 = 121{,}000$$
$$(\Sigma x)(\Sigma y) = 3{,}300$$
$$s_{xx} = 0.20$$
$$s_{xy} = -8.4$$
$$s_{yy} = 452$$

The slope of the regression line is:

$$b = \frac{s_{xy}}{s_{xx}} = \frac{-8.4}{0.20} = -42$$

The y-intercept of the regression line is:

$$a = \frac{\Sigma y - b\Sigma x}{n} = \frac{1100}{10} - \frac{(-42)(3)}{10} = 110 + 12.6 = 122.6$$

Thus the final equation of the least squares regression line is:

$$\hat{y} = a + bx = 122.6 - 42x$$

The total of the y data at each x level is needed to calculate lack of fit and pure error sums of squares. These totals are as follows:

x_i	T_i
0.1	240
0.2	222
0.3	220
0.4	218
0.5	200

There are data values at $k = 5$ different x levels, with $n_i = 2$ values at each level and:

$$\sum_{i=1}^{k} (T_i^2/n_i) = \frac{(240)^2}{2} + \ldots + \frac{(200)^2}{2} = 121{,}404$$

Therefore:

$$SSLF = 121{,}404 - (1{,}100)^2/10 - 352.8 = 51.2$$

and

$$SSPE = 99.2 - 51.2 = 48$$

The mean square values are computed by dividing corresponding sums of squares by their degrees of freedom. The F_1 and F_2 statistics are then calculated as ratios of mean squares. The analysis of variance is shown in Table 9. Using the fitted-regression equation, the following values of $\hat{y}$ can be computed for the values of x listed previously.

x	$\hat{y}$
0.100	118.4
0.200	114.2
0.300	110.0
0.400	105.8
0.500	101.6

The root mean square error is computed as follows:

$$s_y = \sqrt{\frac{\Sigma(y - \hat{y})^2}{n - 2}} = \sqrt{\frac{99.2}{8}} = \sqrt{12.4} = 3.52$$

or

$$s_y = \sqrt{\frac{s_y - b^2 s_{xx}}{n - 2}} = \sqrt{\frac{452 - (-42)^2(0.2)}{8}} = 3.52$$

The correlation coefficient, R^2, is computed as follows:

$$R^2 = \frac{b^2 s_{xx}}{s_{yy}} = \frac{(-42)^2(0.2)}{452} = 0.78$$

Thus, 78% of the variability in the y data about its average is explained by the linear relationship between y and x.

Table 9 Analysis of variance for Example 3

Source of variation	Degrees of freedom, *df*	Sum of squares, *SS*	Mean squares, *MS*	F_{calc}
Regression	1	352.8	352.8	$F_1 = 28.5$
Error	8	99.2	12.4	
Lack of fit	3	51.2	17.07	$F_2 = 1.78$
Pure error	5	48.0	9.6	
Total	9	452.0		

Table 10 ASTM standards related to engineering statistics

ASTM No.	Title
Statistics terminology	
D 4392	"Terminology for Statistically Related Terms"
E 177	"Practice for Use of the Terms Precision and Bias in ASTM Test Methods"
E 456	"Terminology Related to Statistics"
E 1325	"Terminology Relating to Design of Experiments"
E 1402	"Terminology Relating to Sampling"
Statistical conformance with specifications	
D 3244	"Practice for Utilization of Test Data to Determine Conformance with Specifications"
E 122	"Practice for Choice of Sample Size to Estimate a Measure of Quality for a Lot or Process"
Statistical control and comparison of test methods	
D 4356	"Practice for Establishing Consistent Test Method Tolerances"
D 4853	"Guide for Reducing Test Variability"
D 4855	"Practice for Comparing Test Methods"
E 1323	"Guide for Evaluating Laboratory Measurement Practices and the Statistical Analysis of the Resulting Data"
Statistical analysis of test data	
E 178	"Practice for Dealing with Outlying Observations"
E 739	"Practice for Statistical Analysis of Linear or Linearized Stress-Life (S-N) and Strain-Life (ε-N) Fatigue Data"
Statistical guidelines for interlaboratory testing programs	
D 4467	"Practice for Interlaboratory Testing of a Test Method that Produces Non-Normally Distributed Data"
E 691	"Practice for Conducting an Interlaboratory Study to Determine the Precision of a Test Method"
Statistical issues in development of sampling plans	
D 4854	"Guide for Estimating the Magnitude of Variability from Expected Sources in Sampling Plans"
E 105	"Practice for Probability Sampling of Materials"
E 141	"Practice for Acceptance of Evidence Based on the Results of Probability Sampling"

The sum of squares for the regression, total, and error lines are computed as follows:

$$SSR = (-42)^2 (0.20) = 352.8$$

$$SST = 452$$

$$SSE = 452 - 352.8 = 99.2$$

The F_2 value of 1.78 with $k - 2 = 3$ and $n - k = 5$ degrees of freedom is less than the value of 5.41 from Table 8 corresponding to 3 numerator and 5 denominator degrees of freedom. This indicates that lack of fit can be considered insignificant. Thus, it is reasonable to assume that a linear regression adequately describes the data. The F_1 value of 28.5 with 1 and $n - 2 = 8$ degrees of freedom is greater than the value of 5.32 from Table 8 corresponding to 1 numerator and 8 denominator degrees of freedom, so the slope of the regression is found to be significantly different from 0.

Additional details on goodness-of-fit testing and linear-regression procedures can be found in Ref 12 to 15.

Related ASTM Engineering Statistics Standards

Table 10 lists standards published by the American Society for Testing and Materials (ASTM) that address statistical issues relevant to this article. These standards are maintained by ASTM Committee E11 on Statistical Methods. A primary purpose of this group is to advise other ASTM committees in the area of statistics, and to provide information for general application.

REFERENCES

1. C. Lipson and N.J. Sheth, *Statistical Design and Analysis of Engineering Experiments*, McGraw-Hill, 1973
2. *Metallic Materials and Elements for Aerospace Vehicle Structures*, MIL-HDBK-5G, Change Notice 1, 1 Dec 1995
3. "Practice for Use of the Terms Precision and Bias in ASTM Test Methods," E 177, *Annual Book of ASTM Standards*, ASTM
4. "Practice for Conducting an Interlaboratory Study to Determine the Precision of a Test Method," E 691, *Annual Book of ASTM Standards*, ASTM
5. S. Berretta, P. Clerici, and S. Matteazzi, The Effect of Sample Size on the Confidence of Endurance Fatigue Tests, *Fatigue Fract. Eng. Mater. Struct.*, Vol 18 (No.1), 1995, p 129–139
6. C.L. Shen, P.H. Wirsching, and G.T. Cashman, "An Advanced Comprehensive Model for the Satistical Analysis of S-N Fatigue Data," E08, Special Presentation, ASTM, 1994
7. R.C. Rice, K.B. Davies, C.E. Jaske, and C.E. Feddersen, "Consolidation of Fatigue and Fatigue-Crack Propagation Data for Design Use," CR-2586, National Aeronautics and Space Administration, Oct 1975
8. X.L. Zheng, B. Lu, and H. Jiang, Determination of Probability Distribution of Fatigue Strength and Expressions of P-S-N Curves, *Eng. Fract. Mech.*, Vol 50 (No. 4), 1995, p 483–491
9. W. Weibull, A Statistical Representation of Fatigue Failures in Solids, *Acta Polytech., Mech. Eng. Ser.*, Vol 1 (No. 9), 1949
10. K.E. Olsson, Weibull Analysis of Fatigue Test Data, *Quality and Reliability Engineering International*, Vol 10, 1994, p 437–438
11. *Polymer Matrix Composites*, MIL-HDBK-17-1D, 16 Feb 1993
12. M.G. Natrella, *Experimental Statistics*, National Bureau of Standards Handbook 91, 1 Aug 1963
13. J.F. Lawless, *Statistical Models and Methods for Lifetime Data*, John Wiley & Sons, 1982, p 452–460
14. R.B. D'Agostina and M.A. Stephens, *Goodness-of-Fit Techniques*, Marcel Dekker, 1987, p 123
15. D.A. Pierce and K.J. Kopecky, Testing Goodness of Fit for the Distribution of Errors in Regression Models, *Biometrika*, Vol 66, 1979, p 1–5

SELECTED REFERENCES

- F.S. Acton, *Analysis of Straight Line Data*, John Wiley & Sons, 1959
- N.C. Barford, *Experimental Measurements: Precision, Error and Truth*, 2nd ed., John Wiley & Sons, 1985
- A.H. Bowker and G.J. Lieberman, *Engineering Statistics*, Prentice-Hall, 1959
- M.R. Gorman, "Reliability of Structural Systems," Ph.D. thesis, Case Western Reserve University, 1979
- A. Hald, *Statistical Theory with Engineering Applications*, John Wiley & Sons, 1952
- E.B. Haugen, *Probabilistic Approaches to Design*, John Wiley & Sons, 1968
- N.L. Johnson and F.C. Leone, *Statistics and Experimental Design in Engineering and the Physical Sciences*, Volume I, John Wiley & Sons, 1964
- C. Lipson and N.J. Sheth, *Statistical Design and Analysis of Engineering Experiments*, McGraw-Hill, 1973
- M.G. Natrella, *Experimental Statistics*, National Bureau of Standards Handbook 91, U.S. Government Printing Office, 1963

Reliability in Design

Charles O. Smith, Engineering Consultant

RELIABILITY is a measure of the capacity of equipment or systems to operate without failure in the service environment. The National Aeronautics and Space Administration (NASA) defines reliability as the probability of a device performing adequately for the period of time intended under the operating conditions encountered. *Reliability is always a probability*, thus its calculation is one form of applied mathematics. Probabilistical and statistical methods, however, like any other forms of mathematics, are aids to, not substitutes for, logical reasoning.

Reliability is identified with a state of knowledge, not a state of things. Reliability cannot be used to predict discrete, or individually specific, events; only probabilities, that is averages, are predicted. Probability is a measure of what is expected to happen on the average if the given event is repeated a large number of times under identical conditions. Probability serves as a substitute for certainty. A generalization is made from samples, and the conclusions reached cannot be considered to be absolutely correct.

The NASA definition of reliability (and any other definition) includes "adequate performance" of the specific device. There is no general definition of adequate performance. Criteria for adequate performance must be carefully and exactly detailed or specified in advance for each device or system considered. For example, performance that is adequate for a tire of a private automobile would be woefully inadequate for a tire of a race car competing in the Indianapolis 500. The precise definition of adequate performance is completely dependent on the case under consideration.

A reliability of 0.99 implies a probability of 1 failure per 100. A reliability of 0.999 does not imply greater accuracy (one more significant figure) but an order of magnitude difference, that is, a probability of failure of 1 per 1000.

Reliability Tasks

All general definitions, like the NASA one, are qualitative. A quantitative definition of reliability for operating time t is:

$$R(t) = P(T > t) \qquad \text{(Eq 1)}$$

where R is reliability, P is probability, and T is the time to failure of the device. (T itself is a variable.) The device is assumed to be working properly at time $t = 0$. No device can work forever without failure, thus:

$$R(0) = 1 \text{ and } R(\infty) = 0 \qquad \text{(Eq 2)}$$

where $R(t)$ is a nonincreasing (constant, at best; generally decreasing) function between these two extremes. For $t < 0$, reliability has no meaning. It is defined as unity, however, since the probability of failure for $t < 0$ is zero.

The first task in ensuring reliability is to derive and investigate Eq 1. This is first done for a single component. The reliability of an entire system is then determined in keeping with the functions and configuration of the units composing the system. This, in turn, may be a subsystem of a more complex system. The building process continues until the entire system has been treated. The complete system may be so complex that it includes entire organizations, such as maintenance and repair groups with their personnel.

The second task, after the system and its reliability are truly understood, is to find the best way of increasing the reliability. The most important methods for doing this are:

- Reduce the complexity to the minimum necessary for the required functions. Nonessential components and unnecessary complexity increase the probability of system failure, that is, decrease reliability.
- Increase the reliability of components in the system.
- Use parallel redundancy, one or more "hot" spares operate in parallel. If one fails, others still function.
- Use standby redundancy. One or more "cold" spares is switched in to perform the function of a component or subsystem that has failed.
- Employ repair maintenance. Failed components are replaced by a technician rather than switched in as with standby redundancy. Replacement is neither automatic nor necessarily immediate.
- Employ preventive maintenance. Components are replaced periodically by new ones even though they may not have failed prior to the time of replacement.

Combinations of these methods are possible. The first two methods are obviously limited by the current state of technology. The other four methods enable, in principle, the development of systems having reliabilities approaching 100% for all arbitrarily chosen mission times t. This is not always feasible due to various constraints.

The third task is to maximize system reliability for a given weight, size, or cost. Conversely, the task may be to minimize weight, size, cost or other constraints for a given reliability.

Instantaneous Failure Rate and General Reliability Function

The instantaneous failure rate, $\lambda(t)$, that is, the instantaneous probability of failure per component, is assumed to be any variable and integrable function of time. Failure rate is normally expressed as the number of failures per unit time. The most general expression of failure rate is:

$$\lambda(t) = \frac{-1}{R(t)}\left(\frac{dR(t)}{dt}\right) \qquad \text{(Eq 3)}$$

Reliability is also related to the failure density function, or probability density function, $f(t)$:

$$\frac{dR(t)}{dt} = -f(t) \qquad \text{(Eq 4)}$$

Combining Eq 3 and 4 gives an alternative expression for the instantaneous failure rate:

$$\lambda(t) = \frac{f(t)}{R(t)} \qquad \text{(Eq 5)}$$

In other words, the instantaneous failure rate (at time t) equals the failure density function divided by the reliability, with both of the latter evaluated at time t. This is a completely general statement. Rearranging Eq 3 and integrating, knowing that at $t = 0$ and $R(t) = 1$, the general reliability function is:

$$R(t) = e^{-\int_0^t \lambda(t)dt} = \exp\left[-\int_0^t \lambda(t)dt\right] \quad \text{(Eq 6)}$$

This function is independent of the specific failure distribution involved. The case of constant failure rate, that is, $\lambda(t)$ independent of time, has special interest:

$$R(t) = e^{-\lambda t} \quad \text{(Eq 6a)}$$

Graphical Representation of Reliability

The probability of success, $R(t)$, and the probability of failure, $Q(t)$, are mutually exclusive. Thus it is obvious that:

$$R(t) + Q(t) = 1 \quad \text{(Eq 7)}$$

The relationship is shown schematically in Fig. 1. In order to determine the reliability of a system between times t_1 and t_2, after the start of a "mission" at time $t = 0$, first find the probability of failure in that interval, that is:

$$Q_{t_2-t_1} = \int_{t_1}^{t_2} f(t)dt = Q(t)]_{t_1}^{t_2} = Q(t_2) - Q(t_1) \quad \text{(Eq 8)}$$

The probability of nonfailure in the same interval is:

$$R_{t_2-t_1} = 1 - Q_{t_2-t_1} = 1 - [Q(t_2) - Q(t_1)]$$

$$= 1 - R(t_1) + R(t_2) \quad \text{(Eq 9)}$$

This is not, however, the reliability of the system functioning from t_1 to t_2 because the system may fail in the period between $t = 0$ and $t = t_1$. This is the a priori probability that the system will not fail in the specified interval. This is not identical with the probability that the system will operate in that time interval, that is, the system reliability. System reliability is the probability that it will not fail between $t = 0$ and $t = t_1$ and that it will not fail from $t = t_1$ to $t = t_2$. The true reliability is $R(t_2)$.

Mortality Curve

Consider a population of homogeneous components from which a very large sample is taken and placed in operation at time $T = 0$. (T is age, in contrast with t, commonly used for operational life or mission time.) The population will initially show a high failure rate. This decreases rapidly as shown in Fig. 2. This period of decreasing failure rate is called various names, such as early life period, infant mortality period, and shakedown period. Failure occurs due to design or manufacturing weaknesses, that is, weak or substandard components.

When the substandard components have all failed at age T_E, the failure rate stabilizes at an essentially constant rate. This is known as useful life because the components can be used to greatest advantage. Failures during useful life are known as random, chance, or catastrophic because they occur randomly and unpredictably.

When components reach age T_W, the failure rate again increases. Failures begin to occur from degradation due to aging or wear as the components are approaching their rated life.

Early life can be made as short as desired (even eliminated) by proper design, fabrication and assembly, and/or deliberate burn-in periods. For systems that must operate satisfactorily over extended periods, incidence of wear-out can be postponed almost indefinitely by replacing units as they fail during useful life and replacing each unit (even if it has not failed or given any indication of imminent failure) no later than at the end of useful life.

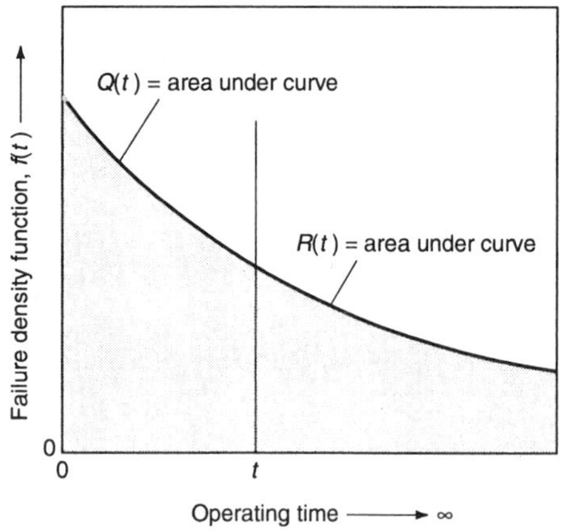

Fig. 1 Schematic representation of relationship between reliability and unreliability. Source: Ref 1

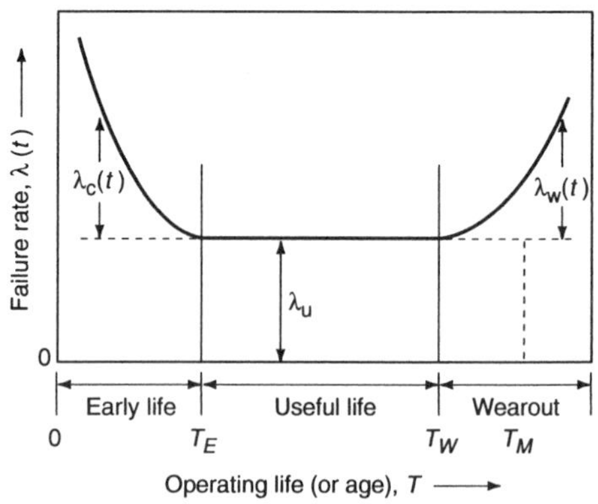

Fig. 2 Mortality curve: failure rate versus age (schematic). Source: Ref 1

Mean Time to Failure

The mean time to failure (MTTF), specifically the mean time to *first* failure, is:

$$\text{MTTF} = \int_0^\infty R(t)dt = \int_0^\infty \left\{\exp\left[-\int_0^t \lambda(\tau)d\tau\right]\right\} dt \quad \text{(Eq 10)}$$

For the special case of useful life with constant failure rate:

$$\text{MTTF} = \int_0^\infty e^{-\lambda t} dt = \frac{1}{\lambda} \quad \text{(Eq 10a)}$$

Strictly speaking, mean time to failure should be used in the case of simple components that are not repaired when they fail but are replaced by good components.

Mean Time between Failures

As indicated MTTF is the mean time to first failure. Mean time between failures (MTBF) is the mean time between two successive component failures. These generally will not be failures of identical components. The relationship between MTBF and MTTF of the individual components is:

$$\frac{1}{\text{MTBF}} = \sum_{j=1}^{m} \frac{1}{(\text{MTTF})_j} \quad \text{(Eq 11)}$$

where the system has m components, all of different ages, each of which is immediately replaced on failure, and MTTF_j is the MTTF of the jth component.

Obviously MTBF is a function of time. When a system is first operated with all new components, the MTTF and the MTBF are identical. After that, the MTBF will fluctuate until, after many failures and replacements, it stabilizes at the value given by Eq 11. For the special case of useful life, both MTTF and MTBF have the same formula given by Eq 10(a). The mathematical equality contributes to confusion between the two despite the clear difference in meaning.

Useful Life

The useful life period is one in which there is constant failure rate (Fig. 2). Equation 6(a) applies, thus:

$$R(t) = \exp\left(-\int_0^t \lambda\, dt\right) = e^{-\lambda t} \quad \text{(Eq 12)}$$

Within useful life, the reliability of a device is the same for all operating periods of the same length; that is, the device is always as "good as new" since the failure rate is constant. In other words, the probability of having a random failure is the same for periods of equal length throughout the entire useful life as failure rate is not affected by time or aging.

The useful life MTBF is often much, much greater than the mean wear-out life T_M. If the failure rate is very small, the MTBF can be many thousands of hours. If a device, however, has an MTBF of 10^6 h, it does not mean the device can be used for 10^6 h. For an operating time equal to the MTBF, there is only about a 37% chance of failureless operation, that is, a reliability of 37%. For MTBF/10, the reliability is about 90%, while

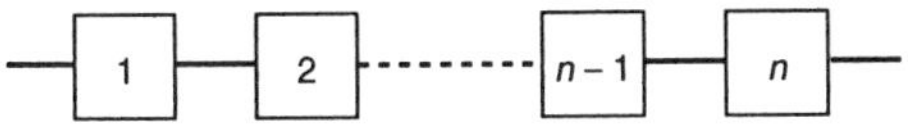

Fig. 3 Schematic system with units in series. Source: Ref 1

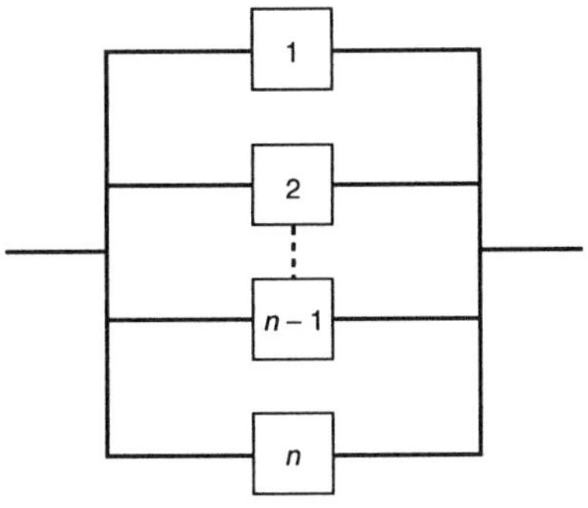

Fig. 4 Schematic system with units in parallel. Source: Ref 1

it is about 95% for MTBF/20 and about 99% for MTBF/100.

Example 1: Estimating Failure and Reliability for a Device. If a device has a failure rate of 0.5×10^{-5} failures/h = 5×10^{-6} failures/h = 0.5%/1000 h, what is its reliability for an operating period of 100 h? If there are 10,000 items in the test, how many failures are expected in 100 h?

$$R(t) = R(100) = e^{-0.5 \times 10^{-5} \times 10^2} = e^{-0.0005}$$
$$R(100) \cong 1 - 0.0005 = 0.9995$$
$$N_s = N_0\, R(t) = 10{,}000 \times 0.9995 = 9995$$
$$N_f = N_0 - N_s = 5 \text{ failures expected}$$

MTTF and MTBF are equal to each other and to $1/\lambda$. For the above device, MTBF is 200,000 h. What is the reliability for an operating time equal to the MTBF? For operating time t = MTBF:

$$R(t) = e^{-t/t} = e^{-1} = 0.3679$$

If the useful life is 100,000 h, what is the reliability for operating over its useful life?

$$R(t) = e^{-0.5 \times 10^{-5} \times 10^5} = e^{-0.5} = 0.6065$$

What is the probability for its surviving the first 100 h? If the device has survived for 99,900 h, what is the probability for its surviving the next 100 h? In both cases, $R(t) = 0.9995$, which is the reliability for any 100 h operating period during useful life.

Series and Parallel Systems

Many systems consist of various components that can be considered to be in series, in parallel, or in combinations of series and parallel. In a series system, all components are so interrelated that the entire system will fail if any one component fails. A parallel system will fail only if all components fail.

Figure 3 shows a schematic series system. Assuming the n components are independent and the system will function for time t if, and only if,

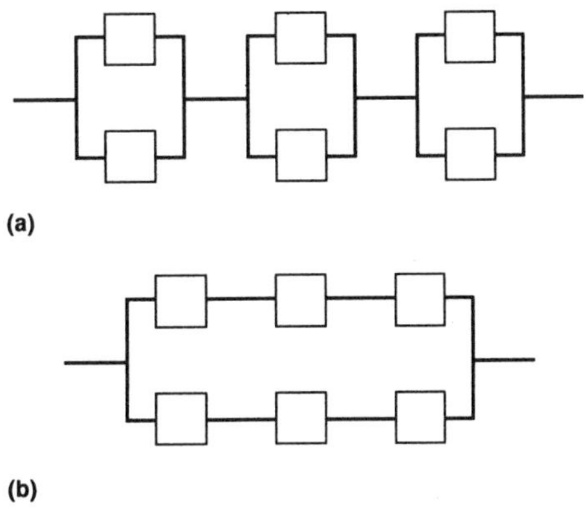

Fig. 5 Configurations of (a) series-parallel and (b) parallel-series systems (logic diagrams). Source: Ref 1

each of the components functions to time t, then the system reliability is:

$$R_{SS}(t) = R_1(t)R_2(t) \ldots R_n(t) = \prod_{i=1}^{n} R_i(t) \qquad \text{(Eq 13)}$$

For useful life with all failure rates constant, the effective failure rate of a system of nonredundant components (not necessarily identical) is equal to the sum of the failure rates of the individual components. The system MTBF can be found by taking the reciprocal of this sum.

A schematic parallel system is shown in Fig. 4. Assuming the n components are independent and the system will function for time t if any one of the components functions (system fails only if all components fail), then the system reliability is:

$$R_{SP}(t) = 1 - \prod_{i=1}^{n} Q_i(t) = \prod_{i=1}^{n} [1 - R_i(t)] \qquad \text{(Eq 14)}$$

The MTBF of a parallel system can be evaluated by using Eq 11. *Reliability in a parallel system cannot be determined by simple substitution of MTBF in Eq 12.* In a parallel system, reliability increases as the number of possible paths (or routes) increases. The rate of increase decreases as the number of paths (n) increases. The greatest gain in reliability comes when a second path is added to a single path.

Complete systems generally consist of several components or units in series. Some of these units may not have sufficient reliability to give the desired system reliability and thus may have parallel units to increase system reliability. This implies use of series-parallel and/or parallel-series arrangements such as shown in Fig. 5. The reliability of a system with n series units with m parallel elements (reliability of a single element is p) in each unit is:

$$\text{series-parallel } R = [1 - (1 - p)^m]^n \qquad \text{(Eq 15)}$$

The reliability of a system with m parallel paths of n elements each (reliability of a single element is p) is:

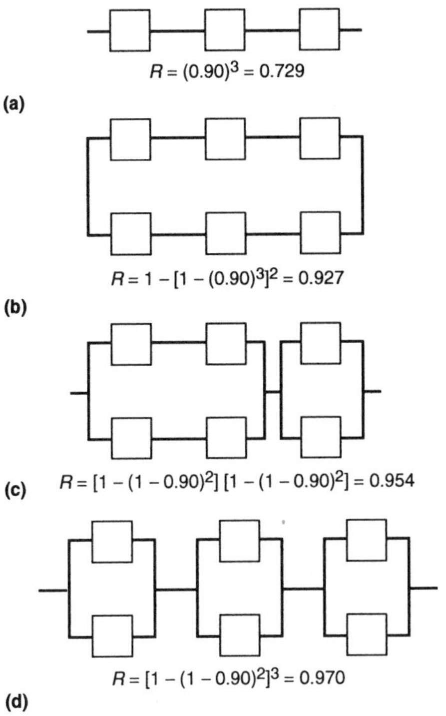

Fig. 6 Reliability calculations for four different arrangements of a three-element system, each element having a reliability of 0.90. (a) Series arrangement. (b) Parallel-series arrangement. (c) Mixed-parallel arrangement. (d) Series-parallel arrangement

$$\text{parallel-series } R = 1 - (1 - p^n)^m \qquad \text{(Eq 16)}$$

Series-parallel arrangements give higher reliability than equivalent parallel-series arrangements. A basic n-element system replicated m times has m^n possible paths for system success for the series-parallel system; there are only m possible paths in the parallel-series system. It follows that the lower the system level at which redundancy is introduced, the more effective redundancy is in increasing reliability. It is also true that as the unit reliability (p) increases, the difference between series-parallel and parallel-series reliabilities decreases.

Example 2: Calculating Reliabilities for Series and Parallel Arrangements of a System. A three-element system, each element having a reliability of 0.90, was considered. Figure 6 shows system reliabilities calculated for four different arrangements of the system.

Logic Diagrams

Use of reliability block diagrams is very helpful when solving reliability problems. A block diagram (or logic diagram) shows the functional relationships among system elements. A schematic physical diagram shows the physical relationships. The logic diagram consists of groups of blocks, signifying system elements, connected together in a manner that traces the system function. In some cases, the two diagrams are identical, but they differ in others.

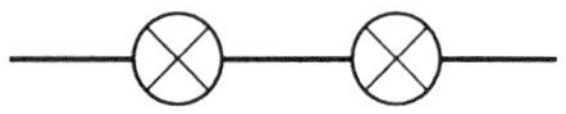

Fig. 7 Schematic diagram of a pipeline with two valves in the line. Source: Ref 1

Example 3: Creating a Logic Diagram for Pipeline Valve Operation. Figure 7 shows the physical diagram of two valves in a pipeline. Should the logic diagram appear as a series or as a parallel system? The answer depends on the criterion of adequate performance. If the two valves are normally shut but expected to open on command to provide flow, this is a series system. If the two valves are normally open but expected to shut on command to stop flow, this is a parallel system.

More Complex Parallel Systems

Equation 14 is based on the criterion for system success requiring successful functioning of only one of the n components in the system. In many systems, there will be a requirement for successful functioning of more than one of the n components. Reliability can be calculated using the binomial distribution. This assumes each component functions properly (p) or it fails (q). From this, the probability of an event not occurring k times in n trials (if P is the probability of the event in each trial), or the probability of the event occurring ($n - k$) times in n trials is:

$$P(\text{exactly } k \text{ failures})\ P_n(k) = C_k^n p^{n-k}(1-p)^k \qquad \text{(Eq 17)}$$

where

$$C_k^n = \frac{n!}{k!\,(n-k)!}$$

The probability of k or fewer failures is:

$$P(k \text{ or fewer failures}) = \sum_{k=0}^{k} P_n(k) \qquad \text{(Eq 18)}$$

Use of the binomial distribution is limited to samples of known size (n is known) with a count of the number of times a certain event is observed.

If the sample size is indefinite, that is, n is not known, the Poisson distribution can be used. In this case, the probability of exactly k failures occurring in time t is:

$$\frac{e^{-\lambda t}(\lambda t)^k}{k!} \qquad \text{(Eq 19)}$$

The probability of k or fewer failures is:

$$\sum_{k=0}^{k} \frac{e^{-\lambda t}(\lambda t)^k}{k!} \qquad \text{(Eq 20)}$$

The Poisson distribution can be used to determine the probability of a specified number of failures in a given mission or to calculate the number of spares required when units operate in useful life.

The binomial distribution, as noted above, applies only to "go-no-go" cases. This is not adequate when a unit can function in more than two ways, that is, function properly and malfunction in two (or more) ways. Such cases can be handled using the multinomial distribution, which is essentially a generalization of the binomial distribution.

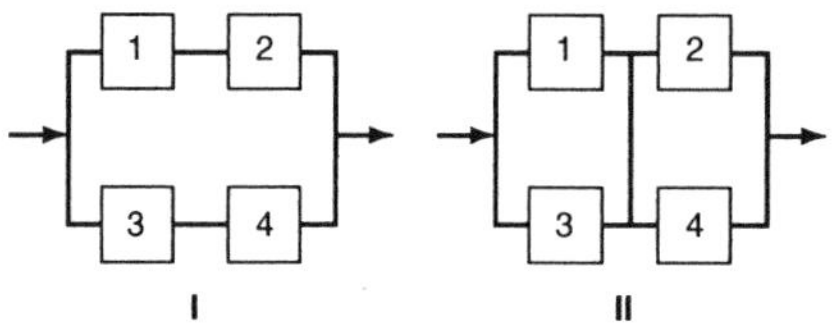

Fig. 8 Diode configurations. See text for discussion.

Example 4: Calculating Reliabilities for Different Arrangements of a Complex System. A diode that can function properly but can malfunction by short circuiting or by open circuiting was considered with these probabilities: p_n is probability of proper operation, p_s is probability of short circuiting, and p_o is probability of open circuiting. (Normally p_n is very close to 1, and p_s and p_o are very small.) The diode was certain to do one of these three; therefore, $p_n + p_s + p_o = 1$.

Four identical diodes were considered for improving the reliability. These were arranged in two possible configurations as shown in Fig. 8. What was the improvement in using a diode quad? (Criterion for system success is flow of a controlled quantity of electrical current.)

Applying the multinomial theorem and expanding:

$$\begin{aligned}(p_n + p_s + p_o)^4 &= p_n^4 + 4p_n^3 p_s + 4p_n^3 p_o \\ &+ 6p_n^2 p_s^2 + 12p_n^2 p_s p_o + 6p_n^2 p_o^2 + 4p_n p_s^3 \\ &+ 12p_n p_s^2 p_o + 12p_n p_s p_o^2 + 4p_n p_o^3 + p_s^4 + 4p_s^3 p_o \\ &+ 6p_s^2 p_o^2 + 4p_s^3 p_o + p_o^4 = 1 + 12p_n p_s^2 p_o\end{aligned}$$

This expansion automatically gives all possible permutations of the four diodes among the three possible modes of function, independently of configuration. The next question was that of satisfactory function of a configuration, each of which must be considered separately. The expansion was examined term by term considering configuration I:

- p_n^4 is all normal. System is a success.
- $4p_n^3 p_s$ is three normal and one short circuited. System is a success. Any single diode can be short circuited with quad operating properly. All four terms apply.
- $4p_n^3 p_o$ is three normal and one open circuited. System is a success. If only one is open circuited, the path containing it is out, but the other path will function properly. All four terms apply.
- $6p_n^2 p_s^2$ is two normal and two short circuited. If diodes 1 and 2 or 3 and 4 are short circuited, there is a direct short circuit, and the quad cannot function. If there is one short circuit in each branch (there are four ways of doing this: 1–3, 1–4, 2–3, and 2–4), then the quad can function. Therefore, four terms are successes, and two terms are failures.

Each remaining term of the expansion was examined, and the following table was developed for configuration I.

Success	Failure
p_n^4	
$4p_n^3 p_s$	
$4p_n^3 p_o$	
$4p_n^2 p_s^2 2$	$2p_n^2 p_s^2$
$12p_n^2 p_s p_o$	
$2p_n^2 p_o^2$	$4p_n^2 p_o^2$
	$4p_n p_s^3$
$8p_n p_s^2 p_o$	$4p_n p_s^2 p_o$
$4p_n p_s p_o^2$	$8p_n p_s p_o^2$
	$4p_n p_o^3$
	p_s^4
	$4p_s^3 p_o$
	$6p_s^2 p_o^2$
	$4p_s p_o^3$
	p_o^4

$$R = \text{system reliability} = \text{probability of quad success}$$

$$Q = \text{system unreliability} = \text{probability of quad failure}$$

$$\begin{aligned}R &= p_n^4 + 4p_n^3 p_s + 4p_n^3 p_o + 12p_n^2 p_s p_o \\ &+ 4p_n^2 p_s^2 + 2p_n^2 p_o^2 + 8p_n p_s^2 p_o + 4p_n p_s p_o^2\end{aligned}$$

$$\begin{aligned}Q &= 2p_n^2 p_s^2 + 4p_n^2 p_o^2 + 4p_n p_s^3 + 4p_n p_s^2 p_o + 8p_n p_s p_o^2 \\ &+ 4p_n p_o^3 + p_s^4 + 4p_s^3 p_o + 6p_s^2 p_o^2 + 4p_s p_o^3 + p_o^4\end{aligned}$$

Eliminating p_n by substitution, that is, $p_n = (1 - p_s - p_o)$ and collecting terms:

$$Q = 2p_s^2 + 4p_o^2 - 4p_o^3 + p_o^4 - p_s^4$$

If p_s and p_o are small, the last three terms can be neglected. Then:

$$Q \cong 2p_s^2 + 4p_o^2$$

The improvement factor is:

$$G_I \equiv \frac{Q \text{ for single}}{Q \text{ for quad}} = \frac{p_s + p_o}{2p_s^2 + 4p_o^2}$$

If $p_s = p_o = 0.01$, then $G_I = 33.33$.

In the above example, quad I was analyzed. What are the reliability and improvement factors if quad II is used? Under what conditions, if any, is quad I preferable? Under what conditions, if any, is quad II preferable?

An analysis for quad II (similar to that for quad I) gives:

$$\begin{aligned}R &= p_n^4 + 4p_n^3 p_s + 4p_n^3 p_o + 2p_n^2 p_s^2 + 12p_n^2 p_s p_o \\ &+ 4p_n^2 p_o^2 + 4p_n p_o p_s^2 + 8p_n p_s p_o^2\end{aligned}$$

$$Q \cong 4p_s^2 + 2p_o^2$$

The improvement factor is:

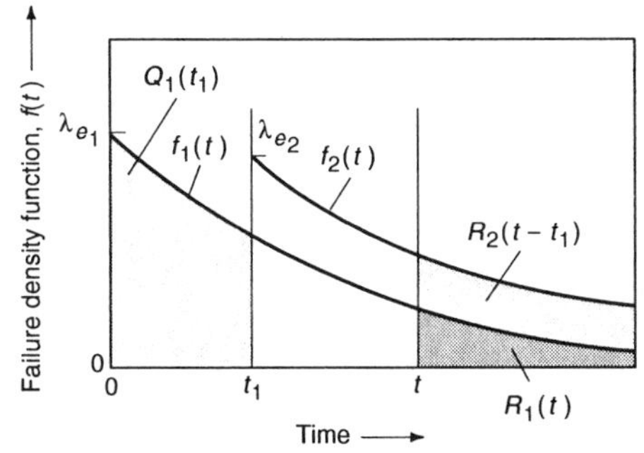

Fig. 9 Graphical representation of terms in Eq 21 in useful life. Source: Ref 1

$$G_{II} = \frac{p_s + p_o}{4p_s^2 + 2p_o^2}$$

if $p_s = p_o$, then there is no difference. If $p_s > p_o$, quad I is preferred. If $p_o > p_s$, quad II is preferred.

Standby Systems

All the previous discussions assumed that the reliabilities of components in a system were independent. This is not always true. For example, if four generators are required to provide the required output, but five are operating in parallel, none will be working at full capacity. Thus they will have a lower failure rate than if operating at full capacity. Reliabilities of individual units in a parallel system, therefore, are generally not independent. When requirements for high reliability make redundancy necessary, a good arrangement has one unit operating until it fails. At that time, a second unit, which has been idly standing by, is switched into the system, commonly by a sensing and switching subsystem.

For a two-unit standby system, the reliability is the probability that unit 1 succeeds for the entire period (t) or that unit 1 fails at some time t_1 prior to t *and* the sensing and switching unit does not fail by t_1 and standby unit 2 does not fail by t_1 *and* successfully functions for the rest of time t. Assuming 100% reliability of the sensing and switching system and unit 2 while idling, the reliability can be written as:

$$R_{SB}(t) = R_1(t) + Q_1(t_1)R_2(t - t_1) \qquad \text{(Eq 21)}$$

where $t_1 < t$. This is shown graphically in Fig. 9. For more complex systems, the above process is expanded, making sure all terms are included.

Example 5: Effect of Standby Arrangements on Reliability. The reliabilities of the four simple systems shown in Fig. 10 were considered. The operating period in each case is 200 h, and the unit failure rate is 2.5×10^{-4} failures/h:

$$\lambda t = (2.5 \times 10^{-4})(200) = 0.050$$

$$\text{Individual unit reliability} = e^{-0.050}$$

$$R = 0.95123$$

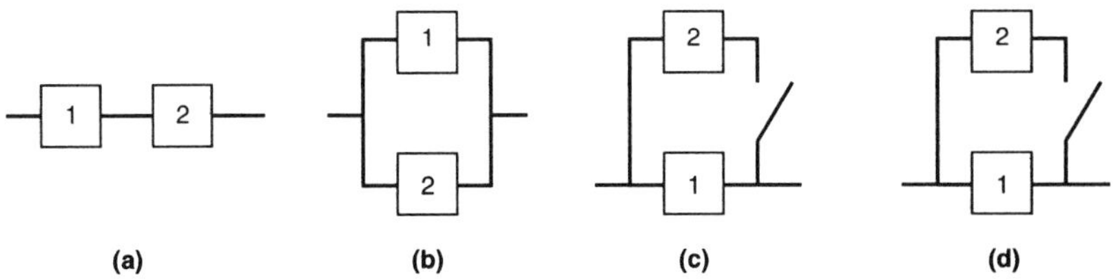

Fig. 10 System configurations for Example 5. See text for discussion.

For the system shown in Fig. 10(a):

$$R_{sys} = (e^{-\lambda t})^2 = 0.90484$$

For the system shown in Fig. 10(b):

$$R_{sys} = 1 - (1 - e^{-\lambda t})^2 = 0.99762$$

For the system shown in Fig. 10(c), the reliability of the sensing switch unit is assumed to be 100%:

$$R_{sys} = e^{-\lambda t} + (\lambda t)e^{-\lambda t} = 0.99879$$

For the system shown in Fig. 10(d), the sensing element failure rate is $\lambda_{se} = 3 \times 10^{-5}$ failures/h, and the switch failure rate is $\lambda_{sw} = 10^{-5}$ failures/cycle:

$$R_{sys} = e^{-\lambda_1 t}$$

$$+ \frac{\lambda_1 e^{-(\lambda_{sw})(1)} e^{-\lambda_2 t}}{\lambda_1 + \lambda_{se} - \lambda_2} [1 - e^{-(\lambda_1 + \lambda_{se} - \lambda_2)t}] = 0.99879$$

If operating time is increased from 200 h, all the calculated reliabilities will decrease.

The calculated reliability of each of the two standby arrangements is slightly better than the arrangement with two units in parallel. This is often the situation when comparing standby arrangements with parallel arrangements. The actuality, however, could be somewhat different. In the parallel arrangement, the two units operate at partial load while the operating unit in either standby arrangement operates at full load.

Also, both standby arrangements have the same calculated reliability. If the failure rates of the sensing and switching units were greater, there would be a difference in the reliability. In general, inclusion of failure rates for sensing and switching is more realistic.

Calculation of reliabilities to five significant figures is questionable when the given data have no more than two significant figures. The use of the greater number of significant figures is for comparison purposes.

Conditional Probability

Many system configurations cannot be reduced to simple series, parallel, standby models, or obvious mixtures of these. In this situation, reliability can be calculated using conditional probabilities. Use of Bayes' theorem requires identification of a keynote component, that is, a component that enhances reliability by its addition but still lets the system operate if it fails. Bayes' theorem then says: The probability of system function is the probability that the system functions given the keynote component is good, multiplied by the probability the keynote component is good *plus* the probability the system functions if the keynote component is bad, multiplied by the probability that the keynote component is bad. In this context, Bayes' theorem can be very helpful (Ref 1-3).

A combination of Smith (Ref 1) and O'Connor (Ref 4, 5, or 6) provides an excellent starting point for anyone who needs a good comprehensive introduction to reliability. Neither book alone is really adequate.

Probabilistic versus Deterministic Design

There has been recognition over the years (however implicit) that demands and capacities do have variations. The statistical nature of design parameters has generally been ignored as shown by efforts to find unique representative values such as minimum guaranteed values, limit loads, etc. The usual approach is to use a safety factor or a worse-case approach. The results have not necessarily been bad. In fact, the most common result has been overdesign, that is, conservative design. It is impossible, however, to determine readily, or even estimate, how conservative the design may be. Nor can marginal or underdesigned situations be identified.

A different approach is to recognize the variations of demand and capacity as indicated schematically in Fig. 11. For this situation, there is a safety factor, based on the mean values. Failure probability is indicated by the extent of overlap of the demand and capability distributions. Using the small standard deviations (A) in Fig. 11, there is very little overlap and thus a very small probability of failure. Using the large standard deviations (C) in Fig. 11, there is much overlap and thus a rather large probability of failure, that is, when a component with low capacity is subjected to a large demand.

Example 6: Probabilistic and Deterministic Design Calculations for a Simple Thin-Wall Cylinder under Internal Pressure (Ref 7). Some data and results for the cylinder are given in Table 1. The safety factor in this example is determined as the ratio of mean yield stress to mean tangential stress. The minimum safety factor is determined as the ratio of minimum yield stress (mean minus 3 standard deviations) to maximum demand (mean plus 3 standard deviations). Cases 1 and 2, using one material but with different standard deviations in applied pressure,

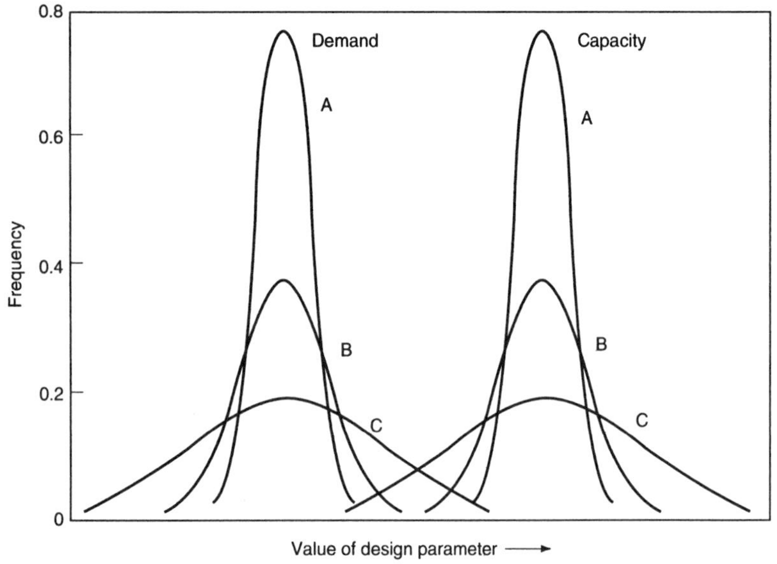

Fig. 11 Schematic distribution of demand (all with same mean values) and capacity (all with same mean values) for three situations: A, small standard deviations; B, moderate standard deviations; and C, large standard deviations. Source: Ref 7

Table 1 Deterministic versus probabilistic calculations for a thin-wall cylinder under internal pressure

See Example 6 in text for discussion.

Case No.	Pressure, MPa	Standard deviation pressure, MPa	Mean yield stress, MPa	Standard deviation yield stress, MPa	Safety factor	Minimum safety factor	Reliability	Predicted No. of failures
1	16.0	0.80	310.0	33.0	1.61	1.85	0.999298	7/10⁴
2	16.0	1.60	310.0	33.0	1.61	2.01	0.998204	18/10⁴
3	16.0	0.80	340.0	21.0	1.77	1.54	>0.9999997	<3/10⁷
4	16.0	1.60	340.0	21.0	1.77	1.68	0.9999987	13/10⁷

Source: Ref 7

have the same safety factor. Case 1 predicts 7 failures in 10^4 while case 2 predicts 18 in 10^4. Cases 3 and 4, with a different material, have the same safety factor. Case 3 predicts less than 3 failures in 10^7 while case 4 predicts 13 in 10^7.

Cases 3 and 4 have a material with a greater mean strength than cases 1 and 2. (These materials values are from actual materials.) The nominal safety factor is increased about 10%. The stronger material, however, has a smaller standard deviation, which results in predictions of far fewer failures, actually by a factor of 1000. This prediction is not remotely indicated by the 10% increase in safety factor.

In cases 1 and 2, the nominal safety factor is less than the minimum safety factor (approximately worst-case treatment). In cases 3 and 4, the nominal safety factor is greater than the minimum safety factor. By inference, comparatively more failures should be expected in cases 1 and 2 than in cases 3 and 4. The probabilistic calculations, however, predict how many failures should be expected in each case. A minimum safety factor cannot be determined without knowledge of ranges of design parameters. If the standard deviations are known, a quantitative prediction can be made of the number of failures instead of relying on an extremely crude prediction from safety factors.

Failure probabilities of 10^{-5}, 10^{-6}, and 10^{-7} may have no real meaning because they are far out of the range that can be verified by experiment or field practice unless one is working with a large number of items. While the lack of verification is real (for small numbers of large items), the probabilistic approach at least provides a scale against which safety of different designs can be compared on a rational basis. Using a safety factor provides no such scale.

Choosing between Probabilistic and Deterministic Approaches. Whether using a probabilistic or deterministic approach, the designer must make a design decision. The deterministic approach can be used with a decision made on the safety factor selected. The probabilistic approach can be used with a decision made on selecting an acceptable number of failures. The latter approach seems more rational.

There is much in the literature regarding the relative merits of probabilistic versus deterministic design. Some representative papers include Smith (Ref 8-12) and Mischke (Ref 13, 14). O'Connor (Ref 5, 6) has a chapter on load-strength interference. In making calculations, common stress (and/or strain) deterministic relationships can be converted to probabilistic relationships by using the "algebra of normal distribution functions" as given in Haugen (Ref 15).

Distributions

There are at least 20 different probability distributions, several of which are very useful in reliability theory and application. Useful life, noted above, uses the (negative) exponential distribution, which has a single parameter. Also as noted above, the binomial and Poisson distributions are very useful in certain situations. The Poisson distribution has a single parameter while the binomial distribution has two parameters.

The Weibull distribution was developed in the late 1940s. It applies to a large number of diverse situations and has found extensive use in general engineering practice. There are three parameters, which are usually determined experimentally from data obtained during testing. All three parameters have physical significance: (a) a lower bound of the variable in question; (b) a characteristic value, related to the median, thus an indicator of central tendency, and (c) Weibull slope, a direct function of the uniformity of the item; the larger the slope, the greater the uniformity. When the Weibull slope is 1, this is a special case, becoming the (negative) exponential distribution. When the Weibull slope is about 3.5, the Weibull distribution is a good approximation to the normal distribution. The Weibull distribution is very good for use with material and mechanical failures and with wear-out (Fig. 2).

Additional information relative to distributions can be found in the section "Statistical Distributions Applied to Design" in the article "Statistical Aspects of Design" in this Volume.

Reliability Testing

Reliability tests measure or demonstrate the ability of equipment to operate satisfactorily for a prescribed time period under specified operating conditions. Assuming a system with ten components in series, which must operate for 1000 h with a reliability of 0.99, then the characteristic life of a component must be about 10^6 h, or about 115 years. To obtain a state of knowledge to determine this requires *more* than 115 years of *failure-free* testing (115 components for 1 year or 1150 components for 6 weeks) for each of 10 components. If the number of parts is several hundred and the desired reliability is 0.999 (not at all uncommon), it is obvious the required number of tests becomes absurdly impossible. Yet, such products are designed, built, and meet guaranteed performance.

Although maximum use is made of past performance data and engineering judgment, it is essential for the designer to know as soon as possible if a design objective for reliability and/or life will be met. To meet a production schedule, it is rarely realistic to test under normal operating conditions. Thus accelerated testing is desired. Three factors control the degree of acceleration: environment, sample size, and testing time. Environment includes any operating condition to which the part is subjected and that may affect its performance and/or durability. Two broad categories of testing are implied: (a) tests to determine which aspects of the environment are truly significant and (b) tests to determine performance as quickly as possible.

Tests for Environmental Factors. Univariate testing (one parameter at a time) is not always useful because there is no indication of any effect of interaction between two or more factors.

Factorial experiments can be used to estimate the main (direct) effect of each "stress," interactions between two (or more) stresses, and experimental error (Ref 16-20). Related information is contained in the articles "Statistical Aspects of Design" and "Robust Design" in this Volume.

To run a complete factorial experiment, especially if one wishes to replicate it, may require a very large number of individual tests or runs. Fortunately, in such a situation, most of the desired information can be obtained by performing only a fraction of the full factorial. In one unusual (and outstanding) situation with 19 variables, a complete full factorial study would have required more than 500,000 samples. A judiciously selected fractional factorial design provided the desired information from a study of only 20 samples (Ref 16-21).

Response surface methodology (RSM) consists of a group of techniques (based on experimental effort as indicated above) for empirical study of relationships between one response (or more responses) and a number of input variables (Ref 16, 22-24). The technique uses a method of steepest ascent to get close to a maximum and then approximates the surface in the vicinity of the maximum with a fitted second-degree equation.

Evolutionary operation (EVOP) is a mode of using a full-scale process so that information on how to improve the process is generated from a simple experiment while the process continues (Ref 16, 25, 26). Evolutionary operation has less applicability in determining effects of environment in reliability than other types of statistical testing, but in some situations it can be of major importance.

Accelerated testing, which is reducing the time required for testing, can be accomplished by: (a) taking a large sample and testing only part of the sample to failure; (b) magnifying the stress; or (c) sudden-death testing. Alternatives (a) and (c) are especially useful in testing a large number of relatively inexpensive components. Sequential testing is more useful when there are only a few relatively expensive units.

If a number of identical units are tested simultaneously, failure will occur in an ordered sequence with the weakest failing first, etc. Such ordering is unique with life testing (Ref 27). The choice of a test involving failure of r units out of n ($n > r$) rather than the choice of a test involving failure of r out of r units will, in general, permit determination of an estimate of characteristic life in a relatively shorter time. For example, if 20 units have to be operated for 200 h to induce all 20 failures, then on the average, only 46 h are needed to observe the first 10 failures in a sample of 20.

Use of magnified loading does reduce testing time and possibly the number of items required for the test. Correlation is a major problem because "normal" must be defined and enough overload data must exist to correlate with normal. Rabinowicz et al. (Ref 28) suggested a technique based on cumulative damage. Test procedure is to run a series of tests to failure at a high stress (S_1) and a second set of tests partly at S_1 and partly at a lower stress (S_2). These two points are plotted with time at S_1 as the ordinate and time at S_2 as the abscissa. A straight line extrapolation between the two points gives the life expected at S_2, which presumably would be the normal operating stress. Caution should be exercised as this is similar to the Palmgren-Miner rule in fatigue failures, which is recognized as a crude approximation in many cases.

Abu Haraz and Ermer (Ref 29) used the cumulative damage theory with some success. They also were successful in using the Arrhenius equation in accelerated testing in a case where the physical law leading to failure was known.

An excellent example of acceleration in fatigue testing was developed by Conover et al. (Ref 30). Intensified loading of programmed fatigue tests developed from field data was used to reduce time to failure in steels for automotive components by a factor of 10. These results were applicable for unnotched specimens, notched specimens, and automotive components.

Sudden-death testing (Ref 1, 17) is especially useful with a large number of relatively inexpensive units. For example, if 50 specimens are available for testing, these could be randomly divided into ten sets of five each. All five are tested simultaneously until one item in the set fails. The rest of the set is removed from the test. Testing in this manner continues until all ten sets have been tested. Then there are 10 numbers, each of which represents the least value in a set of five. These also represent separate estimates of the life of the weakest 12.94% of the population. A Weibull plot gives a straight line describing the distribution of the 12.94% life. The median 50% life point is located, allowing the 12.94% life point to be found on the population line. The population line is drawn through this second point parallel to the first line. (The slopes of the sudden death and population lines are assumed to be identical.) Characteristic life of the population is associated with the 63.2% point on the population line. The use of 50 specimens divided into 10 sets of 5 each is arbitrary for discussion. Any number of specimens can be used divided in whatever manner is desired.

Sequential testing (Ref 1, 17) is especially useful for tests of a small number of relatively expensive units. Specimens are tested in sequence with the test result reviewed after each test is completed. The decision is then made to accept, reject, or continue sampling. This decision is based on predetermined criteria. The sample size is established only after an "accept" or "reject" decision has been made, although an average sample size can be computed once the criteria have been established.

Human Reliability

Human reliability is defined as the probability that a task will be successfully completed by personnel at any required stage in system operation within a required minimum time (if, indeed, a time requirement exists). Successful performance, however, is not errorless performance. It is performance without errors that have a significant effect on correct completion of the task. Related information can be found in the article "Human Factors in Design" in this Volume. Still further information can be found in Ref 31-34.

Availability and Maintainability

The focus of this article is reliability. Other aspects ultimately relate to reliability and operation of reliable systems but cannot be included in this article. These include availability and maintainability.

Availability. Dovich (Ref 2) defines this as a measure of the degree to which an item is in the operable and committable state at the start of the mission when the mission is called for at an unknown (random) time. O'Connor (Ref 6) defines this as the probability that an item will be available when required *or* the proportion of total time that the item is available for use. Total time includes operating time, active repair time, logistic time, and administrative time.

Maintainability. Dovich (Ref 2) defines this as a measure of the ability of an item to be retained in, or restored to, specified condition when maintenance is performed by personnel having specified skill levels, using prescribed procedures and resources, at each prescribed level of maintenance and repair. This also could be stated as a probability.

Reliability Growth

It is common for new products to be less reliable during the early stages of development than later when improvements have been made as a result of testing and/or failures observed and corrected. In addition, items in service often show reliability growth. It is appropriate for designers to track and/or encourage this growth. This was first analyzed by Duane (Ref 35), who observed an improvement of MTBF of a range of items used on aircraft. Duane found that the cumulative MTBF (θ_c), which is total time divided by total failures, gave a straight line when plotted against total time on log-log paper. The slope (α) gave an indication of reliability growth:

$$\log \theta_c = \log \theta_0 + \alpha (\log T - \log T_0)$$

where θ_0 is the cumulative MTBF at the start of the monitoring period T_0. Therefore:

$$\theta_c = \theta_0 \left(\frac{T}{T_0}\right)^{\alpha} \quad \text{(Eq 23)}$$

The slope α gives an indication of the rate of MTBF growth and thus of the effectiveness of the reliability program in correcting failure modes. Duane (Ref 35) observed that α typically ranges between 0.2 and 0.4 and correlates with the intensity of effort on improvement with higher numbers indicating greater intensity. O'Connor (Ref 6) provides a good discussion with an example. Spradlin (Ref 36) provides an excellent example of using the Duane method to improve reliability. Other examples are provided in Ref 37-45.

Conclusions

Determining the reliability of a complex system can be difficult. In principle, proceeding methodically by starting with the simplest units, combining them into subsystems, and then combining the subsystems into the complete system, and determining reliability at each step will lead to the final system reliability. A most important aspect is establishing the criterion of adequate performance of the system. Although difficult, reliability of a system can be established; it is done regularly.

REFERENCES

1. C.O. Smith, *Introduction to Reliability in Design*, McGraw-Hill Publishing Co., 1976
2. R.A. Dovich, *Reliability Statistics*, ASQC Quality Press, 1990
3. H.E. Martz and R.A. Walker, *Bayesian Reliability Analysis*, John Wiley & Sons, Inc., 1982, Reprint, Krieger, 1991
4. P.D.T. O'Connor, *Practical Reliability Engineering*, 3rd ed. rev., John Wiley & Sons, Inc., 1995
5. P.D.T. O'Connor, *Practical Reliability Engineering*, 3rd ed., John Wiley & Sons, Inc., 1991
6. P.D.T. O'Connor, *Practical Reliability Engineering*, 2nd ed., John Wiley & Sons, Inc., 1985
7. C.O. Smith, "Elements of Probabilistic Design," paper presented at International Conference on Engineering Design (ICED), 23–25 Aug 1988 (Budapest, Hungary), available from Heurista (Zurich, Switzerland)
8. C.O. Smith, Design Relationships and Failure Theories in Probabilistic Form, *Nucl. Eng. Des.*, Vol 27, 1974, p 286–292
9. C.O. Smith, "Design of Pressure Vessels to Probabilistic Criteria," Paper M4/3 presented at 1st Intl. Conf. on Structural Mechanics in Reactor Technology, 20–24 Sept 1971 (Berlin, Germany), available from Bundesanstalt für Materialprüfung (BAM) (Berlin, Germany)
10. C.O. Smith, "Design of Rotating Components to Probabilistic Criteria," Paper M5/10 presented at 3rd Intl. Conf. on Structural Mechanics in Reactor Technology, 1–5 Sept 1975 (London, England), available from Bundesanstalt für Materialprüfung (BAM) (Berlin, Germany)
11. C.O. Smith, "Shrink Fit Stresses in Probabilistic Form," ASME Winter Annual Meeting, 10–15 Dec 1978 (San Francisco, CA), *ASME Book No. H00135*, American Society of Mechanical Engineers
12. C.O. Smith, Design of Ellipsoidal and Toroidal Pressure Vessels to Probabilistic Criteria, *J. Mech. Des.*, Vol 102, Oct 1980, p 787–792
13. C.R. Mischke, "A Rationale for Mechanical Design to a Reliability Specification," presented at ASME Design Technology Transfer Conference, 5–9 Oct 1974 (New York, NY)
14. C.R. Mischke, "Implementing Mechanical Design to a Reliability Specification," presented at ASME Design Technology Transfer Conference, 5–9 Oct 1974 (New York, NY)
15. E.B. Haugen, *Probabilistic Mechanical Design*, John Wiley & Sons, Inc., 1980
16. G.E.P. Box, W.G. Hunter, and J.S. Hunter, *Statistics for Experimenters*, John Wiley & Sons, Inc., 1978
17. C. Lipson and N.J. Sheth, *Statistical Design Analysis of Engineering Experiments*, McGraw-Hill Publishing Co., 1973
18. J.D. Hromi, "Some Aspects of Designing Industrial Test Programs," Paper 690022, Society of Automotive Engineers, Jan 1969
19. W.G. Cochran and G.M. Cox, *Experimental Designs*, John Wiley & Sons, Inc., 1950
20. D.R. Cox, *Planning of Experiments*, John Wiley & Sons, Inc., 1958
21. G.E.P. Box and J.S. Hunter, The 2^{k-p} Fractional Factorial Designs, *Technometrics: Part I*, Vol 3 (No. 3), Aug 1961, p 311–351; *Part II*, Vol 3 (No. 4), Nov 1961, p 449–458
22. G.E.P. Box, N.R. Draper, and J.S. Hunter, *Empirical Model-Building and Response Surfaces*, John Wiley & Sons, Inc., 1986
23. W.J. Hill and W.G. Hunter, A Review of Response Surface Methodology: A Literature Survey, *Technometrics*, Vol 8 (No. 4), Nov 1966, p 571–589
24. R. Mead and D.J. Pike, A Review of Response Surface Methodology From a Biometric Point of View, *Biometrics*, Vol 8, 1975, p 803
25. G.E.P. Box and N.R. Draper, *Evolutionary Operation: A Statistical Method for Process Improvement*, John Wiley & Sons, Inc., 1969
26. W.G. Hunter and J.R. Kittrell, "Evolutionary Operation: A Review," *Technometrics*, Vol 8 (No. 3), Aug 1966, p 389–397
27. Epstein and Sobel, Life Testing, *J. American Statistical Association*, Vol 48 (No. 263), Sept 1953
28. E. Rabinowicz, R.H. McEntire, and R. Shiralkar, "A Technique for Accelerated Life Testing," Paper 70-Prod-10, American Society of Mechanical Engineers, April 1970
29. O.B. Abu Haraz and D.S. Ermer, Accelerated Life Tests for Refrigerator Components, *Proceedings, Annual Reliability and Maintainability Symposium*, IEEE, 1980, p 230–234
30. J.C. Conover, H.R. Jaeckel, and W.J. Kippola, "Simulation of Field Loading in Fatigue Testing," Paper 660102, Society of Automotive Engineers, Jan 1966
31. B.A. Sayers, *Human Factors and Decision Making: Their Influence on Safety and Reliability*, Elsevier Science Publishers, 1988
32. K.S. Park, *Human Reliability*, Elsevier Science Publishers, 1987
33. L.S. Mark, J.S. Warren, and R.L. Huston, Ed., *Ergonomics and Human Factors*, Springer-Verlag (New York), 1987
34. B.S. Dhillon, *Human Reliability*, Pergamon Press, 1986
35. J.J. Duane, Learning Curve Approach to Reliability Modeling, *IEEE Transactions, Aerospace*, 2, 1964, p 563–566
36. B.C. Spradlin, Reliability Growth Measurements Applied to ESS, *Annual Reliability and Maintainability Symposium (IEEE)*, Institute of Electrical and Electronics Engineers, 1986, p 97–100
37. E. Demko, True Reliability Growth Measurement, *Annual Reliability and Maintainability Symposium (IEEE)*, Institute of Electrical and Electronics Engineers, 1986, p 92–96
38. J.N. Bower, Reliability Growth During Flight Test, *Annual Reliability and Maintainability Symposium (IEEE)*, Institute of Electrical and Electronics Engineers, 1986, p 101–106
39. C.T. Gray, A Modelling Framework for Reliability Growth, *Annual Reliability and Maintainability Symposium (IEEE)*, Institute of Electrical and Electronics Engineers, 1986, p 107–114
40. L.H. Crow, On the Initial System Reliability, *Annual Reliability and Maintainability Symposium (IEEE)*, Institute of Electrical and Electronics Engineers, 1986, p 115–119
41. J.C. Wronka, Tracking of Reliability Growth in Early Development, *Annual Reliability and Maintainability Symposium (IEEE)*, Institute of Electrical and Electronics Engineers, 1988, p 238–242
42. L.H. Crow, Reliability Growth Estimation With Missing Data—II, *Annual Reliability and Maintainability Symposium (IEEE)*, Institute of Electrical and Electronics Engineers, 1988, p 248–253
43. A.W. Benton and L.H. Crow, Integrated Reliability Growth Testing, *Annual Reliability and Maintainability Symposium (IEEE)*, Institute of Electrical and Electronics Engineers, 1989, p 160–166
44. D.B. Frank, A Corollary to Duane's Postulate on Reliability Growth, *Annual Reliability and Maintainability Symposium (IEEE)*, Institute of Electrical and Electronics Engineers, 1989, p 167–170
45. G.J. Gibson and L.H. Crow, Reliability Fix Effectiveness Factor Estimation, *Annual Reliabil-*

ity and Maintainability Symposium (IEEE), Institute of Electrical and Electronics Engineers, 1989, p 171–177

SELECTED REFERENCES

- W.G. Ireson and G.F. Coombs, Ed., *Handbook of Reliability Engineering and Management*, McGraw-Hill Publishing Co., 1988
- D. Kececioglu, *Reliability and Life Testing Handbook*, Vol 1 and 2, Prentice-Hall, 1993
- D. Kececioglu, *Reliability Handbook*, Vol 1 and 2, Prentice-Hall, 1991
- L.M. Leemis, *Reliability: Probabilistic Models and Statistical Methods*, Prentice-Hall, 1995
- M.O. Locks, *Reliability, Maintainability, and Availability Assessment*, 2nd ed., ASQC, 1995
- *Proceedings, Annual Reliability and Maintainability Symposium*, Institute of Electrical and Electronics Engineers
- P.A. Tobias and D.C. Trindade, *Applied Reliability*, 2nd ed., Van Nostrand Reinhold, 1995

Life-Cycle Engineering and Design

ASM International Materials Life-Cycle Analysis Committee*

ENVIRONMENTAL CONSIDERATIONS play an increasingly important role in design and development efforts of many industries. "Cradle to grave" assessments are being used not only by product designers and manufacturers, but also by product users (and environmentalists) to consider the relative merits of various available products and to improve the environmental acceptability of products.

Life-cycle engineering is a part-, system-, or process-related tool for the investigation of environmental parameters based on technical and economic measures. This article focuses on life-cycle engineering as a method for evaluating impacts, but it should be noted that similar techniques can be used to analyze the life-cycle costs of products (see the article "Techno-Economic Issues in Materials Selection" in this Volume).

Products and services cause different environmental problems during the different stages of their life cycle. Improving the environmental performance of products may require that industry implement engineering, process, and material changes. However a positive change in one environmental aspect of a product (such as recyclability) can influence other aspects negatively (such as energy usage). Therefore a methodology is required to assess trade-offs incurred in making changes. This method is called life-cycle analysis or assessment (LCA).

Life-cycle analysis aims at identifying improvement possibilities of the environmental behavior of systems under consideration by designers and manufacturers. The whole life cycle of a system has to be considered. Therefore it is necessary to systematically collect and interpret material and energy flows for all relevant main and auxiliary processes (Fig. 1).

Life-cycle analysis methods have been developed by governmental, industrial, academic, and environmental professionals in both North America and Europe. Technical documents on conducting LCA have been published by the Society of Environmental Toxicology and Chemistry (SETAC), the U.S. Environmental Protection Agency (EPA), the Canadian Standards Association (CSA), the Society for the Promotion of LCA Development (SPOLD), and various practitioners.

For meaningful comparisons of the life-cycle performance of competing and/or evolving product systems, it is important that associated LCAs be conducted consistently, using the same standards. Although the common methodologies developed by SETAC, EPA, CSA, and SPOLD are a step in that direction, a broad-based international standard is needed. Such an effort is being undertaken by ISO 14000 series (TC207).

Life-cycle thinking and techniques can be applied to products, processes or systems in various ways: it can help assess life-cycle economic costs (LCA_{econ}), social costs (LCA_{soc}) or environmental costs (LCA_{env}).

A primary objective of LCA is to provide a total life-cycle "big-picture" view of the interactions of a human activity (manufacturing of a product) with the environment. Other major goals are to provide greater insight into the overall environmental consequences of industrial activities and to provide decision makers with a quantitative assessment of the environmental consequences of an activity. Such an assessment permits the identification of opportunities for environmental improvement.

Life-Cycle Analysis Process Steps

Life-cycle analysis is a four-step process; each of these steps is described in detail below. The process starts with a definition of the goal and scope of the project; because LCAs usually require extensive resources and time, this first step limits the study to a manageable and practical scope. In the following steps of the study, the environmental burdens (including both consumed energy and resources, as well as generated wastes) associated with a particular product or process are quantitatively inventoried, the environmental impacts of those burdens are assessed, and opportunities to reduce the impacts are identified.

All aspects of the life cycle of the product are considered, including raw-material extraction from the earth, product manufacture, use, recycling, and disposal. In practice, the four steps of an LCA are usually iterative (Fig. 2).

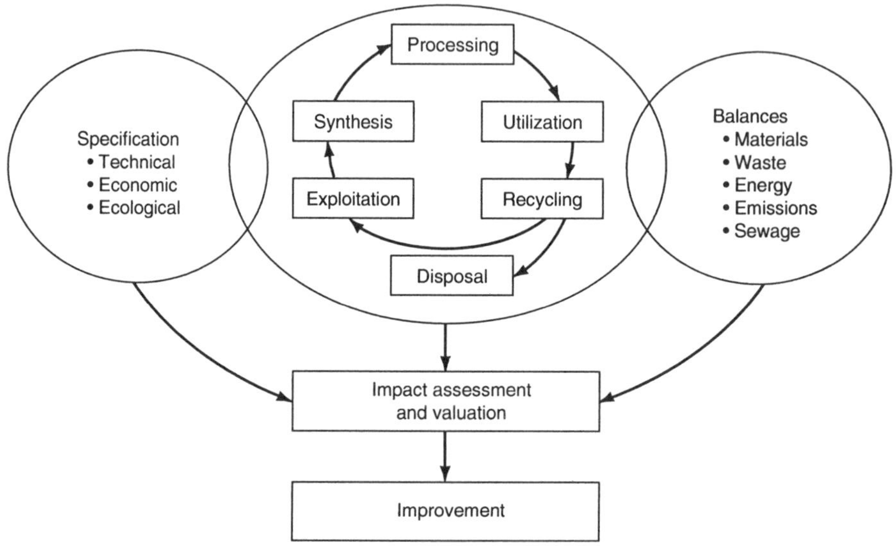

Fig. 1 Factors considered in the life-cycle engineering approach. Source: Ref 1

*This article was prepared by Hans H. Portisch, Krupp VDM Austria GmbH (Committee Chair), with contributions from Steven B. Young, Trent University; John L. Sullivan, Ford Motor Company; Matthias Harsch, Manfred Schuckert, and Peter Eyerer, IKP, University of Stuttgart; and Konrad Saur, PE Product Engineering.

Table 1 Example of a life-cycle inventory for an unspecified product

Substance	Amount	Substance	Amount
Inputs		**Outputs**	
Energy from fuels, MJ		Air emissions, mg	
Coal	2.75	Dust	2,000
Oil	3.07	Carbon monoxide	800
Gas	11.53	Carbon dioxide	11×10^5
Hydro	0.46	Sulfur oxides	7,000
Nuclear	1.53	Nitrogen oxides	11,000
Other	0.14	Hydrogen chloride	60
Total	**19.48**	Hydrogen fluoride	1
Energy from feedstocks, MJ		Hydrocarbons	21,000
Coal	<0.01	Aldehydes	5
Oil	32.75	Other organics	5
Gas	33.59	Metals	1
Other	<0.01	Hydrogen	1
Total feedstock	**66.35**	Solid wastes, mg	
Total energy input, MJ	**85.83**	Mineral waste	3,100
Raw materials, mg		Industrial waste	22,000
Iron ore	200	Slags and ash	7,000
Limestone	150	Toxic chemicals	70
Water	18×10^6	Nontoxic chemicals	2,000
Bauxite	300	Water effluents, mg	
Sodium chloride	7,000	COD	1,000
Clay	20	BOD	150
Ferromanganese	<1	Acid, as H^+	75
		Nitrates	5
		Metals	300
		Ammonium ions	5
		Chloride ions	120
		Dissolved organics	20
		Suspended solids	400
		Oil	100
		Hydrocarbons	100
		Phenol	1
		Dissolved solids	400
		Phosphate	5
		Other nitrogen	10
		Sulfate ions	10

COD, chemical oxygen demand; BOD, bacteriological oxygen demand. Source: Ref 2

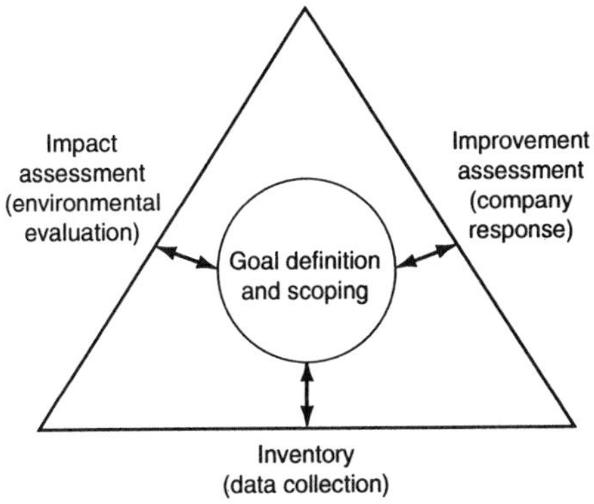

Fig. 2 The life-cycle assessment triangle. Source: Ref 2

Step 1: Goal Definition and Scoping. In the goal definition and scoping stage, the purposes of a study are clearly defined. Subsequently, the scope of the study is developed, which defines the system and its boundaries, the assumptions, and the data requirements needed to satisfy the study purpose. For reasons of economy and brevity, the depth and breadth of the study is adjusted, as required, to address issues regarding the study purpose. Goal definition and project scope may need to be adjusted periodically throughout the course of a study, particularly as the model is refined and data are collected.

Also during this stage, the functional unit is defined. This is an important concept because it defines the performance of a product in measured practical units and acts as a basis for product system analysis and comparison to competing products. For example, the carrying capacity of a grocery bag might be a sensible functional unit.

Finally, the quality of the life-cycle data must be assessed in order to establish their accuracy and reliability. Typically, factors such as data age, content, accuracy, and variation need to be determined. Clearly, data quality affects the level of confidence in decisions that are based on study results.

Step 2: Inventory Analysis. The second stage of LCA is a life-cycle inventory (LCI). It is in this stage that the various inputs and outputs (energy, wastes, resources) are quantified for each phase of the life cycle. As depicted in Fig. 3, systems boundaries are defined in such a way that the various stages of the life cycle of a product can be identified. The separation of burdens (inputs and outputs) for each stage facilitates improvement analysis.

For the purposes of LCI, a "product" should be more correctly designated as a "product system." First, the system is represented by a flowchart that includes all required processes: extracting raw materials, forming them into the product, using the resulting product, and disposing of and/or recycling it. The flowchart is particularly helpful in identifying primary and ancillary materials (such as pallets and glues) that are required for the system. Also identified are the sources of energy, such as coal, oil, gas, or electricity. Feedstock energies, which are defined as carbonaceous materials not used as fuel, are also reported.

After system definition and materials and energy identification, data are collected and model calculations performed. The output of an LCI is typically presented in the form of an inventory table (an example is shown in Table 1), accompanied by statements regarding the effects of data variability, uncertainty, and gaps. Allocation procedures pertaining to co-product generation, recycling, and waste treatment processes are clearly explained.

Step 3: Impact Assessment and Interpretation. Impact assessment is a process by which the environmental burdens identified in the inventory stage of an LCA are quantitatively or qualitatively characterized as to their effects on local and global environments. More specifically, the magnitude of the effects on ecological and human health and on resource reserves are determined.

Life-cycle impact assessment is at this time still in an early phase of development. Although some impact assessment methods have been advanced as either complete or partial approaches, none has been agreed upon. Nevertheless, an approach to impact analysis, known as "less is better," is typically practiced. With this approach, process and product changes are sought that reduce most, if not all, generated wastes and emissions and consumed resources. (Additional information is provided in the article "Design for the Environment" in this Volume.) However, situations in which such reductions are realized are not yet typical. Usually a change in product systems is accompanied by trade-offs between burdens, such as more greenhouse gases for fewer toxins. A fully developed impact analysis methodology would help in the environmental impact assessment of such cases.

As advanced by SETAC, impact analysis comprises three stages: classification, characterization, and valuation.

Classification. In this stage, LCI burdens are placed into the categories of ecological health, human health, and resource depletion. Within each of these categories, the burdens are further partitioned into subcategories, for example, greenhouse gases, acid rain precursors, and toxins of various kinds. Some burdens might fall into several categories, such as sulfur dioxide, which contributes to acid rain, eutrophication, and respiratory-system effects. Environmental burdens are sometimes called "stressors," which are defined as any biological, chemical, or physical entity that causes an impact.

Characterization. In the characterization step of impact assessment, the potential impacts within each subcategory are estimated. Ap-

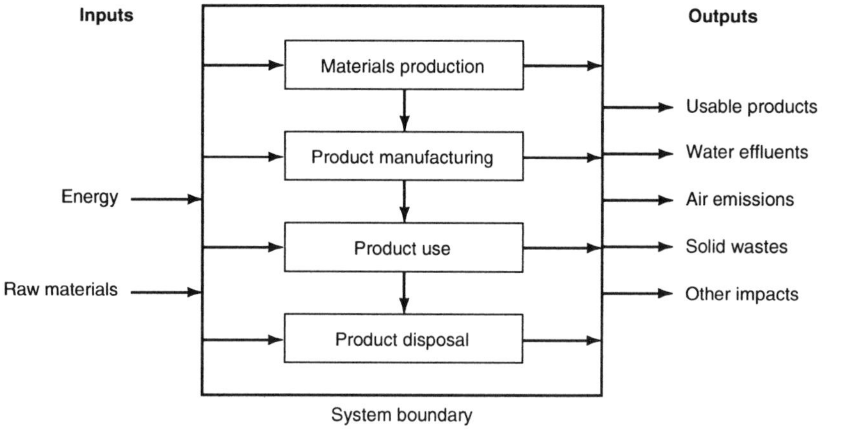

Fig. 3 Generalized system boundaries for a life-cycle inventory of a generic product. Source: Ref 2

proaches to assessing impacts include relating loadings to environmental standards, modeling exposures and effects of the burdens on a site-specific basis, and developing equivalency factors for burdens within an impact subcategory. For example, all gases within the global-warming category can be equated to carbon dioxide, so that a total aggregate "global-warming potential" can be computed.

Valuation. In the valuation step of impact assessment, impacts are weighted and compared to one another. It should be noted that valuation is a highly subjective process with no scientific basis. Further, attaching weighting factors to various potential impacts for comparison purposes is intrinsically difficult. For example, what is more important: the risk of cancer or the depletion of oil reserves? Who would decide this? Because a consensus on the relative importance of different impacts is anticipated to be contentious, a widely accepted valuation methodology is not expected to be adopted in the foreseeable future, if ever.

It is important to recognize that an LCA impact assessment does not measure actual impacts. Rather, an impact in LCA is generally considered to be "a reasonable anticipation of an effect," or an impact potential. The reason for using impact potentials is that it is typically difficult to measure directly an effect resulting from the burdens of a particular product. For example, are the carbon dioxide emissions of any individual's vehicle specifically causing the world to get warmer? It is unlikely that this could ever be shown, though it is reasonable to assume that any individual vehicle contributes its share to the possible effect of global warming caused by human-generated carbon dioxide in proportion to the amount of emissions.

Inventory Interpretation. It is argued by some that, due to the difficulties cited above, the notion of impact assessment should be dropped and replaced by *inventory interpretation*. Classification and characterization could still be used, but all suggestion that environmental effects are assessed is avoided. In comparative assessments, "less is better" is the principle in identifying the environmentally preferable alternative.

Step 4: Improvement Analysis. This step involves identifying chances for environmental improvement and preparing recommendations. Life-cycle assessment improvement analysis is an activity of product-focused pollution prevention and resource conservation. Opportunities for improvement arise throughout an LCA study. Improvement analysis is often associated with design for the environment (DFE) or total quality management (TQM) (see the articles "Design for the Environment" and "Design for Quality" in this Volume). With both of these methodologies, improvement proposals are combined with environmental cost and other performance factors in an appropriate decision framework.

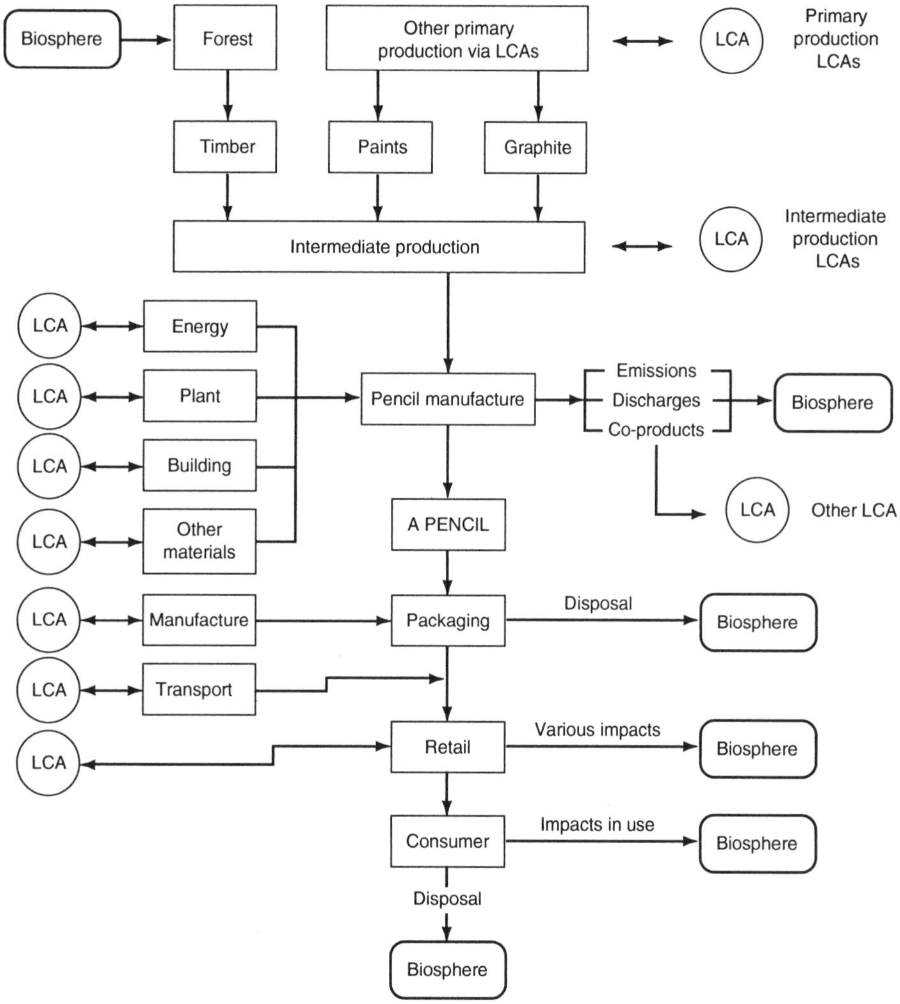

Fig. 4 Simplified life-cycle analysis process for a pencil. Source: Ref 4

Application of Life-Cycle Analysis Results

The results of an LCA can be used by a company internally, to identify improvements in the environmental performance of a product system; and externally, to communicate with regulators, legislators, and the public regarding the environmental performance of a product. For external communications, a rigorous peer-review process is usually required. Virtually all of the peer-reviewed studies conducted to date represent analyses of simple product systems. However, studies for systems as complicated as automobiles are being conducted.

Whether used qualitatively or quantitatively, LCAs often lead to products with improved envi-

	Raw material acquisition	Bulk processing	Engineered materials processing	Assembly and manufacture	Use and service	Retirement	Treatment and disposal
1. Cost							
2. Value/ performance							
3. Environment: compliance							
4. Environment: beyond compliance							
5. Socio-economic							
6. Socio-cultural							

Fig. 5 Matrix showing some possible different approaches to LCA. Source: Ref 4

ronmental performance. In fact, an often-overlooked, important qualitative aspect of LCA is that it engenders a sense of environmental responsibility. Beyond this development within manufacturers, LCA has the potential to become a tool to regulate products, or perhaps even for "eco-labeling." However, such uses are contentious and are expected to remain so.

The bulk of LCA efforts to date have been focused on preparing LCIs, with the impact assessment stage currently seen as the weakest link in the process. Indeed, some companies have even decided to skip this phase of the process altogether, opting to carry out a brief life-cycle review (LCR) before moving straight on to the improvement stage.

Large or small companies and other users will find LCA of value at a number of different levels. Indeed, groups like SETAC and SPOLD now see LCA playing a key role in three main areas:

- *Conceptually*, as a framework for thinking about the options for the design, operation, and improvement of products and systems
- *Methodologically*, as a set of standards and procedures for the assembly of quantitative inventories of environmental releases or burdens—and for assessing their impacts
- *Managerially*, with inventories and—where available—impact assessments serving as a platform on which priorities for improvement can be set

Not surprisingly, perhaps, the bulk of current LCA efforts is devoted to the second of these areas, particularly initiatives such as the 1993 Code of Practice by SETAC (Ref 3). However, the scope of LCA is rapidly spreading to embrace the other two application areas. The "supplier challenges" developed by companies such as Scott Paper, which has incorporated environmental performance standards in its supplier selection process, underscore the very real implications of the managerial phase for suppliers with poor environmental performances. Also, the "integrated substance chain management" approach developed by McKinsey & Company Inc. (Denmark) for VNCI (Association of the Dutch Chemical Industry), covering three chlorine-based products, shows that LCA can produce some fairly pragmatic tools for decision making.

Longer term, the prospects for LCA are exciting. Within a few years, product designers worldwide may be working with "laptop LCAs"—small, powerful systems networked with larger data bases and able to steer users rapidly around the issues related to particular materials, products, or systems. This process would be greatly aided by a widely accepted, commonly understood environmental accounting language.

In the meantime, however, LCA is still quite far from being simple or user-friendly, as is illustrated in the following example.

Example: Life-Cycle Analysis of a Pencil. Anyone who has had even a brief encounter with an LCA project will have seen flow charts rather similar to the one in Fig. 4, which shows the key life-cycle stages for one of the simplest industrial products, a pencil. Most such diagrams are much more complicated, but, as is evident in the figure, even the humble pencil throws an extraordinarily complex environmental shadow.

For example, imagine the flow chart in Fig. 4 is on the pencil maker's PC screen as the computer menu for an electronic information system. When the pencil maker clicks on "Timber," a wealth of data begins to emerge that makes one realize things are not as simple as may have been imagined. Not only is there a potential problem with tropical timber because of the rain forest issue, but the pencil maker now notes that suppliers in the U.S. Pacific Northwest have a problem with the conflict between logging operations and the habitat of the Northern Spotted Owl.

At this point, a pencil maker recognizes the need to examine the LCAs produced by the companies supplying timber, paints, and graphite. Working down the flowchart, the pencil maker sees a total of ten points at which other LCA data should be accessed. This is where complex business life gets seriously complicated. At the same time, however, LCA projects can also be fascinating, fun, and a potential gold mine of new business ideas.

Different Approaches to LCA. As Fig. 5 indicates, the LCA practitioner can look at the life cycle of a product through a number of lenses, focusing down on life-cycle costs or focusing out to the broader sociocultural effects. One example is the Eco-Labeling Scheme (Fig. 6) administered by the European Commission Directorate General XI (Environment, Nuclear Safety, and Civil Protection). This scheme is committed to assessing environmental impacts from cradle to grave.

The sheer variety of data needs, and of data sources, makes it very important for LCA producers and users to keep up to date with the debate and build contacts with other practitioners. Among the biggest problems facing the LCA community today are those associated with the availability of up-to-date data and the transparency of the processes used to generate such data.

Most LCA applications, however, focus—and will continue to focus—on single products and

	Product life cycle				
ENVIRONMENTAL FIELDS	Preproduction	Production	Distribution	Utilization	Disposal
Waste relevance					
Soil pollution and degradation					
Water contamination					
Air contamination					
Noise					
Consumption of energy					
Consumption of natural resources					
Effects on ecosystems					

Fig. 6 The European Community eco-labeling scheme "indicative assessment matrix." Source: Ref 4

on the continuous improvement of their environmental performance. Often, too, significant improvements will be made after a relatively simple cradle-to-grave, or perhaps cradle-to-gate, analysis.

A detergent company, for example, may find that most of the energy consumption associated with a detergent relates to its use, not its manufacture. So instead of just investing in a search for ingredients that require less energy to make, the company may decide to develop a detergent product that gives the same performance at lower wash temperatures.

In short, LCA is not simply a method for calculation, but, potentially, a completely new framework for business thinking.

Case History: LCA of an Automobile Fender

A detailed LCA for an automotive fender as performed by IKP (University of Stuttgart, Germany) and PE Product Engineering (Dettingen/Teck, Germany) is included to illustrate the present status and limitations of this methodology (Ref 1).

Goal and Scope. The specific goal of this investigation was to compare four different fender designs for an average compact class automobile in Germany. The comparison should result in the identification of the best material in terms of resource use, impact on global climate, and recyclability.

The four options were steel sheet; primary aluminum sheet; an injection-molded polymer blend of polyphenylene oxide and nylon (PPO/PA); and sheet molding compound (SMC), a glass-fiber reinforced unsaturated polyester resin. The mechanical requirements for the four fenders were identical; this ensures that the functional unit is well defined and that they are equivalent. Table 2 shows the materials and weights of the four different fender designs.

Data Origin and Collection. *Data* in this context means all pieces of information that might be relevant for the calculation of processes and materials. Such information includes material and energy flows of processes, process descriptions, materials and tools, suppliers, local energy supply, local energy production, production and use of secondary energy carriers (e.g., pressurized air, steam), and location of plants. Which processes are the most relevant and must be considered in more detail depends on the goal and scope of the study. Within this study, the following information (supplier specific, if possible) had to be identified, collected, and examined:

- *Production processes*, with all links in the process chain
- *Primary data* concerning energy and material flow with respect to use of energy carriers (renewable and nonrenewable), use of mineral resources (renewable and nonrenewable), emissions into the air, waterborne emissions, and waste and production residues
- *Coupled and by-products* as well as entries from other process steps (internal loops)
- *Transportation needs* with respect to distance, mode, and average utilization rate
- *Primary energy carriers* and their means of production and distribution
- *Secondary energy carriers* and their means of production and distribution
- *Air and water treatment measures* and disposal of residues

Data collection is not a linear process. Good data collection and evaluation requires iteration steps for identifying relevant flows or additional information, and experience is needed to interpret the collected data. Calculation of modules should be carried out with special regard to the method of data collection (e.g., measured, calculated, or estimated) and the complexity of the system.

Materials Production. The consideration of aluminum shows that not only the main production chain has to be considered, but also the process steps for alumina production (Fig. 7). The steps in electrolysis must be calculated, and the energy use connected with caustic soda and the anode coke has to be examined. The four steps shown in Fig. 7, which must be considered along with a long list of others, demonstrate the difficulty of balancing costs and environmental impacts.

In electrolysis, the source of electric power is important because of the differences in carbon dioxide emissions between plants that are water-power driven and those that burn fossil fuels. Another significant factor is how electrolysis is controlled. Modern plants use technologies that prevent most of the anode effects responsible for the production of fluorocarbon gases, but many older plants emit four or five times as much. This shows the importance of calculating on a site-specific or at least on a country-specific basis.

Because aluminum is globally merchandised, the user frequently does not know the exact source of the metal. The solution to this problem is to calculate the average aluminum import mix. However, this calculation requires detailed information about the different ways aluminum is produced all over the world.

Material Weight. In selection of automotive parts, the usage phase is of great interest. The main environmental factor during this phase is weight difference. Each part contributes to the energy demand for operating an automobile. The share a fender contributes depends only on its mass. However, no data are available for the same car carrying different fenders. Therefore, this study calculated the fuel consumption assuming a steel fender, because average fuel consumption is known for the complete car with the traditional fender. In the same way, possible weight savings are known. Measurements and judgments from all automobile producers show that the assumptions for fuel reduction from weight savings vary within a range of 2.5 to 6% fuel reduction per 10% weight savings. For this study, 4.5% was assumed to be an average value for the kind of cars considered.

Table 2 Material and weight of the different fender designs

Material	Thickness		Weight	
	mm	in.	kg	lb
Steel	0.7	0.0275	5.60	12.35
Sheet molding compound	2.5	0.10	4.97	11.00
Polyphenylene oxide/ polyamide	3.2	0.125	3.35	7.40
Aluminum	1.1	0.043	2.80	6.20

Source: Ref 1

Recycling of the SMC fender shows another weight-related issue. After the useful life of the product, a decision has to be made about whether the part should be dismantled for recycling, or otherwise disposed of. Within this study, the recycling solution was considered because the SMC part can be dismantled easily and ground into granules. Furthermore, SMC can replace virgin material as reinforcement, and granules can be used as filler up to 30%. In addition to the possibility of using recycled material in new parts, the SMC recycling process offers another advantage because the reformulated material has a lower density than the primary material. This means that the use of recycled SMC leads to further weight savings of approximately 8%, while fulfilling the same technical requirements. This example shows that recycling is not only useful for the purpose of resource conservation but may provide other benefits as well. (Unfortunately, at the time this was written the only U.S. company recycling SMC had gone out of busi-

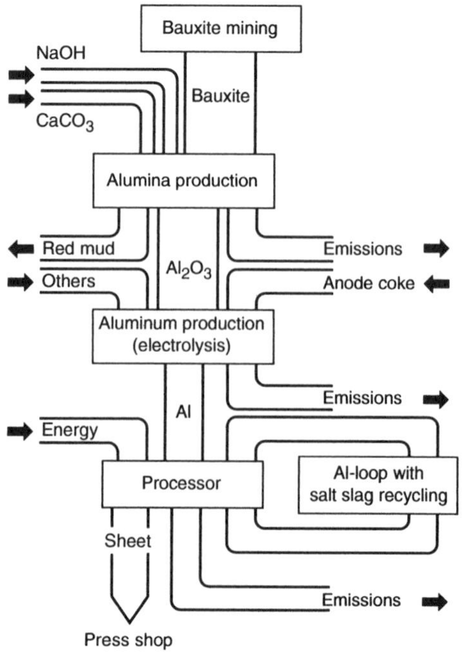

Fig. 7 Main material flow for the production of aluminum sheet parts. Source: Ref 1

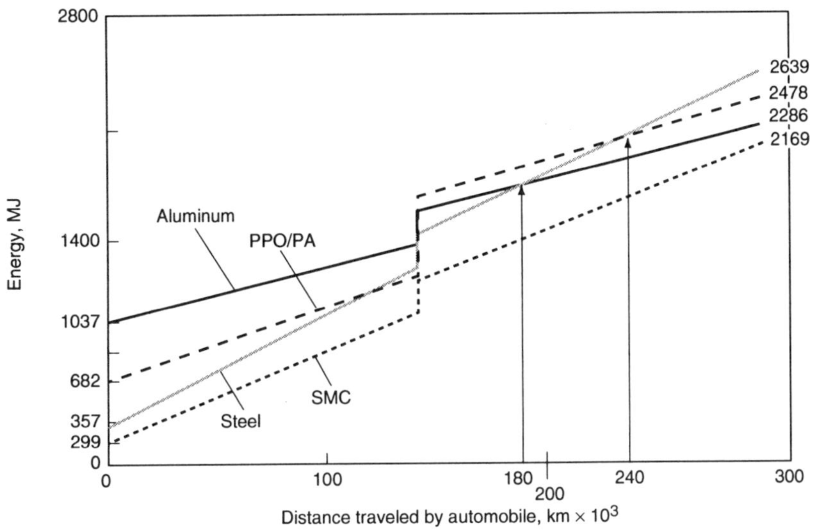

Fig. 8 Energy consumption for the production, use, recycling, and reuse of different fender materials considering the distance traveled by the automobile. Source: Ref 1

ness. This shows that successful recycling requires more than technical feasibility, which is generally achievable—it is highly dependent on viable economics.)

Inventory Results. The discussion of the whole inventory process is not possible here, because it includes up to 30 resource parameters, approximately 80 different emissions into the air, more than 60 water effluents, and many different types of waste. Therefore, this example concentrates on energy demand, selected airborne emissions, and resource use (recyclability).

Energy use is one of the main parameters to consider when selecting automotive parts. It is a reliable basis for judgment because energy use generates waste and emissions, and it requires depletion of resources. Figure 8 shows the energy demand for the different fender materials over two complete usage phases, including production out of raw material and recycling for the second application.

The values at the zero kilometer line represent the energy needed for both material and part production. It is easy to see that aluminum has the highest energy demand of all four materials. This comes mainly from the electrolysis process and the alumina production process.

SMC has the lowest energy demand, needing approximately one-third of the energy required for the aluminum fender. This is due to the fact that SMC is a highly filled material in which the extender is a heavy, relatively inexpensive material. Second best is steel, which requires only a little more energy than SMC. Somewhere in the middle is the PPO/PA blend; the reason for the relatively high energy demand is the feedstock energy of the materials used in polymer production.

The ascending gradients represent the differences arising from the weights of the fenders. The larger the gradient, the higher the weight. It is easy to see that steel, as the heaviest material, loses a lot of its advantage from the production phase. This points out the importance of lightweight designs. The energy demand for the usage phase is approximately four times higher than that required for part production. As a result, the most significant improvements can be made in the usage phase. Nevertheless, SMC still has the lowest energy demand after the first usage phase, and aluminum is still the worst.

Recycling. After the first life cycle of the fender, it is recycled into a new part. The energy needed for recycling of SMC, steel, and aluminum is relatively low; PPO/PA requires much more energy for recycling. The disadvantage of PPO/PA is that although recycling is possible and very energy efficient, the production of the 70% virgin material required in the part is very energy intensive.

Second Use. The second utilization phase shows the same results as the first. In the final analysis, steel turns out to be the most energy-intensive material, followed by the PPO/PA blend. While steel has the disadvantage of its weight, the polymer blend has disadvantages concerning recyclability for external body parts. The situation would be totally different if more material could be recycled, or if the polymer blend could be used more extensively in heavier cars with a longer usage phase. The weight advantage is especially high for aluminum. However, SMC turns out to be the most energy efficient material overall.

Airborne Emissions. Emissions of carbon dioxide, nitrogen oxides, sulfur dioxide, and fluorocarbons were estimated for each material because of their effects on ozone depletion and global warming (Fig. 9). These pollutants were also chosen because they are generated by nearly every manufacturing process, all over the world.

As mentioned before, a high percentage of atmospheric emissions is caused by energy generation. In the case of polymers, emissions are lower than expected because so much energy is stored as material feedstock. Aluminum is the material with the highest energy demand, but emissions are comparatively low because water power is used for a high percentage of aluminum electrolysis. The highest levels of carbon dioxide are emitted during steel production, mainly from the ore reduction process. Carbon dioxide emissions for the production of both polymers are dominated by hydrocarbon processing and refining.

For aluminum, most emissions come from earlier process steps. Alumina is produced mainly in bauxite mining countries, where the least expensive locally available energy is typically generated by burning heavy fuel and coal. Carbon dioxide emissions from aluminum production are dominated by this source, plus the electric power demand of those electrolysis processes that are not based on water power.

Carbon dioxide emissions during usage are directly related to fuel consumption: heavier fenders result in the generation of more carbon dioxide. This is also true for all other emissions considered here. One important approach for a

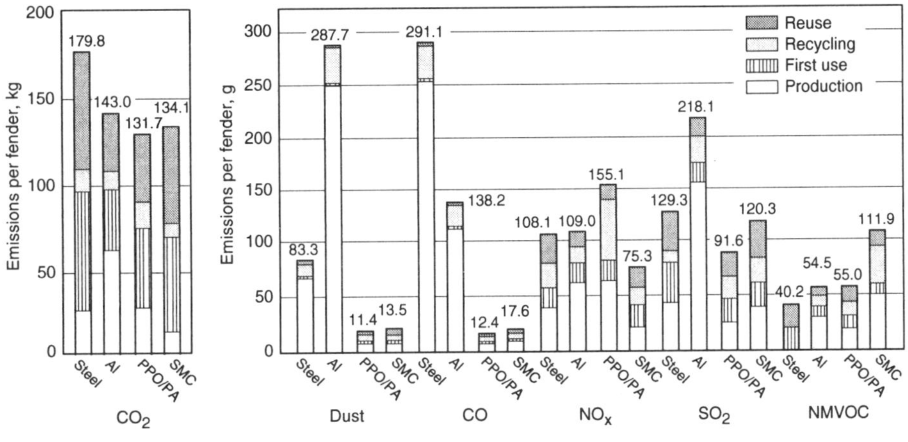

Fig. 9 Selected airborne emissions for the production, use, recycling, and reuse of different fender materials. NMVOC, nonmethane volatile organic compound. Source: Ref 1

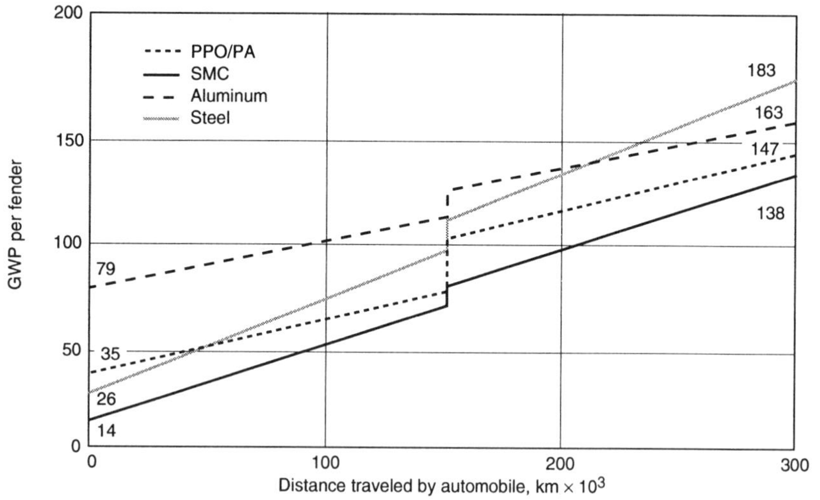

Fig. 10 Calculated contribution to global warming for the production, use, recycling, and reuse of different fender materials considering the distance traveled by the automobile. GWP, greenhouse warming potential (CO_2 equivalents). Source: Ref 1

possible improvement is certainly to reduce this main impact on global warming.

Impact assessment is a special step within the framework of life-cycle assessment. Based on the results of the inventory, conclusions can be drawn, and judgments and valuations are possible. The impact assessment supplies additional information that enables the practitioner to interpret the results from the inventory.

Impact assessment also should allow the practitioner to draw the right conclusions concerning improvement approaches. However, it should be noted that consideration of environmental effects as a consequence of environmental releases is additional information that is not covered by the inventory step. This case history provides only a brief overview.

Impact assessment involves three steps. First is the definition of "environmental problems" or "themes." The problems to be addressed are defined in the scope of the project. Second, emissions are grouped to show their specific contribution to the environmental themes. Third, their shares are calculated. A standard list covers the following themes, which are more or less identical with most of the approaches taken in LCA literature:

- *Global criteria:* Resource use (energy carriers and mineral resources, both renewable and nonrenewable, and water and land use), global warming, ozone depletion, and release of persistent toxic substances
- *Regional criteria:* Acidification and landfill demand
- *Local criteria:* Spread of toxic substances, eutrophication, and formation of photochemicals
- *Others:* Noise, odor, vibration, etc.

In most of the studies conducted by IKP and PE Engineering, resource use and the global climate problems are considered. The methodology for their consideration is broadly accepted. Sometimes acidification or eutrophication is considered as well. All others are more difficult to handle, and appropriate methods are still under discussion.

For the fender example, the contribution to the global-warming problem is calculated by taking into account production, use, recycling, and second use of each material (Fig. 10). The results are mainly influenced by carbon dioxide emissions and energy use and show that lightweight materials have advantages during utilization. However, aluminum is far worse than the others during production because electrolysis is accompanied by fluorocarbon emissions (CF_4 and C_2F_6), which have a very high global warming potential.

Valuation. The second step in the judgment of the environmental impacts is the valuation step. This step may be divided into the normalization process and the final weighing.

Normalization involves scaling absolute contributions to single environmental themes on the same level because absolute numbers may vary within six to ten decades. The effect scores are normalized with the amount of the annual global effect score or the contribution of one process to the theme per year, and so on.

Final weighing involves a personal judgment about the importance of each environmental theme, and the effect of each score on overall impact. This final step is part of the decision-making process. Scientists create tools for this process and help decision makers use and understand them, but the final decisions depend on company policies, not scientific or consultancy work.

Improvement Options. From this study the following conclusions for improvement can be drawn:

- The usage phase is dominated by fuel consumption and the resulting carbon dioxide emissions. For other emissions, the production phase and recycling is also of great importance.
- Reducing part weight may improve energy use and reduce the contribution to global warming. However, reducing part weight may require higher environmental investments during production or recycling. In some cases, these investments are very useful.
- Recycling is more important for expensive and energy-intensive materials.

Experience gained from the evaluation of fender materials shows that the following general conclusions can be made:

- The fuel production has great impact and is not well known today.
- The best basis for decision making is a supplier-specific LCA.
- Close cooperation between producers and suppliers is necessary to find processes that will reduce environmental impacts.

Conclusions

Life-cycle engineering—in particular, life-cycle assessment—is gaining importance for design and materials engineers because environmental considerations are increasingly important factors in design and materials selection. The creation and development of environmental management systems, including extended producer responsibility and product stewardship responsibility, pollution prevention strategies, "green" procurement guidelines, and eco-labeling programs are evidence of the growing importance of life-cycle concerns.

To make a proper assessment the total life cycle of a material, all forms of energy use, waste production, reuse, and recycling have to be considered. Many of these factors are site specific, which complicates calculations and comparisons.

Not all of the steps of a complete LCA are well developed at this time. However, efforts in this direction, particularly regarding standardization of data and methods are in progress. While LCAs for simple products have been performed, more complicated systems are only now being tackled.

Many industry trade organizations have developed or are in the process of developing life-cycle inventory data bases for their products. The Association of Plastics Manufacturers in Europe (APME), the European Aluminum Association (EAA), Finnboard, and the International Iron and Steel Institute are just a few examples. A wide variety of reports and software packages containing inventory data are available. In addition, a large number of national and international database projects exist. A comprehensive listing can be found in Ref 5. Also, Ref 4, while concentrating on Europe, gives an excellent overview and many useful examples and addresses.

Simplification and standardization will lead to more reliable, timely, and cost-effective life-cycle inventories. When consensus about an acceptable impact assessment methodology is reached, life-cycle assessment for simple and then more complex units and systems will be possible.

ACKNOWLEDGMENTS

Portions of this article were adapted from Ref 1 and 2. The authors wish to thank Sustainability Ltd. (United Kingdom) and the Secretariat of SPOLD (Belgium) for allowing the use of some of their information.

REFERENCES

1. M. Harsch et al., Life-Cycle Assessment, *Adv. Mater. Proc.,* June 1996, p 43–46
2. J.L. Sullivan and S.B. Young, Life Cycle Analysis/Assessment, *Adv. Mater. Proc.,* Feb 1995, p 37–40
3. "Guidelines for Life Cycle Assessment: A Code of Practice," Society of Environmental Toxicology and Chemistry (SETAC), Europe (Brussels), 1993
4. *The LCA Sourcebook,* Sustainability Ltd. (London), 1993
5. "Directory of Life Cycle Inventory Data Sources," Society for the Promotion of LCA Development (SPOLD) (Brussels), Nov 1995

SELECTED REFERENCES

- "Life Cycle Assessment: Inventory Guidelines and Principles," U.S. Environmental Protection Agency (EPA), Office of Research and Development (Cincinnati, OH), 1993
- "Life Cycle Design Manual: Environmental Requirements and the Product System," University of Michigan, 1993
- "A Technical Framework for Life Cycle Assessment," SETAC USA (Washington DC), 1991

Design for Quality*

James G. Bralla, Manufacturing Consultant

CONSUMERS have come to expect high quality and dependability in manufactured products. Competitive pressures with respect to quality are stronger than they were in prior years, perhaps thanks to Japanese competition in many product lines (most notably in automobiles). Therefore, designed-in quality is a vital facet to current product design. According to Joseph M. Juran, "One-third of all quality control problems originate in the product's design" (Ref 1).

What Is Quality?

What do customers expect when they purchase products? For both consumer and industrial products, the answers are very nearly the same. Function, performance, and the low price that can result from successful design are important to customers. However, their expectations are not limited to these factors. Customers also want benefits that last as long as the product is owned and used. In the broadest sense of the word, they want products of high lasting quality. In this sense, the word *quality* defines the attributes that an ideal product should have.

D.A. Garvin's classic paper, "What Does Product Quality Really Mean?" (Ref 2) lists "eight dimensions of quality." These form a good starting point for a list of desirable attributes for a product design. Garvin's list includes the following:

- *Performance:* How well the product functions
- *Features:* How many secondary characteristics the product has to enhance its basic function
- *Reliability:* Defined by some as quality in the time dimension; how well the product maintains its quality
- *Conformance:* How well the product conforms to the specifications or standards set for it
- *Durability:* How long the product lasts in use
- *Serviceability:* How easy the product is to maintain
- *Aesthetics:* How attractive the product is
- *Perceived quality:* How high the users believe the quality of the product is, that is, quality reputation of the product

To these desirable attributes, *manufacturability*, how easy and economical the product is to make, should certainly be added. Other desirable characteristics, not mentioned by Garvin, are safety, environmental friendliness, user friendliness or ergonomics, short time to market, and upgradability. Many of these attributes are discussed in other articles in this Section of the Handbook.

Perhaps Garvin's eighth dimension, *perceived quality* is the most important, provided the perception is based on ownership experience. In other words, quality is whatever it is judged to be by the customers of the product in question. Quality is whatever the customer wants. However, this must not be interpreted to mean that quality is whatever sells the product in the store or showroom. It is more a result of how satisfied customers are with the product after they have owned it for some time and have had a chance to weigh its features: ease-of-use, freedom from maintenance, ease of regular service, economy of operation, safety, and other attributes, and, overall, whether the product has met the customers' expectations.

If customers are satisfied with the product after, for example, a year of ownership and at least moderate use and would recommend it to other potential buyers, then perhaps it can be said that the product is of high quality. Other measures, such as whether it conformed to some specifications, whether it had an acceptable reject rate, whether it was made under ISO 9000 conditions, or whether the company producing it got the Malcolm Baldrige award are less meaningful, in the author's opinion, than the customer's evaluation. Customer satisfaction is the prime measure of product quality.

Quality and Robust Design. Taguchi's approach to quality evaluation (see the article "Robust Design" in this Volume) is more quantitative than Garvin's. The highest quality, he contends, is that which minimizes the life-cycle costs of the product. These life-cycle costs include the acquisition cost by the purchaser, which is normally, but not always, closely related to the manufacturing cost. They also include the cost to operate the product, the maintenance expenses for it (including the cost for regular service, repair, and use of a substitute product during maintenance), the cost of any injury resulting from safety flaws, the cost of overcoming any defects it has, including safety defects, and the expense of disposing of it. These life-cycle costs may not all accrue to the same person but, ultimately, are paid by some member or members of society (see the article "Life-Cycle Engineering and Design" in this Volume). In summary, Taguchi's quality cost function measures quality in terms of the cost to any and all members of society who have expenditures resulting from the manufacture, sale, ownership, and disposal of the product. The lower such cost, the higher the product quality. (Taguchi's measure excludes costs due to misuse of the product. For example, an automobile repair due to careless driving is not part of the quality cost of the automobile; an accident due to poor brakes, sloppy steering, or a horn that is awkward to sound would be.)

Phadke's approach (Ref 3) is from a different direction, but perhaps it is not less meaningful. He says that ideal quality means that the product delivers its target performance:

- Each time it is used
- Under all intended operating conditions
- Throughout its intended life
- With no harmful side effects

There may be a conflict between quality and cost, but the conflict is in the initial manufacturing cost, not the life-cycle cost as Taguchi defines it. Many managerial steps taken to enhance quality require a significant initial expense in training, organization, and redirection of systems, procedures, and operating philosophy. Also, corrective action in the product design to solve quality problems often requires an investment in engineering time, new tooling, gaging, or equipment.

Quality and Design for Manufacturability. Many DFM changes, implemented to reduce manufacturing costs, improve quality; however, some may impair quality. One example is the use of free-machining metals for machined parts. They ease and speed up manufacturing operations, but generally have slightly less favorable physical properties than the standard grades so

*Adapted from James G. Bralla, Designing for Higher Quality, Chapter 14, *Design for Excellence*, copyright © 1996 The McGraw-Hill Companies, Inc. Used with permission.

the resulting product may not be quite as strong. Another example is the use of thin walls in injection-molded plastic parts. These speed the molding cycle and save material, but may result in a less-rigid part than one with thicker walls. Another example is the elimination of adjustments, advocated to improve ease of assembly. Such an elimination can have strong beneficial effects on quality, if engineered correctly, because incorrect adjustments are a source of quality defects. If the engineering carried out to eliminate the adjustment is not done soundly, or if specifications on components are not held in production, the lack of capability to adjust may result in a product slightly off in some characteristic, that is, a defective product. Adjustments are normally specified when the designer believes that this is the best way to achieve some precision in dimension or setting as a result of variations in parts or other factors. Eliminating the adjustment may cause the variation to get through to the operation of the product, reducing its quality. Care is required in deciding which approach is best overall.

On the other hand, there are many DFM guidelines that facilitate improved quality. For example, DFM specialists advocate keeping wall thickness in injection-molded plastics parts as *uniform* as possible. This improves the molding operation and also prevents the formation of unsightly sink marks and distortions that impair the fit and quality of plastic parts. In metal stamping, standard DFM guidelines to make bends across the grain of the metal rather than along it and to space punched holes adequately from the edge of a workpiece have the primary purpose of avoiding quality problems.

Additional information about DFM is provided in the article "Design for Manufacture and Assembly" in this Volume.

Management of Quality

Even though initial product design is a strong determinant of eventual product quality, it is far from the only factor. The quality improvement task is dependent on a wide range of factors including company objectives; management and employee attitudes; training; systems and procedures used; the condition of tools, equipment, and facilities; the control exercised by vendors; and many other factors. In short, product quality is heavily dependent on how well the company is managed. J.M. Juran has said, "The most important thing to upgrading quality is not technology, but quality management" (Ref 4). And the management of quality is a broad and complex task.

American industry awoke to the need for improved quality in the 1980s, when Japanese and other international suppliers made large inroads in the U.S. markets for many consumer and industrial products. Analysis showed that *quality cost*, the cost of inspection; scrap; rework; warranties; field service due to quality problems; product call-backs; and most importantly, lost sales due to a poor quality reputation, was a major part of the operating cost of a manufacturing concern. Crosby and others claim that these expenses amount to 25, 30, or even 40% of the cost structure of a company (Ref 5). The corollary is that by spending more money "up front" in quality assurance provisions, manufacturing costs could actually be reduced because scrap, rework, and all the other costs of poor quality would be reduced and sales, market share, and production volumes all would increase when quality was improved.

The up-front costs may be considerable, and the journey from mediocre to superior quality may be a long one. The prioritizing of quality must permeate the whole organization, and considerable careful communication and training will usually be required to obtain it. If all this is done correctly, the savings from reduced quality costs should provide a good return on the up-front investment. The U.S. automobile industry has learned how long it takes to change. Ford Motor Company began its quality improvement program in the early 1980s and, by the 1990s, it is still in progress, not only at Ford but in the rest of the U.S. automobile industry.

A first essential step in managing quality improvement is a firm, sincere commitment by management that quality is a prime priority. The word "sincere" is used advisedly. If management preaches quality, but ships substandard products at the end of the month to meet its monthly billing quota, workers and others in the organization will get the message that quality is not as important to the company as it is touted to be. "Employees are pretty clear on reading signals" (Ref 4). Management must lead the way, but all employees must share a determination to exercise great care in ensuring that all company activities lead to the production of high-quality products.

Statistical Process Control. A prime tool of quality improvement, advocated by Deming and others, is *statistical process control* (SPC) (Ref 6). This is a procedure, using statistical mathematics, which signals that some extraneous factor is affecting the output of a production process. The signal alerts production and quality personnel that some process fault should be looked for and eliminated. In this way, the procedure aids in identifying and correcting the causes of product component defects. Because there are natural random variations in the results of any manufacturing process, the ability to differentiate between these random variations and those caused by some change in process conditions is a critical part of maintaining good control over specified characteristics and dimensions. Broken or worn cutting tools, slipped adjustments, leaks in a pressurized system, and an accidental change to a less-active solder flux are some examples of the kinds of process changes that might otherwise not be noticed, but which may cause a quality deterioration that would be detected by SPC analyses.

Additional information about SPC is provided in *Nondestructive Evaluation and Quality Control*, Volume 17 of *ASM Handbook*.

Total Quality Management. Deming also states that 85% of quality problems are caused by systems, procedures, or management and only 15% by bad workmanship (Ref 6). Blaming workers is not his way to cure quality problems. Incidentally, the 85% attributable to management includes problems traceable to weaknesses or errors in the product design.

Current thinking on the best managerial approaches to control and improve quality involve heavy worker participation in both the monitoring of quality and the corrective actions taken to solve quality problems. One approach that encompasses worker involvement and includes worker empowerment is *total quality management* (TQM). Total quality management is more a broad management philosophy and strategy than a particular technique. Referred to earlier as total quality control, it originated in Japan. It involves:

- A strong orientation toward the customer in matters of quality.
- Emphasis on quality as a total commitment for all employees and all functions including research, development, design, manufacturing, materials, administration, and service. Employee participation in quality matters is standard at all levels. Suppliers also participate.
- A striving for error-free production. Perfection is the goal.
- Use of statistical quality control data and other factual methods rather than intuition to control quality.
- Prevention of defects rather than reaction after they occur.
- Continuous improvement.

Total quality management programs usually stress that quality must be designed into the product rather than tested for at the end of the production process. Additional information is provided in Ref 7.

Quality Function Deployment. Another well-known quality technique is *quality function deployment* (QFD). This is a system that reflects the belief that the customer's viewpoint is the most important element in product quality. Quality function deployment is a technique "for translating customer requirements into appropriate company requirements at each stage—from research and product development through engineering and manufacturing, to marketing, sales, and distribution" (Ref 8). The objective of the approach is to ensure that the customer's preferences are incorporated in all facets of the product. A matrix chart is prepared. Customers' preferences for product attributes (what the customer wants) are listed on the left-hand side of the sheet. Product design features intended to satisfy the customers' requirements are listed across the top of the same sheet. Where a product feature satisfies a customer preference, a mark is placed in the matrix chart. Normally, the mark is coded to indicate the degree to which the customers' preference can be satisfied by the design feature. The objective of this matrix and the whole QFD procedure is to ensure that customers' preferences are satisfied by the product design.

Additional discussion of QFD, including examples of QFD matrices, is provided in the articles "Conceptual and Configuration Design of

Products and Assemblies" and "Concurrent Engineering" in this Volume.

Worker Involvement. The strong worker-participation aspect of TQM is one of its most important components. If workers are given the task of monitoring the quality of their own output, especially by plotting their own SPC charts, and are then encouraged to recommend systems, layout, or workplace changes, quality has the best chance of being improved. Workers know more about the details of their operations than anyone and normally care about the quality of their workmanship. If their knowledge is properly channeled, the best results can be obtained. Utilizing worker suggestions tends to give workers ownership of the quality improvement project and helps to keep them more quality conscious.

Training is a necessary part of a quality improvement effort. The training will encompass not only an appreciation of the quality philosophy but also, for many people, specific statistical and charting know-how. The installation of SPC procedures throughout a factory will take time.

Continuous Improvement. There is much to be said for small-scale incremental improvements in processes, methods, and systems. This is in contrast to the historical pattern in the United States, wherein large-scale, capital-intensive automation projects are used as a means to reduce costs and improve quality. There is nothing wrong with such an approach if the changes are technically and managerially sound and economically justifiable. However, sometimes a series of grass-roots, incremental improvements can yield the same results in the long run with much less investment and upheaval. The *continuous improvement approach*, a major element of TQM, has much to be said for it.

Design of Experiments. Other worthwhile tools for quality improvement are the design of experiments methods discussed in the article "Robust Design" in this Volume. Called *design of experiments* (DOE), *controlled experiments*, *orthogonal arrays*, or *Taguchi methods*, the approaches are most often noted as quality improvement tools, but they are also quite useful for raising process yields and making other manufacturing and product improvements.

Teams. The right way to implement design for quality is through a team approach. To ensure that high product quality is incorporated in the design, an experienced quality person should participate actively in the design process as a member of the project team. This person can supply information about which characteristics, dimensions, and other specifications are likely to be critical to product quality and can make recommendations for testing and testability. Experience with the product line involved or with similar products is obviously important.

Additional information about team-based approaches is provided in the article "Cross-Functional Design Teams" in this Volume.

Principles of quality management can be summarized as follows:

- Management leadership to better quality must be strong and sincere.
- A steady series of small incremental improvements may be preferable to a few major changes.
- Worker involvement is necessary if quality is really going to be improved. In fact, the whole organization must be quality-minded and involved.
- Statistical controls are invaluable in identifying when corrective action needs to be taken.
- Training in statistical methods and quality philosophy are essential elements of a quality improvement program and should be provided.
- Designed experiments are a useful tool, where appropriate.
- Production people should be given the responsibility for quality and the tools and authority to carry out that responsibility.
- It should be remembered that high quality means meeting customer expectations. All

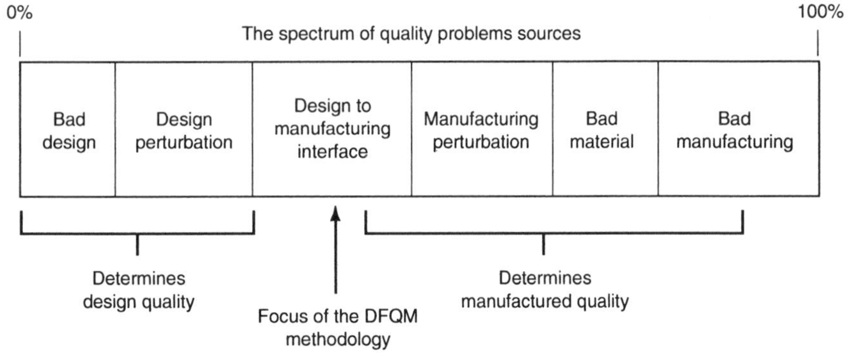

Fig. 1 Summary of the various sources of product quality problems according to the design for quality manufacturability (DFQM) approach. Bad design refers to fundamentally inappropriate design concepts of configurations. Design perturbation refers to minor weaknesses in the design that are capable of correction. Design to manufacturing interface refers to potential sources of quality problems in manufacturing, although the product design is basically sound. Manufacturing perturbation refers to areas where there are weaknesses in the manufacturing process but not full inadequacies. These weaknesses may require improvement to enhance yield, etc. Bad material, perhaps, is more obvious—defects in materials or components purchased. Bad manufacturing refers to errors in workmanship, inadequate training of manufacturing personnel, and defects in equipment and/or tooling due to initial inadequacies or poor maintenance. Source: Ref 11

Matrix chart evaluation of potential quality					
		Ratings alternative		Weighted ratings alternative	
Factor	Weight	No. 1	No. 2	No. 1	No. 2
Does the design reflect customer-indicated preferences for product or component capabilities?	5				
Are proven, existing components and design approaches used in all possible instances?	4				
Do the design specifications and dimensional tolerances conform to the normal capabilities of the process to be used?	3				
Did knowledgeable quality personnel participate in the design process for this item?	4				
Has the item been adequately tested?	3				
Is the item easy to test or inspect for all critical specifications?	3				
Is dimensioning clear and consistent with prior practice?	3				
Are critical dimensions controlled by tooling rather than individual workmanship or machine set up?	4				
Is assembly easy and straightforward with visibility of assembly locations, easy fits, prevention of missing parts, incorrect sequence, or incorrect positioning of parts?	3				
Total weighted ratings:					

Fig. 2 Sample matrix evaluation system for aiding designers in rating the suitability of a product design concept for potential high quality. The component or product with the highest score is deemed to have the best potential quality.

kinds of audit approvals are useless if the customer is not satisfied that the product is good. The customer is king.

- Experienced quality-control analysts and engineers should participate as team members in the design project.
- Be aware of the costs of poor quality including the costs of such items as inspection, screening, rework, scrap, production downtime, delayed deliveries, warranty costs, product returns, lost market share and sales, and lost margin in product pricing.
- Product and process design should take place at the same time.
- Quality faults should be prevented rather than corrected. In other words, quality should be built into manufacturing processes and not achieved by inspection.
- Quality requirements of each product should be well defined.
- If quality problems arise, it is best to concentrate on the most important ones, the ones that Juran calls "the vital few" rather than expending the organization's energy and time on minor problems (Ref 9).
- Sound product design from a quality standpoint must start with some understanding of the actual customer requirements for the product (Ref 10).

How Can Design Unfavorably Affect Product Quality?

As indicated, proper design is a prerequisite for high product quality. Designers must optimize the quality potential of their products. (Proper design doesn't guarantee high quality, however, since poorly managed manufacturing can turn even an optimum design into a defective product.) The following are common causes of inadequate designed-in product quality:

- Separation or isolation of the design activity from production and other support functions
- Failure to consult with or have the participation of experienced quality personnel during the design process
- Failure to address customer wants and needs in the product
- Failure to match the design, particularly the dimensional precision needed, with the capabilities of the manufacturing processes used
- Insufficient thoroughness in initial design efforts, leading to late design changes that tend to cause quality problems in manufacturing
- Insufficient testing of prototypes and pilot production units
- Too great a tendency to reinvent the wheel; that is, not utilizing existing, proven components and designs
- Failure to make the product design simple enough, for example, failure to make the product easy to assemble, potentially leading to assembly and adjustment errors

Evaluating a Product Design for Quality

Granted that design is a major determinant of product quality, how does the designer evaluate a prospective design to judge whether it has the intrinsic high quality that is wanted? Having an objective method of evaluating the product design in terms of quality would be very desirable. Unfortunately, little methodology for this exists as yet. Three limited methods are described below.

U.S. Navy producibility tool No. 2 is a means of evaluating product quality as indicated by the yield of acceptable components from the manufacturing processes that are used to make them. The quality rating of the product (yield) is the product of the yields of each of the parts.

For example, a product composed of five components that have yields of 0.99, 0.98, 0.99, 0.95, and 0.97 would have a yield of $0.99 \times 0.98 \times 0.99 \times 0.95 \times 0.97 = 0.88$. This rating is based on the assumption that the product is defective if any of the five components in it are defective. In other words, the effect of defects in the parts is cumulative. The mathematics applicable is identical to that commonly used to evaluate the reliability of a product in terms of the probability that it will perform for a certain period. If the probabilities for the components operating satisfactorily for the same period are known, the resulting probability of successful operation of the product can be calculated.

The limitation of this method is that it applies only to rejectable defects in components. Sometimes there is a combined effect that is not satisfactory when "good" components are assembled. Usually, components apt to be defective are inspected and sorted before use. Sometimes, also, good components are improperly assembled causing the total product to be defective. The measure also deals with the conformance of characteristics to specifications, not with whether customers accept the product. The mathematics in the system is correct, but the basis for the calculation may not correspond to true quality measurement from the customer's viewpoint. A further limitation, perhaps the most important one, may be the lack of reliable data on the yield of each component of the product, particularly newly designed parts.

Design for Quality Manufacturability Method. S. Das at the New Jersey Institute of Technology (NJIT) is working on another evaluation system that could provide projected quality yield for new parts based on their configuration. The system is still in the initial stages. The fact that product quality is a result of so many factors, many design related but many more manufacturing related, complicates the problem. Figure 1 illustrates Das' interpretation of the spectrum of sources of product quality problems. Das' system also is intended to provide data on assembly as well as individual parts quality. He has analyzed factors that can result in assembly errors, even if the parts assembled do not have defects; for example, part misalignments, misplaced or missing parts, or part interferences. His system is designed to aid the designer in evaluating the potential quality of particular configurations before the design is finalized.

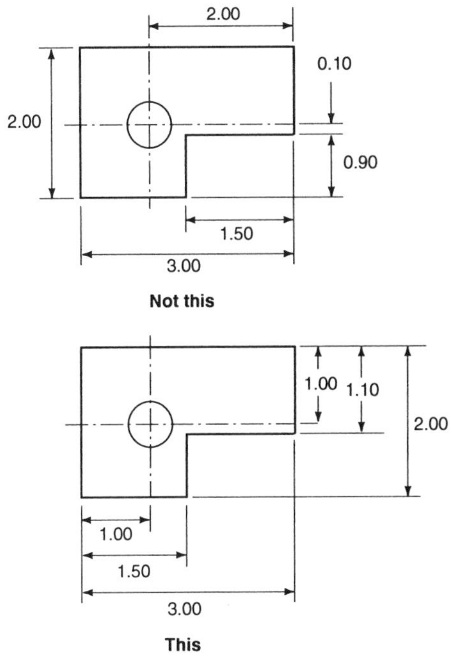

Fig. 3 Dimensioning on engineering drawings. Dimensions should be made from points on the part itself rather than from points in space. It is also preferable to base as many dimensions as possible from the same datum line. These steps help avoid errors when the parts enter production. Source: Ref 12

Matrix Method. The third potential approach for quality evaluation of a proposed design is the matrix method. A matrix for quality evaluation could include managerial as well as technical factors. Figure 2 illustrates a proposed matrix that could aid the designer in ensuring that his or her new component has high quality potential. This kind of evaluation is quite subjective, depending on the knowledge, experience, and judgment of the person making the evaluation. It may not be so suitable for differentiating between subtle differences between two design concepts, an application that is perhaps most important. On the other hand, the procedure lends itself to easy modification so that factors that are particularly important to a particular product line can be included and emphasized, as deemed important.

Guidelines for Promoting Quality

The design guidelines described below are intended to help provide products with a potential for higher levels of quality.

Design the Product, Its Major Subassemblies, and Other Components So That They Can Be Easily Tested. Generally, this means providing space and access for testing devices. The testing should be performable when the component is in process, before its installation in the product. This is when corrective action can be taken most easily and before additional operations or components, not involved in the test, have been added. What kind of test to allow for depends on the product and its specifications. Printed circuit boards often are designed with accessible points for electronic testing. Mechanical components may be tested for the position or adjustment of parts; completeness of assembly; freedom from containing extra, loose parts like dropped fasteners; leaks; actuating force; color, or sound level or other acoustic property. Electronic products are tested primarily for proper function. Testing may be automatic in the case of products manufactured at high production levels. In all cases, the component to be tested must have space for the test device, and the product must be properly supported so that the test can be valid.

Not only must the design provide a product that can be easily tested, the designer must also ensure that there is thorough testing of the design before it is committed to production. Many product quality problems are due to unexpected, unforeseen reactions or interactions or unexpected operating conditions in the product. The more thoroughly the proposed product is tested, the better the chance of detecting potential problems before the product reaches the user.

Utilize Standard, Proven Parts (Ideally, Proven Commercial Parts) Whenever Possible. If standard parts cannot be used, use parts from standard, proven manufacturing processes and proven existing quality control procedures and equipment. If newly designed parts are required, the less the new design departs from existing designs, the less chance there will be for problems that lead to quality deficiencies. Existing, standard mechanisms and circuits should always be employed in favor of new approaches unless there is some specific need for a new approach. In other words, do not reinvent the wheel!

Use Clear, Standardized Dimensioning of Drawings. Dimension as much as possible from the same reference plane. Try to use rectilinear, not angular dimensions. Do not dimension from theoretical points in space but from specific points on the component (Fig. 3).

Design Parts and Set Tolerances To Reduce or Eliminate Adjustments. These, aside from being costly, have been found to be potential sources of quality problems. (This applies to both mechanical and electrical adjustments.) Adjustments also necessitate extra parts in an assembly to provide for both movement and locking. Eliminating the adjustment usually eliminates some parts. Eliminating adjustments normally requires greater precision in some dimension, but often this can be obtained with a process or tooling change. Use of parts with compliance may also, in some cases, eliminate the need for incorporating adjustments in an assembly (Fig. 4).

Design parts so that critical dimensions can be controlled by tooling, rather than by the setup of production equipment or by individual workmanship. (This requires designing for a particular process, always a sound design principle.) Examples of processes in which tooling can be used to control critical dimensions include injection molding, progressive die stamping, die casting, lost-foam casting, and powder metallurgy.

Be Careful of Dimensional Tolerances. The assignment of tolerances can be a critical element that justifies considerable attention from the product designer (see the article "Dimensional Management and Tolerance Analysis" in this Volume). Looser tolerances in component parts almost always result in lower parts cost, but may cause trouble in assembly and in the performance of the finished product if they result in parts fits that are too loose or too tight or in misalignment. Excessively tight tolerances require additional operations or additional care that can dramatically increase costs. However, tight tolerances are generally better from a quality standpoint. Additionally, the Taguchi quality philosophy calls for the closest adherence to nominal dimensions. This implies that close manufacturing control over dimensions is valuable from a total-product-quality standpoint. The designer should specify what dimensions and other specified characteristics are important to the product and should tighten tolerances for these. Noncritical dimensions and other specifications should be more liberally toleranced. Overall, according to Anderson, "The best procedure is to optimize tolerances for a balance of function, quality, safety, and manufacturability" (Ref 1).

Minimize the Number of Different but Similar Part Designs. In other words, standardize on the fewest number of part varieties in order, among other things, to prevent the wrong part from being inadvertently assembled in a product. If this cannot be done, make sure that similar but slightly different parts cannot be accidentally interchanged. Make them very obviously different or, better still, not able to fit into each other's application (Fig. 5).

Use Modular Construction. Modules usually can be tested easily and in other ways have their quality verified.

Thoroughly Analyze Quality Ramifications of Engineering Changes. If engineering changes are made, make sure that their quality ramifications are thoroughly analyzed because quality problems sometimes stem from incompletely engineered design changes. Changes should be clearly and promptly transmitted to manufacturing and promptly implemented (Ref 1). The earlier that the change is made the less chance there will be to encounter quality problems and the lower the cost of the change will be.

Develop More Robust Components and Assemblies. Use Taguchi or other designed experiment methods to develop components and assemblies that are less sensitive to process variations and variations of other conditions.

Design for Ease of Assembly. There are a number of recommendations concerning how parts should be designed to fit together that can have a strong bearing on product quality. Some of these are noted in the article "Design for Manufacture and Assembly" in this Volume. Ease of assembly and freedom from quality problems tend to go together. A simple, easy-to-assemble product design is more apt to provide higher product quality. Some assembly recommendations that bear particularly on product quality can be summarized as follows:

- Design parts so that they can be assembled only in the correct way. This normally involves incorporating some feature that prevents the component from fitting its mating part if it is not oriented correctly. One other possible approach is to make the parts symmetrical, so that there is no feature that can be misplaced (Ref 1).
- Design parts so that if they are omitted in assembly, it will be visually or otherwise obvi-

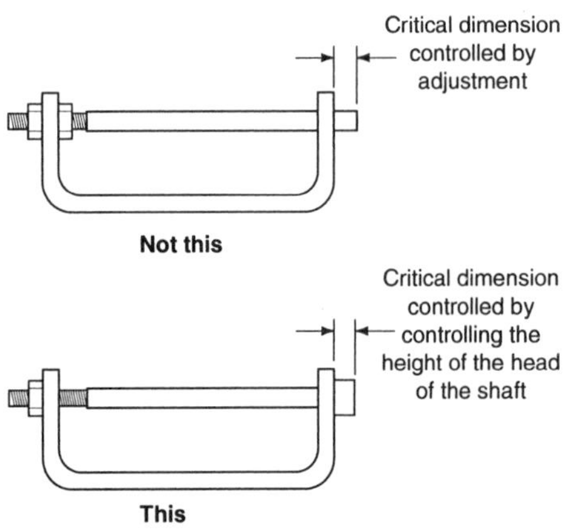

Fig. 4 Example of a design change made to eliminate an adjustment operation. The assembly in the upper sketch is adjusted to set the distance that the pin protrudes from the vertical surface. In the lower view, the adjustment is not needed, but the pin is manufactured with a controlled head height. This design has one less locking nut.

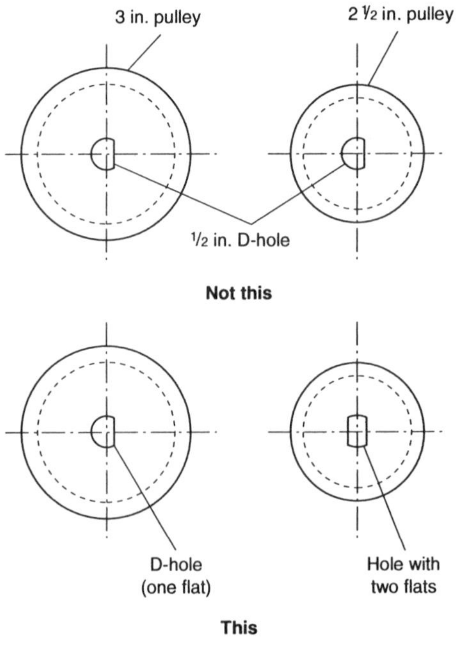

Fig. 5 Two pulleys used in a product, each of a slightly different size. In the upper view, both pulleys use the same design D-hole for mounting on a shaft, and it is possible to put the wrong pulley on a shaft. In the lower view, each pulley has a different mounting hole with different end configurations on the mounting shafts so that the wrong pulley cannot be assembled to each shaft.

ous. (For example, make it a different color than the surrounding parts or design it so that subsequent parts will not fit correctly if it is omitted.)

- Design parts so that they cannot be assembled out of sequence or in the wrong place or so that they can get damaged during assembly. This may involve some change in shape such as an added boss, arm, or other element or a change to make the mounting surface of the part curved or angled.
- Design parts so that they nest into the previously assembled part. This may obviate the need for additional fixtures and will help ensure that parts are assembled correctly.
- Design parts so that access to them in the product and vision of them is unobstructed (Ref 4). (This is a design for service guideline as well.) This will promote correct assembly and will help verify that it is correct. It will facilitate testing and replacement of parts, if necessary.

REFERENCES

1. D.M. Anderson, *Design for Manufacturability,* CIM Press, 1991
2. D.A. Garvin, What Does Product Quality Really Mean? *Sloan Management Review,* 1984, p 25–43
3. M. Phadke, *Designing Robust Products and Processes Using the Taguchi Approach,* video presentation, National Technological University, July 1990
4. Product Quality—Special Report, *Bus. Week,* 8 June 1987
5. P. Crosby, *Quality is Free,* Mentor Books, 1980
6. W.E. Deming, *Out of the Crisis,* Massachusetts Institute of Technology, 1986
7. D.H. Stamatis, *TQM Engineering Handbook,* Marcel Dekker, 1997
8. Using Quality Tools in DFM, *Tool and Manufacturing Engineers Handbook,* Vol 6, *Design for Manufacturability,* Society of Manufacturing Engineers, 1992
9. J.M. Juran, *Juran on Planning for Quality,* Macmillan, 1988
10. Brown, Hale, and Parnaby, An Integrated Approach to Quality Engineering in Support of Design for Manufacture, Chap. 3.3, *Design for Manufacture: Strategies, Principles, and Techniques,* J. Corbett, Ed., Addison-Wesley, 1991
11. S. Das, *Design for Quality Manufacturability,* New Jersey Institute of Technology, 1992
12. J. Bralla, Ed., *Handbook of Product Design for Manufacturing, McGraw-Hill, 1986*

Robust Design*

Lance A. Ealey, McKinsey & Company

ROBUST DESIGN, also known as Taguchi Methods or quality engineering, is an integrated system of tools and techniques that allow engineers and scientists to reduce product or process performance variability while simultaneously guiding that performance toward the best possible level. The techniques, which are useful across a spectrum from the research laboratory to the factory floor and beyond, have been developed and refined by Dr. Genichi Taguchi over the past forty years. Dr. Taguchi, long a respected consultant to Japanese industry, has in the past decade become well known outside of Japan, particularly in the United States and Europe. Thousands of companies routinely apply robust design techniques to achieve rapid, simultaneous improvements in cost and quality, as well as reductions in the time required to develop and optimize products and processes.

Major robust design techniques, tools, and concepts include the quality loss function, parameter design, tolerance design, signal-to noise ratio, technology development, and orthogonal arrays. Each of these is discussed briefly in this article. First, however, this article addresses some of the problems associated with traditional approaches to quality.

The "In Spec" Dilemma

Quality, according to the traditional American manufacturing interpretation, is inexact. Imagine that engineers at a company making axle shafts come up with two tolerance limits, one upper and one lower, for a critical seal fit. If the axle shafts fall within those tolerance limits, they are all thought to be of equally good quality. This is known as conformance to specification. Theoretically, if products conform to specification, they are perfect. Conformance to specification has been characterized as a go/no go or pass/fail type of control: If the part is in specification, it's a "go" or good part; if it doesn't fall within specification limits, it's a "no go" or defective part (Fig. 1). When all axles are "in spec," there are zero defects, and all is right with the world. But is it really? The more uniform the axles are, the more balanced or controllable the next event or process on the assembly line will be. Suppose all the axle shafts conform to specification, although all are grouped near the specification's lower limit—that is, the axle shafts are consistently smaller in diameter than the target specification. Suppose also that the oil seal manufacturer builds parts that conform to specification, although they tend to conform on the high side, consistently greater in diameter than target. The problem begins when the final assembler fits the large-diameter seals on the small-diameter axle shafts. The possibility of leaks occurring in the field is high, because the seals have a looser-than-target fit on the shaft. This phenomenon is known as tolerance stackup.

Tolerance stackup may not be a critical reliability problem when only two variables—the axle shaft and the seal—are involved. Nonetheless, it's what produces variability in final product quality. Tolerance stackup is the reason that one car's transmission shifts so smoothly, while the next works like a stick twirling in a bucket of bolts. In areas where many critical tolerances meet, tolerance stackup can be disastrous.

Example 1: Tolerance Stackup in Automobile Doors. Automobile door closing effort is a famous example of how tolerance stackup can affect a customer's perception of product quality. An automobile door closes over a virtual crossroads of tolerances (Fig. 2). The complexity of this stackup is so great that, even if all door-system components are assembled within specification but not to target, it's sometimes impossible to build a door system that meets all closing-effort and weather-sealing specifications.

In the not too distant past, doors on American cars required a greater effort to close than the doors on Japanese cars, because of tolerance stackup. Customers apparently want a door that closes at a certain number of pounds of pressure because it "feels" right to close the door at that level of force. When American engineers pulled apart Japanese cars to see how they were built, they discovered that, virtually without fail, doors on these cars closed at or around that ideal pres-

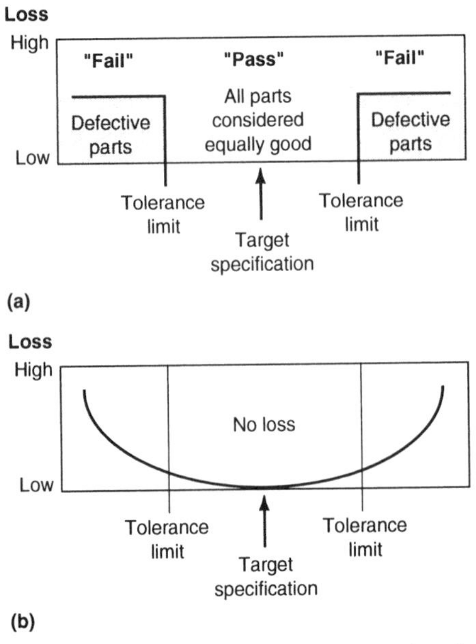

Fig. 1 Two approaches to quality. (a) Traditional conformance to specification (pass/fail) approach. A problem with this approach is that it makes no distinction between on-target performance and marginal, "almost-fail" performance. (b) Taguchi quality loss function approach. Loss increases quadratically as the product or process drifts away from the target specification.

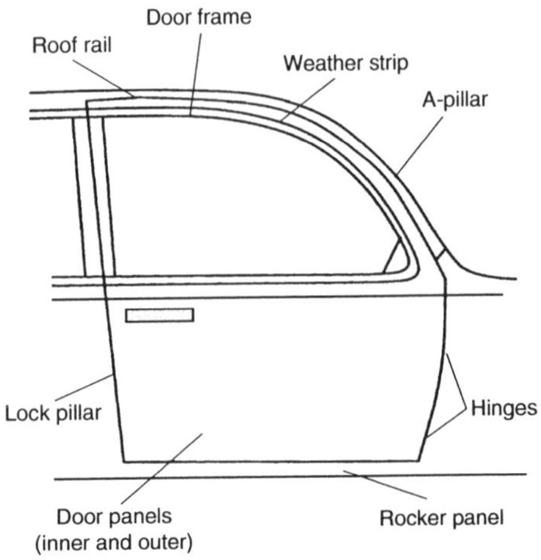

Fig. 2 Automobile door components that can contribute to tolerance stackup. Tolerance stackup may require reworking of parts to achieve proper fit.

*Adapted from Lance A. Ealey, *Quality by Design: Taguchi Methods and US Industry*, 2nd ed., ASI Press and Irwin Professional Publishing, Burr Ridge, IL, 1994. Used with permission. "Taguchi Methods" is a registered trademark of the American Supplier Institute.

sure, about 8 lb of force. The typical closing pressure for an American car was higher, about 12 to 16 lb of force, or varied from car to car. Why? Because all of the tolerances that met at the door were not at or near a target specification, but instead ranged from the specification limit's high to low ends. The result was doors that closed too lightly and thus didn't have watertight fits, or doors that had to be slammed shut.

On many older domestic automobile assembly lines, there are door fit "specialists" who shim hinges and seals or bend metal to make sure the recently attached doors have a watertight and nonbinding fit. As a rule, Japanese automakers don't have many door fit repair personnel on their assembly lines. Some of the latest vehicles from the "Big Three" American manufacturers are following the Japanese model, in which variation-causing tolerance stackup is designed out.

Example 2: Two Approaches to Quality in the Manufacture of Automobile Transmissions. A well-known case pitting conformance to specification against target specification involved an American auto company and a Japanese auto company. The American company asked the Japanese company to build transmissions for a car to be sold in the United States. The American company was building identical transmissions for the same car at an American plant. Both companies were supposed to build the transmissions to identical specifications. The American company, however, was using conformance to specification as its quality standard, while the Japanese company was building every transmission as close to the target specification as it possibly could.

After the transmissions had been on the road for a while, it became apparent that the American-built transmissions were generating higher warranty costs and customer complaints about noise than the Japanese gearboxes. So, the American company disassembled and measured samples of transmissions made by both companies. At first, the American company thought its gages were malfunctioning—it couldn't find any variability from transmission to transmission among the Japanese gearboxes. Instead of using the go/no go conformance to specification approach, the Japanese company had consistently built its transmissions to target specifications.

The American manufacturer discovered that by working to continuously reduce variability around targets, the Japanese manufacturer incurred lower production costs. The American company, on the other hand, was paying for inspection, scrap, rework, and, ultimately, warranty costs. Indeed, the Japanese manufacturer's bid for the work indicated that it planned a total investment two-thirds less than that planned for the American plant. Because the Japanese company worked lean, with very little inventory, it was saving nearly $10 million in its transmission plant, just in material handling equipment. The American company learned this lesson well and has since implemented a very successful policy of continuous improvement around target values as its company-wide goal. In recent years this goal has spread to the company's suppliers through its supplier certification program. In fact, the company's manufacturing productivity today is among the highest in the global auto industry, Japan included.

Cost and Quality. American racing engine builders have always considered uniformity around target specifications to be a higher level of quality than simply staying within specification limits. The whole notion of the desirability of a "blueprinted" engine supports this claim. An engine builder meticulously measures every critical dimension of an engine—bearing clearance, piston fit, and so forth—and removes those parts that aren't at the target specification. The builder then substitutes parts that are built exactly to the target specification and rebuilds the engine. Why? To ensure optimum performance and reliability—that's quality, isn't it? Yes, but inspecting every part and scrapping those that don't measure up is expensive quality.

To a lesser degree, this is the method that American companies have traditionally used to ensure quality, because they did not attach a cost to quality. The Japanese quality surge was keyed to cost reduction, and thus the two terms became interchangeable. This interchangeability is the reason the Japanese are fanatical about quality: not because they care more about their customers than American manufacturers do, but because they care more about cutting costs. This idea is the driving force behind the concept of uniformity at a target specification.

The On-Target Key

Uniformity in and of itself is not the entire answer to having the same quality level from the first part produced to the 10 millionth. The tolerances of those 10 million products must be grouped close to a target specification in order to fulfill the designer's intent, which must be based on customer requirements.

Example 3: Target Specifications for Vinyl Sheet. On many Japanese farms, neat rows of vinyl sheeting are positioned umbrella-like over certain easily damaged crops. These are a low-cost type of greenhouse. The vinyl is used by farmers to protect crops from the elements, thus extending the growing season. Typically, the thickness of the vinyl sheeting is, say, 1.0 mm, and acceptable tolerance limits for the thickness are ±0.2 mm. In other words, sheet that ranges in thickness from 0.8 to 1.2 mm is considered acceptable for use by farmers, according to the Vinyl Sheet Manufacturing Association.

One vinyl sheet manufacturer, through a series of quality improvement efforts, was able to reduce the variation in its processes to a remarkable degree. This manufacturer could hold tolerances of ±0.02 mm, quite an achievement. This new ability to create vinyl sheeting of extremely uniform thickness got the company managers thinking along these lines: If the company could make all its sheeting on the target specification's thin side, say at 0.82 ± 0.02 mm, it could save a considerable amount of money by using less material per square meter of vinyl sheeting. And, it would still be within the association's tolerance limits. Once in the field, however, this uniformly thin vinyl sheet tore easily, and farmers had to constantly repair or replace it. Sheeting that had a manufactured thickness that was uniform around the target thickness of 1.0 mm was thick enough by design to withstand the elements. Because of the increased number of complaints from the field, the association specified that the *average* thickness of the steel must be 1.0 mm and that the thickness variation could range ±0.2 mm.

Despite the fact that the company's sheeting was incredibly uniform in thickness, the technology that made that uniformity possible was not used to gain a quality-driven market advantage. Such an advantage—a sales point—would be increased toughness due to higher uniformity in sheeting thickness, and there would be a resulting increase in the company's reputation among consumers. Instead, the company used its technology to uniformly reduce the robustness of the sheeting, with subsequent unfortunate results for crops, farmers, and, ultimately, consumers.

A Philosophy of Loss

The concept of tying quality to degrees of loss is an important one. Dr. Taguchi defines product and process quality in just those terms. Quality, Taguchi says, is the loss that a product imposes on society after it is shipped. The philosophy behind that statement is elegantly subtle. All products ultimately cause loss: they wear out or break and need to be repaired or replaced. The sign of quality is how much service a product provides before wearing out. A high-quality product causes little loss, because it fulfills its intended function for a period of time that customers consider adequate. A low-quality product, on the other hand, causes loss at every turn.

There are no absolutes in the realm of quality: a superb-quality product of one era may be deemed merely adequate in the next. This is why it is so important for companies to base quality specifications on the end-customer's requirements, not on internal guesses or competitor performance. And customer requirements change constantly. In 1979 in the American automotive market, for example, the American companies were averaging five to eight defects per car produced, while Japanese offerings averaged from two to three. At the time, the Japanese quality levels were considered extraordinary. Today, those 1979 Japanese cars would be considered worse than all but the lowest-quality cars, showing how customer expectations have continued to evolve.

The amount of service a product gives before creating loss has to at least equal the consumer's dollar-value expectation. Ideally, it should exceed this expectation. The only way a manufacturer can ensure this type of quality is by producing products that closely match the original design's blueprint, which in turn must closely match the end-customer's expectations. The manufacturer does this by designing desirable and buildable products and then producing them to these customer-focused target values, rather than by producing them within relatively broad specification limits and then sorting out defective products (which adds cost). This idea represents a dramatic departure from traditional quality/cost thinking, in which quality and cost are thought to be in conflict (the higher the quality, the higher the cost). Taguchi argues that lower cost is the driving force behind higher quality. In a quote from 1984 he noted, "American industry is amazed that Japan can produce high-quality products at such low costs; they're not amazed at Japanese quality. If Japanese automobiles cost two times more, most Americans would stop buying them."

The Quality Loss Function

Taguchi's quality loss function (QLF) allows an engineer to put a dollar value on quality. It provides an evaluation of quality in financial terms—that is, quality loss due to deviation from the target as well as quality improvement. The QLF is perhaps most effective as a way to choose among competing designs for the best, most cost-effective solution, which, in fact, is the primary way many Japanese companies apply it. It can be used both to evaluate loss at the earliest stage of product/process development and to incorporate improvements in quality into management planning. It provides a common language understood by both managers and engineers.

The QLF curve is quadratic in nature: Loss increases by the square of the distance from the target value (Fig. 1b). Thus, if a deviation of 0.02 mm from the target value generates a $0.20 loss, then a deviation of 0.04 mm would cost $0.80, a deviation of 0.06 mm, $1.80, and so forth. In other words, if deviation is doubled, the loss is quadrupled. If deviation is tripled, the loss increases nine times.

The QLF allows a manufacturer to economically justify a practice that has long been a black art in American engineering: the setting of product tolerances. Much traditional tolerancing is done mainly by experience. After designing a product and assigning target specifications for all dimensions, the engineer checks the specifications on the last product the company designed and sets the tolerances accordingly. This method of setting tolerances is usually expensive, because the conscientious engineer will typically set tolerances tighter than probably necessary to reduce the risk of premature failure in the field. Traditionally, tight tolerance limits have been considered more expensive to maintain than looser limits, because they usually require better materials or high-quality manufacturing equipment.

But do these engineering specifications realistically relate to the way customers see quality? Once again, we've come back to the notion of paying for quality. The problem here is that the only real-world feedback comes when warranty costs skyrocket. The engineer then knows that the tolerances weren't correct, but the damage has already been done. On the other hand, if warranty costs do not go up, the engineer has no idea whether the tolerance levels are set too tightly, which could mean the product is overdesigned and thus more expensive than necessary to meet or exceed customer requirements.

By using the QLF, the engineer can set tolerances according to customer sensitivity to loss. Ideally, the manufacturer should set tolerance limits at the breakeven point between the cost to rework a product and the cost of the consumer-absorbed loss.

Example 4: Using QLF to Determine Voltage Tolerances for a Television Set. A television manufacturer builds TVs with a target voltage specification of 115 V. The manufacturer knows from customer research that the consumer will, worst case, tolerate a deviation of ±20 V from this target specification. If the voltage of a TV remains at 115, the consumer experiences very little loss: the TV works fine. If the voltage drops to 95 or less or climbs to 135 or more, most consumers will be dissatisfied with the performance of their TV and will take action. This action will come in the form of demands for the manufacturer to repair or replace the TV, which will cost, say, $100. This tolerance is the absolute limit of acceptance for consumers—many will become dissatisfied long before this limit is reached. In real-world terms, this tolerance usually represents the highest amount of loss possible. Obviously, a manufacturer would not set production tolerances this wide.

This limit of ±20 V is called the functional tolerance, or LD-50 (at least 50% of the consumers are unhappy), in Taguchi terminology. It is not a specification limit; rather, it is a functional limit. Taguchi uses this information and the QLF to develop actual production tolerance limits for the product. Loss projections can be developed for any other deviation from the target voltage if the cost of the loss at the LD-50 has been determined (in this case, $100 ± 20 V).

Now suppose that it is possible to rework the TV at the end of the assembly line so that its voltage is within limits at a cost of $2 per TV. It is now possible to use the QLF to discover at what voltage level the loss to the customer (and possibly to the manufacturer) will be equal to the cost of rework, or $2. In this case, this voltage is ±2.83 V (rounded to ±3.0 V). So, only if the voltage is less than 112 or greater than 118 should the manufacturer spend $2 to fix the TV.

The engineer has economically justified the specification limits: within the range of 115 ±3.0 V, the loss generated by variation equals or is less than the cost of reworking the TV. Had the engineer set the specification limits tighter than ±3 V, the manufacturer would be spending more money than is justifiable to meet the customer's quality needs. If, on the other hand, the engineer had set the tolerance wider than ±3 V, the manufacturer would save money in rework costs, but the cost to the consumer would be a great deal more, due to the nature of the quadratic curve of the QLF. For this reason, Taguchi says a manufacturer who betrays customers in such a fashion is worse than a thief: The manufacturer saves $2 but robs the customer of much more.

In figuring the loss caused by variation, a company can use only concrete dollar items such as warranty or scrap costs or include intangibles such as goodwill or customer inconvenience. If intangibles are included, one must look for a median versus a worst-case example, which could skewer subsequent loss computations.

Justifying Costs

QLF, aside from pegging tolerance limits to real-world effects, also helps a manufacturer in another way. The go/no go approach of conformance to specification makes it impossible for a manager to justify spending money to improve a process if that process is already building parts that conform to specification. Under the company's definition of quality, every part that conforms to specification is perfect.

The manager whose company is instead striving for continuous process improvement realizes that while perfect production (e.g., products that are built exactly to target specification) is impossible to achieve, the philosophy behind the concept provides benefits long after production quality is at world-class levels. Why? Because workers and management have a common goal: continuous improvement. The manager at this company can easily justify the expense by showing that output isn't uniformly grouped close to the target specification but rather is scattered widely across the specification range.

The concept of conformance to specification was an important first step in bringing manufacturing operations into control. But quality is relative. So while traditional American companies continue to measure quality performance in terms of defects per 100 products, world-class western and Japanese companies long ago began thinking in terms of defects per million or even per billion products. The enlightened manager will realize that the competition has moved its manufacturing performance far beyond conformance to specification.

There comes a point of diminishing returns in the quest to build all parts exactly to target specification: Quality becomes so uniformly high that one can't justify spending additional capital to improve it further. Under the concept of continuous improvement, however, workers continue to strive to improve quality through low-cost methods: by developing foolproof tools, refining parts bin positioning, and so forth.

Many American managers have problems with the idea of continuous improvement, in knowing exactly when the point of diminished returns is met. In other words, how do you know when to stop spending cash to improve quality? For the answer to that question, we turn once again to the QLF.

As described earlier, the QLF can be used to set tolerance limits for a product. Knowing the product or process deviation, and the loss due to scrap and rework and current tolerances, the manufacturer can determine the loss per unit for a given performance level of a process. It can then calculate the loss due to variation and decide whether further expenditures for improvements in quality are justified.

It is important to remember that according to the QLF, loss is quadratic, not linear, depending on how far the product is from the target specification. Loss increases dramatically the further the product strays from its target specification. This loss can include everything from the effects of tolerance stackup to the consumer inconvenience caused by products that must be returned for repair. It can also be rooted in strictly accountable costs, such as warranty claims and rework.

Experiment factor level settings

Experiment number	A	B	C	D
1	1	1	1	1
2	1	2	2	2
3	1	3	3	3
4	2	1	2	3
5	2	2	3	1
6	2	3	1	2
7	3	1	3	2
8	3	2	1	3
9	3	3	2	1

The 4 factors being studied in this experiment

The experiments are run across the array

E.g., in experiment 1, all factors are set at their first level; in experiment 2, factor A is set at its first level while B, C, and D are set to level 2

The 3 different settings each factor will be set at during the experiments

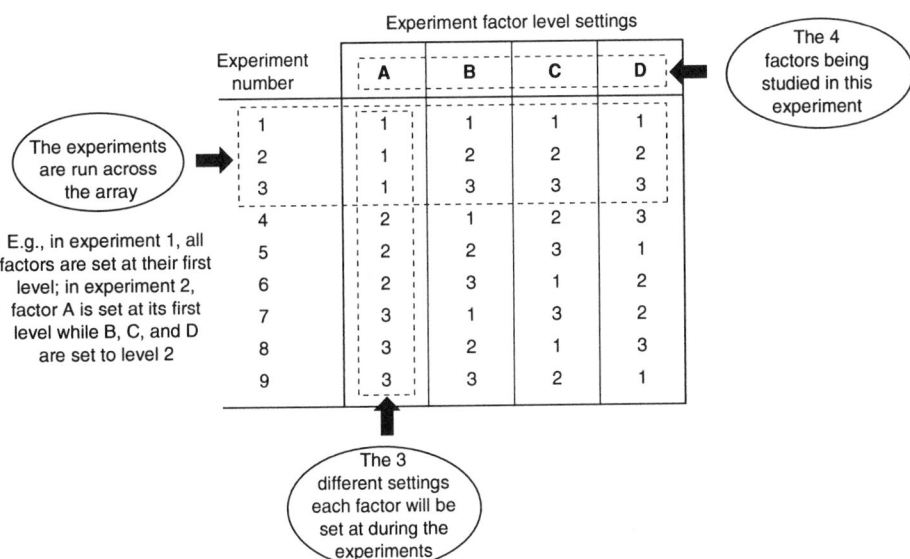

Fig. 3 Example of an orthogonal array. This L9 orthogonal array allows the experimenter to study the effects that four different factors at three different level settings have on the product or process under development. Because all factor levels are exposed to all other factor levels an equal number of times, it becomes possible to isolate the effect that each factor level has on the experiment. This allows the experimenter to identify the combination of factors that should provide optimum performance.

Parameter Design

Taguchi defines the development of a product or a process as a three-stage operation: system design, parameter design, and tolerance design. System design is the primary development stage in which the basic architecture of a product or a process is determined. During system design, the design engineer seeks to create a product or a process with distinct capabilities that will later serve as sales points. The system designer calls upon his or her expertise in the design of similar systems to generate a new system that is superior in some way to previous designs. Robust design usually isn't applied to system design. It's estimated that American engineers focus 70% of their efforts on system design (compared to Japanese engineers' 40%), and only 2% on parameter design. Japanese engineers, on the other hand, apply 40% of their efforts to parameter design, the stage where costs can be most efficiently controlled.

Quality Characteristics. During parameter design, the design or process engineer seeks to optimize system design performance through experimentation that minimizes variation in the face of uncontrollable user and environmental factors. The engineer begins by identifying the quality characteristics that most affect system design performance, especially as it affects the customer. For example, a quality characteristic for a car might be that it starts easily and quickly every time, no matter what the external conditions are (e.g., a temperature range of –30 to +130 °F). Most products will have more than one quality characteristic, and it is important for the engineer to discover which quality characteristics consumers value most highly.

Noise Factors. During parameter design, the engineer seeks out all factors that will have a negative effect on the quality characteristic. In the case of the starting car, the list of factors would include outside temperature, humidity levels, the altitude range at which the car may operate during its lifetime, the fuel grades available in the regions the car may be used in, and so forth.

Notice that the engineer really has no control over these factors. For example, the car company can't tell the car buyer that the car shouldn't be operated when the temperature dips below or climbs above a certain temperature. So the engineer's goal must be to make the car resistant to these uncontrollable factors or, in Taguchi's words, "robust against noise." Taguchi calls these uncontrollable factors *noise factors.* Noise factors usually cause quality characteristics to deviate from target values, which subsequently causes quality loss.

Temperature, altitude, and fuel grade are considered *external noise factors* because they affect the product but are outside the control of the product designer. Two other types of noise factors also exist: *internal* (e.g., deterioration over time as the car battery wears down or as critical engine parts wear out) and *product-to-product* (e.g., one car assembled to target specification always starts while another, plagued by inconsistently made parts, does not).

Once the engineer has identified the various sources of noise that affect the quality characteristic, he or she sets about to create a system design that will allow it to operate consistently over time, regardless of noise. The engineer doesn't attempt to control the noise factors, but instead designs around them: Attempts to control noise usually prove impossible or add cost.

Establishing Design Parameters. The engineer then identifies design elements that need to be robust against the various noise factors. In the case of the car, these would include the drain placed on the battery, the primary ignition voltage, the efficiency of the starter, and the effectiveness of the fuel and air delivery systems. These factors are called parameters: factors that will be optimized to create a robust product (in this case, a car that starts easily, quickly and consistently under all conditions).

Once noise factors and parameters have been identified, the engineer determines multiple settings for each parameter in conjunction with others who have special knowledge regarding the various parameters. For example, starter settings might include turning the engine at three different speeds (i.e., the current speed and two alternatives that the engineers believe, due to their experience, might produce better results).

Experimentation to Determine Optimum Parameter Levels. Once the various settings or levels of operation for each parameter have been determined, the engineer begins to design the series of experiments that will ultimately reveal the optimum level of each parameter. The selection of these various levels is far from random; the engineer must use deep knowledge of the product or process to discover the best alternatives to the current level. With Taguchi's approach to experimental design, the engineer typically needs to run only a portion of the total experiments possible for various factors at their various levels. Rather than focus on the "trivial many," robust design focuses on the "vital few": A large number of factors with different operating levels can be easily assigned with a small number of experiments using orthogonal arrays (described below). However, the objective isn't simply to minimize the number of experiments, but to find reproducible results.

Analyzing Results Using Orthogonal Arrays. The engineer always begins experimentation using the lowest-cost parts and components, with the goal of optimizing them so that the product or process is on target, with little variability, and provides the most stable and reliable performance at the lowest manufacturing cost. At this point, the engineer will use a linear graph to assign the control factors to an *orthogonal array* (Fig. 3). (An *orthogonal array* is a matrix of numbers arranged in rows and columns—each row represents the state of the factors; each column, a specific factor that can be changed from experiment to experiment. The word orthogonal comes from the Latin for "at right angles." Additional information is provided later in this article in the section "Orthogonal Arrays.")

The orthogonal array for the control factors is called the *inner array*. The inner array varies the combinations of the control factors in a systematic fashion: Each factor setting or level of operation is exposed to every other factor level under experimentation the same number of times. This allows the engineer to make a balanced comparison between factor levels under a variety of conditions. A factor that exhibits a consistent effect with various conditions of the other factors will be more likely to reproduce this effect in the field.

Once the inner array is established, the engineer can create an orthogonal array for the noise factors, which is called the *outer array*. Each noise factor is also assigned several levels. The engineer is now ready to begin the series of experiments. Each experiment in the inner array is run several times against the combination of noise factors in the outer array, and each time an experiment is run, the engineer varies the levels of the noise factors. Thus, the engineer exposes the control factors of each experiment to the effects of various combinations of noise factors, which allows the evaluation of the product or process with respect to robustness against noise and the related costs.

The engineer will typically only run a fraction of the total experiments possible for the various parameters at their various levels. Factors that exhibit consistent effects, no matter how many different noise and parameter factors are changed around them, will be significant in increasing the mean result of the experiments or in reducing the deviation from the target specification.

Once all the experiments have been randomly run, the engineer begins to analyze the generated data. Because each experiment was run several times, the engineer has a performance range that shows the average result of the experiment and the performance variation around that average. The engineer must also verify the predicted results of the experimentation by performing a confirmation experiment. If the predicted results are not confirmed, the engineer should redevelop the experiment.

Signal-to-Noise Ratio. The information generated by the experiments is used to calculate a value for the signal-to-noise (S/N) ratio for each experiment. The S/N ratio is in reality a data transformation that expresses the variability and ideal function of the quality characteristic better than the actual data would. The S/N ratio indicates how much variation around the average or mean performance each experiment has generated, projecting field quality performance from the experimental data. The S/N ratio is the operational way of incorporating Taguchi's QLF into experimental design. The larger the S/N ratio, the less variability around the mean, and the smaller the quality loss.

The S/N ratio shows the engineer exactly how robust the product developed from each experiment will be when exposed to a wide range of noise factors. For example, one engine may start well under cold temperatures and at low altitudes but suffer from vapor lock at high temperatures and high altitudes. The engineer is seeking the combination of parameter settings that allows the engine to start easily no matter what the altitude, temperature, and other conditions.

At this point, the engineer could simply choose the experiment that produced the highest S/N ratio as the optimum engine-starting condition. However, the result would probably be far from optimum. Because the experiments were using orthogonal arrays, however, the engineer can go one step further. He or she can conduct other analyses to find the effect of each factor on the average or mean performance of the quality characteristic in the experiment. The engineer is now ready to sort control factors into those that have a strong effect on reducing variability, those that have a strong effect on mean performance, and those that have little effect on either.

Thus, the parameters are classified according to the effect they have on the S/N ratio. Those factors that affect the S/N ratio are used to reduce variability. Those with a strong effect on the mean but little effect on variability are used to adjust the mean performance closer to the target. They are called *adjustment factors*. Those with little effect on either are also important. Here the engineer can select the cheapest levels to reduce cost. They are referred to as *cost-adjustment factors*.

The engineer is thus able to create a product with those factors that can give, in the starting car example, the fastest starting time, those that can reduce the variation in starting time, and those that allow cost to be minimized. The engineer can thus improve quality while reducing cost. From the experimental data, it is possible to choose the parameter settings that will give the fastest starting time with the least amount of variation at the lowest possible cost.

Finally, the engineer combines all of the selected signal, control, and unimportant factors into yet another experiment to check the reproducibility of the results. If the results of this experiment don't come close to yielding the performance specified by the quality characteristic, the engineer must rethink the experiment. Either an important parameter has been missed that directly or indirectly affects starting performance, or the various parameters may have combined to create an interaction that cancels out their combined efficiency, which results when a poor quality characteristic is picked for experimentation. In either case, the engineer will have to redesign the experiment.

If the results of the confirmation experiment do yield an optimum combination that allows consistently fast starts, the engineer checks the results against the desired performance of the original characteristic. If the performance of the optimum combination of factors shows enough improvement in starting-speed consistency, the engineer has successfully completed the experimental-design work.

But what if the improvements in quality and cost still aren't great enough to meet the competition? Suppose, for example, that the starting speed consistency still doesn't meet the quality-characteristic specification. The engineer could try another experiment, or proceed to tolerance design. Improvements made during tolerance design usually require more expensive materials or components and therefore add cost.

Tolerance Design

During tolerance design, the engineer will systematically identify how much the performance levels of certain parameters will have to be increased in order to meet the quality characteristic. The engineer calculates the percentage contribution that each of the factors makes toward achieving the required performance for the quality characteristic. From this information, the engineer decides how much to reduce the tolerance levels of each factor in order to achieve the objective. Tightening the tolerances of the factors almost always involves upgrading with higher-cost parts or components. The QLF is used during tolerance design to find the most cost-effective way of determining which tolerances must be tightened, and which can be left in place or opened up, in order to reduce cost.

For example, if the automobile starter is drawing too much power from the battery, the engineer will upgrade it with a more efficient and expensive unit. By using the QLF, the engineer can calculate just how much better the starter has to be to meet the quality characteristic. Hence, he or she will not upgrade a parameter beyond actual need, which would be more costly than necessary.

At this point, the experimenter can perform another experimental design run to select the most appropriate upgrades for the various factors. The engineer has now completed optimization and can be assured that the starter will perform as intended and under all foreseen conditions. As a final step, the loss function can be used to calculate the savings generated by the optimization.

Technology Development

As Taguchi began to visit more large American companies in the 1980s, he realized that there was a fundamental difference between the ways that American and Japanese companies developed technology. The Americans, he found, do much more "far-upstream" development in technology, which is focused on creating new technologies, whereas the Japanese tend to concentrate much more on "mid-stream" development work, which is typically focused on applying technology at the product and process level. In other words, whereas many American companies develop proprietary technology based on pure research, Japanese companies tend to concentrate their efforts much more on developing technology for specific applications in products and processes.

Because of this, the Japanese company ensures robustness only after the product planning stage, which usually means that every product derived from a specific technology has to be made robust.

This can require significant resources and can take up quite a bit of time, even for companies as efficient at product development as the Japanese.

While many American companies apparently do quite a bit of basic technology research, few ensure robustness at the technology level. Instead, the typical reaction that Taguchi found when talking to research engineers and scientists in American laboratories was that quality assurance was not part of their job: their job was to develop the raw technology itself and pass it on to the product and process development engineers, who would make it work for specific applications.

Thus, both Japanese and American companies have shortcomings in terms of technology development: the Americans do a lot of basic research but do not attempt to make the technology robust, while the Japanese typically work to make technology robust only at the product and process development stage, which tends to be an inefficient use of resources.

Taguchi has said that he considers the western method of research and development to be superior to the traditional Japanese way, but that western companies have to begin to ensure robustness in technology during research and development in order to be successful.

The problem with the Japanese method is simple: when technology is made robust in the context of a specific product application, many times the same technology transferred to another product application will fail miserably, because the robustness achieved earlier was linked directly to the original product application. Instead of this, argues Taguchi, why not make the technology used in the products robust before product planning takes place? By doing this, the company would only have to "tune" the technology to apply it across many different product or process situations. The technology itself would be robust against noise, which in this case could include the product application itself!

In technology development, the engineer works to make the *basic function* of the technology robust, which Taguchi defines as an engineering issue, as opposed to the technology's *objective function*, which is typically a quality characteristic requested by the customer. The basic function can be thought of as the main input/output relationship of a technology. This relationship involves energy transformation. For example, in an automobile engine, the basic function of the engine is to transfer the energy stored in gasoline or diesel fuel into motive power. Objective functions, on the other hand, could include the specific quality characteristics of a specific engine: high miles per gallon, low noise or exhaust emissions, and so forth. But the basic function of the engine remains the same: how efficiently does it transform energy into movement? Taguchi says that once you solve this basic energy transformation question at the technology level, developing the technology for a specific application simply means identifying those factors that will allow you to tune the technology to a specific application.

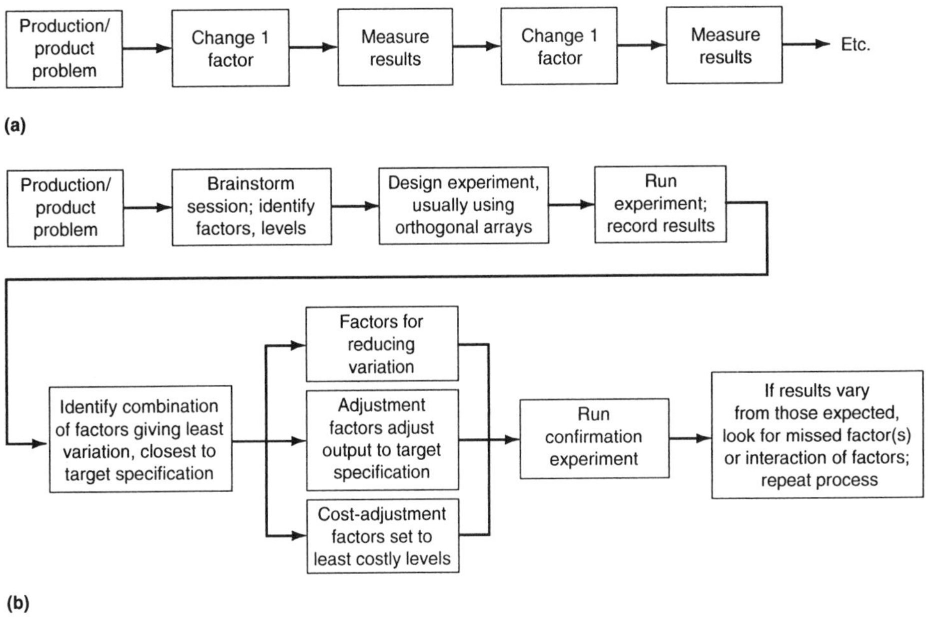

Fig. 4 Two approaches to problem solving during engineering design. (a) Traditional one-factor-at-a-time approach. (b) Taguchi parameter design approach

The engineer has thus made the engine technology itself robust: Developing entire families of engines may be as simple as using the tuning factors to make the technology work across many different applications. The brilliance of technology development is that it removes much of the uncertainty that inevitably surrounds a new technology introduction. Once the technology itself is made robust, application engineering becomes simply a matter of tuning, potentially cutting millions of personnel hours out of product development cycles, allowing a company fast response times to market forces, and cutting development costs at the product level by millions of dollars.

Example 5: Technology Development of a Machining Process. A major Japanese automaker has used technology development to make process technologies robust. The automaker's use of a new heat treatment process raised the hardness of the steel to be machined to such a level that it became very difficult to cut using regular machining technology. The engineering team decided that the energy transformation in this case had to do with the relationship between the dimensions that were programmed into the numerical-control (NC) machine as far as how the steel was to be cut and the actual dimensions of the steel piece after it was cut. The notion was that this should be a perfect proportional relationship.

The goal was to develop the technology itself (i.e., a generic NC machine technology) to such an extent that no matter what hardness of steel, type of part, cutting tool conditions, etc. were involved, the technology would always give this perfectly proportional output to any dimensional programming input. As a result of the study, the standard deviation of the NC machining process was reduced to one-tenth that of prior performance, and this was achieved regardless of what kind of part was to be machined.

The implications of this experiment are enormous. Now, when the company's engineers need to develop a process for machining a part, they can simply tune the NC machining technology that was made robust in the above experiment, cutting the time and effort that have to be expended to a mere fraction of former levels.

Orthogonal Arrays

The concept of the orthogonal array isn't new: The Swiss mathematician Leonhard Euler did considerable research on them in the 1700s. However, orthogonal arrays were little more than academic curiosities until the early to mid-20th century, when researchers led by R.A. Fisher began to exploit their inherent balance when laying out and conducting experimentation. What is new regarding Taguchi's use of orthogonal arrays involves the kinds of information he is looking for when he employs them in experimentation.

Designing Experiments without Using OAs. A typical way of performing industrial research in the United States is known as the "one factor at a time" method (Fig. 4a). An engineer identifies each of the design factors (also known as variables) and its different performance values. During the first experiment, all factors are held at their current values and the result is recorded. In the second experiment, one of the factors is changed to its low or high value. If this causes an improved result compared with the initial experiment, that factor value is adopted and in the next experiment one other factor is changed. The experimenter continues in this manner until the system has been nudged as close as possible to its performance target.

The trouble with results obtained in this manner is that they often have little in common with what happens in real life. In reality, rarely does one variable change while the rest remain fixed. Let's say an engineer is trying to optimize an automobile steering system. The engineer has decided that some 13 factors will be useful in attempting to optimize the steering (e.g., spring stiffness, steering geometry, etc.), and that each factor will be tested at three different settings or levels. The problem with using one factor at a time analysis becomes clear when you realize that, for example, while a certain spring stiffness provides ample performance when the road is dry, it may not work as well if the road is wet or icy, or if tire pressures are too low or too high. Thus, results obtained with the one-factor-at-a-time method may perform brilliantly in the controlled environment of the laboratory, yet fail miserably under the ever-changing conditions of the open road.

What this engineer needs, then, is some way to efficiently compare each of the three performance levels of any given steering factor with all the performance levels of all the other steering factors under test. If the engineer could do this in such a way as to separate, for example, the effect of spring stiffness at its high, low, and current settings on the experimentation as a whole, he or she could select the spring stiffness setting that consistently has the strongest positive effect. Since that spring stiffness setting has performed well, despite being tested in conjunction with every other setting of all 12 other suspension factors, it stands a very good chance of reproducing these positive results in the real world.

One way of doing this would be to perform all 1,594,323 experiments (3 to the 13th power). This would not be very efficient, to say the least. Our engineer needs a method that delivers reproducible results with a minimum of cost and in the shortest time possible. That's where the orthogonal array comes in.

Benefits of Using Orthogonal Arrays. The great power of the orthogonal array lies in its ability to separate the effect of each factor on the average and on the dispersion of the experiment as a whole. By exploiting this ability to sort out individual effects, large numbers of factors can be tracked simultaneously in each experimental run without confusion, thereby obviating the need to perform nearly 1.6 million possible combinations.

The ability to separate these effects is critical to Taguchi's technique (Fig. 4b). As noted above, when factors exhibit a consistent, positive influence on the experiment, no matter how many other factors are changed in the course of experimental runs, the results stand an excellent chance of working effectively under real-world conditions, thus producing a robust product. The ability to generate such reproducible results is the primary reason that Taguchi uses orthogonal arrays in his quality engineering techniques.

The arrays used by Taguchi are balanced, pairwise, across columns. This means that in any pair of orthogonal columns, all combinations of different factor levels occur, and they occur the same number of times.

In the L27 orthogonal array used by our car steering engineer, for example, any column may be compared to any other column, and nine different combinations, (1,1), (1,2), (1,3), (2,1), (2,2), (2,3), (3,1), (3,2), and (3,3), will appear three times each. This means that all three levels of spring stiffness will be exposed to all three levels of bushing stiffness, steering geometry, and so on an equal number of times by the time the final experiment is completed. This systematic exposure of every factor level to every other factor level is what's known as the "balancing" aspect of the orthogonal array.

The orthogonal array thus becomes the framework across which the steering engineer conducts the experimentation. Each of the rows in the array constitutes one experiment, and each column in the array represents a test factor. Column one, for example, could represent spring stiffness. Each of the 13 steering factors in our example will be tested at three different settings (for spring stiffness, these would be the current setting and a stiffer and softer setting, notated 1, 2, and 3 in the array). The engineer would then perform all 27 experiments, adhering to the arrangement of factor levels shown in the OA, and drawing lots to introduce an element of randomness to the experimentation.

It is important to note that the engineer isn't primarily interested in the end results of the 27 experiments in and of themselves. These may not produce large improvements over pre-experiment performance. (In this case, with almost 1.6 million potential combinations, there's a good chance they won't!) Instead, the engineer is mainly interested in exploiting the column-wise orthogonality of the array in order to separate the effect that each of the three spring stiffness settings, for example, has on the outcome.

In Taguchi's system, the orthogonal array allows the experimenter to identify those variables that will reduce performance variation, as well as those that will guide the product back to its performance target once robustness has been achieved. The orthogonal array is also used to find variables that have little effect on robustness or target values. These can and should be set to their least costly levels.

Taguchi often combines a second orthogonal array with the original to introduce "noise" into the experimentation. (The term *noise* in this sense refers to the error-causing factors that are either too expensive or impossible to control. Noise for the steering engineer could include ambient temperature, weather conditions, etc.) This practice can increase the potential robustness of the product to truly formidable levels.

One important final use Taguchi makes of the orthogonal array is as a screen for interactions among the control variables. Such interactions can compromise the reproducibility of the results during the confirmation experiment. If, on running the confirmation experiment, the steering engineer's results fall far short of those anticipated, strong interactions have probably been encountered. The engineer must then rethink the experimentation to find factors or quality characteristics that either do not interact or are robust enough to overcome interactions. In essence, Taguchi harnesses the power of the orthogonal array to provide a "design review" for the product before scale production is begun.

SELECTED REFERENCES

- American Supplier Institute, Inc., *Tenth Annual Taguchi Symposium Case Studies and Tutorials*, Dearborn, MI, ASI Press, 1992
- American Supplier Institute, Inc., *Collection of Publications on Taguchi Methods*, Dearborn, MI, ASI Press, 1984
- Lance A. Ealey, *Quality by Design: Taguchi Methods and US Industry*, 2nd ed., ASI Press and Irwin Professional Publishing, 1994
- R.A. Fisher, *The Design of Experiments*, Edinburgh, Oliver and Boyd, 1960
- M. Phadke, *Quality Engineering Using Robust Design*, Prentice-Hall, 1989
- G. Taguchi, *On-Line Quality Control during Production*, Tokyo, Japanese Standards Association, 1981
- G. Taguchi, *System of Experimental Design*, Quality Resources/Kraus and American Supplier Institute, 1992

Risk and Hazard Analysis in Design

G. Kardos, Carleton University

NO SYSTEM devised or used can be 100% safe or error-free. The objective of risk and hazard analysis is to identify the level of risk and to pinpoint the parts of the system that represent the greatest risk for failure. Then, if the analysis is used properly, steps can be taken to eliminate the cause or reduce the risk to an acceptable minimum. It has been demonstrated that hardware systems approaching a "failure-free" condition can be produced when actions are taken at all levels that are based on:

- Attention to past experiences with similar systems
- Availability of risk information for all project personnel
- A sound, aggressive risk and hazard analysis during all phases
- Development of suitable corrective action and safety programs based on the analysis
- A continuous and searching review of all phases of the program efforts

Rigorous applications of risk and hazard analysis have made difficult technological feats such as landing on the moon relatively accident-free.

The various analysis techniques have grown out of the search for system reliability. Consequently the approach is hardware oriented, with the emphasis on ensuring that hardware is able to perform its intended function. Backup systems and redundancies are used to reduce such risks. Through cost/benefit analysis, the performance of the system will have a computable value that can be compared to the cost of accomplishing the mission.

Risk and hazard analysis tools have been developed to ensure system reliability in critical applications. With the increased emphasis on safety, reliability, and achieving mission objectives, design teams must incorporate risk/hazard considerations in their designs. Figure 1 is a flow chart that shows the integration of risk and hazard analysis in the overall design process. Even if designers or design managers are not directly responsible for carrying out these analyses, they must be familiar with the methodology, so that they understand how they are carried out and how they can respond in terms of design or system changes. Most efforts are best carried out during early design phases, and they can be effectively used during design reviews to provide valuable feedback to the design to avoid failures.

This article outlines the principle methods of risk/hazard analysis as presented in the literature and as practiced by industry. It is intended to provide an appreciation for each method and an insight as to what can and cannot be accomplished with risk/hazard technologies. Related information is provided in the articles "Reliability in Design" and "Safety in Design" in this Volume.

The Nature of Risk

All industrial and human activity involves risk and risk taking. Usually this risk is undertaken unconsciously or with only a vague notion as to the kind and magnitude of the risk. Here, the nature or philosophical basis of risk is explored because all hazard/risk analysis implies the acceptance of some level of risk. All risk analysis methods have as their prime objective the prediction of future risks.

The literature on the prediction of alternate futures is mixed with biases based on personal value systems and assessment of present trends. In truth, most predicted futures tell more about the person making the prediction than about any expected future. Decision makers, such as designers, cannot simply accept the validity of the predictions of the various experts, but must examine the logic, the assumptions and presumptions, to make their own evaluation and conclusions.

To clarify what is acceptable risk, Starr et al. (Ref 2) suggest it is useful to recognize the existence of four different definitions of "risk":

- *Real risk* is determined by actual circumstances as the future unfolds.
- *Statistical risk* is determined from currently available data, based on assumptions that a large number of future systems will, on the average, act the same way as a large number of similar past systems.
- *Predicted risk* is based on analysis of system, models, and historical data.
- *Perceived risk* is risk seen intuitively by individuals or society.

It should be noted that only the "real risk" is a property of the hardware. The other three represent the conception of the risk represented by the system. This conceptual risk can change and the other three forms of risk can change without changing the real risk in any way, simply by a better understanding of the system.

Because of the conceptual nature of risk, any careful analysis, experiment, or historical study may not correspond well with an individual's or society's perception of the risk. Statistical risk

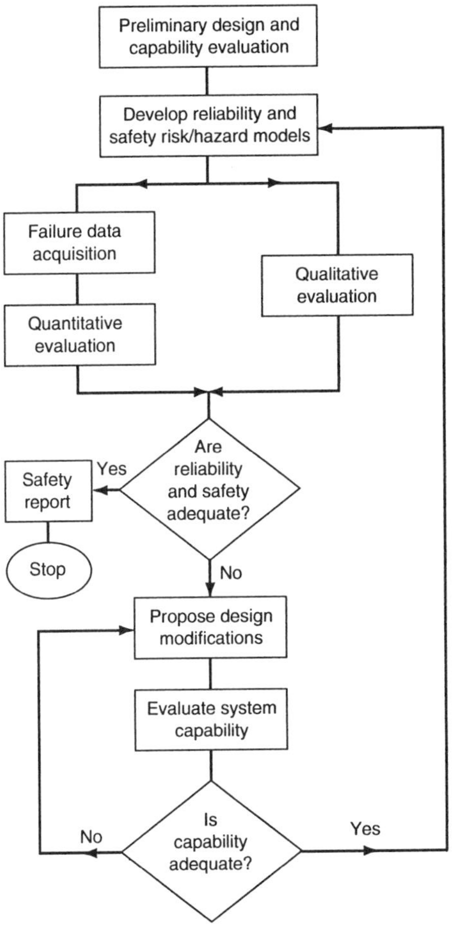

Fig. 1 Flow chart showing the integration of risk and hazard analysis into the design process. Source: adapted from Ref 1

and predicted risk are generally analytical and quantitative, whereas perceived risk is qualitative. Because quantitative evaluations can only deal with those parameters that can be counted or computed, it is bound not to include *all* the parameters that concern an individual.

Perhaps the most important factor not accounted for in analytical methods is risk manageability or controllability. Individuals feel less threatened when they feel they have control over the time, place, and extent of the risk they are exposed to. The same statistical risk may be perceived to be less when individuals have a free choice over whether they will undertake that risk than when they are faced with having to simply "live with it."

Another important factor in the perception of risk is the probable severity of the consequences if an accident were to occur. A complete assessment of risk requires that the potential effects of an accident be somehow integrated with a probability of its occurrence. Unfortunately, the integrating function is not known and must be left to judgment.

Linked closely to severity of consequences in the perception of risk is the episodic nature of consequences. It seems that the size of a potential accident is more important than the probability of occurrence. This truly represents society's value system to a great degree. Activities capable of producing catastrophic accidents, therefore, are seen as a great risk and must be more stringently controlled than high-frequency individual risks.

As with risks, benefits may differ in perception from assessed value. Benefits are surely assessed more positively by those who benefit most directly. As with risks, the more directly the benefit impinges on us personally the more important it is.

It is in this region where gaps between perceived and assessed risks and benefits exist that government agencies are called upon to resolve the conflicts. When there is agreement between public and institutions, unanimity is easily achieved. But when differences in perception occur the resolution becomes difficult.

No general method of dealing with difference exists. The resolution cannot be made by application of technological, probabilistic, or statistical rules. It must be accepted that these differences are real and must be resolved in terms of humanistic and political concepts. The technical methodologies described in this article can identify possible hazards and possibly identify the resources that are necessary to avoid or reduce the risks; however, whether the resources *should* be expended to alleviate the perceived risk remains a political-humanistic-management decision. The justification for undertaking these technological studies is that they give a clearer picture of decisions that must be made. They cannot be expected to provide an automatic decision process.

One method often used for deciding on whether to undertake a project is the risk/benefit analysis (see the section "Risk/Benefit Analysis" in this article). Such an analysis involves two simplifying assumptions: first that risks and benefits can be stated in the same terms (dollars, for example) and second that both the risks and benefits are shared equally by all affected. Both these requirements can easily be satisfied when dealing with acquisition of a piece of hardware to perform a specific narrowly defined function (e.g., military hardware), but in instances of environmental hazards, neither assumption is likely to be satisfied.

The technological methodologies can quantify the inputs independent of the decision process. The value system that establishes the worth of alternatives must be added by all interested parties. In such a situation, clear distinction must be drawn between the scientific estimates of risk and perception of risk.

Generally the scientific-technological community advocates the use of quantitative analysis and risk/benefit assessments as the sole measure of whether to undertake a project. This is attractive because it makes the decision process simpler. But an opposite point of view is advocated by Kletz (Ref 3), who says that if risks exceed a certain level, they should be removed as a matter of priority, even though removal may be expensive; otherwise there is no incentive to look for less expensive solutions.

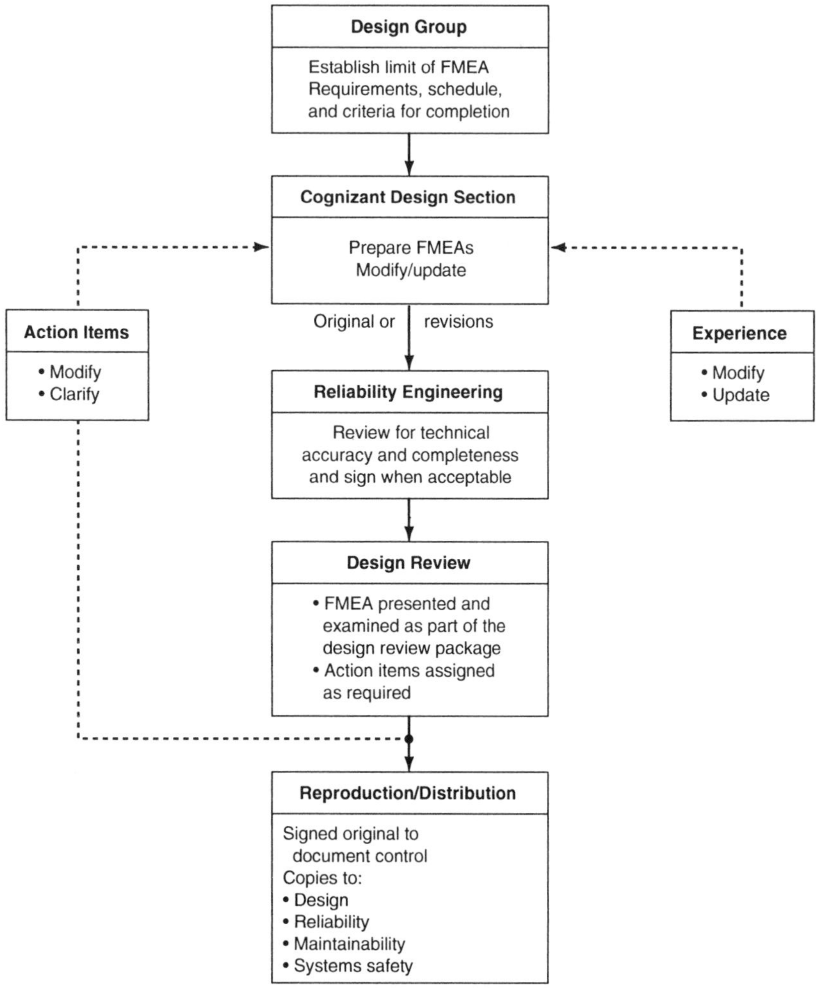

Fig. 2 Failure mode and effect analysis flow chart. Source: Ref 4

Failure Mode and Effect Analysis

Failure mode and effect analysis (FMEA) was originally developed as a tool to review the reliability of systems and components. It has been modified over the years to include criticality analysis (CrA) and failure hazard analysis (FHA) to help determine equipment safety requirements. Its intent is still to ensure reliable system function. The objective of FMEA is to expose all potential element failure modes to scrutiny, to classify and quantify their possible causes with their associated probability of occurrence, and to evaluate their impact on the overall system performance.

Failure mode and effect analysis normally is carried out as part of a comprehensive reliability program. The analysis is performed at increasing levels of complexity as the design progresses, so that the recommendations resulting from the analysis can be implemented early at minimum cost. The results are to be used to guide other activities such as design, purchasing, manufacturing, and quality control.

Failure mode and effect analysis is a systematic, organized procedure for determining, evaluating, and analyzing all potential failures in an operating system; it is used to supplement and support other reliability or safety activities. The

FMEA process involves examining each potential failure mode to assess if the system elements support reliability and to ascertain the consequences of the failure on the entire system. The general process involves the following steps:

1. The probable failure modes of the system are identified, evaluated, assessed, and documented.
2. The documentation of possible failure modes is used to check and verify the safety of the design.
3. Corrective actions are formulated to eliminate or reduce to an acceptable level the consequences of the failure.

To undertake an FMEA, it is essential at the outset to have well-defined and documented criteria for judging performance or safety. What constitutes hazard and at what level must be established at the performance definition stage. To be meaningful, the definition must classify failures in a clear and consistent manner. The *measure of acceptability* must be established before FMEA is undertaken. Similarly, the *conditions* under which the equipment must function throughout its entire life must be specified because the failures have to be analyzed under various operating modes and the transition periods between them.

Failure mode and effect analysis can be carried out at various levels of detail. Failures can be postulated at the component-part level or at subsequent higher subsystem levels. The level of detail at which the analysis is carried out is usually determined by cost constraints and the reliability required of the system. The more detailed the analysis, the greater will be the number of hazards considered, and the greater will be the cost and time required to carry out the analysis. Sometimes a less detailed analysis carried out in time to change the design may, in the long run, be more useful than an in-depth analysis available after completion of the design. The level of detail to which an analysis should be carried can only be based on experience.

Carrying out the FMEA is greatly enhanced by the use of standardized documentation. Failure mode and effect analysis requires an organized systematic approach; standardized documentation is a visible forcing technique that ensures that none of the steps are overlooked. The documentation also provides a historical record of what was considered. The format of the documentation should be arranged to "tell the story" of what happens when an assumed failure occurs.

The formal procedure (Fig. 2) for preparing an FMEA constitutes the following steps:

1. A functional block diagram or logic diagrams are developed for the system or subsystem (Fig. 3). Copies of these diagrams become part of the FMEA documentation. Individual items to be considered are identified by number.
2. All realistic probable failure modes of each element on the diagram are postulated and entered on the FMEA documentation item by item (Fig. 4).

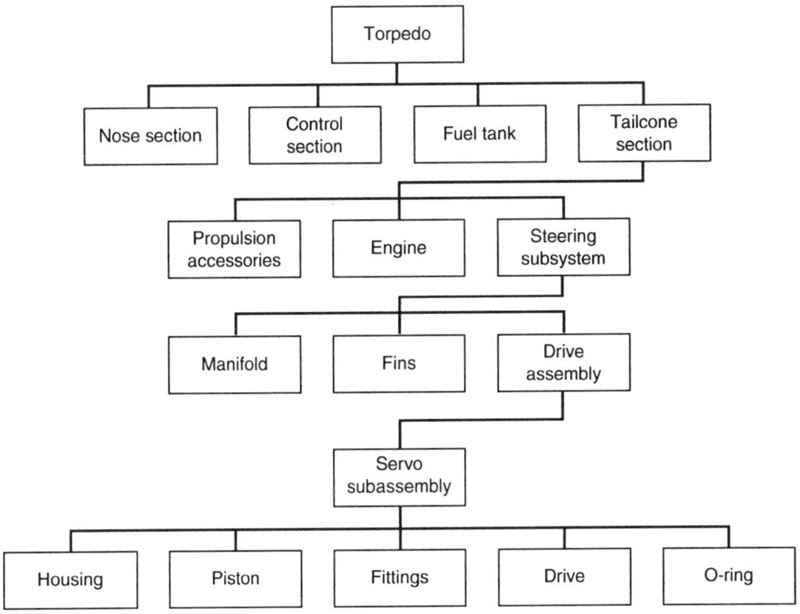

Fig. 3 Block diagram for a torpedo system subassembly. Source: Ref 4

3. The probable causes associated with each failure mode are listed beside each failure mode.
4. The immediate consequences of each assumed failure are described and listed beside the failure mode. Also listed are symptoms that would affect the function.
5. Any internal compensating provisions, within the system, that either circumvents or mitigates the effect of the failure, are evaluated and noted.
6. The effect of the failure on other components or systems is evaluated and documented.

FAILURE MODE AND EFFECTS ANALYSIS

SYSTEM Torpedo — P/N 2793 — PREPARED BY R. Landers — DATE 11/5/96
ASSEMBLY Steering — P/N 9372 — CHECKED BY J. Smith — DATE 11/5/96
SUBASSEMBLY Hyd. Actuator — P/N 2397 — PAGE 5 of 16

FAILURE MODE	FAILURE CAUSE	FAILURE EFFECT	CORRECTIVE ACTION TO MINIMIZE OR ELIMINATE FAILURE/EFFECT	CRITICALITY	FREQUENCY
1. Leakage	1. a. O-ring damaged during assembly. b. O-ring not compatible with skydrol fluid. c. Inadequate design for 3000 psi. d. Poor finish of sliding surfaces. e. Corrosion of surfaces. f. Electrolytic action between parts. g. Damage to O-ring by excess friction from exposed rod.	1. a. Hydraulic fluid will leak into control transformer causing possible damage to electrical wiring. b. Severe leakage will deplete hydraulic supply.	1. a. To assemble, O-ring must pass by sharp corner of orifice opening. Design change will be incorporated to chamfer sharp edges. b. EPR O-ring compatible with skydrol. c. Proven gland standards utilized in design. Qualification program will further prove adequacy. d. 16 finish adequate. e. Exposed rod chrome plated. Skydrol will protect interior surfaces. f. Identical metals used throughout. g. Rod retracted high percentage of operating time. Little opportunity for excessive dryness.	Critical	High
2. Structural failure	2. a. Fatigue. b. Inadequate safety margin.	2. a. Steering assembly will be unable to function within specified limits. Torpedo could go off course.	2. a. Fatigue analysis performed. All corners of adequate radius. b. Stress analysis proves adequate margin of safety. Test programs will prove structural adequacy.	Critical	Low

Fig. 4 Completed FMEA form for a torpedo steering subassembly. Source: Ref 4

7. The level of severity of each mode of failure is evaluated in terms of importance and documented as "critical," "major," or "minor."
8. A quantitative probability of occurrence of the failure mode is made. Where possible the probability should be expressed numerically (Ref 5). In some cases, categorizing may be more meaningful if commonly defined categories are used.
9. Finally, remarks and recommendations are documented. Here the analyst provides any additional data he feels would be meaningful.

An abbreviated example of FMEA documentation is illustrated in Fig. 3 and 4. Figure 3 shows a block diagram for a torpedo system. This system is required to carry out its mission without posing a hazard to its operators or to other equipment. The torpedo must function satisfactorily in an operational environment of 10 to 1000 ft depth of sea water, at –55 to 105 °C, and subject to a launch shock of 10 g. The FMEA is carried out at component subassembly level. The FMEA of steering a hydraulic actuator subassembly is reproduced in Fig. 4. Similar analysis would be generated for each subassembly.

Failure mode and effect analysis procedures may vary in detail from company to company. In fact, it is important to tailor FMEA to the specific needs of the project. However, because of the emphasis on an organized approach, care must be taken that FMEA does not become simply a ritualistic exercise where the form is more important than the content.

The principle benefit achieved through FMEA is the qualitative one of having to scrutinize each item individually with respect to failure. Therefore, FMEA is best carried out as a cooperative effort between the design and the application engineer. The design engineer is best qualified to identify failure modes and effects. The application engineer usually is in the best position to evaluate the consequences and make beneficial recommendations.

When properly carried out, FMEA produces an orderly documentation of possible hazards. This becomes the basis for program planning and a source for design improvements or contingency plans aimed at eliminating or reducing the effects of failure.

Fault Tree Analysis

Fault tree analysis (FTA) is intended to provide a deductive technique of probabilistic safety analysis. Fault trees are a diagram of the Boolean logic that relates the output event (called the TOP event) based on the state of the primary event. The fault tree itself is a graphical representation of the Boolean logic equations.

Boolean logic is made up of a series of simple logic statements called gates, which define the output based on the inputs. Usually, Boolean algebra deals with simple binary statements, true-false, go-no go, and so forth. The most common gates are the AND gate, requiring all inputs to be true for the output to be true, and the OR gate, requiring at least one input to be true for the output to be true. Typical fault tree symbols and organization are shown in Fig. 5.

The versatility in the degree of detail that is reflected in an FTA is a main feature of the fault tree technique. The analyst has options for qualitative or quantitative analysis. The technique provides a basis for analyzing system design, justifying system changes, performing trade-off studies, analyzing common mode failures, and demonstrating compliances with requirements.

The point of view of the FTA is from the TOP event down. Each TOP event failure is analyzed to show what inputs or combinations of inputs from the next level down are necessary to cause the failure. As the fault tree construction proceeds, each fault is developed until basic events are reached. The failed TOP event is thus demonstrated to be a failure situation resulting from the logical interaction of specific basic events. In any one branch the event being developed is called the base event of that branch. A branch of the fault tree is complete only when all events in the branch are developed to basic events. Figure 6 shows a fault tree developed for an automobile braking system. The TOP event is the fault that the brakes fail to operate.

Events that can cause a TOP event are classified as primary faults, secondary faults, primary failures, and secondary failures. A failure is a change in a component that requires repair action; a fault is a change that does not require repair action. Secondary faults and failures are those for which the components are not directly responsible.

The fault tree is constructed from symbols to represent logical outcomes and events that constitute inputs and outputs. A selection of typical symbols used is shown in Fig. 5.

The objective of FTA is to determine the conditions (basic events) that could cause a specific final undesired event (TOP event). Therefore, in constructing the fault tree, only failure-related events are considered. The construction technique involves five steps:

1. The undesired events (TOP events) that will be used to assess system performance are defined. Emphasis is on those overall undesired events that are of primary concern.

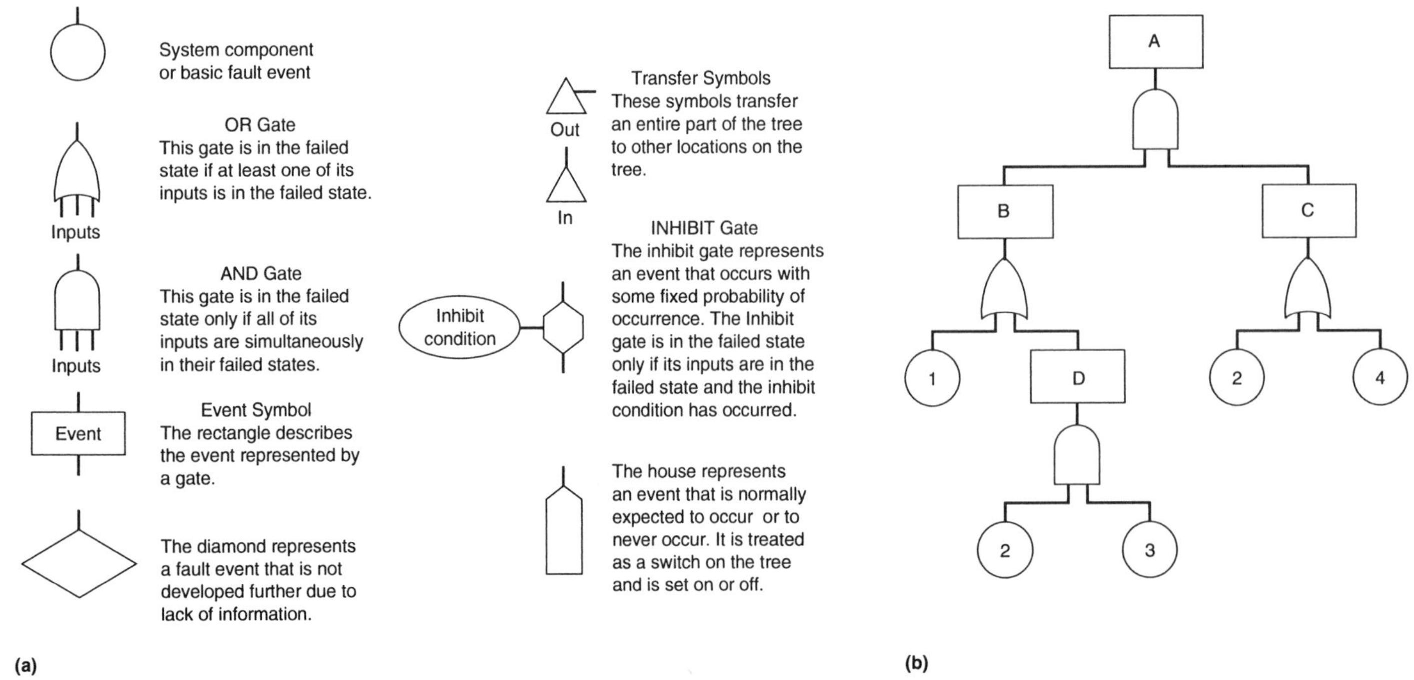

Fig. 5 Fault tree analysis. (a) Symbols. (b) Typical fault tree construction. Source: Ref 1

2. A functional analysis of the system is drawn up. The functional diagram defines the system elements and the interconnections; it represents the physical system.
3. Starting from the TOP event, the system is analyzed to determine the events that can cause the defined system fault.
4. The fault event analysis is continued to the next lower order events that could cause fault. This is done until all events are primary events.
5. Using logical symbols, the logical relationships among fault events are developed. These events are defined in such a manner that probability values can be assigned.

The fault tree presents a logical picture of all the sequences of events that can cause the TOP event fault. This tree is amenable to qualitative and quantitative analysis.

Qualitative analysis includes determining the modes of failure and the components that are responsible for a failure due to a common cause failure. A "minimum cut set" is any mode of failure containing a group of basic events that alone are collectively sufficient to cause the TOP event to occur. A "common cause failure" is a secondary cause of failure that is responsible for more than one component malfunction.

Because the undesired event is the consequence of those basic independent events that singly or in combination terminate direct paths to the top of the tree, often even a qualitative evaluation can identify what corrective actions are necessary to reduce the hazard. Anything that reduces the possibility of the basic events will reduce the hazard.

Quantitative Analysis. It is with respect to quantitative analysis that FTA has received the most attention. By introducing numerical probabilities of the basic events occurring, a probabilistic evaluation of the TOP event occurring can be derived. Considerable effort has been put into developing means of manipulating the probabilistic data with the associated Boolean algebra.

The quantitative analysis is of value for analyzing system design, in demonstrating compliance with safety requirements, and in justifying system changes. If quantitative reliability data are available for individual components and inputs, or can be accurately estimated, the reliability of the system can be calculated.

In systems justifying a thorough component reliability testing program or where extensive reliability history has been documented, quantitative evaluation of the system reliability can be established. However, where such reliability data do not exist the final results can be highly suspect.

Considerable effort has been put into programs for generating and analyzing fault trees by computer. For large systems, this is extremely helpful because otherwise the manpower requirements are excessive. However, much of the benefit derived from FTA is due to the understanding and insight gained while constructing the fault trees. If generation of fault trees is done completely automatically, much of this benefit could be lost.

Advantages and Limitations of FTA. The versatility in the degree of detail that can be

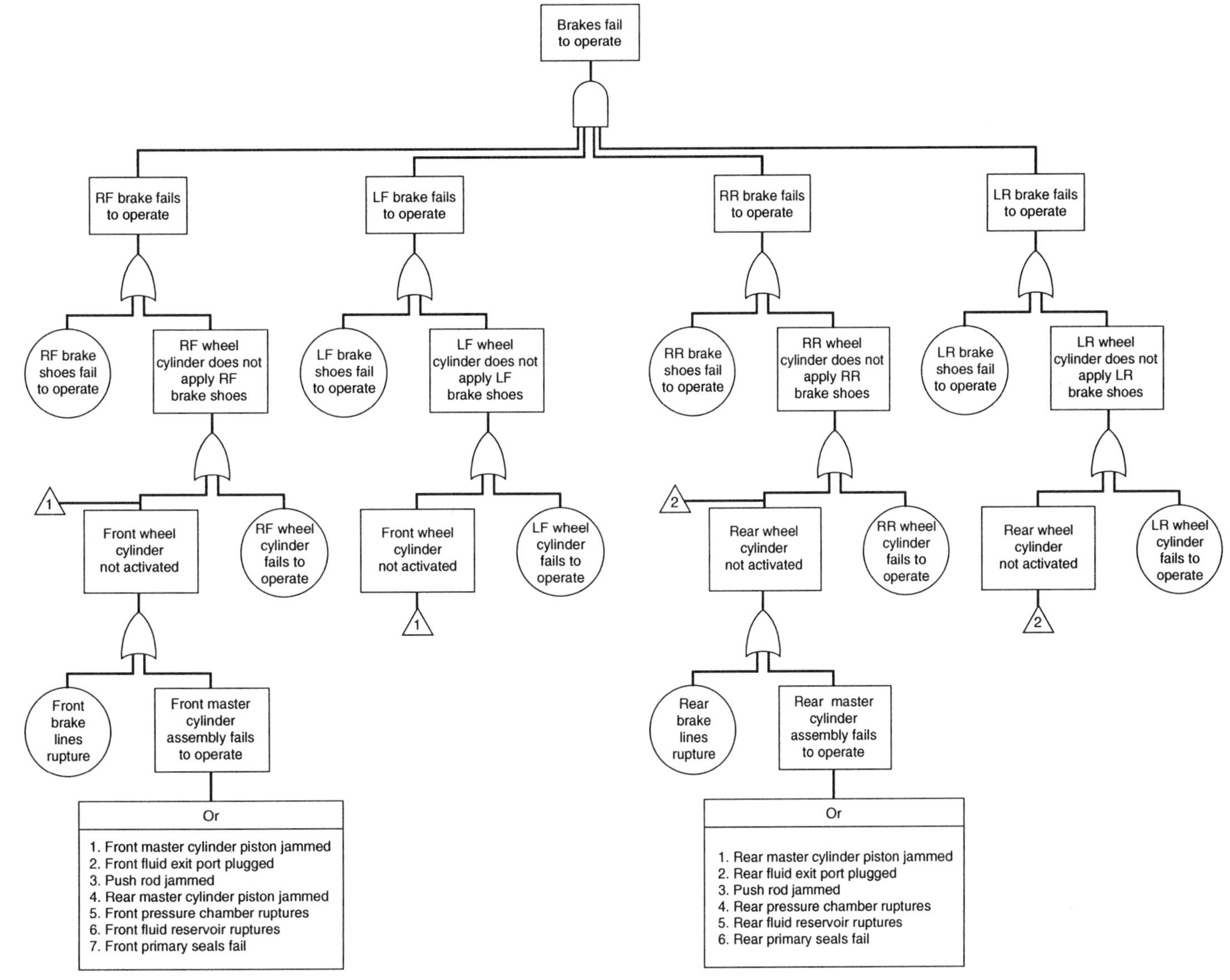

Fig. 6 Fault tree for vehicle braking system. RF, right front; LF, left front, RR, right rear; LR, left rear. Source: Ref 1

reflected in an FTA is a principle feature of the technique. The fault tree is useful in:

- Directing the analyst to ferret out possible failure modes deductively
- Highlighting aspects of the system that are important with respect to the failure of interest
- Providing a graphical aid to those in system management who are removed from system design
- Providing options for qualitative or quantitative system analysis
- Allowing the analyst to concentrate on one mode of failure at a time
- Providing the analyst with insight into the system behavior

Limitations of FTA include:

- In spite of the use of computers, minimum cut sets cannot always be found for large systems because of the large storage required.
- Because all decision equations are binary, some conditions of interest cannot be handled, such as degrading of performance, and standby systems cannot be handled quantitatively without use of Monte Carlo methods (requiring a large number of computer runs).
- Methods for quantitatively dealing with secondary failures do not exist. Nor can the use of repairable systems be dealt with quantitatively.
- Fault tree analysis is severely limited by the shortage of meaningful component reliability data.
- Fault tree analysis is expensive, especially if quantitative results are expected on large systems. The costs of generating suitable and meaningful fault trees for first-time application to a system may be in the same order as the engineering to design the system.

Event Tree Analysis

Event tree analysis (ETA), like FTA, is a binary logic tree. However, it is the reverse of FTA in that it starts with an initiating event (a basic event) and traces all the possible consequences of the event. Possible failure sequences are defined that consist of the initiating event and specific succeeding systems failures. All consequent system states (failure or success) are connected through a branching logic that represents all the specific accident sequences that can arise.

Event tree analysis is an inductive logic approach and is similar to the FMEA. The possible outcome of specific failures are enumerated in the form of sequences of failures that may occur. Physical consequences are determined and associated with each sequence of failures.

Figure 7 is an ETA diagram for a loss-of-coolant accident at a nuclear reactor. In this simplified event tree, the initiating event is assumed to be a pipe break, the probability of which is signified by P_A. The subsequent events depend on whether the break occurs at a time when the electric power is on or off, and a probability P_B is assigned to the failure of electric power at the time of the break. (Because the probability of such a failure is generally less than 0.1, the probability of success, or $1 - P_B$, is always close to 1. Therefore, the probability associated with the upper branches in the tree is assumed to be 1.) One then follows each of the two branches, asking next what the probability of failure of the emergency core-cooling system is, which in turn will depend on the availability or nonavailability of electric power. In the example shown, the probability of failure of the emergency core-cooling system in the absence of electric power is P_{C_2}, and that branch leads farther down the tree. The worst possible case is that given at the bottom, in which the electric power, the emergency core-cooling system, the fission-product removal system and the containment system have all failed, resulting in a very serious accident. The topmost branch illustrates the situation in which the pipe does indeed break, but all the other systems perform flawlessly, so that no risk to the public ensues. The task of ETA is to estimate the probabilities of all intermediate chains of events in such a way that an overall assessment of the risk can be made.

The logic of ETA is quite general and can be used whenever events occur in sequence of logical combination. By superposition of reliability data on the event occurrence and its consequence, a quantitative probabilistic assessment of the likelihood of the system failure can be computed. The initiating event occurrence can usually be derived from historical data. The probability of alternatives occurring can be derived from FTA or FMEA. The mathematical methods developed for FTA can usually be applied to ETA.

Event tree analysis is often used to complement FTA. Whereas most other hazard analysis methods emphasize the inherent reliability of the components and the system, ETA focuses on the consequences of unusual events; such an event could be an input that is out of specification or an unusually large environmental load.

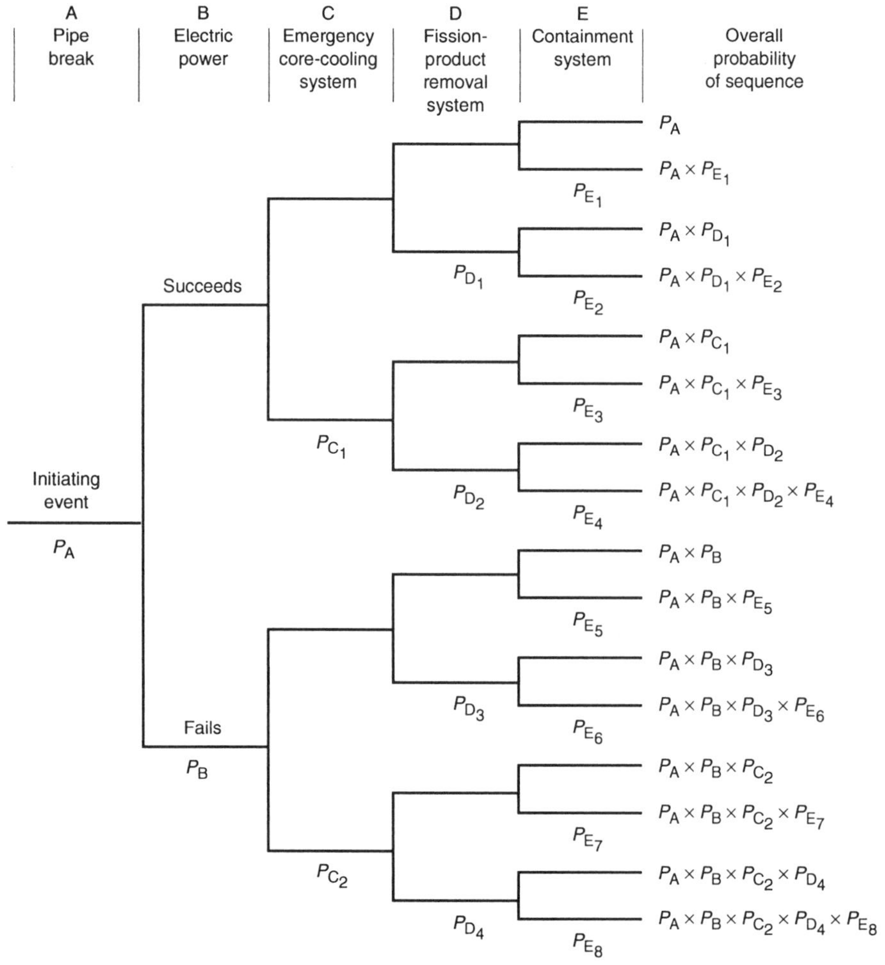

Fig. 7 Event tree analysis for a loss-of-coolant accident at a nuclear reactor. See text for discussion. Source: Ref 6

Risk/Benefit Analysis

Risk/benefit analysis is a management technique to aid in decision making. It is based on the assumption that a *common value scale* can be found for expressing the risks and benefits inherent in a decision. If the common value scale is in dollars, the expected loss index (ELI) is defined as the total dollars being risked times the probability of failure, this is the dollar value of the risk. Then, if an acceptable dollar level of risk can be established, a potential tool for accepting or rejecting any decision can be generated.

By using the dollars at risk as one axis of a graph and the probability of failure as the other axis, a field is defined in which each point represents some specific combination of risk and dollars at risk (Fig. 8). Points with the same expected loss index lie on a straight line at an angle to both axes. By locating the risk versus dollars at risk for any event as a point on the graph it can easily be seen whether the risk is acceptable or not, based on a previously established ELI.

Although this approach is valid for financial considerations, when dealing with safety, it has some serious shortcomings:

- Safety often involves dealing with lives or irreversible damage, which cannot be readily equated to dollars, but has a highly emotional content (see the article "Safety in Design" in this Volume).
- Dollars at risk include elements difficult to isolate or determine: user costs, training costs, cleanup costs, and so forth.
- Reliable event probabilities should be estimates that are based on tests, but in many cases are simply a "best guess" of an expert.

Generally risk/benefit analysis makes sense when dealing with a closed system where only dollars are at stake. Elsewhere risk/benefit analysis must be viewed with caution.

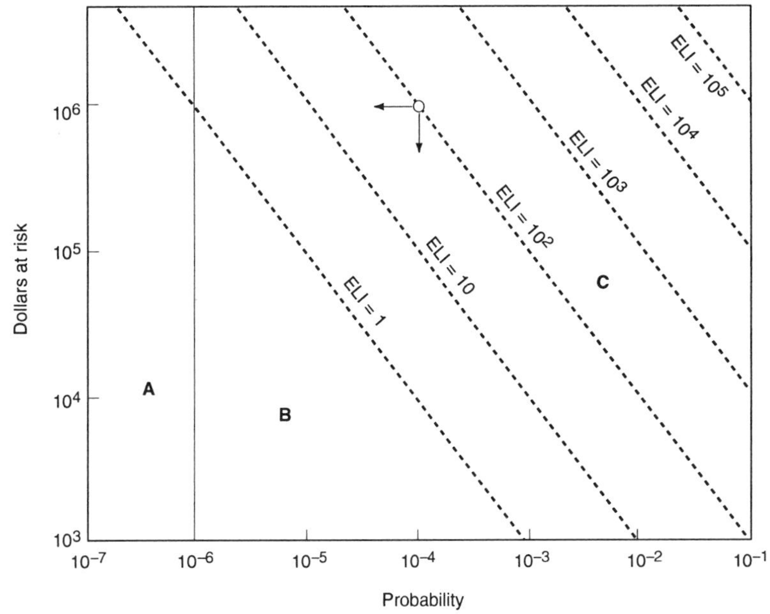

Fig. 8 Typical risk analysis graph. If target ELI is 100, ELI in region C are unacceptable, ELI in regions A and B are acceptable. Source: Ref 7

Safety Analysis

Safety analysis is the synthesis of other risk/hazard analysis techniques to produce a qualitative systematic analysis throughout all phases of the project design, development, and implementation, with the objective of preventing accidents. This includes preventing personnel injuries, damage to equipment, and so forth. The techniques are simple but rigorous and time consuming, and to be effective must be continued throughout the life of the system from conception to shutdown and disposal. The method can be applied to any system but from a different point of view than system analysis. The focus of safety analysis is the prevention of hazardous failure.

The primary tools of safety analysis are worksheets, a hazard log book, a checklist of hazards, and a system description. A basic requirement is that the practitioners have a willingness to challenge the status quo; they cannot be the same as those committed to the design and implementation of the hardware system.

Safety analysis should be started at the same time as the design, by using historical information and knowledge of the system; possible hazards and hazard-causing situations or scenarios should be identified as the design develops. The hazards are categorized according to their potential severity (e.g., catastrophic, critical, controlled, etc.). Worksheets are generated (Fig. 9) that allow each element of the system to be considered and challenged with respect to each potential hazard. The results of these challenges are entered onto the worksheet identifying which components are affected critically and the possible extent of the resultant problem. After the complete system has been challenged with respect to the hazards, the results are simplified by sorting out the "essential hazardous situation" (EHS), that is, the fundamental safety problems to be dealt with. A logbook is established for the system with files to keep track of each essential hazardous situation considered throughout the life of the system. This logbook is used to ensure that all system history from conception to closeout is available for reexamination to anticipate and prevent the essential hazardous situation from occurring.

The identification of the hazards is an initial and continuous activity. Identification initiates an iterative process—the preliminary hazard analysis (Fig. 10) that undertakes to eliminate or reduce the essential hazardous situations to an acceptable minimum. Design changes, safety devices, warning systems, crew procedures, emergency measures, and so forth, are all recommended. These are carried out to control or resolve the hazard situation. This often requires compromises between the designer and the safety analyst, or even changes in the specification governing the functions and use of the system. In any reasonable project, at some logical point before completion, a design review takes place. The safety analyst should become a part of the design review to assess whether the designer has been

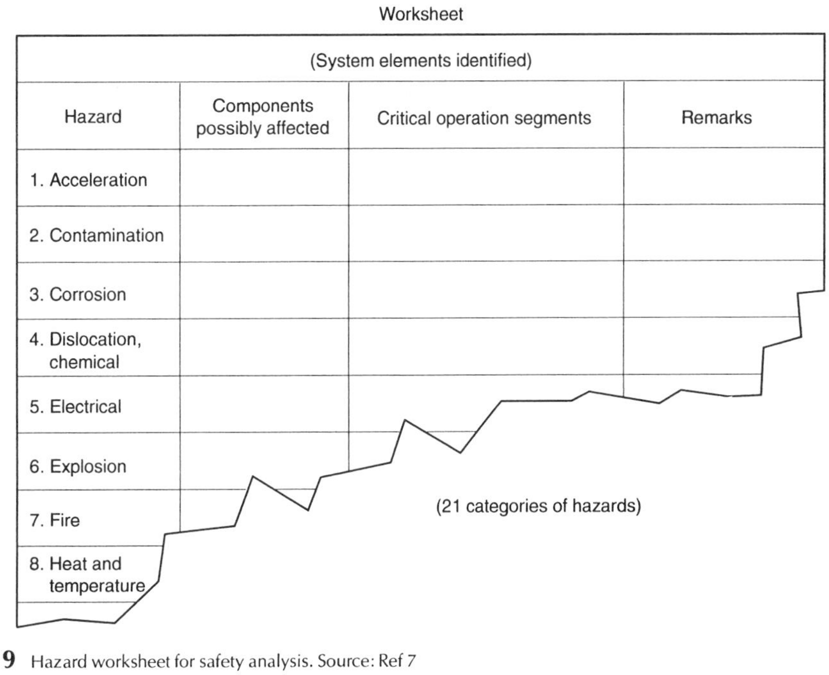

Worksheet
(System elements identified)
Hazard
Components possibly affected
Critical operation segments
Remarks
1. Acceleration
2. Contamination
3. Corrosion
4. Dislocation, chemical
5. Electrical
6. Explosion
7. Fire
8. Heat and temperature
(21 categories of hazards)

Fig. 9 Hazard worksheet for safety analysis. Source: Ref 7

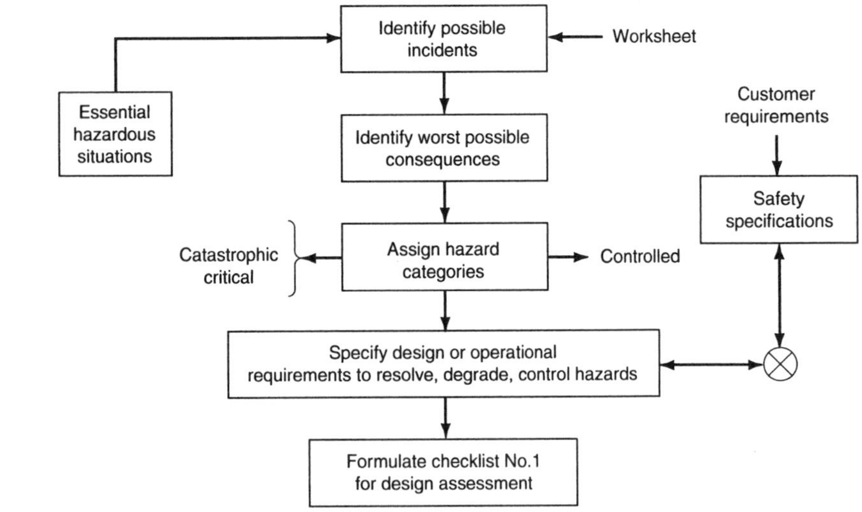

Fig. 10 Safety analysis flow chart for preliminary hazard analysis. Source: Ref 7

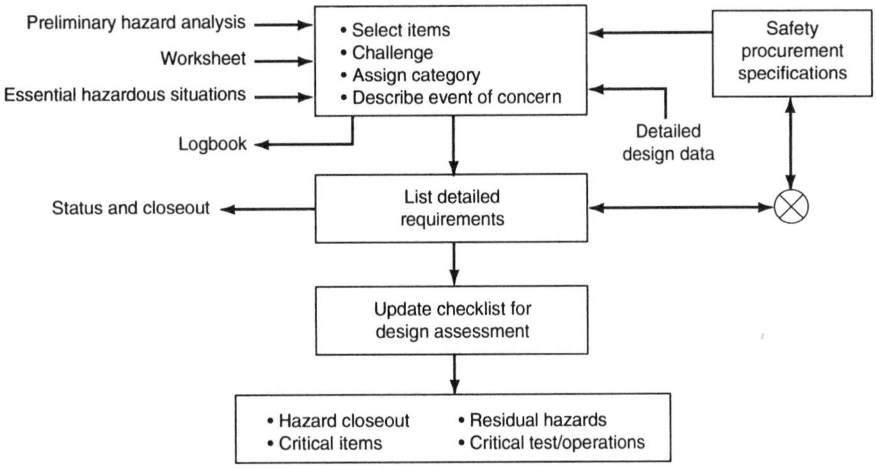

Fig. 11 Safety analysis flow chart for system hazard analysis. Source: Ref 7

cognizant of the essential hazard situations. At this stage, the design has reached a degree of maturity and new hazards may be identified, which may require further action. At the completion of the preliminary hazard analysis:

- The general hazards of concern are identified.
- Additional design and operation requirements have been established to forestall or alleviate anticipated hazards.
- Areas where further improvements with respect to safety are necessary have been identified.
- A databank has been started that can be used to follow and document potentially hazardous situations.

The log book on hazard data is expanded and embellished as the design is refined and as more detailed information becomes available. From this, at any specific point throughout the design, the safety analyst should be able to say what hazards are still possible due to any part of the design and what action is contemplated or necessary to reduce the hazard to an acceptable level. After the system has been completed, has been in service, and is retired, the log book then becomes a valuable historical record on which additional or new systems can draw.

For large complex systems, each part of the system (system, subsystem, assembly, component) is broken down into its basic elements and subjected to its own system hazard analysis (Fig. 11). The complete system hazard analysis is best carried out before commitment to hardware so that the results of its deliberations can be incorporated into the hardware. The techniques used in system hazard analysis are the same as used in the preliminary analysis, but are carried out in greater detail.

Before the design is finalized, a safety review should be held. The purpose is to review all known safety problems and to establish a schedule, budget, and approach to their elimination.

When the design is completed, the data collected on the system safety during the design phase become a source used to determine test requirements, quality control, and operational procedures. Similarly, problems that appear during testing or early operation are fed back into the log book to help formulate future design and safety requirements.

Because no system is perfect, safety analysis cannot stop with the design and production of hardware. The safety analysis evolves into the next stage, operating hazard analysis (OHA) (Fig. 12). This serves to identify solutions where failure that is attributable to operator performance is likely to occur. Human factors studies at this stage are a useful input (see the article "Human Factors in Design" in this Volume). The techniques of carrying out OHA are the same as for system hazard analysis except that the human operator and his limitations must be considered. The outputs from OHA are operational requirements that compensate for or correct a human limitation. The procedures have to either prevent the hazard or confine and minimize its impact.

The outputs from OHA can be in the form of:

- Procedural steps
- Sequence constraints
- Caution or warning notes
- Reference data
- Support or standby equipment

With these various hazard analyses, the equipment and the operating system should be as safe as possible. However, before starting up the facility the system should be subjected to one more review, a readiness review. At this point, all critical personnel are brought together and all safety hazards are reexamined to see whether they have been adequately taken care of. Crew procedures are assessed to ensure that personnel are properly trained, not only for normal operation, but for emergency situations as well. Only when this readiness review panel is satisfied that all necessary actions have been taken should the system be put into service.

The previous discussion presents a description of an ideal safety analysis program. It is doubtful that such an idealized program can be instituted except in the case of small hardware items completely within the control of one firm. However, it is an ideal that can be strived for, especially in the case of large complex systems where the impact of failure can be catastrophic. The institution of a safety analysis program does not exclude the use of other risk/hazard analysis techniques, but instead provides a framework wherein these other techniques can be integrated into design and operating considerations.

Probabilistic Estimates

Most risk/hazard analysis techniques rely on statistical or probabilistic techniques to quantify their results. It should be emphasized that such probabilistic reliability quantities are not properties of the hardware; they represent the state of

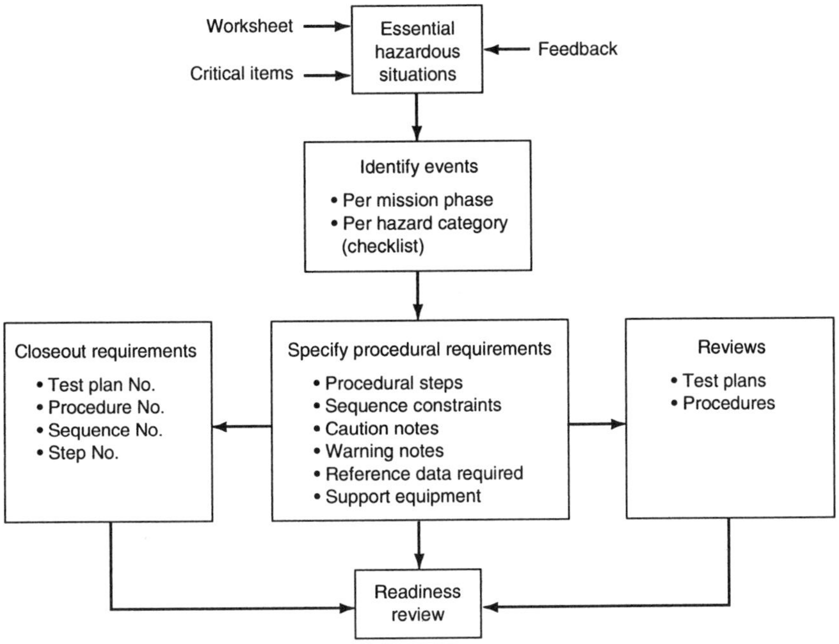

Fig. 12 Safety analysis flow chart for operating hazard analysis. Source: Ref 7

one's understanding of the potential life of the hardware. By increasing the information one has about the system it is possible to change the expressed probability of failure. This will, in no way, affect whether a specific system will fail or not. Probabilistic quantities only deal with one's perception of the world.

Second, statistical and probabilistic quantities, no matter how accurate, do not reflect knowledge about a single specific system. Statistical or probabilistic quantities describe what will happen to a large number of identical systems if they are all asked to perform identical functions under identical circumstances.

Finally, no matter how small the probability of an event may be, there is always a possibility of that event occurring. In fact, a probability specifically predicts that if enough systems are tried, the event *will* occur. Moreover, if probabilistic evaluation shows that an engineered system would only fail catastrophically once in 10^6 years, that does not mean that failure will occur after 10^6 years, it could fail in the next hour. Probabilistic evaluations cannot predict *when* failure will occur.

Consequently, reliance on statistical and probabilistic data for decision making must always be tempered with understanding.

Conclusions

Those who are involved in using risk/hazard analysis methodologies advocate their use, but always with a note of caution. Those who are involved in developing the details of algorithms for manipulating the quantitative aspects of risk/hazard analysis are much more enthusiastic about the methodologies than individuals responsible for applying the methods to hardware.

For the designer, the chief benefits to be gained from risk/hazard analysis are qualitative ones. Because the analyst is required to consider all possible failures and to focus on details, there is less likelihood of anything being overlooked or taken for granted. Corrective action can be taken, because someone has been made responsible for safety.

Although there is great promise in quantified probabilistic hazard assessment, more is in the promise than in the fact. The state-of-the art is not sufficiently developed so that one can realistically expect to specify a number and state that systems that come below that number are satisfactory and systems above that number are unsatisfactory. The only way to eliminate or reduce safety hazards is through a continuous monitoring of the system design and eliminating those problems that may contribute to the hazard.

In spite of the fact that no analysis can produce a simple quantifiable specification requirement, the risk/hazard analysis techniques developed can be used by designers and operators of systems to reduce the hazards inherent in any system.

It is apparent that the various techniques have much in common and that any really effective safety program should use several of the techniques so that they complement one another. There is no doubt that risk/hazard analysis is still an art and not a science. It is expected that two analysts looking at the same problem may generate quite different but equally valid hazard assessments.

To be effective in reducing ultimate hazards, the analysis must be done early and before the design is fixed. Consequently, the risk/hazard analysis must be instituted and carried out by the agency responsible for the design.

ACKNOWLEDGMENT

This article is based on a report prepared by the author entitled "Study of Application and Suitability of Risk/Hazard Analysis Methodology for Environmental Considerations," Environment Canada Contract No. 14SU.KE240-9-0820.

REFERENCES

1. J.B. Fussell and D.P. Wagner, Fault Tree Analysis as a Part of Mechanical Systems Design, *Engineering Design,* Proceedings of the Mechanical Failures Prevention Group, 25th Meeting, NBS 487, National Bureau of Standards, 1976, p 289–308
2. C. Starr et al., The Philosophical Basis for Risk Analysis, *Ann. Rev. of Energy,* Vol 1, 1976, p 629–662
3. T.A. Kletz, The Risk Equations, What Risks Should We Run?, *New Scientist,* 12 May 1977, p 320–322
4. R.R. Landers, A Failure Mode and Effect Analysis Program to Reduce Mechanical Failures, *Engineering Design,* Proceedings of the Mechanical Failures Prevention Group, 25th Meeting, NBS 487, National Bureau of Standards, 1976, p 278–288
5. J.J. Hollenback, "Failure Mode and Effect Analysis," Paper 770740, American Society of Agricultural Engineers, 1976
6. H.W. Lewis, The Safety of Fission Reactors, *Sci. Am.,* March 1980, p 53–65
7. H.D. Wolf, Safety Analysis: Qualitative, Quantitative and Cost Effective, *Engineering Design,* Proceedings of the Mechanical Failures Prevention Group, 25th Meeting, NBS 487, National Bureau of Standards, 1976, p 265–277

SELECTED REFERENCES

- J.D. Andrews and T.R. Moss, *Reliability and Risk Assessment,* John Wiley & Sons, 1993
- J. Bowen, Safety-Critical Systems and the World Wide Web, *Safety Systems: The Safety-Critical Systems Club Newsletter,* Vol 4 (No. 2), Jan 1995. Contains reference to Safety-Critical Systems Archive on World Wide Web, maintained by the Oxford University Computing Laboratory Archive (http://www.comlab.ox.ac.uk/archive/safety.html)
- S. Grey, *Practical Risk Assessment for Project Management,* John Wiley & Sons, 1995
- W.G. Ireson and C.F. Coombs, Ed., *Handbook of Reliability Engineering and Management,* McGraw-Hill, 1988
- A. Kandel and E. Avni, *Engineering Risk and Hazard Assessment,* CRC Press, 1988
- M.M. Neowhouse, The Use of Software Based Risk Assessment Methodologies, *Probabilistic Risk and Hazard Assessment, Proceedings,* A.A. Balkem Pub., 1993
- H.E. Roland and B. Moriarty, *System Safety Engineering and Management,* 2nd ed., John Wiley & Sons, 1990

Human Factors in Design

Charles O. Smith, Engineering Consultant

THE TERMS *human factors*, *ergonomics*, and *human engineering* are often used interchangeably despite some significant differences in the basic meanings of the terms. While one tends to think of human factors as the process of designing products and systems for human use, this is too limited. Human factors also include operational methods and procedures, testing and evaluating these methods and procedures, job design, development of job aids and training materials, and selection and training of people who will be users. This article is primarily concerned with the design of equipment, although work environment and work methods also have significant effects. The central approach of human factors engineering is the systemic application of relevant information about human characteristics and behavior to the design of human-made objects, facilities, and environments that people use. For an acceptable level of human performance, the designer must focus on each of the following elements:

- The state or condition of the *human being*
- The *activity*, including equipment and required tools
- The *context* in which the activity is performed

The Human

The human being is the most complex of the three required elements. Human performance can be affected by a range of conditions that are internal to the individual and have no reference to the activity or context. The designer must understand potential areas of deficiencies in people and account for them when making design decisions.

The major aspects of the human are sensors (vision, hearing, touch, smell, and taste), brain (cognitive) processing (ability to think, reason, and decide), and responders (properly functioning fingers, arms, feet, mouth, etc.). The designer must have a good understanding of how humans function in each of these aspects.

Degraded performance can result if any of the basic capabilities to perform an activity are reduced or lacking. Other factors such as sleep (or lack thereof), fatigue, biological rhythms, illness, drug reactions, lack of basic (or advanced) skills, and/or lack of motivation can influence human performance, usually negatively.

Designers commonly assume that potential users have skills of perception, problem solving, decision making, and movement control. If any of these skills, or any other important skills, is lacking, the designer must provide an efficient way for the user to learn or develop it. This implies an unspoken assumption that the user has the ability to learn the necessary skills in a reasonable time. Some potential users may not have the necessary learning skills, and some of these people will be unable to develop the required skills within an acceptable time. There is a further assumption that potential users are mentally healthy. Designers often assume that all people are much alike and that an individual does not change over time, especially in a very short period. These latter assumptions are seldom justified.

Human beings exhibit a multitude of variations: variations in height, weight, physical strength, visual acuity, hearing, computational capability, intelligence, education, and so on. Designers must consider all these variables, and their ranges, because they recognize that their product will ultimately be used by humans.

Designers seldom have control over the people selected to use the system being designed. The best approach for a designer is to have a clear understanding of the characteristics of the potential user population so that the most appropriate design decisions can be made. Because most products and systems are designed for groups of people, rather than one specific individual, it is necessary to deal with the extremes of strengths and weaknesses of the potential user population. In addition, the designer should document the assumptions about the user population for those who *do* select the users.

The Activity

Cazamian (Ref 1) says: "For a period measured in millennia, work was carried out in the form of crafts. The craftsman or artist was at one and the same time the organizer and executor of his own works. ... Today, we live in a situation where the craftsman's function has, so to speak, split into two parts: those of the organizer (system designer) and those of the executor (systems users). In many, many cases the designer is no longer a user and a user has limited or no input to a designer."

The designer is concerned with the technical and economic aspects of products and systems. The user is much more oriented toward getting through the day, since for many users work is simply a means of existence. Success at work, however, may determine other successes or failures of an individual. A good designer will thoughtfully create the activity to be performed rather than let it evolve.

The designer can control some conditions relating to performance of an activity, so the designer must know what kinds of work people do best, what is best done by machines, how (and what) tasks are best combined into a work module, and what training is required to develop necessary skills for an acceptable level of performance. In many situations, interface requirements must be considered.

The Context

There are two aspects to consider with regard to the context in which a human performs an activity: the *physical context* and the *social context*. In the physical context, noise, lighting, temperature, humidity, vibration, and so on may all have significant effects on the user. In the social context, crowding versus isolation is an example of a factor that may have a significant effect on user performance. The so-called "home court advantage" in athletic events suggests that social context can make a difference in human performance.

Human-Machine Systems

Most systems have some common characteristics. Every system has some purpose or objective that must be clearly understood, including a statement of performance specifications. For the system to fulfill its purpose, specific operational functions must be performed, usually in a definite sequence. Each function must be performed by a person or by a physical component. If the system is a *flow* system, a block diagram can be used for help in allocating each function to

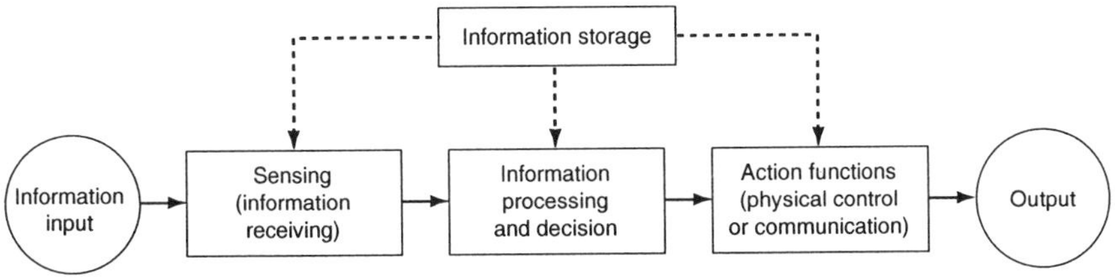

Fig. 1 Schematic representation of basic functions performed by human or machine components of human-machine systems. Source: Ref 2

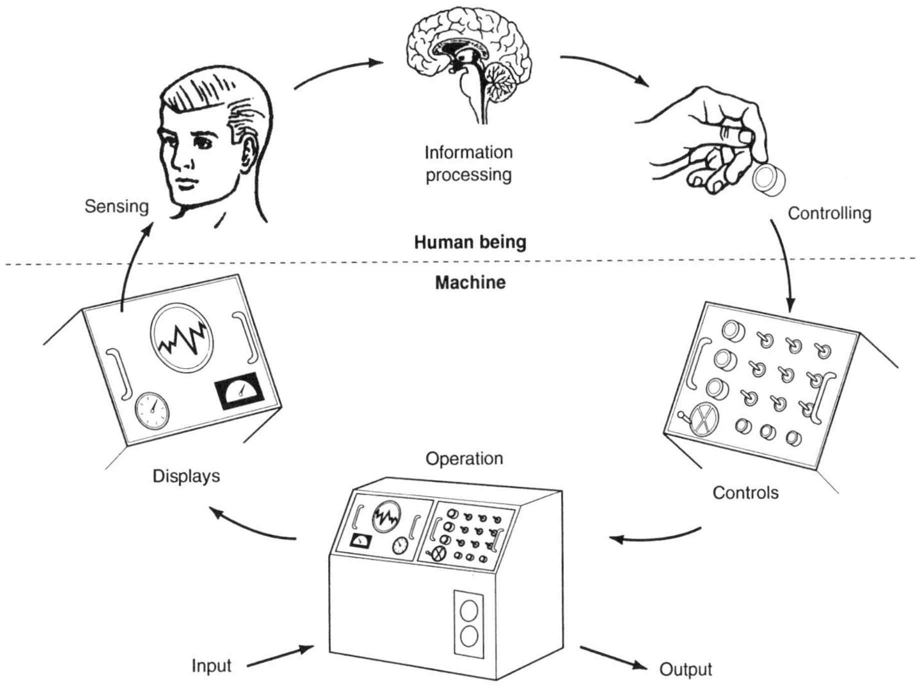

Fig. 2 Schematic representation of a human-machine system showing basic elements and modes of function. Source: Ref 3

a human or to a physical component. In other systems, however, such as driving a motor vehicle, there is no unambiguous way of representing the constantly changing interactions among the driver, the vehicle, and the environment. Execution of any operational function involves a number of basic functions, as indicated schematically in Fig. 1. The basic elements of a human-machine system are shown in Fig. 2.

Human-machine systems can be classified by the degree of human versus machine control:

- *Manual systems* primarily consist of hand tools and other devices that are used together by the human operator, who controls the operation. Examples are a cook plus utensils or a singer plus amplifier.
- *Semiautomatic systems* consist of well-integrated physical components (e.g., powered machine tools) that are generally designed to perform the desired functions with little variation. The machine normally supplies the power, and the human function is primarily one of control, usually through control devices. A motor vehicle is one example.
- *Fully automatic systems* perform all operational functions, including sensing, information processing, decision making, and action. Examples are automatic telephone systems and digital computers.

Distinctions among these three systems are not always readily apparent. Even within any one system, the degree of manual versus automatic aspects can vary among the different components (or subsystems).

In design and development of a new product or system, the majority of the most critical decisions to be made are related to human performance. Informed decisions require the designer to have a good understanding of human engineering, human factors, and ergonomics. Problems are highly likely to develop when humans and machines interact. Human factors engineering attempts to minimize these problems and obtain maximum effectiveness in any human-machine operation by integrating the best capabilities of both.

The designer must avoid any design that expects, or requires, individual operators to exceed their available physical strength, perform too many functions simultaneously, perceive (or detect) and process more information than is possible, perform meticulous tasks under difficult environment conditions, work at peak performance (or capability) for long periods, work with tools in cramped spaces, and so on. Insofar as possible, *the designer should adapt the machine to the human.*

The designer may think in terms of the "typical" or "average" human. This view is much too simple:

- People come in assorted sizes, shapes, capabilities, and varieties. Even when it may be appropriate to design for the average, the designer must remember there is a range of differences from that average.
- Some products are designed for limited groups such as infants, children, teenagers, the elderly, or the infirm. The characteristics of the specific group must be emphasized.
- When designing for the public the designer needs to provide for the entire range of characteristics of people, from babes-in-arms to nonagenarians. For example, doors, ramps, escalators, entries, and so on must be appropriate for the baby in a perambulator, a healthy and active man or woman, and a handicapped or elderly person with a walker or in a wheelchair. The task is not easy.

Broad Design Considerations

The starting point of human factors design is the allocation of functions to the equipment and the operator. Whether a given function should be performed by the equipment, by the operator, or jointly by both depends on which arrangement will give the most effective function in the operational situation. This requires a comparison of the strengths and weaknesses of both. There is no simple answer to the question of best allocation. It has been suggested, however, that if the content of all inputs and outputs of a given function cannot be fully specified, that function should be assigned to the operator, because only humans have the capability and flexibility to make decisions about unexpected events.

If we visualize three stages of system design development (i.e., conceptual, preliminary, and detailed), there are special considerations at each stage.

Conceptual Design. The designer should consider such aspects as:

- *Potential user characteristics* such as size, intelligence, age, reaction time, health, sex, training, and physiological tolerance
- *System features* such as task, operational, and environmental stresses; hazards; and task instructions and procedures
- *Preliminary task descriptions* for users, operators, and maintenance personnel

Preliminary Design. The designer reviews the previous results. In addition, the designer studies or analyzes such things as:

- Human-machine simulations and mockup studies
- Time line analysis (i.e., is the planned use of the operators or maintenance personnel feasible?)
- Task analysis
- Task description
- Link analysis (i.e., is the grouping and sequencing of human tasks logical and manageable?)

Detailed Design. The designer reviews the previous results. In addition, the designer studies such things as:

- Hazard and safety analyses
- Refined time line and link analyses
- Personnel requirements (i.e., how many and what type of people are needed for field operation?)
- Task and training requirements
- Technical literature publications, which must be prepared carefully so that the user can use them effectively

Human-Machine Function Allocation

As noted above, there are some things at which humans are better than machines, but there are also some things at which machines are better than humans. The human factor designer must know which is which. Table 1, while not necessarily complete, gives a substantial number of comparisons of human and machine capability for various functions.

Further Design Guidelines

The designer certainly must consider the hazards in the design when it is used or operated in the intended manner. A hazard is any aspect of technology or activity that produces risk or danger of injury. The designer must also recognize that the product may be used in unintended but foreseeable ways. Protection must be provided against hazards in all uses that can be foreseen by the designer. Unfortunately, even the most diligent search for foreseeable uses may still leave a mode of use undiscovered. In litigation, a key issue often revolves around the question of whether the specific use was foreseeable by a reasonably diligent designer.

Example 1: A Snowmobile. A female teenager and her boyfriend rented a snowmobile for a few hours to ride around on a lake. He was driving; she was a passenger. They collided with another snowmobile rented from the same rental agency. She was thrown on the ice, injuring her left knee. Surgery was performed but left her with permanent injury.

The owner of the agency admitted that the brakes had been disconnected by the previous owner of the agency (while he was an employee) and that he had left them disconnected. The reason given was that renters became confused about the operation of the throttle and the brakes, with the result that they "tore up the equipment." Without brakes, the only mechanism for slowing the snowmobile was to release the throttle and let the engine provide braking. The owner claimed that the throttle had been adjusted to limit the speed to between 20 and 25 mph.

The owner's manual said: "The squeeze-type throttle lever, conveniently located on the right side of the steering handle, enables the operator to control the sled and the engine RPM at the same time." The manual further said: "The hand-operated brake lever is located on the left side of the steering handle. By mounting the brake and

Table 1 Human-machine function comparison

Function	Humans	Machines
Flexibility	Able to handle low-probability alternatives, unexpected events. High flexibility.	Flexibility limited; unexpected events cannot be adequately handled.
Ability to generalize	Able to recognize events as belonging to a given class, to abstract similarities, and to disregard conflicting characteristics.	Zero, or very limited abilities. Can react only to physical properties set up.
Ability to learn	Previous experience can be used to change subsequent behavior.	No change unless equipment configuration is changed.
Sensitivity	Sensitive to wide range of kinds of inputs. Can transduce many kinds of physical stimuli simultaneously. Small amounts of physical energy can be detected.	A given instrument is sensitive to only one kind of input. Thresholds not generally as low as human's. However, can sense energies beyond human spectrum.
Cost	Inexpensive and in good supply. Amount of training required depends on complexity of task.	Cost rises rapidly as complexity increases. Supply limited.
Weight	Light.	To reproduce human functions, would have to be immensely heavy.
Originality	Able to sense and report events and observations not of immediate concern.	No originality in discovering incidental intelligence or the relationships between functions.
Monitoring capacity	Poor.	Good.
Expendability	Nonexpendable; safety considerations have to be imposed.	Expendable.
Manipulability	Ability to perform precise manipulations required in maintenance.	To perform same tasks, machine would be extremely costly and complex.
Reliability	Subject to errors. However, reliability of manual equipment is higher than that of more complex automatic equipment because of smaller number of components.	Makes no errors, but reliability of automatic equipment is lower than that of manual equipment because of increased number of components.
Fluctuation in performance efficiency	Highly variable from day to day.	Minimal variability.
Reaction time	Relatively slow. Shortest human response about 200 ms.	Usually as fast as relay operating time, microsecond lags.
Physical force	Relatively weak unless aided by machinery.	Practically limitless in power.
Boredom or fatigue	A repetitive task results in error. Rest period required. 8 to 10 h maximum efficient work expected.	Capable of many operations without decrement. Only physical limitations, such as heat, corrosion, or wearout, need be considered.
Environmental requirements	Can exist only in a very narrow band of environmental conditions. Physiological maintenance requirements are extensive.	Can tolerate many fluctuations in environment. Restricted only by design specifications.
Tracking ability	Comparatively poor tracking characteristics, although satisfactory in wide range of situations. Can change performance constants to produce best attainable system performance in any situation.	Good tracking characteristics. Considerable complexity needed, however, to track well in all conditions.
Channel capacity	Limited. Has maximum amount of information that can be handled at a given time.	Can be made arbitrarily large. Capacity limited only by design.
Overload operation	May be able to perform better in certain cases of overload than machines; can tolerate temporary overloads without complete disruption.	May break down completely if overloaded; with information-handling capacity fixed, overload leads to system disruption.
Survival	Interested in survival. Situations with possibility of danger are "stress-inducing," resulting in behavior degradation.	Not "conscious" of danger. No behavior decrement in the face of destruction.
Computational ability	Comparatively slow and poor computers.	Excellent and very rapid computers.
Memory storage	Poor short-term storage. Excellent long-term storage.	Excellent short-term storage. Long-term storage very expensive.
Deductive logic	Cannot always be expected to follow optimum strategy. The right premises sometimes lead to wrong conclusions.	Excellent in deductive logic; can exhaust all theorems from a set of postulates. Can store and use optimum strategy for high-probability situations.
Inductive logic	Can go from specific cases to general rules or laws.	No good at inductive logic.
Distraction	Easily distracted by competing stimuli.	Cannot be distracted by competing stimuli.

Source: Ref 4

throttle levers on the steering handle, the operator is able to maintain complete control of his snowmobile using his hands only."

The throttle was opened (to increase speed) by closing (i.e., squeezing) the right-hand lever. The brakes were applied (to decrease speed) by closing (i.e., squeezing) the left-hand lever. When the throttle was released, the engine slowed. At some lower (undetermined) speed the clutch disengaged and the tracks stopped moving, thereby generating significant drag.

As an exercise, this design is analyzed from the human factors viewpoint. If changes would improve the design, what changes would be made?

Most people immediately see this design as very poor from the viewpoint of human factors. Most operators will never see the owner's manual, but in this context that is minor. The natural (automatic) thing for people is that both the left and right hands will do the same thing simultaneously (i.e., both will open or both will squeeze), unless deliberate thought is given to doing otherwise. When a collision is imminent, there is no time for conscious thought, only conditioned reflex. Only well-experienced operators might be expected to have the left hand squeeze while the right hand opens.

It is not necessary, moreover, to have the design that was used. Reversing the action of the brake lever would be a better design (i.e., both hands squeeze to move forward; both hands release to stop). This "deadman" arrangement is common in railway locomotives, power movers, snowblowers, and so on. This arrangement is technically and economically feasible. The argument that keeping the left hand closed is undesirable is weak, because the right hand must be kept closed while the snowmobile is operating.

Another, perhaps superior, alternative is found on many motorcycles and would be familiar to many people. The throttle is operated by rotating a grip on the steering handle. The brake is operated by a squeeze lever on the same side of the steering handle. The two cannot be operated simultaneously. This alternative is technically and economically feasible with no foreseeable adverse effects. Additional information can be found in the section on "Definition of Defects" in the article "Products Liability in Design" in this Volume.

Example 2: A Ladder Label. Figure 3 shows a full-scale copy of a label "permanently attached" (by some sort of adhesive) to the inside of a rail on a fiberglass ladder. The text of the original label had black lettering on a yellow background in keeping with ANSI Z535.1 (Ref 6). The three-line heading was yellow on black. The text print size was 6 points. ("Point" is a printer's measure equivalent to 0.01384 in. or essentially 72 points per in.) ANSI Z535.4 (Ref 7) specifies type (in the text) of 1.5 mm high (minimum), or 5 points. Presumably this label was intended to serve for warnings, directions, or both. *Directions* are intended to *ensure a proper and effective use* of a product. *Warnings* are intended to *ensure a safe use*.

As an exercise, critique this label from the viewpoint of human factors, keeping in mind information given in the articles "Products Liability in Design" and "Safety in Design" in this Volume.

At least two problems deserve comment. One is the content of the text on the label. There are 44 items under five different headings, and many of these items are somewhat ambiguous. For example, in the first three items under "INSPECTION," it is not clear just what one looks for (i.e., there is no definition of terms). Under "SET-UP AND USE," the first item is poorly stated. Engineers have no difficulty in understanding what is intended, but many people besides engineers use ladders. In addition, does any user ever actually measure 75½°, or even "¼ of length being used"? Item 10 seems to conflict with Item 12. Item 11 says: "Always tie top and base to building." Are ladders used only on buildings? Does anyone ever tie a ladder to a building? If so, how? Item 30 says: "Recommend never using if over 65 years of age." Is this the age of the ladder or of the user? And why 65 rather than some other age? One could go on at length about other items. At the bottom the label says: "For additional instructions, see ANSI A14.5." Will the average user have any concept of what this means or where to find a copy? Will any user, even the intelligent engineer, obtain and read ANSI A14.5?

The second problem is the type size. It does conform to ANSI Z535.4. Did you have any difficulty in reading it? Bailey (Ref 8) notes that "type size in books and magazines usually ranges from 7 to 14 points with the majority being about 10 to 11 points. Probably the optimum size is from 9 to 11 points—sizes smaller or larger can slow reading speed."

There was no indication of the location of the label on the ladder rail. If this were an extension ladder, one should find the label at about eye height (5 to 6 ft, 1.52 to 1.83 m) on the base (lower) section. If this were a single ladder, one should find it at the same height on both ends.

How good is this label? How effective is it? Assuming that users do indeed see the label, how many will read it? Of those who read it, how many will really comprehend what the manufacturer is trying to say? It appears that the label was not well thought out, either in content or in phrasing, which is unclear or ambiguous in many places. The label does not provide clear instructions on use or explicitly warn of the dangers. Thus the label is clearly inferior and essentially ineffective. One infers that the manufacturer was trying to cover all possibilities to provide "protection" against products liability suits. It is ineffective in that respect as well.

Anticipating Errors. When humans are involved in the use of a product or system, there will be errors. Some errors are extremely difficult, if not impossible, to anticipate. Also, in many situations, people will abuse equipment. This is commonly a result of poor operating practices or lack of maintenance. In other situations, there may be deliberate action by the user (e.g., trying to fit two components together in a manner

FIBERGLASS
SINGLE & EXTENSION LADDER
FOR SAFETY, READ CAREFULLY

INSPECTION

1. Inspect upon receipt and before use.
2. Never climb a damaged ladder. Return for repair or discard.
3. Check all working parts, rivets, bolts, rope and cable for good working order.
4. Never use ladder with missing parts.
5. Discard if exposed to fire or chemicals.

SELECTION

1. Use 300 lb., and 200 lb. Duty-Rated Ladder for maintenance and heavy-duty work. Never use ladder jacks on 200 lb. or 225 lb. Duty-Rated Ladders.
2. Use ladder with correct duty rating to support combined weight of the user and material. Ladders are available with duty ratings of 200, 225, 250, 300 lb.

SET-UP AND USE

1. Set up ladder at 75½° by placing bottom ¼ of length being used out from vertical resting point.
2. Set ladder on firm level ground. Never lean sideways and never use on ice or snow.
3. Use proper size ladder. Never use temporary supports to increase length or to adjust for uneven surfaces.
4. Keep rungs free from wet paint, mud, snow, grease, or other slippery material.
5. Extend only from ground. Never extend from top or by bouncing.
6. Never walk or jog ladder while on it.
7. Securely engage ladder locks before climbing.
8. Erect ladder with fly (upper) section above and resting on base (lower) section.
9. Each section of a multi-section ladder shall overlap the adjacent section by 3 ft. up to and including 36 ft.; by 4 ft. over 36 ft., up to and including 48 ft.; by 5 ft. over 48 ft., up to and including 60 ft.
10. Always have the four ends of the ladder rails firmly supported.
11. Always tie top and base to building.
12. Project ladder minimum of 3 feet above roof edge.
13. Tie down ladder before stepping onto roof.
14. Never over-reach. Move ladder instead. Keep belt buckle inside ladder side rails.
15. Never use in high winds.
16. Never overload. Ladder designed to support one person when properly used.
17. Never use as a horizontal platform, plank or material hoist.
18. Never use on a scaffold.
19. Never fasten different ladders together to increase length.
20. Never apply a side load to ladder to push or pull anything while on ladder.
21. Never drop or apply impact load to ladder.
22. Never sit on end of ladder rails.
23. When reassembling, properly engage all guide brackets and lock prior to use.
24. Never use in front of unlocked doors.
25. Fly section must have safety shoes if used as a single ladder.
26. Hooks may be attached at or near top for added security.
27. To support the top of a ladder at a window opening, a stabilizer should be attached to span the window.
28. Never use ladder when you are in poor health.
29. Never use if taking drugs or alcoholic beverages.
30. Recommend never using if over 65 years of age.

CLIMBING INSTRUCTIONS

1. Never climb onto ladder from the side or from one ladder to another.
2. Face ladder when ascending or descending. Maintain a firm grip and stand on middle of rung.
3. Never stand above 3rd rung from top.
4. Never climb above support point.

STORAGE

1. Support ladder on racks when stored.
2. Never store material on ladder.
3. Properly support ladder in transit.

FOR ADDITIONAL INSTRUCTIONS, SEE ANSI A14.5

Fig. 3 A black-and-white reproduction of a decalcomania label to be placed on the inside of the side rail of a fiberglass ladder. The heading was yellow lettering on black. The text lettering was black on yellow. Reproduction is 100% of original size.

that is not intended, such as installing thread adapters on pressurized gas containers). There is no question that the designer cannot anticipate all these possibilities and provide protection. Nevertheless, the designer is not relieved of a substantial effort to anticipate such actions and try to thwart them.

How does one proceed? The designer must be well informed on anthropometrics (physical characteristics), how people tend to behave or perform, and how to combine such data to achieve a suitable, effective, and safe design. A wealth of literature is available.

Hunter (Ref 9) includes enough anthropometric data to give insight into the kind of data to expect. He also provides many examples of sources of information. He comments on Department of Defense documents that provide substantial and significant information. The objectives of these various documents can be applied with equal validity to both civilian and military products.

Human behavior is largely a question of psychology, a topic about which most engineers know little. Little information focused for use by engineers seems to be readily available. Possible sources are Ref 10 to 15.

Many publications provide varying degrees of insight and help in applying human factors information to design. Some that may be particularly useful are Ref 4, 16, and 17.

One of the many objectives of the designer is to minimize the probability of "human error," where human error is any personnel action inconsistent with established behavioral patterns considered to be normal, or that differs from prescribed procedures. Predictable errors are those that experience shows will occur repeatedly under similar circumstances. The designer must minimize the possibility of such errors.

People have a strong tendency to follow procedures that require minimum physical and mental effort, discomfort, and/or time. Any task that conflicts with this tendency is highly likely to be modified or ignored by the person who is expected to execute the task.

One of many important considerations in design is to follow common stereotypical expectations as much as possible. Consider a few examples:

- Clockwise rotation of a rotary control (knob) is expected to increase the output.
- Moving a lever forward, upward, or to the right is expected to increase the output.
- On a vertically numbered scale, the higher numbers are expected to be at the top.
- In vehicles, depressing the accelerator is expected to increase speed, and depressing the brake is expected to decrease speed. One expects the right foot to be used to apply force to the accelerator, then moved to the brake pedal. This is true whether one drives on the right or left side of the road.

Smith (Ref 5) tells of a forklift truck that violated the fourth item. The left foot was used to depress a pedal that increased speed, and a brake was applied when the foot was lifted.

Other Information Sources

Hunter (Ref 9) cites an SAE Recommended Practice J833, "Human Physical Dimensions," and other SAE documents. NASA has a three-volume *Anthropometrics Source Book* (Vol 1 has data for the designer, Vol 2 is a handbook of anthropometric data, and Vol 3 is an annotated bibliography) available from the NASA Scientific and Technical Information Office, Yellow Springs, OH. The Department of Defense (DOD) has a basic handbook, *Human Engineering Procedures Guide*, DOD-HDBK-763. One of the basic military specifications is "Human Engineering Design Criteria," MIL-H-1472. DOD documents normally refer to additional references. MIL-H-1472, for example, refers to 54 other documents. All DOD and MIL documents can be obtained from the Standardization Documents Order Desk, Philadelphia, PA.

REFERENCES

1. P. Cazamian, Round Table Discussion on the Social Factors in Ergonomics, Proc. Fourth International Congress on Ergonomics, Strassburg, 1970
2. E.J. McCormick, *Human Factors in Engineering and Design*, 4th ed., McGraw-Hill, 1974
3. M.S. Sanders and E.J. McCormick, *Human Factors in Engineering and Design*, 7th ed., McGraw-Hill, 1993
4. W.G. Ireson, Chap 12, *Reliability Handbook*, McGraw-Hill, 1966
5. C.O. Smith, *Two Industrial Products—Defective Design?*, Paper 93-WA/DE-11, ASME
6. ANSI Z535.1, "American National Standard Safety Color Code," American National Standards Institute, 1991
7. ANSI Z535.4, "American National Standard for Product Safety Signs and Labels," American National Standards Institute, 1991
8. R.W. Bailey, *Human Performance Engineering: A Guide for System Designers*, Prentice-Hall, 1982
9. T.A. Hunter, *Engineering Design for Safety*, McGraw-Hill, 1992
10. E. Grandjean, *Fitting the Task to the Man*, 4th ed., Taylor and Francis, Washington, DC, 1988
11. B.A. Sayers, *Human Factors and Decision Making: Their Influence on Safety and Reliability*, Elsevier Science Publishers, 1988
12. K.S. Park, *Human Reliability*, Elsevier Science Publishers, 1987
13. L.S. Mark, J.S. Warren, and R.L. Huston, Ed., *Ergonomics and Human Factors*, Springer-Verlag, 1987
14. B.S. Dhillon, *Human Reliability*, Pergamon Press, 1986
15. C.D. Wickens, *Engineering Psychology and Human Performance*, 2nd ed., Harper-Collins, 1992
16. W.E. Woodson, *Human Factors Design Handbook*, McGraw-Hill, 1981
17. G. Salvendy, Ed., *Handbook of Human Factors*, Wiley, 1987

SELECTED REFERENCES

- J.A. Adams, *Human Factors Engineering*, Macmillan, 1989
- *Ergonomics: International Journal of Research and Practice in Human Factors and Ergonomics*, Taylor and Francis, Washington, DC
- S.A. Fleger and R.M. Waters, *Human Factors Guidance for Nuclear Safety Analysis Reports at Department of Energy Facilities*, SERA-Vol 4, *Safety and Risk Analysis*, ASME, 1995
- D. Meister, *Conceptual Aspects of Human Factors*, Johns Hopkins Press, 1989
- G. Salvendy, Ed., *Handbook of Industrial Engineering*, 2nd ed., Wiley, 1992
- G. Salvendy and W. Karwowski, *Design of Work and Development of Personnel in Advanced Manufacturing*, Wiley, 1994

Environmental Aspects of Design

Shirley T. Fleischmann, Seymour and Esther Padnos School of Engineering, Grand Valley State University

ONE OF THE HALLMARKS of the late twentieth century has been a growing awareness of the harmful effects on the earth's environment caused by the manufacture and use of a wide variety of products and processes. Ranging from concerns about chlorofluorocarbons (CFCs) and possible destruction of the ozone layer to concerns about shrinking landfill capacity to concerns about pollution of air, water, and land, protection of the earth's environment has become a global issue and an enormous challenge to the engineering community. Engineers stand at the center of what has been called the paradox of technology, namely "that environmental disruption is brought about by the industrial economy, but that advancement of the industrial economy has also been and will be a main route to environmental quality" (Ref 1). This will come about through the development of specific tools that will be used by engineers in environmentally responsible design; it will also require cultural change: "Design should not merely meet environmental regulations; environmental elegance should be a part of the culture of engineering education and practice. Selection and design of manufacturing processes and products should incorporate environmental constraints and objectives at the outset, along with thermodynamic and economic factors" (Ref 1).

The topic of this article is so broad that it is difficult to organize the vast amount of available information into usable form; however, Allenby (Ref 2) has suggested a useful framework. He has identified two streams into which environmental concerns can be divided as engineers develop methodologies for environmentally sound design. Those two streams are generic concerns and specific concerns. While specific concerns often involve product- or process-specific guidelines and techniques intended for use by the design team, generic concerns involve the corporate or management structure that will support specific practices in design. In parallel to total quality management (TQM) methods, which recognize that quality comes only with total and "top-down" commitment, consistently good environmental practice will most likely occur if it is supported throughout a given company—from the top management on down to the hourly workers. The rest of this article follows the two streams suggested by Allenby, first the generic concerns or guidelines that provide the structure in which specific techniques can be developed and used are discussed, then specific tools that have been developed and specific information that engineers can use are presented.

Generic (Structural) Concerns

Consistently good environmental practice will most likely occur within a structure or organization that supports such practice throughout the entire organization. In addition to the required support for the time and material resources that are needed for environmentally responsible design, the design team needs a tremendous flow of information about products and processes at a very detailed level. This information flow begins with those who work most closely with the process or product. Each person in the company has the potential to contribute a different type of information, based on where that person works in the company.

Generic concerns include ethical frameworks within which companies and individuals function. Most professional engineering societies include statements of ethics that place public health and welfare at the center of engineering practice. Many companies have developed mission and values statements as part of a TQM approach. Such statements set the tone for activities within the company, and the inclusion of environmental concerns are a first step to good environmental practice.

General guidelines aid companies in developing a plan for good environmental performance by outlining specific activities to which a company must commit. For example, the *Valdez* principles mandate:

- *Protection of the biosphere:* Companies will minimize the release of any pollutant that may endanger air, water, or earth.
- *Sustainable use of natural resources:* Companies will make sustainable use of renewable natural resources, including the protection of wildlife habitats, open spaces, and wilderness.
- *Reduction and disposal of waste:* Companies will minimize waste and recycle wherever possible.
- *Wise use of energy:* Companies will use environmentally safe energy sources and invest in energy conservation.
- *Risk reduction:* Companies will minimize environmental health risk to employees and local communities.
- *Marketing of safe products and services:* Companies will sell products or services that minimize adverse environmental impacts and are safe for consumer use.
- *Damage compensation:* Companies will take responsibility through cleanup and compensation for environmental harm.
- *Disclosure:* Companies will disclose to employee and community incidents that cause environmental harm or pose health or safety hazards.
- *Environmental directors:* At least one member of a company's board will be qualified to represent environmental interests, and a senior executive for environmental affairs will be appointed.
- *Annual audit:* Companies will conduct an annual self-evaluation of progress in implementing these principles and make results of an independent environmental audit available to the public.

The U.S. Environmental Protection Agency (EPA) certainly has many regulations about environmental performance as do state Departments of Natural Resources. These "command-and-control" guidelines are briefly discussed elsewhere in this article; however, the U.S. EPA also offers more generic EPA Environmental Design Strategies (Ref 3); these strategies emphasize:

- Product system life extension
- Material life extension
- Material selection
- Reduced material intensiveness
- Process management
- Efficient distribution
- Improved management practices

Steps that can be used to standardize an approach to design for the environment (DFE) include (Ref 4):

1. *Initiation:* Define the scope, goal, and system boundaries of the study.

Table 1 Major environmental legislation in the United States

Resource Conservation and Recovery Act, 1976 (RCRA)	Clarified waste disposal issues and established cradle-to-grave control of hazardous waste
Comprehensive Environmental Response, Compensation, and Liability, 1980 (CERCLA)	Reauthorized and amended RCRA, clarified responsibility and liability of parties involved in hazardous material management, established Superfund to pay for remediation (by 1984 the EPA identifies 378,000 sites requiring corrective action)
Hazardous and Solid Waste Amendments, 1984 (HSWA)	Step up national efforts to improve hazardous waste management
Emergency Planning and Community Right to Know Act, 1986 (EPCRA), Also known as Superfund Amendments and Reauthorization Act (SARA)	Section 313 of Title III of this act includes the Toxics Release Inventory (TRI), which makes hazardous waste and toxins a matter of public record. This act sets fines for violations.
Pollution Prevention Act, 1990 (PPA)	Establishes hierarchy of Pollution Prevention (PP or P^2), places source reduction at the head of the list for pollution prevention; firms that must report under TRI must also report level of pollution prevention accomplishments
Clean Air Act, 1970 (CAA)	Controlled chemicals list, allowed EPA to assess risk and act to prevent harm; EPA allowed to enforce law without proof of harm; in 1990 establish Maximum Achievable Control Technology (MACT) instead of previous public health standards; emission levels set by EPA or states (Reauthorized 1996, amendments 1977, 1990, 1995)
Clean Water Act, 1972 (CWA)	Gave EPA authority to control industrial discharge to water by imposing discharge requirements on industry, placing special controls on toxic discharge and requiring a variety of safety and construction measures to reduce spills to waterways
Toxic Substances Control Act, 1976 (TSCA)	All new (toxic) substances or new uses of substances entering the marketplace evaluated for health and environmental effects
Examples of Non-Legislative EPA Measures/Actions	
Environmental auditing, 1986	The EPA attempts to formalize procedures for environmental auditing by developing a generic protocol for environmental auditing.
Industrial Toxics Project (33/50 Program), 1991	Companies with large releases of the 17 chemicals of the TRI (Toxic Release Inventory) reported with largest volume releases asked to target 33% reduction by 1992, 50% reduction by 1995. This is a voluntary measure. There are hopes to foster a pollution prevention ethic in business.

2. *Inventory:* Carry out a life cycle analysis study of the product system.
3. *Impact:* Classify all the relevant environmental data and calculate actual environmental effects.
4. *Improvement:* Having identified the areas for improvement, modify the system specification.

International Standards. ISO 14000, the international standard for Environmental Management Systems (EMS) within a company is another powerful generic guideline. The ISO (International Organization for Standardization) standards are voluntary; however, in a globally competitive marketplace such standards can serve to level the field by defining certain threads that are common to all successful companies. ISO had historically developed specific environmental standards that were oriented toward procedures or specific processes. ISO 14000 is different in that it is a generic standard for environmental management systems, as is discussed later. It is also different from previous standards in that it traces its roots to activities from the United Nations. From the beginning, it has been supported by the international business community as well as by the United Nations. A spirit of cooperation has guided the activities of the ISO Technical Committee 207 (TC 207), which has been charged with the development of this standard. The charge to the committee was to develop a generic standard for the development of environmental management systems that involve good environmental practice, but which also enhance the economic status of the company and encourage trade. Critical organizational performance issues that must be addressed by globally successful companies center on the following five areas: economic, quality, occupational safety and health, environment, and social responsibility.

Registration under ISO 9000 is widely recognized as a core strategic business issue related to quality. It is a generic quality standard that transcends national and cultural boundaries and has, in the past 10 years, swept through the global economic community in a way that any legal or regulatory action could not. ISO 14000 parallels ISO 9000 in that it is a generic standard for environmental management systems rather than quality systems (as for ISO 9000). Like ISO 9000, it will involve third-party registration with regular audits required for renewal of that registration. Because the ISO 14000 standards are voluntary, they provide a mechanism for environmental protection and management that are alternatives to the traditional command-and-control model without infringing on national sovereignty or inhibiting the ability of any company to compete. Environmental responsibility is once again placed within the realm of public trust. Since January of 1993, when the ISO Technical Management Board decided to establish TC 207, and March of that year when Canada was awarded the TC 207 secretariat position, more than 80 countries have participated in numerous working groups on a voluntary basis in order to develop the ISO standards, which are briefly summarized as:

- ISO 14000, environmental management system (EMS)—general guidelines on principles, systems, and supporting techniques
- ISO 14001, EMS—specifications with guidance for use
- ISO 14010, environmental auditing (EA)—general principles
- ISO 14011.1, EA—audit procedures (Part 1): auditing of EMS
- ISO 14012, EA—qualification criteria for environmental auditors
- ISO 14020, environmental labeling (EL)—basic principles
- ISO 14021, EL—self declaration, environmental claims, terms and definitions
- ISO 14022, EL—symbols
- ISO 14023, EL—testing and verification methodologies
- ISO 14024, EL—practitioner programs—guiding principles, practices, and certification procedures of multiple criteria (type 1) programs
- ISO 14031, environmental performance evaluation
- ISO 14040, life-cycle assessment (LCA)—general principles and practices
- ISO 14041, LCA—goal and definition scope and inventory assessment
- ISO 14042, LCA—impact assessment
- ISO 14043, LCA—improvement assessment
- ISO 14050, terms and definitions
- ISO 14060, guide for the inclusion of environmental aspects in product standards

The Regulatory Framework. Environmental regulations often contain elements that are unique to the country or region for which those regulations have been developed. The U.S. EPA oversees environmental regulation. The regulatory framework provided by the EPA is most often the "command-and-control" type of framework, but it offers additional guidance in the development of environmentally responsible design. Major legislation by the EPA is summarized in Table 1. Enforcement of this legislation by the EPA falls under the criminal code of law. States that file a State Implementation Plan (SIP) retain enforcement rights for their own State Department of Natural Resources or Department of Environmental Quality. In addition to EPA regulations, each state has its own regulations and industries in a given state are subject to both the Federal and the State regulations. In either case, violation of environmental law is a violation of criminal law—which is a powerful command and control-driven incentive.

For companies considering new processes or new designs, certain parts of this legislation provide lists of controlled chemicals. Every effort should be made to avoid these chemicals because releases require reporting and legal accountability. For example, the Toxic Release Inventory (TRI), which is a part of the Emergency Planning and Community Right to Know Act (EPCRA) (also known as Superfund Amendments and Reauthorization Act, SARA, Title III, section 313) is a list of restricted toxic chemicals, and under the Clean Air Act (CAA) the National Ambient Air Quality Standards (NAAQS) (section 109 under section 40 of the CAA) and the National Emission Standard for Hazardous Air Pollutants (NESHAPS) (section 112 under section 40 of the CAA) both involve lists of chemicals with standards for discharge. Toxic chemicals with the highest volume releases to the environment, according to the TRI were chosen by the EPA for the 33/50 Program. It was called the 33/50 Program because it sought a 33% voluntary reduction in the emission of these chemicals by 1992 and a 50% reduction by 1995. The 17 toxic chemicals targeted by the program are:

- Benzene
- Cadmium and compounds
- Carbon tetrachloride
- Chloroform
- Chromium and compounds
- Cyanides
- Dichloromethane (methylene chloride)
- Lead and compounds
- Mercury and compounds
- Methyl ethyl ketone
- Methyl isobutyl ketone
- Nickel and compounds
- Tetrachloroethylene
- Toluene
- Trichloroethane
- Trichloroethylene
- Xylene(s)

The chemicals and substances covered by NAAQS include:

- Carbon dioxide
- Hydrocarbons
- Nitrogen oxides
- Suspended particulates
- Photochemical oxidants
- Sulfur oxides
- Lead

NESHAPS is a system of point source air discharge standards, which regulates the amount of chemical that can be emitted, the controls required for managing these emissions, and the reporting of any facility modifications that would effect air emissions. Chemicals included under NESHAPS are:

- Asbestos
- Benzene
- Beryllium
- Coke oven emissions
- Mercury
- Vinyl chloride
- Inorganic arsenic
- Rn-222
- Radionuclides
- Copper
- Nickel
- Phenol
- Zinc and zinc oxide

Lists such as these give the designer specific information about chemicals that are known to be harmful.

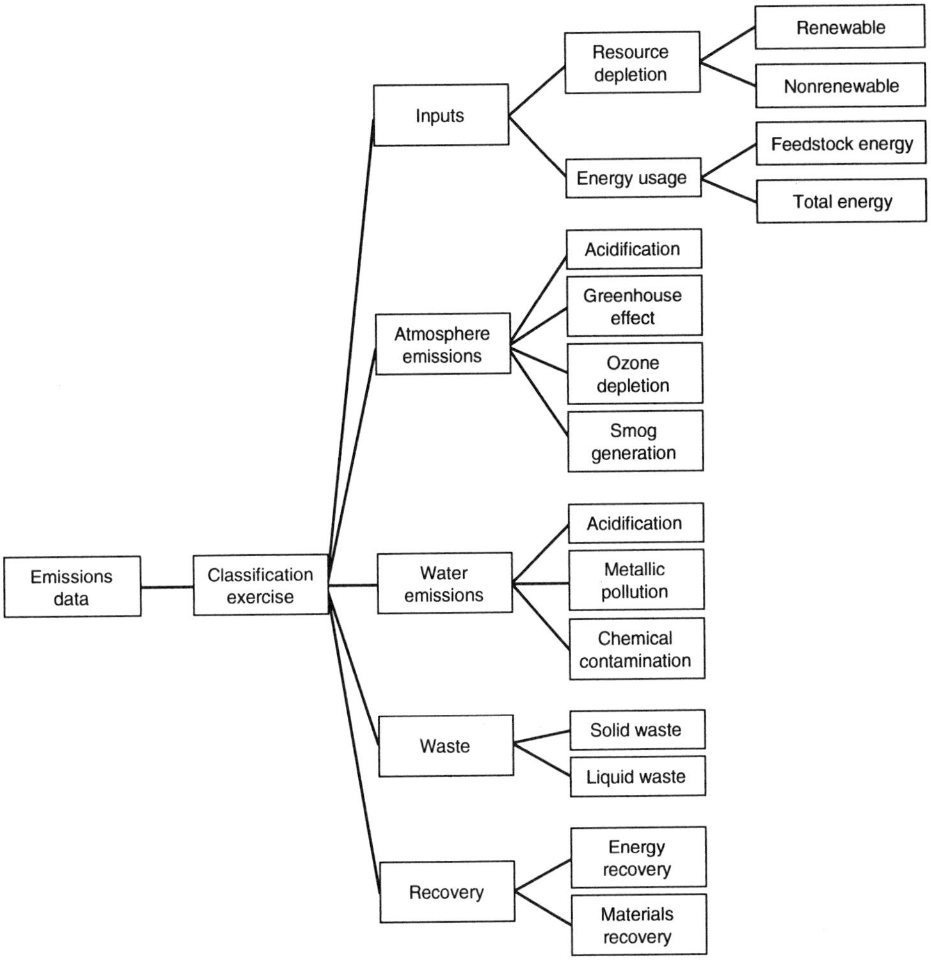

Fig. 1 Schematic showing an environmental impact assessment (EIA) study of emissions data. Source: Ref 5

In addition to its command-and-control approach, the U.S. EPA has also attempted to foster pollution prevention (PP) as a business ethic through various voluntary efforts. The 33/50 Program is one example of these efforts. Pollution prevention roundtables on a national as well as the state level provide forums for the exchange of information. In 1986, the EPA developed a generic protocol for environmental auditing; the use of protocol was/is voluntary, but it does provide a place to begin.

Compliance with environmental regulations on both federal and state level is required and noncompliance can involve substantial fines. The details of the regulatory framework can be overwhelming, and for this reason many companies hire environmental consultants or they designate certain employees to be responsible for environmental compliance. New designs or new processes will have new environmental consequences. For this reason, it is an excellent idea to include someone with knowledge about relevant environmental regulations as part of a design team from the very beginning of the design process. In addition to this designated person(s), it is a good idea to acquaint every engineer involved in design with the basics of the regulatory framework. Note that a company that hopes to be certified under ISO 14000 will most likely undertake these steps as part of its EMS.

Tools for Environmentally Responsible Design

Specific concerns include specific design methods and tools as well as information that is helpful to engineers who hope to complete environmentally sound designs. The various design for "X" (DFX) methodologies, including DFE are discussed in this section. Lists of materials that are frequently recycled and lists of problem materials and products are helpful resources. Organizations such as the Air and Waste Management Association (One Gateway Center, 3rd Floor, Pittsburgh, PA 15222) provide technical manuals and other information exchange. Additionally, trade organizations such as the Institute for Scrap Recycling Industries, Inc. (ISRI) (Suite 1000, 1325 G Street, N.W., Washington, D.C. 20005-3104) can provide specific information. Samples of such information are included. Many professional engineering societies provide journal articles as well as manuals and handbooks—

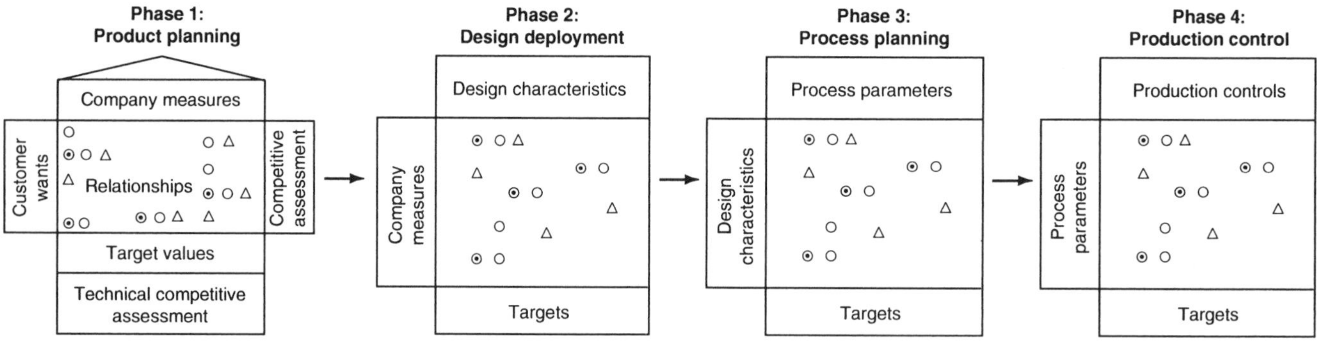

Fig. 2 The four phases of quality function deployment (QFD). Courtesy of the American Supplier Institute

this Volume is just one example—with specific information of value to design engineers. A partial list of software resources is also included.

Life cycle assessment (LCA) is a "cradle-to-grave" assessment of the environmental as well as the energy impacts of a given product design. The article "Life Cycle Engineering and Design" in this Volume provides details about this type of analysis. A full LCA involves a considerable commitment of time and other resources; however, the framework of LCA thinking can be used to prompt the designer to ask questions about the entire product life cycle that might not be asked in a more traditional approach to design. In many cases the primary design focus is on functionality, and environmental performance has not been a significant enough part of the design culture to be considered as part of functionality.

A small example serves to illustrate this point. One company reported providing every employee with a ceramic coffee mug as part of an effort to reduce total discarded paper. (Note that coatings on paper hot beverage containers make them difficult to recycle, so disposal is often the only option.) A narrow concentration on the discarded paper cup waste stream showed a significant decrease in volume; however, life cycle thinking prompted the company to examine the upstream and downstream impacts of this change. They found that many of their employees washed their cups after each use and dried them with paper towels. Not only did they use extra energy by using hot water (which, in turn impacts the environment through discharges from the production of energy), but they also significantly increased the discarded paper towel waste stream. A formal LCA was never done; however, the framework of LCA thinking prompted a much broader and more complete view of the impacts of this change. The habits of the mind that are developed through LCA thinking and that broaden the system boundaries to include cradle-to-grave considerations at the beginning of a design may be the greatest contribution of this type of analysis.

It should also be noted that LCA is a significant part of ISO 14000. There are already examples of standards driving a given design approach; for example, QS 9000 requires failure modes and effects analysis (FMEA). As a result, many automotive suppliers are now including FMEAs as a part of their design process. The same thing may very well happen with LCAs within ISO 14000.

Environmental impact assessment (EIA) can be considered as part of an LCA study; however, it is a separate step that involves its own discipline or approach. Figure 1 shows how an EIA study structures an approach to study the impact of emissions in such a way that major potential problems are identified.

Quality function deployment (QFD) is a structured approach to design that begins with customer requirements ("needs and wants") and deploys those requirements throughout the entire design through production control. QFD consists of four phases: production planning, design deployment, process planning, and production control. The output for each phase feeds into the next phase as shown in Fig. 2. One of the strengths of this approach to design is that environmental requirements can be explicitly entered into phase 1 as customer requirements. Figure 3 shows one way to enter environmental concerns or requirements as a separate environmental block. Another strength of this approach to design is that phase 1 planning requires explicit measures by which the design team will assess the extent to which the design actually meets customer requirements. Designers may enter the design process with good intentions of being environmentally responsible, but unless measures of good environmental performance are specified at the beginning, it may be difficult to ensure that intentions alone will sustain the commitment to environmental performance. By requiring specification of measures at the beginning of this type of a cascaded process, it is far more likely that good environmental practice goals will be met.

The phase 1 chart includes two competitive assessments. On the bottom of the chart, the proposed design can be compared to a benchmark design or a competitive design based on the measures proposed directly above on top of the chart. On the right-hand side, a similar competitive assessment based on customer requirements is included. The competitive assessments presented in this highly visual form make it far easier to determine if improvements must be made from a technical performance or a customer satisfaction point of view. A product with an excellent technical performance and poor customer performance might indicate a need for the marketing department to step in; a product with good customer performance and poor technical performance shows a potential for problems if technical measures are brought into focus.

The "house of quality" at the top of the chart provides a structured approach to identify the trade-offs involved in optimizing the design. Strongly negative interactions in meeting test requirements indicate the areas in which research might be required and in which potential breakthroughs may occur. Outcomes from phase 1 are inputs to phase 2 and so on through the four phases as noted earlier. This structured method provides a way to include and also to track environmental performance throughout the design. Further information about the QFD approach is provided in the articles "Conceptual and Configuration Design of Products and Assemblies" and "Concurrent Engineering" in this Volume.

Design for "X" (DFX) approaches include design for assembly (DFA), design for disassembly (DFD), and so forth. Each approach seeks to optimize design performance with respect to a given variable such as ease of assembly, ease of disassembly, manufacturability, and so forth. While these approaches are discussed in other articles in this Volume, DFD and DFE are appropriate to discuss here. Design for the environment might involve QFD as a design method; it would most likely involve LCA and EIA; that is, it might involve many of the tools discussed to this point. However, the primary focus of the design effort is to optimize environmental performance while also meeting all other requirements.

Failure modes and effects analysis (FMEA) is currently required of many automotive suppliers as part of a QS 9000 (a version of ISO 9000 developed cooperatively by the automotive "Big 3" for automotive supply) quality plan. Like QFD it is a structured approach in which product or process failure modes are identified and ranked. The tool guides the design team to develop methods to avoid these failure modes. While functional failures are often the focus of such an analysis, environmental failure modes such as accidental releases of dangerous chemicals to the environment or inability to recycle and reuse material at the end of the product life and so forth could be included. Additional information about FMEA is provided in the article "Risk and Hazard Analysis in Design" in this Volume.

Software for Environmentally Responsible Design. Boothroyd and Dewhurst, Inc. has developed software to facilitate a number of the DFX approaches, In particular, the company has a DFE product that enables designers to estimate the type of environmental impact that a design will have from the beginning to the end of its useful life. It can assist the designer in assessing costs and benefits directly attributed to disposal, recycling, and reuse of all or part of the product. It is particularly helpful when used in conjunction with another DFX program such as DFA (see the article "Design for Manufacture and Assembly" in this Volume). With little additional effort, the DFA spreadsheets can be imported into a DFE program for a more thorough product analysis.

The Boothroyd-Dewhurst Design for Service (DFS) software enables a designer to estimate ease of product disassembly as well as disassembly time. This is an important part of DFD. Recovery of parts and materials is often very labor intensive in both the disassembly and sorting phases.

AT&T Bell Laboratories is developing (Ref 6) a scoring system that allows the "greenness" of a design and its components to be quantified using factors such as ease of separation, environmental impacts of material choices, relative weight of recyclable to nonrecyclable material, and whether materials are clearly identified. The system advises designers where scoring points are lost so that areas for improvement can be quickly and clearly identified.

Carnegie-Mellon University has developed a computerized DFE tool called ReStar (Ref 7). This program aids the designer in assessing tradeoffs between recycling cost and the benefits derived from reduced environmental distress. For most products, 100% recycling is not economically feasible; instead certain parts or subassemblies might be either recyclable as refurbished (remanufactured) parts or material recovery might be possible. This program serves to aid the designer in deciding how far to go in disassembly. It balances revenue or savings expected from recycling with the costs of recycling.

Design for Disassembly (DFD). If material is to be recovered and reused at the end of the useful life of a product, DFD may be the key to economically feasible recovery. An issue of *Business Week* magazine (Ref 8) featured DFD and a number of new products as well as the companies that led the DFD effort. For example, the BMW Z1 automobile has an all-plastic exterior that is designed to be completely disassembled from the chassis in 20 min. Also, the Polymer Solutions U-Kettle (an electric water heating kettle) can be pulled apart into like-material components after removing a minimal number of screws. Use of two-way pop-in and pop-out fasteners rather than glues or screws and a dramatic decrease in the number of different types of plastic used in a given product are just two of the strategies that were discovered in these early efforts. As noted in the article, "This is not easy stuff. Designing durable goods for disassembly is something of an oxymoron" (David L. Calhoun, G.E. Plastics). For example, Polymer Solutions found that tolerances between parts that snap together had to be very fine in order to eliminate relative motion between parts in use. BMW discovered that they used on the order of 20 different types of plastics in a typical car, and many products are made from resins that are not clearly identified. Because the recovery of materials is typically very labor intensive, the extra time required for the positive identity of plastics when there are literally hundreds of differing formulations on the market can make it economically unfeasible to recycle the product. Guidelines for DFE where many of the suggestions are clearly linked to DFD include (Ref 9):

- Make sure that everyone involved in the product design fully understands DFE principles and design guidelines.
- Avoid the use of toxic materials (as much as possible) in the product and its manufacturing process.
- Design the product and its components to be reusable, refurbishable, or recyclable.
- Minimize the number of parts.
- Minimize the amount of material in the product.
- Avoid (if possible) the use of separate fasteners.
- Utilize the minimum number of screw head types and sizes used in fasteners in one product or portion of the product.
- Use the fewest number of fasteners so as to reduce disassembly time.
- Design parts so that fasteners are easily visible and accessible to aid in disassembly.
- Design the product to be easily disassembled, if possible, even if some parts are corroded.
- Minimize the number of different materials in the product.
- If the number of different materials cannot be reduced for reasons of manufacturing economics or other reasons, choose materials that are compatible and can be recycled together.
- Avoid the use of composite materials such as glass- or metal-reinforced plastics.

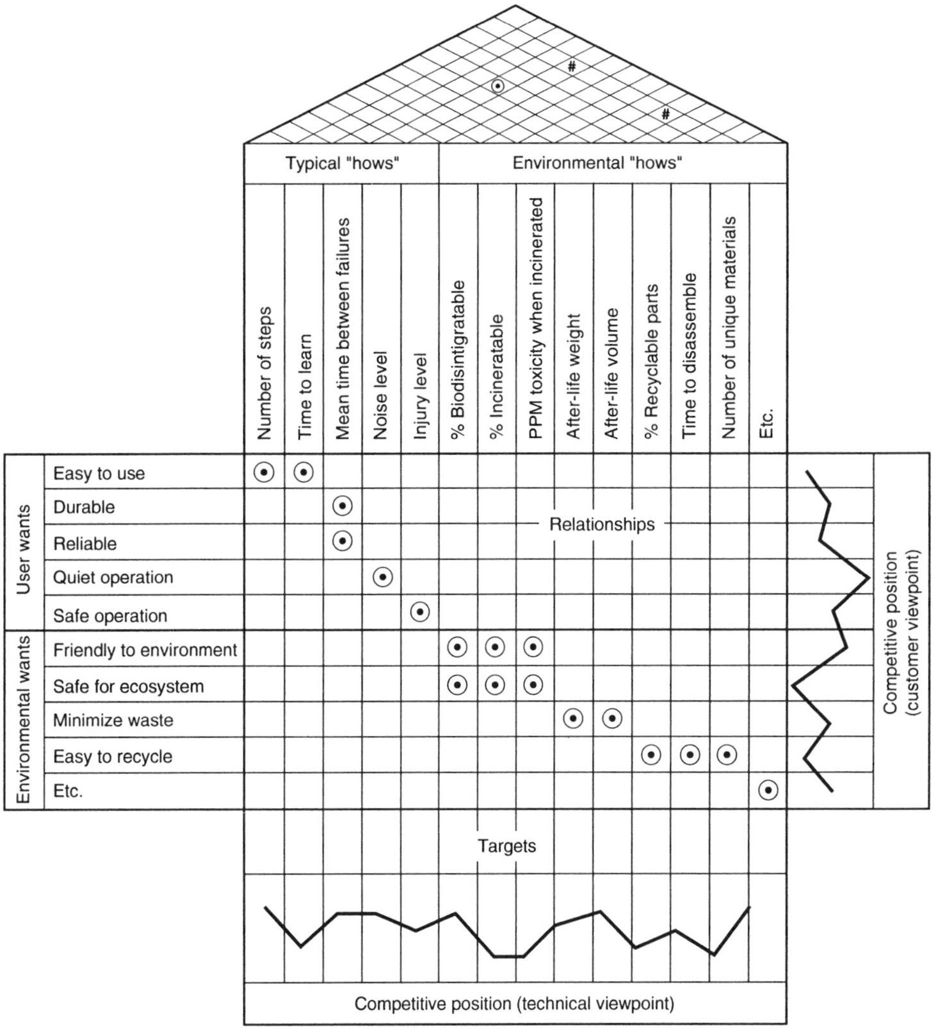

Fig. 3 An example of how environmental concerns can be added to phase 1 of a QFD analysis. Courtesy of the American Supplier Institute

- Standardize components (and subassemblies) to aid in eventual refurbishing of products.
- Use molded-in nomenclature rather than labels or separate nameplates for product identification.
- Use modular designs.
- Wherever feasible (e.g., in molded and cast parts) identify the material from which the part is made right on the part.
- Make separation points between parts as clearly visible as possible.
- Avoid designs requiring spray-painted finishes.
- Provide predetermined break areas if needed.
- Use a woven metal mesh (more easily removable) instead of metal-filled material for welding thermoplastics.
- Design the product to use recycled material from other sources.

Recycling of plastics presents a special problem. To begin with, there are literally hundreds of different types (chemical formulations) of resins—clearly not compatible. Many consumers are familiar with the Society of the Plastics Industry (SPI) coding system found on consumer goods—the molded-in recycling symbol with a number from 1 to 7 in the center. Numbers 1 through 6 are plastics commonly used for packaging; number 7 is designated "other" for all other resins. The Society of Automotive Engineers (SAE) has developed SAE Specification J1344, which is a far more extensive coding system involving many engineering resins. Standards such as ISO 1043, DIN 6120, and VDA 260 are further examples of standards for marking and identification of plastics. Reference 10 contains more information about the identification of plastics.

In addition to the large number of different plastics, plastics are also often lightweight and initially inexpensive. Most secondary materials processors buy and sell material based on weight; a lightweight, high-volume material is more difficult to recycle economically. Because recovery and recycling involve costs of labor and processing and because properties of plastics are often degraded in the recycling process, it is difficult to make recycled material that is cost competitive with often low-cost virgin material. Plastics are also easily contaminated by common chemicals. Additionally, colors that are added to plastics and that eliminate the need for painting in many cases are almost impossible to remove. In addition to sorting by type of plastic, secondary use may also require sorting by color unless the color is not critical. This generally represents more hand sorting and added handling costs. Overall, the market infrastructure for plastics recycling is not nearly as well developed as the secondary materials market for many metals, and partly because of the difficulties in identification and other problems listed above it will be difficult to develop those markets.

Example: Processes and Equipment Used by a Scrap Processor. In order to design products that can be recycled, it is helpful to know how those products are treated in the recycling and material recovery process. This case study is a description of the scrap processing facility owned and operated by the Louis Padnos Iron & Metal Company. The company owns and operates several scrap yards; the Holland, MI (headquarters) yard was chosen because of the variety of processes in a single facility. The ability to move material quickly and efficiently into and out of the facility is extremely important to a scrap processing facility; this particular yard is also interesting because of access by truck, rail, and ship. While the variety of processes described in this study might not be found in a typical facility, the processes and equipment described are typical for the scrap-processing industry.

The Padnos Company processes both postconsumer products and industrial scrap from manufacturing plants. Postconsumer products generally involve mixed materials and possible contamination; separation is required. Industrial scrap is generally a single type of material, that is, turnings from machining operations or clip from stamping operations, which are identified by the company selling the scrap to the company. The metallurgical laboratory is used to verify the identity of incoming material and also to test samples of processed material that is about to be sold.

Materials enter and exit the yard by truck, railcar, or by ship. The latter is least frequent, but the yard is located on Lake Macatawa from which ships have access to the Great Lakes. Railcars and trucks are weighed upon entering and exiting the yard, and they are also scanned for radioactive materials. Such materials are not accepted for processing, and scanning prevents accidental mixing with other materials. Most materials are bought and sold on the basis of weight, so this is a critical operation. In fact, railcars are swept out before loading in order to avoid transportation costs for dirt—a material with no market value.

Briquetter. One of the first operations encountered in the yard is the briquetting machine. Turnings from machining operations are heated (below melting temperature) and pressed into small bars roughly 50 by 150 by 25 mm (2 by 6 by 1 in.) in size. This is a very valuable form of scrap because of its purity and its density. Turnings that arrive from the manufacturer are frequently coated with oil from the machining process. The entire area under the briquetting operation is covered by an underground geomembrane—a thick layer of plastic—which collects the oils that drain off the turnings. The oil and other runoff is pumped to a central collection tank from which oil is recovered. A baghouse air-filtering system traps airborne emissions from the heating and pressing process.

Shear. The 1100 metric ton shear prepares plate metal material for steel melting furnaces. Preprocessing of large products to a size and shape suitable to enter the shear is done by hand—usually using cutting torches. The plate material leaving the shear is then sorted, again by hand, according to size and surface finish. Paint causes problems mainly for furnaces without pollution-control equipment.

The shredder is perhaps the centerpiece of the postconsumer processing equipment in the yard. Automobiles, appliances, and a large number of largely metal consumer goods are processed here. Automobiles have most often been preprocessed by automotive salvagers who remove such valuable parts as radiators (copper), catalytic converters (platinum), working subassemblies such as steering columns, alternators, horns, and so forth, for reconditioning and resale. Airbag deployment

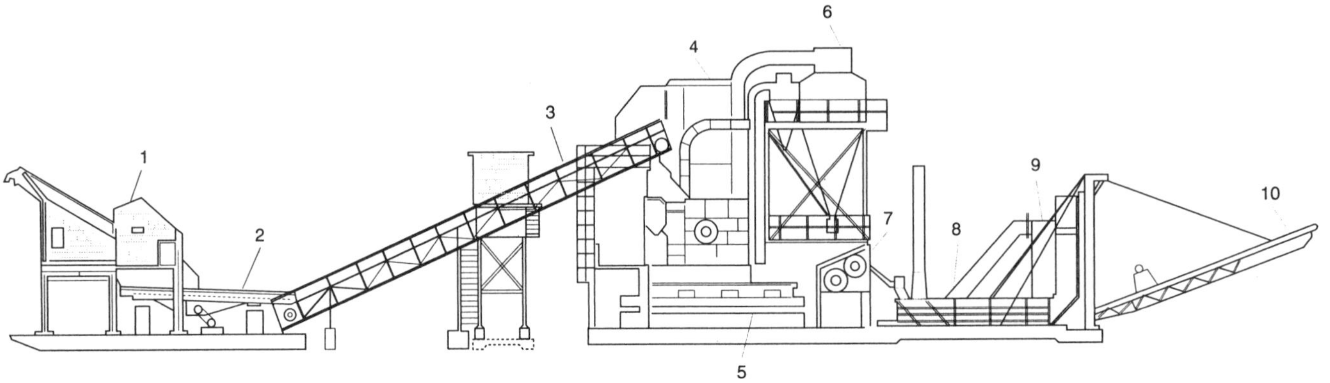

Fig. 4 Schematic of a typical shredder. 1, preshredder; 2, oscillating conveyor; 3, piano hinge conveyor; 4, shredder (ring mill); 5, oscillating conveyor; 6, dust collector unit; 7, magnetic separator; 8, incinerator; 9, smoke scrubber; 10, finished product conveyor

is not required, in part because it is difficult to detect them, but it should be noted that undeployed airbags do pose a safety risk to scrap-processing workers. The Institute of Scrap and Recycling Industries (ISRI) has prepared a pamphlet that explains the industry position on this important safety issue. Fuel tanks must be removed because they are an explosion hazard. Batteries must be removed to avoid lead contamination, and it is preferable to also remove the tires because the tires have little or no value. Engine blocks are removed and treated in a separate part of the yard, and refrigerants must also be removed. The Holland yard does not have a refrigerant recovery facility; however, another scrap processing yard in Grand Rapids (also part of the same company) does have a recovery unit. This is an issue for appliances as well as automobiles.

Figure 4 is a schematic of a typical shredder. Crane-mounted electromagnets or claws lift material into the ripsteel where three slowly counter-rotating wheels literally rip the product apart. The material then enters the hammermill where 36 hammers weighing 350 lb each are attached to a drum which rotates at 893 rpm. The ripped materials are pounded and in the process dissimilar materials are loosened from each other. Material leaving the hammermill drops through a 150 mm (6 in.) grate onto a shaker conveyor. From this point on, the material is separated into three streams: ferrous metals, nonferrous metals, and fluff. Ferrous metals are deposited onto a second conveyor that leads into a large cyclone blower. Light contaminants such as dirts or shreds of plastic are blown off the remaining scrap or "No. 2 frag" is deposited in a large pile. "No. 1 frag" comes from a uniform material source where contamination and mixed material is not a problem. The "No. 2 frag" is more common and comes from a mixed input stream such as has been described (cars, washing machines, etc.). Padnos holds a Q-1 Award from Ford Motor Company for the quality of the ferrous material from the shredder, which Ford uses as feedstock for its steel-making processes. This material can be reused with little or no detriment in its properties. Nonferrous metals are not picked up by the magnets. Large blowers blow the light material such as dirt, plastic, glass, foam, and so forth, away, leaving nonferrous metals for a second output stream. These metals are more difficult to separate and require specialized equipment or hand labor. They are sold to another scrap processor. The third stream, the fluff, which consists of plastic, glass, foam, wood, dirt, carpet, and so forth, is a mixed stream with little value. It is generally landfilled and is therefore a direct cost to the shredding operation.

Note here that the metals (and mainly the ferrous metals) make the shredding operation profitable. As metals usage decreases and plastics and resins are substituted, shredding becomes less profitable. A labeling scheme for the plastics might be proposed, and plastics could be removed before shredding, but such removal and separation often involves hand labor and is both time consuming and costly. Recovery of such materials will occur if it is cost effective; however, consider that a typical American automobile can contain over 200 different resins. Parts not clearly marked or easily separable are most economically landfilled or sent to an incinerator. Finally, the markets for metals are well established, and infrastructure for recovery and reuse of metals exists. Markets for plastics are much more difficult to find, the material has less intrinsic value, and the infrastructure to handle such materials does not exist.

Engine Block Crusher. Engine blocks are placed into a special containment area where fluids are allowed to drain out. Again, this part of the yard has an underground containment system to keep oils and other fluids from draining into the soil and eventually into the watertable. After draining, engine blocks are crushed in a hammermill. The material is washed, and aluminum parts are removed by hand, and the remaining material is further hand sorted by size. Aluminum is melted into large "sows" and sold. Because of its low melting temperature, other metals commingled with aluminum in this process generally do not melt in the furnace and are easily removed in the pouring process. The crushed and cleaned block material is mostly cast iron and can be sold as feedstock to casting furnaces.

Baler. A crane-mounted electromagnet lifts stock into a chamber, and a plunger crushes the material. When the chamber is full of crushed material, metal bands are bound around the completed bale. The stock, which is baled by Padnos in the Holland yard, is typically flat stock that is the scrap from stamping operations. Like the turnings, this material typically comes from a manufacturing plant and is both uniform and the composition is known.

Nonferrous Metals Processing. Padnos also handles nonferrous metals in the Holland yard. Aluminum cans are crushed and baled. Plumbing fixtures, wire, brass stamping scrap, and so forth, are largely processed and sorted by hand.

Paper Processing. Paper is processed in the Holland yard, but Padnos does operate a separate paper-processing plant in Grand Rapids. Paper is separated by hand and baled for sale. The profit margin for paper is small, but with large enough volume and with a ready market for this material it is cost effective to process it. Foil, plastic, nonwater soluble glues, spiral bindings, and so forth, are all contaminants for paper recycling and must be removed and separated in the hand-sorting process.

Conclusions. Understanding how secondary materials are processed is critical to design engineers who wish to design products from which materials will be recovered. It is evident that recovery processes involve a considerable amount of hand labor in the sorting of materials. Strategies such as clearly identifying materials, making a product from the smallest number of different materials possible, making it very easy to disassemble products into marketable parts, and using materials for which a secondary material market exists are all important strategies.

Additional information about recycling of metals is available in the articles "Recycling of Iron, Steel, and Superalloys" in Volume 1 of *ASM Handbook* (Ref 11) and "Recycling of Nonferrous Alloys" in Volume 2 of *ASM Handbook* (Ref 12). Recycling of plastics and recycling of ceramics and glasses are discussed in Ref 13 and 14, respectively.

Additional Environmental Guidelines. What makes a product environmentally unfriendly? Sources of environmentally unfriendliness (Ref 9) are:

- Noxious or poisonous fumes or gases
- Excessive noise
- Hazardous liquids including acids, alkalies, and solvents
- Hazardous solid materials including heavy metals such as mercury, lead, and arsenic
- Safety hazards such as sharp corners or mechanisms that can crush body members or cause electrical shock
- Radioactive materials
- Bacterial contamination of food, drink, or materials that will be used in their preparation/storage

Note that the lists of restricted chemicals from the EPA provide specific information about which materials are hazardous.

Table 2 Materials and products that should be evaluated in terms of environmental risk

Material product	Potential problems
Gas tanks	Explosion, lead from gasoline residue
Batteries	Lead, acid, nickel-cadmium residues
Closed containers and tanks	Explosion, chemical residues
Testing and laboratory equipment, hospital equipment, vials, bottles	Radioactivity, chemical residues, medical and human wastes
Gages and measuring devices	Radioactivity
Compressed gas cylinders	Explosions, hazardous gases
Turnings	Oil, chemical residues
Insulated/coated wire and cable	Asbestos, lead, other coatings or internal chemical residues
Pipe	Asbestos, lead, other coatings or internal chemical residues
Transformers and capacitors	PCBs, oil
Paint/paint thinner cans, buckets	Explosions, lead, cadmium, solvents
Metallic sludges, drosses	Chemical residues
Motor blocks	Oil, solvents, chemical residues
Undeployed automotive airbags	Explosion
Demolition scrap	Radioactivity, asbestos, lead
White goods, appliances	PCBs, CFCs, cadmium, asbestos
Military/government scrap	Explosions, chemical residues
Slag and pit scrap	Chemical residues
Aircraft material	Aluminum,/lithium explosions, chemical residues
Automotive or rail brakes/drums	Asbestos
Waste oil	"Free-flowing," polychlorinated biphenyl (PCB) residues
Plated, coated, painted scrap	Cadmium, lead, zinc

Source: Institute of Scrap Iron and Steel Inc.

Chlorofluorocarbons (CFCs) have been recognized as being environmentally unfriendly due to their potential to destroy the earth's ozone layer. It is now illegal to vent such materials, and care must be taken to recover CFCs from appliances. Alternate materials have been found for decreasing and cleaning, and for many years foams have been made with non-CFC foaming agents. There may be some difficulty with older appliances such as refrigerators with foams that were made with CFC foaming agents. In that case, the CFCs released in crushing the foam may be comparable to the amount released by venting the cooling system. Newer appliances do not have this problem because of design changes made for environmental reasons.

Difficult-to-Recycle Materials. Some materials are simply not economical to recycle, including (Ref 9):

- Laminated materials such as plastics and glass, plastic foam material and vinyl, plastics and metals and dissimilar metals
- Galvanized (zinc-coated) steel
- Thermosetting plastics such as phenolic, urea, or melamine
- Ceramic materials
- Parts with glued or riveted or otherwise fastened identification labels made of a different material than the part (for example, glued paper labels impair the recyclability of glass and plastic containers)
- Subassemblies riveted or otherwise permanently joined dissimilar materials (for example, steel casters on plastic chair bases that cannot be easily extracted)

Difficulty in separation of differing materials and physical properties of materials are frequently the cause of making material recovery uneconomical. For example, thermosetting plastics cannot be melted; when they are formed they are generally made of two or more chemical parts that bond into an almost crystalline lattice structure that must be chemically dissociated. Tires are a thermoset material and also present a large-volume environmental problem, but they also can be considered as a neglected resource (Ref 15). The Institute of Scrap Iron and Steel, Inc. (ISIS) has produced the list shown in Table 2 of problem products. They note that the list is not exhaustive, neither do they suggest that scrap processors refuse to handle these products. They have simply published this list to inform scrap processors about potential problems. The list should be tremendously valuable to designers also.

Conclusions

Changes in the use of international standards, the global marketplace, and the pursuit of quality in manufacturing enterprises have all had a synergistic effect on environmental awareness and environmentally responsible design practice. The specific concerns described in this article can be viewed as the more vertical aspect of design—the specific methods or guidelines used by those designers directly involved in the design of a particular product or process. The generic concerns (organizational structures and guidelines) can be viewed as the horizontal dimension that links design practice into the world in which designs must function. Both the vertical and the horizontal, the generic and the specific concerns, must be addressed if cultural changes are to occur in engineering education and practice.

REFERENCES

1. J.H. Ausubel and H.E. Sladovich, Ed., *Technology and Environment,* National Academy Press, 1989
2. B. Allenby, Industrial Ecology Gets Down to Earth, *IEEE Circuits Devices,* Vol 10 (No. 1), Jan 24–28, 1994
3. G.A. Keoleian and D. Menerey, Environmental Requirements and the Product System, *Life Cycle Design Manual,* U.S. Environmental Protection Agency, Jan 1993
4. C. Fussler, *Life Cycle Assessment—A New Business Planning Tool,* 1993
5. L. Holloway, D. Clegg, I. Tranter, and G. Cockerham, Incorporating Environmental Principles into the Design Process, *Mater. Des.,* Vol 15 (No. 5), 1994
6. W.J. Glantschnig, *Design for Environment (DFE): A Systemic Approach to Green Design in a Concurrent Engineering Environment,* AT&T Bell Laboratories, Princeton, NJ
7. D. Navin-Chandra, *ReStar, a Design for Environment Tool,* Carnegie-Mellon University, Pittsburgh, PA
8. *Business Week,* 17 Sept 1990
9. J.G. Bralla, chapter 18, *Design for Excellence,* McGraw-Hill, 1996
10. *Design for Recycling* (booklet), G.E. Plastics, Pittsfield, MA
11. T.A. Phillips, Recycling of Iron, Steel, and Superalloys, *Properties and Selection: Irons, Steels, and High-Performance Alloys,* Vol 1, *ASM Handbook* (formerly 10th ed. *Metals Handbook*), ASM International, 1990, p 1023–1033
12. D.V. Neff et al., Recycling of Nonferrous Alloys, *Properties and Selection: Nonferrous Alloys and Special-Purpose Materials,* Vol 2, *ASM Handbook* (formerly 10th ed. *Metals Handbook*), ASM International, 1990, p 1205–1232
13. R.L. Walling and R. Ehrig, Recycling of Plastics, *Engineered Materials Handbook Desk Edition,* ASM International, 1995, p 292–296
14. L.M. Sheppard, Recycling and Other Environmental Considerations (for Ceramics and Glasses), *Engineered Materials Handbook Desk Edition,* ASM International, 1995, p 699–706
15. *Decision Maker's Guide for Scrap Tires—The Neglected Resource,* Ohio Council of American Society of Mechanical Engineers, May 1992

SELECTED REFERENCES

- B.R. Allenby and T.E. Graedel, *Industrial Ecology,* Prentice Hall, 1995
- S.B. Billatos and N.A. Basaly, *Green Technology and Design for the Environment,* Taylor & Francis, 1997
- M.R. Block, *Implementing ISO 14001,* American Society for Quality, 1996
- R.E. Cattanach et al., *The Handbook of Environmentally Conscious Manufacturing: From Design and Production to Labeling and Recycling,* Irwin, 1994
- *Design for the Environment: Product Life Cycle Design Guidance Manual,* Government Institute, Mar 1994
- J. Fiksel, Ed., *Design for Environment: Creating Eco-Efficient Products and Processes,* McGraw-Hill, 1996
- H.M. Freeman, Z. Puskas, and R. Olbina, Ed., *Cleaner Technologies and Cleaner Products for Sustainable Development,* NATO ASI Series, Partnership Sub-Series 2, Environment, Vol 2, Springer-Verlag, 1995
- T.E. Graedel and, B.R. Allenby, *Design for Environment,* Prentice Hall, 1997
- *Green Products by Design: Choices for a Cleaner Environment,* Office of Technology Assessment, OTA-E-541, U.S. Government Printing Office, Oct 1992
- M.E. Henstock, *Design for Recyclability,* Institute of Materials/Ashgate, 1988
- D.J. Richards, Ed., *The Industrial Green Game: Implications for Environmental Design and Management,* National Academy Press, 1997

Safety in Design*

Charles O. Smith, Engineering Consultant

THE ASME CODE OF ETHICS says: "Engineers shall hold paramount the safety, health and welfare of the public in the performance of their professional duties." This consideration is not new. Tacitus (Ref 1), in about the first century A.D., said: "The desire for safety lies over and against every great and noble enterprise." Even some 2000 years earlier, the first known written law code (Ref 2), while not specifically mentioning safety, clearly implied a necessity for a builder to consider safety.

The National Safety Council (Ref 3) says: "Each year, accidental deaths and injuries cost our society in excess of $399 billion—in the United States alone. This figure includes lost wages, medical outlays, property damage and other expenses. The cost in human misery is incalculable. Accidents are the fifth leading cause of death. The Council believes that accidents are not just random occurrences, but instead result mostly from poor planning or adverse conditions of the environments in which people live, work, drive and play. In our view, 'accidents' nearly always are preventable—as are many illnesses."

If for no other reason, one should emphasize safety as a matter of enlightened self-interest. Those who design machines and who have an interest in productivity and cost control serve their "customers" well if risks are at a minimum and thus interruptions called accidents are also at a minimum.

Definitions of Safety

One dictionary (Ref 4) definition of safety is: "The quality or condition of being safe; freedom from danger, injury or damage." Most other dictionary definitions are similar. Hammer (Ref 5) says: "Safety is frequently defined as 'freedom from hazards.' However, it is practically impossible to completely eliminate all hazards. Safety is therefore a matter of relative protection from exposure to hazards: the antonym of danger." Lowrance (Ref 6) says: "A thing is safe if its risks are judged to be acceptable." This definition contrasts sharply with the Webster definition (which indicates "zero" risk) and, like Hammer's, implies that nothing is absolutely free of risk. Safety is a relative attribute that changes from time to time and is often judged differently in different contexts. For example, a power saw, lawnmower, or similar powered equipment that may be "safe" for an adult user may not be "safe" in the hands of a child.

Lowrance's definition (Ref 6) emphasizes the relativistic and subjective nature of the concept of safety. It further implies that two very different activities are required for determining how safe a thing is: *measuring risk*, an objective but probabilistic effort; and *judging the acceptability of that risk*, a matter of personal and/or societal value judgment. In addition, the level of acceptable risk involves moral, technical, economic, political, and legal issues.

Technical people are generally qualified to measure risks. The decision whether the general public, with all its individual variations of need, desire, taste, tolerance, and adventurousness, might be (or should be) willing to assume the estimated risks is a value judgment that technical people are no better qualified (and are perhaps less qualified) to make than anyone else.

This matter is explored further in the sections "Paramount Questions" and "Acceptable Level of Danger" of the article "Products Liability and Design" in this Volume.

Hazard, Risk, and Danger

Considerable confusion exists about the meaning of the words "hazard," "risk," and "danger." Products liability litigation in the United States has developed clear distinctions among these three words. See the sections "Hazard, Risk and Danger" and "Definitions of Defects" of the article "Products Liability and Design" in this Volume. "Danger" is the unreasonable or unacceptable combination of hazard and risk. The objective in safe design is to eliminate danger. (Lowrance's use of "risk" seems close to the legal definition of "danger.")

Designer's Obligation

The designer/manufacturer of any product (e.g., consumer product, industrial machinery, tool, system, etc.) has a major obligation to make the product safe (i.e., reduce the risks associated with the product to an acceptable level). In this context, "safe" means a product with an irreducible minimum of danger (as defined in the legal sense), that is, safe not only with regard to intended use (or uses) but also all unintended, but foreseeable, uses. For example, consider the common flat-tang screw driver. Its intended use is well known. Can anyone say that they have never used such a screwdriver for any other purpose? It must be designed and manufactured to be safe in all these uses. That can be done.

There are three aspects, or stages, in designing for safety:

1. Make the product safe (i.e., design all hazards out of the product).
2. If it is impossible to design out all hazards, provide guards that eliminate the danger.
3. If it is impossible to provide proper and complete guarding, provide appropriate directions and warnings.

Make It Safe

In designing any product, there are many considerations, such as function, safety, reliability, producibility, maintainability, environmental impact, quality, unit cost, and so on. In terms of safety, consideration of hazards and their elimination must start with the first concept of the design of the product and be carried through the entire life cycle. As Hunter (Ref 7) says: "This must include hazards which occur during the process of making the product, the hazards which occur during the expected use of the product, the hazards which occur during foreseeable misuse and abuse of the product, hazards occurring during the servicing of the product, and the hazards connected with the disposal of the product after it has worn out."

Because each design is unique, even a modification of an existing product, the designer needs

*Adapted from C.O. Smith, Safety, Chapter 10, *Standard Handbook of Machine Design*, 2nd ed., J.E. Shigley and C.R. Mischke, Ed., copyright © 1996 The McGraw-Hill Companies, Inc. Used with permission.

to give full consideration to safety aspects of the product. Although no fixed, universal set of rules exists that tells the designer how to proceed, some general guidelines are available.

Hazard Recognition

Hazard recognition must start as early as possible in the design stage. Hazard recognition requires much background and experience in identifying accident causation. While academic training is very limited, the National Safety Council and many other organizations publish information on this topic. Any threat to personal safety should be regarded as a hazard and treated as such. These threats come from several sources.

Kinematic/Mechanical Hazards. Any location where moving components come together with resulting possible pinching, cutting, or crushing is a kinematic hazard. Examples are belts and pulleys, sets of gears, mating rollers, shearing operations, and stamping operations with closing forming dies. The author can remember working in a machine shop where individual machines (lathes, grinders, shapers, planers, etc.) were belt driven with power supplied by a large prime mover. Such shops had a great number of nip-point hazards where belts ran onto pulleys and a possible flying object hazard if a belt came apart or slipped off the pulley. Development of low-cost, reliable electric motors for driving individual machines removed the belt-pulley hazards but introduced electrical hazards.

Electrical Hazards. Shock hazard, possibly causing an undesirable involuntary motion, and an electrocution hazard, causing loss of consciousness or death, are the principal electrical hazards for people. Electrical faults (i.e., "short circuits") are the major hazard to property. Massive arcing, cascading sparks, and molten metal often start fires in any nearby combustible material. Any person in the vicinity of a large electrical fault can be severely injured. The danger of electric shock can be reduced by ground fault devices.

Energy Hazards. A potential energy hazard exists if stored energy can be suddenly released in an unexpected manner. Compressed or stretched springs, compressed-gas containers, counterbalancing weights, and electrical capacitors are all energy hazards. Energy hazards are of major importance during servicing of equipment. A designer must develop methods and procedures for placing the product in a "zero energy state" while it is being serviced.

Flywheels, fan blades, loom shuttles, conveyor components, and, in general, any parts with substantial mass that move with significant velocity are kinematic energy hazards that can damage any objects or people who interfere with their motion.

Human Factors/Ergonomic Hazards. All consumer products and most industrial/commercial equipment is intended for use by humans. Ergonomics, defined as the art and science of designing work and products to fit the worker and product user, is a top-priority consideration in the design process.

In many ways a human's capability can exceed that of a machine. The human can adjust to unusual situations; the machine cannot. The human can decide to go over, under, or around an obstacle, and do it; the machine cannot. In an emergency situation, the human can outdo normal performance to a degree that would cause a machine to fail (blow a fuse, pop a gasket, etc.). Unfortunately, humans also make mistakes that can cause accidents.

For additional information, refer to the article "Human Factors in Design" in this Volume.

Environmental Hazards. Internal environmental hazards are things that can damage the product as a result of changes in the surrounding environment. For example, in a water-cooled engine, the water can freeze and rupture the cylinder block if the ambient temperature goes below the freezing point. This problem can be alleviated by using freeze plugs that are forced out of an engine block if the water freezes, adding antifreeze to the cooling water, or using an electrical heating coil that replaces the oil drain plug (standard winter equipment in cities such as Fairbanks, Alaska).

External environmental hazards are adverse effects the product may have on the surrounding environment. These include such effects as noise, vibrations (e.g., from forging and stamping operations), exhaust products from internal combustion engines, various chemicals such as chlorinated fluorocarbons ("Freon") and polychlorinated biphenyls (PCBs), electronic switching devices emitting electromagnetic disturbances, and hot surfaces that can burn a human or cause "thermal pollution."

Hazard Analysis

Hazard analysis is the investigation and evaluation of:

- The interrelationships of primary, initiating, and contributory hazards
- The circumstances, conditions, equipment, personnel, and other factors involved in the safety of a product or the safety of a system and its operation
- Avoiding or eliminating specific hazards by using suitable design, procedures, processes, or materials
- Controls for containing possible hazards and the best methods for incorporating these controls in the product or system
- Damaging effects resulting from lack, or loss, of control of any hazard that cannot be avoided or eliminated
- Safeguards for preventing injury or damage if control of the hazard is lost

References 8 to 12 present typical approaches.

The Consumer Product Safety Commission (CPSC) publishes results of its accident data collections and analyses in the form of Hazard Analyses, Special Studies, and Data Summaries. These identify hazards and report accident patterns by types of products. Information is available from the National Injury Information Clearinghouse, CPSC, 5401 Westbard Avenue, Washington, DC 20207.

Consumer products, as the term implies, are those used by the general public. While service life can be relatively short, some products, such as household refrigerators, stoves, clothes washers, and dryers, often operate for many years, perhaps 20 or more.

In contrast to consumer products, industrial and commercial products are intended to provide revenue for their owners and normally have a relatively long service life. While long life is an advantage from the economic viewpoint, it tends to perpetuate safety design problems for years after safer designs have been developed and distributed in the marketplace. Long service life requires extra care in designing for safety.

Failure Modes and Effects Analysis. Failure modes and effects analysis (FMEA) is performed at the individual component level very early in the design phase to identify all possible ways that components can fail and to determine the effect of such failures on the system (i.e., what the user will experience). FMEA is an inductive process that asks: What if? Additional information can be found in the article "Risk and Hazard Analysis in Design" in this Volume.

Failure Modes and Criticality Analysis. Some components or assemblies in product are especially critical to the product's function and the safety of operators. These should be given special attention with more detailed analysis than others. These more critical components can be identified through experience, engineering judgment, or other analysis. Criticality is rated in more than one way and for more than one purpose. For example, the Society of Automotive Engineers has an Aerospace Recommended Practice (ARP) 926 that establishes four categories of criticality as a function of the seriousness of the consequences of failure. This type of analysis is essentially an extension of FMEA and is designated failure modes, effects, and criticality analysis (FMECA).

Fault Tree Analysis. Fault tree analysis (FTA) differs from FMEA in that it is deductive rather than inductive. FTA starts with what the user experiences and traces back through the system to determine possible alternative causes. The focus is on the product, system, or subsystem as a complete entity. Additional information can be found in the article "Risk and Hazard Analysis in Design" in this Volume.

Fault Hazard Analysis. FMEA and FTA consider only malfunctions. Fault hazard analysis (FHA) has been developed to assess the other categories of hazards. FHA was developed at about the same time as FTA, but it does not use the same logic principles as FTA or the quantitative aspects of FMEA. It was first used by analysts with no knowledge of FTA and by those desiring a tabulated output, which FTA does not provide. FHA is qualitative. It is used mainly as a detailed extension of a preliminary hazard analysis.

Operating Hazards Analysis. FMEA, FMECA, FTA, and FHA are primarily concerned with problems with hardware. In contrast, operating hazards analysis (OHA) investigates the actions of operators involved in activities such as operating, testing, maintaining, repairing, transporting, or handling a product. The primary emphasis is on personnel performing tasks, with secondary consideration of equipment. The end result is usually recommendations for design or operational changes to eliminate hazards or better control them. OHAs should be started early enough to allow time for consideration and incorporation of changes prior to release of a product for production.

Design Review

Design review is an effort, through group examination and discussion, to ensure that a product (and its components) will meet all requirements. In a design of any complexity, there must be a minimum of three reviews: conceptual, interim, and final. Conceptual design reviews have a major impact on the design. As the design becomes fixed, less time is available for major design changes, so interim and final reviews have less impact than the early reviews. It is much easier and much less expensive to design safety in at the beginning of the review process than to include it retroactively. This topic is discussed in greater detail in the section "Design Review" of the article "Products Liability and Design" in this Volume.

Standards

Once a design problem is formulated with the intended function clear, the designer should collect, review, and analyze all pertinent information relative to standards, codes, regulations, industry practices, and so on. From this study, the designer can usually get assistance in hazards analysis and formulate the design constraints resulting from the known requirements. A clear distinction must be made as to those requirements that are voluntary and those that are mandatory. Additional information can be found in the article "Designing to Codes and Standards" in this Volume.

Occupational Safety and Health Administration. The Federal Occupational Safety and Health Act (OSHA) was passed in 1970. One of its goals was "to assure so far as possible every working man and woman in the nation safe and healthful working conditions." These regulations have the force of law, which means employers must provide a workplace with no recognized hazards. Thus employers cannot legally operate equipment that exposes workers to unprotected hazards. Consequently, designers must design hazards out of their products before these products reach the market. Additional information can be found in the article "Designing to Codes and Standards" in this Volume.

Maintenance

Maintenance safety problems can be separated into those that occur during maintenance, from lack of maintenance, or from improper maintenance. Improper maintenance, for example, might be a situation where electrical connections on a metal case were not installed correctly, thus giving a hazardous condition where none existed previously. There seems to be little the designer can do to prevent a lack of maintenance. Much improper maintenance can be avoided by designing products in such a way that incorrect reassembly is essentially impossible.

Equipment of all kinds requires periodic adjustment or replacement of parts. Designers sometimes fail to consider the hazards to which maintenance personnel will be exposed, even in routine maintenance. During maintenance, safety devices must often be disconnected and/or protective guards removed to permit necessary access. In this context, maintenance personnel may need to put parts of their bodies in hazardous locations that were protected by now-bypassed safety devices. It is the responsibility of the designer to provide protection in this situation.

Lockouts, Lockins, and Interlocks. Many injuries and fatalities have occurred when a worker unwittingly started equipment while a maintenance worker was in the equipment. It is necessary to make it impossible for machinery undergoing maintenance to be started by anyone other than the maintenance worker. CFR 1910.147(c)(2)(iii) [OSHA] requires the designer to provide lockout protection.

A lockout prevents an event from occurring or prevents an individual, an object, force, or other factor from entering a dangerous zone. A lockin maintains an event or prohibits an individual, an object, force, or other factor from leaving a safe zone. Locking a switch on a live circuit to prevent current from being shut off is a lockin; a similar lock on a switch on an open current to prevent it from being energized is a lockout. Both lockouts and lockins can be accomplished if each worker has a personal padlock and key (any duplicate key would be in a central office in a locked cabinet). This procedure can mean multiple locks placed on a lockout panel.

Interlocks are provided to ensure that an event does not occur inadvertently or where a sequence of operations is important or necessary and a wrong sequence could cause a mishap. The most common interlock is in the form of an electrical switch that must be in the closed position for power to be supplied to the equipment. If a guard, cover, or similar device is opened, or left open, the machine will not operate. Although interlocks are usually very effective, they can be rather easily bypassed by using some means to keep the switch closed.

Smith (Ref 13) comments on two accidents, one involving a screw auger for mixing core sand in a foundry, the other involving a large batch mixer. In both cases, maintenance workers suffered permanent disabling injuries when another worker negligently switched on the equipment. A lockout or an interlock that functioned when the cover was lifted would have prevented the injuries.

Zero Energy. Many products require storage of energy for operation. For example, energy is stored in any spring that is changed during assembly from its free, unstressed dimensions. This energy storage is also true of cables, cords, and chains that are loaded in tension. Some other sources of stored energy are compressed gases, energized electronic power sources, and lifted counterweights. The zero energy concept requires the designer to provide protection for any operator or maintainer of equipment against the consequences of unanticipated release of stored energy (i.e., neutralize these energy sources in an emergency situation or during maintenance work).

Fail-Safe Designs

A significant fraction of accidents are the result of product failure or malfunction. Fail-safe design seeks to ensure that a failure will either not affect the product or change it to a state in which no injury or damage will occur.

Fail-passive designs reduce the system to its lowest energy level. The product will not operate until corrective action is taken but there will be no further damage from the failure-initiating hazard. Circuit breakers are a good example of fail-passive devices.

Fail-active designs maintain an energized condition that keeps the system in a safe operating mode until corrective action can be taken or the system is replaced. Redundancy using standby equipment is an example of a fail-active system.

Fail-operational designs allow safe continuation of function until corrective action can be taken. Fail-operational is obviously the preferred form, if possible. The ASME requires fail-operational feed-water valves for boilers. Water must first flow under, rather than over, the valve disk. If the disk is detached from the valve stem, water will continue to flow and allow the boiler to function normally.

Designs should be made fail-safe to the greatest degree possible.

General Principles of Designing for Safety

Hunter (Ref 7) gives the following statements as general principles or guidelines for designing safe products:

- Recognize and identify actual or potential hazards, then design them out of the product.
- Thoroughly test and evaluate prototypes of the product to reveal any hazard overlooked in the preliminary design stages.
- Make certain that the product will actually perform its intended function in an acceptable manner so that the user will not be tempted to modify it or need to improvise possibly unsafe methods for using it.
- If field experience reveals a safety problem, determine the root cause, develop a corrective action to eliminate the hazard, and follow up to make certain that the corrective action is successful.
- Design equipment so that it is easier to use safely than unsafely.

- Realize that most product safety problems arise from improper product use rather than product defects.

Safety Checklists

Hammer (Ref 9, 10) and the National Safety Council (Ref 14) provide lists of basic safety requirements for use in developing safe designs. For example, at the top of his list, Hammer (Ref 9, 10) says: "Sharp corners, projections, edges, and rough surfaces which can cause cuts, scratches, or puncture wounds will be eliminated unless required for a specific function." There are 21 more items in the list.

Acceptable Conditions

Hammer (Ref 9, 10) notes that safety engineers (perhaps no one else?) generally consider the following conditions acceptable and indicative of good design.

- Any design that to cause an accident requires at least two *independent* malfunctions, two *independent* errors, or a malfunction and an error that are *independent*
- Any design that positively prevents an error in assembly, installation, connection, or operation that analysis indicates would be safety-critical
- Any design that positively prevents a malfunction of one component (or assembly) from causing other failures that could cause injury or damage (fail-safe)
- Any design that limits and controls the operation, interaction, or sequencing of components (or subassemblies) when an error or malfunction could cause an accident (e.g., when activating switch B before activating switch A could cause damage) (interlock)
- Any design that will safely withstand a release of greater energy than expected, or normally required
- Any design that positively controls buildup of energy to a level that could potentially cause damage (e.g., use of a shear pin to protect a shaft)

Guarding

As indicated above, if it is impossible to design out all hazards, guards must be provided. The basic legal requirements are set forth in CFR 1910.212, *General requirements for all machines* (OSHA), which says:

> (a) Machine guarding. (1) Types of guarding. One or more methods of machine guarding shall be provided to protect the operator and other employees in the machine area from hazards such as those created by point of operation, ingoing nip points, rotating parts, flying chips and sparks. Examples of guarding methods are barrier guards, two-hand tripping devices, electronic safety devices, etc. (2) General requirements for machine guards. Guards shall be affixed to the machine where possible and secured elsewhere if for any reason attachment to the machine is not possible. The guard shall be such that it does not offer an accident hazard in itself.

One should note the key work "all" in the heading. Further, the use of "shall" makes the requirement for guards mandatory.

Most of the dangerous hazards from moving parts of machines occur in three areas:

- *Point of operation*: where the machine works on the workpiece to shape, cut, and so on
- *Power train*: the set of moving parts that delivers power to the point of operation (e.g., shafts, gears, chains, pulleys, cams)
- *Auxiliary components*: such items as feeding mechanisms and other components that move when the machine is in operation

All of these parts have obvious nip points. Less obvious nip points are between an auger screw conveyor and the trough, between a tool rest and a grinding wheel or part being turned on a lathe, between the spokes of a handwheel and the guide or support behind it, between a translating component and a fixed component close to it—in short, a shear of any kind (Ref 15). In general, a nip point occurs when two components are in close proximity with relative motion that reduces the separation between them. There are other hazards, such as potential pressure vessel explosions and bursting flywheels.

The general requirement for a guard is that the point of hazard be substantially enclosed, screened, barricaded, or otherwise protected so that persons, whether workers or bystanders, cannot inadvertently come in contact with the hazard.

Mechanical guards, the most common type, can be fixed, adjustable, or interlocked. Grimaldi and Simonds (Ref 16) give the basic requirements for a mechanical guard as:

- It must be sturdy to prevent damage to the guard from external sources or interference with the operation of the machine. Either of these possibilities would probably cause the operator to remove the guard and not arrange to have it repaired and replaced.
- It must permit maintenance operations without requiring excessive labor for dismantling and reassembling the guard, or else there will be a tendency to omit its installation.
- It must be properly mounted. The mounting must be rigid to prevent objectionable rattles or interference with working parts. The mountings should be strong enough so that they will not fail under use.
- It should be designed so that there are no detachable parts that would reduce its guarding effectiveness if removed and not replaced.
- It should be easy to inspect, and a periodic checkup program, as a part of the maintenance procedure for shop equipment, should be established in order to continue its effectiveness.

Fixed guards should be used wherever possible, because they provide permanent protection against hazardous machinery components. Adjustable guards are used when the mode of operation of the machine is expected to change and adjustment is necessary to accommodate a new set of dimensions. Once adjusted, the guard should then function as a fixed guard. Interlocked guards prevent operation of the machine until the guards have moved into positions that keep the worker out of the hazardous zone. It is essential that the guard put the machine in a safe mode if the guard should fail for any reason (fail-safe).

In using pullbacks, bands are strapped around the operator's wrists, and cords or cables run from the bands to a pulling mechanism synchronized with the down stroke of a power press. If the operator does not remove his or her hands from the hazard area, they are automatically pulled away, even if the press recycles on its own. Pullbacks, however, are not complete protection. The author knows of at least one situation where injury to the worker resulted from an unanticipated recycle. Pullbacks require adjustment to each operator, frequent inspection, and diligent maintenance. Pullbacks are often objectionable to the worker, who feels tied to the machine.

Barrier gates are simple mechanical devices that are opened and shut by machine motion during the operating cycle. This function allows the operator to approach the point of operation (e.g., to feed work stock) but keeps any part of the body from being in the hazard zone when the machine is activated. In most cases, an interlock interrupts power when the gate is open and prevents the operator from opening the gate when the machine is in motion.

Electromechanical Devices. Presence-sensing devices commonly use (a) a light beam and a photoelectric cell ("electric eye") to stop the machine if the light beam is interrupted or (b) a radio frequency electromagnetic field that is disturbed by the capacitance effect of the introducing body.

Distance/Separation Guarding. A logical and effective way of guarding is by separation or distance. The question of location must be considered by the designer. For example, tables of distances and the corresponding openings permitted are given in CFR 1910.217(c)(2)(vi) [OSHA] and in ANSI B11.1.

Input/Output Systems. Systems for feeding stock and ejecting workpieces can provide more safety by using semiautomatic or fully automatic systems. Perhaps the most desirable is a robotic system for mechanical feeding of stock and retrieval of parts. Although more expensive, robots can work where there is a high noise level, can work at a temperature higher than tolerable for most humans, and can perform repetitive monotonous functions indefinitely. One hazard is that the robot may strike a bystander. This hazard, however, can be avoided by barriers or presence sensors.

Auxiliary equipment is generally used in connection with other protective devices to give an additional measure of safety. For example, it is very difficult to provide complete point-of-operation guarding for a band saw, because the saw blade must be exposed in order to accomplish the desired function of cutting material. When small or narrow pieces are being cut, the operator's fingers can get too close unless a push stick or push block is used. The block allows control over the workpiece to get the desired result but keeps the operator's fingers away from the hazard zone. A great variety of pliers, tongs, tweezers, magnetic lifters, suction cup lifters, and so forth are available for use as auxiliary equipment. Such equipment may need to be adjusted for use in different applications.

Controls. Operating controls can be designed to ensure that the operator is out of the hazard zone (point of operation). For example, if only one push button is provided, the other hand could be in the hazard zone. To prevent this, two buttons are provided, far enough apart to require use of both hands, and arranged in series so that both must be pushed to activate the machine. If the stroke time is long enough for the operator to push the buttons and still get a hand into the hazard zone, a requirement can be added that both buttons be held down until the stroke is completed. There is a temptation for workers to tie down one of the buttons, which obviously defeats the two-button safety feature. To circumvent this, a requirement can be added that both buttons must be pressed within a short time period. If the allowable delay is exceeded, the machine will not operate. While most machines should have a two-button control system, there are situations in which a single set of on-jog-off buttons is acceptable because the operator is physically distant from the hazard zone (e.g., control of an overhead crane).

Another aspect of control buttons is that the start, or operate, button (or buttons) should be recessed to reduce the possibility of inadvertent operation. These are also usually green in color. A stop button should have a large, mushroom-shape head that is not recessed and should be easily reached from the normal operating position for use in case of an emergency. The usual color for stop buttons is red.

In cases where a machine runs continuously while the operator is exposed to hazards in any manner, it is necessary to use a control that can immediately stop the machine. The stop button, noted above, is one possibility. In other cases, a trip wire is placed where a worker can easily reach it from any location of the workstation. Pulling on this wire will stop the operation. In one situation (Ref 15), the trip wire was not sufficiently close to be effective when a worker had a hand caught in a shear nip point. In other situations, a force- or pressure-sensitive bar has been used. When the bar is pushed (e.g., if the operator stumbles, loses balance, or is pulled into the machine), the machine will be deactivated. Location is critical. The bar must be located so it will be effective in an emergence but not be inadvertently activated by the material being processed. Presence-sensing devices, "electric eyes," infrared beams, and so forth can also be used to deactivate equipment. Machines that continue to run after power is cut off require a brake for quick stopping.

Data Sources. As noted above, OSHA regulations and ANSI standards are available and can provide much information on guarding. Pertinent data can be found in many other publications (e.g., Ref 7, 16). Information is also available from the National Safety Council (Ref 14, 17, 18), which has videos available for employee training.

Warnings

As noted above, where it is not possible to provide complete and effective guarding, or where such guarding would severely impair the intended function of the product, it is necessary to provide appropriate directions and warnings.

Obviously, eliminating all the potential hazards in a design and/or providing effective guarding is not a simple task. In some cases, it is impossible. Developing a proper, effective warning is generally even more difficult. In large measure, this is because there is hardly consensus, let alone anything approaching unanimity, on a truly adequate and acceptable warning for a given situation. Nonetheless, a full-scale effort must be made.

Directions are instructions intended to *ensure effective use* of a product. *Warnings*, in contrast, are intended to *ensure a safe use* (i.e., to inform of hazards, to inform of improper use, and to instruct how to guard against these, if possible). The distinction is clear in concept, but it is not always possible to tell whether a given statement is a direction or a warning. Lehto and Miller (Ref 19, 20) say: "Perhaps the best way to initially distinguish between warnings and other forms of safety-related information is to state that warnings are specific stimuli which alert a user to the presence of a hazard, thereby triggering the processing of additional information regarding the nature, probability, and magnitude of the hazard. This additional information may be within the user's memory or may be provided by other sources external to the user. Much of the current controversy regarding warnings is actually related to the need for this additional information."

Three criteria must be met for a warning to be fully effective:

- The message must be received.
- The message must be understood.
- The endangered person must act in accordance with the message.

A warning is not effective unless it changes the potential behavior of the endangered individual.

Types of Warnings

Injury or damage can often be avoided by a focus on the existence of a hazard and the need for careful action. Every method for calling attention to a hazard requires communication; each of the human senses, singly or sometimes in concert, have been used for this purpose.

Visual Warnings. It is widely recognized that most information on hazards, perhaps as much as 80%, is visually transmitted to personnel. There are more variations of visual methods than of the other senses. A hazardous area is often more brightly illuminated than other areas in order to focus attention. A piece of equipment can be painted in alternating stripes or in a bright, distinctive color (e.g., fire trucks are now being painted in greenish-yellow rather than red for better visibility). Signal lights are often used (e.g., on emergency vehicles and at railroad crossings). Flags and streamers can be used. Signs are common, such as highway signs.

Auditory warnings may have a shorter range of effectiveness than visual warnings, but their effectiveness may be greater in that short range. Auditory warnings are often coupled with visual warnings, such as on emergency vehicles. Typical devices are sirens, bells, buzzers, and horns (e.g., the intermittent sound of a horn on a truck that is backing up).

Olfactory Warnings. Odorants can be used in some limited, although effective, ways (e.g., addition of small amounts of a gaseous odorant to natural gas to warn of leaks).

Tactile Warnings. Vibration is the major tactile means of warning (e.g., rumble strips on highways). Vibration in machinery may mean the beginning of serious wear or lubrication failure. Temperature sensing, at least an indication of significant temperature change, can also be included in this category.

Tastable warnings have little use in machine applications but have been used with foods and medicines.

Written Warnings

Much confusion exists, especially within the legal system, as to the meaning of the term "warning" when applied to products and their uses. The major reason may be that "warnings" are usually considered synonymous with the explicit "warning labels" that are sometimes placed on products. One consequence is that sources of information that do not explicitly (in words) describe the hazard, specify its intensity, provide instructive countermeasures, and strongly advocate adherence may not be considered adequate warnings. Another reason for the confusion is that society seems to expect warnings to perform multiple functions.

Warnings should supplement the safety-related design features of a product by indicating how to avoid injury or damage from the hazards which could not be feasibly designed out of the product, designed out without seriously compromising utility, or protected against by providing adequate guards. In theory, providing such information will reduce danger by altering the user's behavior or by causing people to avoid using a product. However, from the litigation viewpoint, warnings often perform functions having little to do with either safety or transfer of safety-related information. A manufacturer may view warnings as a defense against litigation. One consequence is extensive use of "warning labels," which sometimes yields no increase in safety. Even more unfortunate is that some manufacturers may use warnings instead of careful design, which is absolutely unacceptable.

As indicated above, for a warning to be effective, the endangered person must receive the message, understand it, and act in accordance with it. The designer/manufacturer obviously has no control over the action, but does have substantial control over sending the warning and making it understandable. Failure on the part of the endangered person to meet any one of the above criteria makes the warning ineffective. The probability that a warning will be effective is certainly no higher than the percentage of users who read the warning. There is general agreement that many people who see a warning label do not read it. Many do not even see the label. This obviously can be discouraging to one trying to develop a proper, effective, warning label. Nonetheless, a major effort must be made.

Every warning, including labels, has an alerting function. The warning label must be prominently located (i.e., in a position such that the user will be virtually sure to see it). The warning label must also be distinctive (i.e., sufficiently different from other labels that there is no question of its identity). Shape has an influence; shapes with rounded or curved boundaries are not as effective in attracting attention as shapes with sharp corners. Rectangles seem to be more effective than squares or triangles. Labels with five or more sides are rarely used on industrial or consumer products.

Three signal words (in relatively large letters) and color combinations are normally used to attract attention:

- *Danger*: The hazard can immediately cause death or severe injury. Letters should be white on a red background.
- *Warning*: The hazard can immediately cause moderate injury or death, or severe injury may eventually result from contact with the source of the hazard. Letters should be black on an orange background.
- *Caution*: The hazard can immediately cause minor injury, or moderate injury may eventually result from contact with the source of the hazard. Letters should be black on a yellow background.

Every warning, especially a label, has a *message*. This message must be clear, simple (unambiguous), succinct, and convincing. Short words are preferred, and as few words as possible. Long sentences with technical terms should be avoided. There are indications in the literature that directions and warnings should be written at the sixth grade level. The use of indices such as the Flesch Reading Ease Formula, Gunning's Fog Index, or McElroy's Fog Count can be helpful. See also Ref 21. For products that will be used only within a country or region in which there is one common language, the choice of language is obvious. For products that will be used in regions with different languages, warning labels must be in those languages. Those who write labels in languages different from that of the manufacturer must be knowledgeable about the linguistic and cultural characteristics of those regions.

A partial alternative to the use of multiple languages is the increasing use of pictographs. A pictograph communicates an idea or concept in one symbol which is universally recognized. For example, there is general recognition that a 45° red diagonal line through an annulus forbids whatever is displayed within the annulus (e.g., a lighted cigarette within the annulus indicates that no smoking is permitted). General guidelines for pictographs are:

- Use a simple design for the symbol.
- Use only one idea per pictograph.
- Use only correct colors and shapes.
- Locate the symbol as close as possible to any related words.

Words on the label must be legible by the average person, some of whom may have uncorrected visual impairment. ANSI Z535.4 (Ref 22) gives requirements for wording and colors to be used, which differ from standards issued by OSHA and the CPSC. ANSI Z535.4 also specifies letter size. Signal words must be at least 3 mm high (9 point type), and the text must be at least 1.5 mm high (5 point minimum). ["Point" is a unit of measure for typeset letters; there are essentially 72 points per inch.] This is a consensus standard and represents the minimal requirements acceptable to those involved in developing the document. Many believe the lettering should be larger. Bailey (Ref 23), for example, notes that the "type size in books and magazines usually ranges from 7 to 14 points with the majority being about 10 to 11 points. Probably the optimum range is from 9 to 11 points—sizes smaller or larger can slow reading speed."

A warning label should be permanent. A warning should not fade or fall off before the end of the service life. Most labels are decalcomanias. Fortunately, they are available with a base of tough wear-resistant material and good adhesive backing. Some products have warnings on stamped or embossed plates permanently secured to the product. Operator manuals and/or maintenance manuals commonly accompany the product as shipped from the manufacturer but do not always find their way to the product in its operational situation. Providing a tough, dirt- and lubrication-resistant envelope that contains the manual and is "permanently" attached to the product (e.g., a power press or similar machinery) by a short chain can be useful for the worker.

CFR 1910.145 (OSHA) specifies requirements for accident prevention signs. By reference, two ANSI standards, Z35.1, "Specifications for Accident Prevention Signs," and Z53.1, "Safety Color Code for Marking Physical Hazards," were incorporated. Designers should consult these as soon as a decision is made to incorporate warnings. It should be noted, however, that in 1979, ANSI Z53 Committee on Safety Colors was combined with ANSI Z35 Committee on Safety Signs to form ANSI Z535 Committee on Safety Signs and Colors. Five subcommittees were formed to update the Z35 and Z53 standards and write two new standards. These are listed in Ref 22 and 24-27. The Society of Automotive Engineers has a recommended practice, J115 (Ref 28), relating to safety signs. This is generally consistent with the ANSI 535 series, but there are some differences. (This situation is an example of old standards still having the force of law in OSHA standards, even though these old standards have been replaced by much more recent standards.)

Smith and Talbot (Ref 29) discuss a label on a 20 ft (6 m) long fiberglass ladder. This is also discussed in some detail in the article "Human Factors in Design" in this Volume (see Fig. 3 and the accompanying discussion in that article).

Although warning that helps prevent an injury may not make great advertising copy, it is a necessity. Warnings are not new. When Samual Jones began manufacturing "Lucifer" matches (smelling of "hellfire and brimstone") in 1829, he printed the following warning on the boxes: "If possible, avoid inhaling the gas that escapes from the combustion of the black composition. Persons whose lungs are delicate should by no means use Lucifers." In terms of the above discussion, this is a relatively good warning.

Sources

There has been much written with regard to warnings in both the technical and legal literature. The best technical source currently available for understanding the nature of warnings and the difficulty in writing them is Lehto and Miller (Ref 19, 20).

Human Factors/Ergonomics

Human beings interact with all products in designing, manufacturing, operating, and maintaining them. Human beings constitute the most complex subsystem in any system because of their abilities and limitations. In addition, the number and variety of actions that a group of persons can take in any situation, either as individuals or as a group, generates a high probability that any deficiency in the system will be linked to, and affected by, personal factors that can generate an accident. In other words, the human being is the most erratic, and the least controllable, parameter in any system.

Additional information can be found in the article "Human Factors in Design" in this Volume.

REFERENCES

1. Publius Cornelius Tacitus, *Annals*, Vol 15
2. *The Code of Hammurabi*, University of Chicago Press, 1904
3. Information Bulletin 000080021, National Safety Council, 1994
4. *Webster's New Twentieth Century Dictionary, Unabridged*, 2nd ed., Simon and Schuster, 1979
5. W. Hammer, *Occupational Safety Management and Engineering*, Prentice-Hall, 1976
6. W.W. Lowrance, *Of Acceptable Risk*, William Kaufman, Inc., Los Altos, CA, 1976

7. T.A. Hunter, *Engineering Design for Safety*, McGraw-Hill, 1992
8. W. Hammer, *Handbook of System and Product Safety*, Prentice-Hall, 1972
9. W. Hammer, *Product Safety Management and Engineering*, Prentice-Hall, 1980
10. W. Hammer, *Product Safety Management and Engineering*, 2nd ed., American Society of Safety Engineers, 1993
11. H.E. Roland and B. Moriarty, *System Safety Engineering and Management*, 2nd ed., 1990
12. J. Stephenson, *Systems Safety 2000*, Van Nostrand Reinhold, 1991
13. C.O. Smith, *Problems in Machine Guarding*, Paper 87-WA/DE-6, ASME
14. *Accident Prevention Manual for Business and Industry*, 10th ed., National Safety Council, 1992
15. C.O. Smith, *System Unsafety in a Transfer Machine*, Proc. Fourth International Conf., System Safety Society, 1979
16. J.V. Grimaldi and R.H. Simonds, *Safety Management*, 5th ed., Irwin, 1989
17. *Safeguarding Concepts Illustrated*, 6th ed., National Safety Council
18. *Power Press Safety Manual*, 4th ed., National Safety Council
19. M.R. Lehto and J.M. Miller, Vol I, *Fundamentals, Design, and Evaluation Methodologies*, Vol I, *Warnings*, Fuller Technical Publications, Ann Arbor, MI, 1986
20. M.R. Lehto and J.M. Miller, *An Annotated Bibliography*, Vol II, *Warnings*, Fuller Technical Publications, Ann Arbor, MI, 1986
21. G.R. Klare, *The Measurement of Readability*, Iowa State University Press, 1963
22. ANSI Z535.4, "American National Standard for Product Safety Signs and Labels," American National Standards Institute, 1991
23. R.W. Bailey, *Human Performance Engineering: A Guide for System Designers*, Prentice-Hall, 1982
24. ANSI Z535.1, "American National Standard Safety Color Code," American National Standards Institute, 1991
25. ANSI Z535.2, "American National Standard for Environmental and Facility Safety Signs," American National Standards Institute, 1991
26. ANSI Z535.3, "Criteria for Safety Symbols," American National Standards Institute, 1991
27. ANSI Z535.5, "Specifications for Accident Prevention Tags," American National Standards Institute, 1991
28. SAE J115, "Safety Signs," SAE Recommended Practice, Society of Automotive Engineers, 1987
29. C.O. Smith and T.F. Talbot, *Product Design and Warnings*, Paper 91-WA/DE-7, ASME

SELECTED REFERENCES

- R.L. Brauer, *Safety and Health for Engineers*, Van Nostrand Reinhold, 1990
- W. Hammer, *Occupational Safety Management and Engineering*, Prentice-Hall, 1989
- Institute for Product Safety, P.O. Box 1931, Durham, NC 27702, various publications
- W.G. Johnson, *MORT Safety Assurance Systems*, Marcel Dekker, 1980
- F.A. Manuele, *On the Practice of Safety*, Van Nostrand Reinhold, 1993
- *Safety, Health and Environmental Resources Catalog*, National Safety Council, annual
- R.A. Wadden and P.A. Scheff, *Engineering Design for the Control of Workplace Hazards*, McGraw-Hill, 1987

Products Liability and Design

Charles O. Smith, Engineering Consultant

PRODUCTS LIABILITY is a legal term for the action whereby an injured party (plaintiff) seeks to recover damages for personal injury or property loss from a producer and/or seller (defendant) when the plaintiff alleges that a defective product caused the injury or loss.

If a products liability suit is entered against a company, the plaintiff's attorney and technical experts will attempt to convince a jury that the manufacturer did not exercise reasonable care in one or more features of design and/or manufacture, and that because the company did not exercise reasonable care, an innocent party was injured. The defendant's team will attempt to convince a jury that the manufacturer was not responsible for the injury.

Products liability is not new. The first law code known to be in writing was established by Hammurabi, King of Babylon, about 4000 years ago, and it contained clauses which clearly relate to products liability (Ref 1).

Who may be a plaintiff? Essentially any consumer, user, or bystander may seek to recover for injury or damages caused by a defective and unreasonably dangerous product.

Who may be a defendant? Any corporation, business organization, or individual who has some degree of responsibility in the "chain of commerce" for a given product, from its inception as an idea or concept to its purchase and use.

The situation is schematically summarized in Fig. 1.

Legal Bases for Products Liability

The three legal theories on which a products liability lawsuit can be based are negligence, breach of warranty, or strict liability. All three are predicated on the fault system (i.e., a person whose conduct causes injury to another is required to fully and fairly compensate the injured party).

The basic method of imposing liability on a defendant requires the plaintiff to prove that the defendant acted in a negligent manner. Under the negligence theory, the plaintiff must essentially establish proof of specific negligence (i.e., prove that the defendant was almost intentionally negligent). Proof of specific negligence is a difficult task.

A user of a product may, as a result of express oral or written statements, or implication, reasonably rely on the manufacturer's express or implied assurance (including advertising material) as to the quality, condition, and merchantability of goods, and as to the safety of using them for their intended purpose and use. If the user relies on these assurances and is injured, suit can be entered on the basis of breach of warranty.

Both negligence and breach of warranty require proof of some fault on the part of the defendant (i.e., the focus is on the action of an individual). Strict liability, however, focuses on the product itself (Ref 2):

> (1) One who sells any product in a defective condition unreasonably dangerous to the user or consumer or to his property is subject to liability for physical harm thereby caused to the ultimate user or consumer, or to his property, if (a) the seller is engaged in the business of selling such a product and (b) it is expected to and does reach

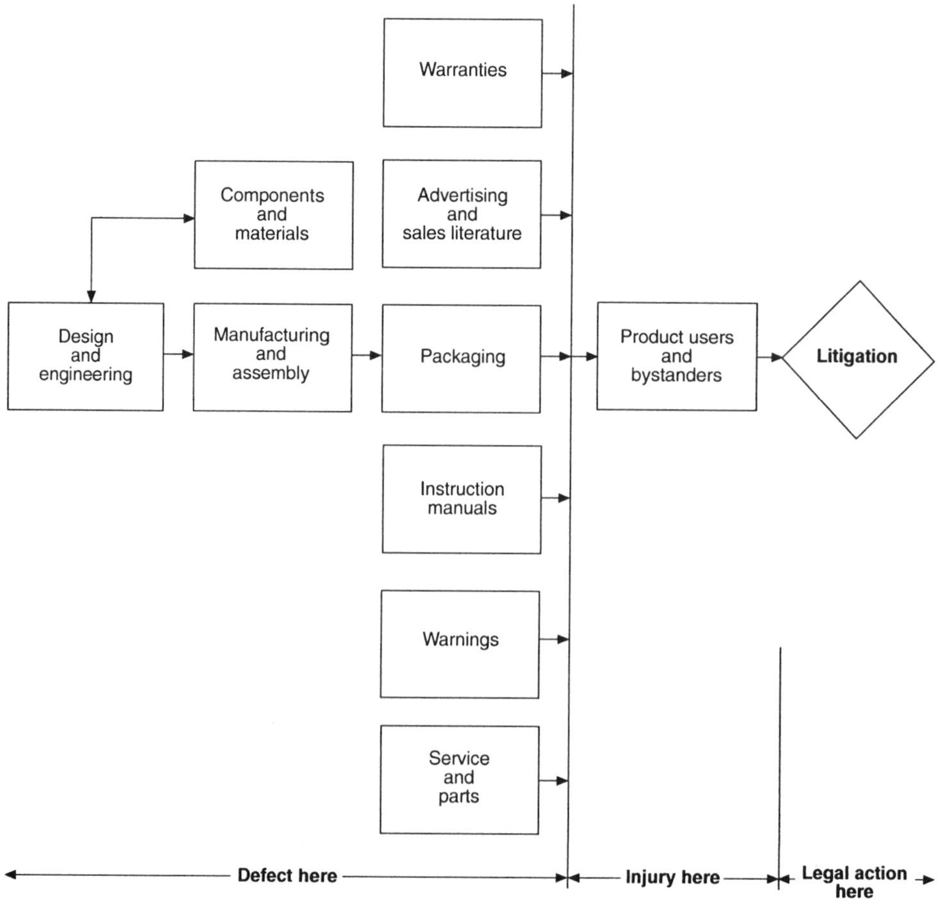

Fig. 1 The essence of products liability

the user or consumer without substantial change in the condition in which it is sold.
(2) The rule stated in Subsection (1) applies although (a) the seller has exercised all possible care in the preparation and sale of his product, and (b) the user or consumer has not bought the product from or entered into any contractual relation with the seller.

Although these three bases apply in general, it should be recognized that there are variations. Interpretation and application may vary greatly depending on the jurisdiction.

Hazard, Risk, and Danger

There is substantial confusion about the meaning of words such as hazard, risk, and danger. Webster (Ref 3) defines danger as: "liability to injury, pain, damage or loss; hazard; peril; risk." Webster makes some distinction by further saying: "Hazard arises from something fortuitous or beyond our control. Risk is doubtful or uncertain danger, often incurred voluntarily."

One can also consider a hazard as any aspect of technology or activity that produces risk; as the potential for harm or damage to people, property, or the environment; and as including the characteristics of things and the actions (or inactions) of individuals. One can also consider risk as a measure of the probability and severity of adverse effects.

With all the products liability litigation in the United States, there has developed a clear distinction among these three words for legal purposes. In this context, a *hazard* is a condition or changing set of circumstances that present an injury potential (e.g., a railroad crossing at grade, a toxic chemical, a sharp knife, the jaws of a power press). *Risk* is the probability of injury and is affected by proximity, exposure, noise, light, experience, attention arresters, intelligence of an involved individual, and so on. Risk (probability of exposure) is obviously much higher with a consumer product than with an industrial product used by trained workers in a shop environment. *Danger* is the unreasonable or unacceptable combination of hazard and risk. The U.S. courts generally hold that any risk that can be eliminated by reasonable accident prevention methods is unreasonable and unacceptable. A high risk of injury could be considered reasonable and acceptable if the injury is minimal and the risk is recognized by the individual concerned.

As might be expected, there is extensive and ongoing debate over the meaning of "reasonable" and "unreasonable." The American Law Institute (Ref 2) says *unreasonably dangerous* means that: "The article sold must be dangerous to an extent beyond that which would be contemplated by the ordinary consumer who purchases it, with the ordinary knowledge common to the community as to its characteristics. Good whiskey is not unreasonably dangerous merely because it will make some people drunk, and is especially dangerous to alcoholics; but bad whiskey, containing a dangerous amount of fusel oil, is unreasonably dangerous." The American Law Institute (Ref 2) further says: "There are some products which, in the present state of human knowledge, are quite incapable of being made safe for their intended and ordinary use. ... Such a product, properly prepared, and accompanied by proper directions and warnings, is not defective, nor is it unreasonably dangerous."

Definitions of Defects

The American Law Institute (Ref 2) says that a product is in a defective condition if "it leaves the seller's hands, in a condition not contemplated by the ultimate user, which will be unreasonably dangerous to him." Peters (Ref 4) indicates that a California Supreme Court decision, *Barker v. Lull* (Ref 5), established a good assessment of "defective condition." This provides three definitions (or criteria) for manufacturing defects and two for design defects:

Manufacturing defects

- Nonconformance with specifications
- Nonsatisfaction of user requirements
- Deviation from the norm

Design defects

- Less safe than expected by ordinary consumer
- Excessive preventable danger

Manufacturing Defects

A failure to conform with stated specifications is an obvious manufacturing defect, and not a new criterion. The aspect of user satisfaction may not be well known, but in the legal context it has long been recognized that a manufacturing defect exists when there is such a departure from some quality characteristic that the product or service does not satisfy user requirements. Under the third criterion (deviation from the norm), added by *Barker*, a manufacturing defect occurs when a product leaves the assembly line in a substandard condition, differs from the manufacturer's intended result, or differs from other, ostensibly identical units of the same product line.

Design Defects

A product may be considered to have a design defect if it fails to perform as safely as an ordinary consumer would expect. This failure to perform safely is interpreted in the context of intended use (or uses) in a reasonably foreseeable manner, where "foreseeable" has the same meaning as "predicted" in failure-modes-and-effects, fault-tree, or hazard analyses. It appears that many "ordinary" consumers would have no concept of how safe a product should, or could, be without the expectations created by statements in sales material, inferences from mass media, general assumptions regarding modern technology, and faith in corporate enterprise.

A design defect also exists if there is excessive preventable danger. The real question is whether the danger outweighs the benefits. A risk-benefit analysis should include at least five factors:

- Gravity of the danger posed by the design (i.e., severity of the consequences in the event of injury or failure)
- Probability (including frequency and exposure of the failure mode) that such a danger will occur
- Technical feasibility of a safer alternative design, including possible remedies or corrective action
- Economic feasibility of these possible alternatives
- Possible adverse consequences to the product and consumer that would result from alternative designs

Additional relevant factors may be included, but design adequacy is evaluated in terms of a balance between benefits from the product and the probability of danger. Quantification of the risk-benefit analysis is not required but may be desirable.

Proving design adequacy places the burden of proof on the defendant. Once the plaintiff proves that the product is a proximate cause of injury, the defendant must prove that the benefits outweighed the risk. Discussion of manufacturing and design defects of various products is given in Ref 6 to 24.

Note: No paper, book, or handbook relative to products liability can be truly current. In addition, there is substantial variation among jurisdictions (federal, state, and local). All cited publications, however, do have something that is currently pertinent.

Other Defects

The engineer must be alert for other possibilities. Smith and Talbot (Ref 25) point out that a marketing defect exists when there is a failure to provide *any* warning of hazard and risk involved with use of a product, provide *adequate* warning of hazard and risk involved with use of a product, or provide appropriate, adequate directions for safe use of a product. In other words, a marketing defect exists when a product, free of design and manufacturing defects, is unreasonably dangerous due to absence of warnings and directions. The designer/manufacturer has control over the directions and warnings provided. The designer/manufacturer is the most knowledgeable about the product and thus presumably the most able to determine the necessary directions and warnings.

Suits against manufacturers often allege a defective product. Careful investigation, however, sometimes shows that the problem is due to improper maintenance (e.g., Ref 26). The designer/manufacturer obviously has no control over the maintenance actually conducted but can try to minimize the possibility of improper practices by providing proper and adequate instructions with the product. In any event, the designer should not overlook the possibility of misuse and improper maintenance on the part of the user.

Example 1: Failure of a High-Speed Steel Twistdrill. A 1.905 cm (3/4 in.) stud broke in the vertical wall of a metalworking machine known as an upsetter. A parallel vertical wall left a lim-

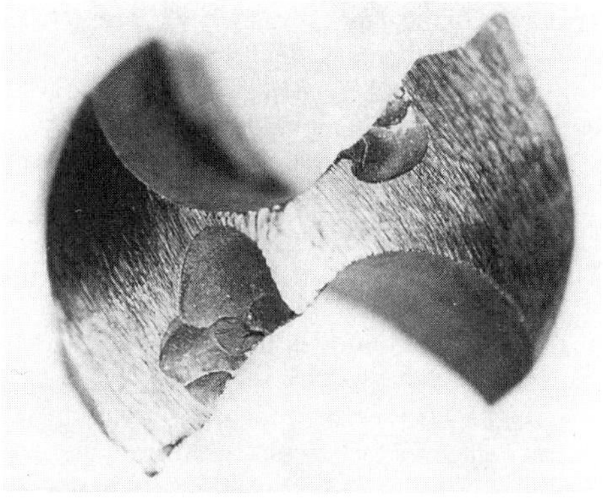

(a)

(b)

Fig. 2 The tip of the broken twistdrill. (a) End view. (b) Top view

ited amount of space in which mechanics could work. A pilot hole was drilled with a 0.476 cm (3/16 in.) drill. The drill was held in a Jacobs chuck in a portable drill press which, in turn, was held to the workpiece by an electromagnet. After the pilot hole was finished, the drill press was removed, the pilot drill was replaced by a 1.587 cm (5/8 in.) high-speed steel drill, and the press was repositioned.

One man was doing the drilling while another man was squirting oil into the hole. When the drill was about 1.27 cm (1/2 in.) to 1.90 cm (3/4 in.) into the pilot hole, there was a "bang." The drill shattered, causing a chip to lodge in the right eye of the oiler, ultimately resulting in loss of vision in that eye. Suit was entered against the drill manufacturer alleging a defective drill.

The plaintiff's attorney retained a metallurgist who examined the fragments. An unetched longitudinal section showed a large nonmetallic inclusion parallel to the axis near the center of the drill. After etching, this section showed carbide bands in a martensitic matrix. Hardness measurements indicated 65 to 66 HRC at the edge of the flute with a bulk hardness of 62 to 64 HRC. The drill tip is shown in Fig. 2.

A second drill from the same lot was also examined. There was carbide banding but to a significantly lesser degree. Hardness was measured as 63 to 64 HRC at both the flute edge and in the bulk of the drill.

Plaintiff's expert concluded that the failed drill was defective while the other drill was satisfactory. He claimed that failure was a cumulative result of the following defective conditions: the steel contained nonmetallic inclusions that were detrimental to the properties of the drill; the carbide segregation was excessive, causing the drill to be brittle; and the cutting edge of the flutes was excessively hard. In his opinion, this high hardness made the edge brittle, so that the edge would chip during drilling. The chips caused the drill to bind and then shatter in a brittle manner because of excessive carbide segregation.

Plaintiff alleged defective design, defective manufacture, unsuitable or defective material, lack of sufficient quality control, and failure to foresee.

The defendant manufacturer believed the twistdrill met all specifications for M1 high-speed steel. Both the supplier and manufacturer inspected for carbide segregation, with the poorest rating being "slight to medium." A "medium" rating was permitted. Heat treatment and nitriding practices were consistent with those published by ASM International. After heat treatment, the drills were within the specified range of 64 to 66 HRC.

Some twenty other inspections for dimensional accuracy, shape, and finish were made after heat treatment. Fourteen drills were given a severe drilling performance test (manufacturer's routine) with no breaking or chipping.

One could argue there was nonsatisfaction of user requirements. The counter argument was that there was too high a demand on the drill.

Conclusions by the plaintiff's experts can be viewed as indicating a deviation from the norm. (Of the 5360 drills made from this one lot of steel, the manufacturer received only this one complaint.) The observed nonmetallic inclusion had a maximum width less than 0.0127 mm (0.0005 in.) and a maximum length less than 0.84 mm (0.033 in.). It was located in the shank, more than 6.35 cm (2 1/2 in.) from the drill tip and along the central axis, which is subjected to essentially none of the bending and twisting loading. While the inclusion is relatively large, it is not likely that it could contribute to the failure.

The largest carbide band was about 8 mm (1 1/3 in.) long and located about one-fourth of the distance from the central axis to the outer edge. It was also some distance from the drill tip. This location implies relatively light loading.

The manufacturer made hardness measurements on the two drills examined by plaintiff's expert. The results are given in Table 1. These indicate no significant difference between the two drills. The higher hardness at the cutting edge is expected and reasonable for a nitrided M1 steel. The hardness of both drills is within normal specification ranges.

The manufacturer examined a number of other M1 drills that had satisfactorily met (corporate) standard drilling tests. One of these had a nonmetallic inclusion 1 1/2 times longer than in the failed drill. Two had edge hardnesses in the 66 to 68 HRC range with carbide banding more pronounced than in the failed drill.

From the viewpoint of design defects, was the drill less safe than expected by the ordinary consumer? Maybe. Presumably the workers did not expect drill failure. It is well known, however, that twist drills do fail, no matter how well designed and manufactured. Using a drill to remove material after drilling a pilot hole is a common practice and clearly foreseeable. It is clearly more hazardous than drilling without a pilot hole. The drill may have been less safe than expected, but it seems more credible that too much was expected.

Table 1 Hardness examination of drills

See Example 1 in text.

Measurement	Average hardness, HRC(a) Bulk	Cutting edge
Plaintiff's measurements		
Broken drill	62-64	65-65
Unbroken drill	63-64	63-64
Defendant's measurements		
Broken drill	64.9, 64.8	65.1(b), 66.5(c)
Unbroken drill	65.0, 65.1	65.6(b), 66.5(c)

(a) Tukon readings (100 g) converted to HRC. (b) At 0.1270 mm (0.005 in.) from the surface. (c) At 0.254 mm (0.001 in.) from the surface

Existence of a design defect related to "excessive preventable danger" seems doubtful. The drill design was highly similar to that used by other manufacturers. All dimensions, tolerances, clearances, and so on were consistent with those used by other manufacturers and were based on years of drill use by a great variety of users. There is no question of potentially severe damage and relatively high probability of exposure. But there are no apparent alternatives that are technically and/or economically feasible.

What is your judgment on the validity of the allegations? How should this litigation be resolved?

Examination of Fig. 2 indicates that one cutting lip is about 0.725 cm (0.286 in.) long while the other is about 0.802 cm (0.316 in.) long, so that the chisel edge is about 2.7 mm (0.018 in.) off center. The shorter lip will contact the work before the longer lip and thus bears all of the initial drilling stresses. The larger of the two chipped areas along the cutting edges in Fig. 2 is on the shorter lip. The broken point also had improper clearance angles (one was close to a negative angle). It was clear that the point of the broken drill was not the original point put on at manufacture but came from regrinding (presumably from "eyeballing" rather than using a jig). The work conditions, including a small pilot hole, a portable drill press, relocation of the press between the two drilling operations, and a questionable supply of coolant, placed abnormal stress on the drill.

The case was eventually settled out of court, with the plaintiff receiving a sum of less than $10,000 at a time when similar injury cases were receiving judgments of $50,000 to $150,000. This was clearly a so-called "nuisance settlement" to get rid of the suit. Much greater detail, both technical and legal, can be obtained by reference to Ref 27 to 29. A side aspect of this case relates to the expert witness. It developed that the plaintiff's expert was not sufficiently knowledgeable about high-speed steels, although he was a competent metallurgist.

Two additional examples arising from litigation are given in the article "Human Factors in Design" in this Volume.

Preventive Measures

What are the implications of the above example for the design engineer? It is necessary to look carefully at the completed design to be sure that it is indeed appropriate and that it does not incorporate problems for which proper technological solutions have existed for some time. (For example, an independent assessment by a design review board, whose members have no parental pride in the design, is highly appropriate.) In addition, there must be recognition that many, perhaps most, consumers have no concept of how safe a product should be. An engineer making a subjective judgment about safety must understand this lack of appreciation of an appropriate safety level.

Acting as a prudent manufacturer is not enough. The focus should be on the product itself, not the reasonableness of a manufacturer's conduct. Obviously, there will be no viable lawsuits if there are no injuries or if there are no violations of the law. Undoubtedly the best practice is to sell a well-designed, well-manufactured product. The manufacturer needs to make certain that all reasonable preventive measures have been used in the design and manufacturing process. Much evidence, however, suggests that one of Casey Stengel's comments applies in the area of preventive measures: "In many areas we have too strong a weakness." While many preventive measures are well known to most design engineers, some comments may be appropriate, even if only in the sense of a checklist of items to be considered.

Design review is an effort, through group examination and discussion, to ensure that a product (and its components) will meet all requirements. In a design of any complexity, there is necessity for a minimum of three reviews: conceptual, interim, and final. Conceptual design reviews have a major impact on the design, while interim and final reviews have relatively less effect as the design becomes more fixed and less time is available for major design changes. *It is much easier and much less expensive to include safety in the initial design than to include it retroactively.*

A more sophisticated product may require several reviews during the design process. These might be: conceptual, definition, preliminary (review of initial design details), critical (or interim review, perhaps several reviews in sequence—review details of progress, safety analyses, progress in hazard elimination, etc.), prototype (review of design before building a prototype), prototype function review, and preproduction review (final review—last complete review before release of the design to production).

These periodic design reviews should review progress of the design, monitor design and development, ensure that all requirements are met, and provide feedback of information to all concerned.

A design review is conducted by an ad hoc design review board composed of materials, mechanical designers, electrical designers, reliability engineers, safety engineers, packaging engineers, various other design engineers as appropriate, a management representative, a sales representative, an insurance consultant, an attorney in products liability, outside "experts" (be sure they are truly expert!), and so on. Members of the design review board should not be direct participants in day-to-day design and development of the product under review, but the engineers should have technical capability at least equal to that of the actual design team. Vendor participation is highly desirable, especially in conceptual and final design reviews.

Design review checklists should be prepared well in advance of actual board meetings. These checklists should cover all aspects of the design and expected performance, plus all phases of production and distribution. A new checklist should be developed for each new product.

It is good practice for a designer/manufacturer to have some sort of permanent review process in addition to the ad hoc board for each individual product. This permanent group should evaluate all new products, reevaluate old products, and keep current with trends, standards, and safety devices.

If properly conducted, a design review can contribute substantially to avoiding serious problems by getting the job done right the first time. Formal design review processes are effective barriers to "quick and dirty" designs based on intuition (or educated guesses) without adequate analyses.

Some Common Procedures. Many engineers and designers are familiar with such techniques and procedures as hazard analysis; failure modes and effects analysis (FMEA); failure modes, effects, and criticality analysis (FMECA); fault tree analysis (FTA); fault hazard analysis (FHA); operating hazard analysis (OHA); use of codes, standards and various regulatory acts, and the Occupational Safety and Health Act (OSHA). These are discussed in the article "Safety in Design" in this Volume. Some other aspects of products liability are perhaps less well known and require some comment.

Prediction methods are necessary in applying FMEA, FTA, and so on. From statistics it is possible to predict performance of a large group of similar products, but it is *not* possible to predict performance of any one individual item of that group. Various statistical and probabilistic techniques can be used to make predictions, but these are predicated on having good data bases.

State of the Art. The meaning of the term *state of the art* should be defined for each specific product. This might be done by comparing the product to those produced by competitors, but this comparison may not be enough. A jury is not bound by negligent practices of a negligent industry, and unfortunately, in some areas industry practices and standards are low-level consensus practices and standards. Being in step with the state of the art may not be enough—one should be ahead of the state of the art (i.e., better than the competitors). It is not enough to explain what was done, because the plaintiff's expert witnesses may point out what could have been done. Purely economic reasons are not a valid defense argument in the courtroom and should be avoided.

Quality Assurance and Testing. A primary function of quality control is to feed back inspection, testing, and other data, showing designers what is happening and revealing any need for design improvement. Manufacturers should test products in various stages of development, including field service, especially if critical components or subassemblies are involved. Final tests are necessary on each individual product or on representative samples of plant output. Care must be taken that quality control is not relaxed, intentionally or unintentionally, for production expediency.

Foreseeability is a factor that requires special attention. It is necessary to determine not only how the product is intended to be used but also every reasonably conceivable way that it can be used and misused. (Who has never used a flat-tang screwdriver for some other purpose?) All reasonable conditions of use, or misuse, that might lead to an accident should be detailed. The designer must conclusively demonstrate that the product cannot be made safer, even to prevent accidents, during use or misuse. The problem of foreseeability is one that seems especially difficult for engineers to accept.

Consumer Complaints. Data on product failures from test facilities, test laboratories, and service personnel are valuable. Each complaint should be quickly, carefully, and thoroughly investigated. An efficient reporting system can result in product corrections before large numbers of the product reach users, or a product recall before there has been a major exposure of the public to an unsafe product.

Warranties and Disclaimers. Warranties and disclaimers are attempts to limit liability. When used, they must be written in clear, simple, and easily understood language. Both should be reviewed by highly competent legal counsel knowledgeable in both the industry and products liability. A copy of the warranty and/or disclaimer must be packaged with the product. All practical means must be used to make the buyer aware of the contents. It must be recognized, however, that warranties and disclaimers, no matter how well written, are an extremely weak defense.

Warnings and Directions. Directions are intended to ensure effective use of a product. Warnings are intended to ensure safe use. Both should be written to help the user understand and appreciate the nature of the product and its dangers. If directions and warnings are inadequate, there is potential liability, because it cannot be said that the user had contributory negligence in failing to appreciate and avoid danger.

The burden of full and effective disclosure is on the manufacturer. Directions and warnings, although essential, do not relieve the manufacturer of the duty to design a safe product. The law will not permit a manufacturer, who knowingly markets a product with a danger that could have been eliminated, to evade liability simply because a warning is placed on the product. *One must design against misuse*.

This topic is discussed in greater detail in the article "Safety in Design" in this Volume. A label is discussed in some detail in the article "Human Factors and Design" in this Volume.

Written Material. All advertising, promotional material, and sales literature must be carefully screened. Warranties can be implied or inferred by the wording on labels, instructions, pamphlets, sales literature, advertising (written and electronic broadcast), and so on, even though no warranty is intended. There must be no exaggeration in such material. The manufacturer must be able to show that the product is properly rated and that the product can safely do what the advertisement says it will do. Additional information on the level of language is given in the article "Safety in Design" in this Volume.

Human Factors. Many products and systems require operation by a human who thereby becomes an integral part of the system. As such, the human can have a very significant effect on system performance. One must recognize that the human being is the greatest, and least controllable, variable in the system. Many attorneys believe that most products liability suits result because someone (usually the designer) did not thoroughly think through how the product interfaced with society.

Additional information can be found in the article "Human Factors in Design" in this Volume.

Products Recall Planning. It is a fact of life that mistakes are sometimes made even by highly experienced professionals exercising utmost care. When such errors occur, a product recall may be necessary. Unless the specific troublesome part can be readily and uniquely identified as to source, production procedure, time of manufacture, and so on, there will be great difficulty in pinpointing the problem within the producing organization. Placing one advertisement for recall purposes in newspaper and magazines (not including TV) throughout the country is very expensive. An obvious economic need, as well as a regulatory requirement, exists for manufacturers (and importers) to have systems in place for expeditious recall of a faulty product.

Records. Once involved in litigation, one of the most powerful defenses that manufacturers and engineers can have is an effective, extensive, and detailed record. Records should document how the design came about, with notes of meetings, assembly drawings (including safety features), checklists, the state of the art at the time, and so on. These records, while no barrier to products liability lawsuits, will go a long way toward convincing a jury that prudent and reasonable care has been taken to produce a safe product.

Paramount Questions

No matter how carefully and thoroughly one executes all possible preventive measures, it is necessary to ask:

- What is the probability of injury?
- Who determines the probability of injury?
- What is an acceptable probability of injury?
- Who determines the acceptable probability of injury?

As Lowrance (Ref 30) suggests, determining the probability of injury is an empirical, scientific activity. It follows that engineers are better qualified by education and experience than most people to determine this probability. Presumably designers will use organized approaches to cope with the complexity. One obvious place for assessing this probability is the design review process. While design review is a most valuable aid for the designer, it is *not* a substitute for adequate design and engineering.

As Lowrance (Ref 30) further suggests, judging the acceptable probability of injury is a normative, political activity. Obviously, assessing the probability of injury is not a simple matter. Assessing the acceptable probability of injury is far more complex and difficult. Use of the word "acceptable" emphasizes that safety decisions are relativistic and judgmental. It implies three questions:

Acceptable in whose view? Acceptable in what terms? Acceptable for whom? This use of "acceptable probability of injury" avoids any implication or inference that safety is an intrinsic, absolute, measurable property.

In assessing acceptable danger, one major task is determining the distribution of danger, benefits, and costs. This determination is both an empirical matter and a political issue. It involves questions such as:

Who will be paying the costs? Will those who benefit be those who pay? Will those endangered be those who benefit? Answers to these questions may be based on quantifiable data but often must be based on estimates or surveys. A related major task is to determine the equity of distribution of danger, benefits, and costs. This asks a question of fairness and social justice for which answers are a matter of personal and societal value judgment.

Who determines the acceptable level of probability of injury? In terms of ability to judge acceptability, designers/engineers are no better qualified than any other group of people and, in general, are less qualified than many others. It is often alleged that engineers (because of their inherent characteristics, education, and experience) are less sensitive to societal influences of their work and products than others. As for most stereotypes, there is some truth in this view. Clemenceau reportedly said: "War is much too serious a matter to be entrusted to the military." Perhaps product design is much too serious a matter to be entrusted solely to designers and (especially) business managers.

Jaeger (Ref 31) has summarized the situation thus:

> Nowadays it seems to me that the risk problem in technology has turned out to become one of the most pressing questions concerning the whole of industrial development. This problem is of fundamental as well as of highly practical importance. The answer to the question "How safe is safe enough?" requires a combination of reflective and mathematical thinking as well as the integration of technological, economic, sociological, psychological and ecological knowledge from a superior point of view.

If the designer cannot adequately make the determination, then who can? Various ideas have been proposed (e.g., Ref 32), but no suggestion yet made is fully satisfactory. The designer/producer must resolve this for each product. References 30 and 33 can be helpful in developing sensitivity to assessing an acceptable probability of injury.

Acceptable Level of Danger

It has been suggested that an acceptable level of danger might be 1 in 4000 per year, or 1 in 10^6 per hour. Statistics indicate that this is about the danger of dying from an automobile accident in the United States. One might infer that U.S. citizens consider this an acceptable level in view of the fact that little apparent effort is expended in trying to decrease the accident rate. The National Highway Traffic Safety Administration indicates that about 50% of fatal traffic accidents in the United States are alcohol related. If there were severe penalties for driving under the influence of alcohol (as there are in some other countries), this danger would presumably decrease to about 1 in 8000 per year. Either level of danger may be rational for the public as a whole (obviously debatable), but it probably is not perceived as such by a bereaved family. Such a rate hardly seems acceptable for consumer products. It certainly is unacceptable for nuclear applications. While the majority of manufactured products have a much lower level of danger than this, many of these products are considered to have a level of danger too high to be acceptable. Juries regularly make this decision in products liability actions.

One aspect of a potentially acceptable level of danger is the manner in which it is stated. Engineers might prefer to state the level in terms of probability. The general public, however, might well prefer it otherwise, or even unstated. The general public must be aware of fatalities from automotive accidents. It is possible that if automobile manufacturers were to point out that there is an annual chance of about 1 in 4000 that an individual will be killed, and a much greater chance of being injured (even seriously, such as spinal injuries, which not only incapacitate the victim but require constant attention by others), the attitude of the public might be different.

It must be recognized that while it is possible to reduce the level of danger to a very small number, *danger cannot be completely eliminated*, no matter how much effort is expended. We do not think there is any one level of acceptable danger. Each situation must be judged independently. The question is not what level of danger the engineer/designer thinks is acceptable for the public but what level the public perceives to be acceptable.

REFERENCES

1. *The Code of Hammurabi*, University of Chicago Press, 1904
2. *Restatement of the Law, Second, Torts*, 2d, Vol 2, American Law Institute Publishers, 1965
3. *Webster's New Twentieth Century Dictionary, Unabridged*, 2nd ed., Simon & Schuster, 1979
4. G.A. Peters, New Product Safety Legal Requirements, *Hazard Prevention*, Sept/Oct 1978, p 21–23
5. *Barker v. Lull Engineering Co.*, 20 C. 3d 413
6. C.O. Smith, Manufacturing/Design Defects, Paper 86-WA/DE-14, ASME
7. C.O. Smith, Mobile Ladder Stand, Paper 87-DE-5, ASME
8. C.O. Smith, Design of a Saw Table, Paper 87-WA/DE-9, ASME
9. C.O. Smith, Coffee Grinder: Safe or Not?, Paper 88-WA/DE-6, ASME
10. C.O. Smith, Collapse of an Office Chair, Paper 89-WA/DE-18, ASME
11. C.O. Smith, Some Subtle (or Not So Subtle?) Product Defects, Paper 90-WA/DE-23, ASME
12. C.O. Smith, A Fatal Helicopter Crash, Paper 91-WA/DE-8, ASME
13. P.D. Beard and T.F. Talbot, What Determines If a Design is Safe, Paper 90-WA/DE-20, ASME
14. T.F. Talbot and C.S. Hartley, Failure of Fastening Devices in Pump Packing Gland Flange, Paper 89-WA/DE-12, ASME
15. T.F. Talbot and M. Crawford, Wire Rope Failures and How to Minimize Their Occurrence, Paper 87-DE-7, ASME
16. T.F. Talbot and J.H. Appleton, Dump Truck Stability, Paper 87-DE-3, ASME
17. T.F. Talbot, Safety for Special Purpose Machines, Paper 87-WA/DE-8, ASME
18. T.F. Talbot, Chain Saw Safety Features, Paper 86-WA/DE-16, ASME
19. T.F. Talbot, Hazards of the Airless Spray Gun, Paper 85-WA/DE-13, ASME
20. T.F. Talbot, Man-Lift Cable Drum Shaft Failure, Paper 87-WA/DE-19, ASME
21. T.F. Talbot, Bolt Failure in an Overhead Hoist, Paper 83-WA/DE-20, ASME
22. W.G. Ovens, Failures in Two Tubular Steel Chairs, Paper 91-WA/DE-9, ASME
23. J.A. Wilson, Log Loader Collapse: Failure Analysis of the Main Support Stem, Paper 89-WA/DE-13, ASME
24. T.A. Hunter, Design Errors and Their Consequences, Paper 89-WA/DE-14, ASME
25. C.O. Smith and T.F. Talbot, Product Design and Warnings, Paper 91-WA/DE-7, ASME
26. C.O. Smith and J.F. Radavich, Failures from Maintenance Miscues, Paper 84-DE-2, ASME
27. C.O. Smith, Failure of a Twistdrill, *J. Eng. Mater. Technol.*, Vol 96 (No. 2), April 1974, p 88–90
28. C.O. Smith, Legal Aspects of a Twistdrill Failure, *J. Prod. Liabil.*, Vol 3, 1979, p 247–258
29. C.O. Smith, "ECL 170, Tortured Twist Drill," Center for Case Studies in Engineering, Rose-Hulman Institute of Technology, Terre Haute, IN
30. W.W. Lowrance, *Of Acceptable Risk*, William Kaufman, Inc., 1976
31. T.A. Jaeger, Das Risikoproblem in der Technik, *Schweizer Archiv fur Angewandte Wissenshafter und Technik*, Vol 36, 1970, p 201–207
32. C.O. Smith, "How Much Danger? Who Decides?" paper presented at ASME Conference "The Worker in Transition: Technological Change," Bethesda, MD, 5-7 April 1989
33. R.A. Schwing and W.A. Albers, *Societal Risk Assessment*, Plenum Press, 1980

SELECTED REFERENCES

- S. Brown, I. LeMay, J. Sweet, and A. Weinstein, Ed., *Product Liability Handbook: Prevention, Risk, Consequence, and Forensics of Product Failure*, Van Nostrand Reinhold, 1990
- V.J. Colangelo and P.A. Thornton, *Engineering Aspects of Product Liability*, American Society for Metals, 1981
- R.A. Epstein, *Modern Products Liability Law: A Legal Revolution*, Quorum Books, Westport, CT, 1980
- P.W. Huber and R.E. Litan, Ed., *The Liability Maze: The Impact of Liability Law on Safety and Innovation*, The Brookings Institute, 1991
- W. Kimble and R.O. Lesher, *Products Liability*, West Publishing Co., 1979
- J. Kolb and S.S. Ross, *Product Safety and Liability*, McGraw-Hill, 1980
- M.S. Madden, *Products Liability*, Vol 1 and 2, West Publishing Co., 1988
- C.O. Smith, *Products Liability: Are You Vulnerable?*, Prentice-Hall, 1981
- J.F. Thorpe and W.H. Middendorf, *What Every Engineer Should Know about Products Liability*, Dekker, 1979

Section 3: Design Tools

Computer-Aided Design

John MacKrell, Principal, CIMdata, Inc.

COMPETITIVE PRESSURES continue to increase in all aspects of the business world of today. For companies to prosper, they must adopt new processes and tools that allow them to work more effectively and efficiently while providing better products to their customers. One manifestation of this is a continuing effort to incorporate concurrent engineering into the product development environment. Integrated product development (IPD) support organizations are beginning to appear in companies that realize the importance of this concept. Another important development is the drive for controlled product development based on a single, master product model concept.

If a master model is to be used as the basis for product development, it must be very rich in intrinsic information. To make the concept viable, the product model must support not only physical design, but analysis, testing, design optimization, simulation, prototyping, manufacturing, maintenance, and many other product development processes. The model has to define all of the product being designed including how parts connect and move with respect to each other. The model must provide an unambiguous representation of the physical product within the product development system. Modern, solids-based computer-aided design/computer-aided manufacturing (CAD/CAM) systems provide a good share of what is needed for companies to develop products using modern methods.

In reality, product design and CAD are processes. Computer-aided design tools can be used to simplify that process, but a CAD tool by itself does not inherently create good design and product development practices. That goal must be accomplished through changes in how individuals and organizations apply CAD/CAM and other tools and methods. In fact, the changes an organization makes in its product development processes are what produce payoffs in faster new product introduction, lower development costs, higher quality, more new products, competitive innovation, and increased profits.

A complete discussion of modern product development processes is a book-length topic of its own. Suffice it to say that the appropriate application of up-to-date technologies can greatly facilitate product development. Modern CAD/CAM systems are one of many technologies that can enable good practices. This article concentrates on describing this important technological area.

Recently, the techniques used to design products have changed dramatically. The advent of parametric, feature-based design creation has caused all of the major CAD vendors to rethink their product offerings and to redesign their CAD systems to present users with a more flexible, easier design process.

Indeed, these changes are having a major impact on the role of CAD in product development. Many companies today are making the transition from two-dimensional drafting and three-dimensional wireframe/surface modeling to complete three-dimensional, solid modeling of new product designs. This leads companies to produce more complete, computer-based product designs that can be used to expand and facilitate more refined product analysis, soft prototyping, computerized simulation, and nonphysical testing. All of these improve productivity, reduce product development time, improve quality, and allow more design creativity.

Because of the close interrelationship of solid modeling with product design in current CAD/CAM environments, additional technologies must be included in any discussion of modern CAD/CAM systems. These are constraint modeling, feature modeling, associativity, assembly design, and design intent. They are described in this article after a brief history of CAD technology. The article closes with a discussion of CAD applications.

Computer-Aided Design: A Brief History

The first commercial CAD systems were introduced in the late 1960s. Two of the earliest, successful products came from Computervision and Applicon. Both companies were heavily influenced by earlier work in the aerospace and automotive industries as well as academic research centers, such as Lincoln Laboratories at the Massachusetts Institute of Technology. In general, these early products basically worked in two dimensions and supported drafting functions. Such so-called electronic drafting boards proved quite efficient when drawings had to be updated and modified frequently. But they did not significantly change the product design process. Through the early 1970s CAD system capabilities were improved to include three-dimensional wireframe and surface design. With these new capabilities, CAD began to be useful for numerical control (NC) programming and some engineering analysis but remained insufficient for real, complex engineering design.

The first commercially available solid modelers appeared in the late 1970s (Euclid from Matra Datavision was the first general CAD system to include solid modeling as its primary design method). By the mid-1980s, virtually all major CAD products had solid modeling capabilities; however, many of them treated solids as an add-on technology, not as their primary modeling method. They arrived to great fanfare and high expectations as the solution to all design problems. The promise was that solids-based CAD would produce large savings throughout design and engineering. However, the reality was quite different: these savings did not materialize because the systems were unnatural and difficult to use, and affordable computing hardware did not provide enough computing power to efficiently create and manipulate the models.

Early solids-based systems relied on Boolean combinations of simple primitive solids (blocks,

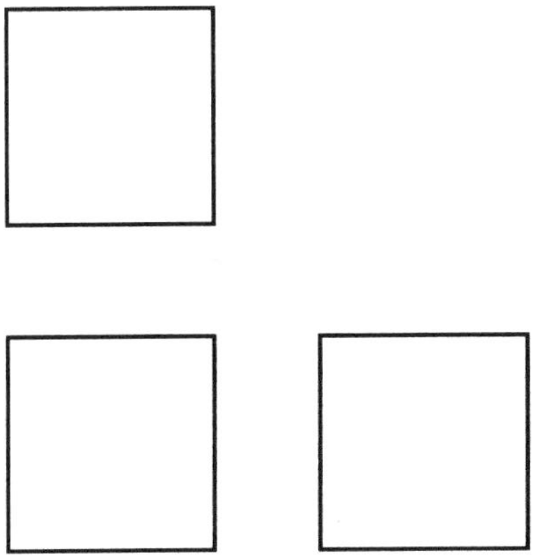

Fig. 1 Cube defined by orthographic projection of two-dimensional geometry

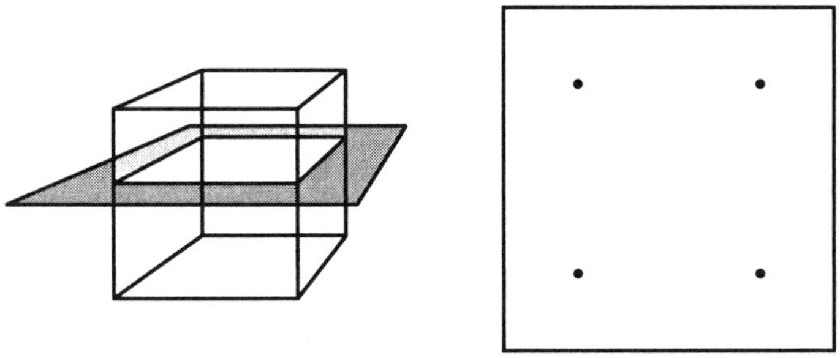

Fig. 2 Cube defined by three-dimensional wireframe geometry

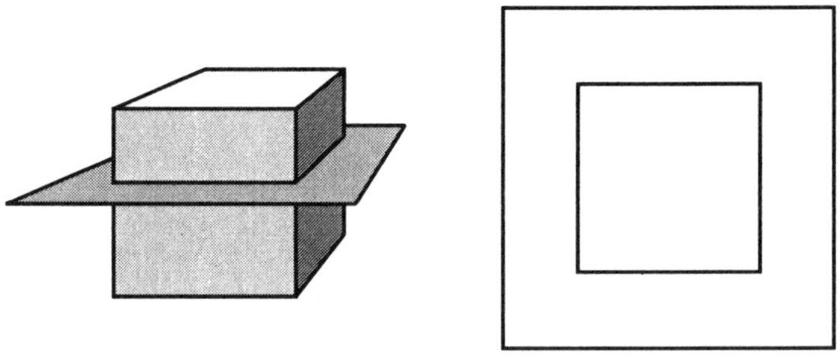

Fig. 3 Cube defined by a surface model

cylinders, spheres, cones, tori, etc.) to build up complex part designs. While Boolean combinations can be used to create very complex designs, they do not match well with the traditional sketching methods that designers have used for most of recent history (several hundred years of formal engineering design, most of it done by creating orthographic-view drawings). The terminology used for these Boolean operations (union, difference, and intersection) is also nontraditional and confused designers and other users, who wanted to make holes, bosses, and other design elements. Boolean methods are also difficult to use when a design has to be changed during the product development process—sometimes requiring parts of the design to be discarded and completely reconstructed.

Sweeping operations (extrusions and rotations) of two-dimensional outlines fit with the designer's sketching methods better, but still require Boolean operations to create details that cannot be included in the swept outline. Examples of this are keyways and bolt holes through a flange.

It was not until 1988 when Parametric Technology Corporation (PTC) introduced its Pro/ENGINEER product with a fundamentally different approach to design modeling that solids became easily accessible and generally productive. The approach established by PTC has three basic components that had been known, but not developed into a coherent, commercial product. These are parametric design, the use of design features, and data associativity. Most other CAD vendors now offer similar capabilities. In the newest constraint-based systems (notably Matra Datavision Euclid Designer, Hewlett-Packard SolidDesigner, SDRC I-DEAS Master Series, Dassault CATIA, and others), the user can create controlling equations and parameters on existing geometric models and make local modifications with ease.

Parametric modeling is a type of constraint modeling, as is variational modeling. Designs modeled in a constraint modeler capture many pieces of information about the shape and size of a design in the form of dimensional and geometric properties. These are related to each other by a set of equations known as relationships or parametric equations. When any dimensional or geometric property is changed, the set of equations can be re-solved to find the effect on all other related dimensions; the geometric description of the design is then updated automatically. This concept is more fully described in the section "Constraint- and Feature-Based Computer-Aided Design" in this article. Simply stated, constraint-based design allows users to make changes easily to their designs that follow their "design intent" without having to re-create or move primitive solids.

Design features are similar to "super" primitives. They are predefined or user-defined groups of geometry with well-defined dimensional attributes and a datum point (or origin) with which they can be easily positioned in the design. Design features include countersunk holes, through holes, tapered holes, bosses, pockets, tabs, draft angles, blends, fillets, shells, and many others. They simplify the design process by combining complicated but easily sized geometry with the appropriate Boolean operation (such as difference to remove a hole) and simplified positioning. They are generally called-out by their common names, such as counterbored hole. Features can also contain other attributes such as the NC instructions required to machine them, or cost-of-manufacturing information.

Parametric Technology Corporation combined parametrically controlled form features with data associativity (further explained below) to create a more easily used design system that produced easily modified designs. Now, most of the other major CAD vendors have copied these techniques or developed similar capabilities.

The current state-of-the-art CAD/CAM systems use constraint-based features along with surface and solid models to create design geometry much more efficiently than was possible with earlier CAD technologies.

Overview of CAD Technology

Modern CAD systems use many forms of geometric modeling. This section is an attempt to provide a very simple, easily understood description of them. In some cases the explanations are oversimplified, but the important points for someone trying to understand how CAD systems work are covered.

The earliest CAD systems used two-dimensional wireframe geometry to create drawings—in much the same way that for the past several hundred years formal drawings have been used to convey and document engineering designs. Due to this long history, two-dimensional drawing is comfortable and natural to use as a design medium, but works best for documenting designs that have already been worked out by manually sketching, making a physical model, or other methods. Many products are too complex to actually design using two-dimensional drawings, and some objects are nearly impossible to define by a set of two-dimensional drawings. For very simple parts and parts that are essentially turned, two-dimensional geometry provides sufficient definition, but for complex parts two-dimensional drawings are just not as complete or efficient as other types of computer-aided design models.

With a two-dimensional wireframe system, the design is created by drawing lines, arcs, circles, and other curves on a plane (i.e., virtual paper) in the computer—much as they would be drawn on a traditional drafting board. Orthographic projections are used to convey the three-dimensional shape of the design. The advantages of CAD over manual drafting methods come in the form of much easier drawing modifications (the designer can erase and redraw and move views around with ease) and in the provision of construction aids of the computer such as selection of existing geometry, construction guides and grids, trimming lines, creating tangents, and many others. However, the design information captured in two-dimensional CAD systems is little better than that developed on a drafting board. The true shape of parts is not determined in a way that helps designers check interferences, develop complex intersections, define free-form surfaces, compute mass properties, create NC programs, perform engineering analyses, or accomplish other common tasks. As is shown in Fig. 1, a cube is defined by two or more views, but the computer cannot automatically determine the volume enclosed by the cube.

A three-dimension wireframe adds a third dimension to two-dimensional geometry, but it does not help very much. With three-dimensional wireframes the geometry of parts is still not fully defined. The surfaces between the wireframe edges are not defined, and the computer is unable to determine what is inside or outside of the part being designed. Many of the problems associated with two-dimensional wireframe designs remain.

For instance, hidden line removal must be done manually. However, three-dimensional wireframe geometry is easy to transfer from one CAD system to another. As shown in Fig. 2, the three-dimensional wireframe cube can be intersected by a plane, but because the surfaces of the cube are not defined, only the four points where the plane intersects the cube's edges can be determined. The computer cannot determine how to connect them into a square.

The next logical step, one that also parallels the evolution of CAD systems, is to define the surfaces that form the "skin" of the design. With the addition of the surfaces, the model becomes much more complex as well as more useful for supporting CAD/CAM applications. The definition of the part is more complete, and the surface geometry can be used to drive NC and finite element modeling. Actual intersections between surfaces can be determined by the computer so that the edges of complex objects can be easily created. However, the computer remains unable to determine the properly bounded intersections between complex parts and cannot automatically insert features such as holes into parts. Surface models do not reliably permit the automatic and correct removal of all hidden lines in a view of a group of parts. Surface models do provide information that can be used to produce NC part programs and rendered, color- shaded images. Other analyses such as finite element modeling and analysis (FEM/FEA) can also be done on surfaced models.

Surface models do not allow any large degree of automation to be applied to engineering analysis and manufacturing applications. The definition is not complete enough (in the vast majority of cases) to allow the computer to determine what is inside and what is outside of the part being designed or to compute mass and other complex properties. Even a completely closed surface model lacks the topological information needed to automatically perform these types of computations. Figure 3 shows that cutting a surface model of the cube provides a square, but does not automatically determine that the inside of the square represents the inside of the part.

Other issues that occur with surface models include:

- Surfaces and patches can be left out of the model creating an incomplete definition of the part, known as an open-surface model.
- Surfaces and patches may overlap or have gaps between them resulting in an erroneous definition of the part geometry, and nonclosed volumes.
- Extra surfaces (inside or outside the part) can be defined, leading to incorrect designs.
- Intersections of surfaces can be imprecise, creating errors in the accuracy of intersection curves
- Nonmanifold conditions can occur at tangencies, edges shared by more than two surfaces, and hanging surfaces, creating parts that cannot be manufactured.

The surface model does not contain information that tells the computer on which side of each surface the material of the object is located. Solid modelers add an ordering to the surfaces and curves found in surface models. This ordering, called topology, allows the computer to determine the volume that is enclosed by the bounding surfaces of the object as opposed to the volume that is outside those surfaces. A solid modeler is able to determine the relationship of any point in space to the solid model—whether it is inside, outside, or lies on the surface of the solid. In Fig. 4, the solid model of the cube, when intersected with the plane, results in not only the square of intersection edges being known, but knowledge that the shaded area is inside the cube.

This is what allows solid modelers to perform automatically Boolean operations between two solid models. Boolean operations are used to combine two solid models into a more complex part. Boolean operations can be used to add material to or remove material from a solid. These operations provide the basis for creating complete and unambiguous descriptions of physical objects within CAD/CAM systems.

The problem with solid modeling has been that Boolean operations are rather difficult to use in many situations. Many designers have found them to be counterintuitive and difficult to control. Form features combined with constraint parameters have greatly simplified how designers work with solid modelers to design parts and assemblies.

Constraint- and Feature-Based Computer-Aided Design

Solid modeling has been around since the early 1970s but it has historically been hard to use, which has delayed its acceptance by many designers. Using Boolean operations and primitives to "sculpt" a solid model has never been an obvious, straightforward concept. Designers have difficulty positioning the primitives, too many operations are required to complete a design, and it is too difficult to change designs. In traditional systems (both manual and CAD) it has been the responsibility of the designer to remember all of the interrelationships among various parts of the design. A simple example of this is that if the diameter of a hole is changed, then the designer must remember to change the diameter of the bolt that passes through the hole. Also, for something as simple as a countersunk hole the problem doubles in complexity—the designer has to remember to change both the diameter of the main hole and the countersink hole as well as the corresponding piece on the mating part. Tracking all of the interrelationships in an assembly can be a daunting task, even in a simple design. Allowing the CAD system to track these relationships automatically translates directly into saved time and fewer design errors. Constraint-based modelers provide a viable solution to these problems.

While constraint modelers provide a more easily used paradigm for sketching and developing modifiable, evolving designs, they also present some problems of their own.

Geometric modeling systems allow designers and others to define the positions of points, lines, curves, and surfaces in two-dimensional and three-dimensional space. The ability of the user to position and move these geometric elements is controlled by the rules embodied in the modeling system being used. For example, in some systems, to move a point that has already been created the point must be deleted and re-created at its new location. This can become very time consuming when modifying a point that happens to be at the ends of several connected lines—each line must be deleted and redrawn to the new location of the point. In some drawing systems, points can be dragged to a new position. The problem then is: which lines, circles, and so forth, that share the point also should drag or change shape, and how?

Constraints are used to create a set of rules that control how changes can be made to a group of geometric elements (lines, arcs, form features, etc.). These rules are typically embodied in a set of equations. The equations can be simple (point A is at location x, y, z) or complex (the length of line B is one-half the length of line A) and can contain geometric as well as numeric positioning information. Four types of constraints are in common use today—numeric, geometric, algebraic, and attributes. Numeric constraints provide positions (x, y, z locations), lengths, diameters, spline parameters, angular values, and other measurable values. Geometric constraints include parallelism, perpendicularity, colinearity, tangency, symmetry, and other nonnumeric parameters that control the positional relationships of one piece

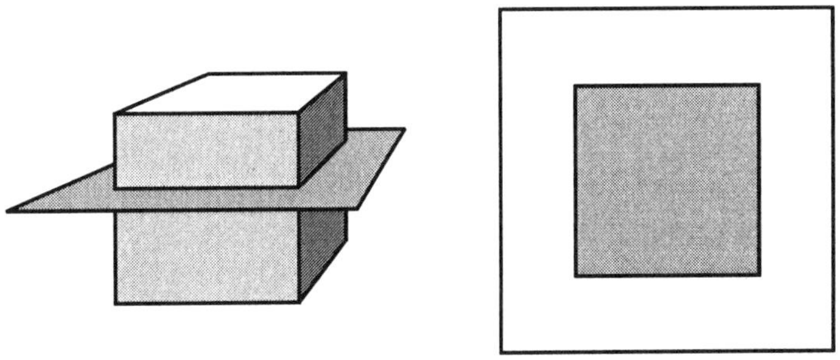

Fig. 4 Cube defined by a solid model

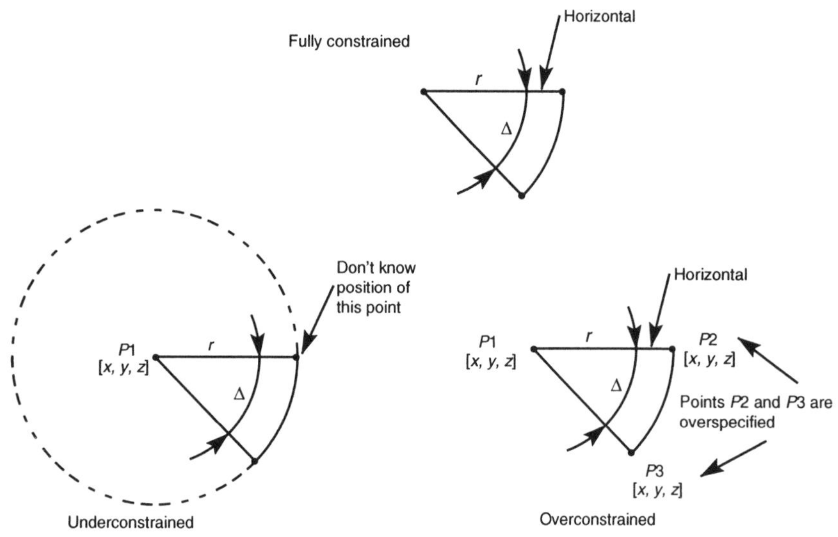

Fig. 5 Simple constraint example

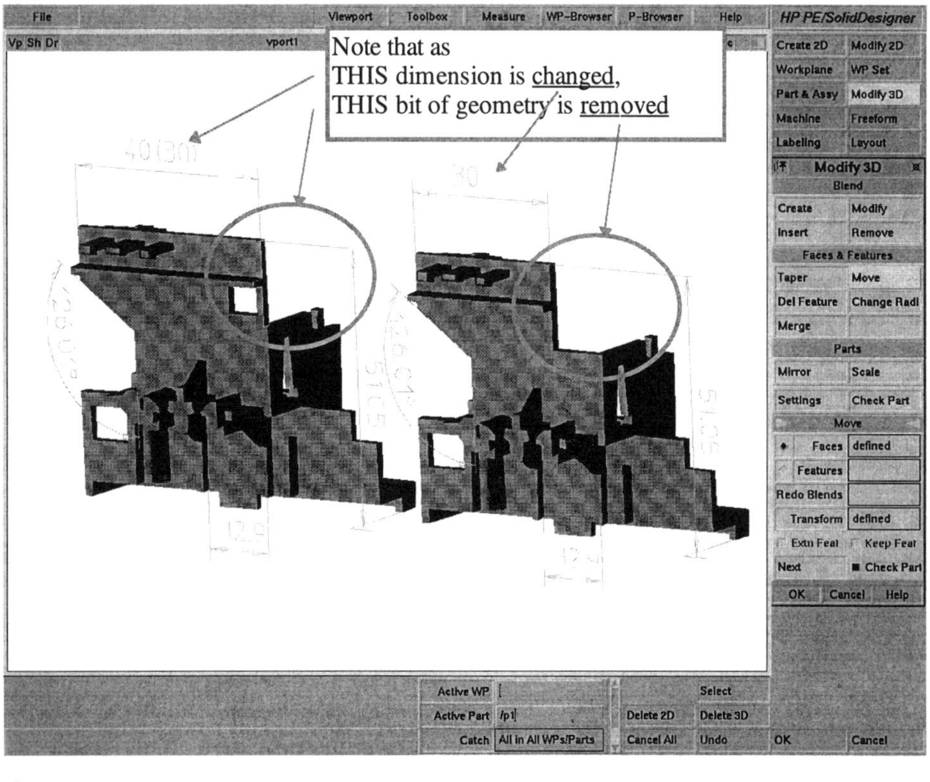

Fig. 6 Object before and after parameters are changed. Courtesy of Hewlett-Packard

of geometry with respect to another. Algebraic constraints combine numeric and geometric constraints in very simple equations (diameter of C = one-half the length of A) or extremely complex sets of equations that include IF-THEN-ELSE branches, inequalities, and calls to external subroutines (e.g., if diameter is less than or equal to 10 then part thickness is 0.125, else part thickness is 0.25). Attribute constraints (e.g., color, material, surface finish, thread type, maximum stress) define other characteristics of the part or its function.

Figure 5 illustrates three typical two-dimensional constraint situations. The pie-shaped piece is fully constrained in the upper view. The center point ($P1$), the radius of the arc (r), the included angle (Δ), and the geometric constraint fixing the top edge as "*horizontal*" completely and unambiguously specify the complete shape and position of the part. In the lower-left example, the "*horizontal*" constraint has been removed. While the remaining constraints still define the shape of the part, its position is no longer specified—it could be oriented at any position around the dotted circle. In the lower-right view, the shape and position of the part are overspecified. Neither $P2$ nor $P3$ add any useful or necessary information about the size or orientation of the piece. Note, however, that $P3$ could be used in place of Δ, and that $P2$ could be used in place of both r and "*horizontal*".

In Fig. 6, note the changes as highlighted in the upper right corner due to the change in the topmost parameter value. In particular, note that the topology of the object has changed, with the square hole being completely suppressed.

Two types of constraint modelers are in general use today: parametric and variational. These types have to do with how the set of constraining equations are defined and solved. In parametric systems all of the constraint equations are captured and solved in the order in which they are created; the design is controlled by a directed graph of operations. The variables used in each equation must have been defined by a previously stated constraint; that is, each geometric element is placed with respect to some previously placed element. However, most systems allow the user to rearrange the order of the equations, thus modifying the implied design intent. In variational systems the set of equations is solved simultaneously so that their ordering is not so important. For instance, if points A and B are located by user-defined dimensions and point C is located by references to the locations of A and B, then three equations define the constraint system:

$$A = \text{point}(x1, y1, z1)$$

$$B = \text{point}(x2, y2, z2)$$

$$C = (A - B)/2$$

In a parametric system, points A and B can be moved and point C will automatically follow. In a variational system any two of the points (for example, A and C) can be moved, and the third point (B) will automatically follow.

Each of these techniques has advantages and disadvantages when compared to the other, but neither holds an overbearing advantage over the other. Whether their differences are important depends on how the design process is to be carried out. The major differences are the order dependence of parametric modeling and the generally slower speed of variational modelers.

Constraint order dependence forces designers to think about the process they will use for a given design and to predetermine the independent parameters that will be used to control changes to the design. The designer may not know enough about the design early in the design process to make valid choices for dependent and independent constraints. In fact, unless the designer devotes considerable preplanning to the design process, the wrong constraints may be selected and the parametric definition will need to be reordered at a later time. Indeed, as the design evolves, unanticipated constraint dependencies are likely to arise; these can force changes to the parametric interdependencies. Changing the parametric ordering can be difficult because the interrelationships among parameters quickly become complex, and even the original designer may not quite understand the nuances of these interrelationships. This is especially true for someone who did not create the design in the first place and has to decipher the meanings and interdependencies of the parameters that have been used by someone else to create the design. Remember that no two designers execute a design problem using exactly the same method.

This problem is alleviated somewhat because most commercially available parametric modelers now allow users to redefine the parameterization and reorder the parametric equations to allow different parameters to control the design. Parametrically defined designs work extremely well for parts and assemblies that have a clearly defined hierarchy of features and components and a few important design parameters that control their overall form. An example of this is families of parts whose sizes are controlled by a few dimensions, such as length and diameter for bolts.

Variational systems require more computing power because they must solve a generally complex set of simultaneous equations (this is precisely the feature that allows them to solve for any variable, not just explicitly defined dependent variables). The method of solution requires numeric approximations that converge on an acceptable result. This can lead to failures in the solution process, although most CAD products appear to handle this convergence quite well. Both of these problems require more computing power than the straightforward, stepwise solution of parametric systems; however, current computer systems are powerful enough to handle substantial systems of equations in "real" time, or so quickly that users do not feel delayed in their work. Variational models work well for designing parts and assemblies where there are no easily defined parameters that control the overall form, or when the design is evolving such that the designer cannot predetermine how the features of the design are going to interrelate. With variational techniques, designers do not have to give a lot of forethought to the hierarchy of the design constraints and can freely change the design without being limited by the order in which constraints were defined and features were added to the design.

Many other factors have an impact on how useful a constraint modeler is. These include the user interface and how well the modeler can deal with a variety of design issues.

Constraints should be applicable to all types of geometry (two-dimensional and three-dimensional curves, surfaces, solid form features, between parts in assemblies, etc.). Some systems have restrictions that allow constraints to be created only during two-dimensional sketching; they allow the sketch to be extruded or rotated to form a solid model, but if the constraints are changed then the solid must be regenerated. In any case, some recomputation of the model is required whenever a constraint is changed. Fortunately, most of the major CAD systems do not restrict constraints to two-dimensional cases, and all of the major CAD vendors have plans to extend their constraint modelers to handle three-dimensional constraints. Surfaces represent another problem area. The issue is whether all of the geometry that controls the definition of a surface can be constrained (control points, slopes, tangency with an adjoining surface, etc.).

How a system handles over- and underconstrained geometry is also an issue. See Fig. 5 for an example of each. Some CAD systems do not allow over- or underconstrained models to be created and stored. They force the user to fully constrain the model. Other systems allow one or both of these conditions to exist. This gives users freedom to work with partially developed designs. In any case, the CAD system should indicate when geometry does not contain a full set of constraints (it is underconstrained) and when it has too many, conflicting constraints (overconstrained). This can be done in many ways—for example, by highlighting geometry that is not fully constrained or that has conflicting constraints, or by reporting the number of degrees of freedom remaining in the model (each degree of freedom corresponds to a missing constraint). In the best case, problems are highlighted on-screen so the user can see exactly what is amiss.

Unconstrained sketching allows the designer to sketch in two-dimensional and expand to three-dimensional without having to supply any dimensions (these can be added later). In this way, a preliminary model of a part can be created very quickly and refined later on.

The vast majority of objects designed in practice are assemblies of interacting parts. A major area of design errors occurs in the interfaces and fit of these parts. So, for designs of assemblies it is imperative that the CAD system allows constraints to be defined among parts of the assembly. When this cannot be done, the designer must remember to update multiple parts whenever the geometry of any of the interfaces of the parts are modified.

Features. All of the major constraint-based CAD systems combine constraints with form features. Form features are a higher-level construct than points, curves, and surfaces. They allow designers to work with a more natural syntax, dealing with slots, bosses, through holes, blind holes, and so forth, rather than Boolean operations on blocks and cylinders.

Form features allow designers to add relatively complex but common shapes to their designs using only a few commands instead of having to create the shape through sweeping or Boolean operations. Features combine geometry with a set of dimensions and with inherent knowledge of how the feature combines with solid objects to which it is applied. The set of dimensions allows the user to size the feature as it is being included in their design and to resize it after it has been

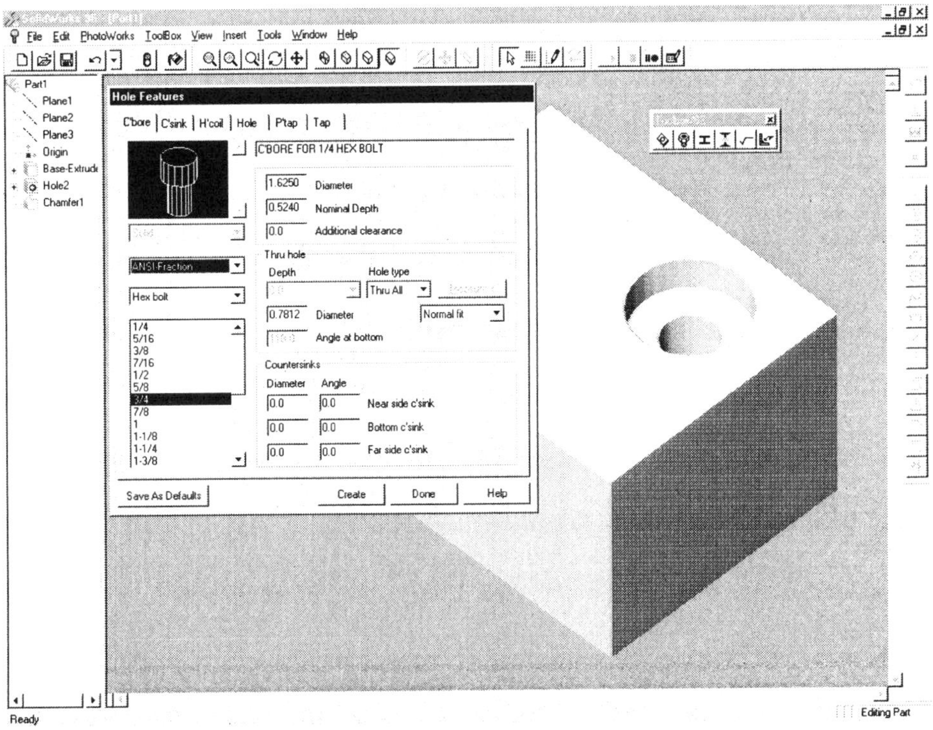

Fig. 7 Dialog for setting parameter values of a counterbored feature. Courtesy of SolidWorks

applied (see Fig. 7 for an example). Features know how they are to be combined with other geometry, for instance a hole feature will automatically be subtracted from the part on which it is applied, while a boss feature will be added (unioned) to the part.

Features are located in the design by positioning them with respect to their own, local coordinate datum. For example, a hole can be positioned by locating the centerpoint of one of its ends at a point on the surface into which the hole passes. The concept of "through" features is very important. A through hole is defined so that it always passes through the object on which it is applied. If the object is made thicker, the through hole does not become a blind hole (as would occur if the hole were made by subtracting a cylinder primitive that has a fixed length), rather, the hole lengthens so that it continues to pass through the part. Variations on through holes include definitions that pass from one face or surface through the next face encountered (but not through any other faces). This prevents the hole from passing through unintended portions of a part. In Fig. 8, a through hole with a fillet is moved to a new location by modifying the positioning parameter of the hole. Not only does the hole get longer, but its fillet changes shape substantially to match the new geometry surrounding the hole.

Most systems provide a library of the most common features. They may also have special sets of features that are tailored for use with specific applications such as sheet metal and mold design. Many systems also allow users to define their own sets of features that are specific to the user's business needs.

Features may contain information in addition to their own geometry and size. For example, manufacturing data such as surface finish and the NC operations required to machine the feature can be stored with the feature. This information can be retrieved by applications to help automate their processes.

Some NC programming systems are beginning to use manufacturing features. These are similar to form features, but contain additional information to help streamline the NC programming process. Note that in many cases the features used by designers are not the same as those required by manufacturing systems. As an example, a designer may design a spoked wheel as a set of boss or spoke features while a manufacturing engineer views the same part as a group of pocket features.

While support for other modeling operations such as Boolean operations and tweaking or local modifications is not always required, these can make the designer's job significantly easier by providing flexible methods for modifying geometries. Local operations include application of draft angles and fillets provided in all of the major CAD systems, but can be extended to other local modifiers such as face lifting and edge dragging. Much greater designing flexibility is gained when these operations work within constraint modeling systems and augment the feature modeling operations typically used in constraint systems.

Data Associativity

Data associativity is the ability of the CAD system to share design information among applications such as design, drafting, CAE, and NC without each application having to translate or transfer the data. Associativity also requires that applications "know" when geometry or other information has been changed so that the application can adjust its own data to suit those changes. Because it helps keep design and other product development modifications synchronized, associativity is an enabler for concurrent engineering. Associativity ensures that as the design changes all of the users of the design data have an up-to-date version of the design. As implemented in CAD systems today, associativity is either unidirectional or bidirectional. Unidirectional associativity provides design changes "downward" to other applications such as drafting and NC. Changes made to the design appear in drawings and NC toolpaths automatically, but changes made in drawings or NC do not change the design. This is also known as downward associativity.

With bidirectional associativity, changes made to any design information in any application (design, analysis, drafting, NC, etc.) are automatically reflected in all other applications of the CAD system. So, when a detailer changes a dimension in a drawing, the geometric model changes as well.

An obvious issue with bidirectional associativity is that people other than designers may be able to change the design. In large product development teams this may violate the team's rules for design control. In small teams, bidirectional associativity can be used to allow highly interactive concurrent engineering. Of course, it should be left to each organization to decide how they want to use or allow associativity.

Most CAD systems provide control over associativity, allowing it to be disabled for certain parts or users, or allowing individual users to decide when associative design changes will affect their own work sets. For instance, an engineer who is working on a finite element model may want to delay accepting design changes that would require remeshing the model until after a preliminary analysis is complete. A product data management (PDM) system can also be used to control who can change designs, drawings, NC part programs, and so forth.

As noted above, bidirectional associativity may or may not be an issue depending on the operating style and rules of the organization. If all team members need to be able to make changes to designs, then bidirectional associativity is a good feature. If some members of your teams are not permitted to make design changes, then either controllable bidirectional associativity or associativity limited to downward changes is needed.

Two types of associativity are commonly in use in current CAD systems, manual and automatic. With manual associativity, the CAD system recognizes that information has changed in one application, but it does not update the related information in other applications until the user tells it to. In automatically associative systems, any change in data in one application (such as the geometric modeler) causes appropriate changes in related data to occur without user intervention in all other applications that use the changed data (such as drafting or NC).

An issue with automatic data associativity is that, in general, the associative update fails when the topology of the model is changed. For example, an NC program may not be able to automatically update a tool path when a feature such as a hole or boss is added to or deleted from the geometric model. In these cases, the user will have to take some action to complete the update.

Assembly Design

Assembly design is the process of creating groups of parts that operate together. Most organizations actually create products that are assemblies as opposed to individual parts. For these groups, it is important that parts can be designed in relation to one another, sharing dimensions and other parameters.

A key element of constraint-driven assembly design is the use of joint or mating conditions. These define how two parts connect with each

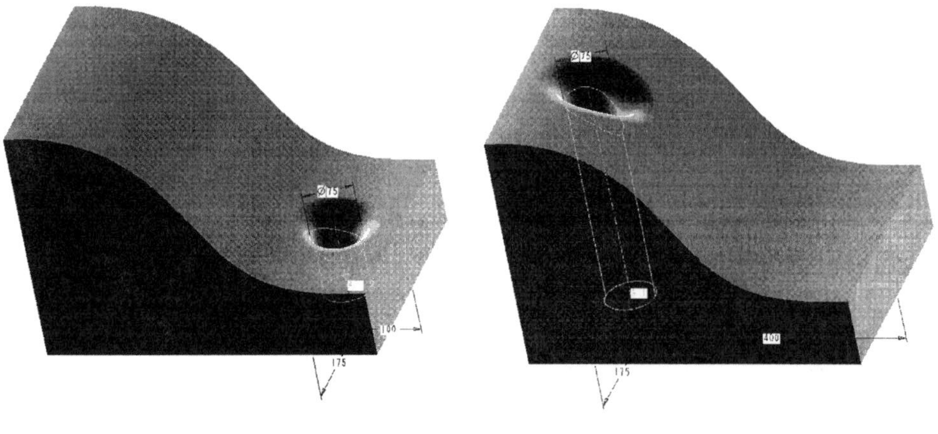

Fig. 8 Through hole feature before and after modification. Courtesy of Parametric Technology

other and how they are allowed to move with relationship to each other. In many CAD systems, the parts of an assembly automatically follow their mating conditions. As one part is moved, the parts that mate with it move so as to preserve the constraints of the mating conditions.

Constraints among parts in assemblies allow designers to control the size and positions of mating features without having to manually update both parts. Constraints among parts are essential to the support of the design of complete assemblies.

In addition, product or assembly structures are a required capability in assembly design systems. They provide a logical model of the relationship among the components (subassemblies and parts) of an assembly. When presented as a graphically displayed tree, they also can be a convenient user interface structure for navigating through complex assemblies (see Fig. 9). In some systems, the product structure can be created and controlled by a PDM system. This allows the structure to be used as a control mechanism for restricting access to certain portions of the assembly to particular product development team members.

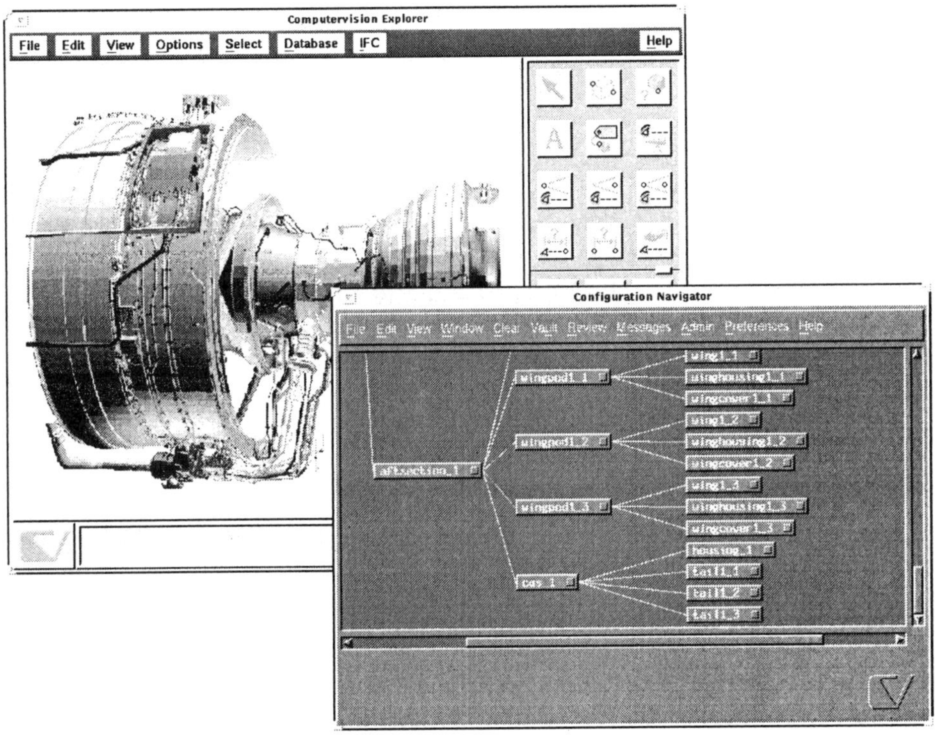

Fig. 9 Complex assembly shown with navigation tree (logical structure). Courtesy of Computervision

Capturing Design Intent

One of the most important advantages of using constraint modelers is that they allow the CAD system to capture some of the designer's intentions. Most designed devices can be characterized by one or more primary and secondary factors that, when changed, alter the size or shape of the device. These might include the length of an engine stroke or the weight of a ship. Each of these is determined by the operational intent of a design, and they may be critical to the success of the device being designed. One of the problems faced by designers is to understand how changes in these factors affect the actual shape, size, and operation of the device being created. Many more detailed forms of design intent exist as well. For example, that two sides of an engine block casting are to be parallel may not affect the operation of the engine, but may be important for manufacturing reasons.

Design history does not necessarily equal design intent. If the intent is simply to get from New York to Washington, whether one flies, drives, or rides a bicycle has little bearing on achieving the goal, but certainly determines the history of events that occur in reaching the goal. Likewise, if three designers are given the same mechanism to design, they are all likely to proceed through very different sets of steps to reach their final designs, although the three designs may result in nearly identical solutions that satisfy the original intent. So, how can design intent be captured?

In addition to the history of steps taken to create a design, the designer must be able to create constraints that embody the functional intent of the device, for example, the relationship between the stroke and bore of a piston determines the displacement of an engine. These controlling parameters are much more important than other constraints (such as the length of the engine block) that may, in fact, be defined earlier during the designer's work within the CAD system. Of course, it is good design practice to think through the design before proceeding with the actual concept development in the CAD system, but it remains extremely difficult to predetermine or even to recognize beforehand every factor that may drive a complex design.

The designer needs the flexibility to proceed with a design according to how creativity drives the process, without having to worry that the CAD system is going to make assumptions about design intent that may be incorrect or difficult to change. Therefore, the CAD system needs to provide mechanisms that allow constraints to be easily reordered, new constraints to be created, and old constraints to be removed.

Because design intent is documented so that the design is easier to update or change, it must represent not only the spatial or geometric relationships among parts of the design, but their operational intent or interrelationships as well. For example, it may be critical to know that a particular joint must be able to both rotate and slide for a mechanism to operate as intended.

The design intent must also be kept in a form that other designers can understand. This requires a capability to document the intent (such as is done in a design notebook). Some systems allow notes to be created and viewed with the design. These notes can contain design parameters, their definitions, present values, and effects. In those systems that allow parametric notes, the value of a design parameter can be changed by selecting it in the note and typing a new value in its place (the design model then is automatically changed to reflect the modified value in the note). However, few of the CAD systems available today provide anything comparable to a traditional design notebook capability.

Applications of CAD Systems

CAD/CAM systems provide much more than the definition of the solid model as a design. The geometric model is only the starting point of a design. It provides the basic geometric definition of the design so that other essential design processes can be performed. Solid models are used to drive engineering analyses of various kinds, drafting and product documentation, prototyping, NC programming, rendering and visualization, manufacturing planning, and other product development activities.

These activities provide the information needed to complete the product development process. Several uses of CAD models are briefly described below; some of these are described in much more detail in other articles that follow in this Section of the Handbook.

Integrated Product Development (IPD). For companies that operate their product development process as a team activity using the concepts of IPD, the need for tightly integrated development tools is apparent. In the past there was a tendency to break the development activity into a number of discreet steps—concept design, detailed design, analysis, drafting, manufacturing engineering, and so forth. So too, the tools used

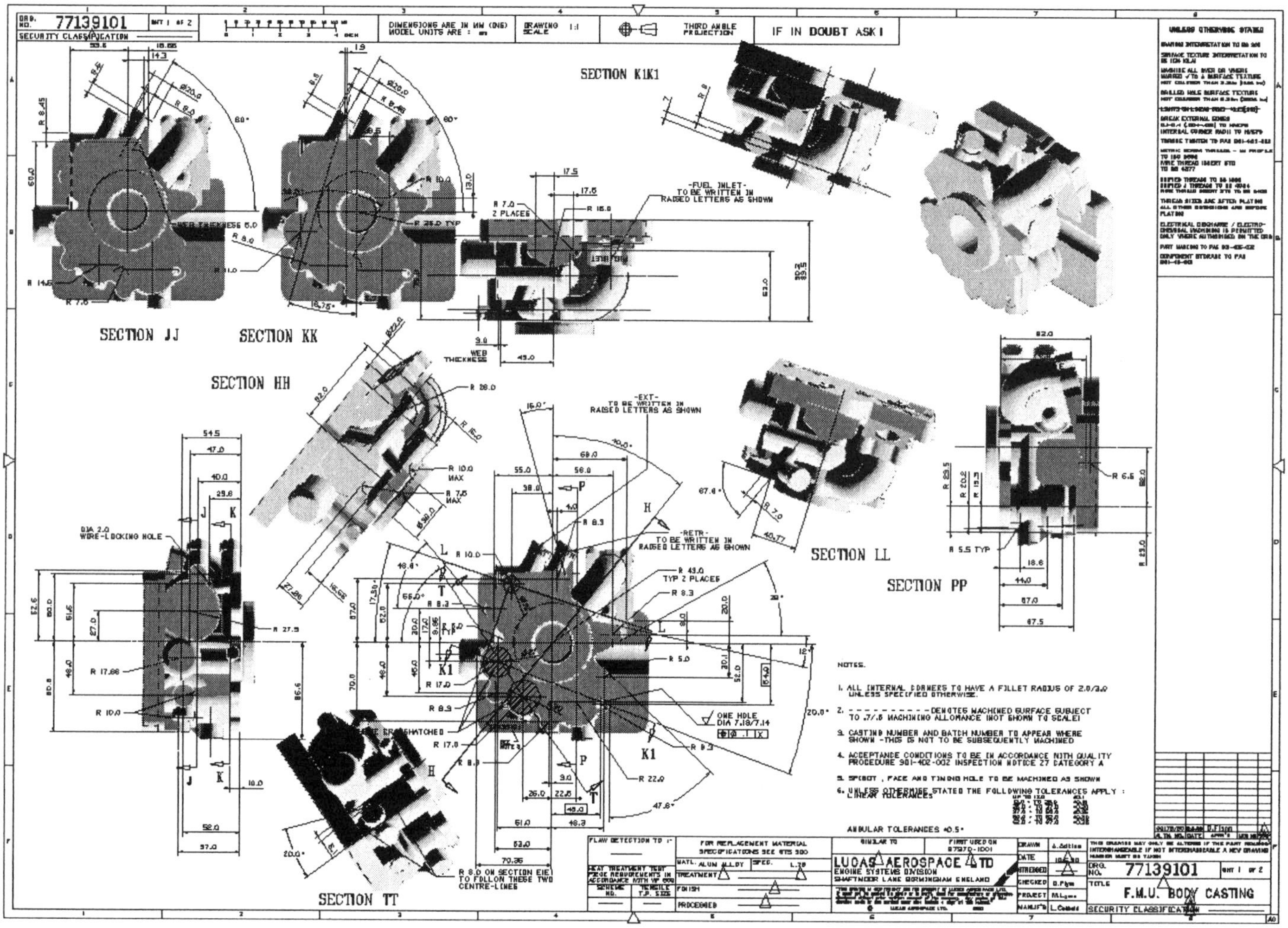

Fig. 10 Detailed drawing with shaded views. Courtesy of Computervision

to develop products were independently developed to support this task-oriented process.

Only recently have CAD vendors begun to deliver highly integrated suites of products that support collaborative, team-oriented product development. An integrated product development process requires several key capabilities. These include:

- Support for assembly modeling that is ubiquitous, that extends throughout the tool set
- A fully associative environment for all design data and activities
- Product data management to control access to, and aid in distribution of product data and changes
- Data exchange versus data translation
- Product simulation within the CAD system (soft prototyping)
- The ability for users to incorporate their own, proprietary tools, data, and processes into the development environment

Several CAD vendors now attempt to fulfill the needs of IPD users. While none have been completely successful, there continues to be progress. As more companies embrace the IPD concept, these systems should continue to grow in capabilities.

Drafting and Product Documentation. Documenting products includes more than drafting (see the article "Documenting and Communicating the Design" in this Volume). In addition to the detailed drawings that have traditionally been used to describe designs, other types of printed and drawn materials are produced. These include specifications, assembly and manufacturing instructions, maintenance manuals, user and instruction manuals, spare parts lists and drawings, and marketing materials among others. Many of these documents require the use of extensive text combined with drawings and artwork of various types.

Solid models can be used to help prepare many types of drawings including exploded assemblies, perspective views, cutaway views, photorealistic renderings, etc. These can be transferred into technical publication, page layout, word processing, and art programs for inclusion in printed and electronic documents. See also the section "Product Visualization" below.

With parametric solid modeling systems, dimensions in drawings can be generated automatically. Solids also provide automatic hidden-line removal, shaded-image views, and section views. Figure 10 illustrates an information-rich alternative to traditional drawings one that is based on having solid models.

Product Visualization. The visualization of the product being designed affects not only how the user sees the design on the CAD system display, but also how pictures of the product can be used in product documentation and other printed materials. Shaded and hidden line displays of solid models on the monitor of the CAD system are common today. In many systems, a shaded image of a part or a complete assembly can be rotated or moved in real time. This is a great help to users in visualizing the shape of the model, manipulating the objects being designed, and simulating (animating) the motions of assemblies. Most systems require advanced graphics hardware to accomplish this level of dynamic display. Other helpful display options include viewing the solid with transparent surfaces so that hidden edges and internal features can be seen and selected, perspective views that look more realistic than isometric projections, dashed hidden lines, and others. All of these can be

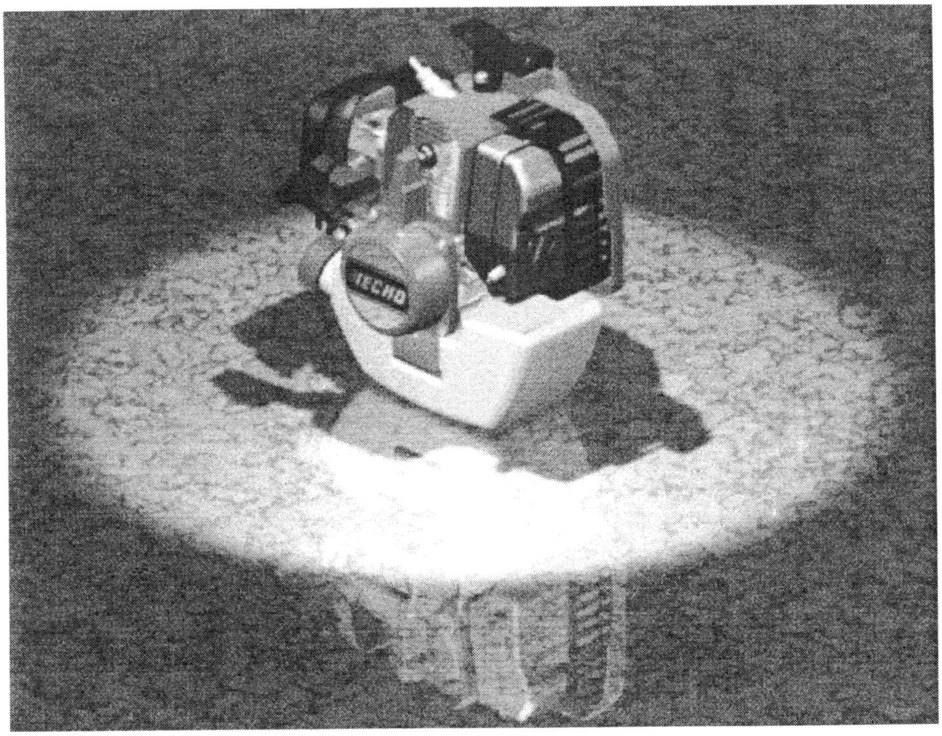

Fig. 11 Rendering of a solid model with textures, reflections and other effects. Courtesy of Matra Datavision

created automatically for solid models, and enhance the usability of the CAD system.

Solids-based CAD systems can provide automatic generation of exploded views in perspective or isometric projection, with hidden lines removed. These provide a significant savings in effort over manual methods of producing these types of drawings. Many CAD systems can provide very realistic renderings of product concept models, as shown in Fig. 11. These can include surface textures, shadows and reflections from multiple light sources, backgrounds, and other realistic effects.

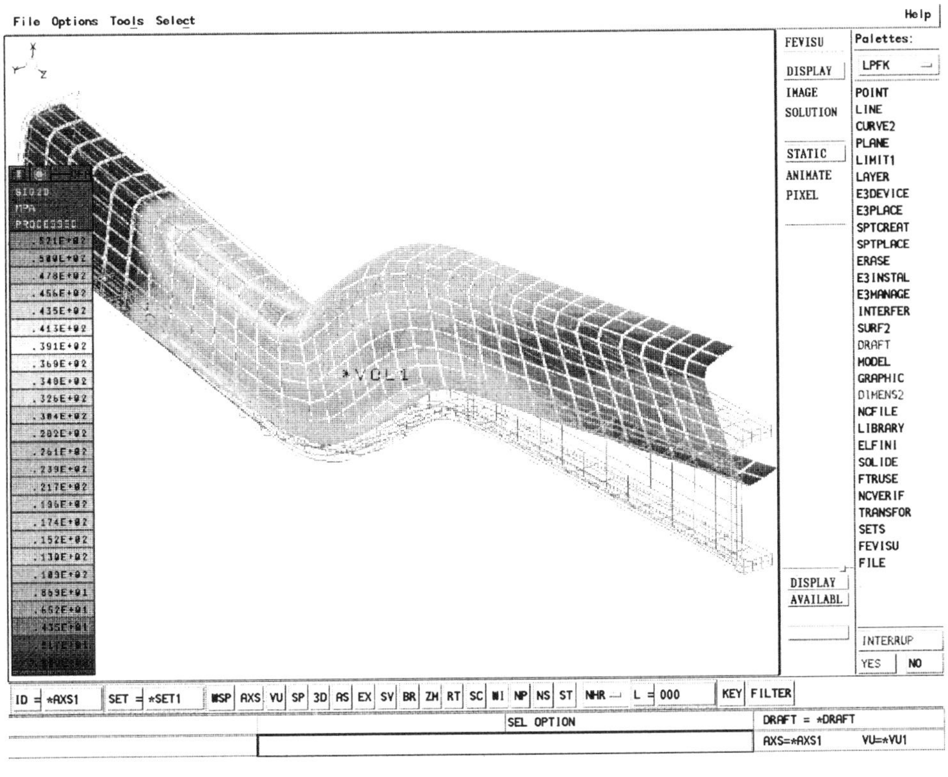

Fig. 12 Display of deflection and stress from FEA. Courtesy of Dassault Systèmes

Structural Analysis. Products that support structural analyses typically use some form of FEM/FEA. This technical area is complex enough, with a number of specialized products, that it would require a book dedicated to it to be adequately covered. This description is deliberately basic. Structural analysis is described in more detail in the article "Finite Element Analysis" in this Volume.

FEM/FEA systems require that a mesh of finite elements be created that covers the model to be analyzed. These elements are typically applied to the surfaces of the model or throughout the solid. Surface element types include triangles, quadrilaterals, and other two-dimensional shapes. Solid elements are tetrahedrons, bricks (or cuboids), or other simple three-dimensional shapes. The elements typically vary in size, with smaller elements used in areas where it is desired to have a refined analysis (such as areas where high stresses are expected, or in areas of fine detail such as around holes). Combined with the elements are loads (forces applied to the model) and constraints (restrictions on how the model can move; anchors, sliding or rotating joints, etc.). Output from the analysis may include plots of stress in the model, deformations, and other parameters (see Fig. 12). Computer-aided design systems may use third-party supplied solvers for FEA or their own, proprietary, built-in solver.

Solid models provide the basis for CAD systems to generate automatically complete surface and solid element meshes. Automatic solid meshing allows these systems to perform design optimization where the FEA analysis is linked to geometric constraint parameters on the model. These parameters can be changed by the system, based on the results of the analysis, thus changing the shape or size of the model. The mesh is regenerated and the analysis re-executed, and so forth, until some user-defined criteria is satisfied. This type of design optimization is impracticable without a solid model.

Mechanisms Analysis. Solid models of mechanisms can be analyzed for their behavior in both static and dynamic conditions. In static cases, the positions of the linkages and parts of the mechanical assembly can be measured and plotted as illustrated in Fig. 13. In dynamic cases, the velocities, accelerations, and other parameters of the actions of the mechanism can be found. In most systems the range of motion of the mechanism can be simulated dynamically on-screen. See the article "Mechanism Dynamics" in this Volume for more information.

Numerical control programming is the process of developing a set of instructions that automatically control a machine tool or other manufacturing device. All machine movements and operations can be controlled by the NC program. The machine tool system consists of the machine itself and the control system that operates it. In NC machining, material is removed

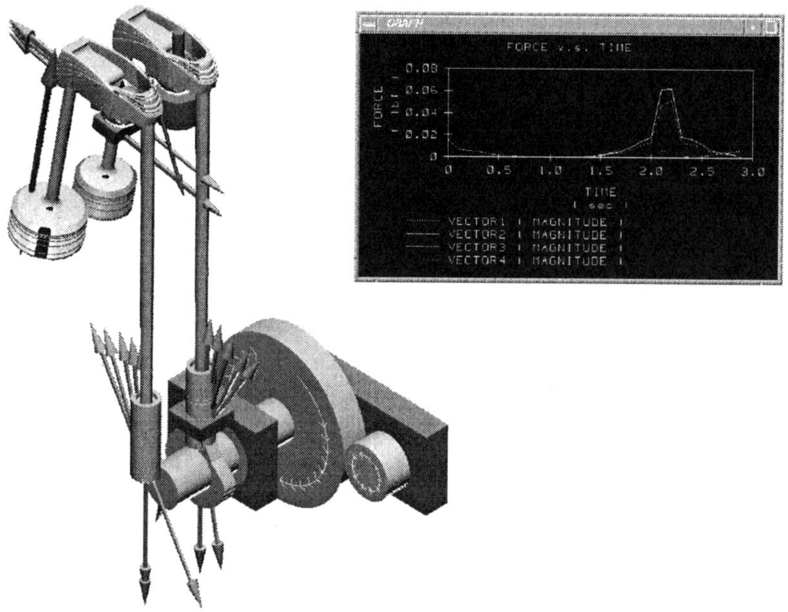

Fig. 13 Mechanism loads and constraints displayed with results graph. Courtesy of Aries Technology

from a piece of stock until the desired shape is achieved.

In computer numerical control (CNC), a dedicated computer is included in the device that controls the machine tool. Computer numerical control machines can be programmed right at the machine or off-line in an NC programming or CAD system. Distributed Numerical Control (DNC) systems receive their NC programs from a centralized computer by means of a network or direct connection. After it receives the NC program, a DNC machine can create parts on its own from the program stored in its local memory. A Direct Numerical Control (also DNC) system uses a central computer to operate directly one or more machine tools, without transferring the NC program to each machine.

Most solids-based CAD systems provide NC programming for a variety of machine tool types. These generally include two- through five-axis milling machines, lathes, punches, laser and other cutters, drills, wire electric discharge machines, and others. High-end systems may also support multifunction machines such as mill-turns and four-axis lathes.

Most NC programming has been done on surface models, applying toolpaths to one surface at a time and manually identifying holes and other nonmachined areas in surfaces. Solid models allow NC systems to produce toolpaths on entire parts with less human intervention, because the relationships among the surfaces of the part can be determined automatically.

Numerical-control programming systems are beginning to take advantage of manufacturing features as well as geometric features. Manufacturing features, while they are similar to form features used in solid modeling, include information that is specific to manufacturing operations (e.g., defining pockets instead of a set of bosses). They can also contain information that is used by the NC programming system to define operations, such as surface finish or type of operation to be used. An entire NC program fragment that machines the particular feature might be included in the feature.

In the NC programming environment, the user develops the machining scenario. The computer can then play the sequence of machining operations so that the user sees a visual simulation of the machining process (see Fig. 14).

Rapid prototyping systems allow a design to be created as a physical object or assembly that can be used for visualization and functional testing. Rapid prototyping systems create the model from plastic, powdered metals, or other materials in a process that builds up the object one layer or slice at a time. The prototyping machines are driven in much the same way as NC machines, using geometry extracted from the solid model. These prototype models can be held, examined for fit and function, used as patterns for molds, and used for other functions. In some cases, rapid prototyping machines are used for pilot and short-run product manufacturing. For more detailed information see the article "Rapid Prototyping" in this Volume.

Wire harness design and layout functions are provided by many CAD systems. These allow users to develop wiring schematics, design connectors, define where wire bundles split and merge, describe wire-routing parameters such as bend and sag, and indicate locations in the mechanical assembly through which the harness may or may not be allowed to pass. Given these types of information, the wire harness package routes the wires from connector to connector. When defined as solid models, additional functions can be performed—such as interference detection of the wire bundle with the other parts of the assembly, bundle volume, and weight.

Harness routings can be controlled by parametric constraints and they may be associative to other geometric data, so that as the shape of the mechanical assembly changes the wire harness changes accordingly. Finally, the wire harness can be laid out to form a manufacturing pin board diagram and for other assembly documentation.

Similar techniques are used for designing tubing, such as hydraulic systems.

Mold, tool, and die design problems can benefit from the application of some specialized functions. For mold design, material flow and cooling analysis, solid mold-base feature and fixture libraries (see Fig. 15), draft features, complex surface design, and materials libraries are available. These help designers analyze mold performance and simplify the design process by providing easy access to information specific to the plastic-molding process.

For tool and die work, libraries of fixtures and tooling are available. Analysis of drawing processes is only beginning to be made available within CAD systems. Analyses include material stretch at each stage of the drawing process, tear analysis, surface curvature analysis, and suggestions for creating progressive dies.

The books and articles listed as "Selected References" contain additional information about solid modeling and CAD/CAM. Many of the books and articles contain further references for additional study.

SELECTED REFERENCES

- A.M. Christman, *NC Software Buyer's Guide*, 4th ed., CIMdata, Ann Arbor, MI, 1996. (This report describes NC programming technology and reviews a number of NC products.)
- J. Encarnação, J., Ed., *Computers and Graphics*, published four times per year. (This journal is a good source for technical information on CAD/CAM and computer graphics topics.)
- I.D. Faux, and M.J. Pratt, *Computational Geometry for Design and Manufacture*, John Wiley, 1979. (This book describes various mathematical concepts for describing geometric models of parts.)
- J.D. Foley, A. Van Dam, S.K. Feiner, and J.F. Hughes, *Computer Graphics, Principles and Practice*, 2nd ed., Addison-Wesley, 1990. (This book is a very comprehensive treatment of computer graphics and user interface design.)
- C.M. Hoffman, *Geometric and Solid Modeling*, Morgan Kaufmann Publishers, Inc., 1989.

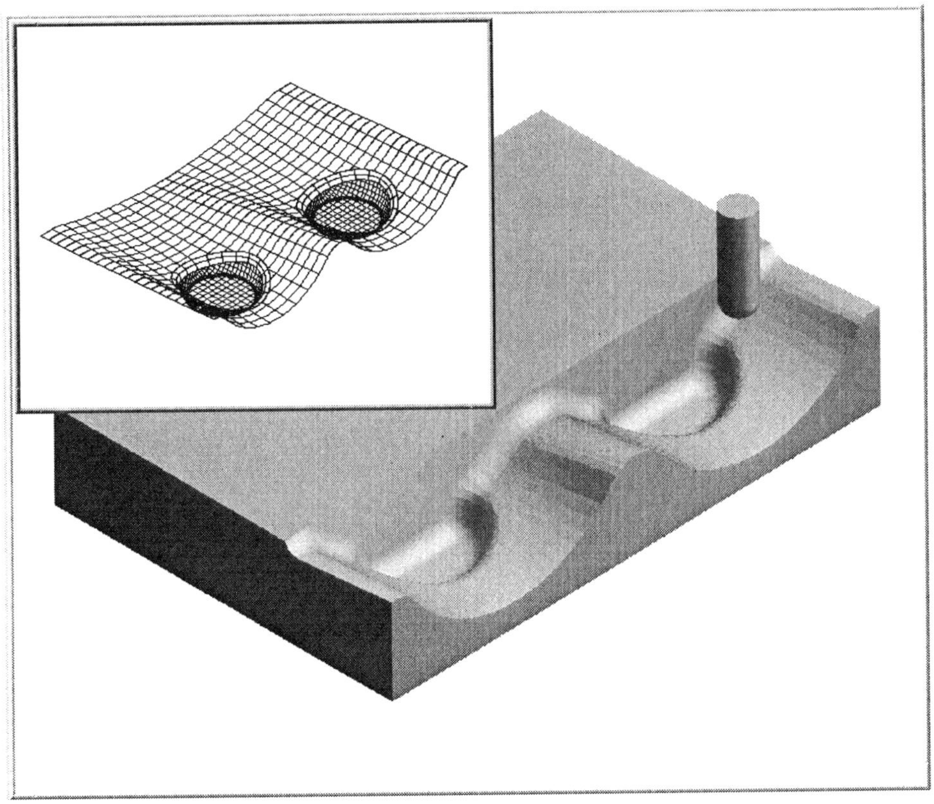

Fig. 14 Numerical-control tool cutting simulation. Courtesy of DP Technology

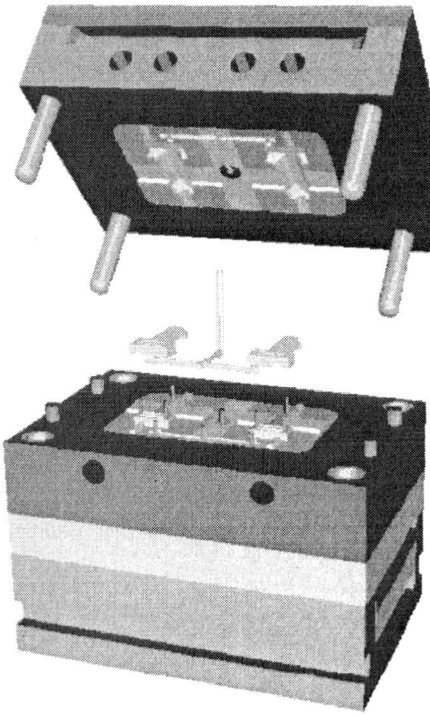

Fig. 15 Mold assembly design. Courtesy of SDRC

(This book is a general overview of solid modeling and its application.)

- W. B. Holtz, *The CAD Rating Guide*, 4th ed., ZEM Press, 1994. (*The CAD Rating Guide* provides information about a large number of CAD/CAM and related software products, ranging from PC-based to high-end CAD systems. It is a good source book, with a several-page section about each product.)
- D. LaCourse, Ed. , *Handbook of Solid Modeling*, McGraw-Hill, 1995. (The essays in this book cover many aspects of CAD/CAM, engineering analysis, and computer graphics.)
- C. Machover, Ed., *The CAD/CAM Handbook*, McGraw-Hill, 1996. (The group of essays in this book cover many aspects of CAD/CAM, engineering analysis, and computer graphics.)
- J.R. MacKrell, *M-CAD Buyer's Guide*, 2nd ed., CIMdata, Ann Arbor, MI, 1994. (This report defines a standard architecture for mechanical CAD/CAM systems and reviews and compares major solids-based CAD products.)
- E.D. Miller, J.R. MacKrell, and A. Mendel, *PDM Buyer's Guide*, 6th ed., CIMdata, Ann Arbor, MI, 1997. (This report describes PDM technology and reviews important PDM products.)
- R. Mills, Solid Modeling Software, *Comput.-Aided Eng.*, Vol 10 (No. 5), May 1991, p 36–58. (This magazine article briefly describes many solid modeling systems.)
- M.S. Pickett, and J.W. Boyse, *Solid Modeling by Computers*, Plenum Press, 1984. (This book is a collection of articles about solid modeling technology that were presented at a symposium sponsored by General Motors.)
- L. Piegl, Ed., *Fundamental Developments of Computer-Aided Geometric Modeling*, Academic Press, 1993. (This book is a compendium of interesting essays by some of the most important early developers of solid modeling and geometric modeling systems.)
- J. R. Rossignac, Issues on Feature-Based Editing and Interrogation of Solid Models, *Comput. Graphics*, Vol 14 (No. 2), 1990, p 149–172. (This article contains a technical description of feature modeling.)
- I. Zeid, *CAD/CAM Theory and Practice*, McGraw-Hill, 1991. (This textbook provides an extensive review of CAD/CAM principles and tools.)

Mechanism Dynamics and Simulation

James E. Crosheck, CADSI

MECHANISM SIMULATION is a subject of such breadth that it is the topic of numerous books and journals and the focus of hundreds of learned papers. The technology is evolving at an ever-increasing rate due to advances in computer technology and numerical analysis techniques. As an overview of this subject as it impacts the rapidly changing design environment, this article offers a series of glimpses into some of the areas of application, providing a direction that the reader may pursue in depth on their own or with the aid of software vendors, consultants, academics, and so forth. A very limited set of references at the end of this article, primarily standard texts, provide the depth that some readers may find valuable. For most readers, the details of the underlying equations are not the immediate concern—rather the questions are related to taking advantage of the technology in an appropriate, cost-effective, and timely manner. Once the direction for solution of the engineering problem is clear, the finer points of the theory and its application can and should be reviewed further to ensure that the solution process properly matches the problem.

Two complementary reasons are usually the impetus for simulation of mechanical systems. First is performance prediction: will the design meet the specifications over the range of conditions in which it is expected to operate? Second is the detailed information that can be extracted from a simulation on component loads, displacements, and accelerations. This information provides the best estimate of the environment of the component, an estimate that is needed for detailed strength analysis using finite element or other techniques.

This article presents an overview of the use of mechanism analysis (kinematics and dynamics) and simulation. It provides indications of the directions in which mechanism simulation is growing and how it is integrated in the evolving computer aided design and computer aided engineering (CAD/CAE) fields. Mechanism simulation is best used as part of a concurrent CAD/CAE approach to design. Current practice seldom utilizes simulation to develop the best design. This article discusses the current state, evolution of, and direction of application for these techniques in a variety of fields. It is presented in the context of material selection and, in turn, component sizing, which is dependent on material selection. The examples presented are not exhaustive; rather, they are used to provide indications of the direction in which simulation is growing. They are intended to illustrate current trends and to stimulate thoughts of opportunities for application by the reader.

Definitions and Basic Concepts

According to Webster's Ninth Collegiate Dictionary, simulation is the "imitative representation of the functioning of one system or process by means of the functioning of another." In current technology, designers typically look to simulation for the solution of questions through computer models that use computational techniques or CAE. The range of techniques varies from queuing theory models used to predict flow of materials through a manufacturing plant to finite element models to predict stresses, heat transfer, or temperature distribution. Of particular interest in this article, and of growing interest in general, is the simulation of the motion and performance of mechanical systems.

This article explores how simulation of nonlinear mechanical systems, which emulates physical systems, is accomplished through the use of representative computer models incorporating a number of interdisciplinary engineering tools. A distinction is made here between linear and nonlinear solutions. Linear equations of motion can be solved using frequency domain analyses. This involves Fourier or Laplace transform techniques that are usually computationally very efficient. Unfortunately, most systems exhibit nonlinear behavior due to their large range of motion, nonlinear stiffnesses, varying mass properties, nonlinear damping, intermittent contact, or excitation from nonlinear forces. Solution techniques for nonlinear systems are demonstrated with examples solved with one of the software packages used in the current engineering environment for the solution of design problems. At the time of this writing, the majority of solutions of mechanism problems are done with one of four commercial systems: DADS from CADSI, Adams from MDI, Applied Motion from PTC, and Working Model from Knowledge Revolution. The examples presented were prepared with a single program, DADS, due to the author's familiarity with it. Similar capability is available from other vendors, and they should be consulted for a description of current features. (The author would highly recommend using a commercial code for simulation of any nontrivial mechanism problem. As recently as the last decade, many engineering analysts wrote their own computer programs to solve specific problems. As the capability of commercial systems has evolved, so has the tendency to write and maintain isolated, specialized programs become less cost effective.)

Before continuing, it is important to provide some definitions for the types of analyses that are commonly performed as part of the general category, mechanical system simulation. A mechanical system is composed of a set of parts or bodies that are connected by joints to form an assembly. The system is forced to move or allowed to move by some form of control system and/or by external forces. For the majority of the problems in mechanical system dynamics, the bodies can be regarded as rigid. However, there are an increasing number of cases in which the flexibility of the part is important in the overall performance of the system. For these cases, some codes will allow the use of information such as mode shapes and frequencies from finite element analyses to calculate small, structural displacements that are superimposed on the large displacements of components of the system. Examples of this are discussed below. Bodies are normally connected by joints, but in many cases the bodies are constrained by elastic elements such as bushings or are confined by intermittent contact with other parts of the system. Again, examples of these real-world situations are briefly described below.

Another term that is useful in the following discussion is "degrees of freedom" (DOF). This quantity is the number of independent directions that a body or system of bodies is free to move. A single body floating in space has six degrees of freedom, three translational and three rotational. If the body is restrained to ground, all six of the DOFs are removed. Suppose two bodies are connected by a rigid hinge joint. The original two bodies had twelve DOFs, but are now reduced to seven. The hinge joint only allows motion in one direction, a rotation of one body relative to another. If one of the bodies is now attached to ground, the complete system only has one DOF left. If the rotation of the hinge is now enforced as a function of time, the system has zero DOF—all independent motion is now defined or restrained. These restraints are algebraic relationships between the original independent DOFs. Note that it is mathematically possible to overconstrain a body. For example, placing two hinge joints between two bodies applies ten algebraic relationships between the motion of the two bodies, but six is the most physically possible. Physically and intuitively, two hinges that are understood to be at arbitrary orientations between the two bodies will lock them together, removing six DOF. Further, if the hinges are aligned as on a door, the system still allows rotation of one body relative to the other. This aspect of the idealization of a system is not followed further. However, it may shed light on some of the art left in simulating a mechanical system (or on some of the skill required in developing computer code to interpret the connections between bodies).

The categories of analyses that are used for mechanical systems include:

- *Kinematics:* Solves an algebraic system of equations with zero DOFs to determine the position, velocity, and acceleration of all bodies at a sequence of user-specified time steps. This means that all motion is prescribed as a function of time. This is often used as a means of checking range of motion and also to verify that the design will allow motion/velocity/acceleration to be in the expected range.
- *Inverse Dynamic:* Solves the same set of equations as a kinematic analysis, but uses the calculated accelerations to find the joint reaction forces. These calculated forces depend on both the external forces generated by the time-dependent, imposed displacements and also the inertial forces.
- *Static Analysis:* Solves a system with a positive number of DOFs to find the position that minimizes all body accelerations subject to external forces and gravity. The resulting position is the equilibrium position for the system. This solution is algebraic, resulting in balanced internal and external forces in the absence of system motion or inertial forces.
- *Quasi-Static:* Solves a sequence of static analyses for different configurations of the system.
- *Assembly:* Solves a set of nonlinear equations to find the position that minimizes the constraint error for all constraints in the model. These constraints are primarily due to alignment of joints between the bodies or placing bodies with enforced displacements. This is sometimes used as a prelude to the actual solution if the system is defined "manually," one part at a time, allowing small errors in locating the bodies. This solution is done or at least checked prior to the start of all analyses.
- *Dynamic:* Solves a system with a positive number of DOFs and numerically integrates the equations to determine the position, velocity, and accelerations of all bodies plus the joint reaction forces.
- *Linearization:* Not really a separate solution procedure, this technique provides information that can be used to communicate with other programs or that can be used for diagnosis of modeling problems. Eigenvalues and the corresponding eigenvectors of the linearized model can be a means of detecting inconsistent data for mass, stiffnesses, or incorrect constraints between parts. Linearized equations for the mechanical (or complete) system allow stability to be easily assessed. A linearization of the model about the current operating condition is also a convenient technique to pass information to control system synthesis and other simulation programs.

Theoretical Background. The theoretical foundations for mechanism kinematics and dynamics can be found in numerous texts and papers. For those interested in a rigorous understanding of the field, a good starting point is texts by Haug, Kane, Greenwood, and Erdman (Ref 1–6).

The basic equations are not complex: a set of second-order differential equations (DEs) that define the force balance on each component of a system plus a set of constraint equations that tie the components together with joints or more general relationships. The resulting equations are normally described as differential-algebraic equations (DAEs). The equations, or one of the variations, can be found in the references listed and are not the primary concern of this article.

However, it should be noted that there are alternative formulations in both academic and commercial software. The simplest formulation uses a Cartesian coordinate system for all forces and displacements. This simplicity also results in a penalty in the form of a larger number of equations in the solution set. Alternatively, relative coordinate formulations create complex relationships in the solution set in exchange for a significant reduction in the number of, or elimination of, constraint equations. Each method has both merits and limitations. For commercial software, the preferences are usually more critical to the software developers than they are to the users of the software. Available commercial codes solve the DAEs or DEs for a wide range of problems with adequate accuracy and in a reasonable amount of computer time and resources. As computer speed and storage increase, the solutions are approaching real time. Potential users of a code should test it on sample problems of their own design and evaluate the tools in conjunction with its own capabilities and its ability to be a part of their product development cycle.

Later sections discuss some further considerations in application of commercial codes, but this is an appropriate point to highlight a technical issue that concerns many application-oriented users. The issue is speed versus accuracy (or better, speed *and* accuracy). Solution of the DAEs requires integration of a set of DEs that are often "stiff" due to physical conditions such as intermittent contact of bodies, stiff springs, compressibility of hydraulic fluid, and so forth. (The definition of "stiff" is not precise, but usually means that the system contains high frequencies and widely spaced eigenvalues.) The accuracy of the solutions varies significantly from code to code. Some implementations feature solutions that track high-frequency components of the solutions at the expense of increased computation time. Others allow the user to tune the integrator to achieve the level of accuracy desired. These options can provide faster solutions if the high-frequency content is not of interest. The critical aspect of this trade-off is recognition by the user that a trade-off is being offered—often this is not clear in the supporting literature. Users of these commercial codes should test the software on problems of the types they expect to routinely solve. Ideally, results should be validated against test data. Unfortunately, time constraints limit the amount of user testing that can be done, resulting in the need for a certain amount of faith in claims of vendors and their demonstrations.

Performance and Function

Some disciplines pose high demands on performance that are difficult to predict without either prototype testing or detailed mechanism dynamic analysis. One example would be a military vehicle that is expected to handle a wide range of loads over a variety of terrain. Maneuvers such as a slalom run on a side slope, obstacle avoidance at highway speeds with full cargo loading, and "high-speed" runs over cross-country terrains need to be reviewed and addressed for the suspension design before the initial prototype is built and tested. In such cases, the loads analysis is secondary to the need to meet stringent performance specifications. The loads are of vital importance for durability and reliability reasons, but only after there is assurance that the concept will meet the performance requirements.

Another example is the antenna deployment of spacecraft. It is difficult to test the lightweight, flexible structures in an earthbound environment. Mechanism analysis, particularly dynamics, is very valuable in checking the functioning of the system before expensive "testing" in actual operation in space. Large numbers of low-frequency modes are often prevalent in these devices (for example, hundreds of modes are predicted to be less than 2 Hz in the Space Station) and the flexibility is often critical to proper performance of the system. Unfortunately, the opposite is also true, that component flexibility may be the source of unexpected behavior that causes problems. Sometimes these problems can be observed with the dynamic simulation prior to actual deployment and corrected prior to deployment. Unfortunately, there will also be a portion of the cases that are not found before actual use of the system. These lead to two observations for users of the technology. First, as many cases as possible should be analyzed before the design is fixed. This increases the possibility of finding as many unexpected phenomena as possible. Second, simulation is still partially an art. If a model is constructed that cannot demonstrate the "problem" behavior, it is a failure of the modeler, not of the technology. For example, if the lockup of a joint is dependent on an over-center latching technique, and it, in turn, only works if a part flexes to allow the latching to occur; a rigid representation of the part is not acceptable. In general, modeling "errors" can usually be traced to assumptions made about the behavior of the system, or parts of the system. These incorrect assumptions are often made to simplify the model, reduce the amount of input data needed, or speed the solution of the simulation. Hindsight is a great teacher in such situations, and each such experience results in a better analyst for the next case!

Example 1: Mechanism Dynamics and Simulation in the Design of Bowling Balls. Rather than delve into a detailed example early in the article, it is appropriate to examine a "simple" problem. Bowling is a deceptively simple game. Use a ball that weighs 16 lb or less and drill some holes in it so it is easy to handle a ball of that weight. Stand a few feet behind a foul line, take three or four steps forward while swinging the ball, release the ball and let it slide or roll down the lane until it hits the set of ten pins 60 ft away. The objective is to knock all ten down with one or two throws. Soon the novice bowler hears advice to throw a "hook" (a curved trajectory) to get more strikes (all ten pins knocked down on the first throw). They also begin to hear debate over the merits of different balls and ways to drill the balls.

Ebonite International has used computer simulation to develop bowling balls that provide a significant improvement in performance over existing balls. Simulation allows Ebonite engineers to design a new ball on the computer and test its performance under controlled conditions in under 30 min. This compares to weeks of time and thousands of dollars previously required to build and test a prototype. Designed using simulation, Ebonite's recently introduced Wolf ball offers a higher level of performance under a broader range of lane conditions.

Ebonite produces bowling balls using the open-casting method. The new Wolf ball has a core made of tungsten graphite, which with a density of 6.0 g/cm^3 is the heaviest material ever used in a bowling ball. Rather than being spherical, the core of the ball is pill-shaped, making it possible to vary the location and degree of the ball's hook so it can be adjusted to a bowler's style and varying lane conditions. In general, bowlers want more hook because it increases the angle at which the ball contacts the pins, which creates a bigger "pocket" within which a strike is most likely.

Although a bowling ball seems to be a simple object, the dynamics of a bowling ball as it travels down the lane are actually quite complex. The ball undergoes gyroscopic influences caused by the core design of the ball and interacts with a lane surface that has varying friction, due to varying levels of oil. The gyroscopic motion also affects the friction by influencing the amount of oil that builds up on the ball. Finally, when the ball reaches the pins, an enormously complex, intermittent contact problem is created, with the ball contacting the pins, the pins contacting other pins, and the pins contacting the lane and gutters.

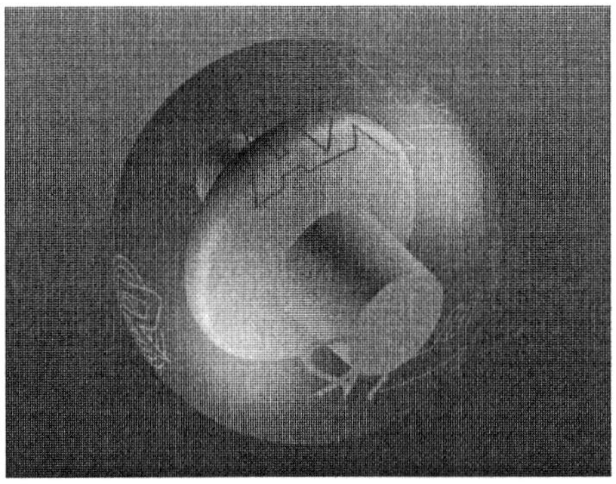

Fig. 1 Modern bowling ball construction with nonspherical, asymmetric cores

The design of bowling balls has gone through a dramatic series of changes. Previously, bowling balls were uniform, or had a spherical core located in the center of the ball. Because the ball had a uniform inertia, it made no difference (from a rotational dynamics standpoint) how the ball was drilled.

The early 1980s saw the introduction of nonspherical, symmetric cores shaped in patterns such as a light bulb or a dumbbell. This type of ball has two principal inertias, one along the axis of the core, and another perpendicular to it. If the axis of the core coincides with the initial rotation of the ball, it will behave in a similar manner to earlier balls, as it will if the initial rotation is in the plane perpendicular to the core axis. However, if the initial rotation of the ball is about any other axis, interesting things start to happen.

First, the rotation axis precesses about the core axis; that is, the motion of the rotation axis describes a cone about the core axis. Generally, increasing the differential between the least moment of inertia axis and the axis of rotation, and judiciously positioning the rotation axis, increases track flare. This in turn increases the amount of friction between the ball and the lane, which allows for a greater hook. The larger hook increases the angle at which the ball hits the pins, increasing the probability of a strike. As a result, nonspherical cores have substantially increased the scores of bowlers that have learned to use them effectively. Nonspherical cores have been adopted by all bowling ball manufacturers and represent the current state-of-the-art. An additional advancement was the introduction of the asymmetric core.

The asymmetric cores used in the most recent bowling balls (Fig. 1) typically are shaped like pills or spheres with opposing sides flattened. The performance of a ball with a conventional symmetric core can be varied only by changing the axis of rotation of the ball in relation to the core axis, while the asymmetric core has three distinct principal inertia axes in relation to which the rotation axis can be positioned. An asymmetric core therefore provides the ball driller with a wide range of track flare potential, which can be used to tailor the ball's performance to a given bowler.

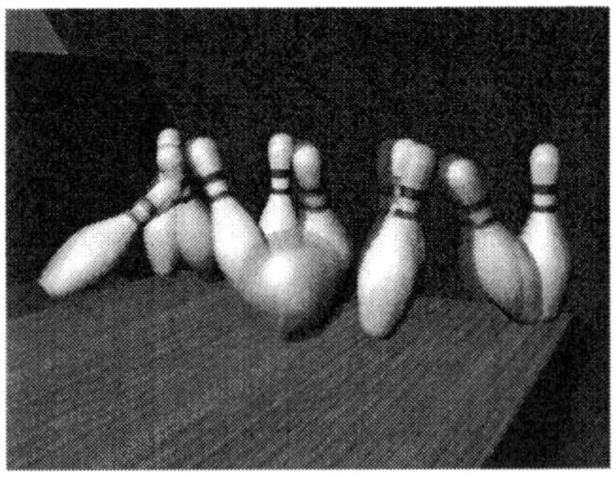

Fig. 2 Accurate simulation of ball performance as it contacts the pins

This new design concept provides a great deal of additional freedom for ball designers, with many more potential ball configurations. However, it also adds difficulty in evaluating all of the possible designs. First of all, there is a much wider range of possibilities for the core shapes. Second, once the core shape is defined, it needs to be drilled in such a way that bowlers can get the full range of performance they desire. Finally, it is up to the manufacturer to precisely determine the performance, in particular the amount of hook, that will be produced by each drilling pattern.

The traditional approach is to build a prototype and test it on the lane; a time-consuming and expensive process. First, it is necessary to build molds for both the core and shell. Then, it has to go through an extensive testing process to objectively determine its performance in relation to other balls. For the new asymmetric core balls, this would have been an overwhelming process. Thus, engineers decided to investigate alternatives using computer simulation.

Engineers begin the simulation process with a Pro/ENGINEER (Pro/E) CAD model of the ball, including its various shells and cores. The geometry and the mass properties (mass and inertias) are extracted from Pro/E and passed to DADS. The lane conditions are described, including oiling levels along and across the lane, and this is converted into a variable friction over the surface of the lane. A series of utilities were created to allow Ebonite to conveniently input ball initial conditions and grip positioning.

After the model is created, the analysis program simulates the movement of the ball down the lane (Fig. 2). Output includes the critical parameters of ball design such as: the reaction length or amount of distance before the ball begins to hook, the backend or sharpness of the hook, and the overall hook radius or magnitude of the hook. The engineers can watch an animation of the throw and can add representations of the track flare, rotation axis motion, and an "inertia map" to enhance understanding of the dynamics. Other utilities can automate changes to the model, allowing the engineers to try dozens of grip drillings, a range of mass properties, or changes in the throw. Tests have shown that the simulation correlates very well with physical testing.

By repetitively performing simulations, engineers were able to hone the performance of the ball to a high level. The documentation that is provided with Ebonite's Wolf ball shows the precise hooking characteristics that can be created by a wide range of drilling positions. By selecting the proper position, bowlers can produce virtually any type of ball performance they desire. Despite the unprecedented number of alternate scenarios that were evaluated, the ball was developed in record time.

By using computer simulation to evaluate alternate design concepts, the design cycle has been compressed to a dramatic degree. The speed and convenience of simulation make it possible to consider many times the number of alternatives that the company was able to evaluate in the past using the build-and-test approach. As a result, the company has been able to make a quantum leap in bowling ball performance while actually reducing the number of engineering hours required.

Load Prediction

Perhaps the longest-running need in engineering has been reasonable estimates of loads to use in sizing components. Not long ago, it was common to put a few strategic accelerometers and strain gages on the mechanism, operate it in a range of conditions including its more severe ones, and then predict the loads on the connection points of components, usually through manual calculations. This process requires that a prototype or a similar mechanism exist for testing. In many cases, test data were not available for a variety of reasons and the loads were based on the equivalent of opinion polls (educated estimates of a group of engineers and technicians). The resulting loads from either incomplete testing or from poor estimating often were seen again later in increased warranty claims. Modern mechanism software allows a more accurate estimate of the loads within a system. This in turn is one of the primary ingredients to successful use of finite element analysis of the components (see the article "Finite Element Analysis" in this Volume).

One area in which mechanism analysis has been used successfully for several years is the suspensions of automobiles and trucks. A typical suspension model (Fig. 3) includes several interconnected components and supporting bushings. The steering input to the front wheels would be run through its range of motion, as would the wheel be run through its vertical range of motion from jounce to rebound (maximum vertical motion upward and downward). Throughout this exercise, characteristics of the suspension such as the caster, camber, and toe-in would be monitored. Forces at individual connections would be plotted. Side forces would be added to the tires to represent the turning forces and the process repeated. Comparison of simulation results with previous designs, successful or not, provides guidance on the quality of the new design concept.

Note that the above process does not really use dynamics in the analysis. Kinematics or quasi-static analysis has been and continues to be an effective tool for suspension design. However, as cars become lighter and more flexible, the chassis becomes more critical in the response of the suspension. Add to that the trend to active or semiactive suspension components and the need for true dynamic analysis of the vehicle becomes important.

The flexibility of the components of a suspension plus the body are also very important to accurate prediction of behavior. Our experience has shown that the flexibility of a truck chassis can reduce peak loads on suspension components by 30 to 50%. If weight and cost are not a concern (not a typical commercial situation), simulation of a suspension that assumes rigid components merely leads to a conservative design. However, the combination of regulatory and economic forces urge designers to be as accurate as possible in their predictions of vehicle behavior. In turn, this pushes designers to incorporate and quantify as many simulation effects, such as flexibility, as they can.

More accurate design loads can be used to reduce cost and improve payload without risking higher failure rates. This implies a need to use dynamic analysis for the fine tuning of the analysis results before the final design decisions are made. The loads can and should be used as input to stress analyses to ensure that the choice of materials for the parts of the suspension are acceptable. Note that the interaction of the other parts of the engineering activity suddenly become more obvious. The properties of a particular material or particular alloy must be balanced against its cost and ease of use in manufacturing. If a change is contemplated, this may imply a need to re-evaluate the mechanism performance due to changes in mass properties of the part. This in turn may require another iteration in the stress analysis. However, loads are only part of the need for prediction of the behavior of a mechanism.

Example 2: Simulation to Predict Loads in Aircraft Landing Gear. The first certification of an American business jet in Russia was accomplished using a mechanism simulation approach that saved three to six months and $100,000. The primary issue in the Russian certification of the Gulfstream IV-SP was the ability of the landing gear of the jet to withstand the rough runways that are common in that country. Conventional methods of validating this capability, developing custom dynamics software or physical testing, would have taken at least six to twelve months. Instead, it took only three months to simulate takeoffs, landings, and taxiing on rough runways using an off-the-shelf program that predicts mechanical performance.

Gulfstream Aerospace is a designer, developer, manufacturer, and marketer of technologically advanced intercontinental business jet aircraft. In 1966, the company created the large cabin, business jet category with the introduction of the Gulfstream II. The Gulfstream IV-SP entered service in 1986 and is still in production, with 305 manufactured to date.

The Gulfstream IV-SP landing gear essentially consists of an oleopneumatic strut. Two chambers, one filled with oil and the other with air, are connected by a bulkhead with a small orifice. The gear is arranged so that when it contacts the ground, oil is forced through the hole into the chamber where it compresses the air. The resis-

Fig. 3 Simulation of suspension systems allowing load predictions to be made

tance of the oil generates a damping force that is proportional to the square of the velocity of the strut. The compression of the air by the oil generates a force that helps to keep the gear extended while the plane is on the ground. The oil plays the same role as an automobile shock absorber while the air is comparable to the springs.

Computer simulation of a landing gear is challenging because of the nonlinearities involved. Geometric nonlinearities arise when the geometry of the landing gear changes over time and changes the very nature of the problem. The translation (extension) of the struts dramatically changes the landing-gear configuration. For example, the swing arm goes through 60° of travel during each landing and takeoff.

Gulfstream worked with CADSI to analyze the main gear and nose gear of the aircraft in landing and taxi simulations (Fig. 4 and 5). The DADS software package allows engineers to develop mechanisms without simplifying assumptions that reduce the accuracy of specialized landing gear code. This software package is also able to incorporate flexibility of components required for accurate simulation of landing-gear struts, and closed-and-open loop control systems needed to simulate the performance of the oleopneumatic strut.

The initial main landing gear model was constructed to simulate a drop test and consisted of eight bodies representing the ground, aircraft, post, piston, cylinder, arm, inboard wheel, and outboard wheel. Eight joints connect the bodies together: bracket (aircraft to post), revolute (post to arm), revolute (arm to inboard wheel), revolute (arm to outboard wheel), spherical (arm to piston), translational (ground to aircraft), translational (cylinder to piston) and universal (post to cylinder). This combination of bodies and restraints leaves each landing gear system with four relative degrees of freedom. One is the vertical translation of the aircraft, another is the motion of the arm and piston relative to the post and cylinder, and the other two are the rotation of the wheels.

The oleopneumatic characteristics of the strut were modeled using DADS control elements. The control system measures the magnitude and velocity of the strut displacement, calculates oil velocity through the orifice and pressure drop across the orifice, and finally determines the force due to the oil flow. The ideal gas law with heat flow involved is used to determine the force due to the air pressure. The total strut force is the summation of the force due to oil flow, the force due to air pressure, and a constant force due to seal friction.

This model was validated by simulating a drop test, the primary method for testing landing gear. A physical drop test involves mounting the gear on a slide attached to a vertical rail. A bucket full of lead bricks is attached to the gear to simulate the weight of the aircraft. The landing gear is instrumented with pressure transducers and accelerometers and dropped to an instrumented steel platform. The simulation results were plotted against physical measurements generated during actual drop tests and found to match well. The next step was enhancing the initial model to include flexibility of key components. The basis of a flexible body in DADS is a mesh of grid points and a set of mode shapes generated by a finite element model of the part in question. MSC/Nastran was used in this case because finite element models had already been prepared for the required components at Gulfstream. The flexibility of the main structural post and trailing arm in the main landing gear were represented with modal coordinates. Each modal coordinate value represents the contribution of that particular mode shape to the overall deformation of the part. Drop-test simulations with flexible bodies added to the model showed near-perfect correlation with physical measurements.

The next step was to add a rigid airframe, aerodynamic equations, variable thrust, and braking. The resulting model was capable of handling realistic landing, takeoff, and taxiing scenarios. The aerodynamic equations calculate lift, drag, and lateral forces, as well as roll, pitch, and yaw moments. These are based on angle of attack, sideslip angle, freestream velocity, dynamic pressure, wing area, mean wing chord, and wing span, as well as many other coefficients that are related to the physical characteristics that determine the aerodynamic characteristics of the aircraft. Logic was added to the model to account for the actions of the pilot such as pulling back on the stick to lift the plane off the runway.

Control elements define the thrust, split it into local X and Z components, and apply the forces to the aircraft. Brake torque is required to simulate a rejected takeoff condition. Brake torque is also applied through control elements. For the rejected takeoff condition, a maximum brake torque is available. The torque is ramped in from zero to the full value over a 2 s time interval. The final modification to the model incorporated a finite element model of the complete airframe. The model is composed primarily of bar elements in a "stick figure" layout along the principle members of the airframe. Also included are the flight control elements with combinations of rigid elements and springs to provide correct stiffness in the model. Concentrated masses are used to define mass and inertia values.

When the Russian certification effort began, the simulation model had already been developed and validated for other purposes. Gulfstream engineers obtained roughness survey data for representative Russian airports from a manufacturer of commercial airliners that had previously gone through the certification process. These survey data were reduced to power spectral density functions that described the runway as a function of roughness frequency. The takeoff, landing, and taxiing simulations were rerun using these data as input to the model.

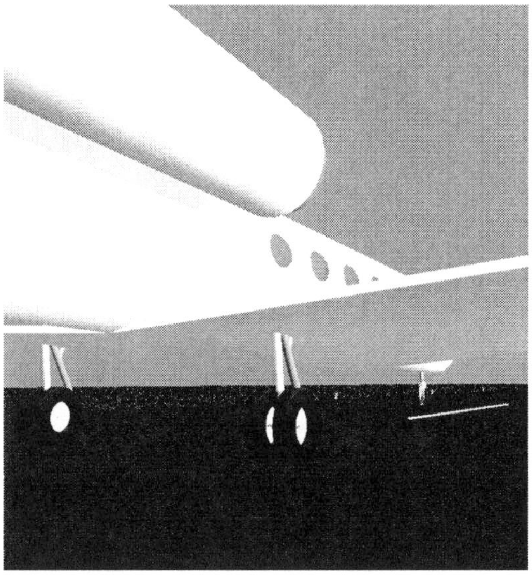

Fig. 4 Taxi simulations that predict loads from the landing gear into the aircraft

Fig. 5 Landing gear mechanism geometry changes significantly during taxi or landing

Gulfstream engineers closely examined the results to determine the response of the structure and the loading on critical components. Both maximum design load conditions and repeated load conditions were investigated. The most sensitive area in the landing gear with respect to runway roughness had previously been shown to be the trunnion where the landing gear attaches to the airframe. The nose landing gear trunnion is the most critical because the nose landing gear has the least stroke and is the most prone to bottoming out. When this occurs, serious damage to the landing gear can result.

The simulation results showed that the aircraft could handle all but the roughest Russian runways without any ill effects. The Russian regulatory authorities were extremely impressed with the analysis results. In particular, the animated simulations provided by the software made it easy for government employees without a strong technical background to easily comprehend effects of landing on runways of various roughness. The certification process proceeded more quickly than expected, allowing Gulfstream to be the first American business jet producer to penetrate the rapidly growing Russian market.

Direct, Concurrent Interfacing of Multiple-Simulation Programs

Most mechanisms are an integral part of a system. Rarely are there simple mechanisms that can be treated in isolation from the controls or the driving forces on the system. In some cases, it is a reasonable approximation of reality to move the mechanism in isolation from the rest of the system. More often, both the large range of motion of a mechanism (which introduces nonlinearity) and the details of its control system must be studied at the same time. For example, the use of kinematics and quasi-static analysis of suspension systems to successfully design their components and review their anticipated behavior has been discussed. However, as designs embrace active components and as the flexibility of the chassis becomes more important in performance (handling and noise creation or transmission), dynamics must become a standard part of the design process.

Software is a transient entity. Codes are becoming more modular, more objectlike, allowing easier coupling and interchange of information. Some mechanism programs now can be linked to traditional control system analysis packages to allow for more complete system simulation. To be accurate, the software linkage must be direct. Intermediate files, which are sometimes used to pass data between "cooperating" programs, indicate that each program is really operating independently of the other and that variables in the two cannot be integrated simultaneously (and accurately). A proper linkage of mechanisms and controls requires integration of the complete set of equations, the DAEs of the mechanism and the state equations in the control system simultaneously. In that environment, the control program can be used in the mode for which it is most powerful, as a control system analysis and synthesis program. Similarly, the mechanism program can be used to determine loads, accelerations, and so forth that are the forte of those codes.

Developing two separate models, one from the controls viewpoint and one from the mechanism side, is one approach that is often used. Normally, depending on the interest and concern of the engineer involved, a detailed model is developed from one vantage point and a cursory model is developed from the other viewpoint. This is simply a reflection of individual interests and expertise, not an indication of deliberate neglect. This also obviously does not allow a robust evaluation from both perspectives, and certainly not of the interaction of the control and mechanical system at all. The best technique is to use two separate programs that complement each other to solve such problems, that is, programs that "co-simulate" on the problem at hand.

Two simulation programs that complement each other as described are DADS and EASY5, by Boeing Computer Services. The details of the mechanism are modeled in DADS and the control system in EASY5 (this interface between the two codes is commercially referred to as DADS/Plant.) To force the concurrency needed, DADS is treated as a module within EASY5. (In control system terminology, the portion of the system that is being controlled is the plant. In this combined system, the mechanical portion of the system is the "plant." Hence, the name, DADS/Plant.) In this mode of operation, EASY5 integrates the equations of motion for the mechanical system as well as the state equations for the control system. DADS determines the velocity, and acceleration information that is needed in EASY5. EASY5 integrates to determine position information plus it predicts forces to be used in DADS. Figure 6 illustrates this interaction.

Note that this is a true direct integration of the programs rather than a process of exporting information from one program to another for use in a separate analysis. Similar techniques are possible with other popular control system analysis programs such as MATLAB/SIMULINK and MATRIX$_x$.

Example 3: Integration of Two Analysis Programs for Simulation of Backhoe Operations. Most backhoes have an open-loop control system actuated by a hydraulic drive system. On the machine, the loop is closed by the operator. In this example, instead of relying on the operator to close the loop, an automatic control system modeled in SIMULINK, is used to control the system (Fig. 7).

Directly linked as a subroutine of SIMULINK as an S-function, DADS/Plant assembles the equations of motion and uses SIMULINK to solve state equations of the mechanical plant model. Simulations can be viewed either graphically as *x-y* plots or with photorealistic animation of the related model geometry. DADS/Plant links the model geometry with the simulation results for interactive animation, which can be used to provide instant feedback to the engineer and other audiences.

The automatic control system allows the bucket to be controlled in order to counteract resonant vehicle motions and provide a smooth ride. This could be desirable to reduce operator fatigue when using the machine to transport material over moderate or long distances.

The first step in this process is to model the mechanism portion of the model in DADS/Plant while the hydraulic system is modeled in SIMULINK. This backhoe example consists of seven bodies connected by a series of joints that produce only four DOFs. These four DOFs allow the operator to position the boom and the tip of the bucket. By modeling and defining geometric relationships of the movable parts and the body properties such as mass, inertia, and joint descriptions, DADS/Plant is able to generate automatically complete nonlinear models.

Four hydraulic actuators are used to maneuver the backhoe. Each actuator is controlled by a spool valve. Each actuator defines two state equations: one for each pressure in the actuator. The pressure state equations are dependent on the volume and the rate of change of volume in the actuator. This requires the stroke of the actuator and the stroke velocity to be fed from the DADS/Plant model into SIMULINK. Further, to close the loop on the controller, the four DOFs of the bucket are needed in the SIMULINK controller.

The mechanism model is included in the block diagram (Fig. 8). A hydraulic system and controller are interconnected to the plant block to form a closed-loop system. The controller modeled in SIMULINK can compute hydraulic forces that are to be applied to the mechanism model to close the feedback loop. The plant elements receive force or torque signals from SIMULINK and, in turn, DADS/Plant passes body positions and velocities back to SIMULINK.

DADS/Plant models can be linearized from SIMULINK. By using the linearized models, users can apply any of the control design tools available with MATLAB, including classical or modern control design methods. The new controller can be incorporated in the SIMULINK block diagram for full nonlinear simulation.

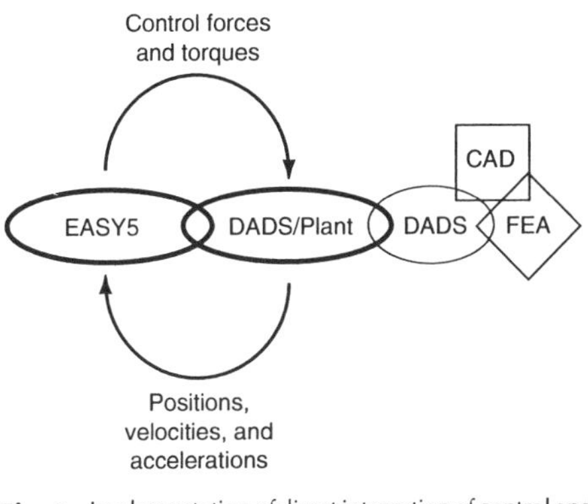

Fig. 6 Implementation of direct integration of control and mechanical system models. This is indicative of the information flow in the DADS/Plant code.

Direct Interfacing of CAD Data and Simulation Programs

For mechanical systems, most of the critical data are tied directly to the information captured in a CAD system. Detailed geometry is certainly contained in the database, but material properties, mass properties of the parts, and locations and types of joints are specified or may be inferred from the geometry. This is especially true if the CAD system is a solid modeling system.

As a simulation specialist attempts to predict the behavior of a mechanical system, the accuracy of the input data is very important. The ease of use issue is also critical in acceptance of simulation in the engineering process.

One of the most common ways in which design accuracy is lost occurs when the analyst is using data that are not current. The simplest solution to maintaining data integrity is the direct incorporation of CAD information from the solid modeler into the simulation model. This direct or integrated approach is both faster and more error-free than indirect procedures that require translation of the geometry to IGES or other representations, followed by translation into the data format of the simulation program. The integrated approach of working on the database in the CAD system guarantees that the same data used by the designer are passed into the simulation program.

Overcoming the ease-of-use issue is addressed by integration of the simulation program within the solid modeler environment. This allows direct data exchange as well as adding menus and operations that produce the simulation model without requiring the user to leave the familiar solid modeling environment.

Example 4: Modeling of an Automobile Door Latch. As an illustration of this modeling process, consider the analysis of the closing and latching of a car door. This model was validated against a set of results from a test fixture at an automotive manufacturer. The latch components were designed and assembled in the Pro/ENGINEER modeling system from Parametric Technology Corporation. The DADS multibody dynamics (MBD) program from CADSI has an integration module that runs within the Pro/E environment. This allows the designer to select parts, joints, constraints, external forces, and so forth, from within a familiar menu environment to quickly generate the model. The resulting model can then be used to predict motion, forces, accelerations, and so forth, to check correlation with test data.

The solid models of the latch mechanism and the complete assembly of parts are shown in Fig. 9. The assembly includes a pawl, a latch, a striker, and other parts including the door frame and door. One of the most critical sets of data is the contact surface shapes. Several parts have intermittent contact and slide against other parts during the closing. This geometric information is simple to extract from the CAD system by picking one segment of the edge that is part of a contact surface. The rest of the edge is then automatically traced and defined as a piecewise curve. Modern contact techniques use the geometry directly from the CAD database and allow Hertzian or other relationships to define the forces and penetration of the mating surfaces.

The analyst can actually run the simulation from the Pro/E environment. Often these simulations require further "tuning" by matching friction and damping coefficients (or other parameters) to correlate with test data. Once acceptable runs are made, the system behavior may be reviewed to determine if it meets the criteria for force levels, bounce, overtravel, and so forth. Both *x-y* plots and realistic animations can be quickly reviewed within this environment to show forces, displacements, accelerations, component interferences, and so forth. If it does not meet all criteria, the design can be quickly updated in the Pro/E model and the process repeated. Note the critical importance of using the CAD data directly; the data should be extracted automatically and accurately and used in a solution procedure that is tested and proven. Speed and accuracy contribute heavily to the value of simulation in the engineering process.

Interactions and Applications with Finite Element Analysis

Another tightly integrated pair of technologies is mechanism analysis with flexible components. Some mechanisms require one or more parts to flex in normal operation. Others have sufficient flexure in parts that it needs to be considered in understanding and predicting behavior. For simple parts, the response can be approximated as a simple spring-mass behavior. For more complex parts, use of the full geometry of the part is the most direct and accurate approach. In these cases, finite element models (e.g., NASTRAN, ANSYS, ABAQUS, etc.) can be created from solid models through CAD systems. These models can then be used to define normal modes and eigenvalues for the constrained part. These can be supplemented with constraint or static modes (deflection shapes due to unit deflections or unit loads at attachment points) to form a set referred to as Craig-Bampton modes. These modes are then used in a technique commonly known as the component mode synthesis (CMS) technique. The procedure uses the set of modes as a basis set for approximation of the displacement of the flexible part and superposes that displacement on the rigid body motion of the part. The result is a highly accurate representation of the overall behavior of the system.

It should be noted that some codes have attempted to include a direct dynamic solution of the finite element model in a dynamic mechanism solution. This results in an increased number of DOFs in the model. Experience to date indicates longer solution times and often less accurate simulations. It is possible that this technique may

Fig. 7 Backhoe system with closed loop to counteract resonant vehicle motion

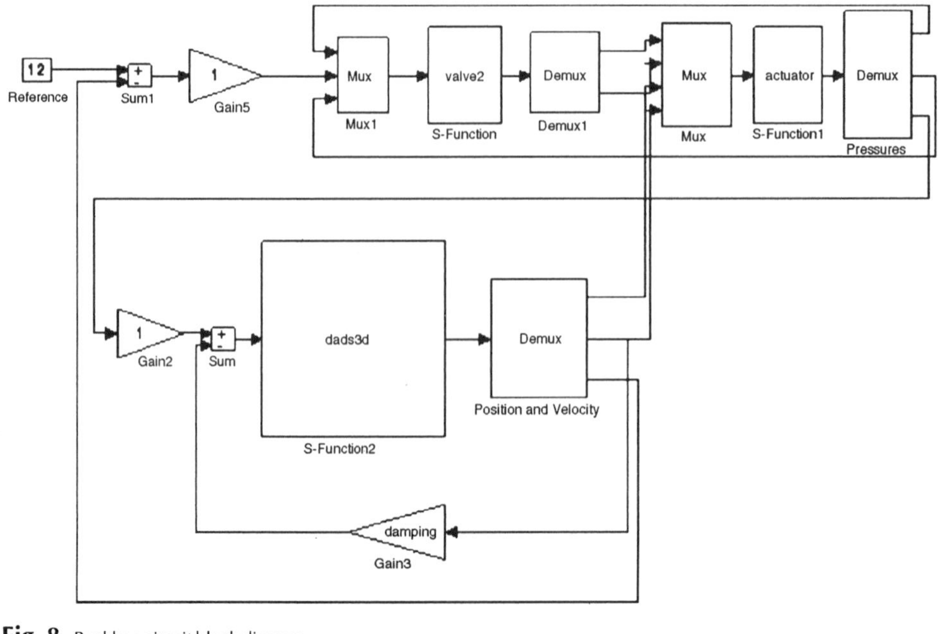

Fig. 8 Backhoe circuit block diagram

have more promise in the future, but at this time a modal superposition solution is more widely accepted as the better solution technique.

Example 5: Integrated Mechanism Dynamics and Finite Element Analysis Approach to Interior Noise in Automobiles. An example that many of us can identify with is the rattle of components in an automobile. As the noise level has been reduced with improved insulation and sound absorption in automotive interiors, sounds that were previously ignored or not otherwise perceptible became unacceptable to many owners. Rattles of cup holders, glovebox doors, windows, latches, and so forth, are now of concern.

The Ford Motor Company has been one of the pioneers in the development of analytic techniques to address these challenging problems before they are manifested in prototype or production vehicles. The following task is one that has served as a test bed for the techniques and is described only briefly, but the concept should be clear.

In rattle cases, the flexibility of one or more of the parts of the systems may be unintentionally "tuned" to the vibrations borne through the structure from the suspension or from the engine. Often, the part has several natural frequencies in the range of the input frequencies, making it difficult to avoid having one or more of these vibration modes respond to the input. In these cases, the amplitude of the response may increase until audible noise is heard. Often an intermittent (certainly nonlinear) restraint force is involved due to contact between a snubber or the latch on a glove box and its supporting structure.

The deflections of the glove box and the supporting structure are small, small enough to be treated as a linear response. The latch components are rigid in comparison to the glove box. These can be modeled very well with a multibody dynamics program. The snubbers between the glove box lid and the supporting structure are highly nonlinear in nature. Further, under rather small loads, the snubbers and springs can be compressed sufficiently to allow the latch and striker to lose contact and then hit again in rebound, creating one cycle in the noise mechanism. The spring holding the latch is set to produce a small load. The spring load cannot be increased significantly or else the glove box will be too difficult for the customer to close. This trade-off is typical of the dilemma confronted by designers.

The latch and striker separation allow two points of view. If the glove box is vibrating, the amplitude, while small, can be large enough to generate loads on the latch that accelerate the lid. When the loads release, the parts strike again. The other view is taken from the "outside," with motion imposed due to road input, engine vibration, and so forth. This input motion of the chassis drives the glove box lid through constraints such as the hinge and the latch and results in vibration of the lid. In this "chicken-and-egg" type of problem, only an integrated solution will yield the accurate answers needed to predict if a design concept reduces rattle.

So how is an integrated solution obtained? Finite element programs cannot easily predict the nonlinear action of the latch. However, the combined use of a finite element and a multibody code can.

In this case, natural frequencies and modes shapes for the glove box lid and some of the supporting structure were determined using MSC/NASTRAN. Frequencies in excess of 50 Hz were needed to cover the frequency range that had been seen experimentally, resulting in use of more than a dozen modes. These modes were then imported into the DADS program directly from the NASTRAN output files to augment the rigid body model of the glove box and latch system (Fig. 10). (Other popular finite element programs, such as ANSYS, PDA/Patran, and I-DEAS, can also serve as the source of modal information used for a flexible component.)

The glove box lid is represented by a combination of a rigid body motion that predicts how its center of gravity (CG) translates and rotates plus a superimposed small displacement motion based on a modal superposition of the modes imported from NASTRAN. This combined motion defines more precisely how the latch moves in space, how the snubbers are compressed, and so forth. This also more accurately predicts when and how much the latch separates from the striker.

This latter information on the intermittent contact can be correlated with noise heard from the latch and glove box. The validated model can now serve as the nominal basis in the search for design changes that will decrease the noise and rattle of a glove box of a passenger vehicle. This search involves iterative changes in the glove box shape, reinforcing ribs, material properties, and so forth, to tune the modes and modify the response. This tight loop from finite element software to multibody dynamic response solution (to test and correlation) can lead to a large improvement in the noise and rattle behavior without the heavy reliance on testing of numerous prototypes.

This example reinforces the point that one approach or technology is often not sufficient to support the engineering process quickly and accurately. In the above case the close link of structural and multibody dynamics to solve the problem is highlighted. While not noted above, it must be recognized that the latch geometry, the input geometry to the FE program, mass properties, and so forth, all came from a CAD system. A

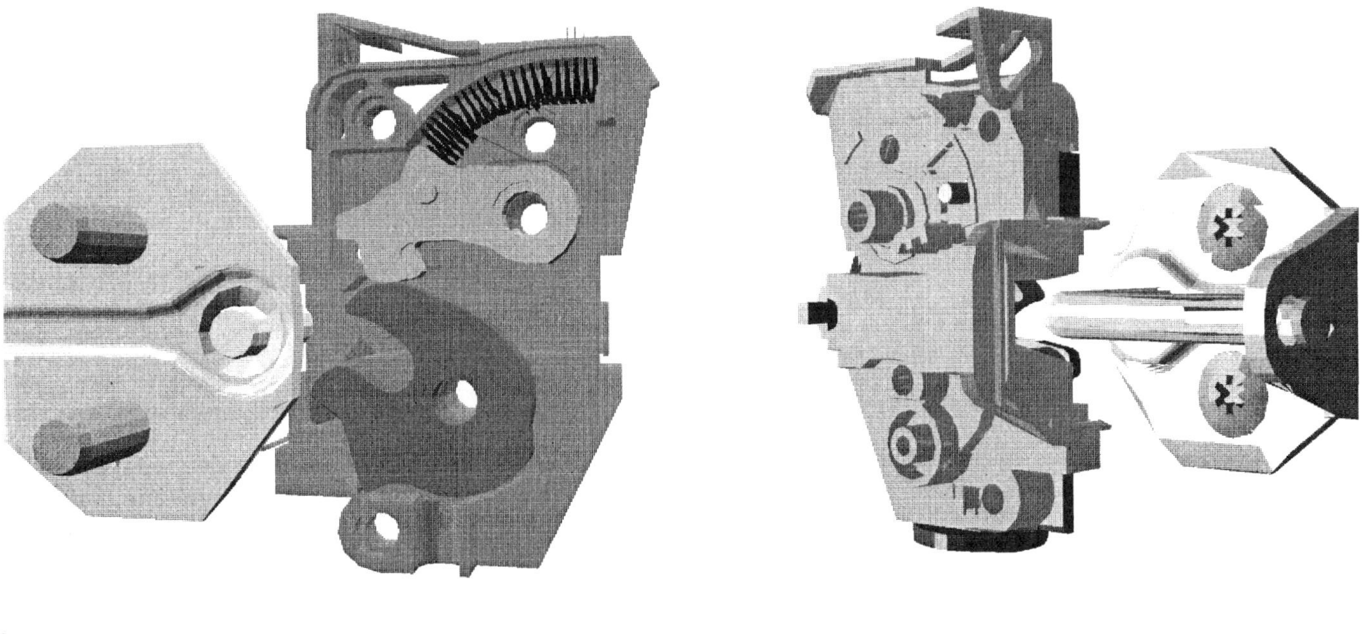

Fig. 9 Two views of an automotive door latch assembly

convenient means to extract that CAD data also was important to the timely and accurate solutions described.

Synthesis, Optimization, Design Guidance, Design Improvement, and Other Factors

Software is evolving into a more synthesis-oriented form, accounting for a broader range of the parameters in a complete design. Much of this is or will be based on the ability to extract design sensitivity (sensitivity of the design's performance, however that is measured, to design parameters) from a given nominal configuration. More and more, it is being recognized that analysis provides too little guidance to design. The results are typically a "go" or "no-go" result in any number of areas (stress, acceleration, force, response time, displacement, etc.). That information is valuable, but does not help pinpoint the design parameters that will improve products, or more important, direct changes. In lieu of this assistance from analysis, numerous trial-and-error attempts are usually made to determine which changes to make. More systematic procedures are occasionally used such as design of experiments (DOE) or Taguchi techniques (see the article "Robust Design" in this Volume and Ref 7) in conjunction with the simulation to guide the design. In the future, there will be sufficient intelligence built into software to provide significant assistance in the definition, as well as the "optimization" of a design.

This author contends that optimization of engineering designs is a misnomer. It is often the focus of papers, books, or lunchtime discussions/arguments. True optimization is seldom possible for real systems. Optimization of idealized, simplified systems is often performed and reported without acknowledging real-world limits such as schedules, manufacturing capabilities, or staff/financial resources. Seldom do the "optimizers" mention that the *design requirements are time dependent,* driven by competitive and customer pressures. Quality improvement relative to current or competitive designs are more typical demands placed on the design process.

Some simulation or analysis software that claim to perform optimization are actually delivering limited parametric variations of a system. Those limitations are usually not understood. The typical approach performs a parameter sweep of one design parameter at a time. Peaks/valleys for each such sweep are then reported as a recommended next trial design. This technique is acceptable if there is no interaction between the design parameters, but that is seldom the case.

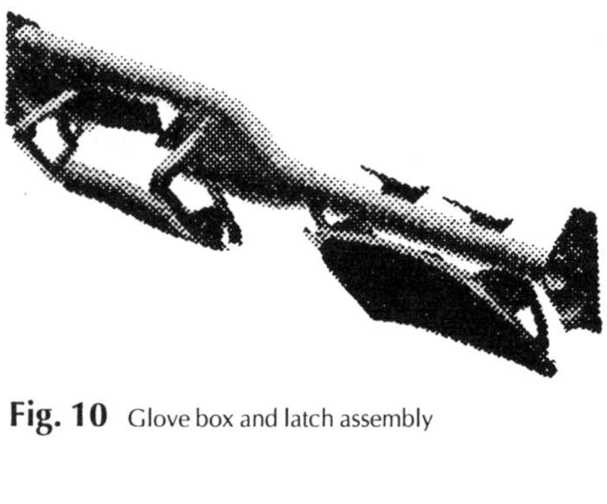

Fig. 10 Glove box and latch assembly

In this section a discussion of a blend of technologies is described that provides a robust solution procedure, is flexible, and is simple to apply. Multiple criteria can be used directly, and trade-off studies may be easily performed. These techniques are based on the statistical procedures of DOE and response surface models (RSMs) (Ref 8).

As a backdrop to this discussion, consider the use of statistical techniques for the creation of models of systems based on experimental tests. The system or process is considered as a "black box" for which a model is needed. The parameters that are thought to be important in describing the behavior of the system are selected and a range of these parameters defined for testing. A form for the model is then selected, usually a simple polynomial in the parameters used in the test. A test plan is created to vary these parameters for a series of tests, and output data are collected to characterize the response of the system. The parameter settings for the tests and the output data can then be used to define specific forms for the model of the response; these resulting models often are referred to as RSMs.

Chemical or manufacturing processes are complex entities that may be described in a narrow range of behavior based on collection of a set of parameters, but the exact mathematical relationship is difficult or impossible to define. In a similar fashion, a complex mechanical system, complete with a control system, can be defined (modeled) mathematically but impossible to comprehend in its entirety. It is attractive to restrict attention to a small subset of the input parameters and to approximate one or more output behaviors of the system as simple functions.

While application of the statistical techniques of DOE and RSM to simulations has been reported for at least two decades, it has only recently become an accepted procedure. The statistical focus has been upon the design of the testing procedure to ensure the maximum resolution of the model in the range of interest. This same "sampling" process can be used to create an accurate, yet simplified, model of the output in terms of the selected design parameters. Once such a model has been created, it may be used to determine how to improve the behavior by adjusting the design parameters. Because realistic designs have multiple-design criteria, multiple-response surfaces must be generated and all (or most) of them used in the design trade-off steps to follow. In principle, these surfaces can be used in an "optimization" scheme to determine the next design iteration. However, the designer usually has sufficient intuition about the design (and cannot explicitly define a mathematical relationship for some of the criteria that determine how good the design will be), thus direct involvement in the trade-off study is vital.

The term design sensitivity was mentioned earlier. What is it and how does it become involved? First, design sensitivity is a numerical measure of how important each parameter is relative to a particular output measure. If a sensitivity coefficient is positive for a given parameter, it implies that a positive increase in the parameter causes a positive change in the output criterion. Conversely, if the coefficient is negative, a positive change causes a decrease in the output. Larger coefficients are more important than small ones. Interaction effects can be identified and quantified relative to the direct effects.

It is not possible to cover the subject of RSM in depth in this article. However, it is important to know that the technique is providing a critical tool in the concurrent engineering suite. Within a single area, it provides a means to determine the importance of a given parameter relative to a specified output. It also provides a simple relationship that can be used to understand trends in design changes. It also provides an important link between technologies. Because several different simulation techniques can provide information regarding a particular performance criteria, all of the sensitivity coefficients that contribute to those effects can be collected together and used in a single trade-off study. A simplistic example might be the vibration felt at the driver's seat. This is influenced by the structural response (finite element analysis), by the suspension behavior (MBD simulation), by the powertrain vibration (MBD or control system simulation), by the seat response (MBD or other programs), by the air flow around the vehicle (computational fluid dynamics codes), and probably by other effects as well. Each of these codes could be separately exercised and the output used to determine the appropriate sensitivity coefficients with combined effects of design changes understood with simple techniques on a spreadsheet program or with more automated procedures.

These examples only sample the possible combinations, the state-of-the-art efforts, in the various fields that support the engineering sciences. However, these cases *do* provide a sense of the direction that must be followed to provide faster and more accurate predictions of system behavior.

Mechanism Dynamics/Simulation and Concurrent Engineering

The focus of this article is on the simulation of mechanical systems, but that cannot be treated as a separate technology. Data integration is becoming both a necessary and a pervasive part of modern engineering. The increasing interrelationship of a variety of technologies, foreshadows the future need for even more interplay of software applications. The integrated exchange of data for models must maintain data integrity, minimize modeling error, and, if successful, will promote concurrent engineering. Products will be of higher quality and be more innovative with the ability to do more "what-if" studies on the computer before committing to hardware.

As an overview of the emphasis on concurrent engineering, consider the current pressure both on decreasing time to market and on increasing quality and how this is reflected in the mechanical engineering process. "Normal" development (only a few years ago) involved preparation of one or multiple physical prototypes of the new product. These prototypes helped find the conflicts in dimensions, incorrect estimates of behavior, inadequate strength, and overall deficiencies in product response to market needs.

Simulation is heavily involved in the evolution of concurrent engineering. Modern CAD systems allow the capture and detailing of design concepts in a short time interval, particularly if the design is evolutionary. With the detailed information represented in CAD, many CAE processes can now be applied in concert to aid in ensuring both quality and speedy design review. This allows full utilization of the valuable modern CAD parametric modeling and visualization tools plus analysis capabilities to determine performance characteristics such as component interference and dynamic system properties.

For example, it is now commonplace to subject a mechanism to a performance review using a multibody dynamics program to predict a variety of information regarding the system. In addition to range of motion studies to determine component interference, important data are provided for loads on the parts of the mechanism. These loads can be used by a finite element program to predict stresses.

If these predicted stresses are too high, design changes are needed to bring the load/stress conditions within performance specifications. The changes in the mechanical system may come in a variety of ways: in its control system, in dimensions of the components of the system, in the manner in which external loads are applied to the parts, or in materials used in the parts. The changes and analyses are an iterative process until the design meets all necessary performance, manufacturing, and cost criteria.

As a result, some companies bypass the prototype stage and go directly to manufacturing, as Boeing did with its 777 airliner. With the amount of regulation in the aerospace field, this was only possible through the use of extensive computer simulations. In the automotive arena, single prototypes are now the norm with few surprises found at late stages in product development.

The above examples illustrate the continuing trend to use more simulation, to use more detailed simulation, and to do it on a shorter timeline. This has forced improvement of the export/import of data from one software package to another—or even better, the integration of one software package into another. Many designers expected to see these data transfers performed in the past through standardized formats such as IGES or STEP or extensions thereof. Instead, most of the really significant progress has been made by individual companies that recognize the need and create a technique for the data exchange or the direct interaction of different software. Fortunately, such state-of-the-art direction and format has often evolved into pseudostandards. Ideally, all CAD vendors and those who derive data from their databases soon will provide convenient access to ISO (STEP) data models that can be directly exchanged.

The current status of mechanism simulation is focused primarily on analysis of a given design or process. There is pressure to use these techniques to provide direction to change designs or processes. Many software packages are now incorporating design sensitivity information calculations. Such directions as sensitivity of measures of quality as a function of design characteristics are being sought. Large companies are cascading sensitivity data from multiple analysis sources to provide a more general model of a product or process. If one simulation yields data used in another simulation, a chain rule procedure (see below) is possible to develop the desired complete model. These models can then be used as a means of "optimizing" the multiple criteria of concern. The chain rule is that used in calculus to determine a total derivative based on a sum of the partial derivatives of the quantity with respect to all its independent variables times the variation in each variable. To illustrate, suppose the designer wants the sensitivity of the peak force at a suspension connection point to changes in the thickness of several body panels. Obviously, the mass will change, but also the natural frequencies of the structure due to the panel thickness changes. The change in the natural frequencies due to the change in thickness can be calculated in standard procedures in programs such as NASTRAN. The total change in a given frequency can be estimated by summing the product of these structural design sensitivities and the respective design changes in the panel thicknesses. The flexible body data needed in a dynamic mechanism simulation are the natural frequencies, but these are now directly related to the panel thicknesses. Hence the thickness data and its sensitivity contributes to the dynamic system sensitivity. This cascading of data from one program to another allows more in-depth analysis of a system than is inherent in a single computer program.

In the future, more detailed simulations will be possible for mechanisms, stress analysis, fluid flow, queuing theory, resource allocation, and so forth, and will be done more reliably and faster due to more automated procedures to collect and distribute data. In addition, more sophisticated techniques will be developed to monitor results and guide design/process revisions, that is, to optimize them. Perhaps it would be just as good to think in terms of continuous product improvement or concurrent engineering or similar business strategies. In conclusion, simulation will become a pervasive activity in engineering, increasingly important for meeting schedules and product goals. Just as most companies have moved to CAD to capture design intent, most companies will use simulation in the future to ensure product quality and viability.

Mechanism analysis is only one portion of the CAD/CAE/CAM process, but is one that can have a significant influence on the final design. Individual tools such as CAD and simulation now allow significant improvement in accuracy and speed relative to their predecessors. Properly orchestrated, these tools can produce products that are closer to optimum with a shorter time-to-market and with reduced production costs. This will be accompanied by increased reliability and maintainability, if these tools are applied as hoped. If instead the tools remain isolated, then the the trends to concurrent engineering and the benefits thereof will not materialize. The extent to which the tools work in concert is only partially the result of efforts by the companies that create the tools. It is also a function of the users, the implementers of the tools, and their ability to change their organizations to take advantage of the new features and functions of the tools.

ACKNOWLEDGMENTS

The author wishes to thank the following people for providing information for this article: Bert Shemwell, Ebonite International; Doug McKissack, Gulfstream Aerospace; Victor Borosky, Ford Motor Company. CADSI and DADS are registered trademarks of Computer Aided Design Software, Inc; DADS/Plant is a trademark; CATIA is registered by Dassault Systemes; PolyFEM and DesignPoint Analysis are registered by IBM Corporation; Pro/ENGINEER is registered by Parametric Technology Corporation; ANSYS is registered by Swanson Analysis Systems, Inc.; MSC/NASTRAN is registered by MacNeal Schwendler; COSMOS/M is registered by Structural Research and Analysis Corporation; PDA/Patran is registered by PDA; EASY5 is registered by Boeing Computer Services; $MATRIX_x$ is a registered trademark of Integrated Systems, Inc.; MATLAB and SIMULINK are registered trademarks of The MathWorks, Inc.; and I-DEAS is trademarked by Structural Dynamics Research Corporation.

REFERENCES

1. E.J. Haug, *Computer-Aided Kinematics and Dynamics of Mechanical Systems,* Vol I, *Basic Methods,* Allyn and Bacon, 1989
2. E.J. Haug, S.S. Kim, and F.F. Tsai, *Computer-Aided Kinematics and Dynamics of Mechanical Systems,* Vol II, *Advanced Methods,* Prentice-Hall, 1992
3. E.J. Haug, *Intermediate Dynamics,* Prentice-Hall, 1992
4. T.R. Kane and D.A. Levinson, *Dynamics: Theory and Applications,* McGraw-Hill, 1985
5. D.T. Greenwood, *Principles of Dynamics,* 2nd ed., Prentice-Hall, 1988
6. A.G. Erdman and G.N. Sandor, *Advanced Mechanism Design: Analysis and Synthesis,* Vol I and II, Prentice-Hall, 1984
7. G.E.P. Box, W.G. Hunter, and J.S. Hunter, *Statistics for Experimenters: An Introduction to Design, Data Analysis, and Model Building,* John Wiley & Sons, 1978
8. R.H. Myers, *Response Surface Methodology,* John Wiley & Sons, 1995

Finite Element Analysis

Donald L. Dewhirst, Ford Motor Company

FINITE ELEMENT ANALYSIS is a computer-based numerical method for solving engineering problems in bodies of user-defined geometry. Interpolation functions are used to reduce the behavior at an infinite field of points to a finite number of points. The interconnectivity of these points is defined by finite elements that fill the appropriate geometry. The genius of the finite element method is that it enables the systematic solution of these field problems with as much fidelity to geometry as needed. The ability to model subtle changes in geometry permits the optimization of size, shape, or form to obtain minimum weight or maximum performance.

Table 1 summarizes the many engineering problems addressed by the finite element method. Because the method was originally developed for structural problems and because most engineers have at least an intuitive understanding of problems of force and displacement, finite elements are discussed here in terms of that class of problems. This permits the discussion to be concrete rather than abstract. The principles are the same regardless of discipline.

Many good academic books have been written on the subject of finite elements (Ref 1-5). Reference 1 is particularly well indexed and treats issues of real concern to the practicing engineer. Reference 2 is unique in its thorough treatment of error estimation and adaptivity.

Table 1 Engineering problems addressed by the finite element method

Discipline	Typical unknown	Possible boundary conditions
Structural	Displacement	Stress or displacement
Thermal	Temperature	Heat flux or convective term or radiative term
Electrical	Voltage	Current source
Magnetic	Electromotive force	Magnetic field source or intensity
Fluid flow	Pressure, velocity	Velocity
Diffusion (Fickian)	Mass concentration	Flux of species
Diffusion (porous media)	Flow velocity	Boundary flow
Corrosion	Anode consumption rate	Electropotential
Crack propagation	Strain energy release rate	Stress
Acoustic noise	Sound pressure level	Velocity

This article introduces the important issues of finite elements (especially accuracy and efficiency) in a nonacademic manner. It does not teach the reader to perform analysis.

A Beam Paradox

An end-loaded aluminum cantilever beam that has a uniform rectangular cross section is analyzed. The designer proposes to add two cylindrical (circular) holes for reasons of access and weight reduction (Fig. 1). This problem could be approached as three-dimensional, but because nothing interesting is happening in the thickness direction, it can be simplified to that of plane stress in two dimensions. In this example, the *finite element mesh* is composed of two-dimensional elements, which are quadrilateral in shape (Fig. 2). There are 16 elements in each finite element model. These are explicitly labeled for the beam without holes; the labels are omitted from the beam with holes. The corners of the quadrilaterals, called *nodes* (sometimes *grids*), are defined by their spatial coordinates. The properties of the aluminum need to be specified. Assuming that the elastic strength of the material will not be exceeded, only Young's modulus, E, and Poisson's ratio, ν, are needed.

Finally, the loads and displacement constraints at the boundaries need to be specified. This is usually the trickiest part of setting up a model. A boundary value problem, a problem of static equilibrium, exists. Something must be known about loads and displacements at the boundaries.

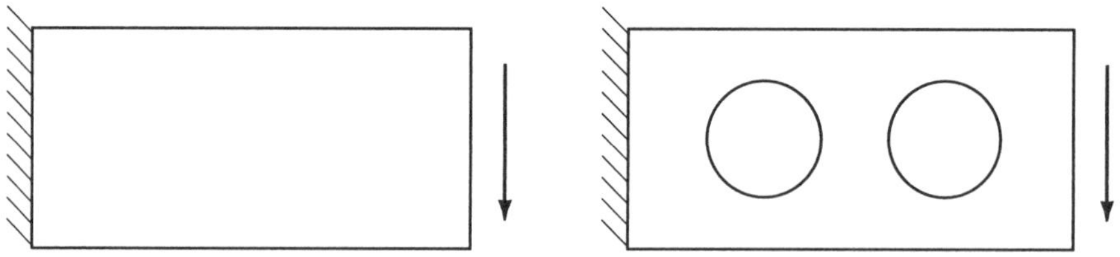

Fig. 1 Cantilever beam with and without access holes

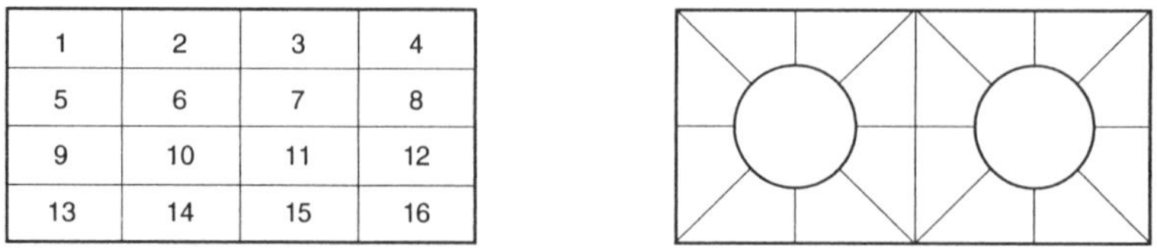

Fig. 2 Finite element meshes of beams of Fig. 1

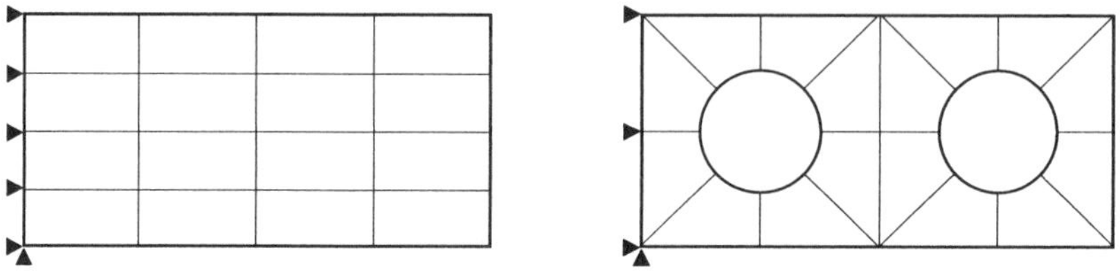

Fig. 3 Possible displacement boundary conditions

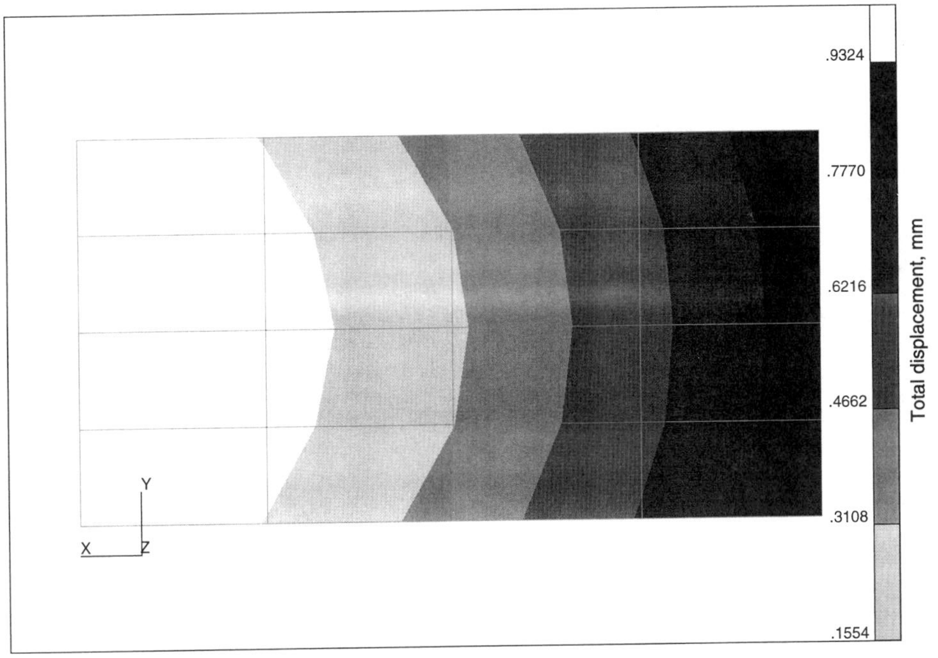

Fig. 4 Displacement contours for a cantilever beam under bending load, peak displacement 0.9324 mm

Generally, the choice is whether to specify loads or displacements or some combination of the two. Care must be taken to prevent rigid-body modes; that is, the body must be anchored in space. A two-dimensional problem, includes three rigid-body modes; a three-dimensional problem includes six rigid-body modes. Figure 3 uses arrowheads to symbolize displacement constraints. In the present model, the entire left-hand boundary is fixed in the *x*-direction, and one node is fixed in the *y*-direction. This is sufficient to remove the rigid body modes and to develop a bending moment at the left end as a reaction to the applied downward shear force (Fig. 1) at the right-hand end. In addition to these displacement constraints, a load set is specified that includes a (parabolically distributed) downward shear load at the right-hand boundary and a similar upward shear load at the left end.

These two problems are solved using a commercial finite element code, and the peak displacements are obtained, as shown in Fig. 4 and 5. There is obviously something wrong with these results. They indicate that the beam *with* the holes is stiffer (has a lower displacement) than the beam *without* the holes. This violates common sense. What is happening? There are two likely answers to the question: not enough elements were used and/or the elements were too distorted to give the appropriate answers. This illustrates one of the fundamental problems facing the finite element analyst: how to design a finite element mesh that is fine enough to give good answers, but coarse enough to be cost effective, bearing in mind that element shape profoundly affects the results. Figure 6 illustrates the typical shape distortions of a quadrilateral element, namely *aspect ratio, skew,* and *taper*. (In three dimensions, one would add the possibility of element *warp*.)

Because this finite element problem is the first one to be approached, the effect of changing the mesh is considered. Three successively finer meshes for the beam with holes are analyzed, and some trends are plotted. The top curve of Fig. 7 shows the peak principal stress in the beam for each of four mesh densities. The principal stress is plotted as a function of the reciprocal of the number of degrees of freedom (DOF). (DOF is the product of the number of nodes multiplied by the number of unknowns per node; the reciprocal of DOF is used for convenience.) As the number of DOF increases, the reciprocal tends to zero. It is generally true that coarse meshes tend to underestimate the stress. Unfortunately, however, a *lower bound answer* is not necessarily guaranteed (Ref 1). Convergence to the appropriate solution by systematically increasing the mesh density is known as *h-convergence*, where the letter

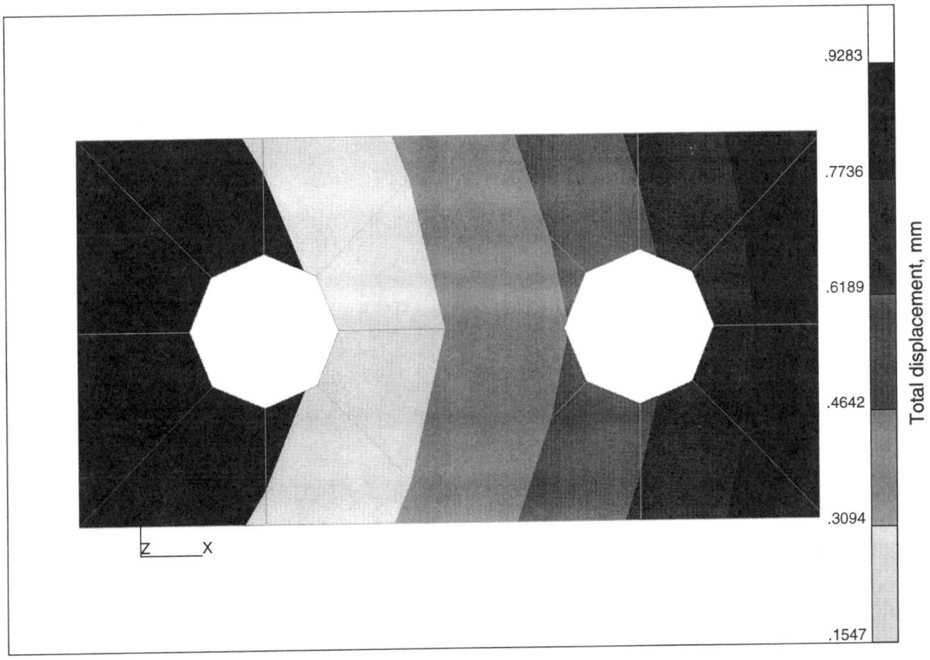

Fig. 5 Displacement contours for a cantilever beam under bending load, peak displacement 0.9283 mm

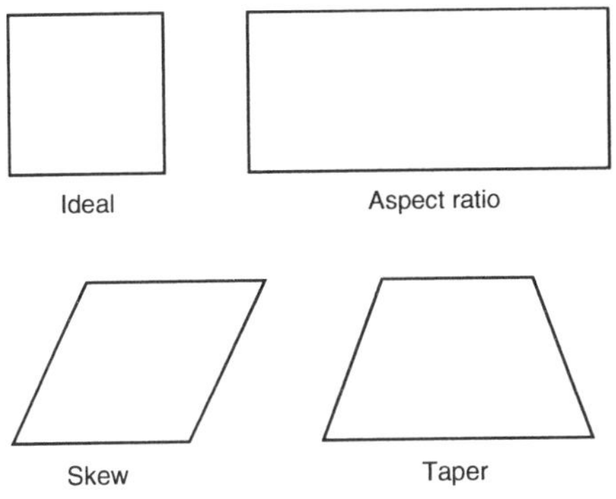

Fig. 6 Types of quadrilateral element distortion

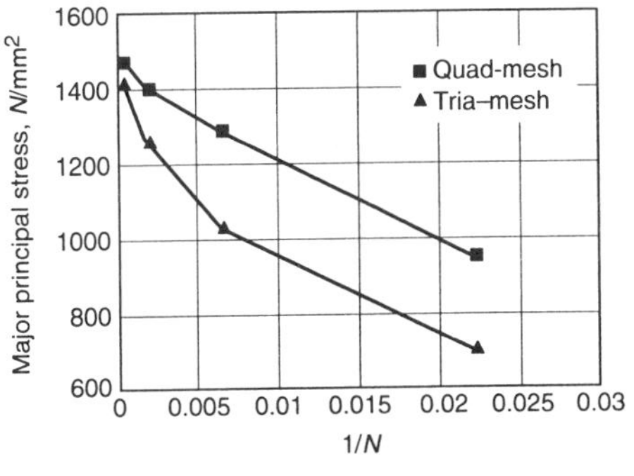

Fig. 7 Peak stress vs. reciprocal number degrees of freedom, quadrilateral and triangular elements

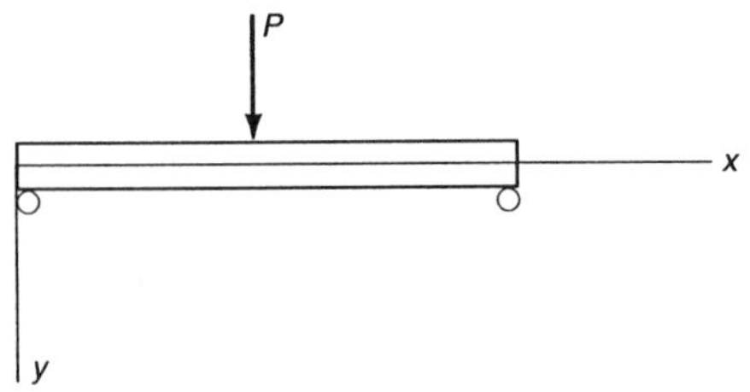

Fig. 8 Simply supported beam with central concentrated load, P

"h" is associated with element size. (The present meshes are uniform in size. It is possible to improve on these meshes by using a nonuniform mesh density. The improved mesh would be finer in regions where stress gradients are highest or where strain energy density is greatest.)

The four-noded quadrilateral is just one of several two-dimensional elements that could have been chosen for this problem. Nodal patterns identical to the mesh of Fig. 2 are used, but the beam is modeled with three-noded triangular elements. See the lower curve of Fig. 7. Both elements converge to the right answer, but the quadrilateral element gives better answers for the same number of nodes. Therefore, the quadrilateral element is more efficient for this class of problem than the triangular element.

The Basis of Finite Elements

To understand why some elements outperform other elements in certain problems, it is appropriate to review the *Rayleigh-Ritz* (Ref 1, 6, 7) procedure on which the finite element method can be based. There are more general and mathematically sophisticated ways to approach finite element theory, but they are less intuitive and less instructive for the present purpose.

The Rayleigh-Ritz procedure is an approximate method for solving structural problems based on the *principle of virtual work*. The principle of virtual work states that the total potential energy of an elastic system is a minimum (or stationary) when the system is in equilibrium. (The total potential energy of the system is the sum of the gravitational potential energy and the elastic strain energy.) The Rayleigh-Ritz method reduces a continuum with infinitely many DOF to a system with finite DOF. It accomplishes this by assuming the displacements of the continuum to be a function of a finite number of undetermined coefficients. The solution for these undetermined coefficients is illustrated below.

The Reduction of an *Infinite* DOF System to a *Finite* DOF System. A continuous system with infinitely many DOF may be approximated by a system with finite DOF. This can be accomplished by assuming a shape function with one or more undetermined coefficients whose values are to be determined.

For example, the simply supported beam of Fig. 8, which is loaded by a central concentrated load, P, is considered. A sinusoidal shape function is assumed, such that:

$$y = A \sin(\pi x/L) \qquad 0 < x < L \qquad \text{(Eq 1)}$$

Here, A is the undetermined coefficient and y is the displaced position of the beam centerline. When $x = 0$ and $x = L$, then $y = 0$. It should also be noted that:

$$d^2y/dx^2 = -A(\pi/L)^2 \sin(\pi x/L)$$

This second derivative, which is proportional to beam curvature and therefore to moment, also vanishes at $x = 0$ and at $x = L$. Therefore, both the deflections and moments are zero at the ends of the beam. These are the appropriate boundary conditions.

The Rayleigh-Ritz procedure can be used to determine the value of the coefficient A. The method consists of finding an expression for the total potential energy of the system in terms of A, differentiating that expression with respect to A, and equating the results to zero. That is, one finds the deformed geometry that provides the minimum value of the total potential energy. In symbols, this is: V = elastic strain energy + gravitational potential energy, or:

$$V = U + \Omega \qquad \text{(Eq 2)}$$

where:

$$U = (1/2)\int EI(d^2y/dx^2)^2 dx \qquad \text{(Eq 3)}$$

E is Young's modulus (material stiffness), I is section modulus (geometric stiffness), and:

$$\Omega = P(-A) \qquad \text{(Eq 4)}$$

For minimum potential energy:

$$dV/dA = 0 \qquad \text{(Eq 5)}$$

so that:

$$dV/dA = d/dA[(\pi^4 EIA^2/4L^3) - PA] = 0 \qquad \text{(Eq 6)}$$

and therefore:

$$A = (2PL^3)/(\pi^4 EI) \qquad \text{(Eq 7)}$$

and:

$$y = (2PL^3)/(\pi^4 EI)\sin(\pi x/L) \qquad \text{(Eq 8)}$$

A comparison of this solution with the known analytical solution shows that the deflections differ by 3% or less. However, the peak stress differs from the accepted solution by 19%.

A better approximation to this beam bending problem should result from adding another degree of freedom, for example:

$$y = A \sin(\pi x/L) + B \sin(3\pi x/L) \qquad \text{(Eq 9)}$$

Following the same procedure as before and using $\partial V/\partial A = 0$ and $\partial V/\partial B = 0$, then:

$$A = (2PL^3)/(\pi^4 EI) \qquad \text{(Eq 10)}$$

and:

$$B = (2PL^3)/(27\pi^4 EI) \qquad \text{(Eq 11)}$$

Comparing this new solution to the analytical solution, deflections are everywhere within a ±1% band of the analytical solution, and peak stress differs from the accepted solution by 8%.

The finite element method can be thought of as an extension to the Rayleigh-Ritz method. There are two major differences:

- In the Rayleigh-Ritz method, the structure is treated in its entirety, as one "element." In the finite element method, multiple elements and nodes are used.
- In conventional finite elements, the values of the nodal displacements and rotations *are* the variables; they *are* the undetermined coefficients. This is advantageous, because it is intuitive to understand and more convenient to specify displacements and rotational constraints at boundaries than to deal with, say, amplitudes of sine waves.

In summary, the finite element method may be thought of as follows. The geometric continuum is divided into a number of elements (e.g., triangles, quadrilaterals, tetrahedra, etc.). The internal displacements of these elements are expressed in terms of the displacements at the nodes of the elements by means of interpolation functions, similar to the sinusoidal approximation (usually, the interpolation functions are polynomials) illustrated in the beam problem. An energy expression is formed and minimized in order to obtain a set of algebraic equations. The solution of these algebraic equations provides the displacements at the nodes. The values of the displacements at each node are analogs to the coefficient A in the example above. Knowing the displacements at each node implicitly gives the displacements and stresses throughout the continuum. As a generalization, the displacements obtained by this method are typically more accurate than the stresses.

Continuum Elements

This section and the one that follows define and contrast two different types of elements: *continuum* and *structural*. A continuum element is one whose geometry is completely defined by its nodal coordinates. Hexahedra, pentahedra, and tetrahedra are the continuum elements commonly used in a three-dimensional domain. Quadrilaterals and triangles are the continuum elements commonly used in a two-dimensional domain.

Tetrahedra, pentahedra, and hexahedra typically have three DOF at each node. The three DOF at each node are the three translations (u, v, w) in the three spatial directions (X, Y, Z or R, Θ, Z or R, Θ, Φ). Figure 9 shows the assumed interpolation function that might be employed for

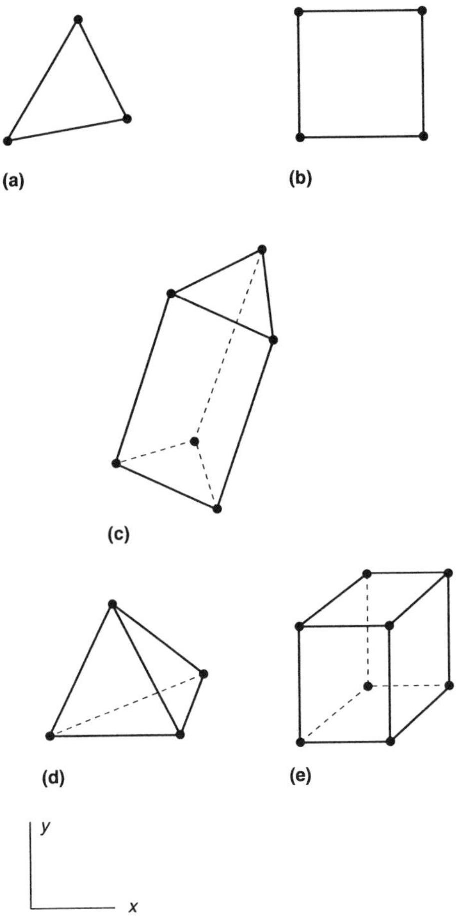

Fig. 9 Continuum elements (lower-order elements). (a) $u = a_o + a_1x + a_2y$; $\partial u/\partial x = a_1$; $\sigma = E\varepsilon_x = E\partial u/\partial x = Ea_1$ = constant. (b) $u = a_o + a_1x + a_2y + a_3xy$; $\partial x/\partial x = a_1 + a_3y$; $\sigma = E\varepsilon_x = E(a_1 + a_3y)$ = linear. (c) u = directionally dependent. (d) $u = a_o + a_1x + a_2y$. (e) $u = a_o + a_1x + a_2y + a_3xy$

each of these fundamental elements. Only the u displacement function is shown; the others have analogous form.

Triangles and quadrilaterals typically have two DOF at each node, the in-plane displacements. Triangles and quadrilaterals can also be structural elements, for example when the bending (rotational) DOF are added. Structural elements are discussed below.

Figure 10 shows the possibility of enhancing continuum elements by adding interior nodes. (A node could also exist at the centroid, or multiple nodes could exist along edges.) The extra nodes supply more independent variables so that the displacement functions can contain more terms. The elements without interior nodes are sometimes called *lower-order elements*, and the ones with interior nodes are sometimes called *higher-order elements*. Most commercial codes permit the selective omission of the interior nodes. This permits joining of lower-order elements to higher-order elements. If interior nodes *are* used, curved edges and faces are possible. When the same interpolation functions are used for both the geometry and the displacements, the elements are known as *isoparametric*.

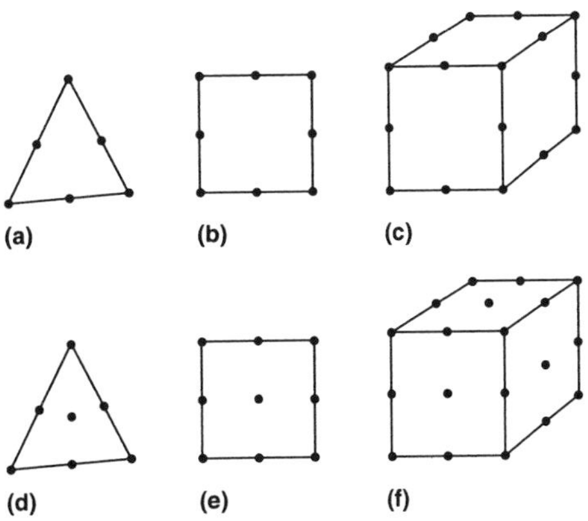

Fig. 10 Continuum elements (higher-order elements). (a) to (c) Sometimes called serendipity elements. (d) to (f) Sometimes called Lagrange elements

The optimum choice of element is code dependent and problem dependent. A generalization that is sometimes made is to use the higher-order elements for "smooth" problems, those relatively free of geometric and load discontinuities. Conversely, the lower-order element is used where there are abrupt changes in geometry, material, or loading.

Structural Elements

In addition to the continuum elements depicted in Fig. 9 and 10, there are a number of elements that behave in accordance with certain structural assumptions that predate finite element technology. As long as these structural assumptions are valid, they provide for maximum solution efficiency (Fig. 11).

Beam elements are typically based on the so-called *Euler-Bernoulli assumptions*. These assumptions are that plane sections of the beam remain plane under deformation and that bending stress is therefore directly proportional to the distance from the bending axis. The through-thickness normal stress is assumed to be zero. Shear deformations are not considered in the Euler-Bernoulli beam. The so-called *Timoshenko beam* adds shear deformation capability and possibly rotary inertia. The shear deformation is assumed to be constant throughout the beam depth, not parabolic. The Timoshenko assumptions tend to "soften" the response of the beam (i.e., displacements are higher and natural frequencies are lower than for the corresponding Euler-Bernoulli beam).

Plate elements are often based on the *Kirchoff assumption* that, after deformation, normals to the midsurface of the element remain normal to that surface. The Kirchoff assumption for plates is analogous to the Euler-Bernoulli assumptions for beams. More sophisticated assumptions are made for the *Mindlin plate element*, which is analogous to the Timoshenko beam. *Shells*, by definition, are curved surfaces. Finite element representation of shells may be as flat facets or as

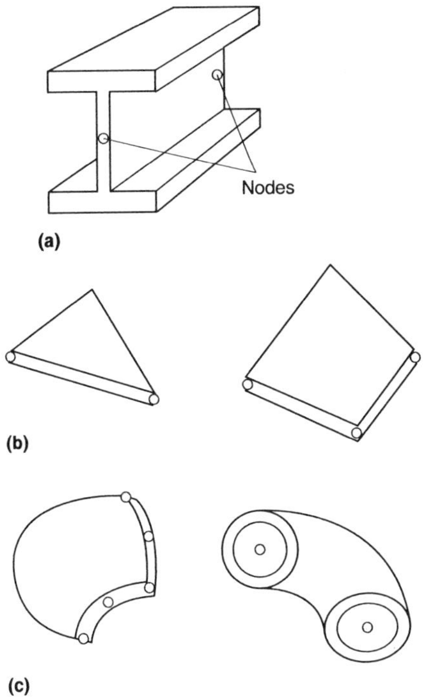

Fig. 11 Structural elements. (a) Beam element. A beam requires two or more nodes, two bending stiffnesses, torsional stiffness, orientation about the line connecting the nodes, material properties, and six degrees of freedom per node. (b) Plate elements. (c) Shell element. (d) Elbow element

curved elements. In either case, the finite elements representing shells must carry membrane as well as bending loads. The use of elements based on these various structural assumptions generally results in a very efficient model, especially for vibration and buckling problems. Accuracy of stresses may suffer at edges or discontinuities in the structure. This could be an important consideration for composite materials, subject to delamination from local though-thickness tensile stresses at the edges.

Multipoint Constraints

In a variety of situations, it is desirable to use multipoint constraints (MPC), that is, to constrain the displacement of a node to that of another node or to a function of the displacements of other nodes. These situations include:

- Joining incompatible elements
- Simulating a rigid body
- Joining structural elements to continuum elements
- Distributing a load in a convenient fashion
- Joining dissimilar materials
- Transition of mesh density

Examples of the first three are shown in Fig. 12. Some comments on these situations are made below.

Joining Incompatible Elements. The eight-noded quadrilateral, 201, is joined to the four-noded quadrilateral, 101, by constraining the mid-side node of 201, node 10, to displace the average of the two corner nodes. In that way, no gap or interpenetration of the two elements can occur along their common edge. The MPC is an effective way to join these two different element types.

Simulating a Rigid Body. Figure 12 shows the simplest form of a rigid body in two dimensions. The node labeled "2" is constrained to the displacements of the node labeled "1."

Joining Structural Elements to Continuum Elements. Shell elements typically have five DOF per node, three translations, and two bending rotations. Three-dimensional continuum elements (i.e., solids) have only the three translational DOF. Joining these two element types will produce a hinge at their interface unless care is taken. Commonly, the hinge is prevented by constraining the interface nodes to immediately adjacent nodes. Reference 8 contains a thorough discussion of this problem.

Linear versus Nonlinear Problems

A majority of the finite element problems solved by engineers use linearizing assumptions. Deflections are assumed to be small, and materials are assumed to be linear elastic. Under these assumptions, solution times are quite predictable. There exists a wide range of problems that are *nonlinear*; that is, the stress and deflection are not directly proportional to the load. These problems usually require iterative solution techniques and additional computer resources (e.g., disk space and solution time).

Problems of large deflection are one class of nonlinear problems. Problems of flat plates, for example, are usually assumed to be linear if the lateral deflection does not exceed the thickness of the plate. After that point, the plate stiffens due to membrane effects. Another type of large deflection problem occurs when the direction of the load follows the deformation of the structure. Still other types occur when a gap is closed or opened, or one component of an assembly strikes another.

Problems of inelastic material behavior are another class of nonlinear problems. Problems of creep, relaxation, hysteresis, phase change, and residual stress fall into this category. Fortunately, most nonlinear material models used in finite element solutions carry over from disciplines that predate finite element technology (e.g., theory of plasticity). However, for problems involving a high rate of deformation (e.g., the forming of metal or the crash of a vehicle), special finite element software based on so-called *explicit integration* techniques is available (Ref 3).

Finally, it is worth mentioning that finite element codes are available that deal with multiple disciplines simultaneously (see Example 3 below).

Solution of the System of Algebraic Equations

For static structural problems, the finite element method results in a system of equations that can be expressed in matrix form as:

$$\{F\} = [K]\{u\} \qquad \text{(Eq 12)}$$

where **{F}** is a column vector of forces, **{u}** is a column vector of displacements, and **[K]** is a square matrix. There will be multiple **{F}** and **{u}** vectors for problems with multiple load cases. The size of the **[K]** matrix is the product of the number of nodes and the number of DOF per node. If all DOF of the problem acted completely independently of each other (e.g., a number of axial springs each separately attached to ground, but not to each other), then the stiffness matrix would be purely *diagonal.* All of the terms off the diagonal would be zero (Fig. 13).

Most structural problems have stiffness terms clustered, or *banded,* around the diagonal. This bandedness may be influenced by either the node numbering strategy or the element numbering strategy. Various solution strategies (solvers) are available in commercial codes, and they exploit (or ignore) bandedness in different ways. Some commercial codes offer only one solver; others may offer a half dozen or more. Be aware that some solvers are sensitive to the node numbering strategy (e.g., sparse solver, skyline solver), while some solvers are sensitive to the element numbering strategy (wavefront, multifront). Iterative solvers, on the other hand, are insensitive to node and element numbering, and are indeed insensitive to bandedness. For this reason, the iterative solvers are excellent for large, chunky continuum problems, the kind of problems with large bandwidth. Iterative solvers are usually well suited to parallel computers (multiple central processing units), whether the computers share memory or not. Iterative solvers, however, are not as efficient for problems with multiple load cases, because each load case requires a repetition of the iteration strategy.

Dynamic structural problems add mass and damping terms to the system of equations:

$$\{F\} = [K]\{u\} + [C]\{du/dt\} + [M]\{d^2u/dt^2\} \qquad \text{(Eq 13)}$$

where **{du/dt}** and **{d²u/dt²}** are column vectors of velocities and accelerations, respectively. The viscous damping matrix **[C]** is usually assumed to be proportional to either the **[K]** matrix or the mass matrix, **[M]**, for computational efficiency. The **[C]** and **[M]** matrices are generally banded. Special techniques of lumping masses at nodes are often used to make the **[M]** matrix diagonal for computational efficiency. Dynamic problems can be solved in the time domain, using time-stepping procedures starting from known initial conditions. Because an additional solution is required for each time increment, this analysis is expensive. Dynamic problems, such as the determination of natural frequencies, can be solved in the frequency domain

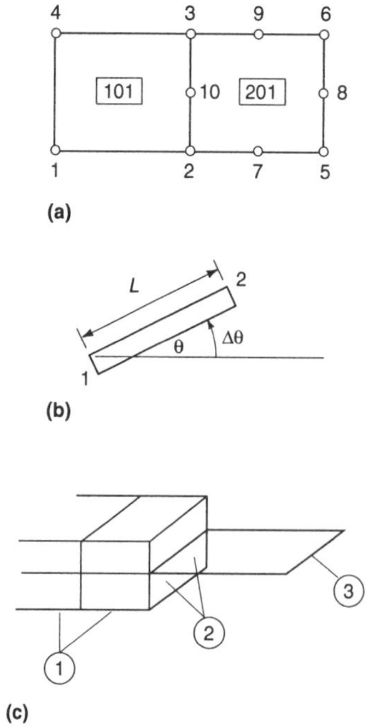

Fig. 12 Multipoint constraints. (a) Joining incompatible elements. If u and v denote displacements in the x and y directions, respectively, and subscripts refer to the nodes, then $u_{10} = (u_2 + u_3)/2$ and $v_{10} = (v_2 + v_3)/2$. This assumes that node 10 is halfway between nodes 2 and 3. (b) Simulating a rigid body. For a rigid body motion $\Delta\theta$, $u_2 = u_1$ $L \sin\theta\,\Delta\theta$ and $v_2 = v_1 + L \cos\theta\Delta\theta$. The motion of node 2 is constrained to that of point 1. (c) Joining structural elements to continuum elements. (1) Solid continuum elements with three translational degrees of freedom per node only. (2) These nodes are slaved to nodes immediately above and below them in order to prevent simulation of the mechanism (hinge). (3) Plate element with bending stiffness (three translational degrees of freedom, two rotational degrees of freedom)

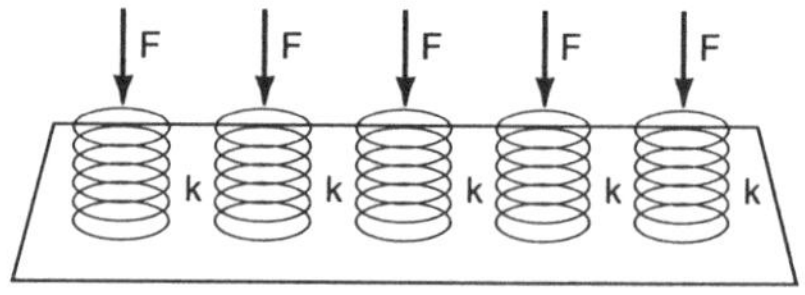

All springs are uncoupled, leading to a diagonal matrix; for unit stiffnesses:

$$[K] = \begin{vmatrix} k & 0 & 0 & 0 & 0 \\ 0 & k & 0 & 0 & 0 \\ 0 & 0 & k & 0 & 0 \\ 0 & 0 & 0 & k & 0 \\ 0 & 0 & 0 & 0 & k \end{vmatrix} = \begin{vmatrix} 1 & 0 & 0 & 0 & 0 \\ 0 & 1 & 0 & 0 & 0 \\ 0 & 0 & 1 & 0 & 0 \\ 0 & 0 & 0 & 1 & 0 \\ 0 & 0 & 0 & 0 & 1 \end{vmatrix}$$

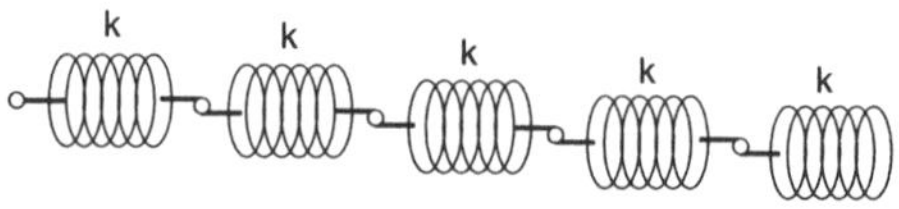

All springs are in series, leading to a banded stiffness matrix; for unit stiffnesses:

$$[K] = \begin{vmatrix} 1 & -1 & 0 & 0 & 0 \\ -1 & 2 & -1 & 0 & 0 \\ 0 & -1 & 2 & -1 & 0 \\ 0 & 0 & -1 & 2 & -1 \\ 0 & 0 & 0 & -1 & 1 \end{vmatrix}$$

Fig. 13 Form of stiffness matrix

as opposed to the time domain. This is accomplished by assuming that the displacement vector is a harmonic function, i.e.,

$$\{\mathbf{u}\} = \{\phi\}e^{i\,\omega\,t} = \{\phi\}(\cos \omega\, t + i \sin \omega\, t) \qquad \text{(Eq 14)}$$

where $i = \sqrt{-1}$, t is time, $\{\phi\}$ is the modal vector (vector of unknown amplitude at the nodes), and ω is the natural frequency associated with the mode shape. This assumption creates an *eigenvalue* problem (Ref 5).

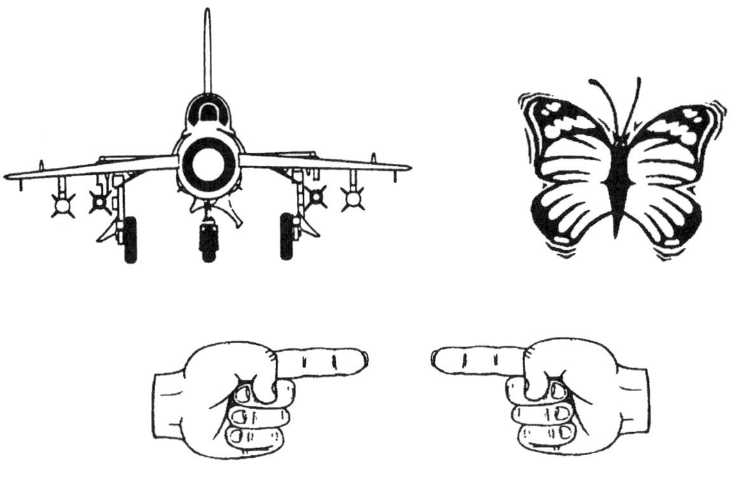

Fig. 14 Objects with bilateral symmetry

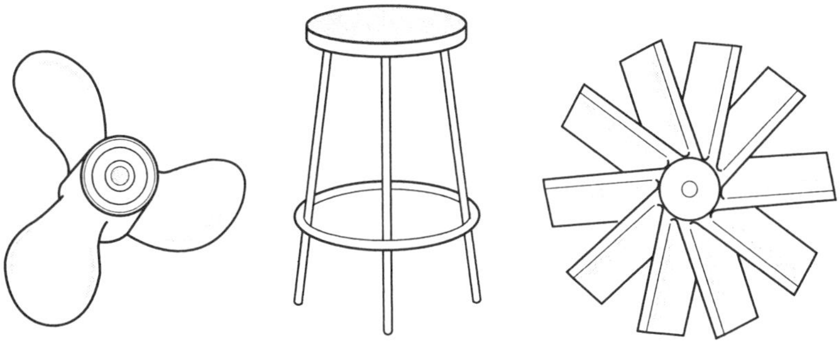

Fig. 15 Objects with rotational symmetry

Symmetry

Symmetry arises when a geometric element is translated, rotated, mirrored, or swept. At least four major types of symmetry arise in finite element problems: bilateral symmetry, rotational symmetry, axisymmetry, and translational symmetry. The exploitation of symmetry can greatly reduce the computer resources required to solve a problem, and can sometimes move problem solution from infeasibility to feasibility.

Bilateral symmetry is synonymous with reflective or mirror symmetry. This is perhaps the simplest, most intuitive type of symmetry. It does carry with it the idea of *handedness*, i.e., right hand versus left hand. If the fundamental region being reflected is itself symmetric, then right hand versus left hand need not be considered. Figure 14 illustrates some examples of bilateral symmetry. Many biological forms have bilateral symmetry, at least in the structural sense. Perhaps none of the objects depicted in Fig. 14 is *perfectly* symmetric. If the small asymmetry is structurally significant, it is sometimes adequate to model the weaker half of the object. An example might be a connecting rod with a small oil-feed hole on one side.

Rotational symmetry exists when a fundamental part of the structure spans 360°/n where n is an integer. Figure 15 shows objects that are rotationally symmetric: a stool, a marine propeller, and a fan blade. *Dihedral symmetry* is a special case of rotational symmetry. It occurs when the fundamental region to be rotated is itself bilaterally symmetric. Note that the stool in Fig. 15 has dihedral symmetry (three fundamental regions, each of which has bilateral symmetry), while the other two objects do not. *Axisymmetry* is another special case of rotational symmetry. Every cross section through the axis of rotation of an axisymmetric object is identical. Axisymmetry can reduce a three-dimensional problem to a two-dimensional one.

Translational symmetry occurs when a fundamental unit is translated one or more times. A prismatic structure is a special case of translational symmetry. Often structures comprise a fundamental element that is both translated and rotated, as shown in Fig. 16. Superelement techniques are sometimes useful for such structures.

Discussion so far has been on symmetry in the context of geometry only. When both the geometry and the loads (including the boundary reactions) are symmetric, it is simple to reduce the problem to that of the fundamental region. Consider the example of a bilaterally symmetric beam with symmetric supports and symmetric loads, as shown in Fig. 17. The behavior of the entire structure can be captured by analyzing one-half of the beam. The midpoint load must be divided by two. The boundary conditions on the cut plane are labeled SYMM. This denotes a specific set of boundary conditions. If one imagines a mirror placed at the cut section in such a way as to appropriately reflect the beam half, then the beam cross section should neither rotate away from the mirror nor interpenetrate the mirror surface as the load is applied. This restriction gives rise to two boundary conditions in a two-dimensional problem, and to three boundary conditions in a three-dimensional problem.

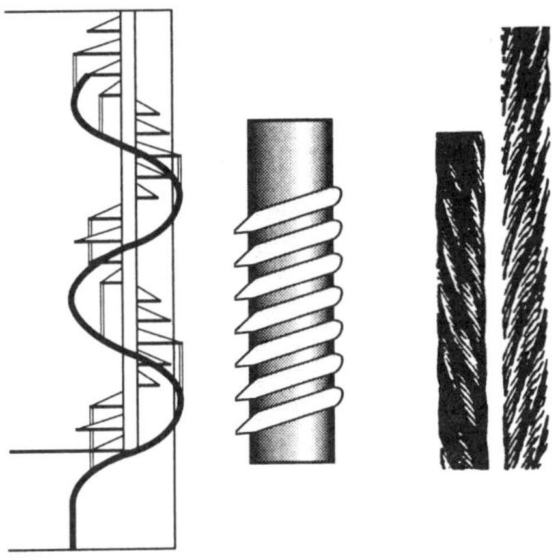

Fig. 16 Objects with translational symmetry combined with rotational symmetry

Antisymmetry refers to symmetry accompanied by a change in some property. For example, a photographic print is antisymmetrical to its negative. In the context of finite element analysis, antisymmetry pertains to a problem whose geometry, loads, and displacement boundary conditions exhibit symmetry, but the sense of whose loads are reversed. In Fig. 18, note that the sense of the forces created by the displacement boundary constraints is also reversed, but because the boundary restraints are bidirectional, it is easier to think of the boundary conditions as symmetric. Note that the boundary conditions at the cut plane are labeled ANTISYMM (antisymmetrical). These boundary conditions are the complement of the SYMM set.

For linear elastic structures, it is valid to superpose or add the results of multiple solutions. This permits decomposing of loads that do not have

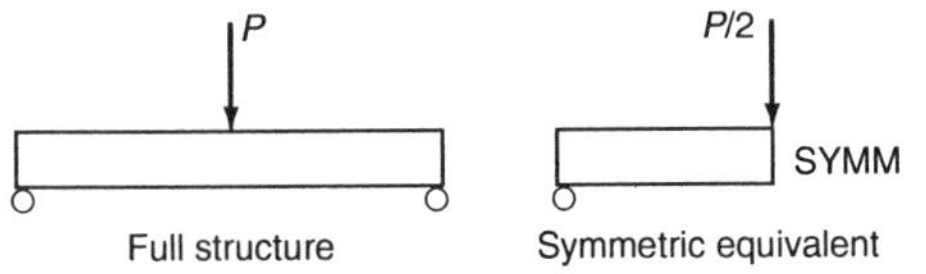

Fig. 17 Symmetric loads. The abbreviation "SYMM" denotes that $u_x = 0$ everywhere on the cross section for continuum elements (e.g., hexagonal and tetragonal elements), or that $u_x = R_y = R_z = 0$ for line elements (e.g., beams).

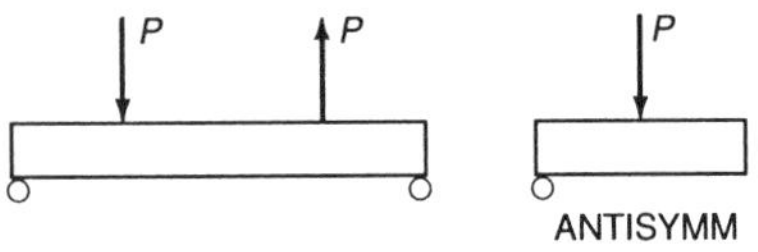

Fig. 18 Antisymmetric loads. The abbreviation "ANTISYMM" denotes boundary conditions that are the complement of the SYMM set, e.g., $u_y = u_z = 0$ everywhere on the cross section for continuum elements.

symmetry into their symmetric and antisymmetric components and superposing these component loads. This allows modeling a structure with geometrical symmetry as a reduced model, even when the loads are not symmetric (Fig. 19).

Superelements (Substructures)

Several standard techniques exist that improve computational efficiency in special situations. The use of the various forms of symmetry has already been discussed. Another of these techniques is variously known as superelement or substructuring technology. The superelements are individually modeled, and the behavior of their interior nodes is determined as a function of the behavior of their external nodes. The assembly of superelements is then analyzed in a separate run. The original motivation for this technique was to permit the solution of problems too large to be accomplished in a single pass. That is, it was an attempt to get around computer resource limitations. With the startling advances in hardware technology, that motivation has largely disappeared. However there are several situations where superelement technology provides important advantages. These are:

- Iterative redesign of a structure will affect only one or a few of the superelements.
- Analysis of assemblies will be done by analysts in disparate locations, perhaps by individual groups on subassemblies (substructures) with results later integrated into the whole.
- Nonlinear analysis is needed in which one or more superelements retain linear behavior.
- Identical units occur repetitively within the structure.
- Mesh redesign is needed in a critical local region for more accurate results.

There are some potential pitfalls with the technology. Two examples are:

- The bandwidth (or wavefront) may be adversely affected by the substructuring process.
- The mass distribution may be adversely affected for dynamic problems, although no loss of accuracy is experienced for static problems. A technique called *component mode synthesis* attempts to compensate for the altered mass distribution (Ref 1).

Finite Element Preprocessing

Figure 20 shows the finite element analysis process. *Finite element preprocessing* comprises geometry definition and manipulation, specification of material properties, generation of the finite element mesh, and definition of generalized loads and displacements both at the boundary and distributed through the body (inertia and gravitational loads). Note that material properties may be specified as constant or variable. When variable, they are often input as tables of properties or as mathematical functions, such as exponentials. Finite element preprocessing may consume as much as 80% of the calendar time in the finite element process.

When mesh is obtained by acting upon a definition of geometry, it is sometimes referred to as *top-down*. When the mesh itself is the only definition of the geometry, it may be referred to as *bottom-up*. Top-down meshing is accomplished by one of two methods: *mapped meshing* or *free meshing*. A mapped mesh requires that the geometry be subdivided into canonical shapes, such as quadrilaterals in two dimensions and cuboids in three dimensions. A mesh is then mapped into these regions. This process generally provides for more user control than free meshing, but it is quite time consuming. Free meshing, by contrast, relies on one of several algorithms to fill any arbitrary geometry with elements. In two dimensions, these elements are usually triangles or quadrilaterals. In three dimensions, robust algorithms exist to fill an arbitrary space with tetrahedra. However, robust algorithms to fill an arbitrary space with *hexahedra* are lacking. One of the contemporary issues in finite elements is the relative performance of these two element types. Because the shape functions of these elements are analogous to those of the triangle and quadrilateral in two dimensions, Fig. 7 implies the superior performance of the low-order hexahedron over the low-order tetrahedron. In practice, high-order tetrahedra are often used to increase accuracy at the expense of increased computational time.

Most finite element preprocessors offer constructs for defining geometry, but it is also desirable that geometry can be imported or exported from/to other software systems. National and international standards organizations have attempted to standardize geometry descriptions toward these ends. Geometry is sometimes described as *clean* or *dirty*, and tools to transform the latter into the former are often needed. Extraneous points or edges must be eliminated. Surfaces or edges smaller than those of the desired finite element size must be blended or eliminated without corrupting the model. Sometimes small features such as fillets and bosses are unimportant for preliminary analyses, but become important as the design matures. For this reason, *feature suppression* is a desirable option.

Another issue to be considered is the number of analysis codes supported. Large organizations often deal with several commercial codes, and the needs of any organization will change with time. Preprocessors are always incomplete with regard to the number of features supported. For example, can seldom-used elements or routines be supported? Can the applied tractions be varied parabolically, sinusoidally, and so on? Can rotation be simulated? Can loads be scaled and superposed? Are contact capabilities supported? Are multiple coordinate systems conveniently supported? Are multipoint constraints supported? What about failure theories? Are composite materials modeled?

It is preferable to define loads and boundary conditions relative to the geometry rather than to the finite element mesh. This permits remeshing of the geometry without the need to redefine loads and boundary conditions. It is also desirable to specify hard points, lines, or surfaces where parts will interact with other parts, so that the interaction (e.g., spot welding, adhesive bonding,

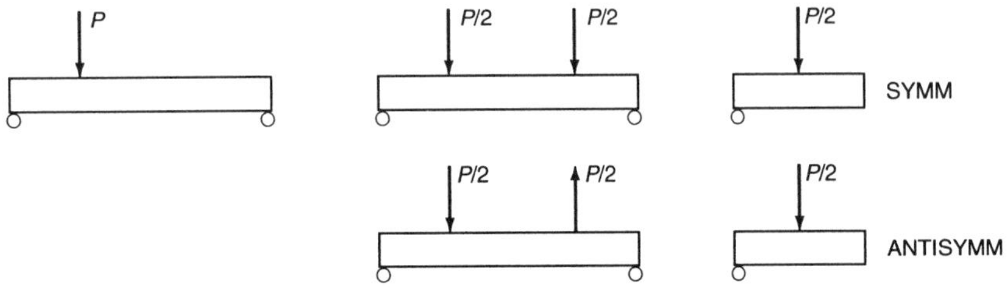

Fig. 19 Superposition of symmetric and antisymmetric loads. The abbreviations "SYMM" and "ANTISYMM" are as defined in Fig. 17 and 18.

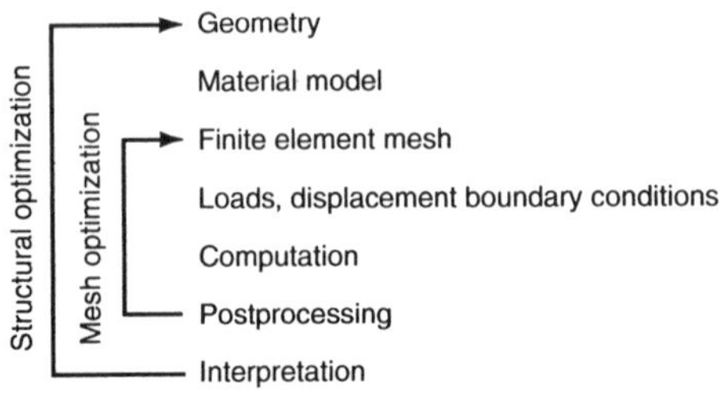

Fig. 20 Typical finite element process

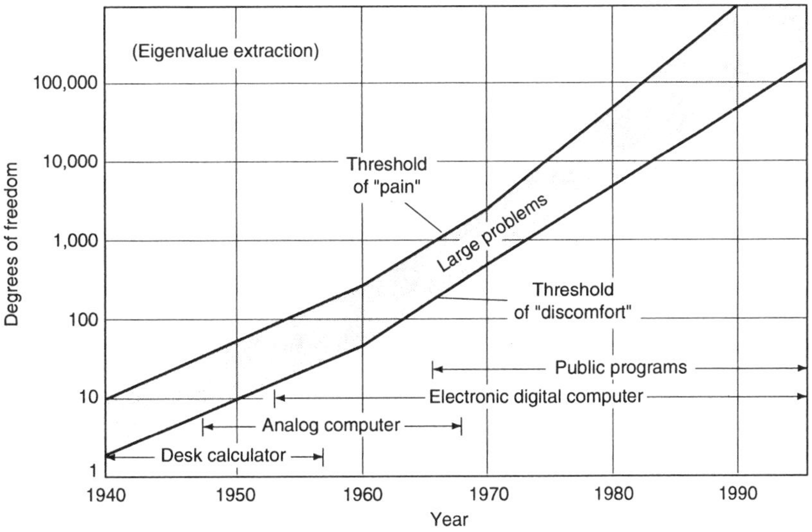

Fig. 21 The changing definition of a large problem. Source: Ref 8

Fig. 22 Solid model of an engine block

etc.) can be properly represented. Those who must create a shell mesh from a solid geometry need a capability called *midsurface extraction* in order to optimally specify the defining surface for plate/shell representation.

Still another function of a preprocessor is the ability to group and categorize various items into convenient sets. Items may be organized geographically (spatially), by material, by a property such as thickness or moment of inertia, by contact pairing, by convection coefficient, or by various other attributes.

Finite element model validation is another capability expected of the finite element preprocessor. Elements need to be investigated for geometric distortion. The traditional criteria (Fig. 6) are aspect ratio, taper, skew, warping, and edge curvature. Because even these traditional metrics are defined differently in different analysis codes, the user needs some control over their definition, their values, or the strategies imposed by the preprocessor for mesh cleanup, mesh smoothing, or element splitting. Still other validation tools are sometimes desirable: the ability to search for duplicate nodes or elements, inconsistent normals in shells, duplicate edges (implying a lack of continuity), missing elements, and so on.

Finally, returning to Fig. 20, it is apparent that the design process is iterative, that both geometry and mesh need to evolve and be updated in an appropriate manner. Macroinstruction capability in a preprocessor can greatly contribute to its usefulness. The ability to be a part of a looping scheme where geometry is modified by the previous iteration in the loop is becoming more and more important.

Finite Element Postprocessing

Finite element solutions comprise hundreds, thousands, or millions of numbers describing calculated field values. These field values might be scalar, vector, or tensor quantities such as temperature, displacement, or stress states, respectively. Usually, some *postprocessing* operations are necessary to interpret these numbers efficiently. Scalars can be plotted as color contours, vectors as arrows denoting magnitude and direction, and tensors as discrete ellipsoids at various positions. Because color contour plots are so intuitive to interpret and so useful, vector and tensor quantities are often plotted as their scalar components.

Many operations can be performed on finite element data. For example, the stresses from two load cases might be averaged and differenced to determine a mean stress and a stress range for fatigue calculations. Alternatively, fatigue damage might be calculated and summed in a manner proportional to the number of cycles predicted for each load. Such fatigue calculations and their display as color contours are a form of postprocessing.

Error indicators or estimators might be calculated and displayed. Optimization theory might be used to drive changes in size (e.g., thickness, moment of inertia), or shape. Other popular post-

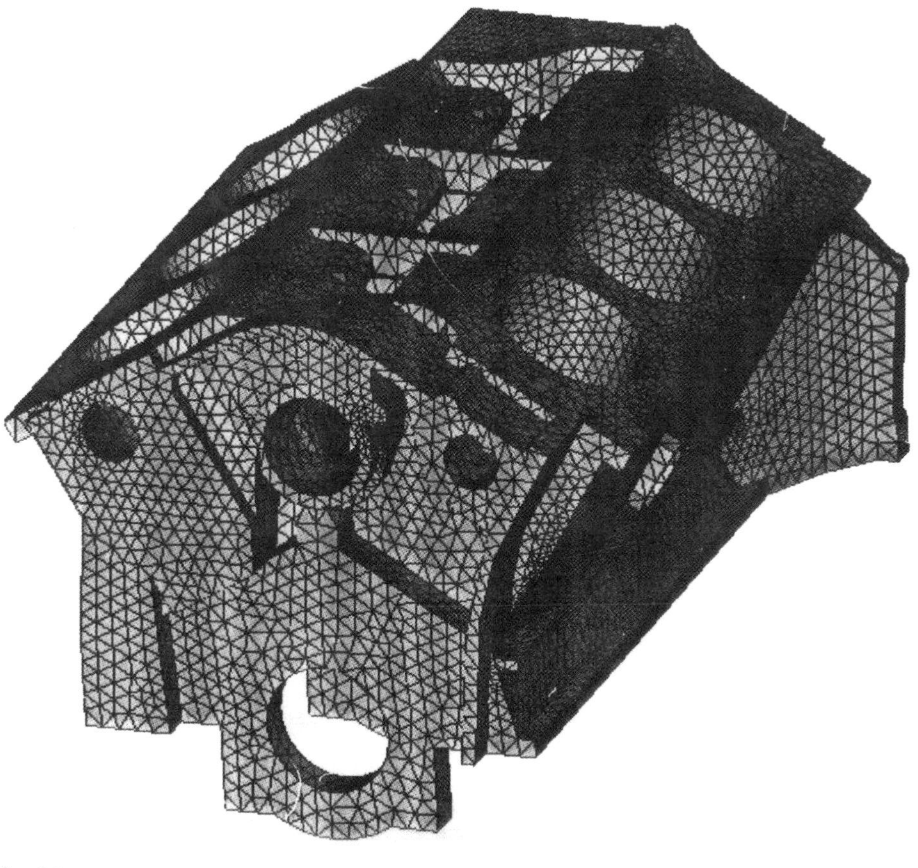

Fig. 23 Finite element model of an engine block

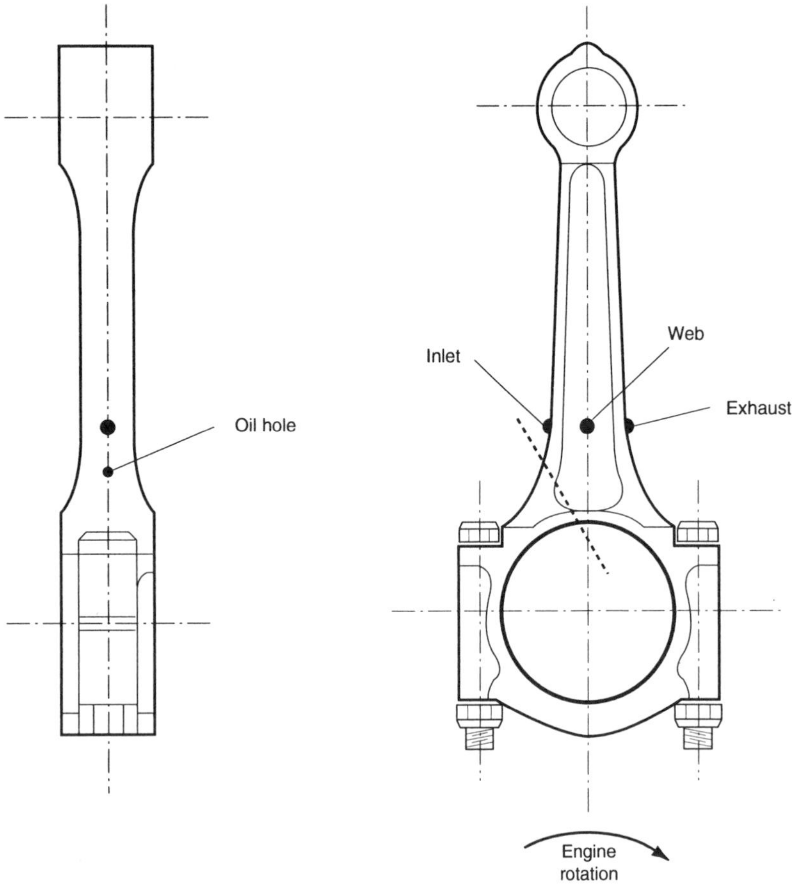

Fig. 24 Automotive connecting rod

processing operations include various *x-y* plots. This might be temperature vs. time, creep displacement vs. time, stress relaxation vs. time, stress vs. load, and so on.

Surprisingly, postprocessing to obtain stress contours is not well standardized. Different schemes exist for where the stresses are calculated, how they are interpolated or extrapolated, and how and whether they are averaged. It is advisable to read your postprocessing manual carefully and/or to compare a postprocessed solution, preferably of a coarse mesh, to a known analytical solution.

Finite Element Analysis Problems and Applications

This section presents three examples to illustrate types of problems that can be addressed and decisions that must be made when using finite element analysis.

Example 1: Finite Element Analysis of an Engine Block (Large Problem). The notion of what constitutes a large problem changes from year to year. MacNeal (Ref 8) has defined a large problem as one that crosses the threshold of "pain," and a medium-sized problem as one that merely crosses the threshold of "discomfort." Figure 21 is his notion of where these two thresholds have moved over the years. Note that a large problem can be large (painful) for a variety of reasons, principally the size of the model, a large number of time-steps needed for successful simulation, and/or nonlinearities such as large deformation, material nonlinearity, or contact.

Figure 22 depicts a solid model of a V-6 automotive engine block. Unimportant geometrical details are intentionally omitted or suppressed. The corresponding finite element model of the block is shown in Fig. 23. The mesh consists of 640,000 nodes and 1.9 million DOF. By 1996 standards, this is clearly a large problem. It required communication with the systems analyst who tends the Cray C90 supercomputer, on which it ran, to obtain access to the appropriate memory and disk resources. Solving for all modes of vibration below 3000 Hz required 23 h on a Cray C90 using a direct (as opposed to an iterative) solver.

This engine block can be modeled by alternative methods. Given sufficient preprocessing time, it can be map meshed with low-order hexahedron elements, which are more efficient than high-order tetrahedra. However, map meshing an engine block requires weeks or months of preparation time. As a compromise, the solid model of the engine block can be regioned into subgeometries such that the complex portions are free meshed with tetrahedra and the simple portions are map meshed with hexahedra. Because these elements are incompatible, constraint equations (multipoint constraints) should be generated to prevent the faces and edges of these elements from separating or interpenetrating. This compromise approach trades off computer resources against human resources.

Assembling this block with its cylinder heads would increase the size of the problem such that superelement or substructuring techniques would likely be necessary. This would be especially recommended if the nonlinear behavior of the gaskets were to be represented. That kind of static analysis is sometimes performed in order to predict cylinder bore distortion as a function of the tightening pattern of the head bolts. A nearly perfect cylinder bore minimizes emissions and maximizes performance.

Example 2: Finite Element Analysis of a Connecting Rod (Levels of Complexity). A typical automotive connecting rod is shown in Fig. 24. A connecting rod is a link that transmits power from the piston to the crankshaft. The rod rotates at its lower end and reciprocates at its upper end. For this reason, engineers are especially concerned to obtain a minimum weight design of the upper end of the rod. The following discussion of its analysis illustrates that finite element models of widely varying fidelity and cost may be appropriate at different stages of analysis and design.

A connecting rod undergoes millions or billions of cycles during its lifetime. Therefore, the problem is not one of determining maximum stress, but of ensuring that the high-cycle fatigue life is sufficient. For fatigue failure analysis of the rod, the *difference* in stress states everywhere in the structure is required. (Von Mises stresses are awkward for this sort of analysis because of the loss of sign; i.e., they are always positive by definition.)

The rod undergoes a continuously varying load during the four-stroke cycle that covers 720° of crankshaft revolution. Therefore, it is sometimes prudent to step the rod through its incremental positions to obtain the neighborhood of those positions in the cycle that produce the maximum and minimum stresses. Furthermore, it may be desirable to consider three or more load cases, say peak torque, peak horsepower, and peak rpm. Because so many analyses are required, a simple two-dimensional model of the rod may be appropriate at the preliminary design stage. The purpose of these analyses is to screen the loads to be used ultimately for the full three-dimensional model.

A second level of modeling might consider the full three-dimensional geometry of the structure, but might apply the loads in some assumed fashion. For example, the pin loading at either end might be arbitrarily spread in a sinusoidal manner over 180° (or less) of arc, at both the upper and lower ends of the rod.

A third level of modeling might explicitly consider all of the contact pairs that occur in the problem. These contact pairs include the two surfaces at the split of the big end of the rod, the crankpin-to-rod interface, the piston-pin-to-rod interface, and the rod bolts at the lower end.

Contact problems are nonlinear and therefore computationally expensive. Additional analysis that should be considered is for the buckling of the rod, or of its flanges, due to peak compressive load. This would typically be done as a linear eigenvalue problem for a rod with low loads and long life. Conversely, for a rod intended to be used in a racing engine, where loads are extremely high, a nonlinear buckling model would be more appropriate.

Example 3: Finite Element Analysis of Brake Assemblies (Multidisciplinary Problem). Brake assemblies for all forms of transportation have been analyzed by finite elements. Assume that the temperature distribution of a brake disk at the end of a panic stop from high speed is to be determined. This problem might be treated strictly as a thermal problem in which the kinetic energy of the vehicle is converted into heat generated by friction at the brake pad/brake disk interface. It might be assumed that the heat input is ramped (decreasing with time) into the disk in a spatially uniform way across the pad. The heat transferred to the atmosphere during the panic braking might be ignored because of the short time duration. If the heat transfer to the atmosphere is to be considered, then a set of values for *convection coefficients* (energy/unit area/unit time) at the disk surface is calculated or assumed. In either case, this would be a transient thermal problem with one degree of freedom, namely temperature, at each node.

However, let us assume that interest is specifically in changing the wheel design to promote brake cooling or in adding an aerodynamic device to augment cooling (as in racing). Here we consider not a single panic stop, but a series of stops or a continuous drag down a grade. A decision might be made that modeling by means of assumed convection coefficients is insufficiently accurate. In that case, a finite element model of both the fluid (i.e., the moving air surrounding the disk) and the solid might be built, wherein the boundary between the solid and the fluid is interior to the model. Analyzing the problem this way is known as *conjugate heat transfer analysis*. This analysis is inherently more costly (many more DOF and many more elements) than the method that employs the assumed convection coefficients. The feasibility of doing this large transient conjugate heat transfer problem might depend upon the ability to invoke a simplifying assumption, as, for example, cyclic symmetry.

Alternatively, if the primary concern is the smoothness of operation of the brake, this problem could be viewed as a coupled mechanical-thermal problem. The heat generated at the surface of the disk is a function of the frictional force and, therefore, the pressure distribution of the brake pad on the disk. Then thermal expansion due to temperature would influence the friction distribution and, therefore, the heat input distribution at the boundary. This coupled problem would have four DOF per node: one temperature and three displacements.

As computers get still faster and cheaper, and as software becomes increasingly sophisticated, multidisciplinary problems will no doubt be attacked more and more frequently.

REFERENCES

1. R.D. Cook, D.S. Malkus, and M.E. Plesha, *Concepts and Applications of Finite Element Analysis*, 3rd ed., John Wiley & Sons, 1989
2. B. Szabo and I. Babuska, *Finite Element Analysis*, John Wiley & Sons, 1991
3. K.J. Bathe, *Finite Element Procedures in Engineering Analysis*, Prentice-Hall, 1982
4. O.C. Zienkiewicz, *The Finite Element Method in Engineering Science*, 4th ed., McGraw-Hill, 1987
5. K.H. Huebner, E.A. Thornton, and T.G. Byrom, *The Finite Element Method for Engineers*, 3rd ed., John Wiley & Sons, 1995
6. H.L. Langhaar, *Energy Methods in Applied Mechanics*, John Wiley & Sons, 1962
7. G. Temple and W. Bickley, *Rayleigh's Principle and Its Application to Engineering, Dover, 1956*
8. R.H. MacNeal, MacNeal Schwendler Corp.

Computational Fluid Dynamics

Peter J. O'Rourke, Los Alamos National Laboratory
Daniel C. Haworth and Raj Ranganathan, General Motors Corporation

COMPUTATIONAL FLUID DYNAMICS (CFD) is one discipline falling under the broad heading of computer-aided engineering (CAE). Computer-aided engineering together with computer-aided design (CAD) and computer-aided manufacturing (CAM) comprise a mathematical-based approach to engineering product and process design, analysis, and fabrication. In this overview of CFD for the design engineer, the purposes are three-fold:

- To define the scope of CFD and motivate its utility for engineering
- To provide a basic technical foundation for CFD
- To convey how CFD is incorporated into engineering product and process design

Introduction

Computational fluid dynamics has as its objective the numerical solution of fluid-flow equations. The calculus problem of solving a coupled system of nonlinear partial differential equations (PDEs) for the variables of interest (e.g., velocity, pressure, and temperature) is transformed into an algebra problem of solving a large system of simultaneous linear equations for discrete unknowns that represent the state of a thermal-fluids system; the latter is amenable to numerical solution on a digital computer.

This is a somewhat abstract description of CFD, but it is necessary to speak in general terms when introducing a subject that encompasses such a wide variety of solution techniques. This overview discusses finite difference, finite volume, finite element, spectral, and some computational particle methods. The emphasis is on the first three, as these are the methods that are primarily used in contemporary CFD codes for engineering design.

In this article, the terminology "computational fluid dynamics" is reserved for computationally intensive three-dimensional simulations of thermal-fluids systems where nonlinear momentum transport plays an important role. It does not encompass all branches of numerical analysis as applied to fluid-dynamics problems. In particular, consideration of zero- or quasi-dimensional analysis of fluids systems (Ref 1, 2) and linear heat conduction or potential flow problems (Ref 3, 4) has been excluded.

The practice of CFD began with the advent of computers; indeed, the first computer was developed, in part, to solve fluid-flow equations. It was recognized by the developers of the atomic bomb at Los Alamos National Laboratory that many fluid-dynamics problems were impossible to solve by analytic means. What was needed was a machine that could perform the massive number of calculations necessary to solve the flow equations by simple finite-difference methods. The ENIAC computer began operating shortly after World War II, and its first calculations were to test various configurations for a hydrogen bomb (Ref 5). Ensuring the safety and reliability of modern nuclear weapons remains a major impetus for the development of more powerful computers and more efficient numerical techniques for solving the fluid-flow equations (Ref 6).

Initially, most numerical solutions were limited to flows that could be approximated as spatially one- or two-dimensional; the time and expense of performing three-dimensional calculations remained prohibitive. Over the last fifteen years, however, CFD calculations of three-dimensional flows have become more common. This has heightened enormously the interest in CFD among engineers, as most real flows are three dimensional. In fact, most fluid-flow problems encountered in industry are so complex that the *only* method of analysis to which they are amenable is CFD. Thus, although use of CFD began only fifty years ago, it is difficult to find problems in fluid dynamics to which computer solution has not been brought to bear.

The capability to perform three-dimensional CFD has resulted primarily from the availability of faster computers with larger memories (Fig. 1) (Ref 7). The development of parallel and massively parallel computers promises to further improve the speed and extend the applicability of CFD. A recent simulation of the oceans (Ref 8, 9) serves to illustrate the problem size and computational requirements that have been realized in modern applications. This problem was run on a 512-node massively parallel computer, requiring ten gigabytes of memory (giga = billion, one byte = eight bits). It ran at a computational speed of four gigaflops ("flops" = floating point operations per second) and required 80 days of computer time. A plot of ocean surface temperature obtained in this simulation is shown in Fig. 2. Computers with maximum performance at one teraflop (tera = trillion) are now becoming available, and petaflop computers (peta = 1000 trillion) are being planned for the next decade (Ref 10).

At the same time, improved numerical methods have yielded higher computational efficiency, that is, fewer operations and/or less memory for a given accuracy. Among the most important of these advances has been the development of faster methods for solving implicit difference approximations (see the following section). A third enabler for three-dimensional CFD has been the formation of improved finite-volume and finite-element methods that better accommodate the complex geometrical boundaries that characterize engineering flows. Examples of engineering applications are given later in this article.

The advent of three-dimensional calculations has increased the engineering relevance of CFD, but many obstacles remain to be overcome before CFD realizes its full potential as an engineering design tool. Foremost among these is spatial resolution. Most flows of practical interest have features whose relevant spatial and temporal scales span many orders of magnitude. For example, in an automotive four-stroke-cycle spark-ignited internal-combustion engine operating at 2000 rev/min, hydrodynamic scales range from 0.01 mm (0.0004 in.) (the turbulence microscale) to 100 mm (4 in.) (the bore diameter); flame thickness (stoichiometric, undiluted reactants) is in the range 0.01 to 0.10 mm (0.0004 to 0.004 in.); and spray droplets issuing from a typical port-fuel injector have diameters as small as 0.10 mm (0.004 in.) (Ref 11). Computers do not exist, and will not exist in the foreseeable future, that can store all the numbers required to fully resolve these phenomena. Thus, the effects of small-scale, unresolvable features on the large-scale, average flow features of interest are "modeled" through modifications to the governing PDEs. Examples of models include turbulence models, combustion models, and multiphase flow models. All models necessarily introduce imprecision, and an ongoing goal of research is to improve the accuracy of these models.

Other issues for three-dimensional engineering CFD include geometry acquisition and grid generation, numerical accuracy, and diagnostics to extract the physical information of interest from the computations. Modeling and other issues are discussed further in subsequent sections.

Fundamentals of Computational Fluid Dynamics

The governing equations of fluid dynamics and an introduction to the CFD techniques for their solution are given in this section. Also, basic terminology used by practitioners of CFD is introduced. Readers who are not interested in the technical foundation of CFD can proceed to the next section in this article, "Computational Fluid Dynamics for Engineering Design."

Governing Equations

The equations of fluid dynamics can be derived from kinetic theory or continuum points of view (Ref 12–14), each of which complements the other. Kinetic theory regards the fluid as made up of molecules subject to collisions and intermolecular forces. Kinetic theory derivations are valid only for dilute gases, but give detailed information about how such transport phenomena as stresses and heat fluxes arise from molecular fluctuations, which in turn are related to the average molecular properties for which the fluid equations solve. Continuum derivations regard the fluid as a continuous medium and show the applicability of the fluid equations to a much broader class of media than dilute gases, but do not give detailed information about transport phenomena.

The Equations of Continuous, Compressible Media. Three basic physical principles, applicable to any continuous medium, are used in continuum derivations:

- Conservation of mass
- Newton's second law that force equals mass times acceleration
- The first law of thermodynamics, which states that total energy, in all its forms, must be conserved

These three principles lead to the following three equations of motion: the mass, or continuity, equation:

$$\frac{\partial \rho}{\partial t} + \frac{\partial \rho u_i}{\partial x_i} = 0 \qquad \text{(Eq 1)}$$

the momentum equation:

$$\frac{\partial \rho u_i}{\partial t} + \frac{\partial (\rho u_i u_j)}{\partial x_j} = \frac{\partial \Sigma_{ij}}{\partial x_j} + \rho F_i \qquad \text{(Eq 2)}$$

the total energy equation:

$$\frac{\partial \rho E}{\partial t} + \frac{\partial (\rho E u_j)}{\partial x_j} = \frac{\partial (\Sigma_{ij} u_j)}{\partial x_i} - \frac{\partial Q_j}{\partial x_j} + \rho F_i u_i \qquad \text{(Eq 3)}$$

These equations are written in Cartesian tensor notation (Ref 15), according to which the subscripts i and j take the values 1, 2, or 3 corresponding to the three Cartesian coordinate directions. A subscript that appears just once in a term takes on one of the three values 1, 2, or 3; repeated subscripts in a term denote a summation of that term over all three coordinate directions. The other notation in Eq 1 to 3 is defined in Table 1.

The total energy E is the sum of the local flow kinetic energy and its internal energy e:

$$E = e + \frac{1}{2} u_i^2 \qquad \text{(Eq 4)}$$

Alternative energy equations for e and for enthalpy $h = e + p / \rho$, where p is the pressure, can easily be derived using Eq 1 to 3. Computational fluid dynamics codes often solve internal energy or enthalpy equations, in place of Eq 3, when calculating compressible flows.

The above equations are expressed in *Eulerian* form; that is, the time derivative is taken at a fixed point in space. This contrasts with the *Lagrangian* form, in which the time derivative is taken following a fluid element (Ref 14). Although the Eulerian form of the equations is most often used in CFD, there are CFD methods that approximate the Lagrangian equations.

When completed with constitutive relations appropriate for fluids, these are the basic equations of compressible fluid dynamics. In practice, one often encounters applications in which extensions of these equations are necessary. Among the most common are extensions to multicomponent and chemically reactive flows (Ref 16), to magnetohydrodynamic flows (Ref 17), and to flows with radiative heat transfer (Ref 18). It is beyond the scope of this overview of CFD to give these extended equations, and the reader is referred to the above references for this information.

Table 1 Computational fluid dynamics nomenclature

Symbol	Definition
C_p	Specific heat at constant pressure
C_v	Specific heat at constant volume
e	Internal energy
E	Total energy
F	Body force per unit mass
h	Enthalpy
K	Turbulent kinetic energy
p	Pressure
Q	Heat flux
R	Universal gas constant
S	Rate of deformation
t	Time
T	Temperature
u	Velocity
W	Molecular weight
x	Spatial location
δ	Kronecker delta function
ε	Dissipation rate of turbulent kinetic energy
κ	Heat conductivity
λ	Second coefficient of viscosity
λ'	Bulk viscosity
μ	First coefficient of viscosity
ρ	Mass density
Σ	Stress

Constitutive Relations of Fluid Flow. To complete these equations the stress Σ_{ij} and heat flux Q_i need to be expressed in terms of known fluid variables and their derivatives. These expressions are known as constitutive relations. A fluid is a medium for which the nonhydrostatic part of its stress depends only on its rate of deformation S_{ij}. The quantity S_{ij} is given by:

$$S_{ij} = \frac{1}{2}\left(\frac{\partial u_i}{\partial x_j} + \frac{\partial u_j}{\partial x_i}\right) \qquad \text{(Eq 5)}$$

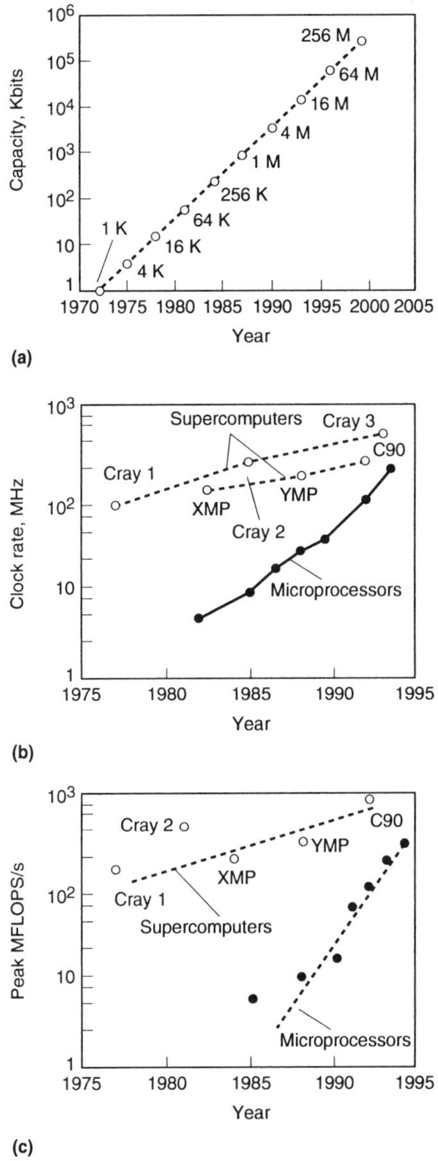

Fig. 1 Growth in computer hardware performance, 1970 to 1995. (a) Memory chip capacity doubles every 1.5 years. (b) Clock rate. (c) Peak single-process megaflops. Source: Ref 7

Thus a fluid has no memory of its previous configurations. This fact, together with the assumption of an isotropic medium in terms of its microstructure, allow the expression (Ref 14) of the full stress as:

$$\Sigma_{ij} = 2\mu S_{ij} + \lambda S_{kk}\delta_{ij} - p\delta_{ij} \quad \text{(Eq 6)}$$

where δ_{ij} is the Kronecker delta function (Ref 15). Fluids with this form of stress tensor are called Newtonian fluids. The thermodynamic pressure p and the first and second coefficients of viscosity μ and λ, depend only on the local thermodynamic state of the fluid. Often the second coefficient of viscosity in Eq 6 is replaced by the bulk viscosity λ', defined by:

$$\lambda' = \lambda + \frac{2}{3}\mu \quad \text{(Eq 7)}$$

The heat flux Q_i depends on gradients in temperature. Again assuming an isotropic fluid Q_i can be written:

$$Q_i = -\kappa \frac{\partial T}{\partial x_i} \quad \text{(Eq 8)}$$

This is Fourier's heat conduction law, and κ is called the heat-conduction coefficient or simply the heat conductivity. Its value depends on the local thermodynamic state of the gas. When one substitutes Eq 6 for the stress tensor and Eq 8 for the heat-flux vector into Eq 1 to 3, the resulting equations are called the compressible Navier-Stokes equations.

The fluid equations are completed by the specification of the quantities p, e, μ, λ, and κ in terms of the local fluid temperature T and density ρ. The equations specifying p and e are referred to as thermal and caloric equations of state, respectively. For a so-called ideal gas, these are given by:

$$p = \rho \frac{R}{W} T \quad \text{(Eq 9)}$$

and

$$e = \int^{T} C_v(T')\, dT' \quad \text{(Eq 10)}$$

where the specific heat at constant volume C_v is a function of temperature. Alternatively, the enthalpy h is given by:

$$h = \int^{T} C_p(T')\, dT' \quad \text{(Eq 11)}$$

where, from the definition of h and the thermal equation of state Eq 9, the specific heat at constant pressure C_p is given by:

$$C_p = C_v + \frac{R}{W} \quad \text{(Eq 12)}$$

Values of C_v and C_p versus temperature are given in Ref 19 and 20.

The quantities μ, λ, and κ are called transport coefficients. How they are related to the local thermodynamic state of the fluid and its molecular properties are given in Ref 21. Given an expression for the viscosity μ, the heat conductivity κ can frequently be approximated by:

$$\kappa = \frac{\mu C_p}{Pr} \quad \text{(Eq 13)}$$

where Pr is the Prandtl number, whose value is nearly constant and often of order unity.

Simplifications of the Fluid-Flow Equations. For certain flow situations, considerable computer time can be saved by solving simplified forms of the compressible flow equations. In this section, the steady-state, inviscid, and incompressible approximations are introduced, and the circumstances under which they may be used are described.

The steady-state approximation is obtained simply by dropping the time-derivative terms in Eq 1 to 3. While solving the steady-state equations can often save computer time, sometimes CFD-solution techniques for the steady-state equations have what are called convergence difficulties, and steady-fluid-flow solutions are more reliably obtained by calculating the long-time limits of solutions to the unsteady equations (Ref 22).

The inviscid, or Euler, equations are obtained by neglecting the viscosity and heat-conduction terms in the preceding equations. A necessary condition for the applicability of the Euler equations is that the Reynolds number Re be much greater than one, where Re is defined by:

$$Re = \frac{\rho u L}{\mu} \quad \text{(Eq 14)}$$

In Eq 14 ρ, u, and μ are characteristic values of the density, velocity, and visocity, respectively, of the fluid, and L is a characteristic distance over which the velocity changes appreciably, also called a gradient length. In a flow to which the Euler equations apply, L is typically the dimension of the apparatus that bounds the flow. The Reynolds number is approximately the ratio of the magnitude of the advective terms to that of the viscous terms in the fluid momentum equation, Eq 2. Thus, when Re is large the viscous terms may sometimes be neglected. When fluid Prandtl numbers are of order unity, smallness of the viscous terms also implies smallness of the heat-conduction terms relative to the advection terms in the energy equation.

There are many high Reynolds number flows, however, where neglect of the viscous and heat-conduction terms is not justified. Sometimes fluid flows have broad regions over which the inviscid equations apply, coupled with thin regions (e.g., boundary layers and shocks) in which the viscous and heat-conduction terms are important. In addition, as Re is increased many flows become turbulent, and the velocity then varies over a range of length scales L. At the smallest of these length scales, Re is of order unity, and viscosity is important because it is responsible for the dissipation of turbulent kinetic energy into heat. (See the section on turbulence below.)

An incompressible flow is one in which the divergence of the velocity field is identically equal to zero:

$$\frac{\partial u_i}{\partial x_i} = S_{ii} = 0 \quad \text{(Eq 15)}$$

A necessary, but not sufficient (see Ref 14), condition that a flow be incompressible is that the Mach number M be much less than one, where M is defined by:

$$M = \frac{u}{c} \quad \text{(Eq 16)}$$

In Eq 16, u and c are characteristic values of the velocity and sound speed of the fluid.

In combination with the continuity equation Eq 1, Eq 15 implies that:

$$\frac{D\rho}{Dt} = \frac{\partial \rho}{\partial t} + u_i \frac{\partial \rho}{\partial x_i} = 0 \quad \text{(Eq 17)}$$

D/Dt is the time derivative following a fluid element, and Eq 17 states that the density of each element of fluid remains a constant along its trajectory. Usually, a more restrictive assumption is made that the density of the whole fluid is equal to a constant ρ_0. In this case, the momentum equation, Eq 2, becomes:

$$\frac{\partial u_i}{\partial t} + \frac{\partial (u_i u_j)}{\partial x_j} = \frac{1}{\rho_0}\frac{\partial \Sigma_{ij}}{\partial x_j} + F_i \quad \text{(Eq 18)}$$

The great simplification of incompressible flow equations is that the energy equation is decoupled from the momentum equation and need not be solved. This is the so-called primitive-variable form of the incompressible flow equations. Another formulation that is used in CFD calculations of two-di-

Fig. 2 Ocean surface temperatures from a recent CFD simulation of the North Atlantic Ocean. Source: Ref 8

mensional, incompressible flows is the stream function and vorticity formulation, which can be found in Ref 3.

Turbulence and Other Models. As has been stated in the introduction, there are many flow situations in which flows have changes in their properties, such as their velocities, with superimposed size scales or time scales that differ by many orders of magnitude. Examples are the seemingly chaotic motions in a turbulent flow or in a multiphase flow such as a liquid spraying into a gas. Classical theories of turbulence predict that the ratios of the largest to the smallest fluctuation length scales of turbulent flows are approximately equal to $Re^{0.75}$, where, in this case, Re is based on the velocity and size scales of the largest turbulent eddies (Ref 23). Even a low value of Re = 10,000 gives fluctuation length scales varying over three orders of magnitude. For such cases it is impossible to resolve the detailed flow fluctuations with CFD methods, and fortunately one usually is not concerned with these detailed fluctuations. The average flow behavior is of interest, however, and it is important to account for the effects of the fluctuations on average flow variables.

There are many ways to define averaged, or filtered, flow variables. In general, space- and time-averages can be defined using a filter function $K(x_i, t)$ whose integral over all space and time is unity. In terms of K, the average of a fluid variable q, denoted by $\bar{q}$, is defined by:

$$\bar{q}(x_i, t) = \iiiint q(y_i, t') K(y_i - x_i, t' - t)\, dy_i\, dt' \quad \text{(Eq 19)}$$

For example, for pure time averaging, one can take $K(x_i, t) = \delta(x_i)\, \Psi_T(t)/T$, where:

$$\Psi_T(t) = \begin{cases} 1 & \text{if } |t| < T/2 \\ 0 & \text{otherwise} \end{cases} \quad \text{(Eq 20)}$$

and $\delta(x_i)$ is the Dirac delta function. Then the average q is defined by:

$$\bar{q}(x_i, t) = \frac{1}{T}\int_{t-T/2}^{t+T/2} q(x_i, t')\, dt' \quad \text{(Eq 21)}$$

and the filter size is said to be T. In addition to space- or time-averaging, one can also use ensemble averaging. This is defined by averaging over an imagined large set of realizations of a fluid experiment. Sometimes ensemble-averaging is combined with space- or time-averaging. In any case, the fluctuation of quantity q from its mean value is denoted by q':

$$q'(x_i, t) = q(x_i, t) - \bar{q}(x_i, t) \quad \text{(Eq 22)}$$

There are two approaches to calculating average flow fields. In the first, called Reynolds averaging because it was first proposed by O. Reynolds (Ref 24), one is interested in predicting *the* average flow field and uses ensemble averaging or a filter size that is large compared to the scales of fluctuations. Thus, the average of the fluctuating part of q is zero:

$$\overline{(q')} = 0 \quad \text{(Reynolds averaging)} \quad \text{(Eq 23)}$$

In contrast, subgrid-scale turbulence models use filters with as small a size as possible, typically comparable to the grid size in the CFD calculation. Thus, one attempts to calculate flow fluctuations with scales larger than the filter, or grid, size and to model only subgrid-scale fluctuations. In a subgrid-scale model, the average of the fluctuation part of q is in general nonzero:

$$\overline{(q')} \neq 0 \quad \text{(subgrid model filtering)} \quad \text{(Eq 24)}$$

Once the method of averaging is chosen, then equations for the averaged flow variables can be obtained by averaging the equations of the preceding sections, or simplified forms of these. In deriving the averaged equations, one finds that the rates of change of average flow variables depend on averages of the products of two fluctuating quantities, also called second-order correlations. The values of these are unknown, and if one tries to close the system of equations by deriving transport equations for the second-order correlations, it is found that these depend on third-order correlations, or averages of three fluctuating quantities. Continuing in this way, one finds that it is impossible to obtain a closed system of equations using either Reynolds-averaging or subgrid-scale-averaging (Ref 25). By using physical and dimensional reasoning and empirical information, the unknown correlation must be expressed (the word "modeled" is also used here) in terms of average flow variables that are known.

A very important example of a second-order correlation and its modeling arises when averaging the incompressible flow momentum equation (Eq 18). The Reynolds-averaged form of this equation is:

$$\rho_0\left(\frac{\partial \bar{u}_i}{\partial t} + \frac{\partial \bar{u}_i \bar{u}_j}{\partial x_j}\right) = \frac{\partial \bar{\Sigma}_{ij}}{\partial x_j} - \frac{\partial}{\partial x_j}\left(\rho_0 \overline{u'_i u'_j}\right) + \rho_0 \bar{F}_i \quad \text{(Eq 25)}$$

The second-order correlation $-\rho_0\overline{u'_i u'_j}$ on the right-hand side of this equation is called the Reynolds stress. The most popular turbulence models in engineering design calculations are the so-called two-equation models, in which the Reynolds stress is, with some theoretical justification (Ref 25, 26), taken to have the form:

$$-\rho_0\, \overline{u'_i u'_j} = 2\mu_T \bar{S}_{ij} - \frac{2}{3}\rho_0 K\, \delta_{ij} \quad \text{(Eq 26)}$$

In this expression μ_T is the turbulent viscosity and K is the turbulent kinetic energy:

$$K = \frac{1}{2}\overline{(u'_i)^2} \quad \text{(Eq 27)}$$

By substituting Eq 26 into Eq 25, one finds that the momentum equation for turbulent flow closely resembles the momentum equation for laminar, or nonturbulent, flow. This is also the case for other averaged fluid equations, and this resemblance allows the same numerical techniques for CFD to be applied to both laminar and turbulent flows.

In two-equation turbulence models, transport equations are solved for the turbulent kinetic energy K and one other scalar that gives a local length or time scale of the turbulence. A popular choice for this second turbulence quantity is the turbulence kinetic energy dissipation rate ε. In terms of K and ε the turbulent viscosity is given by:

$$\mu_T = c_\mu \rho_0 \frac{K^2}{\varepsilon} \quad \text{(Eq 28)}$$

where c_μ is a dimensionless constant. The K-ε turbulence model is described in more detail in Ref 26, which also describes the use of wall functions to calculate wall heat and momentum losses in conjunction with the K-ε model. Descriptions of many two-equation turbulence models, and their relative advantages, can be found in Ref 27.

Numerical Solution of the Fluid-Flow Equations

This section introduces some common techniques for discretizing the fluid-flow equations and methods for solving the discrete equations.

Discretization of the Fluid Equations. In the process of discretization a continuously varying fluid flowfield, which has an infinite number of degrees of freedom, is represented by a finite set of data. This section introduces the discretization techniques used by finite-difference, finite-volume, finite-element, spectral, and some particle methods, and associated concepts of numerical stability and accuracy. The discrete equations of the finite-difference, finite-volume, and finite-element techniques all look similar and are referred to generically as "difference approximations." This introduction to CFD can only scratch the surface of each method. For more in-depth information, the reader should consult Ref 3 and 28 to 32.

Finite-Difference Methods (FDM). In FDMs the entire fluid region of interest is divided into nonoverlapping cells, and approximate values of the fluid variables are stored in each cell. This subdivision is called a grid or a mesh. Derivatives are approximated by taking differences between the variable values in neighboring cells, using the idea of a Taylor-series expansion. Consider the simple one-dimensional example of finite-difference solution of the linear advection equation:

$$\frac{\partial q}{\partial t} + u\frac{\partial q}{\partial x} = 0 \quad \text{(Eq 29)}$$

in the spatial interval $a \leq x \leq b$. This interval is subdivided into cells of equal size $\Delta x = (b - a)/N$, where N is the total number of cells, and q_i^n denotes the approximate value of q at the center of cell i, which lies at the location, or grid point, $x_i = a + (i - ½)\Delta x$, at time $t = n\Delta t$, where Δt is the computational timestep. (In this section, the subscript i represents a cell number, rather than a coordinate direction.) All the values of q_i^n, $1 \leq i \leq N$, for a particular time t =

$n\Delta t$ are stored in computer memory, and values can be computed at time $t = (n + 1)\Delta t$ by using a finite-difference approximation to Eq 29.

To approximate the spatial derivative in Eq 29, q_i^n is considered to be the value at x_i of a differentiable function $q(x, t)$ that can be expanded in a Taylor series about any grid point. Thus the value of q at a neighboring grid point can be expressed in terms of the value of q and its derivatives at grid point i by:

$$q_{i+k}^n = q_i^n + \left(\frac{\partial q}{\partial x}\right)_i^n k\Delta x + \left(\frac{\partial^2 q}{\partial x^2}\right)_i^n \frac{(k\Delta x)^2}{2} + O(\Delta x^3) \quad \text{(Eq 30)}$$

where $O(\Delta x^m)$ represents the fact that the remaining terms in this expansion have as their lowest-order term one in which Δx is raised to the power m. Now the value of the spatial derivative in Eq 29 can be approximated by any finite combination that satisfies:

$$\sum_k a_k q_{i+k}^n = \left(\frac{\partial q}{\partial x}\right)_i^n + O(\Delta x^m) \quad \text{(Eq 31)}$$

when one substitutes from Eq 30 for the q_{i+k}, where a_k are coefficients that depend on Δx. In the approximation Eq 31, m is said to be the order of accuracy of the approximation and all terms containing Δx to some power are said to be truncation errors. As long as $m > 0$, the approximation is said to be consistent. Examples of consistent approximations are the centered-difference approximation:

$$\frac{q_{i+1}^n - q_{i-1}^n}{2\Delta x} = \left(\frac{\partial q}{\partial x}\right)_i^n + O(\Delta x^2) \quad \text{(Eq 32)}$$

which is second-order accurate, and the one-sided approximations:

$$\frac{q_i^n - q_{i-1}^n}{\Delta x} = \left(\frac{\partial q}{\partial x}\right)_i^n + O(\Delta x) \quad \text{(Eq 33a)}$$

and

$$\frac{q_{i+1}^n - q_i^n}{\Delta x} = \left(\frac{\partial q}{\partial x}\right)_i^n + O(\Delta x) \quad \text{(Eq 33b)}$$

which are first-order accurate. If the advection speed u in Eq 29 is positive, then the approximation Eq 33(a) is called an upwind approximation and Eq 33(b) a downwind approximation.

Order of accuracy is one measure of the accuracy of an FDM. To test the accuracy of a finite-difference solution, one can refine the grid by reducing the cell size Δx. When Δx is reduced by a factor of two, numerical errors will be reduced approximately by a factor of four when using a second-order method, but only by a factor of two with a first-order method. It may thus seem to be desirable to use only methods with a very high order of accuracy. In practice, however, it is difficult to define high-order methods near boundaries and often numerical solutions using high-order methods have oscillations in regions of steep gradients. Because of these difficulties, most modern FDMs have second- to fourth-order ac-

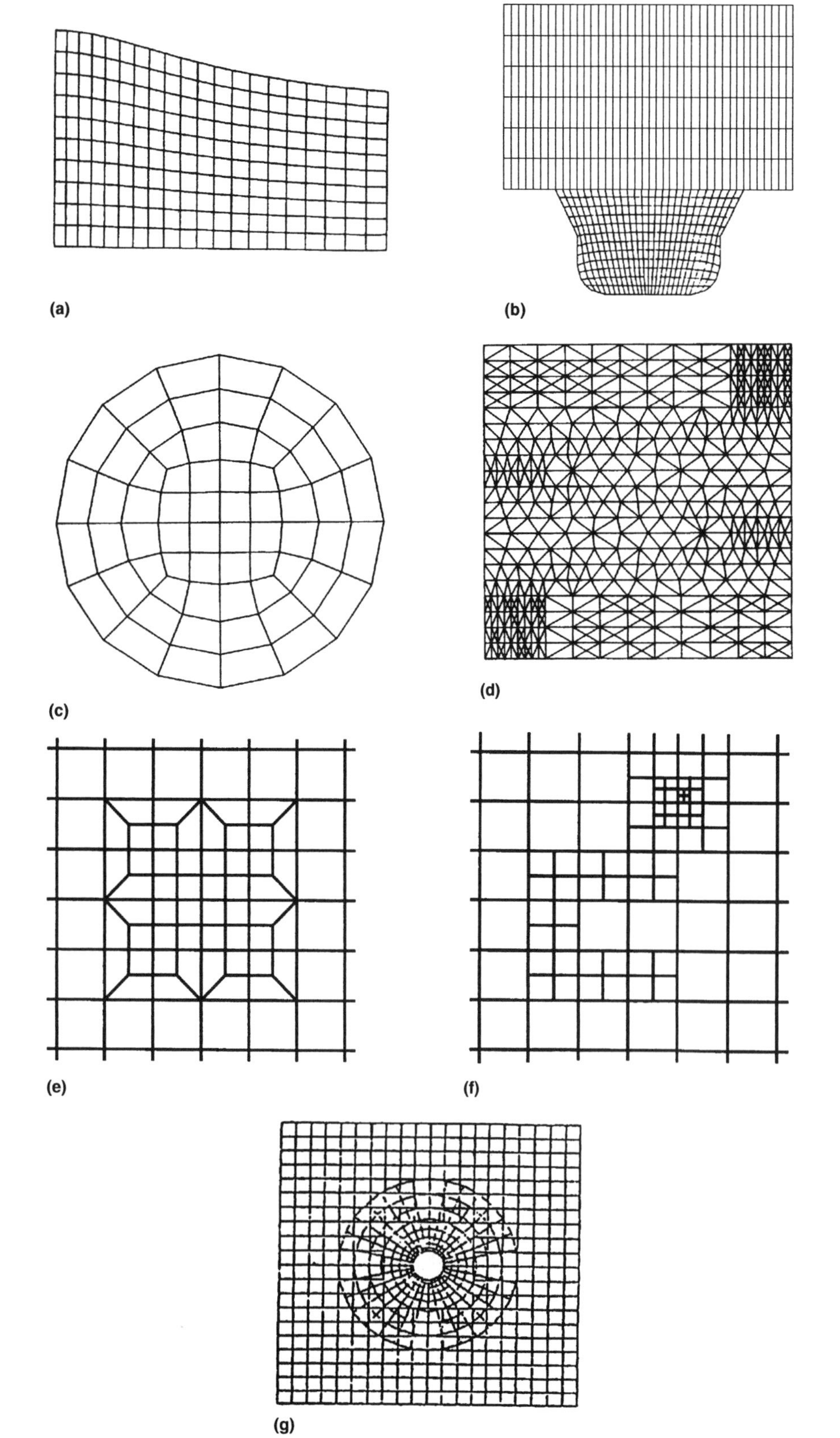

Fig. 3 Examples of grids used in CFD calculations. Two-dimensional examples are shown for clarity. (a) A structured grid. (b) A block-structured grid. (c) An unstructured hexahedral (quadrilateral) grid. (d) An unstructured tetrahedral (triangular) grid. (e) Local mesh refinement via a transition region on an unstructured hexahedral grid. (f) Local mesh refinement via cell splitting on an unstructured hexahedral grid. (g) A chimera grid

curacy and sometimes drop to first-order accuracy in regions of steep gradients.

Returning to the example of the linear advection equation, the time derivative can be approximated in much the same way as the spatial derivative. Because one usually only stores the value of q_i at a single time-level in order to save computer storage, the time derivative is most often approximated by the one-sided finite-difference formula:

$$\frac{q_i^{n+1} - q_i^n}{\Delta t} = \left(\frac{\partial q}{\partial t}\right)_i^n + O(\Delta t) \quad \text{(Eq 34)}$$

When Eq 34 and one of the finite-difference formulas Eq 32, 33(a), or 33(b) is used to approximate the time- and space-derivatives in Eq 29, one obtains a consistent approximation to the linear advection equation that is first-order accurate in time.

When these finite-difference equations are used to advance the numerical solution for q in time, one finds that, in contrast to solutions to the differential equation, solutions to the finite-difference equations using Eq 32 or the downwind approximation Eq 33(b) are subject to catastrophic numerical instabilities, and solutions using Eq 33(a) are only stable if a certain condition is met. This condition, the so-called Courant condition, is that the Courant number $C = (u\Delta t)/\Delta x$ be less than one. The origin of these numerical instabilities was first discovered by J. von Neumann (Ref 33), who devised a method for analyzing the stability of linear finite-difference equations based on examining the behavior of each Fourier component of the solution.

The finite-difference approximations presented so far are explicit in the sense that the solution for q_i^{n+1} can be explicitly found by solving only the finite-difference equation at grid point i. All explicit methods, if they are stable, are subject to Courant conditions to ensure their numerical stability. Intuitively, this condition arises because when using an explicit method, information can only propagate at a speed proportional to $\Delta x/\Delta t$. In order for the numerical solution to approximate the physical solution, the numerical propagation speed must be at least as great as the physical speed. For the simple advection Eq 29, the only physical propagation speed is u. For the fluid equations there are several physical, or characteristic, speeds. The largest of these is $u + c$, where c is the fluid speed of sound, and the Courant condition in explicit CFD calculations is based on the speed $u + c$. To overcome the Courant condition one uses implicit FDMs, in which solution for the value of q_i^{n+1} is implicitly coupled to the solution for q^{n+1} at other grid points. An example of an implicit finite-difference approximation to the linear advection Eq 29 is:

$$\frac{q_i^{n+1} - q_i^n}{\Delta t} = -u\,\frac{q_{i+1}^{n+1} - q_{i-1}^{n+1}}{2\Delta x} \quad \text{(Eq 35)}$$

which can be shown to be unconditionally stable. The disadvantage of implicit methods is that they usually require costly iterative solution. Some iterative solution techniques for implicit equations will be introduced below.

Finite Volume Methods (FVM). As in FDMs, FVMs subdivide the computational region into a mesh of cells, but finite-volume cells can be arbitrary quadrilaterals in two dimensions, hexahedra in three dimensions, or indeed any shape enclosed by a set of corner points. In contrast, FDMs are defined on grids that are obtained using orthogonal curvilinear coordinate systems. Thus, FVMs have much more geometric flexibility than FDMs.

Finite-volume methods approximate forms of the fluid equations that are integrated over these cells, which are also called control volumes. An example is the finite-volume approximation of the integrated form of the mass equation, Eq 1. After integrating Eq 1 over control volume V and applying the Reynolds transport and divergence theorems (Ref 14) one obtains:

$$\frac{d}{dt}\iiint_V \rho\, dv + \iint_S \rho u_i n_i da = 0 \quad \text{(Eq 36)}$$

The quantity $\rho u_i n_i$ is the mass flux (mass per unit area and time) through surface S with unit normal n_i, and Eq 36 is a statement that the time-rate-of-change of the total mass in volume V is equal to the sum of the fluxes, times the areas, through the surface S of the volume. Thus mass is conserved in the sense that there are no internal mass sources. Commonly, the time derivative term in Eq 36 is approximated by:

$$\frac{d}{dt}\iiint_V \rho\, dv \approx V_\nu \frac{\rho_\nu^{n+1} - \rho_\nu^n}{\Delta t} \quad \text{(Eq 37)}$$

where ν is the index of finite-volume cell and V_ν is its volume, and the surface integral is approximated by:

$$\iint_S \rho u_i n_i\, da \approx \sum_\alpha \rho_\alpha (u_i)_\alpha (n_i)_\alpha A_\alpha \quad \text{(Eq 38)}$$

where the sum is over all faces α of control volume ν; ρ_α and $(u_i)_\alpha$ are approximations to ρ and u_i, respectively, on face α; $(n_i)_\alpha$ is an average unit normal vector to face α pointing out of volume V; and A_α is the area of face α.

Using finite-volume methods one can easily construct discrete approximations that have the conservative property; that is, the discrete approximations can mimic the physical laws from which the fluid equations were derived by conserving properties such as computed mass, momentum, and energy. To be more precise, consider the approximation to the mass equation above. A conservative approximation has the property that if ν and μ are two cells that share face α, then when one sums the finite-volume approximations to the change of mass in cells ν and μ, the contributions due to fluxes through common face α cancel each other. This will be true if ρ_α and $(u_i)_\alpha$ are defined the same way in the finite-volume approximations at nodes ν and μ, because the unit outward normal to face α relative to cell ν is minus the outward normal relative to cell μ. Conservative difference approximations have many desirable accuracy properties. For example, it can be shown that difference approximations that conserve mass, momentum, and energy will calculate the correct jump conditions across shocks without having to resolve shock structure (Ref 3).

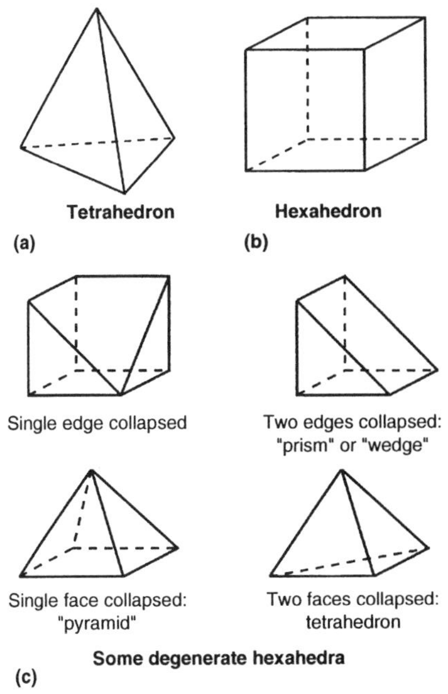

Fig. 4 Principal cell or element types for CFD. (a) Tetrahedron: there are four vertices or nodes, four faces, and six edges for each element. (b) Hexahedron: there are eight vertices or nodes, six faces, and twelve edges for each element. Hexahedral elements generally must remain convex (angles formed by edge and face intersections must remain smaller than 180°). (c) A sampling of possible edge and/or face degeneracies for hexahedral elements

A problem with FVMs is that it is difficult to formulate higher-order FVMs. When a FVM is specialized to a finite-difference grid, the difference approximations look very much like finite-difference approximations, and one can perform Taylor-series expansions and determine the order of accuracy of the method. When more general meshes are used, however, it is unclear whether the same accuracy can be expected.

Finite element methods (FEMs) (Ref 28) use a consistent spatial interpolation when evaluating all the spatial derivative terms in the fluid dynamics equations. These methods have long been popular in stress-analysis problems and have recently been gaining popularity in CFD problems because of advances in the methodology. As in FVMs the computational domain is subdivided into nonoverlapping cells that in three dimensions are either arbitrary hexahedra or tetrahedra (Fig. 3, 4). Finite-element terminology is different, however, in that the cells are called elements, and the vertices of the cells are called nodes. A function $q(x_i,t)$ is represented by an expansion of the form:

$$q(x_i, t) = \sum_{\nu} q_\nu(t) b_\nu(x_i) \qquad \text{(Eq 39)}$$

where the sum is over all the nodes ν in the computational domain. The $b_\nu(x_i)$ are called basis functions and have finite support, meaning that they vanish outside of some neighborhood of the node ν location $(x_i)_\nu$. They also have the properties that:

$$b_\nu((x_i)_\mu) = \delta_{\nu\mu} \qquad \text{(Eq 40)}$$

where $\delta_{\nu\mu}$ is the Kronecker delta function, and:

$$\sum_{\nu} b_\nu(x_i) = 1 \qquad \text{(Eq 41)}$$

for all x_i. Linear (for tetrahedra) or trilinear (for hexahedra) basis functions give rise to second-order numerical methods in the following sense: When the finite-element grid is refined in such a way that the dimensions of elements are reduced by a factor of two, then the difference between the computed and exact solutions, as measured by a global integral of this difference, is reduced by a factor of four. Higher-order FEMs can be constructed by adding midside nodes to the elements and using nonlinear basis functions that have properties Eq 40 and Eq 41 (Ref 28). Because of Eq 40, the coefficient $q_\nu(t)$ is the value of q at location $(x_i)_\nu$ at time t.

Although there are many possibilities for determining $q_\nu(t)$, the most common method is that of the Galerkin finite-element method (GFEM). In GFEM, one substitutes expansions of the form Eq 39 for each function in the fluid equations. To obtain the discrete equations associated with node ν, one multiplies the resulting expanded equations by basis function $b_\nu(x_i)$ and integrates over the entire computational domain. This gives rise to a coupled system of ordinary differential equations for the functions $q_\nu(t)$. Standard numerical methods for ordinary differential equations can then be used to solve for the $q_\nu(t)$. These ordinary differential equations involve coefficients that are integrals of products of the basis functions and their derivatives. Evaluating these coefficients can be a costly step in obtaining a GFEM solution.

Spectral Methods. Like FEMs, spectral methods (Ref 28) represent a function $q(x_i, t)$ by a finite sum:

$$q(x_i, t) = \sum_{n} c_n(t)\, b_n(x_i) \qquad \text{(Eq 42)}$$

but unlike FEMs the basis functions $b_n(x_i)$ are typically orthogonal functions with respect to some weight functions $W(x_i)$; that is:

$$\iiint b_n(x_i)\, b_m(x_i)\, W(x_i)\, dx_i = \delta_{nm} \qquad \text{(Eq 43)}$$

There is no grid in a spectral method. The $c_n(t)$ are no longer the values of q at nodes, but simply the coefficients of the function q in an orthogonal function expansion. Ordinary differential equations for the $c_n(t)$ are obtained by a method that is similar to that of GFEM: one substitutes the expansion Eq 42 into the fluid equations and then multiplies the resulting expanded equation by $b_n(x_i)\, W(x_i)$ and integrates over the computational domain.

Spectral methods are most often used in situations where suitable basis functions can be found that satisfy the boundary conditions of a problem. When this is the case, spectral methods are very efficient for solving fluid-dynamics problems. For example, direct simulations of turbulence with periodic boundary conditions invariably use Fourier series expansions (Ref 34) because of their high accuracy. Because of the difficulty of finding suitable basis functions that satisfy boundary conditions in complex geometries, spectral methods are usually used only for simple geometries.

Computational Particle Methods. Computational particles have long been used for many purposes in CFD calculations (Ref 35). At the simplest level they are used to follow the motion of Lagrangian fluid elements for flow-visualization purposes. At the other extreme, in some particle methods the fluid is completely represented by particles, each of which is endowed with a certain amount of mass, momentum, and energy. This is the case for particle-in-cell (PIC) methods (Ref 36, 37) and for the newer smoothed-particle-hydrodynamics (SPH) methods (Ref 38). The great advantage of the latter two methods is their Lagrangian nature. Because the Lagrangian equations are solved, numerical truncation errors are avoided that arise from finite-difference approximations to the advection terms. These are often the largest errors in approximations of the Eulerian equations. When carefully formulated, PIC and SPH method solutions can also be Galilean invariant and conserve angular momentum (Ref 37).

A disadvantage of particle methods lies in the difficulty of calculating interactions among fluid particles—which give rise, for example, to the pressure gradient terms in the momentum equation. This difficulty manifests itself, particularly in low Mach number calculations, in particle bunching and consequent fluctuations in advective transport. Possibly because of this difficulty, very few commercially available CFD codes use particle methods. An exception is a class of commonly used fluid/particle methods for calculating dispersed, two-phase flows (Ref 39, 40), such as occur when a liquid sprays into a gas. In these methods, computational particles represent the dispersed phase entities and only interact with each other weakly, if they do so at all.

Solution of Implicit Equations. When solving difference approximations to the steady fluid equations or when solving implicit approximations to the unsteady equations, one must solve a large number of coupled algebraic equations for the unknown values of the fluid variables. When the equations are linear, an equation correspond-

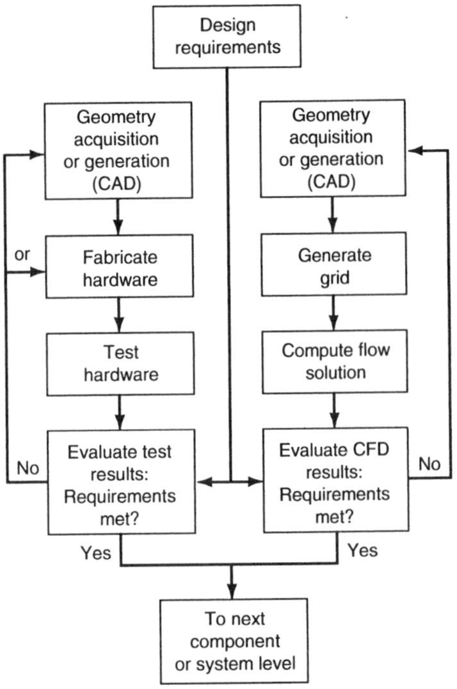

Fig. 5 Engineering component design processes. Left-hand side depicts a hardware-based approach; right-hand side is an analysis- (CFD-) based approach.

ing to the ith cell or node can be written in the form:

$$\sum_{j} a_{ij}\, q_j = s_i \qquad \text{(Eq 44)}$$

where a_{ij} are constant coefficients and s_i is a known source term. The q_j are the unknowns to be solved for; in an unsteady problem $q_j = q_j^{n+1}$. For example, for the implicit approximation Eq 35 to the one-dimensional linear advection equation, one can take $a_{ii} = 1.0$, $a_{ii+1} = u\Delta t/(2\Delta x)$, $a_{ii-1} = -u\Delta t/(2\Delta x)$, and $s_i = q_i^n$. Equation 44 is usually written:

$$\mathbf{Aq} = \mathbf{s} \qquad \text{(Eq 45)}$$

where $\mathbf{A} = (a_{ij})$ is an $N \times N$ matrix of coefficients, N being the number of unknowns; $\mathbf{s} = (s_i)$ is a known source vector; and $\mathbf{q} = (q_i)$ is the vector of unknowns. Because the difference approximation in cell i only depends on the values of q in cell i and a small number of neighbors of cell i, only a small number of elements of the ith row of matrix **A** will be nonzero, and for this reason **A** is referred to as a sparse matrix. The basic problem of implicit fluid dynamics is to solve Eq 45 for the vector of unknowns, given a sparse matrix **A** and source vector **s**.

Only for problems with small N can the matrix problem Eq 45 be solved directly by Gaussian elimination. This is because although the matrix **A** is sparse, and therefore does not require much computer storage for its nonzero elements, when Gaussian elimination is used to solve Eq 45, one finds that it is generally necessary to store in computer memory approximately N^2 nonzero coefficients, which is impossible in problems with a large number of cells.

Thus, interative methods are usually used to solve the matrix problem, Eq 45. Iterative solution methods calculate a sequence of approximations $\mathbf{q}^k$ that converge to the solution $\mathbf{q}$. The exact solution is not obtained, but one stops calculating $\mathbf{q}^k$ when either the difference between successive iterates $\mathbf{q}^{k+1} - \mathbf{q}^k$, or the residual $\mathbf{Aq}^k - \mathbf{s}$, is acceptably small. In the past, popular iterative methods have been point-successive relaxation, line-successive relaxation, and methods based on approximate decomposition of matrix $\mathbf{A}$ into a product of lower and upper triangular matrices that can each be easily inverted (Ref 30). Recently, these methods have largely been supplanted by two methods that have greatly reduced the computer time to solve implicit equations and thereby have made implicit methods more attractive. These more recent methods are conjugate-gradient methods (Ref 41) and multigrid methods (Ref 42).

When nonlinear finite-difference equations are solved, the above iterative methods can be used in conjunction with Newton's method (Ref 43). A nonlinear difference approximation can be written:

$$\mathbf{F}(\mathbf{q}) = \mathbf{O} \qquad \text{(Eq 46)}$$

where $\mathbf{F}$ is a vector-valued function of the vector of unknowns $\mathbf{q}$. If $\mathbf{q}^k$ is the approximation to the solution $\mathbf{q}$ after k Newton-iteration steps, then $\mathbf{q}^{k+1} = \mathbf{q}^k + \boldsymbol{\delta q}$ is obtained by solving the matrix equation:

$$\frac{\partial \mathbf{F}}{\partial \mathbf{q}} \boldsymbol{\delta} \mathbf{q} = - \mathbf{F}(\mathbf{q}^k) \qquad \text{(Eq 47)}$$

The matrix $\partial \mathbf{F}/\partial \mathbf{q}$ is called the Jacobian matrix. Equation 47 is of the form of Eq 45 and can be solved by one of the iterative methods for linear equations. Thus solution for $\mathbf{q}$ involves using an iteration within an iteration. As in the solution of nonlinear equations for single variables, convergence is sometimes accelerated by under-relaxation; that is, one takes $\mathbf{q}^{k+1} = \mathbf{q}^k + \boldsymbol{\lambda\delta q}$ where $\boldsymbol{\delta q}$ is the solution to Eq 47 and $\boldsymbol{\lambda}$ is an underrelaxation factor whose value lies between zero and one.

Newton's method is sometimes used to solve systems of coupled difference equations arising in CFD (Ref 44), but it is often more economical for this purpose to use the simple-implicit method for pressure-linked equations (SIMPLE) method (Ref 45). In the SIMPLE method, a system of coupled implicit equations is solved by associating with each equation an independent solution variable and solving implicitly for the value of the associated solution variable that satisfies the equation, while keeping the other solution variables fixed. As is implied by the acronym SIMPLE, pressure is chosen as an independent variable, and special treatment is used to solve for pressure (Ref 45). The equations are solved sequentially, and repeatedly, until convergence of all the equations is obtained. The SIMPLE method is more efficient if the difference equations are loosely coupled, or if some independent linear combinations of the equations can be found that have little coupling.

Grid Generation for Complex Geometries

Before applying most of the CFD methods outlined above, a computational grid must be generated that fills the flow domain and conforms to its boundaries. For complex domains with curved or moving boundaries, or with embedded subregions that require higher resolution than the remainder of the flowfield, grid generation can be a formidable task requiring more time than the flow solution itself. Two general approaches are available to deal with complex geometries: use of unstructured grids and use of special differencing methods on structured grids.

Unstructured Meshes. Figure 3 shows examples (in two dimensions) of several possible grids arrangements for CFD. In a structured three-dimensional grid (Fig. 3a), one can associate with each computational cell an ordered triple of indices (i, j, k), where each index varies over a fixed range, independently of the values of the other indices, and where neighboring cells have associated indices that differ by ±1. Thus, if N_i, N_j, and N_k are the number of cells in the i-, j-, and k-index directions, respectively, then the number of cells in the entire mesh is $N_i N_j N_k$. Additionally, it is seen that each interior vertex in a structured grid is a vertex of exactly eight neighboring cells.

In an unstructured grid (Fig. 3c and d), on the other hand, a vertex is shared by an arbitrary number of cells. Unstructured grids are further classified according to the allowed cell or element shapes (Fig. 4). In the case of finite-volume methods in particular, an unstructured CFD code may require a mesh of strictly hexahedral cells (Fig. 4b), hexahedral cells with degeneracies (Fig. 4c), strictly tetrahedral cells (Fig. 4a), or may allow for multiple cell types. In any case, the cells cannot be associated with an ordered triple of indices as in a structured mesh.

Intermediate between structured and unstructured meshes are block-structured meshes (Fig. 3b), in which "blocks" of structured grid are pieced together to fill the computational domain.

There are three advantages of unstructured meshes over structured and block-structured meshes. First, unstructured meshes do not require that the computational domain or subdomains be topologically cubic. This flexibility allows one to construct unstructured grids in which the cells are less distorted, and therefore give rise to less numerical inaccuracy, compared to a structured grid. Second, local adaptive mesh refinement (AMR) is naturally accommodated in unstructured meshes by subdividing cells in flow regions where more numerical resolution is required (Fig. 3e, f). Such subdivisions cannot be performed in structured meshes without destroying the logical (i, j, k) indexing. Third, in some cases, particularly when the cells are tetrahedra, unstructured grid generation can be automated with little or no user intervention (Ref 46). Thus, generating unstructured grids can be much faster than generating block-structured grids.

On the other hand, unstructured-mesh CFD codes generally demand higher computational resources. Additional memory is needed to store cell-to-cell and vertex-to-cell pointers on unstructured meshes, while this information is implicit for structured meshes. And, the implied connectivity of structured meshes reduces the number of numerical operations and memory accesses needed to implement a given solution algorithm compared to the indirect addressing required with unstructured meshes.

The relative advantages of hexahedral verses tetrahedral element shapes remain subjects of debate in the CFD community. Tetrahedra have an advantage in grid generation, as any arbitrary three-dimensional domain can be filled with tetrahedra using well-established methodologies (Ref 46). By contrast, it mathematically is not possible to tessellate an arbitrary three-dimensional domain with nondegenerate six-faced convex volume elements. Thus, each of the various automatic hexahedral grid-generation approaches that have been proposed (e.g., Ref 47, 48) either yields occasional degeneracies or shifts the location of boundary nodes, thus compromising the geometry.

Specialized Differencing Techniques. In a second general approach to computing flows in complex geometric configurations, the onus of work is shifted from complexity in grid generation to complexity in the differencing scheme (Ref 49–51). Structured and block-structured grids are used, but one of three numerical strategies is used to extend the applicability of these grids. The first strategy is to use so-called chimera grids (Ref 49) that can overlap in a fairly arbitrary manner (Fig. 3g). Solutions on the multiple grids are coupled by interpolating the solution from each grid to provide the boundary conditions for the grid that overlaps it. This is a very powerful strategy that handles naturally problems in which two flow regions meet at a boundary with a complicated shape or where one object moves relative to another. The second numerical strategy is to use so-called embedded boundaries (Ref 50). Again, structured meshes are used, but the complicated boundary of the computational domain is allowed to cut through computational cells. Special numerical methods are then used in the partial cells that are intersected by the boundary. In the third strategy, local AMR is allowed by using a nested hierarchy of grids (Ref 51). The different grids in the hierarchy are structured and have different cell sizes, but the cells in the more finely resolved grids must subdivide those of the coarser grids.

Although the second general approach affords simplicity in grid generation, it generally is less mature than the various unstructured-mesh approaches. Much development remains before these specialized differencing techniques have the robustness, generality, and efficiency to deal with the variety of problems presented in engineering applications. For the near future, then, the use of various unstructured-mesh approaches is expected to dominate in engineering applications of CFD.

Computational Fluid Dynamics for Engineering Design

This section discusses the process by which the above formalisms are used by the industrial design engineer. Because the use of CFD in engineering design is proliferating rapidly in the 1990s, some of this information, particularly that citing specific software, unavoidably will rapidly become dated. The authors believe that the benefits of providing concrete examples to the reader outweigh the concern of premature obsolescence.

Computational fluid dynamics is one of the tools available to the engineer to understand and predict the performance of thermal-fluids systems. It is used to provide insight into thermal-fluids processes, to interpret experimental measurements, to identify controlling parameters, and to optimize product and process designs. It is the use of CFD as a design tool that is the principal focus here. In the course of a design program, an engineer typically will perform multiple CFD computations to explore the influence of geometry (hardware shape), operating conditions (initial and boundary conditions), and fluid properties. For CFD to be fully integrated into the design process, it must satisfy ever-tightening demands for functionality, accuracy, robustness, speed, and cost.

At present, most engineering CFD using commercially available software can be characterized as having high geometric complexity (domain boundaries are complex three-dimensional surfaces) and moderate physical complexity. The majority of flows considered are steady, incompressible, single-phase, and nonreacting. A common physical complexity encountered in engineering situations is turbulence, as engineering flows typically are characterized by high Reynolds number. Turbulence is modeled using a two-equation model (standard K-ε or variants, Ref 27) in most cases. Applications to transient flows with additional physical complexity and/or more sophisticated models (e.g., compressibility, multiphase, reacting, higher-order turbulence models) are increasing.

The CFD Process

Idealized component design processes are shown schematically in Fig. 5. There the left-hand-side flowchart depicts a hardware-based design process, while the right-hand side represents an analysis- or math-based process. Although CFD is the single analysis tool under consideration here, the right-hand side applies equally well to other mathematical/computational tools (e.g., finite-element structural analysis) that together fall under the heading of CAE.

Both the hardware- and analysis-based processes require the generation or acquisition of geometric data, and the specification of design requirements. Here it is assumed that a three-dimensional CAD geometry model is the preferred method for geometric representation. A hardware approach then proceeds with fabrication of prototypes, followed by testing of prototypes, and evaluation of test results. Design iterations are accomplished either by direct changes to the hardware or by modification of the CAD data set and refabrication, until the design requirements are satisfied. At that point, the original CAD data must be updated (in the case of direct hardware iterations), and the design proceeds to the next component or system level where a similar process is repeated.

Analysis-based design (here, CFD) is not fundamentally different. Mesh generation replaces hardware fabrication, computer simulation substitutes for experimental measurement, and postprocessing diagnostics are needed to extract relevant physical information from the vast quantity of numerical data. To the extent that relatively simple design criteria are available and the component lends itself to a parametric representation, the design-iteration loop can be automated using numerical optimization techniques (Ref 52). Automated computer optimization with three-dimensional CFD remains a subject of research; in most engineering applications, determination of the next design iteration remains largely a subjective, experience-based exercise.

Analysis-based design can be faster and less costly compared to hardware build-and-test. If this is not yet the case in a particular application, it most likely will be true at some point in the future. Thus, analysis affords the opportunity to explore more design possibilities within specified time and budget constraints. Advances in rapid prototyping systems (Ref 53) and other fabrication technology mitigate this advantage to some extent.

A second advantage of analysis is that more extensive information can be extracted compared to experimental measurements. Computational fluid dynamics yields values of the computed dependent variables (e.g., velocity, pressure, temperature) at literally thousands or even millions of discrete points in space and (in time-dependent problems) in time. From this high density of information can be extracted qualitative and quantitative pictures of flow streamlines and three-dimensional isopleths of any computed dependent variable. For time-dependent problems, animation or "movies" reveal the time evolution of physical processes. Application-specific "figures of merit" including total drag force, wall heat flux, or overall pressure drop or rise can be computed. Examples are given in the case studies that follow. Experimental measurements, on the other hand, traditionally have been limited to global quantities or to values of flow variables at a small number of points in space and/or time. Thus in principal, much more complete information is available from CFD to guide the next design iteration. An important caveat is that this additional information is useful only to the extent that it accurately and reliably represents the actual hardware under the desired operation conditions. In most applications of CFD today, there are sufficient sources of uncertainty that abandonment of experimentation is unwarranted. Recent progress in two- and three-dimensional experimental diagnostics (e.g., particle-image velocimetry for velocity fields, Ref 54; laser-induced fluorescence for species concentrations, Ref 55) is enabling higher spatial and/or temporal measurement densities in many applications.

In Fig. 6, the CFD process is modeled as a four-step procedure: (1) geometry acquisition, (2) grid generation and problem specification, (3) flow solution, and (4) postprocessing and synthesis. Depending on the level of integration in the software selected, four (or more) distinct codes may be needed to accomplish these tasks. Some vendors offer fully integrated systems. For the purpose of exposition, we treat each separately.

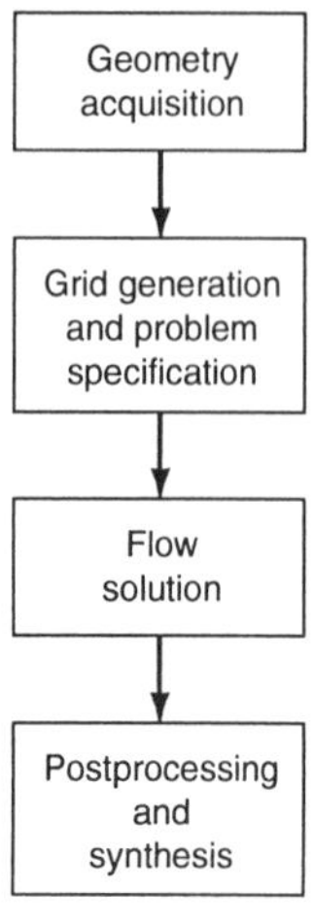

Fig. 6 The CFD process. Examples of available software are given in Table 2.

Table 2 Examples of CFD software available in the United States

This partial listing was extracted from information maintained by several computer hardware and software companies on the Internet early in 1997. Further information on each company and/or code can be found by initiating a network keyword search. Additional information is provided for some companies in Table 3.

Geometry acquisition (CAD)
ICEM CFD
Unigraphics
CATIA
CADDS
I-DEAS
IEMS
Pro-Engineer
Patran
AutoCAD
Grid generation
ICEM CFD
GridGen
Patran
Hexar
CFD-GEOM
Postprocessing (three-dimensional visualization)
ICEM
Patran
Fieldview
Application Visualization System—AVS
Data Visualizer
EnSight
FAST
PLOT3D/TURB3D
MPGS
CFD-VIEW

Geometry Acquisition (CAD). The principal role of CAD software in the CFD process is to provide geometric definition of the bounding surfaces of the three-dimensional computational domain. The computational domain of interest in CFD generally is everything *external* to the solid material; this conveniently might be thought of as the negative of a finite-element structural solid model. Several CAD packages are available commercially; examples are listed in Table 2. These codes are designed primarily with the design and fabrication of three-dimensional solids in mind and have considerable functionality that is not of direct relevance for CFD (Ref 56).

The various CAD packages use different internal representations for curves (one-dimensional objects), surfaces (two-dimensional objects), and solids (three-dimensional objects). The surfaces needed for CFD, for example, may be represented using one of several tensor-product polynomial or spline representations in a two-dimensional parametric space (Ref 57, 58). Any of these representations generally suffice for CFD; most FDM, FVM, and FEM solution methodologies in current engineering CFD codes require at most linear interpolation between the discrete points (nodes or vertices) representing the surface. However, spectral-element methods (Ref 59) and some other high-order orthogonal basis function expansions require a level of surface definition that generally is not available from current commercial CAD systems; this limits the application of such methods to simple geometric configurations at present.

The need to move geometry models among different CAD systems having different internal representations led to the establishment of standards for external geometric data exchange. An early standard supported by most CAD software is the initial graphics exchange specification (IGES, Ref 60). Most CAD-to-CFD interfaces today operate by extracting the outer surfaces and writing an IGES file of "trimmed" B-spline surfaces. Newer standards such as standard for the exchange of products model data (STEP) are merging with IGES and supplanting it; existing standards are evolving rapidly, and new standards are developed as needed. Other external data formats commonly used in the CAD/CAE arena include stereo lithography (STL), where surfaces are processed into a set of triangular facets, cloud-of points (a set of random points in three-dimensional space), and DES (a set of piecewise linear curves describing a surface).

The set of raw surfaces extracted from the CAD model usually requires additional processing before it is suitable for CFD grid generation. The extracted surfaces may not define a closed three-dimensional domain (gaps), there may be more than one surface at a physical location (overlaps), and there simply may be too much geometric detail to be practical for CFD. Modern CAD and grid-generation systems provide fault tolerance and a variety of tools to "clean up" the extracted surfaces prior to grid generation. This cleanup step is labor intensive and often is the single most time-consuming element of the CFD process.

Grid Generation and Problem Specification. The second step in the CFD process is to generate a computational mesh. This might be accomplished using the same software as for geometry acquisition, or a separate code. The grid must satisfy three general requirements:

- It must be compatible with the selected flow solver
- It must be sufficiently fine to satisfy accuracy requirements
- It must be sufficiently coarse to satisfy computational resource limitations

For an unstructured mesh, the minimum information that must be provided from the grid-generation step is the location of each node or vertex, and a description of connectivity among the vertices. A complete problem prescription for CFD requires in addition the specification of initial and boundary conditions for all flow variables (e.g., velocity, pressure, temperature), fluid properties, and any model and numerical parameters. Other code- and application-specific information also may be needed. Because both geometry and grid information are available at the grid-generation stage, this is the most natural time to tag volumes for initial conditions and material properties and surfaces for boundary conditions (e.g., specify which surfaces represent walls, inflow boundaries, etc.). Specific initial values for each dependent variable at each interior cell or vertex, boundary values for each boundary element face or vertex and fluid properties may be set either in the grid-generation software itself or in a separate "preprocessor" provided for the specific CFD code. For present purposes, the preprocessor is considered to be part of the flow solver. Model constants and numerical parameters are specified to the flow solver directly.

Fully automatic tetrahedral-mesh generation is available in a number of commercial and public-domain codes (Ref 46, Table 2). Early generations of automated hexahedral, hexahedral-with-degeneracies, and hybrid hexahedral/tetrahedral strategies (requiring varying levels of manual intervention) also are available at the time of this writing (Ref 47, 48; Table 2). However, a high level of manual intervention still is required to generate high-quality meshes for CFD. This is particularly true in the case of tetrahedral meshes

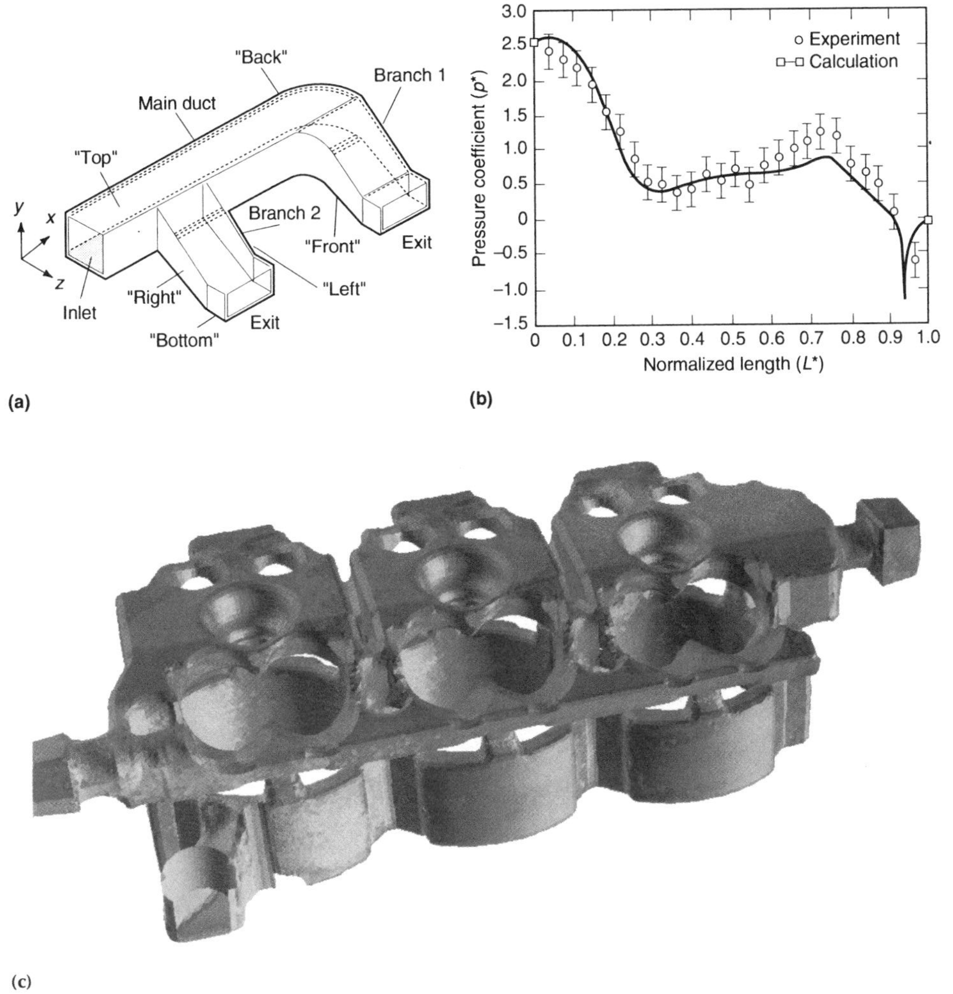

Fig. 7 Examples of internal flow CFD. (a) A simplified automotive HVAC duct (Ref 63). (b) Measured and computed static pressure distributions along the "Top" surface of the main duct and Branch 1 (Ref 63). (c) Computed surface heat transfer coefficients for a production automotive engine block (Ref 64)

in the vicinity of solid walls. A "high-quality" mesh is defined here as one that yields high numerical accuracy for low computational effort (memory and CPU time). This is quantified by performing multiple computations of a single flow configuration using different meshes, and computing the error in each with respect to a benchmark numerical or experimental solution. Discussions of modern mesh-generation techniques for CFD can be found in Ref 32 and 61.

Regardless of the specific methodology used to generate the mesh, it is important that any grid-generation software for CFD maintain separate data structures for geometry definition and for the computational mesh. This ensures that design changes (modifications to CAD surfaces) can be made without redoing the domain decomposition, that boundary conditions can be reset without regenerating the grid, and that mesh density and distribution can be changed independently of the geometry.

Flow Solution. Most contemporary CFD solvers available to the industrial design engineer use either finite-volume or finite-element discretization, with SIMPLE-like iterative pressure-based implicit solution algorithms. Unstructured meshes of primarily hexahedral elements (with limited degeneracies) have been prevalent in most finite-volume formulations to date, although the grid-generation advantages of tetrahedra are leading to an increase in the usage of that element type.

Default or recommended values of numerical parameters are provided by each flow solver. New and/or unusual applications often require experimentation in selecting values of numerical parameters to obtain a stable, converged solution. For the solution methodologies commonly used today, parameters include choice of advection scheme (e.g., the degree of upwinding), convergence criteria for linear equation solvers and pressure iterations, time-step control (for transient problems), mesh adaptation (where available), and other method-specific controls. For this reason, the CFD practitioner needs to have a working knowledge of the information covered in the "Fundamentals" section of this article. With these caveats, flow solution is the step requiring the least manual intervention. The engineer can monitor the solution as it progresses using the available diagnostics, which are discussed next.

Postprocessing and Synthesis. Viewing and making sense of the vast quantities of three-dimensional data that are generated in CFD is a challenging task. Many software packages have been developed for this purpose, both for structured and unstructured meshes (Table 2). All provide considerable flexibility in setting model orientation, in passing cutting planes and/or lines through the computed solution, and in displaying the computed vector and scalar fields. Postprocessors have varying levels of "calculator" capability for computing quantities not supplied directly from the CFD solution, such as vorticity or total pressure. Many allow transient animation to accommodate time-dependent data. Most modern packages provide both a graphics-user interface (GUI) and a save file/read file capability, the latter to allow the user to replicate a particular view of interest for multiple data sets.

Such direct inspection of the computed fields provides detailed insight into flow structure in the same sense as a high-resolution flow visualization experiment. In this respect and others, it had been argued that CFD is more akin to experiment than to theory. Features such as an undesirable flow separation, for example, might provide the engineer with sufficient information to guide a modification to the device geometry for the next design iteration. The connection between device performance or design requirements and the full three-dimensional flow field often is not obvious, however; considerable effort may be required to extract meaningful figures-of-merit from the numerical solution.

Judicious development of diagnostics is necessary to advance CFD from a sophisticated flow-visualization tool to a scientifically based design tool. Quantitative information of direct relevance to the designer is needed to drive design changes toward satisfaction of the design requirements. Such diagnostics are application-specific and have received relatively little attention by CFD researchers and code developers. Examples of diagnostics to extract physical insight and to assess numerical accuracy can be found in Ref 62.

Examples of Engineering CFD

Application areas that have been particularly active in their use of CFD include aircraft and ship design, geophysical fluid flows, and flows in industrial devices that involve energy conversion and utilization. A comprehensive list of the applications of CFD would be difficult to compile, and

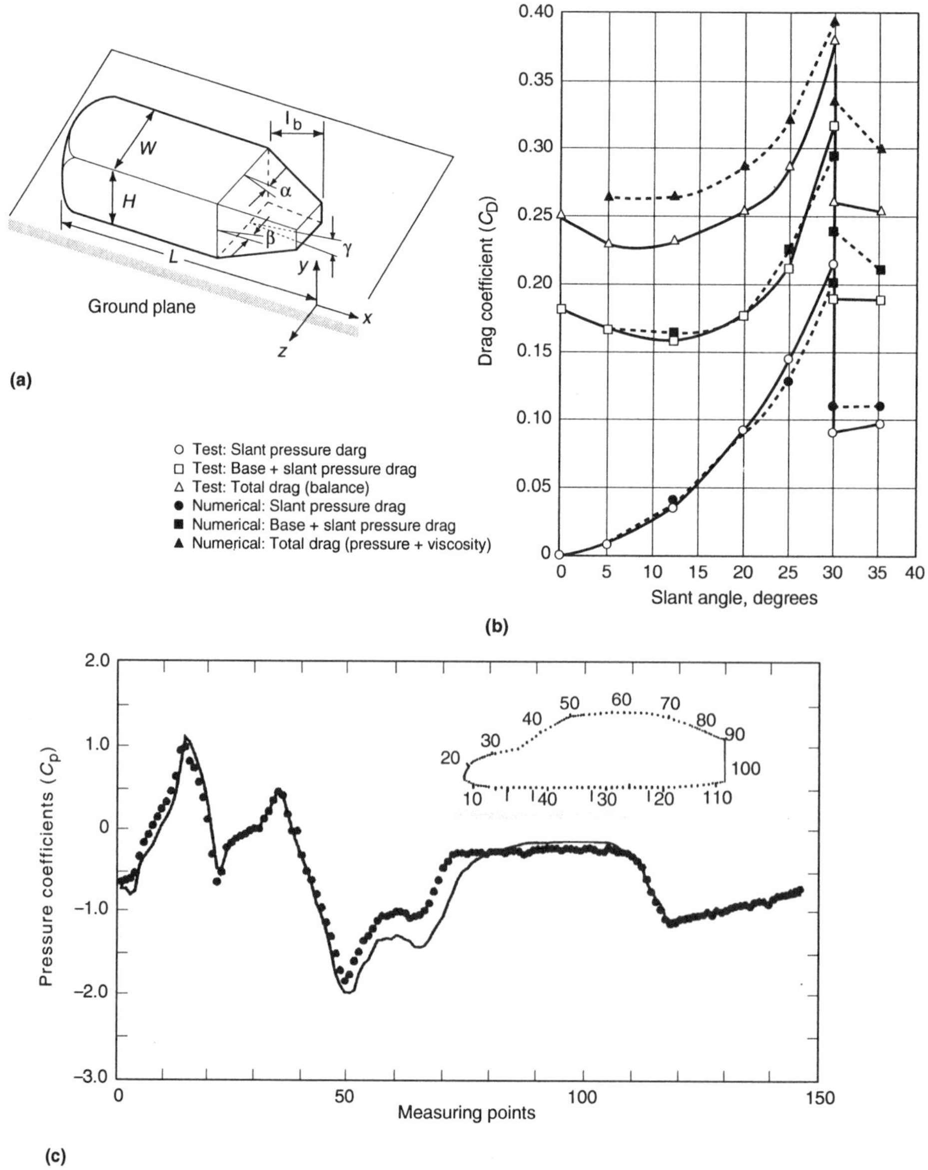

Fig. 8 Examples of external flow CFD. (a) Generic three-dimensional bluff body for validation studies (Ref 65). (b) Computed drag coefficient versus slant angle (angle α) (Ref 66). (c) Measured and computed pressure coefficients along a production car body (Ref 69)

no attempt to do so is made here. Instead, specific case studies are cited with several purposes:

- To illustrate the scope and state-of-the-art in engineering CFD
- To highlight issues that arise in engineering applications of CFD
- To introduce some specific CFD software that is widely used in industry

Internal Duct Flow. Many internal flows of engineering interest can be broadly categorized as complex duct flows. The principle physical complexity is turbulence, particularly as it influences flow separation. A related numerical issue is mesh resolution, especially in the vicinity of walls. Flow losses (pressure drop and separations), flow distribution among multiple branches, mixing, and heat transfer may be important in such configurations.

Two examples of steady, incompressible CFD simulations are given in Fig. 7 (Ref 63, 64). Figures 7(a) and 7(b) show a simplified automotive heating, ventilation, and air-conditioning (HVAC) duct. This is taken from a validation study (Ref 63) where experimental measurements also are available. Results of this kind have allowed engineers to identify flow separations and poor flow distribution among branches; optimized designs for lower pressure drop and more favorable flow distribution are identified using CFD prior to hardware fabrication.

A second internal flow configuration (Fig. 7c) illustrates the geometric complexity that often arises in engineering applications. There, surface heat transfer coefficients from computations of flow in the coolant passages of a production automotive engine block are shown. Such results are used to identify potential "hot spots" and to modify flow passages for more uniform cooling.

External Aerodynamics. External flows comprise a second broad category of engineering interest. This includes flows around immersed bodies such as aircraft, ships, submarines, and automobiles. Bluff-body aerodynamics is particularly challenging; the accurate computation of separation, which may be highly unsteady, is key to predicting lift and drag.

Examples of computations and measurements for idealized three-dimensional bluff bodies are shown in Fig. 8(a) and 8(b) (Ref 65–68). A computational challenge is to capture the sudden drop in drag coefficient at a slant angle of about 30° (Fig. 8b). Computations of flow over realistic vehicle shapes also are feasible using modern CAD/grid generation tools (Fig. 8c) (Ref 69). In all cases shown here, the flows have been computed as steady and incompressible using standard Reynolds-averaged turbulence models to account for unsteadiness.

Manufacturing Processes. Increasing attention is being focused on the design and analysis of engineering processes. Heat transfer accompanied by melting and solidification occurs in manufacturing processes including casting, injection molding, welding, and crystal growth. In such applications, heat conduction in the solid is coupled to convective heat transfer in the fluid.

(a)
(b)
(c)

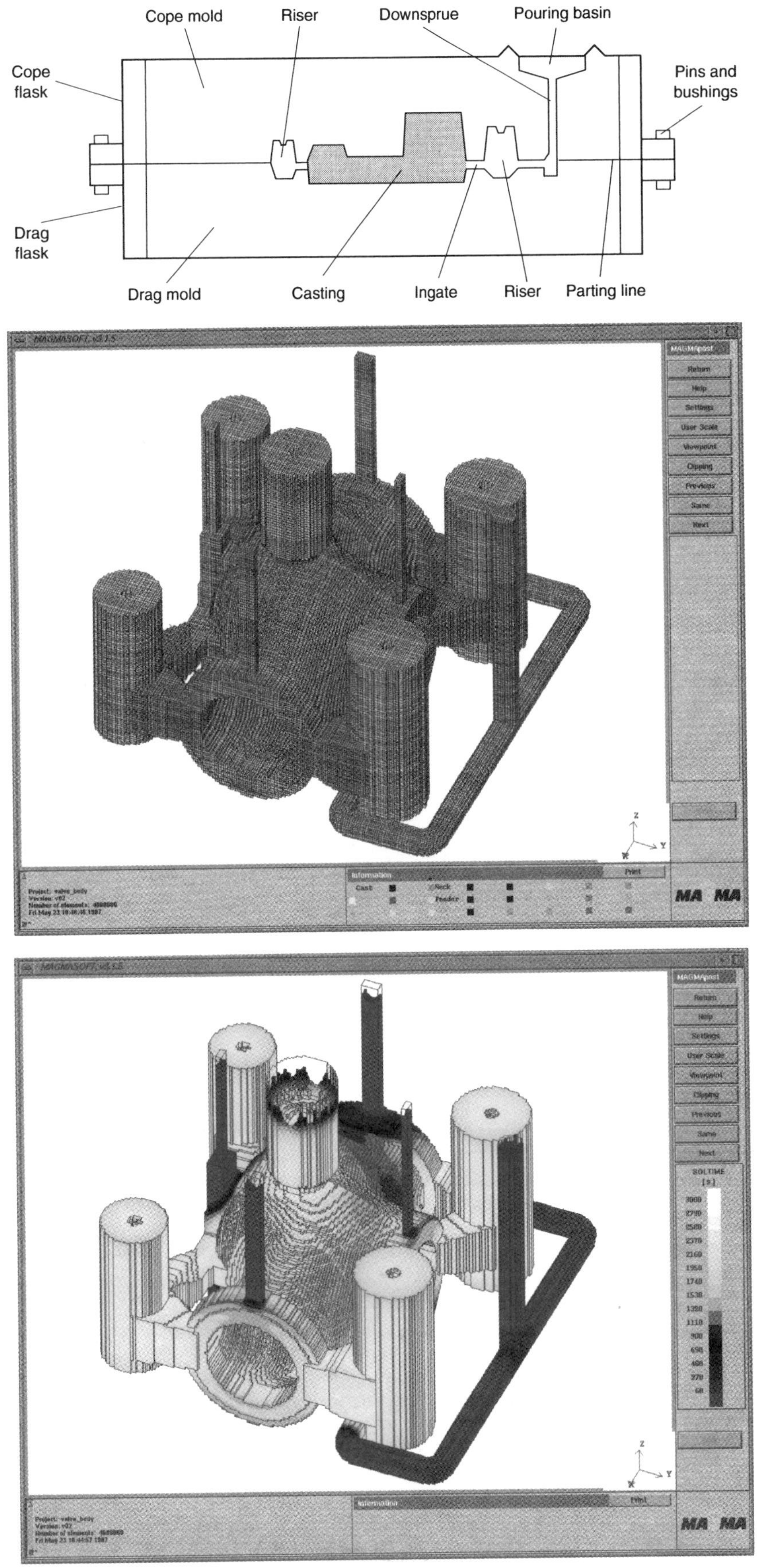

Fig. 9 A metal casting simulation. (a) Typical sand-casting configuration. (b) Automatically generated mesh (five million elements) for casting and cooling channels (Ref 71). (c) Computed local solidification times, which range from 1 to 3000 s (Ref 71)

Table 3 A partial listing of CFD data formats supported by one software vendor

This provides a snapshot in time (late 1996) of the wide variety of commercially available, public domain, and proprietary CFD software used for engineering design and analysis.

Format/code name(s)	Company
FLUENT/UNS, FLUENT-V4, RAMPANT-V2 and V3, NEKTON, TGRID	Fluent Inc., Lebanon, NH
GASP	AeroSoft, Inc., Blacksburg, VA
GMTEC	General Motors Corporation, Warren, MI
HAWK	California Institute of Technology, Pasadena, CA
INCA	Analytical Methods, Inc., Redmond, WA
	Amtec Engineering, Inc., Bellevue, WA
KIVA-3, CHAD	Los Alamos National Laboratory, New Mexico
NASTAR	United Technologies Research Center, East Hartford, CT
NPARC	NPARC Alliance, NASA Lewis Research Center and Arnold Engineering, Cleveland, OH
	Sverdrup Technology, Inc./AEDC Group, Arnold AFB, TN
NS3D	Pratt & Whitney Canada, Longueil, Quebec, Canada
PAB3D	NASA Langley Research Center, Hampton, VA
PARC	NASA Ames Research Center & Boeing Co., NASA Amex Research Center, Moffett Field, CA; Boeing Commercial Airplane Group, Propulsion Research CFD, Seattle, WA
PHOENICS	CHAM Ltd., Wimbledon Village, London, England
POLYFLOW	Polyflow S.A., Louvain-La-Neuve, Belgium
POLY3D	Rheotek, Inc., Montreal, Quebec, Canada
SPECTRUM-CENTRIC	CENTRIC Engineering Systems, Inc., Santa Clara, CA
STARCD	Computational Dynamics Ltd., London, England
TASCflow	Advanced Scientific Computing Ltd., Waterloo, Ontario, Canada
TEAM	Lockheed Aeronautical Systems Co., CA
TNS3Dmb	NASA Langley Research Center, Hampton, VA
TRANAIR	NASA Ames Research Center, Moffett Field, CA
UH3D	Ford Motor Co., Dearborn, MI
USAERO, VSAERO	Analytical Methods, Inc., Redmond, WA
Y237	United Technologies Research Center, East Hartford, CT
ACE-U, CFD-ACE	CFD Research Corporation, Huntsville, AL
AIRFLO3D	Texas Tech University, Department of Mechanical Engineering
ALPHA-FLOW	Fuji Research Institute, Japan
BAGGER	Eglin AFB, FL
CFD++	Metacomp Technologies, Inc., Westlake Village, CA
CFL3D	NASA Langley Research Center, Hampton, VA
CFX	AEA Technology Engineering Software, Inc., Bethel Park, PA
COMCO	The Computational Mechanics Company, Inc., Austin, TX
DSMC-SANDIA	Sandia National Laboratory, Albuquerque, NM
FASTU	NASA Langley Research Center, Hampton, VA
FIDAP	Fluent Inc. (bought FDI Ltd in 1996), Lebanon, NH
	Fluid Dynamics International, Evanston, IL
FIRE	KIT Corporation, St. Paul, MN
FLEX	Sverdrup Technology, Inc., Eglin AFB, FL
FLOTRAN	Swanson Analysis Systems Inc., Houston, PA

Source: Ref 84

The solid-liquid interface moves with time, and its location needs to be tracked as a propagating three-dimensional surface in the CFD solution. Also, fluid properties may be highly temperature dependent and non-Newtonian, including phase changes. Metal casting is cited as one example of such an application.

Casting is a process in which parts are produced by pouring molten metal into a cavity having the shape of the desired product. Figure 9(a) is a schematic of a typical sand-casting configuration. Once the two halves of the mold have been made, they are carefully aligned, one over the other, with the aid of pins and bushings in the sides of the molding boxes, to create the complete mold. Aside from the casting cavity itself, other features are also incorporated into the finished mold such as the pouring basin, downsprue, runners, and ingates that conduct the molten metal into the casting cavity. Risers, or reservoirs of molten metal that remain molten longer than the casting, are needed with most metals and alloys that undergo liquid shrinkage as the casting solidifies. These are placed at critical locations in the mold, generally at heavier sections and areas remote from the ingates. Once the casting has been poured and allowed to cool, and after it has been withdrawn from the sand mold, these appendages are removed before the casting undergoes various finishing operations.

Fluid flow plays two important roles in the casting process. First, and most obviously, the flow of molten metal is necessary to fill the mold. Second, and less obvious, are the effect of convective fluid flow during solidification of the casting. It is the task of the foundry engineer to design grating and riser systems (Fig. 9a) that ensure proper filling and solidification, and CFD is playing an increasingly important role in this field. Proper designs result in lower scrap and less casting repair at the foundry. An example of a computational mesh and computed solidification times is given in Fig. 9(b) and 9(c). One CFD package that has been developed specifically for the modeling of flow and thermal phenomena in casting applications is Magmasoft (Ref 70). Recent references from the literature give ample evidence of the vast amount of CFD activity that is taking place in this area (Ref 71, 72).

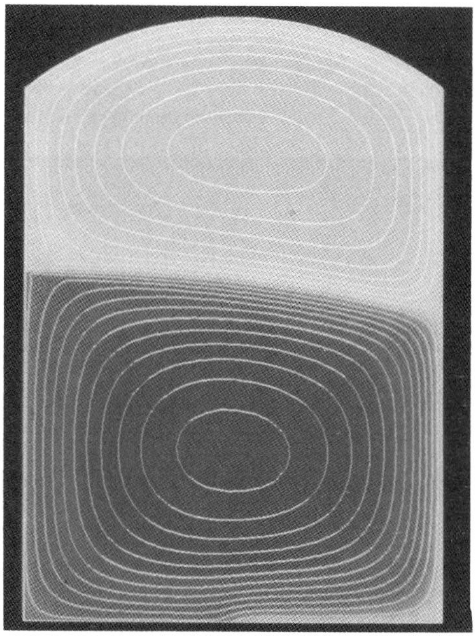

Fig. 10 Flow in the interior of the Sistine Chapel for one possible air-conditioning system configuration. Calculations were done using the FIDAP finite-element CFD code (Ref 73).

Building Interiors. Figure 10 shows an example of CFD applied to building HVAC design. In this case, the geometric configuration is relatively straightforward. The computational domain represents the interior of the Sistine Chapel at the Vatican. The purpose of the analysis was to determine the placement and angles of air-conditioning ducts to minimize deposition of contaminants on the newly restored surface of Michelanglo's frescos. The creation of two separate recirculation cells for the configuration shown in Fig. 10 was deemed to be favorable for isolating traffic-borne particles from chapel visitors in the lower half from the fresco surfaces along the upper walls and ceiling.

Environmental flows include natural phenomena such as atmospheric weather patterns and ocean currents (Fig. 2) and flows of molten rock beneath the crust of the earth. Engineering design issues arise in the extraction of fossil fuels and other materials from the earth, in bridge and building design, and in the treatment and dispersal of wastes from electrical utilities, transportation systems and vehicles, and industrial manufacturing plants. Such problems typically are characterized by a coupling of natural convection (resulting from temperature and/or concentration gradients) with other forces, in many cases including the rotation of the earth.

Internal Combustion Engine. The final example shows a few results from transient computations of flow, fuel spray, and combustion in a reciprocating internal combustion engine (Fig. 11) (Ref 74, 75). This application includes geometric complexity (complex internal flow pas-

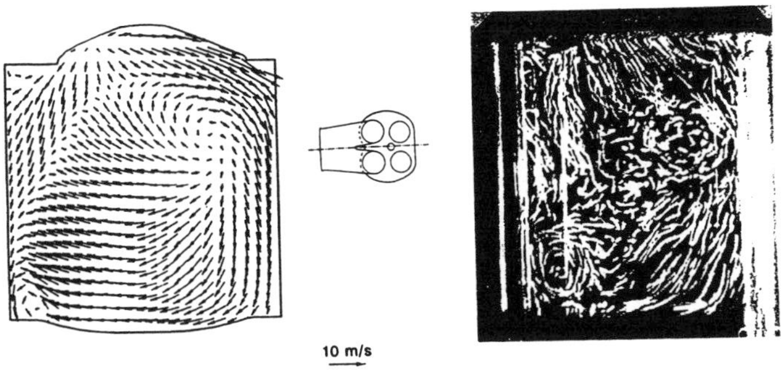

(a)

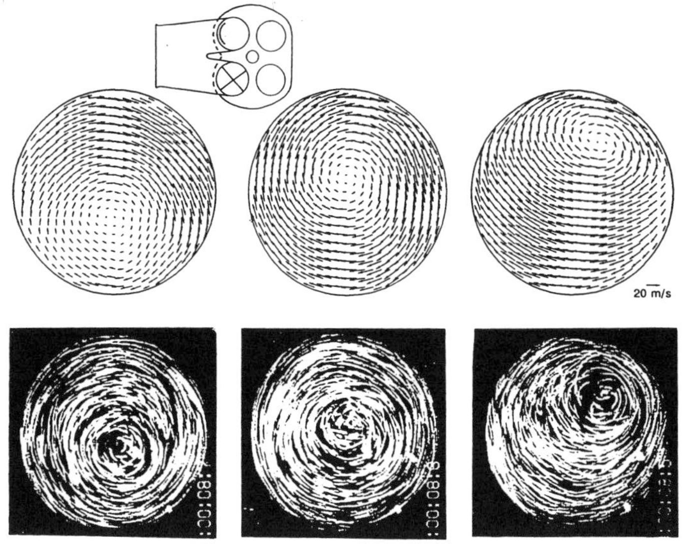

(b)

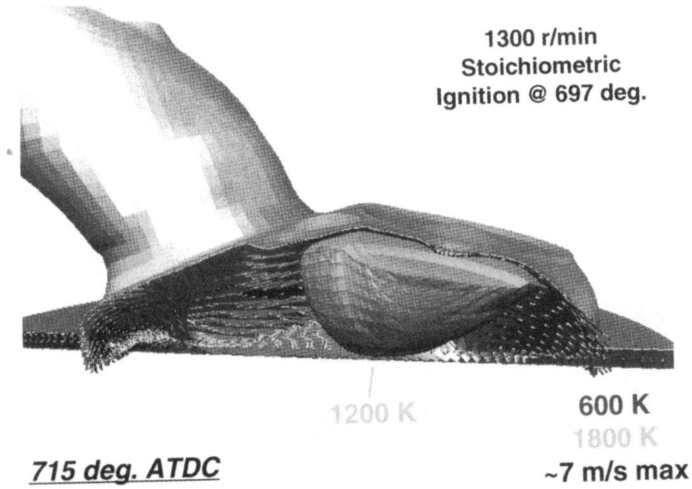

(c)

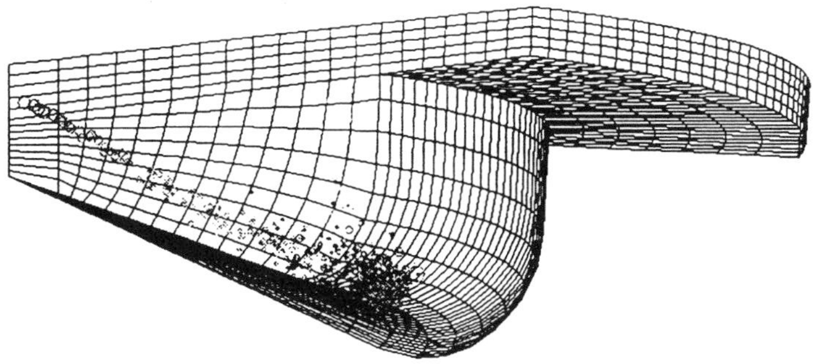

(d)

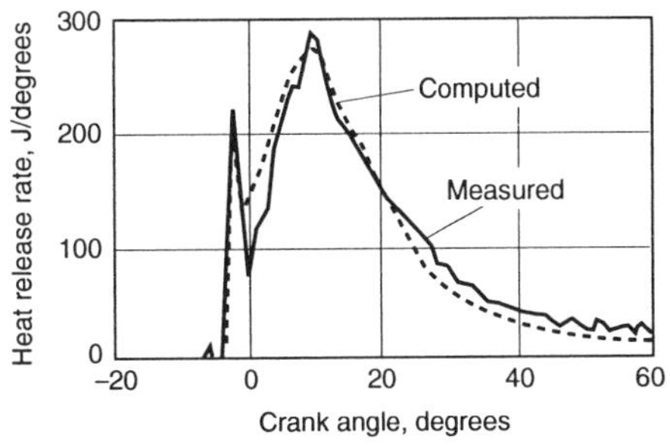

(e)

Fig. 11 Examples of CFD for in-cylinder processes reciprocating IC engines. (a) Instantaneous computed and measured induction flow at piston bottom-dead-center for a port and chamber configuration yielding weakly structured in-cylinder flow (Ref 74). (b) Instantaneous computed and measured induction flow at piston bottom-dead-center for a port and chamber configuration yielding a highly structured in-cylinder flow (Ref 74). (c) Instantaneous computed velocity field and flame propagation near piston top-dead-center for a production four-valve-per-cylinder engine. (d) Instantaneous computed fuel spray for a direct-injection diesel engine (Ref 75). (e) Computed and measured heat release for a direct-injection diesel engine (Ref 75)

sages, moving boundaries—piston and valves), physical complexity (turbulence, combustion, multiphase flow), and numerical challenges (deforming mesh, large density and fluid property variations, coupled Eulerian/Lagrangian algorithms). This represents an application area of CFD that lies at the frontier between research and engineering application.

Of particular interest in a homogeneous-charge spark-ignited engine is the trade-off between flow losses and in-cylinder flow "structure." Flow losses (induction system pressure drop) reduce the quantity of air that can be drawn into the cylinder, lowering engine peak power. A coherent large-scale in-cylinder flow structure tends to yield higher combustion efficiency, but generation of highly structured flow (e.g., a large-scale swirl about the cylinder axis) generally implies a pressure-drop penalty. These trade-offs can be quantified and optimized using CFD (Ref 74). The computations of Fig. 11(a) and 11(b) were performed on unstructured meshes of up to 250,000 predominantly hexahedral cells; computation through one crankshaft revolution required about 150 equivalent single-processor Cray Y-MP CPU hours.

Flame propagation for a production four-valve-per-cylinder automotive engine is shown in Fig. 11(c). Flame shapes and burn rates are tailored by changing the intake port, intake valve, and combustion chamber geometry. A good design generally is one having favorable spark-gap conditions and a flame that propagates uniformly outward to reach all solid walls at the same instant.

Direct-injection diesel and gasoline engines, wherein liquid fuel is injected directly into the combution chamber, are of interest for their high fuel economy potential. Here mixing and fuel stratification are key issues affecting combustion performance; CFD is one tool that is being used to explore the influence of flow structure, injector placement, and injection characteristics on engine combustion performance (Fig. 11d, e) (Ref 75).

Issues and Directions for Engineering CFD

Geometric fidelity between hardware and the computational mesh is crucial to obtaining accurate results. It is characteristic of the highly nonlinear flow equations that small geometric perturbations can result in large changes to the flowfield. One example is shown in Fig. 12 (Ref 76). There, significantly different flow structure and mixing result when the fraction-of-a-millimeter gap between piston and cylinder liner (the "top-ring-land crevice") is included in the mesh compared to when it is ignored. With a top-ring-land crevice, the flow entering the cylinder attaches to the cylinder wall and flows parallel to the wall for an extended time; in the absence of a top-ring-land crevice, the entering flow quickly adopts the port angle on entering the cylinder. This highlights the importance of maintaining a consistent three-dimensional representation of the hardware at all stages of design, analysis, and fabrication. The CFD practitioner should be wary of compromising the geometry in favor of grid-generation expediency, particularly in applications where he or she has little previous experience.

Numerical Inaccuracy. Meshes of hundreds of thousands of computational cells are common in transient engineering applications of CFD today, and several millions of cells are being used in steady-state computations. Even so, numerical inaccuracy remains an issue for three-dimensional CFD. A mesh of 1 million cells corresponds to just 100 nodes in each coordinate direction in a three-dimensional calculation. With the low-order numerics that characterize engineering CFD, this is sufficient to resolve a dynamic range of about one order of magnitude (a factor of 10) in flow scales.

Rapid progress is being made both in discretization schemes for tetrahedral meshes, and in automated grid generation for (primarily) hexahedral meshes; it is unclear at this time which will become dominant in engineering CFD.

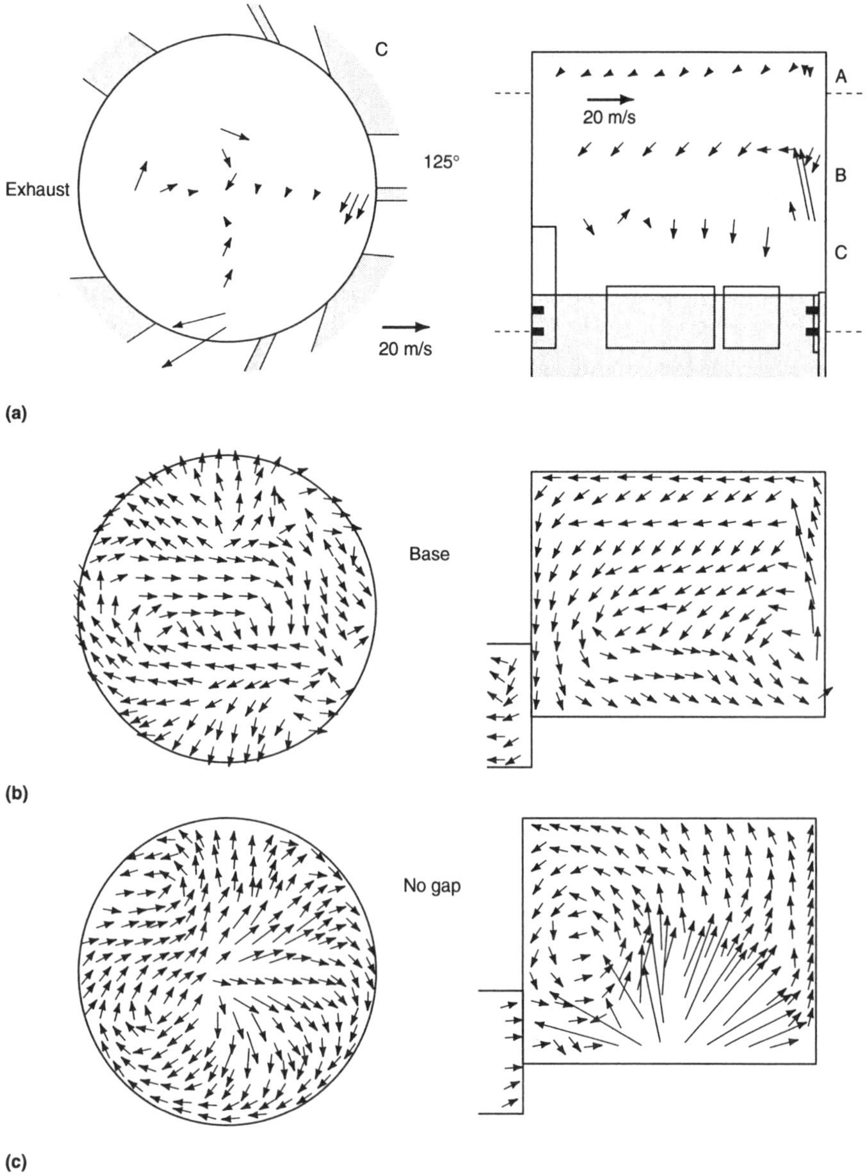

Fig. 12 Computed and measured ensemble-mean velocity fields on two-dimensional cutting planes at 125° after piston top-dead-center for a ported two-stroke-cycle engine. Computational results with and without a top-ring-land crevice are shown. (a) Measured. (b) CFD with top-ring-land crevice. (c) CFD without top-ring-land crevice

The physical models used to represent turbulence, combustion, sprays, and other unresolvable phenomena are a third source of uncertainty in CFD. Turbulence modeling, in particular, is an issue that affects nearly all engineering applications. Research toward improved models continues. Much new physical insight into turbulence is itself being derived from large-scale numerical simulations (Ref 77).

In many high-Reynolds-number engineering applications where the instantaneous flow is highly transient and three-dimensional, turbulence models can be used to reduce the problem to one of steady flow, provided that the *mean* quantities of interest are time independent. This reduces the computational requirements considerably and provides results of acceptable accuracy in many cases. However, as engineering design requirements tighten, there is an increasing number of problems that demand a full three-dimensional transient treatment. Models still are needed to account for scales smaller than those that can be resolved numerically, but subgrid-scale turbulence models are used instead of Reynolds-averaged models. The resulting three-dimensional time-dependent simulations in this case are referred to as large-eddy simulations (LES) (Ref 78). The use of LES in engineering design is expected to proliferate rapidly. Examples of current applications of interest include acoustics and aerodynamic noise (Ref 79) and in-cylinder flows in engines (Ref 80).

Each of these three sources of uncertainty can in principle be isolated and quantified in simple configurations where a second source of data (e.g., experimental measurements) is available. It is more difficult in engineering applications of CFD to isolate and to quantify these errors to obtain meaningful estimates of error bounds. Early in the history of three-dimensional CFD, discrepancies between CFD and experiments generally were attributed to the turbulence model. The importance of the other sources of uncertainty and numerical inaccuracy in particular, has been more widely acknowledged recently (Ref 62, 81, 82). In the authors' experience, most discrepancies between computations and measurements for single-phase nonreacting flows in complex configurations are traced to geometric infidelity or to inadequate mesh resolution (in cases where they have been traced at all).

User Expertise. Computational fluid dynamics codes generally require more experience on the part of the user than other, more mature, CAE tools (e.g., linear FEM structural analysis). "General-purpose" CFD software provides a large number of numerical parameters and problem-specification options. In steady-flow problems, results should be independent of the choice of initial conditions, but different initial conditions may lead to different steady solutions when time-marching to the steady state. The choice of computational domain and specification of boundary conditions always are important, both for steady and time-dependent flows. Minimal user experience may suffice to obtain a reliable solution for steady incompressible flow in a benign geometric configuration, but considerable expertise is needed in problem specification and in results interpretation for complex flows.

CFD and Experimental Measurements. The engineering and scientific community typically accepts measurements from experiments as being more reliable than similar information generated by a CFD calculation. This is the reason for the strong emphasis placed by the profession on "validating" CFD results. While it is true that there are many sources of uncertainty in CFD, the same is true of experiments, particularly for complex systems (e.g., the in-cylinder flow in the last example). When comparing CFD results with measurements for such complex engineering problems, it is more appropriate to approach the exercise as a "reconciliation" rather than a "validation," as the latter implies that the experiment provides the "correct" value.

Interdisciplinary Analysis. In this overview, CFD has been considered as an isolated analysis tool. This is satisfactory only to the extent that one can reasonably prescribe boundary conditions that are independent of the flow solution itself.

For example, in the coolant-flow analysis of Fig. 7(c) temperature boundary conditions might be prescribed from a separate finite-element structural analysis, but the temperature field in the solid depends on the coolant flow itself. One can alternate through a sequence of CFD and thermal structural analyses, taking the most recent boundary conditions available at each step, to obtain a solution that effectively is coupled. A single direct computation of the coupled solution would be more satisfactory, however. In this case, a coupled fluid/heat conduction analysis is feasible as many CFD codes provide conjugate heat transfer capability.

More difficult are cases where fluids and solids interact in a manner that changes the shape of the flow domain. Flow/structure interactions including deformations are important, for example, in some aircraft design problems or in applications where there is significant thermal distortion. Interdisciplinary analysis tools are becoming available for these problems and will see more widespread use in the future.

Future of Engineering CFD. Most contemporary commercial CFD codes start from a discretization of the continuum equations of fluid mechanics and require a computational mesh of discrete cells or elements. An alternative is to approach CFD from a kinetic theory point of view. In Ref 83, for example, an (essentially) grid-free Lagrangian-particle method has been developed and implemented. It is too early, at the time of this writing, to speculate on the future of this approach for engineering design. Computations have been reported for configurations including external flow over simplified and realistic vehicles.

Active research areas for CFD include automated mesh generation, numerical algorithms for parallel computer architectures, linear equation solvers, more accurate and stable discretization schemes, automatic numerical error assessment and correction, improved solution algorithms for coupled nonlinear systems, new and enhanced physical models, more sophisticated diagnostics, interdisciplinary coupled structures/fluids analysis, optimization algorithms, and coupling of three-dimensional CFD into systems-level models.

In the ideal math-based design process, CFD is one part of a multidisciplinary CAE approach, and the full system (versus isolated component) is considered. Grid generation is fully automated to ensure a high-quality (initial) mesh. The flow solver selects all numerical parameters and provides automated solution-adaptive mesh refinement to a specified level of error or allowable computational resource (time or cost). Solution diagnostics provide information of direct relevance to the design requirements. Automated design optimization through modifications to the geometry and/or operating conditions proceeds until design requirements are met.

While much work remains to realize this ideal, CFD already is being used with considerable success in engineering design. Its utility and applicability will increase as the outstanding issues are resolved.

Sources for Further Information. Many references to specific topics have been cited throughout this chapter. General CFD texts include Ref 3, 28, 31, and 32. At all stages of the CFD process (geometry acquisition, grid generation, flow solution, and postprocessing), a broad array of commercial, public-domain, and in-house proprietary codes are being used in engineering design. A small sampling of the software currently available to the design engineer has been mentioned herein. For more comprehensive and up-to-date listings, the reader can consult several sources. Computer hardware companies maintain lists of software that has been ported to their platforms; software vendors maintain lists of codes with which their own products are compatible. Table 3 has been extracted from one such list (Ref 84). General (Ref 85) and industry-specific engineering periodicals often provide reviews of available software. And, a wealth of timely information can be found on the Internet (Ref 86). Given the rapid pace at which CFD technology is evolving, this last source is particularly valuable. In addition to lists and descriptions of the available software, user evaluations and direct comparisons of alternative codes and methodologies can be found there.

CFD is at a relatively early stage of development compared to other areas of CAE such as linear FEM structural analysis. No single code covers all areas of application equally well. While "general-purpose" CFD has been emphasized here, specialized application-specific numerical methods and software often are needed. Specialized experience and expertise can be found within university engineering departments, U.S. national laboratories, and engineering consulting firms; again, the Internet provides a good vehicle for exploring these possibilities.

ACKNOWLEDGMENTS

The authors thank Sherif El Tahry of the General Motors R&D Center for several discussions that helped to shape this article. We are grateful to Diane Poirier of ICEM CFD Engineering for compiling the list of CFD solver software provided in Table 3. We also thank Mathew Maltrud and Richard Smith of Los Alamos National Laboratory for supplying Fig. 2, Norm Carter of General Motors Powertrain Group for supplying Fig. 9(a), and Christopher Rosbrook of MAGMA Foundry Technologies for providing Fig. 9(b) and (c) and the related discussion.

REFERENCES

1. FLOWMASTER USA, Inc. Fluid Dynamics International, Evanston, IL
2. *WAVE Basic Manual: Documentation/User's Manual,* Version 3.4, Ricardo Software, Burr Ridge, IL, Oct 1996
3. P.J. Roache, *Computational Fluid Dynamics,* Hermosa Publishers, 1982
4. N. Grün, "Simulating External Vehicle Aerodynamics with Carflow," Paper 960679, SAE International, 1996
5. W. Aspray, *John von Neumann and the Origins of Modern Computing,* MIT Press, 1990
6. "Accelerated Strategic Computing Initiative (ASCI): Draft Program Plan," Department of Energy Defense Program, May 1996
7. L.H. Turcotte, ICASE/LaRC Industry Roundtable, Oct 3–4, 1994; original source for some data is F. Baskett and J.L. Hennessey, Microprocessors: From Desktops to Supercomputers, *Science,* Vol 261, 13 Aug 1993, p 864–871
8. J.K. Dukowicz and R.D. Smith, *J. Geophys. Res.*, Vol 99, 1994, p 7991–8014
9. R.D. Smith, J.K. Dukowicz, and R.C. Malone, *Physica D,* Vol 60, 1992, p 38–61
10. G. Taubes, Redefining the Supercomputer, *Science*, Vol 273, Sept 1996, p 1655–1657
11. J.B. Heywood, *Internal Combustion Engine Fundamentals*, McGraw-Hill, 1988
12. F.H. Harlow and A.A. Amsden, "Fluid Dynamics," Report LA-4700, Los Alamos Scientific Laboratory, June 1971
13. W.G. Vincenti and C. H. Kruger, *Introduction to Physical Gas Dynamics,* Robert E. Krieger Publishing, 1975
14. P.A.Thompson, *Compressible-Fluid Dynamics,* McGraw-Hill, 1972
15. H. Jeffreys, *Cartesian Tensors,* Cambridge University Press, 1997
16. F.A. Williams, *Combustion Theory,* 2nd ed., Benjamin/Cummings, 1985
17. T.G. Cowling, *Magnetohydrodynamics,* Interscience Tracts on Physics and Astronomy, No. 4, 1957
18. S. Chandrasekhar, *Radiative Transfer,* Dover, 1960

19. D.R. Stull and H. Prophet, *JANAF Thermochemical Tables,* 2nd ed., NSRDS-NBS37, National Bureau of Standards, 1971
20. B. McBride and S. Gordon, "Computer Program for Calculating and Fitting Thermodynamic Functions," NASA-RP-1271, National Aeronautics and Space Administration, 1992
21. R.B. Bird, W. E. Stewart, and E.N. Lightfoot, *Transport Phenomena,* Wiley, 1960
22. L. Crocco, A. Suggestion for the Numerical Solution of the Steady Navier-Strokes Equations, *AIAA J.,* Vol 3 (No. 10), 1965, p 1824–1832
23. L.D. Landau and E.M. Lifshitz, *Fluid Mechanics,* Pergamon Press, 1959
24. O. Reynolds, On the Dynamical Theory of Incompressible Viscous Fluids and the Determination of the Criterion, *Philos. Trans. R. Soc. London,* Series A, Vol 186, 1895, p 123
25. H. Tennekes and J.L. Lumley, *A First Course in Turbulence,* MIT Press, 1972
26. B.E. Launder and D.B. Spalding, *Mathematical Models of Turbulence,* Academic Press, 1972
27. D.C. Wilcox, *Turbulence Modeling for CFD,* DCW Industries, 1993
28. R. Peyret and T.D. Taylor, *Computational Methods for Fluid Flow,* Springer-Verlag, 1983
29. G.D. Smith, *Numerical Solution of Partial Differential Equations,* 2nd ed., Oxford University Press, 1978
30. R.D. Richtmyer and K.W. Morton, *Difference Methods for Initial-Value Problems,* 2nd ed., Interscience Publishers, 1967
31. C.A.J. Fletcher, *Computational Techniques for Fluid Dynamics,* Vol I, *Fundamental and General Techniques,* 2nd ed., Springer-Verlag, 1991
32. C.A.J. Fletcher, *Computational Techniques for Fluid Dynamics,* Vol II, Specific Techniques for Different Flow Categories, 2nd ed., Springer-Verlag, 1991
33. G.G. O'Brien, M.A. Hyman and S. Kaplan, A Study of the Numerical Solution of Partial Differential Equations, *J. Math. Phys.,* Vol 29, 1950, p 223–251
34. M.J. Lee and W.C. Reynolds, "Numerical Experiments on the Structure of Homogeneous Turbulence," Report TF-24, Dept. of Mechanical Engineering, Stanford University, 1985
35. J.U. Brackbill and J.J. Monaghan, Ed., *Proceedings of the Workshop on Particle Methods in Fluid Dynamics and Plasma Physics,* in *Comput. Phys. Commun.*, Vol 48 (No. 1), 1988
36. F.H. Harlow, The Particle-in-Cell Computing Method for Fluid Dynamics, *Fundamental Methods in Hydrodynamics,* B. Alder, S. Fernbach and M. Rotenberg, Ed., Academic Press, 1964
37. J.U. Brackbill and H.M. Ruppel, FLIP: A Method for Adaptively Zoned, Particle-in-Cell Calculations of Fluid Flows in Two Dimensions, *J. Comput. Physics,* Vol 65, 1986, p 314
38. J.J. Monaghan, Particle Methods for Hydrodynamics, *Comput. Phys. Rep.,* Vol 3, 1985, p 71–124
39. J.K. Dukowicz, A Particle-Fluid Numerical Model for Liquid Sprays, *J. Comput. Phys.,* Vol 35 (No. 2), 1980, p 229–253
40. P.J. O'Rourke, "Collective Drop Effects in Vaporizing Liquid Sprays," Ph.D. thesis, Princeton University, 1981
41. Y. Sahd and M. Schultz, Conjugate Gradient-like Algorithms for Solving Non-Symmetric Linear Systems, *Math. Comput.,* Vol 44, 1985, p 417–424
42. W.L. Briggs, *A Multigrid Tutorial,* Society for Industrial and Applied Mathematics (Philadelphia), 1987
43. W.H. Press, B.P. Flannery, S.A.Teukolsky, and W.T. Vettering *Numerical Recipes: The Art of Scientific Computing,* Cambridge University Press, 1987
44. D.A.Knoll and P.R. McHugh, Newton-Krylov Methods Applied to a System of Convection-Diffusion-Reaction Equations, *Compt. Phys. Commun.,* Vol 88, 1995, p 141–160
45. S.V. Patankar, *Numerical Heat Transfer and Fluid Flow,* McGraw-Hill, 1980
46. M.C. Cline, J. K. Dukowicz, and F. L. Addessio, "CAVEAT-GT: A General Topology Version of the CAVEAT Code," report LA-11812-MS, Los Alamos National Laboratory, June 1990
47. HEXAR, Cray Research Inc., 1994
48. G.D. Sjaardeam et al., *CUBIT Mesh Generation Environment,* Vol 1 & 2, SAND94-1100/-1101 Sandia National Laboratories, 1994
49. W.D. Henshaw, A Fourth-Order Accurate Method for the Incompressible Navier-Stokes Equations on Overlapping Grids, *J. Comput. Phys.,* Vol 133, 1994, p 13–25
50. R.B. Pember, et al., "An Embedded Boundary Method for the Modeling of Unsteady Combustion in an Industrial Gas-Fired Furnace," Report UCRL-JC-122177, Lawrence Livermore National Laboratory, Oct 1995
51. J.P. Jessee, et al., "An Adaptive Mesh Refinement Algorithm for the Discrete Ordinates Method," Report LBNL-38800, Lawrence Berkely National Laboratory, March 1996
52. M. Landon and R. Johnson, Idaho National Engineering Laboratory, 1995
53. S. Ashley, Rapid Concept Modelers, *Mech. Eng.,* Vol 118 (No.1), Jan 1996, p 64–66
54. D.L. Reuss, R.J. Adrian, C.C. Landreth, D.T. French, and T.D. Fansler, "Instantaneous Planar Measurements of Velocity and Large-Scale Vorticity and Strain Rate Using Particle Image Velocimetry," Paper 890616, SAE, 1989
55. M.C. Drake, T.D. Fansler, and D.T. French, "Crevice Flow and Combustion Visualization in a Direct-Injection Spark-Ignition Engine Using Laser Imaging Techniques," Paper 952454, SAE International, 1995
56. D. Deitz, Next-Generation CAD Systems, *Mech. Eng.,* Vol 118 (No.8), Aug 1996, p 68–72
57. G. Farin, *Curves and Surfaces for Computer-Aided Geometric Design,* Academic Press, 1990
58. M. Hosaka, *Modeling of Curves and Surfaces in CAD/CAM,* Springer-Verlag, 1992
59. D.C. Chan, A Least Squares Spectral Element Method for Incompressible Flow Simulations, *Proceedings of the Fifteenth International Conference on Numerical Methods in Fluid Dynamics,* Springer-Verlag, 1996
60. "3D Piping IGES Application Protocol Version 1.2," IGES 5.2 Standard, IGES ANS US PRO/IPO-100-1993, U.S. Product Data Association, 1993
61. S. Sengupta, J. Hauser, P.R. Eiseman and J.F. Thompson, Ed. *Numerical Grid Generation in Computational Fluid Dynamics,* Pineridge Press, 1988
62. D.C. Haworth, S.H. El Tahry, and M.S. Huebler, A Global Approach to Error Estimation and Physical Diagnostics in Multidimensional Computational Fluid Dynamics, *Int. J. Numer. Methods Fluids,* Vol 17, 1993, p 75–97
63. C.-H. Lin, T. Han and V. Sumantran, Experimental and Computational Studies of Flow in a Simplified HVAC Duct, *Int. J. Vehicle Design,* Vol 15 (No. 1/2), 1994, p 147–165
64. FLUENT User's Group meeting, FLUENT Inc. (Lebanon, NH), 1995
65. T. Han, D.C. Hammond, and C.J. Sagi, Optimization of Bluff Body for Minimum Drag in Ground Proximity, *AIAA J.,* Vol 30 (No.4). April 1992, p 882–889
66. M.B. Malone, W.P. Dwyer, and D. Crouse, "A 3D Navier-Strokes Analysis of Generic Ground Vehicle Shape," Paper No. AIAA-93-3521-CP, American Institute of Aeronautics and Astronautics, 1993
67. T. Han, Computational Analysis of Three-Dimensional Turbulent Flow around a Bluff Body in Ground Proximity, *AIAA J.,* Vol 27 (No. 9), Sept 1989, p 1213–1219
68. T. Han, V. Sumantran, C. Harris, T. Kuzmanov, M. Huebler, and T. Zak, "Flowfield Simulations of Three Simplified Vehicle Shapes and Comparisons with Experimental Measurements," Paper 960678, SAE International, 1996
69. CFD Research Corporation, Huntsville, AL, 1995
70. MAGMASOFT, developed by MAGMA Giessereitechnologie GmbH, Aachen, Germany; marketed and supported in the U.S. by MAGMA Foundry Technologies, Inc., Arlington Heights, IL
71. R.H. Box and L.H. Kallien, Simulation-Aided Die and Process Design, *Die Cast. Eng.,* Sept/Oct 1994
72. L. Karlsson, Computer Simulation Aids V-Process Steel Casting, *Mod. Cast.*, Feb 1996
73. Fluid Dynamics International, Evanston, IL, 1995
74. B. Khalighi, S.H. El Tahry, D.C. Haworth, and M.S. Huebler, "Computation and Measurement of Flow and Combustion in a Four-Valve Engine with Intake Variations," Paper 950287, SAE International, 1995
75. S. Kong, Z. Han, and R.D. Reitz, "The Development and Application of a Diesel Ignition and Combustion Model for Multidimensional Engine Simulation," Paper 950278, SAE International, 1995
76. D. C. Haworth, M.S. Huebler, S.H. El Tahry, and W. R. Matthes, Multidimensional Calculations for a Two-Stroke-Cycle Engine: a Detailed Scavenging Model Validation, Paper 932712, SAE, 1993

77. P. Moin and J. Kim, Tackling Turbulence with Supercomputers, *Scientific American,* Vol 276 (No. 1), Jan 1997, p 62–68
78. B. Galperin and S.A. Orszag, Ed., *Large-Eddy Simulation of Complex Engineering and Geophysical Flows,* Cambridge University Press, 1993
79. A.R. George, Automobile Aerodynamic Noise, *SAE Trans.,* Vol 99-6, 1990, p 434–457
80. D.C. Haworth and K. Jansen, LES on Unstructured Deforming Meshes: Towards Reciprocating IC Engines, *Proceedings of the 1996 Summer Program,* Stanford University/NASA Ames Center for Turbulence Research, 1996, p 329–346
81. R.W. Johnson and E.D. Hughes, Ed., *Quantification of Uncertainty in Computational Fluid Dynamics,* FED-Vol 213, Fluids Engineering Division, American Society of Mechanical Engineers, 1995
82. I. Celik, C.J. Chen, P.J. Roache, and G. Scheuerer, Ed., *Quantification of Uncertainty in Computational Fluid Dynamics,* FED-Vol 158, Fluids Engineering Division, American Society of Mechanical Engineers, 1993
83. K. Molvig, *Digital Physics: A New Technology for Fluid Simulation,* Exa Corporation, Cambridge, MA, Aug 1993
84. D. Poirier, ICEM CFD Engineering, personal communication, 1996
85. D. Deitz, Designing with CFD, *Mech. Eng.,* Vol 118 (No. 3), March 1996, p 90–94
86. D. Deitz, Engineering Online, *Mech. Eng.,* Vol 118 (No. 1), Jan 1996, p 84–88

Computer-Aided Electrical/Electronic Design

Shaun S. Devlin, Ford Motor Company

DESIGN AND ANALYSIS of electrical/electronic systems and components tends to be different from the corresponding process for most mechanical and hydraulic systems, for one key reason. In most systems, the components interact with each other along a small number of relatively ideal paths (conductors) that have relatively ideal behavior (linear and almost conservative) via two scalar variables (voltage and current). This has led to a greater use of simulation and modeling in all phases of the process and has allowed very complex systems to be modeled, before they were built, with remarkable fidelity. Daniel Whitney (Ref 1) concludes that some of the idealizations that electronic design has employed are becoming less accurate in submicron design. These and other difficulties of the electrical simplifications are described below in contrasting the lumped circuit model and the situations where it is not applicable.

It is useful to think of electrical design following through three overlapping phases: functional, electrical, and physical.

The functional phase involves describing very carefully what the circuit or component is supposed to do (i.e., how do the outputs depend on the inputs). These are the behavioral requirements. If these descriptions can be simulated, we gain increased confidence that the behavioral requirements are correct. Some description techniques support synthesis (nearly automatic translation, usually with the inclusion of library or process technology information) into a full electrical or physical description. This phase of the design is sometimes conducted by the customers (of the design) in order to get the correct requirements in the context of a larger system design.

The ability to synthesize a design from a behavioral description will eliminate many detailed human design steps, thus reducing error, but will require more detail and precision in the behavioral and nonbehavioral requirements. The limitations of the languages and the lack of synthesis tools (except for digital integrated circuits) has limited the automation of this phase. It *is* appropriate to capture at this point any nonelectrical requirements and constraints that the implemented component/system must meet, because the information will be useful in the later implementation stages. Examples are power dissipation (waste energy to be disposed of), size and weight, and any chemical, thermal, or radiation (electromagnetic or nuclear) conditions in the environment that must be met.

The electrical phase is where most designers start describing a design whose desired behavior they often know only informally. The design is often described as a graph of interconnected blocks. The lines or arcs are assumed to be ideal signal conductors and only the blocks have "interesting" explicit behavior. The overall behavior is governed by the individual blocks and their interconnections. The expressive power of these schematic diagrams and their robust mapping to analytical representations such as graph theory and Kirkoff's laws have made these techniques important for many years (Ref 2).

The physical phase is where specific available components are chosen (or requirements for new components are defined and the new component is designed) and descriptions of their electrical and mechanical assembly are elaborated. The electrical interconnections dominate the electrical engineering considerations. The size, strength, reliability, and heat dissipation requirements help in determining the choice between electrically equivalent available parts. The assembly description must provide a geometrical arrangement of conductors (e.g., wires, circuit board traces, silicon metallization), functional elements (e.g., motors, integrated circuit chips, registers), and appropriate physical/geometrical relationships.

Many details of the process, the design tools, and the analysis depend on the physical scale (e.g. chip, board, or systems) and the power levels. These will be discussed at the appropriate places. This article does not discuss any of the commercial tools because of the large number and the high rate of change of these products. The trade publications *Electrical Engineering Times (EETimes)* and *EDN* are recommended for current information about commercial design and analysis tools.

Functional Phase

The process of collecting requirements is essentially informal. There is enormous range of rigor in "writing" the requirements. The more rigorous the expression of the requirements, the more likely that the conformance of the final product to those requirements can be determined with little additional judgment or other human effort. VHDL (Ref 3) and VERILOG (Ref 4) at the behavioral level can be used to express the desired digital behavior. At this time there is no accepted (national, international, or consortium) standard language for describing continuous analog behavior, but there is active work to extend VHDL in this direction. SPICE, which was originally developed at the University of California at Berkeley, is a de facto standard both as a language and a simulator. There are a number of proprietary versions of it (PSPICE, HSPICE) as well as unrelated languages such as the MAST language for Saber (Analogy Corp., Beaverton, OR). Each language is associated with a particular simulator.

Electrical Phase

This discussion begins with the tools that were developed early and are still the most commonly needed and then moves on to more recent and often less needed capabilities. The discussion cannot possibly be complete but will give the reader a feeling for current capabilities, the difficulties in using more than one tool, and the directions of current development.

Schematic Capture

Although there were many design description and analysis tools before graphic computing, the design description techniques (structured text) were not natural for most designers. They were often rigid in syntax and aimed at easy processing into board layout information or simulation. They rarely did both. Graphic schematic capture tools allow designers to draw and interconnect the symbols they are used to drawing on paper. The modification and revision process becomes much easier. Most systems have an internal representation that not only captures the graphics aspect (how the design should be presented on screen or paper) but also "understands" some of the electrical function. If the tool is to be used for simulation, then each symbol must have a simulation model associated with it. The meaning of a line on the screen connecting ports (pins) on two cells (components) has to be stored in such a way that the electrical connection can be understood in the simulator. Alternately, if the design information is to be used for board or chip layout, then the symbol information has to be associated with geometrical information about packages and the physical placement of pins, or (in the case of an integrated circuit) the standard cells (a useful collection of transistors in a controlled library).

To gain user acceptance, the schematic capture tool must provide many of the services of a general-purpose graphic editor and not unduly constrain design creativity with the need to be electrically accurate. For example, many graphic editors force the connection of a line to an adjacent box when the grid feature is turned on. When an electrical schematic editor is used, it should be easy to connect a line to a pin (port) on a symbol, and the editor should not allow a connection at an arbitrary place that does not have electrical significance.

The capability to move symbols around while preserving the connections already drawn ("rubber banding") is very useful. It allows the user to make layout changes to improve the understandability of a diagram after many modifications without altering the essential connectivity. It is usually important to represent a design as either flat (all elements visible on one logical sheet) or hierarchical (blocks or symbols can contain groups of blocks or symbols that may be made visible by some special action on the symbol, often by double clicking). Some tools provide ways of toggling between the two representations.

Even when hierarchical techniques are used, a design may be too large to represent on one sheet of whatever size paper is available. It is then necessary to break the design up into multiple "pages" with indications of how conductivity (lines) on one page are related to connectivity on another. Most systems provide for "off-page connectors" that provide logical linking between pages. These presentation features should not affect the ease or accuracy of simulation. Users must understand that off-page connectors are a notational convenience with no electrical or physical properties, while a real connector certainly has size and weight and may have significant electrical properties.

Figure 1 is a screen image of a session with Viewlogic's Viewdraw, a modern schematic capture package. It can only hint at the capability and usability of such a product.

This is a good time to emphasize the difference between two types of requirements for the outputs of a schematic editor: those that are based on the limitations of human vision and the available printer/plotter sizes, and the need to represent the underlying connectivity that must be manufactured or simulated. There is often a tension between the ease of drawing/annotating and human understandability on one hand and providing a robust "carrier" for the underlying behavioral and/or assembly information. For example, in the author's opinion, a sufficiently complete symbol library with associated simulation models and/or component geometric information is more important than having a variety of fonts or a large number of colors.

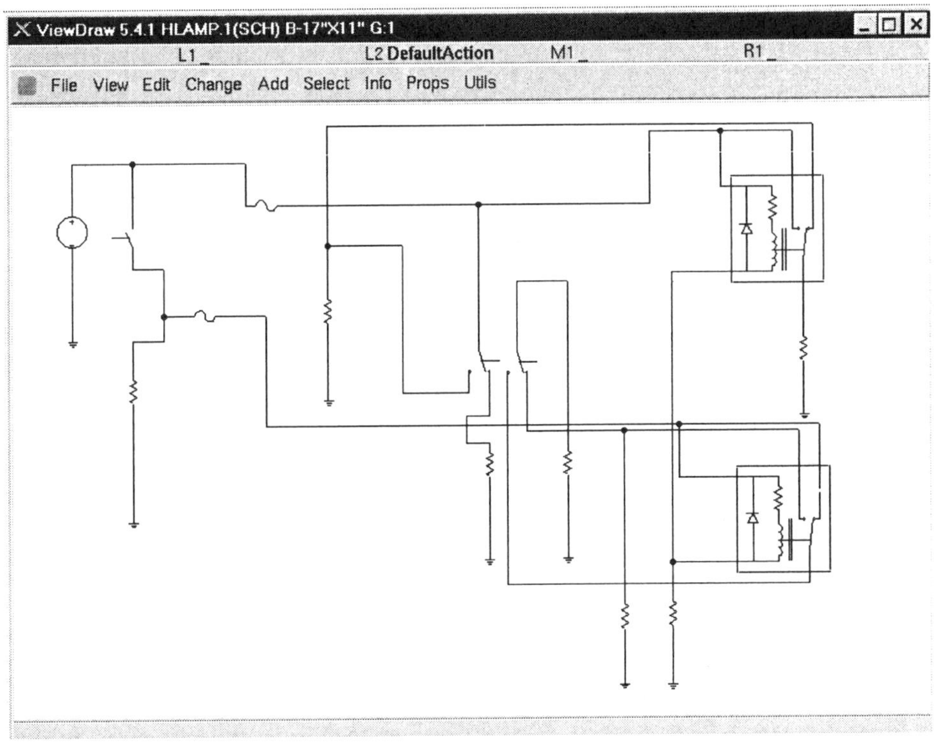

Fig. 1 An electromechanical circuit drawn with a schematic capture tool

Simulation and Testing

Simulation allows the designer to evaluate some of the behavior of a design before any physical parts are built. To perform a simulation the designer must have the design described in a language understandable to the simulation engine ("the simulator"), a simulation engine, and usually a set of time-varying signals (test vectors) that provide inputs to the circuit to be simulated. This set must be carefully chosen to represent the normal and abnormal (but possible) signals that the circuit is likely to experience in practice. These signals can be constructed from recorded data, obtained from simulations of the environment that drives the system under consideration, constructed with the toolkit that is provided with most simulators, or constructed with some combination of these techniques. At the later stages of the design, it is important that these test vectors cause every portion of the design to be exercised. It is also necessary to have a way of recording and displaying the outputs of the design (the results). It is very desirable to have some reasonably concrete ideas about what the "correct" results should be for a certain set of time-varying input signals. Very specific results can be obtained by using the output of a simulation of the behavioral design that has been subjected to the same input sets as are to be used in the physical test.

Simulation engines tend to be optimized for two distinct classes of problems: discrete signal (digital) and continuous signal (analog). The digital simulators are concerned with discrete time changes among a small set of discrete values (low, high, unknown, undetermined, etc.). The simulator is optimized to propagate these changes from the element causing the change to all affected elements and to determine how these changes will generate future changes. In large systems where only a modest fraction of the signals are changing at any given time, minimizing calculations of "no-change" is the best way to improve simulation speed. The currents are rarely evaluated. Analog simulators normally calculate both the voltage and the currents, because the currents drawn by the load often significantly lower the voltages at the driving terminal. In other words, the conservation of current and energy must be considered in the simulation. Because digital circuits are designed to be insensitive to modest changes in the voltage signal and the loads tend to be similar, detailed loading effects are simulated less often and problems are avoided with design rules (e.g., connect fewer than N loads to a source whose "fanout" is N). In contrast, in digital systems the power transmitted between macro design elements is small com-

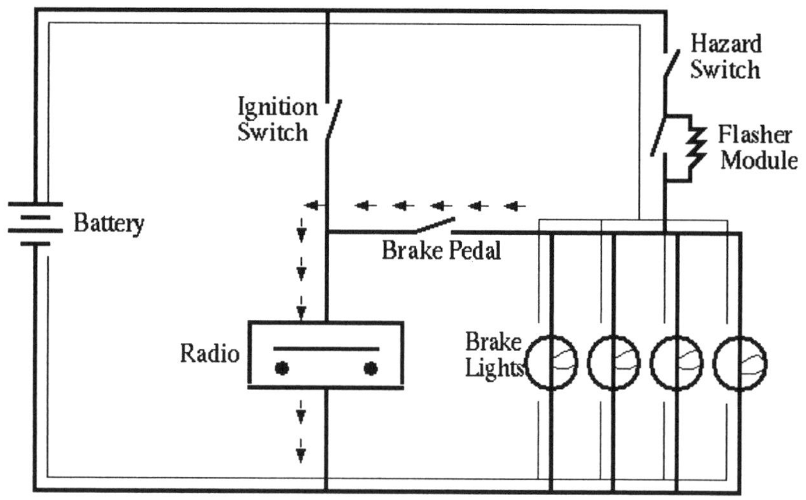

Fig. 2 An automotive example of a "sneak path"

pared to that dissipated in internal switching and is only modeled during design stages where heat removal and packaging are considered.

Analog simulators must evaluate continuous changes in values in continuous time. The analog simulation engine must solve systems of simultaneous (usually nonlinear) differential equations. There will be one equation for each voltage node or current branch in the circuit. The equations represent the application of Kirkoff's laws to circuits with time-varying voltages and currents. The challenge for modern electrical simulators is to choose the form of the mathematical models of each circuit element and a general integration technique such that the time evolution of the circuit voltages and currents can be expressed to suitable accuracy with a minimum number of computational steps. A capability for reducing "waste time" on slowly changing signals can lead to considerable speedup. The SPICE simulator and model description language (Ref 5) pioneered demonstration of these techniques for circuits involving transistors. All the commercially available simulators and languages were influenced by it. The SPICE simulator is in the public domain and is a good starting point, but there are a wide variety of simulators with a broad range of capabilities and costs (e.g., PSPICE, HSPICE, Saber).

Most electrical simulators and models assume that the spatial extent of the component or system does not inherently affect its behavior. More precisely, it is assumed that the delay due to the finite speed of electrical signals in conductors is negligible compared to other delays in the design. An equivalent assumption is that the system is small compared to the electrical wavelength of the frequencies of concern. This is known as the "lumped circuit" approximation, and it allows the dynamics to be described with ordinary differential equations, primarily in time with no spatial definitives, due to the effects of electromagnetic field propagation.

There is a class of problems where this is not a good approximation. Most examples are from high-frequency design (e.g., microwave) where the wavelength of the electrical signals *is* comparable to the dimensions of the circuit elements, including conductors and their spacings. These are best described by partial differential equations in space and time, rather than by the ordinary differential equations in time that are satisfactory for discrete circuits. These partial differential equations cannot be solved analytically for most geometrical situations of interest. Hence, the continuous spatial variation is approximated by three-dimensional meshes of lumped circuit elements whose property variation in space approximates the true variation. This is the finite element method, which has been used in mechanical problems for many years and is now used to calculate the properties of microwave circuit boards, evaluate the magnetic field generated by time-varying currents in motors and transformers, and calculate the electrical response of a piezoresistive strain sensor element to a nonuniform strain field. Of course, within a single transistor, or sometimes between adjacent transistors, the full three-dimensional flow of current carriers must be analyzed, but this is normally the concern of the semiconductor device designer and very specialized tools are used.

The digital simulation engines, in contrast, must solve a set of coupled finite state equations (Ref 6). Each equation predicts the next state (one of a small set of discrete values) as a function of all the current state values and the value of any discrete input signals. In synchronous systems, the prediction is for the next clock signal transition, which occurs at a known time in the future. Hence, for each clock transition, the simulator must obtain the current state variables and the input variables and use them to evaluate the set of state values that will be true at the next clock transition. Since there may be thousands or millions of state variables in a modern digital circuit, the simulator must efficiently evaluate these equations and in particular recognize equations that predict no change. In asynchronous systems, that time interval may be set by any element of the circuit and hence can occur at a time unknown. In a mixed analog digital system, simulation is much more complicated. For example, if an analog integration must start when a particular logical value becomes true, that fact must be communicated from the digital simulator to the analog simulator at the appropriate simulated time. The analog simulator must be prepared to act on this signal, presumably like any other external input. Likewise, if the time at which that integrated value crosses a threshold is an event required in the digital simulation, it must be transmitted back. Few commercial simulators do this well enough to simulate large mixed systems.

The primary initial use of simulation is to help the designer confirm that the design performs its intended function, however completely that is defined. Once the design is frozen, simulation can generate expected results files for test systems. A very wide range of input signal descriptions, designed to exercise all "important" circuits in a design, must be created. Tools exist for creating input signals (test vectors) that can exercise every state that is achievable in a finite number of clock steps. Some states may not be tested in practice because too many steps (and hence prolonged simulation time) are required to reach them. Each organization must make a decision for each product about what fraction of the possible states will not be tested. In systems where 100% test coverage is essential, sometimes extra circuitry (which does not contribute to the base function) is introduced that can read/or set the value of states that would otherwise be costly (in the time sense) to read or set.

Other Analysis

Simulators often perform other analyses on the circuit before the primary simulation begins. In digital simulation, these may include finding circuit elements whose inputs or outputs are unconnected. Additionally, checks may be made to ensure that no output is required to drive more inputs than it is capable of driving (fanout). Checks are also made for locating two or more outputs connected together, which will lead to ambiguous results for most digital technologies. Even in logic families where such connections are allowed and the results are unambiguous, they are normally a small fraction of the connections, and it is useful to highlight them to ensure that they are intentional and not due to a design error.

Another class of errors may occur when designers who are accustomed to conventional electronic logic design use relays or transmission gates to implement logic functions. These errors arise because in conventional electronic logic the signal can only propagate "forward" through gates, while in relays and transmission gates the signal can propagate both ways. Consider these original requirements:

- Load1 (the brake lights) shall be ON if switch1 (ignition switch) and switch2 (brake switch) are both ON.
- Load2 (radio) shall be ON if switch1 (ignition switch) is ON.

Later an additional requirement is added:

- Load1 shall be ON if switch1 (ignition switch) and switch2 (brake switch) are both ON OR if switch3 (hazard switch) AND switch4 (flasher module) is ON.

These are trivial requirements if they arrived at the same time. Let us assume the existing circuit to be modified meets the two original requirements and the designer is asked to make a modification to meet the new requirement. The resulting design, shown in Fig. 2, meets the requirements if it is interpreted colloquially. If the requirement is "Load2 shall be ON *if and only if* switch1 is ON," then the design indicated is wrong because load1 can also be ON if switch2 and switch3 and switch4 are ON. This is a trivial example of an error due to a "sneak path." Today most requirements are not stated with the precision that distinguishes "if then" from "if and only if then." Because the requirement is often ambiguous, the circuit paths that implement the "if then" form can be highlighted by a heuristic "sneak" error detection process or tool as a *possible* error to be examined more closely by the designer. A circuit in Fig. 2 is an example of a common sneak pattern, which occurs when two initially independent circuits of switches and loads are drawn near each other on the page, sharing a common power source at the top and a common power return at the bottom, and then a design change introduces a switched path between these previously independent circuits. This is called an "H" pattern because of its resemblance to a letter "H." Errors like these are easy to make in large electromechanical systems with incremental requirements and numerous switches and relays.

Design errors of this kind can be detected by formal methods where both the requirements and the design are described in languages with semantics that are very well defined. An algorithmic process can then be used to prove whether or not the design requirement is met by the design. This algorithmic process can be automated. There is some progress in this area, but it has been difficult to express requirements in a language that is both precise enough to be mechanically compared to the design and clear enough to discuss with the customer. Likewise, the design language must be precise while not interfering with the creativeness inherent in most good design.

Physical Phase

Printed Circuit Board Layout

The determination of the placement of components and the routing of conductors between them is one of the most important computer-aided capabilities available to the engineer. The algorithms to make the process more automatic have been studied for many years. The process has a blend of electrical and mechanical/geometric aspects, and there are several logical steps.

Choice of Board Material, Outline, and Number of Layers. Most printed circuit boards are made of phenolic. However, unusual requirements of temperature range, thermal dissipation, strength, or dielectric constant may indicate other materials. Although the temperature range may be known a priori, the details of the other considerations may require that the material be chosen before the corresponding analysis is made. For example, the magnitude and location of heat-generating components on a board is known only after placement and simulation. Likewise, the contribution of the conductor to the spreading of heat is determined by the layout. Hence, it may be necessary to run a thermal analysis of the board with all components on it and in an enclosure for several choices of board material. Differential thermal expansion has caused reliability problems, particularly with surface-mounted components. The board size and outline are determined by the size of the circuit to be mounted but are often constrained by the size of the enclosure and the connector arrangement. If the circuit is too large to fit on one board in the chosen enclosure and the partitioning of circuit elements is not obvious (i.e., dictated by a bus architecture), then computer aids are sometime used to allocate the circuit to multiple boards, minimizing the interconnect.

The cost of a board rises steeply with the number of layers. Hence, it is desirable to try to lay out the board with a low number of layers and then increase the number of layers if it is not feasible to route. Feasibility is conditional on how much manual routing is acceptable in the local engineering process. A layout tool that can minimize the need to try several alternatives is very desirable.

Placement of components must account for conductor lengths, thermal dissipation, insertion feasibility, and many other factors. Each of these considerations may be the subject of analysis, and while an optimal set of locations can be calculated for one criterion, no tools take into account multiple considerations. A layout system must first have a library of the "footprints" (a projection of the part outline, including pins on the board, and preferably with recommended pad geometry and drill hole patterns) for each part in the proposed design. A comprehensive layout system provides predesigned drill hole patterns and corresponding electrical pads for each component in the design library. A more comprehensive library will also include the constraints on intercomponent spacing imposed by the insertion process, whether automatic or human.

The library system must certainly allow the addition of new components. It is desirable but much more difficult to be able to add new aspects or attributes of the components when those design attributes become important. For example, if the current library system provides for storing only part shapes that are parameterized by three or four numbers (e.g., cylinders, parallelepipeds, and pyramids), it may be difficult to extend it to be able to describe more complex shapes such as transformers, heat sinks, and other shapes requiring many more parameters, if they can be described parametrically at all. Similarly, a library of component models of geometry for manufacturing and electrical/thermal simulation may be difficult to extend for use with a vibration analysis of a populated board. Hence these possibly future needs should be considered when acquiring a new computer-aided electrical design system.

Routing of Conductors. The conductors must be routed between the appropriate pins on the placed components according to the intent expressed in the netlist (the set of sets of interconnected components and connector pins) but subject to the constraints of the size of the board, minimum widths and spacings, and the number of layers available. Routing of the conductive strips is both the most tedious hand process and the one most investigated theoretically (Ref 7). The netlist provides the connectivity desired. The router attempts to provide that connectivity subject to the constraints of the number of layers and the kind of vias (interlayer connections) that the manufacturing process allows. Interlayer connections should be minimized because they tend to be expensive. Modern routing tools can fully route average boards but may require manual assistance with particularly large designs (large number of component pins per unit area). All tools allow manual routing of difficult or special cases.

Systems Interconnected with Wire and Cable. Electrical/electronic equipment often requires wiring to interconnect circuit boards and other electrical equipment (sensors, switches, motors, etc.). If the equipment in which the electronics is housed is mass produced, the wiring is prebuilt and installed with the electronic boards or modules in the larger system. Examples include a complete computer workstation, a telephone switch, or an automobile. If the product is manufactured in low volume, the wiring is usually installed wire by wire and connectors are attached during the installation process. These two applications have many design similarities but the manufacturing techniques are very different.

Large electrical/electronic systems often consist of printed circuit boards interconnected by more or less organized wiring within a cabinet, vehicle, building, or larger complex. The process of using the netlist as the expression of the desired connectivity is common. The process of determining the routing of the wiring in three-dimensional space is subject to many nonquantifiable constraints related to installation and serviceability. The physical problem of routing individual wires or bundles of wires in three-dimensional space is very similar to the problem of routing piping. In fact, several systems simplify the problem to routing conduits (pipe to protect wiring) or channels (imaginary conduit). Initially only the centerline of the channel path is determined, and the diameter is determined after the number of wires and their diameter are fixed. The capability to use the lengths, spacings, and curvatures defined in the routing process in a calculation of the resulting "parasitic" remittances, capacitances, and inductances is not available in any commercial product in a manner that can be used easily in simulation. In principle, there is an interaction between the routing and the diameter (if there is a voltage drop constraint), but it rarely causes an experimental router to iterate.

Simulation. The metal foil interconnect between a set of pins is not a perfect conductor. Its geometry and the dielectric properties of the board material can lead to (usually) undesired parasitic remittances, and to interconductor capacitance and inductance. For most circuits to operate as intended, these must be below some acceptably low level, determined by the current and frequency levels and the desired function of the circuit. These levels can normally be reached by following the spacing rules that the router uses. Parasitic extraction is the process of calculating the values of these stray parameters from the geometry of the conductors and the parameters of the material and calculating the equivalent lumped circuit elements (resistance, capacitance, inductance) and inserting them in the original netlist. This new actual netlist with the calculated parasitic parameters can now be resimulated to ensure that the circuit meets its requirements. At very high frequencies (>110 MHz), the parasitics are treated not as lumped circuit elements but as transmission lines, and the tailoring of their properties is an essential part of the design process.

Test Design

A stimulus file and the results of a simulation with that stimulus should be usable as an expected results file. The physical design can be verified by subjecting it to a stimulus file in real time and comparing the measured electrical outputs with those predicted by simulation. The detailed physical design files of the unit being tested can be used to help design the mechanical fixtures and electrical probes for the test system.

Standards

It should be clear from this discussion that it is unlikely that a single computer-aided tool (or family of tools) will provide all the electrical/electronic design, manufacturing engineering, and test engineering functions required in a complex enterprise. Standards of representation of the product design and behavior are aimed at providing vendor-neutral file formats or other mechanisms of transferring the required information from one computer system or another. Modern standards of data representation should have a careful definition of the semantics of their terms so that a developer of an application who writes a product description has a clear understanding of the meaning of each term and so that the developer of a reader will have the same understanding. Many standards use the language Express for that purpose (Ref 8). The STEP family of standards developed by ISO TC184-SC4, "Industrial Data," is the most ambitious. It is an evolving series of standards, of which ISO 10303-210, "Electronic Design and Assembly" (AP210) (primarily printed circuit boards and assemblies), and ISO 10303-212 (AP212), "ElectroTechnical Plants," may be of interest to electrical engineers and computer-aided tool developers. The languages VHDL and VERILOG for the description of the behavior and structure of digital systems are now U.S. standards, and VHDL is now an International Electrotechnical Commission (IEC) standard. The EIA/EDIF series of standards (Ref 9) have evolved from a pure netlist standard to include schematics and printed circuit boards with multichip modules.

There is vigorous ongoing work in this area, and it is becoming more important as design and supplier relationships become global and organizations can no longer depend on a single supplier. The process of review required in establishing a national or international standard helps clarify any ambiguities in the meaning of a standard and helps ensure that it will allow description of product aspects that are larger than the scope of any one tool.

REFERENCES

1. D.E. Whitney, Why Mechanical Design Cannot Be Like VLSI Design, *Res. Eng. Des.*, Vol 8, 1996, p 125–139
2. W.-K. Chen, Chap 2, *Applied Graph Theory*, North Holland Publishing Co., 1971
3. *VHDL Language Reference Manual*, IEEE 1076-1993, Institute of Electrical and Electronic Engineers
4. *VERILOG Hardware Description Language Reference Manual*, IEEE 1364–1995, Institute of Electrical and Electronic Engineers
5. L.W. Nagel, SPICE2: "A Computer Program to Simulate Semiconductor Circuits," Electronic Research Laboratory Report ERL-M520, cited in *Semiconductor Device Modeling with SPICE*, P. Antognetti and G. Massobrio, Ed., McGraw-Hill, 1988
6. V.P. Nelson et al., *Digital Logic Circuit Analysis and Design*, Prentice Hall, 1974
7. N. Sherwani, S. Bhingardi, and A. Punyan, *Routing in the Third Dimension*, IEEE Press, 1995
8. "Express I Language Reference," ISO 10303-11: 1994, ISO 10303-11:1994, International Organization for Standardization
9. "Electronic Design Interchange Format," EDIF 200 (EIA 548-1988), EDIF 300 (EIA 618-1994), and EDIF 400 (EIA 682-1996), Electronic Industries Association, Arlington, VA

Design Optimization

Douglas E. Smith, Ford Motor Company

ENGINEERING DESIGN involves the reallocation of materials and energy to improve the quality of life. This occurs in all fields of engineering, including civil, mechanical, electrical, and so forth, and often involves trade-offs based on the requirements of each application. The idea of design optimization suggests that for a given set of possible designs and design criteria, there exists a design that is the best or optimal.

Optimization is a part of everyone's life, either consciously or subconsciously. It is our nature to optimize. Investors want the largest return with the least investment or risk. Marathon runners adjust their pace to achieve the best overall time. This article discusses tools that provide a method for systematic optimization of engineering designs. The primary focus here is on the practical application of optimization technology in a computer-aided engineering (CAE) environment.

The role of the CAE simulation tool is very important in CAE-based design optimization. Computer-aided-engineering-based design optimization does in fact turn CAE *analysis* tools into CAE *design* tools by replacing traditional trial-and-error design approachs with a systematic design-search methodology. Thus, CAE computations that quantify the performance of a particular design are enhanced with information on how to modify the design to better achieve important performance criteria.

It is impossible to cover in detail the broad field of optimal design in this short article. The goal here, therefore, is to acquaint the reader with CAE-based design optimization and to provide direction on where to find additional information on the topic. Although CAE-based optimal design is applicable to a wide array of engineering design problems, much of its development has focused on structural optimization. This fact reflects the greater emphasis devoted to structural optimization in this article. Background in numerical optimization is discussed, and emphasis is placed on identifying specific challenges that are encountered when computing optimal designs with traditional CAE analysis tools. Trends in optimal design for CAE applications are also considered through a discussion of emerging technologies in this area. The interested reader is encouraged to consult the cited and selected references at the end of the article for more information. Other approaches to engineering design that also seek the best design solution can be found elsewhere (see, for example, Taguchi methods in the article "Robust Design" in this Volume).

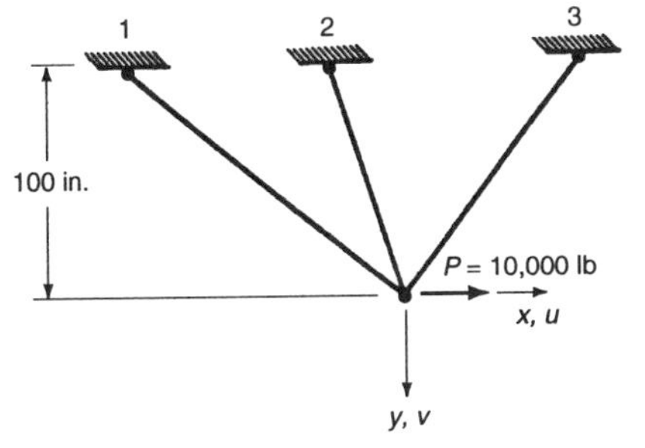

Fig. 1 Three-bar truss example. Source: Ref 1

Numerical Optimization Methods

A key component of CAE-based design optimization is the numerical optimization algorithm. These algorithms solve optimization problems with mathematical programming techniques independent of the physical application. Better designs are computed based on the design definition and the performance measures that evaluate the goodness of a design. This section focuses on numerical optimization algorithms and is intended to provide some background on how these algorithms make decisions when searching for the optimal design. The interested reader is encouraged to find more information in Ref 1 to 5.

The Nonlinear Constrained Optimization Problem

To formulate the design-optimization problem, the notion of having *design parameters* (often referred to as design variables) and *performance measures* is first considered. Design parameters define the process or structure of interest and thus provide a means for changing it to improve its performance. Performance measures that are defined as functions of the design parameters quantify the effectiveness of a given design and enter the optimization problem through the objective function (sometimes referred to as the cost function) and the constraints. The goal when solving an optimization problem is to determine the design parameters that give more desirable objective function and constraint values.

The most general single objective optimization problem is one that minimizes or maximizes an objective function F defined over the N design parameters b_i, $i = 1, 2, \ldots, N$ while satisfying both equality and inequality constraints. In mathematical terms:

$$\begin{array}{lll} \text{Find} & \mathbf{b} = \{b_1, b_2, \ldots, b_N\}^T & \\ \text{To minimize or maximize} & F(\mathbf{b}) & \\ \text{such that} & g_j(\mathbf{b}) \le 0, & j = 1, 2, \ldots, n_g \\ & h_k(\mathbf{b}) = 0, & k = 1, 2, \ldots, n_h \\ & b_i^L \le b_i \le b_i^U & i = 1, 2, \ldots, N \end{array} \quad \text{(Eq 1)}$$

where superscript T indicates the transpose. Constraint functions g and h are defined that mark the boundaries between which designs are allowed and which are infeasible. These constraints are divided into two groups, n_g inequality constraints g_j, $j = 1, 2, \ldots, n_g$, and n_h equality constraints h_k, $k = 1, 2, \ldots, n_h$. The design parameters are assembled in a vector **b**, and initial values are chosen for each component. Side constraints define the upper and lower bounds for each design variable b_i as b_i^U and b_i^L, respectively. The set of all possible designs that can be generated by adjusting the design variables between their respective upper and lower limits is called the *design space.*

A simple structural optimization example is given in Ref 1, where the cross-sectional areas of the truss shown in Fig. 1 are adjusted to obtain a design with minimum mass while satisfying constraints on the maximum allowable tension and compression stresses in each member. In this example, the cross-sectional areas A_1, A_2, and A_3 are the design variables and the mass of the structure is the objective function. Inequality constraints are formed to represent the limits on the maximum stress, and side constraints bound the range of cross-sectional areas to be considered. Truss member stresses are computed from a deformation analysis of the structure under the given loads.

Aspects of Numerical Optimization

Many aspects of the optimization problem play a major role in algorithm selection and performance, and ultimately in the success or failure of the optimization process. Below are some considerations which should be addressed when formulating an optimization problem.

The Nature of the Design Parameters. Design parameters define the structure or process being optimized and thus provide a means to alter or change it. They are often classified as *continuous* or *discrete*. Continuous design parameters are permitted to take any value over a predetermined range. A hole or fillet radius, for example, or the location of a joint in a truss structure can take any value over its permissible range. Discrete or integer variables are restricted to a finite set of values. Discrete design variables are required when design components are limited to sizes that are available, such as sheet metal gage thickness or tubing diameter. Material modulus is also a discrete design variable because there is only a finite set of materials and thus moduli that exist.

Commonly used optimization algorithms accept only continuous design variables. Therefore, the discrete variables are often treated as continuous and the discrete value that is closest to the optimal design is selected. This approach works relatively well when the distance between discrete values is small. Otherwise, *integer programming,* which limits the design variables to the predefined finite set, may be required when the distance between permissible values is large, such as when the variable defines the choice of material (Ref 1).

The Nature of the Performance Measures. Performance measures provide a quantitative measure of the goodness of a design. For each set of design parameters, there is a corresponding set of values, one for each of the performance measures. Performance measures may enter the optimization problem as an objective function or as a constraint. As the objective function, the performance measure is either minimized or maximized. For example, one may wish to minimize the mass of a structure. When the performance measure forms a constraint, it is expressed as a limit on a critical value. In structural optimization, for example, mass is commonly minimized while it is specified that the maximum stress remain below the yield value of the structure's material.

The nature of the performance measures is problem dependent. Objective and constraint functions can be linear or nonlinear functions of the design parameters and can be smooth or even discontinuous functions. General nonlinear mathematical programming algorithms are used when the order and/or smoothness of the performance measures is unknown (Ref 3). Alternatively, special algorithms have been developed when the nature of the objective and constraint functions is known. For example, linear programming methods, though applied to nonlinear problems, are very efficient when solving optimization problems where the objective and constraint functions are linear. Other specialized methods exist to efficiently solve least-squares problems that often arise when analytical models are adjusted to match experimental data.

Local and Global Optimum. The minimization problem in Eq 1 may exhibit a single global minimum and possibly many local minima. The goal of solving Eq 1 is to find the global minimum; however, in practice, in the absence of certain convexity conditions (Ref 3), one can only ensure that the solution is a local minimum. Descent search methods (discussed later in this article) and necessary conditions for an optimum are restricted to values of nearby points, which results in a focus on local or relative minima. In practice, starting with various initial designs alleviates the concern that the global minima is missed; however, local minima are rarely a hindrance to the success of solving practical optimization problems.

Constrained versus Unconstrained Optimization. Optimization problems are classified as constrained or unconstrained. When the range of feasible designs is not restricted, the optimization problem is defined as unconstrained. Alternatively, when limitations are placed on the range of feasible designs through simple bounds on the design variables or with functions of the design variables, the problem is constrained. This distinction plays a significant role in optimization algorithm selection.

Most engineering optimization problems are constrained. Indeed, to minimize the mass of the truss structure in Fig. 1 without constraints on the stress or displacement is of little use. The unconstrained solution is meaningless because all of the areas would simply be reduced to their respective lower bounds. Deformations and stresses in such a design would be well outside their useful ranges, revealing that an ill-defined optimization problem was chosen.

Constraints can significantly affect the computed optimal design because they often force the objective function to assume a higher value. Such constraints are considered to be active because if they were removed, the objective function would decrease in value. It is useful, therefore, to know how a particular constraint influences the value of the optimal objective function value. For smooth, continuous objective functions and constraints, this is accomplished with Lagrange multipliers, which measure the sensitivity of the optimal design to changes in the constraints. Large Lagrange multipliers suggest that even slight changes in the associated constraint limit would result in a significant reduction in the objective function, whereas Lagrange multipliers near zero indicate that the constraint has very little effect on the optimal design.

Multiple- and Single-Objective Optimizations. Many numerical methods have been developed to solve the unconstrained minimization problem given in Eq 1 for a single-objective function F. Often in design, however, there are multiple objectives F_i that may need to be considered. For example, a design may be desired that minimizes both stress *and* weight. Solution techniques for multiple-objective problems, however, have not been developed to the same level as those for single-objective formulations.

Two common methods are used to convert a multiple-objective problem into one that can be solved using single-objective algorithms. The first method defines a new objective function as the weighted sum of each of the individual objectives F_i. Weighting coefficients are selected to reflect the relative importance of each F_i, and care must be taken when using this method because the individual objective with the largest sensitivity always dominates the optimization. The second method chooses the most important F_i as the objective function and defines limits on those remaining, which are then included as constraints in the optimization problem. In the latter, many single-objective optimization problems are typically solved, each with different constraint limits, to understand the behavior of the optimal design.

To a lesser extent, *Edgeworth-Pareto* optimization has been used to solve multiple-objective problems when it is difficult to determine the relative importance of the performance measures (Ref 6, 7). Additionally, compromise programming avoids the sensitivity issues of the weighted objective method by minimizing the difference between each individual objective and its respective target value in a least-squares sense (Ref 8).

Optimization Algorithms

When F is an algebraic function of the design variables, classical methods from elementary calculus can be used to compute the optimal design. For example, when Eq 1 is unconstrained, the design $\mathbf{b}^*$, which satisfies $\nabla F(\mathbf{b}^*) = 0$, and certain criteria on higher-order derivatives comprise the minimum. However, when CAE tools are used to compute the performance of a design, the convenience of having a simple algebraic function is lost because the F is not an explicit function of the design $\mathbf{b}$. Instead, the performance measures are implicitly dependent on $\mathbf{b}$ through a CAE solution. In this case, classical methods may not be applied and iterative schemes that search the design space for the optimal design parameter values must be adopted.

Searching for the Minimum. Most CAE optimal design implementations are based on computationally expensive numerical simulations to evaluate the performance measures and use descent methods to move through the design space. Commonly used descent methods are based on the same underlying structure when systematically adjusting the design variables while searching for a minimum (Ref 3). For unconstrained minimizations, an initial starting design is specified. A search direction is then determined based on some fixed rule, followed by a one-dimensional line search, which minimizes the function along that direction in the design space. This new minimum serves as a starting point for another iteration, and the process is terminated when the objective function cannot be further reduced. The primary difference between descent algorithms is the rule used to define the search direction and the line search minimization technique. Additional distinctions are made between constrained algorithms based on the manner in which they handle the constraints.

Descent methods iteratively update designs as:

$$\mathbf{b}_{I+1} = \mathbf{b}_I + \alpha_I \mathbf{s}_I \quad \text{(Eq 2)}$$

where I is the iteration number and $\mathbf{s}_I$ and α_I are the corresponding search direction and step length, respectively. The only requirement is that a positive movement along $\mathbf{s}_I$, that is, $\alpha_I > 0$, reduces the value of the objective function. Once the search direction $\mathbf{s}_I$ is selected, α_I is computed from a one-dimensional search that minimizes $F(\mathbf{b}_I + \alpha_I \mathbf{s}_I)$.

The method of steepest descent is one of the simplest unconstrained descent algorithms that provides a satisfactory result. This method is rarely used in practical problems because of its poor performance, but it is discussed here to demonstrate the basics of descent algorithms. Furthermore, more advanced descent methods have been motivated by a desire to improve the steepest descent method. The search direction of Eq 2 for the method of steepest descent is the negative of the objective function gradient, that is:

$$\mathbf{s}_I = -\nabla F(\mathbf{b}_I) \quad \text{(Eq 3)}$$

Note that in this case s_I represents the direction of largest decrease in the objective function F. For each iteration, the objective function F and its gradient $\nabla F = -\mathbf{s}_I$ are evaluated. Multiple-function evaluations are then performed during the one-dimensional line search.

More advanced algorithms use higher-order information to compute search directions. Quasi-Newton methods, for example, are popular because they approximate the matrix of second-order sensitivities (the Hessian matrix) with gradient information, thus avoiding its direct computation. As an example, Fig. 2 shows the iterative solution path for the Broyden-Fletcher-Goldfarb-Shanno (BFGS) quasi-Newton algorithm on Rosenbrock's function (Ref 4).

In addition to general-purpose optimization algorithms, efficient techniques of limited scope have also been developed for specific applications. The fully stressed design technique (Ref 1), for example, minimizes the mass of truss structures subject to stress constraints alone. New designs are updated based on optimality criteria, which works well in this case for lightly redundant single-material structures. The limitations of many specific optimization methods render them useless for general applications and thus receive little attention today.

Convergence Criteria. Because numerical optimization is iterative, it is important to know when to stop, that is, when the optimization process has converged to the optimal design. Specifying the maximum number of allowable optimization iterations guarantees that the optimization process terminates; however, it does not ensure convergence is achieved. One convergence criterion is to monitor absolute and relative changes of the objective and constraint functions and the design parameters (Ref 2). Convergence can then be indicated when changes in the performance measures and/or design parameters between successive optimization iterations are within a predefined tolerance. For example, one can choose to terminate an optimization when a new design results in a reduction of mass that is within 1% of the mass for the initial design. Another important convergence criterion is provided by the Kuhn-Tucker necessary conditions for optimality (Ref 1–3). For unconstrained problems, this criterion simply requires that at the optimal design $\mathbf{b}^*$, the objective function gradient $\nabla F(\mathbf{b}^*)$ is less than a small specified constant. The Kuhn-Tucker conditions generalize for constrained optimization problems where a linear combination of the objective function gradient and the constraint gradients are used to indicate convergence (Ref 2, 3).

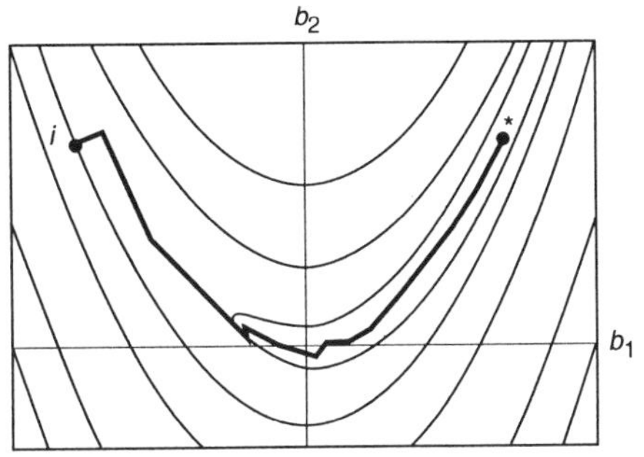

Fig. 2 Unconstrained optimization procedure using BFGS search directions. Shown is the two-dimensional Rosenbrock function $F(\mathbf{b}) = 100(b_2 - b_1^2)^2 + (1 - b_1)^2$, which has a unique minimum at (1,1). *i*, initial design; *, optimal design. Source: Ref 4

Analysis Solutions and Optimization Solutions. The solution of an optimization problem differs significantly from that of a typical CAE simulation. Computer-aided-engineering simulations compute the response or state of a product or process, for example, displacement or temperature; whereas the goal of an optimization solution is to define the product or process itself. Additionally, when analyzing a structure, for example, the displacement solution is almost always guaranteed and under certain conditions, it is unique. On the other hand, the existence and uniqueness of an optimal design is not ensured. Quite possibly, a design may not exist that will merely satisfy the constraints, let alone, be optimal. Furthermore, numerical methods used to solve optimization problems are often sensitive to the initial guess, and solution methods are algorithm dependent. The CAE engineer attempting to optimize his or her design should not be discouraged if the first try is not as successful as expected.

Algorithm Selection. Optimization algorithms are classified by the derivative information that they require to compute $\mathbf{s}_I$ in Eq 2, for example, zero-, first-, and second-order methods. Common unconstrained algorithms include the random search, Powell's conjugate direction, and sequential simplex methods (zero-order); steepest descent, Fletcher-Reeves' conjugate direction, variable metric, Davidon-Fletcher-Powell (DFP), and BFGS methods (first-order); and Newton's method (second-order) (Ref 1–5). Constrained first-order methods include reduced gradient, feasible direction, and sequential linear and quadratic programming methods (Ref 1–5). In CAE-based design optimization, efficient algorithms are desired because each iteration requires one or more computationally expensive numerical simulations. Higher-order algorithms are generally more efficient, that is, they require fewer iterations; however, higher-order derivatives may be impractical to evaluate. First-order methods are typically used in CAE-based design optimization because they require far fewer function evaluations than zero-order methods and avoid the Hessian evaluations required for second-order methods. Reference 4 provides further guidance for algorithm selection when solving unconstrained and linearly and nonlinearly constrained optimization problems.

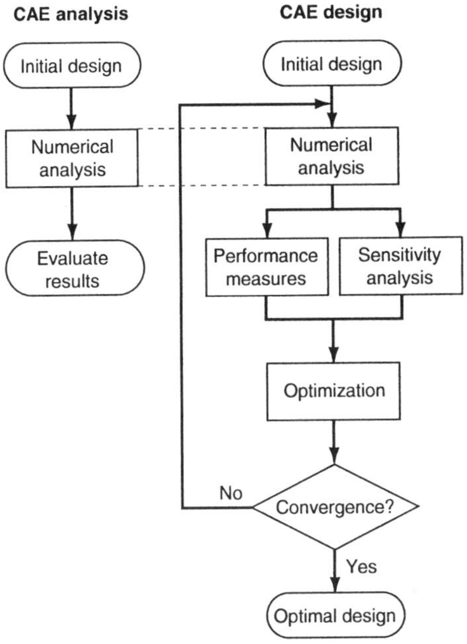

Fig. 3 CAE analysis and CAE design process

Computer-Aided-Engineering-Based Optimal Design

A general framework for CAE-based optimal design is shown in Fig. 3. In CAE analysis, the response of a system (e.g., the displacement in a vehicle structure) is computed using a numerical simulation. While these results are extremely helpful in determining the state of the current design, they do not indicate what changes are required when design criteria are violated.

In CAE design, the simulation software is included in a loop that iteratively updates the initial design to satisfy design criteria. A CAE simulation is performed on the design, which is followed by a computation of the performance measures and the sensitivity of the performance measures with respect to the design parameters.

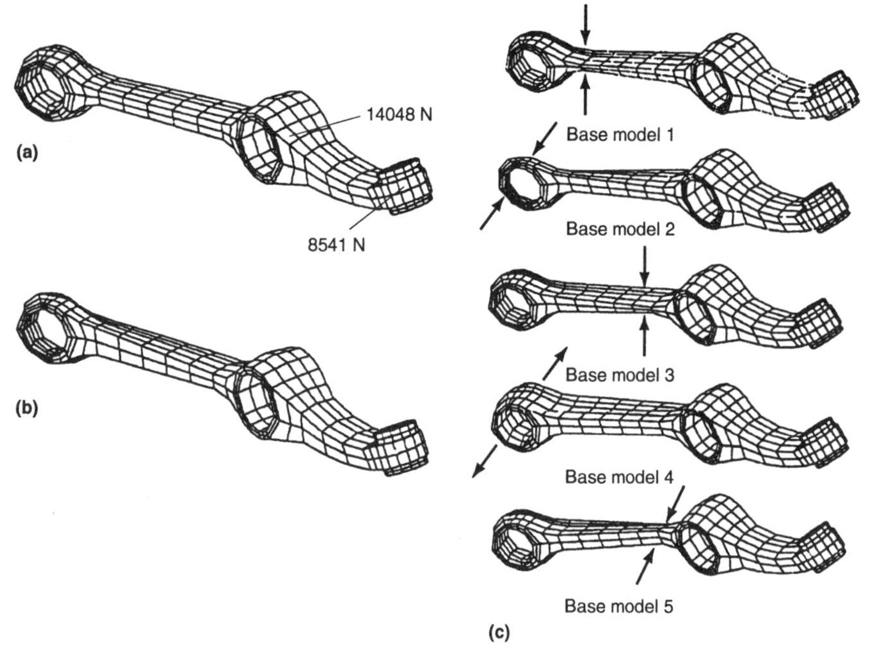

Fig. 4 Basis shapes for structural shape optimization of an automotive lower control arm. (a) Initial design. (b) Optimal design. (c) Five basis shapes of control arm model with arrows that show the location and direction of mesh distortion. Source: Ref 15

The optimization algorithm then computes a new design, and the process is continued. To effectively integrate a CAE simulation package into an optimization environment, particular attention must be given to the simulation program, the design parameterization, describing the desired product or process attributes in terms of mathematical statements that form the optimization problem, selection of an optimization algorithm, and performing efficient and accurate design-sensitivity analyses.

Numerical Simulation

Numerical simulation packages exist that are capable of modeling many physical phenomena involving complex material models, a variety of boundary conditions, and quite arbitrary geometries. These programs take the design definition as input and evaluate the performance measures of interest (see the articles "Mechanism Dynamics and Simulation," "Finite Element Analysis," and "Computational Fluid Dynamics" in this Volume).

Most simulation codes solve partial differential equations by using the finite element method (Ref 9), or some other suitable discretization scheme. In linear structural systems under static loading, for example, the governing differential equations are discretized and assembled to form a system of linear algebraic equations of the form:

$$\mathbf{Ku} = \mathbf{f} \qquad \text{(Eq 4)}$$

where **u** is the displacement at discrete locations in the model (e.g., nodes in a finite element analysis), **K** is the stiffness matrix, and **f** contains the applied forces.

It is of primary importance that the numerical simulation be robust and accurately represent the physical structure or system being designed. Additionally, because the numerical simulation must be fully automated, leading-edge simulation methods that require significant user interaction such as manual mesh adaptations may not be suitable for design optimization. Furthermore, the accuracy of the numerical result must be maintained over the entire design space. Recall that here merely the model is optimized and not the actual physical product. An optimal design based on an inaccurate simulation may not actually be the best design and it may violate important constraints, or both. Computer-aided engineering optimal design is not immune to the old saying "garbage in, garbage out."

Design Parameterization

The design variables in a CAE-based optimization can parameterize any quantity that serves as input to the numerical simulation. This includes material properties such as density, modulus, or viscosity; boundary conditions such as applied loads or displacements; element properties such as plate thicknesses or beam cross sections; and node point locations.

In structural optimization, a distinction is often made between sizing design variables and shape design variables. This distinction has evolved primarily as a result of implementing optimization algorithms with numerical simulation codes and has little to do with the optimization algorithm itself. Recall that optimization algorithms merely take numerical values of the design variables and performance measures as input and generate an updated design, regardless of the design parameter representation in the numerical simulation. This distinction, however, plays a significant role when transforming the design parameter data into inputs for the numerical simulation and when performing design-sensitivity analysis.

Sizing Design Parameters. A sizing design parameter can be thought of as one that does not alter the location of nodal locations in the numerical model. Material properties, boundary conditions, and element properties (such as a bar cross section or plate thickness) are all sizing design parameters. Including these parameters in the optimization process is often quite straightforward because modified values are easily updated in the numerical simulation input files.

Shape Design Parameters. Shape design parameters describe the boundary position in the numerical model and thus define nodal locations. A simple shape parameterization may define the coordinates of a single node point location (which is usually not recommended). Alternatively, adjusting a shape design variable may require that the entire numerical simulation model be remeshed. Implementing shape design variables is often more complex than that for sizing variables because the relationship between a shape parameter and each of the node locations must be specified prior to each optimization iteration. Two approaches are commonly used in shape optimization: geometry-based mesh parameterization and design basis vectors.

Geometry-Based Mesh Parameterization. In this approach, nodal locations are related to higher-level geometry data such as surface-control points or fillet radii through an automatic mesh generator. The approach has been used with mapped meshes and free meshes (Ref 10–12). Geometric-based parameterizations are well suited for integration with parametric solid modelers and are quite attractive because the optimal design is described with respect to realistic physical quantities such as hole diameters or length dimensions. The primary disadvantage is that the mesh generator must be included as part of the function and gradient evaluation process to generate new meshes for each optimization iteration. Additionally, the initial numerical simulation mesh must be generated in terms of the shape parameters, which may be a formidable task for large complex models.

The Reduced-Basis Approach. The reduced-basis method starts with a base configuration with a distinct mesh topology (i.e., the element layout and connectivity), which remains fixed during the optimization (see e.g., Ref 2). The mesh is then distorted during the optimization to form the optimal shape. New nodal coordinates are computed from:

$$\mathbf{X} = \mathbf{X}_o + \sum_{k=1}^{n} b_k \mathbf{V}_k \qquad \text{(Eq 5)}$$

where $\mathbf{X}$ and $\mathbf{X}_o$ are the current and original nodal coordinates, respectively, n is the number of shape parameters, and b_k and $\mathbf{V}_k$ are the kth shape parameter and design basis vector, respectively. Each basis vector, representing the design velocity (Ref 13) $\mathbf{V}_k = \mathbf{X}_k - \mathbf{X}_0$, is of length equal to the number of nodes in the finite element model and relates the motion of each nodal coordinate to the design variable b_k. Note that $b_k = 0$, $k = 1,2,..., n$ gives the initial design. Numerous methods have been used to generate the design basis vectors (Ref 2, 14, 15), and any procedure that produces linearly independent nodal perturbations may be used. New mesh shapes are efficiently computed with the reduced-basis approach; however, care must be taken to ensure that mesh distortion does not degrade the numerical solution. Figure 4 from Ref 15 illustrates the reduced-basis approach for an automotive lower-control arm with five basis shapes. The optimal design is a linear combination of the initial shape and the basis shapes that minimizes mass subject to stress constraints.

Sizing Parameters versus Shape Parameters. From the above definition, it is obvious that material properties and boundary conditions are sizing parameters; however, model geometry variables may or may not be sizing parameters depending on how they are represented in the numerical model. The thickness of a platelike structure, for example, can be parameterized by either a shape or a sizing design parameter. A sizing design parameter can be used to parameterize the element thickness property when plate elements are used in the analysis. Alternatively, when solid brick elements are used to discretize the geometry through the plate thickness, a change in thickness repositions nodal locations. In the latter case, thickness would not be a sizing parameter. The same is true for beam-type structures whose cross-sectional properties are often represented by area or moment of inertia element constants.

Optimization Problem Formulation

The formulation of the optimization problem often defines the success or failure of a CAE-based design optimization. Even with the most accurate numerical simulation and the best optimization algorithm, the success of the optimization may be in jeopardy if Eq 1 is not properly posed. The objective function and constraints must represent the important performance measures in the product or process design. For example, formulating a structural optimization may merely require the minimization of mass subject to stiffness and stress constraints. However, defining the optimization problem to design a casting operation may not be as straightforward. Care must always be taken to guarantee that all important constraints are included in the optimization.

Design Sensitivity Analysis

As mentioned above, descent algorithms use design-sensitivity information to compute the optimal design. The design sensitivities (or design derivatives) quantify the relationships between design parameters and design-performance measures. Computing design sensitivities with respect to each design parameter must be performed accurately and efficiently so that the CAE-optimization process is feasible.

Finite Difference Approximations. Often the design sensitivity of the function F with respect to the design variable b_i is evaluated by the forward finite difference approximation as:

$$\frac{dF}{db_i} \approx \frac{F(\mathbf{b} + \Delta\mathbf{b}_i) - F(\mathbf{b})}{\Delta b_i} \qquad \text{(Eq 6)}$$

where $\Delta\mathbf{b}_i$ is a zero vector with the exception of the ith location that contains Δb_i.

Significant disadvantages of the finite difference method exist when it is used with CAE analysis tools, namely, computational expense and lack of accuracy. Performance measures and thus the system response must be evaluated $N + 1$ times to compute the design sensitivity with respect to each of the N design variables. Computer-aided-engineering models for industrial applications typically have thousands of degrees of freedom, often rendering the finite difference method impractical for optimal design because numerous expensive numerical simulations may be prohibitive. Additionally, when F is nonlinear in the design variable b_i, the finite difference approximation depends on the perturbation size Δb_i. When Δb_i is too large, discretization errors occur, and when Δb_i is too small round-off error due to limits in machine precision corrupts the sensitivity calculation. Small but nonzero sensitivities are particularly susceptible to round-off error. Accuracy may be gained at the expense of computational resources by using double precision calculations or central differencing.

Analytical Design-Sensitivity Analysis. Analytical approaches for sensitivity computations avoid costly perturbation methods by differentiating the governing equations of the system with respect to the design parameters (Ref 13, 16). Here only steady-state solution methods are considered. Note that a performance measure in the optimization problem may be an explicit function of the design parameters $\mathbf{b}$ or it may depend on the response $\mathbf{u}$, which is implicitly dependent on $\mathbf{b}$ through Eq 4. In this case, the performance measure F can be rewritten as:

$$F(\mathbf{b}) = F(\mathbf{u}(\mathbf{b}), \mathbf{b}) \qquad \text{(Eq 7)}$$

where both the implicit and explicit dependence of the performance measure F on the design $\mathbf{b}$ is exposed. Assuming sufficient smoothness, the design sensitivity of F with respect to the design variable b_i, $i = 1,2,..., N$ is calculated from:

$$\frac{dF}{db_i} = \frac{\partial F}{\partial \mathbf{u}} \cdot \frac{d\mathbf{u}}{db_i} + \frac{\partial F}{\partial b_i} \qquad \text{(Eq 8)}$$

where $\nabla F = \{dF/d\mathbf{b}\}^T$. The first term in Eq 8 addresses the implicit dependence of F on the design $\mathbf{b}$ and the last term quantifies the explicit dependence of F on $\mathbf{b}$. For example, the mass of a structure does not depend on the displacement response $\mathbf{u}$ so that only $\partial F/\partial \mathbf{b}$ is nonzero. Alternatively, when the displacement at a node is the performance measure, $\partial F/\partial \mathbf{b} = 0$, and the design sensitivity only has an implicit contribution.

The explicit derivatives $\partial F/\partial \mathbf{u}$ and $\partial F/\partial b_i$ are readily available once the design engineer parameterizes the design and defines the performance measures. The difficulty in evaluating dF/db_i in Eq 8, however, arises from the presence of the implicit response sensitivity $d\mathbf{u}/db_i$, which is defined through the discretized governing equations (see e.g., Eq 4). Therefore, to compute dF/db_i, the implicit response derivative $d\mathbf{u}/db_i$ must be evaluated using the direct differentiation method or eliminated from Eq 8 with the adjoint method (see e.g., Ref 13, 16).

The Direct Differentiation Method. In the direct differentiation method, a pseudo problem is formed for each design parameter by differentiating Eq 4 with respect to each b_i, which after rearranging gives:

$$\mathbf{K}\frac{d\mathbf{u}}{db_i} = -\frac{d\mathbf{K}}{db_i}\mathbf{u} + \frac{d\mathbf{f}}{db_i} \qquad \text{(Eq 9)}$$

Thus computing the response sensitivity $d\mathbf{u}/db_i$ amounts to solving an alternative system of equations that resembles the original analysis of Eq 4. Note that the pseudo problem of Eq 9 must be performed for each b_i where the pseudo load $-d\mathbf{K}/db_i\,\mathbf{u} + d\,\mathbf{f}/db_i$ replaces the load vector $\mathbf{f}$ in Eq 4 and the computed response $\mathbf{u}$ in Eq 4 becomes the pseudo response $d\mathbf{u}/db_i$. The design-sensitivity computation for dF/db_i then follows from Eq 8 where a simple vector dot product and vector addition are performed for any number of performance measures F.

The Adjoint Variable Method. In the adjoint variable method, the implicit response derivative $d\mathbf{u}/db_i$ in Eq 8 is eliminated by first solving the adjoint problem:

$$\mathbf{K}^T\boldsymbol{\lambda} = \left\{\frac{\partial F}{\partial \mathbf{u}}\right\}^T \qquad \text{(Eq 10)}$$

for the adjoint variable vector $\boldsymbol{\lambda}$ and then evaluating the design sensitivity dF/db_i as:

$$\frac{dF}{db_i} = \frac{\partial F}{\partial b_i} - \boldsymbol{\lambda} \cdot \left[\frac{d\mathbf{K}}{db_i}\mathbf{u} - \frac{d\mathbf{f}}{db_i}\right] \qquad \text{(Eq 11)}$$

One adjoint variable vector $\boldsymbol{\lambda}$ is computed for each performance measure F by assembling the adjoint load $\partial F/\partial \mathbf{u}$ and solving the alternative system of Eq 10 that again resembles Eq 4. The design sensitivity dF/db_i is then evaluated with a simple vector dot product computation and a vector addition for each design variable as shown in Eq 11.

Selection and Use of Methods. In contrast to the finite difference method, both the direct differentiation and adjoint methods are efficient and accurate. For example, when the inverted stiffness matrix $\mathbf{K}^{-1}$ (or its transpose $\mathbf{K}^{-T}$) has been stored, the implicit response sensitivity $d\mathbf{u}/db_i$ and the adjoint variable vector $\boldsymbol{\lambda}$ are efficiently computed from Eq 9 and 10, respectively. Furthermore, when the same discretization method is used in the analysis and the sensitivity analysis, the resulting sensitivities are exact for the numerical problem being considered. Additionally, even though the two methods enjoy quite different derivations, they give identical results (Ref 17). In fact, the same explicit sensitivities (i.e., $\partial F/\partial \mathbf{u}$, $\partial F/\partial b_i$, $d\mathbf{K}/db_i$, and $d\mathbf{f}/db_i$) are required for both calculations. The choice of using one method over the other is an efficiency issue that depends on the optimization problem. When the number of performance measures (including the objective function and all of the constraints) exceeds the number of design variables N, the direct differentiation method is more efficient. Conversely, when N exceeds the number of performance measures, the adjoint variable method is preferred (Ref 16).

Analytical approaches to design-sensitivity analysis are far superior to the brute force finite difference method, especially for large models or when the original simulation is either nonlinear or transient. However, they must be fully integrated into the simulation program to be effective, which will likely be a lengthy implementation and require access to the simulation source code. Variations of these analytical approaches exist, for example, natural frequency and mode shape sensitivities, the semianalytical approach, and continuum sensitivity analysis, which may be used to more efficiently compute design sensitivities in other design problems (see e.g., Ref 1, 12, 13, 18). Automatic differentiation methods have also been developed that simplify the design-sensitivity computations by differentiating Fortran programs used in the CAE analysis (Ref 19).

Measuring the Performance of the Optimization

The numerical simulations that evaluate each design in a CAE-based optimization are often complex and computationally expensive. Because the correct optimal design needs to be obtained in a reasonable amount of time, the performance of the optimization process, which includes the algorithm itself and the simulation software, should be considered. The performance of the optimization process may be measured in terms of robustness, accuracy, and efficiency.

The robustness of a CAE-based optimization can be defined as the ability of the code to satisfy the convergence criteria within a reasonable number of optimization iterations (Ref 20). To achieve robustness, the optimization problem must be well posed and must employ a robust optimization algorithm.

The optimization is required not only to converge, but it must converge to the correct design. As discussed previously, the accuracy of the CAE simulation must ensure that the optimal simulation results are the optimal reality. However, an accurate CAE model alone does not guarantee that the desired design has been achieved. Incorrect gradients corrupt the design process and improper optimization problem definitions may miss important constraints. In the latter case, what is asked for is achieved, but it may not be what is wanted. The CAE optimization process should be tested on problems with known solutions, and for more complex problems sound engineering judgement should always be used when assessing an optimal design.

One of the most important performance measures of a CAE optimization is the computational effort required to obtain the optimal design. The total number of function evaluations and the number of optimization iterations are both key indicators regarding the success of the optimization process. The final arbiter that often determines if the optimization is efficient enough, however, is the total amount of time required to obtain an optimal design. Schedules and computer resources may render a CAE-based optimization infeasible even though relatively few function evaluations are needed.

Software Packages for CAE Optimal Design

Commercial programs exist, particularly in structural optimization, which integrate simulation, optimization, and design-sensitivity analysis into a single design environment. It is beyond the scope of this article to discuss details of the optimization programs available today; however, Ref 1 discusses some optimization software, including structural optimization programs. Additionally, an extensive discussion of programs for structural optimization developed in Europe (Ref 21) and by North American government agencies and commercial suppliers (Ref 22) is available elsewhere.

Outside of structural optimization, fully integrated optimization packages are rare. To optimize designs that are governed by other physical phenomena, the CAE designer must integrate the appropriate numerical analysis and optimization software. It is common practice to wrap an optimization algorithm around a numerical analysis package that solves the particular problem of interest. For these applications, it is often not fruitful to set out to develop an optimization algorithm or even to write a program that implements an existing algorithm because general computer codes are available that allow the user a variety of choices (see e.g., Ref 8, 23–25). Function evaluations must avoid user intervention so that the preprocessing, simulation, postprocessing, and design sensitivity analyses must be fully automated. When finite differencing is used for sensitivity evaluation, the integration of one's favorite analysis code with an existing optimization subroutine is straightforward. Alternatively, an extensive implementation may be needed if analytical sensitivities are required.

Approximate Optimization Techniques

When the number of design variables N is small (e.g., less than 10), approximate optimization techniques (also referred to as response surface methods) reduce the number of expensive CAE simulations when compared to direct optimization using finite difference gradients (Ref 2, 26). Approximate techniques use simple expressions such as (Ref 2):

$$F(\mathbf{b}) \approx F(\mathbf{b}^0) + \nabla F(\mathbf{b}^0) \cdot \delta\mathbf{b} + \tfrac{1}{2}\, \delta\mathbf{b} \cdot \mathbf{H}(\mathbf{b}^0)\, \delta\mathbf{b} \qquad \text{(Eq 12)}$$

to approximate the objective function and the constraints about $\mathbf{b}^0$ where F represents any performance measure in the optimization problem of Eq 1. In Eq 12, ∇F is the gradient vector, $\mathbf{H}$ is the Hessian matrix, and $\delta\mathbf{b} = \mathbf{b} - \mathbf{b}^0$. These approximations give an overall view of the design space and can be used to smooth otherwise discontinuous performance measures. Additionally, response surface models simplify the process of integrating several design codes in multidisciplinary optimization (Ref 26, 27).

The approximate optimization process starts by analyzing multiple designs using CAE simulations to generate design sets. Each design set consists of objective and constraint function values corresponding to a particular design $\mathbf{b}$. Approximations such as in Eq 12 are then fit to the available design information, and an optimization is performed using these simple approximate functions rather than expensive CAE simulations. The optimal design for the approximate problem is then evaluated with a CAE simulation, and the approximation is updated using this new response data. The iterative process continues until convergence is achieved.

The application of approximation methods varies based on the choice of optimization algorithm, the method for selecting designs for full CAE simulation, and the sequence of updating terms in Eq 12 (Ref 2, 8, 26, 28). For example, the number of design sets N_d that are required to define Eq 12 is $L = 1 + N + N(N + 1)/2$. For $N_d < L$, only a partial fit of Eq 12 is possible, whereas when $N_d > L$, a least-squares fit is commonly performed to determine the best approximate response surface for the data sets that are available. Weighting is often used to place more emphasis on designs nearest the one with the best overall performance. Designs for CAE simulation may be chosen near the nominal design $\mathbf{b}^0$ using finite difference perturbations that results in a second-order Taylor series expansion for Eq 12. This selection of designs may render an approximation that is only good near $\mathbf{b}^0$ and poorly represents the rest of the design space. Alternatively, designs may be randomly distributed throughout the design space, making Eq 12 merely a quadratic polynomial approximation to the design (Ref 2), which may be good at predicting overall trends but possibly miss local features. Furthermore, care must be taken when selecting candidate designs $\mathbf{b}$ for CAE simulation because linear independence between the design sets must be maintained.

Structural Optimization

Structural optimization enjoys a rich history of advances dating back to the 18th century and possibly further if one considers the beam designs by Galileo Galilei (Ref 29). Early optimization techniques searched for design functions that described the properties of the structure much as the analysis methods of that time solved for displacements and stresses analytically. With the advent of high-speed computers in the late 1950s and early 1960s, numerical solution methods such as the finite element method and numerical optimization algorithms emerged. The analytical solutions and analytical optimizations were replaced by numerical discrete solutions and numerical optimization methods, opening the field of what is now considered structural optimization.

The aircraft, aerospace, and automotive industries have been the primary drivers for structural optimization. Here, the principal focus has been on weight reduction because weight affects many attributes of the design including cost, performance, and fuel economy. Constraints are often imposed to maintain durability and vibration characteristics. Much of the developments in sizing and shape design, design sensitivity analysis (Ref 13), and more recently, topology design (Ref 30) have been pursued to support structural optimization.

A structural optimization example is shown in Fig. 5 (Ref 12). A turbine wheel is designed for minimum mass moment of inertia while satisfying constraints on the maximum von Mises stress. Forces acting on the wheel include centrifugal loads from the turbine blades (not shown) and thermal loads from the hot gases that drive the wheel. By adjusting the shape of the wheel structure in the optimization, the mass moment of inertia and the maximum stress are reduced by 12.5% and 35.0%, respectively.

Emerging Technologies

Computer-aided-engineering-based design optimization has evolved along the classical path of first being researched by universities and government agencies, then demonstrated with special-purpose programs and finally, to some extent, marketed by commercial software vendors. There are still, however, hurdles that must be overcome for widespread use of the technology, and numerous extensions and enhancements are still to be discovered. Below are a few emerging research areas in the optimization community.

Topology Optimization

Topology optimization, a form of structural optimization, computes the best geometric configuration or layout of a structure (Ref 30, 31). It is most beneficial early in the design cycle when least is known about the design and when design changes are easily accommodated. Results from a topology optimization help determine the placement and number of holes and/or stiffening members, and therefore provide a good starting point for further structural refinement via sizing and shape optimization.

The goal of topology optimization is to determine where to place material and where to leave the structure void of material. Two topology optimization approaches have emerged, one based on material homogenization and the other on material density. In the homogenization method (Ref 32), the dimensions and orientation of a void in the material of each element are adjusted as a function of the design variables. Effective material properties are then computed by smearing or homogenizing over each element so that its stiffness and density take on values between those of the void and the solid. The density method (Ref 33, 34) is often considered an engineering approach where the modulus and density of each element are parameterized as functions of the element's design parameter. The element's density is chosen to be a linear function of it's design parameter, and the relationship between it's elastic modulus and this parameter is typically of higher order. This implementation tends to drive the material to become either solid or void by penalizing intermediate designs.

The homogenization method enjoys a rigorous mathematical derivation and has been applied to design composite materials. Homogenization requires three design variables per element for planar structures, which carries with it an undesirable computational burden for practical applications, especially if composites are not to be considered in the final design. The density method works well for isotropic materials, is easily implemented using commercial finite element programs, and accepts multiple objectives and constraints. It also requires only one design variable per element.

Figure 6 (Ref 35) illustrates the use of the homogenization approach of topology design to optimize the material distribution in a frame structure. The first natural frequency is maximized subject to a constraint on the total mass of

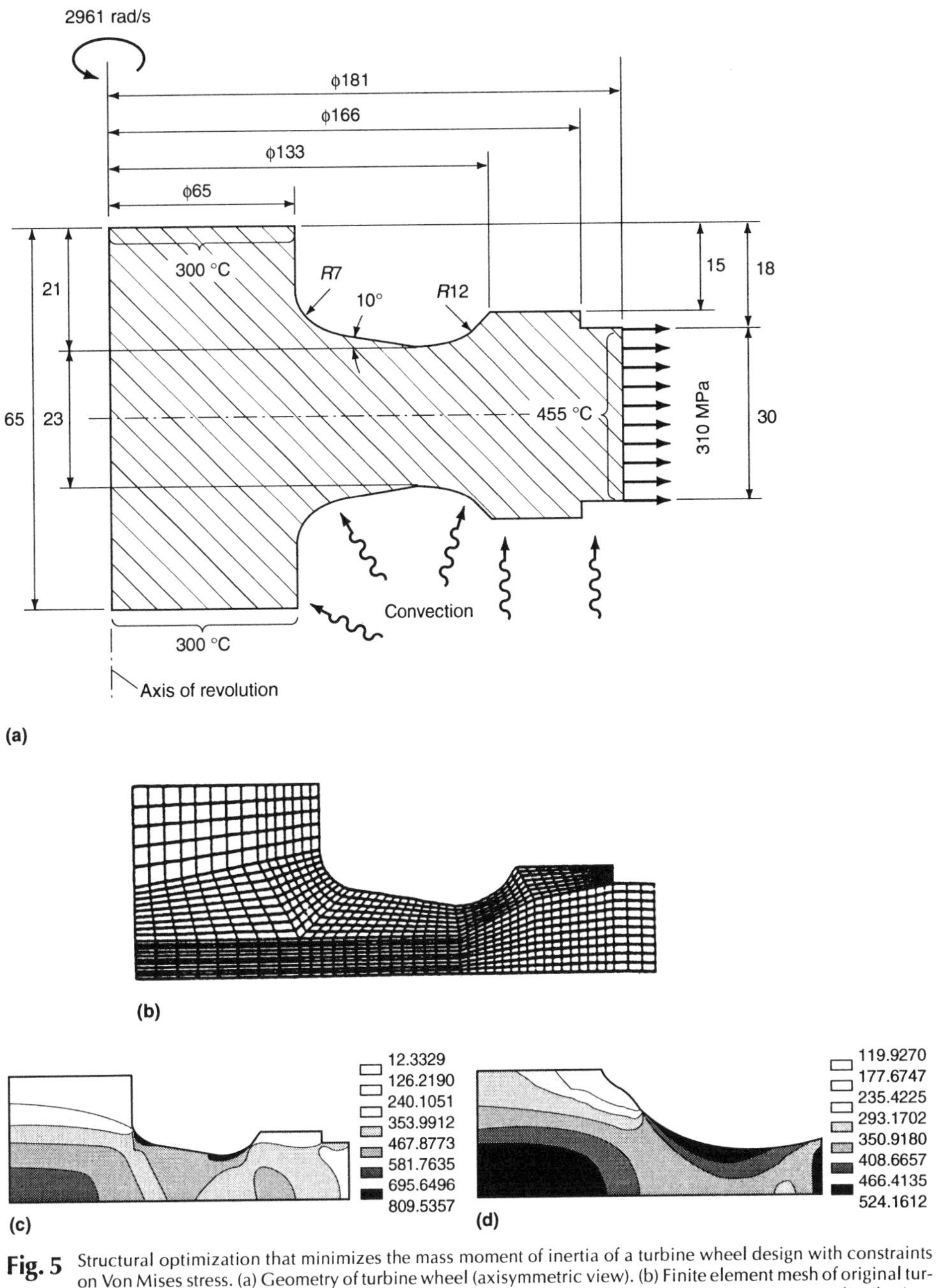

Fig. 5 Structural optimization that minimizes the mass moment of inertia of a turbine wheel design with constraints on Von Mises stress. (a) Geometry of turbine wheel (axisymmetric view). (b) Finite element mesh of original turbine geometry. (c) Initial geometry: Von Mises stress distribution. (d) Optimal geometry: Von Mises stress distribution. Source: Ref 12

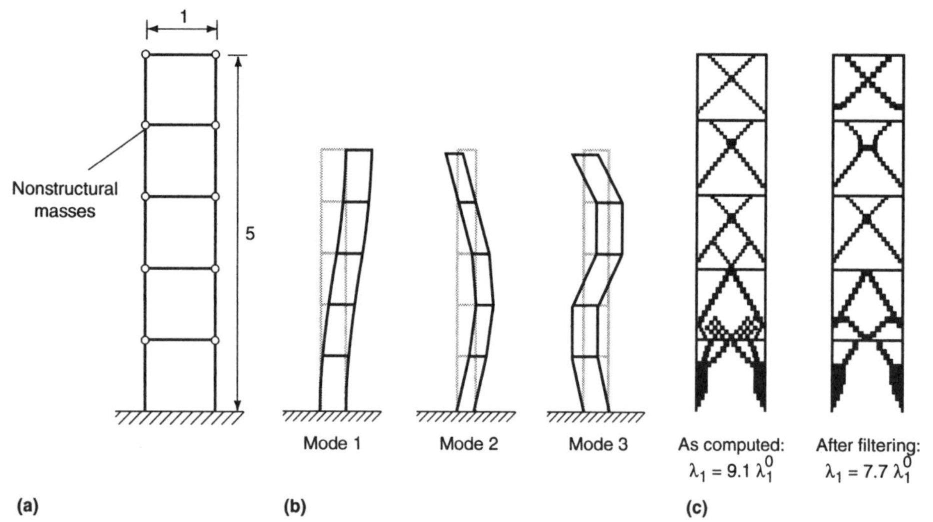

Fig. 6 Topology optimization examples of a frame structure. (a) Initial frame structure showing design domain. (b) First three natural mode shapes. (c) Optimal material distribution from topology optimization as computed and after filtering the topology image to simplify the structural layout. Source: Ref 35

the structure. In this analysis, the first eigenvalue was increased by 910% with the addition of 66.6% of material distributed in an optimal manner within the frame structure.

Materials Processing Optimization

Advances in numerical methods for materials processing analysis have recently made it possible to optimize both a structural component and its manufacturing process. Optimal design has been applied to metal forging (Ref 36), casting (Ref 37), and welding (Ref 38), and to polymer injection molding and sheet extrusion (Ref 39).

Formulation of the optimization problem statement for these applications is critical because all key physical process attributes must be properly represented in the objective function and constraints. Design parameters and performance measures are often unique to the particular processing operation. Furthermore, additional complexity exists because the numerical simulations for these processes are typically nonlinear and transient and often require a coupled simulation to accurately model the interaction between different physical phenomena. Computations can be quite extensive and require accurate and efficient analytical design sensitivities so that the design problem is tractable.

Morthland et al. (Ref 37) optimized the riser design for a hammer casting using design optimization and solidification simulations. Design variables parameterized the shape of the riser shown in Fig. 7, and design sensitivities were computed using a transient direct method. The riser volume was minimized to reduce manufacturing costs. In the same analysis, constraints on the freezing times of the elements labeled in Fig. 7(a) were defined to enforce directional solidification in the section of the hammer casting leading to the riser. The riser volume was decreased by 42% in the optimization, and the liquid region near the end of the solidification process was moved from the hammer to the riser (thus avoiding porosity in the hammer itself).

Multidisciplinary Optimization

Multidisciplinary optimization has recently received much attention in areas such as aerospace, aircraft, and automotive design. Various physics, each of which may require a unique analysis program and, most probably, analytical expertise, must be merged to compute the performance measures and complex couplings that are characteristic of these designs (Ref 40). Aircraft design, for example, must include the interactions between disciplines such as aerodynamics, dynamics, aerolastic stability, structures, controls, and acoustics (Ref 18).

Key issues that must be addressed when performing multidisciplinary design are computational cost, convenience of implementation, data exchange, integration across various design and analysis methods and possibly across engineering organizations, and the use of *black-box* analysis tools. Several approaches have been proposed to solve multidisciplinary optimization problems (Ref 41). *All-at-once* procedures combine two or more disciplines by addressing each design criterion in a single optimization problem statement

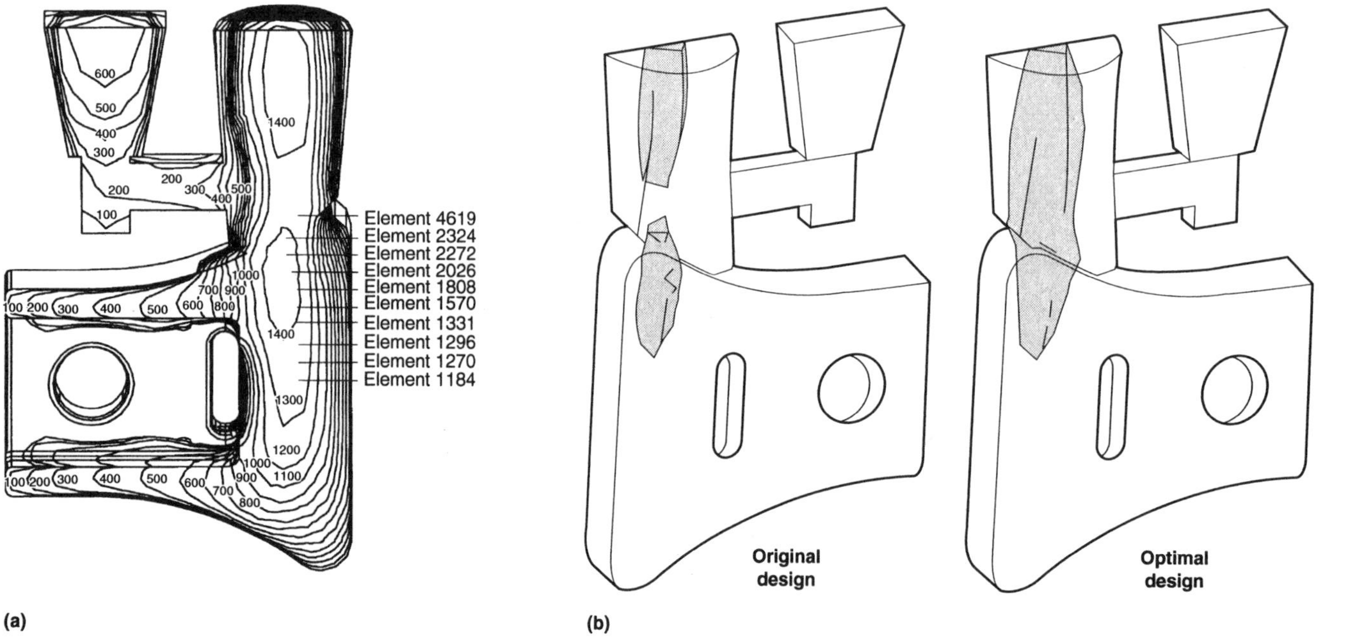

Fig. 7 Optimal riser design for metal casting. (a) Computed solidus isochrones on the symmetry plane for the initial infeasible design. Also shown are the elements specified to enforce directional solidification. (b) Casting regions with liquid after 1565 s for the original (infeasible) and optimal designs. Source: Ref 37

similar to that in Eq 1. This approach may be prohibitive for large-scale applications and is difficult to integrate in moderate or large engineering organizations. Alternatively, multilevel decomposition methods have been developed that replace the large single optimization problem with several subproblems and a coordination problem that is used to maintain the couplings between the subproblems (Ref 42). These methods are designed to promote disciplinary autonomy while achieving interdisciplinary compatibility and are most useful when couplings between the various disciplines can easily be broken or neglected (Ref 41).

One implementation of multidisciplinary optimization for automotive applications employs design sensitivity analysis information to approximate changes in attribute responses (Ref 27). The method considers design attributes such as noise, vibration, and harshness (NVH), durability, safety, vehicle dynamics, and manufacturing, which must all be satisfied in the design process. The method accepts data in the form of design sensitivities and function values from independent analyses. Thus, analysis results from different engineering organizations are easily merged in order to perform the multidiscipline optimization.

Global Optimization Using Stochastic Search Methods

Stochastic search algorithms compute globally optimal designs in a manner that is quite different from the mathematical programming methods discussed previously. Two stochastic search methods, genetic algorithms and simulated annealing, mimic processes found in nature. Genetic algorithms (Ref 43) are based, in principle, on Darwin's theory of survival of the fittest and evolve generations of designs with bias given to the best members in a population. Simulated annealing techniques (Ref 44) are quantitatively based on the behavior of particles in thermal equilibrium where a gradual lowering of the temperature causes atoms to assume a lower, more orderly, energy state analogous to an optimal design.

One advantage of using stochastic search methods is that they are readily adapted to new problems because only function evaluations are required; that is, design sensitivities are not needed. These search algorithms often possess a mechanism for accepting less optimal designs during the search process, which provides a means to escape from a local optimum and find the global minimum. Furthermore, both continuous and discrete design variables can be used in the optimization. The computational requirements for stochastic searches is usually not as great as that for random (zeroth order) searches. However, they often require hundreds of function evaluations making them impractical for optimization problems that rely on computationally expensive CAE simulations. Examples of stochastic searches in structural optimization are given in Ref 45.

Conclusions

Over the past 30 years, numerical optimization and CAE have individually made significant advances and have together been developed to impact the way engineering components and systems are designed. This article has attempted to give a brief overview of current CAE-based optimal design and provide a starting point for further study and/or implementation of these methods. While issues still remain and advanced research and development continue, practical design applications indeed demonstrate that optimization is no longer a subject restricted to the researchers.

REFERENCES

1. R.T. Haftka and Z. Gürdal, *Elements of Structural Optimization,* 3rd ed., Kluwer Academic Publishers, 1992
2. G.N. Vanderplaats, *Numerical Optimization Techniques for Engineering Design: with Applications,* McGraw-Hill, 1984
3. D.G. Luenberger, *Linear and Nonlinear Programming,* 2nd ed., Addison-Wesley, 1984
4. P.E. Gill, W. Murray, and M.H. Wright, *Practical Optimization,* Academic Press, 1981
5. E.J. Haug and J.S. Arora, *Applied Optimal Design,* John Wiley & Sons, 1979
6. W. Stadler, Natural Structural Shapes of Shallow Arches, *J. Appl. Mech. (Trans. ASME),* June 1977, p 291–298
7. S. Adali, Pareto Optimal Design of Beams Subjected to Support Motions, *Comput. Struct.,* Vol 16 (No. 1–4), 1983, p 297–303
8. *DOT v4.20–DOC v1.30 User's Manuals,* Vanderplaats Research and Development, Inc., Colorado Springs, CO, 1995
9. F.L. Stasa, *Applied Finite Element Analysis for Engineers,* CBS College Publishing, 1985
10. J.A. Bennett and M.E. Botkin, Ed., *The Optimal Shape: Automated Structural Design,* Plenum, 1986
11. K.H. Chang and K.K. Choi, A Geometric-Based Parameterization Method for Shape Design of Elastic Solids, *Mech. Struc. Mach.,* Vol 20 (No. 2), 1992, p 215–252
12. N. Olhoff, E. Lund, and J. Rasmussen, Concurrent Engineering Design Optimization in a CAD Environment, *Concurrent Engineering: Tools and Technologies for Mechanical System Design,* E.J. Haug, Ed., Vol F108, NATO ASI Series, Springer-Verlag, 1993
13. E.J. Haug, K.K. Choi, and V. Komkov, *Design Sensitivity Analysis of Structural Systems,* Academic Press, 1986
14. A.D. Belegundu and S.D. Rajan, A Shape Optimization Approach Based on Natural Design Variables and Shape Functions, *Comput. Meth. Appl. Mech. Eng.,* Vol 66, 1988, p 87–106
15. R.J. Yang, A. Lee, and D.T. McGeen, Application of Basis Function Concept to Practical Shape Optimization Problems, *Struct. Optimiz.,* Vol 5, 1992, p 55–63
16. D.A. Tortorelli and P. Michaleris, Design Sensitivity Analysis: Overview and Review, *Inverse Probl. Eng.,* Vol 1 (No. 1), 1993, p 71–105
17. A.D. Belegundu, Lagrangian Approach to Design Sensitivity Analysis, *J. Eng. Mech.,* Vol 111 (No. 5), 1985, p 680–695
18. A. Chattopadhyay and N. Pagaldipti, A Multidisciplinary Optimization Using Semi-Analytical Sensitivity Analysis Procedure and Multilevel Decomposition, *Comp. Math. Appl.,* Vol 29 (No. 7), 1995, p 55–66
19. C. Bischof and A. Griewank, ADIFOR: A Fortran System for Portable Automatic Differentiation, *Fourth AIAA/NASA/ISSMO Symposium on Multidisciplinary Analysis and Optimization,* American Institute of Aeronautics and Astronautics, 1992, p 433–441
20. G. Subbarayan, D.L. Bartel, and D.L. Taylor, A Method for Comparative Performance Evaluation of Structural Optimization Codes: Parts I and II, *Design Engineering Advances in Design Automation,* Vol 14, American Society of Mechanical Engineers, 1988, p 221–232
21. P. Duysinx and C. Fleury, Optimization Software: View from Europe, *Structural Optimization: Status and Promise,* M.P. Kamat, Ed., American Institute of Aeronautics and Astronautics, 1993
22. E.H. Johnson, Tools for Structural Optimization, *Structural Optimization: Status and Promise,* M.P. Kamat, Ed., American Institute of Aeronautics and Astronautics, 1993
23. "Fortran Subroutines for Mathematical Applications," IMSL, Inc., Houston, TX, 1991
24. W.H. Press, S.A. Teukolsky, W.T. Vetterling, and B.P. Flannery, *Numerical Recipes in Fortran,* Cambridge University Press, 1992
25. J.J. Moré and S.J. Wright, *Optimization Software Guide,* Society for Industrial and Applied Mathematics, 1993
26. G. Venter, R.T. Haftka, and J.H. Starnes, Jr., Construction of Response Surfaces for Design Optimization Applications, *Sixth AIAA/NASA/ISSMO Symposium on Multidisciplinary Analysis and Optimization,* American Institute of Aeronautics and Astronautics, 1996, p 548–564
27. K.D. Longacre, J.M. Vance, and R.I. DeVries, A Computer Tool to Facilitate Cross-Attribute Optimization, *Sixth AIAA/NASA/ISSMO Symposium on Multidisciplinary Analysis and Optimization,* American Institute of Aeronautics and Astronautics, 1996, p 1275–1279
28. P. Kohnke, *ANSYS User's Manual for Revision 5.1,* Vol 4, Swanson Analysis Systems, Inc., 1994
29. V.B. Venkayya, Introduction: Historical Perspective and Future Directions, *Structural Optimization: Status and Promise,* M.P. Kamat, Ed., American Institute of Aeronautics and Astronautics, 1993
30. M.P. Bendsøe, *Optimization of Structural Topology, Shape, and Material,* Springer-Verlag, 1995
31. G.I.N. Rozvany, M.P. Bendsøe, and U. Kirsch, Layout Optimization of Structures, *Appl. Mech. Rev.,* Vol 48 (No. 2), Feb 1995, p 41–119
32. M.P. Bendsøe, A. Díaz, and N. Kikuchi, Topology and Generalized Layout Optimization of Elastic Structures, *Topology Design of Struc-*

tures, M.P. Bensøe and C.A. Mota Soares, Ed., Vol 227, NATO ASI Series E: Applied Sciences, Kluwer Academic Publishers, 1993

33. G.I.N. Rozvany, M. Zhou, and T. Birker, Generalized Shape Optimization without Homogenization, *Struct. Optimiz.,* Vol 4, 1992, p 250–252
34. R.J. Yang and C.H. Chuang, Optimal Topology Design Using Linear Programming, *Comput. Struct.,* Vol 52 (No. 2), 1994, p 265–275
35. A. Díaz and N. Kikuchi, Solutions to Shape and Topology Eigenvalue Optimization Problems Using a Homogenization Method, *Int. J. Numer. Methods Eng.,* Vol 35, 1992, p 1487–1502
36. H. Cheng, R.V. Grandhi, and J.C. Malas, Design of Optimal Process Parameters for Non-Isothermal Forging, *Int. J. Numer. Methods Eng.,* Vol 37, 1994, p 155–177
37. T.D. Morthland, P.E. Byrne, D.A. Tortorelli, and J.A. Dantzig, Optimal Riser Design for Metal Castings, *Metall. Mater. Trans.,* Vol 26B (No. 4), Aug 1995, p 871–885
38. P. Michaleris, D.A. Tortorelli, and C.A. Vidal, Analysis and Optimization of Weakly Coupled Thermoelastoplastic Systems with Applications to Weldment Design, *Int. J. Numer. Methods Eng.,* Vol 38, 1995, p 1259–1285
39. D.E. Smith, D.A. Tortorelli, and C.L. Tucker, Optimal Design and Analysis for Polymer Extrusion and Molding, *Sixth AIAA/NASA/ISSMO Symposium on Multidisciplinary Analysis and Optimization,* American Institute of Aeronautics and Astronautics, 1996, p 1019–1024
40. R.J. Balling and J. Sobieszczanski-Sobieski, Optimization of Coupled Systems: A Critical Overview of Approaches, *AIAA J.,* Vol 34 (No. 1), 1996, p 6–17
41. R.T. Haftka, J. Sobieszczanski-Sobieski, and S.L. Padula, On Options for Interdisciplinary Analysis and Design Optimization, *Struct. Optimiz.,* Vol 4, 1992, p 65–74
42. J. Sobieszczanski-Sobieski, Structural Sizing by Generalized, Multilevel Optimization, *AIAA J.,* Vol 25 (No. 1), 1987, p 139–145
43. D.E. Goldberg, *Genetic Algorithms in Search, Optimization and Machine Learning,* Addison-Wesley, 1989
44. S. Kirkpatrick, C.D. Gelatt, and M.P. Vecchi, Optimization by Simulated Annealing, *Science,* Vol 220, 1983, p 671–680
45. P. Hajela, Stochastic Search in Structural Optimization: Genetic Algorithms and Simulated Annealing, *Structural Optimization: Status and Promise,* M.P. Kamat, Ed., American Institute of Aeronautics and Astronautics, 1993

SELECTED REFERENCES

- M.P. Bensøe and C.A. Mota Soares, Ed., *Topology Design of Structures,* Vol 227, NATO ASI Series E, Applied Sciences, Kluwer Academic Publishers, 1993
- J. Cea and E.J. Haug, Ed., *Optimization of Distributed Parameter Structures,* Vol II, NATO ASI, Sijthoff and Noordhoff, 1981
- R.T. Haftka and Z. Gurdal, *Elements of Structural Optimization,* 3rd ed., Kluwer Academic Publishers, 1992
- E.J. Haug, K.K. Choi, and V. Komkov, *Design Sensitivity Analysis of Structural Systems,* Academic Press, 1986
- E.J. Haug, Ed., *Concurrent Engineering: Tools and Technologies for Mechanical System Design,* Vol F108, NATO ASI Series, Springer-Verlag, 1993
- M.P. Kamat, Ed., *Structural Optimization: Status and Promise,* American Institute of Aeronautics and Astronautics, 1993
- C.A. Mota Soares, Ed., *Computer Aided Optimal Design: Structural and Mechanical Systems,* Vol F 27, NATO ASI Series, Springer-Verlag, 1987
- G. Rozvany, Ed., *Structural Optimization,* Springer-Verlag, 1989–date
- G. Rozvany, Ed., *Optimization of Large Structural Systems,* Vol I, II, III, NATO/DFG ASI, 1991
- G.N. Vanderplaats, *Numerical Optimization Techniques for Engineering Design: with Applications,* McGraw-Hill, 1984

Dimensional Management and Tolerance Analysis

Mark Craig, Variation Systems Analysis, Inc.

DIMENSIONAL MANAGEMENT is an engineering methodology combined with computer-simulation tools used to improve quality and reduce cost through controlled variation and robust design. The objective of dimensional management is to create a design and process that "absorbs" as much variation as possible without affecting the function of the product. Dimensional management accomplishes this through optimal selection of datums, feature controls, assembly methods, and assembly sequence.

Companies that design and produce multicomponent assemblies must effectively manage the cost, timing, and quality related to manufacturing and assembly variation if they are to survive in an increasingly competitive world market. Dimensional management differs from the traditional design practice of assigning tolerances to drawings prior to release. In traditional design practice, the design engineer assigns tolerances on component parts just before drawing release. The values of the tolerances might be based on past experience, best guess, or anticipated manufacturing capability. In some cases a one-dimensional tolerance stack-up analysis is performed to determine if an assembly limit would be exceeded when adding the tolerances in any given one-dimensional direction. This approach is still commonly used in many engineering organizations today.

The potential limitations of this traditional approach include:

- The tolerances are assigned at the end of the design cycle where it is too late to make any changes in the design and/or assembly tooling to help desensitize the design and process to variation.
- One-dimensional tolerance analysis does not represent the three-dimensional geometries of the component parts and assemblies.
- The manufacturing, assembly, quality, and supplier team may not be involved during the initial design phase of the product.

The dimensional management process provides an advantage over this traditional approach by combining three-dimensional tolerance analysis and measurement systems within an integrated computer-aided engineering (CAE) system.

Dimensional Management Process

The dimensional management process follows six basic steps as described below.

Step 1: Define Product Dimensional Requirements. The first step in the process is to clearly define the dimensional requirements of the product early in the concept-design phase. This step involves formally documenting the assembly variation targets for the entire product (i.e., assembly specifications). These targets can be identified based on product functional requirements (i.e., seal pressure, leaks, interference concerns, etc.), competitive benchmarks (i.e., fit and finish of competitive products), or quality improvement goals determined from known build problems of an existing, similar product. The product dimensional requirements must be "signed off" by all members of the product team including design, manufacturing, assembly, quality, and suppliers. This process ensures that all members of the product team have a consistent understanding of the product build requirements.

Step 2: Determine Process and Product Requirements. During the design phase of a product, there are only three ways to determine if the product and process, as designed, meets the dimensional product requirements: (1) Make an educated guess. (2) Build many assemblies using production tools and measure the results. (3) Use a computer simulation model to simulate the design and build of the product including the three-dimensional geometry, geometric dimensioning and tolerancing schemes, assembly method variation, assembly sequence, and any known part deflection or distortion.

As previously mentioned, traditional tolerance analysis relies on simple one-dimensional analysis and the "educated guess." Most variation problems are resolved during the prototype building cycle before committing to production tools. This approach lengthens development time.

Simulation of the assembly process is another way to determine if the product and process, as designed, meets the dimensional product requirements. The simulation should predict the amount of assembly variation that is expected to occur and the major contributors to the variation.

Step 3: Ensure Accurate Documentation. Dimensional management product documentation includes: geometric dimensioning and tolerancing schemes, assembly methods, locating schemes, and statistical process control (SPC) checkpoints. The objective is to make sure that the product specifications used as input to the simulation in step 2 are the same specifications documented in step 3 and are fully understood and used by those individuals performing steps 4 to 6.

Step 4: Develop a Measurement Plan that Validates Product Requirements. The simulation performed in step 2 proves that the design, manufacturing, and assembly process as specified meets all dimensional product requirements. The next step in the process is to develop a measurement plan that validates these requirements.

The measurement plan must directly reflect the documented tolerancing schemes and assembly methods represented in the simulation. The features identified as critical in the simulation need to be measured using the same datum reference and feature constraints as defined in the analysis. This approach determines if manufacturing capability achieves actual design intent.

If there is a "disconnect" between the measurement plan and the analysis, the actual measurement data cannot be used to determine if the product variation is acceptable to meet final assembly product objectives. For example, if an organization specifies tolerances as three-dimensional zones, and the assembly is analyzed using one- or two-dimensional tolerance analysis, and then the component parts are checked according to three-dimensional zones, there is no indication if the tolerance zones specified are really required to meet overall product assembly requirements.

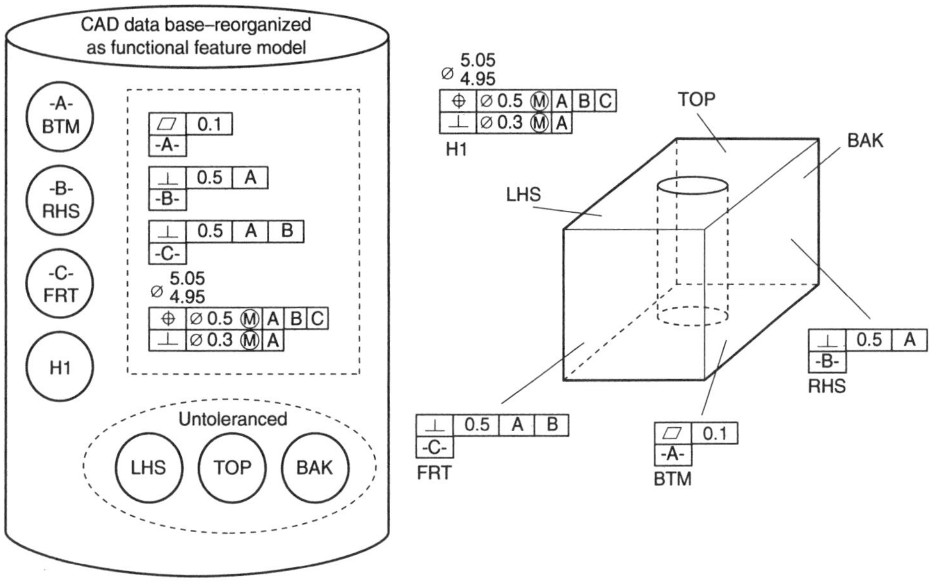

Fig. 1 Functional feature model

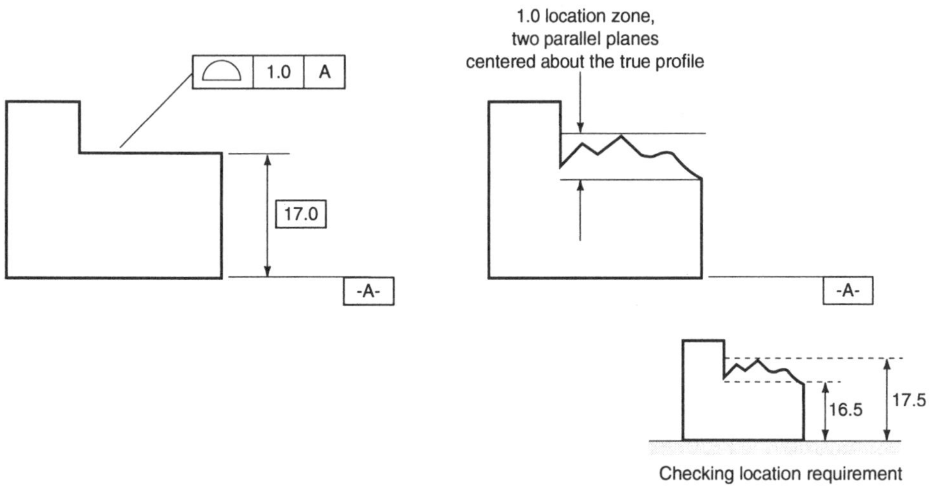

Fig. 2 Component part variation

At the completion of step 4 one should be confident that the product will build within the functional specifications identified in step 1 if:

- The simulation model was created correctly (i.e., no errors).
- The design and process as specified contains all identifiable sources of variation.
- The manufacturing capabilities achieve design intent.

In almost all cases the design and process as specified do not contain all sources of variation. Design and build process documentation typically provides a tolerance specification on component parts and a final build specification as the only sources of variation. Part deflection, weld distortion, gravity effects, fixture variation, and so forth are typically not included in the released design specifications, yet in actual production these variation contributors exist. The dimensional management process helps identify these additional sources of variation and provides a method to capture and quantify their effect on functional requirements.

Step 5: Establish Manufacturing Capabilities versus Design Intent. The measurement plan is next implemented, and capability studies are performed on component parts to ensure component variation meets design intent. Assembly tool validation and verification are also performed to determine if the assembly method variation (between-component variation) meets design intent.

Step 6: Establish Production-to-Design Feedback Loop. In those areas where manufacturing or assembly capability does not meet design intent, the actual production variation data can be input into the simulation model to:

- Determine if the "out-of-spec" conditions adversely affect the overall product function.
- Evaluate several designs or process changes to help reduce the effect that each of the "out-of-spec" conditions has on product function.
- Provide quantification of additional sources of variation that exist in the process that should be specified in the design and process intent documentation.

Since the simulation model comprehends the interactive effects of geometry, assembly methods, and measurement schemes, the model can provide a tool to help ensure that effort is put forth in those areas that will directly improve the overall product.

General Requirements and Simulation

An organizational structure that supports concurrent engineering, computer-aided-engineering simulation tools, and a dimensional management process directly integrated with the existing design-and-build processes are all necessary ingredients. This activity cannot be a part-time job for the design engineer. It requires a dedicated resource to work with the entire product team from concept to production to ensure continuity.

A three-dimensional simulation model is an important consideration in a dimensional management process. Without a simulation model, there seems to be no practical way to determine if the design and process meet the build objectives. Also, just by following the steps necessary to create a simulation model, the product team identifies potential problem areas in the assembly early in the design phase of the product. The creation of a simulation model also forces cross-functional communication as described below.

A functional feature product model contains the three-dimensional surface features on the component parts in an assembly defined by the functional geometric dimensioning and tolerancing (GD&T) scheme (Fig. 1). Features are related to one another according to the GD&T datum references and feature control constraints (i.e., form, orientation, location, and size). These features are the same features that are important for assembly methods, manufacturing process, fixturing, and SPC checking. Identifying them up front in the design process is extremely valuable for the overall product team. (Additional information about GD&T, including a definition of symbols used, is provided in the article "Documenting and Communicating the Design" in this Volume.)

Component part variation is simulated based on the three-dimensional features and constraints defined in a functional feature product model. Each feature is allowed to vary, within its defined zone, according to the feature control constraints established by GD&T. Simulating the variation of the component parts using the same three-dimensional geometry and constraints defined by

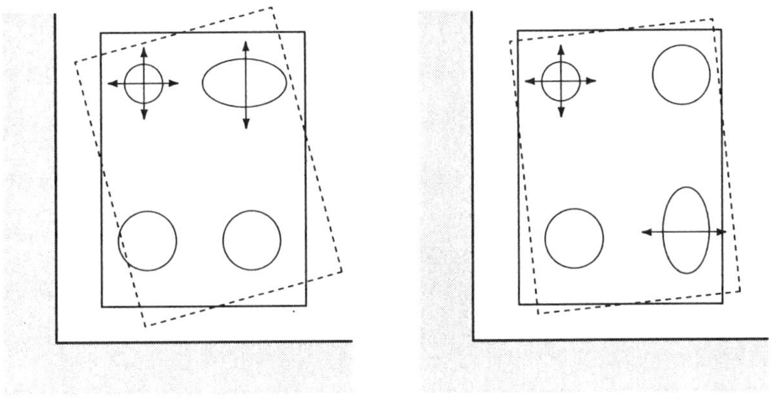

Fig. 3 Assembly method variation

Traditional tolerance analysis focuses on determining if the tolerances assigned on a completed design meets assembly objectives. The tolerances assigned did not relate to the actual manufacturing process, and the tolerance analysis did not reflect the assembly methods.

Dimensional management focuses on creating a robust design that can absorb as much variation as possible without affecting product function. The dimensional management process ensures that the design and process, as specified, meets objectives and that there is a coordinated measurement plan to help resolve any manufacturing build problems during production. Dimensional management is required to achieve continuous improvement in product quality, cost, and design-to-production cycle time. Real concurrent engineering cannot exist without this process.

GD&T provides the link between the component part, downstream measurement scheme, and the assembled product dimensional requirements.

Assembly method variation (or between-component variation) is defined according to how the component parts are assembled (Fig. 3). For example, fixtures, bolt-to-hole clearances, weld sequence, clamp sequence, and gravity effects all should be comprehended in the model and have a major influence on desensitizing the design to variation. The dimensional management process requires the manufacturing assembly engineers to work with the design team, early in the design phase, to determine optimal locating schemes for controlling variation and reducing the cost of manufacturing.

Measurement Schemes. Critical assembly measurements (product dimensional requirements) are defined during concept design. This provides the product team with a well-defined variation goal and provides the quality group with an early indication of what will be required for prototype and production measurement activities.

Assembly sequences are simulated to take into account the effects of subassemblies, assembly method order, and fixtures. Defining the assembly sequence basically defines the process flow through the plant.

Conclusions

A three-dimensional simulation model is often key in dimensional management. Several tolerance analysis simulation packages are commercially available. These packages are integrated with existing CAE systems to provide a direct link between the three-dimensional product design model and the functional feature product model.

SELECTED REFERENCES

- G.P. Dwyer, Applying the Principles of Dimensional Management to Instrument Panel Systems, *International Body Engineering Conference, Interior and Safety Systems Proceedings,* Vol 9, International Body Engineering Conference, Ltd., 1994, p 44–48
- G.P. Dwyer, Driving Product Design Through Early Implementation of 3-D Tolerance Analysis, *International Congress and Exposition, Instrument Panel Design Issues,* SP-1068, 950858, Society of Automotive Engineers, Inc., 1995, p 51–102
- J. Staif, Dimensional Management Process Applied to Body-to-Frame Marriage, *International Body Engineering Conference, Body Design & Engineering Proceedings,* Vol 20, International Body Engineering Conference, Ltd., 1996, p 39–43
- T. Sweder, Driving for Quality, *Assembly,* Chilton Publications, Sept 1995, p 28–33

Documenting and Communicating the Design

Gary Vrsek, Ford Research Laboratory

REDUCING THE TIME TO MARKET for new products is increasingly important. The engineering process can be very complex, involve many people, and require spending large sums of money. Producing a high-quality product initially and maintaining that quality level is more important than ever. The purpose of this article is to describe how documentation supports the process of bringing a product to market, who uses the information, and how it serves as a key form of communication. Volumes have been published on basic drafting principles and techniques, and overall, industry standards have been very effective. Clear and complete documentation is imperative. Properly creating and organizing the necessary documents can greatly facilitate the product development process.

Even the simplest part or product requires geometric definition in some form of drawing and documentation, to bring it from concept to production. However, the documentation needed goes well beyond a few drawings. A company's reputation for delivering a quality product depends on how well the design intent is communicated to the necessary people downstream in the process. It is very important to understand exactly what information is required for individual tasks such as fabrication, machining, or assembly.

An important early step is to define who is involved in the product development effort. Often, the specific configuration of the product will dictate the kinds of disciplines that will be required. For a particular product, the parties involved might include manufacturing, finance, inspection, stamping, assembly, material control, casting, and machining. Each area has its own specific information needs.

Background

The Traditional Design Process. Product design and its accompanying documentation have been approached many different ways through the years. Traditionally, the engineering process has been departmentalized and hierarchical (Fig. 1), with separate entities responsible for product-related engineering, drafting, process engineering, tool engineering, and tool design (Ref 1). A formal request process to initiate a new design or change existing drawings has been typically used to organize and track the work in larger organizations. Even though drafting standards have varied greatly among companies, standards are more important than ever and serve to pass on the experience gained from past projects (see the section "Standards" in this article and the article "Designing to Codes and Standards" in this Volume).

The classic drafting room was usually quite large and full of drawing boards (Ref 2). It was the place where engineers and draftsmen congregated to develop a new product design. The drawings were created using an artistic approach, and the tools consisted of triangles, compasses, and, most important, pencils.

The traditional design process is valid and can result in successful products. However, it has been documented that up to 50% of the drawings created using this approach were flawed (Ref 3). Competitive pressures made it necessary to improve the approach, and new technologies have made this possible. Team work, concurrent engineering philosophies, and communication of the overall scope help all involved ultimately achieve their goals. In industry, all must share the same goals if successful products are to be achieved. The benefit of complete documentation is fewer errors at the hardware stage and less variation in interpretation of the design intent.

The Computer-Aided Design Environment. There has been an evolution in the tools used by designers. The advances in the tools have generated new potential in the way information is formatted and communicated to others. Electronic forms of documentation have become the primary media in recent years. However, changes in the process to maximize the benefits of this technology have not progressed as quickly as many believe they should.

Computer-aided design (CAD) information can take a variety of forms. The key issue to address is what information is required by the various activities involved and how it can best be communicated to those requiring it. Too often, practitioners continue to adhere to the requirements of the historical design process: numerous drawings, prototypes, and physical tests. The acronym "CAD" too often stands for "computer-automated drafting" instead of "computer-aided design" (Ref 4).

A CAD database can contain detailed information about surface topology, contour, size, and location; however, this information generally is not included in a formal drawing format (Ref 5). Generating useful information is a matter of putting oneself in the position of the person who is going to use it. One good example is a machining drawing. Envisioning the information needs of the machinist may help the designer decide how the part should be dimensioned. This outlook will help determine what data should be used and what machines are required to meet the stated tolerances, surface finishes, and so on.

As we examine the documentation issue further, we will better understand the role and benefits of the CAD environment. The two-dimensional (electronic drawings) and three-dimensional (solid models) forms co-exist, with good reason. A mix of information such as wire frame geometry, surface details, solid models, numerical control tool path files, finite element models, and parametric or variational geometry is needed. Spreadsheets, database tools, and file management software also play key roles in the design process.

The Overall Design Process and General Documentation Requirements

The key to a successful design project is to follow a process such as that identified in Fig. 2 (see the article "Overview of the Design Process" in this Volume). Many companies have followed such processes and procedures over the years. While it may seem complex, furnishing complete documentation for such a process is not bureaucratic if the appropriate participants do their part.

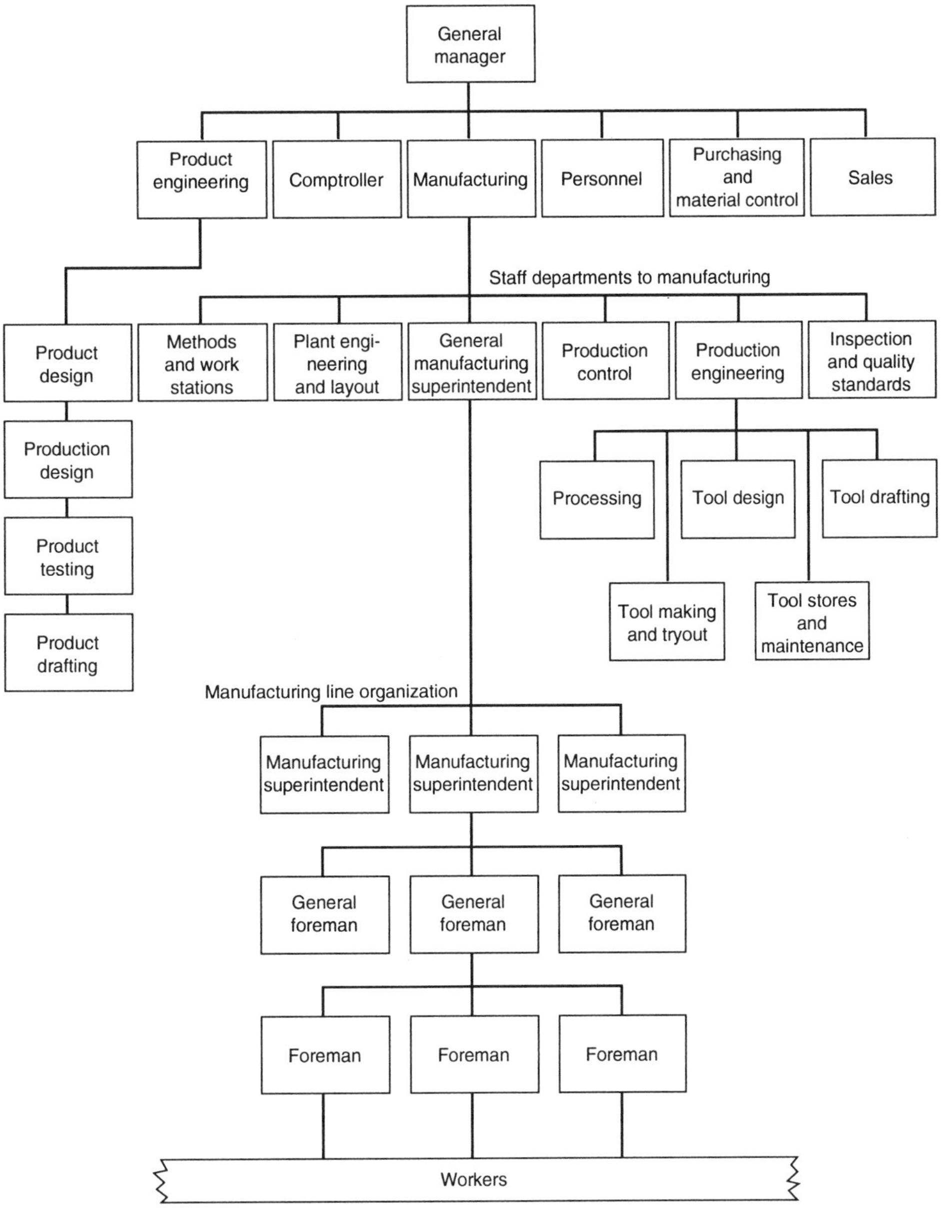

Fig. 1 Traditional organization structure for product design and manufacturing

Often, a product design program has several phases and involves designing, redesigning, and refining the product through the creation of several preliminary prototypes. Sound project management is important to ensure that each phase is started on time and that opportunities to overlap tasks are not missed. It can be argued that extensive documentation is not necessary and is just a sign of unneeded overhead. Also, early in a project, time and resources often are limited, and all aspects of the documentation may not be completed. However, failure to adequately capture a critical new product or technology through documentation can be devastating, and poor documentation can result in safety and liability risks. Following is a brief description of the major design phases and corresponding documentation requirements.

Concept Design. The initial phase of design should focus on the systems-level aspects of the project. The product-related goals must be clearly understood, and a general feeling for the competitive position of the product is needed. Information on the environment is necessary, as well as the definition of key guidelines and performance specifications. Identification of the appropriate level of detail is needed to ensure that ideas can be communicated without spending too much time or effort on parameters that can wait until the detail design phase, such as general tolerances, machining specifications, and so on.

The result of this initial phase is a clear idea of the design, components required, and, generally, the manufacturing process options. Drawings made at this point are the first communication tools used for the project. They must convey the design intent but also expose the design issues and challenges that need to be resolved. A first attempt at determining the cost and competitiveness of the design should also take place at this stage. Because the following phases require more time and labor, it is important to have high confidence in the success of the decisions made in order to continue. The design process tends to filter the various options and refine the chosen concept into a mature design (Ref 6).

Additional information is provided in the articles "Conceptual and Configuration Design of Products and Assemblies" and "Conceptual and Configuration Design of Parts" in this Volume.

Detail design focuses on the individual components. The goal of this phase, historically, has been to generate drawings that can be sent to manufacturing. It is at this point that manufactur-

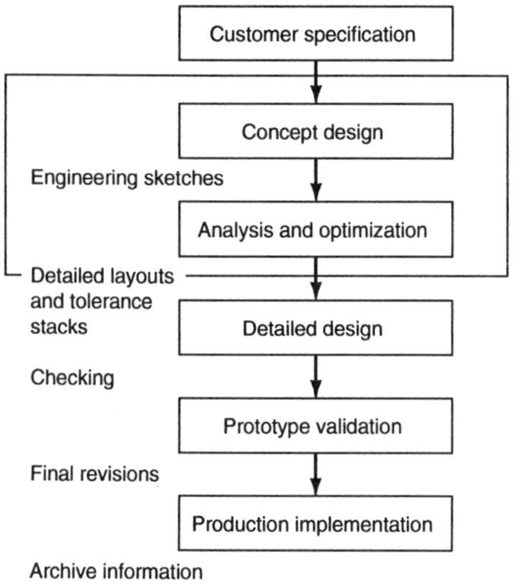

Fig. 2 Overview of the design process and related documentation requirements

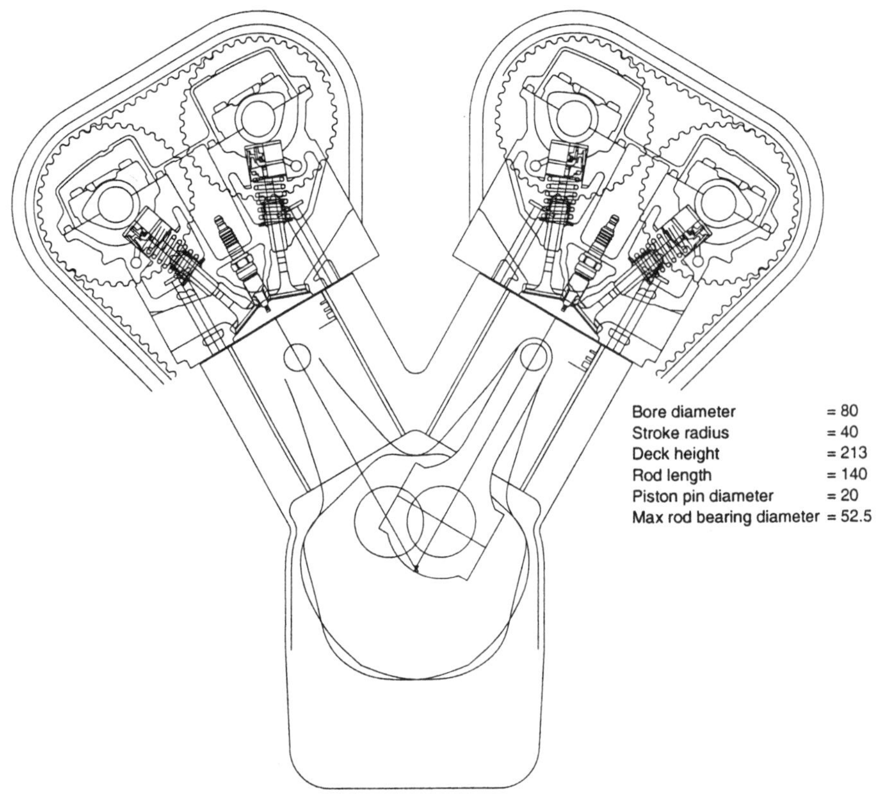

Fig. 3 Example of a drawing produced in the preliminary design phase

ing tolerances and full part definition are determined. The document generated is the primary engineering drawing or database that manufacturing will be working from.

The CAD model is becoming the master for much of the component definition, reducing the amount of information historically found on the detail drawing. This is especially appropriate when dealing with complex surfaces and features that would be difficult to describe dimensionally. Typically, the "detailer" is a novice who is guided by a senior member of the design staff.

Checking is a critical review of the information and documentation generated during the detail design phase. This step ensures the fit and function of the design. Traditionally, checking has been performed by someone not initially involved in the project, who can offer a fresh and objective review of the work done, prior to releasing the job for fabrication. This step is too often passed over to expedite the job, resulting in high cost, last-minute corrections, and modifications to the parts to complete the job. When detailing is done by a less experienced person, checking is especially critical.

Design Approvals and Release. Product design is an evolutionary process, so it is important to record the history of the project and changes along the way. The title block and revision column on a drawing can detail the history of a component. To ensure that changes are properly implemented, they must be communicated to and agreed upon by everyone involved in the processing of the part. A formalized release process is the proven approach to accomplish these goals. It is often necessary to release information early, in order to obtain cost and delivery estimates.

Tracking the document status is important. Some of the things to look for are:

- Key names and dates in steps such as release for quote, release for tooling, product engineering approvals, manufacturing approvals, and materials approval
- Identification of significant or critical characteristics
- Sourcing information (for heat treating, balancing, special tools, etc.)

Filing and Archiving. Ensuring that information is retrievable is extremely important. The issue concerns whether any undocumented changes can be reproduced, or, if the changes did not perform as expected, whether the component can be returned to the original configuration. The process of storing and retrieving information is complicated today. We are faced with many different formats based on whether the data was created manually (on paper) or electronically. If the information is stored electronically, in what format? Can information stored in different electronic formats be communicated across different CAD platforms? In many cases, paper versions or two-dimensional raster images of computer-generated data are stored for easier access. A variety of coding schemes are used in the industry that typically distinguish the various types of information and where to find it. Database technology has enhanced this segment of the business significantly.

Types of Documentation

Documentation must be focused toward explaining a specific task. The techniques presented in this article follow a general design process; differing situations will require adjusting the process accordingly. The following sections identify the key features that most documents must define and what users should be able to determine from these documents.

Historically, the challenge in blueprint reading has been to visualize the part, having only two-dimensional views. Now, electronic tools can represent a three-dimensional image, thereby significantly improving that first step. In either case, drawings are typically done in orthographic projection, with additional views added as required to clearly define the part. The drawing allows people from many different disciplines to understand and contribute to the ultimate goal of producing the part or product. The following sections describe the typical documents created to develop and produce a product.

Product Specifications. Concept design is the most unstructured point in the design process, and initial product specification documentation can come in a wide variety of forms. The simplest documentation can be a typed page of the envisioned product objectives and a list of general parameters, such as cost, size, performance, and/or production volume. The addition of any historical and benchmark information can be helpful. Reinventing the wheel or falling short of the competition are real concerns. Becoming too specific at this stage and requiring too many or unrealistic constraints can cause problems for the project. One must define the project with sufficient detail to enable engineering and design efforts to move forward, but accommodate the project launch as quickly as possible. At this stage,

ENGINE ASSUMPTIONS - DATA SHEET — Pg 1 of 2

SUBSYSTEM NAME						CPSC	Subsystem Affordable Targets (specify abs. or var. from comparator)						
Pistons, Rings, and Connecting Rods						03.11.02		Cost			Mass		Qual/1000
USAGE:	RMC - Rev.	MY	Engine	Vehicle	Applications	Volume		Facilities $	Tooling $	Unit Cost $	(kg)	TGW	12/12 R	3/36 R
New Model			4.0L 5V	Globe	ALL	240,000	Target							
Comparator			4.0L 5V	Star	ALL	20,000	Status							
Cause (CF):			Budget Model:				NVH CRITERIA:							

SUBSYSTEM SPECIFICATION (differences from comparator)

Conn. rod material:	Forged
Conn. rod construction:	I-beam shank;
Conn. rod length:	132.60 mm
Big end ID:	40 mm
Ratio b/a :	1.2 mm
Small end taper:	10 degree
Shank LxWxD:	52 x 22 x 17
Web thickness:	4 mm
Flange Thickness:	2 mm
piston compression height:	40 mm
piston pin assembly:	pressed in conn.rod small end
piston pin OD:	22.00 mm
piston pin ID:	14 mm
piston pin length:	50 mm
Height from pin center to piston top:	40 mm

SUBSYSTEM DESIGN CONCEPT SKETCH

REVISIONS

BENCHMARK (list mfg. and product line):

ENGINEER	PHONE #	SUPERVISOR	MANAGER	DATE

Fig. 4 Example of a project assumptions data sheet

there is often great pressure to cut corners. Examples of preliminary product specifications are provided in the article "Conceptual and Configuration Design of Products and Assemblies" in this Volume.

Engineering sketches are produced by the design engineer or a senior designer/draftsperson. The design will go through several iterations, and the goal of these preliminary drawings is to narrow the options and communicate the objectives of the design (Fig. 3). Frequently, this stage of the design process is carried out in a team environment, depending on the magnitude of the project. The assembly or systems-level aspects are the primary concern of this stage. The engineering sketches serve as important tools for design and redesign.

Product engineering assumptions documents provide a first look at the specifications for the individual components that will be required in the design. These documents should be considered a systems-level communication tool; spreadsheets are well suited for this type of communication (Fig. 4). A preferred format is a list of the basic assumptions and constraints for each component. This document can be used to define the general bounds of the design, and also to clarify peripheral issues such as the need for standard parts, bolts, pins, and other items that will not require redefinition. An assumptions sheet is an important tool for communicating across the organization the scope of the task and objectives for the final product. This document is usually controlled by the responsible engineer and should be considered a working document that will evolve with the design. Its real value comes as a means of sharing the overall design aspects among all of the engineers and designers, keeping them current on issues that might affect their own areas of concern.

Design layouts are produced by the designers assigned to the project; an example is shown in Fig. 5. Once the general design direction is set, the specific aspects of the design and environment need to be defined. Design layouts are very important to design personnel and to other personnel involved with the project, such as manufacturing, finance, quality, and rough forming. They are also useful for designers working on systems or components that are either similar to, or interact with, the ones being designed for the current project. The design layout serves as the main document to keep all participants on the same track.

In the CAD environment, design layouts are critical for ensuring that components fit together properly. These layouts can be very effective with the use of solid models, in either a two-dimensional format for presentation and discussion, or a three-dimensional isometric shaded image that can help others not as closely related with the project to grasp the design fundamentals. Many CAD systems have powerful systems-level tools that allow the user to design and investigate clearances in the same environment.

Detail drawings provide the information that is used to make the parts. All of the information required by a fabricator should be documented in a manner that leaves very little room for misinterpretation. Using CAD, a wealth of information can be extracted from the model. It is important to consider what information might be redundant or counterproductive to add. On the other hand, it is extremely important to ensure that all involved with the detail information can access that data

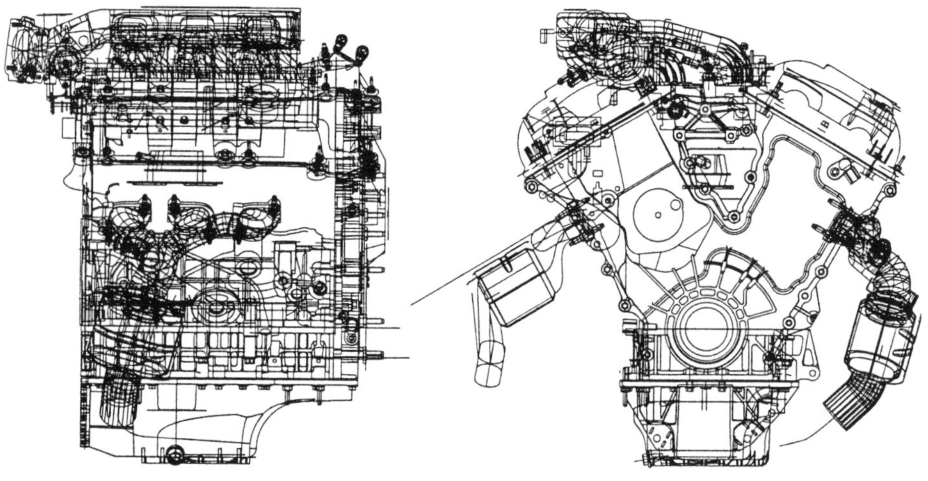

Fig. 5 Example of a design layout

and use it. All of the different areas must be considered; for example, if the quality office needs locations of some critical holes but nothing else, they may want those identified on the drawing even though the machinist did not (Ref 7).

A basic detail drawing includes:

- Standard views (and layers in CAD)—plan, front, side, etc.—the objective is to use conventional views to allow familiarization with the part (Fig. 6)
- Auxiliary views, sections, enlarged views, and isometric views—to aid in the understanding of specific details and clarify the design—will usually have special designations
- Dimensions—for machining information
- Tolerances—dimensional limits—to ensure the appropriate part integration

The approach to the detail drawing can vary (Fig. 7). Some flexibility is necessary to accommodate the unique drawing types. Formats and sizes are typically established by company standards. Other approaches, such as charting dimensions, can benefit organization and consistency (see the section "General Dimensioning Guidelines" in this article). Charting the coordinates as well as the size and tolerance information cleans up the drawing significantly. Charting will probably become increasingly common, because this is an easy step to automate.

Bill of Materials. The entire team is responsible for keeping records that are accurate and current. An important step in the organization of the project is accomplished with a document usually called the "bill of materials" or "parts list" (Fig. 8). This document is used to track the parts in an assembly or project and plays a significant role in information retrieval long after the engineering phases are completed. In its most basic form, the bill of materials lists the part number, part name, and quantity for each part in the complete assembly. Its utility is often increased by adding procurement tracking and special assembly instructions. The generation of this document is an excellent application for spreadsheets; however, for simple assemblies, the bill of materials can be part of the assembly drawing. New aids have been embedded into some CAD packages to allow assembly of the CAD models and to track the components of these assemblies in a bill of materials.

Specialized Drawings. So far, this article has addressed the aspects of the design related to designing the product and machining the part. Other drawings or CAD models are often required to accomplish the task of fully defining the finished product. While it is important to ensure that there is no redundancy in the project, additional forms of information may be required to more clearly communicate the design intent.

A common situation occurs when a casting or forging is required. A "rough form" drawing is helpful for documenting the part prior to any final machining. While it requires additional time to create, it removes possible ambiguity on complex parts, where trying to distinguish between cast and machined features can be difficult. Ideally, a CAD solid model can more completely define the rough form, and computed-aided tools can then be used to finish machine the part.

Assembly drawings are also common tools for clarifying the design intent. These drawings are especially important in cases where the design includes several components and the assembly sequence must be described. Also, where it is difficult to control tolerances, special instructions and part specifications can be identified on the assembly drawing. Examples are select fit parts (batches of like parts with slight size variations) or shims—the use of which is determined at the time of assembly. Additional documents used to clarify the design are worthwhile but will add to the maintenance tasks as the design matures.

Revisions, Record Changes, and Engineering Change Notices. Initiating, tracking, and transmitting changes to the design accurately is very important and can be costly if not handled properly. A formal request system and related documentation is important to ensure that the project flows smoothly and that a change does not negatively affect some unexpected area of the product.

Several issues should be considered when a change to a product is required. On a drawing it is relatively easy to make changes. However, the impact on manufacturing can be significant. Design changes can be categorized as record changes, in-process changes, and replacement or new designs.

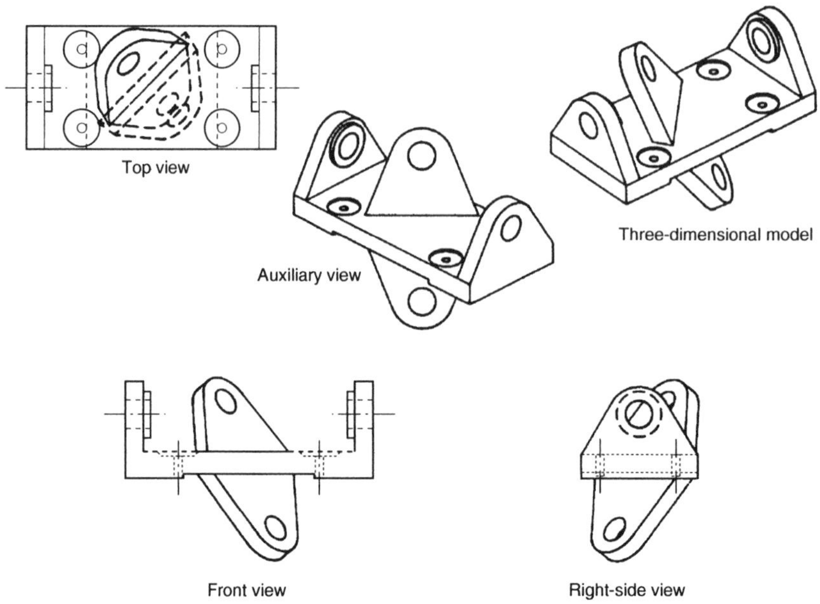

Fig. 6 Examples of standard views in detail design drawings

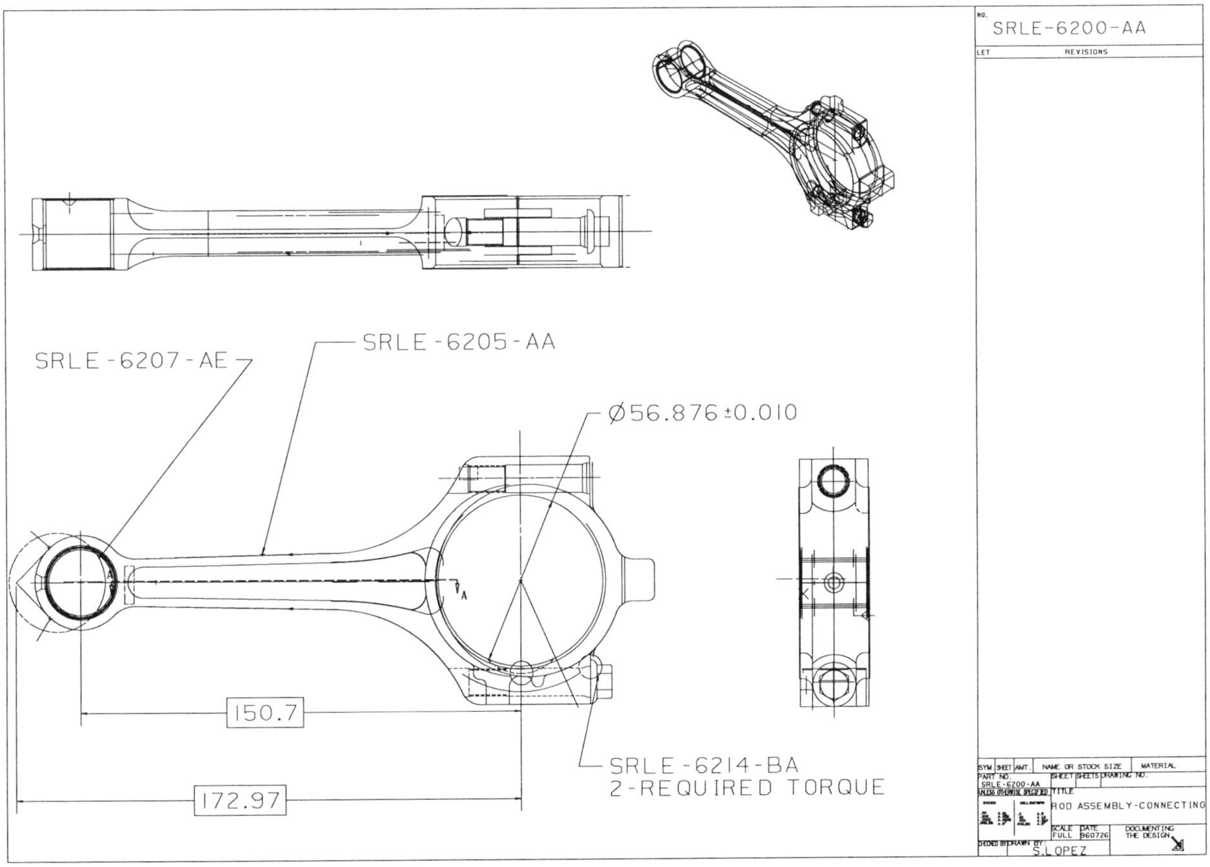

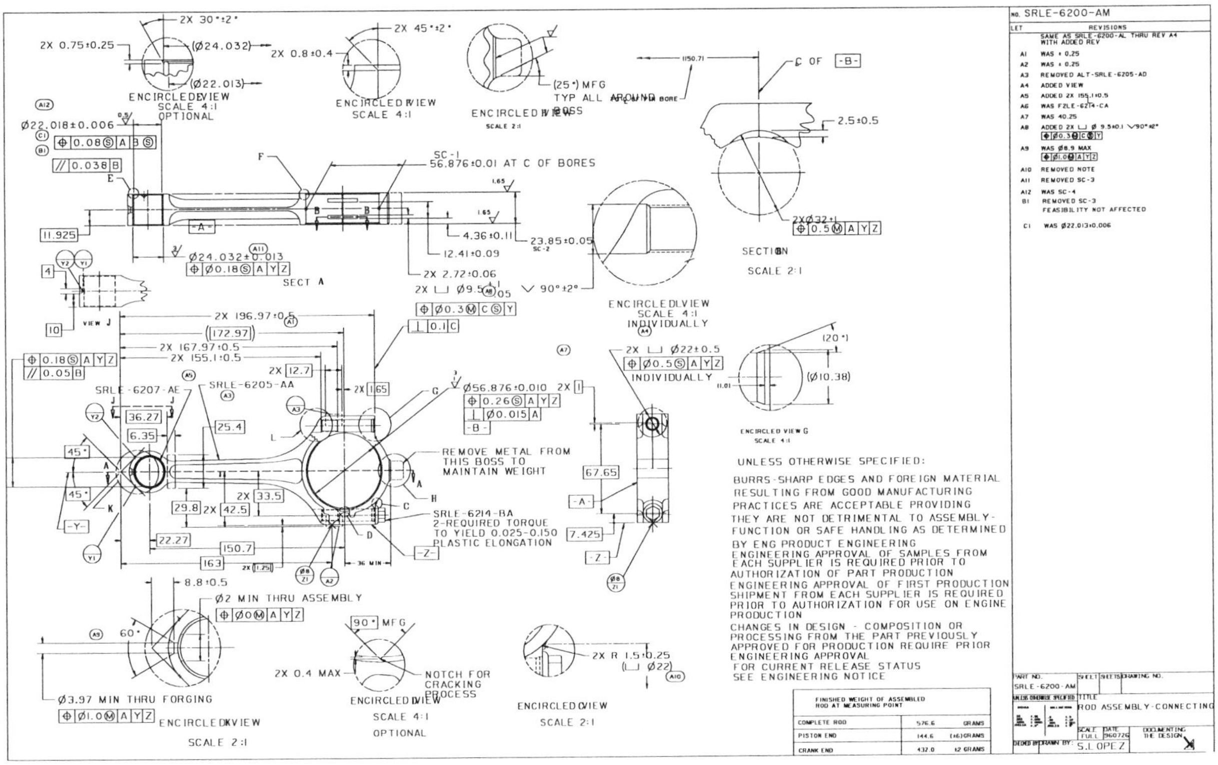

Fig. 7 Examples of detail drawings

Qty / Engine	ENGINE PROGRAM PARTS LIST DOCUMENTING THE DESIGN — PART DESCRIPTION	PART NUMBER Prefix	Base	End	P.O. #	Source	Delivery Date	RESPONSIBILITY Design	Engineer
	PISTON								
6	PISTON (CAST/MACH)	SRLE	6110	24093	RN0694	Ace	11/17/95	S.LOPEZ	M. Mahoney
6	PISTON RING - UP COMPRESSION	SRLE	6150	AC	RN0694	Ace	rec'd FRL	S.LOPEZ	M. Mahoney
6	PISTON RING - LOWER COMPRESSION	SRLE	6152	AC	RN0694	Ace	rec'd FRL	S.LOPEZ	M. Mahoney
12	PISTON RING - SEGMENT OIL CONTROL	SRLE	6159	AC	RN0694	Ace	rec'd FRL	S.LOPEZ	M. Mahoney
6	PISTON RING - SPACER OIL CONTROL	SRLE	6161	AB	RN0694	Ace	rec'd FRL	S.LOPEZ	M. Mahoney
6	PIN - PISTON	SRLE	6135	AA		BN Inc.		S.LOPEZ	M. Mahoney
6	PISTON & CONNECTING ROD ASSY	SRLE	6100	AG				S.LOPEZ	M. Mahoney
6	CONNECTING ROD - FORGING	SRLE	6205	AA		Formall		S.LOPEZ	M. Mahoney
6	CONNECTING ROD ASSY	SRLE	6200	C1		MMR Inc		S.LOPEZ	M. Mahoney
12	BUSHINGS-CONNECTING ROD	SRLE	6207	AE		Bear Inc		S.LOPEZ	M. Mahoney
12	RETAINER - PISTON PIN	SRLE	6140	AC		Spring Co.		S.LOPEZ	M. Mahoney

Fig. 8 Example of a bill of materials

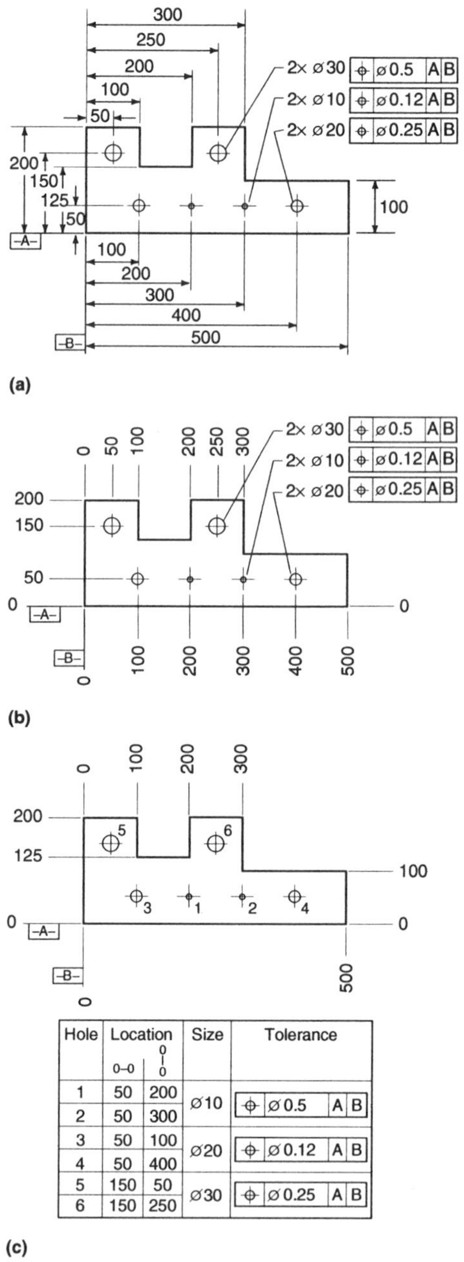

Fig. 9 Approaches to dimensioning on engineering drawings. (a) Linear dimensioning. (b) Zero line dimensioning. (c) Zero line dimensioning with coordinate data and hole specifications

Record changes are those changes that have no manufacturing cost implications. Usually, they are required to correct a mistake on a drawing that was discovered after the drawings were released, but before any parts were made. They are also additions or changes that clarify some aspect of the documented design.

In-process changes are those changes required to solve a product- or manufacturing-related problem. These changes do not affect part interchangeability. Expenses are incurred in the change.

Replacement and new designs are major changes. They generally change product application and interchangeability of parts.

It is important to evaluate how the product will be affected by any changes required. It may be wise to consult manufacturing personnel, asking them to make an assessment and to forge an implementation plan.

Understanding and Using Design Documentation

Layout and assembly drawings promote understanding of the product function and its environment. The detail drawing focuses on construction and manufacturing. It must convey the needs of product engineering, materials engineering, and manufacturing (manufacturing encompasses rough forming, machining, and inspection). This section describes the detailed requirements of engineering and manufacturing and how drawings are used as a communication medium.

Product-Related Information. Product design and systems-level engineers typically focus on the issues that affect the performance and function of the parts. The performance of the final product and aspects related to the assembly will be dictated by the design layouts and detail drawings. Design layouts are extremely important in tracking the environment for the subsystem or assembly. Particularly when dealing with subassemblies and other parts in the environment, issues of interaction are critical. Functional tolerances for fit, durability, cleanliness, and assembly sequence must be clearly conveyed. From the drawings and specifications generated, product engineers must be able to understand the following:

- Size
- Fit (relationships between mating components)
- Function and performance (acceptance test criteria)
- Cost and complexity
- Weight
- Product safety issues
- Inspection requirements

Material specifications affect product function, manufacturing feasibility, and planning. Key issues include basic material properties and factors such as required heat treatment or surface finish. This information is usually contained in the title block or note column. It is important to examine the heat treatment specification, to determine if a specification such as case hardening results in the need for a special machining sequence.

Manufacturing Attributes. It is important for the designer to understand the manufacturing process and include related information in the product definition. The specific information needed varies, depending on the operations used. Especially critical are rough forming operations such as casting, forging, or stamping—operations where the process requirements affect the shape of the part. The most common manufacturing-related features that appear on a drawing are draft and parting line definitions. The information required for all the operations specified will have a significant effect on the product definition. Minimum corners, fillets, and finish machine stock requirements are also specified on the detail drawing.

General Dimensioning Guidelines. Reading a detail drawing can be overwhelming. Among the features found on a drawing are references to milled faces, drilled holes, and threading operations. Viewing the part drawing from the perspective of someone who must perform finishing operations on the rough workpiece can be helpful.

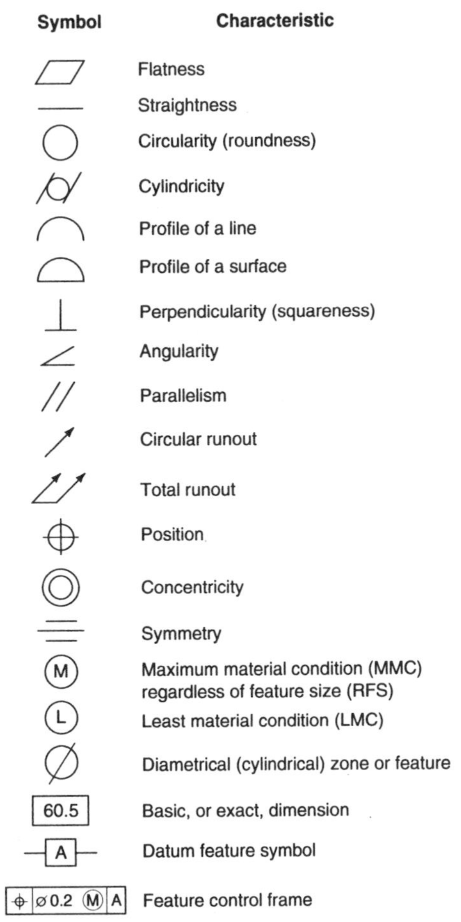
Symbol
Characteristic
Flatness
Straightness
Circularity (roundness)
Cylindricity
Profile of a line
Profile of a surface
Perpendicularity (squareness)
Angularity
Parallelism
Circular runout
Total runout
Position
Concentricity
Symmetry
M
Maximum material condition (MMC)
regardless of feature size (RFS)
L
Least material condition (LMC)
Diametrical (cylindrical) zone or feature
60.5
Basic, or exact, dimension
A
Datum feature symbol
⌀0.2 M A
Feature control frame

Fig. 10 Geometric dimensioning and tolerancing symbols and characteristics. Source: Ref 9

Several manufacturing handbooks give direction in terms of the tolerances to be shown on a drawing and possible manufacturing approaches. Additional features that are typically found on part drawings include special manufacturing locators to set up the part for clamping, manufacturing, and inspection.

There are three primary approaches to dimensioning a component. The traditional approach (Fig. 9a) is to use linear dimensions, with arrowheads and leader lines (Ref 8). This is the preferred approach if manual manufacturing methods are being used, as in prototyping, where features are being addressed one at a time. The approaches shown in Fig. 9(b) and (c) are more appropriate for numerical programming processes. Zero line dimensioning (Fig. 9b), sometimes referred to as datum line dimensioning, has become very popular. This method has occasionally been the source of confusion in the interpretation of the tolerances. The common mistake is to assume that the part tolerances are related to these datum lines, but it is important to remember that zero lines are dimensioning aids only and that datum lines for feature tolerances could be different. The third approach (Fig. 9c) also can utilize the traditional linear dimensions, or zero lines, but it uses tables for the coordinate data and hole specifications (hole charts). Hole charts can be a good way to visually clean up a complex drawing. Reference dimensions, though considered redundant information, are often appropriate in clarifying particular features.

Tolerances and Tolerance Stacks. Tolerances are initially developed to ensure that the part will fit and function as intended. Most often the initial tolerances are selected by referring to past projects of a similar nature, applying the best-known information about that project to the job at hand. Manufacturing engineers use this tolerance information to decide how best to process the job. This includes what machines to use, how to set up and locate the workpiece in the machines, and what sequence of manufacturing operations to use. Tolerances for the part can be found in a variety of locations on the part drawing, including in a general note in the title block, in the notes column, and in the dimensions as a limit or using the ANSI geometric dimensioning and tolerancing (GD&T) notation (Fig. 10).

The variety of approaches to tolerance analysis are beyond the scope of this article (see the article "Dimensional Management and Tolerance Analysis" in this Volume). However, it is important to remember that the aim is to ensure functionality and manufacturability. The tolerance stacks (the cumulative effects of tolerances in an assembly) can provide insight into the product performance and process capability, and all possible sources of variation (dimensional, tooling, heat treatment, surface quality, etc.) must be taken into account.

Standards

Geometric dimensioning and tolerancing has evolved to become the most significant tool for designers to communicate uniformly with the manufacturing world. Throughout this article, the importance of uniformity has been stressed, but it is in the area of dimensioning and tolerancing that it is most difficult to achieve. The completeness of the ASME/ANSI standard has allowed many major companies to abandon the maintenance of separate internal standards, which has proven beneficial to external suppliers as well. Caution must be used to avoid building a specification too complex for the machinist or inspector to understand, too difficult to measure without expensive gaging, or, worse yet, not really applicable to the situation. A company's specific manufacturing practices must dictate the standards that are used by the design group (Ref 10). The GD&T standards define the following:

- *Datum references:* A primary surface or feature, used as a location reference for other features
- *Basic dimensions:* The exact theoretical location for a feature (the tolerance for such a feature is usually in the feature description)
- *Positional tolerances:* Feature tolerances pertaining to location
- *Form tolerances:* Feature tolerances related to the shape, runout, or size variation

The most widely used GD&T standard is ANSI Y14.5M.

ISO 9000. With suppliers playing a key role in the product development and manufacturing processes, standards for passing complete and consistent data is another step toward ensuring the highest possible levels of quality. ISO 9000 is a series of standards that define the criteria for a functional quality assurance system. Maintenance and control of documents is a key component of the ISO 9000 standards and other similar standards.

Local or Corporate Standards. An important goal in design should be to have consistent processes across the company. The ability to attain a level of familiarity with the documentation within the engineering community improves quality and efficiency. Local standards ensure a common language across the various departments in a company that work in related areas, such as powertrains or body structures in an automotive company. Furthermore, the ANSI GD&T specification sometimes allows two or three options in handling certain situations. Local standards can define which option is best suited for the company's manufacturing process.

Typically, key process engineering specifications are required for tests, heat treatment, and screw threads. Local standards for these items help reduce variation within a family of parts, and they also can reduce the amount of written information needed on the part drawing (Ref 1).

Conclusions

The complexity of the various types of engineering documentation can be overwhelming. However, each of the many pieces of information is necessary and reveals its usefulness at some stage of the design process. By the time the detail drawings are prepared, the design has matured and parts are ready to be machined. Below is a step-by-step approach that should be helpful in understanding a design based on its detailed documentation:

1. Visualize the physical part, using the views provided on the detail drawing.
2. Review and understand the correlation between the detail drawings and the parts list.
3. Refer to assembly drawings to understand the intended function of the component and how parts are integrated into the system.
4. Observe the drawing scale used and take time to read the general and/or special notes, instructions, and specifications.
5. Review the material specification and determine the process route for the initial form (e.g., bar stock, casting, forging).

6. Look for the key locating features.
7. Examine dimensions and tolerances to understand the functional requirements and manufacturing strategy.
8. Look for the approvals and revision history.

The key to a successful product development project is communication. All of the parties involved in a project must understand how designs are evolving, and all must be able to make their contributions in a way that brings the product to market quickly. The new electronic environment is helping to generate information more efficiently and more completely. New tools will continue to help designers perform many of these functions concurrently and across greater geographic distances.

ACKNOWLEDGMENTS

The author thanks Sherry Lopez for providing many of the figures and Marshall Mahoney for his technical assistance in preparing this article.

REFERENCES

1. D.F. Eary and G.E. Johnson, *Process Engineering for Manufacturing,* Prentice-Hall, 1962
2. F.E. Giesecke, A. Mitchell, H.C. Spencer, I.L. Hill, and R.O. Loving, *Engineering Graphics,* 2nd ed., Macmillan, 1975
3. A. Krulikowski, GD&T Challenges the Fast Draw, *Manuf. Eng.,* Feb 1994
4. A. Mikulec, J. McCallum, B.A. Shannon, and G.A. Vrsek, Powertrain Engineering Tool, *FISITA 96* (Conference Proceedings), Fédération Internationale des Sociétés d'Ingénieurs des Techniques de l'Automobile, 1996
5. S. Kalpakjian, *Manufacturing Engineering and Technology,* 2nd ed., Addison-Wesley, 1992
6. D.G. Ullman, Issues Critical to the Development of Design History, Design Rationale, and Design Intent Systems, *Design Theory and Methodology,* ASME, 1994
7. L.J. McBee, Creating Views, Sections, and Explosions with CAD/CAM, *Computer-Aided Design, Engineering, and Drafting,* R.M. Dunn and B. Herzog, Ed., Auerbach, 1986
8. W. Hammer, *Blueprint Reading Basics,* 2nd ed., Industrial Press, 1996
9. P.F. Ostwald and J. Muñoz, *Manufacturing Processes and Systems,* 9th ed., John Wiley & Sons, 1997, p 574
10. L.W. Foster, *Geo-Metrics II,* 2nd ed., Addison-Wesley, 1984

Rapid Prototyping

Charles L. Thomas, University of Utah

RAPID PROTOTYPING is a relatively new field in manufacturing that involves techniques/devices that produce prototype parts directly from computer-aided design (CAD) models in a fraction of the time required using traditional techniques. The prototypes are used as form models, to check the touch and feel of the part; as fit models, to verify geometry and alignment of the part in its intended application; and in some cases as function models assembled onto a working mechanism to test the ability of the part under design to perform its intended duty. The prototyped object can also be a mold or a pattern for secondary techniques that produce preproduction or production tooling.

Rapid prototyping techniques generally produce prototypes by decomposing a three-dimensional CAD model into parallel cross sections. Typically, each cross section is constructed atop the previous cross section, building the part layer by layer from layers that are 0.1 to 0.2 mm (0.004 to 0.008 in.) thick. The layers are bonded together either before or after cutting or as a natural consequence of layer formation. The construction materials available for these techniques include photopolymerizable or thermoplastic resin, paper, wax, and metal or ceramic powder. Secondary operations expand the list of available materials to include castable metals and certain forms of composites.

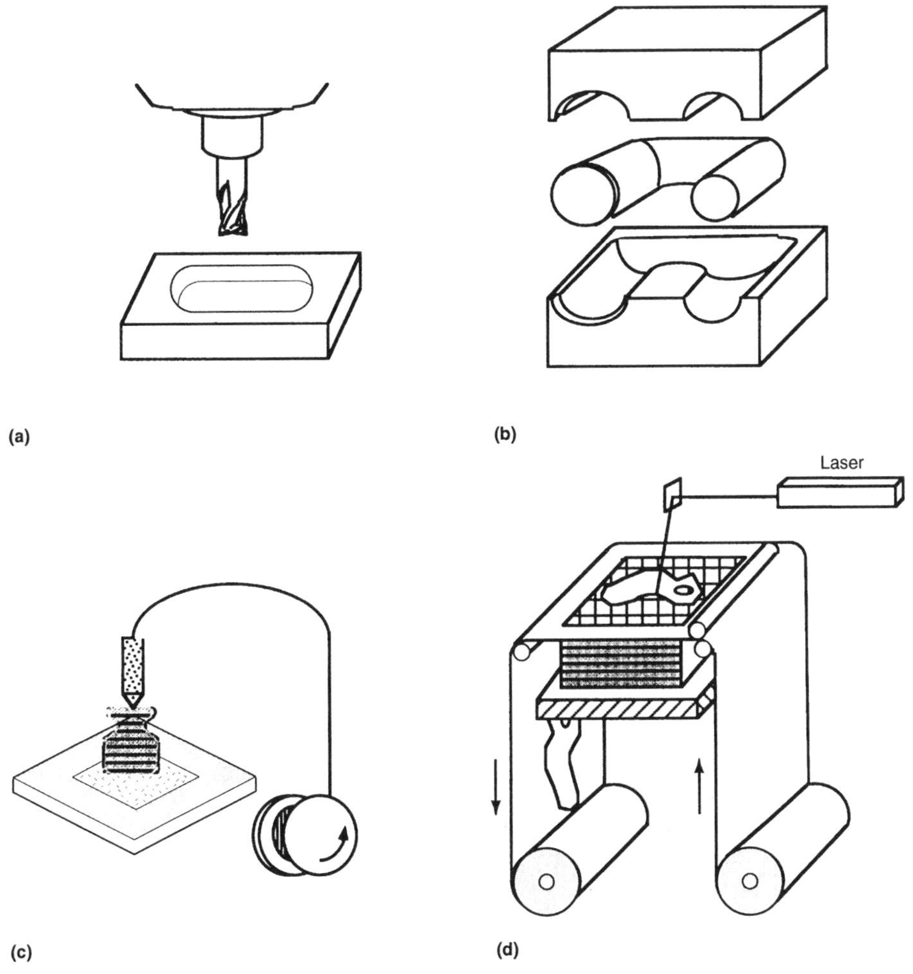

Fig. 1 Four manufacturing paradigms. (a) Subtractive process. (b) Forming process. (c) Additive process. (d) Hybrid process

General Description

The traditional product realization process requires iteration to create a successful design for a manufactured product. Engineers use design and analysis skills to produce an initial design that is then prototyped and tested. If problems are identified during testing, the engineers return to the design and analysis stage. This iterative loop is at the core of successful engineering design.

Using traditional techniques, prototyping and testing are usually expensive and time consuming. As a result, engineers have developed improved modeling and analysis techniques in an effort to catch errors before the physical testing begins. An extreme example of this is seen in the aerospace industry where designs are developed very carefully, using extensive simulation and analysis. The resulting design is often allowed one or no iterations before the design is frozen. When iteration is performed, the engineers often are not allowed to change anything that would require replacing the prototype tooling.

While the value of iteration in the design process is recognized, the cost of iteration (in both time and money) often restricts the number of iterations allowed. This is the impetus for the concept of rapid prototyping (RP). In concept, an RP device is a three-dimensional version of a printer. The designer sends an electronic representation of a three-dimensional object to the device, and an accurate physical representation of the object is created without requiring operator skill or interaction. Actual commercial RP devices typically produce parts with dimensional variations of at least ±0.13 mm (±0.005 in.) and generally require significant skill from the operator to produce prototypes with this accuracy.

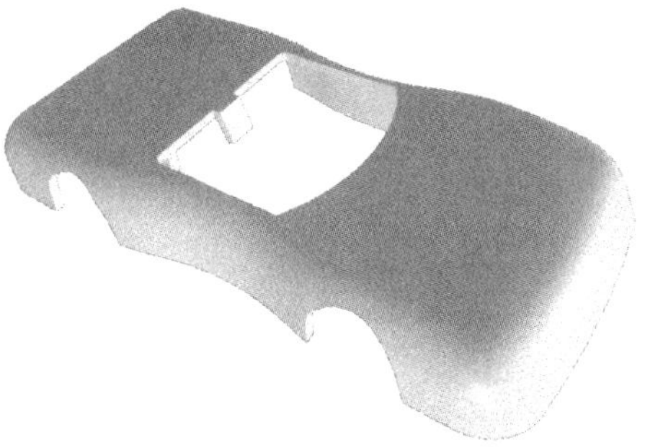

(a)

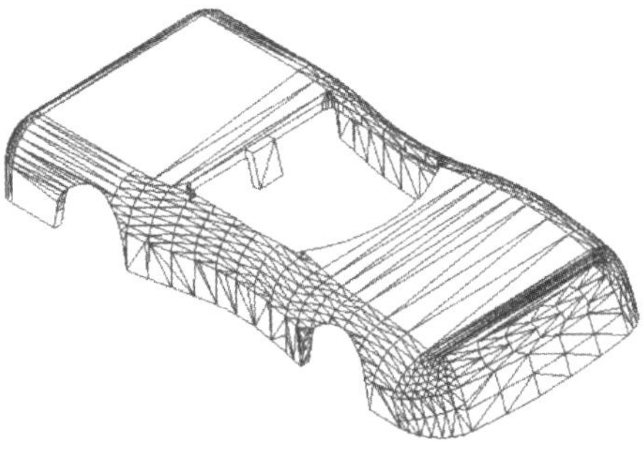

(b)

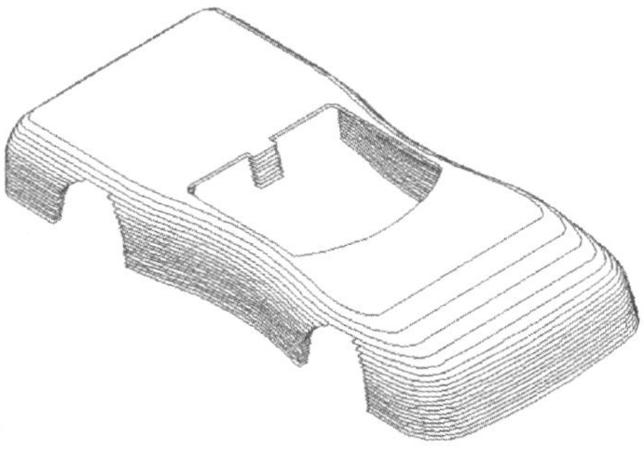

(c)

Fig. 2 Preprocessing for RP begins with the creation of a solid model, which is converted to stereolithography format and then sliced into parallel cross sections. Source: Ref 2

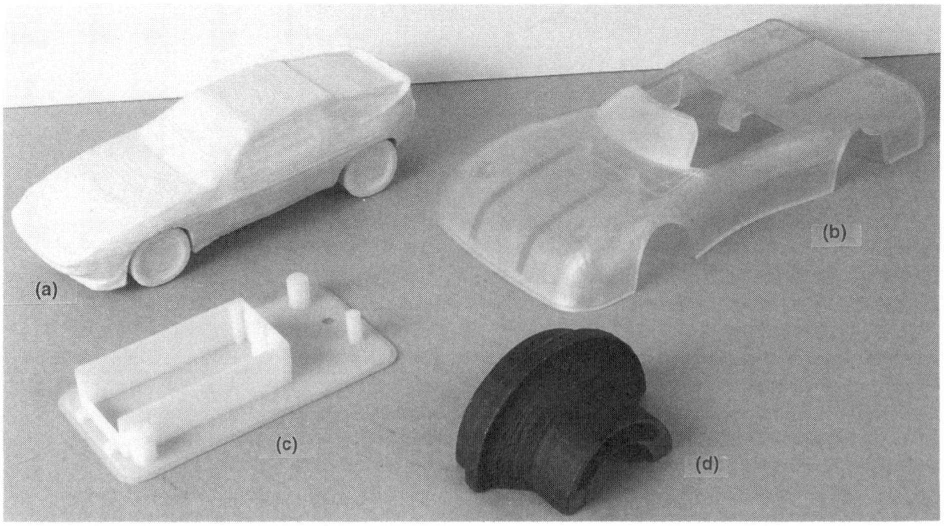

Fig. 3 Parts made by various RP techniques. (a) Car made from layers of paper using a cut-then-stack process. (b) Car cured from liquid photopolymer using stereolithography. (c) Housing base made from thermoplastic polyolefin using FDM. (d) Part investment cast from carbon steel using a foam plastic model made using a cut-then-stack process

Rapid prototyping devices allow the designer to complete a design iteration loop significantly faster than was traditionally possible by reducing the time and cost required to produce a prototype.

Paradigms for Part Creation. Rapid prototyping goes by a variety of names, each preferred by different people with different goals for the process. The term *rapid prototyping* seems to include any method as long as it is fast. Some use the term *solid freeform fabrication* (SFF), which implies that the part is not necessarily a prototype and focuses on the ability of the device to build any desired shape. Numerical control (NC) machining techniques would not be considered SFF, because there are some restrictions on the shapes that can be produced using NC. The term *auto fab* seems to include any automated process, but would not include processes with significant manual components. *Desktop manufacturing* is usually used when referencing small, desktop numerically controlled devices that automate standard machining techniques. *Layered manufacturing* refers to the fact that most techniques discussed here build parts from thin parallel layers.

The search for and discussion of an appropriate term for these techniques is an effort to define a new paradigm for part creation. Although there are many ways to categorize manufacturing techniques, one way they can be discussed is in terms of the mechanics of the part formation process. In these terms, the manufacturing processes fall into four categories (Fig. 1):

- Subtractive processes
- Forming processes
- Additive processes
- Hybrid processes

Most traditional manufacturing processes fall into one of the first two categories. Parts are produced by either removing excess material from a blank (milling, sawing, drilling, etc.), or by forcing raw material to take the shape of a mold (forging, injection molding, casting, etc.).

Most RP techniques fall into one of the last two categories. For additive processes, parts are produced by adding material to create a part. Some of the RP techniques are considered hybrid processes because they require an initial subtractive process to cut each layer from a sheet, but then additively bond the layers together to create the part.

Rapid prototyping devices are generally relatively expensive (often in the range of $60,000 to $500,000). Typically only larger companies can justify the expense of purchasing an RP device. For this reason many prototypes are produced by service bureaus that receive part files by modem and send out the RP prototypes in a day or two. In either case, the time required to produce a prototype is reduced to less than a week. The example later in this section demonstrates the utility and savings that can be achieved through the use of RP.

Typical Rapid Prototyping Process (Stereolithography). The first commercial RP device was introduced in 1987 based on a patent by Charles Hull (Ref 1). The process, termed stereolithography, is described here to outline the general principles of RP. The process can be broken into three subphases:

- Preprocessing
- Building the part
- Postprocessing

Preprocessing includes mathematical manipulation of the electronic model in preparation for the physical construction. Postprocessing covers operations required to produce the finished prototype after the basic shape has been created.

Preprocessing. A designer first develops a three-dimensional solid model of an engineering part. This model may be created in a variety of different CAD softwares (Fig. 2a). The solid model is converted to an industry standard file format called the stereolithography format (or a file extension, *.stl). This conversion is necessary so that the RP device can communicate with the various CAD packages. To convert to the *.stl format, the solid model is tessellated with triangular tiles as shown in Fig. 2(b). Conversion to *.stl format is performed by a module in the CAD software and is a standard function of nearly every CAD software capable of producing solid models. The *.stl file is created and delivered to the RP device. This delivery can be done physically on a disk, over a modem, or through the Internet. The computer controlling the RP device reads the *.stl file and slices the model into a series of parallel cross sections as shown in Fig. 2(c). The cross sections are spaced at a thickness equal to the dimension of the layers that will be created.

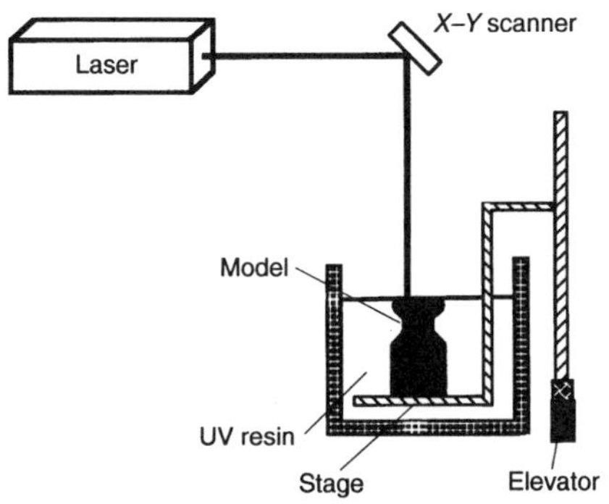

Fig. 4 Schematic of the stereolithography process

Building the Part. At this point the electronic cross-section data must be physically created. While the preprocessing is similar for all RP techniques, the part construction processes are quite varied (Fig. 3). A schematic of the stereolithography process is shown in Fig. 4. Using stereolithography, the part is constructed by curing liquid photopolymer resin in a vat. The photopolymer resin consists of plastic monomer and oligomer combined with a photoinitiator. When light of the appropriate wavelength exposes the liquid, it rapidly polymerizes into a solid. A movable platform immersed in the photopolymer vat is initially positioned just at the liquid surface and then indexed down into the polymer, incrementing after each layer is created. At each increment of depth a liquid layer of resin forms above the part under construction, and the laser scans the liquid surface tracing out the appropriate cross section of the part to be built. Where the light exposes the photopolymer, it rapidly solidifies. In order to create the complete cross section, the laser must scan the periphery of the part and crosshatch the interior solid sections. Once a layer is formed, the build platform drops down and a wiper levels the surface of the liquid. This cycle is repeated for each cross section until the top layer of the part is formed. Finally the platform is raised, and the finished part is removed from the device.

Prototypes do not generally sit directly on the build platform during construction, but are supported by a structure that is built along with the prototype. This support is generated automatically by the preprocessing software that creates the support structure as a hollow rectangular grid. This grid provides an interface between the part and the build platform, it provides support for cantilever sections where there is a significant increase in cross section compared with the previous layer, and it prevents the layers from curling or warping during the building process. While other RP processes do not use these rectangular grids for supports, nearly all processes provide support structures that are automatically generated as the part is constructed.

Figure 3(b) is an example of a part built using stereolithography. This model car body is a thin-shelled structure with stiffening ribs on the interior. The car was built upside down supported by a rectangular grid structure (see Fig. 5a).

Postprocessing. Once the part is removed, it is washed with a solvent to remove excess photopolymer from the surface. The part is placed in a chamber and exposed to ultraviolet (UV) light to ensure that the cure is complete. Any support structures that may have been required are then removed. These operations are specific to stereolithography. Each different RP process has a series of postprocessing operations that are specific to that process.

At this point the part has a stepped surface finish (Fig. 5b). This is not unique to stereolithography, but is a result of the layered construction technique, which makes the surface finish depend on the layer thickness. The part may also have surface imperfections caused by removal of the support structures. Parts are often sanded and/or painted to improve the surface appearance, depending on the wishes of the customer.

Example: Electric Current Sensor Design. This case study, based on the experience of a company that designs and manufactures electric current sensors is given as an example to describe the utility and savings that can be achieved by RP technology. The company (F.W. Bell, a division of Bell Technologies Inc.) is involved in the development of a split core current sensor—a hybrid design using two existing products. The large investment required by such a project could not be justified based only on the two-dimensional drawings supplied by the design team.

Designers at F.W. Bell (with the help of a product engineering and prototyping services bureau) developed a three-dimensional CAD model, taking the ambiguity and miscommunication of part detail typical of two-dimensional drafting out of the design process. In addition, physical models—using a Stratasys Fused Deposition Modeling process (Stratasys Inc., Eden Prairie, MN)—were produced for form, fit, and function evaluation. The models were fabricated using acrylonitrile-butadiene-styrene (ABS) plastic, a material that has properties similar to the polycar-

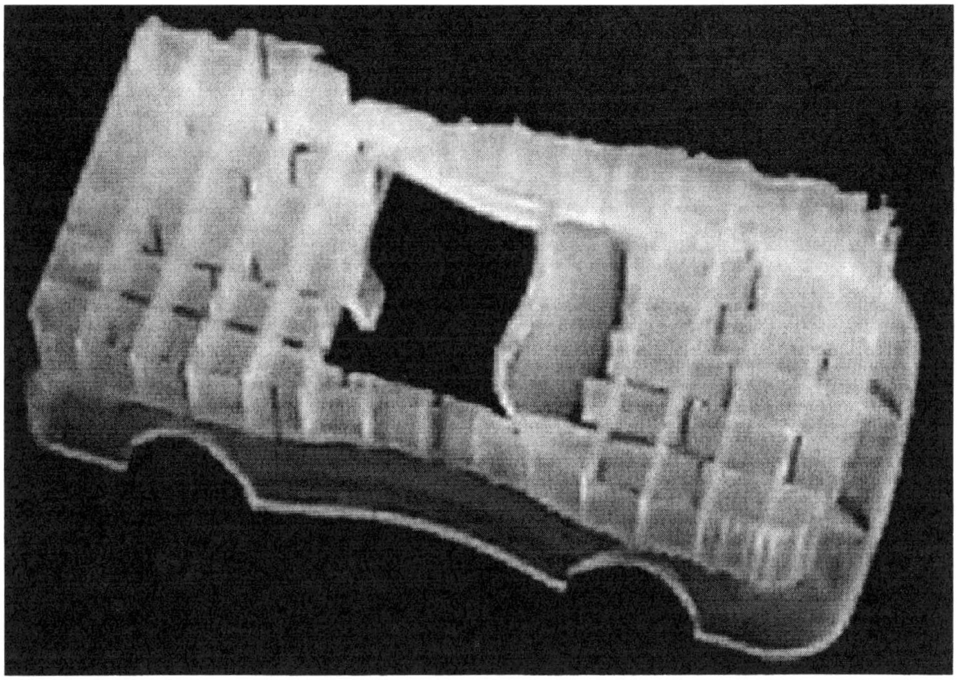

(a)

(b)

Fig. 5 Model car created using stereolithography. (a) Support structures before removal. (b) Surface finish resulting after removal of the supports

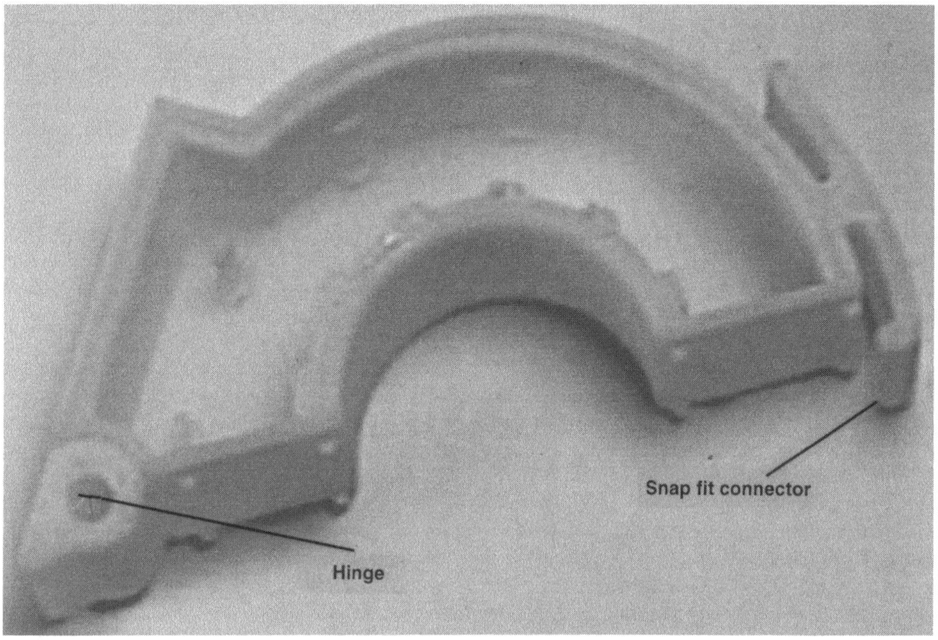

Fig. 6 Housing for a split core electric current sensor was prototyped using FDM. Part is hinged to a mating piece, and the snap fit connector holds the fitting closed.

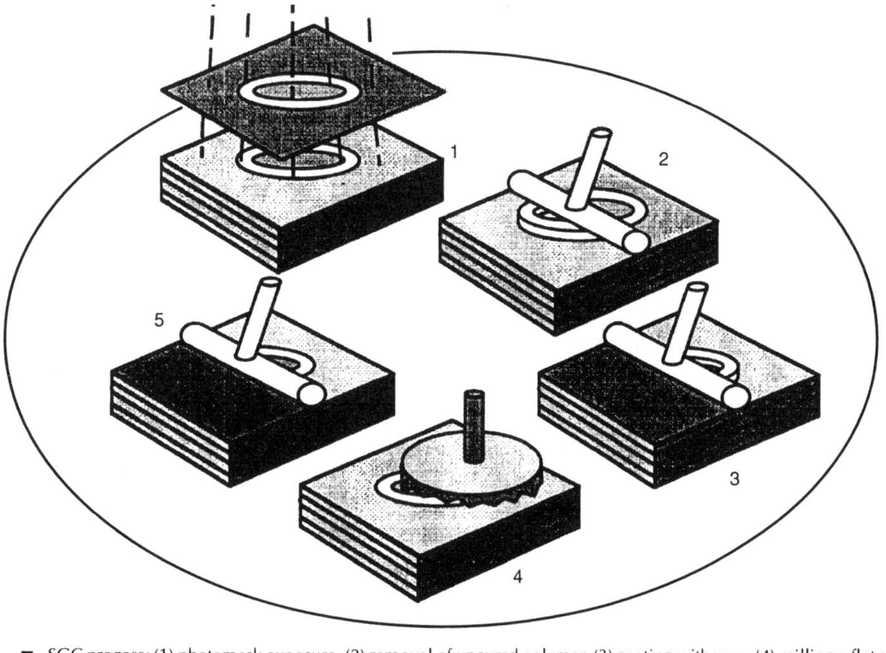

Fig. 7 SGC process: (1) photomask exposure, (2) removal of uncured polymer, (3) coating with wax, (4) milling a flat surface, (5) coating the next polymer layer

bonate considered for production. The design team used the models to identify high-stress areas, to get a concept of the real part, to debug the design, and to perform preliminary electrical tests. A prototype made using fused-deposition modeling (FDM) is shown in Fig. 6.

The snap fits between the components of the split core sensor were an area of high "stress" for the designers. The highly accurate (±0.005 in. tolerances are typical for the FDM process) ABS parts were snapped in place hundreds of times by F.W. Bell's designers, engineers, sales persons, and potential customers. This allowed them to assess not only the design but also the manufacturing processes required to make the parts.

Because the CAD models can be easily modified and the RP service bureau was able to produce the ABS parts literally overnight, several design iterations were produced in a short period of time. This allowed F.W. Bell to try out a multitude of "what-if" scenarios before freezing the design for the development of production tooling. By using ABS models to test design iterations, the company realized cost savings of 60% over the iterative modifications of a production mold, and the time needed to produce the final tooling design was reduced by 70%.

Major Commercial Processes

There are a range of processes currently available commercially, either by purchasing a machine, or through service bureaus. Each process uses a distinctly different method for constructing the part layers and, as a result, each process has a unique set of advantages and disadvantages for producing parts. For discussion here, the various processes have been separated into three categories based on their layer creation method:

- *Selective Cure Layered Processes:* These processes create layers by selectively solidifying portions of a layer of precursor material.
- *Extrusion/Droplet Deposition Processes:* These processes build each cross section by selectively depositing material to create the layer.
- *Sheet Form Fabricators:* Layers are cut from sheet form construction material and laminated together.

Speed and Accuracy. Rapid prototyping techniques in general build parts with an accuracy of approximately ±0.005 in. with closer tolerances possible on small local features. While specific machines may do better or worse than this, it is difficult to tag a machine with a specific number because the tolerance capability generally varies with part geometry, construction material, and operator skill. The build speed for a given machine is also difficult to specify as it is dependent on the part geometry and level of accuracy desired. In general terms, small parts 1 to 3 in.^{3} solid volume, can be built in 1 to 5 h. Larger parts, 60 in.^{3} or more may take several days to complete. In the discussion of the specific commercial machines below, speed and accuracy are not quantified; instead general classes of part geometry and material are discussed that represent the market niche for each device.

Selective Cure Layered Systems

Stereolithography (SLA) was the first commercial system introduced. A basic description of the process is provided in the Section "General Description" in this article. This process has the largest installed user base and the longest development history. Stereolithography builds a part by sequentially solidifying small bits of volume and, as a result, it is most efficient for producing thin-walled structures. Although any part can be built that fits within the 20 by 20 by 23 in. build volume (SLA-500), a large part that requires curing most of the polymer in the build volume would require several days to produce. Parts are produced from a rigid, epoxy-base resin that can be sanded and polished to produce a glossy, optically clear part.

The range of materials available for prototype parts can be increased by prototyping molds instead of parts or by producing patterns for subsequent molding operations. The epoxy-base construction material used in SLA is not directly amenable to investment casting because of thermal expansion of the polymer during the baking of the ceramic shell. To solve this problem, the pattern can be built as a thin-shelled honeycomb structure that will burn out without breaking the ceramic mold. This allows parts to be produced from any metal that can be investment cast.

It is also possible to directly manufacture prototype cavity inserts for injection molding that can be used to produce multiple parts from a select group of injection-moldable polymers and wax (Ref 3). The injection-molded wax patterns can be used to produce multiple metal parts through investment casting. There are restrictions on the geometries that are available using this technique. The prototype part as mold or as pattern can also be used as the starting point for traditional prototyping techniques such as RTV molding.

Solid Ground Curing (SGC). This process involves the curing of each liquid photopolymer layer in a single step using a two-dimensional photomask. The build process results in a polymer part embedded in a solid wax support structure. The surface of each layer is machined to a known thickness, increasing the accuracy of the process. The process steps can be described by the steps shown in Fig. 7.

First, a thin layer of liquid photopolymer is deposited on a flat substrate. A photomask is generated that matches the cross section of the bottom layer of the part, and the photopolymer is exposed through the mask. Uncured polymer is removed, and a layer of wax is deposited, filling in around the cured cross section. The wax is cooled to a solid, and the surface of the layer is machined flat. This process repeats until the complete prototype has been built. Because of the complexity of the process, SGC is one of the more expensive RP processes.

Postprocessing for SGC involves dissolving away the wax supports and any sanding or polishing that might be desired. A postcuring step is not generally needed using SGC. Each layer is cured against a background of solid polymer and wax. Thus, there is no problem with overexposure curing deeper and increasing the layer thickness.

Solid ground curing has several unique advantages. Because the part is completely supported in solid wax, it is not necessary for the software to design support structures. This also implies that there will be no blemishes to remove from the part surface during postfinishing. The solid wax support allows parts to be nested within the build volume, building parts within the hollow regions of other parts, parts on top of parts, and mechanisms consisting of multiple parts to be built as assembled units. While a single part is typically built more rapidly by SLA, SGC can often exceed SLA in total throughput by building many parts simultaneously. This feature is a distinct advantage to service bureaus that typically build multiple parts in a single run. The wax support and machined layers tend to result in accurate parts with little tendency to warp or deform. The wax can be identified as "part" instead of support structure and can be used for investment casting.

Selective laser sintering (SLS) is similar in principle to SLA in that a laser is used to solidify sequential cross sections from thin layers to create a prototype. In SLS the liquid photopolymer resin from SLA is replaced with a fine thermoplastic powder. Following the schematic shown in Fig. 8, the construction platform consists of three cylindrical tanks with pistons mounted to a planar surface. The two outside cylinders supply powder to the surface, and the central cylinder contains the build volume. The build volume piston translates downward a distance equal to one layer thickness, and pistons in the outside cylinders translate upward supplying powder to the construction plane. A roller spreads the powder into a flat layer covering the construction volume. A laser traces the periphery and cross-hatches the solid area of a part cross section on the powder layer sintering the powder into a solid layer. The process repeats until all layers have been created, then the build piston translates upward and the part is removed. To minimize warpage, the entire process is placed in a heated chamber and heated near the melting temperature of the powder. In order to keep the powder chemically stable at these temperatures, the chamber is purged with an inert gas.

Selective laser sintering parts have a surface texture that is somewhat coarse with the surface finish dependent on the particle size of the powder as well as the layer thickness. The parts tend to be somewhat porous. The most common construction materials are nylon and ABS powders; however, the process can theoretically be used with any powder material that can be sintered, which gives SLS a distinct advantage in the range of construction materials.

A process has been recently developed that uses SLS as an initial process in the production of metal injection-molding inserts. This process begins with a steel powder that is coated with a layer of polymer. Using SLS, a "green" part is fabricated by melting the polymer coating on the powder. The green part is placed in an oven with a copper blank. When the temperature is elevated, the polymer burns out of the powder, and the copper melts and infiltrates the powder through capillary action. The result is a composite, steel/copper insert that can be used for low to medium production runs of injection-molded plastic parts. A similar process is under investigation that produces electrical discharge machining (EDM) dies from a composite of ceramic powder and copper (Ref 4).

Extrusion/Droplet Deposition Processes

Fused Deposition Modeling (FDM). This process uses an extruder head attached to a three-axis motion device to generate the part from sequential two-dimensional cross sections (see Fig. 9). Polymer leaves the tip of the extruder at just above the melting temperature of the polymer and rapidly solidifies when it strikes the support substrate or the previous layer. The extruder head acts essentially as a plotter pen, drawing the outline of each cross section and then filling the solid interior. The temperature, plotting speed, and extrusion rate are controlled so that the extruded

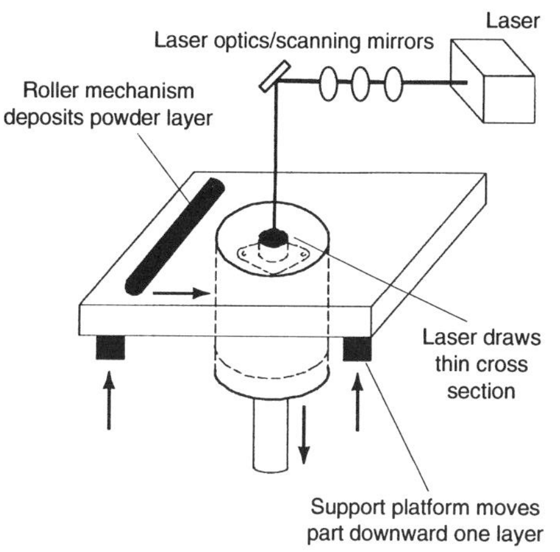

Fig. 8 SLS process. A laser sinters powdered materials, creating the part from thin layers.

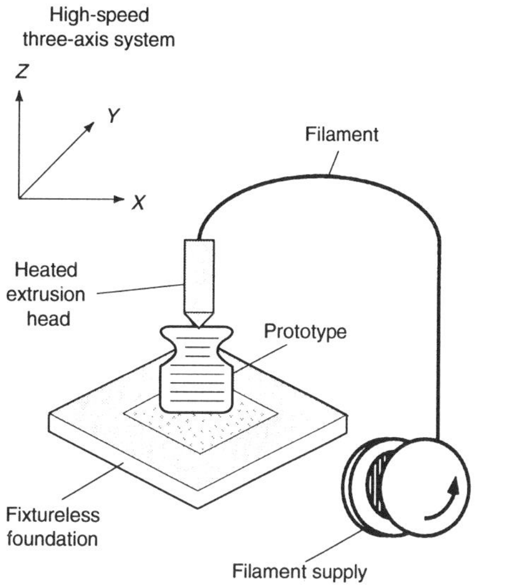

Fig. 9 The FDM process. An extruder head positioned by a three-axis motion controller creates a part from lines of melted polymer.

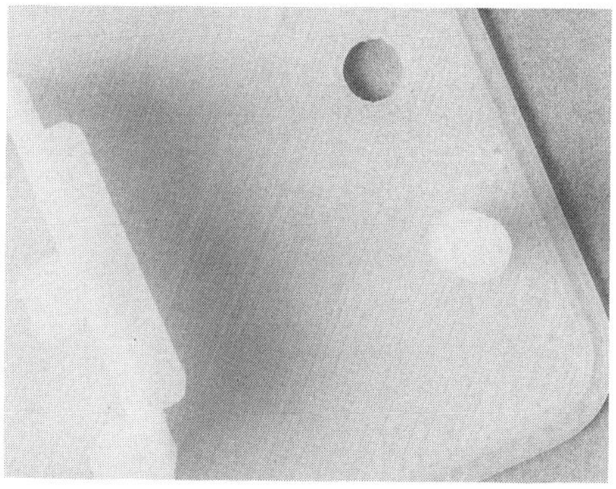

Fig. 10 Porous texture of parts made using FDM caused by adjacent lines of polymer that do not fuse together perfectly

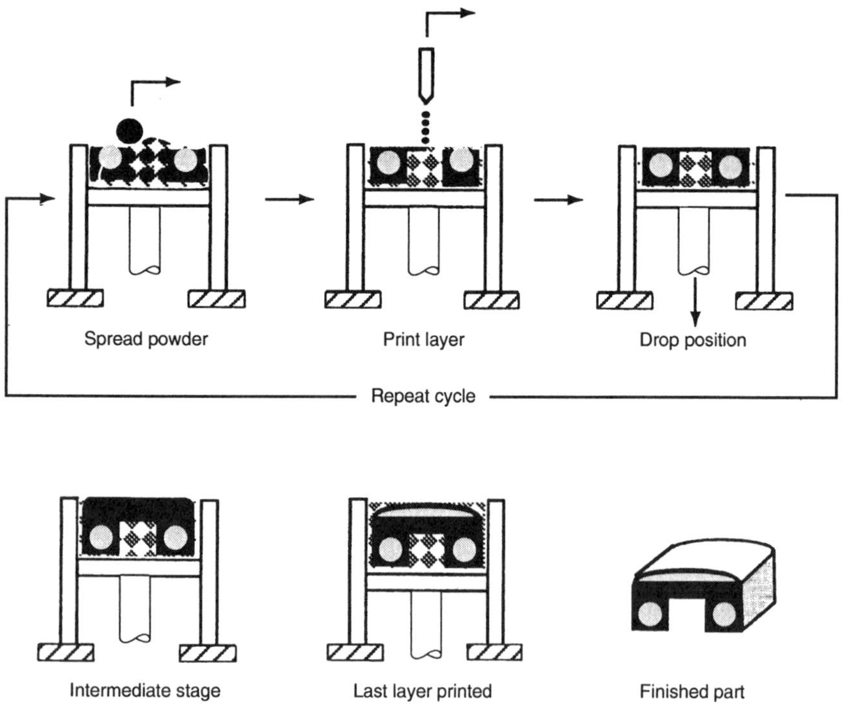

Fig. 11 Three-dimensional printing process solidifying cross sections in layers of powder by depositing a liquid binder

line bonds to the layer below and to some extent to adjacent lines. The direction of the hatching that fills the interior solid area of each layer is varied from layer to layer, which increases the rigidity of the part. The line of extruded polymer is approximately 0.002 to 0.030 in. thick and 0.010 to 0.125 in. wide allowing wall thickness as small as 0.01 in.

The line of extruded polymer requires a base on which to rest, and thus support structures are necessary when cantilever features are created. The supports are created from the polymer construction material and separated from the part with a thin layer of a material with low mechanical properties. This allows the supports to be removed easily. A typical part produced using FDM is shown in Fig. 3(c). Figure 10 shows the surface texture produced using this process. The part is somewhat porous because the lines do not completely fuse to adjacent lines. This effect can be varied somewhat by varying the line spacing.

The FDM process requires very little postprocessing. It is clean, with no chemical fumes or solvents required. The device is relatively small and would be acceptable in an office environment. It is most efficient for thin-walled parts due to the voxel sequential construction process. It is relatively difficult to produce a smooth glossy surface finish even with postprocessing.

Droplet Deposition Processes. Several processes are available that create parts through the deposition of droplets. Two of these processes utilize an inkjet printing head to generate part geometry: one by depositing a liquid binder to solidify a powder and a second that deposits polymer droplets that solidify to form the part. A third process uses a droplet deposition device attached to a five-axis motion controller that allows droplets to be deposited at angles other than vertical.

The three-dimensional printing process developed at the Massachusetts Institute of Technology deposits droplets of a liquid binder material onto thin layers of ceramic powder producing a green ceramic part that is later fired to full strength in a kiln (Fig. 11). The ceramic part is typically used as a casting mold to produce metal parts. Machines using this technique are not commercially available, but investment-cast parts in a variety of castable metals made by this process are available from a single-service bureau.

When the RP processes discussed in this article are used to produce models or molds that are traditionally produced by machining, the models produced must often be considered form prototypes because the material properties and tolerances achieved do not match the requirements for the manufactured parts. However, when using the three-dimensional printing process as a replacement for traditional investment casting, parts can be produced that meet or exceed the material properties and tolerances of the traditional process. Thus, this process is often used for functional prototypes or short-run manufacturing.

Inkjet Droplet Deposition (Polymer). Another class of droplet deposition devices deposit droplets of molten polymer and/or wax. The liquid droplets solidify when they strike the substrate or the previous layer. Two commercial machines are available using this process. The first device uses a relatively large deposition head and creates parts as much as three to five times faster than many competing processes. The accuracy of these models is somewhat reduced.

The second device uses two inkjet heads to deposit droplets of polymer and wax to create a part (Fig. 12). Although it is usually used as support structure, the wax can be used for investment-casting patterns. The commercial device based on this process mills the surface of each layer after the layer is deposited, producing layers as thin as 0.0004 in. thick. Because of the extremely thin layers, parts can be produced with a surface finish that is an order of magnitude smoother than the competing processes. An undesirable result of the thin layers is that the machine is also quite slow. The most appropriate use for the machine is for small models that require excellent surface finish such as jewelry.

Ballistic Particle Manufacturing (BPM). This process is quite similar to polymer droplet deposition. The difference is that BPM allows five-axis control of the droplet deposition head, which allows certain cantilever features to be grown horizontally from the part (eliminating the need for support structures). The commercial BPM device is one of the lowest cost RP devices and uses inexpensive construction materials. It is considered an "office modeler" because the machine is relatively small, quiet, and odor free.

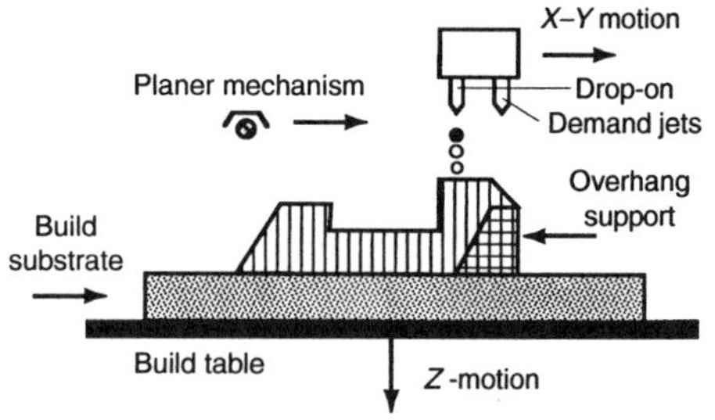

Fig. 12 The inkjet droplet deposition process. Parts are

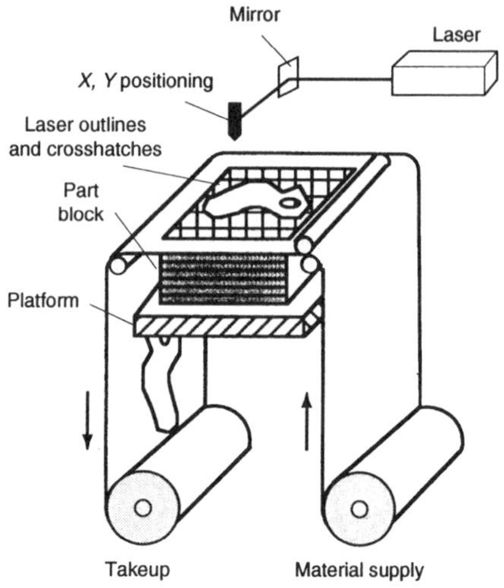

Fig. 13 The laminated object manufacturing process. Parts are built from layers of sheet material (usually paper).

Sheet Form Fabricators

These processes produce parts from thin layers of sheets of construction material. The cross sec-

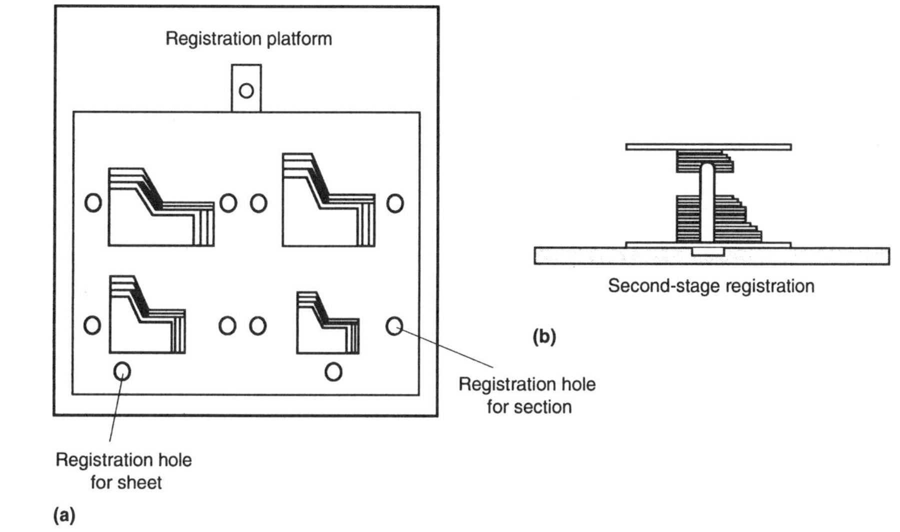

Fig. 14 The two stages of the "cut-then-stack" process. (a) Four subparts are built in parallel. (b) Subparts are stacked to complete the part.

tions are cut from the sheet, stacked, and bonded together to produce the prototypes. There is some variation in the methods used to cut the cross sections from the sheets and in the sequence of operations that gives each process unique advantages and disadvantages.

Stack-Then-Cut (Laminated Object Manufacturing). The laminated object manufacturing (LOM) process follows the schematic shown in Fig. 13. The construction sheet is a roll of paper that has a polyethylene coating on one surface. The paper is unrolled across a flat support, and a heated roller rolls across the build volume, melting the polyethylene and bonding the sheet to the support or to the previous sheet. Next, a laser scans the periphery of the part. The power and speed are adjusted so that the laser burns through only the surface sheet. The laser also cuts a square that defines the edges of the build volume and cuts parting lines and crosshatches the waste material to form small removable cubes. This process is repeated until all layers have been completed, resulting in a rectangular block of construction material with the part hidden inside. Postprocessing consists of manually breaking away the waste material to reveal the part inside and sanding or coating if desired. The finished parts have a texture and consistency similar to wood.

When using LOM, it is necessary to scan only the periphery instead of the entire solid area of each cross section. As a result, this process is most competitive for larger parts with large solid volumes. The build volume of the largest commercial LOM machine available in the United States is 32 by 22 by 20 in. Small features and thin-walled structures must be handled carefully during decubing to avoid breakage. Hollow structures are often built in two pieces so that the interior volume can be decubed and then manually joined during postprocessing. These parts have been used successfully as sand-casting patterns, vacuum-forming patterns, and investment-casting patterns (the paper burns out of the ceramic shell during firing).

Cut-Then-Stack Process. An alternative to the "stack-then-cut" process has been introduced in an inexpensive, partially manual sheet-form prototyping process. The construction sheets consist of an adhesive-backed material attached to a backing sheet. Part cross sections are cut through the adhesive-backed materials—but not through the backing—using a plotter with a knife instead of a pen. Registration and bonding process proceeds manually with an operator registering the sheets on a build platform using registration holes cut through both layers of the construction sheets.

The process uses a parallel decomposition technique that automatically slices small parts into subparts and positions the subparts to cover the entire construction sheet. This is shown in Fig. 14. Using this decomposition a 100-layer part can often be decomposed into ten subparts requiring ten sheets of construction material. The operator performs ten operations registering and bonding the ten sheets resulting in a backing sheet that contains ten, ten-layer thick subparts. The subparts are then stacked, requiring ten more operations.

Common construction materials are 0.1 mm (0.004 in.) thick paper and 1 mm (0.04 in.) thick polystyrene foam. An example paper part is shown in Fig. 3(a). This car, requiring more than 300 layers, was built in two pieces using 155 sheets of construction material. Parts constructed from the foam material can be used as patterns for investment casting. A typical cast part is shown in Fig. 3(d).

The combination of manual registration and bonding, parallel decomposition, and "cut-then-stack" processing results in a unique combination of advantages and disadvantages:

- The machine is an order of magnitude less expensive than most commercial processes.
- The tolerance capability is operator dependent and generally an order of magnitude lower than competing processes.
- Parts that take advantage of the parallel decomposition can often be built four to five times faster than competing processes.

Numerical Control Machining for Prototyping

As techniques for numerical control (NC) machining have improved, it has become possible to create complex parts quite rapidly by machining. Modern CAD tools can interpret the geometric features of a solid model and generate the tool paths necessary to machine the part on an NC machining center. The tool path generation process requires some assistance from a skilled machinist. The machinist must develop a manufacturing plan that may require manual repositioning of the part one or more times during the machining process and may require the use of more than one machine. The machinist is also required for the actual machining.

Numerical control as a prototyping technique has its own advantages and disadvantages. A broad range of construction materials can be used. Once the manufacturing plan is completed and the first part is produced, multiple copies of the part can often be produced very rapidly. For example, the part shown in Fig. 15 was created on a five-axis machining center. The manufacturing plan and initial part required 16 h to develop. Production of subsequent parts required approximately 20 min per part. An added advantage of NC as a prototyping technique is that once the prototype is completed, the designer already has a manufacturing option in place. Numerical control machining techniques are capable of producing very tight tolerances.

Along with these advantages come disadvantages: The level of operator skill required is much higher than for the other techniques discussed here. The range of geometries that can be created using NC is more limited. The ability to create a wide variety of geometries requires a broad range of tools and several machines. Many parts require one or more fixtures to hold the part during machining. This fixture must be designed and built along with the part, requiring added costs and time delay.

Designing with Manufacturable Features. It is interesting to note that the increased flexibility of RP processes over NC machining can be considered both an advantage and a disadvantage. Rapid prototyping techniques put very few limits on the imagination of the part designer. Almost any geometry that can be modeled can be prototyped. However, the designer is generally developing a product that must be manufactured using traditional manufacturing techniques. Using NC as a prototyping tool forces the designer to produce prototypes made up of manufacturable features. Designing with prototypes built using RP techniques can cause the designer to produce a

Fig. 15 Sensor casing for a head position tracking sensor machined from aluminum using an NC machining center. The first copy required 12 h to produce. Duplicates would require 15 min each to produce. Courtesy of S. Drake, University of Utah

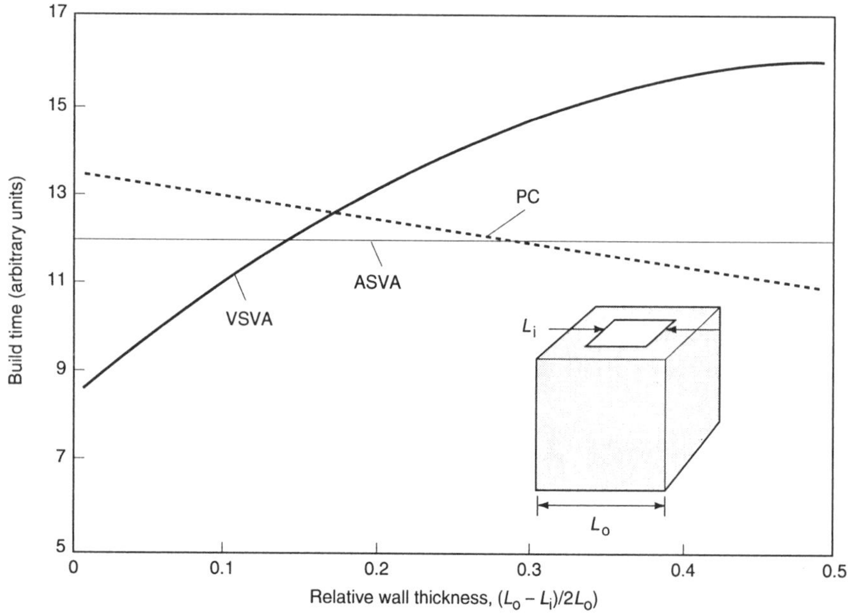

Fig. 16 Variation of build time with wall thickness for three classes of RP processes: voxel sequential volume addition (VSVA), area sequential volume addition (ASVA), and periphery cutting (PC)

design that is impossible to manufacture with any technique except RP.

Analyzing Speed and Accuracy

For each of the processes discussed in this article, the technique used for layer creation produces a definable relationship between part geometry and build speed. Three general classes that appear are voxel sequential volume addition, periphery cutting, and area sequential volume addition.

Voxel Sequential Volume Addition (VSVA). Where a pixel represents the smallest area unit in a two-dimensional picture, the term voxel is used here to represent the smallest volume unit that can be created by an RP device. A voxel is a single droplet for a droplet deposition device or the volume extruded by an FDM device as the extruder head moves a single step of the motion controller. The build time for voxel sequential devices is directly proportional to the solid volume of the part. As a result, these processes create thin-walled structures relatively quickly and thick-bodied structures more slowly. Because they deposit (or cure) only the material needed for the part and required supports, there is little waste.

Periphery Cutting (PC). The sheet-form processes discussed previously in this article create cross sections by cutting the periphery of the part and cutting parting lines to turn the waste material into cubes. In this case, the time required to cut a cross section is proportional to the periphery length. A solid object is constructed faster than if it were hollow because it is not necessary to cut the interior surface. Periphery cutting processes are therefore, optimal for larger, heavy-bodied parts with large volume-to-surface-area ratios. Because the material cut away from each cross section is waste, these processes are relatively wasteful and more so for thin-walled structures.

Area Sequential Volume Addition (ASVA). This class includes processes that create each cross section as a two-dimensional area in a single step. The two-dimensional photomask curing used in the SGC process is an example. For these processes, the build time for the part is only a function of the number of layers in the part. The build times for thin-walled structures and heavy-bodied structures are identical if the vertical height is the same.

These concepts are demonstrated in Fig. 16, in which a generic part is represented by a cube whose wall thickness varies from very thin to a solid cube. The build times for each of the three classes are represented by the equation below:

$$T_B = \sum_{i=1}^{n} (T_R + T_{LCi}) \quad \text{(Eq 1)}$$

where T_B is the time required to build the part, T_R is the time required to produce a new layer of liquid or powder or to apply a new sheet, and T_{LCi} is the time required to create a specific cross section for a part consisting of n layers. For each process T is assumed constant. T for each process is shown below:

For VSVA: $T_{LC} \propto L_o^2 - L_i^2$

For PC: $T_{LC} \propto 4L_o + 4L_i$

For ASVA: $T_{LC} = \text{Constant}$

where L_i and L_o are the inside and outside dimension, respectively, of the cube as shown in Fig. 16.

As the wall thickness increases, the build time for a VSVA process increases while the time actually decreases for the PC process. The build time for the ASVA is unchanged because in this example, the vertical height of the part is unchanged. The exact shape of the curves in Fig. 16 result from the coefficients selected for the equations, the part shape, and the method chosen for varying the geometry. While different assumptions and conditions will change these shapes, the general trends should be preserved.

Current Research Efforts

The discussion so far has been limited to techniques that are available commercially. There are numerous other techniques that are current topics for research. These topics fall into three general classes:

- Improving construction accuracy
- Developing new construction techniques
- Prototyping new materials and complex structures (composites, functionally gradient materials, creating microstructure, embedding components, etc.)

Improving Construction Accuracy. Research in this area proceeds for each of the commercial devices. The work generally revolves around characterizing the behavior of the construction materials during processing (Ref 5, 6), improving the operating characteristics of the machine (Ref 7), or reducing material shrinkage and warp during processing (Ref 8).

Developing New Construction Techniques. The research here is not on processes that would compete with existing processes, but processes that will fill in areas that are not currently represented. A research tool developed for prototyping large objects builds parts from thick sheets of polystyrene foam. This sheet-form process cuts custom-ruled surfaces on the edges of the cross sections, allowing a reasonable representation of the object geometry to be produced from very thick layers. The technique has been used to create models up to 20 ft in length and to produce foam molds for the production of large composite structures (Ref 9). A new technique termed *selective area laser deposition* (SALD) uses laser energy to selectively induce chemical vapor deposition. This process is capable of creating three-dimensional geometric features on the micron scale (Ref 10, 11). An automated version of "cut-then-stack" sheet-form prototyping is under development that will allow layers of a part to be created simultaneously by many machines in parallel, then collated, stacked, and bonded. This has the potential to dramatically increase the speed of prototype production (Ref 12).

Prototyping with New Materials and Complex Structures. Researchers are investigating techniques for directly prototyping fiber-reinforced composites (Ref 13). The three-dimensional printing technique is being used to directly prototype materials with custom microstructures (Ref 14). This allows the potential for creating functionally gradient materials, with properties that vary with position in the prototype. Shape deposition manufacturing allows the construction of prototypes using multiple materials and embedded components (Ref 15). The manual "cut-then-stack" technique is being used to create prototypes that are operational devices by embedding functional components such as motors, switches, and wiring during the stacking process (Ref 12).

ACKNOWLEDGMENTS

Many of the figures were prepared by the students of the Manufacturing Processes Laboratory at the University of Utah: Don Brock, Jen-Ping Mu, Cheol Lee, Andrei Novac, Ravi Vellanki, and Zetian Wang.

REFERENCES

1. M. Burns, *Automated Fabrication: Improving Productivity in Manufacturing,* PTR Prentice Hall, 1993
2. C.L. Thomas, *An Introduction to Rapid Prototyping,* Schroff Development Corp., 1995
3. R. Ponder, personal communication, Department of Mechanical Engineering, Georgia Institute of Technology, Aug 1996
4. K. McAlea and U. Hejmadi, Selective Laser Sintering of Metal Molds: The RapidTool™ Process, *Proc. Solid Free Form Fabrication Symposium,* University of Texas at Austin, 1996, p 97–104
5. J.S. Ullett, S. Rodrigues, and R.P. Chartoff, Characterization of Shrinkage and Stress Build-up during Laser Photopolymerization, *Sixth Int. Conf. Rapid Prototyping,* University of Dayton, 1995, p 57–68
6. T.S. Guess, R.S. Chambers, T.D. Hinnerichs, G.D. McCarty, and R.N. Shagam, Epoxy and Acrylate Stereolithography Resins: In-situ Measurement of Cure Shrinkage and Stress Relaxation, *Sixth Int. Conf. Rapid Prototyping,* University of Dayton, 1995, p 69–80
7. D.C.H. Yang, Y. Juo, T. Kong, J.J. Chaung, and G. Nisnevich, Laser Beam Diameter Compensation for Helysis LOM Machine, *Sixth Int. Conf. Rapid Prototyping,* University of Dayton, 1995, p 171–178
8. J.S. Ullett, R.P. Chartoff, J.W. Schultz, J.C. Bhatt, M. Dotrong, and R.T. Pogue, Low Shrinkage, High T_g Liquid Crystal Resins for Stereolithography, *Proc. Solid Free Form Fabrication Symposium,* University of Texas at Austin, 1996, p 471–480
9. C.L. Thomas, T. Gaffney, S. Kaza, and C. Lee, Rapid Prototyping of Large-Scale Aerospace Structures, *Aerospace Applications Conf. Proc.,* Vol 4, Institute of Electronics and Electronics Engineering, 1996, p 219–230
10. J. Pegna and J.L. Maxwell, The Horton Project: Laser Induced Selective 3 Dimensional Material Deposition for Micromachining and RP of Functional Micro-systems, *Sixth Int. Conf. Rapid Prototyping,* University of Dayton, 1995, p 1–6
11. S. Harrison, J. Crocker, T. Manzur, and H. Marcus, Solid Free Form Fabrication at the University of Connecticut, *Proc. Solid Free Form Fabrication Symposium,* University of Texas at Austin, 1996, p 345–348
12. C.L. Thomas and K. Hayworth, Automating Sheet Based Fabrication, *Proc. Solid Free Form Fabrication Symposium,* University of Texas at Austin, 1996, p 281–289
13. D. Klosterman, R. Chartoff, B. Priore, N. Osborne, G. Graves, A. Lightman, and S. Pak, Structural Composites Via Laminated Object Manufacturing, *Proc. Solid Free Form Fabrication Symposium,* University of Texas at Austin, 1996, p 105–116
14. M.J. Cima, J. Yoo, W. Bae, K. Cho, S. Suresh, and E. Sachs, "Structural Ceramic Components with Computer Derived Microstructure by Three Dimensional Printing," *Solid Free Form Fabrication Symposium* (Austin, TX), University of Texas at Austin, 12–14 August 1996
15. L. Wiess, F. Prinz, G. Neplotnik, P. Padmanabhan, L. Schultz, and R. Merz, Shape Deposition Manufacturing of Wearable Computers, *Proc. Solid Free Form Fabrication Symposium,* University of Texas at Austin, 1996, p 31–38

Section 4: The Materials Selection Process

Overview of the Materials Selection Process

George E. Dieter, University of Maryland

THE SELECTION OF THE CORRECT MATERIAL for a design is a key step in the process because it is the crucial decision that links computer calculations and lines on an engineering drawing with a working design. Materials and the manufacturing processes that convert the material into a useful part underpin all of engineering design. The enormity of the decision task in materials selection is given by the fact that there are well over 100,000 engineering materials from which to choose. On a more practical level, the typical design engineer should have ready access to information on 50 to 80 materials, depending on the range of applications.

The importance of materials selection in design has increased in recent years. The adoption of concurrent engineering methods (see the article "Concurrent Engineering" in this Volume) has brought materials engineers into the design process at an earlier stage, and the importance given to manufacturing in present day product design has reinforced the fact that materials and manufacturing are closely linked in determining final properties. Moreover, world pressures of competitiveness have increased the general level of automation in manufacturing to the point where materials costs comprise 50% or more of the cost for most products. Finally, the great activity in materials science worldwide has created a variety of new materials and focused attention on the competition between six broad classes of materials: metals, polymers, elastomers, ceramics, glasses, and composites. Thus the range of materials available to the engineer is much larger than ever before. This presents the opportunity for innovation in design by utilizing these materials in products that provide greater performance at lower cost. To achieve this requires a more rational process for materials selection.

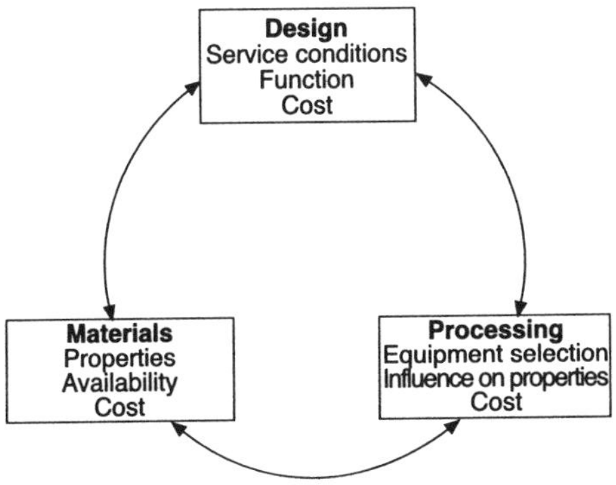

Fig. 1 Interrelationship among design, materials, and processing

Relation of Materials Selection to Design

An incorrectly chosen material can lead not only to failure of the part but also to unnecessary cost. Selecting the best material for a part involves more than selecting a material that has the properties to provide the necessary performance in service; it is also intimately connected with the processing of the material into the finished part (Fig. 1). A poorly chosen material can add to manufacturing cost and unnecessarily increase the cost of the part. Also, the properties of the material can be changed by processing (beneficially or detrimentally), and that may affect the service performance of the part.

With the enormous combination of materials and processes to choose from, the task can be done only by introducing simplification and systemization. Design proceeds from concept design, to embodiment (configuration) design, to detail (parametric) design, and the material and process selection then becomes more detailed as the design progresses through this sequence. The steps in the design process are discussed in the Section "The Design Process" in this Volume. Figure 2 contrasts the design methods and tools

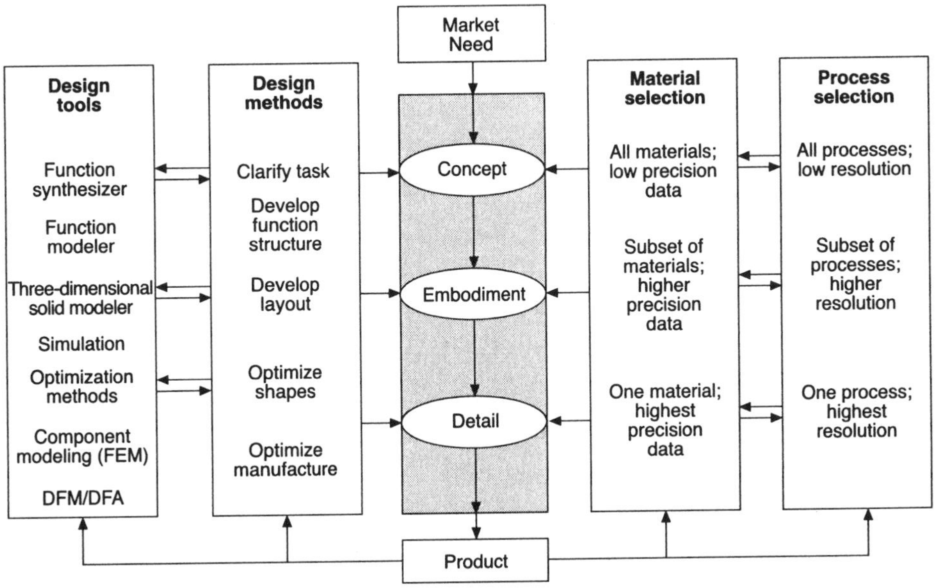

Fig. 2 Schematic of the design process with design tools on the left and materials and process selection on the right. At the concept stage of design, the emphasis is on breadth; in the later stages, it is on precision. FEM, finite element modeling; DFM, design for manufacturing; DFA, design for assembly. Source: Ref 1

used at each stage with the materials and processes selection.

At the concept level of design, essentially all materials and processes are considered rather broadly. The materials selection methodology and charts developed by Ashby (Ref 2) are highly appropriate at this stage (see the articles "Material Property Charts" and "Performance Indices" in this Volume). The decision is to determine whether each design concept will be made from metal, plastics, ceramic, composite, or wood, and to narrow it to a group of materials. The precision of property data needed is rather low. If an innovative choice of material is to be made, it must be done at the conceptual design step because later in the design process too many decisions have been made to allow for a radical change.

At the embodiment or configuration level of design, the emphasis is on determining the shape and approximate size of a part using engineering methods of analysis. Now the designer will have decided on a class of materials and processes, for example a range of aluminum alloys, wrought and cast. The material properties must be known to a greater level of precision.

At the detail or parametric design level, the decision will have narrowed to a single material and only a few manufacturing processes. Here the emphasis will be on deciding on critical tolerances, optimizing for robust design (see the article "Robust Design" in this Volume), and selecting the best manufacturing process using quality engineering and cost modeling methodologies. Depending on the criticality of the part, materials properties may need to be known to a high level of precision. At the extreme, this requires the development of a detailed data base from an extensive materials testing program.

In a more detailed approach to engineering design, Dixon and Poli (Ref 3) suggest a four-level approach to materials selection:

- *Level I.* Based on critical properties, determine whether the part will be made from metal, plastic, ceramic, or composite.
- *Level II.* Determine whether metal parts will be produced by a deformation process (wrought) or a casting process; for plastics, determine whether they will be thermoplastic or thermosetting polymers.
- *Level III.* Narrow options to a broad category of material. Metals can be subdivided into categories such as carbon steel, stainless steel, and copper alloys. Plastics can be subdivided into specific classes of thermoplastics and thermosets such as polycarbonates and polyesters.
- *Level IV.* Select a specific material according to a specific grade or specification.

Thus material and process selection is a progressive process of narrowing from a large universe of possibilities to a specific material and process selection; see the next section "The Process of Materials Selection" in this article. Levels I and II often may suffice for conceptual design. Level III is needed for embodiment (configuration) design and sometimes for conceptual design. Level IV usually can be postponed until detail (parametric) design.

The Process of Materials Selection

A materials selection problem usually involves one of two situations:

- Selection of the materials and the processes for a new product or design
- The evaluation of alternative materials or manufacturing routes for an existing product or design. Such a redesign effort usually is taken to reduce cost, increase reliability, or improve performance. It generally is not possible to realize the full potential of substituting one material for another without fully considering its manufacturing characteristics. In other words, simple substitution of a new material without changing the design rarely provides optimum utilization of the material.

Materials Selection for a New Design. In this situation, these steps must be followed:

1. Define the functions that the design must perform, and translate these into required materials properties such as stiffness, strength, and corrosion resistance, and such business factors as the cost and availability of the material.
2. Define the manufacturing requirements in terms of such parameters as the number of parts required, the size and complexity of the part, its required tolerance and surface finish, general quality level, and overall fabricability of the material.
3. Compare the needed properties and parameters with a large materials property data base (most likely computerized) to select a few materials that look promising for the application. In this initial screening process, it is helpful to establish several screening properties. A screening property is any material property for which an absolute lower (or upper) limit can be established. No trade-off beyond this limit is allowable. It is a go-no go situation. The idea of the screening phase in materials selection is to ask the question: "Should this material be evaluated further for this application?"
4. Investigate the candidate materials in more detail, particularly in terms of trade-offs in product performance, cost, fabricability, and availability in the grades and sizes needed for the application. Material property tests and other testing often is done at this stage. Methods for making this detailed evaluation are discussed in the articles "Performance Indices" and "Use of Decision Matrices in Materials Selection" in this Volume.
5. Develop design data and/or a design specification. Step 4 results in the selection of a single material for the design and a suggested process for manufacturing the part. In most cases, this results in establishing the minimum properties through defining the material with a generic material standard such as those issued by the American Society for Testing and Materials (ASTM), the Society of Automotive Engineers (SAE), the American National Standards Institute (ANSI), and the United States military (MIL specs). For critical parts in sensitive applications, for example in aerospace and nuclear areas, it may be necessary to conduct an extensive testing program to develop design data that are statistically reliable (see the article "Statistical Aspects of Design" in this Volume).

Materials Substitution for an Existing Design. In this situation, the following steps pertain:

1. Characterize the currently used material in terms of performance, manufacturing requirements, and cost.
2. Determine which characteristics must be improved for enhanced product function. Often failure analysis reports play a critical role in this step (see the article "Use of Failure Analysis in Materials Selection" in this Volume).
3. Search for alternative materials and/or manufacturing routes. Use the idea of screening properties to good advantage.
4. Compile a short list of materials and processing routes, and use these to estimate the costs of manufactured parts. A method of engineering analysis called *value engineering* has proven useful for this purpose (see the article "Use of Value Analysis in Materials Selection" in this Volume). Value engineering is a problem-solving methodology that focuses on identifying the key function(s) of a design so that unnecessary costs can be removed without compromising the quality of the design.
5. Evaluate the results in step 4, and make a recommendation for a replacement material. Define the critical properties with specifications or testing as in step 5 of the previous section.

There are two approaches (Ref 3) to determining the material-process combination for a part. In the *material first approach,* the designer begins by selecting a material class and narrowing it down as described above. Then manufacturing processes consistent with the selected material are considered and evaluated. Chief among the factors to consider are production volume and information about the size, shape, and complexity of the part. With the *process first approach,* the designer begins by selecting the manufacturing process, guided by the same factors. Then materials consistent with the selected process are considered and evaluated, guided by the performance requirements of the part. Both approaches end up at the same decision point. Most design engineers and materials engineers instinctively use the materials first approach since it is the method taught in strength of materials and machine design courses. Manufacturing engineers and those heavily involved with process engineering gravitate toward the other approach. No studies have been done to determine which leads to the best results.

Performance Characteristics of Materials

The performance or functional characteristics of a material are expressed chiefly by physical, mechanical, thermal, electrical, magnetic, and optical properties. Material properties are the link between the basic structure and composition of the material and the service performance of the part (Fig. 3). The goal of materials science is to learn how to control the various levels of structure of a material (electronic structure, defect structure, microstructure, macrostructure) so as to predict and improve the properties of a material. Not too long ago metals dominated most of engineering design. Today the range of materials and properties available to the engineer is much larger and growing rapidly. This requires familiarity with a broader range of materials and properties, but it also introduces new opportunities for innovation in product development. Table 1 provides a general comparison of the properties of metals, ceramics, and polymers. A detailed review of fundamental structure-property relationships in engineering materials, and a consideration of the effects of composition, processing, and structure on the properties of steels, nonferrous metallic alloys, ceramics and glasses, engineering plastics, and composites is provided in the Section "Effects of Composition, Processing, and Structure on Materials Properties" in this Volume. Table 2 provides a listing of the broad spectrum of material properties that may be needed.

An important role of the materials engineer is to assist the designer in making meaningful connections between materials properties and the performance of the part or system being designed. For most mechanical systems, performance is limited, not by a single property, but by a combination of them. For example, the materials with the best thermal shock resistance are those with the largest values of $\sigma_f/E\alpha$, where σ_f is the failure stress, E is Young's modulus, and α is the thermal coefficient of expansion. Ashby (Ref 2) showed how to derive these *performance indices* (groupings of material properties that, when maximized, maximize some aspect of performance) and how to use them in conjunction with his materials selection charts (see the article "Performance Indices" in this Volume). Table 3 shows the relationships between standard mechanical properties and the failure modes for materials (Ref 5). For most modes of failure, two or more material properties act to control the material behavior. Also, it must be kept in mind that the service conditions met by materials are in general more complex than the test conditions used to measure material properties. Usually simulated service tests must be devised to screen materials for critical complex service conditions. Finally, the chosen material, or a small group of candidate materials, must be evaluated in prototype tests or field tests to determine their performance under actual service conditions.

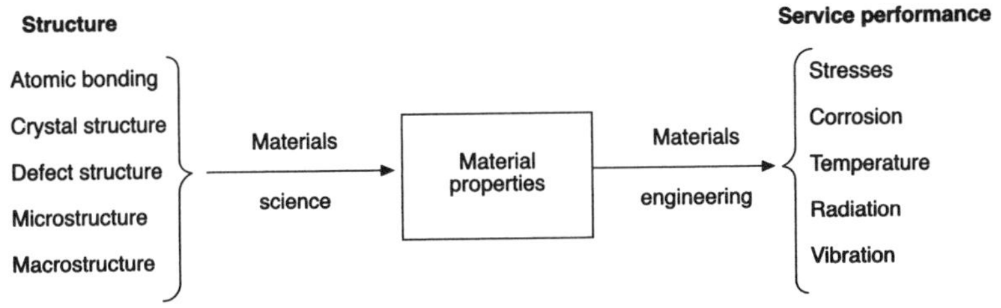

Fig. 3 The role played by material properties in the selection of materials

Table 1 General comparison of properties of metals, ceramics, and polymers

Property (approximate values)	Metals	Ceramics	Polymers
Density, g/cm^3	2 to 22 (average ~ 8)	2 to 19 (average ~ 4)	1 to 2
Melting points	Low (Ga = 29.78 °C, or 85.6 °F) to high (W = 3410 °C, or 6170 °F)	High (up to 4000 °C, or 7230 °F)	Low
Hardness	Medium	High	Low
Machinability	Good	Poor	Good
Tensile strength, MPa (ksi)	Up to 2500 (360)	Up to 400 (58)	Up to 140 (20)
Compressive strength, MPa (ksi)	Up to 2500 (360)	Up to 5000 (725)	Up to 350 (50)
Young's modulus, GPa (10^6 psi)	15 to 400 (2 to 58)	150 to 450 (22 to 65)	0.001 to 10 (0.00015 to 1.45)
High-temperature creep resistance	Poor to medium	Excellent	...
Thermal expansion	Medium to high	Low to medium	Very high
Thermal conductivity	Medium to high	Medium, but often decreases rapidly with temperature	Very low
Thermal shock resistance	Good	Generally poor	...
Electrical characteristics	Conductors	Insulators	Insulators
Chemical resistance	Low to medium	Excellent	Good
Oxidation resistance	Generally poor	Oxides excellent; SiC and Si_3N_4 good	...

Source: Ref 4

Example 1: Materials Selection for an Automotive Exhaust System (Ref 6). The product design specification for the exhaust system must provide for the following functions:

- Conducting engine exhaust gases away from the engine
- Preventing noxious fumes from entering the automobile
- Cooling the exhaust gases
- Reducing the engine noise
- Reducing the exposure of automobile body parts to exhaust gases
- Affecting engine performance as little as possible
- Helping control undesirable exhaust emissions
- Having a service life that is acceptably long
- Having a reasonable cost, both as original equipment and as a replacement part

In its basic form, the exhaust system consists of a series of tubes that collect the gases at the engine and convey them to the rear of the automobile. The size of the tube is determined by the volume of the exhaust gases to be carried away and the extent to which the exhaust system can be permitted to impede the flow of gases from the engine. An additional device, the muffler, is required for noise reduction, and a catalytic converter is required to convert polluting gases to less-harmful emissions. The basic lifetime requirement is that the system must resist the attack of hot, moist exhaust gases for some specified period. In addition, the system must resist attack by the atmosphere, water, mud, and road salt. The location of the exhaust system under the car requires that it be designed as a complex shape that will not interfere with the running gear of the car, road clearance, or the passenger compartment. The large number of automobiles produced each year requires that the material used in exhaust systems be readily available at minimum cost.

This system requires numerous material property requirements. The mechanical property requirements are not overly severe: suitable rigidity to prevent excessive vibration and fatigue plus enough creep resistance to provide adequate service life. Corrosion is the limiting factor on life, especially in the cold end, which includes the resonator, muffler, and tail pipe. Several properties of unique interest, that is, where one or two properties dominate the selection of the material, are found in this system. These pertain to the platinum-base catalyst and the ceramic carrier that supports the catalyst. The majority of the tubes and containers that comprise the exhaust system were for years made of readily formed and welded low-carbon steel, with suitable coatings for corrosion resistance. With the advent of greater emphasis on automotive quality and longer life, the material selection has moved to specially developed stainless steels with improved corrosion and creep properties. Ferritic 11% Cr alloys are used in the cold end components, with 17 to 20% Cr ferritic alloys and austenitic Cr-Ni alloys in the hot end of the system.

Standards and Specifications. Materials properties usually are formalized through standards and specifications. The distinction between these entities is that a standard is intended for use

Table 2 Material performance characteristics

Physical properties	Mechanical properties	Thermal properties
Crystal structure	Hardness	Conductivity
Density	Modulus of elasticity	Specific heat
Melting point	Tension	Coefficient of thermal expansion
Vapor pressure	Compression	Emissivity
Viscosity	Poisson's ratio	Absorptivity
Porosity	Stress-strain curve	Ablation rate
Permeability	Yield strength	Fire resistance
Reflectivity	Tension	
Transparency	Compression	**Chemical properties**
Optical properties	Shear	Position in electromotive series
Dimensional stability	Ultimate strength	Corrosion and degradation
	Tension	Atmospheric
Electrical properties	Shear	Salt water
Conductivity	Bearing	Acids
Dielectric constant	Fatigue properties	Hot gases
Coercive force	Smooth	Ultraviolet
Hysteresis	Notched	Oxidation
	Corrosion fatigue	Thermal stability
Nuclear properties	Rolling contact	Biological stability
Half-life	Fretting	Stress corrosion
Cross section	Charpy transition temperature	Hydrogen embrittlement
Stability	Fracture toughness (K_{Ic})	Hydraulic permeability
	High-temperature behavior	
	Creep	**Fabrication properties**
	Stress rupture	Castability
	Damping properties	Heat treatability
	Wear properties	Hardenability
	Galling	Formability
	Abrasion	Machinability
	Erosion	Weldability
	Cavitation	
	Spalling	
	Ballistic impact	

by as large a body as possible, for example ASTM or ANSI standards, whereas a specification, though dealing with similar technical content, is intended for use by a more limited group, for example a company specification. There are two types of standards or specifications: performance standards and product standards. Performance standards delineate the basic functional requirements of a product and set out the basic parameters from which the design can be developed. Product standards define the conditions under which the components of a design are purchased and manufactured. Materials standards are invariably product standards. They stipulate performance characteristics, quality factors, methods of measurement, tolerances, and dimensions.

Relation of Materials Selection to Manufacturing

The selection of a material must be closely coupled with the selection of a manufacturing process. This is not an easy task for there are many processes that can produce the same part. The goal is to select the material and process that maximizes quality and minimizes the cost of the part. Figure 4 gives a breakdown of manufacturing processes into nine broad classes.

In a very general sense, the selection of the material determines a range of processes that can be used to process parts from the material. Table 4 shows the manufacturing methods used most frequently with different metals and plastics (Ref 7). The material melting point and general level of deformation resistance (hardness) and ductility determine these relationships. The next aspect to consider is the minimum and maximum overall size of the part, often expressed by volume, projected area, or weight. Maximum size often is controlled by equipment considerations. Shape is the next factor to consider. The overall guide should be to select a primary process that makes the part as near to final shape as possible (near-net shape forming) without requiring expensive secondary machining or grinding processes. Sometimes the form of the starting material is important. For example, a hollow shaft can be made best by starting with a tube rather than a solid bar. Shape is often characterized by aspect ratio, the surface-to-volume ratio, or the web thickness-to-depth ratio. Closely related to shape is complexity. Complexity is correlated with lack of symmetry. It also can be measured by the information content of the part, that is, the number of independent dimensions that must be specified to describe the shape. Tolerance is the degree of deviation from ideal that is permitted in the dimensions of a part. Closely related to tolerance is surface finish. Surface finish is measured by the root-mean-square amplitude of the irregularities of the surface. Each manufacturing process has the capability of producing a part with a certain range of tolerance and surface finish (Fig. 5). Polymers are different from metals and ceramics in that they can be processed to a very high surface smoothness, but tight tolerances are seldom possible because of internal stresses left by molding and creep at service temperatures. Manufacturing cost increases exponentially with

Table 3 Relationships between failure modes and material properties

	Material property														
Failure mode	Ultimate tensile strength	Yield strength	Compressive yield strength	Shear yield strength	Fatigue properties	Ductility	Impact energy	Transition temperature	Modulus of elasticity	Creep rate	K_{Ic}(a)	K_{Iscc}(b)	Electrochemical potential	Hardness	Coefficient of expansion
Gross yielding		X		X											
Buckling			X						X						
Creep										X					
Brittle fracture							X	X			X				
Fatigue, low cycle					X	X									
Fatigue, high cycle	X				X										
Contact fatigue			X												
Fretting			X												
Corrosion													X		
Stress-corrosion cracking	X											X	X		
Galvanic corrosion													X		
Hydrogen embrittlement	X												X		
Wear														X	
Thermal fatigue										X					X
Corrosion fatigue					X								X		

An "X" at the intersection of material property and failure mode indicates that a particular material property is influential in controlling a particular failure mode. (a) Plane-strain fracture toughness. (b) Threshold stress intensity to produce stress-corrosion cracking. Source: Ref 5 decreasing dimensional tolerance. Yet another process parameter is surface detail, the smallest radius of curvature at a corner that can be produced. An important practical consideration is the quantity of parts required. For each process, there is a minimum batch size below which it is not economical to go because of costs of tooling, fixtures, and equipment. Also related to part cost is the production rate or the cycle time, the time required to produce one part. The most commonly used manufacturing processes are evaluated with respect to these characteristics in Table 5 (Ref 9).

Ashby (Ref 2) extended his concepts for materials selection for properties at the conceptual design stage to the development of a set of process selection charts that encompass most of the process attributes described above. The approach, as with material selection based on properties, is to search the entire gamut of potential processes to arrive at the small set that should be considered in detail. Figure 6 illustrates the approach. Figure 6(a) shows the relationship between part size and the complexity of the shape. Complexity is expressed in bits of information, and is given by Eq 1:

$$C = n \log_2 \left(\frac{\bar{l}}{\overline{\Delta l}} \right) \qquad \text{(Eq 1)}$$

where C is the part complexity, n is the number of dimensions, $\overline{\Delta l}$ is the geometric mean tolerance, and $\bar{l}$ is the geometric mean dimension.

Simple shapes require only a few bits of information; complex shapes, like integrated circuits, require very many. The casting for an engine block might have 10^3 bits of information, but after machining the complexity increases both by adding new dimensions (n) and improving the precision (reducing $\overline{\Delta l}$). Casting processes can make parts that vary in size from about 1 gram to several hundred tons. Machining adds precision

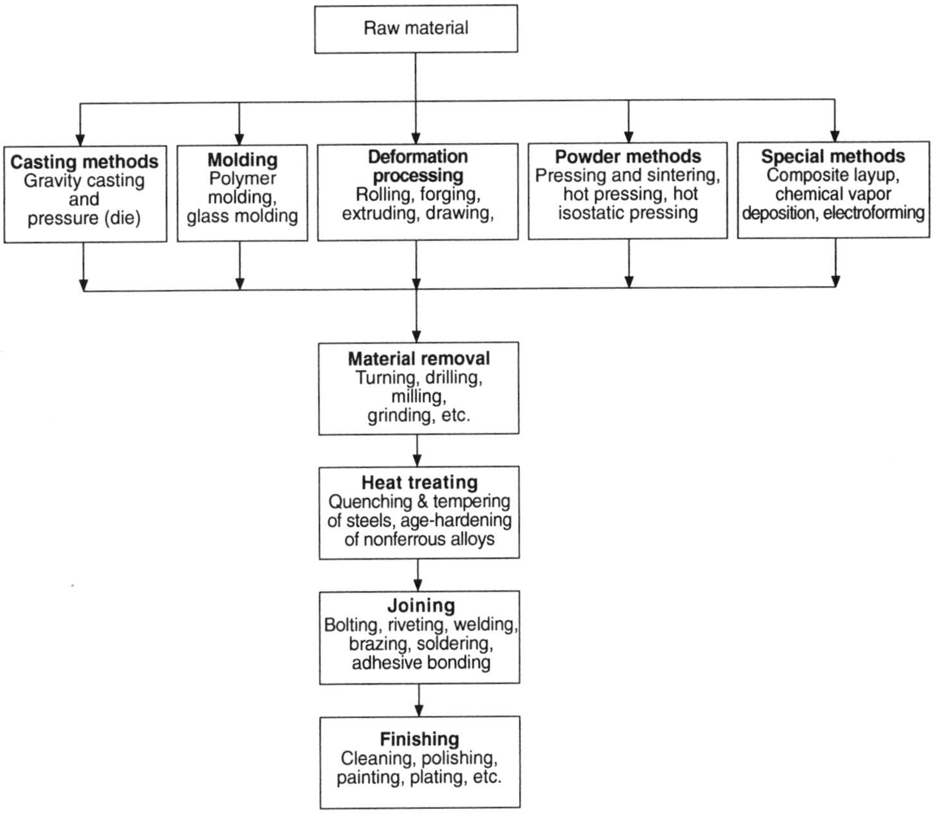

Fig. 4 The nine classes of manufacturing processes. The first row contains the primary forming (shaping) processes. The processes in the vertical column below are the secondary forming and finishing processes. Source: Ref 2

Table 4 Compatibility between materials and manufacturing processes

Process	Cast iron	Carbon steel	Alloy steel	Stainless steel	Aluminum and aluminum alloys	Copper and copper alloys	Zinc and zinc alloys	Magnesium and magnesium alloys	Titanium and titanium alloys	Nickel and nickel alloys	Refractory metals	Thermoplastics	Thermoset plastics
Casting/molding													
Sand casting	•	•	•	•	•	•	—	•	—	•	—	X	X
Investment casting	—	•	•	•	•	•	—	—	—	•	—	X	X
Die casting	X	X	X	X	•	—	•	•	X	X	X	X	X
Injection molding	X	X	X	X	X	X	X	X	X	X	X	•	—
Structural foam molding	X	X	X	X	X	X	X	X	X	X	X	•	X
Blow molding (extrusion)	X	X	X	X	X	X	X	X	X	X	X	•	X
Blow molding (injection)	X	X	X	X	X	X	X	X	X	X	X	•	X
Rotational molding	X	X	X	X	X	X	X	X	X	X	X	•	X
Forging/bulk forming													
Impact extrusion	X	•	•	—	•	•	•	—	X	X	X	X	X
Cold heading	X	•	•	•	•	•	—	—	X	—	X	X	X
Closed die forging	X	•	•	•	•	•	X	•	•	—	—	X	X
Pressing and sintering (P/M)	X	•	•	•	•	•	X	•	—	•	•	X	X
Hot extrusion	X	•	—	—	•	•	X	•	—	—	—	X	X
Rotary swaging	X	•	•	•	•	—	—	•	X	•	•	X	X
Machining													
Machining from stock	•	•	•	•	•	•	•	•	—	—	—	—	—
Electrochemical machining	•	•	•	•	—	—	—	—	•	•	—	X	X
Electrical discharge machining (EDM)	X	•	•	•	•	•	—	—	—	•	—	X	X
Wire EDM	X	•	•	•	•	•	—	—	—	•	—	•	X
Forming													
Sheet metal forming	X	•	•	•	•	•	—	—	—	—	X	X	X
Thermoforming	X	X	X	X	X	X	X	X	X	X	X	•	X
Metal spinning	X	•	—	•	•	•	•	—	—	—	—	X	X

•, normal practice; —, less-common practice; X, not applicable. Source: Adapted from Ref 7

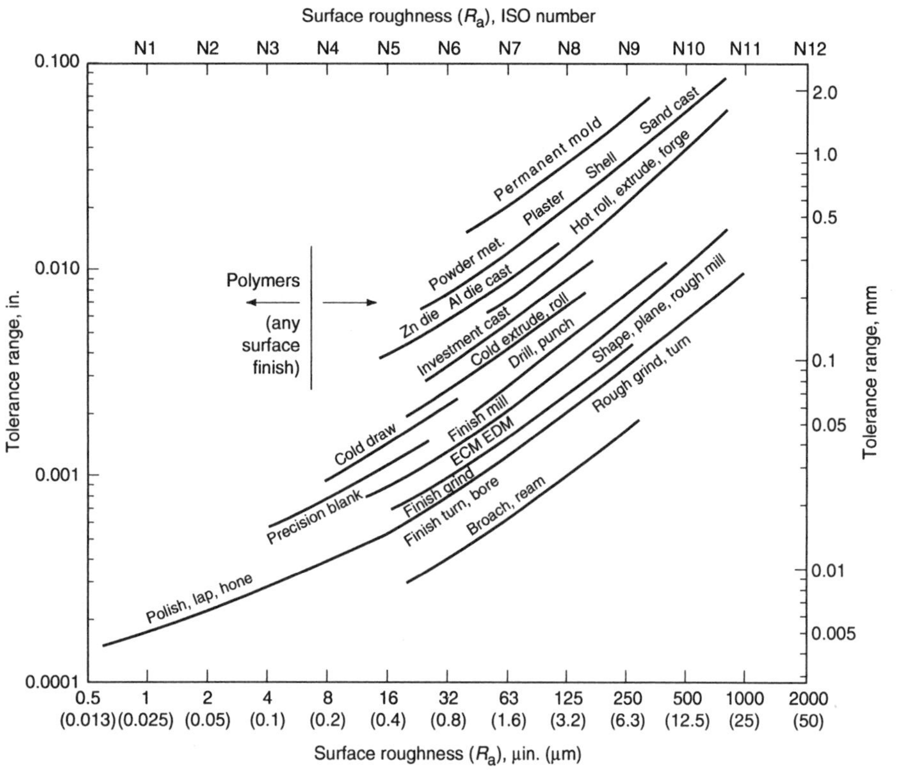

Fig. 5 Approximate values of surface roughness and tolerance on dimensions typically obtained with different manufacturing processes. ECM, electrochemical machining; EDM, electrical discharge machining. Source: Ref 8

by providing parts with a greater range of complexity. Process selection is done by superimposing the envelope of design attributes (Fig. 6b). Sometimes the design sets upper and lower limits on the process attribute, region 1, while other times it imposes only upper limits, region 2. Process selection must be linked with material behavior (Table 4) and constraints imposed by material hardness, melting temperature, and flow and fracture properties.

Much attention has been directed in recent years to developing design rules and computer methods for enhancing manufacturability through proper design. The intent is to reduce product cost and increase product quality by integrating the product and process concepts so that the best match is made between product and process requirements so as to ensure ease of manufacture. This approach is called design for manufacturing (DFM). Detailed information about DFM is provided in the Section "Manufacturing Aspects of Design" in this Volume; additional information on DFM can be found in Ref 7, 10, and 11.

Costs and Related Aspects of Materials Selection

The decision on materials selection ultimately will come down to a trade-off between performance and cost. There is a spectrum of applications varying from those where performance is paramount (defense-related aerospace) to those where cost clearly predominates (household appliances). The total cost of a part includes the cost of the material, the cost of tooling (dies, fixtures), and the processing cost. The unit cost of a part, C, can be expressed by:

$$C = C_m + \frac{C_c}{n} + \frac{C_L}{\dot{n}} \qquad \text{(Eq 2)}$$

where C_m is the material cost; C_c is the capital cost of plant, machinery, and tooling required to make the part; C_L is the labor cost per unit time; n is the batch size; and $\dot{n}$ is the production rate (parts produced per unit time). This cost equation leads to the familiar plot of a break-even point, a batch size (n) beyond which it is more economical to use a manufacturing process that has higher tooling or capital costs (Fig. 7). Below an economical lot size of about 60 pieces, the cost of tooling for making the investment castings cannot be offset by savings in material and machining cost.

Table 5 Manufacturing processes and their attributes

Process	Surface roughness	Dimensional accuracy	Complexity	Production rate	Production run	Relative cost	Size (projected area)
Pressure die casting	L	H	H	H/M	H	H	M/L
Centrifugal casting	M	M	M	L	M/L	H/M	H/M/L
Compression molding	L	H	M	H/M	H/M	H/M	H/M/L
Injection molding	L	H	H	H/M	H/M	H/M/L	M/L
Sand casting	H	M	M	L	H/M/L	H/M/L	H/M/L
Shell mold casting	L	H	H	H/M	H/M	H/M	M/L
Investment casting	L	H	H	L	H/M/L	H/M	M/L
Single point cutting	L	H	M	H/M/L	H/M/L	H/M/L	H/M/L
Milling	L	H	H	M/L	H/M/L	H/M/L	H/M/L
Grinding	L	H	M	L	M/L	H/M	M/L
Electrical discharge machining	L	H	H	L	L	H	M/L
Blow molding	M	M	M	H/M	H/M	H/M/L	M/L
Sheet metal working	L	H	H	H/M	H/M	H/M/L	L
Forging	M	M	M	H/M	H/M	H/M	H/M/L
Rolling	L	M	H	H	H	H/M	H/M
Extrusion	L	H	H	H/M	H/M	H/M	M/L
Powder metallurgy	L	H	H	H/M	H	H/M	L
Key:							
H	>250	<0.005	High	>100	>5000	High	>0.5
M	>63 and <250	>0.005 and <0.05	Medium	>10 and <100	>100 and <5000	Medium	>0.02 and <0.5
L	<63	>0.05	Low	<10	<100	Low	<0.02
Units	μin.	in.		Parts/h	Parts		m^2

Source: Ref 9

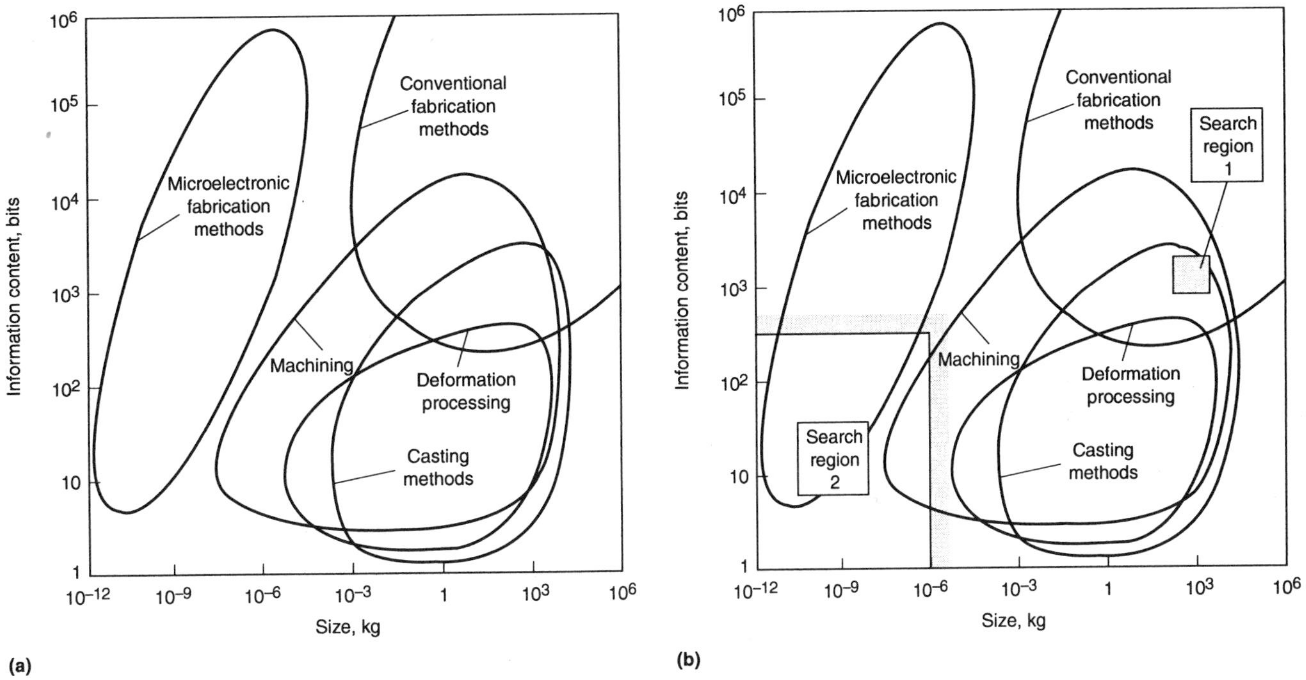

Fig. 6 Ashby process selection charts. Process attributes are given on the axes. A given process maps out a characteristic field. (a) Chart for the relationship between part size and complexity of the shape. (b) Same chart with superimposed regions for attributes of a particular design. The design requirements determine the necessary processing attributes, thus establishing a search area. Processes that overlap the search area become candidates for selection. Source: Ref 2

Prices of plastics are relatively easy to obtain from a trade journal such as *Plastics Technology*. However, prices for metals are more complex. The prices listed in sources such as *American Metal Market/Metalworking News* are nominal, baseline prices to which must be added many extras. For steel products, the following extras may apply (Ref 13):

Metallurgical requirements

- *Grade extra.* Each AISI grade has an extra over the cost of the generic type of steel, that is, hot-rolled bar, hot-rolled plate.
- *Chemistry extra.* Nonstandard chemical composition for the grade of steel
- *Quality extra,* for example, vacuum melting or degassing
- *Inspection and testing.* A charge is made for anything other than routine tensile tests and chemical analysis.
- *Special specifications*

Dimensions

- *Size and form.* Charge for special shapes and sizes
- *Length.* Precise requirements on length are costly.
- *Cutting.* Different charges apply to sheared edge, machined edge, flame cut edge, etc.
- *Tolerances.* Tighter than normal tolerances on outside diameter or thickness cost extra.
- *Metric dimensions.* There may be an extra cost for metric dimensions since United States steel producers produce chiefly to English dimensions.

Processing

- *Special thermal treatments,* for example, normalizing or spheroidizing cost extra.
- *Surface treatment,* for example, pickling or oil dip

Quantity

- *Purchases* in less than heat lots (50 to 300 tons) are an extra cost.

Pack, Mark, and Load

- *Packing.* Special wrapping or boxing
- *Marking.* Charge for other than stamped numbers
- *Loading.* If special blocking is needed in freight cars or trucks

This detailed list shows how inadvertent decisions of the designer can significantly influence cost. Standard chemical composition should be used whenever possible, and the number of alloy grades should be standardized to reduce the cost of stocking many grades of steel. Manufacturers whose production rates do not justify purchasing in heat lots should try to limit their material use to grades that are stocked by local materials service centers. Special section sizes and tolerances should be avoided unless a detailed analysis shows that the extra costs are justified.

To ensure minimum cost in design, every part, every material, and every manufacturing operation should pass the following tests with a negative response (Ref 5).

- Can it be done without it?
- Does it do more than required?
- Does it cost more than it is worth?
- Is there something that does the job better?
- Can it be made by a less-costly method?
- Can a standard item be used?
- Considering the quantities used, could a less costly tooling method be used?
- Can someone else provide it at less cost without affecting dependability; that is, can it be outsourced?

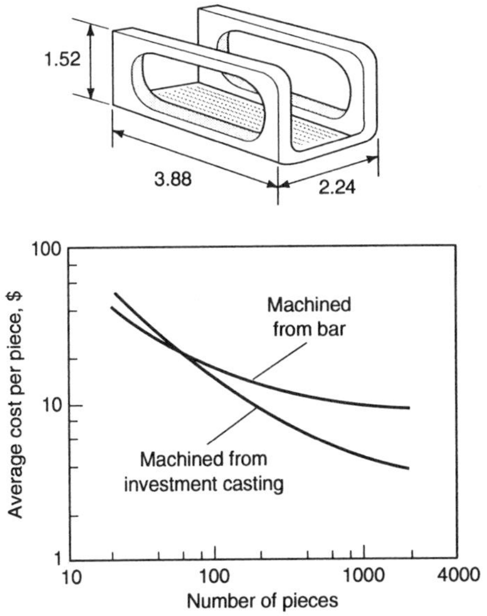

Fig. 7 Unit cost (1961 prices) of roller support made by machining from bar stock compared with machining a blank made by investment casting. Material, AISI 8630 steel. Weight of finished part, 0.5 lb. Part dimensions in inches. Source: Ref 12

- Should the item not be purchased because it costs too much?

These are the basic viewpoints of value analysis, which is described in the article "Use of Value Analysis in Materials Selection" in this Volume.

The chief reason for selecting one material and process combination over another that would satisfy the same functional design requirements is the total cost of the part. It is highly desirable to be able to do this in the conceptual or embodiment phase of design, rather than after most of the design details have been frozen. The development of early cost estimating methods has been actively pursued (Ref 3 and 7), and computer programs are available for this use.

Other factors that affect cost and are key issues in materials selection are:

- *Criticality.* Is this a new material with limited documentation or specifications? Is there more than one supplier, and is it a domestic supplier? Is the lead time for material or tooling acquisition acceptable?
- *Availability.* Is it available in required shapes and sizes?
- *Storage and handling.* Are there any special requirements for the material?
- *Readiness for use.* Can the material be used as is, or does it require a conditioning treatment?
- *Hazardous features.* Are there toxic, allergic, radiological, fire, or explosive factors that must be considered?

More detailed discussion of the selection of materials and processes from the standpoint of cost is provided in the articles "Techno-Economic Issues in Materials Selection" and "Manufacturing Cost Estimating" in this Volume.

Information about Materials Properties

A new material cannot be employed in a design unless the engineer has access to reliable material properties and, hopefully, costs. This illustrates one reason why the tried and true materials are used repeatedly for designs even though better designs could be achieved with newer materials. This article has emphasized how the needs for materials data evolve as the design proceeds from conceptual to detail design. At the start of the design process, low-precision but all inclusive data are needed. At the end of the design process, data are needed for only a single material, but they must be accurate and very detailed. Listed below are readily available sources of materials information useful at various stages of the design process. Much more information can be found in the article "Sources of Materials Properties Data and Information" in this Volume.

Conceptual Design. Useful information sources at this stage in the design process include:

- *Materials Selector,* Penton Publishing Company. A special issue of *Machine Design* magazine giving tabular data for a broad range of metals, ceramics, plastics, and composites
- *The Materials Selector,* 2nd ed., N.A. Waterman and M.F. Ashby, Ed., Chapman & Hall, 1996. A three-volume compilation of data on all materials, with selection and design guide
- *ASM Metals Reference Book,* 3rd ed., M. Bauccio, Ed., ASM International, 1993. A compilation of tabular properties and processing data for ferrous and nonferrous metals
- *ASM Engineered Materials Reference Book,* 2nd ed., M. Bauccio, Ed., ASM International, 1994. A compilation of data for ceramics, polymers, and composites
- *Cambridge Materials Selector, V2.0 for Windows,* Granta Design Limited, Cambridge, UK. Implements the Ashby materials selection scheme for personal computers

Embodiment Design (Configuration Design). At this level, the designer is deciding on the layout and size of the parts and components. The calculations needed for this step require material properties for a narrower class of materials but specific to a particular heat treatment or manufacturing process. These data are typically found in handbooks, computer data bases, and data books published by associations of material producers:

- *Metals Handbook Desk Edition,* ASM International, 1985. Considers all metallic alloys and processing methods
- *Engineered Materials Handbook Desk Edition,* ASM International, 1995. Considers polymers, ceramics, composites, and glasses
- *MAPP,* ESM Software, Hamilton, OH. Materials data bases from ASM International Mat.DB collection, in personal computer format, of properties and processing for metals and some polymers
- *Rover Electronic DataBooks,* William Andrew Inc., Norwich, NY. Materials data bases including ASM International Mat.DB data collection, in personal computer format, of properties and processing for metals and polymers
- *MSC/MVISION,* MacNeal-Schwendler Corp., Los Angeles. A materials data base that is integrated with the Pro/ENGINEER analysis system
- *PERITUS,* Matsel Systems Ltd., Liverpool. A data base for metals, polymers, and ceramics aimed at materials and process selection
- STN International, Chemical Abstracts Services, Columbus, OH. On-line service for metals, ceramics, and polymers data bases developed by the Materials Property Data (MPD) Network

Detail Design. At the detail design stage, very precise data are required. These data are best found in data sheets issued by materials producers. This is particularly true of polymers, whose properties vary considerably depending upon how they are manufactured. For critical parts, tests on the actual material from which the part will be made will be required. There is a very wide range of properties that may be needed at the detail design stage. These go beyond just material properties to include information on manufacturability, cost, experience base in use in other applications, and issues of quality assurance; see Table 6.

Computer Data Bases. The use of computer-aided tools allows the engineer to minimize the materials selection information overload. A computerized materials search can accomplish in minutes what may take hours or days by a manual search. Over 100 materials data bases are available worldwide. However, the data contained in most of them are limited to numerical values and text. Image information (micrographs, fractographs, etc.) can be provided through CD-ROM. All materials property data bases allow the user to search for a material match by comparing four or more property parameters, each of which can be specified as below, above, or within a stated range of values. Some data bases have the ability to weight the significance of the various properties. The most advanced data bases allow the materials property data to be transmitted directly to a design software package, such as finite element analysis, so that the effect of changing material properties on the geometry and dimensions of a part design can be directly observed on the computer monitor. However, this capability is generally limited to a narrow homogeneous group of materials. To compare widely divergent materials, that is, polymers versus aluminum die castings, the *Cambridge Materials Selector* is the only software available.

No materials information data base contains the broad spectrum of information called for in Table 6. Most existing data bases provide numerical material properties. Usually mechanical and corrosion properties are well covered, with less extensive coverage of magnetic, electrical, and thermal properties. Since it is unlikely that any data base will be sufficiently comprehensive for a specific user, it is vital that the system be designed so that users may easily add their own data and subsequently search, manipulate, and compare these values along with the entire collection of data.

The validity of the data in a data base is an important issue. In order to increase the reliability and accuracy of data in a computerized data base, ASTM recommends that the following descriptors be used:

- Source (name of handbook, government agency, producer, etc.)
- Statistical basis of the data
- Status of material (production, experimental, obsolete)
- Evaluation status (if so, by whom and how)
- Validation status (if so, by whom and how)
- Certification status (if so, by whom and how)

Expert systems are computer-based systems with the potential to solve problems that require reasoning. A computerized data base search is only as good as the data in the data base, but a materials selection with an expert system combines qualitative and quantitative information using decision rules in much the same way that

Table 6 Examples of materials information required during detail design

Material identification	Temperature (cryogenic-elevated)	Joining technology applicable
Material class (metal, plastic, ceramic composite)	Tensile strength, yield strength	Fusion
Material subclass	Creep rates, rupture life at elevated temperatures	Adhesive bonding
Material industry designation	Relaxation at elevated temperatures	Fasteners
Material product form	Toughness	Welding parameters
Material condition designation (temper, heat treatment, etc.)	Damage tolerance (if applicable)	Finishing technology applicable
Material specification	Fracture toughness (define test)	Impregnation
Material alternative names	Fatigue crack growth rates (define environment, and load)	Painting
Material component designations (composite/assembly)	Temperature effects	Stability of color
Material production history	Environmental stability	Application history/experience
Manufacturability strengths and limitations	Compatibility data	Successful uses
Material composition(s)	General corrosion resistance	Unsuccessful uses
Material condition (fabrication)	Stress-corrosion cracking resistance	Applications to be avoided
Material assembly technology	Environmental stability	Failure analysis reports
Constitutive equations relating to properties	Toxicity (at all stages of production and operation)	Maximum life service
Material properties and test procedures	Recyclability/disposal	Availability
Density	Material design properties	Multisource? Vendors?
Specific heat	Tension	Sizes
Coefficient of thermal expansion	Compression	Forms
Thermal conductivity	Shear	Cost/cost factors
Tensile strength	Bearing	Raw material
Yield strength	Controlled strain fatigue life	Finished product or require added processing
Elongation	Processability information	Special finishing/protection
Reduction of area	Finishing characteristics	Special tooling/tooling costs
Moduli of elasticity	Weldability/joining technologies	Quality control/assurance issues
Stress-strain curve or equation	Suitability for forging, extrusion, and rolling	Inspectability
Hardness	Formability (finished product)	Repair
Fatigue strength (define test methods, load, and environment)	Castability	Repeatability
	Repairability	
	Flammability	

Source: Ref 14

someone with experience would do. In other words, it provides context dependent selection of a material or a process. A full materials selection expert system is decades away, but successful examples of the method have been developed in narrow areas like the selection of adhesives from a limited set or in designing against corrosion. For more information on expert systems in materials selection and computer-aided materials selection in general, see the article "Computer-Aided Materials Selection" in this Volume.

Methods of Materials Selection

The act of selecting a material is a form of engineering decision making. It has been shown how in the conceptual stage of design it is important to consider a wide spectrum of materials. This list is narrowed considerably to enable embodiment or configuration design to be carried out, and it is narrowed still further to a single material for detail design. Clearly design involves decision making. Many analytical methods have been introduced into engineering design to aid in the evaluation process, which results in a rational decision. These methods are discussed in greater detail in the article "Decision Matrices in Materials Selection" in this Volume and in Ref 15–18.

Cost per Unit Property Method

In many materials selection problems, one property stands out as the most dominant service requirement. In this case, a reasonable screening method is to determine how much a material will cost to provide that requirement, that is, the cost of a function. Cost per unit property is a special type of performance index and is discussed in the article "Performance Indices" in this Volume.

Following the development due to Ashby (Ref 2), it is desired that the material for a solid cylindrical tie-rod of length L to carry a tensile force P with a safety factor S be of a minimum mass. The mass is given by:

$$m = A\,L\,\rho \quad \text{(Eq 3)}$$

where A is the cross-sectional area and ρ is the density of the material from which it is made. The design requires that L and P are specified and cannot be changed; that is, they are constraints. However, the radius of the bar is a free variable so long as the bar supports the load P:

$$P/A = \sigma_f/S \quad \text{(Eq 4)}$$

where σ_f is the failure strength (yield strength). Eliminating A between these two equations results in:

$$m = (S\,P)\,(L)\,(\rho/\sigma_f) \quad \text{(Eq 5)}$$

Equation 5 shows that the performance of the tie-rod is described by three separate factors. The first term in brackets is the functional requirement. The second term is the geometric parameter. The last bracket is the material performance index. For this case of a bar in tension, the performance index M is given by:

$$M = \sigma_f/\rho \quad \text{(Eq 6)}$$

The lightest tie-rod that will safely carry the load P without failing is that made from a material with the largest value of the performance index M. A similar calculation for a light, stiff tie-rod leads to the performance index $M = E/\rho$ where E is Young's modulus. If the design criteria is to minimize cost rather than minimize weight, then the density in Eq 5 is replaced by $C\rho$, where C is the cost per unit mass. Now M is given by:

$$M = \sigma_f\,/\,C\rho \quad \text{(Eq 7)}$$

Other part geometries and types of loading would result in different performance indices. Some of these are given in Table 7.

Weighted Property Index Method

Frequently a design requires that more than one material property or requirement be optimized. The usual approach is to assign a weighting factor to each material requirement or property, depending on its importance. Since this is mostly a subjective process, the weighting must be done with care to prevent bias or getting the answer you intended. The individual weighted property values are then summed to give a comparative weighted property index.

The importance of a screening property for which go-no go limits can be set must be kept in mind. Such parameters like weldability and corrosion resistance must be delineated at the outset, and all materials that do not meet the go-no go criteria are screened out. Other nondiscriminating parameters like availability or fabricability should be used to screen out materials from the weighted property matrix or at least flag them for further deliberation.

The weighted properties index has the drawback of having to combine properties with different units. The best procedure is to normalize these differences by using a scaling factor. Scaling is a simple technique to bring all the different properties within one numerical range. Each property is scaled so that its highest numerical value does not exceed 100. The scaled value of a property, β, is given by:

β = scaled property

$$= \frac{\text{numericalvalue of property(100)}}{\text{largest value in the listd}} \quad \text{(Eq 8)}$$

For properties like cost, corrosion loss, and wear rate, for which a low value is best, the scale factor is formulated as:

β = scaled property

$$= \frac{\text{lowest value in the list (100)}}{\text{numerical value of property}} \quad \text{(Eq 9)}$$

For properties that are not readily expressed in numerical values, for example weldability, some kind of subjective rating is required; see Table 8. Once each material property and requirement is given a scaled value, the weighted property index for the material is simply the sum of the product of the scaled value, β, times the weighting factor, w, for each property:

$$\gamma = \text{weighted property index} = \Sigma \beta_i w_i \quad \text{(Eq 10)}$$

Cost obviously is an important factor in selecting materials. It can be used as one of the properties, but because it is such an important factor, it may over dominate in the selection of weighting factors. One way to prevent this is to use cost as a modifier to the weighted property index instead as one of the properties:

$$\gamma' = \gamma / C\rho \quad \text{(Eq 11)}$$

where C is the material cost per unit mass, and ρ is density.

Example 2: Use of the Weighted Property Index in Materials Selection for a Cryogenic Storage Tank. The use of this index method can be illustrated with the case of materials selection for a large cryogenic storage tank for transporting liquid nitrogen gas (Ref 18). The gas will be stored at −196 °C (−321 °F) so that concern for ductile-to-brittle transition in ferritic steels rules out their use. Also, this rules out most plastics. Since most tanks are fabricated by welding, good weldability is a go-no go requirement. Likewise availability of plate material in the required thickness and size is a nondiscriminating screening factor. With respect to mechanical properties, adequate toughness, high yield strength, and stiffness are important. Stronger material means thinner tank walls, resulting in a lighter tank and lower cool-down losses. Also, a lower density material results in a lighter tank. Lower specific heat reduces cool-down losses while a lower thermal conductivity reduces heat losses. Finally, a lower thermal expansion coefficient reduces thermal stresses. This quick analysis results in seven material properties that must be considered in the material selection. In addition, the go-no go requirements discussed above must be followed.

The first task is to assign weighting factors to each property. The usual procedure is to adjust the weighting factors so that their sum equals unity. This is an area of the analysis wherein bias can easily creep. So that this is avoided, it is good to adopt a formal procedure for assigning weights. Pairwise comparison, sometimes called the digital logic method, works well especially if the decisions are the collective decision of a small group, not just an individual. In pairwise comparison, each property is compared in turn with each of the other properties. Thus with four properties A, B, C, and D, start with A, and ask whether property A or B was the more important. If it is A, then give this a 1, and give B a 0. Then compare A and C. If C is the more important, it gets a 1, and A receives a 0. Finally compare A and D, and find that A gets a 1 and D gets a 0. Now turn to B versus C and B versus D, etc. When all comparisons are made add the ones for each property, and total the results. The weighting factor for A is the sum of the ones it received divided by the total; in this case, 2/6 = 0.33. The total number of decisions to make is given by $N(N-1)/2$.

Table 9 shows the pairwise comparisons made for the cryogenic tank (Ref 18). The resulting weighting factors are given in Table 10. Table 11 shows that seven metallic materials were selected as candidates for constructing the cryogenic tank. Appropriate material properties are given in this table.

These properties have been scaled using Eq 8 and 9 and placed in Table 12. Each scaled property value is then multiplied by the weighting factors in Table 10 and summed according to Eq

Table 7 Performance indices

Part shape and loading for minimum weight	To maximize strength	To maximize stiffness
Tensile tie-rod: load, stiffness, and length are fixed; section area is variable	σ_f / ρ	E/ρ
Torsion bar: torque, stiffness, and length are fixed; section area is variable	$\sigma_f^{2/3}/\rho$	$G^{1/2}/\rho$
Beam: loaded externally or by self weight in bending; stiffness and length fixed; section area free	$\sigma_f^{2/3}/\rho$	$E^{1/2}/\rho$
Plate: loaded externally or by self weight in bending; stiffness, length, and width fixed; thickness free	$\sigma_f^{1/2}/\rho$	$E^{1/3}/\rho$
Cylinder with internal pressure: elastic distortion, pressure, and radius fixed; wall thickness free	σ_f/ρ	E/ρ
Design goal		**Maximize**
Thermal insulation: thickness given; minimum heat flux at steady state		$1/\lambda$
Thermal insulation: thickness given; minimum temperature rise after specified time		$C_p \rho/\lambda$
Thermal storage: maximize energy storage for given temperature and time		$(C_p \rho\lambda)^{1/2}$
Minimum thermal distortion		λ/α
Maximum thermal shock resistance		$\sigma_f/E\alpha$
Pressure vessel: yield before break		K/σ_f

σ_f is failure strength; K is fracture toughness; E is Young's modulus; G is shear modulus; ρ is density; C_p is specific heat capacity; α is thermal expansion coefficient; λ is thermal conductivity. Source: Ref 2

Table 8 Using a subjective value to get a scaled property

	Alternative materials			
Property	A	B	C	D
Weldability	Excellent	Poor	Good	Fair
Relative rating	5	0	3	1
Scaled property	100	0	60	20

Table 9 Use of pairwise comparison in the cryogenic tank example

	Decision number																				
Property	1	2	3	4	5	6	7	8	9	10	11	12	13	14	15	16	17	18	19	20	21
Toughness	1	1	1	1	1	1															
Yield strength	0						1	0	0	1	1										
Young's modulus		0					0					0	0	0	1						
Density			0					1				1				1	1	1			
Thermal expansion				0					1				1			0			1	1	
Thermal conductivity					0					0				1			0		0		0
Specific heat						0					0				0			0		0	1

Source: Ref 18

Table 10 Weighting factors for the cryogenic tank example

Property	Positive decisions	Weighting factors
Toughness	6	0.28
Yield strength	3	0.14
Young's modulus	1	0.05
Density	5	0.24
Thermal expansion coefficient	4	0.19
Thermal conductivity	1	0.05
Specific heat	1	0.05
Total	**21**	**1.00**

Source: Ref 18

Table 11 Properties of candidate materials for the cryogenic tank example

Material	1 Toughness index(a)	2 Yield strength, MPa	3 Young's modulus, GPa	4 Specific gravity	5 Thermal expansion(b)	6 Thermal conductivity(c)	7 Specific heat(d)
Aluminum 2014-T6	75.5	420	74.2	2.8	21.4	0.37	0.16
Aluminum 5052-O	95	91	70	2.68	22.1	0.33	0.16
Stainless steel 301, fully hard	770	1365	189	7.9	16.9	0.04	0.08
Stainless steel 310, ¾ hard	187	1120	210	7.9	14.4	0.03	0.08
Ti-6Al-4V	179	875	112	4.43	9.4	0.016	0.09
Inconel 718	239	1190	217	8.51	11.5	0.31	0.07
70Cu-30Zn	273	200	112	8.53	19.9	0.29	0.06

(a) Toughness index, TI, is based on ultimate tensile strength (UTS), yield strength Y, and ductility e, at −196 °C (−321 °F). TI = (UTS + Y)e/2. (b) Thermal expansion coefficient is given in 10^{-6}/°C. The values are average between room temperature and −196 °C. (c) Thermal conductivity is given in cal/cm^2/cm/°C/s. (d) Specific heat is given in cal/g/°C. The values are average between room temperature and −196 °C. Source: Ref 18

Table 12 Scaled values of properties and calculated weighted property index

Material	Scaled properties 1	2	3	4	5	6	7	Weighted property index
Aluminum 2014-T6	10	30	34	96	44	4.3	38	42.2
Aluminum 5052-O	12	6	32	100	43	4.8	38	40.1
Stainless steel 301, fully hard	100	100	87	34	56	40	75	70.9
Stainless steel 310, ¾ hard	24	82	97	34	65	53	75	50.0
Ti-6Al-4V	23	64	52	60	100	100	67	59.8
Inconel 718	31	87	100	30	82	5.2	86	53.3
70Cu-30Zn	35	15	52	30	47	5.5	100	35.9

Source: Ref 18

Table 13 Relative cost, figure of merit, and ranking of materials

Material	Relative cost(a)	Cost of unit strength × 100	Weighted property index	Figure of merit	Rank
Aluminum 2014-T6	1	0.67	42.2	62.99	2
Aluminum 5052-O	1.05	3.09	40.1	12.98	6
Stainless steel 301, fully hard	1.4	0.81	70.9	87.53	1
Stainless steel 301, ¾ hard	1.5	1.06	50.0	47.17	3
Ti-6Al-4V	6.3	3.20	59.8	18.69	4
Inconel 718	5.0	3.58	53.3	14.89	5
70Cu-30Zn	2.1	8.96	35.9	4.01	7

(a) The costs include stock material and processing cost. The relative cost is obtained by considering the cost of aluminum 2014 as unity and relating the cost of other materials to it. Source: Ref 18

10 to arrive at the weighted property index for each material in Table 11.

So far, cost has not been considered in the material selection analysis. Costs are expressed in Table 13 as relative to the cost of aluminum alloy 2014. The cost of a unit of yield strength is calculated from the reciprocal of Eq 7. This is divided into the weighted property index to give the figure of merit, a cost adjusted weighted property index. It is shown that type 301 stainless steel ranks first but that aluminum alloy 2014 gains several places in rank when adjusted for cost.

While this method of material selection requires considerable calculation and may appear tedious at first reading, the growing use of computerized data bases can greatly facilitate its use. However, a word of warning is in order. It is very easy for engineers to become quickly enamored with a numerical calculation scheme and lose sight of reality. Care must be taken that all nonquantifiable factors have been properly considered in the final decision and that the numerical outcome is not unduly influenced by a hasty choice of weighting factors.

Limits on Properties Method

In this method of materials selection, the performance requirements are divided into three categories:

- Lower limit properties
- Upper limit properties
- Target value properties

For example, if it is required to have a stiff, light material, then a lower limit is put on Young's modulus and an upper limit is put on density. It has already been shown that this is the approach used when screening a large number of materials with a computer data base. After screening, the remaining materials are those whose properties are above the lower limits, below the upper limits, and within the target values of the specified requirements.

To arrive at a merit parameter for each material, the properties are first assigned weighting factors using pairwise comparison. A merit parameter p is then calculated for each material using the relationship:

$$p = \left[\sum_{i=1}^{n_l} w_i \frac{Y_i}{X_i}\right]_l + \left[\sum_{j=1}^{n_u} w_j \frac{X_j}{Y_j}\right]_u + \left[\sum_{k=1}^{n_t} w_k \left|\frac{X_k}{Y_k} - 1\right|\right]_t \quad \text{(Eq 12)}$$

where l, u, and t stand for lower limit, upper limit, and target properties, respectively (Ref 18). Thus n_l, n_u, and n_t are the number of lower limit, upper limit, and target value properties; w_i, w_j, and w_k are the weighting factors on lower limit, upper limit, and target value properties; X_i, X_j, and X_k are the candidate material lower limit, upper limit, and target value properties; and Y_i, Y_j, and Y_k are the specified lower limits, upper limits, and target values. Based on Eq 12, the lower the value of the merit parameter p, the more suitable the material. Cost can be treated as an upper limit property and given the appropriate weight.

Conclusions

The general pattern for material selection has been described in this overview. Many references to more detailed discussions of these topics in this Volume have been given.

A competent job of material selection should include consideration and documentation of the following:

- The problem, design, or redesign objective
- The underlying design criteria (primary, secondary, or supporting criteria; manufacturing method considerations; codes and standards)
- Analysis (preliminary concepts, modeling or simulation, optimization and trade-offs, design reliability, economics and cost)
- Alternatives considered, selection criterion, and decision methodology
- Reasons for the selection of the final material and manufacturing process

REFERENCES

1. M.F. Ashby, Materials, Bicycles, and Design, *Metall. Mater. Trans. A*, Vol 26, Dec 1995, p 3057–3064

2. M.F. Ashby, *Materials Selection in Mechanical Design*, Pergamon Press, 1992
3. J.R. Dixon and C. Poli, *Engineering Design and Design for Manufacturing*, Field Stone Publishers, 1995
4. V. John, *Introduction to Engineering Materials,* 3rd ed., Industrial Press, 1992
5. C.O. Smith and B.E. Boardman, Concepts and Criteria in Materials Engineering, *Metals Handbook,* Vol 3, 9th ed., American Society for Metals, 1980, p 825–834
6. C.O. Smith and B.E. Boardman, Concepts and Criteria in Materials Engineering, *Metals Handbook,* Vol 3, 9th ed., American Society for Metals, 1980, p 826
7. G. Boothroyd, P. Dewhurst, and W. Knight, *Product Design for Manufacture and Assembly,* Marcel Dekker, Inc., 1994
8. J.A. Schey, *Introduction to Manufacturing Processes,* McGraw-Hill Book Co., 1987
9. E.B. Magrab, *Integrated Product and Process Design and Development,* CRC Press, Inc., 1997
10. H.E. Trucks, *Designing for Economical Production,* 2nd ed., Society of Manufacturing Engineers, 1987
11. R. Bakerjian, Ed., *Tool and Manufacturing Engineers Handbook,* Vol 6, *Design for Manufacturability,* 4th ed., Society of Manufacturing Engineers, 1992
12. *Metals Handbook,* Vol 1, 8th ed., American Society for Metals, 1961, p 295
13. R.F. Kern and M.E. Suess, *Steel Selection*, John Wiley & Sons, Inc., 1979
14. *Computer-Aided Materials Selection during Structural Design,* National Academy Press, 1995
15. N. Cross, *Engineering Design Methods,* 2nd ed., John Wiley & Sons, Inc., 1994
16. G.E. Dieter, *Engineering Design: A Materials and Processing Approach,* 2nd ed., McGraw-Hill, 1991
17. W.E. Souder, *Management Decison Methods for Managers of Engineering and Research,* Van Nostrand Reinhold Co., 1980
18. M.M. Farag, *Selection of Materials and Manufacturing Processes for Engineering Design,* Prentice Hall, 1989

SELECTED REFERENCES

- M.F. Ashby, *Materials Selection in Mechanical Design*, Pergamon Press, 1992
- J.A. Charles and F.A.A. Crane, *Selection and Use of Engineering Materials,* 2nd ed., Butterworths, 1989
- G.E. Dieter, *Engineering Design: A Materials and Processing Approach,* 2nd ed., McGraw-Hill, 1991
- M.M. Farag, *Selection of Materials and Manufacturing Processes for Engineering Design,* Prentice Hall, 1989
- G. Lewis, *Selection of Engineering Materials*, Prentice Hall, 1990

Techno-Economic Issues in Materials Selection

Joel P. Clark, Richard Roth, and Frank R. Field III, Massachusetts Institute of Technology

THE SELECTION OF MATERIALS in industrial applications has always been important to product designers. This is a consequence of the fact that the selection of material determines not only the basic physical characteristics of the product, but also the processing technologies that can be employed in its manufacture, the specialized properties that can be developed as a function of that processing, and, ultimately, the cost of the product.

Because of the interplay between material choice, product performance, and manufacturing economics, an effective designer must carefully balance the properties and cost that result from a specific material choice against the willingness of the market to pay the price that such a choice will require. Traditional product design tools employ detailed theoretical and empirical relationships to estimate the engineering performance of a product as a function of its geometry and materials. Through such analyses, the designer attempts to fine tune the design in order to achieve the precise combination of desired characteristics. The economic consequences of this careful design work is rarely treated in such detail. This is a consequence of separation between the design and engineering functions of a firm and the cost accounting part of the firm. While designers rely upon these cost analysts to estimate the costs of their designs, the designer rarely receives the kind of information that would be most helpful in refining the product design; that is, the relationship between specific design elements and cost estimates.

In this article, methods for evaluating materials alternatives on the basis of costs, both direct economic costs and indirect social costs, are presented. The importance of the first of these is obvious. Social costs are presently less crucial because designers rarely consider environmental characteristics in materials selection decisions. Given the interest of government regulators and consumers in the environment, it is expected that environmental performance will increasingly become important as a strategic consideration.

Related information is provided in the articles "Concurrent Engineering," "Life Cycle Engineering and Design," "Design for the Environment," and "Manufacturing Cost Estimating" in this Volume.

Cost Analysis

On the face of it, cost appears to be a relatively simple concept—the financial consequences of product design decisions. However, cost in practice is a difficult metric to determine when material choices are being evaluated. This difficulty arises from the fact that the cost of material choices requires a careful consideration of the material, the process technology to be employed to form the product, and the design of the product itself (Ref 1).

Alternatives for cost analysis can be divided into several categories: rules of thumb, accounting methods, and analytical methods.

Rules of Thumb

One commonly used method of cost estimation is to employ industry rules of thumb (Ref 2). The most typical methods rely upon simple conversions of material cost into total part cost, usually a fixed multiple. For example, a good estimate of the cost of injection molded parts made from commodity plastics is twice the cost of the raw material required. Another example are X-Y plots depicting the combinations of an engineering metric (such as density or stiffness) and an indicative part cost (usually normalized by part weight) that can be achieved with individual materials or material classes (Ref 3). Both the multiplicative factor and the cost-performance plots are not based on detailed technological considerations, but on the past experience of the industry with the material in question. The predictive capabilities of this method can be extremely limited, often inaccurate and overly sensitive to changes in materials costs or to substitution of a more expensive material for a less expensive one. Although the cost-performance plot can be used to make broad conclusions about the suitability (and unsuitability) of particular material classes, the necessary degree of generalization employed to develop the figure necessitates an eventual consideration of cost in more detail.

Rules of thumb are, almost by definition, "back of the envelope" analyses that rely heavily upon the expertise of the person employing them. For example, the injection molding rule of thumb above would not be used by an expert when considering the production of a liquid crystal polymer product, but a novice might easily apply it without recognizing its limitations. In the hands of an expert, rules of thumb combined with years of experience can yield extremely good estimates of the cost of parts that are produced with materials and processes that are within that expert's experience. Unfortunately, that experience is also the greatest limitation of rules of thumb. It is expensive and time consuming to develop, and it can be a hindrance when processes and materials outside that realm of experience are under consideration.

Accounting Methods

Another approach is to examine current accounting data from the manufacturing facility and to try to allocate these costs across all of the products; that is, essentially summing up all the costs and dividing by the number of parts produced (or their mass, etc.) (Ref 4). While this approach is useful from the perspective of the manager interested in the total production cost of a manufacturing facility, it is difficult to allocate cost among the parts. Furthermore, it offers no predictive capabilities. Because this approach relies upon current operating conditions and practice, there is no way to look at the differences in cost, which arise from switching to new materials or technologies.

A variant of this approach is to disaggregate costs between the material content of the product and the time and equipment required to convert the purchased materials into the final part (Ref 4, 5). Accounting information is employed to develop one of two constructs: a machine rental cost or a variable burden rate. In the former, the costs of the capital, energy, and labor required to operate production equipment is summed and then divided by the time they are used to produce parts, yielding a "machine rent," or cost per unit time to operate. In the latter, the capital (and other nonmaterial and nonlabor) costs are distributed over the total number of labor hours (rather than machine hours) that are employed in part production, yielding a "variable burden" rate that essentially can be thought of as a "markup" on labor.

While useful for managerial control, both accounting approaches have important flaws when manufacturing costs of alternative designs are being developed. Most obviously, both approaches are based upon historical data, like the pure accounting method. As such, one is assuming that future costs will be similar to past costs—not a very good assumption when new processes and materials are under explicit consideration! However, the concepts of machine rent and variable burden are apparent attempts to overcome this limitation by introducing parameters that enable consideration of some limited ideas of change; machine rent allows one to consider that the time to produce a part may change, while variable burden allows one to consider that the amount of labor required may change.

Unfortunately, these two concepts still embed strong assumptions that limit their utility for product design. Both the concept of machine rent and variable burden imply that aggregating costs that are otherwise difficult to apportion correctly will still yield cost estimates that can be used to make economic decisions. However, these gross measures of cost essentially assume that all processes are equally capital intensive per unit time, in the case of machine rents, or that all labor requires equal amounts of capital to be effective, in the case of variable burden. Naturally, these measures can be refined by subdividing the accounting data employed, but this approach fails to address the key failure of the underlying assumptions of the method. Each process requires unique combinations of resources, and averaging limits the ability of decision makers to manage these combinations effectively.

In response to this objection, the accounting community has been working to develop a new approach to cost accounting known as activity-based costing (Ref 5). In this approach, the first objective of the cost accountant is to identify each of the activities that the firm undertakes and then to allocate costs according to the activity, rather than the capital or labor within that activity. In this way, managers receive cost information that emphasizes the relationship between what the firm is doing and how it is being achieved, rather than the more gross measures of traditional accounting tools. This approach has been under development for over a decade now, and it has been used to demonstrate that the focus upon business activity, or process, is a key to developing good measures of economic performance.

Analytical Methods

Rules of thumb and traditional accounting techniques offer limited powers for estimating the effect of changes in the material or processing parameters. Activity-based cost accounting is a technique more suited to cost estimation of plants and facilities, not products and processes. As such, it is quite useful in analyzing the cost drivers for a given manufacturing facility, but is incapable of predicting the costs arising from new materials or processes.

However, activity-based costing does demonstrate that consideration of the process is the key factor when attempting to characterize costs. While activity-based costing examines processes from the perspective of how best to manage a process, the designer requires a different emphasis: how does design choice—be it material, processing, or geometry—influence cost, and how is the resulting performance of the product balanced best against that cost. In effect, a more robust, process-based cost-estimation methodology is necessary to analyze these situations. An approach to cost estimation at an early stage in the design process is described in the article "Design for Manufacture and Assembly" in this Volume.

A process-based approach, called technical cost modeling, was developed at MIT in the late 1970s (Ref 6-10). Technical cost modeling requires that the different cost elements for each processing step be estimated separately based on engineering principles, the manufacturing process, and specified economic and accounting assumptions. The relationships between individual elements of cost and the underlying technical and engineering characteristics of each unit operation can be brought into the cost estimate in a way that enables considerable flexibility when exploring the implications of changes in the product design.

Technical Cost Modeling

In the technical cost modeling technique, variable and fixed costs are calculated separately. These categories are further subdivided into the variable costs of material, direct labor, and energy, and the capital costs associated with the main machine, the auxiliary equipment, the tooling, the building, maintenance and overhead on the labor. These individual elements reflect the line items of classical accounting methods in order to maintain consistency with conventional views of cost. However, the elements that underlie these line items are a reflection of how engineering process knowledge influences these conventional cost accounts.

The following paragraphs discuss the basic considerations that go into the development of estimates for each of these elements of cost based on a generic process flow diagram. More detailed descriptions of specific production processes are found in Ref 11-16.

Variable Costs. The terms *variable costs* and *fixed costs* are historically derived and correspond to a simple bifurcation of costs into identifiable categories. Variable costs are those costs that can be directly associated with the production of a unit of output and whose total increases roughly linearly with the total number of units produced. For example, material costs are variable costs since a doubling of the total number of parts produced requires a doubling of the amount of material that will be consumed. Variable costs are contrasted with fixed costs, which are costs that do not increase linearly with total production. For example, the amount of capital equipment required to produce parts does not increase linearly with production. Instead, total capital equipment cost remains fixed until its production capacity is exceeded, whereupon more equipment is required. The perverse consequence of these definitions is that, when considering costs on a per piece basis, the variable costs remain constant as the production volume changes, while the fixed costs change. However, when total costs of producing X parts are considered, the variable costs vary with X, while the fixed costs generally do not vary with X for small changes in X.

The material cost can be estimated as a function of the price of the raw materials, the design of the component, and scrap considerations associated with the process. Essentially, the problem is one of determining the total amount of material actually required by each part, realizing that there will be material losses due to the process and part losses due to quality control. Part mass, process scrap rates, and net production yields can be used to calculate the total material input to the process, while each of these processing parameters can either be a direct input to the model or a more complex function of other engineering parameters.

Direct labor costs are a function of the wages paid (including all costs to the manufacturer of employing a worker), the number of workers necessary to run the process, and the time required to produce the part. Typically, only direct workers are included in this calculation, with the remainder accounted for in the overhead calculation (see below).

Energy consumption can be calculated from the theoretical energy requirements of processing the part. However, it is very difficult to compute this requirement since direct metering of energy consumption is rarely done, and then only for major consumers of energy (furnaces, etc.). Instead, energy costs are often estimated from the consumption requirements of the processing equipment and, again, the time required to produce the part.

Fixed costs are those that are expected to remain essentially constant for a given annual production volume. These include capital expenses for plant, equipment, and tooling, as well as maintenance and overhead labor costs.

Capital costs for machines can be analyzed as a function of factors, such as materials to be processed, the required production volume, and the design of the part (e.g., geometry). Usually, one calculates the number of parallel processing lines needed to manufacture a specified number of parts. Since only an integral number of parallel streams is possible, there are two approaches. One is to round the number of lines to the next highest integer and allocate all of the capital cost of the equipment to that part. This is called the dedicated equipment approach. Often, the machines can be used to manufacture more than one part. For instance, stamping dies can be changed within a matter of minutes to use the stamping press for another part run. In this case, it is appropriate to use a nondedicated assumption, allocating the cost on the basis of the fraction of machine time needed for the annual production volume.

It can be generally assumed that a given level of investment in the primary equipment requires a fixed level of investment in auxiliary equipment. Consequently, in the absence of detailed process information, auxiliary equipment costs can be estimated as a fixed percent of the main machine cost. For certain processes, parameters such as the degree of automation or types of equipment can also be related to the costs of auxiliary equipment.

Tooling cost is possibly the most difficult component of cost to estimate, largely because each set of tools is unique, reflecting design choices and available production equipment. This uniqueness also arises from the fact that tools are largely handmade, requiring considerable craft and special skills that are not widely available. Generally, the total cost of tooling is treated in two parts—the cost of a single set of tools and the number of parts that this tool can be used to produce before it is no longer serviceable. A reasonable approach is to estimate tooling cost based on empirical data that relate various tool types to certain types of components and to previous tooling costs. While such estimates (usually regression models) are imperfect, they can yield good first-order approximations. Tooling costs are always dedicated costs because they apply to specific parts.

Building costs are relatively easy to calculate. Prices per square meter of building space are readily available, and the space requirement can be estimated based on the size of the equipment and conventional practices (e.g., materials handling requirements, safety specifications, etc.).

Maintenance costs are often little understood and would require a complete understanding of the probability of breakdowns occurring on the production line. For the purposes of cost estimation, it is generally sufficient to assume that more expensive equipment requires additional expenditures for maintenance. Consequently, maintenance costs are estimated as a percent of the capital equipment cost.

Overhead labor costs present a similar problem. It is difficult to allocate the costs of supervisors, janitors, etc. across the different aspects of the production facility. Instead this cost is estimated using a burden rate, which is applied to the direct labor requirement. The idea is that a given level of direct labor requires a fixed amount of supervision, janitorial services, and other support staffing. Such a simplistic treatment of maintenance and overhead labor is usually acceptable when the goal is to analyze relative costs of using alternative materials and processes.

Process Engineering Variables. The time to produce a part, or cycle time, is a particularly important feature of technical cost models. Although cycle time is an intermediate variable and does not show up directly in the input or outputs of the models, it is an important determinant of

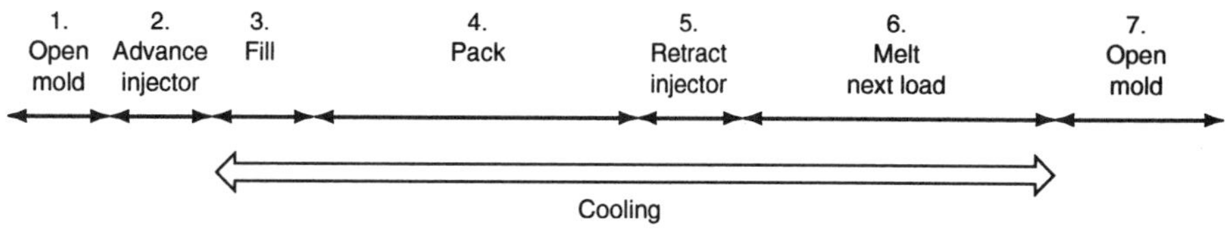

Fig. 1 Plastic injection molding cycle

most of the elements of the manufactured part cost. Fixed costs associated with equipment and tooling are affected because the cycle time influences the number of parallel streams necessary to manufacture a specified production volume. Variable costs are also affected. Labor and energy requirements are dependent on the time it takes to complete the production process.

Example: Estimating Cycle Times for Plastic Injection Molding. In general there are two ways to estimate the cycle time for a given manufacturing process, product design, and materials: theoretical and statistical. For example, consider a relatively simple process, plastic injection molding. The injection molding cycle can be broken into the steps shown in Fig. 1 (Ref 17).

The packing and cooling stages are most suited to theoretical modeling. Packing refers to the injection of additional material into the mold even after the mold is filled. This is done to compensate for shrinkage due to cooling. Packing stops when the material in the injection sprue solidifies, preventing any more material from entering. By computer modeling of the temperature profile in the sprue, one can simulate when this event happens.

A proprietary computer program, MoldFlow (Ref 18), also models the pressure that develops throughout the part during this phase by using a finite element methodology. MoldFlow also models cooling with a finite difference methodology.

If detailed designs (e.g., computer-aided design, CAD, drawings) of the part are available, using theoretical simulation models to simulate the cycle time of the crucial packing and cooling stages is possible. If these designs are not available, it is possible to predict the cycle time using an empirical approach, usually involving statistical estimation.

The statistical approach requires that data be collected from a number of production sites and an equation be fitted to that data using regression analysis. In one study (Ref 19), it was postulated that the cycle time depends primarily on two factors: the time to fill and pack, and the time to cool. Filling and packing are assumed to be a function of the part volume and therefore (through the density) are proportional to the weight of the part.

The form of the equation was:

$$\text{cycle time} = (a + b)(\text{cooling time} + c)(\text{weight})$$

The coefficients (a, b, c) were estimated regression analysis using data from 33 manufacturing operations, representing parts weighing between 5 and 2780 g.

Once the materials and simple part geometry are known, it is easy to calculate the part weight. The cooling time can be estimated using transport theory (Ref 20). In this case, it was assumed that the injection molded parts of interest conform to a slab geometry because they usually have relatively thin and uniform wall thicknesses. The cooling time for a semi-infinite slab is given by the relationship:

$$\text{cooling time} = \left(\frac{d^2 \rho C_p}{\Pi^2 k}\right) \ln \left[\frac{8(T_{\text{melt}} - T_{\text{mold}})}{\Pi^2 (T_{\text{eject}} - T_{\text{mold}})}\right]$$

where d is the nominal wall thickness, ρ is the density, C_p is the specific heat, k is the thermal conductivity, T_{melt} is the process melt temperature, T_{mold} is the temperature at the tool surface, and T_{eject} is the ejection temperature. These data are available in Volumes of the *ASM Handbook* and other materials references. The wall thickness will be known from the part geometry.

The regression estimates cannot be expected to be precise because the estimation of the cooling time represents the minimum time for cooling and does not take into account the information of hot melt during packing or any curved (not slab like) sections in the part. Moreover, each plant is usually operated differently. Nevertheless, the statistical results are relatively good, with the cycle time equation explaining 90% of the variation in the data, and each independent variable statistically significant with a 99% level of confidence. Agreement between the empirical/statistical approach and MoldFlow simulations is within 5% for wall thicknesses between 1 and 5 mm. At wall thicknesses of 10 mm, the approaches result in cooling time estimates that differ by about 20%.

Case Study: Cost Estimation for the Automotive Body-in-White

The cost methods discussed so far are useful for analyzing manufacturing costs. Although manufacturing costs are of greatest importance to product designers, other costs are becoming increasingly relevant. These can be divided into two categories: private costs (those borne by individuals or firms) and social costs (those borne by society). In addition to manufacturing costs, private costs include the costs of using and disposing of (or recycling) the product. Social costs include the cost to society of emissions resulting from extraction, processing, using, and disposing of the materials/products.

These various elements of cost are introduced in the following section with a case study concerned with the manufacture, use, and disposal of the automobile body-in-white. The body-in-white consists of the structural components of the automobile, not including closures (e.g., doors, hood, trunk). Currently a number of designs employing combinations of steel, aluminum, and polymer composites are under consideration by automotive companies.

Technical Cost Modeling of the Body-in-White

The technical cost modeling method has been used extensively to determine manufacturing cost for many production processes, including those involved in the production of automobiles. Body-in-white manufacture begins with the production of individual components, typically metal stampings, but in some cases extrusions and castings are used as well. The parts are assembled using several joining techniques into a complete body-in-white. The choice of joining technique is dependent on the design and material used.

The case study presented here includes several alternative designs; a traditional steel unibody, as well as an aluminum unibody and three aluminum space-frame designs (Ref 16, 21). The unibody designs are based on Ford Taurus. The space-frame designs were chosen to represent a comparable automobile.

In general, an aluminum space-frame design is composed of extrusions, stampings, and castings. In most cases, the extrusions form the predominant load transmitting paths. The frame structure consists of straight and curved extruded profiles, which are primarily joined by cast nodes. These nodes enhance the local rigidity of the structure. Furthermore, castings can be used in areas where the complexity of the design prohibits the use of extrusions, such as in the case of the grill opening. Finally, stamped panels complete the design, in some cases providing some additional structure while in others they are simply "hang-ons." This is entirely different from the unibody design where the stamped panels make up the complete structure of the vehicle.

The availability of part forming operations and joining methods leads to numerous viable design choices for an all aluminum space frame. This case study considered three actual designs developed by aluminum companies for different production volumes (Ref 21). The first design (SF-1) has a high number of aluminum extrusions and no castings. Consequently it is better suited for low production volume applications. The second design (SF-2) consists of a limited number of casting, while the third design (SF-3) uses castings extensively.

Cost models of numerous manufacturing processes were used for this study. All of the designs require the stamping of sheet metal into body panels. Furthermore, in the case of aluminum space frames, many parts are extruded or cast and then machined. Finally, each of the individual components must be assembled using potentially different joining methods into a complete body-in-white.

A typical steel unibody contains more than 100 stampings. It is usually unnecessary to estimate the cost of producing each of these parts. Instead, stampings may be grouped into categories of similar parts, for which materials cost will constitute the same percent of their total costs. This means that it is necessary to model only one part of each type and then to extrapolate the results to all similar components based on the part weight.

Two considerations were used to determine the part classifications: appearance versus nonappearance parts, and complex versus simple geometries. Complex geometries consist of either a deep draw or a large cut out. The roof panel and quarter panel outer were chosen as indicative of the appearance parts. The roof constitutes a simple geometry, while the quarter panel outer is a more complex stamping. The rear floor pan and the quarter panel inner were chosen for the nonappearance parts. The quarter panel inner has a simple geometry, and the rear floor pan has a complex geometry due to the deep draw, which is necessary for the spare tire wheel well (Ref 16).

A metal stamping technical cost model was used to determine the total and material cost of the four characteristic parts. A separate manufacturing model was used to extrapolate these results to all of the parts in the body-in-white. For the aluminum space-frame designs, extrusion, casting, and machining models were used to estimate the costs of parts that were not produced by traditional metal stamping processes.

The body-in-white manufacturing costs were estimated using the results from the various parts fabrication models and the manufacturing extrapolation model in conjunction with a vehicle assembly model, assuming combinations of spot welding and adhesive bonding.

At low production volumes, some of the aluminum space-frame designs could hold a cost advantage over the steel unibody:

Body type	Cost per unit, $, for a total annual production volume of: 20,000	100,000	300,000
Steel unibody	5774	2000	1205
Aluminum unibody	7249	2510	1788
SF-1	4471	...	...
SF-2	...	2926	...
SF-3	6073	2792	2404

Source: Ref 21

Figure 2 shows the effect of production volume on cost for each alternative. In this case, the cost of the three space-frame alternatives is represented as a single function, which includes only the lowest cost design at each production volume.

At high production volumes, the space-frame designs cannot compete on a cost basis with steel (or even with the aluminum unibody). The two monocoque designs are manufactured using stampings and spot welding, both of which are very capital intensive and depend on economies of scale to be cost effective. The main production techniques used for the space-frame designs do

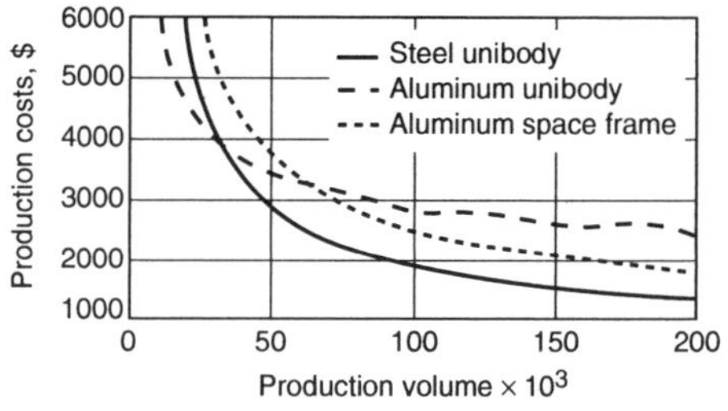

Fig. 2 Production cost comparisons for alternative vehicle designs. Source: Ref 21

not exhibit the same economies of scale and, therefore, are not cost competitive at high volumes.

Life Cycle Cost Analysis of the Body-in-White

In addition to the direct manufacturing costs, it is also possible to analyze the costs borne by the consumer (i.e., the cost of driving and disposal/recycling). This section discusses the life cycle costs of the alternative body-in-white designs. This article also includes a discussion of life cycle analysis (LCA), which is a separate, but related, concept (see the section "Life Cycle Analysis").

The cost of fuel is a strong function of fuel economy. For the steel unibody design, the fuel economy is based on the Ford Taurus: 18 and 26 miles per gallon (mpg) in the city and on the highway, respectively. Using the EPA 55/45 cycle mix of these two driving modes, the steel unibody has a fuel economy of 21.6 mpg (Ref 16). The fuel economy of the aluminum designs is dependent on the weight of the vehicle.

The effects of lightweighting on fuel consumption have been analyzed extensively (Ref 16, 21-24). These studies do not yield identical results. Figure 3 shows that the varying results give a bounded region of possible gains in fuel economy for a given body-in-white weight. For the purposes of this study, an average value was used.

Using the fuel economies as provided by the average line in Fig. 3, it is possible to calculate the cost of powering the vehicle for a given body-in-white weight:

Body type	Lifetime fuel cost, $
Steel unibody	3961
Aluminum unibody	3656
SF-1	3651
SF-2	3630
SF-3	3679

Source: Ref 21

The fuel cost of the vehicle depends on the fuel economy (mpg), the number of miles driven over the lifetime of the vehicle, and the price of gasoline. In the United States, recent Motor Vehicle Manufacturers' Association data show that people drive an average of 10,250 miles per year and cars last an average of 12.5 years (Ref 21). These

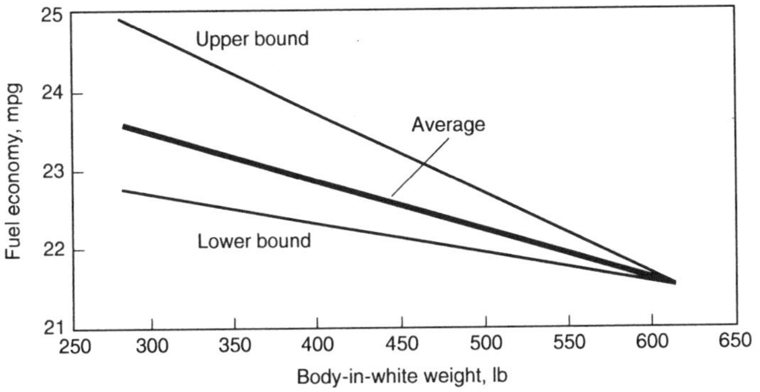

Fig. 3 Effect of vehicle lightweighting on fuel economy

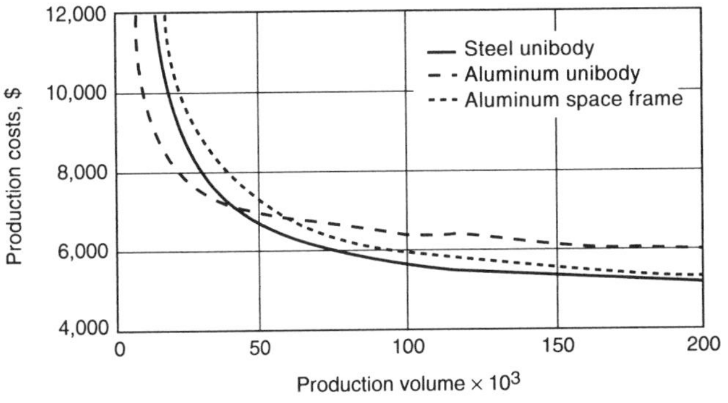

Fig. 4 Life cycle cost comparison for alternative vehicle designs

data were used with a gasoline price of $1.20/gal. A discount rate of 10% was used to convert future dollars to a present value.

Since each of the aluminum designs weighs about 300 lb less than the steel unibody (some more, some less), they offer some cost advantages in the use phase of the life of the vehicle. The vehicle lightweighting achieved in each of the aluminum designs results in use cost savings of approximately $300.

The final phase in the life of the automobile is the disposal stage. At this point, the vehicle has no more transportation value to the owner and has to be discarded. The monetary effects of this stage are restricted to the salvage value of the body-in-white. For this case study, scrap prices were assumed to be $0.05 for steel and $0.30 for aluminum (Ref 16). A discount rate of 10% was used to give the value of the disposed vehicle in present dollars. The net present value (NPV) is:

Body type	NPV, $
Steel unibody	7
Aluminum unibody	20
SF-1	20
SF-2	18
SF-3	22

Source: Ref 21

The aluminum designs have more recoverable value at the end of their useful life. However, in the case of aluminum versus steel, the magnitude of the disposal value is insignificant when compared with the costs arising from the manufacture or use stages.

The total life cycle cost of each design is the sum of its costs through the previously mentioned stages: manufacturing, use, and disposal. Figure 4 shows the effect of production volume on lifetime cost for each alternative. Once again, the cost of the three space-frame alternatives is represented as a single function, which includes only the lowest cost design at each production volume.

At all but the lowest production volumes, the steel unibody is the lowest cost design. However, steel has a diminished cost advantage when compared with the results of the analysis based strictly on manufacturing costs (see Fig. 2). Furthermore, the cross over production volume is shifted up. Based strictly on manufacturing costs, the steel unibody is the low cost design at production volumes down to only 34,000 vehicles per year. However, when total life cycle costs are included, the production volume must rise to above 40,000. Furthermore, the life cycle costs are based on a U.S. gasoline price of $1.20/gal. If this were to rise, or a price more typical of Europe were used, the cost advantage of aluminum in the use phase would be further enhanced, thus increasing its ability to compete at higher production volumes.

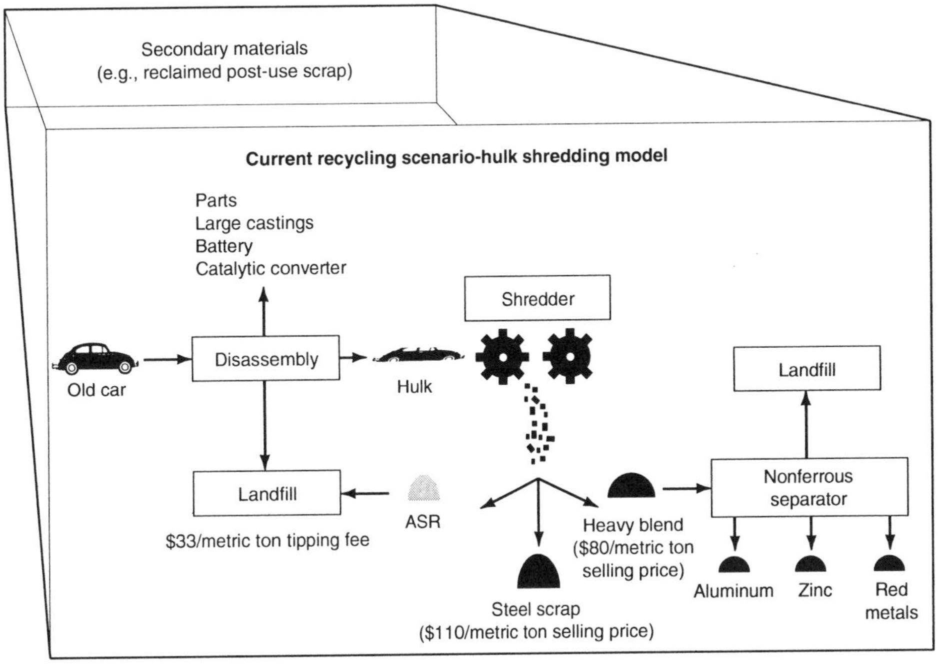

Fig. 5 Current automobile recycling infrastructure. Source: Ref 25

Recycling Economics of the Body-in-White

In the previous analysis, the disposal stage costs of each vehicle were estimated by looking simply at the market for the materials in the body-in-white. However in reality, the economics of recycling is quite a bit more complicated (Ref 25). There are costs associated with the disassembly of the vehicle, shredding of the hulk, separating out the valuable metals, and landfilling the remainder. All of these costs must be considered to get a true understanding of the costs/benefits arising from the disposal stage. Figure 5 shows the typical stages in automotive recycling and disposal.

Figure 5 illustrates just one possible process scenario for reclaiming post-use consumer scrap. In fact, this hulk shredding scheme is currently the dominant recycling procedure. Dismantlers typically pay the last owner of the vehicle $50 or more, depending on its condition. Reusable components and valuable materials are removed prior to shredding. Additional parts, such as tires and fluids, may also be removed, not so much for their resale value, but to ensure that the vehicle hulk is acceptable to the downstream shredder. However, complete vehicle disassembly is not

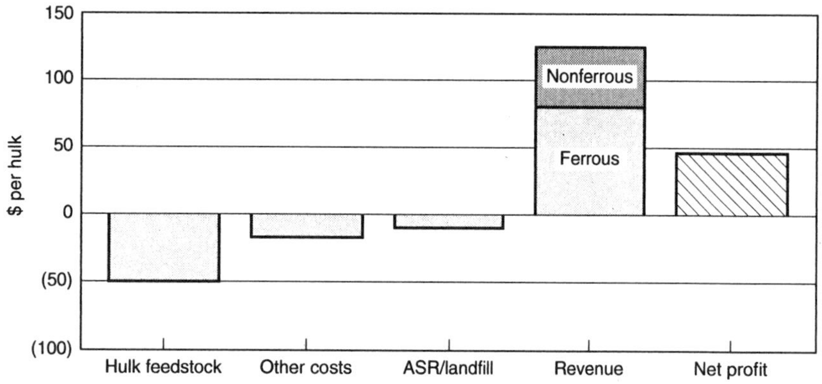

Fig. 6 Cost breakdown for the automobile shredding process (1990 model year). Source: Ref 26

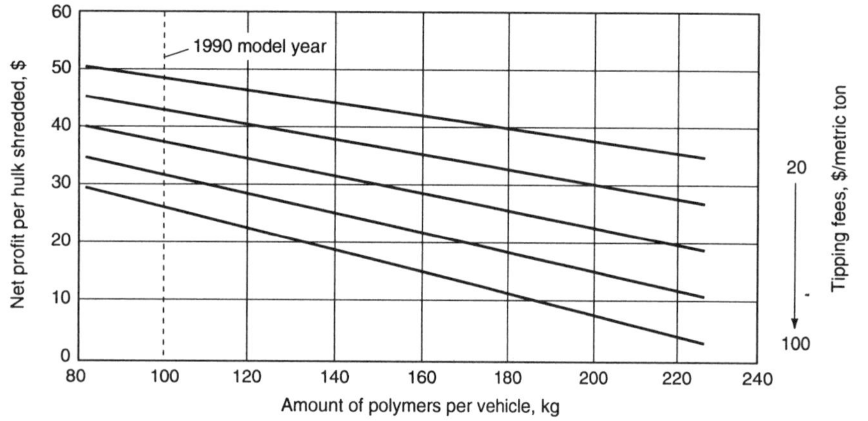

Fig. 7 Effect of polymer content and landfill tipping fee on profit per hulk. Based on price per metric ton of $110 for ferrous scrap and $880 for nonferrous scrap

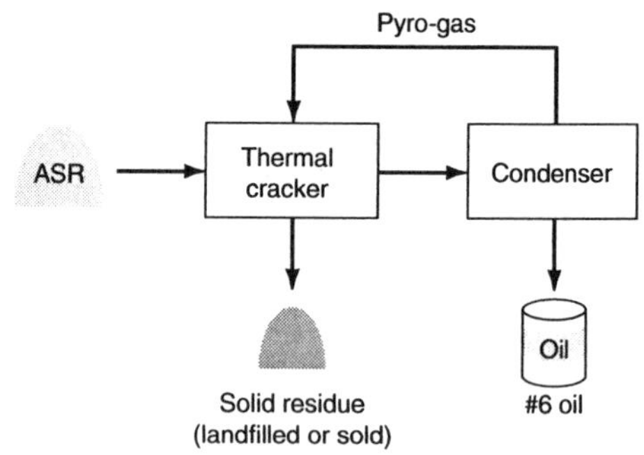

Fig. 8 Schematic diagram of the ASR pyrolysis process

economically advisable since there are diminishing returns associated with the process.

At the shredding facility, hulks are mechanically reduced into fist-sized chunks. Magnetic and specific gravity separation techniques are used to segregate the material into ferrous, nonferrous, and mainly nonmetallic automotive shredder residue fractions. The ferrous fraction is sold to electric arc furnace mini-mills. The nonferrous fraction is sold to specialized shops where aluminum, zinc, and copper can be segregated and resold to their secondary markets. The automotive shredder residue (ASR) is usually sent to a landfill.

Recycling Cost Estimation. The technical cost modeling methodology was applied to the recycling infrastructure to examine the major economic forces at work in this industry (Ref 26, 27). Each process within the recycling infrastructure was divided into its component process steps, and the costs (and revenues) for each of these were estimated. The analysis was based on a polymer intensive vehicle, for which the costs incurred during the disposal phase of the vehicle life may be significant.

Shredder Costs. The first baseline simulation was aimed at understanding the economics of the shredding process. A 1990 model year vehicle was used. In this case shredding resulted in 0.28 metric tons of ASR, 0.66 metric tons of ferrous scrap, and 0.05 metric tons of nonferrous scrap per hulk (Ref 26). The costs estimated with the use of a technical cost model and revenue of this process are shown in Fig. 6.

Given the relatively low disposal tipping fees in the United States, the landfill cost is a small fraction of the total operating cost. The high market value of the metallic fraction, accounting for approximately 71% by weight of the hulk, more than compensates for ASR's relatively small cost liability. The major cost contributor is the purchase of the hulk.

Industry profitability can be affected by changing market conditions such as landfill prices, material content of the hulk, and the feedstock price. Figure 7 demonstrates the extent to which the shredder profit is affected by rising landfill tipping fees and increasing polymer content in the hulk.

Figure 7 shows that only under the extreme conditions of high landfill fees and greatly increased polymer content would the shredder operate at a loss. At tipping fees as high as $100/metric ton, the shredder is still profitable. In fact, net loss for the hulk shredding process will not occur until tipping fees climb to nearly $200/metric ton. Currently, landfill prices in the United States are about $35 to $50 per ton. In Germany, landfill prices are in the $100 to 120/ton range and could rise to $200/ton in the next 5 to 10 years.

Despite the current profitability and robustness of the infrastructure, increasing attention is being focused on looking for alternative recycling schemes. One reason is that the preconditions for collapse of the infrastructure may be realized. Landfill prices are on an upward trend, especially in other countries, and the material content of the vehicle is changing. Furthermore, the hulk transfer price is a key cost to the shredder and is subject to changes in the materials scrap prices and landfill prices. Decreasing metal scrap prices could lower the price the shredder is willing to pay for the hulk. This will have consequences on the dismantler, who may then seek alternative disposal routes for the hulk. The dismantler could in turn transfer this cost on to the last user of the vehicle. This could result in the dismantler charging the vehicle owner for disposal of the car rather than paying for the used vehicle. Vehicle owners may then seek alternate disposal routes, thus bypassing the current recycling infrastructure.

ASR Treatment. Another approach to the vehicle disposal issue involves treatment of the shredder residue. A major advantage of these approaches is their compatibility with the current recycling infrastructure, which is based on shredder processing. The goal of these methods is to recover more of the value imbedded in the ASR, which is currently landfilled, and the concomitant reduction in landfill costs.

Consider two ASR treatment processes, pyrolysis and selective precipitation. Pyrolysis brings the recyclate back to the feedstock stage as petroleum products. Selective precipitation brings the recyclate back to the intermediate product stage as polymer resins (Ref 26, 27).

Pyrolysis is the thermal decomposition of organic materials into oil, gas, and solid residue. Energy is required to start the reaction, but is not necessary to sustain it due to the generation of gas by the process. The resulting oil can be sold, and the solid residue can either be landfilled or sold for its fillers and metal scrap content. A schematic of the ASR pyrolysis process is given in Fig. 8.

The selective precipitation process is aimed at collecting higher valued recyclates. Mechanical separation is used to remove any fluids and obtain both a polymer rich stream and "fines." Fines

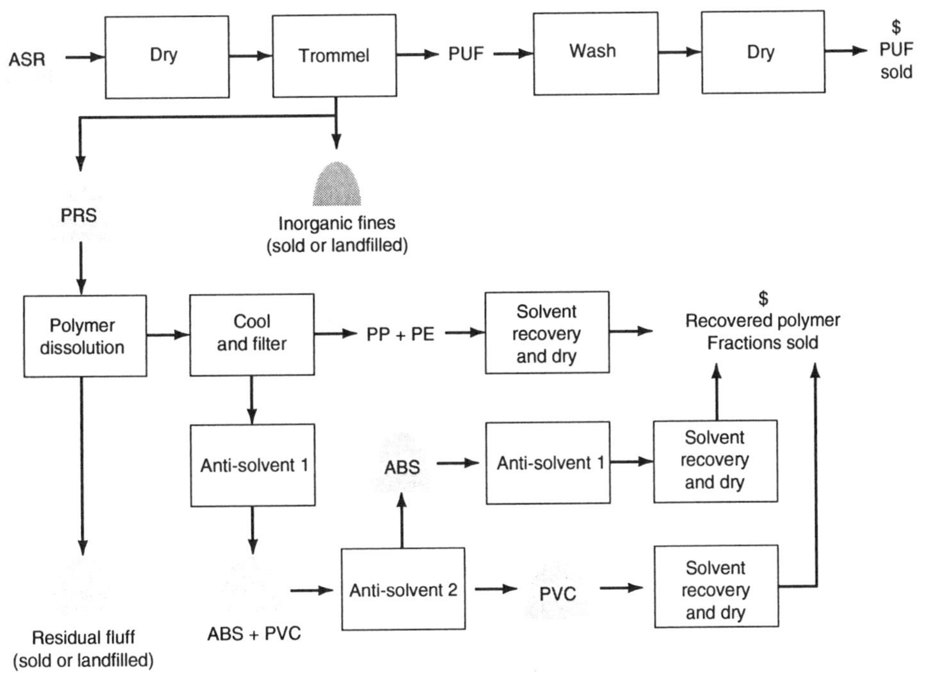

Fig. 9 Flow diagram of the ASR mechanical separation/selective precipitation process. ABS, acrylonitrile-butadiene-styrene; PE, polyethylene; PP, polypropylene; PRS, polymer-rich stream; PUF, polyurethane foam; PVC, polyvinyl chloride. Source: Ref 26

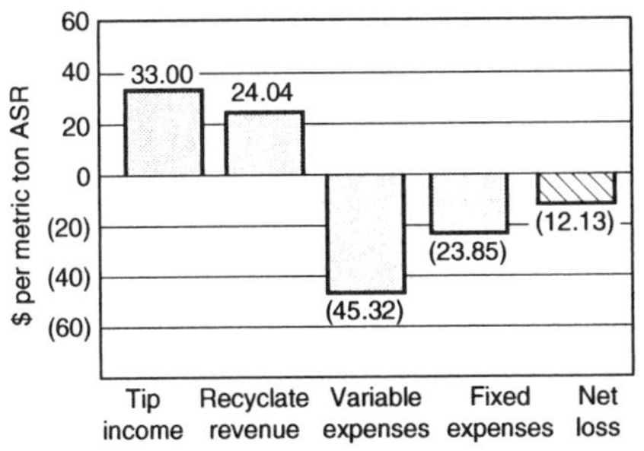

Fig. 10 ASR pyrolysis cost breakdown. Source: Ref 26

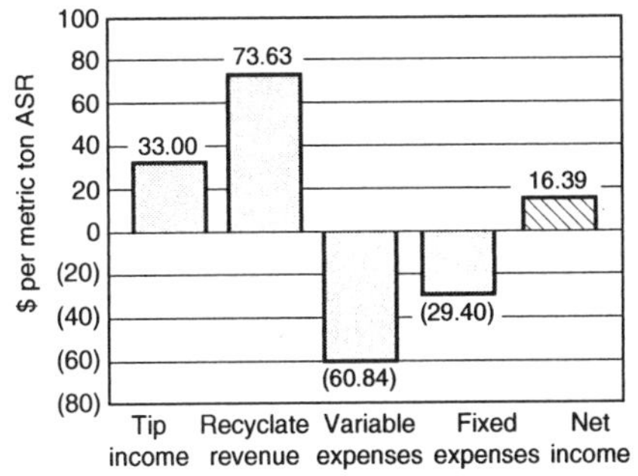

Fig. 11 ASR mechanical separation/selective precipitation cost breakdown. Source: Ref 26

are small particles (less than 0.6 cm in diameter), which are rich in iron oxide and silica. These can be sold as feedstock for cement making. The selective precipitation process then uses extracting solvents at elevated temperatures to remove specific thermoplastic polymers. In this case study, acrylonitrile-butadiene-styrene (ABS), polyvinyl chloride (PVC), and a polyolefin mix of polyethylene (PE) and polypropylene (PP) were considered. Figure 9 shows a schematic of this process, which was developed at Argonne National Laboratory (Ref 26).

Cost models were developed for each of the two ASR treatment processes. The results are shown in Fig. 10 and 11.

The selective precipitation process has a potential profit of \$16/metric ton, while the pyrolysis process has a projected loss of \$12/metric ton of processed ASR. This difference is a consequence of the reclamation of higher valued recyclates in the selective precipitation technique. In fact, these higher valued products more than offset the increased labor and capital cost associated with the process. The recyclate revenues from each process are critical factors, which affect the relative profitability of the processes.

Both of the above two cost projections assume a near 100% capacity utilization. As can be seen, there is a limit to how greater recycling economies of scale can lead to lower recyclate costs. While greater plant utilization certainly leads to a lower per unit fixed cost, the processes modeled still cannot independently achieve profitability. The tipping fee is an important revenue source for either process even at the low values representative of the U.S. prices. In Germany, where tipping fees are about four times those in the United States, landfill prices are the critical variable. Variable costs, such as energy and labor, together with fundamental physical and chemical constraints of materials processing, ultimately impose a financial asymptote.

An additional critical factor for the economics of these processes is the material throughput. Figure 12 shows the net profit of each process for varying levels of ASR throughput.

Regardless of the throughput, pyrolysis results in a net loss, while it is possible for selective precipitation to achieve a profit at high volumes. Another interesting view is based on the availability of shredder fluff in a given region. If less than 30,000 metric tons of ASR (100,000 to 110,000 shredded hulks) can be collected, pyrolysis may be the more economical alternative. There is still a net loss associated with this process, but it is small relative to the selective precipitation technique.

Environmental and Social Costs

Public concern for the environment is increasing, and thus "environmental performance" is becoming the newest strategic objective in mate-

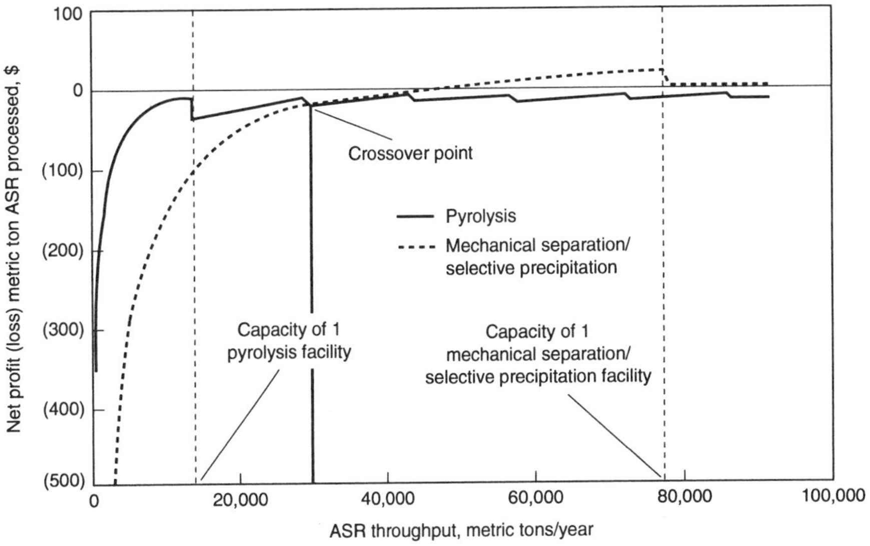

Fig. 12 Throughput implications in alternative recycling processes. Source: Ref 26

Table 1 Percentage of emissions generated during each stage of the vehicle life cycle for a steel unibody and an average aluminum design

Emission type	Emissions produced, %: Mining/refining	Production	Use	Post-use
Steel unibody				
CO_2	2.32	0.05	97.63	~0
HC	0.10	~0	99.90	0
NO_x	2.21	0.04	97.74	~0
CO	0.05	~0	99.95	~0
Particulate	23.30	0.50	76.20	~0
SO_x	97.78	1.97	0	0.25
Aluminum body				
CO_2	8.65	0.09	91.26	0
HC	0.40	~0	99.60	0
NO_x	8.27	0.08	91.64	~0
CO	0.21	0	99.79	0
Particulate	84.10	0.05	15.85	0
SO_x	98.98	0.99	0	0.04

rials selection for automotive design. This objective overlaps with weight reduction, but it is more complex when manufacturing and recycling issues arise. How should designers measure total "environmental performance," and how can they integrate those measures into the whole materials selection problem?

Life Cycle Analysis

Life cycle analysis (LCA) is one of the most actively considered methods for the study and analysis of strategies to meet environmental challenges. Its strengths derive from its roots in traditional process analysis, and from the recognition, implicit in its formulation, that the consequences of technological undertakings are not limited to the performance of a single process or change. Instead, the majority of the effects deriving from an action may only be perceived when the entire range of consequences of that action are taken into account.

The objective of LCA is to provide decision makers, whether consumers, industrialists, or government policy makers, with information that will allow them to understand the environmental impacts of their actions. In order to meet their needs, this tool must be used in alliance with traditional bases for decision making, including economic, engineering, and social objectives.

Life cycle analysis is frequently characterized as a three-step process (Ref 28):

1. *Inventory analysis:* The identification and quantification of energy and resource use and environmental releases to air, water, and land
2. *Impact analysis:* The technical qualitative and quantitative characterization and assessment of the consequences on the environment of these changes
3. *Improvement analysis:* The evaluation and implementation of opportunities to reduce environmental burdens

Inventory. Airborne emissions of carbon dioxide (CO_2), carbon monoxide (CO), particulates, nitrogen oxides (NO_x), sulfur dioxide (SO_2), and other hydrocarbons (HC) are usually considered to be of greatest importance in the automotive body life cycle. The emissions arise from the energy required in each of the stages in the lifetime of the product. In addition, emissions arise from chemical reactions, mainly during the material mining and refining processes. These are strictly related to the material chosen for the production of components for the vehicle. Finally, emissions also arise during the use phase due to the combustion of gasoline.

Table 1 lists the emissions generated for a steel unibody and an average aluminum design over the life cycle of the vehicle. It is important to note that emissions from each of the four aluminum designs do not vary substantially. This is a consequence of the fact that they are similar in weight and have similar emissions released in the use phase. The fact that they are all aluminum designs also means that the emissions generated in the mining and refining stage will also be quite similar.

Results show that the aluminum designs perform better in the pollutant categories that dominate during the use phase (CO_2, hydrocarbons, NO_x, and CO). However, for particulates and sulfur oxides, the steel design is more competitive. Emissions associated with the aluminum design are greater for the mining and refining stage (sulfur oxides and particulates). For the categories that arise predominantly during vehicle use (CO_2, hydrocarbons, NO_x, and CO), the lightweighting achieved in the aluminum designs pays off.

Analysis of the inventory data does not lead to an unambiguous result. On a cost basis, even with a life cycle approach, the steel unibody is most competitive. However, if the goal is to reduce greenhouse gases and smog precursors, one of the aluminum designs may be preferred (Fig. 13).

Impact and Evaluation. Most efforts to develop the LCA technique have focused on constructing a complete set of procedures for the collection and organization of the information that must be developed in the course of a LCA. However, determining what to do with this information, once it is collected, has so far been only imperfectly addressed. Although the reason for employing LCA is to develop activities that reduce environmental impact, establishing how this mass of data informs specific problems has proven to be extremely difficult for all but the simplest of situations.

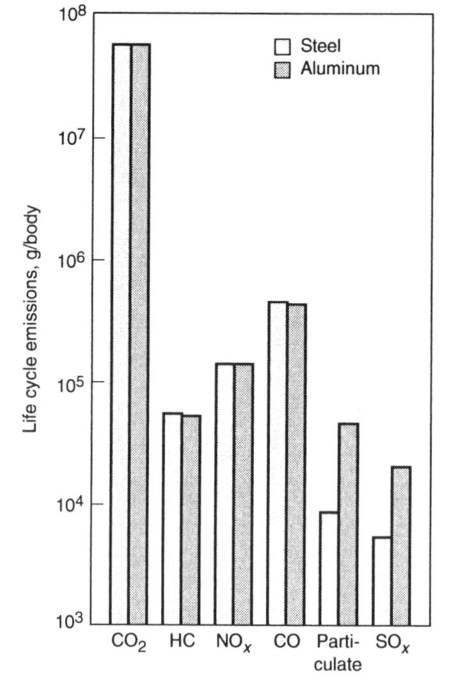

Emission type	Change when using aluminum, %
CO_2	−1.24
HC	−7.71
NO_x	−1.50
CO	−7.50
Particulate	+438
SO_x	+264

Fig. 13 Emissions throughout the entire product life cycle for steel and aluminum automobile bodies

In particular, the most problematic aspect of LCA has been the final, "improvement analysis" component. Improvement analysis implicitly assumes that it is possible to choose (and implement) a "best" action from the set of possible actions, thus yielding improvement. Aside from simple cases where it is possible to find an action that leads to reductions in all impacts on the environment, this choice depends upon the relative importance placed upon each of the possible consequences that are indicated by the analysis. This relative rating of importance is a reflection of the strategic objectives of the user—objectives that are not necessarily shared by all interested stakeholders.

Example: Method for Estimating the "Environmental Load" of Materials. To illustrate the potential and limitations of LCA method, the Swedish Environmental Priority Strategies (EPS), under development by the Swedish Environmental Research Institute, Chalmers Institute of Technology, and the Federation of Swedish Industries, are discussed (Ref 29). EPS translates emissions into a single monetary metric that allows the direct costs of manufacturing, use, and recycling/disposal to be compared with the social costs generated by emissions.

EPS is specifically constructed to associate an "environmental load" with individual activities or processes on a per unit of material consumed or processed basis. For example, EPS might associate X units of environmental load (ELUs) per kilogram of steel produced and Y units of environmental load per kilogram of steel components stamped. Thus, the environmental load of stamping a 5 kg automobile component, requiring 5.3 kg of steel, would be (5.3 X + 5 Y). This load could then be compared to the load associated with a different process stream or with using a different material. The interesting questions are: how are these environmental loads established and what do they mean.

Based on the environmental objectives of the Swedish Parliament, EPS relates all of the physical consequences of the processes under consideration to their impact on five environmental "safeguard subjects": biodiversity, production (i.e., reproduction of biological organisms), human health, resources, and aesthetic values. Because the impacts on any one safeguard subject by a process may take several forms, EPS allows for individual consideration of each of these consequences, called "unit effects." Two criteria are applied when establishing which impacts will become unit effects: the importance of the impact on the sustainability of the environment and the existence of an ability to establish a quantitative value for that impact within traditional economic grounds. Examples of unit effects for human health include: mortality due to increased frequency of cancer; mortality due to increased maximum temperatures; food production decreases (and, hence, increased incidence of starvation) due to global warming.

Once the individual unit effects are established, their value must be determined. This valuation is accomplished by expressing each unit effect in terms of its economic worth and associated risk factors. Formally, the value of each unit effect is set equal to the product of five factors, F1 through F5. F1 is a monetary measure of the total cost of avoiding the unit effect. The extent of affected area (F2), the frequency of unit effect in the affected area (F3), and the duration of the unit effect (F4), represent "risk factors" similar to those employed in toxicological risk evaluations. F5 is a normalizing factor, constructed so that the product F1 × F5 is equal to the cost of avoiding the unit effect that would arise through the use or production of one kilogram of material. The product of all five factors yields the contribution of a particular unit effect to environmental load. Summing the value of each unit effect yields the "environmental load index" (ELI) in units of environmental load per unit of material consumed or processed (ELU/kg). Since these unit effects were specified according to their relevance to the five safeguard subjects, the ELI represents the total environmental load (or impact) of the process.

While this formulation of valuation raises important questions of scientific feasibility (insofar as the ability to characterize fully the unit effects of every process or activity that might be developed is debatable), the crucial valuation questions arise from two other aspects of this scheme: (a) the nature of the economic measures used in calculating the cost of avoiding a unit effect, and (b) the assumption that the value of the total environmental impact of an action (the "environmental load") is equal to the sum of each individual environmental load weighted by the size of each unit effect.

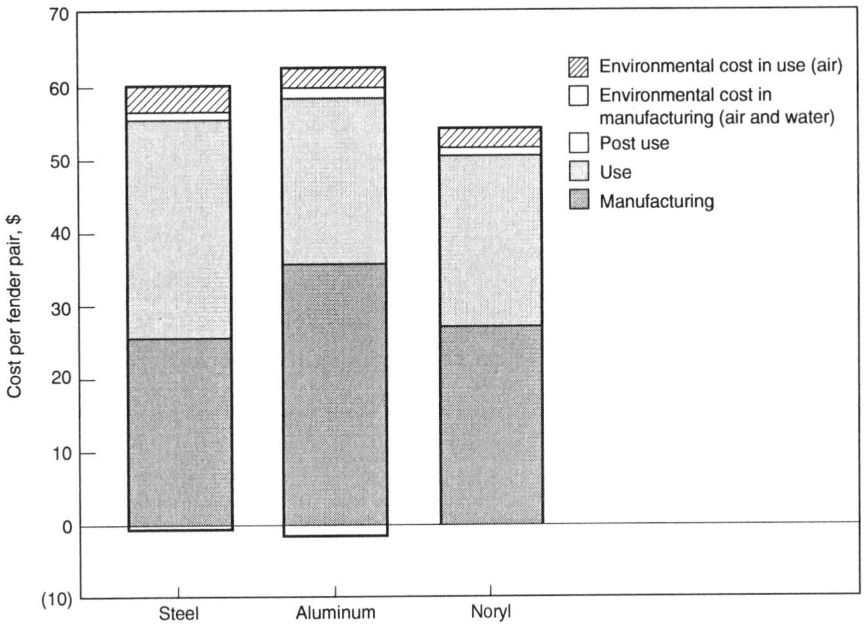

Fig. 14 Estimated life cycle costs by life phase for competing materials for an automobile fender application

The first of these valuation questions relates to the distinction between "cost" and "worth." While the theory of competitive markets argues that prices are the worth of an object, the theory rests upon assumptions that are difficult to support in the case of environment. In the first place, perfect markets assume the availability of perfect information to all participants, which clearly is not the case, or there would be no need to develop life cycle analysis in the first place. Furthermore, the theory of markets routinely discusses "consumer surplus," which can roughly be defined as the difference in the prevailing market price and the higher price that some consumers would have been willing to pay (recall that demand curves slope downward). Finally, there is the critical question how to establish these costs/prices when markets do not exist. While litigators are prepared to place a value on wrongful death or pain and suffering during a civil suit, there are no markets for pain, clean air, or future well-being. Generally, most environmental attributes are "external" to markets; many of the classical examples of market externalities are based on environmental issues.

Where markets exist, EPS uses market prices to establish the costs of avoidance. Where market prices do not exist, EPS relies upon two alternatives. If there are governmental funds allocated to resolve specific problems (e.g., funds to protect a particular species), these funds are normalized and extrapolated to obtain a cost figure (e.g., the value of maintaining biodiversity is established by normalizing the annual budget of the Swedish government for species protection). If relevant financial allocations do not exist, then the method of contingent valuation is employed. This method (or set of methods) is based on direct inquiries of representative populations to determine their willingness to pay to avoid specific effects. As might be expected, this last approach to establishing the appropriate costs of avoidance is somewhat controversial, since it is hard (both conceptually and practically) to design questions that demonstrably extract the "correct" measure of value.

The second of these valuation questions is a reflection of the fact that the mathematical structure of the value function is a consequence of critical assumptions about the nature of the subject's preferences. The valuation employed in the EPS system is an example of a linear, additive preference structure. Each unit effect is reduced to a monetary value, normalized for risk/exposure and for material quantity. Thereafter, the net impact of each increment in unit effect is the same, regardless of how large the effect is, and regardless of the size of any other unit effect. While such value functions are simple to represent and employ (linear combinations of linear functions), it is difficult to argue that they are an accurate, general purpose formulation of value functions for environmental impact. Although the appropriate form of the value function may be linear, EPS does not explicitly make this assumption. Rather, the linearity of EPS valuation is based on the assumption that, because monetiza-

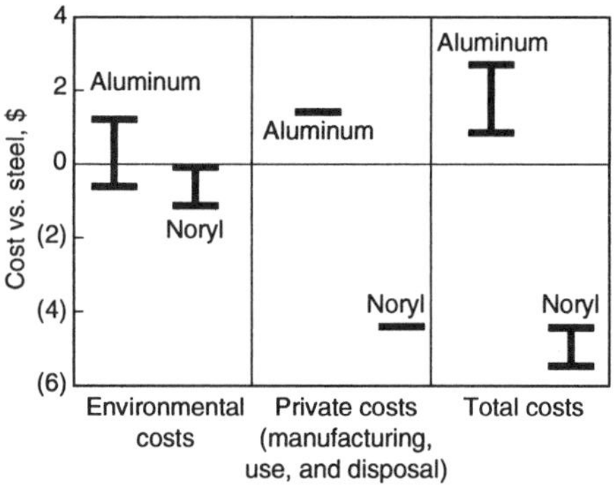

Fig. 15 Total costs relative to steel of competing materials for an automobile fender application

tion reduces all effects to a common metric, the resulting metrics should be additive. In fact, most individuals do not even exhibit linear preferences for money, much less for more subjective attributes. (For example, most individuals would consider paying $0.50 to play a game offering a 50:50 chance of winning $1.00, while rejecting out of hand paying $5,000 to get a 50:50 chance of winning $10,000). In practice, preferences usually reflect nonlinearities in both individual effects and in substitution between effects.

The first two issues (money as a measure of value and linear additive preferences) are not necessarily crippling assumptions when considering the development of value functions for the environment. While difficult, it may be possible for someone to establish the dollar value that exactly offsets a particular unit effect. Similarly, linear additive preferences may be able to model the behavior of an individual over a restricted range. However, it is impossible to state that the same dollar value, or the same linearization of preferences, will be agreeable to every individual in the affected population in the case of environmental considerations. And, if individuals cannot agree on the value or the structure of their preferences, then no single value function can be constructed to represent their wants.

A recent methodology developed at MIT (Ref 30) is similar to EPS, but provides a set of broad ranges of value, in dollars per kilogram of each emission, based on estimates of willingness to pay to avoid the environmental impacts of each pollutant. These ranges reflect scientific uncertainty, variation in context or location, and large variations of possible values for parameters that have a subjective component. The dollars per kilogram ranges can be applied to the life cycle inventories of products to compare material or process alternatives.

The methodology was used to analyze the life cycle costs of three material alternatives for automotive fenders produced at low volumes (60,000/year). The three materials under consideration were steel, aluminum, and Noryl (Noryl is a trademark of General Electric Company for a polyphenylene oxide blend thermoplastic). The results of the base case, employing "best guess" for scientific data and economic valuation, are shown in Fig. 14.

In this scenario, the private costs of manufacturing and use (with German gasoline prices) are significantly greater than the social costs from emissions to the environment. Figure 15 shows the implications of allowing the scientific and economic assumptions to take on the highest and lowest values possible, based on a review of published estimates.

The externalities are the environmental costs of emissions from the extraction to the manufacturing and use stages. The private costs include manufacturing, use, and disposal. The specific assumptions employed in this case study lead to a lower total cost for Noryl, although no clear winner arises under this set of assumptions. Even when no clear choice emerges, the environmental cost drivers can be identified. For instance, the fender case study shows that only 4 or 5 emissions categories account for more than 95% of the total environmental cost for each material.

The EPS system is a commendable attempt at simplifying the enormous detail of inventory data to a representative environmental load. The developers of EPS have pointed out that this system is based on their subjective value judgments, which are not necessarily supportable in all situations worldwide. The ultimate goals for improvement analysis based on life cycle inventories are laudable, but can only be realized by some kind of consensus on the values for avoiding environmental degradation. This suggests that achieving the ultimate stage of LCA will require the development of a basis for devising (and revising) this consensus. In the absence of a common strategic objective, it will be impossible to use LCA to designate ways to achieve environmental improvement beyond straightforward pollution prevention/precautionary principle strategies, because a strategic consensus is required to trade off competing environmental, economic, and engineering goals.

Uses of LCA. In summary, life cycle analysis is a technique that has already shown great promise for improving our understanding of the wider implications and relationships that must be taken into consideration when incorporating environmental concerns into technical decision making. As these concepts diffuse into industrial and technical decision making, LCA will enable industry and government to find ways to be both more efficient and less harmful to the environment.

However, practitioners and proponents must guard against using LCA to determine "best" modes of action when the consequences of the alternatives expose conflicting objectives and values within the group of decision makers. In these cases, no amount of analysis will directly resolve the conflict. Rather, the role of LCA should be to articulate clearly the consequences of each alternative and to provide a framework for the necessary negotiations.

Additional information about LCA is provided in the articles "Life Cycle Engineering and Design" and "Design for the Environment" in this Volume.

Conclusions

There is an ever growing need for consistent methodologies for analyzing the use of new materials, designs, and technologies in many applications. Advances in materials science and in the development of new processing technologies have presented product designers with a wide array of choices previously unavailable to them. This has made the selection of a material for a given application a far more challenging task.

The difficulty confronting designers is compounded by the increasing number of objectives that product designers must satisfy. In the past, the designer simply had to meet a set of performance criteria, at or below a specified cost, from a very limited set of design alternatives. The current situation is much more complicated. In addition to the increasing number of design choices, there are potentially conflicting performance, cost, and environmental characteristics.

Central to all product evaluations is a consideration of the economic consequences of design and materials choice. Cost is one of the key strategic elements of product competitiveness, and an early appreciation of the relationship between major design choices and the cost of the resulting product is a vital element of effective product development. However, cost remains an elusive element of design evaluation. The tools are largely outside the control of the design engineers. The results suggest only a limited number of ways in which cost can be changed, and the costs tend to focus only upon the cost consequences to the firm itself. Unfortunately, designers require a far more comprehensive appreciation of cost, particularly as the number and complexity of design objectives have increased.

The combined technical cost modeling and life cycle analysis methodology offers the product designer a much needed systematic approach for analyzing the trade-offs associated with various choices of materials and technologies. Technical cost modeling enables designers to estimate the manufacturing costs of alternative designs. Its main advantages lie in the fact that it is predictive and allows one to investigate the sensitivity of the outcome to changes in the input parameters. Because it is predictive, it can be used with new processes for which there is no past experience upon which to base cost estimates. The ability to do sensitivity analysis enables the product designer to look at the effects of unknown or uncertain model parameters, capturing the scope and consequences of important processing and market assumptions.

The advantages of the life cycle approach are also two-fold. First, life cycle analysis enables one to look at cost over the entire life of the product, not just the manufacturing phase. For many products, cost can be quite substantial during other parts of the product life, especially the use phase. Second, life cycle analysis is useful for looking at issues relevant to environmental concerns, such as tracking selected emissions throughout the product life. While valuation techniques are rather imperfect, they provide a means for translating these diverse parameters into a common metric, as well as a context for analyzing the implications of distinctions in the strategic objectives of all parties affected by the product and design choice.

The integrated approach provided by technical cost modeling and life cycle analysis is particularly important in industries such as the automotive sector, where both consumer and regulatory pressures are causing the producers to continuously innovate. The combined life cycle cost and emissions methodology offers a systematic and predictive method for addressing some of the fundamental considerations involved in selecting materials and designs for specific products.

REFERENCES

1. R. Roth, F. Field, and J. Clark, Materials Selection and Multi-Attribute Utility Analysis, *J. Computer-Aided Mater. Des.*, Vol 1 (No. 3), ESCOM Science Publishers, Oct 1994
2. J.V. Busch and F.R. Field III, Technical Cost Modeling, *Blow Molding Handbook*, Donald Rosato and Dominick Rosato, Ed., Hanser Publishers, 1988, Ch 24
3. M.F. Ashby, *Materials Selection in Mechanical Design*, Pergamon Press, 1992
4. R. Cooper and P. Kaplan, Measure Costs Right: Make the Right Decisions, *Harvard Business Review*, Sept–Oct 1988
5. "Implementing ABC in the Automobile Industry: Learning from Information Technology Experiences," MIT International Motor Vehicle Program working paper
6. J.F. Elliot, J.J. Tribendis, and J.P. Clark, "Mathematical Modeling of Raw Material and Energy Needs of the Iron and Steel Industry in the USA.," *Final Report to the U.S. Bureau of Mines*, NTIS PB 295-207 (AS), 1978
7. F.E. Katrak, T.B. King, and J.P. Clark, Analysis of the Supply of and Demand for Stainless Steel in the United States, *Mater. Soc.*, Vol 4, 1980
8. P.T. Foley and J.P. Clark, U.S. Copper Supply—An Engineering/Economic Analysis of Cost-Supply Relationships, *Resour. Policy*, Vol 7 (No. 3), 1981
9. J.P. Clark and G.B. Kenney, The Dynamics of International Competition in the Automotive Industry, *Mater. Soc.*, Vol 5 (No. 2), 1981
10. J.P. Clark and M.C. Flemings, Advanced Materials and the Economy, *Sci. American*, Oct 1986
11. Lee Hong Ng and Frank R. Field III, Materials for Printed Circuit Boards: Past Usage and Future Prospects, *Mater. Soc.*, Vol 13 (No. 3), 1989
12. S. Arnold, N. Hendrichs, F.R. Field III, and J.P. Clark, Competition between Polymeric Materials and Steel in Car Body Applications, *Mater. Soc.*, Vol 13 (No. 3), 1989
13. V. Nallicheri, J.P. Clark, and F.R. Field, A Technical & Economic Analysis of Alternative Manufacturing Processes for the Connecting Rod, *Proceedings, International Conference on Powder Metallurgy* (Pittsburgh, PA), Metal Powder Industries Federation, May 1990
14. C. Mangin, J. Neely, and J. Clark, The Potential for Advanced Ceramics in Automotive Engines, *J. Met.*, Vol 45 (No. 6), 1993
15. F.R. Field and J.P. Clark, Automotive Body Materials, *Encyclopedia of Advanced Materials*, R.W. Cahn et al., Ed., Pergamon Press, 1994
16. H. Han and J. Clark, Life Cycle Costing of the Body-in-White: Steel vs. Aluminum, *J. Met.*, May 1995
17. G. Potsch and W. Michaeli, *Injection Molding: An Introduction*, Hanser Publishers, 1995
18. P. Kennedy, *Flow Analysis Reference Manual*, Moldflow Pty. Ltd., Australia, 1993
19. J.V. Busch, "Technical Cost Modeling of Plastics Fabrication Processes," MIT Ph.D. thesis, June 1987
20. G.H. Geiger and D.R. Poirier, *Transport Phenomena in Metallurgy*, Addison-Wesley Publishing Company, 1973
21. D. Politis, "An Economic and Environmental Evaluation of Aluminum Designs for Automotive Structures," MIT S.M. thesis, May 1995
22. M.A. DeLuchi, "Emissions of Greenhouse Gases from the Use of Transportation Fuels and Electricity," Vol 2, U.S. Department of Energy, 1993
23. OECD, Automobile Fuel Consumption in Actual Traffic Conditions, Organization for Economic Co-Operation and Development, Dec 1981
24. SRI International, Potential for Improved Fuel Economy in Passenger Cars and Light Trucks, Prepared for the Motor Vehicle Manufacturers Association, Menlo Park, CA, 1991
25. F.R. Field and J.P. Clark, Recycling Dilemma for Advanced Materials Use: Automotive Materials Substitution, *Mater. Soc.*, Vol 15 (No. 2), 1991
26. A.C. Chen, "A Product Lifecycle Framework for Environmental Management and Policy Analysis: Case Study of Automobile Recycling," MIT Ph.D. thesis, June 1995
27. A.C. Chen, H.N. Han, J.P. Clark, and F.R. Field, A Strategic Framework for Analyzing the Cost Effectiveness of Automobile Recycling, *Proceedings, International Body Engineering Conference* (Detroit), M.N. Uddin, Ed., Society of Automotive Engineers, 1993, p 13–19
28. F.R. Field, J.A. Isaacs, and J.P. Clark, Life Cycle Analysis and Its Role in Product and Process Development, *J. Environmentally Conscious Manufacturing*, 1996
29. B. Steen and S.-O. Ryding, *The EPS Enviro-Accounting Method: An Application of Environmental Accounting Principles for Evaluation and Valuation of Environmental Impact in Production Design*, Swedish Environmental Institute, Dec 1992
30. J. Clark, S. Newell, and F. Field, Life Cycle Analysis Methodology Incorporating Private and Social Costs, in *Life Cycle Engineering of Passenger Cars*, VDI Verlag GmbH, 1996, p 1–19

Material Property Charts

M.F. Ashby, Engineering Design Centre, Cambridge University

MATERIAL PROPERTIES limit performance. However, it is seldom that the performance of a component depends on just one property. Almost always it is a combination (or several combinations) of properties that matter: one thinks, for instance, of the strength-to-weight ratio, σ_f / ρ, or the stiffness-to-weight ratio, E/ρ, which are important in design of lightweight products. This suggests the idea of plotting one property against another, mapping out the fields in property-space occupied by each material class, and the subfields occupied by individual materials.

The resulting charts are helpful in several ways. They condense a large body of information into a compact but accessible form, they reveal correlations between material properties that aid in checking and estimating data, and they lend themselves to a performance-optimizing technique (developed in the article "Performance Indices" following in this Section of the Handbook), which becomes the basis of the selection procedure.

The idea of a materials-selection chart is developed below. Further information about the charts and their uses can be found in Ref 1 to 3 and in the article "Performance Indices."

Displaying Material Properties

Each property of an engineering material has a characteristic range of values. The values are conveniently displayed on materials selection charts, illustrated by Fig. 1. One property (the modulus, E, in this case) is plotted against another (the density, ρ) on logarithmic scales. The range of the axes is chosen to include all materials, from the lightest foams to the heaviest metals. It is then found that data for a given class of materials (polymers for example) cluster together on the chart; the subrange associated with one material class is, in all cases, much smaller than the full range of that property. Data for one class can be enclosed in a property-envelope, as shown in Fig. 1. The envelope encloses all members of the class.

All this is simple enough—just a helpful way of plotting data. However, by choosing the axes and scales appropriately, more can be added. The speed of sound in a solid depends on the modulus, E, and the density, ρ; the longitudinal wave speed v, for instance, is

$$v = \left(\frac{E}{\rho}\right)^{1/2}$$

or (taking logs)

$$\log E = \log \rho + 2 \log v$$

For a fixed value of v, this equation plots as a straight line of slope 1 on Fig. 1. This allows the addition of contours of constant wave velocity to the chart: They are the family of parallel diagonal lines linking materials in which longitudinal waves travel with the same speed. All the charts allow additional fundamental relationships of this sort to be displayed.

A number of mechanical and thermal properties characterize a material and determine its use in engineering design; they include density, modulus, strength, toughness, damping coefficient, thermal conductivity, diffusivity, and expansion. The charts display data for these properties for the nine classes of materials listed in Table 1. Within each class, data are plotted for a representative set of materials, chosen both to span the full range of behavior for the class and to include the most common and most widely used members of it. In this way the envelope for a class encloses data not only for the materials listed in Table 1, but for virtually all other members of the class as well.

The charts show a range of values for each property of each material. Sometimes the range is narrow; the modulus of copper, for instance, varies by only a few percent about its mean value, influenced by purity, texture, and the like. Sometimes the range is wide; the strength of alumina-ceramic can vary by a factor of 100 or more, influenced by porosity, grain size, and so on. Heat treatment and mechanical working have a profound effect on yield strength, damping, and the toughness of metals. Crystallinity and degree of cross-linking greatly influence the modulus of polymers, and so on. These structure-sensitive properties appear as elongated bubbles within the envelopes on the charts. A bubble encloses a typical range for the value of the property for a single material (see Fig. 2). Envelopes (heavier lines) enclose the bubbles for a class.

The data plotted on the charts have been assembled from a variety of sources, the most accessible of which are listed as Ref 4 to 40.

Types of Material Property Charts

The Modulus-Density Chart (Fig. 2). Modulus and density are familiar properties. Steel is stiff, rubber is compliant: these are effects of modulus. Lead is heavy; cork is buoyant: these are effects of density. Figure 2 shows the full range of Young's modulus, E, and density, ρ, for engineering materials. Data for members of a particular class of material cluster together and can be enclosed by an envelope (heavy line). The same class-envelopes appear on all the diagrams, corresponding to the main headings in Table 1.

The density of a solid depends on three factors: the atomic weight of its atoms or ions, their size, and the way they are packed. Metals are dense because they are made of heavy atoms, packed densely; polymers have low densities because they are largely made of carbon (atomic weight:

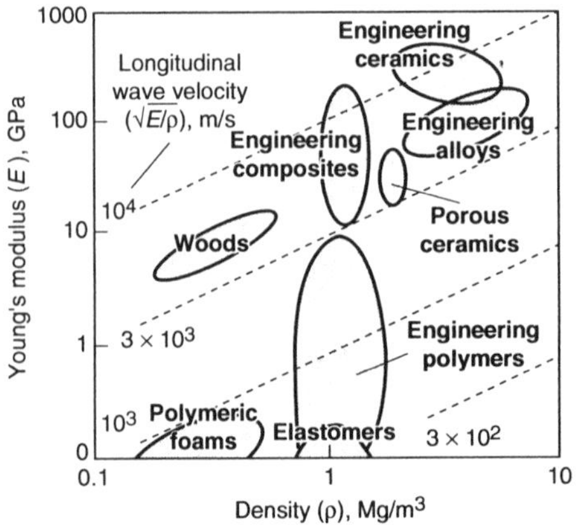

Fig. 1 The idea of a Materials Property Chart: Young's modulus, E, is plotted against the density, ρ, on log scales. Each class of material occupies a characteristic part of the chart. The log scales allow the longitudinal elastic wave velocity $v = (E/\rho)^{1/2}$ to be plotted as a set of parallel contours.

12) and hydrogen in a linear, 2-, or 3-dimensional network. Ceramics, for the most part, have lower densities than metals because they contain light oxygen, nitrogen, or carbon atoms. Even the lightest atoms, packed in the most open way, give solids with a density of around 1 Mg/m^3 (60 lb/ft^3). Materials with lower densities than this are foams—materials made up of cells containing a large fraction of pore space.

The moduli of most materials depend on two factors: bond stiffness, and the density of bonds per unit area. An interatomic bond is like a spring: it has a spring constant, S (units: N/m). Young's modulus, E, is roughly

$$E = \frac{S}{r_o} \qquad \text{(Eq 1)}$$

where r_o is the "atom size" (r_o^3 is the mean atomic or ionic volume). The wide range of moduli is largely caused by the range of values of S. The covalent bond is stiff (S = 20 to 200 N/m, or 0.1 to 1 lb/in.); the metallic and the ionic a little less so (S = 15 to 100 N/m, or 0.075 to 0.5 lb/in.). Diamond has a very high modulus because the carbon atom is small (giving a high bond density), and its atoms are linked by very strong springs (S = 200 N/m, or 1 lb/in.). Metals have high moduli because close packing gives a high bond density and the bonds are strong, though not as strong as those of diamond. Polymers contain both strong diamondlike covalent bonds and weak hydrogen or Van der Waals bonds (S = 0.5 to 2 N/m, or 0.0025 to 0.01 lb/in.); it is the weak bonds that stretch when the polymer is deformed, giving low moduli.

But even large atoms ($r_o = 3 \times 10^{-10}$ m, or 1.2×10^{-8} in.) bonded with weak bonds (S = 0.5 N/m, 0.0025 lb/in.) have a modulus of roughly

$$E = \frac{0.5}{3 \times 10^{-10}} \approx 1 \text{ GPa (200 ksi)} \qquad \text{(Eq 2)}$$

This is the lower limit for true solids. The chart shows that many materials have moduli that are lower than this, but these are not true solids; they are either elastomers or foams. Elastomers have a low E because the weak secondary bonds have melted (their glass transition temperature, T_g, is below room temperature), leaving only the very weak "entropic" restoring force associated with tangled, long-chain molecules. Foams have low moduli because the cell walls bend (allowing large displacements) when the material is loaded.

The chart shows that the modulus of engineering materials spans five decades, from 0.01 GPa (0.5 ksi) (low-density foams) to 1000 GPa (1.5×10^5 ksi) (diamond); the density spans a factor of 2000, from less than 0.1 to 20 Mg/m^3 (6 to 1200 lb/ft^3). At the level of precision of interest here (that required to reveal the relationship between the properties of materials classes) the shear modulus, G, by $3E/8$ and the bulk modulus, K, by E, for all materials except for elastomers (for which $G = E/3$ and $K \gg E$), may be approximated so the G-ρ chart and the K-ρ chart both look almost identical to Fig. 2.

The log scales allow more information to be displayed. The velocity of elastic waves in a material, and the natural vibration frequencies of a component made of it, are proportional to $(E/\rho)^{1/2}$; the quantity $(E/\rho)^{1/2}$ itself is the velocity of longitudinal waves in a thin rod of the material. Contours of constant $(E/\rho)^{1/2}$ are plotted on the chart, labeled with the longitudinal wave speed, which varies from less than 50 m/s (160 ft/s) (soft elastomers) to a little more than 10^4 m/s (33,000 ft/s) (fine ceramics). Note that aluminum and glass, because of their low densities, transmit waves quickly despite their low moduli. One might have expected the sound velocity in foams to be low because of the low modulus; however, the low density almost compensates. The sound velocity in wood is low across the grain, but along the grain, it is high—roughly the same as steel—a fact made use of in the design of musical instruments.

The modulus-density chart helps in the common problem of material selection for applications in which weight must be minimized. Guide lines corresponding to three common geometries of loading are drawn on the diagram; they correspond to the three indices for stiffness-limited minimum-weight design listed in Table 5(a) in the article "Performance Indices" in this Volume, in which their use in selecting materials is explained.

The Strength-Density Chart (Fig. 3). The modulus of a solid is a well-defined quantity with a sharp value. The strength is not. The word "strength" needs definition. For metals and polymers, it is the yield strength, but because the range of materials includes those that have been worked, the range extends from initial yield to ultimate strength; for most practical purposes it is the same in tension and compression. For brittle ceramics, it is the crushing strength in compression, not that in tension which is about 15 times smaller; the envelopes for brittle materials are shown as broken lines as a reminder of this. For elastomers, strength means the tear strength. For composites, it is the tensile failure strength (the compressive strength can be less, because of fiber buckling).

Figure 3 shows these strengths, using the symbol σ_f despite the different failure mechanisms involved, plotted against density, ρ. The considerable vertical extension of the strength bubble for an individual material reflects its wide range, caused by degree of alloying, work hardening, grain size, porosity, and so forth. As before, members of a class cluster together and can be enclosed in an envelope (heavy line). Each envelope occupies a characteristic area of the chart.

The range of strengths for engineering materials, like that of their moduli, spans about five decades: from less than 0.1 MPa (15 psi) (foams, used in packaging and energy-absorbing systems) to 104 MPa (1500 ksi) (the strength of diamond, exploited in the diamond-anvil press). The single most important concept in understanding this wide range is that of the lattice resistance or Peierls stress, which is the intrinsic resistance of the structure to plastic shear. Plastic

Table 1 Engineered material classes included in the material property charts (Fig. 1 to 13)

- Engineering alloys
  - Aluminum (Al) alloys
  - Copper (Cu) alloys
  - Lead (Pb) alloys
  - Magnesium (Mg) alloys
  - Molybdenum (Mo) alloys
  - Nickel (Ni) alloys
  - Steels (MS, mild steels and SS, stainless steels)
  - Cast irons
  - Tin (Sn) alloys
  - Titanium (Ti) alloys
  - Tungsten (W) alloys
  - Zinc (Zn) alloys
  - Beryllium (Be)
  - Boron (B)
  - Germanium (Ge)
  - Silicon (Si)
- Engineering plastics (thermoplastics and thermosets)
  - Epoxies (EP)
  - Melamines (MEL)
  - Polycarbonate (PC)
  - Polyesters (PEST)
  - High-density polyethylene (HDPE)
  - Low-density polyethylene (LDPE)
  - Polyformaldehyde (PF)
  - Polymethyl methacrylate (PMMA)
  - Polypropylene (PP)
  - Polytetrafluoroethylene (PTFE)
  - Polyvinyl chloride (PVC)
  - Polyimides
- Elastomers
  - Natural rubber
  - Hard butyl rubber
  - Polyurethanes (PU)
  - Silicone rubber
  - Soft butyl rubber
- Polymer foams
  - Cork
  - Polyester
  - Polystyrene (PS)
  - Polyurethane (PU)
- Engineering composites (polymer-matrix composites)(a)
  - Carbon-fiber-reinforced polymer (CFRP)
  - Glass-fiber-reinforced polymer (GFRP)
  - Kevlar-fiber-reinforced polymer (KFRP)
- Engineering ceramics
  - Alumina (Al_2O_3)
  - Diamond
  - Sialons
  - Silicon carbide (SiC)
  - Silicon nitride (Si_3N_4)
  - Zirconia (ZrO_2)
  - Beryllia (BeO)
  - Mullite
  - Magnesia (MgO)
- Porous ceramics (traditional ceramics)
  - Brick
  - Cement
  - Common rocks
  - Concrete
  - Porcelain
  - Pottery
- Glasses
  - Borosilicate glass
  - Soda glass
  - Silica (SiO_2)
- Cermets
  - Tungsten carbide/cobalt (WC-Co)
- Woods
  - Ash
  - Balsa
  - Fir
  - Oak
  - Pine
  - Wood products (laminates)

(a) A distinction is drawn in the charts between the properties of uniply and laminated (laminates) composites. (b) Separate property envelopes describe properties of wood parallel, ||, and perpendicular, ⊥, to the grain.

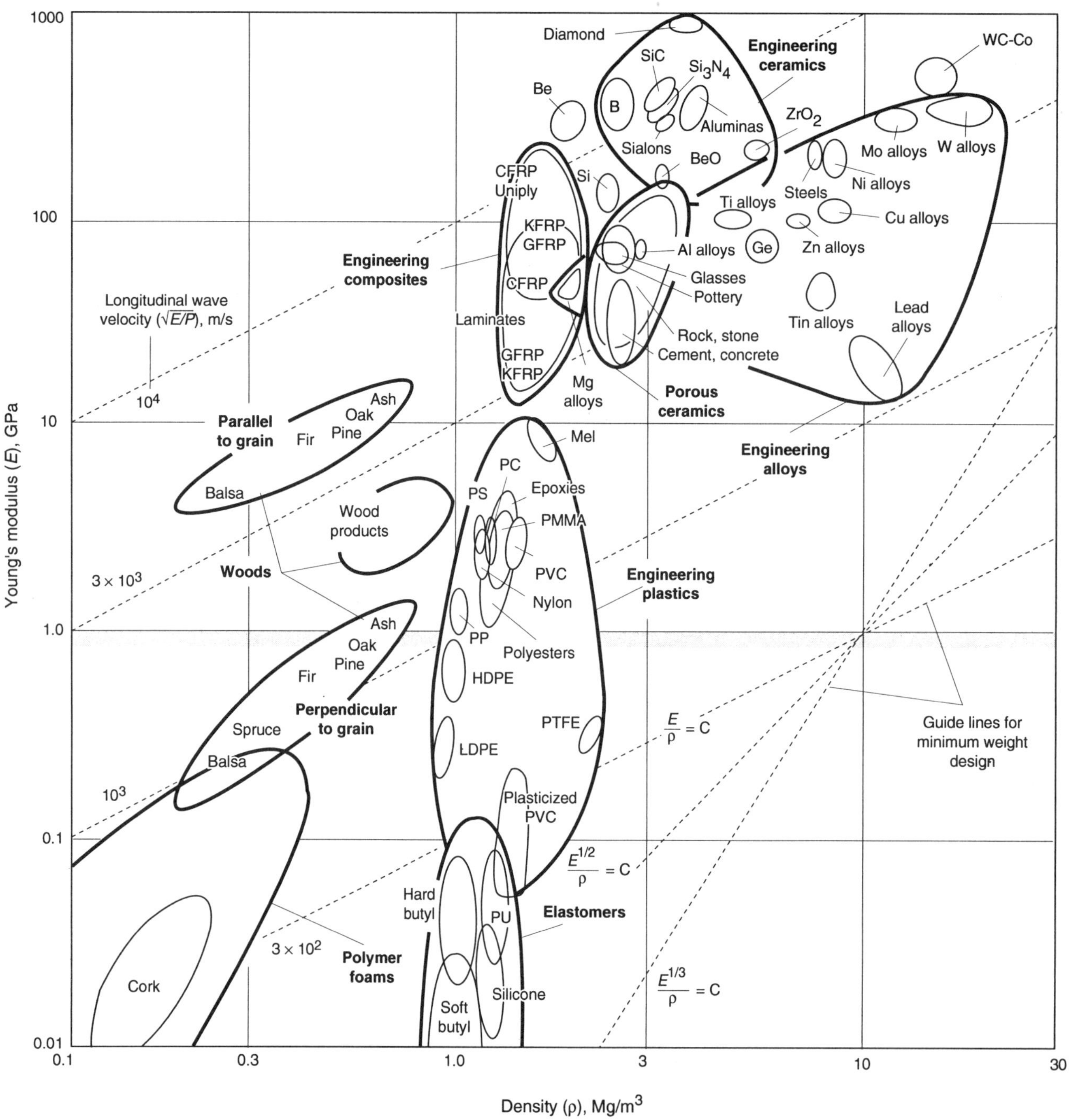

Fig. 2 Young's modulus, E, plotted against density, ρ, for various engineered materials. The heavy envelopes enclose data for a given class of material. The diagonal contours show the longitudinal wave velocity. The guide lines of constant E/ρ, $E^{1/2}/\rho$, and $E^{1/3}/\rho$ allow selection of materials for minimum weight, deflection-limited, design.

shear in a crystal involves the motion of dislocations. Metals are soft because the nonlocalized metallic bond does little to prevent dislocation motion, whereas ceramics are hard because their more localized covalent and ionic bonds, which must be broken and reformed when the structure is sheared, lock the dislocations in place. In noncrystalline solids, on the other hand, the energy is associated with the unit step of the flow process: the relative slippage of two segments of a polymer chain, or the shear of a small molecular cluster in a glass network. Their strength has the same origin as that underlying the lattice resistance: if the unit step involves breaking strong bonds (as in an inorganic glass), the materials will be strong; if it only involves the rupture of weak bonds (the Van der Waals bonds in polymers for example), it will be weak. Materials that fail by fracture do so because the lattice resistance or its amorphous equivalent is so large that fracture happens first.

When the lattice resistance is low, the material can be strengthened by introducing obstacles to slip: in metals, by adding alloying elements, particles, grain boundaries, and even other dislocations ("work hardening"); and in polymers by cross-linking or by orienting the chains so that strong covalent as well as weak Van der Waals bonds are broken. When, on the other hand, the lattice resistance is high, further hardening is superfluous—the problem becomes that of suppressing fracture (see the Section "Fracture Toughness-Density" in this article).

An important use of the strength-density chart is in materials selection in lightweight plastic design. The guide lines performance indices (Table 5b in the article "Performance Indices," which follows in this Section of the Handbook) for materials selection in the minimum-weight

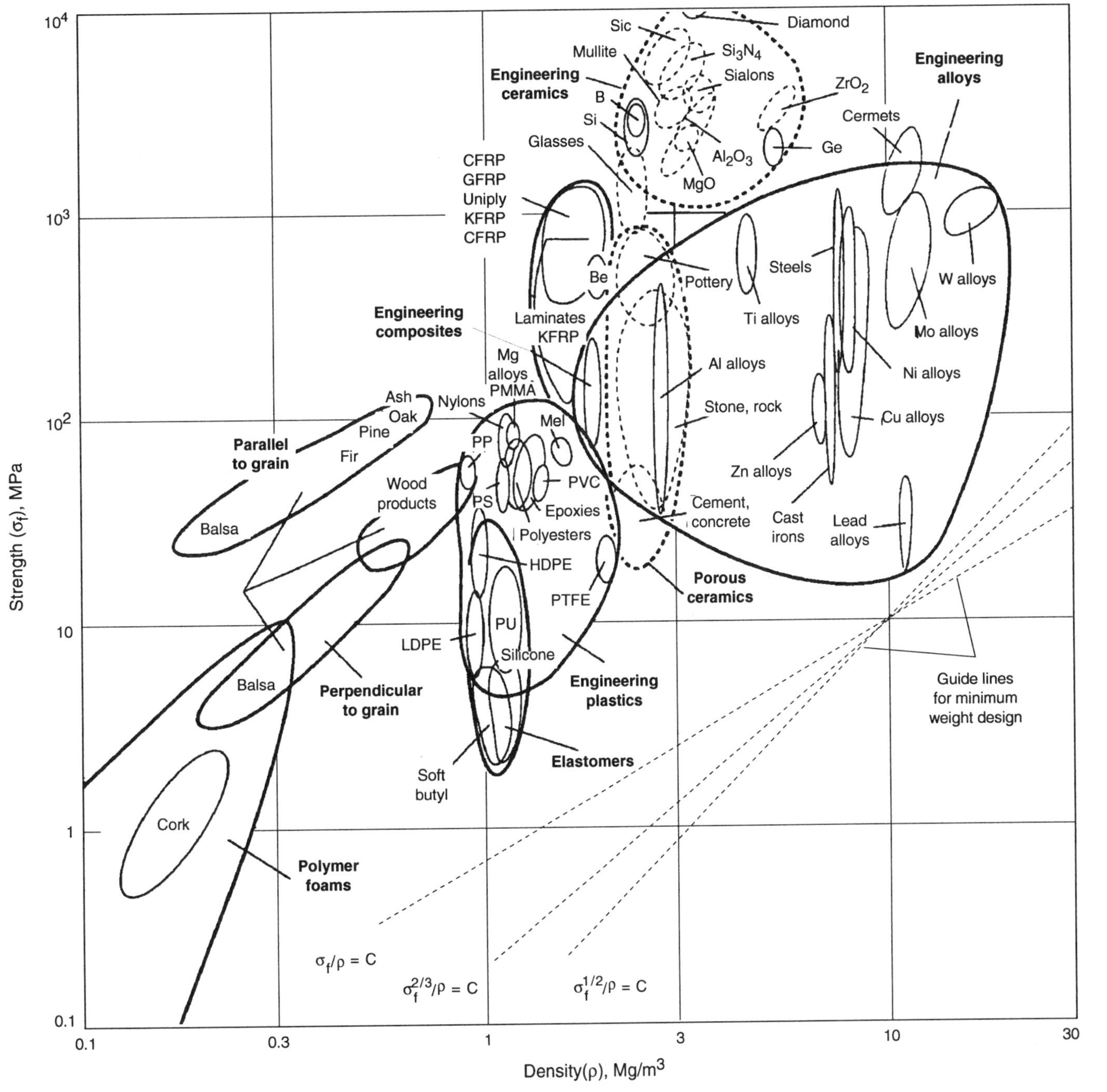

Fig. 3 Strength, σ_f, plotted against density, ρ, for various engineered materials. Strength is yield strength for metals and polymers, compressive strength for ceramics, tear strength for elastomers, and tensile strength for composites. The guide lines of constant σ_f / ρ, $\sigma_f^{2/3}/\rho$, and $\sigma_f^{1/2}/\rho$ are used in minimum weight, yield-limited, design.

design of ties, columns, beams, and plates, and for yield-limited design of moving components in which inertial forces are important.

Aspects of fatigue—the endurance limit, for example—can be displayed in a similar way. Charts relating to fatigue can be found in Ref 41.

The Fracture Toughness-Density Chart (Fig. 4). Increasing the plastic strength of a material is useful only as long as it remains plastic and does not fail by fast fracture. The resistance to the propagation of a crack is measured by the fracture toughness, K_{Ic}. It is plotted against density in Fig. 4. The range is large: from 0.01 to over 100 MPa$\sqrt{m}$ (0.01 to 100 ksi$\sqrt{in.}$). At the lower end of this range are brittle materials that, when loaded, remain elastic until they fracture. For these, linear elastic fracture mechanics works well, and the fracture toughness itself is a well-defined property. At the upper end lie the super-tough materials, all of which show substantial plasticity before they break. For these the values of K_{Ic} are approximate, derived from critical J-integral (J_c) and critical crack-opening displacement (δ_c) measurements (by writing $K_{Ic} = (EJ_c)^{1/2}$, for instance). They are helpful in providing a ranking of materials, but must be used as an indicator only. Guide lines for minimum weight design are based on the indices listed in Table 5(e) in the article "Performance Indices," which follows in this Section of the Handbook. The figure shows one reason for the dominance of metals in engineering; they almost all have values of K_{Ic} above 20 MPa$\sqrt{m}$ (20 ksi$\sqrt{in.}$), a value often quoted as a minimum for conventional design.

The Modulus-Strength Chart (Fig. 5). High tensile steel makes good springs. But so does rubber. How is it that two such different materials are both suited for the same task? This and other

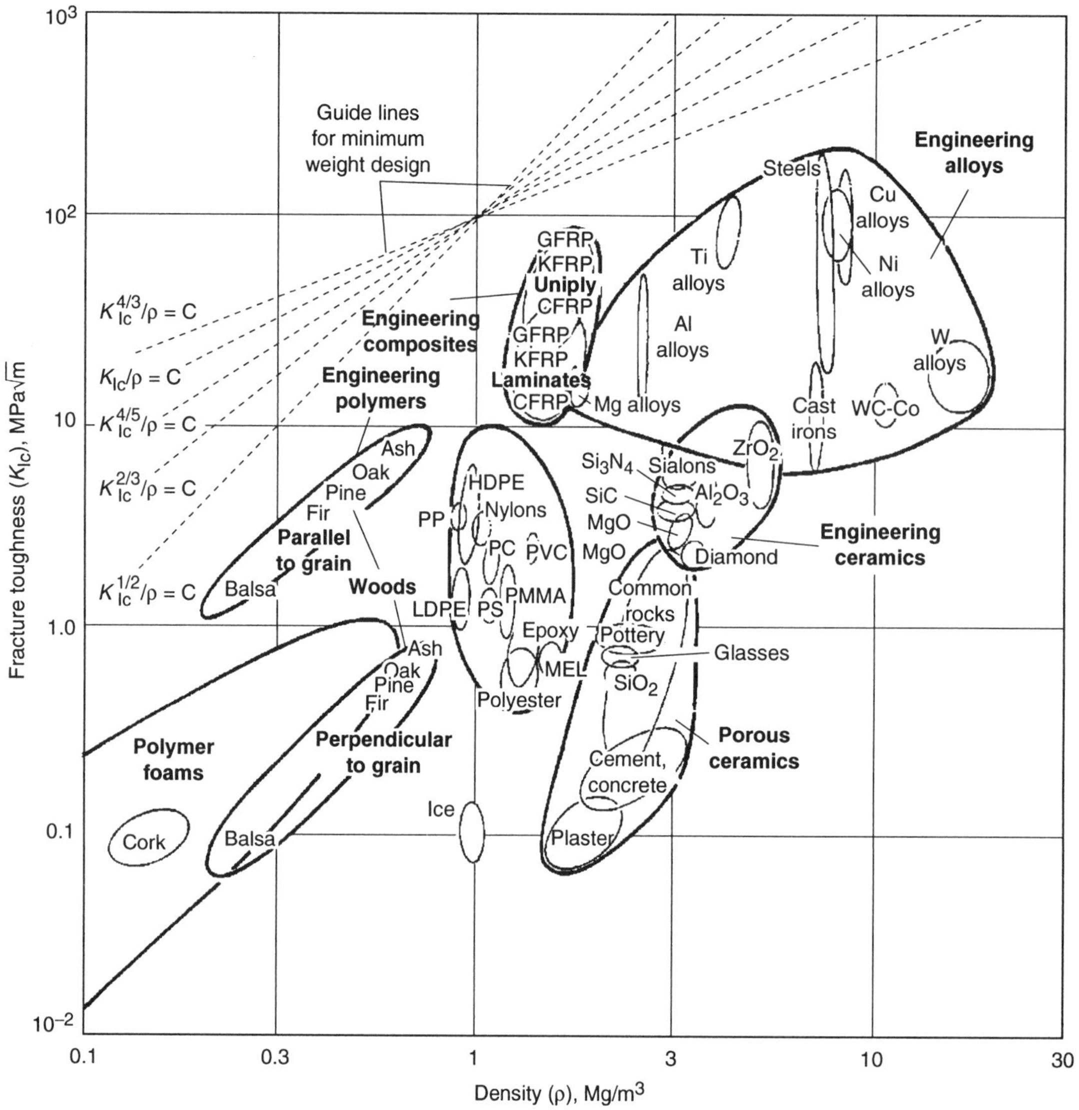

Fig. 4 Fracture toughness, K_{Ic}, plotted against density, ρ. The guide lines of constant K_{Ic}, $K_{Ic}^{2/3}/\rho$, and $K_{Ic}^{1/2}/\rho$, and so forth, help in minimum weight, fracture-limited design. Data for K_{Ic} are valid below 10 MPa√m; data above 10 MPa√m are for ranking only.

questions are answered by Fig. 5, the most useful of all the charts.

It shows Young's modulus, E, plotted against strength, σ_f. The qualifications on "strength" are the same as before: yield strength for metals and polymers, compressive crushing strength for ceramics, tear strength for elastomers, and tensile strength for composites and woods; the symbol σ_f is used for them all. The ranges of the variables, too, are the same. Contours of normalized strength, σ_f / E, appear as a family of straight parallel lines.

Examine these first. Engineering polymers have normalized strengths between 0.01 and 0.1. In this sense they are remarkably strong; the values for metals are at least a factor of 10 smaller. Even ceramics, in compression, are not as strong, and in tension they are far weaker (by a further factor of 15 of so). Composites and woods lie on the 0.01 contour, as good as the best metals. Elastomers, because of their exceptionally low moduli, have values of σ_f / E larger than any other class of material: 0.1 to 10.

The distance over which interatomic forces act is small—a bond is broken if it is stretched to more than about 10% of its original length. So the force needed to break a bond is roughly

$$F = \frac{S r_o}{10} \quad \text{(Eq 3)}$$

where S, as before, is the bond stiffness. If shear breaks bonds, the strength of a solid should be roughly

$$\sigma_y \approx \frac{F}{r_o^2} = \frac{S}{10 r_o} = \frac{E}{10} \quad \text{(Eq 4)}$$

The chart shows that for some polymers it is. Most solids are weaker, for two reasons.

First, nonlocalized bonds (those in which the cohesive energy derives from the interaction of one atom with large number of others, not just with its nearest neighbors) are not broken when the structure is sheared. The metallic bond, and the ionic bond for certain directions of shear, are like this; very pure metals, for example, yield at stresses as low as E/10,000, and strengthening mechanisms are needed to make them useful in engineering. The covalent bond *is* localized, and for this reason covalent solids have yield strengths which, at low temperatures, are as high as E/10. It is hard to measure them (though it can sometimes be done by indentation) because of the second reason for weakness: They generally contain defects—concentrators of stress—from which shear or fracture can propagate, often at stresses well below the "ideal" E/10. Elastomers are anomalous (they have strengths of about E) because the modulus does not derive from bond stretching, but from the change in entropy of the

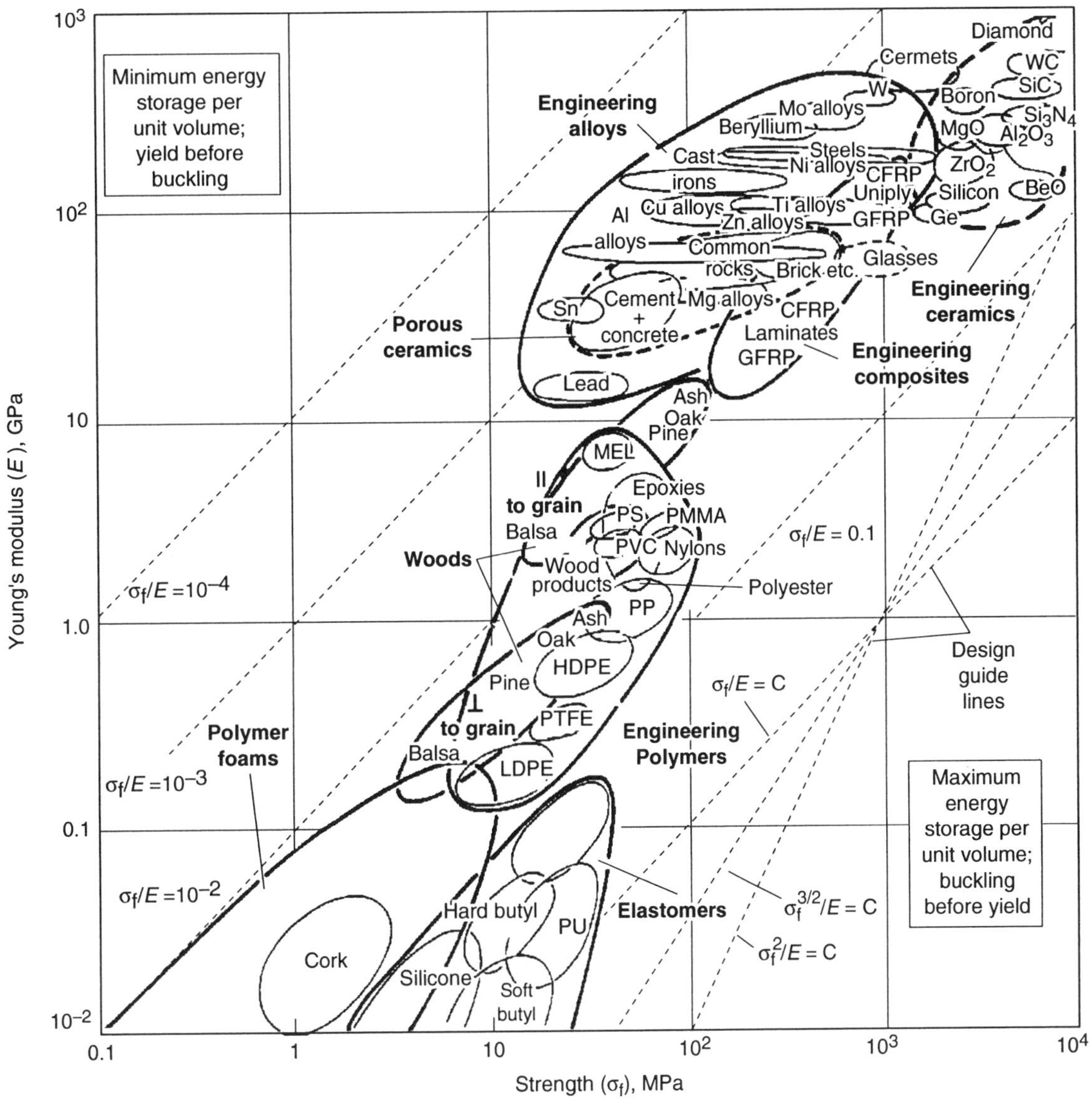

Fig. 5 Young's modulus, E, plotted against strength, σ_f, for various engineered materials. Strength is yield strength for metals and polymers, compressive strength for ceramics, tear strength for elastomers, and tensile strength for composites. The design guide lines help with the selection of materials for springs, pivots, knife edges, diaphragms, and hinges.

tangled molecular chains when the material is deformed.

The performance index for selecting materials for springs (Table 5c in the article "Performance Indices," which follows in this Section of the Handbook) is

$$M = \frac{\sigma_f^2}{E}$$

A guide line for this index is shown on the chart. Using it in the way explained in the article "Performance Indices" reveals that elastomers, high-strength steels, and glass-fiber-reinforced polymer (GFRP) all make good springs.

Equivalent charts for the endurance limit can be found in Ref 41.

The Specific Stiffness-Specific Strength Chart (Fig. 6). Many designs—particularly those for things that move—call for stiffness and strength at minimum weight. To help with this, the data of the modulus-strength chart (Fig. 5) are replotted in Fig. 6 after dividing, for each material, by the density; it shows E/ρ plotted against σ_f / ρ.

Ceramics lie at the top right: they have exceptionally high stiffness and strength per unit weight. The same restrictions on strength apply as before. The data shown here are for compression strengths; the tensile strengths are about 15 times smaller. Composites then emerge as the material class with the most attractive specific properties, one of the reasons for their increasing use in aerospace. Metals are penalized because of their relatively high densities. Polymers, because their densities are low, are favored.

The chart has application in selecting materials for light springs and energy-storage devices (Table 5c in the article "Performance Indices," which follows in this Section of the Handbook). Equivalent charts for the endurance limit are contained in Ref 41.

The Fracture Toughness-Modulus Chart (Fig. 7). As a general rule, the fracture toughness of polymers is less than that of ceramics. Yet polymers are widely used in engineering structures; ceramics, because they are "brittle," are treated with much more caution. Figure 7 helps resolve this apparent contradiction. It shows the fracture toughness, K_{Ic}, plotted against Young's modulus, E. The restrictions described earlier apply to the values of K_{Ic}: When small, they are well defined; when large, they are useful only as a ranking for material selection.

Consider first the question of the necessary condition for fracture. It is that sufficient external work be done, or elastic energy released, to supply the surface energy (2γ per unit area) of the two new surfaces that are created. This is written as:

$$G \geq 2\gamma \qquad \text{(Eq 5)}$$

where G is the elastic energy release rate. Using the standard relation $K \sim (EG)^{1/2}$ between G and stress intensity K, then

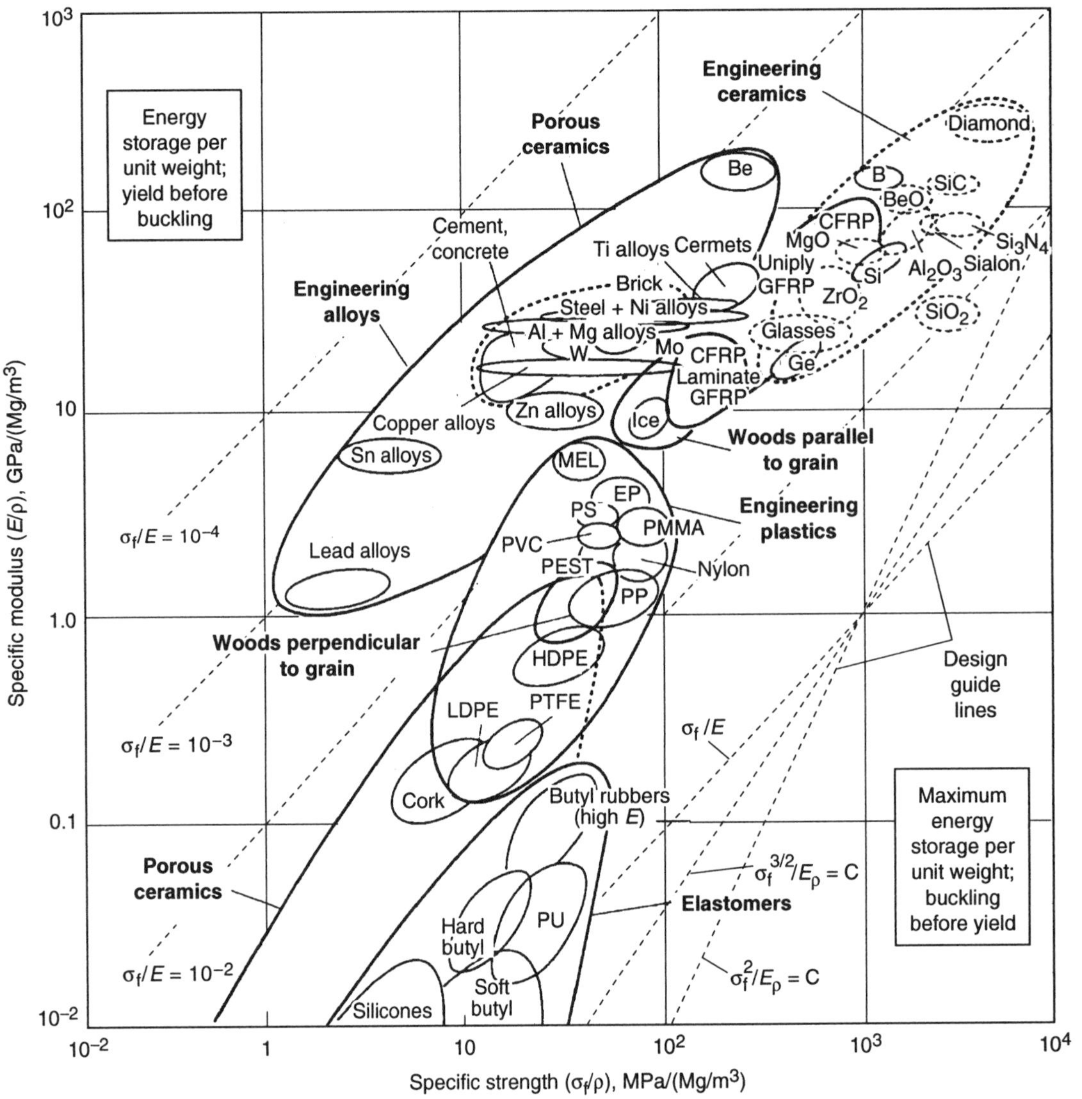

Fig. 6 Specific modulus, E/ρ, plotted against specific strength σ_f/ρ for various engineered materials. Strength is yield strength for metals and polymers, compressive strength for ceramics, tear strength for elastomers, and tensile strength for composites. The design guide lines help with the selection of materials for lightweight springs and energy-storage systems.

$$K \geq (2E\gamma)^{1/2} \quad \text{(Eq 6)}$$

Now the surface energies, γ, of solid materials scale as their moduli; to an adequate approximation $\gamma = Er_0/20$, where r_0 is the atom size, giving

$$K \geq E\left(\frac{r_0}{20}\right)^{1/2} \quad \text{(Eq 7)}$$

The right-hand side of this equation is identified with a lower-limiting value of K_{Ic}, when, taking r_0 as 2×10^{-10} m (8×10^{-9} in.),

$$\frac{(K_{Ic})_{min}}{E} = \left(\frac{r_0}{20}\right)^{1/2} \approx 3\times10^{-6}\sqrt{m}\ (2\times10^{-5}\sqrt{in.}) \quad \text{(Eq 8)}$$

This criterion is plotted on the chart as a shaded, diagonal band near the lower right corner (the width of the band reflects a realistic range of r_0 and of the constant C in $\gamma = r_0/C$). It defines a lower limit on values of K_{Ic}: It cannot be less than this unless some other source of energy (such as a chemical reaction, or the release of elastic energy stored in the special dislocation structures caused by fatigue loading) is available, when it is given a new symbol such as $(K_{Ic})_{scc}$. Note that the most brittle ceramics lie close to the threshold; when they fracture, the energy absorbed is only slightly more than the surface energy. When metals, polymers, and composites fracture, the energy absorbed is vastly greater, usually because of plasticity associated with crack propagation. This is discussed in the following Section of this article.

Plotted on Fig. 7 are contours of toughness, G_{Ic}, a measure of the apparent fracture surface energy ($G_{Ic} \approx K_{Ic}^2/E$). The true surface energies, γ, of solids lie in the range 10^{-4} to 10^{-3} kJ/m^2 (10^{-2} to 10^{-1} ft · lbf/ft^2). The diagram shows that the values of the toughness start at 10^{-3} kJ/m^2 (10^{-1} ft · lbf/ft^2) and range through almost six decades to 10^3 kJ/m^2 (10^5 ft · lbf/ft^2). On this scale, ceramics (10^{-3} to 10^{-1} kJ/m^2, or 10^{-2} to 10 ft · lbf/ft^2) are much lower than polymers (10^{-1} to 10 kJ/m^2, or 10 to 1000 ft · lbf/ft^2); this is part of the reason polymers are more widely used in engineering than ceramics.

The Fracture Toughness-Strength Chart (Fig. 8). The stress concentration at the tip of a crack generates a process zone: a plastic zone in ductile solids, a zone of microcracking in ceramics, a zone of delamination, debonding, and fiber pullout in composites. Within the process zone, work is done against plastic and frictional forces; it is this that accounts for the difference between the measured fracture energy G_{Ic} and the true surface energy 2γ. The amount of energy dissipated must scale roughly with the strength of the material, with the process zone, and with its size, d_y. This size is found by equating the stress field of the crack ($\sigma = K/\sqrt{2\pi r}$) at $r = d_y/2$ to the strength of the material, σ_f, giving

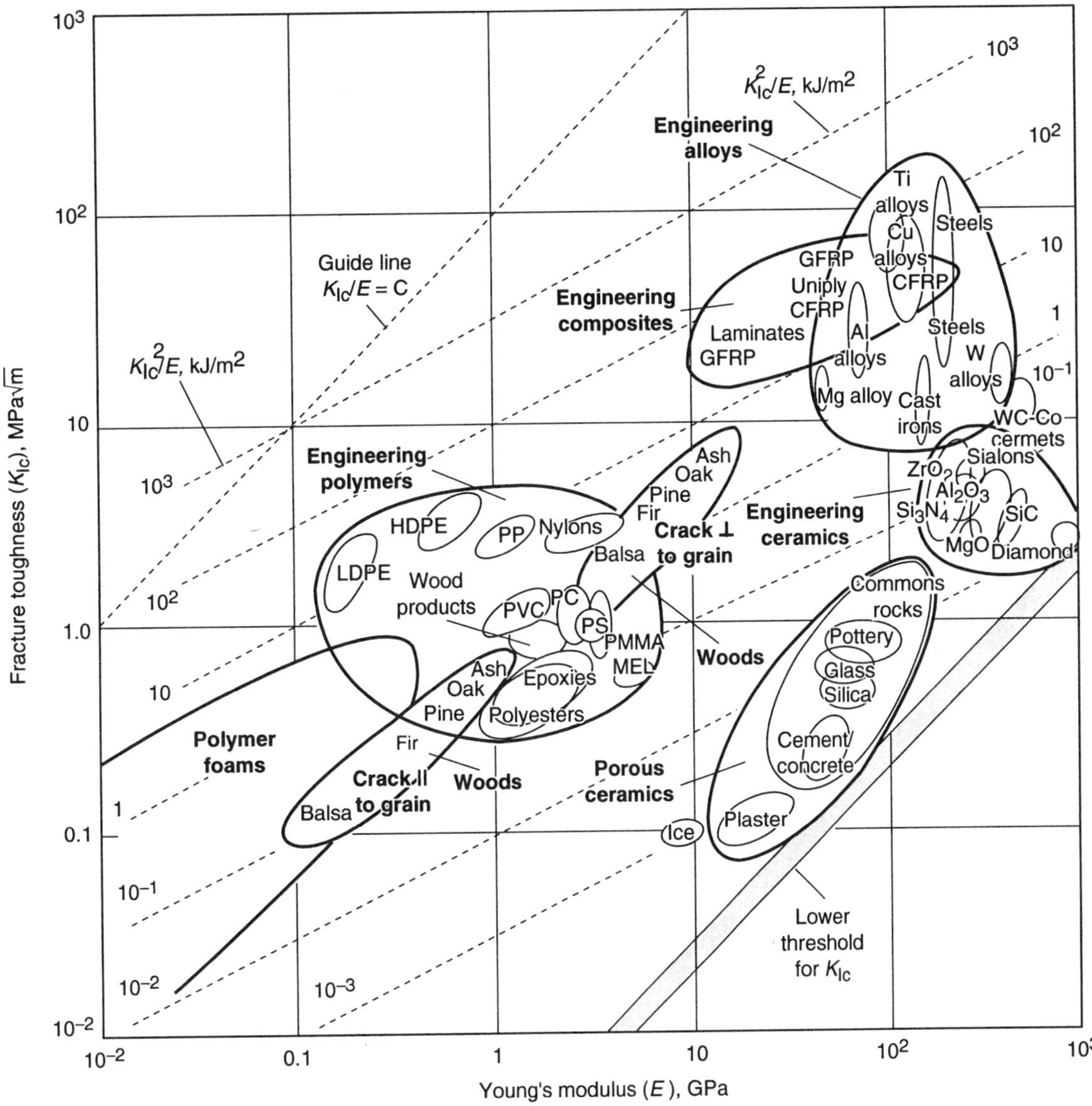

Fig. 7 Fracture toughness, K_{Ic}, plotted against Young's modulus, E. The family of lines are of constant K_{Ic}^2/E (approximately G_{Ic}, the fracture energy). These, and the guide line of constant K_{Ic}/E, help in design against fracture. The shaded band shows the "necessary condition" for fracture. Fracture can, in fact, occur below this limit under conditions of corrosion, or cyclic loading.

$$d_y = \frac{K_{Ic}^2}{\pi \sigma_f^2} \qquad \text{(Eq 9)}$$

Figure 8 (fracture toughness versus strength) shows that the size of the zone, d_y (broken lines) varies enormously, from atomic dimensions for very brittle ceramics and glasses to almost 1 meter for the most ductile of metals. At a constant zone size, fracture toughness tends to increase with strength (as expected); it is this that causes the data plotted in Fig. 8 to be clustered around the diagonal of the chart.

The fracture toughness-strength diagram has application in selecting materials for the safe design of load-bearing structures using the indices described in Table 5(e) in the article "Performance Indices," which follows in this Section of the Handbook.

The Loss Coefficient-Modulus Chart (Fig. 9). Bells are traditionally made of bronze. They can be (and sometimes are) made of glass; and they could (if one could afford it) be made of silicon carbide. Metals, glasses, and ceramics all, under the right circumstances, have low intrinsic damping or "internal friction," an important material property when structures vibrate. Intrinsic damping is measured by the loss coefficient, η, which is plotted in Fig. 9.

The loss coefficient, a dimensionless number, measures the degree to which a material dissipates vibrational energy. If a material is loaded elastically to a stress σ_{max}, it stores an elastic energy

$$U = \int_0^{\sigma_{max}} \sigma d\varepsilon = \frac{1}{2}\frac{\sigma_{max}^2}{E}$$

per unit volume. If it is loaded and then unloaded, it dissipates an energy

$$\Delta U = \oint \sigma d\varepsilon$$

The loss coefficient is

$$\eta = \frac{\Delta U}{2\pi U}$$

The cycle can be applied in many different ways—some fast, some slow. The value of η usually depends on the time scale or frequency of cycling. Other measures of damping include the specific damping capacity, $D = (\Delta U)/U$; the log decrement, Δ (the log of the ratio of successive amplitudes of natural vibrations); the phase lag, δ, between stress and strain; and the "Q"-factor or resonance factor,

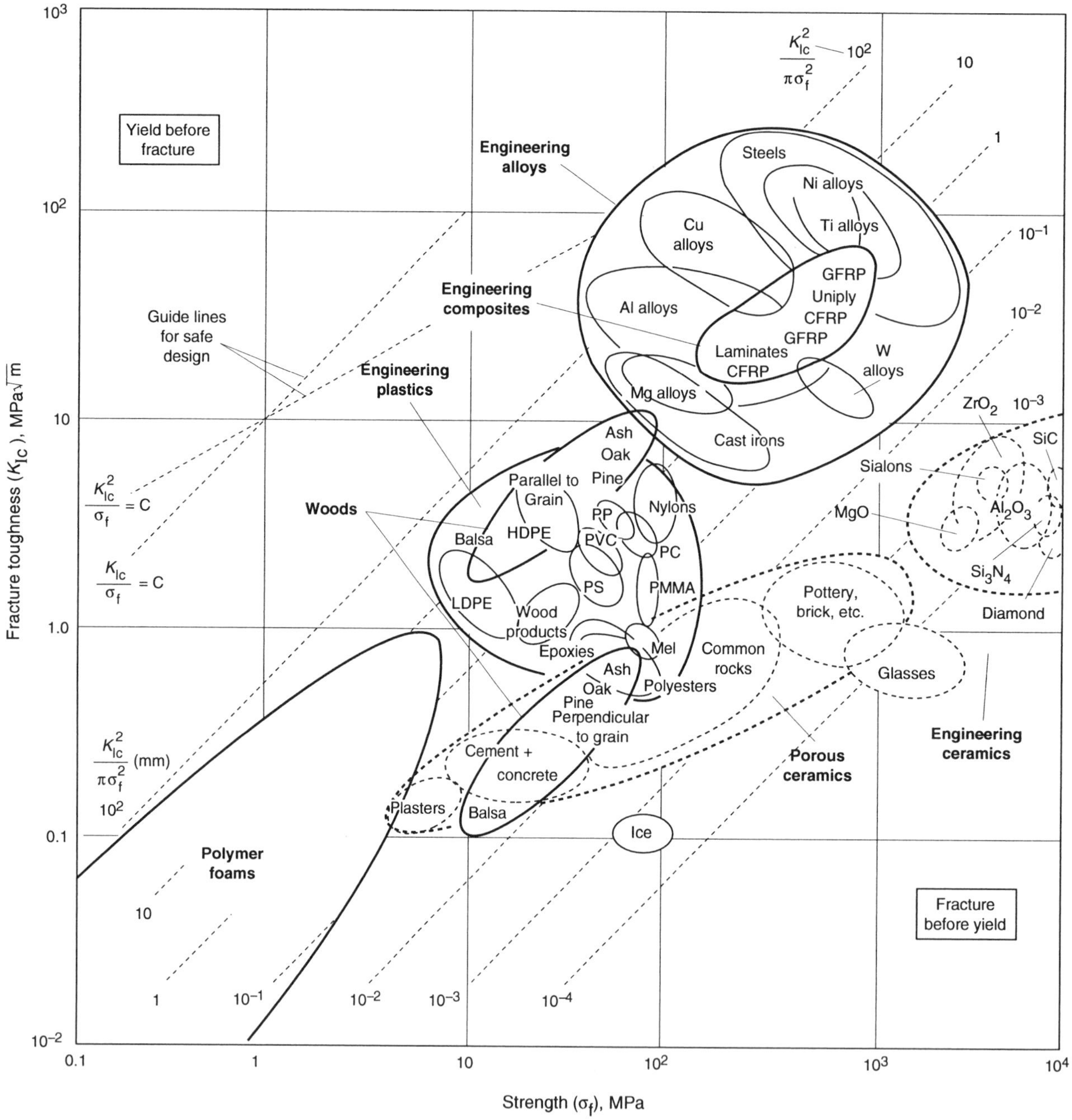

Fig. 8 Fracture toughness, K_{Ic}, plotted against strength, σ_f, for various engineered materials. Strength is yield strength for metals and polymers, compressive strength for ceramics and glasses, and tensile strength for composites. The contours show the value of $K_{Ic}^2/\pi\sigma_f$—roughly, the diameter of the process-zone at a crack tip. The design guide lines are used in selecting materials for damage-tolerant design.

Q. When damping is small ($\eta < 0.01$), these measures are related by

$$\eta = \frac{D}{2\pi} = \frac{\Delta}{\pi} = \tan\delta = \frac{1}{Q}$$

but when damping is large, they are no longer equivalent.

There are many mechanisms of intrinsic damping and hysteresis. Some (the "damping" mechanisms) are associated with a process that has a specific time constant; then the energy loss is centered about a characteristic frequency. Others (the "hysteresis" mechanisms) are associated with time-independent mechanisms; they absorb energy at all frequencies.

In metals a large part of the loss is hysteretic, caused by dislocation movement: it is high in soft metals like lead and pure aluminum. Heavily alloyed metals like bronze and high-carbon steels have low loss because the solute pins the dislocations; these are the materials for bells. Exceptionally high loss is found in the manganese-copper alloys, because of a strain-induced martensite transformation, and in magnesium, perhaps because of reversible twinning. The elongated bubbles for metals span the large range accessible by alloying and working. Engineering ceramics have low damping because the enormous lattice resistance pins dislocations in place at room temperature. Porous ceramics, on the other hand, are filled with cracks, the surfaces of which rub, dissipating energy, when the material is loaded; the high damping of some cast irons has a similar origin. In polymers, chain segments slide against each other when loaded; the relative motion dissipates energy. The ease with which they slide depends on the ratio of the temperature (in this case, room temperature) to the glass transition temperature, T_g, of the polymer. When $T/T_g < 1$, the secondary bonds are "frozen"; the modulus is high and the damping is relatively low.

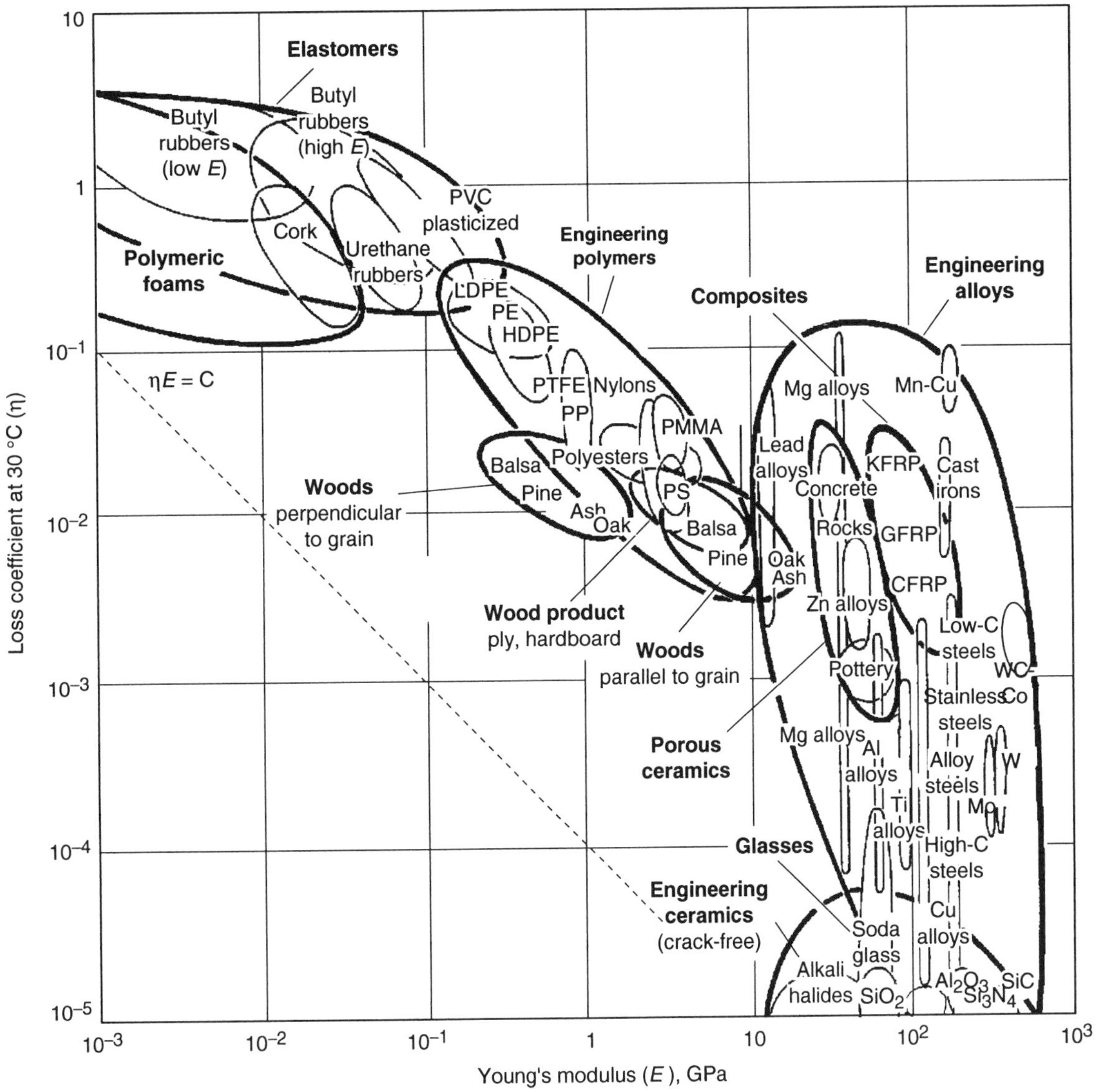

Fig. 9 The loss coefficient, η, plotted against Young's modulus, *E*, for various engineered materials. The guide line corresponds to the condition $\eta = C/E$.

When $T/T_g > 1$, the secondary bonds have melted, allowing easy chain slippage; the modulus is low and the damping is high. This accounts for the obvious inverse dependence of η on *E* for polymers in Fig. 9; indeed, to a first approximation,

$$\eta = \frac{4 \times 10^{-2}}{E} \qquad \text{(Eq 10a)}$$

with *E* in GPa, or

$$\eta = \frac{6}{E} \qquad \text{(Eq 10b)}$$

with *E* in ksi.

The Thermal Conductivity-Thermal Diffusivity Chart (Fig. 10). The material property governing the flow of heat through a material at steady state is the thermal conductivity, λ (units: W/m · K or Btu/h · ft · F); that governing transient heat flow is the thermal diffusivity, *a* (units: m^2/s or $in.^2/s$). They are related by

$$a = \frac{\lambda}{\rho\, C_p} \qquad \text{(Eq 11)}$$

where ρ is the density and C_p the specific heat, measured in J/kg · K; the quantity ρC_p is the volumetric specific heat. Figure 10 relates thermal conductivity, diffusivity, and volumetric specific heat, at room temperature.

The data span almost five decades in λ and *a*. The highest conductivities are those of diamond, silver, copper, and aluminum. The lowest are shown by highly porous materials like firebrick, cork, and foams, in which conductivity is limited by that of the gas in their cells.

Solid materials are strung out along the line

$$\rho C_p \approx 3 \times 10^6\ \text{J/m}^3 \cdot \text{K}\ (45\ \text{Btu/ft}^3 \cdot \text{F}) \qquad \text{(Eq 12)}$$

This can be understood by noting that a solid containing *N* atoms has 3*N* vibrational modes. Each (in the classical approximation) absorbs thermal energy k*T* at the absolute temperature *T*, and the vibrational specific heat is $C_p \sim C_v = 3N\text{k}$ where k is Boltzmann's constant. The volume of *N* atoms is $(N\Omega)\ \text{m}^3$, where Ω is the volume per atom; for almost all solids Ω lies within a factor of two of $1.4 \times 10^{-29}\ \text{m}^3$ ($8.5 \times 10^{-25}\ \text{in.}^3$). The volume specific heat is then (as Fig. 10 shows):

$$\rho\, C_v \cong 3\,N\,\text{k}\,/\,N\,\Omega = \frac{3k}{\Omega}$$

$$= 3 \times 10^6\ \text{J/m}^3 \cdot \text{K}\ (45\ \text{Btu/ft}^3 \cdot \text{F}) \qquad \text{(Eq 13)}$$

Some materials deviate from this rule: they have lower-than-average volumetric specific heat. For a few, such as diamond, it is low because their Debye temperatures lie well above room temperature; then heat absorption is not classical, some modes do not

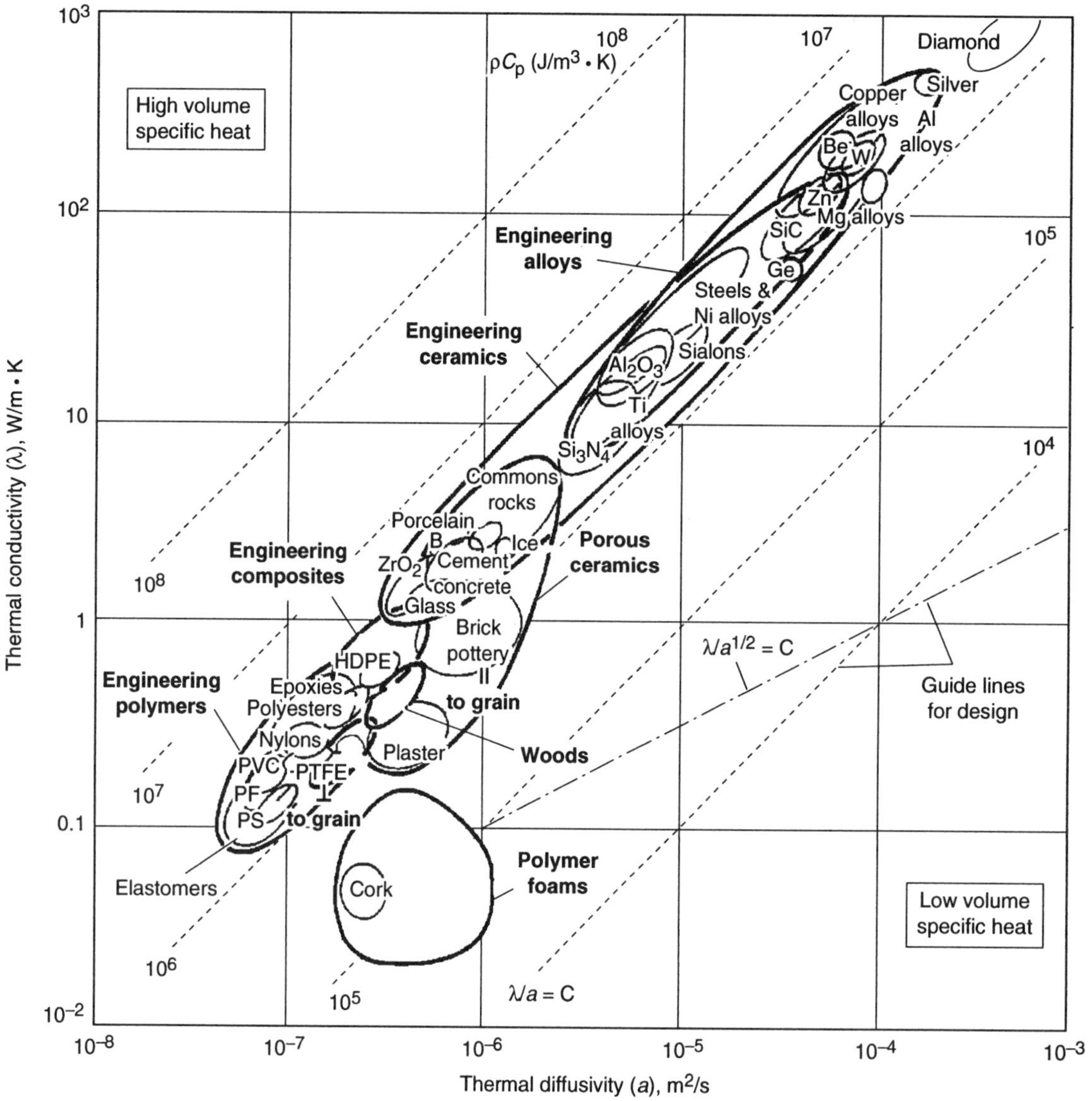

Fig. 10 Thermal conductivity, λ, plotted against thermal diffusivity, a. The contours show the volume specific heat, ρC_p. All three properties vary with temperature; the data here are for room temperature.

absorb kT and the specific heat is less than 3Nk. The largest deviations are shown by porous solids: foams, low density firebrick, woods, and so on. Their low density means that they contain fewer atoms per unit volume and, averaged over the volume of the structure, ρC_v is low. The result is that, although foams have low conductivities (and are widely used for insulation because of this), their thermal diffusivities are not necessarily low: They may not transmit much heat, but they reach a steady state quickly, an important consideration in selecting materials for thermal insulation.

The Thermal Expansion-Thermal Conductivity Chart (Fig. 11). Almost all solids expand on heating. The bond between a pair of atoms behaves like a linear elastic spring when the relative displacement of the atoms is small; but when it is large, the spring is nonlinear. Most bonds become stiffer when the atoms are pushed together and less stiff when they are pulled apart, that is, when they are anharmonic. The thermal vibration of atoms, even at room temperature, involves large displacements; as the temperature is raised, the anharmonicity of the bond pushes the atoms apart, increasing their mean spacing. The effect is measured by the linear expansion coefficient

$$\alpha = \frac{1}{l}\frac{dl}{dT} \quad \text{(Eq 14)}$$

where l is a linear dimension of the body.

The expansion coefficient is plotted against the conductivity in Fig. 11. It shows that polymers have large values of α, roughly 10 times greater than those of metals and almost 100 times greater than those of ceramics. This is because the Van der Waals bonds of the polymer are very anharmonic. Diamond, silicon, and silica (SiO_2) have covalent bonds that have low anharmonicity (that is, they are almost linear-elastic even at large strains), giving them low expansion coefficients. Composites, even though they have polymer matrices, can have low values of α because the reinforcing fibers, particularly carbon, expand very little.

The thermal expansion-thermal conductivity chart shows contours of λ/α, a quantity important in designing against thermal distortion.

The Thermal Expansion-Modulus Chart (Fig. 12). Thermal stress is the stress that appears in a body when it is heated or cooled, but prevented from expanding or contracting. It depends on the expansion coefficient, α, of the material and on its modulus, E. A development of the theory of thermal expansion leads to the relation

$$\alpha = \frac{\gamma_G \rho C_v}{3E} \quad \text{(Eq 15)}$$

where γ_G is Gruneisen's constant; its value ranges between about 0.4 and 4, but for most solids it is

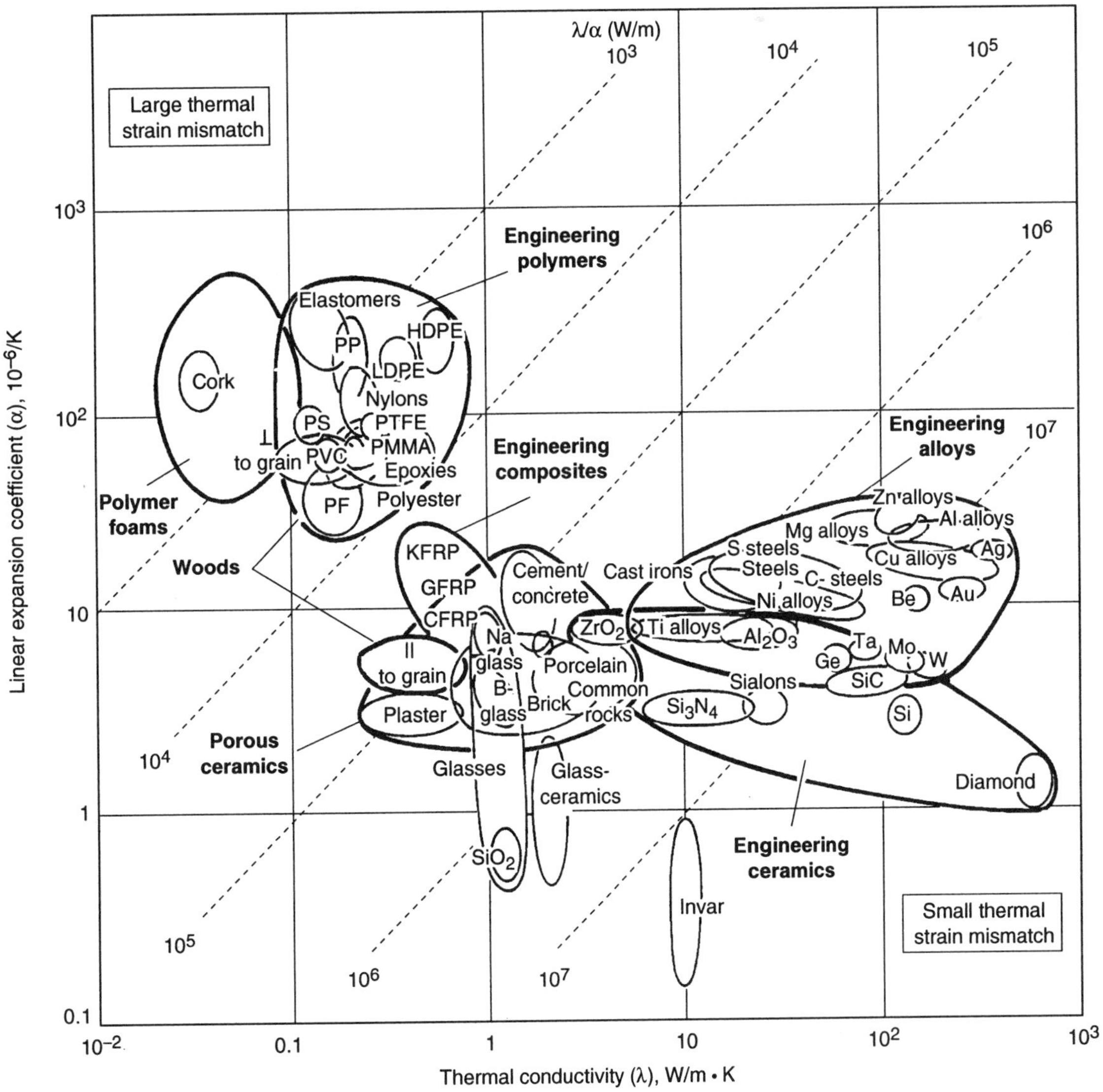

Fig. 11 The linear expansion coefficient, α, plotted against the thermal conductivity, λ. The contours show the thermal distortion parameter λ/α.

near 1. Since ρC_v is almost constant (Eq 12), the equation shows that α is proportional to $1/E$. Figure 12 shows that this is so. Diamond, with the highest modulus, has one of the lowest coefficients of expansion; elastomers with the lowest moduli expand the most. Some materials with a low coordination number (silica and some diamond-cubic or zinc-blend structured materials) can absorb energy preferentially in transverse modes, leading to very small (even a negative) value of γ_G and a low expansion coefficient; that is why SiO_2 is exceptional. Others, like Invar, contract as they lose their ferromagnetism when heated through the Curie temperature and, over a narrow range of temperature, they too show near-zero expansion, which is useful in precision equipment and in glass-metal seals.

One more useful fact: the moduli of materials scale approximately with their melting point, T_m:

$$E \approx \frac{100\,k\,T_m}{\Omega} \quad \text{(Eq 16)}$$

where k is Boltzmann's constant and Ω the volume-per-atom in the structure. Substituting this and Eq 13 for ρC_v into Eq 15 for α gives

$$\alpha = \frac{\gamma_G}{100\,T_m} \quad \text{(Eq 17)}$$

The expansion coefficient varies inversely with the melting point, or (equivalently stated) for all solids the thermal strain, just before they melt, depends only on γ_G, and this is roughly a constant. The result is useful for estimating and checking expansion coefficients.

Whenever the thermal expansion or contraction of a body is prevented, thermal stresses appear that—if large enough—cause yielding, fracture, or elastic collapse (buckling). It is common to distinguish between thermal stress caused by external constraint (a rod, rigidly clamped at both ends, for example) and that which appears without external constraint because of temperature gradients in the body. All scale as the quantity αE, shown as a set of diagonal contours in Fig. 12. More precisely: the stress $\Delta\sigma$ produced by a temperature change of 1 °C in a constrained system, or the stress per °C caused by a sudden change of surface temperature in one that is not constrained, is given by

$$C\,\Delta\sigma = \alpha E \quad \text{(Eq 18)}$$

where $C = 1$ for axial constraint, $(1 - \nu)$ for biaxial constraint or normal quenching, and $(1 - 2\nu)$ for triaxial constraint, where ν is Poisson's ratio. These stresses are large: typically 1 MPa/K; they can cause a material to yield, crack, spall, or buckle when it is suddenly heated or cooled. The resistance of materials to such damage is the subject of the following section.

The Normalized Strength-Thermal Expansion Chart (Fig. 13). The ability of a material to withstand such thermal stress is measured by its

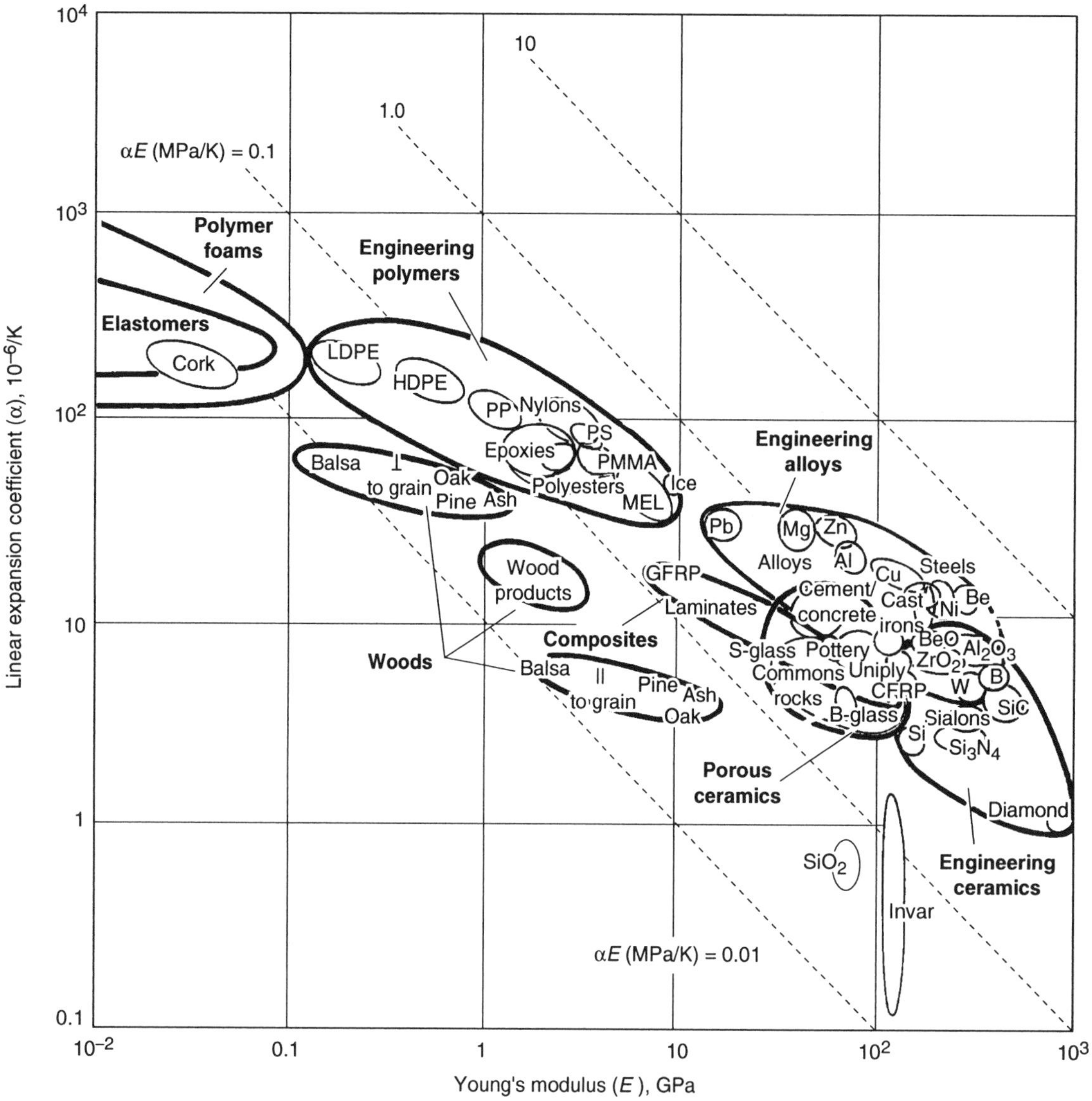

Fig. 12 The linear expansion coefficient, α, plotted against Young's modulus, E. The contours show the thermal stress created by a temperature change of 1 °C if the sample is axially constrained. A correction factor C is applied for biaxial or triaxial constraint (see text).

Table 2 Values for the factor A of Eq 21

Section thickness t = 10 mm

Conditions	Foams	Polymers	Ceramics	Metals
Slow air flow ($h = 10\ W/m^2 \cdot K$)	0.75	0.5	3×10^{-2}	3×10^{-3}
Black body radiation 500 to 0 °C ($h = 40\ W/m^2 \cdot K$)	0.93	0.6	0.12	1.3×10^{-2}
Fast air flow ($h = 10^2\ W/m^2 \cdot K$	1	0.75	0.25	3×10^{-2}
Slow water quench ($h = 10^3\ W/m^2 \cdot K$)	1	1	0.75	0.23
Fast water quench ($h = 10^4\ W/m^2 \cdot K$)	1	1	1	0.1-0.9

thermal shock resistance ΔT_s. It depends on its thermal expansion coefficient, α, and its normalized tensile strength, σ_t/E. They are the axes of Fig. 13, on which contours of constant $\sigma_t/\alpha E$ are plotted. The tensile strength, σ_t, requires definition, just as σ_f did. For brittle solids, it is the tensile fracture strength (roughly equal to the modulus of rupture, or MOR). For ductile metals and polymers, it is the tensile yield strength; for composites it is the stress that first causes permanent damage in the form of delamination, matrix cracking, or fiber debonding.

To use the chart, note that a temperature change of ΔT, applied to a constrained body—or a sudden change ΔT of the surface temperature of a body that is unconstrained—induces a stress

$$\sigma = \frac{E\alpha\Delta T}{C} \qquad \text{(Eq 19)}$$

where C was defined in the previous section. If this stress exceeds the local strength σ_t of the material, yielding or cracking results. Even if it does not cause the component to fail, it weakens it. Then a measure of the thermal shock resistance is given by

$$\frac{\Delta T_s}{C} = \frac{\sigma_t}{\alpha E} \qquad \text{(Eq 20)}$$

This is not quite the whole story. When the constraint is internal, the thermal conductivity of the material becomes important. "Instant" cooling when a body is quenched requires an infinite rate of heat transfer at its surface. Heat transfer rates are measured by the heat transfer coefficient, h, and are never infinite. Water quenching gives a high h, and then the values of ΔT_s calcu-

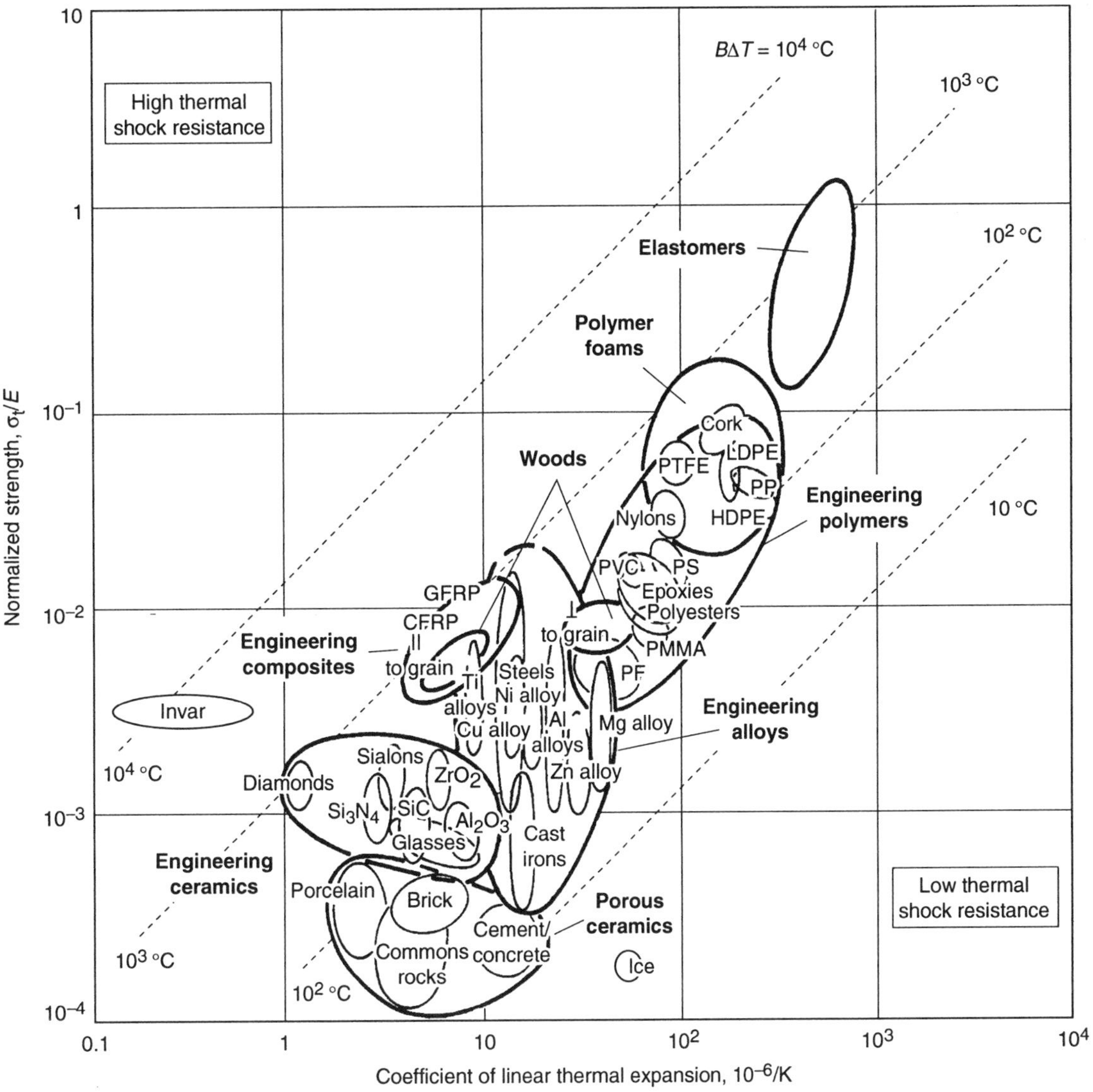

Fig. 13 The normalized tensile strength, σ_t/E, plotted against coefficient of linear thermal expansion, α (see text for strengths). The contours show a measure of the thermal shock resistance, ΔT. Corrections must be applied for constraint and to allow for the effect of thermal conduction during quenching.

lated from Eq 20 give an approximate ranking of thermal shock resistance. However, when heat transfer at the surface is poor and the thermal conductivity of the solid is high (thereby reducing thermal gradients) the thermal stress is less than that given by Eq 20 by a factor A, which, to an adequate approximation, is given by

$$A = \frac{th/\lambda}{1 + th/\lambda} \qquad \text{(Eq 21)}$$

where t is a typical dimension of the sample in the direction of heat flow; the quantity th/λ is usually called the Biot modulus. Table 2 gives typical values of A for each class using a section size t of 10 mm.

The equation defining the thermal shock resistance, ΔT_s, now becomes

$$B\,\Delta T_s = \frac{\sigma_t}{\alpha E} \qquad \text{(Eq 22)}$$

where $B = C/A$. The contours on the diagram are of $B\,\Delta T$. The table shows that, for rapid quenching, A is unity for all materials except the high-conductivity metals, for which the thermal shock resistance is simply read from the contours, with appropriate correction for the constraint (the factor C). For slower quenches, ΔT_s is larger by the factor $1/A$, read from Table 2.

Use of Material Property Charts

The engineering properties of materials are usefully displayed as material selection charts. The charts summarize the information in a compact, easily accessible way, and they show the range of any given property accessible to the designer and identify the material class associated with segments of that range. By choosing the axes in a sensible way, more information can be displayed: A chart of modulus E against density ρ reveals the longitudinal wave velocity $(E/\rho)^{1/2}$; a plot of fracture toughness K_{Ic} against modulus E shows the fracture surface energy G_{Ic}; a diagram of thermal conductivity λ against diffusivity, a, also gives the volume specific heat ρC_v; expansion, α, against normalized strength, σ_t/E, gives thermal shock resistance ΔT_s.

The most striking feature of the charts is the way in which members of a material class cluster together. Despite the wide range of modulus and density associated with metals (as an example), they occupy a field that is distinct from that of polymers, or that of ceramics, or that of composites. The same is true of strength, toughness, thermal conductivity, and the rest: The fields sometimes overlap, but they always have a characteristic place within the whole picture.

The position of the fields and their relationship can be understood in simple physical terms: the nature of the bonding, the packing density, the lattice resistance, and the vibrational modes of the structure (themselves a function of bonding and packing), and so forth. It may seem odd that so little mention has been made of microstructure in determining properties. However, the charts clearly show that the first-order difference between the properties of materials has its origins in the mass of the atoms, the nature of the interatomic forces, and the geometry of packing. Alloying, heat treatment, and mechanical working all influence microstructure and, through this, properties, giving the elongated bubbles shown on many of the charts; yet the magnitude of their effect is less, by factors of 10, than that of bonding and structure.

The charts have numerous applications, among them: data checking, composite design, and identification of applications for new materials (Ref 1). But most important of all, the charts form the basis for a procedure for materials selection. This use is developed further in the following article in this Handbook, "Performance Indices," which contains two case studies to illustrate their use.

ACKNOWLEDGMENTS

The charts reproduced as Fig. 2 to 13 first appeared in Ref 1, where more details about their use can be found.

The author wishes to thank Dr. David Cebon for helpful discussions. The support of the Royal Society, the EPSRC through the Engineering Design Centre at Cambridge, and the Advance Research Project Agency through the University Research Initiative under Office of Naval Research Contract No. N-00014092-J-1808 are gratefully acknowledged.

REFERENCES

1. M.F. Ashby, *Material Selection in Mechanical Design*, Pergamon Press, 1992
2. M.F. Ashby and D. Cebon, *Case Studies in Material Selection*, Granta Design, 1996
3. *CMS Software and Handbooks*, Granta Design, 1995
4. *American Institute of Physics Handbook*, 3rd ed., McGraw-Hill, 1972
5. *Metals Handbook*, 9th ed., and *ASM Handbook*, ASM International
6. *Handbook of Chemistry and Physics*, 52nd ed., The Chemical Rubber Co., Cleveland, OH, 1971
7. *Landolt-Bornstein Tables* , Springer, 1966
8. Materials Selector, *Materials Engineering*, Penton Publishing, 1996
9. C.J. Smithells, *Metals Reference Book*, 7th ed., Butterworths, 1992
10. C.A. Harper, Ed., *Handbook of Plastics and Elastomers*, McGraw-Hill, 1975
11. A.K. Bhowmick and H.L. Stephens, *Handbook of Elastomers*, Marcel Dekker, 1986
12. S.P. Clarke, Jr., Ed., *Handbook of Physical Constants, Memoir 97*, The Geological Society of America, New York, 1966
13. N.A. Waterman and M.F. Ashby, Ed., *The Elsevier Materials Selector*, Elsevier and CRC Press, 1991
14. R. Morrell, *Handbook of Properties of Technical and Engineering Ceramics*, Parts I and II, National Physical Laboratory, London, U.K., 1985 and 1987
15. J.M. Dinwoodie, *Timber, Its Nature and Behaviour*, Van Nostrand-Reinhold, 1981
16. L.J. Gibson and M.F. Ashby, *Cellular Solids, Structure and Properties*, 2nd ed., Cambridge University Press, 1996
17. M.L. Bauccio, Ed., *ASM Engineered Materials Reference Book*, 2nd ed., ASM International, 1994
18. Materials Selector and Design Guide, *Design Engineering*, Morgan-Grampian Ltd, London, 1974
19. *Handbook of Industrial Materials (1992)*, 2nd ed., Elsevier, 1992
20. G.S. Grady and H.R. Clauser, Ed., *Materials Handbook*, 12th ed., McGraw-Hill, 1986
21. A. Goldsmith, T.E. Waterman, and J.J. Hirschhorn, Ed., *Handbook of Thermophysical Properties of Solid Materials*, Macmillan, 1961
22. Colin Robb, Ed., *Metals Databook*, The Institute of Metals, 1990
23. J.E. Bringas, Ed., *The Metals Black Book*, Vol 1, *Steels*, Casti Publishing, 1992
24. J.E. Bringas, Ed., *The Metals Red Book*, Vol 2, *Nonferrous Metals*, Casti Publishing, 1993
25. H. Saechtling, Ed., *International Plastics Handbook*, Macmillan Publishing (English edition), 1983
26. R.B. Seymour, *Polymers for Engineering Applications*, ASM International, 1987
27. *International Plastics Selector, Plastics*, 9th ed., Int. Plastics Selector, San Diego, CA, 1987
28. H. Domininghaus, Ed., *Die Kunststoffe and Ihre Eigenschaften*, VDI Verlag, Dusseldorf, Germany, 1992
29. D.W. van Krevelen, Ed., *Properties of Polymers*, 3rd ed., Elsevier, 1990
30. M.M. Schwartz, Ed., *Handbook of Structural Ceramics*, McGraw-Hill, 1992
31. R.J. Brook, Ed., *Concise Encyclopedia of Advanced Ceramic Materials*, Pergamon Press, 1991
32. N.P. Cheremisinoff, Ed., *Handbook of Ceramics and Composites*, Vol 3, Marcel Dekker, 1990
33. D.W. Richerson, *Modern Ceramic Engineering*, 2nd ed., Marcel Dekker, 1992
34. R. Morrell, *Handbook of Properties of Technical and Engineering Ceramics*, Parts 1 and 2, National Physical Laboratory, Teddington, U.K., 1985
35. W.E.C. Creyke, I.E.J. Sainsbury, and R. Morrell, *Design with Non Ductile Materials*, Applied Science, London, 1982
36. N.P. Bansal and R.H. Doremus, Ed., *Handbook of Glass Properties*, Academic Press, 1966
37. D.S. Oliver, *Engineering Design Guide 05: The Use of Glass in Engineering*, Oxford University Press, 1975
38. S. Musikant, *What Every Engineer Should Know about Ceramics*, Marcel Dekker, 1991
39. J.W. Weeton, D.M. Peters, and K.L. Thomas, Ed., *Engineers Guide to Composite Materials*, ASM International, 1987
40. M.M. Schwartz, Ed., *Composite Materials Handbook*, 2nd ed., McGraw-Hill, 1992
41. N.A. Fleck, K.J. Kang, and M.F. Ashby, The Cyclic Properties of Engineering Materials, *Acta Metall. Mater.*, Vol 42, 1994, p 365–381

Performance Indices

M.F. Ashby, Engineering Design Centre, Cambridge University

ANY ENGINEERING COMPONENT has one or more functions: to carry bending moments, to contain a pressure, to transmit heat, and so forth. In designing the component, the designer has an objective: to make it as inexpensive as possible, or as light, or as safe, perhaps. This must be achieved subject to constraints: that the component must carry the given loads without failure, that certain dimensions are fixed, that it can function in a certain range of temperature and in a given environment. Function, objective, and constraints (Table 1) define the boundary conditions for selecting a material and —in the case of load-bearing components—a shape for its cross section.

From these are derived material property limits and indices that are the key to optimizing the selection. Property limits are bounding values within which certain properties must lie if the material is to be considered further; typically, these are limits on fracture toughness, on maximum service temperature, or on corrosion resistance in a given environment. Performance indices are groupings of material properties which, if maximized, maximize some aspect of the performance of the component. Some are familiar. The specific stiffness, E/ρ, (where E is Young's modulus and ρ is the density) is an index; materials with exceptionally large values of E/ρ are the best candidates for a light, stiff tie-rod. The specific strength, σ_y/ρ, is another; materials with extreme values of σ_y/ρ are good candidates for a light, strong tie-rod. There are many such indices, each characterizing a particular combination of function, objective, and constraint (Fig. 1). Their derivation (Ref 1 to 5) proceeds from a statement of the objective in the form of an objective function, that is, an equation describing the quantity to be maximized or minimized. It contains free variables: parameters of the problem that the designer is free to change, provided the constraints are met. By using the constraint(s) to eliminate the free variable(s), the objective function can be expressed in terms of design-specified quantities and material properties only. The group of material properties is the performance index.

The derivation of indices is described in this article. Their use is illustrated, and a catalog of indices is appended.

Table 1 Questions for determining function, objective, and constraints for a component

Function	"What does component do?"
Objective	"What is to be maximized or minimized?"
Constraints	"What nonnegotiable conditions must be met?"
	"What negotiable but desirable conditions ...?"

Performance Indices and Material Property Charts

Two concepts are used in the "index-and-chart" selection procedure. The first is that of performance indices, which isolate the combination of material properties and shape information that maximize performance; the second is that of material property charts, described more fully in the article "Material Property Charts" in this Volume.

Development of Performance Indices

This article first defines performance indices in a formal way, then specifies how they are derived. The method is illustrated by two examples, one simple, the other—involving section shape as well as material—somewhat more advanced.

The design of a mechanical component is specified by three groups of variables: the functional requirements **F** (the need to carry loads, transmit heat, etc.); the specifications on geometry, **G**; and some combination **M** of the properties p of the material of which it is made (Ref 1 and 2). The performance **P** of the component can be described by an equation with the form

$$\mathbf{P} = f(\mathbf{F}, \mathbf{G}, p) \quad \text{(Eq 1)}$$

where **P** is the quantity for which a maximum or minimum is sought (the mass of the component, or its volume, or cost, or life for example) and f means "a function of." Optimum design can be considered to be selection of the material and geometry that maximize (or minimize) **P**. This optimization is

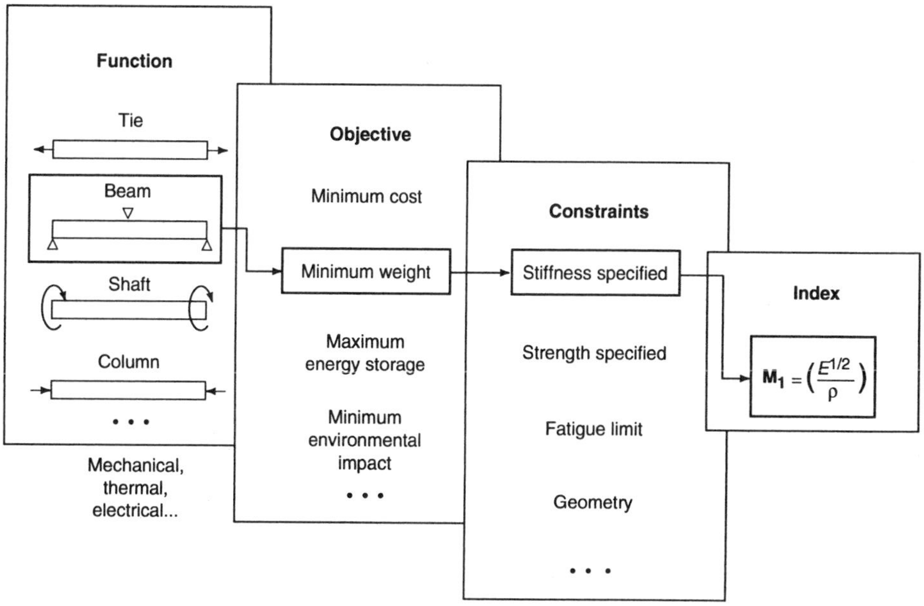

Fig. 1 The specification of function, objective, and constraint leads to a material index, **M**. The combination highlighted by the boxes leads to the index $\mathbf{M}$ - $E^{1/2}/\rho$.

Table 2 Function, objective, and constraints for light, strong tie

Function	"Tie" = carry tensile loads
Objective	"Light" = minimize weight
Constraints	"Strong" = support load F without yield or failure
	"Length l" = length specified (a geometric constraint)

subject to constraints, some of them imposed by the material properties.

The three groups of parameters in Eq 1 are said to be "separable" when the equation can be written

$$\mathbf{P} = f_1(\mathbf{F}) \cdot f_2(\mathbf{G}) \cdot f_3(p) \quad \text{(Eq 2)}$$

where f_1, f_2, and f_3 are functions. When the groups are separable, as they usually are, the optimum choice of material becomes independent of the many of the details of the design: it is the same for all geometries **G** and all values of the functional requirements **F**. Then the optimum material can be identified without solving the complete design problem, or even knowing all the details of **F** and **G**. This enables enormous simplification: the performance for all **F** and **G** is maximized by maximizing $\mathbf{M} = f_3(p)$, which is called the "merit index," or, better, "performance index."

The steps in deriving an index, shown schematically in Fig. 1, are as follows.

Step 1: Identify the *primary function* of the component for which a material is sought. A beam carries bending moments; a heat-exchanger tube transmits heat; a spring stores elastic energy.

Step 2: Write down an equation for the *objective*; it is called the "objective function." The objective is the first and most important quantity to be minimized or maximized. Commonly, it is weight or cost; but it could be energy stored per unit volume or per unit weight (a spring, a flywheel), or depth of dive (a submarine), or energy dissipated in i^2R heating (a bus bar)—it depends on the application.

Step 3: Identify the *constraints*. They are design requirements that must be met and which therefore limit the optimization process of step 2. Commonly these are: a required value for the stiffness S; a required value for the safe load F, moment M, torque T, or pressure p that can be supported; a limit on operating temperature T_{max}; or on resistance to sudden fracture, measured by the fracture toughness K_{Ic}.

It is essential to distinguish between objectives and constraints, and this requires a little thought. For example, in the performance-limited design of a racing bicycle frame, minimizing *weight* might be the objective with stiffness, strength, toughness, and cost as constraints ("as light as possible without costing more than $1000"). But in the design of an inexpensive "shopping" bicycle, minimizing *cost* becomes the objective, and weight becomes a constraint ("as inexpensive as possible, without weighing more than 22 kg").

Lay out the results as in Table 1.

Step 4: Eliminate the free variable(s) in the objective function by using the constraints. Read off the grouping of material properties called the "Performance Index," which maximizes (or minimizes) the value of the objective function. Performance indices are the basis for a method of optimal material selection.

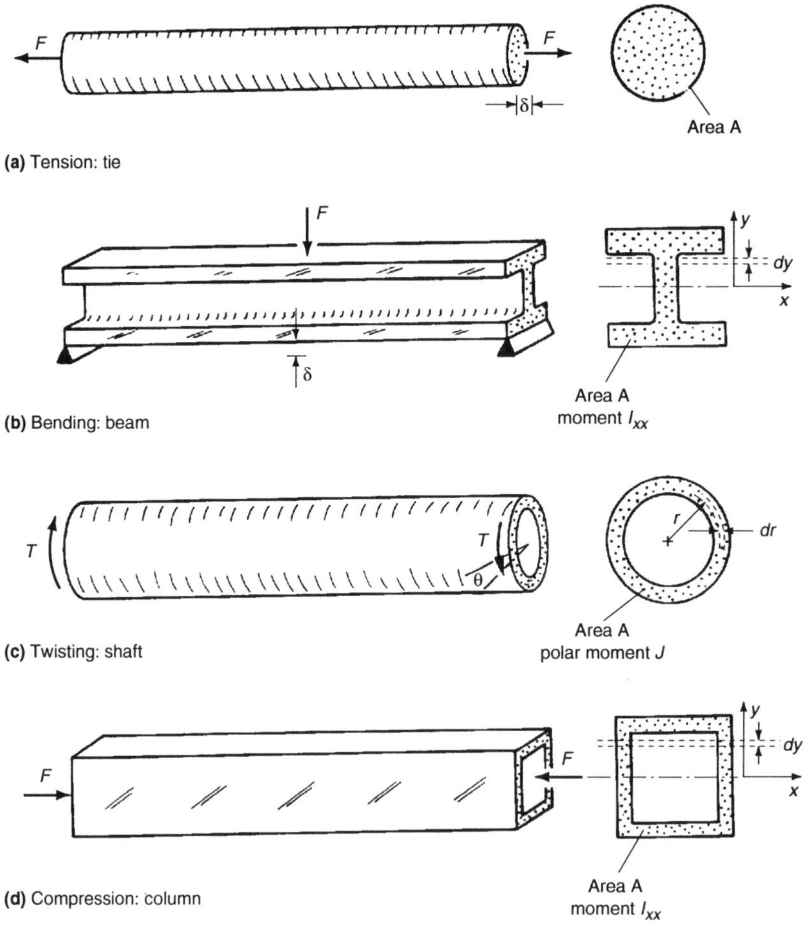

Fig. 2 A tie, a beam, a shaft, and a column. Efficient shapes are shown for each.

Each combination of a function, an objective, and a constraint shown in Fig. 1 leads to a performance index. The index characterizes the combination. The particular combination in the boxes of Fig. 1 describes a light stiff beam; its index (derived below) is:

$$\mathbf{M} = \frac{E^{1/2}}{\rho} \quad \text{(Eq 3)}$$

Indices are always denoted by a bold **M**. There are many others characterizing thermomechanical and electromechanical functions. They are cataloged in Table 5(f) and (g) of this article.

Performance Index for a Light, Strong Tie

The loading on a component can generally be decomposed into some combination of axial tension or compression, bending, and torsion. Almost always, one mode dominates. So common is this that the functional name given to the component describes the way it is loaded: ties carry predominantly tensile loads, beams carry bending moments, shafts carry torques, and columns carry compressive axial loads (Fig. 2).

Consider the first of these. A material is required for a solid cylindrical tie-rod of specified length l, to carry a tensile force F; it is to be of minimum mass. Here, "maximizing performance" means "minimizing the mass while still carrying the load F safely"; thus the design requirements are those specified in Table 2. The mass is

$$m = A\,l\,\rho \quad \text{(Eq 4)}$$

where A is the area of the cross section and ρ is the density of the material of which it is made. Equation 4 is the objective function—the quantity to be minimized. The length l and force F are specified and cannot be changed; the radius r, and thus the cross section, A, of the rod is free. The mass can be reduced by reducing the radius, but there is a constraint: The section area A must be sufficient to carry the tensile load F, requiring that

$$\frac{F}{A} \leq \frac{\sigma_f}{S_f} \quad \text{(Eq 5)}$$

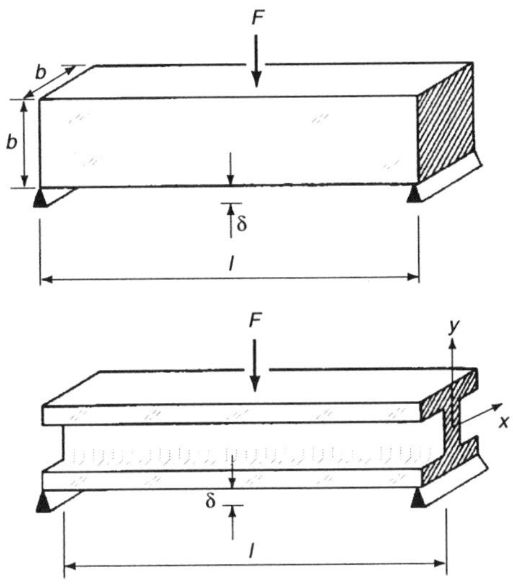

Fig. 3 (a) A solid, square-section beam, loaded in bending. (b) An I-section beam, loaded in bending

where σ_f is the failure strength and S_f is a safety factor, typically in the range 1.2 to 3. Eliminating A between these two equations gives

$$m \geq (S_f F)(l)\left(\frac{\rho}{\sigma_f}\right) \quad \text{(Eq 6)}$$

Note the form of this result. The first bracket contains the functional requirement that the specified load F is safely supported. The second bracket contains the specified geometry (the length l of the tie). The last bracket contains the material properties. The lightest tie-rod that will safely carry the load F without failing is that with the largest value of the material index

$$\mathbf{M} = \frac{\sigma_f}{\rho} \quad \text{(Eq 7)}$$

The convention was adopted that indices are expressed as quantities to be maximized in order to maximize performance; that is why the material properties in Eq 6 are inverted in the definition of **M** in Eq 7. In this example, the index is the "specific strength" mentioned earlier. A similar calculation for a light stiff tie leads to the index

$$\mathbf{M} = \frac{E}{\rho}$$

where E is Young's modulus. This time the index is the "specific stiffness." If the objective is to minimize cost instead of mass, the index becomes

$$\mathbf{M} = \frac{E}{C_m \rho}$$

where C_m is the cost per unit mass of the material.

But things are not always so simple. The text example shows how this comes about.

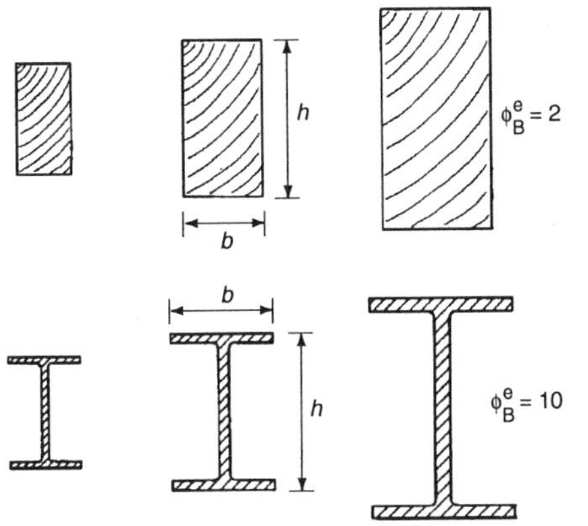

Fig. 4 The shape factor ϕ_B^e characterizes shape and is independent of scale. The shapes shown here have values of ϕ_B^e of 2 and 10. Shape factors as high as 100 are possible.

Performance Index for a Light, Stiff Beam

The mode of loading that most commonly dominates in engineering is not tension, but bending—think of floor joists, of wing spars, of golf-club shafts. Consider, then, a light beam of square, solid section and length l (Fig. 3a) and loaded in bending, which must meet a constraint on its stiffness S, meaning that it must not deflect more than δ under a load F. The words "light, stiff beam" prescribe a function, an objective, and a constraint (Table 3). The stiffness S of an elastic beam is:

$$S = \frac{F}{\delta} \geq \frac{CEI}{l^3} \quad \text{(Eq 8)}$$

where E is Young's modulus, C is a constant that depends on the distribution of load and I is the second moment of the area of the section, which, for a beam of square section is

$$I = \frac{b^4}{12} = \frac{A^2}{12} \quad \text{(Eq 9)}$$

The stiffness S and the length l are specified; the section A is free. The mass of the beam can be reduced by reducing A, but only so far that the stiffness constraint is still met. Using these equations to eliminate A in the objective function, Eq 4 gives

$$m \geq \left(\frac{12S}{C}\right)^{1/2}(l^{5/2})\left(\frac{\rho}{E^{1/2}}\right) \quad \text{(Eq 10)}$$

The brackets are ordered as before: functional requirement, geometry, and material. The best materials for a light, stiff beam are those with large values of the material index

$$\mathbf{M} = \frac{E^{1/2}}{\rho} \quad \text{(Eq 11)}$$

Table 3 Function, objective, and constraints for light, stiff beam

Function	"Beam" = carry bending moments
Objective	"Light" = minimize weight
Constraints	"Stiff" = beam stiffness is specified (a functional constraint)
	"Length l" = length specified (a geometric constraint)

Here, as before, the properties have been inverted; to minimize the mass, **M** must be maximized.

Performance Indices that Include Shape

Wooden beams sometimes have a solid, square section, but they are the exception rather than the rule. If the beam is to be efficient, it must be given an efficient shape: a tube, an I-section, or a box-section, perhaps (Fig. 2 and 3b). All of these are much lighter, for the same stiffness, than a solid, square section. To include shape, Eq 8 is used.

The second moment of area, I, depends on both the shape and scale (size) of the cross section:

$$I_{xx} = \int_{\text{section}} y^2\, dA \quad \text{(Eq 12)}$$

where y is measured normal to the bending axis and dA is the differential element of area at y, as shown in Fig. 2. It is helpful to separate shape from scale (Fig. 4). To do this, the shape factor for elastic bending is defined (Ref 4, 6, and 7):

$$\phi_B^e = \frac{4\pi I}{A^2} \quad \text{(Eq 13)}$$

The second moment, I, has dimensions of $(\text{length})^4$; so does the quantity A^2. The ratio of the two is dimensionless; it describes shape and is independent of scale. Two I-beams that have the same values of ϕ_B^e but that differ in scale have cross sections that look like photographic enlargements or reductions of each other. The factor 4π is inserted to give a solid, circular section a shape factor with a value of 1. Very efficient sections have values of ϕ_B^e as high as 100.

Substituting Eq 13 into Eq 8 gives

$$S = \frac{CE\phi_B^e A^2}{4\pi l^3} \quad \text{(Eq 14)}$$

The free variable, A, in the objective function Eq 4 is now eliminated, giving

$$m = \left(\frac{4\pi S}{C}\right)^{1/2}(l^{5/2}) \cdot \frac{\rho}{(\phi_B^e E)^{1/2}} \quad \text{(Eq 15)}$$

In this equation, the constant C, the stiffness S, and the length l are all specified by the design. For beams of a given, fixed shape, the objective function describing m is minimized by maximizing the same performance index **M** as before—it is that of Eq 11. However, for beams that differ in shape (as aluminum and wood beams generally do), it is minimized by maximizing a modified performance index that now includes the shape factor ϕ_B^e :

$$\mathbf{M} = \frac{E^{1/2}}{\rho} \cdot (\phi_B^e)^{1/2} \quad \text{(Eq 16)}$$

Some materials—steels and aluminum alloys, for instance—are routinely formed to very efficient shapes. Others—wood, for example—are not. Performance indices like that of Eq 16 that include both material and shape allow the coupled selection of both. Further details can be found in Ref 4, 6, and 7.

Other Shape Factors

As has been discussed, a shape factor is a dimensionless number that characterizes the efficiency of a section shape, regardless of scale, in a given mode of loading. Thus, as well as the shape factor, ϕ_B^e , for elastic bending of beams (defined above), there is another, ϕ_T^e , for elastic twisting of shafts (the superscript "e" means "elastic"). These are the appropriate shape factors when design is based on stiffness; when, instead, it is based on strength (i.e., on the first onset of plastic yielding or on fracture) two more shape factors are needed: ϕ_B^f and ϕ_T^f (the superscript "f" meaning "failure"). Finally, there is a shape factor for elastic buckling under axial compression ϕ_E^e . All five shape factors are defined so that they equal 1 for a solid bar with a circular cross section. Each is examined briefly in the following paragraphs.

In the elastic twisting of shafts (Fig. 2c), shape enters through the torsional moment of area, K. For circular sections it is identical with the polar moment of area, J:

$$K = J = \int_{\text{section}} r^2 dA \quad \text{(Eq 17)}$$

where dA is the differential element of area at the radial distance r, measured from the center of the section. For noncircular sections, K is less than J; it is defined such that the angle of twist θ is related to the torque T by

$$\theta = \frac{Tl}{KG}$$

where l is length of the shaft and G the shear modulus of the material of which it is made. Approximate expressions are available to calculate K for common section shapes (Ref 8).

The shape factor for elastic twisting is defined by

$$\phi_T^e = \frac{2\pi K}{A^2} \quad \text{(Eq 18)}$$

It, too, has the value 1 for solid circular shafts and is near 1 for any solid, equiaxed section; but for thin-walled shapes it, too, can be large. As before, sets of sections with the same value of ϕ_T^e differ in size but not shape.

Table 4 Definitions of shape factors

Mode of loading	Constraint	Section moment	Shape factor
Bending	Stiffness	I_{max} = greater of I_{xx} and I_{yy}	$\phi_B^e = \frac{4\pi I_{max}}{A^2}$
Bending	Strength	$Z = I/y_{max}$	$\phi_B^f = \frac{16\pi Z^2}{A^3}$
Torsion	Stiffness	J (circular sections) or K	$\phi_T^e = \frac{2\pi K}{A^2}$
Torsion	Strength	J/r (circular sections) or Q	$\phi_T^f = \frac{4\pi Q^2}{A^3}$
Compression	Buckling	I_{min} = lesser of I_{xx} and I_{yy}	$\phi_E^e = \frac{4\pi I_{min}}{A^2}$

Plasticity starts when the stress, somewhere, first reaches the yield strength, σ_y; fracture occurs when this stress first exceeds the fracture strength, σ_{fr}. Either one of these constitutes failure. The symbol σ_f denotes the failure stress, meaning "the local stress that will first cause yielding or fracture." One shape factor covers both. In bending, the stress σ is largest at the point y_m in the surface of the beam that lies furthest from the neutral axis; it is:

$$\sigma = \frac{M_B y_m}{I} = \frac{M_B}{Z} \quad \text{(Eq 19)}$$

where M_B is the bending moment and Z is the section modulus ($Z = I/y_m$). The shape factor for failure in bending, ϕ_B^f , is defined by

$$\phi_B^f = \frac{16\pi Z^2}{A^3} = \frac{16\pi I^2}{y_m^2 A^3} \quad \text{(Eq 20)}$$

Defined in this way, ϕ_B^f has the value 1 for a solid cylinder. It can be as high as 150 for efficient shapes.

In torsion, the problem is more complicated. For circular tubes or cylinders subjected to a torque T (as in Fig. 2c) the shear stress τ is a maximum at the outer surface, at the radial distance r_m from the axis of bending:

$$\tau = \frac{Tr_m}{J} \quad \text{(Eq 21)}$$

The quantity J/r_m in twisting has the same character as I/y_m in bending. For noncircular sections with ends that are free to warp, the maximum surface stress is given instead by

$$\tau = \frac{T}{Q} \quad \text{(Eq 22)}$$

where Q, with units of m^3, now plays the role in torsion of Z in bending (details in Ref 8). This allows the definition of a shape factor, ϕ_T^f , for failure in torsion:

$$\phi_T^f = \frac{4\pi Q^2}{A^3} \quad \text{(Eq 23)}$$

Like the other shape factors, ϕ_B^f and ϕ_T^f are dimensionless and therefore independent of scale, and both have the value 1 for a solid circular section.

Fully plastic bending or twisting (such that the yield strength is exceeded throughout the section) involve a further pair of shape factors. But, generally speaking, shapes that resist the onset of plasticity also resist full plasticity well. New shape factors for these are not, at this stage, necessary. The shape factor for buckling in compression (Fig. 2d) depends only on I. The appropriate shape factor (ϕ_B^e) has the same form as that for elastic bending (Eq 13) except that the I that enters is the lesser of the principal moments I_{xx} and I_{yy}, whereas in bending the loads are usually applied in such a way that the larger principal moment resists the deformation. The shape factors are summarized in Table 4. Expressions for all expressions of ϕ for common section shapes are listed in Ref 4 and 6.

The Table of Indices (Table 5)

Tables 5(a) to (g) lists some 130 indices, relevant to mechanical, thermomechanical, and electromechanical design. Derivations for almost all of them, following the method outlined here, can be found in Ref 4 and 5. In most selection exercises they are used in the way described below, without consideration of shape. But in mechanical design of components loaded in bending, in torsion, or in compression such that buckling becomes a failure mode, shape and material must be chosen together. The relevant index is then a compound one, like that of Eq 16, containing both material properties and one of the five shape factors of Table 4. Worked examples can be found in Ref 4.

Applying the Indices and Limits: Property Charts

The selection is made by creating materials property charts, onto which performance indices are plotted to give a sequence of selection stages. Figure 5 shows how a single selection stage is performed. Typically, there is a performance index, **M**, made up of two or more properties p_1 and p_2 that characterize them; families of materials appear as larger bubbles. A typical performance index has the form:

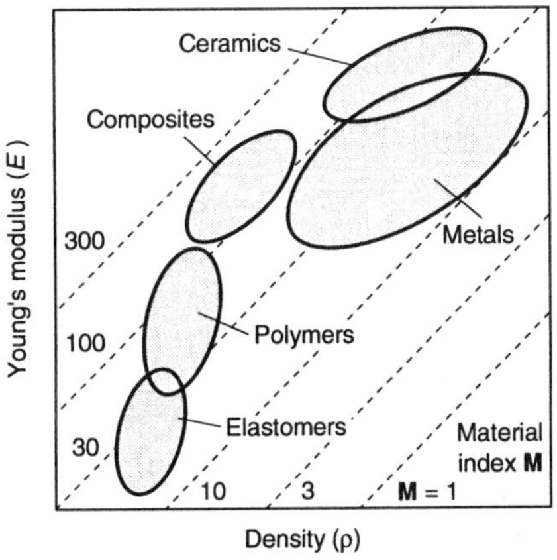

Fig. 5 A schematic E/ρ diagram with contours of the performance index, **M**

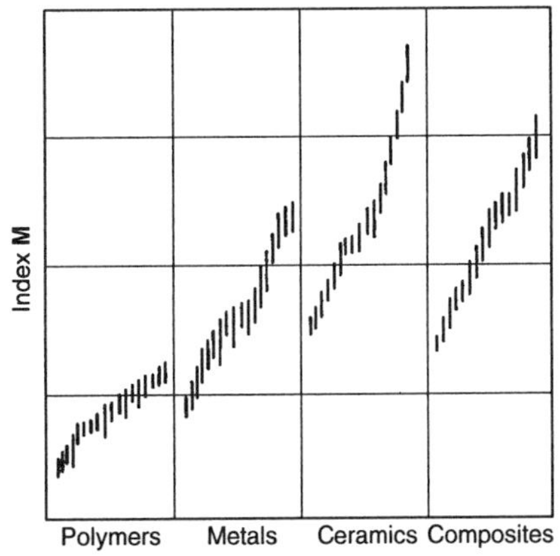

Fig. 6 A bar chart showing the performance index **M**

$$\mathbf{M} = \frac{p_2^n}{p_1} \qquad \text{(Eq 24)}$$

(the index $E^{1/2}/\rho$ is an example). Rearranging this gives:

$$p_2 = (p_1\mathbf{M})^{1/n}$$

Taking logarithms of the last equation gives:

$$\log p_2 = \frac{1}{n}\log p_1 + \frac{1}{n}\log \mathbf{M} \qquad \text{(Eq 25)}$$

Because the axes of the chart are $\log(p_1)$ and $\log(p_2)$, values of **M** can be plotted on it as a family of parallel diagonal lines of slope $1/n$. The selection is optimized by choosing the subset of materials that have the largest values of **M**. In Fig. 5, the materials near the top left are the best choice. Alternatively, the selection can be made as shown in Fig. 6. Here the index itself has been plotted on one axis, giving a bar chart of its values; the materials near the top are the best choice. Shaped materials can be plotted onto charts and selected by a similar procedure, which is illustrated more fully in the case studies that follow.

The method lends itself to implementation in software (Ref 5, 9, and 10). A full selection exercise can involve several stages in which limits are first placed on certain key properties (such as the maximum working temperature, the resistance to corrosion, toughness, or cost), and the subset of materials that pass these limits is then searched using the index-and-chart method to isolate candidates that optimally meet the objective and the constraint.

Table 5(a) Performance indices: stiffness-limited design at minimum mass (cost, energy)

Function and constraints(a)	Maximize(b)
Tie (tensile member)	
Stiffness, length specified; section area free	E/ρ
Shaft (loaded in torsion)	
Stiffness, length, shape specified; section area free	$G^{1/2}/\rho$
Stiffness, length, outer radius specified; wall thickness free	G/ρ
Stiffness, length, wall-thickness specified; outer radius free	$G^{1/3}/\rho$
Beam (loaded in bending)	
Stiffness, length, section shape specified; section area free	$E^{1/2}/\rho$
Stiffness, length, section height specified; width free	E/ρ
Stiffness, length, section width specified; height free	$E^{1/3}/\rho$
Column: compression strut far from buckling limit	
Stiffness and length specified; section area free	E/ρ
Column: compression strut, failure by elastic buckling	
Buckling load, length, section shape specified; section area free	$E^{1/2}/\rho$
Panel (flat plate, loaded in bending)	
Stiffness, length, width specified; thickness free	$E^{1/3}/\rho$
Plate (flat plate, compressed in-plane, buckling failure)	
Collapse load, length, and width specified; thickness free	$E^{1/3}/\rho$
Cylinder with internal pressure	
Elastic distortion, pressure, and radius specified; wall thickness free	E/ρ
Spherical shell with internal pressure	
Elastic distortion, pressure and radius specified; wall thickness free	$E/(1-\nu)\rho$

(a) To minimize cost, use the above criteria for minimum weight, replacing density ρ by $C_m\rho$, where C_m is the material cost per kg. To minimize energy content, use the above criteria for minimum weight replacing density ρ by $q\rho$, where q is the energy content per kg. (b) E = Young's modulus; G = shear modulus; ρ = density

Table 5(b) Performance indices: strength-limited design at minimum mass (cost, energy)

Function and constraints(a)	Maximize(b)
Tie (tensile strut)	
Stiffness, length specified; section area free	σ_f/ρ
Shaft (loaded in torsion)	
Load, length, shape specified; section area free	$\sigma_f^{2/3}/\rho$
Load, length, outer radius specified; wall thickness free	σ_f/ρ
Load, length, wall-thickness specified; outer radius free	$\sigma_f^{1/2}/\rho$
Beam (loaded in bending)	
Load, length, shape specified; section area free	$\sigma_f^{2/3}/\rho$
Load length, height specified; width free	σ_f/ρ
Load, length, width specified; height free	$\sigma_f^{1/2}/\rho$
Column (compression strut)	
Load, length, shape specified; section area free	σ_f/ρ
Panel (flat plate, loaded in bending)	
Stiffness, length, width specified; thickness free	$\sigma_f^{1/2}/\rho$
Plate (flat plate, compressed in-plane, buckling failure)	
Collapse load, length and width specified; thickness free	$\sigma_f^{1/2}/\rho$
Cylinder with internal pressure	
Elastic distortion, pressure and radius specified; wall thickness free	σ_f/ρ
Spherical shell with internal pressure	
Elastic distortion, pressure and radius specified; wall thickness free	σ_f/ρ
Flywheels, rotating disks	
Maximum energy storage per unit volume; given velocity	ρ
Maximum energy storage per unit mass; no failure	σ_f/ρ

(a) To minimize cost, use the above criteria for minimum weight, replacing density ρ by $C_m\rho$, where C_m is the material cost per kg. To minimize energy content, use the above criteria for minimum weight replacing density ρ by $q\rho$, where q is the energy content per kg. For design for infinite fatigue life, replace σ_f by the endurance limit σ_e. (b) σ_f = failure strength (the yield strength for metals and ductile polymers, the tensile strength for ceramics, glasses and brittle polymers); ρ = density

Case Studies in the Use of Performance Indices

The two case studies that follow illustrate the use of material indices, shape factors, and selection charts to select materials.

Example 1: Materials for Springs. Springs come in many shapes (Fig. 7) and have many purposes (Ref 11): One thinks of axial springs (a rubber band, for example), leaf springs, helical springs, spiral springs, and torsion bars. The primary function of a spring is that of storing elastic

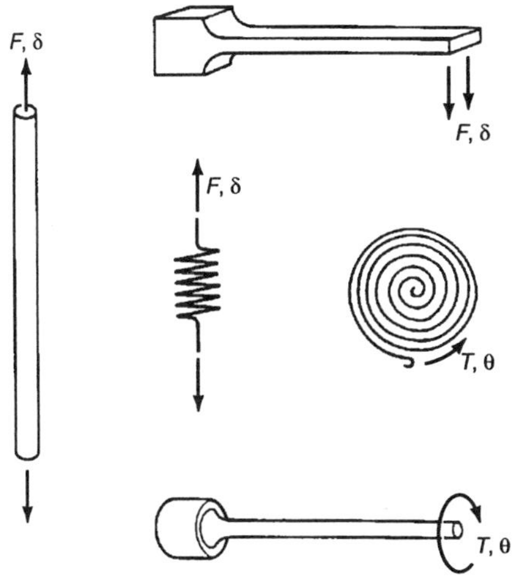

Fig. 7 Springs have many shapes, but all perform the same basic function: that of storing elastic energy.

Table 5(c) Performance indices: strength-limited design for maximum performance of components such as springs and hinges

Function and constraints	Maximize(a)
Springs	
Maximum stored elastic energy per unit volume; no failure	σ_f^2/E
Maximum stored elastic energy per unit mass; no failure	$\sigma_f^2/E\rho$
Elastic hinges	
Radius of bend to be minimized (maximum flexibility without failure)	σ_f/E
Knife edges, pivots	
Minimum contact area, maximum bearing load	σ_f^3/E^2 and H
Compression seals and gaskets	
Maximum conformability; limit on contact pressure	$\sigma_f^{3/2}/E$ and $1/E$
Diaphragms	
Maximum deflection under specified pressure or force	$\sigma_f^{3/2}/E$
Rotating drums and centrifuges	
Maximum angular velocity, radius fixed; wall thickness free	σ_f/ρ

(a) σ_f = failure strength (the yield strength for metals and ductile polymers, the tensile strength for ceramics, glasses, and brittle polymers); H = hardness; ρ = density. For design for infinite life, replace σ_f by the endurance limit σ_e.

Table 5(f) Performance indices: thermal and thermomechanical design

Function and constraints	Maximize(a)
Thermal insulation materials	
Minimum heat flux at steady state; thickness specified	$1/\lambda$
Minimum temp rise in specified time; thickness specified	$1/a = \rho C_p/\lambda$
Minimize total energy consumed in thermal cycle (kilns, etc.)	$\sqrt{a}/\lambda = \sqrt{1/\lambda\rho C_p}$
Thermal storage materials	
Maximum energy stored/unit material cost (storage heaters)	C_p/C_m
Maximize energy stored for given temperature rise and time	$\lambda/\sqrt{a} = \sqrt{\lambda\rho C_p}$
Precision devices	
Minimize thermal distortion for given heat flux	λ/α
Thermal shock resistance	
Maximum change in surface temperature; no failure	$\sigma_f/E\alpha$
Heat sinks	
Maximum heat flux per unit volume; expansion limited	$\lambda/\Delta\alpha$
Maximum heat flux per unit mass; expansion limited	$\lambda/\rho\Delta\alpha$
Heat exchangers (pressure-limited)	
Maximum heat flux per unit area; no failure under Δp	$\lambda\sigma_f$
Maximum heat flux per unit mass; no failure under Δp	$\lambda\sigma_f/\rho$

(a) λ = thermal conductivity; a = thermal diffusivity; C_p = specific heat capacity; C_m = material cost/kg; T_{max} = maximum service temperature; α = thermal expansion coefficient; E = Young's modulus; ρ = density; σ_f = failure strength (the yield strength for metals and ductile polymers, the tensile strength for ceramics, glasses, and brittle polymers)

Table 5(d) Performance indices: vibration-limited design

Function and constraints	Maximize(a)
Ties, columns	
Maximum longitudinal vibration frequencies	E/ρ
Beams	
Maximum flexural vibration frequencies	$E^{1/2}/\rho$
Panels	
Maximum flexural vibration frequencies	$E^{1/3}/\rho$
Ties, columns, beams, panels	
Minimum longitudinal excitation from external drivers, ties	$\eta E/\rho$
Minimum flexural excitation from external drivers, beams	$\eta E^{1/2}/\rho$
Minimum flexural excitation from external drivers, panels	$\eta E^{1/3}/\rho$

(a) E = Young's modulus (the yield strength for metals and ductile polymers, the tensile strength for ceramics, glasses, and brittle polymers); η = damping coefficient; ρ = density. For design for infinite fatigue life, replace σ_f by the endurance limit σ_e.

Table 5(g) Performance indices: electromechanical design

Function and constraints	Maximize(a)
Bus bars	
Minimum life-cost; high-current conductor	$1/\rho_e\rho C_m$
Electromagnet windings	
Maximum short-pulse field; no mechanical failure	σ_y
Maximize field and pulse-length; limit on temperature rise	$C_p\rho/\rho_e$
Windings, high-speed electric motors	
Maximum rotational speed; no fatigue failure	σ_e/ρ_e
Minimum ohmic losses; no fatigue failure	$1/\rho_e$
Relay arms	
Minimum response time; no fatigue failure	$\sigma_e/E\rho_e$
Minimum ohmic losses; no fatigue failure	$\sigma_e^2/E\rho_e$

(a) C_m = material cost/kg; E = Young's modulus; ρ = density; ρ_e = electrical resistivity; σ_y = yield strength; σ_e = endurance limit

Table 5(e) Performance indices: damage-tolerant design

Function and constraints	Maximize(a)
Ties (tensile member)	
Maximize flaw tolerance and strength; load-controlled design	K_{Ic} and σ_f
Maximize flaw tolerance and strength; displacement-control	K_{Ic}/E and σ_f
Maximize flaw tolerance and strength; energy-control	K_{Ic}^2/E and σ_f
Shafts (loaded in torsion)	
Maximize flaw tolerance and strength; load-controlled design	K_{Ic} and σ_f
Maximize flaw tolerance and strength; displacement-control	K_{Ic}/E and σ_f
Maximize flaw tolerance and strength; energy-control	K_{Ic}^2/E and σ_f
Beams (loaded in bending)	
Maximize flaw tolerance and strength; load-controlled design	K_{Ic} and σ_f
Maximize flaw tolerance and strength; displacement-control	K_{Ic}/E and σ_f
Maximize flaw tolerance and strength; energy-control	K_{Ic}^2/E and σ_f
Pressure vessel	
Yield-before-break	K_{Ic}/σ_f
Leak-before-break	K_{Ic}^2/σ_f

(a) K_{Ic} = fracture toughness; E = Young's modulus; σ_f = failure strength (the yield strength for metals and ductile polymers, the tensile strength for ceramics, glasses, and brittle polymers); η = damping coefficient; ρ = density

energy and, when required, releasing it again. The objective, then, is to maximize the energy that can be stored. There is a constraint: The spring, if it is to perform properly, must remain elastic; that is, the stress in it must nowhere exceed the elastic limit or, for cyclic loading, the endurance limit. Table 6 summarizes the design requirements.

The elastic energy W_v stored per unit volume in a block of material of modulus E stressed uniformly to a stress σ is

$$W_v = \frac{1}{2}\frac{\sigma^2}{E}$$

It is this that should be maximized. The spring will be damaged if the stress σ exceeds the elastic limit, yield stress, or failure stress σ_f; the constraint is that $\sigma \le \sigma_f$ with an appropriate safety factor, S_f. So the maximum energy per unit volume that a spring can store is

$$(W_v)_{max} = \frac{1}{2}\frac{\sigma_f^2}{S_f E} \quad \text{(Eq 26a)}$$

Torsion bars and leaf springs are less efficient than axial springs because much of the material is not fully loaded. The material at the neutral axis, for instance, is not loaded at all. For torsion bars:

$$(W_v)_{max} = \frac{1}{3}\frac{\sigma_f^2}{S_f E} \quad \text{(Eq 26b)}$$

and for leaf springs:

$$(W_v)_{max} = \frac{1}{4}\frac{\sigma_f^2}{S_f E} \quad \text{(Eq 26c)}$$

However, as these results show, the mode of loading has no influence on the choice of material. The best materials for compact springs are those with the biggest value of

$$\mathbf{M_1} = \frac{\sigma_f^2}{E} \quad \text{(Eq 27)}$$

If weight, rather than volume, is of concern (light springs), $(W_v)_{max}$ must be divided by the density ρ (giving energy stored per unit weight) and materials must be found with high values of

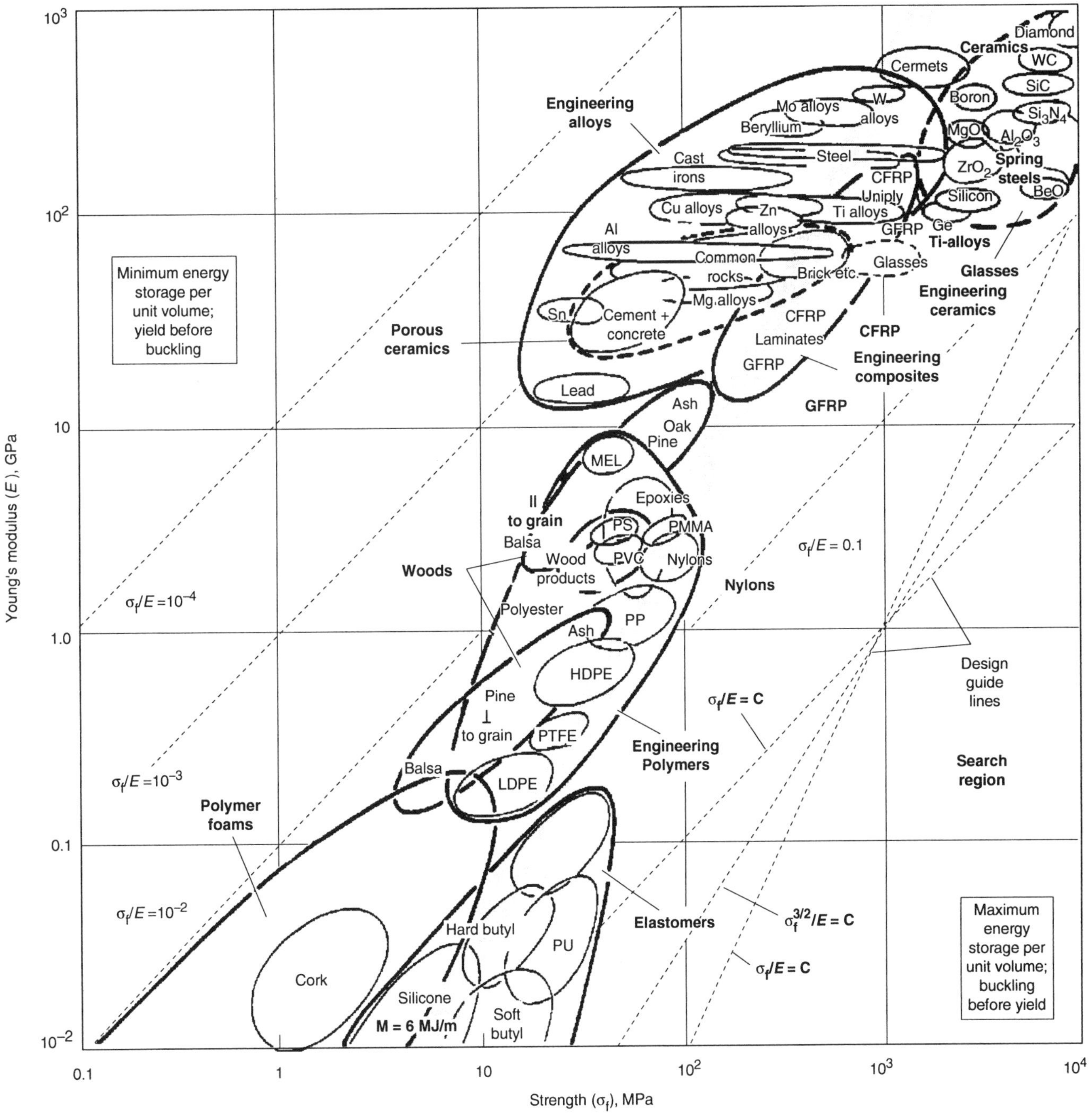

Fig. 8 A chart of Young's modulus, E, plotted against the elastic limit, σ_f. The diagonal line shows $\mathbf{M_1}$.

$$\mathbf{M_2} = \frac{\sigma_f^2}{\rho E} \quad \text{(Eq 28)}$$

Also, if it were the energy per unit cost (inexpensive springs), $(W_v)_{max}$ must be divided by cost per unit volume, $C_m \cdot \rho$ and materials must be found with high values of

$$\mathbf{M_3} = \frac{\sigma_f^2}{C_m \rho E} \quad \text{(Eq 29)}$$

The selection procedure for springs of minimum volume is shown in Fig. 8 (based on Fig. 5 in the article "Material Property Charts" in this Volume), which shows the elastic limit plotted against the modulus. Rearranging Eq 27 gives:

$$E = \frac{\sigma_f^2}{\mathbf{M_1}}$$

or, taking logs,

$$\log(E) = 2\log(\sigma_f - \log(\mathbf{M_1}) \quad \text{(Eq 30)}$$

Family lines of slope 2 link materials with equal values of $\mathbf{M_1} = \sigma_f^2/E$; those with the highest values of $\mathbf{M_1}$ lie toward the bottom right. The heavy line is one of the family; it is positioned so that a subset of materials is left exposed. The best choices are a high-strength steel (spring steel, in fact) lying near the top end of the line, and, at the other end, rubber. But certain other materials are suggested too: GFRP (now used for truck springs), titanium alloys (good but expensive), glass (used in galvanometers), and nylon (children's toys often have nylon springs). Note how the procedure has identified a candidate from almost every class of material: metals, glasses, polymers, elastomers, and composites. They are listed, with commentary, in Table 7.

A similar procedure allows materials selection for light springs using the index $\mathbf{M_2}$ and Fig. 6 of

Table 6 Design requirements for springs

Function	Spring
Objective	1. Maximize the energy stored per unit volume
	2. Maximize the energy stored per unit weight
	3. Maximize the energy stored per unit cost
Constraints	Must not fail by yield, fracture, or fatigue

Table 7 Materials for efficient compact springs

Material	$M_1 = \frac{\sigma_f^2}{E}$ (MJ/m^3)	Comment
Ceramics	(10-100)	Brittle in tension; good only in compression
Spring steel	10	The traditional choice: easily formed and heat treated
Titanium alloys	10	Expensive, corrosion resistant
Carbon-fiber-reinforced polymer (CFRP)	8	Comparable in performance with steel; expensive
Glass-fiber-reinforced polymer (GFRP)	5	Almost as good as CFRP and much less expensive
Glass	10	Brittle in torsion, but excellent if protected against damage; very low loss factor
Nylon	3	The least good; but inexpensive and easily shaped, but high loss factor
Rubber	20	Better than spring steel; but high loss factor

the article "Material Property Charts" in this Volume and for inexpensive springs, using the index $\mathbf{M_3}$ and an appropriate chart (Ref 5).

Other Considerations. Many additional considerations enter the choice of a material for a spring. Springs for vehicle suspensions must resist fatigue and corrosion; internal combustion (IC) valve springs must cope with elevated temperatures. A subtler property is the loss coefficient, shown in Fig. 9 of the article "Material Property Charts" in this Volume. Polymers have a relatively high loss factor and dissipate energy when they vibrate; metals, if strongly hardened, do not. Polymers, because they creep, are unsuitable for springs that carry a steady load, though they are still perfectly good for catches and locating springs which spend most of their time unstressed.

Example 2: Materials for Windsurfer Masts. Windsurfing seems to have its origins in an eccentric idea, published in *Popular Science* in 1965 by an American, Newman Derby. It was that of the "freesail"—a board, steered by back and forward motion of the mast itself. Implementing it successfully took a further idea, that of another American, Jim Drake, who devised a universal joint that allowed a freely swiveling mast to be stuck onto a surfboard, thereby creating a new sport (Ref 12, 13). Since then the equipment (Fig. 9) has evolved, reflecting the technical influence of mechanical and materials engineering. The mast is a key part of the rig and influences the dynamics of the sail. Masts are designed to flex under wind pressure; the flex characteristics of the mast determine sail shape and, through this, sailing characteristics. Thus the first constraint on the mast is that of a specified stiffness. While maintaining stiffness, the designer seeks to limit the outer diameter of the mast in order to reduce its influence on air flow. Also, for reasons of stability, the mast must be as light as possible. Table 8 summarizes the design requirements.

A typical windsurfer mast has a length L of 4.6 m. Its stiffness is characterized by the International Mast Check System (IMCS) number. It is measured by measuring the deflection δ caused by hanging a weight of 30 kg at the midpoint of the mast, supported at its ends; the number is given by

$$\text{IMCS number} = \frac{L}{\delta} \qquad \text{(Eq 31)}$$

The values lie between 20 (a soft mast) and 32 (a hard one). Masts typically weigh between 1.8 and 3 kg. They are tuned by changing the lay-up.

Table 8 Design requirements for windsurfer masts

Function	Windsurfer mast (light, stiff beam)
Objective	Minimize mass, m
Constraints	Stiffness S specified
	Limit on outer diameter of mast

The mast has length L and cross-sectional area A, and for lightness, it is hollow. The objective is to minimize its mass m:

$$m = A\,L\,\rho \qquad \text{(Eq 32)}$$

where ρ is the density of the material of which it is made. There is a constraint on the bending stiffness S of the mast, defined by

$$S = \frac{F}{\delta} = \frac{CEI}{L^3} \qquad \text{(Eq 33)}$$

where E is the (longitudinal) modulus of the material of which it is made, I the second moment of its cross section, and C a constant that depends only on the distribution of load; for three-point bending, as in the ICMS test, $C = 48$. The shape factor for elastic bending, ϕ_B^e, was defined in Eq 13. Because only this shape factor is needed, it shall be denoted by the symbol ϕ alone. Solving Eq 13 for I and substituting in Eq 33 gives an expression for the area A that will just provide the required stiffness S:

$$A = \left(\frac{4\pi S L^3}{CE\phi}\right)^{1/2} \qquad \text{(Eq 34)}$$

Substituting this into Eq 32 gives

$$m = \left(\frac{4\pi S L^5}{C}\right)^{1/2} \left(\frac{\rho}{(\phi E)^{1/2}}\right) \qquad \text{(Eq 35)}$$

the first bracket contains only quantities defined by the design. The second contains only material properties and shape. The mass of the mast is minimized, for a given stiffness, by choosing a material and shape with as large a value as possible of the performance index

$$\mathbf{M} = \frac{(\phi E)^{1/2}}{\rho} \qquad \text{(Eq 36)}$$

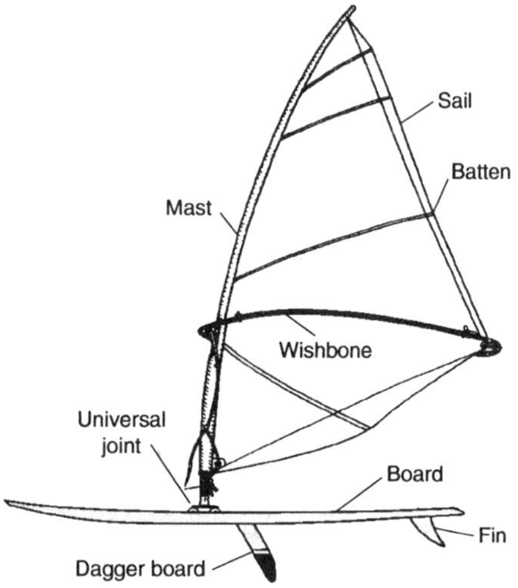

Fig. 9 A windsurfer. The flexure of the mast controls the shape of the sail; its pivoting about the universal joint controls the response of the craft.

There is, additionally, a geometric constraint: The mast must not be too fat. Expressed more formally: The outer radius of the mast must be less than an upper-limiting value:

$$r \leq r_{max} \qquad \text{(Eq 37)}$$

The derivation thus far has been general, applicable to any shape. This new constraint is used by employing results specific to thin-walled tubes. The shape factor for a thin-walled tube is

$$\phi = \frac{4\pi I}{A^2} = \frac{4\pi\,(\pi r^3 t)}{(2\pi r t)^2} = \frac{r}{t} \qquad \text{(Eq 38)}$$

This allows

$$I = \pi r^3 t = \frac{\pi r^4}{\phi} \qquad \text{(Eq 39)}$$

and

$$\phi = \frac{r}{t} \qquad \text{(Eq 40)}$$

Substituting this into Eq 33 gives a lower bound for the quantity E/ϕ:

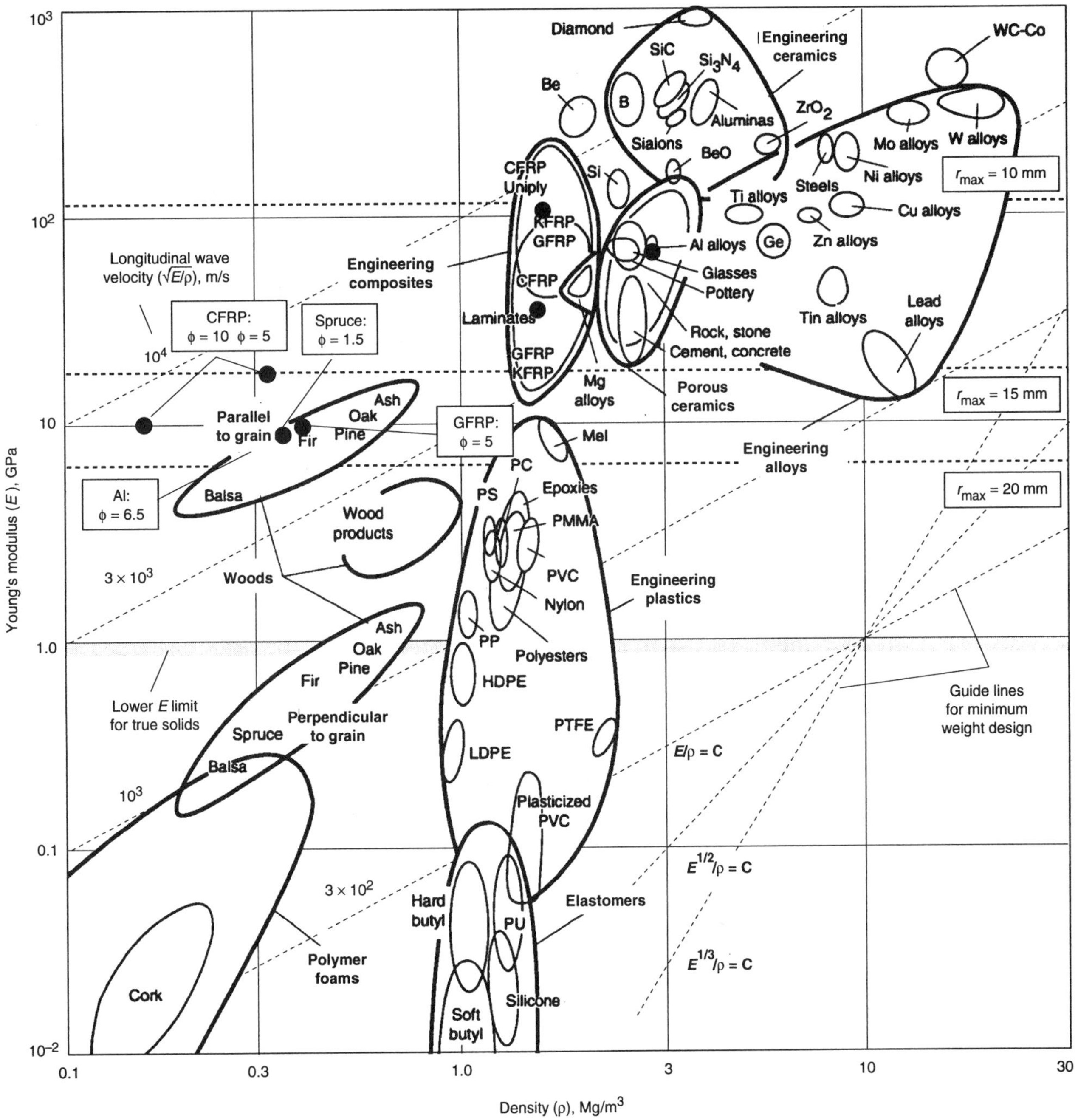

Fig. 10 A chart of Young's modulus, *E*, plotted against density, ρ. Solid materials are plotted at their (*E*, ρ) point; shaping them by a factor φ moves them to the point (*E*/φ, ρ/φ). All the previous selection strategies still apply. The diagonal line shows **M**; the horizontal lines show the limits for *E*/φ.

$$\frac{E}{\phi} \geq \frac{SL^3}{C\pi r_{max}^4} \quad \text{(Eq 41)}$$

To make use of this, the right-hand side must be evaluated. The stiffness $S = F/\delta$ of a "medium" mast with an IMCS number of 26 is 1.7×10^3 N/m. The resulting lower limits for E/ϕ, for $L = 4.6$ m and three values of r_{max}, are listed in Table 9.

Figure 10 shows the selection. The axes are effective modulus E/ϕ and effective density ρ/ϕ. To plot the index **M**, note that

$$\mathbf{M} = \frac{(\phi E)^{1/2}}{\rho} = \frac{(E/\phi)^{1/2}}{\rho/\phi} \quad \text{(Eq 42)}$$

Solving for E/ϕ and taking logs gives

$$\log(E/\phi) = 2\log(\rho/\phi) + 2\log(\mathbf{M}) \quad \text{(Eq 43)}$$

Therefore, the tubes are treated as "materials" with effective modulus E/ϕ and ρ/ϕ, and they are plotted on the conventional E-ρ diagram together with solid materials for which $\phi = 1$. Contours of **M** appear as a family of lines of slope 2. One is shown. The lower limit on effective modulus appears as a horizontal line at the indicated value of r_{max}; only materials

Table 9 Limits on E/ϕ

Maximum radius of mast (r_{max}), mm	Required minimum (E/ϕ), GPa
10	110
15	21
20	6.8

Table 10 Materials for windsurfer masts

Material	Shape factor, $\phi = r/t$	Index M, $GPa^{1/2}/(Mg/m^3)$	Mass (m), kg, for IMCS = 26
r_{max} = 20 mm			
CFRP	14.3	22.9	2.0
GFRP	4.3	6.2	7.3
Aluminum	8.0	8.5	5.3
Wood and (spruce)	1.7	9.0	5.0
r_{max} = 15 mm			
CFRP	4.7	12.9	3.5
GFRP	1.4	3.9	9.2
Aluminum	3.1	4.9	11.6
Wood and (spruce)	Not possible	...	...

above the appropriate line are candidates. Among these, the most attractive are those with the largest value of the index **M**. The selection is listed in Table 10.

Other Considerations. The language of windsurfer design verges on the poetic. The coupled tuning of sail and rig is known in the industry as the "integrated rig concept," a vision that focuses on the "creation of a perfect harmonic functional unit." With harmonic perfection as the criterion of success, the choice of material and overall shape is only the first (though important) step. The distribution of stiffness along the mast now becomes an issue. Masts today are tuned not only to match the weight of the surfer, but also the type of windsurfing (slalom, race, or wave). As in many other sports, the technology of the equipment has become very sophisticated.

ACKNOWLEDGMENTS

The charts shown as Fig. 8 and 10 first appeared in Ref 4, where further details about their use can be found.

The author wishes to thank Dr. David Cebon and Dr. P.M. Weaver for helpful discussions. The support of the Royal Society, the EPSRC through the Engineering Design Centre at Cambridge, and the Advance Research Project Agency through the University Research Initiative under Office of Naval Research Contract No. N-00014092-J-1808 are gratefully acknowledged.

REFERENCES

1. *Metals Handbook*, Vol 1, 8th ed., American Society for Metals, 1961, p 185–187
2. G. Dieter, *Engineering Design, A Materials and Processing Approach*, 2nd ed., McGraw-Hill, 1991, Chapt. 6
3. M.F. Ashby, Materials Selection in Conceptual Design, *Mater. Sci. Technol.*, Vol 5, June 1989, p 517–525
4. M.F. Ashby, *Materials Selection in Mechanical Design*, Pergamon Press, 1992
5. M.F. Ashby and D. Cebon, *Case Studies in Materials Selection*, Granta Design, 1996
6. M.F. Ashby, Materials and Shape, *Acta Metall. Mater.*, Vol 39 (No. 6), June 1991, p 1025–1039
7. P.M. Weaver and M.F. Ashby, The Optimal Selection of Material and Shape, *J. Eng. Des.*, Vol 7 (No. 2), June 1996
8. W.C. Young, *Roark's Formulas for Stress and Strain*, 6th ed., McGraw-Hill, 1989
9. D. Cebon and M.F. Ashby, Computer Aided Material Selection for Mechanical Design, *Met. Mater.*, Jan 1992, p 25–30
10. *CMS Software*, Granta Design, 1995
11. M. Hayes, Materials Update 2: Springs, *Engineering*, May 1990, p 42
12. Farrel O'Shea, *Windsurfing*, Ward Lock, 1991
13. A.J. Barker and J.L. Wearing, Mechanics and Design for Windsurfer Masts, *The Engineering of Sport*, S. Haake, Ed., Balakaema Press, 1996

Decision Matrices in Materials Selection

David L. Bourell, The University of Texas at Austin

THE GOAL of any materials selection endeavor is to choose the best material for a given application. When selection is simple, involving only one or two material demands, the process is relatively straightforward. However, sometimes there are conflicting materials demands, or several individuals or groups want to have input, or the process needs to be documented. In these cases it is helpful to use a decision matrix.

In materials selection, a decision matrix method is any formalized procedure by which materials are ranked prior to a selection decision. The methods described in this article all involve construction of a table, composed of columns of potential materials and rows of important materials functions. A decision matrix is a stimulus to creative work (Ref 1), because it forces the designer to analyze the selection from different angles and to clarify the design issues. Decision matrix methods are sometimes more correctly labeled "evaluation matrix methods."

Advantages of decision matrix methods are:

- They organize and clarify the selection task, providing greater insight into the design situation and ensuring that all appropriate considerations are included.
- They improve understanding of the interactions between proposed solutions, thereby providing a means for new, synergistic solutions. For example, the need to reconcile low cost with corrosion resistance may lead to consideration of plating or surface sheet bonding.
- They provide a written record of the selection process, which can be important if the decision is reviewed or questioned later. In addition, a decision matrix can be a fruitful starting point for future materials selection for a different design with related functions.
- They are versatile, robust, and applicable during all stages of design, including conceptual and configuration design.
- They facilitate participatory decision-making by recording input from various individuals.
- They can be relatively bias-free, thus maximizing creativity and flexibility in materials selection.
- They allow the designer to include or exclude properties or features in excess of the design requirements. For example, consider an application requiring a maximum corrosion rate of 200 mm/year and a minimum yield strength of 200 MPa. Suppose two materials are considered, one with a corrosion rate of 180 mm/year and yield strength of 300 MPa, the other with a corrosion rate of 10 mm/year and yield strength of 210 MPa. Decision matrix methods allow the designer to incorporate into the decision process the relative importance of these "excess" properties and features.

Limitations of decision matrix methods include the following:

- Decision matrix methods are not completely objective. Designers must still make value judgments in comparing materials (e.g., judge the manufacturability of plain carbon steel versus that of low-density polyethylene). Sometimes they must also judge the relative importance of evaluation criteria (e.g., electrical conductivity versus castability versus strength).
- It is important to evaluate the decision matrix method during the design process. However, this "fine tuning" may be misconstrued or abused, so that the inputs are "fudged" in order to obtain a foregone conclusion.
- Often, requisite data are unknown or inaccessible in a timely manner. This is becoming less of a problem with the creation and availability of computer databases (see the article "Computer-Aided Materials Selection" in this Volume).

Alternatives, Criteria, and Weighting Factors

Three important concepts in decision making are alternatives (candidate materials), criteria (objectives), and weighting factors.

Alternatives (candidate materials) are the materials or material systems under consideration. In the conceptual design stage, an alternative might be non-heat-treatable aluminum alloy. In the configuration phase, a specific material, such as ASTM B 209 aluminum alloy 5052-H38, might be an appropriate alternative. Alternatives can be suggested by several sources, including the design engineer's experience, materials databases, materials suppliers, an in-house list of stocked material, and Ashby selection charts (discussed in the article "Material Property Charts" in this Volume).

Criteria (objectives) are the properties or materials features that are being considered in the decision-making process. Criteria should be as specific as the design definition will allow. Early in design, the criteria might be "must be transparent," "should be easily manufactured," "must be tough," or "must have low specific modulus". During configuration design, the criteria might be "toughness exceeding 60 MPa$\sqrt{m}$," "electrical conductivity between 0.1 and 0.4 $\mu\Omega \cdot m$," "hardening without preheat or postheat," or "weldable using gas welding techniques." Criteria generally are derived from the requirements list (also known as the specification sheet or product requirements list) that is developed early in the design process.

Weighting factors are numerical representations of the relative importance of criteria in the selection process. Assuming that all alternatives meet the design demands, weighting factors quantify the relative importance of the criteria (e.g., hardness versus castability versus density) to the overall selection. They can also be used to quantify excess properties. Generally, weighting factors for all criteria sum to unity. All decision matrix methods use alternatives and criteria, but weighting factors may or may not be used.

Decision Matrix Techniques

Numerous decision matrix techniques have been developed (Ref 1–7). The three selection techniques described in this article have been chosen to demonstrate varying degrees of complexity: the Pugh method, the Dominic method, and the Pahl and Beitz method. Their key features are shown in Table 1. The Pugh method is the simplest of the three, with all criteria qualitatively evaluated with equal weight. The Dominic method provides a means for rating the relative importance of criteria by use of weighting factors. The Pahl and Beitz method is a completely numerical approach to materials selection.

Table 1 Comparison of common decision matrix methods

Method	Type	Criteria weighting factors?
Pugh	Qualitative	No
Dominic	Qualitative	Yes
Pahl and Beitz	Quantitative	Yes

The Pugh Method

Description. The Pugh method (Ref 1) is one of the simplest decision matrix methods. Selection is based on qualitative comparison to a datum or reference alternative. The Pugh method is useful early in design, since it requires the least amount of detailed information. It is also useful in redesign, because the current material design serves automatically as the datum.

Method. Alternatives are listed as column headers in the decision matrix. The first column lists criteria. One alternative is selected to be the datum, perhaps the alternative that is best known and understood, or the alternative that is intuitively considered to be the best. Each blank cell is associated with a row (criterion) and a column (alternative). For each blank cell, ask, "Considering this criterion, is this alternative better (+), worse (–), or the same (S) when compared to the datum?" and fill in the appropriate symbol. When there is doubt or disagreement, use "S."

The next step is to create three more rows at the bottom of the matrix with the row headers "Σ+," "Σ–," and "ΣS." For each alternative except the datum, sum the number of "+", "–," and "S" responses in the appropriate blank. Resist the tendency to generate an overall sum for each alternative.

Analyze the results. For each alternative, consider whether some modification might mitigate its weak points relative to the datum. If so, modify the alternative, enter it as a new column on the decision matrix, and evaluate it relative to the datum. Retain the unmodified alternative.

Analyze the criteria. If the method does not adequately distinguish alternatives as strong and weak, perhaps the criteria are overly ambiguous, subject to multiple interpretations, or overlapping. Clarification of criteria may be in order.

Having analyzed the alternatives and criteria, identify the weakest alternatives and eliminate them, thereby shrinking the matrix. If more than a few alternatives appear to be strong, it may be helpful to consider each of them as the final selection, in order to produce additional criteria that will clarify the decision. It is sometimes useful to redo the decision matrix using one of the strongest alternatives as the datum.

At this point, the strongest alternatives may be selected for further development. If the selection process takes place at the final stages of design, the strongest alternative may be selected if one clearly stands apart from the others. Otherwise, it may be prudent to use one of the other, more detailed decision matrix methods.

Example 1: Use of the Pugh Method in Material Selection for Redesign of a Toy Train Spring. The objective in this example was to reconsider selection of a helical steel spring for a child's toy train. The original material was ASTM A 227 class I steel wire, hard drawn for mechanical springs. Three alternatives were considered: a different design geometry of the same material, ASTM A 228 music spring quality steel wire, and ASTM A 229 class I oil-tempered steel wire for mechanical springs. A Pugh decision matrix appears as Table 2. Criteria are grouped by performance and cost factors.

Analysis. Both the music spring quality steel wire and the oil-tempered steel wire appeared to be superior to the current design. The modestly higher ranking of the former arises from consideration of manufacturing cost related to the heat treatment of the latter. Selection of music spring quality steel wire produced an overall cost savings of 33%.

The Dominic Method

Description. The Dominic method (Ref 6) is used to evaluate alternatives using five descriptors: excellent, good, fair, poor, and unacceptable. Further, criteria are rated qualitatively as being high priority, moderate priority, or low priority. The Dominic method, like the Pugh method, is completely qualitative, but it allows for more complexity in the selection process.

Method. A matrix of alternatives and criteria is constructed as described for the Pugh method, but no datum is selected. Alternatives are labeled using capital letters ("A," "B," "C," etc.). Criteria are listed in descending order of importance to the design and are then gathered into three categories: high priority, moderate priority, or low priority (or, alternatively, "crucial," "important," and "desirable"). The matrix is completed by ranking each criterion on an absolute basis as excellent, good, fair, poor, or unacceptable for each alternative.

Once the matrix is complete, the results are tabulated using a Dominic method overall evaluation table (DMOET), Table 3. The rows are the ratings for alternatives and the columns are ratings for criteria. The blanks are filled in with the letters representing alternatives. For example, if alternative B had three "excellent" ratings in high-priority criteria, three B's would be placed in the blank cell for "High" and "Excellent." Once the DMOET is complete, alternatives are given an overall rating. All letters for an alternative must be above an overall rating line to obtain a given rating. The rating lines are shown on the DMOET. To obtain an "excellent" overall rating, the alternative must be rated as excellent in all high-priority criteria, good or excellent in all moderate-priority criteria, and fair or better in the low-priority criteria. The results of the Dominic method are analyzed as described for the Pugh method.

Example 2: Use of the Dominic Method for Material Selection of a Loaded Thermal Conductor in Late Concept Design. The objective in this example was to select a mass-produced, non-heat-treatable cylindrical sheath material to conduct heat efficiently near room temperature and to withstand a static load. A heat transfer application required that a sheet of material be bent around a heat transfer medium. The sheet, whose thickness was defined by the necessary heat transfer conditions, also had to withstand a static compressive load and be resistant to dent-

Table 2 Pugh decision matrix for a toy train helical spring redesign

Criteria	Alternative 1 Present material, hard drawn steel (ASTM A227)	Alternative 2 Hard drawn steel, class I (ASTM A227)	Alternative 3 Music spring quality steel (ASTM A228)	Alternative 4 Oil-tempered steel, class I (ASTM A229)
Wire diameter, mm	1.4	1.2	1.12	1.18
Coil diameter, mm	19	18	18	18
Number of coils	16	12	12	12
Relative material cost (per unit weight)	1	1	2.0	1.3
Tensile strength, MPa	1750	1750	2200	1850
Spring constant	Datum	–	–	–
Durability	Datum	S	+	+
Weight	Datum	+	+	+
Size	Datum	+	+	+
Fatigue resistance	Datum	–	+	S
Stored energy	Datum	–	+	+
Material cost (for one spring)	Datum	+	S	S
Manufacturing cost	Datum	S	+	–
Σ+		3	6	4
ΣS		2	1	2
Σ–		3	1	2

Table 3 Dominic method overall evaluation table (DMOET)

Individual evaluations	Criteria priority High	Moderate	Low
Excellent			
Good			
Fair			
Poor			
Unacceptable			

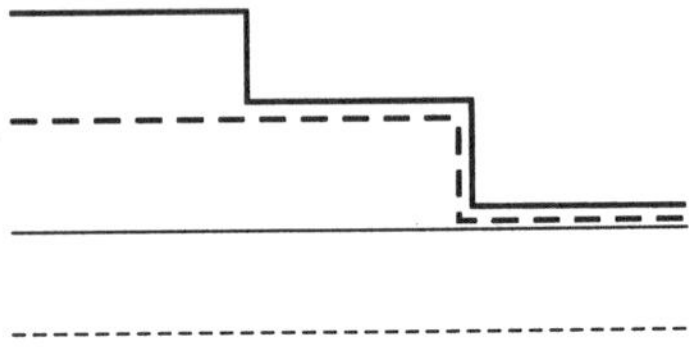

Overall evaluation: all above heavy solid line, excellent; all above heavy dashed line, good; all above light solid line, fair; all above light dotted line, poor ing during handling. Manufacturing issues were tooling cost and springback.

Table 4 lists the materials that were initially considered and their properties. The criteria are listed in Table 5. High-priority criteria were the amount of heat transferred (thermal conductivity, high was desirable), tooling costs (bend force index, low was desirable), and springback (index, low was desirable). Medium-priority criteria were the time to steady-state conduction (thermal diffusivity, high was desirable) and the ability to carry a static load (static load index, high was desirable). The low-priority criterion was the ability to resist denting during handling (hardness, high was desirable). The completed DMOET is shown as Table 6.

Analysis. No material received an overall "excellent" rating. Low-carbon steel, copper, and bronze had "good" ratings. Steel had the highest rating, but steel is limited by its heat transfer characteristics, a difficult parameter to improve on. Aluminum and copper are limited by the high tooling cost associated with their thickness. Because the design stage was late conceptual, it was considered reasonable to continue with low-carbon steel, copper, and bronze into configuration design.

The Pahl and Beitz Method

Description. The Pahl and Beitz method (Ref 7) allows quantitative evaluation of alternatives using weighted criteria. Because of the level of detail associated with its application and the objectivity of numerical approaches, it is particularly well suited for configuration design, using a large number of criteria that require input from technically diverse groups or individuals.

Method. Lists of criteria and alternatives are generated. The criteria are weighted numerically such that the sum of all weighting factors equals 1. (Sometimes 100 is used as the sum.)

An "objective tree" is a useful construction for assigning weighting factors (Fig. 1). The first level represents a simple statement describing the part or component being designed and has an entry value of 1.

The second level is a list of the most basic design requirements, usually arising from the concept design description of functional requirements. The relative value of each is assessed by assigning a fractional weighting factor such that the weighting factor sum of criteria in level 2 equals 1.

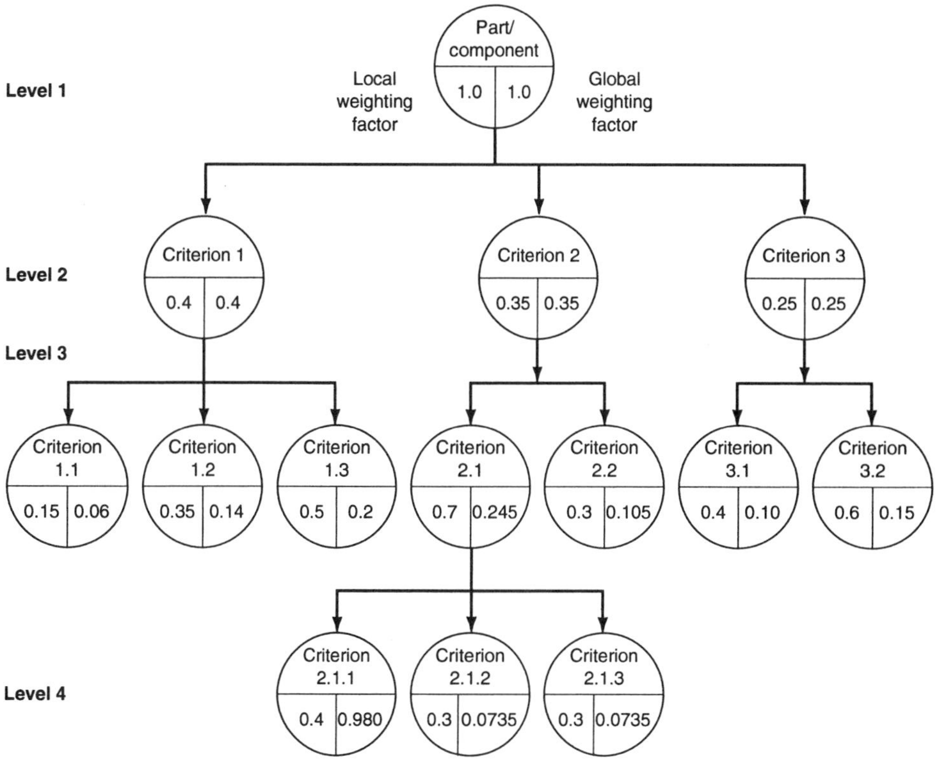

Fig. 1 Pahl and Beitz decision matrix, sample objective tree structure

Level 3 is constructed by further dissection of each level 2 criterion. First, a level 2 criterion is selected. It is dissociated by specification of new criteria with weighting factors such that the sum of the weighting factors equals 1. This continues for each level 2 criterion. The value of these new criteria to the overall design is calculated by multiplying its weighting factor by all "upstream"

Table 4 Properties of candidate materials for the Dominic example

Properties	A 1020 steel (ASTM A611, grade B)	B 304 stainless steel (ASTM A666, grade D)	C 5052-H34 aluminum (ASTM B 209)	D C11000-H04 ETP copper (ASTM B 152)	E C22000-H01 commercial bronze (ASTM B 36)
Thickness (t), in.	0.107	0.107	0.407	0.205	0.205
Thermal conductivity, Btu · ft/h · °F · ft^2	27	9.4	80	224	109
Thermal diffusivity, ft^2/h	909	270	3750	6750	3810
Elastic modulus (E), 10^6 psi	30	28	10.5	17	17
Yield strength (YS), ksi	30	110	26	40	37
Hardness, HB	111	240	68	50	42
Static load index(a), in. · lb	3210	11,770	10,580	8200	7590
Bend force index(a), lb	343	1260	4310	1680	1550
Springback index(a), 10^{-3} in.$^{-1}$	9.35	36.7	6.08	11.5	10.6

(a) Static-load-carrying capability is proportional to YS(t), and large numbers are desirable. Bending force is proportional to YS(t)^2. Extent of springback is proportional to YS/(tE). Small numbers for both indices denote desirable manufacturability.

Table 5 Criteria and ratings for candidate materials for the Dominic example

Criteria	Priority	A 1020 steel (ASTM A611, grade B)	B 304 stainless steel (ASTM A666, grade D)	C 5052-H34 aluminum (ASTM B 209)	D C11000-H04 ETP copper (ASTM B 152)	E C22000-H01 commercial bronze (ASTM B 36)
Heat transfer	High	Good	Fair	Excellent	Excellent	Excellent
Tooling costs	High	Excellent	Good	Fair	Good	Good
Springback	High	Excellent	Fair	Excellent	Excellent	Excellent
Time to steady state	Moderate	Good	Fair	Excellent	Excellent	Excellent
Static load	Moderate	Good	Excellent	Excellent	Excellent	Excellent
Dent resistance	Low	Good	Excellent	Fair	Fair	Fair

Table 6 Dominic method overall evaluation table (DMOET) for the loaded thermal conductor

Individual evaluations	Criteria priority: High	Moderate	Low
Excellent	AACCDDEE	BCCDDEE	B
Good	ABDE	AA	A
Fair	BBC	B	CDE
Poor			
Unacceptable			

Overall evaluation: all above heavy solid line, excellent; all above heavy dashed line, good; all above light solid line, fair; all above light dotted line, poor

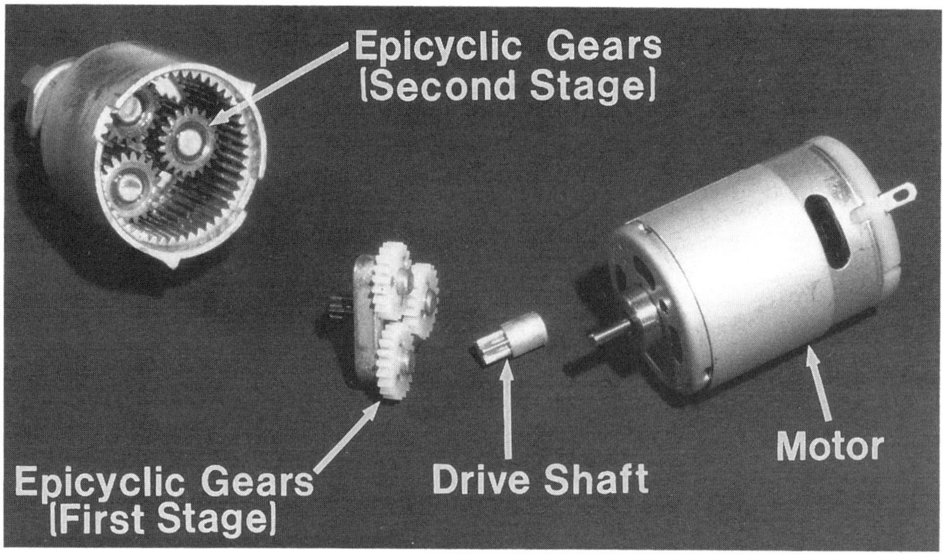

Fig. 2 Epicyclic gear train for a cordless screwdriver

Table 7 Value scales for the Pahl and Beitz method

Value	Meaning
5 value scale	
0	Unsatisfactory
1	Just tolerable
2	Adequate
3	Good
4	Very good
11 value scale	
0	Absolutely useless
1	Very inadequate
2	Weak
3	Tolerable
4	Adequate
5	Satisfactory
6	Good with few drawbacks
7	Good
8	Very good
9	Exceeds requirements
10	Ideal

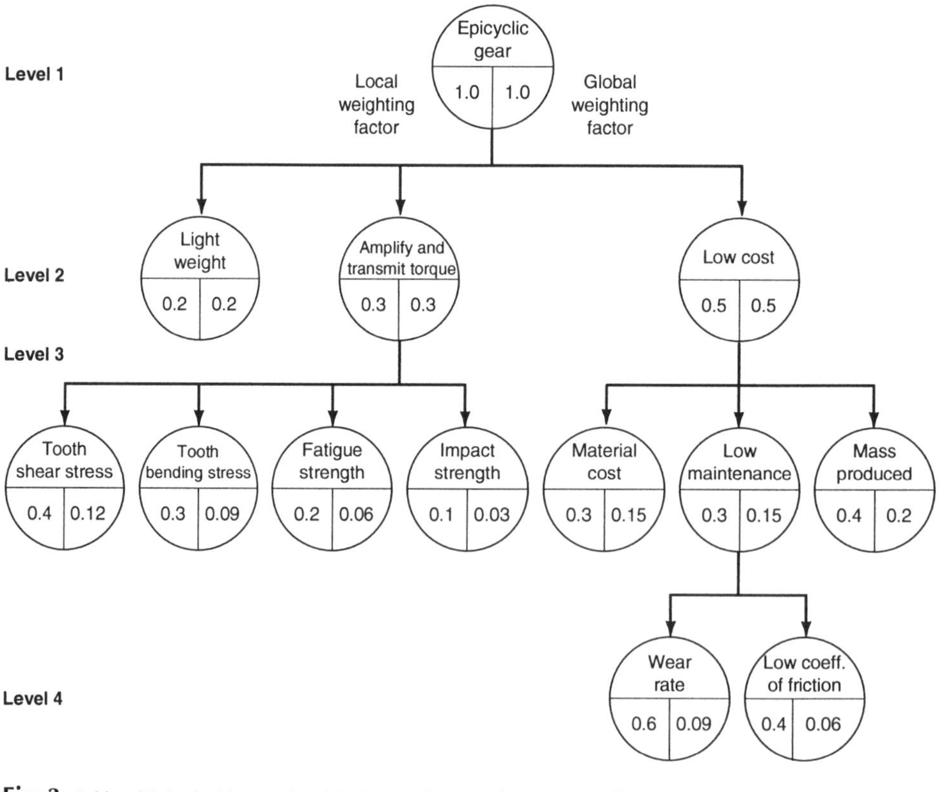

Fig. 3 Pahl and Beitz decision matrix, objective tree for epicyclic gear example

weighting factors in the objective tree. For example, suppose a level 2 criterion with an assigned weighting factor of 0.4 had three level 3 refinements with weighting factors of 0.15, 0.35, and 0.5. The contribution of the first level 3 refinement to the overall design would be (0.15) (0.4) (1) = 0.06. As shown in Fig. 1, both the "local" weighting factor (0.15) and the "global" weighting factor (0.06) are tallied for each criterion.

Further refinement to additional levels may be required, depending on the complexity of the selection process. The final result is a list of the most refined criteria with global weighting factors assigned. These comprise the selection process criteria and associated weighting factors w. These weighting factors sum to 1.

An alternative method of establishing weighting factors is to make a pairwise comparison between all criteria, deciding quantitatively for each pair which is more important (see the article "Overview of Materials Selection" in this Volume).

A matrix of alternatives and criteria is constructed as described for the Pugh and Dominic methods. Suppose there are k alternatives and n criteria. The matrix is completed by assigning a value v to each criterion for each alternative. These values and their meanings come from Table 7 and may be based on either a 5- or 11-value scale (Ref 7, 8). It is sometimes expedient to search for alternatives with extremely good or poor qualities and to assign values there first. Another approach is to convert the property values for each criterion into scaled property values (see the article "Overview of Materials Selection" in this Volume).

The number of values to assign is $(k)(n)$. Each value may be represented by v_{ij}. Once all values are assigned, the $(k)(n)$ weighed values $(wv)_{ij}$ are computed by multiplying each value v_{ij} by the associated criteria weighting factor w_j. Summing all weighted values for alternative i produces the overall weighted value (OWV_i). This is the raw score for a given alternative. It is customary to normalize overall weighted values to 1 by dividing by the maximum value v_{max} (5 or 10, based

Table 8 Pahl and Beitz decision matrix for the selection of an epicyclic gear

Criteria				Alternative 1 30-35% glass-reinforced nylon 6		Alternative 2 40% fiber-glass-reinforced polyester		Alternative 3 Aluminum alloy 380.0-F		Alternative 4 Zinc alloy UNS 33521 (alloy 3)		Alternative 5 Magnesium alloy AZ91B-F	
Primary consideration	No.	Evaluation criterion	Weighting factor (w_j)(a)	Value (v_{1j})	Weighted value ($w_j v_{1j}$)	Value (v_{2j})	Weighted value ($w_j v_{2j}$)	Value (v_{3j})	Weighted value ($w_j v_{3j}$)	Value (v_{4j})	Weighted value ($w_j v_{4j}$)	Value (v_{5j})	Weighted value ($w_j v_{5j}$)
Light weight (0.20)	1	Light weight	0.20	9	1.80	9	1.80	5	1.00	3	0.60	7	1.40
Amplify and transmit torque (0.3)	2	Tooth shear stress	0.12	4	0.48	3	0.36	9	1.08	8	0.96	7	0.84
	3	Tooth bending stress	0.09	4	0.36	3	0.27	8	0.72	7	0.63	6	0.54
	4	Fatigue strength	0.06	6	0.36	6	0.36	7	0.42	6	0.36	7	0.42
	5	Impact strength	0.03	5	0.15	5	0.15	6	0.18	8	0.24	5	0.15
Low cost (0.5)	6	Material cost	0.15	7	1.05	9	1.35	7	1.05	1	0.15	3	0.45
	7	Mass produced	0.20	9	1.80	6	1.20	7	1.40	8	1.60	9	1.80
	8	Wear rate	0.09	5	0.45	6	0.54	8	0.72	7	0.63	6	0.54
	9	Coefficient of friction	0.06	8	0.48	9	0.54	5	0.30	4	0.24	4	0.24
				OWV_1	6.93	OWV_2	6.57	OWV_3	6.87	OWV_4	5.41	OWV_5	6.38
				WR_1	0.693	WR_2	0.657	WR_3	0.687	WR_4	0.541	WR_5	0.638

OWV, overall weighted value; WR, weighted rating. (a) Sum of all weighting factors equals 1.00.

on the value scale chosen in Table 7) and the sum of the weighting factors. The formula for this weighted rating, WR_i (also known as the weighted property index), is:

$$WR_i = \frac{\sum_{j=1}^{n} (w_j \times v_{ij})}{v_{max} \sum_{j=1}^{n} w_j}$$

where normally

$$\sum_{j=1}^{n} w_j = 1$$

The weighted ratings provide an indication of which alternatives are clearly superior or inferior. Ideally, the alternative with the highest rating would be selected. However, as mentioned above, this may be done only after analysis of the data to ensure that other alternatives cannot be improved, and after the criteria and the weighting factors have been reconfirmed to be appropriate in light of the results.

Example 3. Use of the Pahl and Beitz Method in Configuration Design Material Selection of an Epicyclic Gear. The objective in this example was to select a mass-produced, low-cost, lightweight material for a power transmission gear component in a cordless, hand-held power screwdriver. The final design solution was a two-stage epicyclic gear assembly (Fig. 2), necessary to reduce motor rpm by 1:81 and amplify the torque from 50 mJ to over 4 J. The selection was for the three first-stage (low-torque) epicyclic gears. Geometric constraints were a maximum gear diameter of 11 mm and thickness ranging between 1 and 4.5 mm. The torque produced a gear-tooth transmitted shearing/bending force of 35 N. A safety factor of 1.5 was used for all force calculations.

Table 9 Materials properties for the Pahl and Beitz epicyclic gear example

Alternative	Material	Density, g/cm^3	Tensile strength, MPa	Tensile Elongation, %	Impact strength, J/m	Cost, ¢/cm^3
1	Glass-reinforced (30–35%) nylon 6	1.38	165	2.2–3.6	117–181(a)	0.67
2	Fiber-glass-reinforced (40%) polyester	1.3	138	1.4	374(a)	0.32
3	Aluminum alloy 380.0-F	2.71	330	3	400(b)	0.51
4	Zinc alloy UNS 33521 (Alloy 3)	7.1	283	10	5800(c)	1.16
5	Magnesium alloy AZ91B-F	1.81	230	3	270(b)	0.72

(a) Izod. (b) Notched charpy. (c) Unnotched charpy

Three primary considerations for selection were light weight, the ability to transmit and amplify the torque, and low component cost. Table 8 lists the nine grouped criteria based on these considerations. A four-level objective tree was constructed to generate the criteria weighting factors (Fig. 3). These weighting factors are listed in Table 8.

Five alternatives were considered: 30 to 35% glass-fiber-reinforced nylon 6, 40% fiber-glass-reinforced polyester, die-cast aluminum alloy 380.0-F, die-cast zinc alloy UNS 33521 (alloy 3), and die-cast magnesium alloy AZ91B-F. An 11-value scale (Table 7) was used to assign values for each alternative and criterion. Details of the engineering calculations have been omitted for the sake of clarity and brevity. Fundamental property data are compiled in Table 9.

Overall weighted values and weighted ratings for the alternatives were computed and listed at the bottom of Table 8. Nylon and the aluminum alloy had the highest rankings, although the polyester scored well also. After analysis of the selection procedure, nylon was selected for production.

Conclusions

Decision matrix methods in materials selection are formalized procedures by which materials alternatives are ranked. Three common methods, Pugh, Dominic, and Pahl and Beitz, have been presented as examples of decision matrix methods of varying complexity. Factors governing which of these or other decision matrix methods to use include the complexity of the design, when in the design process the method is used, the level of participation involved in the selection process, and the completeness of the knowledge base for the material alternatives being considered. The result of proper use of a decision matrix method is a documented, bias-free materials selection based on adequate consideration of all important issues.

REFERENCES

1. S. Pugh, *Total Design: Integrated Methods for Successful Product Development*, Addison-Wesley, 1991
2. J.R. Dixon and C. Poli, Evaluation and Redesign of Engineering Concepts, *Engineering Design and Design for Manufacturing: A Structured Approach*, Field Stone Publishers, Conway, Mass., 1995
3. K.N. Otto, Measurement Methods for Product Evaluation, *Research in Engineering Design*, Vol 7, 1995, p 86–101
4. M.M. Farag and E. El-Magd, An Integrated Approach to Product Design, Materials Selection and Cost Estimation, *Materials and Design*, Vol 13 (No. 6), 1992, p 323–327
5. N. Cross, *Engineering Design Methods*, 2nd ed., John Wiley, 1994

6. A.E. Howe, P.R. Cohen, J.R. Dixon, and M.K. Simmons, Dominic: A Domain-Independent Program for Mechanical Design, *The International Journal of Artificial Intelligence in Engineering*, Vol 1 (No. 1), 1986
7. G. Pahl and W. Beitz, *Engineering Design: A Systematic Approach*, Springer-Verlag, 1993
8. VDI-Richtlinie 2225, *Technisch-wirtschaftliches Konstruieren* [VDI Guideline 2225, *Technical-Industrial Design*], VDI-Verlag, Düsseldorf

SELECTED REFERENCES

- F.A. Crane and J.A. Charles, The Formalization of Selection Procedures, *Selection and Use of Engineering Materials*, 2nd ed., Butterworths, 1989
- G.E. Dieter, *Engineering Design: A Materials and Processing Approach*, 2nd ed., McGraw-Hill, 1994
- S.S.G. Lee and J. Atkinson, Integration of Materials Selection with Design Analysis, *CME*, 1987, p 35–38
- G. Lewis, Appendix B: Material Property Weighting Factors and Cost Elements, *Selection of Engineering Materials*, Prentice Hall, 1990
- K.N. Otto and K.L. Wood, Estimating Errors in Concept Selection, *1995 Design Engineering Technical Conferences Proceedings*, Vol 83, ASME, 1995, p 397–412

Relationship between Materials Selection and Processing

George E. Dieter, University of Maryland

THE SELECTION OF THE BEST MATERIAL for a particular design is intimately associated with the decisions of how to process the material or manufacture the part. Growing concerns with world economic competitiveness have forced a renewed emphasis on manufacturing as a vital core competency for any industrial enterprise. Major attention is being given to linking product design and manufacturing in a systematic integrated approach called *concurrent engineering*. In concurrent engineering the members of the integrated product design team, from the outset, consider all elements of the product life cycle from concept through disposal, including issues of quality, cost, schedule, and user requirements. (Additional information is provided in the article "Concurrent Engineering" in this Volume.) In view of the increasing recognition of the importance of concurrency in design, there should be no excuse for selection of a material without serious consideration of the alternative manufacturing processes.

As an example, a hydraulic accumulator piston for an automotive transmission was made from a thermoplastic polymer instead of the more conventional aluminum alloy (Ref 1). Use of the glass-filled linear polyphenylene sulfide (PPS) plastic resulted in a 30% reduction in the important functional requirement of weight, coupled with a continuous use temperature of 220 °C (428 °F) and good dimensional stability. The use of precision injection molding permitted as-molded tolerances achieved only in metals with multiple machining operations, and as a bonus, the O-ring was formed into the periphery of the polymer piston. Even though the polymer resin cost more than the aluminum alloy there was a net cost reduction of 10%, due chiefly to the elimination of machining and cleaning steps. This is an example where the selection of a material allowed the use of a manufacturing process with superior capabilities and lower final cost.

As another example, consider a part to be made from an aluminum casting alloy. Three alternative casting processes are sand casting, permanent-mold (gravity-feed) casting, and die casting (see the article "Design for Casting" in this Volume). The chief differentiating factor between these processes is the cooling rate, which varies from about 0.2 °C/s (0.35 °F/s) for sand casting to 500 °C/s (280 °F/s) for die casting. The highest cooling rate results in the finest grain size, which produces a casting with the highest yield strength and the best fatigue properties and wear resistance. Also, die casting produces the best surface finish, the tightest dimensional tolerances, and, because of pressure feeding, the ability to produce wall thicknesses as small as 1.0 to 2.5 mm. On the negative side, the rapid cooling rate may result in a greater degree of porosity. This is a typical example of how the choice of the manufacturing process can significantly influence the properties of the design. Other important factors influencing this decision are discussed below.

Characteristics of Manufacturing Processes

This section describes the basic characteristics of manufacturing processes. Although the level of detail is rather elementary, this section provides sufficient detail for selection of processes at the conceptual or embodiment stages of design. For further detail, see the article "Manufacturing Processes and Their Selection" in this Volume.

Material Factors. In a very general sense the selection of a material determines a range of processes that can be used to produce parts from the material. In the article "Overview of the Materials Selection Process" in this Volume, Fig. 4 shows the common classes of manufacturing processes and Table 4 indicates the manufacturing processes used most frequently with different metals and plastics. The melting point of a material and level of deformation resistance and ductility determine these relationships. Some materials are too brittle to be plastically deformed; others are too reactive to be cast or have poor weldability.

The melting point of a material determines the casting processes that can be employed. Low-melting-point metals can be used with any of a large number of casting processes, but as the melting point of the material rises the number of available processes becomes limited. Ashby (Ref 2), has shown a correlation between the size (weight) of a cast part and the melting point of the material (Fig. 1). This plot shows the regions of size that can be handled by different casting and molding processes and shows how the available number of processes decreases dramatically for materials with high melting point. This is one of a number of process selection charts, dealing with casting, metalworking, polymer processing, power fabrication, and machining, introduced by Ashby in Ref 2 to aid in the selection of a manufacturing process in the conceptual stage of design.

Similarly, yield strength, or hardness, determines the limits in deformation and machining processes. Forging and rolling loads are related to yield strength and tool life (tool load and temperature generation) in machining scales with the hardness of the material being machined. Ultimately the limit of these processes is determined by the workability of the material. See the section "Workability" in this article.

Shape Factors. Each process has associated with it a range of shapes that can be produced. Thus, the first decision in selecting a process is whether it is capable of producing the required shape. A simple classification of shape is as follows:

Two-dimensional (2D)	Profile of product does not change along its length. Examples: wire, pipe, aluminum foil. Many 2D products are used as raw material for processes that make them into 3D shapes.
Three-dimensional (3D)	Profile of the product varies along all three axes. Most products are 3D.
Sheet	Has almost constant section thickness that is small compared with the other dimensions.
Bulk	Has a complex shape, often with little symmetry. Solid: has no significant cavities. Hollow: has significant cavities.

Whether the process can accommodate shapes with undercuts, or reentrant angles, or parts with an element positioned perpendicular to the main die motion also must be determined.

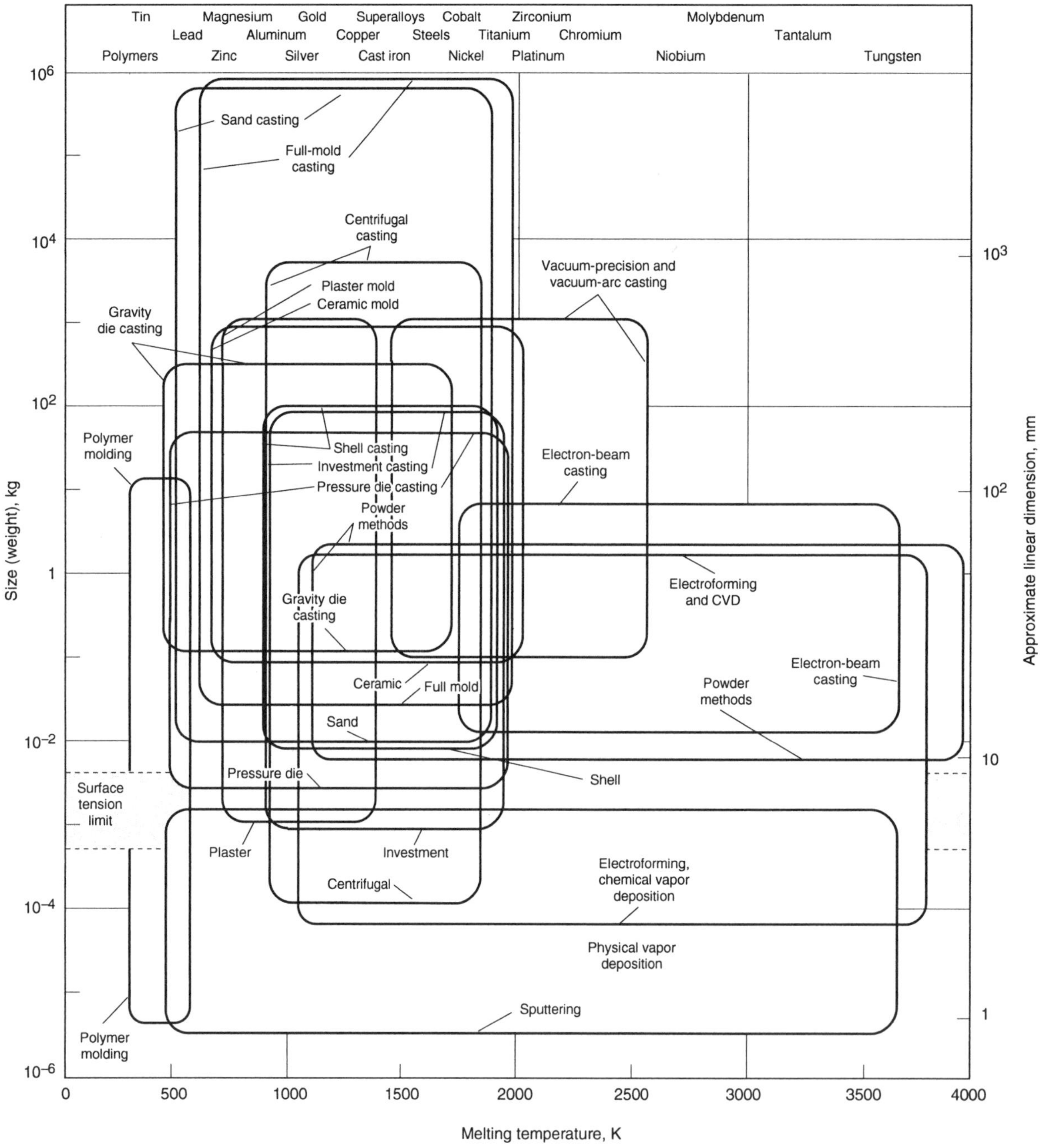

Fig. 1 Ashby process chart for casting. The size in this chart is measured by weight, W. It can be converted to volume via the density, ρ, or to the approximate linear dimension, L, shown on the right-hand-axis via $L = (W/\rho)^{1/3}$. Source: Ref 2

Table 1 Scale for rating manufacturing processes

Rating	Cycle time	Quality	Flexibility	Materials utilization	Operating costs
1	>15 min	Poor quality, average reliability	Changeover extremely difficult	Waste > 100% of finished component	Substantial machine and tooling costs
2	5 to 15 min	Average quality	Slow changeover	Waste 50 to 100%	Tooling and machines costly
3	1 to 5 min	Average to good quality	Average changeover and setup time	Waste 10 to 50%	Tooling and machines relatively inexpensive
4	20 s to 1 min	Good to excellent quality	Fast changeover	Waste < 10% finished part	Tooling costs low/little equipment
5	<20 s	Excellent quality	No setup time	No appreciable waste	No setup costs

Rating scheme: 1, poorest; 5, best. Source: Ref 4

Table 2 Rating of characteristics for common manufacturing processes

Process	Shape	Cycle time	Flexibility	Material utilization	Quality	Equipment tooling costs	Handbook reference
Casting							
Sand casting	3D	2	5	2	2	1	AHB, Vol 15
Evaporative foam	3D	1	5	2	2	4	AHB, Vol 15, p 230
Investment casting	3D	2	4	4	4	3	AHB, Vol 15, p 253
Permanent mold casting	3D	4	2	2	3	2	AHB, Vol 15, p 275
Pressure die casting	3D solid	5	1	4	2	1	AHB, Vol 15, p 285
Squeeze casting	3D	3	1	5	4	1	AHB, Vol 15, p 323
Centrifugal casting	3D hollow	2	3	5	3	3	AHB, Vol 15, p 296
Injection molding	3D	4	1	4	3	1	EMH, Vol 2, p 308
Reaction injection molding (RIM)	3D	3	2	4	2	2	EMH, Vol 2, p 344
Compression molding	3D	3	4	4	2	3	EMH, Vol 2, p 324
Rotational molding	3D hollow	2	4	5	2	4	EMH, Vol 2, p 360
Monomer casting, contact molding	3D	1	4	4	2	4	EMH, Vol 2, p 338
Forming							
Forging, open die	3D solid	2	4	3	2	2	AHB, Vol 14, p 61
Forging, hot closed die	3D solid	4	1	3	3	2	AHB, Vol 14, p 75
Sheet metal forming	3D	3	1	3	4	1	AHB, Vol 14, p 445
Rolling	2D	5	3	4	3	2	AHB, Vol 14, p 343
Extrusion	2D	5	3	4	3	2	AHB, Vol 14, p 315
Superplastic forming	3D	1	1	5	4	1	AHB, Vol 14, p 852
Thermoforming	3D	3	2	3	2	3	EMH, Vol 2, p 399
Blow molding	3D hollow	4	2	4	4	2	EMH, Vol 2, p 352
Pressing and sintering	3D solid	2	2	5	2	2	AHB, Vol 7
Isostatic pressing	3D	1	3	5	2	1	AHB, Vol 7, p 419
Slip casting	3D	1	5	5	2	4	EMH, Vol 4, p 153
Machining							
Single-point cutting	3D	2	5	1	5	5	AHB, Vol 16
Multiple-point cut	3D	3	5	1	5	4	AHB, Vol 16
Grinding	3D	2	5	1	5	4	AHB, Vol 16, p 421
Electrical discharge machining	3D	1	4	1	5	1	AHB, Vol 16, p 557
Joining							
Fusion welding	All	2	5	5	2	4	AHB, Vol 6
Brazing/soldering	All	2	5	5	3	4	AHB, Vol 6, p 929
Adhesive bonding	All	2	5	5	3	5	EMH, Vol 3
Fasteners	3D	4	5	4	4	5	...
Surface treatment							
Shot peening	All	2	5	5	4	5	AHB, Vol 5, p 138
Surface hardening	All	2	4	5	4	4	AHB, Vol 4
CVD/PVD	All	1	5	5	4	3	AHB, Vol 13, p 456

Rating scheme: 1, poorest; 5, best. Ratings from Ref 5. AHB, *ASM Handbook*; EMH, *Engineered Materials Handbook*

Table 3 Comparison of processes for producing automotive bumpers

Process	Cycle time	Flexibility	Materials utilization	Quality	Equipment/ tooling costs
Injection molding	4	1	4	3	1
Reaction injection molding	3	2	4	2	2
Compression molding	3	4	4	2	3
Contact molding	1	4	4	2	4

Rating scheme: 4, best; 1, worst

The factors that contribute to complexity of shape are:

- The requirement to achieve a minimum section thickness
- The presence of undercuts and reentrant angles
- The presence of complex internal cavities

See the article "Manufacturing Processes and Their Selection" in this Volume for more in-depth consideration of shape and its relationship to process selection. The classification of shapes is an active field of research in manufacturing. An important application is *group technology*, which aims at utilizing the similarity of parts based on their geometrical features, because similar shapes tend to be produced by similar processing methods (Ref 3). Application of group technology requires the adoption of detailed numerical coding systems, but the benefits in design and manufacturing can be rather widespread. These include ability to generate quickly designs for similar-shaped parts, control of part proliferation and redundant designs, standardization of design features such as chamfers, corner radii, tolerances, grouping of machine tools into more productive units, development of jigs and fixtures that can accommodate different members of a part family, and the development of computer-based process selection methods.

Process Factors. Key manufacturing process factors include cycle time, quality, flexibility, materials utilization, and operating cost.

Cycle time is the time required to process one unit, once the process has been set up and is operating properly. The inverse of cycle time is production rate.

Quality is an inclusive term describing fitness for use. Included under this characteristic is surface roughness, ability to manufacture parts to a set tolerance, product integrity (freedom from voids, pores, cracks, and inclusions), and ability to produce controlled properties (microstructure control).

Flexibility in manufacturing is the ease with which a process can be adapted to produce different products or variations of the same product. In an era where product customization is becoming more important, this characteristic has gained importance. Flexibility is influenced greatly by the time to change and set up tooling.

Materials utilization measures the amount of material processed in addition to the material required in the product. Most machining operations generate 60 to 80% scrap. Net-shape forging and injection molding are at the other extreme. As materials costs become the greater part of the product cost, materials utilization becomes of greater importance.

Operating cost involves both the capital cost of plant, machinery, and tooling and the labor costs of setting up and running the process. Process selection often is constrained by the available equipment, particularly if the desired process is costly to install. In any manufacturing situation, one available decision is whether to outsource the production to a qualified subcontractor.

A rating system for evaluating these five process characteristics is given in Table 1. This is applied to rating the most common manufacturing processes in Table 2. Another system, using somewhat different process characteristics, is given in Table 5 of the article "Overview of the Materials Selection Process" in this Volume.

Example 1: Selection of Plastics for an Automobile Bumper. This hypothetical, but realistic, example is taken from Ref 4. The functional requirements for a material for a car bumper are:

- Adequate rigidity to maintain dimensional limits of the structure
- Adequate impact resistance down to –30 °C (–22 °F)
- Resistance to degradation from ultraviolet radiation and spillage of gasoline
- Dimensional stability to prevent distortion over the operating temperature range
- Ability for the plastic to be finished to match the adjoining painted metal parts

With these performance requirements of chief importance, four polymeric materials were chosen from the large number of engineering plastics:

Table 4 Cost per part for four processes at different production levels

Process	Cost per part: 1000 parts	10,000 parts	100,000 parts	1,000,000 parts
Injection molding	$451	$46	$5.50	$1.45
Reaction injection molding	$92	$11	$2.90	$2.09
Compression molding	$57	$7.50	$2.55	$2.06
Contact molding	$40	$22	$20.20	$20.02

- Polyester reinforced with chopped-glass fiber to improve toughness
- Polyurethane with glass-flake filler to increase stiffness
- Rubber-modified polypropylene to decrease the ductile-brittle transition to below −30 °C (−22 °F)
- A polymer blend of polyester and polycarbonate to combine the excellent solvent resistance of the former with the high toughness of the latter

It is important to note that, in each case, the plastic was modified in some way to improve one or more critical properties.

The processing characteristics of these polymers and considerations of the shape of the bumpers restrict the choice of manufacturing process to the four processes listed in Table 3. A comparison among the four processes shows that materials utilization and quality are not deciding factors. However, there is a distinct interaction between cycle time (production rate) and flexibility and tooling cost. Deciding on the best manufacturing process requires a more detailed examination of the relationship between production volume and cost.

The four processes differ mainly in the investment in mold costs and in the amount of time required to produce a part. The assumed values for these parameters are:

Process	Mold cost	Labor input/unit
Injection molding	$450,000	3 min = $1
Reaction injection molding	$90,000	6 min = $2
Compression molding	$55,000	6 min = $2
Contact molding	$20,000	1 h = $20

Then the part cost is the sum of the mold cost per part plus the labor input, neglecting material cost. The data in Table 4 indicate that this varies considerably depending on the production volume. At 1000 parts the least expensive process is contact molding. At 10,000 and 100,000 parts it is compression molding, while at 1 million parts it is injection molding. The concept of a break-even point on batch size, or economic lot size (as discussed in the article "Overview of the Materials Selection Process," and shown there in Fig. 7) is exemplified by the data in Table 4.

Additional information on the selection, properties, and processing of engineering plastics is provided in the articles "Overview of the Materials Selection Process," "Effects of Composition, Structure, and Processing on Properties of Engineering Plastics," "Design with Plastics," and "Design for Plastics Processing" in this Volume.

Influence of Materials on Manufacturing Cost

A major shift in manufacturing throughout this century has seen a decrease in the percent of the cost of manufactured goods attributed to direct labor from 50 to 10%, with a corresponding increase in the fraction due to materials from 30 to 60% (Ref 6). As a result, design changes that reduce the amount or cost of materials or purchased components can have a direct and major influence on profits. Figure 2 shows the buildup of cost elements that make up the selling price of a product.

Example 2: Manufacturing Cost Models. Table 5 lists four cost models for producing a product (Ref 7). For simplicity, the selling price per unit is $100. Model A represents the current distribution between the cost elements shown in Fig. 2.

Cost model B shows what would happen if sales were increased by 5%. There would be a 5% increase in the four cost elements, while the unit costs remain unchanged. Costs and profits rise to the same degree as sales.

Cost model C shows what happens with a 5% productivity improvement (5% decrease in direct labor) brought about by a process-improvement program. The small increase in overhead results from the new equipment that was installed to increase productivity. Note that the profit per unit has increased by 10%.

Cost model D shows what happens with a 5% decrease in the cost of material or purchased components. This could also result from a design modification that allows a change to a less-expensive material or eliminates a purchased component. In this case, barring an extensive development program, all of the cost savings goes to the bottom line and results in a 55% increase in the unit profit.

Table 5 Comparison of four product cost models

	Model A	Model B	Model C	Model D
Selling price	$100	$100	$100	$100
Units sold	100	105	100	100
Revenues	$10,000	$10,500	$10,000	$10,000
Direct labor	$1,500	$1,575	$1,425	$1,500
Materials	$5,500	$5,775	$5,500	$5,225
Overhead	$1,500	$1,575	$1,525	$10,500
Sales and distribution	$1,000	$1,050	$1,000	$1,000
Total costs	$9,500	$9,975	$9,450	$9,225
Profit	$500	$525	$550	$775

Source: Ref 7

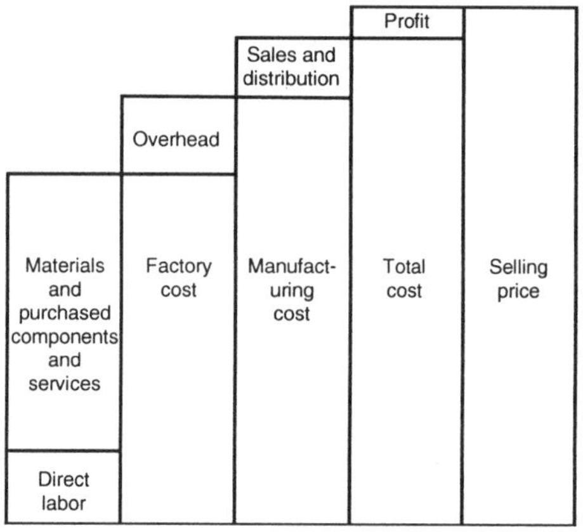

Fig. 2 Buildup of cost elements to establish selling price

Design for Manufacturability

With increased awareness of the importance of the interaction between design and manufacture a new field aimed at formalizing this relationship is evolving, called "design for manufacturability" (DFM). The DFM approach examines the product design in all aspects for ways of integrating the product and its processing so that the best match is made between product and process requirements and that the integrated product/process ensures inherent ease of manufacture. A major objective of DFM is to ensure that the product (including material selection) and the process are designed together. While DFM methods have not given much attention to the materials selection question, the ideas of DFM are important to consider when selecting a material in relationship to possible manufacturing processes. The subject is reviewed briefly here following the approach of Bralla (Ref 8). For a more in-depth consideration of DFM, see the articles "Introduction to Manufacturing and Design," "Design for Manufacture and Assembly," and "Manufacturing Processes and Their Selection" in this Volume.

Minimize Total Number of Parts. Maximum cost savings result when a part is eliminated or combined with another, rather than just being simplified. A part that is eliminated costs nothing to make, assemble, move, store, clean, inspect, rework, or service. A part is a candidate for elimination if there is no need for relative motion between the part and some other element of the design, no need for subsequent adjustment between parts, and no need for the materials to be different. However, part reduction should be taken so far that costs increase because of increased complexity or weight. The best way to achieve part reduction is to establish minimum part count as a functional requirement at the conceptual design stage. Replacing fasteners by snap fits in plastic covers and designing parts to be multifunctional by incorporating hinges, springs, guides, and bearings are good approaches. Plastic parts are particularly well suited for multifunctional integral design (Ref 9).

Standardize. Major cost savings are achieved when the design of products, subassemblies,

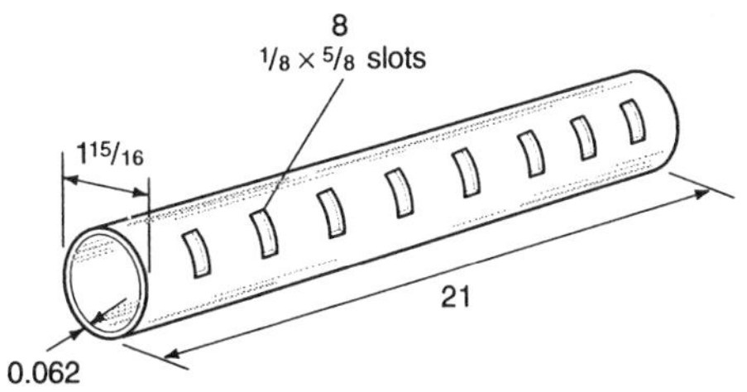

Fig. 3 Slotted tubular part. Dimensions in inches. Source: Ref 13

components, modules, and individual parts are standardized. Attempts should be made to utilize standard parts and components in a line of products. Similar parts should all be shaped and dimensioned the same. Standard catalog components should be used when applicable. The goal should be to not only reduce the number of parts, but to reduce the number of part numbers, that is, the number of distinct part designs needed to produce a line of products.

Use Readily Processed Materials. The topic is addressed later in this article in considerable detail. At this point two cautions are appropriate: (1) never compromise the quality of a part by selecting a more readily processed material, and (2) it is the final part cost that counts, not the initial material cost.

Fit the Design to the Manufacturing Process. The best overall result with respect to cost, function, and quality usually is achieved when the product design is developed for the particular manufacturing process that will be used to make the part. This argues for a process-first, materials-second approach to materials selection. For the many applications where performance is not paramount or difficult to achieve, this is the approach to use. The best manufacturing economy, highest quality, and shortest time-to-market usually are achieved when the product is made using the facilities, equipment, tooling, and know-how that already exist in the company factory or that of a regular vendor.

Design Each Part to be Easy to Make. The inherent capabilities and constraints of each manufacturing process must be known and considered by the design team (Ref 10). These are expressed in the form of design guidelines, design rules, and design standards. The articles in the Section "Manufacturing Aspects of Design" of this Volume that consider each of the manufacturing processes contain much of this type of information. With this knowledge, the designer should minimize needed fixturing, specify tolerances easily met by the process, minimize secondary operations, and design so that finishing operations are not required.

Eliminate Machining and Finishing Operations. Machining and finishing operations are inherently expensive and should be eliminated provided this will not compromise quality. One way to do this is to use as-cast, as-molded, and as-formed surfaces as much as possible instead of machined surfaces. Another approach is to use near-net shape processes such as powder metallurgy, fine blanking, and investment casting.

Factors in Selecting a Material for Production

The ultimate criterion for selection of materials is the cost to produce a quality part. For selection of the best material for producing a part the following factors must be considered:

- Material composition: grade of alloy or plastic
- Form of material: bar, tube, wire, strip, plate, powder, and so forth
- Size: dimensions and tolerances
- Heat-treated condition
- Directionality of mechanical properties (anisotropy)
- Surface finish
- Quality level: control of impurities, inclusions, and microstructure
- Quantity: volume of production (batch size)
- Ease of manufacture: workability, weldability, castability, and so forth
- Ease of recyclability
- Cost of material

Cost is the basis to which the above factors are reduced to reach a decision on material selection. Generally, the least expensive material consistent with achieving the required properties is selected. The selection of the grade of material will establish the composition, which along with the processing, will establish the properties (see the Section "Effects of Composition, Processing, and Structure on Materials Properties" in this Volume). The caveat should be given that it is the final part cost that must be minimized. Thus, it may well be that a more expensive material will result in a less expensive product because it can employ a more economical manufacturing process. This section discusses these kinds of trade-offs. The most common is the decision whether to use material A with process B versus material C with manufacturing process D. Methods for calculating the part or component (assembled parts) cost at an early stage in design have been developed, as described in Ref 11 and 12 and in the article "Design for Manufacture and Assembly" in this Volume. Such methods are essential for widespread adoption of intelligent decision making concerning material selection and process selection.

For geometrically simple parts, such as straight shafts, the most economical raw material form and method of manufacture are readily apparent. As the shape of the part becomes more complex, the applicability of two or more forms and methods of manufacture add complexity to the selection process. For example, a small gear may be completely machined from bar stock. However, it might be more economical to start with a close-tolerance forged gear blank, depending on the total number of parts to be produced. For a small gear the total cost of making 100 pieces might favor machining from bar. At 10,000 or more parts the forged or powder metallurgy gear blank might show an economic advantage. With a larger gear, if only 500 pieces were to be made, it might be more economical to leave the hub solid and drill the bore. At 5000 pieces it would probably pay to have the hub pierced in the forging operation.

Table 6 Range of surface finishes

Surface finish, μin.	Description
250	Results from ordinary machining at medium feeds. Not objectionable in appearance. Can be used for surface of noncritical components.
125	Results from high-quality machining with light cuts, fine feeds, and sharp tools. Can be used for bearing surfaces with light loads and moderately stressed parts. Should not be used for sliding surfaces.
63	A high-quality, smooth machine finish. As smooth a finish as can be produced by turning or milling without a subsequent operation. Suitable for ordinary machine parts where fatigue is not a problem.
32	A fine machine finish produced by careful turning, milling, and drilling followed by grinding and broaching. Normally found in parts subjected to stress concentration and vibration, such as gear teeth and brake drums.
16	This finish is used where the surface finish is of primary importance in the functioning of the part: rapidly rotating shaft bearings, heavily loaded bearings, hydraulic applications. Produced by diamond tools or precision grinding.
8	Scratch-free, close-tolerance finish used for parts like the inside diameter of hydraulic cylinders, pistons, cam faces. Produced by diamond turning or boring, or precision grinding.
4	A mirrorlike surface that is free from visible scratch marks. Typically used in high-quality bearings. Very high cost finish. Produced by lapping, honing, and precision grinding.

Figure 3 shows an example where change in form resulted in economy (Ref 13). A type 321 stainless steel tube was slotted in eight locations. Milling the slots was difficult because the tube was thin, and the labor cost for removing the burrs on the inside was high. The problem was solved by changing the form of the raw material to a type 321 sheet. The slots were punched into the sheet, which was then formed and welded into the required tubular part.

The workpiece should be just enough larger than the final part to allow for removal of surface defects. Thus, extra material in a casting may be necessary to permit machining a critical surface to the needed finish, or a heat-treated steel part may be made oversize to allow the removal of a decarburized layer. Often the manufacturing process dictates the use of extra material, such as sprues and risers in castings and flash in forging and injection molding. Sometimes extra material must be provided for purposes of handling, positioning, or testing the part. Even though extra material removal is costly, it usually is less expensive to purchase a slightly larger workpiece than to pay for a scrapped part.

Most engineering alloys can be purchased from the supplier in a variety of metallurgical conditions other than the annealed state. Examples include quenched-and-tempered steel, solution-treated, cold-worked and aged aluminum alloys, and cold-drawn and stress-relieved brass rod. It may prove to be more economical to have the metallurgical strengthening produced in the

Table 7 Costs of some materials based on mass and volume

Material	Cost per unit mass, $/kg	Material	Cost per unit volume, $/100 cm^3
Germanium	365	Germanium	213
Silver	164	Silver	172
Cobalt	31	Cobalt	27
Polytetrafluoroethylene (PTFE)	12.25	Nickel	7.95
Nickel	8.93	Chromium	5.90
Chromium	8.31	Tin	4.43
Tin	6.06	Brass (sheet)	4.17
Titanium	5.41	Beryllium copper	3.45
Brass (sheet)	5.02	Cadmium	2.94
Aluminum-copper alloy sheet	4.38	Phosphor bronze (ingot)	2.85
Beryllium copper	3.90	18/8 stainless (sheet)	2.71
Polyamide (Nylon 6/6)	3.85	PTFE	2.63
18/8 stainless (sheet)	3.50	Copper (tubing)	2.45
Cadmium	3.40	Titanium	2.43
Phosphor bronze (ingot)	3.24	Copper (grade A ingot)	2.12
Magnesium (ingot)	2.89	Manganese	1.87
Acrylic (polymethyl methacrylate, PMMA)	2.80	Brass (ingot)	1.84
Copper (tubing)	2.75	Aluminum-copper alloy sheet	1.30
Acrylonitrile-butadiene-styrene (ABS)	2.63	Zinc (ingot)	0.81
Manganese	2.52	Lead (ingot)	0.67
Copper (grade A ingot)	2.38	Magnesium (ingot)	0.51
Brass (ingot)	2.18	Mild steel (sheet)	0.47
Amino resin thermoset	1.49	Nylon 6/6	0.44
Aluminum (ingot)	1.37	Aluminum (ingot)	0.37
Phenolic thermoset	1.31	PMMA	0.33
Silicon	1.24	Silicon	0.30
Polystyrene (PS)	1.12	ABS	0.28
Zinc (ingot)	1.12	Mild steel (ingot)	0.25
High-density polyethylene (HDPE)	1.09	Amino resin thermoset	0.23
Polypropylene (PP)	1.00	Cast iron	0.19
Natural rubber	0.98	Phenolic thermoset	0.16
Low-density polyethylene (LDPE)	0.74	PS	0.12
Rigid polyvinyl chloride (PVC)	0.72	Natural rubber	0.12
Mild steel (sheet)	0.60	HDPE	0.11
Lead (ingot)	0.60	Rigid PVC	0.11
Mild steel (ingot)	0.32	PP	0.09
Cast iron	0.26	LDPE	0.07
Portland cement	0.09	Portland cement	0.03
Concrete (ready mixed)	0.04	Concrete (ready mixed)	0.01

Costs are based on July 1991 prices. Source: Ref 14

Table 8 Cost buildup of steel products

Steel product	$/tonne
Pig iron from blast furnace	210
Mild steel ingot	315
Hot-rolled mild steel bar	490
Cold-drawn mild steel bright bar	665
Hot-rolled mild steel sections	498
Hot-rolled mild steel strip	476
Cold-rolled mild steel strip	593
Galvanized mild steel sheet	689
Cold-rolled austenitic stainless sheet	3500

Cost are based on July 1991 prices. Source: Ref 14

workpiece by the material supplier than to heat treat it separately after it has been manufactured. For example, a cap nut was machined from type 1144 steel bar, hot drawn to a yield strength of 825 MPa (120 ksi) and a hardness of 35 HRC. This material replaced 4140 steel, machined and heat treated to 30 to 35 HRC, which had resulted in as much as 40% scrap rate due to distortion (Ref 13).

A metal that has undergone a severe amount of deformation, for example from rolling or wire drawing, develops a preferred orientation, or texture, in which certain crystallographic planes orient themselves in a preferred manner with respect to the direction of maximum strain. A cold-worked metal that has been annealed generally shows a preferred orientation (annealing texture) that is different from and stronger than the texture that existed in the deformed metal. Just as the flow stress of a single crystal varies with orientation, so the tensile properties of a metal sheet with a texture will vary with the direction the axis of the tensile specimen makes with the rolling direction of the sheet. This type of anisotropy is important in producing nonuniform deformation in deep-drawn cups (earing) and in determining the deep-drawing capability of the sheet. It is measured by the *r* value, described in the section "Formability" in this article (see Eq 6).

Another source of the variation of properties with direction of working is mechanical fibering caused by the alignment of inclusions, porosity, and second-phase constituents in the main direction of mechanical working. This kind of anisotropy is important in forgings and rolled plates. The principal direction of working is defined as the longitudinal direction. This is the long axis of a forged bar or the rolling direction of a sheet or plate. The two directions perpendicular to this direction are the transverse directions. Generally, mechanical properties are lowest in the short-transverse direction, for example, the thickness of a plate. Measures of ductility (reduction in area), fatigue, and fracture toughness properties are most affected by mechanical fibering.

In metals there is a general relationship between surface roughness and the tolerance that can be maintained. Moreover, each manufacturing process has a range of surface roughness that can be achieved without extra effort (see Fig. 5 in the article "Overview of the Materials Selection Process" in this Volume). Table 6 describes the range of achievable surface roughness. In polymers, excellent surface finish is obtained with most molding processes, but the dimensional stability (tolerances) depends on the particular polymer and the processing conditions. Whatever the material, tightening the tolerances causes manufacturing cost to increase exponentially.

The cost of the material has a major influence on the cost of a manufactured part. The cost of metals depends on the availability of the ore, the cost of mining, the cost of reduction and refining, and the supply and demand. For polymers, the cost depends on the intermediate chemicals involved in the polymerization. Because metals invariably are used as alloys, and polymers are more frequently used as blends, the costs of the engineering material depend on what materials are combined and how this is achieved. The control of impurities often adds significantly to the material cost. Usually material costs are expressed as cost per unit weight, but often materials serve space-filling functions, so that cost per unit volume would be more appropriate. Table 7 shows a comparison of material costs based on mass and volume. Note how the relative cost of engineering polymers is reduced when expressed on a volume basis. These costs should only be used on a relative basis and then with considerable caution because not all materials are in the same stage of the materials-processing chain. Table 8 illustrates this point by showing the cost buildup in various steel products.

Ease of Manufacture

This section deals with the role of the material in successfully producing a quality part. Such terms as formability, workability, castability, machinability, and weldability are implied by this topic. This section emphasizes defining those material characteristics that aid or hinder the production of a part without defects. Certain characteristics of a material are inherent in the selection of the manufacturing process; for example, a molten metal that reacts strongly with the atmosphere would not be selected for casting in air, but there are more subtle issues and materials selection tests that need to be considered.

Casting of Metals

Castability deals with a complex combination of liquid-metal properties and solidification characteristics that promotes accurate and sound cast-

Table 9 Typical applications for castings and general rating of castability, machinability, and weldability

Casting alloy type	Applications	Castability	Machinability	Weldability
Aluminum	Pistons, clutch housings, exhaust manifolds	1–2	1–2	3
Copper	Pumps, valves, marine propellors	1–3	1–2	3
Gray iron	Engine blocks, gears, brake disks and drums, machine bases	1	2	5
Magnesium	Crankcases, transmission housings	1–2	1	2
Malleable iron	Farm and construction machinery, heavy-duty bearings, railroad rolling stock	2	2	5
Nickel	Gas-turbine blades, pump and valve components for chemical plants	3	3	3
Nodular iron	Crankshafts, heavy-duty gears	2	2	5
Steel (carbon and low alloy)	Die blocks, heavy-duty gear blanks, railroad wheels	3	2–3	1
Steel (high alloy)	Gas-turbine housings, pump and valve components, rock-crusher jaws	3	3	1
White cast iron	Ball mill liners, shot-blasting nozzles, railroad brake shoes, crushers	2	4	4
Zinc	Door handles, radiator grilles, carburetor bodies	1	1	5

1, excellent; 2, good; 3, fair; 4, very poor; 5, difficult. Source: Ref 15

ings. The factors that influence castability are fluidity, shrinkage, and resistance to hot cracking.

Fluidity is the ability of liquid metal to run freely and fill a mold cavity. Because the viscosity of molten metals is very low, fluidity is determined more by the solidification dynamics of the metal and the mold. A high surface tension of the liquid metal reduces fluidity. Oxide films forming on the surface of the molten metal thus have a significantly adverse effect on fluidity. Insoluble inclusions also reduce fluidity. The pattern of solidification of the alloy is important. Fluidity is inversely related to the temperature range over which solidification occurs. Thus, pure metals and eutectic alloys, with short freezing range, have greater fluidity. Fluidity is usually evaluated by pouring the metal into a narrow spiral mold cavity and measuring the length of penetration of the metal before it solidifies.

Shrinkage is an important castability factor in deciding whether the alloy has a high tendency to form shrinkage porosity. Unfortunately, there is no standard test for shrinkage. See the article "Design for Casting" in this Volume for a discussion of how to control shrinkage porosity in casting design.

Resistance to hot cracking is the ability of the alloy to withstand stresses developed by the contraction while cooling from the hot-short temperature range.

Castability is an integrated property that combines these three factors. One way to approach this is to rate the alloys on a scale of 1 to 5 for each of the important factors, use a weighting for the importance of each factor, and use a decision matrix to arrive at a ranking for each casting alloy (see the article "Use of Decision Matrices in Ma-

Table 10 Relative characteristics of aluminum alloys used in sand, permanent mold, and die casting

Rating scheme: 1, best; 5, worst. Individual alloys may have different ratings for other casting processes.

Alloy	Resistance to hot cracking(a)	Pressure tightness	Fluidity(b)	Shrinkage tendency(c)	Corrosion resistance(d)	Machinability(e)	Weldability(f)
Sand casting alloys							
201.0	4	3	3	4	4	1	2
208.0	2	2	2	2	4	3	3
213.0	3	3	2	3	4	2	2
222.0	4	4	3	4	4	1	3
240.0	4	4	3	4	4	3	4
242.0	4	3	4	4	4	2	3
A242.0	4	4	3	4	4	2	3
295.0	4	4	4	3	3	2	2
319.0	2	2	2	2	3	3	2
354.0	1	1	1	1	3	3	2
355.0	1	1	1	1	3	3	2
A356.0	1	1	1	1	2	3	2
357.0	1	1	1	1	2	3	2
359.0	1	1	1	1	2	3	1
A390.0	3	3	3	3	2	4	2
A443.0	1	1	1	1	2	4	4
444.0	1	1	1	1	2	4	1
511.0	4	5	4	5	1	1	4
512.0	3	4	4	4	1	2	4
514.0	4	5	4	5	1	1	4
520.0	2	5	4	5	1	1	5
535.0	4	5	4	5	1	1	3
A535.0	4	5	4	4	1	1	4
B535.0	4	5	4	4	1	1	4
705.0	5	4	4	4	2	1	4
707.0	5	4	4	4	2	1	4
710.0	5	3	4	4	2	1	4
711.0	5	4	5	4	3	1	3
712.0	4	4	3	3	3	1	4
713.0	4	4	3	4	2	1	3
771.0	4	4	3	3	2	1	…
772.0	4	4	3	3	2	1	…
850.0	4	4	4	4	3	1	4
851.0	4	4	4	4	3	1	4
852.0	4	4	4	4	3	1	4
Permanent mold casting alloys							
201.0	4	3	3	4	4	1	2
213.0	3	3	2	3	4	2	2
222.0	4	4	3	4	4	1	3
Permanent mold casting alloys (continued)							
238.0	2	3	2	2	4	2	3
240.0	4	4	3	4	4	3	4
296.0	4	3	4	3	4	3	4
308.0	2	2	2	2	4	3	3
319.0	2	2	2	2	3	3	2
332.0	1	2	1	2	3	4	2
333.0	1	1	2	2	3	3	3
336.0	1	2	2	3	3	4	2
354.0	1	1	1	1	3	3	2
355.0	1	1	1	2	3	3	2
C355.0	1	1	1	2	3	3	2
356	1	1	1	1	2	3	2
A356.0	1	1	1	1	2	3	2
357.0	1	1	1	1	2	3	2
A357.0	1	1	1	1	2	3	2
359.0	1	1	1	1	2	3	1
A390.0	2	2	2	3	2	4	2
443.0	1	1	2	1	2	5	1
A444.0	1	1	1	1	2	3	1
512.0	3	4	4	4	1	2	4
513.0	4	5	4	4	1	1	5
711.0	5	4	5	4	3	1	3
771.0	4	4	3	3	2	1	…
772.0	4	4	3	3	2	1	…
850.0	4	4	4	4	3	1	4
851.0	4	4	4	4	3	1	4
852.0	4	4	4	4	3	1	4
Die casting alloys							
360.0	1	1	2	2	3	4	…
A360.0	1	1	2	2	3	4	…
364.0	2	2	1	3	4	3	…
380.0	2	1	2	5	3	4	…
A380.0	2	2	2	4	3	4	…
384.0	2	2	1	3	3	4	…
390.0	2	2	2	2	4	2	…
413.0	1	2	1	2	4	4	…
C443.0	2	3	3	2	5	4	…
515.0	4	5	5	1	2	4	…
518.0	5	5	5	1	1	4	…

(a) Ability of alloy to withstand stresses from contraction while cooling through hot short or brittle temperature range. (b) Ability of liquid alloy to flow readily in mold and to fill thin sections. (c) Decrease in volume accompanying freezing of alloy and a measure of amount of compensating feed metal required in form of risers. (d) Based on resistance of alloy in standard salt spray test. (e) Composite rating based on ease of cutting, chip characteristics, quality of finish, and tool life. (f) Based on ability of material to be fusion welded with filler rod of same alloy. Source: Ref 16

Table 11 Polymer melt characteristics important in various processes

Process	Viscosity	Thermoset	Thermoplastic	Melt fracture	Die swell	Rate of reaction
Blow molding	Medium		X			
Casting	Very low	X	X			X
Compression molding	High	X				X
Extrusion, film	Medium		X	X	X	
Extrusion, profile	Medium		X	X	X	X
Extrusion, sheet	Medium		X	X	X	X
Filament winding	Medium	X				X
Hand-lay-up	Medium	X				X
Injection molding	Low to medium		X	X	X	
Injection molding foam	Low to medium		X	X		
Injection molding, reactive	Low to medium	X				X
Pultrusion	Medium	X				X
Rotational molding	Medium to low	X	X			X
Thermoforming	Medium		X			
Transfer molding	Medium	X				X

X, important characteristic. Source: Ref 17

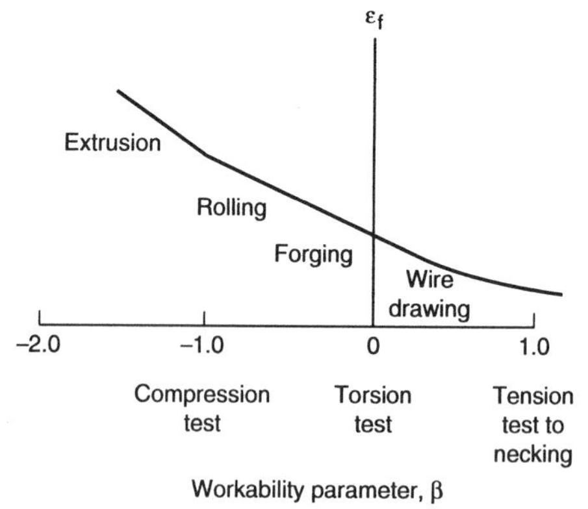

Fig. 4 Influence of stress state on strain to fracture. Source: Ref 18

Table 12 Evaluation of β parameter

Test	Principal stresses	Effective stress	Mean stress	β	Strain-to-fracture measurement
Tension	$\sigma_1{:}\sigma_2=\sigma_3=0$	σ_1	$\frac{\sigma_1}{3}$	1.0	$\varepsilon_f=\ln\frac{A_0}{A_n}$ at necking
Torsion	$\sigma_1=-\sigma_2;\sigma_3=0$	$\sqrt{3}\sigma_1$	0	0	$\varepsilon_f=\frac{r\theta}{\sqrt{3}L}$
Compression	$-\sigma_1;\sigma_2=\sigma_3=0$	σ_1	$-\frac{\sigma_1}{3}$	−1.0	$\varepsilon_f=\ln\frac{A_f}{A_0}$

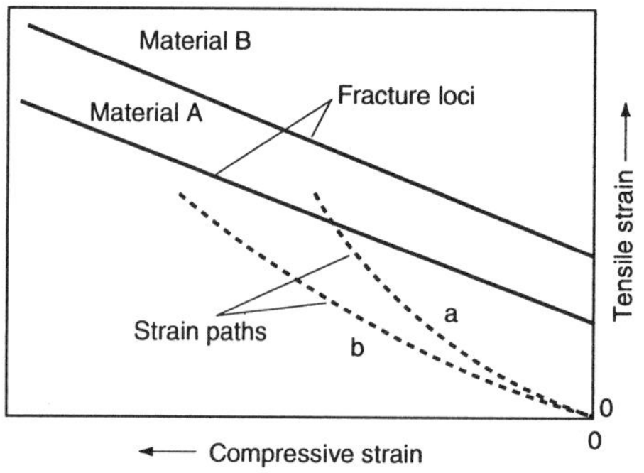

Fig. 5 Schematic workability diagram for bulk deformation processes. Strain path a would lead to failure for material A. Both strain paths can be used for the successful forming of material B. Source: Ref 20

terials Selection" in this Volume). Table 9 shows typical applications for metal alloys and relative ratings with respect to castability, weldability, and machinability. Note that these are general overall ratings for a class of alloys. Within each class, different alloys vary considerably with respect to specific process characteristics (see Table 10).

Polymer Processing

The chief material property influencing the processing of a polymer is its viscosity. Viscosity is a fluid property that relates shear stress experienced during the flow of a fluid to the rate of shear (shear strain rate). This property controls processing because most polymer processes involve the flow of the polymer melt through small channels or die openings. Because of their long-chain molecular structure the viscosity of polymers is many orders of magnitude higher than liquid metals. This means that high pressures are required to force the polymer melt into the mold cavity at a suitable rate. Polymers are non-Newtonian fluids; the viscosity usually decreases with increasing strain rate, as well as with increasing temperature. The fluidity of a polymer melt is often characterized by the melt flow index. This standardized test measures the amount, in grams, of a thermoplastic resin that can be forced through a 2.0955 mm (0.0825 in.) orifice when subjected to 20 N (2160 gf) for 10 min at 190 °C (375 °F). Polymer melts also exhibit viscoelasticity, which results in *die swell* by which the hot polymer expands and changes shape when exiting a die opening or the mold. Table 11 shows whether these and other processing characteristics of the polymer melt are important considerations in selecting a polymer process. The article "Design for Plastics Processing" in this Volume provides a more detailed description of the processes.

Deformation Processing of Metals

Two kinds of plastic deformation processes must be recognized. In three-dimensional, or bulk deformation, processes the material is deformed in all three principal directions. Here, the limit on deformation is concerned chiefly with the fracture of the material, and so ways to describe its workability are of interest. In two-dimensional, or sheet-forming, operations the deformation limit is caused by buckling, excessive thinning, or fracture; therefore, ways to describe the formability of the material are considered.

Workability is a complex technological concept that depends not only on the fracture resistance (ductility) of the material but also on the specific details (stress state) of the deformation process, as determined by die geometry, workpiece geometry, and lubrication conditions. The greater the workability of a material the greater the deformation and/or the more complex the shape that can be produced before fracture occurs. Ease of manufacture also is aided when the material has a low flow stress (yield strength) so that the force that must be applied by the processing equipment and the stresses on the dies are lower. Because flow stress decreases with increasing temperature, much plastic deformation of metals is carried out at elevated temperature (hot working).

Crack formation that limits workability is usually a form of ductile fracture. At temperatures lower than about one-half the melting point (absolute scale) this is the familiar dimple-rupture mode, but at higher temperature fracture initiated by grain-boundary sliding or cavitation can occur (Ref 18). Workability problems can also arise when deformation is localized to a narrow zone. Flow localization is caused by the formation of a dead-metal zone between the workpiece and the tooling, either because of poor lubrication or because the dies are cooler than the workpiece. Localized flow can also result from flow softening due to structural instabilities such as adiabatic heating, grain coarsening, or spheroidization.

The workability of a material is not a fixed material property. The state of stress in the workpiece imposed by the geometry of the deformation process greatly affects the workability. Stress states with high compressive components that impede crack initiation or propagation enhance

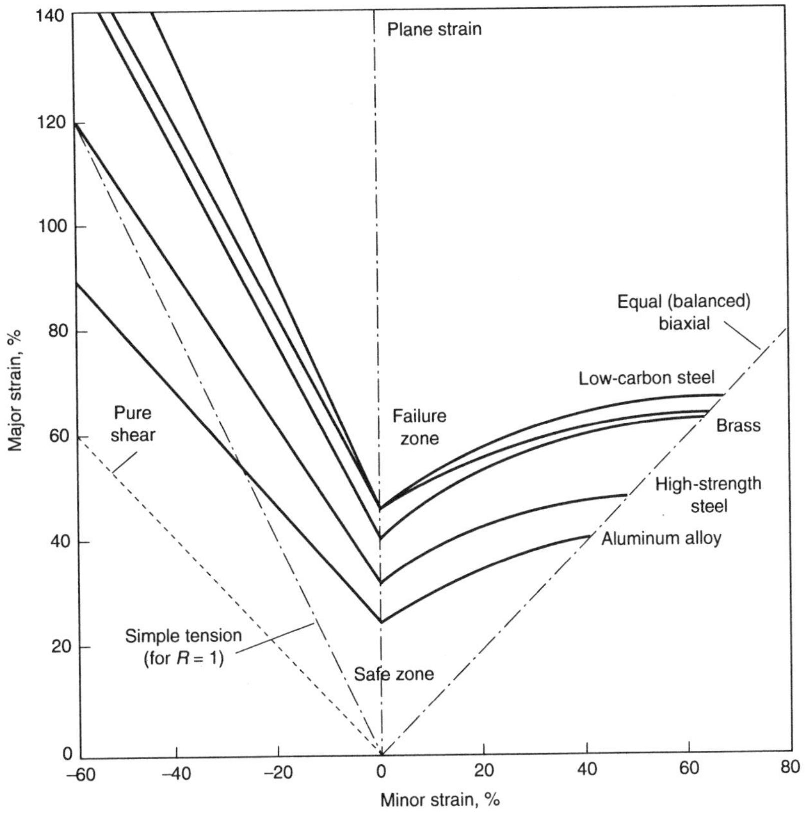

Fig. 6 Typical forming limit diagrams for different sheet metals. To use the diagram, locate the position on the chart for the major strain (always positive) and minor strain that is the most critical combination in the part to be formed. Failure will occur if the point is above the forming limit diagram for the sheet metal. Source: Ref 15

workability. A general workability parameter, β, that corrects for stress state is given as (Ref 19):

$$\beta = \frac{3\sigma_m}{\bar{\sigma}} \quad \text{(Eq 1)}$$

where σ_m is the mean or hydrostatic component of stress:

$$\sigma_m = \frac{1}{3}(\sigma_1 + \sigma_2 + \sigma_3) \quad \text{(Eq 2)}$$

and $\bar{\sigma}$ is the effective stress:

$$\bar{\sigma} = \frac{1}{\sqrt{2}}\sqrt{(\sigma_1 - \sigma_2)^2 + (\sigma_2 - \sigma_3)^2 + (\sigma_3 - \sigma_1)^2} \quad \text{(Eq 3)}$$

Figure 4 shows the strain to fracture plotted against the parameter β. The curve is determined from three basic mechanical property tests: tension, torsion, and compression. The appropriate parameters for each of these tests are given in Table 12. Also shown in Fig. 4 are the location of the most common bulk metal deformation processes in terms of the level of compressive stress state.

One of the most successful and useful design tools to assess workability for free surface fracture is the fracture locus diagram (Fig. 5) (Ref 20). The plot shows the locus of free surface strains (one tensile and the other compressive) that cause fracture. These are obtained from tests on upset compression and bend-test specimens. For many metals the fracture locus is a straight line of slope $-\frac{1}{2}$. To use the fracture diagram, the calculated surface strains estimated to be developed during the deformation process are superimposed on the fracture locus diagram. If the final strain for processing the part lies above the fracture line for the material then part failure is expected. As shown in Fig. 5, one can either change to a material B with better workability or change the process conditions (lubrication, die design, or preform design) to reduce the level of surface tensile strain, for example, strain path b. Additional information is provided in the article "Design for Deformation Processes" in this Volume.

A large number of workability tests have been developed for bulk deformation processes (Ref 21, 22). In addition to the primary tests of tensile, upset compression, bend, and torsion, there are tests such as plane-strain compression, partial-width indentation, ring compression, and forgeability tests such as wedge forging, sidepressing, and the notched-bar upset tests.

Formability. The term formability is commonly used to describe the ability of a sheet metal to maintain its structural integrity while being plastically deformed into a shape. Failure can occur by tearing, buckling and wrinkling, or excessive thinning. An ideal material from the standpoint of formability should:

- Distribute the strain in the sheet uniformly (high m)
- Achieve high strain levels without necking or fracture (high n)
- Withstand in-plane compressive stresses without wrinkling
- Withstand in-plane shear stresses without fracturing
- Retain part shape on removal from the die
- Retain a smooth surface and resist surface damage

The strain distribution is determined by three material properties measured with the tensile test. The *strain-hardening exponent*, also known as the n value, is expressed as:

$$n = \frac{(\ln \sigma)}{d(\ln \varepsilon)} \quad \text{(Eq 4)}$$

where σ is true stress and ε is true strain. The *strain-rate sensitivity*, also known as the m value, is expressed as:

$$m = \frac{d(\ln \sigma)}{d(\ln \dot{\varepsilon})} \quad \text{(Eq 5)}$$

The *plastic strain ratio* (normal anisotropy), or r value, is expressed as:

$$r = \frac{\varepsilon_w}{\varepsilon_t} = \frac{\ln (w/w_0)}{\ln (t/t_0)} \quad \text{(Eq 6)}$$

where ε_w is the true strain in the width direction of the tensile specimen and ε_t is the true strain in its thickness direction.

The r value is a measure of the ability of the material to resist thinning while undergoing width reduction. A high r value is the measure of a material with good deep-drawing characteristics. High values of n and m lead to good formability in stretching operations because they promote uniform strain distribution, but they have little effect on drawability.

A forming limit diagram has been established for sheet metals by subjecting the sheet to various

Table 13 Common machining processes listed in decreasing order of relative difficulty with respect to machinability of the workpiece

Internal broaching
External broaching
Tapping
Generation of gear teeth
Deep drilling
Boring
Screw machining with form tools
High-speed, light-feed screw machine operations
Milling
Shallow drilling
Planing and shaping
Turning with single-point tools
Sawing
Grinding

Source: Ref 24

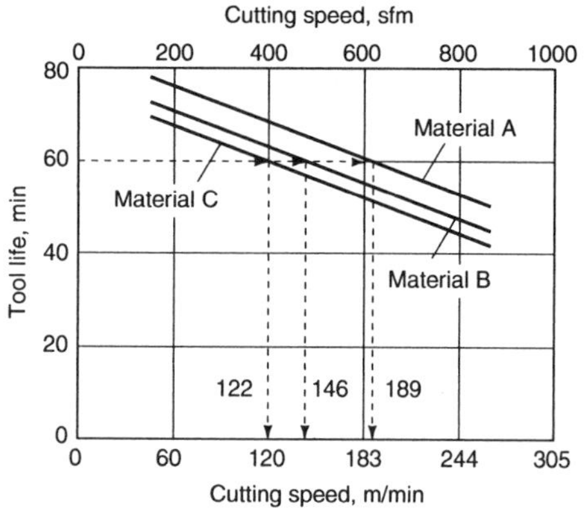

Material	Cutting speed for 60 min tool life		Machinability rating
	m/min	sfm	
A	189	620	100
B	146	480	77 ($^{146}/_{189}$) × 100
C	122	400	65 ($^{122}/_{189}$) × 100

Fig. 7 Relationship between tool life and cutting speed based on Taylor equation. Table shows how the machinability rating would be determined. Source: Ref 25

ratios of major to minor in-plane strains and plotting the locus of strain ratios for which local thinning (necking) and failure occur (Fig. 6). Strain conditions on the left side, where circles distort to ellipses, represent drawing conditions, while the right side where circles distort to larger circles corresponds to stretch conditions. When the minor strain is 0 a plane strain condition is developed. A forming limit diagram for a sheet metal can be used in conjunction with a trial run of the part to determine how close the material is to failure and how lubrication or die parameters should be adjusted to take the material away from a failure condition. The surface of the sheet is covered with a grid of circles produced by electrochemical marking. When the sheet is deformed, the grid of circles distort and the ratio of major to minor strain can be determined at critical points. These strains are plotted in Fig. 6 to determine how close the material is to failure.

Table 14 Approximate energy per unit volume requirements for machining different metals and alloys

Material	Hardness, HB	Unit energy(a), (W/s)/mm^3
Aluminum	50–100	0.55
Aluminum alloys	100–150	0.65
Cast irons	125–175	0.87
	175–250	1.3
Carbon steel	150–200	1.3
	200–250	1.7
	250–300	2.2
Alloy steel	200–250	1.7
	250–300	2.2
	300–350	2.8
	350–400	3.5
Copper	...	1.5
Brass	100–150	1.7
Bronze	100–150	1.7
High-temperature alloys	...	3.3–8.5
Titanium alloys	...	3.0–4.1
Refractory metals	...	3.8–9.3

(a) To convert unit energy to (hp/min)/in.^{3} divide by 2.73. Source: Ref 26, 27

Many simulation tests for the formability of sheet metals have been developed (Ref 23). The Olsen and Erichsen cup tests and the hemispherical dome tests measure stretching, while the Swift cup test measures drawing. The Swift round-bottomed test and the Fukui conical cup tests are involved with combined stretch and drawing.

Machinability

Machinability is the relative ease or difficulty of a material to be machined. A material has good machinability if the tool wear is low, the cutting forces are low, the chips break into small pieces instead of forming long snarls, and the surface finish is acceptable. This illustrates the problem of defining machinability as a characteristic material property. Machinability is more a system property that depends on the workpiece material plus the cutting tool material and its geometry, the type of machining operation, and the operating conditions of speed, feed, depth of cut, cutting tool geometry, nature of the cutting fluid, and the rigidity of the machine tool. The one material generalization that can be applied to machinability is that the higher the hardness of the workpiece material the poorer the machinability. Even this fails at low hardness levels where soft, gummy materials develop chips with built-up edge, and lower machinability. The requirement for the material to have good machinability varies with the machining operation (see Table 13).

The terms machinability index and machinability rating are used as qualitative and relative measures of the machinability of a material under specified cutting conditions. This concept

Table 15 Weldability of specific metals and alloys

Base metals welded	Welding processes: Shielded metal arc	Gas tungsten arc	Plasma arc	Submerged arc	Gas metal arc	Flux-cored arc	Electroslag	Braze	Gas
Aluminums	C	A	A	No	A	No	Exp	B	B
Copper-base alloys									
Brasses	No	C	C	No	C	No	No	A	A
Bronzes	A	A	B	No	A	No	No	A	B
Copper	C	A	A	No	A	No	No	A	A
Copper nickel	B	A	A	No	A	No	No	A	A
Irons									
Cast, malleable nodular	C	No	No	No	C	No	No	A	A
Wrought iron	A	B	B	A	A	A	No	A	A
Lead	No	B	B	No	No	No	No	No	A
Magnesium	No	A	B	No	A	No	No	No	No
Nickel-base alloys									
Inconel	A	A	A	No	A	No	No	A	B
Nickel	A	A	A	C	A	No	No	A	A
Nickel silver	No	C	C	No	C	No	No	A	B
Monel	A	A	A	C	A	No	No	A	A
Precious metals	No	A	A	No	No	No	No	A	B
Steels									
Low-carbon steel	A	A	A	A	A	A	A	A	A
Low-alloy steel	B	B	B	B	B	B	A	A	A
High- and medium-carbon steel	C	C	C	B	C	C	A	A	A
Alloy steel	C	C	C	C	C	C	A	A	A
Stainless steel (austenitic)	A	A	A	A	A	B	A	A	C
Tool steels	No	C	C	No	No	No	No	A	A
Titanium	No	A	A	No	A	No	No	No	No
Tungsten	No	B	A	No	No	No	No	No	No
Zinc	No	C	C	No	No	No	No	No	C

Metal or process rating: A, recommended or easily weldable; B, acceptable but not best selection or weldable with precautions; C, possibly usable but not popular or restricted use or difficult to weld; No, not recommended or not weldable. Source: Ref 29

Table 16 Comparison of weldability tests for fabrication

Test	Fields of use	Controllable variables	Type of data	Specific equipment	Cost
Lehigh restraint test	Weld metal hot and cold cracks, root cracks, HAZ hydrogen cracks, stress-relief cracks	Joint geometry, process, filler metal, restraint level, heat input, preheat, postweld heat treatment	Critical restraint, or % hindered control	None	Costly machining
Slot test	HAZ hydrogen cracks	Filler metal, interpass time, preheat	Time to crack, critical preheat	None	Low cost
Rigid restraint test	Weld metal hot and cold cracks, root cracks, HAZ hydrogen cracks	Joint geometry, process restraint level, filler metal, heat input, preheat	Critical restraint	Restraint jig	Costly machining and setup
Tekken test	Weld metal root cracks, HAZ hydrogen cracks	Joint geometry, process filler metal, heat input, preheat	Critical preheat	None	Low cost
Circular groove test	Weld metal hot and cold cracks, HAZ hydrogen cracks	Process, filler metal, preheat	Go/no-go	None	Costly preparation
Implant test	HAZ hydrogen cracks, stress-relief cracks	Process, filler metal, preheat, postweld heat treatment	Critical fracture stress, critical preheat	Loading jig	Intermediate cost
Tension restraint cracking test	HAZ hydrogen cracks	Process, filler metal, heat input, preheat	Critical fracture stress, critical preheat	Loading jig	Costly machining and setup
Varestraint test	Weld metal and HAZ hot cracks	Process, filler metal, heat input	Crack length, % strain	Loading jig	Costly preparation and analysis
Longitudinal bead-on-plate test	HAZ hydrogen cracks	Electrical type, heat input	% cracking	None	Low cost
Controlled thermal severity test	HAZ hydrogen cracks in fillet welds	Electrical type, cooling rate, preheat	Go/no-go (at two cooling rates)	None	Costly preparation
Cruciform test	HAZ hydrogen cracks, weld metal root cracks	Process, heat input, preheat, filler metal	Go/no-go	None	Costly preparation
Lehigh cantilever test	Lamellar tearing	Process, filler metal, heat input, preheat	Critical restraint stress and strain	Loading jig	Costly specimen preparation
Cranfield test	Lamellar tearing	Filler metal	Number of passes to crack	None	Low cost
Nick bend test	Weld metal soundness	Filler metal	Go/no-go	None	Low cost

Source: Ref 1 in Ref 30

was introduced early in this century when high-speed steel tools dominated the field. The tool life obtained in turning B1112 resulfurized steel with a high-speed steel tool at 180 sfm was assigned a machinability rating of 100. Suppose the tool lasted for 60 min. If in another steel the identical tool lasted for 40 min, then this material would be given a machinability rating of (40/60)100 = 67. Because small changes in cutting speed produce large changes in tool life it is more practical to measure machinability as the cutting speed that causes tool failure in a specified time period, usually 60 min. This makes use of the Taylor tool life equation, that the cutting speed × tool life raised to an exponent equals a constant, (vT^n = C). The plot and method of arriving at the machinability exponent is shown in Fig. 7.

Surface finish is another criterion for evaluating machinability. The better the surface finish under a given set of conditions the higher the machinability of the material. Similarly, cutting force tests to measure the feed force under a given set of conditions are sometimes used. The most fundamental measure of machinability is the energy consumed in machining a unit volume of material (Ref 26). See Table 14 for some values. For details on conducting machinability tests, see Ref 24, 25, and 28.

Microstructure can have an important influence on machinability. Spheroidized carbides are more machinable than pearlite platelets. Control of the inclusions in steel through small additions of sulfur, phosphorus, or calcium can be decisive. The influence of the manufacturing process on machinability is exemplified with the aluminum-silicon casting alloy, A390. Castings made by conventional die casting are more machinable than those produced by sand casting. This is attributable to the microstructure resulting from widely differing cooling rates. The die-casting process produces very fine, evenly distributed silicon particles, while the sand casting has larger silicon particles that are less uniformly distributed.

Finally, it needs to be understood that machinability data show considerable scatter. For B1112 steel, within the composition range permitted by that grade, it was found that the machinability index varied as much as 20% below and 60% above the nominal value of 100 assigned to this grade because of unintentional variations in carbon, sulfur, and silicon content. In general, differences of 5% in machinability ratings are not likely to be significant or reproducible.

Weldability

The weldability of a particular metal combination describes the ease with which the weld can be made and the quality and soundness of that weld. All metals can be welded to themselves by at least one of the welding processes (Table 15); these processes are discussed in the article "Design for Joining" in this Volume. The primary factor that controls weldability is the composition of the base metal and the filler metal. A fusion weld cannot be made between dissimilar metals unless they have metallurgical solubility. Barring this, brittle compounds may be formed. Weldability is affected by such defects as cracking in the weld metal, porosity, or cracking in the heat-affected zone (HAZ). Weld cracking in steel can be caused by the presence of very small amounts of hydrogen, the formation of martensite in the weld, and the presence of high residual stresses. Free-machining steel with high sulfur content cannot be fusion welded without cracking due to hot shortness. Hardenable steels must be welded with caution to avoid brittle martensite formation. Preheating the weld to reduce the quench rate and postheating to temper any martensite should be practiced. To estimate whether hardenability problems are likely to affect weldability of a steel, the carbon equivalent (*CE*) is determined:

$$CE = \%C + \frac{\%Mn}{6} + \frac{\%Ni}{15} + \frac{\%Cu}{15} + \frac{\%Cr}{5} + \frac{\%Mo}{5} + \frac{\%V}{5} \quad \text{(Eq 7)}$$

When the *CE* of a steel is less than 0.45%, weld cracking is unlikely and no heat treatment is required.

The quality of welds is determined from a variety of tests. Tension and bend tests of specimens containing the weld are commonly used. A variety of drop weight and fracture toughness tests are used with steel welds. A large number of tests have been developed for determining the susceptibility of the weld joint to cracking during fabrication. These can be divided into restraint tests, externally loaded tests, underbead cracking tests, and lamellar tearing tests (Ref 30). Table 16 summarizes and compares these tests.

Conclusions

The multitude of factors that go into the selection of the best material *and* process combination for a given design should be apparent from this article. It supports the importance of using an integrated product development team with members having materials and manufacturing expertise to arrive at the final selection. The cost and the quality of the finished part are the most important criteria in this decision. In the space available here only the main factors could be discussed. The reader is directed elsewhere in this Volume for additional details.

There are two generic situations. A material, or small subset of materials, has been selected based on the performance requirements of the design, and it is desired to determine the best manufacturing process. Alternatively, a process is available and well documented as a good way to make a product, and it is desired to determine the range of materials that can be successfully run through the process to make an acceptable product. The first situation is intellectually more challenging and more likely to lead to innovative and successful products. The second situation is more likely to be found with established product lines.

The factors that must be considered fall into three categories: material behavior, technical factors, and business considerations:

Material behavior

- Melting point of material
- Energy required to produce shape (flow stress)
- Workability, formability, machinability, castability, weldability
- Quality of product—freedom from cracks, voids, anisotropy, residual stress

Technical factors

- Availability of needed grade and/or form of material
- Capability to make required shape
- Ability to achieve needed tolerances/surface finish
- Production rate (cycle time)
- Ease of recycling scrap and discarded product
- Environmental impact of process

Business considerations

- Cost of product manufactured
- Availability of process equipment (make or buy decision)
- Flexibility of process; ease of changeover to another product
- Material utilization

REFERENCES

1. *Manuf. Eng.*, June 1996, p 27–29
2. M.F. Ashby, *Materials Selection in Mechanical Design*, Pergamon Press, 1992
3. C.C. Gallagher and W.A. Knight, *Group Technology Production Methods in Manufacture*, John Wiley & Sons, 1986
4. L. Edwards and M. Endean, Ed., *Manufacturing with Materials*, Butterworths, 1990
5. Data Card Index, *Manufacturing with Materials*, The Open University, Milton Keynes, 1990
6. R.S. Nyquist, Paradigm Shifts in Manufacturing Management, *J. Appl. Manuf. Sys.*, Vol 3 (No. 2), Winter 1990
7. A.C. Payne, J.V. Chelson, and L.R.P. Reavill, *Management for Engineers*, John Wiley & Sons, 1996
8. J.G. Bralla, *Design for Excellence*, McGraw-Hill, 1996
9. W. Chow, *Cost Reduction in Product Design*, Van Nostrand Reinhold, 1978
10. J.G. Bralla, Ed., *Handbook of Product Design for Manufacturing*, McGraw-Hill, 1986
11. J.R. Dixon and C. Poli, *Engineering Design and Design for Manufacturing*, Field Stone Publishers, 1995
12. G. Boothroyd, P. Dewhurst, and W. Knight, *Product Design for Manufacture and Assembly*, Marcel Dekker, 1994
13. The Selection of Steel for Economy of Manufacture, *Metals Handbook*, Vol 1, 8th ed., American Society for Metals, 1961, p 290–301
14. V. John, *Introduction to Engineering Materials*, 3rd ed., Industrial Press, 1992
15. S. Kalpakjian, *Manufacturing Processes for Engineering Materials*, 2nd ed., Addison-Wesley, 1991, p 262
16. *Properties and Selection: Nonferrous Alloys and Special-Purpose Materials*, Vol 2, *ASM Handbook*, ASM International, 1990, p 129
17. R.C. Progelhof and J.L. Throne, *Polymer Engineering Principles: Properties, Processes, Tests for Design*, Hanser, 1993
18. G.E. Dieter, Introduction to Workability, *Forming and Forging*, Vol 14, *ASM Handbook*, ASM International, 1988, p 363–372
19. V. Vujovic and A.H. Shabaik, A New Workability Criterion for Ductile Metals, *J. Eng. Mater. Technol. (Trans. ASME)*, Vol 108, 1986, p 245–249
20. H.A. Kuhn, Workability Theory and Application in Bulk Forming Processes, *Forming and Forging*, Vol 14, *ASM Handbook*, ASM International, 1988, p 388
21. G.E. Dieter, Ed., *Workability Testing Techniques*, ASM International, 1984
22. G.E. Dieter, Workability Tests, *Forming and Forging*, Vol 14, *ASM Handbook*, ASM International, 1988, p 378
23. B. Taylor, Formability Testing of Sheet, *Forming and Forging*, Vol 14, *ASM Handbook*, ASM International, 1988, p 877
24. F.W. Boulger, Machinability of Steels, *Properties and Selection: Irons, Steels, and High-Performance Alloys*, Vol 1, *ASM Handbook*, ASM International, 1990, p 591
25. C. Zimmerman, S.P. Boppana, and K. Katbi, Machinability Test Methods, *Machining*, Vol 16, *ASM Handbook*, ASM International, 1989, p 639
26. M.P. Groover, *Fundamentals of Modern Manufacturing*, Prentice-Hall, 1996, p 560
27. S. Kalpakjian, *Manufacturing Processes for Engineering Materials*, 2nd ed., Addison-Wesley, 1991, p 494–496
28. H. Yaguchi, Machinability Testing of Carbon and Alloy Steels, *Machining*, Vol 16, *ASM Handbook*, ASM International, 1989, p 677
29. K.G. Budinski, *Engineering Materials: Properties and Selection*, 4th ed., Prentice-Hall, 1992, p 559
30. S. Liu and J.E. Indacochea, Weldability of Steels, *Properties and Selection: Irons, Steels, and High-Performance Alloys*, Vol 1, *ASM Handbook*, ASM International, 1990, p 603

SELECTED REFERENCES

- G. Boothroyd, P. Dewhurst, and W. Knight, *Product Design for Manufacture and Assembly*, Marcel Dekker, 1994
- G.E. Dieter, *Engineering Design: A Materials and Processing Approach*, 2nd ed., McGraw-Hill, 1991
- L. Edwards and M. Endean, Ed., *Manufacturing with Materials,* Butterworths, 1990
- M.M. Farag, *Selection of Materials and Manufacturing Processes for Engineering Design*, Prentice-Hall, 1990

Computer-Aided Materials Selection

Volker Weiss, Syracuse University

COMPUTERS with elaborate software programs have not only become invaluable tools for computer-aided design and manufacturing (CAD/CAM) processes, but are also emerging rapidly as potentially powerful tools for materials selection. A 1995 report by the National Materials Advisory Board (Ref 1) presented a conceptual architecture of a computer-aided materials selection expert system that is capable of interfacing with a design team and its design-specific software programs (Fig. 1). Basically, such a system consists of three closely integrated parts: databases, knowledge bases, and modeling/analysis capabilities. Databases are discussed extensively elsewhere in this Volume (see the article "Sources of Materials Properties Data and Information"). Knowledge bases include rules for materials selection, design rules, constraints, objects, taxonomies, and lessons learned. The modeling/analysis section generally consists of computer programs for finite element stress analysis, constitutive analysis, life cycle cost analysis, risk analysis, performance simulation, and so forth.

Computerized materials properties databases do exist at present in many companies, and some are commercially available. Much of what is routinely available in materials data handbooks is becoming available for computer use (e.g., on CD-ROMs for use on personal computers). Some computerized materials properties and cost information is beginning to become available on the Internet. The article "Sources of Materials Properties Data and Information" in this Volume provides detailed information on materials properties databases, both print and electronic.

Computerized knowledge bases are less well developed. The subject of potential knowledge bases covers the range of topics of this Section of the Handbook and beyond. The basic nature of the elements of knowledge bases are formulas (such as the calculation of the critical crack length a_{crit} from the yield strength Y and the plane-strain fracture toughness K_{Ic}, namely $a_{crit} = (\frac{1}{2}\pi)(K_{Ic}/Y)^2$), or so-called "if-then rules", or, most usually, company or plant-specific "lessons-learned" files, performance characteristics, and manufacturability information.

At present several company-specific expert systems for materials selection exist, though very few systems are commercially available. Twenty-one commercially available materials databases and selection programs are described briefly in Ref 2. Of these, only two, the Cambridge Materials Selector (CMS 2.0) and Rapa Technology's Plascam software package, a plastics materials selector, are listed as "selectors," and even these are essentially databases, not knowledge bases or expert systems. The artificial intelligence (AI) technology for developing expert systems certainly exists and has been successfully applied to the development of in-house materials selection expert systems (and others, such as failure analysis or materials characterization systems). Some companies offer expert systems for selecting products of theirs that best meet a particular set of requirements, including environmental and chemical environments and geometric compatibility; an example is the bearing design guide developed by Furon, a manufacturer of structural bearings. Generally, little published information is available about many of these excellent in-house systems. Therefore, the following descriptions of various systems for computer-aided materials selection deal primarily with promising prototypes that have emerged for various applications.

Expert Systems: General Description

Expert systems are usually integrations of databases and knowledge bases using search and logic deduction algorithms. Once implemented, the distinction between databases and knowledge bases becomes blurred, because usually they are both treated simply as assertions, for example, "it is the case that the yield strength of 4340 steel, tempered at 400 °F, is 270 ksi." The integration of these information bases is accomplished by inference engines, typically embedded in expert system shells, or by computer languages based on some form of logic search rule (e.g., use of unification and resolution algorithms in Prolog). For all the databases and knowledge bases to be addressable by the inference engine, they have to be in an appropriate format, which is often not the

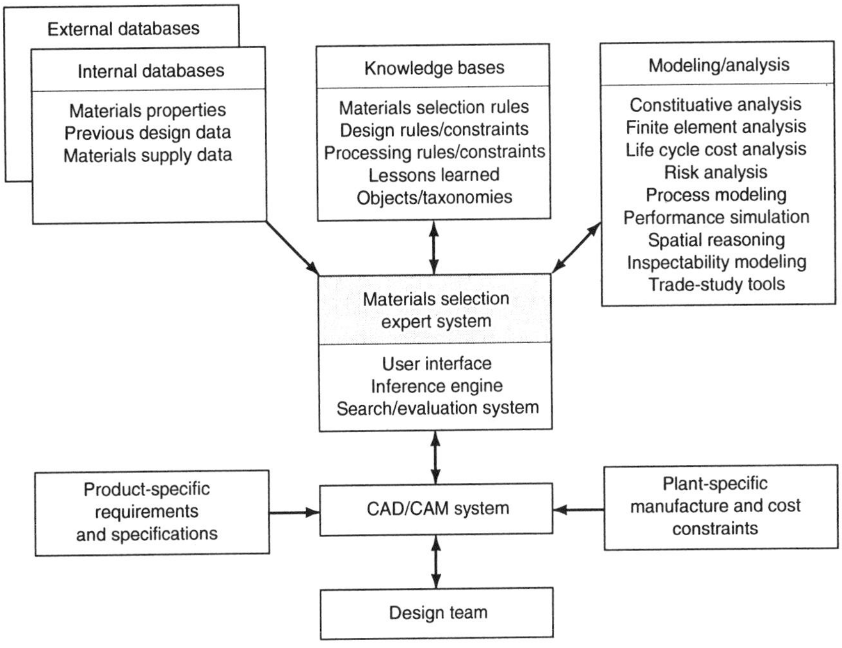

Fig. 1 Conceptual architecture of a computer-assisted materials selection system. Source: Ref 1

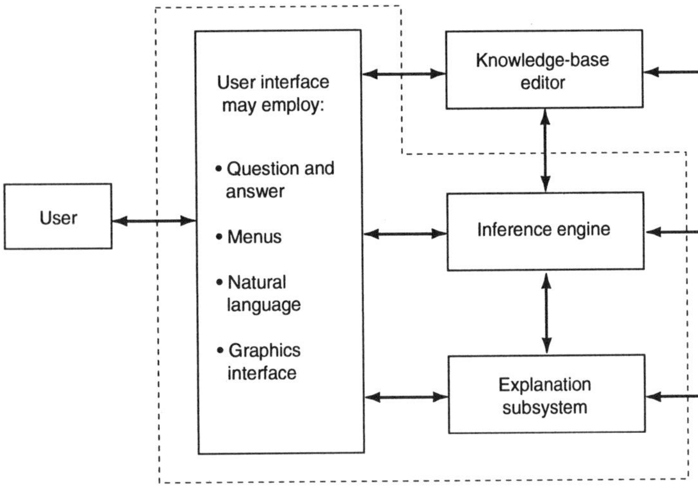

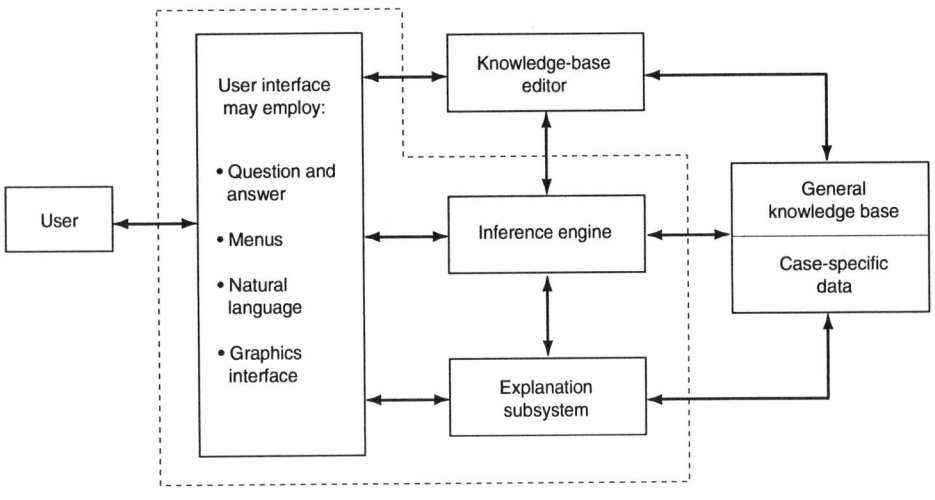

Fig. 2 Architecture of a typical expert system. Source: Ref 4

Table 1 Results of LOG-LISP materials selection query

The conditions were: (1) must safely support a load of 200 ksi, (2) must have a safety factor of 1.75, (3) maximum diameter of 1.5 in., and (4) maximum weight per unit length of 0.5 lb/in.

MAT	COND	YS	DIAM	W/L
ST 4340	QT-400F	262	1.35	0.40
ST 4340	QT-600F	230	1.39	0.43
ST 4130	WQT-400F	212	1.45	0.46
ST 4130	WQT-600F	200	1.49	0.49

MAT, material; ST, steel; COND, condition; QT, quenched and tempered; WQT, water quenched and tempered; YS, yield strength (ksi); DIAM, diameter (in.); W/L, weight per unit length (lb/in.). Source: Ref 6

case, though the situation is improving. Clearly, a fully functional computer aided materials selection system (CAMSS) will assume the character of an expert system. It has also been suggested that the CAMSS will ultimately involve not only a materials selection system, but also will be coupled, perhaps with feedback loops, to the CAD/CAM system, thus forming an integrated set of computer systems that serve the design team.

Expert systems first appeared around 1965; early systems included DENDRAL, MACSYMA, MYCIN, EMYCIN and Prospector (Ref 3). They were generally designed and programmed by one or several "knowledge engineers," who obtained the information from "domain experts" and structured a user interface that was intended to be friendly to the end user, presumably not an expert in artificial intelligence programming. The classical structure of a typical expert system is shown in Fig. 2. Prolog and LISP, and, more recently, object-oriented implementations through OOPS, Smalltalk and Java, are the languages of choice for expert system building. EMYCIN, CLIPS, KEE, ART, and NEXPERT are so-called "expert system shells," that is, empty reasoning systems to which a user can add data and rules to build an expert system. MYCIN was developed for medical diagnostic purposes. EMYCIN is "empty" MYCIN, a shell that is suitable for diagnostic reasoning (such as in failure analysis).

Computer-assisted materials selection, through expert or some other advanced systems, will be primarily of use for the level III and level IV design/materials selection phases (see the article "Overview of Materials Selection" in this Volume). At these stages, a design team may not be aware of, for example, all materials available in a given class (e.g., heat-resistant alloys) that also meet certain specific room-temperature requirements such as minimum fracture toughness and corrosion resistance to sea water after extended high-temperature service.

Quantitative Selection Systems

Any useful quantitatively based expert system for materials selection depends heavily on available data and design rules. The databases must be accessible to the system structure (e.g., languages like Prolog, LISP, or a shell). The data requirement is primarily for well-analyzed and documented data of candidate materials. The application-specific knowledge base will contain the design rules, such as appropriate stress analyses, stress concentration factors, fracture mechanics formulas, creep and fatigue relationships, and so forth. For complicated geometries and critical applications, the application-specific knowledge base might be hybridized to a finite element system to perform the required stress analysis for the geometry under consideration.

The system must have access to, or incorporate, materials properties databases to be used within the reasoning structure of the expert system. For example, a simple expert system for materials selection for a tension bar of minimum weight and a diameter not to exceed a given value and to carry a given load (Ref 5) could have mechanical properties data of materials excerpted or downloaded from a database so that yield strength and density values of a sufficient number of materials are available. Rules such as:

$$\text{Load-carrying capacity} = \text{Yield strength} \times \text{Area} \times \text{Safety factor}$$

and

$$\text{Weight per unit length} = \text{Area} \times \text{Density}$$

and

$$\text{Area} = (\text{Diameter})^2 \times \pi/4$$

need to be invoked in the search through the database to select the materials that meet the requirement. This list can then be sorted in terms of weight per unit length, or yield strength/density, as desired by the designer. Inspection criteria, for example, for cracks, could be added to exclude materials for which the critical crack length is below the inspection limit; this would require fracture toughness values be included for all candidate materials. In addition, exclusion rules could be formulated that remove all materials from consideration that would experience corrosion damage above a specified limit in the service environment; for this purpose, corrosion data for the anticipated service environment would need to be available from the database. In addition, rules for other factors of concern (e.g. cost and appearance, "lessons-learned" rules for similar designs, and "manufacturability" rules) may be formulated and invoked. The result of a query using LOG-LISP (Ref 6) and a database of several hundred materials is shown in Table 1.

This simple example is typical of a wide variety of much more elaborate materials selection problems. Database, knowledge base, and finite element data interaction would come into play for load-carrying parts having complicated geometries. Fatigue and fatigue crack growth knowledge bases and databases will be required for applications with load fluctuations and high-temperature creep data for elevated-temperature applications. For many of these applications, closed-form design formulas will not be available; therefore, the materials selection system will have to be connected to other computer programs (e.g., finite element programs).

Information about the database quality is necessary to decide on the applicability of the materials selection system for preliminary or final design stages. For the former case, typical data, or supplier data, may suffice. For the latter, well-documented and preferably statistically analyzed

Table 2 Results of LOG-LISP materials substitution query

Material and condition to be replaced: ST 4340, QT-800F. Critical properties to be matched or exceeded: TS, YS. Critical alloying elements to be reduced: chromium, nickel

MAT	COND	Cr	Ni	TS	YS
ST 4340	QT-800F	0.90	2.0	213	198
ST 8630	QT-400F	0.60	0.70	238	218
ST 8630	QT-600F	0.60	0.70	215	202
ST 8640	QT-400F	0.60	0.70	270	242
ST 8640	QT-600F	0.60	0.70	240	220
ST 8740	QT-400F	0.60	0.70	290	240
ST 8740	QT-600F	0.60	0.70	240	225

MAT, material; ST, steel; COND, condition; QT, quenched and tempered; TS, tensile strength (ksi); YS, yield strength (ksi). Source: Ref 6

data—of the type and quality characteristic of MIL-HDBK-5 (Ref 7), for example—are required. If data that include statistical information (e.g., standard deviation) are used, the knowledge base will need to have "rules" that deal with that information to determine average and (statistically defined) minimum load-carrying capacities for the structure under consideration. Iterations between design system and materials selection system should be planned for so that the designer can see if design changes also prompt material selection and other (e.g., cost) changes. It should be noted that the rules that manipulate well-defined statistical data are firm, deterministic rules that should not be confused with the application of "fuzzy logic" or "fuzzy numbers," which are addressed in the section "Qualitative and Experiential Systems" in this article.

Another application for a quantitative materials selection system is materials substitution. The query to be answered would be of the type: "find materials that have the same or higher tensile and yield strength as the one currently used (e.g., SAE 4340 steel, quenched and tempered at 800 °F) but contain less chromium and nickel." In addition to calling on mechanical property databases of alloys, a (nominal) chemical composition database must be available to the "inference engine." Table 2 shows the results from the same LOG-LISP prototype expert system used for the tension member design query (Table 1).

The prototype for a materials selection system to meet the special needs of the aerospace industry, the Intelligent Knowledge System for Selection of Materials for Critical Aerospace Applications (IKSMAT), was developed in the late 1980s (Ref 8). The plans were for an elaborate, mostly quantitative system with a front-end program for data entry and a diagnostic program to ensure that all of the important properties and characteristics of all the candidate materials are considered. The model for such a materials selection system is illustrated in Fig. 3. This model was implemented using the system architecture illustrated in Fig. 4. A "master" database facility, containing well documented and evaluated data, mostly of the type found in MIL-HDBK-5 (Ref 7), provided the input through MESSENGER mainframe software. The user interface was planned not only to allow the user to work within the existing knowledge base, but also to add data, rules, new criteria or other information, and to redefine priorities. IKSMAT was turned over to the U.S. Air Force in 1993, and it may have been discontinued due to the expense of maintaining it.

Qualitative and Experiential Selection Systems

Qualitative and experiential materials selection systems are needed for applications that involve considerable uncertainty. In such systems, the computer code seeks to assist or even substitute for design experts. Typical examples are materials selection requirements for service in corrosive environments, for cumulative fatigue damage, and for service under thermal and load cycling. Uncertainties in such applications involve, for example, the effects of combinations of corrosive agents, effects of stress amplitude sequence in cumulative fatigue, and effects of in-phase and out-of-phase load and temperature cycling. Systems that address these problems must be able to deal with approximations, and "fuzzy numbers," and they must compare and manipulate them and present the results of such manipulations with qualifiers, which distinguish them from results obtained from "hard-number" quantitative systems. The system must allow data to be expressed as minima, maxima, ranges, and unknowns, in addition to hard values. Different systems may, based on the underlying "fuzzy-logic" reasoning structure, come to different conclusions. For example, if the corrosion rate for material 1 in polluted harbor water is between 0.04 and 0.38 mm/yr, and that for material 2 between 0.18 and 0.28 mm/yr, which material is more corrosion resistant? Selection based on a "worst-case scenario" would lead to the selection of material 2, while a selection based on average rates would lead to the selection of material 1. For such systems it will be necessary to include the "desired scenario" in the query.

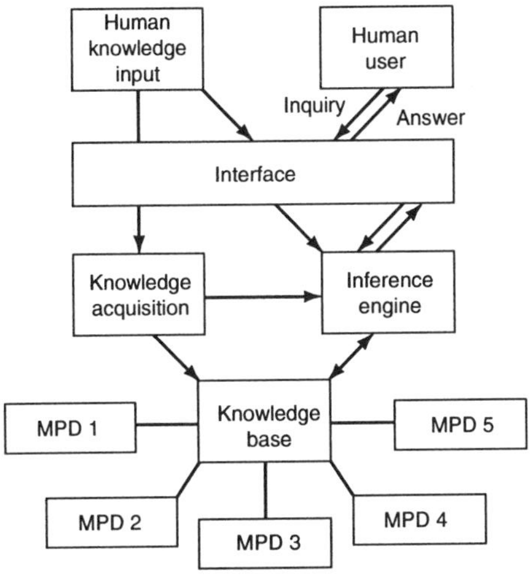

Fig. 3 Model of the IKSMAT system for materials selection in the aerospace industries. MPD, materials properties database. Source: Ref 8

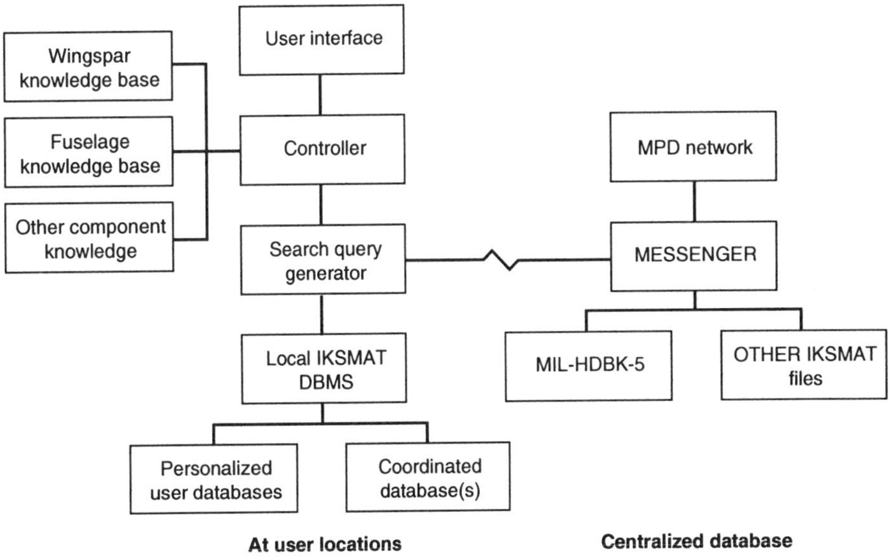

Fig. 4 Specific system architecture for the prototype IKSMAT materials selection system. Source: Ref 8

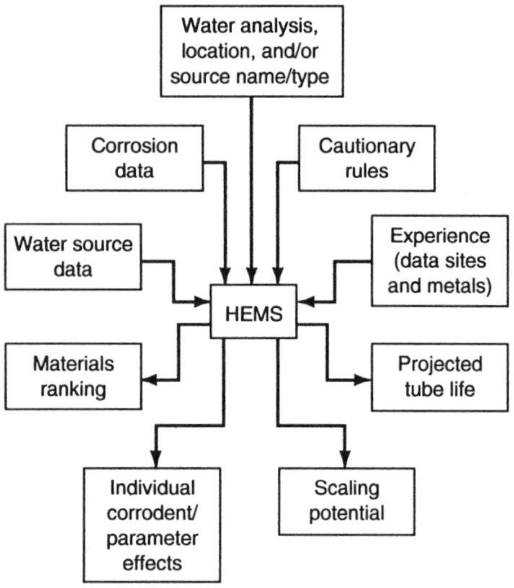

Fig. 5 Structure envisioned for the prototype heat exchanger materials selection (HEMS). Source: Ref 5

Example: Heat Exchanger Materials Selection System. A selection system for materials subjected to service in corrosive environments is typical for this class. Specifically, Syracuse University developed a prototype materials selection system for tube (heat exchanger) applications where corrosion and erosion results from a combined effect of water chemistry and flow rate. The prototype system was implemented in Prolog on a personal computer. The knowledge base for the system consists of six components:

- A corrosion property knowledge base for candidate materials
- A fuzzy number knowledge base for reasoning with uncertain data
- An experiential knowledge base stating conditions for excluding materials
- A location knowledge base containing water analyses and derivable characteristics (e.g., Langelier and Ryzner indices)
- An installation knowledge base providing information on experience of prior installations at the same site
- A control knowledge base containing the rules that govern the selection process

The structure of this prototype heat exchanger materials selection system (HEMS) is shown in Fig. 5. The output of the system, once a location is specified, is the estimated lifetime of heat exchanger tubes, that is, time to first leak. This time is then expressed in "design life fraction" using a company-specified design life, for example, 20 years. The 12 candidate materials are then ranked in order of decreasing life expectancy. A query for tube materials for a hypothetical location (with extremely polluted, sulfide-containing harbor water) resulted in the ranking given in Table 3. It should be noted that there is considerable uncertainty about corrosion rates in the database, especially the effects of combinations of corrosive media. However, a "rule correction module" was incorporated in the knowledge base, which enabled correction of corrosion rate data on the basis of field reports. For the given location, the materials selection system indicates that only titanium tubes are suitable. For other locations, several materials may be adequate, perhaps ranging from design life fractions of 2 to 6. In such a case, the customer must weigh the generally increased cost with increasing design life of the materials, for example, copper, Cu-10Ni, and Cu-30Ni. Such cost information can readily be added to the knowledge base and incorporated in the output.

Table 3 Results of heat exchanger materials selection system query

Query: tube materials suitable for service in extremely polluted, sulfide-containing harbor water

Rank	Material	Design life fraction
1	Titanium	>5.00
2	Type 316 stainless steel	0.35
3	Aluminum bronze	0.26
4	Type 304 stainless steel	0.19
5	Aluminum brass	0.18
6	AISI/SAE 1010 steel	0.13
7	Admiralty brass	0.13
8	Red brass	0.12
9	Copper	0.11
10	Cu-10Ni	0.10
11	Cu-30Ni	0.08
12	Monel alloy	0.07

The development of a user-friendly interface is of great importance when designing expert systems. For HEMS, in the prototype version, developed in Prolog, the user-friendliness was achieved by employing a hypertext link with "buttons" on keywords that activate the appropriate Prolog queries. The following is a short excerpt from the hypertext version of the prototype HEMS system. The words in bold type are "buttons" that activate actions, either display of explanatory text or initiation of a query process, usually through query-specific windows:

> The HEMS Expert System provides help in the selection of tube materials for **centrifugal chillers, reciprocating chillers,** and **other heat exchangers.** The system consists of two major components, an **information base** and a **material selection assistant** expert system. The two components can be used independently. In most instances the user will chose the interactive mode, making use of the local water chemistry database, at least for a start, and of the water chemistry calculation program. The **materials ranking** program, partially implemented here as a prototype system, allows you to select tube materials for typical **water chemistries**.

If the user selects **material selection assistant** or **materials ranking,** windows appear that ask for location, flow rate, and updates on water chemistry and initiates the appropriate queries that result in a ranking of the type given in Table 3. Selecting other buttons provides either information about the subject or connects to databases that provide information such as water chemistries by location and corrosion rates for the candidate materials in various environments. The system was also linked to already existing FORTRAN programs that calculate Langelier and Ryzner indices from the water chemistry data.

Access was provided on two levels, general user and materials specialist. The latter group also had the capability of modifying the system, adding data and programs, and modifying corrosion rate calculations on the basis of new field data. For the materials specialist, who had some familiarity with Prolog programming, ad hoc querying of the knowledge base of the system, beyond the standard prescribed query actions offered through the windows in the general system, could readily be accomplished. To the author's knowledge, the prototype HEMS system was not used commercially.

Object-Oriented Systems

Object-oriented computer languages, such as C++, Java, and so on, gained in popularity in the 1990s, primarily because of the way these languages allow information to be handled in a hierarchical way. Figure 6 shows how the object Al 2024-T3 is treated as a special case of various classes and subclasses. Each instance of an object or class can contain data *and* rules, or methods. Information from general classes is "inherited" to less-general classes and finally to the object. This hierarchy of the knowledge base increases its range of applicability and compactness as compared to a rule-based system operating on relational databases. Design information can be stored in a similar fashion as illustrated in Fig. 7, where the hierarchy for a specific "rod" as a subclass of a solid is presented.

Krishnamurthy and Smith (Ref 10) have developed an experimental quantitative problem solver (QPS) that is applicable to materials selection problems. Instead of asking for materials that satisfy a range of property requirements, one initiates materials selection with QPS by specifying the characteristics of the item being designed and allowing the system to specify the material query.

With increasing computing speed, object-oriented programming has become increasingly practical, even for large systems. Many of the AI languages and expert system shells are now available in object-oriented versions. Interesting new developments are the languages or programs that can interface with the Internet technologies, including the World Wide Web (WWW); two examples are Java and the EXSYS Web runtime engine (EXSYS WREN) (Ref 11). Java is an object-oriented programming language that allows the writing of small programs, called "applets," that can be embedded into a hypertext markup language (HTML) Web page. The HTML can be used to create hypertext documents that can be moved from one platform to another. Embedded applets execute on the local client workstation. Thus it is possible to use Java, or some other similarly capable program, to interact with information on the WWW to provide answers to queries subject to specified rules and constraints. EXSYS WREN (Ref 11) is an expert system development tool that, in connection with rules that can be produced by the compatible EXSYS Rule Book, acts on WWW information to respond to questions. With Java, the rules and other aspects of the applicable knowledge base have to be embedded in the applets. A simple applet may retrieve data on yield strength, modulus, and density from the Web or a database to compute the deflection of a beam of specified dimensions under its own weight. Newer versions of Java offer a new interface and greater database connectivity.

Agents. "Agents" and "agencies," (i.e. collections of "agents") are other developments worthy of consideration for use in constructing systems for materials selection. Agents are "mini expert systems" that have specific limited functions, for example, search-specified databases, or WWW files, for data on a specific subject (such as yield strength and plane strain fracture toughness) to enhance the database of the larger expert system. Agents are usually object-oriented language applications.

Data mining programs also may be useful when building knowledge bases for materials selection expert systems. These programs—some of them using object-oriented languages, some neural networks (Ref 13)—are an extended development of agents. They not only import data of interest, but also are capable of performing statistical and other operations on the data that lets the user make a judgment about their quality or even lead to predictions (perhaps via fuzzy numbers) of outcomes, for example, product demand. They generally can be applied to databases that support open database connectivity (ODBC). Typical commercially available data mining programs are "Alice" (from Isoft), "Profiler" (from Attar Software UK), "Model Quest Miner" (from AbTech Corporation), and "Predict" (from NeuralWare Inc.).

Current Status and Outlook

At present, many components exist that can make up viable computer-based materials selection systems, though no integrated system such as envisioned by the 1995 National Materials Advisory Board study (Ref 1) has been implemented. Several user-specific, smaller-scale, special-purpose materials selector systems have been developed and are in use at various companies and institutions.

Rockwell International developed a "Materials and Process Design Advisor" based on EXSYS. It reportedly consists of expert systems for corrosion protection, adhesive selection, encapsulant selection, conformal coating selection, heat treatment of metallic materials, and selecting soldering processes. The systems are made available to the Rockwell design engineers.

Granta Design Limited is marketing the "Cambridge Materials Selector" (CMS 2.0), which was developed by the Cambridge University Engineering Department. A generic database, containing representatives of virtually all groups of engineering materials, forms the core of the system. After identifying a group of candidate materials, the user can refine the selection with the help of additional databases in the materials group of the candidate materials.

The type of use that is made of these and other emerging systems most critically depends on the quality of data available to the rule base (including design features, environment, etc.) and the inference engine. Material selection for final design of critical components will certainly only be possible if the database consulted consists of well-documented and statistically evaluated data. For materials selection for preliminary or conceptual design stages, databases containing "typical" properties data or information from the material supplier will suffice.

Expert systems for materials selection may even become useful for cases where no commercially available material can satisfy all the performance requirements specified. In such cases, it might be possible to obtain guidance toward potential materials classes by relaxing the requirements until available materials appear as possible candidates. This may lead to identification of materials in research or under development that might eventually help solve the design problem. Ultimately, it is hoped that computing programs will emerge that allow atomic modeling of new materials and their properties. A great number of combinations could be explored inexpensively and in a relatively short time. Leading candidates could then be produced and tested in the laboratory to serve as feedback information for the atomic modeling programs and for further development.

Maintenance of both the databases and the materials selection system is also of critical importance. Initial databases should be designed to allow for efficient updates, deletions, retrieval, summarization, and reporting (see the article "Sources of Materials Properties Data and Information" in this Volume). Only databases that act predictably to queries posed through the materials selection system can be used successfully. In 1969, E.F. Codd developed a relational model and introduced the fundamental concepts for a well-organized, expandable relational database (Ref 14). To deal with these issues and keep an integrated computerized materials selection system current, it is strongly recommended that any such system be under the control of a *human* systems manager, who serves as or directs a *human* database manager specifically assigned to the maintenance, data oversight, and expansion of the knowledge base.

Finally, a note of caution is in order. The range of applicability and usefulness of these computer-aided systems for materials selection will certainly depend on the size of both the database and the knowledge base. Big systems will loose the transparency of small systems, developed by a few company engineers. Retrieval of data from the same original source, but duplicated in numerous tables and handbooks, may give a false sense of statistical security. Rules formulated for a specific case but applied as general rules may lead to wrong conclusions. And, of course, some data may be in error. While several expert systems have programs that explain the logic chain that led to the response for a specific query, these explanations deal with the logic sequences of the search (usually unification and resolution) and not with the accuracy of the assertions in the knowledge base. Therefore, the optimal benefit of these systems is expected when used by or in close association with a materials specialist well informed in the area of application.

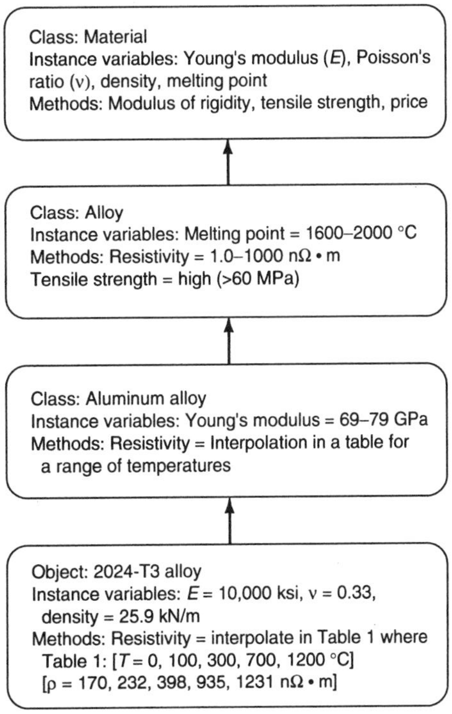

Class: Material
Instance variables: Young's modulus (E), Poisson's ratio (ν), density, melting point
Methods: Modulus of rigidity, tensile strength, price
Class: Alloy
Instance variables: Melting point = 1600–2000 °C
Methods: Resistivity = 1.0–1000 nΩ • m
Tensile strength = high (>60 MPa)
Class: Aluminum alloy
Instance variables: Young's modulus = 69–79 GPa
Methods: Resistivity = Interpolation in a table for a range of temperatures
Object: 2024-T3 alloy
Instance variables: E = 10,000 ksi, ν = 0.33, density = 25.9 kN/m
Methods: Resistivity = interpolate in Table 1 where
Table 1: [T = 0, 100, 300, 700, 1200 °C]
[ρ = 170, 232, 398, 935, 1231 nΩ • m]

Fig. 6 Materials hierarchy in an object-oriented system. Source: Ref 9

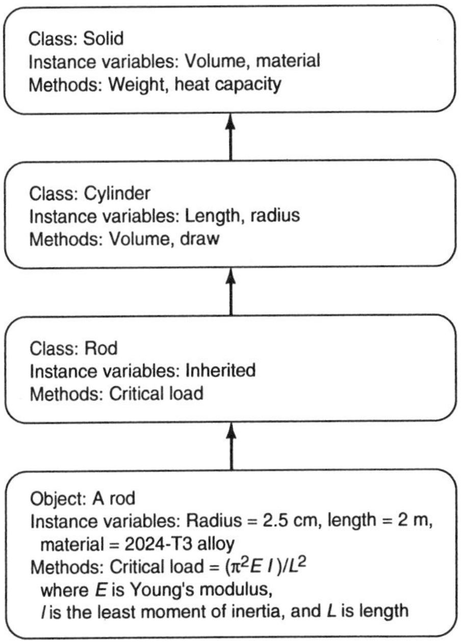

Fig. 7 Geometrical solid hierarchy for a specific rod in an object-oriented system. Source: Ref 9

REFERENCES

1. Committee on Application of Expert Systems to Materials Selections During Structural Design, *Computer-Aided Materials Selection During Structural Design,* Report No. 467, National Materials Advisory Board, National Academy Press, Washington, DC, 1995
2. Software Showcase, *Adv. Mater. Process.,* March 1997
3. P.H. Winston and K.A. Pendergast, Ed., *The AI Business,* MIT Press, 1984
4. G.F. Luger and W.A. Stubblefield, *Artificial Intelligence and the Design of Expert Systems,* Benjamin/Cummings Publishing Co., Redwood, CA, 1989, p 294
5. V. Weiss and K.J. Green, Expert Systems for Materials Selection, *Artificial Intelligence in Minerals and Materials Technology,* U.S. Department of the Interior, Bureau of Mines, Tuscaloosa Research Center, Oct 1987
6. J.A. Robinson and E.E. Sibert, "The LOGLISP User's Manual," Technical Report, School of Computer and Information Sciences, Syracuse University, 1981
7. *Metallic Materials and Elements for Aerospace Vehicle Structures,* MIL-HDBK-5E, *Military Standardization Handbook,* MIL-HDBK-5, Department of Defense, 1986
8. J.G. Kaufman, "An Intelligent Knowledge System for Selection of Materials for Critical Aerospace Applications (IKSMAT)" Final Technical Report, U.S. Air Force Contract F33615-87-C-5305, Department of Defense, 1988
9. F. J. Smith, M.V. Krishnamurthy, S.R. Tripathy, and P. Sage, An Intelligent Object-Oriented Database for Materials Information, *Computerization and Networking of Materials Databases,* C.P. Sturrock and E.F. Begley, Ed., STP 1257, ASTM, 1995
10. M.V. Krishnamurthy and F. J. Smith, Integration of Scientific Data and Formulas in Object-Oriented Knowledge Based Systems, *Knowl. Based Syst. J.,* 1995
11. D. Huntington, Web-Based AI-Expert Systems on the WWW, *PC AI,* March 1997, p 20
12. D.W. Rasmus, Agents—A New Agency, *PC AI,* March 1997, p 37
13. H.M.G. Smets and W.F.L. Bogaerts, Neural Networks for Materials Data Analysis: Development Guidelines, *Computerization and Networking for Materials Databases,* C.P. Sturrock and E.F. Begley, Ed., STP 1257, ASTM, 1995
14. E.F. Codd, A Relational Model for Large Shared Databanks, *Commun. ACM,* Vol 13 (No. 6), 1970, p 377

Value Analysis in Materials Selection and Design

Theodore C. Fowler, Fowler & Whitestone

VALUE ANALYSIS (VA) is a team problem-solving process. Its objective is to improve the value of a product from the viewpoint of the user (customer). Value analysis is applied to problems as diverse as automobile engines and gall bladder operations. It works whenever a product or process performs a function and costs money. While the method is often used to improve on present products, its greatest power is in optimizing a new design.

Throughout this article, the word *product* will be used to describe either a product or a process for which a purchaser, customer, or user can be identified. To be fully effective, VA should be applied to a complete product, not simply to a part or activity or other subelement.

Value Analysis versus Total Quality Management

The past twenty years have seen the development of a number of problem-solving systems. Many of these use teams in the same way VA uses teams. A popular current management philosophy is called total quality management (TQM). Rather than a set of procedures or a methodology, TQM is a state of mind for the modern organization. While it does apply several tools to accomplish its aims, for example, statistical process control (SPC), relations diagrams, quality function deployment (QFD), and policy deployment (PD), it makes primary use of two viewpoints or principles:

- It is customer focused. It concentrates on understanding and satisfying customer needs.
- It treats any endeavor in terms of process (check points) instead of just results (control points). Total quality management adherents follow the precept that approximately 90% of all variation from intended results is caused by problems with the process and only 10% is the result of worker actions. Statistical tools are applied to analyze data with the objective of minimizing variation and improving process.

The two key features distinguishing TQM from VA are:

- TQM presumes that the process under study is needed by the user. Value analysis involves first measuring the user attitude, then defining the user need, and finally applying the VA method.
- The VA procedure starts with the conversion of the product or process into its essential functions. It then establishes the most efficient manner of accomplishing those functions.

The VA process relates the cost of a product to the functions performed by that product. Because product cost is made up of material and labor, the typical ratio being 60 to 40, the VA process places a natural emphasis on the material cost content of a product. (The term material cost as used here includes costs of both raw and fabricated materials.)

Background and General Concepts

Value analysis was developed by Lawrence D. Miles at the General Electric Company in the late 1940s. Most of the newer quality systems and techniques, such as quality circles, poka yoke, grid organization development, Kepner-Tregoe, QFD, brainstorming, continuous systematic improvement, and TQM are team problem-solving systems. What is so special about VA?

The special power of VA comes from a technique called *function analysis*, which dramatically shifts the viewpoint of problem solvers from concrete issues to abstract concepts. When Miles first developed VA, function analysis was its only unique element, and this technique has been the source of all of the breakthroughs that can be attributed to VA.

In addition to function analysis, current VA practice involves several other key attributes, concepts, and activities:

- *Function cost:* A measure of the effort (in dollars, hours, personnel, etc.) spent on each function
- *Function worth:* An analysis of the degree to which a function fulfills the needs and wants of the user
- *Create by function:* A process that forces the problem solver to focus creative effort on abstractly defined customer needs rather than on hardware or activities
- *Value targeting:* A process that directs the problem solver toward areas where function cost does not match what the user/customer says the function is worth
- *Job plan:* A structured approach to implementing VA to maximize output
- *The team:* A multidisciplinary group representing the major aspects of the product under study that carries out the VA activity
- *Ownership:* The concept that each VA proposal must have an owner or "champion" to be successfully implemented

These items are discussed in greater detail in the section "The Value Analysis Process" in this article.

In the fifty years since VA was developed, it has changed mightily. Users have modified the process to fit their requirements. A modified version of the process is in heavy use in the U.S. government, where it is called value engineering (VE). The construction industry has followed their lead with a few further modifications. Their version is also called VE. In the automotive industry, the VA process is sometimes called value management. Each of these names refers to the same basic system. While the fundamental elements of all the systems are similar, there are nevertheless significant differences in focus, procedure, and effectiveness.

Value Engineering in U.S. Government Contracts. For most government products, the design and its related costs are established by contractors outside the government. To apply effective VA to these products, the Department of

Defense created a mechanism called the Value Engineering Change Proposal (VECP).

The VECP is a motivational device. Contractors are not normally inclined to recommend changes in the design of their products. The government specification is commonly rigid and detailed. Though a change to some element of the specification might benefit the product and/or decrease the cost of the product, contractors have discovered through experience that either (1) the government will reject a proposal for change for reasons of varying validity, or (2) the government will accept the proposal and reduce the price of the product. Neither outcome is likely to motivate the contractor to apply the effort required to prepare proposals for change.

A clause is, therefore, inserted in any government contract over $100,000 in total value. This clause encourages submittal of contractor-initiated changes. The contractor is promised a share of the dollar savings for any accepted proposal for change. This motivates the contractor to search out change and motivates the government to submit such proposal to serious evaluation. This VECP clause has been somewhat effective, but it has one very unfortunate unintended effect. It implies that VE is focused only on cost reduction. The "function" element in the value equation is commonly ignored, compromising the VA effort. This misimpression that VA is cost reduction tends to carry over to nongovernment activities and has severely degraded many commercial VA efforts.

Nevertheless, the application of VE in the government is undergoing a rapid expansion since the issuance in 1996 of Public Law 104-106, which requires VE by all units of the U.S. federal administration.

Value Engineering in the Construction Industry. The design of buildings and other civil engineering works follows a rigid and historic contractual path. To accommodate these constraints, VA in construction has also taken a rather different approach.

There has been a drift toward a second-guessing or critique mode. It has proved difficult to reconcile the historic contractual arrangements between owner, architect, engineer, and construction manager with the VA requirement that each team member must be a decision maker whose assignment presently includes responsibilities on the product under study. This has led to a form of VE in which Blue Ribbon Teams of outside experts critique the efforts of the original designers. While this causes some emotional and practical blocks, the process of construction value engineering is developing an impressive record of cost reduction.

Some projects have achieved significant cost reductions (an average of 15 to 20%) and/or significant quality improvements. Material selection in this sector is an important aspect of cost and needs satisfaction and is always analyzed during a VA session. Issued in 1995, ASTM E 1699 provides guidelines for VA as used in construction projects.

The Value Analysis Process

Value analysis begins with implementation planning. This is in keeping with Leonardo da Vinci's famous exhortation, "Think of the end before the beginning." Before starting the study, the team prepares a list of all of the possible areas in which they might anticipate their results to fall. They then list all of the roadblocks that they might expect to face in implementing the results. Next, they list all actions that must be considered during the study to circumvent those roadblocks. These lists are updated throughout the study (a sample is provided in "Example 1: Value Analysis of a Coolant Level Sensor" in this article). This commonly results in very few surprises during the later effort to implement the results of the study. Implementation rates of a properly conducted VA study can approach 100%.

The VA system developed by Miles followed a rigorous procedure he called the *job plan.* Others have varied the job plan to fit their particular constraints. A modern version has eight phases:

1. Preparation
2. Information
3. Analysis
4. Creation
5. Synthesis
6. Development
7. Presentation and report
8. Implementation and follow-up

Each of these phases is described in the following sections.

The Preparation Phase

This phase organizes the study. First, the mandate must be clearly outlined, the objectives set and the logistics established. It is here that the multidisciplinary team is selected. The ideal VA team comprises five experts on the product under study, each from a different discipline. For a complex product, several five-person teams may be appropriate. Each member must be a person whose job includes responsibilities on the product under study. The prime requirement is that each member be a decision maker in the organization. These are referred to as the organization's "movers and shakers." The following capabilities must be included in the team:

- *Design:* Project engineer, chief draftsperson, designer, or materials engineer (ideally, the engineer responsible for design of the product)
- *Operations:* Factory supervisor, industrial engineer, manufacturing engineer, construction supervisor, or methods engineer
- *Cost:* Cost estimator, industrial engineer, or accountant
- *Outreach:* Marketing, sales, field service, purchasing
- *Catalyst:* A constructive troublemaker, possibly an engineer, product manager, marketing person, or construction manager

The team meetings are facilitated by a VA specialist whose role is to ensure that the methodology is followed.

The Information Phase

The information phase includes function-analysis, function-cost, and function-worth activities.

Function Analysis. In the process of function analysis, the product or process under study is first converted into a number of verb-noun pairs. They describe what the product does for the user. The participants in the following phases will then concentrate on those word pairs rather than on the concrete product or process.

The specific form used for these word-pairs is called the *functive.* Each word pair comprises one verb and one noun. Each noun is ideally a parameter or measurable quantity, and each verb is ideally demonstrable on a nonverbal level. One verb and one noun make up one elemental sentence. Therefore, when the key VA question, "What does it do?" is asked of a product or process, the answers become a series of two-word sentences. The number of these sentences could run from forty to eighty for even a simple product or process. In their totality, these two-word statements are a complete and interrelated description of the functions of the product or process. Value analysts call this group of statements a *function analysis.*

By concentrating all their problem-solving effort on these two-word functions, the team minimizes what Miles called "functional fixedness." It is clearly possible to be more creative when tackling a problem defined in terms of improving the function "remove pollutants" than one defined in terms of "make a better catalytic converter." Miles defined this creative focus of function analysis in terms of the technique he called "create-by-function" (see the section "The Creation Phase" in this article).

Function analysis also serves to refocus the team from cost improvement to value improvement. This is accomplished through a fundamental relationship of VA called the *value formula.* It is a simple and elemental relationship and is usually expressed as:

$$\text{Value} = \frac{\text{Function (satisfaction of needs)}}{\text{Cost}}$$

When function is improved, value rises. On the other hand, when cost is reduced, value also rises. If function and cost both increase, value rises if function increases relatively more than cost. This relationship controls all VA and commonly results in designs that optimally satisfy all of the valid needs of the user at minimum cost. This is defined as optimum value.

Function analysis enables the team to capture the essence of a complex product or process on one page of unambiguous statements. Figure 1 is a simplified function diagram of an automobile as developed by a VA team studying the bumpers of a proposed 1992 model of a mid-sized automobile. This structure is called a function analysis system technique (FAST) diagram. Sixty-three functions were identified and arranged in the original diagram, in conformance to a cause/effect logic.

A FAST diagram is a functional model of the product under study. Reading it from left to right takes the reader from the main objective of the product to the details of how the objective is performed. Every function to the right answers the HOW of the function placed to the left of it. Conversely, reading from right to left validates the needs for the detailed functions by answering WHY. For example, in this diagram, one could read:

- HOW do you enhance product? By improving satisfaction.
- HOW do you improve satisfaction? By facilitating maintenance.
- WHY do you facilitate maintenance? To improve satisfaction.
- WHY do you improve satisfaction? To enhance product.

Table 1 Ratings of likes and dislikes of a gas furnace

Example of data collected to establish function worth during value analysis

Likes (features or characteristics)	Importance (most common vote of the users)
Heat exchanger is cast iron for long life	8
Block vent switch is prewired	5
Temperature/pressure gage is easy to read	9
Furnace is low in height	9
Flame rollout location minimizes burnout	5
Dislikes (faults or complaints)	**Seriousness (most common vote of the users)**
Draft hood outlet is too high	8
Pump is in wrong area for transportation	8
Standing pilot rating is not high enough	4
Plastic drain valve handle is too brittle	1

Table 2 Analysis of cost elements from the FAST diagram for an automobile bumper system (Fig. 1)

Priority (according to cost)	Function	Cost, %
1	Please senses	52
	Disguise irregularities	
	Enhance appearance	
	Instill pride	
	Differentiate models	
2	Ensure dependability	42
	Enhance safety	
	Protect pedestrians	
	Increase durability	
	Strengthen structure	
	Resist damage	
	Resist corrosion	
	Reduce repair	
3	Enhance product	6
	Improve satisfaction	
	Facilitate maintenance	

While the focus of a VA study may be only a part of a complete product, for example, a bumper system, the FAST diagram must always describe the complete product, in this instance, the automobile as purchased by a user or customer. This ensures that all evaluation will be related to the wants and needs of the user.

The FAST diagram is a precise analog of the product under study, expressed in the unconstrained language of function. It contains no references whatever to the product hardware or its labor content. It is, thus, the "new viewpoint," which is the underlying basis for all VA.

Function Cost. While the FAST diagram permits team members to create freely in the absence of the constraints represented by drawings, specifications, and bills of material, it is as yet incomplete. The next step turns the costed bill of materials into a costed bill of functions, where costs of items are attached to their respective functions. The procedure is called function-cost allocation.

The VA team allocates the costs of the product or process to the function diagram, resulting in a hierarchical structure called a costed FAST diagram. The team now has a totally new viewpoint, that of function-cost, to replace its customary viewpoint of hardware-cost. Their solutions therefore typically reach far closer to the limits of the envelope of practicality. In the diagram of the motor vehicle, the VA team allocated the cost data to their FAST diagram by assigning each line item of the bill of materials to the function(s) which it performs. The result is a costed bill of functions, the ideal basis for unconstrained, but focused, creativity.

When the VA effort is directed toward the design and development of a new product, a base reference case must first be prepared. This comprises a set of sketches and a pro forma costed bill of materials. The costs are then allocated to the FAST diagram in the same manner as for an existing product.

Function Worth. It is also desirable to add to the FAST diagram data on the needs and desires of the users/customers. To ensure that these data are carefully structured, but unconstrained in viewpoint, they are collected by simply asking a valid sample of users two nonleading, open-ended questions about the product: "What do you like?" and "What do you dislike?" The resulting data are used to assign a function worth to each of the functions.

When the VA effort is being directed toward the design and development of a new product, the collection of user attitude data is focused on currently available versions of the proposed product. In some specific cases, cost and worth in dollars can be replaced by duration, manpower, reliability, or any other criteria worthy of optimization.

Table 1 shows the results of data collected during value analysis of a gas-operated home furnace. The data are a selected sample of the 76 responses collected in a one-day target opportunity panel where 16 users and producers of the home furnace were repeatedly asked, "What do you like about your furnace?" and then, "What do you dislike about your furnace?" The unconstrained responses were rated on a scale of 1 to 10 (10 being highest) on their importance or seriousness. The ratings are of the mode or the "most frequent" value.

The Analysis Phase (Identification of Value Targets)

Knowing the cost and the worth of each function permits the VA team to identify and focus on only those functions where solutions are required, that is, where there is a mismatch between the two parameters of function cost and function worth. In effect, these are functions where cost does not match what the customers say the function is worth to them.

The FAST diagram for the bumper system of a 1992 mid-sized automobile is shown in Fig. 1. The $57.61 total cost is shown allocated to the functions performed by the bumper system. If the team's knowledge is limited to these data—that is, if all they know is cost—they might choose to focus their problem-solving effort simply on the higher cost functions, in accordance with the list shown in Table 2.

A review of this priority list will quickly indicate that the use of cost alone as a criterion would lead the team to apply their redesign effort to functions where the opportunity for increased value is minimal. Fortunately, the VA team had access to data from a target opportunity panel, which shifted their focus from the functions that were simply high cost to those where the function cost failed to match the function worth as measured by the data from the panel. On the basis of these data, they chose to apply their creativity and their synthesis effort to the functions that differentiate models and enhance appearance instead. This demonstrates the importance of good function analysis. Teams often focus on cost reduction only and neglect function enhancement as a method of increasing value.

The Creation Phase

In the creation-phase team session, the focus is on function. This function focus maintains the critical perspective of what it does for the customer, rather than what it is in the mechanical or operational sense. Over the years, the best practitioners of VA have fully understood customer focus and have applied the concept of "create-by-function."

The literature on human creativity comprises hundreds of volumes and thousands of articles in a wide variety of periodicals. The methods and the results described have a common root: brainstorming, a system of creative problem solving developed in the 1940s by Alex Osborn, a New York advertising executive. He successfully used it to generate new approaches to marketing products. When Lawrence Miles was developing VA, he took advantage of Osborn's work and found that the fit between function-oriented problem solving and brainstorming was nearly perfect.

During the creation phase, the VA team simply places the name of a function at the top of an easel pad and calls for an unconstrained outpouring of

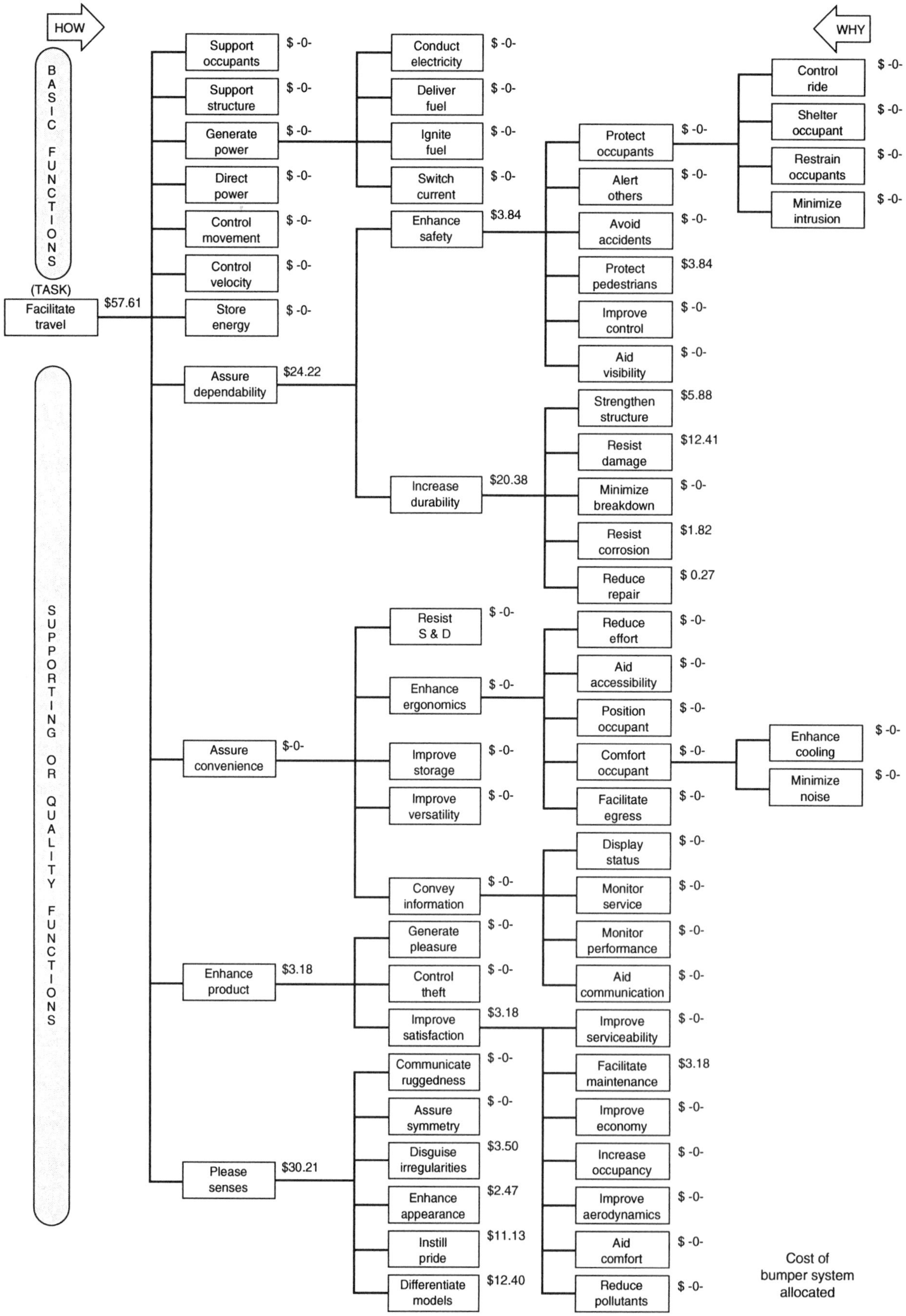

Fig. 1 Simplified function diagram of an automobile

"thoughts and things." The automobile team described above first listed the function *enhance appearance*. More than 100 words and phrases were generated and listed on the easel pad. The rules of brainstorming require the deferring of judgment. As a result, the listed items are seldom solutions, or even ideas. They are merely words, that is, verbalizations of often unconscious mental images. Table 3 shows a partial list of words that were triggered in the minds of the automobile bumper VA team members by the function *enhance appearance*.

Note that most of these words or phrases have no relationship to an automobile or a bumper. This is the secret of effective brainstorming in VA. Team members consciously avoid generating solutions. They play the function-oriented brainstorming game in its pure sense. The list of words/phrases is a result of their focus on the function *enhance appearance* and not on their ultimate task of improving the value of the bumper system.

The secret to effective VA at this stage is to treat a function in its purely semantic sense. The team must rigorously resist reverting to simply creating solutions to the problems of the product under study. Developing this viewpoint is not an easy task and requires much coaching and practice. It is in the later synthesis phase, described below, that the team re-addresses the bumper problem. The unconstrained words/phrases of the creation phase trigger solutions, often unconsciously, in the minds of the expert team members.

Additional information about brainstorming and other methods for spurring creativity is provided in the article "Creative Concept Development" in this Volume.

The Synthesis Phase

A word is not a solution. The words on a brainstorming list are a far leap from a solution that can be implemented. For that reason, VA provides a method for breaking the journey into a series of steps:

Word → Idea → Concept → Solution

The team carefully nurses each word or phrase from its brainstorming list up this ladder of abstraction until an idea, concept, or solution emerges. If the exercise is not successful, that word is deleted and the elevation effort is applied to the next word on the list. The question asked of each word/phrase is "how can I use it?"

At this stage in the VA process, the team has undergone a joint learning process that "loads their minds" with a massive quantity of valid data about the product under study. Each team member is a decision maker in his or her area of specialization, and the information phase has required them to :

- Develop a common and unconstrained language (the FAST diagram)
- Develop an unprecedented understanding of the reason for the existence of each element of material and labor (the function-costing exercise)
- Relate each product function to the needs and desires of the users/customers (the function-worth exercise)

Table 3 Example of creative brainstorming results for the function *enhance appearance*

Blend colors	Electrify
Use makeup	Imply motion
Use translucent materials	Clip fingernails
Illuminate	Coordinate dress
Smooth surfaces	Attract eye
Integrate accessories	Generate emotion
Frame highlights	Round edges
Brush hair	Polish

These three exercises prepare the collective minds of team members for the identification of ideas, concepts, and solutions.

Ownership in the Synthesis Phase. During the synthesis phase of the job plan, ideas and concepts arise. The team leader asks which of the team members will champion the idea or concept. If no one volunteers, the idea or concept is dropped. One who agrees to become a champion is charged with investigating the feasibility and economics of the idea or concept. The result of this phase is, thus, a series of proposals with all the explanations required to be able to select the solutions that are highly likely to be implemented. This includes a description of the present situation and of the proposed alternative, the advantages and disadvantages of the idea, the assumptions used for the calculations, the benefits and costs of the idea, the implementation efforts, and any other information required for a sound decision.

The Development Phase

During the three-week period of the development phase, each champion prepares a rigorously complete set of data on his or her proposals. Included are drawings, cost estimates, engineering calculations, test results, and any other material verifying the efficacy of the proposed change.

The Presentation and Report Phase

The final presentation of results is a brief, but concentrated, report to the management group that authorized the study. Such a presentation has been found to be an essential element in the dynamic of a VA study. It is the focal point toward which all previous activity is directed. It is also framed as "the first day of the implementation phase."

A final report, written by the team members, is issued to each attendee at the final presentation. It contains each of the VA proposal forms, as well as a summary of results and a set of Gantt charts scheduling each activity of the implementation process. This document is, in effect, a "blueprint" for the implementation phase. It should have the power of a sales document because this is the moment at which the champion of each idea will get the go ahead for his or her idea, budget included.

Table 4 Possible areas of change for the coolant level sensor design

See Example 1 in text.

Possible change	Actions required to ensure implementation of change
Probe tip material change	Lab test immersed in coolant at high temperature and pressure
	Verify plan with customer release engineer
Relocation of assembly area	Discuss in advance with Human Resources
	Establish rigorous quality constraints
	Discuss with union
Change in sealing configuration	Accelerated life test to ensure 10 yr mean time between failures
	Qualify to ASME pressure vessel specification

The Implementation and Follow-Up Phase

Without a rigorous system to transform synthesized designs into implemented design changes, an organization stands to waste its problem-solving effort. As noted earlier in this article, VA starts with a process of implementation planning, and this phase is intended to ensure that any proposal reaching this stage has the maximum opportunity for implementation.

A Gantt chart should be prepared describing the activities required to implement each proposal. The implementation plan, as in the example shown in Table 4, serves as a guide to the assignment of a specific person to accomplish each activity. Implementation and follow-up then become simply a matter of carrying out the plan.

Examples

Value analysis techniques are described in this section through excerpts from case studies. Two case studies were selected as best exemplifying the materials-related aspects of the VA process. The first of these studies involves a coolant sensor probe, produced by a U.S. manufacturing company. The second focuses on the development of a helicopter gearbox.

Example 1: Value Analysis of a Coolant Level Sensor. During the implementation planning phase, the team members created a list of

Table 5 Recommended changes to the helicopter gearbox design

See Example 2 in text.

Change	Current situation	Proposed alternative	Advantages	Disadvantages	Weight savings, kg	Weight savings, lb	Savings or (cost), $
1a. Aluminum investment casting	The present gearbox housing is produced via aluminum sand casting with a nominal wall thickness of 0.100 in. and tolerance of 0.020 in. The current material is aluminum alloy AMS 4219, strength class No. 10 to 11.	Produce the housing via aluminum investment casting, which allows for thin-wall castings. It is also suggested that the strength class be increased to No. 12 for improved material properties.	Weight and cost savings. Improved geometric accuracy and consistency. Flexibility of design. Good surface finish for contoured air-passage areas, such as engine air-intakes. Components can be cast with ultrasmooth surfaces. Performance is increased without resorting to expensive secondary operations. Indefinite tool life because investment casting tools are in contact only with wax.	Investment casting reduces the opportunity for concurrent engineering as final detailed drawings are required for tooling definition.	1.4(a)	3.0(a)	1275
1b. Magnesium sand casting	Same as above.	Magnesium would be used to produce sand cast housings. Investment cast magnesium is not economical. However, the switch from aluminum to magnesium sand casting is very feasible and is commonly done in industry.	Magnesium has a density of 0.645 that of aluminum. The weight-to-stiffness ratio favors magnesium at engine operating temperatures. This is particularly important for a casting of a gearbox part. Magnesium has less loss of yield strength with time as temperatures approach operating conditions at 150 °C (302 °F)	Magnesium is prone to corrosion and requires the addition of a thick epoxy coating on the housing. This was estimated at $100. Due to castability issues, wall thickness was increased by 15% over baseline maximum wall thickness.	4.3	9.4	(100)
2a. Thinning of gear webs	On the preliminary design, the nominal thickness of the gear webs of accessory gear trains ranges from 3.05 to 3.81 mm (0.120 to 0.150 in.).	Preliminary stress analysis showed that it would be safe to thin down the webs to 1.91 mm (0.075 in.) for flat gears only. If the gears are thinned down, the pieces may bend, and for this reason, a minimum nominal web thickness of 2.223 mm (0.0875 in.) is assumed. Calculations of savings were made according to this value.	A weight saving of 0.29 kg (0.64 lb). It is unlikely that the removal of the extra 0.8255 to 1.5875 mm (0.0325 to 0.0625 in.) will require an additional cut during machining; therefore, no cost increase is expected to produce the thinner webs.	...	0.3	0.6	0
2b. Cross-drilling of gear webs	The thickness of the gear webs is stated above. They are all full webs.	Cross-drill holes in the gear web in order to remove material and thus weight. Requires extensive analysis because many options are possible. The size, number, and location (balanced, random pattern) of the holes are all variables, but a good disposition must sustain the stresses within a given factor of safety, must not induce vibrations (must not resonate in operating range), etc.	Weight savings. Conservative assumptions were made to give an idea of the possible weight savings, e.g., 3 holes per web, hole diameter of ⅓ of the web's radial length, and accessory gears only (flat and conical).	The cost of cross-drilling three holes in all gears is calculated as: $8.50/hole × 3 holes/gear × 12 gears = $306.	0.75(b)	1.66(b)	(306)(b)
3. Optical torque sensor	The torque being transmitted by the gearbox is measured using a pair of acoustic sensors directed at the intermeshing fingers of two coaxial shafts.	Use an identical configuration, but with photo-electric sensors. The fingers on the shafts are replaced with simple, cross-drilled disks. Each disk has a sensor which "reads" the holes on the disks. Torque is measured as a change of phase between the two signals.	A weight saving is realized by replacing the acoustic sensors with optical ones and the fingers with lightweight disks. Optical sensors are inexpensive, reliable, and require far less machining. The data processing requirements (to interpret the phase, and convert to torque) are essentially unchanged.	...	0.2	0.5	2000
Future option: optical strain gage(c)	Same as above.	An extension of the optical torque sensor proposal. An optic fiber would be mounted on or in the power shaft. A light source is then directed through the fiber, and a detector reads the signal at the fiber exit. As the shaft strains, the fiber is deformed, and the signal altered. Similar applications have been used in many cases, e.g., to measure the strains on the propeller and shaft of an icebreaker. The main requirement is the mounting of the fiber on/in the shaft in a permanent, repeatable manner.	Weight saving. The components used for an optical strain gage are very light. Because the strain is read from the output shaft, the concentric shafts presently employed for this purpose can be removed along with the associated bearings and supports. With the removal of the present torque sensor, the gearbox design can be more compact, thus saving on space and housing size as well. Once calibrated, this system would be simple and inexpensive.	...	...	...	...

(a) Assumes a 2.159 mm (0.085 in.) nominal wall thickness; the actual weight savings achieved will vary depending on the nominal wall thickness selected. (b) Preliminary stress analysis permitted, with reserve, to assume up to 5 holes per gear. In such a case, the weight saved would be 1.22 kg (2.70 lb) at a cost of $510. (c) The solution was thought feasible, but would require some development and thus was recommended to the company for future consideration.

possible areas of change in the coolant probe design, and they updated the list twice during their study. Three typical entries are shown in Table 4.

The coolant sensor contains two conductive probe tips that contact the liquid coolant when it reaches the proper level. This sends a signal to the system indicating that the equipment can be permitted to operate. The remainder of the sensor comprises insulating material and a method of mounting and sealing the assembly.

From the synthesis effort, the team generated several ideas and concepts. This example illustrates only one of them, which was championed by the materials engineer team member. The concept involved the injection molding of the probe body with two probe tips insert-molded in place. It required only a single O-ring. While this new design reduced the parts count from 10 to 4, it left several questions to be resolved in materials selection. Primed by their rigorous function-analysis effort, the full five-person team worked out the design led by the materials engineer.

The team determined that the probe must be conductive and corrosion resistant. Stainless steel (300 series) was chosen, with a cross section to lock into the injected plastic tube and with the required 19 mm (3/4 in.) exposed to match the previous impedance.

The team also determined that the body of the probe must resist the cooling chemicals and withstand temperatures from 1 to 315 °C (33 to 600 °F). The base threads will be molded in. A major concern with the hygroscopicity of nylon was turned into an advantage as a result of the team's focus on required functions and the materials engineer's previous knowledge. Nylon 6/6 was chosen because of its hygroscopic nature. The team eliminated the base O-ring because the nylon body would swell in the presence of water and ensure a seal. For assembly, the unit comes from the press in finished form requiring only deflashing. The seal around the probe tip resists leaks, again because of hygroscopic nature of the nylon.

Results. The new sensor was tooled at a cost of $30,000 and became the new standard for all of the products of the company. It had a failure rate of less than 10% of that of the purchased sensor, with essentially no leakage failures. The unit cost of the new sensor was $1.50, enabling a return on tooling cost in 4.33 months and saving the company $180,000 per year.

Example 2: Value Engineering of a Helicopter Gearbox. The objective of this VE study was to reduce the weight of a new gearbox design for an aerospace company while maintaining or reducing the final cost. The design of the gearbox had not been finalized at the time of the study, so the study focused on weight reduction rather than on detailed design of the gearbox. Material composition and weight considerations were important because the gearbox was to be used in a rotary winged aircraft.

The VE team was composed of four university students along with engineers from the aerospace company. The results and recommendations of this example are presented in some detail to illustrate the level of rigor of the typical value analysis/engineering synthesis and development phases.

Gearbox Function Analysis. While a complete function analysis was done during the study, only the key functions of the gearbox are described below:

Function	Description
Basic functions	
Reduce speed	The gearbox reduces the power turbine speed so that it is suitable for driving the helicopter rotor. In this case, the speed reduction is from approximately 40,000 to 6,000 rpm.
Accommodate accessories	Contained within the gearbox housing are drive gears for mounted accessories that operate at different speeds. These accessories include the alternator, starter, generator, fuel pump, oil pump, hydraulic pump, etc.
Supporting functions	
Transmit signals	The gearbox sends information to the cockpit regarding torque, oil levels, drive speeds, etc.
Detect failure	A gear failure is detected through a metal chip detector installed inside the gearbox housing. This function is for safety only and does not affect the speed-reduction functions.
Simplify maintenance	Appropriate positioning of components facilitates maintenance.
Ensure safety	The gearbox must meet aviation industry standards regarding safety.

The creation, synthesis, and development phases resulted in the proposal and study of many solutions, each of which concentrated on the prime objective of weight reduction through proper material selection and manufacturing methods. A summary of the recommended changes is given in Table 5.

Conclusions of the Helicopter Gearbox Study. The recommendations given in Table 5 considered different materials and casting methods, weight removal through material reduction in gear webs, and the use of an optical torque sensor. It was left to the company to select the most appropriate combination of options given the constraints of the final gearbox design.

A maximum weight saving of 5.2 kg (11.5 lb) with a corresponding cost reduction of $1594 was obtained with the following combination of recommendations:

Recommendation	Weight reduction	
	kg	lb
Magnesium sand casting	4.3	9.4
Cross-drilling of gear webs	0.7	1.6
Using an optical torque sensor	0.2	0.5

A maximum cost reduction of $3275 with a corresponding weight saving of 1.9 kg (4.1 lb) was obtained with the following combination of recommendations:

Recommendation	Cost saving, $
Aluminum investment casting	1275
Thinning of gear webs	0
Using an optical torque sensor	2000

Conclusions

The popularity of the VA process as a tool for design optimization has consistently risen since the early 1960s. Many organizations have applied some version of the process. Wherever the rigor of the process has been maintained, the results are commonly dramatic. This invariably results in the establishment of permanent, institutionalized VA programs.

ACKNOWLEDGMENTS

The author thanks Lucie Parrot, Valorex Inc. and Vince Thomson, McGill University for providing input to a revision of this article.

SELECTED REFERENCES

- A.J. Dell'Isola, *Value Engineering in the Construction Industry,* Construction Publishing, 1974
- C. Fallon, *Value Analysis,* Prentice Hall, 1980
- T.C. Fowler, *Value Analysis in Design,* Van Nostrand Reinhold, 1990
- L.D. Miles, *Techniques of Value Analysis and Value Engineering,* 3rd ed., McGraw-Hill, 1972
- A.E. Mudge, *Value Engineering: A Systematic Approach,* McGraw-Hill, 1971
- M.L. Shillito and D. DeMarle, *Value: Its Measurement, Design and Management,* Rochester Institute of Technology Press, 1991
- T. Snodgrass and M. Kasi, *Function Analysis: The Stepping Stones to Good Value,* University of Wisconsin, 1986

Use of Failure Analysis in Materials Selection

George F. Vander Voort, Buehler Ltd.

MATERIALS SELECTION for parts or components usually occurs under two conditions. The first is when a new part or component is designed, while the second is when an existing part or component needs to be upgraded. Upgrading may be needed due to a variety of reasons. If the part or component is relatively new, upgrading may be needed if part life proves to be marginal or inadequate. Or, a competitor's newly improved part is performing better and sales of your part are dropping. Or, operating conditions may have changed and your part is no longer performing acceptably. Of course, it is possible that parts have failed and the failure study recommended upgrading the material being used, perhaps also with other suggestions regarding the design and manufacturing processes.

Relationship between Failure Analysis and Materials Selection

Several failure analysts have categorized the nature of the failures that their organizations have studied over a long period of time (Ref 1–4). In most such summaries, few, if any, failures are attributed to the wrong material choice. The usual purpose of such surveys is to determine the frequency of failure mechanisms. A survey by Davies (Ref 4) is the exception because it considers mechanisms and causes of failures separately. Table 1 from this survey shows the frequency of failure mechanisms for failed engineering components and aircraft components. As expected, the percentages vary for different types of components. For example, none of the aircraft components failed by brittle fracture, while 16% of the engineering components experienced brittle failure.

While Table 1 is useful and interesting, it tells how components failed but not *why* they failed, which is really of greater importance. Table 2 lists the frequency of the causes of these failures. This information follows from an understanding of the failure mechanism and is vital for prevention of failures of other identical components in service or to be constructed. As for the failure mechanisms, the causes may be quite different for different types of components. Classification by cause will lead to decisions about the need for materials substitution and will provide guidance as to the characteristics that the substitute material must exhibit.

Dolan (Ref 5) described a broad-based failure analysis philosophy aimed at preventing future failures through proper materials selection, the basis of the design and reliance on safety factors, careful fabrication and maintenance with input from failure analysis studies, simulated service tests, and prototype evaluations.

Material selection involves more than merely selecting a particular composition or type of metal, alloy, polymer, ceramic, and so forth. For example, suppose that bearings or races used in a certain application are not lasting long enough. A study reveals that failures always initiated at relatively large, hard oxide inclusions. The composition of the steel used was found to be quite common for this application, but the quality of the material was quite ordinary. In this case, fatigue life could be improved dramatically by purchasing a premium quality level of the same steel. For a few of the more common alloy designations, a manufacturer might produce as many as five to ten different variations within the normal composition range, each designed for a certain balance of properties and quality level for specific applications.

When remedies are needed, the use of a better material may not be considered, or may be considered as a last resort. For example, a failure might be attributed to a problem such as stress-corrosion cracking (SCC). The analyst may have recommended steps to remove, or reduce, the chemical species from the environment that caused SCC, or perhaps the residual or applied stress could be reduced or the stress concentrator at the origin of failure could be avoided. It is also possible that the analyst may recommend switching to a different material that is not susceptible to SCC under the particular conditions. However, this solution inevitably raises the unit cost of the component (as will most solutions), but may not be considered even though a better material might, in the long run, be the lowest-cost, most-reliable solution.

A careful study of the reasons for selecting the original material, or quality level of the material (or both), for an existing product may reveal that the existing material was inadequate. In many cases, the manufacturer may realize that part performance could be substantially improved by changing the material being used, but is reluctant to do so. The larger the company, and the greater

Table 1 Frequency of failure mechanisms

	Percentage of failures	
Mechanism	Engineering components	Aircraft components
Corrosion	29	3
Fatigue	25	61
Brittle fracture	16	...
Overload	11	18
High-temperature corrosion	7	2
SCC/corrosion fatigue/HE	6	8
Creep	3	...
Stress rupture	...	1
Wear, abrasion/erosion	3	7

SCC, stress-corrosion cracking; HE, hydrogen embrittlement. Source: Ref 4

Table 2 Frequency of causes for failure

	Percentage of failures	
Cause	Engineering components	Aircraft components
Improper material selection	38	...
Fabrication imperfections	15	17
Faulty heat treatment	15	...
Design errors	11	16
Unanticipated service conditions	8	10
Uncontrolled environmental conditions	6	...
Inadequate inspection/quality control	5	...
Material mix	2	...
Inadequate maintenance	...	44
Defective material	...	7
Unknown	...	6

Source: Ref 4

the number of parts being made, the harder it is to change the material. The willingness to change is influenced by inventory problems, pricing differences, scheduling problems, document modifications, specifications, and so forth. Furthermore, the certainty that the switch will generate the required degree of performance improvement must be very high, and this is often difficult to predict.

Material selection is primarily performed by manufacturers. Not all university engineering programs offer courses in material selection. This is actually a rather difficult course to teach due to the need for extensive knowledge about the design of components and parts, their manufacture, environmental and usage effects, information not readily available to academics (and rarely published because of competitive issues). Indeed, only a few textbooks (e.g., Ref 6) exist on materials selection, and few academics are considered to be, or claim to be, experts on materials selection. Consequently, students generally get limited training in materials selection at universities. Considerable advances have been made in developing computer software to analyze stress and strain in structures and to predict fatigue life. This information is useful when selecting materials, but it is only part of the overall picture. Materials properties databases are another useful tool in considering alternate materials, although mechanical properties are only part of the overall picture.

The failure analysis methodology generally centers on cause and effect and ignores prevention. Failure studies often do not include recommendations to prevent similar failures in other components in service or to be constructed. In most cases, the analysis concerns only the cause of failure, that is, failure mechanisms, even when the manufacturer's experts perform the analysis. However, after the analysts have determined all of the factors responsible for the failure, they should determine what must be done to prevent similar failures in existing or future components. Prevention should be as important as diagnosis.

The Failure Analysis Process

The basic procedure for performing a failure analysis (Ref 7) involves the following general steps:

- Collecting background information on the component and the incident
- Macro- and microfractography (finding the origin when fracture is involved)
- Verification of materials and properties (compare with specifications)
- Microstructural examination
- Stress and life calculations
- Special measurements, for example, residual stress
- Simulations
- Analysis of the data
- Report preparation
- Follow-up on recommendations

A detailed treatment of the failure analysis process is outside the scope of this article, but can be found in *Failure Analysis and Prevention,* Volume 11 of *ASM Handbook* (Ref 8) and in several textbooks (Ref 9–12 are selected examples).

The failure analysis tends to be influenced strongly by the technical background of the analyst. Mechanical engineers evaluate the design thoroughly and calculate stresses, for example, but tend to ignore the material. On the other hand, metallurgists tend to concentrate heavily on the material and the manufacturing process and do a less thorough study of the stress situation and design. A balanced approach is necessary; this often requires the collaboration of two or more individuals with the required analytical strengths. The trick is knowing when to call in another person with the required engineering background.

Naturally, the primary focus of the failure analysis is the determination of the cause of failure. Indeed, this may be the sole charge given to the analyst. However, in the larger scheme of things, someone must view the reported failure mechanism, material properties, manufacturing procedure, service conditions, and so forth, and decide what must be done if other similar components could fail in the same way. If the failure was a unique case, the material may be quite adequate. However, if the failure indicates the potential for similar failures in other components, either in service or to be built, or if the percentage of failures is higher than acceptable (depending on the degree of risk or potential consequences), or if the component life span is less than desired or required, changes must be made. In any given situation there may be a number of potential ways in which a component can be improved; each must be evaluated as to its potential to cure the problem, its cost, and its potential to produce other problems (which can happen). One option may be the selection of a different material.

Fig. 1 Ductile fracture of a full section X60 grade line pipe tested at 56 °F (13 °C), which is 8 °F above its 50% shear-area DWTT

Fig. 2 Brittle fracture of a full-section X60 grade line pipe tested at −15 °F (−26 °C), well below its 50% shear-area DWTT

Methods for Analyzing Failures to Improve Materials Selection

To evaluate the value of a new, better material, it is necessary to define the required properties of the component. Manufacturing processes can alter the expected properties, as can the environment. Here, the failure study may reveal factors not initially considered, or incorrectly estimated. A host of properties must be considered, as well as their relevance to the part. One must remember that mechanical property tests pertain to idealized situations that have been standardized to reveal comparative information about the material. These data may not be fully relevant to the conditions that exist in the actual component. The uniaxial tension test specimen, toughness tests, such as the Charpy V-notch impact specimen, and hardness tests provide useful comparative data under the test conditions, but the data ranking of materials may not be the same if used in the actual component. Nevertheless, testing by standard procedures is very useful as full-scale testing of components is very difficult and expensive.

Besides these common tests, there are a myriad of other tests and characteristics that can be evaluated for materials under fixed conditions; for example, ductility, fatigue strength, fatigue crack growth rate, notch sensitivity, formability (drawability, stretchability, bendability, etc.), wear resistance (abrasion resistance, adhesion resistance, galling resistance, etc.), machinability, weldability, plus other tests relating to corrosion behavior, thermal expansion/contraction, electrical resistivity, magnetic permeability, and so forth. The failure analysis may show that one or more of these attributes are deficient, and the analyst should be asked to consider these problems, not merely determine the cause of failure.

Once the failure mechanism has been determined, it is important that contributing factors from the design, materials (and quality level), manufacturing and assembly processes, and service conditions be determined. In general, failure analysts tend to do a fairly good job evaluating design and service condition contributions. However, they may overlook the influence of the materials and manufacturing processes unless their assignment includes a specific request to make recommendations for dealing with similar components that may experience the same problems (to prevent failures) or recommendations on avoiding such problems with components to be manufactured. When recommendations are made, they should be carefully reviewed to ensure that new problems are not created by the recommended changes (as was not done in the precipitator wire problem discussed in Example 2 in this article).

Implementing Changes. Once all of the relevant information is assembled, the analyst is faced with the problem of informing those who can implement the required changes. This can be quite frustrating. Some analysts are hired to simply perform the analysis, and their input is not sought beyond the report generation phase. This is often true for independent laboratories and consultants. Their job is done when the report is written, or perhaps after they testify in court. They may not be requested to comment on corrective actions, and the person who hired them may not be concerned about this either.

If a manufacturer or supplier is performing the analysis, they will be more concerned with the work that follows the analysis than with the analysis itself because of their view of the larger picture. Years ago, there was greater resistance to interdepartmental criticism in organizations, even when totally constructive, and resistance to changes required to fix obvious problems with components. Today, most of these barriers have disappeared as companies try to be competitive and to improve products. "Getting the message out" to the required individuals to implement changes, be they in materials or whatever, used to be one of the most difficult tasks of the analyst, but this has changed dramatically over the past decade because of many factors (of which product liability cannot be ignored).

Historical Evolution of Improved Materials

While failures stimulate manufacturers to upgrade their product—through design changes, manufacturing process changes, or materials substitution—they also stimulate research to improve existing alloys and to develop better new alloys. One classic example of this process, the historical evolution of rail steels, is described in the article "Effects of Composition, Processing, and Structure on Properties of Irons and Steels" in this Volume.

Despite some notable historical catastrophes—such as the 15 Jan 1919 failure of a 90 ft diam riveted molasses tank in Boston killing 19 people (Ref 13–18) and the 14 March 1938 failure of the all-welded Vierendeel truss bridge near Hasselt, Belgium (Ref 19–21)—it was not until the failures of welded T-2 tankers and Liberty ships during World War II were analyzed that brittle fracture was demonstrated to occur in ordinary mild steels. This research (Ref 22–26) stimulated not only an understanding of brittle fracture and the measurement of toughness, but also extensive alloy development programs to develop stronger, tougher, weldable carbon- and low-alloy steels in all product forms. These alloys are being continually "tweaked" to further improve properties through advances in steelmaking technology (control of residual elements, gases, inclusion content, segregation, grain size, and microstructure). It is interesting to speculate that, had the RMS Titanic sunk on the night of 14–15 April 1912 in shallow waters so that studies could have discovered the extreme brittleness of the steel used, this failure might have initiated brittle fracture studies three decades earlier.

Failure Analysis Examples

Three examples are provided to illustrate the use of failure analysis in materials selection and materials development/refinement.

Example 1: Use of Failure Analysis Results in the Improvement of Line Pipe Steels. A superb example of continual product refinement, stimulated by product failures, concerns gas transmission line pipe steels. Brittle fracture of some line pipes occurred, and these fast, full-running fractures were spectacular demonstrations of a poor combination of materials, environment, manufacturing and installation problems, and loads (Ref 27, 28). Following previous research on brittle fracture, the initial efforts were to decrease the Charpy ductile-to-brittle transition temperature (DBTT). Other toughness tests, such as full-section drop-weight tests, were used in an effort to make the toughness tests more relevant. Indeed, fracture initiation tests were even performed on actual full lengths of pressurized line pipes (see Ref 29, 30, for example). To illustrate, Fig. 1 shows a full-size section of an X60 grade line pipe that was pressurized internally to 40% of its yield strength and tested at 56 °F (13 °C), 8 °F above its 50% shear-area drop weight transition temperature (DWTT). A 30-grain charge was detonated beneath an 18 in. (45.7 cm) notch cut into the pipe. The crack, moving at 279 ft/s (85 m/s), stopped after a short distance. The fracture was ductile, and the line pipe was tough enough at this temperature to arrest the crack. However, when another pipe was tested in similar fashion at −15 °F (−26 °C), 40 °F below its 50% shear-area DWTT, the crack moved at 2215 ft/s (675 m/s), the fracture was fully brittle, and the line pipe was not ductile enough to stop the crack. Figure 2 shows that the full length of the line pipe opened up in this test.

Despite substantial reductions in the DBTT of line pipe steels, line pipe failures still occurred; this time, however, they occurred under a ductile fracture mode (Ref 31). It was subsequently learned that the absorbed energy on the upper shelf of the Charpy energy-temperature curve was critical for arresting a moving crack. Hence, researchers concentrated on improving the upper-shelf energy. Meanwhile, the strength of these tougher alloys (hot rolled, non-heat treated) was also raised while designers increased the line pipe diameter.

In the 1960s, steelmaking technology was found to be inadequate as pipe sizes increased, stresses increased, and service temperatures decreased. Failure analysis revealed that brittle fracture was the culprit. However, application of the standard approach of reducing the ductile-to-brittle transition temperature of the line pipe steel

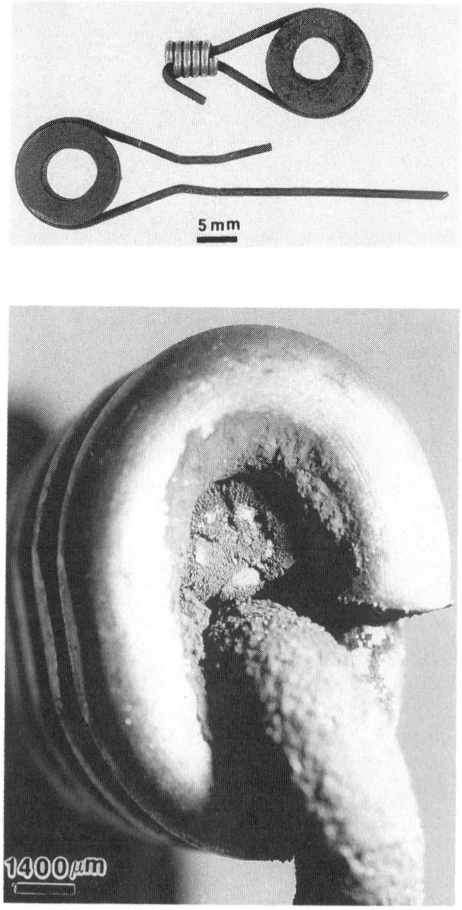

(b)

Fig. 3 Precipitator wires from a basic oxygen furnace. (a) Original AISI 1008 carbon steel wire, wrapped around an insulator spool and fastened with a ferrule made from type 430 ferritic stainless steel. One ferrule has been removed. (b) Close-up view showing the fractured wire face inside the ferrule merely changed the failure mode. Both fracture initiation and fracture propagation had to be controlled. Further studies showed that the upper shelf energy had to be improved in order to stop a crack, once initiated, from propagating. This work led to still further enhancements permitting development of weldable higher-strength, larger-diameter pipe but with satisfactory fracture control. Overall, the changes in steel composition were relatively minor, at least on a total weight percent basis. However, these changes, chiefly in residual gas content, sulfur content, and grain refining additives, required steelmaking technology enhancements. Improved steel processing procedures, chiefly hot-working temperature and deformation control, were also required to optimize microstructure and properties.

Example 2: Failure Analysis Leading to Improved Materials Selection for Precipitator Wires in a Basic Oxygen Furnace. This example of how a component failure stimulated materials selection concerns the failure of wires used in a wet precipitator for cleaning the gases coming off a basic oxygen furnace (BOF). The system consisted of six precipitators, three separate dual units, each composed of four zones. Each zone contained rows of wires suspended between parallel collector plates. The original 0.109 in. (2.77 mm) diam AISI 1008 (for zones 2 to 4 in each precipitator) carbon steel wire was cold drawn to a tensile strength of 100 ksi (690 MPa). One end was looped around an insulator spool at the top and fastened with a ferrule made from AISI 430 stainless steel. The top end of the wire is attached to the insulator on the framework while the bottom end of the wire was attached to a bottle-shaped weight. A potential was placed between the rows of wires and the interspersed parallel collector plates to remove the particulate matter from the gas.

Wires began failing about one year after start-up of the BOF shop. The frequency of failures varied in the different zones. Figure 3 shows an example of an unwrapped 1008 wire and two views of a failed wire. The maintenance metallurgist determined that the 1008 wires failed because of corrosion fatigue. It was decided to replace all of the wires in the two zones with the highest rates of failure with stainless steel wire. Type 304 austenitic stainless steel wire was chosen, and it was ordered cold drawn (150 ksi, or 1034 MPa, tensile strength), mainly because the 1008 wire was cold drawn.

Seven days after the more expensive type 304 wires were installed, the first failed. Thus, switching from inexpensive 1008 carbon steel wire to a much more expensive type 304 austenitic stainless steel wire changed the time to first failure from one year to one week! As a result, further failure analysis was requested.

For the new type 304 wires, ferrules were not used at the ends. Instead, 18 in. (45.7 cm) long cold-drawn 1010 carbon-steel tubes were used.

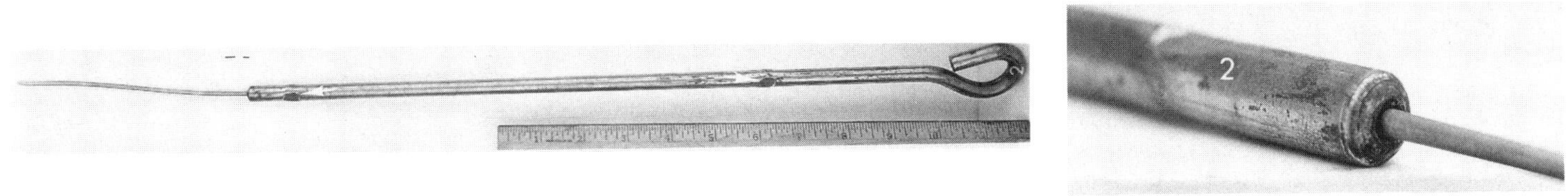

(a) (b)

Fig. 4 Replacement precipitator wires. (a) View of a type 304 replacement precipitator wire and the AISI 1010 tube bent at one end to place over the insulators. The arrows point to the two crimps used to fix the wire in the tube. (b) Close-up view of one of the crimps

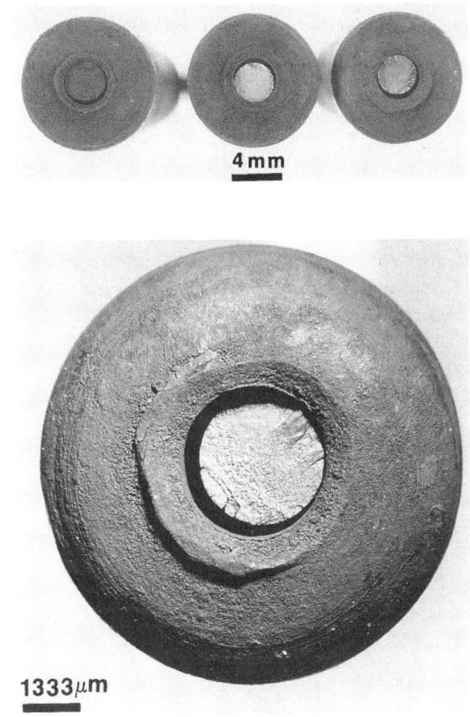

(b)

Fig. 5 Fractured replacement precipitator wires. (a) View of fractured type 304 precipitator wires. (b) Close-up view of one of the wires. Note the deformation at the inside diameter of the tube due to the motion of the wire.

Fig. 6 A 75 in. Morbark total tree harvester

Fig. 7 Chipper knife being installed in a 75 in. Morbark total tree harvester

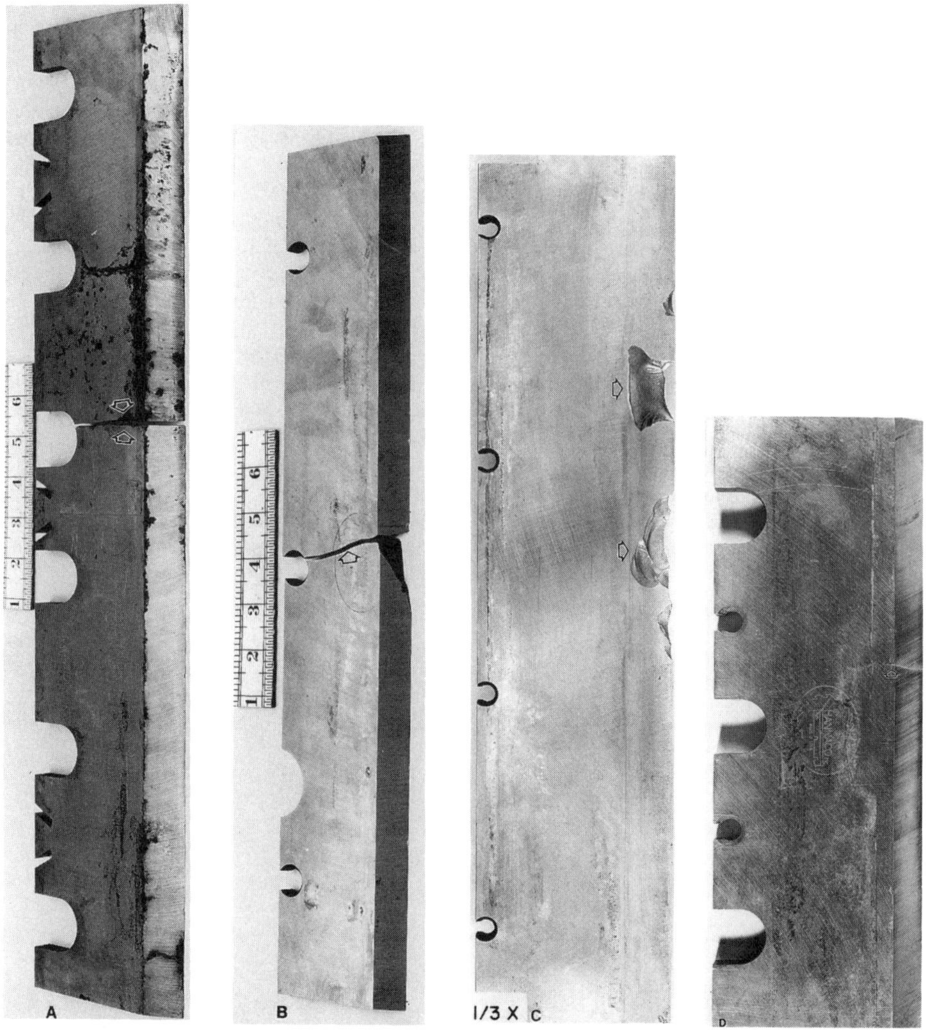

Fig. 8 Examples of a few of the different types of failed chipper knives examined. Arrows point to fractures. (a) Knife from a 75 in. Morbark total tree harvester. (b) Knife from a 66 in. CM&E sawmill chipper. (c) Knife from a Carthage Norman chipper. (d) Knife from a 96 in. Bush chipper

The type 304 wire was inserted into one end of the tube, and the tube was crimped in two places to fix the wire. The other end of the tube was bent so that it could be looped over the insulator on the frame, as shown in Fig. 4. The wires broke at the apex of the tube or just slightly below it, as shown in Fig. 5.

The new type 304 stainless steel wires failed by transgranular SCC. The plant metallurgist was either unaware of the potential for SCC or did not realize that the hot gas from the BOF was cooled with river water before entering the precipitators. Because the river water goes through the plant piping system it was treated three times daily with chlorine to prevent algae growth in condensers, pipes, and so forth. The air temperature within the precipitator varied from about 425 to 200 °F (218 to 93 °C). Thus, the new attachment method produced an ideal stress concentrator, the wire had a high level of residual stresses, the gas contained chlorine ions and the temperature was above 200 °F—ideal conditions for chloride ion SCC of type 304.

Furthermore, the framework holding the wires was vibrated periodically at 60 Hz to dislodge particulate matter from the wires. The bottle weight produced a tensile load of slightly less than 2 ksi (14 MPa). During operation, both ends of the wires were constrained so that the force of the wind through the precipitator concentrated stresses at the tube opening. Calculation of the natural frequency of the wire showed that its third harmonic was almost exactly 60 Hz. Consequently, it was felt that this could be another source of stresses.

To prevent future problems, either from corrosion fatigue or SCC, wires were made from annealed type 430 ferritic stainless steel, which is immune from chloride ion SCC. Attachment was made using type 430 ferrules. The wire diameter and bottle weight were changed so that the natural frequency was about 90 Hz. Over the next few years, most of the precipitator zones (2 to 4, zone 1 used barb wire) were restrung with type 430 wires, and they performed for more than twenty years (except for failures due to electrical short circuits) until the BOF shop was closed down recently.

This is a good example of the problems that can occur in practical failure analyses work. The original carbon steel wire was deemed to be inadequate for the application. The metallurgist decided to replace these wires with a more corrosion-resistant alloy, but did not realize that his choice, under the operating conditions, was unwise. The more expensive type 304 stainless steel wire failed catastrophically after only a week of service time. Use of a ferritic stainless steel solved the problem, but other solutions were also possible.

Example 3: Failure of Chipper Knives. This example describes how a steel research engineer used failure analysis results to identify the need for a better alloy. In this case, a new alloy had to be developed because no suitable composition existed. This example concerns the development of an alloy steel for knives used to chip logs, either hardwoods or softwoods. Traditionally, chipping of logs for making paper, cardboard, or particle board took place in a paper mill using an alloy with a nominal composition of Fe-0.48C-0.30Mn-0.90Si-8.50Cr-1.35Mo-1.20W-0.30V.

This alloy performed well in pulp mill applications where the knives are typically 20 to 30 in. (51 to 76 cm) long, from about 5 to 7 in. (13 to 18 cm) wide, and usually 0.5 to 0.625 in. (13 to 16 mm) thick. One of the long edges is beveled and sharpened.

However, with the development of total tree harvesters, where the chipper is taken out in the field to chip the logs (see Fig. 6), knife failures were frequently encountered due to the less rigid support of the knives and because fewer knives are used (Fig. 7) than in a pulp mill. Typical examples of failed knives are shown in Fig. 8; Fig. 9 shows close-ups of edge damage.

In this case, a tougher alloy steel was needed that would still exhibit all of the other necessary characteristics of a knife steel, for example, edge

Table 3 Results of chipper knife field trials

See Example 3 in text.

Knife material	Average results per run: Run time, min	No. boxes of chips	Resharpening time, min	Stock removal, in.	Stock removal, mm	Production per in. of knife: No. of runs	Run time, min	No. boxes of chips
New	75.0	1.67	29.0	0.036	0.91	27.78	2083.3	46.4
Old A	58.75	1.19	34.4	0.057	1.45	17.54	1030.7	20.9
Old B	53.75	1.17	32.8	0.055	1.40	18.18	977.0	21.37

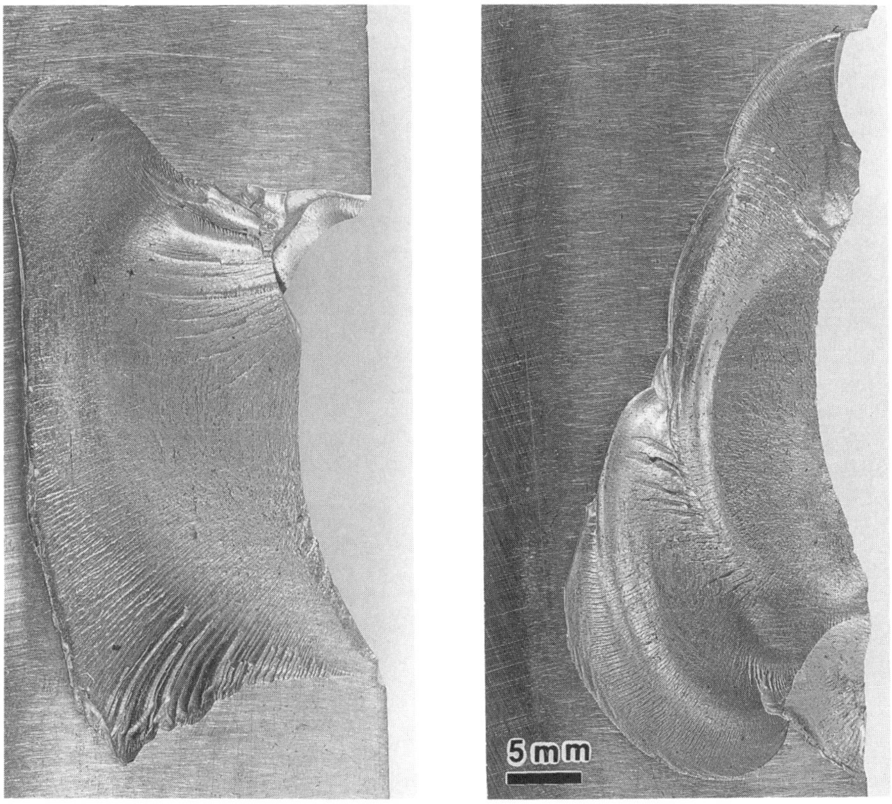

Fig. 9 Close-up of the fracture on the Carthage Norman chipper knife shown in Fig. 8(c)

retention, resistance to softening under frictional heating, wear resistance, ease of heat treatment, dimensional stability in heat treatment, grindability, low alloy cost, and so forth.

The failure analysis revealed longitudinal Charpy V-notch (room temperature) impact strengths of 3 to 5 ft · lbf (4 to 6.8 J), hardnesses from 56 to 59 HRC, and retained austenite contents varying from a trace to 8% in failed knives. The alloy design program was established to produce a lower-cost composition (based on the cost to add various alloying elements) with significantly higher toughness, which is easily heat treated and yields a hardness of at least 58 HRC with as high a tempering temperature as possible. Furthermore, tempering must produce high dimensional stability.

The program developed the nominal composition of Fe-0.50C-0.30Mn-0.40Si-5.00Cr-2.00Mo (Ref 32, 33), which achieved the goals described above. Optimal austenitizing temperature was 1850 °F (1010 °C), and either air or oil quenching could be used. Many manufacturers oil quench using a press to maintain flatness. The optimal tempering temperature was 980 °F (527 °C), followed by a second temper at 940 °F (504 °C), yielding hardness of 59 HRC and absorbed energy of 6.5 to 7.5 ft · lbf (8.8 to 10.2 J) on longitudinal, room-temperature Charpy V-notch specimens. With this tempering practice, all residual retained austenite was eliminated.

Knives were made from the above composition, and blind trials were conducted using a 75 in. Morbark Chip Harvester (Fig. 6 and 7). Knives were also tested from steel of the old composition using two different sources. Knives (B) of the old composition were made by the same company that made the knives from the new composition, while old knives A were made by a competitive knife manufacturer using the former composition. Table 3 presents the results from field trials conducted by a knife user. It is quite obvious that the new composition outperformed the standard composition (and A and B performed similarly) in all areas. Furthermore, in this and in further usage none of the knives made from the new composition broke in service.

This example is typical of many such studies conducted by researchers working for steel and specialty alloy manufacturers. A problem was observed through contacts with manufacturers of knife blades. Knives from a variety of pulp mill, portable tree harvesters, and saw mills were studied to characterize the steel and the reason for their failures. These data then served as the basis for an alloy development program where all of the relevant parameters were evaluated to develop an optimal composition. This was followed by trials in which a knife manufacturer made knives from trial compositions and then evaluated their performance in the field. This information was used to refine the final composition, which was evaluated in blind trials by a disinterested party. The key to the development was a careful study of a number of failed knives with different problems, but chiefly gross fracture, used in different types of operations.

Conclusions

When failures occur in existing products or prototypes, study of the failures provides valuable information to guide the materials selection process, at least in those cases where the material used appears to be part of the problem. Many failures, of course, are not caused by the use of inadequate materials, but a significant percentage of failures do result from use of materials that may not be optimal for the application. Published failure analyses generally concentrate on failure modes and mechanisms and do not always consider if the best material is being used. However, by examining the reasons for the failure and the role of the chosen materials, the appropriateness of the chosen materials can be determined. The analyst should always consider what steps must be taken to prevent such failures in other similar components. The materials development engineers need to be sensitive to the deficiencies of materials under certain operating/environmental conditions. Only by recognizing needs can new and better materials be developed. The materials selection engineer needs to be sensitive to deficiencies in product performance in order to make materials substitutions in a timely manner.

REFERENCES

1. R.K. Penny, Failure Types, Consequences and Possible Remedies, *Int. J. Pressure Vessels Piping*, Vol 61, 1995, p 199–211
2. F.R. Hutchings, The Laboratory Examination of Service Failures, *British Engine Tech. Report*, New Series, Vol III, British Engine Boiler and Electrical Insurance Company, Manchester, England, 1964, p 174–219
3. R.W. Wilson, Diagnosis of Engineering Failures, *Br. Corros. J.*, No. 3, 1974, p 134–146
4. G.J. Davies, Performance in Service, *Essential Metallurgy for Engineers*, E.J. Bradbury, Ed.,

Van Nostrand Reinhold, London, 1985, p 126–155

5. T.J. Dolan, Preclude Failure: A Philosophy for Materials Selection and Simulated Service Testing, *Exp. Mech.*, Jan 1970, p 1–14
6. M.F. Ashby, *Materials Selection in Mechanical Design,* Pergamon Press, 1992
7. G.F. Vander Voort, Conducting the Failure Examination, *Met. Eng. Q.*, Vol 15 (No. 2), May 1975, p 31–36
8. *Failure Analysis and Prevention,* Vol 11, *ASM Handbook* (formerly *Metals Handbook,* 9th ed.), American Society for Metals, 1986
9. V.J. Colangelo and F.A. Heiser, *Analysis of Metallurgical Failures,* John Wiley & Sons, 1974
10. R.D. Barer and B.F. Peters, *Why Metals Fail,* Gordon and Breach Science Publishers, 1970
11. D.J. Wulpi, *Understanding How Components Fail,* American Society for Metals, 1985
12. C.R. Brooks and A. Choudhury, *Metallurgical Failure Analysis,* McGraw-Hill, 1993
13. Disastrous Explosion of a Tank of Molasses, *Sci. Am.*, Vol 120 (No. 5), 1 Feb 1919, p 99
14. Bursting of Molasses Tank in Boston Charged to Bad Design, *Eng. News-Rec.*, Vol 82 (No. 7), 13 Feb 1919, p 353
15. B.S. Brown, Details of the Failure of a 90-Foot Molasses Tank, *Eng. News-Rec.*, Vol 82 (No. 20), 15 May 1919, p 974–976
16. Boston Molasses-Tank Trial: The Case for the Defense, *Eng. News-Rec.*, Vol 85 (No. 15), 7 Oct 1920, p 691–692
17. Experts Deny Bomb Caused Collapse of Boston Molasses Tank, *Eng. News-Rec.*, Vol 87 (No. 9), 1 Sept 1921, p 372–373
18. Bursting of Boston Molasses Tank Found Due to Overstress, *Eng. News-Rec.*, Vol 94 (No. 5), 29 Jan 1925, p 188–189
19. Welded Bridge Failure in Belgium, *Eng. News-Rec.*, Vol 120 (No. 18), 5 May 1938, p 654–655
20. O. Bondy, Brittle Steel a Feature of Belgian Bridge Failure, *Eng. News-Rec.*, Vol 121 (No. 7), 18 August 1938, p 204–206
21. A.M. Portevin, Collapse of the Hasselt Bridge, *Met. Prog.*, Vol 35 (No. 5), May 1939, p 491–492
22. C.F. Tipper, *The Brittle Fracture Story,* Cambridge University Press, 1962
23. M.L. Williams and G.A. Ellinger, Investigation of Structural Failures of Welded Ships, *Weld. J., Res. Suppl.*, Vol 32, Oct 1953, p 498s–537s
24. H.G. Acker, Review of Welded Ship Failures, *Welding Res. Council Bull. Ser.,* No. 19, Nov 1954
25. J. Hodgson and G.M. Boyd, Brittle Fracture in Welded Ships, *Inst. Naval Arch. Q. Trans.*, Vol 100 (No. 3), July 1958, p 141–180
26. M.L. Williams, Correlation of Metallurgical Properties and Service Performance of Steel Plates from Fractured Ships, *Weld. J., Res. Suppl.*, Vol 37, Oct 1958, p 445s–454s
27. G.D. Fearnehough, Fracture Propagation Control in Gas Pipelines: A Survey of Relevant Studies, *Int. J. Pressure Vessels Piping,* Vol 2, 1974, p 257–281
28. J.E. Hood, Fracture of Steel Pipelines, *Int. J. Pressure Vessels Piping*, Vol 2, 1974, p 165–178
29. J.B. Cornish and J.E. Scott, Fracture Study of Gas Transmission Line Pipe, *Mechanical Working & Steel Processing Conf.*, Vol VII, American Institute of Mining, Metallurgical, and Petroleum Engineers, 1969, p 222–239
30. J.F. Kiefner, W.A. Maxey, R.J. Eiber, and A.R. Duffy, Failure Stress Levels of Flaws in Pressurized Cylinders, *Progress in Flaw Growth and Fracture Toughness Testing,* STP 536, ASTM, 1973, p 461–481
31. J.E. Hood and R.M. Jamieson, Ductile Fracture in Large-Diameter Pipe, *J. Iron Steel Inst.*, Vol 211, May 1973, p 369–373
32. G.F. Vander Voort, "Steel Composition for Chipper Knife," U.S. Patent 4,353,743, 12 Oct 1982
33. G.F. Vander Voort, "Method of Heat Treating a Steel Composition for Chipper Knife," U.S. Patent 4,353,756, 12 Oct 1982

Section 5: Effects of Composition, Processing, and Structure on Materials Properties

Introduction

Richard W. Heckel, Michigan Technological University

MATERIALS are selected and used as a result of a match between their properties and the needs dictated by the intended application. In this sense, the properties should be defined broadly and include fabricability, required maintenance, costs associated with the original materials and maintenance, and the behavioral characteristics associated with the function of the item within the overall design (Ref 1). Materials are classified as metals (alloys), ceramics, polymers (plastics), and composites; they are either naturally occurring (e.g., wood, rock, and leather) or manufactured (e.g., organic fibers, aluminum alloys, industrial diamonds, and graphite-reinforced plastics). The rapid increase in the variety of manufactured materials in the last half century and the ranges of material properties thereby available have played a major role in substantial technological progress. In this Section of the Handbook, emphasis is placed on structural materials, that is, those materials whose applications involve primarily their mechanical behavior (e.g., stiffness, strength, ductility, toughness, and hardness) rather than their physical properties (e.g., electrical, thermal, magnetic, and optical behavior). Degradation behavior of structural materials (e.g., corrosion, oxidation, and radiation effects) is also considered where appropriate for specific materials and service conditions.

Composition and Structure Determine Properties

The contemporary view of materials holds that properties are determined by composition and internal structure, with the latter being the result of the processing of a given composition. Simplistically, it is not just the types and amounts of various atoms or ions that constitute the material that alone determine its properties. The arrangements of the atoms or ions and the various types of defects (deviations from perfect atomic arrays) that exist within these arrangements contribute in a major way and positively to the properties of the material. It is the processing of the material during the various stages of its manufacture that determines the atom or ion arrangement and defect structures. Thus, processing parameters control the structure of a given assemblage of atoms or ions; this internal microstructure, along with composition, determines the properties of the material.

The importance of the composition of a material in influencing its properties is readily acceptable to everyone. "What's in it" is important to the taste of our food; the same principle should be applicable (and is) to the properties of materials. The properties of aluminum oxide are expected to differ from those of copper, polyethylene, and graphite because it is known that their compositions vary widely. It is less obvious that internal structural features are important in the determination of properties as well. In fact, a half-century ago it was not uncommon to hear from practitioners that metallic alloys failed by "crystallizing," a statement that implicitly assumed that alloys in service were amorphous and lacked crystalline structure. (It is unfortunate that "failure by crystallization" may yet be heard on rare occasions.) However, experience with food shows that material structure is important (Ref 2). Chefs and lesser cooks know that the texture of their raw materials and the sizes of various components in their recipes must be controlled. Pasta has a wide range of geometric forms, soups contain solids of specific sizes and shapes, bones in broiled fish ruin the entree, and salad dressing can be formulated with two immiscible phases (olive oil and vinegar) to provide two solvents into which a variety of different types of flavorings can be dissolved. Structural features, in addition to composition, are just as important to engineering materials as they are to the preparation of culinary delights even though the raw materials and methods of processing vary substantially.

Most people have experienced examples of the dependence of internal structure on the behavior of materials, but may not have recognized this relationship. A common example is the use of a steel wire coat hanger as the "raw material" in the conversion by hand to another useful form. The bends in the coat hanger that were formed during its factory processing into a coat-hanger shape are difficult to straighten, whereas the already straight sections will readily deform as forces are applied. Clearly, the bent sections are strongest, a fact which indicates that something other than composition controls properties because the coat-hanger wire is of uniform composition (essentially iron with small amounts of carbon, manganese, and silicon). It is now well established that the deformation of the wire during the original production of the coat hanger introduces structural defects (line defects called dislocations) into the otherwise uniform arrangement of the atoms (a body-centered cubic crystalline array for this alloy). These defects increase strength in the regions of the bends, causing the difficulties associated with the straightening of the coat hanger.

A corollary to the above example is the incomplete prediction of properties that usually occurs when composition alone is specified. Chemical analysis of a sample of glass-reinforced plastic composite material (fiberglass) would provide information on the amounts of carbon, hydrogen, oxygen, and silicon plus some additional minor elements in the material. However, a block of material constituted from these elements in the amounts found could behave in a manner entirely apart from glass-reinforced plastic. The keys to the remarkable mechanical behavior of glass-reinforced plastic are the arrangement of the glass fibers in the continuous plastic matrix, the properties of the glass and plastic (due to their individual compositions and structures), and the bond at the interface between the glass and plastic. Thus, the overall composition of the composite, the composition found by the bulk chemical analysis, is determined by the compositions of the individual phases, glass and plastic, and their volume fractions in the composite. The overall composition gives no information on the geometrical features of the glass fibers, their arrangement in the plastic matrix, and the individual compositions of the fiber and matrix phases, all of which contribute to the properties of the composite.

Elements of Structure

Full appreciation of the effects of structure on material properties is impeded by the wide variety of structural elements, their size, and the effects of processing on each of them. In addition, there are a large number of individual properties of interest with each of them being determined typically by more than one structural element. In most instances, more than one property is critical, leading to a large number of critical structural

Dimension, cm	Representative elements	Examination method detectability
10^2	Engineering structures	
	Mechanical test specimens	Human eye radiography ultrasonics
1	Elastic-plastic zones	
	High-strength filament diameter	Penetrants
10^{-2}	Alloy microstructure grain diameter	Optical microscope
10^{-4}	Polymer chain length	Acoustic emission
	Precipitate phases in alloys	Microprobe
10^{-6}	Dislocation spacing	Electron microscope Electropotential measurements
10^{-8}	Crystal lattice spacing Atomic diameter	Field ion microscope

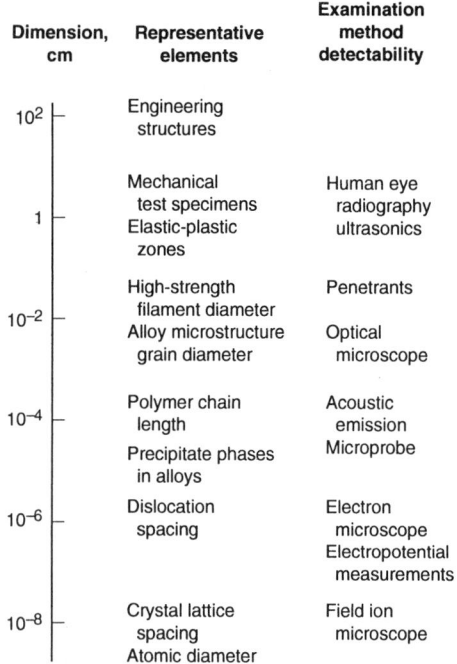

Fig. 1 Comparison of the sizes of microstructural elements and the types of observation techniques. Source: Ref 1

elements to be controlled by processing. Not uncommonly, when several properties are specified, structure/property conflicts occur and the specifications have to be reconsidered because of technological limitations in processing to achieve desired structures and/or the mutual exclusion of specific property combinations.

The common structural elements that are most important in materials include (Ref 3–5):

- Type of atom-to-atom (or ion-to-ion) bonding (primary metallic, ionic and covalent, and secondary)
- Type of interatomic (or interionic) packing: crystalline (crystal structures), amorphous (dimensionality, for example, one-dimensional chains, three-dimensional networks), or mixed (nature of crystalline and amorphous regions and fractions of each)
- Alloying atoms (or ions); these are sometimes classified as "point defects." Crystalline structures either replace host atoms on their crystalline lattice sites (substitutional atoms) or fit into the spaces between host atoms (interstitial atoms). Amorphous structures are atom groups or single atoms substituted into the one-dimensional or three-dimensional arrangements or between adjacent one-dimensional chains causing enhanced interchain bonding.
- Defects in crystalline materials. Point defects are vacant atom or ion positions (the local charge imbalance in the case of an ionic vacancy is compensated by an additional defect). Line defects, or dislocations, are disruptions (displacement of atoms) in the lattice that occur along a single dimension and extend about ten or so atom (or ion) diameters from the core (line) of the defect. Surface defects, or interfaces, are two-dimensional defects between two adjacent crystals of different spatial orientation (defects are termed grain boundaries when the crystals have the same crystal structure as in a two-phase material); free surfaces may be added to this category because they are the termination of the crystal lattices of grains at the surface of the material.
- Microstructure is the arrangement in a material of grains (polycrystalline materials) and discrete regions having different atomic and defect structures; regions with different atomic structure are typically called phases (multiphase materials); often multiphase materials have phases with different chemical compositions as well as different atomic structures. The microstructure is classified by its morphology: the size, shape, amount, and distribution of the discrete regions.
- Macrostructure is, typically, the features of a material that can be observed at very low magnification (or none) such as notches, cracks, and large-scale composition variations (e.g., compositional inhomogeneities in large castings, corrosion products on the surface of a material, and weld-fusion zones).

The dimensions of the defects listed above range from atomic through component size or about ten orders of magnitude. The difficulties associated with the simultaneous visualization of such a range of dimensions further complicates the application of structural concepts in understanding the nature and controlling the behavior of engineering materials using processing. For example, "defects" are universally accepted to be flaws when observing merchandise in stores or showrooms; from a materials structure standpoint, such flaws would be considered as "macrostructure." Defects in the structures of materials, on the other hand, because they may approach atomic or ionic dimensions, are many orders of magnitude smaller in size than what are normally referred to as defects in merchandise and can often confer benefits to the properties.

The small size of many of the critical structural elements has greatly impeded the understanding of composition/structure/processing relationships. Figure 1 (Ref 1) provides a comparison of the sizes of various structural elements in materials along with applicable techniques for their observation. Observation and understanding had to await the development of instrumentation with the resolving power necessary for the study of structure and to correlate such studies with composition and processing conditions and parameters.

Historical Perspective

The Various Materials Ages. Materials have played a dominant role in the continued development of civilization and significant advances have coincided with the development of new materials. Westbrook has provided several useful and detailed reviews (Ref 6–8) of the evolution of various types of materials and their processing over the period of the last 12,000 years. Clearly, a strong linkage has existed and continues to exist between advances in materials and civilization. The evolution of materials over the ages has been summarized by Ashby (Ref 9) as depicted in Fig. 2. For about the first 9000 years of recorded history, essentially all structural materials were naturally occurring. Processing was minimal, and some shaping was possible. Even the copper metal used early in this period was "native copper," a material of high purity found at just a few locations on Earth (e.g., the Upper Peninsula of Michigan, Ref 10) as metallic copper and not as a compound such as copper sulfide that would require extensive refining to yield a useful metal. From the beginning of the Iron Age in 1000 B.C. through most of the 19th century, processing and raw materials development resulted in the advances from iron to cast iron to steels (Ref 11). Major effort was directed toward the development of ferrous production capacity (Ref 12). The transition to steels was accelerated by the expansion of railroads and the industrial revolution. Alloy steels followed in the late 19th and early 20th centuries and were joined by nonferrous alloys based on aluminum, copper (now beginning a growing dependence on sulfide ores rather than "native copper"), nickel, and so forth, in the rapidly growing array of structural materials. As the outset of World War II approached, metallic alloys dominated the overall spectrum of available structural materials in terms of importance to industrial and societal requirements.

The twenty-year period beginning with the onset of World War II clearly marked a transition for materials manufacture, availability, and application as shown in Fig. 2. The trends initiated during this period have continued to the present time (Ref 13). Certainly, the wartime economy of the 1940s and subsequent military requirements greatly accelerated the development of elastomers (artificial rubber) and plastic materials, leading to a wide range of useful properties in these types of materials and production capacities that presently exceed that of steel in the U.S. economy. In addition, the wartime economy expanded the ranges of metallic alloys and their properties. Higher strength and toughness, especially with lower-density and/or higher-temperature capability were in demand and soon provided well-defined needs for the composites segment of the materials field. Ceramic materials were developed in part because of composite materials needs and subsequently have become important in coatings and monolithic components.

More recent societal interests and problems have continued the development and application of materials having increased technological sophistication (traditional materials with improved reliability and new materials with enhanced performance). The new imperatives of increased fuel efficiency, decreased environmental damage, and increased recycling have placed increased emphasis on higher strength-to-weight ratios, higher engine operating temperatures, increased component lifetime, and systems allowing efficient recycling of materials. Furthermore, these advances, to be effective, have carried minimal cost penalties. Thus, thrusts in the development of metals, plastics, composites, and ceramics that began during the 1940 to 1960 transition continue today, driven by expanded sets of societal and industrial goals.

It is meaningful to reflect on the underlying basis for the 12,000 years of materials development depicted by Ashby (Ref 9) in Fig. 2. Following the first 9000 years where practically all materials were naturally occurring (with a minimum amount of shaping), new structural materials became the products of artisans who refined them and then formed them into useful shapes. These growing numbers of artificial materials required carefully selected raw materials and carefully followed recipes for processing because production quality control of chemical composition awaited the 19th century and the understanding of metallic alloy structure was in only rudimentary form in the early 20th century. Arguably, it was (1) the appreciation of the effects of structure, as well as composition, that began to unfold for metals and their alloys in the early 20th century and (2) the recognition in the mid-20th century that processing/structure/property relationships would be important in other types of materials that have been the underlying support for the materials developments to meet industrial and societal needs of the past half-century.

Historians have recognized the importance of materials through the naming of the Stone (prehistoric; prior to 10,000 B.C.), Bronze (4000 to 1000 B.C.), and Iron (1000 B.C. to A.D. 1620) Ages. Another perspective comes from observing the various stages of materials development over the 12,000 years covered by Fig. 2:

- Up to 1000 B.C.: Materials typically used in the "as-found" condition.
- 1000 B.C. to A.D. 1800: Development of new materials where processing altered the composition of the raw material, size was limited, lack of composition and structure analysis limited understanding, quality control based on constant sources of raw materials, and empirically determined processing procedures.
- 1800 to 1880: Development of process mechanization and production quantities (especially in the ferrous industry) and improved compositional control.
- 1880 to 1950: Continuation of mechanization, production capacity development, compositional control; development of new metallic alloys; importance of the structural elements of materials (primarily for metals) began to unfold; and interest in nonmetallic structural materials developed.
- 1950 to present: Emphasis placed on low-cost production of high-quality materials, unified view of materials emerged, importance of structural elements in all classes of materials recognized, new instrumentation providing high-resolution observation of structure developed, detailed understanding of structure/property relationships resulted, and the design *of* materials evolves in partnership with design *with* materials.

Structure and Structure Analysis. Materials science and engineering is certainly an old field and a new one simultaneously. The recognition of the importance of structure in determination of properties is the hallmark of the modern developments in the field. The speculations about the relationships between properties and possible structure changes brought on by processing variations have given way to direct observations of structures developed through controlled processing. A brief history of the development of microstructure analysis techniques is in order. Those interested in detailed treatments should read the presentations of Mehl (Ref 14, 15) and a more recent symposium proceedings that covers advanced microstructural analysis techniques (Ref 16).

Interest in the mechanical behavior of materials preceded an understanding of the causes underlying the variations in material properties. Hooke's law and nonelastic behavior were studied in the 17th century. The mechanics of materials received concentrated attention in the early 1800s by Young, Navier, Cauchy, and Poisson. Mechanical property studies began at this time, and tension test machines were used as early as 1837. Fatigue behavior (*S-N* curves) was studied in the 1860s and mechanical hysteresis (Bauschinger effect) was studied in the 1880s. All of this work on the mechanical behavior of materials predated analysis of the internal structure of materials. Readers desiring detailed information on the subject of mechanical testing may consult Ref 17 and 18.

The modern era of structure analysis had its beginnings in the work of H.C. Sorby (Ref 19, 20), a petrographer at the University of Sheffield, who became interested in the microstructures of metals in the early 1860s. His work necessitated

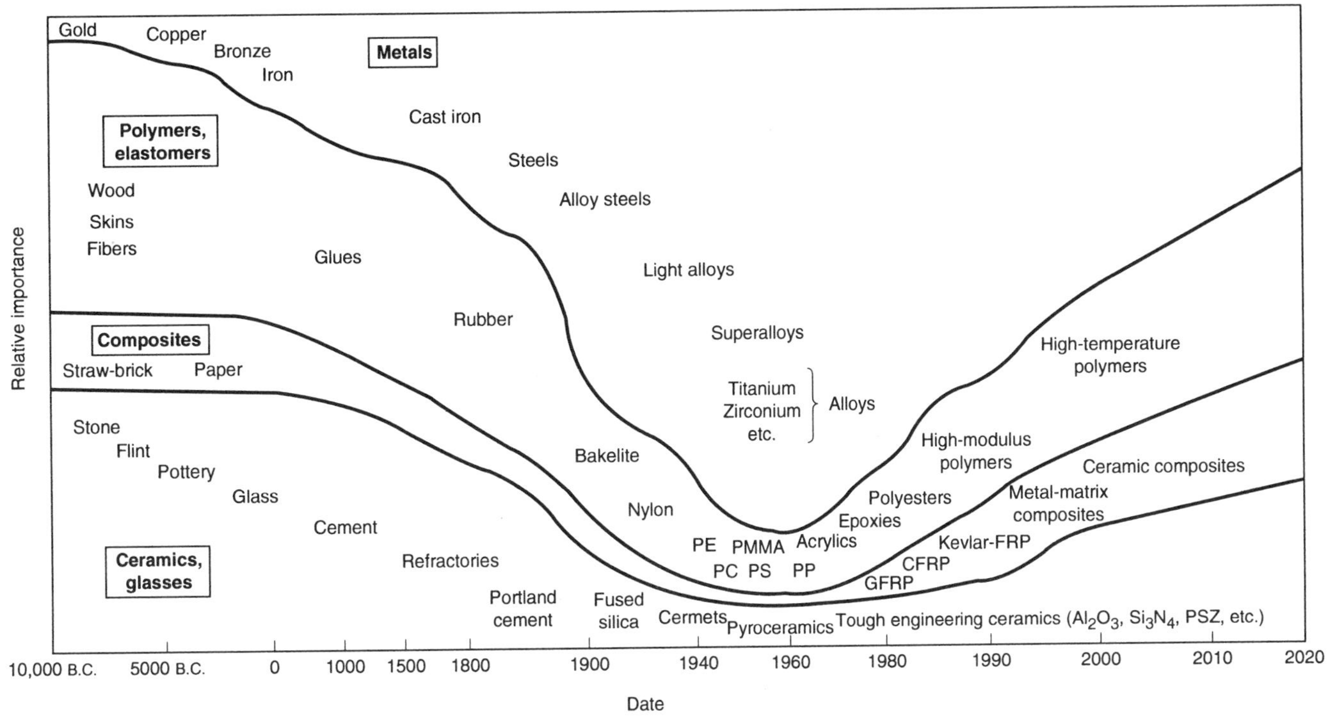

Fig. 2 The evolution of engineering materials. PE, polyethylene; PMMA, polymethylmethacrylate; PC, polycarbonate; PS, polystyrene; PP, polypropylene; CFRP, carbon-fiber-reinforced plastic; GFRP, graphite-fiber-reinforced plastic; PSZ, partially stabilized zirconia. Source: Ref 9

the development of a microscope using reflected light because the transmitted light techniques used for mineral specimens (thin sections) were not appropriate for metals. Sorby studied a piece of steel produced by the now-abandoned Bessemer process and a piece of cast steel. He developed polishing and acid-etching techniques, the latter to provide surface roughening due to preferential attack of certain microstructural features, allowing them to be observed using reflected light at magnifications of 650×. His interest in these microstructures stemmed from his prior interest in metallic meteorites wherein the microstructure was observable (after polishing and etching) with the unaided eye because the microstructural features were coarse. Sorby observed the "pearly" microconstituent in steel (later named pearlite) and correctly concluded it was composed of two phases, essentially pure iron and a hard compound of iron and carbon (later named cementite or iron carbide, Fe_3C). He went on to study quenched steel where he made the seemingly contradictory, but technologically important, observation of the absence of a hard compound phase in spite of the hard condition of the steel. He also studied annealing following cold working and intragranular and transgranular fracture.

Metallography, the subject launched by Sorby's work, is currently the most widely used investigative technique for revealing the microstructures of materials. It is, therefore, remarkable that he spent only about two years in the early 1860s and three months in the mid-1880s studying metals. In fact, the brief work in the 1880s was in preparation for the two presentations to the Iron and Steel Institute in London in 1885, about two decades after his original work (Ref 21). Without these presentations, his subsequent publications (Ref 19, 20), crucial to the subsequent development of a fundamental understanding of materials, may have never been written.

Interest in structural studies using a variety of techniques immediately followed the metallographic work of Sorby. Much of the early work on structure analysis was carried out in Europe; U.S. interests in materials throughout the late 19th and early 20th centuries were directed mainly toward increased steelmaking capacity. From 1890 through 1910, alloy phase equilibrium (phase diagram determination and use) received much attention due, no doubt, as much to the need for practical information on commercial alloys as to the fundamental work of Sorby (Ref 19, 20) and Gibbs (Ref 22). The demonstration by Laue, Friedrich, and Knipping (Ref 23) in 1912 that x-rays would diffract from crystals led to the development and application of a variety of x-ray techniques for crystal structure analysis. The body-centered and face-centered cubic allotropes of iron and the body-centered tetragonal hard phase formed by quenching steel (phase now known as martensite) were determined during the 1920s and elegant studies of crystal structure and lattice parameter variations were undertaken in the 1930s.

Research in the 1930s in the United States provided a new and comprehensive view of steel heat treatment. The understanding of the kinetics of microstructural change benefited significantly (both scientifically and technologically) from the isothermal transformation studies of steels by Davenport and Bain (Ref 24). The presentation of data in the form of isothermal transformation diagrams and the relationships between their findings and the hardenability of steels have scarcely changed since their work. Steel heat treatment, long practiced and seldom understood, had at last been provided with a rational foundation.

Research in the late 1930s contributed insight to the phenomenon of age hardening, which was originally uncovered by Wilm in 1906 in an aluminum alloy. Age hardening could be achieved, even though not understood for several decades. The mystery of the hardening associated with precipitation began to unravel as a result of x-ray diffraction studies; the importance of the nonequilibrium microstructural features was established. Age-hardening alloys began to be exploited in other alloy systems.

The field of materials entered the era of World War II with great anticipation for the unlocking of the potential for property improvement by structure control through processing. The importance of structure was established even if not universally accepted. The missing link in the structure/processing/properties relationship for structural materials was the structure/properties connection. It had been postulated in the mid-1930s and afterward by Orowan (Ref 25), Polanyi (Ref 26), and Taylor (Ref 27) that line defects called dislocations could explain the low strengths of materials relative to their theoretical strengths. The dislocation behavior was suggested as the necessary structural concept to control mechanical properties. Lack of experimental support limited early acceptance of the dislocation hypothesis. Better resolution and higher magnification than available with light microscopy was needed for observation of dislocations and their interactions, both with themselves and with other microstructural features.

Even in the early 1950s, textbooks hardly made mention of dislocations and usually did so in the form of a one-page afterthought. The use of dislocations to understand structure/property relationships in crystalline materials was not found in textbooks until the middle of the 1950s.

The 1950s witnessed the unification of metals, polymers (plastics), ceramics, and composites into a single discipline—materials. The common thread was structure. Not surprisingly, the 1950s also was a period of development and application of instrumentation having resolution high enough to allow structure observation several orders of magnitude above light microscopy. The missing piece of the structure/processing/properties puzzle was at hand. Scanning electron microscopy improved metallographic resolution by an order of magnitude and together with x-ray compositional microanalysis provided information about compositional variations within microstructures. Transmission electron microscopy permitted the direct observation of dislocations and their interactions with each other and other structural elements. As a plethora of other high-resolution instruments continued to evolve, the long-sought-after documentation of processing/structure/property relationships gushed forth. The new, high-resolution techniques were generally applicable to all types of materials. The crystalline material concepts useful in metals were tested on crystalline ceramics; the amorphous material concepts of plastics and glassy ceramics were applied to metallic alloys that previously had been considered to be crystalline if solid. Composite materials took advantage of the understanding developed for metals, plastics, and ceramics. Structure could be characterized, processing could be controlled, material property variations could be understood, and materials could be designed.

REFERENCES

1. G.E. Dieter, *Engineering Design, A Materials and Processing Approach,* 2nd ed., McGraw-Hill, 1991
2. L. Vanasupa, "The Materials Science of Food: or Eating Your Way Through an Introductory Course in Materials Engineering," presented at the 1993 Annual Meeting of the ASEE, University of Illinois, June 1993
3. T.H. Courtney, *Mechanical Behavior of Materials,* McGraw-Hill, 1990
4. J.P. Schaffer, A. Saxena, S.D. Antolovich, T.H. Sanders, and S.B. Warner, *The Science and Design of Engineering Materials,* Irwin, 1995
5. W.D. Callister, *Materials Science and Engineering: An Introduction,* 3rd ed., John Wiley & Sons, 1994
6. J.H. Westbrook, Materials: History Before 1800, *Encyclopedia of Materials Science and Engineering,* M.B. Bever, Ed., Pergamon Press, 1986, p 2816–2827
7. J.H. Westbrook, Materials: History Since 1800, *Encyclopedia of Materials Science and Engineering,* M.B. Bever, Ed., Pergamon Press, 1986, p 2827–2838
8. J.H. Westbrook, *Structural Intermetallics,* R. Darolia, J.J. Lewandowski, C.T. Liu, P.L. Martin, D.B. Miracle, and M.V. Nathal, Ed., The Metallurgical Society, 1993, p 1–15
9. M.F. Ashby, *Materials Selection and Design,* Pergamon Press, 1992
10. A. Murdoch, *Boom Copper,* R.W. Drier and L.G. Koepel (republishers), 1964
11. J.R. Stubbles, *The Original Steelmakers,* Iron and Steel Society, 1984
12. C.D. King, *Seventy-Five Years of Progress in Iron and Steel: Manufacture of Coke, Pig Iron, and Steel Ingots,* American Institute of Mining, Metallurgical, and Petroleum Engineers, 1948
13. Committee on Materials Science and Engineering, *Materials Science and Engineering for the 1990s: Maintaining Competitiveness in the Age*

of Materials, National Research Council, National Academy Press, 1989
14. R.F. Mehl, *A Brief History of the Science of Metals,* American Institute of Mining, Metallurgical, and Petroleum Engineers, 1948
15. R.F. Mehl, *Physical Metallurgy,* R.W. Cahn, Ed., North-Holland, 1965, p 1–31
16. D.B. Williams, A.R. Pelton, and R. Gronsky, Ed., *Images of Materials,* Oxford University Press, 1992
17. I. Curbishley, Ed., *Mechanical Testing,* Institute of Metals, London, 1988
18. *Mechanical Testing,* Vol 8, *ASM Handbook* (formerly *Metals Handbook,* 9th ed.), American Society for Metals, 1985
19. H.C. Sorby, *J. Iron Steel Inst.,* 1886, p 140
20. H.C. Sorby, *J. Iron Steel Inst.,* 1887, p 133
21. C.S. Barrett, *The Sorby Centennial Symposium on the History of Metallurgy,* Gordon and Breach, 1965, p ix–xix
22. J.W. Gibbs, *Trans. Connecticut Academy,* Vol 3, 1876, p 152
23. M. Laue, W. Friedrich, and P. Knipping, *K. Akad. Wiss. Munchen,* 1912, p 303
24. E.S. Davenport and E.C. Bain, *Trans. AIME,* 1930, p 117
25. E. Orowan, *Z. Phys.*, Vol 89, 1934, p 634
26. M. Polanyi, *Z. Phys.,* Vol 89, 1934, p 660
27. G.I. Taylor, *Proc. R. Soc. (London),* Vol 52, 1940, p 23

Fundamental Structure-Property Relationships in Engineering Materials

Thomas H. Courtney, Michigan Technological University

THIS ARTICLE deals with the relationships among material properties (primarily mechanical properties) and material structure. The term *structure* is defined broadly; it relates to factors such as the arrangement of atoms (or ions or molecules) in the solid state. Structure also involves considerations of "defects"—abnormalities—in the idealized atomic arrangements. And structure, too, relates to the collective arrangement of these atoms on a scale much greater than that of an individual atom. In the jargon of the materials engineer, this collective arrangement is called the *microstructure* of the material.

This article considers the several material classes: *metals*, *ceramics*, and *polymers*. All of them are technologically important. Some basic material characteristics (density, elastic modulus, and, to a lesser extent, thermal and electrical conductivity) are determined almost exclusively by material composition. These properties are referred to as *microstructure insensitive*. In contrast, microstructure-sensitive properties depend (sometimes profoundly) on microstructural features. From the standpoint of mechanical behavior, the yield strength and fracture toughness of a material are the most important of the structure-sensitive properties. The yield strength is the applied stress required to initiate material permanent deformation. The fracture toughness is a measure of the resistance to crack propagation (i.e., to fracture). Thus both parameters relate to material "failure," although they represent different kinds of "failure." The yield strength is used in design to prevent (or sometimes, as in metal working, to ensure) plastic deformation; the fracture toughness is utilized to design against material fracture.

Fundamental Characteristics of Metals, Ceramics, and Polymers

Before describing structure-property relationships, it is worthwhile to summarize the essential features of the materials classes.

Metals comprise the vast majority of the elements. Metals more or less occupy the left side of the periodic table and nonmetals the right, although the "dividing line" is not a vertical one. Instead, this line skews from left to right as one looks downward in the table. Metals have few valence electrons. These are collectively shared by the atoms of the solid, giving rise to what is referred to as *metallic bonding*. As a consequence of this bonding, metals are "friendly" at the atomic level. That is, they have a large number (typically 8 to 12) of neighboring atoms in their solid form, and this accounts for the relatively high density of metals. Metallic bonding is also responsible for the high elastic stiffness of metals. Finally, the collective electronic nature of metals explains their high electrical and thermal conductivities.

Metals are distinguished from nonmetals in other ways. In particular, many metals dissolve in each other to form an atomic solution or *alloy*. In the liquid state, this solubility is often complete; that is, the elements dissolve in all proportions just as alcohol and water do. The solubility in the solid state is usually more restricted. Nonetheless, this feature of solid solubility allows materials engineers to manipulate properties of many metals and alloys.

Ceramics. Chemical bonding in ceramics is different than in metals. When elements from the left side of the periodic table (i.e., electropositive elements such as sodium) combine with elements from the right side of the table (electronegative elements such as chlorine), an ionic bond results. Common table salt, NaCl, is a prototypical ionic-bonded ceramic. The coordination number (CN) is the number of cation (positive ion)/anion (negative ion) near neighbors in ionic solids and depends principally upon the relative ionic radii. CNs of ionic solids are, on the average, slightly less than those in metals, and this partially accounts for their slightly lower densities. Ionic-bonded materials have much lower electrical conductivities and, in most cases, much lower thermal conductivities than metals. This is because ions, rather than atoms, are their fundamental units. For example in NaCl, a sodium atom gives up its sole valence electron, and the chlorine ion captures it, to produce ions having full electron orbitals or suborbitals. As a consequence there are no "free" electrons available to conduct electricity. In the liquid state, however, ionic materials often demonstrate respectable electrical conductivities. Flow of electric current in them takes place by ionic motion and is facilitated by the high ionic liquid mobility, which is much greater than the corresponding solid mobility.

Ionic compounds usually form only in stoichiometric proportions (e.g., NaCl, $CaCl_2$, and Al_2O_3). Thus their tolerance for "alloying" is much reduced compared to that of metals, although there are exceptions (some of them notable). "Alloying" that does take place can alter properties considerably. For example, substitution of divalent calcium ions (in amounts on the order of parts per million) for monovalent sodium ions in NaCl increases the electrical conductivity by orders of magnitude, although the resulting conductivity remains much less than that of a typical metal.

Some ceramics are *covalent* solids. Covalent bonding, like metallic bonding, involves electron sharing. However, the sharing is localized in covalent solids. This results in low electrical conductivities; that is, electrons are effectively incapable of detaching themselves from their "parent" atom and moving through the solid in response to an electric field. Another characteristic of the localized electron sharing is a reduced CN. For example, diamond—an elemental covalent solid—has a CN of 4. The relatively low densities of covalent solids are partly a consequence of their low CNs. There are also covalent compounds, such as Si_3N_4, which because of its strength is a potential high-temperature structural material. Both elemental and compound covalent solids are restricted to narrow composition ranges (e.g., pure silicon and stoichiometric Si_3N_4). However, small quantities of "dopants" may be added to, for example, pure silicon, as they are in the semiconductor industry. Such additions—on the order of parts per million—render silicon an

Table 1 Characteristics of the several classes of engineering materials

Material class	Stiffness	Strength	Malleability	Electrical conductivity	Thermal conductivity	Density
Ceramics	High	Very high	Nil	Very low	Low	Medium
Metals	High	High	High	High	High	High
Polymers	Low	Low	High	Very low	Very low	Low

effective engineering *semiconductor*. While the electrical conductivities of doped semiconductors are orders of magnitude below those of metals, their conductivities are much greater, also by orders of magnitude, than those of the elements on which they are based.

Although this discussion has apparently categorized ionic and covalent bonding as distinctly different, the categorization is somewhat arbitrary. Many "ionic" solids, even NaCl, manifest a degree of covalent bonding, and the converse holds for "covalent" compounds. Silica (SiO_2), which exists in a number of structural forms, is an example of a material that manifests mixed ionic and covalent bonding. Materials of this nature are called *polar covalent solids*. Polar covalent solids often demonstrate high melting points and elastic moduli because of this "mixed," strong bonding.

Polymers. Covalent bonding is also dominant in polymers. The resulting low polymeric CNs partly account for their low densities. However, polymers are light primarily because they consist of light atoms (e.g., carbon and hydrogen). As expected, the localized nature of electrons in polymers renders them electrical insulators. They are not good thermal conductors, either.

The molecular architecture of *thermoplastics*, one type of polymer, differs significantly from that of metals and ceramics. The basic atomic or ionic "building blocks" of metals and ceramics usually arrange themselves in a regular three-dimensional pattern, but this is not so with thermoplastics. The basic "molecular unit" in them is a long chain containing many (thousands in some cases) atoms along the chain backbone. Abutting from the chain are "side groups." These range from a single hydrogen atom (as in polyethylene) to a fairly complicated group containing a number of atoms. Bonding between the chains is weak (van der Waals or a variant thereof), in contrast to the strong intrachain covalent bonding. Because of the size and complexity of the individual chains, thermoplastics display a reluctance to assume the long-range atomic ordering that characterizes most ceramics and metals.

Processing Characteristics. The different "chemistries" of the material classes influence the ways in which they are manufactured. Some metals and alloys are melted and then poured directly into a finished shape, as in *castings*. This is possible because liquid metals have high fluidities (conversely, low viscosities), which is a consequence of their small atomic structural units. Other metals are first melted and poured into a large *ingot*, *strand cast billet*, or *slab*. One or more ingot dimensions are then reduced by *deformation processing*, as in rod or wire drawing, plate rolling, forging, or extrusion. The extensive material shape changes accomplished during deformation processing are possible because of the inherent *malleability* of metals. Final shape is imparted to these *wrought* products by a finishing operation, such as machining or stamping.

Although the basic unit of ceramics is a small ion (or group of ions), ceramics are not processed in the ways that metals are. The high melting temperatures of ceramics and/or the tendency of them to react chemically with containers, make it difficult to cast them. Furthermore, ceramics are not malleable. Quite the contrary, they are usually quite brittle and are seldom capable of being extensively deformed. Ceramics are frequently produced by powder processing. Appropriate particulates (made by one of a number of suitable processes) are pressed together, and the resulting compact is heat treated at a high temperature where the powder particles *sinter* together, resulting in partial or complete densification of the powder compact. Ceramics containing residual porosity cannot be used in demanding structural applications, because ceramic fracture resistance is very sensitive to pores. Application of an external pressure during sintering can yield ceramics having minimal porosity. As might be expected, materials made this way cost more. Some metals are also produced by powder processing. For metals, this processing route may be economically viable when the part made is relatively small or has an intricate shape (the near-net-shape powder processing operation is a plus) or when the resulting properties warrant the additional cost.

Thermoplastics are often fabricated in ways similar to those used to fabricate metals. Precursor polymer powders are heated to a temperature at which they flow viscously. Owing to their complex molecular architecture, polymer viscosities are high compared to those of metals. Because of this the polymer can be subjected to mechanical forming while heated. Enclosing dies permit the polymer to be formed directly into a finished shape, such as a soft drink container. It should be noted that the low softening and/or melting temperatures of thermoplastics allow them to be processed at moderate temperatures. Production costs are modest because of this and because the tools used to form polymers have a long life in comparison to tools used to form metals.

This background provides a basis for discussing in more detail the atomic arrangements within the materials classes. The important property differences among the classes are summarized in Table 1, where the classes are roughly categorized in terms of densities, conductivities, malleabilities, and so on.

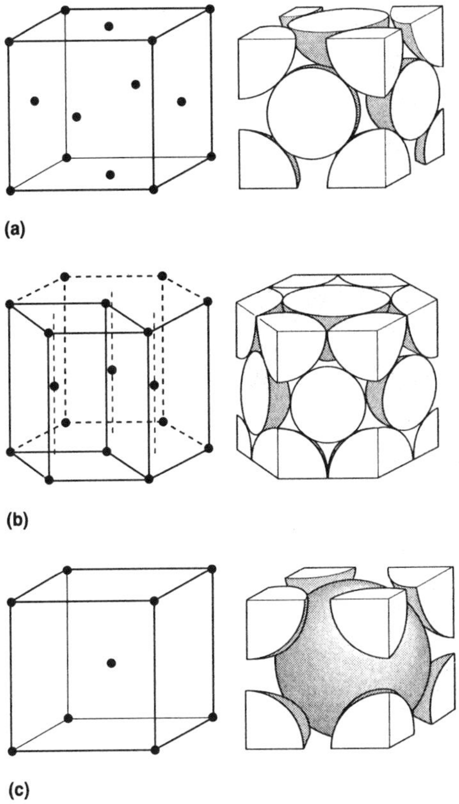

Fig. 1 Representation of several simple unit cells. Points represent positions of atom centers (left), and atoms are represented by spheres or portions of spheres (right). (a) Face-centered cubic unit cell. (b) Hexagonal close-packed unit cell. (c) Body-centered cubic unit cell. Source: Ref 1

Crystal Structure and Atomic Coordination

Basic Concepts. The arrangement of atoms (or molecules or ions) in most solid metals, many solid ceramics, and a few solid polymerics demonstrates a long-range pattern. That is, the atomic packing is repetitive over distances that are large in comparison to the atomic size. Such an arrangement is called *crystalline*, and the repetitive pattern can be described by a fundamental repeating unit or *unit cell*.

Almost all metals crystallize in one of three patterns: face-centered cubic (fcc), hexagonal close-packed (hcp), or body-centered cubic (bcc). The atomic arrangements in these cells are depicted in Fig. 1. The positions of atom centers are noted on the left, and the atoms are represented by spheres (or partial spheres when an atom is shared by adjacent unit cells) on the right. All the arrangements are characterized by efficient atomic packing. Indeed, the fcc array (Fig. 1a) represents the most efficient possible atomic packing, as manifested by the high CN (12) of this structure. Viewing a face of an fcc cell, we see that an atom in a face center is coordinated by four other atoms at cell corners. The distance separating the atom centers is the *atomic diameter*, equal to $a/(2)^{1/2}$ where a is the edge length or *lattice parameter* of the unit cell. However, the

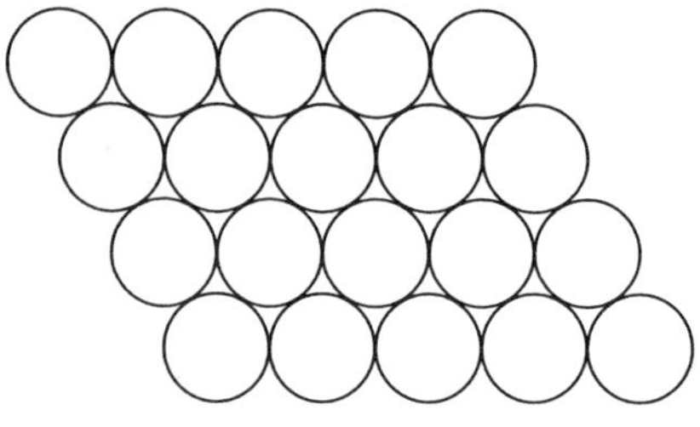

(a)

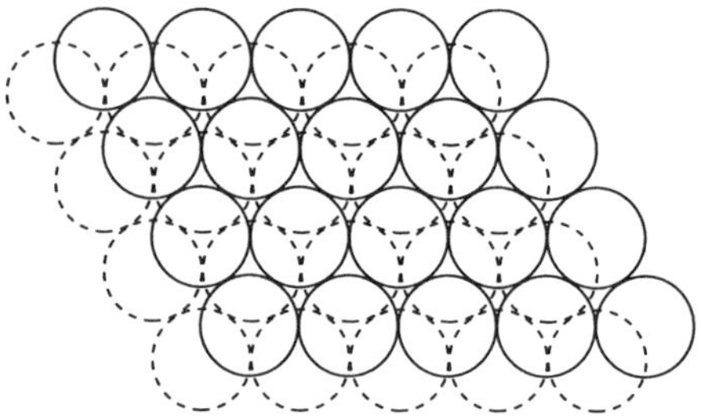

(b)

Fig. 2 (a) Plan view of a close-packed plane in the fcc structure. The directions along which atoms touch are face diagonals. (b) Plan view of two close-packed planes of spheres, with spheres in the top plane (solid circles) situated in interstices in the bottom plane (broken circles). Source: Ref 1

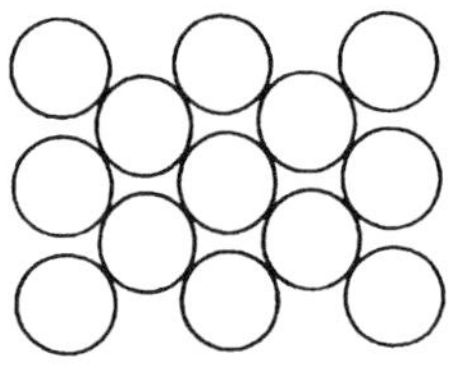

Fig. 3 Plan view of atomic packing in closest-packed bcc plane. Atoms touch along two nonparallel close-packed directions (the cube diagonals). Source: Ref 1

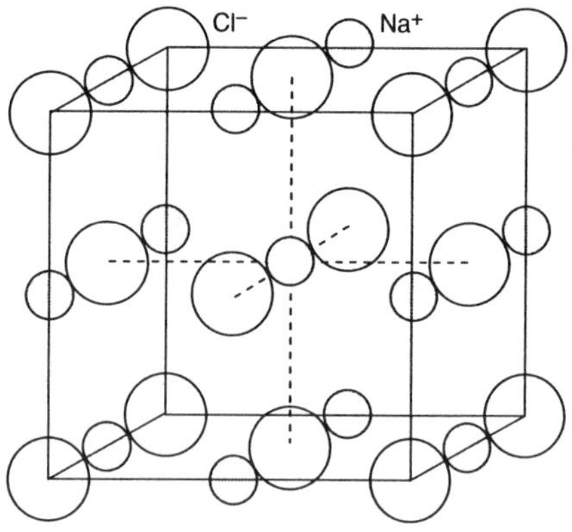

(a)

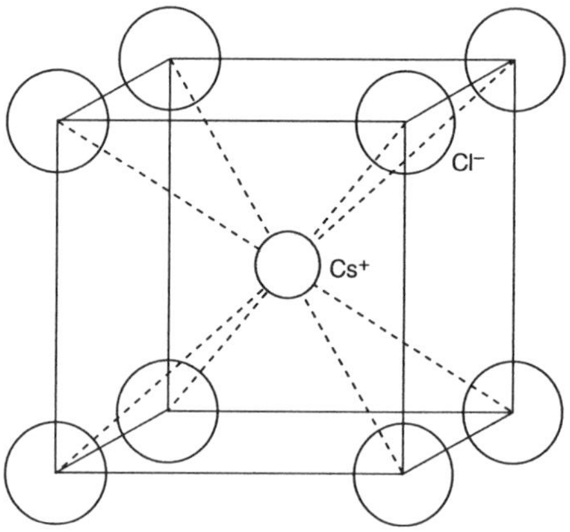

(b)

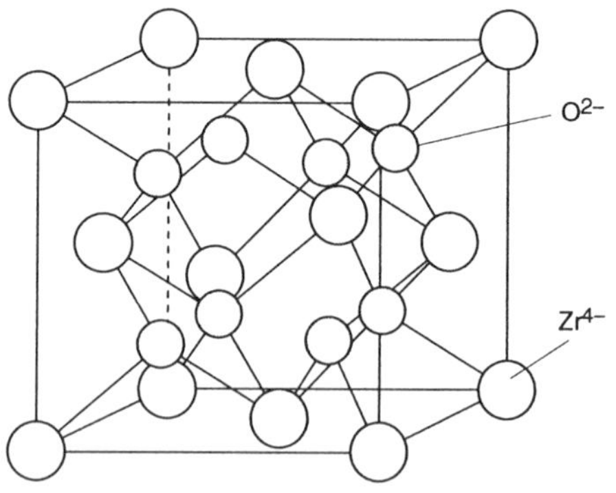

(c)

Fig. 4 Unit cells of several crystalline ceramics. (a) The unit cell of NaCl. Chloride ions assume an fcc array with one sodium ion (displaced by half a lattice parameter along a cube edge) for every chloride ion. Source: Ref 1. (b) The CsCl structure. Chloride ions assume a simple cubic array with a cesium ion in the center of each cell. Source: Ref 1. (c) The structure of ZrO_2. Zirconium ions have an fcc array. The oxygen ions, which take on a simple cubic array, are located within the unit cell. Source: Ref 2

atom at the face center is this same distance from four other atoms on the centers of the four adjoining cell faces. The reference atom is likewise coordinated to four atoms in the centers of adjacent faces in the unit cell directly in front of the unit cell of Fig. 1(a) (not shown). Thus CN = 12 for the fcc structure.

An alternative view of fcc packing permits another way of seeing that it is efficiently packed. Figure 2(a) is a view of a close-packed plane in the fcc structure. A plane is defined by two nonparallel directions; in Fig. 2(a), these are taken as two face diagonals. The atoms in this plane are arranged as billiard balls are in a cue rack. When these atomic planes are stacked vertically and in a direction parallel to the cube diagonal, atoms of one plane lie in the vertices of atoms in the plane beneath (Fig. 2b). Such a stacking pattern generates a close-packed structure. In the fcc pattern, the positions of atom centers repeat every fourth of these planes. That is, atom centers in the fourth plane lie directly above atom centers in the first, atom centers in the fifth plane are directly above those in the second, etc. The stacking is thus described as ABCABC.

The ideal hcp structure (Fig. 1b) is packed as efficiently as the fcc one. Atoms in the close-packed (basal) plane have an atomic arrangement identical to that in a close-packed fcc plane. However in the hcp structure, these planes repeat every other layer (i.e., atom centers in the third layer lie directly above atom centers in the first, atom centers in the fourth layer are directly above atom centers in the second, etc.). This stacking is therefore described as ABAB.

Two lattice parameters (c and a; Fig. 1b) are needed to define the hcp unit cell. An hcp cell has the maximum atomic packing efficiency only when a definite relationship between c and a exists (c/a = 1.63). Few hcp metals exhibit this ratio (most have c/a < 1.63). In these situations, the hcp structure can no longer be viewed as being as efficiently packed as the fcc one.

The CN for the bcc structure (Fig. 1c) is 8. This can be deduced with reference to the atom in the center of the bcc unit cell; it is equidistant from eight atoms at the cell corners. Because the atomic packing is less efficient in bcc, the closest-packed plane in this structure is also less densely packed than in the corresponding fcc plane. A view (Fig. 3) of the closest-packed bcc plane (which is defined by a cell edge and a face diagonal) shows that atoms within this plane touch along the cube diagonals. There are two nonparallel *close-packed directions* of this kind in this plane; the corresponding number for the fcc close-packed plane is 3.

Many metals exist in more than one crystalline form, depending on pressure and temperature. At one atmosphere, for example, iron is bcc at temperatures below 912 °C, is fcc between 912 and 1394 °C, and reverts to the bcc form above 1394 °C until melting at 1538 °C. Titanium, zirconium, and hafnium all exhibit a transition from an hcp structure to bcc on heating. Many other metals (as well as some nonmetals, such as SiO_2) also exhibit such *allotropic* transformations.

The unit cells of three simple ionic solids are depicted in Fig. 4. Chloride ions assume an fcc array in NaCl (Fig. 4a); for each chloride ion there is a sodium ion displaced by half a lattice parameter from the chloride ion along a unit cell edge. In CsCl (Fig. 4b), the chloride ions assume a *simple cubic* array (i.e., they are situated at unit cell corners), and a cesium ion is at the cell center. A more complicated arrangement is found in zirconia, ZrO_2 (Fig. 4c). The structure is not as complex as it might appear at first glance. It is only a variant of simpler structures. The zirconium ions assume an fcc array, with oxygen ions occupying internal cell sites. As indicated in Fig. 4(c), the oxygen ions take on a simple cubic array.

Formal description of ceramic crystalline arrangements often appears cumbersome. However, in many cases the arrangements are only variations of much simpler ones.

Some important ceramics, *glasses*, are noncrystalline. Atomic arrangements in glasses do not repeat over distances that are large in comparison to the atomic size, although there is a definite short-range order to them. Most common glasses are based on polar-covalent-bonded silica. The basic structural unit here (as well as in crystalline silica) is the silicate tetrahedron. As shown in Fig. 5, silicon atoms (ions) are located at tetrahedral centers, and oxygen atoms (ions) are located at their tips. Tetrahedra are then joined tip-to-tip, thereby generating a three-dimensional solid. When the arrangement manifests a long-

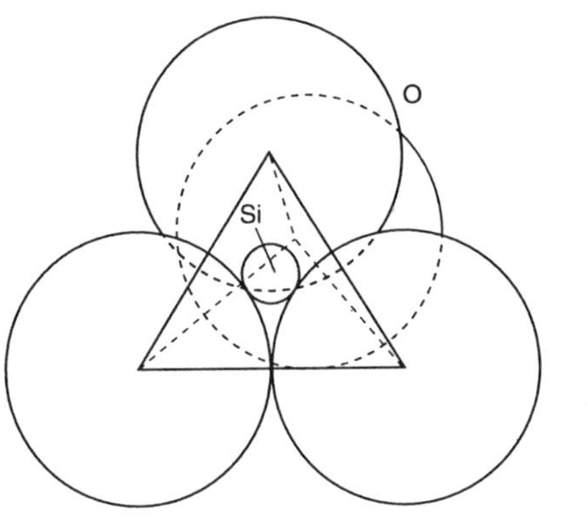

Fig. 5 The basic structural unit in SiO_2 is a tetrahedron in which silicon is located at the center and oxygen at the corners of the tetrahedron. Source: Ref 1

range order, a crystalline form of SiO_2 results; when not, a glassy form is the consequence. The two situations are schematically illustrated in a two-dimensional version in Fig. 6. Liquid SiO_2 has an atomic arrangement essentially the same as that of the more rigid solid glass. Viscosities of silicate glasses are usually much greater than those of metals, even though the basic structural unit (the tetrahedron) is not large. Viscous flow in silicates requires displacement of tetrahedra with respect to each other, and this is difficult because of the strong polar-covalent bonds linking the tetrahedra. Addition of certain impurity ions to silicate glasses reduces their viscosities by orders of magnitude, and this tactic is employed in the economic production of most commercial glassware.

Thermoplastics are seldom crystalline. Polyethylene is an exception, but noncrystalline regions persist in even the most crystalline form of this substance. The structure of most thermoplastics is schematized in Fig. 7, which indicates that the polymeric chains are arranged randomly in three dimensions. As with silicate glasses, the atomic arrangements in "glassy" solid and liquid thermoplastics are similar.

Some Crystallographic Details. Further details of atomic arrangements are described here, because they are relevant to plastic deformation of crystalline materials. As shown in Fig. 2, the close-packed plane in the fcc structure is defined by two face diagonals, and within such a plane there are three nonparallel close-packed directions. In addition, there are four nonparallel planes of this nature in the fcc crystal structure (Fig. 8). (To better illustrate the point, the planes in Fig. 8 are taken from adjacent cells.) There are thus 12 combinations of nonparallel planes and directions (four planes times three directions per plane) in the fcc lattice.

The above is germane because plastic deformation takes place by *slip* (sliding) of close-packed planes over one another. A reason for this slip plane preference is that the separation between close-packed planes is greater than for other crystal planes, and this makes their relative displacement easier. Furthermore, the slip transit direction (or *slip direction*) is a close-packed direction.

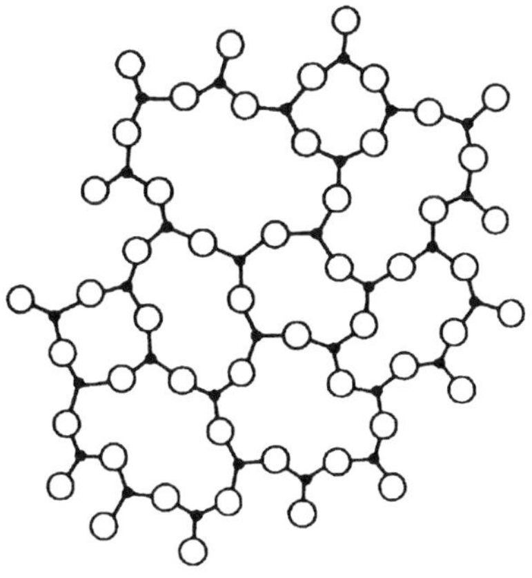

(a)

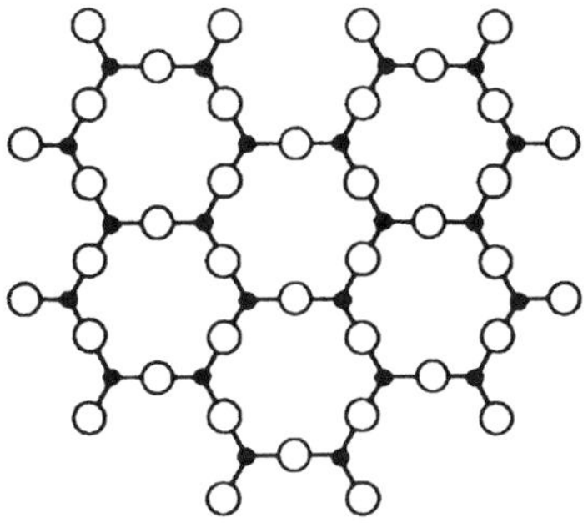

(b)

Fig. 6 Schematic two-dimensional representation of (a) a SiO_2 glass structure and (b) a SiO_2 crystalline structure. Tetrahedra are joined tip to tip in both situations, but in crystalline SiO_2 a long-range pattern to the tetrahedra exists; this is not the case for glassy SiO_2. Source: Ref 1

The combination of planes and directions on which slip takes place (12 for the fcc structure) constitutes the *slip systems* of the material. In polycrystalline materials (defined in the next section), a certain number of slip systems must be available in order for the material to be capable of plastic deformation. Other things being equal, the greater the number of slip systems, the greater the capacity for this deformation. Face-centered cubic metals have a large number of slip systems, and indeed, all of them except two (iridium and rhodium) are capable of moderate to extensive plastic deformation even at temperatures approaching 0 K.

Materials having the bcc structure also often display 12 slip systems, although this number comes about differently than it does for the fcc lattice. A closest-packed bcc plane is defined by a unit cell edge and face diagonal (Fig. 3). There are only two close-packed directions (the cube diagonals) in the closest-packed bcc plane, but there are six nonparallel planes of this type. Over certain temperature ranges, some bcc metals display slip on other than close-packed planes, although the slip direction remains a close-packed one. Thus bcc metals have the requisite number of slip systems to allow for their plastic deformation. That some of them become "brittle" at low temperatures is a result of the strong temperature

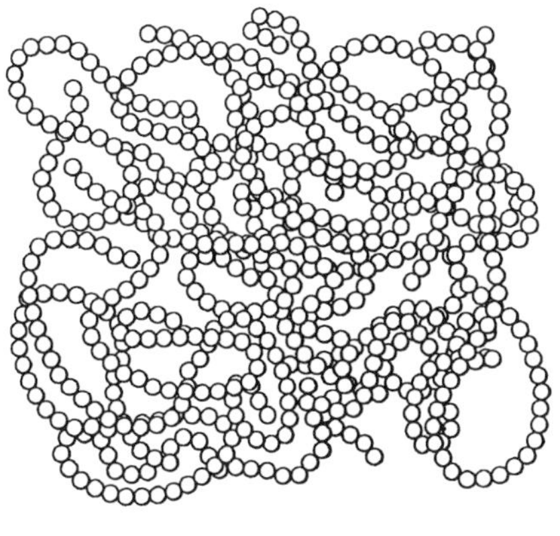

Fig. 7 A schematic representation of a long-chain polymer. The spheres represent repeating units in the polymer chain, not individual atoms. Source: Ref 3

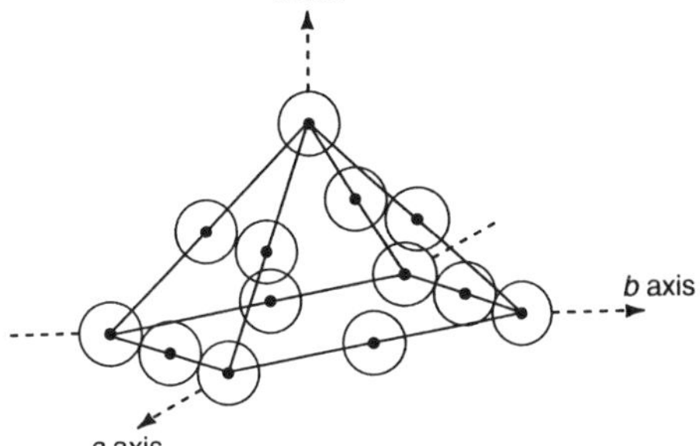

Fig. 8 Four nonparallel close-packed planes characterize the fcc structure. There are three nonparallel close-packed directions within each plane, giving rise to 12 slip systems. Source: Ref 4

sensitivity of their yield strength, which causes them to fracture prior to undergoing significant plastic deformation.

Depending on the *c*/*a* ratio, polycrystalline hcp metals may or may not have the necessary number of slip systems to allow for appreciable plastic deformation. The ideal hcp structure has only three slip systems, as there is only one nonparallel close-packed plane in it (the basal plane, which contains three nonparallel close-packed directions). Three slip systems are insufficient to permit polycrystalline plastic deformation, and so hcp polycrystals for which slip is restricted to the basal plane are not malleable. When *c*/*a* is less than the ideal ratio, basal planes become less widely separated, and other planes compete with them for slip activity. In these instances, the number of slip systems increases, and material ductility is beneficially affected.

The limited ductility of polycrystalline ceramics is often tied to their lack of slip systems. When slip does take place in polycrystalline ionic-bonded ceramics, it does so on planes and in directions that ensure maintenance of local charge neutrality. The number of these slip systems is insufficient to allow significant plastic deformation. At higher temperatures, though, additional slip activity permits some ceramics to demonstrate appreciable ductility.

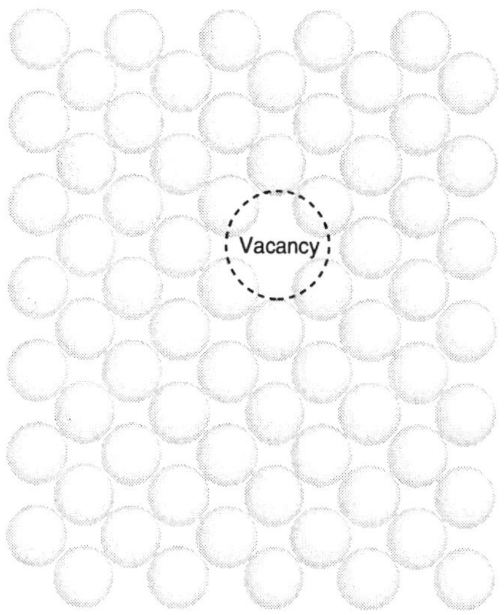

Fig. 9 Two-dimensional representation of a crystal illustrating a vacant lattice site. Source: Ref 3

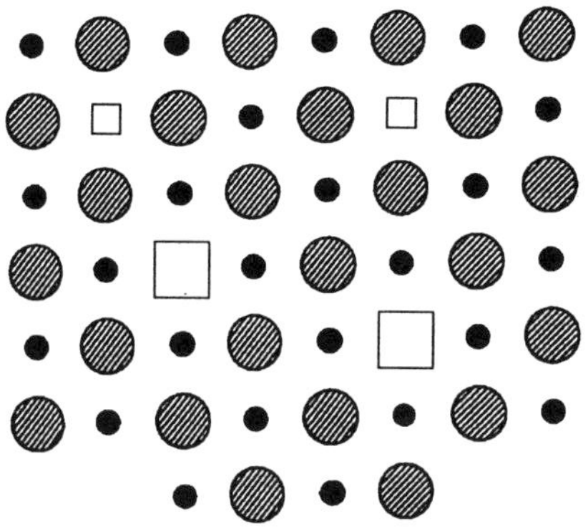

Fig. 10 Schottky defects in an ionic crystal. Cations are represented by solid circles and anions by shaded ones. Equal numbers of cation and anion sites are vacant so as to preserve charge neutrality. Source: Ref 5

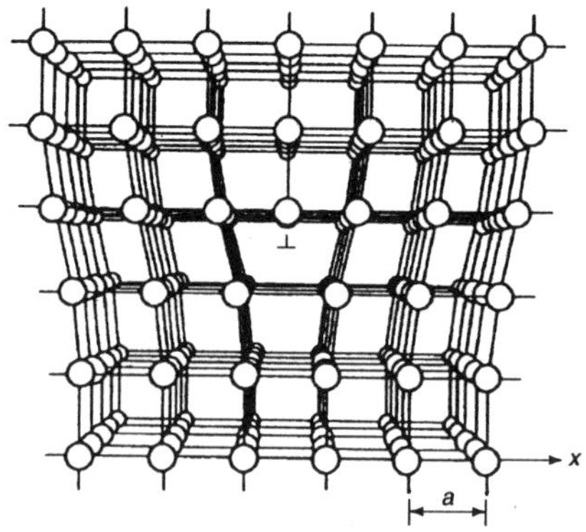

Fig. 11 A schematic of an edge dislocation, represented by a partial atomic plane, in a crystal. The "core" of the dislocation is localized at the partial plane termination. Atomic positions are distorted in region of this core, making slip easier in the vicinity of the dislocation. Source: Ref 6

Crystalline Defects

Atomic arrangements in crystals deviate slightly from the ideal ones described above. Such deviations are called *crystalline defects* (or *imperfections*), although these "defects" often lead to improved material performance. Defects can be classified by their scale or size. The smallest deviation in the ideal crystal arrangement has a volume comparable to that of an atom; such a defect is termed a *point* defect.

Point defects are of two types, *impurity atoms* and *vacancies*. A vacancy is schematically illustrated in Fig. 9; here, rather than having all lattice sites occupied, one site is vacant. Vacancies arise as a result of entropic effects, and the fraction of vacant lattice sites increases with temperature. This fraction is zero at 0 K and is on the order of 10^{-3} for many metals at or close to their melting point. A somewhat different type of "vacancy" found in ionic solids is illustrated in Fig. 10. Here, to maintain electrical neutrality, there must be an equal number of cation and anion vacancies; thus "double" vacancies (*Schottky defects)* are formed.

Vacancies alter properties. Density is (very slightly) decreased by them. Material strength is slightly increased by vacancies (which seems counter-intuitive). Vacancies increase the electrical resistivities of metals, but Schottky defects have an opposite effect in ionic solids. The reason is that conduction in ionic solids results from the motion of ions, and ionic migration is facilitated by vacancies. Vacancies also enhance *atomic diffusion*, the macroscopic atomic mixing that takes place as a result of the motion of many individual atoms. If a layer of copper is placed on one of nickel, for example, and then held at an elevated temperature for a long time, the resultant solid displays a uniform composition as a result of the interdiffusion of copper and nickel atoms.

Impurity atoms are also termed point defects. For example, an fcc unit cell of an alloy of composition 75at.%Cu-25at.%Ni contains, on average, three times as many copper atoms as nickel ones. The substituted nickel atoms are considered defects, since their size differs from that of the host copper atom, and this causes a local distortion of the unit cell. Impurity atoms affect properties, too. Electrical and thermal conductivities in metals are reduced by them. However, metallic strengths are increased by impurities. This *solid solution hardening* is used to strengthen a number of metals. Adding zinc to copper, as in brasses, is a technologically important example.

Small impurity atoms do not substitute for the host atoms, but rather enter into *interstitial* spaces among them and are referred to as interstitials. Typical interstitials in metals are nitrogen, carbon, and oxygen. The sites they occupy in the fcc lattice are the same as those that the smaller sodium ions occupy in the NaCl structure (Fig. 4). Interstitials generally strengthen a metal more than substitutional atoms do, since the interstitials cause more distortion. Carbon atoms in the bcc form of iron are particularly potent hardeners in this respect. The effect is used beneficially in strengthening of *quenched and tempered* steels (see the section "Strengthening of Steels" in this article).

Ionic materials also can have substitutional ions. Potassium, for example, can substitute for sodium in NaCl. When the substituted ion has a different valence than the host, charge balance requirements complicate things. If a calcium ion, with a valence of 2, substitutes for a sodium ion in NaCl, the substitution is accompanied by the formation of a vacant site in the sodium anion array. This satisfies the charge neutrality requirement (two monovalent sodium ions are replaced by a divalent calcium ion and a neutral vacancy). In such situations, the solubility of the replacement ion (i.e., the maximum concentration of impurity that can be dissolved in the host material) is quite restricted. Impurity ions strengthen ionic solids just as impurity atoms do metals. In addition, the generation of excess vacancies, as accompanies the solution of calcium in NaCl, substantially increases electrical conductivity and diffusivity.

Some ionic solids can be viewed as interstitial compounds. Indeed, as alluded to earlier, NaCl can be considered an fcc array of chloride ions, with sodium ions occupying interstitial sites in the lattice.

A *line* defect has two dimensions comparable to an atomic diameter and one dimension that is much greater. An example of a particular line defect, an *edge dislocation*, is shown in Fig. 11. The upper half of the crystal shown contains one more atom column than the lower half of it. The resultant atomic disregistry is centered about a small region. As suggested by Fig. 11, the disregistry is accommodated in an approximately cylindrical volume having a radius comparable to that of an atom and extending along the termination of the atomic column for distances much greater than this. Dislocations are found in all crystalline solids, but the extent to which they exist varies among the material classes. The quantity of dislocations (the *dislocation density*) can be expressed in terms of their number per unit area. With reference to Fig. 11, for example, the dislocation density would be the number of dislocations emerging from a surface divided by the area of the depicted crystal plane. Dislocation densities in metals range from about $10^{10}/m^2$ to $10^{15}/m^2$; in ionic compounds they are usually several orders of magnitude less. Covalent solids have lower dislocation densities still.

Dislocations are important because their motion in response to an applied stress is responsible for plastic deformation in most crystalline solids. As mentioned above, plastic deformation takes place by the relative displacement of atomic planes. This is easier to accomplish when dislocations are present. The atomic disruption in the dislocation vicinity is responsible for the easier slippage of planes on which dislocations are situated. In fact, the stress required to cause dislocations to move is orders of magnitude less than the stress needed to cause slip plane displacement in a "perfect" crystal.

The role of dislocations in plastic flow is verified by the exceptionally high strengths of metal

Fig. 12 Schematic representation of the orientations of individual grains in a polycrystal. Within individual grains, a set of atomic planes has the same orientation in space. At a grain boundary, the orientation changes abruptly. Source: Ref 1

crystals not containing (or containing very few) dislocations. It might be thought that the greater the dislocation density, the lesser the stress required for plastic deformation. This is true for materials containing relatively few dislocations (e.g., about less than $10^8/m^2$). Paradoxically, though, when the dislocation density becomes high enough, the stress required to cause plastic flow increases with dislocation density. This is so because dislocations mutually impede each other's motion. Dislocations in metals also multiply, sometimes substantially, when they are plastically deformed. This is accompanied by an increase in the stress required to continue deformation. This phenomenon of *work hardening* is used to manipulate strengths of a number of metallic materials, including conventional stainless steels and copper and its alloys. The lesser dislocation densities found in ionic and covalent solids is one (but only one) reason that these materials are less malleable than metals.

Crystalline solids also contain internal surface defects. A surface defect has one dimension comparable to the atomic size and two dimensions that are much larger. The most important surface defect is a *grain boundary*. As indicated in Fig. 12, such boundaries separate differently spatially oriented crystals, and the collective aggregate is termed a *polycrystal* (or polycrystalline solid). The average diameter of the individual grains within a polycrystal defines the material *grain size*. Grain sizes in engineered materials vary by quite a bit. They are usually less in nonmetals than in metals, and they can be as fine as 0.1 μm in some ceramics. Metallic grain sizes typically range from several micrometers to, in the case of slowly cooled castings, several centimeters. Some recently developed processes (e.g., *rapid solidification* and *mechanical alloying*) produce materials having grain sizes on the order of nanometers. To put this in perspective, the diameter of a "typical" atom is about 0.25 nm. Thus, grains having a diameter of several nanometers are about ten atoms across.

Grain size affects mechanical properties. The yield strength increases with decreases in grain size, because the distance over which dislocations can move freely is limited to the grain diameter. (Dislocations are restricted from crossing grain boundaries.) Fracture resistance also generally improves with reductions in grain size. This is particularly beneficial, because as discussed below, the general situation is that structural changes that increase yield strength diminish fracture toughness. Fracture resistance improves because the cracks formed during deformation, which are the precursors to those causing fracture, are limited in size to the grain diameter.

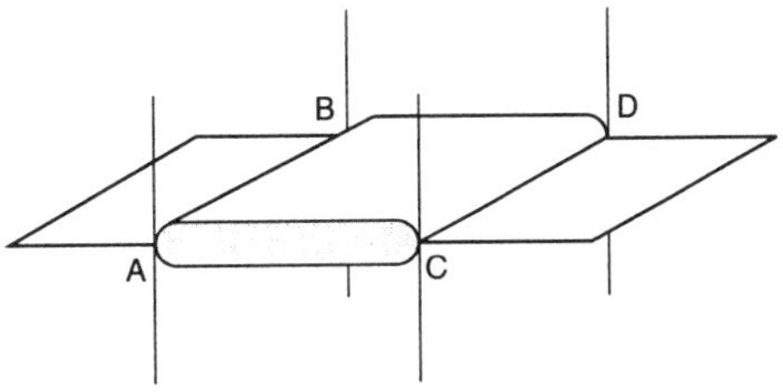

Fig. 13 A three-dimensional sketch of a stacking fault in a fcc crystal. The fault is a narrow ribbon several atomic diameters in thickness. It is bonded by partial dislocations (the lines AB and CD). Source: Ref 4

Stacking faults and *twin boundaries* are other internal surface defects. While found in all crystal structures, they are most easily described with reference to the fcc one. A stacking fault in a fcc lattice corresponds to a "mistake" made in the close-packed plane stacking sequence. Instead of the usual ABCABCABC sequence, an AB-C**ABAB**CAB one is found. The placing of a plane in the A, rather than C, position results in a thin layer of hcp-like material (denoted by **ABAB**). The thickness of this defect is only several atomic diameters in the direction normal to the close-packed planes. Stacking faults in fcc materials generally occur as ribbons (Fig. 13). The fault extends normal to the plane of this figure over distances that are large compared to an atomic size. The ribbon width (the distance between points A and C or B and D in Fig. 13) is highly variable, ranging in size from the order of one to many atomic diameters. Generally, if the energy of the hcp and fcc allotropic forms of the solid are comparable, the width is large, and vice-versa. The boundaries at the edges of the faults (lines AB and CD, Fig. 13) are defined by a special type of dislocation that accommodates the disregistry between the hcp and fcc stacking at the boundaries. Stacking faults play an important role in the work hardening behavior of some fcc metals and alloys. If their width is large, the material work hardens more than if it is small.

The stacking sequence across a twin boundary is ABCA**B**ACBA; the position of the boundary is denoted by **B**. Note that to either side of this boundary the stacking sequence is typical of fcc. (ACBACB represents the same stacking as does ABCABC, in that close-packed layers repeat every fourth layer.) At the twin boundary, a layer of ABA (hcp stacking) exists, so twin boundaries are somewhat akin to stacking faults. However, there are differences between these types of defects. The differences arise from the different positioning of the atoms in the atomic plane twice removed from the respective boundaries. Twins

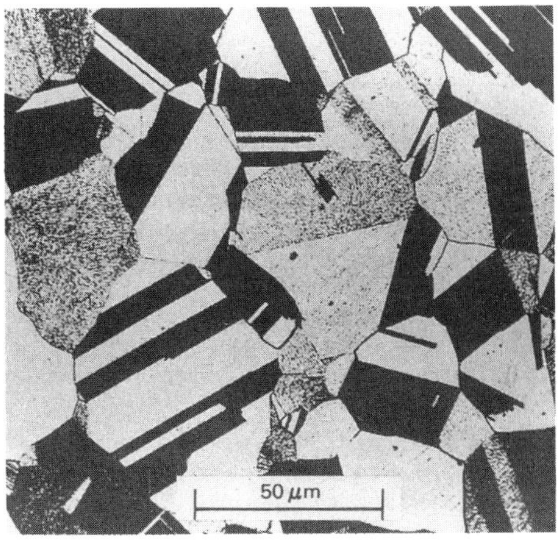

Fig. 14 The microstructure of annealed cartridge brass (70Cu-30Zn), illustrating both grain boundaries and annealing twins. The twins are the regions with parallel sides within the grains. Source: Ref 1

also typically have a width much greater than that of stacking faults. Examples of twins in a copper alloy are shown in Fig. 14. These twins developed in response to heat treatment, and for this reason they are called *annealing twins*. Twins do not affect mechanical behavior to the same degree that stacking faults do (an important exception is low-temperature deformation of bcc metals). Thus, of the several surface defects discussed, grain boundaries play an important role in plastic deformation, stacking faults affect the work hardening behavior of fcc metals, but twins generally play only a minor role in plastic flow.

Volume defects—pores and microcracks—are often present in engineering solids. Volume defects have all three of their dimensions much larger than the atomic size, although the characteristic dimension may still be small (e.g., on the order of 10^{-7} m). Volume defects almost invariably reduce strength and fracture resistance. (An exception is for spherical pores having a radius on the order of nanometers. Such voids are sometimes found in materials exposed to high energy radiation, and a modest increase in strength attends their presence.) The reductions in strength and fracture resistance can be quite substantial, even when the defects constitute only several percentage by volume of the material. In metals, pores are much more likely to be found in cast than in wrought products. The shrinkage accompanying solidification in almost all metals is manifested in microporosity (i.e., pores having diameters on the order of micrometers). The extensive deformation accompanying the production of wrought metals is usually sufficient to "heal" or close this microporosity. Powdered materials, be they metals or ceramics, frequently contain pores. As mentioned, powder products are typically fabricated by a pressing operation followed by a high-temperature heat treatment (sintering) that results in material densification. Full density is difficult to achieve through a "press and sinter" cycle, and thus residual porosity is usually found in the sintered product. Full density is more likely to be obtained when a

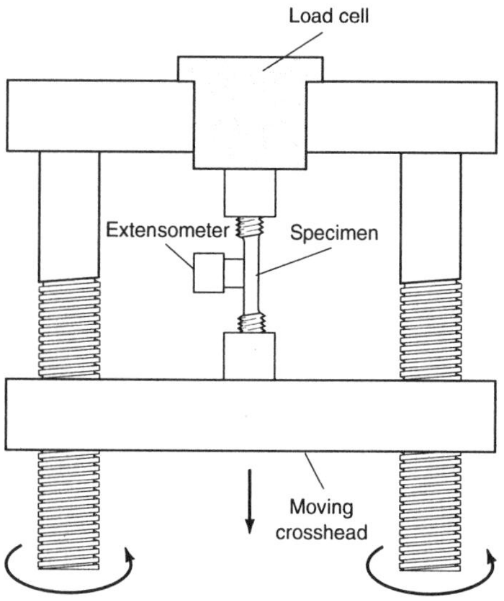

Fig. 15 A schematic of a tensile test. The sample is elongated at a specified rate, and the force required to produce a given elongation is measured by the load cell. Elongation is measured by an extensometer or similar device. Source: Ref 4

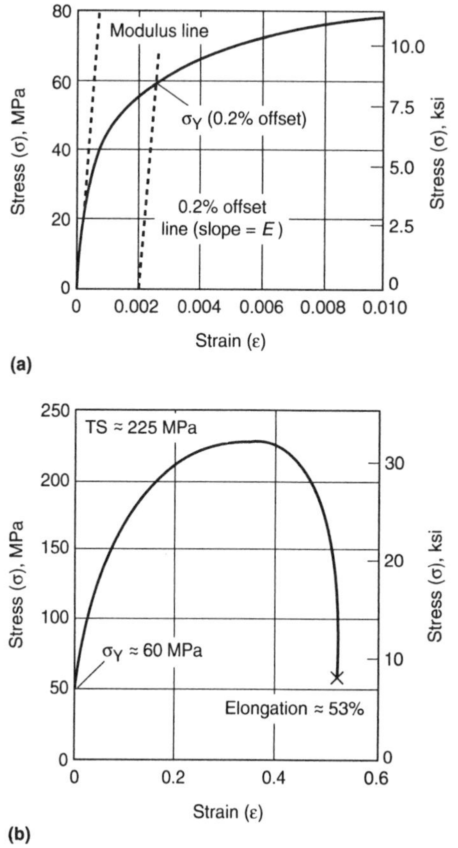

Fig. 16 The engineering stress-strain diagram for soft polycrystalline copper. (a) The low strain region, indicating the initial linear elastic behavior of copper, followed by plastic yielding. The slope of the dotted line defines the elastic modulus (E) and the offset line, with this same slope, is used to determine σ_y. (b) The complete stress-strain curve, indicating the yield strength, the tensile strength (TS), and the percent elongation to fracture. Source: Ref 4

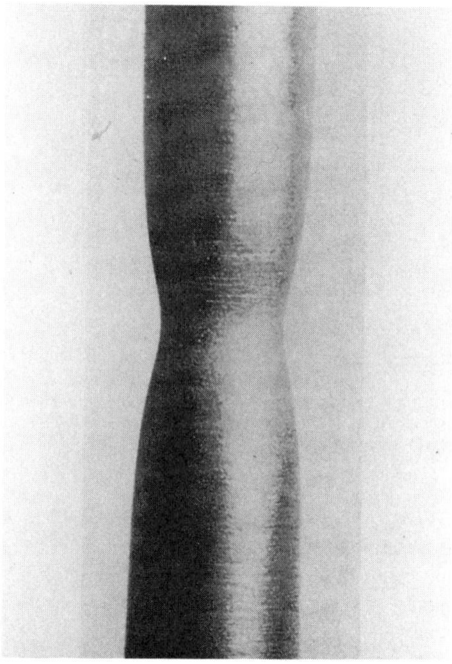

Fig. 17 A neck in a round tensile bar. The neck starts to develop when the tensile strength is reached and becomes more pronounced as the test is continued. Source: Ref 4

stress is applied during sintering (as in hot pressing, in which a uniaxial compressive stress is applied, or hot isostatic pressing, in which the stress state is hydrostatic compression). Pore removal is facilitated by pressure for much the same reason that deformation processing removes pores in the original ingot structure in wrought products.

Evaluation of Mechanical Properties

Since this article is primarily concerned with the relationships among microstructure and mechanical properties, the most common means by which such properties are determined are reviewed here. The relevance of the properties to design is also emphasized.

Tensile Properties. A tensile test (Fig. 15) determines a number of important mechanical properties. The material to be tested, in the form of a cylinder or sheet, is gripped at its ends. A standard length l_0, the sample initial or gage length, is marked on the sample. The initial cross-sectional area (A_0) normal to the sample long axis is also measured. The cross-head of the machine moves downward at a specified rate, thereby elongating the sample. A length measuring device (e.g., an extensometer) measures the instantaneous sample length (l); the change in sample length (the extension, $\delta l = l - l_0$) is then calculated. The load cell simultaneously records the force (F) needed to produce this extension.

Thus, the "raw" data obtained from a tensile test are F and δl. The force depends on the sample cross-sectional area as well as the material. The geometrical dependence is eliminated by dividing F by A_0, and the resulting parameter is termed the engineering stress: $\sigma_E = F/A_0$. The extension is similarly normalized to eliminate sample length effects on test results. This is done by dividing δl by l_0, and the ratio is called the engineering strain: $\varepsilon_E = \delta l/l_0$. Test results, now no longer dependent on sample geometry for the most part, are displayed as a graph of σ_E versus ε_E.

The results of a tensile test on a simple metal, copper, are shown in Fig. 16. Figure 16(a) illustrates the low-strain region of the test, whereas Fig. 16(b) displays the complete test results.

At very low strains (less than about 0.05%), stress and strain are related linearly. This corresponds to *linear elastic deformation*, for which the deformation is elastic or recoverable. That is, when the force is removed the material length returns to its initial length. The proportionality constant relating stress and strain in the elastic region is the *elastic modulus*, E:

$$\sigma = E\varepsilon \qquad \text{(Eq 1)}$$

Values of elastic moduli reflect atomic bond strengths. Moduli are highest in materials with strong bonding (e.g., covalent solids) and are lowest in polymers. The elastic modulus is used in design to limit or control elastic deflection. For a specified stress, a high modulus material deflects less than a low-modulus one.

At a certain stress, the stress and strain no longer relate linearly. This corresponds to the onset of permanent (plastic) deformation. It is difficult to determine the critical stress precisely, because it varies among testing devices and is sensitive to machine "stiffness." An alternative way of defining the approximate onset of plastic deformation is therefore used (Fig. 16a). A specified offset of 0.2% ($\varepsilon = 0.002$) is made on the strain axis. A line parallel to the elastic loading line (of slope E) is then drawn. The intersection of this line with the stress-strain curve defines the 0.2% offset yield strength, σ_y. The procedure ensures that reported yield strengths do not vary from laboratory to laboratory or machine to machine. The physical significance of σ_y is that it is the applied tensile stress producing a permanent strain of 0.2%. The yield strength of a material, unlike its modulus, is structure-sensitive. Some relationships between microstructure and σ_y are discussed in the section "Microstructure and Low-Temperature Strength" in this article. The yield strength is also a design parameter. It is the stress (adjusted by an appropriate safety factor) that engineers use to ensure that plastic deformation does not occur in a structure. It is also the stress that designers of metal-processing equipment consider when developing deformation processing schemes.

For a material like copper, continued plastic deformation beyond yielding is accompanied by an increase in σ_E (Fig. 16b). This phenomenon, work hardening, is a result of dislocation multiplication and a concomitant reduction in disloca-

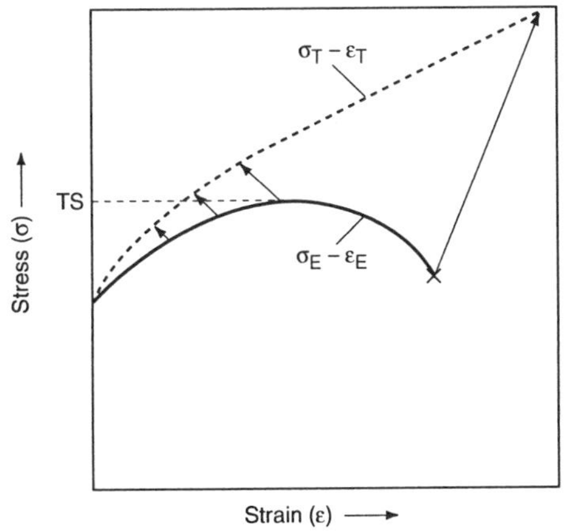

Fig. 18 Schematic illustrating the relationship between the tensile true stress/true strain (dotted line) and engineering stress/engineering strain (solid line). For engineering strains less than the necking strain, the true stress-strain curve is displaced vertically and to the left relative to the engineering stress-strain curve. At some strain beyond the necking strain, the true strain becomes greater than the engineering strain, while the true stress remains greater than the engineering stress. Source: Ref 4

tion mobility, as discussed above. At a certain critical (material-specific) strain, σ_E reaches a maximum value, called the material tensile strength (TS; in Fig. 16b, TS = 225 MPa, or 33 ksi). For strains beyond the tensile strain (0.36 in Fig. 16b), the engineering stress decreases until the material fractures (noted by an X in Fig. 16b).

It might appear that copper "work softens" beyond the tensile strain, but the maximum in σ_E is an artifact. Specifically, the TS corresponds to the onset of *necking* (Fig. 17), wherein a geometrical instability (localized deformation at a specific cross section along the gage length) initiates. Before necking, plastic deformation occurs uniformly along the gage length. That is, the material extends uniformly at all positions along the length, and because material volume remains essentially constant, material diametrical (or thickness for sheet samples) contractions are likewise uniform. However, at necking deformation becomes localized and is restricted to a specific cross-sectional area. Continued straining leads to a further reduction in this area, and the load decreases due to this effect. Since σ_E is calculated on the basis of a constant (initial) cross-sectional area, it also decreases. In other words, although the material continues to work harden, this effect is more than compensated for by the reduction in cross-section in the necked region.

A more precise way of calculating stress following necking is to base the calculation on the cross-sectional area of the neck. In fact, given that the sample cross-sectional area decreases (albeit uniformly) prior to necking, such a procedure is preferable even before necking. To do this a *true stress* is defined ($\sigma_T = F/A_i$, where A_i is the instantaneous sample cross-sectional area). As the gage length increases during deformation, a better way of defining strain is with the *true strain* ($\varepsilon_T = \ln(l_i/l_0)$, where l_i is the instantaneous sample length). Before necking, there are well-defined relations among the respective engineering and true stresses and strains. These come about because material volume remains constant and the initial and instantaneous sample lengths and cross-sectional areas are related by $A_0l_0 = A_il_i$. The following relations then apply:

$$\sigma_T = \sigma_E(1 + \varepsilon_E) \quad \text{(Eq 2a)}$$

$$\varepsilon_T = \ln(1 + \varepsilon_E) \quad \text{(Eq 2b)}$$

It is important to reiterate that Eq 2(a) and (b) hold only for strains prior to necking. The equations show that, for tensile loading, $\sigma_T > \sigma_E$ and $\varepsilon_T < \varepsilon_E$. Thus for strains less than the necking strain, a tensile true stress-strain curve is displaced upward and to the left relative to the tensile engineering stress-strain curve.

To determine the corresponding true stress and strain for strains beyond necking, the neck area (A_{neck}) must be measured. Then σ_T is calculated as $\sigma_T = F/A_{neck}$, and ε_T is determined by $\varepsilon_T = \ln(A_0/A_{neck})$. A schematic comparison of the true stress-strain and engineering stress-strain diagrams is made in Fig. 18. It is clear that the material work hardens throughout plastic deformation (i.e., the slope of the true stress-strain curve is always positive). Furthermore, the necking point cannot be discerned on the true stress-strain curve.

Another useful, but seldom reported, feature that can be obtained from a true stress/true strain curve is the area under this curve. It represents the plastic work per unit volume done on the material through fracture. (Since strain is dimensionless, the units of this area are stress in Pa, or J/m^3.) This area can be considered a measure of the material fracture work. As discussed below, however, there are better measures of resistance to fracture.

To summarize, the important design parameters obtained from a tensile test are E (used in design involving elastic deflection) and σ_y (used in design concerned with plastic flow). The TS is routinely relayed when tensile test results are reported, but TS has no direct bearing in design, although it is a good qualitative measure of the overall strength level of the material. In addition, the percent *elongation to fracture* (53% for the copper of Fig. 16b) is reported. To a degree, this parameter is a measure of material malleability or ductility. However, percent elongation depends on the choice of initial gage length (it is greater the shorter the length). A better parameter to assess ductility is percent *reduction in area*, which is the reduction in sample cross-sectional area of the neck at fracture. An additional parameter useful in metal-forming operations is the work hardening behavior of the material. There are several ways to assess this. All of them relate to the slope of the true stress-strain curve following the onset of plastic deformation; the higher this slope, the greater the work hardening.

Figure 16 illustrates the tensile behavior of a typical ductile metal. Other materials exhibit different behavior, as illustrated in Fig. 19. Brittle ceramics, such as silica glass and alumina (Al_2O_3) at room temperature (Fig. 19a), exhibit linear elastic behavior and undergo no permanent deformation before fracture. The thermoplastic polyethylene displays an extraordinary degree of permanent deformation at room temperature

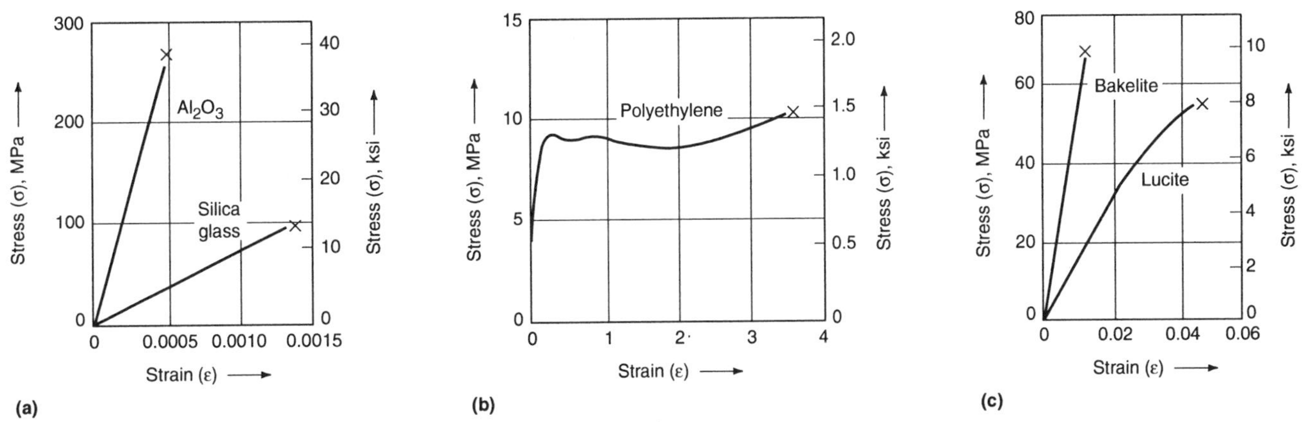

Fig. 19 Room-temperature tensile engineering stress-strain curves for selected materials. (a) The curves for polycrystalline Al_2O_3 and SiO_2 glass are characteristic of brittle solids that only elastically deform prior to fracture. (b) Polyethylene at room temperature demonstrates extensive malleability, coupled with a relatively low strength. (c) Bakelite (a thermosetting polymer) and Lucite (a thermoplastic) are not as "brittle" as Al_2O_3 and silica glass, but they are not malleable materials at room temperature. (Bakelite is never so, but Lucite becomes ductile at a temperature above room temperature.) Source: Ref 4

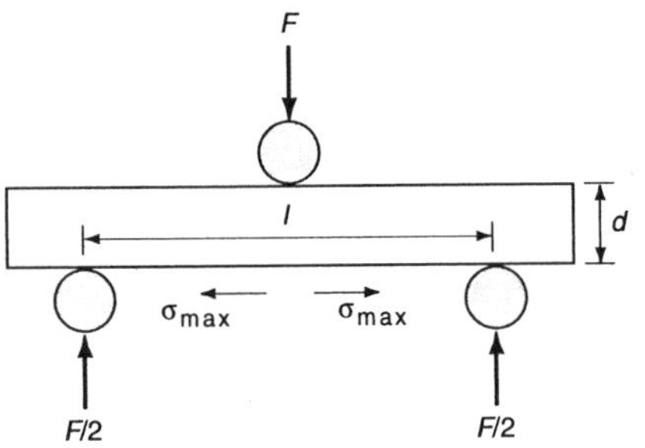

Fig. 20 A bend test eliminates the grip problem in tensile testing ceramics and can be used to measure the material modulus of rupture (Eq 3). Source: Ref 4

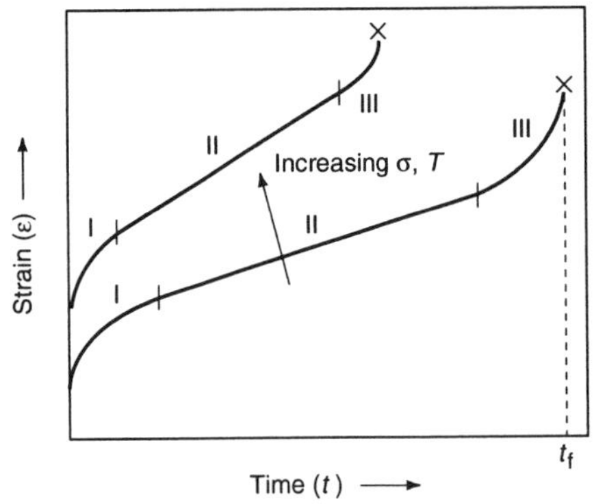

Fig. 21 Schematic results of an elevated-temperature creep test in which a material permanently deforms at a constant stress. As indicated, increasing stress and/or temperature increases the creep strain and the creep rate ($d\varepsilon/dt$). Source: Ref 4

(Fig. 19b), whereas Bakelite (phenol-formaldehyde, a thermosetting polymer incapable of permanent deformation) and Lucite (polymethyl methacrylate, a thermoplastic) exhibit low room-temperature tensile fracture strains (Fig. 19c). Were Lucite tested at a somewhat higher temperature, its stress-strain curve would resemble that of room-temperature polyethylene. Conversely, were polyethylene tested at a low temperature, its tensile behavior would qualitatively resemble that of room-temperature Lucite. Thus material stress-strain behavior is temperature sensitive. For thermoplastic polymers, there is a transition between malleable and "brittle" behavior at a critical temperature close to what polymer chemists call the *glass transition temperature.* Many metals, particularly the bcc transition ones, also exhibit a similar temperature sensitivity of their ductilities.

It is difficult to tensile test ceramics. They are brittle, and gripping them in such a test often leads to their fracturing prematurely. Thus, other tests are often employed to evaluate ceramic mechanical properties. A compression test is one. In such a test a rectangular or circular cross-sectioned sample is placed between two platens, which are displaced so as to reduce sample length, rather than increase it as in a tensile test. For the most part, the platens eliminate the "grip" problem. A compression test can be used to determine the material modulus, yield strength (if it deforms permanently), and compressive fracture stress (if it is brittle). Ceramic compressive fracture stresses are about ten times greater than their tensile fracture stresses. This is so because preexisting flaws, commonplace in ceramics, tend not to extend and grow when subjected to compressive loading, whereas a tensile stress facilitates their growth.

A bend test (Fig. 20) measures something like a tensile fracture stress. During bending, one side of the sample is subjected to tensile loading and the other half to compressive loading. The maximum tensile stress is obtained at the sample surface and, for the configuration of Fig. 20, is given as:

$$\sigma = 6M/bd^2 \qquad \text{(Eq 3)}$$

where M is the applied bending moment (=Fl), b is the sample dimension (width) into the plane of the drawing, and d is the thickness of the sample. The *modulus of rupture*, σ_r, is the stress of Eq 3 evaluated at fracture. Ceramic moduli of rupture are somewhat higher than tensile fracture stresses (although much less than compressive fracture stresses). The reason is that only one-half of a modulus-of-rupture sample experiences a tensile stress, and it is only on the surface of this half that the maximum such stress acts. Since ceramic fracture almost invariably initiates from microscopic flaws, the fraction of the sample experiencing a substantial tensile stress is less in a modulus-of-rupture test than in a tensile test where the material is uniformly loaded. A corollary to this "probability" effect is that fracture stresses of ceramics (whether tensile, compressive or modulus of rupture) are size dependent. Smaller samples display, on the average, higher fracture stresses because the probability of finding large preexisting flaws in them is less. Suitable procedures exist for converting average fracture stresses from one sample size to another. Finally, a fairly large variation in measured fracture stresses of similarly sized samples is also observed in ceramics. This is due to the variations in finding different-sized flaws in a given material volume. Thus, design against ceramic fracture is a "calculated gamble," and this partly accounts for the reluctance to use ceramics in critical applications. Only by processing these materials in a way so as to minimize the presence of microscopic flaws can ceramics be considered for such use. This distinguishes them from metals, which are remarkably flaw tolerant in comparison.

Hardness. A *hardness* test is easy to execute and provides information on the yield strength of a material. In brief, a hard material (the indentor) is placed above the sample surface. A load is then placed on the indentor that penetrates this surface. The size of the resulting indentation, a measure of the permanent deformation the material undergoes in the test, relates to the material hardness; a small indentation indicates high hardness and vice-versa. When the load is divided by the indentation area (sometimes the projected area of the indentation), a hardness number, H, is defined. Hardness has units of stress and, since it relates to permanent deformation, H correlates with σ_y. The stress state arising during a hardness test differs from that in a tensile test, however. This results in H being greater than σ_y. To a good approximation, $H \sim (2.5 \text{ to } 3.0)\sigma_y$; the approximation $H = 3\sigma_y$ is often used.

Strain-Rate Sensitivity. Low-temperature (usually room-temperature) properties of metals and ceramics are not *strain-rate sensitive.* (Polymers are considered later.) That is, σ_y and flow stress depend not at all, or only slightly, on how fast the material is strained. This is not so at elevated temperatures, where materials manifest a higher flow stress the more rapidly they are strained. ("Elevated" here means temperatures typically greater than about 40% of the material's absolute melting temperature.) *Strain rate* is defined as:

$$\dot{\varepsilon} = (1/l)\,(dl/dt) \qquad \text{(Eq 4)}$$

where l is the instantaneous gage length and dl/dt is the rate at which the sample is extended or compressed. At elevated temperatures, the flow stress (at a particular strain) varies with $\dot{\varepsilon}$ according to the empirical equation

$$\sigma = K(\dot{\varepsilon})^m \qquad \text{(Eq 5)}$$

where K is a "strength coefficient" and m is the *strain-rate sensitivity.* Strain-rate sensitivities can be as high as unity, as they are for viscous glasses at high temperature. More typically, high-temperature m values range from about 0.2 to 0.5 for inorganic solids, with m monotonically increasing with temperature.

Creep. A corollary to the above is that at high temperature, materials may permanently deform under a constant stress, contrary to their low-temperature behavior. The phenomenon is called *creep*, and creep deformation, eventually culminating in fracture, is an important high-temperature design consideration. Creep tests assess material suitability for such applications. In these tests the material is subjected to a constant stress (sometimes a constant load) and the extension is measured. A schematic of the results of a creep test is shown in Fig. 21. As indicated, both the strain level and the strain rate increase with stress and/or temperature. Following an initial transient (Stage I), the strain rate attains a minimum, constant value: the steady-state or Stage II creep rate. This creep rate is often a high-temperature design parameter. For a potential application, for example, it might be required that the steady-state creep rate of a material be less than a specified value. Following Stage II creep, the creep rate increases. This Stage III creep is the precursor to material fracture (denoted by X in Fig. 21).

A slightly different test, a stress-rupture test, is also used to assess creep resistance. In this test, the sample is subjected to a constant load, and the time to fracture (also called time to rupture) is measured as it varies with stress and temperature. Stress versus time-to-fracture diagrams are then constructed and used in design. (Extrapolation is required, because anticipated high-temperature material lives can be decades. Such durations exceed by far the times available for laboratory

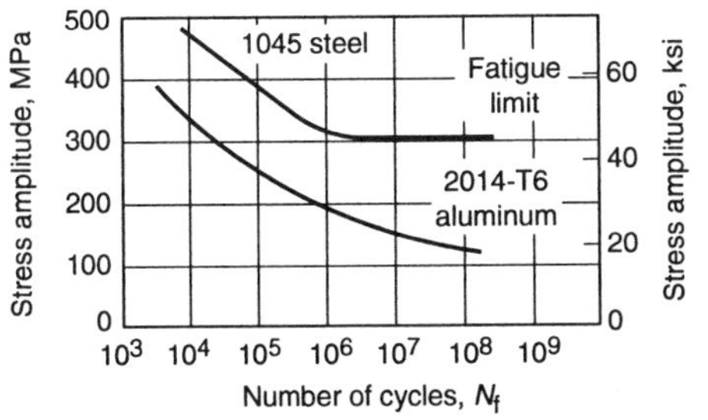

Fig. 22 Maximum stress versus number of fatigue cycles to failure for a plain carbon steel and an age-hardenable aluminum alloy. The steel manifests a fatigue limit, whereas the aluminum alloy does not. Source: Ref 7

testing.) Further information on creep deformation and creep fracture is provided in the article "Design for High-Temperature Applications" in this Volume.

Fatigue and Corrosion. Materials subjected to cyclical loading often fail even when the maximum stresses to which they are subjected are less than their yield strengths. The phenomenon is called *fatigue*. Materials can be assessed for fatigue resistance by subjecting them to cyclical stresses and measuring the number of cycles they sustain before fracturing. One common test applies alternating tensile-compressive stresses of equal magnitude. More specialized tests vary the maximum and minimum stresses, and some tests even stochastically load the material. The results of the first type of test (equal maximum magnitude tensile-compressive stresses) are summarized by plotting the maximum stress versus the logarithm of the number of cycles to failure (Fig. 22). As expected, higher applied stresses result in a lesser number of cycles to failure; the effect is most pronounced at short material lives. At longer lifetimes, material life is not as sensitive to stress. In fact, for the steel whose behavior is illustrated in Fig. 22, there exists a stress level, the *fatigue limit*, below which the material may be assumed not to fail by fatigue, regardless of how many stress cycles it experiences. Such behavior is observed in many ferrous materials, and for them the fatigue limit becomes a design parameter. Other materials do not manifest a fatigue limit (e.g., the aluminum alloy of Fig. 22). For such materials, the stress that causes failure in a very large number of cycles (e.g., 10^8 to 10^9) is the corresponding fatigue design parameter.

Polymers display fatigue behavior somewhat similar to that of metals. However, repeated cycling of polymers leads to their heating (thus test frequency is an important parameter), and this must be taken into account in design. Ceramics, too, are prone to fatigue, but the flaws they contain generally result in short fatigue lives if the applied stress is any appreciable fraction of their fracture strengths.

Corrosive environments often accelerate material failure. The detrimental effect of corrosion and stress acting concurrently can lead to *stress corrosion* fracture. Likewise, gaseous environments may accelerate high-temperature material failure. Creep and fatigue can interact, often deleteriously. Factors like these play an ever more important role as materials are called on to perform in ever more demanding situations. Further information may be found in the articles "Design for Corrosion Resistance," "Design for Fatigue Resistance," and "Design for High-Temperature Applications" in this Volume.

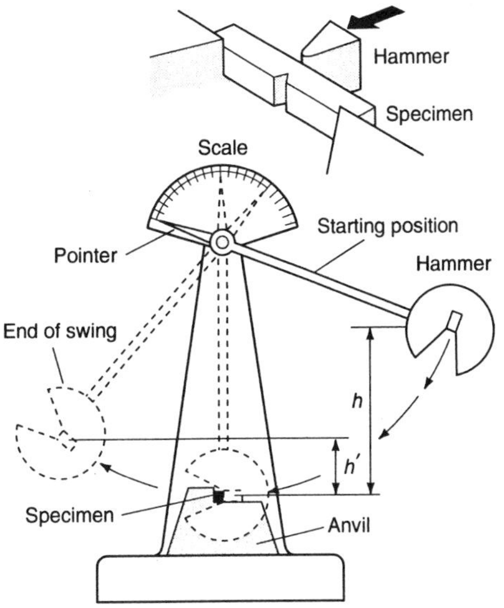

Fig. 23 A Charpy V-notch impact testing apparatus. The hammer is released from a fixed height, strikes the sample, and fractures it. The energy absorbed in fracturing the sample is equal to the loss in potential energy of the hammer. Source: Ref 7

Discussion to this point has focused on failure involving permanent deformation (i.e., by low-temperature yielding, high-temperature creep, or fatigue). We now turn our attention to material fracture.

Fracture Properties. *Fracture mechanics* is the engineering science underlying design against material fracture. Although a relatively new discipline (less than half a century old), fracture mechanics has had a profound impact on design. A fundamental assumption of fracture mechanics is that all structures contain surface and/or interior flaws. It is further assumed that quality control and/or material inspection permit a reasonable estimate of the largest flaw that a material might contain when put into service. These flaws are considered to have a crack-like shape, a shape conducive to producing a high stress concentration in the flaw vicinity. It is the propagation of these flaws, at a stress level that depends upon the flaw size and material, that results in material fracture.

Formal fracture analysis uses mechanics concepts. The stress (σ_{tip}) at the tip of a crack varies with the crack length (c) as:

$$\sigma_{tip} = A\,\sigma_{nom}\,(c)^{1/2} \quad \text{(Eq 6)}$$

where σ_{nom} is the nominal applied stress. The proportionality factor A includes terms such as the radius of curvature of the crack tip. If the material is very brittle (e.g., glass at room temperature) when the stress in front of the crack attains the value required to rupture atomic bonds (the theoretical fracture stress), the crack spontaneously propagates, causing fracture. The nominal stress at which fracture takes place, σ_F, is thus

$$\sigma_F = K/(c)^{1/2} \quad \text{(Eq 7)}$$

where K is a material property (proportional to the theoretical fracture strength in the case of a very brittle material).

Most materials are not as brittle as glass. For them the stress concentration in front of the crack may induce localized plastic deformation there prior to crack propagation. These complexities are taken into account by rewriting Eq 7 as:

$$\sigma_F = K_{Ic}/(\alpha\pi c)^{1/2} \quad \text{(Eq 8)}$$

where K_{Ic} is the *fracture toughness*, a material property. (The π term in Eq 8 is a convention, whereas the α term takes into account details of the crack shape. That is, α is crack-shape dependent, although values of α are always of the order of unity.)

Fracture toughness is straightforwardly measured by any of a number of standard tests. A material containing an intentionally introduced sharp crack (the length of which is measured before testing) is subjected to a tensile stress, σ_F is measured, and Eq 8 permits K_{Ic} to be calculated. For example, a beam with an initial crack on its tensile side can be bent until it breaks, or a plate containing a surface crack can be stressed in tension until the crack propagates.

A hardness test can be used to determine fracture toughnesses of brittle materials. Thus, a hardness test can assess the resistance of a brittle material to both fracture and permanent deformation. To be useful for measuring K_{Ic}, cracks must extend outward from the hardness indentation. K_{Ic} is high when the lengths of these cracks are short, and vice-versa.

As noted, the stress concentration in front of a crack is often sufficient to initiate yielding there. This plastic deformation then accounts for most of the fracture work (the irreversible work accompanying crack propagation). Even very brittle materials (e.g., room-temperature glass) have a work (energy) associated with crack propagation. In the case of glass, this is the energy associated with formation of the fracture surfaces accompanying crack propagation. When a material is capable of plastic deformation, surface energy also accounts for a (very small) portion of the fracture work, but most of this work is taken up in crack-tip plastic deformation. The parameter G_{Ic}, the *toughness*, is a measure of the work of fracture. The dimensions of G_{Ic} are energy/area, and G_{Ic} can thus be thought of as the fracture work per unit area of fracture surface. Toughnesses vary widely. They range from about 1 J/m^2 for brittle materials, such as glass, to something on the order of 10^7 J/m^2 for "tough" steels.

It might be expected that G_{Ic} is related to K_{Ic}; after all, both properties relate to crack propagation and fracture. This is the case, and the relationship is:

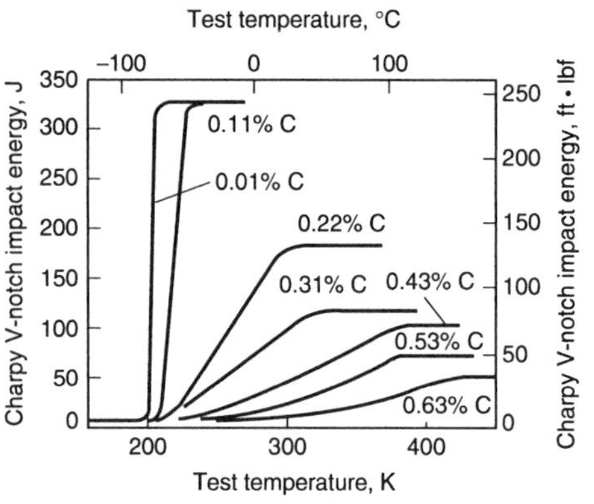

Fig. 24 The temperature variation of impact energy for steels of varying carbon content. For steels with low carbon content, the ductile-to-brittle transition is sharp and clearly identifiable. This transition is more diffuse with higher-carbon content steels. Note that the low-carbon steels are "tougher" at temperatures above the transition temperature. (They are also not as strong as the higher-carbon steels.) Source: Ref 8

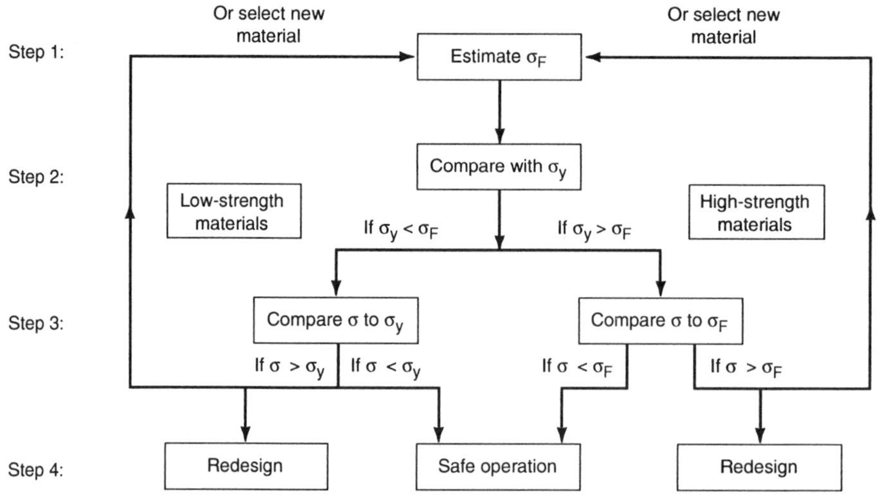

Fig. 25 Schematic of logic employed in design against "failure." Failure can take place by either material permanent deformation or fracture. Thus design entails comparing the 2% offset yield strength (σ_y) to the nominal stress at which fracture occurs (σ_F), and then using the lower of these strengths in design. Source: Ref 4

$$K_{Ic}^{\,2} = EG_{Ic}(1 + \nu) \qquad \text{(Eq 9)}$$

where E is the modulus and ν (about equal to $\frac{1}{3}$ for most metals) is Poisson's ratio.

Impact Properties. While fracture toughness is the most useful material property in design against fracture, there are other tests that provide a measure of fracture resistance. An *impact test*, widely used in the ferrous metals and plastics industries, is one such test (Fig. 23). A notched bar is broken by a falling pendulum whose kinetic energy is partially absorbed in fracturing the bar. The absorbed energy is measured, and the temperature of the sample can also be varied. Graphs of this impact energy versus test temperature for several steels are shown in Fig. 24. Most ferrous metals show an energy absorption/temperature variation similar to that displayed in the figure. The temperature at which the energy decreases rapidly with decreasing temperature relates to the *ductile-to-brittle transition temperature* of the material. Use of the material below this temperature requires some care in preventing fracture. Above this temperature, material "failure" is likely to occur through plastic deformation rather than fracture.

Empirical guides concerning safe low-temperature use of ferrous metals and alloys have been established. For example, the impact energy might be required to be above a specified value, found from experience not to result in unanticipated material fracture. While the results from impact tests are useful, it is worth repeating that fracture mechanics is the preferred tool for design against fracture.

Key Mechanical Properties for Design against Failure. Although a number of mechanical tests have been described above, it is important to recall that the two key material mechanical properties are fracture toughness (K_{Ic}) and yield strength (σ_y). K_{Ic} is used to design against fracture; σ_y is used to design against plastic deformation. In practice, design considers both deformation and fracture, but one factor is often more important. Thus modern design can be simply (but realistically) summarized as shown in Fig. 25.

There is a simple logic to the design process of Fig. 25. After incorporating important safety factors so as to design conservatively, σ_F is compared to σ_y. If σ_y is less than σ_F (generally the situation for low-strength materials), design is conducted to prevent permanent deformation. For high-strength materials (for which σ_F is generally less than σ_y), design is dictated by material fracture. The process requires knowing the appropriate material properties. These are being currently compiled in computer-ready sources, which now complement existing handbooks. Further information is supplied in the article "Sources of Materials Properties Data and Information" in this Volume.

Microstructure and Low-Temperature Strength

The common means for increasing low-temperature strength are described in this section, and brief explanations for their effectiveness are provided. While the various strengthening mechanisms are discussed separately, it is important to realize that often more than one strengthening mechanism is utilized in engineering materials. Practical examples are also provided here. To treat adequately the topics of this section, some "excursions" must be made. In particular, it is difficult to discuss material strengthening without discussing phase diagrams. The phase diagram of a specific system often dictates the means by which it may be strengthened. Space limitations prevent extensive discussion of phase equilibria, so an attempt is made to keep things simple.

Low-Temperature Strengthening of Crystalline Solids

Since dislocations are responsible for crystalline plastic flow, any factor that reduces dislocation mobility increases strength. A number of microstructural features can be used to decrease this mobility. Collectively, they are referred to as *strengthening mechanisms*. This section discusses the most common and important of these: work hardening, solid-solution hardening, grain-size hardening, and particle hardening.

Work Hardening. As noted above, while dislocations are responsible for crystalline plastic flow, there can be "too much of a good thing." If the dislocation density is high, dislocation motion is impeded by the presence of other dislocations. Almost all engineering metals and alloys have relatively high dislocation densities. Furthermore, since the number of dislocations is known to increase during plastic deformation, a material becomes stronger the more it is plastically deformed; the phenomenon is called *work hardening*.

Yield strength is related to dislocation density, ρ, by:

$$\sigma_y = \sigma_0 + \alpha G b (\rho)^{1/2} \qquad \text{(Eq 10)}$$

where G is the material shear modulus, b is the atomic diameter, σ_0 is the yield strength absent work hardening, and α is a material-specific constant. α has a value of about 0.4 for fcc metals and is somewhat higher, about 1.0, for bcc metals.

If dislocation density were to increase linearly with strain, Eq 10 predicts parabolic work hardening (i.e., the flow stress should vary with the square root of the strain). Some metals, notably copper and several of its alloys, exhibit such work hardening characteristics. On the other hand, dislocations are also removed during cold work, albeit not at the same rate they are created. This leads to a less than parabolic work hardening rate. (Under certain conditions some bcc metals linearly harden.) Since it is difficult to theoretically estimate dislocation multiplication and annihilation rates, recourse is made to empirical *constitutive equations* to describe work hardening. One such common expression is:

Table 2 Tensile strengths of copper-zinc alloys

	Tensile strength			
	Soft alloys		Hard alloys	
Alloy composition (wt %)	MPa	ksi	MPa	ksi
Cu	210	30	390	57
Cu-10Zn	250	36	505	73
Cu-20Zn	285	41	610	88
Cu-30Zn	305	44	625	91

Note: Not to be used for design purposes. "Soft" alloys refer to commercially available annealed alloys; "hard" to commercially available heavily cold-worked alloys.

$$\sigma_y = \sigma_0 + K\varepsilon^n \qquad \text{(Eq 11)}$$

where K is a "strength coefficient" and n is the work hardening coefficient. As implied above, n for metals at room temperature is usually less than about 0.5.

Work hardening coefficients can be influenced by other factors affecting strength. These include material composition and whether or not second-phase particles are present in the material. Work hardening is an important consideration in metal forming operations. A high n value indicates a high resistance to neck development, so materials are more formable the more they work harden. Examples of high work hardening materials that are extensively deformed prior to use include brass cartridge shells and aluminum beverage containers. Though most aluminum alloys do not work harden extensively, aluminum beverage container alloys do.

Varying degrees of cold work are used to manipulate the strengths of some alloys. For example, pure, soft copper has a TS of 210 MPa (30 ksi), whereas the TS of extensively cold-worked copper is 390 MPa (57 ksi) (Table 2). The high strength of stainless steel hypodermic needles (required so that they puncture skin with a minimum of pain) derives from their extended drawing during manufacturing. The high yield strength of piano wire (4140 MPa, or 600 ksi) comes primarily from the extensive wire drawing to which it is subjected. To put this strength in perspective, were a 1 in.^{2} section of piano wire available, it could support 300 one-ton automobiles without yielding plastically!

Solid Solution Strengthening. Impurity (solute) atoms strengthen metals and ceramics, sometimes substantially. The impurity atoms have a different size than the host atom and also a different atomic "stiffness" (i.e., atomic bonding is not the same in the solute and solvent). Both effects alter the crystal lattice in the vicinity of a solute atom. As a result, a moving dislocation is either attracted to, or repelled by, the impurity. Either situation results in a strength increase. When the dislocation is attracted to the impurity, the additional force required to pull the dislocation away from it is the cause of the added strength. If the dislocation is repelled by the impurity, an additional force is required to push the dislocation past the solute atom.

Treatments of solid-solution hardening indicate that the hardening depends on the differences in elastic stiffness and atomic size between the solvent and solute. Quantification of the effects is difficult, however, and has only been done in a few cases. Nonetheless, Eq 12 adequately represents the increase in yield strength, $\Delta\sigma_y$, due to solute atoms:

$$\Delta\sigma_y = A(c)^{1/2} \qquad \text{(Eq 12)}$$

where A is a constant that depends upon the solute-solvent combination and c is the solute concentration (in atomic fraction). Equation 12 has been confirmed for a number of materials, with the constant A considered an empirical parameter.

Equation 12 suggests that the maximum strength increase due to solid-solution strengthening depends on the extent to which the solute dissolves in the solvent. Solid solubilities are generally limited. Thus it is seldom that one element dissolves in another to the degree ($c = 0.5$) that the maximum possible strengthening results.

Phase diagrams are useful for indicating solid solubilities; two binary phase diagrams are provided in Fig. 26. A phase diagram is a map in which the "states" of the material are delineated as they vary with temperature and composition. Any combination of temperature and composition that lies within the region designated as "Liquid" in the copper-nickel phase diagram (Fig. 26a) corresponds to a fully liquid state. Likewise, any composition-temperature combination lying within the region marked (Cu, Ni) implies a fully solid state. There is only one solid region in Fig. 26(a), and this indicates that copper and nickel dissolve in all proportions in the solid, as well as the liquid, state. This is unusual; most elemental combinations do not display complete solid solubility. It is possible in the copper-nickel system because the two elements have similar atomic sizes and "chemistry." (They lie beside each other in the periodic table.) The region in the diagram separating the single-phase solid and liquid regions represents temperature-composition combinations for which the state of the material is a mixture of liquid and solid phases. This indicates that copper-nickel alloys do not melt at a specific temperature (as do the pure elements) but rather over a range of temperature. For example, a 50at.%Cu-50at.%Ni alloy begins to melt at about 1240 °C, but it does not become fully liquid until about 1315 °C.

Aluminum is commonly used in applications requiring good strength combined with low density. (The density of aluminum is 2.7 g/cm^3.) Solid aluminum seldom dissolves other elements in large amounts. An example of such limited

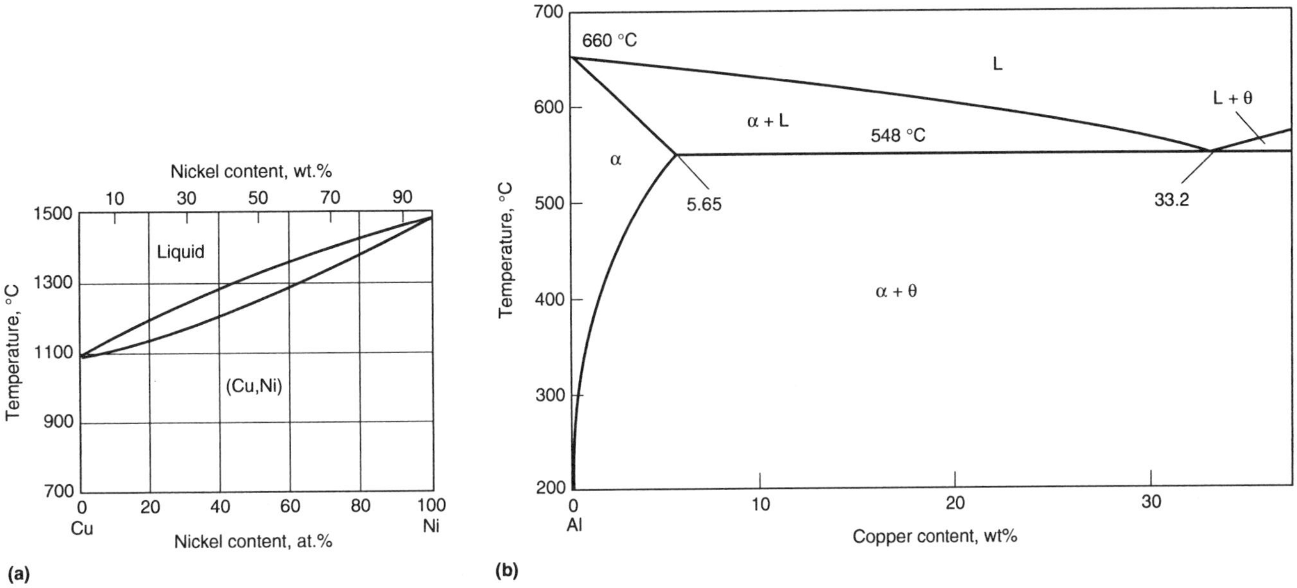

Fig. 26 Two metallic phase diagrams. (a) The copper-nickel diagram. Copper and nickel are completely soluble in both the solid and the liquid states. Source: Ref 9. (b) The aluminum-rich portion of the aluminum-copper diagram. The solubility of copper in solid aluminum is restricted, being a maximum of 5.65 wt% at 548 °C. Source: Ref 4

solid solubility is provided by the aluminum-rich portion of the aluminum-copper phase diagram (Fig. 26b). Here, as in the copper-nickel diagram, there is a single-phase liquid region (abbreviated L in Fig. 26b) and a two-phase (α + L) region. (α refers to a solid phase based on pure aluminum. Greek letters are commonly used in phase diagrams to signify solid phases.) However, there is also a solid two-phase ($\alpha + \theta$) region. The lines separating the single-phase α region and the two-phase regions represent the amount of copper that can dissolve in solid aluminum at different temperatures. The maximum solubility (5.65 wt% Cu) is found at 548 °C, where the boundaries between the lines separating the α region and the different two-phase regions intersect. The solubility of copper in aluminum at room temperature is essentially nil. Figure 26(b) is more or less typical of aluminum phase diagrams in that solid solubilities of most elements in aluminum are limited. As a consequence aluminum is not often strengthened by solid-solution hardening. Rather, the feature of limited solubility allows for hardening by second-phase particles, as discussed below.

The extent of solid-solution strengthening depends on the system-specific value of the constant *A* in Eq 12 as well as the solubility effect. In many cases, the two factors work against each other. That is, high solubility is favored when the atomic sizes and moduli of the solute and solvent do not differ much, yet small differences in these parameters usually correlate with low *A* values. Nonetheless, solid-solution strengthening is utilized for hardening many commercial metals, such as nickel-base superalloys, stainless steels, and brasses. Zinc (the major solute in brass) can dissolve in amounts up to 35 at.% in solid copper. The resulting increases in strength are impressive, as shown in Table 2, which also indicates how solid-solution strengthening and work hardening can be used in tandem. More extended discussion of structure-property-processing relationships in nonferrous alloys can be found in the article "Effects of Composition, Processing, and Structure on Properties of Nonferrous Alloys" in this Volume.

The most technologically important use of solid-solution strengthening is found in iron-carbon *martensites*. As discussed later, carbon does not appreciably dissolve in the low-temperature bcc form of iron. However, appropriate thermal manipulation permits carbon in excess of the equilibrium solubility to be "trapped" in it. The amount of trapped carbon remains low (less than several atomic percent), but significant hardening is provided nonetheless. The reason that carbon is such a potent hardener of iron is that it dissolves interstitially in the bcc lattice, producing a (tetragonal) distortion. The resulting internal stress interacts strongly with moving dislocations, substantially reducing their mobility. More extended discussion of martensite is given in the section "Strengthening of Steels" in this article.

Grain Boundary Strengthening. Dislocations cannot cross grain boundaries, so grain boundaries strengthen a material. A corollary is that the finer the grain size, the higher the yield strength.

Yield strength is found to vary with grain size according to the Hall-Petch relationship:

$$\sigma_y = \sigma_0 + k_y d^{-1/2} \qquad \text{(Eq 13)}$$

where k_y (the Hall-Petch coefficient) is a material constant, *d* is the grain diameter, and σ_0 is the yield strength of an (imaginary) polycrystal having infinite grain size. Hall-Petch coefficient values vary widely; some for metals are given in Table 3, where it is seen that k_y values for fcc metals are usually less than for bcc metals. The values listed in Table 3 can be used to assess how much grain size reductions increase strength. Aluminum, for example, cannot be substantially strengthened by grain size refinements. A reduction in grain size from 100 µm to 1 µm in an aluminum alloy yields a modest increase in σ_y of 61 MPa (9 ksi). On the other hand, an equivalent reduction in grain size in a low-carbon steel increases σ_y by 275 MPa (40 ksi). Thus, grain size reduction can be used to effectively harden iron, but not aluminum, alloys.

As noted, ceramic grain sizes are usually less than metal ones. Moreover, Hall-Petch coefficients are generally greater for ceramics. Thus, grain size effects are important to ceramic strength. And, as noted above, grain size is the one structural feature that both strengthens a material and improves its fracture resistance.

Particle Strengthening. When fine particles are dispersed in a material, strength increases, sometimes appreciably. There are several ways to produce such a dispersion. One involves powder processing, in which small particles of an essentially insoluble solid are mixed with powders of the base material. The mixture is then consolidated and, provided that the dispersoid size is sufficiently fine, effective strengthening can result. Another way to achieve such a dispersion involves selective (internal) oxidation. Aluminum, for example, oxidizes more readily than does copper. If a copper-aluminum solid solution is heat treated in an atmosphere with a suitable oxygen partial pressure, oxygen diffuses into the solid, where it preferentially reacts with aluminum to form minute Al_2O_3 particles that serve as the hardening agent.

Particles can be dispersed in a matrix by other processing routes, as they are for particle-hardened aluminum alloys. Consider again Fig. 26(b). If a 4 wt.% Cu alloy is heated into the single-phase α region (requiring a heat treatment temperature greater than the *solvus* temperature, 500 °C), the structure consists of an aluminum-rich fcc phase containing all the copper in solid solution. On cooling this alloy to a temperature below the *solvus*, the solubility of copper in aluminum is exceeded and θ phase precipitates. (This symbol indicates a phase having the approximate composition $CuAl_2$.) If the θ particles that precipitate are sufficiently fine, appreciable particle hardening is expected.

Precipitation in solids is complex. To form $CuAl_2$ from an aluminum solid solution requires the following events. Copper-rich regions must first form, and this requires diffusion of copper atoms. Then, since the crystal structure of $CuAl_2$ is different from that of aluminum, a cluster, or nucleus, of $CuAl_2$ phase must form. This *nucleation* is not spontaneous, for the nucleus-matrix interface has an associated surface energy (analogous to a grain boundary energy) that impedes nucleation. Following nucleation, the $CuAl_2$ particles grow, and this also requires atomic diffusion, which is temperature sensitive (the higher the temperature, the more capable the atoms are of moving about). Thus, precipitation involves both nucleation and growth of the new phase. Precipitation stops when a certain (equilibrium) amount of precipitate forms, the amount depending on the transformation temperature. The lower it is, the greater the precipitate fraction (the solubility of copper in aluminum decreases with decreasing temperature) and (almost always) the finer the precipitate size. Finally, because of the importance of diffusion, precipitation takes place sluggishly at temperatures below about half of the absolute melting temperature of the material.

Following the above events, the average precipitate particle size increases. Since the precipitate amount remains constant, this particle *coarsening* reduces the precipitate number and increases the spacing between precipitate particles.

While the above describes isothermal precipitation of $CuAl_2$ from supersaturated aluminum, the thermal processing employed to harden commercial aluminum alloys takes a different route. Slow cooling these alloys to room temperature leads to precipitates having too large a size to be effective in hardening. It is feasible to cool the alloy to an intermediate temperature and hold it there for a time that yields an effectively hardening particle dispersion, but there are more convenient ways to achieve the same thing: *quenching* and *aging*. The alloy is first heated into the single-phase α region to *solutionize* it. The alloy is then cooled sufficiently rapidly (quenched) to room temperature so that no precipitation occurs during the operation. Thus, the structure at room temperature is the same (i.e., the solid solution) that it was at the solutionizing temperature.

Following quenching, the alloy is "aged." This involves heating the alloy to a (relatively low) temperature for a specified period. During aging, precipitation takes place and, provided that the proper aging time and temperature are chosen, substantial particle hardening results. Figure 27

Table 3 Values of Hall-Petch coefficients for several metals

Material	Material class	k_y, MPa$\sqrt{m}$	k_y, ksi$\sqrt{in.}$
Low-carbon steel	bcc metal	0.307	1.24
Molybdenum	bcc metal	1.768	7.14
Magnesium	hcp metal	0.279	1.13
Titanium	hcp metal	0.403	1.63
Copper	fcc metal	0.112	0.45
Aluminum	fcc metal	0.068	0.27

Source: Ref 4

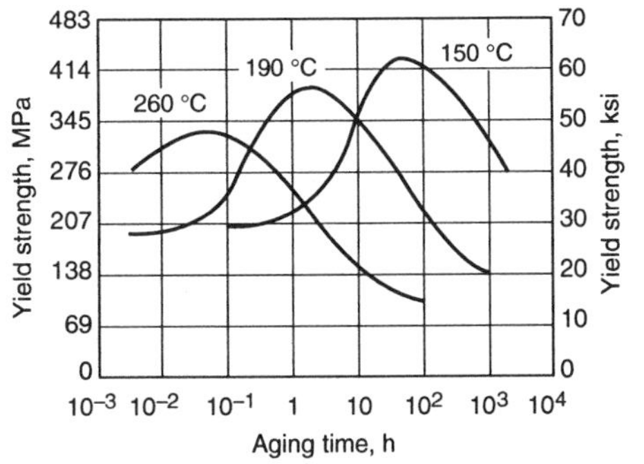

Fig. 27 Room-temperature yield strength vs. aging time at several temperatures for an initially solutionized and quenched Al-4.5wt%Cu alloy. Maximum strength is obtained at lower aging temperatures, but the time required to yield this strength is correspondingly greater.

shows a graph of room-temperature yield strength versus aging time for an Al-4.5wt%Cu alloy aged at different temperatures. Due to precipitation, σ_y initially increases. However, a strength maximum is found at a temperature-dependent aging time; the lower the aging temperature, the longer this optimum time. Moreover, the strength maximum is greater at lower aging temperatures, because a greater precipitate fraction forms at lower temperatures.

Aluminum alloys may be "overaged." That is, if the alloy is heated too long, a strength less than the maximum is obtained. This often results from precipitate coarsening, which generates particles having too large a size for effective hardening.

The above discussion has not specified the hardening precipitate, although it has been implied that it is $CuAl_2$. While this phase does form on aging of aluminum-copper alloys, nonequilibrium (transition) precipitates are precursors to $CuAl_2$, and these are the phases associated with maximum hardening. Although this is so, it does not alter the general description of precipitation given here.

Some factors responsible for particle strengthening can now be considered. The discussion is a little complicated, since several effects can be responsible for the hardening. Some description of the nature of particle-matrix interfaces is first necessary. Precipitation often yields "ordered" particle-matrix interfaces (i.e., atoms to either side of the interface "match up" across the boundary). Due to the different average atomic sizes in the matrix and the precipitate, the crystal lattice is "strained" in the interface vicinity. Yet it is important that atomic continuity persists across the boundary, for it means that dislocations can glide through particles as they wend their way through the material. If the particle-matrix interface is "disordered" (i.e., if atomic matching does not exist across it), dislocations cannot pass through the particle (analogous to the situation pertaining to grain boundaries). Instead, dislocations "bow" around particles with this kind of interface. Disordered particle-matrix interfaces are found, for example, in the powder-processed materials mentioned above.

The "particle hardening" stress depends on whether dislocations pass through or bow around particles. For bowing, it is found that the increase in yield strength due to the presence of the particles is approximately described by the Orowan equation:

$$\sigma_{part} = C[Gb/(\lambda - 2r)] \quad \text{(Eq 14)}$$

where C is a constant of order unity, G is the matrix shear modulus, b is the atomic diameter, r is the particle radius, and λ is the mean center-to-center distance between particles; thus, $(\lambda - 2r)$ is the corresponding mean surface-to-surface distance. (To keep things simple, some liberties have been taken in defining terms. The geometrical distances noted actually apply to the slip plane of the material. The approximations made can be incorporated within the constant C.) Implicit in Eq 14 is that the greater the particle volume fraction (f), the greater the strength. That is, f scales with r/λ and, therefore, for a fixed particle spacing, strength increases with increases in f.

While alloys with disordered particle-matrix interfaces always deform by dislocation bowing, this is also the dominant deformation mechanism in alloys with coarse particle dispersions, regardless of the nature of the particle-matrix interface. This is so because the bowing stress is less than that needed to push dislocations through particles; that is, dislocations take the "easy way out." In fact, the transition from under- to overaging (i.e., the point of maximum strengthening) often corresponds to a transition in dislocation transit mode. Dislocations pass through particles in underaged alloys and bow around them in overaged ones.

When dislocations pass through (shear) particles (as they do for finely dispersed particles with an ordered interface), several factors may contribute to strength. Some of them bear similarity to those leading to solid-solution strengthening. For example, the different atomic bonding and sizes between the precipitate and matrix cause strengthening for much the same reasons that these features result in solid-solution strengthening. However, other factors may also play a role in particle hardening. When a particle is sheared, additional interface area between it and the matrix is generated. Moreover, if the particle is an ordered compound (i.e., if different atoms preferentially occupy certain lattice sites in it), when dislocations pass through the particle they produce atomic disorder. The attendant increase in system energy in both cases is manifested by an increase in strength. The most important strengthening mechanism (or mechanisms) is system specific. While this is so, irrespective of the most important strengthening mechanism, strengths of alloys that deform by particle shearing increase with increased particle fraction.

Multiple Strengthening Mechanisms. Many alloys utilize more than one strengthening mechanism. Work hardening and solid-solution strengthening together impart high strength to cold-worked brasses (Table 2) and stainless steels. Nickel-base superalloys combine solid-solution and particle strengthening, with different kinds of particles used for different purposes. One type of particle, having an ordered interface with the matrix, is located within grains, whereas the other is situated on grain boundaries. Grain boundary hardening is ubiquitous in polycrystals, so such boundaries always strengthen a material (although not always substantially).

The generally accepted view (although with little justification) is that the different strengthening contributions are additive. Given our current understanding, this idea is a useful engineering approximation.

While the basic "physics" of strengthening are known, they are system specific. In addition, it is difficult to calculate a priori the strength increase expected from a specific mechanism for a given system. And, for particle hardening, the most important hardening mechanism can seldom be identified through first principles. So our understanding of strengthening remains semiquantitative. However, knowledge of the parameters affecting strength has been used fruitfully in alloy development. It is therefore appropriate to end this section with some examples. They are taken from ferrous alloys, a good choice because these materials have long been stalwart structural alloys.

Strengthening of Steels. To a metallurgist, the iron-carbon diagram is the eighth wonder of the world. The low-temperature, iron-rich portion of this diagram is shown in Fig. 28. Steels contain less (usually quite a bit less) than about 1.2 wt% C. Steels also almost invariably contain manganese, but steel mechanical properties can be discussed without considering this element.

Thermal treatment is used to vary steel properties. In brief, at some point during processing (which may be concurrent with deformation processing) the steel is heated into the single-phase fcc region (designated γ in Fig. 28; γ is also referred to as austenite, in the jargon of the ferrous industry). We describe the structural changes taking place when a steel is slowly cooled from this region to room temperature; a steel containing 0.4 wt% C is used as an example.

When a steel having this composition is cooled below 780 °C, it enters into a two-phase region of α (also called ferrite, which is bcc) and γ. As it does so, iron-rich α particles precipitate from γ. In many cases, these particles are situated on γ grain boundaries, as indicated in the 780 °C schematic microstructure of Fig. 28. On further cooling, the amount of α increases and, at a temperature slightly above 723 °C, the material is about 50% α and 50% γ. (See the schematic microstructure at this temperature in Fig. 28.) The amount of each phase depends on the steel composition. The 50% ferrite arises here because at a temperature slightly above 723 °C, the composition of a 0.40 wt% C steel lies about halfway between the two boundaries separating the two-phase and single-phase regions. If the composition were increased to, say, 0.6 wt% C (i.e., to a composition closer to the austenite boundary), the amount of α would be reduced to about 25%. Conversely, if

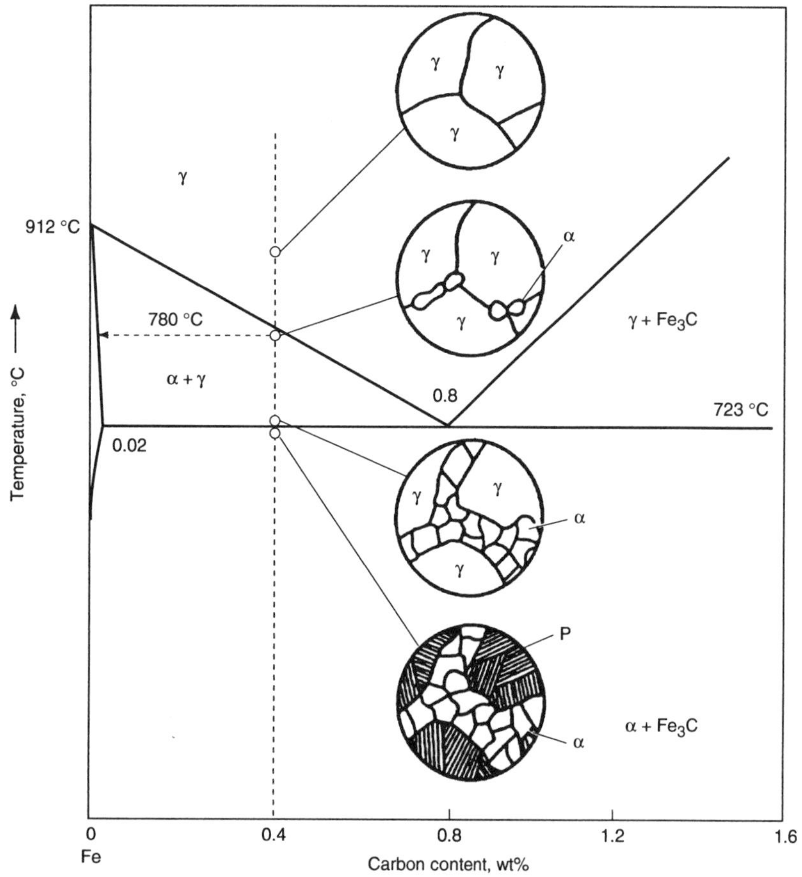

Fig. 28 The low-temperature and iron-rich portion of the iron-carbon phase diagram. The schematic microstructures illustrate the evolution of the structure of a 0.40 wt% C steel upon slowly cooling to room temperature from the γ (austenite) phase field, as described in the text. Source: Ref 3

the composition were 0.2 wt% C, the amount of ferrite would be about 75%.

Further reference to Fig. 28 indicates that something happens on cooling below 723 °C. Specifically, the phases now present are α and Fe_3C (the latter is called cementite). We note that the γ present above 723 °C is not present below this temperature. What has happened is that it has transformed into a mixture of ferrite and Fe_3C; the mixture is called *pearlite*. A schematic of the steel microstructure at a temperature slightly below 723 °C is also provided in Fig. 28. Such a microstructure is representative of "garden variety" or *ferritic-pearlitic* steels.

Properties of ferritic-pearlitic steels depend on composition (i.e., they depend upon how much pearlite the steels contain). The pearlite *microconstituent* of the steels is harder than ferrite, so the strengths of ferritic-pearlitic steels increase with carbon content. A hot-rolled ferritic-pearlitic steel containing about 0.10 wt% C contains about 10% pearlite, and its yield strength is on the order of 275 MPa (25 ksi). By contrast, a 0.80 wt% C hot-rolled steel, which is fully pearlitic, has a yield strength in the range of 450 MPa (65 ksi). Piano wire is a fully pearlitic steel, although its high strength derives primarily from the extensive cold work accompanying its production.

Microalloyed steels are a fairly recent development. They are quite a bit stronger than conventional ferritic-pearlitic steels. Microalloyed steels contain small amounts (about 0.1% by weight) of strong carbide forming elements (e.g., niobium or vanadium). These elements form carbides more readily than iron does; the alloy carbide particles serve to refine ferrite grain size. When processed optimally, a considerable grain size reduction is obtained in microalloyed steels, and since iron can be substantially hardened by grain size reductions, yield strengths of microalloyed steels are high. For example, σ_y of a 0.10 wt% C microalloyed steel is on the order of 550 MPa (80 ksi). This is much higher than the yield strength of a conventional ferritic-pearlitic steel of the same carbon content .

Quenched and tempered steels are very strong. To understand their behavior, we refer to the iron-carbon phase diagram, with a 0.40 wt% C steel again serving as an example. This alloy is heated into the single-phase γ region, as before. However, rather than being slowly cooled, the steel is rapidly *quenched* to room temperature. The quenching prevents the precipitation of α and Fe_3C; the situation is analogous to quenching of solutionized aluminum age-hardenable alloys. There is a fundamental difference with iron-carbon alloys, though. Iron undergoes an allotropic transformation, from fcc to bcc, on cooling. Iron-carbon alloys also can undergo such a transformation. Indeed, when a 0.40 wt% C alloy is quenched rapidly enough to suppress the equilibrium reactions, at a critical undercooling (the temperature of which decreases with increasing carbon content), the fcc structure spontaneously transforms into an essentially bcc one. The transformation takes place by shear absent diffusion. Because of this feature, the transformation is termed a *martensitic* transformation.

Steel martensites are not truly cubic; they are a slightly distorted (tetragonal) form of this structure. However, this alteration is not what causes ferrous martensites to be so strong. One factor is that the carbon content of the martensite is the same as that of its parent austenite, and well in excess of the equilibrium solubility of carbon in bcc iron. This produces substantial solid-solution hardening; the large effect is due to the carbon atoms residing in interstitial sites (this causes the tetragonal distortion), and these atoms interfere markedly with dislocation motion. In addition, a slight volume increase accompanies martensite formation. The accompanying deformation generates a high dislocation density in as-quenched martensite, similar to what is found in cold-worked metals. Finally, twins are present in some quenched martensites. They strengthen the material in a manner similar to (but not as effectively as) grain boundaries. Thus, quenched martensite is strengthened by a combination of solid-solution, dislocation, and twin boundary hardening.

While yield strengths of quenched martensites are high, these materials cannot be effectively used, because they are quite brittle and prone to fracture. Quenched martensites can be rendered more useful by *tempering* them. During tempering, the steel is heat treated in much the same way that aluminum alloys are aged, and somewhat comparable microstructural changes take place during tempering. For example, carbide particles precipitate, and the dislocation density is also reduced. There is an important distinction between aging aluminum alloys and tempering martensite, though. Tempering results in a strength reduction. That is, the strength decrease accompanying the lessened carbon solid-solution hardening is greater than the strength increase arising from carbide precipitation. This is opposite to the effect found on aging aluminum alloys.

Tempering occurs gradually. The matrix composition gradually approaches equilibrium as the carbide particles precipitate, grow, and coarsen. Tempering is usually interrupted before equilibrium is reached, and *tempered martensite* results, which combines high strength (often on the order of 1380 MPa, or 200 ksi, or higher) and reasonable fracture toughness. Tempered martensites find application in highly stressed components, such as automobile drive shafts and aircraft landing gears. Extended tempering of quenched martensite yields a structure called *spheroidite*. The phases present in spheroidite (α and Fe_3C) are the same as in ferritic-pearlitic steels. However, the shape of Fe_3C is spherical in spheroidite. This feature renders spheroidites tougher than ferrite-pearlite mixtures of the same carbon content, but spheroidite is not nearly as strong as tempered martensite.

Quenched martensite is a necessary precursor to tempered martensite. Thus the steel must be rapidly cooled from the γ region if martensite is to form. Martensitic surfaces are almost always obtained by quenching. However, the structure that forms in the interior of the material is another question. Large-diameter sections experience a moderate cooling rate in their interiors, regardless of how rapidly their surfaces might be cooled, and the section interior is thus likely to have a ferritic-pearlitic structure. In fact, with plain carbon steels (i.e., those containing only carbon and manganese as alloying additions), the centers of even fairly thin pieces are ferritic-pearlitic. Alloying elements are added to steels to improve their *hardenability* (i.e., the propensity for forming martensite during cooling). Elements such as chromium, molybdenum, and nickel reduce the rate at which the diffusional transformations from γ to α and/or pearlite take place, but the added elements have minimal effect on the tendency to form martensite. Thus, these elements permit martensite to be formed in sections having respectable diameters, and such alloy steels are the ones used when strength is required throughout large sections. More extended discussion of structure-processing-property relationships in ferrous materials can be found in the article "Effects of Composition, Processing, and Structure on Properties of Irons and Steels" in this Volume.

Low-Temperature Strengthening of Polymers

Long-chain molecules are the basic units in thermoplastics. Intrachain bonding in them is covalent, but interchain bonding is of the weak van der Waals type. As a consequence, thermoplastic permanent deformation takes place by chain sliding. Strengthening of thermoplastics, therefore, comes about by making chain sliding more difficult.

A simple way of strengthening thermoplastics is to make the side groups more "bulky." This produces a "steric" (geometric) hindrance to chain sliding. Thus, in the polymeric series based on the monomer C_2H_3R (where R is a radical), strength increases almost directly with side-group size. Polyethylene (R = H) is the weakest of this series, polypropylene (R = CH_3) is stronger than polyethylene, and polystyrene (R = C_6H_6) is stronger still. Another thermoplastic series, based on monomers with composition $C_2H_2R_1R_2$, is generally stronger than the series having R_1 = H. For example, polyvinylediene chloride ($R_1 = R_2$ = Cl) is stronger than polyvinyl chloride (based on C_2H_3Cl).

Crystalline polymers are stronger than noncrystalline ones. The higher density of the crystalline form signifies that the polymer chains are closer together, and this makes relative chain displacement more difficult. This is put to advantage in strengthening polyethylene by increasing the percentage of crystalline material within it. However, polyethylene is atypical in this respect. Most thermoplastics are difficult to crystallize because of their bulkier side groups and greater viscosities.

Table 4 Strengths of selected filaments and fibers

Material class	Material	Density (ρ), Mg/m^3	Tensile strength (TS)		TS/ρ, MNm/kg
			GPa	ksi	
Metals	Pearlitic steel (piano wire)	7.9	4.2	610	0.53
	Be	1.8	1.3	190	0.72
	W	19.3	3.9	570	0.20
Ceramics	Al_2O_3	3.96	2.0	290	0.51
	Graphite (Kevlar)	1.5	2.8	410	1.87
	S Glass	2.5	6.0	870	2.4
	SiC	2.7	2.8	410	1.04
Polymers	Nylon 66	1.1	1.05	150	0.95
	Polyamide-hydrazide	1.47	2.4	350	1.63
	Copolyhydrazide	1.47	2.7	390	1.84
	Poly (p-phenylene) terephthalamide	1.44	2.8	410	1.94

Source: Ref 4

Some techniques used to strengthen metals also strengthen thermoplastics. When extensively drawn, thermoplastics are strong, remarkably so when taking into account their low densities. Table 4 compares some properties of both organic and inorganic filaments. Unfortunately, such strengths are realized only in filament form; such filaments are the backbone of the carpet and fabric industry. It is an oversimplification to consider polymer filaments as having been "work hardened." During drawing, the long axes of the polymer chains align along the draw direction. Thus when the final filament size is attained, filament strengths reflect the strength of the intrachain covalent bonds and are not a measure of resistance to interchain sliding.

There is minimal polymer analog to solid-solution strengthening, for solid solubilities between polymers are almost always quite limited. (That is one reason polymer recycling is a problem.) An exception occurs when polymers are strengthened by "rigid rod molecules." These molecules, having a stiff "backbone," reinforce the polymer when dispersed in it, in much the same way that fibers reinforce composite materials. In fact, most engineered polymers are one class of composite or another. Metals and ceramics also can be reinforced in composite form.

Composite Materials

Composites are mixtures (blends) of two or more materials having significantly different characteristics. The resulting composite properties are a suitable "average" of the component properties.

Most commercially important composites are polymer-based. The polymer matrix is typically a thermoset, rather than the thermoplastics discussed to this point. Thermosets and thermoplastics differ molecularly. Specifically, thermosets form by a reaction between different chemical species. (The reader may be familiar with setting of epoxy resins, which are thermosets.) The resulting molecular architecture displays a "skeleton-like" array of primary bonds that join the chemical species of the thermoset. Unlike thermoplastics, thermosets are incapable of permanent deformation. Bakelite is probably the most well-known thermosetting plastic.

Thermoset composites are made by adding a reinforcement to the thermoset during or prior to its setting. Strong glass fibers (Table 4) are a common reinforcement. The thermoset serves as a matrix "glue," bonding the fibers together. Glass is also prone to react with water vapor in air. This induces corrosive surface flaws that act much like surface cracks in reducing the glass fracture strength. The polymer matrix also protects the glass-fiber surfaces from this degradation.

Other fibers, even stronger and stiffer, can also reinforce polymers. Various forms of graphite fibers (Table 4) display impressive strengths. Among other uses, graphite-reinforced polymers have application as sporting goods, such as tennis rackets, squash rackets, and golf club shafts. Because of their low densities, polymeric composites have also made inroads in the aerospace industry, where weight considerations are paramount.

As mentioned above, composite properties are some "average" of component properties. Often the average is a volume-weighted one. For example, the volume-fraction rule for the tensile strength of a fiber-reinforced composite is often approximated by:

$$(TS)_c = v_f (TS)_f + (1 - v_f)(TS)_m \quad \text{(Eq 15)}$$

where $(TS)_c$ is the tensile strength of the composite, $(TS)_f$ and $(TS)_m$ are the corresponding fiber and matrix tensile strengths, and v_f and $(1 - v_f)$ are the respective fiber and matrix volume fractions in the composite. Equation 15 applies when the long axes of the fibers are aligned with the stress axis. When the fibers are randomly dispersed, composite strength is less than that given by Eq 15. A lesser strength is also found when the reinforcement shape is particulate, rather than fibrous.

Expressions similar to Eq 15 are used to estimate other composite properties, such as the elastic modulus. However, not all composite properties lend themselves to such a simple formulation. Fracture toughnesses, for example, are not expressed by a volume-fraction rule, although the formulations for toughness do incorporate component toughnesses.

Composites with metal and ceramic matrices are also available, but they are far more expensive than polymeric composite. Metals can be reinforced by ceramic particles or fibers. Examples include aluminum reinforced with SiC particles or graphite fibers. Metal strengths are increased by such reinforcement, but at the expense of fracture toughnesses.

When metals reinforce ceramics, toughness is increased, but usually at the expense of strength. The increased toughness arises because the imbedded metal blunts and/or arrests matrix cracks. Reinforcement shape plays a role in toughening. Fiber- and plate-shaped metal reinforcements generally improve toughness more than equiaxed particles. When cracks approach fibers or plates, especially when the long axis of the fiber or plate is parallel to the stress axis, cracks often deflect along the reinforcement-matrix interface. This provides additional fracture work and a concurrent increase in fracture toughness.

Because of their cost, metal and ceramic composites have not yet made substantial commercial inroads. This especially pertains to composites intended for room-temperature use. However, since polymers have limited high-temperature capacity, metal and ceramic composites might find high-temperature application.

Additional information on structure-property relationships for composite materials is provided in the article "Effects of Composition, Processing, and Structure on Properties of Composites" in this Volume.

Microstructure and High-Temperature Strength

Materials are made stronger at high temperature by increasing their resistance to both time-dependent plastic deformation (creep) and time-independent plastic deformation. For most high-temperature design, preventing or minimizing creep deformation is paramount.

Equation 5, the relation between material flow stress and strain rate, can be rearranged as:

$$\dot{\varepsilon} = K' \sigma^{m'} \quad \text{(Eq 16)}$$

where m' is equal to $1/m$ of Eq 5, and K' is related to K of Eq 5 by $K' = K^{-1/m}$. Equation 16 can be written in a different form, one providing clarity in later discussion:

$$\varepsilon' = AD\,[\sigma/G]^{m''}\,[\sigma\Omega/kT]\,[b/d]^{n'} \quad \text{(Eq 17)}$$

where Ω is the atomic volume (on the order of b^3, where b is the atomic diameter), k is Boltzmann's constant, T is the absolute temperature, A is a material constant, G is the shear modulus, D is a diffusion constant, d is the grain diameter, and n' is an exponent reflecting the effect of grain diameter on creep rate.

Equation 17 represents the creep rate due to a particular creep mechanism. As discussed below, there are several independent creep mechanisms, and the net creep rate is the sum of the creep rates of all of them. However, at a specific stress-temperature combination, one mechanism usually displays a creep rate much greater than the others, and the total creep rate is then well approximated by this maximum rate. However, a different mechanism might dominate at a different stress-temperature combination. The result is that m'' and n' in Eq 17 can change with changes in stress and temperature.

Equation 17 appears complicated at first glance. Application of it can be illustrated with a simple example, viscous flow of glasses. Since glasses are noncrystalline, $n' = 0$. Moreover, $\dot{\varepsilon}$ varies linearly with stress for a viscous glass, so m'' is also zero. The strain rate/stress relationship for a viscous glass is conventionally written as:

$$\dot{\varepsilon} = \sigma/3\eta \quad \text{(Eq 18)}$$

where η is the glass viscosity. Equating Eq 17 and 18, we have:

$$\eta = kT/3AD\Omega \quad \text{(Eq 19)}$$

The relation between viscosity and diffusivity of Eq 19 is consistent with theories of viscosity derived from chemical reaction rate principles.

The appearance of diffusivity in a creep rate equation indicates that high-temperature deformation involves atomic motion. Indeed, for the viscous flow just discussed, deformation comes about solely by diffusion, although to be sure the flow is prompted by an applied stress. Crystalline materials can also diffusionally creep. The situation is illustrated in Fig. 29. The applied stress locally increases the atomic volume in the vicinity of the crystal faces so stressed, resulting in a higher vacancy concentration in the region. The excess vacancies diffuse to the sides of the crystal not stressed, and this is accompanied by diffusion of mass in the opposite direction. Thus mass is transferred from the sides of the crystal grain to the top and bottom crystal faces; the result is that the material extends (creeps) in the direction of the applied stress. While Fig. 29 indicates that diffusion takes place within the crystal volume, the mass flow can also occur along the crystal faces (or the crystal grain boundaries in a polycrystal).

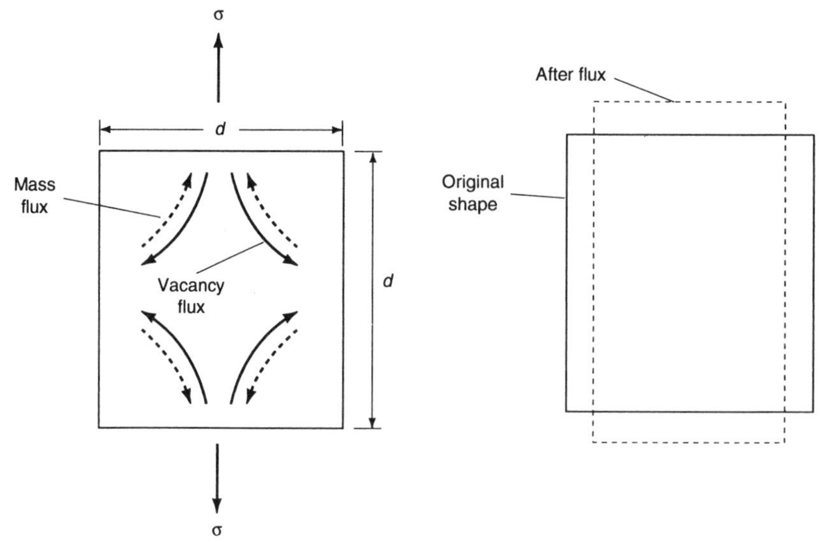

Fig. 29 Diffusional creep results from a higher vacancy concentration in regions of a material that experience a tensile stress compared to regions that do not. This results in a vacancy flux from the former to the latter areas, and a mass flux in the opposite direction. This extends the material, as indicated on the right of this figure. Mass can flow through the grain interior (as indicated here) or along grain boundaries. Source: Ref 4

Grain size plays a role in creep of crystalline solids, at least under stress-temperature combinations at which creep takes place by a mechanism of the kind just discussed. The finer the grain size, the more rapid the mass transport causing permanent deformation. Thus under conditions where creep is solely due to diffusional flow, creep resistance is improved by increasing the grain size of a material. Note that this is different from low-temperature behavior, where a fine grain size is beneficial to strength.

The grain size dependence of diffusional creep depends on whether mass is transported through the crystal volume or along grain boundaries. For the latter (*Coble creep*), n' in Eq 17 is 3. When mass transport is through the crystal volume (*Nabarro-Herring creep*), n' is 2. The different grain size dependencies indicate that, other things being equal, Coble creep should dominate creep in fine-grain materials. The Coble and Nabarro-Herring creep diffusion coefficients also are different, because of the different diffusion paths of the atoms. This results in Nabarro-Herring creep dominating at higher temperatures, and Coble creep dominating at lower temperatures. The temperature at which the transition between the mechanisms takes place is grain size dependent; it is higher the finer the grain size.

Diffusional (Coble and/or Nabarro-Herring) creep dominates creep at relatively low stresses, low enough so that dislocation motion plays no role in creep. At higher stresses, dislocation motion in conjunction with diffusion causes creep.

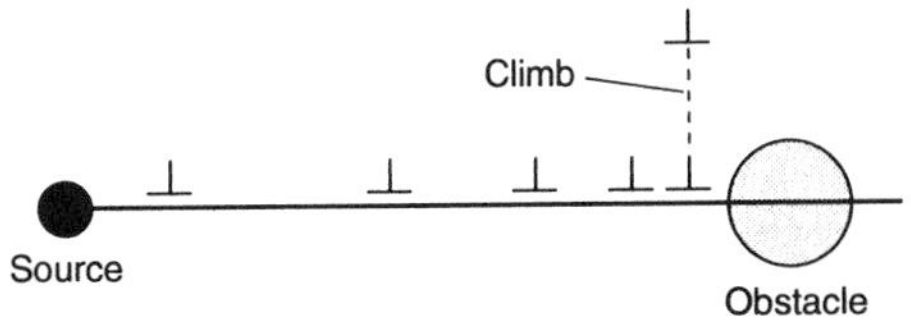

Fig. 30 Schematic representation of the climb of a dislocation that has encountered an obstacle in its slip plane. The "source" noted on the left of the figure represents a region in the material in which dislocations are generated. Source: Ref 10

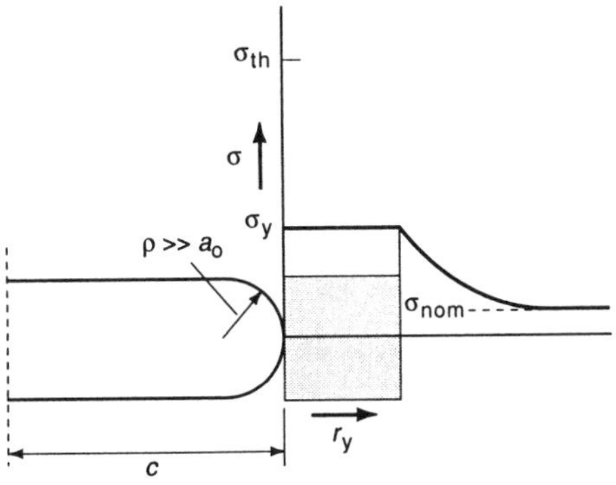

Fig. 31 Schematic of the situation in front of a crack tip in a moderately ductile material. Because of the stress concentration, the material yields plastically in front of the crack tip. Concurrent with yielding, the crack tip is blunted. Source: Ref 4

Figure 30 simply portrays the idea. Here dislocations encounter, and pile up against, a slip-plane obstacle. The applied stress is insufficient to enable the dislocations to bow around, or glide through, the obstacle. However, a dislocation may climb by diffusion to a parallel slip plane. Having climbed, the dislocation proceeds along the new slip plane until it encounters another resistant obstacle, whereupon it climbs (or descends) to another parallel plane and the process repeats. Since dislocation motion depends on both dislocation glide and climb, *climb-glide* creep is used to describe this form of creep. Climb-glide creep depends more strongly on stress than does diffusional creep; m'' ranges from 2 to about 5 in many cases, but sometimes it is substantially more. (The variation arises from the different paths that dislocations take in bypassing obstacles, as well as other microstructural features.) Because of the manner in which the creep rate depends on stress, *power law creep* is another term used to describe this form of creep.

The dominant creep mechanism at a specific stress-temperature combination is found by calculating the creep rate for the several mechanisms and determining the maximum such rate. Although the net creep rate is the sum of the several rates, the maximum one is usually much greater than the others. Deformation mechanism diagrams having axes of stress and temperature can be used to display the results of such calculations. Regions in the diagram delineate stress-temperature combinations for which a given creep mechanism dominates. Of course, to construct the diagrams the appropriate ancillary data (diffusion coefficients, grain size, etc.) must be available.

Design against creep can take several approaches. It makes sense to use a material with a high melting temperature in high-temperature applications, since to a good approximation, diffusion coefficients (on which $\dot{\varepsilon}$ depends) scale with homologous temperature (T/T_m, where T is temperature and T_m is the melting temperature of the material, both in K). There are exceptions to this generality, though. Face-centered cubic metals generally have superior creep resistance to bcc metals at equivalent homologous temperatures, because the slightly more open bcc structure results in slightly greater diffusivities.

Design against creep may also involve first determining the dominant creep mechanism. If diffusional creep dominates, then increasing material grain size is a beneficial structural alteration. If, on the other hand, power law creep dominates, no improvement in material performance is likely to result from such a change. Further, for diffusional creep, improved creep resistance can sometimes be had by placing inert particles on grain boundaries. This increases the effective grain boundary "viscosity," leading to improved creep behavior.

Example: Structure-Property Relationships in Nickel-Base Superalloys for High-Temperature Applications. Nickel-base superalloys find application as turbine blades in engines operating at temperatures as high as 1300 °C (T/T_m = 0.9). The microstructure consists of fine Ni_3Al particles dispersed in an fcc nickel-rich solid solution. These alloys are remarkably resistant to particle coarsening, a big plus, owing to a low matrix-particle surface energy. Solid-solution hardening is also used to strengthen the matrix, and when used in polycrystalline form, inert particles (e.g., carbides) may be dispersed on the material grain boundaries. Increases in engine operating temperatures over the last 50 years have been gradual but significant; from about 400 °C to the above-noted 1300 °C. This increase in operating temperature, of course, translates into improved engine efficiency. Improved high-temperature material properties are largely responsible for the increase in the engine operating temperature. Some improved material performance has come about from minor alterations in alloy chemistry. Some has resulted from processing changes. For example, the major stress axis in a turbine blade is parallel to the blade axis. In a polycrystalline blade, this stress is normal to some grain boundaries, and this causes voids, precursors to fracture, to initiate on the boundaries. Directional solidification (i.e., solidifying the material sequentially from the blade bottom to its top) results in columnar grains having boundaries aligned along the blade axis. This significantly reduces the cavitation problem. Further improvements came with the casting of single-crystal blades. Increasing the superalloy grain size from 100 μm to 1 cm (the order of the thickness of a turbine blade) reduces the creep rate by about six orders of magnitude when Coble creep is the creep mechanism, as it often is under typical blade operating conditions. Such a reduction in creep rate then permits operation at a higher temperature and/or stress.

Microstructure and Low-Temperature Fracture

As described in the section "Evaluation of Mechanical Properties" in this article, design against fracture uses the fracture toughness (K_{Ic}) to determine a safe operating stress. Since K_{Ic} is related to toughness (G_{Ic}), as shown in Eq 9, the features that make a material resistant to crack growth culminating in fracture can be elucidated by discussion of G_{Ic}.

Figure 31 is a sketch of "what happens" in front of a crack. The situation pertains to a material with reasonable capacity for plastic deformation. The stress concentration in front of the crack causes plastic yielding there. In turn this causes crack tip "blunting," such that an initially sharp crack with crack-tip radius of curvature on the order of the atomic size assumes a much greater radius of curvature.

How much work (G_{Ic}) is done during crack propagation? First, there is the work associated with the formation of the fracture surfaces. But in other than very brittle materials, this is much less than that coming from crack-tip plastic deformation. The tensile work to fracture (per unit volume) is given by the integral of the true stress/true strain curve. Thus, we surmise that the corresponding crack-tip work is also likely related to the product of a strength and a fracture strain which, for convenience, we take as the yield strength (σ_y) and the true fracture strain (ε_f). The units for the product of stress and strain are J/m^3. If this product is multiplied by a length, the units are J/m^2, the same as those of G_{Ic}. If ρ is taken as the effective length of the material undergoing crack tip plasticity, we can sensibly write:

$$G_{Ic} = C\sigma_y\varepsilon_f\rho \quad \text{(Eq 20)}$$

The constant C in Eq 20 takes into account that the stress state near the crack tip is not the same as in a tensile test, so that the corresponding yield strength and fracture strain are also not the same as those found in a tensile test. The important point is that toughness scales with flow stress and ductility. In addition, crack-tip blunting (manifested in the length, ρ) is related to ductility. The greater the ductility, the more extensive this blunting.

Material toughnesses vary widely due to variations in material strengths and ductilities. As a rule, increasing σ_y reduces G_{Ic} because of the associated ductility reduction. In terms of toughness, the lessened ductility more than compensates for the strength increase because of the additional effect that ductility has on crack blunting. (However, since K_{Ic} scales with E as well as G_{Ic}, a stiff material will have a greater fracture toughness than a pliable one at an equivalent G_{Ic}.)

Toughnesses range from as high as 10^7 J/m^2 (for tough metals) to as low as about 0.5 J/m^2 (for room-temperature glass). Above their ductile-to-brittle transition temperature, body-centered cubic transition metals and alloys have high toughness. Below this temperature their yield strengths are strongly temperature dependent, and this is accompanied by a substantial ductility reduction. The result is that low-temperature toughnesses of these metals are only on the order of 100 J/m^2. This is still high compared to typical values of G_{Ic} for ceramics (on the order of 30 J/m^2). Since there is usually some (limited) crack-tip deformation in ceramics, their toughnesses exceed that of glass, for which G_{Ic} arises solely from the fracture surface energy. No plastic deformation accompanies room-temperature crack propagation in glass, and the crack-tip radius is on the order of the atomic diameter.

As implied, G_{Ic} (and, therefore, K_{Ic}) is temperature dependent, reflecting the temperature variation of σ_y and ε_f. Yield strengths are not very temperature sensitive for fcc metals and alloys, and G_{Ic} and K_{Ic} are only mildly temperature dependent for this material class as a result. This is not so for structural bcc metals, which manifest a strong temperature dependence of σ_y, as discussed above. Their low-temperature increase in σ_y is accompanied by a heterogeneous pattern of plastic deformation that disproportionately reduces ductility. The result is a sharp temperature sensitivity to G_{Ic} and K_{Ic}. Indeed, the ductile-to-brittle transition in bcc metals often takes place over a narrow temperature range. Results from impact tests mirror this effect. Some ceramics, such as ionic salts, display somewhat similar behavior. Salt transition temperatures, though, (e.g., 650 K for NaCl) are generally higher.

Fracture may take place in essentially "flaw-free" materials; flaw-free meaning here that preexisting crack sizes are too small to cause material fracture. Virgin glass fibers, before they detrimentally interact with the environment, are essentially flaw-free, and their strengths border on the stress needed to rupture atomic bonds. Crystalline solids prone to brittle fracture seldom exhibit such high strengths. When their flaws are of a size less than about the grain diameter, heterogeneous plastic deformation precedes fracture. The "misaccommodation" between plastically deforming and nondeforming regions produces a stress concentration sufficient to nucleate a crack. Lengths of these cracks are on the order of the grain diameter; so this parameter substitutes for crack length in calculation of σ_F (Eq 8). Similar considerations apply to ceramics, but in this instance the misaccommodation often arises from a lack of slip systems (see the section "Crystalline Defects" in this article). For both flaw-free bcc metals and ceramics, σ_F increases with decreases in grain size.

What microstructural alterations might improve fracture toughness? We know that reducing σ_y almost invariably increases K_{Ic}. However, there is a limit to the usefulness of strength reduction for this purpose; too low a value of σ_y means that design is concerned with preventing plastic deformation rather than fracture (Fig. 25). Thus, there are tradeoffs between toughness and strength. Toughness can be improved without significant reduction in strength through processing improvements. Examples include reducing nonmetallic phases (typically oxides, phosphides, etc.) in steels. These brittle inclusions reduce toughness because they fracture in front of the crack tip, causing "leap frogging" of the crack to the inclusion. Since the inclusions do not reduce σ_y, "cleaner" materials are tougher than "dirty" ones at the same strength levels. Furthermore, manipulation of the morphology of the phases in a two-phase material can have a significant effect on fracture toughness.

Ceramics are toughened to an extent by grain size reductions. However, fracture toughnesses of even fine-grained ceramics are only about 5 MPa$\sqrt{m}$ (4.5 ksi$\sqrt{in.}$). This can be compared to a "brittle" metal, for which K_{Ic} might be about 30 MPa$\sqrt{m}$ (27 ksi$\sqrt{in.}$). Low ceramic toughnesses have hampered their use. Recent advances have improved the situation somewhat. The ceramic ZrO_2 has several allotropic forms. When added as a particulate to other ceramics, ZrO_2 can impart toughness in the following way. The stress intensity at a crack tip is sufficient to induce an allotropic, martensitic transformation in the dispersed ZrO_2. The associated transformation work represents additional fracture work, increasing G_{Ic} and K_{Ic}. Toughening by this mechanism has raised K_{Ic} values of certain ceramics to about 16 MPa$\sqrt{m}$ (14 ksi$\sqrt{in.}$). While this is still low, it is approaching a range where, with appropriate care, ceramics could be used in more demanding situations.

Composite toughening has been discussed in the section "Composite Materials" in this article. Only a few additional comments are necessary here. Toughening a ceramic by dispersing a ductile phase within it decreases σ_y while improving fracture resistance. The additional fracture work due to the particles depends on the crack path taken. If the ductile phase fractures in the wake of a propagating crack, the ductile phase fracture work adds to G_{Ic}. If, on the other hand, cracks deflect along the dispersed phase-matrix interfaces, the energy associated with generating these interfaces represents the increment in G_{Ic}.

Toughnesses of thermoplastics are not impressive, even though their malleabilities may be great. This is a result of the comparatively low strengths and moduli of thermoplastics. This does not mean that thermoplastics cannot be used in applications calling for "tough" materials. For example, Lexan (a polycarbonate) is used as protective equipment in contact sports such as football and hockey. Finally, even at temperatures below their glass transition temperatures, thermoplastics have fracture toughnesses much greater than those of inorganic glasses.

Microstructure and High-Temperature Fracture

Creep fracture occurs as follows. Internal voids first nucleate during creep deformation and then grow. When the void volume fraction reaches a critical level, fracture takes place. The voids that are precursors to fracture may reside within grains or, as is more common, along grain boundaries. The accelerating creep rate prior to fracture (Stage III creep; Fig. 21) is often a manifestation of void presence.

Void nucleation is not fully understood. However, it comes about as a result of strain "misaccommodation" of the same nature that causes crack nucleation in some low-temperature fracture. For the high-temperature case, two adjacent grains might have inherently different creep rates. If the rates cannot be made to match, a gap forms between the grains. Higher steady-state creep rates cause voids to form earlier on during creep.

While details of void nucleation remain murky, void growth is not. Voids grow by the same mechanisms that cause creep deformation. Thus an important conclusion: any material alteration that leads to a lower creep rate also improves creep fracture resistance. Note the distinction to low-temperature mechanical behavior, where increases in strength reduce fracture resistance and vice-versa.

One serious issue, extrapolation of short-time experimental results to long-time service situations, arises in high-temperature design. As described above, stress versus time to fracture tests are conducted at stress-temperature combinations for which failure occurs in "reasonable" times. However, intended high-temperature lives are frequently much longer. That is, materials are used at lower stresses and/or temperatures than those employed for laboratory assessment. Suitable means for extrapolating laboratory results must therefore be utilized.

A common extrapolation technique uses the Larson-Miller (LM) parameter (Fig. 32). The data displayed in Fig. 32(a) come from laboratory tests that measured creep fracture times over a range of stress and temperature; they are summarized as a plot of the (logarithm of) applied stress versus (logarithm of) the fracture time. The LM parameter is defined as:

$$LM = T(\log t_f + C) \quad \text{(Eq 21)}$$

where $\log t_f$ is log (to the base 10) of the fracture time (in hours), T is the absolute temperature, and C is a constant having a value close to 20. An LM diagram is constructed by plotting the stress logarithmically versus LM, and the constant C is determined by curve fitting. Results of such a procedure are shown in Fig. 32(b). Note that all data fall within a narrow scatter band, regardless of the test temperature. Curves like that in Fig. 32(b) are then used for extrapolation.

Some caution is required when extrapolating to lower stresses, the usual situation. In particular, there are often changes in dominant creep mechanism (e.g., from power-law to diffusional creep) when stress is lowered. When this is so, fracture times estimated via the LM technique are greater than they actually are. Thus the procedure is not conservative. However, considerable engineering practice, coupled with appropriate safety factors, permits the LM procedure to be usefully employed in high-temperature design. Further information on high-temperature material design is given in the article "Design for High-Temperature Applications" in this Volume.

Microstructure and Fatigue Failure

Fatigue is the most consternating of material failures. It has been estimated that somewhere on the order of 85% of identifiable metallic failures involve fatigue. Design against fatigue is an overriding concern in many aerospace applications, but any component that experiences a time-varying stress is potentially prone to fatigue failure. Examples include automotive drive shafts, springs, and rotating parts. Even bridges fatigue. "Old" bridges were designed to statically support their weight plus that of the bridge traffic. This approach was successful; static collapse of a bridge is rare. Yet the recent spate of bridge failures in the United States shows that the slight load fluctuations on bridge structural components (often in combination with corrosive effects) can cause structural failure after decades of service and many millions of load cycles.

The processes leading to fatigue failure are as follows. Surface or interior imperfections (a machining mark, for example, or perhaps a material inclusion) produce stress concentrations that, under the action of a cyclical stress, nucleate a crack. The crack then slowly grows until it attains a size at which the material "fast fractures." The relation between the maximum applied cyclical stress and the crack length at fracture is given by Eq 8 (i.e., fast fracture is defined by fracture mechanics).

The distance the crack advances per stress cycle during slow growth is related to a parameter ΔK:

$$\Delta K = A(\Delta\sigma)(c)^{1/2} \qquad \text{(Eq 22)}$$

where $\Delta\sigma$ is the algebraic difference between the maximum and minimum applied stress (the stress range) and A is a material constant. Thus, ΔK is to fatigue crack growth as the stress intensity factor is to fast fracture.

Slow crack growth rates are readily measured, and the relationship between these rates and ΔK can thereby be determined. Were the vast majority of material life to lie within the slow crack growth regime, it would then be a simple matter to determine fatigue life. Unfortunately, this situation pertains only to *low-cycle fatigue* (i.e., to fatigue taking place at high stress ranges). For *high-cycle fatigue*, most of the life is spent in crack nucleation. Materials can therefore be made more resistant to high-cycle fatigue by delaying crack nucleation. Since many fatigue cracks initiate on surfaces, surface hardening can be beneficial. Surfaces of ferrous materials can be case hardened (i.e., their surfaces made to contain more carbon than their interiors) or martensitically hardened. Ferrous and other metals can also have their surfaces hardened by cold working.

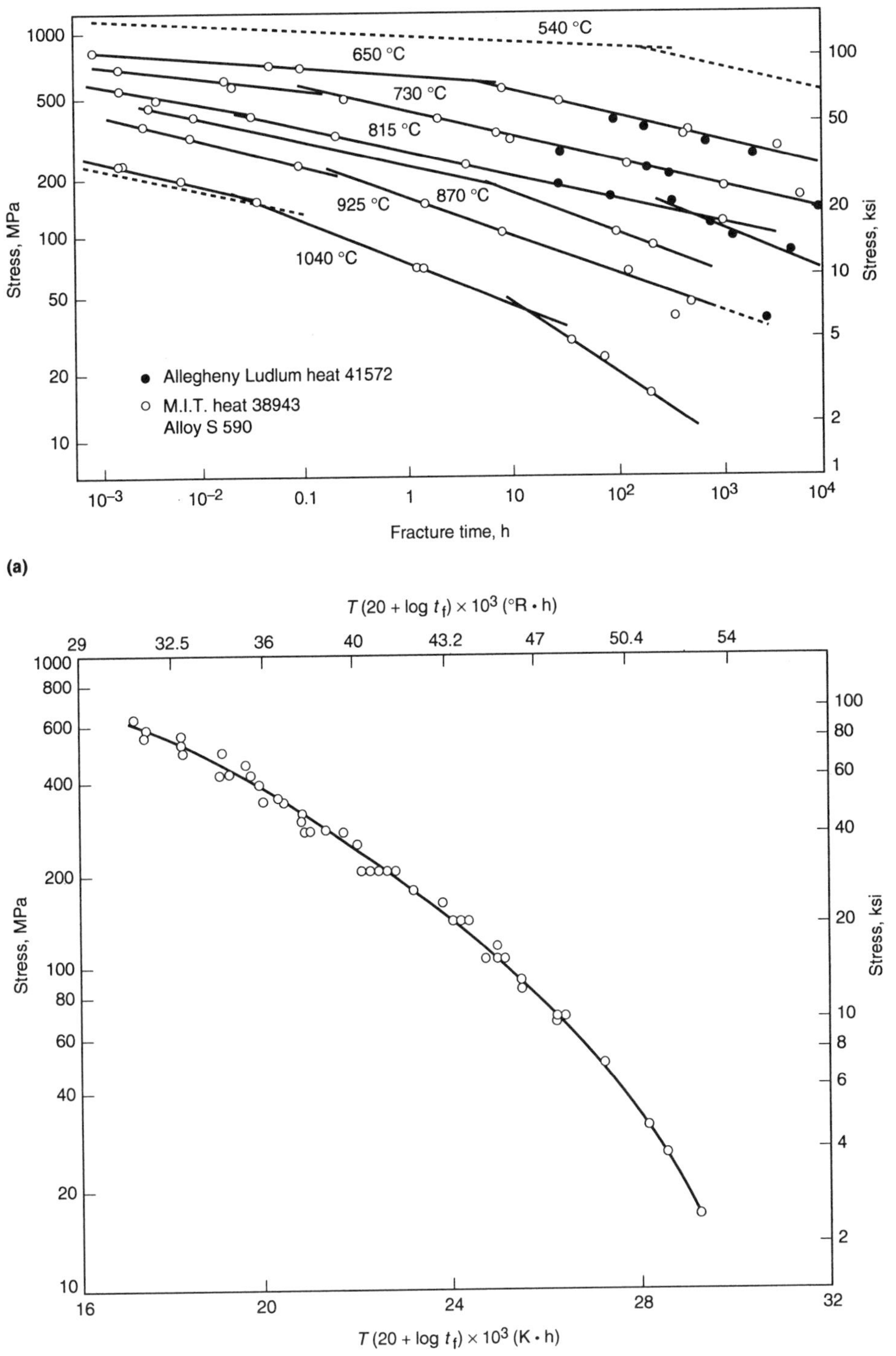

Fig. 32 (a) Stress-creep fracture times for an iron-base alloy at different temperatures. Source: Ref 11. (b) Larson-Miller master plot of the same data. This diagram permits fracture times to be estimated at stress-temperature combinations other than those illustrated in (a). Source: Ref 12

"Shot peening," in which the surface is bombarded by small hardened balls that plastically deform it and additionally induce beneficial residual compressive stresses, is one way of accomplishing this.

As with design against static fracture, fatigue design often assumes that the material contains surface or interior cracks, and that some idea of their size is known a priori. Reasonable estimates on fatigue life can then be made using concepts described above. The difficulty is that if the sizes of fatigue cracks nucleated during service are greater than those to which preexisting cracks have grown at the time of this nucleation, the design process is not conservative. Regular inspection of cyclically stressed parts provides another means of preventing fatigue failures. Further information on design against fatigue is provided in the article "Design for Fatigue Resistance" in this Volume.

REFERENCES

1. K.M. Ralls, T.H. Courtney, and J. Wulff, *Introduction to Materials Science and Engineering,* John Wiley, 1976
2. M.F. Ashby and D.R.H. Jones, *Engineering Materials 1: An Introduction to Their Properties and Applications,* Pergamon Press, 1980
3. W.G. Moffat, G.W. Pearsall, and J. Wulff, *Structure,* Vol I, *The Structure and Properties of Materials,* John Wiley, 1964
4. T.H. Courtney, *Mechanical Behavior of Materials,* McGraw-Hill, 1990
5. W.D. Kingery, H.K. Bowen, and D.R. Uhlmann, *Introduction to Ceramics,* 2nd ed., John Wiley, 1976
6. A.G. Guy and J.J. Hren, *Elementary Physical Metallurgy,* 3rd ed., Addison Wesley, 1974
7. H.W. Hayden, W.G. Moffat, and J. Wulff, *Mechanical Behavior,* Vol III, *The Structure and Properties of Materials,* John Wiley, 1965
8. J.A. Rinebolt and W.J. Harris Jr., *Trans. ASM,* Vol 43, 1951, p 1175
9. M.F. Ashby and D.R.H. Jones, *Engineering Materials 2: An Introduction to Microstructures, Processing and Design,* Pergamon Press, 1986
10. R.W. Evans and B. Wilshire, *Introduction to Creep,* The Institute of Materials, 1993
11. N.J. Grant and A.G. Bucklin, *Trans. ASM,* Vol 42, 1950, p 720
12. R.W. Hertzberg, *Deformation and Fracture Mechanics of Engineering Materials,* John Wiley, 1976

SELECTED REFERENCES

- M.F. Ashby and D.R.H. Jones, *Engineering Materials 1: An Introduction to Their Properties and Applications,* Pergamon Press, 1980
- M.F. Ashby and D.R.H. Jones, *Engineering Materials 2: An Introduction to Microstructures, Processing and Design,* Pergamon Press, 1986
- W.D. Callister, Jr., *Materials Science and Engineering: An Introduction,* 3rd ed., John Wiley, 1994
- A.H. Cottrell, The *Mechanical Properties of Matter,* John Wiley, 1964
- T.H. Courtney, *Mechanical Behavior of Materials,* McGraw-Hill, 1990
- G.E. Dieter, *Mechanical Metallurgy,* 3rd ed., McGraw-Hill, 1986
- A.G. Guy and J.J. Hren, *Elements of Physical Metallurgy,* 3rd ed., Addison-Wesley, 1974
- B. Harris, *Engineering Composite Materials,* The Institute of Metals, London, 1986
- R.W. Hertzberg, *Deformation and Fracture Mechanics of Engineering Materials,* 3rd ed., John Wiley, 1988
- D. Hull, *An Introduction to Composite Materials,* Cambridge University Press, Cambridge, 1981
- W.D. Kingery, H.K. Bowen, and D.R. Uhlmann, *Introduction to Ceramics,* 2nd ed., John Wiley, 1976
- K.M. Ralls, T.H. Courtney, and J. Wulff, *Introduction to Materials Science and Engineering,* John Wiley, 1976
- R.E. Reed-Hill, *Physical Metallurgy Principles,* 2nd ed., D. Van Nostrand, 1973
- R.J. Young, *Introduction to Polymers,* Chapman and Hall, London, 1981

Effects of Composition, Processing, and Structure on Properties of Irons and Steels

Bruce L. Bramfitt, Homer Research Laboratories, Bethlehem Steel Corporation

THE PROPERTIES of irons and steels are linked to the chemical composition, processing path, and resulting microstructure of the material; this correspondence has been known since the early part of the twentieth century. Figure 1 shows the route to attain properties. For a particular iron and steel composition, most properties depend on microstructure. These properties are called structure-sensitive properties, for example, yield strength and hardness. The structure-insensitive properties, for example, electrical conductivity, are not discussed in this chapter. Processing is a means to develop and control microstructure, for example, hot rolling, quenching, and so forth. In this article, the role of these factors is described in both theoretical and practical terms, with particular focus on the role of microstructure.

Basis of Material Selection

In order to select a material for a particular component, the designer must have an intimate knowledge of what properties are required. The composition, mechanical, and physical properties of most irons and steels can be found in the *Properties and Selection: Irons, Steels, and High-Performance Alloys,* Volume 1 of *ASM Handbook*, and in Ref 1. The compositions of selected steels are listed in Tables 1 to 4 of this article. Mechanical properties are listed in Table 5. Con-

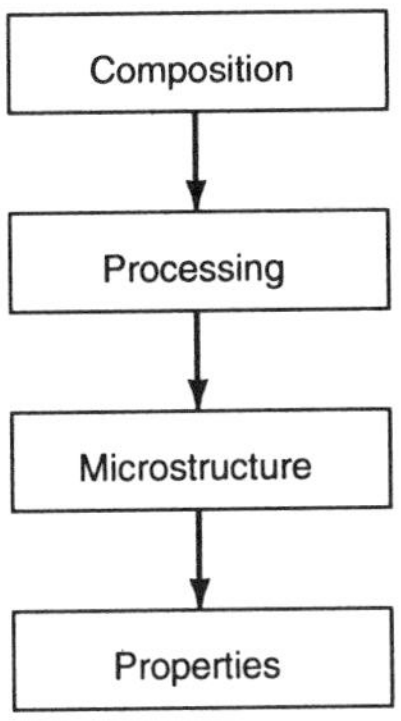

Fig. 1 Relationship among composition, processing, microstructure, and properties

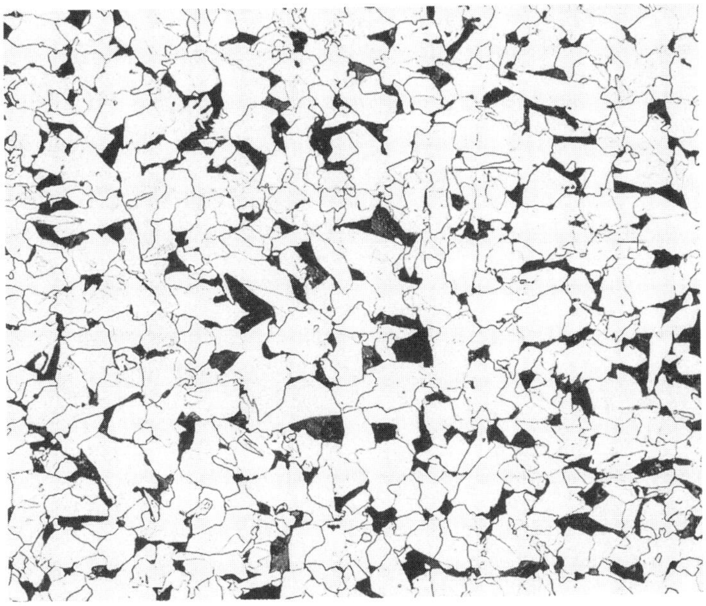

Fig. 2 Microstructure of a typical HSLA structural steel (ASTM A572, grade 50). 2% nital + 4% picral etch. 200×

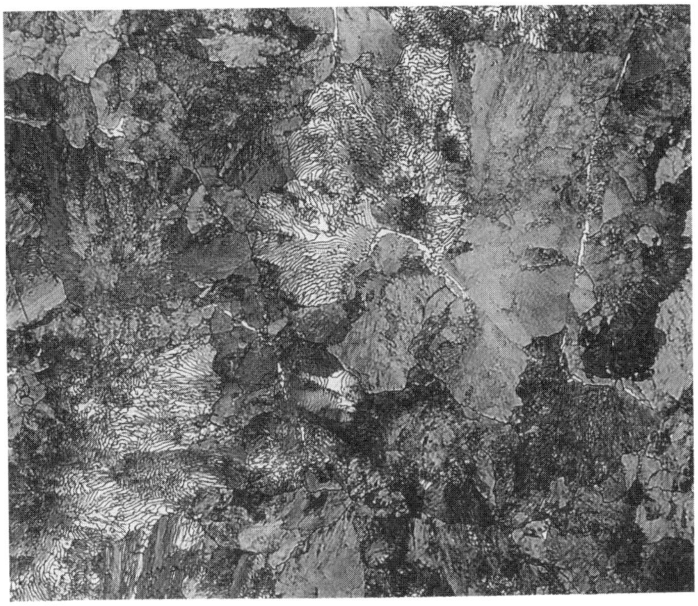

Fig. 3 Microstructure of a typical fully pearlitic rail steel showing the characteristic fine pearlite interlamellar spacing. 2% nital + 4% picral etch. 500×

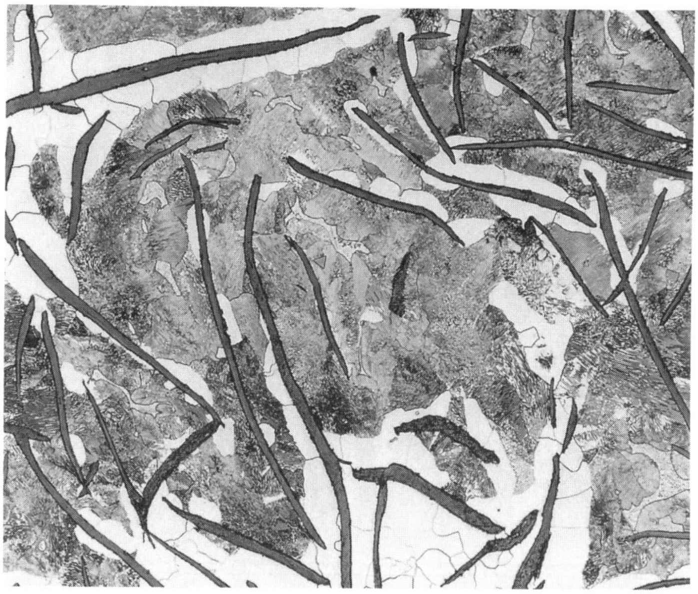

Fig. 4 Microstructure of a gray cast iron with a ferrite-pearlite matrix. 4% picral etch. 320×. Courtesy of A.O. Benscoter, Lehigh University

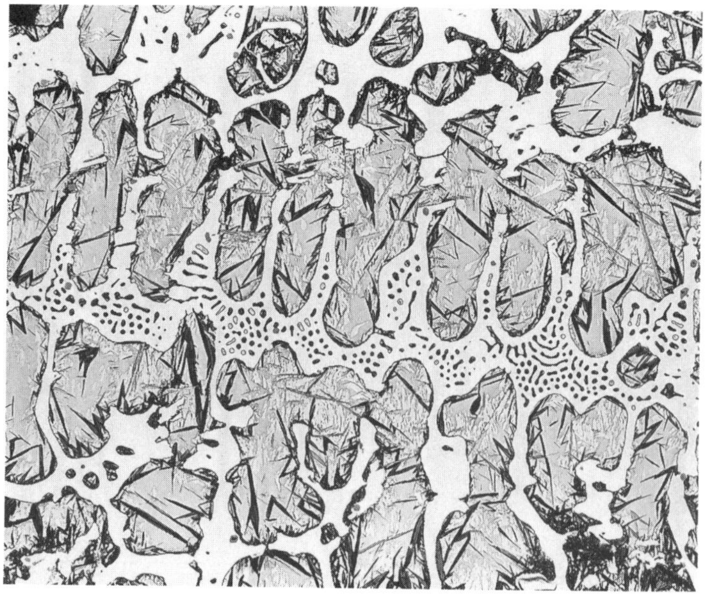

Fig. 5 Microstructure of an alloy white cast iron. White constituent is cementite and the darker constituent is martensite with some retained austenite. 4% picral etch. 250×. Courtesy of A.O. Benscoter, Lehigh University

sideration must be given to the environment (corrosive, high temperature, etc.) and how the component will be fabricated (welded, bolted, etc.). Once these property requirements are established the material selection process can begin. Some of the properties to be considered are:

Mechanical properties	Other properties/ characteristics
Strength	Formability
Tensile strength (ultimate strength)	Drawability
Yield strength	Stretchability
Compressive strength	Bendability
Hardness	Wear resistance
Toughness	Abrasion resistance
Notch toughness	Galling resistance
Fracture toughness	Sliding wear resistance
Ductility	Adhesive wear resistance
Total elongation	Machinability
Reduction in area	Weldability
Fatigue resistance	

In the selection process, what is required for one application may be totally inappropriate for another application. For example, steel beams for a railway bridge require a totally different set of properties than the steel rails that are attached to the wooden ties on the bridge deck. In designing the bridge, the steel must have sufficient strength to withstand substantial applied loads. In fact, the designer will generally select a steel with higher strength than actually required. Also, the designer knows that the steel must have fracture toughness to resist the growth and propagation of cracks and must be capable of being welded so that structural members can be joined without sacrificing strength and toughness. The steel bridge must also be corrosion resistant. This can be provided by a protective layer of paint. If painting is not allowed, small amounts of certain alloying elements such as copper and chromium can be added to the steel to inhibit or reduce corrosion rates. Thus, the steel selected for the bridge would be a high-strength low-alloy (HSLA) structural steel such as ASTM A572, grade 50 or possibly a weathering steel such as ASTM A588. A typical HSLA steel has a ferrite-pearlite microstructure as seen in Fig. 2 and is microalloyed with vanadium and/or niobium for strengthening. (*Microalloying* is a term used to describe the process of using small additions of carbonitride forming elements—titanium, vanadium, and niobium—to strengthen steels by grain refinement and precipitation hardening.)

On the other hand, the steel rails must have high strength coupled with excellent wear resistance. Modern rail steels consist of a fully pearlitic microstructure with a fine pearlite interlamellar spacing, as shown in Fig. 3. Pearlite is unique because it is a lamellar composite consisting of 88% soft, ductile ferrite and 12% hard, brittle cementite (Fe_3C). The hard cementite plates provide excellent wear resistance, especially when embedded in soft ferrite. Pearlitic steels have high strength and are fully adequate to support heavy axle loads of modern locomotives and freight cars. Most of the load is applied in compression. Pearlitic steels also have relatively poor toughness and cannot generally withstand impact loads without failure. The rail steel could not meet the requirements of the bridge builder, and the HSLA structural steel could not meet the requirements of the civil engineer who designed the bridge or the rail system.

A similar case can be made for the selection of cast irons. A cast machine housing on a large lathe requires a material with adequate strength, rigidity, and durability to support the applied load and a certain degree of damping capacity in order to rapidly attenuate (dampen) vibrations from the rotating parts of the lathe. The cast iron jaws of a crusher require a material with substantial wear

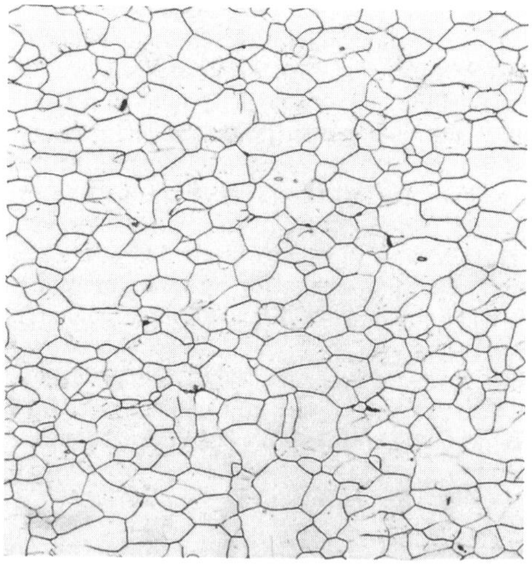

Fig. 6 Microstructure of a fully ferritic, ultralow carbon steel. Marshalls etch + HF, 300×. Courtesy of A.O. Benscoter, Lehigh University

resistance. For this application, a casting is required because wear-resistant steels are very difficult to machine. For the machine housing, gray cast iron is selected because it is relatively inexpensive, can be easily cast, and has the ability to dampen vibrations as a result of the graphite flakes present in its microstructure. These flakes are dispersed throughout the ferrite and pearlite matrix (Fig. 4). The graphite, being a major nonmetallic constituent in the gray iron, provides a tortuous path for sound to travel through the material. With so many flakes, sound waves are easily reflected and the sound dampened over a relatively short distance. However, for the jaw crusher, damping capacity is not a requirement. In this case, an alloy white cast iron is selected because of its high hardness and wear resistance. The white cast iron microstructure shown in Fig. 5

Table 1 Compositions of selected carbon steels

Type No.	UNS No.	Composition(a), % C	Mn	P	S
Nonresulfurized grades, manganese 1.00% max					
1005(b)	G10050	0.06 max	0.35 max	0.040 max	0.050 max
1006(b)	G10060	0.08 max	0.25–0.40	0.040 max	0.050 max
1008	G10080	0.10 max	0.30–0.50	0.040 max	0.050 max
1010	G10100	0.08–0.13	0.30–0.60	0.040 max	0.050 max
1012	G10120	0.10–0.15	0.30–0.60	0.040 max	0.050 max
1015	G10150	0.13–0.18	0.30–0.60	0.040 max	0.050 max
1016	G10160	0.13–0.18	0.60–0.90	0.040 max	0.050 max
1017	G10170	0.15–0.20	0.30–0.60	0.040 max	0.050 max
1018	G10180	0.15–0.20	0.60–0.90	0.040 max	0.050 max
1019	G10190	0.15–0.20	0.70–1.00	0.040 max	0.050 max
1020	G10200	0.18–0.23	0.30–0.60	0.040 max	0.050 max
1021	G10210	0.18–0.23	0.60–0.90	0.040 max	0.050 max
1022	G10220	0.18–0.23	0.70–1.00	0.040 max	0.050 max
1023	G10230	0.20–0.25	0.30–0.60	0.040 max	0.050 max
1025	G10250	0.22–0.28	0.30–0.60	0.040 max	0.050 max
1026	G10260	0.22–0.28	0.60–0.90	0.040 max	0.050 max
1029	G10290	0.25–0.31	0.60–0.90	0.040 max	0.050 max
1030	G10300	0.28–0.34	0.60–0.90	0.040 max	0.050 max
1034	G10340	0.32–0.38	0.50–0.80	0.040 max	0.050 max
1035	G10350	0.32–0.38	0.60–0.90	0.040 max	0.050 max
1037	G10370	0.32–0.38	0.70–1.00	0.040 max	0.050 max
1038	G10380	0.35–0.42	0.60–0.90	0.040 max	0.050 max
1039	G10390	0.37–0.44	0.70–1.00	0.040 max	0.050 max
1040	G10400	0.37–0.44	0.60–0.90	0.040 max	0.050 max
1042	G10420	0.40–0.47	0.60–0.90	0.040 max	0.050 max
1043	G10430	0.40–0.47	0.70–1.00	0.040 max	0.050 max
1044	G10440	0.43–0.50	0.30–0.60	0.040 max	0.050 max
1045	G10450	0.43–0.50	0.60–0.90	0.040 max	0.050 max
1046	G10460	0.43–0.50	0.70–1.00	0.040 max	0.050 max
1049	G10490	0.46–0.53	0.60–0.90	0.040 max	0.050 max
1050	G10500	0.48–0.55	0.60–0.90	0.040 max	0.050 max
1053	G10530	0.48–0.55	0.70–1.00	0.040 max	0.050 max
1055	G10550	0.50–0.60	0.60–0.90	0.040 max	0.050 max
1059(b)	G10590	0.55–0.65	0.50–0.80	0.040 max	0.050 max
1060	G10600	0.55–0.65	0.60–0.90	0.040 max	0.050 max
1064(b)	G10640	0.60–0.70	0.50–0.80	0.040 max	0.050 max
1065(b)	G10650	0.60–0.70	0.60–0.90	0.040 max	0.050 max
1069(b)	G10690	0.65–0.75	0.40–0.70	0.040 max	0.050 max
1070	G10700	0.65–0.75	0.60–0.90	0.040 max	0.050 max
1074	G10740	0.70–0.80	0.50–0.80	0.040 max	0.050 max
1078	G10780	0.72–0.85	0.30–0.60	0.040 max	0.050 max
Nonresulfurized grades, manganese 1.00% max (continued)					
1080	G10800	0.75–0.88	0.60–0.90	0.040 max	0.050 max
1084	G10840	0.80–0.93	0.60–0.90	0.040 max	0.050 max
1086(b)	G10860	0.80–0.93	0.30–0.50	0.040 max	0.050 max
1090	G10900	0.85–0.98	0.60–0.90	0.040 max	0.050 max
1095	G10950	0.90–1.03	0.30–0.50	0.040 max	0.050 max
Nonresulfurized grades, manganese greater than 1.00% max					
1513	G15130	0.10–0.16	1.10–1.40	0.040 max	0.050 max
1522	G15220	0.18–0.24	1.10–1.40	0.040 max	0.050 max
1524	G15240	0.19–0.25	1.35–1.65	0.040 max	0.050 max
1526	G15260	0.22–0.29	1.10–1.40	0.040 max	0.050 max
1527	G15270	0.22–0.29	1.20–1.50	0.040 max	0.050 max
1541	G15410	0.36–0.44	1.35–1.65	0.040 max	0.050 max
1547	G15470	0.43–0.51	1.35–1.65	0.040 max	0.050 max
1548	G15480	0.44–0.52	1.10–1.40	0.040 max	0.050 max
1551	G15510	0.45–0.56	0.85–1.15	0.040 max	0.050 max
1552	G15520	0.47–0.55	1.20–1.50	0.040 max	0.050 max
1561	G15610	0.55–0.65	0.75–1.05	0.040 max	0.050 max
1566	G15660	0.60–0.71	0.85–1.15	0.040 max	0.050 max
Free-machining grades, resulfurized					
1108	G11080	0.08–0.13	0.50–0.80	0.040 max	0.08–0.13
1109	G11090	0.08–0.13	0.60–0.90	0.040 max	0.08–0.13
1110	G11100	0.08–0.13	0.30–0.60	0.040 max	0.08–0.13
1116	G11160	0.14–0.20	1.10–1.40	0.040 max	0.16–0.23
1117	G11170	0.14–0.20	1.00–1.30	0.040 max	0.08–0.13
1118	G11180	0.14–0.20	1.30–1.60	0.040 max	0.08–0.13
1119	G11190	0.14–0.20	1.00–1.30	0.040 max	0.24–0.33
1137	G11370	0.32–0.39	1.35–1.65	0.040 max	0.08–0.13
1139	G11390	0.35–0.43	1.35–1.65	0.040 max	0.13–0.20
1140	G11400	0.37–0.44	0.70–1.00	0.040 max	0.08–0.13
1141	G11410	0.37–0.45	1.35–1.65	0.040 max	0.08–0.13
1144	G11440	0.40–0.48	1.35–1.65	0.040 max	0.24–0.33
1146	G11460	0.42–0.49	0.70–1.00	0.040 max	0.08–0.13
1151	G11510	0.48–0.55	0.70–1.00	0.040 max	0.08–0.13
Free-machining grades, resulfurized and rephosphorized					
1211	G12110	0.13 max	0.60–0.90	0.07–0.12	0.10–0.15
1212	G12120	0.13 max	0.70–1.00	0.07–0.12	0.16–0.23
1213	G12130	0.13 max	0.70–1.00	0.07–0.12	0.24–0.33
1215	G12150	0.09 max	0.75–1.05	0.04–0.09	0.26–0.35
12L14(c)	G12144	0.15 max	0.85–1.15	0.04–0.09	0.26–0.35

(a) The following notes refer to boron, copper, lead, and silicon additions: Boron: standard killed carbon steels, which are generally fine grain, may be produced with a boron treatment addition to improve hardenability. Such steels are produced to a range of 0.0005 to 0.003% B. These steels are identified by inserting the letter B between the second and third numerals of the AISI or SAE number, such as 10B46. Copper: when copper is required, 0.20% min is generally specified. Lead: standard carbon steels can be produced with a lead range of 0.15 to 0.35% to improve machinability. Such steels are identified by inserting the letter L between the second and third numerals of the AISI or SAE number, such as 12L15 and 10L45. Silicon: it is not common practice to produce the 12*xx* series of resulfurized and rephosphorized steels to specified limits for silicon because of its adverse effect on machinability. When silicon ranges or limits are required for resulfurized or nonresulfurized steels, however, these values apply: a range of 0.08% for maximum silicon contents up to 0.15% inclusive, a range of 0.10% for maximum silicon contents over 0.15 to 0.20% inclusive, a range of 0.15% for maximum silicon contents over 0.20 to 0.30% inclusive, and a range of 0.20% for maximum silicon contents over 0.30 to 0.60% inclusive. Example: maximum silicon content is 0.25%, range is 0.10 to 0.25%. (b) Standard grades for wire rod and wire only. (c) 0.15 to 0.35% Pb. Source: Ref 1

is graphite free and consists of martensite in a matrix of cementite. Both of these constituents are very hard and thus provide the required wear resistance. Thus, in this example the gray cast iron would not meet the requirements for the jaws of a crusher and the white cast iron would not meet the requirements for the lathe housing.

Role of Microstructure

In steels and cast irons, the microstructural constituents have the names ferrite, pearlite, bainite, martensite, cementite, and austenite. In most all other metallic systems, the constituents are not named, but are simply referred to by a Greek letter (α, β, γ, etc.) derived from the location of the constituent on a phase diagram. Ferrous alloy constituents, on the other hand, have been widely studied for more than 100 years. In the early days, many of the investigators were petrographers, mining engineers, and geologists. Because minerals have long been named after their discoverer or place of origin, it was natural to similarly name the constituents in steels and cast irons.

It can be seen that the four examples described above have very different microstructures: the structural steel has a ferrite + pearlite microstructure; the rail steel has a fully pearlitic microstructure; the machine housing (lathe) has a ferrite + pearlite matrix with graphite flakes; and the jaw crusher microstructure contains martensite and cementite. In each case, the microstructure plays the primary role in providing the properties desired for each application. From these examples, one can see how material properties can be tailored by microstructural manipulation or alteration. Knowledge about microstructure is thus paramount in component design and alloy development. In this section, each microstructural constituent will be described with particular reference to the properties that can be developed by appropriate manipulation of the microstructure through deformation (e.g., hot and cold rolling) and heat treatment. Further details about these microstructural constituents can be found in Ref 2 to 6.

Ferrite

A wide variety of steels and cast irons fully exploit the properties of ferrite. However, only a few commercial steels are completely ferritic. An

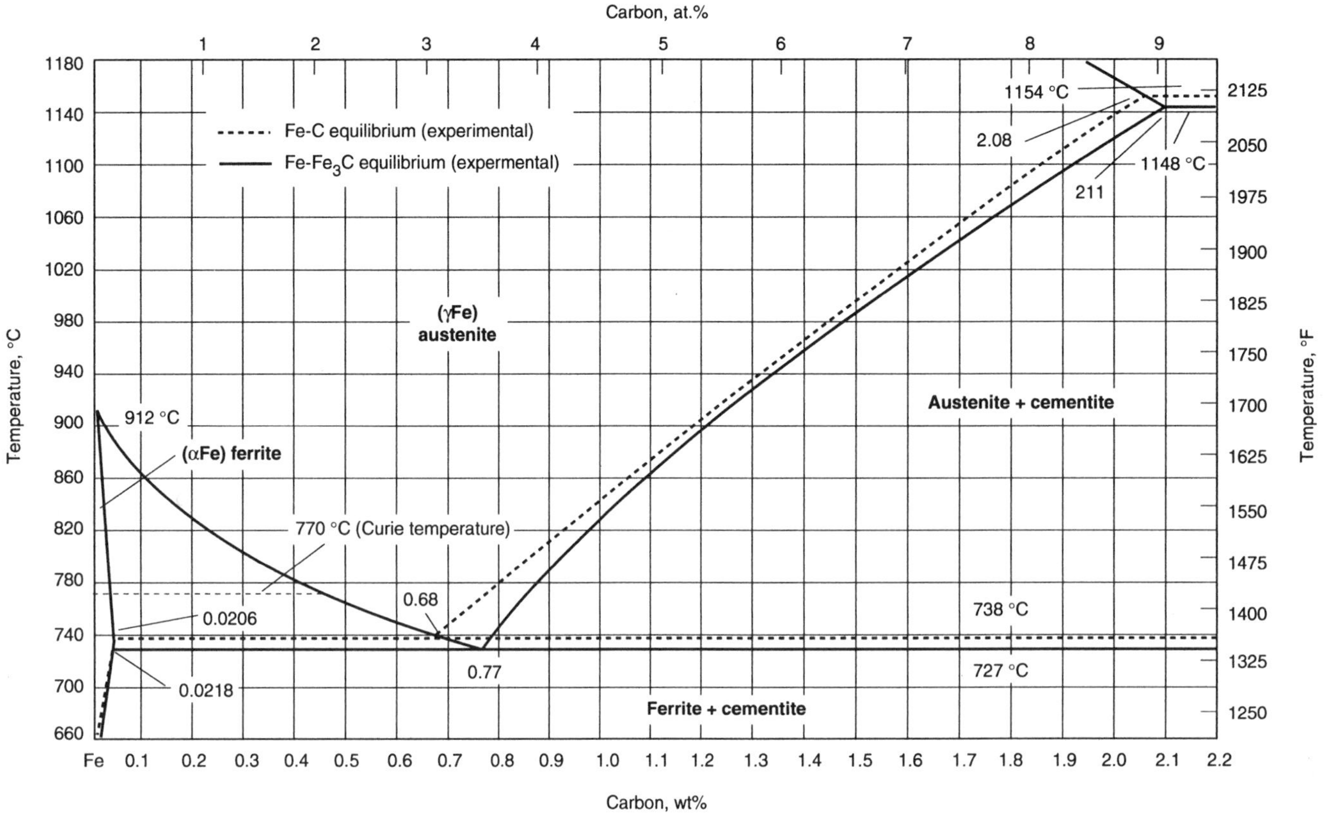

Fig. 7(a) Iron-carbon phase diagram showing the austenite (γFe) and ferrite (αFe) phase regions and eutectoid composition and temperature. Dotted lines represent iron-graphite equilibrium conditions and solid lines represent iron-cementite equilibrium conditions. Only the solid lines are important with respect to steels. Source: Ref 2

example of the microstructure of a fully ferritic, ultralow carbon steel is shown in Fig. 6.

Ferrite is essentially a solid solution of iron containing carbon or one or more alloying elements such as silicon, chromium, manganese, and nickel. There are two types of solid solutions: interstitial and substitutional. In an interstitial solid solution, elements with small atomic diameter, for example, carbon and nitrogen, occupy specific interstitial sites in the body-centered cubic (bcc) iron crystalline lattice. These sites are essentially the open spaces between the larger iron atoms. In a substitutional solid solution, elements of similar atomic diameter replace or substitute for iron atoms. The two types of solid solutions impart different characteristics to ferrite. For example, interstitial elements like carbon and nitrogen can easily diffuse through the open bcc lattice, whereas substitutional elements like manganese and nickel diffuse with great difficulty. Therefore, an interstitial solid solution of iron and carbon responds quickly during heat treatment, whereas substitutional solid solutions behave sluggishly during heat treatment, such as in homogenization.

According to the iron-carbon phase diagram (Fig. 7a), very little carbon (0.022% C) can dissolve in ferrite (αFe), even at the eutectoid temperature of 727 °C (1330 °F). (The iron-carbon phase diagram indicates the phase regions that exist over a wide carbon and temperature range. The diagram represents equilibrium conditions. Figure 7(b) shows an expanded iron-carbon diagram with both the eutectoid and eutectic regions.) At room temperature, the solubility is an order of magnitude less (below 0.005% C). However, even at these small amounts, the addition of carbon to pure iron increases the room-temperature yield strength of iron by more than five times, as seen in Fig. 8. If the carbon content exceeds the solubility limit of 0.022%, the carbon forms another phase called cementite (Fig. 9). Cementite is also a constituent of pearlite, as seen in Fig. 10. The role of cementite and pearlite on the mechanical properties of steel is discussed below.

The influence of solid-solution elements on the yield strength of ferrite is shown in Fig. 11. Here one can clearly see the strong effect of carbon on increasing the strength of ferrite. Nitrogen, also an interstitial element, has a similar effect. Phosphorus is also a ferrite strengthener. In fact, there are commercially available steels containing phosphorus for strengthening. These steels are the rephosphorized steels (type 1211 to 1215 series). Compositions and mechanical property data for these steels can be found in Tables 1 and 5.

In Fig. 11, the substitutional solid solution elements of silicon, copper, manganese, molybdenum, nickel, aluminum, and chromium are shown to have far less effect as ferrite strengtheners than the interstitial elements. In fact, chromium, nickel, and aluminum in solid solution have very little influence on the strength of ferrite.

In addition to carbon (and other solid-solution elements), the strength of a ferritic steel is also determined by its grain size according to the Hall-Petch relationship:

$$\sigma_y = \sigma_o + k_y d^{-1/2} \quad \text{(Eq 1)}$$

where σ_y is the yield strength (in MPa), σ_o is a constant, k_y is a constant, and d is the grain diameter (in mm).

The grain diameter is a measurement of size of the ferrite grains in the microstructure, for example, note the grains in the ultralow carbon steel in Fig. 6. Figure 12 shows the Hall-Petch relationship for a low-carbon fully ferritic steel. This relationship is extremely important for understanding structure-property relationships in steels. Control of grain size through thermomechanical treatment, heat treatment, and/or microalloying is vital to the control of strength and toughness of most steels. The role of grain size is discussed in more detail later in this article.

There is a simple way to stabilize ferrite, thereby expanding the region of ferrite in the iron-carbon phase diagram, namely by the addition of alloying elements such as silicon, chromium, and molybdenum. These elements are called ferrite stabilizers because they stabilize ferrite at room temperature through reducing the amount of γ solid solution (austenite) with the formation of what is called a γ-loop as seen at the far left in Fig. 13. This iron-chromium phase

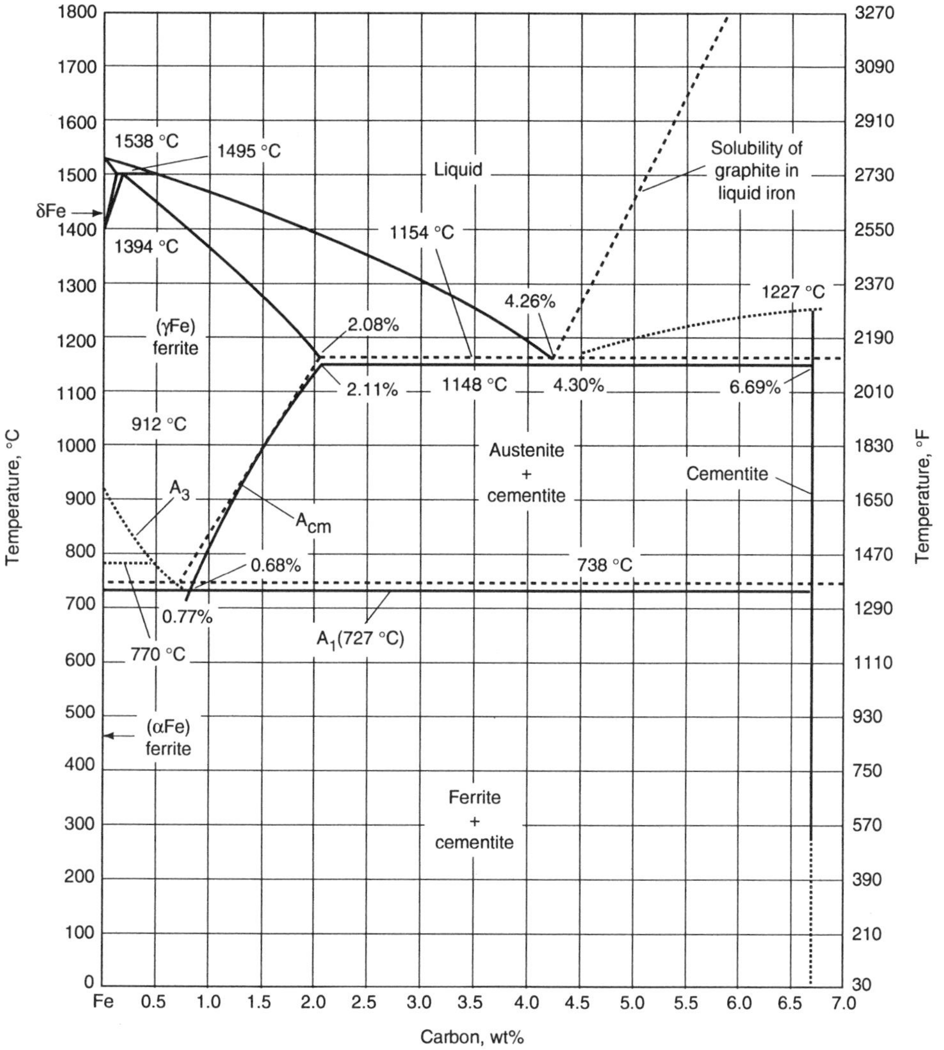

Fig. 7(b) Expanded iron-carbon phase diagram showing both the eutectoid (shown in Fig. 7a) and eutectic regions. Dotted lines represent iron-graphite equilibrium conditions and solid lines represent iron-cementite equilibrium conditions. The solid lines at the eutectic are important to white cast irons and the dotted lines are important to gray cast irons. Source: Ref 2

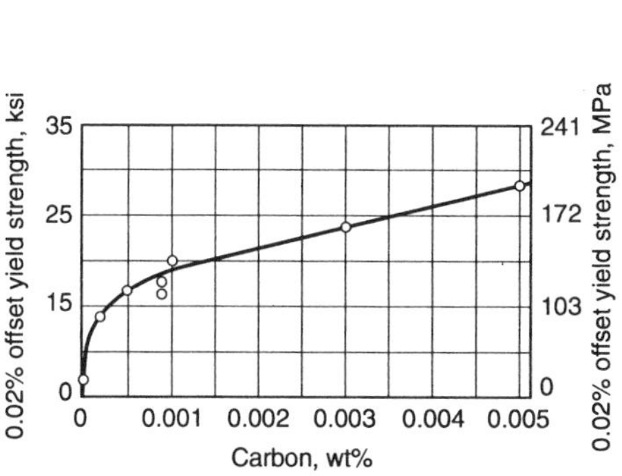

Fig. 8 Increase in room-temperature yield strength of iron with small additions of carbon. Source: Ref 7

diagram shows that ferrite exists up above 12% Cr and is stable up to the melting point (liquidus temperature). An important fully ferritic family of steels is the iron-chromium ferritic stainless steels. These steels are resistant to corrosion, and are classified as type 405, 409, 429, 430, 434, 436, 439, 442, 444, and 446 stainless steels. These steels range in chromium content from 11 to 30%. Additions of molybdenum, silicon, niobium, aluminum, and titanium provide specific properties. Ferritic stainless steels have good ductility (up to 30% total elongation and 60% reduction in area) and formability, but lack strength at elevated temperatures compared with austenitic stainless steels. Room-temperature yield strengths range from 170 to about 440 MPa (25 to 64 ksi), and room-temperature tensile strengths range from 380 to about 550 MPa (55 to 80 ksi). Table 5 lists the mechanical properties of some of the ferritic stainless steels. Type 409 stainless steel is widely used for automotive exhaust systems. Type 430 free-machining stainless steel has the best machinability of all stainless steels other than that of a low-carbon, free-machining martensitic stainless steel (type 416).

Another family of steels utilizing a ferrite stabilizer (γ-loop) are the iron-silicon ferritic alloys containing up to about 6.5% Si (carbon-free). These steels are of commercial importance because they have excellent magnetic permeability and low core loss. High-efficiency motors and transformers are produced from these iron-silicon electrical steels (aluminum can also substitute for silicon in them).

Over the past 20 years or so, a new breed of very-low-carbon fully ferritic sheet steels has emerged for applications requiring exceptional formability (see Fig. 6). These are the interstitial-free (IF) steels for which carbon and nitrogen are reduced in the steelmaking process to very low levels, and any remaining interstitial carbon or nitrogen is tied up with small amounts of alloying elements (e.g., titanium or niobium) that form preferentially carbides and nitrides. These steels have very low strength, but are used to produce components that are difficult or impossible to form from other steels. Very-low-carbon, fully ferritic steels (0.001% C) are now being manufactured for automotive components that harden during the paint-curing cycle. These steels are called bake-hardening steels, and have controlled amounts of carbon and nitrogen that combine with other elements, such as titanium and niobium, during the baking cycle (175 °C, or 350 °F, for 30 min). The process is called aging, and the strength derives from the precipitation of titanium/niobium carbonitrides at the elevated temperature.

Another form of very-low-carbon, fully ferritic steel is motor lamination steel. The carbon is removed from these steels by a process known as decarburization. The decarburized (carbon-free) ferritic steel has good permeability and sufficiently low core loss (not as low as the iron-silicon alloys) to be used for electric motor laminations, that is, the stacked steel layers in the rotor and stator of the motor.

As noted previously, a number of properties are exploited in fully ferritic steels:

- *Iron-silicon steels:* Exceptional electrical properties
- *Iron-chromium steels:* Good corrosion resistance
- *Interstitial-free steels:* Exceptional formability
- *Bake-hardening steels:* Strengthens during paint cure cycle
- *Lamination steels:* Good electrical properties

Pearlite

As the carbon content of steel is increased beyond the solubility limit (0.02% C) on the iron-carbon binary phase diagram, a constituent called pearlite forms. Pearlite is formed by cooling the steel through the eutectoid temperature (the temperature of 727 °C in Fig. 7) by the following reaction:

$$\text{Austenite} \leftrightarrow \text{cementite} + \text{ferrite} \quad \text{(Eq 2)}$$

The cementite and ferrite form as parallel plates called lamellae (Fig. 14). This is essentially a composite microstructure consisting of a very hard carbide phase, cementite, and a very soft and ductile ferrite phase. A fully pearlitic microstructure is

Table 2 Compositions of selected low-alloy steels

Small quantities of certain elements not specified or required are present; maximum contents of incidental elements are: 0.35% Cu, 0.25% Ni, 0.20% Cr, and 0.06% Mo; standard alloy steel may also be produced with 0.15 to 0.35% Pb and are identified with an L inserted between the second and third numerals of the AISI-SAE number (for example, 41L40)

Steel		Composition, %							
Type No.	UNS No.	C	Mn	P max	S max	Si	Ni	Cr	Mo
1330	G13300	0.28–0.33	1.60–1.90	0.035	0.040	0.15–0.35	...	...	...
1335	G13350	0.33–0.38	1.60–1.90	0.035	0.040	0.15–0.35	...	...	...
1340	G13400	0.38–0.43	1.60–1.90	0.035	0.040	0.15–0.35	...	...	...
1345	G13450	0.43–0.48	1.60–1.90	0.035	0.040	0.15–0.35	...	...	...
4023	G40230	0.20–0.25	0.70–0.90	0.035	0.040	0.15–0.35	...	...	0.20–0.30
4024	G40240	0.20–0.25	0.70–0.90	0.035	0.035–0.050	0.15–0.35	...	...	0.20–0.30
4027	G40270	0.25–0.30	0.70–0.90	0.035	0.040	0.15–0.35	...	...	0.20–0.30
4028	G40280	0.25–0.30	0.70–0.90	0.035	0.035–0.050	0.15–0.35	...	...	0.20–0.30
4037	G40370	0.35–0.40	0.70–0.90	0.035	0.040	0.15–0.35	...	...	0.20–0.30
4047	G40470	0.45–0.50	0.70–0.90	0.035	0.040	0.15–0.35	...	...	0.20–0.30
4118	G41180	0.18–0.23	0.70–0.90	0.035	0.040	0.15–0.35	...	0.40–0.60	0.08–0.15
4130	G41300	0.28–0.33	0.40–0.60	0.035	0.040	0.15–0.35	...	0.80–1.10	0.15–0.25
4137	G41370	0.35–0.40	0.70–0.90	0.035	0.040	0.15–0.35	...	0.80–1.10	0.15–0.25
4140	G41400	0.38–0.43	0.75–1.00	0.035	0.040	0.15–0.35	...	0.80–1.10	0.15–0.25
4142	G41420	0.40–0.45	0.75–1.00	0.035	0.040	0.15–0.35	...	0.80–1.10	0.15–0.25
4145	G41450	0.43–0.48	0.75–1.00	0.035	0.040	0.15–0.35	...	0.80–1.10	0.15–0.25
4147	G41470	0.45–0.50	0.75–1.00	0.035	0.040	0.15–0.35	...	0.80–1.10	0.15–0.25
4150	G41500	0.48–0.53	0.75–1.00	0.035	0.040	0.15–0.35	...	0.80–1.10	0.15–0.25
4161	G41610	0.56–0.64	0.75–1.00	0.035	0.040	0.15–0.35	...	0.70–0.90	0.25–0.35
4320	G43200	0.17–0.22	0.45–0.65	0.035	0.040	0.15–0.35	1.65–2.00	0.40–0.60	0.20–0.30
4340	G43400	0.38–0.43	0.60–0.80	0.035	0.040	0.15–0.35	1.65–2.00	0.70–0.90	0.20–0.30
E4340(a)	G43406	0.38–0.43	0.65–0.85	0.025	0.025	0.15–0.35	1.65–2.00	0.70–0.90	0.20–0.30
4615	G46150	0.13–0.18	0.45–0.65	0.035	0.040	0.15–0.35	1.65–2.00	...	0.20–0.30
4620	G46200	0.17–0.22	0.45–0.65	0.035	0.040	0.15–0.35	1.65–2.00	...	0.20–0.30
4626	G46260	0.24–0.29	0.45–0.65	0.035	0.040	0.15–0.35	0.70–1.00	...	0.15–0.25
4720	G47200	0.17–0.22	0.50–0.70	0.035	0.040	0.15–0.35	0.90–1.20	0.35–0.55	0.15–0.25
4815	G48150	0.13–0.18	0.40–0.60	0.035	0.040	0.15–0.35	3.25–3.75	...	0.20–0.30
4817	G48170	0.15–0.20	0.40–0.60	0.035	0.040	0.15–0.35	3.25–3.75	...	0.20–0.30
4820	G48200	0.18–0.23	0.50–0.70	0.035	0.040	0.15–0.35	3.25–3.75	...	0.20–0.30
5117	G51170	0.15–0.20	0.70–0.90	0.035	0.040	0.15–0.35	...	0.70–0.90	...
5120	G51200	0.17–0.22	0.70–0.90	0.035	0.040	0.15–0.35	...	0.70–0.90	...
5130	G51300	0.28–0.33	0.70–0.90	0.035	0.040	0.15–0.35	...	0.80–1.10	...
5132	G51320	0.30–0.35	0.60–0.80	0.035	0.040	0.15–0.35	...	0.75–1.00	...
5135	G51350	0.33–0.38	0.60–0.80	0.035	0.040	0.15–0.35	...	0.80–1.05	...
5140	G51400	0.38–0.43	0.70–0.90	0.035	0.040	0.15–0.35	...	0.70–0.90	...
5150	G51500	0.48–0.53	0.70–0.90	0.035	0.040	0.15–0.35	...	0.70–0.90	...
5155	G51550	0.51–0.59	0.70–0.90	0.035	0.040	0.15–0.35	...	0.70–0.90	...
5160	G51600	0.56–0.64	0.75–1.00	0.035	0.040	0.15–0.35	...	0.70–0.90	...
E51100(a)	G51986	0.98–1.10	0.25–0.45	0.025	0.025	0.15–0.35	...	0.90–1.15	...
E52100(a)	G52986	0.98–1.10	0.25–0.45	0.025	0.025	0.15–0.35	...	1.30–1.60	...
6118(b)	G61180	0.16–0.21	0.50–0.70	0.025	0.040	0.15–0.25	...	0.50–0.70	...
6150(c)	G61500	0.48–0.53	0.70–0.90	0.035	0.040	0.15–0.35	...	0.80–1.10	...
8615	G86150	0.13–0.18	0.70–0.90	0.035	0.040	0.15–0.35	0.40–0.70	0.40–0.60	0.15–0.25
8617	G86170	0.15–0.20	0.70–0.90	0.035	0.040	0.15–0.35	0.40–0.70	0.40–0.60	0.15–0.25
8620	G86200	0.18–0.23	0.70–0.90	0.035	0.040	0.15–0.35	0.40–0.70	0.40–0.60	0.15–0.25
8622	G86220	0.20–0.25	0.70–0.90	0.035	0.040	0.15–0.35	0.40–0.70	0.40–0.60	0.15–0.25
8625	G86250	0.23–0.28	0.70–0.90	0.035	0.040	0.15–0.35	0.40–0.70	0.40–0.60	0.15–0.25
8627	G86270	0.25–0.30	0.70–0.90	0.035	0.040	0.15–0.35	0.40–0.70	0.40–0.60	0.15–0.25
8630	G86300	0.28–0.33	0.70–0.90	0.035	0.040	0.15–0.35	0.40–0.70	0.40–0.60	0.15–0.25
8637	G86370	0.35–0.40	0.75–1.00	0.035	0.040	0.15–0.35	0.40–0.70	0.40–0.60	0.15–0.25
8640	G86400	0.38–0.43	0.75–1.00	0.035	0.040	0.15–0.35	0.40–0.70	0.40–0.60	0.15–0.25
8642	G86420	0.40–0.45	0.75–1.00	0.035	0.040	0.15–0.35	0.40–0.70	0.40–0.60	0.15–0.25
8645	G86450	0.43–0.48	0.75–1.00	0.035	0.040	0.15–0.35	0.40–0.70	0.40–0.60	0.15–0.25
8655	G86550	0.51–0.59	0.75–1.00	0.035	0.040	0.15–0.35	0.40–0.70	0.40–0.60	0.15–0.25
8720	G87200	0.18–0.23	0.70–0.90	0.035	0.040	0.15–0.35	0.40–0.70	0.40–0.60	0.20–0.30
8740	G87400	0.38–0.43	0.75–1.00	0.035	0.040	0.15–0.35	0.40–0.70	0.40–0.60	0.20–0.30
8822	G88220	0.20–0.25	0.75–1.00	0.035	0.040	0.15–0.35	0.40–0.70	0.40–0.60	0.30–0.40
9260	G92600	0.56–0.64	0.75–1.00	0.035	0.040	1.80–2.20	...	...	...
Standard boron grades(d)									
50B44	G50441	0.43–0.48	0.75–1.00	0.035	0.040	0.15–0.35	...	0.40–0.60	...
50B46	G50461	0.44–0.49	0.75–1.00	0.035	0.040	0.15–0.35	...	0.20–0.35	...
50B50	G50501	0.48–0.53	0.75–1.00	0.035	0.040	0.15–0.35	...	0.40–0.60	...
50B60	G51601	0.56–0.64	0.75–1.00	0.035	0.040	0.15–0.35	...	0.40–0.60	...
51B60	G51601	0.56–0.64	0.75–1.00	0.035	0.040	0.15–0.35	...	0.70–0.90	...
81B45	G81451	0.43–0.48	0.75–1.00	0.035	0.040	0.15–0.35	0.20–0.40	0.35–0.55	0.08–0.15
94B17	G94171	0.15–0.20	0.75–1.00	0.035	0.040	0.15–0.35	0.30–0.60	0.30–0.50	0.08–0.15
94B30	G94301	0.28–0.33	0.75–1.00	0.035	0.040	0.15–0.35	0.30–0.60	0.30–0.50	0.08–0.15

(a) Electric furnace steel. (b) Includes 0.10 to 0.15% V. (c) Includes 0.15% minimum V. (d) 0.0005 to 0.003% B. Source: Ref 1

Table 3 Compositions of selected stainless and heat-resisting steels

Compositions are maximum unless otherwise stated.

Steel		Chemical composition, %								
Type No.	UNS No.	C	Mn	P	S	Si	Cr	Ni	Mo	Other elements
201	S20100	0.15	5.50–7.50	0.060	0.030	1.00	16.00–18.00	3.50–5.50	...	N 0.25
202	S20200	0.15	7.50–10.00	0.060	0.030	1.00	17.00–19.00	4.00–6.00	...	N 0.25
205	S20500	0.12–0.25	14.00–15.50	0.060	0.030	1.00	16.50–18.00	1.00–1.75	...	N 0.32–0.40
301	S30100	0.15	2.00	0.045	0.030	1.00	16.00–18.00	6.00–8.00	...	...
302	S30200	0.15	2.00	0.045	0.030	1.00	17.00–19.00	8.00–10.00	...	...
302B	S30215	0.15	2.00	0.045	0.030	2.00–3.00	17.00–19.00	8.00–10.00	...	...
303	S30300	0.15	2.00	0.20	0.15 min	1.00	17.00–19.00	8.00–10.00	0.60(a)	...
303Se	S30323	0.15	2.00	0.20	0.060	1.00	17.00–19.00	8.00–10.00	...	Se 0.15 min
304	S30400	0.08	2.00	0.045	0.030	1.00	18.00–20.00	8.00–10.50	...	...
304L	S30403	0.03	2.00	0.045	0.030	1.00	18.00–20.00	8.00–12.00	...	...
	S30430	0.08	2.00	0.045	0.030	1.00	17.00–19.00	8.00–10.00	...	Cu 3.00–4.00
304N	S30451	0.08	2.00	0.045	0.030	1.00	18.00–20.00	8.00–10.50	...	N 0.10–0.16
305	S30500	0.12	2.00	0.045	0.030	1.00	17.00–19.00	10.50–13.00	...	...
308	S30800	0.08	2.00	0.045	0.030	1.00	19.00–21.00	10.00–12.00	...	...
309	S30900	0.20	2.00	0.045	0.030	1.00	22.00–24.00	12.00–15.00	...	...
309S	S30908	0.08	2.00	0.045	0.030	1.00	22.00–24.00	12.00–15.00	...	...
310	S31000	0.25	2.00	0.045	0.030	1.50	24.00–26.00	19.00–22.00	...	...
310S	S31008	0.08	2.00	0.045	0.030	1.50	24.00–26.00	19.00–22.00	...	...
314	S31400	0.25	2.00	0.045	0.030	1.50–3.00	23.00–26.00	19.00–22.00	...	...
316	S31600	0.08	2.00	0.045	0.030	1.00	16.00–18.00	10.00–14.00	2.00–3.00	...
316F	S31620	0.08	2.00	0.20	0.10 min	1.00	16.00–18.00	10.00–14.00	1.75–2.50	...
316L	S31603	0.03	2.00	0.045	0.030	1.00	16.00–18.00	10.00–14.00	2.00–3.00	...
316N	S31651	0.08	2.00	0.045	0.030	1.00	16.00–18.00	10.00–14.00	2.00–3.00	N 0.10–0.16
317	S31700	0.08	2.00	0.045	0.030	1.00	18.00–20.00	11.00–15.00	3.00–4.00	...
317L	S31703	0.03	2.00	0.045	0.030	1.00	18.00–20.00	11.00–15.00	3.00–4.00	...
321	S32100	0.08	2.00	0.045	0.030	1.00	17.00–19.00	9.00–12.00	...	Ti 5 × C min
329	S32900	0.10	2.00	0.040	0.030	1.00	25.00–30.00	3.00–6.00	1.00–2.00	...
330	N08330	0.08	2.00	0.040	0.030	0.75–1.50	17.00–20.00	34.00–37.00	...	...
347	S34700	0.08	2.00	0.045	0.030	1.00	17.00–19.00	9.00–13.00	...	Nb + Ta 10 × C min
348	S34800	0.08	2.00	0.045	0.030	1.00	17.00–19.00	9.00–13.00	...	Nb + Ta 10 × C min; Ta 0.10 max; Co 0.20 max
384	S38400	0.08	2.00	0.045	0.030	1.00	15.00–17.00	17.00–19.00	...	...
403	S40300	0.15	1.00	0.040	0.030	0.50	11.50–13.00	...	...	...
405	S40500	0.08	1.00	0.040	0.030	1.00	11.50–14.50	...	...	Al 0.10–0.30
409	S40900	0.08	1.00	0.045	0.045	1.00	10.50–11.75	...	...	Ti 6 × C min; 0.75 max
410	S41000	0.15	1.00	0.040	0.030	1.00	11.50–13.50	...	...	...
414	S41400	0.15	1.00	0.040	0.030	1.00	11.50–13.50	1.25–2.50	...	...
416	S41600	0.15	1.25	0.060	0.15 min	1.00	12.00–14.00	...	0.60(a)	...
416Se	S41623	0.15	1.25	0.060	0.060	1.00	12.00–14.00	...	...	Se 0.15 min
420	S42000	>0.15	1.00	0.040	0.030	1.00	12.00–14.00	...	...	...
420F	S42020	>0.15	1.25	0.060	0.15 min	1.00	12.00–14.00	...	0.60(a)	...
422	S42200	0.20–0.25	1.00	0.025	0.025	0.75	11.00–13.00	0.50–1.00	0.75–1.25	V 0.15–0.30; W 0.75–1.25
429	S42900	0.12	1.00	0.040	0.030	1.00	14.00–16.00	...	...	...
430	S43000	0.12	1.00	0.040	0.030	1.00	16.00–18.00	...	...	...
430F	S43020	0.12	1.25	0.060	0.15 min	1.00	16.00–18.00	...	0.60(a)	...
430FSe	S43023	0.12	1.25	0.060	0.060	1.00	16.00–18.00	...	...	Se 0.15 min
431	S43100	0.20	1.00	0.040	0.030	1.00	15.00–17.00	1.25–2.50	...	...
434	S43400	0.12	1.00	0.040	0.030	1.00	16.00–18.00	...	0.75–1.25	...
436	S43600	0.12	1.00	0.040	0.030	1.00	16.00–18.00	...	0.75–1.25	Nb + Ta 5 × C min, 0.70 max
440A	S44002	0.60–0.75	1.00	0.040	0.030	1.00	16.00–18.00	...	0.75	...
440B	S44003	0.75–0.95	1.00	0.040	0.030	1.00	16.00–18.00	...	0.75	...
440C	S44004	0.95–1.20	1.00	0.040	0.030	1.00	16.00–18.00	...	0.75	...
442	S44200	0.20	1.00	0.040	0.030	1.00	18.00–23.00	...	...	...
446	S44600	0.20	1.50	0.040	0.030	1.00	23.00–27.00	...	...	N 0.25
501	S50100	>0.10	1.00	0.040	0.030	1.00	4.00–6.00	...	0.40–0.65	...
502	S50200	0.10	1.00	0.040	0.030	1.00	4.00–6.00	...	0.40–0.65	...
503	S50300	0.15	1.00	0.040	0.040	1.00	6.00–8.00	...	0.45–0.65	...
504	S50400	0.15	1.00	0.040	0.040	1.00	8.00–10.00	...	0.90–1.10	...
	S13800	0.05	0.10	0.01	0.008	0.10	12.25–13.25	7.50–8.50	2.00–2.50	Al 0.90–1.35; N 0.010
...	S15500	0.07	1.00	0.040	0.030	1.00	14.00–15.50	3.50–5.50	...	Cu 2.50–4.50; Nb + Ta 0.15–0.45
...	S17400	0.07	1.00	0.040	0.030	1.00	15.50–17.50	3.00–5.00	...	Cu 3.00–5.00; Nb + Ta 0.15–0.45
...	S17700	0.09	1.00	0.040	0.040	1.00	16.00–18.00	6.50–7.75	...	Al 0.75–1.50

(a) May be added at manufacturer's option. Source: Ref 1

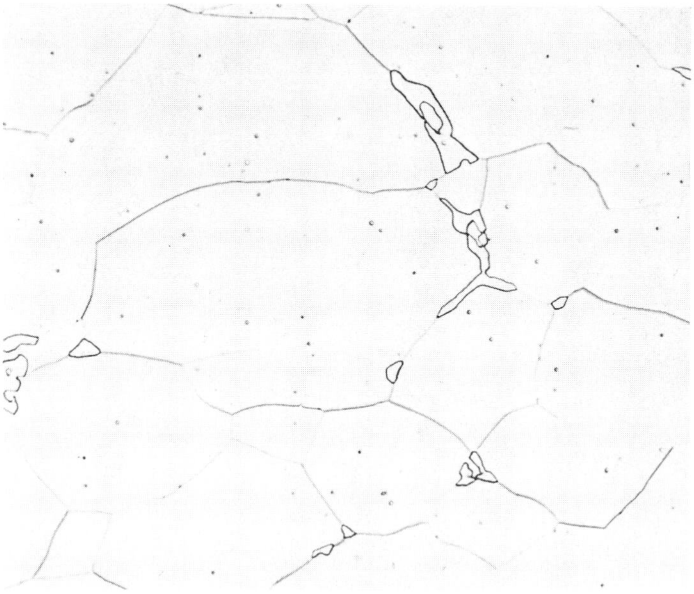

Fig. 9 Photomicrograph of an annealed low-carbon sheet steel with grain-boundary cementite. 2% nital + 4% picral etch. 1000×

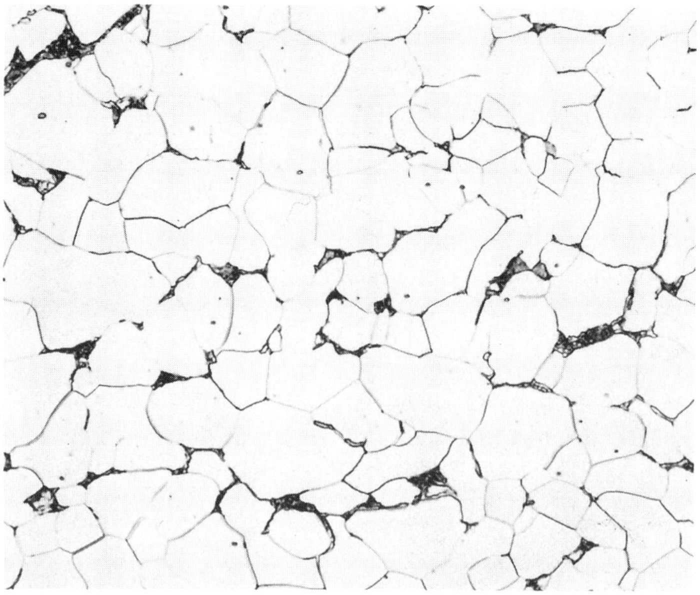

Fig. 10 Photomicrograph of pearlite (dark constituent) in a low-carbon steel sheet. 2% nital + 4% picral etch. 1000×

formed at the eutectoid composition of 0.78% C. As can be seen in Fig. 3 and 14, pearlite forms as colonies where the lamellae are aligned in the same orientation. The properties of fully pearlitic steels are determined by the spacing between the ferrite-cementite lamellae, a dimension called the interlamellar spacing, λ, and the colony size. A simple relationship for yield strength has been developed by Heller (Ref 10) as follows:

$$\sigma_y = -85.9 + 8.3(\lambda^{-1/2}) \quad \text{(Eq 3)}$$

where σ_y is the 0.2% offset yield strength (in MPa) and λ is the interlamellar spacing (in mm). Figure 15 shows Heller's plot of strength versus interlamellar spacing for fully pearlitic eutectoid steels.

It has also been shown by Hyzak and Bernstein (Ref 11) that strength is related to interlamellar spacing, pearlite colony size, and prior-austenite grain size, according to the following relationship:

$$YS = 52.3 + 2.18(\lambda^{-1/2}) - 0.4(d_c^{-1/2}) - 2.88(d^{-1/2}) \quad \text{(Eq 4)}$$

where *YS* is the yield strength (in MPa), d_c is the pearlite colony size (in mm), and *d* is the prior-austenite grain size (in mm). From Eq 3 and 4, it can be seen that the steel composition does not have a major influence on the yield strength of a fully pearlitic eutectoid steel. There is some solid-solution strengthening of the ferrite in the lamellar structure (see Fig. 11).

The thickness of the cementite lamellae can also influence the properties of pearlite. Fine cementite lamellae can be deformed, compared

Table 4 Compositions of selected maraging steels

Grade	Composition, %											
	Ni	Co	Mo	Ti	Al	Si	Mn	C	S	P	Zr	B
18Ni(200)	18.50	8.50	3.25	0.20	0.10	0.10 max	0.10 max	0.03 max	0.01 max	0.01 max	0.01	0.003
18Ni(250)	18.50	7.50	4.80	0.40	0.10	0.10 max	0.10 max	0.03 max	0.01 max	0.01 max	0.01	0.003
18Ni(300)	18.50	9.00	4.80	0.60	0.10	0.10 max	0.10 max	0.03 max	0.01 max	0.01 max	0.01	0.003

Source: Ref 2

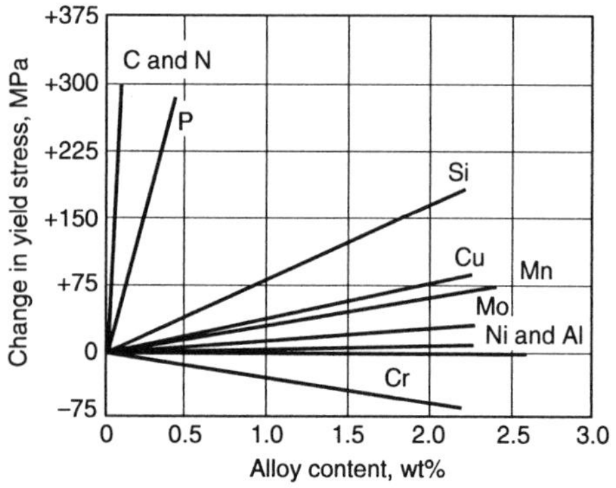

Fig. 11 Influence of solid-solution elements on the changes in yield stress of low-carbon ferritic steels. Source: Ref 5

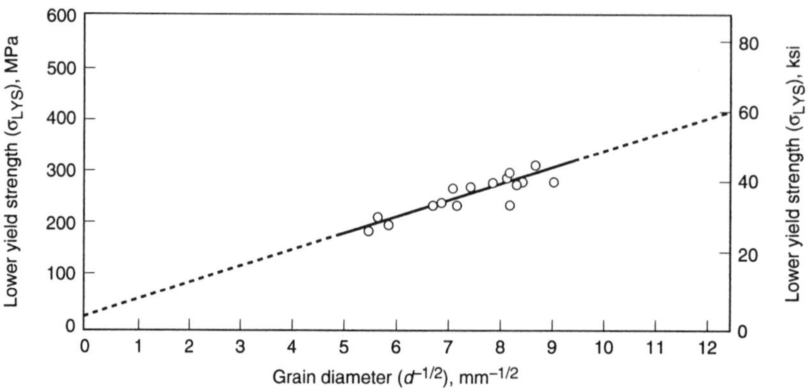

Fig. 12 Hall-Petch relationship in low-carbon ferritic steels. Source: Ref 8

with coarse lamellae, which tend to crack during deformation.

Although fully pearlitic steels have high strength, high hardness, and good wear resistance, they also have poor ductility and toughness. For example, a low-carbon, fully ferritic steel will typically have a total elongation of more than 50%, whereas a fully pearlitic steel (e.g., type 1080) will typically have a total elongation of about 10% (see Table 5). A low-carbon fully ferritic steel will have a room-temperature Charpy V-notch impact energy of about 200 J (150 ft · lbf), whereas a fully pearlitic steel will have room-temperature impact energy of under 10 J (7 ft · lbf). The transition temperature (i.e., the temperature at which a material changes from ductile fracture to brittle fracture) for a fully pearlitic steel can be approximated from the following relationship (Ref 11):

$$TT = 217.84 - 0.83(d_c^{-1/2}) - 2.98(d^{-1/2}) \qquad \text{(Eq 5)}$$

where TT is the transition temperature (in °C).

From Eq 5, one can see that both the prior-austenite grain size and pearlite colony size control the transition temperature of a pearlitic steel. Unfortunately, the transition temperature of a fully pearlitic steel is always well above room temperature. This means that at room temperature the general fracture mode is cleavage, which is associated with brittle fracture. Therefore, fully pearlitic steels should not be used in applications where toughness is important. Also, pearlitic steels with carbon contents slightly or moderately higher than the eutectoid composition (called hypereutectoid steels) have even poorer toughness.

From Eq 4 and 5, one can see that for pearlite, strength is controlled by interlamellar spacing, colony size, and prior-austenite grain size, and toughness is controlled by colony size and prior-austenite grain size.

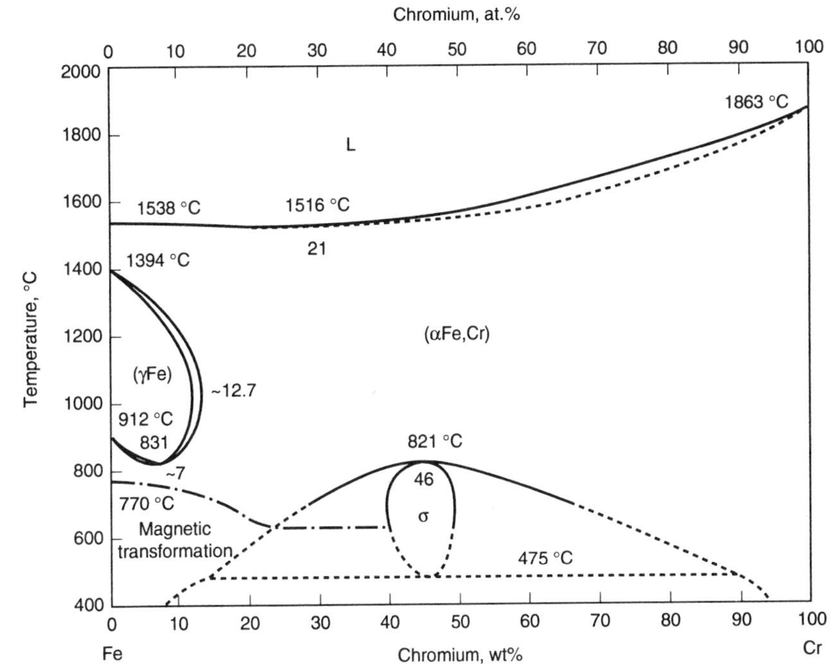

Fig. 13 Iron-chromium phase diagram. Source: Ref 9

Unfortunately, these three factors are rather difficult to measure. To determine interlamellar spacing, a scanning electron microscope (SEM), or a transmission electron microscope (TEM) is needed in order to resolve the spacing. Generally, a magnification of 10,000× is adequate, as seen in Fig. 14. Special statistical procedures have been developed to determine an accurate measurement of the spacing (Ref 12). The colony size and especially the prior austenite grain size are very difficult to measure and require a skilled metallographer using the light microscope or SEM and special etching procedures.

Because of poor ductility/toughness, there are only a few applications for fully pearlitic steels, including railroad rails and wheels and high-strength wire. By far, the largest tonnage application is for rails. A fully pearlitic rail steel provides excellent wear resistance for railroad wheel/rail contact. Rail life is measured in millions of gross tons (MGT) of travel and current rail life easily exceeds 250 MGT. The wear resistance of pearlite arises from the unique morphology of the ferrite-cementite lamellar composite where a hard constituent is embedded into a soft-ductile constituent. This means that the hard cementite plates do not abrade away as easily as the rounded cementite particles found in other steel microstructures, that is, tempered martensite and bainite, which will be discussed later. Wear resistance of a rail steel is directly proportional to hardness. This is shown in Fig. 16, which indi-

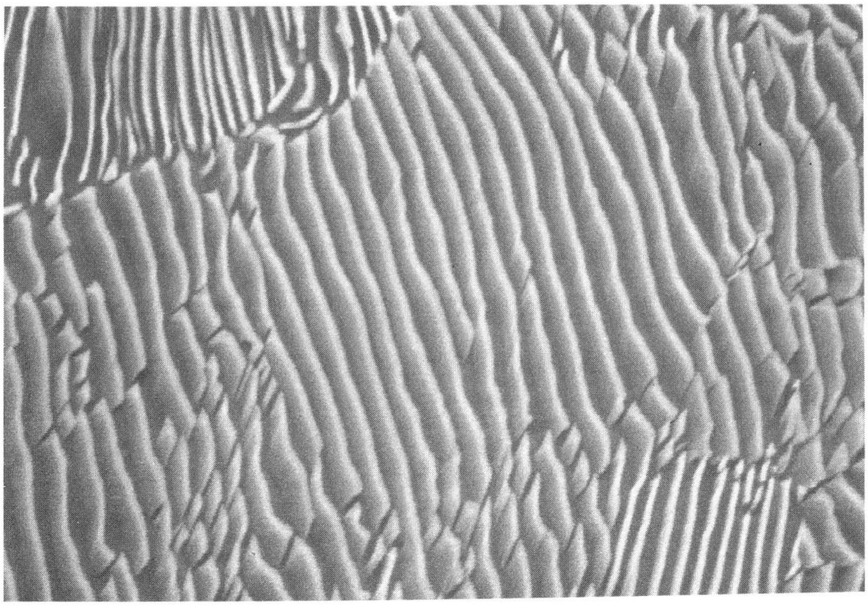

Fig. 14 SEM micrograph of pearlite showing ferrite and cementite lamellae. 4% picral etch. 10,000×

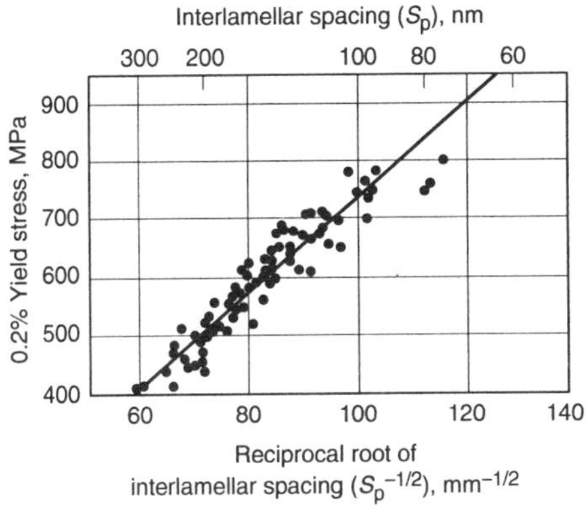

Fig. 15 Relationship between pearlite interlamellar spacing and yield strength for eutectoid steels. Source: Ref 10

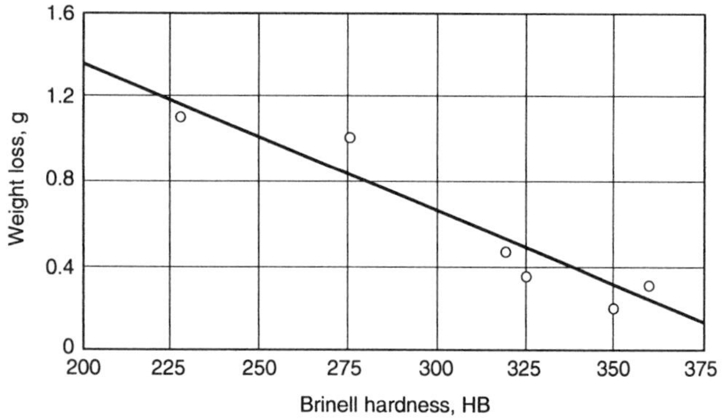

Fig. 16 Relationship between hardness and wear resistance (weight loss) for rail steels. Source: Ref 13

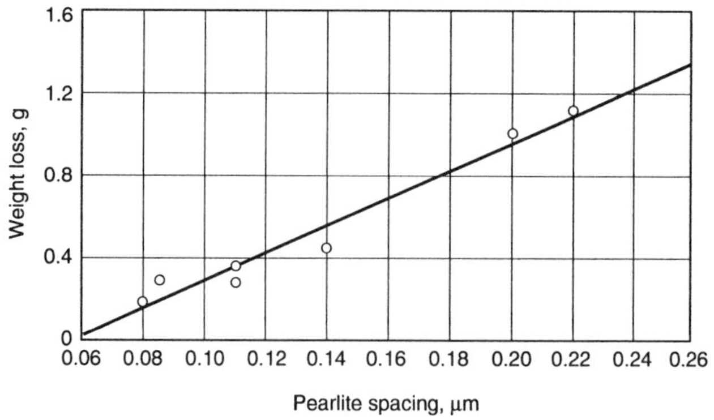

Fig. 17 Relationship between pearlite interlamellar spacing and wear resistance (weight loss) for rail steels. Source: Ref 13

cates less weight loss as hardness increases. Also, wear resistance (less weight loss) increases as interlamellar spacing decreases, as shown in Fig. 17. Thus, the most important microstructural parameter for controlling hardness and wear resistance is the pearlite interlamellar spacing. Fortunately, interlamellar spacing is easy to control and is dependent solely on transformation temperature.

Figure 18 shows a continuous cooling transformation (CCT) diagram for a typical rail steel. A CCT diagram is a time versus temperature plot showing the regions at which various constituents—ferrite, pearlite, bainite, and martensite—form during the continuous cooling of a steel component. Usually several cooling curves are shown with the associated start and finish transformation temperatures of each constituent. These diagrams should not be confused with isothermal transformation (IT or TTT) diagrams, which are derived by rapidly quenching very thin specimens to various temperatures, and maintaining that temperature (isothermal) until the specimens begin to transform, partially transform, and fully transform, at which time they are quenched to room temperature. An IT diagram does not represent the transformation behavior in most processes where steel parts are continuously cooled, that is, air cooled, and so forth.

As shown in Fig. 18, the pearlite transformation temperature (indicated by the pearlite-start curve, P_s) decreases with increasing cooling rate. The hardness of pearlite increases with decreasing transformation temperature. Thus, in order to provide a rail steel with the highest hardness and wear resistance, one must cool the rail from the austenite at the fastest rate possible to obtain the lowest transformation temperature. This is done in practice by a process known as head hardening, which is simply an accelerated cooling process using forced air or water sprays to achieve the desired cooling rate (Ref 15). Because only the head of the rail contacts the wheel of the railway car and locomotive, only the head requires the higher hardness and wear resistance.

Another application for a fully pearlitic steel is high-strength wire (e.g., piano wire). Again, the composite morphology of lamellar ferrite and cementite is exploited, this time during wire drawing. A fully pearlitic steel rod is heat treated by a process known as patenting. During patenting, the rod is transformed at a temperature of about 540 °C (1000 °F) by passing it through a lead or salt bath at this temperature. This develops a microstructure with a very fine pearlite interlamellar spacing because the transformation takes place at the nose of the CCT diagram, that is, at the lowest possible pearlite transformation temperature (see Fig. 18). The rod is then cold drawn to wire. Because of the very fine interlamellar spacing, the ferrite and cementite lamellae become aligned along the wire axis during the deformation process. Also, the fine cementite lamella tend to bend and deform as the wire is elongated during drawing. The resulting wire is one of the strongest commercial products available; for example, a commercial 0.1 mm (0.004 in.) diam wire can have a tensile strength in the range of 3.0 to 3.3 GPa (439 to 485 ksi), and in special cases a tensile strength as high as 4.8 GPa can be obtained. These wires are used in musical instruments because of the sound quality developed from the high tensile stresses applied in stringing a piano and violin and are also used in wire rope cables for suspension bridges.

Ferrite-Pearlite. The most common structural steels produced have a mixed ferrite-pearlite microstructure. Their applications include beams for bridges and high-rise buildings, plates for ships, and reinforcing bars for roadways. These steels are relatively inexpensive and are produced in large tonnages. They also have the advantage of being able to be produced with a wide range of properties. The microstructure of typical ferrite-pearlite steels is shown in Fig. 19.

In most ferrite-pearlite steels, the carbon content and the grain size determine the microstructure and resulting properties. For example, Fig. 20 shows the effect of carbon on tensile and impact properties. The ultimate tensile strength steadily increases with increasing carbon content.

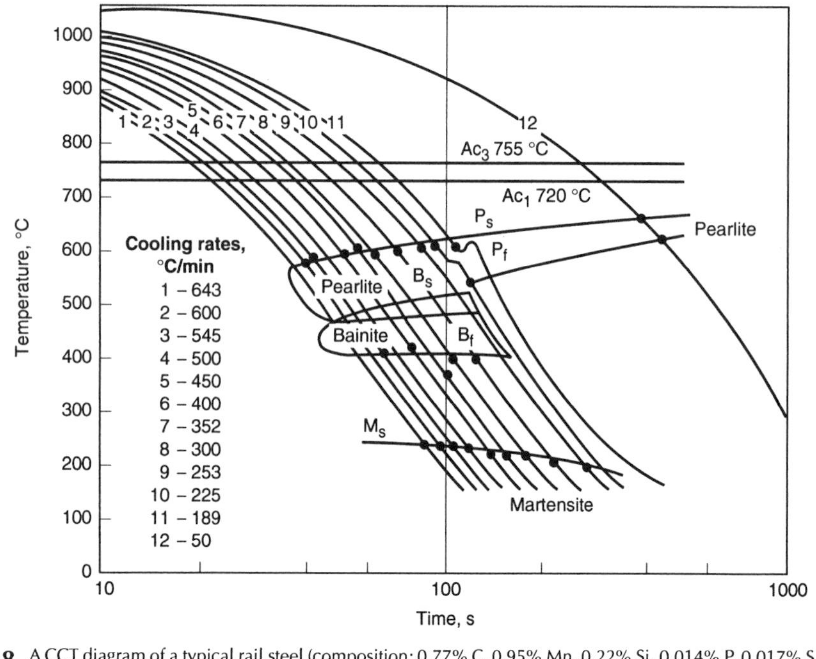

Fig. 18 A CCT diagram of a typical rail steel (composition: 0.77% C, 0.95% Mn, 0.22% Si, 0.014% P, 0.017% S, 0.10% Cr). Source: Ref 14

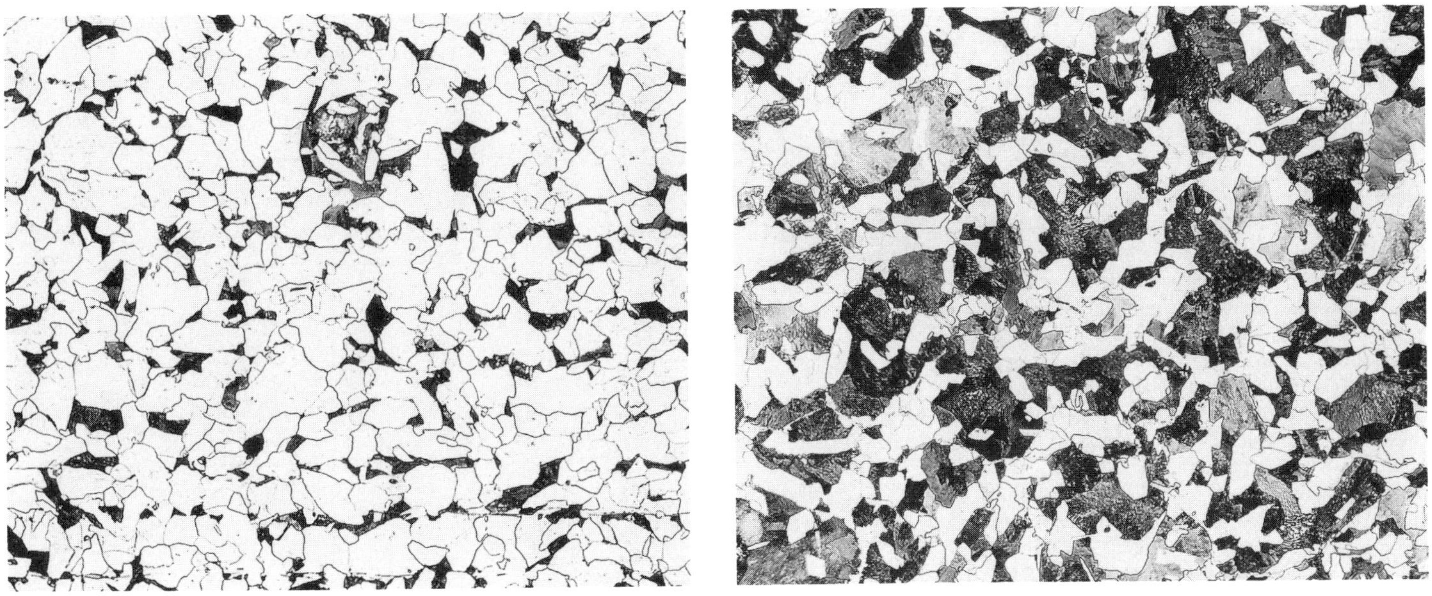

Fig. 19 Microstructure of typical ferrite-pearlite structural steels at two different carbon contents. (a) 0.10% C. (b) 0.25% C. 2% nital + 4% picral etch. 200×

This is caused by the increase in the volume fraction of pearlite in the microstructure, which has a strength much higher than that of ferrite. Thus, increasing the volume fraction of pearlite has a profound effect on increasing tensile strength.

However, as seen in Fig. 20, the yield strength is relatively unaffected by carbon content, rising from about 275 MPa (40 ksi) to about 415 MPa (60 ksi) over the range of carbon content shown. This is because yielding in a ferrite-pearlite steel is controlled by the ferrite matrix, which is generally considered to be the continuous phase (matrix) in the microstructure. Therefore, pearlite plays only a minor role in yielding behavior.

From Fig. 20, one can also see that ductility, as represented by reduction in area, steadily decreases with increasing carbon content. A steel with 0.10% C has a reduction in area of about 75%, whereas a steel with 0.70% C has a reduction in area of only 25%. Percent total elongation would show a similar trend, however, with values much less than percent reduction in area.

Much work has been done to develop empirical equations for ferrite-pearlite steels that relate strength and toughness to microstructural features, for example, grain size and percent of pearlite as well as composition. One such equation for ferrite-pearlite steels under 0.25% C is as follows (Ref 16):

$$YS = 53.9 + 32.34\,(\text{Mn}) + 83.2(\text{Si}) + 354.2(N_f) + 17.4(d^{-1/2}) \quad \text{(Eq 6)}$$

where Mn is the manganese content (%), Si is the silicon content (%), N_f is the free nitrogen content (%), and d is the ferrite grain size (in mm). Equation 6 shows that carbon content (percent pearlite) has no effect on yield strength, whereas the yield strength in Fig. 20 increases somewhat with carbon content. According to Eq 6, manganese, silicon, and nitrogen have a pronounced effect on yield strength, as does grain size. However, in most ferrite-pearlite steels nitrogen is quite low (under 0.010%) and thus has minimal effect on yield strength. In addition, as discussed below, nitrogen has a detrimental effect on impact properties.

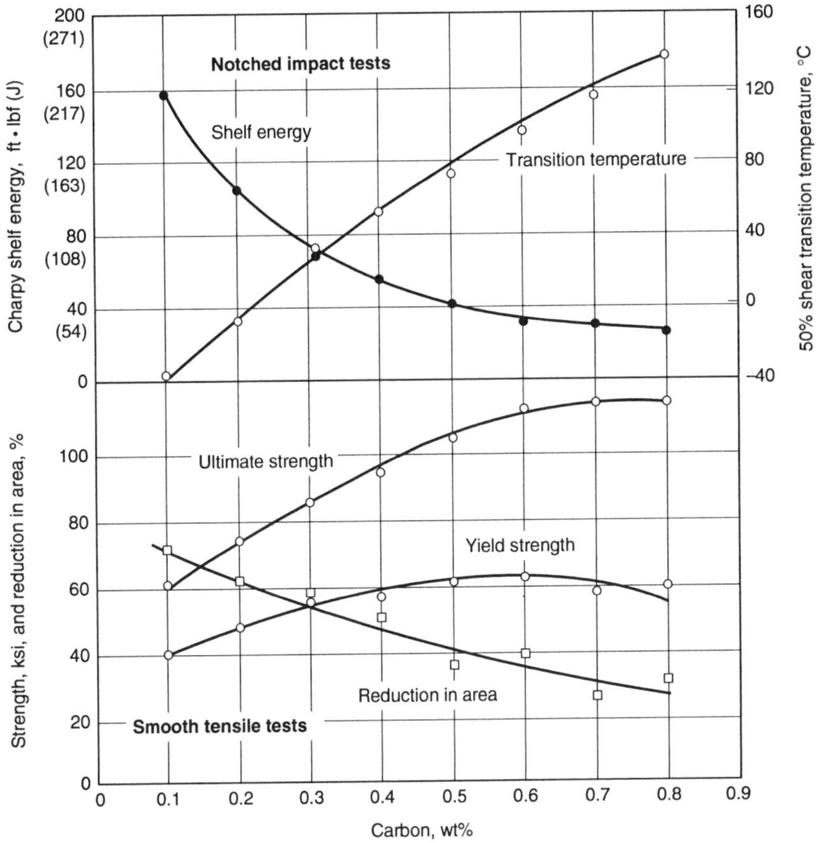

Fig. 20 Mechanical properties of ferrite-pearlite steels as a function of carbon content. Source: Ref 2

The regression equation for tensile strength for the same steels is as follows (Ref 16):

$$TS = 294.1 + 27.7(\text{Mn}) + 83.2(\text{Si}) + 3.9(\text{P}) + 7.7(d^{-1/2}) \quad \text{(Eq 7)}$$

where *TS* is the tensile strength (in MPa) and P is pearlite content (%). Thus, in distinction to yield strength, the percentage of pearlite in the microstructure plays an important role on tensile strength.

Toughness of ferrite-pearlite steels is also an important consideration in their use. It has long been known that the absorbed energy in a Charpy V-notch test is decreased by increasing carbon content, as seen in Fig. 21. In this graph of impact energy versus test temperature, the shelf energy decreases from about 200 J (150 ft · lbf) for a 0.11% C steel to about 35 J (25 ft · lbf) for a 0.80% C steel. Also, the transition temperature increases from about –50 to 150 °C (–60 to 300 °F) over this same range of carbon content. The effect of carbon is due mainly to its effect on the percentage of pearlite in the microstructure. This is reflected in the regression equation for transition temperature below (Ref 16):

$$TT = -19 + 44(\text{Si}) + 700(N_f^{1/2}) + 2.2(\text{P}) - 11.5(d^{-1/2}) \quad \text{(Eq 8)}$$

It can be seen in all these relationships that ferrite grain size is an important parameter in improving both strength and toughness. It can also be seen that while pearlite is beneficial for increasing tensile strength and nitrogen is beneficial for increasing yield strength, both are harmful to toughness. Therefore, methods to control the grain size of ferrite-pearlite steels have rapidly evolved over the past 25 years. The two most important methods to control grain size are controlled rolling and microalloying. In fact, these methods are used in conjunction to produce strong, tough ferrite-pearlite steels.

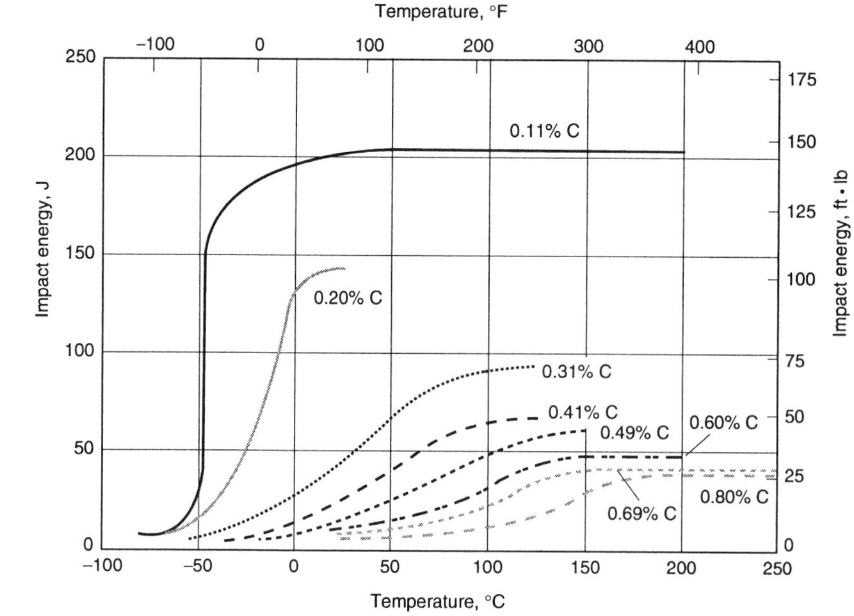

Fig. 21 Effect of carbon content in ferrite-pearlite steels on Charpy V-notch transition temperature and shelf energy. Source: Ref 17

Controlled rolling is a thermomechanical treatment in which steel plates are rolled below the recrystallization temperature of austenite. This process results in elongation of the austenite grains. Upon further rolling and subsequent cooling to room temperature, the austenite-to-ferrite transformation takes place. The ferrite grains are restricted in their growth because of the "pancake" austenite grain morphology. This produces the fine ferrite grain size required for higher strength and toughness.

Microalloying is the term applied to the addition of small amounts of special alloying elements (vanadium, niobium, or titanium) that aid in retarding austenite recrystallization, thus allowing a wide window of rolling temperatures for controlled rolling. Without retarding recrystallization, as in normal hot rolling, the pancake-type grains do not form and a fine grain size cannot be developed. Microalloyed steels are used in a wide

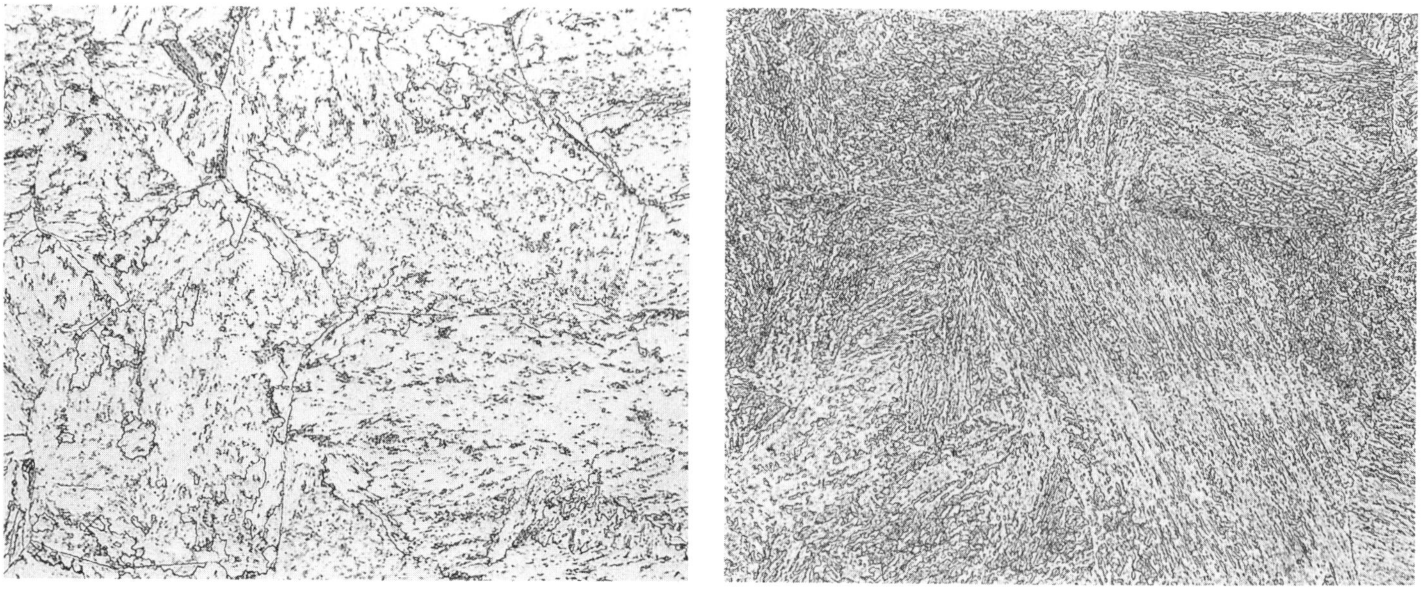

(a) (b)

Fig. 22 Microstructure of (a) upper bainite and (b) lower bainite in a Cr-Mo-V rotor steel. 2% nital + 4% picral etch. 500×

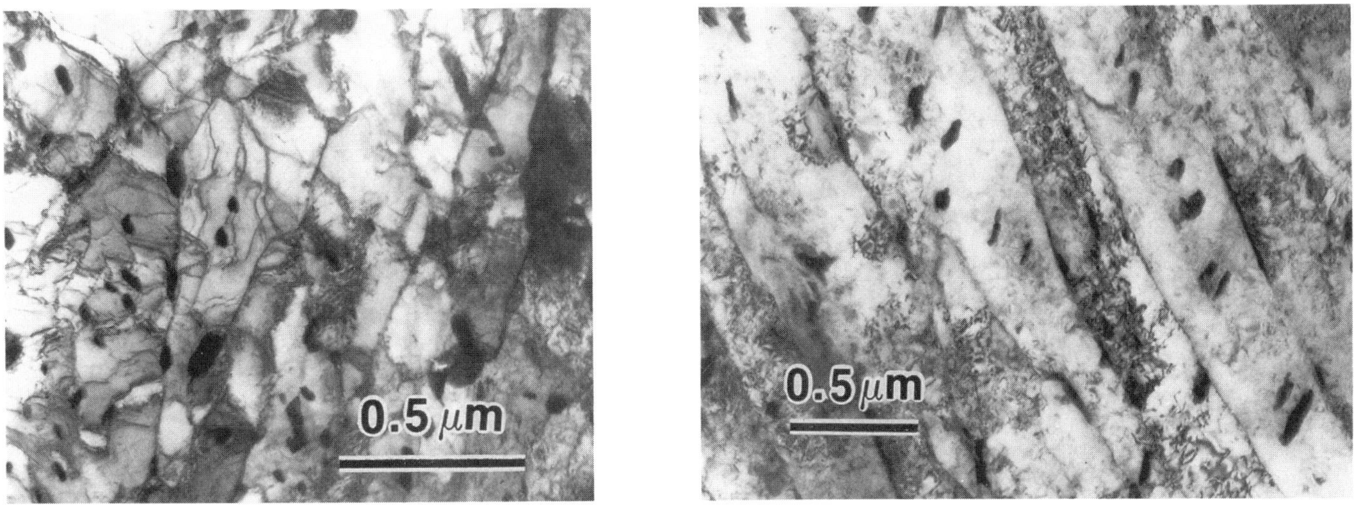

(a) (b)

Fig. 23 TEM micrographs of (a) upper bainite and (b) lower bainite in a Cr-Mo-V rotor steel

variety of high tonnage applications including structural steels for the construction industry (bridges, multistory buildings, etc.), reinforcing bar, pipe for gas transmission, and numerous forging applications.

Bainite

Like pearlite, bainite is a composite of ferrite and cementite. Unlike pearlite, the ferrite has an acicular morphology and the carbides are discrete particles. Because of these morphological differences, bainite has much different property characteristics than pearlite. In general, bainitic steels have high strength coupled with good toughness, whereas pearlitic steels have high strength with poor toughness.

Another difference between bainite and pearlite is the complexity of the bainite morphologies compared with the simple lamellar morphology of pearlite. The morphologies of bainite are still being debated in the literature. For years, since the classic work of Bain and Davenport in the 1930s (Ref 18), there were two classifications of bainite: upper and lower bainite. This nomenclature was derived from the temperature regions at which bainite formed during isothermal (constant temperature) transformation. Upper bainite formed isothermally in the temperature range of 400 to 550 °C (750 to 1020 °F), and lower bainite formed isothermally in the temperature range of 250 to 400 °C (480 to 750 °F). Examples of the microstructure of upper and lower bainite are shown in Fig. 22. One can see that both types of bainite have an acicular morphology, with upper bainite being coarser than lower bainite. The true morphological differences between the microstructures can only be determined by electron microscopy. Transmission electron micrographs of upper and lower bainite are shown in Fig. 23. In upper bainite, the iron carbide phase forms at the lath boundaries, whereas in lower bainite, the carbide phase forms on particular crystal-

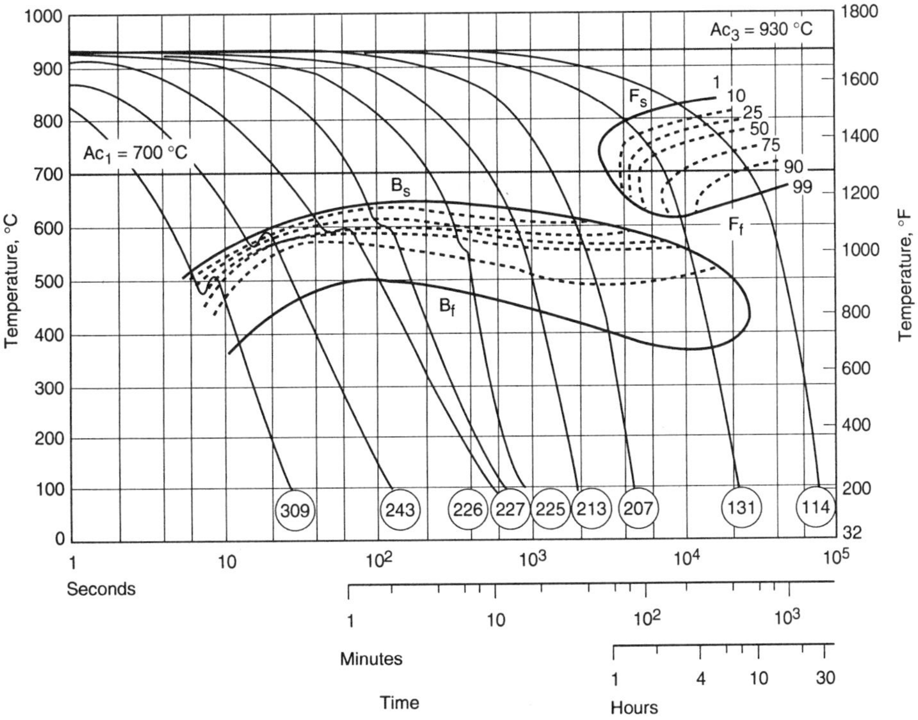

Fig. 24 A CCT diagram of a ½Mo-B steel. Composition: 0.093% C, 0.70% Mn, 0.36% Si, 0.51% Mo, 0.0054% B. Austenitized at Ac_3 + 30 °C for 12 min. B_s, bainite start; B_f, bainite finish; F_s, ferrite start; F_f, ferrite finish. Numbers in circles indicate hardness (HV) after cooling to room temperature. Source: Ref 20

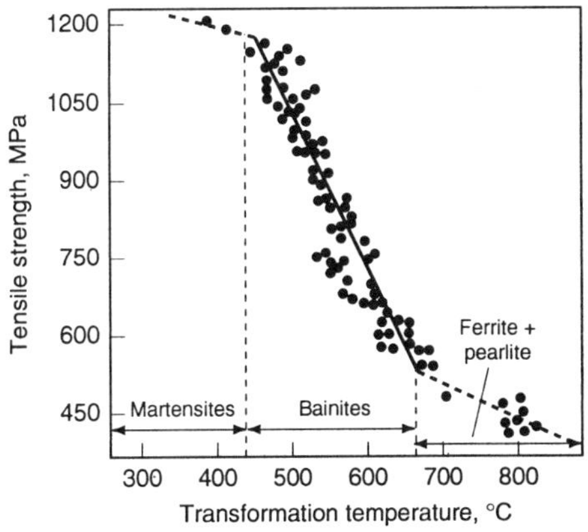

Fig. 25 Relationship between transformation temperature and tensile strength of ferrite-pearlite, bainitic, and martensitic steels. Source: Ref 5

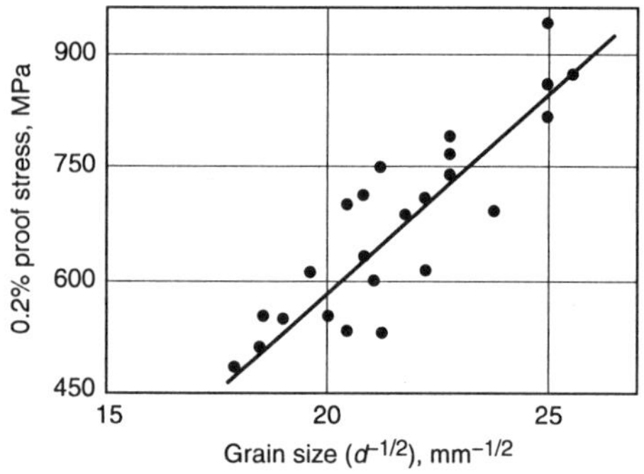

Fig. 26 Relationship between bainite lath width (grain size) and yield strength. Source: Ref 5

lographic habit planes within the laths. Because of these differences in morphology, upper and lower bainite have different mechanical properties. Lower bainite, with a fine acicular structure and carbides within the laths, has higher strength and higher toughness than upper bainite with its coarser structure.

Because during manufacture most steels undergo continuous cooling rather than isothermal holding, the terms upper and lower bainite can become confusing because "upper" and "lower" are no longer an adequate description of morphology. Bainite has recently been reclassified by its morphology, not by the temperature range in which it forms (Ref 19). For example, a recent classification of bainite yields three distinct types of morphology.

- *Class 1* (B_1): Acicular ferrite associated with *intra*lath (plate) iron carbide, that is, cementite (replaces the term "lower bainite")
- *Class 2* (B_2): Acicular ferrite associated with *inter*lath (plate) particles or films of cementite and/or austenite (replaces the term "upper bainite")
- *Class 3* (B_3): Acicular ferrite associated with a constituent consisting of discrete islands of austenite and/or martensite

The bainitic steels have a wide range of mechanical properties depending on the microstructural morphology and composition; for example, yield strength can range from 450 to 950 MPa (65 to 140 ksi), and tensile strength from 530 to 1200 MPa (75 to 175 ksi). Another aspect of a bainitic steel is that a single composition, ½Mo-B steel for example, can yield a bainitic microstructure over a wide range of transformation temperatures. The CCT diagram for this steel is shown in Fig. 24. Note that for this steel the bainite start (B_s) temperature is almost constant at 600 °C (1110 °F). This flat transformation region is important because transformation temperature plays an important role in the development of microstructure. A constant transformation temperature permits the development of a similar microstructure and properties over a wide range of cooling rates. This has many advantages in the manufacturing of bainitic steels and is particularly advantageous in thick sections where a wide range in

Table 5 Mechanical properties of selected steels

Steel	Condition	Tensile strength		Yield strength		Elongation, in 50 mm	Reduction in area, %	Hardness, HB
		MPa	ksi	MPa	ksi	%		
Carbon steel bar(a)								
1006	Hot rolled	295	43	165	24	30	55	86
	Cold drawn	330	48	285	41	20	45	95
1008	Hot rolled	305	44	170	24.5	30	55	86
	Cold drawn	340	49	285	41.5	20	45	95
1010	Hot rolled	325	47	180	26	28	50	95
	Cold drawn	365	53	305	44	20	40	105
1012	Hot rolled	330	48	185	26.5	28	50	95
	Cold drawn	370	54	310	45	19	40	105
1015	Hot rolled	345	50	190	27.5	28	50	101
	Cold drawn	385	56	325	47	18	40	111
1016	Hot rolled	380	55	205	30	25	50	110
	Cold drawn	420	61	350	51	18	40	121
1017	Hot rolled	365	53	200	29	26	50	105
	Cold drawn	405	59	340	49	18	40	116
1018	Hot rolled	400	58	220	32	25	50	116
	Cold drawn	440	64	370	54	15	40	126
1019	Hot rolled	405	59	225	32.5	25	50	116
	Cold drawn	455	66	380	55	15	40	131
1020	Hot rolled	380	55	205	30	25	50	111
	Cold drawn	420	61	350	51	15	40	121
1021	Hot rolled	420	61	230	33	24	48	116
	Cold drawn	470	68	395	57	15	40	131
1022	Hot rolled	425	62	235	34	23	47	121
	Cold drawn	475	69	400	58	15	40	137
1023	Hot rolled	385	56	215	31	25	50	111
	Cold drawn	425	62	360	52.5	15	40	121
1524	Hot rolled	510	74	285	41	20	42	149
	Cold drawn	565	82	475	69	12	35	163
1025	Hot rolled	400	58	220	32	25	50	116
	Cold drawn	440	64	370	54	15	40	126
1026	Hot rolled	440	64	240	35	24	49	126
	Cold drawn	490	71	415	60	15	40	143
1527	Hot rolled	515	75	285	41	18	40	149
	Cold drawn	570	83	485	70	12	35	163
1030	Hot rolled	470	68	260	37.5	20	42	137
	Cold drawn	525	76	440	64	12	35	149
1035	Hot rolled	495	72	270	39.5	18	40	143
	Cold drawn	550	80	460	67	12	35	163
1536	Hot rolled	570	83	315	45.5	16	40	163
	Cold drawn	635	92	535	77.5	12	35	187
1037	Hot rolled	510	74	280	40.5	18	40	143
	Cold drawn	565	82	475	69	12	35	167
1038	Hot rolled	515	75	285	41	18	40	149
	Cold drawn	570	83	485	70	12	35	163
1039	Hot rolled	545	79	300	43.5	16	40	156
	Cold drawn	605	88	510	74	12	35	179
1040	Hot rolled	525	76	290	42	18	40	149
	Cold drawn	585	85	490	71	12	35	170
1541	Hot rolled	635	92	350	51	15	40	187
	Cold drawn	705	102.5	600	87	10	30	207
	Annealed, cold drawn	650	94	550	80	10	45	184
1042	Hot rolled	550	80	305	44	16	40	163
	Cold drawn	615	89	515	75	12	35	179
	Normalized, cold drawn	585	85	505	73	12	45	179
1043	Hot rolled	565	82	310	45	16	40	163
	Cold drawn	625	91	530	77	12	35	179
	Normalized, cold drawn	600	87	515	75	12	45	179
1044	Hot rolled	550	80	305	44	16	40	163
1045	Hot rolled	565	82	310	45	16	40	163
	Cold drawn	625	91	530	77	12	35	179
	Annealed, cold drawn	585	85	505	73	12	45	170
1046	Hot rolled	585	85	325	47	15	40	170
	Cold drawn	650	94	545	79	12	35	187
	Annealed, cold drawn	620	90	515	75	12	45	179
1547	Hot rolled	650	94	360	52	15	30	192
	Cold drawn	710	103	605	88	10	28	207
	Annealed, cold drawn	655	95	585	85	10	35	187
1548	Hot rolled	660	96	365	53	14	33	197
	Cold drawn	735	106.5	615	89.5	10	28	217
	Annealed, cold drawn	645	93.5	540	78.5	10	35	192

(continued)

(a) All values are estimated minimum values; type 1100 series steels are rated on the basis of 0.10% max Si or coarse-grain melting practice; the mechanical properties shown are expected minimums for the sizes ranging from 19 to 31.8 mm (0.75 to 1.25 in.). (b) Most data are for 25 mm (1 in.) diam bar. Source: Ref 1

Table 5 (continued)

Steel	Condition	Tensile strength		Yield strength		Elongation, in 50 mm	Reduction in area, %	Hardness, HB
		MPa	ksi	MPa	ksi	%		
Carbon steel bar(a) (continued)								
1049	Hot rolled	600	87	330	48	15	35	179
	Cold drawn	670	97	560	81.5	10	30	197
	Annealed, cold drawn	635	92	530	77	10	40	187
1050	Hot rolled	620	90	340	49.5	15	35	179
	Cold drawn	690	100	580	84	10	30	197
	Annealed, cold drawn	655	95	550	80	10	40	189
1552	Hot rolled	745	108	410	59.5	12	30	217
	Annealed, cold drawn	675	98	570	83	10	40	193
1055	Hot rolled	650	94	355	51.5	12	30	192
	Annealed, cold drawn	660	96	560	81	10	40	197
1060	Hot rolled	675	98	370	54	12	30	201
	Spheroidized annealed, cold drawn	620	90	485	70	10	45	183
1064	Hot rolled	670	97	370	53.5	12	30	201
	Spheroidized annealed, cold drawn	615	89	475	69	10	45	183
1065	Hot rolled	690	100	380	55	12	30	207
	Spheroidized annealed, cold drawn	635	92	490	71	10	45	187
1070	Hot rolled	705	102	385	56	12	30	212
	Spheroidized annealed, cold drawn	640	93	495	72	10	45	192
1074	Hot rolled	725	105	400	58	12	30	217
	Spheroidized annealed, cold drawn	650	94	505	73	10	40	192
1078	Hot rolled	690	100	380	55	12	30	207
	Spheroidized annealed, cold drawn	650	94	500	72.5	10	40	192
1080	Hot rolled	770	112	425	61.5	10	25	229
	Spheroidized annealed, cold drawn	675	98	515	75	10	40	192
1084	Hot rolled	820	119	450	65.5	10	25	241
	Spheroidized annealed, cold drawn	690	100	530	77	10	40	192
1085	Hot rolled	835	121	460	66.5	10	25	248
	Spheroidized annealed, cold drawn	695	100.5	540	78	10	40	192
1086	Hot rolled	770	112	425	61.5	10	25	229
	Spheroidized annealed, cold drawn	670	97	510	74	10	40	192
1090	Hot rolled	840	122	460	67	10	25	248
	Spheroidized annealed, cold drawn	695	101	540	78	10	40	197
1095	Hot rolled	825	120	455	66	10	25	248
	Spheroidized annealed, cold drawn	680	99	525	76	10	40	197
1211	Hot rolled	380	55	230	33	25	45	121
	Cold drawn	515	75	400	58	10	35	163
1212	Hot rolled	385	56	230	33.5	25	45	121
	Cold drawn	540	78	415	60	10	35	167
1213	Hot rolled	385	56	230	33.5	25	45	121
	Cold drawn	540	78	415	60	10	35	167
12L14	Hot rolled	395	57	235	34	22	45	121
	Cold drawn	540	78	415	60	10	35	163
1108	Hot rolled	345	50	190	27.5	30	50	101
	Cold drawn	385	56	325	47	20	40	121
1109	Hot rolled	345	50	190	27.5	30	50	101
	Cold drawn	385	56	325	47	20	40	121
1117	Hot rolled	425	62	235	34	23	47	121
	Cold drawn	475	69	400	58	15	40	137
1118	Hot rolled	450	65	250	36	23	47	131
	Cold drawn	495	72	420	61	15	40	143
1119	Hot rolled	425	62	235	34	23	47	121
	Cold drawn	475	69	400	58	15	40	137
1132	Hot rolled	570	83	315	45.5	16	40	167
	Cold drawn	635	92	530	77	12	35	183
1137	Hot rolled	605	88	330	48	15	35	179
	Cold drawn	675	98	565	82	10	30	197
1140	Hot rolled	545	79	300	43.5	16	40	156
	Cold drawn	605	88	510	74	12	35	170
1141	Hot rolled	650	94	355	51.5	15	35	187
	Cold drawn	725	105.1	605	88	10	30	212
1144	Hot rolled	670	97	365	53	15	35	197
	Cold drawn	745	108	620	90	10	30	217
1145	Hot rolled	585	85	325	47	15	40	170
	Cold drawn	650	94	550	80	12	35	187
1146	Hot rolled	585	85	325	47	15	40	170
	Cold drawn	650	94	550	80	12	35	187
1151	Hot rolled	635	92	350	50.5	15	35	187
	Cold drawn	705	102	595	86	10	30	207

(continued)

(a) All values are estimated minimum values; type 1100 series steels are rated on the basis of 0.10% max Si or coarse-grain melting practice; the mechanical properties shown are expected minimums for the sizes ranging from 19 to 31.8 mm (0.75 to 1.25 in.). (b) Most data are for 25 mm (1 in.) diam bar. Source: Ref 1

cooling rates is found from the surface to the center of the part.

In designing a bainitic steel with a wide transformation region, it becomes critical that the pearlite and ferrite regions are pushed as far to the right as possible on the CCT diagram; that is, pearlite and ferrite form only at slow cooling rates. Alloying elements such as nickel, chromium, and molybdenum (and manganese) are selected for this purpose.

For low-carbon bainitic steels, the relationship between transformation temperature and tensile strength is shown in Fig. 25 (martensite will be discussed in next section of this article). Note the rapid increase in tensile strength as the transformation temperature decreases. For these steels, a regression equation for tensile strength has been developed as follows (Ref 21):

$$TS = 246.4 + 1925(\mathrm{C}) + 231(\mathrm{Mn} + \mathrm{Cr}) + 185(\mathrm{Mo}) + 92(\mathrm{W}) + 123(\mathrm{Ni}) + 62(\mathrm{Cu}) + 385(\mathrm{V} + \mathrm{Ti}) \quad \text{(Eq 9)}$$

In addition to the elements carbon, nickel, chromium, molybdenum, vanadium, and so forth, it is well known that boron in very small quantities (for example, 0.003%) has a pronounced effect on retarding the ferrite transformation. Thus, in a boron-containing steel (e.g., ½Mo + B), the ferrite nose in the CCT diagram is pushed to slower cooling rates. Boron retards the nucleation of ferrite on the austenite grain boundaries and, in doing so, permits bainite to be formed (Fig. 24). Whenever boron is added to steel, it must be prevented from combining with other elements such as oxygen and nitrogen. Generally, aluminum and titanium are added first in order to lower the oxygen and nitrogen levels of the steel. Even when adequately protected, the effectiveness of boron decreases with increasing carbon content and austenite grain size.

Attempts have been made to quantitatively relate the microstructural features of bainite to mechanical properties. One such relationship is (Ref 22):

$$YS = -194 + 17.4(d^{-1/2}) + 15(n^{1/4}) \quad \text{(Eq 10)}$$

where YS is the 0.2% offset yield strength (in MPa), d is the bainite lath size (mean linear intercept) (in mm), and n is the number of carbides per mm^2 in the plane of section.

With bainitic steels, the lath width of the bainite obeys a Hall-Petch relationship as shown in Fig. 26. The lath size is directly related to the austenite grain size and decreases with decreasing bainite transformation temperature. Because of the fine microstructure of bainite, the measurement of lath size and carbide density can only be done by scanning or transmission electron microscopy (SEM or TEM).

In low-carbon bainitic steels, type B_2 (upper) bainite has inferior toughness to type B_1 (lower) bainite. In both cases, strength increases as the transition temperature decreases. In type B_2 (upper) bainite, the carbides are much coarser than in type B_1 (lower) bainite and have a tendency to crack and initiate cleavage (brittle) fracture. In type B_1 bainite, the small carbides have less tendency to fracture. One can lower the transition temperature in type B_1 bainitic steels by providing a finer austenite grain size through lower-temperature thermomechanical treatment and grain refinement.

Bainitic steels are used in many applications including pressure vessels, backup rolls, turbine rotors, die blocks, die-casting molds, nuclear reactor components, and earthmoving equipment. One major advantage of a bainitic steel is that an optimal strength/toughness combination can be produced without expensive heat treatment, for example, quenching and tempering as in martensitic steels.

Martensite

Martensite is essentially a supersaturated solid solution of carbon in iron. The amount of carbon in martensite far exceeds that found in solid solution in ferrite. Because of this, the normal body-centered cubic (bcc) lattice is distorted in order to accommodate the carbon atoms. The distorted lattice becomes body-centered tetragonal (bct). In plain-carbon and low-alloy steels, this supersaturation is generally produced through very rapid cooling from the austenite phase region (quenching in water, iced-water, brine, iced-brine, oil or aqueous polymer solutions) to avoid forming ferrite, pearlite, and bainite. Some highly alloyed steels can form martensite upon air cooling (see the discussion of maraging steels later in this section). Depending on carbon content, martensite in its quenched state can be very hard and brittle, and, because of this brittleness, martensitic steels are usually tempered to restore some ductility and increase toughness.

Reference to a CCT diagram shows that martensite only forms at high cooling rates in plain-carbon and low-alloy steels. A CCT diagram for type 4340 is shown in Fig. 27, which indicates that martensite forms at cooling rates exceeding about 1000 °C/min. Most commercial martensitic steels contain deliberate alloying additions intended to suppress the formation of other constituents—that is ferrite, pearlite, and bainite—during continuous cooling. This means that these constituents form at slower cooling rates, allowing martensite to form at the faster cooling rates, for example, during oil and water quenching. This concept is called hardenability and is essentially the capacity of a steel to harden by rapid quenching. Most all the conventional alloying elements in steel promote hardenability. For example, type 4340 steel shown in Fig. 27 has significant levels of carbon, manganese, nickel, copper, and molybdenum to promote hardenability. More details about hardenability can be found in Ref 2.

The martensite start temperature (M_s) for type 4340 is 300 °C(570 °F). Carbon lowers the M_s temperature, as shown in Fig. 28, and alloying elements such as carbon, manganese, chromium, nickel, and molybdenum also lower M_s temperature. Many empirical equations have been developed over the past 50 years relating M_s tempera-

Table 5 (continued)

Steel	Condition	Tensile strength, MPa	Tensile strength, ksi	Yield strength, MPa	Yield strength, ksi	Elongation, in 50 mm, %	Reduction in area, %	Hardness, HB
Low-alloy steels(b)								
1340	Normalized at 870 °C (1600 °F)	834	121	558	81	22.0	63	248
	Annealed at 800 °C (1475 °F)	703	102	434	63	25.5	57	207
3140	Normalized at 870 °C (1600 °F)	889	129	600	87	19.7	57	262
	Annealed at 815 °C (1500 °F)	690	100	420	61	24.5	51	197
4130	Normalized at 870 °C (1600 °F)	670	97	435	63	25.5	59.5	197
	Annealed at 865 °C (1585 °F)	560	81	460	67	21.5	59.6	217
	Water quenched from 855 °C (1575 °F) and tempered at 540 °C (1000 °F)	1040	151	979	142	18.1	63.9	302
4140	Normalized at 870 °C (1600 °F)	1020	148	655	95	17.7	46.8	302
	Annealed at 815 °C (1500 °F)	655	95	915	60	25.7	56.9	197
	Water quenched from 845 °C (1550 °F) and tempered at 540 °C (1000 °F)	1075	156	986	143	15.5	56.9	311
4150	Normalized at 870 °C (1600 °F)	1160	168	731	106	11.7	30.8	321
	Annealed at 830 °C (1525 °F)	731	106	380	55	20.2	40.2	197
	Oil quenched from 830 °C (1525 °F) and tempered at 540 °C (1000 °F)	1310	190	1215	176	13.5	47.2	375
4320	Normalized at 895 °C (1640 °F)	793	115	460	67	20.8	51	235
	Annealed at 850 °C (1560 °F)	580	84	425	62	29.0	58	163
4340	Normalized at 870 °C (1600 °F)	1282	186	862	125	12.2	36.3	363
	Annealed at 810 °C (1490 °F)	745	108	470	68	22.0	50.0	217
	Oil quenched from 800 °C (1475 °F) and tempered at 540 °C (1000 °F)	1207	175	1145	166	14.2	45.9	352
4419	Normalized at 955 °C (1750 °F)	515	75	350	51	32.5	69.4	143
	Annealed at 915 °C (1675 °F)	450	65	330	48	31.2	62.8	121
4620	Normalized at 900 °C (1650 °F)	570	83	365	53	29.0	66.7	174
	Annealed at 855 °C (1575 °F)	510	74	370	54	31.3	60.3	149
4820	Normalized at 860 °C (1580 °F)	758	110	485	70	24.0	59.2	229
	Annealed at 815 °C (1500 °F)	685	99	460	67	22.3	58.8	197
5140	Normalized at 870 °C (1600 °F)	793	115	470	68	22.7	59.2	229
	Annealed at 830 °C (1525 °F)	570	83	290	42	28.6	57.3	167
	Oil quenched from 845 °C (1550 °F) and tempered at 540 °C (1000 °F)	972	141	841	122	18.5	58.9	293
5150	Normalized at 870 °C (1600 °F)	869	126	530	77	20.7	58.7	255
	Annealed at 825 °C (1520 °F)	675	98	360	52	22.0	43.7	197
	Oil quenched from 830 °C (1525 °F) and tempered at 540 °C (1000 °F)	1055	159	1000	145	16.4	52.9	311
5160	Normalized at 855 °C (1575 °F)	1025	149	650	94	18.2	50.7	285
	Annealed at 815 °C (1495 °F)	724	105	275	40	17.2	30.6	197
	Oil quenched from 830 °C (1525 °F) and tempered at 540 °C (1000 °F)	1145	166	1005	146	14.5	45.7	341
6150	Normalized at 870 °C (1600 °F)	938	136	615	89	21.8	61.0	269
	Annealed at 815 °C (1500 °F)	670	97	415	60	23.0	48.4	197
	Oil quenched from 845 °C (1550 °F) and tempered at 540 °C (1000 °F)	1200	174	1160	168	14.5	48.2	352
8620	Normalized at 915 °C (1675 °F)	635	92	360	52	26.3	59.7	183
	Annealed at 870 °C (1600 °F)	540	78	385	56	31.3	62.1	149
8630	Normalized at 870 °C (1600 °F)	650	94	425	62	23.5	53.5	187
	Annealed at 845 °C (1550 °F)	565	82	370	54	29.0	58.9	156
	Water quenched from 845 °C (1550 °F) and tempered at 540 °C (1000 °F)	931	135	850	123	18.7	59.6	269
8650	Normalized at 870 °C (1600)	1025	149	690	100	14	45.0	302
	Annealed at 795 °C (1465 °F)	715	104	385	56	22.5	46.0	212
	Oil quenched from 800 °C (1475 °F) and tempered at 540 °C (1000 °F)	1185	172	1105	160	14.5	49.1	352
8740	Normalized at 870 °C (1600 °F)	931	135	605	88	16.0	47.9	269
	Annealed at 815 °C (1500 °F)	696	101	415	60	22.2	46.4	201
	Oil quenched from 830 °C (1525 °F) and tempered at 540 °C (1000 °F)	1225	178	1130	164	16.0	53.0	352
9255	Normalized at 900 °C (1650 °F)	931	135	580	84	19.7	43.4	269
	Annealed at 845 °C (1550 °F)	779	113	485	70	21.7	41.1	229
	Oil quenched from 885 °C (1625 °F) and tempered at 540 °C (1000 °F)	1130	164	924	134	16.7	38.3	321
9310	Normalized at 890 °C (1630 °F)	910	132	570	83	18.8	58.1	269 HRB
	Annealed at 845 °C (1550 °F)	820	119	450	65	17.3	42.1	241 HRB
Ferritic stainless steels(b)								
405	Annealed bar	483	70	276	40	30	60	150
	Cold drawn bar	586	85	483	70	20	60	185
409	Annealed bar	450	65	240	35	25	...	75 HRB
430	Annealed bar	517	75	310	45	30	65	155

(continued)

(a) All values are estimated minimum values; type 1100 series steels are rated on the basis of 0.10% max Si or coarse-grain melting practice; the mechanical properties shown are expected minimums for the sizes ranging from 19 to 31.8 mm (0.75 to 1.25 in.). (b) Most data are for 25 mm (1 in.) diam bar. Source: Ref 1

Table 5 (continued)

Steel	Condition	Tensile strength MPa	Tensile strength ksi	Yield strength MPa	Yield strength ksi	Elongation, in 50 mm %	Reduction in area, %	Hardness, HB
Ferritic stainless steels(b) (continued)								
430 (con't)	Annealed and cold drawn	586	85	483	70	20	65	185
442	Annealed bar	515	75	310	45	30	50	160
	Annealed at 815 °C (1500 °F) and cold worked	545	79	427	62	35.5	79	92 HRC
446	Annealed bar	550	80	345	50	25	45	86 HRB
	Annealed at 815 °C (1500 °F) and cold drawn	607	88	462	67	26	64	96 HRB
Martensitic stainless steels(b)								
403	Annealed bar	515	75	275	40	35	70	82 HRB
	Tempered bar	765	111	585	85	23	67	97 HRB
410	Oil quenched from 980 °C (1800 °F); tempered at 540 °C (1000 °F); 16 mm (0.625 in.) bar	1085	158	1005	146	13	70	...
	Oil quenched from 980 °C (1800 °F); tempered at 40 °C (104 °F); 16 mm (0.625 in.) bar	1525	221	1225	178	15	64	45 HRB
414	Annealed bar	795	115	620	90	20	60	235
	Cold drawn bar	895	130	795	115	15	58	270
	Oil quenched from 980 °C (1800 °F); tempered at 650 °C (1200 °F)	1005	146	800	116	19	58	...
420	Annealed bar	655	95	345	50	25	55	195
	Annealed and cold drawn	760	110	690	100	14	40	228
431	Annealed bar	860	125	655	95	20	55	260
	Annealed and cold drawn	895	130	760	110	15	35	270
	Oil quenched from 980 °C (1800 °F); tempered at 650 °C (1200 °F)	831	121	738	107	20	64	...
	Oil quenched from 980 °C (1800 °F); tempered at 40 °C (104 °F)	1435	208	1140	166	17	59	45 HRC
440C	Annealed bar	760	110	450	65	14	25	97 HRB
	Annealed and cold drawn bar	860	125	690	100	7	20	260
	Hardened and tempered at 315 °C (600 °F)	1970	285	1900	275	2	10	580
Austenitic stainless steels(b)								
201	Annealed	760	110	380	55	52	...	87 HRB
	50% hard	1035	150	760	110	12	...	32 HRC
	Full hard	1275	185	965	140	8	...	41 HRC
	Extra hard	1550	225	1480	215	1	...	43 HRC
202	Annealed bar	515	75	275	40	40	...	...
	Annealed sheet	655	95	310	45	40	...	...
	50% hard sheet	1030	150	760	110	10	...	...
301	Annealed	725	105	275	40	60	70	...
	50% hard	1035	150	655	95	54	61	...
	Full hard	1415	205	1330	193	6	...	...
302	Annealed strip	620	90	275	40	55	...	80 HRB
	25% hard strip	860	125	515	75	12	...	25 HRC
	Annealed bar	585	85	240	35	60	70	80 HRB
303	Annealed bar	620	90	240	35	50	55	160
	Cold drawn	690	100	415	60	40	53	228
304	Annealed bar	585	85	235	34	60	70	149
	Annealed and cold drawn	690	100	415	60	45	...	212
	Cold-drawn high tensile	860	125	655	95	25	...	275
305	Annealed sheet	585	85	260	38	50	...	80 HRB
308	Annealed bar	585	85	205	30	55	65	150
309	Annealed bar	655	95	275	40	45	65	83 HRB
310	Annealed sheet	620	90	310	45	45	...	85 HRB
	Annealed bar	655	95	275	40	45	65	160
314	Annealed bar	689	100	345	50	45	60	180
316	Annealed sheet	580	84	290	42	50	...	79 HRB
	Annealed bar	550	80	240	35	60	70	149
	Annealed and cold-drawn bar	620	90	415	60	45	65	190
317	Annealed sheet	620	90	275	40	45	...	85 HRB
	Annealed bar	585	85	275	40	50	...	160
321	Annealed sheet	620	90	240	35	45	...	80 HRB
	Annealed bar	585	85	240	35	55	65	150
	Annealed and cold-drawn bar	655	95	415	60	40	60	185
330	Annealed sheet	550	80	260	38	40	...	...
	Annealed bar	585	85	290	42	45	...	80 HRB
347	Annealed sheet	655	95	275	40	45	...	85 HRB
	Annealed bar	620	90	240	35	50	65	160

(continued)

(a) All values are estimated minimum values; type 1100 series steels are rated on the basis of 0.10% max Si or coarse-grain melting practice; the mechanical properties shown are expected minimums for the sizes ranging from 19 to 31.8 mm (0.75 to 1.25 in.). (b) Most data are for 25 mm (1 in.) diam bar. Source: Ref 1

ture to composition. One recent equation by Andrews (Ref 24) is:

$$M_s\,(°C) = 539 - 423(C) - 30.4(Mn) - 12.1(Cr) - 17.7(Ni) - 7.5(Mo) \quad \text{(Eq 11)}$$

With sufficient alloy content, the M_s temperature can be below room temperature, which means that the transformation is incomplete and retained austenite can be present in the steel.

The microstructure of martensitic steels can be generally classed as either lath martensite, plate martensite, or mixed lath and plate martensite. In plain carbon steels, this classification is related to carbon content, as shown in Fig. 28. Lath martensite forms at carbon contents up to about 0.6%, plate martensite is found at carbon contents greater than 1.0%, and a mixed martensite microstructure forms for carbon contents between 0.6 and 1.0%. An example of lath martensite is shown in Fig. 29 and plate martensite in Fig. 30. Generally, plate martensite can be distinguished from lath martensite by its plate morphology with a central mid-rib. Also, plate martensite may contain numerous microcracks, as shown in Fig. 31. These form during transformation when a growing plate impinges on an existing plate. Because of these microcracks, plate martensite is generally avoided in most applications. The important microstructural units measured in lath martensite are lath width and packet size. A packet is a grouping of laths having a common orientation.

Plain-carbon and low-alloy martensitic steels are rarely used in the as-quenched state because of poor ductility. To increase ductility, these martensitic steels are tempered (reheated) to a temperature below 650 °C (1200 °F). During tempering, the carbon that is in supersaturated solid solution precipitates on preferred crystallographic planes (usually the octahedral {111} planes) of the martensitic lattice. Because of the preferred orientation, the carbides in a tempered martensite have a characteristic arrangement as seen in Fig. 32.

Tempered martensite has similar morphological features to type B_1 (lower) bainite. However, a distinction can be made in terms of the orientation differences of the carbide precipitates. This can be seen by comparing type B_1 bainite in Fig. 23 with tempered martensite in Fig. 32. However, unless the carbide morphology is observed it is very difficult to distinguish between B_1 bainite and tempered martensite.

The hardness of martensite is determined by its carbon content, as shown in Fig. 33. Martensite attains a maximum hardness of 66 HRC at carbon contents of 0.8 to 1.0%. The reason that the hardness does not monotonically increase with carbon is that retained austenite is found when the carbon content is above about 0.4% (austenite is much softer than martensite). Figure 34 shows the increase in volume percent retained austenite with increasing carbon content. Yield strength also increases with increasing carbon content as seen in Fig. 35. This empirical relationship between the yield strength and carbon content for untempered low-carbon martensite is (Ref 25):

Table 5 (continued)

Steel	Condition	Tensile strength MPa	Tensile strength ksi	Yield strength MPa	Yield strength ksi	Elongation, in 50 mm %	Reduction in area, %	Hardness, HB
Austenitic stainless steels(b) (continued)								
347 (con't)	Annealed and cold drawn bar	690	100	450	65	40	60	212
384	Annealed wire 1040 °C (1900 °F)	515	75	240	35	55	72	70 HRB
Maraging steels(b)								
18Ni(250)	Annealed	965	140	655	95	17	75	30 HRC
	Aged bar 32 mm (1.25 in.)	1844	269	1784	259	11	56.5	51.8 HRC
	Aged sheet 6 mm (0.25 in.)	1874	272	1832	266	8	40.8	50.6 HRC
18Ni(300)	Annealed	1034	150	758	110	18	72	32 HRC
	Aged bar 32 mm (1.25 in.)	2041	296	2020	293	11.6	55.8	54.7 HRC
	Aged sheet 6 mm (0.25 in.)	2169	315	2135	310	7.7	35	55.1 HRC
18Ni(350)	Annealed	1140	165	827	120	18	70	35 HRC
	Aged bar 32 mm (1.25 in.)	2391	347	2348	341	7.6	33.8	58.4 HRC
	Aged sheet 6 mm (0.25 in.)	2451	356	2395	347	3	15.4	57.7 HRC

(a) All values are estimated minimum values; type 1100 series steels are rated on the basis of 0.10% max Si or coarse-grain melting practice; the mechanical properties shown are expected minimums for the sizes ranging from 19 to 31.8 mm (0.75 to 1.25 in.). (b) Most data are for 25 mm (1 in.) diam bar. Source: Ref 1

$$YS\,(\text{MPa}) = 413 + 17.2 \times 10^5 (C^{1/2}) \qquad \text{(Eq 12)}$$

Lath martensite packet size also has an influence on the yield strength, as shown in Fig. 36. The linear behavior follows a Hall-Petch type relationship of ($d^{-1/2}$).

Most martensitic steels are used in the tempered condition where the steel is reheated after quenching to a temperature less than the lower critical temperature (Ac_1). Figure 37 shows the decrease in hardness with tempering temperature for a number of carbon levels. Plain-carbon or low-alloy martensitic steels can be tempered in lower or higher temperature ranges, depending upon the balance of properties required. Tempering between 150 and 200 °C (300 and 390 °F) will maintain much of the hardness and strength of the quenched martensite and provide a small improvement in ductility and toughness (Ref 26). This treatment can be used for bearings and gears that are subjected to compression loading. Tempering above 425 °C (796 °F) significantly improves ductility and toughness but at the expense of hardness and strength. The effect of tempering temperature on the tensile properties of a typical oil-quenched low-alloy steel (type 4340) is shown in Fig. 38. These data are for a 13.5 mm (0.53 in.) diam rod quenched in oil. The as-quenched rod has a hardness of 601 HB. Note that by tempering at 650 °C (1200 °F), the hardness (see *x*-axis) decreased to 293 HB; or to less than half the as-quenched hardness. The tensile strength has decreased from 1960 MPa (285 ksi) at a 200 °C (400 °F) tempering temperature to 965 MPa (141 ksi) at a 650 °C (1200 °F) tempering temperature. However, the ductility, represented by total elongation and reduction in area, increases dramatically. The tempering process can be retarded by the addition of certain alloying elements such as vanadium, molybdenum, manganese, chromium, and silicon. Also, for tempering, temperature is much more important than time at temperature.

Temper embrittlement is possible during the tempering of alloy and low-alloy steels. This embrittlement occurs when quenched-and-tempered steels are heated in, or slow cooled through the 340 to 565 °C (650 to 1050 °F) temperature range. Embrittlement occurs when the embrittling elements, antimony, tin, and phosphorus, concentrate at the austenite grain boundaries and create intergranular segregation that leads to intergranular fracture. The element molybdenum has been shown to be beneficial in preventing temper embrittlement.

The large variation in mechanical properties of quenched-and-tempered martensitic steels provides the structural designer with a large number of property combinations. Data, like that shown in Fig. 38, are available in Volume 1 of *ASM Handbook* as well as from other sources. Hardnesses of quenched-and-tempered steels can be estimated by a method established by Grange, et al. (Ref 27). The general equation for hardness is:

$$HV = HV_C + \Delta HV_{Mn} + \Delta HV_P + \Delta HV_{Si} + \Delta HV_{Ni} + \Delta HV_{Cr} + \Delta HV_{Mo} + \Delta HV_V \qquad \text{(Eq 13)}$$

where HV is the estimated hardness value (Vickers).

In order to use this relationship, one must determine the hardness value of carbon (HV_C) from Fig. 39. For example, if one assumes that a tempering temperature of 540 °C (1000 °F) is used and the carbon content of the steel is 0.2% C, the HV_C value after tempering will be 180 HV. Second, the effect of each alloying element must be determined from a figure such as Fig. 40. This graph represents a tempering temperature of 540 °C (1000 °F). Graphs representing other tempering temperatures can be found in Ref 27.

To illustrate the use of the Grange, et al. method, the same type 4340 steel shown in Fig. 38 is used. The composition of the steel is 0.41% C, 0.67% Mn, 0.023% P, 0.018% S, 0.26% Si, 1.77% Ni, 0.78% Cr, and 0.26% Mo. Assuming a

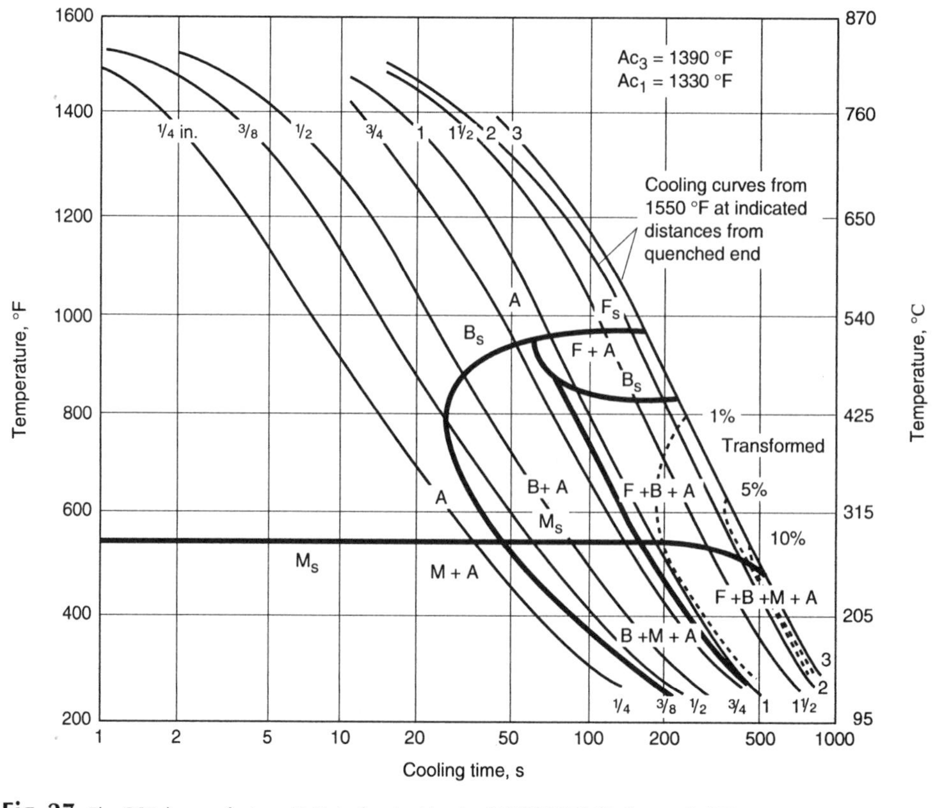

Fig. 27 The CCT diagram for type 4340 steel austenitized at 845 °C (1550 °F). Source: Ref 22

540 °C (1000 °F) tempering temperature, the estimated hardness value for carbon is 210 HV. From Fig. 39, the hardness values for each of the other alloying elements are:

Element	Content, %	Hardness, HV
C	0.41	210
Mn	0.67	38
P	0.023	7
Si	0.26	15
Ni	1.77	12
Cr	0.78	43
Mo	0.26	55
Total hardness		**380**

According to Fig. 38, the hardness value after tempering at 540 °C (1000 °F) was 363 HB (see Brinell hardness values along *x*-axis). From the ASTM E 48 conversion table (included in *Mechanical Testing*, Volume 8 of *ASM Handbook*), a Brinell hardness of 363 HB equates to a Vickers hardness of 383 HV. The calculated value of 380 HV (in the table above) is very close to the actual measured value of 383 HV. Thus, this method can be used to estimate a specific hardness value after a quenching-and-tempering heat treatment for a low-alloy steel. Also, as a rough approximation, the derived Brinell hardness value can be used to estimate tensile strength by the following equation (calculated from ASTM E 48 conversion table):

$$TS\,(\text{MPa}) = -42.3 + 3.6\,\text{HB} \quad \text{(Eq 14)}$$

For the above example, a type 4340 quenched-and-tempered (540 °C, or 1000 °F) steel with a calculated hardness of 363 HB would have an estimated tensile strength from Eq 14 of 1265 MPa (183 ksi). From Table 5, this measured tensile strength of a type 4340 quenched-and-tempered (540 °C, or 1000 °F) steel is 1255 MPa (182 ksi).

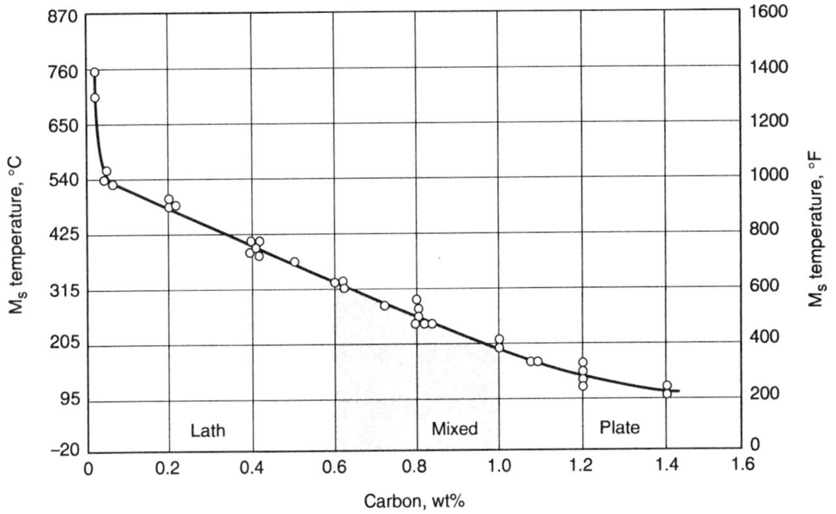

Fig. 28 Effect of carbon content on M_s temperature in steels. Source: Ref 6

It is seen that quenched-and-tempered martensitic steels provide a wide range of properties. The design engineer can choose from a large number of plain-carbon and low-alloy steels (Tables 1 and 2). In addition to this large list of steels, there are two other commercially important categories of fully martensitic steels, namely, martensitic stainless steels (Table 3) and maraging steels (Table 4).

Like the ferritic stainless steels, martensitic stainless steels (e.g., type 403, 410, 414, 416, 420, 422, 431, and 440) are high chromium-iron alloys (12 to 18% Cr), but with deliberate additions of carbon (0.12 to 1.2% C). These steels use carbon in order to stabilize austenite in iron-chromium alloys (Fig. 13). The expanded region of austenite is called the γ-loop. In the Fe-Cr phase diagram (without C), the γ-loop extends to about 12% Cr (see Fig. 13). With carbon additions, austenite can exist up to 25% Cr. These steels can be heat treated much like those of the low-alloy steels. However, martensitic stainless steels, with such high chromium contents, can form martensite on air cooling, even in thick sections. Martensitic stainless steels are considered high-strength stainless steels because they can be treated to achieve a yield strength between 550 MPa (80 ksi) and 1.7 GPa (250 ksi). On the other hand, ferritic stainless steels, which do not contain carbon, are not considered high-strength steels because their yield strength range is only 170 to 450 MPa (25 to 64 ksi). Because of their high strength and hardness, coupled with corrosion resistance, martensitic stainless steels are used for knives and other applications requiring a cutting edge as well as some tool steel applications.

Maraging steels are a separate class of martensitic steels and are considered ultrahigh-strength

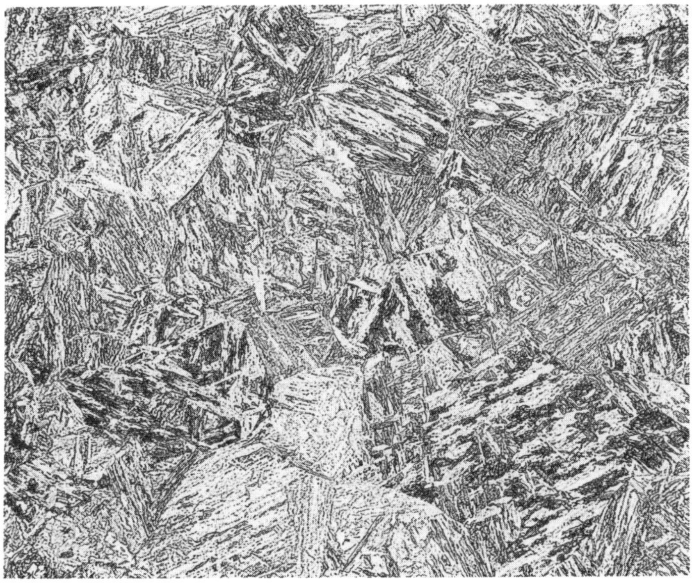

Fig. 29 Microstructure of a typical lath martensite. 4% picral + HCl. 200×

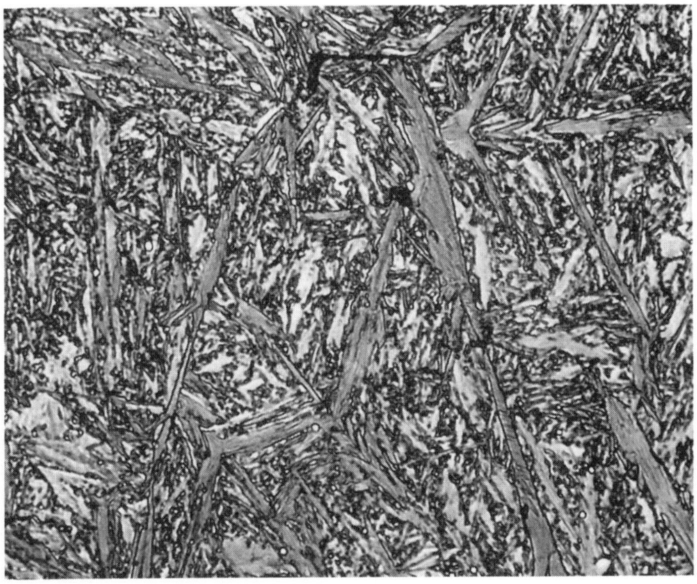

Fig. 30 Microstructure of a typical plate martensite. 4% picral + HCl. 1000×

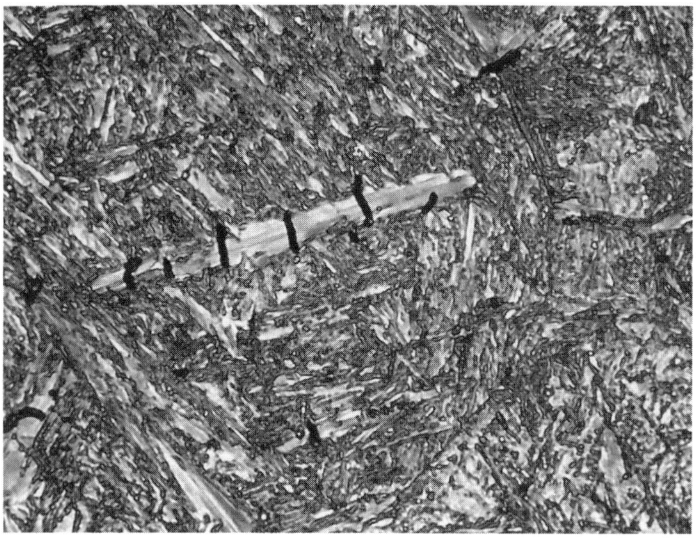

Fig. 31 Microcracks formed in plate martensite. 4% picral + HCl/sodium metabisulfite etch. 1000×

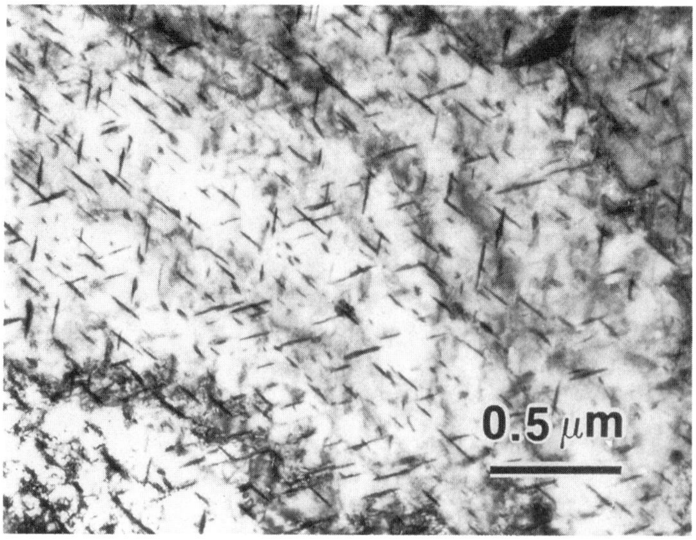

Fig. 32 TEM micrograph showing carbide morphology in tempered martensite

steels with yield strength levels as high as 2.5 GPa (360 ksi), as seen in Table 5. In addition to extremely high strength, the maraging steels have excellent ductility and toughness. These very-low carbon steels contain 17.5 to 18% Ni, 8.5 to 12.5% Co, 4 to 5% Mo, 0.20 to 1.8% Ti, and 0.10 to 0.15% Al. Because of the high alloy content, especially the cobalt addition, they are very expensive. Their high strength is developed by austenitizing at 850 °C (1560 °F), followed by air cooling to room temperature to form lath martensite. However, the martensitic constituent in maraging steels is relatively soft—28 to 35 HRC—which is an advantage because the component can be machined to final form directly upon cooling. The final stage of strengthening is through an aging process, carried out at 480 °C (900 °F) for 3 h. During aging, the hardness increases to about 51 to 58 HRC depending on the grade of maraging steel. The aging treatment promotes the precipitation of a rodlike intermetallic compound Ni_3Mo. These precipitates can only be observed at high magnification (e.g., by TEM). The precipitates strengthen the surrounding matrix as they form during aging. Full hardening can be developed, even in very thick sections. Maraging steels are used for die-casting molds and aluminum hot-forging dies as well as numerous aircraft and missile components.

Austenite

Austenite does not exist at room temperature in plain-carbon and low-alloy steels, other than as small amounts of retained austenite that did not transform during rapid cooling. However, in certain high-alloy steels, such as the austenitic stainless steels and Hadfield austenitic manganese steel, austenite is the microstructure. In these steels, sufficient quantities of alloying elements that stabilize austenite at room temperature are present (e.g., manganese and nickel). The crystal structure of austenite is face-centered cubic (fcc) as compared to ferrite, which has a (bcc) lattice. A fcc alloy has certain desirable characteristics; for example, it has low-temperature toughness, excellent weldability, and is nonmagnetic. Because of their high alloy content, austenitic steels are usually corrosion resistant. Disadvantages are their expense (because of the alloying elements), their susceptibility to stress-corrosion cracking (certain austenitic steels), their relatively low yield strength, and the fact that they cannot be strengthened other than by cold working, interstitial solid-solution strengthening, or precipitation hardening.

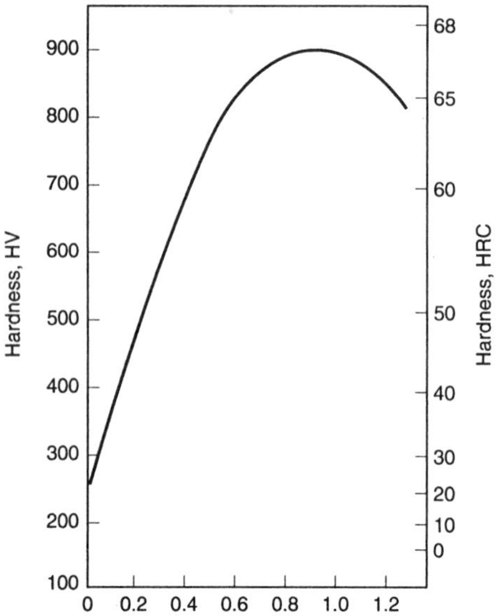

Fig. 33 Effect of carbon content on the hardness of martensite. Source: Ref 4

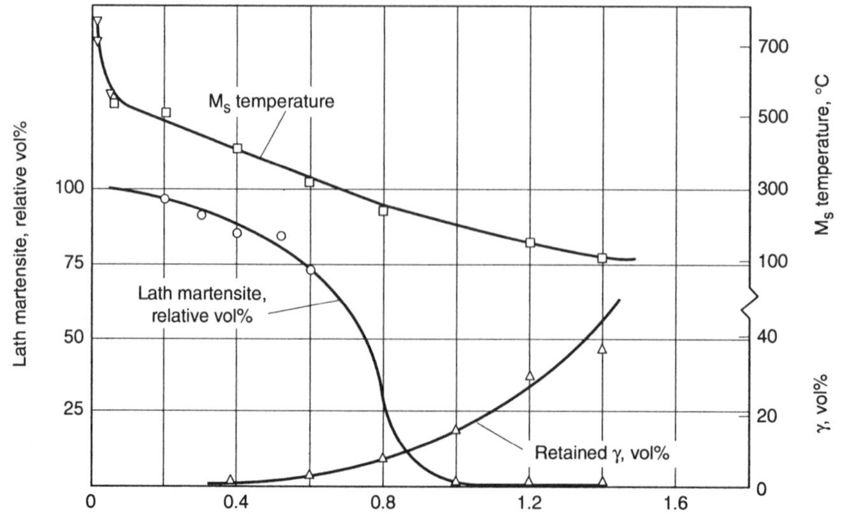

Fig. 34 Effect of carbon content on the volume percent of retained austenite (γ) in as-quenched martensite. Source: Ref 4

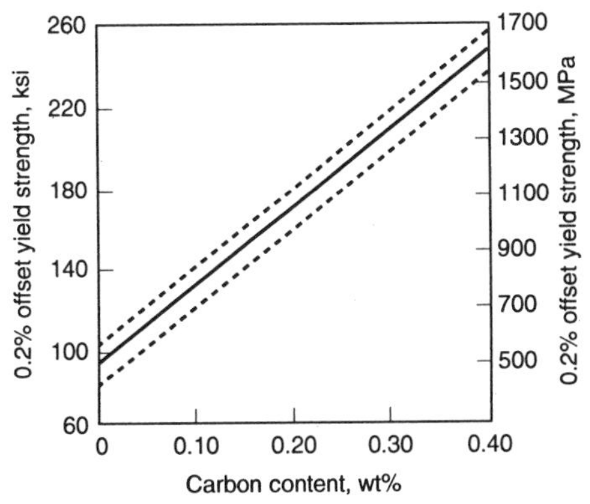

Fig. 35 Relationship between carbon content and the yield strength of martensite. Source: Ref 4

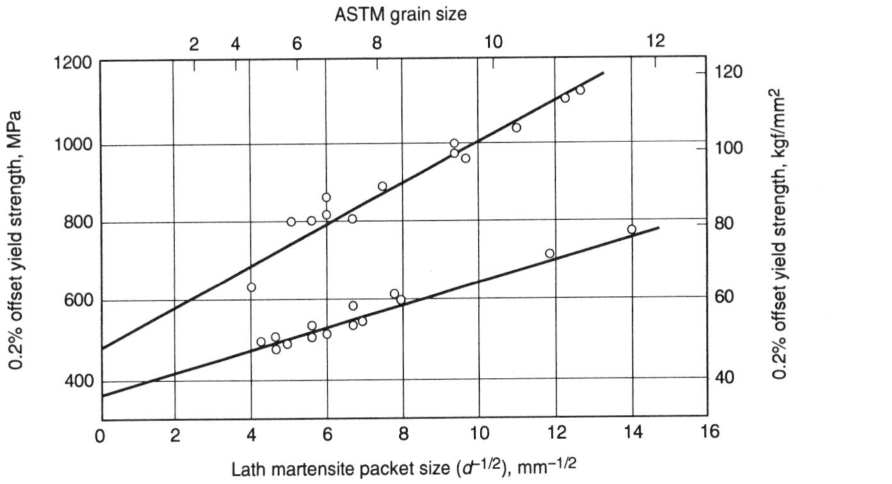

Fig. 36 Relationship between lath martensite packet size (*d*) and yield strength of Fe-0.2%C (upper line) and Fe-Mn (lower line) martensites. Source: Ref 2

The austenitic stainless steels (e.g., type 301, 302, 303, 304, 305, 308, 309, 310, 314, 316, 317, 321, 330, 347, 348, and 384) generally contain from 6 to 22% Ni to stabilize the austenite at room temperature. They also contain other alloying elements, such as chromium (16 to 26%) for corrosion resistance, and smaller amounts of manganese and molybdenum. The widely used type 304 stainless steel contains 18 to 20% Cr and 8 to 10.5% Ni, and is also called 18-8 stainless steel. From Table 5, the yield strength of annealed type 304 stainless steel is 290 MPa (40 ksi), with a tensile strength of about 580 MPa (84 ksi). However, both yield and tensile strength can be substantially increased by cold working as shown in Fig. 41 (see Table 5). However, the increase in strength is offset by a substantial decrease in ductility, for example, from about 55% elongation in the annealed condition to about 25% elongation after cold working.

Some austenitic stainless steels (type 200, 201, 202, and 205) employ interstitial solid-solution strengthening with nitrogen addition. Austenite, like ferrite, can be strengthened by interstitial elements such as carbon and nitrogen. However, carbon is usually excluded because of the deleterious effect associated with precipitation of chromium carbides on austenite grain boundaries (a process called sensitization). These chromium carbides deplete the grain-boundary regions of chromium, and the denuded boundaries are extremely susceptible to corrosion. Such steels can be desensitized by heating to high temperature to dissolve the carbides and place the chromium back into solution in the austenite. Nitrogen, on the other hand, is soluble in austenite and is added for strengthening. To prevent nitrogen from forming deleterious nitrides, manganese is added to lower the activity of nitrogen in the austenite, as well as to stabilize the austenite. For example, type 201 stainless steel has composition ranges of 5.5 to 7.5% Mn, 16 to 18% Cr, 3.5 to 5.5% Ni, and 0.25% N. The other type 2*xx* series of steels contain from 0.25 to 0.40% N.

Another important austenitic steel is austenitic manganese steel. Developed by Sir Robert Hadfield in the late 1890s, these steels remain austenitic after water quenching and have considerable strength and toughness. A typical Hadfield

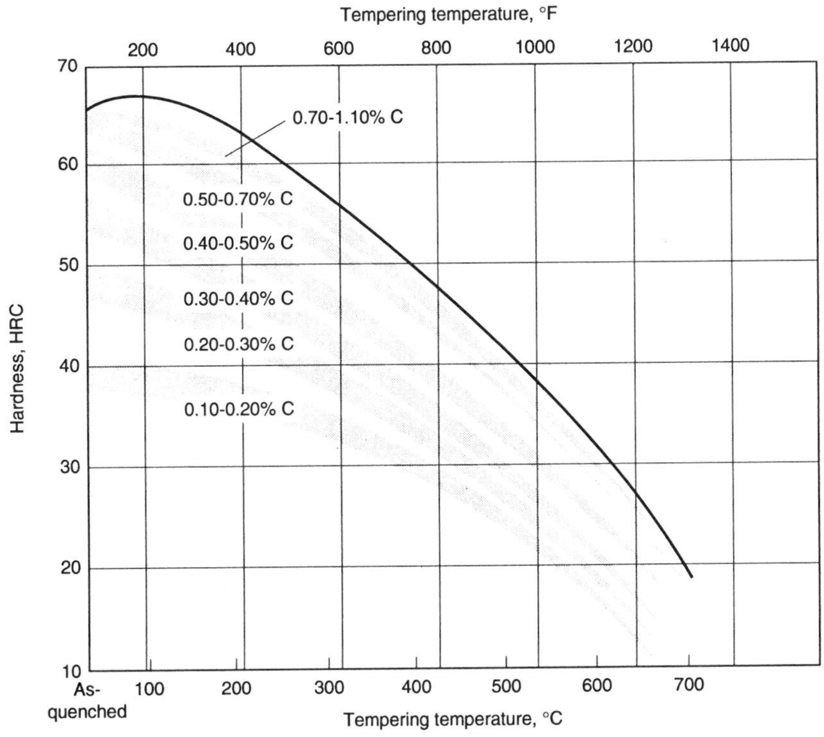

Fig. 37 Decrease in the hardness of martensite with tempering temperature for various carbon contents. Source: Ref 2

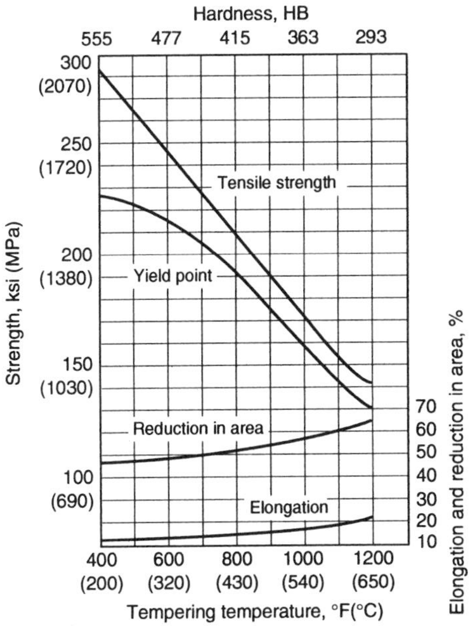

Fig. 38 Effect of tempering temperature on the mechanical properties of type 4340 steel. Source: Ref 2

manganese steel will contain 10 to 14% Mn, 0.95 to 1.4% C, and 0.3 to 1% Si. Solution annealing is necessary to suppress the formation of iron carbides. The carbon must be in solid solution to stabilize the austenite. When completely austenitic, these steels can be work hardened to provide higher hardness and wear resistance. A work-hardened Hadfield manganese steel has excellent resistance to abrasive wear under heavy loading. Because of this characteristic, these steels are ideal for jaw crushers and other crushing and grinding components in the mining industry. Also, Hadfield manganese steels have long been used for railway frogs (components used at the junction point of two railroad lines).

Ferrite-Cementite

When plain-carbon steels are heated to temperatures just below the lower critical temperature (Ac_1), the process of spheroidization takes place. Figure 42 shows a fully spheroidized steel microstructure. The microstructure before spheroidization is pearlite. During spheroidization, the cementite lamellae of the pearlite must change morphology to form spheroids. The process is controlled by the diffusion rate of carbon and portions of the lamellae must "pinch-off" (dissolve), and that dissolved carbon must diffuse to form a spheroid from the remaining portions of lamellae. This process takes several hours. Spheroidization takes place in less time when the starting microstructure is martensite or tempered martensite. In this process, the spheroidized carbides are formed by growth of carbides formed during tempering.

A fully spheroidized structure leads to improved machinability. A steel in its fully spheroidized state is in its softest possible condition. Some steels, such as type 1020, are spheroidized before cold forming into tubing because spheroidized steels have excellent formability.

Ordinary low-carbon, cold-rolled, and annealed sheet steels have ferritic microstructures with a small amount of grain-boundary cementite, as shown in Fig. 9. These carbides nucleate and grow on the ferrite grain boundaries during the annealing process, which takes place in the lower portion of the intercritical temperature region (i.e., the region between the A_3 and A_1 temperatures shown in the iron-carbon diagram, Fig. 7). Many modern-day automotive sheet steels are produced with very low carbon levels to avoid these grain-boundary carbides because they degrade formability.

Ferrite-Martensite

A relatively new family of steels called dual-phase steels consists of a microstructure of about 15 to 20% martensite in a matrix of ferrite. The microstructure of a typical dual-phase steel is shown in Fig. 43. In most plain-carbon and low-alloy steels, the presence of martensite in the microstructure is normally avoided because of the deleterious effect that martensite has on ductility and toughness. However, when the martensite is embedded in a matrix of ferrite, it imparts desirable characteristics. One desirable characteristic is that dual-phase steels do not exhibit a yield point. Figure 44 compares the stress-strain behavior of four steels: plain carbon, SAE 950X, and SAE 980X, which exhibit a yield point with the fourth, a dual-phase steel (GM 980X). This means that the cosmetically unappealing Lüders bands that form during the discontinuous yielding (i.e., yield point) are absent in a dual-phase steel. Also note in Fig. 44 that the dual-phase steel has much more elongation than the SAE 980X of similar tensile strength. These characteristics are especially important in formability.

A unique characteristic of a ferrite-martensite dual-phase steel is its substantial work hardening capacity. This allows the steel to strengthen while being deformed. By proper design of the stamping dies, this behavior can be exploited to produce a high-strength component. Most conventional high-strength steels have limited formability because their high strength is developed prior to the forming process.

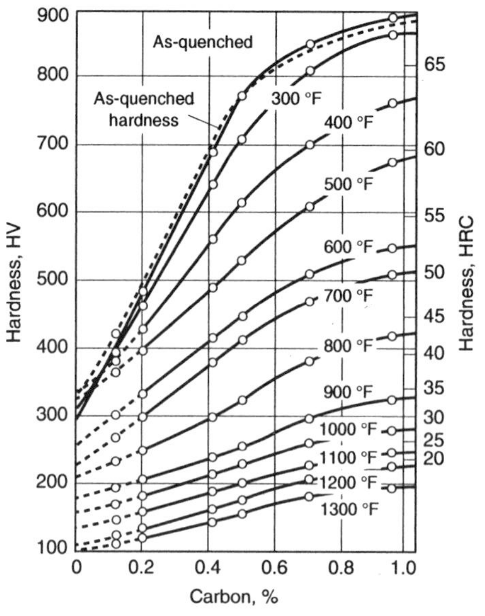

Fig. 39 Relationship between hardness of tempered martensite with carbon content at various tempering temperatures. Source: Ref 2

Graphite

When carbon contents of iron-carbon alloys exceed about 2%, there is a tendency for graphite to form (see Fe-C diagram in Fig. 7b). This is especially true in gray cast iron in which graphite flakes are a predominant microstructural feature (Fig. 4). Gray cast iron has been used for centuries because it melts at a lower temperature than steel and is easy to cast into various shapes. Also, the graphite flakes impart good machinability, acting as chip breakers, and they also provide excellent damping capacity. Damping capacity is important in machines that are subject to vibration. However, gray cast iron is limited to applications that do not require toughness or ductility,

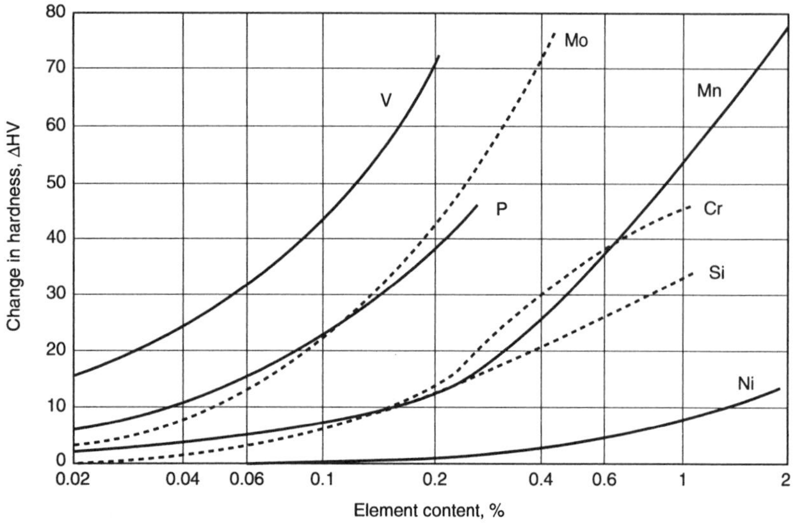

Fig. 40 Effect of alloying elements on the retardation or softening during tempering at 540 °C (1000 °F) relative to iron-carbon alloys. Source: Ref 2

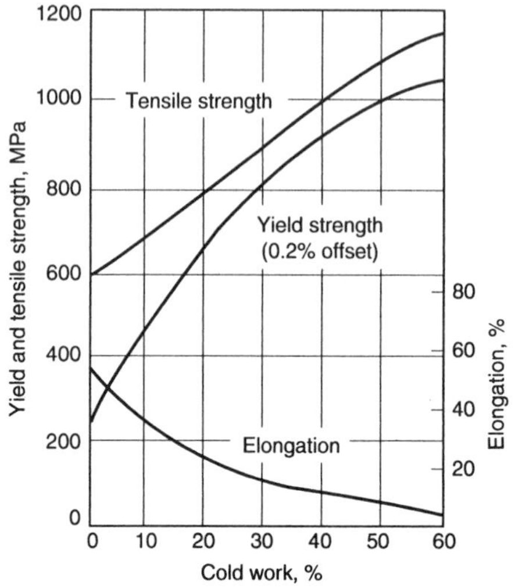

Fig. 41 Influence of cold work on mechanical properties of type 304 stainless steel. Source: Ref 4

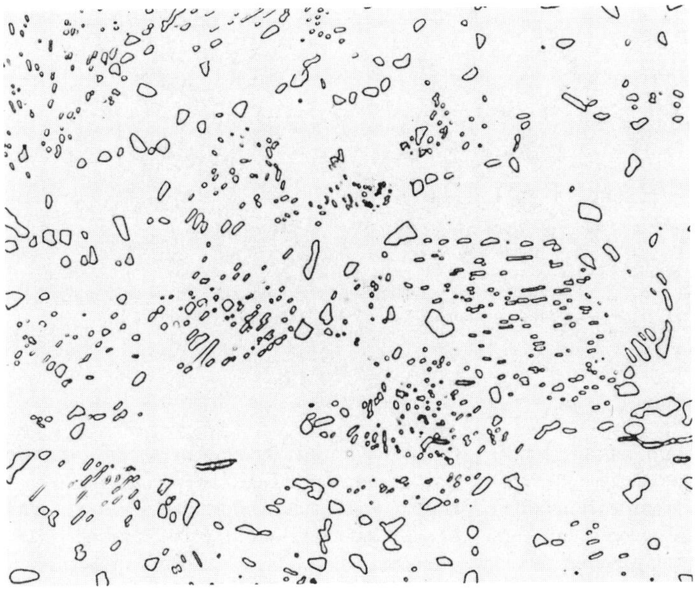

Fig. 42 Microstructure of a fully spheroidized steel. 4% picral etch. 1000×

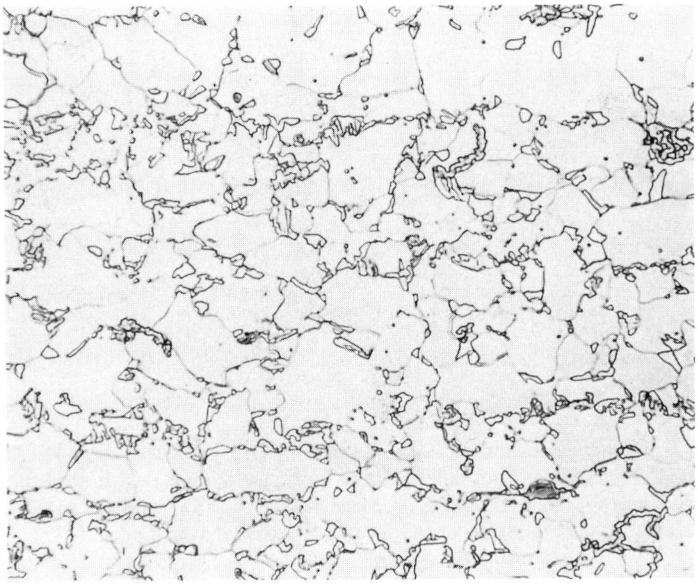

Fig. 43 Microstructure of a typical dual-phase steel. 2% nital etch. 250×

for example, total elongation of less than 1%. The flake morphology of the graphite provides for easy crack propagation under applied stress.

Gray cast irons usually contain 2.5 to 4% C, 1 to 3% Si, and 0.1 to 1.2% Mn. The graphite flakes can be present in five different morphologies as seen in Fig. 45. Type A, because of its random orientation and distribution, is preferred in many applications, for example, cylinders of internal combustion engines. The matrix of a typical gray cast iron is usually pearlite. However, ferrite-pearlite or martensitic microstructures can be developed by special heat treatments. As a structural material, gray cast iron is selected for its high compressive strength, which ranges from 572 to 1293 MPa (83 to 188 ksi), although tensile strengths of gray iron range only from 152 to 431 MPa (22 to 63 ksi). Gray cast irons are used in a wide variety of applications, including automotive cylinder blocks, cylinder heads and brake drums, ingot molds, machine housings, pipe, pipe fittings, manifolds, compressors, and pumps.

Another form of graphite in cast iron is spheroidal graphite found in ductile cast irons (also called nodular cast irons). The microstructure of a typical ductile cast iron is shown in Fig. 46. This form of graphite is produced by a process called inoculation, in which a magnesium or cerium alloy is thrust into molten cast iron immediately prior to the casting operation. These elements form intermetallic compounds that act as a nucleating surface for graphite. With a spherical morphology, the graphite no longer renders the cast iron brittle as do graphite flakes in gray cast iron. Ductile irons have much higher ductility and toughness than gray iron and thus expand the use of this type of ferrous alloy. Most ductile iron castings are used in the as-cast form. However, heat treatment can be employed to alter the matrix microstructure to obtain desired properties. The matrix can be fully ferritic, fully pearlitic, fully martensitic, or fully bainitic, depending on composition and heat treatment. The yield strength of typical ductile cast irons ranges from 276 to 621 MPa (46 to 76 ksi), and their tensile strengths range from 414 to 827 MPa (60 to 120 ksi). Total elongation ranges from about 3 to 18%. Heat treated, austempered ductile irons have yield strengths ranging from 505 to 950 MPa (80 to 138 ksi), tensile strengths ranging from 860 to 1200 MPa (125 to 174 ksi), and total elongations ranging from 1 to 10%. Uses for ductile iron include gears, crankshafts, paper-mill dryer rolls, valve and pump bodies, steering knuckles, rocker arms, and various machine components.

Malleable cast iron encompasses yet another form of graphite called temper carbon. The microstructure of a typical malleable cast iron is shown in Fig. 47. This form of graphite is produced by the heat treatment of white cast iron, which does not contain graphite, but does contain a high percentage of cementite. When a white cast iron is heated for an extended period of time (about 60 h) at a temperature of 960 °C (1760 °F), the cementite decomposes into austenite and graphite. By slow cooling from 960 °C (1760 °F), the austenite transforms into ferrite or pearlite, depending on the cooling rate and the diffusion rate of carbon. The ductility and toughness of malleable iron falls between that of ductile cast iron and gray cast iron. Because white iron can only be produced in cast sections up to about 100 mm (4 in.) thick, malleable iron is thus limited in section size. The yield strength of typical malleable cast irons ranges from 207 to 621 MPa (30 to 70 ksi), and the tensile strength ranges from 276 to 724 MPa (40 to 105 ksi). Total elongation ranges from 1 to 18%. Applications for malleable iron include pipe fittings, valves, crankshafts, transmission gears, and connecting rods.

Cementite

A major microstructural constituent in white cast iron is cementite. The microstructure of a typical white cast iron is shown in Fig. 48. The cementite forms by a eutectic reaction during solidification:

$$\text{Liquid} \leftrightarrow \text{Cementite} + \text{Austenite} \qquad \text{(Eq 15)}$$

The eutectic constituent in white cast iron is called ledeburite and has a two-phase morphology shown as the smaller particles in the white matrix in Fig. 49. The eutectic is shown in the Fe-C binary diagram in Fig. 7(b). The austenite in the eutectic (as well as the austenite in the primary phase) transforms to pearlite, ferrite-pearlite, or martensite, depending on cooling rate and composition (e.g., in Fig. 47 the

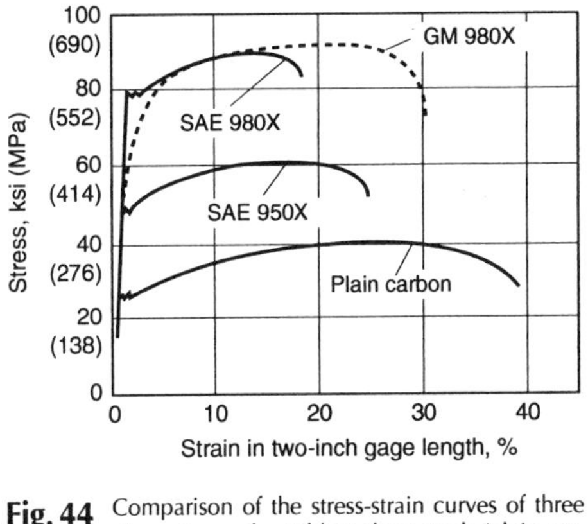

Fig. 44 Comparison of the stress-strain curves of three discontinuously yielding sheet steels (plain carbon, SAE 950X, and SAE 980X) and a dual-phase steel (GM 980X). In addition to the differences in yielding behavior, note the higher percentage of uniform elongation in the dual-phase steel compared with the conventional SAE 980X of similar tensile strength. Source: Ref 2

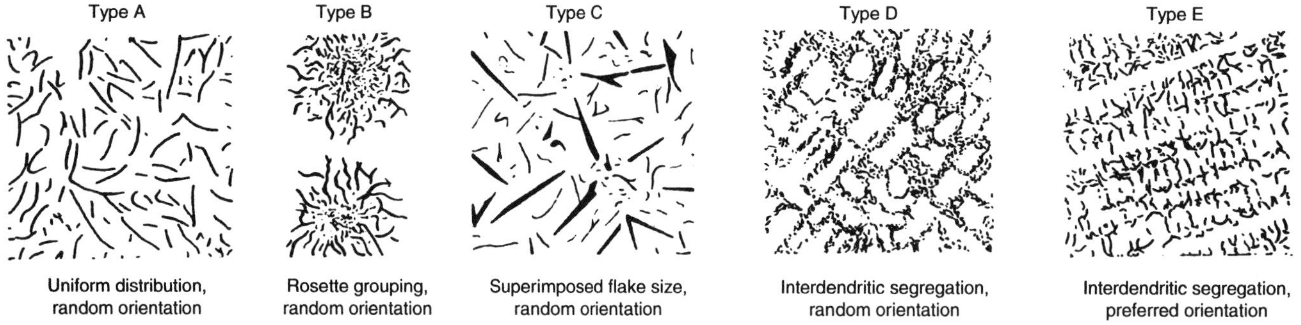

Fig. 45 Classification of different graphite flake morphology

austenite transformed to pearlite). Because of the high percentages of cementite, white cast irons are used in applications requiring excellent wear and abrasion resistance. These irons contain high levels of silicon, chromium, nickel, and molybdenum and are termed alloy cast irons. Such applications include steel mill rolls, grinding mills, and jaw crushers for the mining industry. Hardness is the primary mechanical property of white cast iron and ranges from 321 to 400 HB for pearlitic white iron and 400 to 800 HB for alloy (martensitic) white irons.

Evolution of Microstructural Change in Steel Products

It is interesting to study how microstructural and compositional changes have evolved in certain steel products. In some products, these changes have taken place over a 150-year span, and in other products over just a few decades. Examples of long-time evolution are rail steel and ductile/malleable cast iron, and an example of a short-term evolution is automotive sheet steel.

Evolution of Rail Steels. Up until the 1860s, most rails in the United States were either cast iron, wrought iron, or plain carbon steel imported from England. The steel rails were much preferred because the cast iron rails were brittle and the wrought iron rails were soft and had poor wear resistance. In 1865, the first steel rails were produced in the United States and since then, steel has become the dominant rail material.

Early steel rails had carbon contents much below the current eutectoid carbon level. This is because higher carbon content rails had a tendency to fracture and cause serious derailments. Not until the 1930s did railroad engineers realize that the problem was not the higher carbon content, but hydrogen. Laboratory and commercial studies proved that the problem of premature failure was caused by hydrogen flakes, or what the rail engineers called shatter cracks. Practices were put in place in the late 1930s to remove hydrogen by slow cooling of the rails during the manufacturing process in order to allow the hydrogen to diffuse out of the rails.

With this problem solved, the carbon content of rail steel quickly rose to eutectoid levels (0.76 to 0.82%). With this increase in carbon, the wear resistance increased and the railroads could haul heavier axle loads. Unfortunately, as axle load steadily increased from the 1940s to the 1980s, even the fully pearlitic rails began to wear more rapidly. During this time, it was discovered that if one produced a finer pearlite interlamellar spacing, the hardness of the rail increased as did the wear resistance. This led to manufacturing processes in which accelerated cooling was employed to attain a finer pearlite interlamellar spacing. These processes included oil quenching (in the 1940s), forced-air cooling, water-spray cooling, and aqueous-polymer quenching (in the 1980s).

Large improvements in wear resistance were obtained by these new processes, and it was soon discovered that rail life was no longer determined by the rail wearing out. Indeed, rail life in the 1990s is now determined by fatigue life and fracture toughness. The fatigue life could be improved by cleaner steels (fewer inclusions), but fracture toughness could not be improved sufficiently in a fully pearlitic steel, even by finer interlamellar spacings. That is, the pearlitic steel has essentially been pushed to its limit as serving

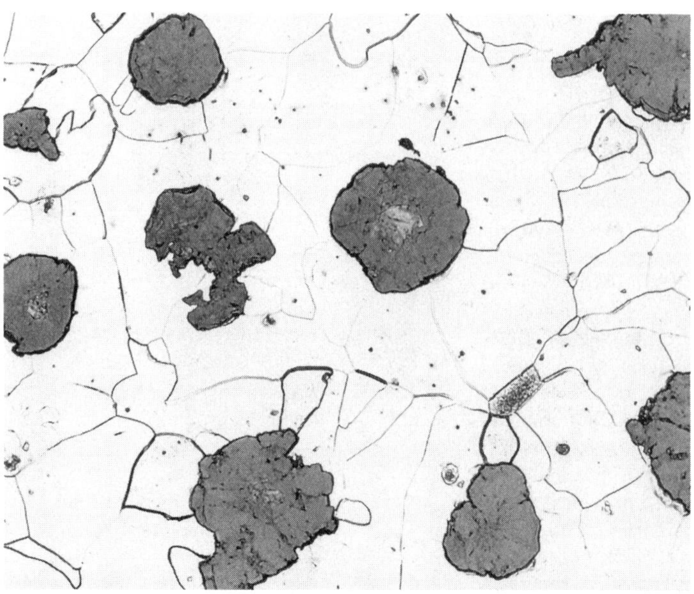

Fig. 46 Microstructure of a typical ductile (nodular) cast iron showing graphite in the form of spheroids. 2% nital etch. 200×. Courtesy of A.O. Benscoter, Lehigh University

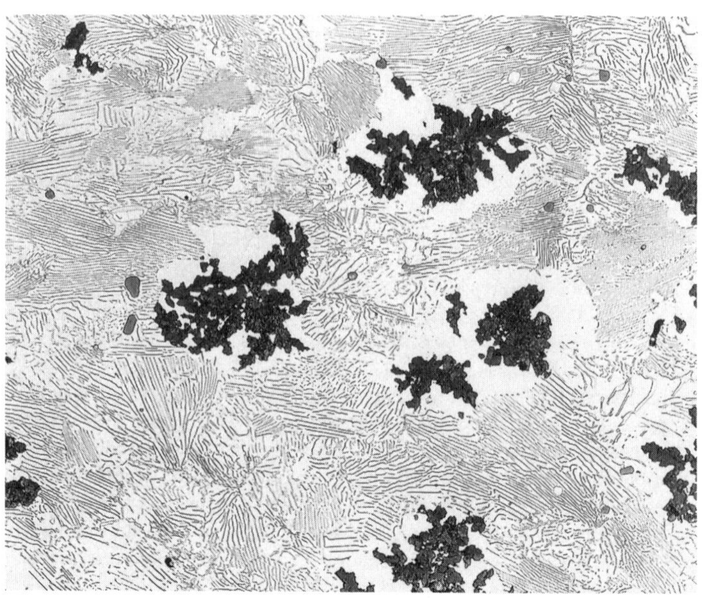

Fig. 47 Microstructure of a typical malleable cast iron showing graphite in the form of temper carbon. 4% picral etch. 250×. Courtesy of A.O. Benscoter, Lehigh University

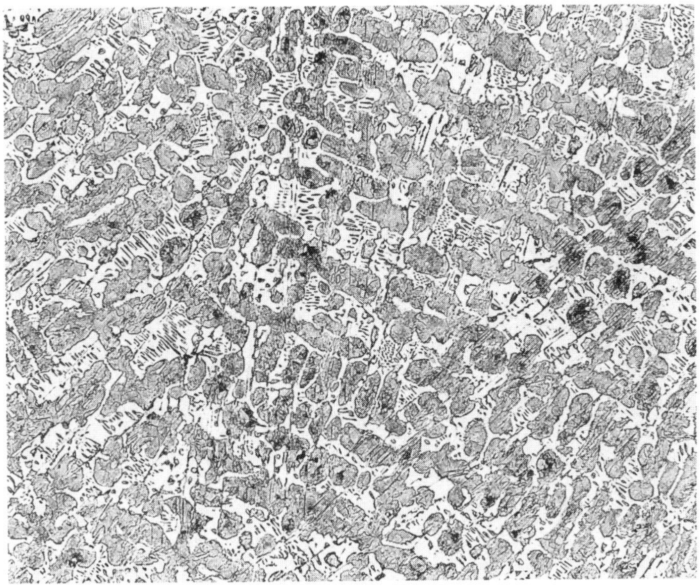

Fig. 48 Microstructure of a typical white cast iron. 4% picral etch. 100×. Courtesy of A.O. Benscoter, Lehigh University

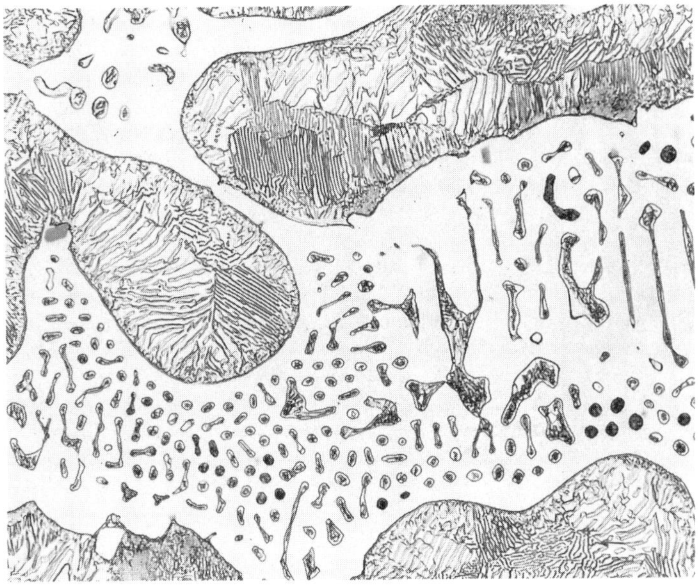

Fig. 49 Microstructure of the eutectic constituent ledeburite in a typical white cast iron. 4% picral etch. 500×. Courtesy of A.O. Benscoter, Lehigh University

as a rail steel. Thus, a revolutionary change must take place in rail steel technology. One such possible change is to completely switch from a fully pearlitic microstructure to a fully bainitic microstructure. Rail producers around the world are currently investigating high-strength bainitic rail steels. A German rail manufacturer has an experimental bainitic steel rail in track.

This example is used to illustrate how the demands in the field produce evolutionary microstructural, property, and processing changes to keep pace. However, as these demands increase, there comes a point where a revolutionary change must take place because the useful properties attributed to the original microstructure have been fully exhausted.

Evolution of Cast Iron. Another historic evolutionary change in microstructure took place in the manufacture of cast iron. Both gray and white cast iron, being among the earliest ferrous materials, were brittle and could not be fabricated into shapes other than those made by the casting process. Although quite useful, gray and white cast iron were relegated to simple shapes such as cooking pots, cannon balls, window sashes, stove panels, and radiators that did not demand a material with ductility or malleability. The reason for the lack of ductility is the flake-shaped graphite in gray cast iron and the cementite networks in white cast irons. The only way to attain ductility and malleability and therefore expand the usefulness of these irons was to alter the microstructure. The first commercial success was that of R.A.F. de Réaumer, a French metallurgist, in 1720. Réaumer used the process of decarburization of white cast iron by packing the castings in iron ore and heating the material to "bright redness" for several days. The lengthy heat treatment allowed the cementite to decompose into iron and carbon. The iron ore provided oxygen for decarburization. The final casting was completely free of cementite and the microstructure was fully ferritic or a mixture of ferrite and pearlite. Because the process depended on decarburization (and thus carbon diffusion), only thin castings could be treated in this way. In larger castings, cementite remained in the thicker sections, thus limiting their usefulness.

About 110 years later, S. Boyden of the United States invented what is known today as malleable iron. The advantage of malleable iron was that the process did not depend on days and weeks of decarburization, but depended on the decomposition of the cementite in white cast iron into iron and "free-carbon." This free-carbon formed as particles of carbon called "temper carbon," (see Fig. 47). The Boyden process involved a heat treatment of only 30 h and large castings could be heat treated. This process of microstructure alteration greatly expanded the usefulness of cast iron for more than a century and a half.

Another microstructural alteration process was invented in the 1960s. Here liquid cast iron is inoculated with a magnesium alloy, resulting in spheroids, rather than flakes, of graphite forming during solidification. The cast iron is called ductile iron and has a microstructure as shown in Fig. 46. The matrix in ductile iron can consist of ferrite, pearlite, ferrite-pearlite, martensite, or bainite, depending on alloy composition and heat treatment. Currently, austempered ductile irons, with a bainitic matrix, provide high strength and toughness heretofore not achieved in a cast iron product. Ductile iron castings now replace many components that were historically produced from plain-carbon and low-alloy steels. This evolutionary path of microstructural alteration in cast iron is an important example of the role microstructure plays in the development of structure-sensitive mechanical properties.

Evolution of Steel Sheet. An example of rapid-paced evolutionary change is in sheet steel for the automotive industry. For years, automobiles were produced using inexpensive, low-carbon sheet steel. However, during the oil crisis in 1974, the drastic increase in the price of gasoline and the Western world's almost total dependence on imported oil, mandated improved fuel economy for automobiles. This translated into production of lighter weight automobiles; this, in turn, created a revolution in sheet steel metallurgy. Since the oil crisis, ordinary low-strength, low-carbon sheet steel has been replaced by a number of higher-strength sheet steels requiring new process technology. These new steels include the high-strength, precipitation-strengthened steels, the dual-phase and tri-phase steels, and the bake-hardenable steels. Also, new coating techniques have been developed to protect these new steels from corrosion. This rapid change in automobile sheet steel is one example of the importance of basic understanding of metallurgy and the application of physical metallurgy to engineering product design.

REFERENCES

1. *Engineering Properties of Steel*, P.D. Harvey, Ed., American Society for Metals, 1982
2. G. Krauss, *Principles of the Heat Treatment of Steel*, American Society for Metals, 1980
3. R.W.K. Honeycombe, *Steels—Microstructure and Properties*, American Society for Metals, 1982
4. W.C. Leslie, *The Physical Metallurgy of Steels*, McGraw-Hill, 1981
5. F.B. Pickering, *Physical Metallurgy and the Design of Steels*, Applied Science, 1978
6. G. Krauss, Microstructures, Processing, and Properties of Steels, *Properties and Selection: Irons, Steels, and High-Performance Alloys*, Vol 1, *ASM Handbook*, 1990, p 126

7. E.C. Bain and H.W. Paxton, *Alloying Elements in Steel*, 2nd ed., American Society for Metals, 1961, p 62
8. *Microalloying 75*, Conference Proceedings (Washington, D.C., Oct 1975), Union Carbide Corporation, 1977, p 5
9. T.B. Massalski, J.L. Murray, L.H. Bennett, and H. Baker, Ed., *Binary Alloy Phase Diagrams*, Vol 1, American Society for Metals, 1986, p 822
10. W. Heller, R. Schweitzer, and L. Weber, *Can. Metall. Q.*, Vol 21 (No. 1), 1982, p 3
11. J.M. Hyzak and I.M. Bernstein, *Metall. Trans. A*, Vol 7A, 1976, p 1217
12. G.F. Vander Voort and A. Roósz, *Metallography*, Vol 17 (No. 1), 1984, p 1
13. H. Ichinose et al., paper I.3, *Proc. First Int. Heavy Hauls Railway Conf.*, Association of American Railroads, 1978, p 1
14. *Atlas of Time-Temperature Diagrams for Irons and Steels*, G.F. Vander Voort, Ed., ASM International, 1991, p 570
15. B.L. Bramfitt, *Proc. 32nd Mechanical Working and Steel Processing Conference*, Vol 28, ISS-AIME, 1990, p 485
16. F.B. Pickering, *Towards Improved Toughness and Ductility*, Climax Molybdenum Co., 1971, p 9
17. G.J. Roe and B.L. Bramfitt, Notch Toughness of Steels, *Properties and Selection: Irons, Steels, and High-Performance Alloys*, Vol 1, *ASM Handbook*, ASM International, 1990, p 739
18. E.C. Bain, *The Sorby Centennial Symposium on the History of Metallurgy*, TMS-AIME, 1963, p 121
19. B.L. Bramfitt and J.G. Speer, *Metall. Trans. A*, Vol 21A, 1990, p 817
20. *Atlas of Time-Temperature Diagrams for Irons and Steels*, G.F. Vander Voort, Ed., ASM International, 1991, p 249
21. W. Steven and A. G. Haynes, *J. Iron Steel Inst.*, Vol 183, 1956, p 349
22. R.W.K. Honeycombe and F.B. Pickering, *Metall. Trans. A*, Vol 3A, 1972, p 1099
23. *Atlas of Time-Temperature Diagrams for Irons and Steels*, G.F. Vander Voort, Ed., ASM International, 1991, p 544
24. K.W. Andrews, *J. Iron Steel Inst.*, Vol 203, 1965, p 271
25. G.R. Speich and H. Warlimont, *J. Iron Steel Inst.*, Vol 206, 1968, p 385
26. G. Krauss, *J. Iron Steel Inst. Jpn., Int.*, Vol 35 (No. 4), 1995, p 349
27. R.A. Grange, C.R. Hribal, and C.F. Porter, *Metall. Trans. A*, Vol 8A, 1977, p 1775

Effects of Composition, Processing, and Structure on Properties of Nonferrous Alloys

Ronald N. Caron, Olin Corporation
James T. Staley, Alcoa Technical Center

A BROAD RANGE OF PROPERTIES are available with commercially available nonferrous alloys; this articles provides an overview of these alloys and describes the specific microstructure/property relationships that are used to make specific properties available to designers of structural applications. Control of the microstructure during processing is as important for fabricability as it is for providing the desired engineering properties in the final product. Manufacturing alloy components with the desired structural properties requires applying the proper choice of processing method to the alloy composition capable of delivering those desired properties. Alloy and process development have sought to improve upon the properties available with metals by adjusting alloy additions along with processing parameters to meet the requirements of new technology. This process is ongoing today with the base alloys available for centuries (iron, copper, tin) as well as with the newer base alloys currently recruited to meet modern needs with unique property requirements (titanium, tungsten, niobium). When currently available alloys fall short of meeting the needs of the newest structural designs, the solution will be found as it has in the past by cooperative alloy/process development efforts with the supplier of the particular base-alloy system that has the properties closest to the particular design need.

No metal or alloy is entirely unique; similarities in process and property characteristics do exist among the wide variety of commercially available nonferrous alloys. However, each metal and alloy offers unique combinations of useful physical, chemical, and structural properties that are made available by its particular composition and the proper choice of processing method. This article focuses on the monolithic form of nonferrous alloys. However, it should be recognized that these alloys are also used in structural combinations, either as simple or complex composites, or with coatings in order to provide even more unique, economical combinations of useful properties. The importance of availability and cost are generally ignored in the following discussion except where these attributes are significant.

Detailed information and additional references to the literature on processing and property characteristics of nonferrous alloys are provided in *Properties and Selection: Nonferrous Alloys and Special-Purpose Materials,* Volume 2 of *ASM Handbook* (Ref 1). Basic data on nonferrous alloy phase relationships are found in *Alloy Phase Diagrams,* Volume 3 of *ASM Handbook* (Ref 2).

Aluminum and Aluminum Alloys

Aluminum alloys are second in use only to steel as structural metals. More than 500 alloys are registered with the Aluminum Association. Typical tensile strengths of aluminum alloy products range from 45 MPa (6.5 ksi) for 1199-O sheet to almost 700 MPa (100 ksi) for 7055-T77511 extruded products. The low density combined with high strength have made aluminum alloys the standard material for applications such as aircraft, where specific strength (strength-to-weight ratio) is a major design consideration. Because of their corrosion resistance, moderate strength, and good ductility, they are also used for a wide variety of applications including beverage cans; building and construction; marine, rail, truck, and auto transportation; and tool and jig plate.

Wrought aluminum products are produced using all available metalworking techniques, and castings are made using all standard solidification processes. Their properties depend on a complex interaction of chemical composition and microstructural features developed during solidification, thermal treatments, and (for wrought products) deformation processing. Although pure aluminum is very resistant to corrosion because of the presence of a film of aluminum oxide, corrosion resistance generally decreases with increasing alloy content, so tempers have been developed to improve the corrosion resistance of the highly alloyed materials.

Comprehensive information about the properties, processing, and applications of aluminum alloys is provided in the *ASM Specialty Handbook: Aluminum and Aluminum Alloys* (Ref 3). Other useful reference sources are listed as Ref 4 to 17.

Aluminum Alloy Phase Diagrams

Although few products are sold and used in their equilibrium condition, equilibrium phase diagrams are an essential tool in understanding effects of composition and both solidification and solid-state thermal processing on microstructure. For aluminum alloys, phase diagrams are used to determine solidification and melting temperatures, the solidification path, and the equilibrium phases that form and their dissolution temperatures. In addition to determining appropriate temperatures for casting and thermal treatments, phase diagrams are used to determine the maximum levels for ancillary element additions of certain elements to prevent the crystallization of coarse primary particles. The most important liquid-to-solid transformations for aluminum alloys are the eutectic and the peritectic. Examples of phase diagrams illustrating eutectic and peritectic reactions are discussed in the following paragraphs, and phase diagrams for aluminum alloys can be found in Ref 2, 3, 10, 11, and 17.

The eutectic reaction is illustrated by the aluminum-copper system (Fig. 1). When the liquidus temperature of aluminum-rich alloys is reached during solidification, the liquid begins to solidify into a solid solution of copper in aluminum (α-aluminum). As temperature approaches the solidus, the α-aluminum becomes more enriched with copper. When the temperature falls below the solidus temperature in alloys containing less

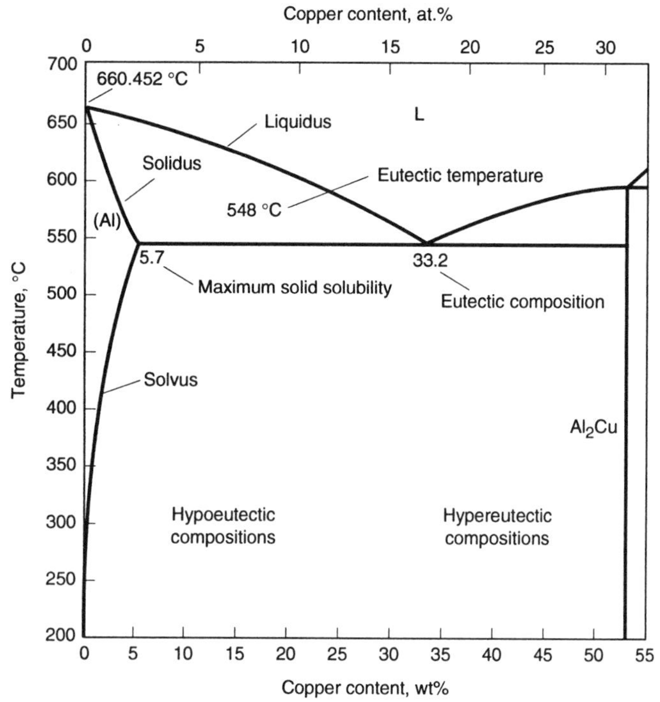

Fig. 1 Aluminum-copper phase diagram illustrating the eutectic reaction

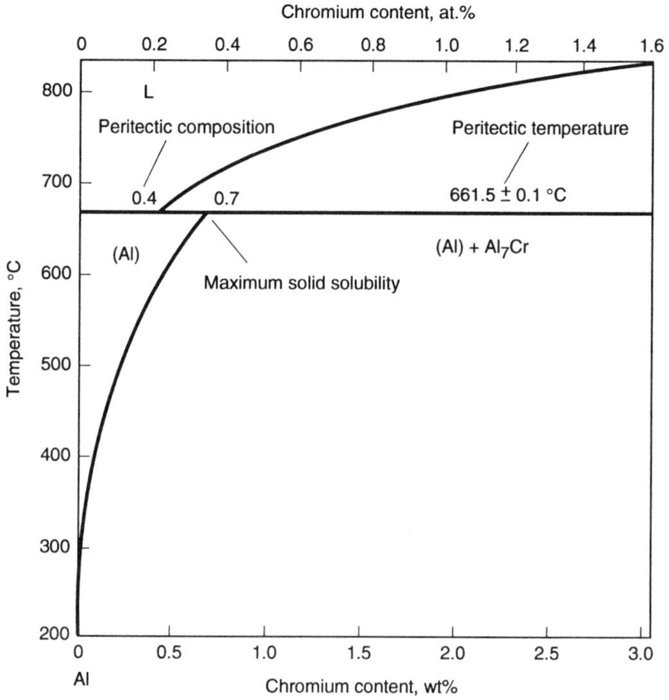

Fig. 2 Aluminum-chromium phase diagram illustrating the peritectic reaction

than the maximum solubility, 5.7% Cu, solidification is complete. At temperatures below the solvus, Al_2Cu particles precipitate, depleting the α-aluminum of copper. When cooled to room temperature under near-equilibrium conditions, the α-aluminum contains little copper, so strength is low. To increase strength, the material must be solution heat treated, quenched, and aged to develop metastable precipitates as described later in this section. In alloys containing more than 5.7% Cu, some liquid remains when the eutectic temperature is reached. This liquid solidifies at this temperature by a eutectic reaction to α-aluminum and Al_2Cu intermetallic particles. On cooling below the eutectic temperature, the α-aluminum rejects copper as Al_2Cu precipitates. It is important to realize that the eutectic reaction can occur in alloys containing less than the maximum solid solubility under commercial casting conditions, even though the equilibrium phase diagram does not predict that. Consequently, Al_2Cu particles form during solidification of most aluminum alloy ingots and shaped castings. Therefore, they are "preheated" or homogenized to dissolve the intermetallic particles.

The peritectic reaction in aluminum alloys is typified by the aluminum-chromium system (Fig. 2). During equilibrium solidification of alloys containing more than the peritectic composition, 0.41% Cr, but less than the maximum solid solubility of 0.77%, an intermetallic compound, Al_7Cr, forms when the liquidus temperature is reached. When the temperature falls to the peritectic temperature, 661 °C, the remaining liquid along with the Al_7Cr transforms to α-aluminum. Under commercial solidification conditions, however, the primary particles of Al_7Cr would not have the opportunity to transform to α-aluminum, so they would remain. Consequently, maximum chromium limits are established so that all of the chromium remains in supersaturated solid solution in the ingot. It precipitates as chromium-bearing dispersoids during ingot preheat.

Aluminum Alloy Impurity and Alloying Elements

All commercial aluminum alloys contain iron and silicon as well as two or more elements intentionally added to enhance properties. The phases formed and the function of the alloying elements are described.

Iron. Virtually all aluminum alloys contain some iron that is an impurity remaining after refining bauxite and smelting. The phase diagram predicts that during solidification of an aluminum-iron alloy containing a few tenths of a percent of iron, most of the iron remains in the liquid phase until a eutectic of solid solution plus Al_3Fe intermetallic constituent particles having a monoclinic crystal structure freezes. Depending on solidification rate and on the presence of other elements such as manganese, constituent particles of the metastable orthorhombic Al_6Fe phase can form instead of the equilibrium Al_3Fe. The maximum solid solubility of iron in aluminum is 0.05%, but the solubility is much lower in most structural alloys.

Silicon. This element is also a ubiquitous impurity in commercial aluminum alloys. Two ternary phases, cubic α-$Al_{12}Fe_3Si$ and monoclinic β-$Al_9Fe_2Si_2$, form by a eutectic reaction. At low silicon contents, almost all of the iron is present as Al_3Fe. With increasing silicon contents, first the α- then the β-Al-Fe-Si phases appear. Phases in commercial products may not be those predicted by the equilibrium phase diagrams because of the long times at high temperatures required to approach equilibrium. In large amounts, silicon improves castability and fluidity. Consequently, it is used in 4*xxx* brazing sheet and in 3*xx.x* and 4*xx.x* casting alloys. Silicon ranges from about 5 to 20% in casting alloys. Hypereutectic alloys (those containing >12.6% Si, the eutectic composition) are used for engine blocks because the primary silicon particles are wear resistant. Some 3*xx.x* casting alloys contain small additions of magnesium to render them capable of being age hardened.

Silicon is deliberately added to some alloys containing magnesium to provide precipitation hardening. The Al-Mg-Si system is the basis for the 6*xxx* alloys. At low magnesium contents, elemental silicon may be present as second-phase particles. As magnesium increases, both silicon particles and equilibrium hexagonal Mg_2Si constituents may be present. At higher magnesium contents, only Mg_2Si is present. Ternary alloys are strengthened by precipitation of metastable precursors to Mg_2Si. With the addition of copper, a complex quaternary $Al_4CuMg_5Si_4Q$ phase can form. A precursor to this quaternary phase strengthens Al-Cu-Mg-Si alloys.

Manganese. The aluminum-manganese system is the basis for the oldest aluminum alloys. Such alloys, known as 3*xxx*, are the most widely used wrought alloys because of their excellent formability and resistance to corrosion. Commercial aluminum-manganese alloys contain both iron and silicon. During solidification of commercial size ingots, some of the manganese forms

Table 1 Designation system for wrought aluminum alloys

Aluminum, at least 99.00%	1*xxx*
Aluminum alloys grouped by major alloying elements:	
Copper	2*xxx*
Manganese	3*xxx*
Silicon	4*xxx*
Magnesium	5*xxx*
Magnesium and silicon	6*xxx*
Zinc	7*xxx*
Other element	8*xxx*

The last two digits indicate the aluminum purity or identify the aluminum alloy. The second digit indicates modifications of the original alloy or impurity limits.

Table 2 Designation system for cast aluminum alloys

Aluminum, at least 99.00%	1*xx.x*
Aluminum alloys grouped by major alloying elements:	
Copper	2*xx.x*
Silicon with added copper or manganese	3*xx.x*
Silicon	4*xx.x*
Magnesium	5*xx.x*
Zinc	7*xx.x*
Tin	8*xx.x*
Other element	9*xx.x*

The second two digits indicate the aluminum purity or identify the aluminum alloy. The last digit indicates the product form, castings (*xxx*.0) or foundry ingot (*xxx*.1 and *xxx*.2). A modification of the original alloy or impurity limits is designated by a serial letter before the numerical designation. The series 6*xx.x* is unused.

Table 3 Nominal chemical compositions of selected 3*xxx* and 5*xxx* aluminum alloys

	Alloying element content, wt %			
Alloy	Cu	Mn	Mg	Cr
3003	0.12	1.2	...	...
3004	...	1.2	1.0	...
3005	...	1.2	0.4	...
3105	...	0.6	0.5	...
5005	...	...	0.8	...
5050	...	...	1.4	...
5052	...	...	2.5	0.25
5252	...	...	2.5	...
5154	...	...	3.5	0.25
5454	...	0.8	2.7	0.12
5056	...	0.12	5.0	0.12
5456	...	0.8	5.1	0.12
5182	...	0.35	4.5	...
5083	...	0.7	4.4	0.15
5086	...	0.45	4.0	0.15

All contain iron and silicon as impurities.

Al_6(Mn,Fe) and cubic Al_{12}(Fe,Mn) Si by eutectic reactions. The remaining manganese remains in solution and is precipitated during the ingot preheat as Al_{12}(Mn,Fe)Si and Al_6(Mn,Fe) dispersoids. These dispersoids strengthen the material and control recrystallized grain size. In alloys containing copper, manganese precipitates as $Al_{20}Cu_2Mn_3$ dispersoid particles. Effects on strength are minor, but the dispersoids aid in grain size control after solution heat treatment.

Magnesium. The aluminum-magnesium system is the basis for the wrought 5*xxx* and cast 5*xx.x* non-heat-treatable aluminum alloys, which provide excellent combinations of strength and corrosion resistance by solid-solution strengthening and work hardening. Although in principle this phase diagram exhibits a positively sloping solvus, a necessary condition for a precipitation-hardening system, difficulty in nucleating the face-centered cubic (fcc) Al_3Mg_2 precipitates has precluded commercialization of heat-treatable aluminum-magnesium alloys, unless they contain enough silicon, copper, or zinc to form Mg_2Si, Al-Cu-Mg, or Al-Zn-Mg precipitates.

Copper. The aluminum-copper system is the basis for the wrought 2*xxx* and cast 2*xx.x* alloys, and many other heat-treatable alloys contain copper. In commercial aluminum-copper alloys, some of the copper chemically combines with aluminum and iron to form either tetragonal Al_7Cu_2Fe or orthorhombic α(Al,Cu,Fe) constituent particles during solidification. These constituents cannot be dissolved during subsequent thermal treatments, but one can transform to the other during thermal treatments of ingots or castings. During heat treatment of aluminum-copper alloys containing little magnesium, Al_2Cu precipitates as the strengthening phase.

Adding magnesium to aluminum-rich aluminum-copper alloys results in the formation of the Al_2CuMg phase by eutectic decomposition. Metastable precursors to face-centered orthorhombic Al_2CuMg precipitates are used to strengthen several structural alloys used in the aerospace industry because they confer a desirable combination of strength, fracture toughness, and resistance to the growth of fatigue cracks.

Zinc. This element confers little solid-solution strengthening or work hardening to aluminum, but Al-Zn-Mg precipitates provide the basis for the 7*xxx* wrought alloys and the 7*xx.x* cast alloys. Two phases can form by eutectic decomposition in commercial Al-Zn-Mg alloys: hexagonal $MgZn_2$ and body-centered cubic (bcc) $Al_2Mg_3Zn_3$. Depending on the zinc/magnesium ratio, copper-free alloys are strengthened by metastable precursors to either $MgZn_2$ or $Al_2Mg_3Zn_3$. In Al-Zn-Mg-Cu alloys, copper and aluminum substitute for zinc in $MgZn_2$ to form $Mg(Zn,Cu,Al)_2$. Al_2CuMg particles can also form in these alloys by eutectic decomposition and solid-state precipitation.

Chromium. In commercial alloys, the solubility can be reduced to such an extent that Al_7Cr primary particles can form by a peritectic reaction at chromium contents lower than that indicated by the binary aluminum-chromium phase diagram. Because coarse primary particles are harmful to ductility, fatigue, and fracture toughness, the upper limits of chromium depend on the amount and nature of the other alloying and impurity elements. In 5*xxx* alloys, fcc cubic $Al_{18}Mg_3Cr_2$ dispersoids precipitate during ingot preheating. In 7*xxx* alloys, the composition of the dispersoids is closer to $Al_{12}Mg_2Cr$. Chromium dispersoids contribute to strength in non-heat-treatable alloys and control grain size and degree of recrystallization in heat-treatable alloy products.

Zirconium. This element also forms a peritectic with aluminum. The phase diagram predicts that the equilibrium Al_3Zr phase is tetragonal, but fine dispersoids of metastable cubic Al_3Zr form during ingot preheating treatments. Most 7*xxx* and some 6*xxx* and 5*xxx* alloys developed since the 1960s contain small amounts of zirconium, usually less than 0.15%, to form Al_3Zr dispersoids for recrystallization control.

Lithium. This element reduces the density and increases the modulus of aluminum alloys. In binary alloys it forms metastable Al_3Li precipitates and combines with aluminum and copper in Al-Cu-Li alloys to form a large number of Al-Cu-

Table 4 Nominal chemical compositions of selected 2*xxx*, 6*xxx*, and 7*xxx* aluminum alloys

	Alloying element content, wt %							
Alloy	Fe	Si	Cu	Mn	Mg	Cr	Zn	Zr
2008	0.40(a)	0.65	0.9	0.3(a)	0.4	...	...	...
2219(b)	0.30(a)	0.20(a)	6.3	0.3	...	...	...	0.18
2519(b)	0.39(a)(c)	0.30(a)(c)	5.8	0.3	0.25	...	...	0.18
2014	0.7(a)	0.8	4.4	0.8	0.5	...	...	...
2024	0.50(a)	0.50(a)	4.4	0.6	1.5	...	...	...
2124	0.30(a)	0.20(a)	4.4	0.6	1.5	...	...	...
2224	0.15(a)	0.12(a)	4.4	0.6	1.5	...	...	...
2324	0.12(a)	0.10(a)	4.4	0.6	1.5	...	...	...
2524	0.12(a)	0.06(a)	4.25	0.6	1.4	...	...	...
2036	0.50(a)	0.50(a)	2.6	0.25	0.45	...	...	...
6009	0.50(a)	0.8	0.4	0.5	0.6	...	...	...
6061	0.7(a)	0.6	0.3	...	1.0	0.2	...	...
6063	0.50(a)	0.4	...	...	0.7	...	...	...
6111	0.4(a)	0.9	0.7	0.3	0.8	...	...	...
7005	0.40(a)	0.35(a)	...	0.45	1.4	0.13	4.5	0.14
7049	0.35(a)	0.25(a)	1.6	...	2.4	0.16	7.7	...
7050	0.15(a)	0.12(a)	2.3	...	2.2	...	6.2	0.12
7150	0.15(a)	0.10(a)	2.2	...	2.4	...	5.4	0.12
7055	0.15(a)	0.10(a)	2.3	...	2.1	...	8.0	0.12
7075	0.50(a)	0.40(a)	1.6	...	2.5	0.25	5.6	...
7475	0.12(a)	0.10(a)	1.6	...	2.2	0.20	5.7	...

(a) Maximum allowable amount. (b) 2219 and 2519 also contain 0.10% V and 0.06% Ti. (c) 0.40% max Fe plus Si

Li phases. Because of its high cost relative to other alloying elements, lithium alloys have been found to be cost effective thus far only in space and military applications.

Aluminum Alloy Designation System

The alloy designation system classifies alloys by their composition. In the United States, aluminum alloys are designated by two distinct four-digit systems, one for wrought aluminum (Table 1) and one for cast products (Table 2). The system for designating wrought alloys has been adopted worldwide. No worldwide system for designating aluminum casting alloys exists; European manufacturers often use tradenames as identification.

Nominal compositions for several wrought structural alloys are presented in Tables 3 and 4. Nominal compositions of several casting alloys are presented in Table 5.

Aluminum Alloy Temper Designation System

The temper designation system (Table 6) classifies tempers by certain thermal and mechanical processes that control strength and other characteristics of both wrought and cast aluminum alloy products. In general, wrought alloys in the 2*xxx*, 6*xxx*, and 7*xxx* series and cast alloys in the 2*xx.x*, 3*xx.x*, and 7*xx.x* series are referred to as heat-treatable alloys because they can be precipitation hardened and are used in T tempers. Typical tensile properties of heat-treatable wrought and cast alloy products are presented in Tables 7 and 8, respectively. The remaining are referred to as non-heat-treatable alloys and are strengthened by solid-solution strengthening, dispersion strengthening and, for wrought alloys, work hardening. They are used in either O or H tempers. Typical tensile properties of wrought non-heat-treatable alloy sheet products are presented in Table 9.

O Temper. Ductility of all aluminum products is highest in this temper. Factors determining properties of annealed products depend on the alloy system. The annealed strength of unalloyed aluminum, 1*xxx* series, generally increases and ductility decreases with increasing impurity level, and the amounts of magnesium and manganese largely determine the bulk properties of 3*xxx* and 5*xxx* products. Strength increases as magnesium (solid-solution strengthening) and manganese contents (dispersion and solid-solution strengthening) increase. Ultimate tensile strength increases more significantly with increasing magnesium content than does tensile yield strength because of the potent effect of magnesium on work hardening. Parts of heat-treatable alloy products that are difficult to form are often formed in the O temper, then heat treated to a final temper.

H Tempers. Cold working of annealed material to H1 tempers increases the dislocation density. This increases strength, particularly yield strength, and decreases ductility. In unalloyed aluminum and in alloys containing little magnesium, cold working produces cells that have walls containing a high density of dislocations enclosing a volume of relatively strain-free material. In alloys containing sufficient amounts of magnesium, however, the dislocations form a tangled forest. In highly worked aluminum-magnesium alloys, rearrangement of the dislocation structure occurs over long times at room temperature. Stabilizing treatments, H3 tempers, prevent loss of strength in certain 3*xxx* and 5*xxx* alloys during subsequent long-time exposure. During partial annealing treatments, H2 tempers, cell walls either form or become more perfect, and dislocations within the cell migrate to cell boundaries. If the temperature exceeds a critical level, which depends on alloy content and strain, the cold-worked product will either partially or completely recrystallize. Materials in H2 tempers provide a combination of strength and ductility generally superior to that of material in H1 tempers. Cold-worked alloys containing above approximately 3.5% Mg and annealed alloys containing above approximately 4.5% Mg can also suffer a degradation in corrosion resistance caused by precipitation of a continuous film of Al_3Mg_2 on grain boundaries at temperatures between ambient and approximately 205 °C (400 °F). Special H116 and H321 controlled hot-rolling tempers have been developed that either avoid precipitation of Al_3Mg_2 on grain boundaries or agglomerate the precipitate to increase corrosion resistance. For a particular strength level, a higher resistance to stress corrosion is

Table 5 Nominal chemical compositions of selected aluminum casting alloys

Alloy	Alloying element content, wt %				
	Fe	Si	Cu	Mn	Mg
354.0	0.2(a)	9.0	1.8	...	0.5
A356.0	0.2(a)	7.0	0.1(a)	0.1(a)	0.3
365.0	0.15(a)	10.5	...	0.65	0.3

(a) Maximum allowable amount

Table 6 Temper designation system for aluminum alloys

Designation	Processing	Comments
F	As fabricated	Products of shaping processes in which no special control over thermal conditions or work hardening is employed.
O	Annealed	Wrought products that are annealed to obtain the lowest strength temper and to cast products that are annealed to improve ductility and dimensional stability.
H	Work hardened (wrought products only)	Products that have their strength increased by work hardening, with or without supplementary treatments to produce some reduction in strength. The H is always followed by two or more digits.
H1	Work hardened only	The number following this designation indicates the degree of work hardening.
H2	Work hardened and partially annealed	Products that are work hardened more than the desired final amount and then reduced in strength by partial annealing. The number following this designation indicates the degree of work hardening after the product has been partially annealed.
H3	Work hardened and stabilized	Products that are work hardened and whose properties are stabilized either by a low-temperature thermal treatment or as a result of heat introduced during fabrication. The number following this designation indicates the degree of work hardening remaining after the stabilization procedure.
W	Solution heat treated	An unstable temper applicable only to alloys that spontaneously age at room temperature after solution heat treatment. This designation is specific only when the period of natural aging is indicated. For example, W ½ h
T	Thermally treated to produce stable properties other than F, O, or H	Products that are thermally treated, with or without supplementary work hardening, to produce stable tempers. The T is always followed by one or more digits.
T3	Solution treated, quenched, cold worked, and naturally aged to a substantially stable condition	Products that are cold worked after solution treatment and the effect of cold work is recognized in mechanical property limits.
T4	Solution treated, quenched, and naturally aged to a substantially stable condition	Products that are either not cold worked after solution heat treatment or are cold worked and the effects of cold work might not be recognized in mechanical property limits.
T5	Quenched from an elevated-temperature shaping process and then artificially aged	...
T6	Solution heat treated, quenched, and artificially aged	Products that are either not cold worked after solution heat treatment or are cold worked and the effects of cold work might not be recognized in mechanical property limits.
T7	Solution treated, quenched, and overaged/stabilized	Wrought products that are overaged to carry them beyond a point of maximum strength to provide control of some other characteristic, usually corrosion resistance. Applies to cast products that are artificially aged after solution heat treatment to provide dimensional and strength stability.
T8	Solution treated, quenched, cold worked, and artificially aged	Products that are cold worked after solution treatment and the effect of cold work is recognized in mechanical property limits.
T9	Solution treated, quenched, artificially aged, and then cold worked	Products that are cold worked to improve strength.

obtained by increasing magnesium and manganese rather than by increasing work hardening.

W and T Tempers. The highest strengths are obtained by precipitation hardening. The material is held for a sufficient time above the solvus to dissolve essentially all of the major alloying elements, quenched at a rate to retain most of these elements in solid solution, then aged either at room temperature (natural aging) or at a modestly elevated temperature (artificial aging). The highest-strength alloys contain the largest concentration of the major alloying elements. For a particular alloy system, strength typically increases with increasing alloy content. Most 2*xxx* and 6*xxx* wrought alloys and 2*xx.x* and 3*xx.x* cast alloys are strengthened during natural aging by Guinier-Preston (G-P) zones, which are precursors to Al_2Cu, Al_2CuMg, Mg_2Si, or $Al_4CuMg_5Si_4$ phases. Strength of these materials increases for about 4 days, then stabilizes (T4 temper). In contrast, in 7*xxx* alloys containing G-P zone precursors to phases such as $MgZn_2$, strength continues to increase indefinitely at room temperature (W temper). The ductility in the freshly quenched (W <½ h temper) condition is high enough for many forming operations. Consequently, many parts are formed shortly after quenching from the solution-heat-treatment temperature. To prevent the formation of large grains during the solution treatment of formed parts, a critical amount of strain must be avoided. Although this critical strain is alloy dependent, strains near 10% are particularly troublesome for most alloys. In addition, ductility in the T4 or T3 tempers is sufficiently high that some parts can be formed successfully in this condition. Strength, particularly yield strength, increases substantially with artificial aging (T6 temper). This increase is accompanied by a loss in ductility. Strength of materials hardened by Al_2CuMg, Al_2Cu, or Al_2CuLi precipitates may be increased by cold work prior to artificial aging, T8, treatments. The increase in strength of these materials is attributed to a refinement of Al_2CuLi and of the metastable precursors to Al_2CuMg and Al_2Cu. Additions of silicon and other alloying elements can also serve to refine the size of precipitates in certain 2*xxx* alloys. Cold-finishing rod and bar products after artificial aging increases their strength (T9 temper).

The solution treatment, in most cases, is a separate operation. In particular circumstances, however, the heat from a shaping process may be sufficient to provide solution treatment. These products can be cooled after the shaping process and subsequently aged to develop useful properties (T5 temper). Some 6*xxx* alloys attain the same specified properties whether furnace solution heat treated or cooled from an elevated-temperature shaping process at a rate rapid enough to maintain sufficient silicon and magnesium in solution. In such cases, the T6 temper designation may be used.

Table 7 Typical tensile properties of selected 2*xxx*, 6*xxx*, and 7*xxx* aluminum alloy products

Alloy	Temper	Product(a)	Tensile strength		Yield strength		Elongation in 50 mm, %
			MPa	ksi	MPa	ksi	
2008	T4	Sheet	250	36	125	18	28
	T6	Sheet	300	44	240	35	13
2014	T6, T651	Plate, forging	485	70	415	60	13
2024	T3, T351	Sheet, plate	450	65	310	45	18
	T361	Sheet, plate	495	72	395	57	13
	T81, T851	Sheet, plate	485	70	450	65	6
	T861	Sheet, plate	515	75	490	71	6
2224	T3511	Extrusion	530	77	400	58	16
2324	T39	Plate	505	73	415	60	12
2524	T3, T351	Sheet, plate	450	65	310	45	21
2036	T4	Sheet	340	49	195	28	24
2219	T81, T851	Sheet, plate	455	66	350	51	10
	T87	Sheet, plate	475	69	395	57	10
2519	T87	Plate	490	71	430	62	10
6009	T4	Sheet	220	32	125	18	25
	T62	Sheet	300	44	260	38	11
6111	T4	Sheet	285	41	165	24	25
	T6	Sheet	350	51	310	45	10
6061	T6, T6511	Sheet, plate, extrusion, forging	310	45	275	40	12
	T9	Extruded rod	405	59	395	57	12
6063	T5	Extrusion	185	27	145	21	12
	T6	Extrusion	240	35	215	31	12
7005	T5	Extrusion	350	51	290	42	13
7049	T73	Forging	540	78	475	69	10
7050	T74, T745X	Plate, forging, extrusion	510	74	450	65	13
7150	T651, T6151	Plate	600	87	560	81	11
	T77511	Extrusion	650	94	615	89	12
7055	T7751	Plate	640	93	615	89	10
	T77511	Extrusion	670	97	655	95	11
7075	T6, T651	Sheet, plate	570	83	505	73	11
	T73, T735X	Plate, forging	505	73	435	63	13
7475	T7351	Plate	505	73	435	63	15
	T7651	Plate	455	66	390	57	15

(a) Properties of sheet and plate are for the long-transverse direction, and those of extrusions and forgings are for the longitudinal direction.

Table 8 Typical tensile properties of selected aluminum cast products used in the automotive industry

Alloy	Temper	Casting process	Tensile strength		Yield strength		Elongation in 20–25 mm, %
			MPa	ksi	MPa	ksi	
354.0	T6	Permanent mold	372	54	255	57	8
A356.0	T6	Sand	276	40	207	30	6
		Permanent mold	283	41	207	30	12
365.0	T6	Vacuum die	206	30	139	20	17

The mechanical properties of a casting can vary due to composition, geometry, and fabrication process, which are selected based on design requirements. Properties in certain areas can be adjusted by process and tooling design.

Table 9 Typical tensile properties of selected 3*xxx* and 5*xxx* aluminum alloy sheet products

Alloy	Temper	Tensile strength		Yield strength		Elongation in 50 mm, %
		MPa	ksi	MPa	ksi	
3003	O	100	16	40	6	30
	H14	125	18	115	17	9
	H18	165	24	150	22	5
3004	O	180	26	70	10	20
	H34	240	35	200	29	9
	H38	285	41	250	36	5
	H19	295	43	285	41	2
3005	O	130	19	55	8	25
	H14	180	26	165	24	7
	H18	240	35	225	33	4
3105	O	115	17	55	8	24
	H25	180	26	160	23	8
	H18	215	31	195	28	3
5005	O	125	18	40	6	25
	H34	160	23	140	20	8
	H38	200	29	185	27	5
5050	O	145	21	55	8	24
	H34	190	28	165	24	8
	H38	220	32	200	29	6
5052	O	195	28	90	13	25
	H34	260	38	215	31	10
	H38	290	42	255	37	7
5252	O	180	26	85	12	23
	H25	235	34	170	25	11
	H28	285	41	240	35	5
5154	O	240	35	115	17	27
	H34	290	42	230	33	13
	H38	330	48	270	39	10
	H112	240	35	115	17	25
5454	O	250	36	115	17	22
	H34	305	44	240	35	10
	H111	250	36	125	18	18
	H112	250	36	125	18	18
5056	O	290	42	150	22	35
	H18	435	63	405	59	10
	H38	415	60	345	50	15
5456	O	310	45	160	23	24
	H112	310	45	165	24	22
	H116	350	51	255	37	16
5182	O	275	40	130	19	21

Aluminum Alloy Microstructural Features Not Inferred from the Alloy-Temper Designation Systems

The alloy designation system defines the alloy content, and the temper-designation system identifies many of the thermal and mechanical processes that control the microstructure and, hence, the bulk properties of aluminum alloy products. Nevertheless, many metallurgical features are not specified by these systems. The features include nonmetallic inclusions, porosity, second-phase particles, grain and dislocation structure, and crystallographic texture.

Inclusions are typically oxides of aluminum and magnesium including spinel, $MgAl_2O_4$. Oxides form on the surface of molten aluminum and become entrapped when turbulent flow forces them below the surface. Filtration of the molten metal is used to control inclusions. Inclusions can give rise to problems ranging from pinholes in foil to reduced fatigue life in structural wrought products and castings.

Porosity reduces ductility and increases susceptibility to the initiation of fatigue cracks. Porosity may arise from either shrinkage during solidification or from hydrogen. Hydrogen control during solidification is extremely important because of the ten-fold decrease in the solubility of hydrogen in aluminum as it solidifies. Hydrogen-induced porosity can also occur in solid aluminum products when they are heated to high temperatures in humid environments. Provided that the hydrogen content is low enough, most of the porosity can be closed by thermomechanical treatments. Isostatic pressure can be used to close the pores in castings, and conventional forging and extrusion are effective in healing ingot porosity. Porosity in thick-rolled products is particularly difficult to close; tensile stresses in the short-transverse direction may arise during the initial rolling of thick plate because the amount of deformation per pass is limited. This stress causes pores to enlarge. With additional rolling to thinner plate, the pores heal.

Second-phase particles are divided into four classes based on their mode of formation and their ability to be dissolved: primary particles, constituents, dispersoids, and precipitates.

Primary Particles. These particles form when some phase other than aluminum solid solution separates first from the melt. Primary silicon particles form in castings when hypereutectic aluminum-silicon alloys solidify by eutectic decomposition. Ductility decreases with increasing size of the silicon particles, so size control is important. The coarse, faceted primary silicon particles are refined to a fine spherulitic structure using additives containing phosphorus. In certain casting alloys and 8*xxx* wrought alloys, primary iron-bearing constituents can form if the alloying content is such that the alloy is hypereutectic. In wrought alloys, macroscopically large, undesirable primary particles of Al_7Cr, Al_3Ti, or Al_3Zr can form by a peritectic reaction if chemical composition is not closely controlled.

Constituents. These particles may be either intermetallic compounds or essentially pure silicon that forms during solidification of hypoeutectic aluminum-silicon alloys. They range in size from a few micrometers to tens of micrometers. Constituents can be classified either as virtually insoluble or soluble. Because the low maximum solid solubility of iron in aluminum is further reduced by other alloying elements to 0.01 wt% or less, constituents containing iron are insoluble. Iron-free constituents containing silicon can be either soluble or insoluble depending on the chemical composition of the alloy. Major alloying elements can combine either with each other or with aluminum to form soluble constituent particles. Most of these soluble constituents dissolve either during ingot preheating prior to deformation processing or during the solution heat treatment of cast shapes or wrought products. Constituent size decreases with increasing solidification rate. In hypoeutectic 3*xx*.0 and 4*xx*.0 castings, modification by elements such as strontium significantly refine the flake structure of the silicon particles to a finer fibrous morphology.

Constituent particles are generally not beneficial and are detrimental to the fatigue resistance and fracture toughness of high-strength alloy products. These particles fracture at relatively low plastic strains and provide low-energy sites for the initiation and growth of cracks. Several high-purity (low iron and silicon) versions of 2024 and 7075 have been commercialized, and the maximum allowable impurity levels of all modern high-strength alloys are significantly lower than those of older alloys. Despite the harmful effects of constituents in high-strength alloys, the ability of alloy 3004-H19 to make commercially successful beverage containers relies on careful control of size, volume fraction, and distribution of $Al_{12}(Fe,Mn)Si$ constituent particles. These constituent particles serve to "scour" the die during the drawing operation so that galling is minimized. Attempts to produce can stock from roll-cast sheet have generally not been successful because the particle size distribution in roll-cast sheet is not as effective in minimizing galling.

Dispersoids. These form by solid-state precipitation, either during ingot preheating or during the thermal heat treatment of cast shapes, of slow-diffusing supersaturated elements that are soluble in molten aluminum but which have limited solubility in solid aluminum. Manganese, chromium, or zirconium are typical dispersoid-forming elements. Unlike the precipitates that confer precipitation hardening, dispersoids are virtually impossible to dissolve completely, once precipitated. In addition to providing dispersion strengthening, the size distribution of dispersoids in wrought alloys are a key factor in controlling degree of recrystallization, recrystallized grain size, and crystallographic texture. Dispersoids in non-heat-treatable alloys also stabilize the deformation substructure during elevated-temperature exposures, for example, during paint baking.

In contrast to the commercially significant dispersion strengthening provided by dispersoids in 3*xxx* and 5*xxx* alloys, the level of dispersion strengthening afforded by dispersoids in wrought heat-treatable alloys is trivial. In 2*x*24 alloys, $Al_{20}Cu_2Mn_3$ dispersoids nucleate dislocations at the particle-matrix interface during the quench. These dislocations serve as nucleation sites for subsequent precipitation. The newer 7*xxx* alloys contain zirconium, which forms coherent Al_3Zr dispersoids while most of the older 7*xxx* alloys contain $Al_{12}Mg_2Cr$ dispersoids which exhibit incoherent interfaces. The incoherent interfaces serve to nucleate $MgZn_2$ precipitates during the quench, so alloys containing these precipitates lose a great deal of their potential to develop high strength after slow quenching (quench sensitivity). Nucleation is difficult on coherent interfaces, so the newer alloys are less quench sensitive. A number of casting alloys, and some wrought alloys, contain elements that can form either constituents or dispersoids depending on the solidification rate.

Precipitates can form during any thermal operation below the solvus. In properly solution-heat-treated products, all precipitates dissolve during the solution-heat-treatment operation. Depending on quench rate and alloy, precipitates can form during the quench from the solution-heat-treatment temperature at grain and subgrain boundaries and at particle-matrix interfaces. These coarse precipitates do not contribute to age hardening and can serve to reduce properties such as ductility, fracture toughness, and resistance to intergranular corrosion. After the quench, G-P zones form at ambient temperature (natural aging). These are agglomerates of atoms of the major solute elements with a diffuse, coherent boundary between the G-P zone and the matrix. During elevated-temperature precipitation heat treatments (artificial aging) G-P zones may either nucleate metastable precipitates or they may dissolve, and metastable precipitates nucleate separately. Cold working subsequent to quenching introduces dislocations that may serve to nucleate metastable or equilibrium precipitates. With prolonged artificial aging, equilibrium precipitates may form. Coarse equilibrium precipitates form during annealing treatments of heat-treatable alloy products, O temper. They also form during most thermomechanical treatments prior to solution heat treatment.

Grain Structure. The grain size of aluminum alloy ingots and castings is typically controlled by the introduction of inoculants that form intermetallic compounds containing titanium and/or boron. During deformation processing, the grain structure becomes modified. Most aluminum alloy products undergo dynamic recovery during hot working as the dislocations form networks of subgrains. New dislocation-free grains may form between and following rolling passes (static recrystallization) or during deformation processing (dynamic recrystallization). During deformation, the crystal lattice of the aluminum matrix rotates at its interfaces between constituent and coarse precipitate particles. These high-energy sites serve to nucleate recrystallization. This process is termed particle-stimulated nucleation and is an important mechanism in the recrystallization process of aluminum. The particle size that will serve as a nucleus decreases as deformation temperature decreases and strain and strain rate increase. Dispersoid particles retard the movement of high-angle grain boundaries. Consequently, hot-worked structures are resistant to recrystallization and often retain the dynamically recovered subgrain structure in the interiors of elongated cast grain boundaries. In heat-treated products containing a sufficient quantity of dispersoids—the unrecrystallized structure of hot-worked-plate—forgings and extrusions can be retained after solution heat treatment.

Degree of recrystallization of hot-worked products has an effect on fracture toughness. Unrecrystallized products develop higher toughness than do products that are either partially or completely recrystallized. This behavior is attributed to precipitation on the recrystallized high-angle grain boundaries during the quench. These particles increase the tendency for low-energy intergranular fracture. Products such as sheet, rods, and tubing that are cold rolled invariably recrystallize during solution heat treatment or annealing to O temper.

Decreasing the grain size can increase strength of 5*xxx* alloy products in the O temper by 7 to 28 MPa (1 to 4 ksi), but grain size is not a major factor in increasing strength of other aluminum alloy products. Several measures of formability are influenced by grain size, however, so grain size is controlled for this reason. One particular use of grain size control is to produce stable, fine grains, which are essential in developing superplastic behavior in aluminum alloy sheet.

Crystallographic Texture. Cast aluminum ingots and shapes generally have a random crystallographic texture; the orientation of the unit cells comprising each grain are not aligned. With deformation, however, certain preferred crystallographic orientations develop. Many of the grains rotate and assume certain orientations with respect to the direction of deformation. For flat-rolled products and extrusions having a high aspect ratio of width to thickness, the deformation texture is similar to that in pure fcc metals. These orientations are described by using the Miller indices of the planes {nnn} in the grains parallel to the plane of the worked product and directions [nnn] parallel to the working direction. The predominant textures are {110}[112], {123}[634], and {112}[111]. During recrystallization, a high concentration of grains in the {001}[100] or {011}[100] orientations may develop. Alternatively, if particle-stimulated nucleation is present to a large extent, the recrystallized texture will be random. Control of crystallographic texture is particularly important for non-heat-treatable sheet that will be drawn. If texture is not random, ears form during the drawing process. In extruded or drawn rod or bar, the texture is a dual-fiber texture in which almost all grains are aligned so that the grain directions are either [001] or [111]. In heat-treatable alloys, texture has the most potent effect on the properties of extrusions that have the dual-fiber texture. Strengthening by this process is so potent that the longitudinal yield strengths of extruded products exhibiting this texture are about 70 MPa (10 ksi) higher than strength in the transverse direction. If this dual-fiber texture is lost by recrystallization, strength in the longitudinal direction decreases to that in the transverse directions.

Copper and Copper Alloys

After iron and aluminum, copper is the third most-prominent commercial metal because of its availability and attractive properties: excellent malleability (or formability), good strength, excellent electrical and thermal conductivity, and superior corrosion resistance (Ref 18–25). Copper offers the designer moderate levels of density (8.94 g/cm^3, or 0.323 lb/in.^{3}), elastic modulus (115 GPa, or 17 × 10^6 psi), and melting temperature (1083 °C, or 1981 °F). It forms many useful alloys to provide a wide variety of engineering property combinations and is not unduly sensitive to most impurity elements. The electrical conductivity of commercially available pure copper, about 101% IACS (International Annealed Copper Standard), is second only to that of commercially pure silver (about 103% IACS). Standard commercial copper is available with higher purity and, therefore, higher conductivity than what was available when its electrical resistivity value at 20 °C (70 °F) was picked to define the 100% level on the IACS scale in 1913. The thermal conductivity for copper is also high, 391 W/m · K (226 Btu/ft · h · °F), being directly related to the electrical conductivity through the Wiedemann-Franz relationship.

Copper and the majority of its alloys are highly workable hot or cold, making them readily commercially available in various wrought forms: forgings, bar, wire, tube, sheet, and foil. In 1995, copper used in wire and cable represented about 50% of U.S. production and in flat products of various thickness another 15%, rod and bar about 14%, tube about 14.5%, with foundries using about 5% for cast products, and metal powder manufacturers about 0.6%. Besides the more familiar copper wire, copper and its alloys are used in electrical and electronic connectors and components, heat-exchanger tubing, plumbing fixtures, hardware, bearings, and coinage.

As with other metal systems, copper is intentionally alloyed to improve its strength without unduly degrading ductility or workability. However, it should be recognized that additions of alloying elements also degrade electrical and thermal conductivity by various amounts depending on the alloying element, its concentration and location in the microstructure (solid solution or dispersoid). The choice of alloy and condition is most often based on the trade-off between strength and conductivity. Alloying also changes the color from reddish brown to yellow (with zinc, as in brasses) and to metallic white or "silver" (with nickel, as in U.S. cupronickel coinage).

Copper and its alloys are readily cast into cake, billet, rod, or plate—suitable for subsequent hot or cold processing into plate, sheet, rod, wire, or tube—via all the standard rolling, drawing, extrusion, forging, machining, and joining methods. Copper and copper alloy tubing can be made by the standard methods of piercing and tube drawing as well as by the continuous induction welding of strip. Copper is hot worked over the temperature range 750 to 875 °C (1400 to 1600 °F), annealed between cold working steps over the temperature range 375 to 650 °C (700 to 1200 °F), and is thermally stress relieved usually between 200 and 350 °C (390 and 660 °F). Copper and its alloys owe their excellent fabricability to the face-centered cubic crystal structure and its twelve available dislocation slip systems. Many of the applications of copper and its alloys take advantage of the work-hardening capability of the material, with the cold processing deformation of the final forming steps providing the required strength/ductility for direct use or for subsequent forming of stamped components. Copper is easily processible to more than 95% reduction in area. The amount of cold deformation between softening anneals is usually restricted to 90% maximum to avoid excessive crystallographic texturing, especially in rolling of sheet and strip.

Although copper obeys the Hall-Petch relationship and grain size can be readily controlled by processing parameters, work hardening is the only strengthening mechanism used with pure copper. Whether applied by processing to shape and thickness, as a rolled strip or drawn wire, or by forming into the finish component, as an electrical connector, the amount of work hardening applied is limited by the amount of ductility required by the application. Worked copper can be recrystallized by annealing at temperatures as low as 250 °C (480 °F), depending on prior degree of cold work and time at temperature. While this facilitates processing, it also means that softening resistance during long-time exposures at moderately elevated temperatures can be a concern, especially in electrical and electronic applications where I^2R heating is a factor. For applications above room temperature, but at temperatures lower than those inducing recrystallization in commercial heat treatments, thermal softening can occur over extended periods and characteristics such as the half-softening temperature should be considered; that is, the temperature for which the worked metal softens to half its original hardness after a specific exposure time, usually 1 h.

A more useful engineering property for many electrical contact applications is stress-relaxation resistance, the property that characterizes the decrease in contact load supported by a mechanical contact over time at a given temperature, typically measured at room temperature between exposures to elevated temperature (Ref 20). Figure 3 illustrates the characteristics of the tensile-stress-relaxation property of drawn (worked) copper wire; the degree of relaxation increases with temperature and time. It also increases with the initial temper or degree of cold work in the material. The mechanism is the thermally activated and applied-stress directed motion of crystal lattice defects, such as point defects and dislocations. Consequently, the application of a

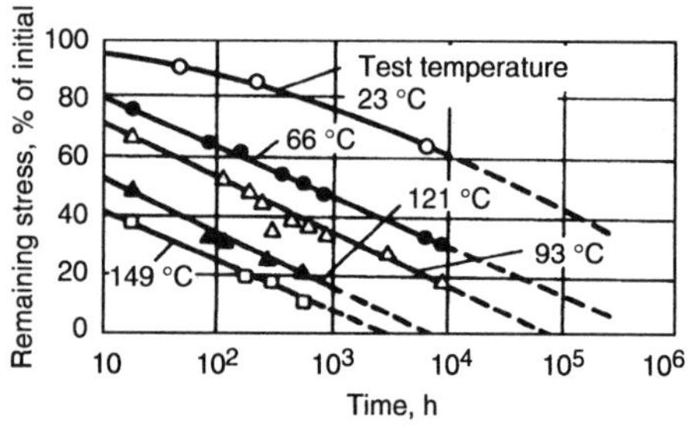

Fig. 3 Tensile-stress-relaxation characteristics of copper alloy C11000. Data are for tinned 30 AWG (0.25 mm diam) annealed ETP copper wire; initial elastic stress, 89 MPa (13 ksi).

thermal heat treatment (stabilization anneal) to induce recovery mechanisms to tie up mobile components of dislocations will improve the stress-relaxation resistance. Alloying elements also restrict dislocation motion and provide a more potent remedy for improving stress-relaxation resistance of cold-worked metal in service. For example, the improvement in stress-relaxation resistance obtainable by alloying copper with 5% Sn (alloy C51000) in combination with a low-temperature stabilization heat treatment and as a function of sheet orientation is illustrated by comparing the alloy data at 93 °C (200 °F) in Fig. 4 with those for copper in Fig. 3.

Wrought Copper Alloys

The purpose of adding alloying elements to copper is to optimize the strength, ductility (formability), and thermal stability, without inducing unacceptable loss in fabricability, electrical/thermal conductivity, or corrosion resistance. A list of selected wrought copper alloy compositions and their properties is given in Table 10. In this table, the alloys are arranged in their common alloy group: the coppers (99.3% min Cu), the high-coppers (94% min Cu), brasses (copper-zinc), bronzes (copper-tin, or copper-aluminum, or copper-silicon), copper-nickels, and the nickel silvers (Cu-Ni-Zn). Composition and property data are given by the Copper Development Association (CDA) and are incorporated in the ASTM numbering system, wherein alloys numbered by the designations (now UNS) C10100 to C79900 cover wrought alloys and C80100 to C99900 apply to cast alloys. Copper alloys show excellent hot and cold ductility, although usually not to the same degree as the unalloyed parent metal. Even alloys with large amounts of solution-hardening elements—zinc, aluminum, tin, silicon—that show rapid work hardening are readily commercially processed beyond 50% cold work before a softening anneal is required to permit additional processing. The amount of cold working and the annealing parameters must be balanced to control grain size and crystallographic texturing. These two parameters are controlled to provide annealed strip products at finish gage that have the formability needed in the severe forming and deep drawing commonly done in commercial production of copper, brass, and other copper alloy hardware and cylindrical tubular products.

The pure copper alloys, also called the coppers (C10100 to C15900), are melted and cast in inert atmosphere from the highest-purity copper in order to maintain high electrical conductivity (oxygen-free, or OF, copper, C10200). Copper is more commonly cast with a controlled oxygen content (0.04% O as in electrolytic tough pitch, or ETP, copper, C11000) to refine out impurity elements from solution by oxidation. Included in this group are the alloys that are deoxidized with small addition of various elements such as phosphorus (C12200, Cu-0.03P) and the alloys that use minor amounts of alloy additions to greatly improve softening resistance, such as the silver-bearing copper alloys (C10500, Cu-0.034 min Ag) and the zirconium-bearing alloys (C15000 and C15100, Cu-0.1Zr).

High-copper alloys (C16000 to C19900) are designed to maintain high conductivity while using dispersions and precipitates to increase strength and softening resistance: iron dispersions in Cu-(1.0–2.5)Fe alloys (C19200, C19400), chromium precipitates in Cu-1Cr (C18200), and the coherent precipitates in the Cu-(0.3–2.0)Be-Co,Ni age-hardening alloys (C17200, C17410, and C17500).

Brass alloys are a rather large family of copper-zinc alloys. A significant number of these are binary copper-zinc alloys (C20500 to C28000), utilizing the extensive region of solid solution up to 35% Zn, offering excellent formability with good work-hardening strength at reasonable cost. The alloys below 15% Zn have good corrosion and stress-corrosion resistance. Alloys above 15% Zn need a stress-relieving heat treatment to avoid stress corrosion and, under certain conditions, can be susceptible to dezincification. Alloys at the higher zinc levels of 35 to 40% Zn contain the bcc beta phase, especially at elevated temperatures, making them hot extrudeable and forgeable (alloy C28000 with Cu-40Zn, for example). The beta alloys are also capable of being hot worked while containing additions of 1 to 4% Pb, or more recently bismuth, elements added to provide the dispersion of coarse particles that promote excellent machinability characteristics available with various commercial Cu-Zn-Pb alloys (C31200 to C38500). The tin-brasses (C40400 to C49000) contain various tin additions from 0.3 to 3.0% to enhance corrosion resistance and strength in brass alloys. Besides improving corrosion-resistance properties in copper-zinc tube alloys, such as C44300 (Cu-30Zn-1Sn), the tin addition also provides for good combinations of strength, formability, and electrical conductivity required by various electrical connectors, such as C42500 (Cu-10Zn-2Sn). A set of miscellaneous copper-zinc alloys (C66400 to C69900) provide improved strength and corrosion resistance through solution hardening with aluminum, silicon, and manganese, as well as dispersion hardening with iron additions.

Bronze alloys consist of several families named for the principal solid-solution alloying element. The familiar tin-bronzes (C50100 to C54400) comprise a set of good work-hardening,

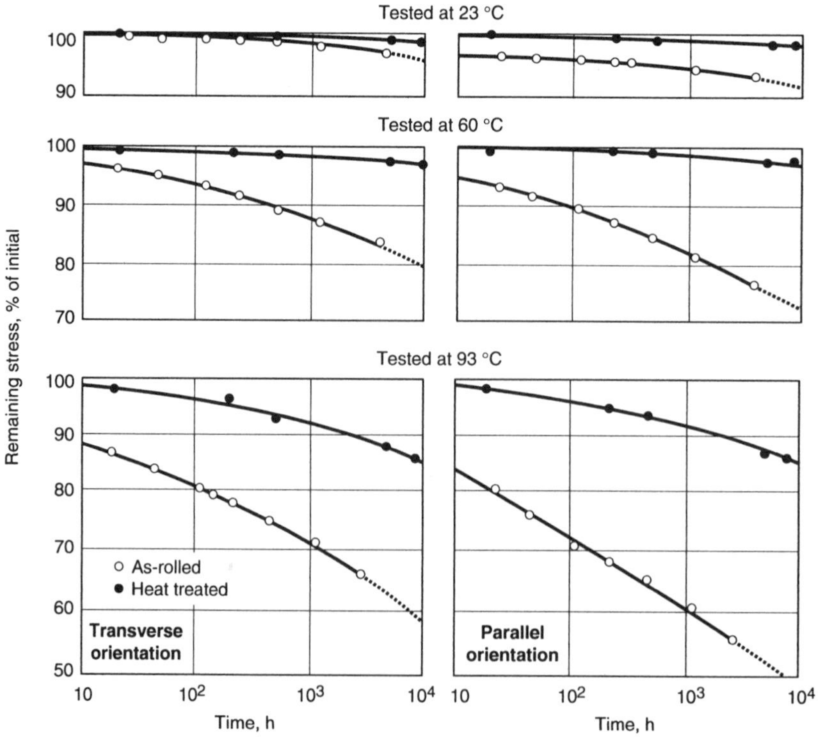

Fig. 4 Anisotropic stress-relaxation behavior in bending for highly cold-worked C51000 copper alloy strip. Data are for 5% Sn phosphor bronze cold rolled 93% (reduction in area) to 0.25 mm (0.01 in.) and heat treated 2 h at 260 °C (500 °F). Graphs at left are for stress relaxation transverse to the rolling direction; graphs at right, for stress relaxation parallel to the rolling direction. Initial stresses: as rolled, parallel orientation, 607 MPa (88 ksi); as rolled, transverse orientation, 634 MPa (92 ksi); heat treated, parallel orientation, 641 MPa (93 ksi); heat treated, transverse orientation, 738 MPa (107 ksi)

solid-solution alloys containing from nominally 0.8% Sn (C50100) to 10% Sn (C52400), usually with a small addition of phosphorus for deoxidation. These alloys provide an excellent combination of strength, formability, softening resistance, electrical conductivity, and corrosion resistance. The aluminum-bronze alloys contain 2 to 15% Al (C60800 to C64200), an element adding good solid-solution strengthening and work hardening, as well as corrosion resistance. The aluminum-bronzes usually contain 1 to 5% Fe, providing elemental dispersions to promote dispersion strengthening and grain size control. The silicon-bronze alloys (C64700 to C66100) generally offer good strength through solution- and work-hardening characteristics, enhanced in some cases with a tin addition, as well as excellent resistance to stress corrosion and general corrosion.

Cupronickels are copper-nickel alloys (C70100 to C72900) that utilize the complete solid solubility that copper has for nickel to provide a range of single-phase alloys (C70600 with Cu-10Ni-1.5Fe, and C71500 with Cu-30Ni-0.8Fe, for example) that offer excellent corrosion resistance and strength. The family of copper-nickel alloys also includes various dispersion- and precipitation-hardening alloys due to the formation of hardening phases with third elements, such as Ni_2Si in C70250 (Cu-3Ni-0.7Si-0.15Mg) and the spinodal hardening obtainable in the Cu-Ni-Sn alloys (C72700 with Cu-10Ni-8Sn, for example).

Copper-nickel-zinc alloys, also called nickel-silvers, are a family of solid-solution-strengthening and work-hardening alloys with various nickel-zinc levels in the Cu-(4–26)Ni-(3–30)Zn ternary alloy system valued for their strength, formability, and corrosion and tarnish resistance, and, for some applications, metallic white color.

Strengthening Mechanisms for Wrought Copper Alloys

Solution Hardening. Copper can be hardened by the various common methods without unduly impairing ductility or electrical conductivity. The metallurgy of copper alloys is suited for using, singly or in combination, the various common strengthening mechanisms: solid solution and work hardening, as well as dispersed particle and precipitation hardening. The commonly used solid-solution hardening elements are zinc, nickel, manganese, aluminum, tin, and silicon, listed in approximate order of increasing effectiveness. Commercial alloys represent the entire range of available solid-solution compositions of each element: up to 35% Zn, and up to (and even beyond) 50% Ni, 50% Mn, 9% Al, 11% Sn, and 4% Si. The relative amount of solution strengthening obtained from each element or particular combination of elements is determined by the ability of the solute to interfere with dislocation motion and is reflected in the work-hardening rate starting with the annealed condition, as illustrated by the increase in tensile strength with cold work shown in Fig. 5 and also Table 10.

Work hardening is the principal hardening mechanism applied to most copper alloys, the degree of which depends on the type and amount of alloying element and whether the alloying element remains in solid solution or forms a dispersoid or precipitate phase. Even those alloys that are commercially age hardenable are often provided in the mill hardened tempers; that is, they have been processed with cold work preceding and/or following an age-hardening heat treatment. For the leaner alloys (below about 12% Zn, or about 3% Al, for example), processing generates dislocations that develop into entanglements and into cells, with some narrow shear band formation beyond about 65% cold reduction in thickness. After about 90% cold work, the distinct "copper" or "metal" deformation crystallographic texture begins to develop. With the richer solid-solution alloys that lower the stacking-fault energy, planar slip is the dominant dislocation mechanism, with associated higher work hardening. Beyond about 40% cold work in these richer alloys, stacking faults, shear banding, and

Table 10 Compositions and properties of selected wrought copper alloys

Alloy	UNS No.	Nominal composition	Treatment	Tensile strength		Yield strength		Elongation, %	Rockwell hardness
				MPa	ksi	MPa	ksi		
Pure copper									
OFHC	C10200	99.95 Cu	...	221-455	33-66	69-365	10-53	55-4	...
High-copper alloys									
Beryllium-copper	C17200	97.9Cu-1.9Be-0.2Ni or Co	Annealed	490	71	...	...	35	60 HRB
			Hardened	1400	203	1050	152	2	42 HRC
Brass									
Gilding, 95%	C21000	95Cu-5Zn	Annealed	245	36	77	11	45	52 HRF
			Hard	392	57	350	51	5	64 HRB
Red brass, 85%	C23000	85Cu-15Zn	Annealed	280	41	91	13	47	64 HRF
			Hard	434	63	406	59	5	73 HRB
Cartridge brass, 70%	C26000	70Cu-30Zn	Annealed	357	52	133	19	55	72 HRF
			Hard	532	77	441	64	8	82 HRB
Muntz metal	C28000	60Cu-40Zn	Annealed	378	55	119	17	45	80 HRF
			Half-hard	490	71	350	51	15	75 HRB
High lead brass	C35300	62Cu-36Zn-2Pb	Annealed	350	51	119	17	52	68 HRF
			Hard	420	61	318	46	7	80 HRB
Bronze									
Phosphor bronze, 5%	C51000	95Cu-5Sn	Annealed	350	51	175	25	55	40 HRB
			Hard	588	85	581	84	9	90 HRB
Phosphor bronze, 10%	C52400	90Cu-10Sn	Annealed	483	70	250	36	63	62 HRB
			Hard	707	103	658	95	16	96 HRB
Aluminum bronze	C60800	95Cu-5Al	Annealed	420	61	175	25	66	49 HRB
			Cold rolled	700	102	441	64	8	94 HRB
Aluminum bronze	C63000	81.5Cu-9.5Al-5Ni-2.5Fe-1Mn	Extruded	690	100	414	60	15	96 HRB
			Half-hard	814	118	517	75	15	98 HRB
High-silicon bronze	C65500	96Cu-3Si-1Mn	Annealed	441	64	210	31	55	66 HRB
			Hard	658	95	406	59	8	95 HRB
Copper nickel									
Cupronickel, 30%	C71500	70Cu-30Ni	Annealed	385	56	126	18	36	40 HRB
			Cold rolled	588	85	553	80	3	86 HRB
Nickel silver									
Nickel silver	C75700	65Cu-23Zn-12Ni	Annealed	427	62	196	28	35	55 HRB
			Hard	595	86	525	76	4	89 HRB

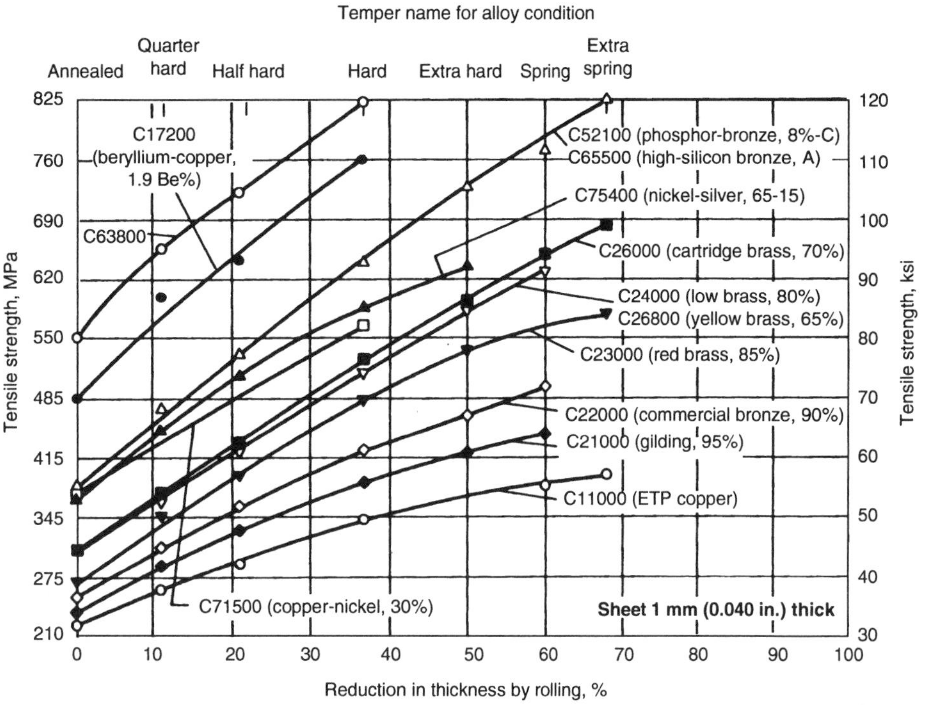

Fig. 5 Tensile strength of single-phase copper alloys as affected by percentage reduction in thickness by rolling (temper). Curves of lesser slope indicate a low rate of work hardening and a higher capacity for redrawing. ETP, electrolytic tough pitch

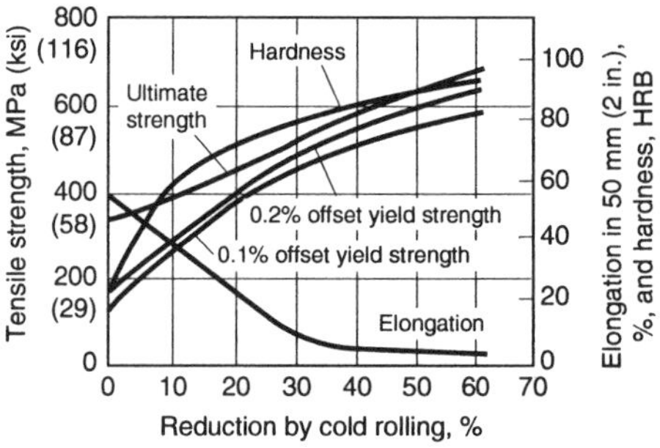

Fig. 6 The effect of cold rolling on the strength, hardness, and ductility of annealed copper alloy C26000 when it is cold rolled in varying amounts up to 62% reduction in thickness

deformation twinning become important deformation mechanisms that, beyond 90% cold work, lead to the "brass" or "alloy" type of crystallographic deformation texture and accompanying anisotropy of properties. The variation in tensile properties with cold working of an annealed Cu-30 Zn alloy (C26000) is shown in Fig. 6. The degree of work hardening seen with cold working several selected single-phase copper alloys is illustrated by the cold-rolling curves in Fig. 5. Many copper alloys are used in wrought forms in a worked temper, chosen for the desired combination of work-hardened strength and formability, either for direct use in service or for subsequent component fabrication.

Dispersion strengthening is used in copper alloys for hardening, controlling grain size, and providing softening resistance, as exemplified by iron particles in copper-iron alloys, C19200 or C19400, and in aluminum bronzes, C61300 or C63380. Cobalt silicide particles in alloy C63800 (Cu-2.8Al-1.8Si-0.4Co), for example, provide fine-grain control and dispersion hardening to give this alloy high strength with reasonably good formability. Alloy C63800 offers an annealed tensile strength of 570 MPa (82 ksi) and rolled temper tensile strengths of 660 to 900 MPa (96 to130 ksi). Alloys offering exceptionally good thermal stability have been developed using powder metallurgy (P/M) techniques to incorporate dispersions of fine Al_2O_3 particles (3 to 12 nm in size) in a basically copper matrix, which is finish processed to rod, wire, or strip products. This family of alloys, C15715 to C15760, can resist softening up to and above 800 °C (1472 °F).

Precipitation Hardening. Age-hardening mechanisms are used in those few but important copper systems that offer a decreasing solubility for hardening phases. The beryllium-copper system offers a series of wrought and cast age-hardening alloys, UNS C17000 to C17530 and C82000 to C82800. The wrought alloys contain 0.2 to 2.0% Be and 0.3 to 2.7% Co (or up to 2.2% Ni). They are solution heat treated in the range 760 to 955 °C (1400 to 1750 °F) and age hardened to produce the beryllium-rich coherent precipitates when aged in the range 260 to 565 °C (500 to 1050 °F), the specific temperature being chosen for the particular alloy and desired property combination (Fig. 7). The precipitation sequence during aging consists of the formation of solute-rich G-P zones, followed in sequence by coherent platelets of the metastable intermediate phases γ' and γ''. Overaging is marked by the appearance of the B2 ordered equilibrium γ-BeCu phase as particles within grains and along grain boundaries, large enough to be seen in the light microscope. The cobalt and nickel additions form dispersoids of equilibrium (Cu, Co, or Ni)Be that restrict grain growth during solution annealing in the two-phase field at elevated temperatures (Fig. 7b). A cold-working step following solution annealing is often used to increase the age-hardening response. Alloy C17200 (Cu-1.8Be-0.4Co), for example, can be processed to reach high strength: that is, tensile strengths after solutionization (470 MPa, or 68 ksi), after cold rolling to the hard temper (755 MPa, or 110 ksi), and after aging (1415 MPa, or 205 ksi). While they are commercially available in the heat-treatble (solutionized) condition, the beryllium-copper alloys are commonly provided in the mill-hardened temper with the optimal strength/ductility/conductivity combination suitable for the application.

Other age-hardening copper alloys include the chromium-coppers, which contain 0.4 to 1.2% Cr (C18100, C18200, and C18400); these alloys produce arrays of pure chromium precipitates and dispersoid particles when aged. The Cu-Ni-Si alloys, C64700 and C70250, age harden by precipitating the Ni_2Si intermetallic phase (Fig. 8). Compositions in the Cu-Ni-Sn system, C71900 and C72700, are hardenable by spinodal decomposition, a mechanism that provides high strength and good ductility through the formation of a periodic array of coherent, fcc solid-solution phases that require the electron microscope to be seen. Each of these alloys, including the beryllium-coppers can be thermomechanically processed to provide unique combinations of strength, formability, electrical conductivity, softening resistance, and stress-relaxation resistance.

Copper Casting Alloys

The copper casting alloys, numbered UNS C80100 to C99900, are available as sand, continuous, centrifugal, permanent mold, and some die castings (Ref 22, 23). They are generally similar to the wrought counterparts, but they do offer their own unique composition/property characteristics. For example, they do offer the opportunity to add lead to levels of 25% that could not be easily made by wrought techniques in order to provide compositions in which dispersions of lead particles are useful for preventing galling in bearing applications. The copper casting alloys are used for their corrosion resistance and their high thermal and electrical conductivity. The most common alloys are the general-purpose Cu-5Sn-5Pb-5Zn alloy (C83600), used for valves and plumbing hardware, and C84400, widely used for cast plumbing system components. C83600 contains lead particles dispersed about the single-phase matrix and offers good machinability, with moderate levels of corrosion resistance, tensile strength (240 MPa, or 35 ksi), ductility, and conductivity (15% IACS).

While the Cu-Sn-Pb-(and/or Zn) casting alloys have only moderate strength, the cast manganese

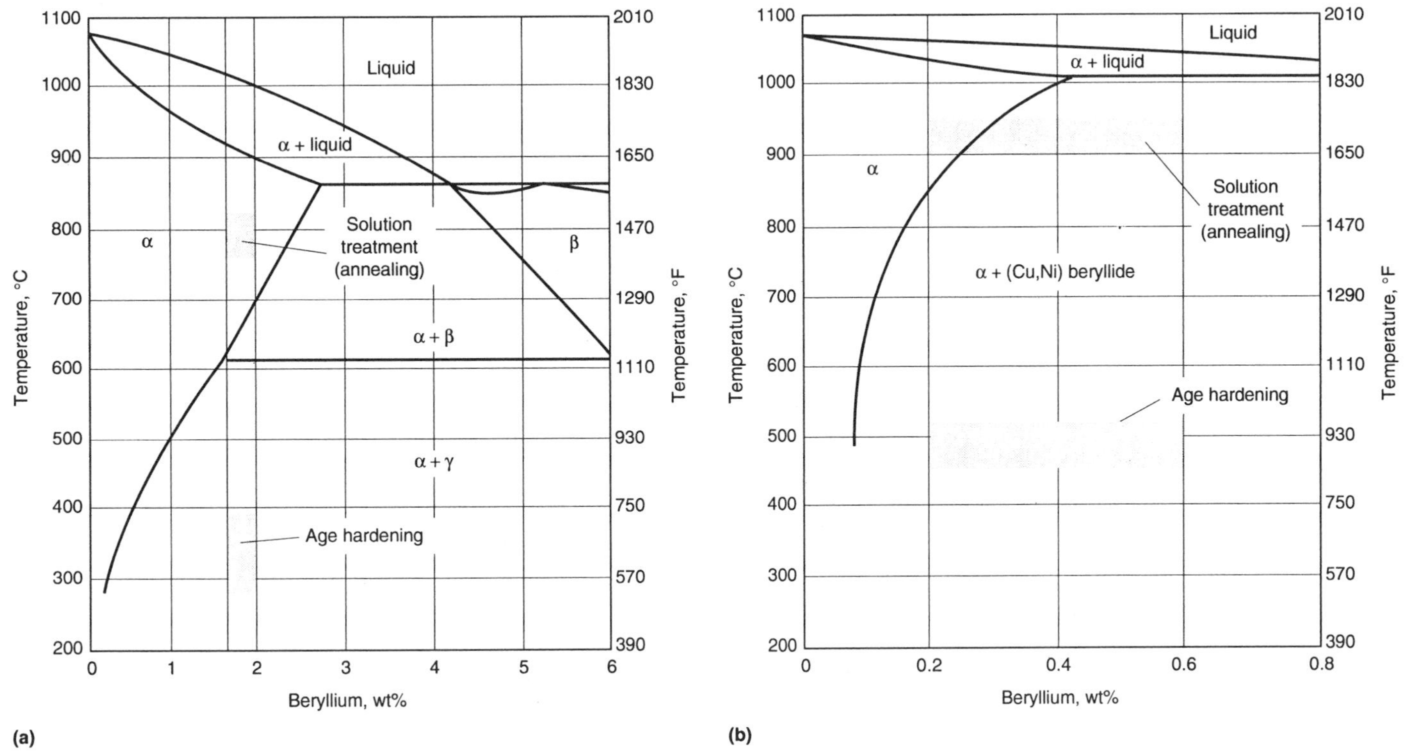

Fig. 7 Phase diagrams for beryllium-copper alloys. (a) Binary composition for high-strength alloys such as C17200. (b) Pseudobinary composition for C17510, a high-conductivity alloy containing Cu-1.8Ni-0.4Be

and aluminum bronzes offer higher tensile strengths, 450 to 900 MPa (65 to 130 ksi). As with the wrought alloys, the cast aluminum-bronze alloys commonly contain an iron addition (0.8 to 5.0%) to provide iron-rich particles for grain refinement and added strength. In addition, at aluminum levels in the range 9.5 to 10.5% (or 8.0 to 9.5% Al with nickel or manganese additions) the alloys are heat treatable for added strength. Depending on the section thickness and cooling rate of the casting, as well as the alloy composition and heat treatments, the microstructures can be rather complex. The aluminum-bronzes can be annealed completely or partially in the β field and quenched to form β martensite with α needles. Aging these alloys will temper the martensite by precipitation fine α needles. One of the aluminum-bronze alloys, Cu-10.5Al-5Fe-5Ni, for example, is used for its combination of high strength and good corrosion resistance. Through heat treatment, the intermetallic κ-phase, with its complex composition (Fe,Ni,Cu)Al and CsCl crystal structure, provides a strengthening component in any of its morphologies: as globular particles, fine precipitates, or as a component of cellular eutectoid colonies.

Copper Alloy Powders

Unique structural components are commercially made of copper and its alloys by P/M methods. Copper and prealloyed powders are made by reduction of oxides, or by atomization, wherein the solidification of liquid droplets from a pour stream is broken up by an impinging jet of a liquid or gas. Self-lubricating sintered bronze bearings are deliberately not pressed and sintered to 100% density in order to maintain an interconnected porosity to serve as an oil reservoir. The P/M technique is uniquely suited to permit the addition of up to about 1.5% graphite in these bearings. Likewise, various multiphase P/M copper alloys containing a mixture of hard and soft phases in a copper matrix (Cu-7Sn-3Fe-6Pb-6 graphite-3SiO_2, for example) are made for friction materials. The combination of thermal stability, wear resistance, and sliding friction properties make these materials suitable for use in clutch plates, and so forth. Various structural parts are made of bronze, brass, and nickel-silver alloys. In addition, P/M techniques are used to prepare the initial stages of the oxide-dispersion-strengthened (ODS) copper alloys, which are fabricated into finished forms by standard wrought methods to provide good softening resistance with excellent thermal or electrical conductivity. These ODS materials are prepared by internal oxidation of powder copper-aluminum alloy to form a dispersion of fine Al_2O_3 particles, about 3 to 12 nm in size.

Nickel and Nickel Alloys

Nickel and nickel alloys are used in the chemical processing, pollution control, power generation, electronic, and aerospace industries, taking advantage of their excellent corrosion, oxidation, and heat resistance (Ref 18, 26–32). Nickel is ductile and can be made by the conventional processing methods into cast, P/M, and various wrought products: bar/wire, plate/sheet, and tube. Commercially pure nickel has moderately high values of melting temperature (1453 °C, or 2647 °F), density (8.902 g/cm^3, or 0.322 lb/in.^{3}), and elastic modulus (204 GPa, or 30 × 10^6 psi). It is ferromagnetic, with a Curie temperature of 358 °C (676 °F) and good electrical (25% IACS) and thermal conductivity (82.9 W/m · K, or 48 Btu/ft · h · °F). Elemental nickel is used principally as an alloying element to increase the corrosion resistance of commercial iron and copper alloys; only about 13% of annual consumption is used in nickel-base alloys. Approximately 60% is used in stainless steel production, with another 10% in alloy steels and 2.5% in copper alloys. Nickel is also used in special-purpose alloys: controlled expansion, electrical resistance, magnetic, and shape memory alloys.

Effects of Alloying Elements in Nickel Alloys

Nickel has an fcc crystal structure, to which it owes its excellent ductility and toughness. Because nickel has extensive solid solubility for many alloying elements, the microstructure of nickel alloys consists of the fcc solid-solution austenite (γ) in which dispersoid and precipitate particles can form. Nickel forms a complete solid solution with copper and has nearly complete solubility with iron. It can dissolve about 35% Cr, about 20% each of molybdenum and tungsten, and about 5 to 10% each of aluminum, titanium, manganese, and vanadium. Thus, the tough, ductile fcc matrix can dissolve extensive amounts of elements in various combinations to provide so-

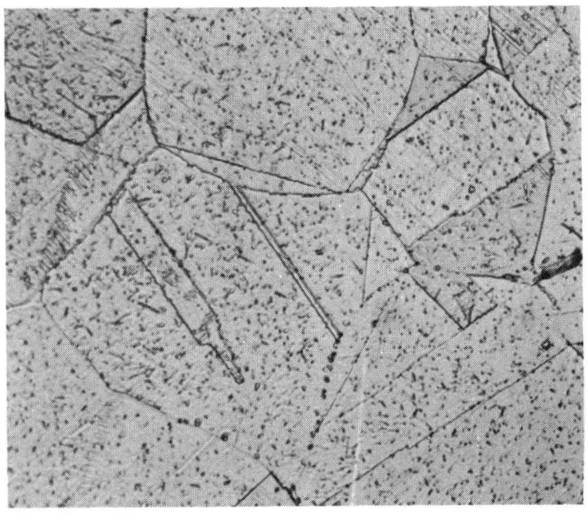

Fig. 8 Photomicrograph showing the dispersion of Ni_2Si precipitates in the quenched and aged condition of copper alloy C64700, Cu-2Ni-0.7Si. Magnification: 500×

Fig. 9 Replica electron micrograph of the nickel alloy Udimet 700 (Ni-15Cr-17Co-5Mo-3.5Ti-4Al-0.06C) in the solution-annealed and aged condition, showing precipitation of carbide at grain boundaries and arrays of γ' within grains of the γ solid-solution matrix. 4500×

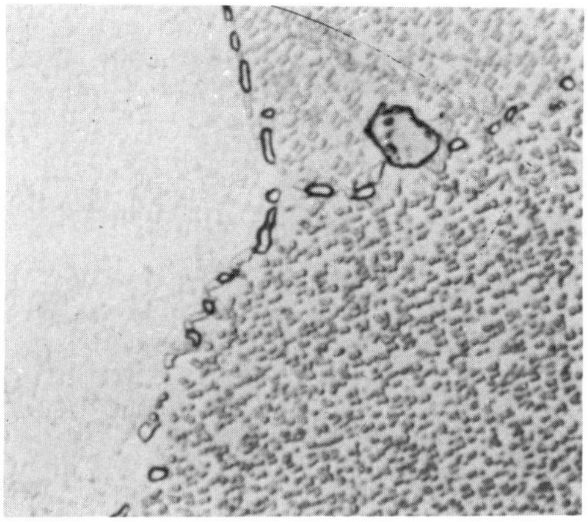

Fig. 10 Photomicrograph of the nickel alloy Udimet 700 (Ni-15Cr-17Co-5Mo-3.5Ti-4Al-0.06C in the solution-annealed and aged condition, showing precipitation of $M_{23}C_6$ carbide at grain boundaries and arrays of γ' within grains of the γ solid-solution matrix. 1000×

lution hardening as well as improved corrosion and oxidation resistance. The degree of solution hardening has been related to the atomic size difference between nickel and the alloying element, and therefore the ability of the solute to interfere with dislocation motion. Tungsten, molybdenum, niobium, tantalum, and aluminum, when aluminum is left in solution, are strong solution hardeners, with tungsten, niobium, tantalum, and molybdenum also being effective at temperatures above 0.6 T_m (T_m = melting temperature), where diffusion-controlled creep strength is important. Iron, cobalt, titanium, chromium, and vanadium are weaker solution-hardening elements. Aluminum and titanium are usually added together to form the age-hardening precipitate, $Ni_3(Al,Ti)$.

Gamma Prime (γ') Precipitation. Gamma-prime (γ'), $Ni_3(Al,Ti)$, and the closely related γ'', Ni_3Nb, are the major precipitation-hardening phases in nickel alloys. These precipitates are based on the intermetallic compound, Ni_3Al, which has an fcc $L1_2$ ordered crystal structure with a lattice parameter differing from the nickel austenite (γ) matrix by ≤1%. This misfit allows the homogeneous nucleation and growth of rather stable arrays of coherent precipitates (Fig. 9). Strengthening is provided in part by the hindrance to dislocations moving across the γ-γ' interface and, more importantly, as they cut across the ordered precipitate, where they must split into partials to maintain the ordered crystal structure. Moreover, Ni_3Al is one of the unique phases that show a significant increase in flow stress with temperature. In particular, the yield strength increases over the range 300 °C (572 °F) to above 900 °C (1650 °F), showing a broad peak at about 600 °C (1110 °F).

In addition, some alloying elements can partition to γ', affecting the interface mismatch and precipitate-coarsening kinetics as well as contributing a solution-hardening component to strength, with titanium being the most effective at room and elevated temperatures. However, titanium, niobium, and tantalum can influence mechanical properties still further by encouraging the formation of other similar types of precipitates. With higher titanium content, γ' will transform to the hexagonal close-packed (hcp) η-phase, Ni_3Ti, which has an acicular or cellular morphology. With increased amounts of niobium, γ' transforms to the commercially important metastable body-centered tetragonal (bct) phase γ''. A decrease in hardening will result if the equilibrium orthorhombic phase, Ni_3Nb, is allowed to form. The actual phases precipitated and their effectiveness in hardening the microstructure are dependent on the alloy composition, the applied heat treatments, the resulting precipitate volume fraction, and the service conditions.

Carbides. Although not a carbide former, nickel dissolves many elements that readily form the carbides seen in nickel alloys (MC, M_6C, M_7C_3, $M_{23}C_6$). The MC carbides (where M = W, Ta, Ti, Mo, Nb) are usually large, blocky, and undesirable. The M_6C carbides (M = Mo, W) can precipitate as small platelets in the grains or as blocky particles in boundaries useful for grain control, but deleterious for ductility and stress rupture properties. The M_7C_3 (M = Cr) can be useful when precipitated as discrete particles, but more so are grain boundary particles of $M_{23}C_6$ (M = Cr, Mo, W), where they can enhance creep rupture properties (Fig. 10). If carbides are allowed to agglomerate or form grain-boundary films during heat treatment or in service at elevated temperatures, they can seriously impair ductility and cause embrittlement. As in stainless steels, precipitation of chromium carbides at boundaries can lead to intergranular corrosion due to the chromium-depleted zone alongside the grain boundary becoming anodic to the rest of the grains. This grain-boundary sensitization is controlled in several ways: (1) by avoiding the chromium-carbide aging temperature range (425 to 760 °C, or 800 to 1400 °F) during processing, (2) with stabilization heat treatments to tie up carbon with more stable carbide formers (niobium, tantalum, titanium), and (3) by reducing the carbon level in the base alloy.

Nickel Alloys

Nickel is alloyed to extend the good corrosion resistance and good heat resistance of elemental nickel. Even with extensive amounts of alloying elements, the tough, ductile fcc austenitic matrix is preserved. It is convenient to describe nickel alloys by grouping them into their two broad application areas: corrosion resistance, especially in aqueous environments, and heat resistance. Naturally, this artificial separation should not be considered a rigid barrier as the corrosion-resistant alloys have good strength above room temperature and the heat-resistant alloys have good corrosion resistance. The unique, special-property alloys, many of which are also used for their good corrosion and heat resistance as well as high strength, are described separately.

Corrosion-Resistant Nickel Alloys. A list of selected corrosion-resistant nickel alloys with nominal values of mechanical properties is given in Table 11. The commercially pure nickel grades, Nickel 200 to 205, are highly resistant to many corrosive media, especially in reducing environments, but also in oxidizing environments where they can maintain the passive nickel oxide surface film. They are used in the chemical processing and electronics industries. They are hot worked at 650 to 1230 °C (1200 to 2250 °F), annealed at 700 to 925 °C (1300 to 1700 °F), and are hardened by cold working. For processed sheet, for example, the tensile properties in the annealed condition (460 MPa, or 67 ksi, tensile strength; 148 MPa, or 22 ksi, yield strength; and 47% elongation) can be increased by cold rolling up to 760 MPa (110 ksi) tensile strength, 635 MPa (92 ksi) yield strength, and 8% elongation. Because of its nominal 0.08% C content (0.15% max), Nickel alloy 200 (UNS No 2200) should not be used above 315 °C (600 °F), because embrittlement results from the precipitation of graphite in the temperature range 425 to 650 °C (800 to 1200 °F). The more widely used low-carbon alloy Nickel 201 (UNS No 2201), with 0.02% max C, can be used at temperatures above 290 °C (550 °F). Higher-purity nickel is commercially available for various electrical applications.

The low-alloy nickels contain 94% min Ni. The 5% Mn solid-solution addition in Nickel 211 protects against sulfur in service environments. As little as 0.005% S can cause liquid embrittlement at unalloyed nickel grain boundaries in the range between 640 and 740 °C (1185 and 1365 °F). Duranickel, alloy 301 (Ni-4.5Al-0.6Ti), offers the corrosion resistance of commercially pure nickel with the strengthening provided by the precipitation of γ′. There is sufficient alloying additions in alloy 301 to lower the Curie temperature, making the alloy weakly ferromagnetic at room temperature.

The nickel-copper alloys are strong and tough, offering corrosion resistance in various environments, including brine and sulfuric and other acids, and showing immunity to chloride-ion stress corrosion. They are used in chemical processing and pollution control equipment. Capable of precipitating γ′, $Ni_3(Al,Ti)$, with its 2.7Al-0.6Ti alloy addition, alloy K-500 adds an age-hardening component to the good solution strengthening and work-hardening characteristics already available with the nominal 30% Cu in alloy 400. The composition of these alloys can be adjusted to decrease the Curie temperature to below room temperature.

The Ni-Cr-Fe(-Mo) alloys might simply be thought of as nickel-base analogs of the iron-base austenitic stainless steel alloys, with an interchange of the iron and nickel contents. In these commercially important alloys the chromium content in general ranges from 14 to 30% and iron from 3 to 20%. With a well-maintained Cr_2O_3 surface film, these alloys offer excellent corrosion resistance in many severe environments, showing immunity to chloride-ion stress-corrosion cracking. They also offer good oxidation and sulfidation resistance with good strength at elevated temperatures. These nickel-rich Ni-Cr-Fe alloys have maximum operating temperatures in the neighborhood of 1200 °C (2200 °F). Alloy 600 (UNS N06600, with Ni-15Cr-8Fe) is a single-phase alloy that can be used at temperatures from cryogenic to 1093 °C (2000 °F). The modest yield strength of strip in the annealed condition (207 to 310 MPa, or 30 to 45 ksi) can be readily work hardened by cold rolling to reach yield strengths of 827 to 1100 MPa (120 to 160 ksi) and can retain most of this strength up to about 540 °C (1000 °F).

The Ni-Cr-(Fe)-Mo alloys consist of a large family of alloys that are used in the chemical processing, pollution control, and waste treatment industries to utilize their excellent heat and corrosion resistance. Alloys in this commercially important family, such as C-276 and alloy 625, are made even more versatile by their excellent welding characteristics and the corrosion resistance of welded structures. The molybdenum additions to these alloys improve resistance to pitting and crevice corrosion. Aluminum improves the protective surface oxide film, and the carbide formers titanium and niobium are used to stabilize the alloys against chromium-carbide sensitization. Even with the low-level additions of aluminum and titanium to alloy 800, for example, small amounts of γ′ can form in service during exposure to elevated temperatures. The high molybdenum and silicon additions in Hastelloy B and D promote good corrosion resistance in the presence of hydrochloric and sulfuric acids.

Table 11 Compositions and properties of selected corrosion-resistant nickel-base alloys

Alloy	Nominal composition, wt %	Ultimate tensile strength, MPa	Ultimate tensile strength, ksi	Yield strength (0.2% offset), MPa	Yield strength (0.2% offset), ksi	Elongation in 50 mm (2 in.), %	Rockwell hardness
Commercially pure and low-alloy nickels							
Nickel 200	99.0% Ni	462	67	148	21.5	47	109 HB
Nickel 201	99.0% Ni	403	58.5	103	15	50	129 HB
Nickel 211	Ni-4.75Mn-0.75Fe	530	77	240	35	40	...
Duranickel 301	Ni-4.5Al-0.5Ti	1170	170	862	125	25	30–40 HRC
Nickel-copper alloys							
Alloy 400	Ni-31Cu-2.5Fe	550	80	240	35	40	110-150 HB
Alloy K-500	Ni-30Cu-2Fe-1.5Mn-2.7Al-0.6Ti	1100	160	790	115	20	300 HB
Nickel-molybdenum and nickel-silicon alloys							
Hastelloy B	Ni-28Mo-5.5Fe-2.5Co						
Sheet		834	121	386	56	63	92 HRB
Investment cast		586	85	345	50	10	93 HRB
Hastelloy D	Ni-9.25Si-3Cu-1.5Co	793	115	...	...	...	30–39 HRC
Nickel-chromium-iron alloys							
Alloy 600	Ni-15Cr-8Fe	655	95	310	45	40	75 HRB
Alloy 800	Ni-21Cr-39.5Fe-0.4Ti-0.4Al	600	87	295	43	44	138 HB
Alloy 617	Ni-22Cr-3Fe-12Co-9Mo-1Al	755	110	350	51	58	173 HB
Alloy 690	Ni-29Cr-9Fe	725	105	348	50	41	88 HRB
Alloy 751	Ni-15Cr-7Fe-1Nb-2Ti	1310	190	976	142	22	352 HB
Nickel-chromium-molybdenum alloys							
Alloy C-276	Ni-15.5Cr-16Mo-5.5Fe-3.75W-1.25Co + V	785	114	372	54	62	209 HB
Alloy 625	Ni-21.5Cr-9Mo-3.65Nb + Ta-2.5Fe	930	135	517	75	42.5	190 HB
Nickel-chromium-iron-molybdenum-copper alloys							
Hasteloy G	Ni-22.25 Cr-19.5Fe-6.5Mo-2Cu + Co,Nb,Ta	690	100	320	47	50	79 HRB
Alloy 825	Ni-21.5Cr-30Fe-3Mo-2.25Cu + Al	690	100	310	45	45	...

Heat-Resistant Nickel Alloys. Chemical compositions of selected heat-resistant superalloys are given in Table 12. A glance at this list reveals that these nickel-containing materials include nickel-, iron-nickel-, or cobalt-base alloys. They can be made by wrought and P/M methods, and also with castings produced with carefully controlled conditions to provide the desired polycrystal, or elongated (directionally solidified), or single-crystal grain structure for improved elevated-temperature mechanical properties. The majority of the nickel-base superalloys utilize the combined strengthening of a solution-hardened austenite matrix with γ′ precipitation. The niobium-rich, age-hardening precipitate, γ″, offers the ease of heat treatment and weldability that has made alloy 718 the most important nickel-base superalloy for aerospace and nuclear structural applications. Alloy 718 is a high-strength, corrosion-resistant alloy that is used at temperatures from –250 to 700 °C (–423 to 1300 °F). Some of the alloys, Hastelloy X for example, obtain additional strengthening from carbide precipitation instead of γ′. Others, MA 754 for example, utilize P/M techniques involving mechanical alloying (Ref 27) to achieve a dispersion of about 1 vol% of very fine (25 nm) inert oxide particles, such as Y_2O_3 , to promote higher elevated-temperature tensile and stress-rupture strength.

The iron-base Fe-Ni-Cr heat-resistant alloys are extensions of the iron-base stainless steels with higher nickel and additions of other alloying elements. Retaining the fcc iron-nickel austenite matrix, these alloys (alloys A-286 and 901, for example) are workable into various wrought forms and are capable of precipitation hardening with γ′. Alloys 903 and 909 are controlled thermal expansion Fe-Ni-Co-base alloys that are capable of age hardening with $Ni_3(Nb,Ti)$ precipitation and are designed to have high strength and low coefficient of thermal expansion for applications in gas turbine rings and seals up to 650 °C (1200 °F) (Ref 26). These alloys are hot worked at about 870 to 1120 °C (1600 to 2050 °F) and solution heat treated at 815 to 980 °C (1500 to 1800 °F). The standard aging treatment consists of 720 °C (1325 °F) for 8 h, furnace cool at 55 °C (100 °F)/h to 620 °C (1150 °F) for 8 h, followed by air cooling. Alloy 909 in the as-hardened condition, for example, retains much of its room-temperature yield strength (1070 MPa, or 155 ksi) at 540 °C (1000 °F), namely, 895 MPa (130 ksi) (Ref 30).

Specialty Nickel Alloys. Unique combinations of properties are available with other nickel-base alloys for special applications. While some of these properties are also available to some extent with alloys described above, the alloys described below were developed to promote their rather unique properties.

There are many electrical resistance alloys used for resistance heating elements. They can contain 35 to 95% Ni, but invariably contain

Table 12 Chemical compositions of selected superalloys

Alloy (a)	Composition, wt %													
	Ni	Cr	Fe	Co	Mo	W	Nb	Ti	Al	C	Mn	Si	B	Other
Wrought alloys														
Waspaloy	58.0	19.5	...	13.5	4.3	...	...	3.0	1.3	0.08	...	...	0.006	...
Udimet 700	55.0	15.0	...	17.0	5.0	...	...	3.5	4.0	0.06	...	...	0.030	...
Hastelloy X	47.0	22.0	18.5	1.5	9.0	0.6	...	...	...	0.10	0.5	0.5	...	...
A-286	26.0	15.0	54.0	...	1.3	...	...	2.0	0.2	0.05	1.3	0.5	0.015	...
Incoloy 901	42.5	12.5	36.0	...	5.7	...	...	2.8	0.2	0.05	0.1	0.1	0.015	...
Inconel 718	52.5	19.0	18.5	...	3.0	...	5.1	0.9	0.5	0.04	0.2	0.2	...	...
Incoloy 903	38.0	...	42.0	15.0	...	...	3.0	1.4	0.9	...	0.8	1.0	...	...
Incoloy 909	38.0	...	42.0	13.0	...	...	4.7	1.5	0.03	...	...	0.4	...	...
Haynes 188	22.0	22.0	3.0	bal	...	14.0	...	...	...	0.10	1.25	0.35	...	...
Haynes 25 (L-605)	10.0	20.0	3.0	bal	...	15.0	...	...	...	0.10	1.5	1.0	...	0.05 La
Cast alloys														
MAR-M-200(P)	bal	9.0	...	10.0	...	12.5	1.8	2.0	5.0	0.15	...	...	0.015	0.05 Zr
B-1900 (P)	bal	8.0	...	10.0	6.0	...	...	1.0	6.0	0.10	...	...	0.015	4.3Ta, 0.08 Zr
Nimocast 90 (P)	bal	19.5	1.5	18.0	...	...	...	2.4	1.4	0.06	...	...	...	...
MAR-M-509 (P)	10.0	24.0	1.0	bal	...	7.0	...	0.2	...	0.60	...	...	...	7.5 Ta
FSX-414 (P)	10.5	29.5	2.0	bal	...	7.0	...	...	...	0.25	...	...	0.012	...
MAR-M-247 (DS)	bal	8.0	...	10.0	0.6	10.0	...	1.0	5.5	0.15	...	...	0.015	3.0 Ta, 0.03 Zr, 1.5 Hf
CM 247 LC (DS)	bal	8.0	...	9.0	0.5	10.0	...	0.7	5.6	0.07	...	...	0.015	3.2 Ta, 0.010 Zr, 1.4 Hf
PWA 1484 (SC)	bal	5.0	...	10.0	2.0	6.0	...	...	5.6	...	...	...	...	9.0 Ta, 3.0 Re, 0.1 Hf
CMSX-4	bal	6.0	...	9.0	0.6	6.0	...	1.0	5.6	...	...	...	...	7 Ta, 3 Re, 0.1 Hf
Powder metallurgy alloys														
MA754	bal	20.0	...	...	...	...	...	0.5	0.3	0.05	...	...	...	0.6 Y_2O_3
MA 6000	bal	15.0	...	2.0	2.0	4.0	...	2.5	4.5	0.05	...	...	0.1	2Ta
MERL 76	bal	12.2	...	18.2	3.2	...	1.3	4.3	5.0	0.025	...	...	0.02	0.3 Hf, 0.06 Zr
Rene′ 95	bal	12.8	...	8.1	3.6	3.6	3.6	2.6	3.6	0.08	...	...	0.01	0.053 Zr

(a) P, polycrystalline casting; DS, directionally solidified casting; SC, single-crystal casting; bal, balance

greater than 15% Cr to form an adherent surface oxide to protect against oxidation and carburization at temperatures up to 1000 to 1200 °C (1850 to 2200 °F) in air. Examples are Ni-20Cr (UNS N06003), Ni-15Cr-25Fe (UNS N06004), and Ni-20Cr-3Al-3Fe. These alloys are single-phase austenite and have the needed properties for heating elements: desirably high, reproducible electrical resistance; low thermal expansion to minimize thermal fatigue and shock; good creep strength; strong and ductile for fabrication (Ref 27, 28).

The ferromagnetic characteristics of nickel allow formulation of nickel-base alloys for corrosion-resistant soft magnets for a variety of applications, typified by Ni-5Mo-16Fe. Low thermal expansion characteristics are shown by Fe-(36-52)Ni-(0-17)Co alloys, making these materials useful for glass-to-metal sealing and containment equipment for liquefied natural gas, for example. The controlled thermal expansion alloys, typified by alloy 903 (Ni-42Fe-15Co + Nb,Al,Ti), are also γ'-precipitation hardenable, offering high strength and low, relatively constant thermal expansion coefficient for applications up to about 650 °C (1200 °F). With nearly 50-50 at.%, nickel forms a shape memory intermetallic alloy with titanium, which offers 8% of reversible strain via a thermoelastic martensitic transformation, along with good ductility and corrosion resistance.

The $L1_2$ intermetallic compound, Ni_3Al, has been the focus of development work to create a strong, corrosion-resistant material for elevated-temperature applications. Wrought and cast beryllium-nickel alloys are commercially available (UNS N03360 with Ni-2Be-0.5Ti, for example) and respond to processing and age-hardening heat treatments as readily as the beryllium-copper alloys, but offer higher strength with better resistance to thermal softening and stress relaxation (Ref 25, 29–32).

Cobalt and Cobalt Alloys

Falling between iron and nickel in the periodic table, cobalt has many similar properties as these other two more familiar transition metals (Ref 18, 33). Its melting temperature (1493 °C, or 2719 °F), density (8.85 g/cm^3, or 0.322 lb/in.^{3}), thermal expansion coefficient (13.8/K, or 7.66/°F), thermal conductivity (69.0 W/m · K), and elastic modulus (210 GPa, or 30×10^6 psi) are all rather similar to the respective values of iron and nickel. All three are ferromagnetic, but the Curie temperature of cobalt, 1123 °C (2050 °F), is significantly higher than that of iron (770 °C, or 1418 °F) or nickel (358 °C, or 676 °F). The crystal structure and chemical and mechanical properties differ enough to give cobalt a viable commercial life of its own.

Cobalt is not as readily geologically available as its two companion metals are, making cobalt about 100 times more costly than iron and about 8 times more costly than nickel. Cobalt is heavily used as an alloying element in nickel-base heat-resistant alloys and as the ductile binder phase (3 to 25%) for tungsten carbide particles in cemented carbides. Nonetheless, the excellent wear resistance, elevated-temperature hardness, and corrosion resistance of cobalt alloys have been commercially utilized in gas turbine engines, earthmoving equipment, and as bearing materials. Cobalt alloys are most often used for their wear resistance either in solid forms or as welded or thermally sprayed overlays to hardface other structural materials. They are also available as P/M products and in wrought sheet, bar, and tube forms.

Elemental cobalt has an hcp crystal structure (ε-Co) at room temperature, transforming to fcc (α-Co) at 417 °C (783 °F). Although the principal alloying elements affect the temperature of this transition (chromium, tungsten, and molybdenum stabilize the hcp phase, and iron and nickel stabilize the fcc structure), the fcc-to-hcp transformation is notably sluggish especially in alloyed cobalt. The alloys usually go into service at room temperature in the metastable fcc form. The α-to-ε transformation usually occurs by the strain-induced martensitic (or shear) reaction, which also contributes to the high work-hardening rates generally seen with cobalt alloys. Carbon, one of the principal alloying elements, has a profound influence on hardness, elevated-temperature strength, and creep resistance, as well as resistance to abrasive wear through formation of carbide phases.

Effects of Alloying Elements in Cobalt Alloys

The traditional cobalt alloys are known as Stellite alloys and were originally formulated around 1900 for their wear and hardness properties. They are composed of significant amounts of chromium, tungsten, molybdenum, iron, nickel, and carbon. The ferromagnetism of cobalt is suppressed by the heavy additions of chromium and

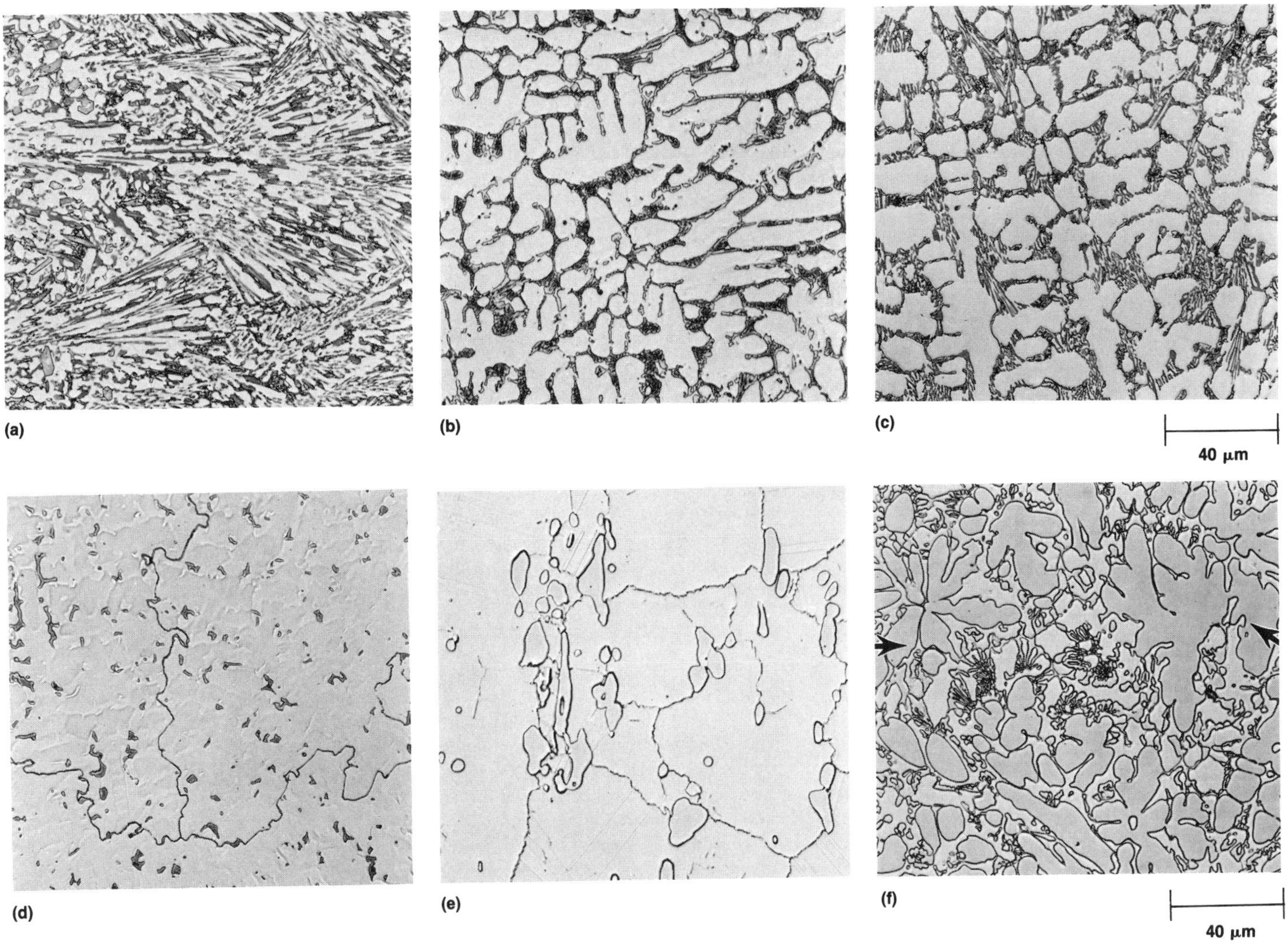

Fig. 11 Microstructures of various cobalt-base wear-resistant alloys. (a) Stellite 1, two-layer GTAW deposit. (b) Stellite 6, two-layer GTAW deposit. (c) Stellite 12, two-layer GTAW deposit. (d) Stellite 21, two-layer GTAW deposit. (e) Haynes alloy 6B, 13 mm (0.5 in.) plate. (f) Tribaloy alloy (T-800) showing the Laves precipitates (the largest continuous precipitate, some of which are indicated with arrows). All 500×

retained as expected by additions of iron and nickel. Chromium is restricted to less than 25% to avoid precipitation of the chromium-rich σ-phase. The microstructures consist of hard complex alloy carbides (M_7C_3, M_6C, and $M_{23}C_6$) in a tough solid-solution alloy matrix. The morphology and volume fraction of the carbide phase are a function of alloy composition and fabrication method. Cobalt-base alloys can be grouped into their three major property application areas: wear resistance, elevated-temperature strength, and corrosion resistance. A representative sampling of the many commercially available cobalt-base alloys is listed in Table 13. In most applications, the cobalt alloy must also exhibit good properties in all three general property areas.

Wear-Resistant Cobalt Alloys

Cobalt alloys are used mostly for their excellent wear resistance (Ref 33, 34). The Co-Cr-W-C Stellite wear-resistant alloys contain generous amounts of carbide-forming elements chromium, tungsten, and molybdenum, with carbon varying from 0.1 to 3.3% to encourage the formation of hard carbide particles. The microstructures generally consist of chromium-rich M_7C_3 carbides and, in the higher tungsten alloys, the tungsten-rich M_6C type of carbide, with enough chromium and tungsten alloying element left in solid solution to provide a tough, solution-hardened matrix. Stellite 6, for example, was reported (Ref 34) to contain about 12% Cr_7C_3 in a solid-solution matrix of about 58Co-18Cr-4W. The somewhat richer alloy Stellite 1 contains about 27% M_7C_3 and 1.5% W_6C, in a matrix of 45Co-11Cr-10W-0.3C. In general, alloys with more than about 1.3% C can be made only by casting because of limited ductility. These higher-carbon alloys are applied as hardfacings onto other structural metals using weld overlays or spray coatings. The solidification microstructures (Fig. 11), resulting from multipass weld overlays using the gas tungsten arc welding (GTAW) process, for example, contain arrays of eutectic carbides (dark etching phases) between alloy-rich dendrites (white etching phases in Fig. 11a to d).

Haynes alloy 6B is available in plate, sheet, and bar wrought forms, with a more homogeneous microstructure, containing globular carbides in a worked or an annealed matrix (Fig. 11e). In wrought form, alloy 6B offers tensile properties of 619 MPa (90 ksi) yield strength, 998 MPa (145 ksi) tensile strength, and 11% elongation. On the other hand, the commercial alloy Tribaloy T-800 contains large additions of molybdenum and silicon to produce a large volume fraction (about 50%) of a hard, corrosion-resistant Laves phase, an intermetallic compound of the type $MgZn_2$ (Fig. 11f). Depending on the cobalt alloy, this phase can have a complex composition beyond a simple AB_2 stoichiometry (CoSiMo, or Co_2W). Although the massive amounts of hard second-phase particles limit room-temperature ductility in the Tribaloy alloys, they are readily formed as thermally sprayed or as P/M components to provide exceptionally good wear resistance.

Abrasion Resistance. The characteristics of both the carbide and the matrix give cobalt alloys their good resistance to the various types of wear encountered in service. The hard second-phase carbide particles in cobalt alloys provide the resistance to abrasive wear whether of the high-stress type, where the abrasive medium is crushed after entrapment between metallic surfaces, or of the low-stress type, where the moving surfaces come into contact with packed abrasive particles such as soil and sand. The resistance to the cutting/plowing abrasive wear mechanisms shown by cobalt-base alloys generally depends on the hardness of the solid-solution matrix as well as the hardness, volume fraction, and distribution of the carbide or Laves phases.

Sliding Wear Resistance. Sliding wear, where two surfaces are forced together and move relative to one another especially without lubricant, consists of a complex set of mechanisms that can occur singly or in combination. The three mechanisms involve, first, generation of localized high temperatures (at low contact forces) to form adherent oxide glazes that can reduce wear rates or, by spalling of the oxide glaze, create oxide debris that combines abrasion with sliding wear, called fretting. The second type is related to high stress metal-to-metal contact (at high contact forces) resulting in cold welding and fracture of small surface pieces, transferring them from one surface to another (called galling). The third mechanism is subsurface initiation and growth of fatigue cracks. The solid-solution matrix of cobalt-base alloys gives the good resistance to sliding wear through control of the oxidation behavior, the inherent resistance to deformation, fatigue, and fracture at elevated temperatures, independent of the presence of hard second-phase particles.

Erosion Resistance. Cobalt alloys are resistant to the several types of erosive wear. The ductility and toughness of the solid-solution matrix are important in resisting the cutting/plowing action of solid-particle erosion, that is, impingement of small, solid particles carried by a gas or by a liquid in a slurry against the metal surface. Resistance to liquid-droplet erosion and cavitation erosion, however, requires the ability to absorb shock (stress) waves without microstructural fracture. While liquid-droplet and cavitation erosion have the same effect on the surface, they are caused by different mechanisms. Both result in a succession of shock (or stress) waves induced by liquid drops hitting the surface. In cavitation erosion the metal surface is in contact with a liquid undergoing pressure changes. The collapse of bubbles, momentarily formed by localized low-pressure areas or by evaporated pockets of the liquid, causes surface damage by the impact of liquid jets from the bubble implosion. The good resistance to cavitation erosion shown by cobalt alloys is attributable to its tough matrix having the ability to absorb the shock waves through work hardening and the energy-absorbing, strain-induced α to ε martensitic phase transformation. The important role that the strain-induced phase transformation plays in the good erosion wear resistance shown by cobalt alloys is indicated by the detection of an increase in ε-cobalt on the alloy surface and the identification of this phase in the wear debris in wear test experiments with Stellite alloy 6.

Heat-Resistant Cobalt Alloys

The high-temperature resistance of cobalt alloys is due to the stability of both the arrays of carbide particles and the matrix. The hardness and creep rupture strengths are stable to about 800 °C (1472 °F). The carbide particles pin grain boundaries to suppress grain growth (hardness) and grain-boundary sliding (creep resistance). Moreover, the room-temperature hardness is essentially unchanged after exposure to relatively high elevated temperatures. Cobalt alloys are used in aircraft gas turbines because they show good elevated-temperature strength, and resistance to thermal fatigue and thermal shock as well as good oxidation resistance. The alloys also have the good resistance to sulfidation needed in land-base turbines that use lower-quality fuels.

Both Haynes alloy 25 (L605) and alloy 188 are available in wrought forms as sheet, plate, bar, or pipe. In either alloy, the chromium and tungsten provide solution hardening and nickel stabilizes the α-phase for improved ductility. Both alloys work harden rapidly and offer a small aging response, with good ductility shown after the aging/stabilization heat treatment. For example, the strength of alloy 25 in the annealed condition—476 MPa (69 ksi) yield strength, 917 MPa (133 ksi) tensile strength, 41% elongation—will increase after a 30% cold-working step to 1000 MPa (145 ksi) yield strength, 1345 MPa (195 ksi) tensile strength, 16% elongation. When quenched from its annealing temperature (1175 to 1230 °C, or 2150 to 2250 °F), the alloy shows a small increase in hardness accompanied by a small decrease in ductility when aged or exposed to 480 to 650 °C (900 to 1200 °F). Alloy 25 retains good strength at elevated temperatures. At 982 °C (1800 °F), for example, the annealed temper offers 165 MPa (24 ksi) yield strength, 255 MPa (37 ksi) tensile strength, and 72% elongation. In Haynes 188, the carbides are precipitated during the aging treatment in the range 650 to 1175 °C (1200 to 2150 °F), with M_6C forming at the higher aging temperature and $M_{23}C_6$ at the lower end of the range. A small lanthanum addition gives this alloy its excellent oxidation resistance by its making the oxide scale more adherent and more impermeable to diffusion and further growth.

MAR-M Alloy 509 is capable of being cast into complex shapes as an investment casting alloy that offers high strength, high creep strength, as well as good resistance to oxidation, sulfidation, and thermal shock. This alloy is used in the cast condition at temperatures up to about 955 °C (1750 °F). In this type of alloy, chromium and tungsten are the main solid solution strengthening elements, nickel stabilizes the fcc phase to improve ductility, and the tantalum, titanium, zir-

Table 13 Nominal compositions of selected cobalt-base alloys

Alloy tradename	Nominal composition, %									
	Co	Cr	W	Mo	C	Fe	Ni	Si	Mn	Others
Cobalt-base wear-resistant alloys										
Stellite 1	bal	31	12.5	1 (max)	2.4	3 (max)	3 (max)	2 (max)	1 (max)	...
Stellite 6	bal	28	4.5	1 (max)	1.2	3 (max)	3 (max)	2 (max)	1 (max)	...
Stellite 12	bal	30	8.3	1 (max)	1.4	3 (max)	3 (max)	2 (max)	1 (max)	...
Stellite 21	bal	28	...	5.5	0.25	2 (max)	2.5	2 (max)	1 (max)	...
Haynes alloy 6B	bal	30	4	1	1.1	3 (max)	2.5	0.7	1.5	...
Tribaloy T-800	bal	17.5	...	29	0.08 (max)	...	...	3.5	...	...
Stellite F	bal	25	12.3	1 (max)	1.75	3 (max)	22	2 (max)	1 (max)	...
Stellite 4	bal	30	14.0	1 (max)	0.57	3 (max)	3 (max)	2 (max)	1 (max)	...
Stellite 190	bal	26	14.5	1 (max)	3.3	3 (max)	3 (max)	2 (max)	1 (max)	...
Stellite 306	bal	25	2.0	...	0.4	...	5	...	...	6 Nb
Stellite 6K	bal	31	4.5	1.5 (max)	1.6	3 (max)	3 (max)	2 (max)	2 (max)	...
Cobalt-base high-temperature alloys										
Haynes alloy 25 (L605)	bal	20	15	...	0.10	3 (max)	10	1 (max)	1.5	...
Haynes alloy 188	bal	22	14	...	0.10	3 (max)	22	0.35	1.25	0.05 La
MAR-M alloy 509	bal	22.5	7	...	0.60	1.5 (max)	10	0.4 (max)	0.1 (max)	3.5 Ta, 0.2 Ti, 0.5 Zr
Cobalt-base corrosion-resistant alloys										
MP35N, Multiphase alloy	bal	20	...	10	...	...	35	...	...	...
Haynes alloy 1233	bal	25.5	2	5	0.08 (max)	3	9	...	...	0.1N (max)

bal, balance

conium additions form MC type carbides. Its room-temperature tensile properties are typically 586 MPa (85 ksi) yield strength, 780 MPa (113 ksi) tensile strength, and 4% elongation. Good strength is retained even at temperatures as high as 982 °C (1800 °F), where typical tensile properties are 190 MPa (28 ksi) yield strength, 248 MPa (36 ksi) tensile strength, and 26% elongation.

Corrosion-Resistant Cobalt Alloys

Several cobalt-chromium-nickel-molybdenum alloys have been developed for providing excellent corrosion resistance along with high strength and toughness. In these alloys, the carbon is minimized to avoid carbide precipitation along grain boundaries, maximizing corrosion resistance. The presence of chromium in cobalt alloys permits them to passivate by forming adherent films of Cr_2O_3 as do stainless steels. These alloys are provided in various wrought forms, in the work-hardened or work-plus-age-hardened condition. The alloys work harden rapidly due to the strain-induced transformation that provides a dispersion of fine hcp platelets. The tensile strength of the annealed condition (937 MPa, or 130 ksi) can be increased to 1585 MPa (230 ksi), with about 50% cold reduction and raised still further (2000 MPa, or 290 ksi) after a subsequent aging treatment at about 530 to 593 °C (1000 to 1100 °F) for 4 h. In this aged condition the alloy retains favorable levels of ductility and toughness, showing around 10% elongation and 46% reduction in area. Good strength and ductility are available at elevated temperatures as well.

Because of their biocompatibility, nonmagnetic cobalt-chromium-molybdenum alloys, such as Co-28.5Cr-6Mo (ASTM F 75 and F 799), have been used for orthopedic implants. When made by investment casting, the alloys are strengthened by carbide particles. When made by forging or hot isostatic pressing, the alloys are hardened by a combination of nitrogen and carbide strengthening mechanisms. In addition to the excellent corrosion resistance for these applications, these alloys also have the required high-cycle fatigue resistance as well as strength, ductility, and wear resistance.

Titanium and Titanium Alloys

The good strength, low density (4.5 g/cm^3, or 0.16 lb/in.^{3}), relatively high melting point (1668 °C, or 3034 °F), excellent corrosion resistance, and good heat-transfer properties of titanium and its alloys have made them attractive to structural designers for use primarily in aerospace and chemical industries where the combination of unique properties can justify the cost (Ref 18, 35–39). Titanium is also seeing increased use for sporting goods, such as golf clubs. The alloys are available as castings, P/M products, and in basically all wrought plate, sheet, tube, forging, bar, and wire forms. Titanium exists in two crystallographic forms: the room-temperature hcp α-phase transforms to the bcc β-phase at 883 °C (1620 °F). Manipulating the morphology of these two allotropic phases through alloy additions and thermomechanical processing is commonly done to provide a wide range of useful mechanical property combinations. The coefficient of thermal expansion, 8.41 μm/m · K, is lower than that of steel or aluminum, while its elastic modulus (110 GPa, or 16.2 × 10^6 psi), falls between the values of steel and aluminum. Titanium can be used for cryogenic applications because it has no debilitating ductile-brittle transition temperature (DBTT), although decreases in toughness below room temperature can be seen in β alloys. The maximum useful temperature range for structural applications is 425 to 580 °C (800 to 1100 °F), depending on the alloy and condition.

Titanium readily forms a stable adherent oxide layer, TiO_2, that passivates the metal and provides resistance to attack by most mineral acids and chlorides. Titanium is nontoxic and resists human body fluids, making it biocompatible for use in the biomaterials field. In an anhydrous or nonoxygen environment, however, the oxide film will not re-form if damaged, making titanium susceptible to crevice corrosion. Titanium is also susceptible to hydrogen embrittlement, through its ability to form hydride by absorbing hydrogen in pickling solutions at room temperature or from reducing atmospheres at elevated temperatures. Because of its excellent corrosion resistance and good heat-transfer characteristics, unalloyed (commercially pure) titanium is used in heat exchangers, condensers, jet engine shrouds, submarine components, reactor vessels, and storage tanks.

Titanium is made by double, or triple melting in vacuum electric arc furnaces, with the ingots from the first melt used as consumable electrodes for subsequent melts. In addition to homogenization, the purpose of this melt practice is to reduce the number of hard, embrittling oxygen or nitrogen-rich inclusions and to remove volatile elements especially hydrogen. The impurity elements carbon, nitrogen, silicon, iron, and oxygen raise strength and lower ductility. In particular, the iron and oxygen impurity content control the strength of unalloyed titanium, while the carbon and nitrogen impurities must be kept to a minimum to avoid embrittlement. Extra-low-interstitial (ELI) grades are commercially available that maximize ductility and toughness.

Effects of Alloying Elements in Titanium Alloys

The relatively low strength of the commercially pure grades of titanium are readily improved with the addition of alloying elements in conjunction with the application of thermomechanical processing during fabrication into final products (Table 14). The alloying elements used in titanium are classified by their individual effects on the phase diagram: whether they stabi-

Table 14 Compositions and properties of selected wrought titanium and titanium alloys

	Tensile strength (min)		0.2% yield strength (min)		Impurity limits, wt%					Nominal composition, wt %				
Designation	MPa	ksi	MPa	ksi	N (max)	C (max)	H (max)	Fe (max)	O (max)	Al	Sn	Zr	Mo	Other
Unalloyed grades														
ASTM grade 1	240	35	170	25	0.03	0.10	0.015	0.20	0.18	...	...	...	...	...
ASTM grade 4	550	80	480	70	0.05	0.10	0.015	0.50	0.40	...	...	...	...	...
Alpha and near-alpha alloys														
Ti-5Al-2.5Sn	790	115	760	110	0.05	0.08	0.02	0.50	0.20	5	2.5	...	...	...
Ti-8Al-1Mo-1V	900	130	830	120	0.05	0.08	0.015	0.30	0.12	8	...	...	1	1V
Ti-2.25Al-11Sn-5Zr-1Mo	1000	145	900	130	0.04	0.04	0.008	0.12	0.17	2.25	11.0	5.0	1.0	0.2 Si
Ti-6Al-2Sn-4Zr-2Mo	900	130	830	120	0.05	0.05	0.0125	0.25	0.15	6.0	2.0	4.0	2.0	0.08 Si
Alpha-beta alloys														
Ti-6Al-4V(A)	900	130	830	120	0.05	0.10	0.0125	0.30	0.20	6.0	...	...	...	4.0V
Ti-6Al-2Sn-4Zr-6Mo(b)	1170	170	1100	160	0.04	0.04	0.0125	0.15	0.15	6.0	2.0	4.0	6.0	...
Ti-3Al-2.5V(c)	620	90	520	75	0.015	0.05	0.015	0.30	0.12	3.0	...	...	...	2.5V
Beta alloys														
Ti-3Al-8V-6Cr-4Mo-4Zr(a)	900	130	830	120	0.03	0.05	0.020	0.25	0.12	3.0	...	4.0	4.0	6.0 Cr, 8.0 V
Ti-15V-3Cr-3Al-3Sn(b)	1000	145	965	140	0.05	0.05	0.015	0.25	0.13	3.0	3.0	...	...	3.0 Cr, 15 V
Ti-10V-2Fe-3Al	1170	170	1100	160	0.05	0.05	0.015	2.2	0.13	3.0	...	...	...	10.0 V

(a) Mechanical properties given for annealed condition; may be solution treated and aged to increase strength. (b) Mechanical properties given for solution treated and aged condition; alloy not normally applied in annealed condition. Properties may be sensitive to section size and processing. (c) Primarily a tubing alloy; may be cold drawn to increase strength

Table 15 Ranges and effects of some alloying elements used in titanium

Alloying element	Range (approx), wt %	Effect on structure
Aluminum	3–8	α stabilizer
Tin	2–4	α stabilizer
Vanadium	2–15	β stabilizer
Molybdenum	2–15	β stabilizer
Chromium	2–12	β stabilizer
Copper	~2	β stabilizer
Zirconium	2–5	α and β strengthener
Silicon	0.05–0.5	Improves creep resistance

lize the α- or the β-titanium phase (Table 15). Titanium and its alloys are processed at temperatures both above and below the β transus, the temperature above which the alloy is 100% β, in order to manipulate the amount and morphology of the α and β phases during processing as well as the relative amounts retained in the final product. The base composition and the processing temperatures dictate the microstructural constitution of the alloys, which are classified according to the dominant phase in the alloy. Alpha alloys are predominately α, usually with minor amounts of β present. Alpha-beta alloys will obviously contain both phases, with more β than the α -alloys. The β content at room temperature may be as low as 5 to 10%, and greater. Beta alloys have sufficient β stabilizer content that the alloys can be solution treated above the β transus, water quenched, and retain 100% β. The commercial alloys are normally metastable and are aged to precipitate α to increase strength. The microstructure will thus consist of α in a β matrix.

Aluminum in solution hardens the α phase, but at levels >6 wt% it can form the embrittling intermetallic phase Ti_3Al. Zirconium and tin solution harden both the α and the β phases, and both elements retard the rates of phase transformation, permitting greater control of microstructures during heat treatment. Niobium is added to titanium to improve oxidation resistance at elevated temperatures. Chromium additions enhance corrosion resistance.

Titanium Alloy Microstructure/Property Relationships

Alpha Phase. The α and near-α titanium alloys are processed in the β or α/β phase fields at elevated temperatures to take advantage of the greater processibility of the bcc β phase than the α phase. Beta is strain-rate sensitive and is processed at low strain rates to reduce the resistance to deformation. Beta phase responds well to solutionizing (hardenability) and age hardening treatments. The alloys are processed in the β or the α/β two-phase field, cooled to room temperature, and heat treated to form controlled amounts and morphology of β-to-α phase transformation products. For example, when the near-α alloy, Ti-8Al-1Mo-1V, is forged at the appropriate elevated temperature within the α/β two-phase field, the resulting microstructure (Fig. 12b), consists of relatively fine equiaxed α grains retained from their equilibrium presence at the processing temperature (primary α) in a matrix comprising acicular α grains that transformed from the β grains that existed at the elevated temperature. It is this microstructure (Fig. 12b), that provides the best combination of strength and ductility for this alloy. When this alloy is hot forged at a temperature much lower in the α/β two-phase field, the microstructure at room temperature consists of nearly all α grains (Fig. 12a), resulting in lower strength than the alloy is capable of providing. This microstructure could also be achieved by processing at the same temperature as in Fig. 12(b), but with slow cooling from the processing (or solution-treatment) temperature. This microstructure in Fig. 12(a) has better toughness than that in Fig. 12(b). In contrast, hot forging of this alloy at a temperature in the all-β regime will result in the relatively coarse array of acicular α transformation products produced from a coarse grain β parent, Fig. 12(c). This structure provides the maximum fracture toughness, with a notable reduction in ductility and fatigue strength.

Alpha-Beta Structures. The α-β titanium alloys offer similar microstructures as the α alloys, but with a greater amount of retained β, which affects the properties. For example, when the commercially important α-β alloy, Ti-6Al-4V, is cooled from an elevated-temperature processing operation or heat treatment, the β phase can transform to the coarse acicular structure seen in Fig. 12(c) or to a fine acicular α structure via a diffusion-controlled nucleation and growth mechanism, or to needlelike α′ via a martensitic reaction as it cools below the martensite start temperature, M_s (Fig. 13). The coarseness of the transformed structure is controlled by cooling rate. The martensite reaction produces the α′ phase from the β by a diffusionless or shear movement of the product grain boundary, mechanistically similar to what occurs during the quenching of steels. The martensitic product has an hcp crystal structure and is in a supersaturated state with respect to the alloy content because it inherits the composition of the parent β. A second type of martensitic transformation product, the orthorhombic α″, can occur either upon cooling (athermal) or during cold working (strain induced). When aged, both metastable martensites precipitate equilibrium amounts of α + β.

The fine acicular α may be difficult to distinguish from the martensite as illustrated in Fig. 13. In general, acicular (or lamellar) α improves creep resistance, fracture toughness, and crack growth resistance, but at a sacrifice in ductility and fatigue (with a slight drop in strength). The coarser the transformed-β structure, the greater the improvement in toughness and crack-growth resistance, and the greater the loss in ductility and fatigue strength.

Typical microstructures representative of those most commonly used for α/β alloys are shown in Fig. 14. Proceeding from Fig. 14(a) to (d) will generally result in progressively decreased tensile and fatigue strengths, with increasing improvements in damage tolerance type properties. The difference in microstructure between 14(a) and (b) is due to the differences in processing history. The temperature during sheet

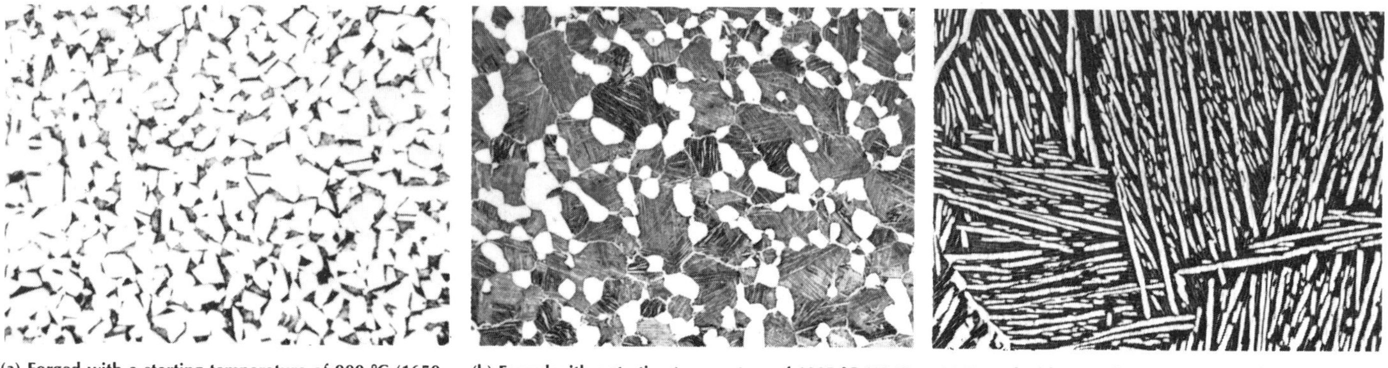

(a) Forged with a starting temperature of 900 °C (1650 °F), which is below the normal temperature range for forging Ti-8Al-1Mo-1V

(b) Forged with a starting temperature of 1005 °C (1840 °F), which is within the normal range, and air cooled

(c) Forged with a starting temperature of 1093 °C (2000 °F), which is above the beta transus temperature, and rapidly air cooled after finish forging

Fig. 12 Microstructures of near-α alloy Ti-8Al-1Mo-1V after forging with different starting temperatures. (a) Equiaxed α grains (light) in a matrix of α and β (dark). (b) Equiaxed grains of primary α (light) in a matrix of transformed β (dark) containing fine acicular α. (c) Transformed β containing coarse and fine acicular α (light). Etchant: Kroll's reagent (192). All micrographs at 250×

rolling decreases as rolling proceeds, and the final rolling temperature is significantly lower than the final forging temperature. Thus, there is less retained β at the final working temperature for the sheet, and a predominantly "globular"-type α microstructure (the black features in Fig. 14a are retained or transformed β). The final forging temperature is significantly higher, with more retained β, accounting for the higher amount of lamellar transformed β microstructure. The slow cooling of the recrystallized annealed structure permits the primary α to grow during cooling, consuming most of the β. Retained β is observed at some α-α boundaries and triple points. The solution-treated and aged condition is not commonly used for Ti-6Al-4V, but it is the standard heat treatment for aerospace fasteners.

Beta Phase. The β alloys are rich enough in β-stabilizing elements (Tables 14 and 15) that the beta phase can be retained at room temperature with appropriate cooling rates. The martensite reaction does not occur because the M_s is below room temperature. In this metastable condition, the retained β is hardenable by precipitation of α given the proper aging heat treatment. The family of β alloys are classified as "lean" and "rich" β alloys, a useful oversimplification in spite of the continuum of phase changes that occur with increasing alloy content (Fig. 15). The lean β alloys are prone to form the embrittling, metastable ω phase during cooling or aging, while the rich β alloys are stable enough to avoid formation of ω phase during processing. In practice, this is not a problem as these alloys are normally aged at a temperature above the ω solvus so α and β are the only phases present. However, this would preclude using β alloys in the solution-treated condition for elevated-temperature application below about 425 °C (800 °F) where ω precipitation will occur. Using aged β alloys below this temperature should not be a problem as the aging stabilizes the β sufficiently to prevent ω formation. The formation of ω phase can also be effectively suppressed by cold working the parent β phase before aging to encourage extensive heterogeneous nucleation of the desired equilibrium α phase.

Besides shaping the alloy, the thermomechanical working is designed to provide the microstructure for the optimal combination of properties required by the application. The morphology (shape/size/volume fraction) of the α phase determines structural properties. For the near-α and the α-β alloys, in general, the equiaxed α morphology promotes higher ductility and formability, higher strength, better low-cycle fatigue, and higher threshold stress for hot-salt stress corrosion compared to the acicular α morphology. However, when processed with a predominately acicular morphology, the alloy has superior creep resistance, higher fracture toughness, superior stress-corrosion resistance, and lower crack-propagation rates. A general comparison of the structural property combinations available with either the β or the α-β processing mode applied

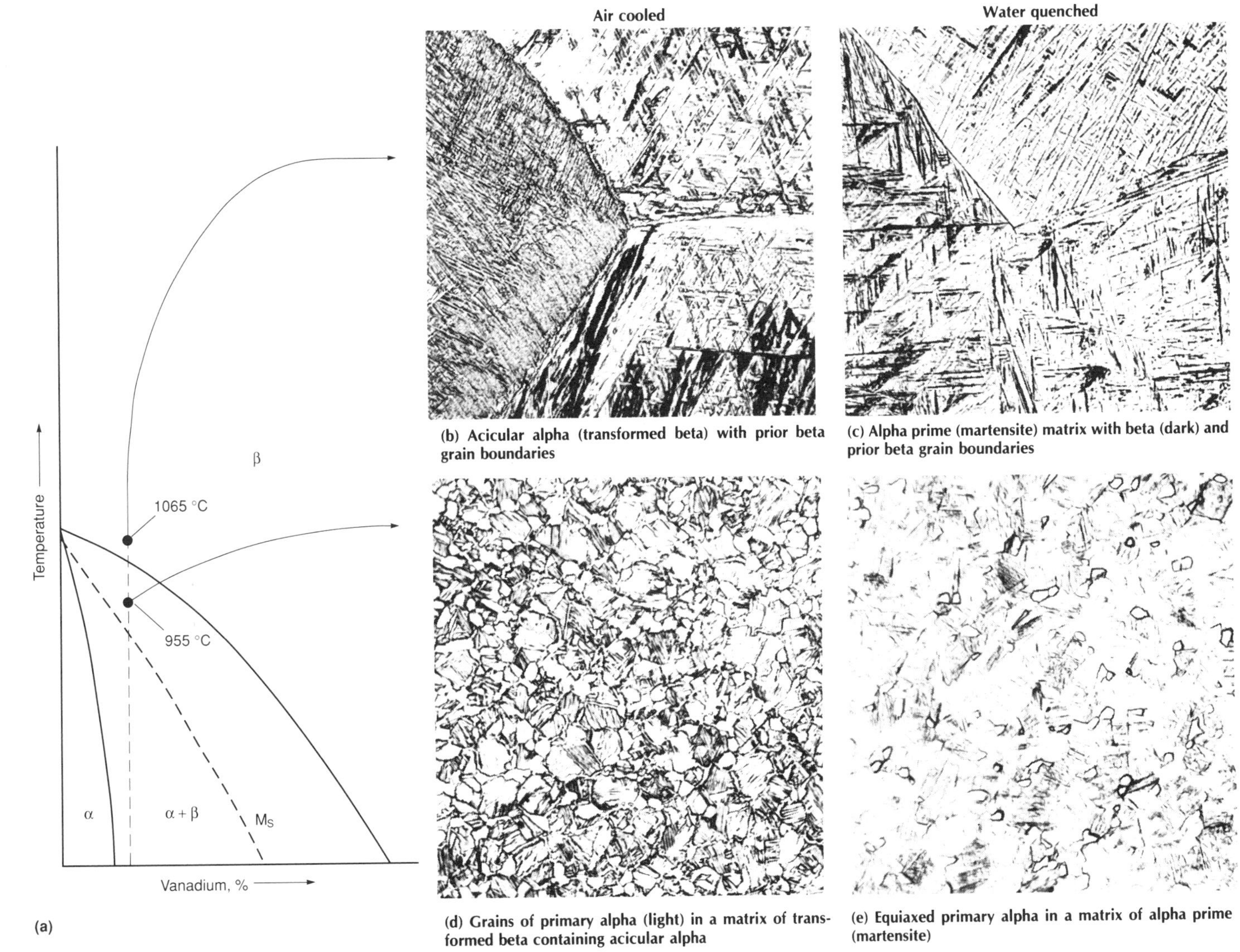

(b) Acicular alpha (transformed beta) with prior beta grain boundaries

(c) Alpha prime (martensite) matrix with beta (dark) and prior beta grain boundaries

(d) Grains of primary alpha (light) in a matrix of transformed beta containing acicular alpha

(e) Equiaxed primary alpha in a matrix of alpha prime (martensite)

Fig. 13 Microstructures of alloy Ti-6Al-4V after cooling from different areas of the phase field shown in (a). The specimens represented in micrograph (e) provided the best combination of strength and ductility after aging. Etchant: 10 HF, 5 HNO_3, 85 H_2O. All micrographs at 250×

to the α-β alloys are qualitatively described in Table 16.

Wrought Titanium Alloys

Unalloyed titanium is all α phase at room temperature. The acicular form is readily obtained as the principal β transformation product, but the equiaxed α morphology is obtainable only by applying a recrystallizing heat treatment to a cold-worked condition. The commercially pure grades have relatively low strength (Table 14), but with good control of impurities it has good toughness, offering impact toughness values equivalent to quenched-and-tempered steels. There is good creep resistance up to about 315 °C (600 °F). Because of the good corrosion resistance and the ability to maintain a clean surface in service, the base titanium metal/oxide system offers good heat-transfer properties making commercially pure grades particularly useful for heat-exchanger applications. Commercially pure grades are also available with minor alloy elements. For example, a small solid-solution addition of 0.2 Pd improves corrosion resistance still further. The alloy Ti-0.3Mo-0.8Ni (UNS R53400) offers higher strength with corrosion resistance between that of the unalloyed and the titanium-palladium grades.

Titanium and its alloys are processed at elevated temperatures to take advantage of the better forgeability of the bcc β phase. It also results in less crystallographic texturing and more uniform properties than does processing the hcp α phase. In the initial ingot breakdown operations, the metal is forged in the β field just above the β transus. However, as the metal is taken down through the secondary processing steps, such as

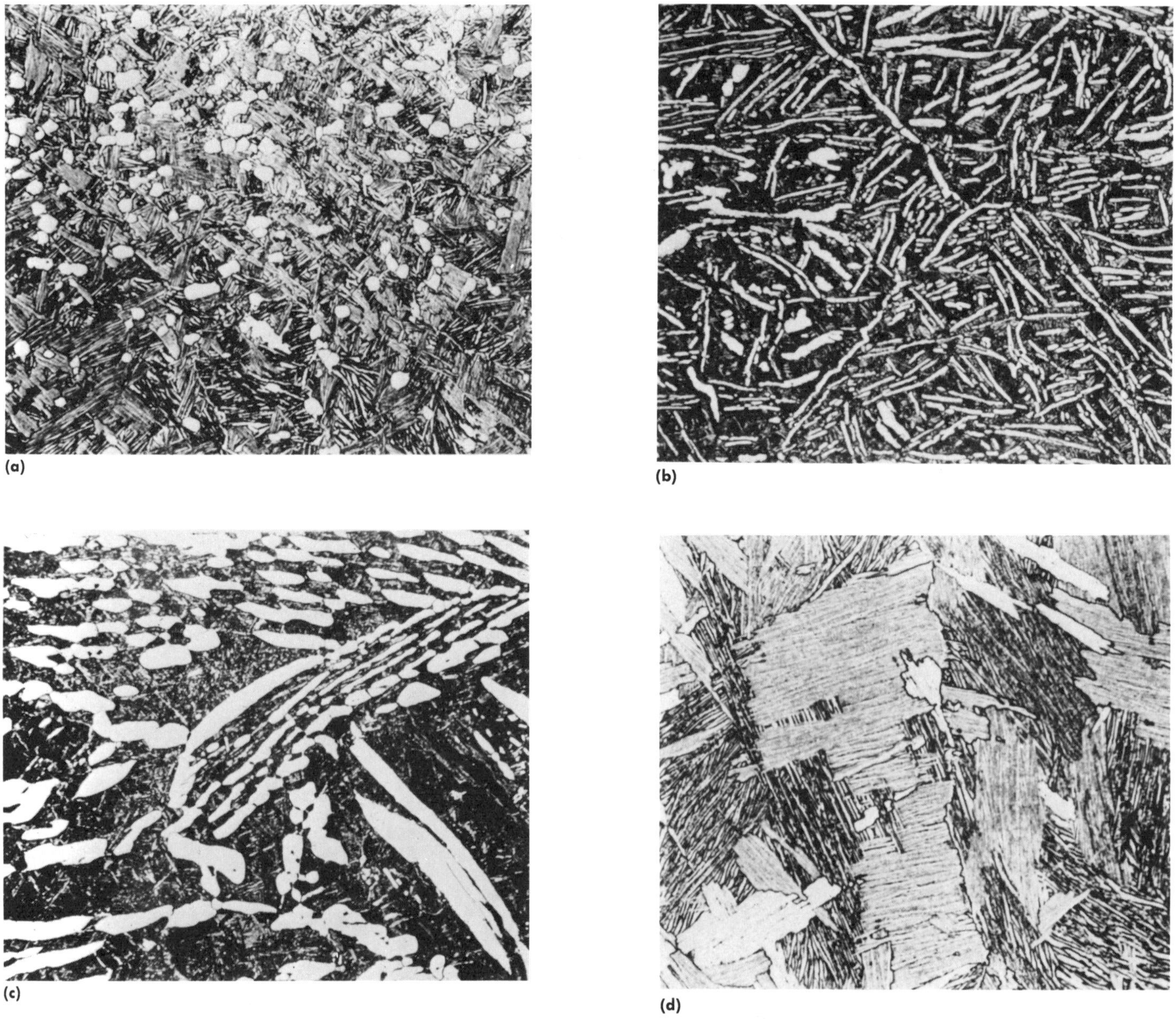

Fig. 14 Microstructures corresponding to different combinations of properties in Ti-6Al-4V forgings. (a) 6% equiaxed primary α plus fine platelet α in Ti-6Al-4V α-β forged, then annealed 2 h at 705 °C (1300 °F) and air cooled. (b) 23% elongated, partly broken up α plus grain-boundary α in Ti-6Al-4V, forged and water quenched, then annealed 2 h at 705 °C and air cooled. (c) 25% blocky (spaghetti) α plates plus very fine platelet α in Ti-6Al-4V forged from spaghetti-α starting structure, then solution treated 1 h at 955 °C (1750 °F) and reannealed 2 h at 705 °C. (d) 92% α basketweave structure in Ti-6Al-4V β forged and slow cooled, then annealed 2 h at 705 °C. Structures in (a) and (b) produced excellent combinations of tensile properties, fatigue strengths and fracture toughness. Structure in (c) produced very poor combinations of mechanical properties. Structure in (d) produced good fracture toughness, but poor tensile properties and fatigue resistance.

rolling, it is processed high in the α/β field just below the β transus to avoid excessive β grain growth.

Many titanium alloys are placed in service in heat-treated conditions. After being shaped into final form or before machining, the alloys are commonly given softening, stress-relieving, or solutionizing and aging thermal treatments. The alloys are commonly given stress-relief anneals without adversely affecting strength or ductility in order to relieve the stresses induced by prior thermomechanical processing and forming operations. However, care is taken to choose temperatures that are high enough to accomplish the stress relieving and low enough to avoid overaging (for the solution treated and aged condition) or strain aging, or to avoid recrystallizing of a cold-worked condition. Cooling rates from stress relieving or aging treatments must be sufficiently slow and uniform to avoid reintroducing thermal stresses. Aging treatments are carried out at a temperature in the range 425 to 650 °C (800 to 1200 °F) as appropriate for the specific alloy and desired properties.

Heating rates to the aging temperature of β alloys must be rapid enough to avoid precipitation of the ω phase. The α and α-β alloys are solution treated just below the β transus for optimal ductility, toughness, and creep strength of the aged condition. On the other hand, annealing or processing is done well below the β transus to encourage a fine equiaxed α in the final product for high strength, good ductility, and resistance to fatigue crack initiation.

The annealing treatments need to be carried out in inert atmosphere or in vacuum to avoid excessive oxidation of the surface. When an oxygen-enriched layer is formed it is called "α case", and it must be chemically or mechanically removed before being put into service. Titanium should not be heat treated in a hydrogen atmosphere; the hydrogen content is limited to a maximum of 125 to 150 ppm to avoid embrittlement, which manifests itself as reduced impact strength, low notched tensile strength, and delayed cracking.

Titanium alloys in general can be made to exhibit superplasticity under the right conditions, but the α-β alloys show the highest degree of this phenomenon. Superplasticity is exceptionally high ductility when processed or tensile tested with very fine α-β structures, with a primary α grain size of about 10 μm (400 μin.) at very high temperature and with controlled, (slow) strain rates. The α-β alloy, Ti-6Al-4V, for example, has been superplastically formed commercially into complex parts at 870 to 925 °C (1600 to 1700 °F), under an argon atmosphere, and at strain rates in the range 1.3×10^{-4} to 1.0×10^{-3}/s.

Alpha Alloys. The α-titanium alloys are slightly less corrosion resistant but offer higher strength than unalloyed titanium. The α alloy Ti-5Al-2.5Sn (UNS R54520), for example, shows good ductility at a tensile strength level of 790 MPa (115 ksi) (Table 14). However, outside of grain size control of strength, it cannot be strengthened by heat treatment. Its primary use is for cryogenic applications.

The near-α titanium alloys utilize a small amount of β-stabilizer elements to retain some β for additional microstructural and property control. These alloys, Ti-8Al-1Mo-1V (UNS R54810) and Ti-6Al-2Sn-4Zr-2Mo, for example, are processed high in the α/β field to restrict β grain growth. The microstructures can range from equiaxed to acicular α (Fig. 12a to c). The latter alloy is used primarily for elevated-temperature applications, up to 540 °C (1000 °F).

Alpha-Beta Alloys. The α-β titanium alloys contain one or more α with one or more β-stabilizing elements, offering increased strength and a wider range of properties, with the usual trade-off in strength versus ductility. These alloys can be strengthened by heat treatment or by thermomechanical processing to produce the wide variety of microstructures described above (Fig. 13 and 14). These alloys can be aged, with better aging response shown when the β is more rapidly cooled to room temperature. The most commonly used alloy in this family, Ti-6Al-4V (UNS R56400), can be processed to provide a wide range of tensile properties. The alloy exhibits yield strengths in the range of 830 to 970 MPa (120 to 140 ksi), tensile strengths of 900 to 1070 MPa (130 to 155 ksi), and elongations of about 10 to 15% in the annealed condition. In the solution treated and aged condition the properties range from 1000 to 1100 MPa (145 to 160 ksi) yield strength, 1070 to 1170 MPa (155 to 170 ksi) tensile strength, and elongations of 5 to 8%. The aged properties are strongly dependent on section thickness.

Beta Alloys. With the β titanium alloys, it is difficult to obtain a great variety of useful property combinations by processing, but they do offer better fracture toughness and better room-temperature forming characteristics than the α-β alloys. The β alloys also provide high strength where yield strength instead of creep strength is important. The beta alloys are more hardenable and better able to retain the desired as-quenched condition in heavier sections than the α-β alloys. Aging treatments are done at 450 to 650 °C (850 to 1200 °F) producing dispersions of fine alpha particles in the retained beta matrix. Yield strengths approaching 1380 MPa (200 ksi) are available with β alloys in the solution treated and aged condition. The β alloy Ti-3Al-8V-6Cr-4Mo-4Zr (UNS R58640) exhibits cold-drawn and aged tensile strengths of 1310 to 1450 MPa (190 to 210 ksi) with a minimum of 10% elongation for wire.

Titanium Casting Alloys

The titanium casting alloys comprise less than 2% of total mill products and have been based on traditional wrought alloy compositions, with Ti-6Al-4V being used for about 85% of the total. Casting permits the manufacture of increasingly complex parts and have been used primarily in pumps and valves for marine and chemical plant applications as well as for air frames, gas turbine engine components, and surgical implant prostheses. The cast microstructure invariably consists of β-transformation products formed during cooling of the casting: α platelets, grain boundary α, and α colonies. The inevitable shrinkage porosity seen in the as-cast microstructure has been successfully eliminated in commercial practice by applying subsequent hot isostatic pressing (HIP) operations. With the application of pressure under argon of 103 MPa (15 psi) at temperatures of 815 to 980 °C (1500 to 1800 °F), the shrinkage porosity collapses and the interfaces diffusion bond to provide a high-quality product. Stress-relief annealing is done at lower temperatures (730 to 845 °C, or 1350 to 1550 °F) and in vacuum to remove hydrogen and protect surfaces from oxidation.

Table 16 The effect of processing mode on the properties of α-β titanium alloys

Property	β processed	α/β processed
Tensile strength	Moderate	Good
Creep strength	Good	Poor
Fatigue strength	Moderate	Good
Fracture toughness	Good	Poor
Crack growth rate	Good	Moderate
Grain size	Large	Small

Industrial and marine applications for the cast alloys are mostly for corrosion resistance. The α-β alloy Ti-6Al-4V is used in the annealed condition with tensile properties of 855 MPa (124 ksi) yield strength, 930 MPa (135 ksi) tensile strength, and 12% elongation. The α-β alloy Ti-6Al-2Sn-4Zr-2Mo, making up of about 6% of casting products, is employed at higher operating temperature where its good creep resistance up to 500 to 600 °C (930 to 1110 °F) is needed. The β alloy Ti-15V-3Al-3Cr-3Sn, developed as a cold-formable, age-hardenable sheet alloy, is also heat treatable as a casting alloy, where it has tensile strengths of 1200 MPa (175 ksi).

The inherently large β grains resulting from casting (0.5 to 5 mm, or 0.02 to 0.2 in.) are beneficial for fracture toughness, creep resistance, and fatigue crack propagation resistance, but they are harmful to fatigue strength and tensile elongation. Heat treatments are applied to

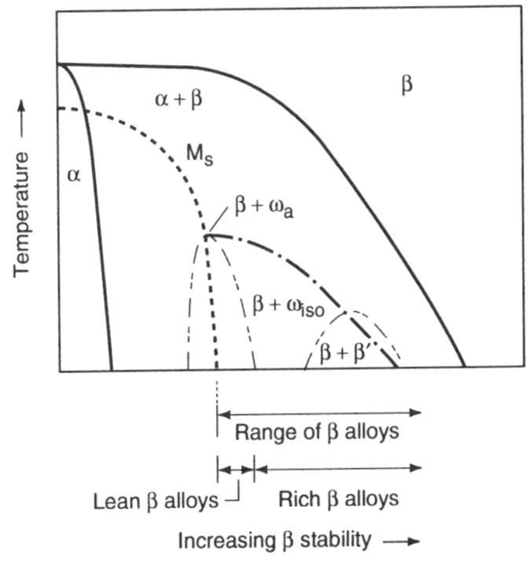

Fig. 15 Schematic phase diagram of a β–stabilized titanium system, indicating the compositional range that would be considered β alloys and the subdivision of this range into the lean and rich β alloys

minimize the amount of large boundary α for improved fatigue strength.

Titanium Alloy Powders

Titanium and its alloys, particularly the α-β alloy Ti-6Al-4V, have been fabricated by P/M techniques to take advantage of the net-shape manufacture and of the complex configurations that are possible with these methods. Blended elemental powder components are cold pressed and sintered to achieve densities on the order of 95 to 99% of theoretical and can be hot isostatically pressed to higher densities. Full densities are not achieved unless low Cl^- powder is used. Prealloyed powder is directly hot isostatically pressed to the desired shape. Unlike the P/M products of some other metals, titanium cannot tolerate the least amount of porosity lest it degrade fatigue and fracture properties. For this reason, contamination in the original powder must be avoided and hot isostatic pressing (or other hot processing) steps are used after sintering to ensure high-integrity P/M products.

When powdered fines from the sponge manufacturing method are used, the sintering is done in the β field at temperatures in the range 1150 to 1315 °C (2100 to 2400 °F). The resulting P/M product usually comprises relatively coarse α platelets in colonies of similar orientations. Prealloyed powder products are used for more demanding applications and processing to achieve 100% density is a requirement. Post-sintering thermal or thermomechanical treatments are applied to the compacts, and powder with a low Cl^- content must be used to ensure 100% density and to refine the α microstructure for improved strength and fatigue resistance.

Ti-6Al-4V products made by the prealloyed powder route offer respectable tensile properties for the pressed-and-sintered condition, nominally 847 MPa (123 ksi) yield strength, 930 MPa (135 ksi) tensile strength, and 14% elongation. When the sintered product is forged, the tensile properties are 951 MPa (138 ksi) yield strength, 1027 MPa (149 ksi) tensile strength, and 9% elongation, with improved fatigue properties. These properties are comparable with what is available with wrought products of similar microstructures. Beta alloys are also successfully made by P/M methods. After thermal treatment, the alloy Ti-8V-1.3Al-5Fe has shown yield and tensile strengths as high as 1461 MPa (212 ksi) and 1516 MPa (220 ksi), respectively, with 8% elongation. An additional benefit of P/M is its inherent effectiveness at providing a homogeneous composition; castings of a composition with such a high iron content as the latter alloy would have segregation problems that would degrade properties if made in cast or wrought products.

Zinc and Zinc Alloys

Table 17 Composition and properties of selected zinc alloys

UNS No.	Common name	Nominal composition, %	Tensile strength MPa	Tensile strength ksi	Elongation, %	Hardness, HB(a)
Die-casting alloys						
Z35541	No. 2, AC43A	Zn-4Al-2.5Cu-0.04Mg	359	52	7	100
Z33520	No. 3, AG40A	Zn-4Al-0.04 Mg	283	41	10	82
Z 35531	No. 5, AC41A	Zn-4Al-1Cu-0.05Mg	329	48	7	91
Z 33523	No. 7, AG40B	Zn-4Al-0.015Mg-0.012Ni	283	41	14	76
Z 35635	ZA-8	Zn-8Al-1Cu-0.025 Mg	374	54	8	103
Z 35630	ZA-12	Zn-11Al-1Cu-0.025Mg	404	58	5	100
Z 35840	ZA-27	Zn-27Al-2Cu-0.015Mg	426	62	2	119
Wrought alloys (hot-rolled condition)						
Z 21220	...	Zn-0.06Pb-0.06Cd	150–170	21–25	52–30	43
Z 44330	...	Zn-1Cu	170–210	24–30	50–35	52
Z 41320	...	Zn-0.8Cu-0.15Ti	221–290	32–42	38–21	61

(a) Test duration, 30 s

Zinc is the fourth most used industrial metal (after iron, aluminum, and copper). It is principally used in galvanized coatings and anodes for corrosion protection of iron and steels, zinc casting alloys (die, permanent, sand mold), copper (brasses), aluminum, and magnesium alloys, wrought zinc alloys, and zinc chemicals (zinc oxide) (Ref 18, 40, 41). In the United States during 1994, zinc usage was about 54% in galvanizing, 21% in zinc-base alloys, 15% in brass and bronze, and 10% in other uses. As a structural material zinc offers good strength at a moderate density (7.133 g/cm^3, or 0.258 lb/in.^{3}), and good electrical conductivity (28% IACS) and thermal conductivity (113 W/m · K). Its relatively low boiling point (906 °C, or 1663 °F) and its relatively low melting point (420 °C, or 788 °F), make it easy to purify by distillation and to cast into useful products. The high fluidity of the liquid contributes to its good castability, permitting the casting of thin-wall components. Wrought zinc and zinc alloys are available in flat-rolled, wire-drawn, and extruded and forged shapes. Zinc and its alloys are widely used in the automotive, hardware, home appliance, and electrical components industries.

Zinc and its alloys show excellent bearing and wear properties because of its self-lubricative surface characteristics resulting in a low coefficient of sliding friction and resistance to galling. Zinc also offers good spark resistance and good machining characteristics. Its excellent sound and vibration-damping properties, that increase with temperature, are similar to those of cast irons. Zinc can be soldered, but solder alloys based on lead, cadmium, or tin cannot be used, because these elements promote intergranular corrosion. In addition to its attractive physical and mechanical properties, zinc and its alloys have a relatively low cost and are environmentally friendly.

Because of its highly anodic position (second to magnesium) in the electromotive series of metals, zinc is widely used as a sacrificial anode for protection of iron and steels as galvanized coatings. Zinc structural components have good corrosion resistance provided impurity limits are not exceeded and provided that the component is a stand-alone or is otherwise designed properly to avoid galvanic corrosion by contact with other metals. The thin adherent surface film, zinc "patina" ($ZnCO_3 \cdot 3Zn(OH)_2$) readily forms in air, giving zinc alloys their resistance to atmospheric corrosion. However, this film is attacked by acids and does not stand up well in continuous contact with moisture. Zinc responds well to a wide variety of finishing and coating treatments, such as buffing, plating, chromating, painting, anodizing, plastic coating, and phosphating. Chromating and anodizing treatments also provide good corrosion protection.

Although adequate for its many room-temperature applications, commercially pure zinc has poor creep resistance at modestly elevated temperatures, which can be improved upon somewhat by alloying. With such a low melting temperature (420 °C, or 788 °F), one would expect high diffusion rates, natural aging reactions, and active creep mechanisms leading to microstructural and dimensional instability at and just above room temperature (homologous temperature of 0.43 T_m). The modulus of elasticity of zinc is listed as a range (70 to 140 GPa, or 10 to 20 × 10^6 psi) because there is no region of strict proportionality in the tensile stress-strain curve in polycrystalline zinc. For this reason, yield strength values are not listed for zinc or its leaner alloys. Alloying beyond 4% Al adds stiffening second phases to allow yield strength and the elastic modulus to be measured in standard tensile tests. But even so, tensile and other mechanical properties are affected by operating temperature and time. For hardness data, the test duration, usually 30 s, should also be specified along with test load and penetrator shape. The mechanical property and dimensional changes shown by a rapidly cooled alloy casting can be countered by a post solidification stabilization anneal, typically 100 °C (212 °F) for 3 to 6 h followed by air cooling. At 100 °C (212 °F), tensile strengths are about 65 to 75% of the room-temperature values.

Unalloyed zinc is available in three basic grades differing by impurity content: special high grade, high grade, and prime western. Impurities, especially lead, cadmium, and tin, must be tightly controlled to avoid intergranular and subsurface corrosion. The use of high-purity zinc with minute amounts of a specific element, such as 0.02% Mg or 0.01% Ni, that counteract impurities via mechanisms that are not yet clear, greatly reduces the corrosion susceptibility. A "steam test" consisting of exposure to saturated water vapor at 95 °C (203 °F) for 10 days correlates well with the corrosion behavior of zinc-aluminum alloys in humid atmospheres. Iron and silicon impurities are also controlled to minimize the formation of hard particles that harm machinability.

The poor room-temperature working characteristics and directional properties of zinc are a direct consequence of its hcp crystal structure. Of the twenty-two elements with an hcp structure, only zinc and cadmium exhibit a c/a ratio greater than the ideal close-packed value of 1.633. Not only is slip difficult because of the limited number of available slip systems, but cleavage readily occurs along the basal plane. Zinc has a ductile-brittle transition temperature (DBTT) just below room temperature and sharp radii must be avoided to reduce its notch sensitivity. Much of the useful deformation strain in processing is by twinning. In a single crystal, the interatomic bonds in the hexagonal layer, or basal plane, are significantly stronger than those between the layers, or normal to the basal plane. This anisotropy in atom-bonding strength leads to significantly larger values of compressibility and coefficient of thermal expansion along the direction of the c-axis. For example, in the basal plane along the distance of closest atomic approach, the thermal expansion coefficient is 15 ppm/°C (8.3 ppm/°F), while normal to the basal plane it is as high as 62 ppm/°C (34 ppm/°F). This anisotropy is inherent in wrought products that have crystallographic texture due to processing.

Effects of Alloying Elements in Zinc Alloys

Of the eleven elements exhibiting a reasonable level of solubility in zinc, silver and gold are too costly, while cadmium, mercury, and as little as 0.08% Mg lead to hot shortness. It is aluminum and copper, with minute amounts of magnesium (<0.08%), and more recently titanium, that are mainly added to casting and wrought zinc alloys. The principal role of the alloy elements is, in general, to provide second-phase particles or precipitates to increase mechanical strength, while restraining recrystallization, grain growth, and creep processes. The compositions of the various casting alloys are distinguishable from wrought, as seen from the selected examples listed in Table 17. The rationale for these differences is discussed below.

Zinc forms a eutectic with aluminum at 5% Al and at 382 °C (720 °F). The maximum solid solubility of aluminum in zinc is about 1% which decreases to 0.05% at room temperature. Consequently, the microstructure of zinc-aluminum alloys consists of a mixture of aluminum-rich and zinc-rich phases. The aluminum phase is solution hardened by zinc in solution and by precipitation of zinc particles; the zinc-rich phase is hardened by precipitation of the aluminum phase particle, with little aluminum in solution at equilibrium at room temperature (Fig. 16). The precipitation of the less dense aluminum phase contributes to the dimensional changes with time after zinc is rapidly cooled after casting or heat treatment. It is the primary function of the stabilization anneal to provide a more equilibrium volume fraction of the aluminum phase to stabilize the microstructure.

Zinc Casting Alloys

Pure zinc is coarse grained, attacks iron when molten, and exhibits property anisotropy when cast and processed. Because molten zinc does not dissolve hydrogen, there is no need to flux and cast products are generally sound. The addition of aluminum greatly reduces the iron dissolution rate, permitting the use of ferrous die-casting equipment. Thus, aluminum is the principal alloying element in the cast zinc alloys, providing good castability and increased cast strength. The casting alloys can be divided into two principal classes based on aluminum content: the hypoeutectic alloy with about 4% Al for optimal mechanical properties and those with much greater levels of aluminum, as seen in Table 17. The cast microstructure of the 4% Al alloys consists of zinc-rich dendrites, with interdendritic eutectic arrays of zinc and aluminum (Fig. 16). Alloys with >5% Al are hypereutectic and have aluminum-rich dendrites, with interdendritic aluminum/zinc eutectic structures. Aluminum additions above 4% add strength, hardness, and stiffness, but at the price of lower ductility. Because the electrical conductivity of aluminum (higher than zinc) is lowered by dissolved zinc and alloying elements have little solubility in the zinc phase, electrical and thermal conductivity values are rather constant across the range of alloys.

Copper, magnesium, and titanium are the other common alloying elements. Copper is added for strength, hardness, and creep resistance. However, with increasing amounts of copper, ductility decreases and the degree of dimensional growth with time increases. Where it is an important design criterion, dimensional stability can be a problem especially if moderately elevated temperatures are expected in service. For example, after die casting, Zn-4Al shows an initial shrinkage that decreases with time. The effect of copper in this alloy was quantified in a laboratory experiment where after the initial shrinkage after casting, exposure for 1 year in dry air at 95 °C (205 °F) resulted in dimensional growth of 0.0002 in./in. at 0% Cu and 0.002 in./in. for a Zn-4Al-1.5Cu alloy. The amount of growth seen at the 1.25% Cu level is considered tolerable for many applications. Originally, copper was also added to counteract the harmful effects of impurity elements, a role taken over by very low, but deliberate additions of magnesium. It is still unclear how the impurities causing intragranular corrosion are neutralized by copper or magnesium. Nickel works as magnesium, but it is harder to control. Magnesium also has the added benefit of lowering the DBTT or embrittlement temperature. However, exceeding about 0.08% Mg leads to hot shortness during cooling of the casting. Exceeding the strict impurity limits in general will lead to intragranular corrosive attack and warping and cracking with time after casting.

The die-casting alloys (No. 2 to 7 in Table 17) are basically a family of Zn-4Al alloys because this composition provides excellent casting characteristics and optimal strength and ductility. The copper addition provides strength, hardness, and creep resistance, but it is limited by the expansion or swelling with time it induces. The expansion with time shown by alloy No. 2 (2.5% Cu) might be too excessive for some applications. On the other hand, alloy No. 3 (0% Cu) has low creep resistance and should not be exposed to temperature above 50 °C (120 °F) under load.

The gravity-fed castings (sand and permanent mold) are made of higher aluminum-containing alloys, ZA-8, ZA-12, and ZA-27, to take advantage of the higher strength and creep and fatigue resistance than the 4% Al alloys. The improved creep resistance is directly related to the coarser

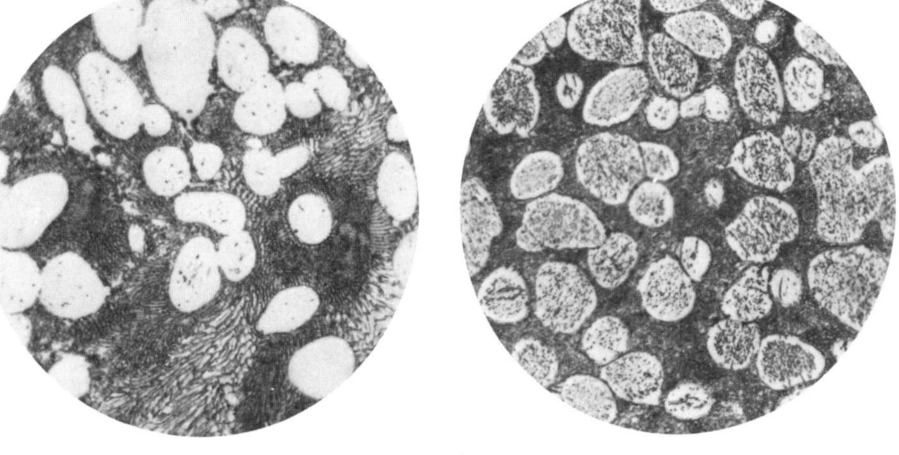

(a) (b)

Fig. 16 Photomicrographs of zinc die-casting alloy No. 5 (UNS Z35531), Zn-4Al-1Cu-0.05Mg, showing (a) the zinc solid solution surrounded by eutectic in the as-die-cast microstructure and (b) the precipitation of aluminum-rich precipitates in the zinc phase after aging 10 days at 95 °C (205 °F). 1000×

cast grain structure. These alloys are also more dimensionally stable and have lower densities than the 4% Al alloys. ZA-27 is the lowest density (5.00 g/cm^3, or 0.181 lb/in.^{3}), the strongest, and the hardest zinc-casting alloy, but it suffers from low ductility and impact strength.

Wrought Zinc Alloys

The wrought zinc alloys are fabricated as flat-rolled, wire-drawn, and extruded and forged products. The flat-rolled products are processed from cast slabs 25 to 100 mm (1 to 4 in.) thick to gages as thin as foil, 0.025 mm (0.001 in.). Zinc alloys are hot worked at 120 to 150 °C (250 to 300 °F), but even at room temperature, the heat generated during an individual processing or forming step is generally sufficient to cause recrystallization, precluding the need for annealing.

The wrought alloys are relatively lean in alloy content and the strict impurity control noted for the casting alloys also applies to the wrought alloys. Owing to the crystallographic texture derived from working of this hcp material, the mechanical and thermal expansion properties are anisotropic (directional) in the rolled product, with the orientation transverse to the rolling direction showing higher tensile strength and lower thermal expansion than the longitudinal orientation (rolling direction), as illustrated by the properties for one of the rolled zinc alloys (Zn-0.06Pb-0.06Cd), listed in Table 18.

In addition to pure zinc, there are three commercial rolled zinc alloys: Zn-0.08Pb (UNS Z21210), Zn-0.06Pb-0.06Cd (UNS Z21220), and Zn-0.3Pb-0.03Cd (UNS Z21540) that are deep drawn into battery cans, eyelets, and grommets. They are readily formable provided that the temperature is kept above 21 °C (70 °F) and are generally hot worked at 120 to 275 °C (250 to 525 °F). Higher strength and creep resistance are obtainable with the alloys in which copper is used as a hardening agent, through the precipitation of the ε-copper-zinc phase: Zn-1.0Cu (UNS Z44330), Zn-0.8Cu-0.010Mg (UNS 45330), and Zn-0.8Cu-0.15Ti (UNS Z41320). The last named alloy offers higher room-temperature tensile strength (200 to 262 MPa, or 29 to 38 ksi) and creep resistance than the other wrought alloys provided that the alloy is heat treated after rolling (typically at 250 °C, or 480 °F, for 45 min) to precipitate the intermetallic compounds that zinc forms with copper and titanium. In this latter alloy, the precipitation of $TiZn_{15}$ at grain boundaries restrains grain growth (Fig. 17).

One of the most-studied superplastic materials is the Zn-22Al eutectoid alloy, to which 0.5% Cu and 0.02% Mg can be added for enhanced creep strength. When solutionized, quenched, and annealed, this alloy forms a microstructure comprising small equiaxed grains of zinc and aluminum terminal solid solution phases (Fig. 18). It is this fine-grain structure that exhibits a tensile elongation in excess of 2500% when tensile tested in the superplastic regime at 250 °C (480 °F) and is easily formed into complex shapes at 250 to 270 °C (480 to 520 °F). When heated above 275 °C (527 °F) and slowly cooled to room temperature, it loses its superplastic properties. The alloy has the directionality in properties of a wrought product, with tensile properties in the rolling direction of 255 MPa (37 ksi) yield strength, 310 MPa (45 ksi) tensile strength, and 27% elongation; in the orientation transverse to the rolling direction, 295 MPa (43 ksi) yield strength, 380 MPa (55 ksi) tensile strength, and 25% elongation. Forgings and extruded products are made primarily of either the zinc-aluminum alloy, Zn-14.5Al-0.75Cu-0.02Mg, or the zinc-titanium alloy, Zn-1.0Cu-0.01Ti, with the former showing good tensile strength with higher impact strength at low temperatures.

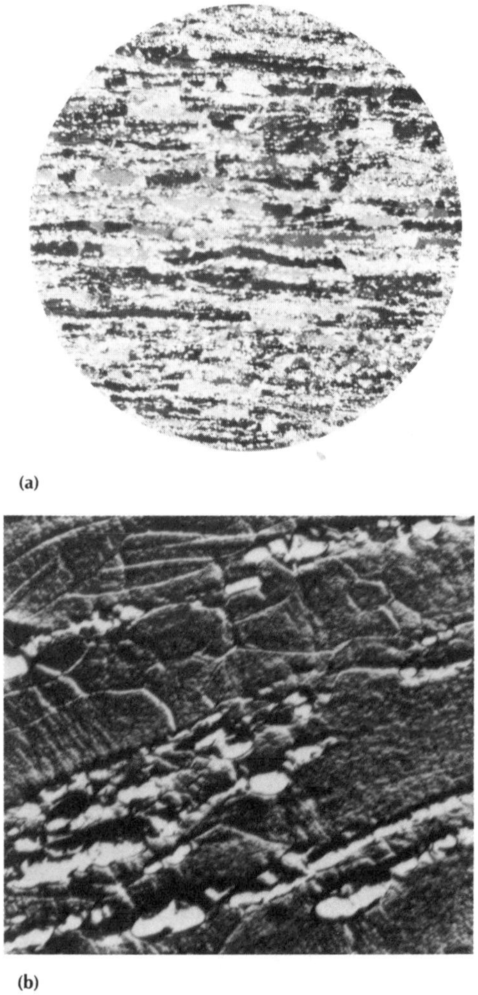

Fig. 17 Photomicrographs of hot-rolled zinc alloy Zn-0.6Cu-0.14Ti. (a) Photographed under polarized light to show the elongated grains (parallel to the direction of rolling) restrained by stringers of the $TiZn_{15}$ particles. 250×. (b) Replica electron micrograph showing the morphology and distribution of the $TiZn_{15}$ particles (white) along the grain boundaries. 4400×

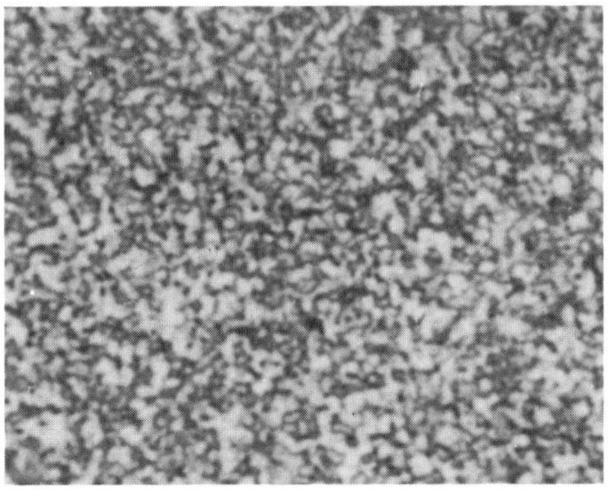

Fig. 18 Photomicrograph of Zn-22Al alloy (eutectoid composition) showing superplastic, fine-grained structure obtained by annealing at 350 °C (660 °F) and water quenching. 2500×

Magnesium and Magnesium Alloys

Magnesium and its alloys have been employed in a wide variety of structural applications because of their favorable combination of tensile strength (140 to 365 MPa, or 20 to 53 ksi), elastic modulus (45 GPa, or 6.5 × 10^6 psi), and low density (1.74 g/cm^3, or 0.0628 lb/in.^{3}). Magnesium alloys have high strength-to-weight ratios (tensile strength/density), comparable to those of other structural metals (Ref 18, 42, 43). Approximately 15% of annual production is used in structural applications, primarily as die castings (11%), gravity-fed castings (1%), and wrought mill products (3%). The bulk of the remainder is used as alloying additions to aluminum alloys (54%), to desulfurize steel (11%), and to nodularize cast iron (4%). The wrought products start out as cast ingots and are processed into extruded bars, billets, wire, plate, sheet, or general forged shapes. Its rather low melting point (650 °C, or (1202 °F), facilitates its use in die casting, permanent-die, and gravity-feed sand-casting operations, provided the melt is protected from oxidation with inert atmospheres. Magnesium has relatively good electrical (38% IACS) and thermal conductivity (418 Wm · K, or 240 Btu/ft · h · °F) values. It also has a remarkably high damping capacity, that is, the ability to absorb elastic vibrations.

Magnesium falls at the highest anodic position in the galvanic series of metals, making it ideally suited as a sacrificial anode material to protect underwater or underground structures, such as piping, ship hulls, and ballast tanks. On the other hand, this position in the galvanic series also requires proper design of structural applications of this metal to avoid unwanted galvanic corrosion in service. Proper design may involve nothing more than ensuring that the magnesium component be insulated from other metals lower in the galvanic series or be protected from the envi-

Table 18 Property directionality in the rolled zinc alloy Zn-0.06Pb-0.06Cd

	Direction of measurement	
Property	Longitudinal (rolling direction)	Transverse to the rolling direction
Tensile strength, MPa (ksi)	150 (22)	170 (25)
Coefficient of thermal expansion, µm/m · K (µin./in. · °F)	32.5 (18.1)	23.0 (12.7)

ronment with a well-maintained chromate coating among others.

Magnesium and its alloys are used in a wide variety of structural applications in automotive, industrial, materials-handling, commercial, and aerospace equipment. They are used in support brackets, various types of automotive housings, grain shovels, hand-held tools, luggage frames, computer housings, and ladders. Magnesium and its alloys are quite easily machined, permitting dry cutting at high speeds and great cutting depths, requiring caution to prevent fine machining chips from heating to near the melting point and catching fire. Magnesium alloys can be welded by gas metal arc and by resistance spot welding, but it is also often joined by adhesive bonding.

The primary grade of unalloyed magnesium commercially available is grade 9980A (ASTM B 92) with a nominal purity of 99.8%, which is adequate for most uses. The available higher-purity grades must be specified when improved corrosion resistance is needed. The elements nickel, iron, and copper, in particular, must be kept below restricted limits (typically, 0.005% Ni, 0.005% Fe, 0.05% Cu for some alloys) or corrosion resistance rapidly degrades due to the formation of second-phase particles that are cathodic to magnesium: Mg_2Ni, $FeAl_3$, Mg_2Cu, $MgCu_2$, Fe. A minimum level of manganese is sometimes specified to ensure a critical iron/manganese ratio for tying up iron. One important consequence of the low melting point (650 °C, or 1202 °F) of magnesium is its susceptibility to creep under load at moderately elevated temperatures. Magnesium is alloyed to improve its creep performance to acceptable levels by controlling grain-boundary sliding and migration, the primary creep mechanism in this metal.

Magnesium has poor ductility at room temperature and below. This hcp metal is more easily worked at temperatures above 225 °C (440 °F), usually in the range 345 to 510 °C (650 to 950 °F), where additional hcp slip systems (mainly pyramidal) and twinning become active, augmenting the primary basal slip and twinning deformation modes available at lower temperature. Another consequence of the predominant basal plane slip at room temperature is the mechanical property anisotropy (or directionality) in cold-rolled sheet due to crystallographic texture. For example, a mildly rolled sheet has a tensile strength and tensile elongation in the rolling direction (241 MPa, or 35 ksi, and 2% elongation) that differ from those transverse to the rolling direction (262 MPa, or 38 ksi, and 8% elongation). Moreover, the hcp crystal structure combined with coarse solidification grain structure also leads to tensile properties that vary with location and direction, especially in large castings. The mechanical property anisotropy in wrought products is also manifested by the observation that the 0.2% offset yield strength in compression is equal to about 40 to 70% of the corresponding yield strength in tension. In the case of cast products, however, the yield strength in compression is roughly equal to the tensile yield strength.

Magnesium Alloys

There are two basic families of magnesium alloys: alloys with 2 to 10% Al with additions of zinc and manganese, and alloys containing zinc with additions of rare earth (RE) metals, thorium, silver, zirconium for strength, creep resistance, and grain refinement. Because the rare earth metals behave similarly, they are alloyed in a group by adding mischmetal, which naturally contains about 50% Ce. The maximum solid solubility of aluminum in magnesium is 12.7% at the eutectic temperature, 437 °C (819 °F), which drops to about 1% at 100 °C (212 °F). The maximum solubility of zinc in magnesium is 6.2% at its eutectic temperature (340 °C, or 644 °F), which drops to 2.0% at 100 °C (212 °F). Thus, alloys containing aluminum and/or zinc are capable of age hardening to combine particle strengthening with solution hardening by aging to form metastable precipitates or the equilibrium phases, $Mg_{17}Al_{12}$ or MgZn. An alloy containing 9Al-1Zn, for example, is precipitation hardenable either naturally or artificially in a furnace. After a solutionizing treatment above 370 °C (700 °F), aging above 205 °C (400 °F) will precipitate $Mg_{17}Al_{12}$ in a continuous array of Widmanstätten particles, whereas aging at lower temperatures will form this phase in a discontinuous cellular morphology. The other alloying elements commonly used in magnesium—zirconium, manganese, thorium, and cerium—have <1.0% maximum solubility. The magnesium alloy designation code identifies the main alloying elements and the amount added. For example, alloy AZ31B contains 3Al-1Zn and is the second alloy (B) in the family sequence. Each of the standard alloying elements are indicated by code, the most important of which are: A (Al), C (copper), E (rare earth), H (thorium), K (zirconium), L (lithium), M (manganese), Q (silver), S (silicon), T (tin), Z (zinc). Magnesium and its alloys are available in the as-cast or the as-fabricated (F), the annealed (O), or the worked (H24, H26) tempers.

The properties of a group of selected commercially available magnesium alloys are given in Table 19. The tensile properties available with magnesium alloys generally range from 70 to 340 MPa (10 to 49 ksi) yield strength, 140 to 365 MPa (20 to 53 ksi) tensile strength, and 1 to 15% elongation. The principal magnesium die-casting alloys are used directly in the as-cast (F) condition and contain Al-Zn-Mn, Al-Mn, or Al-Si-Mn. As aluminum is added to magnesium, the strength continuously rises to 10% Al, but with tensile elongation peaking at about 3% Al. Even though alloys with ≥4% Al are age hardening, aluminum is generally not effective at improving creep resistance. Magnesium alloys with 3% Al have optimal ductility, those with 9% Al offer the best strength, but those with about 6% provide the best combination of strength and ductility. Zinc as an alloying element can be viewed similarly; that is, ductility is optimum for a 3% addition and a good combination of strength and ductility is available with 5% Zn.

The remarkable effectiveness of zirconium in grain-refining cast magnesium has been explained by the similarities in crystal structure and lattice parameters of the two elements. The respective distance of closest atomic approach between them differs by about 0.75%, allowing magnesium grains to nucleate epitaxially on the (0001) basal planes of hcp α-zirconium phase crystals, which are first to separate during cooling of the casting. The zirconium addition is usually kept below 0.8%. It readily forms compounds with iron, aluminum, silicon, carbon, oxygen, and nitrogen, and it reacts with hydrogen to form a hydride that is insoluble in magnesium. Additions of thorium and the rare earth metals reduce microporosity and hot cracking, while improving strength and creep resistance. Manganese is added for corrosion stability in Mg-Al-Zn alloys, and yttrium is added for the same purpose in the aluminum-free and the zinc-free magnesium al-

Table 19 Compositions and properties of selected cast and wrought magnesium alloys

Alloy and temper	Nominal composition, wt %	Tensile strength, MPa	Tensile strength, ksi	Yield strength, MPa	Yield strength, ksi	Elongation in 50 mm (2 in.), %	Hardness, HRB
Sand and permanent-mold castings							
AM100A-T61	Mg-10Al-0.1Mn	275	40	150	22	1.0	69
AZ63A-T6	Mg-6Al-3.0Zn-0.13Mn	275	40	130	22	5.0	73
EZ233A-T5	Mg-2.7Zn-0.6Zr-3.3RE(a)	160	23	110	16	2.0	50
HK31A-T6	Mg-3.3Th-0.7Zr	220	32	105	15	8.0	55
ZC63A-T6	Mg-0.5Mn-6Zn-2.7Cu	210	30	125	18	3.5	62
ZK61A-T6	Mg-6Zn-0.7Zr	310	45	195	28	10.0	70
Die castings							
AS41A-F	Mg-43Al-0.35Mn-1.0Si	220	32	150	22	4.0	...
AZ91A, B, and D-F	Mg-9Al-0.13Mn-0.7Zn	230	33	150	22	3.0	63
Extruded bars and shapes							
AZ61A-F	Mg-6.5Al-1.0Zn	310	45	230	33	16.0	60
ZK60A-T5	Mg-5.5Zn-0.45Zr	365	53	305	44	11.0	88
Sheet and plate							
AZ31B-H24	Mg-3Al-1.0Zn	290	42	220	32	15.0	73
HK31A-H24	Mg-3Th-0.6Zr	255	37	200	29	9.0	68

(a) RE, rare earth addition

loys. Manganese forms elemental manganese particles in aluminum-free magnesium alloys, but forms $MnAl_n$ (n = 1, 4, or 6) in aluminum-containing alloys. Magnesium alloys are not susceptible to stress corrosion, except for alloys containing aluminum above 1.5%, which may require a stress-relief anneal, especially after welding, before being placed in service to help reduce susceptibility to stress corrosion.

Besides being readily die cast directly from a liquid melt, magnesium alloys are also formed from semisolid slugs. The most common die-casting alloy for applications at room temperature is AZ91D, containing Mg-9Al-1Zn-0.13 min Mn. The alloy shows excellent castability from melt temperatures in the range 625 to 700 °C (1157 to 1292 °F). It is prone to hot shortness at 400 °C (752 °F), even though its solidus and liquidus occur at 470 and 595 °C (878 and 1103 °F), respectively. The alloy shows a good combination of mechanical properties. The higher-purity version AZ91E also offers good resistance to saltwater corrosion, as long as good control of iron, nickel, and copper impurities is maintained. A good dispersion of $Mg_{17}Al_{12}$ particles obtainable in die-cast AZ91D is illustrated by Fig. 19. As described above, this phase can be solutionized and precipitated throughout the microstructure with appropriate heat treatments. The higher-purity alloy AM60B, containing Mg-6Al-0.1Mn, shows greater ductility and toughness and is used for automobile wheels and for sports equipment. Alloys with silicon, such as AS41A (Mg-4.3Al-0.35Mn-1.0Si) are more resistant to creep at temperatures up to 175 °C (347 °F). The sand-casting alloy EZ33A, Mg-3RE-3Zn-0.6Zr, is used for its good strength and creep resistance at temperatures up to 260 °C (500 °F) and where there is a special need for pressure tightness (soundness) and good damping capacity. It is also readily welded via a gas shielded arc using welding rods of the same alloy.

Increased creep resistance is available with the alloy HK31A (Mg-3Th-0.7Zr), which is available as either a casting or as a wrought alloy. The alloying elements provide particles to pin and restrict the sliding and migration of grain boundaries at elevated temperatures, which help retain usable strength up to 315 to 345 °C (600 to 650 °F). As a sheet and plate alloy, HK31A also has good formability and weldability. The alloy ZK60A, containing 5.5Zn-0.45Zr, offers still higher strength and good ductility. This alloy is hot worked at 315 to 400 °C (600 to 752 °F) and can be age hardened at 150 to 200 °C (302 to 392 °F). Although the solidus and liquidus for this alloy are 520 and 635 °C (968 and 1175 °F), respectively, the as-cast condition is susceptible to hot shortness as low as 315 °C (600 °F); the wrought product experiences hot shortness at 510 °C (950 °F).

AZ31B, containing Mg-3Al-1Zn, is a widely used sheet and plate alloy, which offers moderate strength with good ductility and formability. In the annealed (O) condition it offers a yield strength of 150 MPa (22 ksi), a tensile strength of 255 MPa (37 ksi), and 21% elongation. In the rolled/relief annealed temper (H24) this alloy shows tensile properties of 220 MPa (32 ksi) yield strength, 290 MPa (42 ksi) tensile strength, and 15% elongation. The fine, elongated grain morphology seen in the warm-worked microstructure is illustrated in Fig. 20. The alloy can be used up to 100 °C (212 °F) in either temper. The alloying additions lower electrical conductivity (18.6% IACS) and thermal conductivity (96 W/m · K) while providing solution hardening. The good formability is obtainable through hot working at 230 to 425 °C (450 to 800 °F). As little as 15% cold deformation will be recrystallized by a 1 h exposure to 205 °C (400 °F). As noted previously, still higher strength is available with higher aluminum content, such as in the wrought alloy AZ61A. This alloy is capable of being used up to 200 °C (392 °F) and is readily hot worked at 230 to 400 °C (446 to 752 °F). However, it is also hot short at 415 °C (780 °F), a temperature significantly below its solidus, 525 °C (977 °F). The strength level obtainable and the difference in tensile and compressive yield strength seen with this type of alloy are indicated by the tensile properties obtainable with processed sheet: 305 MPa (44 ksi) tensile strength, 220 MPa (32 ksi) tensile yield strength, 8% elongation, and 150 MPa (22 ksi) compressive yield strength.

Alloys with very low density and good formability have been made by adding lithium to magnesium. LA141A, for example, has sufficient levels of lithium (14%, or 36 at.%) to put the alloy into the bcc-lithium phase field. Experimental alloys containing Mg-37.6Li-5Al have been made with a density of 0.95 g/cm^3 (0.034 lb/in.^{3}), a value lower than that of water. This level of lithium in magnesium is well within the single bcc-lithium phase field and has a moderately low solidus, (340 °C, or 644 °F), and liquidus (425 °C, or 797 °F).

Beryllium and Beryllium Alloys

Because of its unusual combination of physical and chemical properties, beryllium finds its way into unique structural applications when its relatively high cost can be justified (Ref 18, 44, 45). It has a very low density (1.848 g/cm^3, or 0.067 lb/in.^{3}), a moderately high melting point (1290 °C, or 2341 °F), and a high elastic modulus (303 GPa, or 44 × 10^6 psi). It has good electrical conductivity (40% IACS) and thermal conductivity (190 W/m · K, or 109 Btu/ft · h · F). Beryllium is an important alloying element in copper and nickel to produce commercially important age-hardening alloys that are used in electrical contacts, springs, spot welding electrodes, and nonsparking tools. It is also added to aluminum and magnesium for grain refinement and oxidation resistance.

While it can be melted and cast, the resulting casting has coarse grains (>100 μm) that are difficult to process and attempts to refine them by alloying have been unsuccessful. The fine grain size is produced primarily by P/M techniques, from which it inherits the fine grain size (5 to 15 μm) from the powder processing. Powder consolidation is done by vacuum hot pressing and has also been done by the hot or cold isostatic pressing methods in argon atmosphere. It shows only modest ductility (≤3% tensile elongation) at room temperature, which is attributed to a large covalent component in its atomic bonding in the c-axis direction and to its hcp crystal structure, which is limited at room temperature to only one slip direction on two crystal planes, basal and prism. This low ductility is not improved even when impurities are reduced to levels as low as ≈10 ppm. However, beryllium does show improved ductility at temperatures above 200 to 250 °C (392 to 482 °F), where more crystal deformation modes are available. At 400 °C (752 °F), for example, it shows a tensile elongation as high as 50%. Beryllium transforms to a bcc phase at 1270 °C (2318 °F), only 13 °C below the melting point.

Beryllium is flat rolled, extruded into shapes, and forged at elevated temperatures. It is worked warm; the handbooks suggest a temperature range 800 to 1100 °C (1470 to 2110 °F) for hot working. Beryllium is commercially available as

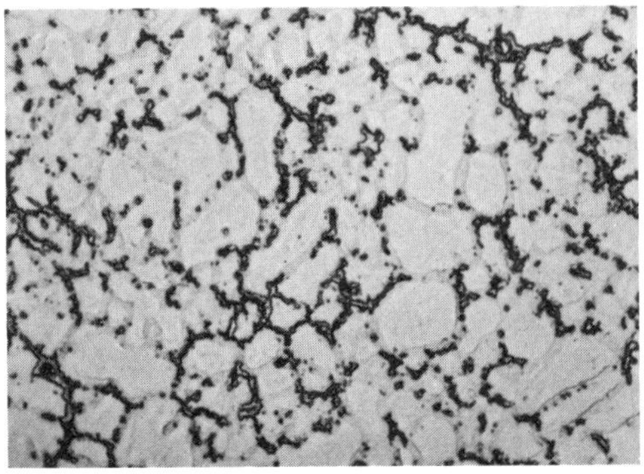

Fig. 19 Photomicrograph of the ZA91A-F die casting alloy, Mg-9Al-0.13Mn-0.7Zn, showing the particles of $Mg_{17}Al_{12}$ at the boundaries of small, cored grains. Segregation (coring) in the grains and absence of precipitated discontinuous $Mg_{17}Al_{12}$ are results of the rapid cooling rate of die castings. 500×

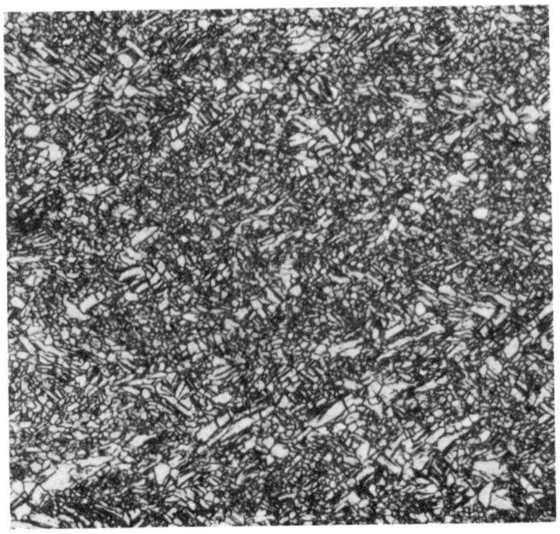

Fig. 20 Photomicrograph of the longitudinal edge view of worked sheet of the magnesium alloy AZ31B-H24 (Mg-3Al-1Zn), showing fine, elongated grains resulting from warm working of this sheet material. 250×

tube shapes and as plate, sheet, and foil in thickness ranging from 0.013 to 15.2 mm (0.0005 to 0.6 in.). Recrystallization annealing is done at 725 to 900 °C (1340 to 1650 °F). As with other hcp metals, beryllium displays anisotropic mechanical properties in wrought forms because of crystallographic texture and grain morphology. The traditional ball-milling method of making powder produces flat, crystallographically oriented powder particles that naturally leads to a crystallographic texture in the initial pressing stage. Newer powder preparation techniques, involving impact-grinding or atomization methods, produce a more equiaxed product, avoiding the oriented flat powders produced by ball milling and reducing the amount of crystallographic texture during the initial consolidation operation. Even so, care is taken to fabricate beryllium with a fine grain and as random a crystallographic orientation as possible. The room-temperature tensile properties of cross-rolled sheet are 345 to 414 MPa (50 to 60 ksi) yield strength, 483 to 621 MPa (70 to 90 ksi) tensile strength, 10 to 40% elongation. Both the tensile and the yield strength have been shown to obey the Hall-Petch relationship between yield strength and grain size.

Beryllium forms a thin protective oxide (BeO) coating that provides good corrosion resistance to the atmosphere at room temperature and giving it excellent resistance to pure water up to 300 °C (570 °F). However, corrosion resistance is impaired by impurities. The presence of carbides or chlorides, in particular, can cause a reaction with moist air at room temperature. Above 760 °C (1400 °F), beryllium reacts with oxygen and nitrogen, but not at all with hydrogen at any temperature. It is attacked by many dilute common acids at room temperature. It can be anodized or chromate coated for improved corrosion resistance. Beryllium shows the machining characteristics of cast iron when the required carbide tools are used. It is commonly brazed and adhesive bonded and with care can be welded.

Unalloyed beryllium is used in weapons, spacecraft, rocket nozzles, structural tubing, optical components, and precision instruments. The specific modulus, that is the ratio of elastic modulus to density, is higher than that of aluminum, magnesium, or titanium. The high specific modulus and the low thermal expansion translates into good physical stability and low distortion during inertial or thermal stressing, making beryllium useful for critical aerospace components. Its exceptionally good dimensional stability and its ability to reflect infrared radiation makes beryllium highly suited for mirrors, either as polished or as a plated substrate. The near transparency of beryllium to x-rays and other high-energy electromagnetic radiation is responsible for its widespread use in foil gage as windows in x-ray tubes and radiation-detection devices.

Beryllium Structural Materials

Commercially available grades of beryllium are distinguished by impurity levels and BeO content. The metal, in fact, normally contains 0.7 to 4.25% of BeO, which is unavoidable and increases with the fineness of the beryllium powder used in the consolidation process. It has extremely low solubility for oxygen and readily forms oxide particles that are situated at grain boundaries, which help to control grain growth without being harmful to ductility below 1.2% BeO. Of the structural grades, indicated by the S in the designation, S-65B, with 99.0% min Be and 0.7% BeO, offers the most ductility at room temperature: 207 MPa (30 ksi) yield strength, 290 MPa (42 ksi) tensile strength, and 3% min elongation. Grade S-200F, with 98.5% min Be and 1.5% BeO is stronger, but less ductile, at room temperature: 240 MPa (35 ksi) yield strength, 325 MPa (47 ksi) tensile strength, and 2.0% elongation. This grade has usable strength and modulus up to 600 to 650 °C (1110 to 1200 °F), at which temperature the tensile properties are: 100 MPa (15 ksi) yield strength, 130 MPa (19 ksi) tensile strength, and 10% elongation.

Beryllium is almost too reactive to accept alloying elements; it reacts to form stable compounds with most of the other common metals. While beryllium shows some solid solubility for copper, iron, nickel, cobalt, palladium, platinum, gold, and to a lesser extent, <1.0%, for chromium and cerium, compounds formed with these metals tend to cause embrittlement. For example, the inherent impurity levels of iron and aluminum in commercial beryllium have been singled out as being particularly harmful to ductility: iron either in solid solution or as $FeBe_{11}$ particles causes embrittlement, and aluminum segregated to grain boundaries leads to hot shortness. These two elements are controlled through the appropriate processing to form discrete grain-boundary precipitates of $AlFeBe_4$.

Beryllium is mutually insoluble with and forms binary eutectic phase diagrams with silicon and aluminum. Alloys in each of these systems have recently been developed to take advantage of the properties made available by creating intimate mixtures of the respective two metals. One of these alloys, for example, is cast at the eutectic composition, Be-46Si, and contains a mixture of beryllium and silicon grains to provide a material with thermal expansion characteristics that are useful for mirror substrate application. Be-(38-50)Al alloys can be made by rapid-solidification techniques, using liquid metastable miscibility gap to separate Al + Be liquid phases into a fine, interconnected morphology after solidification. The structural properties obtainable are determined by rule-of-mixtures of the two metals and combine the ductility of aluminum (and potential for age hardening) with the strength, stiffness, and low density of beryllium (Ref 46).

Inhalation of respirable beryllium and its compounds should be avoided. Users should comply with occupational safety and health standards applicable to beryllium in Title 29, Part 1910, Code of Federal Regulations.

Refractory Metals and Alloys

The refractory metals and their alloys have been so named for their heat resistance because of their characteristic high melting points, good elevated-temperature strength and mechanical stability, and good resistance to softening. They also have high densities, low vapor pressure, and relatively low thermal expansion coefficients compared to the other common structural metals and alloys (Ref 47). The unique physical properties available with the refractory metals are compared to those of iron, aluminum, and copper in Table 20. It is natural to find the refractory metals in elevated-temperature applications. While these metals generally have high chemical reactivity and oxidation tendencies, some of these metals (tantalum and niobium) show exceptionally good resistance to certain corrosion environments, making them uniquely suitable for containing and protecting against particular corrosive media in the petrochemical and aerospace industries.

Compared to aluminum and copper, the refractory alloys are generally difficult to cast and to process at either room or elevated temperatures because of their high melting temperatures and sensitivity to impurities. Besides requiring high processing temperatures, the refractory metals generally require protection from oxygen. These bcc crystal structure alloys show DBTTs at and above room temperature. Moreover, ductility is sensitive to the presence of impurity elements segregated to recrystallized grain boundaries. However, these materials are being fabricated into useful forms by good control of composition and processing parameters.

Table 20 Properties of the refractory metals compared to those of iron, copper, and aluminum

Metal	Crystal structure	Specific gravity	Melting point		Elastic modulus		Thermal expansion coefficient at room temperature	
			°C	°F	GPa	10^6 psi	ppm/°C	ppm/°F
Refractory metals								
Niobium	bcc	8.57	2468	4474	102	15	7.3	4.1
Tantalum	bcc	16.6	2996	5425	186	27	6.5	3.6
Molybdenum	bcc	10.2	2610	4730	324	47	4.9	2.7
Tungsten	bcc	19.2	3410	6170	400	58	4.6	2.6
Rhenium	hcp	21.0	3180	5755	460	67	6.7	3.7
Metals for comparison								
Iron	bcc	7.87	1538	2798	208	30	11.8	6.5
Copper	fcc	8.93	1085	1981	115	17	16.5	9.2
Aluminum	fcc	2.70	660	1220	62	9	23.6	13.1

Table 21 Typical tensile properties of selected niobium alloys in the annealed condition

Alloy	Test temperature	Tensile strength		Yield strength		Elongation, %
		MPa	ksi	MPa	ksi	
Unalloyed niobium	Room temperature	207	30	103	15	35
	1093 °C (2000 °F)	45	6	...	...	...
Nb-1Zr	Room temperature	240	35	138	20	20
	1093 °C (2000 °F)	138	20	69	10	40
C-103 (10Hf-1Ti)	Room temperature	414	60	310	45	25
	1093 °C (2000 °F)	207	30	138	20	40
C-129Y (10W-10Hf-0.1Y)	Room temperature	586	85	448	65	25
	1093 °C (2000 °F)	275	40	207	30	40

Source: Ref 18 (p 1145), Ref 47 (p 565–571), and trade literature from Teledyne Wah Chang, Albany, OR

Niobium and its Alloys

Niobium (Nb), formerly called columbium (Cb) in the United States, has a lower density (8.57 g/cm^3, or 0.310 lb/in.^{3}), a lower melting point (2468 °C, or 4474 °F) and a lower modulus of elasticity (102 GPa, or 15×10^6 psi) than the other refractory metals. Most of the production of niobium is used as an alloying element in steels and nickel alloys; only about 6% of its annual use is made into niobium and niobium alloy mill products. Niobium and its alloys can be made by P/M techniques, but are usually made by consumable electrode vacuum arc melting and electron beam melting ingot techniques.

After careful surface preparation to remove defects, the materials are hot worked at 800 to 1200 °C (1470 to 2190 °F) and cold worked and formed at 200 to 300 °C (390 to 570 °F). The alloys of niobium are recrystallized at 1000 to 1300 °C (1830 to 2370 °F) and stress relieved at 900 to 1000 °C (1650 to 1830 °F), in vacuum or inert atmospheres. With good control of impurity elements and processing variables, niobium can be hot worked and exhibits excellent cold-working characteristics at room temperature in either the stress relieved or the recrystallized conditions. Tight control of impurities must be maintained during manufacture of niobium and its alloys to guarantee processing capability. Contamination by interstitial impurity elements impairs room-temperature ductility. The DBTT of this bcc material is comfortably below room temperature (–125 °C, or –195 °F) for the pure metal, but the DBTT increases with higher impurity content. Because of its high fabricability, niobium is commercially available in plate, sheet, wire, and tube forms. Owing to its adherent oxide, niobium is highly resistant to corrosion in most aqueous media and to liquid metal systems (lithium and sodium-potassium alloys) up to 1000 °C (1830 °F). Niobium and its alloys are gas tungsten arc weldable provided that great care is taken to clean the surface before and after welding and to shield the metal from the atmosphere.

Niobium offers reasonably good strength at elevated temperatures, but must be protected against hydrogen, nitrogen, and oxygen pickup during manufacture and in service. The oxides, nitrides, and hydrides formed with niobium by these gaseous elements are detrimental to ductility. Niobium has a high solubility for oxygen and absorbs it internally from the surface oxide that readily forms at temperatures above about 400 °C (752 °F). It will absorb hydrogen in the temperature range 250 to 950 °C (480 to 1740 °F), and it readily forms an embrittling hydride at 500 to 600 °C (932 to 1110 °F). Formation of hydride precipitates is in fact deliberately used to embrittle niobium to facilitate the grinding steps used in commercial powder preparation. Niobium and its alloys are protected from these atmospheric gases by melting and casting in vacuum, or by processing in sacrificial metallic containers made of molybdenum. Low levels of zirconium and yttrium are added as alloying elements to tie up oxygen internally as oxide particles to promote better ductility. It is not uncommon to see ZrO_2 particles scattered about the microstructure. Before placing niobium products in service, they can be protectively coated with aluminide or silicide compounds, such as hafnium-silicide, or applied as a mixture of Si-20Cr-20Fe.

Unalloyed niobium is basically soft and malleable. It can withstand extremely high amounts of cold working between anneals (whether stress relieving or recrystallizing). Unalloyed niobium recrystallizes at 1090 °C (2000 °F) for 70 to 80% cold working, but it can recrystallize at temperatures as low as 815 °C (1500 °F). Obtaining a uniform, fine recrystallized grain requires the proper balance between the degree and type of work and the temperature/time parameters of the anneal. The specification limits for the troublesome impurity elements set for good commercial quality niobium are 100, 200, 100, and 15 ppm max for each of carbon, oxygen, nitrogen, and hydrogen, respectively. Carbon is kept low (below 0.072%) to avoid poor ductility caused by niobium carbide formation at grain boundaries and within grains. Niobium has usable strength at 1000 to 1200 °C (1830 to 2190 °F), but the environment and mechanical loading will dictate whether and how it can be designed into the application.

The alloy Nb-1Zr, one of the first commercially available, was developed for the nuclear industry to take advantage of its low thermal nuclear capture cross section and improved creep strength over unalloyed niobium. The alloy can be used over the temperature range 980 to 1200 °C (1800 to 2200 °F). It has excellent machinability (80% on a scale where free-machining brass is defined as 100%).

Niobium shares complete solid solubility with each of tantalum, tungsten, vanadium, and molybdenum, and it is capable of dissolving significant amounts of hafnium and titanium as well as zirconium. The solubility of zirconium in niobium is complete above 970 °C (1778 °F), but it drops with temperature from the 9% value at the monotectoid reaction at 620 °C (1148 °F). Higher strength and higher application temperatures are available with significant additions of these solution-hardening elements. Tantalum, vanadium, and titanium increase ductility along with the higher strength by tying up oxygen as dispersoids of oxide particles. Alloy C103 is one such solution-hardening composition containing Nb-10Hf-1Ti. It offers higher strength and higher application temperature range, making it useful for aerospace applications, especially in rocket propulsion systems, such as rocket nozzles. The microstructure utilizes oxide particles of the alloying elements for grain size control and dispersion strengthening in an otherwise solid-solution matrix. C103 alloy is easily fabricated into sheet and rod products, and it can be welded with the proper precautions. Its service temperature range is 980 to 1315 °C (1800 to 2400 °F).

The more heavily alloyed C129Y, which contains Nb-10W-10Hf-0.1Y, 0.5 Ta max, 0.5 Zr max and can be solution annealed (at 1850 °C, or 3362 °F) and aged (at about 1200 °C, or 2190 °F, for 1 h) to form dispersoids of hafnium, zirconium, or yttrium oxides. While this alloy is somewhat more difficult to fabricate and weld than unalloyed niobium and alloy C103, it offers higher strength and extends the service temperature range of niobium to 1650 °C (3000 °F) (Table 21).

Alloy Cb-752, containing Nb-10W-2.5Zr, also offers higher strength and modulus of elasticity. It is given an aging treatment to precipitate compounds of zirconium with interstitial impurity elements. There are a number of other commercial niobium alloys available that combine a dispersion of ZrO_2 particles in a solution-hardened matrix, namely, FS-48 (Nb-15W-5Mo-1Zr), FS-82 (Nb-33Ta-1Zr), FS-85 (Nb-28Ta-10W-1Zr), and B-66 (Nb-5Mo-5V-1Zr). A series of niobium alloys that use carbide precipitation for increased strength and creep resistance include D43 (Nb-10W-1Zr-0.1C), D31 (Nb-10Mo-10Ti-0.1C), and C3015 (Nb-29Hf-14.5W-2Ta-1.5Zr-0.5Ti-0.2C). The processing temperatures are high. Alloy D43, for example, is solution treated at 1649 °C (3000 °F) and aged at 1316 °C (2400 °F). Still other alloys have been made with TiO_2 dispersoids in niobium using P/M techniques. The alloy Nb-(45-55)Ti is a metal superconductor at low temperatures. It is readily fabricated into fine wires, but it needs to be quenched from elevated temperatures to avoid the titanium β-α phase transformation.

Tantalum and Its Alloys

Tantalum melts at 2996 °C (5425 °F) and has a density of 16.6 g/cm^3 (0.60 lb/in.^{3}). Its excellent cold fabricability is unique among the refractory metals. Tantalum has an extremely low DBTT for a bcc metal, falling as low as –260 °C (–452 °F) for the fully annealed condition and not seen with the cold-worked temper. About 25% of annual production appears as mill products in the form of plate/sheet, rod/bar, and tube. Because of its relatively good thermal conductivity and its remarkably good corrosion resistance to many industrial chemicals and body fluids, tantalum mill products are used in the chemical process industry as heat exchangers and tank liners as well as for prosthetic devices. It is clad to copper, or aluminum, or steel to take advantage of the combination of properties available with a composite, including lower cost. In elevated-temperature service, tantalum and its alloys are used in various aerospace applications and in heating elements in vacuum furnaces to take advantage of the good forming and welding capability as well as its good elevated-temperature strength. Its high density and excellent formability are called upon for shape-charge liner components of armor penetrators. About 50% of annual tantalum production is used in the electronics industry as capacitors to take advantage of the uniquely high value and temperature stability (over the range –55 to 125 °C, or –67 to 257 °F) of the dielectric constant of its oxide.

The tough, impermeable nature of the natural oxide film on tantalum is responsible for its excellent corrosion resistance to nitric, hydrochloric, bromic, and sulfuric acids. It is also resistant to the liquid metals bismuth, lead, lithium, magnesium, mercury, potassium, sodium, and sodium-potassium alloys. However, when exposed to air or an oxidizing atmosphere above 260 °C (500 °F) tantalum readily oxidizes and must be protected during processing and while in service above this temperature. Moreover, as with the other refractory metals, the properties are sensitive to impurity levels of carbon, oxygen, nitrogen, and hydrogen. When these elements exceed specified minima (100 to 200 ppm O, 50 to 75 C, 50 to 75 N, and 10 H) oxides, carbides, nitrides, and hydrides are formed, raising the strength and lowering ductility.

Tantalum is consolidated and refined in electron beam furnaces and in vacuum arc remelting furnaces, with some products made using P/M methods. Its excellent ductility and low work-hardening coefficient make tantalum and its alloys readily cold processible, permitting cold reductions of 75% or higher between intermediate process annealing steps. Tantalum and its alloys are rolled at 260 to 370 °C (500 to 700 °F) to take advantage of the increased plasticity made manifest by the more rapid dropoff in yield strength relative to the tensile strength at temperatures above 50 °C (122 °F). Recrystallization annealing of tantalum is done at temperatures in the range 1200 to 1400 °C (2200 to 2550 °F), in vacuum or inert atmospheres. Tantalum is weldable by gas tungsten arc or metal arc methods as well as by resistance and electron beam techniques, provided that good protection from the atmosphere is used. It has been successfully brazed with silver, copper, and specially designed refractory metal alloys. Tantalum has good machining and stamping characteristics, matching that of copper, but tantalum tends to gall and to smear processing tools.

The high-purity, unalloyed tantalum products are made using electron beam melting (UNS R05200) or sintered P/M products (UNS R05400). Tantalum has mechanical properties and forming characteristics similar to low-carbon steel, with a lower elastic modulus (186 GPa, or 26 × 10^6 psi). The room-temperature annealed tensile strength of 210 to 390 MPa (30 to 55 ksi) can reach 1400 MPa (200 ksi) with work hardening. While it has good elevated-temperature stability, tantalum must be alloyed to improve its low strength at elevated temperatures.

The doped tantalum alloys contain small additions (<100 ppm) of yttria, thoria, silicon, or yttrium, that provide fine particles to resist grain growth at elevated temperature. They are used for capacitor lead wires, electronic filaments and supports, and furnace parts.

Tantalum shows complete solid solubility for tungsten, niobium, and molybdenum and rather extensive solubility for hafnium, zirconium, rhenium, and vanadium, giving a wide choice of alloying elements available for solution hardening of the base metal. Ta-2.5W (63 Metal), for example, is used in heat exchangers and in welded-tube applications requiring high formability. The tungsten addition adds solid-solution hardening, increasing the room-temperature tensile strength by about 25%, or by 35 to 70 MPa (5 to 10 ksi), but doubling the elevated-temperature yield strength at 200 °C (390 °F) over that of unalloyed tantalum. At 1000 °C (1830 °F), this alloy has tensile properties of 69 MPa (10 ksi) yield strength, 124 MPa (18 ksi) tensile strength, and 20% elongation. The alloy Ta-10W (UNS R05255) offers still higher room-temperature tensile strength (482 MPa, or 70 ksi, min) and a higher elastic modulus (207 GPa, or 30 × 10^6 psi) than unalloyed tantalum. This alloy is used in aerospace applications and can be used at temperatures up to 2480 °C (4500 °F).

Still other tantalum alloys are available that combine solution-hardening elements, such as alloy T-111, containing 8W-2Hf, and alloy T-222, containing 10W-2.5Hf-0.01C. In addition, the solid-solution alloy Ta-40Nb is regarded as an inexpensive substitute for alloys richer in tantalum. This solution-hardened alloy combines the tensile strength of Ta-2.5W with a lower density (12.1 g/cm^3, or 0.437 lb/in.^{3}) and elastic modulus (152 GPa, or 22 × 10^6 psi), falling proportionally between the values of the pure metals, as one would expect for a solid solution.

Molybdenum and Its Alloys

Molybdenum has a density of 10.22 g/cm^3 (0.369 lb/in.^{3}) and a melting point of 2610 °C (4730 °F). Much of its annual production is used as alloying additions in other metal systems. Less than 5% of annual production is in the form of molybdenum mill products. The element has a bcc crystal structure and has a rather inconvenient DBTT of 7 °C (45 °F) even for extremely high-purity material (1 at. ppm impurity level). Molybdenum offers a rather high strength at elevated temperatures, a high elastic modulus (324 GPa, or 47 × 10^6 psi), a low coefficient of thermal expansion, a low vapor pressure at elevated temperatures, and good fabricability above the DBTT. It has a reasonably good thermal and electrical conductivity at room temperature (31% IACS). A large market for molybdenum is electronics, where the material, in monolithic form or as clad with copper or nickel, is used as heat sinks for silicon devices and packaging components, taking advantage of its good thermal conductivity and close thermal expansion match to silicon.

Molybdenum is extremely sensitive to impurities at grain boundaries and exhibits intergranular fractures when in the fully recrystallized condition. It is processed as a P/M product or starting from arc-melting processes. Molybdenum is cold worked at elevated temperatures below the recrystallization temperature (about 1010 °C, or 1850 °F) and can be processed to foil gages (0.012 mm, or 0.0005 in.) with interpass process anneals done at stress-relieving (recovery) temperatures. Molybdenum is generally commercially available in the worked/stress-relief-annealed condition. The microstructure of wrought molybdenum products consists of elongated, interlocked cold-worked grains, a structure that exhibits a DBTT of –20 to 40 °C (–4 to 104 °F) and directional mechanical properties. For example, molybdenum sheet can be readily bent in the direction of rolling, but it is difficult to make a bend at 90° to the rolling direction because of the alignment of the worked, interlocked boundaries aligned in the rolling or process direction. Deep-drawing quality is cross rolled to reduce the sheet anisotropy. Should the wrought/stress-relief-annealed product be heated above its recrystallization temperature in service, the DBTT rises to 40 to 80 °C (100 to 175 °F), making the metal brittle when cooled to room temperature.

When heated in atmospheres containing oxygen above 425 °C (800 °F), molybdenum forms a volatile, external oxide, MoO_3, that appears as a white odorless smoke, necessitating the use of protective coatings for applications above 500 °C (932 °F) to avoid undesirable thinning of the metal. Because of the extremely low solubility of oxygen in molybdenum, oxygen does not diffuse into molybdenum as it does in niobium. Molybdenum is inert to hydrogen, carbon dioxide (below 1205 °C, or 2200 °F), and ammonia gases, but molybdenum will readily react with ceramics, carbon, hydrocarbons, graphite, and silicon carbide. Molybdenum does not react with Al_2O_3 up to 1705 °C (3100 °F), making the latter the refractory of choice where one is needed in proximity to molybdenum in vacuum and inert atmospheres. Molybdenum is resistant to mineral acids without oxidizing agents and is resistant to liquid lithium, sodium, potassium, sodium-potassium, magnesium, and bismuth, but it is not resistant to molten tin, aluminum, iron, or cobalt.

For structural applications, molybdenum is limited to about 1650 °C (3000 °F). The tensile strength of molybdenum at 1205 °C (2200 °F) is about 96 MPa (14 ksi). Its high thermal conduc-

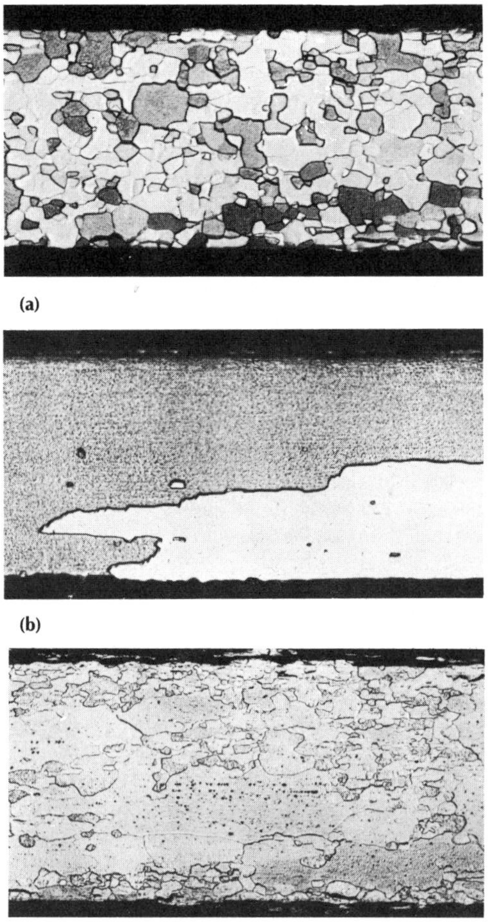

(a) (b) (c)

Fig. 21 Photomicrographs comparing (a) the fully recrystallized, equiaxed grains in undoped tungsten wire to (b) the "interlocked" recrystallized grain structure of doped tungsten wire and to (c) the recrystallized structure of mixed grain size due to ThO_2 particles (black dots) in thoriated tungsten wire. Wire diam., 0.2 mm (0.007 in.); annealed at 2700 °C (4890 °F) for 5 min. 200×

tivity and low thermal-expansion characteristics make it resistive to cracking on quenching and less susceptible to distortion in service, making molybdenum suitable for use as furnace resistance elements and heat-treating baskets. Taking advantage of its ability to resist elevated-temperature corrosion and abrasive conditions, molybdenum and its alloys are used for tools and dies for elevated-temperature processing of other metals. The low vapor pressure and structural stability at elevated temperatures make molybdenum suitable for elevated-temperature structural components in vacuum applications up to 1200 to 1300 °C (2190 to 2370 °F). Molybdenum is weldable, but success depends on recognizing the need for protection from the atmosphere and that the weld zone will be brittle because of its coarse grain.

Limited in number, the alloys of molybdenum deliver higher elevated-temperature strength and creep resistance than unalloyed molybdenum. They are typically made using P/M techniques. Because the recrystallization temperature effectively places the upper temperature limit of wrought molybdenum at about 1010 °C (1850 °F), the HT grade of molybdenum uses nonmetallic elements at dopant levels to stabilize the interlocked wrought microstructure to above 1480 °C (2700 °F). An additional benefit is that the DBTT is below room temperature (–70 to –40 °C, or –94 to –40 °F) for either the wrought annealed or the recrystallized condition. These doped alloys require highly worked structures (>95% reduction in area) with the work being done in a single direction to develop optimal properties. Also processed this way are several oxide-doped alloys, such as TEM (Mo-0.2La_2O_3), MLR (Mo-0.7La_2O_3), and Z-6 (Mo-0.2ZrO_2).

TZM, Mo-0.5Ti-0.1Zr-0.03C (UNS R03630, arc cast, and UNS R03640, P/M product) is the most common commercial alloy and offers high-temperature stability and strength owing to its high recrystallization temperature of 1370 °C (2500 °F). However, it too is embrittled when recrystallized in service. The higher strength and creep resistance make this alloy suitable for heat engines, heat exchangers, nuclear reactors, radiation shields, and extrusion dies.

The three commercial alloys—TZC (Mo-1Ti-0.3Zr-0.15C), MHC (Mo-1.2Hf-0.05C), and ZHM (Mo-0.4Zr-1.2Hf-0.12C)—are carbide-strengthened alloys like TZM, but with richer alloy additions to develop greater strength at temperature and increased resistance to recrystallization. The Mo-30W alloy, available in arc-cast and P/M grades, is used specifically for tooling and components handling molten zinc. Finally, there are several molybdenum-rhenium alloys available (25, 41, 47.5% Re) where rhenium is added to improve the DBTT behavior and as a potent strengthener, greatly increasing the work-hardening rate of molybdenum.

Tungsten and Its Alloys

The key characteristics and properties of tungsten are its high melting point (3410 °C, or 6170 °F), high stability (strength and softening resistance, low vapor pressure) at elevated temperatures, high elastic modulus (400 GPa, or 58×10^6 psi), good electrical and thermal conductivity, and a high density (19.2 g/cm^3, or 0.697 lb/in.^{3}). The metal suffers from brittleness and notch sensitivity, with a DBTT falling above room temperature (150 to 300 °C, or 302 to 572 °F, for annealed unalloyed tungsten). As seen with the other refractory metals, tungsten is sensitive to impurity elements segregated to recrystallized grain boundaries (especially the interstitial elements oxygen, carbon, and phosphorus). Tungsten alloys are used in radiation shields, as counterweights and inertial weights, kinetic energy penetrators, heavy-duty electrical contacts, and targets for x-ray tubes. Its low thermal expansion coefficient is compatible with glasses over a broader temperature range than seen with iron-nickel Invar alloys, for example, making tungsten suitable as a glass-sealing material. Tungsten can be used at temperatures as high as 2500 °C (4530 °F).

About 25% of the annual production of tungsten is made into mill products; another 60% is used in cutting tools in the form of tungsten carbide. Tungsten is manufactured primarily by P/M techniques using powders in sizes ranging 1 to 10 μm (39 to 390 μin.) made from hydrogen-reduced tungsten oxide. Relatively thick-walled products are made by chemical vapor deposition (CVD) techniques by reacting WF_6 with H_2. Proper control of the process can produce fine equiaxed grains instead of the coarse columnar grain structure that is naturally favored by CVD.

Basic commercial processing involves extrusion and forging with rolling and drawing used as follow-up for making the finish forms. The alloys are "warm" worked at about 1200 °C (2190 °F), that is, at temperatures above the DBTT (about 300 °C, or 572 °F, for unalloyed tungsten) and below the recrystallization temperature (about 1370 °C, or 2500 °F) in order to avoid embrittlement caused by the crack sensitivity of recrystallized grain boundaries. Processing sequences of warm working at temperatures above about 650 °C (1200 °F) with interpass softening anneals (recovered, not recrystallized) intentionally develops a microstructure with elongated, cold-worked grains, with interlocked boundaries and with a recovered dislocation substructure, providing alternate, less-embrittling sites for interstitial impurity segregation. In this condition the material can be further worked or can be used in applications requiring hardness and strength at moderate, subrecrystallization temperatures. Such a cold-worked structure exhibits property anisotropy in sheet form. For higher-temperature use, the more stable, recrystallized grain structure needed to provide creep resistance and mechanical stability is obtained by alloy-doping techniques.

Tungsten has a low resistance to oxidation at temperatures greater than 1120 °C (2050 °F), primarily because of cracking in the oxide layer allowing further oxide penetration; it must be protected against oxidation during fabrication and use. It is common to process within a protective container or cladding made of a sacrificial alloy (molybdenum, niobium, stainless steel cans, etc.). Tungsten does not react with H_2, N_2, or NH_3; it will not form an embrittling hydride. Tungsten also resists attack from liquid metals: mercury, potassium, sodium-potassium, magnesium, bismuth, and zinc; however, it is attacked by aluminum. While tungsten is designed for use at elevated temperatures, care needs be taken to retain ductility at room temperature. Residual stresses must be avoided during manufacture, even while applying the frequently needed stress-relief annealing treatments. While tungsten is machinable, it must be kept cool during grinding operations to avoid cracking due to residual stresses.

In unalloyed tungsten, with low impurity content (99.95% pure), cold working lowers the DBTT to below room temperature. Notch sensitivity is lower with a retained cold-worked structure, with increasing grain size up to 1 mm (0.039 in.), with even lower impurity content, and with even very low porosity inherited from incomplete sintering. But, unalloyed tungsten can show progressive grain growth with heating in applications such as heating elements for vacuum furnaces. Worked sheet can show tensile strengths of 965 to 1720 MPa (140 to 250 ksi) at room temperature and 21 MPa (3 ksi) at 2000 °C (3630 °F). Drawn fine wire can reach tensile strengths as high as 5860 MPa (850 ksi).

The doped tungsten alloys, or AKS alloys, are made by adding aluminum, potassium, and silicon compounds at dopant levels to the tungsten powder before pressing. During the sintering treatment at about 3000 °C (5430 °F) in hydrogen, the oxides are reduced and the dopants volatilized. Potassium, in particular, at residual levels of about 85 wt ppm (400 at. ppm) is entrapped in stringers of nanometer size bubbles that become aligned along grain boundaries, creating jagged boundaries during grain growth during in-process recrystallization anneals (Fig. 21). The resulting interlocking of recrystallized grain boundaries promotes good creep resistance, providing a more stable structure that shows less deflection and distortion during repeated heating cycles at high temperatures, making them suitable for common electric lamp filaments that can reach temperatures as high as 2500 °C (4530 °F) in service.

Thoriated or zirconiated tungsten alloys have been made by adding thorium or zirconium to form 1 or 2% of their respective oxide particles (ThO_2 and ZrO_2), which contribute dispersion hardening and grain-growth restraint at elevated temperatures (Fig. 21). These alloys are more machinable than unalloyed tungsten and are suitable for electronic filament and heater wire applications, as well as for gas tungsten arc welding electrodes. The alloy W-3.6Re-1ThO_2 combines solution hardening (rhenium) with dispersion hardening (ThO_2) for imparting even higher strength at higher temperatures. Some manufacturers have replaced thoria with rare earth oxides such as lanthana, yttria, and so forth.

The tungsten-rhenium and tungsten-molybdenum alloys are important solid-solution alloys that provide more ductility and fabricability than unalloyed tungsten. They also exhibit higher recrystallization temperatures and low DBTT temperatures. While they work harden more rapidly, they also exhibit lower strength than unalloyed tungsten, explained by the phenomenon of "solution softening." Alloys commercially developed for optimal properties are W-25Re, and W-25Re(or 30Re)-30 at.% Mo.

The heavy-metal tungsten alloys, also called composite alloys, comprise a high volume fraction (about 85 vol%) of hard/strong, spheroidal tungsten grains in a continuous matrix of a ductile/tough nickel-base solid-solution alloy, nickel-copper, nickel-iron, or nickel-cobalt (Fig. 22). These alloys are made by powder metallurgy techniques starting with about 80 to 98% W powder and the remainder nickel alloy powder. These alloys are sintered at 1480 °C (2700 °F), that is, at temperatures above the melting point of the nickel alloy ("liquid metal sintering"), keeping to a minimum both residual porosity and any continuity of tungsten grains for improved ductility. Cracks can initiate at tungsten-tungsten grain interfaces with as little as 3 to 5% strain. During sintering, tungsten diffuses into the nickel phase where it provides added solution hardening. The strength and ductility are dependent on the strength of the nickel-alloy matrix. The alloys are annealed in vacuum or inert atmosphere, followed by rapid cooling to retain impurities in solution, to minimize impurity segregation to boundaries, and to minimize intermetallic phase precipitation. In the commercial W-Ni-Fe alloys the nickel/iron ratio is controlled to optimize mechanical properties by avoiding μ-phase (Fe_7W_6) seen with nickel/iron ratios less than 1 and Ni_4W precipitates seen with nickel/iron ratios greater than 4 to 1.

These composite alloys continue to be developed for improved ductility and toughness of tungsten while preserving the desired combination of good strength and high density, for use in counterweights, electrical contacts, and kinetic-energy penetrators. Other alloying elements such as tantalum, molybdenum, and rhenium are added to solution harden the nickel phase and for either increasing (with rhenium) or decreasing (with molybdenum) the density. A similar microstructure is made when tungsten is alloyed with silver (W-25Ag) or copper (W-30Cu) using liquid metal sintering as above or using liquid copper or silver infiltration of a porous tungsten powder compact. These latter alloys are used as electrical contacts, offering high conductivity and high arc and welding resistance.

Other alloying elements, such as aluminum, yttrium, zirconium, and boron, are added to scavenge oxygen, nitrogen, and phosphorus impurity elements. During aging, the alloying elements niobium, tantalum, hafnium, and zirconium as well as tungsten can form carbide precipitates for dispersion hardening, such as is used in W-4Re-0.3HfC. Tungsten carbide precipitates, however, will coarsen with time at temperatures above 425 °C (800 °F); however, the other refractory metal carbides are resistant to coarsening at temperatures as high as 1650 °C (3000 °F).

Coatings of metal or refractory oxide or compound coatings are required in many applications to protect the base tungsten alloy against oxidation. Coatings protective against oxidation are applied in the form of metals, or oxides (zirconia), or as other compounds (carbides, nitrides, borides, and silicides). The metal coating systems include hafnium-tantalum, nickel-chromium (good to 1000 to 1400 °C, or 1830 to 2550 °F), platinum, and rhodium (good to 1650 °C, or 3000 °F), and usually are applied over a diffusion-barrier layer between the tungsten alloy and the coating.

Rhenium and Its Alloys

Rhenium is a strong, ductile, refractory metal with an hcp crystal structure. It has a very high density (21.0 g/cm^3, or 0.760 lb/in.^{3}) and melting

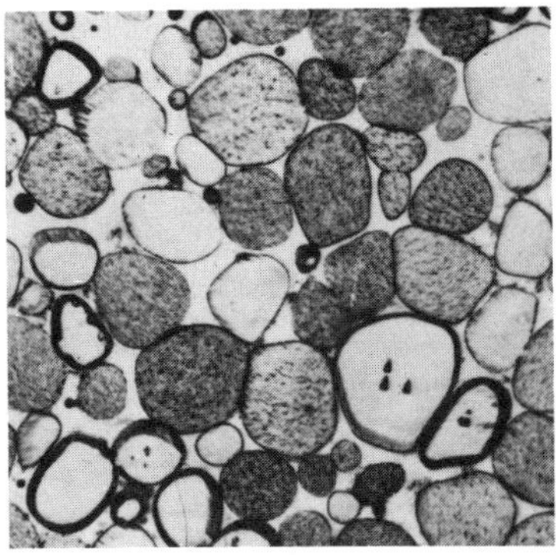

Fig. 22 Photomicrograph of the as-sintered condition of the tungsten heavy-metal alloy 90W-6Ni-4Cu, showing the spheroidal particles of tungsten (white and gray) embedded in a matrix of copper-nickel solid solution (white). 200×

point (3180 °C, or 5755 °F). It has good mechanical stability at elevated temperatures, offering good resistance to thermal shock and wear, and higher creep resistance and strength than the other refractory elements. The annealed condition, showing a room-temperature tensile strength of 1170 MPa (170 ksi), still has a tensile strength of about 48 MPa (7 ksi) at a temperature as high as 2710 °C (4910 °F). Rhenium is used for electrical contacts, thermocouples, filaments for electrical devices, including large-diameter lamp filaments. Because it is in short supply, it is costly and used mostly as an alloying addition to the other refractory alloys.

Rhenium is primarily made using P/M techniques with some also made by arc melting in an inert atmosphere. It has high cold ductility, but because of its very high work-hardening rate it requires the use of light deformation passes with frequent intermediate anneals, either stress relieving or recrystallizing at 1225 to 1625 °C (2237 to 2960 °F) in vacuum or dry H_2 or H_2-N_2 mixtures. Hot deformation must be carried out in vacuum or hydrogen to prevent hot cracking caused by formation of the low-melting-point oxide that penetrates grain boundaries during hot working in air. This is one case where the metal has a higher temperature stability than its oxide: the metal catastrophically oxidizes in air at moderately elevated temperatures, forming Re_2O_7, which melts at 297 °C (567 °F) and boils at 363 °C (685 °F), billowing off as a white cloud. Thus, the metal must be protected from oxidation during processing or while in service. A coating of iridium has been used. Because rhenium does not form a carbide, it is resistant to carbonaceous environments and is suitable for use in contact with graphite.

The high work-hardening rate of rhenium at room temperature translates into rapid increases in strength with cold work. For an annealed condition starting with a tensile strength of 1050 MPa (152 ksi), for example, cold working only 10% increases the tensile strength to 1900 MPa (276 ksi), and 20% to 2000 MPa (291 ksi), and with 40% cold work still further to 2670 MPa

(388 ksi). The metal can be processed, but it must frequently be given interpass anneals. Crystallographic texturing of the hcp crystal structure and the high work-hardening rate lead to mechanical property anisotropy in processed sheet.

Rhenium is an important alloying element for tungsten and molybdenum, forming useful solid-solution alloys, W-(10–26)Re and Mo-(11–50)Re, with good combinations of strength and ductility over those of the unalloyed metals, reportedly by adding a twinning mode of crystal plastic deformation to the basic dislocation slip mechanism. The rhenium addition also lowers the DBTT and reduces the susceptibility to impurity embrittlement at recrystallized grain boundaries. This enhanced ductility with rhenium in solid solution has been called the "rhenium effect."

REFERENCES

1. *Properties and Selection: Nonferrous Alloys and Special-Purpose Materials*, Vol 2, *ASM Handbook* (formerly 10th ed., *Metals Handbook*), ASM International, 1990
2. H. Baker, Ed., *Alloy Phase Diagrams,* Vol 3, *ASM Handbook,* ASM International, 1992
3. J.R. Davis, Ed., *ASM Specialty Handbook: Aluminum and Aluminum Alloys,* ASM International, 1993
4. D. Altenpohl, *Aluminum Viewed from Within, an Introduction to the Metallurgy of Aluminum Fabrication*, Aluminium-Verlag, Dusseldorf, 1982
5. *Aluminum Standards and Data*, The Aluminum Association, 1993
6. C. Brooks, *Heat Treatment, Structure and Properties of Nonferrous Alloys*, American Society for Metals, 1982
7. J. Hatch, Ed., *Aluminum: Properties and Physical Metallurgy,* American Society for Metals, 1984
8. W.E. Haupin and J.T. Staley, Aluminum and Aluminum Alloys, *Encyclopedia of Chemical Technology*, 1992
9. Heat Treating of Aluminum Alloys, *Heat Treating*, Vol 4, *ASM Handbook,* ASM International, 1991, p 841–879
10. W. Petzow and G. Effenberg, Ed., *Ternary Alloys: A Comprehensive Compendium of Evaluated Constitutional Data and Phase Diagrams*, VCH Verlagsgesellschaft, Weinheim, Germany, 1990
11. H.W.L. Phillips, *Equilibrium Diagrams of Aluminium Alloy Systems*, Aluminum Development Association, 1961
12. I.J. Polmear, Light Alloys, *Metallurgy of the Light Metals*, 3rd ed., Arnold, 1995
13. R.E. Sanders, Jr., S.F. Baumann, and H. Stumpf, Non-Heat-Treatable Aluminum Alloys, *Aluminum Alloys, Their Physical and Mechanical Properties*, Engineering Materials Advisory Services Ltd, 1986, p 1441–1484
14. T.H. Sanders, Jr., and J.T. Staley, Review of Fatigue and Fracture Research on High-Strength Aluminum Alloys, *Fatigue and Microstructure*, American Society for Metals, 1979, p 467–522
15. J.T. Staley, Metallurgical Factors Affecting Strength of High Strength Alloy Products, *Proceedings of Fourth International Conference on Aluminum Alloys*, Norwegian Institute of Technology, Department of Metallurgy and SINTEF Metallurgy, 1994
16. E.A. Starke, Jr., and J.T. Staley, Application of Modern Aluminum Alloys to Aircraft, *Progr. Aerosp. Sci.*, Vol 32 (No. 2–3), 1996, p 131–172
17. K.R. Van Horn, Ed., *Aluminum*, Vol I, *Properties, Physical Metallurgy and Phase Diagrams,* American Society for Metals, 1967
18. Properties of Pure Metals, *Properties and Selection: Nonferrous Alloys and Special-Purpose Materials,* Vol 2, *ASM Handbook,* ASM International, 1990, p 1099–1201
19. D.E. Tyler and W.T. Black, Introduction to Copper and Copper Alloys, *Properties and Selection: Nonferrous Alloys and Special-Purpose Materials,* Vol 2, *ASM Handbook,* ASM International, 1990, p 216–240
20. D.E. Tyler, Wrought Copper and Copper Alloy Products, *Properties and Selection: Nonferrous Alloys and Special-Purpose Materials,* Vol 2, *ASM Handbook,* ASM International, 1990, p 241–264
21. P. Robinson, Properties of Wrought Coppers and Copper Alloys, *Properties and Selection: Nonferrous Alloys and Special-Purpose Materials,* Vol 2, *ASM Handbook,* ASM International, 1990, p 265–345
22. R.F. Schmidt and D.G. Schmidt, Selection and Application of Copper Alloy Castings, *Properties and Selection: Nonferrous Alloys and Special-Purpose Materials,* Vol 2, *ASM Handbook,* ASM International, 1990, p 346–355
23. A. Cohen, Properties of Cast Copper Alloys, *Properties and Selection: Nonferrous Alloys and Special-Purpose Materials,* Vol 2, *ASM Handbook,* ASM International, 1990, p 356–391
24. E. Klar and D.F. Berry, Copper P/M Products, *Properties and Selection: Nonferrous Alloys and Special-Purpose Materials,* Vol 2, *ASM Handbook,* ASM International, 1990, p 392–402
25. J.C. Harkness, W.D. Speigelberg, and W.R. Cribb, Beryllium-Copper and Other Beryllium-Containing Alloys, *Properties and Selection: Nonferrous Alloys and Special-Purpose Materials,* Vol 2, *ASM Handbook,* ASM International, 1990, p 403–427
26. W.L. Mankins and S. Lamb, Nickel and Nickel Alloys, *Properties and Selection: Nonferrous Alloys and Special-Purpose Materials,* Vol 2, *ASM Handbook,* ASM International, 1990, p 428–445
27. J.J. deBarbadillo and J.J. Fischer, Dispersion-Strengthened Nickel-Base and Iron-Base Alloys, *Properties and Selection: Nonferrous Alloys and Special-Purpose Materials,* Vol 2, *ASM Handbook,* ASM International, 1990, p 943–949
28. R.A. Watson et al., Electrical Resistance Alloys, *Properties and Selection: Nonferrous Alloys and Special-Purpose Materials,* Vol 2, *ASM Handbook,* ASM International, 1990, p 822–839
29. D.W. Dietrich, Magnetically Soft Materials, *Properties and Selection: Nonferrous Alloys and Special-Purpose Materials,* Vol 2, *ASM Handbook,* ASM International, 1990, p 761–781
30. E.L. Frantz, Low-Expansion Alloys, *Properties and Selection: Nonferrous Alloys and Special-Purpose Materials,* Vol 2, *ASM Handbook,* ASM International, 1990, p 889–896
31. D.E Hodgson, M.H. Wu, and R.J. Biermann, Shape Memory Alloys, *Properties and Selection: Nonferrous Alloys and Special-Purpose Materials,* Vol 2, *ASM Handbook,* ASM International, 1990, p 897–902
32. C.T. Liu, J.O. Stiegler, and F.H. Froes, Ordered Intermetallics, *Properties and Selection: Nonferrous Alloys and Special-Purpose Materials,* Vol 2, *ASM Handbook,* ASM International, 1990, p 913–942
33. P. Crook, Cobalt and Cobalt Alloys, *Properties and Selection: Nonferrous Alloys and Special-Purpose Materials,* Vol 2, *ASM Handbook,* ASM International, 1990, 446–454
34. K.C. Antony, Wear Resistant Cobalt-Base Alloys, *J. Met.*, Vol 35, 1983, p 52–60
35. J.D. Destefani, Introduction to Titanium and Titanium Alloys, *Properties and Selection: Nonferrous Alloys and Special-Purpose Materials,* Vol 2, *ASM Handbook,* ASM International, 1990, p 586–591
36. S. Lampman, Wrought Titanium and Titanium Alloys, *Properties and Selection: Nonferrous Alloys and Special-Purpose Materials,* Vol 2, *ASM Handbook,* ASM International, 1990, p 592–633
37. D. Eylon, J.R. Newman, and J.K. Thorne, Titanium and Titanium Alloy Castings, *Properties and Selection: Nonferrous Alloys and Special-Purpose Materials,* Vol 2, *ASM Handbook,* ASM International, 1990, p 634–646
38. D. Eylon and F.H. Froes, Titanium P/M Products, *Properties and Selection: Nonferrous Alloys and Special-Purpose Materials,* Vol 2, *ASM Handbook,* ASM International, 1990, p 647–660
39. R Boyer, G. Welsch, and E.W. Collings, Ed., *Materials Properties Handbook: Titanium Alloys,* ASM International, 1994
40. R.J. Barnhurst, Zinc and Zinc Alloys, *Properties and Selection: Nonferrous Alloys and Special-Purpose Materials,* Vol 2, *ASM Handbook,* ASM International, 1990, p 527–542
41. *Engineering Properties of Zinc Alloys*, 2nd ed., International Lead Zinc Research Organization, 1981
42. S. Housh, B. Mikucki, and A. Stevenson, Selection and Application of Magnesium and Magnesium Alloys, *Properties and Selection: Nonferrous Alloys and Special-Purpose Materials,* Vol 2, *ASM Handbook,* ASM International, 1990, p 455–479
43. S. Housh, B. Mikucki, and A. Stevenson, Properties of Magnesium Alloys, *Properties and*

Selection: Nonferrous Alloys and Special-Purpose Materials, Vol 2, *ASM Handbook,* ASM International, 1990, p 480–516
44. A.J. Stonehouse and J.M. Marder, Beryllium, *Properties and Selection: Nonferrous Alloys and Special-Purpose Materials,* Vol 2, *ASM Handbook,* ASM International, 1990, p 683–687
45. A.J. Stonehouse, Physics and Chemistry of Beryllium, *J. Vac. Sci. Technol., A*, Vol 4 (No. 3), 1986, p 1163–1170
46. D.H. Carter et al., Age Hardening in Beryllium-Aluminum-Silver Alloys, *Acta Mater.*, Vol 44, 1996, p 4311–4315
47. J.B. Lambert, Refractory Metals and Alloys, *Properties and Selection: Nonferrous Alloys and Special-Purpose Materials,* Vol 2, *ASM Handbook,* ASM International, 1990, p 557–585

SELECTED REFERENCES

- J.R. Davis, Guide to Materials Selection: Nonferrous Alloys, *Engineered Materials Handbook Desk Edition*, ASM International, 1995, p 119–127
- *Metallography and Microstructures*, Vol 9, *ASM Handbook* (formerly 9th ed. *Metals Handbook*), ASM International, 1985

Effects of Composition, Processing, and Structure on Properties of Ceramics and Glasses

Victor A. Greenhut, Rutgers—The State University of New Jersey

CERAMICS are most commonly defined as man-made, nonmetallic, inorganic materials. Common inorganic glasses should be considered a subcategory of ceramics. (Amorphous plastics and metals are technically also classed as glasses, but in this article the common understanding of "glass" is used; silicate glasses receive major emphasis because they represent the "ceramic" glasses most commonly used in consumer and engineering applications.) Ceramics are composed of crystals that have a long-range periodic atomic arrangement. Glasses have short-range order, but possess no long-range periodic crystal structure. Many ceramics can exist in both states, as for example silicon dioxide, which can be crystalline quartz or glassy, fused silica. Glass ceramics and vitrified bodies may be thought of as intermediate combinations of crystalline and glassy constituents (they are discussed later in this article).

Ceramics and glasses are usually composed of oxides, carbides, borides, or nitrides and show a number of common features attributable to their covalent/ionic bonding. Some ceramics are relatively more covalent in nature; these include silicon nitride and silicon carbide used for various high-temperature mechanical structures and wear parts. Other materials are almost purely ionic in bonding, including halides such as magnesium fluoride used for infrared transmitting windows and optical glass fibers. Most ceramics show some mix of covalent and ionic bonding, yielding a wide range of performance, but with common trends.

At all but extremely high temperatures, ceramics and glasses are mechanically hard and brittle with a potentially high strength limited chiefly by flaws and microstructure (see the article "Design with Brittle Materials" in this Volume). Most ceramics are quite refractory and allow application at quite high temperatures in many oxidizing and reducing environments. Ceramics are usually thermal and electrical insulators, but unlike metals these properties can often be varied independently with composition and microstructure control. Materials such as aluminum nitride and beryllia can show metallike thermal conductivity while functioning as good electrical insulators. Capacitive properties can be designed with appropriate additives and microstructure to vary greatly, either increasing or decreasing with temperature and applied voltage. Ceramics and glasses are generally transparent to light as individual crystals or particles, but can be made translucent or opaque by light scattering from interfaces of particle aggregates. These materials are also very resistant to reactive environments and thus are used for containment of chemicals and industrial reactants, including molten metals and glasses at elevated temperatures. This chemical resistance is called "durability"; the term is used, in particular, to describe resistance to water and relative inertness to aqueous solutions, both acid and basic. Because of the variety of properties that can be produced through appropriate processing of controlled chemistries and microstructures, the dollar value of ceramics used worldwide (based on total sales) is similar to that of metals.

Ceramics are commonly used as a polycrystalline aggregate most often processed at elevated temperature to yield a solid. Properties in the individual ceramic crystals are usually quite directional, and the random or preferred crystal orientation can lead to major differences in properties as a function of processing. Increased strength due to resistance to flaw propagation and directional piezoelectric properties are just two examples of how anisotropy in microstructure and properties can be manipulated to obtain desired performance. Many ceramics are multicomponent systems and are affected by the specific scale, shape, and microstructural arrangement of the various phases. In addition, most ceramics contain some level of porosity and the pore structure, whether continuous or isolated, can have profound effects on properties and performance. A great deal is understood about how chemical composition and the structure of starting materials can be manipulated by processing to yield microstructures with particular properties and performance. However, a full understanding of these relationships is still elusive. This article presents general principles and trends, but the selected references at the end of this article and the experience of experts should be used to reliably extend what is presented.

The various broad classes of ceramic materials are reviewed in this article, relating composition and structure to properties. General processing variables that can affect structure and compositional homogeneity are discussed and related to properties. Glasses are treated first because they play a role in other ceramic composition/structure-property relationships and are probably the most completely understood in terms of these relationships. This is followed by glass ceramics, which are usually derived from glass by thermal treatment. Traditional ceramics usually pertaining to clay-based systems follow. Technical ceramics, which include both oxide and nonoxide ceramics, conclude this article. Technical ceramics are often divided into engineering ceramics and advanced ceramics to distinguish more-developed materials from new, very high performance systems. It should be noted that the divisions and subdivisions used herein are those commonly followed by ceramists and that many materials may fall into two or several classifications depending on which compositional, microstructural, property, or performance features are being considered. The emphasis of this article is on mechanical and related properties; however, electrical, electronic, and magnetic properties are also very dependent on composition, processing, and structure.

Glass

Glasses are amorphous materials that exhibit a glass transition, that is, a temperature at which the amorphous solid exhibits a sudden rate of change in thermodynamic quantities (discontinuous slope) with temperature (e.g., heat capacity and thermal expansion) without a first-order phase change. Glasses are usually products of fusion (liquids) that cool to a rigid state without crystallizing, yielding a rigid elastic mechanical solid. Some glasses such as sol-gel and vapor-derived glasses are made without melting of starting materials. Such materials, as well as nonoxide (ceramic) glasses, have growing commercial significance.

Traditional Glasses

Oxide glasses have as their chief component glass formers such as SiO_2, Pb_2O_5, and B_2O_3. Glass formers form the three-dimensional network of the glass structure. Glass modifiers enter the network structure at interstitial sites and modify the properties of formers. Alkaline oxides (Li_2O, Na_2O, K_2O, etc.) as modifiers tend to increase glass fluidity and lower the forming temperature of the "molten" glass. With such additions, glass that results has increasing thermal expansion and solubility in aqueous environments. The smaller the ionic radius (Li), the more effective are these modifier effects. The alkaline earth oxides (MgO, CaO, etc.) are sometimes termed stabilizers because they tend to restore durability while raising the working temperature of glasses containing substantial amounts of alkaline modifiers. Alkaline earth oxides also tend to promote crystallization of the glass as it is held at elevated temperature. The most commonly used glasses today are soda-lime-silica glasses (Table 1) based on the composition 74% SiO_2, 16% Na_2O, 10% CaO (wt%). These provide workability and glass fluidity over an extended range of moderate temperatures, with reasonable environmental durability. These glasses are the most inexpensive and are used for windows, containers, drinking glasses, plates, light bulbs, and so forth. Other modifiers (BaO, ZnO, etc.) may increase durability, workability, or some other property of the resulting glass. Glass intermediates such as Al_2O_3 and PbO can act as both glass formers and modifiers.

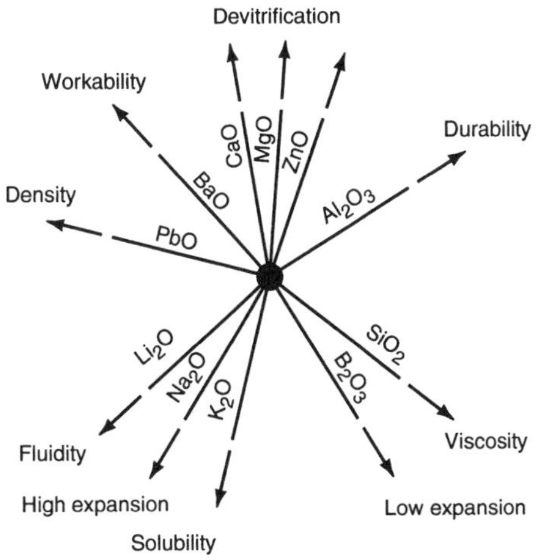

Fig. 1 Effect of glass composition on properties

Trace additions are made to glass to provide color. Oxidized iron provides blue, green, amber or yellow depending on its oxidation state, while copper may yield green or blue and cadmium sulfide provides yellow, orange, or red. Other transition metal oxides yield such colors as blue (cobalt), green (chromium), and violet (nickel). Colloidal metals can also provide color; colloidal gold or copper can yield brilliant reds. Transparent colloidal silver halide particles incorporated in a glass decompose under ultraviolet light to yield opaque colloidal silver particles, which darken the glass. The effect is reversible (when the ultraviolet light is removed), producing photochromic glass. The ultraviolet absorption of glass is increased with additions of CeO_2, TiO_2, Fe_2O_3, V_2O_5, or CrO_3. Infrared absorption is increased by additions of FeO and CuO. Glass can be opacified with CaF_2, NaF, ZnS, or $Ca_2(PO_4)_3$, which can give an opalescent appearance to the glass. Trace iron in the raw materials is often masked with decolorizers such as MnO, Se^{2-}, NiO, or Co_3O_4.

The general attributes of the various major glass ingredients are shown in Fig. 1. Each ingredient can produce multiple effects so that the arrows indicate the tendency for each additive.

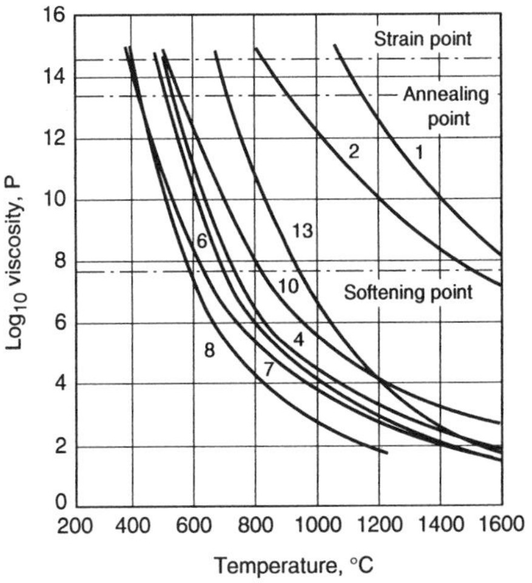

Standard viscosity points	Dynamic viscosity, P
Melting point	10^2
Gob temperature	10^3
Working temperature	10^4
Flow point	10^5
Softening point	$10^{7.6}$
Annealing point	10^{13}
Strain point	$10^{14.5}$

Fig. 2 Relationship between temperature and viscosity of glass. The numbers refer to glasses listed in Table 2.

Thus, PbO is most effective in increasing density while improving workability by extending and reducing the forming temperature range of a more fluid glass. Lead glasses also have a high index of refraction, making them useful for optical applications such as lenses. Their extended working range and brilliant light reflection makes them useful for such aesthetic applications as handcrafted "crystal" and glazes for fine china. However, a high PbO addition particularly in the presence of alkaline oxide decreases the glass durability. Silica glasses are very viscous at their high working temperatures (over 2000 °C), which makes them costly to produce. Very high silica glasses are among the most chemically du-

Table 1 Normal compositions of commercial glass by application

Oxide	Content, %																
	Optical (vitreous silica)	High silica (Vycor)	Plate	Window	Container	Light bulb	Tubing	Lime tableware	Low-expansion borosilicate	Thermometer	Borosilicate crown	Lead tableware	Halogen lamp	Textile fiber (E-glass)	S-glass	Optic flint
SiO_2	100.0	94.0	72.7	72.0	74.0	73.6	72.1	74.0	81.0	72.9	69.6	67.0	60.0	52.9	65.0	49.8
Al_2O_3	...	...	0.5	0.6	1.0	1.0	1.6	0.5	2.0	6.2	...	0.4	14.3	14.5	25.0	0.1
B_2O_3	...	5.0	...	...	...	...	...	...	12.0	10.4	9.9	...	...	9.2	...	...
SO_3	...	...	0.5	0.7	Trace	5.2	...	...	...	...	...	...	0.3	...	...	...
CaO	...	...	13.0	10.0	5.4	3.6	5.6	7.5	...	0.4	...	...	6.5	17.4	...	...
MgO	...	...	...	2.5	3.7	...	3.4	...	...	0.2	...	...	...	4.4	10.0	...
BaO	...	...	...	...	Trace	...	...	...	...	...	2.5	...	18.3	...	...	13.4
PbO	...	...	...	...	...	...	...	...	...	...	...	17.0	...	...	...	18.7
Na_2O	...	1.0	13.2	14.2	15.3	16.0	16.3	18.0	4.5	9.8	8.4	6.0	0.01	...	...	1.2
K_2O	...	...	...	...	0.6	0.6	1.0	...	...	0.1	8.4	9.6	Trace	1.0	...	8.2
ZnO	...	...	...	...	...	...	...	...	...	...	...	...	...	...	...	8.0
As_2O_3	...	...	Trace	Trace	Trace	Trace	...	Trace	...	Trace	0.3	Trace	...	...	...	0.4

Source: Ref 1

rable and lowest-expansion glasses, making them desirable for high-temperature chemical processing, ultraviolet lamps, optical fiber, and solid-state electronic processing. Borosilicate glasses (sold under the tradename Pyrex) provide a compromise of these properties with much lower forming temperatures and associated fabrication costs. Aluminosilicate glasses have relatively low expansion, high-temperature resistance, and chemical durability, but are less difficult to produce than high-silica glasses. They are used in fiberglass, electronics, and high-temperature laboratory equipment. Table 1 gives nominal compositions of many common commercial glasses.

A summary of properties for various commercial glasses is given in Table 2. The viscosity characteristics with temperature are defined in Fig. 2 (curve numbers refer to Table 2). The strain point is the temperature (viscosity) below which the glass acts like a perfectly elastic solid, for all practical purposes. That is, at any industrially practical strain rate the glass would not accumulate residual stresses, but would deform to accommodate the stress. Glass is held above the annealing point and slowly cooled below the strain point to remove any residual stresses introduced during production. As the temperature is raised, molecular mobility decreases the viscosity of the glass so that it flows more readily. The working temperature defines the approximate temperature at which glass is formed or shaped. In the case of blow-molded glass, this may be from a small piece of glass, a "gob," sheared from a larger mass at a higher temperature.

Below the glass transition temperature (T_g), glass behaves as a perfectly elastic, brittle solid. The strength of glass is limited by flaws such as scratches, inclusions, small bubbles (seeds), and small inclusions (stones). Optical fiber produced without flaws and immediately coated to prevent surface abrasion and water attack approaches theoretical strength levels of about 17 GPa (2.5×10^6 psi). Commercial fiber has a strength about an order of magnitude lower, while blown glass has strengths up to 700 MPa (100 ksi). A typical strength for abraded glass is only about 10 MPa (5 ksi), but after exposure to atmospheric water this may decrease to a few ksi, particularly if a tensile stress is applied simultaneously. The nominal design strength is usually about 10 MPa (1 ksi).

Glass can be considerably strengthened by thermal tempering, rapidly cooling the glass from just above the softening point. The surface will be placed in a state of compression, typically about 70 to 200 MPa (10 to 30 ksi), which must be overcome to induce failure. As a result, abraded, tempered glass has an improved strength of about 100 to 250 MPa (15 to 35 ksi). Only simple shapes such as plate glass are amenable to thermal tempering. Tempered glass provides safety on failure as the release of balanced surface compression and interior tensile stresses will shatter the glass into many fine pieces less sharp than typical glass fragments.

An even higher surface compressive stress up to 700 MPa (100 psi) can be produced by "chemical tempering." The glass is immersed in a bath of molten alkali salt (usually KNO_3 for a sodium-modified glass) that "stuffs" the near-surface layers by ion exchange via the larger ion. This has the advantage of tempering the surface of complex shapes, but long treatment times are needed to produce a depth that will resist surface scratches.

Glass is often considered a very environmentally inert material, but even the most durable glasses are attacked by strongly acidic or basic solutions. Soda-lime-silica glass (and other high alkali-containing glasses) can be permanently damaged within the short period of a month if exposed to high humidity or if moisture is held on the surface by a dirt film. Moisture causes sodium ion exchange from the glass. The high-pH surface film that results causes accelerated attack of the glass. Mechanical strength can also be reduced by the attack of specific species in an aqueous environment. Glasses high in SiO_2, Al_2O_3, and CaO are usually most durable. Surface treatments with SnO_2, TiO_2, and SO_2 can also be used to increase durability.

Specialty Glasses

Solder (sealing) glasses are often based principally on PbO to give workability at low temperature. SiO_2 can also be used in these formulations. Such glasses have very low softening and working temperatures and are used in a variety of joining applications including glass joining of television picture tube plate/funnel assemblies. Other applications include electronic package sealing, hermetic feedthroughs, thick film materials, and component mounting material. The thermal-expansion characteristics are often adjusted by composition control for the specific glass/ceramic/metal assembly.

Chalcogenide glasses are based on the arsenic-selenium family and show controlled electrical conductivity and photoconductivity. A major application is the photosensitive coating of xerography drums, which is applied by vacuum deposition or reactive sputtering. Other applications include infrared transmitting fiber-optic waveguides and infrared lenses. Inability to achieve sufficient purity and homogeneity has limited fiber lengths, to date. These materials are also being studied for solar-cell applications.

Heavy metal fluoride glasses (HMFG) are candidates for repeaterless transoceanic communication links because of their low theoretical light attenuation (10^{-3} dB/km at 3.5 μm). Oxygen impurities and crystallization have limited realization of HMFG potential. Increased chemical durability to water is also required for practical application. Some of these glasses are commercially available for short transmission distance infrared applications.

Sol-gel glasses are chemically synthesized from a sol or solution using hydrolizable alkoxides. The final material is often chemically similar to materials produced by more conventional melt processes. In the case of a solution, the material converts to an alcogel, which may be dried by natural evaporation yielding an xerogel or by hypercritical extraction in a pressure vessel yielding an aerogel. A sol will form a hydrogel or precipitate resulting in turn in a cake or powder.

Table 2 Properties of selected commercial glasses

No.(a)	Designation	Viscosity data: Strain point, °C	Annealing point, °C	Softening point, °C	Flow point, °C	Coefficient of linear thermal expansion (0 to 300 °C), 10^{-7}/°C	Specific gravity	Refractive index at Na D line	Volume resistivity (log scale), Ω · cm: 250 °C	350 °C	Dielectric properties at 1 MHz and 20 °C: Power factor	Dielectric constant	Young's modulus: GPa	10^6 psi
1	Silica glass (fused silica)	1070	1140	1667	...	5.5	2.20	1.458	12.0	9.7	0.0002	3.78	69	10
2	96% silica glass	820	910	1500	...	8	2.18	1.458	9.7	8.1	0.0005	3.8	67	9.7
3	Soda-lime—window sheet	505	548	730	920	85	2.46–2.49	1.510–1.520	6.5–7.0	5.2–5.8	0.004–0.011	7.0–7.6	69	10
4	Soda-lime—plate glass	510	553	735	920	87	2.46–2.49	1.510–1.520	6.5–7.0	5.2–5.8	0.004–0.011	7.0–7.6	69	10
5	Soda-lime—containers	505	548	730	920	85	2.46–2.49	1.510–1.520	6.5–7.0	5.2–5.8	0.004–0.011	7.0–7.6	69	10
6	Soda-lime—electric lamp bulbs	470	510	696	880	92	2.47	1.512	6.4	5.1	0.009	7.2	68	9.8
7	Lead-alkali silicate—electrical	395	435	626	850	91	2.85	1.539	8.9	7.0	0.0016	6.6	62	9.0
8	Lead-alkali silicate—high-lead	395	430	580	720	91	4.28	1.639	11.8	9.7	0.0009	9.5	52	7.6
9	Aluminoborosilicate—apparatus	540	580	795	...	49	2.36	1.49	6.9	5.6	0.010	5.6	...	...
10	Borosilicate—low-expansion	520	565	820	1075	32	2.23	1.474	8.1	6.6	0.0046	4.6	68	9.8
11	Borosilicate—low-electrical loss	455	495	...	910	32	2.13	1.469	11.2	9.1	0.0006	4.0	47	6.8
12	Borosilicate—tungsten sealant	460	500	703	900	46	2.25	1.479	8.8	7.2	0.0033	4.9	...	...
13	Aluminosilicate	670	715	915	1090	42	2.53	1.534	11.4	9.4	0.0037	6.3	88	12.7

(a) See Fig. 2

Crystalline materials can also be produced by these routes. These techniques can lead to glasses of exceptional purity and homogeneity based on the purity of starting components. The product material typically has a very high nanoscale porosity that may be useful for high surface area applications such as catalyst substrates or where low density is required. The material can be consolidated at moderate temperatures. Chief applications are as films and coatings with some fibers and powders produced by this technology. Bulk shapes can be made, but the high cost of starting materials and material processing limit bulk applications.

Glass Ceramics

Glass ceramics are produced by melting a glass composition that is then formed into the desired shape by conventional high-speed glass manufacturing methods. The shape is then subjected to a multiple-temperature crystallization heat treatment so that up to 99% of the final material is fine-grained crystals (about 1 μm) in a continuous glassy matrix with the original external shape maintained. The fine grain size, fully dense structure yields materials with mechanical properties and high-temperature resistance generally superior to the glasses from which they are derived. The glassy phase will usually limit performance at very high temperatures. These materials are often superior in strength to conventionally processed ceramics containing porosity and show superior toughness because crack fronts deflect around the fine crystals. If the index of the crystalline and glassy phases do not match, the glass ceramic will usually be white due to diffuse reflection of light. Matched index yields transparent glass ceramics.

Thermal expansion can be tailored by selection of the initial glass composition. The common compositions are based on low-expansion borosilicate glass chemistry. As with glasses generally, low modifier content yields low expansion. This low expansion, coupled with good strength and toughness, has led to applications in cookware and stove tops, which require thermal shock resistance. Near zero-thermal expansion coefficient can be designed over a temperature range useful for telescope mirrors and heat exchangers.

Machinable glass ceramics can be made by incorporating crack-arresting mica flakes in the structure, either by crystallization or by mechanical blending with frit. The resulting glass ceramics can be shaped economically in limited production by high-speed machine tools. Photomachinable glass ceramics include a photoactive element such as silver so that portions of the glass precursor exposed to light form crystals on heating. The crystallized portion is then etched away chemically. This allows complex shapes to be produced by techniques resembling photolithography. The photomachining process has been used to produce electronic and optical components as well as artware such as Christmas tree ornaments.

Glazes and Enamels

Glaze is a continuous adherent layer of glass applied to the surface of a ceramic body. It is applied to the surface of a ceramic as a suspension of ingredients (as indicated above for bulk glasses). After drying, the glaze is fired to yield a hard, nonabsorbent surface layer. The thermal expansion is usually selected by compositional control (by varying glass formers and modifiers as for bulk glass) to be slightly less than the ceramic body. Upon cooling, the slightly lower expansion results in a surface residual compression that increases the strength of the product. The flow temperature is adjusted by composition control to provide a firing maturation with the body, after bisque (partial) or full firing of a ware. Some glazes contain a crystalline constituent or partially devitrify upon cooling, providing opacity. Colorants are included for decoration. Properties of the glaze are similar to that of their bulk glass or glass ceramic composition.

Porcelain enamels are chemically durable and physically strong glasses, usually alkali-borosilicate glasses, applied to a metal either wet or nominally dry and fused to the surface at elevated temperature. As with glazes, they may contain crystalline phases for opacity and glass colorants. They are also tailored compositionally to yield favorable thermal expansion as well as to bond to the metal they coat. Properties can be inferred from bulk glass or glass ceramic behavior, as in the case of glazes.

Traditional Ceramics

Traditional ceramics usually use clay as a major component of the batch. Most are clay-based bodies containing one or more clays such as kaolin (china) and ball clays combined with other minerals in the form of particulate powder. Some ingredients such as feldspar and nepheline syenite act as fluxes lowering the temperature at which glassy phase forms so that the ceramic can be fired at temperatures typically from 1000 to 1300 °C (1830 to 2370 °F). Fillers are added to assist in forming and firing and yield desired properties for the fired product. Silica is the most common filler and may melt partially into the glassy phase during firing. Alumina is used as a filler principally to impart strength to the final product. The combination of clay, flux, and filler of controlled particle sizes together with added and/or native organic matter and water provide a plastic material termed a "green" ceramic. This can be formed into the object by various mechanical and slip-casting methods followed by the firing step. During firing, glass forms that binds the particulate material together during the process of vitrification, sometimes termed vitreous sintering. There is considerable shrinkage of the body during this process, and some of the glass may crystallize (devitrify) upon cooling.

A solid ceramic results with some features common to all traditional ceramics. Grains of filler and crystallized material are held together in a continuous glassy phase. Pores are a feature that may be regarded as a third phase. Pores may be isolated, interconnected, or a combination. The extent of these three general microstructural features—crystal grains, glass, and porosity—determines the classification and properties of traditional ceramics, as shown in Fig. 3. The size, shape orientation, composition, and distribution of these features determines the properties of the final product. A first major classification is into fine particle ware (particles less than 0.2 mm, or 8 mils) and coarse ware (chiefly larger particles, but less than 8 mm, or 0.3 in., for the coarsest structures). Further subdivision is based on the open porosity, as determined by water absorption. Color and application determine the further subdivisions shown in Fig. 3.

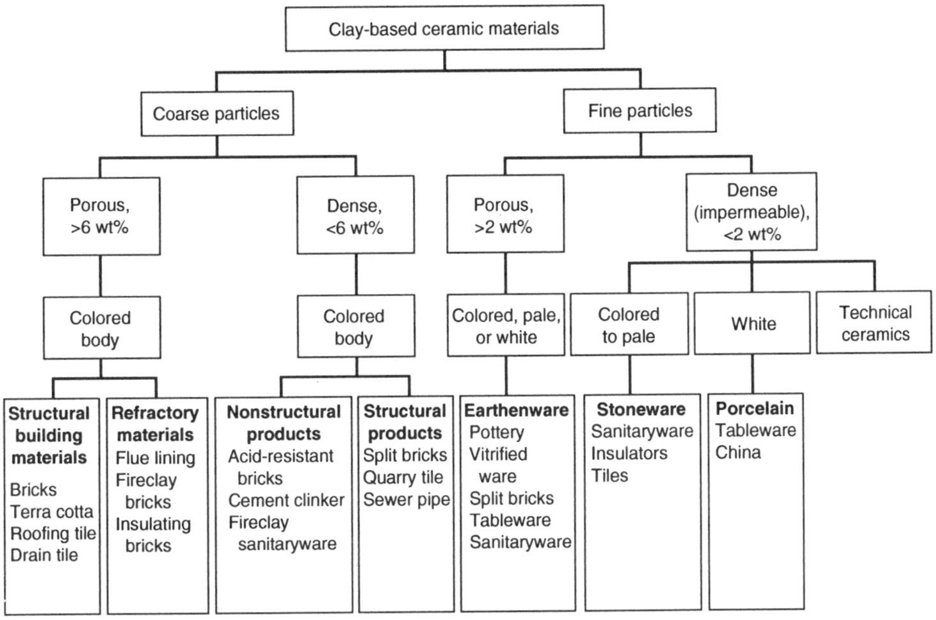

Fig. 3 Classification of clay-based ceramics. Weight percent values represent water absorptivity.

Table 3 Ceramic whiteware classes and products

Class and type	Absorption, %	Products
Earthenware		
Natural	+15	Tableware, artware, tiles
Fine	10–15	Tableware, artware, kitchenware, tiles
Talc	10–20	Artware, tiles, ovenware
Semivitreous	4–9	Tableware, artware
Stoneware		
Natural	0–5	Drain pipe, kitchenware, artware, tiles
Fine	0–5	Cookware, artware, tableware
Technical vitreous	0–0.2	Chemical ware
Jasper	0–1	Artware
Basalt	0–1	Artware
China		
Hotel	0.1–0.3	Tableware
Bone	0.3–2	Tableware, artware
Frit	0–0.5	Tableware, artware
Vitreous plumbing fixtures	0.1–0.3	Lavatories, closet bowls, flush tanks, urinals
Cookware	1–5	Ovenware, stoneware
Porcelain		
Hard	0–0.5	Tableware, artware
Technical vitreous	0–0.2	Chemical ware, ball mill balls and linings
Triaxial electrical	0–0.2	Low-frequency insulators
High-strength electrical	0–0.2	Low-frequency insulators
Dental	0–0.1	Dentures
Technical ceramics		
Steatite	0–0.5	High-frequency insulators, low-loss dielectrics
Electrical porcelains	0–0.2	Low-frequency insulators

Table 4 Typical ceramic body compositions

Ceramic	Content, %: China clay	Ball clay	Feldspar	Flint	Other	Firing range, PCE(a)
Fine whiteware						
White body, low porosity	30.6	14.4	27.5	27.5	...	8
Ivory body, high workability	18	38	32	12	...	9
Earthenware, wall tile	22	25	3	25	10 talc, 15 pyrophyllite	5–6
Vitreous floor tile	32	...	58	8	2 talc	9
Chemical stoneware	55	...	30	14	1 whiting ($CaCO_3$)	14
Tender porcelain	30	...	30	35–37	3–5 whiting ($CaCO_3$)	8–9
Hard porcelain	50.4	...	17.1	32.3	...	14
Electrical porcelain						
High voltage	20	25	35	...	20 alumina	10–11
Low voltage	17	35	32	16	...	10–11
Steatite porcelain	7	...	6	...	87 talc	
Bone china	37.5	...	20	5	37.5 bone ash ($[Ca_3(PO_4)]_2$)	8–10
Hotel china						
Standard	36.8	6.2	21.7	35.2	...	...
High-strength body	34.7	7.1	18.3	27.5	12 alumina	10–12
Cookware	62	...	10	7	21 $MgCo_3$	14
Vitreous china, sanitaryware	22.3	26.7	33.7	17.3	...	9

(a) PCE, pyrometric cone equivalent

Compositions that contain more fluxing raw material tend to produce more glassy phase at lower temperatures, as does firing to higher temperature or for longer times. A higher glass content tends to vitreous sinter to a greater extent, yielding a body with higher density and lower porosity. A fully vitrified body may show no continuous porosity and zero water absorption. The extent of closed porosity depends on the specific composition and firing cycle selected. Usually, strength, environmental resistance, density, and translucency increase with vitrification because the greater amount of glass eliminates porosity and more fully binds together the crystalline particles of the microstructure. Glassy phases that form at higher temperatures tend to cause greater strength, lower thermal expansion, and improved resistance to water and chemical attack, if fired at higher temperature to achieve an equivalent structure to that of lower-firing body formulations. These whiteware properties follow the general behavior of glasses as discussed in the section "Glass" above. Fine-grained microstructures tend to show greater strength if other factors are held constant. Because of the wide combination of raw materials, particle sizes, and firing cycles it is difficult to establish a systematic description of resulting microstructures and related properties.

Several raw material and processing "errors" can result in undesirable structure. If, due to fast firing, the surface closes over, water and decomposing organic materials can cause bloating in which the body swells. There are limited applications in which this is used to make a very open pore structure for lightweight material or thermal insulation. If the surface closes after water is removed the decomposition can lead to "black core," a discolored and often cracked, swollen center. Inhomogeneity, differential heating, or friction with the furnace setter (ware support material) can result in ware warping. A few included large grains can act as defects, as can large pores resulting from particle agglomerates, and inhomogeneous organic matter. These defects act as strength-limiting intrinsic defects. A thermal expansion mismatch between filler grains and glassy phase as well as the quartz inversion (transformation of high-temperature silicates back to quartz) can introduce internal cracks.

Whitewares (Fine Ceramics)

The fine particle branch of Fig. 3, fine ceramics, is often termed whitewares or whiteware ceramics. Most are made using triaxial (clay/feldspar/flint) bodies as the basis for formulation. Sometimes earthenware is distinguished as a separate group because of its greater water absorption (>2%) and range of color to browns and

Table 5 Mechanical properties of ceramic whiteware

Ceramic	Tensile strength, MPa	Tensile strength, ksi	Compressive strength, MPa	Compressive strength, ksi	Flexural strength, MPa	Flexural strength, ksi	Modulus of elasticity, GPa	Modulus of elasticity, 10^6 psi	Impact strength(a), J	Impact strength(a), ft · lbf
Low-voltage porcelain	10–17	1.5–2.5	172–345	25–50	24–41	3.5–6	48–69	7–10	0.3–0.4	0.2–0.3
Cordierite refractory	7–24	1.0–3.5	138–310	20–45	10–48	1.5–7	13.8–34	2–5	0.3	0.2
Alumina, aluminosilicate refractories	5–21	0.7–3.0	90–414	13–60	10–41	1.5–6	14–34	2–5	0.3	0.2
Magnesium silicate	17	2.5	138–207	20–30	48–62	7–9	28–34	4–5	0.3	0.2
High-voltage protection ceramic	21–55	3–8	172–345	25–50	62–103	9–15	48–97	7–14	0.3–0.4	0.2–0.3
Alumina porcelain	55–207	8–30	55–172	8–25	138–310	20–45	103–359	15–52	0.7–0.9	0.5–0.7
Steatite	55–69	8–10	448–896	65–130	110–165	16–24	90–103	13–15	0.4–0.5	0.3–0.4
Forsterite	55–69	8–10	414–690	60–100	124–138	18–20	90–103	13–15	0.4–0.5	0.3–0.4
Zirconia porcelain	69–103	10–15	551–1034	80–150	138–241	20–35	138–207	20–30	0.5–0.7	0.4–0.5
Lithia porcelain	...	...	414	60	55	8	...	...	0.4	0.3
Titania, titanate ceramics	28–69	4–10	276–827	40–120	69–152	10–22	2.1–3.4	0.3–0.5	0.4–0.7	0.3–0.5

(a) 13 mm (0.5 in.) rod

Table 6 Physical properties of ceramic whiteware

Ceramic	Specific gravity	Coefficient of linear thermal expansion at 20–700 °C, μm/m · °C	Maximum safe operating temperature, °C	Thermal conductivity, cal/cm^2 · s · °C	Thermal shock resistance	Dielectric strength(a), V/mil	Resistivity at 22 °C, Ω · cm	Power factor at 1 MHz	Dielectric constant
Low-voltage porcelain	2.2–2.4	5.0–6.5	900	0.004–0.005	Moderate	40–100	10^2–10^4	0.010–0.020	6.0–7.0
Cordierite refractory	1.6–2.1	2.5–3.0	1250	0.003–0.004	Excellent	40–100	10^2–10^4	0.004–0.010	4.5–5.5
Alumina, aluminosilicate refractories	2.2–2.4	5.0–7.0	1300–1700	0.004–0.005	Excellent	40–100	10^2–10^4	0.0002–0.010	4.5–6.5
Magnesium silicate	2.3–2.8	11.5	1200	0.003–0.005	Good	80–100	10^2–10^5	0.0008–0.010	5.0–6.0
High-voltage protection ceramic	2.3–2.5	5.0–6.8	1000	0.002–0.005	Moderate to good	250–400	10^{12}–10^{14}	0.006–0.010	6.0–7.0
Alumina porcelain	3.1–3.9	5.5–8.1	1350–1500	0.007–0.05	Excellent	250–400	10^{14}–10^{15}	0.001–0.002	8.0–9.0
Steatite	2.5–2.7	8.6–10.5	1000–1100	0.006–0.006	Moderate	200–350	10^{13}–10^{15}	0.008–0.0035	5.5–7.5
Forsterite	2.7–2.9	11	1000–1100	0.005–0.010	Poor	200–300	10^{13}–10^{15}	0.0003	6.2
Zirconia porcelain	3.5–3.8	3.5–5.5	1000–1200	0.010–0.015	Good	250–350	10^{13}–10^{15}	0.0006–0.0020	8.0–9.0
Lithia porcelain	2.3	1	1000	...	Excellent	200–300	...	0.05	5.6
Titania, titanate ceramics	3.5–5.5	7.0–10.0	...	0.008–0.01	Poor	50–300	10^8–10^{15}	0.0002–0.050	15–10,000

(a) 6 mm (0.25 in.) specimen

reds as distinct from the white, cream, ivory, or pale gray appearance often associated with the denser, more impermeable whitewares. Generally, proceeding from left to right among the fine ceramics in Fig. 3, the porosity and crystal grain size decrease while strength and durability (environmental resistance) increase. Aesthetic properties such as whiteness and translucency also improve. Definitions of whiteware bodies in terms of water absorption and typical applications are given in Table 3. The name of the subclass often helps to identify its source or microstructural characteristics. The typical formulations of a variety of whitewares are given in Table 4, and typical properties are listed in Tables 5 and 6. Vitreous bodies have no open porosity and are translucent in thin sections. With correct, defect-free manufacture, these are premier materials within each class, known both for their aesthetic beauty and performance in terms of strength and durability. Nonvitreous bodies have a relatively large water absorption due to open porosity and correspondingly poorer properties. The various classes and subclasses overlap significantly in structure and properties so that strict division and distinction by structure, chemistry, and/or properties may not be possible.

Earthenware is a low-temperature-fired nonvitreous clay-based body usually with relatively high porosity and water absorption (10–20%), except semivitreous earthenware, which is fired to medium porosity and water absorption (4–9%). The typical ware is opaque and generally has the lowest strength and resistance to water of whitewares because of its high porosity level. Semivitreous earthenware products have somewhat superior properties at higher firing cost. Other subclasses are natural earthenware, made from a single unbeneficiated clay; fine earthenware, made from a triaxial (clay/feldspar/flint) body with beneficiated clay and talc earthenware, formulated with a high talc level and showing greatest porosity levels. Typical bodies are red because of high iron oxide levels in the clays, but range to white for triaxial and talc formulations. Ware may be used glazed or unglazed.

Earthenware is used for a variety of household tableware, containers, and oven cookware. The water absorption of this red-colored pottery is often used to advantage in applications such as flower pots or other vessels that cool by evaporation. This type of ware is common in less-developed societies and is frequently used for aesthetic reasons in industrially developed countries.

Stoneware is intermediate in structure, properties, and firing between earthenware and porcelain ceramics (Fig. 3). Stoneware bodies are fine-textured, semivitreous, or vitreous bodies with moderate strength and environmental resistance attributable to an intermediate porosity level (water absorption of 0 to 5%). It is rather durable, but lacks the whiteness and translucency of china. Stoneware is also frequently colored as a result of impurities such as iron in the clay. Stoneware is used for tile and drain pipe as well as less demanding dinnerware and artware. Natural stoneware is made from a single unbeneficiated clay, while fine stoneware is made from beneficiated clay raw materials and nonplastic minerals. Basalt stoneware contains relatively high iron levels and is typically red or brown. Jasper stoneware is formulated with high barium levels. Basalt and jasper stoneware generally have improved strength and water absorption properties because their porosity is lower than that of natural and fine stoneware. Technical stoneware has the lowest porosity of this class of ceramics (0–0.2%) and superior properties. It is made from beneficiated materials fired to higher temperatures to achieve very low porosity suitable to application as chemical ware.

China is a term frequently used as synonymous with soft paste (tender) porcelain. The term porcelain is often used inclusively for china, porcelain tableware/artware, and technical porcelains. However, true porcelain is typically stronger but more brittle and more translucent than china, and it is blue-white in color, whereas china is cream-white. Microstructurally, porcelain contains more crystalline and less amorphous component than china. Typical high-fired china has very low continuous porosity and water absorption (0.1 – 0.5%) and may be used glazed or unglazed. While china bodies can be made from simple clay/flux/silica formulations, they frequently contain varying amounts of powders of alumina, bone, ash, frit (glass particulate), lithium minerals and/or cordierite (low thermal expansion). These additives alter the green and fired properties of the china as well as the appearance. China bodies and glaze are fired in a one-step process in which body and glaze mature at the same time, except when decoration is applied or very high luster is desired as in the case of tableware. Hotel china, for commercial food handling, and vitreous china, used for sanitary plumbing fixtures (sanitaryware), are developed with low porosity for low water absorption and superior impact resistance, high strength, and aqueous chemical durability. Fine china, usually considered the finest grade of china, is a relatively thin, translucent body made from bone (usually bovine), frit (Belleck, named for the famous Irish ware that uses crushed glass), or feldspar china. It is usually fired to mature the body and again to develop a very high gloss glaze. The term cookware is used across all whitewares, but most often for china. It refers to a body with a smooth glaze (produced in a single firing) that is resistant to contact by food acids, bases, and abrasives. Such ware is strong and resistant to chipping and thermal shock.

Porcelain is made from a triaxial composition (see the article, "Design for Ceramic Processing" in this Volume). Porcelains may have alumina or zircon substituted for silica in order to increase strength or be high in talc and low in alkali (steatite porcelain) for electrical and strength applications. Compositions based on magnesium aluminosilicates are often used for electronic applications (such as direct current power transmission) to avoid electrically mobile alkali ions. Porcelains have very low water absorption (less than 0.5%). Glaze may be applied after a partial firing (bisque fire) to allow volatiles to burn off. Hard porcelain is used for tableware and artware, while most other porcelain compositions are designed for specific application as technical ceramics.

Technical Ceramics. Whitewares, chiefly chinas and porcelains, are also used as technical

Table 7 Structural ceramic products and related ASTM standards

Product	ASTM standard
Facing materials	
Face brick	C 216
Terra-cotta	...
Thin brick veneer	C 1088
Sculptured brick	...
Special shapes	...
Glazed brick	C 126
Glazed structural tile	C 126
Load-bearing units	
Building brick	C 62
Hollow brick	C 652
Face brick	C 216
Structural tile facing	C 212
Structural tile floor	C 57
Structural tile wall	C 34
Paving units	
Light traffic pavers	C 902
Quarry tile	...
Paving brick	...
Chemically resistant units	
Chemically resistant masonry units	C 279
Industrial floor brick	C 410
Chemical stoneware	...
Sewer pipe	C 700
Chimney brick	C 980
Filter block	C 159
Miscellaneous	
Roofing tile	...
Flue lining	C 315
Lightweight aggregate	C 331
Drain tile	C 4

ceramics for applications such as chemical ware, dental prostheses, electrical insulators, ball mills, refractories, and structural shapes. They are usually extremely low in porosity with near-zero water absorption. Strength, abrasion resistance, chemical durability, electrical properties, and/or high-temperature resistance are adjusted by compositional control of the glassy and crystalline phases(s). Other technical ceramics are made with little or no clay and are discussed in the section "Engineering Ceramics" in this article.

Structural Clay Products

Clay-based ceramic materials with relatively coarse particle sizes (left branch of Fig. 3) are used in various construction applications including brick, roofing and drain tile, terra cotta, and pipe. Tile is the sole exception, having a fine grain size that classifies it with other whiteware ceramics. Tile is classified by the American Standards Institute (ANSI A137.1) and European standards into categories such as ceramic mosaic tile, decorative wall tile, paver tile, porcelain tile, quarry tile, and wall tile. Various applications and their ASTM designations are given in Table 7. Raw materials are clays and/or shales with some fluxing agent and an amount of filler (usually silica) that depends on the application. Most compositions use 35 to 55% clay, 25 to 55% fluxing material, and the remainder filler material, usually silica. As with whitewares a portion of the constituents fuse and form a glassy bond on cooling. Structural clay ceramics show the existence of at least a minimal fired bond.

Table 8 The general range of property values for commercial brick used in the United States

Property	Property value
Compressive strength	
Extruded brick, MPa (psi)	41–110 (6,000–16,000)
Molded brick, MPa (psi)	21–45 (3,000–6,500)
Modulus of rupture, MPa (psi)	4.8–27.6 (700–4,000)
Modulus of elasticity, GPa (10^6 psi)	9.7–34.5 (1.4–5.0)
Bulk density of solid volume, g/cm^3	1.65–2.08
Absorption for room-temperature water, %	0.5–10.0
Saturation coefficient	0.6–0.9
Initial rate of absorption, kg/m^2 · min (g/min · 30 in.^{2})	0.05–3.6 (1.70)
Coefficient of linear thermal expansion, μm/m · K (μin./in. · °F)	4.5–9.0 (2.5–4.5)
Moisture expansion, %	0.02–0.09
Shrinkage in service, %	0
Corrosion resistance	Can contain all acids hot or cold except hydrofluoric; resistant to all alkalies
Thermal conductivity, W/m · K (Btu · in./ft^2 · h · °F)	0.43–1.44 (3–10)

The most important properties of structural clay products are low density (which affects transportation cost), water absorption rate and water capacity (which depend chiefly on pore structure), and compressive strength (which depends on the fired bond and pore structure). Load bearing brick and tile (floor tile, paving brick, bearing wall brick) have strengths from about 17 to 140 MPa (2.5 to 20 ksi) with a relatively low porosity and higher fired bond. Decorative brick and tile are typically less dense, lowering bearing weight as well as shipping and material costs. Strength is lower, 3.5 to 70 MPa (0.5 to 10 ksi), due to the greater porosity. Drain tile for pipe are intermediate in strength (10 to 75 MPa, or 1.5 to 11 ksi). Thermal conductivity also depends on pore structure, with denser material being more thermally conductive. Frost damage and efflorescence can occur from porous structural clay products exposed to water, depending on the specific porosity and chemistry of the product.

The open pore structure is fundamental to the performance of structural clay products. A greater quantity of pores, as defined by the weight percent of water absorption, decreases strength, resistance to cyclic freezing of water-saturated material, insulating value, and corrosion resistance of product. Frost resistance and minimum capillarity is provided when the preponderant size distribution of pores is substantially greater than 1 μm in size for the minimum pore dimension. Often, maximum damage occurs when the mean pore size is in the range 3 to 0.9 μm. Interconnection of pores and pore shape influences fluid permeability of the pores. The path between pores can be tortuous or sealed against fluid penetration, thereby restricting water flow. This would be desirable for water containment (pipe) or preventing water from entering a structure (brick, tile). Alternatively, high permeability is the engineered structure for drainage tile.

When increased compressive structural strength is desired, a material is fired at higher temperatures and/or longer times to achieve a greater fired bond. A body fired at higher temperatures contains more molten material, drawing particles together and furnishing a denser material. This yields fewer undesirable small pores from the standpoint of frost damage. The molten material solidifies on cooling, chiefly into glassy material. This binding phase provides mechanical strength, particularly in compression. As with whitewares, a large amount of "fused" (glassy) material is termed a vitreous product and those with a lesser amount semivitreous. A wide variation in classification and porosity exists specific to particular product type. Chemically resistant vitreous brick may have a porosity of about 0.5%, while vitreous sewer pipe may have a 6% porosity.

The range of properties that can be found within a structural clay product can be seen from the example of bearing-wall brick as given in Table 8. It can be seen that several mechanical properties such as compressive strength, bend strength (modulus of rupture), and elastic modulus can vary by several times. There is a parallel difference in pore structure shown in the table by the various water absorption characteristics. Differences in the fused bond can be inferred from the previous discussion.

Refractories

Refractories are materials that maintain their structural strength and physical integrity at the elevated temperatures of a furnace environment. They must also resist thermal shock, thermal cycling, mechanical fatigue, and physical wear. Many are resistant to attack by aggressive environments of chemical processing, exhaust vapors, molten metals, cement clinker production, and fused glass (including slags) at extreme temperatures. The most demanding applications are containment refractories, which are nominally fully dense (near-zero porosity) and designed in terms of chemistry and phase structure to resist the particular environment and conditions of a manufacturing process. Chrome magnesite materials are used to resist molten iron and steel and their slags. Zirconia-base ceramics are employed for nonreactivity with cobalt- and nickel-base superalloys. Alumina and zirconia grain in an aluminosilicate matrix, AZS refractories, resist attack by molten soda lime silica glass and high alkali borosilicate glass used for fiberglass. Common refractories are made from natural materials, but high-performance refractories are often composed of more purely refined or chemically derived ceramics starting materials. Most refractories are metal oxide ceramics, silicon-carbide-base materials or graphite. Newer carbides, borides, nitrides, and silicides find use where their higher expense is justified.

A refractory usually has a designated upper-use temperature determined by the softening or melting point of constituent oxides. This temperature is usually a range because the refractory is a complex mixture of phases. A combination of temperature, time, load, and environment all interactively determine the actual utilization limits. Sometimes the pyrometric cone equivalent (PCE) is used to quantify refractoriness. This is a measure of the heat content that a refractory can withstand without softening as determined by slumping of the pyrometric cone (actually a tall, narrow-based, tilted tetrahedral pyramid). This can better reflect service because it combines temperature and time factors. Another approach is to determine the pyroplastic sag under its own weight of a narrow bar of refractory supported near its ends. The hot modulus of rupture (HMOR) is determined by bend testing at elevated temperature. The performance of a refractory depends on its phase distribution, particle size distribution, and porosity as determined by starting materials, processing, and firing conditions. The oxidizing or reducing nature of firing and use atmosphere can affect structure and properties. This is sometimes determined with a carbon monoxide resistance test, which evaluates the change in HMOR and other properties as well as microstructural alteration with prolonged high-temperature exposure to a carbon monoxide atmosphere. Refractories are termed low, medium, or high duty depending on their relative refractoriness within or between refractory types.

Refractories must also resist *spalling,* cracking or flaking that results from thermal cycling and thermal shock, differential expansion of microstructural constituents, temperature-dependent phase changes, and thermal gradients. The ability to insulate or conduct heat depends on thermal conductivity (Fig. 4). Glass tends to lower thermal conductivity, but conductivity cannot be predicted based only on glass and crystalline content. Phase distribution, particle size, and porosity are also very significant factors. Dimensional stability is also important. Reversible expansion during heating and cooling (Fig. 5) must either be accommodated in design or minimized. If a refractory is not fully dense, further densification can occur during use, resulting in unacceptable dimensional change. Under load (including self-load) this dimensional change (often called "subsidence") can lead to system failure. Minimization of open and closed porosity can avoid this effect. Porosity is closely controlled in refractory manufacture because it leads to a loss of mechanical strength and allows for penetrating chemical attack by gases and liquids. Internal pores do decrease thermal conductivity and can modestly improve fracture toughness (mechanical and thermal shock resistance). Improved strength, strength retention with thermal cycling, and blunting of cracks have all been observed in well-designed porous refractories.

The most common refractories are based on oxides of SiO_2, Al_2O_3, MgO, CaO, Cr_2O_3, and ZrO_2. All are considered "basic" except for SiO_2 and ZrO_2, which are considered "acidic." This distinction is based on historical concepts of the chemistry and chemical interaction of refractories with molten slags or fundamental chemical relationships. While these concepts have not proved fully accurate, the division is nonetheless useful. Oxide refractories are also divided into clay and nonclay materials. The composition and properties of selected basic refractories and high-duty refractories are given in Tables 8 and 9. The properties of selected refractory brick are given in Table 10, thermal conductivity with temperature in Fig. 4, and thermal expansion with temperature in Fig. 5.

The majority (about two-thirds) of refractories are used in the form of bricks, but a wide variety of other shapes and forms are available. For some critical applications such as high-attack zones in glass tanks, the ceramic is fusion cast into large structural shapes and precision cut to obtain maximum density and minimize brick junctures as zones of localized attack. Such melt casting is used for alumina, alumina-zirconia-silica (AZS), alumina-silica, magnesite, chrome-magnesite, zircon, and spinel. Monolithic refractories—castables, plastics (plastic-working consistency), mortars, gunning mixes, and ramming mixes—are placed or cast directly in the furnace and fired in place. Mortars can also be used to lay ceramic brick and repair worn or cracked refractories. Many of these materials involve cementitious reactions based on calcium aluminate and to a lesser extent portland cement (see the section "Portland Cement and Concretes" in this article). Frequently, users erroneously omit the required firing of the material after the cementitious reaction. The unfired material lacks strength and water resistance. Other systems involve phosphate bond and boron-base binder materials. The drying and firing steps are important in all these materials in order to cause the phase transformations that fully develop the desired phase structure and low porosity required for good strength, abrasion resistance, thermomechanical stability, environmental durability, and resistance to water. Some systems are also used to produce a porous insulating material by evolving gas during setting of the directly applied or cast monolithic.

Clay refractories include fireclay and alumina materials (see Tables 9 and 10). Fireclay refractories are usually made from kaolinite-base clay with minor amounts of other clays. The refractoriness increases as alumina levels are increased from about 25 to 45%. Alkali and iron impurities decrease the maximum use temperature because they modify the glassy bond. Fireclay provides low thermal expansion and intermediate insulation with moderate resistance to thermally in-

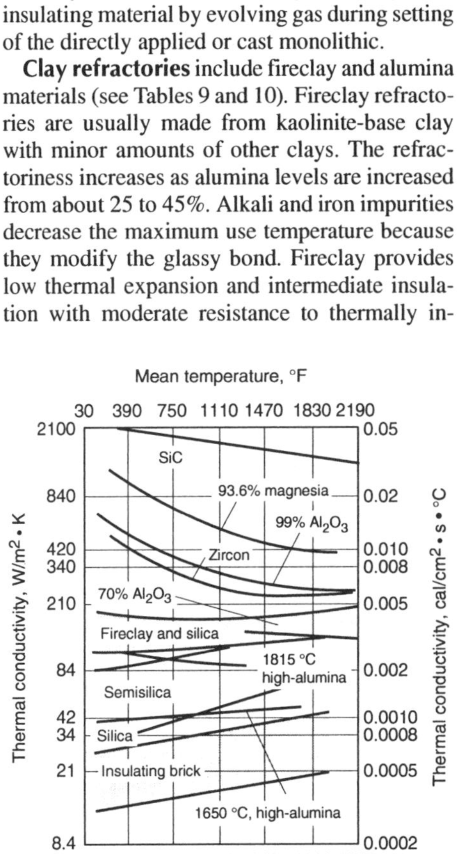

Fig. 4 Thermal conductivity of various types of refractory brick

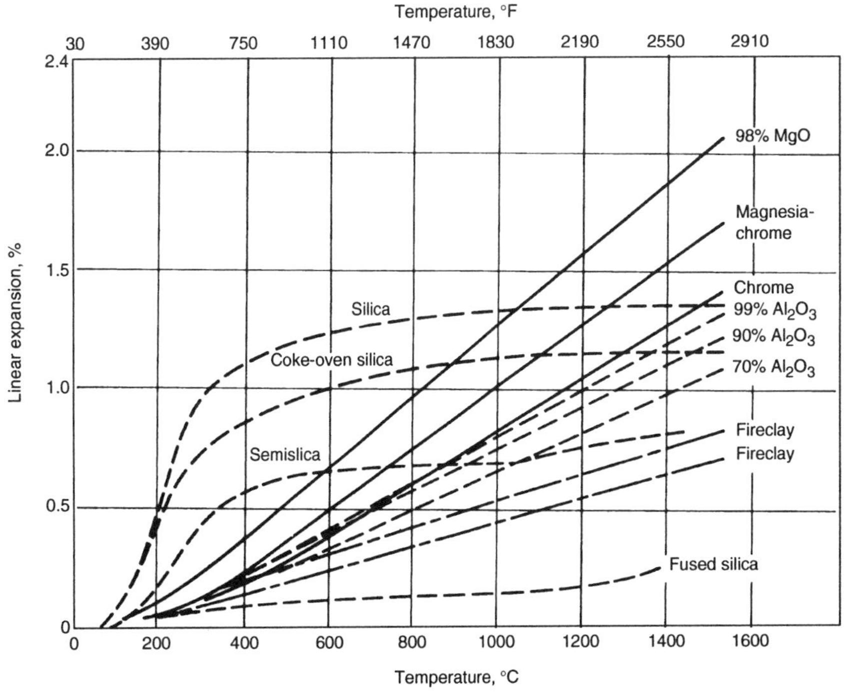

Fig. 5 Linear thermal expansion of magnesia and chrome brick (solid lines) and silica-alumina brick ranging in composition from 96% to 99% alumina (dashed lines)

duced stress. They provide poor resistance to bases at elevated temperature, but can perform adequately in acidic environments. Fireclay is used in steel pouring pit applications and as a general cost-effective refractory. Alumina refractories (Tables 5, 6, and 11) and are made from bauxite or other raw materials that yield a refractory with 50 to 87.5% alumina. They have superior volume stability and spall resistance as compared to other clay refractories. They are more resistant to chemical attack than fireclay and are used in more severe steelmaking applications. They are used in lower service parts of glass-melting furnaces and steel ladles. Phosphate-bonded and alumina-carbon brick are often included in this category.

Basic refractories are produced from various combinations of magnesite (dead-burned or fused), dolomite, chrome ore, and carbon (see Tables 9 and 10). Magnesia (magnesite) brick consists chiefly of periclase (MgO) in fired, pitch-bonded (tar-impregnated) or chemically bonded forms. For resistance to chemical attack and refractoriness, 80 to 95% MgO is used. Dolomite brick (CaO) is made from highly calcined dolomite ore ($CaCO_3 \cdot MgCO_3$) usually bonded with pitch or tar to extend service life by preventing hydration of the lime component. Dolomite brick is used in rotary cement kilns and steel ladles. Chrome (chromite) brick is made from natural chrome ore and is a complex combination of spinel minerals based on chrome, iron, magnesium, and aluminum oxides. It contains 30 to 45% chrome oxide. Some silicate phases are also present. They have excellent resistance to basic conditions and moderate resistance to acidic environments. Chrome-magnesite brick has outstanding resistance to steelmaking slags and is used in severe service areas of basic oxygen furnace vessels and electric arc furnaces. Thermal expansion is reduced by using a larger proportion of chrome ore than magnesite. High-purity magnesite often derived from seawater or brine wells provides superior resistance to attack. Pitch and tar-bonded materials and those containing flake graphite are often used. Very low porosity and high homogeneity can be achieved by melting the raw materials with an electric arc and casting large shapes (fusion cast) or by using fusion cast grain as a starting material for brick manufacture (rebonded fused grain, RFG).

High and extra-high alumina refractories are compositions containing more than about 85% alumina (see Tables 10 and 11). They are made from alumina (or bauxite for lower alumina levels) and are densely sintered or fusion cast. Usually, greater refractoriness is accomplished by increasing alumina content—nominally 100% alumina may be used to 1800 °C (3300 °F). Higher alumina levels also result in greater strength and resistance to attack. Induction furnaces and gates for continuous casting often use 90% alumina. Phosphate-bonded alumina is used in molten aluminum manufacture because of its relative resistance to attack and its suitable refractoriness.

Mullite refractories ($3Al_2O_3 \cdot 2SiO_2$) are similar in performance to high alumina refractories (see Tables 10 and 11). The aluminosilicate raw materials must be sintered at very high temperatures or electric arc melted to produce the mullite phase. The thermal conductivity of mul-

Table 9 Composition and selected properties of basic refractory materials

		Maximum use temperature in oxygen		Thermal conductivity, kcal/min. · °C			Refractoriness under load of 197 kPa (28.5 psi)	
Type	Composition	°C	°F	At 300 °C (570 °F)	At 800 °C (1470 °F)	At 1200 °C (2190 °F)	°C	°F
Silica	93–96% SiO_3	1700	3090	0.8–1.0	1.2–1.4	1.6–1.8	1650–1700	3000–3090
Fireclay	15–45% Al_2O3, 55–80% SiO_2	1300–1450	2370–2640	0.8–0.9	1.0–1.2	2.5–2.8	1250–1450	2280–2640
Magnesite	80–95% MgO, Fe_2O_3,Al_2O_3	1800	3270	3.8–9.7	2.8–4.7	2.5–2.8	1500–1700	2730–3090
Chromite	30–45% Cr_2O_3, 14–19% MgO, 10–17% Fe_2O_3, 15–33% Al_2O_3	1700	3090	1.3	1.6	1.8	1400–1450	2550–2640
Chrome magnesite	>60% MgO, Fe_2O_3, Al_2O_3	1800	3270	1.9–3.5	1.4–2.5	1.8	1500–1600	2730–2910

Table 10 Physical properties of some typical fired refractory brick

Property	Magnesia (95% MgO)	Chrome (30% Cr_2O_3)	90% alumina	70% alumina	Zircon	Fireclay (Missouri superduty)	Silicon carbide	Silica (superduty)
Bulk density, g/cm^3	2805–2950	3060–3140	2900–2965	2530–2600	3605–3720	2310–2370	2565–2660	1780–1875
Porosity,%	15–19	16–20	14–18	17.5–21.5	19–23	11–14	11–15	20–24
Cold crushing strength, MPa (ksi)	48–70 (7–10)	35–55 (5–8)	62–95 (9–14)	27–48 (4–7)	48–76 (7–11)	12–21 (1.7–.3)	69–83 (10–12)	27–41 (4–6)
Modulus of rupture, MPa (ksi)	17–24 (2.5–3.5)	14–21 (2–3)	17–21 (2.5–3)	7.6–11 (1.1–1.6)	15–23 (2.2–3.3)	4.8–6.9 (0.7–.1)	21–24 (3–3.5)	4.1–6.9 (0.6–1)
Reheat test, % permanent linear change after heating to:								
1600 °C (2910 °F)	...	...	...	3.5 to 6.0	...	0 to 0.9	...	...
1650 °C (3000 °F)	...	...	...	...	0	...	...	...
1725 °C (3135 °F)	–0.2 to 1.0	...	0.1 to 1.0	...	...	...	–0.1 to 0.1	...
Load test at 170 kPa (25 ksi); withstands load temperature, °C (°F)	1620 (2950)	1400 (2550)	1760 (3200)	1450 (2640)	1600 (2910)	1450 (2640)	1650 (3000)	1680 (3055)

Table 11 Composition and selected properties of high–duty refractory oxides

		Melting point		Maximum use temperature in oxygen		Thermal conductivity, kcal/min °C				Refractoriness under load of 196 kPa (28.4 psi)	
Type	Composition	°C	°F	°C	°F	At 100 °C (212 °F)	At 500 °C (930 °F)	At 1000 °C (1830 °F)	At 1500 °C (2730 °F)	°C	°F
Aluminum oxide	100% Al_2O_3	2015	3660	1950	3540	26.0	9.4	5.3	5.0	2000	3630
Beryllium oxide	100% BeO	2550	4620	2400	4350	189.0	56.3	17.5	13.5	2000	3630
Magnesium oxide	100% MgO	2800	5070	2400	4350	31.0	12.0	6.0	5.4	2000	3630
Silicon dioxide	100% SiO_2	...	...	1200	2190	0.8	1.4	1.8	...	...	...
Mullite	72% Al_2O_3, 28% SiO_2	1830(a)	3325(a)	1850	3362	5.3	3.8	3.4	...	...	...

(a) Incongruent

lite is significantly lower than alumina, particularly at temperatures below 1000 °C (1830 °F). This provides better insulation than alumina.

Silica refractories are listed in Tables 9 and 10. In Fig. 5 it can be seen that the thermal expansion of crystalline silica (quartz) is very high to about 500 °C (900 °F). In order to prevent thermal shock failure, the heating and cooling rates of silica refractories must be extremely slow through this range. Silica may be classed as superduty with very low alumina and alkali, regular (conventional duty), and coke-oven quality (free of iron spots, chips, and other defects). There are two corresponding ASTM designations based on adding the alumina content and twice the alkali level to yield a flux factor. Type A has a flux factor less than and type B greater than 0.5. Both must have a modulus of rupture greater than 3.5 MPa (500 psi). The fused (glassy) form shows very low expansion and has been developed for applications such as hot patching as well as shrouds, glass tank refractories, and linings for chemical reactors.

Silicon carbide (SiC) refractories have good load-bearing capability at very high temperature, excellent thermal shock resistance, and extremely high thermal conductivity (see Table 9 and Fig. 4). It is increasingly used for kiln furniture supporting and separating whitewares, particularly as ceramic fast firing has become more prevalent. Electric heating elements are also made of SiC. Silicon carbide refractories are produced by electric furnace reaction of sand and coke. Nitride-bonded SiC is used to process molten aluminum.

Zirconia is used as an additive to increase thermal shock resistance and lower corrosive attack in a variety of refractories (AZS, for example). As a refractory (Table 10), it is usually magnesia or calcia stabilized to prevent disruptive phase transformations on heating and cooling. Zirconia refractories are extremely resistant to attack by molten metals, glasses, and slags; have relatively low expansion; and possess high use temperature because of their very high melting point. Applications include the melting and casting of various specialty alloys.

Carbon and graphite can only be used in low oxygen or reducing environments. These refractories are resistant to many molten glasses, ceramics, and metals. Carbon-base refractories have high thermal conductivity and very high use temperatures.

Insulating refractories have a structure and composition to optimize insulating efficiency, maximum-use temperature, and subsidence under compressive load. When a containment re-

Table 12 Approximate composition and fineness ranges for ASTM standard types of portland cement

Compound/property	Type I	Type II	Type III	Type IV	Type V
	Composition, wt %/property value				
C_3S	42–65	35–60	45–70	20–30	40–60
C_2S	10–30	15–35	10–30	50–55	15–40
C_3A	0–17	0–8	0–15	3–6	0–5
C_4AF	6–18	6–18	6–18	8–15	10–18
CSH_2	3–6	3–6	3–6	3–6	3–6
Specific surface area, m^2/kg (ft^2/lb)	300–400 (1465–1955)	280–380 (1365–1855)	450–600 (2195–2930)	280–320 (1365–1560)	290–350 (1415–1710)

Source: Ref 2

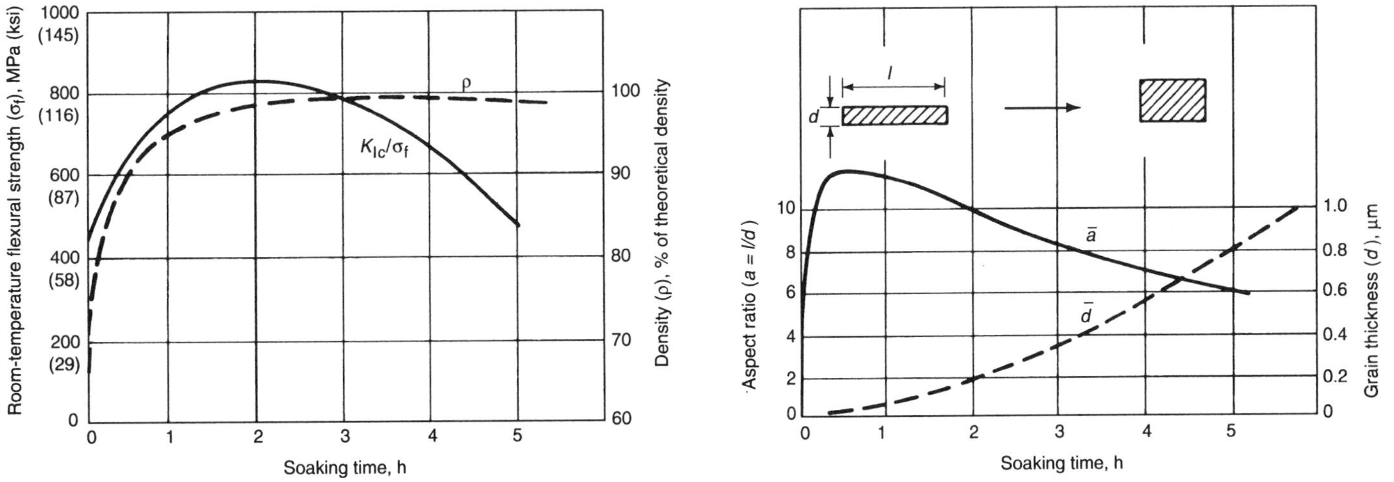

Fig. 6 Changes in flexural strength (σ_{Ic}) and fracture toughness (K_{Ic}) of silicon nitride as a function of aspect ratio and diameter of β-grains. Values of (K_{Ic}) and σ_f vary directly with the microstructure.

Table 13 Compressive strength data for typical cements versus time

Cement type	Description	Applications	1-day strength MPa	1-day strength ksi	3-day strength MPa	3-day strength ksi	7-day strength MPa	7-day strength ksi	28-day strength MPa	28-day strength ksi	91-day strength MPa	91-day strength ksi
Type I	General use	Pavement, sidewalks, buildings, bridges, other structures, tanks	9.3	1.3	22.5	3.3	32.0	4.6	42.0	6.1	50.5	7.3
Type II	Moderate heat of hydration	Piers, heavy retaining walls	14.0	2.0	27.0	3.9	36.6	5.3	46.3	6.7	52.5	7.6
Type III	High early strength	Structures put into service quickly	21.0	3.0	37.5	5.4	44.2	6.4	52.3	7.6	56.0	8.1
Type IV	Low heat of hydration	Large gravity dams	...	...	9.6	1.4	13.9	2.0	34.3	5.0	...	...
Type V	Sulfate-resistant	Structures in areas with high sulfate levels	...	...	22.1	3.2	29.5	4.3	41.3	6.0	...	...
White portland	Limited iron content	Architectural applications	...	...	26.5	3.8	36.2	5.3	46.5	6.7	...	...
Portland/blastfurnace	Contains blast furnace slag	General construction applications	8.6	1.2	13.1	1.9	25.3	3.7	45.0	6.5	53.3	7.7

Table 14 Mechanical properties of cements

Material	Strength, MPa	Fracture toughness, MPa
Conventional cement paste	7–10	0.2–0.4
Conventional concrete	2–3	0.2
Conventional fiber reinforced concrete	6–13	5–21
High-strength concrete	8–11	0.7–1.5
DSP cement	15–25	0.4–0.5
MDF cement	150–200	3
Glass fiber reinforced MDF cement	70–124	1.2–2.5
Polymer fiber reinforced MDF cement	54–120	1.7–2.8
Kevlar fiber reinforced MDF cement	58–128	2.1–5.3
Carbon fiber reinforced MDF cement	91–123	2.7–3.2
Aluminum fiber reinforced MDF cement	94–127	2.5–2.8

Source: Ref 3–5

fractory does not line the furnace, the inner insulating refractory must also be chemically resistant to the furnace environment—usually furnace gas and evolved by-products. Many insulating refractories are not mechanically resistant or chemically durable (resistant) in a water accident. The best insulating behavior is provided by maximizing the porosity (lightweight brick). The pores should be at a medium to fine level and tortuous to prevent air (gas) circulation. Additional benefit can be derived from a lower thermal conductivity and higher heat reflectivity for the ceramic. Thus, light-colored oxide ceramics are preferred with silica-base materials (fireclay, silica) providing better insulating power than alumina (high alumina, alumina). High porosity levels detract from two additional properties: crush strength and maximum-use temperature. Compressive strength decreases with porosity, but can be improved by selecting a refractory material with sufficient high-temperature strength and a pore structure that provides some mechanical rigidity due to the structural arrangement of the material. At high temperatures the insulating refractory continues to sinter or deform by viscous flow under load. This process proceeds more rapidly with a greater starting porosity or higher temperature. In glass-bonded material, glass modifiers—particularly alkaline oxides—will accelerate high-temperature subsidence. Formulations with higher temperature ratings, such as alumina, can be selected.

Because strength and subsidence requirements counter insulating power, a layered structure is usually used in furnace construction. The inner material is denser (or may be a containment refractory) designed to resist the highest temperature and the furnace environment. Successive layers are more porous, with greater insulating power, because they need only resist a lower temperature and are not exposed to attack. The outermost layers can be refractory fire-board or fiber-blanket materials with high loft (open volume) to provide maximum insulating power at lower temperatures.

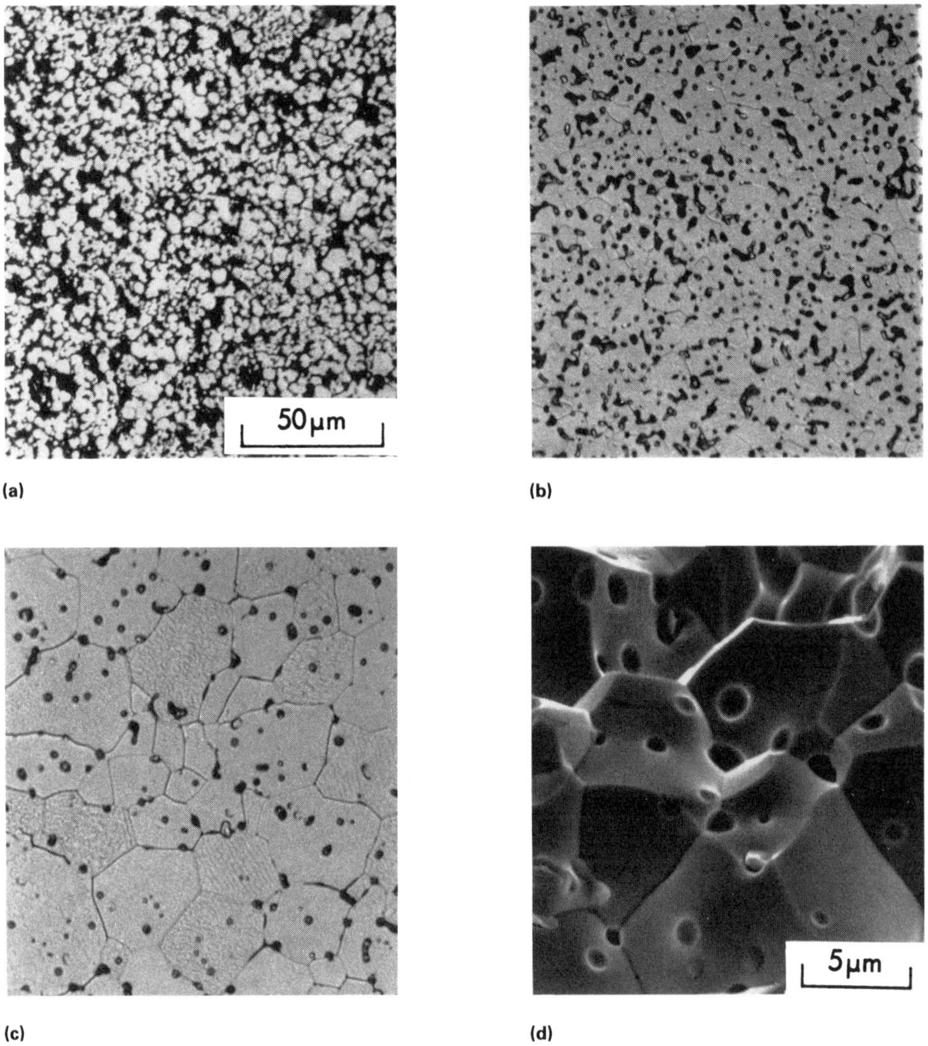

Fig. 7 Progressive densification and grain growth at several stages of sintering: (a) initial stage, (b) intermediate stage, (c) final stage, and (d) fracture surface. The fracture surface micrograph shows the desirable placement of spherical pores on grain boundaries in the final stage of sintering.

Portland Cements and Concrete

Portland cements are composed chiefly of calcium silicate powders, which when mixed with water to form a paste, react slowly under ambient conditions to yield a hard mass used chiefly for structural applications. The mass is a porous material composed primarily of poorly crystallized calcium silicate hydrate. When mixed with a fine aggregate less than 2 mm (0.1 in.) in size, such as sand, it is termed a mortar. A typical cement-to-aggregate-to-water ratio is 1 to 3 to 0.5 by weight. When the material contains a coarse aggregate greater than 5 mm (0.2 in.) in size, such as crushed rock or river gravel, it is termed a concrete. A typical cement-to-fine aggregate-to-coarse aggregate-to-water ratio is 1 to 2 to 3 to 0.5 by weight. The cementitious portion of the mixture may also contain chemical and/or mineral admixtures (discussed below). Both mortars and cements can be used as castable or moldable materials fabricated in a plant or on-site with steel or fiber reinforcement added for more demanding mechanical performance.

Blended cements are calcined combinations of raw materials such as limestone, low alkali clay, iron ore, and gypsum furnishing the common oxides ingredients: CaO (C), SiO_2 (S), Al_2O_3 (A), Fe_2O_3 (F), FeO (f), and S or SO_3 ($\bar{S}$) (the single letters are the usual industry designation). About 5% gypsum or other calcium sulfate is added during grinding of the fired clinker to prevent rapid setting. Most cements contain various proportions of alite (C_3S), belite (C_2S), aluminate phase (C_3A), ferrite phase (C_4AF), and gypsum ($C\bar{S}H_2$). Table 12 lists the compositions of ASTM standard portland cement types, and Table 13 gives applications and compressive strengths.

The amount of MgO (periclase) in the cement as well as the sodium and potassium levels must be controlled to preserve water and chemical durability. Insoluble silica residue and hydration

Table 15 Mechanical properties of selected advanced ceramics

Material	Crystal structure	Theoretical density, g/cm^3	Knoop or Vickers hardness, GPa	Knoop or Vickers hardness, 10^6 psi	Transverse rupture strength, MPa	Transverse rupture strength, ksi	Fracture toughness, $MPa\sqrt{m}$	Fracture toughness, $ksi\sqrt{in.}$	Young's modulus, GPa	Young's modulus, 10^6 psi	Poisson's ratio
Glass-ceramics	Variable	2.4–5.9	6–7	0.9–1.0	70–350	10–51	2.4	2.2	83–138	12–20	0.24
Pyrex glass	Amorphous	2.52	5	0.7	69	10	0.75	0.7	70	10	0.2
TiO_2	Rutile tetragonal	4.25	7–11	1.0–1.6	69–103	10–15	2.5	2.3	283	41	0.28
	Anatase tetragonal	3.84	...	...	...	...	...	...	...	...	...
	Brookite orthorhombic	4.17	...	...	...	...	...	...	...	...	...
Al_2O_3	Hexagonal	3.97	18–23	2.6–3.3	276–1034	40–150	2.7–4.2	2.5–3.8	380	55	0.26
Cr_2O_3	Hexagonal	5.21	29	4.2	>262	>38	3.9	3.5	>103	>15	
Mullite	Orthorhombic	2.8	...	...	185	27	2.2	2.0	145	21	0.25
Partially stabilized ZrO_2	Cubic, monoclinic, tetragonal	5.70–5.75	10–11	1.5–1.6	600–700	87–102	(a)	(a)	205	30	0.23
Fully stabilized ZrO_2	Cubic	5.56–6.1	10–15	1.5–2.2	245	36	2.8	2.5	97–207	14–30	0.23–0.32
Plasma-sprayed ZrO_2	Cubic, monoclinic, tetragonal	5.6–5.7	...	...	6–80	0.9–12	1.3–3.2	1.2–2.9	48(b)	7(b)	0.25
CeO_2	Cubic	7.28	...	...	...	...	...	...	172	25	0.27–0.31
TiB_2	Hexagonal	4.5–4.54	15–45	1.5–6.5	700–1000	102–145	6–8	5.5–7.3	514–574	75–83	0.09–0.13
TiC	Cubic	4.92	28–35	4.0–5.1	241–276	35–40	...	...	430	62	0.19
TaC	Cubic	14.4–14.5	16–24	2.3–3.5	97–290	14–42	...	...	285	41	0.24
Cr_3C_2	Orthorhombic	6.70	10–18	1.5–2.6	49	7.1	...	...	373	54	...
Cemented carbides	Variable	5.8–15.2	8–20	1.2–2.9	758–3275	110–475	5–18	4.6–16.4	396–654	57–95	0.2–0.29
SiC	α, hexagonal	3.21	20–30	2.9–4.4	(c)	(c)	(d)	(d)	207–483	30–70	0.19
	β, cubic	3.21	...	...	...	...	...	...	...	...	...
SiC (CVD)	β, cubic	3.21	28–44	4.1–6.4	(e)	(e)	5–7	4.6–6.4	415–441	60–64	...
Si_3N_4	α, hexagonal	3.18	8–19	1.2–2.8	(f)	(f)	(g)	(g)	304	44	0.24
	β, hexagonal	3.19	...	...	...	...	...	...	...	...	...
TiN	Cubic	5.43–5.44	16–20	2.3–2.9	...	...	...	...	251	36	...
BeO	...	2.8–2.9	11–14	1.6–2.0	...	...	5	4.6	340	49	...
MgO	...	3.5	5–6	0.7–0.9	...	...	3–5	2.7–4.6	300	44	...
$MgAl_2O_4$	...	3.2	12–15	1.7–2.2	...	...	2–5	1.8–4.6	260	38	...
$ZrO_2 \cdot SiO_2$	...	4.25	...	...	...	...	2–4	1.8–3.6	160	23	...
B_4C	...	2.3–2.5	27–31	3.9–4.5	...	...	...	...	450	65	...
BN	...	2.0–2.1	(h)	(h)	...	...	...	...	20–100(i)	3–15(i)	...
Graphite (C)	...	1.9	(j)	(j)	...	...	...	...	3–15(i)	0.4–2(i)	...
WC	...	15	13–16	1.9–2.3	...	...	(k)	(k)	600	87	...

(a) 8–9 (7.3–8.2) at 293 K, 6–6.5 (5.5–5.9) at 723 K, and 5 (4.6) at 1073 K, in units of $MPa\sqrt{m}$ ($ksi\sqrt{in.}$). (b) 21 (3) at 1373 K, GPa (10^6 psi). (c) Sintered: 96–520 (14–75) at 300 K, and 250 (36) at 1273 K. Hot pressed: 230–825 (33–120) at 300 K, and 398–743 (58–108) at 1273 K, MPa (ksi). (d) Sintered: 4.8 (4.4) at 300 K, and 2.6–5.0 (2.4–4.6) at 1273 K. Hot pressed: 4.8–6.1 (4.4–5.6) at 300 K, and 4.1–5.0 (3.7–4.6) at 1273 K, $MPa\sqrt{m}$ ($ksi\sqrt{in.}$). (e) 1034–1380 (150–200) at 300 K, and 2060–2400 (300–350) at 1473 K, MPa (ksi). (f) Sintered: 414–650 (60–94). Hot pressed: 700–1000 (100–145). Reaction bonded: 250–345 (36–50), MPa (ksi). (g) Sintered: 5.3 (4.8). Hot pressed: 4.1–6.0 (3.7–5.5). Reaction bonded: 3.6 (3.3), $MPa\sqrt{m}$ ($ksi\sqrt{in.}$). (h) Soft anisotropic. (i) Anisotropic. (j) Soft. (k) Up to 20 $MPa\sqrt{m}$ (18 $ksi\sqrt{in.}$) with cobalt additions

and carbonation products must also be kept within limits. Impurities can result in additional cement phases or different crystal forms not discussed above. Variations in raw materials, raw material processing, and kiln construction and operation can cause different clinker microstructure with resulting differences in performance.

Admixtures, finely divided materials blended with portland cement, can have significant effects on performance. Pozzolanic mineral additives are most common. These are siliceous or aluminosilicate materials that react with calcium hydroxide and water to form cementitious compounds at ambient temperatures. From 5 to 30% by weight may be added to the cement. Chemical admixtures are water-soluble organic compounds or inorganic salts added in small quantities (0.1 to 2%). Admixtures can significantly alter set time and/or reduce the amount of water required.

Concretes are composed of cement (with or without admixtures) and inert aggregate and reinforcement intended to maintain their form and mechanical properties in the hardened cement. They are formulated to provide suitable workability prior to setting and a required minimum compressive strength with time. To minimize cost, aggregate should be maximized. Greatest final environmental durability is often achieved with a minimum water-to-cement ratio and excess cement to ensure that all spaces between aggregates are filled. Reinforcing steel can be incorporated to provide strength, particularly in tension, and prevent catastrophic failure.

A number of new cements have been developed with significant strength and toughness enhancement (Table 14). These materials have low water-to-cement ratios, and low porosity is an important factor in their relatively high compressive and flexural strength. Special processing also prevents macrodefects such as bubbles, which decrease strength. Densified-with-small-particle (DSP) cements contain fly ash, silica fume, and superplasticizer in their formulation. Macro-defect-free (MDF) cements incorporate a small amount of polymeric material and one of a number of reinforcement fibers. Calcium-aluminate-base cements can also show superior properties to portland cement and are particularly useful for high-temperature applications. They can also show significant durability to water if fired to sufficient temperature to promote full phase development.

Engineering Ceramics

Ceramics that exhibit superior properties for demanding engineering applications are called *engineering ceramics.* If they are established materials, they are often referred to as *technical ceramics. Advanced ceramics* are relatively new materials or materials that show exceptional properties at the performance edge of ceramic technology. These terms are used interchangeably here, and mechanical performance is stressed. Several examples of engineering ceramics are presented earlier in this article. A major proportion of refractories are used for demanding engineering purposes and can be classed as technical ceramics. Porcelains and chinas used for electrical and chemical ware applications also could be classed as technical ceramics. Some of the new high-performance cements and concretes can be regarded as advanced materials. The discussion centers on newer monolithic materials with higher performance. A summary of properties of engineering ceramics is given in Tables 15 to 17.

Most engineering ceramics are made from relatively pure powders of well-characterized components with little batch-to-batch variability.

Table 16 Thermal properties of selected advanced ceramics

Material	Crystal structure	Coefficient of linear thermal expansion, 10^{-6}/K	Thermal conductivity, W/m · K	Specific heat, J/kg · K	Emittance(a)	Thermal shock resistance parameter(b)
Glass ceramics	Variable	5–17	2.0–5.4 at 400 K 2.7–3.0 at 1200 K	795–1298	0.9 at 300 K (T)	1.2(c)
Pyrex glass	Amorphous	4.6	1.3 at 400 K 1.7 at 800 K	335 at 100 K 1170 at 700 K	0.85 at 100 K (N) 0.85 at 900 K (N) 0.75 at 1100 K (N)	0.2
TiO_2	Rutile tetragonal Anatase tetragonal Brookite orthorhombic	9.4	8.8 at 400 K 3.3 at 1400 K	799 at 400 K 920 at 1700 K	0.83 at 450 K (T) 0.89 at 1300 K (T)	0.2
Al_2O_3	Hexagonal	7.2–8.6	27.2 at 400 K 5.8 at 1400 K	1088	0.75 at 100 K (N) 0.53 at 1000 K (N) 0.41 at 1600 K (N)	6.5
Cr_2O_3	Hexagonal	7.5	10–33 at 350 K	670 at 300 K 837 at 1000 K 879 at 1600 K	0.69 (N) 0.91 (N)	2.7
Mullite	Orthorhombic	5.7	5.2 at 400 K 3.3 at 1400 K	1046	0.5 at 1200 K (N) 0.65 at 1550 K (N)	0.9
Partially stabilized ZrO_2	Cubic, monoclinic, tetragonal	8.9–10.6	1.8–2.2	400	...	0.5
Fully stabilized ZrO_2	Cubic	13.5	1.7 at 400 K 1.9 at 1600 K	502 at 400 K 669 at 2400 K	0.82 at 0 K (N) 0.4 at 1200 K (N) 0.5 at 2000 K (N)	0.8
Plasma-sprayed ZrO_2	Cubic, monoclinic, tetragonal	7.6–10.5	0.69–2.4		0.61–0.68 at 700 K (T) 0.25–0.4 at 2800 K (T)	0.2
CeO_2	Cubic	13	9.6 at 400 K 1.2 at 1400 K	370 at 300 K 520 at 1200 K	0.65 at 1300 K (T) 0.45 at 1550 K (T) 0.40 at 1800 K (T)	...
TiB_2	Hexagonal	8.1	65–120 at 300 K 33–80 at 1100 K 54–122 at 2300 K	632 at 300 K 1155 at 1400 K	0.8 at 1000 K (N) 0.85 at 1400 K (N) 0.4 at 2800 K (N)	21
TiC	Cubic	7.4–8.6	33 at 400 K 43 at 1400 K	544 at 293 K 1046 at 1366 K	0.5 at 800 K (N) 0.85 at 1500 K (N) 0.38 at 2800 K (N)	2.2
TaC	Cubic	6.7	32 at 400 K 40 at 1400 K	167 at 273 K 293 at 1366 K	0.2 at 1600 K (N) 0.33 at 3000 K (N)	3.7
Cr_3C_2	Orthorhombic	9.8	19	502 at 273 K 837 at 811 K	...	0.2
Cemented carbides	Variable	4.0–8.3	14.3–119	197–544	...	13(d)
SiC	α hexagonal β cubic	4.3–5.6	63–155 at 400 K 21–33 at 1400 K	620–1046	0.85 at 400 K (N) 0.80 at 1800 K (N)	31
SiC (CVD)	β cubic	5.5	121 at 400 K 34.6 at 1600 K	837 at 400 K 1464 at 2000 K	...	...
Si_3N_4	α hexagonal β hexagonal	3.0	9–30 at 400 K	400–1600	0.9 at 600 K (N) 0.8 at 1300 K (N)	16
TiN	Cubic	8.0	24 at 400 K 67.8 at 1773 K 56.9 at 2573 K	628 at 273 K 1046 at 1366 K	0.4 at 800 K (N) 0.8 at 1400 K (N) 0.5 at 2100 K (N) 0.33 at 3000 K (N)	...

(a) N, normal; T, total hemispherical. (b) Calculated using $R = k\sigma(1 - \mu)/E\alpha$. (c) Corning grade 9606. (d) For Kennametal grade 701 tungsten carbide-cobalt

Controlled powder-processing techniques and consolidation are employed. Powder agglomeration is avoided as it can lead to inhomogeneity and large pores, which can act as strength-limiting intrinsic flaws and deteriorate other properties. Most advanced ceramics are fired as close to full density as is possible. With rare exceptions, porosity detracts from performance. Development of improved materials concentrates on relationships among composition, processing, microstructure, properties, and performance. Considerable understanding and new basic principles have been achieved during the past two decades and are emphasized in this section. A few examples of important new or developing materials are presented.

Structural Ceramics

Structural ceramics are intended to resist significant loads, be fracture tough, provide beneficial tribological properties, and/or exhibit wear resistance, often taking advantage of the good performance of ceramics at elevated temperature and in corrosive environments. Ceramics also provide high stiffness and low weight. As indicated, microstructural uniformity and low porosity are very important to reliable, high performance. Applications include wear materials, bearings, cutting tools, automotive parts, power-generation equipment, biomedical prostheses, and various aerospace/military uses. Many of the newly developed materials show exciting performance (Table 15) and are receiving increasing industrial acceptance.

Optimization of strength and toughness are often at odds in terms of microstructure. Usually a fine grain size is required for strength, while toughness is enhanced by elongated grain structures that are quite large. Well-engineered efforts to introduce elongated grains can result in an optimal elongated grain microstructure for strength and toughness. Figure 6 shows such an optimum for silicon nitride, an important advanced ceramic for heat engine applications. Phases with elongated grains or whiskers may be added as a reinforcement in the starting batch material. When elongated grains are grown in situ during sintering and/or subsequent firing, the re-

Table 17 Engineering properties of α-Al_2O_3

Property	Single crystal	99.9%	99.9%	99.5%	96%	90%
Physical						
Density, g/cm^3	...	3.99	3.96	3.87	3.72	3.60
Grain size, μm (mil)	...	15–45 (0.6–1.8)	1–6 (0.04–0.24)	5–50 (0.2–2)	2–20 (0.08–8)	2–10 (0.08–0.4)
Surface finish, (authentic average), μm (mil)	...	62 (2.5)	50 (2)	87 (3.5)	162 (6.5)	162 (6.5)
Color	...	Translucent white	Ivory	Ivory	White	White
Mechanical						
Modulus of elasticity, GPa (10^6 psi)	434 (63)	393 (57)	366 (53)	372 (54)	303 (44)	275 (40)
Modulus of rigidity, GPa (10^6 psi)	...	162 (24)	158 (23)	151 (22)	124 (18)	117 (17)
Bulk modulus, GPa (10^6 psi)	...	234 (34)	227 (33)	227 (33)	172 (25)	158 (23)
Poisson's ratio	...	0.22	0.22	0.22	0.21	0.22
Flexural strength, MPa (ksi)						
At 25 °C (77 °F)	634 (92)	282 (41)	551 (80)	379 (55)	358 (52)	337 (49)
At 1000 °C (1830 °F)	413 (60)	172 (25)	413 (60)	...	172 (25)	...
Compressive strength, MPa (ksi)						
At 25 °C (77 °F)	...	2549 (370)	3790 (550)	2618 (380)	2067 (300)	2480 (360)
At 1000 °C (1830 °F)	...	482 (70)	1929 (280)	...	...	...
Tensile strength, MPa (ksi)						
At 25 °C (77 °F)	...	206 (30)	310 (45)	262 (38)	193 (28)	220 (32)
At 1000 °C (1830 °F)	...	103 (15)	220 (32)	...	96 (14)	108 (16)
Transverse sonic velocity, 10^3 m/s	...	9.9	9.9	9.8	9.1	8.8
Hardess (R45N)	...	85	90	83	78	79
Coefficient of thermal expansion, 10^{-6}/K						
From 25–400 °C (77–750 °F)	...	7.4	7.4	7.6	7.4	7.0
From 25–1000 °C (77–1830 F)	...	8.3	8.3	8.3	8.2	8.1
Thermal conductivity, W/cm · K						
At 20 °C (68 °F)	0.43	0.39	0.39	0.35	0.24	0.16
At 100 °C (212 °F)	...	0.28	0.27	0.26	0.19	0.13
At 400 °C (750 °F)	...	0.13	0.13	0.12	0.10	0.08
Dielectric constant at 25 °C (77 °F)						
At 1 kHz	...	10.1	9.9	9.8	9.0	8.8
At 1 MHz	...	10.1	9.8	9.7	9.0	8.8
At 10 GHz	...	10.1	9.8	9.7	8.9	8.7
Dissipation factor at 25 °C (77 °F)						
At 1 kHz	...	0.00050	0.0020	0.0002	0.0011	0.0006
At 1 MHz	...	0.00004	0.0002	0.0003	0.0001	0.0004
At 10 GHz	...	0.00009	0.0050	0.0002	0.0006	0.0009
Loss factor at 25 °C (77 °F)						
At 1 kHz	...	0.0050	0.020	0.002	0.010	0.005
At 1 MHz	...	0.0004	0.002	0.003	0.001	0.004
At 10 GHz	...	0.0010	0.005	0.002	0.005	0.008
Dielectric strength, AC, kV/cm (average root mean square values at 60 Hz AC)						
0.63 cm thick	...	90.5	94.5	86.6	82.6	92.5
0.13 cm thick	...	200.7	181.1	169.3	145.6	177.1
0.02 cm thick	...	...	314.9	330.7	228.3	299.2
Volume resistivity, Ω · cm^2/cm						
At 25 °C (77 °F)	...	...	$>10^{15}$	$>10^{14}$	$>10^{14}$	$>10^{14}$
At 500 °C (930 °F)	...	...	3.3×10^{12}	...	4.0×10^{9}	2.8×10^{8}
At 1000 °C (1830 °F)	...	...	1.1×10^{7}	...	1.0×10^{6}	8.6×10^{5}

Source: Ref 6

sulting material is said to be self-reinforced. These reinforcements may be distributed isotropically or anisotropically depending on the production method.

The grain structure that results from sintering depends strongly on the grain (powder) size and grain size distribution of the starting material as well as minor additive sintering aids and grain growth inhibitors. One approach to achieving fine ultimate grain size and high strength is to use very fine, uniform starting powder grain size. Some success has been found using uniform nanoscale powders, although such powders do complicate microstructural (final property) control during the materials handling and forming stages. Pore structure and grain growth interact and must be controlled; this can be achieved with selected additives and controlled firing time/temperature (Fig. 7) to minimize porosity and control grain growth. The existence of a few larger starting particles coupled with the effects of additives can cause exaggerated grain growth (Fig. 8). Such grains can act as strength, inhibiting flaws whether on the surface or in the interior of a ceramic. This can also be a method for introducing elongated particles for toughening.

The sintering additives and impurities can significantly affect final properties. Frequently, they form a glassy phase on grain boundaries that limits the strength and the high-temperature strength of a structural ceramic. Figure 9 shows that hot-pressed silicon nitride (HPSN) and hot isostatically pressed silicon nitride (HIPSN) have significantly higher strength than silicon nitride materials with more grain-boundary phase. The drop-off in strength above about 1100 °C (1850 °F) can be attributed to softening of a thin grain-boundary layer in the higher-performance materials. Another view is that the grain-boundary phase may provide some toughening at elevated temperature.

While texture—directional grain structure and/or crystal orientation—exists in many ceramics, its effect on mechanical behavior is not well documented. Some hot-pressed carbide and nitride ceramics have been seen to have as much as a 50% difference in strength relative to hot-pressing direction. The honeycomb monolith for catalytic converters is extruded green with grain orientation of the calcined kaolinite, talc, and alumina, which are its main constituents. After vitrification and cooling, cordierite forms with an orientation that has high strength and low thermal expansion in the long direction of the catalyst support. This is necessary to resist thermal shock failure of the part. In addition to controlling texture, the fine pore structure must maximize surface area for platinum catalyst deposition.

Alumina (Al_2O_3) ceramics are transition materials between technical and advanced ceramics. Alumina is used for such diverse application as electronic substrates, spark plugs, containment and insulating refractories, bearings, chemical containment vessels, water faucet valves, medical prostheses, and halogen lamp enclosures. It is also used for windows (infrared and nonscratch watch crystal) in its single crystal form, sapphire. It displays good strength, chemical inertness, wear resistance, and refractoriness. Alumina ceramics are usually classified according to their nominal alumina content, for example 96% alumina and 99.5% alumina. These are not specific materials because the percentage refers to the weight percent of alumina raw material in the body and many different grades of alumina can be used. The remaining percentage can be any additive, although similar additives are used for most materials. The higher level aluminas, such as 99.5% alumina, are usually made from a high quality technical grade of alumina with about 0.5% magnesia addition. There are usually low level impurities in the alumina which should be summed with the additive quantity to get the overall composition. The additive plus impurities form glass grain boundary phases, which have an important effect on properties. Some alumina dissolves into the glassy phase so that the amount of glass can be several times the amount of additive. For 96% alumina, an additive of silica and alkaline earth oxides is usually part of the formulation. The author has seen this composition vary widely among manufacturers—particularly among small job shops. Switching from one manufacturer to another can result in a rather different product.

A "higher" alumina, one with a greater alumina content, will generally show superior mechanical, thermal, and chemical resistance properties at a greater product cost for the same density (porosity). However, as more additive is incorporated in the alumina ceramic, firing to nominally full density can be done at lower temperature and in shorter time. Higher density results in superior performance, with the exception of thermal insulation, for which porosity is desired. The relationship among the amount of grain boundary glass, the fired density, and the firing cost must be engineered for the particular application.

The specific additives also have a major effect on properties. For example, a more soluble grain boundary phase (more glass modifier ions) can severely impair chemical resistance of even a very high alumina. Alumina grains may also be pulled out of the surface of a weak grain boundary glass under frictional loading. Thus, a valve used for dispensing water-based chemicals may deteriorate rapidly if the wrong additives and starting alumina grade are used. The observed differences in metal joining to alumina result in part because the grain boundary glass composition can affect metallization of alumina electronic substrates. Translucent alumina used for halogen lamp enclosures uses grain boundary glasses that match the refractive index of alumina so that light is not reflected and dispersed from alumina grain boundaries. The author and others have found that careful selection of the grain boundary glass and high microstructural uniformity can lead to major strength and toughness advantages. That is, a major composite strengthening (about double) and toughening effect relative to 99.5% alumina can be engineered by "adding" a major amount of suitable grain boundary glassy phase through the selection of a suitable composition and processing route. It has also been found that crystallizing the grain boundary by heat treatment can lead to considerable strengthening. The concepts of composite formulation and grain boundary heat treatment are now being applied to some proprietary advanced alumina (and other ceramic) formulations.

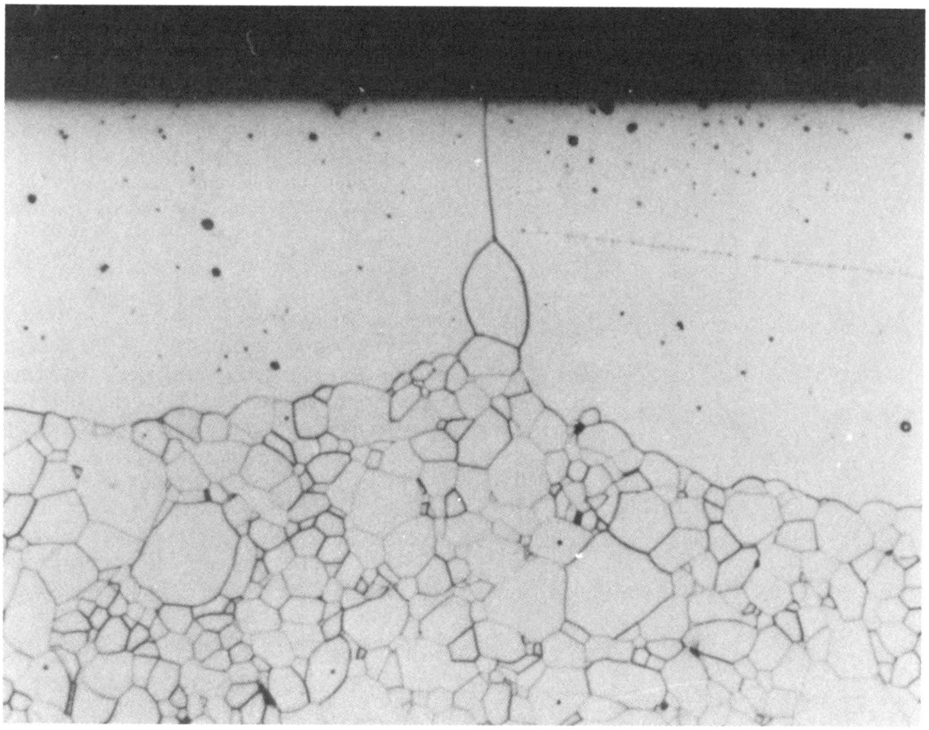

Fig. 8 Evaporation of MgO from the surface region of sintered MgO-doped Al_2O_3 that led to exaggerated grain growth. Source: Ref 7

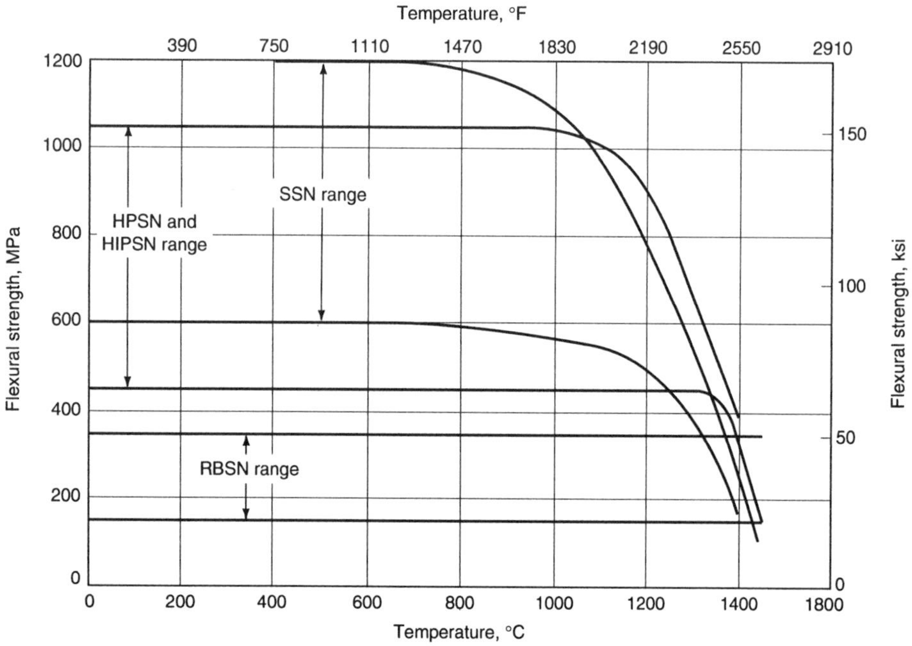

Fig. 9 Variation of flexural strength with temperature for various types of silicon nitride ceramics. SSN, sintered silicon nitride; HPSN, hot-pressed silicon nitride; HIPSN, hot isostatically pressed silicon nitride; RBSN, reaction-bonded silicon nitride

Transformation-Toughened Zirconia. One remarkable method of microstructural control to yield improved properties is transformation (martensitic) toughening, first observed in zirconia (ZrO_2) ceramics. Zirconia exhibits phase transformations from cubic to triclinic to monoclinic as its temperature is lowered. If the zirconia is stabilized with such materials as yttria, calcia, or magnesia, the triclinic phase can be preserved in a metastable state at ambient (or other temperature) dependent on the amount of additive, heat treatment, cooling rate, grain size, and grain size distribution. As shown in Fig. 10, the tetragonal zirconia can transform to monoclinic under the influence of a crack tip stress. The transformation may cause the crack tip to deflect due to the shear stress (about 1 to 7%) or to be impeded by the local compressive stress (>3%). A very significant toughening effect results (see Fig. 11), which can as much as triple K_{Ic}. A fracture toughness similar to high-strength aluminum alloys results, which makes the ceramic very resistant to impact failure. Flexural strengths as great as 2000 MPa (300 ksi) have been observed. It was found that these (and other structural ceramics) exhibit *R*-curve behavior in which the K_{Ic} increases as a crack grows in slow crack growth.

Zirconia-toughened ceramics also show remarkable resistance to abrasion and wear because surface layers transform under stress (Fig. 12), creating a highly compressive surface layer. This layer resists abrasion and the state of compression will renew if it is worn away. Zirconia cutting tools, thread guides, and wear surfaces have been found to last ten or more times longer than their tool steel counterparts in industrial applications. Some consumer applications such as household knives and scissors as well as break-

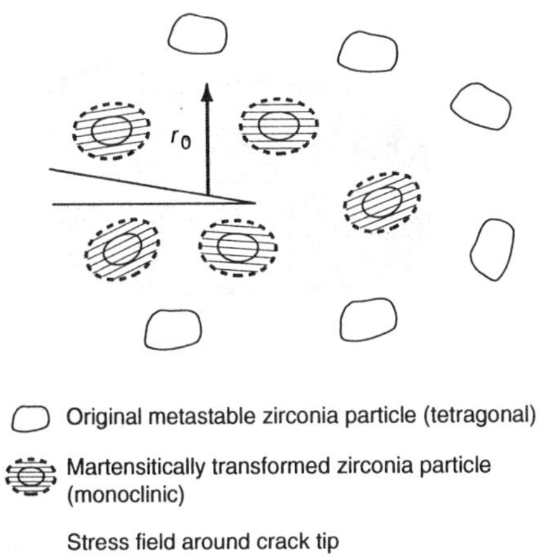

Fig. 10 Stress-induced transformation of metastable ZrO_2 particles in the elastic stress field of a crack

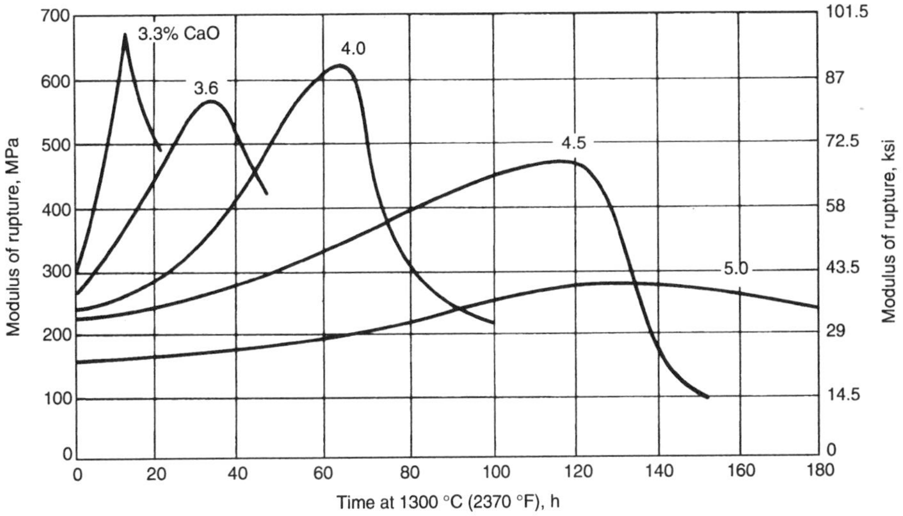

Fig. 11 Strength/aging curves obtained by heat treatment at 1300 °C (2370 °F) for various compositions of CaO-PSZ materials

proof buttons have been commercialized. Metastable tetragonal zirconia particles can also be added to various engineering ceramics and provide a substantial toughening effect. This is shown for alumina, silicon nitride, and silicon carbide in Fig. 13.

The strength and toughness properties of the transformation-toughened materials are useful only to slightly elevated temperature. Applications include industrial knives and cutting tools as well as consumer knives for home and sports use. Ceramic cutting tool blades retain their edge many times longer than the best metal knives—indefinitely in kitchen use. The material is also used for guides and wear surfaces in highly abrasive and corrosive environments. An intriguing application is for nonsparking hammers for flammable environments. The toughened materials are also being used in a variety of tool and die applications. A major success in biomedical applications has been it use as the ball for ball-and-socket joints such as the hip. The chemical inertness of zirconia coupled with the abrasion resistance instilled by transformation toughening have resulted in commercial development of zirconia valves for severe chemical and mechanical environments. Another novel application is for shirt buttons that do not break during aggressive laundering and pressing.

R-curve behavior is possibly the most exciting mechanical material discovery in recent history. It was first observed in zirconia ceramics, but has since been seen for alumina, reinforced materials, composites, and nonoxide ceramics. These materials show "*R*-curve" behavior when the progressive resistance to crack propagation is measured. This is shown schematically in Fig. 14. As a crack propagates, increased resistance to propagation occurs. *R*-curve behavior depends on a well-engineered microstructure. A variety of fiber, particulate, grain boundary and second-phase reinforcements impede or resist crack propagation. A number of systems have been developed, including the zirconia based system mentioned above, with toughness enhancements of a factor of two or more as compared to monolithic ceramics. The fracture toughness is not a single value as reinforcements behind the crack bridge it and hold it together. A microscopic crack could grow several centimeters before the maximum toughness value is achieved. This is something quite new in mechanical behavior of materials and suggests a new approach for relatively brittle materials—a margin of safety can be engineered into the system by employing the increasing resistance to crack propagation over a range of crack sizes. This could imply immunity to failure under conditions of controlled use.

Nonoxide ceramics such as silicon nitride (Si_3N_4), silicon carbide (SiC), and boron nitride (BN) show significant strength and toughness at ambient and high (over 1300 °C) temperatures. They are also relatively stable in sever chemical environments because of the covalent nature of their bond structure. The advanced nonoxide ceramics are often called "covalent ceramics." These materials also show a rather low friction coefficient and good wear resistance.

Silicon carbide and silicon nitride have been shown able to sustain high tensile loads at temperature in excess of 1500 °C (2730 °F) and 1300 °C (2370 °F), respectively. They are being implemented for next generation of auxiliary aircraft and turbines for electric power generation and rocket engines. The increased material operating temperature can result in a simpler and more efficient part, without the cooling passages required for a conventional metal turbine blade. For many high-temperature applications, most structural metals show poor creep resistance. These ceramic materials have relatively high fracture toughness and low thermal expansion, making them resistant to thermal shock and mechanical stress at elevated and room temperature. Flexural strengths exceed 700 MPa (100 ksi) for silicon nitride and 400 MPa (60 ksi) for silicon carbide. An example of the strength of these materials with temperature is shown in Fig. 15. These materials are being used in high-performance automotive applications as turbocharger rotors, valves, valve seats, and other parts. They are used regularly for the precombustion chamber and valve parts in truck applications.

The low friction and wear characteristics of nonoxide ceramics have resulted in a variety of bearing applications for severe environments and elevated temperatures. Other applications include wear plates for grinding and milling machinery, water jet nozzles, and high performance cutting tools. The inertness of the covalent ceramics has been applied to containment of chemically active materials, particularly at elevated temperature.

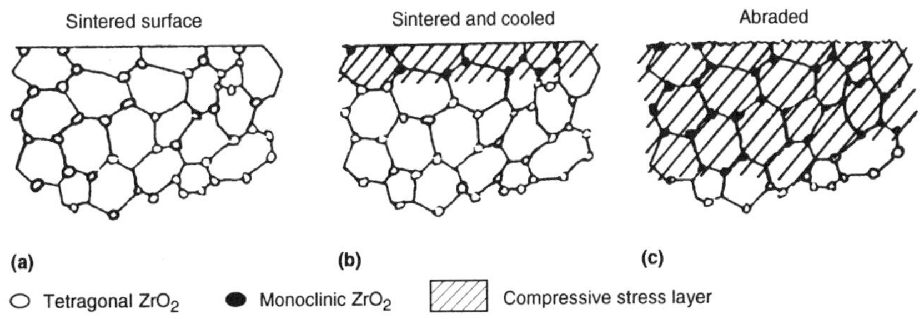

Fig. 12 Diagram of a section through a free surface at (a) the sintering temperature. On cooling, particles of ZrO_2 near the surface (b) transform due to reduced constraint, developing a compressive stress in the matrix. The thickness of this compressively stressed layer can be increased (c) by abrasion or machining.

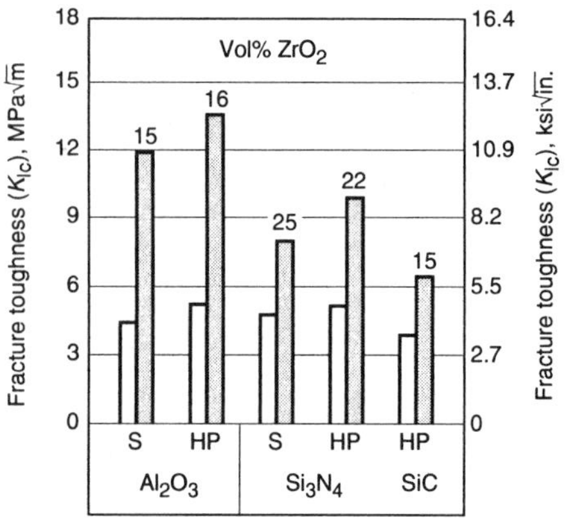

Fig. 13 Increase in fracture toughness (K_{Ic}) observed upon inclusion of zirconia particles in the ceramic matrix. The volume of zirconia added is shown in the shaded histogram. The toughness of the matrix material is shown in the adjacent white histogram bars. S, sintered. HP, hot pressed

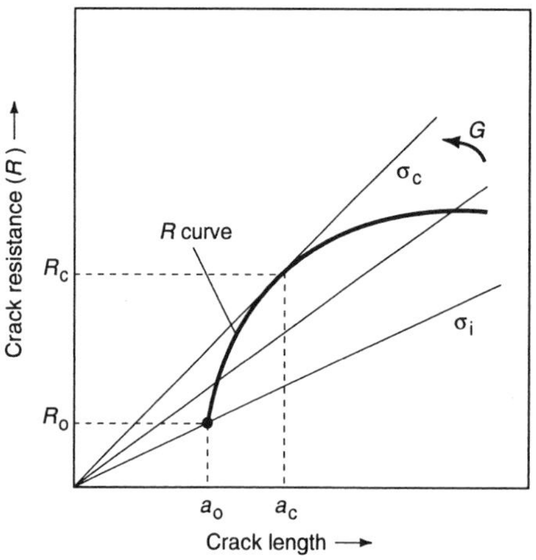

Fig. 14 Schematic representation of *R*-curve behavior. Crack resistance increases with crack length, so catastrophic failure will occur only when the strain energy release rate (*G*) exceeds the critical crack resistance value (R_c). a_o, initial crack size; a_c, critical crack size

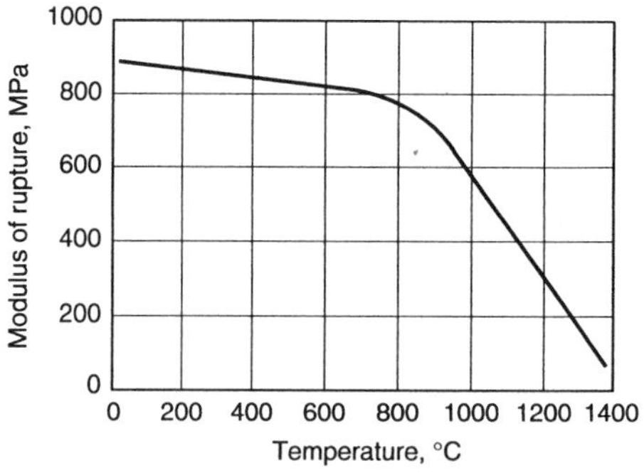

Fig. 15 Bend strength of a typical commercial monolithic silicon nitride as a function of temperature. Higher strength with temperature can be achieved with control of composition and processing, and by the incorporation of reinforcements.

Grain boundary phases—resulting from additives, oxidized material present on raw material particle surfaces, and oxygen introduced during processing—have a strong effect on properties. When suitable in chemistry and distribution, these secondary phases provide a composite strengthening effect, high temperature resistance, and fracture toughness. However, the reverse effects on properties can occur without sophisticated engineering of the chemistry and microstructure of the nonoxide ceramic. Many of the nonoxide ceramics have been engineered to contain randomly oriented, elongated second-phase particles, which can be "grown" during firing or mechanically mixed in the starting powders. Some materials, such as silicon nitride, have also been "self-reinforced" with elongated grains of the same material, which grows during the sintering process. These materials can provide superior mechanical strength and toughness at ambient and high temperature. Several commercial materials of this type display strong R-curve behavior, which may account for their superior strength and impact resistance in application.

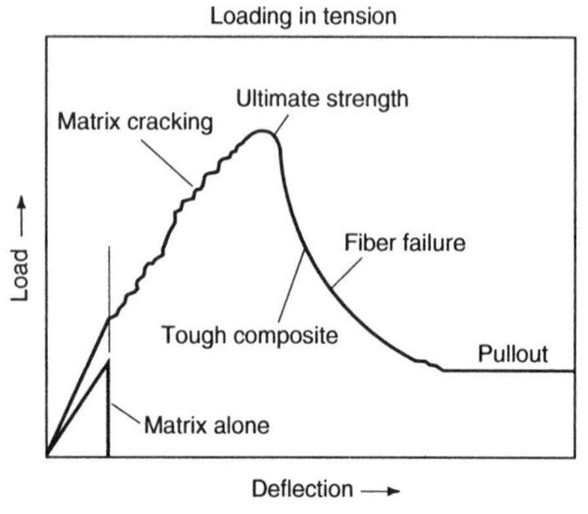

Fig. 16 Schematic load-deflection curve for a fiber-reinforced ceramic-ceramic composite. The increase in elastic modulus (initial slope) is imparted by the high-modulus fiber. An increase in strength occurs to the start of matrix cracking. The ultimate strength is significantly greater than that of the matrix alone. The crack aperture continues to be held together by fiber until fibers pull out of the matrix.

Continuous (ceramic) fiber-reinforced ceramic matrix composites (CFCMCs) show a unique resistance to conventional brittle failure. They also demonstrate a moderate increase in elastic modulus because of the incorporation of high-modulus fiber. The composite strength can be increased significantly, and the material demonstrates a unique "graceful" failure. The ceramic or glass matrix of the composite will fail at a stress higher than the monolithic ceramic, as shown in Fig. 16. Even when the matrix has failed completely the composite does not break. Instead, fibers bridge the crack and hold the material together. In order for matrix cracks to deflect around the fiber the fiber matrix bond cannot be too strong. This is accomplished by providing an interfacial debonding material between fiber and matrix. If the fibers and matrix are strongly bonded, failure cracks proceed straight through the fiber and little strengthening is achieved. For the weakly bonded interface case the entire matrix can crack without fiber failure. Such mechanisms lead to a substantial increase (3 to 4 times is common) in ultimate strength (Fig. 16). A noncatastrophic failure (ultimate strength in Fig. 16) results in which further energy is absorbed in fiber fracture and pull out of the fiber from the matrix. In a tensile test, the sample can be visibly two pieces, with a visible gap, and yet hold together. The result is a new stress-strain behavior, for which the correct design criteria are not yet fully determined. A noncatastrophic failure is possible, and there is some indication that partial failure in the form of matrix cracking can be repaired by a "crack healing" technique. This behavior requires a controlled bond between the fiber and matrix. This bond cannot be too strong in order to achieve the property enhancements. The distribution of fiber has a strong influence on strength and work of fracture (area under a stress-strain curve).

The first matrices for successful ceramic fiber reinforced composites were glasses such as lithium aluminosilicates (LAS glass). These materials were limited to temperatures of several hundred degrees Celsius. Newer, high temperature ceramic matrix composites have been developed that can be used at temperatures well over 1000 °C. Not only the thermal and oxidation resistance of the matrix is important. The environmental resistance of the fiber is also a matter of concern. If the matrix cracks and the fiber is susceptible to oxidation, as, for example, in the case of common silicon carbide reinforcements, the fiber will deteriorate with exposure. Most such fiber is derived from the pyrolysis of organic precursors, and residual oxygen can also detract from high-temperature fiber strength. The temperature resistance has been improved for silicon carbide fiber by improved process control in its manufacture. More oxidation resistant alumina and aluminosilicate fibers have been developed but are still relatively developmental.

The cost of both the fiber and production of a finished composite has prevented widespread application. High-quality silicon carbide fiber costs hundreds of dollars per pound while special fiber may cost as much as several thousand dollars a pound. Highest temperature resistance can be achieved by carbon-carbon composites, but these require proprietary protective coatings to keep from burning in air. Relatively high composite processing costs must be added to these raw material expenses. Much of the activity in this area has been fostered by military and aerospace applications where devices are already being tested and/or used. One application with some promise is electric power generation turbines. An example of a potential commercial application (is for turbine combustor linings. In such an application a 1.25% decrease in pressure drop could be achieved with a 0.75% increase in power and a 0.4% increase in efficiency. This would yield fuel savings of $370,000 per year in a typical turbine as well as significantly reduced emissions.

One new extension of the composite technology is the area of biomimetics. Composites can be designed which mimic the structure of biological systems such as wood, bone and shells. Such materials are of interest because they are often uniquely tough and strong. Composite principles can be extended to these materials. A wide variety of forming methods are used to produce these and other ceramic-fiber ceramic-matrix composites.

Substantial improvements have also been made in ceramic-fiber ceramic-matrix composites containing long fiber ceramic reinforcement. Both increases in strength and work of fracture can be achieved relative to the matrix material alone. Complete failure of the matrix can occur, while fibers still hold together the crack aperture. This implies both safety from catastrophic failure and potential for repairability.

Electronic and Magnetic Ceramics

Ceramics serve in many electronically passive and active roles. Electronic substrates and packages for hybrid microelectronics are made of alumina, beryllia, aluminum nitride, and boron nitride. Transducers and actuators are made from various formulations based on lead zirconate titanate (PZT), lead magnesium niobate (PMN), and lead magnesium titanium niobate (PMTN). Oxygen sensors are made from zirconia and temperature sensors from nickel and iron oxides and pyroelectrics ($LiTaO_3$). Barium titanate and other related materials are used for small, high-reliability ceramic capacitors. Semiconducting ceramics such as silicon carbide show promise for elevated-temperature applications, while high-temperature (high T_c) superconductors based on $YBa_2Cu_3O_{7-x}$ are already being used in some devices such as superconducting quantum interference devices (SQUIDs).

Hard and soft magnetic ceramics can be made from a variety of ferrites or garnets and magnetoplumbites or γ–Fe_2O_3. Applications include recording tape, high-speed switching (miniature) transformers, inductors, microwave generators, cryogenic refrigerators, and recording heads.

The properties of both electronic and magnetic ceramics depend very strongly on composition, including trace chemistry, and structure of raw materials as well as the processing and firing route chosen. Subtle changes in microstructure, chemistry, chemical distribution, grain-boundary material, and oxidation state can have profound effects on properties. It is beyond the scope of this article—which has focused on mechanical performance—to develop these issues further, but many of the considerations presented for mechanical applications can be extended to electronic and magnetic materials.

REFERENCES

1. A.K. Varshneya, *Fundamentals of Inorganic Glasses,* Academic Press, 1994
2. "Guide to the Selection and Use of Hydraulic Cements," ACI 225R-85, American Concrete Institute, 1985
3. J.F. Young, Macro-Defect-Free Cement: A Review, *Specialty Cements with Advanced Properties,* Vol 179, *MRS Symposium Proceedings,* B. Scheetz, A. Landers, I. Odler and H. Jennings, Ed., Materials Research Society, 1991
4. L. Stuble, E. Garbrozi, and J. Clifton, Durability of High-Performance Cement-Based Materials, *Advanced Cementitious Systems: Mechanisms and Properties,* Vol 245, *MRS Symposium Proceedings,* Materials Research Society, 1992, p 324–340
5. M.R. Sisbee, D.M. Ray, and M. Perez-Pena, Recent Developments in MDF Cement Materials: An Overview, *Advances in Cementitious Materials,* Vol 16, *Ceramics Transactions,* American Ceramic Society, 190, p 395–411
6. C.T. Lynch, Ed., *Handbook of Materials Science,* Vol 2, CRC Press, 1974, p 360–362
7. J.E. Burke, K.W. Lay, and S. Prochazka, The Effect of MgO on the Mobility of Grain Boundaries and Pores in Aluminum Oxides, *Materials Science Research,* Vol 13, G.C. Kuczynski, Ed., Plenum Press, 1980, p 417–425

SELECTED REFERENCES

- *American Ceramic Society Bulletin,* American Ceramic Society, Westerville, OH
- *Ceramic Forum International,* Bauverlag GmbH, Weisbaden, Germany
- *Ceramic Industry,* Business News Publishing Co., Troy, MI
- *Ceramic Science and Engineering Proceedings,* American Ceramic Society, Westerville, OH
- *Composites,* Butterworth-Heinemann US, Stoneham, MA
- *Composites Science and Technology,* Elsevier Science Publishers LTD, United Kingdom
- R.H. Doremus, *Glass Science,* 2nd ed., John Wiley & Sons, 1994
- *Engineered Materials Handbook Desk Edition,* ASM International, 1995
- *Engineered Materials Handbook,* Vol 4, *Ceramics and Glasses,* S.J. Schneider, Ed., ASM International, 1991
- I. Fanderlik, *Optical Properties of Glass,* Elsevier, 1983
- V.E. Henkes, G.Y. Onoda, and W.M. Carty, *Science of Whitewares,* American Ceramic Society, 1996
- J.T. Jones and M.F. Berard, *Ceramics—Industrial Processing and Testing,* Iowa State University Press, 1972
- *Journal of Composites Technology and Research,* ASTM, West Conshohocken, PA
- *Journal of the American Ceramic Society,* American Ceramic Society, Westerville, OH
- *Journal of the European Ceramic Society,* Elsevier Science Publishers LTD, United Kingdom
- W.D. Kingery, H.K. Bowen, and D.R. Uhlmann, *Introduction to Ceramics,* 2nd ed., John Wiley & Sons, 1976
- R.L. Lehman, et al., Materials, Chapter 12, *CRC Mechanical Engineering Handbook,* CRC Press, 1997
- J.S. Reed, *Principles of Ceramic Processing,* 2nd ed., John Wiley & Sons, 1995
- D.W. Richerson, *Modern Ceramic Engineering,* 2nd ed., Marcel Dekker, 1992
- E.B. Shand, *Glass Engineering Handbook,* 2nd ed., McGraw–Hill, 1982
- F.V. Tooley, *Handbook on Glass Manufacturing,* Ashlee Publishing, 1988
- *Uhlmann's Encyclopedia of Industrial Chemistry,* 5th ed., VCH, 1986

Effects of Composition, Processing, and Structure on Properties of Engineering Plastics

A.-M.M. Baker and C.M.F. Barry, University of Massachusetts Lowell

PLASTICS are so prevalent in our lives that it is easy to overlook the vast differences in their properties and how specialized many polymers have become. Consider the differences between aramid bulletproof vests and the polyurethane foam used in pillows. Why can plates made of crystallized polyethylene terephthalate be microwaved successfully while plastic film wrap (polyvinylidene chloride) has poor elevated-temperature properties? Consider how different polycarbonate is from plastic foam (expanded polystyrene); why is one plastic suitable for motorcycle helmets and the other for disposable coffee cups?

The answers to these questions lie in the chemical nature of the plastic and in the morphology assumed by the polymer chains. Plastics are polymeric materials that have been compounded with fillers, colorants, reinforcing agents, thermal stabilizers, plasticizers, and other modifying agents or additives. Table 1 shows the structures and transition temperatures of selected polymers. This article clarifies the importance of chemical composition and morphology to mechanical properties and reviews basic plastic processing techniques.

The difference between plastics and metals or ceramics is that plastics can be melted at relatively low temperatures and formed into a variety of shapes. Advantage can be taken of their non-Newtonian flow behavior in selecting a suitable molding or finishing process. Atoms can be specifically selected to design a polymer with the desired properties through a fundamental understanding of how submolecular, molecular, intermolecular, and supermolecular forces behave. A polymer scientist can custom polymerize a plastic to meet specific application requirements. This article starts at the most fundamental building-block level, atomic, then expands to a discussion of molecular considerations, intermolecular structures, and finally supermolecular issues. An explanation of important physical properties, many of which are unique to polymers, follows, and the final section discusses processing techniques.

More detailed information about properties and performance of polymers is provided in *Engineering Plastics,* Volume 2 of the *Engineered Materials Handbook,* and in the *Engineered Material Handbook Desk Edition,* both published by ASM International. Related coverage in this Volume is contained in the articles "Design with Plastics" and "Design for Plastics Processing."

Composition: Submolecular Structure

Polymers are organic molecules, with the carbon atom playing a critical role in developing their final properties. Hydrogen, oxygen, nitrogen, fluorine, and chlorine are among the many atoms that are built into polymer structures in order to tailor specific properties. Table 2 lists common atoms found in plastics and gives both the electronegativity (relative tendency to attract electrons) of the atom, as well as the number of unpaired electrons present in the outer shell. The number of unpaired electrons governs the number of covalent bonds the atom will form. The electronegativities of the constituent atoms that make up the polymer control its polarity. This, in turn, regulates the ability of the polymer to form the secondary bonds (e.g., hydrogen bonds) that have marked effects on the final thermomechanical properties.

Carbon is of fundamental importance as the most basic building block of all polymers. Carbon contains six valence electrons, two of which are located in the inner, most protected orbital, and all or four of which are in the outer orbital. It is the presence of four outer orbital electrons (exactly halfway between zero and eight) that causes carbon to be a neutral atom. Consequently, the electronegativity of carbon is 2.5. Metal atoms tend to be large, with a propensity to lose electrons when forming bonds; thus, their electronegativities are lower than 2.5. Elements that tend to gain electrons have electronegativities greater than 2.5. Carbon atoms share electrons when forming bonds with other carbons and, while the resultant materials can vary dramatically from diamond to graphite to hydrocarbon polymers such as polyethylene, the neutral carbon–carbon covalent bonds are stable to heat and ultraviolet (UV) light exposure.

Because carbon can form four bonds, it may bond more than once with other carbon atoms. As shown in Fig. 1, carbon–carbon single bonds are relatively stable. While carbon–carbon double bonds are shorter (as evidenced by their greater bond dissociation energy), they are more subject to attack by atmospheric oxygen. Consequently, polymers, such as polyisoprene and polybutadiene, are usually compounded with antioxidants. Carbon–carbon triple bonds are even more sensitive to oxygen attack. Although these are rarely found individually in commercial polymers, alternating triple and single bonds (called conjugated triple bonds) impart electrical conductivity to polymers, such as polyacetylene. Conjugated double bonds are more rigid. Rings of carbon–carbon single bonds, such as found in cyclohexane, assume nonplanar configurations. In contrast, rings of conjugated carbon–carbon double bonds, which occur in benzene, phenyl groups, and phenylene groups, are rigid and planar. As is discussed later in this article, these groups impart rigidity to polymers such as polystyrene (PS) and polycarbonate (PC).

Attaching other elements to a carbon atom introduces polarity, which changes the balance of the electron cloud. This can be regarded as either reducing the stability of an all-carbon material, or increasing its reactivity. Introducing polarity to the molecule through electronegativity differences between atoms has significant effects on thermal properties such as melting temperature (T_m) and mechanical properties such as Young's modulus (E). The presence of polar bonds produces higher thermal and mechanical properties

Table 1 Structures and properties of selected commodity and engineering plastics

Common name	Tensile strength, MPa	Glass transition temperature (T_g), °C	Melting temperature (T_m), °C	Repeating unit
Low-density polyethylene (LDPE)	10–12	−120	110	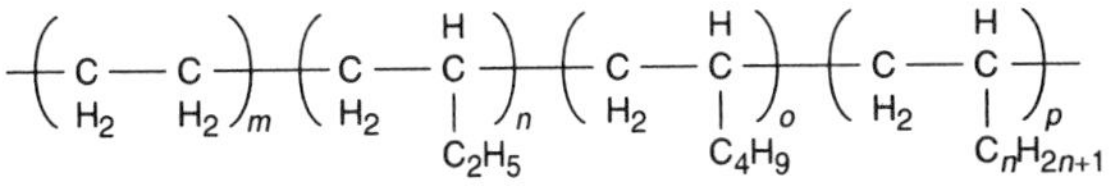
High-density polyethylene (HDPE)	26–33	−120	135	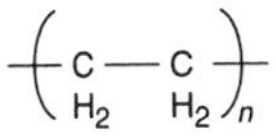
Linear low-density polyethylene (LLDPE)	15–32	−120	125	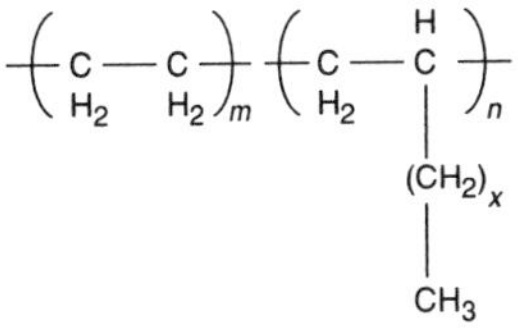 where x = 1 to 5
Isotactic polypropylene (PP or i-PP)	31–37	−10	165	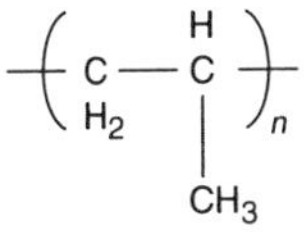
Cis-1,4-polyisoprene, natural rubber	...	−67	15–50	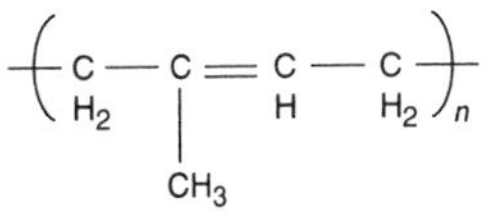
Trans-1,4-polyisoprene, gutta percha or balata	...	−71	56–65	
Polybutadiene:				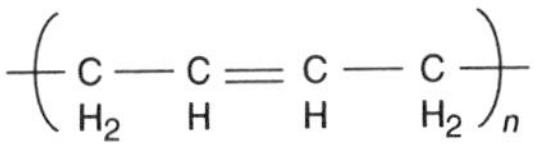
1,4-cis	21(a)	−102	...	
1,4-trans	14(a)	−107	...	
1,2-isotactic	10(a)	−15	...	
1,2-syndiotactic	11(a)	−15	90	
Poly-(4-methyl-1-pentene) (TPX)	28	55	245	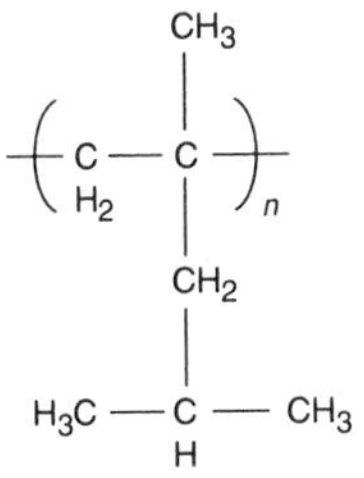
Atactic-polystyrene (PS or a-PS)	50	100	...	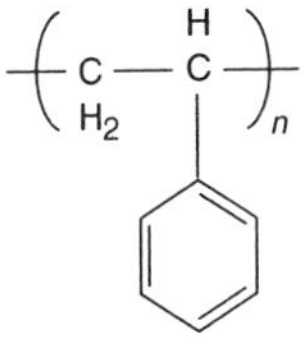
Syndiotactic-polystyrene (s-PS)	41	100	270	
Polymethylacrylate	...	0	...	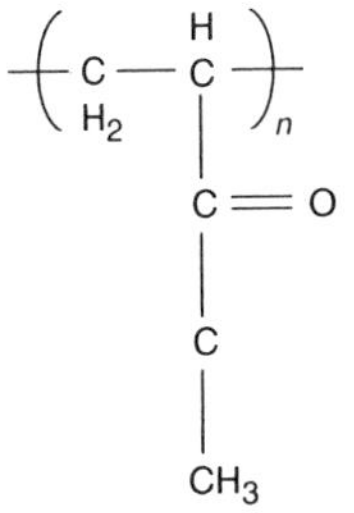

(a) When vulcanized. Source: Ref 1–6

(continued)

Table 1 (continued)

Common name	Tensile strength, MPa	Glass transition temperature (T_g), °C	Melting temperature (T_m), °C	Repeating unit
Polymethyl methacrylate				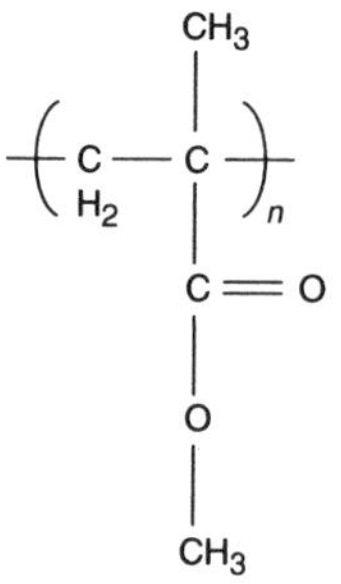
PMMA	70	100, 105	...	
i-PMMA	...	45	160	
Polyvinyl chloride (PVC)	55	80, 87	212	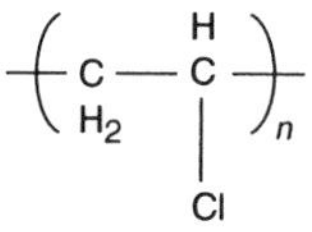
Polyvinylidene chloride (PVDC)	...	−17	198	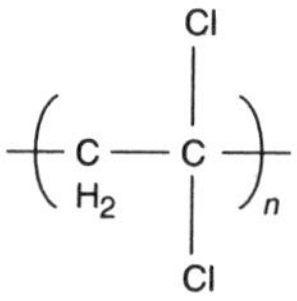
Polyvinyl fluoride (PVF)	66–131	−20	200	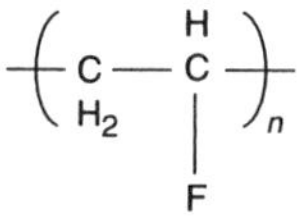
Polyvinylidene fluoride (PVDF or PVF_2)	48	−35	171	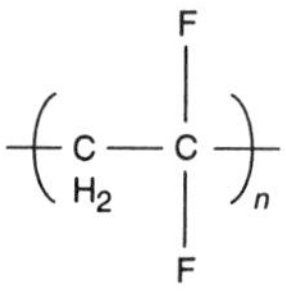
Polychlorotrifluoro-ethylene (PCTFE)	30–39	45, 52	220	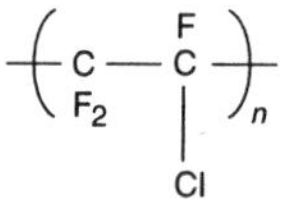
Polytetrafluoroethylene (PTFE)	17–21	126	327	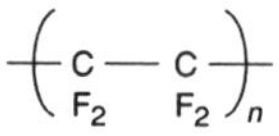
Polyvinyl acetate (PVAC)	Soft	29	...	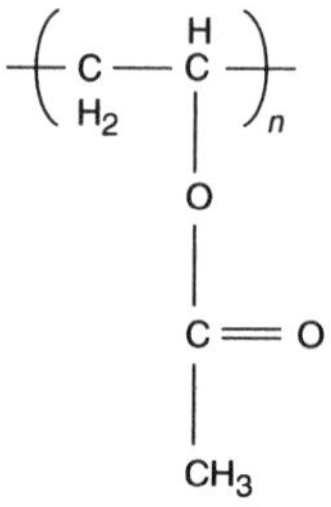
Polyvinyl alcohol (PVOH)	83–152	85	$T_d < T_m$	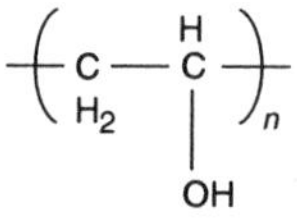
Polyacrylonitrile (PAN)	...	104	$T_d < T_m$	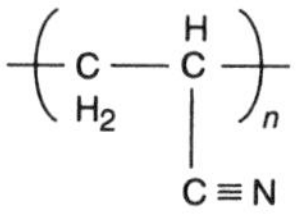

(a) When vulcanized. Source: Ref 1–6

(continued)

Table 1 (continued)

Common name	Tensile strength, MPa	Glass transition temperature (T_g), °C	Melting temperature (T_m), °C	Repeating unit
Polyoxymethylene (POM or polyacetal)	70	–50	175	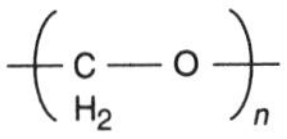
Polyethylene oxide (PEO)	13–22	–55	66	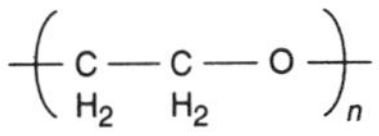
Polypropylene oxide	...	–62	65	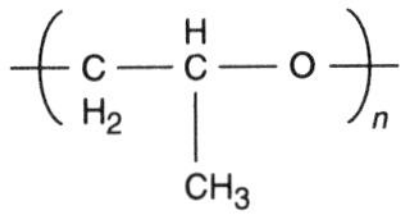
Polyamide 11 (nylon 11)	38	...	185	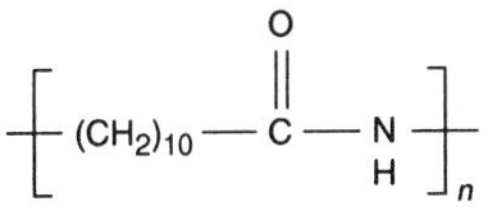
Polyamide 12 (nylon 12)	45	...	175	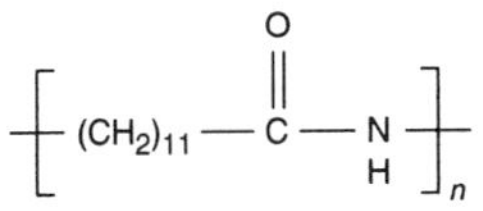
Polyamide 4/6 (nylon 4/6)	100	...	295	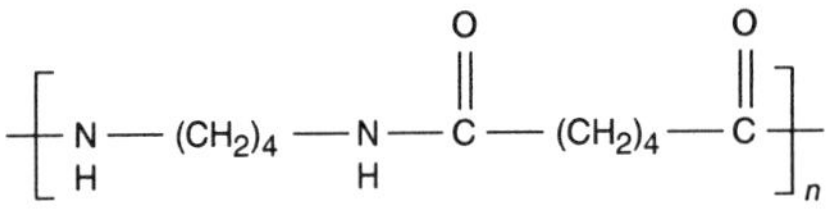
Polyamide 6/6 (nylon 6/6)	80	60	264	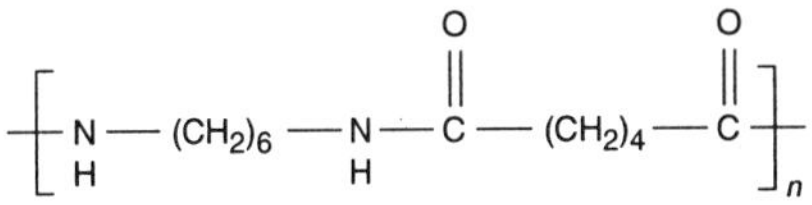
Polyamide 6/10 (nylon 6/10)	55	40	215	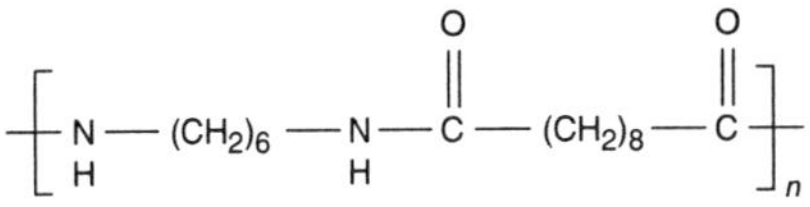
Polycarbonate (PC)	62	150	...	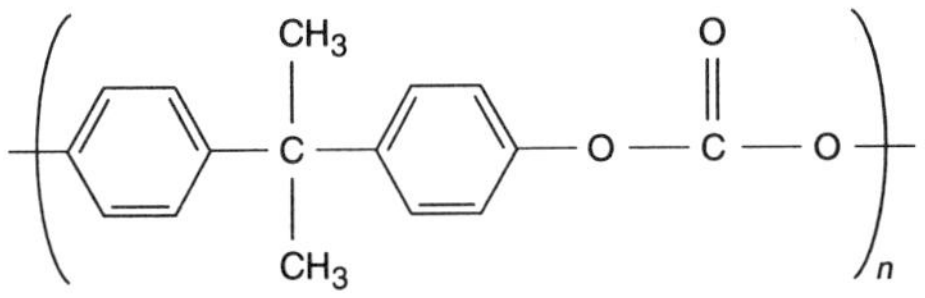
Polyethylene terephthalate (PET)	72	69	265	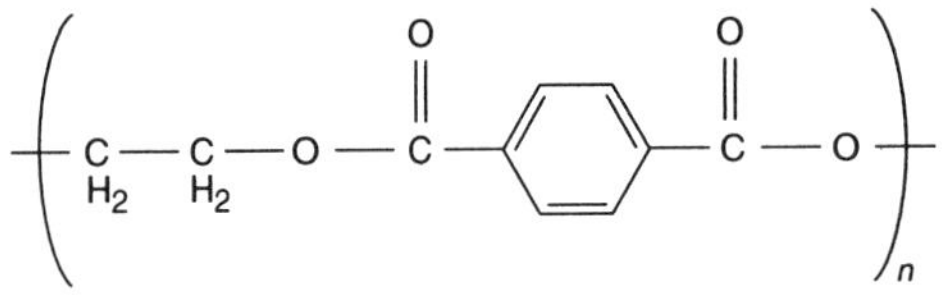
Polybutylene terephthalate (PBT)	52	60	232	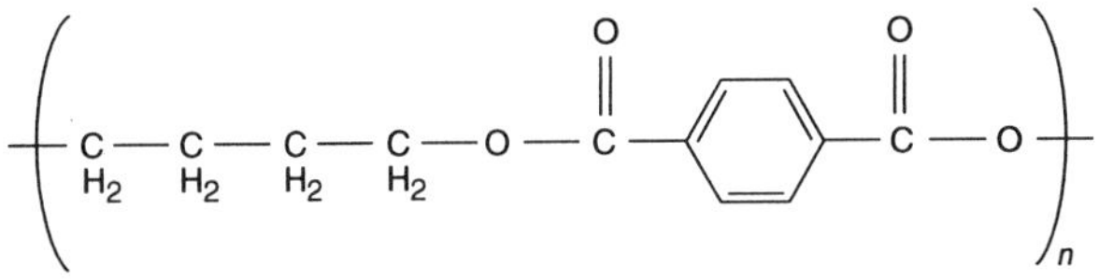
Polyetherimide (PEI)	105	215	...	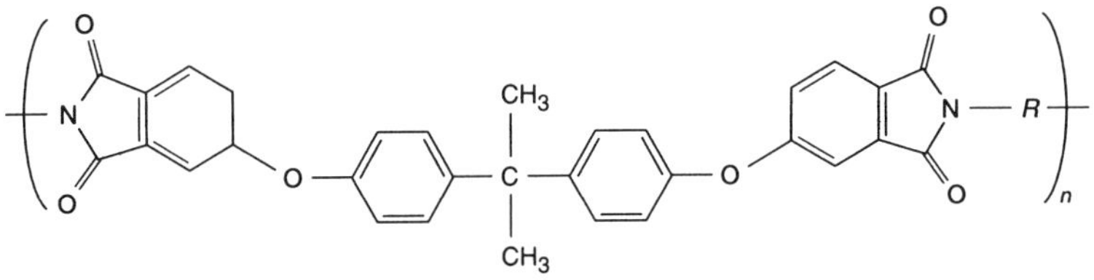

(a) When vulcanized. Source: Ref 1–6

(continued)

Table 1 (continued)

Common name	Tensile strength, MPa	Glass transition temperature (T_g), °C	Melting temperature (T_m), °C	Repeating unit
Polyamide-imide (PAI)	152	275	...	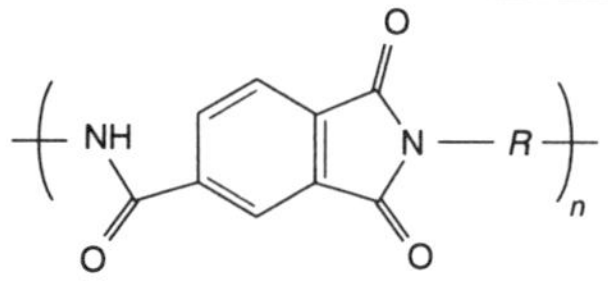
Polyimide (PI)	72–118	310–365	...	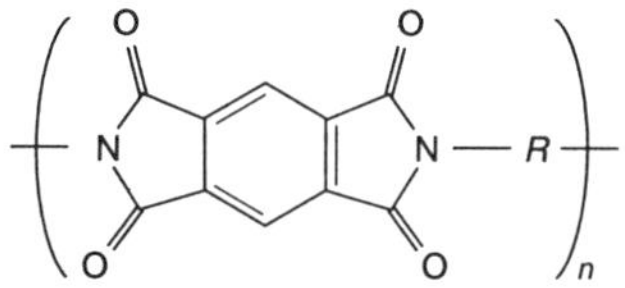
Polysulfone (PSU or PSF)	70	195	...	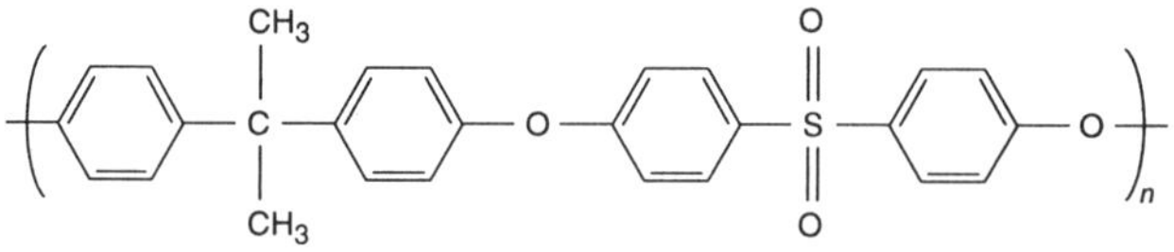
Polyarylether sulfone (PAS)	70	220	...	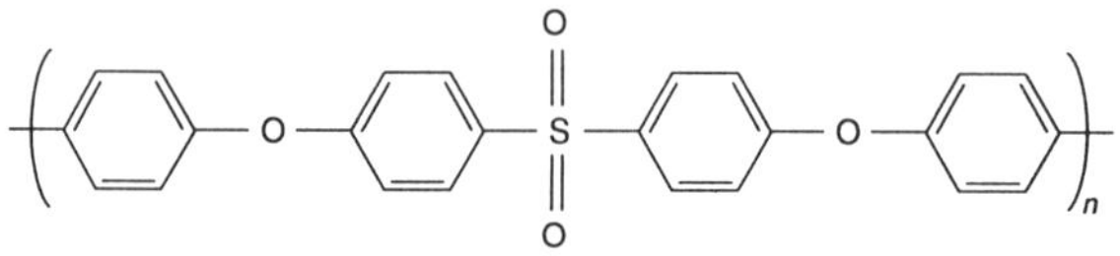
Polyether sulfone (PES)	90	230	...	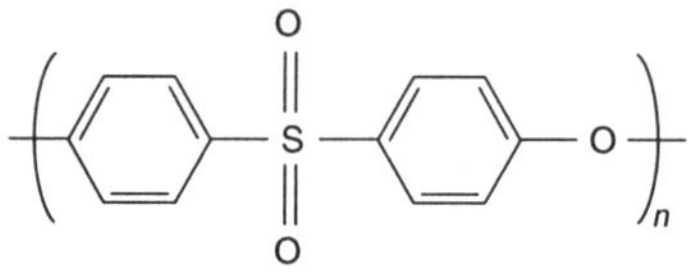
Polyphenylene sulfide (PPS)	70	85	288	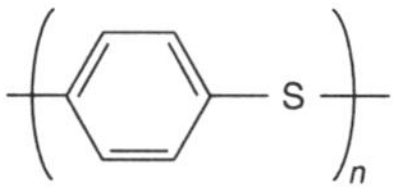
Polyetherketone (PEK)	110	155	365	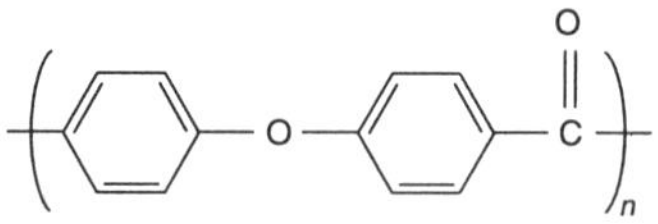
Polyetheretherketone (PEEK)	92	143	334	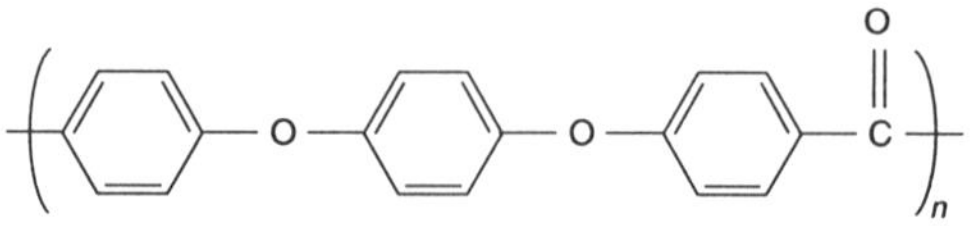
Polyetherketoneketone (PEKK)	102	156	338	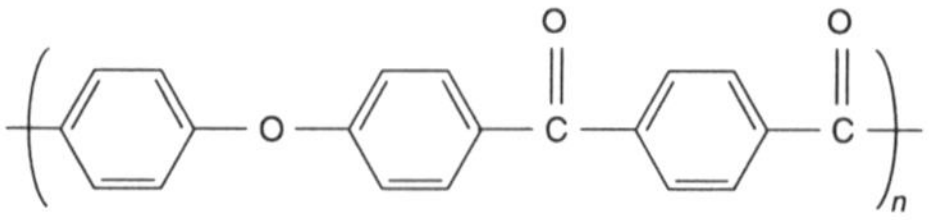
Polyetheretherketone-ketone (PEEKK)	100	167	360	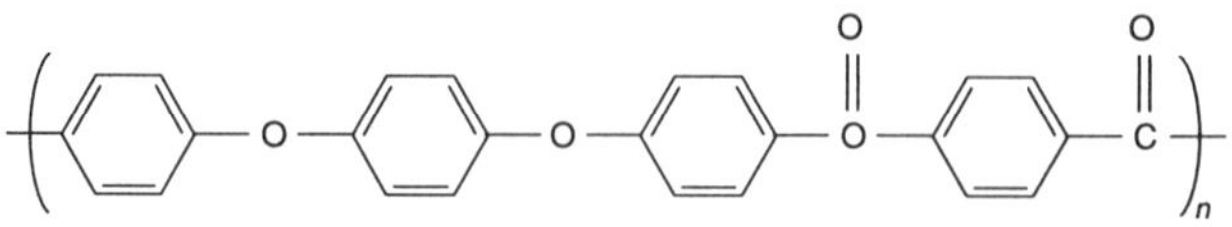
Polyetherketone-etherketone-ketone (PEKEKK)	118	170	381	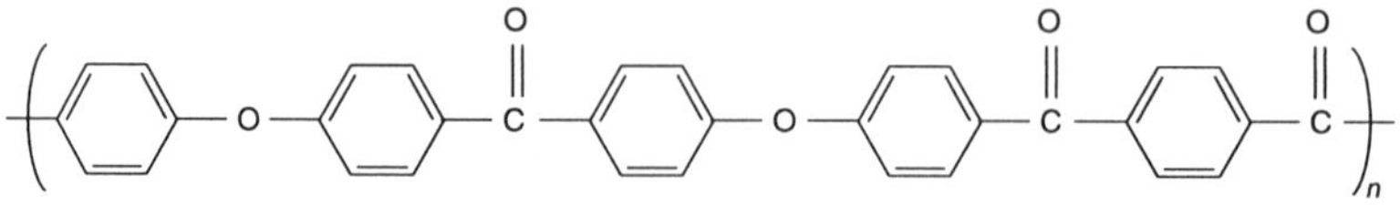
Polyphenylene oxide (PPO)	72	220	...	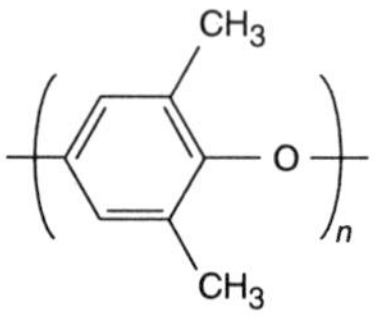

(a) When vulcanized. Source: Ref 1–6

(continued)

Table 1 (continued)

Common name	Tensile strength, MPa	Glass transition temperature (T_g), °C	Melting temperature (T_m), °C	Repeating unit
Modified polyphenylene oxide (PPO/PS)	55	140	...	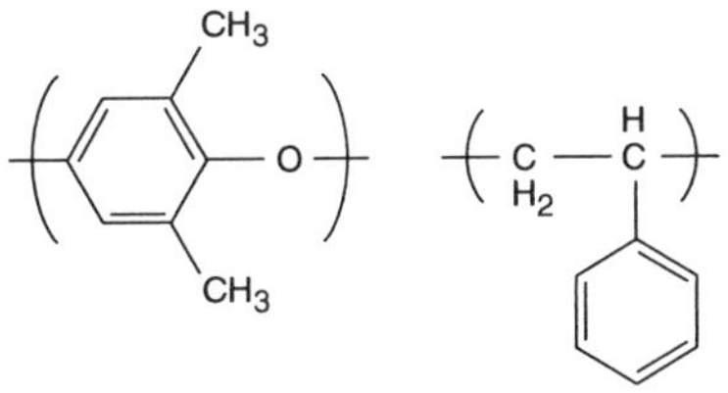
Polydimethylsiloxane (PDMS)	...	–123	–85 to –65	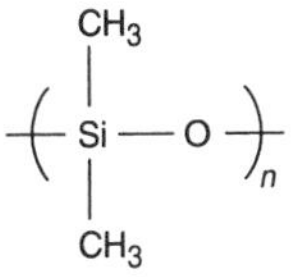

(a) When vulcanized. Source: Ref 1–6

engineering plastics. Figure 1 presents chemical groups commonly found in plastics and the bond dissociation energies (E_d) for selected groups.

Hydrogen. Because the electronegativity of hydrogen, 2.1, is only slightly more electropositive than carbon, carbon–hydrogen bonds are almost as stable as carbon–carbon bonds. In the absence of atmospheric oxygen, carbon–hydrogen bonds have good thermal and UV stability. Materials containing aliphatic (i.e., noncyclic) carbon–hydrogen bonds, such as polyethylene (PE) and polypropylene (PP), are marked by low surface energies, low adhesion, and low coefficients of friction. This makes PP automobile bumpers difficult to paint and is the why printing inks adhere poorly to untreated PE bags. Aromatic carbon–hydrogen bonds (for example, in a benzene ring) are stabilized by resonance and are more stable than aliphatic carbon–hydrogen bonds.

Hydrogen can also bond to elements other than carbon, such as oxygen, in the case of the common hydroxyl group, –OH. Due to the electronegativity of oxygen, the hydroxyl group is more polar and less balanced, making this bond more highly reactive than the previous bonds considered.

Table 2 Number of covalent bonds formed and electronegativities of atoms commonly found in plastics

Atom	Total number of electrons	Number of unpaired electrons	Number of covalent bonds formed	Electro-negativity(a)
H	1	1	1	2.1
C	6	4	4	2.5
N	7	3	3	3.0
O	8	2	2	3.5
F	9	1	1	4.0
Si	14	4	4	1.8
P	15	3	3 or 5	2.1
S	16	2	2 or 6	2.5
Cl	17	1	1	3.0
Br	35	1	1	2.8

(a) Electronegativity data from Ref 7

Oxygen. With an electronegativity of 3.5, oxygen introduces significant polarity to polymers. It is a unique atom in that it has two pairs of readily available unbonded electrons that can form fairly strong hydrogen bonds with neighboring molecules. These unbonded electron pairs also impart high surface energy to oxygen-containing polymers. Thus, such polymers have higher mechanical properties and provide better adhesion than nonpolar hydrocarbon polymers. Hydrogen bonds are further discussed in the section on intermolecular arrangements.

Carbon and oxygen are the components of several major functional groups shown in Fig. 1. The stability of the –C–O–C– ether bond is dependent on attached groups. Because aromatic ethers have a resonating system that includes the two electron pairs from the oxygen, the larger extended structure is stabilized through resonance. This contributes to the high thermal stability and high heat-distortion temperatures of engineering plastics such as polysulfones (PSUs) and polyetherketones (PEKs). In contrast, the bond of a hydrogen to an atom adjacent to the oxygen in an aliphatic ether (referred to as the α-hydrogen) is destabilized in the presence of the oxygen. Thus, polymers—such as polyvinyl acetals and cellulosics—exhibit instability because their $-O-CH_2-O-$ linkages are particularly sensitive to acid hydrolysis.

The carbonyl group of ketones, esters, and carbonates (shown in Fig. 1) strongly absorbs UV light in the 2800 to 3200 Å range, thus leading to polymer instability and poor outdoor aging characteristics. The ester group may hydrolyze and degrade upon exposure to water; manufacturers capitalize on this reactivity to produce polyvinyl alcohol (PVOH) from polyvinyl acetate (PVAC). Polyvinyl alcohol is a water-soluble, film-forming polymer that finds extensive use in applications ranging from photographic film to packaging. Polyvinyl acetate is not water soluble and is used in adhesives, textile applications, and latex paint.

Nitrogen, with an electronegativity of 3.0, generally forms strong bonds with carbon and, as in the case of oxygen, the unbonded electron pair generates a highly polar molecule available to form secondary bonds. The presence of both oxygen and nitrogen in the amide, urea, and urethane groups leads to strong hydrogen bonding and high sensitivities to water in the corresponding polymers. An alternative bond that nitrogen can form with carbon is an extremely rigid triple bond. This nitrile group is instrumental in generating high-modulus, heat-resistant engineering plastics such as styrene-acrylonitrile (SAN) copolymers and acrylonitrile-butadiene-styrene (ABS).

Fluorine is the most electronegative of all elements, with an electronegativity of 4.0. Its small atomic radius means that the carbon–fluorine bond length is very short. The strong bonds it forms with carbon impart low surface energy to fluoropolymers and allow them to be used for nonwetting applications such as nonstick cookware. The carbon–fluorine bond is also low in friction, which is suitable for high-lubricity applications such as mold lubricants and self-lubricating gears and bearings. This bond is extremely stable to heat, UV light, and chemical exposure making it appropriate for high-temperature plastics and elastomers. Table 3 highlights the effects of different degrees of fluorination on maximum-use temperature from PE to polytetrafluoroethylene (PTFE). It is evident that the reduction in fluorine content generates thermal instability, but does result in a more easily processed polymer. Highly fluorinated plastics such as PTFE are not melt processible by traditional methods.

Chlorine. While chlorine has seven valence electrons like fluorine, its larger atomic radius reduces its electronegativity to 3.0. Thus, chlorine bonds less strongly to carbon than does fluorine. The presence of such a large and electronegative atom generates polarity that has a marked effect on mechanical properties such as stiffness. A nonpolar molecule, PE, has a tensile modulus of 175 to 280 MPa and a T_m of 105 to 110 °C. Polyvinyl chloride (PVC), which substitutes a single chlorine atom onto the PE structure, has a tensile modulus of 2400 to 6500 MPa and a glass transition temperature (T_g) (amorphous) of 75 to 105 °C.

Composition: Molecular Structure

Polymer molecules contain multiple repeat units called mers. The number of repeat units can be varied and this strongly affects the thermal, mechanical, and rheological properties of plastics as shown in Table 4. Polymer size is quantified primarily by molecular weight (MW), molecular-weight distribution (MWD), and branching.

Molecular weight is generally defined as either number average ($\overline{M}_n$) or weight average ($\overline{M}_w$) depending on whether the length of each molecule is averaged according to numbers of molecules present at that length (as in the case of $\overline{M}_n$) or whether large molecules are more heavily considered (as in the case of $\overline{M}_w$). Equations 1 and 2 define $\overline{M}_n$ and $\overline{M}_w$, respectively, as:

$$\overline{M}_n \equiv \frac{\sum_{i=1}^{\infty} w_i}{\sum_{i=1}^{\infty} N_i} = \frac{\sum_{i=1}^{\infty} M_i N_i}{\sum_{i=1}^{\infty} N_i} \quad \text{(Eq 1)}$$

$$\overline{M}_w \equiv \frac{\sum_{i=1}^{\infty} M_i w_i}{\sum_{i=1}^{\infty} w_i} = \frac{\sum_{i=1}^{\infty} M_i^2 N_i}{\sum_{i=1}^{\infty} M_i N_i} \quad \text{(Eq 2)}$$

where w_i is the weight of polymer species i, N_i is the number of moles of species i, and M_i is the molecular weight of that species. If a polymer system has 7 moles of 20,000 MW species and 5 moles of 60,000 MW species, then the $\overline{M}_n$ and the $\overline{M}_w$ can be calculated as follows, according to Eq 3 and 4, respectively:

$$\overline{M}_n = \frac{(7 \cdot 20{,}000) + (5 \cdot 60{,}000)}{(7+5)} \cong 37{,}000 \quad \text{(Eq 3)}$$

$$\overline{M}_w = \frac{(7 \cdot 20{,}000^2) + (5 \cdot 60{,}000^2)}{(7 \cdot 20{,}000) + (5 \cdot 60{,}000)} \cong 47{,}000 \quad \text{(Eq 4)}$$

Because in the case of $\overline{M}_w$ the higher MW fractions of a polymer contribute more heavily, $\overline{M}_w$ is always greater than or equal to $\overline{M}_n$. $\overline{M}_n$ can be measured by methods that depend on end-group analysis or colligative properties such as osmotic pressure, boiling-point elevation, or freezing-point depression. $\overline{M}_w$ can be measured by light-scattering techniques or ultracentrifugation, both of which depend on the mass of species present (Ref 12).

As shown in Fig. 2, many physical and mechanical properties vary significantly as a function of MW, up to a threshold value, whereupon they level off asymptotically at higher MWs. Molecular entanglement can be dramatically demonstrated by the relationship of melt viscosity, η, to $\overline{M}_w$; melt viscosity being a measure of the tendency of the material to resist flow. Below a critical $\overline{M}_w$, denoted as M_c, there is little chain entanglement, and the melt viscosity increases linearly with $\overline{M}_w$ until it reaches the M_c threshold. At this point the melt viscosity increases as an exponential function of $\overline{M}_w$, with the exponent approximating 3.4 for many polymers, as shown in Fig. 3. In the elevated-slope region, molecular entanglements inhibit molecular slippage. The increased occurrence of physical chain entanglements associated with higher MWs accounts for the elevation of melt viscosity.

Below M_c the chains are short enough to align in the direction of flow and to slip past each other with relative ease. Once the critical length has been achieved, entangled polymers offer more resistance to the stresses inducing flow. This property, associated with the high MWs of engineering plastics, dramatically distinguishes them from Newtonian rheological behavior as is further explored in the sections on thermal and mechanical properties in this article.

This concept of M_c can be related to mechanical properties intuitively. The degree of intermolecular attractive forces is limited by the chain length. For example, at low MWs (below M_c), chain disentanglement can occur. Above a certain size (greater than M_c), the system is highly entangled and has maximized its intermolecular bonding such that it is now limited by the strength of the chain backbone. Most industrial engineering plastics have MWs well above M_c so that moderate changes in MW will not appreciably affect properties such as yield stress or modulus.

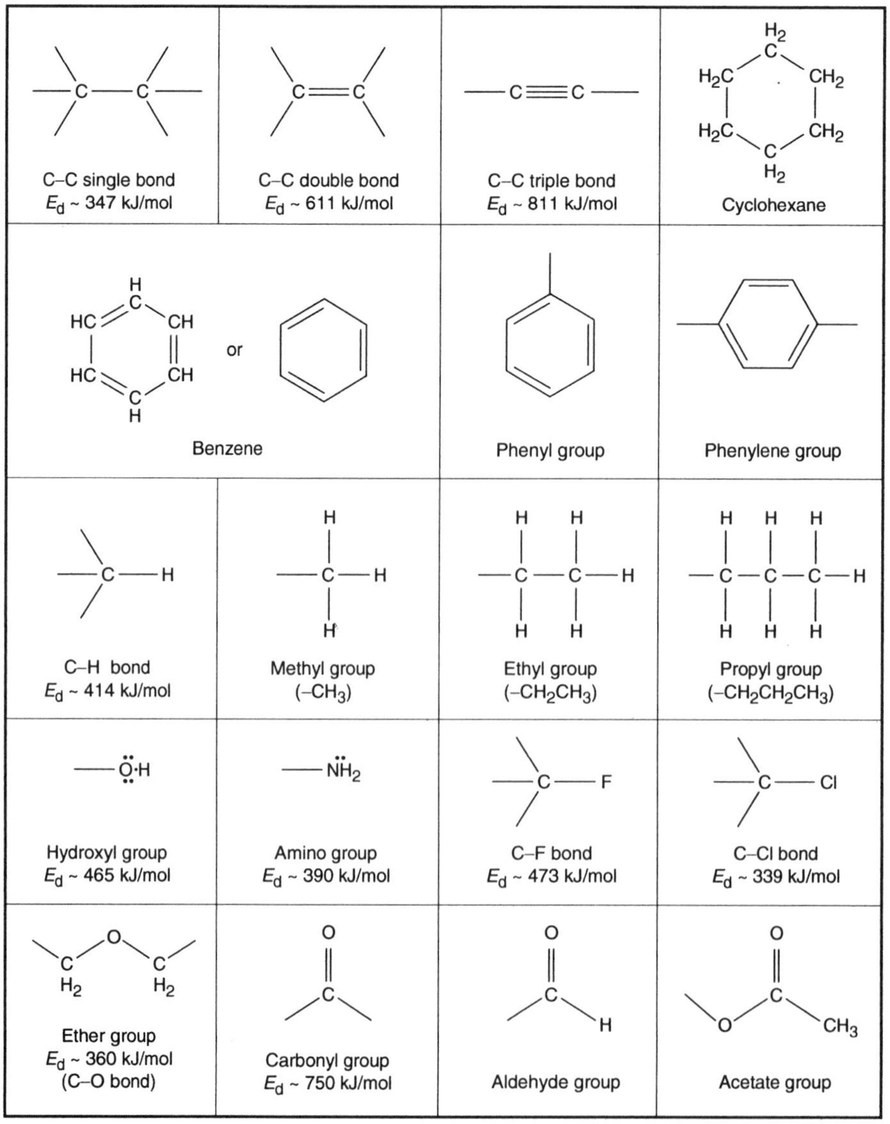

Fig. 1 Chemical groups and some bond dissociation energies (E_d) used in plastics. Adapted from Ref 8; dissociation energies from Ref 9

$\overline{M}_n$ finds relevance in relating properties that depend on small molecules (like environmental stress cracking resistance), while $\overline{M}_w$ is well suited for relating properties that depend on intermolecular attractions, because as chain length increases, the number of intermolecular bonds per molecule also increases. This is important when the property of interest measures the ability of a material to disentangle chains.

The breadth of MW range in a sample can be represented by a polydispersity index, which is equal to the ratio of $\overline{M}_w$ to $\overline{M}_n$. A material with a broad range of MWs (i.e., a high polydispersity index or broad MWD) will melt at lower temperatures than the equivalent material with a narrow range of MWs because the components with lowest MW will melt first.

Recent use of metallocene catalysts during polymerization has resulted in greater control over MWD. The narrow MWD linear low-density polyethylenes (LLDPEs) have better strength and heat-sealing properties because the lower MW components are no longer present. However, the lack of shorter polymer chains increases melt viscosity to such a degree that processing problems are often encountered. The use of blends of high- and low-MW LLDPE generates a bimodal MWD that produces a balance of good strength and ease of processing.

Chain branching also has a significant effect on flow properties. For a polymer of a given MW, the more highly branched the structure is, the lower its density will be and the lower the degree of entanglement. Moreover, for any given polymer, the lower its MW, the more flexible it will be as there are a greater number of chain ends per unit volume for short chain species. Chain ends reduce packing efficiency, and the additional free volume available offers sites into which the polymer can be displaced under stress. Once the MW is greater than the M_c the end-group concentration change is insignificant for further MW increases, and the mechanical properties plateau when the total intermolecular attractions are greater than the strength of the polymer backbone.

In addition to MW and chain branching, repeat units can be added in either a random or ordered fashion. In atactic polymers, such as PS and polymethyl methacrylate (PMMA), the mers are added randomly. In contrast, the repeat units of isotactic and syndiotactic polymers are ordered. Because the side chains of atatic polymers are randomly oriented as shown in Fig. 4(a), they inhibit crystallization (as is discussed later in this article). In isotactic polymers (Fig. 4b) the side chains all extend from the same side of the backbone, while in syndiotactic polymers they alternate sides (Fig. 4c). This regularity facilitates crystallization.

Inherent Flexibility. Before expanding the scope of consideration to include interactions between neighboring molecules, it is important to appreciate the inherent flexibility of the backbone of any given molecule. In this discussion it is first assumed that every carbon–carbon bond segment is completely free to assume any position as long as the equilibrium requirement that the carbon–carbon bond angle be maintained at 109° is met. A random conformation that might occur can be shown as:

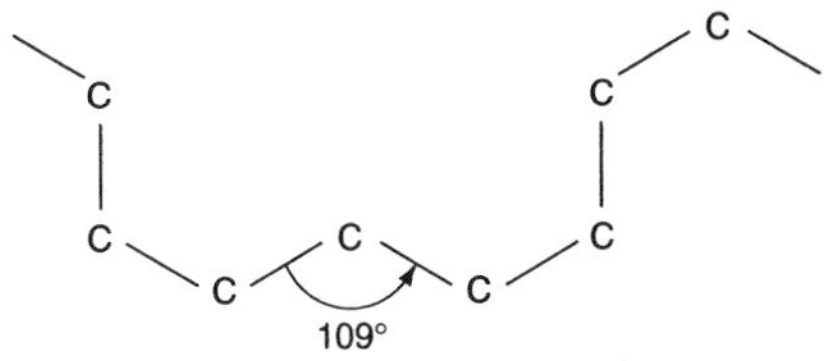

Inclusion of the hydrogen atoms (which fill the valence electron requirements of carbon) in the spatial consideration introduces limitations to the flexibility. The hydrogen atoms impose restrictions on the number of energetically viable positions that the chain can assume. Figure 5 plots an example of different energetically favorable conformations for the case of ethane (C_2H_6). It considers what happens as one carbon is rotated around the carbon–carbon bond and demonstrates the effects of trying to force the hydrogen atoms of one carbon atom to be spatially close to the hydrogen atoms of an adjacent carbon atom.

Figure 6 dramatically demonstrates the effects of the replacement of two hydrogens by carbon–carbon triple bonds. For example, there are fewer

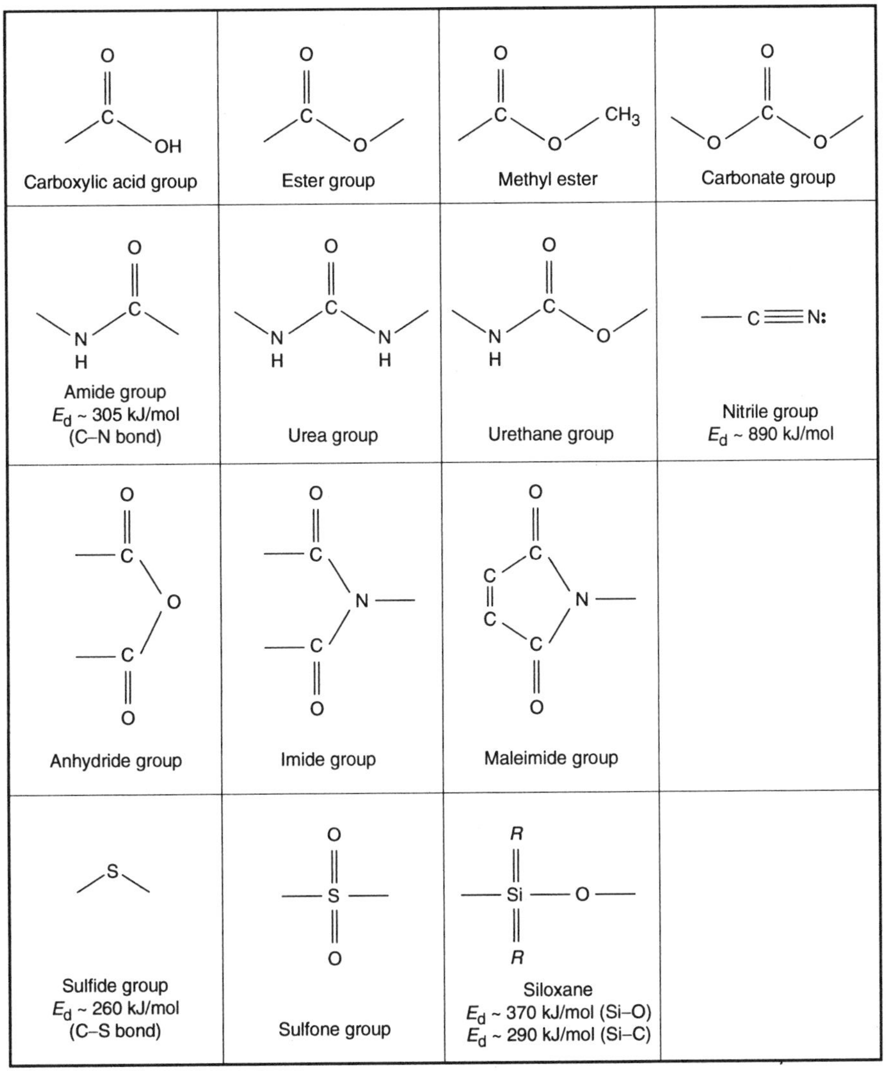

Fig. 1 (continued) Chemical groups and some bond dissociation energies (E_d) used in plastics. Adapted from Ref 8; dissociation energies from Ref 9

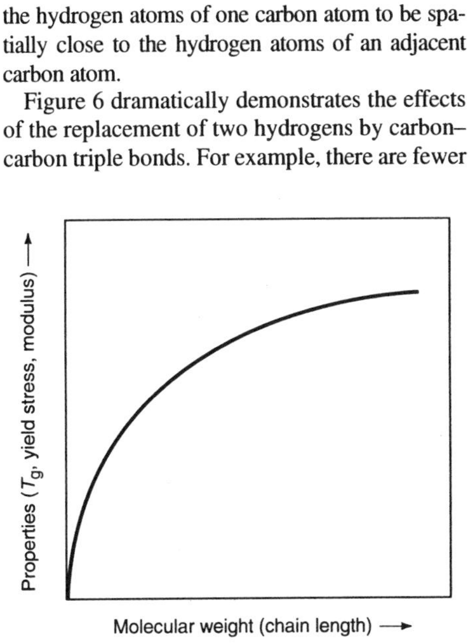

Fig. 2 General influence of molecular weight on polymer properties. Source: Ref 13

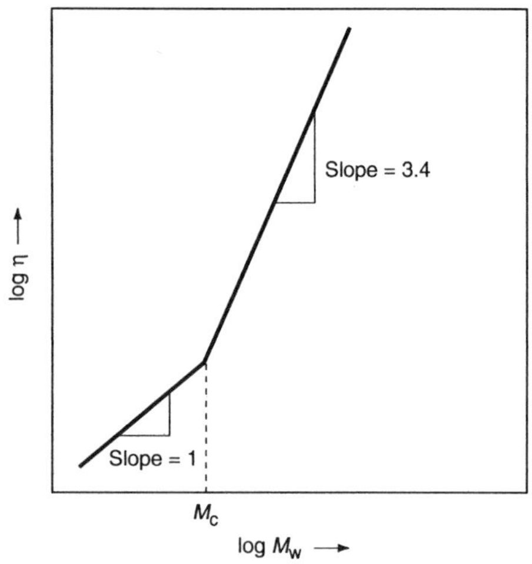

Fig. 3 Viscosity dependence on molecular weight exhibiting M_c. Source: Ref 14

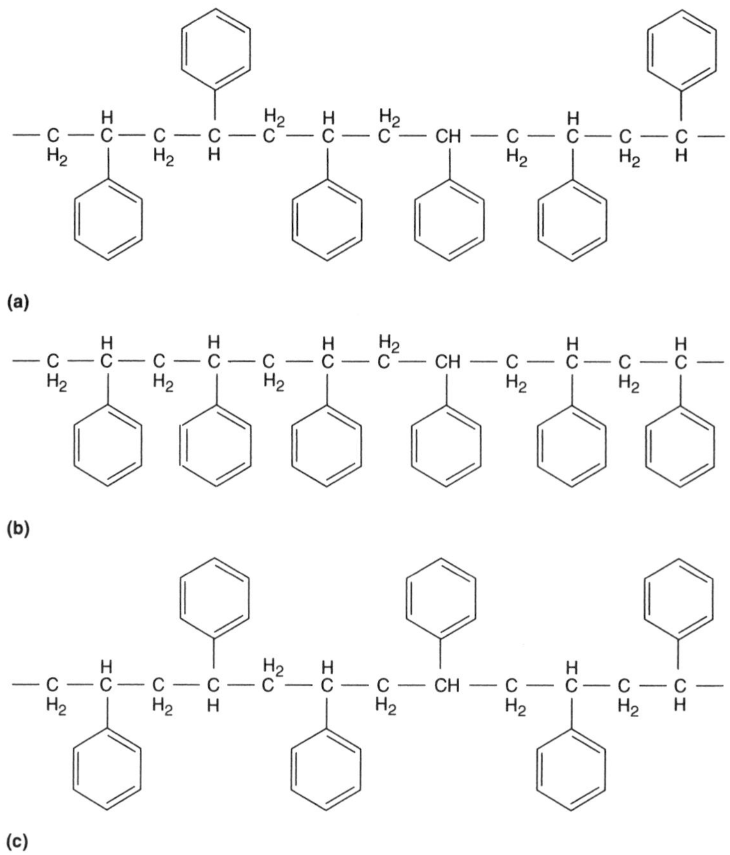

Fig. 4 Tacticity in polymers as shown by (a) atactic, (b) isotactic, and (c) syndiotactic polystyrene

atoms surrounding the central carbon of methylacetylene, which allows greater freedom of rotation for the carbon–carbon single bond. Of course, the greater electron density of the carbon–carbon triple bond does restrict the motion of that bond. Consideration of neopentane shows the resulting reduction of degrees of freedom when substituent hydrogens are replaced by the considerably more bulky methyl ($-CH_3$) groups. Introduction of the electronegative oxygen-containing side groups further increases stiffness of the backbone by reducing flexibility not only due to the size of this side group but because of electrical repulsion as well.

This concept accounts for the flexibility of rubbers, such as cis-1,4-polybutadiene (Table 1), that have double bonds on their main chain. The double bond eliminates two hydrogen atoms, and the additional free volume results in additional flexibility. One of the most flexible polymers, polydimethylsiloxane (PDMS), has a flexible ether linkage on the main chain and nonpolar side groups, which accounts for its lack of rigidity.

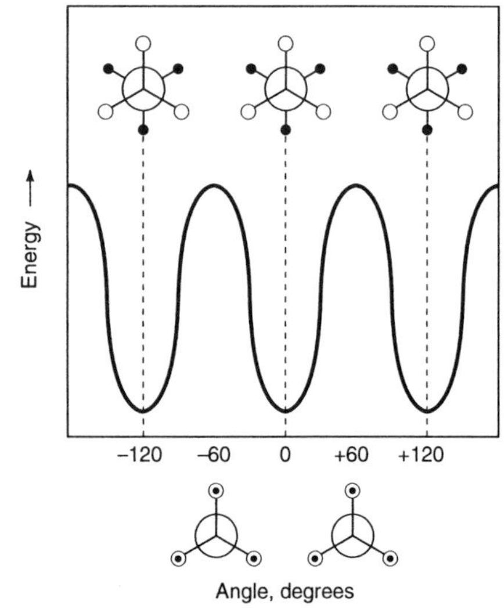

Fig. 5 Steric hindrance of ethane. Source: Ref 15

The ether oxygen only forms bonds with two carbon atoms, and the lack of hydrogen atoms means that ether linkages are surrounded by ample free volume. This promotes ease of rotation. Flexibility is also introduced by ether linkages due to the smaller atomic radius of oxygen atoms compared to those of carbon atoms. The structure of PDMS, one of the few commercially significant polymers without carbon on its backbone, is shown in Table 1. Cis-1,4-polybutadiene is used for flexible hose, gasketing, and rubber footwear application, while PDMS is used for embedding electrical components, high-temperature gaskets, and rubber-covered rollers for laminators.

Ring structures on the backbone reduce flexibility. The presence of a phenylene group combined with the resonance among the adjacent oxygen structures of polyethylene terephthalate (PET) explains its rigidity. This is one reason why PET is well suited for the manufacture of thin-walled soda bottles. Materials such as polyacrylonitrile (PAN) and PVC are all rigid due to electrical repulsion (the nitrile group is highly polar) as well as steric hindrance (chlorine is a large atom). The repeating unit structures for PET, PAN, and PVC are shown in Table 1.

These molecular factors account in part for the elevated T_g and elastic moduli values of engineering polymers; these are discussed later in the section "Thermal and Mechanical Properties of Solid Engineering Plastics" in this article. Polyacrylonitrile is used principally for synthetic (acrylic) fibers because its rodlike molecules form highly crystalline bundles and the high degree of hydrogen bonding provides high mechanical, thermal, and chemical resistance. With its low cost and relatively high modulus, PVC is used for water and gas pipes, window frames, siding, gutters, and identification and credit cards.

Main chain restrictions to rotation are important when considering inherent flexibility, but side chains and their morphology also play an important role. Side-chain contributions to molecular flexibility are affected by three characteristics:

- Presence of branching in the side chain
- Length of the side chain
- Polarity of the side chain

Branched side chains are even bulkier than their linear counterparts and offer greater steric hindrance. Steric hindrance is the restriction of free rotation due to spatial limitations imposed by the presence of atoms. This reduced flexibility is manifested as higher gas transition and wetting temperatures. Cyclic side groups stiffen the molecule, although this stiffening effect is diminished as the cyclic group occurs further from the main backbone. This can be seen when considering the phenyl group, which introduces significant steric hindrance, is present on increasingly

Table 3 Continuous service temperature as a function of degrees of fluorine substitution on polyethylene

Name	Continuous service temperature, °C	Repeat structure
PE	60–75	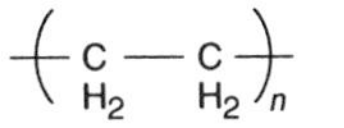
PVF	100–120	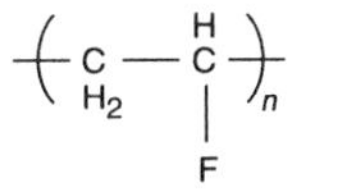
PVDF	150	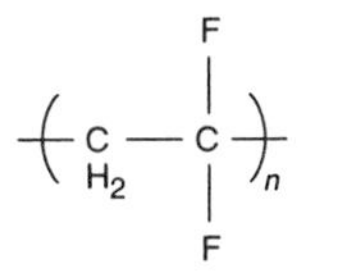
PTFE	250	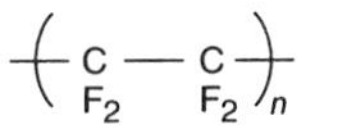

Source: Ref 10

Table 4 Effect of molecular weight on polyethylene

Number of $-CH_2-CH_2-$ units	Molecular weight (MW)	Softening temperature, °C	Character of polymer at 25 °C
1	30	−169(a)	Gas
6	170	−12(a)	Liquid
35	1,000	37	Grease
140	4,000	93	Wax
250	7,000	98	Hard wax
430	12,000	104	Plastic
750	21,000	110	Plastic
1,350	38,000	112	Plastic

(a) Melting point. Source: Ref 11

Table 5 Effect of side-chain length on glass transition and melting temperatures

Side chain structure	Glass transition temperature (T_g), °C	Melting temperature (T_m), °C
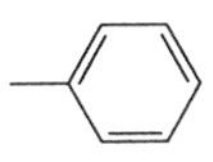	83	240
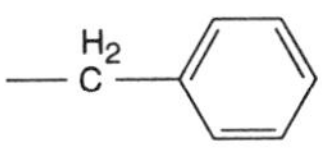	60	208
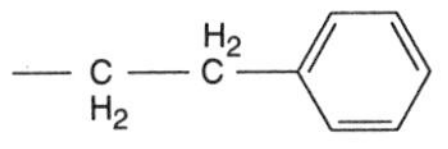	10	160

Source: Ref 17

long side chains. Table 5 indicates the effect of length of side chain on thermal transitions. When the phenyl group is pendant to the main chain, as in the case of PS, the T_m is 240 °C. Locating further away by one carbon atom reduces T_m to 208 °C, and when it is two carbon units away the T_m is only 160 °C.

There is an interesting limitation to the lowering of thermal transition temperatures by increasing the length of purely aliphatic (linear chains, without rings) side chains. It occurs when the reduction of stiffening through increased intermolecular distance is offset by side-chain crystallization. After the aliphatic side chain reaches eight to ten carbon units in length, side-chain crystallization can occur which again increases T_g and T_m. This is demonstrated in Table 6 and Fig. 7 for a series of polyolefins (saturated polymers containing only carbon and hydrogen). While PE has a T_m of 137 °C, introduction of a one-carbon side chain in PP increases the T_m to 176 °C due to limited chain mobility. However, longer side chains increase the free volume enough to reduce T_m until the side-chain length exceeds eight to ten carbons. These long side chains then have sufficient mobility to crystallize and again increase T_m.

Electrical repulsion between polar side chains disrupts random coil formation of the backbone and imposes what is known as "rigid-rod" conformation. This occurs in engineering thermoplastics such as PTFE and PAN.

Composition: Intermolecular Considerations

Intermolecular arrangements are governed by both spatial considerations (such as order and distance to neighboring molecules) and by the presence of attractive forces between molecules. Intermolecular order is defined as either amorphous, crystalline, or oriented, as shown in Fig. 8.

Amorphous Versus Semicrystalline. While amorphous materials assume random, three-dimensional structures, semicrystalline polymers have very ordered, tightly packed three-dimensional arrangements connected by amorphous regions. In the melt or solution, the chains of all polymers, except liquid crystalline polymers (LCPs), exhibit random or amorphous configurations. Liquid crystalline polymers form randomly arranged rodlike bundles. Upon cooling of the melt or evaporation of the solvent, some polymers remain amorphous whereas others crystallize. The state is determined by the regularity and flexibility of the polymer structure and the rate at which the melt is cooled or the solvent evaporated.

Polymers, such as atactic PS, atactic PMMA, atactic PP, and PVC, have large side chains or pendant groups added at irregular intervals. Because these groups prevent such polymers from forming crystalline regions, polymers with irregular structures are usually amorphous. When the pendant group or side chain is small enough, such as in PVOH and PAN, the side group can be tucked into ordered structures resulting in polymers that are semicrystalline. Moreover, regular addition of even large side groups permits the formation of tightly packed regions. Consequently, isotactic PP and syndiotactic PS are semicrystalline polymers, whereas the atactic forms are amorphous.

Table 6 Effect of length of aliphatic side chain on glass transition and melting temperatures of polyolefins

Olefin	Number of carbons in side chain	Glass transition temperature (T_g), °C	Melting temperature (T_m), °C
PE	0	−122	137
PP	1	−19	176
Poly-(1-butene)	2	−24	120
Poly-(1-pentene)	3	−47	70
Poly-(1-hexene)	4	−50	−55
Poly-(1-heptene)	5	...	−40
Poly-(1-octene)	6	−60	−38
Poly-(1-dodecene)	10	...	45
Poly-(1-octadecene)	16	...	70

Source: Ref 18

Because chain mobility is required to form ordered structures, polymers with regular, but rigid, structures cannot crystallize under normal processing conditions. Polycarbonate can crystallize if annealed at sufficiently high temperatures for long periods of time; however, under typical processing conditions PC is amorphous. In contrast, the structure of PE is so flexible that crystallization occurs even when the polymer melt is quenched (cooled rapidly).

Amorphous polymers exhibit a T_g that is the temperature at which the amorphous regions become mobile. In contrast, semicrystalline polymers exhibit both a T_g and a T_m. At this latter temperature, the ordered crystalline regions melt and become disordered random coils. While the magnitude of the T_g of a polymer depends only on the inherent flexibility of the polymer chain, the magnitude of T_m is also a function of the attractive forces between chains.

Although the degree of crystallinity in a given polymer varies with the processing conditions, the maximum degree of crystallinity depends on the polymer structure. Polymers such as PE, PP, polyoxymethylene (POM), and nylon 6/6 have regular, flexible structures that permit high levels of crystallization. As indicated in Table 7, increased branching that reduces the regularity of the polymer structure and its density, also decreases the degree of crystallinity. The molecular architecture of these grades, shown in Fig. 9,

Table 7 Properties of polyethylenes of varying degrees of crystallinity

Property	Low density	Medium density	High density
Density range, g/cm^3	0.910–0.925	0.926–0.940	0.941–0.965
Crystallinity, approximate %	42–53	54–63	64–80
Melting temperature (T_m), °C	110–120	120–130	130–136
Hardness, Shore D	41–46	50–60	60–70
Tensile modulus, MPa	97–260	170–380	410–1240

Source: Ref 19

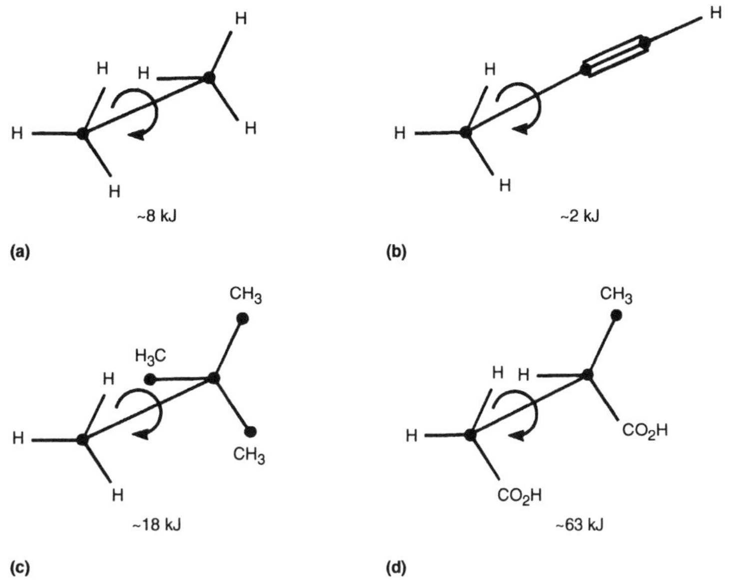

Fig. 6 Rotational energy barriers as a function of substitution. (a) Ethane. (b) Methylacetylene. (c) Neopentane. (d) Methylsuccinic acid. Source: Ref 16

explains why HDPE can achieve the highest level of crystallinity. Because the linear molecule is unimpeded by the random branches found in low-density polyethylene (LDPE), it can assume a tightly packed crystalline form.

The influence of crystallinity is best illustrated through the properties presented in Table 7 for PEs of various degrees of crystallinity. As shown in Table 7, the T_m, modulus, and hardness increase with crystallinity.

Orientation. Oriented polymers are often confused with semicrystalline polymers. In the case of oriented polymers, localized regularity is induced by mechanical deformation and is limited to small areas. Straining of polymers can result in stretched areas of parallel, linear, partially ordered structures as shown in Fig. 8(c). This uniaxial orientation results from forming processes such as fiber spinning, pipe and profile extrusion, and flat-film extrusion. The polymer chains can also be aligned parallel and perpendicular (transverse) to the primary direction of flow as shown in Fig. 8(d). Blown-film extrusion and blow molding inherently produce this biaxial orientation. In contrast, in the production of PET sheet, uniaxial orientation occurs during extrusion while the biaxial orientation is induced during a secondary stretching operation. Biaxial orientation is also the underlying concept of shrink-wrap films that revert to their amorphous conformations when enough heat is applied to reverse the induced orientation. Rotomolding and other low-shear processes produce little orientation.

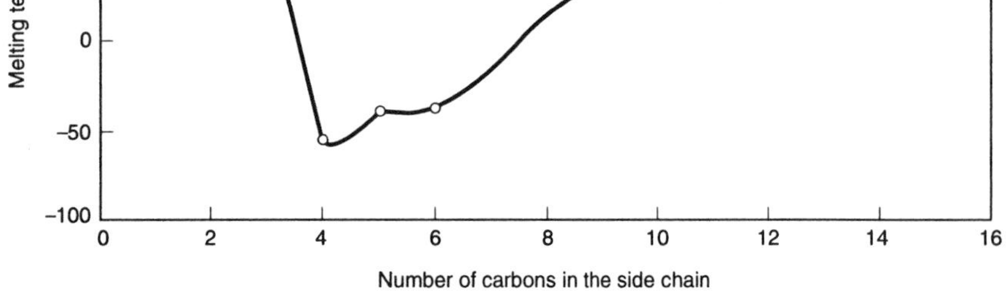

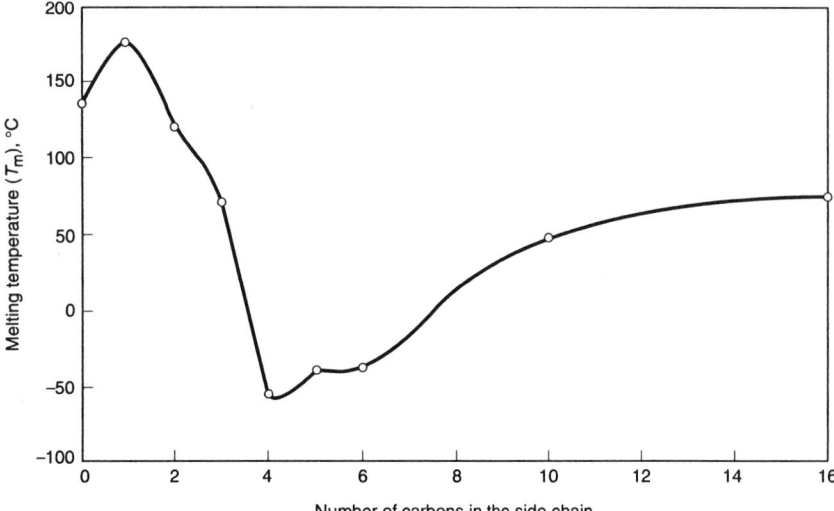

Fig. 7 The effect of aliphatic side chain on the melting temperature of polyolefins

Intermolecular Attractions. Secondary intermolecular attractive forces that promote crystallinity include London dispersion forces, dipole forces (either induced or permanent), hydrogen bonding, and ionic bonding. These secondary bonds do not actually connect two atoms through equally shared electrons the way that a primary covalent bond does; therefore, the energy required to break secondary bonds is less than the 300 to 420 kJ/mol (Ref 21) strength of covalent bonds. The interatomic distance of covalent bonds is quite short, generally between 1 and 2 Å (Ref 21). When primary, or covalent, bonds join adjacent polymer chains, the polymer is cross-linked.

London dispersion forces are the weakest of the secondary bonds with energies of 4 to 8 kJ/mol and an intermolecular distance of 3 to 5 Å (Ref 21). They are the only secondary interactions in linear, nonpolar hydrocarbons and fluoropolymers. The mobility of the valence electron clouds in these polymers results in transient states of electrical imbalance, and this momentary polarity draws two molecules together. London dispersion forces also provide significant intermolecular attractions in polar polymers, such as PVC, nylons, and PET, which form other secondary bonds.

Dipole Forces. In the presence of a polar molecule, an induced dipole can be set up in a neighboring molecule. Dipoles are the result of a covalent bond between atoms of differing electronegativities, and the resulting polarity accounts in large part for high thermal and mechanical properties of polar polymers such as PVC. These forces result in an intermolecular attraction of 4 to 21 kJ/mol (Ref 21) and often control solubility.

Hydrogen bonding occurs when the electron pair of an electronegative atom is shared by a hydrogen. The typical length of these bonds is 3 Å, with strengths of 6 to 25 kJ/mol (Ref 21). The degree to which hydrogen bonding occurs is re-

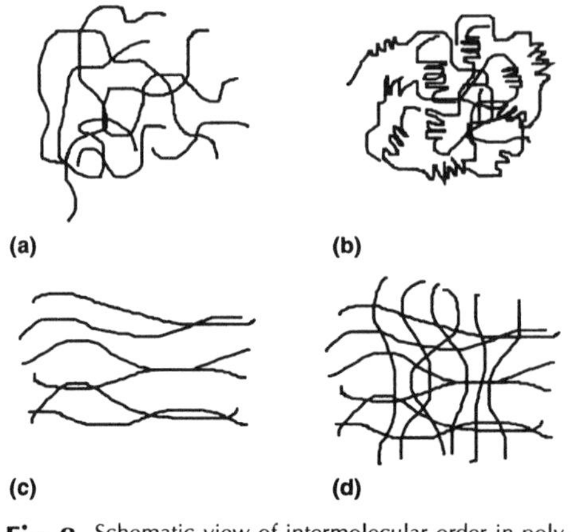

Fig. 8 Schematic view of intermolecular order in polymers. (a) Amorphous. (b) Semicrystalline. (c) Uniaxial orientation. (d) Biaxial orientation

lated to the number of hydrogen bonding sites available, which in turn is related to the MW of the molecule. The facility with which they are formed is aided by regular, crystalline structures. Hydrogen bonding accounts for the high strengths of aliphatic polyamides such as nylon 6/6, and is so strong in aromatic polyamides, such as aramid fibers, that the polymers degrade before they melt.

Ionic bonding is less common than hydrogen bonding. Ionic bonding is the binding force that results from the electrostatic attraction of positively and negatively charged ions. Ionomers, with their equal numbers of positively and negatively charged ions, have high T_m and moduli. Overall, they are electrically neutral. Bond strengths are of the order of 42 to 84 kJ/mol (Ref 21). These bonds are easily eliminated by polar liquids (of high dipole moments) such as water because surface ions are readily extricated when in contact with these liquids.

Cross-linking is the creation of a three-dimensional network by forming covalent bonds between polymer chains as shown in Fig. 10. While the degree of cross-linking can vary, highly thermoset systems are typically rigid. Upon exposure to elevated temperatures, cross-linked polymers cannot melt and flow. The covalent bonds that form the three-dimensional network prevent melting and also do not permit dissolution in solvents. Thermoset systems, such as unsaturated polyester, epoxy, thermoset polyurethanes, polyureas, phenol formaldehyde, and melamine formaldehyde, are shaped and cross-linked during processing. Representative structures are shown in Table 8.

Normally thermoplastic resins, such as PE, can also be cross-linked after the shaping operation. Cross-linking of PE does not introduce many cross-links because PE is quite unreactive. The few cross-links that form actually reduce regularity and therefore crystallinity. Thus, the modulus of semicrystalline thermoplastics is not increased upon cross-linking, although hot creep is reduced. Hot creep is the deformation of plastics exposed to stress and elevated temperatures for prolonged periods. Polyethylenes used in wire coating are frequently cross-linked for this reason.

Composition: Supermolecular Considerations

Supermolecular considerations include copolymerization, polymer blends, plasticization, incorporation of additives, and foaming.

Copolymerization. Copolymers are polymer molecules that contain several different repeat units. Usually two monomers are polymerized into one of four different configurations: random, alternating, block, or graft (Fig. 11). In random copolymers the units are distributed randomly along the polymer chains, whereas with alternating copolymers every second repeat unit is the same. Block copolymers also contain alternating segments of each monomer, but the segments are usually several repeat units long. Graft copolymers consist of a main chain composed of only one repeat unit with side chains of the second monomer.

The properties and processing characteristics of copolymers are often very different from those of the component polymers. Processing and properties can also vary with the ratio of the components and their arrangement within the copolymer. Block and graft copolymers can form two-phase systems similar to those observed with immiscible polymer blends.

Examples of random copolymers are ethylene propylene rubber (EPR), polystyrene-co-acrylonitrile (SAN), and fluorinated ethylene propylene (FEP). Ethylene propylene rubber is an amor-

Table 8 Representative structures of thermoset plastics

Common name	Representative polymer structure
Phenol-formaldehyde, phenolic	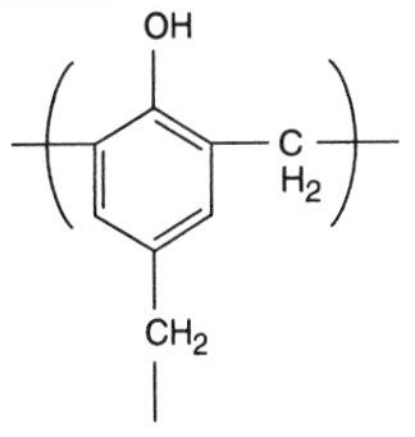
Melamine-formaldehyde, melamine	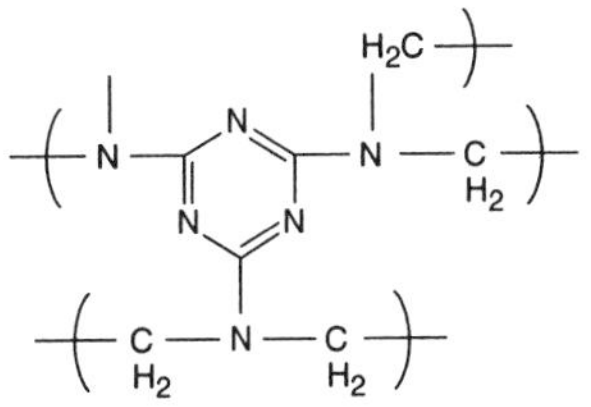
Urea-formaldehyde	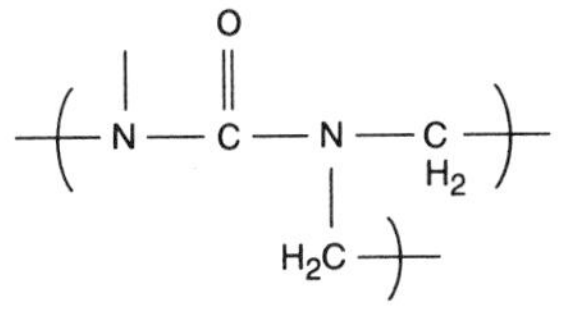
Epoxy	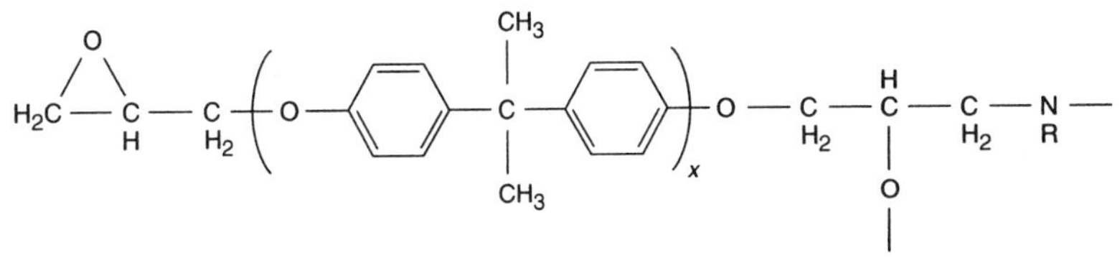
Unsaturated polyester	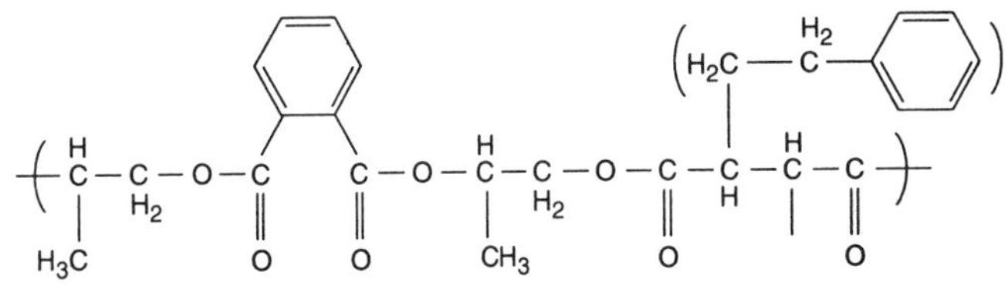
Polyurethane	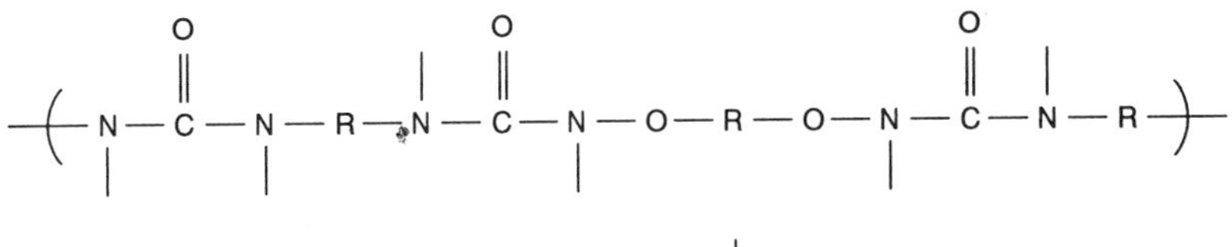
Silicone	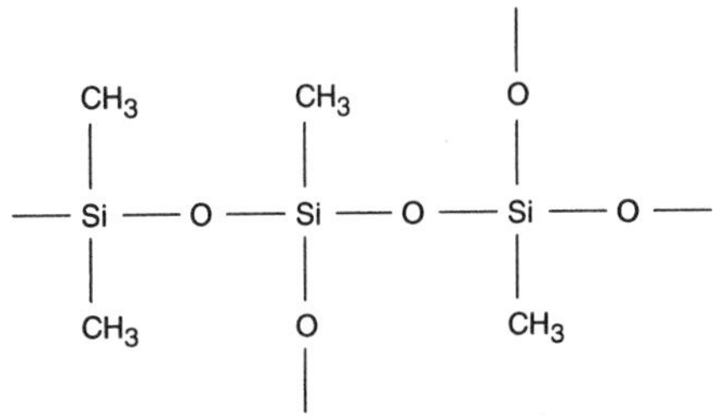

Source: Ref 8

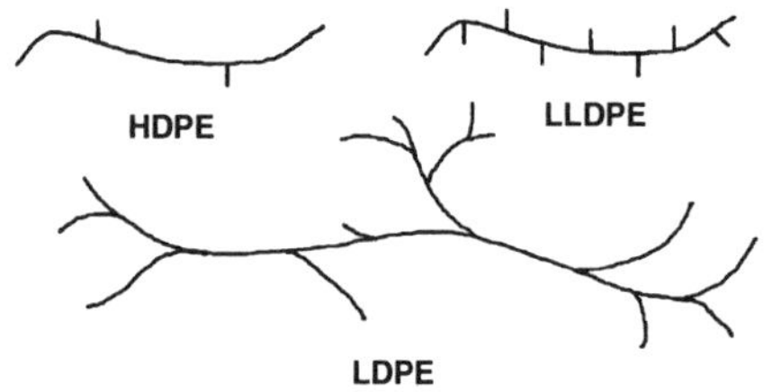

Fig. 9 Molecular architecture of high-density, low-density, and linear low-density polyethylenes. Source: Ref 20

phous elastomer, whereas PE and PP are semicrystalline plastics. Because acrylonitrile (which as PAN is difficult to process) is the minor component of SAN, it increases the melt temperature and stiffness of the PS without affecting its processibility. Fluorinated ethylene propylene is a melt-processible copolymer, while its major component, PTFE, is not.

Typical block copolymers are polyetheramides, hard segment-soft segment polyurethanes, and "styrenic" elastomers (for example, styrene-butadiene-styrene, or SBS, and styrene-ethylene-butylene-styrene, or SEBS). Graft copolymers are present in impact-modified polystyrene (HIPS) and ABS terpolymers. Alternating copolymers have, until recently, been laboratory curiosities.

Polymer Blends. While copolymers are mixtures of monomers that were joined together during polymerization, polymer blends are mixtures of polymer chains. The component polymers may be miscible, immiscible, or partially miscible. In the case of miscible blends, the polymers mix on a molecular level to produce a single phase. The most prominent example of this is modified polyphenylene oxide, which is a blend of polyphenylene oxide (PPO) with either PS or HIPS. Such systems exhibit a single T_g, and the mechanical properties are not affected by processing any differently than homopolymers.

With immiscible blends, the polymers cannot mix on a molecular level and therefore separate into two phases that exhibit the transition temperatures of the component polymers. In partially miscible blends, intermolecular attractions between the component polymers produce two phases that are not as sharply separated as those of immiscible blends. These blends exhibit transition temperatures that are shifted from those of the component polymers. The properties of both immiscible and partially miscible blends are sensitive to composition and processing conditions.

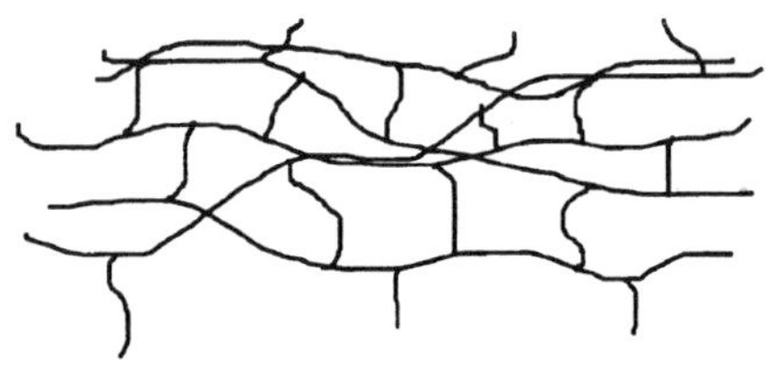

Fig. 10 Schematic of a cross-linked polymer

For immiscible latex systems, such as HIPS and ABS, the size of the rubbery phases and the degree of grafting between the rigid and rubbery phases is determined during the polymerization process. However, for mechanically blended systems, the morphology is determined during the blending process and can be altered during injection molding.

Immiscible and partially miscible blends can be made compatible to provide better adhesion between the two phases. Typically, a third component, such as a block copolymer or reactive copolymer, is added to the blend to form a link between the phases.

Plasticizers are small molecules that are added to plastics to reduce viscosity during processing and to increase the flexibility of the finished product. Plasticizers such as phthalates are typically incorporated into vinyl compositions to produce flexible PVC automotive upholstery, raincoats, and luggage. Water and solvents are used as temporary plasticizers during the processing of polymers such as cellulosics and PAN.

Additives can produce significant changes in the properties and processibility of polymers. Some additives such as colorants, antioxidants, and thermal stabilizers, do not affect the mechanical properties, but may influence viscosity during processing. In contrast, mineral fillers and glass or carbon fibers affect both mechanical properties and processibility. Fillers such as talc, calcium carbonate, and silica often reduce cost and increase the modulus, melt viscosity, and the deflection temperature under load. While fibers can significantly improve mechanical properties, their performance depends on orientation and fiber length, both of which can be affected by processing.

Foams. In foamed plastics a dispersed gaseous phase is incorporated into the plastic from the physical introduction of air or nitrogen, the degradation of chemical blowing agents or the addition of microballoons (hollow glass or plastic microspheres) to the polymer. This gas phase reduces the weight and thermal conductivity of the plastic. While the resulting foams are classified many ways, they can generally be divided into open-cell and closed-cell foams. The individual cells (gas phases) of closed-cell foams are separated, whereas in open-cell foams these cells interconnect. Consequently, closed-cell foams are typically buoyant and are frequently used for life jackets, buoys, and other flotation devices. Foamed plastics can be made from either thermoplastic or thermoset polymers, and the modulus of the base polymer determines the flexibility of the foam. Because the walls of flexible foams collapse when pressure is applied, these materials easily dissipate mechanical and acoustic energy. This makes flexible foams particularly suitable for packaging, cushions, padding, and related applications. In contrast, high-modulus polymers produce rigid foams with a high ratio of load-bearing strength to weight. These foams typically find applications in airplane wings and automotive parts.

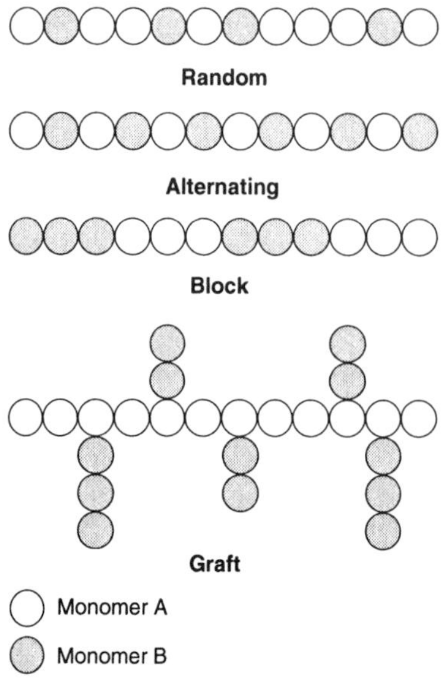

Fig. 11 Copolymer configurations. Source: Ref 22

Thermal and Mechanical Properties of Solid Engineering Plastics

A typical plot of stress versus strain behavior for an engineering thermoplastic is shown in Fig. 12. This classic relationship is characterized by a linear region (shown as segment AB), which is called the linear viscoelastic region. In this region the polymer chains stretch and disentangle in response to the stress being imposed. The ratio of stress to strain (the slope) is known as either Young's modulus or the elastic modulus. Behavior in this region is like that of a purely elastic, ideal solid, governed by Hooke's law:

$$\sigma = E\varepsilon \qquad \text{(Eq 5)}$$

where σ is stress, ε is strain, and the proportionality constant E is known as the spring constant or as stated earlier, Young's modulus or the elastic modulus.

Beyond this point, known as the yield point (shown as point B in Fig. 12), increased strain can be achieved with reduced stress. Secondary bonds are broken, and the strain is now irreversible. Permanent deformations such as necking begin to occur. Prior to point B, removal of

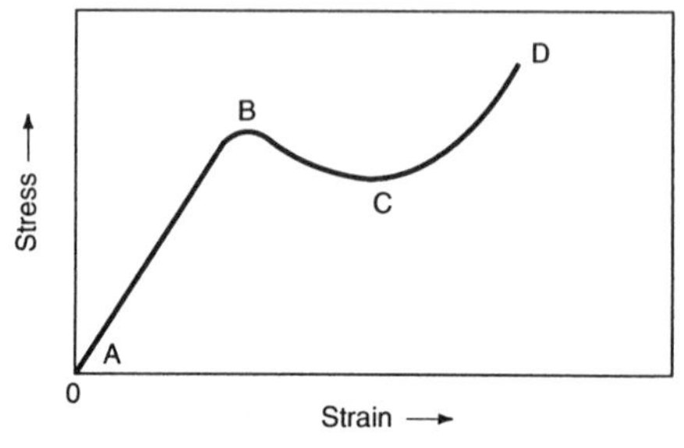

Fig. 12 Typical stress-strain curve for a polymer

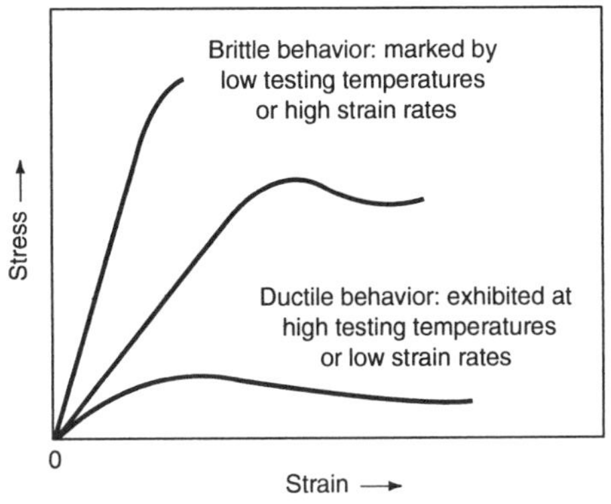

Fig. 13 Mechanical behavior of a plastic tested under different temperatures and strain rates

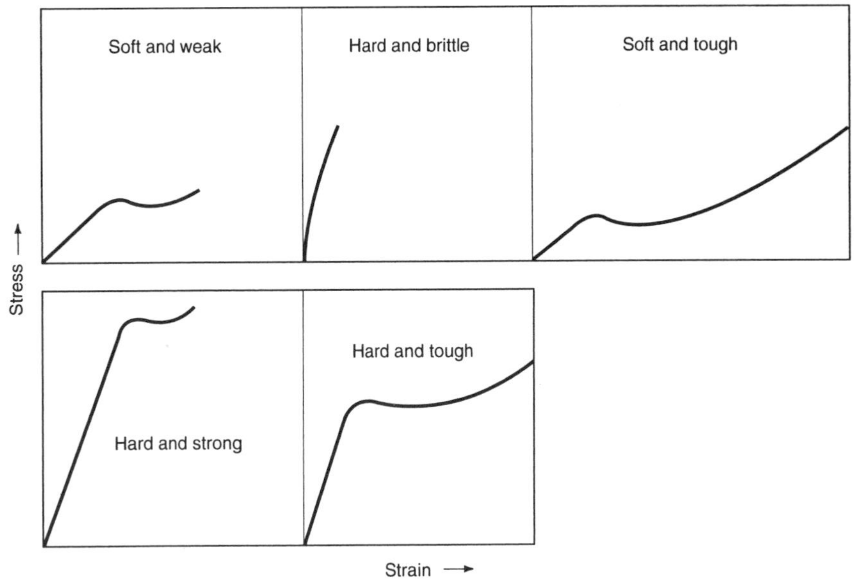

Fig. 14 Tensile stress-strain curves for several types of polymeric materials. Source: Ref 23

stress allows the material to recover its original dimensions.

Eventually, at point C, the slope increases due to mechanically induced orientation of the polymer chains. This orientation in the direction of the imposed stress effectively increases the strength of the material. Finally, the breaking point (D) is achieved where the ultimate, or breaking, stress and strain are defined. For tensile properties, the stress is often referred to as tensile strength whereas the strain is elongation.

The stress-strain behavior presented in Fig. 12 varies strongly as a function of both strain rate and temperature. At very high strain rates, the molecules do not have adequate time to disentangle from each other and physically respond to the imposed stress. High-speed testing, known as impact testing, yields a high modulus response and low ultimate strains. In cases where the stress is imposed very slowly, the polymer chains have adequate time to disentangle and deform.

Temperature also plays an important role. At very low temperatures, polymer molecules do not have much thermal energy or mobility. Therefore, they exhibit higher moduli and lower ultimate strains than at higher temperatures. At elevated temperatures, the molecules are more flexible and can distort and orient in response to the stress imposed by testing. Figure 13 shows the response of the same engineering plastic to different strain rates and different temperatures.

Figure 14 highlights the mechanical behavior of different plastics. "Strong" and "weak" are distinguished by differences in ultimate stress values, while "hard" and "soft" are differentiated by Young's moduli differences (the slope of the linear region). "Brittle" refers to a low ultimate strain, and "tough" is generally related to a large area under the stress-versus-strain curve. This definition of tough can be misleading because reinforced plastics have low ultimate strains, but are almost unbreakable.

The classic relationship of elastic modulus to temperature for polymers is presented in Fig. 15. The glassy state is characterized by limited motion of small segments of the molecule, one to four atoms in length. Behavior in this region is like that of a purely elastic, ideal solid. In the leathery region, the modulus decreases by up to three orders of magnitude for amorphous polymers. The temperature at which the polymer behavior changes from glassy to leathery is known as the T_g. This corresponds to approximately 2.5% free volume, which is the unoccupied space between molecules. The rubbery plateau has a relatively stable modulus. As temperature is further increased, rubbery flow begins, but motion does not yet involve entire molecules. In this region, deformations begin to become nonrecoverable as permanent set takes place. There is little elastic recovery in the liquid flow region, and these viscous materials, if ideal, would obey Newton's law:

$$\sigma = \eta\dot{\varepsilon} \quad \text{(Eq 6)}$$

where σ is stress, $\dot{\varepsilon}$ is strain rate, and the proportionality constant η is referred to as viscosity. The transition from the rubbery plateau to liquid flow occurs at the T_m. At this temperature, entire molecules are in motion.

Effects of Structure on Thermal and Mechanical Properties. Because free volume is

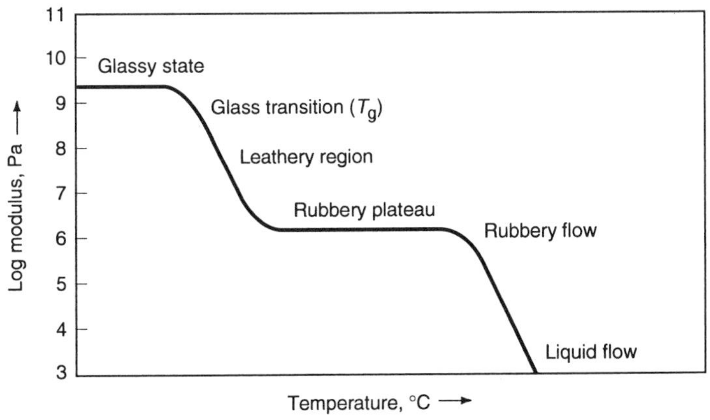

Fig. 15 Thermal dependence of elastic modulus for a typical polymer. Source: Ref 24

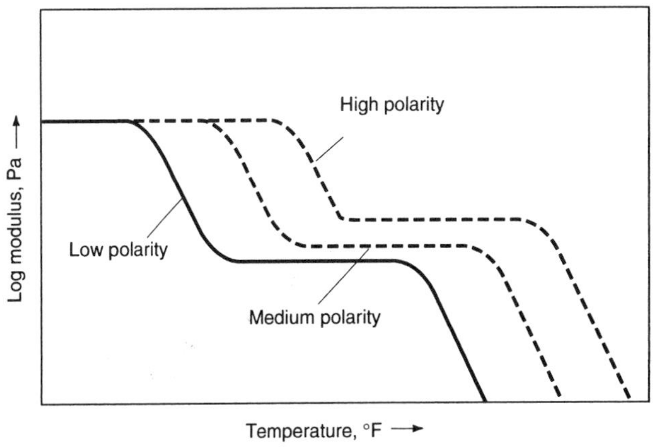

Fig. 16 Effect of temperature on modulus for polymers with different polarities. Source: Ref 25

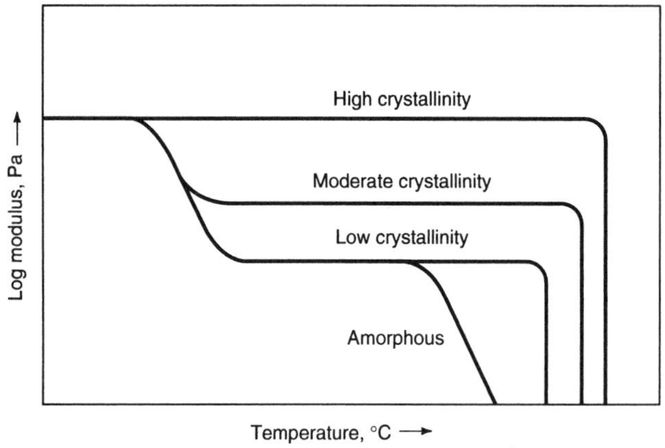

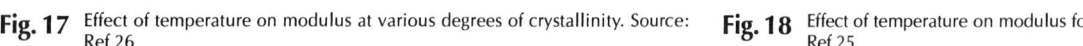

Fig. 17 Effect of temperature on modulus at various degrees of crystallinity. Source: Ref 26

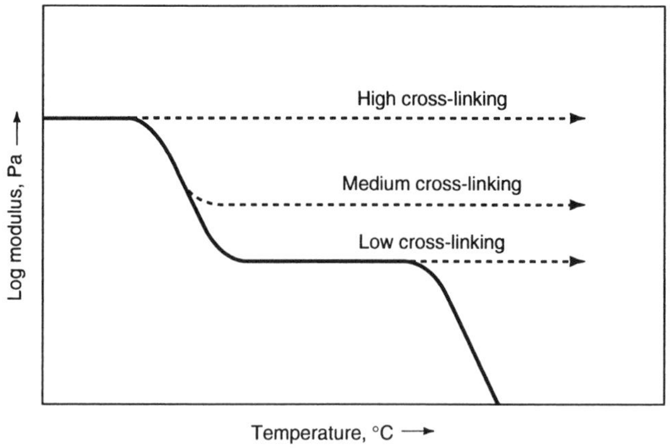

Fig. 18 Effect of temperature on modulus for different degrees of cross-linking. Source: Ref 25

generally associated with end-group concentration, T_g is a function of MW, particularly $\overline{M}_n$. Higher MWs mean longer chains, typically reduced relative concentration of end groups, and a reduction in the associated free volume. This leads to greater opportunity for molecular entanglements, which behave as physical (albeit temporary) cross-links and thus drives the onset of T_g to higher temperatures. Addition of plasticizer is a means of reducing the overall "effective" MW through the incorporation of typically low MW entities into the plastic. While unimolecular plasticizers provide significant increases of free volume, which allows for enhanced rotational degrees of freedom for the plasticized polymer, more permanent polymeric plasticizers with their greater MW and internal plasticizers (flexible segments incorporated into the polymer) permit far less mobility. Consequently, the latter two must be added in larger amounts to achieve the same effects as produced with unimolecular plasticizers.

Increasing polarity in the polymer produces stronger attractive forces between molecules. As shown in Fig. 16, this so stiffens the polymer that the onset of T_g can be delayed. Because more thermal energy is required to overcome the stronger polar attractive forces of the molecules, T_m is increased. Thus, the nonpolar high-density polyethylene (HDPE), the moderately polar POM, and the highly polar nylon 6/6 exhibit T_g of –120, –50, and 60 °C, respectively, while their T_m increase as 135, 175, and 264 °C, respectively.

Figure 17 presents the effect of crystallinity on the modulus-temperature relationship. At T_m the crystal structure is overwhelmed by thermal motion of the chains, and flow occurs. Increasing the degree of crystallinity does not affect the T_g, which involves much smaller structural components than the crystal lattice. However, polymers with higher degrees of crystallinity do require higher temperatures in order to melt. Higher degrees of crystallinity lead to higher T_m and rubbery plateaus, which occur at higher moduli.

High MWs extend the rubbery region as increased entanglements serve to postpone flow or deformation. In the extreme case of numerous covalent bonds linking molecules together, cross-linked polymers never exhibit the transition from the rubbery plateau into the flow regime. The covalent bond cross-links preclude flow, and the rubbery plateau simply extends until the decomposition temperature, at which point the covalent bonds are broken down. As the degree of cross-linking is increased, the onset of the rubbery plateau occurs at increasingly higher moduli as shown in Fig. 18.

Thermal and Mechanical Properties of Molten Engineering Plastics

Newton's law, given by Eq 6, only applies to ideal, viscous materials. A plot of log viscosity versus log shear rate for polymer melts (Fig. 19) exhibits three different regions of behavior. Regions A and C are Newtonian in that the viscosity is invariant with shear rate. Region A is often referred to as the "lower Newtonian plateau" and represents the viscosity at rates of shear that are low enough to allow the molecules to remain randomly tangled. As the rate of shear increases, region B is entered where molecules are now starting to align in the direction of flow. Because these aligned molecules offer less resistance to flow, viscosity is reduced. Finally, in region C, often referred to as the "upper Newtonian plateau," the molecules are aligned as much as possible and further increases in shear rate are no longer able to further reduce resistance to flow. This is the minimum viscosity that the molecules can achieve at a given temperature.

Processes such as extrusion and injection molding generate shear rates that are within region B, where the viscosity versus shear rate relationship is often approximated by the power law, given in Eq 7:

$$\eta = k\dot{\gamma}^{n-1} \quad \text{(Eq 7)}$$

where η is the viscosity, $\dot{\gamma}$ is shear rate, k is a material constant called the consistency index, and n is a constant called the power-law index. Power-law indices approximate the shear sensitivity of a polymer; values for common polymers are given in Table 9. Polymers that have very stiff backbones, such as PC and PS, tend to exhibit lower Newtonian plateaus that extend to shear rates of 1000 s^{-1} or more. Consequently, as is discussed in the section "Processing" in this article, shear thinning does not often reduce the viscosity of these polymers during extrusion.

Viscoelasticity

Mechanical analogs to purely elastic Hookean solid behavior and purely viscous Newtonian melt behavior help describe why polymers have intermediate (viscoelastic) properties, which are time dependent. Most commonly, a spring is used to model Hookean behavior, and a dashpot (representing a piston in a viscous "hydraulic fluid"-like material) represents viscous behavior. These models, and their concomitant stress and strain behaviors are shown in Fig. 20(a) and (b).

Application of a deforming force (i.e., pulling) on the spring results in an immediate stretching and thus an immediate strain. Once the force is released, the spring immediately recovers its initial length. Pulling with twice the force results linearly in twice the strain. The case of the dash-

Table 9 Sample power-law indices (*n*) for common plastics

Polymer	*n*
LDPE	0.35
LLDPE	0.60
HDPE	0.50
PP	0.35
PS	0.30
ABS	0.25
PMMA	0.25
PVC	0.30
PC	0.70
PET	0.60
PBT	0.60
Nylon 6	0.70
Nylon 6/6	0.75

Source: Ref 28

pot, however, is significantly different. When the "piston" has a force applied to it, it slowly starts to move (no instant displacement as in the case of the spring), and when the force is released, the dashpot stays in its new conformation. Once a force causes an ideal viscous polymer melt to flow, it remains in its new position.

Two models, combining the spring and the dashpot either in series or parallel, have been developed that attempt to better describe real polymer flow behavior. These models, Maxwell and Voigt, are named after their creators and are shown in Fig. 20(c) and (d). Figure 21, very similar to Fig. 15, shows which mechanical analogs model different regions of the log modulus versus temperature curve. The behavior shown in the Voigt model helps to explain the action known as creep. Creep occurs when, under a static load for extended periods of time, increased strain levels slowly develop, as in the case of a refrigerator that after many years distorts a linoleum floor. The Maxwell model describes stress relaxation, which occurs when polymers are subjected to a constant strain environment. Over time, the molecules relax and orient themselves to the strained position, thereby relieving stress. This occurs in applications such as threaded metal inserts into plastic parts and threaded plastic bottle caps.

Properties of Engineering Plastics and Commodity Plastics

Engineering plastics generally offer higher moduli and elevated-service temperatures compared to the lower-cost, high-volume, commodity plastics such as PE, PP, and PVC. These improved properties are due to chemical substituents, inherently rigid backbones, and the presence of secondary attractive forces as discussed earlier in this article. Engineering thermoplastics (e.g., POM, PC, PET, and polyether-imide, or PEI) are polymerized from more expensive raw materials, and their processing requires higher energy input compared to that of commodity plastics, which is why the engineering thermoplastics are more expensive.

Structures of Commodity Plastics. It is interesting to note the T_m elevation of HDPE from LDPE. The effect of the branched structure on density and morphology enables the high-density version to form more tightly packed crystalline regions that require more thermal energy to overcome the cohesive forces keeping the plastic from melting. Substituting a methyl group in place of a hydrogen, in the case of PP, increases T_m and tensile strength further above that of HDPE. In this case, steric hindrance due to the additional size of the methyl group stiffens the chain and restricts rotation. The substitution of a large and highly electronegative chlorine atom in PVC prevents crystallization and also increases the onset of T_g, both due to steric hindrance effects and to the attractive polar forces generated. Polar attractive forces are so extensive that the tensile strength can be seen to increase to 55 MPa. Polystyrene is amorphous and transparent due to the atactic positioning of the pendant phenyl group, whose randomness destroys crystallinity. The tensile strength of PS is less than that of PVC due to the lack of the highly polar pendant group.

Structures of Engineering Plastics. Phenylene and other ring structures (Table 1) attached directly into the backbone often stiffen the polymer significantly, imparting elevated-thermal properties and higher mechanical properties such as increased strength. Polyoxymethylene is essentially PE with an ether substitution, but it has a much higher T_m (200 °C versus 135 °C for HDPE) because of its polarity. Both of these features promote a highly crystalline morphology. The high dimensional stability, good friction and abrasion characteristics, and ease of processing of this polymer make it a popular engineering plastic for precision parts.

Polycarbonate has an extended resonating structure because of the carbonate linkage. It has such a stiff backbone that crystallization is impeded, and the resultant amorphous structure is transparent, much like PET. Physical properties of PET, however, depend strongly both on its degree of crystallinity, which is governed by degree of orientation imparted during processing, and on its annealing history. The high strength, ease of processing, and clarity of PET make it

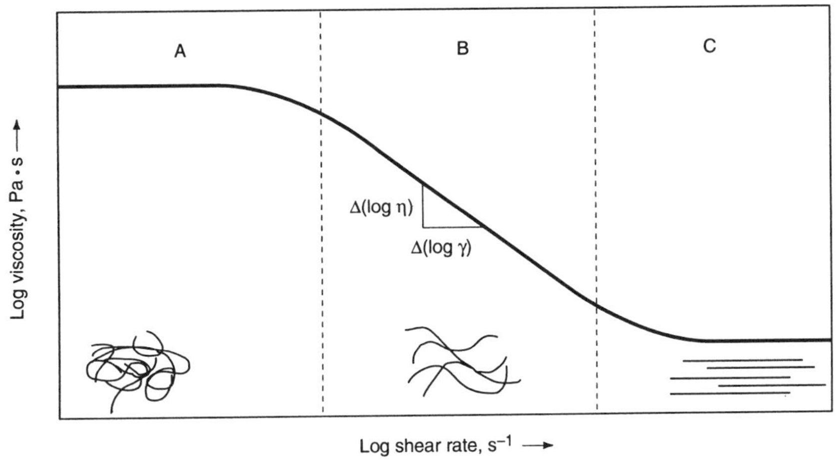

Fig. 19 General pseudoplastic behavior. Source: Ref 27

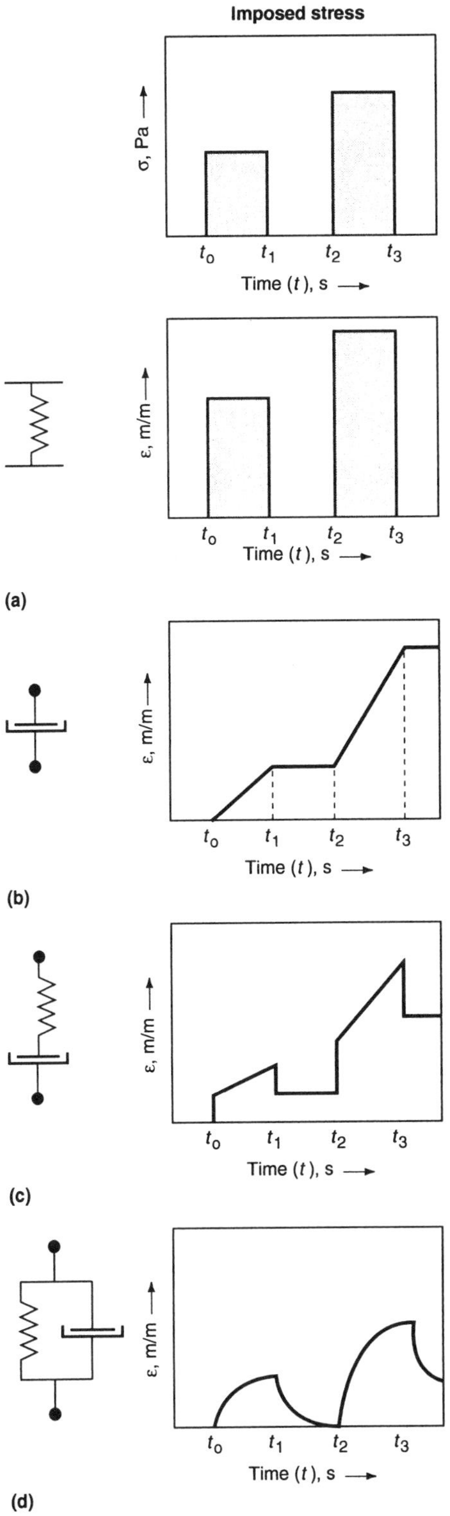

Fig. 20 Mechanical models and typical behavior. (a) Ideal Hookean solid ($\sigma = E\varepsilon$; spring model; elastic response). (b) Ideal viscous Newtonian liquid ($\sigma = \eta\dot{\varepsilon}$; dashpot model). (c) Maxwell's mechanical model for a viscoelastic material. (d) Voigt's mechanical model for a viscoelastic material. Source: Ref 29

ideal for soda bottles and polyester fibers. Polycarbonate has high strength, stiffness, hardness, and toughness over a range of –150 to 135 °C and can be reinforced with glass fibers to extend elevated-temperature mechanical properties. The high impact strength of high-MW PC makes it suitable for applications such as motorcycle helmets. The carbonate linkage of PC causes a susceptibility to stress cracking.

Polyetherimide has both imide groups and flexible ether groups, resulting in high mechanical properties but with enough flexibility to allow processing. Its highly aromatic (presence of benzene rings) structure allows it to be used for specialty applications.

Polyetheretherketone (PEEK), PPO, and PPS also rely on backbone benzene rings to yield high mechanical properties at elevated temperatures. Both sulfur and oxygen are electronegative atoms, creating dipole moments that promote intermolecular attractions and thus favorably affect elevated-temperature properties.

While the composition of thermoset plastics vary widely, the three-dimensional structure produced by cross-linking prevents melting and hinders creep. Overall properties such as stiffness and strength are determined by the flexibility of the polymer structure and the number of cross-links (cross-link density). Because epoxies, phenolics, and melamine formaldehyde contain aromatic rings, they are typically rigid and hard. Epoxies are used for adhesives, assorted electronics applications, sporting goods such as skis and hockey sticks, and prototype tooling for injection molding and thermoforming. Melamine formaldehyde is easily colored and so is often found in household and kitchen equipment, electronic housings, and switches. In contrast, phenolics are naturally dark colored and are limited to electronic and related applications where aesthetics are less important. Silicones with their flexible ether linkages are softer and often used as caulking and gasket materials. Thermoset polyurethanes vary widely from flexible to relatively rigid depending on the chemical structure between urethane groups. Unsaturated polyesters are used for potting and encapsulating compounds for electronics and in glass-fiber-reinforced molding compounds.

This discussion of the major commodity and engineering plastics is by no means complete. It is meant rather to include concepts touched on earlier in evaluating structures in relation to their resultant properties.

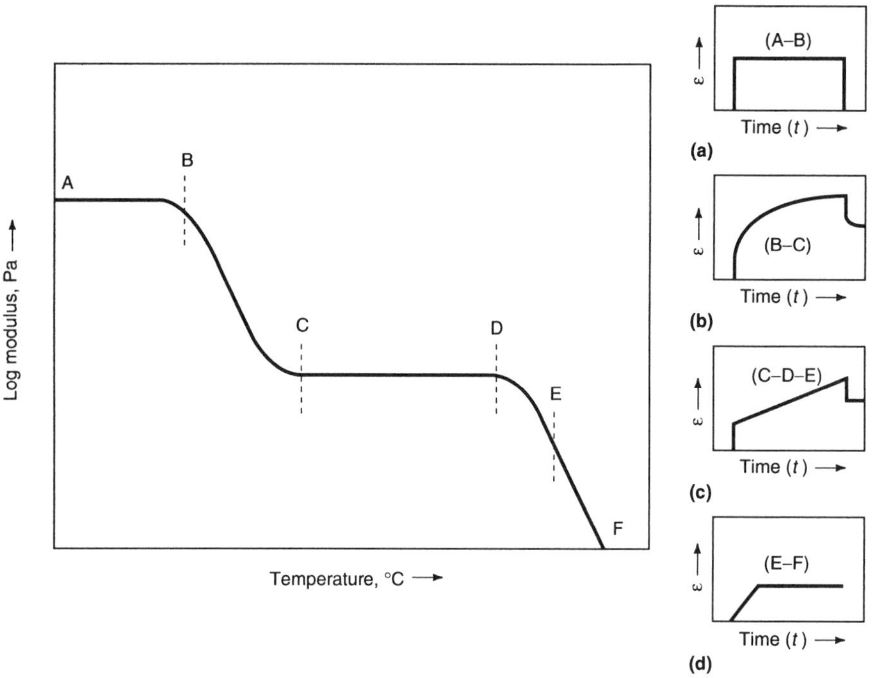

Fig. 21 Thermal dependence of elastic modulus for polystyrene. (a) Glassy region corresponding to Hookean solid behavior. (b) Leathery region corresponding to Voigt model behavior. (c) Rubbery plateau region corresponding to Maxwell model behavior. (d) Liquid flow region corresponding to Newtonian liquid behavior. Source: Ref 30

Table 10 Electrical properties of selected plastics

Plastic	Surface resistivity, Ω	Volume resistivity, Ω · cm	Dielectric strength, kV/mm	Dielectric constant		Dissipation factor	
				At 50 Hz	At 10^6 Hz	At 50 Hz	At 10^6 Hz
LDPE	10^{13}	$>10^{16}$	>70	2.3	2.3	2×10^{-4}	2×10^{-4}
PTFE	10^{17}	$>10^{18}$	60–80	2.1	2.1	2×10^{-4}	2×10^{-4}
PS	10^{14}	…	…	2.6	…	0.5×10^{-4}	2.5×10^{-4}
PMMA	5×10^{13}	$>10^{15}$	30	3.7	2.6	0.060	0.015
PVC	…	$>10^{15}$	20–40	3.5	2.7	0.003	0.002
Plasticized PVC	…	10^{15}	28	6.9	3.6	…	…
POM	10^{13}	10^{15}	70	…	3.7	0.0015	0.0055
Nylon 6/6	…	10^{15} (dry) 10^{11} (wet)	40 (dry)	4.0 (dry) 6.0 (wet)	3.4	0.02 (dry) 0.20 (wet)	…
PET	6×10^{14}	2×10^{14}	60	3.4	3.2	0.002	0.021
PBT	5×10^{13}	5×10^{13}	>45	3.0	2.8	0.001	0.017
PC	$>10^{15}$	$>10^{16}$	>80	3.0	2.9	0.900	11
Modified PPO	10^{14}	$>10^{15}$	22	2.7	2.6	4×10^{-4}	9×10^{-4}
PAI	5×10^{18}	2×10^{15}	23	…	3.9	…	0.030
PEI	…	7×10^{15}	24	3.15	3.05	0.0015	0.0064
PSU	3×10^{16}	5×10^{16}	20	3.15	3.10	0.001	0.005
PEEK	…	5×10^{16}	19	3.20	…	0.003	…

Source: Ref 4

Electrical Properties

Volume and/or surface resistivity, the dielectric constant, dissipation factor, dielectric strength, and arc or tracking resistance are considered important electrical properties for design. These properties relate to structural considerations such as polarity, molecular flexibility, and the presence of ionic impurities, which may result from the polymerization process, contaminants, or plasticizing additives. Table 10 shows some typical electrical property values for selected plastic materials.

Volume resistivity is a measure of the resistance of an insulator to conduction of current. Most neat polymers have a very high resistance to flow of direct current, usually 10^{15} to 10^{20} Ω · cm compared to 10^{-6} Ω · cm for copper. Electrical conductivity in normally insulating polymers results from the migration of ionic impurities and is affected by the mobility of these ionic species. Generally, plasticizers with their increased mobility and high relative concentration of end groups reduce resistivity and therefore increase electrical conductivity. Because absorption of water increases the mobility of ionic species, this also reduces volume resistivity. Thus, the volume resistivity of nylon 6/6 is reduced by four decades when the polymer absorbs water at ambient conditions. Addition of antistatic agents decrease surface resistivity because the polar additives migrate to the surface of the polymer and absorb humidity. In contrast, conductive fillers, such as carbon black powders and aluminum flake, can form three-dimensional pathways for conduction through insulating polymer matrices. Finally, highly conjugated polymers such as polyacetylene and polyaniline provide sufficient electron movement to reach semiconductor conductivity. For full conductivity, they rely on dopants.

Dielectric Constant and Dissipation Factor. In the presence of an electric field, polymer molecules will attempt to align in that field. The dielectric constant (or permittivity), ε or ε′, is a measure of this polarization. While the dielectric constant varies from 1 for a vacuum (where nothing can align) to 80 for water, the values for polymers (shown in Table 10) are generally so low that most polymers are insulators. The dielectric constant also varies with temperature, rate or frequency of measurement, polymer structure and morphology, and the presence of other materials in the plastic. The dielectric constant of polymers typically peaks at the major thermal transition temperature (T_g and/or T_m) and then decreases because of random thermal motions in the melt. As shown in Fig. 22(a), the dielectric constant decreases abruptly as frequency increases. This occurs between 1 Hz and 1 MHz and is a result of the inability of the dipoles to align with the high-frequency electric fields. The dielectric loss, ε″, is a measure of the energy lost to internal motions of the material, and as shown in Fig. 22(b), peaks where the dielectric constant changes abruptly. The dissipation factor, tan δ, which is given by:

$$\tan \delta = \frac{\varepsilon''}{\varepsilon'} \quad \text{(Eq 8)}$$

is a measure of the internal heating of plastics. Thus, little heating should occur in insulators ($\tan \delta < 10^{-3}$), whereas high-frequency welding necessitates that tan δ be much greater (Ref 32).

Because polymer molecules are typically too long and entangled to align in electric fields, the dielectric constant usually arises from shifting of the electron shell of the polymer and/or alignment of its dipoles in the field. For nonpolar polymers, such as PTFE and PE, only electron polarization occurs and the dielectric constant can be approximated by:

$$\varepsilon = n^2 \quad \text{(Eq 9)}$$

where *n* is the optical refractive index of the polymer. These values vary little with frequency, and changes occurring with increased temperatures are caused by changes in free volume of the polymer. In contrast, the dielectric constants of polar polymers, such as PVC and PMMA, are greater than n^2 and change substantially with temperature and frequency. Backbone flexibility or ease of rotation of polar side groups allows some polymers to orient quickly and easily. If the electric field alternates slowly enough, the molecule may be able to align or orient in the field depending upon its flexibility and mobility. Consequently, relatively flexible polymers, such as PVC and PMMA, exhibit greater decreases in dielectric constant with increased frequency than polymers, such as PEI and PSU, that have rigid backbones. The additional free volume and mobility of the plasticized PVC allows the molecules to align with minimal delay; as shown in Table 10, this doubles the dielectric constant at low frequencies.

Dielectric Strength. As the electric field applied to a plastic is increased, the polymer will eventually break down due to the formation of a conductive carbon track through the plastic. The voltage at which this occurs is the breakdown voltage, and the dielectric strength is this voltage divided by the thickness of the plastic. The dielectric strength decreases with the thickness of the insulator because this prevents loss of internal heat to the environment. Dielectric strength is increased by the absence of flaws.

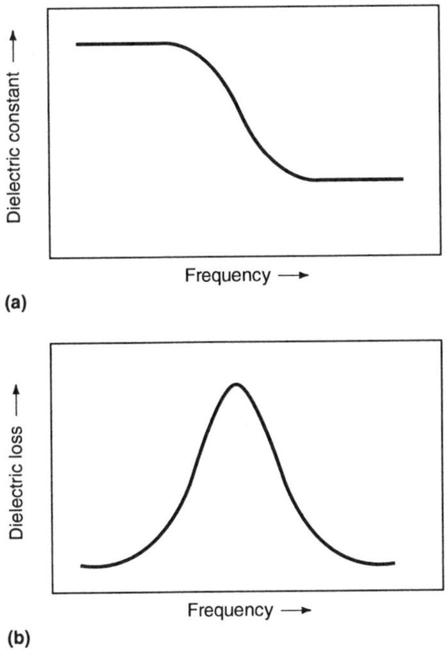

Fig. 22 Frequency dependence of the (a) dielectric constant and (b) dielectric loss. Source: Ref 31

Arc Resistance. In contrast to the dielectric strength, arc resistance is the ability of a polymer to resist forming a carbon tracking on the surface of the polymer sample. Because these tracks usually emanate from impurities surrounding electrical connections, arc resistance is measured by the track times. Polymers, such as PC, PS, PVC, and epoxies (which have aromatic rings, easily oxidized pendant groups, or high surface energies), are prone to tracking (Ref 33) and exhibit typical track times of 10 to 150 s (Ref 34). However, polyesters may have better tracking resistance than phenolics because of the heteroatomic backbone that disrupts the carbon track. Nonpolar aliphatic compounds or those with strongly bound pendant groups usually have better arc resistance; thus, the tracking times for PTFE, PP, PMMA, and PE are greater than 1000 s (Ref 33).

Optical Properties

Transparency, opacity, haze, and color are all important characteristics of plastics. Optical clarity is achieved when light is able to pass relatively unimpeded through a polymer sample. This is usually defined by the refractive index, *n*, which is shown in Fig. 23 and given by:

$$n = \frac{\sin \alpha}{\sin \beta} \quad \text{(Eq 10)}$$

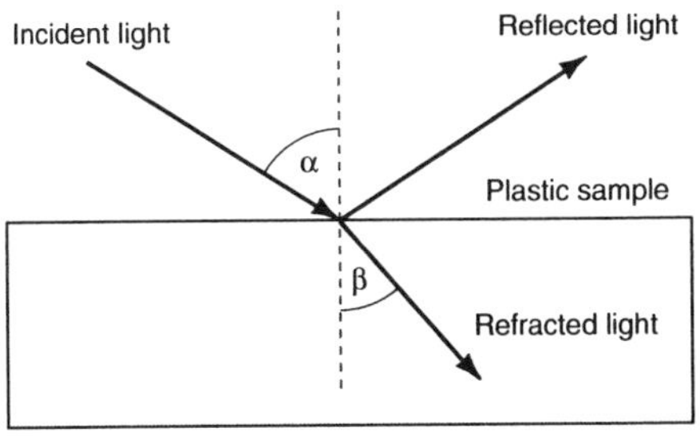

Fig. 23 Light refracted by a plastic sample

where α is the angle of incident light and β is the angle of refracted light. While *n* for most polymers is 1.40 to 1.70, it increases with the density of the polymer and varies with temperature. In order for a material to be clear, light has to be transmitted with minimal refraction. Unstressed, homogeneous, amorphous polymers, such as PS, PMMA, and PC, exhibit a single refractive index and thus are optically clear. However, when these polymers are severely oriented, and therefore stressed, the areas with different refractive indices produce birefringence in the molded products. Because amorphous, but heterogeneous, systems, such as the immiscible polymer blends ABS and HIPS, typically exhibit a refractive index for each polymer phase, they are usually opaque or translucent. Semicrystalline polymers, such as HDPE and nylon-6/6, effectively have two phases, the amorphous and crystalline regions. Consequently, semicrystalline polymers are usually not transparent. Finally, introduction of any nonpolymeric phases, such as fillers or fibers, into the plastic material induces opacity because these phases have their own refractive indices.

Optical clarity can also be controlled by polymerization techniques. When the refractive indices of multiphase systems are matched, these plastics can be optically clear, but usually only over narrow temperature ranges. Neat poly-(4-methyl-1-pentene) (TPX) is clear because the bulky side chains produce similar densities (0.83 g/cm^3), and thus similar refractive indices, in the amorphous and crystalline regions of the polymer. Matching of refractive indices of PVC and its impact modifier is often used in transparent films for food packaging. Domains (second phases) that are smaller than the 400 to 700 nm wavelengths of visible light will not scatter visible light, and thus do not reduce clarity. In impact-modified polymers, the minor rubbery phase is usually dispersed as particles with diameters greater than 400 nm, so most of them are opaque. However, when the domains have diameters less than 400 nm or when the two phases form concentric rings whose width is too narrow to scatter visible light, the blends are clear.

When crystals are smaller than the wavelength of visible light, they will also not scatter light and the plastic will be optically clear or translucent. These crystal sizes can be controlled by quenching, use of nucleating agents, stretching, and copolymerization. In quenching, the plastic melt is rapidly cooled below the transition temperature of the polymer. The resultant reduction in thermal mobility of the polymer molecules limits crystal growth because the molecules are not able to form ordered structures. While quenching is more easily accomplished with thin parts and films, nucleating agents can reduce crystal size in a wider range of parts. The agents are small particles at which the crystallization process can begin. Consequently, many such sites competing for polymer chains will reduce the average crystal size. Stretching also promotes clarity because the mechanical stretching can break up large crystals, and the resultant thinner films are more liable to transmit light without refraction. Finally, copolymerization can reduce the regularity of the polymer structure enough to inhibit formation of large crystals. As discussed earlier, structural regularity is required of a polymer is to pack into tightly order crystallites, and randomization of the structure results in smaller areas capable of being packed together.

The surface character of processed parts also controls optical properties. Smooth surfaces reflect and transmit light at limited angles, whereas rough surfaces scatter the light. Consequently, smooth surfaces produce clear and glossy products while rough surfaces appear dull and hazy. Because surface character is usually controlled by processing, it is discussed in the next section.

Unmodified polymers are usually clear to yellowish in color. Other colors are produced by dispersing pigments or dyes uniformly within the plastic. Poor dispersion can produce the marbled or speckled appearances favored for cosmetic cases. However, degradation of polymers will produce yellowing or browning of the plastic. Polymers such as PVC, which are particularly subject to degradation, are also discussed in the section "Processing" in this article.

Chemical Properties

Solubility is the ease with which polymer chains go into solution and is a measure of the attraction of the polymer to solvent molecules. The old adage of "like dissolves like" can be explained by considering the balance of forces that occur during dissolution of the polymer. Solubility is determined by the relative attraction of polymer chains for other polymer chains and polymer chains for solvent molecules. If the polymer-solvent interactions are strong enough to overcome polymer-polymer interactions, dissolution occurs; otherwise, the polymer remains insoluble. Swelling can be considered as partial solubility because the solvent molecules penetrate the polymer, but they cannot completely separate the chains.

When solvents and polymers have similar polarities, the polymer will dissolve in or be swollen by the solvent. Because longer chains are more entangled, higher MW hinders dissolution. Semicrystalline polymers are much harder to dissolve than similar amorphous materials. The tightly packed crystalline regions are not easily penetrated because the solvent molecules must overcome the intermolecular attractions. Elevated temperatures, which increase the mobility of solvent molecules and polymer chains, facilitate dissolution. The presence of cross-links completely prevent dissolution, and such polymers merely swell in solvents.

Plasticizers must be soluble in the polymer to prevent migration to the surface (blooming) and extraction by solvents. Consequently, the relatively expensive primary plasticizers for PVC closely match the solubility of the polymer, while less expensive secondary plasticizers are less compatible with the PVC.

Permeability is a measure of the ease with which molecules diffuse through a polymer sample. The low densities of polymers compared with metals and ceramics allow enhanced permeation of species such as water, oxygen, and carbon dioxide. If there are strong interactions between the polymer and the migrating species, adsorption will be high, but permeation may be low as the migrating species is delayed from diffusing. For example, the electronegative chlorine atoms substitution in polyvinylidene chloride (PVDC) enhances adsorption of oxygen, nitrogen, carbon dioxide, and water while its tightly packed chain arrangement restricts diffusion of these species. Thus, PVDC films (commonly used as plastic wrap) are extremely valuable in food packaging operations. As shown in Fig. 24, permeability can also be inhibited by the addition of platelike fillers, which increase the distance that water must travel in order to pass completely through the plastic.

Environmental stress cracking occurs when a stressed plastic part is exposed to a weak solvent, often moisture. The stress imparts strain to the polymer, which allows the solvent to penetrate and either extract small molecules of low $\overline{M}_n$, or to plasticize and weaken the polymer. The stress then causes fracture at these weak areas. Polymers which are exposed to UV light are particularly susceptible to environmental stress cracking. Resistance is enhanced when the permeability of the polymer to water is low.

Processing

Most thermoplastic processing operations involve heating, forming, and then cooling the polymer into the desired shape. This section briefly outlines the most common plastics manufacturing processes. The factors that must be considered when processing engineering thermoplastics are also discussed. These include melt viscosity and melt strength; crystallization; orientation, die swell, shrinkage, and molded-in stress; polymer degradation; and polymer blends.

Overview of the Major Thermoplastics Processing Operations. Although there are a number of variants, the major thermoplastics processing operations are extrusion, injection molding, blow molding, calendering, thermoforming, and rotational molding. Characteristics of each of these processes are described briefly below. Additional information is provided in the article "Design for Plastics Processing" in this Volume.

Extrusion is a continuous process used to manufacture plastics film, fiber, pipe, and profiles. The single-screw extruder is most commonly used. In this extruder, a hopper funnels plastic pellets into the channel formed between the helical screw and the inner wall of the barrel that contains the screw. The extruder screw typically consists of three regions: a feed zone, a transition or compression zone, and a metering or conveying zone (see Fig. 10 in the article "Design for Plastics Processing" in this Volume). The feed zone compacts the solid plastic pellets so that they move forward as the solid mass. As the screw channel depth is reduced in the transition zone, a combination of shear heating and conduction from the heated barrel begins to melt the pellets. The fraction of unmelted pellets is reduced until finally in the metering zone a homogeneous melt has been created. The continuous rotation of the screw pumps the plastic melt through a die to form the desired shape.

The die and ancillary equipment produce different extrusion processes. With blown-film extrusion, air introduced through the center of an annular die produces a bubble of polymer film; this bubble is later collapsed and wound on a roll. In contrast, flat film is produced by forcing the polymer melt through a wide rectangular die and onto a series of smooth cooled rollers. Pipes and profiles are extruded through dies of the proper shape and held in that form until the plastic is cooled. Fibers are formed when polymer melt is forced through the many fine, cylindrical openings of spinneret dies and then drawn (stretched) by ancillary equipment. In extrusion coating, low-viscosity polymer melt from a flat-film die flows onto a plastic, paper, or metallic substrate. However, in wire coating, wire is fed through the die and enters the center of the melt stream before or just after exiting the die. Finally, coextrusion involves two or more single-screw extruders that separately feed polymer streams into a single die

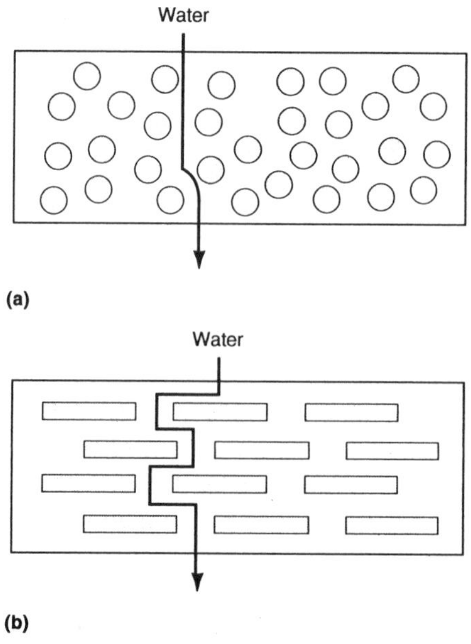

Fig. 24 Barrier pigment effect. Water passes relatively unobstructed through a polymer with spherical additives (a), but must travel around platelike fillers (b). Source: Ref 35

assembly to form laminates of the polymers. Typical extrusion pressures range from 1.5 to 35 MPa.

While single-screw extruders provide high shear and poor mixing capabilities, they produce the high pressures needed for processes such as blown and flat-film extrusion. Screw designs are changed to improve mixing, to shear gel (unmelted polymer) particles, and to provide more efficient melting. The latter designs are particularly critical to the extrusion of PE films where partially melted polymer particles are not desirable.

In addition to single-screw extruders, twin-screw extruders are available. While twin-screw extruders use two screws to convey the polymer to a die, the configuration of the screws produce different conveyance mechanisms. Intermeshing twin-screw extruders transfer the polymer from channel to channel, whereas nonintermeshing twin-screw extruders—like single-screw extruders—push the polymer down the barrel walls. In addition, intermeshing corotating twin-screw extruders tend to move the polymer in a figure-eight pattern around the two screws. Because this produces more shear and better mixing, corotating twin-screw extruders are well suited to mixing and compounding applications. Intermeshing counterrotating twin-screw extruders channel the polymer between the two screws. Twin-screw extruders also permit tighter control of shear because twin screws are usually not a single piece of metal, but two rods on which component elements are placed. Consequently, screw profiles can be "programmed" to impart specific levels of shear.

In contrast to the single- and twin-screw extruders, ram extruders have no screw, but merely use a high-pressure ram to force the polymer through a die. This provides for minimal shear and much higher pressures than available in single-screw extruder. However, ram extrusion is a batch operation, not a continuous operation.

Injection molding is a batch operation used to rapidly produce complicated parts. Plastic pellets are fed through a hopper into the feed zone of a screw and melted in much the same way as occurs in a single-screw or ram extruder. However, rather than being forced through a die, in an injection-molding machine the melt is accumulated and subsequently forced under pressure into a mold by axial motion of the screw. This pressure is typically quite high and for rapid injection and/or thin-walled parts can exceed 100 MPa. Once the part has cooled sufficiently, the mold is opened, the part ejected, and the cycle recommences. The use of multiple-cavity molds allows for simultaneous production of a large number of parts, and often little finishing of the final part is required. Polymer from multiple plasticating units (extruders) can also be injected sequentially into the same mold to form "coinjected" parts. In gas-assisted injection molding, gas is injected into the melt stream and accumulates in thicker sections of the part, whereas in foam processes the introduced gas forms small pockets (cells) throughout the melt.

Blow molding operations generate hollow products, such as soda bottles and automobile fuel tanks. The three basic processes are continuous extrusion, intermittent extrusion, and injection blow molding. In continuous-extrusion blow molding, a tube of polymer is continuously extruded. Pieces of this tube (called parisons) are cut off, inserted into the mold, and stretched into the cavity of the blow mold by air pressure. Although intermittent extrusion blow molding is similar, the tube of plastic is injected from the extruder rather than continuously extruded. In the injection blow molding process. a plastic preform, which for bottles resembles a test tube with threads, is injected molded. Then this preform is brought to the forming temperature (either as part of the cooling from injection molding or after being reheated) and expanded into the blow mold. Stretch blow molding is a variant of the blow-molding process, in which the preform is stretched axially by mechanical action and then expanded in the transverse direction to contact the walls of the mold.

Calendering uses highly polished precision chromium rolls to transform molten plastic continuously into sheet (>0.25 mm) or film (≤0.25 mm) for floor coverings. This process can also be used to coat a substrate, for example, cords coated with rubber for automotive tire use (Ref 36). Usually an extruder provides a reservoir of plastic melt, which is then passed between two to four calender rolls whose gap thickness and pressure profiles determine the final gage of the sheet being formed. Chill rolls are used to reduce the sheet temperature, and a windup station is generally required to collect the sheet product.

Thermoforming operations are used to produce refrigerator liners, computer housings, food containers, blister packaging, and other items that benefit from its low tooling costs and high output rates. In this process, infrared or convection ovens heat an extruded or calendered sheet to its rubbery state. Mechanical action, vacuum and/or air pressure force the heated sheet into complete contact with cavity of the thermoforming mold.

Rotational molding, or rotomolding, involves charging a polymeric powder or liquid into a hollow mold. The mold is heated, and then cooled, while being rotated on two axes. This causes the polymer to coat the inside of the mold. Because rotomolding produces hollow parts with low molded-in stresses, it is often used for chemical containers and related products where environmental stress crack resistance is required. It can also be used for hollow parts with complicated geometries that cannot be produced by blow molding.

Melt viscosity and melt strength are major factors to be considered when choosing a resin and a processing operation. While flexible polymers are generally less viscous than polymers with more rigid structures, MW, MWD, and additives are used to tailor plastics for specific processes. Resins are typically rated by their melt index, which is the flow of the melt (in grams per 10 min) through a geometry and under a load specified by ASTM D 1238 (Ref 37). Although this generates the flow at very low shear rates, it is an indication of the melt viscosity of the plastic. Extrusion blow molding processes require that the melt index be below 2 g per 10 min, whereas other extrusion processes require somewhat greater flow. In contrast, high-melt-index resins (6 to 60 g per 10 min) are necessary in extrusion coating, injection molding, and injection blow molding.

Low-viscosity polymers such as nylon 6/6 tend to leak (drool) from the nozzles of injection-molding machines, so they require special nozzles for injection molding. Aliphatic nylons exhibit narrow melting ranges and so need special screws in which the transition zone is relatively short, typically two or three turns (flights). Molecular weight distribution also factors into the extrusion of relatively low-viscosity polymers such as PEs. A wider MWD provides easier processing, but is detrimental to final properties such as strength and heat sealing. Narrower MWDs, particularly with linear polymers such as HDPE and LLDPE, often necessitate changes to extruder.

High-viscosity polymers, such as PC and PSU, typically require high injection pressures and clamping tonnages. If, however, the pressure required to fill the cavity exceeds the maximum injection pressure for the press, then the cavity is underfilled. When the injection pressure is greater than clamp pressure (tonnage), then the melt can force its way through the parting line (where the mold opens to eject the finished part) and damage the mold. The former problem is common in high-speed or thin-wall injection molding of PC and other high-viscosity resins. While increasing processing temperatures does decrease the melt viscosity, increased plasticating (screw) speeds do not reduce viscosity much due to the rigid backbones of PC and PSU, which extend the lower Newtonian plateau beyond the shear rates typical of plasticating units. However, high shear is still produced during injection and can break the polymer chains, which lowers mechanical properties, such as the impact strength of PC. High-flow resins (melt index > 40 g per 10 min) are available, but these generally exhibit lower MWs with the corresponding changes in properties. Other high-flow resins, which are usually immiscible blends of the primary polymer with a higher-flow plastic or additive, also affect final thermomechanical properties.

Very-high MW or very rigid structures produce polymers that are not truly melt processible. In high-MW materials such as ultrahigh-molecular-weight polyethylene (UHMWPE) and PTFE, the intermolecular attraction and excessive chain length do not allow the materials to melt. Heat will soften these polymers, but they are usually processed as slurries in which a solvent or oil carries the unmolten polymer particles. Because this requires excessive pressure, PTFE is often processed using a ram extruder. Ultrahigh-molecular-weight polyethylene needs less pressure, but is also processed on ram or twin-screw extruders to prevent excessive shearing (as is discussed later in this article). The high MW ($\sim 10^6$

Daltons, Ref 38) of the PMMA used for Plexiglas (trademark of Rohm and Haas Corp.) sheet does not permit melt processing, but rather the sheet is cast (polymerized) from the monomer (molding grade PMMA resins have MWs in the range of 60,000 Daltons, Ref 38).

The very inflexible structures of polyimides and aromatic polyamides do not permit melt processing. While polyimides are cast, more flexible variations, such as PEI and polyamide-imide (PAI) are melt processible. Similarly, copolymers and other variants of PTFE are melt processible. In both cases, the properties of the melt-processible polymers are less than those of the originals. Polyphenyl oxide is barely processible. However, blends of PPO with PS or HIPS are.

Additives such as processing aids and colorants can severely alter the viscosity of a polymer. It is not unusual for the same polymer compounded in different colors to have very different flow characteristics. Fillers and fibers typically increase melt viscosity. High loadings of fine particulate fillers, such as carbon black and titanium dioxide, can alter the low shear-rate behavior of the plastic; because these materials exhibit yield stresses, more force or pressure is required to initiate movement of the molten polymer. Regrind (processed polymer from runners and sprues) is often recombined with the virgin resin. However, because the regrind usually has a lower MW than the virgin resin, the flow characteristics of the mixture differ from those of the neat polymer.

Control of viscosity is critical in several processes. In coextrusion, the polymers must form layers and not mix with each other. Thus, the maximum viscosity difference for multimanifold dies is 400 to 1, whereas it is 2 or 3 to 1 for feed blocks where the molten layers are in contact longer. In gas-assisted injection molding, the polymer viscosity determines where the bubble will form. Viscosity also allows the polymer flow in rotary molding and extrusion coating.

Melt strength is the ability of the molten polymer to hold its shape for a period of time. Because long entangled polymer chains produce melt strength, these resins are high-MW polymers (with the related low-melt index values). However, polymers, such as PS, PET, and some nylons, which do not permit sufficient entanglement, always have low melt strength. Consequently, the processing equipment must accommodate this. Fiber extrusion lines usually place the extruder two or three floors above the windup units and draw the low-melt-strength fibers with gravity. This technique has also been used in blown-film extrusion of nylons. Polystyrene and PET are generally processed using flat-film extrusion so that the melt flows from the die to chill rollers that support the melt. As discussed earlier, biaxially oriented PET films are then produced by heating the flat film to its rubbery state and stretching it on a center frame. Low-melt-strength polymers must always be injection blow molded.

Sheet materials used for thermoforming require hot strength to prevent excessive sagging of the rubbery polymeric sheet during heating. While this strength is also related to the MW and MWD, it incorporates the transition temperatures of the polymer. Because amorphous polymers exhibit broad transitions from their T_g to the molten state, they are easily thermoformed. The sharper melting transitions of polymers, such as PP, PET, and nylons, provide narrow processing temperature ranges and tend to be either too solid to form or too molten and sag. Broadening of the MWD of PP and copolymerization of PET have produced grades of these resins suitable for thermoforming. There are also special techniques that use the ductility of PP to thermoform parts.

Crystallization has two components: nucleation and crystal growth. Nucleation is the initiation of crystallization at impurities in the polymer melt and is enhanced by rapid cooling rates and nucleating agents. Crystal growth is favored by slower cooling rates (which allows the molecules enough thermally induced mobility to assume a crystalline structure). Although the maximum crystallinity occurs if the polymer is held at 0.9 T_m (K), the degree of crystallinity developed is a function of the temperatures achieved and how long the molten plastic is kept warm. Consequently, because rapid cooling produces no crystallinity or many small crystallites, it is used to produce optically clear PE-blown film and blow-molded PET bottles. Slower cooling or annealing—which produces fewer, but larger, crystals—is not always favored because mechanical properties such as impact strength are adversely affected. Moreover, while the intermolecular bonding that occurs in a crystalline polymer results in improved mechanical and thermal properties, the desire for crystalline, stress-annealed parts is balanced by economics, which usually dictate that plastics be cooled as rapidly as possible to reduce production time.

The volumetric changes (tight molecular packing) associated with crystallization produce shrinkage in plastics products. Consequently, the semicrystalline plastics shrink far more than amorphous plastics, and the degree of shrinkage varies with the cooling rate. Typical shrinkage values are presented in Table 11, but the incorporation of additives—such as fillers and glass fibers, which interrupt or enhance crystallinity—can affect shrinkage. Because flexible polymers, such as aliphatic nylons and PP, exhibit high levels of shrinkage, particularly in thick cross sections, they reduce shrinkage during extrusion by utilizing the high pressures of ram extruders to process the polymers slightly below their melting temperatures.

Crystallinity can also vary through the thickness of a part with the rapidly cooled outside surfaces and the slowly cooled core having different levels of crystallinity. This effect, which varies with polymer type and processing conditions, can alter plastic properties. With flexible polymers, such as PP, crystallization occurs throughout the thickness. However, at relatively slow injection speeds and low mold temperatures, relatively rigid polymers, such as syndiotactic PS, polyphenyl sulfide (PPS), and polyketones, produce layers of amorphous polymer at the surface and core of the part with a semicrystalline region between these layers (Ref 40). At high temperatures, these polymers behave more like PP.

Table 11 Typical shrinkage values for selected polymers

Polymer	Shrinkage, mm/mm: Polymer	Shrinkage, mm/mm: Polymer with 30% glass fiber
HDPE	0.015–0.040	0.002–0.004
PP	0.010–0.025	0.002–0.005
PS	0.004–0.007	...
ABS	0.004–0.009	0.002–0.003
POM	0.018–0.025	0.003–0.009
Nylon 6/6	0.007–0.018	0.003
PET	0.020–0.025	0.002–0.009
PBT	0.009–0.022	0.002–0.008
PC	0.005–0.007	0.001–0.002
PSU	0.007	0.001–0.003
PPS	0.006–0.014	0.002–0.005

Source: Ref 39

Orientation. Different levels of orientation—and the related phenomena of die swell, shrinkage, and molded-in stress—are introduced during processing. Because gravity is the only force acting on the melt during rotational molding, very little orientation occurs in this process. Uniaxial orientation results from pipe, profile, flat-film and fiber extrusion, and calendering, whereas blow molding and blown-film extrusion induce biaxial orientation. While the actual orientation in injection molding varies with the mold design, the high flow rates generally align the polymer molecules in the direction of flow. Thermoforming also orients the polymer chains according to the design of the product.

Die swell is the expansion of the polymer melt that occurs as the extruded melt exits the die. This occurs when the aligned polymer chains escape the confines of the die and return to their random coil configuration. Die swell is dependent on processing conditions, die design, and polymer structure. It typically increases with screw speed (output rate) and decreases with higher melt temperatures and longer die land lengths. Increased MW, which produces more entanglement, also increases die swell.

Melt Fracture. At high extrusion rates, the polymer surface may also exhibit sharkskin or melt fracture. When the shear stress during extrusion exceeds the critical shear stress for the polymer, a repeating wavy pattern known as sharkskin occurs. In high-MW polyolefins this may disappear as the shear rate reaches the stick/slip region where the defect is present, but not visible. At even higher speeds, the polymer surface breaks up again in the defect known as melt fracture. This is particularly important in continuous and intermittent extrusion blow molding where these high-MW polymers are used; the output rates for continuous extrusion blow molding are typically below the critical shear rate, while those for intermittent extrusion blow molding place the process in the stick/slip region.

Table 12 Water absorption, processing temperatures, and maximum shear conditions for selected polymers

Polymer	Water absorption, %	Processing temperatures, °C	Maximum shear stress, MPa	Maximum shear rate, $10^3\ s^{-1}$
HDPE	<0.01	180–240	0.20	40
PP	0.01–0.03	200–260	0.25	100
PMMA	0.10–0.40	240–260	0.40	40
PVC, rigid	0.04–0.40	140–200	0.20	20
ABS	0.20–0.45	200–260	0.30	50
POM	0.25–0.40	190–230	0.45	40
Nylon 6/6	1.00–2.80	270–320	0.50	60
PET	0.10–0.20	280–310	0.50	...
PBT	0.08–0.09	220–260	0.40	50
PC	0.15	280–320	0.50	40
PS	0.30	310–340	0.50	50

Source: Ref 8, 39

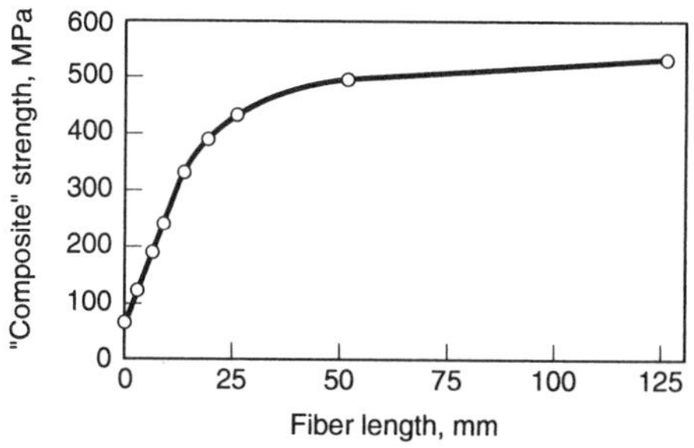

Fig. 25 The effect of fiber length on material strength. Source: Ref 41

Shrinkage. Although shrinkage results from the volumetric contraction of the polymer during cooling, it is influenced by the relaxation of oriented polymer molecules. During processing the polymers align in the direction of flow, and their relaxation causes swelling perpendicular to this direction. Consequently, shrinkage in the direction of flow is usually much greater than transverse to flow. Addition of fillers and fibers, which also align in the flow, reduces shrinkage because they prevent the aligned molecules from relaxing. While rapid cooling can prevent the aligned polymer chains from relaxing, these chains contribute to molded-in stress.

Molded-in stress is the worst in regions where the polymer chains are highly aligned and not allowed to relax. Thus, processes with high levels of orientation produce the greatest molded-in stress. The stressed areas are points of attack for chemicals and sources of future breaks and cracks. Annealing will remove some of these stresses and is routinely required for some polymers such as PSUs. Because processes such as thermoforming and injection blow molding do not actually melt the plastic, but shape it at lower temperatures, the stretching produces high levels of molded-in stress. Usually the gate region of an injection-molded part will have the highest stresses, and consequently gate location is an important consideration in part design and failure analysis.

Polymer Degradation. Polyvinyl chloride, other chlorine-containing polymers, fluoropolymers, and POM tend to degrade under normal processing conditions. The dehydrochlorination of PVC occurs relatively easily and requires tightly controlled processing conditions. Hydrochloric acid formed during the degradation of PVC is not only corrosive to the equipment, but it catalyzes further degradation. The remaining polymer becomes increasingly rigid and discolored due to the formation of conjugated carbon–carbon double bonds. A similar reaction occurring in fluoropolymers produces the equally corrosive hydrofluoric acid. In contrast, POM depolymerizes from the ends of the polymer in an action called "unzipping"; this produces formaldehyde, which further catalyzes the depolymerization. To prevent or minimize degradation of PVC (or other chloropolymers and fluoropolymers), stabilizers are added to the plastic. With POM, copolymerization with cyclic ethers (such as ethylene oxide) or incorporation of blocking groups at the ends of the polymers (end capping) prevents unzipping.

Because many engineering polymers were produced by condensing two components to produce water, the presence of water during melt processing reverses this reaction. Thus, chains are broken, the MW is reduced, and properties decrease. In addition, water migrates to the surface of the part, resulting in the visual defect known as splay. While water uptake varies with the polarity and storage conditions of the plastic, most engineering plastics require drying before processing. Of the polymers shown in Table 12, only HDPE, PP, and rigid PVC are usually processed without some drying. While undried ABS and PMMA will not exhibit chain scission, they are typically dried to prevent splay. The remaining polymers in Table 12 are subject to chain scission and visual defects. Control of the water content in PET is of major importance for clarity of blow-molded bottles.

The combination of temperature and shear can also degrade plastics. The long entangled polymer chains of UHMWPE are easily severed in single-screw extruders. Heat-sensitive polymers such as PVC also degrade when the viscous dissipation from shear raises the melt temperature above the degradation temperature. Because counterrotating twin-screw extruders have positive material conveying characteristics, uniform residence time, and uniform temperature distributions, they are used for extruding materials such as rigid PVC. Ultrahigh-molecular-weight polyethylene is often processed on twin-screw extruders or ram extruders (which have little shearing action). While shear can be a problem in extrusion processes, it is usually greatest in injection molding where polymer is forced at high velocities through small orifices. As indicated in Table 12, the processing temperatures and maximum shear conditions vary from polymer to polymer. However, as mentioned previously, when forcing highly viscous melts through thin channels, these maximum values are easily exceeded. Excess shear rates produce chain scission, whereas excess shear stress tends to produce cracking and related defects in the plastics product.

When continuous-glass fibers or glass mats are processed using traditional thermoset processing techniques, the glass fibers usually remain unbroken. However, the discontinuous glass fibers commonly added to engineering resins are often broken during plastication and molding. As shown in Fig. 25, the fiber length is critical to the strength of the "composite." Reduction of the fiber length below a critical value results in a rapid decrease in strength. Consequently, glass fibers are often compounded into polymers using the controlled shear of twin-screw extruders. Special nonreturn valves (at the end of screws in injection-molding machines) also minimize fiber degradation.

Blends. The properties of immiscible and partially miscible blends depend on their processing conditions. Some are engineered so that one phase migrates to the air interface and governs surface properties. In immiscible polyblends, morphology is very sensitive to temperature and shear. These determine the size of the domains and whether the domains are spherical, elongated, or laminar. Phases may elongate in the flow direction.

ACKNOWLEDGMENTS

The authors would like to acknowledge Dr. Rudolph Deanin of University of Massachusetts at Lowell's Plastics Engineering Department, whose seminal text, *Polymer Structure, Properties and Applications,* is a major source of information for this article.

REFERENCES

1. J.A. Brydson, *Plastics Materials,* 5th ed., Butterworths, 1989
2. R.J. Cotter, *Engineering Plastics Handbook of Polyarylethers,* Gordon and Breach, 1995
3. R.D. Deanin, *Polymer Structure, Properties and Applications,* Cahners Books, 1972
4. H. Dominghaus, *Plastics for Engineers: Materials, Properties, and Applications,* Hanser Publishers, 1988
5. F. Rodriguez, *Principles of Polymer Systems,* 3rd ed., Hemisphere Publishing, 1989
6. J.H. Schut, Why Syndiotactic PS Is Hot, *Plast. Technol.,* Feb 1993, p 26–30

7. R.D. Deanin, *Polymer Structure, Properties and Applications,* Cahners Books, 1972, p 27
8. L.L. Clements, Polymer Science for Engineers, *Engineering Plastics,* Vol 2, *Engineered Materials Handbook,* ASM International, 1988, p 56–57
9. F. Rodriguez, *Principles of Polymer Systems,* 3rd ed., Hemisphere Publishing, 1989, p 23
10. H. Dominghaus, *Plastics for Engineers: Materials, Properties, and Applications,* Hanser Publishers, 1988, p 34, 347
11. R.D. Deanin, *Polymer Structure, Properties and Applications,* Cahners Books, 1972, p 54
12. S.L. Rosen, *Fundamental Principles of Polymeric Materials,* 2nd ed., John Wiley & Sons, 1993, p 53, 54, 59
13. R.D. Deanin, *Polymer Structure, Properties and Applications,* Cahners Books, 1972, p 55
14. J.M. Dealy and K.F. Wissbrun, *Melt Rheology and Its Role in Plastics Processing; Theory and Applications,* Van Nostrand Reinhold, 1990, p 369
15. R.D. Deanin, *Polymer Structure, Properties and Applications,* Cahners Books, 1972, p 130
16. J.A. Brydson, *Plastics Materials,* 5th ed., Butterworths, 1989, p 58
17. R.D. Deanin, *Polymer Structure, Properties and Applications,* Cahners Books, 1972, p 141
18. R.D. Deanin, *Polymer Structure, Properties and Applications,* Cahners Books, 1972, p 138
19. S.L. Rosen, *Fundamental Principles of Polymeric Materials,* 2nd ed., John Wiley & Sons, 1993, p 45
20. S.L. Rosen, *Fundamental Principles of Polymeric Materials,* 2nd ed., John Wiley & Sons, 1993, p 46
21. F. Rodriguez, *Principles of Polymer Systems,* 3rd ed., Hemisphere Publishing, 1989, p 23–24
22. W. Michaeli, *Plastics Processing, an Introduction,* Hanser Publishing, 1992, p 19
23. C.C. Winding and G.D. Hiatt, *Polymeric Materials,* McGraw-Hill, 1961
24. R.D. Deanin, *Polymer Structure, Properties and Applications,* Cahners Books, 1972, p 89
25. R.D. Deanin, *Polymer Structure, Properties and Applications,* Cahners Books, 1972, p 342
26. R.D. Deanin, *Polymer Structure, Properties and Applications,* Cahners Books, 1972, p 240
27. C. Rauwendaal, *Polymer Extrusion,* 2nd ed., Hanser Publishers, 1990, p 182
28. C. Rauwendaal, *Polymer Extrusion,* 2nd ed., Hanser Publishers, 1990, p 218
29. M.M. McKelvey, *Polymer Processing,* John Wiley & Sons, 1962, p 26, 30
30. J.M.G. Cowie, *Polymers: Chemistry & Physics of Modern Materials,* 2nd ed., Blackie Academic and Professional, 1991, p 248
31. R.D. Deanin, *Polymer Structure, Properties and Applications,* Cahners Books, 1972, p 109
32. W. Michaeli, *Plastics Processing, an Introduction,* Hanser Publishing, 1992, p 59
33. C.C. Ku and R. Liepins, *Electrical Properties of Polymers: Chemical Principles,* Hanser Publishers, 1987, p 181–182
34. A.B. Strong, *Plastics: Materials and Processing,* Prentice-Hall, 1996, p 144
35. M.J. Austin, Inorganic Anti-Corrosive Pigments, *Paint and Coating Testing Manual,* J.V. Koleste, Ed., ASTM, 1995, p 239
36. W. Michaeli, *Plastics Processing, an Introduction,* Hanser Publishing, 1992, p 159
37. ASTM D 1238, *Annual Book of ASTM Standards,* Vol 08.01, ASTM
38. J.A. Brydson, *Plastics Materials,* 5th ed., Butterworths, 1989, p 382
39. *Modern Plastics Encyclopedia '92,* McGraw-Hill, 1992, p 378–428
40. Y. Ulcer, M. Cakmak, J. Miao, and C.M. Hsiung, Structural Gradients Developed in Injection Molded Syndiotactic Polystyrene (S-PS), *Annual Technical Conference of the Society of Plastics Engineers,* 1995, p 1788
41. P.K. Mallick, *Fiber-Reinforced Composites,* Marcel Dekker, 1988, p 83

Effects of Composition, Processing, and Structure on Properties of Composites

R. Laramee, Intermountain Design Inc.

COMPOSITES fabricated with fiber reinforcement and a resin, carbon, or metal matrix are versatile materials that offer several advantages for today's innovative and demanding designs. In general, composites are lightweight, strong, and impact and fatigue resistant. They can be cost competitive, and are adaptable to many applications. Composites can be readily tailored in composition and manufacturing processing to meet specific engineering-design applications and loading conditions.

This article describes the interaction of composition, manufacturing process and composite properties and how variations in the composition, manufacturing, shop process instructions, and loading/environmental conditions can affect the use of a composite product in a performance/service life operation.

For composite matrix and reinforcement systems, the reinforcement type and orientation will in most instances be the dominant contributor to properties. With the use of good coupling agents, the reinforcement and matrix can be more effective in working together to resist all loading conditions. Fillers can be used effectively to lower density and cost, change strength properties, and facilitate the manufacturing process.

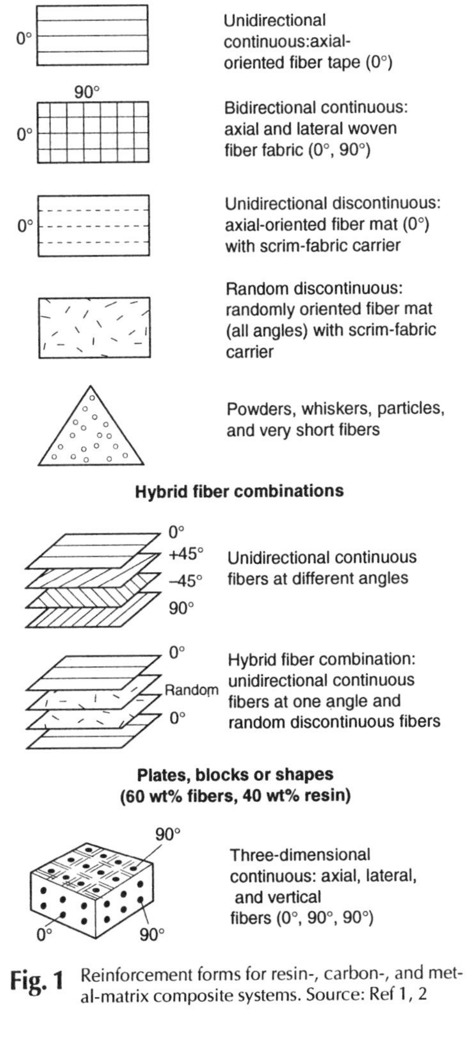

Fig. 1 Reinforcement forms for resin-, carbon-, and metal-matrix composite systems. Source: Ref 1, 2

Table 1 Types of materials used in composites

Fiber reinforcements	**Matrix materials (continued)**
Inorganic	Metal
Glass	Stainless steel alloy
Boron/tungsten wire	Aluminum alloy
Silicon carbide	Titanium alloy
Organic	Carbon
Aramid (Kevlar)	Carbonized resin
Carbon	CVD carbon or graphite deposition
Graphite	Carbon powder
Matrix materials	Filler
Resin	Powder
Thermoplastic	Silica
Polyester	Carbon
Polyamide	Microballoon
Polysulfone	Phenolic
Thermoset (virgin or carbonized)	Carbon
Epoxy	Glass
Phenolic	Solid particles
Polyester	Carbon
Polyimide	Silicon carbide
Bismaleimide	Ceramic
Pitch	**Resin-fiber coupling agents**
	Silane

Source: Ref 1–3

Generally, in manufacturing, a longer process time with a higher pressure and temperature will result in higher properties and an improved product. Knowledge of the environment and the loading conditions will enable the design/manufacturing product team to specify any necessary coatings, optimal structural cross sections, a maintenance schedule, and inspection criteria that will anticipate possible problems.

However, in the actual application of composites to primary and secondary structures—a trade-off among weight, cost, size, and stress/deflections with other materials—composites may not be the answer for every problem. As existing and future designs for mass vehicles such as automotive, trains, ships, and aircraft are evaluated, a hybrid mixture of metal alloys and composites may provide the optimal solutions. An open, creative, and aggressive mind plus good material databases and unique design concepts will go far to provide a balanced materials application to a broadening field of problem solutions.

The design properties of resin-matrix composites are based on a standard composition and a standard processing cure, with planned variations to meet specific design applications. Standard compositions range from a 33 to 66% matrix material plus a 33 to 66% reinforcing fiber. These percentages hold true for resin-, carbon-, and metal-matrix compositions. For resin-matrix composites, the standard processing cures include pressures from 0 to 1700 kPa (0 to 250 psi), temperatures from room temperature to 177 °C (350 °F), and time at maximum temperature of 1 h/in. of component thickness.

Databases of mechanical and thermal properties versus temperature exist for standard materials, as supplied by prepreg suppliers such as Thiokol Inc. and Fiberite Inc. for resin-matrix composites, and as developed in-house for carbon- and metal-matrix products by other manufacturers, for various processing cycles. The following discussion provides data on matrix composition, manufacturing, and mechanical properties.

Composition of Composites

For a standard composite panel or structural shape, the ratio of matrix material (resin, metal, or carbon) to the fiber reinforcement ranges approximately from 1:2 to 2:1. The fiber is the main tailoring element for design properties, while fiber orientation and fillers can provide secondary fine tuning for the product application.

The resin, carbon, or metal matrix provides (1) stable dimensional control to the fiber laminate, (2) a small participating component for properties, and (3) a shear resistance between reinforcing fibers. A coupling agent enhances resin matrix-to-fiber bonding, while the filler can fine tune such properties as density, cost/pound, processing viscosity, strength, and flame-retardant characteristics.

Reinforcing-fiber characteristics such as density; fiber diameter, strength, and modulus; fiber-filament bundle size; and woven-fiber fabric type or chopped-fiber form are initially optimized approximately by the "rule of mixtures" to help meet the composite properties needed for the design application engineering criteria for product operation and performance.

In addition, the selection of the ratio of matrix to reinforcement constituents is influenced by the loading patterns to the product, environmental operating conditions, the standard manufacturing-processing methods, reinforcement forms, costs, and completion time for the particular company and industry.

Reinforcement forms for the various matrix systems are shown in Fig. 1. Table 1 identifies a short list of current reinforcements, matrix materials, and coupling agents. Most frequently used reinforcements include:

- Uniaxial continuous fiber for end-tape filament winding, braiding, or pultrusion

Table 2 Reinforcements for metal- and carbon-matrix composites

Fiber	Density gm/cm³	Diameter µm	Diameter µin	Tensile strength GPa	Tensile strength 10^6 psi	Tensile modulus GPa	Tensile modulus 10^6 psi
Metal-matrix reinforcements: boron and alumina fibers							
Boron-tungsten	2.6	100–200	3950–7850	5.5–7.0	0.80–1.0	400	58
Boron-carbon	2.3	100–200	3950–7850	5.0	0.73	400	58
α-alumina(a)	3.95	20	790	14(b)	2.0(b)	390(b)	57(b)
				19(c)	2.8(c)	390(c)	57(c)

Material	Specific gravity	Melting point °C	Melting point °F	Tensile strength MPa	Tensile strength ksi	Young's modulus of elasticity GPa	Young's modulus of elasticity 10^6 psi	Coefficient of thermal expansion, 10^{-6}/K
Metal-matrix reinforcements: metallic wires								
Aluminum	2.71	660	1220	290	40	68.9	10.0	23.6
Beryllium	1.85	1350	2460	1100	160	310	45.0	11.6
Copper	8.90	1083	1980	413	60	124	18.0	16.5
Tungsten	19.3	3410	6170	2890	130	345	50.0	4.6
Austenitic stainless steel	7.9	1539	2800	2390	350	200	29.0	8.5
Molybdenum	10.2	2625	4750	2200	320	331	48.0	...

Material	Density, g/cm³	Tensile strength GPa	Tensile strength 10^6 psi	Tensile modulus GPa	Tensile modulus 10^6 psi
Metal-matrix reinforcements: short fibers and whiskers					
Alumina					
Whiskers	4.0	10–20	1–3	700–1500	100–220
Sintered fibers	<4.0	0.2–0.7	0.030–0.10	140–300	20–40
Boron, thermally formed fibers	2.3	2.75	0.400	400	60
Boron nitride, fibers	1.8–2.0	0.3–1.4	0.045–0.20	28–80	4–10
Silicon nitride, whiskers	3.2	5–7	0.75–1.0	350–380	50–55
Carbon-matrix reinforcements					
Carbon whiskers	>2.0	...	...	700	100
Carbon fibers	1.8–2.0	2–3	0.29–0.44	230–550	35–80

(a) Slurry-spun continuous fiber. (b) Uncoated. (c) Silicon carbide-coated. Source: Ref 4

Table 3 Common fillers for resin matrices

Category	Fillers
Lubricating fillers	Molybdenum disulfide, graphite, teflon
Magnetic fillers	Black iron oxide, powdered iron or shot, barium and strontium ferrites
Lower-density fillers	Microballoons: expanded perlite, glass nodules
Reinforcement, viscosity control, decorative fillers	Cotton flock, α-cellulose fibers
Lower compression, sound absorption fillers	Powdered cork, protein
Pigment, packing fillers	Carbon blacks, powdered bituminous coal
Electrical insulation	Fused silica
Lowers composite expansion	Beta encryptite (lithium aluminosilicate)
Electrical conduction	Carbon powder or fibers

Source: Ref 5

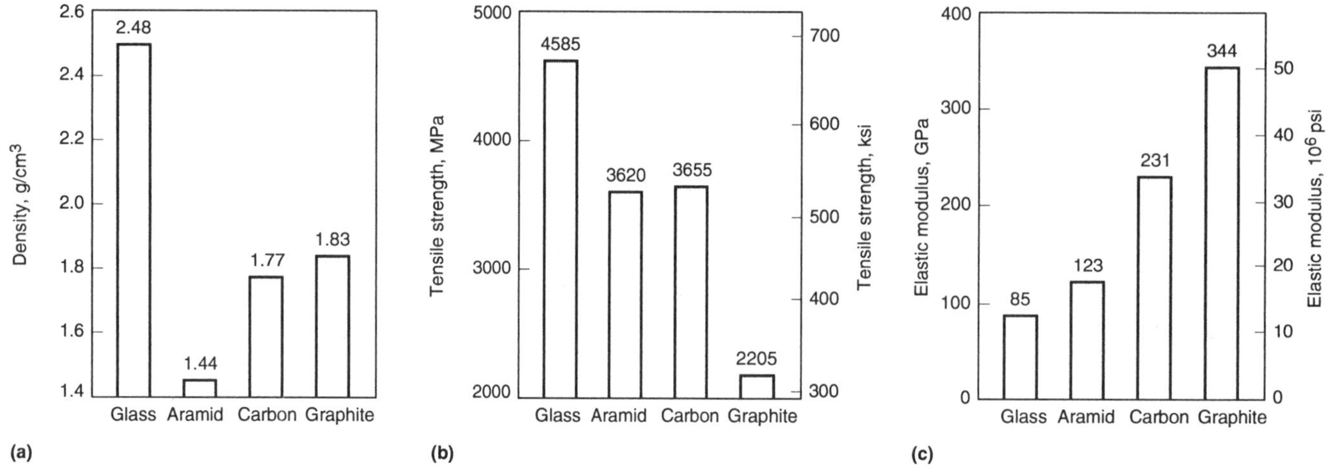

Fig. 2 Mechanical properties and service temperatures for selected reinforcement fibers. Inorganic fibers: glass (maximum temperature 970 °C, or 1780 °F) and aramid (maximum temperature 500 °C, or 930 °F). Organic fibers: carbon (maximum temperature 2500 °C, or 4500 °F) and graphite (maximum temperature 3000 °C, or 5500 °F). (a) Density. (b) Tensile strength. (c) Elastic modulus. Source: Ref 4

- Fabric (warp and fill continuous fiber) for tape wrapping and lay-up fabrication processes
- Chopped, continuous short-fiber reinforcement for molding compounds, injection or resin-transfer molding, and bulk- or sheet-molding compound as a charge for compression molding

Coupling agents are added to assist the binding of organic and inorganic fibers to the resin matrix.

Resin, carbon, or metal matrix, with or without fillers, can be added to the reinforcement fibers or woven fabric as a partially staged resin (partially solidified, stabilized, or cured liquid resin) or as a metal sheet or foil or powder form, or matrix

Table 4 Resin-matrix composite closed-mold processing characteristics

Process	Resin	Reinforcement	Filler	Thermoplastics	Thermoset	Fiber orientation	Cure operation	
							Pressure(a)	Temperature(b)
Preform molding								
Transfer mold	Prepreg, preform, BMC	Short fibers	Yes	Yes	Yes	Parallel to mold surfaces	High	Low
Compression mold	Prepreg, preform, BMC-SMC	Short fibers	Yes	Yes	Yes	Parallel to mold surfaces	High	Low
Resin-transfer mold	Pressurized liquid resin injection	Preform or mat, fabric lay-up, 2D or 3D preform	Yes	Yes	Yes	Parallel to mold surfaces and reinforcement surface	Medium	Medium
Preform flow-die molding								
Injection mold	Resin, liquid, powder, pellets, crystals, prepreg	Short fibers	Yes	Yes	Yes	Parallel to mold surfaces	High	High
Extrusion	Resin, liquid, powder, pellets, crystals, prepreg	Short fibers	Yes	Yes	Yes	Parallel to orifice die inside surface	High	High
Reaction injection molding	Liquid resin reactants, curing agents	Short fibers, powder, whiskers	Yes	Yes	Yes	Parallel to mold surfaces	High	Medium
Thin-shell molding								
Blow molding	Hollow resin, tube heated	Powder	Yes	Yes	Yes	Parallel to mold surfaces	High	Medium
Rotational molding	Liquid resin	Powder, whiskers	Yes	Yes	No	Parallel to mold surfaces	Medium	Medium
Slip casting	Semiliquid resin	Powder, whiskers	Yes	Yes	No	Parallel to mold surfaces	Low	Low
Miscellaneous								
Foam molding	Liquid resin plus blowing agent	Short fibers	Yes	Yes	Yes	Parallel to mold surfaces	Low/medium	Low
Lost-core molding	Liquid resin, molten wax, ceramic spray	Powder, whiskers	Yes	Yes	No	Parallel to mold surfaces	Low	Medium/very high
Thermoforming	Resin, prepreg, SMC	Short fibers, cut continuous fibers	Yes	Yes	Yes	Parallel to mold surfaces	Low	Medium

BMC, bulk molding compound; SMC, sheet molding compound; 2D, two-dimensional; 3D, three-dimensional. (a) Pressure ranges: low, ≤100 kPa (≤15 psi); medium, 100–1725 kPa (15–250 psi); high, up to 100 MPa (15 ksi). (b) Temperature ranges: low, room temperature to 165 °C (330 °F); medium, 165–190 °C (330–370 °F); high, 190–205 °C (370–400 °F); very high, 205–815 °C (400–1500 °F). Source: Ref 6–11

Table 5 Resin-matrix composite open-mold processing characteristics

Process	Resin	Reinforcement	Filler	Thermoplastics	Thermosets	Fiber orientation	Cure operation	
							Pressure(a)	Temperature(b)
Moving-mold mandrel								
Filament winding and fiber	Wet fiber, winding, prepreg tape	Undirectional continuous fiber tape	Yes	Yes	Yes	Multidirectional parallel to mandrel	Low	Low
Braiding	Wet resin impregnation after winding	Continuous fiber	Yes	yes	Yes	Parallel to braid mold surface, multidirectional fibers	Medium/high	Medium
Fabric tape wrapping	Prepreg tape	Continuous fabric (tape width)	Yes	Yes	Yes	Multidirectional, parallel to mandrel surface	High	Low
Stationary-mold mandrel								
Hand lay-up	Prepreg liquid	Fiber, fabric mat, short continuous	Yes	Yes	Yes	Random, parallel to mold surface	Low	Low
Gun spray-up	Liquid	Chopped short fiber	Yes	Yes	Yes	Random, parallel to mold surface	Low	Low
Fiber-tape lay-down	Prepreg tape, wet tape	Continuous unidirectional fiber tape	Yes	Yes	Yes	Multidirectional, parallel to mandrel	Medium	Low
Pultrusion	Wet resin impregnation of fibers during pull through die	Continuous fiber, fiber mat	Yes	Yes	Yes	Parallel to laminate exterior surface and mold surface	Medium	Medium
Laminated plates, tubes	Prepreg	Fiber mat, fabric (unidirectional, fiber oriented)	Yes	Yes	Yes	Parallel to mold plates	High	Low
Miscellaneous								
Casting neat resin	Liquid resin	Powder, short fiber	Yes	Yes	Yes	Random, parallel to mold surface	Low	Low
Calendering	Semiliquid resin, catalyst, cure agent	Fiber, powder	Yes	Yes	No	Parallel to sheet surface	Medium	Low

(a) Pressure ranges: low, ≤100 kPa (≤15 psi); medium, 100–1725 kPa (15–250 psi); high, up to 100 MPa (15 ksi). (b) Temperature ranges: low, room temperature to 165 °C (330 °F); medium, 165–190 °C (330–370 °F); high, 190–205 °C (370–400 °F); very high, 205–815 °C (400–1500 °F). Source: Ref 6–11 material can be added to the fibers (in situ) during placement onto or into the open- or closed-mold surface. The matrix can be in the form of a liquid, powder, particles, foil, sheet, or fiber.

Fillers or additives can replace up to 33% of the weight of a resin matrix to tailor composite properties for such characteristics as density, wear resistance, color, ductility, flame-smoke retardation, moisture resistance, lubricity, and dimensional stability. The components added to the resin matrix will also change the cost per pound, the softening and gelation of the final cure process and a change of the service temperature.

Currently, twelve or more fiber-reinforcement systems are available for resin-matrix-composite fabrication into industrial, commercial, and aerospace products. However, this article concentrates mainly on four fiber types: glass, aramid, carbon, and graphite. Figure 2 shows a comparison of the fiber characteristics and properties. Density ranges from 1.44 to 2.48 g/cm^3, strength from a minimum of 2200 MPa (320 ksi) to a maximum of 4585 MPa (665 ksi), and modulus from 85 to 345 GPa (12 to 50 × 10^6 psi), and service-temperature capability in inert atmospheres from 500 to 3040 °C (930 to 5500 °F). Reinforcements for metal- and carbon-matrix composites are shown in Table 2. Typical property variations for three reinforcements are:

Reinforcement	Density, g/cm^3	Tensile strength MPa	Tensile strength ksi	Elastic modulus GPa	Elastic modulus 10^6 psi
Boron metal deposited on a carbon fiber	2.3	5030	730	400	58
Alumina powder pressed, sintered, and formed into fiber	<4.0	450	65	205	30
Tungsten powder pressed, sintered, and formed into fiber	19.3	900	130	345	50

Matrix materials include resin, carbon, and metal. Resin systems include at least 36 types or hybrid combinations, divided into thermoplastics (heat affected) or thermoset (heat permanently cured) divisions. The baseline room-temperature properties and service temperatures of three thermoplastic and three thermosetting resins in Fig. 3 show the ranges of density from 1.14 to 1.43 g/cm^3, tensile strength from 53 to 112 MPa (7.7 to 16.2 ksi), tensile modulus from 875 to 4135 MPa (127 to 600 ksi), and service temperature from 130 to 370 °C (266 to 700 °F).

A common metal-matrix material is aluminum alloy, with properties that include low liquid temperature of 650 °C (1200 °F) (which aids fabrication), a low density (2.75 g/cm^3), a good elastic modulus (70.3 GPa, or 10.2 × 10^6 psi), and an excellent heat-treated yield strength (minimum 276 MPa, or 40 ksi).

For resin-matrix composites, most materials commonly used are preformulated at the supplier for the preimpregnated fiber, resin, curing agent, and coupling system. However, the manufacturer will often add small amounts of a filler to aid in the processing; change surface texture, color, and thermal and electrical conductivity; provide moisture resistance; and serve as an antioxidant. A short list of such fillers for resin matrices is given in Table 3. Modifications to the composition, however, must be tested for compatibility and successful cure processing to achieve properties tailored to meet specific product performance and service operation requirements.

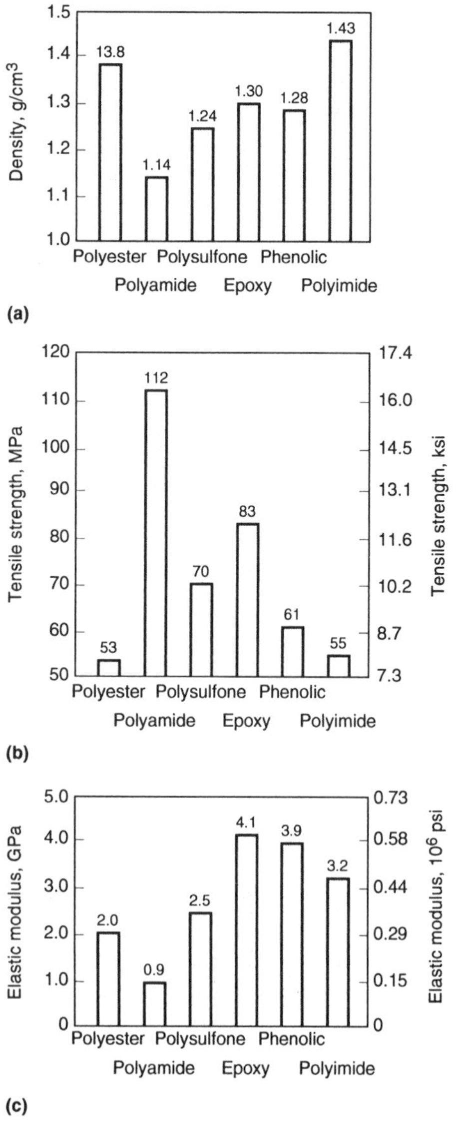

Fig. 3 Mechanical properties and service temperatures for selected matrix resins. Thermoplastics: polyester (unfilled; maximum temperature 140 °C, or 284 °F), polyamide (nylon 6/6, unfilled; maximum temperature 130 °C, or 266 °F), and polysulfane (standard; maximum temperature 160 °C, or 320 °F). Thermosets: epoxy (unfilled; maximum temperature 260 °C, or 500 °F), phenolic (unfilled; maximum temperature 230 °C, or 450 °F), and polyimide (unfilled; maximum temperature 370 °C, or 700 °F). (a) Density. (b) Tensile strength. (c) Elastic modulus. Source: Ref 4

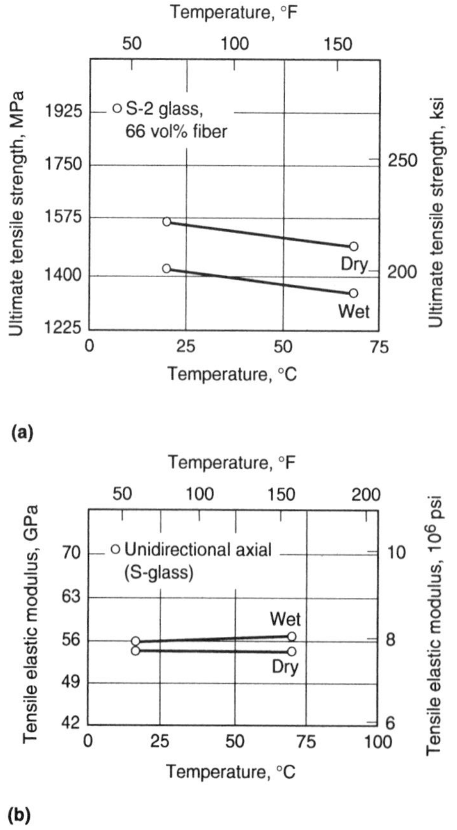

Fig. 4 Effect of temperature on the strength of S-glass-fiber/epoxy-matrix composites. (a) Tensile strength. (b) Elastic modulus. Source: Ref 4

Manufacturing of Composites

The manufacturing process shapes the composite laminate through a cure, or solidification, cycle consisting of pressure and temperature plus time at maximum temperature. Higher temperatures and pressures produce thinner parts with higher density and higher mechanical properties. In addition, the open or closed mold surface(s) defines the laminate shape, dimensional stability, and final fiber orientation. The range of part properties depends on the matrix material, the material and type of fiber reinforcement, and any fillers and coupling agents used.

Manufacturing processes include sixteen for fabricating resin-matrix composites in closed and open molds or dies (Tables 4, 5), six for metal-matrix composites (Table 6), and five for carbon-carbon matrix types (Table 7), for a minimum of 27 methods of manufacturing. Each table identifies the state of the fiber and matrix at the start of the process, the fiber orientation, cure or solidification process parameters, and the resin-, carbon-, or metal-matrix material used in the process.

Resin-Matrix Processing

Closed mold methods of fabrication are grouped into four families having similar characteristics (Ref 6–9, 16).

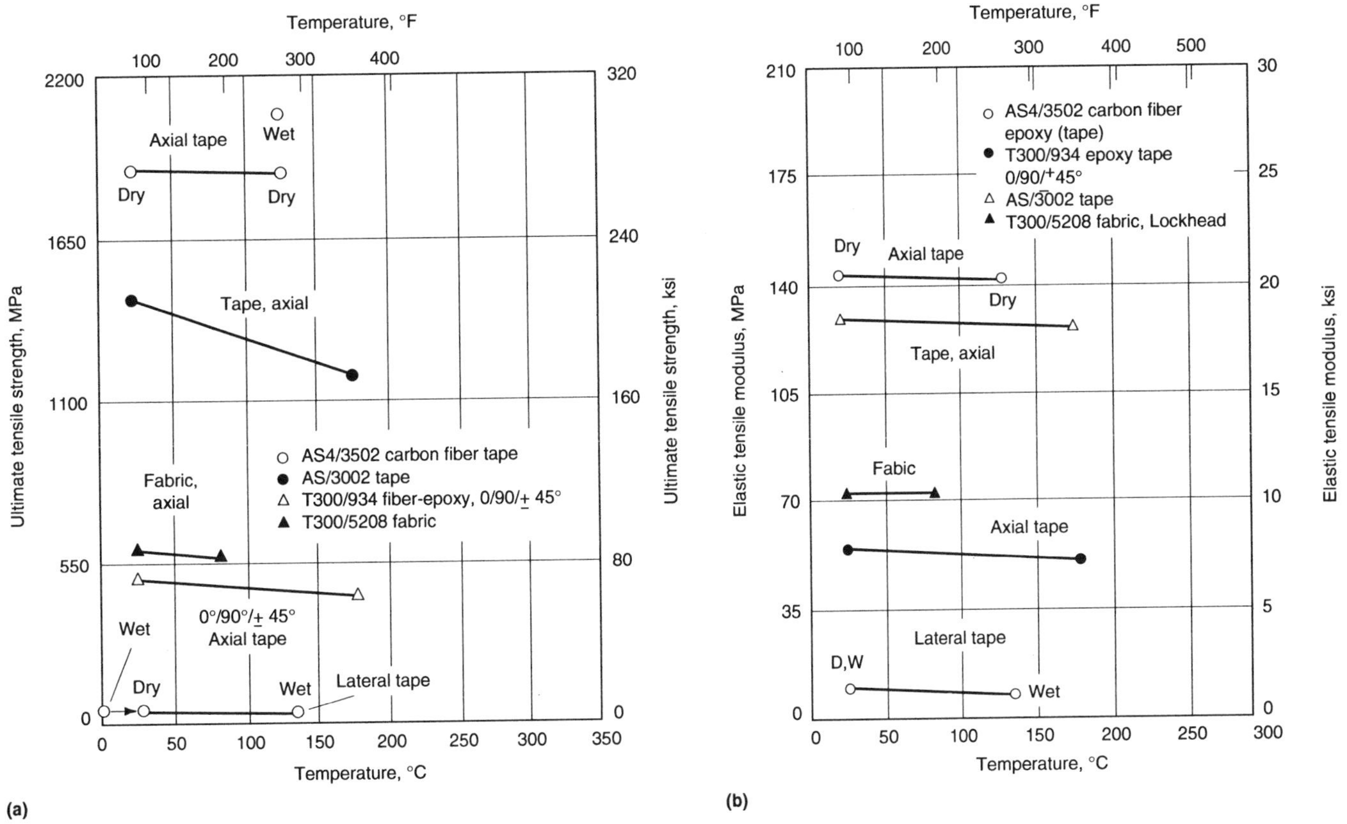

Fig. 5 Effect of temperature on the strength of carbon-fiber/epoxy-matrix composites. (a) Tensile strength. (b) Elastic modulus. Source: Ref 4

Preform molding in transfer, compression, or resin-transfer molds uses a preform of fiber or fabric in a resin matrix. Fiber orientation is parallel to the mold surfaces.

Preform flow-die molding by injection, extrusion, or reaction injection uses a preform of fiber or fabric in a resin matrix that is forced through a forming die at high pressure and high temperature and then is cured. Fiber orientation is parallel to the mold centerline or to the mold surfaces.

Thin-shell molding by blow molding, rotational molding, or slip casting uses a liquid resin plus a preform of fiber in a resin matrix that is cured with a temperature, pressure, time cure cycle. Fiber orientation is parallel to the mold surfaces.

Miscellaneous molding processes such as foam, lost core, and thermoforming use short fibers in a resin matrix that is cured using a cure cycle. Fiber orientation is parallel to the mold surfaces.

Open-mold methods are grouped into the following three families by their characteristics (Ref 6–9).

Moving-mold mandrel processes (filament winding, braiding, and fabric-tape wrapping) use a continuous fiber or fabric with a resin matrix that is wound on a moving mandrel and is subsequently cured. Fiber orientation is parallel to the mandrel surface.

Stationary-mold mandrel processes (fabric hand lay-up, chopped-fiber gun lay-up, fiber-tape lay-down, fiber pultrusion through a die, fabric structural-shape lamination) use a resin matrix that is cured after the required thickness is achieved. Fiber orientation is parallel to the mandrel surface.

Miscellaneous molding processes (such as casting and calendering) use a liquid resin, with or without a fiber reinforcement, shaped to the mold surface by the mold cavity or rollers and cured.

Whereas as the closed-mold tooling includes the capacity for pressure and temperature control

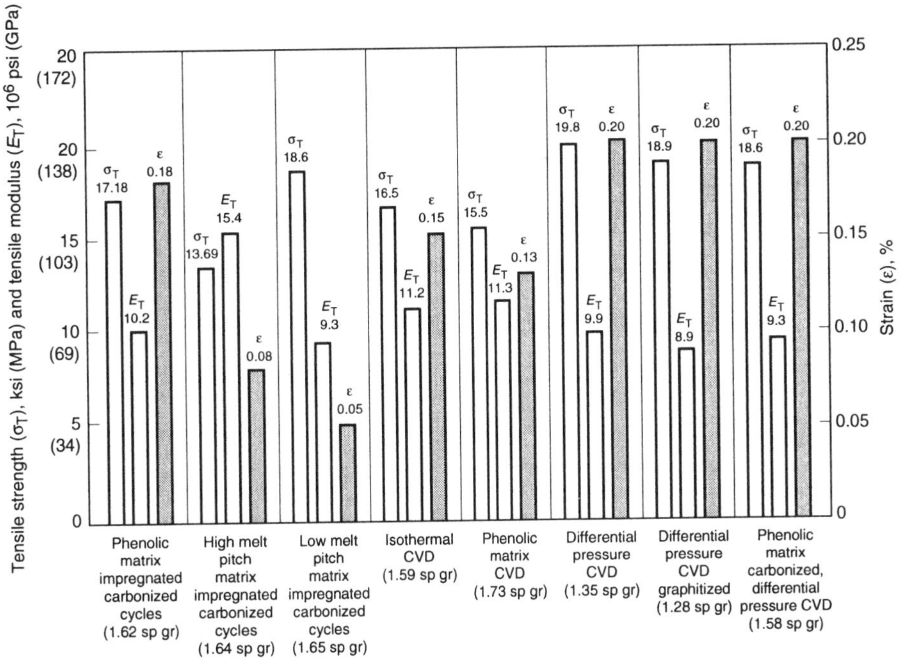

Fig. 6 Room-temperature mechanical properties of carbon-fiber/carbon-matrix (carbonized resin/CVD carbon) composites (tensile hoop rings). Source: Ref 15

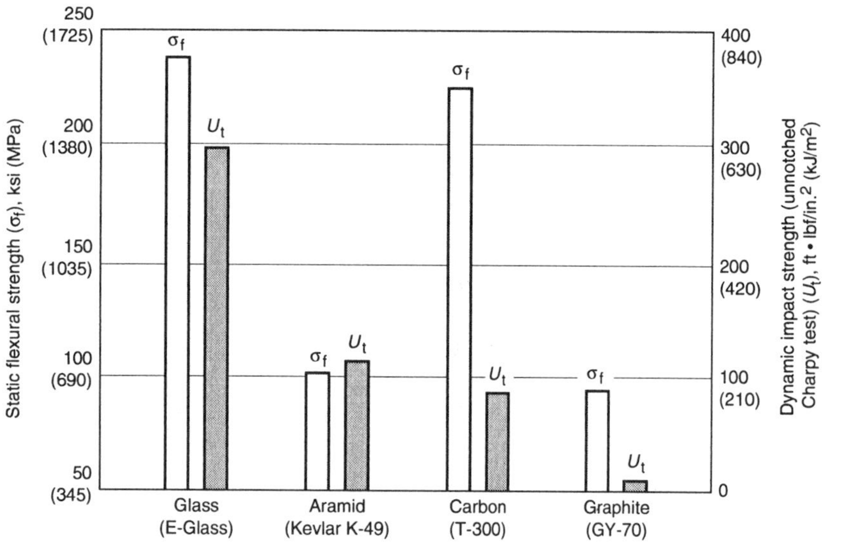

Fig. 7 Effect of fiber type on the flexure strength and impact strength of fiber/epoxy composites. Source: Ref 1

over time, open-mold processes require additional equipment for the resin-cure cycle such as:

- Ovens
- Autoclaves (heated pressurized air chambers)
- Vacuum bags and bleeder/release materials
- Hydroclaves (heated pressurized water and component rubber bags)
- Furnaces and induction heaters
- Electron-beam or ultraviolet light resin-cure systems

The additional curing equipment ensures a uniform distribution of the resin matrix, a highly uniform density, and a high-quality component.

Metal-Matrix Processing

Metal-matrix processing involves higher temperatures and pressures for laminate metal-matrix solidification than resin-matrix processing does. Both open- and closed-mold presses are used in conjunction with plasma-arc metal-spray equipment. The family of processes used for metal-matrix composites are listed below (Ref 4, 12, 13).

Powder Metallurgy. Aluminum, titanium, or stainless steel powder plus fiber reinforcement is compacted at a pressure of 21 to 28 MPa (3 to 4 ksi) and sintered at temperatures up to 1760 °C (3200 °F) to form solid right-cylinder billets in closed molds. Subsequent forming to the final shape is by extrusion, rolling, or forging.

Rotating-mandrel processes are used to deposit the following material forms:

- Plasma-arc spraying of the metal matrix and gun-spraying of fiber reinforcement for hollow structural shapes
- Filament winding of continuous reinforcement fiber and plasma arc spraying of the metal matrix (aluminum, titanium, or stainless steel) for plates and shapes
- Filament winding of continuous fiber reinforcement that has been coated with molten metal matrix for plates and shapes

The material is processed on the open mandrel under a low pressure (up to 100 kPa, or 15 psi) and at low temperature (up to 165 °C, or 330 °F) for final solidification. The metal mandrel may or may not be removed, depending on the design concept.

Table 6 Metal-matrix composite processing

Manufacturing process	Matrix material	Reinforcement material	Densification method	Final shape operations
Powder metallurgy				
P/M process	Aluminum, titanium, or stainless steel powder	Silicon carbide powder	Temperature, pressure, and time sintering	Extrusion Roll sheet, plate Forging
Rotating mandrels				
Cospray of molten matrix and SiC whiskers on rotating mandrel	Aluminum, titanium, or stainless steel melt	Silicon-carbide whiskers/powder Carbon, boron	Spray pressure	Mandrel rotation Machine finish
Filament wind with fiber and resin/powder matrix	Aluminum, titanium, or stainless steel powder Resin (thermoplastic, thermoset)	Metal fiber (aluminum, titanium, or stainless steel) prepregged with resin, powder matrix	Pressure and temperature on mandrel	As wound and surface coated
Filament wind with fiber and spray with molten metal matrix (plasma-arc spray)	Aluminum, titanium, or stainless steel powder pressure molten spray between fiber layers	Metal fiber or man-made fiber (boron, SiC, alumina glass, carbon) filament wound on mandrel	Pressure temperature consolidated Removed from mandrel in sheet form and molded to structural shape	As molded and surface coated
Hot press/diffusion bond				
Hot, mold of metal/reinforced fiber	Aluminum, titanium, or stainless steel foil or thin sheet	Glass, carbon, boron, graphite depending on metal melt temperature	Mold temperature pressure/inside sealed vacuum retort	Machine and finish
Diffusion bond	Aluminum, titanium, or stainless steel thin foil	Metal fiber or man-made fiber (boron, alumina glass, carbon)	Wound on steel tube, sealed and evacuated with metal bag Isostatic pressure in furnace	Machine and finish
Multiple hot press and diffusion bond	Aluminum, titanium, or stainless steel thin metal foil	Metal fiber or man-made fiber (boron, alumina glass, carbon)	Thin-pressed metal-fiber sheets diffusion bonded Sheets superplastic formed to shape with temperature, pressure	Machine and finish
Compressed preform densified				
Compressed preform in CVD furnace	CVD pyrolytic carbon infiltrated into preform in CVD furnace	Preform reinforcement of ceramic, metal, or man-made fibers, in particle, whisker, short- or long-fiber form	Vacuum, inert gas purge temperature conversion of organic gas into carbon deposition and H_2	Machine and finish
Compressed preform in hot-melt metal matrix	Molten metal pressed, vacuumed, or wicked into preform reinforcement	Preform reinforcement of ceramic, metal, or man-made fibers, in particle, whisker, short- or long-fiber form	Temperature, pressure, time consolidation	Forge, extrude, or roll Machine and finish

CVD, chemical vapor deposition. Source: Ref 4, 12–14

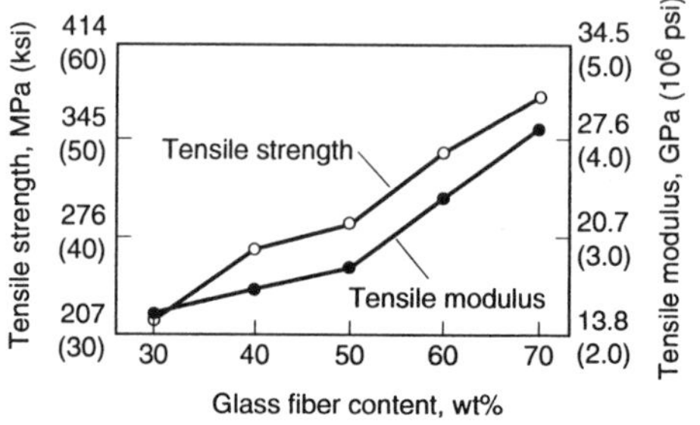

Fig. 8 Effect of glass fiber content on the longitudinal mechanical properties of pultruded E-glass/polyester composite sheets. Source: Ref 1

Hot Pressing and Diffusion Bonding. A matrix of carbon, boron, silicon carbide, stainless steel, or titanium in the form of thin metal foil is reinforced with continuous fiber to form a flat sheet preform, which is subsequently densified by pressure and temperature diffusion bonding. High pressure (up to 100 MPa, or 15 ksi) and ultrahigh metal melt temperature (up to 2760 °C, or 5000 °F) is required for solidification to final shape. Shapes are then machined and final finished. The temperature used depends on the metal matrix and fiber types used.

Compressed Reinforcement Preform with Matrix Infiltrations. The preform reinforcement used is usually a metal or organic short fiber pressed into a sheet or solid billet preform, with infiltration of molten aluminum, titanium, or stainless steel by pressure impregnation or carbon infiltration by chemical vapor deposition (CVD). The pressure used may vary from low (0 to 100 kPa, or 0 to 15 psi) for CVD, to high (up to 100 MPa, or 15 ksi) for molten metal, and temperature is very high (up to 2760 °C, or 5000 °F). The temperature used again depends on the metal matrix and fiber types used.

Additional process equipment for the metal-matrix composite fabrication includes:

- CVD furnaces
- Closed molds with temperature, time, and pressure controls

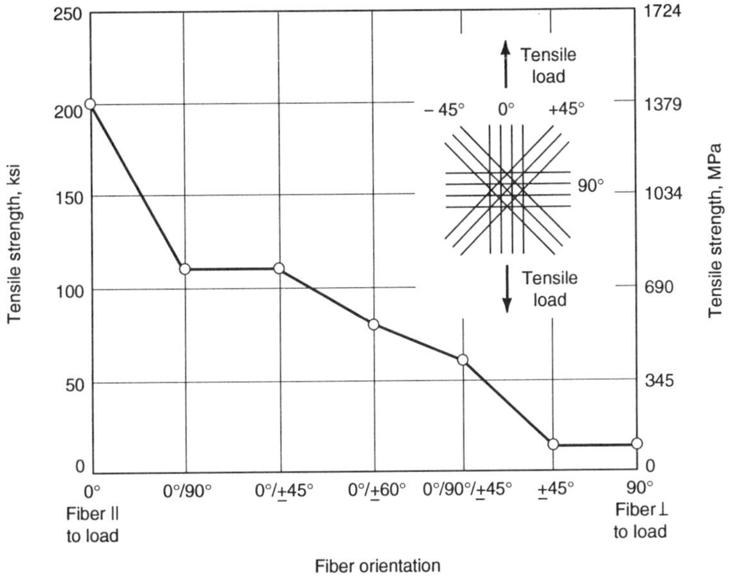

Fig. 9 Effect of fiber orientation on the strength of carbon-fiber/epoxy composites. Source: Ref 1

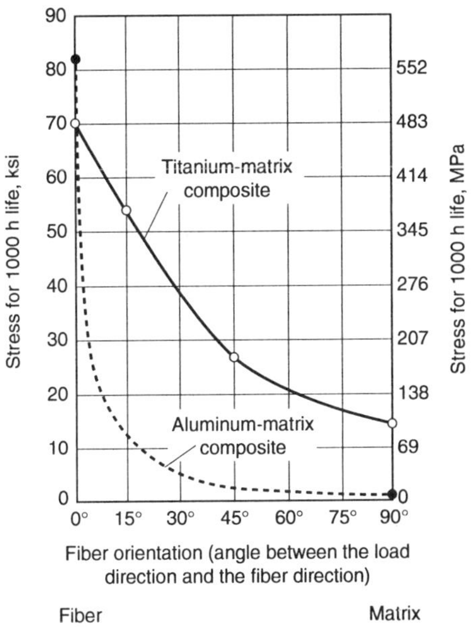

Fig. 10 Effect of fiber orientation on the creep strength of two metal-matrix composites reinforced with boron/SiC fiber. Titanium-matrix composite has matrix of Ti-6Al-4V; test temperature 425 °C (800 °F). Aluminum-matrix composite has matrix of 6061 aluminum; test temperature 300 °C (575 °F). Source: Ref 17

Table 7 Carbon-reinforcement/carbon-matrix processing

Manufacturing process	Matrix material	Reinforcement material	Densification method	Final shape operations
Rotating mandrels				
Three-dimensional woven/filament wound carbon/carbon (also braided)	Phenolic, pitch, and furan resins convertible with inert atmosphere heat to densified carbon char	Carbon (rayon, pan, pitch base), graphite fiber prior processed to 1650–2760 °C (3000–5000 °F) for fiber weaving dry on a rotating bulk graphite mandrel with radial fibers. Filament-wound axial, hoop fibers	Vacuum pressure impregnation with pitch resin 1650 °C (3000 °F) carbonized in inert vacuum. Repeat cycle five times to specific gravity = 1.90 max. Optional final graphitization at 2760 °C (5000 °F). Densification cycles may require up to 1 week	Machine and finish after each impregnation and carbonization cycle to open voids for the next densification cycle
Two-dimensional fabric tape-wound carbon/carbon	Same as three dimensional	Prepreg carbon phenolic fabric laid up, vacuum bagged, and cured at 165 °C (325 °F) and 1720 kPa (250 psi) for up to 18 hours. Heat-up, cure, cool down	Carbonize billet, vacuum pressure impregnate with pitch resin and cure. Repeat cycle six times to specific gravity = 1.80 max. Final graphitization at 2760 °C (5000 °F) is optional	Machine and finish after each impregnation and carbonization cycle may be necessary to open voids for next impregnation
Two-dimensional filament wound CVD infiltrated (also braided)	CVD of pyrolytic carbon into filament wound fiber preform; more than one densification cycle may be required	Carbon or graphite fiber filament wound shape with helical, polar, or hoop windings with or without coupling agent and binder resin	CVD of pyrolytic carbon into filament wound fiber preform may take one to five cycles depending on woven preform, density desired, and the surface buildup and penetration. Final graphitization at 2760 °C (5000 °F) is optional	Machine and finish after each impregnation may be necessary to open voids for infiltration
Static (male/female) open/closed mold mandrel				
Two-dimensional felt preform CVD infiltrated or resin impregnated; or fabric-sheet lay-up	CVD of pyrolytic carbon into fiber felt preform; more than one densification may be required	Carbon or graphite short fibers are felted into fiber preform	Same as two-dimensional filament wound for rotating mandrels	Same as two-dimensional filament wound for rotating mandrels
Compressed fiber preform powder, whisker particles	Resin or CVD impregnation or both	Bulk graphite or graphite particles preform	Multiple resin/CVD densification cycles	Machine and shape per densification cycle

Source: Ref 2, 4, 14, 15

Table 8 Typical mechanical properties of metal-matrix composites

Matrix material (a)	Reinforcement			Tensile strength(b)		Tensile modulus(b)	
	Material	Form	Content, vol%	MPa	ksi	GPa	10^6 psi
Aluminum (6061-T6)	None	...	...	306	44	70	10
	T-300 carbon	Fiber	35–40	1034–1276 (L)	150–185 (L)	110–138 (L)	16–20 (L)
	Boron	Fiber	60	1490 (L)	216 (L)	214 (L)	31 (L)
				138 (T)	20 (T)	138 (T)	20 (T)
	SiC	Powder	20	552	80	119	17
Aluminum (201)	GY-70 carbon	Fiber	37.5	793 (L)	115 (L)	207 (L)	30 (L)
	Al_2O_3	Fiber	60	690 (L)	100 (L)	262 (L)	38 (L)
				172–207 (T)	25–30 (T)	152 (T)	22 (T)
Titanium (Ti-6Al-4V)	None	...	...	890	129	120	17
	SiC	Fiber	35–40	820 (L)	119 (L)	225 (L)	33 (L)
				380 (T)	55 (T)	...	...
	SCS-6(c)	Fiber	35–40	1455 (L)	211 (L)	240 (L)	35 (L)
				340 (T)	49 (T)	...	...

(a) Matrix material baseline is aluminum or titanium alloy in sheet, foil, melt, powder, or spray form and solidified with a temperature, pressure, and time cycle. Reinforcement fiber or powder is processed to higher temperatures than the metal matrix, thus the solidification cycles have little or no effect on the fiber. (b) For fiber-reinforced materials, properties are reported in the direction of the fiber (L) or transverse to the fiber (T). (c) Coated SiC fiber. Source: Ref 1

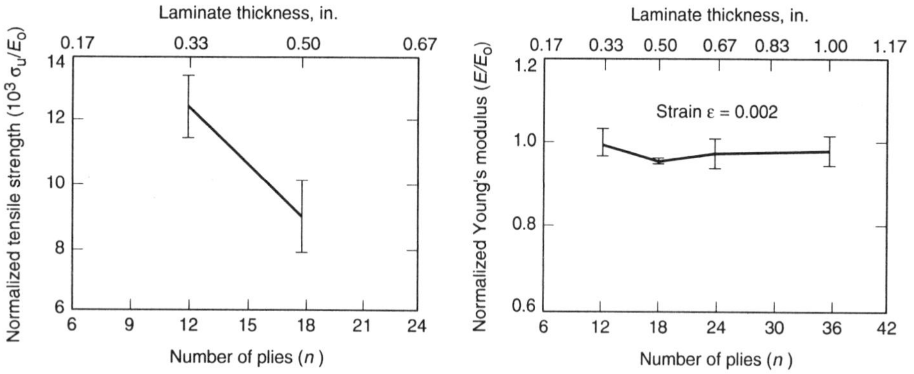

Fig. 11 Effect of laminate thickness on the mechanical properties of unidirectional AS4 carbon-tape/Epon 828 epoxy composites processed at 100 psi and 170 °C (275 °F) for 2 h. 1 ply = 0.7 mm (0.028 in.). Source: Ref 18

- Hot-vacuum, closed-mold presses
- Plasma-arc metal spray and powder-gun equipment

This equipment ensures a matrix laminate having a uniform density with a minimum of subsurface discontinuities.

Carbon/Carbon Matrix Processing

Carbon/carbon processing methods involve the use of the highest temperature and pressure open molds and vacuum/inert heating chambers for the multiple-resin pressure-impregnation cycles and the subsequent multiple carbonization/graphitization of the resin and/or CVD densification cycles (Ref 2, 4, 15).

Rotating mandrels of bulk graphite are used to deposit a fiber (two- or three-dimensional) lay-down of a reinforcement preform that is subsequently pressure impregnated with a resin, cured, and carbonized. The resin/carbonization cycles may be repeated up to eight times to achieve the density goals. Chemical vapor deposited carbon may also be infiltrated into preform in the late densification cycles as a surface and subsurface strengthening agent and as an oxidation-resistant material.

Filament winding, braiding, or fabric-tape wrapping is used to produce two-dimensional fiber preforms or a three-dimensional fiber shape having radial in-wound or drilled/bonded-in-place radial-rod reinforcements. The fiber or fabric can be preimpregnated with resin or post-wind pressure impregnated with resin and then cured. The preform is subsequently carbonized and/or carbon (CVD) densified. This cycle is repeated three to eight times and may be graphitized to a final density of 1.45 to 1.95 g/cm^3.

Static (Male/Female) Open/Closed Mold Mandrels. A fiber mat or a fabric-sheet pattern lay-up is resin transfer molded with resin or carbon CVD densification that is carbonized/graphitized and repeat cycled to the density goal.

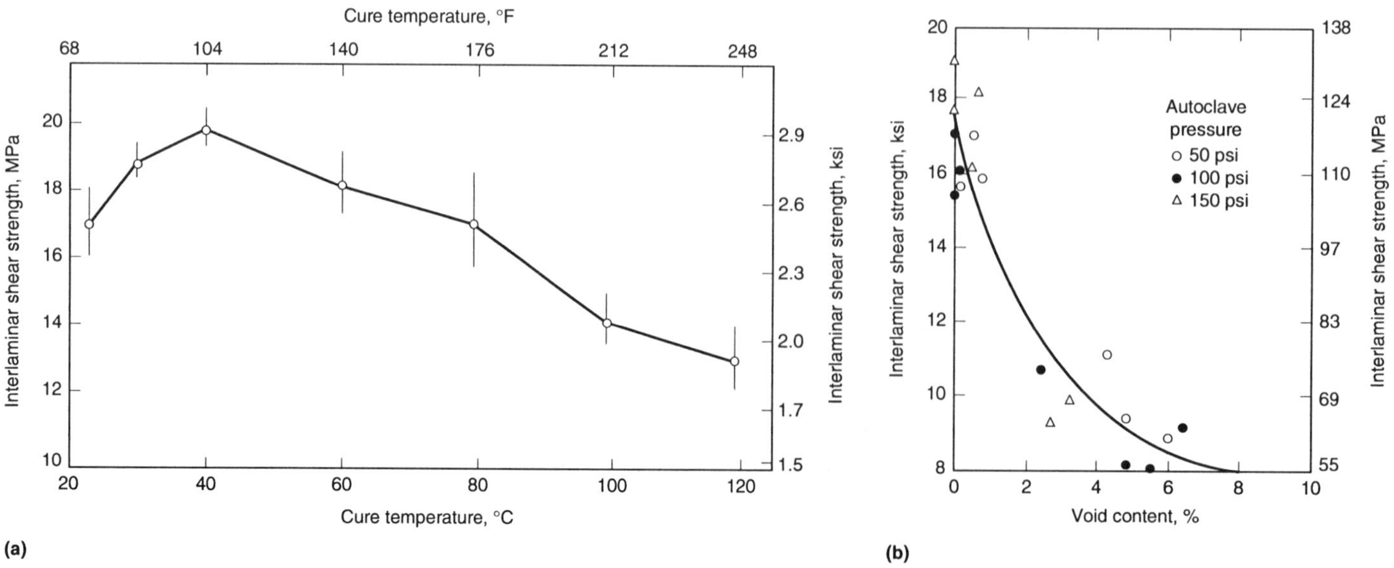

Fig. 12 Effect of interlaminar shear strength of (a) post-curing temperature for a glass/phenolic composite and (b) void content for a carbon/epoxy composite. Source: Ref 1, 7

Table 9 Effect of chopped-fiber length on the mechanical and thermal properties of E-glass/phenolic composites

Fiber length		Notched izod impact strength		Tensile strength		Flexural strength		Flexural modulus		Compressive strength		Coefficient of thermal expansion (23–250 °C), 10^{-4} in./in./°C
mm	in.	J/m	ft · lbf/in.	MPa	ksi	MPa	ksi	GPa	10^6 psi	MPa	ksi	
<6	<¼	30–65	0.6–1.2	55–83	8–12	69–103	10–15	13.8–24.1	2.0–3.5	241–276	35–40	1.5–2.5
6	¼	320–426	6–8	83–103	12–15	103–138	15–20	13.8–17.2	2.0–2.5	241–276	35–40	1.5–2.5
13	½	800–1065	15–20	83–110	12–16	138–207	20–30	17.2–20.7	2.5–3.0	207–241	30–35	1.0–1.5
25	1	960–1335	18–25	103–138	15–20	172–241	25–35	20.7–24.1	3.0–3.5	172–207	25–30	1.0–1.5

E-glass content, 30 to 40 wt%. Source: Ref 7

Alternatively, a fiber preform is placed in a closed chamber and vacuum-pressure resin impregnated, cured, carbonized three to eight times, and graphitized. A variation of this second process is to carbon CVD infiltrate the preform after 2 to 4 cycles of resin char processing. The CVD process is continued for 2 to 3 cycles, with machining/cleaning of the outside surfaces after each cycle to open the preform-billet pores and void areas for further impregnation and densification.

Carbon-fiber/carbon-resin-char-matrix processing is accomplished first by a resin-cure cycle at 163 °C (325 °F) and a pressure of 1725 to 6900 kPa (250 to 1000 psi) in a vacuum-bag enclosure. Second, the cured resin is carbonized and graphitized in inert vacuum chambers. Carbonization requires a 1370 to 1930 °C (2500 to 3500 °F) temperature and a time of approximately one week, and graphitization requires a temperature of 2480 to 3040 °C (4500 to 5500 °F) and a one-week period. The reimpregnation of the pores and void cavities of the charred component, after machining the surfaces and cleaning, is at a pressure of 10.3 to 17.2 MPa (1500 to 2500 psi) in an inert vacuum chamber heated to 540 to 1095 °C (1000 to 2000 °F) for less than one week. An occasional carbon (CVD) infiltration into the carbonized preform is performed by processing in an inert vacuum furnace at 1370 to 1930 °C (2500 to 3500 °F) for less than one week.

Equipment in addition to the open graphite mandrels for carbon/carbon matrix processing includes carbonization and graphitization furnaces and carbon (CVD) infiltration chambers. In addition, resin-impregnation pots are used to fill the fiber preforms with a pitch or phenolic-resin system. Lathes or large turning machines are used after each densification cycle to prepare a fresh surface for the next resin-impregnation/char-densification cycle and to ensure final dimensional control. The open graphite mandrel is used for fabrication of the carbon/carbon fiber filament-wound preform, while closed chambers are used for the resin impregnation, carbonization, graphitization, and carbon (CVD) infiltration densification cycles.

Additional information about processing of composites is provided in the article "Design for Composite Manufacture" in this Volume.

Table 10 Effect of silane coupling agents on the strength of E-glass fiber-reinforced polyester rods

Treatment	Dry strength		Wet strength(a)	
	MPa	ksi	MPa	ksi
No silane	916	133	240	35
Vinyl silane	740	107	285	41
Glycidyl silane	990	144	380	55
Methacryl silane	1100	160	720	104

(a) After boiling water at 100 °C (212 °F) for 72 h. Source: Ref 1

Table 11 Effect of processing-cure parameters on the compressive properties of graphite-fabric/epoxy composites

Cure process	Material preheat temperature		Cure cycle pressure		Vacuum pressure duration into cure cycle, h	Degassing/ bleeder plies	Prepreg mold stiffener plate, plies	Post cure	Compressive strength		Compressive modulus	
	°C	°F	kPa	psi					MPa	ksi	GPa	10^6
Conventional	None	None	607	88	1	2/none	None	Yes	270	39.3	42.7	6.2
Optimized	65	149	1014	147	2	None/2	4	No	290	42.2	48.3	7.0

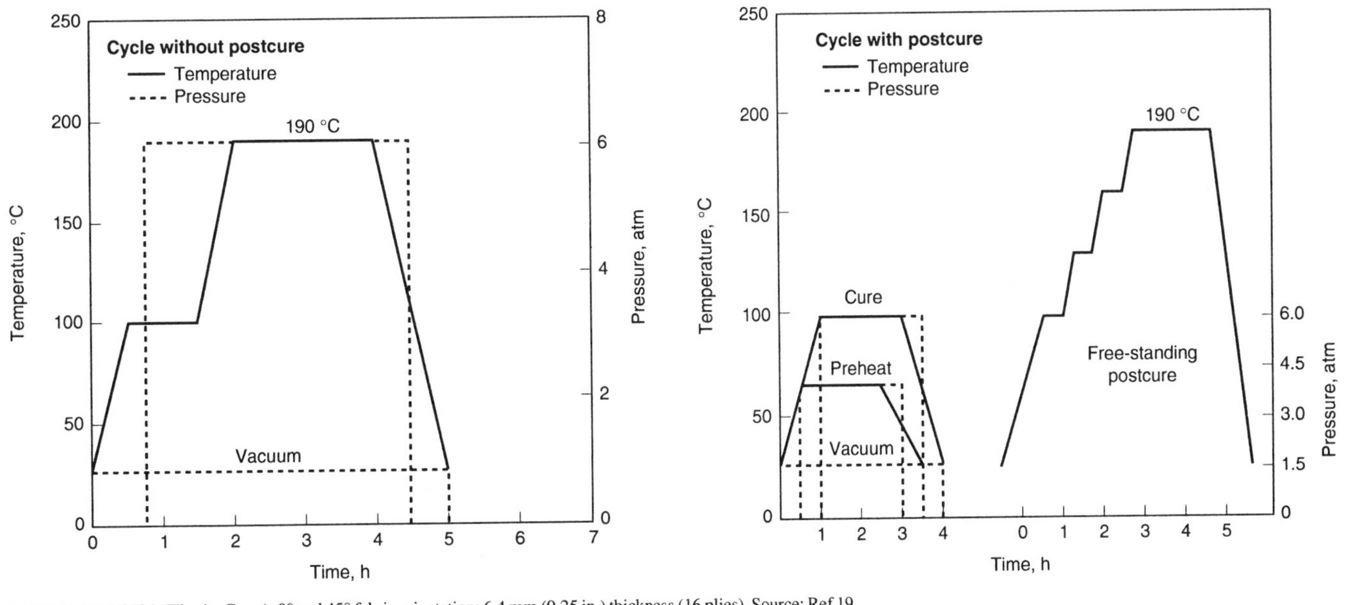

Material: MXG 7620/2534 (Fiberite Corp.); 0° and 45° fabric orientation; 6.4 mm (0.25 in.) thickness (16 plies). Source: Ref 19

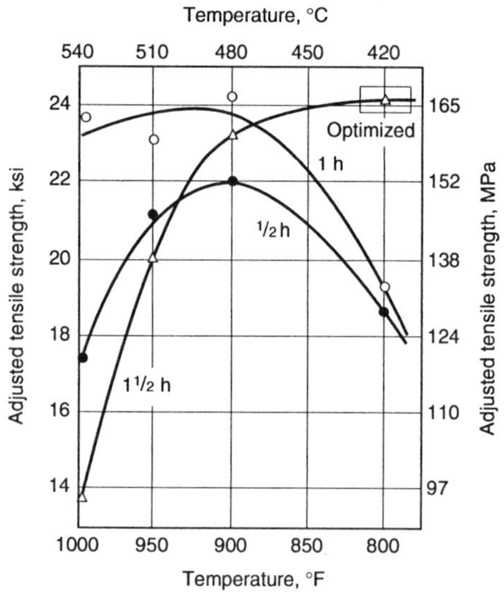

Fig. 13 Effect of bond time and temperature on the longitudinal strength of boron-fiber/aluminum-matrix composites. Fiber content, 3.3 vol%; bonding pressure, 345 MPa (50 ksi). Source: Ref 12

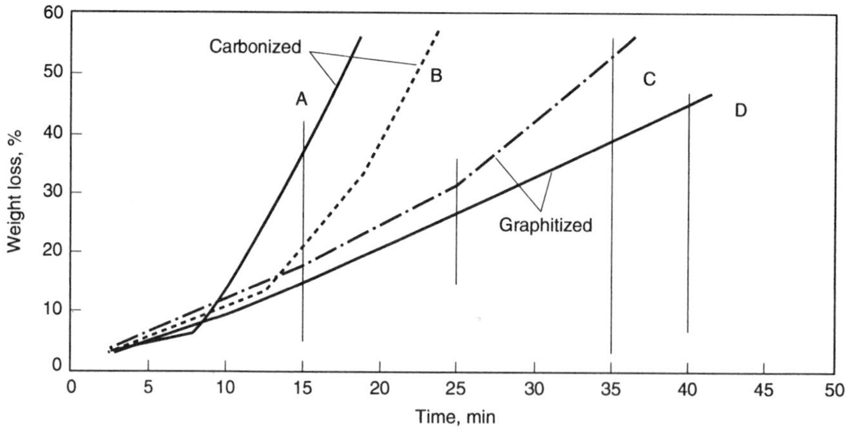

Material (a)	Heat treat temperature °C	Heat treat temperature °F	Oxidation temperature °C	Oxidation temperature °F
A	1500	2732	630	1166
B	1800	3272	698	1288
C	2000	3632	721	1330
D	2400	4352	771	1420

(a) Material A and B heat treated at carbonizing temperatures; material C and D heat treated at graphitizing temperatures

Fig. 14 Effect of final processing temperature on the oxidation weight loss of carbon/carbon composites. Source: Ref 15

Mechanical Properties of Composites

Within a composite laminate consisting of a reinforcing fiber/fabric, fillers, coupling agent and resin, metal, or carbon-matrix system, the greatest influence on mechanical properties is the reinforcement type and its percentage of the total constituents. That is, among composite components that have the same fiber orientation in the laminates and the same matrix material, the component having the highest-strength fiber and the greatest percentage (by weight) of fiber in the laminate will exhibit the greatest strength. Likewise, in the component, the highest strength exists in the planes with the highest percentage of fiber (Fig. 1). For a three-dimensional block, strength is greatest in the vertical and axial fiber directions. For a two-dimensional block, strength is greatest in the axial direction.

The decrease in laminate strength with an increase in temperature, as shown in Fig. 4 and 5 for glass and carbon fibers in an epoxy-resin matrix, is caused by the softening and a weight loss of water and solvent in the resin system. Resin weight loss begins at 120 to 175 °C (250 to 350 °F), and resin conversion to a porous char state begins at 315 to 760 °C (600 to 1400 °F). Resin matrix carbon retention for carbon matrix conversion at 760 °C (1400 °F) is 55% for phenolic, 17% for polyester, and 10% for epoxy (novolac). Fiber strength does not degrade significantly until the following temperatures are reached:

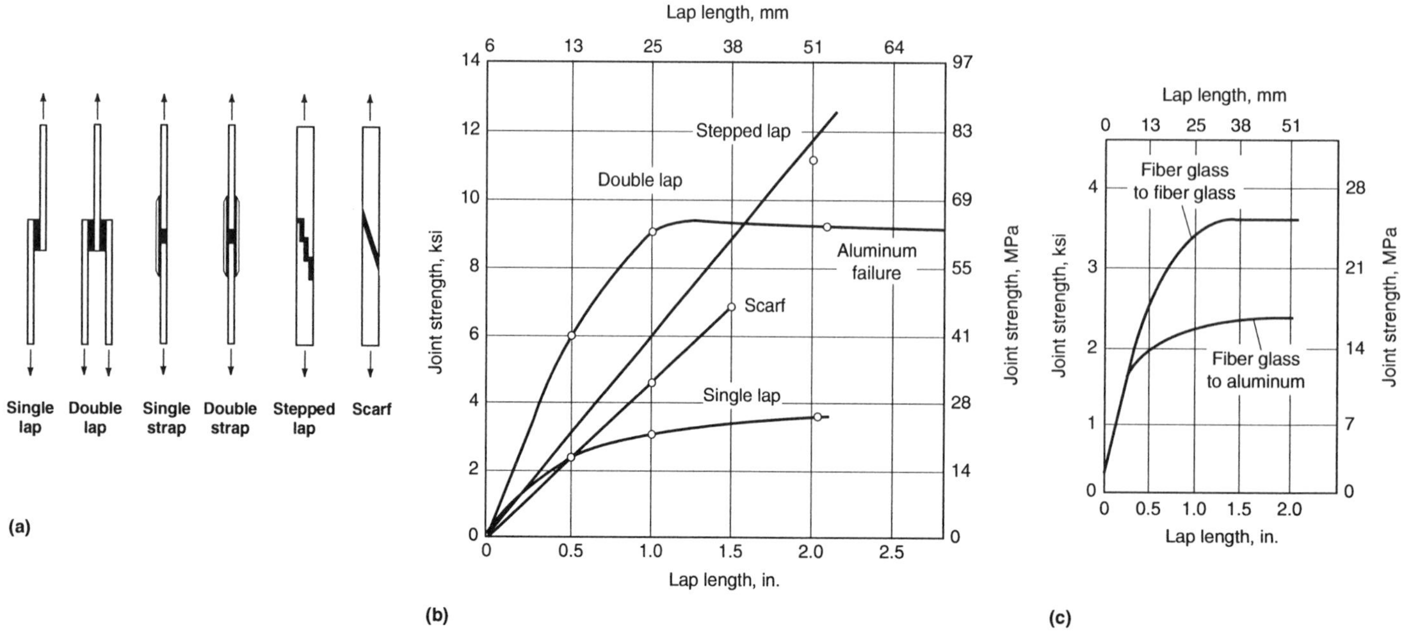

Fig. 15 Effect of lap length on strength of the adhesive bond. (a) Joint configurations. (b) Boron/epoxy composites epoxy bonded to aluminum. (c) Fiber glass bonded to fiberglass and to aluminum (single-lap bonds; epoxy adhesive). Source: Ref 1

Fiber	Temperature °C	Temperature °F
Kevlar	160	320
Glass	430	800
Al_2O_3	1200	2200
Boron	1650	3000
SiC	1650	3000
Carbon, graphite	1930	3500

Resin-matrix composites (as well as metal- and carbon-matrix composites) are used in many aerospace and military applications at temperatures below −20 °C (0 °F) and above 760 °C (1400 °F). The resin-matrix composite having the best elevated-temperature properties is carbon fiber (highest degradation temperature) in phenolic resin (highest resin matrix carbon retention).

The strengths of metal-matrix composites using three different nonferrous alloys as the matrix materials are shown in Table 8. As can be seen, the titanium alloy provides higher strength and modulus (and higher density and cost) than the aluminum-alloy matrix. Of the available reinforcing fibers, boron, glass, and carbon provide the highest composite strengths, while graphite, silicon carbide (SiC), and aluminum oxide (Al_2O_3) provide the highest composite modulus.

A variety of carbon-reinforcing-material types (fiber, fabric, felt, particles) can be bonded together with a carbon or graphite form of a phenolic or pitch resin-matrix system or a CVD carbon- or graphite-matrix system to form carbon/carbon composite shapes. At least six types (having specific gravities of 1.30 to 2.00) exist for the fabrication of laminates, rings, and cones by molding, tapewrapping, lay-up, or filament winding. Figure 6 shows the properties of eight such types. Of these, the material that uses a carbon fiber in a CVD carbon matrix has the highest tensile strength and percent strain, while the material that uses carbon fiber in a high-melt-temperature pitch-resin matrix exhibits the highest modulus.

Effects of Composition Variations on Properties

A composite material can be "tailored" to a product application by varying the composition and/or percentage of the total composition weight or volume of any or all parts of a composite composition, such as the matrix, fiber reinforcement, and coupling agent.

An S or E glass fiber reinforcement will give the highest strength at the least cost, while a graphite fiber provides the highest modulus at the highest cost. Fiber content in a matrix can vary from 33 to 66% by weight, can be a unidirectional continuous fiber or a two-directional woven fiber fabric, and the angle orientation between successive plys of material can be changed for further strengthening. A fiber/matrix coupling agent at 1 to 2% (by weight) on the fibers can help the laminate become stable, uniform, continuous, low in void content, and homogenous, thereby enhancing properties up to 25%. In addition, a filler can modify the modulus, strength, resin flow (viscosity), uniformity, ease of processing, and the density of a resin-matrix material. Hollow microspheres of carbon in the resin matrix can lower the composite density by 5 to 15% and decrease strength levels by 15 to 25% with an addition of only 1 to 10% by weight of spheres. Fillers in the resin matrix are usually kept to a low percentage (by weight), but may increase up to 20 to 30% when used in powder or whisker form.

It should be noted that two or more different fibers in a resin, metal, or carbon matrix can be formulated successfully as long as they are cure or bond-matrix compatible and their service performance is satisfactory.

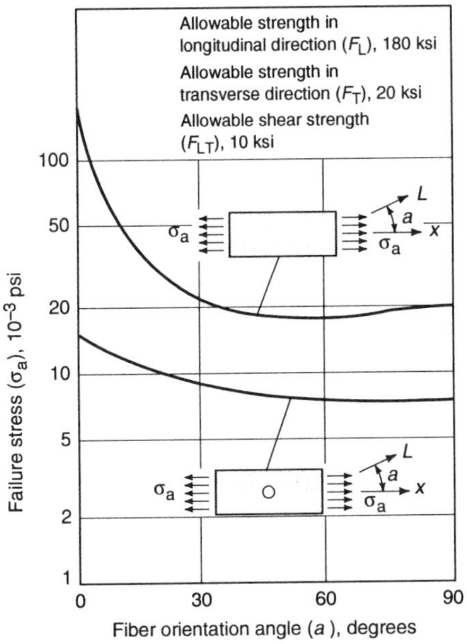

Fig. 16 Effects of fiber orientation and cut-out on the failure stress of boron/epoxy composite plates. Source: Ref 20

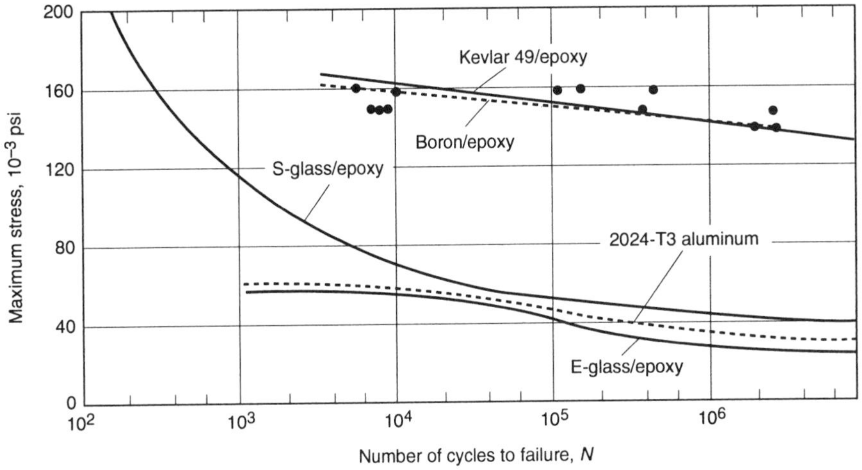

Fig. 17 Comparison of the fatigue strength (tension-tension loading) of an aluminum alloy and several unidirectional resin-matrix composite materials. Source: Ref 17

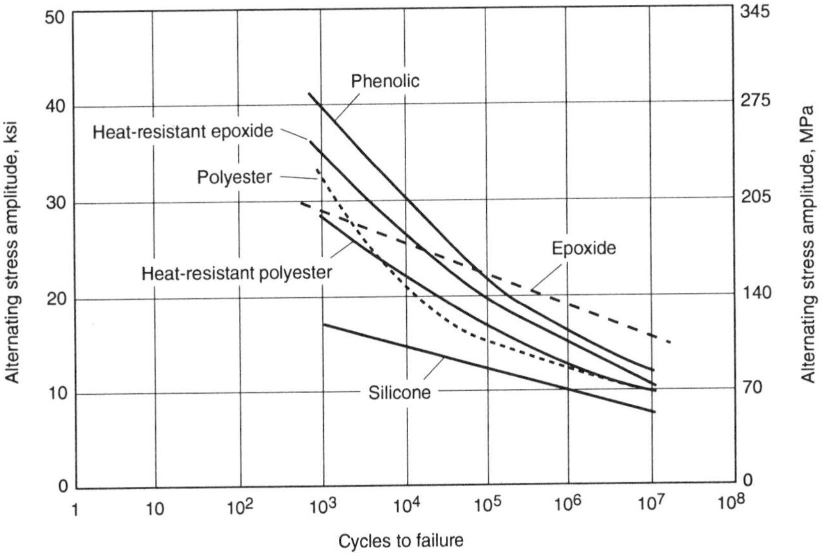

Fig. 18 Effect of type of resin matrix material on the fatigue strength of glass-fabric/resin composites. Source: Ref 20

Table 12 Weather serviceability data for a glass/phenolic composite

Property	Original	Stored 3 yr	Weathered 3 yr Florida	Stress weathered 3 yr Florida
Tensile strength, MPa (ksi)	408 (59.2)	401 (58.2)	367 (53.2)	349 (50.6)
Compression strength, MPa (ksi)	278 (40.3)	291 (42.2)	254 (36.8)	201 (29.1)
Flexure strength, MPa (ksi)	421 (61.0)	408 (59.2)	356 (51.6)	333 (48.3)
Modulus of elasticity, GPa (10^6 psi)	24.8 (3.6)	24.7 (3.58)	22.1 (3.2)	18.6 (2.7)

Material: heat-resistant phenolic glass fabric composite conforming to MIL-R-9299 (outdoor Florida weathering). Source: Ref 7

Table 13 Effect of exposure to boiling water on interlaminar shear strengths of type P1 glass phenolic laminate

Exposure time, h	Interlaminar shear strength MPa	psi
4	13.6 ± 1.6	1972 ± 232
8	13.4 ± 1.6	1943 ± 232
16	12.9 ± 0.8	1871 ± 116
24	12.8 ± 1.6	1856 ± 232
48	11.8 ± 1.1	1711 ± 160

Source: Ref 7

Table 14 Effect of dry and wet conditioning on the interlaminar shear strengths of glass/phenolic composites

Panel condition	Interlaminar shear strength MPa	psi
As-received	25.17 ± 2.58	3650 ± 374
Dry	29.14 ± 5.89	4225 ± 854
Wet	26.42 ± 3.01	3831 ± 436

Source: Ref 7

Variations in composition can include the following significant factors:

- Fiber type
- Fiber content, %
- Fiber length
- Fiber orientation
- Laminate thickness
- Fiber/resin coupling agent
- Filler types

A primary composition decision is the fiber type to be selected. Among four of the common fibers—glass, aramid, carbon, and graphite—glass fiber offers high strength and impact resistance (see Fig. 7) with good damping characteristics and low cost but a high density (1.90–2.10 g/cm^3). Graphite offers lower strength and impact resistance; it also exhibits a high modulus, thermal stability, good oxidation resistance at temperature, and a light weight (1.32–1.90 g/cm^3), but at a higher cost. Density is an important factor in part-quality inspection and achieving system weight requirements.

Figure 8 shows the increase in composite strength and modulus with an increase in the percent of glass fiber in a fiber/polyester laminate; this same trend also exists for carbon felt in a CVD carbon- or graphite-matrix material (Ref 12). The use of longer-length reinforcing fibers in a resin-matrix composite also provides higher strength than shorter fibers (Table 9).

As shown in Fig. 9 (for a resin-matrix composite) and Fig. 10 (for two metal-matrix composites), strength of a composite is also highest when the fiber direction is parallel to the loading direction (0°) and the lowest when the fiber is oriented perpendicular to the loading direction (90°) (where strength is a measure of resin strength only).

Another factor that affects mechanical properties is laminate thickness. As shown in Fig. 11, as the plate thickness of carbon/fiber epoxy resin-matrix composite increases from 12 to 18 plies (13 to 22 mm, or 0.5 to 1.0 in.), strength and modulus decrease as more void areas and/or resin-rich areas of lower local properties are introduced into the laminate.

Coupling agents such as silane compounds placed on the reinforcing fiber prior to bonding to the resin matrix will mostly increase the tensile strength of the composite in wet or dry environmental conditions (Table 10). The strength increases will depend on the type of coupling agent and its compatibility with the resin system.

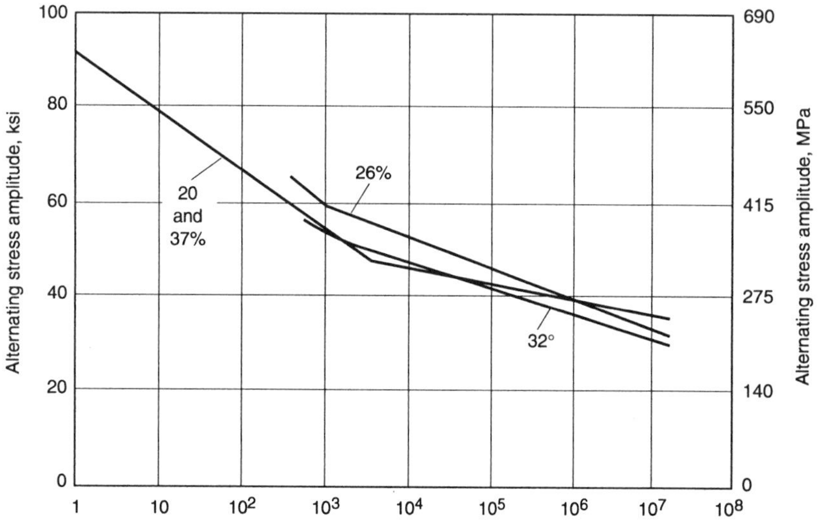

Fig. 19 Effect of resin matrix material content on the fatigue strength of ±5° glass-fiber/epoxy composites. Source: Ref 20

Effects of Processing on Properties

For resin-, metal-, and carbon-matrix composites, the basic parameters of temperature, time, and pressure are the key variables of the processing cure, the solidification bonding, or the conversion of resin to carbon char.

The cure process used for the resin matrix, with and without a post-cure cycle, can make a difference in the compressive strength and modulus of the matrix material. The conventional and optimized cure processes are listed in Table 11. A higher compressive strength and modulus result from the optimized process. This is caused by the higher cure pressure and preheat temperature used, plus the longer time used in the vacuum-bag application.

Generally, use of a post cure for laminates or structural shapes will lower the percent of residual volatiles and increase the properties (interlaminar shear strength), provided the temperature and time is optimized. Some post-cure temperatures and duration times can actually lower the properties of the original composite, if caution is not used, as shown in Fig. 12(a). The premise of a lower void content enhancing properties is shown in Fig. 12(b); plus, a higher autoclave pressure (150 psi) provides a higher shear strength at 0% void content.

In metal-matrix processing, the layers of reinforcing fiber in a metal matrix are diffusion bonded with pressure, temperature, and time until the bonding operation is complete. However, the right combination of diffusion-bond parameters, depending on the thickness, size, shape, and complexity, needs to be optimized (Fig. 13). In this example, 1½ h at 425 °C (800 °F) and a 345 MPa (50 ksi) bond pressure provides an optimized tensile strength of 167 MPa (24.2 ksi).

The matrix processing of carbon/carbon composites also involves temperature, pressure, and time parameters. In this instance, the parameters are those required to complete the conversion of the phenolic, pitch, or furane resin systems to a charred carbon matrix and its eventual densification through multiple impregnation and char conversion cycles. After final densification, the composite shape is brought to the full carbonization (1320 to 1870 °C, or 2400 to 3400 °F) or graphitization (1980 to 2980 °C, or 3600 to 5400 °F) temperature to finish the process. Carbonization provides a higher final strength, a higher density, and a higher weight loss due to oxidation than does graphitization.

A high graphitization temperature (2000 to 2400 °C, or 3630 to 4350 °F) provides a microstructure having a smaller separation (in angstrom units) between carbon platelets, resulting in a form of carbon/carbon composite that is more heat stable and exhibits a higher thermal conductivity, lower thermal expansion, lower density, and a higher oxidation resistance to weight loss. Thus, the A and B carbonization composites in Fig. 14 display a higher weight loss earlier than the graphitized composites, and the graphitized composites took 20 min longer to reach the same weight loss as the carbonized composites.

Effects of Design Loadings and Environmental Conditions on Composite Properties

The types of loadings and environmental conditions existing for an applied composite product design must be considered in advance for the composite to successfully meet the required performance life and operating conditions of the system.

Adequate bonding of joints between composites or between a composite and a metal requires a minimum of 1 in. or more lap to develop a usable strength level for single- and double-lap joints and from 1.5 to 2.0 in. for strap, step, and scarf joints (Fig. 15).

The presence of a hole in a flat laminate will lower the failure stress by an average factor of three to five, due to the stress concentration caused by the hole. This is shown in Fig. 16, which also shows the effect of fiber angle in respect to the load.

Environmental conditions also affect the mechanical properties of composites. Repeated tension, flexural, or compression dynamic loadings (fatigue) will lower the allowable strength levels for resin-matrix composites. Reduction factors from 2 to 4 are common between static (no cycles) and dynamic (10 million cycles) allowable stresses (Fig. 17–19). The same pattern for allowable strength versus fatigue number of cycles also holds true for carbon- and metal-matrix composites.

A glass/phenolic exterior panel stored, weathered, or stress weathered in Florida could lose 15% or more in strength after 3 years (Table 12). Table 10 indicates the tensile strength of dry and wet glass/polyester composites. The effects of exposure to room-temperature water, boiling water, salt, and acid are given in Table 13 and 14 and Fig. 20 and 21. As the resin-matrix composite gains weight in boiling water, it experiences a corresponding loss of properties.

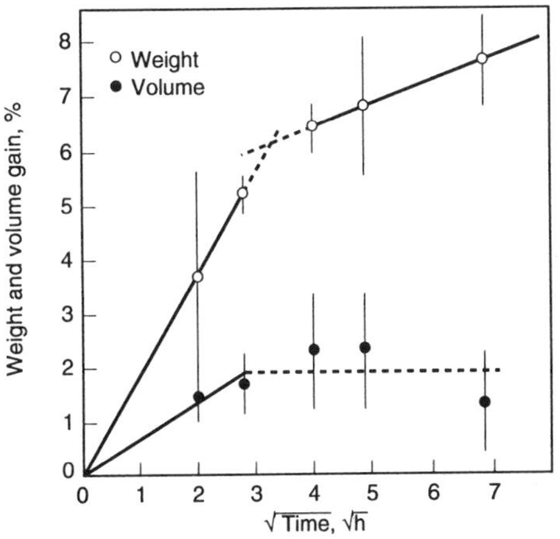

Fig. 20 Effect of time in boiling water on the weight and volume gain of type P1 glass/phenolic composites. Source: Ref 7

Conclusions

Composite material development design concepts should be tested for confirmation of material supplier test properties and design loaded with test fixture criteria for subscale and full-scale application. Existing materials and fabrication methods for composite design should be used wherever possible to control the unknown factors involved in advanced-concept applications.

It is important to have a thorough understanding of the following factors for composite material applications:

- The material composition and the percentage (by weight) of all constituents and the storing, shipping, and handling requirements.
- The cure/forming cycle suggested by the material supplier and the preferred machining, surface treating, and finishing operations (including the bands of acceptability for expected variations).
- The cured properties of the composite for the range of operating loads and environments to be expected in the performance service life.

If changes to the baseline composition, cure/forming cycle, and properties are required, it is important to stay in close contact with the material supplier's technical representatives and laboratory scientists and their corresponding equivalents within your company and your industry. The first few components fabricated should be expected to change in material composition, processing, properties, shape, fabrication step sequencing, and specification criteria. However, with a sound engineering approach over the life of the program, the necessary adjustments usually can be made to produce a high-quality product.

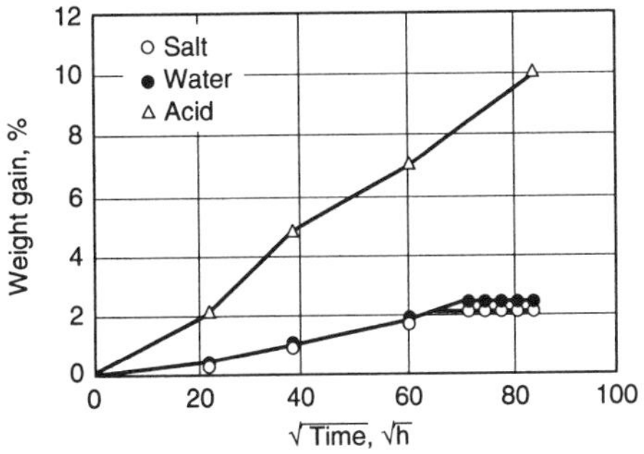

Fig. 21 Effect of time on the moisture absorption of glass/phenolic composites exposed to salt, acid, and water. Source: Ref 7

REFERENCES

1. P.K. Mallick, Materials Manufacturing and Design, *Fiber Reinforced Composites,* 2nd ed., Marcel Dekker, 1993, p 16, 71, 213, 289, 301, 373, 390, 476–478, 533
2. *Engineered Materials Handbook Desk Edition,* ASM International, 1995, p 477, 532–582, 1057–1094
3. *Modern Plastics Encyclopedia,* McGraw-Hill, 1994, p 233
4. *Composites,* Vol 1, *Engineered Materials Handbook,* ASM International, 1987, p 118, 119, 355, 360, 363, 373, 381, 405, 410, 861, 862
5. H.S. Katz, and J.V. Milewski, *Handbook of Fillers for Plastics,* Van Nostrand, 1987, p 56, 57, 75
6. R. Flinn and P. Trojan, *Engineering Materials and Applications,* Houghton-Mifflin, 1990, p 618, 619
7. *Handbook of Advanced Material Testing,* Marcel Dekker, 1995, p 945–948, 950, 955
8. E.P. DeGamo, *Materials and Processes in Manufacturing,* 4th ed., Macmillan, 1974, p 190–212
9. C.A. Harper, Ed., *Handbook of Plastics, Elastomers and Composites,* 2nd ed., McGraw-Hill, 1992, sections 1.5, 4.4, 5.31, 11.2
10. R.B. Seymour, *Reinforced Plastics,* ASM International, 1991, p 9, 51
11. "Fiberglass Reinforced Plastics," Owens-Corning Corp., 1964, p 24–30
12. C.T. Lynch and J.P. Kershaw, *Metal Matrix Composites,* CRC Press, 1972, p 16, 17, 51
13. K.A. Lucas and H. Clarke, *Corrosion of Alumina Based Metal Matrix Composites,* John Wiley & Sons, 1993, p 17, 20, 21
14. *Carbon Composite and Metal Composite Systems,* Vol 7, Technomic
15. G. Savage, *Carbon-Carbon-Composites,* Chapman Hall, 1993, p 95, 97, 98, 101, 125, 126, 129, 200
16. L. Edwards and M. Endean, *Manufacturing with Materials,* Butterworths, 1990, p 73, 145
17. J.A. Lee and D.L. Mykkanen, *Metal and Polymer Matrix Composites,* Noyes Data Corp., 1987, p 113, 150
18. *Design and Manufacturing—Advanced Composites*, Fifth Conference, ASM International, 1989, p 89, 94–95, 143–145, 164–165
19. S.Y. Yang, C.K. Huang, and C.B. Wu, Influence of Processing on Quality of Advanced Composite Tools, *SAMPE J.,* April, 1996, p 37
20. *Composites Materials—Testing and Design,* STP497, ASTM, p 98, 146, 378, 507, 509–510, 556, 559, 588–600

Effects of Surface Treatments on Materials Performance

Arnold R. Marder, Lehigh University

SURFACE ENGINEERING is a multidisciplinary activity intended to tailor the properties of the surfaces of engineering components so that their function and serviceability can be improved. The *ASM Handbook* defines surface engineering as "treatment of the surface and near-surface regions of a material to allow the surface to perform functions that are distinct from those functions demanded from the bulk of the material" (Ref 1). Of concern to the design engineer is the availability of surface-specific properties for the component that can provide:

- Protection in a desired environment
- Improved mechanical properties
- Electronic or electrical properties
- Desired appearance

Of further concern to the designer is the availability of economical processes to produce the required properties. These processes include solidification treatments such as hot dip coatings, weld-overlay coatings, and thermal spray surfaces; deposition surface treatments such as electrodeposition, chemical vapor deposition (CVD), and physical vapor deposition (PVD); and heat treatment coatings such as diffusion coatings and surface hardening.

Surface treatments are used in a variety of ways to improve the material properties of the component. Coating mechanical properties, for example, hardness, strength, and toughness, can improve the component wear, fatigue, and erosion properties, respectively. Electrical properties in circuit design are dependent on the surface-deposition processes. Similarly, environmental properties such as resistance to aqueous corrosion and high-temperature oxidation and sulfidation can be improved by selective surface treatments.

The bulk of the material or substrate cannot be considered totally independent of the surface treatment. Most surface processes are not limited to the immediate region of the surface, but can involve the substrate by exposure to either a thermal cycle or a mechanical stress. For example, diffusion coatings often have high-temperature thermal cycles that may subject the substrate to temperatures that can cause phase transformations and thus property changes, or shot-peening treatments that deliberately strain the substrate surface to induce improved fatigue properties. It is the purpose of this article to review information on surface treatments that improve service performance so that the design engineer may consider surface-engineered components as an alternative to more costly materials.

Related information is provided in the article "Design for Surface Finishing" in this Volume. More detailed information about the processes described in this article can be found in *Surface Engineering,* Volume 5 of *ASM Handbook* (Ref 1).

Solidification Surface Treatments

Solidification surface treatments include hot dip coatings, weld overlays, and thermal spray coatings.

Hot Dip Coatings

Hot dip coatings are predominantly used to improve the aqueous corrosion of steel. Processing of hot dip coatings involve either batch or continuous processing. The continuous process is more advantageous for sheet steels, whereas the batch process is normally used for individual parts. Details of the processing techniques are outlined in Ref 1. In the batch galvanizing process, the two types of conventional practices used at the present time are the wet process and the dry process (Ref 2). The wet process involves a flux blanket on the top of the molten zinc bath to remove impurities from the surface of the steel and also to keep that portion of the surface of the zinc bath, through which the steel is immersed, free from oxides. In the dry process the steel is usually cleaned, treated with an aqueous solution, dried, and then dipped in the molten zinc bath. The molten zinc bath is maintained at temperatures between 445 and 455 °C (830 and 850 °F) and times in the range of 3 to 6 min. The time of immersion is used to control the thickness of the coating, which consists of iron-zinc alloy phases at the interface along with a top coat of pure zinc. Good cooling control is necessary because the zinc can continue to react with the substrate to produce further alloying and detrimentally affect the properties of the coating such as the spangle finish (or grain size).

In continuous hot dip processing, welded coils of steel are coated at speeds of 200 m/min. The flux or Cook-Norteman line is similar to the batch process in that the sheet is cleaned and fluxed in line prior to immersion. The hot-processed continuous line is more complex in that the steel sheet is first cleaned at temperature in a reducing environment, annealed above the recrystallization temperature of about 700 °C (1290 °F) and then immersed in the molten bath. As the strip exits the bath, the thickness of the molten metal film is controlled by gas wiping dies that remove excess coating metal. After coating, the sheet is either cooled by forced air or subjected to an in-line heat treatment, called galvannealing, before being rewound into coil or sheared into cut lengths at the exit of the line.

In general, the coating microstructure consists of the substrate, the interfacial alloy layer, and the overlay cast structure. Depending on the type of coating, the microstructure and composition of these constituents changes. As expected, the substrate plays a major role in the type of coating obtained, and substrate composition can affect growth kinetics of the phases formed. For example, if the substrate contains silicon the well-known Sandelin Effect can influence the iron-zinc phase reaction and consequently the thickness of the coating (Ref 3). Similarly, alloy additions to the steel to improve sheet formability, for example, interstitial-free (IF) steels with titanium, titanium/niobium, and phosphorus, can influence the microstructure of the iron-zinc phases in galvanized and galvannealed steel (Ref 4). Substrate grain size has also been shown to greatly affect the nucleation of the iron-zinc phases. In aluminum-containing baths, the structure formed first is an inhibition layer that is dependent on bath composition and prevents further alloying for a certain short time before the inhibition layer becomes unstable (Ref 5).

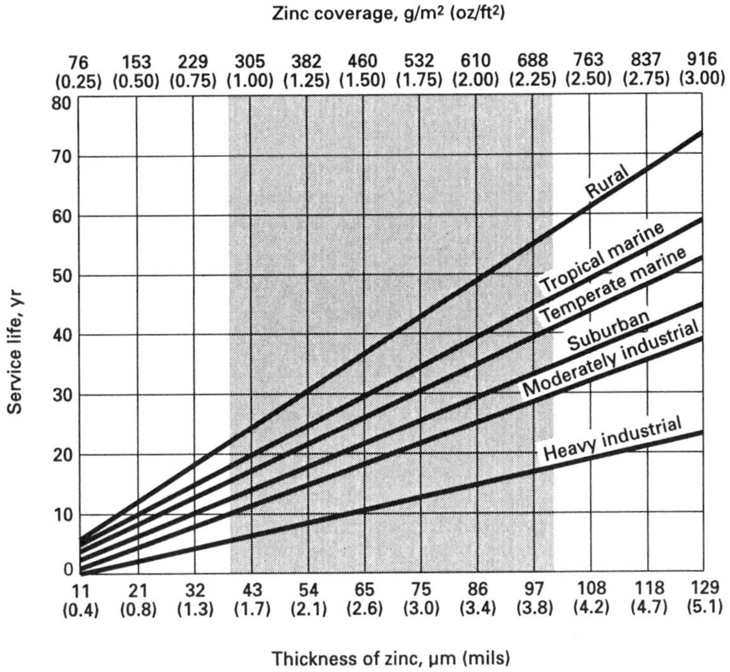

Fig. 1 Service life (time to 5% rusting of steel surface) versus thickness of zinc for selected atmospheres. Shaded area is thickness range based on minimum thicknesses for all grades, classes, and so forth, encompassed by ASTM A 123 and A 153. Source: Ref 10

When the zinc galvanizing bath contains only a trace of aluminum, zinc attack of the substrate is uniform and the phases that form are governed by the iron-zinc binary phase diagram. In zinc baths containing aluminum, the stability of the inhibition layer governs the amount of iron-zinc phases formed. Once the inhibition layer is no longer stable, outbursts or rapid growth of iron-zinc phases occur during hot dipping (Ref 6). During the thermal cycle of the galvannealed process, the inhibition layer dissolves and iron-zinc phase layer growth occurs in a controlled manner until the entire coating is made up of iron-zinc phases (Ref 7). Both galvanized and galvannealed alloy phase growth are determined by Fe-Al-Zn ternary diffusion, and the overlay cast microstructure greatly depends on aluminum content of the bath. The pure-zinc and low-aluminum coatings form an overlay of pure Zn (η) phase. Zn-5wt%Al (Galfan) solidifies as eutectic microstructure, and the Zn-55 wt%Al (Galvalume and Zincalume) solidifies as aluminum dendrites with zinc-rich interdendritic regions. The aluminum coatings (Type I and Type II) either form overlays of aluminum-silicon or aluminum alloy, respectively.

Galvanized coatings are commonly characterized by surface spangles. In cross section, an Fe_2Al_5(Zn) inhibition layer develops first, preventing any iron-zinc intermetallic phase formation. The overlay layer is made up of dendrites of pure Zn (η) phase and appears as a polycrystalline structure. The three surface finishes commonly produced are:

- Regular spangle, where the coating solidifies from the dipping temperature by air cooling
- Minimum spangle, where the coating is quenched using water, steam, chemical solutions, or by zinc powder spraying
- Extra-smooth temper roll finish carried out as an additional operation with regular and minimum spangle material

Aluminum is probably the most important alloying element added to the hot dip galvanizing bath, with different levels required to produce different properties in the bath (Ref 8). Aluminum levels of 0.005 to 0.02 wt% are added to brighten the initial coating surface. The effect is related to the formation of a continuous alumina (Al_2O_3) layer on the coating surface that inhibits further oxidation by acting as a protective barrier layer. This effect is also responsible for the reduced atmospheric oxidation of the zinc bath. In addition, aluminum in the range of 0.1 to 0.3 wt% is added to the zinc bath to suppress the growth of brittle iron-zinc intermetallic phases at the steel coating interface by forming the Fe_2Al_5 (Zn) inhibition layer. The end of this incubation period is marked by the disruption of the initial layer, followed by rapid attack of the substrate steel. An increase in the incubation period depends on increased aluminum bath composition using a low bath temperature, having low bath iron content, agitation, and the presence of solute additions in the steel. Thus, during commercial production, the immersion time is kept below the incubation period in order to obtain a highly ductile product.

Zinc coatings add corrosion resistance to steel in several ways. As a barrier layer, a continuous zinc coating separates the steel from the corrosive environment. By galvanic protection, zinc acts as a sacrificial anode to protect the underlying steel

Table 1 Effects of hot dip coatings on threshold voltages for cratering of cathodic electrophoretic primer

Type of surface	Cratering threshold, V
Uncoated bare steel	>400
Zinc	275
Zinc-iron	225
Zn-55Al	375
Aluminum	>400

Source: Ref 11

at voids, scratches, and cut edges of the sheet. The sacrificial properties of zinc can be seen in a galvanic series where the potential of zinc is less noble than steel in most environments at ambient temperatures. In addition, after dissolution of the zinc metal, zinc hydroxide can precipitate at the cathodic areas of the exposed steel, forming a secondary barrier layer. Zinc corrodes at a slower rate than the steel substrate, although the corrosion rate of zinc varies depending on the atmosphere to which it is exposed (Ref 9), as shown in Fig. 1.

During forming, especially stretch-forming operations, increased friction of the zinc can result in less total stretch before fracture. In severe forming operations, galling and coating pickoff can also occur. Furthermore, coating particulate buildup on die surfaces can lead to poor surface appearance of formed parts. Proper lubrication is essential in the design of any forming process, especially when forming zinc-coated parts. Weldability of zinc coatings is also an important property of the coating. Spot weldability properties are particularly important because most galvanized product is joined in this manner. Zinc coatings reduce the life of welding electrodes because of the copper electrode alloys with zinc. This effect leads to higher resistance, localized heating, and increased pitting and erosion of the electrode tip. As a result, manufacturing costs increase because lower tip life reduces productivity due to frequent downtime in the welding operation to redress tips.

Although zinc coatings are often used in the as-coated state, some applications call for a painted surface, and therefore paintability is an important design property of the coating. It has been shown that large-spangle material is difficult to paint; therefore, most painted products are either minimum spangle or temper rolled. It is usually necessary to pretreat a hot dip galvanized coating with a zinc phosphate or complex oxide thin coating before prepainting. In the automobile industry, following the pretreatment most automobile bodies are primed with an electrophoretic paint (e-coat), and, as a result, resistance to e-coat cratering is an important property. At high e-coat voltages, sparking as a result of exceeding the dielectric properties of the deposited paint film cases localized heat generation, film disruption, and premature curing of the paint. After paint curing, these sparked areas form pinpoint craters that result in a paint surface with a detrimental appearance. Therefore, resistance to e-coat cra-

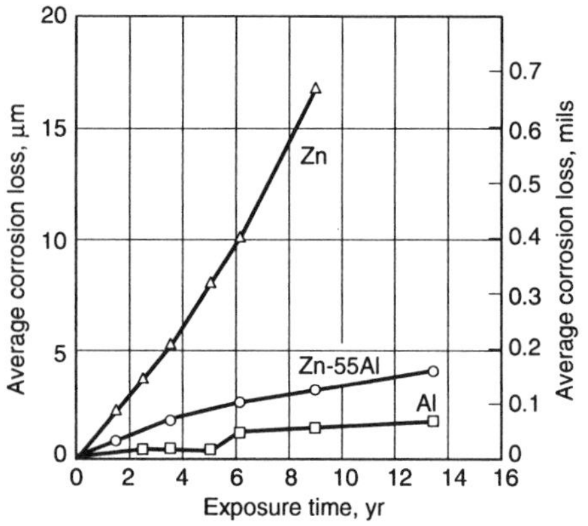

Fig. 2 Corrosion losses of hot dip coatings in the industrial environment of Bethlehem, PA. Source: Ref 13

Table 2 Coating thickness losses for galvanized steel and type 2 aluminized steel in atmospheric exposure

Years exposed	Middletown, OH				Kure Beach, NC			
	G90(a)		Type 2		G90		Type 2	
	µm	mils	µm	mils	µm	mils	µm	mils
1	2.6	0.1	0.5	0.02	7.0	0.28	1.4	0.04
2	5.2	0.2	0.7	0.028	8.6	0.34	2.4	0.09
4	9.3	0.37	1.2	0.047	12.7	0.5	3.8	0.15
6	14.5	0.57	3.1	0.12	16.2	0.64	4.5	0.18
10	24.4	0.96	2.9	0.11	23.5	0.93	6.0	0.24
15	...	...	5.3	0.21	26.0	1.02	6.7	0.26

(a) G90 galvanized steel has a coating weight of 0.90 oz/ft^2 (270 g/m^2). Source: Ref 14

tering, expressed in cratering threshold voltage, is an essential paintability property (Table 1).

Galvanneal coatings are essentially diffusion coatings that expose the zinc galvanized steel to an annealing temperature around 500 °C (930 °F) to produce a fully alloyed coating containing iron-zinc intermetallic phases. This is accomplished by inserting heating and cooling capacity above the liquid zinc pot in order for the galvannealing process to be continuous. Good process control requires that the effects of heating rate, hold temperature and time, and cooling rate on the iron-zinc reaction kinetics be well understood. Galvanneal coatings have been classified as (Ref 7):

- *Type 0:* Underalloyed coating containing predominantly ζ-phase.
- *Type 1:* Optimal alloyed coating with less than a 1 µm interfacial γ-layer and a top layer containing δ-phase interspersed with a small amount of ζ-phase
- *Type 2:* Overalloyed coating with a γ-layer more than 1 µm thick and an overlay of δ-phase containing basal plane cracks.

Formability is an important property in galvanneal coatings because iron-zinc intermetallic phases are considered brittle. As a result, powdering and flaking of the coating can occur during the forming operation, resulting in reduced corrosion resistance and impaired paintability. The type 1 coating was found to have the best formability properties (Ref 12), but as in most forming operations lubrication to improve metal flow is essential. Spot weldability of galvanneal coatings are improved over galvanized coatings because it is more difficult for these iron-zinc phases to alloy with the copper electrode. Paintability is also better than that of galvanized coatings because of the microscopically rough surface formed as a result of the iron-zinc alloy phases throughout the coating. However, galvanneal coatings are more prone to cratering during e-coating (Table 1). Conversely, corrosion resistance can be slightly reduced because of the increased iron in the coating from the iron-zinc phases; the galvanic potential is not as great as it is for pure zinc.

Zn-5Al alloy coating (Galfan) is near the eutectic point in the aluminum-zinc equilibrium phase diagram. Two compositions have been reported based on additions to the eutectic composition: small (up to about 0.5%) mischmetal additions containing lanthanum and cerium and additions of 0.5% Mg. These additions are made to improve the wettability and suppress bare spot formation as well as to produce a typical "minimized spangle" structure. The microstructure of Galfan is characterized by a two-phase structure, a zinc-rich proeutectoid η-phase surrounded by eutectic phase consisting of lamellae of α-aluminum and zinc-rich η-phase. However, the microstructure can be varied depending on the cooling rate. In the range of normal bath temperatures, 420 to 440 °C (790 to 825 °F) there is no visible intermetallic layer or at least an extremely thin layer (<0.5 µm) at the interface between the steel substrate and the overlay coating. Thus, Galfan coatings have excellent formability and cut-edge corrosion protection.

Zn-55Al alloy coating (marketed under the tradename Galvalume) contains about 1.5% Si added for the purpose of preventing an exothermic reaction at the coating overlay/substrate steel interface. As a result, the coating contains α-aluminum dendrites, zinc-rich interdendritic regions, and a fine dispersion of silicon particles, along with a prominent Fe-Al-Zn intermetallic alloy layer at the interface between the steel substrate and the overlay coating. The surface of the coating contains characteristic spangles that consist of aluminum dendrites with a clearly measurable dendrite arm spacing. Cooling rate after dipping can significantly refine the microstructure of the coating, increasing the number of silicon particles and constraining the growth of aluminum dendrites.

Initially, the atmospheric corrosion of the Zn-55Al coating takes place in the zinc-rich interdendritic regions, enabling the coating to exhibit galvanic protection. As the coating continues to corrode, the zinc corrosion products become trapped in the interdendritic regions and act as a further barrier to corrosion. Eventually, the aluminum dendrites, which also acted as a barrier layer, add to the corrosion protection, as does the Fe-Al-Zn intermetallic alloy layer. This results in a parabolic type of corrosion as evidenced in Fig. 2. Although its galvanic protection is less than that provided by galvanized coatings, Zn-55Al is generally adequate to protect against rust staining at scratches and cut edges of the steel sheet.

Aluminum coatings are produced as type 1 coating, a thin (20 to 25 µm) aluminum-silicon alloy coating, and type 2, a thicker (30 to 50 µm) pure aluminum coating. Silicon is present in type 1 coatings in the range of 5 to 11 wt% to prevent formation of a thick iron-aluminum intermetallic layer at the coating/steel substrate interface. Instead, a thin Fe-Al-Si intermetallic layer is formed, allowing for good formability and coating adherence. These coatings are intended primarily for applications requiring improved appearance, good formability, and resistance to high temperatures, as in automobile exhaust components. The type 2 coating has a microstructure containing a pure aluminum overlay and a thick iron-aluminum intermetallic alloy layer. Thus, the formability and adhesion of this coating is limited by the poor ductility of the alloy layer. Nevertheless, the coating is used for outdoor construction applications (e.g., roofs, culverts, etc.) that require resistance to atmospheric corrosion (Table 2). The aluminum outerlayer offers excellent corrosion resistance because of the good barrier properties provided by the increased thickness of the coating (Fig. 2).

Weld-Overlay Coatings

Welding is a solidification method for applying coatings with corrosion, wear, and erosion resistance. Weld-overlay coatings, sometimes referred to as hardfacing, offer unique advantages over other coating systems in that the overlay/substrate weld provides a metallurgical bond that is not susceptible to spallation and can easily be applied free of porosity or other defects. Welded deposits of surface alloys can be applied in thicknesses greater than most other techniques, typically in the range of 3 to 10 mm. Most welding processes are used for application of surface coatings and on-site deposition can be more easily carried out, particularly for repair purposes. Weld overlays are very versatile because a large number of commercially available alloys can be selected to provide protection from a wide range of environmental degradation mechanisms.

During weld-overlay surfacing, the coating material is raised to its melting point and then solidified on the surface of the substrate, which

Table 3 Weld surfacing processes

Process	Approximate deposit thickness (min), mm	Deposition rate, kg/h	Dilution single layer, %	Typical uses
Oxyacetylene (OAW)	1.5	≤1	1–5	Small area deposits on light sections
Powder weld (PW)	0.1	0.2–1		Small area deposits on light sections
Shielded metal arc (SMAW)	3	1–4	15–30	Multilayers on heavier sections
Gas tungsten arc (GTAW)	1.5	≤2	5–10	High-quality low-dilution work
Plasma transferred arc (PAW)	2	≤10	2–10	High-quality lowest-dilution work
Gas metal arc (GMAW)	2	3–6	10–30	Faster than SMAW, no stub-end loss; positional work possible
Flux-cored arc (FCAW)	2	3–6	15–30	Similar to GMAW. Mainly for iron-base alloys for high abrasion resistance
Submerged arc (SAW)				
Wire	3	10–30	15–30	Heavy section work; higher-quality deposits than FCAW
Strip	4	10–40	10–25	Corrosion-resistant cladding of large areas
Bulk	...	...	...	Similar to SAW wire but other alloys possible
Electroslag (ESW)	4	15–35	5–20	High-quality deposits at higher deposit rates than SAW. Limited alloy range

Source: Ref 16

means that metals and alloys used for this purpose must have melting points similar to or less than the substrate material. The effectiveness of the weld-overlay coating depends mainly on the welding process and the overlay alloy composition. The welding process must be selected and optimized to apply protective overlays at high deposition rates and thermal efficiency, with good control over the overlay/substrate dilution and coating thickness. The overlay alloy composition must be selected to provide the required properties to prevent coating degradation, and the alloy composition must be readily weldable.

A number of welding processes are available for applying protective weld overlays, and many welding parameters must be considered when attempting to optimize a particular process for a given application. The process principles and their characteristics for some processes are summarized for comparison purposes in Table 3 and are described in Ref 15 and 16. The processes can be grouped as torch processes, arc welding processes, and high-energy-beam techniques. The torch process, oxyacetylene welding (OAW), is the oldest and simplest hardfacing process and involves simply heating the substrate with the flame and then melting the filler rod to get the hardfacing to melt. High-energy-beam techniques use laser beam welding (LBW) or electron beam welding (EBW) to alloy the surface by adding alloy powders to the weld pool.

In arc welding, the heat is generated by an arc between an electrode and the workpiece. Arc welding processes can be grouped into nonconsumable electrode processes and consumable electrode processes. Nonconsumable electrode processes, gas tungsten arc welding (GTAW) and plasma arc welding (PAW), both involve a tungsten electrode and the introduction of the filler metal (in the form of rod or wire in GTAW and powder in PAW). The arc melts the filler metal to form a molten pool that is protected from the atmosphere by an inert gas shield. In plasma arc welding, an additional inert gas flows through a constricted electric arc in the welding torch to form the plasma. In general, for consumable electrode processes, the arc is maintained between the consumable electrode and the workpiece. In shielded metal arc welding (SMAW), the electrode consists of a core wire surrounded by a flux covering, that upon melting forms a liquid slag and gas to protect the molten metal pool. In flux core arc welding (FCAW), the flux is contained in the core of the metallic tubular electrode, whereas in gas metal arc welding (GMAW) the consumable wire electrode and substrate metal is protected from the atmosphere by a gas fed axially with the wire through the welding gun nozzle. In submerged arc welding (SAW), the arc, which is submerged beneath a covering of flux dispensed from a hopper, melts the electrode, the surface of the workpiece, and some of the flux that protects the molten pool from oxidation. Electroslag welding (ESW) uses equipment similar to SAW for strip cladding.

There are a large number of processing parameters that must be considered when attempting to optimize welding processes for surface application:

All processes

Voltage across the arc
Current through the arc
Current polarity
Current pulsing parameters
Travel speed of heat source
Shielding gas type (except SAW)

Consumable processes

Filler metal feed rate
Electrode diameter
Electrode extension ("stick-out" length)

Nonconsumable processes

GTAW electrode tip angle (vertex angle)
PAW plasma gas flow rate

However, the important factors considered in terms of arc welding overlay parameter optimization and process performance include arc efficiency, melting efficiency, deposition rate, dilution, and coating thickness (Ref 17). Arc efficiency is only a function of the arc welding process; melting efficiency increases with increasing arc power and travel speed, and the maximum deposition rate is directly related to both the arc and melting efficiency. During the deposition of the weld-overlay coating, the base metal and the filler metal are melted and mixed in the liquid state to form a fusion bond. Depending

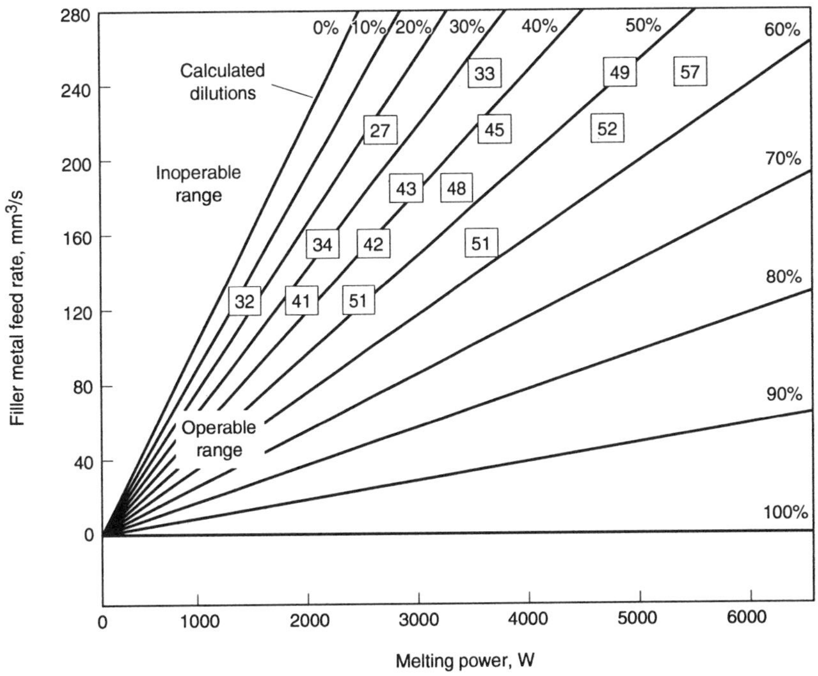

Fig. 3 Effect of processing parameters on dilution with experimental data plotted for SAW process. Source: Ref 18

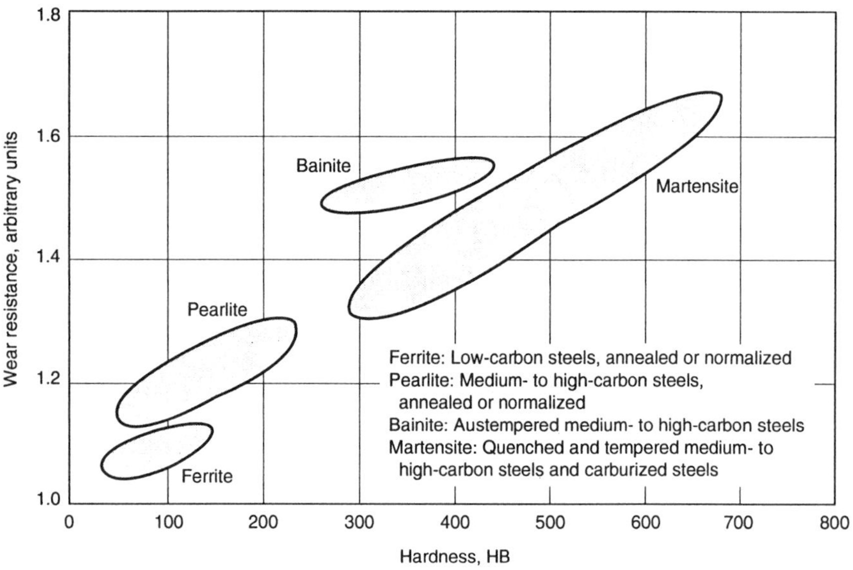

Fig. 4 Effect of structure and hardness on abrasion resistance. Source: adapted from Ref 21

on the weld-overlay coating thickness, if a large portion of the substrate is melted and allowed to mix appreciably with the weld overlay, dilution can cause the overall alloy content of the coating to be significantly reduced. The level of mixing is quantified as the dilution ratio and is one of the most important parameters in a surface application because the original filler metal mechanical and corrosion properties can be altered. The extent to which dilution occurs depends on the surfacing and substrate materials used, the welding process chosen, and the parameters employed. Table 3 indicates the range for dilution expected for the various processes employed. Figure 3 is a surfacing diagram that relates dilution for various arc welding processes according to filler metal feed rate and melting power (a function of arc and melting efficiency) and can be used to facilitate process selection and parameter optimization in weld-overlay applications (Ref 18).

During welding, the base metal is subjected to peak temperatures that are at least as high as the melting temperature of the substrate. The properties of the weld and the adjacent heat-affected zone (HAZ) strongly depend on the thermal history as dictated by the heat input. Preheating the part may be a necessary step in reducing the residual stress and distortion associated with welding. Preheat and maintenance of a specific minimum temperature during the welding cycle can also reduce the cooling rate to prevent the formation of a detrimental transformation region in the HAZ of ferrous alloys. Interpass temperature is another important factor needed to be controlled in order to prevent increased dilution and HAZ grain growth at high temperatures. Postweld heat treatment can take many forms, depending on whether the weld-overlay coating needs to be stress-relief annealed or must be heat treated for specified properties.

Excellent reviews of hardfacing metallurgy and the application of weld-overlay consumables are found in Ref 16 and 19. For overlay coatings, components are designed to provide resistance to various forms of wear, erosion, and corrosion over a large temperature range. Thus, properties such as hardness, microstructure, and corrosion resistance are more important for the coating than tensile strength and elongation, which are usually provided by the substrate material.

Generally, coatings selected for wear resistance require high hardness as a characteristic, thus the term "hardfacing." It is believed that most hardfacing alloys develop their wear resistance by virtue of wear-resistance carbides (Ref 20). Almost all hardfacing alloys can be separated into two major groups based on chemical compositions of the primary solidified hard phases:

- *Carbide hardening alloys*, including cobalt-base/carbide (WC-Co) and some iron-base superalloys
- *Intermetallic hardening alloys,* for example, nickel-base superalloys, austenitic stainless steels, and iron-aluminides

However, although increased hardness generally increases wear resistance, different microstructures containing the same carbide type can also have significant effect on wear resistance (Fig. 4).

Erosion resistance of materials is very dependent on the erosion conditions, the effects of which are dominated by a number of variables including particle size, shape, composition, and velocity; angle of incidence; and temperature. Unlike wear properties, the erosion rate of weld-overlay coatings generally increases with increasing hardness (Fig. 5). However, the erosion resistance of weld-overlay alloys depends on whether the coating can be classified as a brittle or ductile material (Ref 22). Those materials that can be deformed plastically (ductile) produce a large plastic zone beneath the eroded surface, and the increased plastic zone size can be directly correlated to an improved steady-state erosion resistance. For those materials that cannot deform plastically (brittle), an increase in coating hardness sometimes may lead to a decrease in volumetric erosion rate. Thus, materials that can dissipate particle impact energy through plastic deformation (plastic zone) exhibit low erosion rates. However, for materials that do not deform plastically (no plastic zone) and do not undergo plastic deformation, the ability to resist brittle fracture (i.e., cracking) becomes a major factor that can control the erosion resistance.

The corrosion resistance of weld-overlay coatings follows the corrosion-resistant properties of the bulk materials and is also dependent on the corrosive environment. Weld-overlay coatings provide resistance to oxidation and sulfidation. Dilution, as discussed previously, can be expected to modify the behavior of the coating alloy from the properties quoted for the undiluted bulk materials. In weld-overlay coatings such as austenitic steels, dilution can affect corrosion resistance because of a reduction in the effective chromium content or an increase in carbon content through carbon pickup from the substrate steel. Iron aluminides appear to be potentially important weld-overlay coatings for sulfidation environments. Figure 6 shows isothermal weight gain studies for a number of weld-overlay coatings exposed to H_2S-H_2-H_2O-Ar gas mixtures at 800 °C (1470 °F) (Ref 23). This work showed that compositions containing at least 30% Al and ≥2% Cr had excellent sulfidation resistance and, at increased chromium levels, corrosion rates increased but were still superior to other alloy classes such as stainless steels.

Detailed information about weld overlay coatings is available in the article "Hardfacing, Weld Cladding, and Dissimilar Metal Joining," in *Welding, Brazing, and Soldering,* Volume 6 of *ASM Handbook* (Ref 24).

Thermal Spray Coatings

Thermal spraying is a generic term for a group of processes that apply a consumable in the form of a spray of finely divided molten or semi-molten droplets to produce a coating. A number of extensive reviews of the topic can be found in Ref 25 to 29. The characteristics that distinguish thermal spray processes from weld-overlay coatings are indicated as follows (Ref 29):

- Substrate adhesion, or bond strength, is dependent on the materials and their properties and generally is characterized as a mechanical bond between the coating and the substrate, unlike the metallurgical bond found in weld-overlay coatings.
- Spray deposits can be applied in thinner layers than welded coating, but thick deposits are also possible.
- Provided there is a stable phase, almost all material compositions can be deposited, including metals, cermets, ceramics, and plastics.

- Thermal spray processes are usually used on cold substrates, preventing distortion, dilution, or metallurgical degradation of the substrate.
- Thermal spray processes are line-of-sight limited, but the spray plume often can be manipulated for complete coverage of the substrate.

Processes for thermal spray coatings can be classified into two categories, arc processes and gas combustion processes, depending on the means of achieving the heat for melting of the consumable material during the spraying operation.

In the lower-energy electric arc (wire arc) spray process, heating and melting occur when two electrically opposed charged wires, comprising the spray material, are fed together to produce a controlled arc at the intersection. The molten material on the wire tips is atomized and propelled onto the substrate by a stream of gas (usually air) from a high-pressure gas jet. The highest spray rates are obtained with this process, allowing for cost-effective spraying of aluminum and zinc for the marine industry. In the higher-energy plasma arc spray process, injected gas is heated in an electric arc and converted into a high-temperature plasma that propels the coating powder onto the substrate at very high velocities. This process can take place in air with air plasma spraying (APS), or in a vacuum with vacuum plasma spray (VPS) or low-pressure plasma spraying (LPPS).

For gas combustion processes, the lower-energy flame spray process uses oxyfuel combustible gas as a heat source to melt the coating material, which may be in the form of rod, wire, or powder. In the higher-energy, high-velocity, oxyfuel combustion spray (HVOF) technique, internal combustion of oxygen and fuel gas occurs to produce a high-velocity plume capable of accelerating powders at supersonic speeds and lower temperatures than the plasma processes. Continuous combustion occurs in most commercial processes, whereas the proprietary detonation gun (D-gun) process uses a spark discharge to propel powder in a repeated operating cycle to produce a continuous deposit.

In the lower-energy processes, electric arc (wire arc) spray and flame spray processes, adhesion to the substrate is predominantly mechanical and is dependent on the workpiece being perfectly clean and suitably rough. Some porosity is always present in these coatings, which may present problems in both corrosion and erosion. The higher-energy processes—APS, VPS, LPPS, and HVOF processes—were developed to reduce porosity and improve adhesion to the substrate. In addition, these processes are capable of spraying materials with higher melting points, thus widening the range of applications to include high-temperature coatings and thermal and mechanical shock-resistant coatings. With these higher-energy processes, bond strengths are higher because of the possible breakup of any oxide films present on the particles or the workpiece surface, allowing for some diffusion bonding to take place (Ref 29). Typical design features of the various thermal spray processes are listed in Table 4.

The variations in oxide content and porosity, as well as the chemical composition of the coating, greatly affect the properties of the deposit and, in the case of corrosion, the underlying substrate. The splat morphology and, more importantly, the splat/splat and splat/substrate interface are critical to properties such as bond strength, wear, erosion, and corrosion. The mechanical properties of thermal spray coatings are not well documented except for their hardness and bond strength. Table 5 contains typical mechanical properties data for a large range of plasma sprayed materials; however, the sensitivity of the properties of the coatings to specific deposition parameters makes universal cataloging of properties by process and composition "virtually meaningless" (Ref 26).

One of the most extensive uses for thermal spray coatings is in wear applications. Generally,

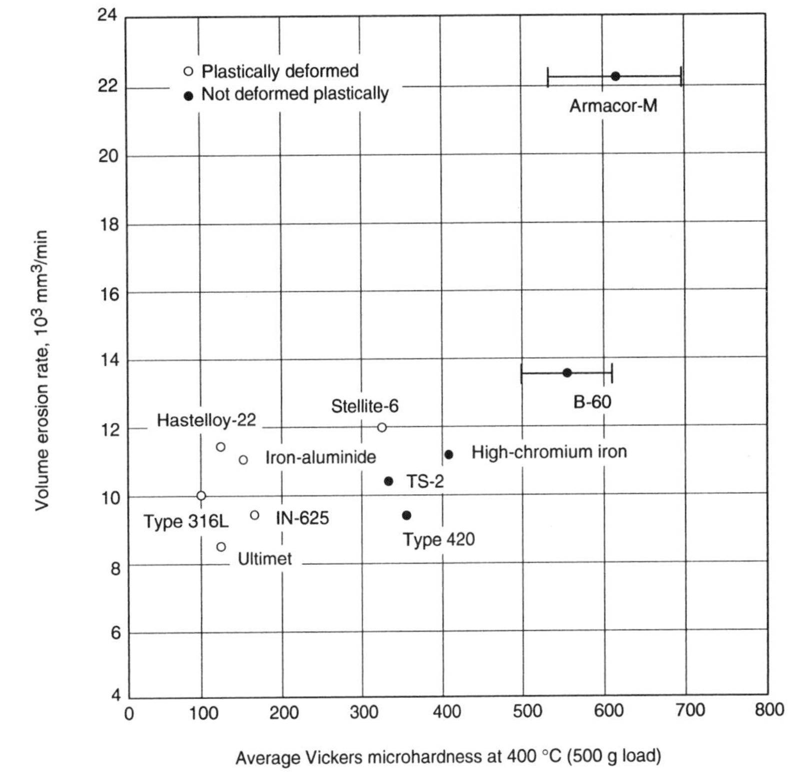

Fig. 5 Volume steady-state erosion rates of weld-overlay coatings at 400 °C (750 °F) as a function of average microhardness at 400 °C (90° impact angle; alumina erodent). Source: Ref 22

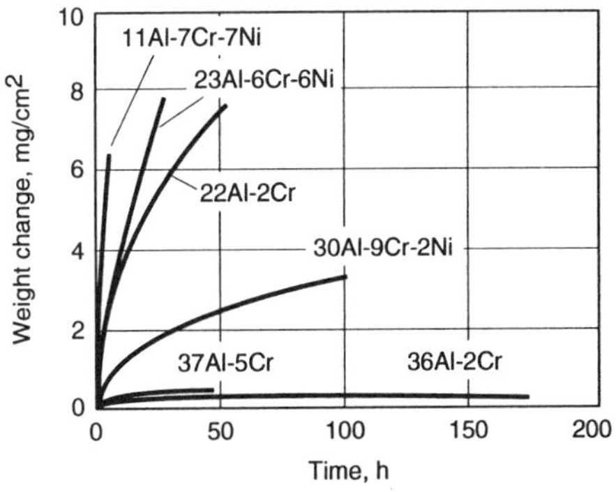

Fig. 6 Weight change versus time for specimens cut from iron-aluminide weld overlays and isothermally exposed to H_2S-H_2-H_2O-Ar at 800 °C (1470 °F). The elemental concentrations shown are in at.%; the balance is iron. Source: Ref 23

Table 4 Typical design characteristics of thermal spray processes

Process	Gas temperature, °C	Particle velocity, m/s	Adhesion, MPa	Oxide content, %	Porosity, %	Spray rate, kg/h	Relative cost, low = 1	Typical deposit thickness, mm
Flame	3,000	40	8	10–15	10–15	2–6	1	0.1–15
Arc wire	NA	100	12	10–20	10	12	2	0.1 to >50
High velocity oxyfuel (Jet Kote)	3,000	800	>70	1–5	1–2	2–4	3	0.1 to >2
Detonation gun	4,000	800	>70	1–5	1–2	0.5	NA	0.05–0.3
Air plasma spray	12,000	200–400	4 to >70	1–3	1–5	4–9	4	0.1–1
Vacuum plasma spray	12,000	400–600	>70	ppm	<0.5	4–9	5	0.1–1

NA, not applicable. Source: Ref 29

the wear resistance of coatings increases with their density and cohesive strength, so that HVOF coatings provide the best wear resistance in contrast to plasma spray coatings (Table 6). Carbide cermets were found to be good for both wear and erosion environments, and the optimal amount of hard phase (oxide and carbide) has been determined for erosion resistance. A comparison between the presprayed powder and the actual coating microstructure showed that the retention of the FeCrAlY matrix is much better than the chromium carbide particles, which can form oxides during HVOF thermal spraying (Ref 30). Erosion tests show that both carbides and oxides increase the erosion rate of the coating (Fig. 7) and that low amounts of hard constituents are preferable for erosion resistance.

Thermal spray zinc, aluminum, and zinc-aluminum alloys are used for sacrificial galvanic protection for corrosion resistance on bridges, ships, and other large structures. Other corrosion-resistant applications for thermal spray coatings include oxidation and sulfidation resistance in power boilers and other high-temperature uses. In a comparison of several different thermal spray techniques, it was found that high-temperature corrosion-resistant coatings must have compositions that promote the formation of protective oxides at splat boundaries, be dense enough so that protective oxides can form within and fill voids, and be thick enough to postpone the diffusion of corrosive species to the substrate material along the fast diffusion paths of the coating (Ref 31). It has been found that corrosion attack of the substrate generally decreases as the free path to the substrate (that is, the diffusion path the corrosive species takes to the substrate) increases. Thus, as in erosion and wear, the splat/splat and splat/substrate boundaries are critical to properties of the thermal spray coatings.

Table 5 Typical mechanical properties of plasma sprayed coatings

Material	Bond tensile strength (a) MPa	Bond tensile strength (a) ksi	Rockwell macro/micro hardness	Density g/cm^3	Density lb/ft^3
Pure metals					
Aluminum	8.3	1.2	45/58 HRH	2.48	155
Copper	21.4	3.1	65/142 HRB	7.20	449
Molybdenum (fine)	57.2	8.3	70/1450 HR15N	9.90	618
Molybdenum (coarse)	55.2	8.0	65/1448 HRA	8.96	559
Nickel (fine)	23.4	3.4	84/... HR15T	7.95	496
Nickel (coarse)	33.1	4.8	81/... HR15T	7.48	467
Niobium	54.5	7.9	61/1344 HRC	7.06	441
Tantalum	46.9	6.8	65/1585 HRA	14.15	883
Titanium	41.4	6.0	78/... HR15N	4.17	260
Tungsten	40.0	5.8	50/500 HRA	16.90	1055
Alloy metals					
304 stainless	17.6	2.55	88/... HR15T	7.22	451
316 stainless	23.4	3.4	70/... HR30T	6.80	425
431 stainless	31.0	4.5	35/... HRC	6.25	390
80Ni-20Cr (fine)	31.0	4.5	90/... HR15	7.48	467
80Ni-20Cr (coarse)	29.0	4.2	90/... HR15T	7.19	449
40Ni-60Cu	24.1	3.5	72/... HRB	7.89	493
35Ni-5In-60Cu	24.1	3.5	83/... HR15T	7.94	496
10Al-90Cu (fine)	28.3	4.1	88/... HR15T	6.73	418
10Al-90Cu (coarse)	22.1	3.2	81/... HR15T	6.30	393
Hastelloy 31 (fine)	41.4	6.0	79/... HR15T	7.65	478
Hastelloy 31 (coarse)	23.4	3.4	79/... RH15T	7.83	489
5Al-95Ni	68.3	9.9	80/490 HRB	7.51	469
20Al-80Ni	47.6	6.9	80/510 HRB	6.92	432
6Al-19Cr-75Ni	49.6	7.2	90/250 HRB	7.51	469
12Si-88Al	16.5	2.4	78/60 HR15T	2.49	155
5Al-5Mo-90Ni	37.9	5.5	80/200 HRB	7.43	464
Hastelloy X	42.7	6.2	89/... HR15T	7.65	478
Hastelloy C	42.1	6.1	90/... HR15T	8.25	515
420 stainless	22.1	3.2	70/... HR15N	7.10	443
0.9C stainless	33.8	4.9	35/... HRC	7.05	440
Cast iron	35.9	5.2	28/... HRC	7.00	437
Ti-6Al-4V	33.1	4.8	35/... HRC	4.30	268
Monel	44.8	6.5	35/... HR15N	8.50	531
0.2C steel	22.1	3.2	95/... HRB	6.90	431
Metal composites					
95Ni-5Al	33.8	4.9	80/500 HR15T	7.39	461
80Ni-20Al	32.4	4.7	86/500 HR15T	7.02	438
65Ni-35Ti	32.1	4.65	72/660 HR15N	6.62	413
75Ni-19Cr-6Al	42.7	6.2	92/250 HR15T	7.71	481
75Ni-9Cr-7Al-5Mo-5Fe	27.6	4.0	80/250 HRB	6.90	431
90Ni-5Al-5Mo	48.3	7.0	80/200 HRB	7.40	462
Carbide powders and blends					
88WC-12Co (cast, fine)	44.8	6.5	88/... HR15N	13.75	858
88WC-12Co (cast, coarse)	44.8	6.5	81/... HR15N	12.41	775
88WC-12Co (sintered)	55.2	8.0	85/... HR15N	14.55	908
83WC-17Co	68.9	10.0	85/950 HR15N	11.10	693
$75Cr_3$-C_2-25NiCr (fine)	41.4	6.0	84/950 HR15N	6.41	400
$75Cr_3$-C_2-25NiCr (coarse)	34.5	5.0	80/1850 HR15N	6.23	389
$75Cr_3$-C_2-25NiCr (composite)	...	...	.../1850 HR15N	...	...
$85Cr_3$-C_2-15NiCr	...	...	80/1850 HR15N	5.80	362
Ceramic oxides					
ZrO_2 (calcinated)	44.8	6.5	70/... HR15N	5.30	331
Chromium oxide	44.8	6.5	90/... HR15N	4.80	300
$80ZrO_2$-20yttria	15.2	2.2	80/... HR15N	5.00	312
TiO_2	...	...	87/... HR15N	4.10	256
Al_2O_3 (white)	44.8	6.5		...	...
$87Al_2O_3$-$13TiO_2$	15.5	2.25	90/... HR15N	3.50	218
$60Al_2O_3$-$40TiO_2$	27.6	4.0	90/850 HR15N	3.50	218
$50Al_2O_3$-$50TiO_2$	...	...	85/... HR15N	4.0	250
Al_2O_3-gray (fine)	6.9	1.0	87/193 HR15N	3.30	187
Al_2O_3-gray (coarse)	...	...	85/... HR15N	3.30	187
Magnesium zirconate	17.2	2.5	75/... HR15N	4.20	262

(a) Over a grit-blasted surface roughened to 2.5 to 4.1 μm (100 to 160 μin.) AA (arithmetic average). Source: Ref 28

Deposition Surface Treatments

Deposition surface treatments include electrochemical plating, chemical vapor deposition, and physical vapor deposition processes.

Electrochemical Deposition

Electrochemical methods are well-established processes for applying metal coatings for improved surface properties of materials. Electrodeposition or electroplating is defined as the deposition of a coating by electrolysis, that is, depositing a substance on an electrode immersed in an electrolyte by passing electric current through the electrolyte. The process can take place in an aqueous electrolyte near ambient tem-

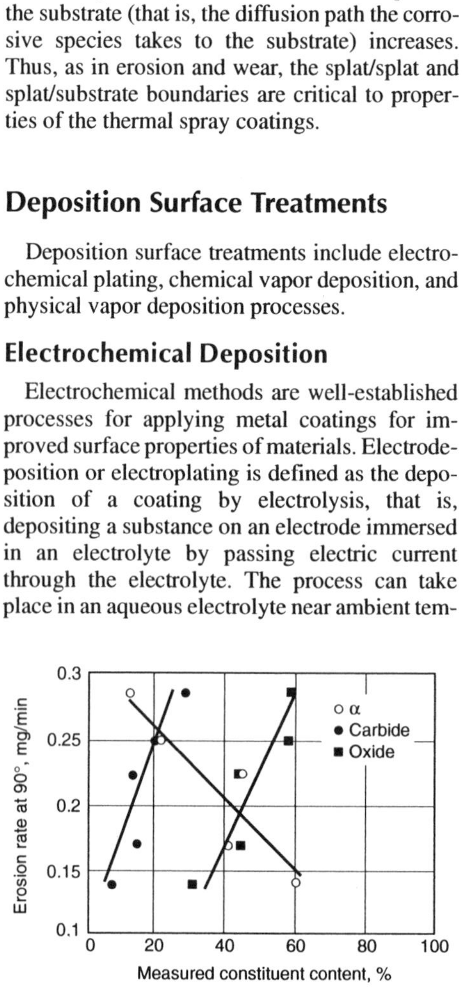

Fig. 7 Steady-state erosion rates versus constituent composition. Source: Ref 30

peratures (called aqueous solution electroplating) or in a fused metal salt at high temperatures (called metalliding or fused salt electroplating). Modifications of the electroplating process include occlusion or composite deposition plating. Excellent reviews of electroplating can be found in Ref 32 and 33. Wear- and corrosion-resistance data for selected electrochemical coating methods are summarized in Table 7.

Aqueous solution electroplating provides decorative and protective finishes for use at ambient temperatures and in a variety of environments. A main drawback in electroplating is the inability of achieving uniform deposition, which is related to the throwing power of the electrolyte. Throwing power is substrate-shape dependent and also depends on the anode/cathode configuration and the current density, as well as on the composition and conductivity of the electrolyte. A second difficulty is that not all metallic elements can be deposited. Another problem with electroplating is evolution of hydrogen at the electrodes when the cathode efficiency is less than 100%. If a ferrous substrate is to be plated, absorbed atomic hydrogen can cause embrittlement of the part. Unless the parts are heat treated to remove the absorbed hydrogen, they will be brittle and unusable for any application involving elastic strain.

Substrate preparation for plating is critical to ensure good adhesion and surface quality. Maximum adhesion depends on both the elimination of surface contaminants in order to induce a metallurgical bond and the generation of a completely active surface to initiate plating on all areas. The cleaning steps for steel substrates usually involve precleaning, intermediate alkaline cleaning, electrocleaning, acid treatments, and anodic desmutting (Ref 35). Electrodeposited metals can have a very good bond to the substrate, but that bond will never be the same as a fusion bond, and poor bonds can go undetected unless techniques are used to test the actual bond strength. The electrodeposited coating usually ends up with a surface texture that is the same as the starting substrate surface texture, unless an intermediate leveling coating (such as copper) is used.

Nevertheless, the electroplating process produces a coating with distinct advantages (Ref 36):

- The workpiece will not undergo distortion or metallurgical changes because the operating temperature of the bath does not exceed 100 °C (212 °F).
- Plating conditions can be adjusted to modify hardness, internal stress, and metallurgical characteristics of the coating.
- Coatings are dense and adherent to the substrate.
- The thickness of the coating is proportional to the current density and length of time of the deposition.
- Although deposition rate seldom exceeds 75 μm/h (3 mils/h), it can be accelerated by forced circulation of the electrolyte, and can be as high as 150 or 200 μm/min (6 or 8 mils/h) for some metals in high-speed plating.
- There is no technical limit to the thickness of electrodeposits.
- Application of coatings is not confined to the line of sight. Although throwing power may be limited, the freedom of anode design and location is helpful.
- Areas not requiring deposition can be masked.
- Only the tank size of the bath limits the dimensions of the part, although large parts such as gun barrels can be the tank itself; as another alternative, brush plating can be used.
- The process is suitable for automation and has economic advantages over other coating processes.

In general, aqueous electroplating has minimal effect on substrate properties (apart from hydrogen embrittlement). Coated substrates can also be heat treated to promote interdiffusion, although this may result in concentration of elements at grain boundaries, causing embrittlement. Specific elemental electrodeposition processes and properties are reviewed in Ref 37; some examples are given here.

Nickel plating is widely used for a corrosion- and wear-resistant finish. Typical applications, with a thin top coat of electrodeposited chromium, are decorative trim for automotive and

Table 6 Abrasive wear data for selected thermal spray coatings

Material	Type	Wear rate, mm^3/1000 rev
Carballoy 883	Sintered	1.2
WC-Co	Detonation gun	0.8
WC-Co	Plasma spray	16.0
WC-Co	Super D-gun	0.7
WC-Co	High-velocity oxyfuel	0.9

ASTM G 65 dry sand/rubber wheel test. 50/70 mesh Ottawa silica. 200 rpm. 30 lb load. 3000-revolution test duration. Source: Ref 26

Table 7 Characteristics of wear- and corrosion-resistant electrochemical finishes for engineering components

Characteristic	Copper	Electroplated nickel	Electroless nickel	Chromium	Electroless nickel + chromium	Electroplated nickel + chromium	Comment
Maximum working temperature, °C	50	650	550	650	550	650	Short times at higher temperatures possible
Nontoxicity	Excellent	Very good	Very good	Very good	Very good	Very good	...
Covering complex shapes	Medium	Medium	Excellent	Very poor	Very good	Medium	...
Thickness range, μm	12.5–500	12.5–500	12.5–500	12.5–500	12.5–500 + 25–50	12.5–500 + 25–50	Grinding needed over 200 μm
Wear							
Hardness, HV	60–150	200–300	450–500 (900–1000 after heat treatment)	850–950	850–950	850–950	Indication of abrasive-wear resistance
Low friction, antistick	Poor	Poor	Medium	Excellent	Excellent	Excellent	Indication of adhesive-wear resistance
Resistance to impact	Medium	Very good	Medium	Medium	Medium	Very Good	Thin coats and soft substrates prone to damage
Corrosion resistance	Very poor	Very good	Medium	Poor	Very good	Very good	Nickel at least 50 μm for corrosive environments
Typical applications	Buildup; lubricant in forming; heat sink; selective case hardening	Buildup under or instead of chromium in corrosive conditions; printing surfaces	PVC molding tools and dies; moving parts in process industries; glass and rubber molds	Molds, tools, valves, rams, pistons, shafts, gauges, dies, saw blades	High-temperature, antiseize bolting; ball valves; shafts	Marine crane rams and hydraulics, mine roof supports, print rolls	...

Source: Ref 34

consumer products and office furniture. Nickel deposits are also used for nondecorative purposes for improved wear resistance, for example, on pistons, cylinder walls, ball studs, and so forth.

Chromium electroplating is also used as decorative and hard coatings. Colored and tarnish-resistant chromium decorative coatings are produced over a base deposit of copper and/or nickel for applications such as those noted above for nickel. Hard chromium coatings are used for hydraulic pistons and cylinders, piston rings, aircraft engine parts, and plastic molds, where resistance to wear, heat abrasion, and/or corrosion are required.

Cadmium and zinc electroplating provides galvanic corrosion protection when coated on steel. Deposit thickness can vary between 5 and 25 μm (0.2 to 1 mil), and typical applications for both coatings are found in Table 8. Cadmium is preferred for the protection of steel in marine environments, whereas zinc is preferred in industrial environments. Cadmium is also preferred for fastening hardware and connectors because its coefficient of friction is less than zinc. Cadmium is toxic and should not be used in parts that will have contact with food. Precautions for minimizing hydrogen embrittlement should be taken because cadmium plating is more susceptible to such embrittlement than any other plated metal. The development of continuous electrogalvanizing lines has produced a thin, formable coating that is ideal for deep drawing or painting. Automotive body panels are typically new applications for electrogalvanized zinc, zinc-nickel, and zinc-iron alloys. Processing details applicable to electrogalvanizing can be found in Ref 39.

Tinplate is another continuous electrolytic plating process that has been used for the past 200 years to make containers for the long-term storage of food (Ref 40). The typical tinplate product consists of five layers: an innermost layer of steel sheet, a tin-iron intermetallic compound layer, a free-tin layer, a thin passivation layer based on chromium oxide, and a top layer of oil film for lubrication. The corrosion characteristics of tinplate are documented in the literature (Ref 41).

Metalliding, or fused-salt electroplating, is a process for surface modification and surface hardening by electrodeposition from fused-salt electrolytes. Two unique aspects of this electrodeposition process are: (1) elements that cannot be plated by conventional processes may plate by fused-salt electrodeposition and (2) if the deposition rate is controlled to match the diffusion rate of the deposition species in the substrate at the fused-salt temperatures 400 to 900 °C (750 to 1650 °F), the substrate will develop a diffusion coating, termed metalliding. In electroplating (Ref 42), the molten-salt medium in which the anode and cathode are immersed, consists of a soluble form of the metal to be plated dissolved in a molten-salt solvent, such as an alkali metal halide, that does not participate in the plating process. Coating thickness is determined by the electrical charge, and a sharp interface between the coating and substrate is maintained. At the higher-temperature plating range of the bath, some coating/substrate interdiffusion can occur. In metalliding (Ref 43), the element to be diffused is made the anode of the molten-salt electrochemical cell (usually fluoride) and the substrate is the cathode. A more electrochemically active anode diffuses into the cathode when the electrodes are connected. Because the process is diffusion controlled, a sharp coating/substrate interface does not exist; instead a diffusion gradient in the substrate occurs.

Although fused-salt electroplating has only found limited application for refractory metals and ceramic coatings, some success has been obtained with the platinum-group metals. The process conditions for this technology are too stringent and economically unfeasible. Nevertheless, the process requirements for the electrodes, the melt, and cell operation have been outlined (Ref 42). On the other hand, metalliding is a unique electrodeposition process for applying elements that are difficult to electrodeposit on substrates that usually cannot be plated. It is an important process for improving the surface hardness and corrosion resistance of metals without producing significant dimensional changes. Small-scale, small-size, specialized or strategically important components can be considered for technological development for coating by metalliding if the cost justification can be made, in view of the fact that a one-step diffusion process can be achieved more easily.

Precious Metal Plating. Silver, gold, and the platinum metal groups are electroplated by either aqueous solution electroplating or fused-salt electrodeposition. Both silver and gold are used for decorative purposes as well as industrial uses; the aqueous plating process is reviewed in Ref 44 and 45. Decorative applications of both elements still predominate, but silver has been successfully substituted for gold in some functional uses in electronics. Silver is used on metallic leadframes, the device that supports the majority of silicon chips. New silicon-to-silver bonding techniques have been used to replace the more expensive gold. However, in electrical contact applications, where long-term surface integrity is important, silver has not been able to replace gold because of its tendency to oxidize or sulfidize on the surface, increasing the contact resistance of the component.

Both aqueous and fused-salt electrolytes have been used for plating the platinum group elements. Platinum has been used as a diffusion barrier layer in aluminiding nickel-base alloys and MCrAlY coatings. Platinum from the aqueous electrolyte is highly stressed unlike the fused-

Table 8 Recommended minimum thicknesses and typical applications for zinc and cadmium coatings electrodeposited on iron and steel

Service conditions	Coating thickness, μm	Coating thickness, mils	Chromate finish	Time to white corrosion in salt spray, h	Typical applications
Electrodeposited zinc					
Mild (indoor atmosphere; minimum wear and abrasion)	5	0.2	None	...	Screws, nuts and bolts, wire goods, fasteners
			Clear	12–24	
			Iridescent	24–72	
			Olive drab	72–100	
Moderate (mostly dry, indoor atmosphere; occasional condensation, wear, and abrasion)	8	0.3	None	...	Tools, zipper pulls, shelves, machine parts
			Clear	12–24	
			Iridescent	24–72	
			Olive drab	72–100	
Severe (exposure to condensation; infrequent wetting by rain and cleaners)	13	0.5	None	...	Tubular furniture, window screens, window fittings, builders' hardware, military hardware, appliance parts, bicycle parts
			Clear	12–24	
			Iridescent	24–72	
			Olive drab	72–100	
Very severe (exposure to bold atmospheric conditions; frequent exposure to moisture, cleaners, and saline solutions; likely damage by abrasion or wear)	25	1	None	...	Plumbing fixtures, pole line hardware
Electrodeposited cadmium					
Mild (see above)	5	0.2	None	...	Springs, lock washers, fasteners, tools, electronic and electrical components
			Clear	12–24	
			Iridescent	24–72	
			Olive drab	72–100	
Moderate (see above)	8	0.3	None	...	Television and radio chassis, threaded parts, screws, bolts, radio parts, instruments
			Clear	12–24	
			Iridescent	24–72	
			Olive drab	72–100	
Severe (see above)	13	0.5	None	...	Appliance parts, military hardware, electronic parts for tropical service
			Clear	12–24	
			Iridescent	24–72	
			Olive drab	72–100	
Very severe (see above)(b)	25	1	None	...	...
			Clear	24	
			Iridescent	24–72	
			Olive drab	72–100	

(a) Thickness specified is after chromate coating, if used. (b) There are some applications for cadmium coatings in this environment; however, these are normally satisfied by hot dipped or sprayed coatings. Source: Ref 38 salt deposit. Substances particularly considered for platinum-group metal coating are the refractory group alloys of molybdenum, tungsten, tantalum, niobium, and vanadium, which tend to form volatile oxides at high temperatures thus reducing their usefulness as corrosion-resistant materials. However, cost is still a major factor, and these metals can be used for diffusion barrier layers only if components are small and strategically important.

Electroless plating baths have been developed for copper, nickel, silver, and a number of other materials, but the system with the most importance for corrosion and wear applications is the nickel/phosphorus system. Electroless nickel plating is used to deposit nickel without the use of an electric current; thus it is sometimes called autocatalytic plating. In this process the part is immersed in a aqueous solution containing metal salts, a reducing agent, and other chemicals that control the pH and reaction rates. The part acts as a catalyst for the reduction of the nickel ions by the reducing agent. The reducing agent causes the metal ion reduction and the nickel coating on the part continues to act as a catalyst as the plating process continues, unlike in electroplating where the ions pick up electrons from the cathode. When the process takes place using a hypophosphite-reducing agent, the finished nickel coating is not pure nickel, but contains phosphorus inclusions. Phosphorus content can be as high as 13% (Ref 46).

As applied, nickel-phosphorus coatings are uniform, hard, relatively brittle, lubricious, easily solderable, and highly corrosion resistant (Ref 47). Wear resistance equivalent to hard chromium coatings can be obtained when the coating is heat treated at low temperatures to produce a very hard precipitation-hardened structure. As applied, most of these coatings are amorphous metal glasses that when heated first form nickel phosphite (Ni_3P) particles; at temperatures above 320 °C (610 °F), the deposit crystallizes. Internal stresses are primarily a function of coating composition, and coating thickness uniformity can be easily controlled. Adhesion to most metals is excellent, and frictional properties are also excellent and similar to chromium.

As deposited, the microhardness of electroless nickel coating is about 500 to 600 HVN (48 to 50 HRC), equivalent to many hardened steels. After precipitation hardening, hardness values as high as 110 HVN are reported, which is equivalent to commercial hard-chromium coatings. Because of their high hardness, electroless nickel coatings have excellent wear and abrasion resistance in both the as-deposited and hardened condition.

Electroless nickel coatings can be easily soldered and are used in electronic applications to facilitate soldering of light metals such as aluminum. Electroless nickel is often used as a barrier coating; to be effective, the deposit must be free of pores and defects. In the as-deposited amorphous state, the coating corrosion resistance is excellent (Table 9), and in many environments is superior to that of pure nickel or chromium al-

Table 9 Corrosion of electroless nickel coatings in various environments

Environment	Temperature		Corrosion rate: Electroless nickel-phosphorus(a)		Corrosion rate: Electroless nickel-boron(b)	
	°C	°F	μm/yr	mil/yr	μm/yr	mil/yr
Acetic acid, glacial	20	68	0.8	0.03	84	3.3
Acetone	20	68	0.08	0.003	Nil	Nil
Aluminum sulfate, 27%	20	68	5	0.2	...	...
Ammonia, 25%	20	68	16	0.6	40	1.6
Ammonia nitrate, 20%	20	68	15	0.6	(c)	(c)
Ammonium sulfate, saturated	20	68	3	0.1	3.5	0.14
Benzene	20	68	Nil	Nil	Nil	Nil
Brine, 3.5% salt, CO_2 saturated	95	205	5	0.2	...	...
Brine, 3.5% salt, H_2S saturated	95	205	Nil	Nil	...	...
Calcium chloride, 42%	20	68	0.2	0.008	...	...
Carbon tetrachloride	20	68	Nil	Nil	Nil	Nil
Citric acid, saturated	20	68	7	0.3	42	1.7
Cupric chloride, 5%	20	68	25	1	...	...
Ethylene glycol	20	68	0.6	0.02	0.2	0.008
Ferric chloride, 1%	20	68	200	8	...	...
Formic acid, 88%	20	68	13	0.5	90	3.5
Hydrochloric acid, 5%	20	68	24	0.9	...	...
Hydrochloric acid, 2%	20	68	27	1.1	...	...
Lactic acid, 85%	20	68	1	0.04	...	...
Lead acetate, 36%	20	68	0.2	0.008	...	...
Nitric acid, 1%	20	68	25	2	...	...
Oxalic acid, 10%	20	68	3	0.1	...	...
Phenol, 90%	20	68	0.2	0.008	Nil	Nil
Phosphoric acid, 85%	20	68	3	0.1	(c)	(c)
Potassium hydroxide, 50%	20	68	Nil	Nil	Nil	Nil
Sodium carbonate, saturated	20	68	1	0.04	Nil	Nil
Sodium hydroxide, 45%	20	68	Nil	Nil	Nil	Nil
Sodium hydroxide, 50%	95	205	0.2	0.008	...	...
Sodium sulfate, 10%	20	68	0.8	0.03	11	0.4
Sulfuric acid, 65%	20	68	9	0.4	...	...
Water, acid mine, 3.3 pH	20	68	7	0.3	...	...
Water, distilled, N_2 deaerated	100	212	Nil	Nil	Nil	Nil
Water, distilled, O_2 saturated	95	205	Nil	Nil	Nil	Nil
Water, sea (3.5% salt)	95	205	Nil	Nil	...	...

(a) Hypophosphite-reduced electroless nickel containing approximately 10.5% P. (b) Borohydride-reduced electroless nickel containing approximately 5% B. (c) Very rapid. Specimen dissolved during test. Source: Ref 47

Table 10 Mechanical properties of electrodeposited cermets

Cermet	Hardness, HV: As plated	Hardness, HV: Annealed	Yield strength(a), MPa	Elongation(a), %
Ni	187	118	93.0	6.26
Ni-2.02Al_2O_3	275.4	247	68.4(b) 67(a), 68.4(b)	...
Ni-3.33TiO_2	354	254	222.4	3.75
Ni-6.80Cr_2O_3	409	295	284.0	1.50
Ni-3.6(TiO_2 + $CrSi_2$)	283	209	198.5	2.60
Ni-24Co	280	150	120.6	2.3
Ni-23.4Co-3TiO_2	383	219	206.6	2.0
Ni-22.5Co-6.07Cr_2O_3	462	285	264	1.2
Ni-23.7Co-3.8$CrSi_2$	302	207	196.4	1.6
Ni-23.7Co-3.4(TiO_2+$CrSi_2$)	359	211	196.5	...

(a) Annealed. (b) As plated. Source: Ref 34

Table 11 Typical deposition temperatures for thermal and plasma CVD

Material	Deposition temperature: Thermal CVD		Deposition temperature: Plasma CVD	
	°C	°F	°C	°F
Silicon nitride	900	1650	300	570
Silicon dioxide	800–1100	1470–2010	300	570
Titanium carbide	900–1100	1650–2010	500	930
Titanium nitride	900–1100	1650–2010	500	930
Tungsten carbide	1000	1830	325–525	615–975

Source: Ref 50

Table 12 Selected wear and corrosion properties of CVD coating materials

Material	Hardness, GPa	Hardness, 10^6 psi	Thermal conductivity, W/m · K	Coefficient of thermal expansion at 25 °C (77 °F) 10^{-6}/K	Remarks
Titanium carbide	31.4	4.5	17	7.6	High wear and abrasion resistance, low friction
Titanium nitride	20.6	3.0	33	9.5	High lubricity; stable and inert
Titanium carbonitride	24.5–29.4	3.5–4.3	20–30	8	Stable lubricant
Chromium carbide	22.1	3.2	11	10	Resists oxidation to 900 °C (1650 °F)
Silicon carbide	27.4	4.0	125	3.9	High conductivity, shock resistant
Titanium diboride	33.0	4.7	25	6.6	High hardness, high wear resistance
Alumina	18.8	2.7	34	8.3	Oxidation resistant, very stable
Diamondlike carbon	29–49	4.2–7.1	200	...	Very hard, high thermal conductivity
Diamond	98	14.2	180	2.9	Extreme hardness and high thermal conductivity

Source: Ref 50

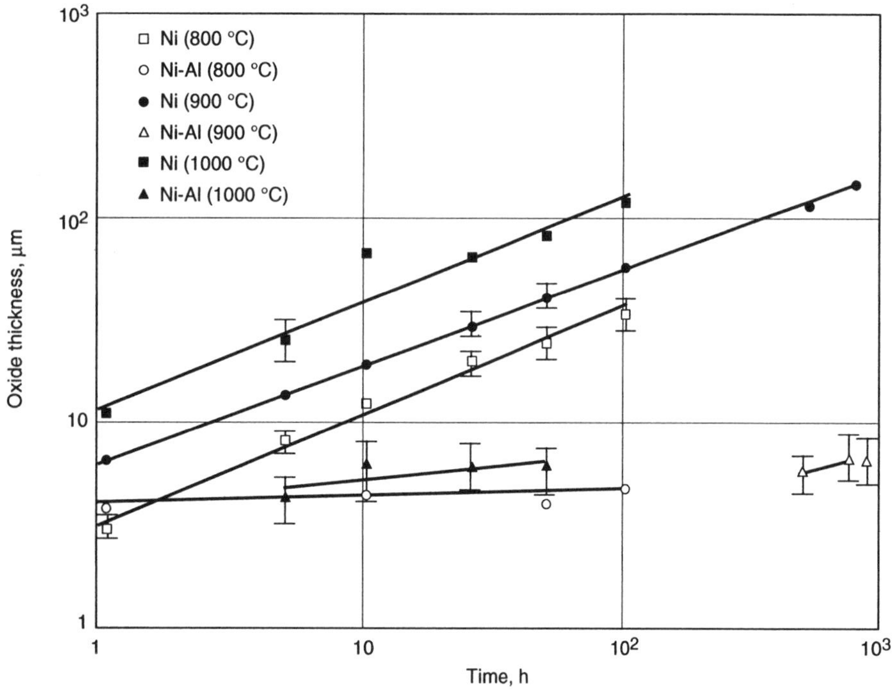

Fig. 8 Oxide depth versus time plotted on log-log scale for pure nickel and nickel-alloy coatings exposed in air at 800, 900, and 1000 °C. Source: Ref 49

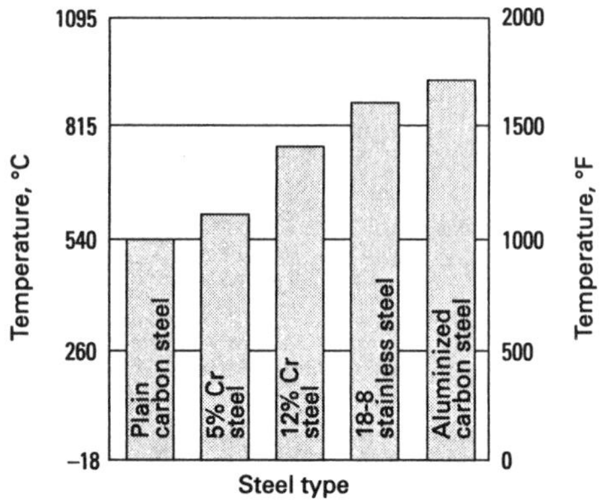

Fig. 9 Oxidation of steels in air at the temperature at which scaling is less than 10 mg/cm^2. Source: Ref 58

loys. However, after heat treatment the corrosion resistance can deteriorate.

Composite deposition plating is a further extension of aqueous solution electroplating or electroless coatings in that particles or fibers are suspended in the electrolyte, then occluded in the deposit. Oxides, carbides, silicides, refractory powder, metallic powder, and organic powder can be introduced into the electrolyte. The most widely used electrodeposited composites are cermet coatings, with alumina (Al_2O_3), zirconia (ZrO_2), titania (TiO_2), and silicon carbide (SiC) added to increase strength, hardness, and wear resistance (Table 10). The amount of ceramic particles incorporated in the coating depends on the current density and the bath loading, that is, the amount of particulate in the suspension. It has been shown that coatings up to 40 vol% Al_2O_3 were produced at 0.5 A/dm^2 and 5.3 vol% bath loading. The amount of Al_2O_3 incorporated into the coating was seen to decrease with increasing current density and decreasing bath loading (Ref 48). Hardness values ranged from approximately 250 to 580 HVN, depending on the amount of Al_2O_3 incorporated.

Metallic particles such as chromium can be introduced into a metal plating electrolyte (for example, nickel and cobalt), and the deposited composite can be subsequently heat treated to form high-temperature oxidation-resistant alloys. MCrAlY composites have been made by depositing 10 µm CrAlY powder in a cobalt or nickel matrix. Heat treatment bonds the coating to the substrate and interdiffuses the cobalt and nickel with the CrAlY particles. The CoCrAlY coatings produced have been shown to be superior to some plasma sprayed CoCrAlY and pack aluminized coatings (Ref 34). Aluminum particles have been codeposited in a nickel matrix and subsequently heat treated to produce a nickel-aluminum intermetallic coating with exceptional oxidation resistance (Ref 49) (Fig. 8).

Polytetrafluoroethylene (PTFE), diamond, and SiC particles can also be incorporated into a nickel electroless plating for improved properties. Diamond and silicon carbide are used to enhance abrasion resistance; the surface hardness of these composites is reported to be 1300 HVN. Polytetrafluoroethylene is added to the electroless nickel bath to provide a composite coating with enhanced lubrication. Almost any particulate material can be deposited in a metallic matrix, provided the particles are sufficiently small to remain suspended in the bath and that the particles do not react chemically with the bath during electrodeposition.

Chemical Vapor Deposition and Related Processes

Chemical vapor deposition (CVD) involves the formation of a coating by the reaction of the coating substance with the substrate. The coating species can come from a gas or gases or from contact with a solid as in the pack cementation process. The process is more precisely defined as the deposition of a solid on a heated surface by a chemical reaction from the vapor or gas phase (Ref 50). In general, three processing steps are involved in any CVD reaction: (1) the production of a volatile carrier compound, (2) the transport of the gas to the deposition site without decomposition, and (3) the chemical reaction necessary to produce the coating on the substrate. The numerous chemical reactions used in CVD include thermal decomposition (pyrolysis), reduction, hydrolysis, disproportionation, oxidation, carburization, and nitridation. These reactions take place singly or in combination and are controlled by thermodynamics, kinetics, mass transport, chemistry of the reaction, and processing parameters of temperature, pressure, and chemical activity.

Chemical vapor deposition processes can be classified as either open reactor systems, including thermal CVD and plasma CVD, or as a closed-reactor system, as in pack cementation. In thermal CVD, reactions usually take place above 900 °C (1650 °F), whereas plasma CVD usually operates at temperature between 300 and 700 °C (570 and 1290 °F) (Table 11). Using the lower-reaction-temperature plasma CVD enables coatings to be produced on substrates with low melting points or that otherwise would undergo solid-state transformations over the range of deposition temperatures. Furthermore, the low deposition temperature of plasma CVD coatings limits the stresses due to the large mismatches in thermal expansion that can lead to cracking and delamination of the coating.

Materials that cannot ordinarily be deposited by electrodeposition—for example, the refractory metals tungsten, molybdenum, rhenium, niobium, tantalum, zirconium, hafnium, and so forth—are deposited using CVD processes. Typical products produced are crucibles, rocket nozzles, and other high-temperature components; linings for chemical vessels; and coatings for electronic components. These refractory metals are deposited at temperatures far below their melting points or sintering temperatures, and coatings can be produced with a preferred grain size and grain orientation. For example, tungsten that is deposited by the hydrogen reduction of the halide and deposition at a lower temperature (500 °C, or 930 °F) gives a finer grain size with higher strength (83 MPa, or 12 ksi) than deposition at a higher temperature (700 °C, or 1290 °F) (Ref 50).

The deposition of ceramics involves many different compounds, for example, titanium carbide (TiC), titanium nitride (TiN), Al_2O_3, diamond, and so forth. Table 12 lists the properties of typical CVD coating materials for wear and corrosion resistance. Coatings for the cutting tool industry utilize CVD processes, particularly TiC coatings for cemented tungsten carbide tools and TiN and carbonitride coatings for high-speed tool steels and cemented carbide tools (Ref 50). Nearly all coatings are multilayer systems that combine TiN for lubricity and galling resistance, Al_2O_3 for chemical inertness and thermal insulation, and TiC and carbonitride for abrasion resistance. A comparison of coatings for cutting tools is given in Table 13.

Diamond films grown by CVD exhibit outstanding properties approaching natural diamond, such as high electrical resistivity, high optical transparency, extreme hardness, high refractive index, and chemical inertness. Different film deposition techniques and system configurations result in films with different characteristics. Diamond films can be grown using processing variables of different concentrations of methane in methane-hydrogen gas mixtures and flow rates (Ref 52). The CVD of diamond requires the presence of atomic hydrogen, which selectively removes graphite and activates and stabilizes the diamond structure. The basic reaction involves the decomposition of methane, which can be activated by microwave plasma, thermal means (hot filament), plasma arc, or laser.

The pack cementation process originally involved pack carburizing, which is the process of diffusing carbon into the surface of iron or low-carbon steel by heating in a closed container filled with activated charcoal. The simplest and oldest carburizing process involves filling a welded sheet metal or plate box with granular charcoal that is activated with chemicals such as barium carbonate to assist the formation of carbon monoxide (CO). In the heated box, charcoal forms carbon dioxide (CO_2), which converts to CO in an environment with an excess of carbon. The CO then forms atomic carbon at the component surface and diffuses into the part. The pack carburization process is of little commercial importance, although it is still done occasionally. It has given rise to other CVD pack diffusion processes including aluminizing, siliconizing, chromizing, and boronizing.

Pack cementation is a modified CVD deposition batch process that involves heating a closed/vented pack to an elevated temperature (e.g., 1050 °C, or 1920 °F) for a given time (e.g., 16 h) during which a diffusional coating is produced (Ref 53). The traditional pack consists of four components: the substrate or part to be coated, the master alloy (i.e., a powder of the element or elements to be deposited on the surface of the part), a halide salt activator, and relatively inert filler powder. The master alloy, the filler, and halide activator are thoroughly mixed together, and the part to be coated is buried in this mixture in a retort (Ref 54). When the mixture is heated, the activator reacts to produce an atmosphere of source element(s) halides that diffuse into the pack and transfer the source element(s) to the substrate on which the coating is formed (Ref 55). A summary of CVD pack diffusion process characteristics is given in Table 14.

Aluminizing. An aluminizing pack cementation process is commercially practiced for a range

Table 13 Criteria for selecting coating materials for cutting tools

Property	Best materials(a)
Oxidation and corrosion resistance; high-temperature stability	Al_2O_3, TiN, TiC
Crater-wear resistance	Al_2O_3, TiN, TiC
Hardness and edge retention	TiC, TiN, Al_2O_3
Abrasion resistance and flank wear	Al_2O_3, TiC, TiN
Low coefficient of friction and high lubricity	TiN, Al_2O_3, TiC
Fine grain size	TiN, TiC, Al_2O_3

(a) For each property, best material is identified first. Source: Ref 51

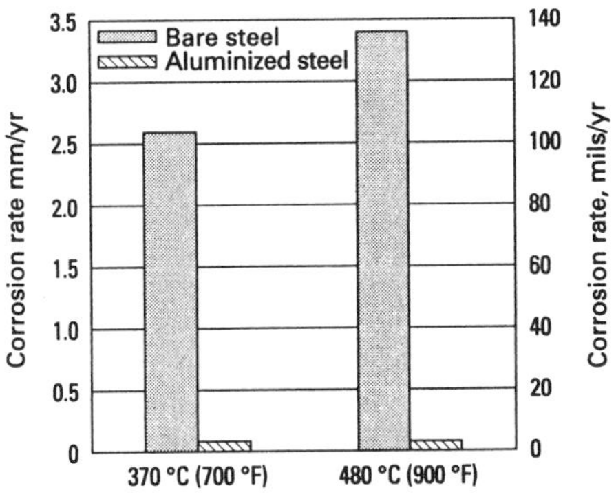

Fig. 10 Relative corrosion rates of 9Cr-1Mo alloy steel in 5 mol% H_2S at 3550 kPa (515 psi) for 300 h. Source: Ref 59

Table 14 Typical characteristics of pack cementation processes

Process	Nature of case	Process temperature, °C (°F)	Typical case depth	Case hardness, HRC	Typical base metals	Process characteristics
Aluminizing (pack)	Diffused aluminum	870–980 (1600–1800)	25 μm–1 mm (1–40 mils)	<20	Low-carbon steels	Diffused coating for oxidation resistance at elevated temperatures
Siliconizing by chemical vapor deposition	Diffused silicon	925–1040 (1700–1900)	25 μm–1 mm (1–40 mils)	30–50	Low-carbon steels	For corrosion and wear resistance, atmosphere control is critical
Chromizing by chemical vapor deposition	Diffused chromium	980–1090 (1800–2000)	25–50 μm (1–2 mils)	Low-carbon steel, <30; high-carbon steel, 50–60	High- and low-carbon steels	Chromized low-carbon steels yield a low-cost stainless steel; high-carbon steels develop a hard corrosion-resistant case
Titanium carbide	Diffused carbon and titanium, TiC compound	900–1010 (1650–1850)	2.5–12.5 μm (0.1–0.5 mil)	>70(a)	Alloy steels, tools steels	Produces a thin carbide (TiC) case for high resistance wear; temperature may cause distortion
Boriding	Diffused boron, boron compounds	400–1150 (750–2100)	12.5–50 μm (0.5–2 mils)	40–70	Alloy steels, tool steels, cobalt and nickel alloys	Produces a hard compound layer, mostly applied over hardened tool steels; high process temperature can cause distortion

(a) Requires quench from austenizing temperature. Source: Ref 56

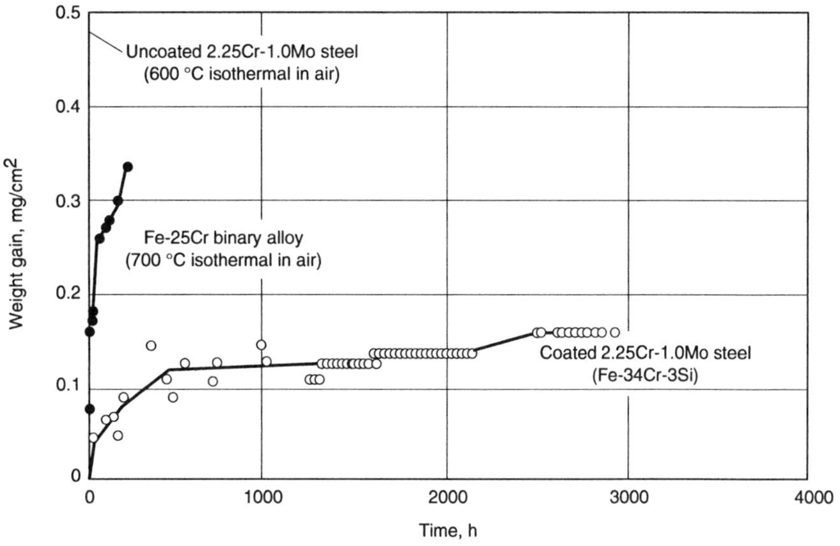

Fig. 11 A plot of weight gain versus time for the cyclic oxidation in static air at 700 °C (1290 °F) of coated samples with surface compositions of Fe-34 Cr-3 Si. Source: Ref 63

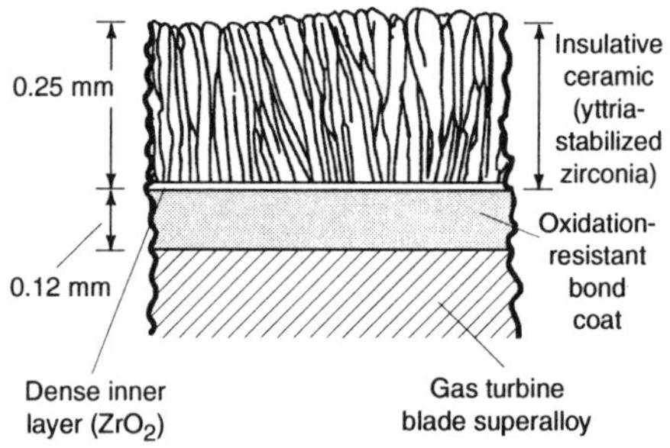

Fig. 12 Cross section illustrating the strain-tolerant columnar ZrO_2 microstructure of EB/PVD zirconia thermal barrier coatings. Source: Ref 69

of alloys, including nickel- and cobalt-base superalloys, steels, and copper. Simple aluminide coatings resist high-temperature oxidation by the formation of an alumina protective layer and can be used up to about 1150 °C (2100 °F), but the coating can degrade by spallation of the oxide during thermal cycling. For extended periods of time at temperatures in excess of 1000 °C (1830 °F), interdiffusion of the coating will cause further degradation, and therefore practical coating life is limited to operating temperatures of 870 to 980 °C (1600 to 1800 °F). Pack compositions, process temperatures, and process times depend on the type of base material to be aluminized and fall into the following classifications (Ref 57):

Class	Alloy
I	Carbon and low-alloy steels
II	Ferritic and martensitic stainless steels
III	Austenitic stainless steels with 21–40% Ni and iron-base superalloys
IV	Nickel- and cobalt-base superalloys

As a general rule, overall aluminum diffusion is slowed as the nickel, chromium, and cobalt contents increase. Thus, higher temperatures and longer processing times are required to produce greater aluminum diffusion thicknesses as the base material increases in alloy content.

Stainless steels are oxidation resistant as a result of the formation of a thin chromium-rich oxide on the component surface. A similar reaction occurs in aluminized steels in which a thin slower-growing aluminum-rich oxide forms. Unlike chromium oxide, Al_2O_3 does not exhibit volatility in the presence of oxygen above 927 °C (1700 °F). Figure 9 compares an aluminized carbon steel with several alloys at a temperature in which scaling remains less than 10 mg/cm^2 for oxidation in air. In sulfidizing environments, pack aluminized coatings have excellent resistance to corrosive attack. In contrast to stainless steels, the aluminum-rich surface (50% Al) and diffusion zone (20% Al min) of the coating is far more resistant than chromium to sulfidation corrosion. Figure 10 compares the corrosion rates of bare and aluminized 9Cr-1Mo steel in a hydrogen sulfide (H_2S) environment.

Siliconizing, the diffusion of silicon into steel, occurs similarly to aluminizing. There are pack and retort processes in which parts are subjected to gas atmospheres that react with the heated part surface to produce nascent silicon that diffuses into the substrate to be coated. In a pure silicon pack that is activated with NH_4Cl, $SiCl_4$ and $SiHCl_3$ gases form, which are reduced by hydrogen gas to deposit elemental silicon on the surface of the part (Ref 55). Another process involves tumbling parts in a retort with SiC. When a temperature of 1010 °C (1850 °F) is reached, silicon tetrachloride gas is introduced, which reacts with the part and the SiC particles to produce a concentration gradient of silicon on the part surface as the silicon diffuses into the substrate. The process normally takes place on low-carbon steels, and these steels develop case depths up to 1 mm (0.040 in.) with a silicon content of 13 wt% (Ref 60). Case depths developed on these siliconized steels have hardnesses of about 50 HRC and therefore can be used for wear resistance. The presence of silicon on the surface allows for the formation of a stable silicon dioxide (SiO_2) phase in oxidizing environments and excellent corrosion resistance.

Chromizing. Chromium can be applied in the same manner as aluminum and silicon to produce a chromium-rich coating, and many of the same principles of aluminizing packs apply to chromizing packs. Parts are packed in chromium powder with an inert filler such as aluminum oxide. A halide salt activator is added that changes to the vapor phase at the processing temperature and serves as a carrier gas to bring chromium to the surface of the part. Diffusion coatings can be formed on nickel-base superalloys by pack cementation using ammonium chloride as a chromium-alumina activator. These coatings usually contain 20 to 25 wt% Cr at the outer surface and involve approximately equal rates of interdiffusion of chromium and nickel. Significant depletion of aluminum and titanium from the alloy surface occurs, thus producing a coating that is a solid solution of the chromium in the remaining nickel-base superalloy. The deposited coating is usually overlaid with a thin layer of α-chromium, which must be removed chemically (Ref 55).

In low-alloy steels, it has been shown that chromizing is much more complex, leading to microstructures that may behave detrimentally in some environments (Ref 61). In a chromized 2.25Cr-1Mo alloy, the coating contains a thin outer layer (~5 μm) of mostly chromium (>80 wt%), which is essential for corrosion protection. Large columnar ferritic grains, containing between 30 and 15 wt% Cr, are found beneath the outer layer. The columnar grain boundaries, as well as the boundary between the outer chromium-rich layer and the columnar grains, are decorated with chromium carbides that were found to contribute to coating degradation. A layer of Kirkendall voids (also decorated by carbides), iron carbides at the coating/substrate interface, and a large decarburized zone in the substrate are also produced by the process.

Evaluation of samples exposed to a fossil-fired boiler up to two years revealed two degradation mechanisms: cracking and sulfidation corrosion. Cracking of the outer coating layer allowed ingress of sulfur, resulting in intergranular sulfidation corrosion attack. Once the outer protective chromium layer has been breached, the columnar grain boundary orientation promotes crack initiation and propagation along the carbides when the tube is subjected to axial thermal loading. However, it should be noted that chromized coatings have been used up to 10 years in some fossil-fired boilers. Therefore, stress and environmental conditions are critical to the successful use of these pack cementation coatings, as long as the effect of the processing thermal cycle on the coating and substrate morphology is understood.

Boriding, or boronizing, is a thermochemical surface-hardening process that can be applied to a wide variety of ferrous, nonferrous, and cermet materials. The boronizing pack process is similar

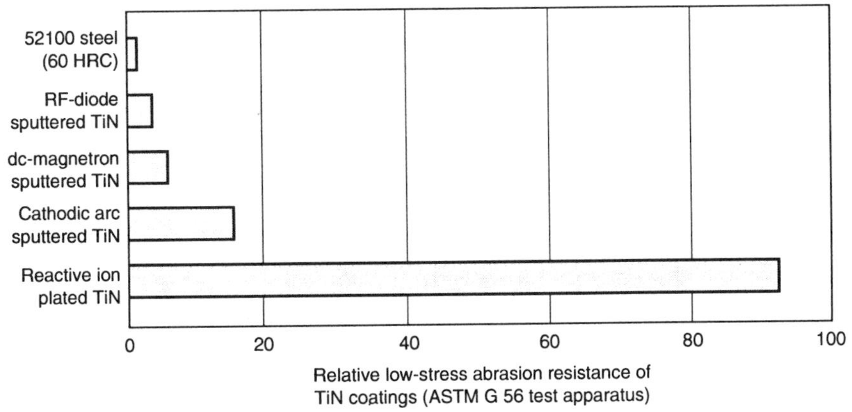

Fig. 13 Effect of coating technique on the relative abrasion resistance of TiN on hardened steel applied by various processes. Source: Ref 71

to pack carburizing with the parts to be coated being packed with a boron-containing compound such as boron powder or ferroboron. Activators such as chlorine and fluorine compounds are added to enhance the production of the boron-rich gas at the part surface. Processing of high-speed tool steels that were previously quenched hardened is accomplished at 540 °C (1000 °F). Boronizing at higher temperatures up to 1090 °C (2000 °F) causes diffusion rates to increase, thus reducing the process time. The boron case does not have to be quenched to obtain its high hardness, but tool steels processed in the austenitizing temperature range need to be quenched from the coating temperature to harden the substrate.

Boronizing is most often applied to tool steels or other substrates that are already hardened by heat treatment. The thin (12 to 15 μm) boride compound surfaces provide even greater hardness, improving wear service life. Distortion from the high processing temperatures is a major problem for boronized coatings. Finished parts that are able to tolerate a few thousandths of an inch (75 μm) distortion are better suited for this process sequence because the thin coating cannot be finish ground (Ref 60).

Multicomponent Coatings. Multicomponent pack cementation coatings have been developed from chromium/aluminum and chromium/silicon processing for high-temperature corrosion resistance. In general, simple binary alloys—for example, iron-chromium, nickel-aluminum, and so forth—are not as effective for oxidation resistance as a ternary alloy using the interaction of two oxidation-resistant elements. Considering an Fe-20Cr-5Al alloy exposed to a high-temperature environment, a continuous adherent chromia (Cr_2O_3) scale initially is formed, which prevents the rapid oxidation of the iron (Ref 62). The Cr_2O_3 scale also prevents inward diffusion of the oxygen, negating internal oxidation of aluminum, and allows for the formation of a much more stable but slower-growing and protective Al_2O_3. Similar results were found for an Fe-30Cr-3Si alloy in which a Cr_2O_3 scale formed initially during exposure to an oxidizing environment, preventing the rapid oxidation of iron and the internal oxidation of silicon and allowing a very slow protective SiO_2 film to form beneath the Cr_2O_3 (Ref 62).

The codeposition of two or more elements in a halide-activated cementation pack is inherently difficult because of the large differences in the thermodynamic stabilities of their volatile halides. A variety of alloys have been coated with chromium/aluminum (low-carbon and alloy steels; types 410, 304, and 316 stainless steels; and nickel-base superalloys) and chromium/silicon (low-carbon and low-alloy steels, types 304 and 409 stainless steels, Incoloy 800). Each alloy requires a specific pack chemistry to obtain optimal coating composition for superior oxidation resistance. Details of the required pack compositions for each coating/alloy combination are reviewed in Ref 63.

During chromizing/aluminizing of low-alloy steels, a ferrite layer is initially formed on the steel allowing for subsequent diffusion of chromium and aluminum into the ferrite matrix substrate. Rejection of carbon into the ferrite/austenite interface at the reaction temperature minimized decarburization and prevents the formation of chromium carbides at the surface. The resulting chromium/aluminum coating provides excellent isothermal and cyclic oxidation resistance at 700 °C (1290 °F) by the formation of an Al_2O_3 scale. Similar results were found for a chromizing/silicon process in which improvement over an uncoated steel was significant (Fig. 11). Even in austenitic stainless steels and superalloys the key element in the reduction of oxidation kinetics is the presence of ferrite at the surface, with its higher diffusion coefficients. The high diffusivity for ferrite promotes the early formation and retention of the thermodynamically more stable oxide scales in chromium/silicon coatings, that is, chromia rather than iron oxide at an early stage of oxidation and silica rather than chromia at steady state.

Physical Vapor Deposition Processes

Physical vapor deposition (PVD) processes involve the formation of a coating on a substrate by physical deposition of atoms, ions, or molecules of the coating species (Ref 64). There are three main techniques for applying PVD coatings: thermal evaporation, sputtering, and ion plating. Thermal evaporation involves heating of the material until it forms a vapor that condenses on a substrate to form a coating. Sputtering involves the electrical generation of a plasma between the

Table 15 Comparison of PVD process characteristics

Process	Processing temperature, °C	Throwing power	Coating materials	Coating applications and special features
Vacuum evaporation	RT–700, usually <200	Line of sight	Chiefly metal, especially Al (a few simple alloys/a few simple compounds)	Electronic, optical, decorative, simple masking
Ion implantation	200–400, best <250 for N	Line of sight	Usually N (B, C)	Wear resistance for tools, dies, etc. Effect much deeper than original implantation depth. Precise area treatment, excellent process control
Ion plating, ARE	RT–0.7 T_m of coating. Best at elevated temperatures	Moderate to good	Ion plating: Al, other metals (few alloys) ARE: TiN and other compounds	Electronic, optical, decorative. Corrosion and wear resistance. Dry lubricants. Thicker engineering coatings
Sputtering	RT–0.7 T_m of metal coatings. Best >200 for nonmetals	Line of sight	Metals, alloys, glasses, oxides. TiN and other compounds(a)	Electronic, optical, wear resistance. Architectural (decorative). Generally thin coatings. Excellent process control
CVD	300–2000, usually 600–1200	Very good	Metals, especially refractory TiN and other compounds(a), pyrolytic BN	Thin, wear-resistant films on metal and carbide dies, tools, etc. Free-standing bodies of refractory metals and pyrolytic C or BN

RT, room temperature; ARE, activated reactive evaporation; T_m, absolute melting temperature. (a) Compounds: oxides, nitrides, carbides, silicides, and borides of Al, B, Cr, Hf, Mo, Nb, Ni, Re, Si, Ta, Ti, V, W, Zr. Source: Ref 65 coating species and the substrate. Ion plating is essentially a combination of these two processes. A fourth type of process is ion implantation, which does not produce a coating; instead, high energy ions are made to penetrate into and modify the substrate surface. A comparison of the process characteristics is provided in Table 15. Reviews of these processes can be found in the literature (for example, Ref 64, 66, and 67).

Originally PVD was used to deposit single metal elements by transport of a vapor in a vacuum without involving a chemical reaction. Today, PVD technology has evolved so that a wide array of inorganic materials (including metals, alloys, compounds, or their mixtures) and organic compounds can be deposited. The PVD process occurs in a vacuum chamber and involves a vapor source and the substrate on which deposition occurs. Different techniques arise because of variations in atmospheres, vapor source heating method, and electrical voltage of the substrate, all of which contribute to the structure, properties, and deposition rate of the coating (Ref 67). The steps in deposition occur as follows:

1. Synthesis of the material deposited (transition from a condensed state, solid or liquid, to the vapor phase, or, for deposition of compounds, reaction between the components of the compound, some of which may be introduced into the chamber as a gas or vapor)
2. Vapor transport from the source to the substrate
3. Condensation of the vapors followed by film nucleation and growth

The PVD process produces coatings for a range of applications including electronics, optics, decoration, and corrosion and wear prevention. Only engineering uses, that is, corrosion- and wear-resistant coatings, are discussed below. The coatings used for wear applications are usually hard compounds, and, from the designers point of view, thin-film wear coatings can be used for the same type of applications as chromium electroplate. Physical vapor deposition coatings have hardnesses greater than any metal and are used in systems that cannot tolerate even microscopic wear losses. Most processes are operated on a batch basis, and the component size is limited by the size of the vacuum chamber. Provided that the substrate can be manipulated to face the coating source, the size and shape of objects are limited by the capital and operating expenditures involved rather than by the fundamental characteristics of the process. Furthermore, cleanliness of the substrate is critical and far exceeds surface preparation requirements for other coatings.

Thermal evaporation is the oldest and probably the most widely used PVD technique. It accounts for the major proportion of both the equipment in use and the area coated. Thermal evaporation occurs in a hard vacuum of 0.1 to 10 mPa, at which pressures the mean free path of a gas atom, that is, the average distance the atom travels before colliding with another atom, is greater than the chamber dimensions. An atom evaporating from a source travels in a straight line; thus the process is line-of-sight limited, and coating around corners or reentrant angles is not possible without substrate manipulation. A description of the process and the equipment is found in Ref 68.

Aluminum and chromium coatings for automotive trim are probably the largest application of this process. Hard coatings of chromium and Al_2O_3 are sometimes deposited on steel or tungsten carbide tools, but this application has been replaced by sputter coatings. Aerospace applications use aluminum and nickel-chromium for corrosion protection and aluminum and silver as solid lubricants. Electron beam/physical vapor deposition (EB/PVD) is widely used in the gas turbine industry for applying MCrAlY metallic coatings on turbine blades and vanes for oxidation and corrosion protection (Ref 69). Zirconia thermal barrier coatings (TBCs) can also be deposited using this technique. Bond coats for EB/PVD TBCs are normally MCrAlY type coatings, similarly processed. The major advantage of EB/PVD TBCs is the columnar outer structure of the ZrO_2 (Fig. 12), which reduces the stress buildup within the body of the coating. These TBCs have superior degradation resistance that has been confirmed in gas turbine flight tests. For example, EB/PVD zirconia was found to be better than plasma sprayed zirconia or metallic MCrAlY after 4200 h tests of first-stage blades.

Table 16 Research and development applications for ion implantation

Surface properties modified	Substrates studied	Ions species used	Comments
Wear	Steels, WC, Ti, Co/Cr alloys, TiN coatings, electroplated Cr	N, C 10–20 at.% $\geq 10^{17}$ ions/cm^2	Ti,Co/Cr alloys largest use commercially in orthopedic devices
Friction	Steels	Ti plus C implants $\geq 10^{17}$ ions/cm^2	Dual implants give amorphous surface layer
Fatigue	Ti alloys, steels	N, C $\geq 10^{17}$ ions/cm^2	Implantation effective for surface initiated fatigue
Fracture toughness	Ceramics: Al_2O_3, TiN	Ar 10^{15}–10^{17} ions/cm^2	Radiation damage critical; ion-induced compressive stress helpful
Aqueous corrosion catalysis	Steels, Ti alloys, Pt	Cr, Ta, Cr+P $\geq 10^{17}$ ions/cm^2	Ion implant can mimic "normal" alloys; amorphous and unique surface alloys possible
Oxidation	Superalloys	Y, Ce $\geq 10^{15}$ ions/cm^2	Low effective doses; implanted species stay at metal-oxide interface
Electrical conductivity	Polymers	Ar, F 10^{15}–10^{17} ions/cm^2	Permits chain scissoning, doping; conductivity approaches disordered metal levels
Optical: refractive index	Glasses, electrooptics	Li, Ar 10^{15}–10^{17} ions/cm^2	Chemical doping and lattice disorder both important

Source: Ref 73

Table 17 Response of steels and cast irons to flame hardening

Material	Typical hardness, HRC, as affected by quenchant		
	Air(a)	Oil(b)	Water(b)
Plain carbon steels			
1025–1035	...	...	33–50
1040–1050	...	52–58	55–60
1055–1075	50–60	58–62	60–63
1080–1095	55–62	58–62	62–65
1125–1137	...	...	45–55
1138–1144	45–55	52–57(c)	55–62
1146–1151	50–55	55–60	58–64
Carburized grades of plain carbon steels(d)			
1010–1020	50–60	58–62	62–65
1108–1120	50–60	60–63	62–65
Alloy steels			
1340–1345	45–55	52–57(c)	55–62
3140–3145	50–60	55–60	60–64
3350	55–60	58–62	63–65
4063	55–60	61–63	63–65
4130–4135	...	50–55	55–60
4140–4145	52–56	52–56	55–60
4147–4150	58–62	58–62	62–65
4337–4340	53–57	53–57	60–63
4347	56–60	56–60	62–65
4640	52–56	52–56	60–63
52100	55–60	55–60	62–64
6150	...	52–60	55–60
8630–8640	48–53	52–57	58–62
8642–8660	55–63	55–63	62–64
Carburized grades of alloy steels(d)			
3310	55–60	58–62	63–65
4615–4620	58–62	62–65	64–66
8615–8620	...	58–62	62–65
Martensitic stainless steels			
410, 416	41–44	41–44	...
414, 431	42–47	42–47	...
420	49–56	49–56	...
440 (typical)	55–59	55–59	...
Cast irons (ASTM classes)			
Class 30	...	43–48	43–48
Class 40	...	48–52	48–52
Class 45010	...	35–43	35–45
50007, 53004, 60003	...	52–56	55–60
Class 80002	52–56	56–59	56–61
Class 60-45-15	...	...	35–45
Class 80-60-03	...	52–56	55–60

(a) To obtain the hardness results indicated, those areas not directly heated must be kept relatively cool during the heating process. (b) Thin sections are susceptible to cracking when quenched with oil or water. (c) Hardness is slightly lower for material heated by spinning or combination progressive-spinning methods than it is for material heated by progressive or stationary methods. (d) Hardness values of carburized cases containing 0.90 to 1.10% C. Source: Ref 75

Sputter coating is a vacuum process that involves the use of ions from a gas-generated plasma to dislodge coating atoms or molecules from a target made of the material that will become the coating. The plasma is established between the target and the substrate by the application of a direct current potential or an alternating potential (radio frequency). An inert gas is introduced into the chamber to form the glow discharge plasma between the electrodes. The materials that can be sputter coated are pure metals, alloys, inorganic compounds, and some polymeric materials. A major restriction to be considered for the substrate material is the temperature of the process, which can range from 260 to 540 °C (500 to 1000 °F). Details of the process and equipment have been reviewed (Ref 70). Sputtering is often used for depositing compounds and materials that are difficult to coat by thermal evaporation techniques. Engineering applications of sputter coatings include:

- Corrosion and oxidation resistance, for example, nickel-chromium, MCrAlY, and polymers
- Lubrication, for example, silver, indium, MoS_2, PTFE, selenides, silicides, and tellurides
- Wear resistance, for example, TiN, other nitrides, tungsten, molybdenum, carbides, borides, and diamondlike carbon

Titanium nitride coatings are generally used for wear resistance, and Fig. 13 shows that TiN coatings increase the abrasion resistance of a hardened steel (Ref 64).

Ion plating is a vacuum coating process in which a portion of the coating species impinges on the substrate in ionic form (Ref 64). The process is a hybrid of the thermal evaporation process and sputtering with the evaporation rate being maintained at a higher rate than the atoms that can be sputtered from the substrate. Some evaporant atoms pass through the plasma in atomic form, while some atoms collide with electrons from the substrate and become ions. They impinge on the substrate in ionic form, pick up electrons and return to the atomic state, forming the coating. A detailed description of the process and equipment is provided in Ref 72.

A variant process is reactive ion plating in which the metallic constituent (titanium) of the compound (TiN) is evaporated into the reactive gas mixture of argon and nitrogen that is enhanced by the glow discharge, depositing a golden-colored TiN coating on the substrate. Films of TiN are applied to a wide range of tools such as bits, punches, dies, taps, and so forth, to improve tool life by three to ten times (Ref 69). Figure 13 shows the improvement in abrasion resistance of reactive ion plating over other PVD sputtering processes. The results show that the higher-energy reactive ion plating process had more than an order of magnitude improvement over the radio-frequency diode sputtered coatings (Ref 64). These results indicate that application technique should be made a coating selection factor.

Ion implantation involves the bombardment of a solid material with medium- to-high-energy ionized atoms and offers the ability to alloy virtually any elemental species into the near-surface region of any substrate (Ref 73). The advantage of such a process is that it produces improved surface properties without the limitations of dimensional changes or delamination found in conventional coatings. During implantation, ions come to rest beneath the surface in less than 10 to 12 s, producing a very fast quench rate and allowing the development of nonequilibrium surface alloys or compounds. Details of the process and associated equipment are documented in Ref 73. Ion implantation is commercially applied to various steels, tungsten carbide/cobalt materials, and alloys of titanium, nickel, cobalt, aluminum and chromium, although applications are restricted to temperatures below 250 °C (480 °F) for steels and 450 °C (840 °F) for carbides.

Table 16 lists some of the applications for the ion-implantation process. Ion-implantation surfaces produce exceptional results in reducing wear, friction, and corrosion. Commercial applications involve tooling, bearings, and biomedical components. Nitrogen implantation, especially in alloy surfaces containing elements forming stable nitrides, has found use in tools and dies such as cobalt-cemented tungsten carbide wire drawing inserts. Nitrogen implantation has been especially successful in increasing the life (up to 20 times) of tools and parts used in the manufacture of injection-molded plastics. Titanium and cobalt-chromium alloy orthopedic prostheses for hip and knee joints are among the most successful commercial applications for ion-implantation components for wear resistance.

Heat Treatment Coatings

Several types of coatings involve a heat treatment either after processing or as part of the coating process. These processes involve (1) surface hardening in steels, where localized heat treatment of the surface component occurs, and (2) diffusion coatings, in which the surface composition is modified. These composition modifications can be produced in processes involving carburizing and/or nitriding of steel component surfaces when interstitial atoms of carbon or nitrogen are diffused into the steel surface at high temperature in order to increase the surface hardness, strength, and fatigue resistance.

Surface Hardening

Surface hardening in a general sense involves many processes that improve the wear resistance of parts while utilizing the tough interior properties of the steel component. In this case, surface hardening is limited to localized heat-treating processes that produce a hard quenched surface without introducing additional alloying species. This approach consists of hardening the surface by flame, induction, laser beam, or electron beam heating. More detailed information on surface hardening of steels can be found in Ref 74.

Flame hardening consists of austenitizing the surface of steel by heating with an oxyacetylene or oxyhydrogen torch and immediately quenching with water (Ref 74). After quenching, the microstructure of the surface layer consists of hard martensite over a lower-strength interior core of other steel morphologies such as ferrite and pearlite. A prerequisite for proper flame hardening is that the steel must have adequate carbon and other alloy additions to produce the desired hardness, because there is no change in composition. Flame-hardening equipment utilizes direct impingement of a high-temperature flame or high-velocity combustion product gases to austenitize the component surface and quickly cool the surface faster than the critical cooling

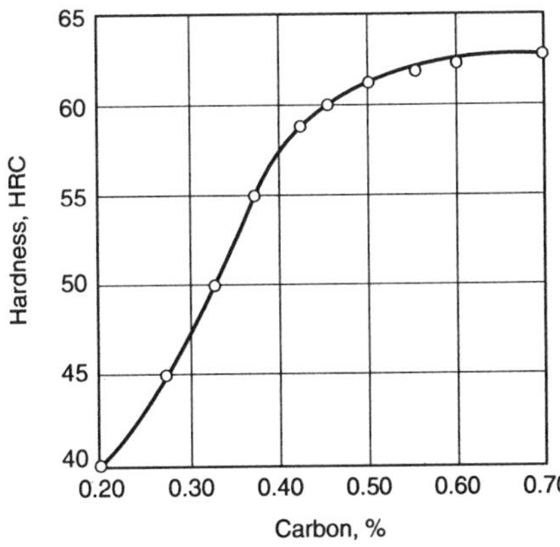

Fig. 14 Relationship of carbon content to minimum surface hardness attainable by flame or induction heating and water quenching. Practical minimum carbon contents can be determined from this curve. Source: Ref 75

Table 18 Comparison of flame- and induction-hardening processes

Characteristic	Flame	Induction
Equipment	Oxyfuel torch, special head quench system	Power supply, inductor, quench system
Applicable material	Ferrous alloys, carbon steels, alloy steels, cast irons	Same
Speed of heating	Few seconds to few minutes	1 to 10 s
Depth of hardening	1.2 to 6.2 mm (0.050 to 0.250 in.)	0.4 to 1.5 mm (0.015 to 0.060 in.); 0.1 mm (0.004 in.) for impulse
Processing	One part at a time	Same
Part size	No limit	Must fit in coil
Tempering	Required	Same
Can be automated	Yes	Yes
Operator skills	Significant skill required	Little skill required after setup
Control of process	Attention required	Very precise
Operator comfort	Hot, eye protection required	Can be done in suit
Cost		
Equipment	Low	High
Per piece	Best for large work	Best for small work

Source: Ref 76

rate to produce martensite in the steel. This is necessary because the hardenability of the component is fixed by the original composition of the steel. Thus, equipment design is critical to success of the operation. Flame-heating equipment may be a single torch with a specially designed head or an elaborate apparatus that automatically indexes, heats, and quenches parts. With improvements in gas-mixing equipment, infrared temperature measurement and control, and burner rig design, flame hardening has been accepted as a reliable heat-treating process that is adaptable to general or localized surface hardening for small or medium-to-high production requirements (Ref 35).

The flame-hardening process is used for a wide variety of applications (Ref 74). These include (1) parts that are so large that conventional furnace treatments are impractical or uneconomical, (2) prevention of detrimental treatment of the entire component when only small segments of the part require heat treatment, and (3) use of less costly material to obtain the desired surface properties where alloyed steels would be normally applied. Flame hardening is limited to hardenable steels (wrought or cast) and cast iron. Typical hardnesses obtained for the flame-hardened grades depend on the quench media (Table 17). The practical level of minimum surface hardness attainable with water quenching for various carbon contents is shown in Fig. 14.

Induction hardening is a versatile heating method that involves placing a steel part in the magnetic field generated by high-frequency alternating current passing through an inductor, usually a water-cooled copper coil. The depth of hardening increases as the frequency of the alternating current decreases. Other variables important to the process include the coil current, heating time, and the coil design (Ref 76). The specific ferrous alloys that are commonly used in induction surface hardening are the same as those used in flame hardening (Table 17). The minimum carbon contents to obtain specific surface hardness are also shown in Fig. 14. Electrical properties of the alloy are an important consideration when selecting induction treatment as a surface-hardening technique. In induction hardening the electrical resistivity and magnetic properties of the alloy can produce significant differences in heating characteristics. Thus, different steels require differing induction-heating parameters. Table 18 compares the flame- and induction-hardening processes.

Electron- and laser-beam methods use high-energy beams to heat treat the surface of hardenable steel. The EB heat treating process uses a concentrated beam of high-velocity electrons as an energy source to heat selected parts of the steel component. In laser heat treatment a laser beam is used to harden localized areas of ferrous parts. These processes are similar to flame and induction hardening, except that the need for quenchants is eliminated as long as a sufficient size workpiece is being used.

Electron-Beam Hardening. In electron-beam (EB) hardening, the surface of the hardenable steel is heated rapidly to the austenitizing temperature, usually with a defocused electron beam to prevent melting. The mass of the workpiece conducts the heat away from the treated surface at a rate that is rapid enough to produce hardening. Materials for application of EB hardening must contain sufficient carbon and alloy content to produce martensite. With the rapid heating associated with this process, the carbon and alloy content should be in a form that will quickly allow complete solid solution in the austenite at the temperatures produced by the electron beam. In addition, the mass of the workpiece should be sufficient to allow proper quenching; for example, the part thickness must be at least ten times the depth of hardening, and hardened areas must be properly spaced to prevent tempering of pre-

Table 19 Typical characteristics of carburizing, nitriding, and carbonitriding diffusion treatments

Process	Name of case	Process temperature, °C (°F)	Case hardness, Typical case depth	Hardness, HRC	Typical base metals	Process characteristics
Carburizing						
Pack	Diffused carbon	815–1090 (1500–2000)	125 μm–1.5 mm (5–60 mils)	50–63(a)	Low-carbon steels, low-carbon alloy steels	Low equipment costs, difficult to control case depth accurately
Gas	Diffused carbon	815–980 (1500–1800)	75 μm–1.5 mm (3–60 mils)	50–63(a)	Low-carbon steels, low-carbon alloy steels	Good control of case depth, suitable for continuous operation, good gas controls required, can be dangerous
Liquid	Diffused carbon and possibly nitrogen	815–980 (1500–1800)	50 μm–1.5 mm (2–60 mils)	50–65(a)	Low-carbon steels, low-carbon alloy steels	Faster than pack and gas processes, can pose salt disposal problem, salt baths require frequent maintenance
Vacuum	Diffused carbon	815–1090 (1500–2000)	75 μm–1.5 mm (3–60 mils)	50–63(a)	Low-carbon steels, low-carbon alloy steels	Excellent process control, bright parts, faster than gas carburizing, high equipment costs
Nitriding						
Gas	Diffused nitrogen, nitrogen compounds	480–590 (900–1100)	125 μm–0.75 mm (5–30 mils)	50–70	Alloy steels, nitriding steels, stainless steels	Hardest cases from nitriding steels, quenching not required, low distortion, process is slow, is usually a batch process
Salt	Diffused nitrogen, nitrogen compounds	510–565 (950–1050)	2.5 μm–0.75 mm (0.1–30 mils)	50–70	Most ferrous metals including cast iron	Usually used for thin hard cases <25 μm (<1 mil), no white layer, most are proprietary processes
Ion	Diffused nitrogen, nitrogen compounds	340–565 (650–1050)	75 μm–0.75 mm (3–30 mils)	50–70	Alloy steels, nitriding steels, stainless steels	Faster than gas nitriding, no white layer, high equipment costs, close case control
Carbonitriding						
Gas	Diffused carbon and nitrogen	760–870 (1400–1600)	75 μm–0.75 mm (3–30 mils)	50–65(a)	Low-carbon steels, low-carbon alloy steels, stainless steels	Lower temperature than carburizing (less distortion), slightly harder case than carburizing, gas control critical
Liquid (cyaniding)	Diffused carbon and nitrogen	760–870 (1400–1600)	2.5–125 μm (0.1–5 mils)	50–65(a)	Low-carbon steels	Good for thin cases on noncritical parts, batch process, salt disposal problems
Ferritic nitrocarburizing	Diffused carbon and nitrogen	480–590 (900–1090)	2.5–25 μm (0.1–1 mil)	40–60(a)	Low-carbon steels	Low-distortion process for thin case on low-carbon steel, most processes are proprietary

(a) Requires quench from austenitizing temperature. Source: Ref 56 viously hardened areas. The most suitable materials for EB hardening are the same steels used in flame hardening (Ref 76):

- 1045 to 1080 carbon steels
- Medium- to high-carbon alloy steels (4140, 4340, 8645, 52100, and so forth)
- Pearlitic matrix cast irons
- W1, W2, O1, O2, L2, L6, S1, S2 tool steels

There are two basic types of EB systems: stationary or movable. In the movable process, the workpiece is fixed and the gun is moved to produce heating for the hardening. Stationary guns require manipulation of the workpiece under the beam. However, in both cases the area to be hardened on the workpiece must be in a line of sight with the beam. Gun movement or workpiece manipulation is accomplished by computer control to produce any desired pattern, and the beam can be oscillated or pulsed by standard controls. To produce an electron beam, a high vacuum of 10^{-5} torr (10^{-3} Pa) is required in the region where the electrons are emitted and accelerated. This vacuum environment protects the emitter from oxidizing and avoids scattering of the electrons while they are still traveling at a relatively low velocity. Electron-beam hardening in hard vacuum units require that the part be placed in a chamber that is sufficiently large to manipulate the gun or the workpiece. Out-of-vacuum units usually involve shrouding the workpiece; a partial vacuum (10^{-2} torr, or 13 Pa) is obtained in the work area by mechanical pumps.

Laser-Beam Hardening. Lasers can be used to perform selective hardening with hardening depths and material constraints similar to those of EB hardening. Laser heat treatment produces thin surface zones, which are heated and cooled rapidly, resulting in very fine martensitic microstructures, even in steels with relatively low hardenability. This process produces typical case depths for steel ranging from 0.75 to 1.3 mm (0.030 to 0.050 in.) depending on the laser power range, and hardness values as high as 60 HRC (Ref 35). Laser processing has advantages over EB hardening in that laser hardening does not require a vacuum, wider hardening profiles are possible, and there can be greater accessibility to hard-to-get areas with the flexibility of optical manipulation of light energy. A major disadvantage of lasers is the need to use surface treatments to prevent reflectivity of the laser beam.

Diffusion Coating (Carburizing/Nitriding)

Diffusion coatings (Table 19) involve heat treating processes that cause some elemental species to diffuse into the surface of the part to alter the surface properties. These coatings are developed by either heating the component in a gaseous atmosphere of the coating material (carburizing, nitriding, etc.) or by two processes discussed earlier: metalliding and pack cementation. Diffusion processes depend on the concentration gradient of the diffusing species, the diffusivity of the atomic species in the host material, and the time and temperature at which the process takes place.

All carburizing and nitriding processes increase the surface carbon or nitrogen content of the alloy to allow the surface to respond to quench hardening. The heat treater usually relies on empirical data to determine how long to expose the part to achieve the desired carbon or nitrogen diffusion. The term used for the entire field of surface-hardening processes is case hardening, and the case indicates the depth of hardening below the surface. Although the depth of hardening decreases gradually because the diffused species does not stop abruptly, the effective case depth is considered to be the depth at which the hardness falls below 50 HRC.

Carburizing is the addition of carbon to the surface of low-carbon steels at temperatures (generally between 850 and 950 °C, or 1560 and 1740 °F) at which austenite, with its high solubility for carbon, is the stable crystal structure (Ref 74). Hardening of the component is accomplished by removing the part and quenching or allowing the part to slowly cool and then reheating to the austenitizing temperature to maintain the very hard surface property. On quenching, a good wear- and fatigue-resistant high-carbon martensitic case is superimposed on a tough, low-carbon steel core. Carburized steels used in case hardening usually have base carbon contents of about 0.2 wt%, with the carbon content of the carburized layer being fixed between 0.8 and 1.0 wt% (Ref 77). Carburizing methods include gas carburizing, vacuum carburizing, plasma carburizing, salt bath carburizing, and pack carburizing. These methods introduce carbon by use of an atmosphere (atmospheric gas, plasma, and vacuum), liquids (salt bath), or solid compounds (pack). The vast majority of carburized parts are processed by gas carburizing, using natural gas, propane, or butane. Vacuum and plasma carburizing are useful because of the absence of oxygen in the furnace atmosphere. Salt bath and pack carburizing have little commercial importance, but are still done occasionally.

Gas carburizing can be run as a batch or a continuous process. Furnace atmospheres consist of a carrier gas and an enriching gas. The carrier gas is supplied at a high flow rate to ensure a positive furnace pressure, minimizing air entry into the furnace. The type of carrier gas affects the rate of carburization. Carburization by methane is slower than by the decomposition of CO. The enriching gas provides the source of carbon and is supplied at a rate necessary to satisfy the carbon demand of the work load. Most gas carburizing is done under conditions of controlled carbon potential by measurement of the CO and CO_2 content. The objective of the control is to maintain a constant carbon potential by matching the loss in carbon to the workpiece with the supply of enriching gas. The carburization process is complex, and a comprehensive model of carburization requires algorithms that describe the various steps in the process, including carbon diffusion, kinetics of the surface reaction, kinetics of the reaction between the endogas and enriching gas, purging (for batch processes), and the atmospheric control system. Possible models of each of these steps have been outlined (Ref 78).

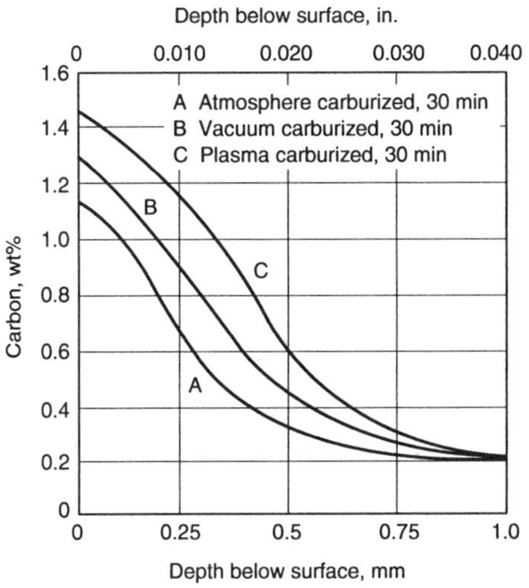

Fig. 15 Carbon gradient profile of atmosphere, vacuum, and plasma carburizing of AISI 8620 steel at 980 °C (1800 °F) saturation conditions for 30 min and followed by direct oil quenching. Source: Ref 81

Vacuum carburizing is a nonequilibrium, boost-diffusion-type carburizing process in which austenizing takes place in a rough vacuum, followed by carburization in a partial pressure of hydrocarbon gas, diffusion in a rough vacuum, and then quenching in either oil or gas (Ref 79). Vacuum carburizing offers the advantages of excellent uniformity and reproducibility because of the improved process control with vacuum furnaces, improved mechanical properties due to the lack of intergranular oxidation, and reduced cycle time. The disadvantages of vacuum carburizing are predominantly related to equipment costs and throughput.

Plasma carburizing is basically a vacuum process utilizing glow-discharge technology to introduce carbon-bearing ions to the steel surface for subsequent diffusion (Ref 80). This process is effective in increasing carburization rates because the process bypasses several dissociation steps that produce active soluble carbon. For example, because of the ionizing effect of the plasmas, active carbon for adsorption can be formed directly from methane (CH_4) gas. High temperatures can be used in plasma carburizing because the process takes place in an oxygen-free vacuum, thus producing a greater carburized case than both atmospheric gas and vacuum carburizing (Fig. 15).

Nitriding is a process similar to carburizing, in which nitrogen is diffused into the surface of a ferrous product to produce a hard case. Unlike carburizing, nitrogen is introduced between 500 and 550 °C (930 and 1020 °F), which is below the austenite formation temperature (Ac_1) for ferritic steels, and quenching is not required. As a result of not austenitizing and quenching to form martensite, nitriding results in minimum distor-

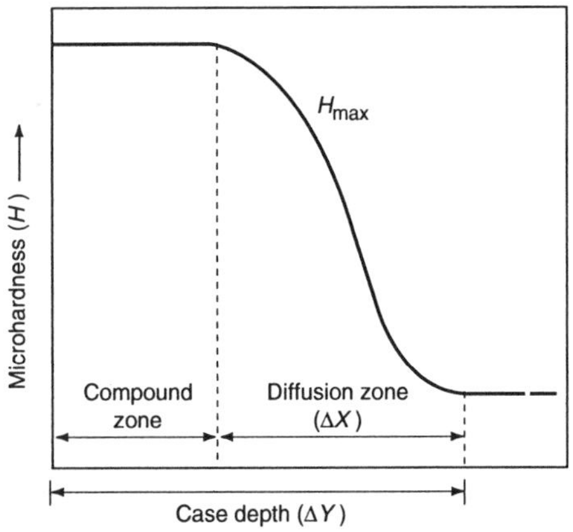

Fig. 16 Factors affecting the microhardness profile of a nitrided steel. The hardness of the compound zone is unaffected by alloy content, while the hardness of the diffusion zone is determined by nitride-forming elements (Al, Cr, Mo, Ti, V, Mn). ΔX is influenced by the type and concentration of alloying elements; ΔY increases with temperature and decreases with alloy concentration. Source: Ref 82

tion and excellent control. The various nitriding processes (Table 19) include: gas nitriding, liquid nitriding, and plasma (ion) nitriding.

All hardenable steels must be quenched and tempered prior to nitriding. The nitriding process is used to obtain a high surface hardness, improve wear resistance, increase fatigue resistance, and improve corrosion resistance (except for stainless steel). The case structure of a nitrided steel, containing a diffusion zone with or without a compound zone (Fig. 16), depends on the type and concentration of alloying elements and the time-temperature exposure of a particular nitriding treatment (Ref 82). The diffusion zone is the original core microstructure with the addition of nitride precipitates and nitrogen solid solution. The compound zone is the region where γ' (Fe_4N) and ε ($Fe_{2-3}N$) intermetallics are formed. Commercial steels containing aluminum, chromium, vanadium, tungsten, and molybdenum are most suitable for nitriding because they readily form nitrides that are stable at the nitriding temperatures (Ref 83). The following steels can be nitrided for specific applications:

- Aluminum-containing low-alloy steels: Nitralloys
- Medium-carbon, chromium-containing low-alloy steels: 4100, 4300, 5100, 6100, 8600, 8700, and 9800 series
- Low-carbon, chromium-containing low-alloy steels: 3300, 8600, and 9300 series
- Hot-working die steels containing 5% Cr: H11, H12, and H13
- Air-hardenable tool steels: A2, A6, D2, D3, and S7
- High-speed tool steels: M2 and M4
- Nitronic stainless steels: 30, 40, 50, and 60
- Ferritic and martensitic stainless steels: 400 and 500 series
- Austenitic stainless steels: 200 and 300 series
- Precipitation-hardened stainless steels: 13-8 PH, 15-5 PH, 17-4 PH, 17-7 PH, A-286, AM 350 (Ref 83), and AM 355

Gas nitriding (Ref 83) is a case-hardening process that takes place in the presence of ammonia gas. Either a single-stage or a double-stage process can be used when nitriding with anhydrous ammonia. The single-stage process, in which a temperature of 495 to 525 °C (925 to 975 °F) is used, produces the brittle nitrogen-rich compound zone known as the white nitride layer at the surface of the nitrided case. The double-stage process, or Floe process, has the advantage of reducing the white nitrided layer thickness. After the first stage, a second stage is added which either by continuing at the first-stage temperature or increasing the temperature to 550 to 565 °C (1025 to 1050 °F). The use of the higher-temperature second stage lowers the case hardness and increases the case depth.

Liquid nitriding (nitriding in a molten salt bath) uses similar temperatures as in gas nitriding and a case-hardening medium of molten, nitrogen-bearing, fused-salt bath containing either cyanides or cyanates (Ref 84). Similar to salt bath carburizing, liquid nitriding has the advantage of processing finished parts because dimensional stability can be maintained due to the subcritical temperatures used in the process. Furthermore, at the lower nitriding temperatures, liquid nitriding adds more nitrogen and less carbon to ferrous materials than that obtained with high-temperature treatments because ferrite has a much greater solubility for nitrogen (0.4% max) than carbon (0.02% max).

Plasma (ion) nitriding is a method of surface hardening using glow-discharge technology to introduce nascent (elemental) nitrogen to the surface of a metal part for subsequent diffusion into the material (Ref 82). The process is similar to plasma carburizing in that a plasma is formed in a vacuum using high-voltage electrical energy and the nitrogen ions are accelerated toward the workpiece. The ion bombardment heats the part, cleans the surface, and provides active nitrogen. The process provides better control of case chemistry, case uniformity, and lower part distortion than gas nitriding.

Carbonitriding introduces both carbon and nitrogen into the austenite of the steel. The process is similar to carburizing in that the austenite composition is enhanced and the high surface hardness is produced by quenching to form martensite. This process is a modified form of gas carburizing in which ammonia is introduced into the gas carburizing atmosphere (Ref 85). As in gas nitriding, elemental nitrogen forms at the workpiece surface and diffuses along with carbon into the steel. Typically, carbonitriding takes place at a lower temperature and a shorter time than gas carburizing, producing a shallower case. Steels with carbon contents up to 0.25% are commonly carbonitrided; these include 1000, 1100, 1200, 1300, 1500, 4000, 4100, 4600, 5100, 6100, 8600, and 8700 series.

Ferritic nitrocarburizing is a subcritical heat treatment process, carried out by either gaseous or plasma techniques, and involves the diffusion of carbon and nitrogen into the ferritic phase. The process results in the formation of a thin white layer or compound layer, with an underlying diffusion zone of dissolved nitrogen in iron, or alloy nitrides (Ref 86). The white layer improves surface resistance to wear, and the diffusion zone increases the fatigue endurance limit, especially in carbon and low-alloy steels. Alloy steels, cast irons, and some stainless steels can be treated. The process is used to produce a thin, hard skin, usually less than 25 μm (0.001 in.) thick, on low-carbon steels in the form of sheet metal parts, powder metallurgy parts, small shaft sprockets, and so forth.

REFERENCES

1. Surface Engineering, Vol 5, *ASM Handbook,* ASM International, 1994
2. D. Wetzel, Batch Hot Dip Galvanized Coatings, *Surface Engineering,* Vol 5, *ASM Handbook,* ASM International, 1994, p 360–371
3. J. Foct, The Morphology of Zinc Coatings, *The Physical Metallurgy of Zinc Coated Steel,* A.R. Marder, Ed., TMS, 1994, p 1
4. C.E. Jordan and A.R. Marder, Alloy Layer Growth During Hot-Dip Galvanizing at 450 °C, *Galvatech '95,* Iron and Steel Society, 1995, p 319
5. A.R.P. Ghuman and J.I. Goldstein, Reaction Mechanisms for the Coatings Formed During Hot Dipping of Iron in 0 to 10 Pct Al-Zn Baths at 450–700 °C, *Metall. Trans.,* Vol 2A, 1971, p 2903
6. Y. Hisamatsu, Science and Technology of Zinc and Zinc Alloyed Coated Steel Sheet, *Galvatech '89,* Iron and Steel Institute Japan, 1989, p 3
7. C. E. Jordan and A.R. Marder, Morphology Development in Hot-Dip Galvanneal Coatings, *Metall. Mater. Trans.,* Vol 25A, 1994, p 937
8. A.R. Marder, Microstructural Characterization of Zinc Coatings, *Zinc-Based Steel Coating System: Metallurgy and Performance,* TMS, 1990, p 55
9. D.C.H. Nevinson, Corrosion of Zinc, *Corrosion,* Vol 13, *ASM Handbook,* ASM International, 1987, p 755–769
10. *Hot Dip Galvanizing for Corrosion Protection of Steel Products,* American Galvanizers Association, 1989
11. H.E. Townsend, Continuous Hot Dip Coatings, *Surface Engineering,* Vol 5, *ASM Handbook,* ASM International, 1994, p 339–348
12. C.E. Jordan, K. Goggins, and A.R. Marder, Interfacial Layer Development in Hot-Dip Galvanneal Coatings on Interstitial Free (IF) Steel, *Metall. Mater. Trans.,* Vol 25A (No. 10), 1994, p 2101–2109
13. H.E. Townsend and J.C. Zoccola, *Mater. Perform.,* Vol 10, 1979, p 13–20
14. H.F. Graff, Aluminized Steel, *Encyclopedia of Materials Science and Engineering,* Pergamon Press, 1986, p 138–141

15. K.G. Budinski, *Surface Engineering for Wear Resistance,* Prentice-Hall, 1988, p 209
16. *Engineering Coatings, Design and Application,* S. Grainger, Ed., Abington Publishing, 1989, p 33
17. J. N. DuPont and A.R. Marder, Thermal Efficiency of Arc Welding Processes, *Weld. J.,* Dec 1995, p 406-s
18. J.N. DuPont and A.R. Marder, Dilution in Single Pass Arc Welds, *Metall. Mater. Trans.,* Vol 27B, 1996, p 481
19. K.G. Budinski, *Surface Engineering for Wear Resistance,* Prentice-Hall, 1988, p 242
20. M. Scholl, R. Devanathan, and P. Clayton, Abrasive and Dry Sliding Wear Resistance of Iron-Molybdenum-Nickel-Silicon-Carbon Weld Hardfacing Alloys, *Wear,* Vol 135 (No. 2), 1990, p 355
21. K.H. Zum Gahr, How Microstructure Affects Abrasive Wear Resistance, *Met. Prog.,* Sept 1971, p 46–49
22. B.F. Levin, J.N. DuPont, and A.R. Marder, Weld Overlay Coatings for Erosion Control, *Wear,* Vol 181–183, 1995, p 810
23. P.F. Tortorelli, I.G. Wright, G.M. Goodwin, and M. Howell, High-Temperature Oxidation/Sulfidation Resistance of Iron-Aluminide Coatings, *Elevated Temperature Coatings: Science and Technology II,* N.B. Dahorte and J.M. Hampikian, Ed., TMS, 1996, p 175
24. J.R. Davis, Hardfacing, Weld Cladding, and Dissimilar Metal Joining, *Welding, Brazing, and Soldering,* Vol 6, *ASM Handbook,* 1993, p 789–829
25. K.G. Budinski, *Surface Engineering for Wear Resistance,* Prentice-Hall, 1988, p 219
26. R.C. Tucker, Jr., Thermal Spray Coatings, *Surface Engineering,* Vol 5, *ASM Handbook,* ASM International, 1994, p 497–509
27. H. Herman and S. Sampath, Thermal Spray Coatings, *Metallurgical and Ceramic Protective Coatings,* K. Stern, Ed., Chapman and Hall, 1996, p 261
28. *Thermal Spraying: Practice, Theory, and Application,* American Welding Society, 1985
29. *Engineering Coatings: Design and Application,* S. Grainger, Ed., Abington Publishing, 1989, p 77
30. K.J. Stein, B.S. Schorr, and A.R. Marder, Erosion of Thermal Spray FeCrAlY-Cr_3C_2 Cermet Coatings, *Elevated Temperature Coatings: Science and Technology II,* N.B. Dahorte and J.M. Hampikian, Ed., TMS, 1996, p 99
31. S.T. Bluni and A.R. Marder, Effects of Thermal Spray Coating Composition and Microstructure on Coating Response and Substrate Protection at High Temperatures, *Corrosion,* Vol 52, 1996, p 213
32. F. Lowenheim, *Modern Electroplating,* 3rd ed., John Wiley, 1974
33. J.W. Dini, *Electrodeposition: The Materials Science of Coatings and Substrate,* Noyes, 1993
34. M.G. Hocking, V. Vasatasree, and P.S. Sidky, *Metallic and Ceramic Coatings: Production, High Temperature Properties and Applications,* John Wiley & Sons, 1989, p 206
35. J.R. Davis, Surface Engineering of Carbon and Alloy Steels, *Surface Engineering,* Vol 5, *ASM Handbook,* ASM International, OH, 1994, p 701–740
36. *Engineering Coatings: Design and Application,* S. Grainger, Ed., Abington Publishing, 1989, p 101
37. M.E. Browning, Section Ed., Plating and Electroplating, *Surface Engineering,* Vol 5, *ASM Handbook,* ASM International, 1994, p 165–332
38. *Quality Metal Finishing Guide,* Metal Finishing Suppliers' Association
39. S.G. Fountoulakis, Continuous Electrodeposited Coatings for Steel Strip, *Surface Engineering,* Vol 5, *ASM Handbook,* ASM International, 1994, p 349–359
40. A.J. Killmeyer, Tin Plating, *Surface Engineering,* Vol 5, *ASM Handbook,* ASM International, 1994, p 239–241
41. D.J. Maykuth and W.B. Hampshire, Corrosion of Tin and Tin Alloys, *Corrosion,* Vol 13, *ASM Handbook,* ASM International, 1987, p 770–783
42. K.H. Stern, Electrodeposition of Refractory Metals from Molten Salts, *Metallurgical and Ceramic Protective Coatings,* K. Stern, Ed., Chapman and Hall, London, 1996, p 9
43. K.H. Stern, Metalliding, *Metallurgical and Ceramic Protective Coatings,* K. Stern, Ed., Chapman and Hall, London, 1996, p 38
44. A. Blair, Silver Plating, *Surface Engineering,* Vol 5, *ASM Handbook,* ASM International, 1994, p 245–246
45. A.M. Weisberg, Gold Plating, *Surface Engineering,* Vol 5, *ASM Handbook,* ASM International, 1994, p 247–250
46. K.G. Budinski, *Surface Engineering for Wear Resistance,* Prentice-Hall, 1988, p 52
47. D.W. Baudrand, Electroless Nickel Plating, *Surface Engineering,* Vol 5, *ASM Handbook,* ASM International, 1994, p 290–310
48. K. Barmak, S.W. Banovic, C.M. Petronis, D.F. Susan, and A.R. Marder, Structure of Electrodeposited Graded Composite Coatings of Ni-Al-Al_2O_3, *J. Microsc.,* Vol 185, part 2, Feb 1997, p 265
49. D.F. Susan, K. Barmak, and A.R. Marder, Diffusion and Oxidation Behavior of Electrodeposited Ni Al Particle Composite Coatings, *Materials, Coatings and Processes for Improved Reliability of High Temperature Components,* N.S. Cheruvu and K. Dannemann, Ed., to be published 1997
50. H.O. Pierson, Chemical Vapor Deposition of Semiconductor Materials, *Surface Engineering,* Vol 5, *ASM Handbook,* ASM International, 1994, p 510–516
51. D.G. Bhat and P.F. Woerner, Coatings for Cutting Tools, *J. Met.,* Feb 1986, p 68
52. D.R. Chopra, A.R. Chourasia, M. Green, R.C. Hyer, K.K. Mishra, and S.C. Sharma, Diamond and Amorphous Films, *Surface Modification Technologies IV,* T.S. Sudarshan, D.G. Bhat, and M. Jeandin, TMS, 1991, p 583
53. M.A. Harper and R.A. Rapp, Codeposition of Chromium and Silicon in Diffusion Coatings for Iron-Base Alloys Using Pack Cementation, *Surface Modification Technologies IV,* T.S. Sudarshan, D.G. Bhat and M. Jeandin, TMS, 1991, p 415
54. R. Bianco, M.A. Harper, and R.A. Rapp, Codepositing Elements by Halide-Activated Pack Cementation, *JOM,* Nov 1991, p 68
55. G.W. Goward and L.L. Seigle, Diffusion Coatings for Gas Turbine Engine Hot Sections, *Surface Engineering,* Vol 5, *ASM Handbook,* ASM International, 1994, p 611–617
56. K.G. Budinski, *Surface Engineering for Wear Resistance,* Prentice-Hall, 1988, p 116–117
57. L.K. Bennett and G. T. Bayer, Pack Cementation Aluminizing of Steels, *Surface Engineering,* Vol 5, *ASM Handbook,* ASM International, 1994, p 617–620
58. W. Beck, "Comparison of Carbon Steel, Alonized Type 304 for Use as Dummy Slabs in Reheat Furnace Operation," Alon Processing, Inc. Tarentum, PA
59. T. Perng, "A Fundamental Study of the Noxso NO_x/SO_2 Flue Gas Treatment," Noxso, 1984
60. K.G. Budinski, *Surface Engineering for Wear Resistance,* Prentice-Hall, 1988, p 111
61. B.J. Smith and A.R. Marder, Characterization of Chromium Diffusion (Chromize) Coatings in a High Temperature Coal Combustion Atmosphere, *Surface Modification Technologies IV,* T.S. Sudarshan, D.G. Bhat and M. Jeandin, TMS, 1991, p 471
62. C. Wagner, *Corros. Sci.,* Vol 5, 1965, p 751
63. R. Bianco and R.A. Rapp, Pack Cementation Diffusion Coatings, *Metallurgical and Ceramic Protective Coatings,* K. Stern, Ed., Chapman and Hall, 1996, p 236
64. K.G. Budinski, *Surface Engineering for Wear Resistance,* Prentice-Hall, 1988, p 138
65. *Engineering Coatings, Design and Application,* S. Grainger, Ed., Abington Publishing, 1989, p 119
66. D.M. Mattox, Section Ed., Vacuum and Controlled-Atmosphere Coating and Surface Modification Processes, *Surface Engineering,* Vol 5, *ASM Handbook,* ASM International, 1994, p 495–626
67. M.G. Hocking, V. Vasatasree, and P.S. Sidky, *Metallic & Ceramic Coatings: Production, High Temperature Properties & Applications,* J. Wiley & Sons, 1989, p 49
68. D.M. Mattox, Vacuum Deposition, Reactive Evaporation and Gas Evaporation, *Surface Engineering,* Vol 5, *ASM Handbook,* ASM International, 1994, p 556–572
69. R.L. Jones, Thermal Barrier Coatings, *Metallurgical and Ceramic Protective Coatings,* K. Stern, Ed., Chapman and Hall, 1996, p 194
70. S.L. Rhode, Sputter Deposition, *Surface Engineering,* Vol 5, *ASM Handbook,* ASM International, 1994, p 573–581
71. E.J. Lee and R.G. Bayer, Tribological Characteristics of Titanium Nitride Thin Coatings, *Met. Finish.,* July 1985, p 39–42
72. D.M. Mattox, Ion Plating, *Surface Engineering,* Vol 5, *ASM Handbook,* ASM International, 1994, p 582–592
73. J.K. Hirvonen and B.D. Sartwell, Ion Implantation, *Surface Engineering,* Vol 5, *ASM Handbook,* ASM International, 1994, p 605–610

74. S. Lampman, Introduction to Surface Hardening of Steels, *Heat Treating,* Vol 4, *ASM Handbook,* ASM International, 1991, p 260–267
75. T. Ruglic, Flame Hardening, *Heat Treating,* Vol 4, *ASM Handbook,* ASM International, 1991, p 268–285
76. K.G. Budinski, *Surface Engineering for Wear Resistance,* Prentice-Hall, 1988, p 120
77. G. Krauss, *Steels: Heat Treatment and Processing Principles,* ASM International, 1990, p 286
78. C.A. Stickels and C.M. Mack, Overview of Carburizing Processes and Modeling, *Carburizing Processing and Performance,* G. Krauss, Ed., ASM International, 1989, p 1
79. J. St. Pierre, Vacuum Carburizing, *Heat Treating,* Vol 4, *ASM Handbook,* ASM International, 1991, p 348
80. W.L. Grube and S. Verhoff, Plasma (Ion) Carburizing, *Heat Treating,* Vol 4, *ASM Handbook,* ASM International, 1991, p 352–362
81. S.H. Verhoff, *Ind. Heat.,* March 1986, p 22–24
82. J.M. O'Brien and D. Goodman, Plasma (Ion) Nitriding, *Heat Treating,* Vol 4, *ASM Handbook,* ASM International, 1991, p 420–424
83. C.H. Knerr, T.C. Rose, and J.H. Filkowski, Gas Nitriding, *Heat Treating,* Vol 4, *ASM Handbook,* ASM International, 1991, p 387–409
84. Q.D. Mehrkam, J.R. Easterday, B.R. Payne, R.W. Foreman, D. Vukovich, and A.D. Godding, Liquid Nitriding, *Heat Treating,* Vol 4, *ASM Handbook,* ASM International, 1991, p 410–419
85. J. Dossett, Carbonitriding, *Heat Treating,* Vol 4, *ASM Handbook,* ASM International, 1991, p 376–386
86. T. Bell, Gaseous and Plasma Nitrocarburizing, *Heat Treating,* Vol 4, *ASM Handbook,* ASM International, 1991, p 425–436

Sources of Materials Property Data and Information

Jack H. Westbrook, Brookline Technologies

ONLY TWO GENERATIONS AGO the information needs of a materials engineer were rather simply served. In the first place, because there was little intersubstitution among the major classes of materials and because the interdisciplinary field of materials science was not yet born, the engineer was, most likely, a metallurgist with little concern for materials other than metals. For such an individual the one-volume *Metals Handbook* largely filled his needs for data and tutorial reviews.

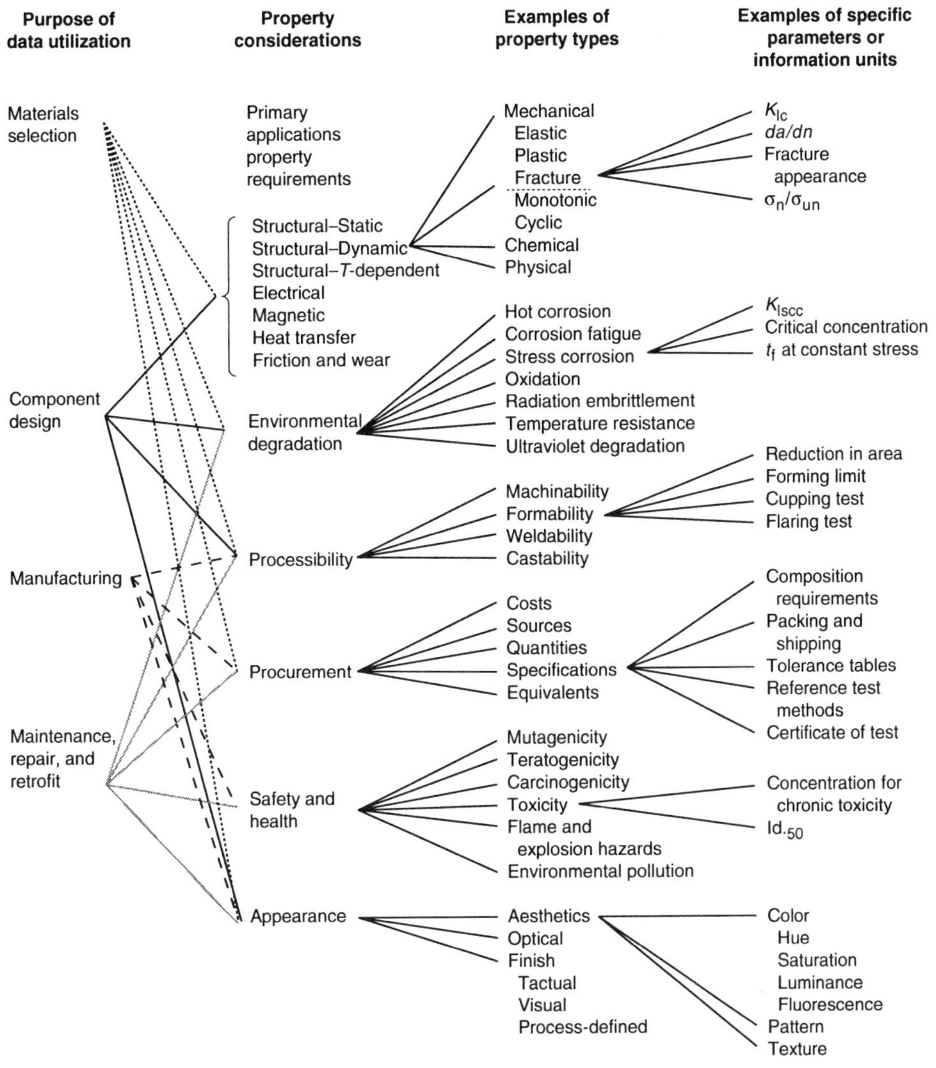

Fig. 1 Taxonomy of materials information. Source: Ref 1

How different is the world of today! There are now of the order of 100,000 different commercially available materials from which to choose. Increasingly, in many applications, plastics, ceramics, and glasses compete directly with metals; processing techniques once regarded as unique to a certain materials class are being adapted to other quite different classes (e.g., slip casting or vacuum forming of metals); and the development and proliferation of composites further blur the boundaries between the subfields of materials. Thus, the scope of materials for which information is required has been greatly broadened.

In addition, the sheer volume of information has increased exponentially. Not only has there been a general explosion in the primary journal publications, but another exponentially growing collection (more difficult to be aware of and to access) has developed in the report literature. The *ASM Handbook* has now grown to twenty volumes and yet is still challenged to do as complete a job of summarizing current knowledge as did its one-volume predecessor of the 1940s. Furthermore, an increasing proportion of materials data is not in hard-copy print form but in electronic formats, available on floppy disks, tapes, CD-ROMs, or on-line. Finally, new social and environmental concerns and increased world trade have demanded the availability of new types of materials information. Health, safety, and environmental issues require—for the application, storage, and disposal of materials—types of information unheard of a few decades ago. Worldwide sourcing of materials, multinational operations of many companies, and proliferation of the use of the Système International d'Unités (SI) (the "metric system") place new demands on the searcher, compiler, and disseminator of materials information.

It is also important to emphasize that the materials information needs of engineers in industry are more demanding and difficult to satisfy than those of materials scientists. Engineers must not only have accessible the best existing value for

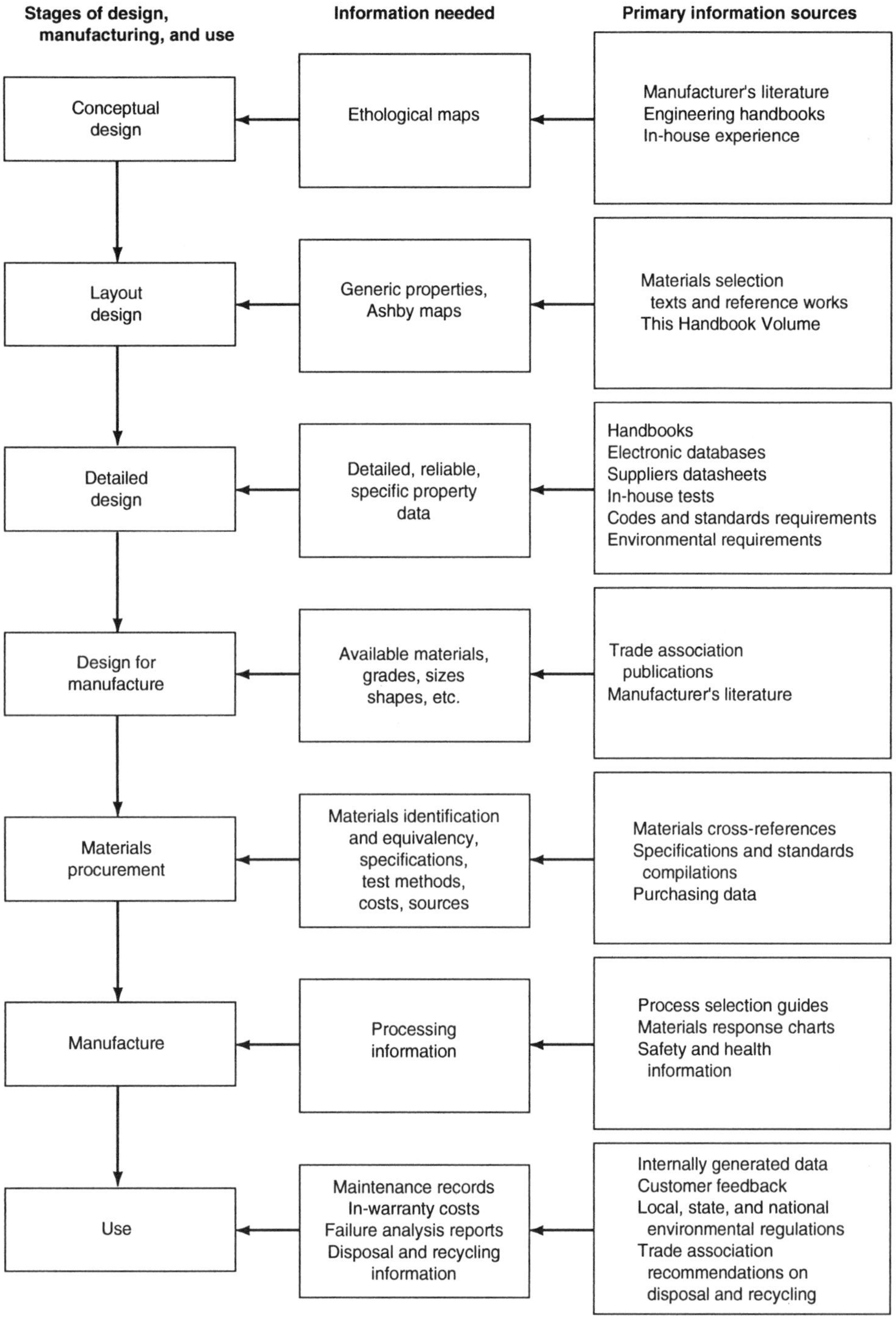

Fig. 2 Information needs and sources at various steps in the design-manufacture-use cycle

the property in question, but they must also know the limits of uncertainty for the value in order to estimate the reliability of a design. Rather than working solely with elemental metals or individual chemical compounds, engineers are frequently confronted with complex alloys, clads, and other composites whose properties and behavior must also be known or reliably estimated. Furthermore, as noted above, many properties of commercial materials are not fixed values, but depend on processing history and resulting structure and effect of external variables. In addition, engineers frequently require data that cannot be expressed in terms of a single property or combination of properties, but must be related to a performance test or service application.

Fewer sources of compiled and evaluated data cater to the needs of industry as compared to scientists, and, under business pressures, engineers in industry have less time available to search for answers to their questions or to reconcile conflicting data from diverse sources. Furthermore, existing materials data compilations, both print and electronic, display a disconcerting variety of terminologies, test methods, property units, and so forth, making integration difficult at best, a situation exacerbated by the great disparity of materials classes to be considered. The available data compilations may be from a publisher, from a materials supplier, or result from an in-house testing program. Endorsement of reliability of the information may be keyed to the individual editor/compiler, a professional association, a trade association, the data generator, or the distribution agency. Examples of each are presented later in this article.

Property data and information are needed at each stage in the sequence: design, manufacture, application and recycling of industrial or consumer goods. The number and diversity of the specific data required are staggering. Not only do each of these functional needs subsume a variety of specific types of data, but also within each type there may be a multiplicity of individual parameters or information units to characterize the attribute in question. The main features of this taxonomy are shown in Fig. 1. Only selected items have been expanded for illustration; the reader may readily detail others appropriate to individual cases. It must also be remarked that most of the parameters listed at the far right in Fig. 1 are not fixed values determined solely by the composition of the materials selected, but will vary as a function of structure, processing, or the effect of external variables such as temperature, section size, or surface finish. The particular property and materials classes important for application vary among industrial groups as indicated in Table 1, but the materials information needs for each are always quite broad.

Information Needs

This section examines more closely the information needs at various points in the design, manufacture, and use cycle. These points are laid out schematically in Fig. 2 with notation as to the types of information needed at each stage. Actual practice is even more complex than shown with numerous reiteration loops that refine the design, materials selection, and process selection and provide feedback of application experience that impacts the next generation of product. These matters are discussed in more detail in other articles in this Volume as well as in the work by Waterman and Ashby (Ref 2).

A few words follow that more fully characterize the various sources of information and differentiate the quality of that information at different points in the design-manufacture cycle, even when the nominal identification of property needed is the same. Finally, examples are given of specific sources of information needed at each stage. (Detailed source listings are provided in the section "Specific Sources of Data" in this article.)

Requirements definition, the first step in particularizing a conceptual design, is based in part on ethological maps, two- or three-dimensional graphs showing the relationship of important parameters of existing products. Such information

Table 1 Materials information needs by industrial groupings

Industry group	Materials classes									Properties groups					
	Ferrous metals	Nonferrous metals	Refractory metals and superalloys	Ceramics and glasses	Composites	Inorganic compounds	Plastics	Semiconductors	Wood	Mechanical	Thermal	Electrical, electronic, and magnetic	Other physical	Corrosion and oxidation	Processibility
Civil engineering	X	X		X			X		X	X	X			X	X
Transportation	X	X	X		X		X			X	X			X	X
Power generation	X	X	X							X	X	X		X	X
Aerospace	X	X	X	X	X			X		X	X	X	X	X	X
Defense	X	X	X	X	X	X				X	X	X	X	X	X
Chemical	X	X		X		X				X	X			X	
Oil and gas	X		X							X	X			X	
Electronics and communication		X		X				X				X	X		X
Materials producers	X	X	X	X			X		X	X	X	X	X	X	X
Consumer products	X	X		X	X		X	X		X	X	X		X	X
Other industrial products	X	X					X			X	X	X			X

Source: Ref 1

is rarely available already compiled, but must be assembled on an ad hoc basis from the generic sources shown. This step focuses attention on which properties are likely to be critical for various components.

Layout design (also called embodiment design) presents a general arrangement of the conceptual design with appropriate dimensions. Ashby maps can be used to facilitate consideration of all possible generic classes of materials at this very early stage in the design process. Ashby maps are available in print form (Ref 2) or electronically via the *Cambridge Materials Selector,* a "Windows"-based PC tool-kit (available from Granta Design, Ltd., Cambridge, United Kingdom); examples of Ashby maps are provided in the article "Material Property Charts" in this Volume. Another approach using formal logic trees has been developed by Allen et al. (Ref 3) at Brigham Young University. At this stage the aim is to view potential materials as widely as possible and with a focus on just a few critical properties. Nominal values of key properties are sufficient, but costs should also be considered.

Detailed design is the stage in which a fully dimensioned preferred layout design is developed. The layout out design includes dimensional tolerances, surface finishes, and determination of the actual specifications of materials to be used.

Table 2 Contrasting features of databases for materials selection and design

Feature	Selection	Design
Scope	Broad	Narrow
No. of properties	Few	Many
No. of materials	Many	Few
Multifaceted properties (composition, processing, time, and temperature dependent)	NA	Key requirement
Data quality	Nominal	Precise
Cost of material	Important	NA
Manipulation capability of database	NA	Essential

NA, not applicable

Table 3 Directories and cross references to materials standards and specifications

(C) *Alloy Finder CD-ROM,* 2nd ed., ASM International. Incorporates *Woldman's Engineering Alloys,* 8th ed., 1994; *Worldwide Guide to Equivalent Irons and Steels,* 3rd ed., 1993; and *Worldwide Guide to Equivalent Nonferrous Metals and Alloys,* 3rd ed., 1996

(P) *British and Foreign Specifications for Steel Castings,* Steel Casting Research and Trade Association, Sheffield, UK, 1980

(P) *Department of Defense Index of Specifications and Standards,* Part 1, Alphabetical Listing; Part 2, Numerical Listing. U.S. Government Printing Office, Published annually

(C) EQUIST 2.0, Hungarian database with comparisons and equivalencies for 6500 standard steels from 18 countries

(P) European Committee for Iron and Steel Standardization (ECISS), *Index of Standards,* PSP 2, British Steel, plc., Head Office Standards, London, United Kingdom, 1986

(P/C) J.P. Frick, Ed., *Woldman's Engineering Alloys,* 8th ed., ASM International, 1994. Lists compositions, selected properties, and manufacturers for 55,000 alloys. Available in print (1400 pp) and on diskette; data also included in *Alloy Finder CD-ROM* (see separate listing)

(P) *German Standards,* Beuth, Germany. English translation of more than 4750 popular DIN standards plus 1500 DIN-EN, DIN-EC, and DIN-ISO standards

(C) *Global Standards Network,* A directory on the World Wide Web of >100,000 regional, national, and international standards, produced by ANSI and NIST (http: //www.nssn.org)

(P) *Handbook of Comparative World Steel Standards,* DS 67, ASTM, 1996, 552 pp

(P) *Handbook of International Alloy Compositions and Designations,* Vol 1R, Titanium, Metals and Ceramics Information Center, Battelle Columbus Laboratories, 1990

(P) W. Hufnagel, Ed., *Key to Aluminium,* 4th ed., Aluminium AG, Dusseldorf, Germany, 1991

(P) *Index of Aerospace Materials Specifications,* Society of Automotive Engineers, 1986, 224 pp

(P) *International Trade-names Directory,* Gale Research, 1989

(P) *Iron and Steel Specifications,* 7th ed., British Steel, London, 1989, 346 pp. Steel grades to BS970 together with French, German, Japanese, and Swedish standards

(P) *ISR/VAMAS Unified Classification Scheme for Advanced Ceramics,* ASTM, 1993

(P) M. Kehler, *Aluminium,* Vol 3, *Handbook of International Alloy Compositions and Designations,* Aluminium Verlag, Heyden, Germany, 1981, 859 pp

(P) *Materials,* Vol 4, *World Standards Speedy Finder,* The International Technical Information Institute of Japan, Tokyo

(P) *Metals and Alloys in the Unified Numbering System,* 7th ed., DS 56F, ASTM, 1996, 450 pp

(C) *Metals Infodisk,* A CD-ROM database integrating data on tens of thousands of metal grades, over 5000 worldwide standards, a directory of leading suppliers, and limited property information. Also accessible via a Web site. ILI Infodisk, Inc., Paramus, NJ

(P/C) D.L. Potts and J.G. Gensure, *International Metallic Materials Cross-Reference,* 3rd ed., Genium Publishing, 1988. Print (699 pp) and diskette

(P) R.B. Ross, Ed., *Metallic Materials Specifications Handbook,* 4th ed., Chapman and Hall, 1992, 830 pp

(P/C) H. Schmitz, *Stahl-Eisen-Listen/Steel-Iron Lists,* 7th ed., VDEh, Dusseldorf, Germany, 1987. 2500 German and 1000 foreign materials covered; also in a computerized database

(P/C) *Stahlschlüssel (Key to Steel),* 17th ed., Verlag Stahlschlüssel, 1995. Print (580 pp) and CD-ROM

(C) *Standards Infodisk,* A CD-ROM database covering >300,000 U.S. and foreign standards issued by 42 U.S. and 28 international or European organizations, ILI Infodisk, Inc., Paramus, NJ

(P) *Supplement to Dechema Corrosion Handbook—Concordance of U.S. and German Designations for Metallic Materials,* VCH, Weiheim, Germany, 1992, 254 pp

(P) *Technical Data,* Technical Indexes, Ltd., Bracknall, UK, updated monthly. >100 indexed microfilm or microfiche files covering 10^7 pp of technical information, 1000 pp printed index

(P) F.T. Traceski, *Specifications and Standards for Plastics and Composites,* ASM International, 1990, 224 pp

(C) *UNSearch,* a computerized form of the book *Metals and Alloys in the Unified Numbering System,* ASTM/SAE

(P) M.J. Wahl, *Handbook of Superalloys,* International Alloy Compositions and Designations Series, Springer Verlag, 1982, 258 pp

(P) W.A. Woishnis, Ed., *Engineering Plastics and Composites,* 2nd ed., ASM International, 1993, 590 pp. Identifies tradenames and manufacturer designations for polymeric materials

(D) World Metal Index, Sheffield City Libraries, Sheffield, United Kingdom. A service covering 70,000 standard grades of metallic materials worldwide

(P/C) *Worldwide Guide to Equivalent Irons and Steels,* 3rd ed., ASM International, 1992. Available in print (600 pp) and on diskette; data also included in *Alloy Finder CD-ROM* (see separate listing)

(P/C) *Worldwide Guide to Equivalent Nonferrous Metals and Alloys,* 3rd ed., W. Mack, Ed., ASM International, 1995. Available in print (500 pp) and on diskette; data also included in *Alloy Finder CD-ROM* (see separate listing)

C, computer readable; D, data center; P, print

Table 4 Guides and directories to sources of materials data and information

General

L. Anthony, Ed., *Information Sources in Engineering,* 2nd ed., H. Zell Publications, 1985, 560 pp

H. Behrens and G. Ebel, *Physik Daten,* Vol 3-1 to 3-4, Fachinformationszentrum, Karlsruhe, Germany, 1976–1979

Guide to American Scientific and Technical Directories, 3rd ed., Todd Publications, 1993, 33 pp

C.D. Hurt, Ed., *Information Sources in Science and Technology,* Libraries Unlimited, 1988, 400 pp

R.S. Marvin and G.B. Sherwood, A Guide to Sources of Information on Materials, *Handbook of Materials Science,* Vol 3, CRC Press, 1980, p 603–627

M.N. Patten, Ed., *Information Sources in Metallic Materials,* Bowker-Saur, 1990, 415 pp

H. Wawrousek, J.H. Westbrook, and W. Grattidge, Data Sources of Mechanical and Physical Properties of Engineering Materials, Second International Symposium, *Computerization and Networking of Materials Databases,* STP 1106, J.G. Kaufman and J.S. Glazman, Ed., ASTM, 1991, p 142

H. Wawrousek, J.H. Westbrook, and W. Grattidge, Data Sources of Mechanical and Physical Properties of Engineering Materials, *Physik Daten,* No. 30-1, Fachinformationszentrum, Karlsruhe, Germany, 1989

J.H. Westbrook, Materials Information Sources, *Encyclopedia of Materials Science and Engineering,* M.B. Bever, Ed., Pergamon, 1986, p 527

J.H. Westbrook and K.W. Reynard, Sources of Information for Materials Economics, Policy, and Management, *Concise Encyclopedia of Materials Economics, Policy, and Management,* M.B. Bever, Ed., Pergamon, 1993, p 35

By country

Asia/Oceania, *The CODATA Directory of Data Sources for Science and Technology in Asia-Oceania Countries,* CODATA, Paris, 1994, 188 pp

France, *Banques del Donnés Factuelles sur les Matériaux,* B. Marx, Ed., CODATA France, Committee on Data for Science and Technology, Paris, 1987

Germany, *Quellenverzeichnis Werkstoffdaten,* Gessellschaft für Information and Dokumentation mbH, Heidelberg, Germany, 1987

Israel, *Directory of Data Bases in Israel,* E. Hoffmann, Ed., National Center of Scientific and Technological Information, Tel Aviv, Israel, 1986

Japan, *Directory of Data for Science and Technology in Japan,* Japanese National Committee for CODATA, Tokyo, 1987

Nordic Countries, *Directory of On-line Databases in the Nordic Countries,* E. Miskos, Ed., Fabrittius, Oslo, NORDINFO Publication 12, 1988, 152 pp

Russia/Eastern Europe, Material and Substance Data Banks in COMECON Countries and in the USSR, *The Provision of Materials Property Data via Computerized Systems,* A.D. Kozlov, Ed., Report on a CODATA Symposium held at the Siberian Division of the USSR Academy of Sciences; *CODATA Bull.,* Vol 22, 1990

United Kingdom

- Materials Data Sources, P.T. Houldcroft, Ed., Institution of Mechanical Engineers, London, 1987, 111 pp
- UK Materials Information Sources—1992, K.W. Reynard, Ed., The Design Council, London, 1992

Guides to on-line or machine-readable sources

F.C. Allan and W.R. Ferrell, Numeric Databases in Science and Technology: An Overview, *Database,* Vol 12 (No. 3), 1989, p 50–58

C.J. Armstrong, Ed., *World Databases in Physics and Mathematical Science,* Bowker-Saur, London, 1995

A.J. Barrett, Ed., *International Register of Materials Database Managers,* Special Report No. 14, 2nd ed., CODATA, 1993

The CD-ROM Directory 1993, TFPL Publishing, Washington, DC, 1993. Available in print and on CD-ROM

CD-ROMs in Print, Meckler, 1993. Available in print and on CD-ROM

Guides to on-line or machine-readable sources (continued)

The CODATA Referral Database (CRD), based on the 1988 revision of the UNESCO *Inventory of Data Referral Sources of Science and Technology,* J.H. Westbrook and W. Grattidge, Ed. Available only on diskette from CODATA, Paris, France

Data Base Directory 1991–1992, Knowledge Industry Publications/ASIS, 1991, 456 pp

A Directory of Databases Available Through SearchMAESTRO, Defense Technical Information Center, Cameron Station, Alexandria, VA, 1987

Directory of Scientific Databases, 4th ed., Longman Group, Ltd., 1986, 232 pp

Gale Directory of Databases, Vol 1, *On-Line Sources;* Vol 2, *CD-ROM, Diskette, Magnetic Tape, Handheld and Batch Access Database Products;* Gale Research International, 1993

S.I. Gurke and J.G. Kaufman, *A Computerized Directory of Directories of Scientific and Technical Databases,* Report to USNC of CODATA, Oct 1991

NASA On-Line Directory of Numeric Databases, NASA Langley Research Center, Hampton, VA

Non-bibliographic Data Banks in Science and Technology, S. Schwarz, D.G. Watson, and O. Alvfeldt, Ed., ICSU Press, 1985

NTIS Directory of Computerized Data Files, NTIS PR-629, National Technical Information Service. Lists 1300 databases for 50 Federal agencies

Numeric Databases—A Directory, ICSTI, Paris, France, 1991

Online Database Selection: A Users Guide to the Directory of On-line Databases, Cuadra/Elsevier, 1989

J. Rumble, J. Sauerwein, and S. Pennell, *Scientific and Technical Factual Databases for Energy Research and Development,* DOE Report DE87001518, U.S. Dept. of Energy, 1986, available from NTIS

J. Schmittroth and D.M. Maxfield, *On-Line Database Search Services,* Gale Research, 1984

M.E. Williams, *Computer-Readable Databases—A Directory and Data Sourcebook in Science, Technology, and Medicine,* American Library Association, 1985

Examples of topical guides

S. Clark-Kerwien and G. Papatrefon, Guide to Information Sources and Standards, *Ceramics and Glasses,* Vol 4, *Engineered Materials Handbook,* ASM International, 1991, p 38–40

Directory of Resources for the Reinforced Plastics/Composites Industry, Composites Institute, Society of the Plastics Industry, 1986

C.M. Hudson and S.K. Seward, A Compendium of Sources of Fracture Toughness and Fatigue-Crack Growth Data for Metallic Alloys, *Int. J. Fract.,* Vol 14, 1978, R51–R184

Information Sources in Polymers and Plastics, R.T. Adkins, Ed., Bowker-Saur, UK, 1989, 313 pp

S.R. Kaplan, Ed., *A Guide to Information Sources in Mining, Minerals, and Geosciences,* Interscience, 1965, 599 pp

A.H. Landrock, Guide to General Information Sources, *Adhesives and Sealants,* Vol 3, *Engineered Materials Handbook,* ASM International, 1990, p 65–72

A.H. Landrock, Guide to General Information Sources, *Engineering Plastics,* Vol 2, *Engineered Materials Handbook,* ASM International, 1988, p 92–95

B.G.R. Logan, Ed., *International Ceramic Directory,* London and Sheffield Publishing Co., Ltd., 1984

Minerals Data Source Directory, U.S. Bureau of Mines Information Circular 8881, U.S. Government Printing Office, 1982, 296 pp

R. Serjantson, Ed., *Ferro-Alloy Directory,* 3rd ed., Metal Bulletin Books, 1992

F.T. Traceski, Guide to General Information Sources, *Composites,* Vol 1, *Engineered Materials Handbook,* ASM International, 1987, p 40–42

United Nations Industrial Development Organization, Wein, Austria. Several published guides to information sources in specific topical areas, including iron and steel, ceramics, glass, pulp and paper, and cement and concrete

Calculational activities include finite element analysis; failure modes and effects analysis; effects of notches, stress concentrations, and other fracture-mechanics-related factors; effects of mode of loading (fatigue and impact); and effects of environment.

At this point, nominal values of properties no longer suffice. Probabilistic design approaches require still more detailed information. Property values must be as precise as possible with effects of all known variables considered, and their reliability or statistical variability well documented. The contrast between the features of information needs for preliminary materials selection and detailed component design are highlighted in Table 2. A common attribute of materials property information sources, either print or electronic, is that they often are not clearly differentiated and documented as to whether they are intended for preliminary materials selection or for detailed design. It is also often the case that publicly available information does not include effects of the known important variables or the expected statistical spread of the property concerned. In such cases, resort must be made to in-house testing, following accepted standard test methods, and a proprietary handbook built up with the necessary information. Sometimes such programs are sponsored by a consortium of interested parties who cooperatively fund the work, carry out and analyze the tests, and report the data to their membership. Examples of this kind include the work of the Materials Property Council, the U.S. Department of Defense Military Handbook Committees, trade associations (e.g., Aluminum Association, Copper Development Association, American Iron and Steel Institute), and user groups (e.g., the Government-Industry Data Exchange Program). Finally, designers at this stage must also be cognizant of any requirements imposed by codes and standards groups (Ref 4) or by environmental considerations. Representative articles describing the uses of information in materials selection and design are given in Ref 5 to 8, and this topic is discussed in several of the articles in the Sections "The Design Process," "Criteria and Concepts in Design," and "The Materials Selection Process" in this Volume.

Design for Manufacture. This stage comprises analysis and possibly modification of the detailed design to establish its amenability to low-cost manufacturing and assembly with an acceptably small (or negligible) downgrading of product performance. The first stage in this process requires assessment of commercial availability of the material in the grades, sizes, and shapes determined by prior design work. Such information is usually available in publications of trade associations such as those previously cited.

Materials Procurement. The first step in materials procurement is to learn the most common designation for the material—international, national, trade association, or other—and its equivalence in other designation systems. Various directories and cross-reference guides have been developed to aid in this activity (see Table 3).

Next, it must be determined what specifications shall be invoked to ensure that the materials acquired will behave as expected. A specification has been defined as "a document intended primarily for use in procurement that clearly and accurately describes the essential technical requirements for items, materials, or services, including the procedures by which it will be determined that the requirements have been met" (Ref 4). The principal content elements of a materials specification are: title, statement of scope, requirements, quality-assurance provisions, applicable reference documents, preparations for delivery, notes, and definitions.

The most effective specification is that which accomplishes the desired result with the fewest requirements. In writing a new specification, properties and performance should be emphasized rather than how the objectives are to be achieved. Effort should be made by a company to avoid "specials" and to try whenever possible to adopt a standard or specification already established by an organization at a higher hierarchical level, that is, a trade association, or national or international standard. Sources of standards and specifications have been compiled in directories (see Table 3) and are discussed in Ref 9 (see also the article "Designing to Codes and Standards" in this Volume). Whenever necessary to be included in a specification, the test method or quantitative inspection technique cited should be one recognized nationally, for example, an ASTM standard. Finally, for actual procurement of the material it is necessary to have information on sources (vendors) and costs. Again, a variety of directories and periodicals are available.

Manufacture. Information is also needed during manufacture itself for several purposes: selection of the most economic process (may impact back on the component design), to understand the response of the material to processing variables, and to observe best practices with respect to safety and health in the workplace.

Use. When the component made from the material selected is placed in service, it now generates over its lifetime information that may be profitably used, either in redesign of the component or in the next generation of a product to serve the same function. This category includes both internally generated data and that resulting from customer feedback.

Locating Sources of Data

A persistent problem for seekers of materials information, whether printed or electronic, is the lack of adequate guides and directories. The inadequacies of existing directories arise from poor currency, narrowness of focus (by country, by materials class, by properties class, by format, etc.), inadequate descriptions of database or print sources, and poor characterization of the reliability of the information contained. References 1 and 10 are overviews of materials information sources. The first covers guides to technical infor-

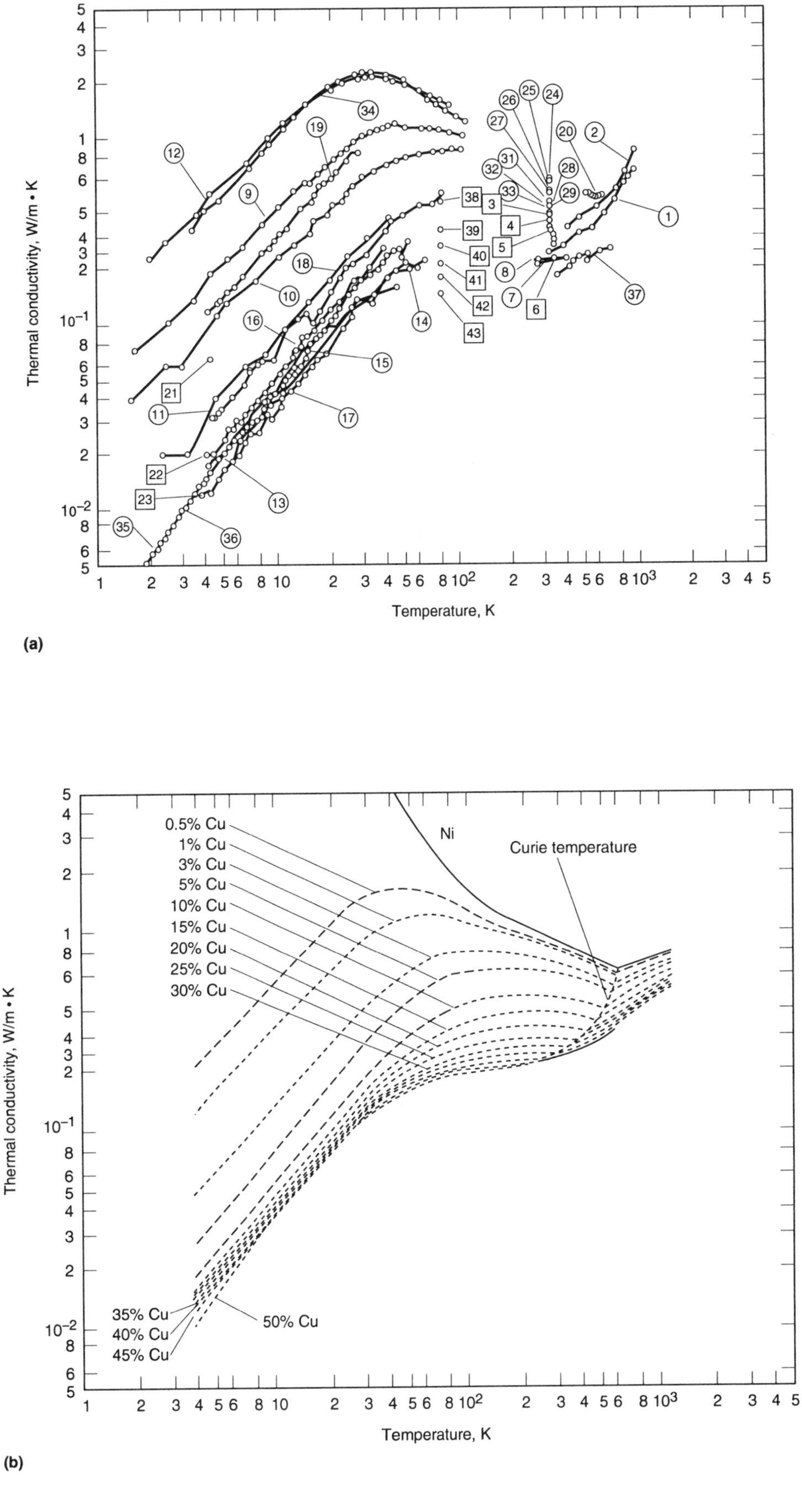

Fig. 3 Thermal conductivity data for nickel-copper alloys. (a) Experimental data from the literature. (b) Following evaluation, the recommended synthesized data. Source: After Ho and Touloukin (Ref 24)

Table 5 Sources of materials data

General

(D) Advanced Materials and Processes Technical Information and Analysis Center (AMPTIAC), operated by the Illinois Institute of Technology Research Institute (IITRI), 201 Mill St., Rome NY 13440. Tel: (315) 339-7117. Incorporates the Metals and Ceramics Information Center (MCIC), the High-Temperature Materials Information Analysis Center (HTMIAC), the Ceramics Information Analysis Center (CIAC), and the Metal-Matrix Information Analysis Center (MMIAC). These encompass the following subjects: ceramics and ceramic composites; organic structural materials and matrix composites; monolithic metals, alloys and metal-matrix composites; electronic, optical and photonic materials; and environmental protection and special function materials

(C) AEA Metals Databases (*Ni-DATA, Ti-DATA, Al-DATA, and Fe-DATA*), AEA Technology, Engineering Software Inc., Bethel Park, PA. Databases designed to work with thermochemical modeling software *MTDATA* or *Thermo-calc* for predicting phase behavior and properties

(P) *Aerospace Structural Metals Handbook,* 5 vol with quarterly updates, Mechanical Properties Data Center, Battelle Columbus Laboratories, over 4000 pp. Referenced alloy data of special interest to the aerospace designer. The data, currently covering 200 alloys, are carefully characterized as to source, product form, size, and pertinent details of thermal-mechanical processing and testing history

(P/C) *Alloy Digest,* (1952–present) ASM International. Approximately 3500 data sheets on commercially available metals and alloys; eight new data sheets appear each month. Now also available on CD-ROM

(D) American Bureau of Metal Statistics (ABMS), 400 Plaza Drive, Harmon Meadow, P.O. Box 1405, Secaucus, NJ 07094-0405. Tel: (201) 863-6900

(P/C) *Annual Book of ASTM Standards,* ASTM, West Conshohocken, PA. Approximately 60 vol per year. A comprehensive compilation of standards, tests, and specifications. Each volume published separately throughout the year, covering a specific topic. Available in electronic formats

(D) ASM International, Materials Park, OH 44073. Tel: (440) 338-5151

(P) M. Bauccio, Ed., *ASM Engineered Materials Reference Book,* 2nd ed., 1994, 580 pp. A compilation of data for ceramics, polymers, and composites

(P) M. Bauccio, Ed., *ASM Metals Reference Book,* 3rd ed., ASM International, 1993, 614 pp. A compilation of tabular properties and processing data for ferrous and nonferrous metals

(P) C.F. Beaton and G.F. Hewitt, *Physical Property Data for the Design Engineer,* Hemisphere, 1988

(P) M.B. Bever, Ed., *Encyclopedia of Materials Science and Engineering,* MIT Press, 1986, 8 volumes

(C) *Binary Alloy Phase Diagrams CD-ROM,* ASM International. Collection of 4700 binary metallic phase diagrams derived from the classic 3-vol handbook edited by T. Massalski

(P) A. Blake, Ed., *Handbook of Mechanics, Materials, and Structures,* Wiley, 1985, 736 pp

(P) D. Bloor et al., Ed., *Encyclopedia of Advanced Materials,* Pergamon, 1994, 4 vol

(P) G.S. Brady, H.R. Clauser, and J.A. Vaccari, Ed., *Materials Handbook,* 14th ed., McGraw-Hill, 1996, 1072 pp. Has 800 A-to-Z entries providing descriptions and data for more than 13,000 materials and tradenames

(C) *Cambridge Materials Selector, V2.0 for Windows,* Granta Design Limited, Cambridge, UK. Implements the Ashby materials selection scheme for personal computers

(C) *CenBASE/Materials on CD-ROM,* CenTOR Software Corp., Garden Grove, CA. Detailed information on over 30,000 grades of thermoplastics, thermosets, elastomers, rubbers, adhesives, composites, fibers, ceramics, and metals. Also available via the World Wide Web

(D) Center for Information and Numerical Data Analysis and Synthesis (CINDAS), Purdue University, 2595 Yeager Rd., West Lafayette, IN 47906. Tel: (765) 494-9393. Thermal, electrical, mechanical, physical, and chemical properties of more than 3000 materials, including ferrous and nonferrous metals, organic and inorganic materials, glasses, ceramics, polymers, and composites

(C) *CETIM-Materiaux* (common properties of engineering materials used in materials selection), Centre Technique des Industries Mechaniques (CETIM). CETIM, 74 route de la Jonoliere, B.P. 957, 44076 Nantes Cedex, France

(P) N.P. Cheremisinoff, *Materials Selection Deskbook,* Noyes, 1997, 207 pp

(C) *DASMAT* (Data Sheet system of Materials), Ishikawajima Haruna Heavy Industries Co. Ltd., 3-1-15 Toyosu, Koto-ku, Tokyo 135, Japan

(P) *Engineered Materials Handbook Desk Edition,* ASM International, 1995, 1300 pp. Considers polymers, composites, ceramics, and glasses

(D) Engineering Sciences Data Unit (ESDU), 27 Corsham St., London N1 5UA, United Kingdom. Tel: 44 171 490-5151. Critically evaluated engineering design data on physical properties, heat transfer, stress and strength, fluid mechanics, fatigue, structures, and aerodynamics; also publishes for Ministry of Defense a handbook of data on aerospace materials

(P) D.J. Fisher, *Rules of Thumb for the Physical Scientist,* Trans Tech, Soloturn, Switzerland, 1988, 320 pp. Collects many simple, empirical correlations observed between diverse properties; helpful in estimating unknown properties and in forecasting likely properties of unknown systems

(P) *Handbook of Industrial Materials,* 2nd ed., Elsevier, 1992, 803 pp. Covers ferrous metals, nonferrous metals and alloys, nonmetallic materials, thermoplastics, thermoset plastics (processed), plastis, materials for electrical applications, joining materials

(D) Information Center on Advanced Materials. Institute of Materials Science (ICAM-RIMSE) Tsinghua University (TH Univ), Main Building of Tsinghua University, Beijing, China 10084. Tel: 2824541, ext. 2758

(D) Information System on Measurement of Mechanical Quantities, Bundesanstalt für Materialforschung und Prüfung (BAM), Berlin, Unter den Eichen 87, D-12200 Berlin 45, 3615, Germany. Tel: 49 30 8104 3615

(P) *Key Engineering Materials,* Mechanical Properties Research, Trans Tech, 1990

(P) C.T. Lynch, Ed., *Practical Handbook of Materials Science,* CRC Press, 1989, 636 pp. Selected reference data for the physical properties of solid-state and structural materials in tabular form

General (continued)

(C) *MAPP,* ESM Software, Hamilton, OH. Materials databases from ASM International Mat.DB collection, in personal computer format, of properties and processing for metals and some polymers

(C) *MATADOR* (nonferrous and ferrous metals, numerical data for physical and mechanical properties). Motoren und Turbinen Union München GmbH, Postfach 50 06 40. Tel: 49 89 1489 3323

(D) Materials Research Society (MRS), 506 Keystone Dr., Warrendale, PA 15086. Tel: (412) 779-3003

(D) The Materials Properties Council, Inc. (MPC), 345 East 47th St., New York, NY 10017. Tel: (212) 705-7693

(P) *Materials Selector,* Penton Publishing Co., a special issue of *Machine Design* magazine giving tabular data for a broad range of metals, ceramics, plastics, and composites

(P) *Metals Handbook Desk Edition,* American Society for Metals, 1985, 1376 pp. Considers all metallic alloys and processing methods

(C) *Metals Infodisk,* A CD-ROM database integrating data on tens of thousands of metal grades, over 5000 worldwide standards, a directory of leading suppliers, and limited property information. Also accessible via a Web site. ILI Infodisk, Inc., Paramus, NJ

(D) The Minerals, Metals & Materials Society (TMS), 420 Commonwealth Dr., Warrendale, PA 15086. Tel: (412) 776-9000

(C) *MSC/MVISION,* MacNeal-Schwendler Corp., Los Angeles. A materials database that is integrated with the Pro/ENGINEER analysis system

(P) C.E. Nunez, R.A. Nunez, and A.C. Mann, *CenBASE Materials in Print, 1990,* 4 vol, J. Wiley, 1990, 4492 pp

(C) *PERITUS,* Matsel Systems Ltd., Liverpool. A database for metals, polymers, and ceramics aimed at materials and process selection

(P) D.D. Pollock, *Physical Properties of Materials for Engineers,* Vol 1, 1981, 224 pp; Vol 2, 1982, 312 pp; CRC Press

(D) Research into Artifacts, Center for Engineering, University of Tokyo. 2-11-16 Yayoi, Bunkyo-ku, Tokyo 113, Japan. Tel: 81-3-3812-2111. Generic data on all engineering materials, nuclear fuel data, fusion reactor materials data, industrial standards for metals and alloys, high-temperature superconducting materials

(P) C. Robb, *Metals Databook,* Institute of Metals, London, 1990. Chemical, physical, and mechanical properties of commercially available metallic materials

(P) R.B. Ross, Ed., *Handbook of Metal Treatments and Testing,* 2nd ed., Wiley, 1988,

(C) *Rover Electronic DataBooks,* William Andrew Inc., Norwich, NY. Materials databases including ASM International Mat.DB data collection, in personal computer format, of properties and processing for metals and polymers

(P) J.F. Shackleford, W. Alexander, and J.S. Park, Ed., *CRC Materials Science and Engineering Handbook,* 2nd ed., CRC Press, 1994, 1532 pp

(P) C.J. Smithells, Ed., *Smithells Metals Reference Book,* 7th ed., Butterworth, 1991

(C) STN International, Chemical Abstracts Services, Columbus, OH. On-line service for metals, ceramics, and polymers databases developed by the Materials Property Data (MPD) Network

(P) *Structural Alloys Handbook,* 3 vol with annual updates, CINDAS/Purdue University. Vol 1, *Wrought Iron and Cast Iron;* Vol 2, *Wrought Stainless Steel, Cast Steel, and Cast Stainless Steel;* Vol 3, *Structural Steel, Wrought and Cast Aluminum, Copper, Brass, Bronze, Magnesium, and Titanium*

(C) Supplier's Data on a Full Spectrum of Materials: Plastics, Metals, Ceramics and Other Nonmetallics (MATUS). Engineering Information Company, Ltd., 30 Hampton House, 15/17 Ingate Place, London SW8 3NS, United Kingdom. Tel: 44 71-662-8155

(D) TNO Metal Institute. Netherlands Organization for Applied Scientific Research, Laan Van Westenenk 501, Postbus 541, 7300 AM Apeldorn, Netherlands. Tel: (055) 73344

(P/C) P. Villars, A. Prince, and H. Okamoto, Ed., *Handbook of Ternary Alloy Phase Diagrams,* ASM International. >15,000 diagrams from 7380 systems. 10-vol set in print, or full set CD-ROM, or 7 subsets in CD-ROM (industrial/heat resistant alloys; solders, brazes, and copper alloys; light metal structural alloys; electronic materials/semiconductors; precious metals; rare earth/actinides; carbides/nitrides)

(P) N.A. Waterman and M.F. Ashby, Ed., *The Materials Selector,* 2nd ed., Chapman and Hall, 1997. A three-volume compilation of data on all materials, with selection and design guide

(D) World Bureau of Metal Statistics, 27a High St., Ware, Herts SG12 9BA, United Kingdom. Tel: 44 0920 461 274

(D) World Metal Index, Sheffield Library and Information Service, Central Library, Surrey St., Sheffield, S. Yorkshire S1 1XZ, United Kingdom. Tel: 44 114 273 4744 and 44 114 273 4714. A service with coverage of over 70,000 grades of ferrous and nonferrous materials, including all British Standards and most United States, European, and National Standards. Also features a large collection of metallurgical and engineering journals and literature

Ferrous metals

(D) American Iron and Steel Institute (AISI), 1101 17th St., N.W., Suite 1300, Washington, DC 20036-4700. Tel: (202) 452-7100

(P) H.T. Angus, *Cast Iron: Physical and Engineering Properties,* 2nd ed., Butterworths, 1976, 542 pp

(D) British Cast Iron Research Association (BCIRA), Alvechurch, Birmingham, Warwickshire B48 7QB, United Kingdom. Tel: 44 527 66414. Databases on ductile and malleable iron and on gray cast iron

(P) J.R. Davis, Ed., *ASM Specialty Handbook: Carbon and Alloy Steels,* ASM International, 1996, 731 pp

(P) J.R. Davis, Ed., *ASM Specialty Handbook: Cast Irons,* ASM International, 1996, 450 pp

(P) J.R. Davis, Ed., *ASM Specialty Handbook: Stainless Steels,* ASM International, 1994, 577 pp

(continued)

C, computer readable; D, data center; P, print

Table 5 (continued)

Ferrous metals (continued)

(P) J.R. Davis, Ed., *ASM Specialty Handbook: Tool Materials,* ASM International, 1995, 501 pp
(P) P. Harvey, Ed., *Engineering Properties of Steel,* American Society for Metals, 1982, 527 pp
(D) International Iron & Steel Institute, Rue Colonel Bourg 120, B-1140 Brussels, Belgium. Tel: 32-2-735-90-75
(D) Iron & Steel Society (ISS), 410 Commonwealth Dr., Warrendale, PA 15086. Tel: (412) 776-1535
(C) Material Strength Database for Engineering Steels and Alloys (ENSTAL), National Research Institute for Metals—Japan Information Center for Science and Technology, Tsukuba, Japan
(P/C) *The Metals Black Book,* Vol 1, *Ferrous Metals,* 2nd ed., J.E. Bringas, Casti Publishing, Edmonton, Alberta, Canada, 1996, 740 pp
(C) Minitech Computerized Alloy Steel Information System (CASIS). Hardenability and related properties including as-rolled, quenched and tempered, and carburized properties. Minitech Ltd., Box 5185, Station E, Hamilton, Ontario, Canada L8S 4L3. Tel: (416) 523-7721
(P) F.B. Pickering, Ed., *Constitution and Properties of Steels,* Vol 7, *Materials Science and Technology—A Comprehensive Treatment,* R.W. Cahn, P. Haasen, and E.J. Kramer, Ed., VCH Publishers, 1991, 824 pp
(P) *Properties and Selection: Irons, Steels, and High-Performance Alloys,* Vol 1, *ASM Handbook* (formerly *Metals Handbook,* 10th ed.), ASM International, 1990, 1063 pp
(C) Properties of Iron Castings, Institüt für Leichtbau, Dresden, Germany. Database containing physicochemical properties, data on processing and reprocessing of materials, information on the form and production technology of 200 types of iron casting materials selected by application with reference to their properties, testing conditions, and treatment
(C) Properties of Steel, Institüt für Leichtbau, Dresden, Germany. Database containing physicochemical properties, information on processing and reprocessing, on the form and production technology of 300 types of German steels together with their specific application areas and test conditions
(P) R. Serjantson, *Ferro-Alloy Directory and Databook,* 3rd ed., Metal Bulletin, 1992
(P) Y.S. Touloukian and C.Y. Ho, Ed., *Properties of Selected Ferrous Alloying Elements,* Hemisphere, 1989, 269 pp
(P) C.W. Wegst, Ed., *Stahlschlüssel (Key to Steel),* 17th ed., Verlag Stahlschlüssel GmbH, Germany, 1995
(C) Werkstoffdatenbank (WDB), Kraftwerk Union AG-Erlangen (KWU), Werkstoffabteilung, Postfach 32 20, Hammerbacherstr. 12 + 14, 8520, Erlangen, Germany. Tel: 49 9131 18 5785. Data on the technical and mechanical properties of steels and iron alloys

Nonferrous metals

(D) Aluminum Association, Inc., 900 19th St., NW, Suite 300, Washington DC 20006. Tel: (202) 862-5100
(P) *Aluminum: Properties and Physical Metallurgy,* J.E. Hatch, Ed., American Society for Metals, 1984, 424 pp
(P) J.R. Davis, Ed., *ASM Specialty Handbook: Aluminum and Aluminum Alloys,* ASM International, 1993, 784 pp
(P) B.T. Barry and C.G. Thwaites, *Tin and Its Alloys and Compounds,* John Wiley & Sons, 1983, 268 pp
(P) W. Betteridge, *Nickel and Its Alloys,* John Wiley & Sons, 1984, 211 pp
(P) R. Boyer, E.W. Collings, and G. Welsch, Ed., *Materials Properties Handbook: Titanium Alloys,* ASM International, 1994, 1169 pp
(P) J.E. Bringas, Ed., Vol 2, *The Metals Red Book, Non-ferrous Metals,* Casti Publishing, Edmonton, Alberta, Canada, 1993, 600 pp
(D/C) Copper Development Association Inc., 260 Madison Ave., New York, NY 10016. Tel: (212) 251-7200. Technical data and information on the selection, fabrication, and use of copper and its alloys. Operates the Copper Data Center (an on-line bibliographic database of the world literature on copper and copper technology) in cooperation with the International Copper Association
(P) W. Hofmann, *Lead and Lead Alloys—Properties and Technology,* Springer-Verlag, 1970, 551 pp
(D) International Copper Association Ltd., 260 Madison Ave., New York, NY 10016. Tel: (212) 251-7240
(D) International Lead Zinc Research Organization (ILZRO), 2525 Meridian Parkway, P.O. Box 12036, Research Triangle Park, NC 27709. Tel: (919) 361-4647
(D) International Magnesium Association (IMA), 1303 Vincent Place, Suite 1, McLean, VA 22101. Tel: (703) 442-8888
(D) International Precious Metals Institute (IPMI), 4905 Tilgman St., Suite 160, Allentown, PA 18104. Tel: (610) 395-9700
(D) International Titanium Association (ITA), 1871 Folsom St., Suite 100, Boulder, CO 30302. Tel: (303) 443-7515. Formerly the Titanium Development Association (TDA)
(D) Lead Development Association, 42 Weymouth St., London W1N 3LQ, United Kingdom. Tel: 071 499-8422
(D) Lead Industries Association, Inc., 295 Madison Ave., New York, NY 10017. Tel: (212) 578-4750
(P) K.H. Matucha, *Structure and Properties of Non-Ferrous Alloys,* Vol 8, *Materials Science and Technology—A Comprehensive Treatment,* R.W. Cahn, P. Haasen, and E. Kramer, Ed., VCH Publishers, 1995
(D) Nickel Development Institute (NiDI), 214 King St. W., No. 510, Toronto, Ontario M5H 3S6, Canada. Tel: (416) 591-7999
(P) F. Porter, *Zinc Handbook—Properties, Processing and Use in Design,* Marcel Dekker, 1991, 648 pp
(P) *Properties and Selection: Nonferrous Alloys and Special-Purpose Materials,* Vol 2, *ASM Handbook* (formerly *Metals Handbook,* 10th ed.), ASM International, 1991, 1328 pp
(P) *The Properties of Aluminium and its Alloys,* 8th (revised) ed., Aluminium Federation, ALFED, Birmingham, United Kingdom, 1983, 88 pp

Nonferrous metals (continued)

(D) Refractory Metals Association, 105 College Rd., East, Princeton, NJ 08540. Tel: (609) 452-7700
(P) E.M. Savitskii, Ed. (A. Prince, Ed., English ed.), *Handbook of Precious Metals,* Taylor and Francis, 1990, 600 pp
(P) E.A. Starke, Jr. and T.H. Sanders, Jr., Ed., *Aluminum Alloys: Their Physical and Mechanical Properties,* Engineering Materials Advisory Services Ltd., West Midlands, United Kingdom, 1986, 1700 pp
(P) E.G. West, *Copper and Its Alloys,* Ellis Horwood Industrial Metals Series, John Wiley & Sons, 1982, 241 pp
(P) S.W.H. Yik and C.T. Wang, *Tungsten—Sources, Metallurgy, Properties and Applications,* Plenum, 1979, 516 pp
(D) Zinc Development Association, 42 Weymouth St., London W1N 3LQ, United Kingdom. Tel: (071) 499-6636

Plastics

(P) M.B. Ash and I.A. Ash, Ed., *Handbook of Plastic Compounds, Elastomers, and Resins: An International Guide by Category, Tradename, Composition, and Supplier,* VCH, 1992, 872 pp. Information on over 15,000 chemical trademark products currently sold throughout the world, including tradename by category reference, tradename cross reference, chemical component cross reference, and chemical manufacturer directory
(D) Association of Plastics Manufacturers in Europe, 250 Avenue Louise, Box 73, 1050 Brussels, Belgium. Tel: 02 640 28 50
(P) M.L. Berins, Ed., *Plastics Engineering Handbook of the Society of the Plastics Industry Inc.,* 5th ed., Van Nostrand Reinhold, 1991, 900 pp
(P) J. Brandrup and E.H. Immergut, Ed., *Polymer Handbook,* 3rd ed., John Wiley & Sons, 1989
(C) *CAMPUS,* M-Base GmbH, Aachen, Germany. A database of engineering polymers compiled from data submitted by the manufacturers
(C) *CETIM-ASTAR* (properties and calculations for 340 thermoplastics). Centre Technique des Industries Mecaniques (CETIM-Nantes) 74 route de la Joneliers, 44076 Nantes, France. Tel: (33) 40 37 36 35
(P) N.P. Cheremisinoff, *Application and Processing Operations,* Vol 3, *Handbook of Polymer Science and Technology,* Marcel Dekker, 1989, 680 pp
(P) N.P. Cheremisinoff, *Composites and Specialty Applications,* Vol 4, *Handbook of Polymer Science and Technology,* Marcel Dekker, 1989, 760 pp
(P) N.P. Cheremisinoff, *Performance Properties of Plastics and Elastomers,* Vol 2, *Handbook of Polymer Science and Technology,* Marcel Dekker, 1989, 758 pp
(P) N.P. Cheremisinoff, *Synthesis and Properties,* Vol 1, *Handbook of Polymer Science and Technology,* Marcel Dekker, 1989, 798 pp
(C) Commercially Available Plastics and Elastomers (*PLASCAMS-DB*). RAPRA Technology Ltd., Shropshire, United Kingdom. Tel: (44) 030-250383
(C) Commercially Available Plastics and Their Properties (*PLASTI-SERV*). Mechanical, physical, electrical, and thermal properties of 9000 plastics. Plastiserv, P.O. Box 20933, Columbus, OH 43220. Tel: (614) 457-8126
(P) P.J. Cornish, Ed., *Concise Encyclopedia of Polymer Processing and Applications,* Pergamon, 1992, 771 pp
(P) C.A. Daniels, *Polymers: Structure and Properties,* Technomic Publishing, 1989, 113 pp
(P) *Effect of Temperature and Other Factors on Plastics,* Plastics Design Laboratory, 1990
(P) *Effect of UV Light and Weather on Plastics and Elastomers,* Plastics Design Laboratory, 1994
(P) *Engineering Plastics,* Vol 2, *Engineered Materials Handbook,* ASM International, 1988, 883 pp
(C) *ENGINPLAST* (guide to the selection of engineering plastics). ENGINPLAST, 148 rue d'Aulnay, 92290 Chatenay Malabry, France. Tel: (33) 46 60 17 55
(P) C.A. Harper, Ed., *Handbook of Plastics, Elastomers, and Composites,* 3rd ed., McGraw-Hill, 1996, 848 pp
(P/C) *International Plastics Selector, Plastics,* 9th ed., International Plastics Selector, San Diego, CA, 1987. Also two computer databases: *PLASTICS* (thermoplastics and thermosets) and *SELECTOR* (adhesives, sealants, composites, and laminates)
(C) *IPS Plastics and Adhesives,* International Plastics Selector, 15 Iverness Way East, Englewood, CO 80112. Tel: (303) 397-2472. Plastics properties for >15,500 thermoplastics and thermosets from >230 manufacturers. Adhesives properties for >7800 adhesives, sealants, and primers from >230 manufacturers
(P) J.I. Kroschwitz, Ed., *Concise Encyclopedia of Polymer Science and Engineering,* Wiley, 1990, 1341 pp
(P/C) J.E. Kroschwitz, Ed., *Encyclopedia of Polymer Science and Engineering,* 2nd ed., Wiley, 1985–1990, 23 vol. Coverage includes natural and synthetic polymers, plastics, fibers, elastomers, computer topics, and processing. Also available on CD-ROM and on-line
(P) S. Levy and J.H. Dubois, *Plastic Product Design Engineering Handbook,* 2nd ed., Chapman and Hall, 1985, 376 pp
(P) J.M. Margolis, Ed., *Conductive Polymers and Plastics Handbook,* Chapman and Hall, 1989, 208 pp
(P) E. Miller, Ed., *Plastic Product Design Handbook,* part B, Marcel Dekker, 1983, 392 pp
(P) P. Mitchell, Ed., *Plastic Part Manufacturing,* Vol 8, *Tool and Manufacturing Engineers Handbook,* Society of Manufacturing Engineers, 1996
(P) *Modern Plastics Encyclopedia,* McGraw-Hill, Annual. A special issue each year of the magazine *Modern Plastics.* Both an encyclopedia and a directory of information about plastics. Covers resins and compounds, chemicals and additives, reinforced plastics and compounds, primary processing, auxiliary equipment and components, and fabricating and finishing. Also included are sections on engineering data, a buyer's guide, and a directory of tradenames

(continued)

C, computer readable; D, data center; P, print

Table 5 (continued)

Plastics (continued)

(C) *Plaspec,* Mechanical, physical, optical, and electrical properties of more than 4200 plastics and polymers. Plaspec, 633 Third Ave., 32nd Floor, New York, NY 10017
(D/C) Plastics and Rubber Advisory Service, 6 Bath Place, Rivington St., London EC2A 3JE, United Kingdom. Tel: 44 171 457 5000. Backed by computer- and text-based databases
(C) *POLYMAT,* DKI Kunstsoff Datenbank (Suppliers data on commercially available plastics), Deutsches Kunststoff-Institut (DKI) Abteilung Dokumentation, Schlossartenstr. 6R, 6100 Darmstadt, Germany. Tel: 49 (06151) 16-2106
(P) *Polymer Engineering Principles: Properties, Process, and Tests for Design,* Hanser Gardner Publications, 1993
(C) *PROCOP-M* (physical, thermodynamic and rheological data on thermoplastics for injection molding). Available through Cisigraph, 76 rue des Gemeaux SILIC 413, 94573 Rungis Cedex, France. Tel: (33) 14 687 22 16
(P) I.I. Rubin, Ed., *Handbook of Plastic Materials and Technology,* Wiley, 1990, 1792 pp. Comprises 119 chapters on plastic materials, properties, processes, and industry practices
(P) H.J. Saechtling, Ed., *International Plastics Handbook,* Macmillan, London, 1983 (English ed.)
(C) *SciPolymer,* SciVision Inc., Lexington MA. A program for de novo design of polymers or estimation of polymer properties. Includes a database of 650 polymers, 200 common monomers, and 150 common substituent groups
(P) R.B. Seymour, *Polymers for Engineering Applications,* ASM International, 1987, 198 pp
(C) *SPAO* (computer-assisted selection of polymers). Provides information on 4000 to 5000 commercial grades of polymers on the European market: manufacturer, applicable standards and test methods, forming characteristics, engineering properties, and applications. Laboratoire National d'Essais, 1 rue Gaston Boissier, 75015 Paris, France. Tel: (33) 145 32 29 89
(P) F.M. Sweeney, Ed., *Polymer Blends and Alloys—Guidebook to Commercial Products,* Technomic Publishing, 1988, 568 pp
(P) D.W. van Krevelen, *Properties of Polymers,* 3rd ed., Elsevier, Amsterdam, 1990
(P) W.A. Woishnis, Ed., *Engineering Plastics and Composites,* 2nd ed., ASM International, 1993

Elastomers

(P) A.K. Bhowmick and H.L. Stephens, *Handbook of Elastomers,* Marcel Dekker, 1988, 816 pp
(P) N. Cheremisinoff, Ed., *Elastomer Technology Handbook,* CRC Press, 1993
(P) R. Crawford, *Plastics and Rubber—Engineering Design and Applications,* Mechanical Engineering Publications, London, 1985, 316 pp. Properties of plastics and rubbers, processing and fabrication, designing with plastics and rubbers, design limitations
(P) Engineering Data Sheets, Malaysian Rubber Producers' Research Association (MRPRA), Hertford, United Kingdom
(P) I. Frantz, *Elastomers and Rubber Compounding Materials: Manufacture, Properties, and Applications,* Elsevier, 1990, 580 pp
(P) W. Hofmann, *Rubber Technology Handbook,* Oxford University Press, 1989
(P) W.A. Lee and R.A. Rutherford, Fluoropolymer Transition Temperature Data Sheets, Rubber and Plastics Research Association of Great Britian, (RAPRA), United Kingdom
(P) P.B. Lindley, "Engineering Design with Natural Rubber," NR Technical Bulletin, Malaysian Rubber Producers' Research Association (MRPRA), Brickendonbury, Hertford SG13 8NL, United Kingdom, 1978
(P) B.M. Walker and C.P. Rader, *Handbook of Thermoplastic Elastomers,* 2nd ed., Van Nostrand Reinhold, 472 pp

Ceramics and glasses

(C) *Advanced Ceramics Data Base (ACDB),* Tsinghua University, 321 Main Bldg., Tsinghua University, Beijing, China. Tel: +86 10 625 957 67
(D) American Ceramic Society, Inc., 735 Ceramic Place, Westerville, OH 43081. Tel: (614) 890-4700
(P) R.J. Brook, Ed., *Concise Encyclopedia of Advanced Ceramic Materials,* MIT Press, 1990, 400 pp
(C) *Ceramic Phase Diagrams,* NIST/American Ceramic Society. A CD-ROM collection of >13,000 ceramic phase diagrams
(P) *Ceramics and Glasses,* Vol 4, *Engineered Materials Handbook,* ASM International, 1991, 1217 pp
(D) Ceramics Information Analysis Center (CIAC), Advanced Materials and Processes Technical Information and Analysis Center, 201 Mill St., Rome NY 13440. Tel: (315) 339-7117
(P) N.P. Cheremisinoff, Ed., *Handbook of Ceramics and Composites,* 3 vol, Marcel Dekker, 1990
(C) *INTERGLAD.* A CD-ROM based database for glass compositions and properties, developed in Japan. Described by I. Yasui and F. Utsuno, *Bull. Am. Cer. Soc.,* Vol 72 (No. 10), 1993, p 65–71
(D) KERAMIK, Institut fur Nichtmetallische Werkstoffe der TU Berlin, Englische Strasse 20, 1000 Berlin 12, Germany. Tel: (030) 314-3425
(P) V.K. Marghussian, *Thermo-Mechanical Properties of Ceramic Fibers,* Parthenon Press, 1986, 152 pp
(P/C) O. Mazurin, et al., *Handbook of Glass Data,* 5 vol, Elsevier Science, 1985–1992. Electronic version available on diskettes
(P) R. Morrell, *Handbook of Properties of Technical and Engineering Ceramics,* Parts I and II, National Physical Laboratory, Her Majesty's Stationery Office, London, 1985 and 1987
(C) *NIST Structural Ceramics Database,* Ceramics Division, National Institute of Standards and Technology, Gaithersburg, MD 20899. Tel: (301) 975-6127. Materials specifications, thermal and mechanical properties, and corrosion data
(C) *NIST Tribo-Ceramic Materials Database,* Standard Reference Data Program, National Institute of Standards and Technology, Gaithersburg, MD 20899. Tel: (301) 975-2200. Friction, lubrication, and wear data for industrially important ceramic materials
(P) D.S. Oliver, *Engineering Design Guide 05: The Use of Glass in Engineering,* Oxford University Press, Oxford, United Kingdom, 1975

Ceramics and glasses (continued)

(P) G.C. Phillips, *A Concise Introduction to Ceramics,* Van Nostrand Reinhold, 1991
(P) H.O. Pierson, Ed., *Handbook of Refractory Carbides and Nitrides: Properties, Characteristics, Processing, and Applications,* Noyes, 1997, 362 pp
(C) *SG PROMAT* (mechanical, thermal, physical, and thermodynamic properties of glass and related materials). Saint Gobain Recherche, 39 quai Lucien Lefranc, 93304 Aubervilliers, France
(P) H. Rawson, *Properties and Applications of Glass,* Glass Science and Technology Series, Vol 3, Elsevier, 1980, 318 pp
(P) J.S. Reed, *Principles of Ceramics Processing,* 2nd ed., Wiley, 1995
(P) S. Saito et al., Ed., *Ceramics Databook,* Gordon and Breach, 1987, 739 pp. Translated from the Japanese edition (Tokyo, 1986) by International Research Foundation
(P) H. Scholze, *Glass: Nature, Structure and Properties,* Springer Verlag, Berlin, 1991
(P) M.M. Schwartz, *Ceramic Joining,* ASM International, 1993, 196 pp
(P) M.M. Schwartz, *Handbook of Structural Ceramics,* McGraw-Hill, 1992
(C) *SciGlass,* SciVision Inc., Lexington, MA. A comprehensive glass property information system in a PC-based format, including property data for 85,000 glasses, calculation formulae, and patent and standards information
(P) R.A. Terpstra, P.P.A.C. Pex, and A.H. DeVries, Ed., *Ceramic Processing,* Chapman and Hall, 1995
(P) D.R. Uhlman and N.J. Kreidel, Ed., *Elasticity and Strength in Glass,* Vol 5, *Glass Science and Technology,* Academic Press, 1980
(P) D.R. Uhlman and N.J. Kreidel, Ed., *Structure, Microstructure, and Properties,* Vol 4a, *Glass Science and Technology,* Academic Press, 1989
(P) J.B. Wachtman, *Mechanical Properties of Ceramics,* Wiley-Interscience, 1996, 448 pp

Composites

(C) *Advanced Composite Materials (ACM),* Industrial Products Research Institute AIST, MITI, Higashi 1-1-4, Yatabe-machi, Tsukuba-gun, Ibaraki 305, Japan. Tel: (81) 0298-54-6701
(P) K.J. Bowman, S.K. El-Rahaiby, and J.B. Wachtman, Jr., Ed., *Handbook on Discontinuously Reinforced Ceramic Matrix Composites,* American Ceramic Society/CINDAS, 1995, 115 pp
(P) C.X. Campbell and S.K. El-Rahaiby, Ed., *Databook on Mechanical and Thermophysical Properties of Particulate-Reinforced Ceramic Matrix Composites,* American Ceramic Society, 1995, 511 pp
(P) C.X. Campbell and S.K. El-Rahaiby, Ed., *Databook on Mechanical and Thermophysical Properties of Whisker-Reinforced Ceramic Matrix Composites,* American Ceramic Society, 1995, 590 pp
(C) *CETIM-FIMAC* (properties of 6000 composite materials), Centre Technique des Industries Mecaniques (CETIM-Nantes), 74 route de la Joneliere, 44076 Nantes, France. Tel: (33) 40 37 36 35
(P) K.K. Chawla, *Ceramic Matrix Composites,* Chapman and Hall, 1993
(P) *Composites,* Vol 1, *Engineered Materials Handbook,* ASM International, 1987, 983 pp
(P) *Composites and Laminates,* D.A.T.A. Inc., 1987, 550 pp
(C) Data Base of Composite Materials for Future Industries (both resin and metallic matrices), R&D Institute of Metals & Composites for Future Industries, Agency of Industrial Science and Technology, MITI, 1-7-2, Nishi-shinbashi, Minatoku, Tokyo 105, Japan. Tel: (81) 03-595-2405
(P) J.-B. Donnet and R.C. Bansal, *Carbon Fibers,* 2nd ed., Marcel Dekker, 1989, 496 pp
(P) M. Grayson, Ed., *Encyclopedia of Composite Materials and Components,* Wiley, 1983, 1161 pp
(P) *Handbook of Composites,* Chapman and Hall, 1997
(P) N.L. Hancox and R.M. Mayer, *Design Data for Reinforced Plastics: A Guide for Engineers and Designers,* Chapman and Hall, 1994, 326 pp
(P) A.R. Ibbotson and N.J. Lindsay, "Further Development of an Engineering Database on Selected Metal-based Composites," DERA Report DRA/SMC/CR 954018, 1995, 38 pp
(D) Information Center on Advanced Materials, Institute of Materials Science (ICAM-RIMSE), Tsinghua University, Beijing, China 10084. Tel: 2824541, ext. 2758
(P) A. Kelly, Ed., *Concise Encyclopedia of Composite Materials,* rev. ed., Pergamon, 1994, 378 pp
(P) S.M. Lee, Ed., *International Encyclopedia of Composites,* VCH, 1990–1992, 6 vol
(P) R.L. Lehman, S.K. El-Rahaiby, and J.B. Wachtman, Jr., Ed., *Handbook on Continuous Fiber-Reinforced Ceramic Matrix Composites,* American Ceramic Society, 1995, 601 pp
(P) P.K. Mallick, *Composites Engineering Handbook,* Marcel Dekker, 1997, 1248 pp
(P) P.K. Mallick, *Fiber-Reinforced Composites: Materials, Manufacturing, and Design,* Marcel Dekker, 1993, 584 pp
(P) J.V. Milewski, *Handbook of Reinforcements for Plastics,* Van Nostrand Reinhold, 1988
(P) S. Peters, Ed., *Lubin's Handbook of Composites II,* Chapman and Hall, 1995
(P) M.M. Schwartz, *Composite Materials Handbook,* 2nd ed., McGraw-Hill, 1992
(D) Society for the Advancement of Material and Process Engineering (SAMPE), 1055 W. San Bernadino Rd., Covina, CA 91722. Tel: (818) 331-0616
(D) Suppliers of Advanced Composite Materials Association (SACMA), 1600 Wilson Blvd., Suite 1008, Arlington, VA 22209. Tel: (703) 841-1556
(P) J.W. Weeton, D.M. Peters, and K.L. Thomas, *Engineers Guide to Composite Materials,* ASM International, 1987, 397 pp

High-temperature materials

(P) *Aerospace Structural Metals Handbook,* 5 vol updated quarterly. Battelle-Columbus Laboratories
(P) H.E. Boyer, Ed., *Atlas of Creep and Stress-Rupture Curves,* ASM International, 1988, 663 pp
(P) J.R. Davis, Ed., *ASM Specialty Handbook: Heat-Resistant Materials,* ASM International, 1997
(C) *DIWA,* Databank information system for materials behavior (especially time- and temperature-dependent properties which can be displayed graphically), Staatliche Materialprufungsanstalt, Universitat, Stuttgart, Pfaffenwaldring 32, 7000 Stuttgart 80, Germany. Tel: 49 (0711) 685-3034, -3031
(P) R.D. Dudley and P.D. Desai, Ed., *Beryllides, and Miscellaneous Alloys,* Vol III, *Properties of Intermetallic Alloys,* CINDAS/Purdue University, 1994

(continued)

C, computer readable; D, data center; P, print

Table 5 (continued)

High-temperature materials (continued)

(P) B.F. Gilp and P.D. Desai, Ed., *Silicides*, Vol II, *Properties of Intermetallic Alloys*, CINDAS/Purdue University, 1994

(C) *HIGHTEMP*, Swedish Institute of Metals Research, Drottning Kristinas Väg 48, S-11428, Stockholm, Sweden. Tel: 46-8-24-3330

(C) *High Temperature Materials Data Bank (HTM-DB)*. On-line access to mechanical properties of Alloy 800, an iron-base superalloy, and 31 other high-temperature alloys. Commission of the European Communities, DG XII, JRC, NL 1755 ZG Petten, The Netherlands. Tel: 31 2246-5208

(D) High-Temperature Materials Information Analysis Center (HTMIAC), Advanced Materials and Processes Technical Information and Analysis Center, 201 Mill St., Rome NY 13440. Tel: (315) 339-7117

(P) T.Ya. Kosolapova, Ed., *Handbook of High Temperature Compounds: Properties, Production, Applications*, Hemisphere, 1990, 933 pp

(D) *Mechanical Properties of Metals at Elevated Temperatures*, The Iron and Steel Institute of Japan (ISIJ), Keidanren Kaikan (3rd floor), No. 9-4, Otemachi, 1-Chome, Chiyoda ku, Tokyo 100, Japan. Tel: 81 3 3279-6021

(P) *Metallic Materials and Elements for Aerospace Vehicle Structures*, MIL-HDBK-5D, 1986. Updated periodically, 2 vol. Available from the Superintendent of Documents, U.S. Government Printing Office, Washington, DC 20401

(P) J.E. Payne and P.D. Desai, Ed., *Aluminides*, Vol I, *Properties of Intermetallic Alloys*, CINDAS/Purdue University, 1994

(P) M.F. Rothman, Ed., *High Temperature Property Data: Ferrous Alloys*, ASM International, 1987, 550 pp

(P) G. Samsonov and I.M. Vinitskii, *Handbook of Refractory Compounds*, Plenum Publishing, 1980, 567 pp

(P) C.P. Sullivan, M.J. Donachie, and F.R. Morral, *Cobalt-Base Superalloys*, Centre d'Information du Cobalt, Brussels, 1970

(D) Ultra-refractories Laboratory, BP 5, Odeillo, 66120 Font Romeu, France. Tel: 33 68 30 10 24

(P) V.E. Zinov'yev, *Standard Handbook of (Physical) Properties of Metals at High Temperatures*, Hemisphere, 1990, 258 pp

Electrical, electronic, magnetic, and optical materials

(P) D. Bimberg, R. Blachik, et al., *Semiconductors*, Vol III/17, Subvol a–i (1982–1985); III/22 Subvol a (1987) and b (1989). Landolt-Bornstein. *Numerical Data and Functional Relationships in Science and Technology*, New Series, K.H. Hellwege, Ed., Springer, Berlin

(D) *Colorimetry and Optical Properties of Materials*, Bundesanstalt für Materialsforschung und Prüfung (BAM), Berlin, Unter den Eichen 87, D-1000 Berlin 45, Germany. Tel: 49 30 8104 5400

(C) *Compound Semiconductor Databases*, A, *Crystal Growth of 4–6 Solid Solution Semiconductors;* B, *Optical Data on 2–6 Compound Semiconductors;* C, *Structures and Properties of 2–4 Intermetallic Compounds;* D, *Crystal Data of Intermetallic Compounds*, Electronics Materials Lab., Dept. of Material Science, Faculty of Engineering, Tohoku University, Aramaki, Aoba, Sendai 980, Japan. Tel: (81) 022-222-1800, ext. 4463

(P) *The Databook of Materials for Infrared Detectors*, Japanese Electronics Industry Development Association, 1979

(P) A.S. Diamond, *Handbook of Imaging Materials*, Marcel Dekker, 1991, 608 pp

(P/D) Electronic Materials Information Service (EMIS). Data Reviews Series comprises books on semiconductors and other electronic materials. Twelve compilations have been published: *Properties of Wide Bandgap II–VI Semiconductors*, 1997, 250 pp; *Properties of Gallium Arsenide*, 3rd ed., 1996, 1012 pp; *Properties of III–V Quantum Wells and Superlattices*, 1996, 400 pp; *Properties of Metal Silicides*, 1995, 300 pp; *Properties of Silicon Carbide*, 1995, 300 pp; *Properties of Strained and Relaxed Silicon Germanium*, 1995, 200 pp; *Properties of Group III Nitrides*, 1994, 280 pp; *Properties of Narrow Gap Cadmium-Based Compounds*, 1994, 600 pp; *Properties of Growth of Diamond*, 1994, 437 pp; *Properties of Lattice-Matched and Strained Indium Gallium Arsenide*, 1993, 317 pp; *Properties of Aluminum Gallium Arsenide*, 1993, 345 pp; *Properties of Indium Phosphide*, 1991, 495 pp. Institution of Electrical Engineers, Herts, United Kingdom. Available in the Americas from INSPEC Department, IEEE Operations Center, Piscataway, NJ

(C) *Electronic Materials Properties Database (EMPDb)* at the Institute of Inorganic Chemistry of the Russian Academy of Sciences, Novosibirsk. Tel: 7-3832-355-950. Three subsidiary databases: thermodynamic properties; structural properties; and physical properties; supplemented by numerous application programs for searching, organizing files, and calculations

(D/C) Electronic Properties Information Center (EPIC), CINDAS, Purdue University, 3595 Yeager Rd., W. Lafayette, IN 47906. Tel: (765) 494-9393. Electrical, electronic, magnetic, and optical properties of electrical and electronic materials

(P) J.A. King, *Materials Handbook for Microelectronics*, Artech House, 1988

(P) P. Klocek, *Handbook of Infrared Optical Materials*, Marcel Dekker, 1991, 624 pp

(C) *MAGNELEC* (magnetic and electric properties of metals), Ecole Superieure d'Electricite, Gif, France

(P) F.F. Mazda, *Electronic Engineer's Reference Book*, Butterworths, 1989

(P) T.W. McDaniel and R.H. Victora, Ed., *Handbook of Magneto-Optical Data Recording Materials, Sub-Systems, and Techniques*, Noyes, 1997, 967 pp

(P) M.L. Minges, Ed., *Packaging*, Vol 1, *Electronic Materials Handbook*, ASM International, 1989, 1224 pp

(C) *NIST High Temperature Superconductors Database*, Ceramics Division, National Institute of Standards and Technology, Gaithersburg, MD 20899. Tel: (301) 975-6127. Evaluated property data for oxide superconductors

Electrical, electronic, magnetic, and optical materials (continued)

(C) *Optical Properties of Solids and Liquids (OPTROP)*. Sandia National Laboratory, Division 7243, P.O. Box 5800, Albuquerque, NM 87185. Tel: (505) 844-6242

(C) *OPTIMATR*, Applied Research Corporation, Landover, MD. Software for the calculation of optical properties of materials. The underlying database contains data for >120 crystals, dielectrics, semiconductors, metals, and glasses

(P) E.D. Palik, Ed., *Handbook of Optical Constants of Solids*, 2nd ed., Academic Press, 1991, 1096 pp

(P) K. Schroder, Ed., *Electronic, Magnetic and Thermal Properties of Solid Materials*, Vol 8, Electrical Engineering and Electronics Series, Marcel Dekker, 1978

(P) H.P.J. Wijn, Ed., *Magnetic Properties of Metals: D-elements, Alloys and Compounds*, Springer Verlag, Berlin, 1991, 200 pp

(P) E.P. Wohlfarth, Ed., *Handbook on Ferromagnetic Materials: A Handbook on the Properties of Magnetically Ordered Substances*, 5 vol, Elsevier, 1980–1990

(P) M. Yobe, *Compilation of Materials Data for Development of Specific Electronic Devices*, Proc. Ninth International CODATA Conf. (Jerusalem), 1984

Cement and concrete

(D) American Concrete Institute, P.O. Box 19150, Detroit, MI 48219. Tel: (313) 532-2600

(P) P. Barnes, *Structure and Performance of Cements*, Elsevier, 1983, 558 pp

(D) British Cement Association, Century House, Telford Ave., Crowthorne, Berkshire RG45 6YS, United Kingdom. Tel: 44 1344 762 676

(P) G.C. Bye, *Portland Cement: Composition, Production and Properties*, Pergamon, 1983, 149 pp

(P) *Cement-Manual*, 47th ed., Bauverlag, Wiesbaden, Germany, 1978/1980, 594 pp

(D) Hydraulic and Structural Research Centre: Service for Testing Materials CRIS-SM, Italian Electricity Board (ENEL) (Soil, rock, concrete, bituminous concrete, and geotextiles), Via Ornato 90/14, 20162 Milano, Italy. Tel: 39 2 88471

(D) Measurement of Mechanical Quantities, Bundesanstalt für Materialforschung und Prufung (BAM), Unter den Eichen 87, D-1000 Berlin 45, Germany. Tel: 49 30 8104 6102

(P) A.M. Nevill, *Properties of Concrete*, 3rd ed., Halstead Press, 1986, 792 pp

(P) V.S. Ramachandran, *Concrete Admixtures Handbook: Properties*, Science & Technology, Noyes Publications, 1984, 626 pp

Wood and paper

(P) J.A. Bristow and P. Kolseth, *Paper Structure and Properties*, Marcel Dekker, 1986, 416 pp

(D) Forest Products Laboratory, One Gifford Pinchot Dr., Madison WI 53705-2398. Tel: (608) 231-9200

(D) Fraunhofer-Institute for Wood Research, Fraunhofer-Institut fur Holzforschung (WKI), Fachbereich Anorganisch gebundene Holzwerkstoffe, Bienroder Weg 54E, 3300 Braunschweig, Germany. Tel: (0531) 3090-0

(P) *Handbook of Wood Technology and House Construction*, Research and Education Association, 1981, 864 pp

(P) *Handbook of Wood and Wood-Based Materials*, Forest Products Laboratory, Hemisphere, 1989, 416 pp

(P) M. Lewin and I.S. Goldstein, Ed., *Wood Structure and Composition*, Marcel Dekker, 1991, 512 pp

(P) R.E. Mork, Ed., *Handbook of Physical and Mechanical Testing of Paper and Paperboard*, Marcel Dekker, Vol 1, 1983, 662 pp; Vol 2, 1984, 528 pp

(D) PAPERFACTS, Papiertechnische Stiftung fur Papiererzeugung und Papierverarbeitung (PTS), Hessstr. 130a, D8000 Munchen 40, Germany. Tel: 49 (89) 12601-46

(D) Technical Association of the Pulp and Paper Industry (TAPPI), Technology Park/Atlanta, P.O. Box 105113, Atlanta, GA 30348. Tel: (770) 446-1400

(P) F.F. Wangaard, Ed., *Wood: Its Structure and Properties*, Materials Research Laboratory, The Pennsylvania State University, 1979, 407 pp

(P) *Wood Engineering Handbook*, U.S. Forest Products Laboratory, Prentice-Hall, 1990, 480 pp

(P) *Wood Handbook: Wood as an Engineering Material*, U.S. Dept. of Agriculture, *Forest Service Agriculture Handbook 72*, U.S. Government Printing Office, 1987

Adhesives

(P) *Adhesives and Sealants*, Vol 3, *Engineered Materials Handbook*, ASM International, 1990, 893 pp

(P) E.W. Flick, *Handbook of Adhesives Raw Materials*, Noyes, 1989

(P) I. Skeist, Ed., *Handbook of Adhesives*, 3rd ed., Van Nostrand Reinhold, 1990, 779 pp

(P) A.H. Landrock, *Adhesives Technology Handbook*, Noyes, 1985, 445 pp

(C) *PAL*, Permabond Division of National Starch and Chemical, Eastleigh, Hants, United Kingdom. An expert system and database for adhesive selection. Described in W.A. Lees and P.J. Selby, Adhesive Selection by Means of the PAL Program, *Computerization and Networking of Materials Databases*, ASTM STP 1140, T.I. Barry and K.W. Reynard, Ed., ASTM, 1992, p 301–317

(P) D. Satas, *Handbook of Pressure Sensitive Adhesive Technology*, 2nd ed., Van Nostrand Reinhold, 1989, 768 pp

(P) J. Shields, *Adhesives Handbook*, 3rd ed., rev. 1985, Butterworths, 384 pp

(P) R.L. Patrick, Ed. (Vol 1–6), J.D. Minford, Ed. (Vol 7), *Treatise on Adhesion and Adhesives*, Marcel Dekker, 1967–1991

Nuclear applications

(D) Atomic Energy Research Establishment (AERE), Building 393, Harwell, Oxfordshire OX11 ORA, United Kingdom. Tel: 44 235 24141. Data on metals and alloys, ceramics, polymers, cement, and composites

(C) *Atomic and Molecular Data for Fusion (ALLADIN)*, International Atomic Energy Agency, Vienna, Austria. Numerical data of atomic and plasma-material interaction data for fusion

(continued)

C, computer readable; D, data center; P, print

Table 5 (continued)

Nuclear applications (continued)

(C) *Atomic and Molecular Data Information System (AMDIS),* International Atomic Energy Agency, Vienna, Austria. Atomic, molecular, plasma-material interaction, and materials properties data
(P) K.H. Bockhoff, *Nuclear Data for Science and Technology,* Kluwer, 1983
(P) E.H.P. Cordfunke and R.J.M. Konings, Ed., *Thermochemical Data for Reactor Materials and Fission Products,* North Holland, Amsterdam, 1990, 696 pp
(C) *Fusion Materials Data Base—J (Irradiation Creep and Swelling Data),* Physical Metallurgy Lab., Japan Atomic Energy Research Institute (JAERI), Tokai-Mura, Naka-gun, Ibaraki, Japan 319-11. Tel: (81) 0292-82-5391
(C) *Fusion Reactor Materials Database, (FRMDB),* University of Tokyo, Dept. of Nuclear Engineering, 1984. Cites primarily the physical properties, mechanical properties and radiation effects
(P) C.K. Gupta, *Materials in Nuclear Energy Applications,* CRC Press, 2 vol, 1989, 512 pp
(C) *Material Database for Nuclear Application (Creep of Hastelloy XR),* Materials Engineering Lab., Dept. of Fuels and Materials Research, Tokai Div., Japan Atomic Energy Research Institute (JAERI), Rokai-mure, Naka-gun, Ibaraki, Japan 319-11. Tel: (81) 0292-82-5381
(C) *Material Database for Nuclear Applications (Stress Corrosion Cracking of Ni-Mo and Ni-Cr-Mo-V Steels for Pressure Vessels),* Materials Engineering Lab., Dept. of Fuels and Materials Research, Tokai Div., Japan Atomic Energy Research Institute (JAERI), Rokai-mura, Naka-gun, Ibaraki, Japan 319-11. Tel: 81 0292-82-5381
(C) *Pressure Vessel Steels (PVS),* Materials Research and Computer Simulation Corp., 5266 Hollister Ave., Suite 323, Santa Barbara, CA 93111. Tel: (805) 964-8781. Tensile and toughness-related properties of approximately 200 structural steels
(C) *Structural Materials for Fast Breeder Reactors,* Materials Development Section, Oarai Engineering Center, Power Reactor and Nuclear Fuel Development Corporation, Tokyo, Japan. Tel: (81) 0292-67-4141

Miscellaneous materials

(P) W. Archer, *Industrial Solvents Handbook,* Marcel Dekker, 1996, 316 pp
(P) K.J.A. Brookes, *World Directory and Handbook of Hardmetals,* International Carbide Data, 4th ed., 1987, 490 pp
(P) G. Davies, Ed., *Properties and Growth of Diamond,* EMIS Data Reviews Series, No. 9, 437 pp
(P) D.L. Dreifus, A. Collins, T. Humphreys, K. Das, and P. Pehrsson, *Diamond for Electronic Applications,* Materials Research Society, 1996, 477 pp
(P) M.D. Drory, D.B. Bogy, M.S. Donley, and J.E. Field, *Mechanical Behavior of Diamond and Other Forms of Carbon,* Materials Research Society, 1995, 474 pp
(P) J.E. Field, Ed., *Properties of Natural and Synthetic Diamond,* Academic Press, 1992, 728 pp
(P) E.W. Flick, *Handbook of Paint Raw Materials,* 2nd ed., Noyes, 1988

Miscellaneous materials (continued)

(C) *Foamed Materials Database,* Granta Design Ltd. Database for use with *Cambridge Materials Selector.* Contains information on 118 open and closed cellular solids, both commercial and natural materials, their properties, uses, tradenames, and suppliers
(P) K.A. Gschneidner and L. Eyring, Ed., *Handbook on the Physics and Chemistry of Rare-Earths,* 6 vol, North-Holland, 1979–1983
(P) G.L. Harris, *Properties of Silicides,* EMIS Data Review Series, No. 13, 1995, 300 pp
(P) C. Hornbostel, *Construction Materials: Types, Uses and Applications,* 2nd ed., Wiley, 1991, 1023 pp
(P) T. Iida and R.I.L. Guthrie, Ed., *The Physical Properties of Liquid Metals,* Oxford University Press, 1988, 288 pp
(P) Y. Kawazoe et al., Ed., *Ternary Amorphous Alloy Factual Database,* Springer-Verlag, 1997
(P) M. Lewin and S.B. Sello, Ed., *Handbook of Fiber Science and Technology,* 3 vol, Marcel Dekker, 1983-1985
(P) K. Maex and M. van Russum, Ed., *Properties of Metal Silicides,* EMIS Data Review Series, No. 14, 1995, 300 pp
(P) L. Michaels and S.S. Chissick, Ed., *Asbestos: Properties, Applications, and Hazards,* Vol 1, 1978, 572 pp; Vol 2, 1983, 666 pp; Books on Demand
(D) Minerals Information Office, U.S. Geological Survey, Reston VA. Tel: (703) 648-7712 (asphalt information); (703) 648-7726 (asbestos)
(P) U. Mizutani, Y. Hoshino, and Y. Yamada, Ed., *The Handbook of Formation of Binary Amorphous Alloys by Rapid Quenching Methods* (in Japanese), Agne Publishing Co., Tokyo, 1986
(P) F. Moavenzadeh, Ed., *Concise Encyclopedia of Building and Construction Materials,* Pergamon, 1990, 682 pp. A comprehensive reference work covering all aspects of building and construction materials
(P) H.O. Pierson, *Handbook of Carbon, Graphite, Diamond and Fullerenes: Properties, Processing and Applications,* Noyes, 1993, 405 pp
(P) M.A. Prelas, G. Popovici, and L.K. Bigelow, Ed., *Handbook of Industrial Diamonds and Diamond Films,* Marcel Dekker, 1997, 1240 pp
(D) Rare-Earth Information Center, Institute for Physical Research and Technology, Iowa State University, Ames Laboratory, Ames, IA 50011-3020. Tel: (515) 294-2272
(P) *Rare Metals Handbook,* 2nd ed., C.A. Hampel, Ed., Krieger, 1971, 715 pp. Availability, processing, and physical properties of rare metals
(P) R.P. Reed, F.R. Fickett, L.T. Summers, and M. Stieg, Ed., *Advances in Cryogenic Engineering Materials,* Vol 40A&B, Plenum Press, 1994, 1490 pp
(C) *TAPP,* ESM Software, Hamilton, OH. Materials property and phase diagram database covering properties of 17,000 solids, 14,000 fluids, and thermodynamic properties of >10,000 compound phases and solutions
(P) R.T. Tung, K. Maex, P.W. Pellegrini, and L.H. Allen, Ed., *Silicide Thin Films: Fabrication, Properties, and Applications,* Materials Research Society, 1996, 648 pp

C, computer readable; D, data center; P, print

mation sources: encyclopedias; dictionaries, thesauri, and glossaries; numeric, graphical, and pictorial data sources; auxiliary information sources; the primary literature; reviews; and special topics. The second treats data and information sources relevant to materials economics, policy, and management. A comprehensive materials information directory that is most relevant in the present context has been compiled by Wawrousek et al. (Ref 11); it is an indexed catalog of some 1250 different sources of data on the mechanical and physical properties of engineering materials. Each identified source was categorized by type (computer-readable, data center, printed handbook, etc.) and located in one or several of the approximately 1100 cells of a materials-property matrix. The resource is described in more detail in an ASTM publication (Ref 12). Unfortunately, this directory is almost 10 years old at the time of this writing and is not accessible for on-line computer searching.

Other useful guides to materials information sources are the series of critical surveys of data sources in a variety of fields (Ref 13–16), sponsored by the National Bureau of Standards (now the National Institute of Standards and Technology). For each selected data source covered, the editors defined the publisher or custodian; the scope (the materials/properties covered); the sources of the data; the size of the data bank; the means of data storage and search; the selectivity, timeliness, and availability of the data; the cost of access; and general comments. These surveys are more than 20 years old at the time of this writing and do not extend to all materials and properties of current interest. Other guides and directories are listed in Table 4.

Table 6 Sources of materials purchasing information

American Metal Market (5 per week) Fairchild, NY
D.C. Basil, et al., Ed., *Purchasing Information Sources,* Gale Research, Detroit, MI, 1977, 256 pp
J.C. Bittence, Ed., *Guide to Engineering Materials Producers,* ASM International, 1993, 392 pp
G.S. Brady, H.R. Clauser, and J.A. Vaccari, *Materials Handbook: An Encyclopedia for Managers, Technical Professionals, Purchasing and Production Managers, Technicians, and Supervisors,* 14th ed., McGraw-Hill, 1996, 1072 pp
Ceramic Sources (annual), American Ceramic Society, Westerville, OH
Chemical Economics Handbook, Stanford Research Institute, Palo Alto, CA, 1982–1984
J. Frick, Ed., *Woldman's Engineering Alloys,* 8th ed., ASM International, 1994, 1400 pp
A.H. Grant, *Guide to Buying Castings,* State Mutual, New York
M.R. Leenders and H.E. Fearon, Ed., *Purchasing and Materials Management,* 10th ed., Irwin, Homewood, *1992*
New Trade Names in the Rubber and Plastics Industry, RAPRA Technology, Ltd., Shropshire, United Kingdom. 53,000 trade names with product description and supplier's name and address, available both in print and on CD-ROM
Product Source (annual), *R&D Magazine,* Cahners, Newton, MA
Roskill's Metals Databook, 7th ed., TMS-AIME, 1991. Prices, production, and producers of 32 metals worldwide
R. Serjeanson, Ed., *Metal Bulletin's Prices and Data,* 6th ed., Metal Bulletin Journals Ltd., 1982, 423 pp

Suppliers of Databases (Hosts). The organizations listed below each supply numerous databases (bibliographic as well as numeric or factual) relevant to materials. A given database may be available from several suppliers:

- DIALOG and DataStar, Knight-Ridder Information, Inc., 2440 W. El Camino Real, Mountain View, CA 94040. Tel: (800) 334-2564
- Questel-Orbit, Inc., France Telecom Group, 8000 Westpark Drive, Suite 130, McLean, VA 22102. Tel: (800) 456-7248
- STN International—The Scientific and Technical Information Network, operated cooperatively by Fachinformationszentrum (FIZ) Karlsruhe, Chemical Abstracts Service (CAS), and The Japan Science and Technology Corporation Information Center for Science and Technology (JICST), c/o Chemical Abstracts Service, 2540 Olentangy River Road, Columbus, OH 43210. Tel: (614) 447-3600

For more information about database suppliers, see the sources by Allan and Ferrell and by Schmittroth and Maxfield listed in Table 4.

The Internet is rapidly developing as a source of information on a vast array of topics. Unfortunately as of this writing, there is relatively little information available of materials interest. Refer-

Table 7 Sources of specific types of materials properties and performance data

Mechanical properties

(D) ALUFAT, Aluminum Fatigue Data Base, Fachhochschule Wilhelmshaven, Fachbereich Maschinenbau, Friedrich-Paffrath-Strasse 101, 2940 Wilhelmshaven, Germany. Tel: (0442) 8041, 501638

(C) *Bolting Database (BOLTS),* Materials Research and Computer Simulation Corp., 5266 Hollister Ave., Suite 323, Santa Barbara, CA 93111. Tel: (805) 964-8781. Mechanical properties of 80 base metals

(P) E.R. Booser, *Tribology Data Handbook,* CRC Press, 1997, 1264 pp

(P) H.E. Boyer, Ed., *Atlas of Creep and Stress-Rupture Curves,* ASM International, 1988, 663 pp

(P) H.E. Boyer, *Atlas of Fatigue Curves,* American Society for Metals, 1986, 518 pp

(P) H.E. Boyer, *Atlas of Stress-Strain Curves,* American Society for Metals, 1986, 640 pp

(D) Characteristic Values for Fatigue of Technical Materials, Fachbereich 12-Werkstofftechnik der Universitat (Gesamthochschule) Essen, Postfach 6843, Universitatsstr. 15, 4300 Essen 1, Germany. Tel: (0201) 183-2916, -2621

(D) CYMAP, Cyclic Material Properties, Fachgebiet Werkstoffmechanik der TH Darmstadt, Petersentr. 13, Germany. Tel: (06151) 16-3645

(C) *Data Base on Fatigue Strength of Metallic Materials, Data Base on Fatigue Crack Growth Rates of Metallic Materials,* The Society of Materials Science, Japan Committee on Fatigue Strength of Materials, 1-101 Yoshidaizumidono-cho, Sakyo-ku, Kyoto, Japan 606. Tel: (81) 075-761-5321

(D) Elastic Constants at Low Temperatures for Metals and Alloys and Other Materials, National Institute of Standards and Technology, MC 853, 325 Broadway, Boulder, CO 80303. Tel: (303) 497-3443

(P) *ESDU Engineering Sciences Data Items on Fatigue-Endurance Data,* 8 vol; *Fatigue-Fracture Mechanics,* 4 vol, ESDU International, 1992 (updated frequently)

(C) *FATDAT* (fatigue data on materials), INSA-Lyon 20 rue Albert Einstein, 69621 Ville-urbanne Cedex, France. Tel: (33) 78 94 80 46

(P) *Fatigue Data Book: Light Structural Alloys,* ASM International, 1995, 397 pp

(C) *Fatigue of Aluminum Alloy Weldments,* Engineering Research Institute, Iowa State University, Ames, Iowa 55011. Tel: (515) 294-6979

(D) Fatigue Testing Division, National Research Institute for Metals, Tsukuba, Japan

(P) *Fatigue and Tribological Properties of Plastics and Elastomers,* Plastics Design Library, 1995, 475 pp

(C) *Fracture Toughness of Aluminum Alloys,* Alcoa Technical Center, Alcoa Center, PA 15069. Tel: (412) 339-6651. Data also accessible on STN International on-line service as ALFRAC

(P) H.J. Frost and M.F. Ashby, *Deformation-Mechanism Maps: The Plasticity and Creep of Metals and Ceramics,* Pergamon, Oxford, 1982, 184 pp

(C) *Kawasaki Fatigue Strength Analysis System (KAFSAS),* Strength Research Laboratory, Technical Institute, Kawasaki Heavy Ind. Co., 1-1 Kawasaki-cho, Akashi-city 673, Japan. Tel: (81) 078-921-1626. 400 steels, 40 stainless steels, 60 nonferrous alloys

(P) W.T. Matthews, *Plain Strain Fracture Toughness (K_{Ic}) Data Handbook for Metals,* Army Materials and Mechanics Research Center, 1974, 86 pp. Available from NTIS, Springfield, VA

(D) *Mechanical Properties of Metals at Elevated Temperatures,* The Iron and Steel Institute of Japan (ISIJ), Keidanren Kaikan (3rd floor), No. 9-4, Otemachi, 1-Chome, Chiyoda ku, Tokyo 100, Japan. Tel: 81 3 3279-6021

(P) H. Nakazawa, Ed., *Data Book on Fatigue Crack Growth Rates of Metallic Materials,* 2 vol, Society of Materials Science, Japan, 1983

(C) *NIST Tribomaterials I (ACTIS) Database,* Standard Reference Data Program, National Institute of Standards and Technology, Gaithersburg, MD 20899. Tel: (301) 975-2200. Property data for materials commonly used in tribology applications

(P) M.F. Rothman, Ed., *High Temperature Property Data: Ferrous Alloys,* ASM International, 1987, 550 pp

(P) R.P. Skelton, *High Temperature Fatigue: Properties and Prediction,* Elsevier, 1987

(P) Y. Tanaka, Ed., *Data Book on Fatigue Strength of Metallic Materials,* 3 vol, Society of Materials Science, Japan, 1982

(P/C) D. Taylor, *A Compilation of Fatigue Thresholds and Crack Growth Rates,* EMAS Publishers Ltd., Warley, West Midland, United Kingdom, 1986

(P) H. Zahoor, *Ductile Fracture Handbook,* Electric Power Research Institute, 1989

Corrosion and oxidation behavior

(C) *ACHILLES* (a database and expert system covering corrosion and protection of metals), Atomic Energy Research Establishment, Bldg. 393, AERE, Harwell, Oxfordshire OX11 ORA, United Kingdom. Tel: (44) 235 24141

(P) *Chemical Resistance Guide for Elastomers II,* Compass, 1996, 440 pp

(P) *Chemical Resistance Guide for Metals and Alloys,* Compass, 1996, 436 pp

(P) D.E. Clark et al., *Corrosion of Glass,* Asklee Publishing Co., 1979, 75 pp

(P) *Compass Corrosion Guide II,* Compass, 1996, 115 pp

(C) *CORROSION,* SDC Information Services, 2500 Colorado Ave., Santa Monica, CA 90406. Tel: (213) 820-4111. Carries data on the effects of over 600 agents on 20 metals, carbon, and glass, 26 plastics, and 13 rubbers over a temperature range of 60 to 560 °F. Database corresponding to the Marcel Dekker publication, *Corrosion Resistance Tables*

(P) *Corrosion,* Vol 13, *ASM Handbook,* (formerly *Metals Handbook,* 9th ed.), ASM International, 1987, 1432 pp

Corrosion and oxidation behavior (continued)

(D) Corrosion Data Center, National Institute of Standards and Technology, Materials Bldg., Room B266, Gaithersburg, MD 20899. Tel: (301) 975-4679

(P/C) B.D. Craig and D. Anderson, Ed., *Handbook of Corrosion Data,* 2nd ed., ASM International, 1995, 998 pp. Data also available on diskette

(P) *DECHEMA Corrosion Handbook: Corrosive Agents and Their Interaction with Materials,* D. Behrens, Ed., Vol 1–9; and G. Kreysa and R. Eckermann, Ed., Vol 10–12, VCH, 1987–1993. A series of 12 vol describing the corrosion behavior of technically important materials and prevention methods

(D) *DECOR, DECHEMA'S Materials and Corrosion Data,* Deutsche Gesellschaft für chemische Apparatewesen, chemische Technik und Biotechnologie e.V., I&D-Informationssysteme und Datenbanken, Postfach 970146, Theodor-Heuss-Allee 25, 6000 Frankfurt a.M. 97, Germany. Tel: (069) 7564-0

(P) D.J. De Renzo, Ed., *Corrosion-Resistant Materials Handbook,* 4th ed., Noyes, 1985, 965 pp

(P) W.Z. Friend, *Corrosion of Nickel and Nickel-Base Alloys,* Wiley-Interscience, 1980, 459 pp

(P) D.L. Graver, *Corrosion Data Survey—Metals Section,* 6th ed., NACE, 1986, 192 pp

(P) N.E. Hammer, *Corrosion Data Survey—Nonmetals Section,* 5th ed., NACE, 1975, 494 pp

(C) Information System on High Temperature Corrosion; Corrosion of High Cr Alloys and Structural Materials by H_2/H_2S Gases at High Temperature, Royal Military College, Canada

(P) C.E. Jaske, et al., *Corrosion Fatigue of Metals in Marine Environments, Metals and Ceramics Information Center Series,* Battelle, 1981, 245 pp

(P) K. Komai, Ed., *Data Book on Stress Corrosion and Corrosion Fatigue Properties of Metallic Materials,* 4 vol, Society of Materials Science, Japan, 1987

(P) G.Y. Lai, *High Temperature Corrosion of Engineering Alloys,* ASM International, 1990, 231 pp

(P) L. Lay, *Corrosion Resistance of Technical Ceramics,* Her Majesty's Stationary Office, United Kingdom, 1984, 147 pp

(C) *Materials Deterioration Data Program (MADD)* (formerly DETER), Plastics Technical Evaluation Center (PLASTEC), USA Armament R&D Center, Building 351N, Dover, NJ 07801. Atmospheric chemical resistance data on metals and nonmetals (plastics, elastomers, inorganics)

(P) A.J. McEvily, Ed., *Atlas of Stress-Corrosion and Corrosion-Fatigue Curves,* ASM International, 1990, 541 pp

(C) *NACE/NIST Corrosion Performance Databases,* Corrosion Data Center, National Institute of Standards and Technology, Gaithersburg, MD 20899. Tel: (301) 975-6027. Reference data on the performance of materials in corrosive environments

(P) P.A. Schweitzer, Ed., *Corrosion and Corrosion Protection Handbook,* 2nd ed., Marcel Dekker, 1989, 660 pp

(P) P.A. Schweitzer, Ed., *Corrosion Resistance Tables,* 4th ed., Marcel Dekker, 3 vol, 1995, 3256 pp

(P) J.F. Stringer, *High Temperature Corrosion of Aerospace Alloys,* AGARD-AG 200, 1975

(P) R.S. Treseder, R. Baboian, and C.G. Munger, Ed., *NACE Corrosion Engineer's Reference Book,* 2nd ed., NACE, 1991

(P) H.H. Uhlig and R.W. Revie, *Corrosion and Corrosion Control: An Introduction to Corrosion Science and Engineering,* 3rd ed., Wiley, 1985, 441 pp

Thermal and thermophysical properties

(P) F.J. Blatt, P.A. Schroeder, C.L. Foiles, and D. Greig, *Thermoelectric Power of Metals,* Plenum, 1976, 264 pp

(P) Yu.V. Boyko et al., Ed., *Handbook of Thermodynamic and Optical Properties of Plasmas of Metals and Dielectrics,* Hemisphere, 1991, 424 pp

(P) N.P. Cheremisinoff, *Handbook of Heat and Mass Transfer,* 3 vol, Gulf Publishing, 1986–1989

(P) V.P. Glushko and V.A. Medvedev, *Thermal Constants of Substances,* 10 vol, Hemisphere, 1984, 2000 pp

(C) *JICST Thermophysical and Thermochemical Property Database* (TH System). Research Institute for Production Development, Thermophysical and Thermochemical Properties DB Working Group, Japan Information Center of Science and Technology, 15 Morimoto-cho, Skimogamo Sakyo-Ku Kyoto, Japan. PVT properties, critical density and volume, sound velocity, entropy and enthalpy of organic and inorganic substances

(P) R.E. Krzhizhanovskii, *Thermophysical Properties of Metals and Alloys,* Metallurgia Publishing, Moscow, 1970, 500 pp

(P) K. Schroder, Ed., *Electronic, Magnetic and Thermal Properties of Solid Materials,* Vol 8, Electrical-Engineering and Electronics Series, Marcel Dekker, 1978

(P) P.A. Schroder, *Handbook of Electrical Resistivities of Binary Metallic Alloys,* CRC Press, 1983

(P) E.U. Schunder et al., Ed., *Heat Exchanger Design Handbook,* 5 vol, Hemisphere, 1984–1991

(D) Thermal Systems Division, National Engineering Laboratory, East Kilbridge, Glasgow G75 OQU, United Kingdom. Tel: 44 1355 272 072

(C/D) Thermophysical Properties Engineering Knowledge Base, National Center for Excellence in Metalworking Technology, Concurrent Technologies Corp., 1450 Scalp Ave., Johnstown, PA 15905. Tel: (814) 269-2731. Thermophysical properties data for use in modeling, control, and optimization of metalworking processes

(D) Thermophysical Properties Research Center (TPRC), CINDAS, Purdue University, 3595 Yeager Rd., W. Lafayette, IN 47906. Tel: (765) 494-9393

(P) E. Yu Tonkov, Ed., *High Pressure Phase Transformations: A Handbook,* Gordon and Breach, 1996, 197 pp

Table 8 Sources of materials processing data

Machining

(C) Banque de Donnees Materiaux a Utiliser en decolletage et leur traitements (C.T. Dec-Doc) now known as VULCHAIN-BDM. Centre Technique De L'Industrie du Decolletage (C.T. Dec), BP65, 74301 Cluses, Cedex, France. Tel: (33) 50 98 20 44. Machining data for several thousand metal alloys including ferrous and nonferrous alloys

(P) K.J.A. Brookes, *World Directory and Handbook of Hardmetals,* 4th ed., International Carbide Data, 1987, 490 pp

(C) Informationszentrum für Schnittwerte (INFOS), Laboratorium für Werkzeugmaschinen und Betriebslehre der RWTH EXAPT eV, Postfach 5 87, 5100 Aachen, Germany. Tel: 49 0241 25607

(D/C) Machinability Data Center, Institute for Advanced Manufacturing Sciences, Inc., 1111 Edison Dr., Cincinnati, OH 45216. Tel: (513) 948-2000. Operates CUTDATA machinability database with data for ferrous, nonferrous, and nonmetallic materials

(P) *Machining,* Vol 16, *ASM Handbook* (formerly *Metals Handbook,* 9th ed.), ASM International, 1989, 944 pp

(C) *Machining Data Base* (MDB), Technical Information Service Section, Technical Research Institute of Japan Society for the Promotion of Machine Industry, 1-1-12 Hachiman-cho, Higashikurume-city, Tokyo 203, Japan. Tel: (81) 0424-75-1177. Machining data and related technical information

(P) *Machining Data Handbook,* 3rd ed., 2 vol, Metcut Research Associates, 1980. Available from Machinability Data Center, Institute for Advanced Manufacturing Sciences, Inc., 1111 Edison Dr., Cincinnati, OH 45216. Tel: (513) 948-2000

Heat treating

(P) H. Chandler, Ed., *Heat Treater's Guide: Practices and Procedures for Irons and Steels,* 2nd ed., ASM International, 1995, 904 pp

(P) H. Chandler, Ed., *Heat Treater's Guide: Practices and Procedures for Nonferrous Alloys,* ASM International, 1995, 669 pp

(P) *Heat Treating,* Vol 4, *ASM Handbook,* ASM International, 1991, 1012 pp

(D) Heat Treating Network (HTN), 16600 West Sprague Rd., Suite 445, Cleveland, OH 44130. Tel: (216) 243-8990

(D) Heat Treating Society, ASM International, Materials Park, OH 44073. Tel: (440) 338-5151

(P) *Heat Treatment Reference Book,* C. Learch, Wild Barfield, Imperial Way, Watford, Herts. WD2 4QQ, United Kingdom

(C) *Minitech Computerized Alloy Steel Information System (CASIS),* Minitech Ltd., Box 5185, Station E, Hamilton, Ontario, Canada L8S 4L3. Tel: (416) 523-7721. Hardenability and related properties including as-rolled, quenched-and-tempered, and carburized properties

(P) G.E. Totten and M.A.H. Howes, Ed., *Steel Heat Treatment Handbook,* Marcel Dekker, 1997, 1224 pp

(P) G. Vander Voort, *Atlas of Time-Temperature Diagrams for Irons and Steels,* ASM International, 1991, 804 pp

(P) G. Vander Voort, Ed., *Atlas of Time-Temperature Diagrams for Nonferrous Alloys,* ASM International, 1991, 474 pp

Casting

(D) American Foundrymen's Society, Inc. (AFS), 505 State St., Des Plaines, IL 60016. Tel: (847) 824-0181

(C/D) British Cast Iron Research Association (BCIRA), Alvechurch, Birmingham, Warwickshire B48 7QB, United Kingdom. Tel: 44 527 66414. Databases on ductile and malleable iron and on gray cast iron

(D) British Investment Casting Trade Association (BICTA), The Holloway, Alvechurch, Birmingham, B48 7QA, United Kingdom. Tel: 44 1527 584 770

(C) *CADMOULD/CADFORM,* Institut for Kunststoffverarbeitung (IKV), Ponstr. 49, 5100 Aachen, Germany. Tel: (0241) 80-3806

(P) *Casting,* Vol 15, *ASM Handbook* (formerly *Metals Handbook,* 9th ed.), ASM International, 1988, 937 pp

(P) E. Hermann and D. Hoffman, *Handbook on Continuous Casting,* Aluminum, Germany, Heyden, 1980

(D) Steel Founders' Society of America (SFSA), 455 State St., Des Plaines, IL 60016. Tel: (847) 299-9160

Deformation processing

(P) *Atlas of Cold Working Properties,* Vol 1, *Aluminum Materials;* Vol 2, *Copper Materials;* Vol 3, *Precious Metals, Ni and Ni Alloys, Zn and Zn Alloys, Other Nonferrous Metals,* German Society of Metallurgy, 1987

(C) *Atlas of Formability Engineering Knowledge Base,* National Center for Excellence in Metalworking Technology (NCEMT), Concurrent Technologies Corp., 1450 Scalp Ave., Johnstown, PA 15905. Tel: (814) 269-2731

(P) *Atlas of Hot Working Properties of Non-Ferrous Metals,* Vol 1 & 2, Deutsche Gesellschaft für Metallkunde, Oberursel, Germany, 1979

(P) T.T.G. Byrer, Ed., *Forging Handbook,* Forging Industry Association and American Society for Metals, 1985, 296 pp

(D) Forging Industry Association (FIA), 25 Prospect Ave. West, Suite 300, Cleveland, OH 44115. Tel: (216) 781-6260

(P) *Forming and Forging,* Vol 14, *ASM Handbook* (formerly *Metals Handbook,* 9th ed.), ASM International, 1988, 978 pp

(P) H.J. Frost and M.F. Ashby, *Deformation-Mechanism Maps: The Plasticity and Creep of Metals and Ceramics,* Pergamon, Oxford, 1982, 184 pp

(D) National Center for Excellence in Metalworking Technology (NCEMT), Concurrent Technologies Corp., 1450 Scalp Ave., Johnstown, PA 15905. Tel: (814) 269-2731

(P) Y.V.R.K. Prasad and S. Sasidhara, Ed., *Hot Working Guide: A Compendium of Processing Maps,* ASM International, 1997, 545 pp

(C/D) Thermophysical Properties Engineering Knowledge Base, National Center for Excellence in Metalworking Technology, Concurrent Technologies Corp., 1450 Scalp Ave., Johnstown, PA 15905. Tel: (814) 269-2731. Thermophysical properties data for use in modeling, control, and optimization of metalworking processes

Joining

(D) American Welding Society (AWS), 550 N.W. LeJeune Rd., Miami FL 33126. Tel: (800) 443-9353. Offers Teleweld information service

(D) Edison Welding Institute (EWI), 1250 Arthur E. Adams Dr., Columbus, OH 43221. Tel: (614) 688-5000

(P) *Handbook of Plastics Joining: A Practical Guide,* Plastics Design Library, 1996, 485 pp

(D) Information Systems on Materials Engineering—Joining and Nondestructive Testing, Bundesanstalt für Materialforschung und Prüfung (BAM), D-12200 Berlin, Germany. Tel: 49 30 8104 1555

(C) *LOTPROSA,* Soldering Database for Microelectronics, Lehrstuhl für Metallurgie und Metallkunde der TU Munchen, Postfach 202420, Arcisstr. 21, 8000 Munchen 2, Germany. Tel: (089) 2105-3544

(P/C) *The Metals Blue Book: Welding Filler Metals,* American Welding Society and Casti Publishing, 1995, 417 pp

(P) R.O. Parmley, Ed., *Standard Handbook of Fastening and Joining,* 3rd ed., McGraw-Hill, 1997

(P) *Welding, Brazing, and Soldering,* Vol 6, *ASM Handbook,* ASM International, 1993, 1299 pp

(P) *Welding Handbook,* 8th ed., American Welding Society: *Welding Technology,* Vol 1, 1991, 638 pp; *Welding Processes,* Vol 2, 1991, 955 pp; *Materials and Applications,* Part I, Vol 3, 1996, 526 pp

(D) *Welding of Materials,* Schweisstechnische Lehr- und Versuchsanstalt Hannover e.V. (SLV), Am Lindener Hafen 1, 3000 Hannover 91, Germany. Tel: (0511) 21962-0

(C) *ZUSMAT,* Data on welding additives and auxiliary materials, Rheinisch-Westfalischer Technischer Uberwachung-Verein e.V., Zentralabteilung Informationssysteme und Statistik, Postfach 103261, Kurfurstenstr, 58, 4300 Essen, Germany. Tel: (0201) 825-2624

Recycling

(D) Institute of Scrap Recycling Industries Inc. (ISRI), 1325 G. Street N.W., Suite 1000, Washington, DC 20005. Tel: (202) 466-4050

(C) Polydata GmbH, Frankenbergerstrasse 30, D-5100, Aachen, Germany. Tel: 49 241 502 023. Database of technical support for dismantling and recycling

C, computer readable; D, data center; P, print

ences 17 and 18 are two general guides to science and technology on the Internet. References 19 and 20 are introductions to the Internet for materials scientists and engineers. E-mail, PC software, and mathematical routines are typical examples of available general-interest applications on the Internet. Newsgroups such as "sci.engr.metallurgy" and "sci.materials" can provide free information (from the personal knowledge of individuals) on the existence of databases covering particular materials topics. Because such a vast quantity of information is available on the Internet, considerable time or specialized browsing software is needed to deal with it. Unfortunately, there is no index via browser programs such as Gopher or World Wide Web (WWW) for materials-related archives at the time of this writing. Many materials databases are accessible through the Internet, for example, those on STN International (Ref 21), which has an Internet link. King et al. (Ref 22) have developed an inventory of materials databases to promote knowledge of available (not necessarily via the Internet) materials and chemical property databases. Links to materials-related Web sites are provided in many places on the WWW (for example, at sites sponsored by professional societies, government agencies, and college/university materials science departments).

Specific Sources of Data

In addition to the materials information sources already cited directly, many others are listed in Tables 5 to 9 to give the reader some indication of what is available, but without implication that those listed are the most current, most authoritative, or most important. Thus, the listings in these tables are *representative,* not exhaustive. Many more can be found by perusal of Ref 11. In selecting sources for inclusion, emphasis was placed on sources containing substantial amounts of quantitative information on engineering properties of materials, rather that those of a descriptive or tutorial character in their coverage of particular materials classes, properties, phenomena, or application areas. In contrast, sources of data on crystallography, thermodynamics, and analytical information receive scant attention. Three different types of source are coded as C, computer readable; D, data center; and P, print:

- *Computer readable* comprises all electronic formats: floppy disk, tape, CD-ROM, and online. (The state-of-the-art with respect to electronic access to materials information as of 1995 is reviewed in Ref 23.)
- *Data centers* are organizations that maintain both print and electronic files of data in one or more fields. Their staffs often are not only information specialists, but also individuals with expertise in the subject field of their collections and who usually have experience in the evaluation of the data in the collection. Both direct and indirect access to the files is usually possible as well as consultative assistance.

Table 9 Sources of materials safety and health information

(P) *Chemical Safety Data Sheets,* Vol 1, *Solvents,* 1989, 300 pp; Vol 2, *Main Group Metals and Their Compounds,* 1989, 500 pp; Vol 3, *Corrosives and Irritants,* 1990, 300 pp, Royal Society of Chemistry Publications, CRC Press

(P) H.E. Christensen and E.J. Fairchild, Ed., *Registry of Toxic Effects of Chemical Substances,* U.S. Dept. of Health, Education, and Welfare, Government Printing Office, 1986, 1245 pp

(P) G.S. Dominguez and K.G. Bartlett, *Hazardous Waste Management,* Vol I, CRC Press, 1986, 272 pp

(P) L. Friberg, G.F. Nordberg, and V.B. Vouk, Ed., *Handbook on the Toxicology of Metals,* 2nd ed., Elsevier, 1986

(P) R.A. Goyer, Toxicity of Metals, *Properties and Selection: Nonferrous Alloys and Special-Purpose Materials,* Vol 2, *ASM Handbook* (formerly *Metals Handbook,* 10th ed.), ASM International, 1990, p 1233–1269

(P) P. Gross and D.C. Brown, *Toxic and Biomedical Effects of Fibers; Asbestos, Talc, Inorganic Fibers, Man-Made Vitreous Fibers and Organic Fibers,* Noyes, 1984, 257 pp

(D/C) Hazardous Materials Service, c/o Atomic Energy Authority Technology PLC, Harwell, Didcot, Oxfordshire OX11 ORA, United Kingdom. Tel: 44 1235 821111. A service for information on hazardous materials; operates *CHEMDATA,* a computerized database of chemical products

(P) *International Register of Potentially Toxic Chemicals (IRPTC),* United Nations Environment Programme, Pavillon du Petit Sacannex, 16 Ave Jean Trembly, 1209 Geneva, Switzerland

(P/C) *Materials Safety Data Sheet Collection,* 3 vol, Genium Publishing, Schenectady, NY. Covers 1019 substances; also available on CD-ROM

(C) National Library of Medicine, 8600 Rockville Pike, Bethesda, MD 20894. Tel: (301) 496-6095, (800) 638-8480, or (301) 496-6308. Database on toxicity of substances, and physical and chemical properties

(P) W.B. Neely and G.E. Blau, *Environmental Exposure from Chemicals,* Vol I, 256 pp; Vol II, 192 pp, CRC Press, 1985

(P) J.L. O'Donoghue, *Neurotoxicity of Industrial and Commercial Chemicals,* Vol I, CRC Press, 1985, 232 pp; Vol II, 1985, 224 pp

(C) *Oil and Hazardous Materials Technical Assistance Data System (OHM-TADS),* U.S. Environmental Protection Agency. Automated information retrieval file for data on 1402 hazardous substances

(P) N.R. Sax and R.J. Lewis, *Dangerous Properties of Industrial Materials,* 7th ed., 3 vol, Van Nostrand Reinhold, 1984, 4000 pp

(P) H.G. Seiler and H. Sigel, *Handbook on Toxicity of Inorganic Compounds,* Marcel Dekker, 1987, 1025 pp

(P) V.O. Sheftel, *Toxic Properties of Polymers and Additives,* RAPRA Technology, Shrewsbury, UK, 1990

(P) M. Sittig, *Handbook of Toxic and Hazardous Chemicals,* Noyes, 1981, 729 pp

(P) J.V. Soderman, Ed., *CRC Handbook of Identified Carcinogens and Non-Carcinogens: Carcinogenicity—Mutagenicity Database,* Vol I, *Chemical Class File,* 1982, 696 pp, Vol II, *Target Organ File,* CRC Press, 1982, 640 pp

(P) *Toxic and Hazardous Industrial Chemicals Safety Manual,* The International Technical Information Institute, Tokyo, Japan, 1984, 591 pp

(D) Toxic Materials Information Center (TMIC), Environmental Information System Office, Oak Ridge National Laboratory, P.O. Box X, Oak Ridge, TN 37830. Tel: (423) 574-1145

(D) Toxicology Data Network and Environmental Health Information Program, National Library of Medicine, Specialized Information Services Division, 8600 Rockville Pike, Bethesda, MD 20894. Tel: (301) 496-6531

(D) Toxicology Information Center, Environmental Studies and Toxicology Board, National Academy of Sciences, 2101 Constitution Ave., Washington, DC 20418. Tel: (202) 334-2387

C, computer readable; D, data center; P, print

Table 10 Types of metadata associated with materials properties data

Completeness of description of material

- Producer
- Heat/lot identification
- Status of material (commercial or experimental)
- Name/UNS No.
- Specification(s)
- Condition/temper/grade/class
- Product form
- Chemical analysis
- Any special melting practice or source material, for example, vacuum cast, powder metallurgy product, etc.
- Process/thermal history
- Microstructure
- Background properties (e.g., tensile properties)

Completeness of description of test method

- Test standard
- Specimen type, size, shape, finish
- Specimen location and orientation
- Specimen relation to end-use situation
- Loading rate
- Temperature and method of measurement
- Environment and method of monitoring
- Statements of precision and accuracy

Completeness of reporting of test data

- Report format
- Coverage of logical variables
- Tests run to completion according to test plan
- Replications
- Documentation of units and conversions
- Consistency of results
- Failure type and description

Source: Ref 29

- *Print* includes handbooks, review articles, data sheets, and reports. The information provided may be numeric tables, graphs, images, or other pictorial formats.

Evaluation and Interpretation of Data

Having located one or more pieces of data relevant to his or her problem, the scientist or engineer must usually answer additional questions before the information can be used. If the data found are for conditions not precisely those pertinent to the problem situation, how would the requisite interpolation or extrapolation be performed? If two or more sets of data are not in agreement with each other, which should be chosen; or should some sort of averaging technique be applied? Is the functional dependence of the data on measured variables such as temperature or composition in such accord with theoretical expectations as to give credence to their validity? What does the internal scatter of the data or the disparity between these data and those from another source imply about their reliability? Can a self-consistent data set of stated reliability be synthesized from limited, fragmentary, and conflicting experimental data? Such questions have been dealt with in a general way by Ho and Touloukian (Ref 24), Lide and Paul (Ref 25), and Barrett (Ref 26–28), among others.

Development of answers to the questions just posed involves a sequence of processes in the analysis of the available data. Briefly, these comprise:

- *Reduction of data:* A process by which physical measurements of observations are subjected to a series of adjustments (including such steps as the correction of the measured results for known errors of the instruments, the conversion to standard units and, perhaps, the transformation to standard conditions appropriate to the type of experiment or observation), so as to permit the results of observations made at different times or places to be compared and combined more readily.
- *Homogenization:* The process by which comparable data sets arising from dissimilar sources, possibly with differing nomenclatures and associated units are converted to a harmonized terminology and unit system applicable to a local standard.
- *Evaluation:* The process of establishing the accuracy and integrity of property data. Evaluation involves examination and appraisal of the data presented, assessment of experimental technique and associated errors, consistency checks for allowed values and units, comparison with other experimental or theoretical values, reanalysis and recalculation of derived quantities as required, selection of best values, and assignment of probable error or reliability.

A typical result of such an analysis is shown in Fig. 3. It must be emphasized that these are not clerical activities, but ones that require substantial expertise by individuals experienced in the subject area of the data but also in the techniques of data analysis. These persons usually are relatively senior and require extensive bibliographic and computer support; hence the data analysis process can be quite expensive.

In selecting data for use or for incorporation in an authoritative reference work or electronic database, there are two overriding considerations to be applied, even before the analytical processes just reviewed are undertaken. First, can an explicit trail be documented to the original source of the raw data; and second, are the metadata associated with the number or numbers to be used adequately complete? Metadata are the descriptive data about the material reported on, about the test methods used to determine the properties cited, and about the reporting of the test data. They have been more fully detailed by Kaufman (Ref 29) as listed in Table 10.

Data may also be considered for use that do not derive from physical tests, but result from some theoretical prediction or are estimated from values for similar materials. Here, too, metadata are required that describe the theoretical techniques, the basis for the estimate, the estimation procedures and reference cases, and so forth. Other references (Ref 30–33) also examine the importance of metadata and the multiple roles metadata can play.

Table 11 ASTM E-49 materials database standards

ASTM No.	Topic
Materials designations	
E 1308	Polymers
E 1338	Metals and alloys
E 1339	Aluminum alloys
Data recording formats	
E 1313	Material properties data
E 1723	Test data for plastics
E 1761	Fatigue and fracture data for metals
Codes for use in databases	
E 1722	Materials and chemical property data codes
Terminology	
E 1314	Structuring terminological records
E 1443	Terminology for materials and chemical databases
E 1013	Terminology for computerized systems
Data and database quality	
E 1407	Materials database management
E 1484	Quality indicators for data and databases
E 1485	Material and chemical property database descriptors
Chemical structural information	
E 1586	Computerization of chemical structural information
Laboratory information management systems	
E 1578	Guide for laboratory information management systems
E 622	Guide for development of computerized systems
E 625	Guide for training users
E 1340	Guide for rapid prototyping
E 1283	Guide for procurement of computer integrated manufacturing systems
E 1034	Guide for classifying industrial robots
E 627	Guide for documenting computerized systems
E 919	Specification for software documentation of computerized systems
E 731	Guide for selection and acquisition of commercially available computer systems
E 1206	Guide for computerization of existing equipment

The various data evaluation methods and procedures to be used have been reviewed in some detail for the general case by Barrett (Ref 28) and need not be recapitulated here. However, some additional comments are important. First, while the procedures Barrett describes can be generally applied, more specific techniques have been examined for important individual mechanical property classes such as creep, fatigue, and stress rupture (Ref 34–38). Assessment of large data sets can be eased in cases where some constitutive law, theoretical or empirical, is applicable. A familiar example is the so-called Larson-Miller analysis of stress-rupture data as shown in Fig. 4. Such an approach is useful for identification of a discordant data set or for interpolation within the range of the actual test parameters; it should not be used for extrapolation. A long-range goal is to have the capability for prediction of material behavior in more realistic situations where two or more processes are combined, for example, creep and fatigue, or where the stress and/or temperature history is complex, even within a single deformation regime. For recent progress in this area see the chapters by Jenkins, Evans, and McLean in Ref 38.

Table 12 Standards developed by ASTM E-49 in cooperation with, but maintained by, other committees or organizations

No.	Topic	Jurisdictional body
D 5592	Guide for materials properties needed in engineering design using plastics	ASTM D-20
E 1309	Identification of composite materials	ASTM D-30
E 1434	Recording mechanical property test data for high-modulus fiber-reinforced composites	ASTM D-30
E 1471	Identification of fibers, fillers, and core materials	ASTM D-30
G 107	Recording corrosion data for metals	ASTM G-1
E 1454	Recording computerized ultrasonic test data	ASTM E-7
E 1475	Recording computerized radiological test data	ASTM E-7
A9.1	Recording data descriptive of arc welds	AWS
A9.2	Recording arc weld property and inspection data	AWS

Obtaining and Reporting Test Data

Of all test data on materials, those most readily available and most used in materials selection and design activities are mechanical properties data. The subsequent discussion relates specifically to that class of properties, but the principles outlined can be extended to other families of properties.

The utility of test data is not simply enhanced when standard tests are used, but, as previously discussed, may only be *permitted* when:

- Standard tests are used
- The test equipment has been calibrated using well-defined, certified reference materials
- Standards for reporting of test data have been fully met
- Terms used, descriptive of the material and the test, are recorded and defined

The accelerating use of computers to store and analyze test data has forced increasing rigor, especially with respect to the last two points. Without the meeting of these conditions, it would be difficult to search a database or to properly merge or compare independent data sets. For these reasons ASTM E-49 (Committee on Computerization of Material and Chemical Property Data) has been active in recent years developing standards for describing materials and recording test results and properties. Notwithstanding the desirability of using standardized test methods whenever possible, prototype service testing can often yield crucial information to aid a design decision.

The materials property database standards issued by ASTM E-49 as of 1996 (Table 11) include guides for identification of a material under discussion (Ref 40) and for data recording (Ref 41). Specific recommendations are made for bearing tests, plane strain fracture toughness, tension, compression, and notch-bar impact test data

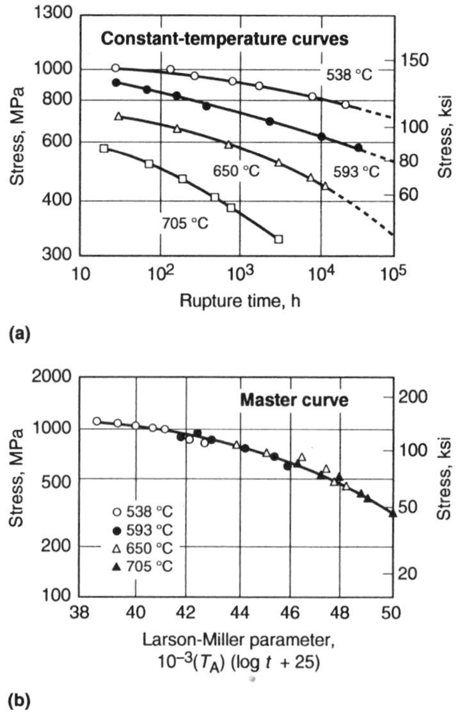

Fig. 4 Example of the application of a "constitutional law" to aggregate and harmonize data. (a) Stress-rupture data for Inconel 718. (b) Larson-Miller analysis of the data. Source: Ref 39

for metals and plastics as well as for arc-welded properties and corrosion data. Related work includes standards for codes, terminology, quality indicators, chemical structure information, and laboratory management information systems as summarized in Table 11. Other work by ASTM E-49 concerns standards developed in collaboration with other professional organizations but maintained by them as summarized in Table 12. Still more work of this kind is also going forward for other types of tests and other classes of materials, but has not yet resulted in a formal standard.

The work of ASTM was preceded by a general reconnaissance of standards and metadata requirements for computerization of selected mechanical properties of metallic materials (Ref 31). The specific properties covered were tensile behavior, hardness numbers, notch-bar impact test parameters, and fatigue properties. While test methods for the properties examined were considered to be generally adequate, improved standards for data recording were recommended as well as the availability of standard reference materials certified as to their mechanical properties.

The standardization of industrial data by and for ISO is reviewed in Ref 42, which describes three ISO-developed standards: STEP (standard for exchange of product model data), P-LIB (parts libraries), and MANDATE (manufacturing data exchange).

Other useful references on data development, recording, and use may also be cited. The *ASM Handbook* includes volumes that cover mechanical testing (Ref 43), corrosion (Ref 44), nondestructive testing (Ref 45), tribological behavior (Ref 46), and fatigue and fracture (Ref 47). Other ASM reference works are devoted specifically to tensile testing (Ref 48) and hardness testing (Ref 49). Two British works considered high-temperature mechanical properties, specifically, Loveday et al. (Ref 50) and Curbishley (Ref 51). The latter work is directed at the nonspecialist who needs to know what types of tests are available and how to select the most appropriate test for a particular purpose, and who needs to have an appreciation of the methods for data determination and analysis and a knowledge of applicable standards. The emphasis is on metals, but the principles are pertinent to ceramics, polymers, and composites as well. The particular properties addressed include: tensile, creep, fatigue crack growth, low-cycle fatigue, high-cycle fatigue, and fracture toughness. Unfortunately, this work fails to show, even by reference, data compilations, print or computerized, in any of the subject fields examined.

For ceramics, a two-part handbook for the engineer and designer has been prepared (Ref 52, 53). The first part presents an overview of the properties of technical and engineering ceramics (not just mechanical properties, but thermal, electrical, optical, and chemical as well), and the second part is intended to cover data sheets for specific classes of commercially available ceramic materials (oxides, silicates, silicon carbide, silicon nitride, nonoxide ceramics, halides, glasses, etc.). To date, only the volume on high-alumina ceramics has appeared (Ref 53).

Conclusions

Great progress has been made in the materials information field in the second half of the 20th century. Enormous problems still remain, however. Adequate directories of information sources are needed; much of the accessible data is of indeterminate reliability; gaps exist in the coverage of the materials-properties matrix; most existing data sources are mutually incompatible and cannot easily be used together; only a fraction of all needed data is in electronic form; and there has been a general failure to appreciate how expensive, yet rewarding, are data compilation and evaluation activities. This situation has been analyzed in some detail by Westbrook and Kaufman (Ref 54).

REFERENCES

1. J.H. Westbrook, Material Information Sources, *Encyclopedia of Materials Science and Engineering,* M.B. Bever, Ed., Pergamon, 1986, p 527
2. N.A. Waterman and M.F. Ashby, Ed., *The Materials Selector,* 2nd ed., 3 Vol, Chapman and Hall, 1997
3. D.K. Allen and P.R. Smith, *Engineering Materials Taxonomy,* Monograph No. 4, Computer-Aided Manufacturing Laboratory, Brigham Young University, 1982
4. N.E. Promisel, R.J.H. Bollard, R.E. Harmon, S.A. Hoenig, I. Katz, M.F. McFadden, M.A. Steinberg, E.I. Shobert, and C.R. Sutton, "Materials and Processes Specifications and Standards," NMAB Report 330, National Academy of Sciences, 1977
5. P. Sargent, *Materials Information for CAD/CAM,* Butterworths-Heinemann, Oxford, England, 1991
6. E.F. Begley and R.G.Munro, Issues in Development of an Advanced Ceramic Materials Selector System, *Computerization and Networking of Materials Databases,* Vol 3, STP 1140, T.I. Barry and K.W. Reynard, Ed., ASTM, 1992, p 272
7. "Automating Materials Selection During Structural Design," NMAB 467, National Materials Advisory Board, 1995
8. J. Datsko, *Materials Selection for Design and Manufacturing: Theory and Practice,* Marcel Dekker, 1997, 320 pp
9. J.H. Westbrook, Materials Standards and Specifications, *Kirk-Othmer Encyclopedia of Chemical Technology,* Vol 16, 4th ed., John Wiley & Sons, 1995, p 33–67
10. J.H. Westbrook and K.W. Reynard, Sources of Information for Materials Economics, Policy, and Management, *Concise Encyclopedia of Materials Economics, Policy, and Management,* M.B. Bever, Ed., Pergamon Press, Oxford, 1993, p 35
11. H. Wawrousek, J.H. Westbrook, and W. Grattidge, Data Sources of Mechanical and Physical Properties of Engineering Materials, *Physik Daten,* No. 30-1, Fachinformationszentrum, Karlsruhe, Germany, 1989, 257 pp
12. H. Wawrousek, J.H. Westbrook, and W. Grattidge, Data Sources of Mechanical and Physical Properties of Engineering Materials, Second International Symposium, *Computerization and Networking of Materials Databases,* STP 1106, J.G. Kaufman and J.S. Glazman, Ed., ASTM, 1991, p 142
13. R. B. Gavert, R.L. Moore, and J.H. Westbrook, "Critical Surveys of Data Sources: Mechanical Properties of Metals," Special Publ. 396-1, National Bureau of Standards, 1974, 81 pp
14. D.M. Johnson and J.F. Lynch, "Critical Surveys of Data Sources: Properties of Ceramics," Special Publ. 396-2 , National Bureau of Standards, 1975, 47 pp
15. R.B. Diegle and W.K. Boyd, "Critical Surveys of Data Sources: Corrosion of Metals," Special Publ. 396-3, National Bureau of Standards, 1976, 29 pp
16. M.J. Carr, R.B. Gavert, R.L. Moore, H. Wawrousek, and J.H. Westbrook, "Critical Surveys of Data Sources: Electrical and Magnetic Properties of Metals," Special Publ. 396-4, National Bureau of Standards, 1976, 83 pp
17. G.P. Clement, "Science and Technology on the Internet PLUS," Laboratory Solutions, San Carlos, CA, 1995, 264 pp + diskettes
18. E.J. Renehan, Jr., *Science on the Web,* Springer, 1996, 382 pp
19. K.J. Meltsner, Understanding the INTERNET: A Guide for Materials Scientist and Engineers, *J. Met.,* Vol 47 (No. 4), 1995, p 9
20. B.J. Thomas, *The Internet for Scientists and Engineers: Online Tools and Resources,* International Society for Optical Engineering, 1996, 445 pp; also available from TMS
21. V.J. Drago and J.G. Kaufman, Technical Features of the Chemical and Materials Property Data Network Service, *J. Chem. Info. Comput. Sci.,* Vol 33 (No. 1), 1993, p 32
22. T.M. King, Y. Monma, and S. Nishijima, Promoting Knowledge of Materials and Chemical Property Databases, *Proc. 15th Int. CODATA Conf.* (Tsukuba, Japan), Oct 1996
23. J.H. Westbrook, J.G. Kaufman, and F. Cverna, Electronic Access to Factual Materials Information: The State of the Art, *MRS Bull.,* Vol 20 (No. 8), 1995, p 40–48
24. C.Y. Ho and Y.S. Touloukian, Methodology in the Generation of Critically Evaluated, Analyzed and Synthesized Thermal, Electrical and Optical Properties Data for Solid Materials, *Proc. Fifth Int. CODATA Conf.* (Boulder, CO), 1976, p 615
25. D.R. Lide and M.A. Paul, Ed., *Critical Evaluation of Chemical and Physical Structural Information,* National Academy of Sciences, 1974, 628 pp
26. A.J. Barrett, The Evaluation/Validation Process—Practical Considerations and Methodology, *Development and Use of Numerical and Factual Databases,* AGARD LS-130, NATO Advisory Group for Aerospace Research and Development, Paris, 1983, p 5.1–5.14
27. A.J. Barrett, On the Evaluation and Validation of Engineering Data, *Computer Handling and Dissemination of Data,* P.S. Glaeser, Ed., Elsevier, 1987, p 124
28. A.J. Barrett, Data Evaluation, Validation, and Quality, *ASTM Manual on the Building of Materials Databases,* C.H. Newton, Ed., ASTM, 1993, 107 pp
29. J.G. Kaufman, Standards for Computerized Material Property Data, ASTM Committee E-49, *Computerization and Networking of Materials Databases,* STP 1017, J.S. Glazman and J.R. Rumble, Jr., Ed. ASTM, 1989, p 7–22
30. J.L. McCarthy, Metadata Management for Large Statistical Databases, *Proc. Eighth Int. Conf. on Very Large Databases* (Mexico City), 1982, p 234–243
31. J.H. Westbrook, "Standards and Metadata Requirements for Computerization of Selected Mechanical Properties of Metallic Materials," Special Publ. 702, National Bureau of Standards, 1985
32. J.L. Dolby, The Role of Description in the Evaluation of Data Quality, *The Role of Data in Scientific Progress,* P.S. Gaeser, Ed., Elsevier/CODATA, 1985, p 357
33. J.H. Westbrook and W. Grattidge, The Role of Metadata in the Design and Operation of a Materials Database, *Computerization and Networking of Materials Databases,* Vol 2, STP 1106, ASTM, 1991 p 84–102

34. "Statistical Analysis of Fatigue Test Results," Fatigue Endurance Data Items 68012–68017, ESDU International, London, 1968
35. M.T. Yu, F.J. Lu, T.H. Topper, and P. Irving, Fatigue Data Pooling and Probabilistic Design, *Computerization and Networking of Materials Databases,* Vol 2, STP 1106, J.G. Kaufman and J.S. Glazman, Ed., ASTM, 1991, p 197–213
36. S. Nishijima, Y. Monma, and K. Kanazawa, "Significance of Data Evaluation Models in Materials Databases," VAMAS Technical Report No. 6, National Research Institute for Metals, Tokyo, 1990
37. Y. Monma, K. Kanazawa, and S. Nishijima, "Computational Models for Creep and Fatigue Data Analysis," VAMAS Technical Report No. 7, National Research Institute for Metals, Tokyo, 1990
38. P. Spilling, Ed., *Characterization of High Temperature Materials: Numerical Techniques,* Institute of Metals, 1989, 226 pp
39. H.R. Voorhees, Assessment and Use of Creep-Rupture Properties, *Mechanical Testing,* Vol 8, *ASM Handbook* (formerly 9th ed. *Metals Handbook),* 1985, p 329–342
40. B. Moniz, Nomenclature and Current Standards for Identification of Engineering Materials, *ASTM Manual on the Building of Materials Databases,* C.H. Newton, Ed., MNL 19, ASTM, 1993, 107 pp
41. M.W. Wardle, Nomenclature and Current Standards for Recording of Test Results and Properties, *ASTM Manual on the Building of Materials Databases,* C.H. Newton, Ed., MNL 19, ASTM, 1993, 107 pp
42. A.-F. Cutting-Decelle and A.-M. Dubois, Standardization of Industrial Data: The ISO Approach Through the Product Data Technology-Based Standards: STEP, P-LIB, and MANDATE—Some Industrial Applications, *Proc. 1996 International CODATA Conference,* Tsukuba, Japan, 1996
43. *Mechanical Testing,* Vol 8, *ASM Handbook* (formerly 9th ed. *Metals Handbook*), American Society for Metals, 1985, 778 pp
44. *Corrosion,* Vol 13, *ASM Handbook* (formerly 9th ed. *Metals Handbook*), ASM International, 1987, 1432 pp
45. *Nondestructive Evaluation and Quality Control,* Vol 17, *ASM Handbook* (formerly 9th ed. *Metals Handbook*), ASM International, 1989, 795 pp
46. *Friction, Lubrication, and Wear Technology,* Vol 18, *ASM Handbook,* ASM International, 1992, 942 pp
47. *Fatigue and Fracture,* Vol 19, *ASM Handbook,* ASM International, 1996, 950 pp
48. P. Han, Ed., *Tensile Testing,* ASM International, 1992, 207 pp
49. Hardness Testing, ASM International, 1987, 188 pp
50. M.S. Loveday, M.F. Day, and B.F. Dyson, *Measurement of High Temperature Properties of Materials,* Her Majesty's Stationery Office, London, 1982
51. I. Curbishley, Ed., *Mechanical Testing,* Vol 3, *Characterization of High Temperature Materials,* The Institute of Metals, London, 1988, 357 pp
52. R. Morrell, *An Introduction for the Engineer and Designer,* Part 1 of *Handbook of Properties of Technical and Engineering Ceramics,* Her Majesty's Stationery Office, London, 1985, 348 pp
53. *Data Reviews,* Section 1, *High Alumina Ceramics,* Part 2 of *Handbook of Properties of Technical and Engineering Ceramics,* Her Majesty's Stationery Office, London, 1987, 255 pp
54. J.H. Westbrook and J.G. Kaufman, Impediments to an Elusive Dream: Computer Access to Numeric Data for Engineering Materials, *Modeling Complex Data for Creating Information,* J.-E. Dubois and N. Gershon, Ed., Springer Verlag, Berlin, 1996, p 125–132

Section 6: Properties versus Performance of Materials

Properties Needed for the Design of Static Structures

Mahmoud M. Farag, The American University in Cairo

ENGINEERING DESIGN can be defined as the creation of a product that satisfies a certain need. A good design should result in a product that performs its function efficiently and economically within the prevailing legal, social, safety, and reliability requirements. In order to satisfy such requirements, the design engineer has to take into consideration a large number of diverse factors:

- *Function and consumer requirements,* such as capacity, size, weight, safety, design codes, expected service life, reliability, maintenance, ease of operation, ease of repair, frequency of failure, initial cost, operating cost, styling, human factors, noise level, pollution, intended service environment, and possibility of use after retirement
- *Material-related factors,* such as strength, ductility, toughness, stiffness, density, corrosion resistance, wear resistance, friction coefficient, melting point, thermal and electrical conductivity, processibility, possibility of recycling, cost, available stock size, and delivery time
- *Manufacturing-related factors,* such as available fabrication processes, accuracy, surface finish, shape, size, required quantity, delivery time, cost, and required quality

Figure 1 illustrates the relationship among the above three groups. The figure also shows that there are other secondary relationships between material properties and manufacturing processes, between function and manufacturing processes, and between function and material properties.

The relationship between design and material properties is complex because the behavior of the material in the finished product can be quite different from that of the stock material used in making it. This point is illustrated in Fig. 2, which shows that in addition to stock material properties, production method and component geometry have direct influence on the behavior of materials in the finished component. The figure also shows that secondary relationships exist between geometry and production method, between stock material and production method, and stock material and component geometry. The effect of component geometry on the behavior of materials is discussed in the following section.

Effect of Component Geometry

In almost all cases, engineering components and machine elements have to incorporate design features that introduce changes in their cross section. For example, shafts must have shoulders to take thrust loads at the bearings and must have keyways or splines to transmit torques to or from pulleys and gears mounted on them. Under load, such changes cause localized stresses that are higher than those based on the nominal cross section of the part. The severity of the stress concentration depends on the geometry of the discontinuity and the nature of the material. A geometric, or theoretical, stress concentration factor, K_t, is usually used to relate the maximum stress, S_{max}, at the discontinuity to the nominal stress, S_{av}, according to the relationship:

$$K_t = S_{max}/S_{av} \quad \text{(Eq 1)}$$

The value of K_t depends on the geometry of the part and can be determined from stress concentration charts, such as those given in Ref 2 and 3. Other methods of estimating K_t for a certain geometry include photoelasticity, brittle coatings, and finite element techniques. Table 1 gives some typical values of K_t.

Experience shows that, under static loading, K_t gives an upper limit to the stress concentration value and applies it to high-strength low-ductility materials. With more ductile materials, local yielding in the very small area of maximum stress causes some relief in the stress concentration. Generally, the following design guidelines should be observed if the deleterious effects of stress concentration are to be kept to a minimum:

- Abrupt changes in cross section should be avoided. If they are necessary, generous fillet radii or stress-relieving grooves should be provided (Fig. 3a).
- Slots and grooves should be provided with generous run-out radii and with fillet radii in all corners (Fig. 3b).
- Stress-relieving grooves or undercuts should be provided at the end of threads and splines (Fig. 3c).
- Sharp internal corners and external edges should be avoided.
- Oil holes and similar features should be chamfered and the bore should be smooth.

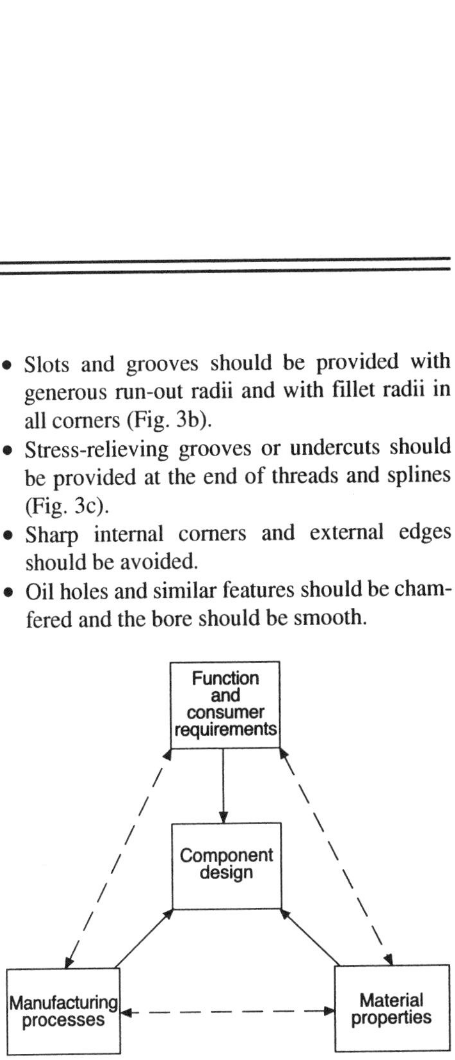

Fig. 1 Factors that should be considered in component design. Source: Ref 1

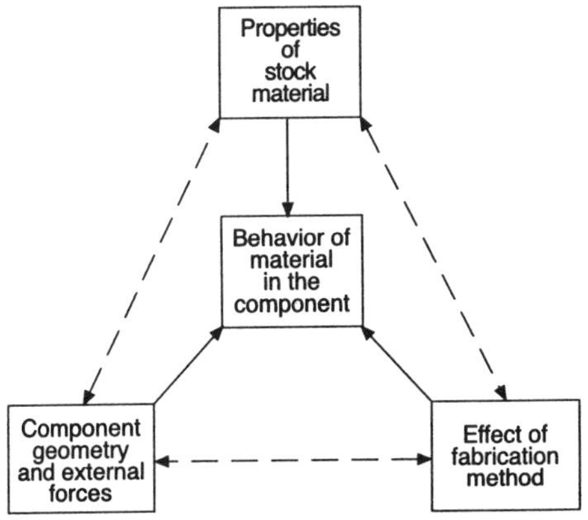

Fig. 2 Factors that should be considered in anticipating the behavior of material in the component. Source: Ref 1

- Weakening features like bolt and oil holes, identification marks, and part numbers should not be located in highly stressed areas.
- Weakening features should be staggered to avoid the addition of their stress concentration effects (Fig. 3d).

Factor of Safety

The term factor of safety is applied to the factor used in designing a component to ensure that it will satisfactorily perform its intended function. The main parameters that affect the value of the factor of safety, which is always greater than unity, can be grouped into:

- Uncertainties associated with material properties due to variations in composition, heat treatment, and processing conditions as well as environmental variables such as temperature, time, humidity, and ambient chemicals. Manufacturing processes also contribute to these uncertainties as a result of variations in surface roughness, internal stresses, sharp corners, and other stress raisers.
- Uncertainties in loading and service conditions

Generally, ductile materials that are produced in large quantities show fewer property variations than less ductile and advanced materials that are produced by small batch processes. Parts manufactured by casting, forging, and cold forming are known to have variations in properties from point to point.

To account for uncertainties in material properties, the factor of safety is used to divide into the nominal strength (S) of the material to obtain the allowable stress (S_a) as follows:

$$S_a = S/n_s \quad \text{(Eq 2)}$$

where n_s is the material factor of safety.

In simple components, S_a in the above equation can be viewed as the minimum allowable strength of the material. However there is some danger involved in this use, especially in the cases where the load-carrying capacity of a component is not directly related to the strength of the material used in making it. Examples include long compression members, which could fail as a result of buckling, and components of complex shapes, which could fail as a result of stress concentration. Under such conditions it is better to consider S_a as the load-carrying capacity that is a function of both material properties and geometry of the component.

In assessing the uncertainties in loading, two types of service conditions have to be considered:

- *Normal working conditions,* which the component has to endure during its intended service life
- *Limited working conditions,* such as overloading, which the component is only intended to endure on exceptional occasions, and which if repeated frequently could cause premature failure of the component

Table 1 Values of the stress concentration factor K_t

Component shape		Value of critical parameter, K_t	
Round shaft with transverse hole			
Bending	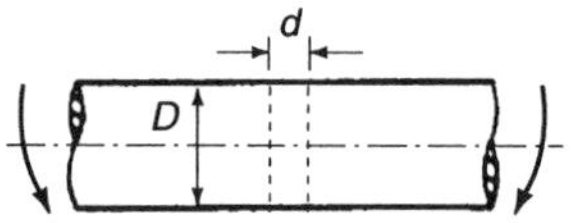	$d/D = 0.025$	2.65
		$= 0.05$	2.50
		$= 0.10$	2.25
		$= 0.20$	2.00
Torsion	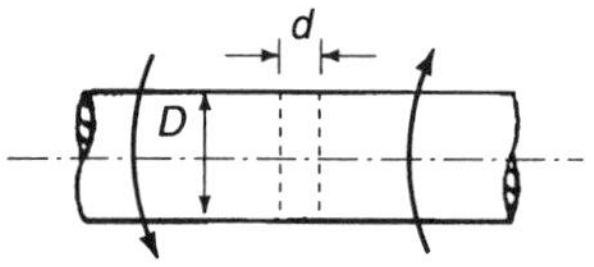	$d/D = 0.025$	3.7
		$= 0.05$	3.6
		$= 0.10$	3.3
		$= 0.20$	3.0
Round shaft with shoulder			
Tension	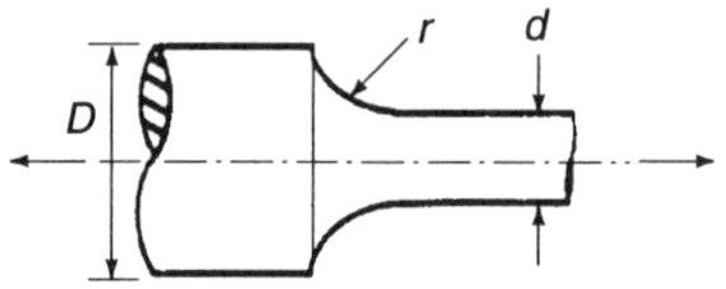	$d/D = 1.5$, $r/d = 0.05$	2.4
		$r/d = 0.10$	1.9
		$r/d = 0.20$	1.55
		$d/D = 1.1$, $r/d = 0.05$	1.9
		$= 0.10$	1.6
		$= 0.20$	1.35
Bending	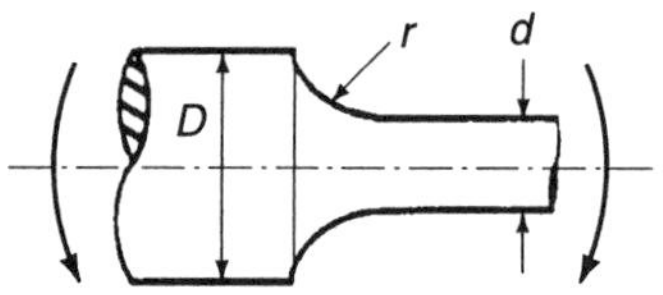	$d/D = 1.5$, $r/d = 0.05$	2.05
		$r/d = 0.10$	1.7
		$r/d = 0.20$	1.4
		$d/D = 1.1$, $r/d = 0.05$	1.9
		$r/d = 0.10$	1.6
		$r/d = 0.20$	1.35
Torsion	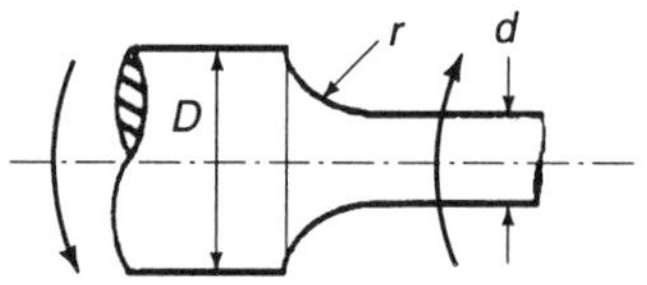	$d/D = 1.5$, $r/d = 0.05$	1.7
		$r/d = 0.10$	1.45
		$r/d = 0.20$	1.25
		$d/D = 1.1$, $r/d = 0.05$	1.25
		$r/d = 0.10$	1.15
		$r/d = 0.20$	1.1
Grooved round bar			
Tension	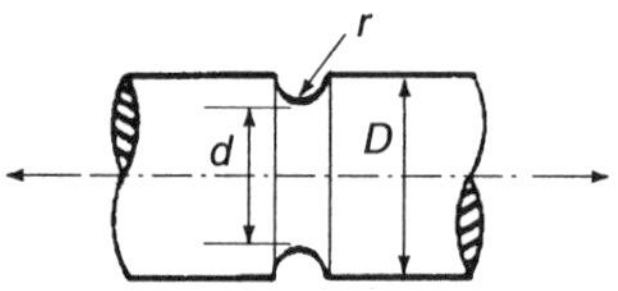	$d/D = 1.1$, $r/d = 0.05$	2.35
		$r/d = 0.10$	2.0
		$r/d = 0.20$	1.6
Bending	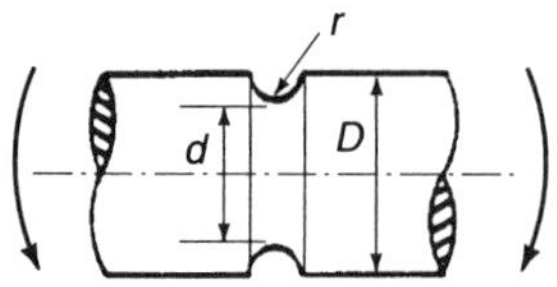	$d/D = 1.1$, $r/d = 0.05$	2.35
		$r/d = 0.10$	1.9
		$r/d = 0.20$	1.5
Torsion	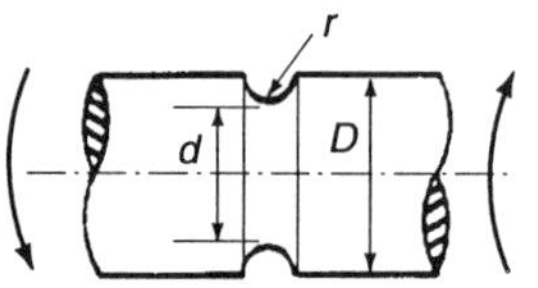	$d/D = 1.1$, $r/d = 0.05$	1.65
		$r/d = 0.10$	1.4
		$r/d = 0.20$	1.25

Source: Ref 1

In a mechanically loaded component, the stress levels corresponding to both normal and limited working conditions can be determined from a duty cycle. The normal duty cycle for an airframe, for example, includes towing and ground handling, engine run, take-off, climb, normal gust

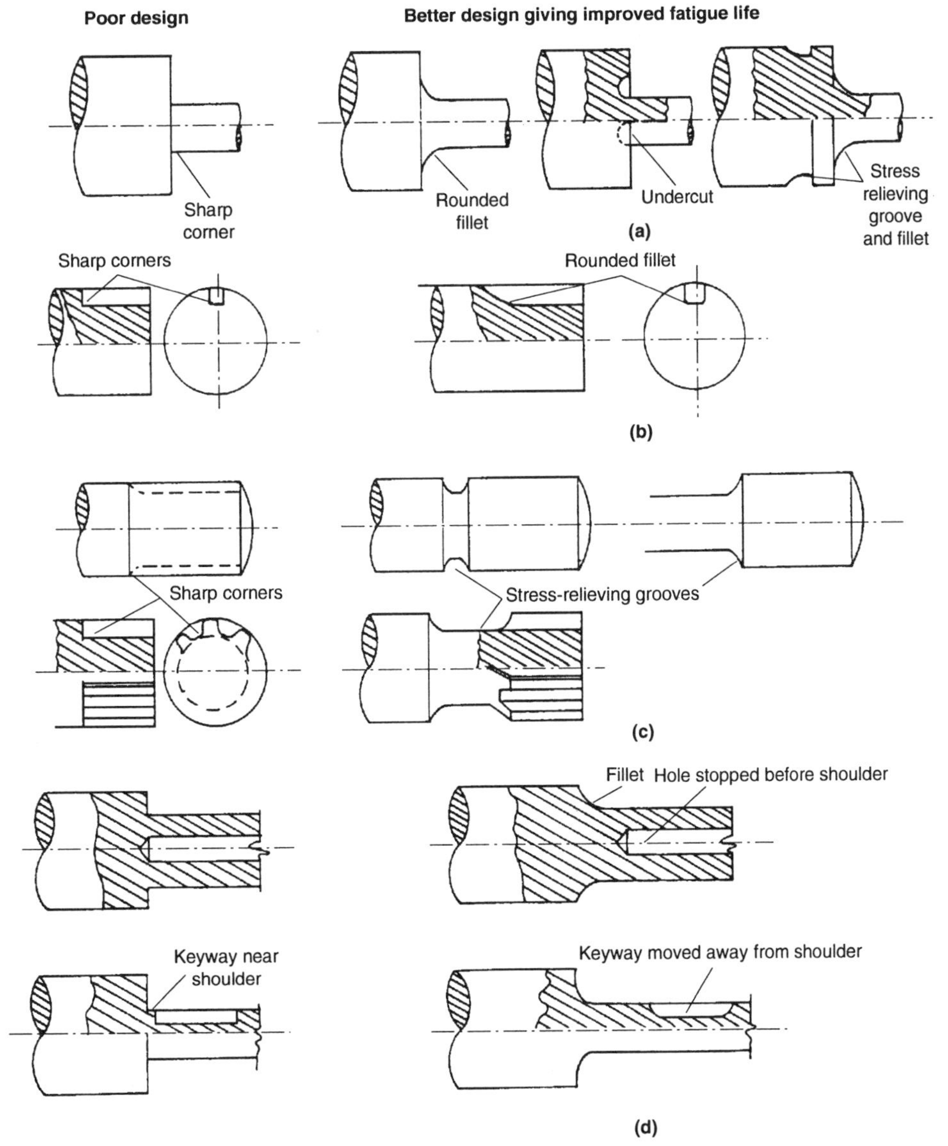

Fig. 3 Design guidelines for reducing the deleterious effects of stress concentration. See text for discussion. Source: Ref 1

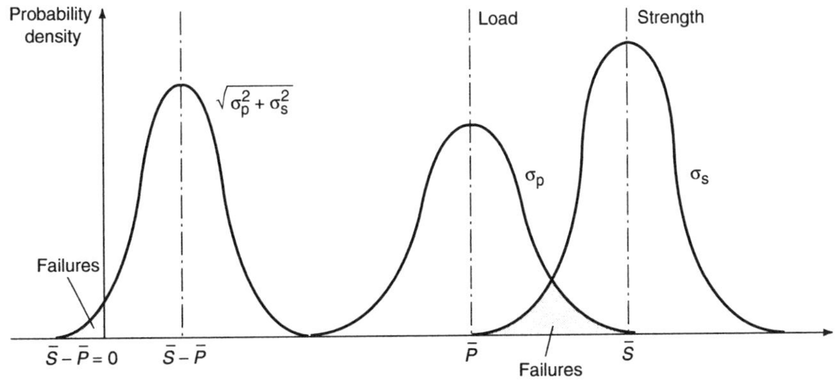

Fig. 4 Effect of variations in load and strength on the failure of components. Source: Ref 1

loadings at different altitudes, kinetic and solar heating, descent, and normal landing. Limited conditions can be encountered in abnormally high gust loadings or emergency landings. Analyses of the different loading conditions in the duty cycle lead to determination of the maximum load that will act on the component. This maximum load can be used to determine the maximum stress, or damaging stress, which if exceeded would render the component unfit for service before the end of its normal expected life. The load factor of safety (n_l) in this case can be taken as:

$$n_l = P/P_a \quad \text{(Eq 3)}$$

where P is the maximum load and P_a is normal load.

The total or overall factor of safety (n) that combines the uncertainties in material properties and external loading conditions can be calculated as:

$$n = n_s \cdot n_l \quad \text{(Eq 4)}$$

Factors of safety ranging from 1.1 to 20 are known, but common values range from 1.5 to 10.

In some applications a designer is required to follow established codes when designing certain components, for example, pressure vessels, piping systems, and so forth. Under these conditions, the factors of safety used by the writers of the codes may not be specifically stated, but an allowable working stress is given instead.

Probability of Failure

As discussed earlier, the actual strength of the material in a component could vary from one point to another and from one component to another. In addition, it is usually difficult to precisely predict the external loads acting on the component under actual service conditions. To account for these variations and uncertainties, both the load-carrying capacity S and the externally applied load P can be expressed in statistical terms. As both S and P depend on many independent factors, it would be reasonable to assume that they can be described by normal distribution curves. Consider that the load-carrying capacity of the population of components has an average of $\bar{S}$ and a standard deviation σ_S while the externally applied load has an average of $\bar{P}$ and a standard deviation σ_P. The relationship between the two distribution curves is important in determining the factor of safety and reliability of a given design. Figure 4 shows that failure takes place in all the components that fall in the area of overlap of the two curves, that is, when the load-carrying capacity is less than the external load. This is described by the negative part of the $(S - P)$ curve of Fig. 4. Transforming the distribution $(S - P)$ to the standard normal deviate z, the following equation is obtained:

$$z = [(S - P) - (\bar{S} - \bar{P})]/[(\sigma_S)^2 + (\sigma_P)^2]^{1/2} \quad \text{(Eq 5)}$$

From Fig. 4, the value of z at which failure occurs is:

$$z = -(S - P)/[(\sigma_S)^2 + (\sigma_P)^2]^{1/2} \quad \text{(Eq 6)}$$

Table 2 Values of *z* and corresponding levels of reliability and probability of failure

z	Reliability	Probability of failure
−1.00	0.8413	0.1587
−1.28	0.9000	0.1000
−2.33	0.9900	0.0100
−3.09	0.9990	0.0010
−3.72	0.9999	0.0001
−4.26	0.99999	0.00001
−4.75	0.999999	0.000001

For a given reliability, or allowable probability of failure, the value of z can be determined from cumulative distribution function for the standard normal distribution. Table 2 gives some selected values of z that will result in different values of probabilities of failure.

Knowing σ_S, σ_P, and the expected $\bar{P}$, the value of $\bar{S}$ can be determined for a given reliability level. As defined earlier, the factor of safety in the present case is simply S/P. The following example illustrates the use of the above concepts in design; additional discussion of statistical methods is provided in the article "Statistical Aspects of Design" in this Volume.

Example 1: Estimating Probability of Failure. A structural element is made of a material with an average tensile strength of 2100 MPa. The element is subjected to a static tensile stress of an average value of 1600 MPa. If the variations in material quality and load cause the strength and stress to vary according to normal distributions with standard deviations of $\sigma_S = 400$ and $\sigma_P = 300$, respectively, what is the probability of failure of the structural element? The solution can be derived as follows: From Fig. 4, $(\bar{S} - \bar{P})$ = 2100 − 1600 = 500 MPa, standard deviation of the curve $(\bar{S} - \bar{P}) = [(400)^2 + (300)^2]^{1/2} = 500$; from Eq 6, $z = -500/500 = -1$. Thus, from Table 2, the probability of failure of the structural element is 0.1587 (15.87%), which is too high for many practical applications.

One solution to reduce the probability of failure is to impose better quality measures on the production of the material and thus reduce the standard deviation of the strength. Another solution is to increase the cross-sectional area of the element in order to reduce the stress. For example, if the standard deviation of the strength is reduced to $\sigma_S = 200$, the standard deviation of the curve $(\bar{S} - \bar{P})$ will be $[(200)^2 + (300)^2]^{1/2} = 360$, $z = -500/360 = -1.4$, which, according to Table 2, gives a more acceptable probability of failure value of 0.08 (8%).

Alternatively, if the average stress is reduced to 1400 MPa, $(\bar{S} - \bar{P}) = 700$ MPa, $z = -700/500 = -1.4$, with a similar probability of failure as the first solution.

Experimental Methods. As the above discussion shows, statistical analysis allows the generation of data on the probability of failure and reliability, which is not possible when a deterministic safety factor is used. One of the difficulties with this statistical approach, however, is that material properties are not usually available as statistical quantities. In such cases, the following approximate method can be used.

In the case where the experimental data are obtained from a reasonably large number of samples, more than 100, it is possible to estimate statistical data from nonstatistical sources that only give ranges or tolerance limits. In this case, the standard deviation σ_S is approximately given by:

$$\sigma_S = (\text{maximum value of property} - \text{minimum value})/6 \qquad \text{(Eq 7)}$$

This procedure is based on the assumption that the given limits are bounded between plus and minus three standard deviations.

If the results are obtained from a sample of about 25 tests, it may be better to divide by 4 in Eq 7 instead of 6. With a sample of about 5, it is better to divide by 2.

In the cases where only the average value of strength is given, the following values of coefficient of variation, which is defined as $\nu' = \sigma_S/S$ can be taken as typical for metallic materials: $\nu' = 0.05$ for ultimate tensile strength, and $\nu' = 0.07$ for yield strength.

Designing for Static Strength

The design and materials selection of a component or structure under static loading can be based on static strength and/or stiffness depending on the service conditions and the intended function.

In the case of ductile materials, designs based on the static strength usually aim at avoiding yielding of the component. The manner in which the yield strength of the material affects the design depends on the loading conditions and type of component, as illustrated in the following cases.

Designing for Axial Loading. When the component is subjected to uniaxial stress, yielding takes place when the local stress reaches the yield strength of the material. The critical cross-sectional area, A, of such a component can be estimated as:

$$A = K_t n P/YS \qquad \text{(Eq 8)}$$

where K_t is the stress concentration factor (described in the section "Effect of Component Geometry" in this article), P is the applied load, n is the factor of safety, described in the section "Factor of Safety" in this article), and YS is the yield strength of the material.

Designing for Torsional Loading. The critical cross-sectional area of a circular shaft subjected to torsional loading can be determined from the relationship:

$$2 I_p/d = K_t n T/\tau_{max} \qquad \text{(Eq 9)}$$

where d is the shaft diameter at the critical cross section, τ_{max} is the shear strength of the material, T is the transmitted torque, and I_p is the polar moment of inertia of the cross section. ($I_p = \pi d^4/32$ for a solid circular shaft; $I_p = \pi (d_o^4 - d_i^4)/32$ for a hollow circular shaft of inner diameter d_i and outer diameter d_o.)

While Eq 9 gives a single value for the diameter of a solid shaft, a large combination of inner and outer diameters can satisfy the relationship in the case of a hollow shaft. Under such conditions, either one of the diameters or the required thickness has to be specified in order to calculate the other dimension.

The ASTM code of recommended practice for transmission shafting gives an allowable value of shear stress of 0.3 of the yield or 0.18 of the ultimate tensile strength, whichever is smaller. With shafts containing keyways, ASTM recommends a reduction of 25% of the allowable shear strength to compensate for stress concentration and reduction in cross-sectional area.

Designing for Bending. When a relatively long beam is subjected to bending, the bending moment, the maximum allowable stress, and dimensions of the cross section are related by:

$$Z = (n M)/YS \qquad \text{(Eq 10)}$$

where M is the bending moment and Z is the section modulus ($Z = I/c$, where I is the moment of inertia of the cross section with respect to the neutral axis normal to the direction of the load, and c is the distance from center of gravity of the cross section to the outermost fiber). Bending moment can be determined by consulting a standard reference on the strength of materials, such as Ref 4.

Designing for Stiffness

In addition to being strong enough to resist the expected service loads, there may also be the added requirement of stiffness to ensure that deflections do not exceed certain limits. Stiffness is important in applications such as machine elements to avoid misalignment and to maintain dimensional accuracy of machine parts.

Design of Beams. The deflection of a beam under load can be taken as a measure of its stiffness, and this depends on the position of the load, the type of beam, and the type of supports. For example, a beam that is simply supported at both ends suffers maximum deflection (y) in its middle when subjected to a concentrated central load (P). In this case the maximum deflection, y, is a function of both E and I, as follows:

$$y = (P \cdot L^3)/(48 \cdot E \cdot I) \qquad \text{(Eq 11)}$$

where L is the length of the beam, E is Young's modulus of the beam material, and I is the second moment of area of the beam cross section with respect to the neutral axis.

Design of Columns. Columns and struts, which are long slender parts, are subject to failure by elastic instability, or buckling, if the applied axial compressive load exceeds a certain critical value, P_{cr}. The Euler column formula is usually used to calculate the value of P_{cr}, which is a function of the material, geometry of the column,

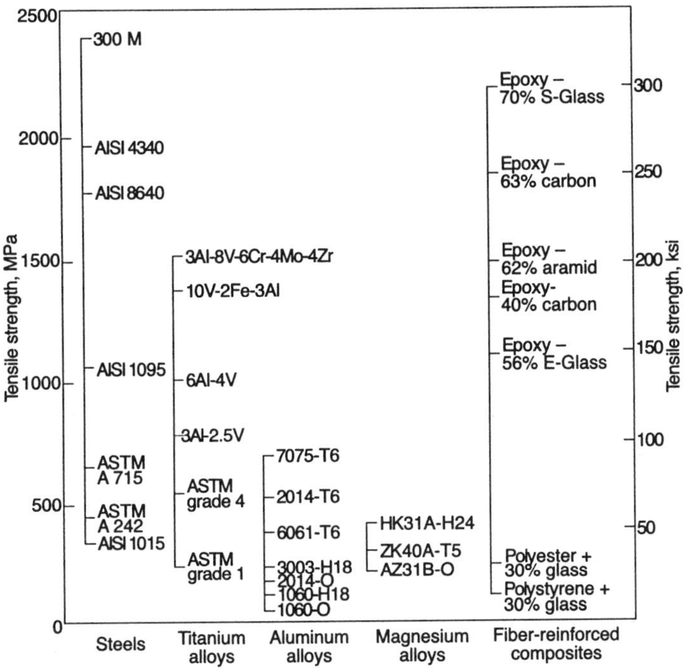

Fig. 5 Comparison of various engineering materials on the basis of tensile strength. Source: Ref 1

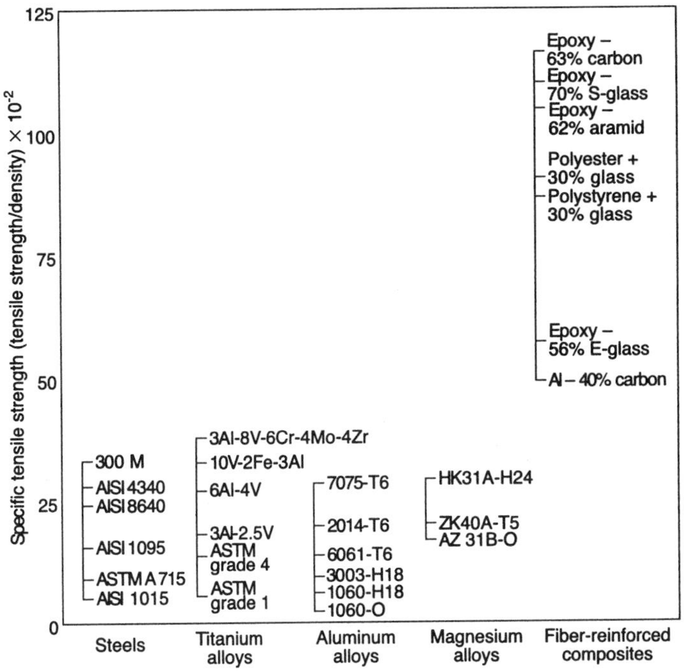

Fig. 6 Comparison of various engineering materials on the basis of specific tensile strength. Source: Ref 1

and restraint at the ends. For the fundamental case of a pin-ended column, that is, ends are free to rotate around frictionless pins, P_{cr} is given as:

$$P_{cr} = \pi^2 EI/L^2 \quad \text{(Eq 12)}$$

where I is the least moment of inertia of the cross-sectional area of the column, and L is the length of the column.

The above equation can be modified to allow for end conditions other than the pinned ends. The value of P_{cr} for a column with both ends fixed—built-in as part of the structure—is four times the value given by Eq 12. On the other hand, the critical load for a free-standing column—one end is fixed and the other free as in a cantilever—P_{cr} is only one-quarter of the value given by Eq 12.

The Euler column formula given above shows that the critical load for a given column is only a function of E and I and is independent of the compressive strength of the material. This means that resistance to buckling of a column of a given material and a given cross-sectional area can be increased by distributing the material as far as possible from the principal axes of the cross section to increase I. Hence, tubular sections are preferable to solid sections. Reducing the wall thickness of such sections and increasing the transverse dimensions increases the stability of the column. However, there is a lower limit for the wall thickness below which the wall itself becomes unstable and causes local buckling.

Experience shows that the values of P_{cr} calculated according to Eq 12 are higher than the buckling loads observed in practice. The discrepancy is usually attributed to manufacturing imperfections, such as lack of straightness of the column and lack of alignment between the direction of the compressive load and the axis of the column. This discrepancy can be accounted for by using an appropriate imperfection parameter or a factor of safety. For normal structural work, a factor of safety of 2.5 is usually used. As the extent of the above imperfections is expected to increase with increasing slenderness of the column, it is suggested that the factor of safety be increased accordingly. A factor of safety of 3.5 is recommended for columns with $[L(A/I)^{1/2}] > 100$, where A is cross-sectional area.

Equation 12 shows that the value of P_{cr} increases rapidly as the length of the column, L, decreases. For a short enough column, P_{cr} becomes equal to the load required for yielding or crushing of the material in simple compression. Such a case represents the limit of applicability of the Euler formula as failure takes place by yielding or fracture rather than elastic instability. Such short columns are designed according to the procedure described for simple axial loading.

Design of Shafts. The torsional rigidity of a component is usually measured by the angle of twist, θ, per unit length. For a circular shaft, θ is given in radians by:

$$\theta = T/G\,I_p \quad \text{(Eq 13)}$$

where G is the modulus of elasticity in shear, and:

$$G = E/[2(1 + \nu)] \quad \text{(Eq 14)}$$

where ν is Poisson's ratio.

The usual practice is to limit the angular deflection in shafts to about 1° (π/180 radians) in a length of 20 times the diameter.

Selection of Materials for Static Strength

Static Strength and Isotropy. The resistance to static loading is usually measured in terms of yield strength, ultimate tensile strength, and compressive strength. When the material does not exhibit a well-defined yield point, the stress required to cause 0.1 or 0.2% plastic strain, the proof stress, is used instead. (This is usually called the 0.2% offset yield strength.) For most ductile wrought metallic materials, the tensile and compressive strengths are very close, and in most cases only the tensile strength is given. However, brittle materials like gray cast iron and ceramics are generally stronger in compression than in tension. In such cases, both properties are usually given. For polymeric materials, which usually do not have a linear stress-strain curve, and whose static properties are very temperature dependent, other design methods must be used; additional information is provided in the article "Designing with Plastics" in this Volume.

Although many engineering materials are almost isotropic, there are important cases where significant anisotropy exists. In the latter case, the strength depends on the direction in which it is measured. The degree of anisotropy depends on the nature of the material and its manufacturing history. Anisotropy in wrought metallic materials is more pronounced when they contain elongated inclusions and when processing consists of repeated deformation in the same direction. Composites reinforced with unidirectional fibers also exhibit pronounced anisotropy. Anisotropy can be useful if the principal external stress acts along the direction of highest strength.

The level of strength in engineering materials may be viewed either in absolute terms or relative to similar materials. For example, it is generally understood that high-strength steels have tensile strength values in excess of 1400 MPa (200 ksi), which is also high strength in absolute terms. Relative to light alloys, however, an aluminum alloy with a strength of 500 MPa (72 ksi) would also be designated a high-strength alloy even though this level of strength is low for steels.

From the design point of view, it is more convenient to consider the strength of materials in absolute terms. From the materials and manufacturing point of view, however, it is important to consider the strength as an indication of the degree of development of the material concerned, that is, relative to similar materials. This is because highly developed materials are often complex, more difficult to process, and relatively more expensive. Figure 5 gives the strength of some materials both in absolute terms and relative to similar materials. In a given group of materials, the medium-strength members are usually more widely used because they generally combine optimal strength, ease of manufacture, and economy. The most-developed members in a given group of materials are usually highly specialized, and as a result they are produced in much lower quantities. The low-strength members of a given group are usually used to meet requirements other than strength. For example, electrical and thermal conductivities, formability, or corrosion resistance may be more important than strength in some applications.

Frequently, higher-strength members of a given group of materials are more expensive. However, using a stronger but more expensive material could result in a reduction of the total cost of the finished component. This is because less material would be used, and consequently processing cost could also be less.

Weight and Space Limitations. The load-carrying capacity of a given component is a function of both the strength of the material used in making it and its geometry and dimensions. This means that a lower-strength material can be used in making a component to bear a certain load, provided that its cross-sectional area is increased proportionally. However, the designer is not usually completely free in choosing the strength level of the material selected for a given part. Other factors such as space and weight limitations could limit the choice.

Weight limitations are encountered with many applications including aerospace, transport, construction, and portable appliances. In such cases, the strength/density, or specific strength, becomes an important basis for comparing the different materials. Figure 6 compares the materials of Fig. 5 on the basis of specific strength, which is the tensile strength of a material divided by its density. The figure shows a clear advantage of the fiber-reinforced composites over other materials.

Using stronger material will allow smaller cross-sectional area and smaller total volume of the component. It should be noted, however, that reducing the cross-sectional area below a certain limit could cause failure by buckling due to increased slenderness of the part.

Table 3 Comparison of materials considered for a cylindrical compression element

See Example 2 in text.

Material	Strength, MPa	Elastic modulus, GPa	Specific gravity	Diameter based on strength, mm	Diameter based on buckling, mm	Mass based on larger diam, kg	Remarks
Steels							
ASTM A675, grade 45	155	211	7.8	20.3	15.75	...	Reject(a)
ASTM A675, grade 80	275	211	7.8	15.2	15.75	0.3	Reject(b)
ASTM A715, grade 80	550	211	7.8	10.8	15.75	0.3	Reject(b)
Aluminum							
2014-T6	420	70.8	2.7	12.3	20.7	...	Reject(a)
Plastics and composites							
Nylon 6/6	84	3.3	1.14	27.5	44.6	...	Reject(a)
Epoxy-70% glass	2100	62.3	2.11	5.5	21.4	...	Reject(a)
Epoxy-62% Kevlar	1311	82.8	1.38	7.0	19.9	0.086	Accepted

(a) Material is rejected because it violates the limits on diameter. (b) Material is rejected because it violates the limits on weight. Source: Ref 1

Example 2: Materials Selection for a Cylindrical Compression Element. A load of 50 kN is to be supported on a cylindrical compression element of 200 mm length. As the compression element has to fit with other parts of the structure, its diameter should not exceed 20 mm. Weight limitations are such that the mass of the element should not exceed 0.25 kg. Table 3 shows the calculated diameter of the compression element when made of different materials. The diameter is calculated on the basis of strength and on the basis of buckling. The larger value for a given material is used to calculate the mass of the element. The results in Table 3 show that only epoxy-62% Kevlar satisfies both the diameter and weight limits for the compression element.

Selection of Materials for Stiffness

Deflection under Load. As discussed in the section "Design of Beams" in this article, the stiffness of a component may be increased by increasing its second moment of area, which is computed from the cross-sectional dimensions, and/or by selecting a high-modulus material for its manufacture.

An important characteristic of metallic materials is that their elastic moduli are very difficult to change by changing the composition or heat treatment. Using high-strength materials in attempts to reduce weight usually comes at the expense of reduced cross-sectional area and reduced second moment of area. This could adversely affect stiffness of the component if the elastic constant of the new strong material does not compensate for the reduced second moment of area.

Selecting materials with higher elastic constant and efficient disposition of material in the cross section are essential in designing beams for stiffness. Placing material as far as possible from the neutral axis of bending is generally an effective means of increasing I for a given area of cross section. See the discussion of shape factor in the article "Performance Indices" in this Volume.

When designing with plastics, whose elastic modulus is 10 to 100 times less than that of metals, stiffness must be given special consideration. This drawback can usually be overcome by making some design adjustments. These usually include increasing the second moment of area of the critical cross section, as shown in the following example.

Example 3: Design Changes Required for Materials Substitution. This example considers the design changes required when substituting high-density polyethylene (HDPE) for stainless steel in making a fork for a picnic set while maintaining similar stiffness. The narrowest cross section of the original stainless steel fork is rectangular with an area of 0.6 by 5 mm.

Analysis:

- E for stainless steel = 210 GPa.
- E for HDPE = 1.1 GPa.
- I for the stainless steel section = $5 \times (0.6)^3/12 = 0.09$ mm^4.
- From Eq 11, EI should be kept constant for equal deflection under load.
- EI for stainless steel = $210 \times 0.09 = 18.9$.
- EI for HDPE design = $1.1 \times I$.
- I for HDPE design = 17.2 mm^4. Taking a channel section of thickness 0.5 mm, web height 4 mm, and width 8 mm, $I = [8 \times (4)^3 - 7 \times (3.5)^3]/12 = 17.7$ mm^4, which meets the required value.
- Area of the stainless steel section = 3 mm^2.
- Area of the HDPE section = 7.5 mm^2.
- The specific gravity of stainless steel is 7.8 and that of HDPE is 0.96.
- Relative weight of HDPE/stainless steel = $(7.5 \times 0.96)/(3 \times 7.8) = 0.3$.

Weight Limitations. In applications where both the stiffness and weight of a structure are important, it becomes necessary to consider the stiffness/weight (specific stiffness), of the structure. In the simple case of a structural member

Table 4 Comparison of the stiffness of selected engineering materials

Material	Modulus of elasticity (E), GPa	Density (ρ), mg/m³	E/ρ $\times 10^{-5}$	$E^{1/2}/\rho$ $\times 10^{-2}$	$E^{1/3}/\rho$
Steel (carbon and low alloy)	207	7.825	26.5	5.8	35.1
Aluminum alloys (average)	71	2.7	26.3	9.9	71.2
Magnesium alloys (average)	40	1.8	22.2	11.1	88.2
Titanium alloys (average)	120	4.5	26.7	7.7	50.9
Epoxy-73% E-glass fibers	55.9	2.17	25.8	10.9	81.8
Epoxy-70% S-glass fibers	62.3	2.11	29.5	11.8	87.2
Epoxy-63% carbon fibers	158.7	1.61	98.6	24.7	156.1
Epoxy-62% aramid fibers	82.8	1.38	60	20.6	146.6

Source: Ref 1

under tensile or compressive load, the specific stiffness is given by E/ρ, where ρ is the density of the material. In such cases, the weight of a member of a given stiffness can be easily shown to be proportional to ρ/E and can be reduced by selecting a material with lower density or higher elastic modulus. When the component is subjected to bending, the dependence of the weight on ρ and E is not as simple. From Eq 11 it can be shown that the weight w of a simply supported beam of square cross-sectional area is given by:

$$w = L \cdot b^2 \cdot \rho = \frac{L^{5/2}}{2} \cdot \left(\frac{P}{y}\right)^{1/2} \cdot \frac{\rho}{E^{1/2}} \quad \text{(Eq 15)}$$

Related information is provided in the article "Performance Indices" in this Volume.

Equation 15 shows that for a given deflection y under load P, the weight of the beam is proportional to $(\rho/E^{1/2})$. As E in this case is present as the square root, it is not as effective as ρ in controlling the weight of the beam. It can be similarly shown that the weight of the beam in the case of a rectangular cross section is proportional to $(\rho/E^{1/3})$, which is even less sensitive to variations in E. This change in the effectiveness of E in affecting the specific stiffness of structures as the mode of loading and shape change, is shown in Table 4.

Buckling Strength. Another selection criterion that is also related to the elastic modulus of the material and cross-sectional dimensions is buckling under compressive loading. The compressive load, P_b, that can cause buckling of a strut is given by Euler formula (Eq 12).

Equation 12 shows that increasing E and I will increase the load-carrying capacity of the strut. For an axially symmetric cross section, the weight of a strut, w, is given by:

$$w = L \cdot \frac{\pi \cdot D^2}{4} \cdot \rho = \frac{2 \cdot L^2 \cdot P_b^{1/2}}{\pi^{1/2}} \left(\frac{\rho}{E^{1/2}}\right) \quad \text{(Eq 16)}$$

Equation 16 shows that the weight of an axisymmetric strut can be reduced by reducing ρ or by increasing E of the material, or both. However, reducing ρ is more effective, as E is present as the square root. In the case of a panel subjected to buckling, it can be shown that the weight is proportional to $(\rho/E^{1/3})$.

Types of Mechanical Failure under Static Loading at Normal Temperatures

Generally, a component can be considered to have failed when it does not perform its intended function with the required efficiency. The general types of mechanical failure encountered in practice are:

- *Yielding of the component material under static loading.* Yielding causes permanent deformation that could result in misalignment or hindrance to mechanical movement.
- *Buckling.* This type of failure takes place in slender columns when they are subjected to compressive loading, or in thin-walled tubes when subjected to torsional loading.
- *Failure by fracture due to static overload.* This type of failure can be considered an advanced stage of failure by yielding. Fracture can be either ductile or brittle.
- *Failure due to the combined effect of stresses and corrosion.* This usually takes place by fracture due to cracks at stress concentration points, for example, caustic cracking around rivet holes in boilers.

Of the above types of mechanical failure, the first two do not usually involve actual fracture, and the component is considered to have failed when its performance is below acceptable levels. On the other hand, the latter two types involve actual fracture of the component, and this could lead to unplanned load transfer to other components and perhaps other failures.

Causes of Failure of Engineering Components

As discussed earlier, the behavior of a material in service is governed not only by its inherent properties, but also by the stress system acting on it and the environment in which it is operating. Causes of failure of engineering components can be classified into the following main categories:

- *Design deficiencies.* Failure to evaluate working conditions correctly due to the lack of reliable information on loads and service conditions is a major cause of inadequate design. Incorrect stress analysis, especially near notches, and complex changes in shape could also be a contributing factor.
- *Poor selection of materials.* Failure to identify clearly the functional requirements of a component could lead to the selection of a material that only partially satisfies these requirements. As an example, a material can have adequate strength to support the mechanical loads, but its corrosion resistance is insufficient for the application.
- *Manufacturing defects.* Incorrect manufacturing could lead to the degradation of an otherwise satisfactory material. Examples are decarburization and internal stresses in a heat-treated component. Poor surface finish, burrs, identification marks, and deep scratches due to mishandling could lead to failure under fatigue loading.
- *Exceeding design limits and overloading.* If the load, temperature, speed, and so forth, are increased beyond the limits allowed by the factor of safety in design, the component is likely to fail. Subjecting the equipment to environmental conditions for which it was not designed also falls under this category. An example here is using a freshwater pump for pumping seawater.
- *Inadequate maintenance and repair.* When maintenance schedules are ignored and repairs are poorly carried out, service life is expected to be shorter than anticipated in the design.

As this article has described, various material properties influence the design of components. The type of property and the sensitivity of the design to variations in this property depend on the component geometry and type of load. Underestimation of the load and/or overestimation of the material property and service conditions could lead to failure in service.

REFERENCES

1. M.M. Farag, *Selection of Materials and Manufacturing Processes for Engineering Design,* Prentice Hall, London, 1989
2. R.E. Peterson, *Stress-Concentration Design Factors,* John Wiley and Sons, 1974
3. J.E. Shigley and L.D. Mitchell, *Mechanical Engineering Design,* 4th ed., McGraw-Hill, 1983
4. W.C. Young, *Roark's Formulas for Stress and Strain,* 6th ed., McGraw-Hill, 1989

SELECTED REFERENCES

- V.J. Colangelo and F.A. Heiser, *Analysis of Metallurgical Failures,* John Wiley and Sons, 1987
- N.H. Cook, *Mechanics and Materials for Design,* McGraw-Hill, 1985
- F.A.A. Crane and J.A. Charles, *Selection and Use of Engineering Materials,* Butterworths, 1984
- M.M. Farag, *Materials Selection for Engineering Design,* Prentice Hall, London, 1997

Design for Fatigue Resistance

Erhard Krempl, Rensselaer Polytechnic Institute

FATIGUE is a gradual process caused by repeated application of loads, such that each application of stress causes some degradation or damage of the material or component. An appropriate measure of deterioration or damage has not been found, and this fact makes fatigue design difficult.

The design of components against fatigue failure may involve several considerations of irregular loading, variable temperature, and environment. In this article, the effect of environment is excluded. The effects of environment on material performance are considered elsewhere in this Volume.

The main objective here is the discussion of design considerations against fatigue related to material performance under mechanical loading at constant temperature (isothermal fatigue, or simply fatigue). In this article, periodic loading of specimens is considered, and the material properties related to fatigue derived from these tests are discussed.

Design methods considering the irregular nature of actual load applications, which requires the statistical treatment of the load histories and their translation into load spectra, cycle counting methods, and damage accumulation will not be discussed. These topics are addressed in more detail in *Fatigue and Fracture,* Volume 19 of *ASM Handbook* (Ref 1).

This article reviews "traditional" methods of fatigue design. In recent years, the fracture mechanics approach to crack propagation has gained acceptance in the prediction of fatigue life. In this approach, crack initiation is neglected and crack propagation to final failure is considered. This method can give a conservative estimate of fatigue life. There has been considerable success with this method. However, the treatment of short cracks, the propagation of cracks for negative *R*-ratios, and for irregular loading, crack closure effects and multiaxial loadings provide significant challenges. Reference 2 gives an introduction to the current research in this area.

In addition, the design methods reviewed in this article focus principally on smooth and notched components. Mechanically fastened joints and welded joints require special attention. In particular, such joints tend to negate alloy and composition effects. More detailed information on fatigue of welds and mechanical joints is contained in Ref 1, 3, and 4.

The Fatigue Process

The fatigue process consists of a crack initiation and a crack propagation phase. The demarcation between these two phases is, however, not clearly defined. There is no general agreement when (or at what crack size) the crack initiation process ends, and the crack growth process begins. Nonetheless, the separation of the fatigue process in initiation and propagation phases has been developing during the last forty years. The tendency is now to investigate crack growth of preexisting flaws and to neglect the crack initiation process. This method has been driven by the development of the fracture mechanics approach. Previously, the cycles-to-failure needed to completely separate the test specimens were reported, and no separation in crack initiation and crack growth life had been made.

For metals and alloys, two regimes of the fatigue phenomenon are generally considered—high-cycle fatigue and low-cycle fatigue.

High-cycle fatigue involves nominally linear elastic behavior and causes failure after more than approximately 100,000 cycles. As the loading amplitude is decreased, the cycles-to-failure increase. For many alloys, a fatigue (endurance) limit exists beyond 10^6 cycles. The endurance or fatigue limit represents a stress level below which fatigue life *appears* to be infinite upon extrapolation. However, fatigue limits from controlled laboratory tests (often limited from 10^6 to 10^8 cycles) are not necessarily valid in application environments. For example, a fatigue limit can be eradicated by one large overload or the onset of additional crack-initiation mechanisms such as corrosion at the surface. Moreover, even though testing ceases at 10^6 or 10^8 cycles for practical reasons, it is known that fatigue limits cannot be extrapolated to infinity. In some cases, there may be a change in failure mechanisms at lives above 10^6 or 10^8 cycles.

Due to the nominally elastic behavior, dissipation is small and high-frequency testing at small amplitudes can be performed without causing self-heating of the specimen. In mechanical testing machines, frequencies up to 300 Hz are possible. Usual testing speeds, however, are below 100 Hz. At this frequency, approximately 8.6×10^6 cycles are accumulated per day. To impose 8.6×10^8 cycles, more than a hundred days of testing time at 100 Hz are required. It is therefore no surprise that fatigue data extending to more than 10^8 cycles are hard to find.

Stresses involving inelastic deformation so that a significant stress-strain hysteresis loop develops during cyclic loading lead to low-cycle fatigue failure, usually in less than 100,000 cycles. Cyclic inelastic deformation causes dissipation of energy, which can lead to significant self-heating of a specimen if a high frequency is used for testing. This is one reason why low-cycle fatigue testing is usually performed at frequencies below 1 Hz. Low-cycle fatigue investigations started in the 1950s in response to failures that were found in power-generation equipment. The problems were caused by frequent start/stop operations and thermal stresses induced by temperature changes.

The simultaneous advent of servocontrolled testing machines, using feedback control, clip-on extensometers, and computer control made testing relevant to the engineering problem at hand and established low-cycle fatigue testing as a separate discipline. Complex waveforms can be imposed in both displacement (strain) or load (stress) control; no backlash exists upon zero-load crossing with modern machines. Reliable stress-strain data are generated. Computer control and computer analysis of data permit a detailed correlation between deformation behavior and fatigue life.

Because of the nominally elastic behavior during high-cycle fatigue loading, cracks will initiate from defects such as inclusions, second-phase particles, and other stress concentrators, and inelastic deformation is restricted to these sites. Persistent slip bands, which cover a large area of the specimen gage section volume, are not to be expected because the nominal behavior is elastic in high-cycle fatigue loading. Therefore, initiation from persistent slip bands, which is the predominant mechanism for low-cycle fatigue crack initiation, can only occur in very few, highly stressed grains. These sites will become less frequent as the load amplitude is decreased. Defects are normally randomly distributed in a material and vary in their size and distribution from specimen to specimen. Therefore, crack initiation and propagation are not going to be identical in different specimens. A comparatively large scatter of high-cycle fatigue data is to be expected. The nominally inelastic deformation in low-cycle fatigue loading causes many persistent slip bands to develop. Crack initiation and propagation can be more uniformly distributed than in high-cycle fatigue loading. Therefore, the low-cycle-fatigue data scatter is expected to be less than that of the high-cycle-fatigue data.

Nomenclature. This article uses, unless stated otherwise, engineering stress σ and strain ε and cycles to failure *N*, where failure denotes separation of the test specimens. Results from constant amplitude, periodic loadings are the basis of most of the discussions.

A periodic loading of stress (sinusoidal, triangular, or other) is imposed, and test results are reported in terms of stress. In reference to Fig. 1, the maximum and minimum stress is designated by σ_{max} and σ_{min}, respectively. The mean stress σ_{mean} and the stress range $\Delta\sigma$ are given by (a tensile stress is introduced as a positive quantity and a compressive stress is defined as a negative quantity):

$$\sigma_{mean} = (\sigma_{max} + \sigma_{min})/2 \text{ and } \Delta\sigma = (\sigma_{max} - \sigma_{min}) \quad \text{(Eq 1)}$$

respectively. The stress amplitude or alternating stress is $\sigma_a = \Delta\sigma/2$. The stress range and the stress amplitude are always positive; the mean stress can change sign. A completely reversed loading has zero mean stress.

In addition to these quantities the *R*-ratio:

$$R = \sigma_{min}/\sigma_{max} \quad \text{(Eq 2)}$$

and the *A*-ratio:

$$A = \sigma_a/\sigma_{mean} = (\sigma_{max} - \sigma_{min})\,(\sigma_{max} + \sigma_{min}) \quad \text{(Eq 3)}$$

are used. A simple calculation shows that:

$$A = (1 - R)/(1 + R) \quad \text{(Eq 4)}$$

and that:

$$R = (1 - A)/(1 + A) \quad \text{(Eq 5)}$$

Either one of the three expressions σ_{mean}, *A*, or *R* can be used to describe the loading. For completely reversed loading $\sigma_{mean} = 0$, $R = -1$, and $A = \infty$; for tension/tension loading $\sigma_{mean} > 0$, $0 < R < 1$, and $A > 0$. Similar values hold for other types of loading that are less frequently employed.

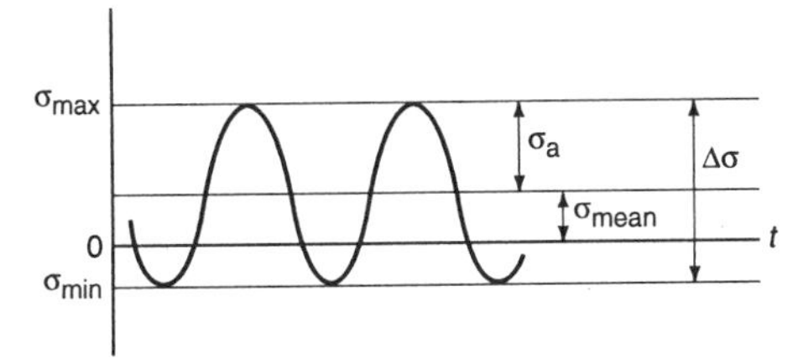

Fig. 1 Schematic showing the imposed periodic stress and the definition of terms

High-Cycle Fatigue

To establish a fatigue curve several identical specimens are needed. The first specimen is subjected to a given loading intended to result in a number of cycles to failure. For the next specimen, the loading is either increased or decreased, and the number of cycles is observed again. In such a way the fatigue or endurance curves, also referred to as *S-N* curves, shown in Fig. 2, are established. The data are for steels and other metallic alloys subjected to completely reversed loading. The decrease of the maximum strength with cycles is evident for every alloy. For steels, the endurance limit, or fatigue limit—the stress below which no fatigue failure is expected no matter how many cycles are applied—is well pronounced.

Under certain conditions, an endurance limit may be observed in steels at ambient temperature, but it may not be present at elevated temperatures, or may be eradicated with an overload or the onset of corrosion. In other alloys, such as age-hardening aluminum alloys for example, endurance limits at 10^8 cycles, are not observed. Thus, the endurance limit is not an inherent property of metallic alloys.

Fatigue Strength and Tensile Strength. Figure 2 clearly demonstrates that the fatigue performance increases with an increase in tensile strength. The increase of the fatigue strength with tensile strength (Fig. 3) is true for specimens with good surface finish and without stress concentrators and only up to a certain hardness where flaws do not govern behavior. In the presence of notches or of corrosive environment, the fatigue strength does not improve substantially with an increase of tensile strength. Notches and stress raisers are likely going to be present in actual components, and it may not be possible to achieve the desired fatigue strength by selecting an alloy with increased tensile strength without changing the geometry.

Figure 4(a) shows the relation between tensile strength and the fatigue strength for wrought steels, and it is seen that the endurance ratio

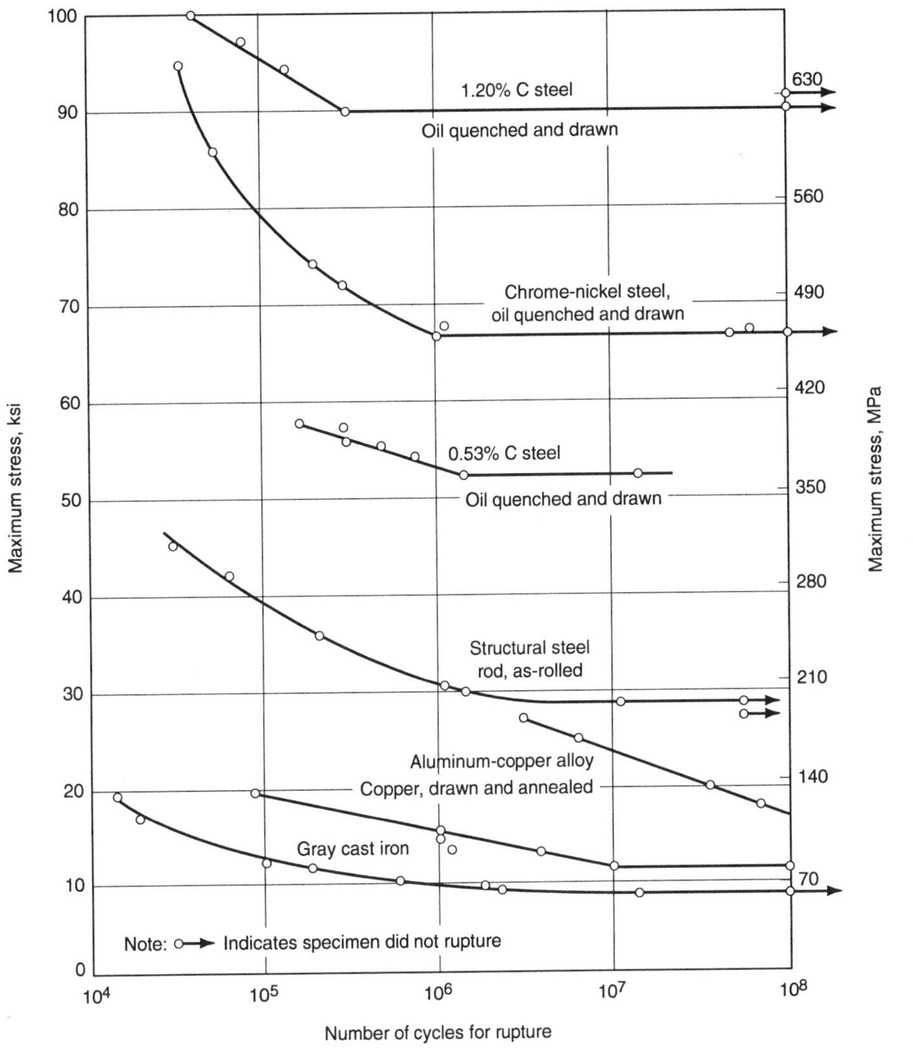

Fig. 2 *S-N* diagram for various alloys subjected to completely reversed loading at ambient temperature. The decrease in fatigue strength with cycles and the endurance limit of some steels is shown. Source: Ref 6

(fatigue strength at the endurance limit or at a given number of cycles/tensile strength) is between 0.6 and 0.35. A comparable relation holds for aluminum alloys (see Fig. 4b) with the endurance ratio between 0.35 and 0.5.

For steels, the fatigue strength at high number of cycles is approximately 50% of the tensile strength.

Data Scatter. For most of the fatigue curves in Fig. 2, only one test per stress level was performed and the fatigue (S-N) curve was drawn through these data points, giving the impression that there is a unique relationship between cycles to failure and fatigue strength. If more than one specimen is tested at the same stress level, so that the loading and testing conditions are duplicated within experimental accuracy, different cycles to failure will in general be found. It is then possible to plot a histogram of number of failed specimens in a certain cycle interval, as shown in Fig. 5. Normalization of the histogram is accomplished by dividing the ordinate by the total number of specimens, 57 in this case. Then the ordinate gives the percentage of specimens that fail at a given number of cycles to failure. This method of testing and plotting reveals that fatigue is associated with scatter. The scatter has been found in carefully performed tests and has been established as a basic property of the fatigue strength. (Note also the difference in the histogram when plotted on a linear and a log scale.)

To appreciate the effects of scatter it is advantageous to keep the relation shown in the schematic of Fig. 6 in mind. In this figure, a normalized histogram as shown in Fig. 5 is superposed on the fatigue curve at three different stress levels. The ordinates of the bell-shaped curves give the percentage of specimens expected to fail at a given number of cycles. The peaks of the distributions can be connected by a fatigue curve. This curve would indicate the number of cycles to failure at which the largest percentage of specimens are expected to fail. If only a few tests are run, the results are expected to be close to this curve. The width of the band in Fig. 6 indicates that all specimens are expected to fail in this interval. The distributions become broader as the stress levels decrease, indicating an increase in scatter. The reason for this observation has been discussed in the section "The Fatigue Process" in this article.

It is possible to treat fatigue as a probabilistic process and to use methods of probability theory for design. However, a probabilistic design of components is prevalent if a large number of components are involved and is then part of a reliability analysis. In the majority of cases, a probabilistic fatigue design is not performed because of the expense and time involved in getting the data. However, it is always important to keep the nondeterministic nature of fatigue strength in mind. A discussion of probabilistic design methods for fatigue can be found in Ref 12 to 14. Statistical analysis of fatigue data is described in Ref 15 and 16.

Mean Stress Effects. Fatigue life is affected by the presence of a mean stress as shown schematically in Fig. 7. A tensile mean stress reduces the

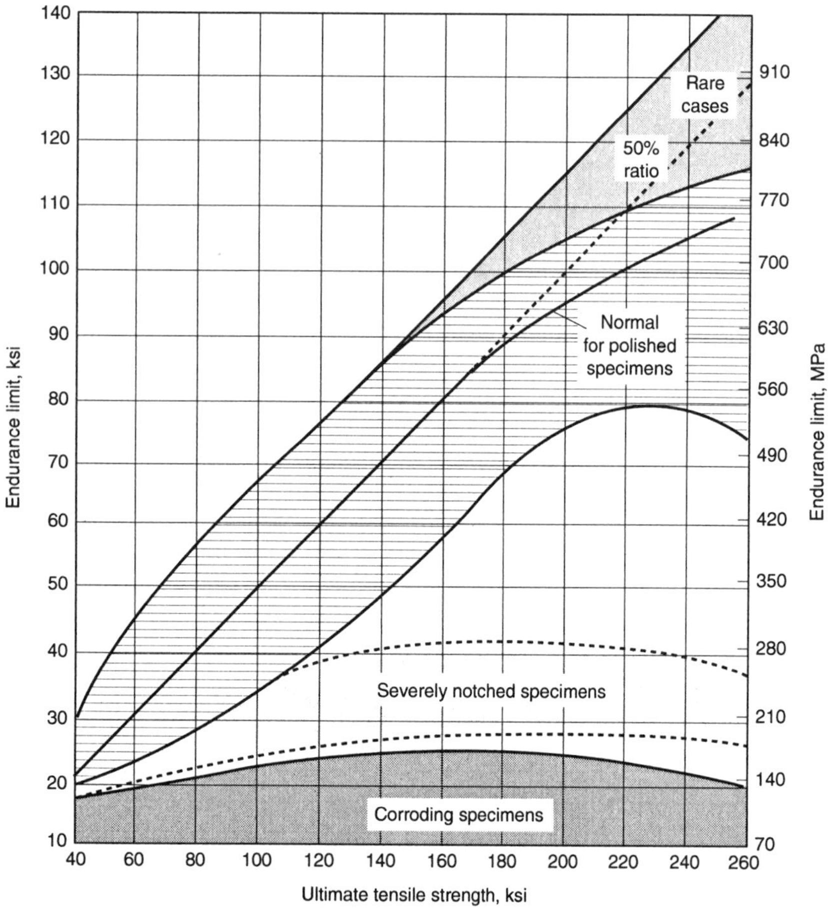

Fig. 3 The relation between fatigue strength and tensile strength of polished, notched specimens and of specimens subjected to a corrosive environment. Source: Ref 7

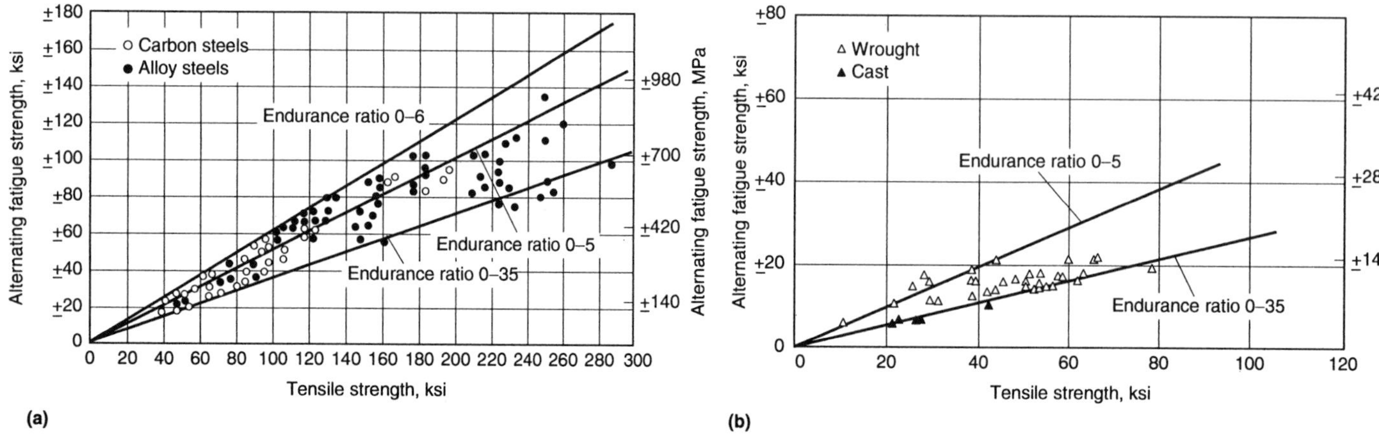

Fig. 4 Alternating fatigue strength in rotating bending (stress amplitude) and tensile strength. Wrought steels, fatigue strength between 10^7 to 10^8 cycles. Source: Ref 8

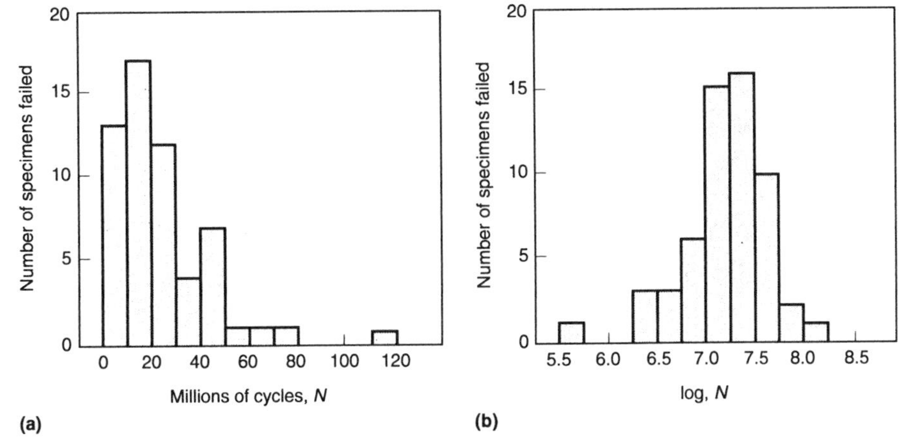

Fig. 5 Histograms showing fatigue-life distribution for 57 specimens of a 75S-T6 aluminum alloy tested at 30 ksi. Note the influence of a linear (a) or logarithmic (b) plot of cycles to failure N on the shape of the histogram. Source: Ref 10

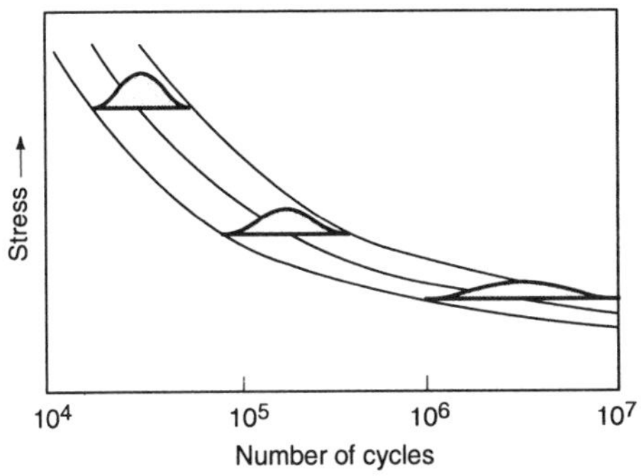

Fig. 6 Schematic showing the fatigue curve with the distribution of lives at three different stress levels. Note that the distribution widens as the stress level decreases. Source: Ref 11

life at a given amplitude and a compressive mean stress increases it.

Test data, which most likely pertain to endurance or to a life around the endurance limit, for example, 10^8 cycles, are shown for steels and aluminum alloys in Fig. 8. The beneficial effects of a compressive mean stress and the deleterious effect of a tensile mean stress are evident. It is interesting to observe that in this normalized plot very little difference exists between steels and aluminum alloys.

Figure 9 shows constant fatigue-life diagrams from 10^4 to 10^7 cycles for smooth specimens (solid lines) and for notched specimens (dashed lines) for an age-hardened aluminum alloy. Both maximum/minimum stress coordinates and alternating/mean stress coordinates are used for plotting. In addition, lines of constant A and R ratios are also entered. The lines for the smooth specimens end at mean stress of 82 ksi, which corresponds to the ultimate strength of the aluminum alloy. In plotting the data, the authors have used the relationships given in Eq 1 through 5. Although the diagram is very busy, it shows, in addition to the fatigue information, how the same data can be plotted differently.

Figure 9 demonstrates again that for a given fatigue life the allowable amplitude decreases as the mean stress increases.

The influence of mean stress in high-cycle fatigue at the endurance limit is often predicted using only the engineering ultimate tensile or yield strength as shown schematically in Fig. 10. The fatigue strength amplitude for $R = -1$ is plotted on the ordinate and the ultimate stress and the yield strength are marked on the abscissa. Straight lines are then drawn to the endurance at zero mean stress. These lines are known as Soderberg and modified Goodman lines as seen in Fig. 10. The modified Goodman line is sometimes too conservative, and therefore the Gerber parabola is drawn as shown. Forrest (Ref 18) shows that the data fall largely between the Gerber parabola and the modified Goodman line. This empirical construction gives a good prediction of the mean stress effect for positive mean stresses.

Stress Concentration. The equilibrium conditions of mechanics result in a constant stress for a bar of uniform cross section and loaded in the direction of the axis of the bar. In bending or torsion, a linear distribution of stress across the section is computed under the assumption of linear elastic behavior. For a real specimen, the actual distribution of stresses varies from grain to grain and at the grain boundaries and cannot be determined accurately, if at all, without knowing the exact location, the individual properties, and the orientations of the grains.

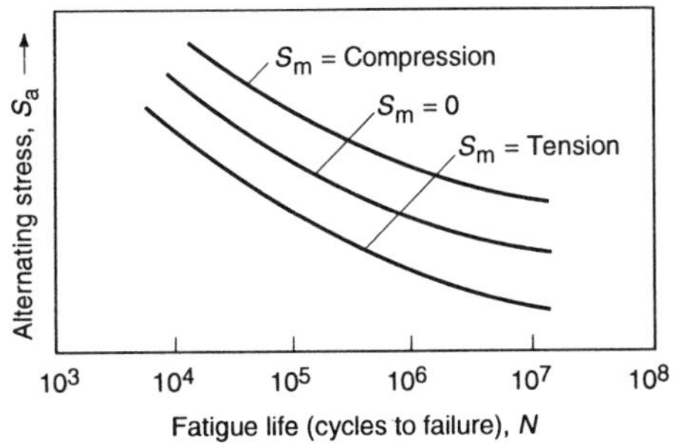

Fig. 7 Schematic showing the influence of mean stress on fatigue life. S_m, mean stress. Source: Ref 9

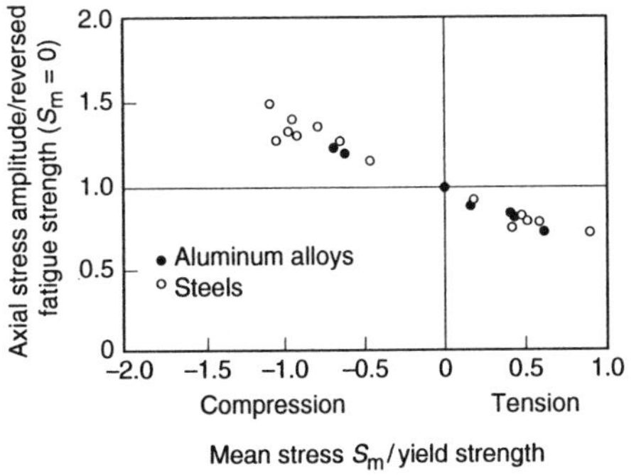

Fig. 8 The influence of compressive and tensile mean stresses on the fatigue strength amplitude of steels and aluminum and alloys. Source: Ref 17

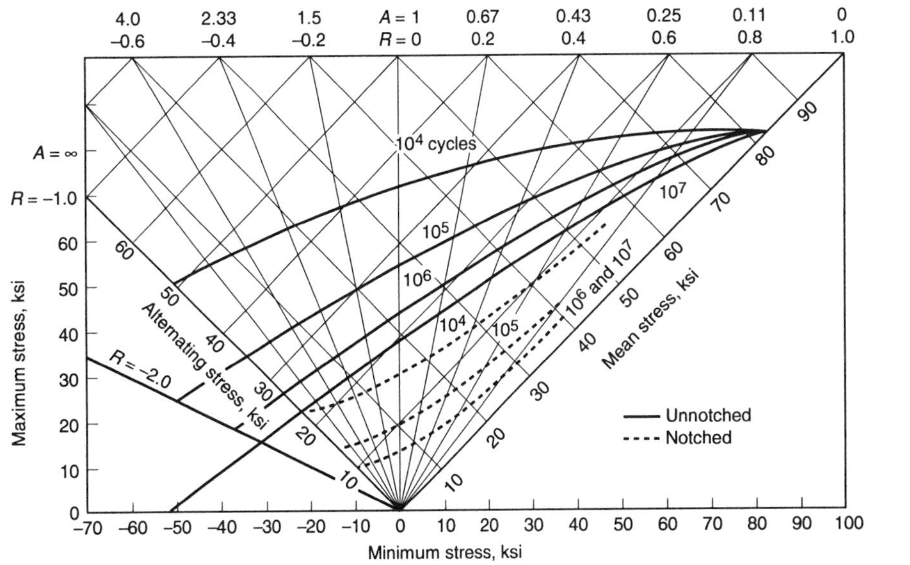

Fig. 9 Constant-life fatigue diagrams for unnotched (solid lines) and notched (dashed lines) 7075-T6 specimens with an ultimate tensile strength of 82 ksi. The stress concentration factor for the notched specimens is $K_t = 3.4$. Data were obtained at 2000 cycles/min. 1 ksi = 6.895 MPa. Source: Ref 17

The average of the actual distribution must be equal to the stress derived from continuum analysis, and this average stress is used in calculations.

If changes in cross section occur in a component, the stress distribution calculated by elastic stress analysis is no longer uniform and no longer one-dimensional. An example of an actually calculated linear elastic stress distribution for a deeply notched bar under axial load P is given in Fig. 11. The graph is normalized and is valid for any material that exhibits linear elastic behavior.

The principal normal stresses plotted are the longitudinal stress S_l, the circumferential stress S_c, and the radial stress S_r. The variation of these stresses are shown as a function of the radius. The distribution is symmetric to the center line at r = 0, and no shear stresses are present due to symmetry. The radial stress S_r is 0 at the notch root, and the state of stress is two-dimensional there but changes to a three-dimensional one in the interior. The longitudinal stress reaches its highest value in the notch root, which is 2.44 times the nominal stress S_n based on the minimum cross section. This geometry is said to have a stress concentration factor of 2.44. (The solution used in Fig. 11 is for a deep notch; the outer diameter is very large and need not be specified.)

The stress concentration factor K_t is defined as the ratio of the highest longitudinal stress over the net section nominal stress (sometimes the nominal stress is based on the unnotched section). It simply measures how many times the notch-root longitudinal stress is greater than the nominal stress. Effects of multiaxiality are not included.

For many geometries and simple loading cases the stress concentration factor has been determined, for example, in Ref 20 and 21. If in addition to the stress concentration factor the stress distribution is desired, it can be determined by experiments or by analysis, or, of course, by finite-element calculations.

The stress concentration factor does not give any information on the fatigue strength of a notched member. To find out how weakening the notch is going to be in cyclic loading, a fatigue test must be performed. As in the case of smooth specimens a fatigue curve is obtained. Using the smooth and the notched specimen data obtained under the same loading conditions a fatigue strength reduction factor K_f can be calculated by:

$$K_f = \left(\frac{\text{Smooth specimen strength}}{\text{Notched specimen strength}}\right)_{\text{at a given life}} \quad \text{(Eq 6)}$$

where the notched specimen strength is based on the nominal stress usually referred to the minimum cross section. Most of the tests are done for completely reversed conditions, but results with mean stress can also be analyzed as long as the conditions for the smooth and the notched specimens are the same. In contrast to the stress concentration factor K_t, the fatigue reduction factor K_f is material specific.

The results shown in Fig. 9 permit the calculation of the fatigue strength reduction factor K_f based on stress amplitude (alternating stress) for several values of the mean stress at different cycles to failure. It can be seen that:

$$K_f \leq K_t \quad \text{(Eq 7)}$$

always. This has been found true in many other studies and a "fatigue notch sensitivity index" q is in use:

$$q = (K_f - 1)/(K_t - 1) \quad \text{(Eq 8)}$$

The index varies from 0 for a material with no notch sensitivity to full sensitivity if $q = 1$.

A conservative estimate for the fatigue strength reduction factor K_f is the assumption that it is equal to the stress concentration factor K_t or that $q = 1$.

This estimate is too conservative in some cases and does not consider the multiaxial state of stress in the notch root. Other, more complicated methods of predicting fatigue strength of notched

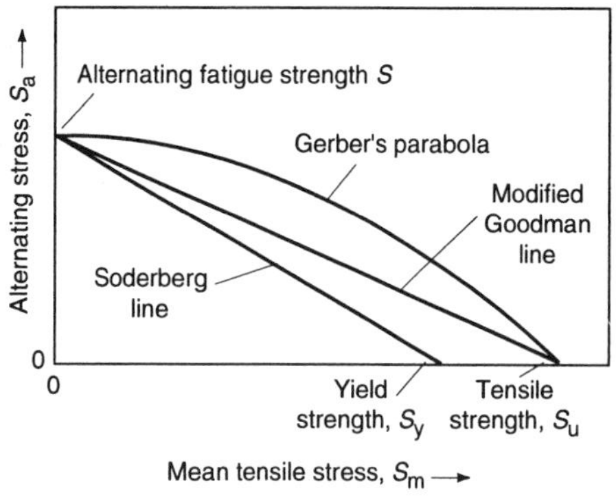

Fig. 10 Schematic showing the construction of Soderberg and modified Goodman lines and of the Gerber parabola. Note that this construction is a prediction of the mean stress influence on fatigue. Source: Ref 18

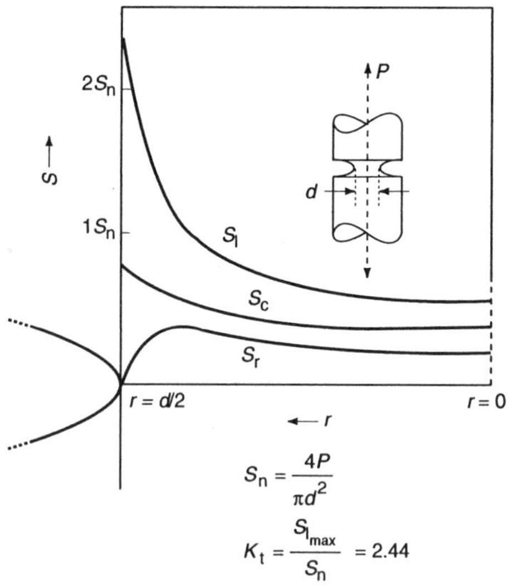

Fig. 11 Stress distribution in a deeply notched bar. S_l, S_c, and S_r are the longitudinal, the circumferential, and the radial stress, respectively. The nominal stress is given by $S_n = (4P)/(\pi d^2)$. Source: Ref 19

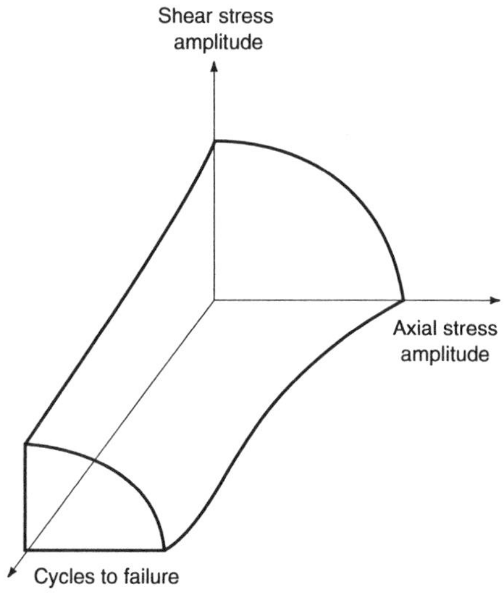

Fig. 12(a) Schematic showing *S-N* curves for multiaxial loading and surfaces of equal cycles to failure

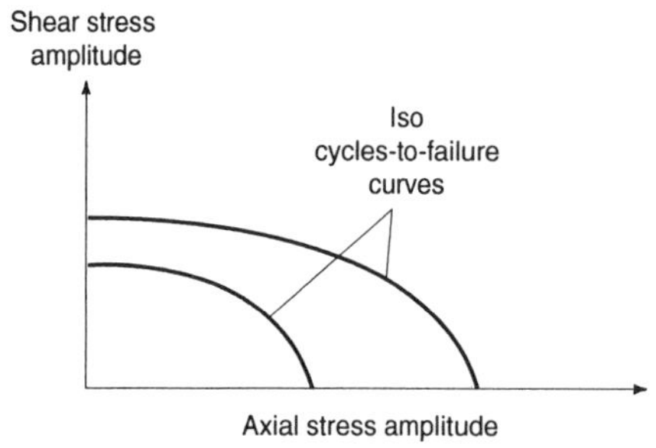

Fig. 12(b) The results of Fig. 12(a) replotted as iso-cycles-to-failure curves

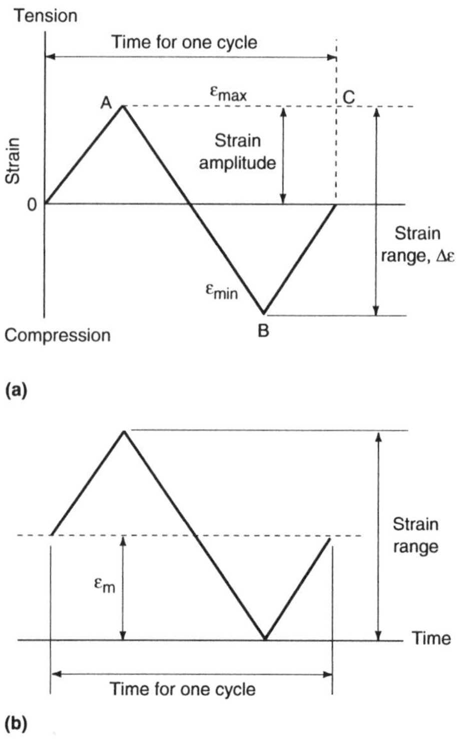

Fig. 13 Definition of terms in low-cycle fatigue testing. Source: Ref 31

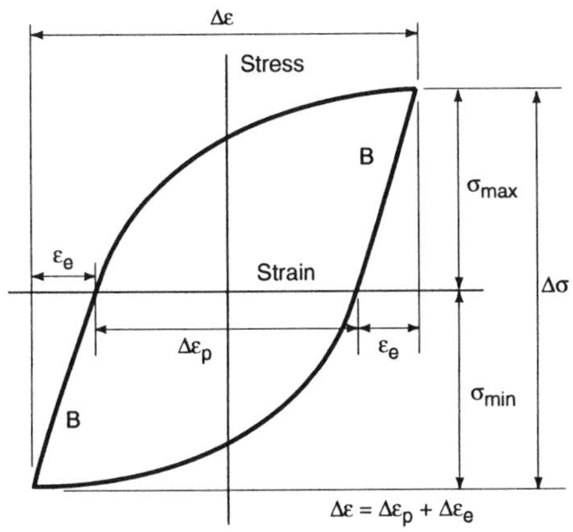

Fig. 14 Definition of terms related to the stable hysteresis loop. Source: Ref 32

members have been proposed: see Ref 3 and 22 to 24.

Multiaxial Fatigue. Components can be subjected to complex loadings in the presence of multiaxial states of stress and design methods must be developed for these conditions. Multiaxial fatigue tests are performed to elucidate the basic fatigue properties from which life-prediction methods can be derived. A thin-walled tube subjected to axial and torsional loading and bars subjected to bending and torsion are examples of possible specimens and loading conditions. As a first step, it is desirable to have a measure of stress that can correlate the data obtained under different conditions.

A thin-walled tube subjected to axial and torsional loading is considered to elucidate the approach. For the tube, the torque generates shear stresses τ, and the axial forces induces a normal stress σ in planes perpendicular to the axis. For a sinusoidal loading:

$$\sigma = \sigma_0 \sin \omega t$$

$$\tau = \tau_0 \sin(\omega t + \varphi) \quad \text{(Eq 9)}$$

where the subscript 0 indicates amplitude and where ω and φ denote the frequency and the phase shift, respectively.

If the phase shift φ is 0, proportional loading is imposed. Fatigue curves can be obtained for axial and torsional loading and combined loading. Figure 12(a) shows a schematic plotting multiaxial fatigue results for constant amplitude, completely reversed loading. The usual *S-N* curves are obtained for axial and torsion loading. Iso cycles-to-failure curves are obtained for $N = 10^4$ and $N = 10^8$ cycles. They represent the combination of axial and shear stress that lead to failure in the stated number of cycles. These curves are replotted in Fig. 12(b).

A multiaxial stress criterion should be able to reproduce the iso cycles-to-failure curves shown in Fig. 12(b).

The von Mises or effective stress criterion is in indicial notation:

$$\bar{\sigma}^2[N] = \frac{3}{2}(\sigma_{ij} - \frac{1}{3}\sigma_{kk}\delta_{ij})(\sigma_{ij} - \frac{1}{3}\sigma_{kk}\delta_{ij}) \quad \text{(Eq 10)}$$

where σ_{ij} is the stress matrix, δ_{ij} is the Kronecker delta, and $\bar{\sigma}$ is the effective stress, which is a function of cycles to failure *N*.

If Eq 10 is specialized for the axial-torsion case, the equation of iso cycles to failure curves is given by:

$$\bar{\sigma}^2[N] = \sigma_0^2 + 3\tau_0^2 \quad \text{(Eq 11)}$$

It is seen that the von Mises criterion requires that the iso cycles to failure curves are ellipses that vary as a function of *N*. Other criteria, the Tresca criterion for example, result in different iso-cycle curves.

The von Mises criterion has been used as a model of a yield surface in plasticity. Plastic deformation is known to be insensitive to superposed pressure up to large pressures. As a consequence, Eq 10 would predict a 0 right-hand side for pure hydrostatic loading. Fatigue on the other hand has been shown to be sensitive to superposed hydrostatic pressure; see Ref 25 for a summary. Therefore, the von Mises criterion and other yield criteria are not suitable for correlating fatigue data on principal grounds. It is therefore not surprising that the use of yield criteria for correlating fatigue data has not always shown

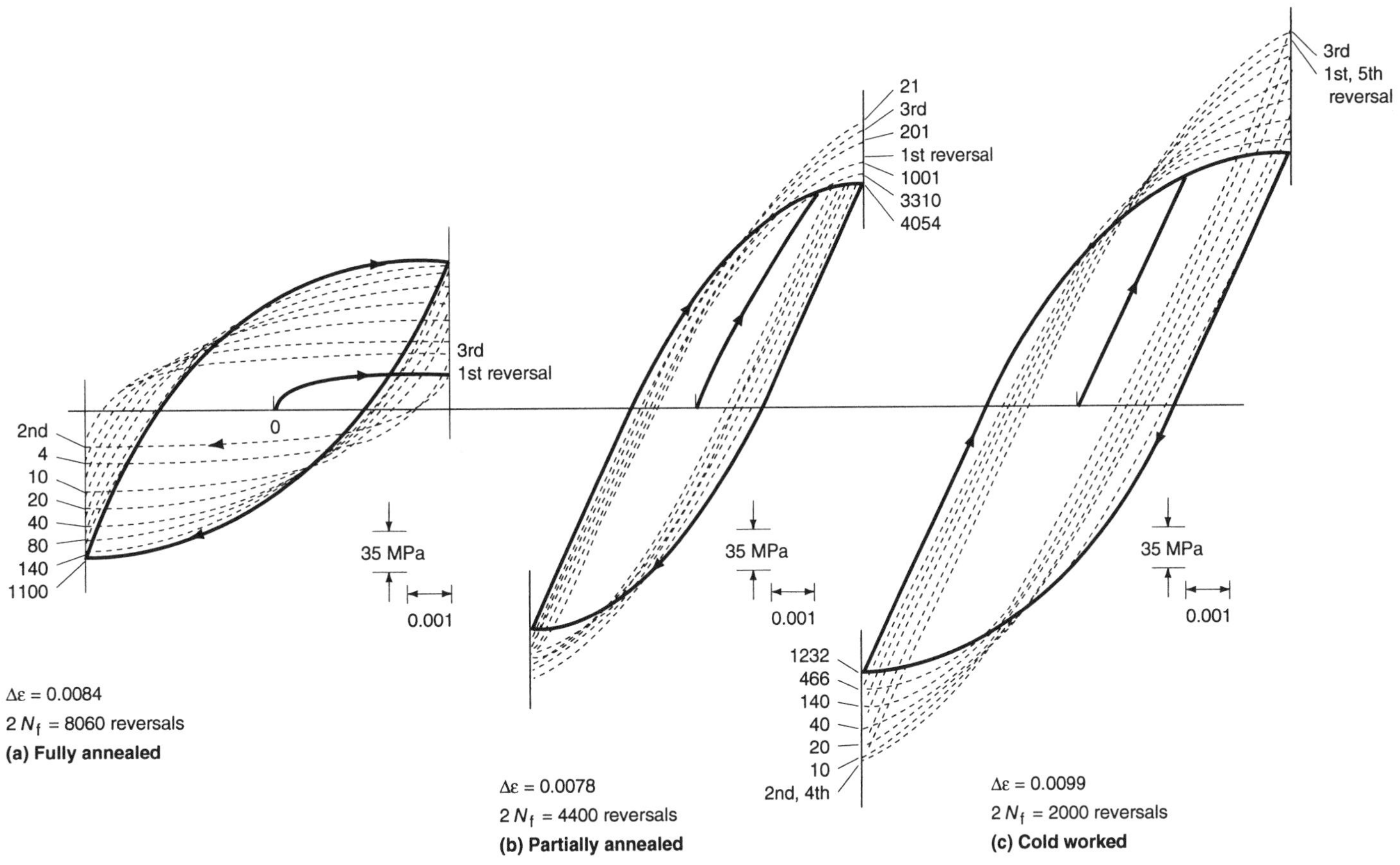

Fig. 15 Hysteresis loops for copper with varying degree of prior cold work. Source: Ref 33

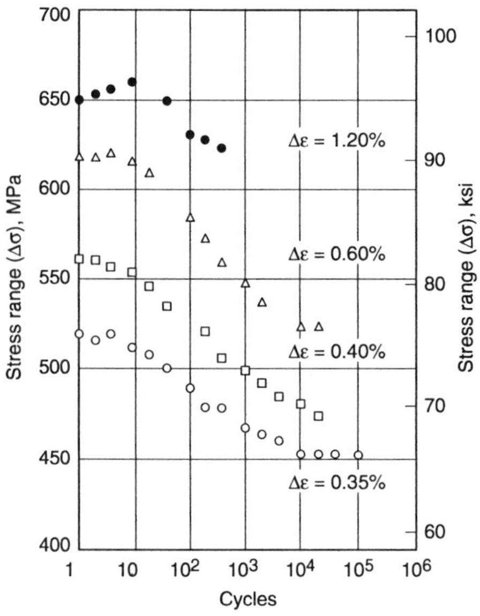

Fig. 16 Stress range versus cycles for type 304 stainless steel showing initial hardening or neutral behavior followed by softening. Source: Ref 34

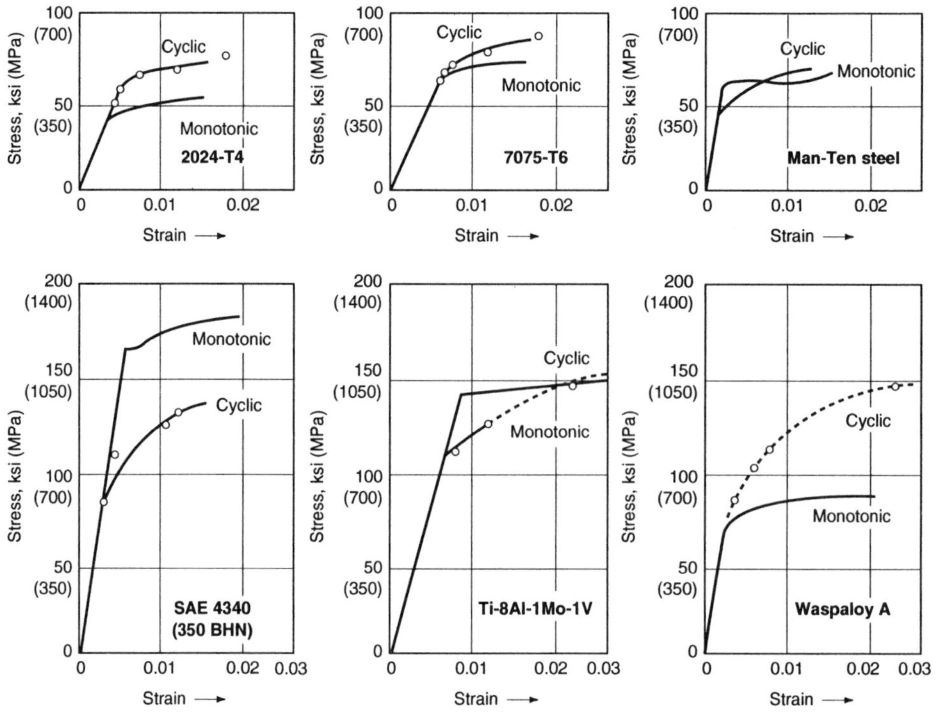

Fig. 17 Monotonic and cyclic stress-strain diagrams for six different engineering alloys. O, companion specimens. ———, incremental step. Source: Ref 35

good results (see Ref 25). Despite these fundamental problems, the yield criteria continue to be used in design codes.

Loading with constant normal and shear stress amplitudes is assumed. Proportional loading is obtained for zero phase shift, $\varphi = 0$. It can be shown easily that in this case the planes of maximum shear stress amplitude are stationary. If, on the other hand, a phase shift between axial and shear loading exists, $\varphi \neq 0$, nonproportional loading takes place and the planes of maximum shear stress amplitude change during a cycle.

Because of the rotation of the planes of maximum shear stress in out-of-phase loading, $\varphi \neq 0$, there are more potential initiation sites in nonproportional than in proportional loading. Different fatigue lives are to be expected, even if axial and shear stress amplitudes are the same. This is indeed the case.

These and other observations and considerations have led to "critical plane approaches" (see Ref 26).

Additional overviews of multiaxial fatigue are given in Ref 27 to 29. A review paper on multiaxial fatigue and notches is given in Ref 30.

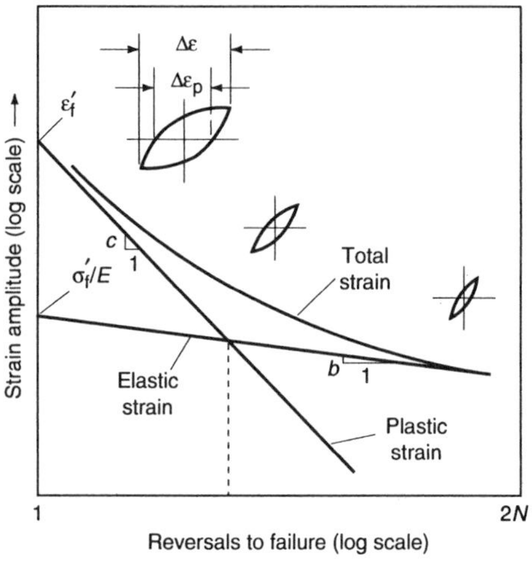

Fig. 18 Strain amplitude versus cycles to failure showing elastic and plastic contributions. Source: Ref 40

Low-Cycle Fatigue

Low-cycle fatigue loading usually involves bulk inelastic deformation behavior, and a mechanical hysteresis loop develops between stress and strain. Low-cycle fatigue testing is performed at low frequencies, usually below 1 Hz. The demands of the application and the development of clip-on axial and diametral extensometers and of servocontrolled testing machines have established strain control as the standard practice in low-cycle fatigue.

Low-cycle fatigue investigation was in part prompted by component failures in the power-generation industry. These failures at stress risers (notches, fillets) were caused by few, repeated thermal stresses due to startup and shutdown operations. The elastically stressed neighborhood limited the strains of the inelastic deformation region associated with stress riser. The plastic zone is considered to be under strain control by the surrounding elastic material.

In testing, strain control eliminates potentially large strains caused by small errors in controlling the load and by the small tangent modulus in the plastic range. Strain-controlled testing limits the strains automatically.

Figure 13 explains the terminology used in low-cycle fatigue testing. The terms are essentially the same as shown in Fig. 1, except that strain is replaced by stress. As in the case of high-cycle fatigue, completely reversed testing is done in the majority of cases. In contrast to high-cycle fatigue where the deformation behavior is nominally linearly elastic, nonlinear, inelastic behavior is encountered in low-cycle fatigue testing. A hysteresis loop develops, and terms related to the stable hysteresis loop are explained in Fig. 14.

Deformation Behavior

Figure 15 shows hysteresis loops for copper at three different levels of cold work. The curves on the left pertain to a fully annealed material (no prior cold work), and the stress range increases considerably. The increase of stress range diminishes from cycle to cycle, and finally the hysteresis loop is traced over and over again (not shown). A stable state, shakedown, a cyclic steady state, or cyclic neutral behavior has been reached.

The hysteresis loops of the partially annealed material reach a higher stress amplitude than the ones for the fully annealed material. Hardening continues and is followed by softening, a decrease in stress range (see the cycle numbers written on top right; the stress level at first reversal is higher than that at the 4054th). There may

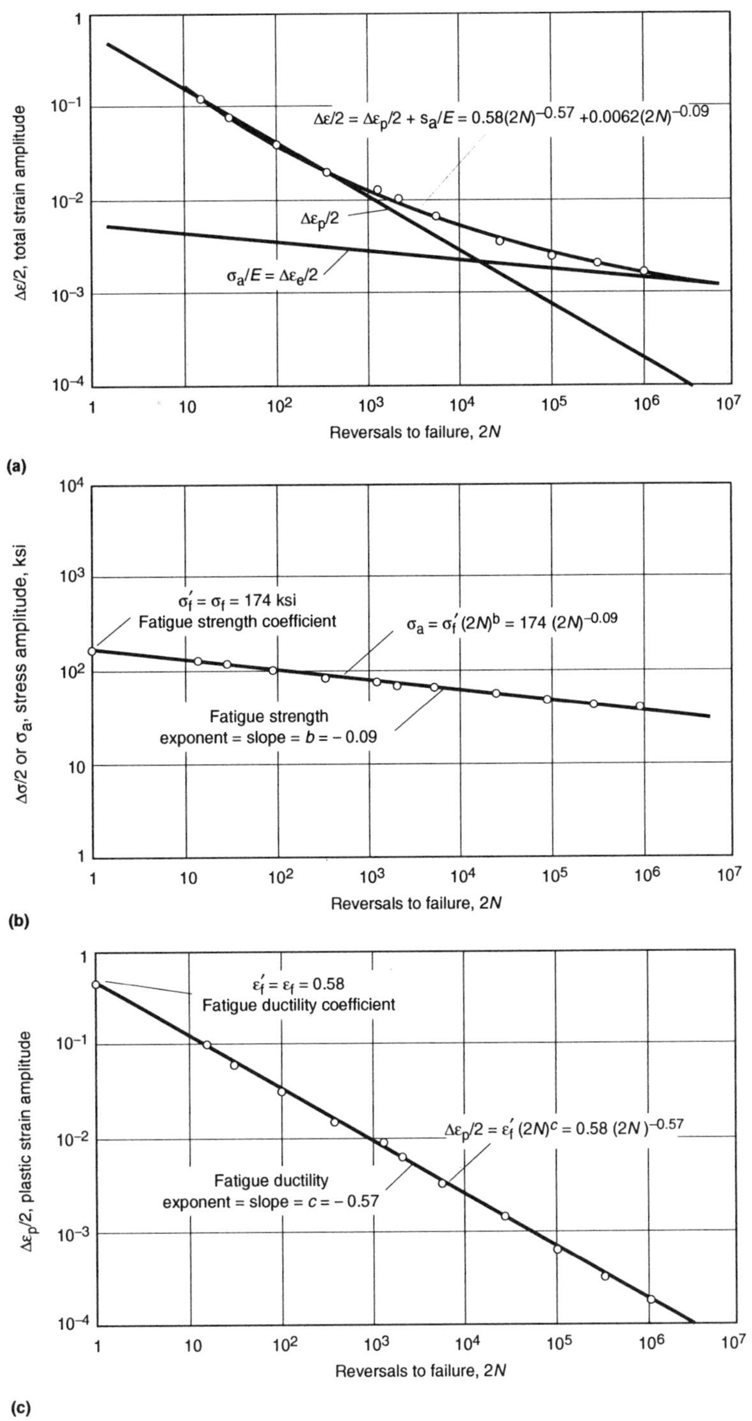

Fig. 19 Determination of the constants for the fatigue life curve for annealed 4340 steel. Source: Ref 42

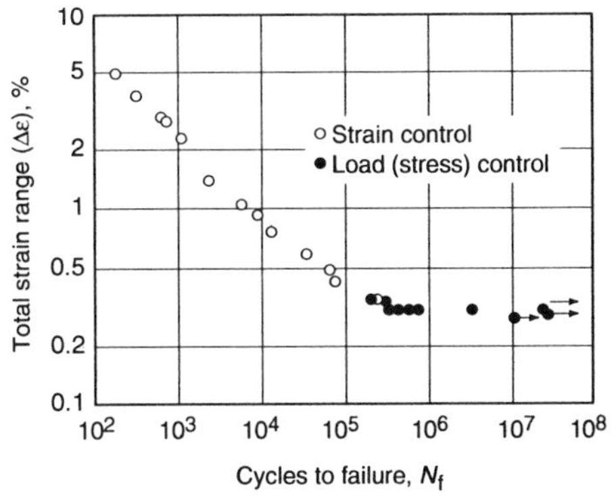

Fig. 20 Transition from low-cycle fatigue to high-cycle fatigue for carbon steel plate. Source: Ref 44

have been a quasi-steady-state between the 2000th and 4000th cycle, where the maximum stress changed very little when measured on a per cycle basis.

The fully cold-worked specimen, hysteresis loops on the right, cyclically softens, and the stress range decreases. The tests depicted in Fig. 15 were done in strain control. The strain amplitude was kept constant. All other quantities, the stresses, especially the stress range, and the plastic strain range (the width of the hysteresis loop at zero stress) can change with cycles. Figure 15 shows that these changes can be significant.

All dependent variables (all those that are not enforced, or controlled) can change in response to the internal microstructural rearrangements and are macroscopic indicators of such changes. If the hysteresis loop changes from cycle to cycle, a new material forms at each cycle, in principle. While the changes and therefore the differences in the material from cycle to cycle are small, they can be significant between the initial condition and steady-state condition.

Figure 15 illustrates a general trend found in cyclic inelastic deformation. Annealed materials tend to harden cyclically. In contrast, cyclic softening is observed in cold-worked materials. The trend is not always uniform; sometimes cyclic hardening (softening) is followed by softening (hardening). An example is given in Fig. 15 for the partially annealed specimen and in Fig. 16 for type 304 stainless steel. At the high strain ranges, cyclic hardening is observed followed by significant softening. At the low strain ranges the initial behavior is cyclic neutral and is followed by softening. It should be noted that the initial behavior is for a small number of cycles only and that the drops in stress range appear exaggerated due to the vertical scale starting at 400 MPa.

Designers allow for the difference of a cyclically loaded material and the material when cycling started by constructing the so-called cyclic stress-strain diagrams. Here the stress amplitude at cyclic steady state, or, if such a steady state does not exist, the stress amplitude at half the number of cycles to failure is determined. The corresponding strain amplitude and stress amplitude are then entered into a stress versus strain diagram. Only a few points (around 5 points usually) are obtained this way, but they suffice in most cases to draw a cyclic stress-strain diagram. No significant microstructural changes take place for elastic cycling, and therefore the monotonic and the cyclic stress-strain diagram have the same elastic slope. Differences appear at strain amplitudes beyond the elastic range as shown in Fig. 17, where the monotonic and the cyclic stress-strain diagrams for six different engineering alloys are presented. When the cyclic curve is above (below) the monotonic curve, cyclic hardening (softening) is achieved. For many of the alloys, the cyclic curve is equal or above the monotonic curve.

For an analytical representation, the Ramberg-Osgood formula is used for both the cyclic and

Table 1 Monotonic and cyclic fatigue properties of selected engineering alloys

Material	Process description	S_u, MPa	S_y/S_y', MPa/MPa	K/K', MPa/MPa	n/n'	$\varepsilon_f/\varepsilon_f'$	σ_f/σ_f', MPa/MPa	b	c	S_f ($2N = 10^7$), MPa	S_f/S_u
Steel											
1005–1009	HR sheet	345	262/228	531/462	0.16/0.12	0.16/0.10	848/641	–0.109	–0.39	148	0.43
1005–1009	CD sheet	414	400/248	524/290	0.049/0.11	1.02/0.11	841/538	–0.073	–0.41	195	0.47
1020	HR sheet	441	262/241	738/772	0.19/0.18	0.96/0.41	710/896	–0.12	–0.51	152	0.34
0030(c)	Cast steel	496	303/317	...	0.30/0.13	0.62/0.28	750/653	–0.082	–0.51	190	0.38
Man-Ten	HR sheet	510	393/372	.../786	0.20/0.11	1.02/0.86	814/807	–0.071	–0.65	262	0.51
1040	As forged	621	345/386	...	0.22/0.18	0.93/0.61	1050/1540	–0.14	–0.57	173	0.28
RQC-100	HR sheet	931	883/600	1172/1434	0.06/0.14	1.02/0.66	1330/1240	–0.07	–0.69	403	0.43
4142	Drawn at temperature	1062	1048/745	...	.../0.18	0.35/0.22	1115/1450	–0.10	–0.51	310	0.28
4142	Q&T	1413	1379/827	...	0.051/0.17	0.66/0.45	1825/1825	–0.08	–0.75	503	0.36
4142	Q&T	1931	1724/1344	...	0.048/0.13	0.43/0.09	2170/2170	–0.081	–0.61	589	0.31
4340	HR and annealed	827	634/455	...	.../0.18	0.57/0.45	1090/1200	–0.095	–0.54	274	0.33
4340	Q&T	1241	1172/758	1579/...	0.066/0.14	0.84/0.73	1655/1655	–0.076	–0.62	492	0.40
4340	Q&T	1469	1372/827	...	.../0.15	0.48/0.48	1560/2000	–0.091	–0.60	467	0.32
9262	Annealed	924	455/524	1744/1379	0.22/0.15	0.16/0.16	1046/1046	–0.071	–0.47	348	0.38
9262	Q&T	1000	786/648	.../1358	0.14/0.12	0.41/0.41	1220/1220	–0.073	–0.60	381	0.38
Aluminum											
1100-0	As received	110	97/62	...	.../0.15	2.09/1.8	.../193	–0.106	–0.69	37	0.33
2024-T3	...	469	379/427	455/655	0.032/0.065	0.28/0.22	558/1100	–0.124	–0.59	151	0.32
2024-T4	...	476	303/441	807/...	0.20/0.08	0.43/0.21	634/1015	–0.11	–0.52	175	0.37
5456-H3	...	400	234/359	...	.../0.16	0.42/0.46	524/725	–0.11	–0.67	124	0.31
7075-T6	...	579	469/524	827/...	0.11/0.146	0.41/0.19	745/1315	–0.126	–0.52	176	0.30

These values do not represent final fatigue design properties. HR, hot rolled; CD, cold drawn; Q&T, quenched and tempered. Ultimate engineering strength, S_u; engineering yield strength, S_y; fatigue strength, S_f'. All other quantities are defined in Eq 15 and 16. Source: Ref 36, 37

the monotonic stress strain diagram. For the monotonic case:

$$\varepsilon = \frac{\sigma}{E} + \left(\frac{\sigma}{K}\right)^{1/n} \quad \text{(Eq 12)}$$

and for the cyclic condition:

$$\frac{\Delta\varepsilon}{2} = \frac{\Delta\sigma}{2E} + \left(\frac{\Delta\sigma}{2K'}\right)^{1/n'} \quad \text{(Eq 13)}$$

where E and K are the elastic modulus and strength constant with dimensions of stress, respectively, and n is the dimensionless strain-hardening exponent. Primed quantities refer to the cyclic stress-strain diagram, and $\Delta\varepsilon$ and $\Delta\sigma$ denote the strain and stress range, respectively. Table 1, from Ref 38, gives typical data for some steels and some aluminum alloys.

Equations 12 and 13 are the simplest representation of stress-strain diagrams. They are only valid for monotonic loading. If the hysteresis loop and its evolution have to be described and modeled, incremental plasticity or unified state variable theories are needed (see Ref 39).

Fatigue Life

For design purposes, the strain amplitude or the strain range are plotted versus cycles to failure as shown in Fig. 18. Log-log coordinates are used, and the elastic and plastic contributions are shown. In addition, hysteresis loops are also drawn to show the nature of the deformation.

In high-cycle fatigue, cycles-to-failure referred to complete separation of a specimen. In low-cycle fatigue, it is customary to plot reversals to failure ($2N$) on the abscissa. In addition, cycles to failure does not designate separation of the specimen. In low-cycle fatigue, strain control is used. As the crack develops (it is assumed to occur within the extensometer gage length), the specimen becomes more and more compliant, and less and less load is required to reach the strain limits. With a large crack, the specimen could be cycled for a long time at very low stresses before complete separation, if any, is experienced.

Because of this uncertainty, cycles to failure are generally determined with regard to the stress response. When the maximum stress or sometimes the stress range is reduced by a certain percentage, usually 5 or 10%, the test is terminated and the specimen is declared to have failed.

The life-time curve is based on strain. It is described by postulating that the total strain range is the sum of the elastic and the plastic strain ranges:

$$\Delta\varepsilon = \Delta\varepsilon^{el} + \Delta\varepsilon^{pl} \quad \text{(Eq 14)}$$

and that each "component" is linearly related to the reversals-to failure on log-log coordinates as shown in Fig. 18. When these relations are substituted:

$$\frac{\Delta\varepsilon}{2} = \frac{\Delta\varepsilon^{el}}{2} + \frac{\Delta\varepsilon^{pl}}{2} = \frac{\sigma_f'}{E}(2N)^b + \varepsilon_f'(2N)^c \quad \text{(Eq 15)}$$

The quantities σ_f' and ε_f' are called fatigue strength coefficient and fatigue ductility coefficient, respectively, and are usually determined from tests, as are the dimensionless powers b and c.

Equation 15 was derived from the Coffin-Manson relationship, and the constants can be estimated from tensile tests, a test with one reversal (Ref 41). Table 1 gives the constants for selected steels and aluminum alloys. Extensive tabulations of these coefficients and exponents are given in Ref 43.

Figure 19 shows the details of the construction for a 4340 steel. The number of cycles to failure at the intersection of the elastic and plastic line is called the transition fatigue life.

A rule of thumb for ductile alloys at room temperature is that a fatigue life of 1000 cycles is reached at a strain range of 1%.

The transition from low-cycle to high-cycle fatigue is illustrated in Fig. 20 for specimens made from a carbon steel plate (Ref 44). Note that the data points at high number of cycles to failure are filled. This indicates that these tests were run in load control. They blend smoothly with the strain-controlled tests, and a fatigue curve ranging from 10^2 cycles to the endurance limit is obtained. (Clip-on extensometers for high strain ranges do not have the resolution to accurately measure and control small strain ranges. They also limit the frequency at which cycling can be done. After an initial adjustment period that determines the load range, the extensometer is taken off and the test is continued in load control.)

Notched Members

A first hypothesis for the calculation of the life to crack initiation in notched members is to assume that the maximum strain range in the root of the notch governs fatigue life. This hypothesis is called the strain approach to life prediction. It is widely accepted and used (see, for example, Ref 45).

Plastic deformation is involved, and the elastically calculated stress (strain) distributions are no longer valid. For the notch root, the axial strain (the strain in the direction of loading) increases whereas the axial stress decreases when inelastic deformation sets in. This is demonstrated in Fig. 21, where the plastic stress and strain distributions (labeled actual) for a given applied load are plotted. Note that the integral of the axial stresses over the cross section must equal the applied load to satisfy equilibrium. The actual stresses in the interior are higher than the elastic ones to compensate for the drop at the notch root (see Fig. 21). In contrast to elasticity where a single normalized stress distribution exists, the stress distribution varies with load in plasticity.

In elasticity, there is only one concentration factor and it is the same for every load in the elastic range. For plasticity, there is a strain and a stress concentration factor and they change with the applied load. Formally:

$$K_\sigma = \frac{\sigma_{max}}{\sigma_N} \quad \text{(Eq 16)}$$

and:

$$K_\varepsilon = \frac{\varepsilon_{max}}{\varepsilon_N} \quad \text{(Eq 17)}$$

where the subscripts max and N denote the value of the maximum axial component in the notch root and the nominal value (usually based on the minimum cross section), respectively.

Because of the redistribution of stress and strain, the strain concentration factor K_ε increases with nominal stress, whereas the stress concentration factor K_σ decreases (Fig. 22). The bifurcation at a nominal stress of about 15 ksi (103 MPa) is very well marked. A relationship similar to Fig. 22, but for an elastic stress concentration factor of 6 is given in Fig. 8.6 of Ref 28.

Two simplified procedures to determine the maximum strain in the notch root are given in Fig. 23. In each case, the cyclic stress-strain diagram and the smooth bar low-cycle fatigue curve are needed. Depending on the approach, the stress or the strain concentration factor must also be determined. The strain approach is mostly pursued in the published papers (for a review see Ref 3, 28, and 49).

For the determination of the stress (strain) concentration factors the empirical Stowell-Ohman-Hardrath Method (Ref 49), or the approximate Neuber Method and variants are used (Ref 24, 28). The latter is preferred in applications.

The Stowell-Ohman-Hardrath Method yields (Ref 28):

$$K_\sigma = 1 + (K_t - 1)\frac{E_s}{E_N} \quad \text{(Eq 18)}$$

where E_s and E_N are the secant moduli in a uniaxial stress-strain diagram evaluated at the maximum stress and strain and at the nominal stress and strain, respectively. They are given by $E_s = \sigma_{max}/\varepsilon_{max}$ and by $E_N = \sigma_N/\varepsilon_N$. From the definition of the stress and strain concentration factors, $K_\sigma/K_\varepsilon = E_s/E_N$. With this relation, the strain concentration factor is:

$$K_\varepsilon = K_t - 1 + E_N/E_s \quad \text{(Eq 19)}$$

The strain concentration factor can be determined by an iterative solution using Eq 12 (or a stress-strain diagram) and Eq 19 (Ref 50).

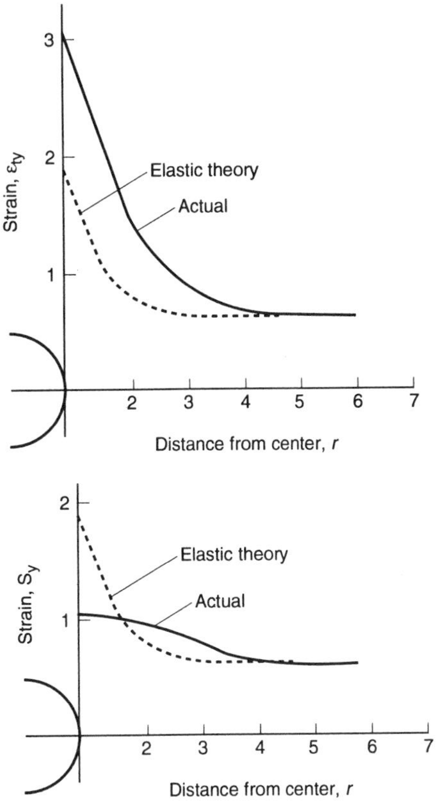

Fig. 21 Changes in the axial strain and stress distribution for elastic and plastic behavior. ε_{ty}, total yield strain; S_y, yield stress. Source: Ref 46

The starting point for the Neuber Method is again the approximate relation:

$$K_t^2 = K_\sigma K_\varepsilon \quad \text{(Eq 20)}$$

This expression is freely applied to strain and stress ranges and to the cyclic stress-strain diagram to determine the notch root strain in a procedure similar to the one discussed above (see Ref 24).

These methods are all approximate and derived from empirical relations. Plasticity involves nonlinear behavior, and analytical solutions are only available for simple geometries. Elastoplastic finite-element methods are now available to calculate the maximum strain in the notch strain, analyses can be performed using the cyclic stress strain diagram.

For additional methods of the analysis of notched members and their life prediction, see Ref 24, 28, and 30.

Multiaxial Low-Cycle Fatigue

Deformation Behavior. The cyclic-hardening (softening) phenomena experienced in uniaxial loading are also encountered in multiaxial loading. While in uniaxial loading, the planes of maximum shear stress are stationary; this need not be the case for biaxial or multiaxial loading. This causes additional features of hardening to appear. Included are "extra" hardening in out-of-phase loading and sequence effects.

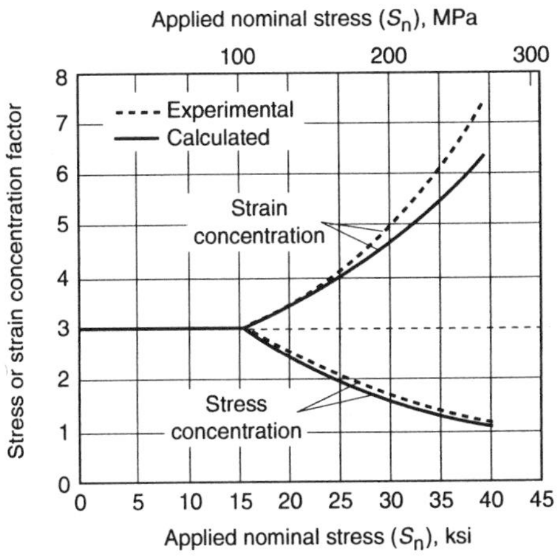

Fig. 22 The variation of the stress (strain) concentration factors with applied nominal stress S_n. Note that initially the stress and strain concentration factors are equal. They diverge as the nominal stress increases. Source: Ref 47

To demonstrate the latter tubular specimens of annealed type 304 stainless steel were cycled at an effective von Mises strain amplitude of 0.008 ($\varepsilon_e = (\varepsilon^2 + \gamma^2/3)^{1/2}$), where γ is the engineering shear strain) at a constant effective strain rate of 3×10^{-4} 1/s. Pure axial cycling and pure torsional cycling tests were performed in the first and second test, respectively. In the third test, torsion and axial loading alternated. After each cycle, the stress and strain were returned to zero. The results were plotted using the von Mises effective stress:

$$\sigma_e = (\sigma^2 + 3\tau^2)^{1/2}$$

and the accumulated effective inelastic strain:

$$p = \int_0^t [(\dot{\varepsilon} - \dot{\sigma}/E)^2 + (\dot{\gamma} - \dot{\tau}/G)^2]^{1/2}\, d\eta$$

where τ is the shear stress, G is the shear modulus, and η is the integration variable. While the continuous cycling responses are reasonably correlated by the effective quantities, the intermittent axial/torsion test exhibits considerably more hardening, as seen in Fig. 24. Extrapolation of the hardening curves would indicate that the saturation stress level will be different for continuous and intermittent cycling.

The hardening is not only sequence dependent but also amplitude dependent as shown in Fig. 25. By plotting the results of uniaxial and 90° out-of-phase tests performed at different effective strain amplitudes, it is demonstrated that the hardening depends not only on the accumulated inelastic strain, but also on the amplitude. It is also observed that the 90° out-of-phase tests harden considerably more than the uniaxial tests. It is most interesting that the true effective stress in a tensile test hardens even more than the cyclic tests. In this case, the strain rate was cycled between 1.5×10^{-3} and 1.5×10^{-5} 1/s, which explains the steps in the graph.

Finally, Fig. 26 demonstrates the path dependence of cyclic hardening for two materials. The top row has icons for the paths executed in strain control at the same effective stress amplitude as indicated in the left column. It is seen that the hardening as measured by the steady-state stress range is path dependent and that the 90° out-of-phase tests show the strongest hardening. The hardening that exceeds the one obtained in proportional loading is termed "extra hardening."

The "extra hardening" is observed in annealed copper and stainless steels, but not all alloys investigated so far exhibit this effect. It appears that cyclic neutral materials do not (Ref 53). The almost cyclic neutral, age-hardenable, isotropic superalloy B1900 + Hf did not exhibit extra hardening at two elevated temperatures (Ref 54). Also, aluminum is reported to exhibit no extra hardening (Ref 55). This author is not aware of any investigation using cyclic-softening alloys in multiaxial loadings involving different strain paths.

A review of the microstructural investigations associated with multiaxial loading is presented in Ref 56. They report on the microstructural reasons of the "extra hardening" referred to as overstrengthening.

It is comparatively easy to model the effects of biaxial loading in the absence of extra hardening (Ref 57). However, the modeling of extra hardening has not been satisfactorily accomplished. For a review, see Ref 58. An energy-based approach is championed in Ref 28.

Even if finite-element calculations are used to calculate the cyclic response, an accurate modeling of these extra-hardening effects cannot be expected because a completely satisfactory constitutive model is not available. (For a finite-element model to be realistic, the appropriate material description has to be implemented into the program.)

Fatigue Life. Because out-of-phase loading can reduce the fatigue life by more than a factor of ten (Ref 59) and because it is assumed that the strain history determines fatigue performance, it is important to capture the actual strain history in a component or specimen. Increasingly, finite-element analyses are used to obtain the desired strain ranges and mean strain and stresses.

As in the uniaxial case, strain-based approaches are preferred in multiaxial low-cycle fatigue-life prediction. Rather than relying on strain invariants, such as effective strain or maximum shear strain, for correlation and prediction of multiaxial low-cycle fatigue life, the trend in the 1970s shifted to strain-based "critical-plane" approaches, which according to Ref 60, made the connections between fatigue crack initiation on specific planes on the surface of the material and the maximum shear strain range and/or normal strains on these planes. An overview is given in Ref 61 for constant amplitude loading. This approach deals in essence with the life to crack initiation, and crack growth is not included. Increasingly, the activity shifts toward inclusion of crack propagation in life prediction (Ref 62, 63).

According to Ref 64, multiaxial fatigue is best understood by examining crack propagation mechanics, because it is the physical growth of cracks that controls damage accumulation. This trend will undoubtedly continue, and the crack initiation and crack-propagation methods will have to be merged to arrive at a "complete" life-prediction method.

Thermal Fatigue

Many components in the power-generation and propulsion industry are subjected to the simultaneous action of mechanical loads and changing temperature. Examples are turbine blades and their coatings, first-stage shells in turbines and brakes of airplanes, cars, and railroads. In recent years failures of solder joints of electronic packages have been reported that can be attributed to thermal fatigue and are caused by the changing temperature in the devices.

Thermal cycles occur less than a few ten thousand times during the lifetime of the components. Frequently, the thermal stresses set up by these temperature changes can exceed the elastic region and inelastic deformation takes place. (At high temperatures, the linear elastic region may not exist and inelasticity can always be present.) After return to ambient temperature, residual stresses are present. Due to repeated cycles and due to the inelastic deformation, a low-cycle fatigue situation exists and cracking is experienced,

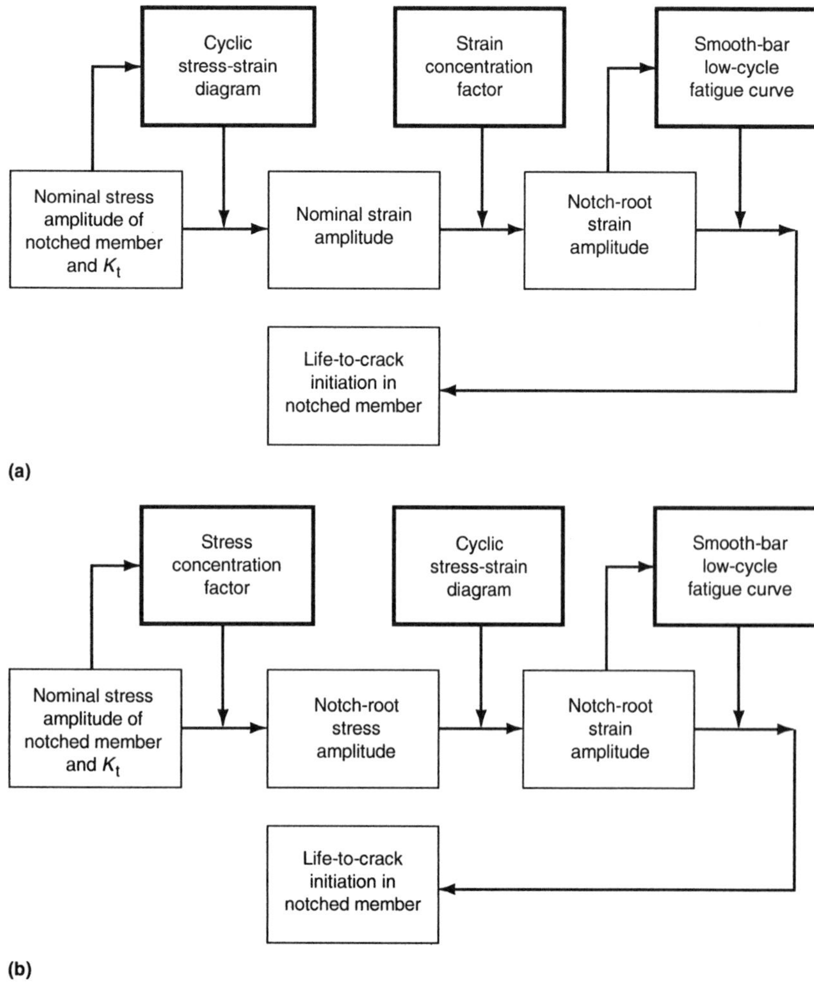

Fig. 23 Schematic showing how the life to crack initiation for notched members can be determined. (a) Strain approach. (b) Stress approach. Source: Ref 48

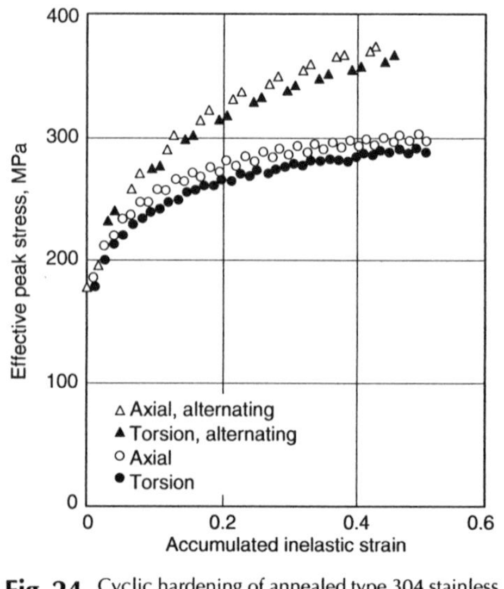

Fig. 24 Cyclic hardening of annealed type 304 stainless steel: Effective peak stress versus accumulated inelastic strain for pure axial and torsional cycling and intermittent axial/torsion cycling at an effective strain amplitude of 0.008. Source: Ref 51

especially at stress raisers, within the design lifetime of the component.

In many cases, the assumption is made that the life under thermal and mechanical cycling can be predicted by using the isothermal low-cycle fatigue life at the maximum temperature of the cycle. There is, however, no guarantee that this method leads to a conservative prediction.

Although thermal fatigue testing is more elaborate than isothermal low-cycle fatigue investigations, it can be performed using the test setup used in isothermal low-cycle fatigue tests. Suitable heating (induction heating is used mostly) and cooling facilities (mostly pressurized air) must be added. It should be mentioned that thermal fatigue investigations preceded isothermal low-cycle fatigue.

The lifetime analysis is also based on a strain approach as in low-cycle fatigue, and the methods of life prediction are similar.

Further details can be found in Ref 65, a recent, comprehensive overview.

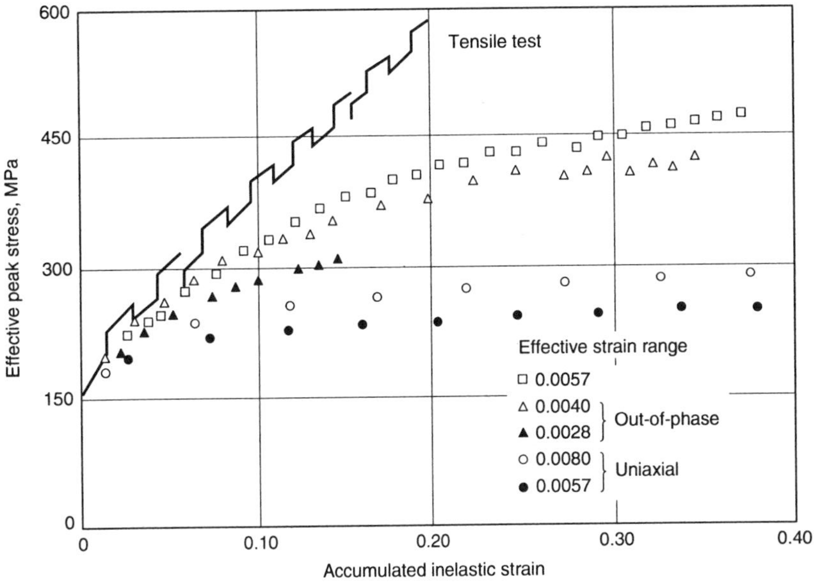

Fig. 25 Cyclic hardening of annealed type 304 stainless steel: True effective stress (for uniaxial monotonic loading with strain rate cycling between 1.5×10^{-3} and 1.5×10^{-5} 1/s), effective stress (uniaxial completely reversed strain controlled at two amplitudes and effective stress in 90° out-of-phase cycling (three amplitudes) versus accumulated inelastic strain. The effective strain rate is 3×10^{-4} 1/s in all cyclic tests. Source: Ref 51

High-Temperature Behavior

It is best to use homologous temperature θ (the ratio of test temperature over melting temperature, both measured in degrees Kelvin). For structural applications, θ is below 0.5. Higher homologous temperatures are used in processing.

High-Cycle Fatigue

In general, high-cycle fatigue strength is decreased as the temperature increases. This is demonstrated in Fig. 27, where most of the data show a decrease in fatigue strength with increasing temperature. There are, however, exceptions. The fatigue strength of 0.17% C steel has a maximum at about 350 °C. High-strength cast iron also exhibits an increase of fatigue strength with temperature before the decrease begins.

This anomalous behavior is most likely caused by dynamic strain aging. According to Ref 66, when solute atoms are sufficiently mobile to catch up with mobile dislocations, a particular dynamic interaction takes place. Whenever this phenomenon is present, it tends to produce behavior that is opposite to the one expected; for example, the fatigue strength increases with a temperature increase. Strain aging is confined to certain temperature ranges that depend on the alloy, and "normal" behavior is frequently found outside these temperature ranges.

The results presented in Fig. 27 are for completely reversed bending. There is a mean stress effect at high temperatures that can deviate from the one shown in Fig. 8 through 10. An example of such behavior is given in Fig. 28 for S-816 unnotched and notched specimens at three different temperatures at 2.16×10^7 cycles or 100 h life.

When a static stress is applied at high homologous temperature, creep and creep rupture can be experienced at stress levels below yield. This is not the case at low homologous temperature. Creep must be considered as an additional failure mechanism at high homologous temperature. To construct the diagrams showing the influence of mean stress, the following practice is adopted. An iso-time procedure is followed. The creep rupture strength corresponding to the time it takes (in this case 100 h) to complete the specified number of cycles to failure (in this case, 2.16×10^7 cycles) is entered on the abscissa if it is less than the yield strength. The abscissa then represents the creep rupture strength in most cases.

It is seen in Fig. 28 that the iso-cycles curve have a tendency to bow out; the presence of an alternating stress can allow for an increased mean stress when compared to pure static loading. This behavior is frequently observed at elevated temperature. (Compare with Fig. 9 to see the difference.)

When the notched and unnotched results are compared, it is seen that the rule of thumb expressed in Eq 7 is still valid. This result corresponds to other findings.

Low-Cycle Fatigue

Deformation behavior is characterized by the presence of rate (time)-dependent effects that are found in addition to plasticity. The majority of tests are conducted in strain control to simulate the effect of elastic stress fields surrounding inelastically deformed regions in the vicinity of stress raisers.

After initial changes from idle to operating conditions with attendant increases in temperature, steady-state conditions are reached for power-generation equipment, jet engines, and

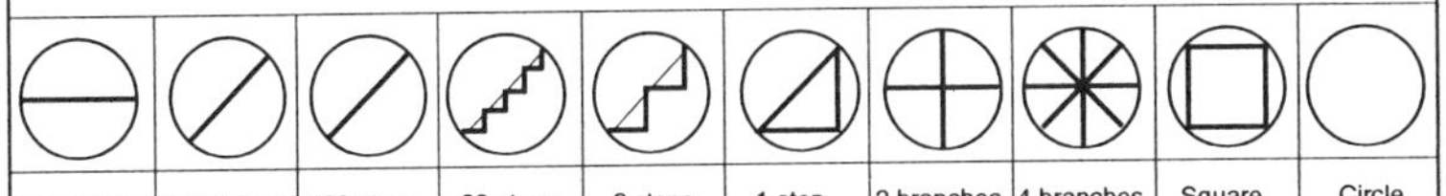

Material and maximal equivalent strain over a cycle	Cyclic strain paths ε, γ/√3 plane: Proportional	Proportional	128 steps	32 steps	2 steps	1 step	2 branches	4 branches	Square	Circle
2024 Al alloy Δε = 0.75%	330	330	330	340	380	420	...	...	...	...
2024 Al alloy Δε = 0.50%	270	270	270	290	320	330	...	...	350	370
A316 stainless steel Δε = 0.50%	350	350	360	370	430	500	510	530	570	590
	Hardening at the steady state, MPa									

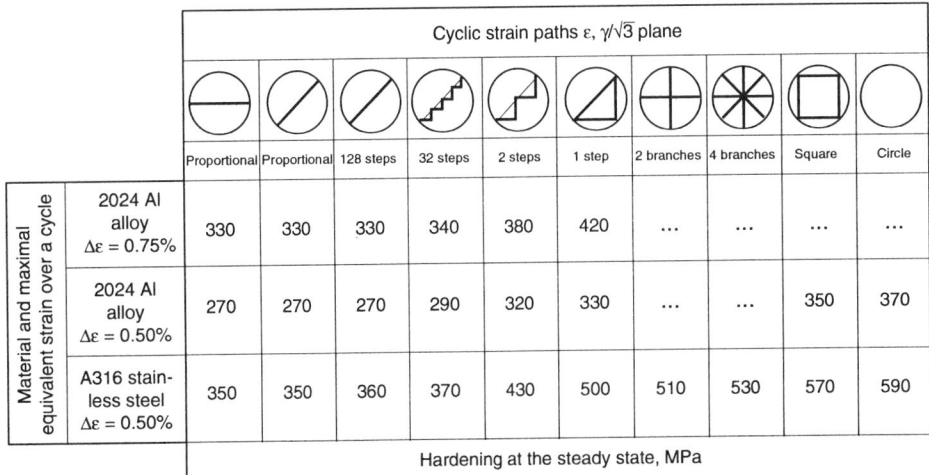

Fig. 26 The path dependence of cyclic hardening of type 316 stainless steel and of 2024 aluminum alloy. It is seen that the hardening is path dependent and that the 90° out-of-phase loading generates the highest steady stress range, the highest extra hardening. Source: Ref 52

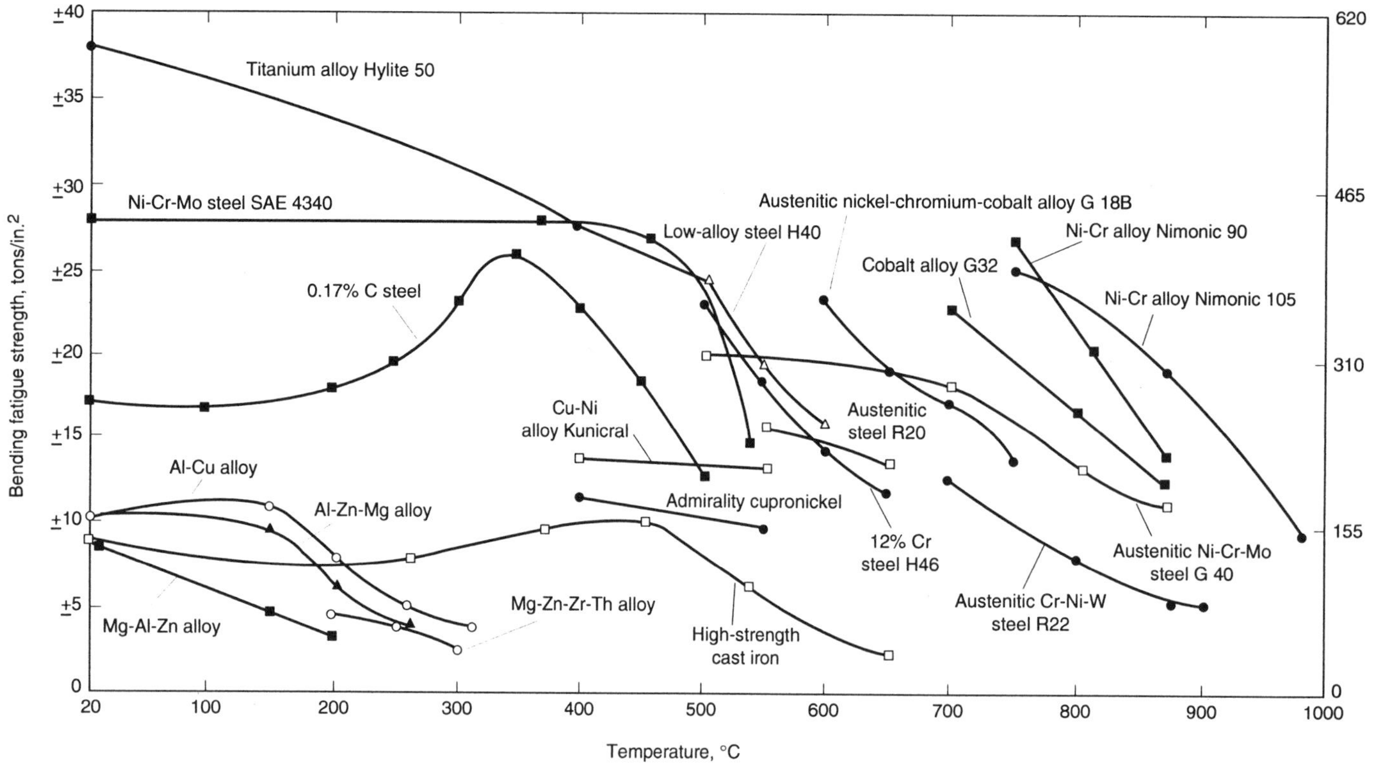

Fig. 27 The influence of temperature on the 10^7 cycle fatigue strength in bending. Source: Ref 22

other high-temperature components. The power output and the temperature are nearly constant. During these times, creep processes will be operative that contribute to the damage inflicted on the material.

Hold or dwell times at constant strain are introduced in testing to simulate these periods of nearly constant power output. In strain control, relaxation sets in and the stress decreases. The amount of stress decrease changes from cycle to cycle and may not reach a steady-state condition.

A schematic showing the imposed strain conditions, the resultant stress response, and the cross-plot of input and output—the hysteresis curve—is shown in Fig. 29.

The variation of the peak stress with cycles for 316 stainless steel for three different hold times at a strain range of 3% is shown in Fig. 30. The tests show cyclic hardening at each hold time followed by a steady-state range before the incipient crack decreases the peak stress. The curve for the longest hold time of 480 min has the smallest peak stress, and the highest peak stress is obtained for

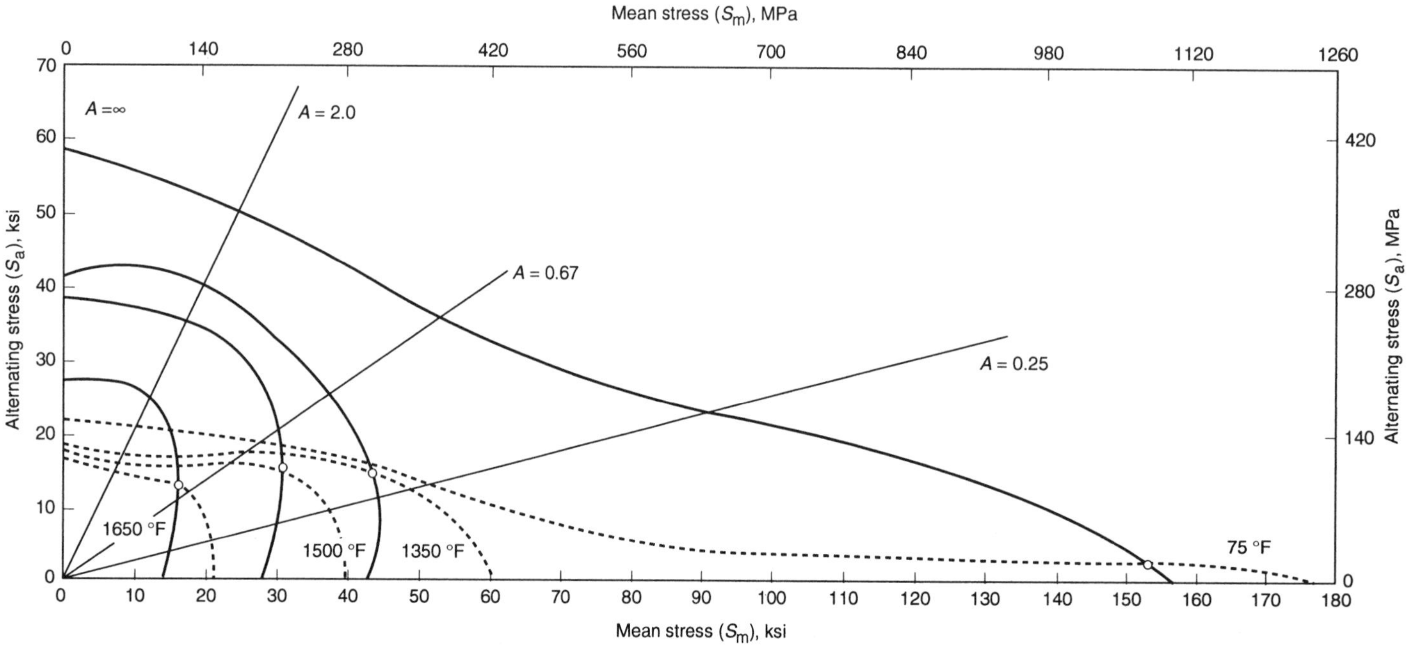

Fig. 28 Mean stress effects at high temperature for notched K_t = 3.4 (dashed lines) and unnotched (solid lines) specimens of S-816 alloy for 2.16×10^7 cycles or 100 h. Source: Ref 67

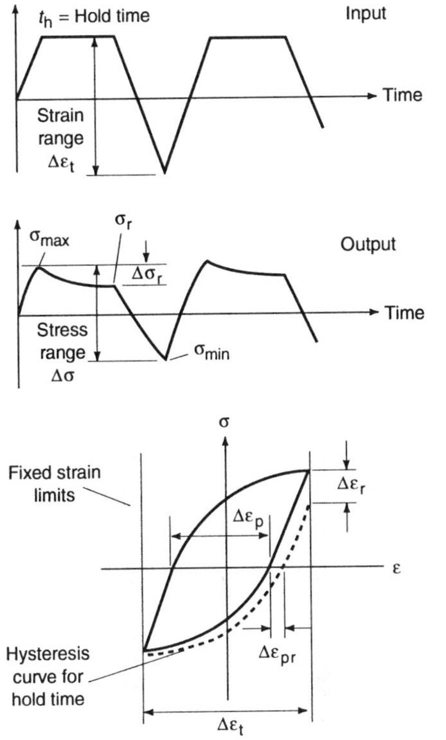

Fig. 29 Schematic showing imposed strain history, stress response, and hysteresis loop. Note that the drop in relaxation adds additional plastic strain labeled $\Delta\varepsilon_{pr}$ to the width of the loop at zero stress. Source: Ref 68

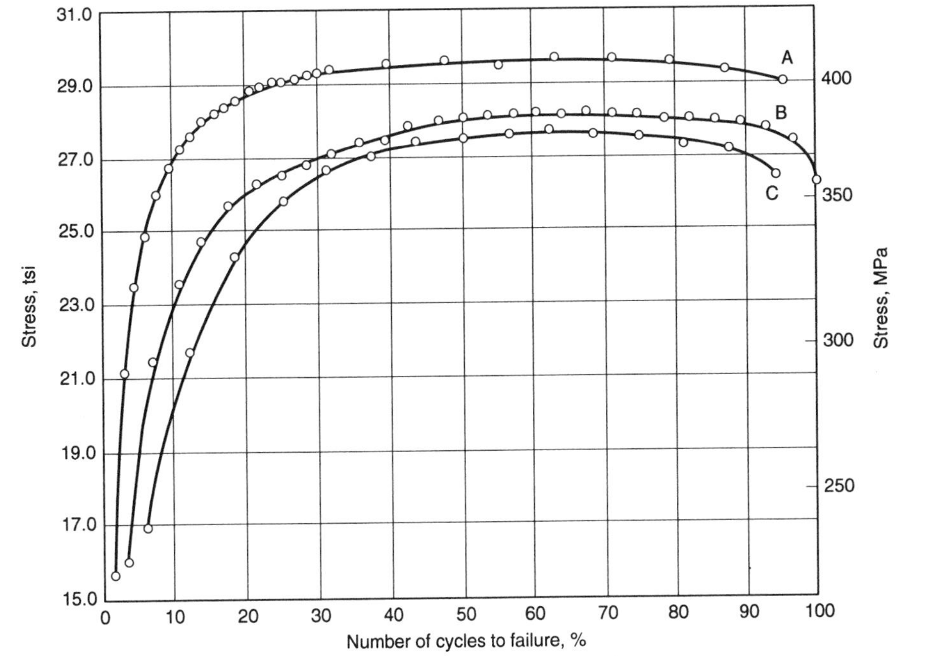

Fig. 30 The influence of hold time on the variation of the peak stress with cycles for 316 stainless steel at a strain range of 3% at 600 °C. Material A8; push-pull test. Dwell time: A, 63 cycles at 30 min; B, 28 cycles at 100 min; C, 16 cycles at 480 min. $E_t = \pm 1.5\%$. Source: Ref 69

the shortest hold time of 30 min. (Note that the hardening appears exaggerated due to method of plotting; also the strain range is quite high.)

Figure 31 shows that the relaxation drop during hold time increases with cycles by more than a factor of two. As in Fig. 30, the strain range is quite high, and this fact tends to exaggerate the change in the relaxation drop with cycles. At a reduced strain range, the change in drop may not be as large.

When frequency is changed from test to test, the stress range for the highest frequency will be the highest if normal deformation behavior prevails (no strain aging). Reduced frequencies will have reduced stress ranges. Frequencies must be changed by at least an order of magnitude to have a noticeable effect on the stress range.

During relaxation the strain is constant and the inelastic strain is "traded" for elastic strain, and in this process damage is introduced.

If the tests are run in load control then creep periods are introduced directly as shown in Fig. 32. During each hold time, creep strain is accumulated, and the specimen can fail by excessive deformation before fatigue occurs. This phenomenon is known as ratcheting. It cannot be found in strain control.

For design purposes, the deformation behavior illustrated above in a cursory way must be captured in suitable material models, the constitutive equations. It is recognized that the changing internal state due to inelastic deformation must be accounted for in a realistic model. To this end, state variable models have been developed where the state variables represent the changing internal structure and where all inelastic deformation is

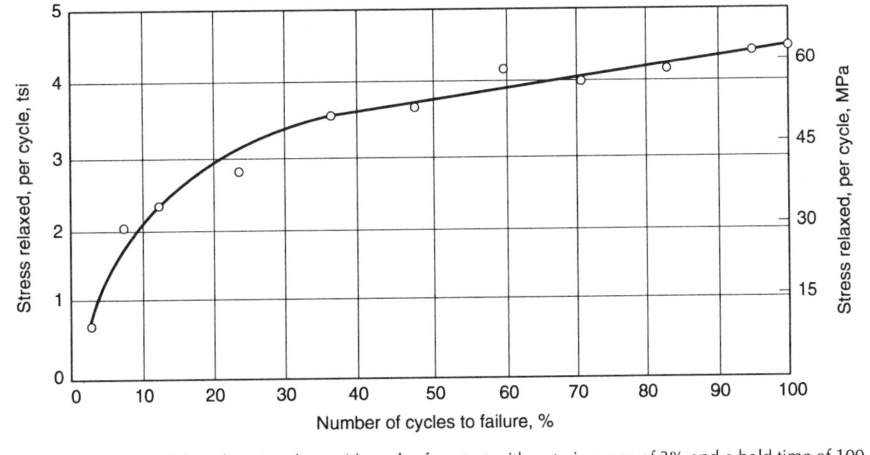

Fig. 31 The variation of the relaxation drop with cycles for a test with a strain range of 3% and a hold time of 100 min. Material A6; push-pull test. Dwell time: 42 cycles at 100 min. $E_t = 1.49\%$. Source: Ref 70

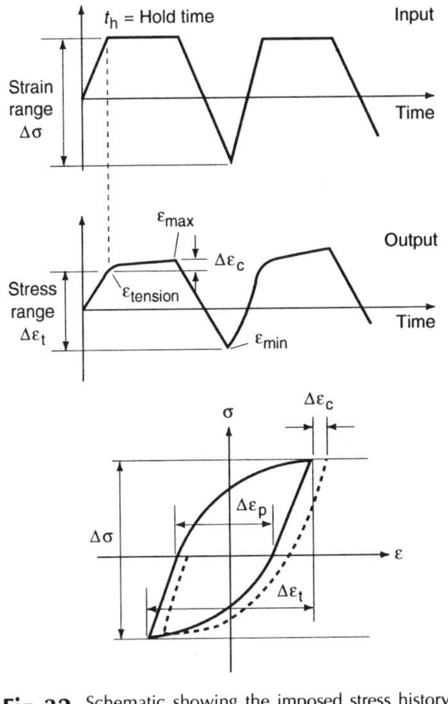

Fig. 32 Schematic showing the imposed stress history, the strain response, and the hysteresis loop. Note that the creep strain accumulated in each cycle can lead to failure due to excessive elongation (ratcheting) before failure due to fatigue. Source: Ref 68

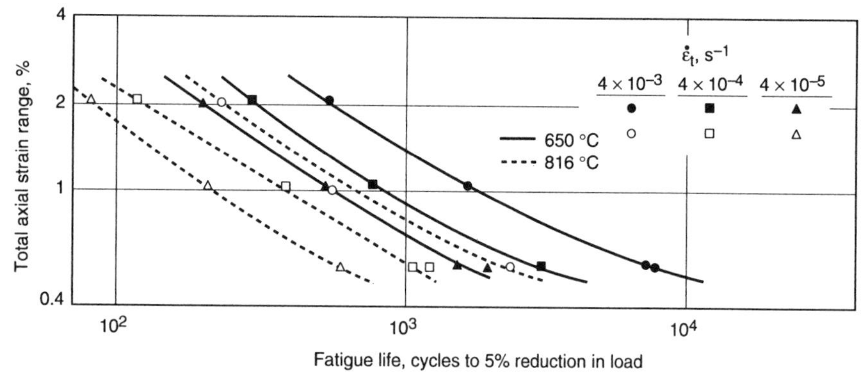

Fig. 33 The influence of strain rate on cycles to failure for type 304 stainless steel at 650 and 816 °C. Source: Ref 72

considered rate (time)-dependent (see Ref 39). The state variable theories use incremental formulation so that they can be integrated for any loading history. The differential equations are nonlinear, and a numerical integration scheme is needed. The recent advances in computing power make this easily possible.

The old design approach combined rate (time)-independent plasticity with separately formulated creep equations. But this formulation has led to many contradictions and is no longer used in advanced work. State variable constitutive equations were considered for design approaches, and their predictions were compared with experiments (Ref 71).

Lifetime. Similar to the deformation behavior, cycles to failure are also influenced by loading rate and hold time. In general, cycles to failure are determined as explained above by percentage of load drop, but early investigations may have adopted a different definition, especially when tests were run in bending.

When the strain rate is increased, fatigue life increases as seen in Fig. 33 for type 304 stainless steel. Note that the strain rate is changed by an order of magnitude at a time. The fatigue life changes by less than an order of magnitude.

An increase in dwell time decreases fatigue life, as shown in Fig. 34 where the inelastic strain range is plotted versus cycles to failure. It is seen that the life reduction of tensile hold time is significant. At the low strain range, a 30 min tensile hold time can reduce the life by one order of magnitude. It is also seen that the introduction of hold times at other positions does not have the life-reducing effect as do tensile hold times.

Life-prediction methods have to model these effects, and it is clear that the methods given previously for room-temperature applications are not sufficient.

Detailed discussions of the methods applicable to high temperature including the frequency-modified strain range methods and strain-range partitioning methods are in use (see Ref 74–76).

These methods are all aimed at a typical cycle, and it is presumed that this cycle is representative of the actual loading. In high-temperature fatigue testing various wave forms are imposed (see Fig. 29 and 32, and Fig. 1.3 in Ref 75). It is then very difficult to find a representative cycle. Moreover in real applications, variable amplitude loadings are experienced, and a typical cycle is hard to define.

Following the developments of incremental constitutive equations that are in principle applicable to any history, incremental life-prediction laws have been developed (Ref 77, 78). These laws need to be integrated for each history to determine the lifetime and are ideally suited for a joint use with incremental constitutive laws.

In hold-time and slow-frequency tests, creep and fatigue damage interact as evidenced by the change from transgranular to intergranular cracking as the hold time increases. The subject of high-temperature fatigue and microstructure is discussed in depth in Ref 79.

Recently high-temperature fatigue considerations of electronic components have become important, especially for solder joints. These joints operate at high homologous temperature, and frequency and hold time have an important effect on lifetime. The reader is referred to Ref 80.

Effect of Environment. All the discussions so far were for test results obtained in laboratory air. It has been found out that other environments can have a significant influence on fatigue performance. Discussions of the effects of environment are given in Ref 81 and 82, to name just two sources.

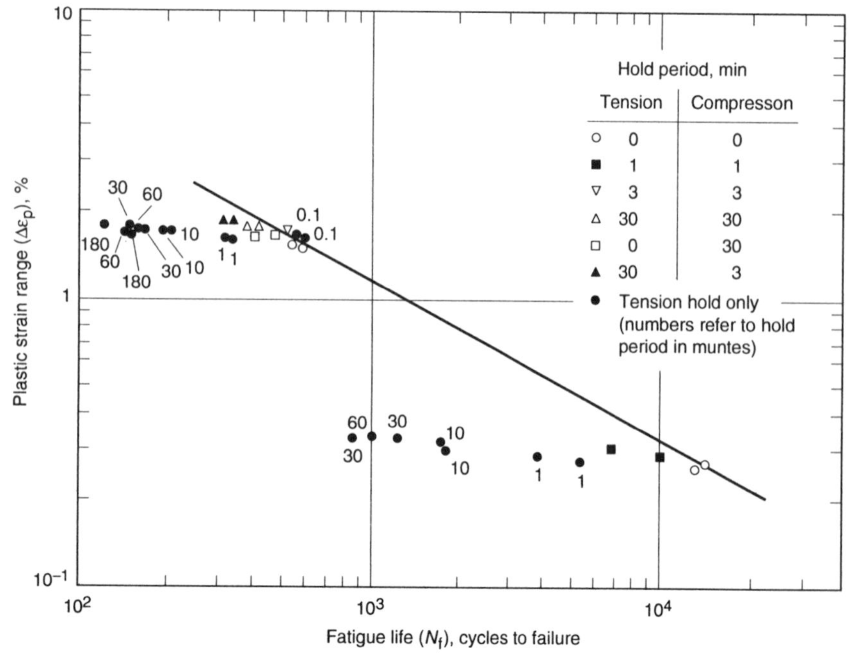

Fig. 34 Effect of hold period and position of hold period on fatigue life for type 304 stainless steel at 1200 °F and a strain rate of 4×10^{-3} 1/s. Source: Ref 73

REFERENCES

1. *Fatigue and Fracture,* Vol 19, *ASM Handbook,* ASM International, 1996
2. M.R. Mitchell and R.W. Landgraf, Ed., *Advances in Fatigue Lifetime Predictive Techniques,* STP 1122, ASTM, 1992
3. H.O. Fuchs and R.I. Stephens, *Metal Fatigue in Engineering,* John Wiley & Sons, 1980
4. M.L. Sharp and G.E. Nordmark, *Fatigue Design of Aluminum Components and Structures,* McGraw-Hill, 1996
5. S. Suresh, *Fatigue of Materials,* Cambridge University Press, 1994
6. H.E. Boyer, Ed., *Atlas of Fatigue Curves,* American Society for Metals, 1986, p 30
7. P.G. Forrest, *Fatigue of Metals,* Pergamon Press, 1962, p 59
8. P.G. Forrest, *Fatigue of Metals,* Pergamon Press, 1962, p 58
9. P.G. Forrest, *Fatigue of Metals,* Pergamon Press, 1962, p 73
10. P.G. Forrest, *Fatigue of Metals,* Pergamon Press, 1962, p 47
11. P.E.K. Frost, J. Marsh, and L.P. Cook, *Metal Fatigue,* Clarendon Press, 1974, p 241

12. P.E.K. Frost, J. Marsh, and L.P. Cook, *Metal Fatigue,* Clarendon Press, 1974, p 417–427
13. H.O. Fuchs and R.I. Stephens, *Metal Fatigue in Engineering,* John Wiley & Sons, 1980, p 94–98
14. H.O. Fuchs and R.I. Stephens, *Metal Fatigue in Engineering,* John Wiley & Sons, 1980, p 304–308
15. P.S. Veers, Statistical Considerations in Fatigue, *Fatigue and Fracture,* Vol 19, *ASM Handbook,* ASM International, 1996, p 295–302
16. R.C. Rice, Fatigue Data Analysis, *Mechanical Testing,* Vol 8, 9th ed., *Metals Handbook,* 1985, p 695–719
17. H.O. Fuchs and R.I. Stephens, *Metal Fatigue in Engineering,* John Wiley & Sons, 1980, p 75
18. P.G. Forrest, *Fatigue of Metals,* Pergamon Press, 1962, p 94–102
19. H.J. Grover, "Fatigue of Aircraft Structures," NAVAIR 01 1A 13, Naval Air Systems Command, U.S. Government Printing Office, 1966, p 59
20. W.D. Pilkey, *Peterson's Stress Concentration Factors,* 2nd ed., John Wiley & Sons, 1997
21. E.A. Avallone and T. Baumeister III, Ed., *Mark's Standard Handbook for Mechanical Engineers,* 10th ed., McGraw-Hill, 1996
22. P.G. Forrest, *Fatigue of Metals,* Pergamon Press, 1962
23. H.J. Grover, "Fatigue of Aircraft Structures," NAVAIR 01 1A 13, Naval Air Systems Command, U.S. Government Printing Office, 1966
24. N.E. Dowling, *Mechanical Behavior of Materials: Engineering Methods for Deformation, Fracture, and Fatigue,* Prentice-Hall, 1993
25. E. Krempl, *The Influence of State of Stress on Low-Cycle Fatigue of Structural Materials. A Literature Survey and Interpretive Report,* STP 549, ASTM, 1974
26. M.W. Brown and K.J. Miller, A Theory of Fatigue under Multiaxial Stress-Strain Conditions, *Proc. Inst. Mech. Eng.,* Vol 187, 1978, p 745–755
27. Y.S. Garud, Multiaxial Fatigue: A Survey of the State of the Art, *J. Test. Eval.,* Vol 9, 1981, p 165–178
28. F. Ellyin, *Fatigue Damage, Crack Growth and Life Prediction,* Chapman & Hall, 1997
29. D.L. McDowell, Multiaxial Fatigue Strength, *Fatigue and Fracture,* Vol 19, *ASM Handbook,* ASM International, 1996, p 263–273
30. S.M. Tipton and D.V. Nelson, Advances in Multiaxial Fatigue for Components with Stress Concentrations, *Int. J. Fatigue,* in press
31. J.B. Conway and L.H. Sjodahl, *Analysis and Representation of Fatigue Data,* ASM International, 1991, p 2
32. J.B. Conway and L.H. Sjodahl, *Analysis and Representation of Fatigue Data,* ASM International, 1991, p 5
33. H.O. Fuchs and R.I. Stephens, *Metal Fatigue in Engineering,* John Wiley & Sons, 1980, p 25
34. J.B. Conway and L.H. Sjodahl, *Analysis and Representation of Fatigue Data,* ASM International, 1991, p 9
35. R.W. Hertzberg, *Deformation and Fracture Mechanics of Engineering Materials,* 4th ed., John Wiley & Sons, 1996, p 262
36. Technical Report on Fatigue Properties, *SAE J.,* Vol 1099, Feb 1975
37. R.I. Stephens, G. Mauritzson, P.H. Benner, and D.R. Galliart, *J. Steel Casting Res.,* No. 83, July 1978
38. H.O. Fuchs and R.I. Stephens, *Metal Fatigue in Engineering,* John Wiley & Sons, 1980, p 298
39. A.S. Krausz and K. Krausz, Ed., *Unified Constitutive Laws of Plastic Deformation,* Academic Press, 1996
40. H.O. Fuchs and R.I. Stephens, *Metal Fatigue in Engineering,* John Wiley & Sons, 1980, p 77
41. H.O. Fuchs and R.I. Stephens, *Metal Fatigue in Engineering,* John Wiley & Sons, 1980, p 78
42. H.O. Fuchs and R.I. Stephens, *Metal Fatigue in Engineering,* John Wiley & Sons, 1980, p 79
43. Parameters for Estimating Fatigue Life, *Fatigue and Fracture,* Vol 19, *ASM Handbook,* ASM International, 1996, p 963–979
44. J.B. Conway and L.H. Sjodahl, *Analysis and Representation of Fatigue Data,* ASM International, 1991, p 22
45. N.E. Dowling, Estimating Fatigue Life, *Fatigue and Fracture,* Vol 19, *ASM Handbook,* ASM International, 1996, p 250
46. H.J. Grover, "Fatigue of Aircraft Structures," NAVAIR 01 1A 13, Naval Air Systems Command, U.S. Government Printing Office, 1966, p 66
47. H.J. Grover, "Fatigue of Aircraft Structures," NAVAIR 01 1A 13, Naval Air Systems Command, U.S. Government Printing Office, 1966, p 67
48. E. Krempl, "The Effect of Stress Concentration on the Low-Cycle Fatigue of Three Low-Strength Structural Steels at Room Temperature and at 550 °F," GEAP-1 0170, AEC Research and Development Report, General Electric Company, March 1970,
49. B.M. Wundt, *Effects of Notches on Low-Cycle Fatigue. A Literature Survey,* STP 490, ASTM, 1972
50. B.M. Wundt, *Effects of Notches on Low-Cycle Fatigue. A Literature Survey,* STP 490, ASTM, 1972, p 15
51. E. Krempl and H. Lu, The Path and Amplitude Dependence of Cyclic Hardening of Type 304 Stainless Steel at Room Temperature, *Biaxial and Multiaxial Fatigue,* EGF 3, M.W. Brown and K.J. Miller, Ed., Mechanical Engineering Publications, 1989, p 89–106
52. A. Benallal, P. LeGallo, and D. Marquis, An Experimental Investigation of Cyclic Hardening of 316 Stainless Steel and of 2024 Aluminium Alloy under Multiaxial Loadings, *Nucl. Eng. Des.,* Vol 114, 1989, p 345–353
53. E. Krempl and H. Lu, Comparison of the Stress Responses of an Aluminum Alloy to Proportional and Alternate Axial and Shear Strain Paths at Room Temperature, *Mech. Mater.,* Vol 2, 1983, p 183–192
54. U.S. Lindholm, K.S. Chan, S.R. Bodner, R.M. Webster, K.P. Walker, and B.N. Cassenti, "Constitutive Modeling for Isotropic Materials (HOST)," Second Annual Status Report, CR 174980 (SwRI-7576/30) National Aeronautics and Space Administration, July 1985
55. M. Clavel and X. Feaugas, Micromechanisms of Plasticity under Multiaxial Cyclic Loading, *Fourth Int. Conf. Biaxial/Multiaxial Fatigue,* Vol I and II, Societé Française de Metallurgie et de Materiaux (SF2M), Sixth Int. Spring Meeting, European Structural Integrity Society (ESIS), 1994, p 18
56. M. Clavel and X. Feaugas, Micromechanisms of Plasticity under Multiaxial Cyclic Loading, *Fourth Int. Conf. Biaxial/Multiaxial Fatigue,* Vol I and II, Societé Française de Metallurgie et de Materiaux (SF2M), Sixth Int. Spring Meeting, European Structural Integrity Society (ESIS), 1994
57. D. Yao and E. Krempl, "Viscoplasticity Theory Based on Overstress: The Prediction of Monotonic and Cyclic Proportional and Nonproportional Loading Paths of an Aluminum Alloy, *Int. J. Plast.,* Vol 1, 1985, p 259–274
58. N. Ohno, Recent Topics in Constitutive Modeling of Cyclic Plasticity and Viscoplasticity, *Appl. Mech. Rev.,* Vol 43, 1990, p 283–295
59. D. Socie, Critical Plane Approaches for Multiaxial Fatigue Damage Assessment, *Advances in Multiaxial Fatigue,* D.L. McDowell and R. Ellis, Ed., STP 1191, ASTM 1993, p 31
60. D.L. McDowell and R. Ellis, Ed., *Advances in Multiaxial Fatigue,* STP 1191, ASTM, 1993, p 1
61. D. Socie, Critical Plane Approaches for Multiaxial Fatigue Damage Assessment, *Advances in Multiaxial Fatigue,* D.L. McDowell and R. Ellis, Ed., STP 1191, ASTM 1993, p 7–36
62. T. Ogata, A. Nitta, and J.J. Blass, Propagation Behavior of Small Cracks in 304 Stainless Steel under Biaxial Low-Cycle Fatigue at Elevated Temperature, *Advances in Multiaxial Fatigue,* STP 1191, D.L. McDowell and R. Ellis, Ed., ASTM, 1993, p 313–325
63. J.-Y. Berard, D.L. McDowell, and S.D. Antolovich, Damage Observations of a Low Carbon Steel under Tension-Torsion Low-Cycle Fatigue, *Advances in Multiaxial Fatigue,* STP 1191, D.L. McDowell and R. Ellis, Ed., ASTM, 1993, p 326–344
64. M.W.K. Brown, J. Miller. U.S. Fernando, J.R. Yates, and D.K. Suker, Aspects of Multiaxial Fatigue Crack Propagation, *Fourth Int. Conference on Biaxial/Multiaxial Fatigue,* Vol I and II, Societé Française de Metallurgie et de Materiaux (SF2M), Sixth Int. Spring Meeting, European Structural Integrity Society (ESIS), p 3–16
65. H. Sehitoglu, Thermal and Thermomechanical Fatigue of Structural Alloys, *Fatigue and Fracture,* Vol 19, *ASM Handbook,* ASM International, 1996, p 527
66. Y. Estrin, Dislocation-Density-Related Constitutive Model, *Unified Constitutive Laws of Plastic Deformation,* A.S. Krausz and K. Krausz, Ed., Academic Press, 1996, p 69–106
67. H.O. Fuchs and R.I. Stephens, *Metal Fatigue in Engineering,* John Wiley & Sons, 1980, p 245

68. E. Krempl and B.M. Wundt, *Hold-Time Effects in High-Temperature Low-Cycle Fatigue: A Literature Survey and Interpretive Report,* STP 489, ASTM, Sept 1971, p 4
69. E. Krempl and B.M. Wundt, *Hold-Time Effects in High-Temperature Low-Cycle Fatigue: A Literature Survey and Interpretive Report,* STP 489, ASTM, Sept 1971, p 9
70. E. Krempl and B.M. Wundt, *Hold-Time Effects in High-Temperature Low-Cycle Fatigue: A Literature Survey and Interpretive Report,* STP 489, ASTM, Sept 1971, p 10
71. T. Inoue, N. Ohno, A. Susuki, and T. Igari, Evaluation of Inelastic Constitutive Models under Plasticity-Creep Interaction for 2 1/2 Cr 1 Mo Steel at 600 °C, *Nucl. Eng. Des.,* Vol 114, p 295–309
72. J.B. Conway and L.H. Sjodahl, *Analysis and Representation of Fatigue Data,* ASM International, 1991, p 26
73. E. Krempl and B.M. Wundt, *Hold-Time Effects in High-Temperature Low-Cycle Fatigue: A Literature Survey and Interpretive Report,* STP 489, ASTM, Sept 1971, p 12
74. A.D. Batte, Creep-Fatigue Life Prediction, *Fatigue at High Temperature,* R.P. Skelton, Ed., Applied Science, 1983, p 365–401
75. J.B. Conway and L.H. Sjodahl, *Analysis and Representation of Fatigue Data,* ASM International, 1991
76. M.A. McGaw, Cumulative Creep-Fatigue Damage Evolution in an Austenitic Stainless Steel, *Advances in Fatigue Life Predictive Techniques,* M.R. Mitchell and R.W. Landgraf, Ed., STP 1122, ASTM, 1992, p 84–106
77. S. Majumdar, Designing Against Low-Cycle Fatigue at Elevated Temperature, *Nucl. Eng. Des.,* Vol 63, 1981, p 121–135
78. N.-M. Yeh and E. Krempl, An Incremental Life Prediction Law for Multiaxial Creep-Fatigue Interaction and Thermomechanical Loading, *Advances in Multiaxial Fatigue,* D.L. McDowell and R. Ellis, Ed., STP 1191, ASTM 1993, p 107–119
79. A. Pineau, High Temperature Fatigue Behaviour of Engineering Materials in Relation to Microstructure, *Fatigue at High Temperature,* R.P. Skelton, Ed., Applied Science, 1983, p 305–364
80. S.A. Schroeder and M.R. Mitchell, Ed., *Fatigue of Electronic Materials,* STP 1153, ASTM, 1994
81. P. Marshall, The Effect of Environment on Fatigue, *Fatigue at High Temperature,* R.P. Skelton, Ed., Applied Science, 1983, p 259–303
82. T. Sugiura, A. Ishikawa, T. Nakamura, and Y. Asada, Formulation on Air Environmental Effect on Creep-Fatigue Interaction, *Nucl. Eng. Des.,* Vol 153, 1994, p 87-95

Design for Fracture Toughness

M.P. Blinn and R.A. Williams, Materials Characterization Laboratory

MODERN TECHNOLOGY has many challenges for the scientist, engineer, or designer of the machinery used by the people of the world. Because mechanical equipment is often used at or near design limitations, great care must be employed in selecting the proper materials to use for a particular design application. The need for high-performance materials in such industries as aerospace and power generation has advanced the use of design parameters in the evaluation of material behavior. In particular, fracture toughness is used as a design parameter to establish safe, efficient performance levels for mechanical equipment.

Fracture toughness, in the most general of definitions, is the ability of a material to withstand fracture in the presence of cracks. The American Society for Testing and Materials (ASTM) has numerous commonly accepted standards for fracture testing, and it defines fracture toughness as "a generic term for measures of resistance to crack extension" (Ref 1). In design, fracture toughness is used as a parameter to avoid the catastrophic failures often associated with fracture. When employing fracture toughness as a design criterion, the user must be well aware that historically, catastrophic fracture failures have occurred at nominal stresses far below the design stresses of the structures. To corroborate this point, the literature contains numerous examples of unanticipated mechanical and structural failures due to fracture (Ref 2–5).

In the design process, the scientist, engineer, or designer should recognize that cracks, discontinuities, heterogeneities, or numerous other anomalies may exist in any "homogeneous" material used in manufacturing a product (Ref 6, 7). These cracks can exist at any stage in the life of the structure or machine component, and they should be accounted for not only after a part has been in service for a length of time, but also during the entire manufacturing process. Fracture mechanics provides a useful tool in planning for the occurrence of cracking; in particular, the use of fracture toughness as a design parameter (or limit) can alleviate the occurrence of fracture.

The focus of this article is on the use of fracture toughness as a parameter for engineering and design purposes. Although the subject of fracture mechanics is discussed in a general manner, most of the information in this article is related to the fracture behavior of metals. Both linear elastic and elastic-plastic fracture mechanics concepts are reviewed as they relate to fracture toughness and the design process. The article explores the use of plane strain fracture toughness (K_{Ic}), crack-tip opening displacement, and the *J*-integral as criteria for the design and safe operation of structures and mechanical components. In particular, the use of fracture toughness in damage-tolerant methods in aircraft and pressure vessel design are reviewed.

A Brief History of Fracture Mechanics

It is generally accepted that the study of fracture mechanics was pioneered by A.A. Griffith in the first half of the 20th century, although much earlier fracture studies have been attributed to da Vinci (Ref 3). Regardless of who was the earliest fracture experimentalist, the importance of Griffith's contributions to engineering was most pronounced in the new paradigm that he introduced to the world: a quantitative relationship for fracture by the use of fracture mechanics. The history of fracture mechanics is well documented in the literature (Ref 5).

In his seminal work in the 1920s on fracture mechanics (Ref 8), Griffith tested glass specimens that were scratched and unscratched and compared their respective breaking strengths. Griffith developed the following expression for the fracture strength of solids that met stringent boundary conditions:

$$\sigma_c = \sqrt{\frac{2\gamma E}{\pi a}} \quad \text{(Eq 1)}$$

where σ_c is the critical stress, E is the elastic modulus, γ is the surface energy, and a is the crack length.

From his findings, Griffith showed that assuming a material to be a homogeneous solid could lead to grave miscalculations in design. Further, Griffith advocated the use of more refined materials to increase strength, but cautioned that discrepancies in strength between scratched and unscractched specimens would still exist, regardless of how well the material was refined. In essence, Griffith established the concept of designing for fracture; however, it took years for the engineering and scientific community to fully appreciate (or understand) his findings.

Fracture mechanics became a more workable topic for most engineers due to the studies of Irwin in the late 1940s and 1950s (Ref 5, 9). Irwin proposed an energy approach to describe fracture mechanics, based on the work of Griffith. Irwin defined an expression for the energy release rate as:

$$G = -\frac{d\Pi}{dA} \quad \text{(Eq 2)}$$

where G is the energy release rate, $d\Pi$ is the change in potential energy of the cracked body, and dA is the change in crack-surface area.

Taking this equation out further, the elastic energy release rate of a plate can be described as:

$$G = \frac{\pi\sigma^2 a}{E} \quad \text{(Eq 3)}$$

where σ is the applied stress.

In order to obtain more workable engineering units, Eq 3 can be rearranged and rewritten as:

$$K = f\left(\frac{a}{W}\right)\sigma\sqrt{\pi a} \quad \text{(Eq 4)}$$

where K is the stress intensity factor (in general form), which is a function of geometry ($f(a/W)$), stress, and crack length. The stress intensity factor is generally expressed in units of MPa$\sqrt{\text{m}}$ (ksi$\sqrt{\text{in.}}$).

Currently, it is the stress intensity factor that is most commonly used as a basis for the design and analysis of structures or mechanical components that contain cracks. The stress intensity factor is essentially a pseudostress; that is, it is calculated and used in design much the same as a stress, only the crack is now considered in the equation. Although very useful in design, the stress intensity factor as represented in Eq 4 is limited to elastic material behavior. It should be clearly understood that the stress intensity factor (K or K_I, as will be discussed later) described in this article is not the same as the stress concentration factor (also denoted by K in many texts). A stress concentration

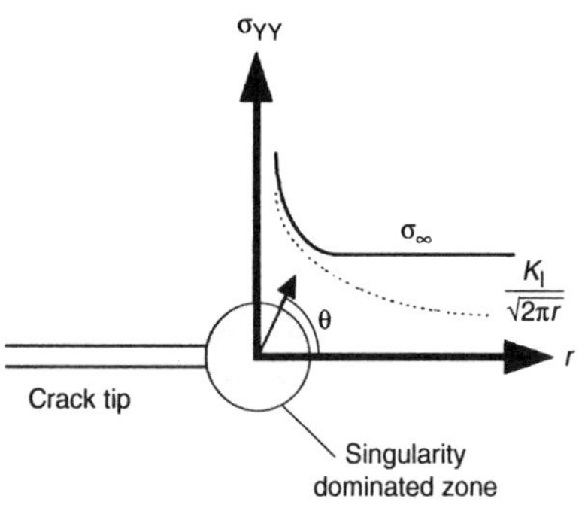

Fig. 1 Stress field normal to the crack plane in mode I, where $\theta = 0°$. The far field stress is σ and the stress intensity factor is K_I. Source: Ref 5

factor is generally used as a stress multiplier, applied to the calculated stress of the structure or component. For example, where there is a change in the component shape (geometry), such as at a sharp edge or hole, a stress concentration factor is applied to calculated stress to account for the geometric constraint.

The challenge in fracture mechanics in the 1960s and 1970s was to establish working models for materials that were too tough to be characterized by the Griffith-Irwin fracture mechanics approach, which was somewhat limiting to elastically behaving materials. In the early 1960s, Wells proposed a crack-tip opening displacement method (CTOD) based on his observations that the degree of crack-tip blunting was in proportion to the toughness of the material (Ref 5, 10). From this work, the following expression is one form that relates the elastic component of CTOD (δ) to the stress intensity factor (K_I):

$$\delta = \frac{4K_I^2}{\pi \sigma_{YS} E} \qquad \text{(Eq 5)}$$

where σ_{YS} is the yield stress.

To develop an alternative means to evaluate the fracture resistance of high-toughness materials, the J-integral was first proposed by Rice in 1968 (Ref 11) as a path-independent integral for measuring the stress and strain field ahead of cracks. The J-integral is expressed mathematically as:

$$J = \int_\Gamma \left(W dy - T_i \frac{\partial u_i}{\partial x} ds \right) \qquad \text{(Eq 6)}$$

where J is the contour integral, W is the strain energy density, T_i is the traction vector outward normal to the arc Γ, $T_i = \sigma_{ij} n_j$, u_i is the displacement vector, and s is the arc length along Γ.

Early experimental work by Begley and Landes in the 1970s (Ref 12–14) developed a multispecimen method in the laboratory, with a proposal to use the J-integral to evaluate the onset of crack initiation (J_{Ic}). The early form of the J-integral expression that Begley and Landes worked with was:

$$J = -\frac{1}{B} \left(\frac{\partial U}{\partial a} \right) \qquad \text{(Eq 7)}$$

where U is the potential energy and B is the specimen thickness.

Adding to this early laboratory work, Andrews et al. (Ref 15) developed an unloading compliance method for a single-specimen test. Additionally, numerous investigators continued exploring mathematical expressions for the J-integral (Ref 5, 16, 17). Further discussion on the J-integral and elastic-plastic fracture mechanics is presented in the next section of this article.

Categories of Fracture Mechanics

Fracture mechanics is a broad subject, but there are fundamentally two categories: linear elastic fracture mechanics (LEFM) and elastic-plastic fracture mechanics (EPFM). LEFM, the more mature of the two categories, studies the behavior of materials where cracking is presumed to occur under essentially elastic conditions. Additionally, LEFM assumes that the crack tip is sharp, with a limited amount of plasticity. Some materials that are designed using LEFM concepts are high-strength steels, titanium, and aluminum alloys.

Although LEFM has been used in design and analysis with a measure of success, there are limitations to its applications. Lower-strength steels can behave in a relatively plastic manner, so they are not easily characterized by LEFM. To account for materials that do not fall within the limitations of LEFM, research has been actively pursued from the 1960s through the present to determine and refine different methods of characterizing the crack tip and the fracture process. For the study of more ductile materials, EPFM evolved from LEFM as a somewhat more complex science. In contrast to LEFM, EPFM assumes that the crack tip is not sharp and that there is a degree of crack tip plasticity (or blunting). Some materials that are designed using EPFM concepts are common structural materials, such as lower-strength, higher-toughness steels.

Linear Elastic Fracture Mechanics. Although some of the background work previously discussed in this article touched on the stress intensity factor (K_I), the concept was not fully developed. The stress intensity factor is at the heart of characterizing LEFM, and a general example is reviewed below. As further reference material for the solid mechanics approach to the stress intensity factor, the literature contains numerous texts (some older, but excellent background material) on LEFM theory (Ref 3–5, 9, 18, 19).

Consider a crack in a material as shown schematically in Fig. 1. The stress field ahead of the crack tip can be described by the following equations, assuming a linear elastic, isotropic material:

$$\sigma_{XX} = \frac{K_I}{\sqrt{2\pi r}} \cos\left(\frac{\theta}{2}\right)\left[1 - \sin\left(\frac{\theta}{2}\right)\sin\left(\frac{3\theta}{2}\right)\right] \qquad \text{(Eq 8)}$$

$$\sigma_{YY} = \frac{K_I}{\sqrt{2\pi r}} \cos\left(\frac{\theta}{2}\right)\left[1 + \sin\left(\frac{\theta}{2}\right)\sin\left(\frac{3\theta}{2}\right)\right] \qquad \text{(Eq 9)}$$

$$\pi_{XY} = \frac{K_I}{\sqrt{2\pi r}} \cos\left(\frac{\theta}{2}\right)\sin\left(\frac{\theta}{2}\right)\cos\left(\frac{3\theta}{2}\right) \qquad \text{(Eq 10)}$$

$$\sigma_{ZZ} = \nu\left(\sigma_{XX} + \sigma_{YY}\right) \quad \text{(plane strain)} \qquad \text{(Eq 11)}$$

$$\sigma_{ZZ} = 0 \quad \text{(plane stress)} \qquad \text{(Eq 12)}$$

$$\tau_{XZ} = \tau_{YZ} \qquad \text{(Eq 13)}$$

where K_I is the mode I stress intensity factor (explained below), τ is the shear stress, ν is the Poisson's ratio, and r is the radius.

It is important to note the subscript used on the stress intensity factors, which refers to the loading mode. There are essentially three loading modes discussed in the literature that can be applied to a crack developing in a material, as shown schematically in Fig. 2. Mode I, the opening mode of loading, is by far the most commonly studied. Modes II and III are the in-plane and the out-of-plane shear modes, respectively, and are not further discussed in this article. These loading modes are less significant in fracture toughness

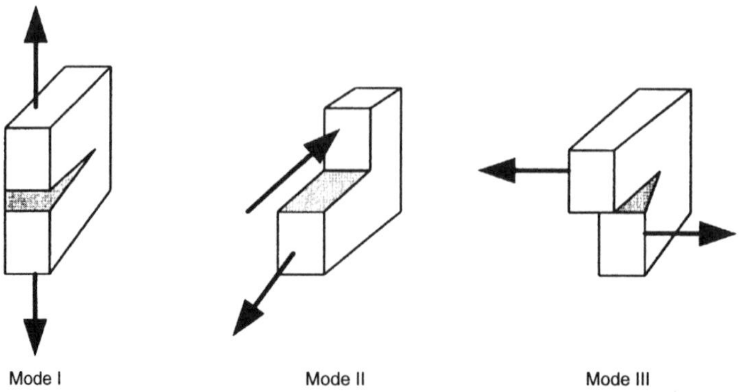

Fig. 2 The three modes of loading that can be applied to a crack. Mode I is the opening mode, mode II is the in-plane shear mode, and mode III is the out-of-plane shear mode. Source: Ref 5

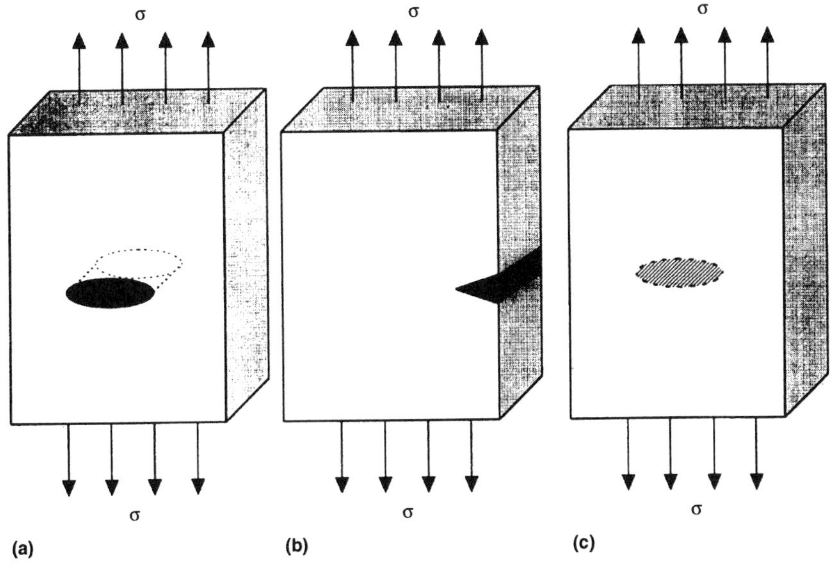

Fig. 3 Schematic of the three load-crack situations that have closed-form solutions in fracture mechanics. (a) The through crack in an infinite plate subjected to a remote tensile stress. (b) The semi-infinite plate with an edge crack. (c) The penny-shaped circular crack embedded in a solid. Source: Ref 5

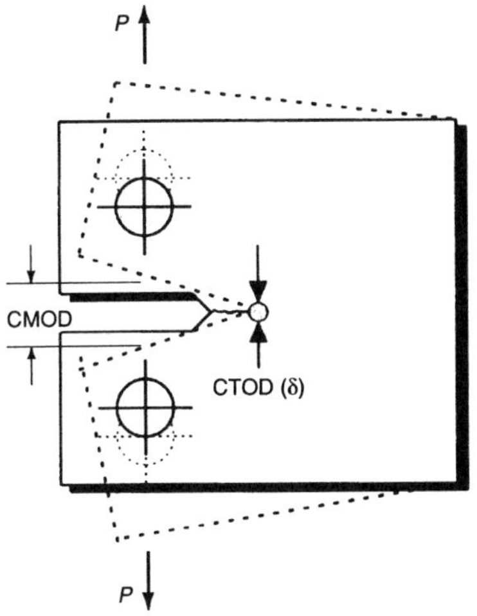

Fig. 4 Schematic of a sample specimen showing the definition of crack-mouth opening displacement (CMOD) and crack-tip opening displacement (CTOD). CTOD is the diameter of the circular arc at the blunted crack tip and should not be confused with the plastic zone. Source: Ref 16

testing, but they are described in ASTM E 616 and E 399 (Ref 1, 20).

Closed-form solutions exist for many stress intensity factor load-crack situations, and some of the more common expressions are given in Eq 14 to 16 (Ref 5). These three stress intensity solutions are perhaps the most simplistic of the closed-form solutions. To accompany these equations, various schematics are shown in Fig. 3(a) to (c), each picture representing one of the three stress intensity factor load-crack situations. For the through crack in an infinite plate subjected to a remote tensile stress (Fig. 3a), the stress intensity factor is given as:

$$K_I = \sigma\sqrt{\pi a} \quad \text{(Eq 14)}$$

For the semi-infinite plate with an edge crack (Fig. 3b), the stress intensity factor is given as:

$$K_I = 1.12\sigma\sqrt{\pi a} \quad \text{(Eq 15)}$$

where 1.12 is a correction factor to account for the different boundary condition.

For the "embedded penny" crack in an infinite medium (Fig. 3c), the stress intensity factor is given as:

$$K_I = \frac{2}{\pi}\sigma\sqrt{\pi a} \quad \text{(Eq 16)}$$

where $2/\pi$ accounts for a radial crack.

Having described the stress intensity factor (K or K_I) used for LEFM, it is perhaps appropriate to define plane-strain fracture toughness (K_{Ic}), which is the material response characteristic that is often used in designing against fracture. K_{Ic} is a limiting value; that is, it can be used in static fracture mechanics design in much the same way as the yield strength (σ_Y) is used in static stress analysis. As the subscript of K_{Ic} implies, it is a critical stress intensity factor for mode I cracking and is determined for a material by following a rigorous test procedure, as prescribed in ASTM E 399 (Ref 20). More detailed descriptions of K_{Ic} and ASTM E 399 are given in later sections of this article.

Elastic-Plastic Fracture Mechanics. Different approaches are used for evaluating materials that fracture (or behave) more plastically than in LEFM. In evaluating EPFM, the CTOD and J-integral methods are most often used to characterize fracture at the crack tip. The more current literature contains the best sources for additional review of EPFM (Ref 5, 16, 21–23). Closed-form CTOD and J-integral solutions are also available in these references.

To evaluate the use of the CTOD in essentially the elastic case, consider the through crack in an infinite plate subject to a remote tensile stress (Fig. 3a). The most general form of this load-crack situation is expressed as follows (Ref 5):

$$\delta = \frac{K_I^2}{m\sigma_{YS}E} = \frac{G}{m\sigma_{YS}} \quad \text{(Eq 17)}$$

where δ is the CTOD, and m is a constant (1.0 for plane stress, 2.0 for plane strain).

The concept of the CTOD and the crack-mouth opening displacement (CMOD) is shown schematically in Fig. 4, which shows a sample specimen before and after (hidden lines) deformation (Ref 16). Note that the CMOD is evaluated at the load line (centerline of the loading) and the CTOD is evaluated at the crack tip. Some test methods used for evaluating the CTOD are British Standard 7448, Part 1 and ASTM E 1290 (Ref 24, 25).

The J-integral method is another means of characterizing the crack tip in an elastic-plastic type of material. The mathematical definition of the J-integral was previously given in Eq 6; the alternative definition is "a mathematical expression, a line or surface integral that encloses the crack front from one crack surface to the other, used to characterize the local stress-strain field around the crack front" (Ref 1).

To expand on the background work discussed above and summarize the link between the J-integral and Griffith-Irwin fracture concepts, a general (and simplistic) form of the J-integral may be given as follows (Ref 17):

$$J = J_{el} + J_{pl} \quad \text{(Eq 18)}$$

$$J_{el} = \frac{K_I^2(1-\nu^2)}{E} \quad \text{(Eq 19)}$$

$$J_{pl} = \frac{\eta A_{pl}}{B_N B_0} \quad \text{(Eq 20)}$$

where J_{el} is the elastic portion of the J-integral, J_{pl} is the plastic portion of the J-integral, η is a factor based on the type of specimen used, A_{pl} is the area under the load against displacement plot, B_N is the specimen thickness, and B_0 is the uncracked portion of the specimen (ligament).

Note that Eq 18 to 20 apply to specific specimen geometries and load-crack situations, as described in the ASTM test standards for fracture toughness testing (Ref 17, 25, 26). These expressions were given only as an example to show the link between J and K, and to further show how the J-integral is expressed in two parts: the elastic and the plastic components. Finally, Fig. 5 is presented to show schematically how the plane-strain fracture toughness (J_{Ic}) is obtained after testing. Once a candidate J value (J_Q) is determined, it must pass through numerous validity

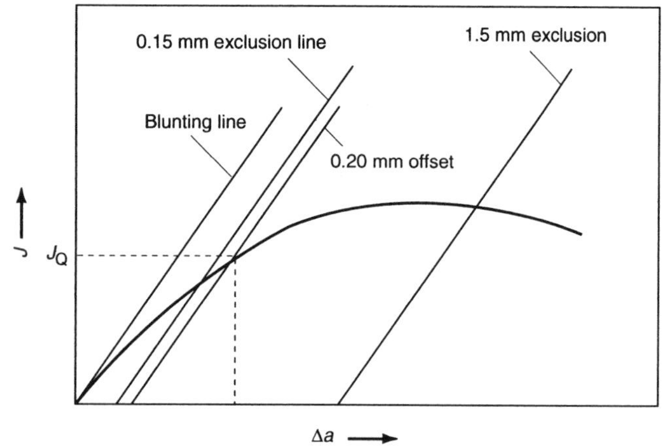

Fig. 5 Schematic of a J-Δa plot showing the point at which J_Q is determined for evaluation as J_{Ic}. These plots are generated by first obtaining the plot of load against crack-mouth opening displacement (CMOD), and then determining the J and Δa values accordingly. Only for compact tension specimens is J estimated from some function of CMOD. The blunting line approximates the apparent crack advance, due to crack-tip blunting in the absence of slow stable crack tearing (Ref 1). The offset and exclusion lines are used in data qualification in the determination of J_{Ic}. Source: Ref 20

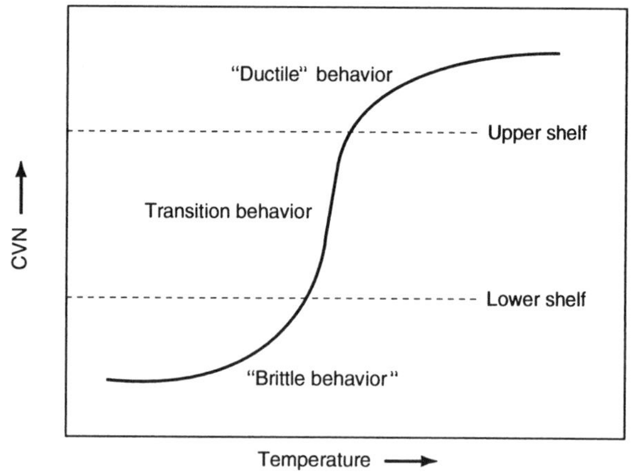

Fig. 6 Charpy V-notch energy (CVN) plotted against temperature, showing the three regions of material behavior ("brittle," transition, and "ductile"). This schematic approximates the S-shape of the curves for low-strength steel; other materials, such as aluminum alloys and high-strength steels, show a flat-shaped curve.

criteria before it can be considered as a fracture toughness value to use in design.

Dynamic and Impact Fracture Mechanics. Although dynamic and impact test methods, such as the Charpy and Izod tests (Ref 27), have been essentially overshadowed by LEFM and EPFM in the past few decades, they do have historical significance in fracture mechanics. In particular, the concept of transition temperature behavior in fracture is well grounded in impact testing (Ref 3–5).

The transition temperature is shown schematically in Fig. 6, where the Charpy V-notch energy (CVN) is plotted against temperature. Generally, numerous specimens are tested at various temperatures to establish plots that may reveal temperature effects in a material. Lower-strength steels tend to give an S-shaped plot, as shown in Fig. 6; higher-strength steels and aluminum alloys will give a flat, lower-energy plot (Ref 3). The lower, flat portion of the plot is generally referred to as the "lower shelf" region; the higher, flat portion of the plot is the "upper shelf" region. Additionally, there is a "transition" region between the upper and lower shelves. Metallurgically, early investigators studied the apparent "brittle" behavior of steels at the lower shelf and the "ductile" behavior at the upper shelf.

To elaborate on the use of a plot of CVN against temperature, it should be noted that in essence this plot is a toughness against temperature plot. For impact tests, Charpy energy can be replaced by percent shear fraction or percent lateral expansion as alternative measures of toughness. Additionally, plots of CTOD or J against temperature can be employed in similar fashion, to serve the same purpose. Results of these S-shaped plots can be translated from the left or right by varying the strain rate, thickness, cold forming, welding, or geometries associated with testing. That is to say, "ductile" (or tough) materials may behave in a "brittle" (or less tough) manner if any (or all) of these parameters are varied. Conversely, a less tough material can be used successfully by careful attention to these parameters.

In tandem with the transition temperature effects, many early investigators proposed correlations between fracture toughness and CVN values (Ref 3, 4, 28). This information was especially important in the upper shelf region, where LEFM methods to determine fracture toughness become impractical. One such correlation was reported by Begley and Logsdon (Ref 3, 29) who proposed a three-point correlation for rotor steel alloys (NiCrMoV, NiMoV, CrMoV) of the entire transition temperature regime with the following three equations:

$$(K_{Ic})_1 = 0.45\sigma_{YS} \quad \text{(at 0\% shear fracture)} \qquad \text{(Eq 21)}$$

$$(K_{Ic})_2 = \sigma_{YS}\left[5\left(\frac{\text{CVN}}{\sigma_{YS}} - 0.05\right)\right]^{1/2}$$
(at 100% shear fracture) (Eq 22)

$$(K_{Ic})_3 = \frac{1}{2}\left[(K_{Ic})_1 + (K_{Ic})_2\right]$$
(at 50% shear fracture) (Eq 23)

Equations 21 to 23 are not intended to be used for all materials (or design applications); they apply only to the rotor steels described in the literature (Ref 3, 29). No CVN-K_{Ic} correlation is universal; it may vary with product form, temper, and numerous other material parameters. Additionally, the variability in CVN-K_{Ic} correlations can be quite large, so these equations should be used with caution.

Variables Affecting Fracture Toughness

Fracture toughness is highly dependent on many variables, many of which can dramatically alter the fracture resistance characteristics of a structure, mechanical component, or test specimen. During analysis and design, consideration of yield strength, loading rate, temperature, thickness, and orientation should all be addressed. When fracture toughness data are generated, laboratory conditions for testing should model actual service conditions, as practically as possible. Although not a complete list, some of the more important variables that affect fracture toughness are presented in this section. These variables can be studied in greater detail in some of the numerous texts on fracture mechanics (e.g., Ref 3–5, 9, 19).

Yield Strength. A general phenomenon with many (but not all) steels and high-strength alloys is that: increasing the fracture toughness of the material lowers the yield strength. Conversely, increasing the yield strength of the material lowers the fracture toughness. This consideration of yield strength, in tandem with fracture toughness, is especially important in design applications where the material will be taken to the upper limits of stress. Table 1 presents a compilation of yield strength and corresponding fracture toughness values for various aircraft structural materials (Ref 18).

The general relationship between toughness and yield strength may not be the case for structural steels used in various other applications. Modern steelmaking techniques can routinely increase the yield strength of the material, with little or no sacrifice in toughness. In some cases, an increase in toughness may be achieved in tandem with an increase in yield strength. The important point is that the user of a material should be careful not to enhance one material quality (such as yield strength) at the expense of another material quality (such as toughness).

Loading Rate. There is a tendency for many materials, such as steels, to exhibit different values of fracture toughness at different loading rates. In terms of stress intensification rates ($\dot{K}$),

Table 1 Yield strength (σ_{YS}) and fracture toughness (K_{Ic}) of various plate materials tested at room temperature

Material identification	Process description	σ_{YS} MPa	σ_{YS} ksi	K_{Ic} MPa$\sqrt{m}$	K_{Ic} ksi$\sqrt{in.}$
Steel					
4340	260 °C temper	1495–1640	217–238	50–63	45–57
D6AC	540 °C temper	1495	217	102	93
HP 9-4-20	550 °C temper	1280–1310	186–190	132–154	120–140
HP 9-4-30	540 °C temper	1320–1420	192–206	90–115	82–105
10 Ni (vim)	510 °C temper	1770	257	54–56	49–51
18 Ni (200)	Marage	1450	210	110	100
18 Ni (250)	Marage	1785	259	88–97	80–88
18 Ni (300)	Marage	1905	277	50–64	45–58
Aluminum					
2014-T651	...	435–470	63–68	23–27	21–25
2020-T651	...	525–540	76–78	22–27	20–25
2024-T351	...	370–385	54–56	31–44	28–40
2024-T851	...	450	65	23–28	21–25
2124-T851	...	440–460	64–67	27–36	25–33
2219-T851	...	345–360	50–52	36–41	33–37
7050-T73651	...	460–510	67–74	33–41	30–37
7050-T73651	...	515–560	75–81	27–31	25–28
7050-T73651	...	400–455	58–66	31–35	28–32
7050-T73651	...	525–540	76–78	29–33	26–30
7050-T73651	...	560	81	26–30	24–27
Titanium					
T1-6Al-4V	Mill annealed	875	127	123	112
T1-6Al-4V	Recrystallized annealed	815–835	118–121	85–107	77–9

Note: These values are dependent on other factors, such as the actual plate thickness that specimens are cut from. Source: Ref 18

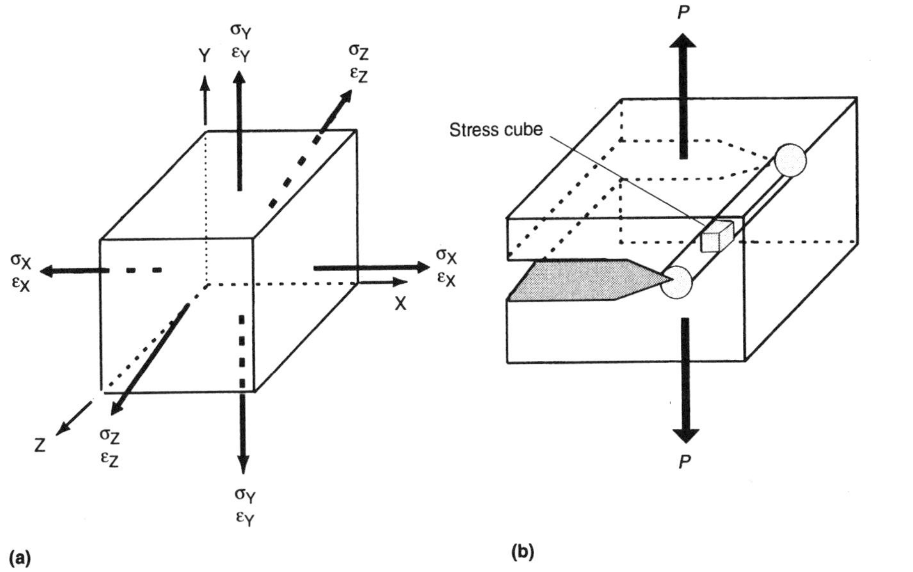

Fig. 7 Solid mechanics type approach to the description of stress and strain using a stress cube. (a) An expanded stress cube, with the normal stress (σ) and strains (ε) depicted. (b) Schematic of a plate with a crack, loaded in tension. The gray circular "cylinder" in the plate represents the plastic zone, which will vary in size and shape with thickness (it is not a perfect cylinder as shown). The radius of the plastic zone decreases along the *z*-axis from the outside surface to the center of the specimen.

impact tests are done at relatively fast rates in the range of 10 to 10^4 MPa$\sqrt{m}$/s (9.1 to 9.1 × 10^3 ksi$\sqrt{in.}$/s), as opposed to the quasistatic fracture test rates of less than 2.75 MPa$\sqrt{m}$/s (2.5 ksi$\sqrt{in.}$/s) (Ref 30). In general, impact test fracture toughness results tend to be lower (more conservative) than fracture toughness results determined with a test done according to ASTM E 399 or E 813.

Impact loading and rapid loading rate are not the same thing; in fact, a rapid loading rate can have strain rates an order of magnitude less than those of an impact test. Some examples of rapid loading are dynamic bridge loading and ship slamming, although they are often thought to be impact loading situations. This misunderstanding arose from early observations of steam locomotives crossing bridges and feeling the "hammering" of the counterweights on the driving wheels.

Temperature. Fracture toughness is highly dependent on temperature, and an effort should be made early in the design process to establish characteristic temperatures that model the service applications of the structure or mechanical component under consideration. Many design applications, such as aircraft, pressure vessels, and turbines, run at varying temperatures, and it may be necessary to generate a spectrum of fracture toughness information regarding temperature. Various articles and texts relate the issues of testing at different temperatures (Ref 3, 4, 21, 31) and discuss the transition temperature phenomenon, as well as the use (and misuse) of terminology such as "ductile" and "brittle" behavior in fracture.

The greatest challenge in assessing fracture toughness at a range of temperatures is determining what test method is appropriate. Modern structural steels used in land and marine vehicles and structures (high-strength low-alloy steel, quenched and tempered steel, carbon steel, etc.) exhibit significant plasticity to invalidate plane-strain toughness testing at any temperature. For other materials, such as the higher-strength materials used in aerospace applications, plane-strain fracture toughness testing may be appropriate at lower to moderate temperatures, but not for higher service temperatures. Methods such as *J*-integral, CTOD, and even CVN testing are available for higher-temperature testing, but the test method should be carefully chosen.

As a general description of temperature effects in fracture of steels, it is instructive to paraphrase from Ewalds and Wanhill (Ref 9). In general, at low temperatures the crack tip is sharp, the material yield stress is high, and at failure the plastic zone size is small (as is K_{Ic}). Conversely, at higher temperatures the crack tip blunts, the material yield stress decreases, and there is a larger plastic zone (K_{Ic} is larger too).

Material Thickness (Plane Stress and Plane Strain). Material thickness is an especially important consideration in the design for fracture, as thicker sections can lead to a decrease in the fracture toughness of the structure or component part under consideration. What this means is that as the thickness of a plate (or section) is increased, the fracture toughness value can decrease, at least to a limiting value. This is important for the designer to note, as a design "fix" to a highly stressed component may be to increase the cross-sectional area of the part, thus reducing the nominal stress. However, the fracture toughness of the component may be decreased by the change in geometry, and it should be reevaluated in terms of adequacy for design.

Material thickness is linked with plane-stress and plane-strain behavior in fracture mechanics, due to the dependency of the stress state at the

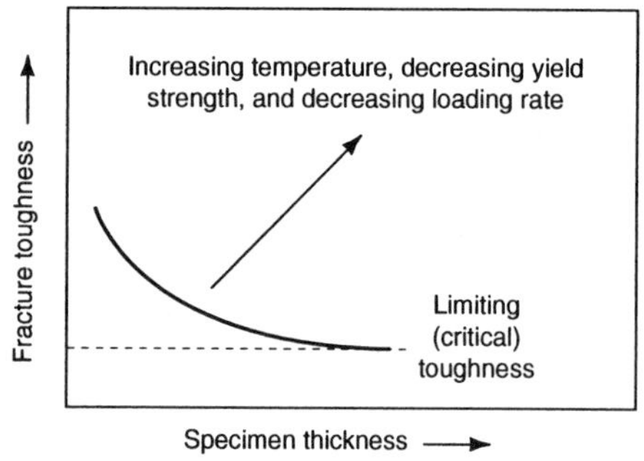

Fig. 8 Effect of specimen thickness, loading rate, and temperature on fracture toughness. In general, increasing the specimen thickness will decrease the fracture toughness to a limiting (or critical) value. Additionally, increasing the temperature and decreasing the loading rate will shift upward the curve for fracture toughness vs. specimen thickness.

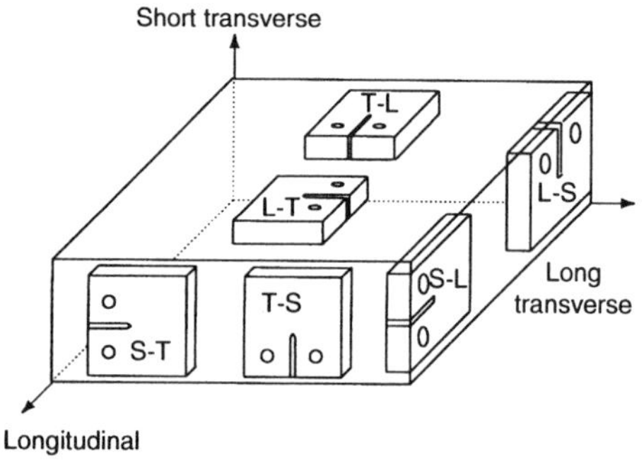

Fig. 9 Specimen orientation scheme showing the longitudinal, long transverse, and short transverse directions. There are six possible specimen designations: L-S, L-T, S-L, S-T, T-L, and T-S. The first letter denotes the direction of the applied load; the second letter denotes the direction of crack growth. Source: Ref 20

Table 2 Fracture toughness (K_{Ic}) values for aluminum and steel alloys

Material	Product form	Thickness range	K_{Ic}, MPa$\sqrt{m}$ (ksi$\sqrt{in.}$) L-T	T-L	S-L
Aluminum					
2014-T651	Plate	0.0–1.0	24 (22)	23 (21)	20 (18)
2024-T351	Plate	0.5–2.0	34 (31)	32 (29)	24 (22)
2024-T851	Plate	0.5–1.5	25 (23)	22 (20)	19 (17)
7050-T7451	...	0.5–2.0	36 (33)	31 (28)	28 (25)
7050-T7452	Hand forging	1.5	37 (34)	24 (22)	24 (22)
7075-T6	Extrusion	0.5–0.75	32 (29)	23 (21)	21 (19)
7075-T651	Plate	0.5–2.0	29 (26)	24 (22)	20 (18)
7075-T7351	Plate	0.5–2.0	33 (30)	33 (30)	25 (23)
7075-T7651	Plate	0.5–2.0	30 (27)	25 (23)	21 (19)
7178-T651	Plate	0.5–1.0	26 (24)	22 (20)	17 (15)
7178-T7651	Plate	0.5–1.0	31 (28)	26 (24)	19 (17)
Steel					
300M (235 ksi)	Plate	0.5	47 (43)	...	...
H-11 (220 ksi)	Plate	0.5	37 (34)	35 (32)	...
17-4 PH (175 ksi)	Plate	0.5	46 (42)	42 (38)	...
D6AC (210 ksi)	Plate	0.5–1.0	40 (36)	...	...

Note: These values are for information only. Source: Ref 33

Table 3 Typical "defects" in alloys

Metallurgical microstructure	Processing defects	Operations defects
Inclusions (oxides, sulfides, constituents)	Casting pores	Corrosion pits
Large precipitates (carbides, intermetallics, dispersoids)	Powder contaminants (hairs, ceramics, other metals)	Wear/fretting damage
Clusters or bands (inclusions, precipitates)	Weld defects	Surface oxidation and carburization
Brittle surface coatings	Forging/extrusion "cracking" and geometry discontinuities	Hydrogen attack
Local "soft spots" (precipitate-free zones)	Tool marks and dents	Creep voiding/cracking
Local "hard spots" (solute-enriched phases)	...	...

Note: In this table, the term "defect" is defined as "a feature, sized between 1 μm and 1 mm, that impairs the mechanical integrity and performance of a component." Source: Ref 59

crack tip on the thickness. When a component (or test specimen) is thin, there is essentially no lateral constraint (in the direction of the thickness), and plane-stress conditions will dominate (Ref 4). Conversely, when a component is very thick, there is a great deal of lateral constraint, and plane-strain conditions can prevail.

The effects of plane stress and plane strain can be explained by the use of a stress cube and sample specimen, as shown schematically in Fig. 7. If the thickness of the specimen is minimal, then σ_z will be essentially zero, with σ_x, σ_y, ε_x, ε_y, and ε_z all having a magnitude (value). This situation is considered to be plane stress (σ_x and σ_y have a magnitude and are "in-plane"). Conversely, if the thickness of the specimen is increased substantially, then σ_z, σ_x, σ_y, ε_x, and ε_y all will have magnitudes and ε_z will be essentially zero. This situation is considered to be plane strain (ε_x and ε_y have a magnitude and are "in-plane"). Plane stress and plane strain also affect the plastic zone size; the plane-stress and plane-strain plastic zones can be simply modeled (along the *x*-*y* plane) as:

$$r_y = \frac{1}{2\pi}\frac{K^2}{\sigma_{YS}^2} \quad \text{(plane stress)} \qquad \text{(Eq 24)}$$

$$r_y = \frac{1}{6\pi}\frac{K^2}{\sigma_{YS}^2} \quad \text{(plane strain)} \qquad \text{(Eq 25)}$$

where r_y is the radius of the plastic zone. It should be noted that r_y decreases along the *z*-axis from the outside surface to the center of the specimen.

Figure 8 summarizes the effects of loading rate, temperature, and material (specimen) thickness in a somewhat simplistic plot. In general, increasing the specimen thickness will decrease the fracture toughness value to a limiting (critical) value. Additionally, increasing the temperature and decreasing the loading rate will shift upward the curve of fracture toughness versus specimen thickness.

Material Orientation and Anisotropy. The forming process of a material, such as drawing, extruding, rolling, or forging, can have a great effect on the microstructure and texture (preferred grain orientation) of the material. Additionally, the fracture toughness of a material is affected by the microstructural and mechanical changes a material undergoes through processes such as heat treating. These changes in the material fracture toughness properties can be advantageous to the scientist, engineer, or designer who is aware of their discrepancies. For the unaware and unknowing, these changes in material fracture toughness behavior can have disastrous consequences.

Material anisotropy (or simply anisotropy) is a term used to describe the dependence of material properties, such as fracture toughness, yield strength, and elastic modulus, to the texture. When a material has mechanical properties that are independent of direction, it is said to be *isotropic*. Conversely, if the material has mechanical properties that depend on a particular (or preferred) orientation(s), it is said to be *anisotropic* (Ref 32).

To provide a common scheme for describing material anisotropy, ASTM has standardized the following 6 orientations: L-S, L-T, S-L, S-T, T-L, and T-S (Ref 1). As shown in Fig. 9, the first letter denotes the direction of the applied load; the second letter denotes the direction of crack growth. In designing for fracture toughness, consideration of anisotropy is very important, as different orientations can result in widely differing fracture toughness values. Table 2 shows how fracture toughness can vary among materials, as well as orientation (Ref 33).

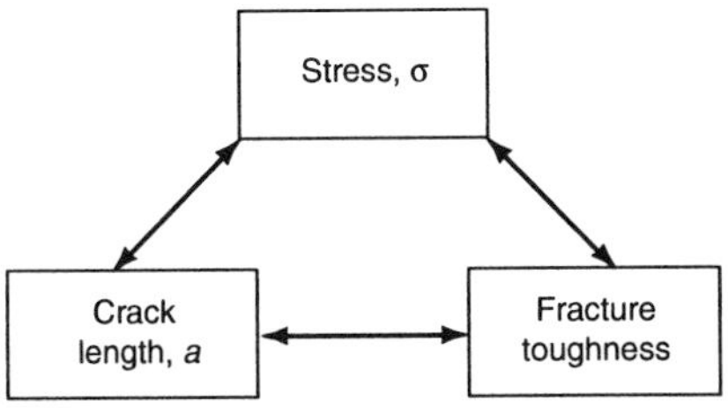

Fig. 10 The fundamental relationship between applied stress, crack length, and fracture toughness in fracture mechanics design

Table 4 Inspection techniques used for cracks

Technique	Principles	Applications and notes
Direct		
Visual	Naked eye, assisted by magnifying glass, low-power microscope, lamps, mirrors	Only at places easily accessible. Detection of small cracks requires much experience.
Penetrant	Colored liquid penetrant is brushed on material and allowed to penetrate into cracks. Penetrant is washed off and a developer is applied. Remnants of penetrant in crack are extracted by developer and give colored line.	Only at places easily accessible. Sensitivity on the same order as visual inspection.
Magnetic particles	Part to be inspected is covered with a layer of a fluorescent liquid containing iron powder. Part is placed in a strong magnetic field and is observed under ultraviolet light. At cracks, the magnetic field lines are disturbed.	Only applicable to magnetic materials. Parts have to be dismounted and inspected in a special cabin. Notches and other irregularities give indications. Sensitive method.
X-ray	X-rays emitted by portable x-ray tube pass through structure and are caught on film. Cracks, absorbing less x-rays than surrounding material, are delineated by black line on film.	Has great versatility and sensitivity. Interpretation problems arise if cracks occur in fillets or at edge of reinforcement. Small surface flaws in thick plates are difficult to detect.
Indirect		
Ultrasonic	Probe (piezoelectric crystal) transmits high-frequency wave into material. The wave is reflected at the ends and also at a crack. The input-pulse and the reflections are displayed on an oscilloscope. Distance between first pulse and reflection indicates position of crack. Interpretation: Reflections of crack disappear upon change of direction of wave.	Universal method since a variety of probes and input pulses can be selected. Information about the size and nature of the defect (which may not be a crack) are difficult to obtain.
Eddy current	Coil induces eddy current in the metal. In turn this induces a current in the coil. Under the presence of a crack the induction changes; the current in the coil is a measure for the surface condition.	Cheap method (no expensive equipment) and easy to apply. Coils can be made small enough to fit into holes. Sensitive method when applied by skilled person. Provides little or no information about nature and size of defect.
Acoustic emission	Measurement of the intensity of stress waves emitted inside the material as a result of plastic deformation at crack tip, and as a result of crack growth.	Inspection must be done while structure is under load. Continuous surveillance is possible. Expensive equipment is required. Interpretation of signals is difficult.

Source: Ref 19

Analysis, Detection, and Design Criterion

The fundamental basis for designing a structure or mechanical component to be fracture resistant is to explore the interrelationship among the important elements in fracture. The most basic fracture prevention model contains three elements, as shown in Fig. 10: applied stress, crack length, and fracture toughness. Stated another way, the basic fracture prevention model should account for the applied stress (analysis), crack length (detection), and fracture toughness (design criterion).

Analytical methods for fracture mechanics have been thoroughly studied in both LEFM and EPFM, and many closed-form solutions can be found in fracture mechanics textbooks (e.g., Ref 3–5, 9, 19, 34–48). With regard to stress intensity factors, small samples of some of these closed-form solutions were presented in Eq 14 to 16. Additionally, some handbooks and codes show K and J solutions that are used in various industries, such as the military aircraft and nuclear pressure vessel industries (Ref 49–58). Finite element methods in fracture mechanics, which came into vogue in the 1980s, are most commonly found in journal articles. Several popular fracture mechanics journals are good sources for all types of analytic procedures: *Engineering Fracture Mechanics*, *International Journal of Fracture Mechanics*, and the ASTM *Special Technical Publications*.

Cracks and Methods of Detecting Cracks. It is not an overstatement to say that defining and detecting cracks is just as important as computing the stress intensity factor. What has to be resolved in the initial phase of a fracture-resistant design is what defines the crack, either in terms of constitution, geometric size, or measurability (what is practically detectable). In the literature, various terms or phrases are used to describe a crack (or nucleation site of a crack), such as flaw, defect, discontinuity, heterogeneity, or anomaly, to name just a few. Regardless of which term is used, what is important to note goes back to Griffith (as paraphrased from Hoeppner, Ref 6): "all materials contain inherent discontinuities and heterogeneities and at times they can be crack-like." Building on this thought, it is instructive to look at what types of cracks may exist in a component or structure and where these cracks might nucleate, at least from a metallurgical viewpoint. Table 3 contains a list of typical defects in alloys and shows various means by which a crack can nucleate in a material (Ref 59).

As defined in Table 3, a defect may be too small to be detected, especially in a large structure such as an aircraft. It is the role of the scientist, engineer, and designer to determine the practical limitations of the available detection methods, both physically and statistically. Further, the scientist, engineer, and designer must judiciously employ these detection methods, in agreement with the fracture strength of the material (fracture toughness), for the safe operation of structures and mechanical components (Ref 60). Table 4 provides a sample of some of the crack detection methods that are commonly used by industry (Ref 19).

Design Criterion—Determination of Fracture Toughness Values. The fracture toughness value of a material is determined in the laboratory, with the most generally accepted test methods following standards published by ASTM. Currently, ASTM has standards E 23 for notch toughness testing, E 399 for plane-strain fracture toughness testing, and E 813, E 1737, and E 1290 for elastic-plastic fracture toughness testing (Ref 17, 20, 25–27). Other standard fracture test methods have been developed, such as British Standard 7448, Part 1 (Ref 24).

The importance of laboratory testing in determining material design parameters, such as fracture toughness, should never be underestimated. The users of a material in a design application must communicate with the testing laboratory in order to choose the appropriate test for the specific application. A structure or mechanical component may encounter a plethora of variables that affect fracture toughness, besides those discussed above. Further, any number of possible combinations of these variables can occur, leading to potentially synergistic effects.

Laboratory testing for fracture toughness is well documented in the literature, and the literature provides an excellent supplement to the test descriptions provided in the ASTM standards (Ref 3–5, 61–63). A summary of fracture tests and their comparisons is presented in Table 5 (Ref 16).

Charpy Impact Testing. Although perhaps more of an impact energy test than a true fracture toughness test, the Charpy impact test is often used to obtain fracture toughness results. The history and early uses of this easy-to-use and relatively inexpensive test method are well described in the literature (Ref 3–5, 28). The Charpy impact test as described in ASTM E 23 (Ref 27) uses a hammer on the end of an arm to strike a CVN specimen, supported on an anvil. The pendulum motion of the hammer striking the CVN specimen results in an energy transfer to the specimen, which is measured with a visual indi-

Table 5 Fracture toughness tests

Method	Advantages	Disadvantages
K_{Ic}, ASTM E 399	This method is the most reliable to get fracture toughness values at lower temperatures. The success of all other methods is based on their ability to give data comparable to this method.	The high cost of testing the large specimens required for higher temperature tends to reduce the number of data points. Linear extrapolation from valid K_{Ic} at lower temperature to higher temperature produces conservatism. No valid K_{Ic} values at higher temperature.
J-integral, ASTM E 813	Provides fracture toughness values that agree with K_{Ic} method. Yields realistic fracture toughness data at higher temperature. Has the advantage (over CTOD) of a sound theoretical basis, which permits evaluation of stable crack growth. Determination of *dJ/da* is a measure of the resistance to continued crack propagation. Testing many small *J* specimens provides an indication of material toughness variation.	Not able to evaluate irregular crack propagation due to residual stress or at HAZ near welds. Not accurate enough at low temperatures. Measurements are inaccurate due to irregular crack fronts. Not valid for thin materials where K_J is 2.5 K_{Ic}. When heat tinting is used, the additional number of specimens adds to testing costs.
CTOD, BS 7448, ASK-AAN 220	Provides fracture toughness values that agree with ASTM K_{Ic} method. Yields realistic fracture toughness data at higher temperatures. CTOD results have shown good consistency and comparability with toughness values using other methods. Simultaneous measurement of CTOD and *J*-int. is possible for a minor extra cost.	Variations in the measurement of δ results in variations of K_{Ic} of up to a factor of 2. This method restricted to temperatures above –60 °C.
Simple equal energy	Provides fracture toughness values that agree with ASTM K_{Ic} method. Yields realistic fracture toughness data at higher temperatures. Toughness data are identical or closely similar to *J*-int. data.	Limitations similar to those of the *J*-int. method. This method is more empirical in nature, so *J*-int. testing is preferred.
Instrumented Charpy testing	Requires small specimens. Practically suited for determination of toughness variations in small regions of complex parts, in HAZ of welds, and in other locally embrittled zones. Error in K_{Ic} is small (in comparison to ASTM K_{Ic} method) for predominantly brittle failure.	Can provide very pessimistic values, particularly at higher temperatures. K_{Ic} is slightly underestimated at low temperatures, but considerable scatter of measurements exists above the brittle-transition temperature within a factor of 3 due to small size of specimens. Difficulty in separating the crack-initiation and crack propagation components of fracture.
Empirical methods per Begley and Logsdon	Requires small specimens. Offers a rapid and inexpensive technique to estimate K_{Ic} for wrought ferritic steels. This method indicates that Charpy K_{Ic} values are scattered and lie entirely below ASTM K_{Ic} data. Conservative by a factor of up to 3. K_{Ic} by this method provides narrow scatter band with the results below ASTM K_{Ic} by a factor of 2.	Can provide very pessimistic values, particularly at higher temperatures. Cannot give information relevant to small regions such as HAZ at welds, castings, or materials other than the ferritic steels.

CTOD, crack-tip opening displacement; HAZ, heat-affected zones. Source: Ref 16

cator. Charpy energy, percent shear fracture, or percent elongation is generally plotted against a range of temperatures, as shown in Fig. 6.

Compliance-Based Fracture Toughness Testing. Laboratory testing for fracture toughness has become increasingly reliant on servohydraulic equipment, and a synthesis of mechanical test apparatus with sophisticated computer data acquisition and controls is becoming the fracture toughness test standard. Compliance-based fracture testing employs a displacement (CMOD) gage. DC signals are amplified and conditioned to control and monitor the test, as shown schematically in Fig. 11. The load is generally monitored by the use of a load cell, mounted within the test frame in the load train.

Compliance-based fracture testing employs the relationship between compliance, which is the reciprocal of the load-displacement plot generated during testing, and the crack length. For example, the equation for the crack mouth opening compliance of a compact tension specimen for LEFM (following ASTM E 399) can be expressed as follows (Ref 20):

$$C = \frac{EvB}{P} \quad \text{(Eq 26)}$$

where C is the compliance, v is the displacement of the clip gage, B is the specimen thickness, and P is the load.

The relationship between compliance and crack length has been verified in the literature (Ref 64), and these relationships are generally good for specific specimen geometries, such as compact tension (C(T)) and single-edged notch bend specimens (SE(B)), as found in ASTM E 399, E 813, and E 1737 (Ref 20, 26, 17). With reference to Fig. 12, and using the C(T) specimen for an example, the elastic compliance relationship (following ASTM designation E 399) works as follows:

1. The clip gage records a displacement and the load cell records the load, both values of which are input into the compliance equation (Eq 26).
2. The compliance is input into an equation of the form:

$$u_x = \left\{\sqrt{C} + 1\right\}^{-1} \quad \text{(Eq 27)}$$

where u_x is a value to use for a polynomial expression.

3. Continuing, u_x is input into a fifth-order polynomial expression with various constants as follows:

$$\frac{a}{W} = c_0 + c_1u_x + c_2u_x^2 + +c_3u_x^3 + c_4u_x^4 + c_5u_x^5 \quad \text{(Eq 28)}$$

where W is the specimen width and c_i are various constants. Multiplying both sides of the equation by W obtains the crack length.

4. Finally, the stress intensity factor can be determined by the following expression:

$$K = \left(\frac{P}{B\sqrt{W}}\right) f\left(\frac{a}{W}\right) \quad \text{(Eq 29)}$$

It should be noted that the compliance-based procedure just developed describes only LEFM testing in detail. For EPFM testing the procedure is essentially the same, except that plasticity must be accounted for in the determination of J and CMOD. For information on that topic, see ASTM E 813 and E 1737, and some of the more current fracture mechanics texts (Ref 5, 17, 23, 26, 44, 45).

The direct current electric potential (DCEP) technique for fracture toughness testing is being recognized as a viable alternative to compliance-based testing, and it has recently been incorporated into ASTM E 1737 as Appendix A (Ref 17). Higher-temperature applications are well suited for the DCEP technique, as the electrical wiring involved is less cumbersome than CMOD gages, especially in an environmental chamber, and more cost effective. The DCEP technique is not particularly well suited for relatively large or thick specimens.

The DCEP technique is relatively simple, generally consisting of a direct current source, voltmeter, amplifier, and recording device. A constant direct current is passed through the specimen, which acts as a resistor in the circuit. During testing, the crack in the fracture specimen increases, reducing the remaining uncracked portion of the specimen (the ligament). Reduction in the cross-sectional area of the specimen ligament constricts the current flow and increases the resistance of the specimen. Hence, during a DCEP fracture toughness test, the electric circuit can be modeled by the following simple equation:

$$V = IR \quad \text{(Eq 30)}$$

where V is the electric potential, I is the direct current, and R is the specimen resistance.

A reference voltage (U_0) is established at the start of the test, with the subsequent voltages (U_i) recorded at predetermined intervals. Crack length is determined using a calibration procedure, as described in ASTM E 1737, and the ratio of U_0 and U_i is an important input in the crack

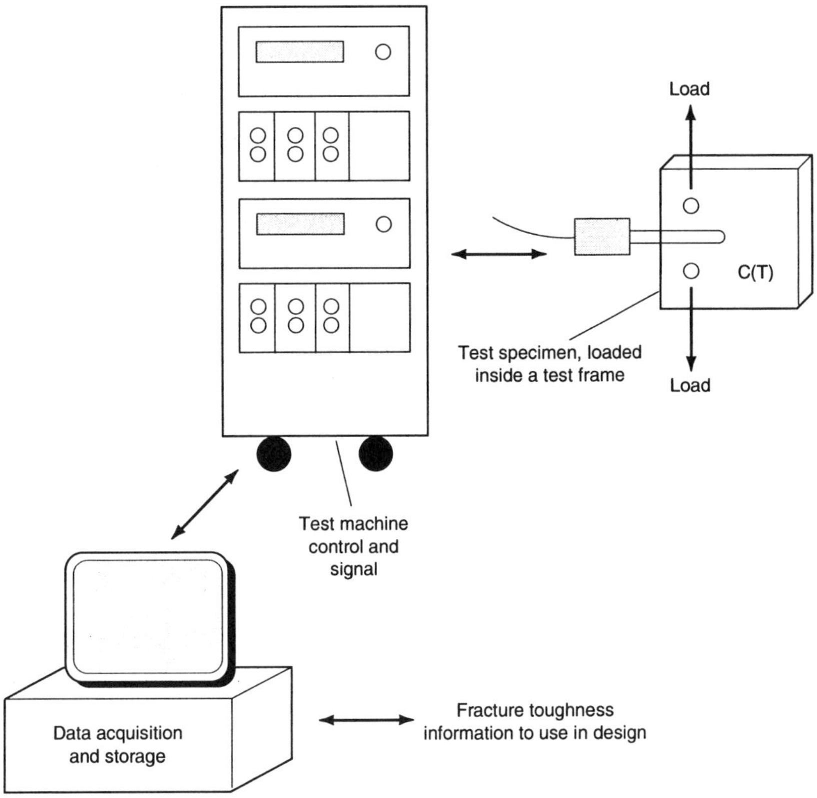

Fig. 11 Common fracture toughness testing setup, showing the interaction of the specimen tested to the control and data acquisition. Current systems generally employ servohydraulic test systems; this schematic shows a gage of crack-mouth opening displacement mounted in the compact tension (C(T)) specimen.

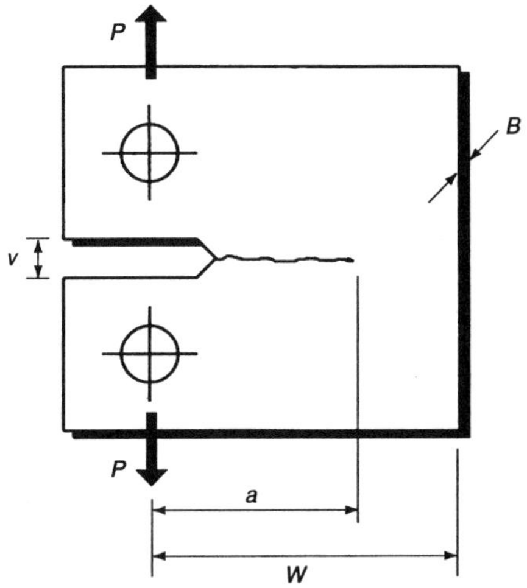

Fig. 12 Schematic of a compact tension specimen used for fracture testing. *P*, load; *W*, width; *B*, thickness; *v*, displacement; *a*, crack length

length calculations. The calculations of the stress intensity factors are the same as in the unloading compliance method; it is the method of establishing the crack length that differs.

Fracture Toughness and Design Philosophies

It was not until the second half of the 20th century that scientists and the engineering community at large embraced fracture mechanics for design. Slightly earlier to (and nearly concurrent with) the advances made in fracture design was the progress in the study of fatigue. It is interesting to note that just as a great deal of fracture failures nucleate from fatigue cracks, fracture design evolved from fatigue design. To fully appreciate the use of fracture toughness in design, it is instructive to follow the evolution of fatigue and fracture in design, from infinite-life design to modern damage-tolerant design philosophies.

Infinite-life design is perhaps the oldest fatigue design philosophy, where the design stress of the structure or component is established below the fatigue limit of the material. This philosophy relies on stress-life plots of uncracked specimens to determine the fatigue limit of a material. It is still in use, although more contemporary fatigue and fracture design philosophies predominate. What can be called into question about this design philosophy is the initial premise that the material is crack free; Griffith disproved this premise long ago.

Safe-life design developed as a logical progression away from the inherent conservatism of infinite-life design. Many structures and mechanical components are intended to be in service for a specified length of time, rather than indefinitely. Hence, the restrictions of a safe-life design can be relaxed by the use of "finite-life" design. Instead of establishing the design stress below the fatigue limit, the design stress is based on the stress associated with the design number of cycles, with an appropriate safety factor applied. This design philosophy is the basis for some of the ASME boiler and pressure vessel codes and the ground vehicle industry. In modern aircraft design, the safe-life design approach is used for components whose failure would result in the loss of the aircraft in service, such as the landing gear (Ref 18).

Safe-life design assumes a macrocrack-free structure, and the structure (or component) under consideration must remain macrocrack-free during service. A macrocrack-free component is not the same as a crack-free component; a macrocrack can be detected during inspection.

Fail-safe design ensures that the structure will support load after any single member fails, or after there is partial damage to an extensive structure. While there is nothing implicit about the presence or absence of cracks, some fail-safe designs account for the "possibility" of cracks by incorporating crack arresters, crack stoppers, and crack deflectors. In aircraft design, a complete fail-safe design provides for failure of a single component or for partial damage to the structure, while maintaining adequate stiffness in the structure for control.

Damage tolerance is the ability of a structure to sustain anticipated loads in the presence of fatigue, corrosion, or accidental damage until such damage is detected through inspections or malfunctions and repaired. Damage tolerance techniques can be applied to safe-life and fail-safe design. However, with respect to fracture mechanics, damage tolerance is a more contemporary approach to the design of structures and mechanical components, where the initial basis for design is the belief that cracks "will" exist in the structure or mechanical component. Taken a step further, the "complete" damage-tolerant design assumes that cracks may occur at any stage in the manufacturing process or service life of a product. Further, damage-tolerant design is intended to be intimately linked with periodic inspections of the structure or mechanical component under consideration, ensuring that repair or replacement of a damaged item is performed in a safe, timely manner.

As a summary of the different fatigue and fracture mechanics design philosophies and their relationship with fracture toughness, an example from aircraft design is shown in Table 6 (Ref 33). To show their significance in design, items related to fracture toughness are shown in bold print.

Examples of Fracture Toughness in Design

The use of fracture toughness in design is extensively covered in numerous articles, textbooks, and handbooks related to aircraft design and damage tolerance (Ref 65–71). Additionally, there are various articles, texts, handbooks, and codes related to pressure vessel design (Ref 72,

Table 6 Design criteria for sizing aircraft, showing how fracture toughness is used in damage-tolerant design

Mode of failure	Design criteria	Allowable data
Static strength of undamaged structure	Structure must support ultimate loads without failure for 3 s	Static properties
Deformation of undamaged structure	Deformation of the structure at limit loads may not interfere with safe operation	Static properties and creep properties for elevated-temperature conditions
Fatigue crack initiation of undamaged structure	Fail-safe structure must meet customer service life requirements for operational loading conditions. Safe-life components must remain crack-free in service. Replacement times must be specified for limited life components.	Fatigue properties
Residual static strength of damaged structure	**Fail-safe structure must support 80-100% limit loads without catastrophic failure, a single member failed in redundant structure or partial failure in monolithic structure.**	Static properties and **fracture toughness properties**
Crack growth life of damaged structure	**For fail-safe structure, inspection techniques and frequency must be specified to minimize risk of catastrophic failures. For safe-life structure, must define inspection techniques and frequencies and replacement times, so that probability of failure due to fatigue cracking is extremely remote.**	Crack growth properties and **fracture toughness properties**

Note: Bold text shows items related to fracture toughness. Source: Ref 33

Table 7 Fracture properties of the pressure vessel liner material in Example 1

Specimen	Test temperature °C	Test temperature °F	Fracture toughness MPa$\sqrt{m}$	Fracture toughness ksi$\sqrt{in.}$	Microstructure	Validity
A-4	25	77	99.4	90.4	F/B	No
A-2	25	77	117.3	106.7	F/B	No
B-1	25	77	99.4	90.4	P	Marginal
B-4	25	77	90.1	82.0	P	Marginal
B-2	−40	−40	69.4	63.1	P	Yes
B-3	−40	−40	64.7	58.9	P	Yes
A-1	−40	−40	119.7	108.9	F/B	No
A-3	−40	−40	98.6	89.7	F/B	Marginal

F/B, ferrite bainite; P, pearlite. Source: Ref 81

73). Fracture mechanics is pertinent to other disciplines as well, such as the automotive and aerospace industries (Ref 74, 75).

Other excellent sources of literature on fracture mechanics are failure analysis studies. Some of these are contained in fracture texts (Ref 4); others are found in complete texts covering the subject of failure analysis (Ref 76, 77). A rigorous failure analysis gives good insight into how a fracture failure can occur and what the scientist, engineer, and designer should consider for a damage-tolerant structure or material component.

The reader also should be aware of the use of various engineering "tools" to assist in LEFM and especially EPFM analysis of structures or components. Some of these tools are the Welding Institute CTOD design curve, the failure assessment diagram (FAD), and the deformation plasticity failure assessment diagram (DPFAD). The use of these approaches in the design against fracture is well documented, and some excellent articles related to their use (including case studies) may be found in the literature (Ref 78–80). Another general reference book on fracture mechanics is *Fatigue and Fracture*, Volume 19 of *ASM Handbook*.

Example 1: LEFM Analysis of a Cracked Jacketed Pressure Vessel. The following example summarizes the use of LEFM in the analysis of a jacketed thick-walled pressure vessel, subjected to high pressures above 69 MPa (10 ksi). This case study, originally published as a journal article (Ref 81), is a straightforward example of the use of fracture toughness and damage tolerant methods in design. What is especially interesting is the magnitude of the pressures involved, and the application of a unique technology in controlling these pressures: high-pressure engineering.

Many large, vintage 1945 pressure vessels were returned to service in the late 1980s and were subjected to repeated pressurizations. It was observed that some of these pressure vessels developed long, shallow cracks at the inside diameter (ID) in the longitudinal direction. A damage tolerance assessment was performed to determine three important quantities: the fracture toughness of the material, the applied stresses during service (stress intensities in service), and the critical flaw size (crack length). Note that these are the same quantities discussed in the fundamental fracture mechanics relationship diagram (Fig. 10).

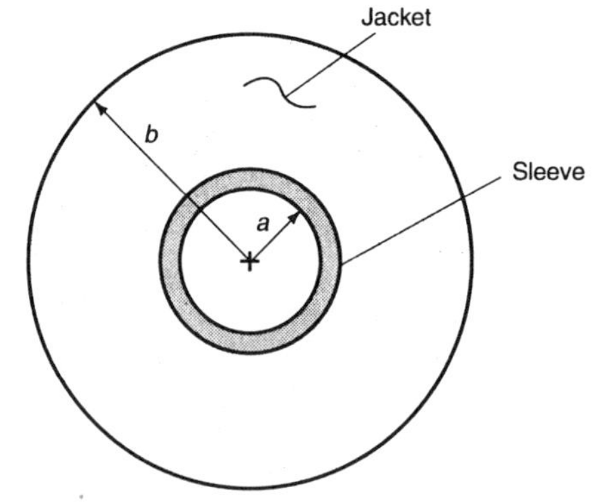

Fig. 13 Schematic of the pressure vessel example, showing the pertinent dimensions for analysis. Source: Ref 81

The geometry of the pressure vessel is shown schematically in Fig. 13, where a is the inside radius (IR) of the liner, b is the outside radius (OR) of the liner (and the IR of the jacket), and c is the OR of the jacket. The dimensions of the pressure vessel were as follows: a is 20.32 cm (8.00 in.), b is 25.81 cm (10.16 in.), and c is 62.23 cm (24.50 in.). The k ratio (ratio of c/a), an important measure in high-pressure engineering (Ref 82), was calculated as 3.06. It is interesting to note, from the standpoint of damage tolerance, that the liner in a pressure vessel served as a crack arrestor.

Metallurgical examination of the liners revealed two microstructures to consider in a damage tolerance analysis. Plane-strain fracture toughness tests were conducted according to ASTM E 399 on two different materials, one containing a pearlitic microstructure and one containing a ferrite/bainite microstructure. Further, to evaluate temperature effects, two temperatures were selected: −40 and +25 °C (−40 and +70 °F). The results of the fracture toughness tests, presented in Table 7, determined the first part of the damage tolerance assessment.

The stress analysis was performed using finite element models and evaluating the effects of shrink fitting the liner into a much thicker jacket. This was the second step of the damage tolerance assessment. The basic tangential stresses (σ_θ) at the ID and the stress intensity factor used in the fracture toughness analysis were calculated as follows (Ref 82):

$$\sigma_\theta = \frac{2pk^2}{k^2 - 1} \quad \text{(internal pressure)} \qquad \text{(Eq 31)}$$

$$\sigma_\theta = \frac{-2b^2}{b^2 - a^2}\,\frac{E}{\delta}\left(\frac{(c^2 - b^2)(b^2 - a^2)}{2b^2\,(c^2 - a^2)}\right)$$

(shrink fit) (Eq 32)

$$f\left(\frac{l}{W}\right) = 1.12 - 0.175\left(\frac{l}{W}\right) - 0.515\left(\frac{l}{W}\right)^2 + 1.43\left(\frac{l}{W}\right)^3$$

(internal pressure) (Eq 33)

Table 8 Calculated value of the critical crack length in Example 1

Internal pressure σ_θ (ID)	Critical crack length at indicated fracture toughness (K_{Ic}) value				
	137.8 MPa (20 ksi)	275.6 MPa (40 ksi)	413.4 MPa (60 ksi)	551.2 MPa (80 ksi)	689.0 MPa (100 ksi)
296.5 MPa (43.0 ksi)	0.5 mm (0.02 in.)	2.0 mm (0.08 in.)	5.3 mm (0.21 in.)	4.57 mm (0.36 in.)	...
342.7 MPa (49.7 ksi)	0.5 mm (0.02 in.)	1.5 mm (0.06 in.)	3.6 mm (0.14 in.)	6.4 mm (0.25 in.)	9.9 mm (0.39 in.)

$$f\left(\frac{l}{W}\right) = 1.12 - 0.217\left(\frac{l}{W}\right) - 0.268\left(\frac{l}{W}\right)^2 + 0.943\left(\frac{l}{W}\right)^3$$

(shrink fit) (Eq 34)

$$K = \sigma_\theta \text{ (ID) } \sqrt{\pi l}\, f\left(\frac{l}{W}\right)$$

(internal pressure and shrink fit) (Eq 35)

where σ_θ (ID) is the stress at the ID, l is the crack length, and $f(l/W)$ is the geometric shape factor.

Finally, by establishing the relationship among internal pressure, interference fit, tangential stress, and fracture toughness, a critical crack length was determined by graphical analysis. Essentially, the equation for the stress intensity factor (Eq 35) was rearranged for both the internal pressure and shrink fit cases and plotted as K against crack length. From these plots, a third plot, the algebraic sum of the two plots, determined the combined solution. For this analysis, the critical crack lengths were evaluated at two internal pressures, 296.5 and 342.7 MPa (43 and 49.7 ksi). Table 8 shows estimates of the critical crack sizes for both internal pressure cases, evaluated at various fracture toughness values. For example, a high-pressure vessel loaded at 342.7 MPa (49.7 ksi) internal pressure and having a material fracture toughness of 551.2 MPa (80 ksi) would have a critical crack length of 6.4 mm (0.25 in.). This is the final step in this basic damage tolerance model, and from the critical crack size, service life and inspection techniques can be reviewed.

The Future Direction of Fracture Toughness in Design

Although fracture mechanics is a relatively mature science, there is still a great deal of refinement that can be made to the use of fracture mechanics in design applications. In the 1960s and 1970s, the focus was on LEFM analysis and testing methods; from the 1970s to the present, EPFM analysis and testing methods have been of greatest interest. LEFM and EPFM will be the paradigm for fracture mechanics into the next century, and they will surely be improved upon. Additionally, in the next few decades, with the trend towards maintaining longer service lives of structures such as military and commercial aircraft, there will undoubtedly be improvements in crack (flaw) detection.

Finally, there will be constant challenges for the scientist, engineer, and designer of damage-tolerant structures. Materials will be taken to their limits, and each new prototypic structure will contain some "exotic" or modified material(s) that will have to be characterized for fracture resistance. Another development, advocated by Griffith as early as the 1920s, will be the refinement of engineering materials to improve fracture strength. This approach should be viewed in terms of microdiscontinuities (which can be addressed by advancements in material refining techniques) and macrodiscontinuities (which can be alleviated by careful attention to the manufacturing and fabrication processes). The metallurgist may have a role in developing an engineering material for a component or structure; the mechanical, civil, or structural engineer may have a role in designing a workable component or structure using this material. Regardless on what level people work, everyone in the manufacturing and design process should always look towards improving the product.

REFERENCES

1. "Standard Terminology Relating to Fracture Testing," E 616-89, *Annual Book of ASTM Standards*, ASTM
2. J.A. Bennett and H. Mindlin, Metallurgical Aspects of the Failure of the Point Pleasant Bridge, *J. Test. Eval.*, Vol 1 (No. 2), March 1973, p 152–161
3. S.T. Rolfe and J.M. Barsom, *Fracture and Fatigue Control in Structures*, 4th ed., Prentice-Hall, Inc., 1996
4. R.W. Hertzberg, *Deformation and Fracture Mechanics of Engineering Materials*, 3rd ed., John Wiley & Sons, Inc., 1983
5. T.L. Anderson, *Fracture Mechanics—Fundamentals and Applications*, 2nd ed., CRC Press, 1995
6. D.W. Hoeppner, Parameters That Input to Application of Damage Tolerance Concepts to Critical Engine Components, *Damage Tolerance Concepts for Critical Engine Components*, Conference Proceeding 393, Advisory Group for Aerospace Research and Development, 1985
7. D.W. MacFerran, "Towards a Postulate-Based Methodology for Developing Specifications and Failure Criteria," Ph.D. dissertation, University of Utah, 1994
8. A.A. Griffith, The Phenomenon of Rupture and Flow in Solids, *Philosophical Transactions*, Royal Society of London, Vol 221A (No. 163), 1920, p 861–906
9. H.L. Ewalds and R.J.H. Wanhill, *Fracture Mechanics*, E. Arnold Publishing Co., 1984
10. A.A. Wells, "Unstable Crack Propagation in Metals: Cleavage and Fast Fracture," Paper 84, *Proc. Crack Propagation Symp.*, Vol 1, 1961
11. J.R. Rice, A Path Independent Integral and the Approximate Analysis of Strain Concentration by Notches and Cracks, *J. Appl. Mech.*, Vol 35, 1968, p 379–386
12. J.A. Begley and J.D. Landes, The J-Integral as a Fracture Criterion, *Fracture Toughness*, ASTM STP 514, 1972, p 1–20
13. J.D. Landes and J.A. Begley, The Effect of Geometry on J_{IC}, *Fracture Toughness*, ASTM STP 514, 1972, p 24–29
14. J.D. Landes and J.A. Begley, Test Results from J-Integral Studies: An Attempt to Establish a J_{IC} Testing Procedure, in ASTM STP 560, 1975, p 170–186
15. W.R. Andrews, G.A. Clark, P.C. Paris, and D.W. Schmidt, Single Specimen Tests for J_{IC} Determination, in ASTM STP 590, 1976, p 27–42
16. R.D. Venter and D.W. Hoeppner, "Crack and Fracture Behavior in Tough Ductile Materials," report submitted to the Atomic Energy Control Board of Canada, Oct 1985
17. "Test Method for J-Integral Characterization of Fracture Toughness," E 1737-96, *Annual Book of ASTM Standards*, ASTM
18. H.O. Fuchs and R.I. Stephens, *Metal Fatigue in Engineering*, John Wiley & Sons, Inc., 1980
19. D. Broek, *Elementary Engineering Fracture Mechanics*, 4th ed., Martinus Nijhoff, 1986
20. "Test Method for Plane-Strain Fracture Toughness of Metallic Materials," E 399-90, *Annual Book of ASTM Standards*, ASTM
21. R.B. Scarlin and M. Shakeshaft, Limitations of Some Methods of Establishing Fracture-Toughness Data, *Met. Technol.*, Vol 8, Part 1, Jan 1981, p 1–9
22. H.S. Reemsnyder, "Fatigue and Fracture of Ship Structures," paper presented at the Symposium and Workshop on the Prevention of Fracture in Ship Structures, National Academy of Science, Washington, D.C., March 1995
23. J.A. Joyce, *Manual on Elastic-Plastic Fracture: Laboratory Test Procedures*, MNL 27, ASTM, 1996
24. "Fracture Mechanics Tests, Part 1: Method for Determination of K_{IC}, Critical CTOD and Critical J Values of Metallic Materials," BS 7448: Part 1, The British Standards Institution, 1991
25. "Standard Test Method for Crack Tip Opening Displacement (CTOD) Fracture Toughness Measurement," E 1290-93, *Annual Book of ASTM Standards*, ASTM
26. "Test Method for J_{IC}, A Measure of Fracture Toughness," E 813-89, *Annual Book of ASTM Standards*, ASTM
27. "Test Methods for Notched Bar Impact Testing of Metallic Materials," E 23-96, *Annual Book of ASTM Standards*, ASTM
28. D.A. Shockey, Charpy Impact Testing, *Mechanical Testing*, Vol 8, *ASM Handbook*, American Society for Metals, 1985, p 261–268
29. J.A. Begley and W.A. Logsdon, "Correlation of Fracture Toughness and Charpy Properties for Rotor Steels," WRL Scientific Paper 71-1E7-MSLRF-P1, Westinghouse Research Laboratory, Pittsburgh, July 1971

30. D.A. Shockey, Dynamic Fracture Testing, *Mechanical Testing*, Vol 8, *ASM Handbook*, American Society for Metals, 1985, p 259
31. F.A. Nichols, How Brittle is Brittle Fracture, *Engineering Fracture Mechanics*, Vol 12, 1979, p 307–316
32. M.P. Blinn, "The Effects of Frequency and Anisotropy as Related to Corrosion Fatigue Crack Propagation in Aluminum Alloy 2224-T3511," Ph.D. dissertation, University of Utah, 1995
33. M.C.Y. Niu, *Airframe Structural Design*, Conmilit Press, Ltd., 1988
34. J.F. Knott, *Fundamentals of Fracture Mechanics*, Butterworths, London, 1973
35. A.S. Kobayashi, Ed., "Experimental Methods in Fracture Mechanics," SESA Monograph 1, Society for Experimental Stress Analysis, 1973
36. A.S. Kobayashi, Ed., "Experimental Methods in Fracture Mechanics, 2," SESA Monograph 2, Society for Experimental Stress Analysis, Westport, Conn., 1975
37. G.C. Sih, H.C. Van Elst, and D. Broek, *Prospects in Fracture Mechanics*, Noordhoff International, Leyden, The Netherlands, 1974
38. G.P. Cherepanov, *Mechanics of Brittle Fracture*, McGraw-Hill, 1979
39. J.J. Burke and V. Weiss, *Applications of Fracture Mechanics to Design*, Plenum Press, 1979
40. L.H. Larsson, Ed., *Advances in Elastic-Plastic Fracture Mechanics*, Applied Science Publishers, London, 1980
41. K. Hellan, *Introduction to Fracture Mechanics*, McGraw-Hill, 1984
42. D.R.J. Owen and A.J. Fawkes, *Engineering Fracture Mechanics: Numerical Methods and Applications*, Pineridge Press, Ltd., 1983
43. A.P. Parker, *The Mechanics of Fracture and Fatigue*, E. & F.N. Spoon Ltd., London, 1981
44. M.F. Kanninen and C.H. Popelar, *Advanced Fracture Mechanics*, Oxford University Press, 1985
45. A.G. Atkins and Y.W. Mai, *Elastic and Plastic Fracture*, John Wiley & Sons, Inc., 1985
46. D. Broek, *The Practical Use of Fracture Mechanics*, Kluwer Academic Publishers, Boston, 1988
47. D. Taylor, *Fatigue Thresholds*, Butterworths, London, 1989
48. R.N.L. Smith, *Basic Fracture Mechanics*, Butterworth-Heinemann Ltd., Oxford, 1991
49. H. Tada, P. Paris, and G. Irwin, *The Stress Analysis of Cracks Handbook*, 2nd ed., Paris Productions, Inc., St. Louis, Mo., 1985
50. *Fracture Mechanics Design Handbook*, AD-A038 457, U.S. Army Missile Command, Redstone Arsenal, Alabama
51. G.C. Shih, *Handbook of Stress-Intensity Factors*, Institute of Fracture and Solid Mechanics, Lehigh University, 1973
52. "Flaw Evaluation Procedures: ASME Section XI," Report EPRI NP-719-SR, Electric Power Research Institute, Aug 1978
53. *Rules for Inservice Inspection of Nuclear Plant Components*, ASME Boiler and Pressure Vessel Code, Section XI, 1996
54. V. Kumar, M.D. German, and C.F. Shih, "An Engineering Approach for Elastic-Plastic Fracture Analysis," Report EPRI NP-1931, Project 1237-1, General Electric Co., R&D Center, Schenectady, N.Y., July 1981
55. V. Kumar, W.W. Wilkening, W.R. Andrews, M.D. German, H.G. DeLorenzi, and D.F. Mowbray, "Estimation Technique for the Prediction of Elastic-Plastic Fracture of Structural Components of Nuclear Systems," Report EPRI RP1237-1, General Electric Co., R&D Center, Schenectady, N.Y., 31 Jan 1982
56. V. Kumar, M.D. German, W.W. Wilkening, W.R. Andrews, H.G. DeLorenzi, and D.F. Mowbray, "Advances in Elastic-Plastic Fracture Analysis," Report EPRI NP-1931, Project 1237-5, General Electric Co., R&D Center, Schenectady, N.Y., June 1987
57. V. Kumar and M.D. German, "Further Developments in Elastic-Plastic Fracture Analysis of Through-Wall Flaws and Surface Flaws in Cylinders," Report 87-SED-017, Project 1237-5, General Electric Co., R&D Center, Schenectady, N.Y., June 1987
58. *Ductile Fracture Handbook*, Novetech Corp., Gaithersburg, Md.
59. R. Gangloff, Effects of Defects in Metals, *Fracture Mechanics and Fatigue: A Modern View with Applications to Engineering Design*, Section 16, Union College Summer Technical Institute, July 1992
60. D.E. Pettit and W.E. Krupp, The Role of Nondestructive Testing in Fracture Mechanics Applications, *Fracture Prevention and Control*, American Society for Metals, 1973, p 53–69
61. G.R. Irwin, Fracture Mechanics, *Mechanical Testing*, Vol 8, *ASM Handbook*, American Society for Metals, 1985, p 439–464
62. G.E. Dieter, *Mechanical Metallurgy*, 3rd ed., McGraw-Hill, Inc., 1986
63. R.E. Reed-Hill and R. Abbaschian, *Physical Metallurgy Principles*, 3rd ed., PWS-Kent Publishing Co., 1992
64. A. Saxena and S.J. Hudak, Jr., Review and Extension of Compliance Information for Common Crack Growth Specimens, *Int. J. Fract.*, Vol 14 (No. 5), Oct 1978, p 453–468
65. W.E. Krupp and D.W. Hoeppner, Fracture Mechanics Applications in Materials Selection, Fabrication Sequencing and Inspection, *J. Aircr.*, Vol 10 (No. 11), 1973, p 682–688
66. U.G. Goranson, J. Hall, J.R. Maclin, and R.T. Watanabe, Long-Life Damage Tolerant Jet Transport Structures, *Design of Fatigue and Fracture Resistant Structures*, ASTM STP 761, 1982, p 47–90
67. J. Hall and U.G. Goranson, "Principles of Achieving Damage Tolerance with Flexible Maintenance Programs for New and Aging Aircraft," paper presented at the 13th Congress of the International Council of the Aeronautical Sciences/AIAA Aircraft Systems and Technologies Conferences (Seattle, WA), Aug 1982
68. U.G. Goranson, Elements of Structural Integrity Assurance, *Fatigue*, Vol 16 (No. 1), 1994, p 43–65
69. T. Swift, Damage Tolerance Capability, *Fatigue*, Vol 16 (No. 1), 1994, p 75–94
70. J.W. Lincoln, Challenges for the Aircraft Structural Integrity Program, *FAA/NASA International Symposium on Advanced Structural Integrity Methods for Airframe Durability and Damage Tolerance*, NASA Conference Publication 3274, Part 1, 1994, p 409–424
71. *Metallic Materials and Elements for Flight Vehicle Structures*, MIL-HDBK-5F, U.S. Govt. Printing Office, Washington, D.C., 1990
72. R.A. Ainsworth, I. Milne, A.R. Dowling, and A.T. Stewart, Assessing the Integrity of Structures Containing Defects by the Failure Assessment Diagram Approach of the CEGB, *Fatigue and Fracture Assessment by Analysis and Testing*, Vol 103, ASME PVP, 1986, p 123–129
73. R.L. Jones, T.U. Marston, S.W. Tagart, D.M. Norris, and R.E. Nickell, Applications of Fatigue and Fracture Damage Tolerant Design Concepts in the Nuclear Power Industry, *Design of Fatigue and Fracture Resistant Structures*, ASTM STP 761, 1982, p 28–46
74. R.C. Rice and C.E. Smith, Fatigue and Fracture Tolerance Evaluation of Tall Loran Tower Eyebolts, *Design of Fatigue and Fracture Resistant Structures*, ASTM STP 761, 1982, p 424–444
75. S.H. Smith, N.D. Ghadiali, A. Zahoor, and M.R. Wilson, Fracture Tolerance Analysis of the Solid Rocket Booster Servo-Actuator for the Space Shuttle, *Design of Fatigue and Fracture Resistant Structures*, ASTM STP 761, 1982, p 445–476
76. D.J. Wulpi, *Understanding How Components Fail*, American Society for Metals, 1985
77. C.R. Brooks and A. Choudhury, *Metallurgical Failure Analysis*, McGraw-Hill, Inc., 1993
78. H.S. Reemsnyder, "Structural Integrity Analysis of a Heavy Rolled Structural Section," paper presented at the AWS\IIW Conference on Fitness for Purpose of Welded Structures, Key Biscayne, Florida, October 24, 1991
79. H. Reemsnyder, "Engineering Approaches to Elastic/Plastic Failure Analysis, " Bethlehem Steel Corp., 7 Nov 1994
80. H. Reemsnyder, Fatigue and Fracture of Ship Structures, *Prevention of Fracture in Structure*, Ship Structures Committee, Washington, D.C., 1997
81. J.A. Kapp, M.D. Witherell, R.J. Fujczak, T.M. Hickey, and J.J. Zalinka, "Fracture Mechanics Assessment of a Cracked 16 Inch ID Vintage 1945 Jacketed Pressure Vessel," *High Pressure Technology: Material, Design, Stress Analysis, and Applications*, Vol 148, ASME PVP, 1988, p 83–89
82. T.E. Davidson and D.P. Kendall, in *The Mechanical Behavior of Materials Under Pressure*, H.L.I.D. Pugh, Ed., Elsevier Publishing, 1970

Design for Corrosion Resistance

F. Peter Ford and Peter L. Andresen, General Electric Corporate Research and Development Center
Peter Elliott, Corrosion and Materials Consultancy, Inc.

THE ANNUAL COST OF CORROSION in the United States in 1949 (Ref 1) was 2.1% of the gross national product (GNP), or $5 billion. Subsequent analysis (Ref 2) in 1975 indicated an increase to 4.2% GNP ($70 billion), with an estimate of $126 billion by 1982. Similar values have been quoted in other countries, including the United Kingdom (Ref 3) and Taiwan (Ref 4). These figures are associated with the direct cost of replacement of the corroded structure; additional costs are associated with maintance, increased regulatory demand (e.g., environment control), and lost production.

Not only is there a direct and indirect financial burden, but corrosion also has a considerable societal impact. Crucial industries such as energy, aerospace, transportation, food, agriculture, electronics, marine, and petrochemical rely on the safety and availability of their infrastructures. In many instances, these requirements have been compromised by corrosion events that have led to loss of life, environmental pollution, and loss of power to industry. However, between 25 and 50% of the economic impact (depending on the specific industry) could have been prevented by the use of well-accepted materials selection and corrosion prevention measures. The purpose of this article is to give the reader a concept of the principles of corrosion and the basis of the various prevention measures that can be taken for the different corrosion modes. Reference may be made to numerous textbooks, handbooks, and articles for specific material/environment systems for more detailed discussions (see Ref 5–28 and the "Selected References" listed at the end of this article).

Basic Principles of Aqueous Corrosion

Aqueous corrosion phenomena are largely understood in terms of the electrochemical reactions that occur at the metal-environment interface. The metal surface can be in elemental form (e.g., copper or gold), as an alloy perhaps containing a mixture of phases, but more often the metal surface will have an oxide, compound, or salt layer between it and the environment. The role of this layer is discussed later in this section. The environment is usually water and its dissociation products H^+ (or H_3O^+) and OH^-, plus ionic impurities such as Na^+, Cl^-, or SO_4^{2-} and dissolved gases (e.g., oxygen, carbon dioxide). The amounts of these impurities can range from parts per billion to highly concentrated conditions; even at the lowest levels, however, these impurity levels can significantly affect certain corrosion events.

Metal atom (M) "removal" from the surface occurs primarily by reactions of the general type

$$M + zH_2O \rightarrow M^{n+} \cdot zH_2O + ne^- \quad \text{(Eq 1)}$$

$$xM + yH_2O \rightarrow M_xO_y + 2yH^+ + 2ye^- \quad \text{(Eq 2)}$$

where n is the number of electrons involved in the overall oxidation of an atom of metal and z is the hydration number.

For instance, bare surface corrosion occurs by a series of individual atomic steps that may be described overall by Eq 1. In essence, a metal atom combines with z solvating water molecules (zH_2O) and goes into solution as a solvated cation ($M^{n+} \cdot zH_2O$), liberating n electrons into the underlying metal substrate. Alternatively, the metal atom may interact with the water molecules to form a metal oxide (M_xO_y) on the surface by the solid-state reaction described overall by Eq 2. Because both of these reactions involve the liberation of electrons, they are known in electrochemical terms as "oxidation" reactions that occur at "anodic" sites on the metal.

Under natural corrosion conditions the electrons liberated by these oxidation reactions will be consumed by "reduction" reactions occurring at adjacent "cathodic" sites (Fig. 1). Such reactions may be of the type

$$2H^+ + 2e^- \rightarrow 2H_{ads} \rightarrow H_2\uparrow \quad \text{(Eq 3)}$$

where, in acid solutions, hydrogen ions are reduced to form adsorbed hydrogen atoms (H^+) on the surface. These adsorbed hydrogen atoms may then either be absorbed into the metal (discussed below) or, as shown in Eq 3, combine with adjacent atoms to form hydrogen gas, which may bubble off as gas, allowing corrosion to proceed, or remain as a film that eventually slows or halts corrosion. In more neutral or alkaline solutions, the reduction reaction (because of the dissociation relationships between H_2O, H^+, and OH^-) is of the form

$$2H_2O + 2e^- \rightarrow H_2 + 2OH^- \quad \text{(Eq 4)}$$

In solutions containing dissolved oxygen, the electron consumption at the cathodic site in alkaline solutions is usually achieved by the reduction reaction

$$O_2 + 2H_2O + 4e^- \rightarrow 4OH^- \quad \text{(Eq 5a)}$$

or in acid solutions by:

$$O_2 + 4H^+ + 4e^- \rightarrow 2H_2O \quad \text{(Eq 5b)}$$

In other instances where reducible cations are present, the cathodic reaction may be:

$$Fe^{3+} + e^- \rightarrow Fe^{2+} \quad \text{(Eq 6)}$$

or

$$Cu^{2+} + e^- \rightarrow Cu^+ \quad \text{(Eq 7)}$$

Thus, in summary, the metal corrosion (or oxidation) reactions are always associated with one

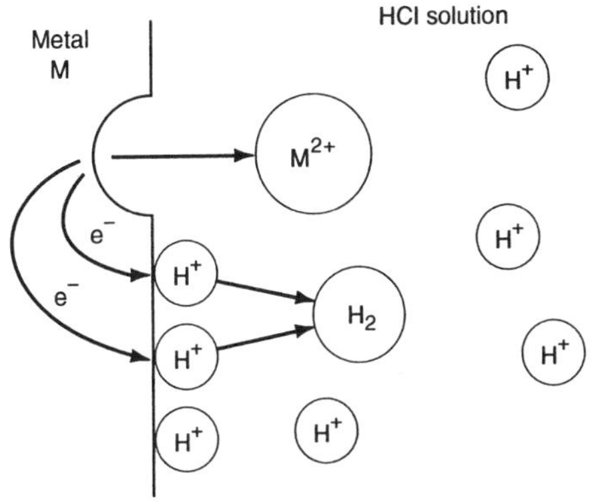

Fig. 1 Schematic representation of oxidation and reduction reactions occurring on a corroding metal surface. Source: Ref 5

or more reduction reactions. In order to conserve charge, the sum of these oxidation and reduction rates must be equal. Thus, in simplistic terms, the corrosion of, say, zinc in a deaerated acid solution would be described by Eq 1 and 3 to give:

$$Zn + 2H^+ \rightarrow Zn^{2+} + H_2 \quad \text{(Eq 8)}$$

that is, zinc dissolution with hydrogen evolution.

Because these oxidation and reduction reactions involve a transfer of charged species adjacent to the surface, an equivalent capacitor is formed for which there is an associated electrode potential across the metal/solution interface (Fig. 2). Under "equilibrium" (reversible) conditions, the oxidation and reduction reactions shown in Eq 9, which are associated with metal dissolution (the forward reaction) and metal deposition (the backward reaction), occur at equal rates, and the equilibrium potential is given by a function of the free energies of formation of the relevant species and their activities (this relationship is discussed later in this section):

$$M + zH_2O \leftrightarrow M^{n+} \cdot zH_2O + ne^- \quad \text{(Eq 9)}$$

Note that, if the metal/solution potential is moved either way from the equilibrium potential, there will be a net increase in oxidation rate (i.e., dissolution) or reduction rate (metal deposition). Because these reactions involve electron movement, the rates of reaction may be quantified in terms of current that can often be measured. Because the reaction rates occur over a specific area, the unit of practical significance is the current density (i). Faraday's law relates this current density to the amount of metal being uniformly dissolved or reduced, in unit time:

$$CR = \frac{M}{n\rho F} i_a \quad \text{(Eq 10)}$$

where CR is the corrosion rate (cm/s), i_a is the oxidation current density (amps/cm^2), n is the number of electrons (or "equivalents") involved (e.g., in the oxidation reaction of Eq 1), ρ is the density of corroding metal (g/cm^3), F is the Faraday constant (96,500 coulombs/"equivalent"), and M is the atomic weight of corroding metal (g).

As a point of reference, an anodic current density of 1 μA/cm^2 corresponds to approximately 0.4 to 0.5 mils/year (mpy) metal removal rate for many ferrous, aluminum, copper, and nickel-base alloys.

The relationships between the rates of oxidation and reduction reactions shown in Eq 9 and the electrode potential (known as "polarization" relationships) are shown schematically in Fig. 3. The most fundamental relationship between the oxidation reaction rate for Eq 9 and the overpotential (η_a), which is defined as the difference between the electrode potential and the equilibrium potential, is given by:

$$\eta_a = \beta_a \log \frac{i_a}{i_o} \quad \text{(Eq 11)}$$

where i_o is known as the "exchange current density" and β_a is the Tafel coefficient for the anodic reaction, which is the slope of η_a vs. log i_a. The oxidation (or reduction) rate may rise exponentially with increasing overpotential (Fig. 3a). This increase is not endless, however, for eventually the reaction rate will become controlled by the diffusion of reactants to, or the reaction products from, the metal/solution interface region. This mass transfer will be dependent on convection considerations as well as on the bulk concentration of the relevant dissolved species, and it will be evidenced by a reaction rate that is not directly dependent on potential (Fig. 3b). In systems where an oxide can form on the reacting surface in certain potential ranges (e.g., Eq 2), the dissolution rate (i_a) may abruptly drop due to "passivation" to an extremely low value (Fig. 3c); however, at higher potentials the passive film may lose its protective nature because of its dissolution (or "transpassivity") or to localized degradation due to aggressive anions (e.g., pitting in saline solutions). Thus, the polarization curves may change with different alloy/environment conditions, and, as will be seen later, this behavior will give the possibility of a large combination of corrosion responses.

Bearing in mind that several reactions may be possible at a given metal/solution interface, a composite reaction rate diagram can be developed. For instance, the two-reaction system shown in Fig. 4 illustrates the oxidation reaction associated with the corrosion of a metal (Eq 1) and the corresponding reduction of H^+ ions in an acid environment (Eq 3). Based on this reaction rate construction (known as an Evans diagram), and bearing in mind that the net oxidation and net reduction reaction rates must be equal, the corrosion current density (i_{corr}) can then be predicted, and this result can be converted to uniform dimensional loss by Eq 10.

As indicated in Fig. 4 for a naturally corroding system, the potential between the metal and solution will have a value known as the "corrosion potential" (E_{corr}), defined by the balance of the net oxidation and reduction reaction rates. The relative difference between E_{corr} and the equilibrium potentials for the various possible reactions is of importance in determining potential corro-

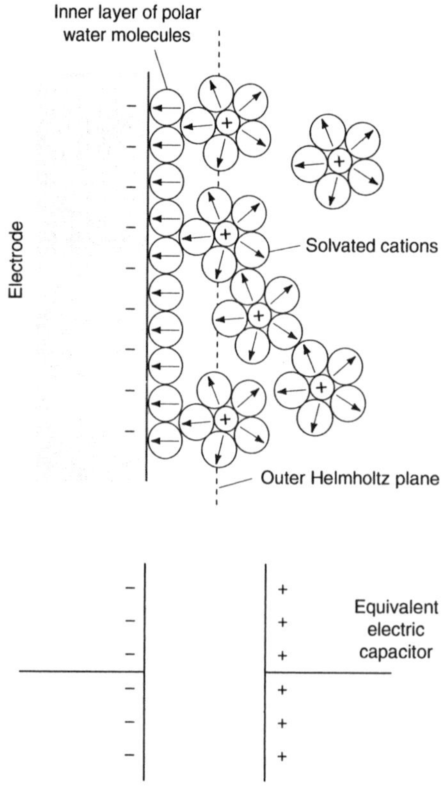

Fig. 2 Schematic electrode surface structure with equivalent electric capacitor. The circles containing arrows represent water molecules; the direction of arrows indicates the orientation of the hydrogen atoms in the molecule. Source: Ref 5

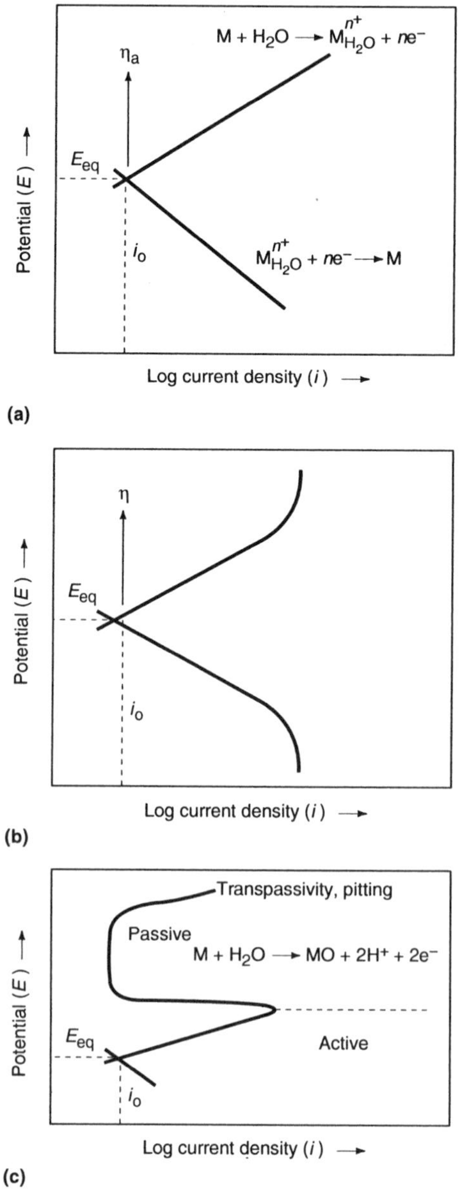

Fig. 3 Schematic Evans diagrams showing polarization relationships between electrode potential (E) and current density (i) indicating: (a) activation control, (b) onset of mass-transport (diffusion) control at higher overpotentials, and (c) active-to-passive transitions due to formation of surface oxide, followed by breakdown or dissolution of oxide at higher overpotentials. E_{eq}, equilibrium potential

sion mitigation actions. Thus, a brief review of the equilibrium thermodynamics at the metal/solution interface is given below, because it is relevant to the arguments that are given in a later section of this article, "Forms of Corrosion and General Mitigation Approaches."

When computing the metal/solution potential associated with the various half-cell reactions (e.g., Eq 1–7), it is conventional to formulate the reactions as reduction processes, for example:

$$Ni^{2+} + 2e^{-} \rightarrow Ni \qquad \text{(Eq 12)}$$

The "reduction" potential (E) associated with the metal/solution interface (Fig. 2) will be related to the free-energy changes (ΔG) in the reaction shown in Eq 12:

$$E = -\frac{\Delta G}{n\text{F}} \qquad \text{(Eq 13)}$$

Since ΔG of a reaction can be expressed as RT ln (products/reactants) and a ΔG^0 (and, therefore, E^0) is used to define ΔG under standard conditions, an overall relationship, known as the Nernst equation, can be formulated for the Ni^{2+}/Ni reaction:

$$E = E^0 + \frac{\text{RT}}{n\text{F}} \ln (a_{Ni^{2+}}) = E^0 + 0.03 \log (a_{Ni^{2+}}) \qquad \text{(Eq 14)}$$

where E^0 is the "standard potential" for Ni^{2+}/Ni, and $a_{Ni^{2+}}$ is the chemical activity of Ni^{2+} cations in the solution. Standard potentials are those realized when the relevant species are in their "standard" states of unit activity (for dissolved species) or unit fugacity (for gaseous species) usually at 25 °C.

These metal/solution potentials cannot be directly measured experimentally (because measurement automatically requires a second metal/solution half-cell electrode); therefore, they are defined in relation to the standard H^+/H_2 half-cell reaction (often measured on inert platinum):

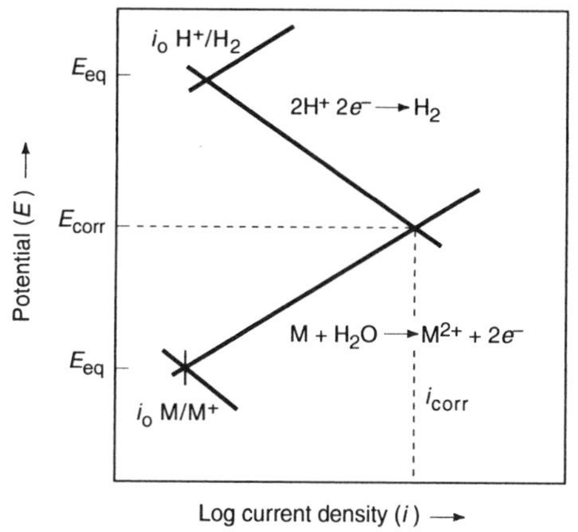

Fig. 4 Schematic Evans diagram showing oxidation and reduction reactions associated with, for instance, $M \rightarrow M^+$ and $H^+ \rightarrow H_2$ reactions, illustrating the fact that when the relevant oxidation and reduction reaction rates are in dynamic equilibrium, the metal/solution interface will attain the measured corrosion potential (E_{corr}) and the metal will corrode at a rate corresponding to i_{corr}

$$2H^+_{\text{(unit activity)}} + 2e^- \leftrightarrow H_{2\,\text{(unit fugacity)}} \qquad \text{(Eq 15)}$$

whose standard potential is, by convention, defined as zero at all temperatures. Thus, all metal/solution potentials are measured with respect to this standard hydrogen electrode (SHE) or to an equivalent, well-defined half-cell reaction (reference electrode) that has been calibrated against the SHE. Table 1 lists the standard (reduction) potentials for several metals and nonmetals with respect to their dissolved form. This table illustrates (bearing in mind that the potential is related to free energy changes) that the stability of the elements with respect to their dissolved cations ranges from the very stable noble metals (gold, platinum) to the very unstable elements (sodium, potassium). However, it is important to recognize that Table 1 does not *necessarily* give the relative ranking of metal corrosion resistance, because this ranking will depend on the specifics of the environment, the extent of elemental alloying, and the formation of surface oxides. These specific factors are covered in the section "Forms of Corrosion and General Mitigation Approaches" in this article.

Thus, the equilibrium potential for, say, a Ni^{2+}/Ni half-cell reaction shown in Eq 12 can be defined as:

$$E = -0.25 + 0.03 \log (a_{Ni^{2+}}) \quad \text{volts (SHE)} \qquad \text{(Eq 16)}$$

This particular equilibrium potential relationship is independent of the activity of H^+ ions (i.e., the pH). However, this is not the situation for other technically important reactions. For instance, the formation of an oxide by a solid-state reaction (e.g., Eq 2) will be dependent on the pH of the solution, e.g.:

$$Ni + H_2O \leftrightarrow NiO + 2H^+ + 2e^- \qquad \text{(Eq 17)}$$

and the equilibrium potential will be defined by $E = +0.11 + 0.03 \log (a_{H^+})^2$, and because pH = –log (a_{H^+}):

$$E = +0.11 - 0.06\,\text{pH} \qquad \text{(Eq 18)}$$

Alternately, there will be an equilibrium between the activity of the solvated metal cations ($Ni^{2+} \cdot zH_2O$) and the precipitation of an oxide or hydroxide in the solution, due merely to the fact that the solubility of the oxide/hydroxide has been exceeded. In such instances, depending on the specific solid-phase components being formed, the equilibrium dissolved metal activity may be independent of potential and be solely dependent on pH. For example, for the Ni^{2+}/NiO equilibrium:

$$Ni^{2+} + H_2O \leftrightarrow NiO + 2H^+ \qquad \text{(Eq 19)}$$

the equilibrium relationship will be given by:

$$\log (a_{Ni^{2+}}) = 12.18 - 2\,\text{pH} \qquad \text{(Eq 20)}$$

Note that the dependence of a reaction on pH is explicitly shown by the presence of H^+ or OH^- in the reaction, and its dependence on potential is shown by the presence of electrons.

Based on such various equilibria considerations, it is possible to define the relative stabilities of the metal, its dissolved species (e.g., Ni^{2+}, $HNiO_2^-$), solid-state oxides (NiO, Ni_3O_4, Ni_2O_3, NiO_2) and hydroxides $Ni(OH)_2$ as a function of potential and pH. Such a composite potential/pH diagram, known as a Pourbaix diagram, is shown in Fig. 5 for Ni/H_2O. Superimposed on these stability regions are the equilibrium relationships for the H_2O/H_2 (Eq 4) and O_2/H_2O reactions (Eq 5), illustrated by the lines marked "a" and "b," respectively. The insertion of these latter relationships onto the Pourbaix diagram are important for corrosion assessments, as described below.

If we arbitrarily assume that "corrosion" is defined by dissolution in solutions containing an activity of dissolved metal cations of 10^{-6} moles/L (i.e., ~0.06 mg/L (ppm) for iron, copper, or zinc), then the Pourbaix diagram can be divided into regions denoted as "corrosion immunity" (where, thermodynamically, the metal cannot undergo net dissolution), "corrosion" (where the metal will be oxidized to a soluble species), and "passivation." In this latter region, an oxide can form at the metal/solution surface, and, depending on its properties, it *may* confer protection against corrosion. Regions of "corrosion immunity," "corrosion," and "passivation" for Ni/H_2O are shown in Fig. 5.

There is, as was pointed out earlier, a further criterion for corrosion to occur, and that is the fact

Table 1 Standard electromotive force potentials (reduction potentials)

Reaction	Standard potential, E^0 (volts vs. SHE)
Noble (more cathodic)	
$Au^{3+} + 3e^- = Au$	+1.498
$Cl_2 + 2e^- = 2Cl^-$	+1.358
$O_2 + 4H^+ + 4e^- = 2H_2O$ (pH 0)	+1.299
$Pt^{2+} + 3e^- = Pt$	+1.118
$NO_3^- + 4H^+ + 3e^- = NO + 2H_2O$	+0.957
$O_2 + 2H_2O + 4e^- = 4OH^-$ (pH 7)	+0.82
$Ag^+ + e^- = Ag$	+0.799
$Hg_2^{2+} + 2e^- = 2Hg$	+0.799
$Fe^{3+} + e^- = Fe^{2+}$	+0.771
$O_2 + 2H_2O + 4e^- = 4OH^-$ (pH 14)	+0.401
$Cu^{2+} + 2e^- = Cu$	+0.342
$Sn^{4+} + 2e^- = Sn^{2+}$	+0.15
$2H^+ + 2e^- = H_2$	0.000
$Pb^{2+} + 2e^- = Pb$	–0.126
$Sn^{2+} + 2e^- = Sn$	–0.138
$Ni^{2+} + 2e^- = Ni$	–0.250
$Co^{2+} + 2e^- = Co$	–0.277
$Cd^{2+} + 2e^- = Cd$	–0.403
$2H_2O + 2e^- = H_2 + 2OH^-$ (pH 7)	–0.413
$Fe^{2+} + 2e^- = Fe$	–0.447
$Ce^{3+} + 3e^- = Cr$	–0.744
$Zn^{2+} + 2e^- = Zn$	–0.762
$2H_2O + 2e^- = H_2 + 2OH^-$ (pH 14)	–0.828
$Al^{3+} + 3e^- = Al$	–1.662
$Mg^{2+} + 2e^- = Mg$	–2.372
$Na^+ + e^- = Na$	–2.71
$K^+ + e^- = K$	–2.931
Active (more anodic)	

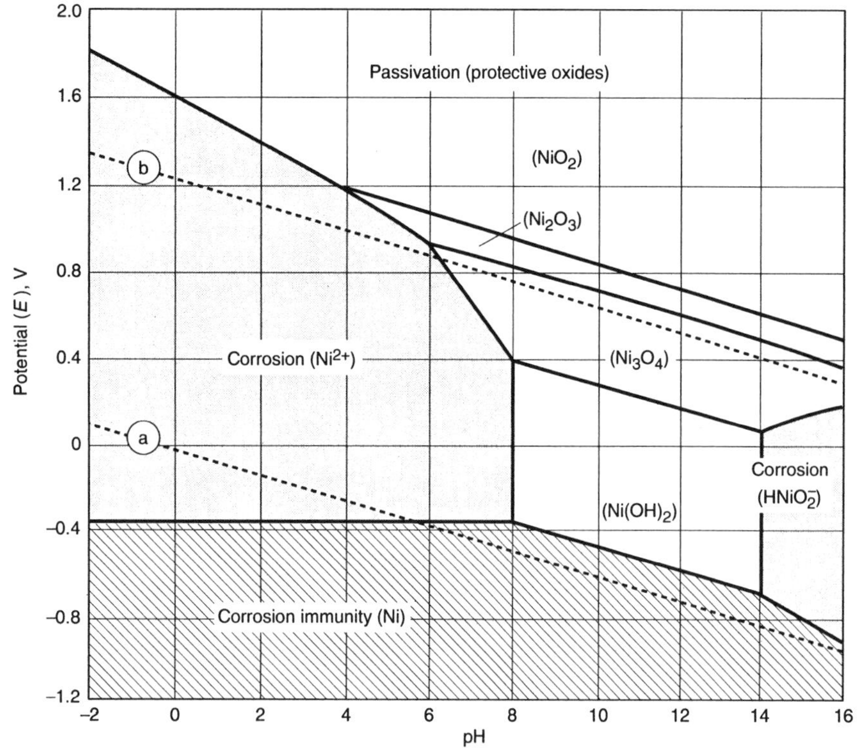

Fig. 5 Pourbaix diagram for the nickel/water system, indicating the *E*/pH regions of thermodynamic stability of various dissolved and solid species and the regions of "corrosion immunity," "passivation," and "active corrosion." Dashed lines show the equilibrium potentials for the (a) H_2/H_2O and (b) O_2/H_2O systems

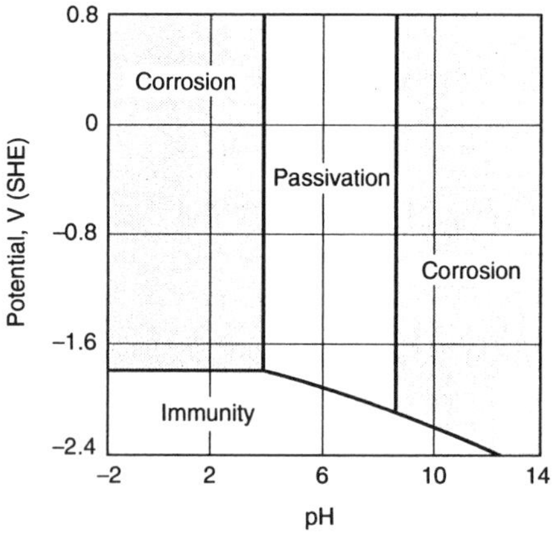

Fig. 6 Pourbaix diagram for aluminum/water

that a corresponding reduction reaction must occur. In water, these reactions are often associated with water reduction (Eq 4) and oxygen reduction (Eq 5). Thus, referring to Fig. 5, corrosion of nickel in deaerated acids will be confined thermodynamically to the *E*/pH "space" between line "a" and the line described by Eq 12 and 16 (at about –0.35 V_{SHE} in Fig. 5), whereas in aerated acids the corrosion region will be extended up to line "b." Thus, the Pourbaix diagrams are important tools for defining the *E*/pH regimes where corrosion *might* occur and the regions where, thermodynamically, corrosion *cannot* occur. It is important to recognize, however, that Pourbaix diagrams do not give information on the rates of corrosion.

Forms of Corrosion and General Mitigation Approaches

From the previous section, it is seen that there are four conjoint requirements for corrosion of metals in aqueous environments to occur. They are:

- A metal oxidation (or "corrosion") reaction is thermodynamically possible under the given potential and pH conditions (Fig. 5).
- A corresponding reduction reaction is thermodynamically possible given the presence of various reducible species (Eq 3–7) under given potential and pH conditions.
- There is an electronic path in the metal between the anodic and cathodic sites where the oxidation and reduction reactions are occurring (compare with Fig. 1).
- There is an ionic path in the environment between the anodic and cathodic sites.

With this background, the specific forms of corrosion will now be discussed.

General Corrosion

General corrosion occurs more or less uniformly over the whole structure area. This generally implies that the surface is fairly uniform metallurgically, and that the environment is relatively constant throughout the system. General corrosion often results from atmospheric exposure (especially polluted industrial environments), marine exposure, or exposure in chemical plants involving relatively concentrated aqueous environments. The mechanisms of degradation are reasonably well understood, and the problem usually occurs because of poor materials choice or changes in process chemistry.

Alloy Choice and Design. There is no question that, from a thermodynamic viewpoint, materials can often be selected that can withstand even highly aggressive conditions. For instance, the noble metals (e.g., gold, silver, and platinum) are resistant to most concentrated acids. These metals, however, are hardly materials of widespread use, and for most engineering materials, general corrosion protection is achieved by the presence of a surface film or oxide that has very limited solubility in the environment. Thus, material choices have to be based on the specific environment under consideration.

Alloying is generally needed for engineering-strength purposes, and knowledge of corrosion mechanisms can facilitate the design of alloys that also have adequate corrosion resistance. For instance, based solely on examination of the Pourbaix diagram, nickel and its alloys are normally a good choice for use in alkaline environments, given the fact that this metal has a relatively small stability region for the dissolved $HNiO_2^-$ anion in high-pH solutions; hence, the use of nickel electrodes in alkaline batteries (Fig. 5). Conversely, aluminum and its alloys would *not* be good candidates for alkaline environments, given the relatively wide stability range for the AlO_2^- anion at pH values greater than ~8.5 (Fig. 6).

The general corrosion resistance of many alloys is increased by elements that enhance the passivation region on the Pourbaix diagram due to the formation of mixed surface oxides:

Base metal	Alloying element	Alloy
Fe	Cr	400-series ferritic stainless steels (11–23% Cr)
Fe	Cr, Ni	300-series austenitic stainless steels (16–26% Cr, 6–22% Ni)
Ni	Cr	Alloy 600 (16% Cr, 0.2% Cu, 8% Fe) Nimonic 75 (20% Cr)
Ni	Mo, Cr	Hastelloy B (28% Mo, 1.5% Fe, 1.5% Cr) Hastelloy C (17% Mo, 5% Fe, 15% Cr), Hastelloy C-22 Hastelloy C-276 (16% Mo, 16% Fe, 15% Cr)
Ni	Si	Hastelloy D (10% Si, 3% Cu)
Ti	Mo, Ta, Al	3–6% Al, 2.5–4% V
Zn	Al	35–50% Al
Co	Cr	Vitallium and Stellite orthopedic inserts

The increase in the corrosion resistance associated with alloying may be indicated by the Pourbaix diagrams for the major elements in the alloy. For instance, the superposition of the Cr_2O_3 stability domains onto the Pourbaix diagram for

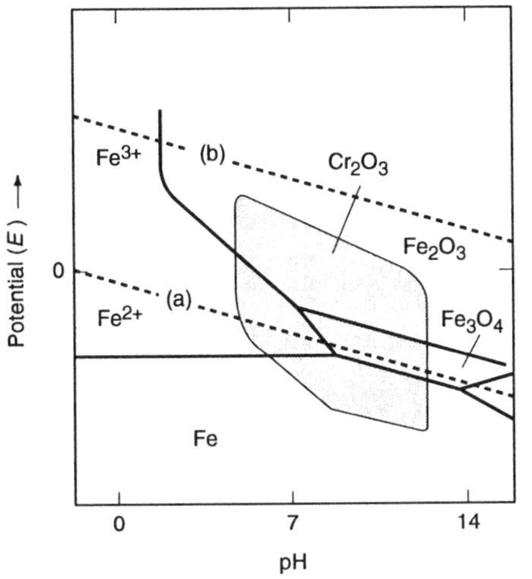

Fig. 7 Superposition of the Cr_2O_3 stability regions (shaded) on the iron/water Pourbaix diagram, to illustrate the *E*/pH regions where chromium alloying might confer added corrosion resistance. Dashed lines show the equilibrium potentials for (a) H_2/H_2O and (b) O_2/H_2O

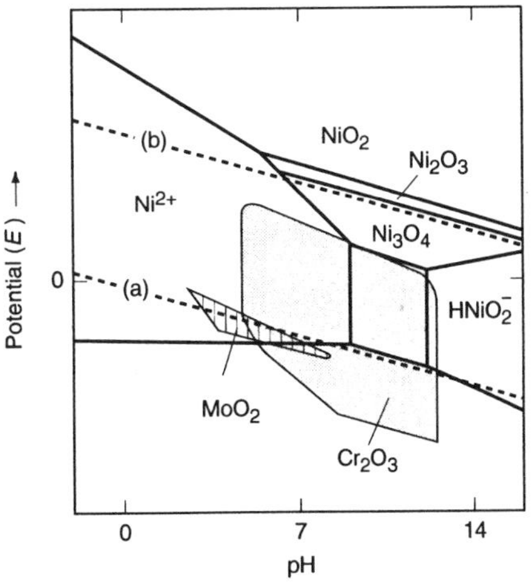

Fig. 8 Superposition of the MoO_4 and Cr_2O_3 stability regions (shaded) on the nickel/water Pourbaix diagram, to illustrate the thermodynamic reason for the improved corrosion resistance of various nickel alloys in neutral and acid environments

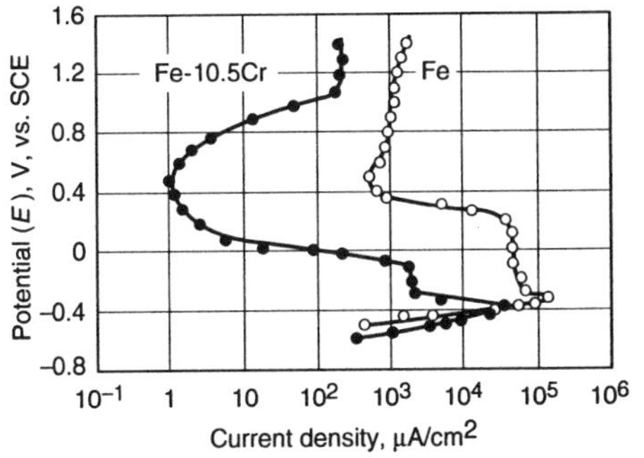

Fig. 9 Evans diagram showing the potentiostatic anodic polarization of pure iron and Fe-10.5Cr alloy in H_2SO_4. Compare with Fig. 3(c). Source: Ref 26

iron-water (Fig. 7) or the MoO_4 and Cr_2O_3 stability domains onto the Pourbaix diagram for nickel-water (Fig. 8) indicate that, in both instances, the "passivation" domains are reinforced in neutral/alkaline environments and are increased in acidic environments. This behavior gives some rationale for the corrosion resistance improvement observed in, for instance, the ferritic and austenitic stainless steels and in the Nichrome and Hastelloy types of nickel-base alloys.

It should be emphasized again that such material-design analysis based on thermodynamic considerations does not give a direct prediction of the corrosion rates. However, when considering the anodic polarization curves for iron and iron-chromium alloys in deaerated sulfuric acid (Fig. 9), it is not surprising that, at low overpotentials, the dissolution rate increases with increasing polarization (compare with the schematic Evans diagram, Fig. 3a), followed by a decrease at a potential associated with the formation of a protective "passivating" film (Fig. 3c). Moreover, the marked decrease in corrosion rate due to the chromium alloying and the resultant formation of a low-solubility iron-chromium spinel oxide is understandable. Similar polarization curves for nickel-base alloys in deaerated sulfuric acid are shown in Fig. 10 to illustrate the combined effects of different alloying additions. As would be expected from Fig. 8, molybdenum has only a mild effect on the passivation characteristics in acidic conditions and, hence, alloying nickel primarily with molybdenum in Hastelloy B does not significantly reduce the dissolution kinetics under these particular environmental conditions. However, Fig. 8 also indicates that the passivation domain for nickel is considerably increased in acidic environments by further alloying with chromium, and the effect of this observation is reflected in Fig. 10 by the vastly reduced oxidation rates for the chromium-containing Hastelloy C family of alloys. Thus, under deaerated acid conditions, where the corrosion potential will be about −0.250 V (versus the standard calomel electrode, or SCE) in Fig. 10, there is little difference in corrosion current densities (~10 μA/cm^2) between the three alloys. Under aerated or oxidizing acid conditions, however, when the corrosion potential will be at more positive potentials, there is a considerable benefit in choosing the chromium-containing (and more expensive) Hastelloy C grades. The combination of the high molybdenum and chromium alloying contents also confers enhanced corrosion resistance in environments other than oxidizing acids. For instance, the increased general and pitting resistance of the molybdenum- and chromium-containing nickel-base alloys allow their satisfactory long-term use in desalination plant and seawater heat exchangers.

The increase in corrosion resistance due to alloying elements need not, however, necessarily be confined to increasing the passivation region on the Pourbaix diagram. This behavior is illustrated by the copper-base alloys, where the corrosion resistance in aqueous environments at room temperature is primarily due to the cuprous oxide (Cu_2O) film. This oxide is a *p*-type semiconductor and, therefore, its growth by an oxidation reaction similar to Eq 2 depends on the transport of copper cations and electrons through the oxide.

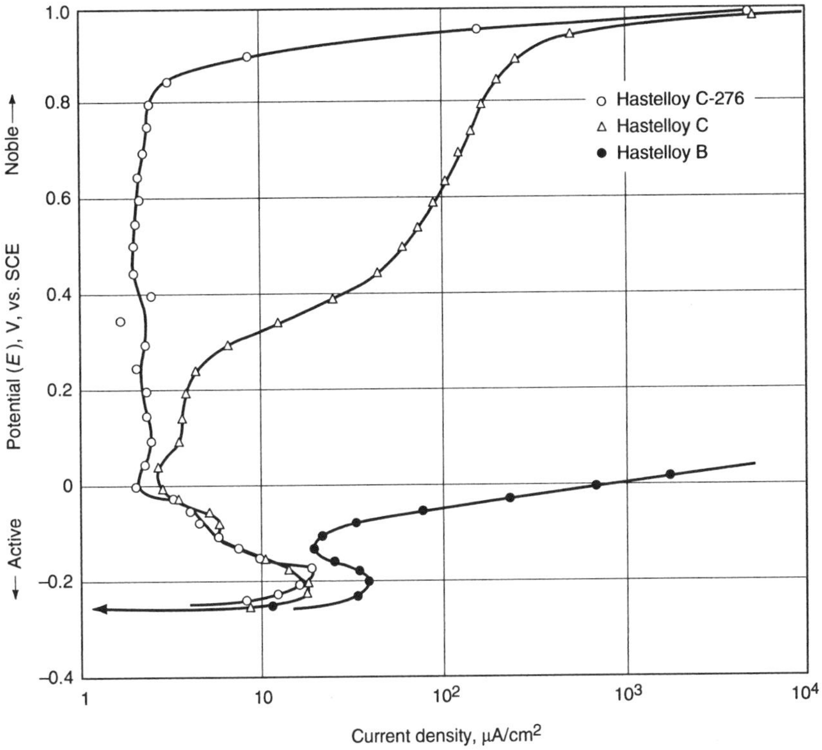

Fig. 10 Evans diagram showing the comparison of potentiostatic anodic polarization of nickel alloys in H_2SO_4 at ambient temperature. Source: Ref 5, with courtesy of F.G. Hodge, Haynes International

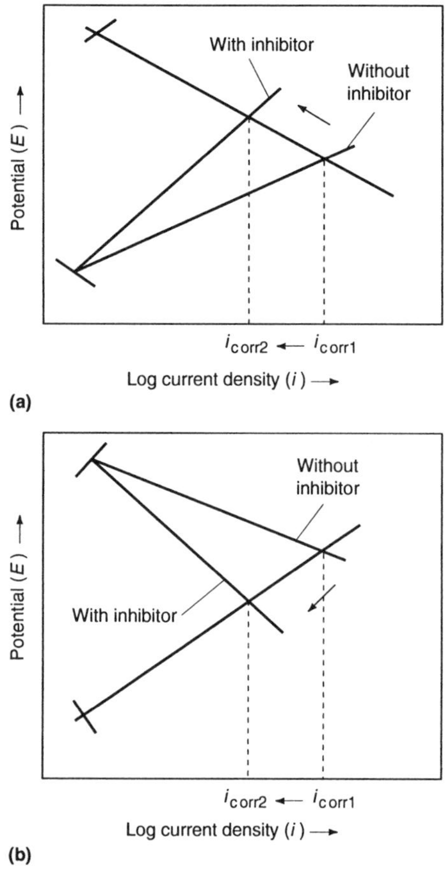

Fig. 11 Schematic Evans diagrams similar to that in Fig. 4, illustrating (a) anodic inhibition and (b) cathodic inhibition where changes in the oxidation or reduction reaction kinetics alter the corrosion potential (E_{corr}) and corrosion rate (i_{corr})

Table 2 Organic coating classifications and characteristics

Resin type	Estimated chemical and weather resistance: Acid	Alkali	Solvent	Water	Weather	Comments
One-component systems						
Alkyd (oil base)	F	P	P	F	VG	Excellent adhesion to poorly prepared surfaces. Adequate for mild chemical fumes. Not chemically resistant.
Modified alkyd						
Silicon	F	F	F	F	VG	Improved durability, gloss retention, heat resistance.
Amino	F	F	F	G	VG	High humidity resistance.
Phenolic	F	F	F	VG	VG	Suitable for immersion service.
Epoxy ester	F	F	P	F–G	P–F	Similar to alkyds. Somewhat better chemical resistance. More expensive.
Vinyl	VG	VG	P	VG	G	Widely used industrial coating. Easily recoated. Low toxicity, tasteless. High volatiles content—possible fire hazard.
Chlorinated rubber	G	G	P	G	VG	Good vapor barrier. Poor resistance to sunlight.
Acrylics (solvent base)	F	G	G	G	VG	Automotive topcoats.
Two-component systems						
Phenolic	F	F	G	VG	VG	Immersion service.
Polyamide	F	VG	VG	P	G	Tough, flexible, abrasion resistant. Difficult to topcoat.
Coal tar epoxy	VG	VG	G	VG	G	Can be applied without primer. Very adherent. Excellent chemical resistance.
Polyester urethane	F–G	G	VG	F–G	VG	Can be designed for wide range of service depending on modifying groups.

VG, very good; G, good; F, fair; P, poor

Thus, alloying elements that form divalent or trivalent cations will reduce the ionic and electronic conductivity of the oxide and decrease the oxidation kinetics. This result is the fundamental basis for the development of aluminum bronze (5–12% Al), brasses (30% Zn), cupronickels (10–30% Ni), and the mixed brasses and bronzes (admiralty and naval brass, Cu-Zn-Sn; aluminum-brass, Cu-Zn-Al; phosphor bronze, Cu-Sn-P).

Inhibitors. Mitigation of general corrosion by adding inhibitors to the aqueous environment relies on an alteration of the oxidation or reduction kinetics and/or the creation of an adsorbed film between the metal and the corrosive environment. This approach is practically limited to closed environment systems that have good circulation, so that an adequate and controlled concentration of inhibitor is ensured. Such conditions can be met, for instance, in industrial cooling systems, oil production, the refining industry, and acid pickling of steel components.

In their simplest form, inhibitors may be large organic molecules, such as ureas, mercaptans, pyridenes, and benzoates, which chemisorb on the metal surface and thereby hinder the access of the reactants to the corrosion process. In many instances, however, effective inhibition of the corrosion process is achieved by the use of either "anodic" inhibitors or "cathodic" inhibitors that quite specifically decrease the kinetics of the oxidation and reduction processes, respectively, and thereby (Fig. 11) decrease the corrosion current density.

Examples of anodic inhibitors include chromates, nitrites, molybdates, and phosphates. For instance, chromates derive their effectiveness in protecting ferrous and nonferrous metals by forming protective mixed oxides at the anodic sites. It is important to recognize, however, that because these inhibitors rely on the creation of a thermodynamically stable oxide or compound, their use might be dependent on the presence or absence of additional species. For instance, nitrites require deaerated conditions, or else the nitrite is oxidized to nitrate at the more positive potentials in aerated solutions. Conversely, molybdates, phosphates, and borates all require the presence of an oxidant, usually oxygen, in order to ensure that the metal/solution interface is at a potential where the protective film can form. Furthermore, the critical concentration of anodic inhibitor that is required may depend on the presence of other corrosive anions (such as chloride) that may degrade the passivity of the protective film (discussed below). If this critical inhibitor concentration is not maintained, then excessive localized corrosion attack can occur. Moreover, because the efficiency of this type of inhibitor depends on a specific type of protective film, the choice of a given inhibitor might be good for one alloy system but not for another. For instance, nitrites can be extremely effective for mitigating general corrosion in low-carbon steels, but they are ineffective on zinc and are deleterious for copper and for lead-tin solders. Thus, it is common to have quite complicated inhibitor mixtures for mixed alloy systems.

Thus, "anodic" inhibitors can be classified as dangerous if there is not continual monitoring of water composition (below a minimal concentration of inhibitor, the corrosion rate is dramatically increased), whereas "cathodic" inhibitors are generally regarded as safer, if somewhat less powerful. An obvious method of inhibiting the reduction reactions occurring at the cathodic sites is to lower the activity of the reactant. Thus, sulfites and hydrazine are both commonly used cathodic inhibitors in industrial boilers because they remove oxygen (Eq 5). For example:

$$N_2H_4 + O_2 \rightarrow 2H_2O + N_2 \quad \text{(Eq 21)}$$

Similarly, excessive corrosion in acid systems (e.g., pickling baths) can be controlled by the use of additives such as arsenic that poison the adsorbed hydrogen atom recombination step in Eq 3 and hence lower the overall reduction reaction rate (e.g., Fig. 11b).

Inhibitors may also rely on the alkalinity produced at cathodic sites in aerated environments due to the reactions in Eq 5. For instance, the addition of calcium bicarbonate will cause a reaction with the hydroxide produced during the reduction reaction to precipitate calcium carbonate, which then prevents further oxygen diffusion to the cathodic site. Similar cathodic inhibition due to precipitate formation is achieved by various zinc salts, where the inhibiting efficiency is increased over that due to zinc hydroxide by com-

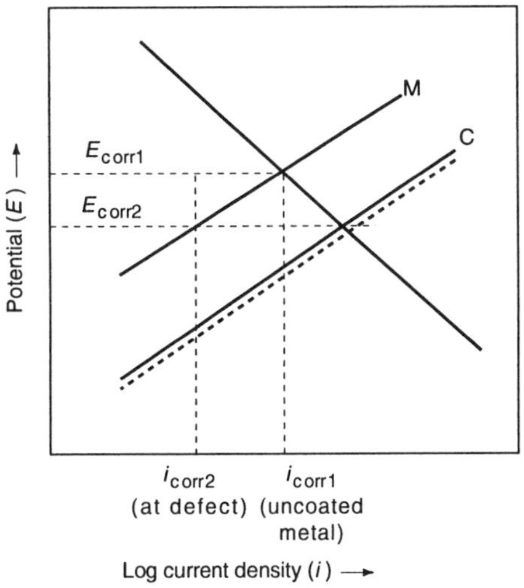

Fig. 12 Schematic Evans diagram illustrating the effect of a metallic coating (C) in galvanically protecting a metal (M) that is exposed at a defect in the coating. Mechanism relies on the effective drop in corrosion potential on the exposed metal from E_{corr1} to E_{corr2}, with a resulting drop in current density i_{corr1} to i_{corr2} and a corresponding drop in corrosion rate.

binations of zinc chromates, phosphates, or polyphosphates.

Coatings. Paint or metallic coatings can be applied to a metal surface to act as a physical barrier between the metal and an aqueous environment and/or have a further role in mitigating electrochemical corrosion. In contrast to inhibitors, however, where the coating may be of molecular dimensions, paint and metallic coatings often range from 25 to 100 μm to millimeters in thickness.

Paint. The primary role of paint coatings is to act as a physical barrier to the diffusion of water, oxygen, and aggressive anions (Cl^-, SO_4^{2-}, etc.) to the metal surface. However, very few, if any, paint systems are completely impermeable, and thus some further means of mitigating electrochemical corrosion is necessary for longer-term protection. There are numerous organic paint systems available, each with very specific applications (Table 2). In general, however, the paint will comprise three components: the "vehicle," such as linseed oil, which confers fluidity during application; a "pigment," generally a solid suspension of 30 to 90 vol%, which not only gives color but, more importantly, imparts extra corrosion resistance via inhibition or sacrificial mechanisms; and an "additive" that accelerates the drying process by evaporation, polymerization, or oxidation to give a binder that holds the pigment to the surface. The entire paint system may comprise three or more coatings, with the first (primer) coat applied to the clean, grease-free surface. The primer coat provides corrosion suppression due to the presence of, for instance, sacrificial zinc or aluminum metallic particles, or zinc chromate or phosphate inhibitors. The outer (top-coat) paint layer may contain additives for specialty purposes such as quartz for wear resistance or copper for antifouling purposes in marine environments.

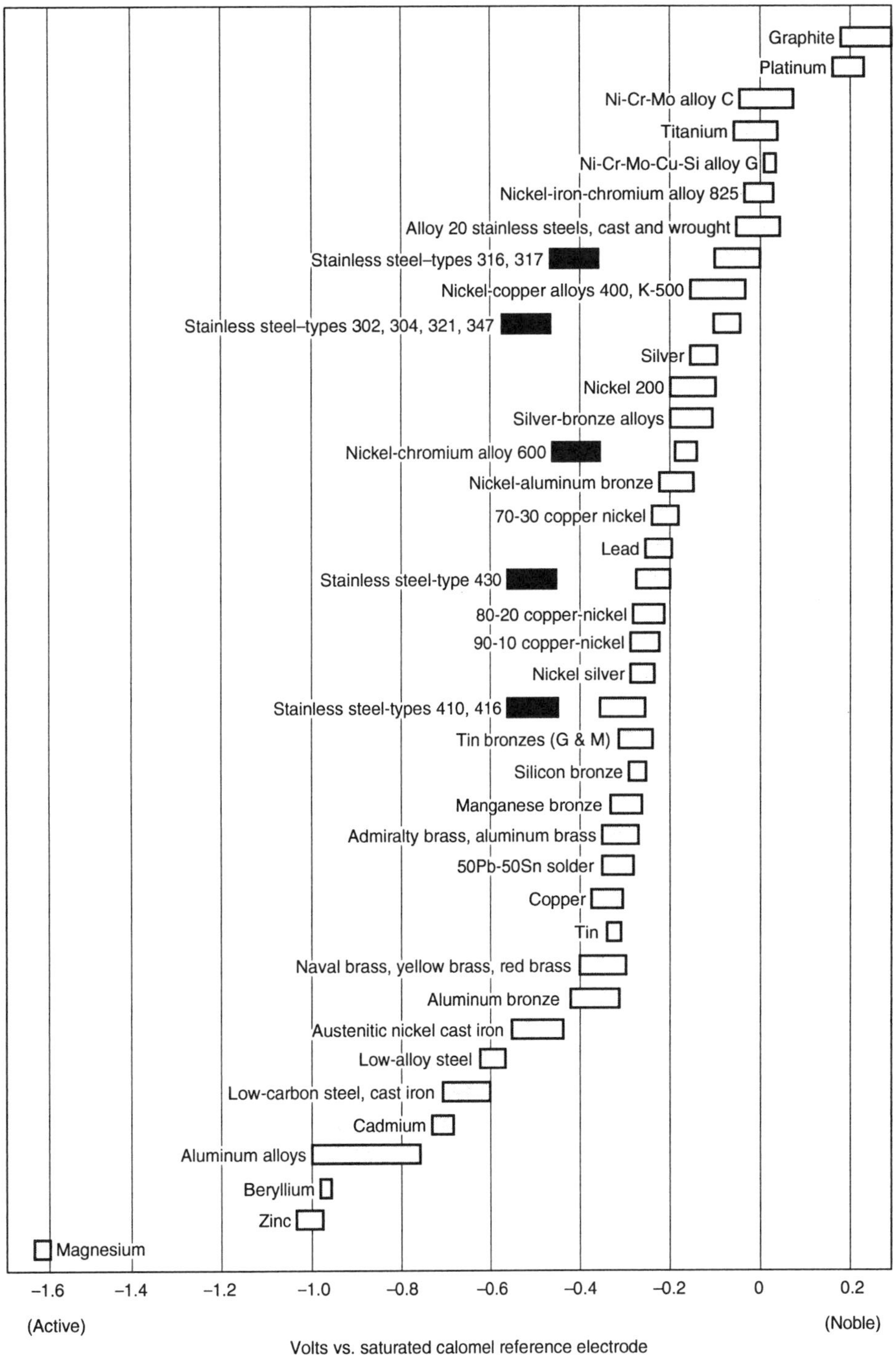

Fig. 13 Galvanic series for seawater. Dark boxes indicate active behavior for alloys that exhibit both active and passive behavior.

Numerous organic compounds can be used in the paint formulation, such as oil-base alkyds, silicon-modified alkyds, epoxy esters, acrylics, and polyurethanes. More recently, solvent-free paints have been developed to conform to environmental control regulations. These latter paints are basically coatings of nylon, polyvinyl chloride, or polytetrafluoroethylene, that are deposited by dipping, electrostatic or airless spraying, or, more recently, by thermal spraying. The corrosion protection of the various paints is due to the formation of a plastic insulating barrier between the metal and the aqueous environment.

Metal Coatings. Corrosion protection can also be obtained via metallic coatings deposited by electroplating, hot dipping, thermal spraying, overlay cladding (explosive, rolling, welding), or diffusion coating. The utility of most such coatings is due to the fact that they are relatively impermeable (compared to paints) and may protect the underlying metal even if the coating is mechanically ruptured. The mechanism of this latter property relates to the fact that the corrosion potential of any corroding system is determined by the dynamic equilibrium between all the oxidation and reduction reactions (e.g., Fig. 4). Thus, referring to the schematic Evans diagram in Fig. 12, it is seen that an uncoated metal (M) in, for example, an aerated environment where the reduction reaction is given by Eq 5, will be corroding at a rate controlled by the current density, i_{corr1}, and the corrosion potential will be E_{corr1}. If the surface is completely covered with a metallic coating (C), then the corrosion potential on the coating surface will be E_{corr2}. If the coating is now ruptured, then (because of the ratio of the areas of the coating to the exposed metallic substrate) the corrosion potential of the exposed metallic substrate will be decreased from E_{corr1} to E_{corr2} and the current density at the exposed metal substrate will be correspondingly decreased from i_{corr1} to i_{corr2}. In other words, the underlying metal surface will be protected even if the outer metallic coating is ruptured. The area of the surface defect that can be protected will be dependent on the conductivity of the environment. For instance, in distilled water the maximum defect dimension that can be protected may be <3 mm, whereas in seawater these dimensions may be an order of magnitude greater.

Specific protective metallic coatings for a given metallic substrate can now be designed based on this understanding. For instance, the metallic coating must have a corrosion potential more negative than the substrate to be protected. An indication of whether this criterion can be met for protection of metals in seawater can be gained from the galvanic series (Fig. 13), which indicates the corrosion potentials of various alloys in

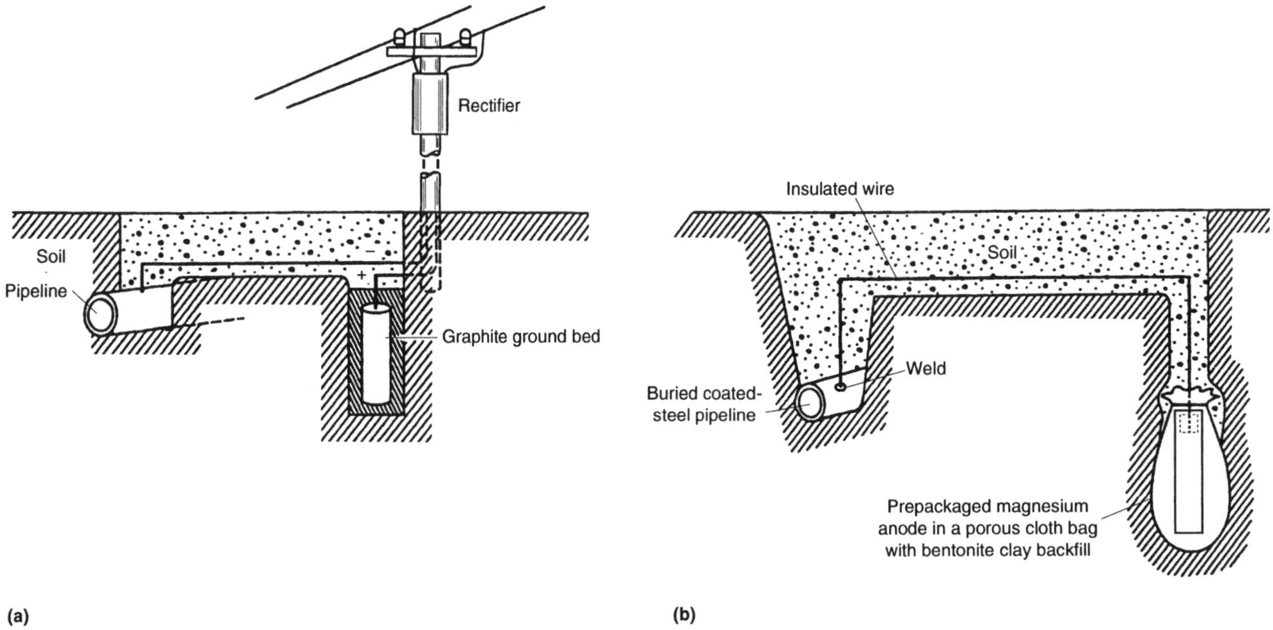

Fig. 14 Two methods of cathodic protection: (a) by impressed current and (b) by a sacrificial anode. Source: Ref 27

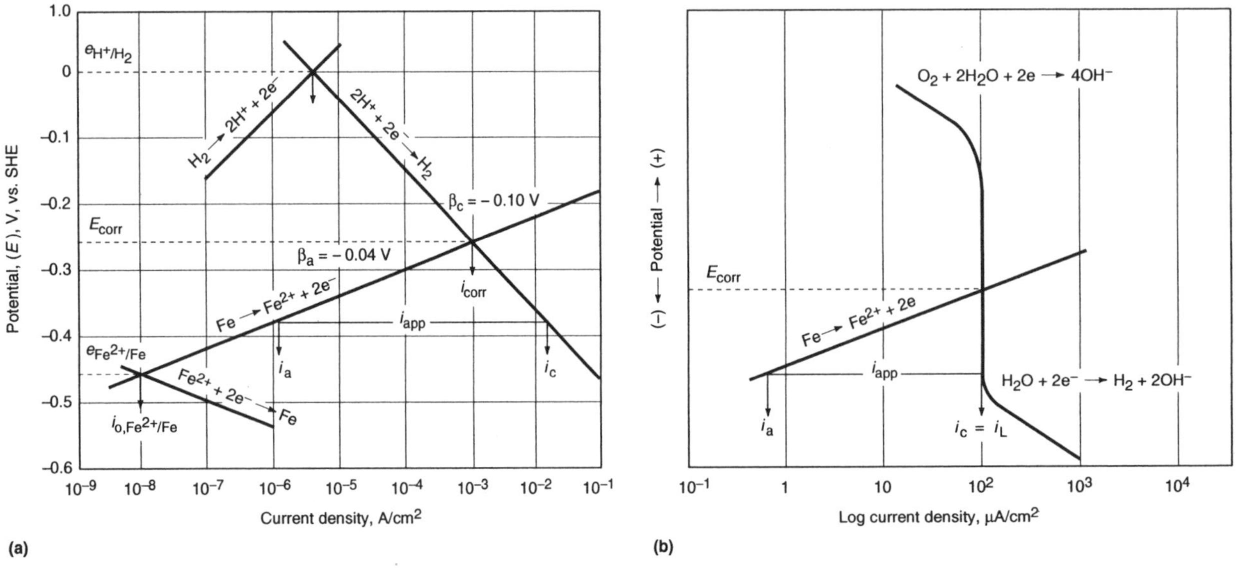

Fig. 15 Cathodic protection with an impressed current density i_{app} for (a) steel in acid solution and (b) steel in neutral aerated water. Source: Ref 5

flowing seawater. Thus, low-carbon steel with a corrosion potential of about −650 mV (vs. SCE) will be protected by, for instance, cadmium-, aluminum-, zinc- or magnesium-base coatings, even if there is a defect in the coating that exposes the underlying low-carbon steel to the seawater. Examination of the galvanic series indicates that tin would not be expected to provide galvanic protection for low-carbon steel, and this situation is true for flowing, aerated environments. However, this is not true in *deaerated* solutions where the relative positioning of the corrosion potential for tin and low-carbon steel are reversed. This combination gives the rationale for the use of tin as the protective coating for the inside of sealed beverage and food containers, and it explains why protection is lost once the container is opened and the inside of the can is exposed to air.

Because the coating is the main component to corrode, its effective life as a protection to the underlying metal (even if that is exposed at defects in the coating) depends on the thickness of the coating and the environment. For instance a 100 μm thick zinc coating on low-carbon steel might offer protection in seawater for 5 years, but its lifetime in a nonpolluted inland environment might be 35 years. Conversely, the useful lifetime in seawater might be extended to 15 years by increasing the zinc coating thickness to 300 μm or replacing the coating with 55% Al-Zn.

Other metallic coatings might not give galvanic protection to the underlying metal substrate, and excessive attack at defects in the coating can occur unless other design features are employed. A good example of this is chromium plating on low-carbon steel, which has decorative purposes as well as being hard and wear resistant; however, the chromium plating can easily crack and thereby expose the underlying steel to the environment. Because of the very large area of "noble" chromium surface, excessive attack will occur on the small area of low-carbon steel exposed at a crack in the coating. The solution to this situation is to place two intermediate layers of nickel and nickel-sulfur alloy between the chromium coating and the steel. The nickel is thus potentially attacked, although slowly, and the integrity of the steel substructure is maintained.

Cathodic and Anodic Protection. Examination of the Pourbaix diagrams for, for instance, iron (Fig. 7) or nickel (Fig. 5) indicates that, in mildly acidic or neutral environments, corrosion protection can be attained either by ensuring that the metal/solution potential is depressed below the equilibrium potential for Fe^{2+}/Fe or Ni^{2+}/Ni, or by elevating the potential to a value where a protective passivating oxide is formed. Such corrosion mitigation actions are known respectively as "cathodic" and "anodic" protection.

Cathodic protection can theoretically be applied to virtually all metal/water systems, but it has the widest commercial use for steels in aerated soil or seawater (e.g., the oil/petrochemical and marine industries). There are two approaches to lowering the metal/solution potential, involving the use of either an impressed current or a sacrificial anode. In the first method, a direct current (such as from a rectified ac power source or solar power cells) is passed between the structure to be protected and an adjacent inert anode (Fig. 14a). This anode can be inert graphite, high-silicon iron, platinized titanium, lead-platinum alloy, or lead-silver alloy, depending on the environment. In the second method (Fig. 14b), the structure to be protected is directly connected to a sacrificial anode, such as magnesium, zinc, aluminum, or their alloys. The fundamentals of these approaches are clear by examination of the relevant Evans polarization diagram (Ref 5). For instance, for cathodic protection by impressed current of low-carbon steel in acidic solutions, the relevant reduction reaction is H^+/H_2 (Eq 3) and the oxidation reaction is:

$$Fe \rightarrow Fe^{2+} + 2e^- \qquad \text{(Eq 22)}$$

It is seen in Fig. 15(a) that the corrosion rate of the unprotected steel is approximately 1 mA/cm^2 (~460 mils/year). If a direct current is applied between the mild steel structure and an inert electrode (Fig. 14a), the corrosion current will drop as the steel is cathodically polarized, and, if an acceptable corrosion rate is 10^{-6} A/cm^2 (or 0.46 mils/year), then a current density of ~10^{-2} A/cm^2 would have to be applied. However, this practice would be an impractical solution to this particular problem, for it would require excessively high currents (e.g., for the protection of 1 m^2 exposed area, >100 A would be required), and there would be a safety problem of dealing with the excessive amounts of hydrogen evolved at the inert anode. A more practical use for this impressed current approach to cathodic protection is for low-carbon steels in aerated, neutral environments (such as seawater). In this situation, the relevant reduction reaction is O_2/OH^- (Eq 5a), whose rate is limited by oxygen diffusion control (compare with Fig. 3b) at the high cathodic overpotentials. Thus, as shown in Fig. 15(b), the applied current necessary to reduce the oxidation current for the Fe/Fe^{2+} reaction to 1 μA/cm^2 is only 100 μA/cm^2 (i.e., markedly less than in acidic environments).

The principles of the sacrificial-anode approach follow those principles outlined above for sacrificial metallic coatings. In this situation, however, the sacrificial anodes are physically separated from the protected structure (Fig. 14b). The extent of the physical separation of the sacrificial anodes is dictated by the conductivity of the environment (e.g., the soil and its backfill) and

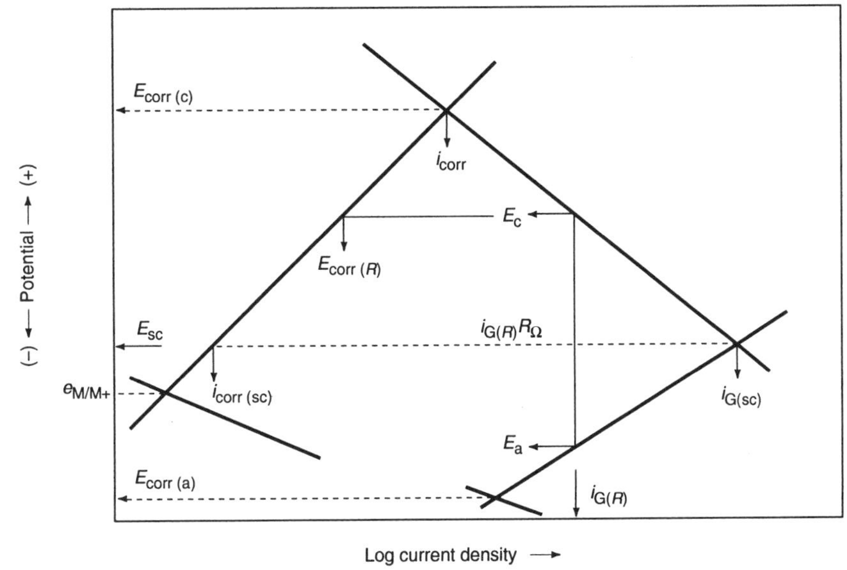

Fig. 16 Schematic polarization diagram for a sacrificial anode, G, coupled to a cathodically protected metal structure, i_{corr}, is reduced to $i_{corr(sc)}$ by the galvanic current $i_{G(sc)}$. Note that the presence of a higher-resistivity environment between anode and structure leads to an iR potential drop, which increases the metallic structure corrosion current to $i_{G(R)}$, separates anode and cathode by a potential $i_{G(R)} R_{\Omega}$, and reduces i_{corr} to $i_{corr(R)}$. Source: Ref 5

Table 3 Cathode materials for anodic protection

Cathode material	Environment
Platinum on brass	Various
Steel	Kraft digester liquid
Illium G	Sulfuric acid (78–105%)
Silicon cast iron	Sulfuric acid (89–105%)
Copper	Hydroxylamine sulfate
Stainless steel	Liquid fertilizers (nitrate solutions)
Nickel-plated steel	Chemical nickel plating solutions
Hastelloy C	Liquid fertilizers (nitrate solutions)
	Sulfuric acid
	Kraft digester liquid

Table 4 Reference electrodes for use in anodic protection

Electrode	Solution
Calomel	Sulfuric acid
Silver-silver chloride	Sulfuric acid, Kraft solutions, fertilizer solutions
$Mo\text{-}MoO_3$	Sodium carbonate solutions
Bismuth	Ammonium hydroxide solutions
Type 316 stainless steel	Fertilizer solutions, oleum
$Hg\text{-}HgSO_4$	Sulfuric acid, hydroxylamine sulfate
Pt-PtO	Sulfuric acid

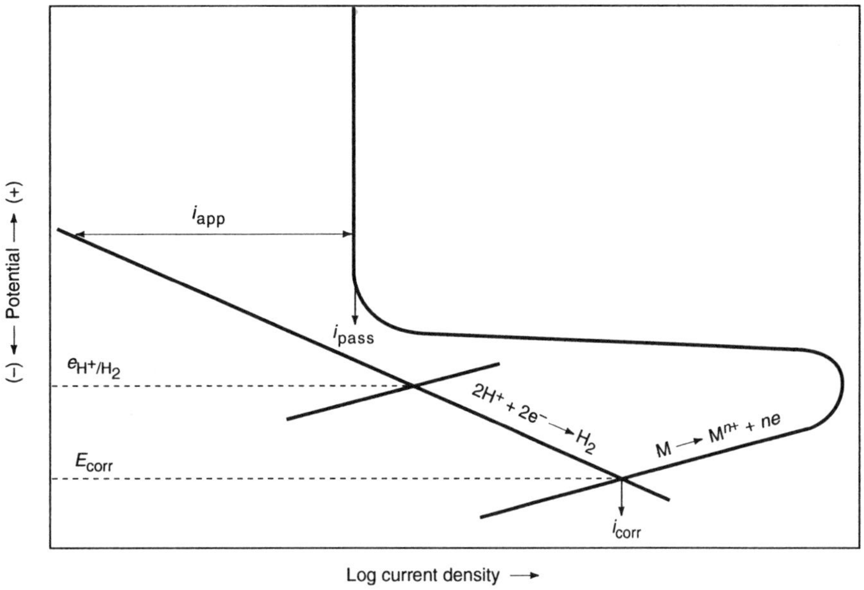

Fig. 17 Potential control in the passive region for anodic protection. Source: Ref 5

the acceptable corrosion rates of the protected structure and the sacrificial anode. For instance, it is seen in the Evans diagram in Fig. 16 that the corrosion rate of the metal structure can be reduced from i_{corr} to a miniscule rate $i_{corr(sc)}$ by coupling it electrically to a sacrificial anode, G. However, the corrosion rate on the sacrificial anode can be significant, given the fact that the cell current is concentrated on the relatively small area of the anode; this high sacrificial dissolution rate will reduce the design life of the protection system. This arrangement can be modified by changing the anode material (e.g., substitute a 55Al-4Zn alloy rather than use pure zinc) to one that is less active. In this way, the corrosion rate of the main metal structure may be slightly increased from $i_{corr(sc)}$, but it is still acceptable and the design life of the system is increased. An alternative approach would be to decrease the conductivity of the backfill (assuming that this can be done) so that the *i*R potential drop between anode and cathode is increased (Fig. 16), thereby increasing the potential of the metal, decreasing the potential of the sacrificial anode, and leading to compromise corrosion rates of the metal and sacrificial anode.

Anodic protection is applicable only to alloys that exhibit an active/passive transition in their anodic polarization characteristics (refer to Fig. 3c, 9, 10) and is generally confined to stainless steels, nickel and its alloys, and titanium in relatively acidic solutions. The application method is similar to cathodic protection by impressed current, except that the structure is *anodically* polarized so that the metal potential is in the passive regime. The cathodes are, for instance, nickel-plated steel, platinum, brass, silicon cast iron, or copper, depending on the environment (Tables 3, 4). Constant monitoring of the potential of the metal structure is needed, because if the potential falls from the desired passivation potential range into the active range, excessive corrosion can occur (Fig. 17).

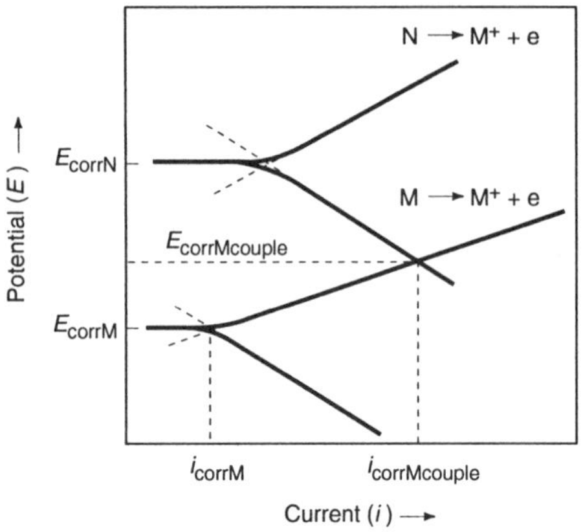

Fig. 18 Schematic potential/current curves for metals M and N, illustrating the fact that the corrosion current on M (i_{corrM}) increased to $i_{corrMcouple}$ when connected to metal N. This current is determined by the criterion that the total oxidation and reduction currents be equal.

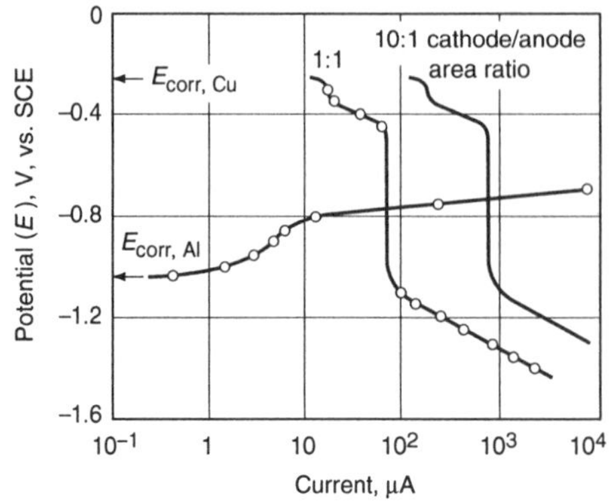

Fig. 19 Prediction of galvanic corrosion rates of aluminum/copper couples and effect of aluminum/copper surface area ratio. Source: Ref 5

Localized Attack

Dissimilar-Metal Corrosion. If two metals, say aluminum and copper, are immersed in aerated chloride solutions, they will individually attain corrosion potentials of about −1.0 and −0.25 V (vs. SCE), respectively, as indicated in the galvanic series (Fig. 13). If these metals are electrically connected, then electrons will flow through the connection from the aluminum to the copper, and the aluminum will become anodically polarized and copper will be cathodically polarized with respect to their original (unconnected) corrosion potentials. The extent of this polarization and the resultant corrosion rates on the two metals will be dependent on the relative areas of the two electrodes and the conductivity of the solution. For instance, referring to the schematic Evans polarization diagrams for two metals M and N (Fig. 18), it is seen that their corrosion potentials change from the "unconnected" values E_{corrM} and E_{corrN} toward a common value $E_{corr\text{-}couple}$, which is determined by the criterion that the total reduction current is balanced by the total oxidation current at the various parts of both surfaces. The result is that the corrosion current on the more active metal is increased from I_{corrM} to $I_{corrMcouple}$. The increase depends on the relative areas of the two metals; this important aspect can be illustrated by considering the aluminum/copper couple in aerated chloride solutions (Fig. 19). If it is assumed that the area of the two metals are equal (say 1 cm^2), then in the uncoupled condition the corrosion rate on the aluminum is about 0.7 μA/cm^2 (0.3 mil/year), but when connected to the copper, the corrosion potential will increase from −1.05 V (vs. SCE) to approximately −0.8 V (vs. SCE), and the corrosion rate on the aluminum will increase dramatically to ~70 μA/cm^2 (30 mils/year). However, if the area of the copper is increased by a factor of ten, then the oxygen reduction current on the copper will increase by a factor of ten, and consequently, the current density on the original 1 cm^2 of aluminum will be increased to a catastrophic 700 μA/cm^2 (300 mils/year).

The other important factor in this phenomenon is the ionic and electronic resistivity of the overall circuit. For instance, as the ionic resistance increases, the resultant "*i*R" drop in the solution will eventually cause the corrosion potentials to revert to the uncoupled values. However, it is incorrect to assume that all areas of a given metal

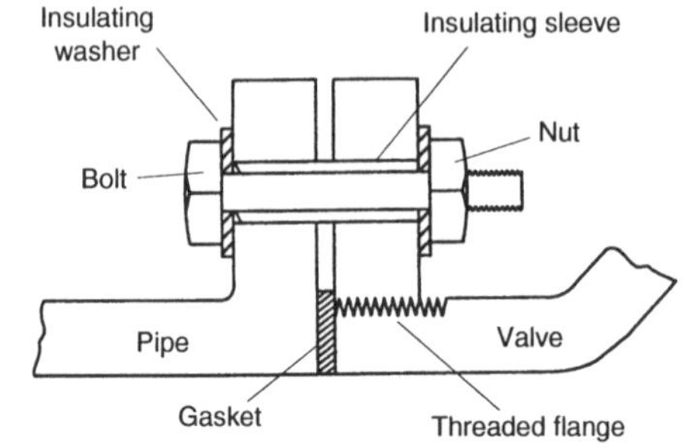

Fig. 20 Flange insulated to eliminate a galvanic couple. Source: Ref 10

surface are at the same corrosion potential (which is created by the *local* metal-environment interface) because of local polarization (e.g., areas of the aluminum adjacent to the copper) or chemistry variations (e.g., oxygen consumption in crevices and cracks). Thus, if the metals are physically in contact, the area immediately adjacent to the joint will be polarized, causing the attack at dissimilar-metal joints to be localized near the joint itself.

The obvious design factors in systems involving dissimilar metals are to: (a) select metals with the least difference in "uncoupled" corrosion potentials, or (b) minimize the area of the more noble metal with respect to that of the active metal, or (c) insert an electrically insulating gasket between the two dissimilar metals (Fig. 20).

Intergranular Attack. Metals discussed so far in this article have been regarded as homogeneous, but in fact, most engineering alloys are multiphased or contain distributions of solid-solution elements that have different chemical activities. Such metallurgical inhomogeneities occur especially at grain boundaries, either due to grain-boundary segregation or intermetallic precipitation, and these can give rise to localized intergranular corrosion.

Intergranular corrosion of austenitic Fe-Cr-Ni stainless steels offers an ideal example of the phenomenon and design methods to counteract it. In this system, chromium carbide can precipitate at the grain boundary by a classical thermally activated nucleation and growth process during heat treatment or welding. This carbide precipitation, called *sensitization,* occurs at about 510 to 790 °C (950 to 1450 °F) and is accompanied by chromium depletion in the adjacent metal matrix such that the chromium content can fall from 18% to less than 10% Cr in a band up to 10 μm from the grain boundary. This depleted zone will have markedly different corrosion properties from the adjacent high-chromium matrix. For instance, it is seen (Fig. 21) that if the chromium content falls much below 12%, then the corrosion rate in acidic solutions rises markedly in a given oxidizing potential range, and preferential corrosion will occur. This behavior will be aggravated by the fact that this narrow depleted zone will have a lower corrosion potential than the larger area of passivated high-chromium alloy connected to it, and hence further corrosion will occur due to galvanic effects, as explained in the previous section, "Dissimilar-Metal Corrosion."

This type of attack depends on two conjoint conditions, and the problem can be resolved by attention to only one of these conditions. First, grain-boundary chromium-carbide precipitates and their associated chromium-depleted zone are necessary. These conditions can be avoided by attending to such factors as the temperature/time requirements for such "sensitization," by annealing the structure to solutionize the carbide, and by making alloy compositional changes such as low carbon content (e.g., L-grade stainless steels with <0.03% carbon) or the addition of elements such as niobium or titanium, which form more stable carbides than chromium carbide. Second, the intergranular corrosion will be observed only under oxidizing potential conditions; thus, avoidance of oxidizers such as oxygen, Fe^{3+} or Cu^{2+} (e.g., Eq 5–7) may well alleviate the problem.

Similar intergranular attack phenomena are seen in other passive systems, where the localized attack is associated with either active depleted zones (e.g., the copper-depleted zones in Al-Cu or Al-Zn-Mg-Cu alloys or the molybdenum-depleted zones in Ni-Cr-Mo alloys) or with active precipitates (e.g., Mg_2Al_3 in the Al-Mg alloys or $MgZn_2$ in Al-Zn-Mg alloys). In all instances, however, mitigation can be obtained by attention to the metallurgical conditions that give rise to the precipitation or to the specific environments in which galvanic attack occurs.

Dealloying corrosion is associated with alloys whose constituents are elements having very different electrochemical activities. For instance, zinc-copper, gray cast iron, and aluminum-tin alloys give rise to phenomena classified as "dezincification," "graphitic corrosion," and "dealuminification," respectively. In two-phase structures such as α/β brasses and graphitic iron, the mechanism is partially galvanic (or dissimilar-metal) attack, but it is believed that in some systems, the mechanism is enhanced by combined dissolution of both elemental components and then reprecipitation of the more noble element. The resultant damage is a friable surface with large amounts of porosity. Given these component parts to the mechanism, it is apparent that potential changes can either exacerbate or mitigate the problem.

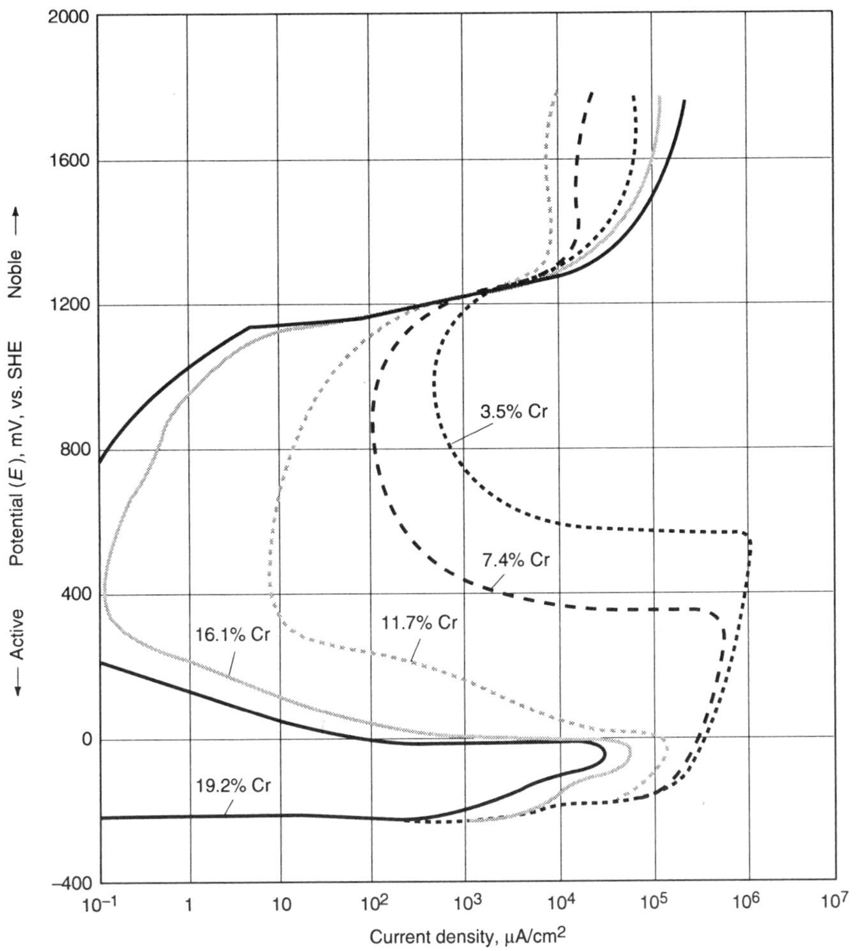

Fig. 21 Effect of chromium content on anodic polarization of Fe-Ni alloys of 8.3 to 9.8% Ni in 2N H_2SO_4 at 90 °C. Source: Ref 106

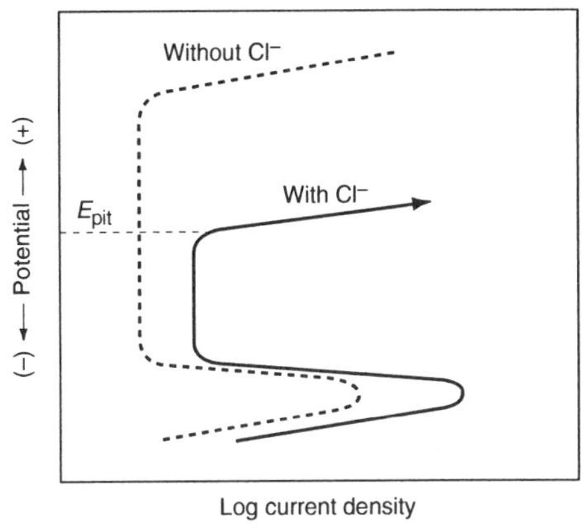

Fig. 22 Schematic representation of critical pitting potential, E_{pit}, due to anodic polarization. Source: Ref 5

Pitting, Crevice Corrosion, and Differential Aeration. Pitting and crevice corrosion arise from the creation of a localized aggressive environment that breaks down the normally corrosion-resistant passivated surface of the metal. This localized environment normally contains halide anions (e.g., chlorides) and is generally created because of differential aeration, which creates potential drops between most of the surface and occluded regions (e.g., pits, crevices, and inclusions) that concentrate the halide at discrete locations.

In pitting, this localization may begin at microscopic heterogeneities such as scratches and inclusions (e.g., sulfides). Above a given potential, negatively charged anions (e.g., Cl^-) accumulate on the metal surface and can cause breakdown of the protective oxide. The breakdown mechanism continues to be a topic of research (Ref 28). Catastrophic localized breakdown occurs at a specific corrosion potential, E_{pit}, that is a function of the material, chloride concentration, pH, and temperature (Fig. 22). Once this breakdown occurs, pit propagation can progress rapidly, because:

- The environment within the pit is deaerated (i.e., at low potential), thereby setting up a potential drop between the pit and the higher-potential surface surrounding the pit. This potential drop concentrates the aggressive anion in the pit.
- Electroneutrality considerations dictate that the increase in negatively charged non-OH^- anions be counterbalanced by an increase in cations, and these usually are hydrogen ions (i.e., the pH decreases).
- The combination of the two factors above leads to an increased metal dissolution rate within the pit.
- The high concentration of metal cations produced in the pit leads to the precipitation of, for example, metal hydroxides near the mouth of the pit, which helps ensure that the severe localized environment is contained (Fig. 23).
- The precipitation of noble ions on the metal surface, or within the alloy as second phases, further accelerates the corrosion rate.

A very similar sequence of events occurs in crevice corrosion, the main difference being that the initiation event is associated with the creation of a localized aggressive environment in a macroscopic (often a designed-in, geometric) crevice. As with pitting, however, the mechanism of this localization is associated with deaeration (low potential) inside the crevice, coupled with an aerated (high-potential) environment outside.

Environmentally Assisted Cracking

Environmentally assisted cracking describes the initiation and subcritical crack propagation in structural metals due to the combined action of tensile stress, material microstructure, and environment. The conditions of these three parameters may be specific to a given alloy/environment system, and if one of these conditions is not met, the problem does not occur.

The topic is made more difficult, however, by the superposition of various mechanisms (slip dissolution, hydrogen embrittlement) and phenomena (stress corrosion, corrosion fatigue, hydrogen embrittlement). Part of this problem is addressed in the section "Life Prediction and Management" in this article. The rest of this section addresses the understanding of the mechanisms of environmentally assisted cracking and how this can be used qualitatively to control cracking.

Candidate Crack-Propagation Models. The basic premise for all of the proposed crack-propagation mechanisms for ductile alloys in aqueous solutions is that the crack tip must propagate faster than the corrosion rate on the unstrained crack sides. If this were not true, the crack would degrade into a blunt notch (Ref 29, 30). Indeed, the suppression of both stress corrosion and corrosion fatigue in many systems can be explained in terms of blunting of cracks during the early propagation stage. For instance, low-alloy steels will not exhibit stress corrosion in acidic or concentrated chloride solutions unless the general corrosion/blunting effect is counteracted with chromium or nickel alloying additions (Ref 31, 32). Similar blunting explanations can be proposed for the case of corrosion-fatigue crack initiation of aluminum in chlorides in comparison to hydroxides (Ref 33).

Numerous crack propagation mechanisms were proposed in the period 1965–1979 (Ref 34–44). With the advent in recent years of more sensitive analytical capabilities, however, many of the earlier cracking hypotheses have been shown to be untenable. The candidate mechanisms for environmentally assisted crack propagation (for both stress corrosion and corrosion fatigue) have been narrowed down to slip dissolution, film-induced cleavage, and hydrogen embrittlement.

Qualitative Prediction Methods for Environmentally Assisted Cracking in Ductile Alloy/Aqueous Environments. Qualitative predictions of cracking have centered around the observation that the rate-determining step in all of the above cracking mechanisms is not necessarily the atom-atom rupture process itself, but is one (or a combination) of mass transport of species to and from the crack tip, passivation reactions, and the dynamic strain processes at the crack tip (Ref 29, 30). Thus, changes in cracking susceptibility for most ductile alloy/aqueous environment systems with, for instance, changes in temperature, electrode potential, stressing mode, or environmental composition, can be explained logically (Ref 29, 30) using a reaction-rate surface (Fig. 24), regardless of the specific atom-

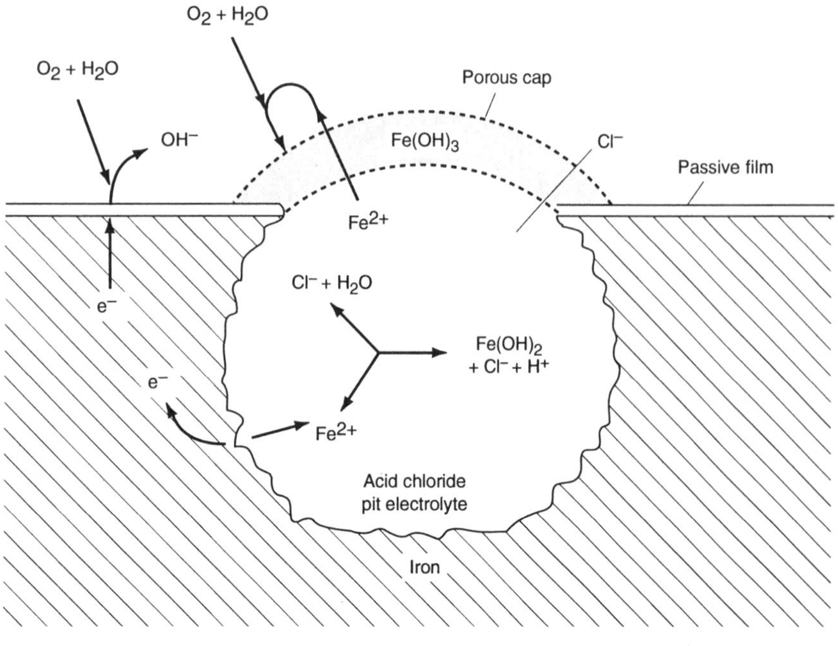

Fig. 23 Schematic representation of processes occurring in an actively growing pit in iron. Source: Ref 5

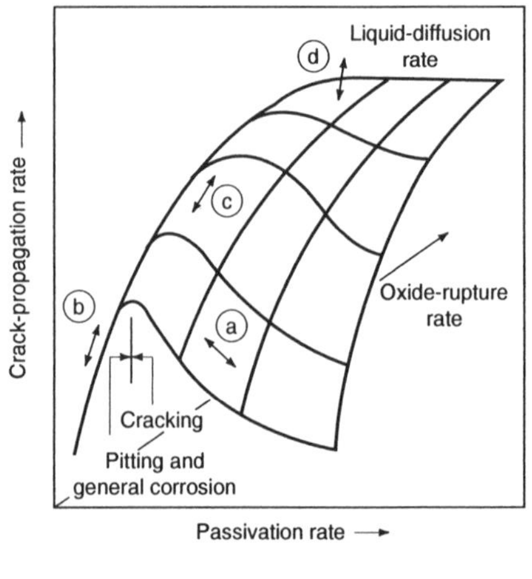

Fig. 24 Schematic reaction-rate surface illustrating the variation in crack-propagation rate with the rate-controlling parameters in the slip-dissolution, film-induced cleavage and hydrogen-embrittlement mechanisms for environmentally assisted cracking in ductile alloy/aqueous environment systems. Source: Ref 29, 30

atom rupture mechanism at the crack tip. This fact can be reinterpreted in terms of the crack propagation rate/stress intensity (Fig. 25) (Ref 29) relationship for a given alloy/environment system subjected to different loading histories in which the limiting and rate-controlling reactions can be defined (Ref 29).

The importance of passivation kinetics on crack propagation is well recognized (Ref 29, 45–49). Very slow passivation rates at the crack tip will promote crack blunting due to excessive dissolution on the crack sides, whereas very fast rates will minimize the amount of crack-tip penetration per oxide-rupture event. Maximum susceptibility, with high-aspect-ratio cracks, will occur at intermediate passivation rates or in regimes of barely stable passivity (e.g., near the passivation potential). The effects of potential, anion (or cation) content, and alloying addition on cracking susceptibility can be quantitatively understood by this simple concept, regardless of whether the advancement mechanism is slip dissolution or hydrogen embrittlement. For instance, cracking susceptibility in poorly passivating systems (e.g., austenitic stainless steel in caustic at high temperature, low-carbon steel in caustic or phosphate) will be increased by actions that promote passivation. Thus in these systems, cracking susceptibility will be greatest in potential ranges adjacent to active/passive transitions on a polarization curve (Fig. 26). In contrast, systems that exhibit strongly passivating behavior (e.g., aluminum alloys, austenitic stainless steels in neutral solutions) will crack most severely under potential conditions where incipient passivity breakdown occurs due to the presence of aggressive anions (e.g., chloride) (Fig. 26b).

The fundamental importance of passivation on the crack-propagation process in ductile alloy/aqueous environment systems also indicates an analytical method of determining the potential ranges where cracking susceptibility may be severe. An example is the direct measurement of passivation rate (Ref 50), by comparing the bare surface and fully passivated dissolution rates to determine whether a high-aspect-ratio crack is possible, or by performing potentiodynamic scans at various rates. Such rapid prediction capabilities are of use in preliminary failure analyses or risk assessments, but this usefulness should be tempered by the realization that these techniques will indicate only the possibility of severe susceptibility.

As indicated in Fig. 24 and 25, the rate-determining step can change as the system parameters (e.g., corrosion potential, stressing frequency, temperature) lead to an increase in crack-propagation rate. Ultimately the rate-determining step will often be liquid diffusion, either of solvating water molecules, anions, or solvated cations to and from the crack tip. Under these conditions, the propagation rate will become independent of stress intensity (i.e., the stage II region in Fig. 25) and will exhibit a temperature dependence associated with an activation enthalpy of ~4 kcals/(g mole) associated with liquid diffusion. It is important from a practical viewpoint to realize that this is a limiting condition, and that the activation enthalpy can change continuously between 4 and ~30 kcals/(g mole) (symptomatic of passivation control) with corresponding changes in, for instance, loading rate (Ref 29), and temperature (Ref 29). Thus, mechanistic analyses based on a specific value of activation enthalpy must be treated with caution, unless it is determined that a limiting value is being measured.

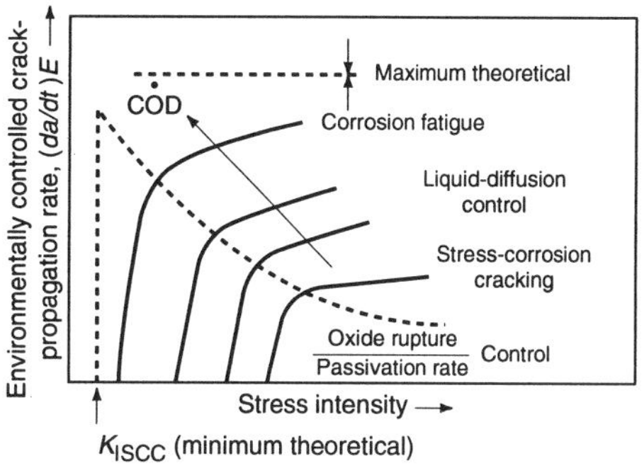

Fig. 25 Suggested variation (Ref 29) in environmentally controlled crack-propagation rate with stress intensity for various crack-tip deformation rates COD. Note the suggested rate-controlling parameters and the fact that these relationships should be bounded by a maximum crack propagation and a minimum theoretical K_{Iscc} or K_{TH}.

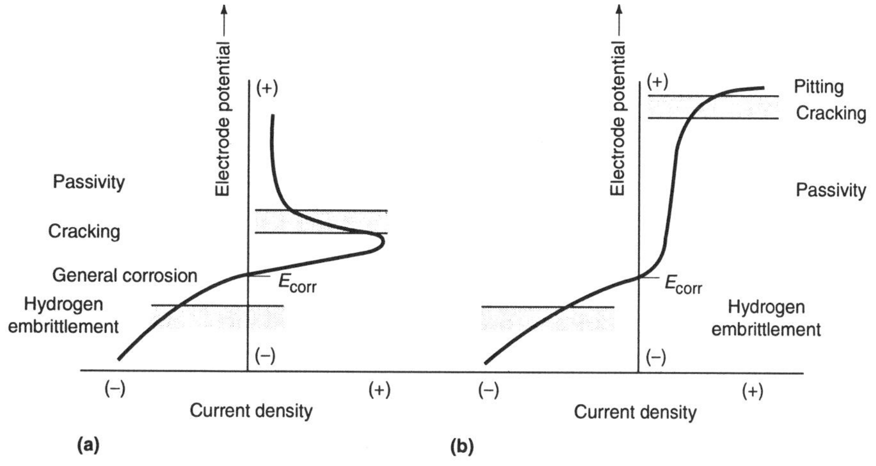

Fig. 26 Schematic electrode-potential/current-density relationship for (a) "poorly" passivating and (b) "strongly" passivating systems, indicating where severe cracking susceptibility in ductile alloy/aqueous environment systems is commonly encountered

Engineering Design Principles

The earlier sections of this article provide a background about the principles of corrosion. This section focuses on engineering aspects of design that can, without due care and attention, precipitate unexpected premature failure. More extensive texts relating specifically to design are available (Ref 51–54), as are guides from various material suppliers and promoters (Ref 55–57). However, it is the fine details of engineering design, often compounded by human errors or poor communication (Ref 22, 58–64), that account for many unexpected failures, at times sig-

Table 5 Design aspects for reliability

Approach	Factors to be taken into account (typical)
"Avoid failure" approach	
Control product duty	Fail safe. Visible instrumentation.
Keep product simple	Fewer functions. Fewer joints.
Predict reliability	Maintain in fixed limits. Overdesign (thicker), etc.
Study variable interfaces	Contact faces, bearings. Lubrication (wear, friction). Seals. Consult experts and test work.
Instill care for familiar items or designs	Complacency. Negligence.
Study innovation	Consult experts and test work. Records of operation. Records of failure mode/frequency.
Redundant items	Stand-by units in good repair. Storage. Handling. Vibration, impact, wear, etc.
Recognize human limits	Misreading of instructions. Sequence of controls (hypnosis?). Adequate housekeeping. User attitudes.
Test product design	Proving the design. Variables. Safety limits. Pilot plant and real-life testing.
Control malfunctions	Human. Automatic. Mixture. Important for software/hardware.
"Keep it working" approach	
Maintainability	Access.
Short-term items	Replace routinely before failure. Regular inspection. Monitoring.
Spares	Identity correctly coded. Location known. Easy access replacement. Avoid identical mating systems.
Built-in adjustments	Corrects for progressive deterioration. Manual/automatic.
"Let it fail then repair" approach	
Defects	Subtle or catastrophic.
Spares	Need quantified reliability (difficult for corrosion).
Minimize off period	Keep repair time low. Speedy fault diagnosis. Speedy removal of failed part(s). Speedy replacement. Speedy check of assembly. Experts, inspectors.

Source: Ref 66

nificant. On occasion a poor design can cause premature failure of the most advanced corrosion-resistant materials.

Design Considerations

Designing for corrosion control can only be effective if it is part of the overall design philosophy. However, a designer is seldom a corrosion engineer, so it is necessary to convey corrosion knowledge to the designer. Unlike conventional engineering, the basic difficulty is that corrosion is not a tangible property; it is more a *behavioral* pattern. Thus, to realize safe, reliable designs, it is essential that there be a rigid control on materials and fabrication and an extensive effort to eliminate human errors or misunderstandings that result from poor communication.

The results of a survey of chemical-process plants (Ref 65) showed that design faults ranked highest (58%) in the reasons for failure. Of almost equal ranking was the incorrect application of protective treatments (55%), followed by categories that demonstrate a lack of knowledge about the operating conditions (52%), lack of process control (35%), and an unawareness that there was actually a corrosion risk (25%).

In an ideal world, designers would call for some corrosion assessment *prior* to preparing the detailed engineering design. Typically, schemes would permit some form of evaluation with respect to both function and the necessary action, for example from the proposal-to-production planning stages (Fig. 27) (Ref 65). In the practical "real" world, however, communication of "agreed" reasons for failures may not always reach the designer. Indeed, communication to contractors, who are closest to the application, is even poorer (Ref 65, 66). Studies have shown that, while management is always informed of the reasons for failure in the chemical-processing industry, site personnel are informed only 77% of the time, designers 55%, material suppliers 37%, and contractors only 11% of the time.

"Draftsmen's Delusions." A further complication in designing against corrosion relates to the general interpretation of design drawings in what has been referred to as "draftsmen's delusions." For example, the draftsman might be considering a certain piece of equipment without knowledge of the fact that there may be unusual shapes, moving parts, or environmental issues. A lack of attention to design detail causes many premature failures by corrosion-related processes.

All too often, the designer will have in mind one thing, which in reality becomes totally different. For example, a simple cross-over line between two reactor vessels might, in practice, become an extended line with several turns, merely to position a shut-off valve at a more convenient and accessible position closer to ground level. There are countless examples of this situation in

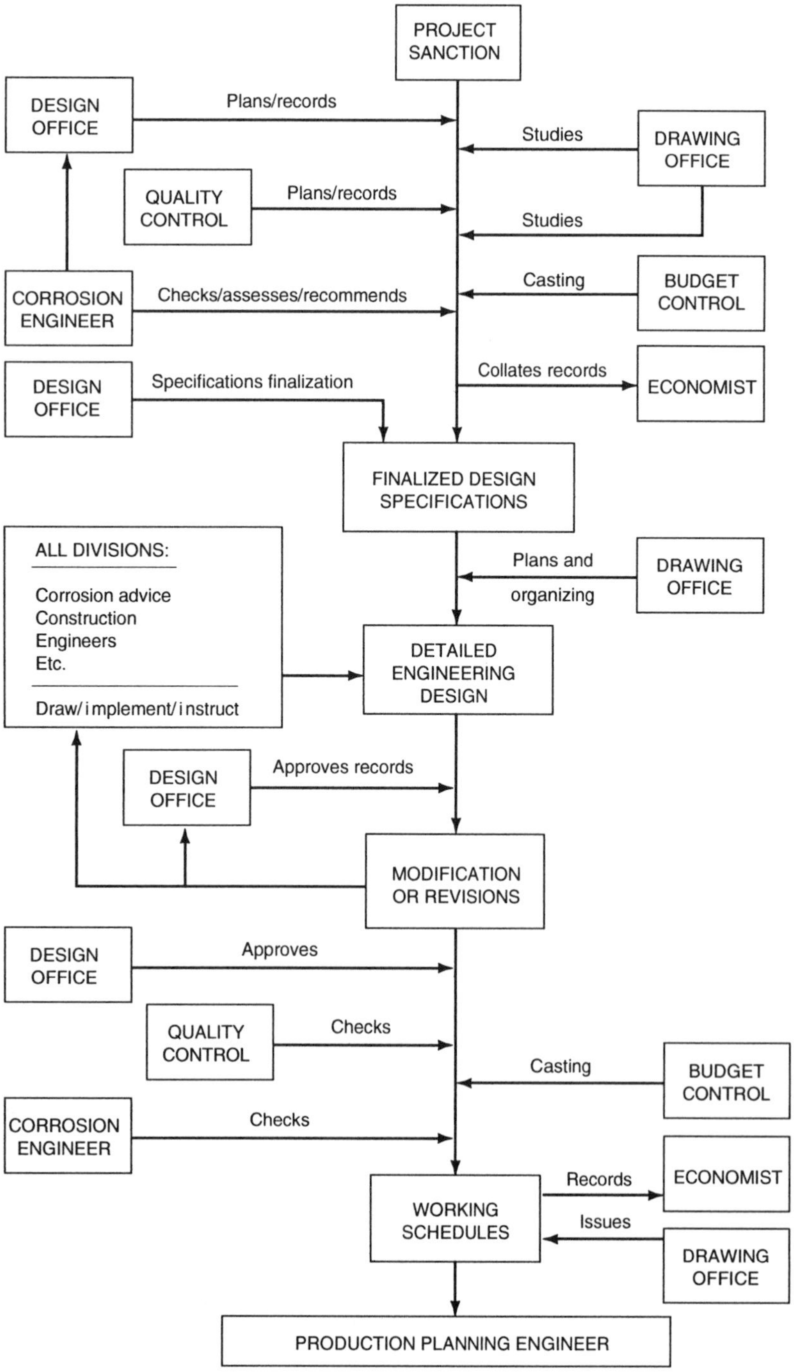

Fig. 27 Action steps during design. Source: Ref 64

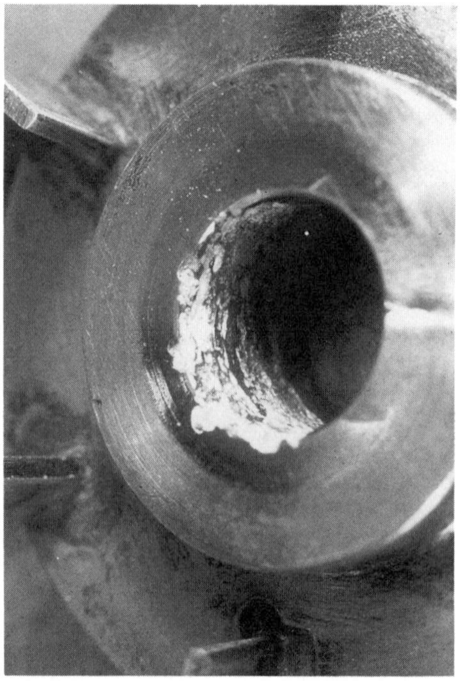

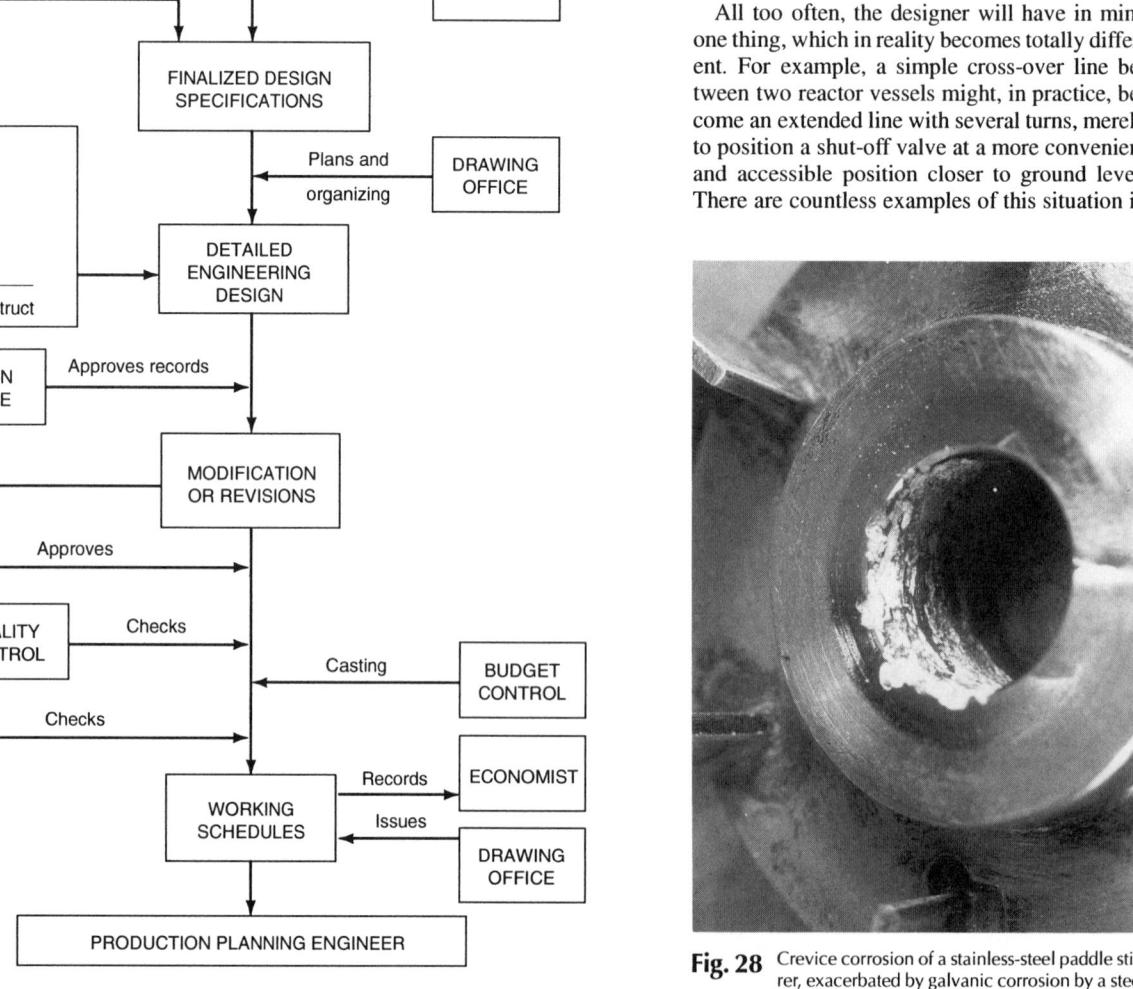

Fig. 28 Crevice corrosion of a stainless-steel paddle stirrer, exacerbated by galvanic corrosion by a steel retaining pin

"real-world" failure analysis, where the actual designs were not those originally intended; examples will be provided below.

Quality assurance and control usually ensure that the requested material is what it should be, given rigorous inspection. However, material controls at fabrication are sometimes less than perfect. In one case, an inspector noticed a blemish on the outside of a steel pressure vessel. Upon closer scrutiny (from the vessel interior) the problem was found to be a steel bolt that had been inadvertently rolled into the steel plate during fabrication.

While considering materials, it is important to avoid nonspecific descriptions or terms in reference to design drawings and specifications. There are many instances where generic terms, such as "stainless steel," bronze," "Hastelloy," or "Inconel," are too vague and the ultimate choice is far from what was expected and required. Wherever possible, and notably in high-risk areas, materials should be selected and tested according to code requirements (Ref 67, 68). Substitutes, if requested, should be properly evaluated before use.

Reliability Engineering

The designer should play a significant role in reliability management. The communication chain, or the "reliability loop" from the designer,

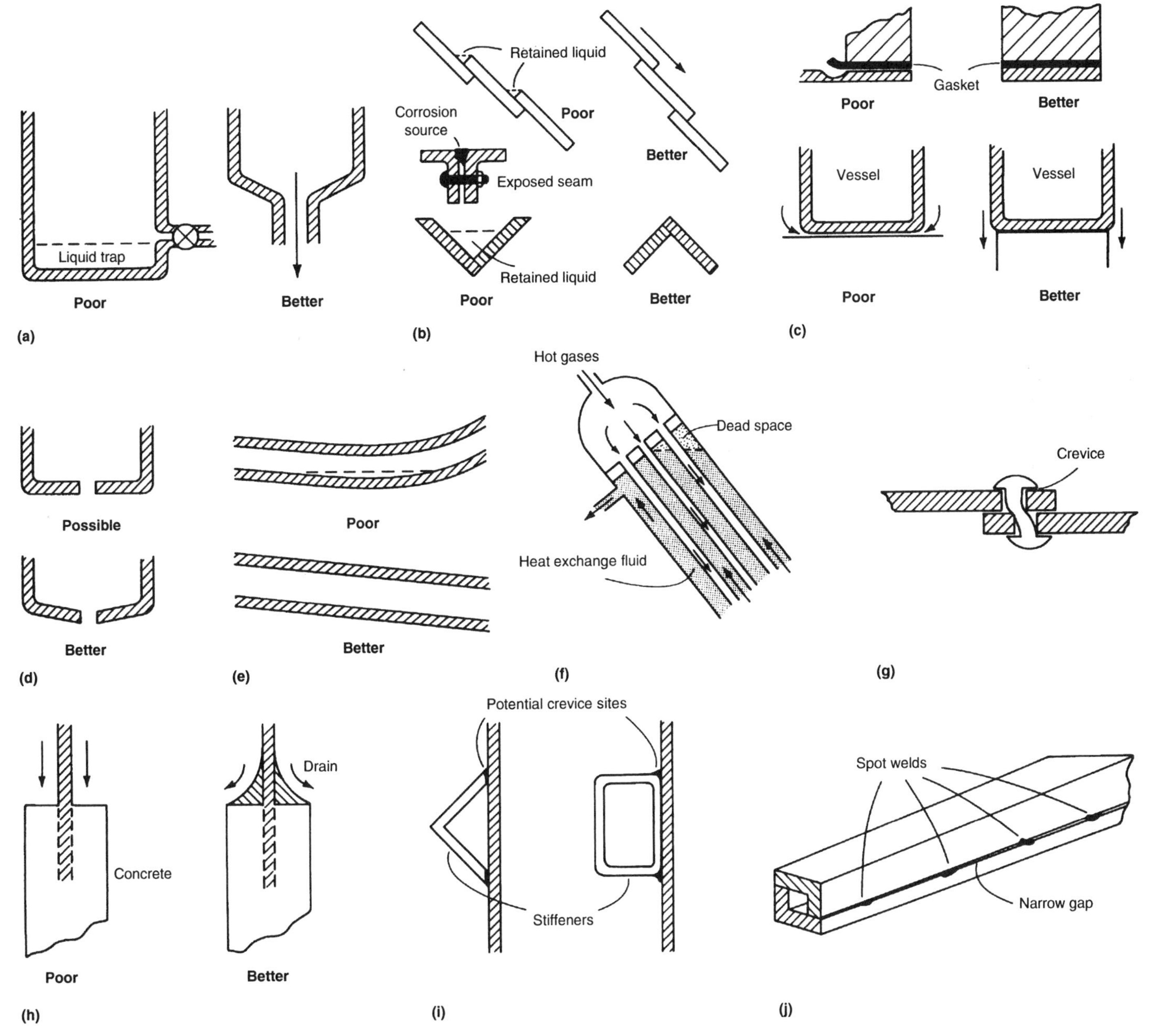

Fig. 29 Examples of how design and assembly can affect localized corrosion by creating crevices and traps where corrosive liquids can accumulate. (a) Storage containers or vessels should allow complete drainage; otherwise, corrosive species can concentrate in vessel bottom, and debris may accumulate if the vessel is open to the atmosphere. (b) Structural members should be designed to avoid retention of liquids; L-shaped sections should be used with open side down, and exposed seams should be avoided. (c) Incorrect trimming or poor design of seals and gaskets can create crevice sites. (d) Drain valves should be designed with sloping bottoms to avoid pitting of the base of the valve. (e) Nonhorizontal tubing can leave pools of liquid at shutdown. (f) to (j) Examples of poor assembly that can lead to premature corrosion problems. (f) Nonvertical heat exchanger assembly permits dead space that is prone to overheating if very hot gases are involved. (g) Nonaligned assembly distorts the fastener, creating a crevice that can result in a loose fit and contribute to vibration, fretting and wear. (h) Structural supports should allow good drainage; use of a slope at the bottom of the member allows liquid to run off, rather than impinge directly on the concrete support. (i) Continuous weld for horizontal stiffeners prevents traps and crevices from forming. (j) Square sections formed from two L-shape members need to be continuously welded to seal out the external environment.

to the manufacturer, then to the user and back to the designer, is a key factor (Ref 66, 69). When a risk is well documented, it should be possible to overdesign or at least isolate the area to minimize risk to users. Where a risk of failure is high, the emphasis should be toward a "fail-safe" or "no-fail/replace" procedure. Failures vary considerably: the design function can be partially or totally affected; the onset of failure can be gradual or sudden. The combination of sudden and total failure represents the worst catastrophic situation (e.g., explosions, fires, and total structural collapse), many of which can be attributed to "a small design detail."

Corrosion Awareness

This article is intended to improve corrosion awareness, but clearly this is only a starting point. To be effective, the user must also be willing to take action and the designer should insist on appropriate codes and/or recommended working practices. Whether the necessary action will be taken is affected by financial, technical, safety, social, and/or political issues (Ref 69).

To prevent corrosion/degradation, the designer can:

- Avoid obvious design weaknesses (see examples below).
- Use more reliable materials, even if this entails greater cost.
- Introduce additional precautions (inhibitors, cathodic protection, coatings).
- Establish efficient maintenance/repair teams having detailed procedures and including qualified surveyors, inspectors, and supervisors.
- Ensure that standby products are available, fully labeled, and properly stored (using dessicants and noncorrosive packaging)

As noted above, the planned approach for reliable engineering design should include corrosion control and/or preventive measures, for which standards and specifications are available. Actual (real-world) approaches vary from plant to plant (or component to component) and include the:

- "Avoid failure" approach
- "Keep-it-working" approach
- "Let-it-fail-then-repair" approach

Aspects germane to these approaches are summarized in Table 5 (Ref 66, 69).

New plants and process equipment are commonly designed to prevent or reduce problems that occurred previously. Updated and improved procedures are of little use, however, if no action is made until after the equipment fails (Ref 70). Situations have occurred where cathodic protection was ineffective because the anode material was not connected to the structure it was protecting; a monitoring signal (a bell) was ignored because of noise; online monitoring data were estimated (guessed at) so that the inspector could avoid travel in blizzard conditions; and a sentinel hole (weep hole) leaked for weeks without any investigation as to cause until the vessel exploded.

Why Failures Occur

In the context of design, there are several factors that relate to material/component failure:

- *Overload* suggests a weakness in plant control instrumentation or operation.
- *Abnormal conditions* can result from a lack of process control or variations in raw materials (alloy type or chemical inhibitors, etc.).
- *Poor fabrication* may relate to inadequate instructions or inspection (e.g., excessive cold work, overmachining, flame cutting, or excessive torque loading).
- *Poor handling.* Scratches or machine marks can result from poor detailing or poor instruction. Not to be excluded are identity marks (incised codes) and inspection stamps.
- *Assembly,* if incorrect (e.g., welds and fasteners) can seriously influence stress, flow, and compatibility.
- *Storage and transportation* can significantly influence material performance, especially for items shipped to/from tropical, humid climates where heavy rains, violent seas, storms, and cargo sweat may each contribute to material degradation unless adequate precautions are taken (Ref 71). Proper design and effective corrosion control management normally should accommodate these aspects of material handling.

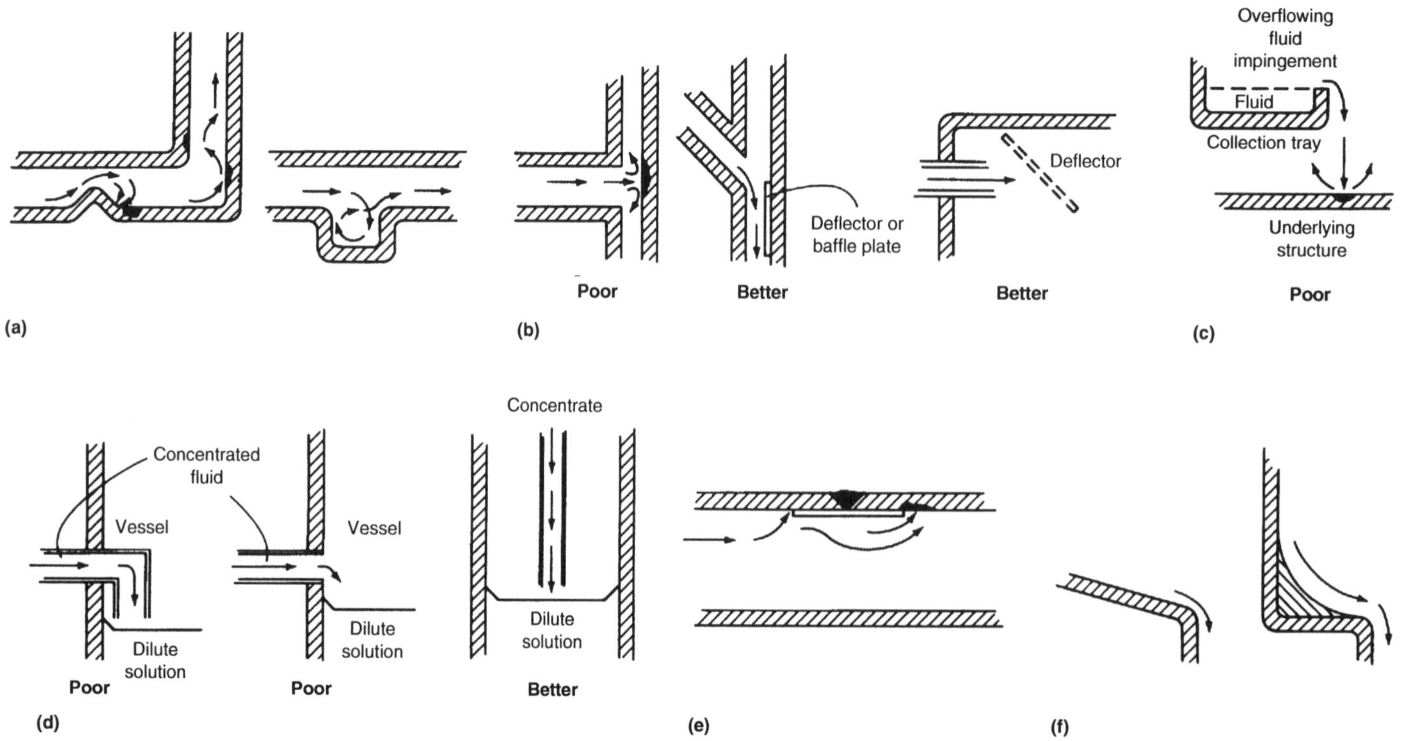

Fig. 30 Effect of design features on flow. (a) Disturbances to flow can create turbulence and cause impingement damage. (b) Direct impingement should be avoided; deflectors or baffle plates can be beneficial. (c) Impingement from fluid overflowing from a collection tray can be avoided by relocating the structure, increasing the depth of the tray, or using a deflector. (d) Splashing of concentrated fluid on container walls should be avoided. (e) Weld backing plates or rings can create local turbulence and crevices. (f) Slope or modified profiles should be provided to permit flow and minimize fluid retention.

- *Accessibility.* Some structures or components may not be accessible for remedial work, even if a corrosion risk is recognized. Buried structures can be affected where soil and bacterial corrosion might apply (Ref 72).

Design and Materials Selection

Corrosion control measures are best initiated at the design stage (Fig. 27). Materials are usually selected to perform a basic function or to provide a functional requirement (see the article "Materials Selection" in Ref 22). Therefore, in many instances the material choice is dictated not by corrosion, but by characteristics such as strength, reflectivity, wear resistance, and dimensional stability. In some situations a corrosion-resistant alloy may not be satisfactory.

High-Temperature Service. The subject of designing for high-temperature service is outside the scope of this article (see the article "Design for High-Temperature Applications" in this Volume and Ref 58, 73, and 74); however, high temperatures always accelerate corrosion processes, and certain gases or liquids, which are considered innocuous under ambient conditions, become aggressive to materials when hot. A tenfold change in corrosion rate is not uncommon for a temperature change of 30 °C under aqueous conditions. The same tenfold change (or considerably worse) can occur with a 20 °C change under high-temperature conditions. If temperatures are too high, the material might oxidize (i.e., scale). Thick scales and metal loss result from overheating (loss of water cooling, absence of insulation, etc.). Heat-transfer contributions will increase as the scale/deposit thicknesses increase.

Candidate high-temperature materials need to be strong and resistant to oxidation or to other corrosion processes that might involve complex multioxidant environments having highly volatile phases and molten salts. It is important for designers and others to recognize that several corrosion elements might simultaneously be involved in an application (e.g., oxygen, halogen, and sulfur) (Table 6). Material selection for high-temperature service needs to be reviewed for each individual part and application. Alloy steels and more sophisticated alloys based on nickel and cobalt are most commonly used, in which key elements for high-temperature corrosion resistance include chromium, aluminum, silicon, and rare-earth additions for scale retention.

Different behavior can arise at similar temperatures, depending on the source of heat, such as electrical heating elements; fuel combustion, flue gases and deposits; flame impingement; friction and wear. When hot gases cool, condensation can

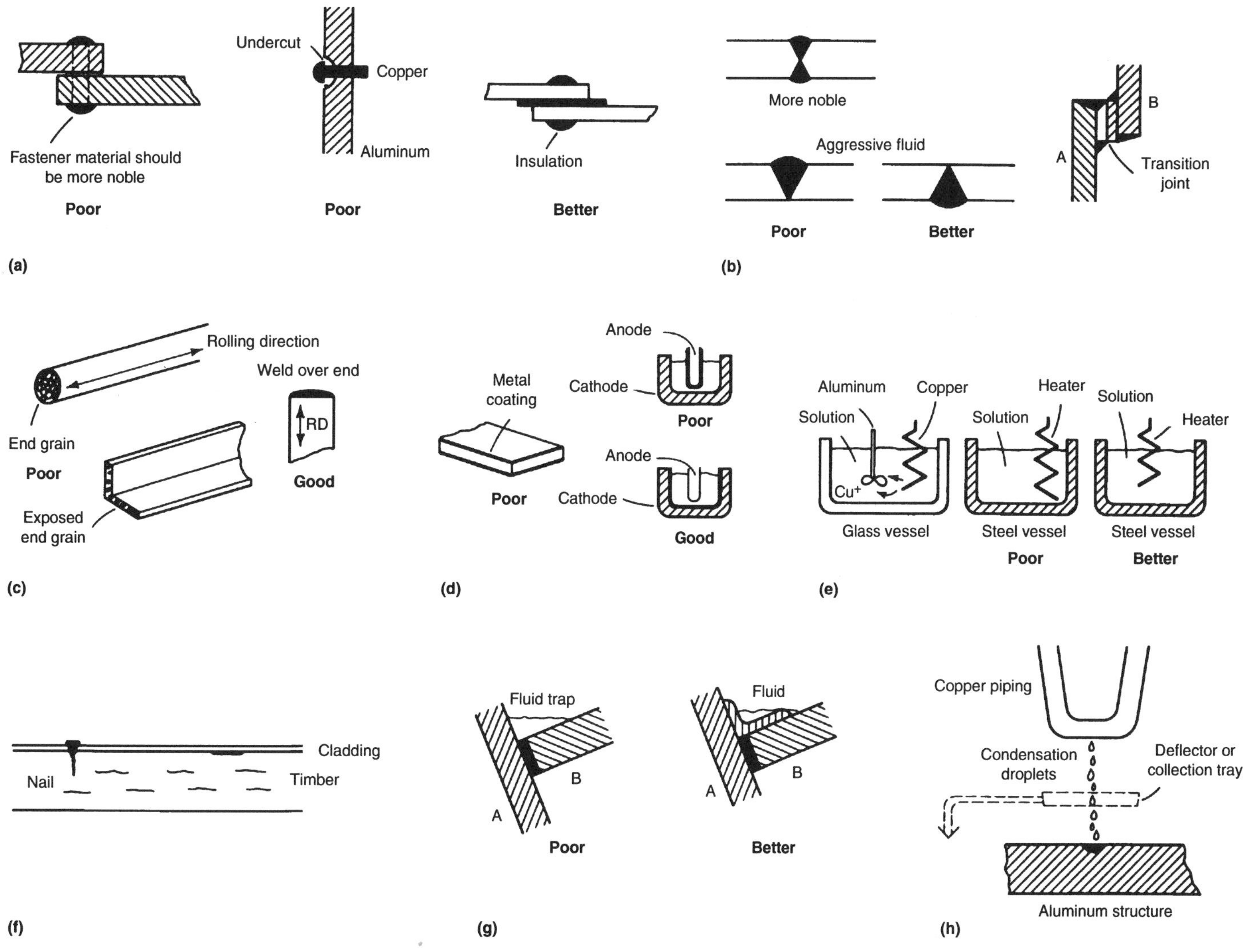

Fig. 31 Design details that can affect galvanic corrosion. (a) Fasteners should be more noble than the components being fastened; undercuts should be avoided, and insulating washers should be used. (b) Weld filler metals should be more noble than base metals. Transition joints can be used when a galvanic couple is anticipated at the design stage, and weld beads should be properly oriented to minimize galvanic effects. (c) Local damage can result from cuts across heavily worked areas. End grains should not be left exposed. (d) Galvanic corrosion is possible if a coated component is cut. When necessary, the cathodic component of a couple should be coated. (e) Ion transfer through a fluid can result in galvanic attack of less noble metals. In the example shown at left, copper ions from the copper heater coil could deposit on the aluminum stirrer. A nonmetallic stirrer would be better. At right, the distance from a metal container to a heater coil should be increased to minimize ion transfer. (f) Wood treated with copper preservatives can be corrosive to certain nails, especially those with nobility different from that of copper. Aluminum cladding can also be at risk. (g) Contact of two metals through a fluid trap can be avoided by using a collection tray or a deflector.

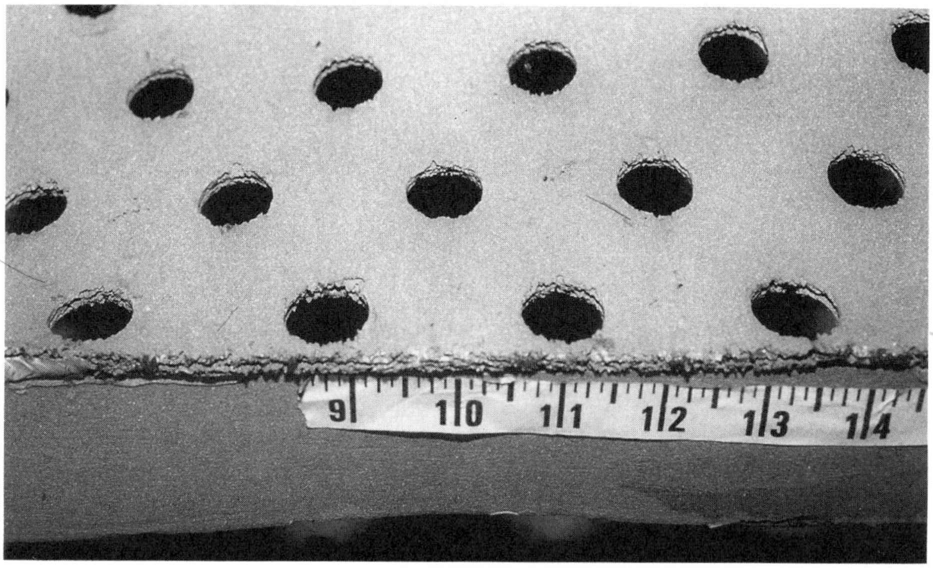

Fig. 32 End-grain corrosion along cut edges and punched holes in a reactor tray made from type 316 stainless steel

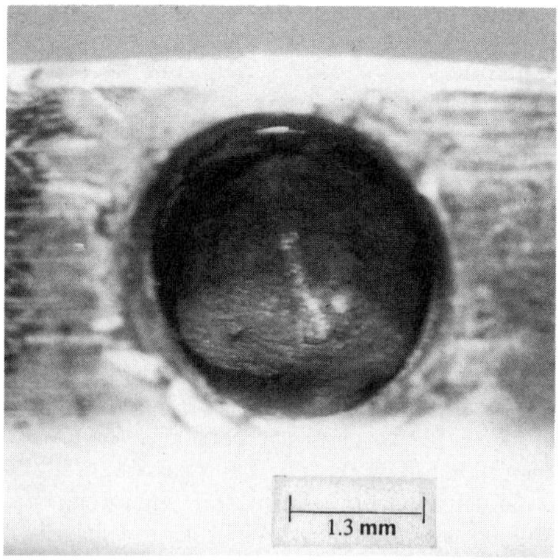

Fig. 33 The incorrect choice of a carbon-steel retaining bolt for a stainless-steel spindle resulted in localized galvanic corrosion of the paddle-stirrer assembly (see Fig. 28)

cause acid dewpoint conditions, thereby changing the material choice to a corrosion-resistant alloy.

Design Factors That Influence Corrosion. The basic factors that most influence design for corrosion resistance are summarized in Table 7 (Ref 58). Each factor plays a unique yet not always unrelated role with other factors. For example, localized corrosion damage adjacent to the spindle support of a stainless steel paddle stirrer (Fig. 28) resulted not only from crevice corrosion (oxygen differential cell) but also because of galvanic corrosion, caused by a small steel retaining screw (anodic) that had been used inadvertently for assembly. Ultimately, the stirrer support loosened, which allowed further deterioration by fretting.

Design Details. The following sections demonstrate aspects of design detail that can accelerate corrosion.

Location. Exposure to winds and airborne particulates can lead to deterioration of structures. Designs that leave structures exposed to the elements should be carefully reviewed, because atmospheric corrosion is significantly affected by temperature, relative humidity, rainfall, and pollutants. Also important are the season and location of on-site fabrication, assembly, and painting. Codes of practice must be adapted to the location and the season.

Shape. Geometrical form is basic to design. The objective is to minimize or avoid situations that worsen corrosion. These situations can range from stagnation (e.g., retained fluids and/or solids; contaminated water used for hydrotesting) to sustained fluid flow (e.g., erosion/cavitation in components moving in or contacted by fluids, as well as splashing or droplet impingement).

Common examples of stagnation include nondraining structures, dead ends, badly located components, and poor assembly or maintenance practices (Fig. 29). General problems include localized corrosion associated with differential aeration (oxygen concentration cells), crevice corrosion, and deposit corrosion.

Movement. Fluid movement need not be excessive to damage a material. Much depends on the nature of the fluid and the hardness of the material. A geometric shape may create a sustained delivery of fluid or may locally disturb a laminar stream and lead to turbulence. Replaceable baffle plates or deflectors are beneficial where circumstance permit their use; they eliminate the problem of impingement damage to the structurally significant component.

Careful fabrication and inspection should eliminate or reduce poor profiles (e.g., welds, rivets, bolts), rubbing surfaces (e.g., wear, fretting), and galvanic effects due to the assembly of incompatible components. Figure 30 shows typical situations in which geometric details influence flow.

Compatibility. In plant environments, it is often necessary to use different materials in close proximity. Sometimes, components that were designed in isolation can end up in direct contact in the plant (Fig. 31). In such instances, the ideals of a total design concept become especially apparent, but usually in hindsight. Direct contact of dissimilar metals introduces the possibility of galvanic corrosion, and small anodic (corroding) areas should be avoided wherever this contact is apparent.

Galvanic corrosion resulting from metallurgical sources is well documented. Problems such as weld decay and sensitization can generally be avoided by material selection or suitable fabrication techniques. Less obvious instances of local-

Table 6 Types of corrosion and corrodents encountered in high-temperature processes or components

Process components	Temperature, °C	Types of corrosion or corrodent
Chemical/petrochemical		
Ethylene steam cracking furnace tubes	to 1000	Carburization, oxidation
Steam reforming tubes	to 1000	Oxidation, carburization
Vinyl chloride crackers	to 650	Halide gas
Hydrocracking heaters, reactors	to 550	H_2S and H_2
Petroleum coke calcining recuperators	816	Oxidation, sulfidation
Cat cracking regenerators	to 800	Oxidation
Flare stack tips	950–1090	Oxidation, thermal fatigue, sulfidation, chlorination, dewpoint
Carbon disulfide furnace tubes	850	Sulfidation, carburization, deposits
Melamine production (urea)-reactors	450–500	Nitriding
Other processes		
Titanium production reactor vessels	900	Oxidation, chlorination
Nitric acid—catalyst grid	930	Oxidation, nitriding, sulfidation
Nuclear reprocessing reactors	750–800	Oxidation (steam), fluorination (HF)
Oil-fired boiler superheaters	850–900	Fuel ash corrosion
Gas turbine blades corrosion	to 950	Sulfates, chlorides, oxidation, ash
Waste incinerators—superheaters	480	Chlorination, sulfidation, oxidation, molten salts
Fiberglass manufacturing recuperators	1090	Oxidation, sulfidation, molten salts

Source: Ref 58

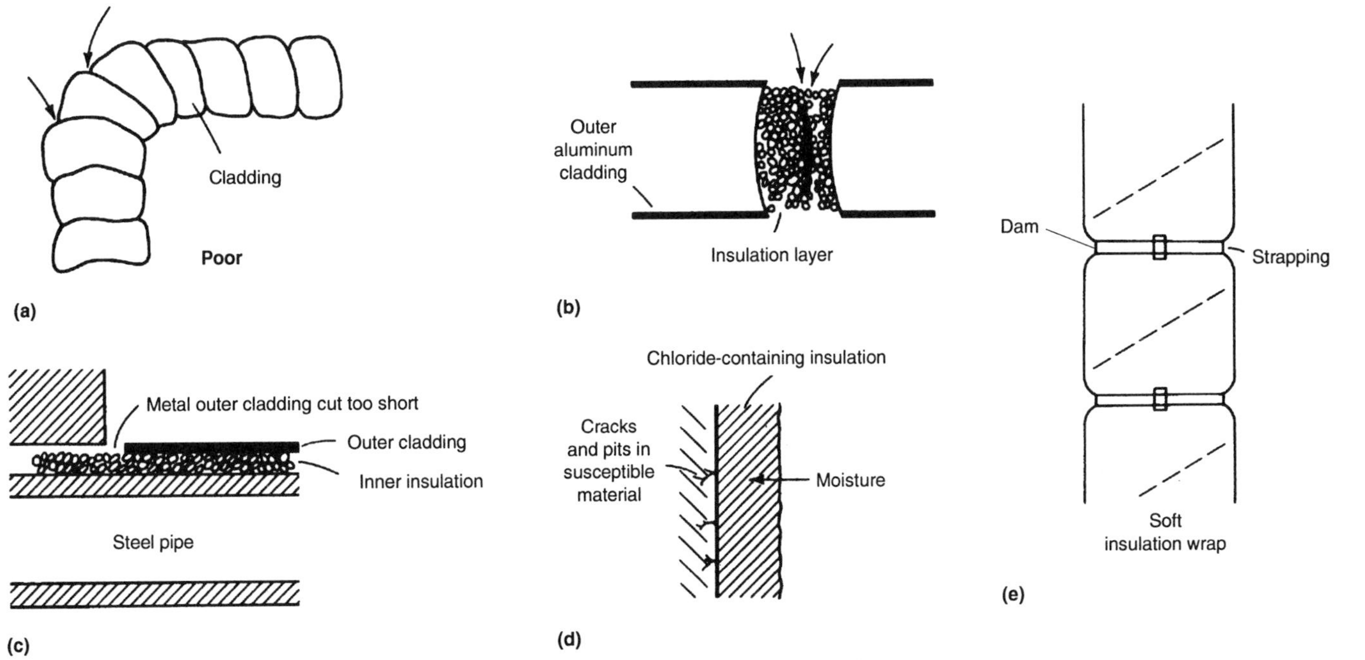

Fig. 34 Corrosion problems associated with improper use of insulation and lagging. (a) Incorrect overlap in lobster-back cladding does not allow fluid runoff. (b) Poor installation left a gap in the insulation that allows easy access to the elements. (c) Outer metal cladding was cut too short, leaving a gap with the inner insulation exposed. (d) Insufficient insulation can allow water to enter; chloride in some insulation can result in pitting or stress-corrosion cracking of susceptible materials. (e) Overtightened strapping can damage the insulation layer and cause fluid "dams" on vertical runs

ized attack occur because of end-grain attack and stray-current effects, which can render designs ineffective.

End-grain attack, or preferential attack of grains exposed by cross-cutting through a metal plate or rod (Fig. 31c), occurs in many corrosive fluids. An example is the cut edges and punched-out "holes" in a stainless steel reactor tray (Fig. 32).

Stray-current effects are common on underground iron or steel pipelines that are located close to electrical supply lines, or where stray currents can cause active corrosion at preferred sites. For example, stray-current attack caused a titanium flange spacer to become anodic "during service" due to inappropriate grounding of welding equipment on an adjacent part of the plant structure.

Designers, when aware of compatibility effects, need to exercise their ingenuity to minimize the conditions that most favor increased corrosion currents. Table 8 provides some suggestions.

The most common design details relating to galvanic corrosion include jointed assemblies (Fig. 29, 31). Where dissimilar metals are to be used, some consideration should be given to compatible materials known to have similar potentials (Fig. 13). Care should be exercised because the galvanic series is limited and refers to specific environments, usually seawater.

Where noncompatible materials are to be joined, it is advisable to use a more noble metal in a joint (Fig. 31b). Effective insulation can be useful if it does not lead to crevice corrosion. Some difficulties arise in the use of adhesives, which may not be sealants.

The relative surface areas of anodic and cathodic surfaces should not be underestimated, because corrosion at a small anodic zone can be several hundred times greater than that for the same bimetallic components of similar area. As noted already, synergistic effects must also be recognized, such as failure resulting from a combination of galvanic and crevice attack. This example shows the unfortunate choice of a carbon steel bolt to tighten the spindle in a stainless steel paddle stirrer (Fig. 33). The result was crevice corrosion at the stirrer support (Fig. 28), which was exacerbated by the galvanic action caused by the carbon steel/stainless steel metal-to-metal contact.

Less obvious examples of galvanic corrosion occur when ion transfer results in the deposition of active and noncompatible deposits on a metal surface. For example, an aluminum stirrer plate used in water was extensively pitted because the water bath was heated by a copper heater coil (Fig. 31e). The pits resulted from deposition of copper ions from the heater element. More rapid, but similar, damage occurred when a dental aspirator (Teflon-coated aluminum) was attacked by mercury from a tooth filling. These two metals

Table 7 Corrosion factors that can influence design considerations

Environment	Natural Chemical Storage/transit
Stress	Residual stress from fabrication Operating stress—static, variable, alternating
Shape	Joints, flanges Crevices, deposits Liquid containment and entrapment
Compatibility	Metals with metals Metals with other materials Quality control
Movement	Flowing fluids Parts moving in fluids Two-phase fluids
Temperature	Oxidation, scales Heat-transfer effects Molten deposits Condensation and dewpoint
Control	Surface cleaning and preparation Coatings Cathodic protection Inhibitors Inspection Planned maintenance

Source: Ref 57

Table 8 Sources of increased corrosion currents and related design considerations

Source	Design considerations
Metallurgical sources (both within the metal and for relative contact between dissimilar metals)	Difference in potential of dissimilar materials; distance apart; relative areas of anode and cathode; geometry (fluid retention); mechanical factors (e.g., cold work, plastic deformation, sensitization)
Environmental sources	Conductivity and resistivity of fluid; changes in temperature; velocity and direction of fluid flow; aeration; ambient environment (seasonal changes, etc.)
Miscellaneous sources	Stray currents; conductive paths; composites (e.g., concrete rebars)

rank as a "high risk" combination for galvanic corrosion (see Fig. 13). The aluminum section was rapidly pitted once the Teflon had first been worn away by sharp fragments of tooth enamel.

Anodic components can on occasion be overdesigned (made thicker) to allow for the anticipated corrosion loss. In other instances, easy replacement is a cost-effective option (Table 5).

Where metallic coatings are used, there may be a risk of galvanic corrosion, especially along the cut edges. Rounded profiles and effective sealants or coatings are beneficial. Transition joints can be introduced when different metals are to be in close proximity. These and other situations are illustrated in Fig. 31. Another possibility is coating of the cathodic material for corrosion control. Ineffective painting of an anode in an assembly can significantly reduce the desired service lifetime, because local defects (anodes) effectively multiply the risk of localized corrosion.

Insulation represents another area for potential corrosion attack, although most problems arise because of poor installation. Insulation types and properties vary considerably, and expert advice from suppliers is recommended. The most common corrosion problems include crevice corrosion (where insulation and/or adhesives are tightly held against a metal surface, for example when straps or ties are too tight) and pitting corrosion (where moisture condenses on the metal, usually because the insulation barrier was too thin or was improperly installed). Moisture-absorbing tendencies vary from one insulation to another (Ref 75).

Wet-dry cycling has been known to lead to concentration effects (e.g., chloride ions from calcium-silicate insulation). There have been reported instances of chloride stress-corrosion cracking (SCC) in certain stainless steel pipes and vessels, or pitting of these and other materials, such as aluminum, when contacted by insulation. The early instances of SCC failure were mainly attributable to high chloride levels (500–1500 ppm) associated with asbestos-type materials. The chloride levels have been significantly reduced in recent years to a level that is not expected to cause SCC. Standards are now available, as are tests to evaluate insulation materials (Ref 76). A parallel German standard calls for zero nitrite content and <0.2% ammonia levels in elastomeric insulation that is used for copper and copper-alloy piping to reduce the risk of SCC (Ref 77).

Figure 34 shows some typical examples in which design and installation procedures could have been improved. Other problems occur when insulation is torn or joints are misaligned or incorrectly sealed with duct tape or similar bandaging, none of which is recommended by insulation suppliers.

Stress. From a general design philosophy, environments that promote metal dissolution can be considered more damaging if stresses are also involved. In such circumstances, materials can fail catastrophically and unexpectedly. Safety and health may also be significantly affected.

A classic example of chloride SCC occurred in a type 304 (UNS S30400) stainless steel vessel (Fig. 35). The stress-corrosion cracks extended radially over the area where a new flanged outlet was welded into the vessel. Residual stresses (from flame cutting) and the fluids inside the vessel (acidic with chlorides) were sufficient to cause this failure in a matter of weeks.

Figure 36 shows examples of using design detail to minimize stress. Perfection is rarely attained in general practice, and some compromise on materials limitation, both chemical and mechanical, is necessary. Mechanical loads can contribute to corrosion, and corrosion (as a corrosive environment) can initiate or trigger mechanical failure. Designs that introduce local stress concentrations directly or as a consequence of fabrication should be carefully considered.

Of particular importance in design are stress levels for the selected material: the influence of tensile, compressive, or shear stressing; alternating stresses; vibration or shock loading; service temperatures (thermal stressing); fatigue; and wear (fretting, friction). Profiles and shapes contribute to stress-related corrosion, especially if material selection dictates the use of materials that are susceptible to failure by SCC or corrosion fatigue (Ref 10, 22, 30, 78).

Materials selection is especially important wherever critical components are used. Also important is the need for correct procedures at all stages of operation, including fabrication, transport, storage, startup, shutdown, and normal operation.

Surfaces. Corrosion is a surface phenomenon, and the effects of poorly prepared surfaces, rough textures, and complex shapes and profiles can be expected to be deleterious (Ref 79). Figure 37 shows some examples in which design details could have considerably reduced the onset of corrosive damage resulting from ineffective cleaning or painting.

Designs should provide for surfaces that are free from deposits; access to remove retained soluble salts before painting; free-draining as-

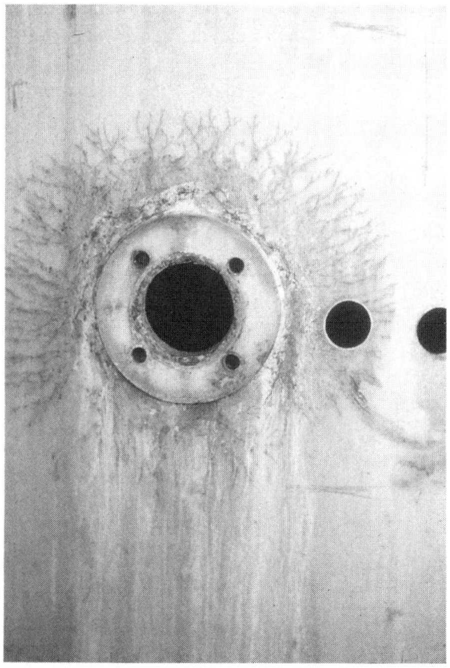

Fig. 35 Chloride SCC in a type 304 stainless-steel vessel after a new flange connection was welded into place

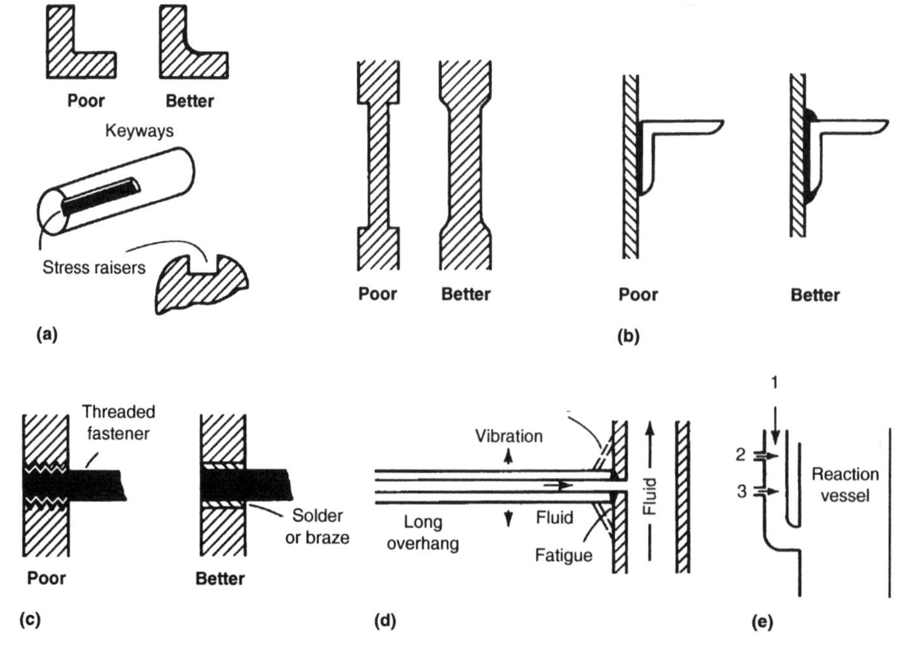

Fig. 36 Design details that can minimize local stress concentrations. (a) Corners should be given a generous radius. (b) Welds should be continuous to minimize sharp contours. (c) Sharp profiles can be avoided by using alternative fastening systems. (d) Too long an overhang without a support can lead to fatigue at the junction. Flexible hose may help alleviate this situation. (e) Side-supply pipework may be too rigid to sustain thermal shock from a recurring sequence that involves (1) air under pressure, (2) steam, and (3) cold water.

semblies; proper handling of components to minimize distortion, scratches, and dents; and properly located components relative to adjacent equipment (to avoid carryover and spillages). Other recommended procedures for coating constructional materials are shown in Fig. 38 (Ref 18).

Neglect and poor (or no) maintenance caused localized pitting on the underside of a type 304 stainless steel vessel lid that was exposed to high humidity, steam, and chloride vapors. Access in this example was possible, but not used. Common engineering structural steelwork requires regular preventive maintenance, and restricted access makes this impossible. Figure 37 shows situations in which surface cleaning and/or painting is difficult or impossible. Condensation in critical areas can also contribute to corrosion. Typical structures susceptible to this phenomenon include automobile exhaust systems and chimneys or exhausts from high-temperature plants, such as boilers, kilns, furnaces, or incinerators.

Painting and surface coating techniques have advanced in recent years and have provided sophisticated products that require careful mixing and application. Maintenance procedures frequently require field application where some control (use of trained inspectors) is essential, as in offshore oil and gas rigs. Inspection codes and procedures are available and total design should incorporate these wherever possible. In critical areas, design for online monitoring and inspection will also be important. The human factor in maintenance procedures is often questionable. Adequate training and motivation are of primary importance in ensuring that design details are appreciated and implemented.

Life Prediction and Management

As pointed out in previous sections, corrosion degradation is largely understood mechanistically, and logical mitigation actions and design decisions can be formulated with a reasonable scientific basis. However, in recent years there has been a further requirement in some industries that the lifetime of various components be predicted and that the technical (and economic) benefits of life extension actions be quantitatively defined. Such further design requirements for life management are especially seen, for instance, in the aerospace and power industries.

To meet these particular design criteria, it is necessary to derive accurate and verifiable life prediction methods. An example is given below of such a derivation for the example of environ-

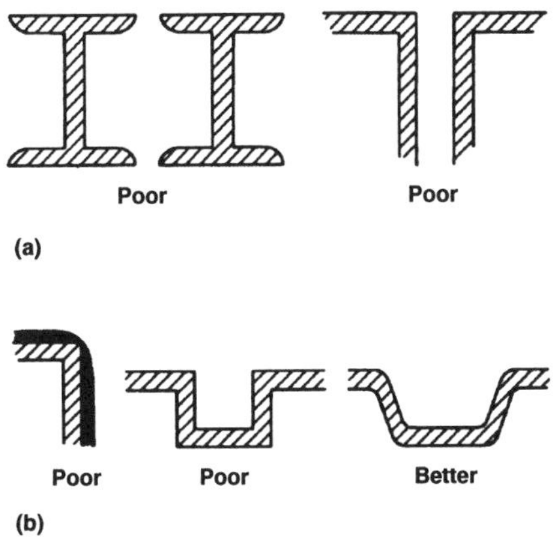

Fig. 37 Effects of design on effectiveness of cleaning or painting. (a) Poor access in some structures makes surface preparation and painting difficult; access to the types of areas shown should be maintained at a minimum of 45 mm (1¾ in.), or one-third of the height of the structure. (b) Sharp corners and profiles should be avoided if the structure is to be painted or coated.

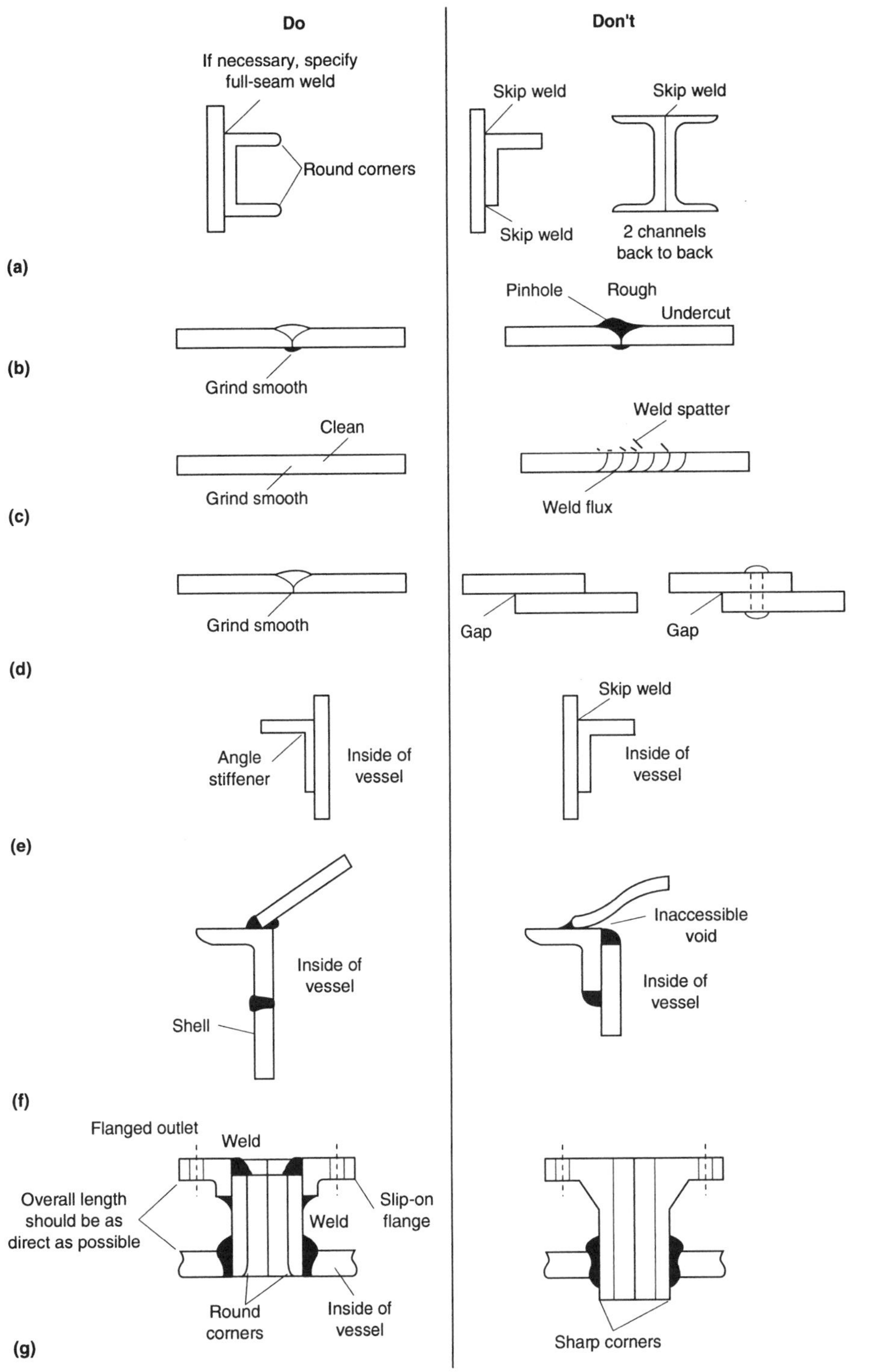

Fig. 38 Suggestions for steel construction to be coated. (a) Avoid pockets or crevices that do not drain or cannot be cleaned or coated properly. (b) Joints should be continuous and solidly welded. (c) Remove weld spatter. (d) Use butt welds rather than lap welds or rivet joints. (e) Keep stiffeners to outside of tank or vessel. (f) Eliminate crevice (void) at roof-to-shell interface in nonpressure vessel. (g) Outlets should be flanged or pad type, not threaded. Where pressure limits allow, slip-on flanges are preferred, because the inside surface of the attaching weld is readily available for radiusing and grinding. Source: Ref 80

mental degradation in light-water nuclear reactors (LWRs). The specific corrosion phenomena in these structures have been widely publicized and discussed, especially in a series of conferences (Ref 81–90) organized by the National Association of Corrosion Engineers (NACE, now known as NACE International), the American Institute of Mining, Metallurgical, and Petroleum Engineers (AIME), and the American Nuclear Society (ANS), and in books (Ref 91, 92). Moreover, this particular topic of corrosion in nuclear reactors has been addressed in the *ASM Handbook* (Ref 93). These incidences have covered all forms of corrosion, including erosion corrosion, galvanic attack, general corrosion, intergranular corrosion, SCC, and corrosion-fatigue cracking in, primarily, carbon/low-alloy steels, austenitic stainless steels, nickel-base alloys, and zirconium-base alloys. To one extent or another, these phenomena have affected the integrity of most of the components in reactors, including irradiated core structures, piping, tubing, pressure vessel, service water systems, and heat exchangers. Moreover, the publications have addressed the whole spectrum, from the mechanistic aspects of degradation in a particular alloy/environment system to the adequacy of the existing design and life prediction methods.

The objective of this section is to focus on one form of degradation, environmentally assisted cracking, not only because it is one of the severest forms of life-limiting phenomena, but also because it represents an area where "fundamental" approaches to resolving the design/prediction problems are most advanced.

There have been well-documented instances of environmentally assisted cracking (e.g., stress-corrosion and strain-assisted cracking and corrosion fatigue) in high-temperature water of austenitic stainless steels, nickel-base alloys, low-alloy and carbon steels, and their weld metals in various subcomponents of LWRs. These events have had a considerable economic impact (Ref 81) and are occurring at a time when competitive and deregulation pressures are forcing nuclear utilities (at least in the United States) to reduce costs (Ref 82). Thus, there is an incentive to develop validated life prediction methods to (a) evaluate the extent of a problem (i.e., is an incident an isolated occurrence or is it the precursor to a more widespread generic problem?) and (b) quantitatively define the benefit of a specific mitigation action and evaluate its cost-effectiveness.

Unfortunately, the scatter in data from which such design or life prediction codes can be formulated is extreme. This scatter applies to data obtained from both field experience (Ref 94) and "well-controlled" laboratory experiments. This latter observation is illustrated in Fig. 39 by crack-propagation rate/stress-intensity relationships for stainless steels (Ref 96). Similar data scatter has been noted for nickel-base alloys (Ref 97) and low-alloy steels (Ref 97) in 288 °C high-purity water. In other instances, a large database does not exist, so it is impossible to define a usable life prediction algorithm. An example of this is irradiation-assisted stress-corrosion cracking of stainless steels in reactor cores.

Consequently, the current life prediction codes for environmentally assisted cracking of structural materials in 274 to 345 °C water, repre-

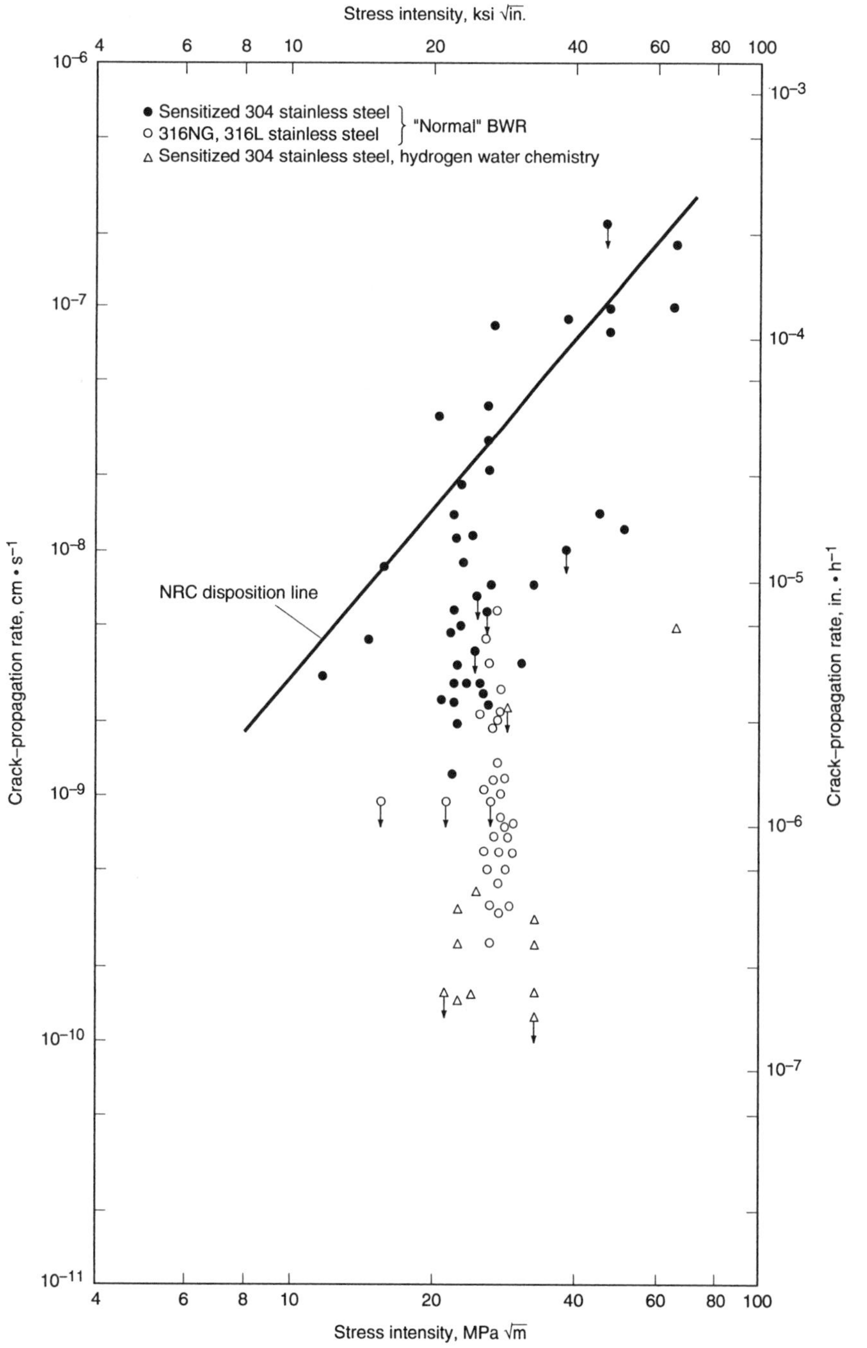

Fig. 39 Crack-propagation rate vs. stress-intensity data for stainless steel in 288 °C water. Also shown is the Nuclear Regulatory Commission disposition line. Source: Ref 95

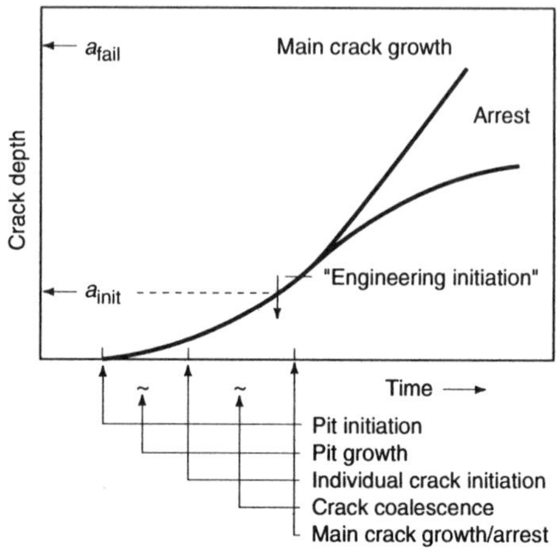

Fig. 40 Proposed sequence of crack initiation, coalescence, and growth for steels undergoing subcritical cracking in aqueous environments

sentative of the coolant, used in boiling-water reactors (BWRs) and pressurized-water reactors (PWRs), usually represent upper bounds of the database available at a given time. Moreover, a single crack-propagation-rate/stress-intensity relationship (e.g., ASME Section XI for life prediction/evaluation) or cyclic-stress/cycles-to-crack-initiation relationship (e.g., ASME Section III for fatigue design) is usually quoted with no account of the different cracking susceptibilities that are associated with the wide range of material/environment conditions in LWR systems. This conservative "upper bound" scenario can put unreasonable constraints on the continued operation of specific plants that have been operating under system conditions far better than those on which the design or life evaluation code was originally based.

Various statistical (extreme-value, pattern-recognition) or neural-network approaches can be applied to the available cracking database to develop an algorithm for component life, or crack depth, as a function of all the reactor operating parameters. This "classical" approach is limited, however, because there are comparatively few datasets for crack propagation, for example, that have been obtained under adequate control of system conditions (e.g., water purity, loading) and using sensitive enough techniques that continuously monitor propagation rates that are relevant to design lives of the order of 40 to 60 years for LWRs.

This situation is offset by the increase in mechanistic understanding of environmentally assisted cracking, which has accelerated in the last 20 years due to the availability of experimental and analytical procedures that allow quantification and validation of various cracking hypotheses. Because the precise shape of the crack depth/operational time relationship must be a function of the specific material, environmental, and stressing conditions, it follows that if there is a range in the actual system operating conditions, then there will be a predictable range in the observed cracking susceptibility, with the distribution of the range in cracking mirroring the distribution of system conditions that affect the propagation process. This predictable range in observed cracking susceptibility provides a bridge between deterministic and statistical/probabilistic life prediction methods (Ref 96, 99).

Historically, environmentally assisted cracking has been divided into the "initiation" and "propagation" periods. To a large extent, this division is arbitrary, because in most investigations, "initiation" is defined as the time at which a crack is detected, or when the load has relaxed a specific amount (in a strain-controlled situation). Such a definition of initiation, however, can correspond to a crack depth of significant metallurgical dimensions (e.g., >2 mm). Thus, for the purpose of lifetime modeling, it is proposed (Fig. 40) that, phenomenologically, initiation is associated with microscopic crack formation at localized corrosion or mechanical-defect sites and is generally related to pitting, intergranular attack, scratches, weld defects, or design notches. It is further proposed that the probability is high of such microscopic initiation sites existing or developing relatively early in the life of the component. Then the problem of life prediction devolves to understanding the growth of small cracks from these

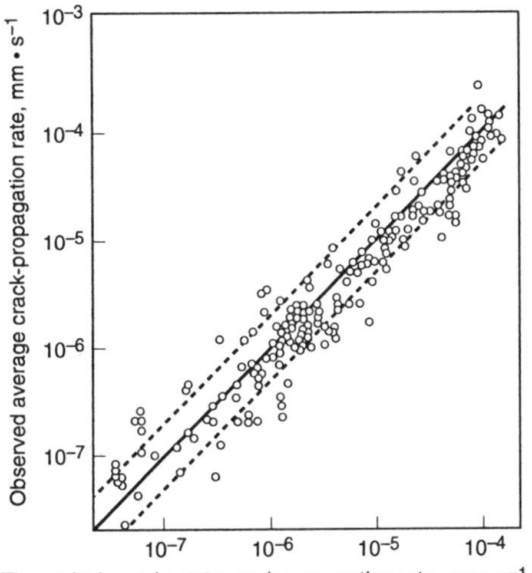

Fig. 41 Comparisons between observed and theoretical crack-propagation rates for type 304/316 stainless steels in 288 °C water. This database represents a wide combination of stressing material and environmental conditions. Source: Ref 96

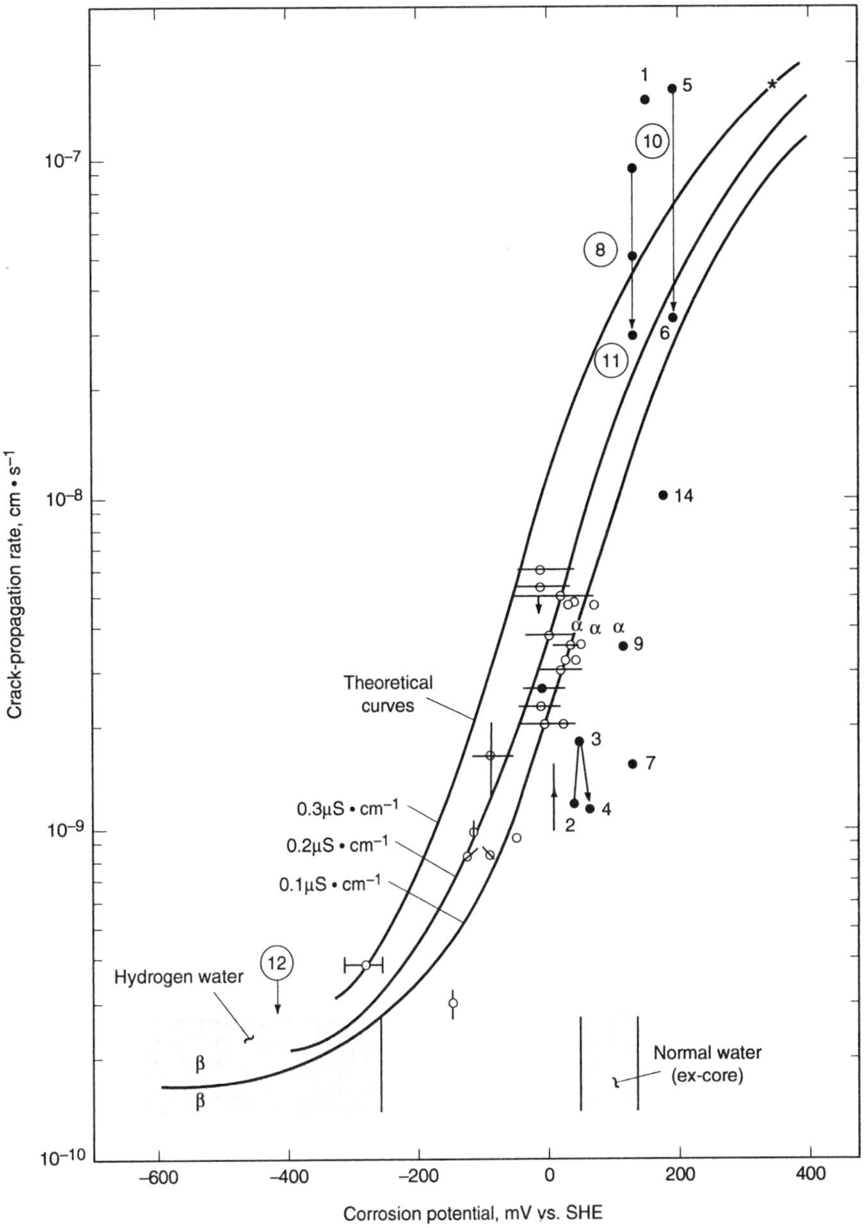

Fig. 42 Observed and predicted sensitivity of stress-corrosion-cracking sensitivity to corrosion potential for sensitized type 304 stainless steel in 288 °C water. The data points are measurements made in the laboratory or in reactors. The curves are the predicted relationships for the indicated conductivities. The numbered data points were obtained at the Harwell variable-energy cyclotron. The circled numbers were with the proton irradiation turned on, and the uncircled numbers were with the irradiation off. Similarly the data point * was obtained under fast neutron irradiation in a boiling-water-reactor core.

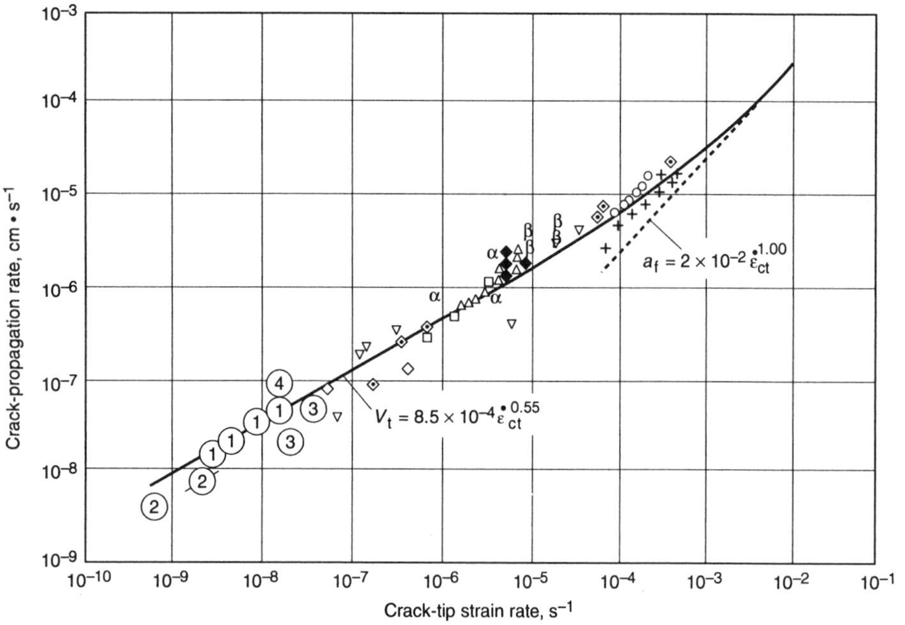

Fig. 43 Predicted and observed crack-propagation rate/crack-tip strain-rate relationships for sensitized type 304 stainless steel in 8 ppm oxygenated, 0.5 μS · cm^{-1} purity water at 288 °C

geometrically separated initiation sites, and the coalescence of these small cracks to form a major crack, which may then accelerate or arrest, depending on the specific material, environment, and stress conditions.

Attention has been focused primarily on the propagation rates for "deep" cracks (e.g., approximately one grain diameter and greater), recognizing that the resultant life prediction can be conservative because the microscopic crack initiation and crack coalescence periods are not being accounted for.

Numerous mechanisms have been advanced for crack propagation in ductile alloy/aqueous environment systems. The usual procedure in developing a quantitative predictive model for crack propagation is to advance a hypothesis for the cracking mechanism and then independently quantify the relevant parameters in that mechanism. The resultant prediction algorithms are then validated by comparison with observed crack-propagation data. Once validated, these algorithms are used for design-life-extension decisions. This process is described very briefly below for stainless steels in 288 °C water (the boiling-water reactor coolant), where the hypothesis is made that the cracking mechanism is slip dissolution (Ref 96, 100, 101).

Quantification of the Slip-Oxidation Mechanism for Stainless Steels in 288 °C High-Purity Water. Various theories have been proposed to relate crack propagation in various ductile alloys, including stainless steels, to oxidation rates and the stress/strain conditions at the crack tip. These theories were supported by an early correlation between the average oxidation current density on a straining surface and the crack-propagation rate for several systems (Ref 101, 102). Experimentally validated elements of these earlier proposals have been combined such that crack propagation in many systems can be correlated with the oxidation that occurs when the protective film at the crack tip is ruptured (Ref 103, 104).

Consider the change in oxidation charge density with time following the rupture of a protective film at the crack tip. Initially, the oxidation rate and, hence, crack-advance rate are rapid and are typically controlled by activation or liquid diffusion kinetics as the bare metal dissolves. In most LWR cracking systems, a protective oxide rapidly reforms at the bared surface and the total oxidation rate (and crack-tip advance) will slow with time. Thus, crack advance can be maintained only if the film rupture process is repetitive. Therefore, for a given crack-tip environment and material condition, the crack-propagation rate will be controlled by the change in oxidation-charge density with time and the frequency of film rupture at the strained crack tip. This latter parameter is determined by the fracture strain of the film, ε_f, and the strain rate at the crack tip, $\dot{\varepsilon}_{ct}$. By invoking Faraday's law, the average environmentally controlled crack-propagation rate, $\overline{V}_t$, can be related to the oxidation-charge density (metal oxidation current integrated over time) passed between film-rupture events, Q_f, and the strain rate at the crack tip, $\dot{\varepsilon}_{ct}$:

$$\overline{V}_t = \frac{M}{z\rho F} \frac{Q_f}{\varepsilon_f} \dot{\varepsilon}_{ct} \qquad \text{(Eq 23)}$$

where M and ρ are the atomic weight and density of the crack-tip metal, respectively.

Because the oxidation-charge density on a bare surface varies with time at a rate that is dependent on the material and environment compositions, Eq 23 can be reformulated in terms of a power-law relationship:

$$\overline{V}_t = A\,(\dot{\varepsilon}_{ct})^n \qquad \text{(Eq 24)}$$

where A and n are constants that depend on the material and environment compositions at the crack tip, and are related to the oxidation reaction rates, or current densities, in the specific crack-tip material/environment system (Ref 102).

Under constant load or displacement conditions, the crack-tip strain rate can be related to the creep processes at a moving crack tip, and under monotonically increasing or cyclically changing bulk strain conditions, ε_{ct} can be related to an applied strain rate, $\dot{\varepsilon}_{app}$ (Ref 96, 102). Thus, the slip-oxidation mechanism can be applied not only to stress corrosion but also to strain-induced cracking (Ref 105) and corrosion fatigue. Under cyclic loading conditions, however, the crack is also moving forward by irreversible cyclic plastic deformation (e.g., fatigue-striation formation). Because this mechanical crack advance occurs independently of the crack advance by oxidation processes, these two crack-advance mechanisms (fatigue-striation formation and corrosion oxidation) are considered additive.

Thus, the starting point in the development of a quantitative prediction method for environmentally assisted cracking is Eq 24. However, to develop a *useful* methodology, it is necessary to redefine this fundamental equation in terms of measurable engineering or operational parameters. This redefinition involves: (a) defining the crack-tip alloy/environment composition (e.g., in terms of bulk alloy composition, anionic concentration or solution conductivity, dissolved-oxygen content or corrosion potential); (b) measuring the reaction rates for the crack-tip alloy/environment system that corresponds to the "engineering" system; and (c) defining the crack-tip strain rate in terms of continuum parameters such as stress, stress intensity, and loading frequency. Extensive work has been conducted in these areas, which has been reviewed elsewhere (Ref 10).

As a result of these examinations of the crack-tip metallurgical, chemical, and stressing conditions, practical crack-propagation-rate algorithms of the following form have been developed for stainless steels in 288 °C BWR water:

$$V_t = 7.8 \times 10^{-3} n^{3.6} (\dot{\varepsilon}_{ct})^n \qquad \text{(Eq 25)}$$

$$n = f(\kappa, \text{EPR}, \phi_c) \qquad \text{(Eq 26)}$$

$$\dot{\varepsilon}_{ct} = 6 \times 10^{-14} K^4 \quad \text{for constant load} \qquad \text{(Eq 27)}$$

$$\dot{\varepsilon}_{ct} = 5\dot{\varepsilon}_{app} \quad \text{for monotonically increasing strain} \qquad \text{(Eq 28)}$$

$$\dot{\varepsilon}_{ct} = 100\dot{\nu}A_R\,\Delta K^4 \quad \text{for cyclic loading} \qquad \text{(Eq 29)}$$

where κ is the conductivity of coolant ($\mu S \cdot cm^{-1}$), ϕ_c is the corrosion potential of the steel (mV_{SHE}), EPR is the measurement of grain-boundary chromium depletion due to heat treatment annealing or welding, K is the stress intensity (ksi$\sqrt{in.}$), $\dot{\varepsilon}_{app}$ is the applied strain rate (s^{-1}), $\dot{v}$ is the cyclic-loading frequency (s^{-1}), ΔK is the stress-intensity amplitude under cyclic loading, and A_R is a parameter that is a function of the mean stress under cyclic loading.

Validation of Life-Prediction Algorithms and Their Application. The overall comparison between the observed and theoretical crack-propagation rates in type 304/316 stainless steels in 288 °C water is shown in Fig. 41. The laboratory database upon which this comparison was made was obtained under a wide range of stressing (static, monotonically increasing, and cyclic load), material (solution annealed vs. various degrees of sensitization) and water composition (<10 ppb O_2 to >8 ppm O_2, <0.1 to 10 $\mu S \cdot cm^{-1}$). It is seen that there is a reasonable agreement between observation and prediction.

Changes in corrosion potential within the range expected in BWRs can have a significant effect on the cracking susceptibility of type 304/316 stainless steels, especially under constant-load conditions. This predicted and observed effect is illustrated in Fig. 42 for furnace-sensitized type 304 stainless steel under constant stress intensity (25 ksi$\sqrt{in.}$) in water with the conductivity in the range 0.1 to 0.3 $\mu S \cdot cm^{-1}$. It is seen that over the corrosion potential range −550 mV_{SHE} to +250 mV_{SHE} (spanning "hydrogen-water" conditions to those under "normal" core conditions) the crack-propagation rate can change three orders of magnitude. From an operational design viewpoint, therefore, it is seen that considerable benefit may be predicted by developing actions that lower the corrosion potential of the stainless steel structures, thereby highlighting remedial actions that lower the effective concentration of oxidants (oxygen, hydrogen peroxide) in the coolant. Solution conductivity is also predicted to have an effect on the cracking susceptibility, as indicated by the three theoretical relationships shown in Fig. 42, thereby highlighting the quantitative value of maintaining water-purity control.

So far, the comparisons between observation and theory have centered on material/environment systems variables that affect n in Eq 25 and 26. The effect of stressing/straining conditions on the cracking susceptibility occur primarily through their effect on the crack-tip strain rate in Eq 27–29. It follows that because the crack tip does not recognize *how* the strain rate is maintained, the cracking susceptibility for a given material/environment condition should adhere to the same crack-propagation rate/crack-tip strain-rate relationship, regardless of the stressing/straining mode. The truth to this statement is illustrated in Fig. 43, which shows the theoretical and observed crack-propagation rate strain-rate relationship for a severely sensitized type 304 stainless steel in 8 ppm O_2, 0.5 $\mu S \cdot cm^{-1}$ water. Movement along the strain-rate axis has been achieved by increasing stress intensity under constant-load conditions, increasing applied strain rate under monotonically increasing strain conditions, or cyclic loading under a variety of stress-intensity amplitude, mean stress, and loading frequency conditions. The single theoretical relationship line in Fig. 43 adequately predicts the cracking under this wide range of loading modes, indicating that the prediction method applies to stress-corrosion cracking (SCC), strain-induced cracking (SIC), and corrosion fatigue (CF).

The old lore that these types of cracking (SCC, SIC, CF) are separate phenomena with, by implication, different mitigation or design modification needs is probably incorrect. For instance, it follows from Eq 25 that the sensitivity of the cracking susceptibility to the crack-tip strain rate will be a function of the material/environment conditions that affect n (Eq 26). Thus, the slope of the crack-propagation-rate/strain-rate relationship will be relatively shallow for severe environ-

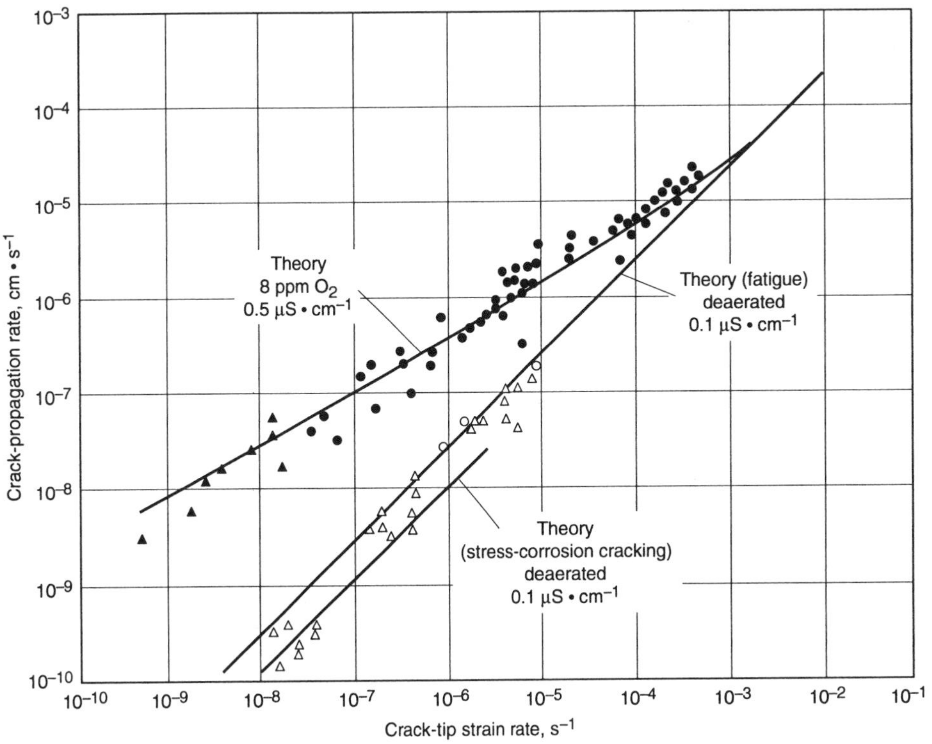

Fig. 44 Predicted and observed crack-propagation rate/crack-tip strain-rate relationships for stainless steels in a variety of material/environment systems

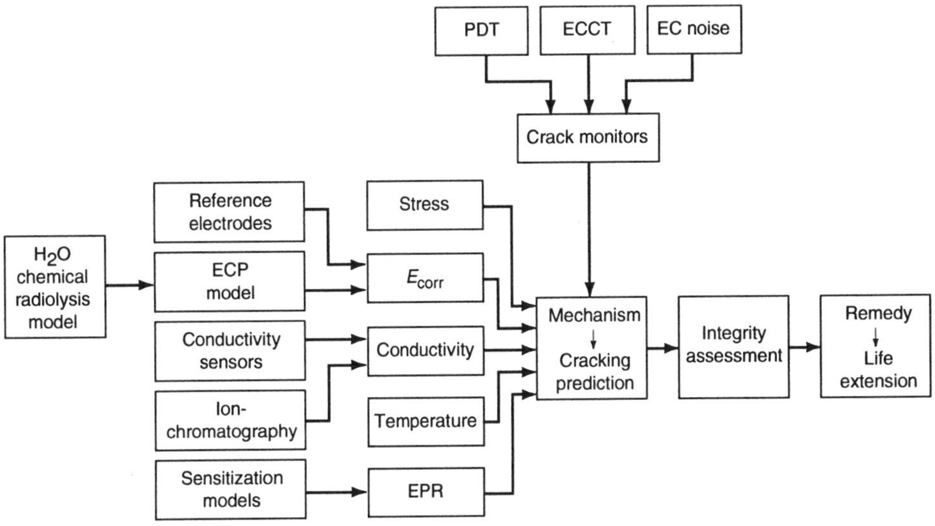

Fig. 45 The integration of system monitors, sensors, and environmental/material models as inputs to a crack-propagation-rate model

mental and material conditions (e.g., high dissolved oxygen, impure water, and high degrees of grain-boundary sensitization), and the relationship will be steep for less severe material/environmental conditions. This predicted and observed (Fig. 44) change in propagation-rate/strain-rate dependency with system conditions is significant when evaluating the validity of accelerated tests that are often used for development of design codes. For instance, increasing the crack-tip strain rate, and hence cracking susceptibility, by using the "slow-strain-rate test" is a valid test acceleration procedure (because it is accelerating one of the rate-determining steps in the cracking mechanism), but the factor of improvement between a reference condition and a proposed mitigation condition will be less in this test than at the lower stressing or strain-rate conditions expected in the operating plant. The relationship (i.e., Fig. 44) also gives an explanation for the lore that the cracking susceptibility is more dependent on the specific environmental conditions under constant-load stress-corrosion conditions than under corrosion-fatigue conditions.

In summary, therefore, it is apparent that the crack-prediction algorithms are able to quantitatively explain the changes in crack-propagation rates for type 304/316 stainless steel in water at 288 °C for a wide combination of water composition (corrosion, potential, conductivity), material sensitization, and stressing (constant load/displacement, cyclic load) conditions. It follows, however, that because the cracking response is so sensitive to changes in combinations of system conditions, it is necessary to combine the predictive method with system-defining sensors and models (Fig. 45). Provided this combining is done, it is then possible to make predictions of the extent of cracking in specific plant components (Fig. 46) and the increase in life associated with specific system changes (Fig. 47).

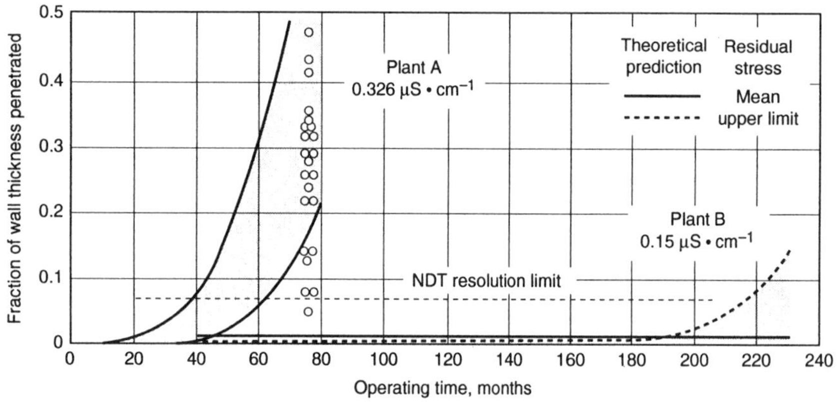

Fig. 46 Theoretical and observed intergranular stress corrosion crackdepth vs. operational-time relationships for 28 in. diameter schedule 80 type 304 stainless steel piping for two boiling-water reactors operating at different mean coolant conductivities. Note the bracketing of the maximum crack depth in the lower-purity plant by the predicted curve, which is based on the maximum residual-stress profile and the predicted absence of observable cracking in the higher-purity plant (in 240 operating months).

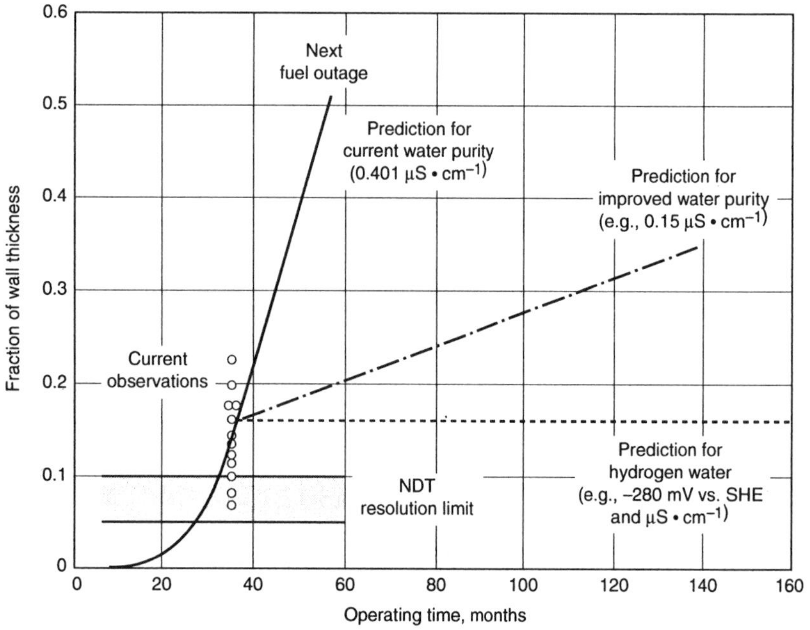

Fig. 47 Predicted crack depth vs. time response for defected 28 in. diameter schedule 80 recirculation piping in a given boiling-water reactor to defined changes in water purity. Also shown is the crack-depth limit that can be resolved by nondestructive testing (NDT).

REFERENCES

1. H. Uhlig, *Chemical and Engineering News,* Vol 97, 1949, p 2764
2. Editorial, *Corrosion Prevention and Control,* Vol 27, 1980, p 1
3. T.P. Hoar, *Report of the Committee on Corrosion and Protection,* Her Majesty's Stationery Office, London, 1971
4. *Proc. 1986 Joint Chinese-American Corrosion Workshop,* Industrial Technology Research Institute, Hsinchu, Taiwan, Dec 1986
5. D.A. Jones, *Principles and Prevention of Corrosion,* 2nd ed., Prentice Hall, 1996
6. K.R. Trethewey and J. Chamberlain, *Corrosion for Science and Engineering,* 2nd ed., Longman, 1995
7. P. Marcus and J. Oudar, *Corrosion Mechanisms in Theory and Practice,* Marcel Dekker, Inc., 1995
8. C.P. Dillon, *Corrosion Resistance of Stainless Steels,* Marcel Dekker, Inc., 1995
9. B.D. Craig, *Fundamental Aspects of Corrosion Films in Corrosion Science,* Plenum Press, 1991
10. M.G. Fontana, *Corrosion Engineering,* 3rd ed., McGraw-Hill Book Co., 1986
11. W.W. Kirk and H.H. Lawson, *Atmospheric Corrosion,* ASTM, 1995
12. J.C. Scully, *The Fundamentals of Corrosion,* Pergamon Press, 1975
13. H.P. Hack, *Galvanic Corrosion,* ASTM, 1988
14. S.L. Chawla and R.K. Gupta, *Materials Selection for Corrosion Control,* ASM International, 1993
15. P.A. Schweitzer, *Corrosion and Corrosion Protection Handbook,* 2nd ed., Marcel Dekker, 1989
16. G. Moran and P. Labine, *Corrosion Monitoring in Industrial Plants Using Nondestructive Testing and Electrochemical Methods,* ASTM, 1986
17. D.O. Northwood, W.E. White, and G.F. Vander Voort, *Corrosion, Microstructure, and Metallography,* American Society for Metals, 1985
18. R.S. Treseder, R. Baboian, and C.G. Munger, Ed., *NACE Corrosion Engineer's Reference Book,* 2nd ed., NACE, 1991
19. R.B. Seymour, *Plastics vs. Corrosives,* John Wiley & Sons, 1982
20. M. Henthorne, *Localized Corrosion—Cause of Metal Failure,* ASTM, 1972
21. R. Baboian, *Electrochemical Techniques for Corrosion Engineering,* NACE, 1985
22. *Corrosion,* Vol 13, *ASM Handbook* (formerly *Metals Handbook,* 9th ed.), ASM International, 1987
23. S.K. Coburn, *Corrosion Source Book,* American Society for Metals, 1984

24. A.J. McEvily, Jr., *Atlas of Stress-Corrosion and Corrosion Fatigue Curves,* ASM International, 1990
25. L.L. Shreir, R.A. Jaman, and G.T. Burstein, *Corrosion Metal/Environment Reactions,* Butterworth Heinenmann, Ltd., 1994
26. R.F. Steigerwald and N.D. Greene, *J. Electrochem. Soc.,* Vol 109, 1962, p 1026
27. H.H. Uhlig and R.W. Rene, *Corrosion and Corrosion Control,* 3rd ed., John Wiley & Sons, 1985, p 217
28. Z. Szklarska-Smialawska, *Pitting Corrosion of Metals,* NACE, 1986
29. F.P. Ford, "Mechanisms of Environmental Cracking Peculiar to the Power Generation Industry," Report NP2589, EPRI, 1982
30. F.P. Ford, Stress Corrosion Cracking, *Corrosion Processes,* R.N. Parkins, Ed., Applied Science, 1982
31. R.N. Parkins, N.J.H. Holroyd, and R.R. Fessler, *Corrosion,* Vol 34, 1978, p 253
32. B. Poulson and R. Robinson, *Corr. Sci.,* Vol 20, 1980, p 707
33. J. Congleton, "Some Aspects of Crack Initiation in Stress Corrosion and Corrosion Fatigue," paper presented at Corrosion 88, NACE, St. Louis, 21–25 March 1988
34. *Conf. Proc., Environmental-Sensitive Mechanical Behavior* (Baltimore, MD, June 1965), A.R.C. Westwood and N.S. Stoloff, Ed., Gordon and Breach, 1966
35. R.W. Staehle, A.J. Forty, and D. Van Rooyen, Ed., *The Fundamental Aspects of Stress-Corrosion Cracking,* Ohio State University, Sept 1967
36. J.C. Scully, Ed., *Theory of Stress Corrosion Cracking,* NATO, Brussels, March 1971
37. O. Devereaux, A.J. McEvily, and R.W. Staehle, Ed., *Corrosion Fatigue Chemistry, Mechanics and Microstructure,* University of Connecticut, Storrs, June 1971
38. M.P. Bastein, Ed., *L'Hydrogene dans les Metaux,* Science et Industrie, Paris, 1972
39. L.M. Bernstein and A.W. Thompson, Ed., *Hydrogen in Metals, L,* American Society for Metals, 1973
40. R.W. Staehle, J. Hochmann, R.D. McCright, and J.E. Slater, Ed., Stress-Corrosion Cracking and Hydrogen Embrittlement of Iron-Base Alloys, NACE, 1977
41. A.W. Thompson and I.M. Bernstein, Ed., *Proc. Effect of Hydrogen on Behavior of Materials* (Jackson Lake, WY, Sept 1975), TMS, 1976
42. R.M. Latanision and J.T. Fourie, Ed., *Surface Effects on Crystal Plasticity (*Hohegeiss, Germany, 1975), Noordhof-Leyden, 1977
43. P.R. Swann, F.P. Ford, and A.R.C. Westwood, Ed., *Mechanisms of Environment Sensitive Cracking of Materials,* The Metals Society, April 1977
44. Corrosion Fatigue, *Met. Sci.,* Vol 13, 1979
45. T.R. Beck, *Corrosion,* Vol 30, 1974, p 408
46. R.W. Staehle, in *Theory of Stress Corrosion Cracking,* J.C. Scully, Ed., NATO, Brussels, March 1971
47. J.C. Scully, *Corros. Sci.,* Vol 8, 1968, p 771
48. D.J. Lees, F.P. Ford, and T.P. Hoar, *Met. Mater.,* Vol 7, 1973, p 5
49. J.R. Ambrose and J. Kruger, *J. Electrochem. Soc.,* Vol 121, p 1974, p 599
50. F.P. Ford and M. Silverman, *Corrosion,* Vol 36, 1980, p 558
51. V.R. Pludek, *Design and Corrosion Control,* MacMillan, 1977
52. R.J. Landrum, *Fundamentals of Designing for Corrosion Control,* NACE International, 1989
53. R.N. Parkins and K.A. Chandler, *Corrosion Control in Engineering Design,* Department of Industry, Her Majesty's Stationery Office, London, 1978
54. L.D. Perrigo and G.A. Jensen, Fundamentals of Corrosion Control Design, *The Northern Engineer,* Vol 13 (No. 4), 1982, p 16
55. *Designer Handbooks,* Specialty Steel Industry of North America, Washington, D.C.; also publications relative to design, Nickel Development Institute, Toronto, Canada
56. *Guides to Practice in Corrosion Control,* Department of Industry, Her Majesty's Stationery Office, London, 1979–1986
57. *Engineering Design Guides,* Design Council, British Standards Institute, Council of Engineering Institutions, Oxford University Press, 1975–1979
58. P. Elliott and J.S. Llewyn-Leach, *Corrosion Control Checklist for Design Offices,* Department of Industry, Her Majesty's Stationery Office, London, 1981
59. P. Elliott, *Corrosion Control in Engineering Design,* audiovisual for Department of Industry, United Kingdom, 1981
60. O.W. Siebert, Classic Blunders in Corrosion Protection, *Mater. Perform.,* Vol 17 (No. 4), 1978, p 33 and Vol 22 (No. 10), 1983
61. T.F. Degnan, *Mater. Perform.* Vol 26 (No. 1), 1987, p 11
62. P. Elliott, Why Must History Repeat Itself?, *Ind. Corros.,* Feb/March 1991, p 8
63. P. Elliott, Process Plant Corrosion—Recognizing the Threat, *Process Eng.,* Vol 65 (No. 11), 1984, p 43
64. P. Elliott, Understanding Corrosion Attack, *Plant Eng.,* Oct 1993, p 68
65. P. Elliott, *Corrosion Survey,* Supplement to *Chem. Eng.,* Sept 1973
66. P. Elliott, Catch 22 and the UCS Factor—Why Must History Repeat Itself?, *Mater. Perform.,* Vol 28 (No. 7), 1989, p 70 and Vol 28 (No. 8), 1989, p 75
67. *Standards for Corrosion Testing of Metals,* ASTM, 1990
68. R. Baboian, Ed., *Corrosion Tests and Standards: Applications and Interpretation,* ASTM Manual Series, MNL-20, 1995
69. H.J.H. Wassell, *Reliability of Engineered Products,* Engineering Design Guide, Design Council, Oxford University, 1980
70. P. Elliott, We Never get Corrosion Problems, *Super News,* 1974, p 70
71. A. Sparks, *Steel—Carriage by Sea,* 2nd ed., Lloyd's of London Press, 1995
72. G. Kobrin, Ed., *Microbiologically Influenced Corrosion,* NACE International, 1993
73. P. Elliott, *Practical Guide to High Temperature Alloys, Mater. Perform.,* Vol 28, 1989, p 57
74. G.Y. Lai, *High Temperature Corrosion of Engineering Alloys,* ASM International, 1990
75. W. Pollock, *Corrosion under Wet Insulation,* NACE International, 1988
76. "Specification for Wicking-Type Thermal Insulation for Use Over Austenitic Stainless Steel," C 795, *Annual Book of ASTM Standards,* ASTM
77. "Codes of Practice for Drinking Water Installations (TRWI)," 628.1.033:696.11:620.193, DIN, Teil 7, 1988
78. H.H. Uhlig, *Corrosion and Corrosion Control,* 2nd ed., John Wiley & Sons, 1971, p 314
79. C.G. Munger, *Corrosion Prevention by Protective Coatings,* NACE International, 1984
80. P.E. Weaver, "Industrial Maintenance Painting," RP0178, NACE International, 1973, p 2
81. R.L. Jones, "Corrosion Experience in U.S. Light Water Reactors—NACE 50th Anniversary Perspective," Paper 168, presented at Corrosion 93, NACE, 1993
82. R.L. Jones, "Critical Corrosion Issues and Mitigation Strategies Impacting the Operability of LWRs," Paper 103, presented at Corrosion 96, NACE, 1996
83. Conf. Proc., *Environmental Degradation of Materials in Nuclear Systems—Light Water Reactors,* J. Roberts and W. Berry, Ed., NACE, 1983
84. Conf. Proc., *Environmental Degradation of Materials in Nuclear Systems—Light Water Reactors,* J. Roberts and J. Weeks, Ed., ANS, 1985
85. Conf. Proc., *Environmental Degradation of Materials in Nuclear Systems—Light Water Reactors,* J. Weeks and G. Theus, Ed., TMS, 1987
86. Conf. Proc., *Environmental Degradation of Materials in Nuclear Systems—Light Water Reactors,* G. Theus and D. Cubicciotti, Ed., NACE, 1989
87. Conf. Proc., *Environmental Degradation of Materials in Nuclear Systems—Light Water Reactors,* D. Cubicciotti and E. Simonen, Ed., ANS, 1991
88. Conf. Proc., *Environmental Degradation of Materials in Nuclear Systems—Light Water Reactors,* R. Gold and E. Simonen, Ed., TMS, 1993
89. Conf. Proc., *Environmental Degradation of Materials in Nuclear Systems—Light Water Reactors,* R. Gold and E. McIlree, Ed., NACE, 1995
90. H. Okada and R. Staehle, Ed., *Predictive Methods for Assessing Corrosion Damage to BWR Piping and PWR Steam Generators,* NACE, 1982
91. D.D. MacDonald and G.A. Cragnolino, Corrosion of Steam Cycle Materials, *ASME Handbook on Water Technology for Thermal Power Systems,* P. Cohen, Ed., ASME, 1979
92. J.T.A. Roberts, *Structural Materials in Nuclear Power Systems,* Plenum Press, 1981
93. J.C. Danko, Corrosion in the Nuclear Power Industry, *Corrosion,* Vol 13, *ASM Handbook,* ASM International, 1987
94. D.A. Hale, C.W. Jewett, and C.S. O'Toole, "BWR Coolant Impurities Program," First An-

nual Progress Report, Report NP2293, EPRI, Nov 1985

95. W.S. Hazelton, "Technical Report on Materials Selection and Processing Guidelines for BWR Coolant Pressure Boundary Piping," Draft report NUREG 0313 Rev. 2, U.S. Nuclear Regulatory Commission, 1978
96. F.P. Ford, D.F. Taylor, P.L. Andresen, and R.G. Ballinger, "Corrosion Assisted Cracking of Stainless Steel and Low Alloy Steels in LWR Environments," Report NP5064S, EPRI, Feb 1987
97. P.L. Andresen, *Corrosion 47,* NACE, 1991, p 917–938
98. F.P. Ford, "Environmentally Assisted Cracking of Low Alloy Steels," Final Report of Contract C102-1, Report NP7473-L, EPRI, Jan 1992
99. F.P. Ford, P.L. Andresen, M.G. Benz, and D. Weinstein, On-Line BWR Materials Monitoring and Plant Component Lifetime Prediction, *Proc. Nuclear Power Plant Life Extension,* American Nuclear Society, Vol 1, June 1988, p 355–366
100. F.P. Ford, "Mechanisms of Environmental Cracking Peculiar to the Power Generation Industry," Report NP2589, EPRI, Sept 1982
101. F.P. Ford, Stress Corrosion Cracking, *Corrosion Processes,* R.N. Parkins, Ed., Applied Science, 1982
102. F.P. Ford, The Crack Tip System and its Relevance to the Prediction of Environmentally Assisted Cracking, *Proc. First International Conf. Environment Induced Cracking of Metals,* NACE, Oct 1988, p 139–166
103. R.N. Parkins, Environment Sensitive Fracture—Controlling Parameters, *Proc. Third International Conf. Mechanical Behavior of Materials,* K.J. Miller and R.F. Smith, Ed., Pergamon, Vol 1, 1980, p 139–164
104. T.R. Beck, *Corrosion 30,* NACE, 1974, p 408
105. J. Hickling, "Strain Induced Corrosion Cracking: Relationship to Stress Corrosion Cracking/Corrosion Fatigue and Importance for Nuclear Plant Service Life, paper presented at Third IAEA Specialists Meeting on Subcritical Crack Growth, Moscow, May 1990
106. K. Osozaawa and H.J. Engell, *Corros. Sci.,* Vol 6, 1966, p 389

SELECTED REFERENCES*

- V.A. Ashworth and P. Elliot, Guide to the Corrosion Resistance of Metals, *Metals Reference Book,* 5th ed., C.J. Smithells and E.A. Brandes, Ed., Butterworths, 1976, p 1460
- B.D. Craig and D. Anderson, Ed., *Handbook of Corrosion Data,* 2nd ed., ASM International, 1995
- *Corrosion Data Survey: Metals Section,* 6th ed., NACE, 1985
- *Corrosion Data Survey: Nonmetals Section,* 5th ed., NACE, 1975
- D.J. De Renzo, Ed., *Corrosion-Resistant Materials Handbook,* 4th ed., Noyes, 1985
- *DECHEMA Corrosion Handbook: Corrosive Agents and Their Interaction with Materials,* D. Behrens (Vol 1-9) and G. Kreysa and R. Eckermann (Vol 10-12), Ed., VCH, 1987-1993
- *NACE/NIST Corrosion Performance Databases,* Corrosion Data Center, National Institute of Standards and Technology, Gaithersburg, MD
- P.A. Schweitzer, Ed., *Corrosion Resistance Tables,* 3 vol, 4th ed., Marcel Dekker, 1995
- H.H. Uhlig, *Corrosion Handbook,* John Wiley & Sons, 1948
- H.H. Uhlig and R.W. Revie, *Corrosion and Corrosion Control: An Introduction to Corrosion Science and Engineering,* 3rd ed., Wiley, 1985

*See also Ref 5–28, 68, and 79 in the list of numbered references.

Design for High-Temperature Applications

David A. Woodford, Materials Performance analysis, Inc.

APART FROM nineteenth-century steam boilers, machines and equipment for high-temperature operation have been developed principally in the present century. Energy conversion systems based on steam turbines, gas turbines, high-performance automobile engines, and jet engines provide the technological foundation for modern society. All of these machines have in common the use of metallic materials at temperatures where time-dependent deformation and fracture processes must be considered in their design. The single valued time-invariant strain associated with elastic or plastic design analysis in low-temperature applications is not applicable, nor is there in most situations a unique value of fracture toughness that may be used as a limiting condition for part failure. In addition to the phenomenological complexities of time-dependent behavior, there is now convincing evidence that the synergism associated with gaseous environmental interactions may have a major effect, in particular on high-temperature fracture.

This article reviews the basic mechanisms of elevated-temperature behavior and associated design considerations with emphasis on metals. Subsequently, the engineering analysis will be confined to presenting data in the form that a designer might use, with emphasis on design principles rather than detailed design analysis. Thus, multiaxial stresses, part analysis, and creep-fatigue interaction are not formally treated. However, remaining life assessment and the effect of nonsteady stresses are covered. A broader treatment of most of these aspects can be found in other articles that appear in the *ASM Specialty Handbook: Heat-Resistant Materials* and in *Mechanical Testing,* Volume 8 of the *ASM Handbook.* Emphasis here is placed on developing an appreciation of the uses (and abuses) of creep and rupture testing, data presentation, data analysis, limitations of long-time tests, and alternative approaches to high-temperature design. The objective is to provide a solid foundation for design principles from a materials performance perspective.

Historical Development of Creep Deformation Analysis

The phenomenon of time-dependent deformation was referred to as slow stretch by Philips (Ref 1) and as viscous flow by Andrade (Ref 2) at the beginning of this century and subsequently became known as creep. There were several seminal ideas in the Andrade work that have had a lasting impact on scientific studies and engineering dogma. The initial work was primarily on lead wires at room temperature (a high temperature relative to the melting point for lead) with some additional experiments on a 78.5% Sn 21.5% Pb alloy and copper. Andrade noted that after applying a fixed load the rate of extension initially decreased then became constant for a time, but finally increased and continued increasing until failure. He recognized that as the wire stretched, the load per unit area increased. Subsequently, he devised a scheme to compensate for this and maintain a constant stress on the wire. As a result of this, the extent of viscous flow, that is, extension linearly dependent on time, increased as shown in Fig. 1. Andrade also recognized that the length of wire being experimented on at any time is increasing and thus used the concept of true strain. He derived a formula to describe the observed deformation:

$$l = l_0 (1 + \beta t^{1/3})\, e^{kt} \qquad \text{(Eq 1)}$$

where l and l_0 are the current and initial specimen lengths, t is the time, and β and k are constants. The initial transient strain (later to be called primary creep) was referred to as beta creep and followed a time to the one-third law, the viscous region (later to be called steady-state creep) was proportional to time, and the accelerating strain region leading to fracture, which was not specifically treated by Andrade, later became known as tertiary creep. Much later, in a comprehensive study of creep in copper and aluminum, Wyatt (Ref 3) concluded that there are two types of transient creep in metals:

- At higher temperatures, beta creep predominates as in Andrade's experiments.
- At lower temperatures, the strain is proportional to log(time), and the flow is referred to as alpha creep.

From this early work, subsequent studies diverged into two investigative paths. The first sought understanding of creep deformation micromechanisms in pure metals and solid-solution alloys in relatively short-term tests, accepted the concept of steady-state creep (although testing was more often conducted at constant load rather than constant stress), and often assumed implicitly that viscous flow was history independent. This means that not only is there a steady creep rate associated with a given applied stress, but that this rate is obtained despite previous deformation at different stresses and temperatures. Although this might be a reasonable approximation for pure metals, it is manifestly wrong for most engineering alloys.

The second investigative path concentrated on generating long-time creep data on engineering materials. The testing was invariably at constant load, and data extracted included times for specific creep strains, minimum creep rates (al-

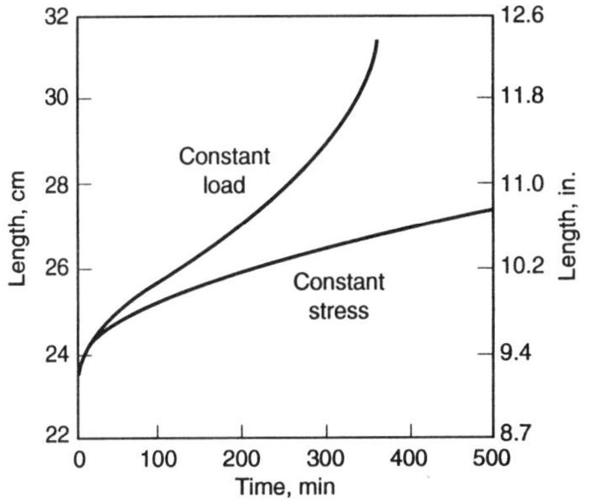

Fig. 1 Creep tests on lead wire. In both tests, initial lengths and initial loads were the same. Source: Ref 2

though the term steady state was often used despite the fact that constant rates cannot be expected when the stress is changing), and time to failure (often referred to as rupture life). This latter measurement was of special significance because it became a basis for design against part failure, and later as a basis for estimating remaining life of operating components.

There thus emerged a framework for design against both creep deformation and fracture using a single testing procedure. It formed a basis for what might be called an uncracked body analysis and comprises the major part of this article. Analysis of cracked bodies involving fracture mechanics concepts as applied to creeping structures is not covered although some reference is made as appropriate. In particular, the importance of fatigue loading is emphasized in the article "Creep-Fatigue Interaction" in the *ASM Specialty Handbook: Heat-Resistant Materials.*

Until the last quarter century, virtually all creep and creep fracture studies were on metallic materials. However, as early as 1903, Philips (Ref 1) recognized that the phenomenon was not unique to the metallic bond and that materials with covalent and ionic bonds showed similar effects. In fact, creep of polymers is now of considerable importance in plastic automobile components and gas lines, and creep of ceramics is of interest in aerospace applications.

Basic Concepts of Elevated-Temperature Design

Time-dependent deformation and fracture of structural materials at elevated temperatures are among the most challenging engineering problems faced by materials engineers. In order to develop an improved design methodology for machines and equipment operating at high temperatures, several key concepts and their synergism must be understood. As is described in this section, these include:

- Plastic instability at elevated temperatures
- Deformation mechanisms and strain components associated with creep processes
- Stress and temperature dependence
- Fracture at elevated temperatures
- Environmental effects

Design Phenomenology

The issues of interest from a design basis are the nature of primary creep, the validity of the concept of viscous steady-state creep, and the dependence of deformation on both temperature and stress. The simplest and most pervasive idea in creep of metals is an approach to an equilibrium microstructural and mechanical state. Thus a hardening associated with dislocation generation and interaction is countered by a dynamic microstructural recovery or softening. This process proceeds during primary creep and culminates in a steady-state situation. The idea was first presented by Bailey (Ref 4) and subsequently in the following mathematical form by Orowan (Ref 5):

$$d\sigma = \frac{\partial \sigma}{\partial \varepsilon} d\varepsilon + \frac{\partial \sigma}{\partial t} dt \qquad \text{(Eq 2)}$$

where $d\sigma$ represents the change in flow stress, $\partial\sigma/\partial\varepsilon$ represents the hardening that results from an increment of plastic strain $d\varepsilon$, and $\partial\sigma/\partial t$ represents the softening due to recovery in a time increment dt.

At constant stress (and temperature), the steady-state creep rate is given by:

$$\dot{\varepsilon} = -\frac{\partial\sigma/\partial t}{\partial\sigma/\partial\varepsilon} \qquad \text{(Eq 3)}$$

The numerator is frequently given the symbol r as the recovery rate associated with thermal softening, and the denominator is referred to as the strain-hardening coefficient, h. Although there is evidence that both hardening and softening processes occur during creep, and despite the fact that numerous studies have attempted to quantify Eq 2, it is in fact incorrect. As pointed out by McCartney (Ref 6), Eq 2 implies that an equation of state exists of the form:

$$\sigma = \sigma(\varepsilon, t) \qquad \text{(Eq 4)}$$

Equation 2 is the differential form of Eq 4, which assumes that the variables ε and t are independent. Since measured strain is in fact a function of elapsed time t, it follows that the partial derivatives have no meaning. Lloyd and McElroy (Ref 7) also concluded that the concept required a history-dependent term and the idea had serious deficiencies. They further concluded that the related concept of the applied stress being the sum of an internal stress opposing dislocation motion and an effective stress as the driving force for motion was inconsistent with real behavior. Their alternative theory draws on the observation of anelastic phenomena, which is considered in a subsequent section.

The concept of steady-state creep has been addressed rigorously in very few publications (Ref 8). From these limited studies, however, it can be stated that a constant creep rate cannot occur during the changing stress conditions of the common constant load test (or, if the creep rate appears constant, it cannot be steady state). Further, it can be said that most engineering alloys undergo purely time-dependent changes at temperature associated with an approach to thermodynamic equilibrium, such as precipitate coarsening. With the additional complication of strain-induced changes, it is unlikely that a steady state could be established. It is especially improbable that such a state could be history independent. Any search for a true steady state should, therefore, be limited to pure metals or solid-solution alloys and would require constant stress testing and true-strain plotting.

Plastic Instability

A major issue in the tensile creep test is the role of plastic instability in leading to tertiary creep. Understanding of the nature of plastic instability for time-dependent flow has depended on the theory of Hart (Ref 9). He showed that the condition for stable deformation is:

$$\gamma + m \geq 1 \qquad \text{(Eq 5)}$$

where m, which equals $[(\partial \ln \sigma)/(\partial \ln \dot{\varepsilon})]_{\varepsilon}$, is the strain-rate sensitivity, and γ, which equals $[(\partial \ln \sigma)/(\partial \varepsilon)]_{\dot{\varepsilon}}$, is a measure of the strain-hardening rate. For steady-state flow, γ is equal to 0. For constant stress tests, Burke and Nix (Ref 10) concluded that flow must be unstable when steady state is reached according to Hart's criterion but that macroscopic necking is insignificant and that the flow remains essentially homogeneous. They concluded that a true steady state does exist. Hart himself questioned the conclusions based on their analysis but did not rule out the possibility of a steady state for pure metals (Ref 8). In a very careful experimental analysis, Wray and Richmond (Ref 11) concluded that the concept of a family of steady states is valid. They advocated tests in which two of the basic parameters (stress, strain rate, and temperature) are held constant. However, they reported the intrusion of nonuniform deformation before the steady state was reached. They also pointed out the complexities associated with uncontrolled and often unmeasured loading paths, which produce different structures at the beginning of the constant stress or constant strain rate portions of the test. For constant stress tests in pure metals, although the concept of steady state (viscous flow in Andrade's terminology) is appealing, it appears not yet to have been rigorously demonstrated.

In constant load tests, steady-state behavior would of course result in an increasing creep rate after the minimum, as the true stress increases. As such, the test is inappropriate to evaluate the concept. However, it is by far the most common type of creep test and can be analyzed for instability (Ref 12). The condition for instability may be stated:

$$\ddot{A} \geq 0 \qquad \text{(Eq 6)}$$

where $\ddot{A}$ is the second derivative of specimen cross-sectional area with respect to time. This in turn leads to a point of instability expressed in terms of gage length:

$$\frac{\ddot{A}}{A} = \frac{\ddot{L}}{L} - \frac{2\dot{L}}{L} = 0 \qquad \text{(Eq 7)}$$

This criterion is shown in Fig. 2 for constant load tests on nickel. The instability criterion is fulfilled at a strain very close to that of the minimum creep rate. However, the value of this criterion remains low up to 20 to 25% strain, at which separate measurements of specimen profiles indicate that macroscopic necking occurs. In this respect, the results are similar to constant stress results (Ref 10) in that although

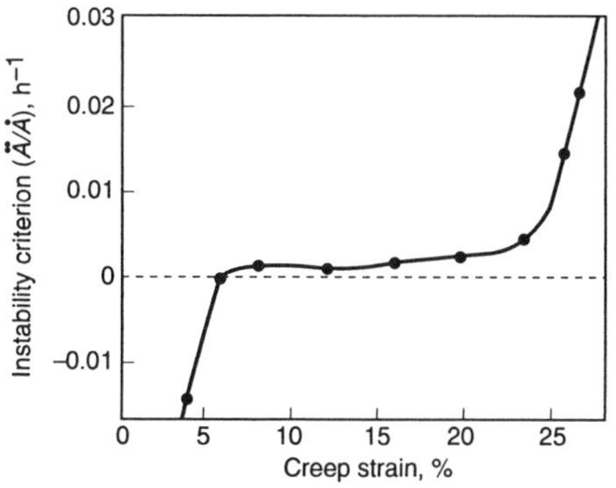

Fig. 2 Change in the parameter $\ddot{A}/\dot{A}$ with creep strain in nickel at 525 °C (980 °F) and 138 MPa (20 ksi). Source: Ref 12

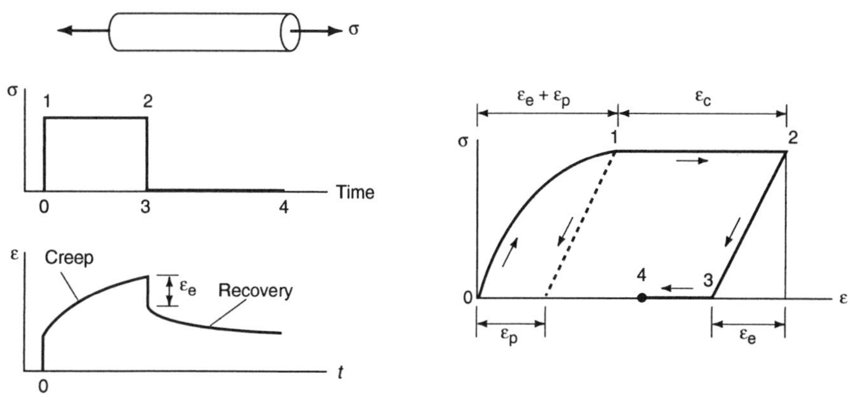

Fig. 3 Stress-time step applied to a material exhibiting strain response that includes time-independent elastic, time-independent plastic, time-dependent creep, and time-dependent anelastic (creep-recovery) components. Source: Ref 20

deformation is potentially unstable at the end of the primary stage, it is not grossly so.

Creep Processes

Creep behavior can be characterized either in terms of deformation mechanisms or in terms of strain constituents.

Deformation Mechanisms. Creep of metals is primarily a result of the motion of dislocations, but is distinct from time-independent behavior in that flow continues as obstacles, which may be dislocation tangles or precipitate particles, are progressively overcome. The rate-controlling step involves diffusion to allow climb of edge dislocations or cross slip of screw dislocations around obstacles. In steady-state theory, there is a balance between the hardening associated with this dislocation motion and interaction, and a dynamic recovery associated with the development of a dislocation substructure. Theory for such a process predicts a power-law dependence of creep rate on applied stress. For example, climb-controlled dislocation creep gives an exponent $n = 4$ in the following equation (Ref 13):

$$\dot{\varepsilon} = C\sigma^n \quad \text{(Eq 8)}$$

where C is a constant. Nothing in this or similar theories allows for history effects, and although the power function connection may be applicable, the value of n is not only invariably higher, but strongly history dependent in structural alloys (see the following section).

At very high homologous temperatures (T/T_m) and low stresses, creep may occur in both metals and ceramics by mass transport involving stress-directed flow of atoms from regions under compression to regions under tension. In this case, theory indicates that there is a stress dependence of unity and that the process is controlled either by bulk diffusion (Ref 14, 15) or by grain-boundary diffusion (Ref 16). These various processes of creep (dislocation controlled as well as diffusion controlled) may be represented on a deformation mechanism map to highlight regimes of stress and temperature where each mechanism, based on current theories, may be operating (Ref 17). However, such maps are only as good as the theories on which they are based and give no guidance on deformation path dependence.

Another important deformation process in metallic and ionic polycrystals at high temperature and low stresses is grain-boundary sliding (Ref 18). The resistance to sliding is determined by the mobility of grain-boundary dislocations and by the presence of hard particles at the boundary. This sliding leads to stress concentrations at grain junctions, which are important in nucleating cracks. In ductile materials, these stress concentrations may be relieved by creep and stress relaxation in the matrix or by grain-boundary migration (Ref 19).

Strain Components. There are several different sources of strain at high temperature in response to an applied stress. The elastic strain is directly proportional to stress, and a modulus that is temperature dependent can be determined. For metallic materials and ceramics, although there is a strain-rate dependence of elastic modulus, it is small and often ignored. For polymers, by contrast, the elastic modulus is ill defined because of viscoelasticity.

Plastic strain for all materials may be treated as three separate constituents:

- Time-independent nonrecoverable, which may be thought of as an instantaneous deformation
- Time-dependent nonrecoverable, which may involve any or all of the micromechanisms described above
- Time-dependent recoverable

The first of these is unlikely to be significant in practical applications except in the region of stress concentrations since loading is normally well below the macroscopic yield stress. The second is the major source of creep in normal laboratory testing. The third constituent is not widely studied or analyzed, but may become very important at low stresses and under nonsteady conditions, that is, high-temperature service. It leads to what has been termed creep recovery and anelasticity.

At high temperatures, the application of a stress leads to creep deformation resulting from the motion of dislocations, mass transport by diffusion, or grain-boundary sliding. These processes in turn lead to a distribution of internal stresses that may relax on removal of the stress. This relaxation leads to a time-dependent contraction in addition to the elastic contraction and results in the phenomenon of creep recovery illustrated in Fig. 3. In polymers this phenomenon, which may account for nearly all the nonelastic strain, is termed viscoelastic recovery and is associated with the viscous sliding and unkinking of long molecular chains (Ref 21). In metals it is associated with the unbowing of pinned dislocations (Ref 7), rearrangement of dislocation networks (Ref 22), and local grain-boundary motion (Ref 23). In ceramics it appears to be primarily a grain-boundary phenomenon (Ref 24).

Whereas the importance of creep recovery is well recognized in polymer design, it has often been ignored in design of metallic and ceramic materials. A few extensive studies have been reported on metals (Ref 25–27) that have led to several broad conclusions:

- Creep-recovery strain increases linearly with stress for a fixed time at a given temperature, but is dependent on prestrain.
- The rate of creep recovery increases with increasing temperature.
- When the stress is low enough, essentially all transient creep is linear with stress and recoverable.
- Mathematically, the recovery may be described by a spectrum of spring dashpot combinations with a wide range of relaxation times.

Assuming that the measured recovery strain after unloading had made an equivalent contribution to forward creep (Ref 28), it was possible in these studies to separate the anelastic and plastic creep components as shown in Fig. 4. Because the anelastic component is linear with stress and the plastic component is a power function of stress (for the same time), at very low stresses the strain is entirely anelastic. This observation led to the definition of a plastic creep limit that was time dependent. For times up to 100 h in a low-alloy

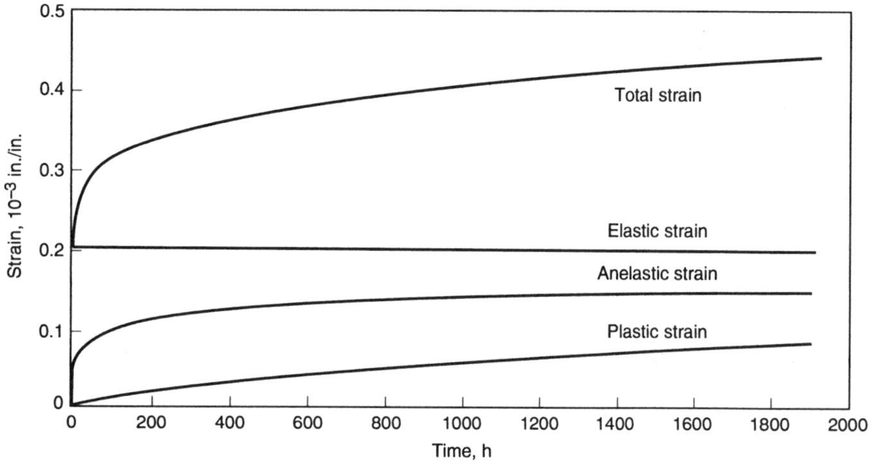

Fig. 4 The separation of strain components for a creep test on Cr-Mo-V steel at 538 °C (1000 °F) and 35 MPa (5 ksi). Source: Ref 27

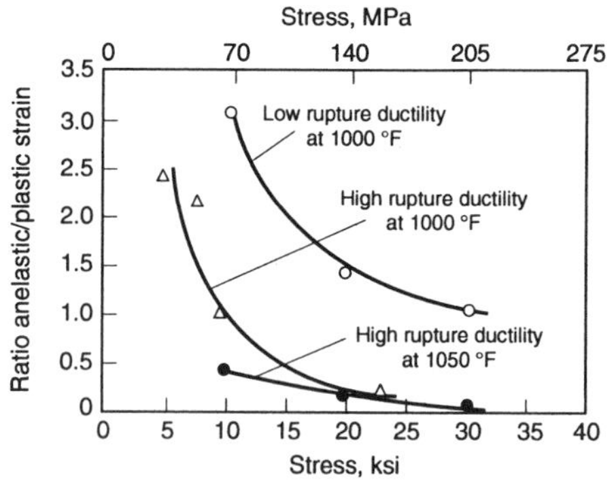

Fig. 5 Effect of ductility on recoverable creep strain for Cr-Mo-V steel after 1000 h creep exposure. Source: Ref 27

steel tested at 425 °C (800 °F), Lubahn (Ref 26) found this limit to be 140 MPa (20 ksi) (Ref 5); all creep below this stress was fully recoverable. In tests on a similar alloy at 538 °C (1000 °F), Goldhoff (Ref 27) found that the creep limit ranged from 150 MPa (22 ksi) for 1 h to zero at 5000 h. By plotting the ratio of anelastic to plastic strain for a fixed time (1000 h) as a function of stress (Fig. 5), Goldhoff (Ref 27) showed how the former became dominant at low stresses. Figure 5 also shows that a heat treatment that produces low ductility leads to higher ratios, suggesting a link between anelastic deformation and intergranular fracture, which was consistent with microstructural observations of fracture in this alloy.

There have been even fewer systematic studies of creep recovery in ceramics, but silicon carbide fibers have been shown to recover fully their creep strain between 1000 and 1400 °C (1830 and 2550 °F) (Ref 24). Additionally, provided an appropriate period was allowed for recovery after each stress cycle, tension-tension fatigue resulted in zero cumulative creep strain. This indicates the potential importance of anelastic phenomena in damage accumulation for nonsteady conditions. Very recent work on large specimens of silicon nitride have shown recovery of most of the accumulated strain after unloading from stress-relaxation tests (Fig. 6).

There are strong indications that anelastic phenomena should be included in design considerations. Anelastic contraction as well as extension can occur depending on whether the stress is decreased or increased, whereas plastic shortening never occurs. Although several authors have pointed out that, because of the linear stress dependence the analysis should be much simpler than for plastic creep analysis (Ref 7, 27), accurate measurements at the low stresses of interest for service applications are difficult. The possible link with fracture processes is also of great interest, but neither consideration has influenced design practice.

Stress and Temperature Dependence

The minimum creep rate in both constant load and constant stress tests is normally represented by a power function of stress (Eq 8), and the temperature by an Arrhenius expression including an activation energy term (Q) derived from chemical reaction rate theory (Ref 29):

$$\dot{\varepsilon} = S\sigma^n e^{-Q/RT} \quad \text{(Eq 9)}$$

where S, which is a constant, depends on structure. Although an exponential or hyperbolic sine stress function may provide a better fit in some cases, the power function has generally prevailed and has become strongly linked with mechanistic treatments. In pure metals, early studies indicated a stress exponent on the order of four and an activation energy close to that for self-diffusion (Ref 13, 29, 30). For engineering alloys, the stress exponents are generally higher and may not be constant (Ref 31), and the value of the activation energy may be much higher than that for the alloy matrix self-diffusion and may be sensitive to test temperature.

Because the basic formulation of Eq 9 is used to correlate much engineering data and is used in creep analysis of components, it is useful to examine critically some of the limitations in this analysis as they apply to engineering alloys. It was first shown by Lubahn (Ref 32) that, because of the rapidly decreasing creep rate in the primary stage, a strain-time plot of a portion of this stage always appears to show approximately constant rates at the longest times. This has led to many errors in the literature with false minimum creep rates. Some of these errors may lead to apparent n values close to one and consequent speculations about Newtonian viscous creep (Ref 33). Figure 7 shows results for minimum creep rates in a Cr-Mo-V steel in tests lasting up to 50,000 h. Also included are plots where time restrictions on the measurements were imposed to illustrate this potential for error. Nevertheless, the true minimum data points indicate n values ranging from 3.3 to 12.

As pointed out by Woodford (Ref 33), the curvature indicates that Eq 9 with S as a constant does not apply over the stress range, and it is meaningless to consider both n and S changing. In fact, the slope at any point has no clear physical significance because the structural state at the minimum creep rate is different for each test because of the different deformation history. To

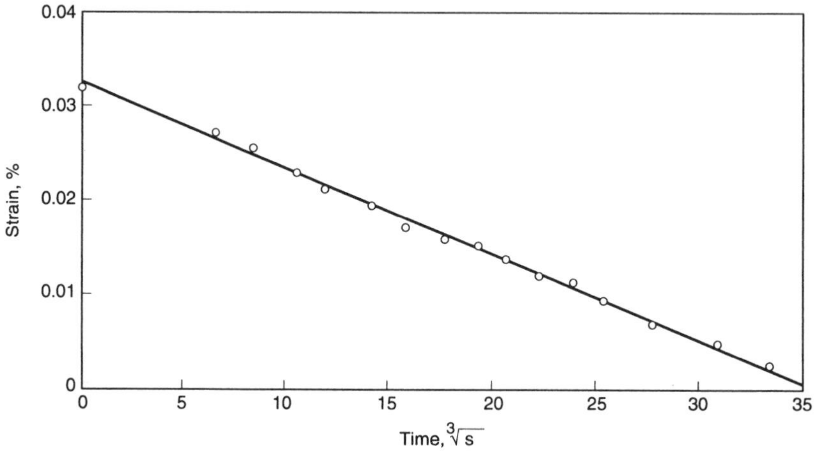

Fig. 6 Recovery of creep strain in silicon nitride at 1200 °C (2190 °F) after unloading from a stress-relaxation test started at 300 MPa (43.5 ksi), showing a time to the one-third dependence

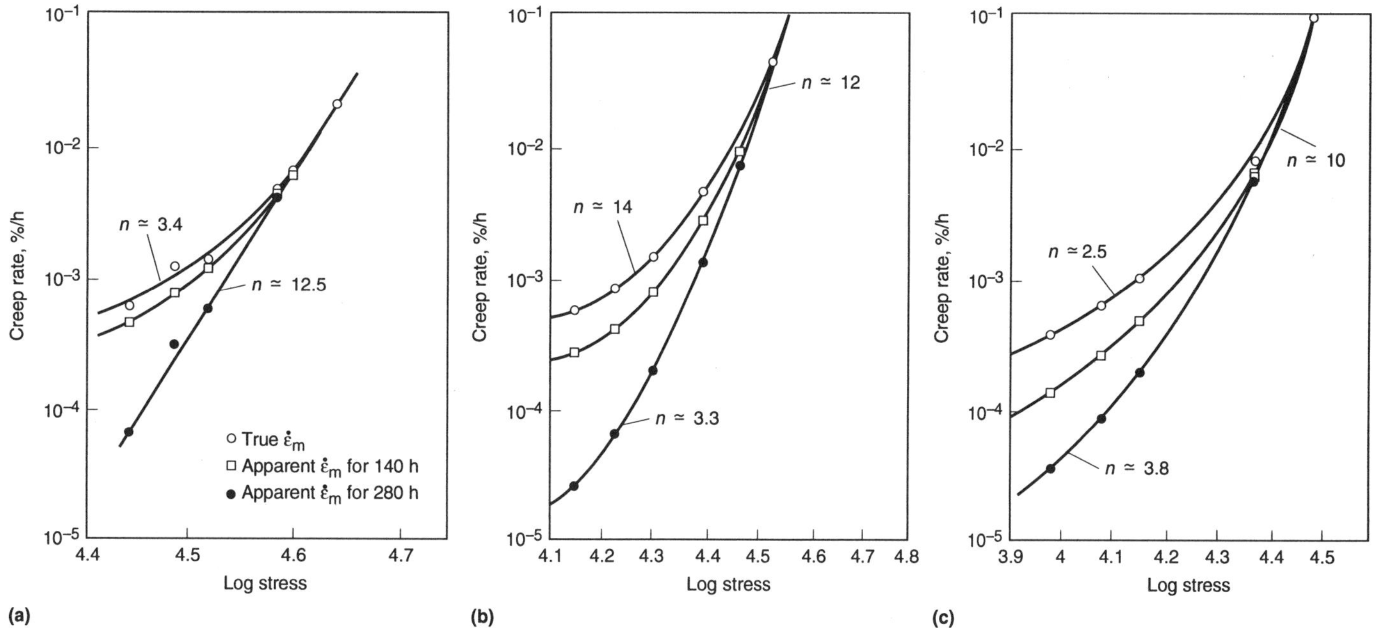

Fig. 7 Effect of test time restrictions on the apparent stress sensitivity of creep rate for a chromium-molybdenum steel at temperatures of (a) 510 °C (950 °F), (b) 565 °C (1050 °F), and (c) 593 °C (1100 °F). Source: Ref 33

approximate a constant structure determination, creep rates have been measured under decreasing stress either by discrete stress drops (Ref 34) or during stress relaxation (Ref 35). The stress exponents measured from these data are much higher than those obtained from the minimum creep-rate data, but have clear physical significance because they relate to an approximately constant structure. An alternative approach for measuring stress dependence at close to constant structure is to monitor the creep rate and corresponding stress increase in constant load tests at strains beyond those corresponding to the minimum creep rates (Ref 33, 36). In this method $n = d \log \dot{\varepsilon}/d \log \sigma_0 (1 + \varepsilon)$ where σ_0 is the initial stress and ε the nominal strain. Results for the steel data are shown in Fig. 8 giving n values for individual tests between 30 and 100, which are much higher than values estimated from the slopes of the lines drawn through the minimum creep rates. It has been shown that, as in the stress-decrement measurements, the values of n may be related to a particular structural state. The reciprocal of these values gives a measure of strain-rate sensitivity and correlates well with elongation at fracture (Ref 33, 36, 37).

Although the representation of creep data in the engineering literature has been strongly influenced by the simple correlations reported for short-time tests for pure metals, it is clear that any physical significance is lost for most structural materials. The stress dependence of creep determined from the slope of a line drawn through minimum creep-rate data is expected to be quite different from that determined for a stress change on an individual specimen. The importance of deformation history is again apparent. Likewise,

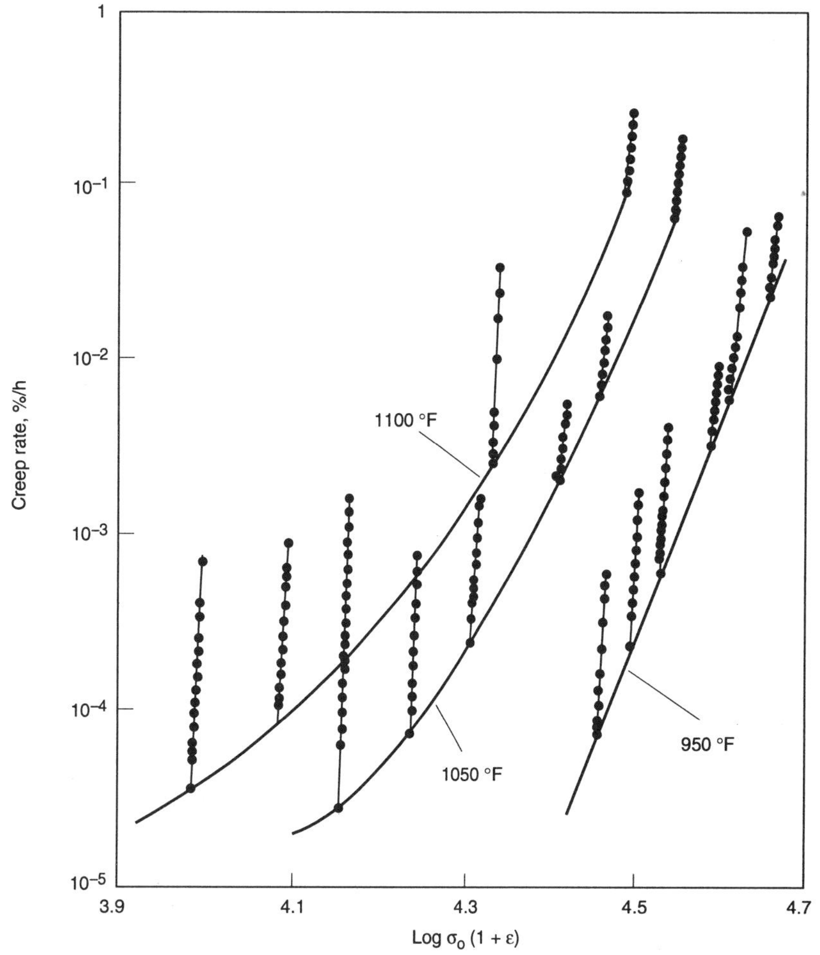

Fig. 8 Creep rate for a chromium-molybdenum steel as a function of the true stress showing that the stress sensitivity measured in a single test is different from that measured in separate tests. Source: Ref 33

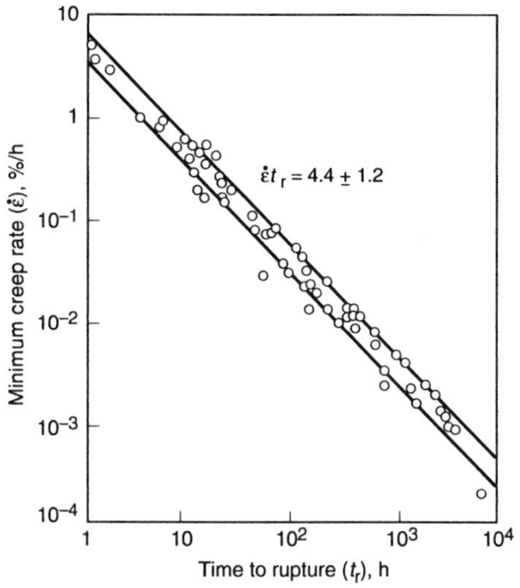

Fig. 9 Monkman-Grant relationship between minimum creep rate and time to rupture for a 2¼Cr-1Mo steel. Source: Ref 39

an exponential temperature dependence of minimum creep rates should be viewed as an empirical correlation. Temperature change experiments on a single specimen usually do not give the same activation energy, and because the structural state changes with temperature, a temperature change sequence effect on the apparent activation energy is also to be expected.

Fracture at Elevated Temperatures

As indicated previously, the constant load creep rupture test is the basis for design data for both creep strength (minimum creep rate or time to a specific creep strain) and failure (time to rupture). The various ways in which such data are presented, correlated, and extrapolated are addressed in subsequent sections. However, it is useful to note here that the well-known Monkman-Grant relationship (Ref 38) shown in Fig. 9 indicates that the time to rupture is reciprocally related to the minimum creep rate. This relationship is commonly observed in ductile materials and has been used to predict one property from the other. However, the true significance of the correlation is that the rupture life is principally a measure of creep strength rather than fracture resistance. This leads to a number of inconsistencies in design procedures that are discussed later in this article.

At this point, it is appropriate to consider the processes leading to fracture. Plastic instability in ductile materials has already been reviewed. This process may lead directly to fracture in pure metals and contribute significantly to fracture in engineering materials at moderately high stresses. However, of much greater concern are the processes leading to intergranular fracture with reduced ductility at low stresses and high temperatures. Here again, many of the basic studies have been conducted on pure metals and solid-solution alloys.

Crack Nucleation and Morphology. Two types of cracking have been identified: wedge-shaped cracks emanating from grain-boundary triple points and the formation of cavities or voids on grain-boundary facets often oriented perpendicular to the applied tensile stress (Ref 40). An example of creep cracks in nickel that appears to show both forms is given in Fig. 10. A fractographic study of creep cavities in tungsten concluded that the different crack morphologies actually reflected differences in growth rate. At low growth rates, surface diffusion allowed the cavities to reduce their surface tension by assuming nearly equiaxed polyhedral shapes. At higher growth rates, irregular two-dimensional cracks developed that on sectioning appeared as wedge cracks (Ref 42).

Although much work continues to model the nucleation and growth of these cracks and cavities (Ref 43), there are uncertainties in the mechanism of nucleation and in the identification of a failure criterion. For example, McLean has shown that a stress concentration up to 1000 is needed to nucleate a hole unless it is stabilized by internal pressure (Ref 44). As a consequence, the nucleation stage has been treated with less enthusiasm than has the modeling of growth. This issue may well be resolved on the basis of environmental interaction (see the section "Environmental Effects" in this article). Another major problem is the effect of temperature and stress on the extent of cracking at failure. Most theories assume that failure occurs at some critical cavity distribution or crack size. However, it has been shown that the extent of cavitation at failure or at any given fraction of the failure life is very sensitive to the test conditions (Ref 45, 46). Thus cavitation damage at failure at a high stress may be comparable to damage in the very early stage of a test at low stress. For stress-change experiments, there is therefore a loading sequence effect on rupture life, which is discussed later in this article, for engineering alloys.

Embrittlement Phenomena. As pointed out previously, rupture life is primarily a measure of creep strength; fracture resistance would be identified better with a separate measure that reflects the concern with embrittlement phenomena that may lead to component failure. Most engineering alloys lose ductility during high-temperature service. This has been shown to be a function of temperature and strain rate (Ref 47) so that there is a critical regime for maximum embrittlement. At a fixed strain rate, for example, ductility first decreases with increasing temperature. This is believed to be caused by grain boundaries playing an increasing role in the deformation process leading to the nucleation of intergranular cracks. At still higher temperatures, processes of recovery and relaxation at local stress concentrations lead to an improvement in ductility. Figure 11 is an example of a ductility contour map for a low-alloy steel based on measurements of reduction of area (RA) of long-term rupture tests (Ref 48).

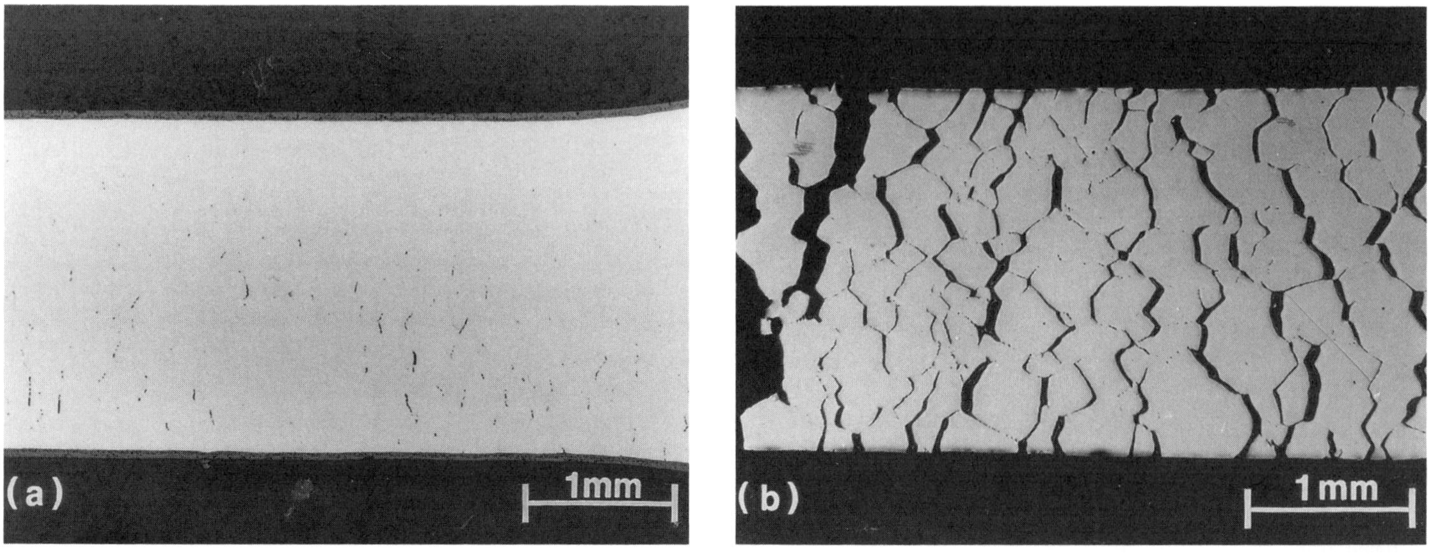

Fig. 10 Unetched microstructure of nickel samples after air testing at 15.8 MPa (2.3 ksi) and 800 °C (1470 °F). (a) Low-carbon Ni270 unloaded after 500 h slight cavitation. (b) Standard Ni270 after failure in 23 h. Source: Ref 41

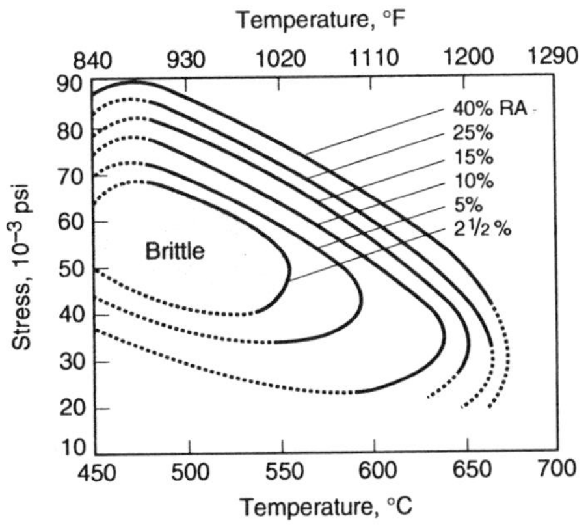

Fig. 11 Ductility contour map in stress/temperature space for a Cr-Mo-V steel. RA, reduction in area. Source: Ref 48

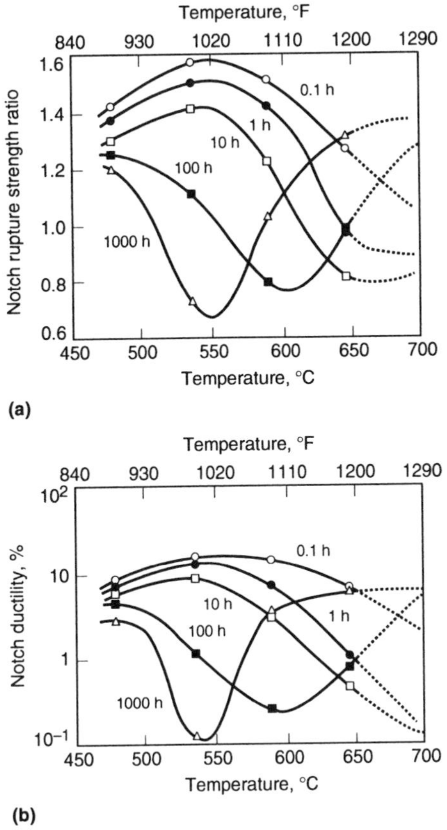

Fig. 12 Ratio of notched bar strength to smooth bar strength (a) and notch ductility (b) as a function of test temperature for various rupture times in a Cr-Mo-V steel. Source: Ref 49

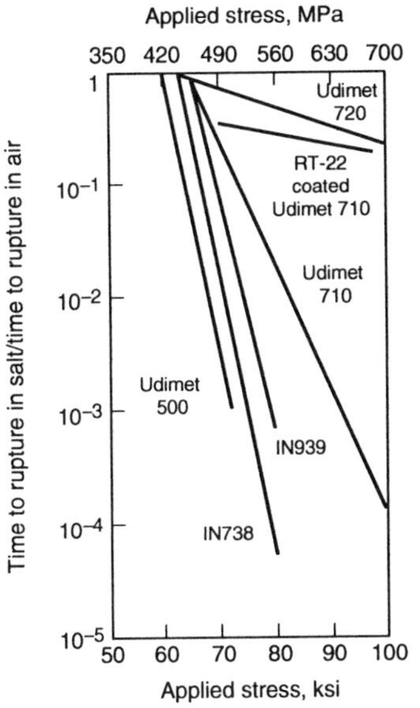

Fig. 13 Relative reductions in rupture life due to sulfate/chloride salt at 705 °C (1300 °F) for several superalloys. For RT-22 coated Udimet 710, rupture time in salt for coated alloy divided by time in air for uncoated alloy. Source: Ref 57

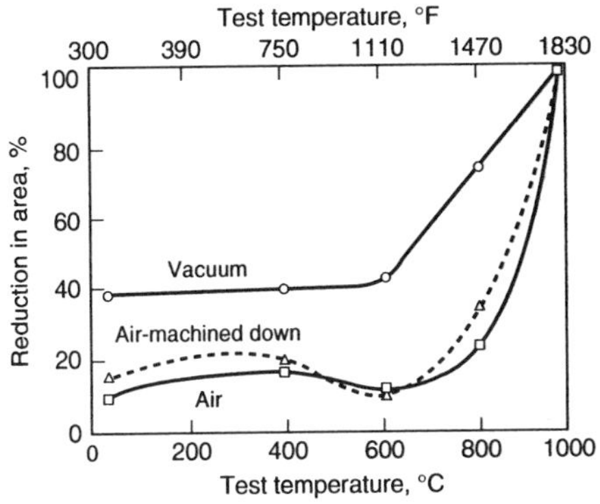

Fig. 14 Tensile ductility of IN903A after air and vacuum exposures at 1000 °C (1830 °F) for 100 h as a function of test temperature in vacuum tests. Embrittlement remained after reducing to half the initial diameter. Source: Ref 41

Maximum embrittlement occurred in a critical range of temperature and stress (or strain rate). This type of embrittlement generally coincides with a sensitivity to notches, which emphasizes its practical significance. For example, Fig. 12 shows that the ratio of notch strength to smooth strength for various test times passes through minima corresponding to ductility minima (Ref 49) based on reduction in area at failure in the notch. This tendency to develop so-called notch weakening at temperature is of great concern in selecting alloys and monitoring their service performance.

The embrittlement phenomena may be associated with precipitation in the alloy that interferes with the ability to accommodate stress concentrations at grain boundaries (Ref 48), with segregation of embrittling species from the grain interior to the grain boundaries (Ref 50, 51), or with intergranular penetration of embrittling species from the environment (Ref 41).

Environmental Effects

It has long been known that test environment may affect creep-rupture behavior. Until quite recently, however, the work has been largely empirical with creep tests being conducted in various atmospheres and differences noted in creep rates and rupture lives (Ref 52). The effect on rupture life, in particular, was often less than a factor of ten in environments such as oxygen, hydrogen, nitrogen, carbon dioxide, and impure helium compared with vacuum. In many cases, it was not clear how inert the vacuum was, and little account was taken of specimen thickness. Often, effects on ductility were not reported, and there were very few studies of crack propagation.

A renewed interest developed in the 1970s as a result of observations of a dominant role played by the environment in high-temperature fatigue crack growth of superalloys (Ref 53). There were subsequent studies of sustained-load crack propagation that also showed very strong effects (Ref 54–56). In some cases at high stresses, the test environment was so severe, as in the case of sulfur, that profound changes were seen in smooth bar rupture life. Such an example is shown in Fig. 13, in which the time to rupture in common superalloys was reduced by several orders of magnitude in tests in a sulfate/chloride mixture at 705 °C (1300 °F) (Ref 57). These results were explained in terms of grain-boundary penetration of sulfur, which leads to rapid crack propagation. The coated specimen was far less susceptible, and the addition of a grain-boundary modifier (in this case, boron) in Udimet 720 gave an enormous improvement relative to Udimet 710.

Embrittling Effects of Oxygen. At about the same time that the ideas on environmental attack at an intergranular crack tip were being developed, it was also shown that short-term prior exposure in air at high temperature (greater than about 900 °C, or 1650 °F) could lead to profound embrittlement at intermediate temperatures (700 to 800 °C, or 1290 to 1470 °F) (Ref 41, 58–61). This was shown to be caused by intergranular diffusion of oxygen that penetrated on the order of millimeters in a few hours at 1000 °C (1830 °F). The embrittlement was monitored using measurements of tensile ductility at intermediate temperatures in iron-, nickel-, and cobalt-base alloys (Ref 41). An example of the results for the Fe-Ni-Co alloy, IN903A, is shown in Fig. 14, which also confirms the extent of damage penetration from tests in which the specimen diameter of 2.54 mm (0.1 in.) was reduced by half. Post-exposure tests on cast alloys showed that this embrittlement could also lead to a reduction in rupture life of several orders of magnitude. An example for alloy IN738 is shown in Fig. 15.

Using model alloys based on nickel, it was shown that oxygen in the elemental form in high-purity nickel did not embrittle; a chemical reaction was necessary (Ref 41). Three embrittling reactions were confirmed: a reaction with carbon to form carbon dioxide gas bubbles; a reaction with sulfides on grain boundaries to release sulfur, which does embrittle in the elemental form; and a reaction with oxide formers to form fine oxides that act to pin grain boundaries. These phenomena are believed to be the same processes that serve to embrittle the region ahead of a crack tip. Thus, oxygen attack may occur dynamically to account for the accelerated advance of a crack in air tests compared with inert environment tests, and it may occur during higher-temperature ex-

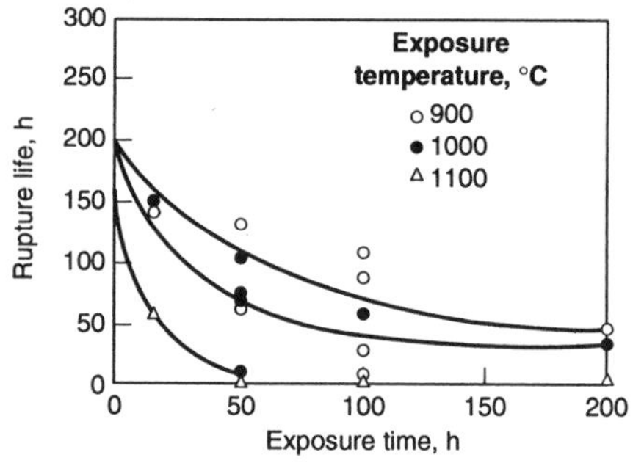

Fig. 15 Effect of exposure in air at various temperatures on stress-rupture life of IN738 at 800 °C (1470 °F) and 400 MPa (58 ksi). Source: Ref 59

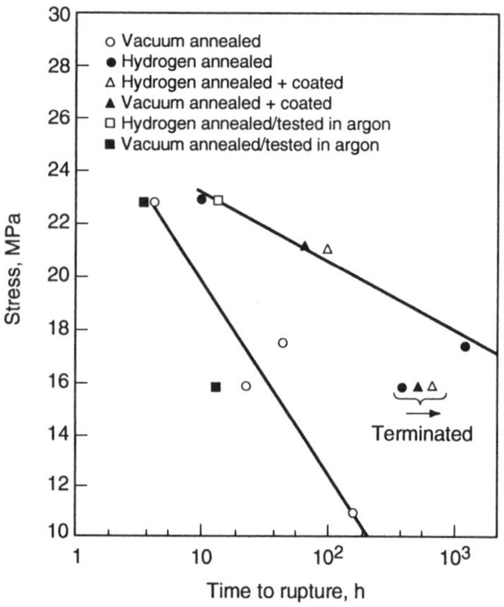

Fig. 16 Effect of environmental interaction on rupture life of Ni270 at 800 °C (1470 °F). Longer lives are obtained by preventing cavitation nucleation from carbon dioxide gas formation. This is achieved by decarburizing to eliminate carbon or by coating to prevent oxygen penetration. Source: Ref 63

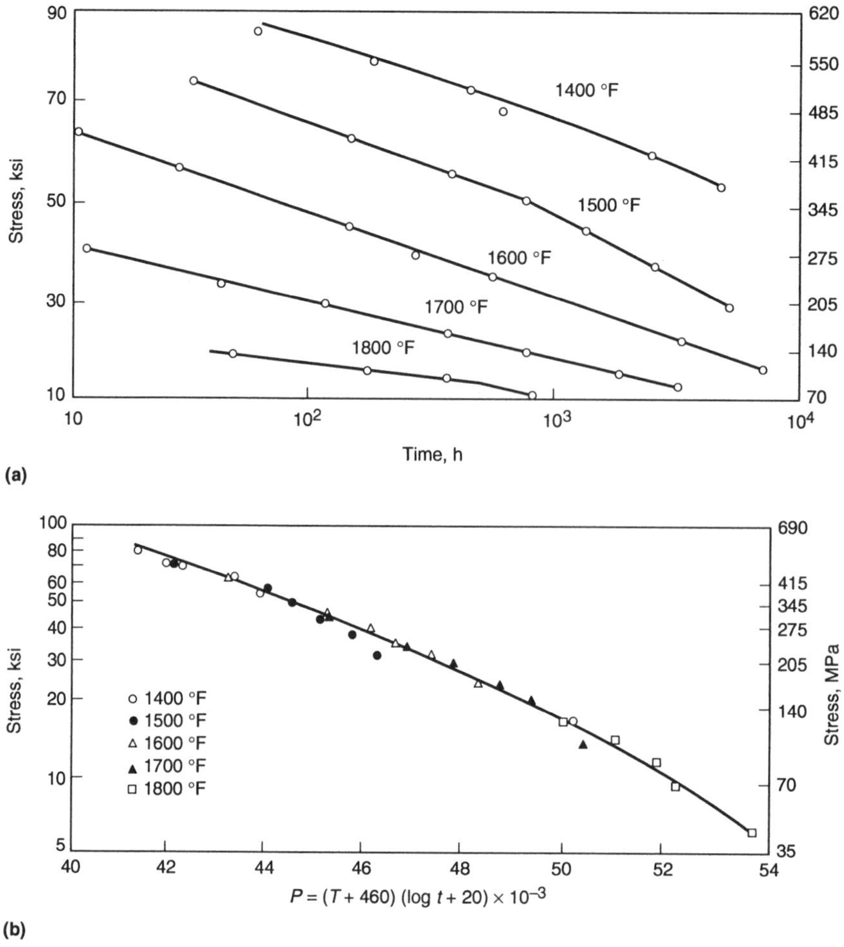

Fig. 17 Stress-rupture behavior of Astroloy. (a) Stress versus time curves. (b) Larson-Miller plot. Source: Ref 69

posure with or without an applied stress to set up an embrittlement situation. Thermal fatigue in combustion turbines is a particularly challenging situation for oxygen attack since maximum strains develop at intermediate temperatures in the cycle, but holding may be at the maximum temperature (Ref 62).

Combined Effects of Oxygen and Carbon. Of special interest relative to the previous discussion of creep cavitation is the reaction between diffusing oxygen and carbon. In nickel, it was found that if this reaction were prevented, creep cavitation could not develop during creep tests. Prevention was achieved either by removing the carbon (decarburizing) or by applying an environmental protective diffusion or overlay coating (Ref 63). Air tests at 800 °C (1470 °F) and at various stresses showed an enormous increase in rupture life (Fig. 16) if the gas bubble formation did not occur. Since nickel has been used as an archetypical metal for the study of creep cavitation, the confirmation that the cavities are nucleated as gas bubbles now solves the problem of nucleation (Ref 44). The observation that even in superalloys gas bubbles are frequently nucleated at carbides, which may serve as cavity nuclei (Ref 59), and that cavitation during creep is often concentrated near the specimen surface (Ref 64), points to the likelihood that oxygen attack may be invariably associated with creep cavitation. The gas pressures developed in the bubbles appear to be quite adequate for nucleation (Ref 65).

Effect of Other Gaseous Elements. Hydrogen, chlorine, and sulfur may also embrittle as a result of penetration. Sulfur is particularly aggressive in that it diffuses more rapidly and embrittles more severely than does oxygen (Ref 66). It is also frequently found in coal gasification and oil-refining processes as well as industrial gas turbines operating on impure fuel.

Design Methodology

This section describes the basic presentation and analysis methods for creep rupture that are currently widely used. In addition to the application of these methods to materials selection and the setting of basic design rules, some consideration will be given to their application to remaining life assessment of operating components. The interaction with fatigue will not be included since some discussion of that complex topic is covered in the article "Creep-Fatigue Interaction" in the *ASM Specialty Handbook: Heat-Resistant Materials*. Also omitted from this discussion are multiaxial stress effects and sustained load (or creep) crack growth. However, it should be recognized that components in service normally operate under multiaxial stress systems, and detailed procedures are used for analysis that are based on effective stresses and strains. The considerable amount of work conducted in recent years on sustained-load crack growth, and the other topics alluded to, are reviewed in several recent texts (Ref 20, 39, 67) and in *Mechanical Testing*, Volume 8 of the *ASM Handbook*.

Creep Rupture Data Presentation. Laboratory creep tests are typically run between 100 and 10,000 h, although a few are run for shorter times (for example, for acceptance tests), and occasionally some testing is conducted for longer times. Since most high-temperature components are expected to last ten years or more, service stresses are obviously lower than those used in the longest creep tests to generate data for most of the alloys used. Therefore, to provide data for creep rates and rupture lives that are appropriate for the setting of design stresses, it became necessary to

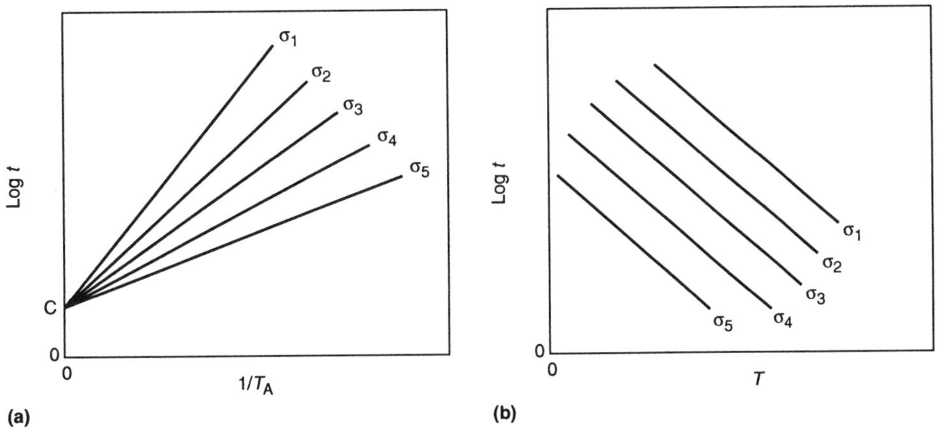

Fig. 18 Two common time-temperature parameters for rupture life. (a) Larson-Miller parameter. $f(\sigma) = T_A$ (lot t + C). (b) Manson-Succop parameter. $f(\sigma) = \log t - BT$. Source: Ref 69

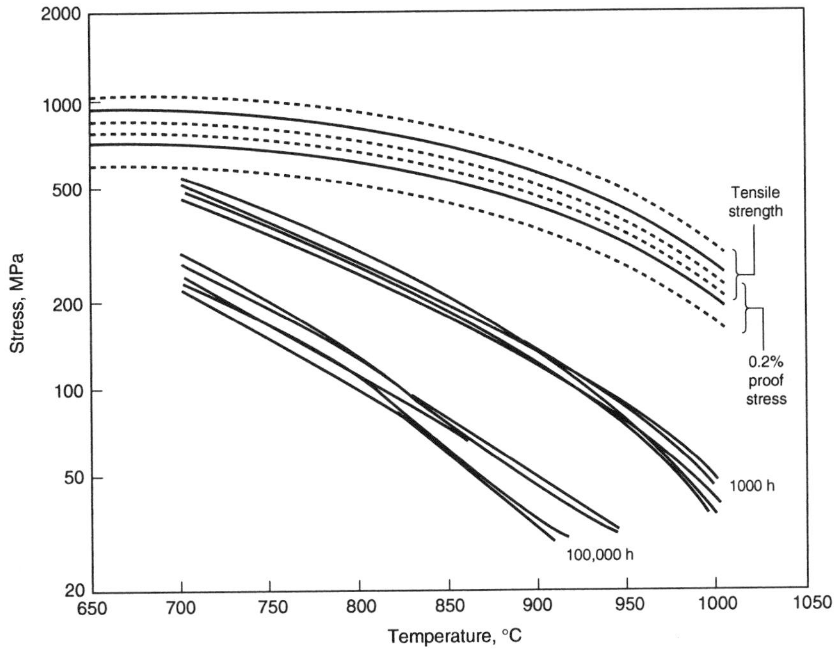

Fig. 19 Temperature dependence of 0.2% yield stress, tensile strength, and creep rupture strength at 1000 and 100,000 h for a nickel-base superalloy casting. Source: Ref 73

develop methods for extrapolation. Over the years, a tremendous amount of effort has gone into optimizing methods of data extrapolation (Ref 68, 69).

One of the major considerations in such procedures must be statistical issues, such as the best estimate of the stress associated with a given median life or creep rate, the use of stress or time as the dependent variable in the data fitting, the treatment of variability among heats of the same alloy, and the analysis of data with run-outs. All of these issues have been treated with considerable rigor and shown to be important relative not only to the proper interpretation of data, but to the proper design of experiments (Ref 69). In addition, there are different practices among testing laboratories that may have appreciable effects on results. These include specimen geometry, loading procedure, specimen alignment, furnace type, and temperature control.

Despite all these concerns regarding proper statistical treatment of data, a methodology has been developed based on time-temperature parameters that is now in widespread use. The approach may be used to achieve the following major design objectives:

- It allows the representation of creep rupture (or creep) data in a compact form, allowing interpolation of results that are not experimentally determined.
- It provides a simple basis for comparison and ranking of different alloys.
- Extrapolation to time ranges beyond those normally reached is straightforward.

Based on the Arrhenius rate equation and a previous tempering parameter (Ref 70), Larson and Miller (Ref 71) developed the most commonly used parameter:

$$P = f(\sigma) = T(C + \log t_r) \quad \text{(Eq 10)}$$

where σ is stress, T is absolute temperature, and t_r (or t_R) is time to rupture. In their original paper, the constant C was set equal to 20, which provided a good fit for a variety of alloys, and is still widely used today. A set of σ versus log t_r data for the alloy Astroloy and the parametric master curve derived from it is shown in Fig. 17.

Many other parametric forms have been developed (Ref 68, 69), the most common of which are described in the article "Assessment and Use of Creep-Rupture Data" in the *ASM Specialty Handbook: Heat-Resistant Materials*. The time-temperature parameter provides the means for interchanging time and temperature so that a long time may be computed from a short-time test at higher temperature. Figure 18 provides two parametric forms, the shape of which depend on whether isostress lines are parallel or converge on plots of log t versus $1/T$ (the Larson-Miller parameter in Fig. 18a) or T (the Manson-Succop parameter in Fig. 18b). The Manson-Succop parameter (Ref 72) has been used extensively by the Japanese National Research Institute for Metals in compiling data sheets on a wide range of alloys. For example, the data shown in Fig. 19 are for Ni-19Cr-18Co-4Mo-3Ti-3Al-B superalloy castings (Ref 73). The rupture data are statistically treated based on a Manson-Succop correlation, and the tensile data show 95% prediction intervals (dashed lines).

At least two approaches have been used to allow a given data set to determine objectively the best of the common parameters or to provide new parametric forms. The minimum commitment method (MCM) (Ref 74) and the graphical optimization procedure (GOP) (Ref 75) provide optimal data presentation but lose the advantage of representing different alloys on the same parametric plot. The GOP method uses the fact that all parameters are in the form of one variable expressed as the product of functions of two others, for example, $t_r = H(\sigma)Q(T)$ or $T = H(\sigma)C(t_r)$. These functional forms are solved graphically to optimize the parameter. For example, Fig. 20(a) shows data on Cr-Mo-V steel plotted according to a Larson-Miller parameter. The curves at each temperature are clearly ordered, whereas Fig. 20(b) shows the same data using a parameter determined using the GOP method, which includes the optimized values of the $C(t_r)$ function. There is in this case no separation of the isothermal segments. The optimized value of the Larson-Miller "constant" actually varied from 28.5 at 10 h to 13.6 at 100,000 h (Ref 76) for this data set.

Design rules for high-temperature time-dependent deformation and fracture may be established based on formal codes or on proprietary manufacturers specifications. For example, an ASME code (Ref 77) is used for the design of fossil-fuel boilers and for pressure vessel and piping systems in the petroleum and chemical process industries. The allowable stresses are to be no higher than the lowest of:

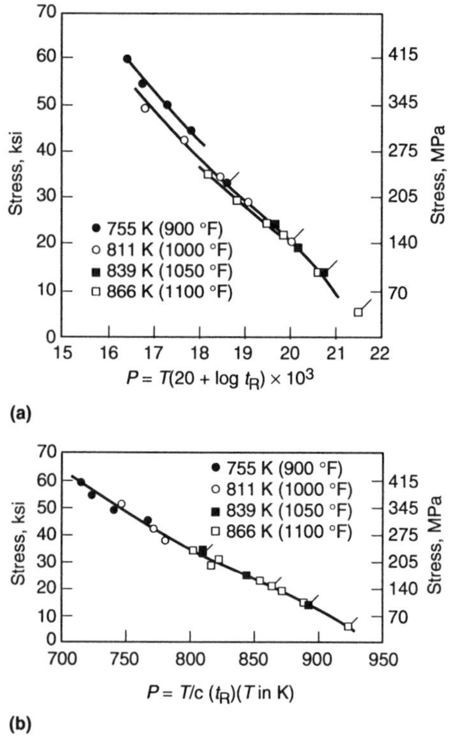

Fig. 20 Creep parameters for a Cr-Mo-V steel. (a) Larson-Miller plot using a constant of 20 showing segmenting of the data. (b) Same data using an optimized parameter based on the graphical optimization procedure (GOP) method. Source: Ref 76

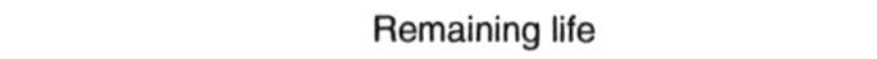

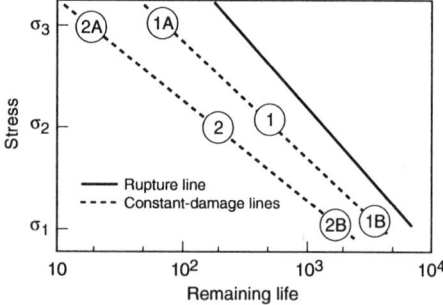

Fig. 21 Schematic plot illustrating construction of constant damage curves in terms of remaining life for different stresses. Source: Ref 83

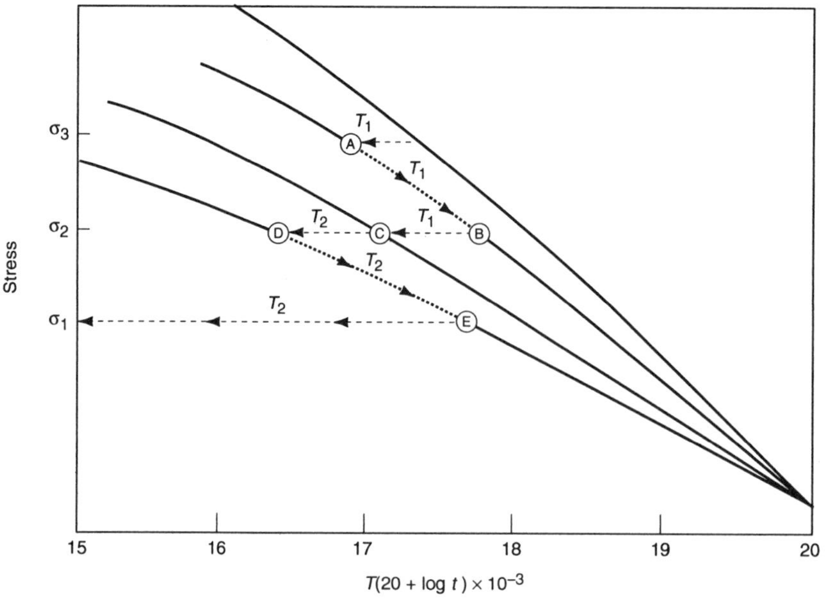

Fig. 22 Schematic plot illustrating construction of constant damage curves in terms of the Larson-Miller parameter. The sequence ABCDE is an example of a particular stress-temperature path. Source: Ref 83

- 100% of the stress to produce a creep rate of 0.01% in 1000 h
- 67% of the average stress to produce rupture in 100,000 h
- 80% of the minimum stress to produce rupture in 100,000 h

These stresses may be determined from parametric plots or from derived curves, such as those shown in Fig. 19. It is of interest to note that it is implicit in these rules that the rupture life provides a measure of creep strength. The connection between creep rate and rupture life may be made through the Monkman-Grant relationship (Ref 38) or through the Gill-Goldhoff correlation (Ref 78), which relates stress for rupture with stress for creep to a specific strain for a fixed time and temperature. Both of these methods are described in the article "Assessment and Use of Creep-Rupture Data" in the *ASM Specialty Handbook: Heat-Resistant Materials.*

With nickel-base superalloys, it has been found that surface cracks related to environmental attack may develop at strains as low as 0.5% (Ref 79). Since these cracks result in severe loss in fatigue life, this is an appropriate failure criterion rather than rupture life. Gas turbine blades may therefore be designed on the basis of time to 0.5% creep with a suitable safety factor on stress.

Damage Accumulation and Life Prediction

Engineering procedures for life management of operating components assume that the material is progressively degraded or damaged as creep strain increases and operating time accumulates. Damage may be in the form of precipitate changes that may result in softening (overaging) and reduced creep strength, or embrittlement and reduced resistance to fracture. The embrittlement may be due to segregation of harmful species, either from the interior or from the external environment, to interfaces, especially grain boundaries. Damage may also occur as a result of progressive intergranular cavitation and cracking, as previously described. Some of this damage may be reversible by suitable heat treatment or by hot isostatic pressing and may allow the possibility of component rejuvenation. However, for the purposes of component life management, which allows decisions to be made regarding part replacement, repair, or rejuvenation, there is a critical need to quantify the accumulation of damage as a function of operating conditions.

There are two basic approaches to using the concept of damage accumulation for life assessment:

- Based on a detailed knowledge of the operating conditions, including temperature and stress changes, the remaining life is estimated from the known original properties of the material of construction.
- Remaining life estimates are made using post-exposure measurements of microstructural changes, intergranular cavitation, or mechanical properties such as hardness, impact energy, or stress-rupture life.

Creep under Nonsteady Stress and Temperature. There are two common approaches to analyzing creep when the temperature and/or the stress change. The strain-hardening law assumes an equation of state of the form:

$$\dot{\varepsilon} = f_1\,(\sigma, T, \varepsilon_c) \quad \text{(Eq 11)}$$

which defines the state of the material in terms of creep strain (ε_c). The time-hardening law:

$$\dot{\varepsilon} = f_2\,(\sigma, T, t) \quad \text{(Eq 12)}$$

defines the state in terms of time. It should be clear that neither of these is physically rigorous. However, they provide analytical options, for example, when a stress change calls for a transfer to a new creep curve at the same strain or time, respectively. Although strain is intuitively a better compromise as a state variable, the time-hardening law is often easier to handle analytically for complex sequences of stress and temperature changes. Moreover, if the sequences involve multiple increases and decreases in stress and temperature, the two approaches may produce similar results. Modifications of the laws including normalizing the strain with the fracture strain, normalizing the time with the failure time, or using a mean function have also been used.

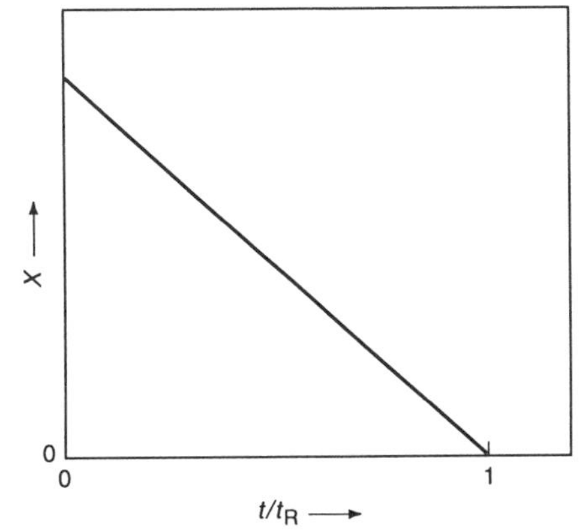

Fig. 23 Ideal behavior of a property suitable for damage monitoring as a function of life fraction

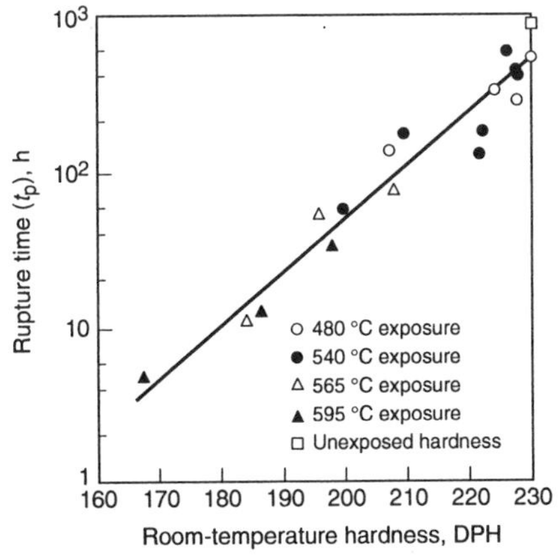

Fig. 24 Correlation between post-exposure rupture life for a Cr-Mo-V steel tested at 240 MPa (35 ksi) and 540 °C (1000 °F) and room temperature hardness. DPH, diamond pyramid hardness. Source: Ref 85

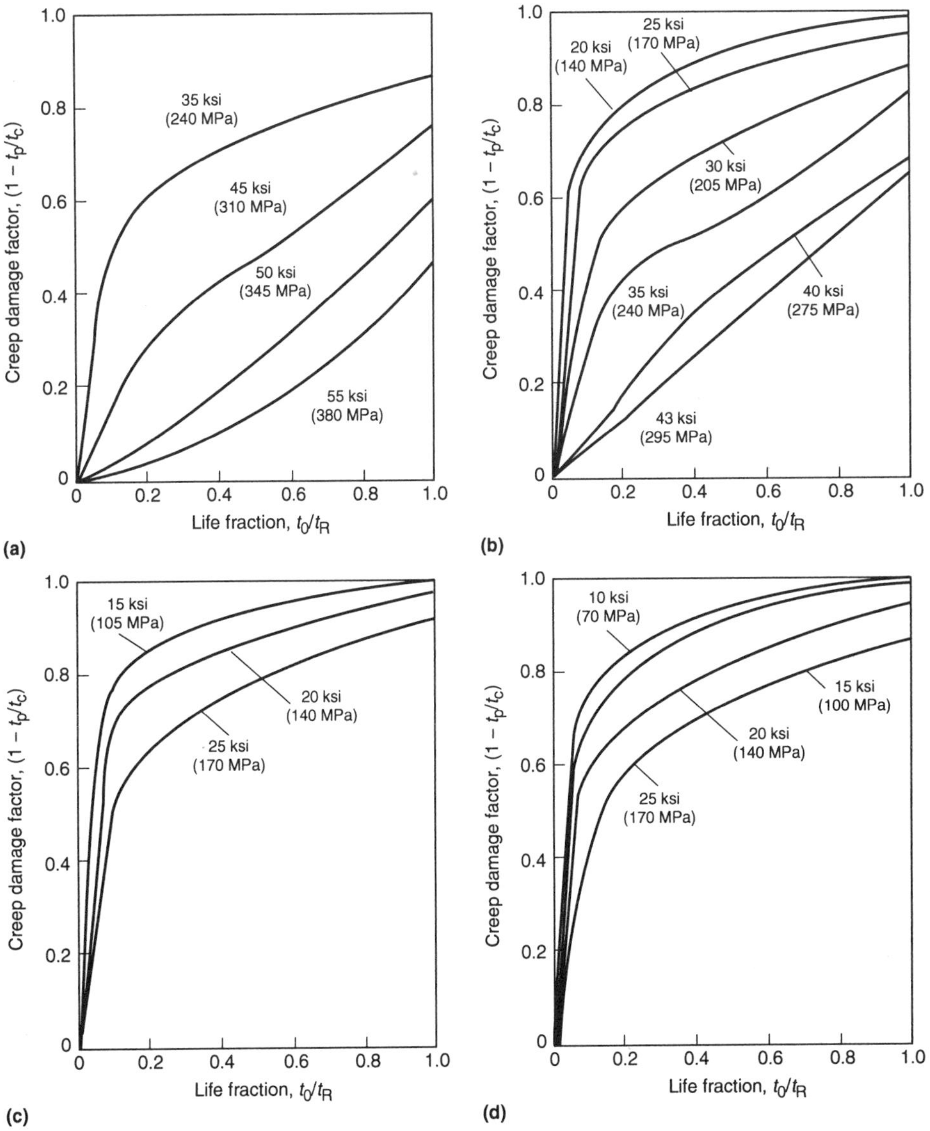

Fig. 25 Graphical description of used life as a function of life-fraction exposure for Cr-Mo-V steel at various stresses and temperatures. (a) At 480 °C (900 °F). (b) At 540 °C (1000 °F). (c) At 565 °C (1050 °F). (d) At 595 °C (1100 °F). Source: Ref 85

These same ideas have been extended to predict failure either in terms of strain fractions or time fractions. In this case, however, by far the most extensively used in design analysis and life prediction has been the life-fraction approach. This concept was first introduced by Robinson (Ref 80, 81) as an accounting system for materials in which the rupture time is a power function of stress and the log rupture time is a linear function of temperature. The analytical solutions for such a material subjected to various temperature and stress cycles were given by Robinson (Ref 81), but the concept is now generally presented in the form:

$$\sum_{i=1}^{n} \frac{\Delta t_i}{t_{Ri}} = 1 \quad \text{(Eq 13)}$$

where t is time spent at temperature T or stress σ, t_R is rupture time at temperature T or stress, and n is the number of temperature or stress changes.

Each fractional expenditure of life, or accumulated damage, is considered to be independent of all others. Failure is predicted when the sum of the fractions of life equals unity. This hypothesis remains a widely used tool for life assessment.

Considerable experimental work has attempted to confirm or refute the rule. However, when the results of multiple-stress or temperature-change experiments are analyzed in terms of the life-fraction rule, it is only possible to determine whether the rule is appropriate, but not why it fails. Only when damage is defined in terms of remaining life rather than used life is it possible to formulate the appropriate damage law (Ref 82). Figure 21 illustrates the procedure schematically for a stress-change experiment where the abscissa is the logarithm of the remaining life. The continuous line represents remaining life for the virgin material, that is, the stress-rupture data for the initial condition. Point 1 is the remaining life on two specimens exposed at stress σ_2. One of these specimens is then tested to failure at σ_3, and the measured remaining life is represented by point 1A. The stress on the second specimen is then reduced to σ_1, resulting in a measured remaining life indicated by 1B. The dashed line drawn through the points 1A, 1, and 1B is then defined as a constant damage line in terms of remaining life. Clearly, any number of lines such as 2A, 2, 2B may be constructed.

A necessary condition for the life-fraction rule to apply is that the constant damage lines should be displaced horizontally by a constant distance when remaining life is plotted on a logarithmic scale for both temperature and stress changes. For low-alloy steels, the life-fraction rule applies quite well for temperature changes but not for stress changes. For stress changes, the curves often converge at low stresses, leading to a loading sequence effect. For example, a stress increment leads to a life fraction at failure less than one and a stress decrement leads to a life fraction at failure greater than one. This behavior may be

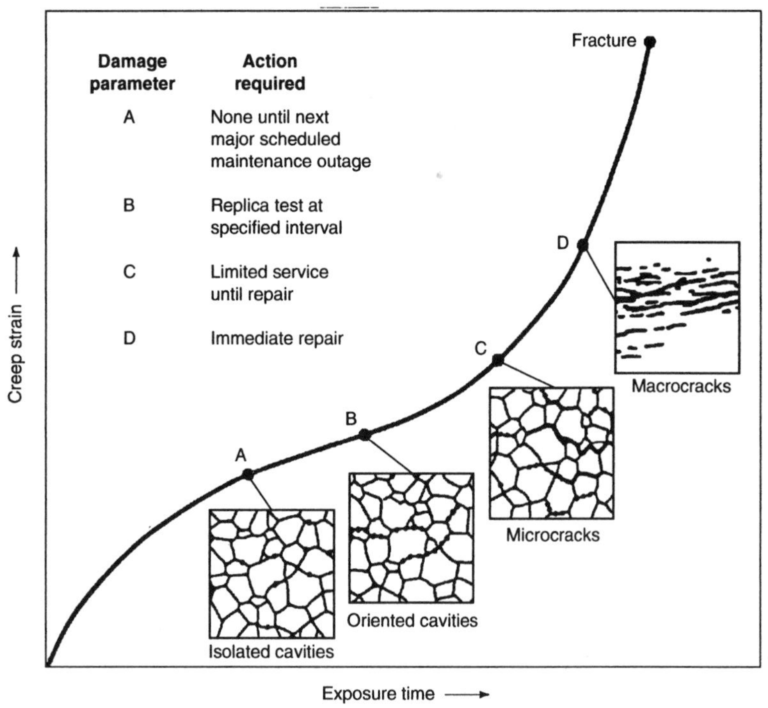

Fig. 26 Creep life assessment based on cavity classification in boiler steels. Source: Ref 87

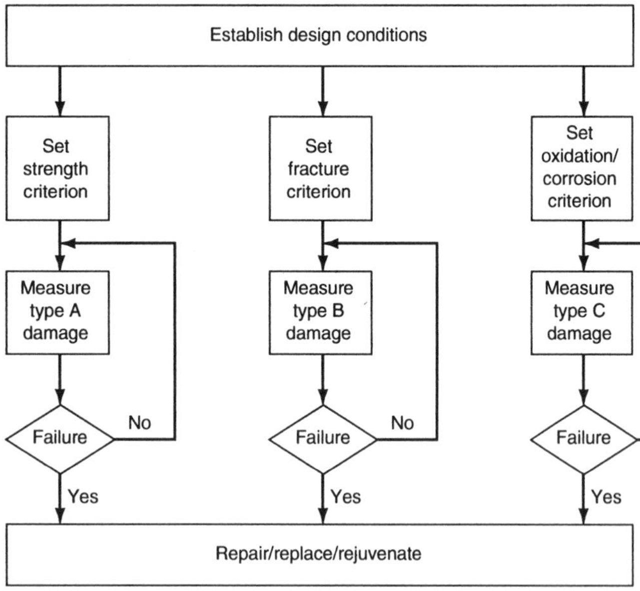

Fig. 27 The design-for-performance concept. Source: Ref 92

combined on a parametric plot (Fig. 22), where a sequence of both temperature and stress changes may be represented. For detailed discussion on the construction and use of the remaining life curves for nonsteady conditions and component life assessment, see Ref 82 to 84.

Post-Exposure Evaluation. Ideally, a microstructural characterization or mechanical property should change linearly with life fraction and independently of the test conditions, as shown in Fig. 23. In fact, the changes are invariably nonlinear and nonunique with life fraction. Both cavitation changes and creep strength changes are dependent on the test conditions (Ref 45, 46, 83). Thus, the amount of damage at failure or at a given life fraction is a strong function of the stress in particular.

A comprehensive study of a low-alloy steel in which miniature stress-rupture specimens were taken from large specimens, which had previously been exposed to various strains and times up to 60,000 h at four temperatures over a range of stresses, confirmed this complexity (Ref 85). The initial exposure led to progressive softening, which correlated well with the post-exposure rupture life (Fig. 24) in a standard test at 540 °C (1000 °F) and 240 MPa (35 ksi). Using parametric correlations with the original exposures, a series of damage curves were generated (Fig. 25) in which the ordinate is one minus the ratio of the post-exposure rupture life (t_p) divided by the mean rupture life of the unexposed material (t_c) under the same test conditions. This definition of creep damage increases to a limiting value of unity as t_p approaches zero.

It is readily apparent from these curves that (1) the damage accumulates in a nonlinear manner, (2) the damage at failure is dependent on the test conditions, and (3) the damage at a constant life fraction is strongly stress dependent but quite insensitive to temperature for a given stress. These points are consistent with the previous commentary on the limitations of the life-fraction rule and with the general applicability of time-temperature parameters.

Despite similar concerns with the use of microstructural observations of cavitation to predict remaining life, there have been extensive reported studies of such measurements (Ref 39, 86). The first detailed attempt to relate remaining life of power plant to cavitation observed on replicas was that of Neubauer and Wedel (Ref 87), who set up recommended procedures based on the observation of four levels of cavitation severity in steam pipes (Fig. 26). There were subsequent attempts to quantify the cavitation in terms of mechanistic concepts (Ref 88). Other studies have examined the cavitation in nickel-base alloys for gas turbines (Ref 89) and set up similar criteria for end of useful life.

The replica studies in particular have usually been limited to surface observations so that the possibility of environmental interaction is strong (see the section "Environmental Effects" in this article). Also, many of these studies have recently been criticized on the basis that the observation of cavities is very sensitive to preparation techniques (Ref 90).

Limitations and Alternative Design Approaches

In the increasingly competitive global market for manufactured components operating at high temperatures, there are three primary technical objectives:

- The development cycle for new materials and new designs must be reduced.
- The designs must be optimized for efficiency and performance.
- Procedures must be developed for component life management to allow timely repair or replacement.

Current Limitations. The primary design of most high-temperature components involves the application of some proportion of the stress for failure in a given time (usually in the range of 20,000 to 100,000 h) calculated from extrapolated constant load stress-rupture tests. It might appear, therefore, that the most direct method of measuring remaining life after extended service would be to compute it based on the current remaining life of a sample taken from the component. Accordingly, methods are being developed to take miniature stress-rupture samples from components and assess remaining life based on extrapolation from such data. There are a number of reasons why this approach may be unsound:

- Component failure is often localized with little or no material degradation or damage remote from the failure.
- Cracks frequently initiate from the surface so that any post-exposure property measurement on material taken from the interior has limited value.
- Interactions with the operating gaseous environments, which may have profound effects on crack initiation and propagation, are generally ignored.
- The changing stress in a constant load test is not normally accounted for in summing life fractions to predict remaining life.
- The sources of scatter in experimental property measurement (for example, alignment, temperature control, precision of stress and strain measurement, specimen geometry) bear little connection to the sources of scatter in service.
- Time to rupture in an unnotched ductile alloy is principally a measure of deformation behavior rather than fracture resistance.

The failure to establish a clear separation of a strength requirement from a failure criterion (an appropriate analogy might be between yield strength and fracture toughness for many low-temperature components) leads to a paradox (Ref 91). This may be stated: When a component fails, the material of the component has a finite life, sometimes approaching the original design life, at the operating conditions of the component. It thus follows that a remaining life estimated from a sample taken from an operating component may bear no relationship to the actual component remaining life.

Low-temperature design is dominated by properties that uniquely characterize the mechanical state in terms of stiffness (modulus), strength (yield), and fracture resistance (K_c). If service-induced changes in state occur, for example, radiation hardening and embrittlement, these changes are monitored in terms of their effects on changes in the same short-time properties. Thus life management decisions are based on the same performance criteria as in the original design: There is no remaining life paradox in this case.

The basis for current methods of high-temperature design is different in that the objective is to incorporate time-dependent changes in the test methodology. The creep rates and rupture lives relate to the starting material only in terms of the specific deformation path being imposed, for example, a variable strain rate or a variable true stress test with an arbitrary interaction between creep deformation and fracture processes. This history is quite different from any real service history. To measure these properties, unlike in the low-temperature situation, the structural and mechanical states must be changed. As an example, the minimum creep rate at a given temperature and stress is a measure of the creep strength of the material in its current state; it relates to its initial state only for the particular deformation history. The designer may claim that long-time data are required for component creep strength evaluation, but in reality what is needed is low strain rate data. It will be shown that such data may be obtained with precision in short-time tests.

Simulating service complexity in material testing involves a hierarchy of increasing complexity and expense. Thus long-term creep tests are inadequate; nonsteady test conditions, complex stresses, cyclic stresses, environmental effects, and the synergism among them must be taken into account. Even with the most complex test plan, most service operations cannot be simulated.

Alternative Design Approaches and Tests. An alternative approach is to simplify the test methodology and develop tests to measure separately the high-temperature creep strength and fracture resistance, ideally to evaluate the current state in terms of these properties. The consequences of microstructural evolution, induced in service or in laboratory simulations, can then be assessed using the same short-time tests. Design is then based on minimum acceptable performance levels (Ref 92).

The concept is illustrated in Fig. 27 for three fundamental properties, including an environmental damage criterion, which could be critical in a number of high-temperature applications. Additional criteria could be added. However, it might not be necessary to include some important failure processes, such as high-cycle fatigue and thermal fatigue, because these in principle are either derivable from, or correlatable with, the other properties or are too complex and parameter sensitive to be assigned minimum performance values.

The creep strength criterion for high-temperature applications may be developed from a series of carefully conducted stress-relaxation tests (SRT). The stress versus time response is converted to a stress versus creep strain-rate response (Ref 92, 93). This is, in effect, a self-programmed

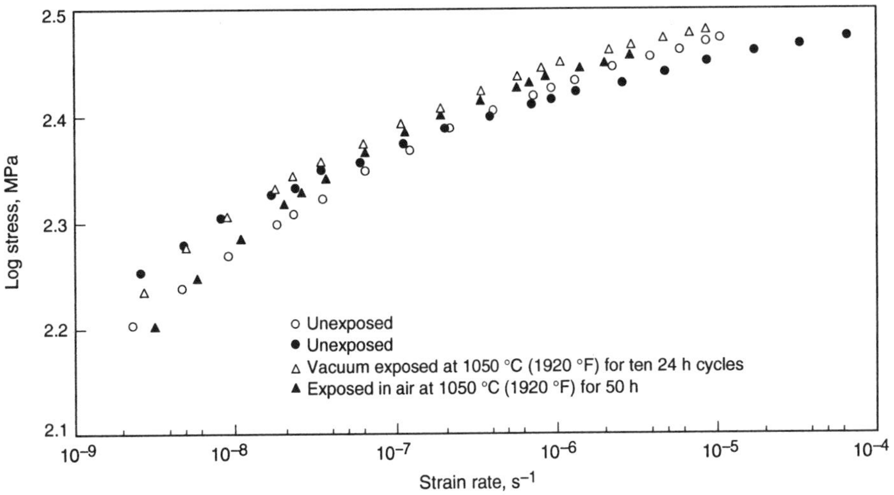

Fig. 28 Insensitivity of creep strength of IN738 to various thermal exposures as determined from stress versus creep-rate behavior calculated from stress-relaxation tests. Source: Ref 92

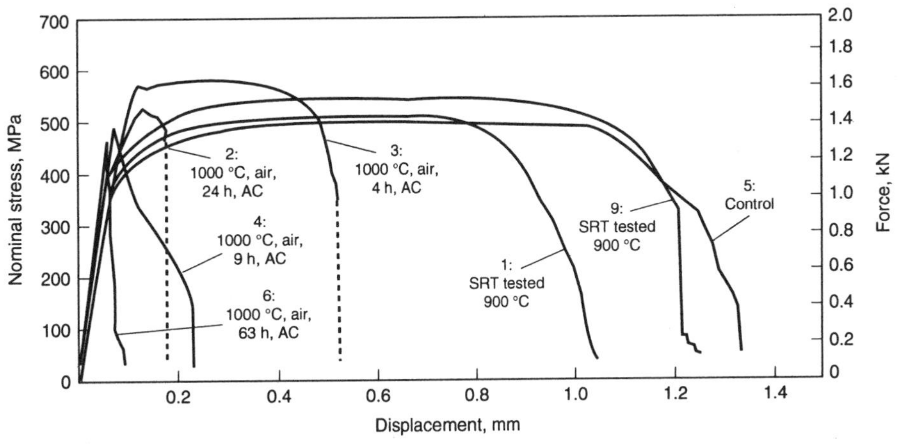

Fig. 29 Embrittlement of IN738 with increasing severity of exposure in air demonstrated in constant displacement rate (4 × 10^{-5} mm/s) tensile tests at 800 °C (1470 °F). SRT, stress-relaxation tests. AC, air cooled. Source: Ref 95

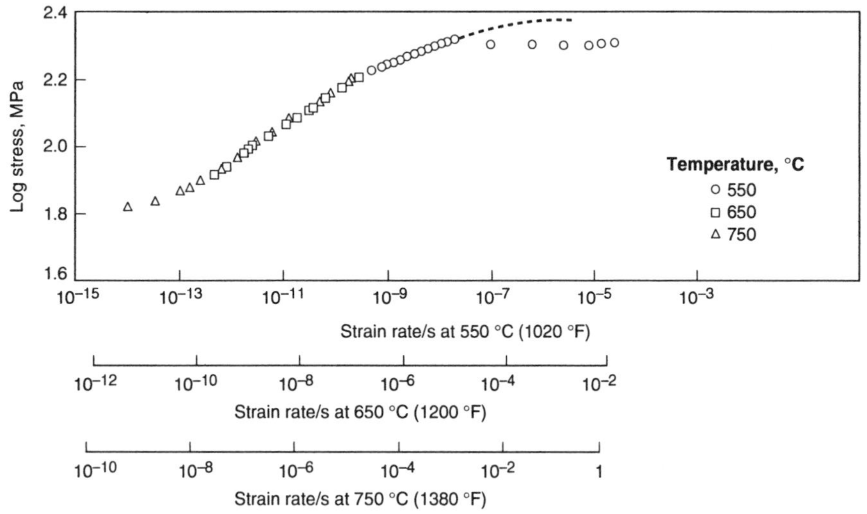

Fig. 30 Master stress versus creep-rate curve for Incoloy 800H between 550 and 750 °C (1020 and 1380 °F) achieved by horizontal translation of data generated on one specimen. Source: Ref 95

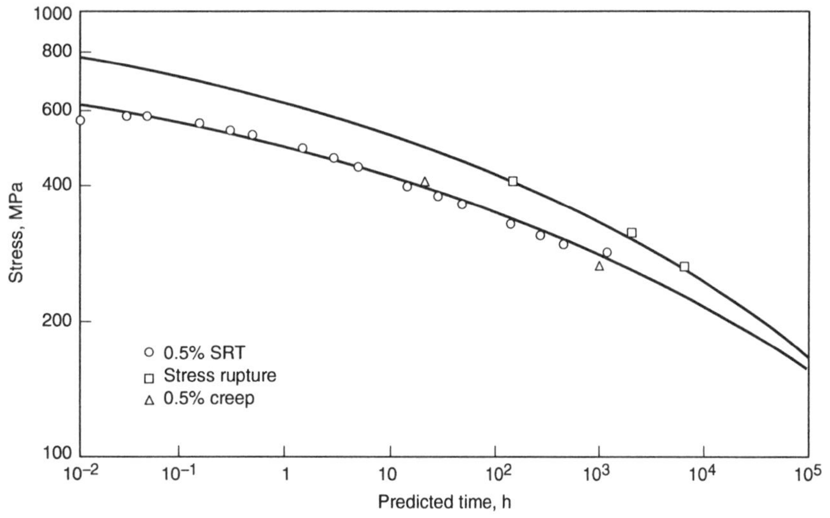

Fig. 31 Comparison of predicted time to 0.5% creep based on stress-relaxation measurements at 650 °C (1200 °F) for an austenitic iron-base alloy with measurements made on conventional creep tests. SRT, stress-relaxation tests

variable stress creep test. Typically, a test lasting less than one day may cover five decades in creep rate. The accumulated nonelastic strain is usually less than 0.1%, so that several relaxation runs at different temperatures and from different stresses can be made on a single specimen with minimal change in the mechanical state. Thus an enormous amount of creep data can be generated in a short time.

A constant displacement rate (CDR) test (Ref 92, 94), in which a crack is enabled to initiate naturally at a grain boundary and propagate slowly under closed-loop displacement control, was developed as a separate basis to evaluate the fracture resistance. Both types of tests were first performed on large specimens to establish the broad principles of the methodology and then compared with data taken from miniature specimens machined from actual gas turbine blades (Ref 95). It was established that environmental interaction, specifically gas phase embrittlement (GPE) resulting from intergranular penetration of oxygen, could drastically reduce the fracture resistance with little or no effect on the creep strength (Ref 92, 95).

Figure 28 is an example of four relaxation runs at 900 °C (1650 °F) showing no appreciable effect of severe high-temperature exposure in air or vacuum on the stress versus creep-rate response for a cast nickel-base gas turbine blade made of alloy IN738. By contrast, Fig. 29 shows that there is a profound degradation in the CDR fracture resistance for high-temperature air exposures. This approach is clearly valuable for assessing performance and continued operation when embrittlement occurs with minimal effect on creep strength. It can also be used directly as a basis for creep design. For example, Fig. 30 shows how relaxation runs at different temperatures can be fitted by horizontal displacement of the strain-rate scales to allow stresses corresponding to very low creep rates to be estimated. In fact, this procedure allows design creep rates to be attained based on several one-day tests (Ref 92). It also allows comprehensive determination of the effect of prior exposure, either from laboratory simulations or component service, to be evaluated.

Recently, it has been shown how creep data obtained from stress-relaxation testing can be compared directly with long-time creep data in terms of time to specific creep strains. Since the two approaches use different deformation histories, neither of which is fundamentally preferred compared with component deformation, precise conformity should not be anticipated or required. However, for a number of alloys (Ref 95) and polymers (Ref 96), the agreement is surprisingly good. Figure 31 shows such a comparison for an iron-base superalloy.

Thus the approach offers accelerated creep strength evaluation, which may be used either directly or through calibration with long-time tests, and may be used to assess the creep strength of operating components. In particular, by recognizing high-temperature fracture as a property that is not uniquely connected to creep strength, the approach offers a basis to understand why some components fail very early in their design lives and why others appear undamaged after exceeding their design lives.

Conclusions

Time-dependent deformation and fracture of structural materials are among the most challenging engineering problems faced by materials engineers. The critical role played by high-temperature energy-conversion machines in modern society attests to the remarkable success of the design methodology developed during the present century. However, modern design needs, including accelerated evaluation and development of advanced materials, and improved remaining life assessment methods for operating equipment have identified some ways in which the methodology might be improved. It has been pointed out in this article that it is desirable to decouple the creep strength and fracture resistance criteria. It has also been shown how this could lead to new accelerated short-time testing in which the objective is not to attempt to incorporate microstructural evolution and damage-development in the test, as in the traditional long-time creep-to-rupture test. Rather, the accelerated test may be used to measure separately the consequences of these changes on creep strength and fracture resistance.

The generally neglected anelastic or time-dependent recoverable component of creep may be the dominant strain component in many service situations at low stresses and needs to be incorporated in design analysis. This is true for ceramics and metals as well as polymers. It may also provide, in some cases, a critical link between deformation and fracture.

ACKNOWLEDGMENTS

Of all the people who have influenced his thinking over the years, the author would like to give special thanks to the late Robert Goldhoff, who directed him into this field of study and research; Louis Coffin, who has provided direct and indirect influence over the years; Edward Hart, a rigorous thinker and writer; Roger Bricknell, an intellectual partner; Michael Henry, a creative materials engineer; the late Chester Sims, a force in superalloys; Joanne Beckman, his first Ph.D. student; Donald Van Steele, an outstanding experimentalist; and David Stiles, a supporter of new ideas.

REFERENCES

1. F. Philips, The Slow Stretch in India Rubber, Glass and Metal Wire When Subjected to a Constant Pull, *Philos. Mag.*, Vol 9, 1905, p 513
2. E.N. da C. Andrade, The Viscous Flow in Metals and Allied Phenomena, *Proc. R. Soc.*, Vol A84, 1910, p 1–13
3. O.H. Wyatt, Transient Creep in Pure Metals, *Proc. Phys. Soc.*, Vol 66B, 1953, p 459–480
4. R.W. Bailey, *J. Inst. Met.*, Vol 35, 1926, p 27
5. E. Orowan, *J. West Scot. Iron and Steel Inst.*, Vol 54, 1946–1947, p 45
6. L.N. McCartney, No Time—Gentlemen Please, *Philos. Mag.*, Vol 33 (No. 4), 1976, p 689–695
7. G.J. Lloyd and R.J. McElroy, On the Anelastic Contribution to Creep, *Acta Metall.*, Vol 22, 1974, p 339–348
8. E.W. Hart, A Critical Examination of Steady State Creep, *Creep in Structures*, A.R.S. Ponter and D.R. Hayhurst, Ed., Springer-Verlag, 1981, p 90–102
9. E.W. Hart, Theory of the Tensile Test, *Acta Metall.*, Vol 15, 1967, p 351
10. M.A. Burke and W.D. Nix, *Acta Metall.*, Vol 23, 1975, p 793
11. P.J. Wray and O. Richmond, Experimental Approach to a Theory of Plasticity at Elevated Temperatures, *J. Appl. Physics*, Vol 39 (No. 12), 1968, p 5754–5761
12. D.A. Woodford, Creep Ductility and Dimensional Instability of Nickel at 500 and 525 °C, *Trans. ASM*, Vol 59, 1966, p 398–410
13. J. Weertman, *J. Appl. Phys.*, Vol 26, 1955, p 1213–1217
14. F.R.N. Nabarro, *Proc. Conf. Strength of Solids*, Physical Society, London, 1948, p 75
15. C. Herring, *J. Appl. Phys.*, Vol 21, 1950, p 437–445
16. R.L. Coble, *J. Appl. Phys.*, Vol 34, 1963, p 1679–1684
17. M.F. Ashby, *Acta Metall.*, Vol 20, 1572, p 887–897
18. R.L. Bell and T.G. Langdon, Grain Boundary Sliding, *Interfaces Conf.*, R.C. Gifkins, Ed., Butterworths, 1969, p 115–137
19. J.L. Walter and H.E. Cline, Grain Boundary Sliding, Migration, and Deformation in High-Purity Aluminum, *Trans. AIME*, Vol 242, 1968, p 1823–1830
20. N.E. Dowling, *Mechanical Behavior of Materials*, Prentice Hall, 1993
21. R.J. Crawford, *Plastics Engineering*, Pergamon Press, 1987, p 222
22. J.C. Gibeling and W.D. Nix, Observations of Anelastic Backflow Following Stress Reductions During Creep of Pure Metals, *Acta Metall.*, Vol 29, 1981, p 1769–1784
23. C.M. Zener, *Elasticity and Anelasticity of Metals*, University of Chicago, 1948
24. J.A. DiCarlo, *J. Mater. Sci.*, Vol 21, 1986, p 217–224
25. J. Henderson and J.D. Snedden, Creep Recovery of Aluminum Alloy DTD 2142, *Appl. Mater. Res.*, Vol 4 (No. 3), 1965, p 148–168
26. J.D. Lubahn, The Role of Anelasticity in Creep, Tension and Relaxation Behavior, *Trans. ASM*, Vol 45, 1953, p 787–838
27. R.M. Goldhoff, Creep Recovery in Heat Resistant Steels, *Advances in Creep Design*, A.I. Smith and A.M. Nicolson, Ed., John Wiley & Sons, 1971, p 81–109
28. T.S. Ke, Experimental Evidence of the Viscous Behavior of Grain Boundaries in Metals, *Phys. Rev.*, Vol 71, 1947, p 533
29. J.E. Dorn, Some Fundamental Experiments on High-Temperature Creep, *J. Mech. Phys. Solids*, Vol 3, 1954, p 85–116
30. O.D. Sherby, Factors Affecting the High Temperature Strength of Polycrystalline Solids, *Acta Metall.*, Vol 10, 1962, p 135–147
31. R. Viswanathan, The Effect of Stress and Temperature on the Creep and Rupture Behavior of a 1.25pct. Chromium-0.5 pct. Molybdenum Steel, *Metall. Trans. A*, Vol 8, 1977, p 877–993
32. J.D. Lubahn, "Creep of Metals," Symposium on Cold Working of Metals, American Society for Metals, 1949
33. D.A. Woodford, Measurement and Interpretation of the Stress Dependence of Creep at Low Stresses, *Mater. Sci. Eng.*, Vol 4, 1969, p 146–154
34. S.K. Mitra and D. McLean, Cold Work and Recovery in Creep at Ostensibly Constant Structure, *Met. Sci.*, Vol 1, 1967, p 192
35. E.W. Hart, "Phenomenological Theory: A Guide to Constitutive Relations and Fundamental Deformation Properties, *Constitutive Equations in Plasticity*, A.S. Argon, Ed., MIT Press, 1975
36. D.A. Woodford, Analysis of Creep Curves for a Magnesium-Zirconium Alloy, *J. Inst. Met.*, Vol 96, 1968, p 371–374
37. D.A. Woodford, Strain Rate Sensitivity as a Measure of Ductility, *Trans. ASM*, Vol 62, 1969, p 291–293
38. F.C. Monkman and N.J. Grant, *Proc. ASTM*, Vol 56, 1956, p 595
39. R. Viswanathan, *Damage Mechanisms and Life Assessment of High Temperature Components*, ASM International, 1989, p 82
40. F. Garofalo, *Fundamentals of Creep and Creep Rupture*, MacMillan, 1965
41. D.A. Woodford and R.H. Bricknell, Environmental Embrittlement of High Temperature Alloys by Oxygen, *Embrittlement of Engineering Alloys*, C.L. Briant and S.K. Banerji, Ed., Academic Press, 1983, p 157
42. J.O. Steigler, K. Farrell, B.T.M. Loh, and H.E. McCoy, Creep Cavitation in Tungsten, *Trans. ASM*, Vol 60, 1967, p 494
43. A.C.F. Cocks and M.F. Ashby, On Creep Fracture by Void Growth, *Prog. Mater. Sci.*, Vol 27, 1982, p 189–244
44. D. McLean, The Physics of High Temperature Creep in Metals, *Rept. Prog. Phys.*, Vol 29, 1966, p 1–33
45. D.A. Woodford, A Parametric Approach to Creep Damage, *Met. Sci. J.*, Vol 3, 1969, p 50–53
46. D.A. Woodford, "Density Changes During Creep in Nickel," *Met. Sci. J.*, Vol 3, 1969, p 234–240
47. F.N. Rhines and P.J. Wray, Investigation of the Intermediate Temperature Ductility Minimum in Metals, *Trans. ASM*, Vol 54, 1961, p 117–128
48. D.A. Woodford and R.M. Goldhoff, An Approach to the Understanding of Brittle Behavior of Steel at Elevated Temperatures, *Mater. Sci. Eng.*, Vol 5, 1970, p 303–324
49. W.E. Brown, M.H. Jones, and D.P. Newman, *Symp. on Strength and Ductility of Metals at Elevated Temperatures*, STP 128, ASTM, 1952, p 25
50. M.P. Seah, Grain Boundary Segregation, *Metal Phys.*, Vol 10, 1980, p 1043–1064
51. E.P. George, P.L. Li, and D.P. Pope, Creep Cavitation in Iron—Sulfides and Carbides as Nucleation Sites, *Acta Metall.*, Vol 35 (No. 10), 1987, p 2471–2486
52. R.H. Cook and R.P. Skelton, The Influence of Environment on High Temperature Mechanical Properties of Metals and Alloys, *Int. Met. Rev.*, Vol 19, 1974, p 199
53. L.F. Coffin, Fatigue at High Temperature, *Fatigue at Elevated Temperature*, A.E. Carden, A.J. McEvily, and C.H. Wells, Ed., STP 520, ASTM, 1972, p 5–36
54. K. Sadananda and P. Shahinian, The Effect of Environment on the Creep Crack Growth Behavior of Several Structural Alloys, *Mater. Sci. Eng.*, Vol 43, 1980, p 159–168
55. K.R. Bain and R.M. Pelloux, Effect of Environment on Creep Crack Growth in PM/HIP René 95, *Metall. Trans. A*, Vol 15, 1984, p 381–388
56. S. Floreen and R.H. Kane, Investigation of the Creep-Fatigue-Environment Interaction in a Nickel Base Superalloy, *Fatigue Eng. Mater. Struct.*, Vol 2, 1980, p 401
57. G.A. Whitlow, C.G. Beck, R. Viswanathan, and E.A. Crombie, The Effects of a Liquid Sulfate/Chloride Environment on Superalloys, *Metall. Trans. A*, Vol 15, 1984, p 23–28
58. W.H. Chang, *Proc. Conf. Superalloys—Processing*, Section V, MCIC-7210, AIME, 1972
59. D.A. Woodford, Environmental Damage of a Cast Nickel Base Superalloy, *Metall. Trans. A*, Vol 12, 1981, p 299–308
60. R.H. Bricknell and O.A. Woodford, The Embrittlement of Nickel Following High Temperature Air Exposure, *Metall. Trans. A*, Vol 12, 1981, p 425–433

61. M.C. Pandey, B.F. Dyson, and D.M.R. Taplin, Environmental, Stress-State and Section-Size Synergisms During Creep, *Proc. R. Soc. London A,* Vol 393, 1984, p 117–131
62. D.A. Woodford and D.F. Mowbray, Effect of Material Characteristics and Test Variables on Thermal Fatigue of Cast Superalloys, *Mater. Sci. Eng.,* Vol 16, 1974, p 5
63. R.H. Bricknell and D.A. Woodford, Cavitation in Nickel during Oxidation and Creep, *Int. Conf. on Creep and Fracture of Engineering Materials and Structures,* B. Wilshire and R.W. Evans, Ed., Pineridge Press, Inst. of Metals, 1991, p 249–262
64. E.C. Scaife and P.L. James, *Met. Sci. J.,* Vol 2, 1968, p 217
65. H. Reidel, Fracture at High Temperatures, Springer-Verlag, 1987
66. J.P. Beckman and D.A. Woodford, Gas Phase Embrittlement of Nickel by Sulfur, *Metall. Trans. A,* Vol 21, 1990, p 3049–3061
67. G.A. Webster and R.A. Ainsworth, *High Temperature Component Life Assessment,* Chapman and Hall, 1994
68. J.S. Conway, *Stress Rupture Parameters: Origin, Calculations and Use,* Gordon and Breach, 1969
69. R.M. Goldhoff, "Development of a Standard Methodology for the Correlation and Extrapolation of Elevated Temperature," EPRI FP-1062, Electric Power Research Institute, 1979
70. J.H. Holloman and L.C. Jaffe, Time-Temperature Relations in Tempering Steel, *Trans. AIME,* Vol 162, 1945, p 223–249
71. F.R. Larson and J. Miller, A Time-Temperature Relationship for Rupture and Creep Stresses, *Trans. ASME,* Vol 74, 1952, p 765–775
72. S.S. Manson and G. Succop, *Stress Rupture Properties of Inconel 700 and Correlation on the Basis of Several Time-Temperature Parameters,* ASTM, STP 174, 1956
73. NRIM Creep Data Sheet, No. 34B, National Research Institute for Metals, Tokyo, Japan, 1975
74. S.S. Manson and C.R. Ensign, "A Specialized Model for Analysis of Creep Rupture Data by the Minimum Commitment Method," Tech. Memo TMX 52999, National Aeronautics and Space Administration, 1971
75. D.A. Woodford, A Graphical Optimization Procedure for Time-Temperature Rupture Parameters, *Mater. Sci. Eng.,* Vol 15, 1974, p 169–175
76. D.A. Woodford, Perspectives in Creep and Stress Rupture, *Int. Conf. on Creep,* Tokyo, JSME, IMechE, ASME, ASTM, 1986, p 11–20
77. ASME Boiler and Pressure Vessel Code, Section 1, ASME
78. R.M. Goldhoff and R.F. Gill, A Method for Predicting Creep Data for Commercial Alloys on a Correlation between Creep Strength and Rupture Strength, *ASME J. Basic Eng.,* Vol 94, Series D, No. 1, 1972, p 1–6
79. W.L. Chambers, W.J. Ostergren, and J.H. Wood, Creep Failure Criteria for High Temperature Alloys, *J. Eng. Mater. Technol.,* Vol 101, 1979, p 374–379
80. E.L. Robinson, Effect of Temperature Variation on the Creep Strength of Steels, *Trans. ASME,* Vol 60, 1938, p 253–259
81. E.L. Robinson, Effect of Temperature Variation on the Long-Time Rupture Strength of Steels, *Trans. ASME,* Vol 74, 1952, p 777–781
82. D.A. Woodford, A Critical Assessment of the Life Fraction Rule for Creep-Rupture under Nonsteady Stress or Temperature, Paper No. 180.1, *Int. Conf. on Creep and Fatigue in Elevated Temperature Applications,* Inst. Mech. Eng., 1973-1974, p 1–6
83. D.A. Woodford, Creep Damage and the Remaining Life Concept, *J. Eng. Mater. Technol.,* Vol 101, 1979, p 311–316
84. R.V. Hart, Assessment of Remaining Creep Life Using Accelerated Stress Rupture Tests, *Met. Technol.,* Vol 13, 1976, p 1
85. R.M. Goldhoff and D.A. Woodford, *The Evaluation of Creep Damage in a Cr-Mo-V Steel,* STP 515, ASTM, 1972, p 89–106
86. R. Viswanathan and S.M. Gehl, Life-Assessment Technology for Power-Plant Components, *JOM,* Vol 44 (No. 2), 1992, p 34–42
87. B. Neubauer and U. Wedel, Restlife Estimation of Creeping Components by Means of Replicas, *Advances in Life Prediction Methods,* ASME, 1983, p 307–314
88. B.J. Cane and M.S. Shammas, "A Method for Remnant Life Estimation by Quantitative Assessment of Creep Cavitation on Plant," Report TPRD/U2645/N84 CEGB, Leatherhead Lab., 1984
89. S.A. Karllsson, C. Persson, and P.O. Persson, Metallographic Approach to Turbine Blade Lifetime Prediction, *Baltica III Conf. on Plant Condition and Life Management,* Helsinki, 1995, p 333–348
90. I. LeMay, T.L. da Silveira, and S.K.P. Cheung-Mak, Uncertainties in the Evaluation of High Temperature Damage in Power Stations and Petrochemical Plant, *Int. J. Press. Vess. Piping,* Vol 59, 1994, p 335–342
91. D.A. Woodford, The Remaining Life Paradox, *Int. Conf. on Fossil Power Plant Rehabilitation,* ASM International, 1989, p 149
92. D.A. Woodford, Test Methods for Accelerated Development, Design, and Life Assessment of High Temperature Materials, *Mater. Design,* Vol 14 (No. 4), 1993, p 231–242
93. E.W. Hart and H.D. Solomon, Stress Relaxation Testing of Aluminum, *Acta Metall.,* Vol 21, 1973, p 295
94. J.J. Pepe and D.C. Gonyea, Constant Displacement Rate Testing at Elevated Temperatures, *Int. Conf. on Fossil Power Plant Rehabilitation,* ASM International, 1989, p 39
95. D.A. Woodford, The Design for Performance Concept Applied to Life Management of Gas Turbine Blades, *Baltica III, Int. Conf. on Plant Condition and Life Management,* Helsinki, 1995, p 319–332
96. S.K. Reif, K.J. Amberge, and D.A. Woodford, Creep Design Analysis for a Thermoplastic from Stress Relaxation Measurements, *Mater. Design,* Vol 16 (No. 1), 1995, p 15–21

Design for Oxidation Resistance

James L. Smialek and Charles A. Barrett, NASA Lewis Research Center
Jon C. Schaeffer, General Electric Aircraft Engines

ALLOYS intended for use in high-temperature environments rely on the formation of a continuous, compact, slow-growing oxide layer for oxidation and hot corrosion resistance. To be protective, this oxide layer must be chemically, thermodynamically, and mechanically stable. Successful alloy design for oxidative environments is best achieved by developing alloys that are capable of forming adherent scales of either alumina (Al_2O_3), chromia (Cr_2O_3), or silica (SiO_2).

In this article, emphasis has been placed on the issues related to high-temperature oxidation of superalloys used in gas turbine engine applications. Despite the complexity of these alloys, optimal performance has been associated with protective alumina scale formation. As will be described below, both compositional makeup and protective coatings play key roles in providing oxidation protection. Other high-temperature materials described include nickel and titanium aluminide intermetallics, refractory metals, and ceramics. Additional information on the oxidation resistance of other structural alloys, including chromia-forming ferrous alloys for industrial applications can be found in the many references provided and in the *ASM Specialty Handbook: Heat-Resistant Materials.*

The Oxidation Process

Except for a few noble metals, such as gold, platinum, and palladium, all metals will react with oxygen when exposed to air at high temperatures. This is due to basic thermodynamics, that is, highly negative free energies for the general oxidation reaction:

$$aM + \tfrac{1}{2}bO_2 = M_aO_b \quad \text{(Eq 1)}$$

where M is the reacting metal, M_aO_b is its oxide and a and b are the moles of the metal and oxygen, respectively, in 1 mol of the oxide. This fact can be readily assessed from the Ellingham diagram of the free energies of formation of the oxidation product. In general, the most reactive or oxygen active (least noble) metals are found in the alkali and alkaline earth (groups IA and IIA of the periodic table), the scandium group IIIA (including the lanthanide and actinide series), and the titanium group IVA metals. The common high-temperature structural base metals (iron, cobalt, and nickel group VIIIA) are more noble, but still oxidize readily in typical oxygen-containing environments. The partial pressure of oxygen below which oxidation will not occur is defined by the thermodynamic relationship for Eq 1 as:

$$\Delta G = -RT \ln (1/p^b_{O_2}) \quad \text{(Eq 2)}$$

where ΔG represents the Gibbs energy change, R is the gas constant, and T is the absolute temperature, and where the metal and oxide are in pure, standard states.

Consumption of the metal and loss of its engineering properties are determined by the oxidation rate. The rate of oxidation is determined by the slowest step in the oxidation process. In some less-common cases a volatile, gaseous oxide, or highly porous solid oxide scale may form where the consumption rate may be controlled by the metal-oxygen reaction or by diffusion through a gaseous boundary layer. However in the vast majority of applications of high-temperature materials, a relatively dense external solid scale forms. The rate of oxidation in this case is controlled by counter-current diffusion of oxygen and metal ions through an ionic oxide layer. As is the case with most diffusion-controlled processes, the instantaneous rate is inversely proportional to the thickness of scale that must be penetrated to sustain the process. This leads to the classic parabolic growth rate:

$$x^2 = k_p t \quad \text{(Eq 3)}$$

where x is the oxide scale thickness or mass (weight) gain per unit area, k_p is the parabolic growth rate constant with the units cm^2/s for scale thickness measurements and the units $g^2/cm^4 \cdot s$ for mass gain measurements, and t is time.

Thus, slow-growing scales are made up of oxides with low diffusion constants. It can be seen from Fig. 1 that, over the temperature regime of most practical interest, scales based on Al_2O_3, Cr_2O_3, or SiO_2 produce the most desirable rates (Ref 1). (BeO and B_2O_3 have also been shown to exhibit low growth rates, but are subject to special problems, such as moisture-enhanced growth or toxicity, and are not widely employed as protective oxides.) Accordingly, high-temperature alloy design for oxidation resistance is based on the ability to form these three simple elemental oxide scales. Although more complex oxide phases, such as $NiAl_2O_4$ and $NiTa_2O_6$ may form on many commercial alloys, none are known to possess lower diffusion and growth rates than the single Al_2O_3, Cr_2O_3, or SiO_2 oxides. However, the rate and specific type of ionic diffusion can be altered to varying degrees by small amounts of dopants of secondary cations, especially at the oxide grain boundaries.

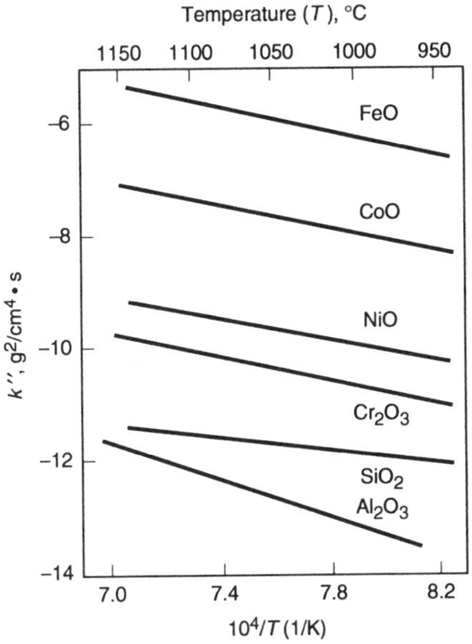

Fig. 1 Parabolic oxidation rate of various oxide scales (Arrhenius plot). The relationship between k'' and k_p (parabolic rate constant described by Eq 3) is $k'' = (8/V)^3 k_p$, where V is the equivalent volume of the oxide. Source: Ref 1

General Methodologies

The basic test methods to evaluate oxidation resistance are to expose the material, monitor the kinetics, and subsequently characterize the oxidation products and substrate material. Exposures are most often done in air, but can be in any controlled mix of corrosive gases, such as H_2/H_2S, CO/CO_2, or O_2/SO_2. Oxidation kinetics are usually recorded as weight changes, but may also be followed as changes in scale thickness. Continuous weight change is usually obtained by thermogravimetric analysis (TGA), in which the sample is suspended from the bottom of an analytical balance into the hot zone of a vertical tube furnace. Continuous oxidation kinetics can also be observed by a manometric apparatus in which the pressure drop caused by oxygen consumption in a closed system is monitored and converted to weight of oxygen gained per unit area of sample. Cyclic oxidation is widely used to characterize the resistance of the scale to spallation. Here automated retraction of samples from the hot zone (by means of timers, counters, switches, and motors or pneumatic actuators) can easily expose multiple samples for thousands of cycles of a specified duration. The most simple test is that of interrupted oxidation, where samples are oxidized continuously, except for the limited cycle that occurs during the periodic withdrawal needed for weighing and examination.

More complex rigs are used at dedicated research laboratories if oxidation under more realistic conditions is the primary objective of the test. Thus, rigs with natural-gas or jet-fuel burners are common in the land-based and aircraft gas turbine industries. Petrochemical reaction chambers, exhaust flue gas, carburizing or nitriding heat-treating furnaces, and heat exchangers for coal-fired power stations are other examples of special cases where testing should simulate the actual environment to obtain a faithful representation of the corrosion process.

Weight change data provide a very useful measure of the amount of metal converted to oxide and provide a succinct, helpful comparison among alloys. However, the ultimate engineering

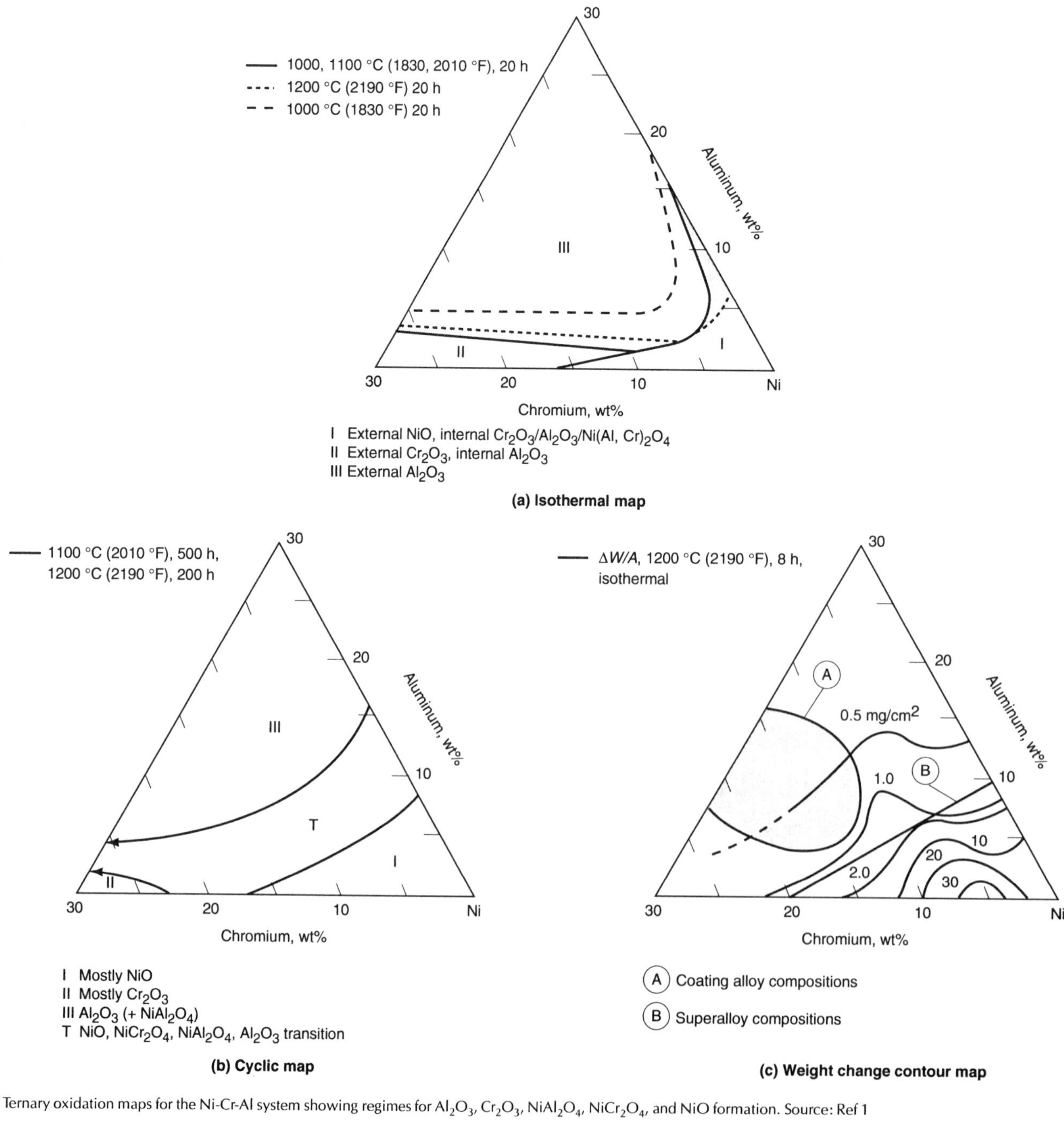

Fig. 2 Ternary oxidation maps for the Ni-Cr-Al system showing regimes for Al_2O_3, Cr_2O_3, $NiAl_2O_4$, $NiCr_2O_4$, and NiO formation. Source: Ref 1

criteria of the oxidation resistance of a material is retention of mechanical properties after exposure. This depends not only on the reduction of a load-bearing section from oxidation, but also on the morphology of the oxidation product. Consequently, environmental attack is further categorized by a thorough characterization of the corrosion products in conjunction with weight change data. For example, a simple metallographic cross section can illustrate the remaining unaffected metal core of the specimen. The nature of the attack (irregular intrusions, preferential phase attack, intergranular oxidation, internal oxidation, uniform planar external scales, and depletion zones) can also be readily distinguished. Scanning electron microscopy/energy dispersive spectroscopy (SEM/EDS) provides the next level of detail, with higher magnification and chemical analysis capability. This tool is especially useful in characterizing the morphology and distribution of features on the external surfaces. It is also used to determine the grain size of the oxide scales, the growth rate of which is often controlled by short-circuit grain-boundary diffusion. X-ray diffraction (XRD) of the external surface is widely used to determine the scale phases, and electron microprobe analysis of cross sections can accurately determine scale compositions and alloy-depletion profiles. More elaborate studies may employ transmission electron microscopy/scanning transmission electron microscopy (TEM/STEM), Auger spectroscopy, x-ray photoelectron spectroscopy, or chemical analyses of stripped corrosion products.

Alloy Design for Optimal Performance

The most general rule followed in alloy design for oxidation is that the alloys must form scales composed primarily of either Al_2O_3, Cr_2O_3, or SiO_2. For model iron-, cobalt-, or nickel-base binary alloys, this occurs at approximately 15 wt% Al, 20 wt% Cr, or 5 wt% Si. Unfortunately, these levels of aluminum, chromium, and silicon result in brittle alloys, and—in the case of silicon—very low melting temperatures. Silicon then is only used in small (1–3 wt%) additions to help stabilize Al_2O_3 or Cr_2O_3 scales. Many high-temperature alloys are, however, based on M-chromium systems having a good compromise of high-temperature mechanical properties and oxidation resistance. Small additions of manganese and silicon are widely used to stabilize Cr_2O_3 scales at concentrations below 20 wt% Cr. Similarly, chromium is widely used at 5 to 15 wt% levels to help stabilize the formation of Al_2O_3 scales at lower aluminum levels near 5 wt%.

The rationale for these secondary additions derives from the opposing trends of transient versus selective oxidation. Here, the initial oxidation process has been described as concurrent oxidation of all the major alloying elements present on the surface, even though one oxide is usually greatly favored from a thermodynamic standpoint. This proceeds until the initial rapid kinetics have transitioned to lower rates and the p_{O_2} at the oxide-metal interface has been lowered, both of which favor the formation of a healing layer of the most stable oxide. If silicon or chromium are added to a nickel-aluminum alloy, the contributions of slow growing SiO_2 or Cr_2O_3 nuclei during this transient period allow a complete transition to an Al_2O_3 inner healing layer at lower amounts of aluminum. The reduced aluminum level due to chromium additions can be seen in the Ni-Cr-Al ternary oxidation map of Fig. 2 that indicates which compositions form protective Al_2O_3 scales.

Another important consideration is that of scale adhesion during thermal cycling. The thermal expansion of oxides is much lower than that of most engineering metals such that large compressive stresses are built up in the scale upon cooling. As the scale thickens, these stresses eventually cause some form of scale fracture and spallation. For a strongly adherent scale, spallation often takes the form of microscopic segments in outer portions of the scale, thus relieving stresses without severe degradation. However, a weakly bonded scale can spall at the oxide/metal interface, exposing bare metal upon reheating. This results in more rapid consumption of the critical scale-forming elements, as indicated by Eq 3. Also, depletion zones are more likely to form, and subsequent oxidation products can now include more of the nonprotective transition scales, such as NiO.

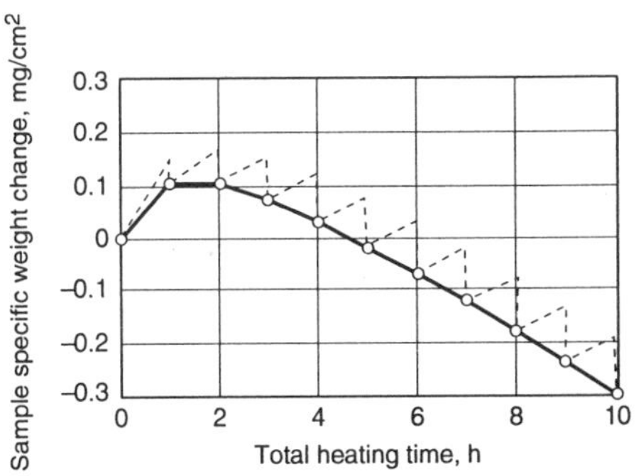

Fig. 3 Idealized cyclic oxidation weight change curve showing parabolic growth legs, punctuated by fractional spall events upon cooldowns. Source: Ref 2

The spalling process is of course directly reflected in the weight-change curve (Fig. 3), where each heating cycle is associated with a weight gain, and each cooling cycle produces a superimposed weight loss. For many alloys, the amount of scale spalled has been found to be a function of the (scale thickness)^2 times a spalling constant, Q_o. This process has been modeled in great detail, such that gravimetric curves can be predicted for various k_p growth rates and Q_o (or k_s) spall constants as shown in Fig. 4 (Ref 2–5).

Adherence is dramatically improved in alloys and coatings by the addition of about 0.01 to 0.1 wt% Y. An example of this effect is shown in Fig. 5. An explanation of the detailed mechanism is still evolving, but many critical experiments have indicated that interfacial segregation of sulfur is associated with scale debonding, presumably by weakening the interfacial bonds (Ref 6, 7). Conversely, the absence of sulfur segregation in desulfurized alloys is associated with good adhesion, even without the use of yttrium (Ref 8). When yttrium is added, it is believed to tie up the sulfur, present as a 1 to 10 ppm impurity in the bulk, by forming stable yttrium-sulfur sulfides or complexes. Other sulfur active elements, such as

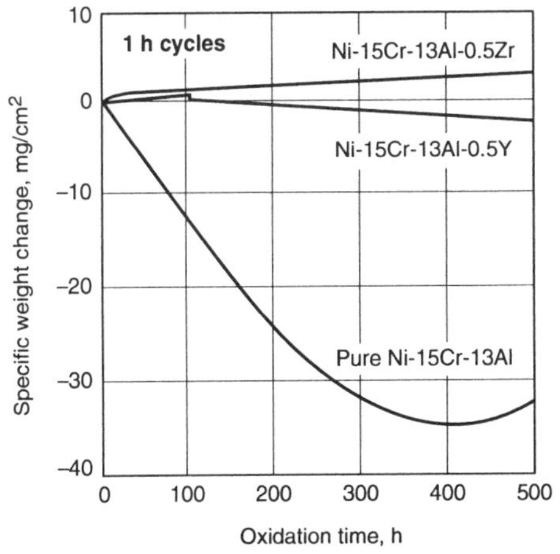

Fig. 5 The effect of a small addition of reactive elements (yttrium or zirconium) on scale adhesion and the 1100 °C (2010 °F) cyclic oxidation resistance of a model Ni-15Cr-13Al (wt%) coating alloy

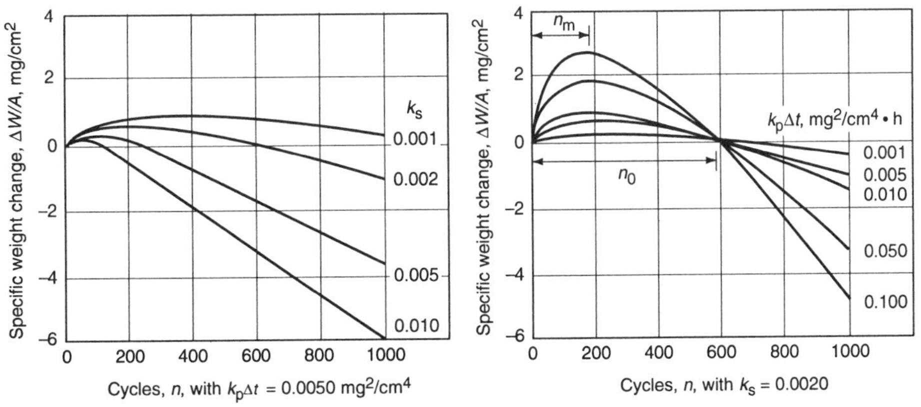

Fig. 4 Family of model cyclic oxidation curves for various parabolic rate constants, k_p, and spall constants, k_s. Source: Ref 5

Table 1 Ranking of 34 superalloys by the 1100 °C (2010 °F), 200 h cyclic oxidation attack parameter, K_a, and primary composition makeup

Also shown are model K_a values calculated from Eq 5 (see text), final weight changes, $\Delta W/A$, given in mg/cm^2, standard deviations, Σ, and number of duplicates, n.

Alloy	K_a (avg)	K_a (calc)	K_a (Σ)	$\Delta W/A$ (avg)	$\Delta W/A$ (Σ)	n	Composition, wt% (a) Cr	Al	Ta	Mo	W	Nb	Ti	Zr	Co	Other	½Cr + Al
TRW-R	0.11	0.54	0.01	−0.96	0.16	2	8.0	5.3	6.0	3.0	4.0	0.3	0.8	0.1	8.0	1.0 Hf	9.3
B-1900	0.19	0.31	0.07	−1.40	0.54	7	8.0	6.0	4.3	6.0	0.1	0.1	1.0	0.1	10.0	...	10.0
NASA-VIA	0.33	0.35	0.10	−1.48	0.83	6	6.1	5.4	9.0	2.0	5.8	0.5	1.0	0.1	7.5	0.4 Hf, 0.5 Re	8.5
MAR-M-247	0.51	0.77	0.08	−4.22	1.17	5	8.2	5.5	3.0	0.6	10.0	0	1.0	0.1	10.0	1.50 Hf	9.6
TAZ-8A	0.61	0.52	0.43	0.56	2.93	11	6.0	6.0	8.0	4.0	4.0	2.5	0	1.0	0	...	9.0
IN-713 LC	0.72	0.94	...	−6.20	...	1	12.0	5.9	0	4.5	0	2.0	0.6	0.1	0	...	11.9
B-1900+Hf	0.72	0.33	0.89	−1.68	0.25	3	8.0	6.0	4.3	6.0	0.1	0.1	1.0	0.1	10.0	1.0 Hf	10.0
TRW-1800	0.73	0.87	...	−8.65	...	1	13.0	6.0	0	0	9.0	1.5	0.6	0.1	0.6	...	12.5
MAR-M-246	1.55	0.84	...	−24.44	...	1	11.0	5.0	2.0	0	0	0	1.5	0	11.0	0.1 Cu	10.5
René-125	3.02	2.06	2.46	−21.22	17.42	3	9.0	5.0	3.8	2.0	7.0	0	2.5	0.1	10.0	1.5 Hf	9.5
Astroloy	3.24	9.14	...	−30.25	...	1	15.0	4.4	0	5.3	4.0	0	3.5	0.1	15.0	...	11.9
NX-188	3.45	2.28	0.28	−48.76	13.71	2	0(b)	8.0	0	18.0(b)	0	0	0	0	0	...	8.0
René-120	6.85	8.86	...	−38.57	...	1	9.0	4.3	3.8	2.0	7.0	0	4.0	0.1	10.0	...	8.8
U-700	6.96	5.42	9.20	−55.21	66.80	27	15.0	4.3	0	4.5	0	0	3.5	0.1	18.5	...	11.8
Mar-M-200	8.21	14.35	5.26	−53.59	3.95	3	9.0	5.0	0	0	12.5	2.7	2.0	0.1	10.0	...	9.5
Mar-M-421	9.53	8.64	...	−74.11	...	1	15.8	4.3	0	2.0	3.8	2.0	1.8	0.1	9.5	...	12.2
Waspaloy	9.63	15.18	12.38	−79.29	113.00	4	19.5	1.3	0	4.3	0	0	3.0	0.1	13.5	...	11.1
Mar-M-200+Hf	17.31	16.18	3.94	−89.80	28.36	6	9.0	5.0	0	0	11.5	1.0	2.0	0.1	10.0	1.5 Hf	9.5
WAZ-20	21.15	15.09	9.40	−198.05	60.17	2	0	6.5	0	0	18.5(b)	0	0	0.2	0	...	6.5
IN-792	22.55	19.20	4.69	−166.26	17.64	7	12.7	3.2	3.9	2.0	3.9	0	4.2	0.1	9.0	0.8 Hf	9.6
Mar-M-509	25.43	25.67	4.11	−174.15	52.40	2	23.5	0	3.5	0	7.0	0	0.2	0.5	54.7	...	11.8
IN-718	28.57	29.09	...	−284.60	...	1	18.0	0.4	0	3.1	0	5.0	0.9	0	0	18.5 Fe	9.4
IN-625	28.72	11.28	...	−293.20	...	1	22.5	0.2	1.9	9.0	0	1.8	0.2	0	0	...	11.5
IN-738	29.33	19.60	9.56	−232.45	110.83	10	16.0	3.4	1.8	1.8	2.6	0.9	3.4	0.1	8.5	...	11.4
U-520	31.65	17.26	...	−172.80	...	1	19.0	2.0	0	6.0	1.0	0	3.0	0	12.0	...	11.5
U-720	32.34	19.29	...	−313.50	...	1	18.0	2.5	0	3.0	1.2	0	5.0	0	15.0	...	11.5
IN-939	32.58	30.14	...	−227.60	...	1	22.0	2.0	1.5	0	2.0	1.0	3.6	0.1	19.0	...	13.0
U-710	33.76	20.21	...	−270.20	...	1	18.0	2.5	0	3.0	1.5	0	5.0	0	15.0	...	11.5
X-40	35.57	24.46	...	−206.30	...	1	25.5	0	0	0	7.5	0	0	0	56.0	...	12.8
René-80	37.40	20.00	3.47	−330.35	135.84	2	14.0	3.0	0	4.0	0	0	5.0	0	9.5	...	10.0
R-150-SX	45.01	68.24	...	−596.40	...	1	5.0	5.5	6.0	1.0	5.0	0	0	0	12.0	3.0 Re, 2.2 V	8.0
W-152	45.29	54.96	42.86	−569.30	14.57	2	21.0	0	0	0	11.0	2.0	0	0	65.6	2.0 Fe	10.5
IN-100	46.06	24.31	70.06	−180.33	245.46	3	10.0	5.5	0	3.0	0	0	5.5	0.1	15.0	1.0 V	10.5
Mar-M-211	73.46	11.60	102.19	−269.76	360.82	2	9.0	5.0	0	2.5	5.0	2.7	2.0	0.1	10.0	...	9.5
						Range	**5–25.5**	**0–8**	**0–9**	**0–9**	**0–12**	**0–5**	**0–5.5**	**0–1**	...	...	**6.50–13**

(a) Ni = balance. (b) Rare exceptions out of the typical composition range. Source: Ref 18

scandium, lanthanum, zirconium, and hafnium, have also shown this effect and are occasionally present in engineering alloys or coatings. The phenomenon is especially important for single-crystal superalloys, which have the potential of being very oxidation resistant, but are not easily manufactured with yttrium.

While the adhesion effect is prominent for Al_2O_3 and Cr_2O_3 scales, it has not been well documented for SiO_2 scales. Cr_2O_3 scales are also subject to fast outward growth when the alloy is not doped with reactive elements. Furthermore, all Cr_2O_3 scales are subject to degradation by formation of a volatile CrO_3 gaseous species and are therefore less suitable for high-velocity applications. Many other phenomena may be encountered in the oxidation of high-temperature alloys, such as the preferential oxidation of carbide phases or grain boundaries. The fundamentals of these phenomena are not widely reported nor easily generalized. However, in practice some alloy-specific process may override otherwise protective behavior, such that performance cannot be projected without consulting test data of the alloy in question.

The preceding is a brief introduction to some general aspects of oxidation that apply to typical high-temperature alloys. They apply equally well to intermetallic compounds, aluminide and MCrAlY coatings, and bond coats for thermal barrier coatings. Excellent treatments of the fundamental aspects can be found in reference books by Kofstad (Ref 9) or Birks and Meier (Ref 10). More detailed information on a wide range of commercial alloys in a variety of environments (carburizing, nitriding, halogens, salts, liquid metals) can be found in Lai (Ref 11). Supplementary information on superalloy oxidation, corrosion, and coatings is present in Ref 1 and 12 to 16 and in the review of high-temperature oxidation of iron-, nickel-, and cobalt-base alloys in Ref 17.

Performance Characteristics of Superalloys

Cyclic Oxidation. The oxidation resistance of commercial superalloys can vary dramatically with alloy type, temperature, and specific environment. Thus, although many of the mechanisms referred to in the previous section are usually applicable, it is difficult to generalize the complex oxidation behavior inherent to these multielement systems. One can obtain a better appreciation of this fact by examining the behavior of a large number of commercial superalloys tested in one study (Ref 18). Here, 34 alloys were tested in cyclic oxidation in air at 1100 °C (2010 °F) for 200 h, often with multiple samples, and characterized by weight change and scale phases. Selected alloys were also tested at 1000 °C (1830 °F) for 500 h and at 1150 °C (2100 °F) for 100 h. The result is a database allowing alloy comparisons on a one-to-one basis as well as an illustration of broad, overall compositional effects.

The weight change curves were described by paralinear or linear kinetics, which had been previously correlated to metal loss by means of an attack parameter, K_a (Ref 4). This parameter defines the combined effect of parabolic-scale growth rates and spalling rates on the consumption of metal. It can therefore be used as an engineering design figure-of-merit related to the residual load-bearing cross section of a component.

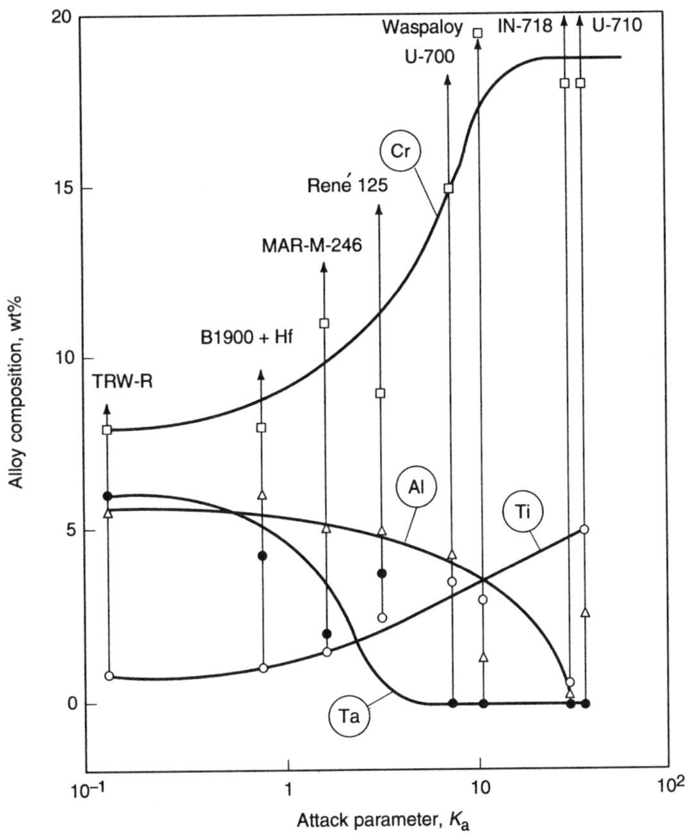

Fig. 6 Compositional trends of eight commercial superalloys with their 1100 °C (2010 °F) cyclic oxidation attack parameter. Source: Ref 18

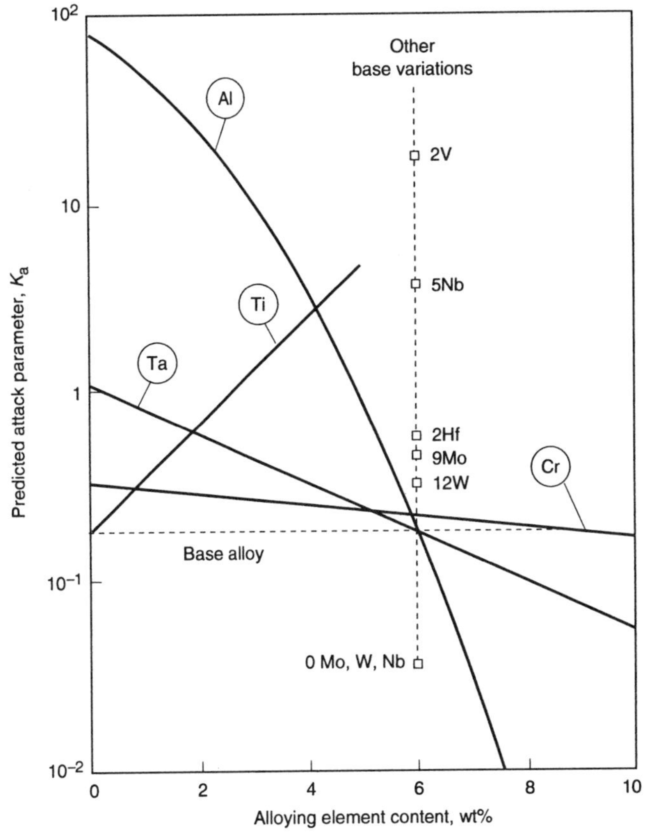

Fig. 7 Modeled (individual element) compositional effects on the 1100 °C (2010 °F) cyclic oxidation attack parameter for a Ni-8Cr-6Al-6Ta-4Mo-4W-1Nb base alloy

The results are summarized in a ranking table (Table 1), which orders the alloys according to K_a, along with the final weight change, number of duplicates, and standard deviations. For the most part, weight change and K_a, track each other, to the extent that the 200 h weight loss at 1100 °C (2010 °F) equals 7.6 K_a, in mg/cm^2, from regression analysis. Exceptions occur when an unusually large spread in the data occur for duplicate samples. These spreads may bias the average performance parameters differently than weight change.

The compositions of the 34 alloys are also listed in Table 1. In most cases, multiple elemental changes occur with progression down the table, thus obscuring any simple compositional effects on performance. However, many broad general trends can be discerned. For example, fairly consistent compositional blocks have been drawn in. These show consistently high aluminum (>5%), high tantalum (>3%), and low titanium (<1%) for the top-performing alloys. Conversely, lower aluminum (<5%), lower tantalum (<2%), and higher titanium (3–5%) occur more commonly for the poorer-performing alloy groupings. The highest levels of chromium (20–25%) did not necessarily ensure good performance, but 5 to 15% Cr was present in all the best-performing alloys. The very high chromium alloys (20%) did not rank very highly because they also typically contained low aluminum (0–2%). In fact, it has been pointed out that chromium and aluminum contents are usually inversely correlated in commercial superalloys (Ref 1), such that:

$$\tfrac{1}{2}\,\mathrm{Cr} + \mathrm{Al} = 10 \qquad \text{(Eq 4)}$$

The value of this compositional parameter is also listed in Table 1 and can be seen to vary primarily between the narrow range of 8 to 13.

A graphical illustration of the points made above can be seen in Fig. 6. Here the chromium, aluminum, tantalum, and titanium contents of a select group of eight alloys is plotted against the 1100 °C (2010 °F) cyclic oxidation attack parameter. The broad trend of improved performance (low K_a) with higher aluminum, higher tantalum, and lower titanium contents is apparent. It also appears that good performance is associated with a low chromium content. However, because chromium is inversely correlated with aluminum content so strongly, a separable chromium effect cannot be claimed from this figure alone.

These eight alloys in Fig. 6 were selected from the master database in Table 1 because their chromium, aluminum, tantalum, and titanium contents all varied in a more or less monotonic fashion with increasing K_a. There were nonsystematic variations in other alloying elements, such as molybdenum, tungsten, and niobium, which are considered to have only secondary effects. Within reason, some compositional trends can be concluded from this simple two-dimensional plot. However, the complexity and interactions of systems containing ten or more alloying elements often obscures individual elemental effects and compositional dependencies. As a result, multiple linear regression was appropriately used to extract more precise compositional parameters affecting performance (Ref 18). An equation thus results that gives the best overall fit ($R^2 = 0.84$) and accounts for the simultaneous variations of 11 elements and temperature:

$$\log_{10} K_a = x_1(1/T_K) + x_2(\mathrm{Cr}/T_K) + x_3(\mathrm{Cr}) + x_4(\mathrm{CrW}) + x_5(\mathrm{Al}^2) + x_6(\mathrm{AlTa}) + x_7(\mathrm{AlMo}) + x_8(\mathrm{AlV}) + x_9(\mathrm{TaNb}) + x_{10}(\mathrm{TaTi}) + x_{11}(\mathrm{NbTi}) + x_{12}(\mathrm{NbHf}) + x_{13}(\mathrm{Ti}) + x_{14}(\mathrm{Re}) + x_{15} \qquad \text{(Eq 5)}$$

where:

$x_1 = -28{,}734$	$x_9 = 0.05346$
$x_2 = 924.75$	$x_{10} = 0.01932$
$x_3 = -0.71874$	$x_{11} = 0.08140$
$x_4 = 0.00373$	$x_{12} = 0.24155$
$x_5 = -0.05162$	$x_{13} = 0.08345$
$x_6 = -0.03008$	$x_{14} = 0.21293$
$x_7 = 0.01273$	$x_{15} = 22.756$
$x_8 = 0.16396$	

Equation 5 can be used to show the relative importance of various elemental changes to a hypothetical superalloy, where, for illustration purposes, only one element was varied at a time. For example, this technique allows the effects of aluminum and chromium to be decoupled. A base alloy was chosen near the composition of the best-performing group of alloys (i.e., Ni-8Cr-6Al-6Ta-4Mo-4W-1Nb-0Ti-0Hf-0V). The variations covered the compositional ranges typical for most of the superalloys in Table 1. Attack parameters predicted by Eq 5 for 1100 °C (2010 °F) oxidation are shown in Fig. 7. The performance of the base alloy is shown as the horizontal dashed line and its intersection with each of the elemental curves. It can be seen that aluminum additions are the most effective in reducing K_a. Tantalum and now chromium can be seen to be beneficial, in that order. Vanadium and titanium were the most deleterious alloying elements. Detrimental effects were also predicted for separately adding the upper limit of the compositional range for niobium, hafnium, tungsten, or molybdenum. Alternatively, improved performance is predicted for removing the base levels of 4Mo, 4W, and 1Nb. No significant trend with cobalt content was found. It should be noted that although some of the modeled behavior exhibits K_a <0.1, these low values were not experimentally observed in this database. The intent of this plot is to address broad trends rather than pinpoint the precise performance of alloys not included in the database.

A simplified equation was also obtained from a regression model that did not use any cross-term interactions. The resulting equation ($R^2 = 0.80$) thus extracts coefficients that apply only to individual elements:

$$\log_{10} K_a = \Sigma x_i(y_i) + x_{12} \qquad \text{(Eq 6)}$$

where:

y_i =	x_i =
$1/T_k$	−17,305
Chromium	−0.08308
Aluminum	−0.33925
Tantalum	−0.15488
Titanium	0.26408
Molybdenum	0.04527
Niobium	0.24172
Hafnium	0.17782
Zirconium	0.37654
Rhenium	0.87295
Carbon	2.00000
x_{12} =	14.77172

Again, the relative benefits of aluminum, tantalum, and chromium and the detriment of titanium are evident. Negative effects of niobium, hafnium, zirconium, rhenium, and carbon are also indicated. It should be noted that of these only niobium and rhenium are typically added in excess of 1 wt%. Low hafnium additions (0.1 wt%) are generally considered beneficial to scale adhesion. However, "overdoping" with up to 2 wt% Hf to prevent grain-boundary cracking in columnar-grained superalloys can be detrimental to overall oxidation performance.

Finally, the 1100 and 1150 °C (2010 and 2100 °F) data, along with that from the few alloys tested at 1000 °C (1830 °F), was also incorporated by the model to obtain temperature effects. Thus, for the base alloy Eq 5 produced an Arrhenius relationship for the attack parameter, namely:

$$K_a = 6.5 \times 10^{14} \exp\{(-97{,}570\ \text{cal/mole})/RT\} \qquad \text{(Eq 7)}$$

The apparent activation energy, 98 kcal/mole, is quite in line with those generally observed for diffusion in, and growth of nickel-, chromium-, and aluminum-base oxide scales (i.e., 50–125 kcal/mole).

Interrupted Long-Term Oxidation. Another large database was developed for a wider representation of high-temperature alloys (Ref 19). This study tested 68 commercial alloys, including iron-base ferritic and austenitic stainless steels, at 982 °C (1800 °F) for 10,000 h in air, cycled 10 times for 1000 h each. The results are summarized in Tables 2(a) through 2(d). These tables list the performance (final weight change) and major chemistry by 4 categories:

- Iron-base alloys (Table 2a)
- Cobalt-base alloys (Table 2b)
- Nickel-base sheet alloys (Table 2c)
- Nickel-base superalloys (Table 2d)

The best-performing group of alloys was the ferritic alloys with aluminum. These alloys typically contained from 2 to 32% Al and 0 to 18% Cr. They gained between just 1 to 4 mg/cm² and formed primarily protective Al_2O_3 scales, as expected from the composition (FeCrAl, FeAl, or Fe_3Al). Conversely, the ferritic alloys without aluminum (Croloys and ferritic stainless steels) oxidized the most as a group, gaining between 100 to 300 mg/cm² for all but one sample that lost

Table 2(a) Ranking of iron-base ferritic alloys and austenitic stainless steels in a 10,000 h, 982 °C (1800 °F), interrupted oxidation test

See Tables 2(b), 2(c), and 2(d) for the rankings of cobalt- and nickel-base alloys that were also tested. A total of 68 commercial high-temperature alloys are ranked.

				Composition, wt%								
Alloy	Type	$\Delta W/A$ (max)	Rank	Ni	Fe	Cr	Al	Ti	Mo	Mn	Si	Other
RA-310	F	0.50	2	20.0	52.9	25.0	0	0	0	1.5	0.5	...
HOS-875	F	0.83	7	0	71.4	22.5	5.5	0	0	0	0.5	...
NASA-18T	F	1.25	9	0	76.5	18.0	2.0	0.5	0	0.4	1.3	1.3 Ta
TRW Valve	F	1.47	10	0.1	66.8	0	32.0	0	0.1	0.1	0.4	0.5 W
Thermenol	F	1.70	12	0	80.7	0	16.0	0	3.3	0	0	...
18SR	F	3.47	18	0.3	77.9	18.0	2.0	0.4	0	0.4	1	...
IN-800	A	−9.67	26	32.0	45.3	20.5	0.4	0.4	0	0.8	0.4	...
RA-330	A	−9.72	27	35.0	43.2	19.0	0	0	0	1.5	1.3	...
RA-26-1	F	−18.66	36	0.2	71.6	26.0	0	0.5	1.0	0.3	0.3	...
316SS	A	−51.31	41	12.0	65.4	17.0	0	0	2.5	2	1	...
T439SS	F	59.98	42	0.5	79.3	18.3	0.2	1.0	0	0.4	0.4	...
334SS	A	63.04	43	20.0	58.4	19.0	0.3	0.4	0	1	0.3	...
310SS	A	−92.36	46	20.5	50.8	25.0	0	0	0	2	1.5	...
321SS	A	112.18	51	10.5	69.0	17.0	0	0.4	0	2	1	...
430SS	F	−164.65	53	0	81.9	17.0	0	0	0	0	1	...
RA-309	A	−176.65	54	14.0	60.6	23.0	0	0	0	1.5	0.8	...
347SS	A	178.17	55	11.0	67.1	18.0	0	0	0	2	1	0.4 Nb, 0.4 Ta
409SS	F	184.59	57	0	88.5	11.0	0	0.5	0	0	0	...
309SS	A	−187.09	58	13.5	60.3	23.0	0	0	0	2	1	...
410SS	F	208.56	59	0	85.4	12.5	0	0	0	1	1	...
304SS	A	−227.65	60	10.0	67.9	19.0	0	0	0	2	1	...
Croloy 9	F	308.47	62	0	88.5	9.0	0	0	1.4	0.5	0.5	...
Croloy 5	F	310.63	63	0	93.7	5.0	0	0	0.6	0	0.5	...
Croloy 7	F	315.07	64	0	91.1	7.0	0	0	0.6	0.5	0.8	...
Multimet	A	−320.82	65	20.0	30.2	21.0	0	0	3.0	1.5	0.5	20 Co, 2.5 W, 0.5 Nb, 0.5 Ta

SS, stainless steel; A, austenitic; F, ferritic. Source: Ref 19

165 mg/cm^2. The major scales phases were Fe_2O_3 and occasionally Cr_2O_3 or $(Ni,Fe)Cr_2O_4$ spinel. The austenitic stainless steels performed midway between these groups, with weight changes of 19 to 228 mg/cm^2. One exception was alloy RA-310, which ranked among the better alloys with a gain of only 0.4 mg/cm^2. The major scale phases among this group were $(Ni,Fe)Cr_2O_4$ spinel and Cr_2O_3 or Fe_2O_3.

The cobalt-base alloys exhibited weight changes between a gain of 1 mg/cm^2 and a loss of 10 mg/cm^2 with two exceptions: L-605 samples lost 70 to 100 mg/cm^2 and WI-52 lost 414 mg/cm^2. All alloys exhibited MCr_2O_4 spinel and Cr_2O_3 as the primary scale phases, with no correlation with performance.

The nickel-base sheet alloys with aluminum were often among the best-performing alloys, with maximum weight changes less than 3 mg/cm^2.

The nickel-base alloys with iron exhibited a wide span of performance, with gains as low as 3 mg/cm^2 for IN-601 or losses ranging from 7 mg/cm^2 for Hastelloy X to 510 mg/cm^2 for Hastelloy N. The primary scale phases were usually $(Ni,Fe)Cr_2O_4$ and Cr_2O_3, except for the $NiMoO_4$ and MoO_4 observed on Hastelloy N.

The nickel-base alloys without iron also exhibited widely varying behavior, with weight gains ranging from 0.5 mg/cm^2 for U-700 to 178 mg/cm^2 for WAZ-20, and with weight losses ranging from 1.7 mg/cm^2 for Hastelloy S to −300 mg/cm^2 for IN-100. The scale phases were $Ni(Al,Cr)_2O_4$, NiO, Cr_2O_3, Al_2O_3, or $Cr(Ta,Mo,W)O_4//Ni(Ta,Mo,W)_2O_6$ tri-rutile phases, without an overall correlation with performance. However Al_2O_3-forming alloys did the best within this group.

Table 2(b) Ranking of cobalt-base alloys in a 10,000 h, 982 °C (1800 °F), interrupted oxidation test

See Tables 2(a), 2(c), and 2(d) for the rankings of iron- and nickel-base alloys that were also tested. A total of 68 commercial high-temperature alloys are ranked.

Alloy	ΔW/A (max)	Rank	Composition, wt %									
			Ni	Co	Fe	Cr	Al	Ti	W	Nb	Ta	Other
X-40	−0.79	5	10.5	56.0	0	25.5	0	0	7.5	0	0	...
Belgian P-3	1.10	8	11.6	32.9	29.5	25.0	0	0	0	0	0	...
MAR-M-509	1.53	11	10.0	54.7	0	23.5	0	0.2	7.0	0	3.5	0.5 Zr
H-150	−4.41	20	3.0	47.6	20.0	28.0	0	0	0	0	0	...
HA-188	−4.87	21	22.0	39.4	1.5	22.0	0	0	14.0	0	0	...
Belgian S-57	−7.99	24	10.0	56.5	0	25.0	3.0	0	0	0	5.0	0.5 Y
L-605	−101.81	50	10.0	52.9	0	20.0	0	0	15.0	0	0	...
WI-52	−413.72	67	0	65.6	0	21.0	0	0	11.0	2.0	0	...

Source: Ref 19

Table 2(c) Ranking of nickel-base sheet alloys in a 10,000 h, 982 °C (1800 °F), interrupted oxidation test

See Tables 2(a), 2(b), and 2(d) for the rankings of iron- and cobalt-base alloys and nickel-base superalloys. A total of 68 commercial high-temperature alloys are ranked.

Alloy	ΔW/A (max)	Rank	Composition, wt %				
			Ni	Fe	Cr	Al	Other
DH-245, 0.74	4	74.1	0.7	20.0	3.5	...	
IN-702	0.80	6	80.5	0.5	15.4	3.1	0.4 Ti
IN-601	2.71	16	60.3	14.1	23.0	1.4	...
Chromel A	−5.10	22	78.0	0.4	20.0	0	...
Chromel C	−11.65	29	59.3	24.0	15.0	0	...
DH-241	−12.07	30	76.3	0.3	20.0	0	...
Tophet 30	−12.21	31	68.6	0	30.0	0	1.4 Si
Chromel AA	−14.70	34	67.8	10.0	20.0	0	...
DH-242	−17.78	35	77.8	1.0	20.0	0	
Ni-40Cr	−21.90	38	60.0	0	40.0	0	...
Chromel P	75.70	44	89.4	0.2	10.0	0	...
Pure Ni-270	114.0	52	100.0	0	0	0	...

Source: Ref 19

Table 2(d) Ranking of nickel-base superalloys in a 10,000 h, 982 °C (1800 °F), interrupted oxidation test

See Tables 2(a), 2(b), and 2(c) for the rankings of iron- and cobalt-base alloys and nickel-base sheet alloys. A total of 68 commercial high-temperature alloys are ranked.

Alloy	ΔW/A (max)	Rank	Composition, wt %										
			Ni	Co	Fe	Cr	Al	Ti	Mo	W	Nb	Ta	Other
NASA-VIA	−0.08	1	62	7.5	0	6.1	5.4	1	2	5.8	0.5	9	0.4Hf, 0.5Re
U-700	0.63	3	54	18.5	0	15	4.3	3.5	4.5	0	0	0	...
Hastelloy S	−1.72	13	67.4	0	1	15.5	0.2	0	15.5	0	0	0	...
B-1900	−1.83	14	64.3	10	0	8	6	1	6	0.1	0.1	4.3	...
René-120	2.05	15	59.8	10	0	9	4.3	4	2	7	0	3.8	...
IN-617	−3.23	17	55.4	12.5	0	22	1	0	9	0	0	0	...
TAZ-8A	−4.11	19	68.4	0	0	6	6	0	4	4	2.5	8	1.0Zr
Hastelloy X	−7.00	23	46.3	1.5	18.5	22	0	0	9	0.6	0	0	...
IN-713LC	−8.67	25	74.8	0	0	12	5.9	0.6	4.5	0	2	0	...
IN-671	−11.02	28	51.6	0	0	48	0	0.4	0	0	0	0	...
RA-333	−12.35	32	45.2	3	18	25	0	0	3	3	0	0	...
MAR-M-200	−14.23	33	58.6	10	0	9	5	2	0	12.5	2.7	0	...
IN-706	−19.74	37	77.5	1	0	16	0.4	1.8	0	0	2.9	0	...
IN-600	−29.33	39	76.5	0	7.2	15.8	0	0	0	0	0	0	...
Hastelloy C-276	−34.36	40	55.4	2.5	5.5	15.5	0	0	16	3.8	0	0	0.35V
IN-750X	−80.82	45	73.6	0	6.8	15	0	2.5	0	0	0.9	0	...
IN-804	−93.81	47	42.6	0	25.4	29.5	0.3	0.4	0	0	0	0	...
Hastelloy G	−96.20	48	45.8	0	19.5	22	0	0	6.5	0.5	1	1	...
IN-738X	−96.51	49	61.4	8.5	0	16	3.4	3.4	1.8	2.6	0.9	1.8	...
WAZ-20	178.29	56	74.7	0	0	0	6.5	0	0	18.5	0	0	...
René-80	−239.97	61	64.3	9.5	0	14	3	5	4	0	0	0	...
IN-100	362.88	66	59.8	15	0	10	5.5	5.5	3	0	0	0	1.0V
Hastelloy N	−511.92	68	69.9	0.2	5	7	0	0	16.5	0	0	0	...

Source: Ref 19

The general picture obtained from this work is that Al_2O_3 formers and some Cr_2O_3 formers can be very oxidation resistant at 982 °C (1800 °F) for 10,000 h. The iron-base materials (ferritic and austenitic stainless steels) containing no aluminum were the least-protective classes of alloys, with a wide range of intermediate to excellent behavior exhibited by the nickel- and cobalt-base alloys.

Similar ranking behavior was observed for the same test conducted at 815 °C (1500 °F) for the majority of alloys (Ref 20). However, now a considerable number of chromia-formers performed almost as well as the alumina-formers. Especially notable improvements, on a relative basis, were observed for Hastelloy C-276, IN-600, IN-706, and type 304 stainless steel when tested at 815 °C (1500 °F) as compared to 982 °C (1800 °F).

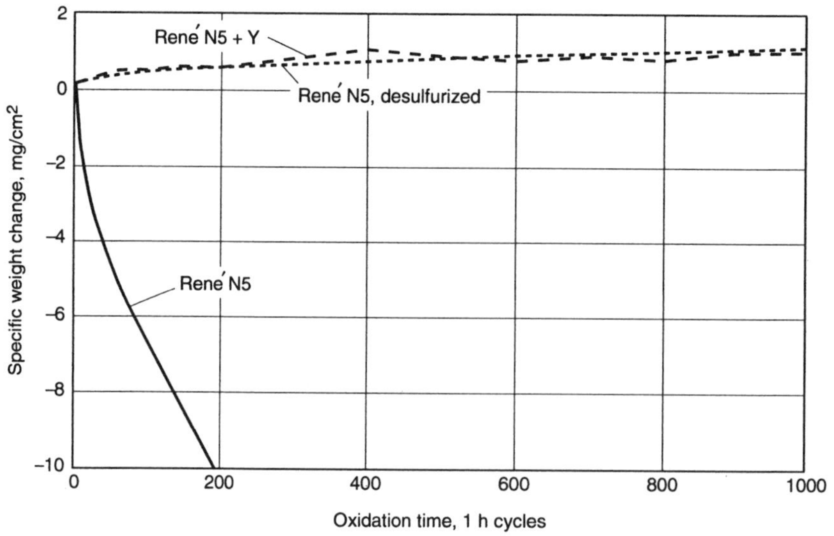

Fig. 8 The effect of desulfurization (1280 °C, or 2335 °F, 100 h hydrogen anneal) or yttrium additions (100–150 ppm) on the 1150 °C (2100 °F) cyclic oxidation behavior of René N5. Source: Ref 8

Performance Characteristics of Single-Crystal Superalloys

A great deal of current interest in the field of oxidation centers around advanced nickel-base single-crystal alloys. These alloys are used in the most demanding high-temperature applications of gas turbines. Some of the most common of these alloys are PWA 1480, 1484, and 1487 (Pratt and Whitney), René N4, N5, and N6 (General Electric), CMSX-4, -6, and -10 (Cannon-Muskegon), and SRR 99 and RR 2000 (Rolls-Royce). Most of these alloys fall within the following compositional range:

Element	wt%
Chromium	5–10
Aluminum	5–6
Tantalum	2–8
Molybdenum	1–3
Tungsten	4–6
Niobium	0–1
Rhenium	0–6
Titanium	0–5
Hafnium	0–0.3
Cobalt	5–10

These levels make possible, first and foremost, a continuous Al_2O_3 film, upon which excellent oxidation resistance depends. Many of the other elements also participate in scale formation as $Ni(Al,Cr)_2O_4$, $(Al,Cr)TaO_4$, and $NiTa_2O_6$ oxides. These phases, when located as an outer layer, may take part in a relatively innocuous transient growth process. Other nonprotective scales—(NiO, $NiTiO_3$, Cr_2O_3, Ta_2O_5, or $(Mo,W)O_3$—can form after an interval of cyclic oxidation if scale adhesion is lacking.

While this alloy class was not included in the 34 alloy database described earlier (see Table 1), Eq 5 can be used to predict attack parameters. For example, the 1100 °C (2010 °F) K_a for René N5 was calculated to be 0.22, which is comparable to the low value of 0.19 obtained for the base alloy in Fig. 7. Furthermore, the oxidation resistance of these alloys can be dramatically improved by maximizing Al_2O_3-scale adhesion. This is accomplished by the addition of reactive elements or by desulfurization by melt purification and/or hydrogen annealing (Ref 21–25). Examples of these effects can be seen in Fig. 8 to 10 for René N5, CMSX-4, and PWA 1480, respectively. Yttrium is difficult to homogeneously incorporate at high levels, but can be effective at just 15 ppm when the sulfur content is low (<2 ppm) (Ref 24). Alternatively, good adhesion is obtained without yttrium if the sulfur content is extremely low (<0.5 ppm).

Although some of the best oxidation resistance has been documented for single-crystal alloys, it is not due to the unique monocrystalline structure. Rather, the primary oxidation response stems from the high aluminum and tantalum contents, dopant levels of yttrium and hafnium, low sulfur, and the remainder of the compositional makeup. The absence of grain-boundary carbide strengtheners may also contribute to a more continuous Al_2O_3 scale.

The previous discussion has outlined some of the oxidation issues applying to bare (uncoated) alloys. High-temperature test data have been used for the purpose of discriminating compositional effects. It should be noted that many of the alloys will exhibit acceptable performance for intermediate- to low-temperature applications. But, for the most demanding applications, oxidation-resistant coatings combined with the best alloys become the most practical defense against aggressive environments.

Coating Concepts for Superalloys

Aluminides. The most widely used coating in gas turbines is the intermetallic compound NiAl. The monolithic material has been extensively studied and shown to form protective Al_2O_3 scales, even to the lower aluminum limits of the single-phase stoichiometry. However, in a cyclic mode, the hypostoichiometric compositions have been shown to exhibit degradation because the aluminum depletion from scale spallation can result in depletion zones with reduced oxidation resistance (Ref 26, 27). The use of reactive elements such as zirconium or yttrium have been shown to reduce interfacial spalling and produce some of the highest oxidation resistance ever measured for nickel-base material (Ref 28–30).

The coatings are generally produced by chemical vapor deposition (CVD) processes (aluminum halide gas or aluminum powder pack) by diffusional reaction with the superalloy substrates. Depending on the temperature and the aluminum activity of the pack, the coatings may be formed by outward nickel or inward aluminum diffusion (Ref 31). Inward diffusion coatings may require a higher temperature anneal to convert a very brittle Ni_2Al_3 phase into NiAl. Inward coatings also have a higher level of substrate elements present. At or below 1000 °C (1830 °F), very long oxidative lifetimes can be expected due to the low growth rate of Al_2O_3 scales and limited spalling. As temperature exceeds 1100 °C (1830 °F), interdiffusion with the substance becomes a major mode of aluminum loss, which then triggers initial coating degradation (Ref 27, 32). Once this dilution takes place, less-protective scales may form with excessive spalling and eventual failure. Interdiffusion is also responsible for potentially embrittling carbide or σ-Widmanstätten phases in the coating/substrate interdiffusion zone.

Dopant elements have been found to beneficially affect aluminide coating performance. Hafnium, zirconium and yttrium increase alumina scale adhesion. Silicon modifications can serve in this regard, but they are especially effective in stabilizing Al_2O_3-scale formation at very low aluminum concentrations. The most widespread aluminide modification is platinum. For example, a 20 μm layer of platinum is first electroplated onto the superalloy and then aluminized to form a $PtAl_2$-(Ni,Pt)Al two-phase mixture or layered structure. Platinum serves many purposes: it stabilizes an aluminum-rich surface phase for continued Al_2O_3 formation; it improves Al_2O_3-scale adhesion; and it improves the hot corrosion resistance of NiAl (Ref 33).

An example of a platinum-aluminide coating on a single-crystal superalloy, exposed in a high by-pass turbofan engine, is shown in Fig. 11. The outer layer is depleted NiAl, with particles of Ni_3Al forming at grain boundaries. The inner layer is an interdiffusion zone, enriched in alloying elements due to nickel dilution in forming the outer layer. Finally, a modified substrate structure is formed beneath this, with topologically closed-packed chromium-rich phases (σ).

Ni(Co)CrAlY. M-(10-30)Cr-(5-20)Al-(0.1-1)Y coatings are also widely used to solve the oxidation/corrosion problems of high-temperature turbine materials. They are generally applied by physical vapor deposition (PVD) or low-pressure (vacuum) plasma spraying (LPPS or VPS). These coatings are also based on excellent Al_2O_3-forming capability (see Fig. 2) but at lower aluminum levels and with an increase in low-temperature ductility. The presence of chromium gives an added defense against sodium sulfate hot corrosion, for which Al_2O_3 formers are not as resistant as Cr_2O_3-formers. Because of the "overlay" coating process and the compositional similarity to the superalloy substrates, these coatings are much less susceptible to diffusional wearout than the aluminide conversion coatings. However, there is a reduced aluminum reservoir to sustain Al_2O_3-scale growth or reformation after spallation.

Thermal barrier coatings (TBCs) enable a large improvement in turbine durability or performance by reducing the operating temperature of the superalloy surface or by allowing hotter gas temperatures. The insulating layer is generally about 125 to 250 μm of 6 to 8% Y_2O_3-ZrO_2, generally referred to as yttria partially stabilized zirconia (YSZ). It is usually applied by plasma spraying (PS) or electron beam (EB)-PVD. A preliminary bond coat is necessary for optimal adhesion. For PS-YSZ, an LPPS NiCoCrAlY type of bond coat is conventionally used, whereas for EB-PVD coatings, a platinum aluminide diffusion coating is used.

The failure mechanism of the TBC in clean environments is a combination of bond-coat oxidation and thermal fatigue (Ref 34). As the Al_2O_3 scale thickens at the YSZ-bond coat interface, it increases the stresses on the adjacent zirconia. During cooldown, the higher-expansion substrate puts large compressive thermal stresses on the zirconia, encouraging it to buckle and pull the thermally grown Al_2O_3 scale off the bond coat. Continued cycling results in crack growth until a critical length is achieved and macrospalling of the ceramic coating occurs (failure).

The surface roughness of the plasma-sprayed system enhances the bond strength by the increased contact area and mechanical interlocking. However, this roughness also produces tensile stress concentrations at the tips of the bond coat asperities and thus initiates interfacial cracking. The porosity and weak, microcracked splat boundaries in the PS-YSZ accommodates much of this stress. In contrast, the EB-PVD coatings with aluminide bond coats have a much smoother interface and less stress concentration, albeit with less mechanical bonding. The thermal stress is accommodated by the columnar structure of the EB-PVD zirconia, which demonstrates a propensity to crack along column boundaries, rather than buckle and debond at the oxide/metal interface. An example of an EB-PVD coating applied to a platinum-aluminide-coated René N5 alloy is shown in Fig. 12. Both systems are expected to benefit by improved Al_2O_3-scale adhesion. This can be accomplished by low-sulfur alloys, low-sulfur coatings, and by reactive element dopants to both.

For more corrosive environments, such as shipboard marine-power turbines, low-melting $NaVO_3$ deposits may destabilize YSZ by leaching out the yttrium to form YVO_4. Destabilization allows a martensitic transformation to monoclinic zirconia accompanied by a large volume expansion and powdering of the coating. Scandia (+india)-stabilized zirconia has been found to be resistant to vanadate corrosion and may provide

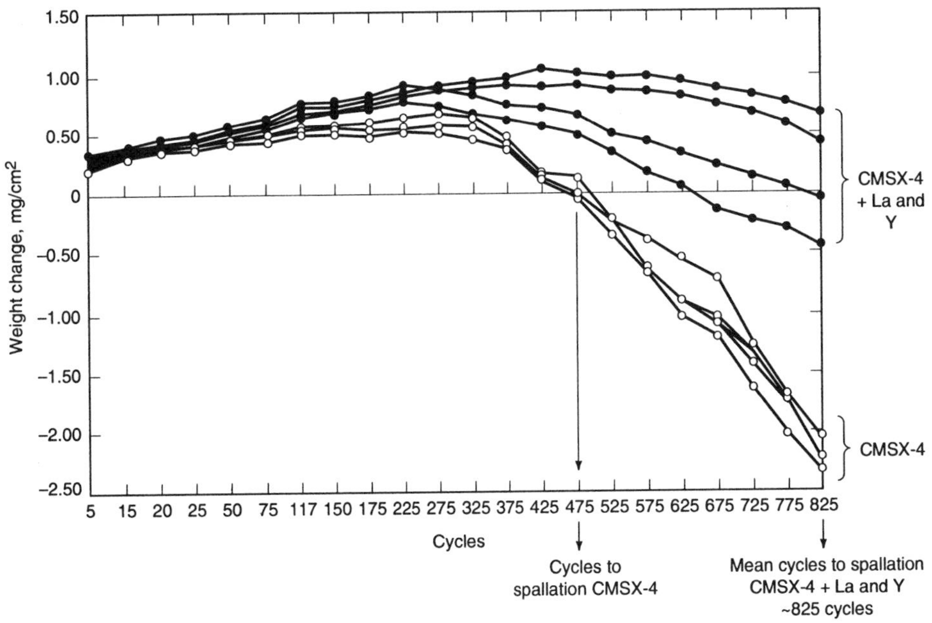

Fig. 9 The effect of 15 to 20 ppm of lanthanum and yttrium on the 1177 °C (2150 °F) cyclic oxidation behavior of low sulfur (<2 ppm) CMSX-4. Source: Ref 24

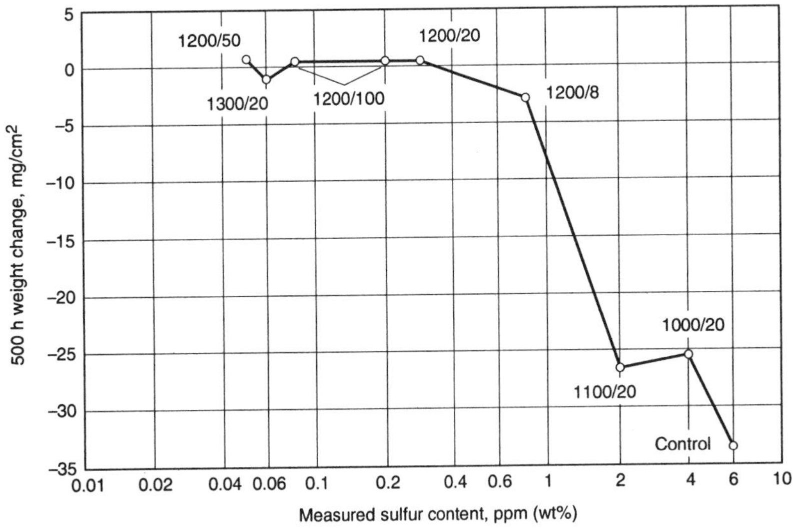

Fig. 10 The effect of sulfur content on the 500 h (cycles), 1100 °C (2010 °F) cyclic oxidation weight change of hydrogen-annealed PWA 1480. The values given inside the figure represent the annealing temperature/time in °C/h. Source: Ref 25

a benefit in the presence of sea salt and vanadium impurities in the fuel (Ref 35).

The state-of-the-art of processing, stability, service experience, thermal and stress modeling and design, and advanced concepts have been summarized in a recent NASA workshop on TBCs (Ref 36). In addition to aeroturbine components, the applications of this concept to diesel engine and land-based turbines has been described.

Chromizing. In the latter stages of the turbine (low-pressure turbine), the temperatures are reduced to below 900 °C (1650 °F). Thus, high-temperature oxidation is much less of a concern, but the conditions are optimal for Na_2SO_4 deposition and high-temperature corrosion (see the section on "Hot Corrosion" in this article). Chromizing is the CVD or pack deposition of elemental chromium, followed by a diffusion anneal to a nickel-chromium outer layer. This enrichment of chromium provides protection against the scale-fluxing attack that occurs under a molten sulfate deposit.

Performance Characteristics of Other High-Temperature Materials

Nickel and Titanium Aluminides. The oxidation resistance of NiAl has been actively studied since 1970. The fundamental mechanisms were described by Pettit, and regimes of protective behavior were mapped over the equilibrium diagram (Ref 37). Al_2O_3 was found to be stable over the entire temperature range for aluminum concentrations above about 30 at.%. Cyclic oxidation showed rapid spalling and depletion at 1100 °C (2010 °F) for compositions below about 40 at.% (Ref 27). Zirconium and other reactive element additions produced large improvements in scale adhesion, allowing protective behavior at 1200 °C (2190 °F) for 3000 h as shown in Fig. 13 (Ref 28, 29). Reactive element oxide dispersoids were also effective in this regard (Ref 30). Long-term durability is directly keyed to the large aluminum reservoirs in Ni-50Al-base alloys and the ability to supply aluminum to the surface with minimal depletion-zone formation until the bulk composition has dropped to 40 at.% (Ref 26). Impressive gains in low-temperature ductility and high-temperature strength were made for microalloyed single-crystal NiAl alloys (Ref 38, 39). However, widespread use as a free-standing structural material has not yet been realized.

Ni_3Al has been ductilized (by 0.1% B additions) to a much greater extent and has moderate oxidation resistance below 1000 °C (1830 °F) (Ref 40). However, an intermediate temperature embrittlement phenomenon is observed. Similarly, Fe_3Al and FeAl alloys have been engineered with moderate ductility (1–5%) and moderate oxidation resistance (Ref 41, 42). Some of the iron aluminide alloys have exhibited exceptional sulfidation resistance in recuperator environments from coal gasification plants and are also a low-cost alternative to nickel-base candidates.

Various titanium aluminides are being actively researched and considered for the latter stages of the compressor or turbine sections and have also found application in automotive valves and turbochargers. Ti-24Al-11Nb (at.%), Ti-22Al-24Nb (at.%), and Ti-48Al-2Cr-2Nb (at.%) (called the α_2-, orthorhombic-, and γ-phase alloys, respectively) have very attractive specific strength values because of their very low densities. The oxidation behavior has proved to be very complex, in that layered, intermixed $TiO_2 + Al_2O_3$ scales are the rule, with possible beneficial or detrimental effects of air versus oxygen environments (due to titanium nitride) (Ref 43–45). In general, the oxidation rate becomes excessive above 900 °C (1650 °F) for the α_2 and orthorhombic alloys. Furthermore, at 700 °C (1290 °F), surface hardening (embrittlement) of these two alloys has been observed due to oxygen interstitial diffusion (Ref 45). An effective coating would be very desirable; however, most attempts at aluminide, silicide, or MCrAlY coatings have only resulted in lowering fatigue lives.

The oxidation rate of γ-phase TiAl-base alloys is much more acceptable, and the embrittlement phenomena are much less prominent, requiring exposure above 800 °C (1470 °F) to achieve measurable hardening (Ref 45). However, TiAl alloys are inherently brittle, with maximum tensile ductilities on the order of just 1 to 2%. A predominance of Al_2O_3 in the scale on γ-phase alloys results in moderate oxidation resistance at 800 °C (1470 °F) as shown in Fig. 14. Considerable improvements are indicated by some experimental coatings that are based on the TiAl + Cr alloys (Fig. 15).

The oxidation behavior of nickel, iron, and titanium aluminides, as well as a number of other developmental intermetallic compounds, have been extensively reviewed. Recently published reviews can be found in Ref 46 to 49.

Refractory metals such as niobium, molybdenum, tantalum, tungsten, and rhenium are natural candidates for high-temperature applications because of their extremely high melting points (2250 to 3400 °C, or 4080 to 6150 °F). However, they exhibit some of the highest oxidation rates of structural metals to the extent that they are rarely used in oxygen-containing environments without coatings. Even with coatings, the potential for catastrophic failure generally limits their applications to less-critical components or short missions (e.g., rocket nozzles or exhaust nozzles of military turbines).

Coatings for refractory metals and their alloys are typically based on disilicides (MSi_2) that

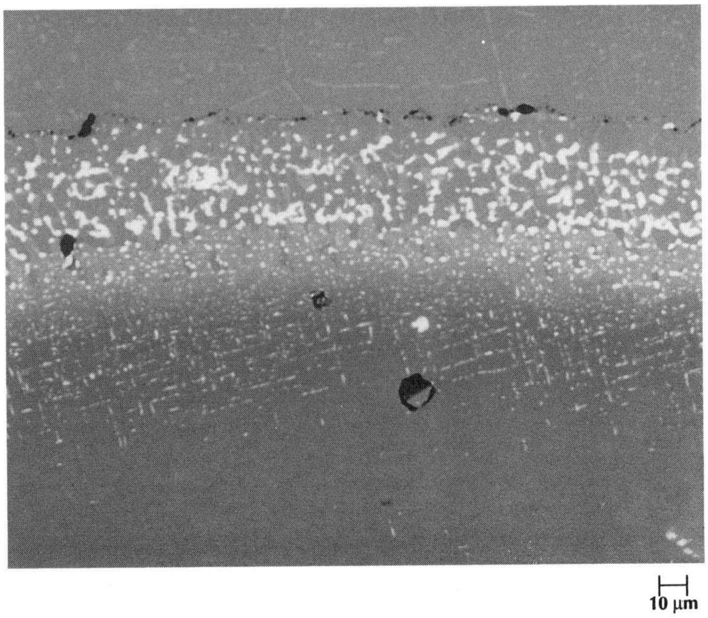

Fig. 11 Microstructure of a platinum-modified pack aluminide coating on a single-crystal superalloy after engine exposure. 600×

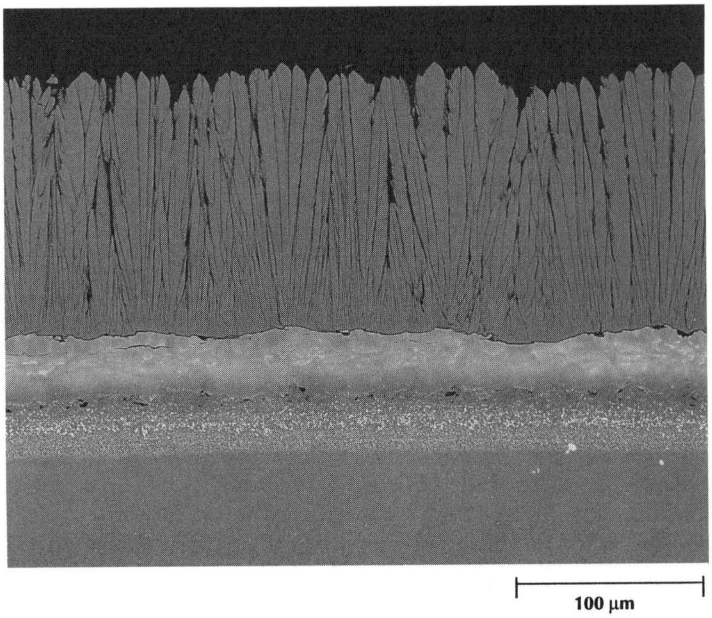

Fig. 12 Columnar structure of an EB-PVD 7% yttria-zirconia thermal barrier coating on a platinum-aluminide-coated single-crystal superalloy. 250×

form protective SiO_2 scales. Some additives enable lower-melting silicates to flow and seal through-cracks that inevitably form because of thermal expansion mismatch stresses with the substrate. The coatings are generally applied by slurry spraying followed by vacuum sintering, where other additives are used to lower the processing fusion temperature of the coating. Typical components include molybdenum, niobium, titanium, iron, cobalt, chromium, and silicon (Ref 50).

While maximum lifetimes of hundreds of hours at 1500 °C (2730 °F) may be expected in an isothermal exposure for, say, a defect-free $MoSi_2$ coating, repeated cycling will trigger degradation by cracking. For example, bars of FS85 (Nb-28Ta-10W-1Zr) were slurry fusion coated with a commercial R512E coating (20Cr-20Fe-60Si) and tested in a 30 min cycle Mach 0.3 burner rig (Ref 51). This exposure produced coating breaches (defined by –5 mg/cm^2 weight losses) at 170, >200, 120, and 110 cycles for 982, 1093, 1260, and 1371 °C (1800, 2000, 2300, and 2500 °F) exposures, respectively. Except for the 1093 °C (2000 °F) exposure, all samples showed catastrophic oxidation pits or massive substrate attack at the specimen edges. Oxidation products were friable mixtures of Nb_2O_5, $FeNbO_4$, SiO_2, and amorphous silicates.

$MoSi_2$ has also been developed as a free-standing structural material. Its most common application is as ultrahigh temperature heating elements (SuperKanthal) for furnaces (up to 1700 °C, or 3090 °F). Its oxidation resistance is derived from extremely slow-growing SiO_2 scales (Fig. 1). However, at low temperatures (500 to 700 °C, or 930 to 1290 °F), $MoSi_2$ exhibits a classic "pest" catastrophic oxidation mechanism, where a sample literally turns to dust in a short amount of time. This mechanism has been shown by TEM techniques to be caused by stresses from a voluminous oxidation product of an inward-growing SiO_2 and MoO_x microfilament mixture (Ref 52, 53). Rapid inward diffusion of oxygen is allowed by the MoO_x microfilaments. In a defect-free and stress-free material, pest can be avoided or at least delayed. In practice, pest is prevented by

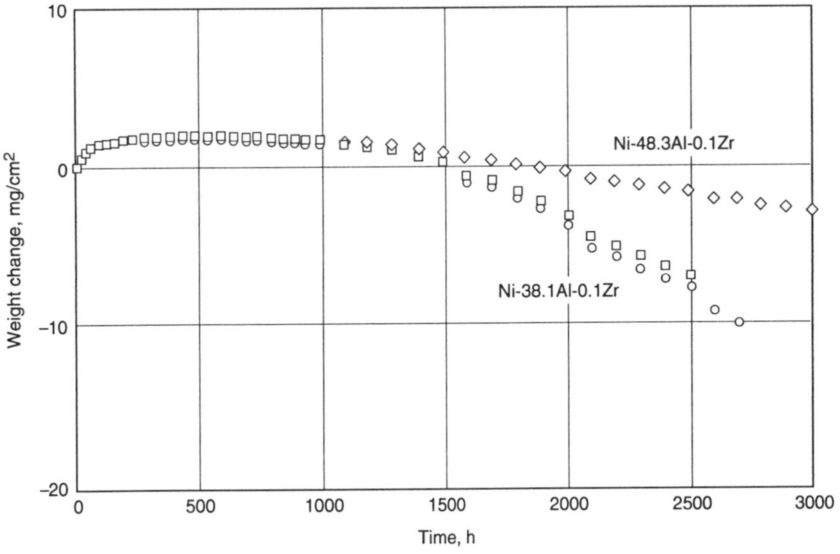

Fig. 13 Long-term protective 1200 °C (2190 °F) cyclic oxidation behavior of NiAl + 0.1 at.% Zr. Source: Ref 28

compositing with silica-base glass, boron compounds, or silicon nitride (Ref 54). At high temperature, MoO_3 is very volatile and leaves the scale by vaporization, with little effect on the silica growth rate. More detailed information on high-temperature silicides can be found in Ref 55.

Ceramics also have very high melting (or decomposition) temperatures and are candidate materials for high-temperature applications. The primary corrosion and oxidation mechanisms and considerations have been reviewed in a number of publications by Jacobson (Ref 56, 57). From the standpoint of oxide ceramics, the conventional oxidation reaction is precluded. In actual combustion environments, however, there may be reactions with combustion products leading to volatile species and surface recession. For example, the reaction of water vapor with SiO_2 and some silicates leads to gaseous $Si(OH)_4$, with loss rates that could measurably decrease component thickness in high-temperature, high-velocity, high-pressure environments. Conversely, the reactivity of Al_2O_3 in a combustion environment is very limited, without any serious conversions to gaseous products, and zirconia reactions are virtually nonexistent.

The nonoxide ceramics, silicon carbide (SiC) and silicon nitride (Si_3N_4), are very refractory, oxidation-resistant materials that have been widely studied for high-temperature applications. In pure oxygen, these materials, like $MoSi_2$, are among the most oxidation-resistant (nonoxide) materials known. The low scale-growth rate is due to slow-growing SiO_2 scales, which may be amorphous below about 1300 °C (2370 °F) or for short oxidation times. At extremely high tem-

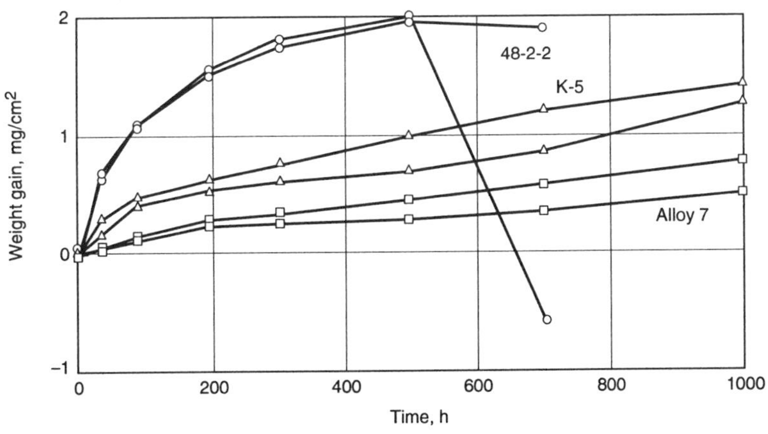

Fig. 14 Interrupted oxidation of structural γ-TiAl alloys at 800 °C (1470 °F) in air. Alloys tested include: Ti-48Al-2Cr-2Nb (48-2-2); Ti-46-5Al-3Nb-2Cr-0.2W (K-5); and Ti-46Al-5Nb-1W (Alloy 7). Source: Ref 45

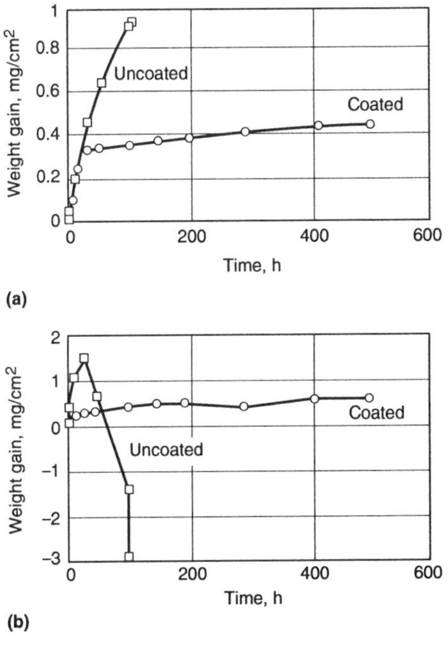

Fig. 15 The effect of a low-pressure plasma sprayed Ti-51Al-12Cr coating on the (a) 800 °C (1470 °F) and (b) 1000 °C (1830 °F) interrupted oxidation behavior of a Ti-48Al-2Cr-2Nb γ-TiAl alloy in air. Source: Ref 45

peratures (1700 °C, or 3090 °F) the evolution of CO or N_2 reaction products may produce large bubbles that could disrupt the protective scales (Ref 58). In addition, at low p_{O_2} and high temperatures, a passive-to-active oxidation transition occurs in which the reaction product is now a nonprotective SiO gas. These silica-formers are also subject to the volatility issues in water vapor discussed above.

Other corrosive reactions with ceramics are not widely encountered. However, if alkali-containing deposits appear on silicon-base systems, a particularly damaging reaction may occur (Ref 56). For example, the catalyst for sea salt hot corrosion in turbine alloys, Na_2SO_4, has been shown to react with SiO_2 scales formed on SiC or Si_3N_4. A molten sodium-silicate product rapidly results, with bubble and pit formation in the substrate and fracture strength reductions of up to 50% (Ref 59, 60).

Degradation Due to Hot Corrosion and Particulate Ingestion

Most of the preceding discussions have evaluated alloy and coating performance on the basis of oxidation resistance and the appropriate mechanisms. However, there are other common environmental durability issues that in some cases supersede oxidation as the primary mode of attack of gas turbine components. As described below, these include hot corrosion of superalloys and airfoil degradation due to deposits resulting from ingested particles or sand.

Hot Corrosion

The subject of hot corrosion of superalloys is extremely complex and to a certain degree controversial. It entails the detailed chemical reactions involved with the effects of a molten Na_2SO_4 deposit on the oxidation process and the dissolution of oxide scales coupled with the formation of sulfides in the alloy. Na_2SO_4 forms from the reaction of ingested NaCl salt with the sulfur impurity in the fuel. Because deposition is not favored at high temperatures (>1000 °C, or 1830 °F), hot corrosion is generally limited to lower temperatures. Increased corrosion results from higher deposition rates (higher sodium and sulfur impurity levels, higher pressure, and lower temperature).

The propagation modes for hot corrosion are intimately related to the reactions between the molten deposits and the alloys. In particular, the deposits cause nonprotective reaction products to be formed. As will be briefly described below, the nonprotective reaction product is formed because of a "fluxing" action of the molten deposit. A more detailed review of the overall formalisms of hot corrosion is provided by Pettit and Giggins (Ref 14).

The thermodynamics of Na_2SO_4 dissociation define an SO_3 pressure and oxide ion (O^{2-}) activity product (equilibrium constant). This equilibrium is used to assess whether certain oxides will dissolve in the molten deposit. The term basic fluxing describes oxide dissolution when O^{2-} combines with the scale to form a soluble MO^{2-} radical. It is generally associated with temperatures >900 °C (1650 °F) and is so termed high-temperature hot corrosion (type I). Sulfide formation and alloy depletion ahead of the corrosion front is a typical characteristic.

The term acidic fluxing defines dissolution when a soluble metal M^{2+} ion and O^{2-} are produced from the scale. Acidic fluxing is generally associated with a temperature range of 650 to 800 °C (1200 to 1470 °F) and is thus termed low-temperature hot corrosion (type II). However, the presence of molybdenum, tungsten, or vanadium in the alloy can also induce these soluble M^{2+} ions to form, even at higher temperatures. Type II corrosion does not generally exhibit a depletion zone or sulfides ahead of the corrosion front.

For protection against hot corrosion, a high chromium content is the best alloying strategy under all conditions, unlike the recommendations outlined for simple oxidation. For basic fluxing, a sufficiently high aluminum content (10%) to ensure continuous Al_2O_3 formation can physically inhibit corrosion, but is not as chemically resistant as Cr_2O_3. Chromium also decreases the O^{2-} concentration and prevents NiO basic dissolution. Thus high chromium can protect both physically and chemically. For acidic fluxing, a high aluminum content is not effective in preventing corrosion. High molybdenum, tungsten, or vanadium contents exacerbate acidic corrosion, while chromium decreases the rate of acidic corrosion.

Hot Corrosion Data. A large database of alloy hot-corrosion resistance has been developed (Ref 61). Here 96 commercial and laboratory superalloys were exposed to a Mach 0.3 burner rig test, at 900 °C (1650 °F) for 300 h in 1-h cycles, using 0.5 ppm NaCl in the combustion air and jet A-1 fuel with 0.05 to 0.07 wt% S. The extent of corrosion was monitored by the percentage of metal cross section consumed, as determined by quantitative metallography.

The top 12 alloys exhibited less than 2.0% of consumed cross-sectional area. The only compositional distinction was that the chromium content varied from 12.6 to 17.9% and the aluminum content varied from 2.1 to 6.9%. Conversely, the 12 poorest-performing alloys contained 4.0 to 10.7% Cr and 4.9 to 6.5% Al. Thus, improved performance was definitely biased toward high-chromium alloys. High-aluminum/low-chromium alloys were not especially corrosion resistant. No distinction in the titanium, tungsten, molybdenum, tantalum, niobium, cobalt, or zirconium content was apparent in these two extreme groups.

Sulfur-Induced Hot Corrosion. Sulfidation is another corrosion mechanism that is of great concern in coal-fired power systems (boiler tubes, exhaust gas heat exchangers, etc.). Basically, nickel-, cobalt-, and iron-base alloys are not sulfidation resistant in high p_{S_2} or p_{SO_2} (low p_{O_2}) environments because diffusion is very rapid in the sulfide scales that form on these alloys. Nickel sulfide has a parabolic rate constant seven orders of magnitude higher than that of nickel oxide. These sulfides have a very defective (nonstoichiometric) lattice, allowing fast cation transport. Low-melting sulfides are common, which only intensifies the problem. The addition of chromium (>50 at.%) will decrease the nickel sulfidation rate by three orders of magnitude, but this rate is still a factor of 10^5 times that of Cr_2O_3 growth. A similar effect occurs with the addition of 20 wt% Al to Fe-20Cr, where the resulting improvement is vastly inferior to that produced in oxidation environments.

The corrosion products and mechanisms are much dependent on the partial pressures of the sulfur-containing gas, for example, H_2S/H_2, S_2/Ar, SO_2/O_2, as well as on temperature. Preoxidation has been found to delay the onset of sulfidation, especially with Al_2O_3 scales; however, the oxides of nickel, cobalt, iron, chromium, and aluminum have all been found to allow sulfur penetration. Thus, the design of materials for ultimate sulfidation resistance is lacking. Alloy design for immediate levels of sulfur-containing gases has been fruitful (Ref 40). More detailed information on sulfidation can be found in Ref 9 to 11.

Airfoil Degradation Caused by Particulate Deposits

In particularly dusty or sandy terrain (the Middle Eastern countries and the southwestern United States), substantial amounts of particulates can be ingested from airborne particles. These have a geological connection with the region and have been identified as silica sand, complex silicates, dolomite/calcium sulfates, and salts. The chemistries vary from pure SiO_2 sand to Ca-Al-Fe-Mg silicates, $(Ca,Mg)CO_3$, $CaSO_4$, and NaCl (Ref 62, 63). When passed through the combustor, these particles melt and are splat deposited on the surfaces of airfoils. The deposits have chemistries similar to the fines analyzed from sand samples. However, the carbon content is reduced because of gaseous CO_2 formation and the sulfur content is increased because of calcium and magnesium reactions with the sulfur impurity in the fuel which leads to sulfate deposits.

Although the calcium-magnesium-aluminum-silicate (CMAS) deposits are relatively refractory, they easily cover over cooling holes and decrease cooling effectiveness at the outer airfoil surfaces. The consequent temperature increase has an obvious structural debit, but it also increases the oxidation rate of the alloy. In extreme cases, burn-through was apparent in helicopter turbine airfoils. Furthermore, it poses the probability of accelerated release and diffusion of foreign elements from the deposit into the protective scales and alloy surfaces. Thus, sulfide formation or very low melting calcium-aluminum or magnesium-aluminum eutectics might result. For TBC-coated airfoils, these deposits have been found to infiltrate the porous zirconia and degrade resilience by filling beneficial microcracks and by increased high-temperature sintering (Ref 63). Any defense against this type of attack, however, is difficult to imagine from the standpoint of improved alloy/coating design. Filters for helicopter turbines and frequent inspections or washdowns would be more appropriate for known high-risk service routes or land-based turbine locations.

Limitations: Testing Techniques and Life Prediction

Isothermal thermogravimetric tests and cyclic furnace tests are relatively standard throughout the industry such that the measured k_p or weight change curves from different laboratories can be compared. Some ramifications occur because of variations in the length of the tests or the cycle duration. Burner rig testing becomes more problematic in that different rigs may operate at different velocities, may be pressurized, or may be operated at various levels of impurity dopants (salts) to the fuel. The most common Mach 0.3 atmospheric rig, with 1 h cycles is probably the closest standard oxidation rig test. However, to screen systems for corrosion resistance, individual laboratory preferences for salt level and cycle profile vary and depend on proprietary correlations with engine experience or models of atmospheric salt ingestion. Furnace corrosion tests are almost exclusively targeted toward basic understanding because of the high levels of salt deposits and unrealistically short incubation times. Thermal barrier coating performance is also difficult to faithfully simulate because the failure modes are stress related and the true stress state on a cooled airfoil is substantially different than that on a furnace coupon (no thermal gradient) or uncooled burner rig bar. Mechanical degradation of stressed high-temperature components (turbine airfoils) may also have an oxidation component, but the synergistic mechanisms are not as widely studied or recognized.

REFERENCES

1. J.L. Smialek and G.M. Meier, High-Temperature Oxidation, *Superalloys II,* C.T. Sims, N.S. Stoloff, and W.C. Hagel, Ed., John Wiley & Sons, 1987, p 293–323
2. C.E. Lowell, C.A. Barrett, R.W. Palmer, J.V. Auping, and H.B. Probst, COSP: A Computer Model of Cyclic Oxidation, *Oxid. Met.,* Vol 36 (No. 1/2), 1991, p 81–112
3. C.E. Lowell, J.L. Smialek, and C.A. Barrett, Cyclic Oxidation of Superalloys, *High Temperature Corrosion,* NACE-6, R.A. Rapp, Ed., National Association of Corrosion Engineers, 1983, p 219–261
4. C.A. Barrett and C.E. Lowell, Resistance of Ni-Cr-Al Alloys to Cyclic Oxidation at 1100 and 1200 °C, *Oxid. Met.* Vol 11 (No. 4), 1977, p 199–223
5. J.L. Smialek, Oxide Morphology and Spalling Model for NiAl, *Metall. Trans.,* Vol 9A, 1978, p 308
6. J.L. Smialek and R. Browning, Current Viewpoints on Oxide Adherence Mechanisms, *Electrochemical Society Symposium Proceedings on High Temperature Materials Chemistry III,* 1986, p 259–271
7. J.G. Smeggil, A.W. Funkenbusch, and N.S. Bornstein, *Metall. Trans.* A, Vol 17A, 1986, p 923–932
8. J.L. Smialek, D.T. Jayne, J.C. Schaeffer, and W.H. Murphy, Effects of Hydrogen Annealing, Sulfur Segregation and Diffusion on the Cyclic Oxidation Resistance of Superalloys: A Review, ICMC, *Thin Solid Films,* Vol 253, 1994, p 285–292
9. P. Kofstad, *High Temperature Corrosion,* Elsevier Applied Science, 1988
10. N. Birks and G.H. Meier, *Introduction of High Temperature Oxidation of Metals,* Arnold, 1983
11. G.Y. Lai, *High-Temperature Corrosion of Engineering Alloys,* ASM International, 1990
12. R. Wasielewski and R. Rapp, High Temperature Oxidation, *Superalloys,* C.T. Sims and W. Hagel, Ed., John Wiley & Sons, 1972
13. S.J. Grisaffe, Protective Coatings, *Superalloys,* C.T. Sims and W. Hagel, Ed., John Wiley & Sons, 1972
14. F.S. Pettit and C.S. Giggins, Hot Corrosion, *Superalloys II,* C.T. Sims, N.S. Stoloff, and W.C. Hagel, Ed., John Wiley & Sons, 1987, p 327–358
15. J.H. Wood and E. Goldman, Protective Coatings, *Superalloys II,* C.T. Sims, N.S. Stoloff, and W.C. Hagel, Ed., John Wiley & Sons, 1987, p 359–384
16. N. Birks, G.H. Meier, and F.S. Pettit, High Temperature Corrosion, *Superalloys, Supercomposites and Superceramics,* J.K. Tien and T. Caulfield, Ed., Academic Press, 1989, p 439–489
17. I.G. Wright, "Oxidation of Iron-, Nickel-, and Cobalt-Base Alloys," MCIC 72-07, Metals and Ceramics Information Center, June 1972
18. C.A. Barrett, "A Statistical Analysis of Elevated Temperature Gravimetric Cyclic Oxidation Data of Ni- and Co-Base Superalloys Based on an Oxidation Attack Parameter," TM-105934, NASA Lewis Research Center, Dec 1992
19. C.A. Barrett, "10,000-Hour Cyclic Oxidation Behavior at 982 °C (1800 °F) of 68 High-Temperature Co-, Fe-, and Ni-Base Alloys," TM107394, NASA Lewis Research Center, June 1987
20. C.A. Barrett, 10,000-Hour Cyclic Oxidation Behavior at 815 °C (1500 °F) of 33 High-Temperature Alloys, *Environmental Degradation of Engineering Materials,* College of Engineering, Virginia Tech, Blacksburg, VA, 10–12 Oct 1977, p 319–327
21. J.L. Smialek and B.K. Tubbs, Effect of Sulfur Removal on Scale Adhesion to PWA 1480, *Metall. Mater. Trans.,* Vol 26A, 1995, p 427–435
22. M.A. Smith, W.E. Frazier, and B.A. Pregger, Effect of Sulfur on the Cyclic Oxidation Behavior of a Single Crystalline Nickel-Based Superalloy, *Mater. Sci. Eng.,* Vol 203, p 388–398
23. M. Gobel, A. Rahmel, and M. Schutze, The Cyclic-Oxidation Behavior of Several Nickel-Base Single-Crystal Superalloys Without and With Coatings, *Oxid. Met.,* Vol 41 (No. 3/4), 1994, p 271–300
24. R.W. Broomfield, D.A. Ford, H.R. Bhangu, M.C. Thomas, D.J. Fraisier, P.S. Burkholder, K. Harris, G.L. Erikson, and J.B. Wahl, "Development and Turbine Engine Performance of Three Advanced Rhenium Containing Superalloys for Single Crystal and Directionally Solidified Blades and Vanes," Paper No. 97-GT-117, Presented at the ASME International Gas Turbine and Aeroengine Congress and Exhibition (Orlando, FL), June 1997
25. J.L. Smialek, "Oxidation Resistance and Critical Sulfur Content of Single Crystal Superalloys," Paper No. 96-GT-519, Presented at the ASME International Gas Turbine and Aeroengine Congress and Exhibition (Brussels, Belgium), June 1990
26. J.A. Nesbitt and E.J. Vinarcik, *Damage and Oxidation in High Temperature Composites,* G.K. Haritos and O.O. Ochoa, Ed., American Society of Mechanical Engineers, 1991, p 9–22
27. J.L. Smialek and C.E. Lowell, Effects of Diffusion on Aluminum Depletion and Degradation of NiAl Coatings, *J. Electrochem. Soc.,* Vol 121, 1974, p 80
28. C.A. Barrett, The Effect of 0.1 Atomic Percent Zirconium on the Cyclic Oxidation Behavior of β-NiAl for 3000 Hours at 1200 °C, *Oxidation of High-Temperature Intermetallics,* T. Grobstein and J. Doychak, Ed., The Minerals, Metals, & Materials Society, 1988, p 67–82
29. J. Doychak, J.L. Smialek, and C.A. Barrett, The Oxidation of Ni-Rich Ni-Al Intermetallics, *Oxidation of High-Temperature Intermetallics,* T. Grobstein and J. Doychak, Ed., The Minerals, Metals & Materials Society, 1988, p 41
30. B.A. Pint, The Oxidation Behavior of Oxide-Dispersed β-NiAl: I. Short-Term Performance at 1200 °C and II. Long-Term Performance at 1200 °C, Submitted to *Oxidation of Metals,* 1997
31. G.W. Goward and D.H. Boone, Mechanisms of Formation of Diffusion Aluminide Coatings on Nickel Base Superalloys, *Oxid. Met.,* Vol 3, 1971, p 475–495
32. M. Goebel, A. Rahmel, M. Schutze, M. Schorr, and W.T. Wu, Interdiffusion Between the Platinum-Modified Aluminide Coating RT22 and Nickel-Based Single-Crystal Superalloys at 1000 and 1200 °C, *Mater. High Temp.,* Vol 12 (No. 4), 1994
33. J.S. Smith and D.H. Boone, "Platinum Modified Aluminides—Present Status," Paper No. 90-GT-319, Presented at the ASME International Gas Turbine and Aeroengine Congress and Exhibition (Brussels, Belgium), June 1990
34. R.A. Miller and C.E. Lowell, Failure Mechanisms of Thermal Barrier Coatings Exposed to Elevated Temperatures, *Thin Solid Films,* Vol 67, 1984, p 517–521
35. R.L. Jones, Thermogravimetric Study of the 800 °C Reaction of Zirconia Stabilizing Oxides with SO_3-$NaVO_3$, *J. Electrochem. Soc.,* Vol 139 (No. 10), 1992, p 2794–2799

36. "First Thermal Barrier Coating Workshop," CP-3312, W.J. Brindley, Ed., NASA Lewis Research Center, Oct 1995
37. F.S. Pettit, Oxidation Mechanisms for Nickel-Aluminum Alloys at Temperatures Between 900 and 1300 °C, *Trans. TMS-AIME,* Vol 239, 1967, p 1296–1305
38. R. Darolia, NiAl Alloys for High-Temperature Structural Applications, *J. Met.,* Vol 43 (No. 3), 1991, p 44–49
39. R.D. Noebe, R.R. Bowman, and M.V. Nathal, "Review of the Physical and Mechanical Properties and Potential Applications of the B2 Compound NiAl," TM 105598, NASA Lewis Research Center, April 1992
40. J.H. DeVan and C.A. Hippsley, Oxidation of Ni_3Al Below 850 °C and Its Effect on Fracture Behavior, *Oxidation of High-Temperature Intermetallics,* T. Grobstein and J. Doychak, Ed., The Minerals, Metals, & Materials Society, 1988, p 31–40
41. J. H. DeVan, Oxidation Behavior of Fe_3Al and Derivative Alloys, *Oxidation of High-Temperature Intermetallics,* T. Grobstein and J. Doychak, Ed., The Minerals, Metals, & Materials Society, 1988, p 107–116
42. J.L. Corkum and W.W. Smeltzer, The Synergistic Effect of Aluminum and Silicon on the Oxidation Resistance of Iron Alloys, *Oxidation of High-Temperature Intermetallics,* T. Grobstein and J. Doychak, Ed., The Minerals, Metals, & Materials Society, 1988, p 97–106
43. A. Rahmel, W.J. Quadakkers, and M. Schutze, Fundamentals of TiAl Oxidation—A Critical Review, *Mater. Corros.,* Vol 46, 1995, p 271–285
44. G.H. Meier, D. Appalonia, R.A. Perkins, and K.T. Chiang, Oxidation of Ti-Base Alloys, *Oxidation of High-Temperature Intermetallics,* T. Grobstein and J. Doychak, Ed., The Minerals, Metals, & Materials Society, 1988, p 185–194
45. M.P. Brady, W.J. Brindley, J.L. Smialek, and I.E. Locci, The Oxidation and Protection of Gamma Titanium Aluminides, *JOM,* Vol 48 (No. 11), Nov 1996, p 46–50
46. Oxidation of High-Temperature Intermetallics, T. Grobstein and J. Doychak, Ed., The Minerals, Metals, and Materials Society, 1988
47. J. Doychak, Oxidation Behavior of High-Temperature Intermetallics, *Intermetallic Compounds,* J.H. Westbrook and R.L. Fleischer, Ed., John Wiley & Sons, 1994, p 977
48. G. Welsch, J.L. Smialek, J. Doychak, J. Waldman, and N.S. Jacobson, High Temperature Oxidation and Properties, *Oxidation and Corrosion of Intermetallic Alloys,* G. Welsh and P.D. Desai, Ed., Purdue Research Foundation, 1996, p 121–266
49. J.L. Smialek, J.A. Nesbitt, W.J. Brindley, M.P. Brady, J. Doychak, R.M. Dickerson, and D.R. Hull, Service Limitations for Oxidation Resistant Intermetallic Compounds, *Mater. Res. Soc. Symp.,* Vol 364, 1995, 1273–1284
50. C.M. Packer, Overview of Silicide Coatings for Refractory Metals, *Oxidation of High-Temperature Intermetallics,* T. Grobstein and J. Doychak, Ed., The Minerals, Metals, & Materials Society, 1988, p 235–244
51. J.L Smialek, M.D. Cuy, and D. Petrarca, unpublished research, NASA Lewis Research Center, 1992
52. J. Doychak, R.R. Dickerson, D. Hull, and M. Maloney, unpublished research, NASA Lewis Research Center, 1993
53. D. Berztiss, R.R. Cerchiara, E.A. Gulbransen, F.S. Pettit, and G.H. Meier, *Mater. Sci. Eng.,* Vol A155, 1992, p 164–181
54. M.G. Hebsur, Pest Resistant and Low CTE $MoSi_2$-Matrix for High Temperature Structural Applications, *Mater. Res. Soc. Symp. Proc.,* Vol 350, 1994, p 177–182
55. A.K. Vasudevan and J.J. Petrovic, High Temperature Structural Silicides, Proceedings of the First High Temperature Structural Silicides Workshop, *Mater. Sci. Eng.,* Vol 155, 1992
56. N.S. Jacobson, Corrosion of Silicon-Based Ceramics in Combustion Environments, *J. Am. Ceram. Soc.,* Vol 76 (No. 1) 1993, p 3–28
57. N.S. Jacobson, J.L. Smialek, and D.S. Fox, Molten Salt Corrosion of Ceramics, *Corrosion of Advanced Ceramics,* K.G. Nickel, Ed., Kluwer Academic Publishers, 1996, p 205–222
58. G.H. Schiroky, Oxidation Behavior of Chemically Vapor-Deposited Silicon Carbide, *Adv. Ceram. Mater.,* Vol 2 (No. 2), 1987, p 137–141
59. J.L. Smialek and N.S. Jacobson, Mechanisms of Strength Degradation for Hot Corrosion of α-SiC, *J. Am. Ceram. Soc.,* Vol 69, 1986, p 741–752
60. D.S. Fox and J.L. Smialek, Burner Rig Hot Corrosion of Silicon Carbide and Silicon Nitride, *J. Am. Ceram. Soc.,* Vol 73, 1990, p 303–311
61. C.A. Stearns, D.L. Deadmore, an C.A. Barrett, Effect of Alloy Composition on the Sodium-Sulfate Induced Hot Corrosion Attack of Cast Nickel-Base Superalloys at 900 °C, *Alternate Alloying for Environmental Resistance,* The Metallurgical Society, 1987, p 131–144
62. J.L. Smialek, F.A. Archer, and R.G. Garlick, Turbine Airfoil Degradation in the Persion Gulf War, *JOM,* Vol 46 (No. 12), Dec 1994, p 39–41
63. M.P. Borom, C.A. Johnson, and L.A. Peluso, "A Role of Environmental Deposits in Spallation of Thermal Barrier Coatings on Aeroengine and Land-Based Gas Turbine Hardware." Paper No. 96-GT-285, Presented at the ASME International Gas Turbine and Aeroengine Congress & Exhibition (Birmingham, UK), June 1996

Design for Wear Resistance

Raymond G. Bayer, Tribology Consultant

WEAR is damage to a solid surface as a result of relative motion between it and another surface or substance (Ref 1). The damage usually results in the progressive loss of material. The scientific measure used for wear is volume loss. However, in engineering the concern with wear is usually associated with dimensional or appearance changes that eventually affect performance and not with volume loss. As a result other measures are often used in practice, such as depth of the wear scar on a mechanical component or the degree of haze with optical components.

For any material, wear can occur by a variety of mechanisms, depending on the properties of the material and the situation in which it is being used. Wear resistance is, therefore, not an intrinsic material property like hardness or elastic modulus. Both wear and wear resistance are system properties or responses.

Collection of all the mechanical, chemical, and environmental elements that can affect wear and wear behavior is referred to as the tribosystem. Typical factors that can affect wear behavior are the properties of the materials, the nature of the relative motion, the nature of the loading, the shape of the surface(s), the surface roughness, the ambient temperature, and the composition of the environment in which the wear occurs. Tribosystem design parameters are those parameters that affect wear and that the designer can specify and alter. Designing for the control of wear involves selecting values for tribosystem design parameters in order to obtain acceptable wear behavior or life. The process for doing this is called wear design.

Fundamentally, wear design consists of identifying those design factors that can affect wear and then determining values for them on the basis of their effect on wear rate. The aim is to achieve an acceptably low wear rate. Common tribological design parameters are materials, surface contours, lubrication, and roughness. However, these are not the only ones that can be considered design parameters. For example, loading, type of relative motion, and various environmental parameters may be utilized as design parameters in some situations.

Different design parameters influence wear rate in different ways. There are four fundamental ways to reduce wear rates: by modifying the surface to make it more wear resistant, by using a more wear-resistant material or material pair, by increasing the separation between the surfaces, and by reducing the severity of the contact (i.e., by modifying those features of the wear situation that tend to increase wear rate). Often, the wear rate reduction associated with a design parameter comes from a combination of these methods.

For example, one common method for reducing wear rate is to use a lubricant. A lubricant can reduce wear rate by reducing surface shear forces, by reacting with the surface to form a more wear-resistant surface layer, or by acting as an interposing layer that decreases the amount of contact between the contacting surfaces. In most situations all three elements are present. A lubricant can also conduct heat away from the contact region. In situations where temperature of the contacting surfaces is a factor in wear behavior, this is another way by which a lubricant can affect wear.

Wear design methodology consists of four general activities: system analysis, model selection, data selection, and verification. In order to perform these activities successfully, it is necessary to have a basic understanding of wear behavior and a familiarity with engineering models for wear. Consequently, this article provides a brief overview of these topics in the sections "Wear Behavior," "Lubrication," "Material Selection for Wear Applications," and "Wear Models." Wear design principles and practices are addressed in the sections "Wear Design" and "Methods for Wear Design." Additional comprehensive information on tribology is provided in *Friction, Lubrication, and Wear Technology*, Volume 18 of the *ASM Handbook*.

Wear Behavior

Three general types of wear situations are common. The first occurs when two solid bodies are in contact and move relative to one another. This first situation is generally subdivided by the predominant nature of the relative motion, such as sliding, rolling, or impact. The second situation occurs when the wear is caused by a liquid moving relative to a solid surface. Wear in that situation is often called erosion or erosive wear. The third situation, which is generally referred to as abrasive wear, occurs when the wear is caused by hard particles. Erosion and abrasive wear situations can also be subdivided into more specific categories. Examples of these are cavitation erosion, solid particle erosion, gouging abrasion, and slurry erosion.

Wear Mechanisms

There are four general ways by which a material can wear in the aforementioned situations: adhesive processes, abrasive or deformation processes, fatigue or fatigue-like processes, and oxidative or corrosive processes.

Adhesive Wear. With adhesive processes, wear occurs as a result of the bonding that takes place between two surfaces in contact. With subsequent separation of the two surfaces, material from either surface may be pulled out, resulting in wear.

Abrasive wear processes are those fracture, cutting, and plastic deformation processes that can occur when a harder surface engages a softer surface. These mechanisms tend to produce machining-chip-like debris.

Fatigue or fatigue-like wear processes are those associated with crack initiation and propagation or progressive deformation as a result of repeated contact, such as the ratchet mechanism proposed by K. Johnson (Ref 2, 3).

Corrosive wear processes are those associated with the loss of wear of in situ formed reaction product (e.g., oxide layers).

Interaction among Wear Mechanisms. The foregoing mechanisms are not mutually exclusive. They can coexist and interact to form more complex wear processes.

When worn surfaces are examined, features indicative of more than one mechanism are usually found. However, in most tribosystems one type of mechanism tends to predominate and ultimately be the controlling one. Except for adhesive wear processes, these processes can result from sliding, rolling, and normal impact motions. Adhesive wear processes normally occur only with sliding. However, adhesive wear processes can occur under nominal impact and rolling conditions, because of the slip that is often present in those situations.

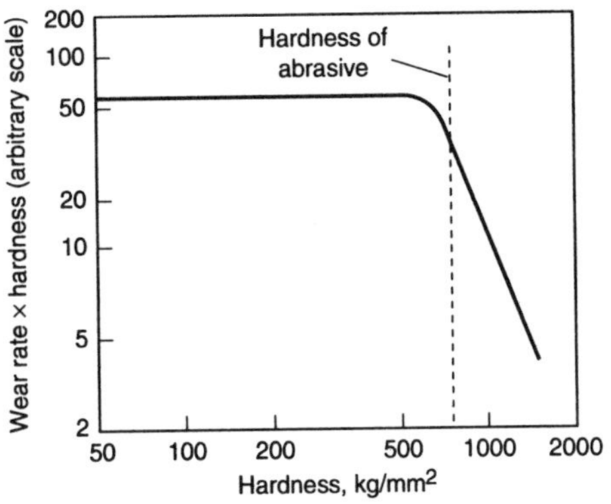

Fig. 1 Abrasive wear behavior as a function of surface hardness. Source: Ref 4

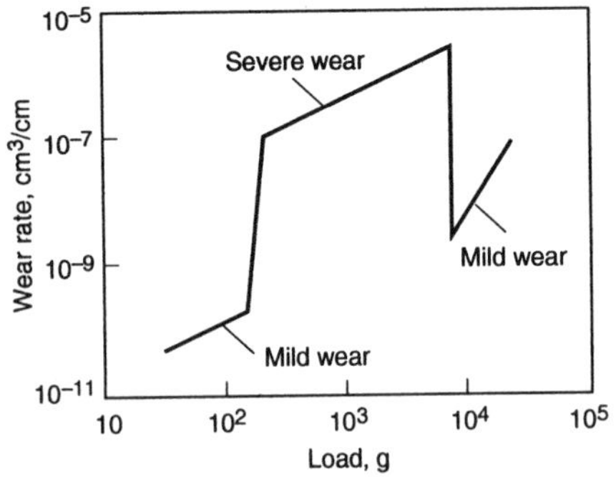

Fig. 2 Mild to severe wear transition in the case of unlubricated sliding between two steel surfaces. Transitions occur as a result of changes in the nature of the oxide formed in each region. Source: Ref 6

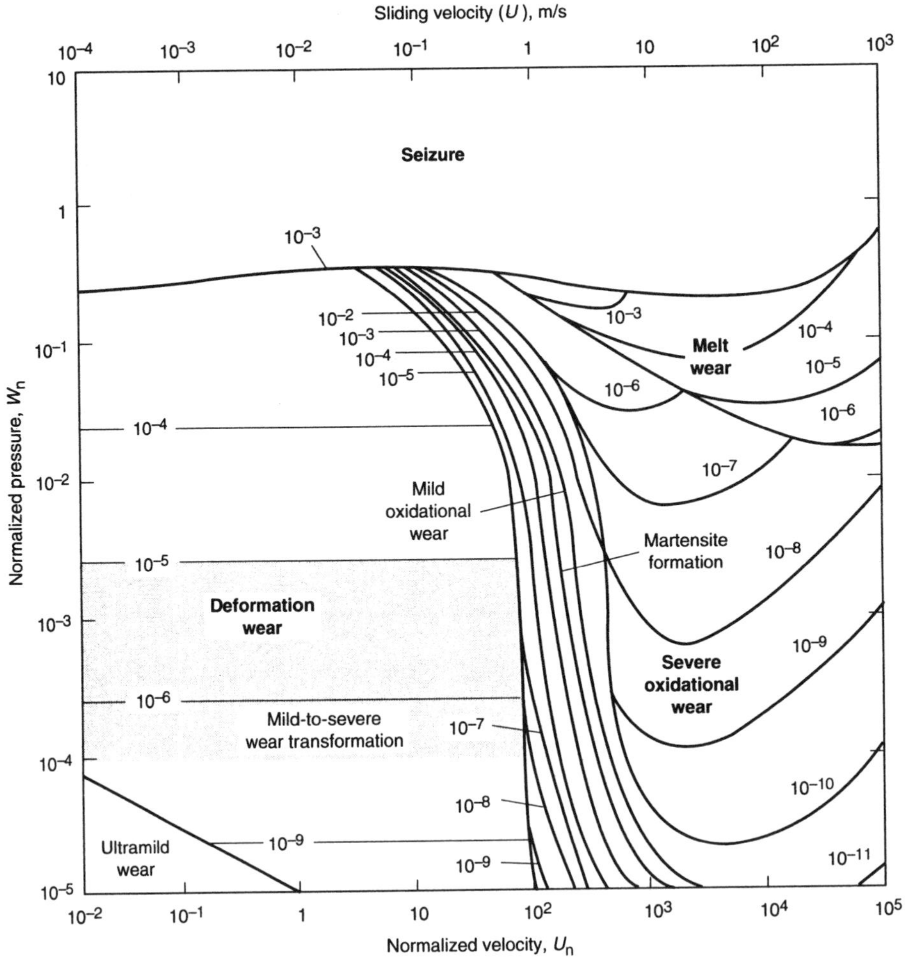

Fig. 3 Wear map for unlubricated sliding between two steel surfaces. Source: Ref 7

Table 1 Sliding wear coefficients

Wear mechanism	Sliding wear coefficient(a)
Adhesive wear	10^{-7} to 10^{-1}
Abrasive wear	10^{-6} to 10^{-1}
Corrosive wear	10^{-7} to 10^{-2}
Fatigue wear	$\leq 10^{-6}$

(a) $(V \times H)/(P \times S)$, where V is the volume of wear, H is the hardness (load per unit area), P is the applied load, and S is the sliding distance

The relative wear rates of these mechanisms can differ by several orders of magnitude, depending on the wear situation. Table 1 shows the observed range of a dimensionless wear coefficient for each of these mechanisms under sliding conditions. This coefficient is obtained by dividing the amount of wear by the distance of sliding and the load and multiplying by the hardness of the wearing surface. The value for each of the mechanisms is based on data obtained from tribosystems in which that mechanism is considered to be the predominant one. The higher the value, the more severe the wear situation. Generally, engineering applications require that this coefficient be in the range of 10^{-5} or less. In some cases, very small values are required. Automotive engine cylinder applications often require values in the range of 10^{-8}. Table 1 shows that the most severe forms of wear are associated with adhesive and abrasive wear mechanisms.

Control Methods. Fortunately, the severity of these mechanisms can be controlled and reduced to acceptable levels by the appropriate choice of various design parameters.

Adhesive Wear Control. The severity of adhesive wear mechanisms can be reduced by the appropriate selection of material pairs and by the use of lubricants, as shown by the range in data in Table 1. Studies have indicated that adhesive wear requires a minimum stress to operate. This stress is typically a small fraction of the materials hardness. Consequently, the severity of adhesive wear can also be reduced by reducing the contact stress.

Abrasive Wear Control. The severity of abrasive wear mechanisms caused by the roughness of a counterface can be reduced by reducing the roughness of that surface. With wear by particles, the severity can be reduced by reducing the number and size of the particles causing the wear. This is shown by the range in abrasive wear coefficients in Table 1. For abrasive or deformation wear processes to occur, the wearing surface must be softer than the particle or asperity causing the wear (Fig. 1). Consequently, these mechanisms can be eliminated by making the wearing surface harder than the particles causing the wear. However, this does not mean that these particles can no longer cause wear. Wear is still possible by fatigue and fatigue-like processes, which are generally milder.

Tribofilms. In addition to these material wear processes, there are system wear processes associated with the formation of tribofilms. Tribofilms are films composed of wear debris that form on wear surfaces. If the layer is entirely composed of material from the counterface, it is called a transfer film. If it is composed of material from both surfaces, it is called a third-body film. (The term *third-body film* is also used to describe any material layer that accommodates velocity changes between two surfaces and those films resulting from environmental interactions.) With respect to wear, these films tend to act as lubricants by separating the two surfaces. However, these films do not necessarily result in reduced friction. The relative motion between the two surfaces is accommodated by relative motion in the film or at a film and surface interface. When there is sufficient coverage, the wear rates of the surfaces are controlled by the rate at which material is lost from the tribofilms. Under these conditions, normalized wear rates tend to be in the range of 10^{-5} or less. Tribofilm formation is a

Table 2 Operational classification of wear situations

Motion	Environment	Mechanisms
Two-body contact		
Rolling		
With slip	Without particles	Fatigue Adhesive
	With particles	Fatigue Adhesive Abrasive
Without slip		Fatigue
Trends: Wear increases with increasing slip and presence of particles. Adhesive and abrasive wear modes can predominate but with pure rolling fatigue modes predominate. Mildest wear situation. Smooth surfaces preferred		
Impact		
With stationary body		Fatigue (elastic or plastic)
With moving body	Dry, without particles	Fatigue (elastic or plastic) Adhesive
	Dry, with particles	Fatigue (elastic or plastic) Adhesive Abrasive
	Fluid, without particles	Fatigue (elastic or plastic) Adhesive
	Fluid, with particles	Fatigue (elastic or plastic) Adhesive Abrasive
Trends: With stationary body, induced vibrations and misalignment can cause fretting, which tends to increase wear. Plastic deformation generally unacceptable except in short life applications. For lives greater than 10^6, contact stresses need to be in the elastic range. With moving object, wear increases with the amount of sliding and sliding effects can predominate. Fluid lubrication effects can be very significant. With particles, wear tends to increase.		
Sliding		
Unidirectional	Dry	Fatigue Adhesive
	Fluid	Fatigue Adhesive
	Particles	Fatigue Adhesive Abrasive
Sliding (continued)		
Cyclic	Large amplitude, dry	Fatigue Adhesive
	Large amplitude, fluid	Fatigue Adhesive
	Large amplitude, particles	Fatigue Adhesive Abrasive
	Small amplitude, dry	Fatigue Adhesive
	Small amplitude, fluid	Fatigue Adhesive
	Small amplitude, particles	Fatigue Adhesive Abrasive
Trends: More than one type of mechanism involved. Fatigue mechanism is mildest and conditions that minimize adhesion and abrasion preferred. Low contact stresses preferred. Mild to severe wear transitions often associated with the transition from elastic to plastic deformation. Predominant mechanisms(s) can change with wear. In mild wear situations, terminal mode often fatigue. With the presence of particles, abrasion tends to become predominant mode. Nature of contact shape, such as large area, conforming, line, and point, can be significant in wear behavior.		
One-body contact with fluid		
Impingement		
Low angle	Without particles	Fatigue
	With particles	Abrasive (cutting)
High angle	Without particles	Fatigue
	With particles	Fatigue Abrasive (deformation)
Flow		
Streamline	Without particles	None
	With particles	Abrasive (cutting)
Turbulent	Without particles	Fatigue
	With particles	Fatigue Abrasive
Trends: In flow situations, without particles, cavitation wear mechanism. Impinging fluid droplets act like particles. Corrosion effects often present		

General comments: With fluids other than lubricants, synergistic effects between corrosion and wear often occur. In all situations, oxidative wear effects are probable with metals and ceramics. Operational characterization may be different for different locations on a part. Source: Ref 8

major factor in the wear behavior when self-lubricating materials are used, such as the polytetrafluoroethylene-filled plastics, but the effect is not limited to those types of materials. Tribofilms have also been observed to occur with metals, plastics, and ceramics. Because lubrication tends to inhibit the formation of these films, the effect of these films tends to be more significant in unlubricated systems than lubricated ones.

Wear-In. Initial wear behavior of a tribosystem is often different from long-term behavior. There are two reasons for this. One is that wear and wear processes modify surfaces and it may take some time before stable surface or contact conditions are established. Another is that the relative mix of mechanisms can change, either as a result of the changes to the surfaces and the interface, or because some mechanisms, such as fatigue, may require some time to become significant. The most significant changes occur during the initial period, which is often called the break-in or wear-in period. During this period, conformity between surfaces is improved, roughness is modified, stable oxides and tribofilms are formed, and work hardening and thermal softening can occur. Wear rates are generally higher during this period, as well. While long-term wear rates tend to be lower than initial wear rates, they are not necessarily constant. Depending on the tribosystem, long-term wear rates may change as a result of the nature of the dominant wear mode. For example, fatigue wear modes are frequently stress dependent. For non-conformal contacts at constant load, wear rates would continue to decrease as the wear results in increased contact area and, hence, lower contact stress.

Severe versus Mild Wear. There are two broad, qualitative classifications for wear behavior: severe and mild, Wear is generally classified as severe in those situations where the features of the worn surfaces are relatively coarse and have high normalized wear coefficients. For sliding, severe wear behavior is generally associated with coefficients greater than 10^{-4}. With mild wear, the features of the worn surfaces are finer. Wear coefficients are also smaller but not insignificant. Severe wear behavior cannot be tolerated in most applications. All materials can exhibit both mild and severe wear characteristics, depending on the tribosystem in which they are used. Transitions between these two modes are often related to such parameters as load, speed, stress, and amount of lubrication, but they can also be related to other parameters (Ref 5). For example, changes in relative humidity or the type of abrasive particles encountered can also result in mild-severe wear transitions. Often these types of transitions are sharp (Fig. 2). While mild-severe wear transitions are generally the most significant in engineering, transitions can also occur in mild and severe wear behavior. Wear maps are a graphical representation of transition boundaries (Ref 7). This can be seen in Fig. 3, which shows a wear map for unlubricated sliding of a steel couple.

The relationship between wear or wear rate and design parameters depends on the nature of the wear mechanisms involved and, consequently, on the tribosystem. These relationships can be linear or nonlinear. For example, in the case of adhesive wear, simple theoretical models indicate that wear volume is proportional to load. On the other hand, models used to describe fatigue wear typically indicate that wear volume is proportional to some high power of the load. As a result, if adhesive wear is the predominant mechanism, the relationship between wear volume and load would be linear. If fatigue is the predominant mode for the tribosystem, a non-

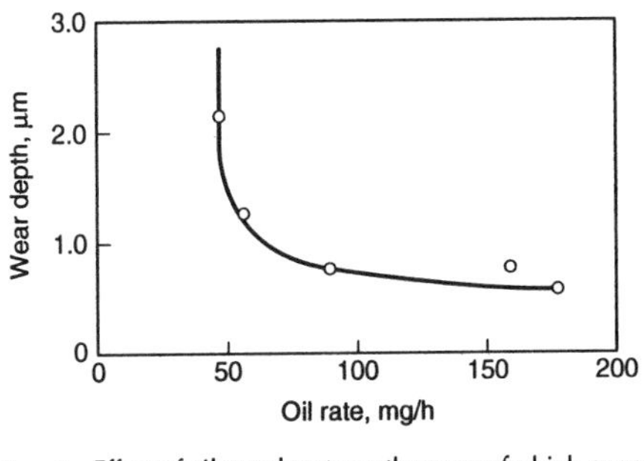

Fig. 4 Effect of oil supply rate on the wear of a high-speed printer component. Wear occurred at the interface between a pivoting type element and a type carrier backstop. The materials were hardened steel. The wear resulted from a combination of impact and fretting. Source: Ref 12

Table 3 Wear models for design

Sliding	
General model	$V = kL^m S^n$
K factor model	$V = KLS$
Zero wear model (conditions for zero wear)	$\tau_{max} \le 0.54\tau_y$
	$2000(\gamma_R \tau_y)^9 = (S/W)(\tau_{max})^9$
Measurable wear model(a)	
Variable energy mode	$d[Q/(\tau_{max} W)^{4.5}] = Cd\,(S/W)$
Constant energy mode	$dQ = Cd\,(S/W)$
Impact	
Percussive impact model	$V = kv^m N$
Zero wear model (condition for zero wear)(b)	$N_0 = \frac{2000}{1+\beta}\left(\Gamma_R \frac{\sigma}{\sigma_y}\right)^9$
Measurable wear model(c)	$dV = \frac{V}{N} dN + g \frac{9V}{\sigma} d\sigma$
Rolling	
General model	$V = kL^m N$
Surface endurance model (condition for surface cracks)	$N_1 \sigma_1^m = N_2 \sigma_2^m$
Load-stress factor model	$L_e = K_1 \frac{w}{1/R_1 + 1/R_2}$
	$\log K_1 = \frac{B - \log N}{A}$
Abrasion	
General model	$V = kLS$
Erosion	
General model(d)	$e = KAI$
Liquid drops model	$e = K \sin^n \alpha M v^m$
Particles model	$e = [K_d v^n \cos^n \alpha \sin(\pi\alpha/\beta) + K_b v^m \sin^m \alpha] M$
Vibration-induced cavitation model	$e = K\delta^n$
Jet-induced cavitation model	$e = Kv_j^n$
Key to symbols	
A	Function of attack angle
e	Erosion rate
I	Function of stream intensity
L	Load
L_e	Endurance load
M	Erodent rate
N	Number of impacts
N_0	Zero wear life (impacts)
Q	Cross sectional area of scar
S	Sliding distance
v	Velocity of impact (impact model); velocity of drops or particles (erosion model)
V	Volume of wear
v_j	Velocity of jet
W	Length of contact area
α	Angle of incidence
β	Ratio of surface damage to subsurface damage in compound impact situations (impact model); tribosystem empirical coefficient (erosion model)
δ	Amplitude of vibration
σ	Peak contact pressure
σ_y	Yield point in tension
τ_{max}	Maximum shear stress
τ_y	Yield point in shear
$C, k, K, K_b, K_d, m, n, \gamma_R, \Gamma_R$	Tribosystem empirical coefficients

(a) Zero wear model for sliding can be used to determine C. (b) $\beta = 0$ for pure impact. (c) For constant-energy mode, $g = 0$. For variable-energy mode, $g = 1$. (d) Where K is a function of time, A is a function of angle, and I is a function of stream intensity. Source: Ref 8–10

Table 4 Effect of different lubricants on friction and wear for reciprocating sliding in a ball-plane

Oil	Steel/steel Coefficient of friction	Steel/steel Depth of wear, μm	Stainless steel/steel Coefficient of friction	Stainless steel/steel Depth of wear, μm
A	0.14	1.93	0.16	0.30
B	0.24	0.65	0.13	0
C	0.38	0.95	0.18	0.13
D	0.14	0	0.13	0.45
E	...	...	0.16	0.23

Source: Ref 14

linear relationship would exist. This is also illustrated by the different dependencies on parameters such as load, usage, and hardness found in empirical studies of different wear situations.

Table 2 provides an operational classification of tribosystems, identifying the type of wear mechanisms involved and controlling factors. Relationships that can be used to describe wear in these situations are shown in Table 3.

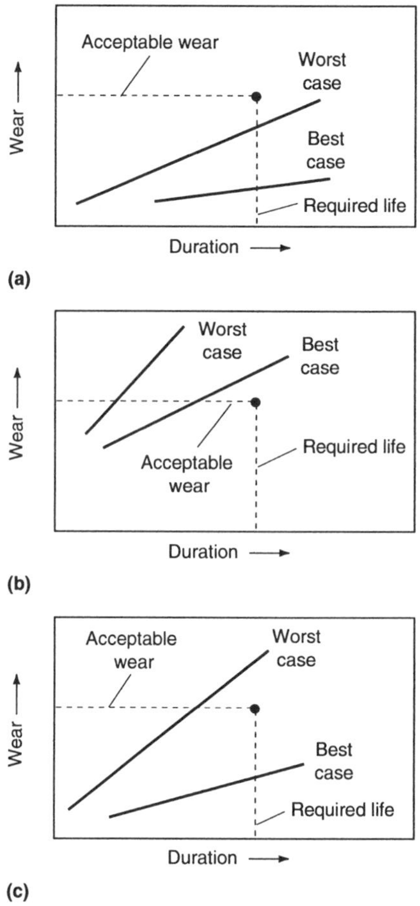

Fig. 5 Bracketing analysis. (a) The design is satisfactory and no further modifications are required. (b) A design change is needed. (c) Further work is needed to identify satisfactory materials and , if applicable, the correct model for the wear situation.

Lubrication

Friction is the resistance to relative motion between two bodies in contact. A lubricant is defined as any substance that reduces the friction between two surfaces. The use of these materials is one of the principal ways of reducing wear and extending the life of mechanical equipment. Lubricants provide a low shear interface between surfaces by physically separating those surfaces and by allowing the formation or modification of surface films on those surfaces. While the principal effect of lubrication on wear behavior is associated with the reduction of adhesive wear, the other types of mechanisms can also be affected by the presence of a lubricant. For example, mechanisms that are influenced by shear stresses, such as some deformation and fatigue mechanisms, can be affected by changes in traction caused by the use of a lubricant. Oxidation wear mechanisms can be affected by the changes in surface films resulting from the use of a lubricant. Lubricants can also inhibit or modify the formation of tribofilms.

In general, lubricants tend to reduce wear. However, there are some situations in which they can increase wear. For example, they may inhibit the formation of a beneficial tribofilm without providing adequate lubrication. This is often the case in tribosystems in which self-lubricating materials are used. Another example is abrasive wear situations, where abrasive particles tend to agglomerate or where wear debris can clog an abrasive counterface. These actions tend to reduce abrasive wear rates, but in these situations, lubricants tend to prevent the agglomeration or clogging, which leads to a higher wear rate. In abrasive wear situations a lubricant can also reduce the critical angle for cutting, which also tends to increase wear.

Lubrication can also affect wear behavior in other ways. In circulating systems, lubricants can provide cooling to the interface and remove wear debris. In non-circulating systems, oil and grease films tend to hold abrasive particles in dirty environments, which can lead to increased wear.

Except for abrasive and erosive wear situations, the effective application of lubrication generally results in mild wear behavior. Severe wear behavior in a lubricated situation is often an indication of lubricant breakdown, improper lubricant selection, or an inadequate supply. Even though one of the ways a lubricant works is to separate the surfaces, thick lubricant layers are not required to obtain significant improvements in wear performance. Thin films, even down to monomolecular layers, can have significant effects provided that they are maintained.

Lubrication by thin films is often referred to as boundary lubrication. Its effectiveness primarily depends on the ability of the lubricant to coat or react with the surface. With fluids it is possible to generate thicker films as a result of squeeze film effects in the fluid. As a result of these effects it is sometimes possible to achieve complete separation of the surfaces, which virtually eliminates wear. This is referred to as fluid lubrication. (De-

Table 5 Typical wear applications for selected engineering materials

Materials	Sliding: Unlubricated wear	Sliding: Lubricated wear	Sliding: Abrasion	Rolling: Unlubricated wear	Rolling: Lubricated wear	Rolling: Abrasion	Impact wear	Three-body abrasion	Fluid erosion	Cavitation	Drop erosion	Particle erosion
Structural alloys												
Surface treatments	X	X	X			X		X				
Hard surfacing	X	X	X			X	X	X			X	X
Soft coatings	X	X										X
Alloy steels		X	X	X	X	X	X	X				X
Tool steels	X	X	X	X	X	X	X	X		X		
Stainless steels												
Precipitation hardened			X	X				X				
Martensitic		X	X	X	X			X	X			
Cast irons												
Graphitic	X	X	X	X		X		X				
White	X		X			X		X				
High-temperature alloys												
Refractory materials	X	X					X	X				
Superalloys		X		X						X	X	X
Copper-base alloys												
Bronze		X									X	
Beryllium copper		X					X					
Soft bearing alloys (babbitts)		X										
Carbides	X	X	X	X				X			X	X
Ceramics	X		X					X	X		X	X
Polymers												
Thermosets	X											
Thermoplastics	X	X									X	
Elastomers				X	X		X	X		X	X	X
Carbons	X											
Lubricating composites	X	X										

Source: Ref 8

Table 6 Application wear models

AFBMA model for ball bearings
AFBMA model for roller bearings
Taylor model for tool wear
PV model for plastic bearings
K factor model for journal bearings
Zero/measurable wear model for journal bearings
Impact wear model for elastomers
Type wear in impact printers

AFBMA, Antifriction Bearing Manufacturers' Association. Source: Ref 8–11

sign procedures for lubrication are beyond the scope of this article and can be found in books on bearing design and fluid lubrication.) The state of lubrication between boundary and fluid lubrication is called mixed lubrication. Wear rates are lower with mixed lubrication than with boundary lubrication (Ref 11).

By using almost any lubricant, wear rates are generally reduced by one to two orders of magnitude. However, there can be significant differences in the lubricating abilities of different lubricants, so it is often possible to obtain larger reductions (e.g., several orders of magnitude) by optimizing the selection. Lubricant selection for a wear application often involves two other elements. One is supply of the lubricant. If there is an inadequate supply, wear rate will increase, as shown in Fig. 4. The other is the chemical stability of the lubricant in the application. Generally, when a lubricant degrades, its ability to protect the surface is decreased and wear rates increase.

Types of Lubricants. Lubricants are either fluids or solids. The fluid category includes gases, liquids, and greases. Common solid lubricants are molybdenum disulfide, polytetrafluoroethylene, graphite, and soft metals. Because fluids tend to be displaced easily, other materials are frequently added to enhance their boundary lubrication characteristics. Solid lubricants and reactive compounds are often used in greases and oils for this reason. These types of additives are generally called extreme pressure (EP) additives. Solid lubricants are also used as fillers in plastics to make self-lubricating materials. With these materials the solid lubricants provide lubrication by forming tribofilms on the rubbing surfaces. While fluid films tend to be more easily displaced than solid lubricant films, they have the ability to self-heal. Solid lubricant layers do not. However, this limitation of solid lubricants is removed when they are used as additives and fillers (Ref 13).

While lubrication is used to reduce both friction and wear, the effects that a lubricant has on each of these phenomena can be different. As a consequence, the best lubricant for friction reduction is not necessarily the best for wear reduction. Often the coefficients of friction can be similar for different lubricants, while wear rates can differ by orders of magnitude. Table 4 contains data that illustrate these two points.

Material Selection for Wear Applications

Some fundamental criteria can be applied in the selection of a material for wear applications. The primary criterion is that the material remain chemically, mechanically, and thermally stable under the operating conditions. A secondary criterion is that the nominal contact stresses be within the elastic range of the material. If either of these criteria is not met, it is likely that severe wear behavior and unacceptably high wear rates will result. For long life under sliding it is generally desirable to have a 2-to-1 or larger ratio between yield point and nominal contact stress. For impact and rolling this ratio can be smaller, approaching 1, and still be acceptable, In abrasive wear situations, it is generally desirable to have the material harder than the abrasives present, or at least of comparable hardness, to minimize wear and obtain long life. A corollary is that if a particular material must be used, the conditions of use need to be changed so that these criteria are satisfied.

As stated above, material wear resistance is not an intrinsic property, like elastic modulus or density. It tends to vary with the wear situation and is best viewed as a system response. While there is a general trend for wear to decrease with increasing hardness, there is considerable scatter about that trend. As a consequence, material hardness is generally not a sufficient indicator of wear resistance or wear performance in specific situations. It is often necessary to consider other properties of the material as well. Basically this is because hardness is not the only material property that is associated with wear behavior, and because the differences between materials are not limited to hardness, particularly in the case of different types or classes of materials. Because of the many factors associated with wear behavior, different types of materials tend to be used for different

Table 7 Design rules for wear applications

- Reliance on analytical design procedures increases the degree of conservatism that should be used.
- Wear is a system property; utilize all the parameters that influence wear.
- Design with the limits and characteristics of the materials in mind.
- Design so that a mild wear condition exists.
- Minimize exposure to abrasive particles.
- Optimize contact to minimize stresses.
  - Ensure good alignment.
  - Round corners and edges.
- Use a lubricant whenever possible.
- Use dissimilar materials.
- To increase system life (reduce system wear), it is sometimes necessary to increase hardness of both members.
- Rolling is preferred over sliding.
- Sliding or fretting motions should be eliminated in impact wear situations.
- Impacts should be avoided in sliding contacts.
- Elastomers frequently out perform harder materials in impact situations.
- Thickness of conventional coatings generally should be greater than 100 μm.
- Use moderate surface roughness.
- Avoid the use of stainless steel shafts with impregnated sintered bronze.
- When molded, filled plastics tend to exhibit significant difference between initial and long-term wear behavior.
- When glass or other hard fillers are used, the hardness of the counterface should be equal to or greater than that of the filler. For glass, this hardness should be >60 HRC.
- The tendency for galling can be reduced by using dissimilar and hard materials of low ductility, lubricating, and reducing contact stresses; stress level above threshold levels for galling should be avoided.
- Avoid designs in which fretting motions can occur.
- When fretting motions are present, design for optimum sliding wear life and to minimize abrasive wear.
- Sacrificial wear design should be considered when satisfactory life cannot be achieved by other means.
- Conform to vendor recommendations for optimum wear performance.
- Changes associated with design modifications or new applications should be reviewed carefully with respect to their affect on potential wear behavior.

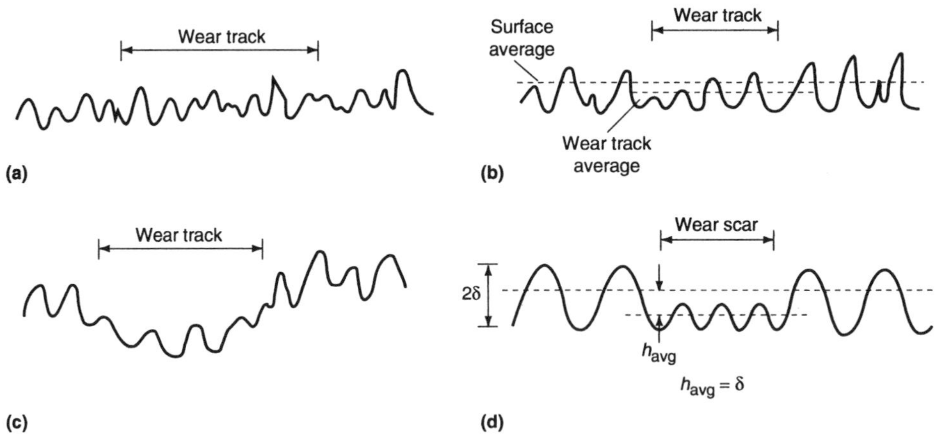

Fig. 6 Illustration of zero wear concept. (a) Surface profile when wear is less than the zero wear criterion. (b) Near the zero wear criterion. (c) Above the zero wear criterion. (d) Definition of the zero wear criterion

wear situations (Ref 15). Table 5 provides an overview of typical wear applications for different classes of materials.

Because of the system nature of wear, rankings of materials in terms of their wear resistance often change with the nature of the wear applications or the nature of the wear test used to determine the rankings. For example, rankings obtained from abrasive wear tests are typically not the same as those obtained from nonabrasive wear tests. Similarly, different rankings tend to be obtained in lubricated and unlubricated tests or in high- and low-speed tests, among others. Because of this, wear tests used to rank or evaluate materials for use in specific situations need to simulate the wear application in all essential details (Ref 8, 16–21). The basic elements that need to be considered in simulation are loading conditions, contact geometry, motions involved, and environmental conditions.

Highly wear-resistant materials are often materials that maintain average wear characteristics under extreme conditions, such as high ambient temperatures or a corrosive environment. As a result, these kinds of materials may provide no better and sometimes poorer wear performance than other materials in situations where these extreme conditions do not exist.

Wear Models

Because of the complex nature of wear behavior, there is no universal wear model that is applicable to all situations. However, there are wear models that can be used for design for specific situations (Ref 8–10, 22). There are models for generic wear situations, such as rolling and sliding, as well as models for specific devices, such as journal and roller bearings. Table 3 contains a list of models used for sliding, rolling, impact, and erosion. Table 6 lists some application models. These models provide relationships between wear and design parameters in a number of forms. In some cases the relationships are for the amount of wear, others are for wear and erosion rates, and still others are for equivalent wear conditions (i.e., those combinations of load and usage that result in the same amount of wear).

All these models involve one or more empirical coefficients, which are material and environment dependent. While empirical, they tend to be heuristically related to or based on a variety of general physical concepts and mechanisms. These heuristic concepts can often be used as an aid in the application of these models to particular situations. For example, they are often of use in estimating the values for the empirical coefficients, in evaluating the applicability of a model to a given situation, or in extending a model. In-depth treatment of these models and their use can be found elsewhere (Ref 8–10).

While there are a number of different models available for design use, most wear situations encountered in design can be adequately covered with the use of relatively few. The zero wear model for sliding (Ref 8, 23), the measurable wear model for sliding (Ref 8, 24), the zero wear model for impact (Ref 8, 25), the measurable wear model for impact (Ref 8, 26), and the surface endurance model for rolling (Ref 8, 27), summarized in Table 3, are applicable in most engineering situations where abrasion is not predominant. In those situations in which abrasion predominates, the abrasive wear model given in Table 3 provides good correlation with performance.

The coefficients of the models are determined from wear tests that match the conditions required by the model. In most cases not all of the coefficients of the model can be determined simply by measuring wear after a certain amount of time, sliding distance, or number of operations. It

Table 8 Key to symbols used in Fig. 7 to 9

Symbol	Description
a	The radius of that circle which best fits the contour of the edge in the region of contact, in.
E	Young's modulus of elasticity, psi
H_m	Microhardness, kg/mm^2
K	Stress-concentration factor used in cases where the contact between bodies terminates abruptly
L	Length of contact in cases involving cylinders and planes, in.
L'	Width of contact in cases of flat-on-flat conforming geometry, in.
N	Number of passes
n_o	Frequency of oscillation, oscillations/min
n_r	Angular velocity of a shaft, rpm
P	Normal force between two contacting surfaces, lb
P_A	Allowable normal force for a specific geometry, material, lubrication, and life, lb
R	Principal radius of curvature, in.
x	Operation time, h/month
y	Machine life, months
z	Total linear travel for one oscillation cycle, in./cycle
α	Angle of contact in a shaft-in-hole conforming geometry, degree
γ_R	Zero-wear factor
μ	Coefficient of friction
ν	Poisson's ratio
σ_y	Tensile yield stress, psi
τ_{max}	Maximum shear stress in contact region, psi
τ_y	Shearing yield stress, psi ($\tau_y = \sigma_y/2$)
ϕ	Total angular travel for one oscillation cycle, degree/cycle

Note: Subscripts 1 and 2 of these variables apply to Body 1 and Body 2 as shown in the sketches and equations in Fig. 7 to 9.

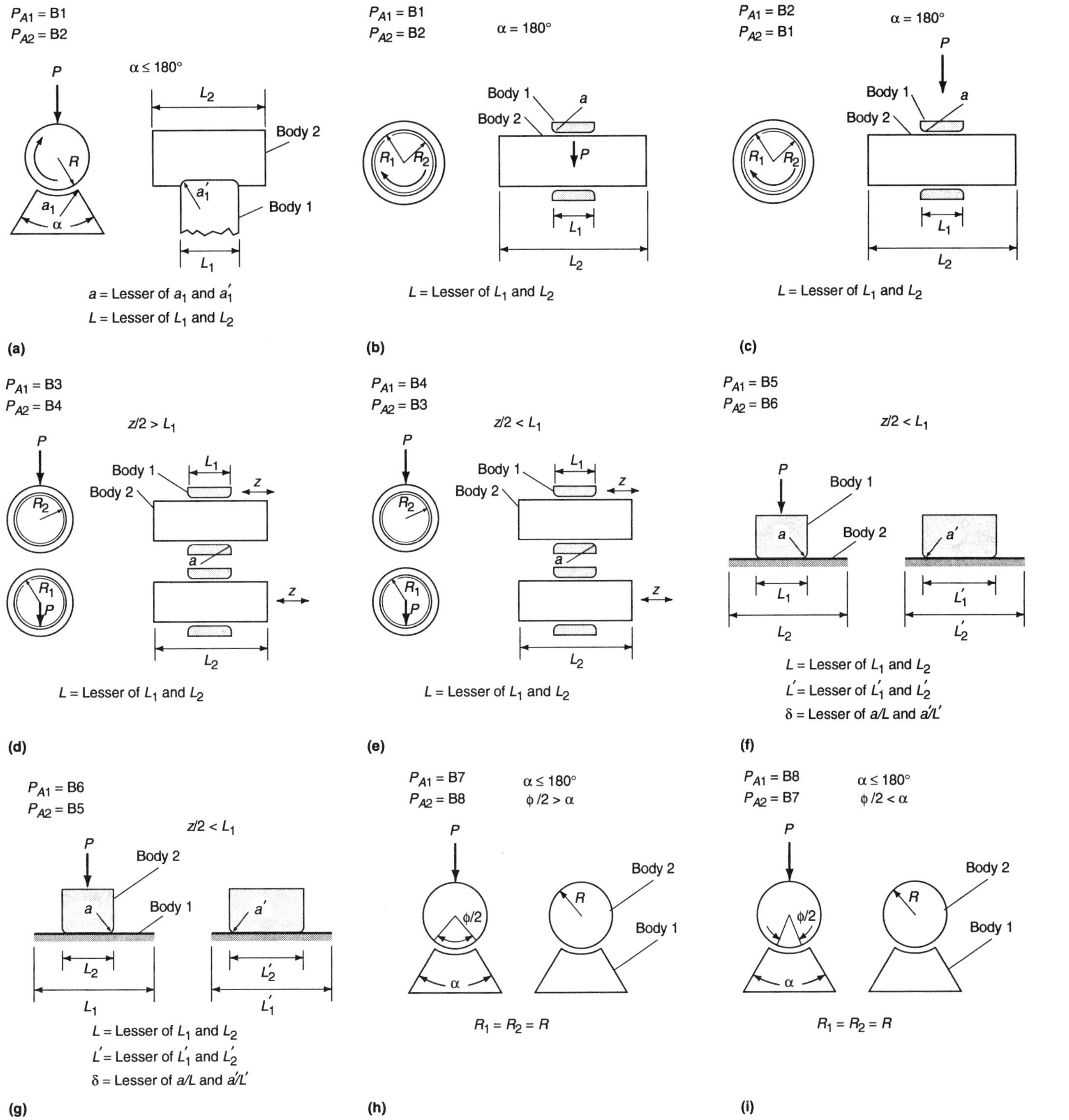

Fig. 7 Allowable load relationships for area-contact sliding mechanisms. (a) Cylinder in hole. Rotating cylinder; fixed partial hole; $R_1 = R_2 = R$. (b) Cylinder in hole. Rotating cylinder; fixed full hole; $R_1 = R_2 = R$. (c) Cylinder in hole. Fixed cylinder; rotating hole; $R_1 = R_2 = R$. (d) Cylinder in hole. Linear oscillation; either member fixed; travel ($z/2$) greater than contact length; $R_1 = R_2 = R$. (e) Cylinder in hole. Linear oscillation; either member fixed; travel ($z/2$) less than contact length; $R_1 = R_2 = R$. (f) Plane on plane. Linear translation; travel ($z/2$) greater than contact length. (g) Plane on plane. Linear translation; travel ($z/2$) less than contact length. (h) Sphere in socket. Partial socket (<180°); either member fixed; oscillation in one plane; travel ($\phi/2$) greater than contact angle. (i) Sphere in socket. Partial socket (<180°); either member fixed; oscillation in one plane; travel ($\phi/2$) less than contact angle. Source: Ref 28

is generally necessary to develop a wear curve or series of wear curves that can be analyzed to determine the coefficients for the model. A wear curve is a plot of wear versus time, sliding distance, or number of operations. While laboratory tests are often used to determine values for different materials and environmental conditions, it is sometimes possible to analyze existing hardware data to determine the coefficients. In situations where there is neither available data for the specific materials or environmental conditions involved in the application nor the ability to perform the appropriate tests, it is generally possible to estimate the values, based on published data regarding these models. Such information can be found in Ref 8 to 10 and 15.

Wear Design

Wear design involves four elements (Ref 8): system analysis, modeling, data gathering, and verification.

System analysis is the starting point of a wear design. It begins with the examination of the design and the identification of possible wear

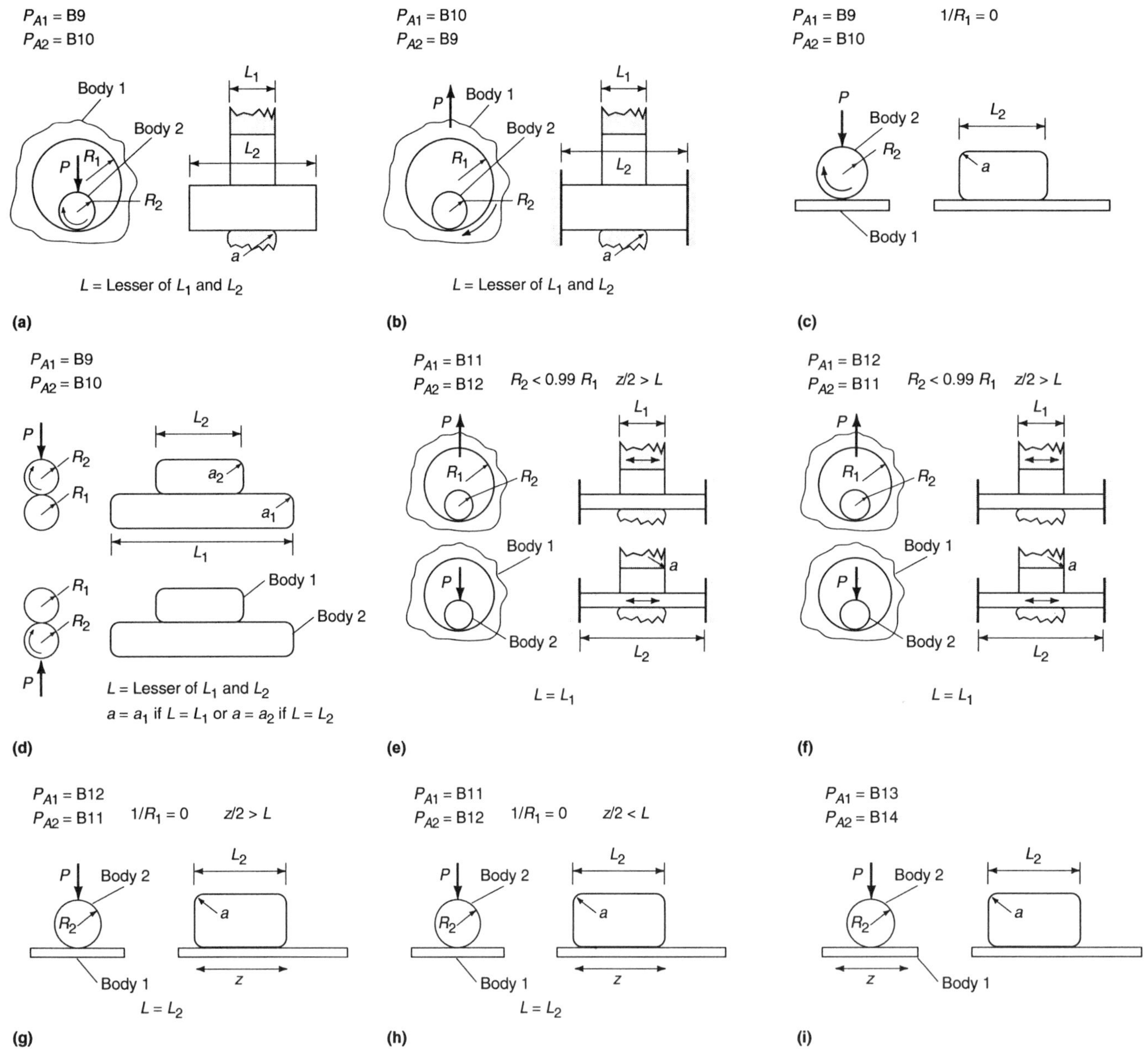

Fig. 8 Allowable load relationships for "line"-contact sliding mechanisms. (a) Cylinder in hole. Rotating cylinder; fixed hole; $R_2 < 0.99R_1$. (b) Cylinder in hole. Rotating hole; fixed cylinder; $R_2 < 0.99R_1$. (c) Cylinder on plane. Rotating cylinder; fixed plane. (d) Cylinder on cylinder. One cylinder fixed; one rotating. (*Note:* Reverse sign of R_1 in B9 and B10.) (e) Cylinder in hole. Linear oscillation; either member fixed; travel ($z/2$) greater than contact length. (f) Cylinder in hole. Linear oscillation; either member fixed; travel ($z/2$) less than contact length. (g) Cylinder on plane. Linear oscillation parallel to cylinder axis; either member fixed; travel ($z/2$) greater than contact length. (h) Cylinder on plane. Linear oscillation parallel to cylinder axis; either member fixed; travel ($z/2$) less than contact length. (i) Cylinder on plane. Linear oscillation perpendicular to cylinder axis; either member fixed. Source: Ref 28

points or concerns. It then involves the characterization of the tribosystem associated with each one. Initially this characterization may be very general and more qualitative than quantitative. It might simply be the determination of the general nature of the wear situation, such as lubricated sliding at low or moderate temperatures or a particle erosion situation with possible corrosion. As the wear design proceeds, system analysis involves such elements as the determination of loads and contact stresses, detailed characterization of the environment, detailed characterization of the motions involved, and determination of the factors that affect these variables. System analysis also involves the determination of how much wear can be allowed and the establishment of a failure criterion. In general it consists of all those elements needed to implement the modeling and data gathering elements.

Modeling. The modeling phase of a wear design approach involves the selection of a model, which provides a basis for the determination of the design. Model selection is basically done by matching the characteristics of the tribosystem to the descriptions of the various wear models and selecting the most appropriate one. Once this is done, the model is then used to determine the values of the design parameters necessary to obtain the desired wear life or performance.

Data Gathering. In order to use models in the fashion described above, it is generally necessary to determine the values of one or more empirical coefficients, which generally are material and environment dependent. Existing data may be used for this purpose if they were obtained for conditions that match or simulate the conditions of the current application. If not, appropriate wear tests need to be done to determine those coefficients. Estimates for these coefficients based on theoretical considerations or extrapolation of existing data can also be used, but these are generally less accurate then those obtained from wear tests that simulate the wear situation.

In some situations it may be necessary to consider more than one model. This could be because there is inadequate information available to differentiate between models, or because the wear situation is so complex that several different con-

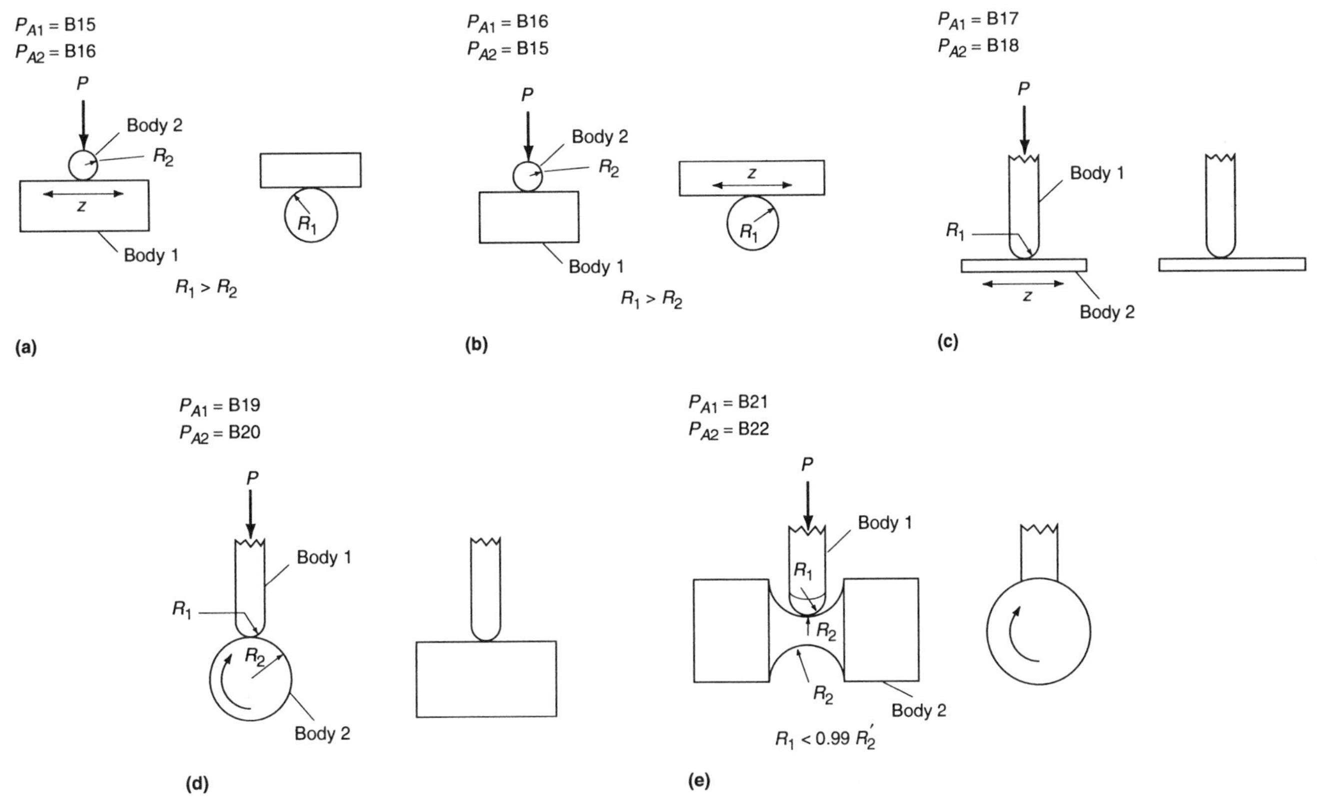

Fig. 9 Allowable load relationships for "point"-contact sliding mechanisms. (a) Crossed cylinders. Unequal cylinder diameters; linear oscillation parallel to axis of larger-diameter cylinder. (b) Crossed cylinders. Unequal cylinder diameters; linear oscillation parallel to axis of smaller-diameter cylinder. (c) Sphere on plane. Linear oscillation; either member fixed. (d) Sphere on cylinder. Rotating cylinder; fixed sphere. (e) Sphere on cylinder. Rotating grooved cylinder; fixed sphere. Source: Ref 28

ditions need to be considered. In this case the design parameters should be selected so that adequate wear performance is predicted by all the models. Alternatively, this complexity may be eliminated by doing further system analysis, doing some tests to identify the appropriate model, or introducing elements in the design to eliminate some of the possibilities. For example, damping may be introduced to eliminate possible fretting motions that may contribute to the wear, or seals may be used to eliminate the possibility of abrasive particles in the contact region.

Verification. In addition to the normal verification that the design works, it is necessary to verify the validity of different assumptions made in the other three phases. These include examinations to verify that the characteristics of the wear are consistent with the modeling (e.g., correct location and appearance of wear scar).

Theoretical versus Empirical Wear Design Approaches. In practice, wear design approaches can be completely theoretical, semi-empirical, or completely empirical. In the completely theoretical approach the basis for selection of a model is a description of the tribosystem. Also, existing empirical coefficients or estimates based on them are used to predict wear behavior using those models. In the semi-empirical approach, some testing is done to determine values for the coefficients or to verify the applicability of a model that was selected on a theoretical basis. In the completely empirical approach the model and coefficients are determined empirically. An example of this might be the use of regression analysis to determine a suitable model. While the completely theoretical approach is most desirable from a design standpoint, it tends to be the least accurate and to be associated with a higher degree of risk. As a result, such approaches should be more conservative and use larger safety factors for establishing designs than those involving some experimental elements.

Designing for Preferred Modes of Wear. An important factor in wear design is the recognition that the selection of design parameters and the overall nature of the design affects not only the wear rate but also the wear modes and behavior. Preferred modes of wear can be ensured by proper design. The primary criterion is that the design be selected to ensure mild wear behavior. In two-body, nonabrasive wear situations this generally means that contact stresses should be in the elastic range. In the case of sliding, contact stresses should be a small fraction of the yield strength, generally less than 0.5 and less than 0.2 for very low wear rates. Some form of lubrication should also be used. For rolling and impact, the stresses can be significantly higher (i.e., greater than 0.5) for long life, provided that sliding is not involved. In abrasive wear situations, materials similar in hardness to the abrasives or, preferably, harder than the abrasives should be used. Materials should be compatible with the environment in which they are to be used. A list of other rules for wear design is given in Table 7.

Bracketing. One technique that is often helpful in wear design is called bracketing, which is illustrated in Fig. 5. This involves the development of two theoretical wear projections for a design. Different models may be used for each projection. For example, in the case of sliding, one projection might be based on the K-factor model and the other on the combination of the sliding zero wear and measurable wear models. One projection is an optimistic projection, using the most favorable values for the coefficients that have been reported for the most favorable model, if two or more models are potentially applicable. The other is a pessimistic or worst-case projection, using the least favorable values for the least favorable model. If the optimistic projection exceeds the allowable wear, it is generally advisable to change the design concept. If the pessimistic projection gives acceptable wear, the proposed design is acceptable. If the two curves bracket the required wear behavior, the design is considered to be feasible but further work is required to establish a satisfactory design. This further work generally requires the determination of the coefficients for different materials. In cases where more than one model was used, this generally requires some testing to determine which model is the correct one.

Design Modifications. While the focus of the wear design methodology is the avoidance of

Table 9 Allowable loads for mechanisms shown in Fig. 7

$$B1 = 1.535\,\alpha^{1/9}\left(\frac{RL(\sin\alpha/2)}{K}\right)\left(\frac{\gamma_R\tau_{y1}}{\sqrt{0.25+\mu^2}}\right)\left(\frac{1}{xyn_r}\right)^{1/9}$$

$$B2 = 2.953\left(\frac{RL(\sin\alpha/2)}{K}\right)\left(\frac{\gamma_R\tau_{y2}}{\sqrt{0.25+\mu^2}}\right)\left(\frac{1}{xyn_r}\right)^{1/9}$$

$$B3 = 2.953\,L^{10/9}\left(\frac{R}{K}\right)\left(\frac{\gamma_R\tau_{y1}}{\sqrt{0.25+\mu^2}}\right)\left(\frac{1}{xyzn_o}\right)^{1/9}$$

$$B4 = 2.734\frac{RL}{K}\left(\frac{\gamma_R\tau_{y2}}{\sqrt{0.25+\mu^2}}\right)\left(\frac{1}{xyn_o}\right)^{1/9}$$

$$B5 = 1.476\,L^{10/9}\left(\frac{L'}{K}\right)\left(\frac{\gamma_R\tau_{y1}}{\sqrt{0.25+\mu^2}}\right)\left(\frac{1}{xyzn_o}\right)^{1/9}$$

$$B6 = 1.367\frac{LL'}{K}\left(\frac{\gamma_R\tau_{y2}}{\sqrt{0.25+\mu^2}}\right)\left(\frac{1}{xyn_o}\right)^{1/9}$$

$$B7 = 4.638\,\alpha^{1/9}\left(\frac{R^2\sin^2\alpha/2}{K}\right)\left(\frac{\gamma_R\tau_{y1}}{\sqrt{0.25+\mu^2}}\right)\left(\frac{1}{xyn_o}\right)^{1/9}$$

$$B8 = 4.294\left(\frac{R^2\sin^2\alpha/2}{K}\right)\left(\frac{\gamma_R\tau_{y2}}{\sqrt{0.25+\mu^2}}\right)\left(\frac{1}{xyn_o}\right)^{1/9}$$

Stress-concentration factor for B1-B4:

$$K = \frac{0.6318}{\left[\left(\frac{1-\nu_1^2}{E_1}+\frac{1-\nu_2^2}{E_2}\right)\frac{aP}{RL^2\sin\alpha/2}\right]^{1/3}}$$

Stress-concentration factor for B5 and B6:

$$K = \frac{0.4778}{\left[\left(\frac{1-\nu_1^2}{E_1}+\frac{1-\nu_2^2}{E_2}\right)\frac{\delta P}{LL'}\right]^{1/3}}$$

Stress-concentration factor for B7 and B8:

$$K = \frac{0.181}{\left[\left(\frac{1-\nu_1^2}{E_1}+\frac{1-\nu_2^2}{E_2}\right)\frac{aP}{\alpha R^3}\right]^{1/3}}$$

Table 10 Allowable loads for mechanisms shown in Fig. 8

$$B9 = 32.07\left(\frac{1-\nu_1^2}{E_1}+\frac{1-\nu_2^2}{E_2}\right)^{5/4}\left(\frac{L}{R_2^{1/4}}\left[\frac{1}{1/R_2-1/R_1}\right]^{5/4}\right)\left(\frac{\gamma_R\tau_{y1}}{K(1+\mu)}\right)^{9/4}\left(\frac{1}{xyn_r}\right)^{1/4}$$

$$B10 = 27.39\left(\frac{1-\nu_1^2}{E_1}+\frac{1-\nu_2^2}{E_2}\right)\left(\frac{L}{1/R_2-1/R_1}\right)\left(\frac{\gamma_R\tau_{y2}}{K(1+\mu)}\right)^{2}\left(\frac{1}{xyn_r}\right)^{2/9}$$

$$B11 = 6.848\left(\frac{1-\nu_1^2}{E_1}+\frac{1-\nu_2^2}{E_2}\right)\left(\frac{1}{0.25+\mu^2}\right)\left(L\right)^{11/9}\left(\frac{1}{1/R_2-1/R_1}\right)\left(\frac{\gamma_R\tau_{y1}}{K}\right)^{2}\left(\frac{1}{xyzn_o}\right)^{2/9}$$

$$B12 = 5.870\left(\frac{1-\nu_1^2}{E_1}+\frac{1-\nu_2^2}{E_2}\right)\left(\frac{1}{0.25+\mu^2}\right)\left(\frac{L}{1/R_2-1/R_1}\right)\left(\frac{\gamma_R\tau_{y2}}{K}\right)^{2}\left(\frac{1}{xyn_o}\right)^{2/9}$$

$$B13 = 23.48\left(\frac{\gamma_R\tau_{y1}}{(1+\mu)}\right)^{2}\left(\frac{L_2R_2}{K^2}\right)\left(\frac{1-\nu_1^2}{E_1}+\frac{1-\nu_2^2}{E_2}\right)\left(\frac{1}{xyn_o}\right)^{2/9}$$

$$B14 = 50.86\left(\frac{\gamma_R\tau_{y2}}{(1+\mu)}\right)^{9/4}\left(\frac{L_2^4R_2^5}{K^9}\right)^{1/4}\left(\frac{1-\nu_1^2}{E_1}+\frac{1-\nu_2^2}{E_2}\right)^{5/4}\left(\frac{1}{xyzn_o}\right)^{1/4}$$

Stress-concentration factor for B9-B14:

$$K = \left[1+\frac{1/a}{1/R_2-1/R_1}\right]\left[\frac{2}{m^2n}\right]$$

$$\cos\theta = \frac{1/a+1/R_1-1/R_2}{1/a-1/R_1+1/R_2}$$

wear problems in new designs, it can also be applied to the resolution of existing wear problems. It provides a methodology for establishing corrective actions when a design modification is required. In these situations the existence of wear data and worn hardware is often valuable. The failure analysis and hypothesis development activity that is normally associated with problem-solving efforts provides valuable input to the system analysis and modeling required for wear design.

Methods for Wear Design

In most cases a wear design starts out with a general design outline based on function. This design outline defines the generic shapes, motions, and environment. It may also provide some limits on the type of materials and method of lubrication that can be used. Based on this input a wear model is selected, then used to determine the specifics of the design based on wear considerations. This may take the form of sequentially evaluating designs, where specific materials, dimensions, tolerances, and roughnesses are assumed, to see if any of the combinations will provide adequate wear behavior. It may also take the form of determining what material properties, dimensions, and so on are required in order to obtain adequate wear performance. In practice, the former is often tried first and if a successful design cannot be found, the latter approach is incorporated into the analysis. If a successful design cannot be obtained within this framework, it is then usually necessary to see if other general designs for that function can provide adequate wear performance. It is in these cases that the bracketing technique is often useful.

Example: Design Approach for Low-Wear Computer Peripherals. The details of wear design methods tend to vary not only with the type of wear but also with the needs of the organization. However, a design approach used for low-wear applications in computer peripherals can be used to illustrate the general nature of a wear design approach (Ref 28, 29). Other examples of specific approaches can often be found in wear and design publications, such as *Machine Design*, *Wear*, and *Tribology International*. A number of wear design examples and methods can also be found in Ref 8.

The design approach used for computer peripherals is primarily a theoretical approach, which is based on the zero wear model for sliding. In this model zero wear is defined as a wear scar whose average depth is the center-line average (CLA) roughness of the surface, as illustrated in Fig. 6. This model relates material properties and stress levels so that a zero wear condition will not be exceeded. It can be used to determine the maximum load that can be applied to a design for a given lifetime and not have the depth of wear exceed the surface roughness level. Expressions for the allowable load were developed for a number of different geometries and motions, as shown and described in Fig. 7 to 10 and Table 8 to 11.

The wear design method is to evaluate a proposed design by using the relationships shown in Fig. 7 to 9 to determine an allowable load. The allowable load is then compared with an estimate of the actual load. If the allowable load is significantly higher than the estimated actual operating load, the design is accepted and is given further

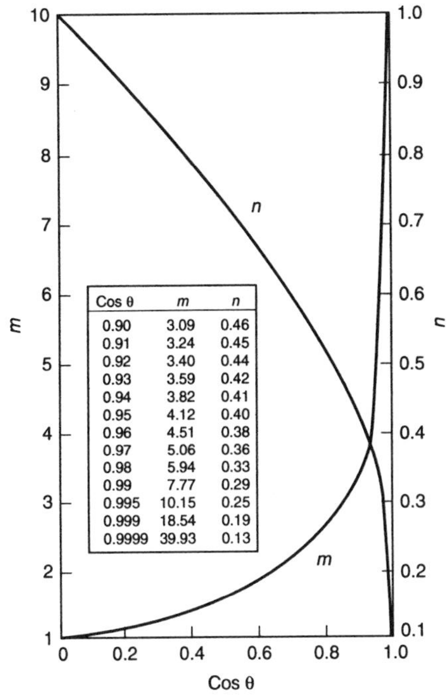

m
n
Cos θ
Cos θ | m | n
0.90 3.09 0.46
0.91 3.24 0.45
0.92 3.40 0.44
0.93 3.59 0.42
0.94 3.82 0.41
0.95 4.12 0.40
0.96 4.51 0.38
0.97 5.06 0.36
0.98 5.94 0.33
0.99 7.77 0.29
0.995 10.15 0.25
0.999 18.54 0.19
0.9999 39.93 0.13

Fig. 10 Plot of m and n vs. cos θ. Values for m and n appear in the stress-concentration factor (K) expression in Fig. 8 and in some of the allowable-load expressions in Fig. 9. Source: Ref 28

consideration only if problems develop in subsequent functional tests.

If the allowable and estimated loads are comparable, changes to the design are considered to improve the safety margin. If this cannot be done, the wear of the device is closely monitored in life and functional tests to verify adequate performance.

If the allowable load is less than the estimated load, design changes are explored to obtain a condition in which the allowable load exceeded the estimated load. Typical changes that are considered in this approach are dimensional and material changes, as well as any changes that would affect the loading situation.

This wear design method is a conservative approach, because most applications can tolerate wear depths one to two orders of magnitude above the CLA value. In situations where an acceptable "zero wear design" cannot be found because of cost or technical reasons, a less conservative approach is then used. In this situation a more realistic failure criterion is used for the amount of wear that can be tolerated. Generally this is the amount of wear that would cause functional problems. A combination of the zero wear model for sliding and the measurable wear model for sliding is then used to project wear and evaluate potential designs. Examples of this method can be found in Ref 21 and 30 to 32.

Table 11 Allowable loads for mechanisms shown in Fig. 9

$$B15 = 52.80\left(\frac{\gamma_R \tau_{y1}}{\beta}\right)^3 \left(\frac{m^3 n^3 R_1^2 R_2^2}{(R_1+R_2)^2}\right)\left(\frac{1-\nu_1^2}{E_1}+\frac{1-\nu_2^2}{E_2}\right)^2 \left(\frac{1}{xyn_o}\right)^{1/3}$$

$$B16 = 153.4\left(\frac{\gamma_R \tau_{y2}}{\beta}\right)^{27/8} (m^{27/8} n^{15/4})^{3/8} \left(\frac{R_1 R_2}{R_1+R_2}\right)^{19/8} \left(\frac{1-\nu_1^2}{E_1}+\frac{1-\nu_2^2}{E_2}\right)^{19/8} \left(\frac{1}{xyzn_o}\right)^{3/8}$$

$$B17 = 29.57\left(\frac{\gamma_R \tau_{y1}}{\beta}\right)^{27/8} (R_1)^{19/8} \left(\frac{1-\nu_1^2}{E_1}+\frac{1-\nu_2^2}{E_2{}^{19/8}}\left(\frac{1}{xyzn_o}\right)\right)^{3/8}$$

$$B18 = 13.20\left(\frac{\gamma_R \tau_{y2}}{\beta}\right)^3 (R_1)^2 \left(\frac{\gamma-\nu_1^2}{E_1}+\frac{1-\nu_2^2}{E_2}\right)^2 \left(\frac{1}{xyn_o}\right)^{1/3}$$

$$B19 = 76.99\left(\frac{\gamma_R \tau_{y1}}{\beta}\right)^{27/8} (m^{27/8} n^{15/4})\, R_2^2 \left(\frac{R_1}{2R_2+R_1}\right)^{19/8} \left(\frac{1-\nu_1^2}{E_1}+\frac{1-\nu_2^2}{E_2}\right)^{19/8} \left(\frac{1}{xyn_r}\right)^{3/8}$$

$$B20 = 66.53\left(\frac{\gamma_R \tau_{y2}}{\beta}\right)^3 (mn)^3 \left(\frac{R_1 R_2}{2R_2+R_1}\right)^2 \left(\frac{1-\nu_1^2}{E_1}+\frac{1-\nu_2^2}{E_2}\right)^2 \left(\frac{1}{xyn_r}\right)^{1/3}$$

$$B21 = 77.03\left(\frac{\gamma_R \tau_{y1}}{\beta}\right)^{27/8} (m^{27/8} n^{15/4}) (R_2)^2 \left(\frac{R_1 R'_2}{2R_2 R'_2 + R_1 R'_2 - R_1 R_2}\right)^{19/8} \left(\frac{1-\nu_1^2}{E_1}+\frac{1-\nu_2^2}{E_2}\right)^{19/8} \left(\frac{1}{xyn_r}\right)^{3/8}$$

$$B22 = 66.56\left(\frac{\gamma_R \tau_{y2}}{\beta}\right)^3 (mn)^3 \left(\frac{R_1 R_2 R'_2}{2R_2 R'_2 + R_1 R'_2 - R_1 R_2}\right)^2 \left(\frac{1-\nu_1^2}{E_1}+\frac{1-\nu_2^2}{E_2}\right)^2 \left(\frac{1}{xyn_r}\right)^{1/3}$$

For B15-B22, β = the greater of μ or 0.31

For B15 and B16, $\cos\theta = \frac{R_1 - R_2}{R_1 + R_2}$

For B17-B20, $\cos\theta = \frac{R_1}{2R_2 + R_1}$

For B21 and B22, $\cos\theta = \frac{R_1 (R_1 + R_2)}{2R_2 R'_2 + R_1 R'_2 - R_1 R_2}$

Note: The relationship between cos θ and values for m and n is shown in Fig. 10.

REFERENCES

1. ASTM G 40-92, "Standard Terminology Relating to Wear and Erosion"
2. A.F. Bower and K.L. Johnson, *J. Mech. Phys. Solids,* Vol 37 (No. 4), 1989, p 471–493
3. K. Johnson, *Proc. 20th Leeds-Lyon Symp. Tribology,* Elsevier, 1994, p 21
4. F. Aleinikov, *Soviet Phys.-Tech. Phys.,* Vol 2, 1957, p 505, 2529
5. P. Blau, *Friction and Wear Transitions of Materials,* Noyes Publications, Park Ridge, NJ, 1989
6. N.C. Welsh, *Proc. Royal Society,* Vol A257, 1965, p 31
7. S.C. Lim and M.F. Ashby, *Acta Metall.,* Vol 35, 1987, p 1-24
8. R.G. Bayer, *Mechanical Wear Prediction and Prevention,* Marcel Dekker, 1994
9. P. Blau, Ed., *Friction, Lubrication, and Wear Technology,* Vol 18, *ASM Handbook,* ASM International, 1992
10. M. Peterson and W. Winer, Ed., *Wear Control Handbook,* American Society of Mechanical Engineers, 1980
11. E.R. Booser, Ed., *Handbook of Lubrication,* Vol II, CRC Press, 1984
12. R. Bayer, *Wear,* Vol 35, 1975, p 35–40
13. E.R. Booser, Ed., *Handbook of Lubrication,* Vol III, CRC Press, 1994
14. IBM General Products Division Technical Report TR 01.17.142.678, 10 April 1962
15. B. Bhushan and B.K. Gupta, *Handbook of Tribology,* McGraw-Hill, 1991
16. R.G. Bayer, Ed., *Selection and Use of Wear Tests for Metals,* STP 615, American Society for Testing and Materials, 1976
17. R.G. Bayer, Ed., *Wear Tests for Plastics: Selection and Use,* STP 701, American Society for Testing and Materials, 1979
18. R.G. Bayer, Ed., *Selection and Use of Wear Tests for Coatings,* STP 769, American Society for Testing and Materials, 1982

19. C.S. Yust and R.G. Bayer, Ed., *Selection and Use of Wear Tests for Ceramics,* STP 1010, American Society for Testing and Materials, 1988,
20. A.W. Ruff and R.G. Bayer, Ed., *Tribology: Wear Test Selection for Design and Application,* STP 1199, American Society for Testing and Materials, 1993
21. *Mechanical Testing,* Vol 8, *Metals Handbook,* 9th ed., American Society for Metals, 1985
22. K. Ludema and R.G. Bayer, Ed., *Tribological Modeling for Mechanical Designers,* STP 1105, American Society for Testing and Materials, 1991
23. R.G. Bayer, W.C. Clinton, C.W. Nelson, and R.A. Schumacher, *Wear,* Vol 5, 1962, p 378–391
24. R.G. Bayer, *Wear,* Vol 11, 1968, p 319-332
25. P. Engel, T. Lyons, and J. Sirico, *Wear,* Vol 23, 1973, p 185
26. P. Engel and R.G. Bayer, *J. Lubr. Technol.,* Oct 1974, p 595
27. G. Talbourdet, Paper 54-Lub-14, American Society of Mechanical Engineers, 1954
28. R.G. Bayer, A.T. Shalkey, and A.R. Wayson, *Mach. Des.,* 9 Jan 1969, p 142–151
29. R.G. Bayer and A.R. Wayson, *Mach. Des.,* 7 Aug 1969, p 118–127
30. R.G. Bayer, *Proc. First European Tribology Congress,* The Institution of Mechanical Engineers, 1973, p 79–84
31. R. Bayer, *Standardization News,* American Society for Testing and Materials, 9 Feb 1974, p 29–32, 57
32. P. Engel et al., *J. Lubr. Technol.,* Vol 100, 1978, p 189–195

Properties Needed for Electronic and Magnetic Applications

Eugene J. Rymaszewski, Rensselaer Polytechnic Institute

PROPERTY can be defined as a quantitative response to a stimulus. In the context of electronic and magnetic applications, such a response usually has two aspects: another stimuluslike parameter and one or, more likely, several attributes of a physical body responding to the stimulus. For example, a voltage V applied to both ends of a wire causes an electrical current I to flow through the wire. The value of the current can be expressed as (Ohm's Law):

$$I = \frac{V}{R} \quad \text{(Eq 1)}$$

$$V = IR \quad \text{(Eq 2)}$$

$$R = \frac{\rho l}{A} \quad \text{(Eq 3)}$$

$$\rho = \frac{RA}{l} \quad \text{(Eq 4)}$$

where R is the electrical (Ohmic) resistance of the wire, in Ω; ρ is the electrical resistivity of the wire material, in $\Omega \cdot$m; A is the cross-sectional area of the wire, in m^2; and l is the wire length, in m. Note that A and l are the dimensions—structural parameters—of the wire and ρ is the property of the wire material. The wire is directly involved in the voltage/current interaction, as opposed to a structure supporting the wire, which may be involved indirectly.

The physical dimensions of this structure and its mechanical and thermal properties may also be relevant, albeit indirectly, to the numerical determination of ρ. For example, sufficiently high values of V and I generate enough heat to raise the wire temperature and, consequently, alter the value of R and, therefore, ρ. The temperature rise will depend on the thermal path (cooling) between the wire and its environment determined by the structure. The temperature rise causes expansion in most materials that can lead to mechanical stresses in the wire and/or its support structure. Thus, there are mechanical and thermal interactions to be considered. Such interactions, if not properly considered, may lead to wrong conclusions. For example, if one were to measure the voltage or current dependency of ρ and not factor in the temperature rise and its effect, then an excessively high positive value of $\Delta\rho/\Delta V$ would be "measured." In reality, it may be zero (or even negative); or the presumed resistor stability, which neglects its thermal coefficient of resistivity (TCR) and temperature rise, will not occur, and the design may fail to perform as expected.

The nonelectromagnetic parameters fall into two distinctly different categories: those involved in fabrication (melting solder, making thermocompression bond, sputtering or evaporating film, diffusing impurities into semiconductors, sintering ceramics), and those involved in operation of the equipment, such as the wire example described above. This article focuses, as its name indicates, on this second category of properties.

Background Information

Classes of Materials. The value of electrical resistivity of materials, ρ, and its inverse, the electrical conductivity, σ, vary over almost 24 orders of magnitude at or near room temperature. Its subranges classify the materials as conductors, semiconductors, and insulators. Conductors are typically metals with ρ ranging from 10^{-8} to 10^{-4} $\Omega \cdot$m. The electrical conductivity results from the presence of "free" electrons, that is, free to move in response to the presence of an electric field.

No "free" electrons are present in the insulators, and consequently, their ρ is from 10^4 to 10^{16} $\Omega \cdot$ m (Ref 1). Insulating materials fall into two major categories, the organic (polymers)—pure, as in polymers, or composite, as in epoxy-impregnated fiberglass cloth—and inorganic, usually ceramics.

The gap in the values of ρ between the conductors and insulators is filled by the semiconductors whose electrical conductivity mechanism is best explained by the band-gap theory (Ref 2). The ρ of semiconductors is in the 10^{-4} to 10^4 $\Omega \cdot$ m range. The semiconductors are either group IV elements, silicon (Si) and germanium (Ge), or composite—formed by two or more group III–V or group II–VI materials, such as GaAs (Ref 3).

The magnetic materials can be metals, alloys, and ceramics. They all have the ability to carry high magnetic flux in response to applied magnetizing force. Their broad classification is based on the atomic structure and values of relative permeability. The high-permeability materials fall into two distinct groups, the magnetically hard and magnetically soft. The hard materials have high coercivity; once magnetized, they resist demagnetizing forces caused by any applied or stray magnetic fields (Ref 4).

The electronic and magnetic applications cover a very wide span of generating and processing information: from placing a telephone call (in a house, on the road, in an airplane) to watching TV from the air, cable, or satellite, to using computers of various sizes and performance capabilities, to remotely interacting with a space ship, and so forth. Many, probably most, of these applications entail generation, processing, transmission, and storage of information, as well as using information to initiate an action—creation of sound, light, or mechanical force, for example. All of the above material categories are used in the electronics hardware in various structures, with widely varying structural dimensions and demands on their properties. As a rule, mutually compatible sets of materials must be employed, placing further demands on their properties.

Small, inexpensive items, such as handheld calculators, are designed for the lowest manufacturing cost, which often precludes repair (a "throwaway" product). The more expensive, and more complex, products usually consist of several subassemblies that are semipermanently joined. These joints are either solder or connectors, a technology in itself to provide satisfactory performance and reliability at affordable cost (Ref 5).

Electronic Applications. The semiconductors play a pivotal role in electronics for processing and storing data and in interfacing the electronic circuits with their exterior, for example, fiberoptic interconnections, optical scanners, dis-

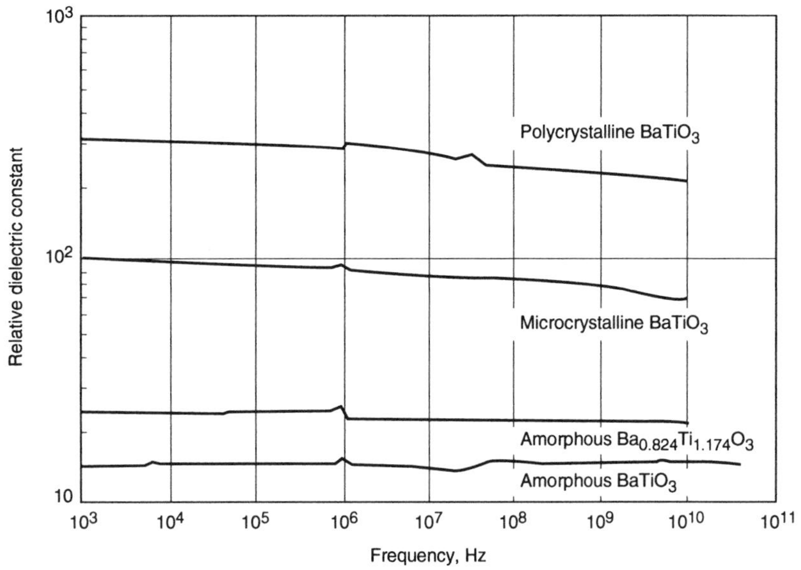

Fig. 1 Frequency dependency of dielectric constant for selected dielectric materials

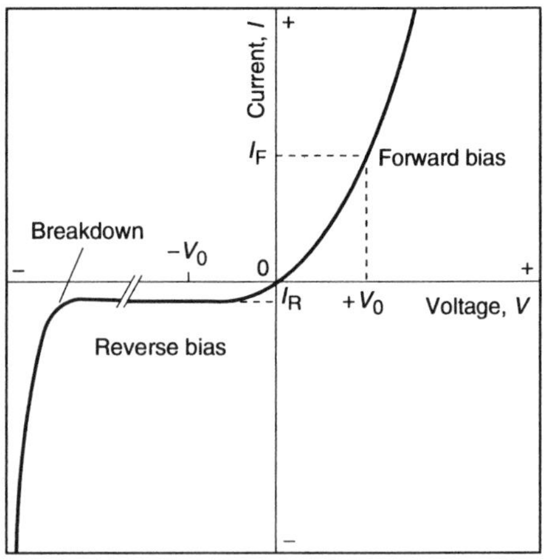

Fig. 2 Voltage-current characteristic of a semiconductor junction. Source: Ref 9

plays and, ultimately, humans. The metals and insulating materials (organic and/or inorganic) are directly involved in electronic circuit embodiments by providing interconnections. They also provide mechanical support, environmental protection, and facilitate heat transfer (Ref 5).

Magnetic materials play a key role in permanent storage and retrieval of information. Furthermore, their use in transformers enables more than just changing magnitude of the voltage. A current transformer reduces high current values to make them more suitable for an ammeter. Transformers (both voltage and current) also isolate the output circuit(s) from the input,—which is important for a multitude of reasons, such as keeping away very high powerline voltages from the instrument panel or separating the telephone line from the interior circuitry of a modem.

The materials properties, such as ρ, are customarily given for the "bulk" materials—those having similar order-of-magnitude cross-sectional dimensions, on a macroscopic scale. The numerical values of these properties may change when the dimensions are reduced to submicron range, as is the case with thin films which see an increasing use in electronic applications (Ref 6). Another factor influencing the numerical values is the microstructure of the material: totally amorphous, crystalline regions within an amorphous matrix, polycrystalline, or monocrystalline. Chemical composition of some amorphous structures may deviate from the stoichiometric, with the resultant alterations in their properties. Figure 1 shows an example of near-zero dispersion of amorphous $Ba_xTi_{2-x}O_y$ thin films and still very low dispersion of partially and fully crystallized barium-titanate films (Ref 7, 8). Properties of the crystalline materials may be direction dependent. As already indicated above, a materials property is often temperature dependent.

The Ohm's law equations (Eq 1–4) indicate that either a voltage V or a current I can be the stimulus that, through the geometry and resistivity/conductivity of the material will determine the value of the response. An underlying assumption is the independence of $\rho = 1/\sigma$ on either voltage or current. This is not always the case, even at a constant temperature. Perhaps the most striking example is the *V-I* (voltage-current) characteristic (interdependence) of a semiconductor diode, formed at the junction of two oppositely doped regions. In addition to being highly nonlinear, it is very strongly polarity dependent, as shown in Fig. 2 (Ref 9). Such nonlinear response is pivotal to the very existence of all digital electronic circuits.

Information-handling applications require periodic or, at least, occasional changes in the values of voltages and currents. Consequently, dependency of the materials properties on such changes must be known and considered for a particular application. For example, the resistivity equations (Eq 1–4) imply uniform current density throughout the cross-sectional area A. However, as the frequency of the current increases, the current penetration depth decreases (skin effect), causing the resistivity to increase. In the time domain, the skin effect manifests itself as time-dependent resistance: high shortly after start of the current flow and decreasing to its DC value. The time it takes depends on the dimensions of the conductor (Ref 10).

Materials may be stressed beyond their mechanical limits, with the resultant permanent deformation. An equivalent situation may occur with the excessive electric stresses that can affect insulators (dielectric breakdown) and conductors (melting, electromigration, void migration) (Ref 11, 12).

Properties of Interest. In this section, an overview of the electric and magnetic parameters is followed by detailed discussion of the significance of these parameters for electronic applications. Many applications prescribe particular values and tolerances of such parameters, and significant efforts are still underway to achieve a particular combination of values of several properties in a particular material. Increasingly, they have to be engineered because such combinations of properties are not available in the existing materials.

Some of these parameter values or properties are not desired because they have deleterious impacts. In such cases they are considered parasitic and, consequently, attempts are made to keep them small, below or within acceptable limits. For example, a particular resistor value is required in a circuit. By contrast, resistance of an interconnection should be low to avoid excessive values of voltage drop. Also, resistance of an insulator, for example, dielectric of a capacitor, must be high enough to prevent excessive leakage currents.

As is shown at the end of the section "Power and Energy Conversion, Storage, and Transmission," some sets of values cannot be altered beyond certain limits for a given set of structural dimensions regardless of the materials properties, only appropriate dimensional changes will produce the desired results. The most notable example is the product of inductance and capacitance of an interconnection (a pair of electrodes, such as wires). This product reaches a minimum for nonmagnetic materials ($\mu = 1$) with low dielectric constants $\varepsilon_r \approx 1$ (e.g., air). Any further reduction is possible only with a shorter length of the interconnection. Such reduction can easily lead to a complete structural redesign.

An analogous situation may occur when the heat generated by electronic circuits is within a small area, causing high values of heat flux, often expressed in W/cm^2. The heat flowing to its environment encounters thermal resistance, causing the temperature to rise. The thermal conductivity of the materials in the heat-flow path and their structural geometries (cross-sectional area and path length) control the thermal resistance. As shown in Fig. 3, the power densities on the order of 100 W/cm^2 and higher—not unusual in some aggressive initial electronics designs—are likely

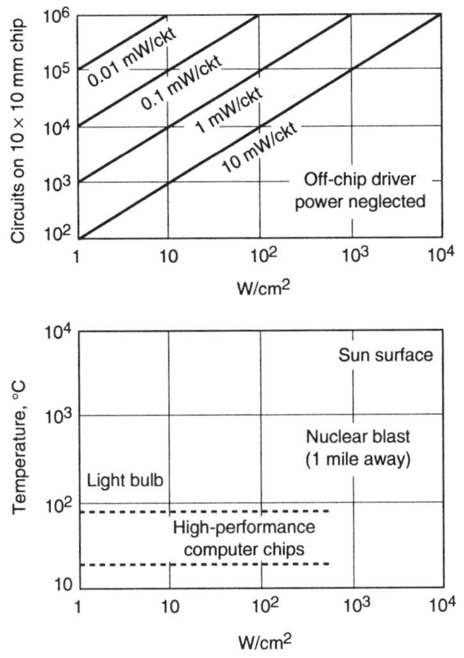

Fig. 3 Temperature versus power density. Source: Ref 5

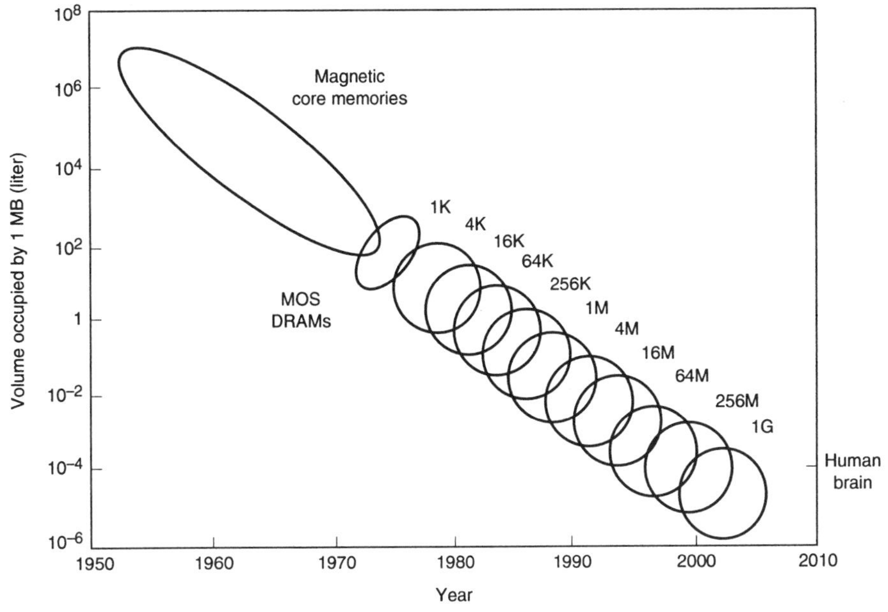

Fig. 4 Memory density trends. Source: Ref 5

to cause excessively high operating temperatures. The limited thermal conductivities of materials used often require close attention to the structural dimensions. Some relief may soon be provided from development of affordable exotic materials with high thermal conductivity, for example, diamond films. An alternative is the use of Peltier-junction based (Ref 13) thermoelectric coolers.

Most electronic structures, whether integrated circuit (IC) interconnections or packaging, are multilayer structures, with conducting layers alternating with the insulating layers. All layers are appropriately patterned and bonded by a variety of processing techniques. The interlayer bonds are part chemical, part mechanical. The bond strength is often of concern, as delamination can have disastrous reliability consequences. Inadequate bond strength and excessive shear forces (caused, for example, by mismatched thermal coefficients of expansion and large temperature variations) are of prime concern.

Important Trends and Considerations Not Addressed in Detail. The trend in electronic applications is toward increasing density—more functions per unit area or per unit volume. Figure 4 shows dramatic reductions in the volume required for 1 MB (megabyte) of electronic memory storage from 1950 to 2010, a sixty-year interval during which the technology changed from small magnetic (ferrite) cores to IC chips (Ref 13). Packaging had evolved as well and contributed to the volumetric improvements beyond those that would have been possible with just the magnetic core and semiconductor chip technologies which are basically planar, two dimensional.

At the same time, the trend is often toward increasing the area of the IC chips, multichip modules (MCMs) and printed wiring boards (PWBs). Thus, the number of pixels per each subassembly is steadily increasing (Ref 14). This trend taxes the capabilities of design support systems and, especially, the defect-free manufacturing processes. There is a strong competitive pressure to balance the design of leading-edge products against the expense of establishing, maintaining, and improving defect-free manufacturing environment. Consequently, a certain percentage of the product, especially during its early manufacturing phase, is likely to be defective. It has to be weeded out prior to the final assembly, or even sooner. Sophisticated testing facilitates the weeding out. However, nonkiller defects escape detection, are shipped, and contribute to failures early in the life of the products. A prestress process, known as "burn-in," endeavors to enhance the weeding-out of defects (Ref 11, 15).

Overview of Electric and Magnetic Parameters and Materials Properties

Electric voltage, current, and power are fundamental electrical units—*V*, *I*, and *P*. They are measured in volts (V), amperes (A), and watts (W), respectively. A potential difference, or electromotive force, of 1 V exists between two points of a conductor carrying a constant current of 1 A when the power dissipated between these points equals 1 W (Ref 16). Note that implicit to this definition is the electrical resistance of 1 Ohm (Ω) present between these points.

Power and Energy Conversion, Storage, and Transmission. The very definition of fundamental electrical units invokes power dissipation, that is, the conversion of the electrical energy into thermal. The energy can also be temporarily stored, either in the electric field associated with a voltage between two or more conductors, or in the magnetic field caused by and associated with the current flowing through a conductor. Changes in the field energies are linked with the corresponding temporary flow of current and appearance/application of voltage. Electric permittivity, a materials property, is the proportionality constant that directly relates the voltage and structural geometry to the electrical field energy. Magnetic permeability is its magnetic field energy counterpart.

The static (constant with time) magnetic and electric fields may and do exist independent of each other. The dynamic (changing with time) fields do not: a charge must be transported to alter the electrical potential between any two (or more) electrodes, such as capacitor plates or electrical connections. Note that $I(t) = dQ/dt = C(dV/dt)$. Such current $I(t)$ creates a magnetic field associated with it. This current will in turn cause a voltage $V = L(dI/dt)$. Thus, the dynamic voltages and currents are interdependent and interlinked in space. An interconnection consisting of two wires of uniform cross-sectional geometry has an inductance per unit length, L_0, and a capacitance per unit length, C_0.

The ratio of the above *V* and *I* on an infinitely long interconnection (no reflections) with a uniform cross-sectional geometry is referred to as the characteristic impedance Z_0:

$$Z_0 = \frac{V}{I} = \sqrt{\frac{L}{C}} = \sqrt{\frac{L_0}{C_0}}$$

The value of Z_0 and, therefore, the ratio $L/C = L_0/C_0$ depend on cross-sectional geometry, magnetic permeability, and electric permittivity of the material(s) surrounding the interconnection conductors. The geometry dependence for some selected geometries is shown in Fig. 5; this dependency as well as very detailed data are presented in Ref 17. The detailed theory is discussed in many texts, such as Ref 18 and 19.

The velocity of propagation v of the "interlocked" voltage $V(t)$ and current $I(t)$ waves does not depend on the cross-sectional geometry, only on the permeability μ and permittivity ε of the insulating material between the conductors,

$$v = \sqrt{L_0 C_0} = \frac{c}{\sqrt{\mu\varepsilon}}$$

where c is the velocity of light in free space. Thus, *for a given length, the product LC is fixed* and reaches a minimum for $\mu = \varepsilon = 1$. To reduce the value of LC, the interconnection length must be reduced—often a difficult task.

Capacitance and Dielectric Materials. An ideal capacitor has no electrical conduction between its plates; it only stores any amount of electrical energy and returns it in full. The dielectric material between the capacitor plates largely determines the degree to which these ideal characteristics are met. It has four key properties—the dielectric constant, the dielectric leakage, the dielectric breakdown, and the dielectric loss—which limit the ideal behavior of a capacitor. The leakage current drains the charge and may be very detrimental in applications in which the capacitor is expected to separate (block) dc circuits or to maintain its charge, as in dynamic random access memory (DRAM) chips.

The dielectric leakage depends on the electric field strength. It is commonly normalized to the electrodes area. Its metric is then the current density J, often given in μA/cm^2. In many applications, it is very important to keep the capacitor area A to a minimum. Two factors determine the capacitance density (for example, expressed in nF/cm^2): the dielectric constant ε and the thickness of the dielectric, d. Thinner dielectric produces higher electric field $E = V/d$, often measured in MV/cm, for the same applied voltage V. Usually, it increases the value of J. Figure 6 shows examples of the leakage currents J versus capacitance densities for some recently developed thin films, with the values of d well into the submicron range (Ref 13).

There is another limitation to the voltage between the capacitor plates—the dielectric breakdown, a situation caused by excessive electric field strength and manifested by avalanching current, often producing permanent destruction (Ref 10). The breakdown field strength is inversely dependent on the dielectric constant, as illustrated in Fig. 7, breakdown field versus permittivity (Ref 13). It must be noted here that very thin films may well have disproportionately low breakdown voltages (and average fields) because the dielectric will first break down at locations of reduced thickness caused, for example, by microroughness of the metals onto which the dielectric was deposited or by presence of small particles.

Unlike the leakage currents and breakdowns that occur at dc, the dielectric losses are an ac phenomenon. In an ideal capacitor, there is a 90° phase shift at the angular frequency $\omega = 2\pi f$ between the applied ac voltage vector V and the vector I of the ac current flowing through the capacitor C; $I = V \times j2\pi fC = V \times j\omega C$, with $C = \varepsilon A/d$. With all real materials, the dielectric constant is a complex number $\varepsilon = \varepsilon' + j\varepsilon''$. As a rule $\varepsilon' \gg \varepsilon''$. The ratio of ε'' to ε' is often referred to as the "loss-angle tangent," tan δ. For the most insulating materials, it remains relatively constant and small at all frequencies at which ε' remains constant. A change (reduction) in ε' with frequency is known as dispersion and is associated with rising ε'' and, therefore, a higher value of tan δ that fairly accurately tracks $d\varepsilon'/df$ ($d\varepsilon'/d\omega$) (Ref 4, 10, 12, 20).

The onset of dispersion occurs at lower frequencies for the higher values of ε. This is another factor to be considered, in addition to the lower breakdown field strength shown in Fig. 7. As shown in Fig. 1, minimum dispersion, if any, occurs with the amorphous thin films. Another property of higher dielectric constant materials to consider is the temperature dependency of ε.

Inductance is conceptually related to the resistance in the sense that a change in the current flowing through it is resisted by appearance of voltage $V = L(dI/dt)$ or, in case of a steady alternating current of frequency f, $V = I \times j\omega L$. The term ωL is called reactance, the positive counterpart to reactance of a capacitor $1/j\omega C$ or $-j/\omega C$. Again, an ideal inductor temporarily stores energy in its magnetic field and returns all of it. In the real world, this is never fully the case (except in superconductors) because of the Ohmic wire resistance and, if magnetic materials are used, because of their losses caused by the remanence (area within the hysteresis loop).

Signal Transmission. Unlike water flow from a hose or light emitted by its source, the electric current can flow only in a circuit (containing resistances and reactances) that forms a closed loop, that is, with connections to both terminals of the source. It is common among electronics engineers to imply presence of a "return" current path, without being explicit about it. In most situations, it works well and justifies the simplification. The return current path is usually formed by the "ground": the body (chassis) of equipment or planes used to deliver the power (see the next section "Mutual Inductance"). However, such a design has a poorly controlled return-current path, which can cause performance problems (Ref 5). Characteristic impedance and signal velocity of propagation along such lines are briefly covered in the section "Power and Energy Conversion, Storage, and Transmission." Ideal transmission lines preserve the amplitude and transition time of the signal. The inevitable losses occur due to the interconnection resistance (particularly troublesome in the small cross-section lines of high density ICs and packages) and, to a much lesser degree, in the insulating dielectrics.

Mutual Inductance. If a second conductor is present in the magnetic field associated with the current through the first conductor, it induces a voltage proportional to the ratio of the field common to both conductors to the total field of the first conductor. This phenomenon is the base for various transformer designs which, as a rule, seek "tight coupling" between the first (primary) and second (secondary) wires. This coupling is further enhanced by use of ferromagnetic flux concentrators. A second, less desirable, effect occurs when two or more signal transmission lines are in proximity of each other. The mutual inductance between them is, as a rule, accompanied by mutual capacitance. The combined effect is known as "crosstalk." Typically, it is very small (close to zero) at the far end (away from the driver of the first line) and sizeable at the near end. A "directional coupler" utilizing this phenomenon permits separation of two waves propagating in opposite directions along the first line (Ref 5).

Overview of Parameters and Materials Properties Other than Electric and Magnetic

As already briefly indicated, the temperature dependence of many properties, singly and in the context of an electronic structure, is of paramount

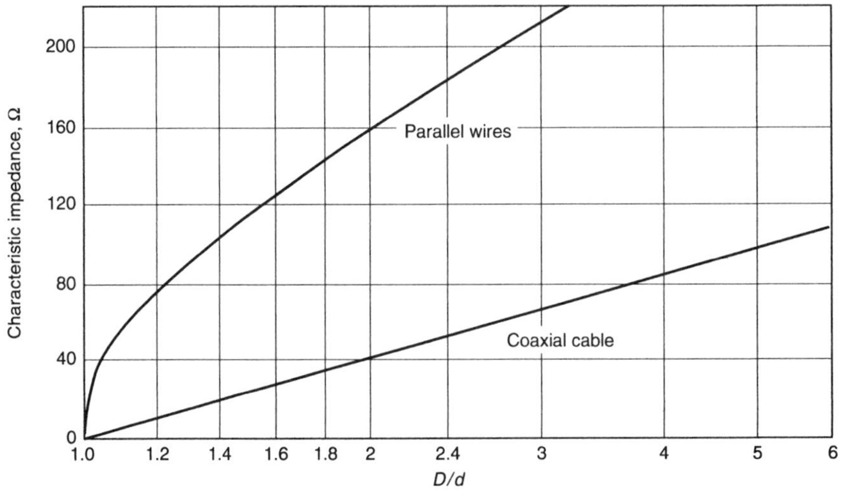

Fig. 5 Characteristic impedance versus cross-sectional geometries of two parallel wires and coaxial cable

importance to proper functioning of devices, circuits, and equipment. The malfunctions can be immediate, for example, caused by the drift of device characteristics outside of design limits, or they can be the result of failures caused by time or cycle-dependent disruptions of the structural integrity, for example, cracking of the solder joints. The mechanical stresses can also originate from a hostile operating environment: shocks and vibrations being the major ones. A hostile operating environment (salt water, for example) may chemically attack the electronic components.

Temperature and Temperature Dependencies. An important consideration, especially in semiconductor-based components and ICs is the strong temperature dependency of the conduction mechanism; it often limits the temperatures at which a circuit functions reliably. There are several other temperature considerations, temperature values, and their effects:

- Temperatures required for processing (manufacturing) of the semiconductor devices and IC chips, such as diffusion or sputtering
- Temperatures required for packaging (assembly) of the electronic equipment, for example, solder
- Environmental temperatures encountered during the product life, for example, transportation in an unheated compartment of a jet plane, products under the hood of a vehicle in tropic and arctic environments
- Temperature rise above the environment caused by power dissipation

The effects of these temperature variations are many:

- Shift in the electrical characteristics already mentioned
- Shift in the device switching speed (bipolar digital circuits are faster at higher temperatures, complementary metal-on-silicon, or CMOS, circuits are slower)
- Excessive mechanical stresses caused by mismatches in the thermal coefficients of expansion
- Instability of the materials such as decomposition of polymers and melting of solder

The designs for temperature control become increasingly difficult with wider range of operating temperatures (e.g., in automotive applications from arctic to tropic) and greater power densities. Thermoelectric coolers/heaters may be required. They increasingly employ Peltier-effect junctions.

Interface Adhesion. Essentially all electronic structures have thin-film layers of insulators and conductors. The interlayer adhesion is, as already mentioned, an extremely important design consideration that often requires compromises in the choice of materials, their deposition/lamination techniques, and surfaces treatments to facilitate the adhesion.

An additional consideration is the quality and integrity of the vertical electrical connections between two or more conducting layers. Such connections require creation of apertures in the insulating layer, commonly called vias or via holes, and their subsequent filling with conducting materials. Additional challenge results from diminishing dimensions, currently in the submicron range. The integrity of the via connections requires sufficient removal of the insulating material and metal deposition that has a good bond to the conductor at the via hole bottom as well as to the via hole sides. This integrity is threatened during the subsequent product manufacturing and through its life in the field, mainly by mechanical stresses.

Stress-Strain, Creep, and Fatigue. There is an unavoidable mismatch in the thermal coefficients of expansion (TCE) between the semiconductor IC bodies, the aluminum and insulator thin films on them, and the packaging to which the chips are attached. The need for low thermal resistance further compounds the challenge of providing adequate stress relief. The packaging structures usually employ copper conductors and various organic and inorganic insulators, each with their own challenges. Creep is often relied upon to provide stress relaxation. However, repeated temperature cycling fatigues the bonds and, thus, limits the product life. The mechanism is sufficiently understood (Ref 5) to offer design guidance. The failures may result from structural damage that disrupts either electrical or thermal conduction paths.

Electronic Applications

The electronic applications (as opposed to electric power) are still growing rapidly, doubling in dollar volume every decade. They seem to penetrate every appliance and gadget, and even go beyond dreams of science-fiction writers; just think of cell phones, personal digital assistants, and small, easily portable geopositioning equipment. Yet, many of these applications share the same basic technologies: semiconductor IC chips, packaging (supporting chips and other components, such as resistors, capacitors, and inductors), source(s) and distribution of electric power, and means of interfacing with humans and other equipment. This section deals with the components of the electronic circuits—analog and digital. The next section discusses the augmenting technologies: magnetic and special technologies such as electrooptical.

Semiconductors. Understanding of the operational principles of a semiconductor junction is barely half a century old; the IC technology is two decades younger. Yet the progress has been spectacular, as demonstrated in Fig. 4. The fabrication of ICs is a highly specialized, still rapidly evolving field (Ref 3, 21, 22). The IC chips consist of two major domains: the semiconductor devices and the thin-film interconnections (conductors, insulators, and interlayer connections) for signal and power distribution. An integration of IC chips with the rest of the components occurs in the packaging, the task of which is to provide me-

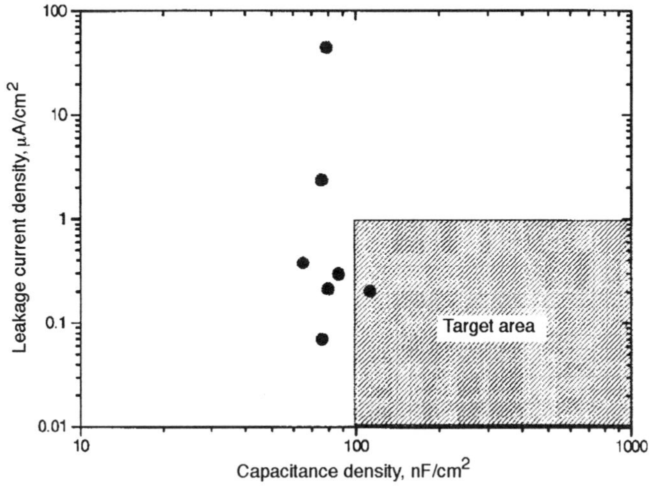

Fig. 6 Leakage current densities versus capacitance densities of new dielectric thin

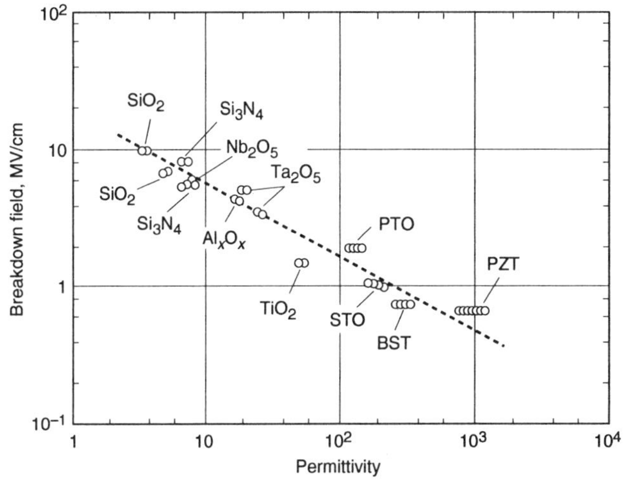

Fig. 7 Breakdown field dependency on the dielectric constant

chanical support and protection, signal and power connections, and to conduct away the inevitable heat. Thus, sufficient electrical, thermal, and mechanical interfaces must be established during manufacture and assembly and maintained through the useful product life. Depending on the application, there may be a need to disassemble the package in order to remove and replace one or several of the failed chips.

Resistors. The very definition of electrical parameters (Ohm's law) includes the resistor. Yet, the real-life realization of just about any resistor has several challenges, made especially severe in demanding applications. The thermal coefficient of resistivity had been already mentioned, the value and sign of which are controlled by the composition of materials and by the physical dimensions. The range of resistor values of interest extends over at least 6 to10 orders of magnitude. The additional considerations are the variation of the value when first manufactured and within the operating temperature and frequency ranges, and the power-handling capability. The very recent challenges are presented by the quest for the lowest manufacturing and assembly costs and smallest area or volume. Such challenges drive development of "embedded" resistors, as well as capacitors and inductors, labeled imbedded passives (Ref 13). They are produced with thin films of proper composition, dimensions, and processing. In addition to determining the power-handling capabilities, the structural dimensions of resistors control their high-frequency characteristics. The impedance of low-value resistors (below 100 Ω) has the tendency to acquire a sizeable series inductance, and high-value resistors (above 1 kΩ) tend to have a parallel capacitor component. The skin effect is less pronounced in thin-film resistors, but its effect should at least be considered.

Capacitors. The range of capacitance value is even greater than those of resistors, from fraction of a picofarad (pF) to one or several Farads, or at least 12 orders of magnitude. Some of the applications require tight tolerances (initial value, throughout the life, versus temperature and frequency) while others are open-ended toward higher values requiring only some minimum value. The operating and breakdown voltage and leakage currents are another important consideration that may dictate the choice of materials, as, for example, shown in Fig. 7. The embedded capacitors often require compatibility with organic materials used in packaging structures. This need rules out many excellent processing techniques that require high temperatures.

The series inductance of the capacitor with its connections, L_s, affects the high-frequency behavior in a peculiar way— there is an apparent increase in the capacitor value because total $X_c = (1/\omega C - \omega L_s) < 1/\omega C$. To make matters worse, above the resonance frequency, at which $1/\omega C = \omega L_s$, the inductance dominates the total impedance (plus the losses represented by a series resistance R_s). Just as with the resistors, careful control of geometry is required to push the resonance frequency to values higher than the highest operating frequency (Ref 23).

Signal Transmission. Transmission of electrical signals tolerates only relatively minor distortion, be it injected noise or signal amplitude degradation. The signal delay is unavoidable and is factored into designs, but it is often required to be as small as achievable as well as to remain constant. The signal transition time limits the maximum transmittable data rate. It is controlled by the losses in the conductors (series resistance) and insulators (dielectric losses and, in extreme situations, leakage currents). The quest for higher densities (Ref 14) drives cross-sectional geometries toward smaller values, requiring low-loss materials. For example, aluminum served the IC industry well since its inception a third of a century ago. While copper offers lower losses, and will eventually replace aluminum, the IC industry will use it with greatest reluctance. By contrast, the electronic packaging industry has been copper based from the start. Both are seeking dielectric materials with permittivities below 2 to minimize signal delay time and degradation of the signal transition time. Nonisometric values of permittivity are of concern.

Power Distribution. The power distribution system resides between the source of power—a battery or power supply circuits that convert the power line ac voltage into appropriate value dc voltage—and the power users: a multitude of circuits connected to the distribution at a multitude of locations. Ideally, all power users should have the same voltage (within tight tolerance) free of superimposed ac noise. The conductivity is, again, a key parameter, along with proper geometries. Suppression of the ac noise (decoupling) is usually accomplished with discrete capacitors whose high-frequency behavior sets the limits (Ref 5, 23).

Magnetic Applications

Some of these applications are as old as the electromagnetic technology, or even its precursor, the electricity and magnetism branch of physics. The unwanted (parasitic) effects of magnetic field energy associated with a current flow are described elsewhere in this section. The desired effects can be classified into several major groups: those requiring flux enhancers—such as ferromagnetic cores—and those without, sometimes referred to as having an air core. Some applications require conversion of the magnetic energy into energy other than electrical, for example, mechanical or thermal.

Inductors and transformers require no energy conversion. Some conversion is unavoidable and is considered parasitic. The inductors have only one set of terminals and one coil associated with the magnetic flux. The transformers have at least two coils (primary, to which the input voltage or current are applied, and the secondary) that provide isolation and, depending on the ratio of the turns in the coils, a voltage or current transformation. The volume of magnetic core is determined by the transmitted power.

Electromagnets, relays, motors, and activators utilize the magnetic field(s) to exert mechanical force either by interaction with the field of a permanent magnet, or between two electromagnets or (in most relays) by the attractive force between a solenoid core and the structure carrying contacts.

Magnetic storage requires initial magnetization and subsequent interrogation (readout) that causes minimum alteration of the magnetization to enable multiple readouts. The main advantages of the magnetic storage are the permanence—ability to store without continued application of electric power—and high storage capacity and essentially an unlimited number of the write/read cycles. The main disadvantage is the need to move the storage elements with respect to the readout head, by either a moving tape or a spinning disk. Another disadvantage is a relatively long read-out time.

Special Applications

These applications augment the mainstream electronic and magnetic applications by utilizing materials properties such as piezoelectric, optoelectric, optomagnetic, superconducting, and so forth.

Piezoelectric applications are similar to magnetic transducers, sometimes serving the same purpose, for example, in phonographic pickups. The amount of mechanical displacement is very small—an advantage when minute, well-controlled movements are the goal, as in microscopic positioning mechanisms. The piezoelectric crystals have very constant frequencies of mechanical oscillation. They are popular in applications requiring generation of accurate ac signals, such as watches and tuning circuits for radio receivers and transmitters.

Optoelectronics are gaining in popularity because modulated coherent light enables very high data transfer rates over large distances. The transmission medium is a fiber light guide whose refraction coefficient confines the light beam to the interior and transmits it with very low losses. However, generation and modulation of light at the sending end still poses challenges, and these challenges are greater than those at the receiver: the wide-band demodulation of light and conversion to electrical signals.

Display panels use two types of technology: the well-established cathode-ray tubes (CRTs) used in most television sets and desktop computers, and flat-panel displays, which begin to challenge the bulky and heavy CRTs in area and are far superior in total volume and mass.

Optoelectronic storage applications are best known by the compact disc (CD), its bigger cousin, laser disc (LD), and the emerging DVD, which originally stood for a digital video disc. All offer a very low-cost, high-capacity storage, mostly for read-only use—the information is permanently embedded during the manufacturing. These discs supplement the magnetic storage and, to a degree, compete. Write/read variants are also available, but at high cost.

REFERENCES

1. J.F. Shackelford, *Introduction to Materials Science for Engineers*, 3rd ed., Macmillan, 1992, Fig. 11.7.1
2. L. Solymar and D. Walsh, *Lectures on the Electrical Properties of Materials,* 3rd ed., Oxford University Press, 1984
3. S.K. Ghandhi, *VLSI Fabrication Principles,* 2nd ed., John Wiley & Sons, 1994
4. P.S. Neelakanta, *Handbook of Electromagnetic Materials*, CRC Press, 1995
5. R. Tummala, E.J. Rymaszewski, and A.G. Klopfenstein, Ed., *Microelectronics Packaging Handbook*, Part 1, 2nd ed., Chapman & Hall, 1997
6. S.P. Murarka and M.C. Peckerar, *Electronic Materials Science and Technology*, Academic Press, 1989, p 272
7. E.J. Rymaszewski, High-Dielectric Constant Thin Films for High-Performance Applications, *Proc. DARPA Physical Electronic Packaging Program Review*, Washington, DC, March 1993
8. W.-T. Liu, S. Cochrane, P. Beckage, D.B. Knorr, T.-M. Lu, J.M. Borrego, and E.J. Rymaszewski, Deposition, Structural Characterization, and Broadband (1 kHz–40 GHz) Dielectric Behavior of $Ba_xTi_{2-x}O_y$ Thin Films, *Proc. Mater. Res. Soc.,* Spring, 1993.
9. W.D. Callister, Jr., *Materials Science and Engineering: An Introduction,* 3rd ed., John Wiley & Sons, 1994, Fig. 19.20
10. S.O. Kasap, *Principles of Electrical Engineering Materials and Devices,* McGraw-Hill, 1997
11. R. Tummala, E.J. Rymaszewski, and A.G. Klopfenstein, Ed., *Microelectronics Packaging Handbook,* Part 2, *Semiconductor Packaging,* 2nd ed., Chapman & Hall, 1997
12. D.D. Pollock, *Physical Properties of Materials for Engineers*, 2nd ed., CRC Press, 1993
13. E.J. Rymaszewski, Revolution in Packaged Electronics, *MRS 1996 Fall Meeting Tutorial Program*, Symposium J, 2 Dec 1996
14. E.J. Rymaszewski, Dense, Denser, Densest…, *J. Electron. Mater.,* Vol 18 (No. 2), 1989
15. R. Tummala, E.J. Rymaszewski, and A.G. Klopfenstein, Ed., *Microelectronics Packaging Handbook,* Part 3, *Subsystem Packaging,* 2nd ed., Chapman & Hall, 1997
16. R.C. Weast, D.R. Lide, M.J. Astle, and W.H. Reyer, Ed., *CRC Handbook of Chemistry and Physics*, 70th ed., CRC Press, 1989–1990
17. *Reference Data for Radio Engineers,* 4th ed., International Telephone and Telegraph Corp., 1956, Chapter 20
18. S. Ramo and J.R. Whinnery, *Fields and Waves in Modern Radio,* 2nd ed., John Wiley and Sons, Chapman & Hall, 1953
19. D.K. Chang, *Field and Wave Electromagnetics,* 2nd ed., Addison-Wesley, 1990
20. J.P. Schaffer, A. Saxena, S.A. Antolovich, T.H. Sanders, and S.B. Warner, *The Science and Design of Engineering Materials*, Irwin, 1995
21. S.A. Campbell, *The Science and Engineering of Microelectronic Fabrication,* Oxford University Press, 1996
22. J.W. Mayer and S.S. Lau, *Electronic Materials Science: For Integrated Circuits in Si and GaAs,* Macmillan, 1990
23. J.N. Humenik, J.M. Oberschmidt, L.L. Wu, and S.G. Paul, Low-Inductance Decoupling Capacitor for the Thermal Conduction Modules of the IBM Enterprise System/9000 Processors, *IBM J. Res. Dev.,* Vol 36 (No. 5), Sept 1992, p 935–942

Design with Brittle Materials

Stephen F. Duffy, Cleveland State University
Lesley A. Janosik, NASA Lewis Research Center

BRITTLE MATERIALS (e.g., ceramics, intermetallics, and graphites) are increasingly being used in the fabrication of lightweight components. From a design engineer's perspective, brittle materials often exhibit attractive high-strength properties at service temperatures that are well beyond use temperatures of conventional ductile materials. For advanced diesel and turbine engines, ceramic components have already demonstrated functional abilities at temperatures reaching 1370 °C (2500 °F), which is well beyond the operational limits of most conventional metal alloys. However, a penalty is paid in that these materials typically exhibit low fracture toughness, which is usually defined by a critical stress intensity factor, and typically quantified by K_{Ic}. This inherent undesirable property must be considered when designing components. Lack of ductility (i.e., lack of fracture toughness) leads to low strain tolerance and large variations in observed fracture strength. When a load is applied, the absence of significant plastic deformation or microcracking causes large stress concentrations to occur at microscopic flaws. These flaws are unavoidably present as a result of fabrication or in-service environmental factors. Note that nondestructive evaluation (NDE) inspection programs cannot be successfully implemented during fabrication. The combination of high strength and low fracture toughness leads to relatively small critical defect sizes that cannot be detected by current NDE methods. As a result, components with a distribution of defects (characterized by various sizes and orientations) are produced, which leads to an observed scatter in component strength. Catastrophic crack growth for brittle materials occurs when the crack driving force or energy release rate reaches a critical value and the resulting component failure proceeds in a catastrophic manner.

The emphasis in this article is placed on design methodologies and characterization of certain material properties. Of particular interest to the design engineer is the inherent scatter in strength noted above. Accounting for this phenomenon requires a change in philosophy on the design engineer's part that leads to a reduced focus on the use of safety factors in favor of reliability analyses. If a brittle material with an obvious scatter in tensile strength is selected for its high-strength attributes, or inert behavior, then components should be designed using an appropriate design methodology rooted in statistical analysis. However, the reliability approach presented in this chapter demands that the design engineer must tolerate a finite risk of unacceptable performance. This risk of unacceptable performance is identified as the probability of failure of a component (or alternatively, component reliability). The primary concern of the engineer is minimizing this risk in an economical manner.

This article presents fundamental concepts and models associated with performing time-independent and time-dependent reliability analyses for brittle materials exhibiting scatter in ultimate strength. However, the discussion contained within this article is not limited to materials exposed to elevated service temperatures. The concepts can be easily extended to more mundane applications where brittle materials such as glass or cements are used. Specific applications that have utilized ceramic materials at near-ambient temperatures include wear parts (nozzles, valves, seals, etc.), cutting tools, grinding wheels, bearings, coatings, electronics, and human prostheses. Other brittle materials, such as glass and graphite materials, have been used in the fabrication of infrared transmission windows, glass skyscraper panels, television cathode ray tubes (CRTs), and high-temperature graphite bearings. Thus, in this article the design methodologies used to analyze these types of components, as well as components exposed to elevated service temperatures, are presented. Reliability algorithms are outlined, and several applications are presented to further illustrate the utilization of these reliability algorithms in structural applications. For further background material on statistical methods, see the article "Statistical Aspects of Design" in this Volume.

Time-Independent Reliability Analyses

An engineer is trained to quantify component failure through the use of a safety factor. By definition, the safety factor for a component subjected to a single load L is given by the ratio:

$$\text{Safety factor} = \frac{R}{L} \qquad \text{(Eq 1)}$$

where R is the resistance (or strength) of the material from which the component is fabricated. Making use of the concept of a safety factor, the probability of failure (P_f) for the component where a single load is applied is given by the expression:

$$P_f = \text{Probability}\left(\frac{R}{L} \geq 1\right) \qquad \text{(Eq 2)}$$

In making the transition from a deterministic safety factor for a component to a probability of failure, for the most general case, the assumption is made that both R and L are random variables. Under this assumption P_f is the product of two finite probabilities summed over all possible outcomes. Both probabilities are associated with an event and a random variable.

The first event is defined by the random variable L taking on a value in the range:

$$\left(x - \frac{dx}{2}\right) \leq L \leq \left(x + \frac{dx}{2}\right) \qquad \text{(Eq 3)}$$

The probability associated with this event is the area under the probability density function (PDF) for the load random variable (f_L) over this interval, i.e.,

$$P_1 = f_L(x)\, dx \qquad \text{(Eq 4)}$$

The second event is associated with the probability that the random variable R is less than or equal to x. This is the area under the probability density function for the resistance random variable (f_R) from $-\infty$ (or an appropriate lower limit defined by the range of the resistance random variable) to x. This second probability is given by the cumulative distribution function (CDF) for the resistance random variable (F_R) evaluated at x, that is:

$$P_2 = F_R(x) \qquad \text{(Eq 5)}$$

With the probability of failure defined as the product of these two probabilities, summed over all possible values of x, then:

$$P_f = P_1 P_2 = \int_{-\infty}^{+\infty} F_R(x) f_L(x) dx \quad \text{(Eq 6)}$$

To interpret this integral expression, consider the graphs in Fig. 1. In this figure, the graph of an arbitrary PDF for the resistance random variable is superimposed on the graph of an arbitrary PDF for the load random variable. Note that R and L must have the same dimensional units (e.g., force or stress) to superimpose their graphs in the same figure. A common misconception is that P_f is the area of overlap encompassed by the two probability density functions. Scrutiny of Eq 6 leads to the appropriate conclusion that the probability of failure is really the area under the composite function:

$$g_{RL}(x) = F_R(x) f_L(x) \quad \text{(Eq 7)}$$

which is also illustrated in Fig. 1.

Next, consider the situation where the load random variable has very little scatter relative to the resistance random variable. For example, if a number of test specimens were fabricated from a brittle material (a monolithic ceramic), the ultimate tensile strength can easily vary by more than 100%. That is, the highest strength value in the group tested can easily be twice as large as the lowest value. Variations of this magnitude are not typical for the load design variable, and the engineer could easily conclude that load is a deterministic design variable while strength is a random design variable. This assumption can be accommodated in this development by allowing the PDF for the load random variable to be defined by the expression:

$$f_L(x) = \delta(x - x_o) \quad \text{(Eq 8)}$$

Here δ is the Dirac delta function defined as:

$$\delta(x - x_o) = \begin{cases} \infty & x = x_o \\ 0 & x \neq x_o \end{cases} \quad \text{(Eq 9)}$$

Note that the Dirac delta function satisfies the definition for a PDF; that is, the area under the curve is

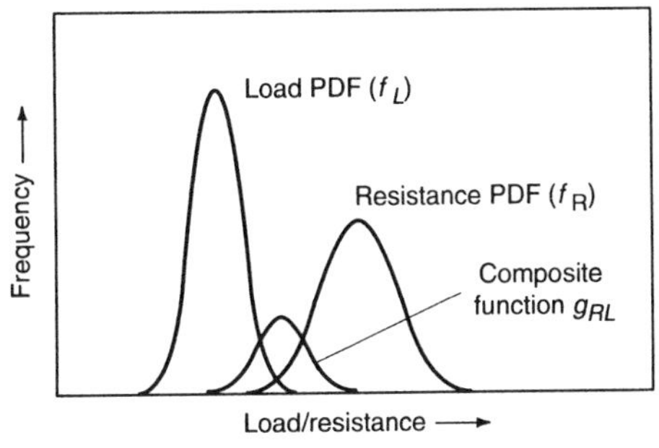

Fig. 1 Interference plot for load and resistance random variables

equal to 1, and the function is greater than or equal to 0 for all values of x. The Dirac delta function represents the scenario where the standard deviation of a random variable approaches 0 in the limit, and the random variable takes on a single value, that is, the central value identified here as x_o. Because the Dirac delta function is being used to represent the load random variable, then x_o represents the deterministic magnitude of the applied load. Keep in mind that the applied load can have units of force or stress. However, load and resistance are commonly represented with units of stress. Thus x_o is replaced with σ, an applied stress, and the probability of failure is given by the expression:

$$P_f = \int_{-\infty}^{+\infty} F_R(x)\delta(x - \sigma) dx \quad \text{(Eq 10)}$$

However, with the Dirac delta function embedded in the integral expression, the probability of failure simplifies to:

$$P_f = F_R(\sigma) \quad \text{(Eq 11)}$$

Thus the probability of failure is equal to the CDF of the resistance random variable evaluated at the applied load, σ. The use of the Dirac delta function in representing the load design variable provides justification for the use of the Weibull CDF (or a similarly skewed distribution) in quantifying the probability of failure for components fabricated from ceramics or glass.

System Reliability

A unique property of most brittle materials is an apparent decrease in tensile strength as the size of the component increases. This is the so-called size effect. As an example, consider a simple component such as a uniaxial tensile specimen. Now suppose that two groups of these simple components have been fabricated. Each group is identical with the exception that the size of the specimens in the first group is uniformly smaller than the specimens in the second group. The mean sample strength from the first group would be consistently and distinctly larger in a manner that cannot be accounted for by randomness. Thus Eq 11 must be transformed in some manner to admit a size dependence. This is accomplished through the use of system reliability concepts. (See the article "Reliability in Design" in this Volume for details on formulating the basic equations for system reliability.) After the following discussion the reader should be cognizant that the expression given in Eq 11 represents the probability of failure for a specified set of boundary conditions. If the boundary conditions are modified in any way, Eq 11 is no longer valid. To account for size effects and to deal with the probability of failure for a component in a general manner, the component should be treated as a system, and the focus must be directed on the probability of failure of the system.

The typical approach to designing structural components with varying stress fields involves discretizing the component in order to characterize the stress field using finite element methods. Because component failure may initiate in any of the discrete elements, it is convenient to consider a component as a system and utilize system reliability theories. A component is a series system if it fails when one discrete element fails. This type of failure can be modeled using weakest-link reliability theories. A component is a parallel system when failure of a single element does not cause the component to fail. In this case, the remaining elements sustain load through redistribution. This type of failure can be modeled with what has been referred to in the literature as "bundle theories." Weakest-link theories and bundle theories represent the extremes of failure behavior modeled by reliability analysis. They suggest more complex systems such as "r out of n" systems. Here a component (system) of n elements functions if at least r elements have not failed. This type of system model has not found widespread application in structural reliability analysis. The assumption in this article is that the failure behavior of the brittle materials is sudden and catastrophic. This type of behavior fits within the description of a series system, thus a weakest-link reliability system is adopted.

Now the probability of failure of a discrete element must be related to the overall probability of failure of the component. If the failure of an individual element is considered a statistical event, and if these events are independent, then the probability of failure of a discretized component that acts as a series system is given by the expression:

$$P_f = 1 - \prod_{i=1}^{N} (1 - p_i) \quad \text{(Eq 12)}$$

where N is the number of finite elements for a given component analysis. Here p_i is the probability of failure of the ith discrete element.

In the next section an expression is specified for the probability of failure (or alternatively, the reliability) of the ith discrete element for a simplified state of stress, that is, a uniaxial tensile stress. This expression allows the introduction of size scaling. Once size-scaling relationships are established for a simple state of stress, the relationships are extended to multiaxial states of stress.

Two-Parameter Weibull Distribution and Size Effects

In the ceramic and glass industry the Weibull distribution is universally accepted as the distribution of choice in representing the underlying PDF for tensile strength. A two-parameter formulation and a three-parameter formulation are available for the Weibull distribution. However, the two-parameter formulation usually leads to a more conservative estimate for the component probability of failure. The two-parameter Weibull PDF for a continuous random strength variable, denoted as Σ, is given by the expression:

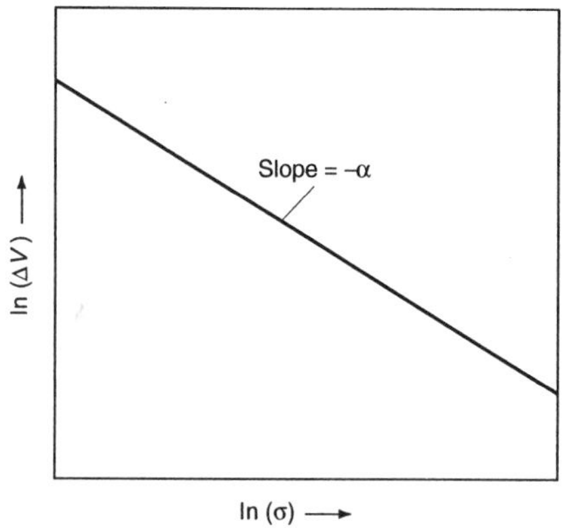

Fig. 2 Specimen gage volume plotted as a function of failure stress

$$f_\Sigma(\sigma) = \left(\frac{\alpha}{\beta}\right)\left(\frac{\sigma}{\beta}\right)^{(\alpha-1)} \exp\left[-\left(\frac{\sigma}{\beta}\right)^{\alpha}\right] \quad \text{(Eq 13)}$$

for $\sigma > 0$, and

$$f_\Sigma(\sigma) = 0 \quad \text{(Eq 14)}$$

for $\sigma \le 0$. The cumulative distribution is given by the expression:

$$F_\Sigma(\sigma) = 1 - \exp\left[-\left(\frac{\sigma}{\beta}\right)^{\alpha}\right] \quad \text{(Eq 15)}$$

for $\sigma > 0$, and

$$F_\Sigma(\sigma) = 0 \quad \text{(Eq 16)}$$

for $\sigma \le 0$. Here α (a scatter parameter, or Weibull modulus) and β (a central location parameter, or typically referred to as the Weibull scale parameter) are distribution parameters that define the Weibull distribution in much the same way as the mean (a central location parameter) and standard deviation (a scatter parameter) are parameters that define the Gaussian (normal) distribution. Note that in the ceramics and glass literature when the two-parameter Weibull formulation is adopted then "m" is used for the Weibull modulus α, and either σ_0 or σ_θ (see the discussion in the parameter estimation section regarding the difference between σ_0 and σ_θ) is used for the Weibull scale parameter. In this article, the (α, β) notation is used exclusively and reference is made to the typical notation adopted in the ceramics literature. The reason for this is the tendency to overuse the "σ" symbol (e.g., σ_θ, σ_0, σ_i-failure observation, and σ_t-threshold stress, etc.). Throughout this discussion the symbol "σ" implies applied stress.

If the random variable representing uniaxial tensile strength of an advanced ceramic is characterized by a two-parameter Weibull distribution, that is, the random strength parameter is governed by Eq 13 and 14, then the probability that a uniaxial test specimen fabricated from an advanced ceramic will fail can be expressed by the CDF:

$$P_f = 1 - \exp\left[-\left(\frac{\sigma_{max}}{\beta_\theta}\right)^{\alpha}\right] \quad \text{(Eq 17)}$$

Note that σ_{max} is the maximum normal stress in the component. When used in the context of characterizing the strength of ceramics and glasses, the central location parameter is referred to as the Weibull characteristic strength (β_θ). In the ceramic literature, this parameter can either be identified as the Weibull characteristic strength or the Weibull scale parameter. Because tensile strength is the random variable of interest, this parameter is referred to as a strength parameter throughout the rest of this article. The characteristic strength is dependent on the uniaxial test specimen (tensile, flexural, pressurized ring, etc.) used to generate the failure data. For a given material, this parameter will change in magnitude with specimen geometry (the so-called size effect alluded to earlier). The Weibull characteristic strength typically has units of stress. The scatter parameter α is dimensionless.

With the tensile strength characterized by the two-parameter Weibull distribution, the discussion returns to the weakest-link expression for component probability of failure defined by Eq 12. Let $\Re_i$ represent the reliability of the ith continuum element where:

$$\Re_i = 1 - p_i \quad \text{(Eq 18)}$$

The reliability of this continuum element is then governed by the following expression:

$$\Re_i = \exp\left(-\left(\frac{\sigma}{\beta_0}\right)^{\alpha} \Delta V\right) \quad \text{(Eq 19)}$$

where σ is the principal tensile stress applied to the continuum element. The volume of this arbitrary continuum element is identified by ΔV. In this expression β_0 is the Weibull material scale parameter and can be described as the Weibull characteristic strength of a specimen with unit volume loaded in uniform uniaxial tension. This is a material specific parameter that is utilized in the component reliability analyses that follow. The dimensions of this parameter are stress × (volume)$^{1/\alpha}$.

The requisite size scaling discussed earlier is introduced by Eq 19. To demonstrate this, take the natural logarithm of Eq 19 twice, that is:

$$\ln\ln(\Re_i) = \ln\left(-\left(\frac{\sigma}{\beta_0}\right)^{\alpha} \Delta V\right) \quad \text{(Eq 20)}$$

Manipulation of Eq 20 yields:

$$\ln(\Delta V) = -\alpha \ln(\sigma) + \ln\ln(\Re_i) + \alpha \ln(-\beta_0) \quad \text{(Eq 21)}$$

with

$$y = \ln(\Delta V) \quad \text{(Eq 22)}$$

$$x = \ln(\sigma) \quad \text{(Eq 23)}$$

$$m = -\alpha \quad \text{(Eq 24)}$$

and

$$b = \ln\ln(\Re_i) + \alpha \ln(-\beta_0) \quad \text{(Eq 25)}$$

then it is apparent that Eq 21 has the form of a straight line, that is, $y = mx + b$.

Once again consider the two groups of test specimens fabricated from the same material mentioned at the beginning of the section on system reliability. Recall that the specimens in each group are identical with each other, but the two groups have different gage sections such that ΔV (which is now identified as the gage section volume) is different for each group. Estimate Weibull parameters α and β_0 from the failure data obtained from either group (parameter estimation is discussed in detail in a following section). After the Weibull parameters are estimated the straight line in Fig. 2 is located by setting $\Re_i$ equal to 0.5 (i.e., the 50th percentile) in Eq 21. This value for $\Re_i$ should establish a line that correlates well with the median values in each group.

Now return to the data sets mentioned above and establish the stress value associated with the median in each group. Plot the gage volumes (ΔV) of each group as a function of the median stress values in Fig. 2. If no size effect is present, the median failure strengths of the groups will fall close to a horizontal line. This would indicate no correlation between gage volume and the median strength value. Keep in mind that the discussion here could proceed using any percentile value, not just the 50th percentile. A systematic variation away from a horizontal line indicates a size effect exists that must be considered in engineering design. If the median values for each group follows the trend indicated by the solid line in Fig. 2 the design engineer should have no apprehensions using Weibull analysis with size scaling. Figures 1 and 3 in Ref 1 are two excellent examples of these types of graphs with actual data.

The ability to account for size effects of individual elements is introduced through the expression for $\Re_i$ given by Eq 19. A rational approach for justifying this expression is outlined above. Now a general expression for the probability of failure for a component (i.e., a general form for Eq 17) is derived based on Eq 19. Under the assumptions that the component consists of an infinite number of elements (i.e., the continuum assumption) and that the component is best represented by a series system, then:

$$P_f = 1 - \lim_{k\to\infty}\left(\prod_{i=1}^{k} \Re_i\right) \quad \text{(Eq 26)}$$

Substituting for $\Re_i$ yields:

$$P_f = 1 - \exp\left(-\lim_{k\to\infty}\sum_{i=1}^{k}\left(\left(\frac{\sigma}{\beta_0}\right)^{\alpha}\Delta V\right)_i\right) \quad \text{(Eq 27)}$$

Here ΔV once again represents the volume of an element. The limit inside the bracket is a Riemann sum. Thus:

$$P_f = 1 - \exp\left[-\int\left(\frac{\sigma}{\beta_0}\right)^{\alpha} dV\right] \quad \text{(Eq 28)}$$

Weibull (Ref 2) first proposed this integral representation for the probability of failure. The expression is integrated over all tensile regions of the specimen volume if the strength-controlling flaws are randomly distributed through the volume of the material, or over all tensile regions of the specimen area if flaws are restricted to the specimen surface. For failures caused by surface defects, the probability of failure is given by the expression:

$$P_f = 1 - \exp\left[-\int\left(\frac{\sigma}{\beta_0}\right)^{\alpha} dA\right] \quad \text{(Eq 29)}$$

which is derived in a manner similar to Eq 28. The segregation of defect populations into volume and surface-distributed flaws hints at the possibility of multiple defect populations. Reference 3 presents an in-depth treatment of this topic as it relates to ceramic materials.

The Weibull material scale parameter β_0 has units of stress × $(\text{volume})^{1/\alpha}$. If the strength-controlling flaws are restricted to the surface of the specimens in a sample, then the Weibull material scale parameter has units of stress × $(\text{area})^{1/\alpha}$. For a given specimen geometry, Eq 17 and 28 can be equated, yielding an expression relating β_0 and β_θ. Methods for converting β_θ to an equivalent β_0 value are addressed in ASTM Standard Practice C 1239-95.

Three-Parameter Weibull Distribution

The three-parameter Weibull PDF for a continuous random strength variable, denoted as Σ, is given by the expression:

$$f_\Sigma(\sigma) = \left(\frac{\alpha}{\beta}\right)\left(\frac{\sigma-\gamma}{\beta}\right)^{(\alpha-1)} \exp\left[-\left(\frac{\sigma-\gamma}{\beta}\right)^{\alpha}\right] \quad \text{(Eq 30)}$$

for $\sigma > \gamma$, and

$$f_\Sigma(\sigma) = 0 \quad \text{(Eq 31)}$$

for $\sigma \le \gamma$. In Eq 30 α is once again the Weibull modulus (or the shape parameter), β is the Weibull scale parameter, and γ is a threshold parameter. The cumulative distribution is given by the expression

$$F_\Sigma(\sigma) = 1 - \exp\left[-\left(\frac{\sigma-\gamma}{\beta}\right)^{\alpha}\right] \quad \text{(Eq 32)}$$

for $\sigma > \gamma$, and

$$F_\Sigma(\sigma) = 0 \quad \text{(Eq 33)}$$

for $\sigma \le \gamma$. The same reasoning presented in the previous section on size scaling utilizing a two-parameter formulation can be applied using the three-parameter formulation. The resulting expression for the probability of failure of a component subjected to a single applied stress σ is:

$$P_f = 1 - \exp\left[-\int\left(\frac{\sigma-\gamma}{\beta_0}\right)^{\alpha} dV\right] \quad \text{(Eq 34)}$$

if the defect population is spatially distributed throughout the volume. A similar expression exists for failures caused by area defects. The focus of the discussion in the next section turns to accommodating multiaxial stress states in Eq 28 and Eq 34. This involves the development of multiaxial reliability models.

The approach outlined in this section and previous sections to account for the scatter in failure strength and the size effect of brittle materials was first introduced by Weibull (Ref 2 and 4). The concepts were based on the principles of weakest-link theory presented earlier. A number of authors including Pierce (Ref 5), Kontorova (Ref 6), as well as Frenkel and Kontorova (Ref 7) have made contributions in this area. In fact, Pierce first proposed the weakest-link concept while modeling yarn failure. However, Pierce assumed a Gaussian distribution for the strength random variable of yarn, and Weibull developed the unique PDF for his work that now bears his name. Hu (Ref 8) explored the difficulties associated with parameter estimation when a Gaussian or log normal distribution is adopted for the strength random variable. Shih (Ref 9) has shown that the three-parameter Weibull distribution is a more accurate approximation of brittle material behavior (specifically monolithic ceramics) than the Gaussian or other distributions. However, most analyses incorporate a two-parameter Weibull PDF where the threshold stress (the value of applied stress below which the failure probability is 0) is taken as 0. The reliability predictions obtained using the two-parameter function are more conservative than those obtained with the three-parameter model.

Multiaxial Reliability Models

Over the years a number of reliability models have been presented that extend the uniaxial format of Eq 28 and 34 to multiaxial states of stress. Only models associated with isotropic brittle materials are presented here. Anisotropic reliability models are beyond the scope of this article. References 10 and 11 contain information pertaining to reliability models for brittle composites. The monolithic models highlighted here include the principle of independent action (PIA) model, the normal stress averaging (NSA) model, and Batdorf's model. A brief discussion is presented for each. A detailed development is omitted for the sake of brevity. In order to simplify the presentation of each model, recast Eq 28 as:

$$P_f = 1 - \exp\left[-\int \psi\, dV\right] \quad \text{(Eq 35)}$$

where ψ is identified as a failure function per unit volume. What remains is the specification of the failure function ψ for each reliability model.

Phenomenological Models (NSA and PIA). To predict the time-independent (also referred to as fast-fracture) material response under multiaxial stress states Weibull (Ref 2) proposed calculating a failure function per unit volume (Weibull identified the function as the risk of rupture) by averaging the tensile normal stress raised to an exponent in all directions over the area of a unit radius sphere for volume flaws. This is known as the NSA model where:

$$\psi = k\,\overline{\sigma_n^{\alpha}} \quad \text{(Eq 36)}$$

where

$$\overline{\sigma_n^{\alpha}} = \frac{\int_A \sigma_n^{\alpha}\, dA}{\int_A dA} \quad \text{(Eq 37)}$$

and

$$k = \frac{(2\alpha+1)}{(\beta_0)^{\alpha}} \quad \text{(Eq 38)}$$

The area integration in Eq 37 is performed over the region of a unit sphere where σ_n (the Cauchy normal stress) is tensile. The reader is directed to Ref 12 for an in-depth explanation of the constants appearing in the equations above. Gross (Ref 13) demonstrated that for surface flaws this same averaging technique can be executed over the contour of a circle with a unit radius. Although the surface flaw technique is intuitively plausible for the NSA model, the approach is somewhat arbitrary. In addition, it lacks a closed-form solution, and therefore requires computationally intensive numerical modeling.

Barnett et al. (Ref 14) and Freudenthal (Ref 15) proposed an alternative approach usually referred to as the PIA model. Here:

$$\psi = \left(\frac{\sigma_1}{\beta_0}\right)^{\alpha} + \left(\frac{\sigma_2}{\beta_0}\right)^{\alpha} + \left(\frac{\sigma_3}{\beta_0}\right)^{\alpha} \quad \text{(Eq 39)}$$

where σ_1, σ_2, and σ_3 are the three principal stresses at a given point. The PIA model is the probabilistic equivalent to the deterministic maximum stress failure theory.

The NSA model, and in particular the PIA model, have been widely applied in brittle material design. The reader is directed to Ref 16 to 20 for a more in-depth development and discussion of the merits of these two models. Historically, the NSA and the PIA models have been popular methods for multiaxial stress state analysis. However, the NSA and PIA models are in essence phenomenological and do not specify the nature of the defect causing failure. As a consequence, no foundation exists for extrapolating predictions made by these models to conditions different from the original test conditions. Other models that are rooted in the principles of fracture mechanics are discussed in the next section.

Batdorf's Theory—Mechanistic Model. The concepts proposed by Batdorf (Ref 21), and later refined by Batdorf and Crose (Ref 22), are important in that the approach incorporates a mechanistic basis for the effect of multiaxial states of stress into the weakest-link theory. Here material defects distributed throughout the volume (and/or over the surface) are assumed to have a random orientation. In addition, the defects are assumed to be noninteracting discontinuities (cracks) with an assumed regular geometry. Failure is assumed to occur when a far-field effective stress associated with the weakest flaw reaches a critical level. The effective stress is a predefined combination of the far-field normal stress and the far-field shear stress. It is also a function of the assumed crack configuration, the existing stress state, and the fracture criterion employed (hence the claim that the approach captures the physics of fracture). Accounting for the presence of a far-field shear stress reduces the far-field normal stress needed for fracture. This model is identified by taking:

$$\psi = \alpha\, k_B \int_0^{(\sigma_e)_{max}} \frac{\Phi\,(\Sigma, \sigma_{cr})}{4\pi}\, \sigma_{cr}^{\alpha-1}\, d\sigma_{cr} \qquad \text{(Eq 40)}$$

where Φ is a solid angle that is dependent on the fracture criterion selected, the crack configuration, and the applied stress state. The maximum effective stress $(\sigma_e)_{max}$ is defined as an equivalent mode I fracture stress for mixed-mode loading. The crack-density coefficient k_B is obtained from the following expression:

$$k_B = \frac{\eta_v\,(\sigma_{cr})}{(\sigma_{cr})^{\alpha}} \qquad \text{(Eq 41)}$$

Here σ_{cr} is defined as the critical far-field normal stress for a given crack configuration under mode I loading. Once again Ref 12 can provide a detailed interpretation of the parameters appearing in Eq 40. For the most part, the Batdorf model yields more accurate reliability analyses than those produced by either the NSA or PIA models.

Numerous authors have discussed the stress distribution around cracks of various types under different loading conditions and proposed numerous criteria to describe impending fracture. Specifically, investigators such as Giovan and Sines (Ref 23), Batdorf (Ref 21), Stout and Petrovic (Ref 24), as well as Petrovic and Stout (Ref 25) have compared results from the most widely accepted mixed-mode fracture criteria with each other and with selected experimental data. The semiempirical equation developed by Palaniswamy and Knauss (Ref 26) and Shetty (Ref 27) seemingly provides enough flexibility to fit to experimental data. In addition, Shetty's criterion can account for the out-of-plane crack growth that is observed under mixed-mode loadings. However, several issues must be noted. No prevailing consensus has emerged regarding a best probabilistic fracture theory. Most of the available criteria predict somewhat similar results, despite the divergence of initial assumptions. Moreover, one must approach the mechanistic models with some caution. The reliability models based on fracture mechanics incorporate the assumptions made in developing the fracture models on which they are based. One of the fundamental assumptions made in the derivation of fracture mechanics criteria is that the crack length is much larger than the characteristic length of the microstructure. This is sometimes referred to as the continuum principle in engineering mechanics. For the brittle materials discussed here, that characteristic length is the grain size (or diameter). If one contemplates the fact that most brittle materials are high strength with an attending low fracture toughness, then the critical defect size can be quite small. If the critical defect size approaches the grain size of the material, then the phenomenological models discussed above may be more appropriate than the mechanistic models.

Parameter Estimation

As indicated earlier, the distribution of choice for characterizing the tensile strength of brittle materials is the Weibull distribution. One fundamental reason for this choice goes beyond the fact that the Weibull distribution usually provides a good fit to the data. While the log-normal distribution often provides an adequate fit, it precludes any accounting of size effects. Reference 8 provides a detailed discussion on this matter. As it turns out, once a conscious choice is made to utilize the Weibull distribution, Eq 17 provides a convenient formulation for parameter estimation. However, one cannot extract the fundamental distribution parameters needed for general component analysis from this expression, unless the test specimen has the same geometry and applied loads as the component. The fundamental distribution parameters (identified previously as material specific parameters) were embedded in Eq 28. Thus, together Eq 17 and 28 provide a convenient method for extracting material specific parameters from failure data.

Tensile strength measurements are taken for one of two reasons: either for a comparison of the relative quality of two materials or for the prediction of the failure probability for a structural component. The latter is the focus of this article, although the analytical details provided here allow for either. To obtain point estimates of the unknown Weibull distribution parameters, well-defined functions are utilized that incorporate the failure data and specimen geometry. These functions are referred to as estimators. It is desirable that an estimator be consistent and efficient. In addition, the estimator should produce unique, unbiased estimates of the distribution parameters. Different types of estimators exist, including: moment estimators, least squares estimators, and maximum likelihood estimators. This discussion initially focuses on maximum likelihood estimators (MLE) due to the efficiency and the ease of application when censored failure populations are encountered. The likelihood estimators are used to compute parameters from failure populations characterized by a two-parameter Weibull distribution. Alternatively, nonlinear regression estimators (discussed later) are utilized to calculate unknown distribution parameters for a three-parameter Weibull distribution.

Many factors affect the estimates of the distribution parameters. The total number of test specimens plays a significant role. Initially, the uncertainty associated with parameter estimates decreases significantly as the number of test specimens increases. However, a point of diminishing returns occurs when the cost associated with performing additional strength tests may not be justified by improvements in the estimated values of the distribution parameters. This suggests that a practical number of strength tests should be performed to obtain a desired level of confidence associated with a parameter estimate. This point cannot be overemphasized. However, quite often 30 specimens (a widely cited rule-of-thumb) is deemed a sufficient quantity of test specimens when estimating Weibull parameters. One should immediately ask why 29 specimens would not suffice. Or more importantly, why is 30 specimens sufficient? The answer to this is addressed in ASTM Standard Practice C 1239-95 where the details of computing confidence bounds for the maximum likelihood estimates (these bounds are directly related to the precision of the estimate) are presented. Duffy et al. (Ref 28) discusses the reasons why these same confidence bounds are not available for the nonlinear regression estimators.

Tensile and flexural specimens are the most commonly used test configurations in determining ultimate strength values for brittle materials. However, as noted earlier, most brittle material systems exhibit a decreasing trend in material strength as the test specimen geometry is increased. Thus, the observed strength values are dependent on specimen size and geometry. Parameter estimates can be computed based on a given specimen geometry; however, the parameter estimates should be transformed and utilized in a component reliability analysis as material-specific parameters. The procedure for transforming parameter estimates for the typical specimen geometries just cited is outlined in ASTM Standard Practice C 1239-95. The reader should be aware that the parameters estimated using nonlinear regression estimators are material-specific parameters. Therefore, no transformation is necessary after these parameters have been estimated.

Brittle materials can easily contain two or more active flaw distributions (e.g., failures due to inclusions or machining damage) and each will have its own strength distribution parameters. The censoring techniques for the two-parameter Weibull distribution require positive confirmation of multiple-flaw distributions, which neces-

Table 1 Alumina fracture stress data

Specimen No.	Stress, MPa	Specimen No.	Stress, MPa	Specimen No.	Stress, MPa
1	307	13	347	25	376
2	308	14	350	26	376
3	322	15	352	27	381
4	328	16	353	28	385
5	328	17	355	29	388
6	329	18	356	30	395
7	331	19	357	31	402
8	332	20	364	32	411
9	335	21	371	33	413
10	337	22	373	34	415
11	343	23	374	35	456
12	345	24	375		

sitates fractographic examination to characterize the fracture origin in each specimen. Multiple-flaw distributions may also be indicated by a deviation from the linearity of the data from a single Weibull distribution (see Fig. 3). However, observations of approximately linear behavior should not be considered a sufficient reason to conclude a single flaw distribution is active. The reader is strongly encouraged to integrate mechanical failure data and fractographic analysis.

As was just noted, discrete fracture origins are quite often grouped by flaw distributions. The data for each flaw distribution can also be screened for outliers. An outlying observation is one that deviates significantly from other observations in the sample. However, an apparent outlying observation may be an extreme manifestation of the variability in strength. If this is the case, the data point should be retained and treated as any other observation in the failure sample. Yet the outlying observation can be the result of a gross deviation from prescribed experimental procedure, or possibly an error in calculating or recording the numerical value of the data point in question. When the experimentalist is clearly aware that either of these situations has occurred, the outlying observation may be discarded, unless the observation (i.e., the strength value) can be corrected in a rational manner. For the sake of brevity, this discussion omits any discussion on the performance of fractographic analyses and omits any discussion concerning outlier tests.

Two-Parameter MLEs. With the above discussion serving as background, attention is now focused on obtaining estimated values of the Weibull parameters α and β_θ. This discussion focuses on MLEs because of their efficiency and ease of application when censored failure populations are encountered. When a sample containing ultimate strength observations yields two or more distinct flaw distributions, the sample is said to contain censored data. The maximum likelihood methodology accounts for censored data in a rational, straightforward manner. Other estimation techniques (specifically linear regression estimators) must appeal to ad hoc reranking schemes in the presence of censored data.

Johnson and Tucker (Ref 1), as well as others, have shown that the MLE method is more efficient in estimating parameters. Here, efficiency is measured through the use of confidence bounds.

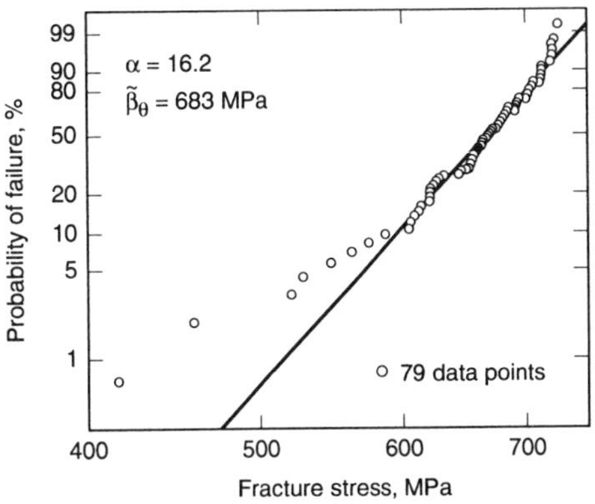

Fig. 3 Sample with multiple failure populations

For an equivalent confidence level, the authors of these works have demonstrated that the confidence bounds for an MLE is always smaller than the confidence bound obtained using linear regression. For this reason the likelihood estimators should be used to compute parameters from failure populations characterized by a two-parameter Weibull distribution.

The parameter estimates obtained using the maximum likelihood technique are unique (for a two-parameter Weibull distribution), and as the size of the sample increases, the estimates statistically approach the expected values of the true population parameter. Let $\sigma_1, \sigma_2, \ldots, \sigma_N$ represent realizations of the ultimate tensile strength (a random variable) in a given sample, where it is assumed that the ultimate tensile strength is characterized by the two-parameter Weibull distribution. The likelihood function associated with this sample is the joint probability density of the N random variables and thus is a function of the unknown Weibull distribution parameters (α, β_θ). The likelihood function for an uncensored sample under these assumptions is given by the expression:

$$L = \prod_{i=1}^{N} \left(\frac{\tilde{\alpha}}{\tilde{\beta}_\theta}\right)\left(\frac{\sigma_i}{\tilde{\beta}_\theta}\right)^{\tilde{\alpha}-1} \exp\left[-\left(\frac{\sigma_i}{\tilde{\beta}_\theta}\right)^{\tilde{\alpha}}\right] \quad \text{(Eq 42)}$$

The parameter estimates (the Weibull modulus $\tilde{\alpha}$ and the characteristic strength $\tilde{\beta}_\theta$) are determined by taking the partial derivatives of the logarithm of the likelihood function with respect to $\tilde{\alpha}$ and $\tilde{\beta}_\theta$, and equating the resulting expressions to 0. Note that the tildes distinguish a parameter estimate from its corresponding true value. The system of equations obtained by differentiating the log likelihood function for a censored sample is given by:

$$\frac{\sum_{i=1}^{N} (\sigma_i)^{\tilde{\alpha}} \ln(\sigma_i)}{\sum_{i=1}^{N} (\sigma_i)^{\tilde{\alpha}}} - \frac{1}{N}\sum_{i=1}^{N} \ln(\sigma_i) - \frac{1}{\tilde{\alpha}} = 0 \quad \text{(Eq 43)}$$

and

$$\tilde{\beta}_\theta = \left[\left(\sum_{i=1}^{N} (\sigma_i)^{\tilde{\alpha}}\right)\frac{1}{N}\right]^{1/\tilde{\alpha}} \quad \text{(Eq 44)}$$

Equation 43 is solved numerically, because a closed-form solution for $\tilde{\alpha}$ cannot be obtained from this expression. Once $\tilde{\alpha}$ is determined this value is inserted into Eq 44 and $\tilde{\beta}_\theta$ is calculated directly. The

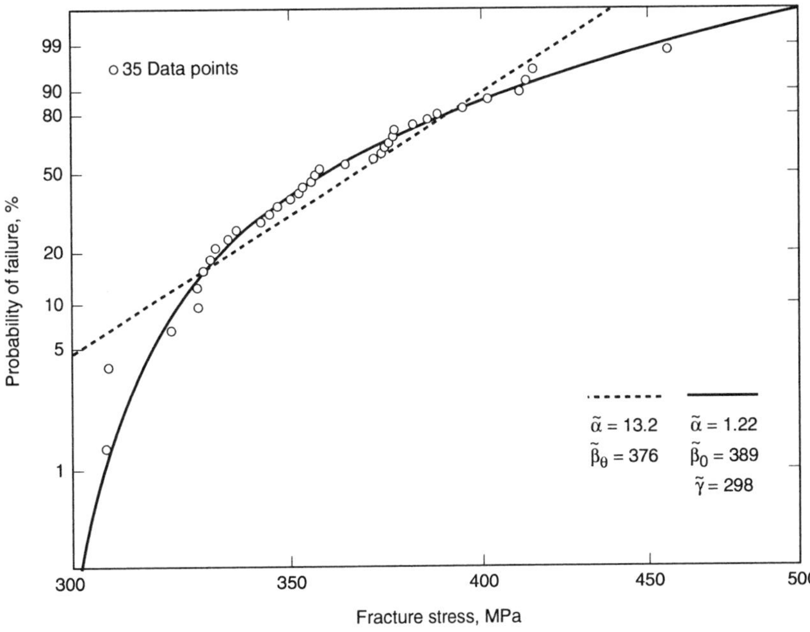

Fig. 4 Alumina failure data (see Table 1) and probability of failure curves based on estimated parameters for the two- and three-parameter Weibull distributions

reader is once again directed to ASTM Standard Practice C 1239-95 for the expressions corresponding to samples with censored data.

Three-Parameter Linear Regression. To date, most reliability analyses performed on structural components fabricated from ceramic materials have utilized the two-parameter form of the Weibull distribution. The use of a two-parameter Weibull distribution to characterize the random nature of material strength implies a nonzero probability of failure for the full range of applied tensile stress. This represents a conservative design assumption when analyzing structural components. The three-parameter form of the Weibull distribution was presented earlier in Eq 30 and 31. The additional parameter is a threshold stress (γ) that allows for zero probability of failure when the applied stress is at or below the threshold value. Certain monolithic ceramics have exhibited threshold behavior. The reader is directed to an extensive database assembled in Ref 29, the silicon nitride data in Ref 30, as well as data (with supporting fractography) presented in Ref 31 that was analyzed later in Ref 28.

When strength data indicates the existence of a threshold stress, a three-parameter Weibull distribution should be employed in the stochastic failure analysis of structural components. By employing the concept of a threshold stress, an engineer can effectively tailor the design of a component to optimize structural reliability. To illustrate the approach, Duffy et al. (Ref 28) embedded the three-parameter Weibull distribution in a reliability model that utilized PIA. Analysis of a space-shuttle main engine (SSME) turbopump blade predicted a substantial improvement in component reliability when the three-parameter Weibull distribution was utilized in place of the two-parameter Weibull distribution. Note that the three-parameter form of the Weibull distribution can easily be extended to Batdorf's (Ref 21, 22) model, reliability models proposed for ceramic composites by Duffy et al. (Ref 32), or Thomas and Wetherhold (Ref 33), as well as the interactive reliability models proposed by Palko (Ref 34).

The nonlinear regression method presented here was first proposed by Margetson and Cooper (Ref 35). However, these estimators maintain certain disadvantages relative to bias and invariance, and these issues were explored numerically in Ref 28. The Monte Carlo simulations in Ref 28 demonstrated that the functions proposed in Ref 35 are neither invariant nor unbiased. However, they are asymptotically well behaved in that bias decreases and confidence intervals contract as the sample size increases. Thus, even though bias and confidence bounds may never be quantified using these nonlinear regression techniques, the user is guaranteed that estimated values improve as the sample size is increased.

Regression analysis postulates a relationship between two variables. In an experiment typically one variable can be controlled (the independent variable), while the response variable (or dependent variable) is not. In simple failure experiments the material dictates the strength at failure, indicating that the failure stress is the response variable. The ranked probability of failure (P_i) can be controlled by the experimentalist, because it is functionally dependent on the sample size (N). After arranging the observed failure stresses ($\sigma_1, \sigma_2, \sigma_3, \ldots, \sigma_N$) in ascending order, and specifying:

$$P_i = \frac{(i - 0.5)}{N} \tag{Eq 45}$$

then clearly the ranked probability of failure for a given stress level can be influenced by increasing or decreasing the sample size. The procedure proposed in Ref 35 adopts this philosophy. They assume that the specimen failure stress is the dependent variable, and the associated ranked probability of failure becomes the independent variable.

Using the three-parameter version of Eq 34, an expression can be obtained relating the ranked probability of failure (P_i), to an estimate of the failure strength ($\tilde{\sigma}_i$). Assuming uniaxial stress conditions in a test specimen with a unit volume, Eq 34 yields:

$$\tilde{\sigma}_i = \tilde{\gamma} + \tilde{\beta}_0 \left[\ln \left(\frac{1}{1 - P_i} \right) \right]^{1/\tilde{\alpha}} \tag{Eq 46}$$

where $\tilde{\alpha}$, $\tilde{\beta}_0$, and $\tilde{\gamma}$ are estimates of the shape parameter (α), the scale parameter (β_0), and threshold parameter (γ), respectively. Expressions for the evaluation of these parameters for a test specimen subjected to pure bending are found in Ref 28. Defining the residual as:

$$\delta_i = \tilde{\sigma}_i - \sigma_i \tag{Eq 47}$$

where σ_i is the ith ranked failure stress obtained from actual test data, then the sum of the squared residuals is expressed as:

$$\sum_{i=1}^{N} (\delta_i)^2 = \sum_{i=1}^{N} \left[\tilde{\gamma} + \tilde{\beta}_0 (W_i)^{1/\tilde{\alpha}} - \sigma_i \right]^2 \tag{Eq 48}$$

Here the notation of Ref 35 has been adopted where:

$$W_i = \ln \left(\frac{1}{1 - P_i} \right) \tag{Eq 49}$$

Note that the forms of $\tilde{\sigma}_i$ and W_i change with specimen geometry. This is discussed in more detail in Ref 28.

It should be apparent that the objective of this method is to obtain parameter estimates that minimize the sum of the squared residuals. Setting the partial derivatives of the sum of the squared residuals with respect to $\tilde{\alpha}$, $\tilde{\beta}_0$, and $\tilde{\gamma}$ equal to zero yields the following three expressions:

$$\tilde{\beta}_0 = \frac{N \left[\sum_{i=1}^{N} \sigma_i (W_i)^{1/\hat{\alpha}} \right] - \left[\sum_{i=1}^{N} \sigma_i \right] \left[\sum_{i=1}^{N} (W_i)^{1/\hat{\alpha}} \right]}{N \sum_{i=1}^{N} (W_i)^{2/\hat{\alpha}} - \left[\sum_{i=1}^{N} (W_i)^{1/\hat{\alpha}} \right] \left[\sum_{i=1}^{N} (W_i)^{1/\hat{\alpha}} \right]} \tag{Eq 50}$$

$$\tilde{\gamma} = \left\{ \left[\sum_{i=1}^{N} \sigma_i \right] \left[\sum_{i=1}^{N} (W_i)^{2/\hat{\alpha}} \right] - \left[\sum_{i=1}^{N} \sigma_i (W_i)^{1/\hat{\alpha}} \right] \left[\sum_{i=1}^{N} (W_i)^{1/\hat{\alpha}} \right] \right\} \times \left\{ N \sum_{i=1}^{N} (W_i)^{2/\hat{\alpha}} - \left[\sum_{i=1}^{N} (W_i)^{1/\hat{\alpha}} \right] \left[\sum_{i=1}^{N} (W_i)^{1/\hat{\alpha}} \right] \right\}^{-1} \tag{Eq 51}$$

and

$$\left| \sum_{i=1}^{N} \sigma_i (W_i)^{1/\tilde{\alpha}} \ln (W_i) - \tilde{\gamma} \sum_{i=1}^{N} \sigma_i (W_i)^{1/\tilde{\alpha}} \ln (W_i) - \tilde{\beta}_0 \sum_{i=1}^{N} \sigma_i (W_i)^{2/\tilde{\alpha}} \ln (W_i) \right| \leq \kappa_{conv} \tag{Eq 52}$$

in terms of the parameter estimates. The solution of this system of equations is iterative, where the third expression is used to check convergence of an iteration. The initial solution vector for this system is determined after assuming a convenient value for $\tilde{\alpha}$, say $\tilde{\alpha} = 1$. Then $\tilde{\beta}_0$ is computed from Eq 50 and $\tilde{\gamma}$ is calculated from Eq 51. The values of these parameter estimates are then inserted into Eq 52 to determine if the convergence criterion is satisfied to within some predetermined tolerance (κ_{conv}). If this expression is not satisfied, $\tilde{\alpha}$ is updated and a new iteration is conducted. This procedure continues until a set of parameter estimates is determined that satisfy Eq 52.

The estimators perform reasonably well in comparison to estimates of the two-parameter Weibull distribution for the alumina data found in Table 1. Figure 4 is a plot of probability of failure versus failure stress for this data. The straight line represents the two-parameter fit to the data where $\tilde{\alpha} = 143.2$, $\tilde{\beta}_\theta = 395$ ($\tilde{\gamma} \equiv 0$) using values from Ref 29 for the shape and scale parameters. The nonlinear curve represents the three-parameter fit to the data where $\tilde{\alpha} = 1.22$, $\tilde{\beta}_0 = 389$, and $\tilde{\gamma} = 298$. Note that the three-parameter distribution appears more efficient in predicting the failure data in the high-reliability region of the graph. This is a qualitative assessment. Goodness-of-fit statistics such as the Kolmogorov-Smirnov statistic, the Anderson-Darling statistic, and likelihood ratio tests could provide quantitative measures to establish which form of the Weibull distribution would best fit the experimental data. These statistics are utilized in conjunction with hypothesis testing to assess the significance level at which the null hypothesis can be rejected. Comparisons can then be made based on the value of the significance level.

Time-Independent Reliability Algorithms

After a reliability model has been adopted and the failure function ψ has been specified, the primary task is the evaluation of the integral given in Eq 35. Closed-form solutions exist for only the simplest of component geometries and boundary conditions. Therefore, integrated computer algorithms have been developed that enable the design engineer to predict the time-independent (fast-fracture) reliability of components subjected to thermomechanical loading. Two algorithms are discussed here. One algorithm has been developed at the NASA Lewis Research Center and has been given the acronym CARES (Ceramics Analysis and Reliability Evaluation of Structures). This algorithm is widely discussed in Ref 12 and 36 to 38. The second computer algorithm, given the acronym ERICA, was developed by AlliedSignal (Ref 39, 40) with funding provided by the U.S. Department of Energy. Both algorithms are discussed briefly, and design examples are illustrated.

CARES Algorithm. The NASA Lewis Research Center CARES algorithm couples commercially available finite element programs, such as MSC/NASTRAN, ANSYS, or ABAQUS, with the probabilistic design models discussed previously. The algorithm contains three software modules that:

- Perform parameter estimation using experimental data obtained from standard laboratory specimens
- Generate a neutral database from MSC/NASTRAN, ABAQUS, and ANSYS finite element results files
- Evaluate the reliability of thermomechanically loaded components

Heat-transfer and linear-elastic finite element analyses are used to determine the temperature field and stress field. The component reliability analysis module of CARES uses the thermoelastic or isothermal elastostatic results to calculate the time-independent reliability for each element using a specified reliability model. Each element can be made arbitrarily small, such that the stress field in an element can be approximated as constant throughout the element (or subelement). The algorithm is compatible with most (but not all) two-dimensional elements, three-dimensional elements, axisymmetric elements, and shell elements for the commercial finite element algorithms mentioned above. Reliability calculations are performed at the Gaussian integration points of the element or, optionally, at the element centroid. Using the element integration points enables the element to be divided into subelements, where integration point subvolumes, subareas, and subtemperatures are calculated. The location of the Gaussian integration point in the finite element and the corresponding weight functions are considered when the subelement volume and/or area is calculated. The number of subelements in each element depends on the integration order chosen and the element type. If the probability of survival for each element is assumed to be a mutually exclusive event, the overall component reliability is the product of all the calculated element (or subelement) survival probabilities. The CARES algorithm produces an optional PATRAN file containing risk-of-rupture intensities (a local measure of reliability) for graphical rendering of the critical regions of the structure.

ERICA Algorithm. Unlike CARES, the AlliedSignal algorithm ERICA has a software architecture with a single module. Currently, only one finite element program interface exists for the algorithm, that is, an interface with the ANSYS finite element program. Once again stress and temperature information from the solution of a discretized component are used in conjunction with a specified reliability model to assess component reliability. ERICA admits multiple flaw distributions that can be spatially distributed through the volume, along the surface, and along the edges of a component. Both isotropic material behavior, and to a limited extent, anisotropic material behavior (for surface calculations) are taken into account. This anisotropic surface option allows the user to account for various types of surface finish on a component (e.g., ground, as fired, etc.). The ERICA algorithm can function on any platform that supports ANSYS. A limited number of element types are supported that offer the user some flexibility in modeling a component. Note that neither CARES nor ERICA support a full suite of elements for any of the commercial finite element algorithms.

Time-Independent Design Examples

Reliability analyses are typically segregated into two categories: time-independent and time-dependent. This classification is rooted both in the historic development of the reliability models presented here and also in a practical approach to the analysis of a component. Yet in many instances, a component must perform in an adequate fashion over a predetermined service life. To accomplish this design goal, the component must survive the initial load cycle. Thus, the calculated time-independent reliability value is used as a screening criterion and can also be used as an initial value for the time-dependent analyses discussed later. A fundamental premise of probabilistic analysis dictates that if the reliability of a component varies with time then it should never exceed the initial value (unless there exists some physical mechanism such as flaw healing that can account for this phenomenon). Typically, materials deteriorate with time, and this assumption is incorporated throughout this chapter. From historical perspective, the authors simply point out that the time-independent models were developed first (hence they are presented first here). In addition, the time-independent approach has been rigorously exercised over the years. Extensive design experience and databases have been established prior to proposal of the time-dependent modeling efforts outlined later in this chapter.

Both the CARES and ERICA reliability algorithms have been used in the design and analysis of numerous structural components. Of the two, the NASA CARES algorithm has been more widely utilized for proprietary reasons. The CARES reliability algorithm has been used to design glass and ceramic parts for a wide range of applications. These include hot section components for turbine and internal combustion engines, bearings, laser windows on test rigs, radomes, radiant heater tubes, spacecraft activation valves and platforms, CRTs, rocket launcher tubes, and ceramic packaging for microprocessors. Illustrated below are some typical design and analysis applications that have utilized the CARES software. In the interest of brevity, a complete example problem cannot be included in

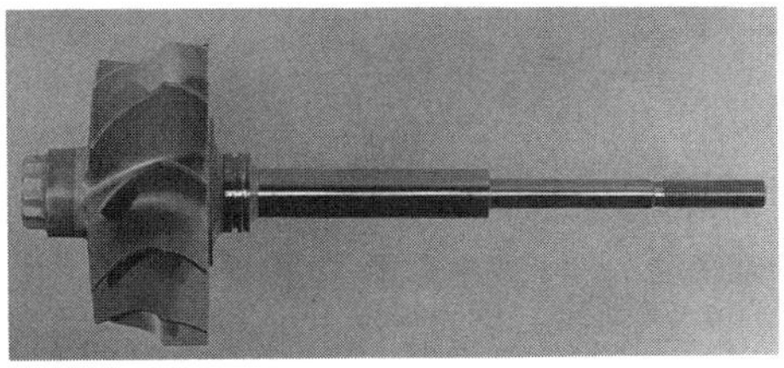

Fig. 5 Ceramic automotive turbocharger wheel. Courtesy of AlliedSignal Turbocharging and Truck Brake Systems

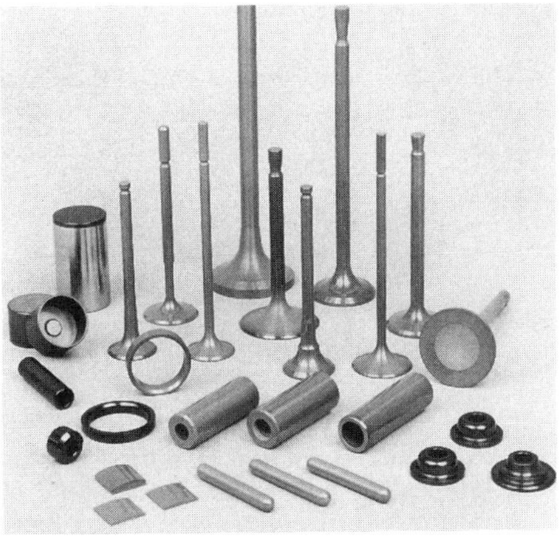

Fig. 6 Automotive valves and engine components. Courtesy of TRW Automotive Valve Division

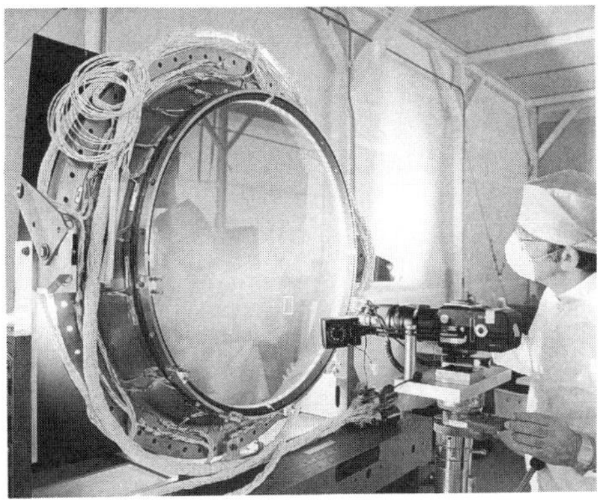

Fig. 7 The largest known ZnSe window manufactured for a cryogenic vacuum chamber. Courtesy of Hughes Danbury Optical Systems

this article. For a complete step-by-step procedure on conducting a time-independent component reliability analysis, the reader is directed to Ref 12.

The CARES algorithm has been successfully used in the development of ceramic turbocharger wheels (Ref 41). Specifically, the CARES algorithm was utilized to design the CTV7301 silicon nitride turbocharger rotor, depicted in Fig. 5, which was implemented in the Caterpillar 3406E diesel engine. The reduced rotational inertia of the silicon nitride ceramic rotor compared to a metallic rotor significantly enhanced the turbocharger transient performance and reduced emissions. Note that this was a joint effort involving AlliedSignal and Caterpillar and represents the first design and large-scale deployment of ceramic turbochargers in the United States. More than 1700 units have been supplied to Caterpillar Tractor Company for on-highway truck engines. These units together have accumulated a total of over 120 million miles of service.

Extensive work has been performed to analyze graphite and ceramic structural components such as high-temperature valves, test fixtures, and turbine wheels using CARES. A silicon nitride turbine wheel has been designed as a retrofit to replace components fabricated from Waspaloy in a military cartridge-mode air turbine starter (Ref 42). The silicon nitride component reduced cost and weight while increasing resistance to temperature, erosion, and corrosion.

The CARES algorithm has been used to analyze a ceramic-to-metal brazed joint for automotive gas turbine engines (Ref 43, 44). A major design hurdle in ceramic-to-metal joining is the thermal expansion mismatch between the two different materials. This results in high residual stresses that increase the likelihood of ceramic failure. One of the goals of this work was to improve the capability of the metal shaft to transmit power by reducing concentrated tensile stresses. The results confirmed the importance of probabilistic failure analysis for assessing the performance of various brazed joint designs.

A monolithic graphite spacecraft activation valve was designed (Ref 45) to direct reaction control gases for fine tuning the trajectory of a high-performance kinetic energy kill vehicle during the last 9 s of flight. Utilizing the CARES software, the valve was designed to withstand a gas pressure of 11.4 MPa (1.6 ksi) at 1930 °C (3506 °F).

A design study (Ref 46) demonstrated the viability of an uncooled silicon nitride combustor for commercial application in a 300 kW engine with a turbine inlet temperature of 1370 °C (2498 °F). Using the CARES algorithm, an analysis identified the most severe transient thermal stress in an emergency shutdown. The most critical area was found to be around the dilution port.

Ceramic poppet valves for spark ignition engines have been designed (Ref 47). These parts, depicted with other engine components in Fig. 6, have been field tested in passenger cars with excellent results. Potential advantages offered by these valves include reduced seat insert and valve guide wear, improved valve train dynamics, increased engine output, and reduced friction loss using lower spring loads.

The largest known zinc-selenide (ZnSe) containment window (depicted in Fig. 7) was designed using the CARES algorithm. The window formed a pressure barrier between a cryogenic vacuum chamber containing optical equipment and a sensor chamber. The window measured 79 cm (31 in.) in diameter by 2.5 cm (1 in.) thick and was used in a test facility for long-range infrared sensors.

The previous examples cited successful applications of the reliability algorithms in the design and analysis of commercial applications. In many cases, the algorithms have been an integral component of research and development efforts in government-supported programs. A specific example of this is the use of the CARES algorithm by participating organizations in the Advanced Turbine Technology Applications Program (ATTAP) to determine the reliability of structural component designs. The ATTAP program (Ref 48) is intended to advance the technological readiness of the ceramic automotive gas turbine engine. Structural ceramic components represent the greatest technical challenge facing the commercialization of such an engine and are thus the prime project focus. Cooperative efforts have been developed between industry, key national facilities, and academia to capitalize on the unique capabilities and facilities developed for ceramic materials characterization and processing technology. Figure 8 depicts engine components, including structural, combustion, regeneration, and insulation applications designed using the NASA-developed CARES software.

Life Prediction Using Reliability Analyses

The discussions in the previous sections assumed all failures were independent of time and history of previous thermomechanical loadings. However, as design protocols emerge for brittle material systems, designers must be aware of several innate characteristics exhibited by these materials. When subjected to elevated service temperatures, they exhibit complex thermomechanical behavior that is both inherently time dependent and hereditary in the sense that current behavior depends not only on current conditions, but also on thermomechanical history. The design engineer must also be cognizant that the ability of a component to sustain load degrades over time due to a variety of effects such as oxidation, creep, stress corrosion, and cyclic fatigue. Stress corrosion and cyclic fatigue result

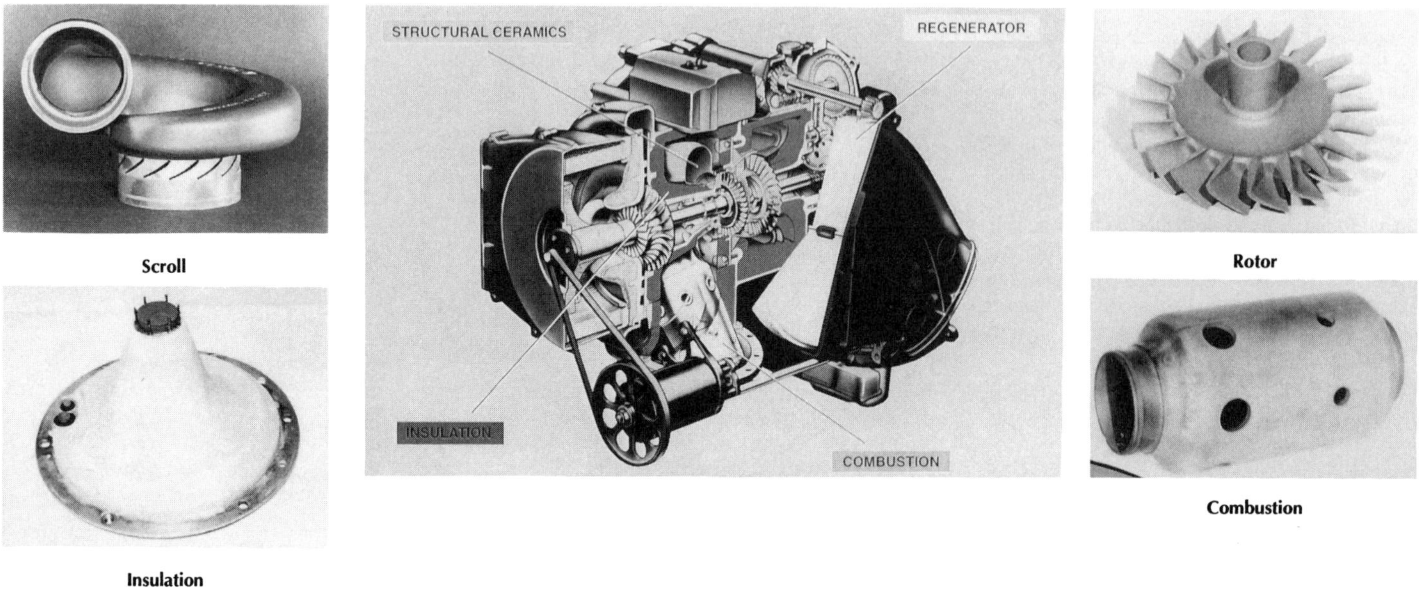

Fig. 8 Gas turbine engine and components. Scroll and rotor are made from structural ceramics. Courtesy of Allison Engine Company

in a phenomenon called subcritical crack growth (SCG). This failure mechanism initiates at a preexisting flaw and continues until a critical length is attained. At that point, the crack grows in an unstable fashion leading to catastrophic failure. The SCG failure mechanism is a time-dependent, load-induced phenomenon. Time-dependent crack growth can also be a function of chemical reaction, environment, debris wedging near the crack tip, and deterioration of bridging ligaments. Fracture mechanism maps, such as the one developed for ceramic materials (Ref 49) depicted in Fig. 9, help illustrate the relative contribution of various failure modes as a function of temperature and stress.

In addition to the determination of the Weibull shape and scale parameters discussed previously, analysis of time-dependent reliability in brittle materials necessitates accurate stress field information, as well as evaluation of distinct parameters reflecting material, microstructural, and/or environmental conditions. Predicted lifetime reliability of brittle material components depends on Weibull and fatigue parameters estimated from rupture data obtained from widely used tests involving flexural or tensile specimens. Fatigue parameter estimates are obtained from naturally flawed specimens ruptured under static (creep), cyclic, or dynamic (constant stress rate) loading. For other specimen geometries, a finite element model of the specimen is also required when estimating these parameters. For a more detailed discussion of time-dependent parameter estimation, the reader is directed to the *CARES/Life (CARES/Life Prediction Program) Users and Programmers Manual* (Ref 50). This information can then be combined with stochastic modeling approaches and incorporated into integrated design algorithms (computer software) in a manner similar to that presented previously for time-independent models. The theoretical concepts upon which these time-dependent algorithms have been constructed and the effects of time-dependent mechanisms, most notably subcritical crack growth and creep, are addressed in the remaining sections of this article.

Although it is not discussed in detail here, one approach to improve the confidence in component reliability predictions is to subject the component to proof testing prior to placing it in service. Ideally, the boundary conditions applied to a component under proof testing simulate those conditions the component would be subjected to in service, and the proof test loads are appropriately greater in magnitude over a fixed time interval. This form of testing eliminates the weakest components and, thus, truncates the tail of the strength distribution curve. After proof testing, surviving components can be placed in service with greater confidence in their integrity and a predictable minimum service life.

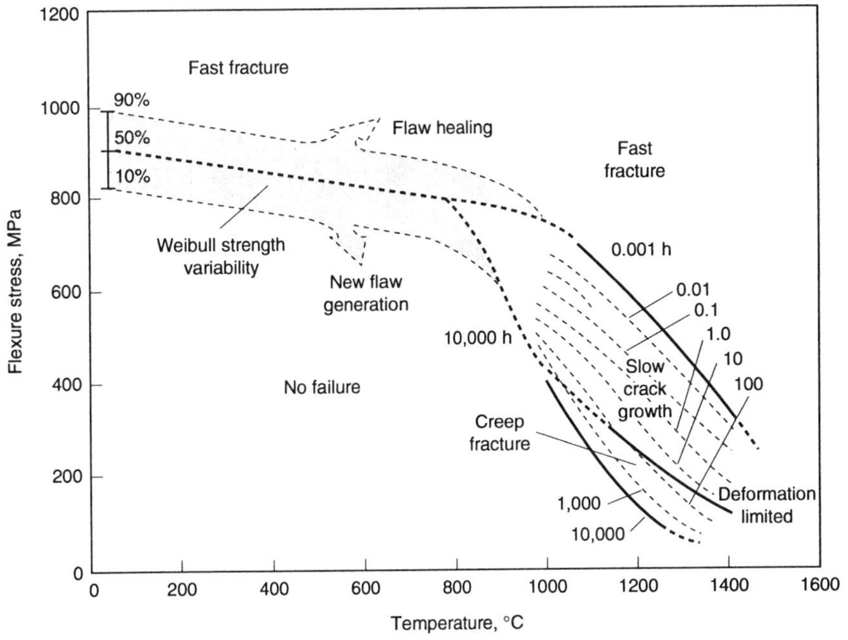

Fig. 9 Fracture mechanism map for hot-pressed silicon nitride flexure bars. Fracture mechanism maps help illustrate the relative contribution of various failure modes as a function of temperature and stress. Source: Ref 49

Need for Correct Stress State

With increasing use of brittle materials in high-temperature structural applications, the need arises to accurately predict thermomechanical behavior. Most current analytical methods for both subcritical crack growth and creep models use elastic stress fields in predicting the time-dependent reliability response of components subjected to elevated service temperatures. Inelastic response at high temperature has been well documented in the materials science literature for these material systems, but this issue has been ignored by the engineering design community. However, the authors wish to emphasize that accurate predictions of time-dependent reliability demand accurate stress-field information. From a design engineer's perspective, it is imperative that the inaccuracies of making time-dependent reliability predictions based on elastic stress fields are taken into consideration. This section addresses this issue by presenting a recent formulation of a viscoplastic constitutive theory to model the inelastic deformation behavior of brittle materials at high temperatures.

Early work in the field of metal plasticity indicated that inelastic deformations are essentially unaffected by hydrostatic stress. This is not the case for brittle (e.g., ceramic-based) material systems, unless the material is fully dense. The theory presented here allows for fully dense material behavior as a limiting case. In addition, as pointed out in Ref 51, these materials exhibit different time-dependent behavior in tension and compression. Thus, inelastic deformation models for these materials must be constructed in a manner that admits sensitivity to hydrostatic stress and differing behavior in tension and compression.

A number of constitutive theories for materials that exhibit sensitivity to the hydrostatic component of stress have been proposed that characterize deformation using time-independent classical plasticity as a foundation. Corapcioglu and Uz (Ref 52) reviewed several of these theories by focusing on the proposed form of the individual yield function. The review includes the works of Kuhn and Downey (Ref 53), Shima and Oyane (Ref 54), and Green (Ref 55). Not included is the work by Gurson (Ref 56), who not only developed a yield criteria and flow rule, but also discussed the role of void nucleation. Subsequent work by Mear and Hutchinson (Ref 57) extended Gurson's work to include kinematic hardening of the yield surfaces.

Although the previously mentioned theories admit a dependence on the hydrostatic component of stress, none of these theories allows different behavior in tension and compression. In addition, the aforementioned theories are somewhat lacking in that they are unable to capture creep, relaxation, and rate-sensitive phenomena exhibited by brittle materials at high temperature. Noted exceptions are the recent work by Ding et al. (Ref 58) and the work by White and Hazime (Ref 59). Another exception is an article by Liu et al. (Ref 60), which is an extension of the work presented by Ding and coworkers. As these authors point out, when subjected to elevated service temperatures, brittle materials exhibit complex thermomechanical behavior that is inherently time dependent and hereditary in the sense that current behavior depends not only on current conditions, but also on thermomechanical history.

The macroscopic continuum theory formulated in the remainder of this section captures these time-dependent phenomena by developing an extension of a J_2 plasticity model first proposed by Robinson (Ref 61) and later extended to sintered powder metals by Duffy (Ref 62). Although the viscoplastic model presented by Duffy (Ref 62) admitted a sensitivity to hydrostatic stress, it did not allow for different material behavior in tension and compression.

Willam and Warnke (Ref 63) proposed a yield criterion for concrete that admits a dependence on the hydrostatic component of stress and explicitly allows different material responses in tension and compression. Several formulations of their model exist, that is, a three-parameter formulation and a five-parameter formulation. For simplicity, the overview of the multiaxial derivation of the viscoplastic constitutive model presented here builds on the three-parameter formulation. The attending geometrical implications have been presented elsewhere (Ref 64, 65). A quantitative assessment has yet to be conducted because the material constants have not been suitably characterized for a specific material. The quantitative assessment could easily dovetail with the nascent efforts of White and coworkers (Ref 59).

The complete theory is derivable from a scalar dissipative potential function identified here as Ω. Under isothermal conditions, this function is dependent on the applied stress σ_{ij} and internal state variable α_{ij}:

$$\Omega = \Omega(\sigma_{ij}, \alpha_{ij}) \qquad \text{(Eq 53)}$$

The stress dependence for a J_2 plasticity model or a J_2 viscoplasticity model is usually stipulated in terms of the deviatoric components of the applied stress, $S_{ij} = \sigma_{ij} - (1/3)\,\sigma_{kk}\delta_{ij}$, and a deviatoric state variable, $a_{ij} = \alpha_{ij} - (1/3)\,\alpha_{kk}\delta_{ij}$. For the viscoplasticity model presented here, these deviatoric tensors are incorporated along with the effective stress, $\eta_{ij} = \sigma_{ij} - \alpha_{ij}$, and an effective deviatoric stress, identified as $\Sigma_{ij} = S_{ij} - a_{ij}$. Both tensors, that is, η_{ij} and Σ_{ij}, are utilized for notational convenience.

The potential nature of Ω is exhibited by the manner in which the flow and evolutionary laws are derived. The flow law is derived from Ω by taking the partial derivative with respect to the applied stress:

$$\dot{\varepsilon}_{ij} = \frac{\partial \Omega}{\partial \sigma_{ij}} \qquad \text{(Eq 54)}$$

The adoption of a flow potential and the concept of normality, as expressed in Eq 54, were introduced by Rice (Ref 66). In his work, the above relationship was established using thermodynamic arguments. The authors wish to point out that Eq 54 holds for each individual inelastic state.

The evolutionary law is similarly derived from the flow potential. The rate of change of the internal stress is expressed as:

$$\dot{\alpha}_{ij} = -h\,\frac{\partial \Omega}{\partial \alpha_{ij}} \qquad \text{(Eq 55)}$$

where h is a scalar function of the inelastic state variable (i.e., the internal stress) only. Using arguments similar to Rice's, Ponter, and Leckie (Ref 67) have demonstrated the appropriateness of this type of evolutionary law.

To give the flow potential a specific form, the following integral format proposed by Robinson (Ref 61) is adopted:

$$\Omega = K^2\left[\left(\frac{1}{2\mu}\right)\int F^n\, dF + \left(\frac{R}{H}\right)\int G^m\, dG\right] \qquad \text{(Eq 56)}$$

where μ, R, H, and K are material constants. In this formulation μ is a viscosity constant, H is a hardening constant, n and m are unitless exponents, and R is associated with recovery. The octahedral threshold shear stress K appearing in Eq 56 is generally considered a scalar state variable that accounts for isotropic hardening (or softening). However, because isotropic hardening is often negligible at high homologous temperatures ($T/T_m \geq 0.5$), to a first approximation K is taken to be a constant for metals. This assumption is adopted in the present work for brittle materials. The reader is directed to Ref 68 for specific details regarding the experimental test matrix needed to characterize these parameters.

The dependence on the effective stress Σ_{ij} and the deviatoric internal stress a_{ij} is introduced through the scalar functions $F = F(\Sigma_{ij}, \eta_{ij})$ and $G = G(a_{ij}, \alpha_{ij})$. Inclusion of η_{ij} and α_{ij} will account for sensitivity to hydrostatic stress. The concept of a threshold function was introduced by Bingham (Ref 69) and later generalized by Hohenemser and Prager (Ref 70). Correspondingly, F is referred to as a Bingham-Prager threshold function. Inelastic deformation occurs only for those stress states where $F(\Sigma_{ij}, \eta_{ij}) > 0$.

For frame indifference, the scalar functions F and G (and hence Ω) must be form invariant under all proper orthogonal transformations. This condition is ensured if the functions depend only on the principal invariants of Σ_{ij}, a_{ij} η_{ij}, and α_{ij}; that is, $F = F(\tilde{I}_1, \tilde{J}_2, \tilde{J}_3)$, where $\tilde{I}_1 = \eta_{ii}$, $\tilde{J}_2 = (1/2)\,\Sigma_{ij}\,\Sigma_{ij}$, $\tilde{J}_3 = (1/3)\,\Sigma_{ij}\,\Sigma_{jk}\,\Sigma_{ki}$, and $G = G(\hat{I}_1, \hat{J}_2, \hat{J}_3)$, where $\hat{I}_1 = a_{ii}$, $\hat{J}_2 = (1/2) a_{ij}\, a_{ij}$, $\hat{J}_3 = (1/3)\, a_{ij}\, a_{jk}\, a_{ki}$. These scalar quantities are elements of what is known in invariant theory as an integrity basis for the functions F and G.

A three-parameter flow criterion proposed by Willam and Warnke (Ref 63) serves as the Bingham-Prager threshold function, F. The William-Warnke criterion uses the previously mentioned stress invariants to define the functional dependence on the Cauchy stress (σ_{ij}) and internal state variable (α_{ij}). In general, this flow criterion can be constructed from the following general polynomial:

$$F = \lambda\left(\frac{\sqrt{\tilde{J}_2}}{\sigma_c}\right) + B\left(\frac{\tilde{I}_1}{\sigma_c}\right) - 1 \qquad \text{(Eq 57)}$$

where σ_c is the uniaxial threshold flow stress in compression and B is a constant determined by considering homogeneously stressed elements in the virgin inelastic state $\alpha_{ij} = 0$.

Note that a threshold flow stress is similar in nature to a yield stress in classical plasticity. In addition, λ is a function dependent on the invariant J_3 and other threshold stress parameters that are defined momentarily. The specific details in deriving the final form of the function F can be found in Willam and Warnke (Ref 63), and this final formulation is stated here as:

$$F(\tilde{I}_1, \tilde{J}_2, \tilde{J}_3) = \frac{1}{\sigma_c}\left[\frac{1}{r(\tilde{\theta})}\right]\left[\frac{2\tilde{J}_2}{5}\right]^{1/2} + \frac{\tilde{I}_1}{3\rho\sigma_c} - 1 \qquad \text{(Eq 58)}$$

for brevity. The invariant $\tilde{I}_1$ in Eq 58 admits a sensitivity to hydrostatic stress. The function F is implicitly dependent on $\tilde{J}_3$ through the function $r(\tilde{\theta})$, where the angle of similitude, θ, is defined by the expression:

$$\cos(3\tilde{\theta}) = \frac{(3\sqrt{3})\,\tilde{J}_3}{2(\tilde{J}_2)^{3/2}} \qquad \text{(Eq 59)}$$

The invariant $\tilde{J}_3$ accounts for different behavior in tension and compression, because this invariant changes sign when the direction of a stress component is reversed. The parameter ρ characterizes the tensile hydrostatic threshold flow stress. For the Willam-Warnke three-parameter formulation, the model parameters include σ_t, the tensile uniaxial threshold stress, σ_c, the compressive uniaxial threshold stress, and σ_{bc}, the equal biaxial compressive threshold stress.

A similar functional form is adopted for the scalar state function G. However, this formulation assumes a threshold does not exist for the scalar function G and follows the framework of previously proposed constitutive models based on Robinson's viscoplastic law (Ref 61).

Employing the chain rule for differentiation and evaluating the partial derivative of Ω with respect to σ_{ij}, and then with respect to α_{ij}, as indicated in Eq 54 and 55, yields the flow law and the evolutionary law, respectively. These expressions are dependent on the principal invariants (i.e., $\tilde{I}_1$, $\tilde{J}_2$, $\tilde{J}_3$, $\hat{I}_1$, $\hat{J}_2$, and $\hat{J}_3$) the three Willam-Warnke threshold parameters (i.e., σ_t, σ_c, and σ_{bc}), and the flow potential parameters utilized in Eq 56 (i.e., μ, R, H, K, n, and m). These expressions constitute a multiaxial statement of a constitutive theory for isotropic materials and serve as an inelastic deformation model for ceramic materials.

The overview presented in this section is intended to provide a qualitative assessment of the capabilities of this viscoplastic model in capturing the complex thermomechanical behavior exhibited by brittle materials at elevated service temperatures. Constitutive equations for the flow law (strain rate) and evolutionary law have been formulated based on a threshold function that exhibits a sensitivity to hydrostatic stress and allows different behavior in tension and compression. Furthermore, inelastic deformation is treated as inherently time dependent. A rate of inelastic strain is associated with every state of stress. As a result, creep, stress relaxation, and rate sensitivity are phenomena resulting from applied boundary conditions and are not treated separately in an ad hoc fashion. Incorporating this model into a nonlinear finite element code would provide a tool for the design engineer to simulate numerically the inherently time-dependent and hereditary phenomena exhibited by these materials in service.

Life Prediction Reliability Models

Using a time-dependent reliability model such as those discussed in the following section, and the results obtained from a finite element analysis, the life of a component with complex geometry and loading can be predicted. This life is interpreted as the reliability of a component as a function of time. When the component reliability falls below a predetermined value, the associated point in time at which this occurs is assigned the life of the component. This design methodology presented herein combines the statistical nature of strength-controlling flaws with the mechanics of crack growth to allow for multiaxial stress states, concurrent (simultaneously occurring) flaw populations, and scaling effects. With this type of integrated design tool, a design engineer can make appropriate design changes until an acceptable time to failure has been reached. In the sections that follow, only creep rupture and fatigue failure mechanisms are discussed. Although models that account for subcritical crack growth and creep rupture are presented, the reader is cautioned that currently available creep models for advanced ceramics have limited applicability because of the phenomenological nature of the models. There is a considerable need to develop models incorporating both the ceramic material behavior and microstructural events.

Subcritical Crack Growth. A wide variety of brittle materials, including ceramics and glasses, exhibit the phenomenon of delayed fracture or fatigue. Under the application of a loading function of magnitude smaller than that which induces short-term failure, there is a regime where subcritical crack growth occurs and this can lead to eventual component failure in service. Subcritical crack growth is a complex process involving a combination of simultaneous and synergistic failure mechanisms. These can be grouped into two categories: (1) crack growth due to corrosion and (2) crack growth due to mechanical effects arising from cyclic loading. Stress corrosion reflects a stress-dependent chemical interaction between the material and its environment. Water, for example, has a pronounced deleterious effect on the strength of glass and alumina. In addition, higher temperatures also tend to accelerate this process. Mechanically induced cyclic fatigue is dependent only on the number of load cycles and not on the duration of the cycle. This phenomenon can be caused by a variety of effects, such as debris wedging or the degradation of bridging ligaments, but essentially it is based on the accumulation of some type of irreversible damage that tends to enhance crack growth. Service environment, material composition, and material microstructure determine if a brittle material will display some combination of these fatigue mechanisms.

Lifetime reliability analysis accounting for SCG under cyclic and/or sustained loads is essential for the safe and efficient utilization of brittle materials in structural design. Because of the complex nature of SCG, models that have been developed tend to be semiempirical and approximate the behavior of SCG phenomenologically. Theoretical and experimental work in this area has demonstrated that lifetime failure characteristics can be described by consideration of the crack growth rate versus the stress intensity factor (or the range in the stress intensity factor). This is graphically depicted (see Fig. 10) as the logarithm of crack growth rate versus the logarithm of the mode I stress intensity factor. Curves of experimental data show three distinct regimes or regions of growth. The first region (denoted by I in Fig. 10) indicates threshold behavior of the crack, where below a certain value of stress intensity the crack growth is zero. The second region (denoted by II in Fig. 10) shows an approximately linear relationship of stable crack growth. The third region (denoted by III in Fig. 10) indicates unstable crack growth as the materials critical stress intensity factor is approached. For the stress-corrosion failure mechanism, these curves are material and environment sensitive. This SCG model, using conventional fracture mechanics relationships, satisfactorily describes the failure mechanisms in materials where at high temperatures, plastic deformations and creep behave in a linear viscoelastic manner (Ref 71). In general, at high temperatures and low levels of stress, failure is best described by creep rupture, which generates new cracks (Ref 72). The creep rupture process is discussed further in the next section.

The most-often-cited models in the literature regarding SCG are based on power-law formulations. Other theories, most notably Wiederhorn's (Ref 73), have not achieved such widespread usage, although they may also have a reasonable physical foundation. Power-law formulations are used to model both the stress-corrosion phenomenon and the cyclic fatigue phenomenon. This modeling flexibility, coupled with their widespread acceptance, make these formulations the most attractive candidates to incorporate into a design methodology. A power-law formulation is obtained by assuming the second crack growth region is linear and that it dominates the other regions. Three power-law formulations are useful for modeling brittle materials: the power law, the Paris law, and the Walker equation. The power law (Ref 71, 74) describes the crack velocity as a function of the stress intensity factor and implies that the crack growth is due to stress corrosion. For cyclic fatigue, either the Paris law (Ref 75) or Walker's (Ref 76, 77) modified formulation of the Paris law is used to model the SCG. The Paris law describes the crack growth per load cycle as a function of the range in the stress intensity factor. The Walker equation relates the crack growth per load cycle to both the range in the crack tip stress intensity factor and the maximum applied crack tip stress intensity factor. It is useful for predicting the effect of the *R*-ratio (the ratio of the minimum cyclic stress to the maximum cyclic stress) on the material strength degradation.

Expressions for time-dependent reliability are usually formulated based on the mode I equivalent stress distribution transformed to its equivalent stress distribution at time $t = 0$. Investigations of mode I crack extension (Ref 78) have resulted in the following relationship for the equivalent mode I stress intensity factor:

$$K_{\mathrm{Ieq}}(\Psi, t) = \sigma_{\mathrm{Ieq}}(\Psi, t)\, Y \sqrt{a(\Psi, t)} \qquad \text{(Eq 60)}$$

where $\sigma_{\mathrm{Ieq}}(\Psi, t)$ is the equivalent mode I stress on the crack, Y is a function of crack geometry, $a(\Psi, t)$ is the appropriate crack length, and Ψ represents a spatial location within the body and the orientation of the crack. In some models (such as the phenomenological Weibull NSA and the PIA models), Ψ represents a location only. Y is a function of crack geometry; however, herein it is assumed constant with subcritical crack growth. Crack growth as a function of the equivalent mode I stress intensity

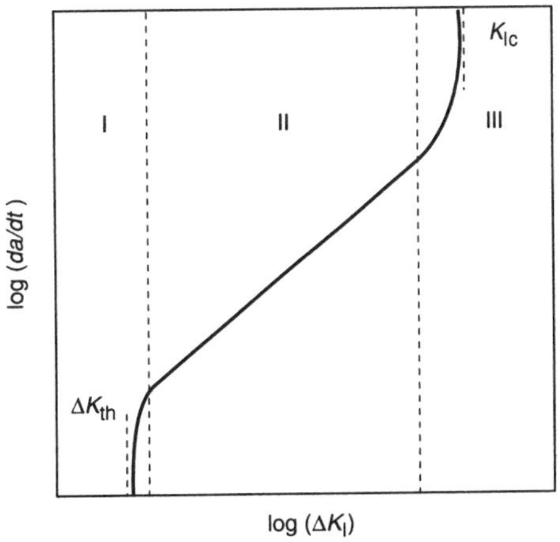

Fig. 10 Schematic illustrating three different regimes of crack growth

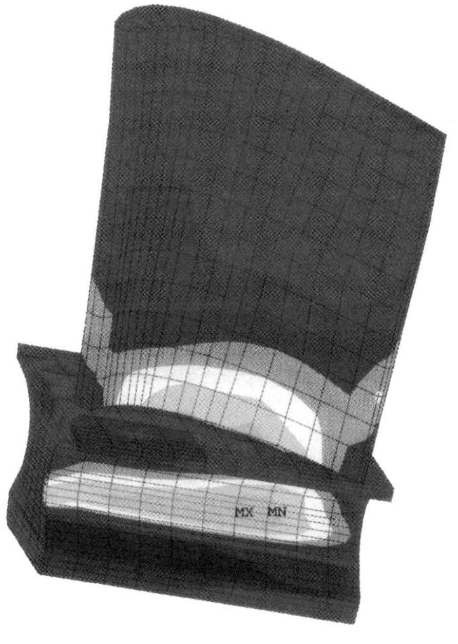

Fig. 11 Stress contour plot of first-stage silicon nitride turbine rotor blade for a natural-gas-fired industrial turbine engine for cogeneration. The blade is rotating at 14,950 rpm. Courtesy of Solar Turbines Inc.

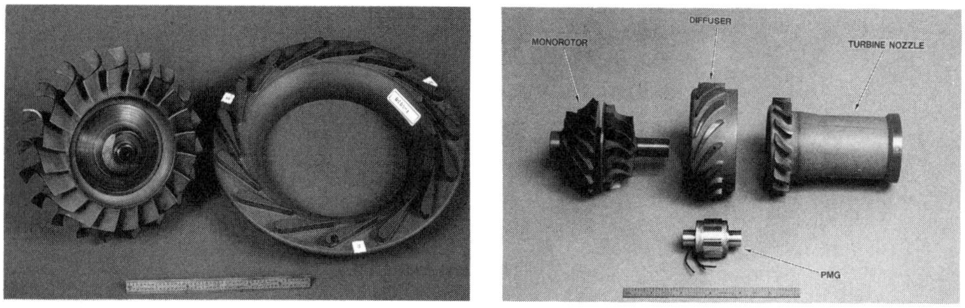

(a) (b)

Fig. 12 (a) Ceramic turbine wheel and nozzle for advanced auxiliary power unit. (b) Ceramic components for small expendable turbojet. Courtesy of Sundstrand Aerospace Corporation

factor is assumed to follow a power-law relationship:

$$\frac{da\,(\Psi, t)}{dt} = A\,K_{\mathrm{Ieq}}^{N}\,(\Psi, t) \qquad \text{(Eq 61)}$$

where A and N are material/environmental constants. The transformation of the equivalent stress distribution at the time of failure, $t = t_f$, to its critical effective stress distribution at time $t = 0$ is expressed (Ref 79, 80):

$$\sigma_{\mathrm{Ieq},0}\,(\Psi, t_f) = \left[\frac{\int_0^{t_f} \sigma_{\mathrm{Ieq}}^{N}\,(\Psi, t)\,dt}{B} + \sigma_{\mathrm{Ieq}}^{N-2}\,(\Psi, t_f)\right]^{1/(N-2)} \qquad \text{(Eq 62)}$$

where

$$B = \frac{2}{A\,Y^2\,K_{\mathrm{Ic}}^{N-2}\,(N-2)} \qquad \text{(Eq 63)}$$

is a material/environmental fatigue parameter, K_{Ic} is the critical stress intensity factor, and $\sigma_{\mathrm{Ieq}}\,(\Psi, t_f)$ is the equivalent stress distribution in the component at time $t = t_f$. The dimensionless fatigue parameter N is independent of fracture criterion. B is adjusted to satisfy the requirement that for a uniaxial stress state, all models produce the same probability of failure. The parameter B has units of stress^2 × time.

Because SCG assumes flaws exist in a material, the weakest-link statistical theories discussed previously are required to predict the time-dependent lifetime reliability for brittle materials. An SCG model (e.g., the previously discussed power law, Paris law, or Walker equation) is combined with either the two- or three-parameter Weibull cumulative distribution function to characterize the component failure probability as a function of service lifetime. The effects of multiaxial stresses are considered by using the PIA model, the Weibull NSA method, or the Batdorf theory. These multiaxial reliability expressions were outlined in the previous section on time-independent reliability analysis models, and, for brevity, are not repeated here. The reader is directed to see the previous section or, for more complete details, to consult Ref 50.

Creep Rupture. For brittle materials, the term creep can infer two different issues. The first relates to catastrophic failure of a component from a defect that has been nucleated and propagates to critical size. This is known as creep rupture to the design engineer. Here, it is assumed that failure does not occur from a defect in the original flaw population. Unlike SCG, which is assumed to begin at preexisting flaws in a component and continue until the crack reaches a critical length, creep rupture typically entails the nucleation, growth, and coalescence of voids which eventually form macrocracks, which then propagate to failure. The second issue related to creep reflects back on SCG as well as creep rupture, that is, creep deformation. This section focuses on the former, while the latter (i.e., creep deformation) is discussed in a previous section.

Currently, most approaches to predict brittle material component lifetime due to creep rupture employ deterministic methodologies. Stochastic methodologies for predicting creep life in brittle material components have not reached a level of maturity comparable to those developed for predicting fast-fracture and SCG reliability. One such theory is based on the premise that both creep and SCG failure modes act simultaneously (Ref 81). Another alternative method for characterizing creep rupture in ceramics was developed by Duffy and Gyekenyesi, (Ref 82), who developed a time-dependent reliability model that integrates continuum damage mechanics principles and Weibull analysis. This particular approach assumes that the failure processes for SCG and creep are distinct and separable mechanisms.

The remainder of this section outlines this approach, highlighting creep rupture with the intent to provide the design engineer with a method to determine an allowable stress for a given component lifetime and reliability. This is accomplished by coupling Weibull theory with the principles of continuum damage mechanics, which was originally developed by Kachanov (Ref 83) to account for tertiary creep and creep fracture of ductile metal alloys.

Ideally, any theory that predicts the behavior of a material should incorporate parameters that are relevant to its microstructure (grain size, void spacing, etc.). However, this would require a determination of volume-averaged effects of microstructural phenomena reflecting nucleation, growth, and coalescence of microdefects that in many instances interact. This approach is difficult even under strongly simplifying assumptions. In this respect, Leckie (Ref 84) points out that the difference between the materials scientists and the engineer is one of scale. He notes the materials scientist is interested in mechanisms of deformation and failure at the microstructural level and the engineer focuses on these issues at the component level. Thus, the former designs the material and the latter designs the component. Here, the engineer's viewpoint is adopted, and readers should note from the outset that continuum damage mechanics does not focus attention on microstructural events, yet this logical first approach does provide a practical model, which macroscopically captures the changes induced by the evolution of voids and defects.

This method uses a continuum-damage approach where a continuity function, ϕ, is coupled with Weibull theory to render a time-dependent damage model for ceramic materials. The continuity function is given by the expression:

$$\phi = [1 - b\,(\sigma_0)^m\,(m+1)t]^{(1/(m+1))} \qquad \text{(Eq 64)}$$

where b and m are material constants, σ_0 is the applied uniaxial stress on a unit volume, and t is time. From this, an expression for a time to failure, t_f, can be obtained by noting that when $t = t_f$, $\phi = 0$. This results in the following:

$$t_f = \frac{1}{b\,(\sigma_0)^m\,(m+1)} \qquad \text{(Eq 65)}$$

which leads to the simplification of ϕ as follows:

$$\phi = [1 - (t/t_f)]^{(1/(m+1))} \qquad \text{(Eq 66)}$$

The above equations are then coupled with an expression for reliability to develop the time-dependent model. The expression for reliability for a uniaxial specimen is:

$$R = \exp\,[-V\,(\sigma/\beta)^\alpha] \qquad \text{(Eq 67)}$$

where V is the volume of the specimen, α is the Weibull shape parameter, and β is the Weibull scale parameter. Incorporating the continuity function into the reliability equation and assuming a unit volume yields:

$$R = \exp\,[-(\sigma_0/\phi\beta)^\alpha] \qquad \text{(Eq 68)}$$

Substituting for ϕ in terms of the time to failure results in the time-dependent expression for reliability:

$$R = \exp\left\{-\left(\frac{\sigma_0}{\beta}\right)^\alpha \left[1 - \frac{t}{t_f}\right]^{-(\alpha/m+1)}\right\} \qquad \text{(Eq 69)}$$

This model has been presented in a qualitative fashion, intending to provide the design engineer with a reliability theory that incorporates the expected lifetime of a brittle material component undergoing damage in the creep rupture regime. The predictive capability of this approach depends on how well the macroscopic state variable ϕ captures the growth of grain-boundary microdefects. Finally, note that the kinetics of damage also depend significantly on the direction of the applied stress. In the development described previously, it was expedient from a theoretical and computational standpoint to use a scalar state variable for damage because only uniaxial loading conditions were considered. The incorporation of a continuum-damage approach within a multiaxial Weibull analysis necessitates the description of oriented damage by a second-order tensor.

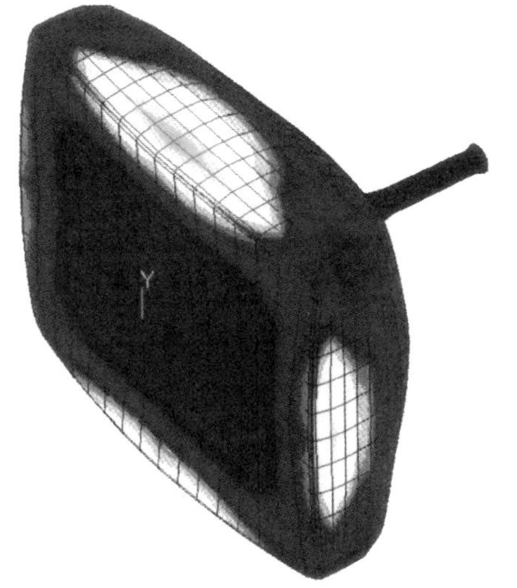

Fig. 13 Stress plot of an evacuated 68 cm (27 in.) diagonal CRT. The probability of failure calculated with CARES/Life was less than 5.0×10^{-3}. Courtesy of Philips Display Components Company

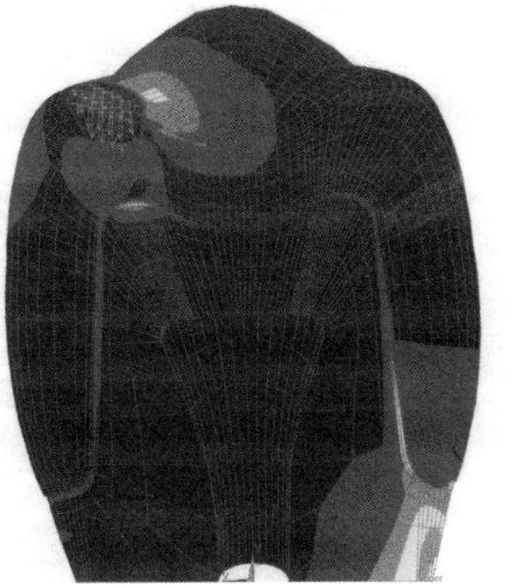

Fig. 14 Stress contour plot of ceramic dental crown, resulting from a 600 N biting force. Courtesy of University of Florida College of Dentistry

Life-Prediction Reliability Algorithms

The NASA-developed computer program CARES/*Life* (Ceramics Analysis and Reliability Evaluation of Structures/Life-Prediction program) and the AlliedSignal algorithm ERICA have the capability to evaluate the time-dependent reliability of monolithic ceramic components subjected to thermomechanical and/or proof test loading. The reader is directed to Ref 39 and Ref 40 for a detailed discussion of the life-prediction capabilities of the ERICA algorithm. The CARES/*Life* program is an extension of the previously discussed CARES program, which predicted the fast-fracture (time-independent) reliability of monolithic ceramic components. CARES/*Life* retains all of the fast-fracture capabilities of the CARES program and also includes the ability to perform time-dependent reliability analysis due to SCG. CARES/*Life* accounts for the phenomenon of SCG by utilizing the power law, Paris law, or Walker equation. The Weibull cumulative distribution function is used to characterize the variation in component strength. The probabilistic nature of material strength and the effects of multiaxial stresses are modeled using either the PIA, the Weibull NSA, or the Batdorf theory. Parameter estimation routines are available for obtaining inert strength and fatigue parameters from rupture strength data of naturally flawed specimens loaded in static, dynamic, or cyclic fatigue. Fatigue parameters can be calculated using either the median value technique (Ref 85), a least squares regression technique, or a median deviation regression method that is somewhat similar to trivariant regression (Ref 85). In addition, CARES/*Life* can predict the effect of proof testing on component service probability of failure. Creep and material healing mechanisms are not addressed in the CARES/*Life* code.

Life-Prediction Design Examples

Once again, because of the proprietary nature of the ERICA algorithm, the life-prediction examples presented in this section are all based on design applications where the NASA CARES/*Life* algorithm was utilized. Either algorithm should predict the same results cited here. However, at this point in time comparative studies utilizing both algorithms for the same analysis are not available in the open literature. The primary thrust behind CARES/*Life* is the support and development of advanced heat engines and related ceramics technology infrastructure. This U.S. Department of Energy (DOE), and Oak Ridge National Laboratory (ORNL) have several ongoing programs such as the Advanced Turbine Technology Applications Project (ATTAP) (Ref 48, 86) for automotive gas turbine development, the Heavy Duty Transport Program for low-heat-rejection heavy-duty diesel engine development, and the Ceramic Stationary Gas Turbine (CSGT) program for electric power cogeneration. Both CARES/*Life* and the previously discussed CARES program are used in these projects to design stationary and rotating equipment, including turbine rotors, vanes, scrolls, combustors, insulating rings, and seals. These programs are also integrated with the DOE/ORNL Ceramic Technology Project (CTP) (Ref 87) characterization and life prediction efforts (Ref 88, 89).

The CARES/*Life* program has been used to design hot-section turbine parts for the CSGT development program (Ref 90) sponsored by the DOE Office of Industrial Technology. This project seeks to replace metallic hot-section parts with uncooled ceramic components in an existing design for a natural-gas-fired industrial turbine engine operating at a turbine rotor inlet temperature of 1120 °C (2048 °F). At least one stage of blades (Fig. 11) and vanes, as well as the combustor liner, will be replaced with ceramic parts. Ultimately, demonstration of the technology will be proved with a 4000 h engine field test.

Ceramic pistons for a constant-speed drive are being developed. Constant-speed drives are used to convert variable engine speed to a constant output speed for aircraft electrical generators. The calculated probability of failure of the piston is less than 0.2×10^{-8} under the most severe limit-load condition. This program is sponsored by the U.S. Navy and ARPA (Advanced Research Projects Agency, formerly DARPA, Defense Advanced Research Projects Agency). As depicted in Fig. 12, ceramic components have been designed for a number of other applications, most notably for aircraft auxiliary power units.

Glass components behave in a similar manner as ceramics and must be designed using reliability evaluation techniques. The possibility of alkali strontium silicate glass CRTs spontaneously imploding has been analyzed (Ref 91). Cathode ray tubes are under a constant static load due to the pressure forces placed on the outside of the evacuated tube. A 68 cm (27 in.) diagonal tube was analyzed with and without an implosion protection band. The implosion protection band reduces the overall stresses in the tube and, in the event of an implosion, also contains the glass particles within the enclosure. Stress analysis (Fig. 13) showed compressive stresses on the front face and tensile stresses on the sides of the tube. The implosion band reduced the maximum principal stress by 20%. Reliability analysis with CARES/*Life* showed that the implosion protection band significantly reduced the probability of failure to about 5×10^{-5}.

The structural integrity of a silicon carbide convection air heater for use in an advanced power-generation system has been assessed by ORNL and the NASA Lewis Research Center. The design used a finned tube arrangement 1.8 m (70.9 in.) in length with 2.5 cm (1 in.) diam tubes. Incoming air was to be heated from 390 to 700 °C (734 to 1292 °F). The hot gas flow across the tubes was at 980 °C (1796 °F). Heat transfer and stress analyses revealed that maximum stress gradients across the tube wall nearest the incoming air would be the most likely source of failure.

Probabilistic design techniques are being applied to dental ceramic crowns, as illustrated in Fig. 14. Frequent failure of some ceramic crowns (e.g., 35% failure of molar crowns after three years), which occurs because of residual and functional stresses, necessitates design modifications and improvement of these restorations. Thermal tempering treatment is being investigated as a means of introducing compressive stresses on the surface of dental ceramics to improve the resistance to failure (Ref 92). Evaluation of the risk of material failure must be considered not only for the service environment, but also from the tempering process.

REFERENCES

1. C.A. Johnson and W.T. Tucker, Advanced Statistical Concepts of Fracture in Brittle Materials, Vol 4, *Engineered Materials Handbook*, ASM International, *Ceramics and Glasses,* 1991, p 708–715
2. W.A. Weibull, The Phenomenon of Rupture in Solids, *Ing. Veten. Akad. Handl.,* No. 153, 1939
3. C.A. Johnson, Fracture Statistics of Multiple Flaw Populations, *Fracture Mechanics of Ceramics,* Vol 5, R.C. Bradt, A.G. Evans, D.P.H. Hasselman, and F.F. Lange, Ed., Plenum Press, 1983, p 365–386
4. W.A. Weibull, A Statistical Distribution Function of Wide Applicability, *J. Appl. Mech.,* Vol 18 (No. 3), Sept 1951, p 293–297
5. F.T. Pierce, "The Weakest Link" Theorems of the Strength of Long and of Composite Specimens, *Text. Inst. J.,* Vol 17, 1926, p T355–T368
6. T.A. Kontorova, A Statistical Theory of Mechanical Strength, *J. Tech. Phys. (USSR),* Vol 10, 1940, p 886–890
7. J.I. Frenkel and T.A. Kontorova, A Statistical Theory of the Brittle Strength of Real Crystals, *J. Phys. USSR,* Vol 7 (No. 3), 1943, p 108–114
8. J. Hu, "Modeling Size Effects and Numerical Techniques in Structural Reliability Analysis," Master's thesis, Cleveland State University, Cleveland, OH, 1994
9. T.T. Shih, An Evaluation of the Probabilistic Approach to Brittle Design, *Eng. Fract. Mech.,* Vol 13 (No. 2), 1980, p 257–271
10. S.F. Duffy and S.M. Arnold, Noninteractive Macroscopic Statistical Failure Theory for Whisker Reinforced Ceramic Composites, *J. Compos. Mater.,* Vol 24 (No. 3), 1990, p 293–308
11. S.F. Duffy and J.M. Manderscheid, Noninteractive Macroscopic Reliability Model for Ceramic Matrix Composites with Orthotropic Material Symmetry, *J. Eng. Gas Turbines Power (Trans. ASME),* Vol 112 (No. 4), 1990, p 507–511
12. N.N. Nemeth, J.M. Manderscheid, and J.P. Gyekenyesi, "Ceramics Analysis and Reliability Evaluation of Structures (CARES) Users and Programmers Manual," TP-2916, National Aeronautics and Space Administration, 1990
13. B. Gross and J.P. Gyekenyesi, Weibull Crack Density Coefficient for Polydimensional Stress States, *J. Am. Ceram. Soc.,* Vol 72 (No. 3), 1989, p 506–507
14. R.L. Barnett, C.L. Connors, P.C. Hermann, and J.R. Wingfield, "Fracture of Brittle Materials under Transient Mechanical and Thermal Loading," AFFDL-TR-66-220, U.S. Air Force Flight Dynamics Laboratory, March 1967
15. A.M. Freudenthal, Statistical Approach to Brittle Fracture. Fracture, An Advanced Treatise, *Mathematical Fundamentals,* Vol 2, H. Liebowitz, Ed., Academic Press, 1968, p 591–619
16. J. Margetson, "A Statistical Theory of Brittle Failure for an Anisotropic Structure Subjected to a Multiaxial Stress State, Paper 76-632, American Institute of Aeronautics and Astronautics, July 1976
17. A. Paluszny and W. Wu, Probabilistic Aspects of Designing with Ceramics, *J. Eng. Power,* Vol 99 (No. 4), Oct 1977, p 617–630
18. G.J. DeSalvo, "Theory and Structural Design Application of Weibull Statistics," WANL-TME-2688, Westinghouse Astronuclear Laboratory, 1970
19. J.L. Wertz and P.W. Heitman, "Predicting the Reliability of Ceramic Turbine Components. Advanced Gas Turbine Systems for Automobiles," SP-465, Society of Automotive Engineers, 1980, p 69–77
20. W.H. Dukes, *Handbook of Brittle Material Design Technology,* AGARDograph 152, AGARD, Paris, France, 1971
21. S.B. Batdorf, Fundamentals of the Statistical Theory of Fracture, *Fracture Mechanics of Ceramics,* Vol 3, R.C. Bradt, D.P.H. Hasselman, and F.F. Lange, Plenum Press, 1978, p 1–30
22. S.B. Batdorf and J.G. Crose, A Statistical Theory for the Fracture of Brittle Structures Subjected to Nonuniform Polyaxial Stresses, *J. Appl. Mech.,* Vol 41 (No. 2), June 1974, p 459–464
23. M. Giovan and G. Sines, Biaxial and Uniaxial Data for Statistical Comparison of a Ceramic's Strength, *J. Am. Ceram. Soc.,* Vol 62 (No. 9), Sept 1979, p 510–515
24. M.G. Stout and J.J. Petrovic, Multiaxial Loading Fracture of Al_2O_3 Tubes: I, Experiments, *J. Am. Ceram. Soc.,* Vol 67 (No. 1), Jan 1984, p 14–18
25. J.J. Petrovic and M.G. Stout, Multiaxial Loading Fracture of Al_2O_3 Tubes: II, Weibull Theory and Analysis, *J. Am. Ceram. Soc.,* Vol 67 (No. 1), Jan 1984, p 18–23
26. K. Palaniswamy and W.G. Knauss, On the Problem of Crack Extension in Brittle Solids Under General Loading, *Mech. Today,* Vol 4, 1978, p 87–148
27. D.K. Shetty, Mixed-Mode Fracture Criteria for Reliability Analysis and Design with Structural Ceramics, *J. Eng. Gas Turbines Power (Trans. ASME),* Vol 109 (No. 3), July 1987, p 282–289
28. S.F. Duffy, L.M. Powers, and A. Starlinger, Reliability Analysis of Structural Components Fabricated from Ceramic Materials Using a Three-Parameter Weibull Distribution, *J. Eng. Gas Turbines Power (Trans. ASME),* Vol 115 (No. 1), Jan 1993, p 109–116
29. G.D. Quinn, "Flexure Strength of Advanced Ceramics—A Round Robin Exercise," Materials Technology Laboratory TR-89-62 (Available from the National Technical Information Service, AD-A212101, 1989)
30. M.R. Foley, V.K. Pujari, L.C. Sales, and D.M. Tracey, Silicon Nitride Tensile Strength Data Base from Ceramic Technology Program for Reliability Project, *Life Prediction Methodologies and Data for Ceramic Materials,* C.R. Brinkman and S.F. Duffy, Ed., ASTM, to be published
31. L.-Y. Chao and D.K. Shetty, Reliability Analysis of Structural Ceramics Subjected to Biaxial Flexure, *J. Am. Ceram. Soc.,* Vol 74 (No. 2), 1991, p 333–344
32. S.F. Duffy, J.L. Palko, and J.P. Gyekenyesi, Structural Reliability of Laminated CMC Components, *J. Eng. Gas Turbines and Power (Trans. ASME),* Vol 115 (No. 1), 1993, p 103–108
33. D.J. Thomas and R.C. Wetherhold, Reliability of Continuous Fiber Composite Laminates, *Comput. Struct.,* Vol 17, 1991, p 277–293
34. J.L. Palko, "An Interactive Reliability Model for Whisker-Toughened Ceramics," Master's thesis, Cleveland State University, Cleveland, OH, 1992
35. J. Margetson and N.R. Cooper, Brittle Material Design Using Three Parameter Weibull Distributions, *Proceedings of the IUTAM Symposium on Probabilistic Methods in the Mechanics of Solids and Structures,* S. Eggwertz and N.C. Lind, Ed., Springer-Verlag, 1984, p 253–262
36. S.S. Pai and J.P. Gyekenyesi, "Calculation of the Weibull Strength Parameters and Batdorf Flaw Density Constants for Volume and Surface-Flaw-Induced Fracture in Ceramics," TM-100890, National Aeronautics and Space Administration, 1988
37. J.P. Gyekenyesi and N.N. Nemeth, Surface Flaw Reliability Analysis of Ceramic Components with the SCARE Finite Element Postprocessor Program, *J. Eng. Gas Turbines Power (Trans. ASME),* Vol 109 (No. 3), July 1987, p 274–281
38. J.P. Gyekenyesi, SCARE: A Postprocessor Program to MSC/NASTRAN for the Reliability Analysis of Structural Ceramic Components, *J. Eng. Gas Turbines Power (Trans. ASME),* Vol 108 (No. 3), July 1986, p 540–546
39. J.C. Cuccio, P. Brehm, H.T. Fang, J. Hartman, W. Meade, M.N. Menon, A. Peralta, J.Z. Song, T. Strangman, J. Wade, J. Wimmer, and D.C. Wu, "Life Prediction Methodology for Ceramic Components of Advanced Heat Engines, Phase I," ORNL/Sub/89-SC674/1/V1, Vol 1, Final Report, Oak Ridge National Laboratory, March 1995
40. J.C. Cuccio, P. Brehm, H.T. Fang, J. Hartman, W. Meade, M.N. Menon, A. Peralta, J.Z. Song, T. Strangman, J. Wade, J. Wimmer, and D.C. Wu, "Life Prediction Methodology for Ceramic Components of Advanced Heat Engines, Phase I," ORNL/Sub/89-SC674/1/V2, Vol 2, Final Report, Oak Ridge National Laboratory, March 1995
41. C. Baker and D. Baker, Design Practices for Structural Ceramics in Automotive Turbocharger Wheels, *Ceramics and Glasses,* Vol 4, *Engineered Materials Handbook,* ASM International, 1991, p 722–727

42. C.J. Poplawsky, L. Lindberg, S. Robb, and J. Roundy, "Development of an Advanced Ceramic Turbine Wheel for an Air Turbine Starter," Paper 921945, presented at Aerotech '92, Anaheim, CA, Society of Automotive Engineers, 5–8 Oct 1992
43. J.H. Selverian, D. O'Neil, and S. Kang, Ceramic-to-Metal Joints: Part I-Joint Design, *Am. Ceram. Soc. Bull.*, Vol 71 (No. 9), 1992, p 1403–1409
44. J.H. Selverian and S. Kang, Ceramic-to-Metal Joints: Part II-Performance and Strength Prediction, *Am. Ceram. Soc. Bull.*, Vol 71 (No. 10), 1992, p 1511–1520
45. C.L. Snydar, "Reliability Analysis of a Monolithic Graphite Valve," presented at the 15th Annual Conference on Composites, Materials, and Structures (Cocoa Beach, FL), American Ceramic Society, 1991
46. J.A. Salem, J.M. Manderscheid, M.R. Freedman, and J.P. Gyekenyesi, "Reliability Analysis of a Structural Ceramic Combustion Chamber," Paper 91-GT-155, presented at the International Gas Turbine and Aeroengine Congress and Exposition, Orlando, FL, 3–6 June 1991
47. R.R. Wills and R.E. Southam, Ceramic Engine Valves, *J. Am. Ceram. Soc.*, Vol 72 (No. 7), 1989, p 1261–1264
48. J.R. Smyth, R.E. Morey, and R.W. Schultz, "Ceramic Gas Turbine Technology Development and Applications," Paper 93-GT-361, presented at the International Gas Turbine and Aeroengine Congress and Exposition (Cincinnati, OH), 24–27 May 1993
49. G.D. Quinn, Fracture Mechanism Maps for Advanced Structural Ceramics: Part 1; Methodology and Hot-Pressed Silicon Nitride Results, *J. Mater. Sci.*, Vol 25, 1990, p 4361–4376
50. N.N. Nemeth, L.M. Powers, L.A. Janosik, and J.P. Gyekenyesi, "CARES/*Life* Prediction Program (CARES/Life) Users and Programmers Manual," TM-106316, to be published
51. T.-J. Chuang, and S.F. Duffy, A Methodology to Predict Creep Life for Advanced Ceramics Using Continuum Damage Mechanics, *Life Prediction Methodologies and Data for Ceramic Materials,* STP 1201, C.R. Brinkman and S.F. Duffy, Ed., ASTM, 1994, p 207-227
52. Y. Corapcioglu and T. Uz, Constitutive Equations for Plastic Deformation of Porous Materials, *Powder Technol.,* Vol 21, 1978, p 269–274
53. H.A. Kuhn and C.L. Downey, Deformation Characteristics and Plasticity Theory of Sintered Powder Metals, *Int. J. Powder Metall.,* Vol 7, 1971, p 15–25
54. S. Shima and M. Oyane, Plasticity Theory for Porous Metals, *Int. J. Mech. Sci.,* Vol 18, 1976, p 285
55. R.J. Green, A Plasticity Theory for Porous Solids, *Int. J. Mech. Sci.,* Vol 14, 1972, p 215
56. A.L. Gurson, Continuum Theory of Ductile Rupture by Void Nucleation and Growth: Part I¾ Yield Criteria and Flow Rules for Porous Ductile Media, *J. Eng. Mater. Technol.,* Vol 99, 1977, p 2–15
57. M.E. Mear and J.W. Hutchinson, Influence of Yield Surface Curvature on Flow Localization in Dilatant Plasticity, *Mech. Mater.,* Vol 4, 1985, p 395–407
58. J.-L. Ding, K.C. Liu, and C.R. Brinkman, A Comparative Study of Existing and Newly Proposed Models for Creep Deformation and Life Prediction of Si3N4, *Life Prediction Methodologies and Data for Ceramic Materials,* STP 1201, C.R. Brinkman and S.F. Duffy, Ed., ASTM, 1994, p 62–83
59. C.S. White, and R.M. Hazime, Internal Variable Modeling of the Creep of Monolithic Ceramics, *proceedings of the 11th Biennial Conference on Reliability, Stress Analysis, and Failure Prevention,* O. Jadaan, Ed., American Society of Mechanical Engineers, 1995
60. K.C. Liu, C.R. Brinkman, J.-L. Ding, and S. Liu, Predictions of Tensile Behavior and Strengths of Si_3N_4 Ceramic at High Temperatures Based on a Viscoplastic Model, *ASME Trans.,* 95-GT-388, 1995
61. D.N. Robinson, "A Unified Creep-Plasticity Model for Structural Metals at High Temperature," ORNL/TM 5969, Oak Ridge National Laboratory, 1978
62. S.F. Duffy, A Unified Inelastic Constitutive Theory for Sintered Powder Metals, *Mech. Mater.,* Vol 7, 1988, p 245–254
63. K.J. Willam and E.P. Warnke, Constitutive Model for the Triaxial Behaviour of Concrete, *Int. Assoc. Bridge Struct. Eng. Proc.,* Vol 19, 1975, p 1–30
64. L.A. Janosik and S.F. Duffy, A Viscoplastic Constitutive Theory for Monolithic Ceramic Materials—I, paper 15, *Proceedings of the Physics and Process Modeling (PPM) and Other Propulsion R&T Conference,* Vol I, *Materials Processing, Characterization, and Modeling; Lifing Models,* CP-10193, National Aeronautics and Space Administration, 1997
65. L.A. Janosik and S.F. Duffy, A Viscoplastic Constitutive Theory for Monolithic Ceramics—I, paper No. 96-GT-368, International Gas Turbine Congress, Exposition, and Users' Symposium (Birmingham, UK), American Society of Mechanical Engineers, 10–13 June 1996
66. J.R. Rice, On the Structure of Stress-Strain Relations for Time-Dependent Plastic Deformation in Metals, *J. Appl. Mech.,* Vol 37, 1970, p 728
67. A.R.S. Ponter and F.A. Leckie, Constitutive Relationships for Time-Dependent Deformation of Metals, *J. Eng. Mater. Technol. (Trans. ASME),* Vol 98, 1976
68. L.A. Janosik, "A Unified Viscoplastic Constitutive Theory for Monolithic Ceramics," Master's thesis, Cleveland State University, Cleveland, OH, 1997, to be published
69. E.C. Bingham, *Fluidity and Plasticity,* McGraw-Hill, 1922
70. K. Hohenemser and W. Prager, Ueber die Ansaetze der Mechanik Isotroper Kontinua, *Z. Angewandte Mathemat. Mech.,* Vol 12, 1932 (in German)
71. A.G. Evans and S.M. Wiederhorn, Crack Propagation and Failure Prediction in Silicon Nitride at Elevated Temperatures, *J. Mater. Sci.,* Vol 9, 1974, p 270–278
72. S.M. Wiederhorn and E.R. Fuller, Jr., Structural Reliability of Ceramic Materials, *Mater. Sci. Eng.,* Vol 71, 1985, p 169–186
73. S.M. Wiederhorn, E.R. Fuller, and R. Thomson, Micromechanisms of Crack Growth in Ceramics and Glasses in Corrosive Environments, *Met. Sci.,* Aug–Sept 1980, p 450–458
74. S.M. Wiederhorn, *Fracture Mechanics of Ceramics,* R.C. Bradt, D.P. Hasselman, and F.F. Lange, Ed., Plenum, 1974, p 613–646
75. P. Paris and F. Erdogan, A Critical Analysis of Crack Propagation Laws, *J. Basic Eng.,* Vol 85, 1963, p 528–534
76. K. Walker, The Effect of Stress Ratio During Crack Propagation and Fatigue for 2024-T3 and 7075-T6 Aluminum, *Effects of Environment and Complex Load History on Fatigue Life,* STP 462, ASTM, 1970, p 1–14
77. R.H. Dauskardt, M.R. James, J.R. Porter, and R.O. Ritchie, Cyclic Fatigue Crack Growth in SiC-Whisker-Reinforced Alumina Ceramic Composite: Long and Small Crack Behavior, *J. Am. Ceram. Soc.,* Vol 75 (No. 4), 1992, p 759–771
78. P.C. Paris and G.C. Sih, Stress Analysis of Cracks, *Fracture Toughness Testing and Its Applications,* STP 381, ASTM, 1965, p 30–83
79. T. Thiemeier, "Lebensdauervorhersage fun Keramische Bauteile Unter Mehrachsiger Beanspruchung," Ph.D. dissertation, University of Karlesruhe, Germany, 1989 (in German)
80. G. Sturmer, A. Schulz, and S. Wittig, "Lifetime Prediction for Ceramic Gas Turbine Components," Preprint 91-GT-96, American Society of Mechanical Engineers, 3–6 June 1991
81. F. Lange, Interrelations Between Creep and Slow Crack Growth for Tensile Loading Conditions, *Int. J. Fract.,* Vol 12, 1976, p 739–744
82. S.F. Duffy and J.P. Gyekenyesi, "Time Dependent Reliability Model Incorporating Continuum Damage Mechanics for High-Temperature Ceramics," TM-102046, National Aeronautics and Space Administration, May 1989
83. L.M. Kachanov, Time of the Rupture Process Under Creep Conditions, *Izv. Akad. Nauk. SSR, Otd Tekh. Nauk,* Vol 8, 1958, p 26
84. F.A. Leckie, Advances in Creep Mechanics, *Creep in Structures,* A.R.S. Ponter and D.R. Hayhurst, Ed., Springer-Verlag, 1981, p 13
85. K. Jakus, D.C. Coyne, and J.E. Ritter, Analysis of Fatigue Data for Lifetime Predictions for Ceramic Materials, *J. Mater. Sci.,* Vol 13, 1978, p 2071–2080
86. S.G. Berenyi, S.J. Hilpisch, and L.E. Groseclose, "Advanced Turbine Technology Applications Project (ATTAP)," Proceedings of the Annual Automotive Technology Development Contractor's Coordination Meeting (Dearborn, MI), 18–21 Oct 1993, SAE International
87. D.R. Johnson and R.B. Schultz, "The Ceramic Technology Project: Ten Years of Progress," Paper 93-GT-417, presented at the International Gas Turbine and Aeroengine Congress and Exposition (Cincinnati, OH), 24–27 May 1993, American Society of Mechanical Engineers
88. J. Cuccio, "Life Prediction Methodology for Ceramic Components of Advanced Heat En-

gines," Proceedings of the Annual Automotive Technology Development Contractor's Coordination Meeting (Dearborn, MI), 18–21 Oct 1993

89. P.K. Khandelwal, N.J. Provenzano, and W.E. Schneider, "Life Prediction Methodology for Ceramic Components of Advanced Vehicular Engines," Proceedings of the Annual Automotive Technology Development Contractor's Coordination Meeting (Dearborn, MI), 18–21 Oct 1993
90. M. van Roode, W.D. Brentnall, P.F. Norton, and G.P. Pytanowski, "Ceramic Stationary Gas Turbine Development," Paper 93-GT-309, presented at the International Gas Turbine and Aeroengine Congress and Exposition (Cincinnati, OH), 24–27 May 1993, American Society of Mechanical Engineers, 24–27 May 1993
91. A. Ghosh, C.Y. Cha, W. Bozek, and S. Vaidyanathan, Structural Reliability Analysis of CRTs, *Society for Information Display International Symposium Digest of Technical Papers,* Vol XXIII, 17–22 May 1992, Society of Information Display, Playa Del Ray, CA, p 508–510
92. B. Hojjatie, Thermal Tempering of Dental Ceramics, *Proceedings of the ANSYS Conference and Exhibition,* Vol 1, Swanson Analysis Systems Inc., Houston, PA, 1992, p I.73–I.91

Design with Plastics

G.G. Trantina, General Electric Corporate Research and Development

THE KEY to any successful part development is the proper choice of material, process, and design matched to the part performance requirements. The ability to design plastic parts requires knowledge of material properties—performance indicators that are not design or geometry dependent—rather than material comparators that apply only to a specific geometry and loading. Understanding the true effects of time, temperature, and rate of loading on material performance can make the difference between a successful application and catastrophic failure. Examples of reliable material performance indicators and common practices to avoid are presented in this article. Simple tools and techniques for predicting part performance (stiffness, strength/impact, creep/stress relaxation and fatigue) integrated with manufacturing concerns (flow length and cycle time) are demonstrated for design and material selection.

Engineering plastics are now used in applications where their mechanical performance must meet increasingly demanding requirements. Because the marketplace is more competitive, companies cannot afford overdesigned parts or lengthy, iterative product-development cycles. Therefore, engineers must have design technologies that allow them to create productively the most cost-effective design with the optimal material and process selection.

The design-engineering process involves meeting end-use requirements with the lowest cost, design, material, and process combination (Fig. 1). Design activities include creating geometries and performing engineering analysis to predict part performance. Material characterization provides engineering design data, and process selection includes process/design interaction knowledge. In general, the challenge in designing with structural plastics is to develop an understanding not only of design techniques, but also of manufacturing and material behavior.

Engineering thermoplastics exhibit complex behavior when subjected to mechanical loads. Standard data sheets provide overly simplified, single-point data that are either ignored or, if used, are probably misleading. Some databases provide engineering data (Ref 1) over a range of application conditions and knowledge-based material selection programs have been written (Ref 2). A methodology for optimal selection of materials and manufacturing conditions to meet part performance needs is being developed. This methodology is clarified in this article by describing and demonstrating simple tools and techniques for the initial prediction of part performance, leading to the optimal selection of materials and process conditions. Related coverage is provided in the articles "Effects of Composition, Processing, and Structure on Properties of Engineering Plastics" and "Design for Plastics Processing" in this Volume. Detailed information about many of the concepts described in this article can be found in *Engineering Plastics*, Volume 2 of the *Engineered Materials Handbook* published by ASM International.

Mechanical Part Performance

There are a wide variety of part performance requirements. Some, such as flammability, transparency, ultraviolet stability, electrical, moisture, and chemical compatibility, as well as agency approvals, are specified as absolute values or simplified choices. However, mechanical requirements such as stiffness, strength, impact, and temperature resistance cannot be specified as absolute values. For example, a part may be required to have a certain stiffness—maximum deflection for a given loading condition. The part geometry (design) and the material stiffness combine to produce the part stiffness. Thus, it is impossible to select a material without some knowledge of the part design. Similarly, the part may be required to survive a certain drop test and/or a certain temperature/time/loading condition. Again, it is impossible to select a material or design a part by using traditional, inadequate, single-point data such as notched Izod or heat distortion temperature (HDT). In addition, it is important to consider the effects of the design and material selection of a part on its fabrication (see the article "Relationship between Materials Selection and Processing" in this Volume). Considerations such as flow and cycle time should be quantitatively included in the design and material-selection process. Simple yet extremely useful tools and techniques for the initial prediction of part performance are presented here.

The design process for thermoplastic part performance can be divided into two categories based on time-independent and time-dependent material behavior (Fig. 2). For time-independent material behavior, elastic material response is used to predict the displacement of a part under load. The maximum load occurs when the strength of the material is reached as fully plastic yielding for ductile materials or brittle failure for glass-filled materials. Time-dependent material behavior becomes important for three types of loading: monotonic loading at a given strain-rate until failure occurs, constant load for a period of time, or cyclic load. In the first case strain-rate-dependent material behavior becomes important; for constant load or displacement, time-dependent deformation or stress relaxation becomes an important design consideration; for cyclic loading, fatigue failure is an important consideration. In the next five sections, stiffness, strength, im-

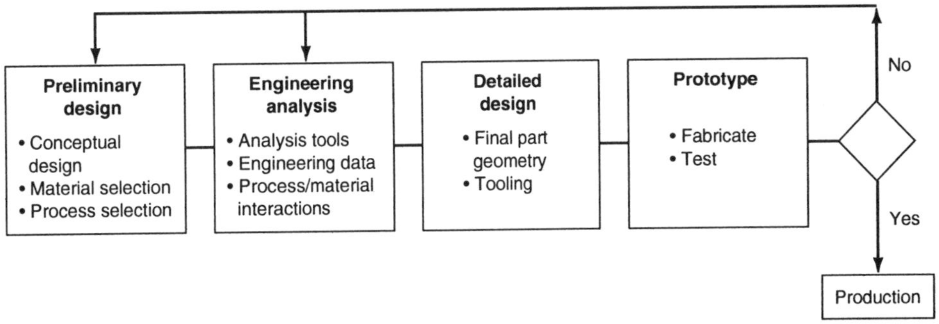

Fig. 1 Design-engineering process. The goal is to meet the end-use requirements the first time with the lowest cost.

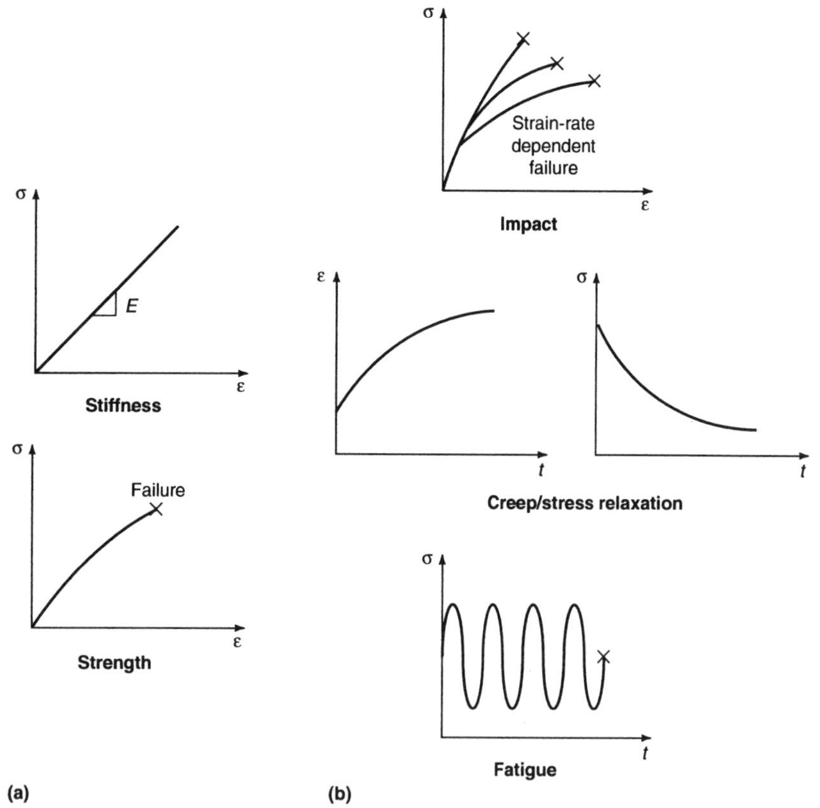

Fig. 2 Design for thermoplastic part performance. (a) Time-independent. (b) Time-dependent

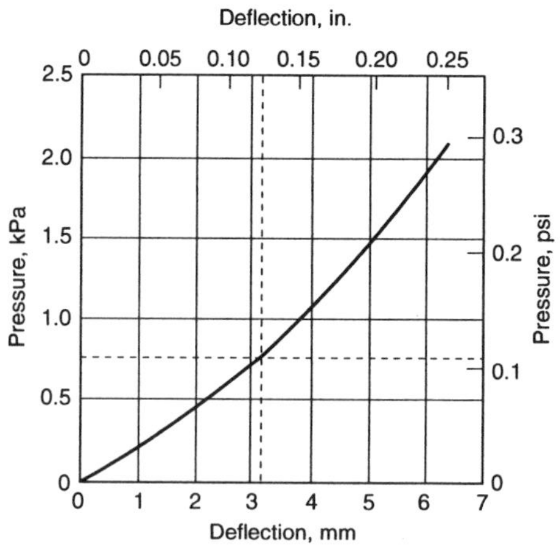

Fig. 3 Nonlinear pressure-deflection response for a 254 by 254 mm (10 by 10 in.) plate with a thickness of 2.5 mm (0.1 in.) and a material with a modulus of 2350 MPa (340 ksi)

pact, creep/stress relaxation, and fatigue behavior will be related to part performance. More details of these important design issues can be found in Ref 3.

Part Stiffness. Many thermoplastic parts are platelike structures that can be treated as a simply supported plate, possibly reinforced with ribs. A procedure intended to provide quick, approximate solutions for the stiffness of laterally loaded rib-stiffened plates has been developed (Ref 4). The computer program employs the Rayleigh-Ritz energy method and is capable of including the geometric nonlinearities associated with the large-displacement response typical of low-modulus materials such as thermoplastics. The program allows the user to input the important parameters of specific plate structures (length, width, thickness, number of ribs, rib geometry), the boundary conditions (simply supported, clamped, point supported), and the loading (central point, uniform pressure, torsion loading). With the capability of multiple rib pattern definitions, the user can quickly determine the load-deflection response for different designs to select the one that is most effective for the specific application. This tool has been validated with finite element results. An example demonstrating the prediction of the nonlinear load displacement response is shown in Fig. 3.

Strength and Stiffness of Glass-Filled Plastic Parts. An accurate characterization of the strength and stiffness of glass-filled thermoplastics is necessary to predict the strength and stiffness of components that are injection molded with these materials. The mechanical properties of glass-reinforced thermoplastics are generally measured in tension using end-gated, injection-molded ASTM type I (dog-bone) specimens (Ref 3). However, the gating and the direction of loading of these molded specimens yields nonconservative stiffness and strength results caused by the highly axial orientation of glass that occurs in the direction of flow (and loading) during molding.

Previous studies (Ref 5) have shown that injection-molded, glass-reinforced thermoplastics are anisotropic; that is, stiffness and strength values in the cross-flow direction are substantially lower than in the flow direction. The tensile stiffness and strength were measured by using dog-bone specimens that were cut in both the flow and cross-flow direction from edge-gated plaques of

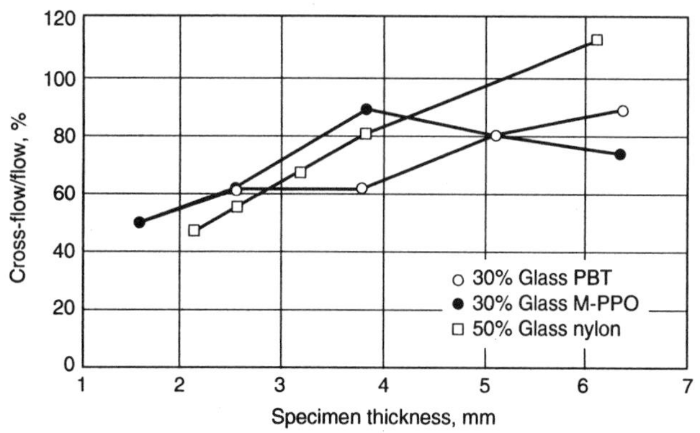

Fig. 4 Ratio of cross-flow/flow tensile modulus as a function of specimen thickness. PBT, polybutylene terephthalate; M-PPO, modified polyphenylene oxide

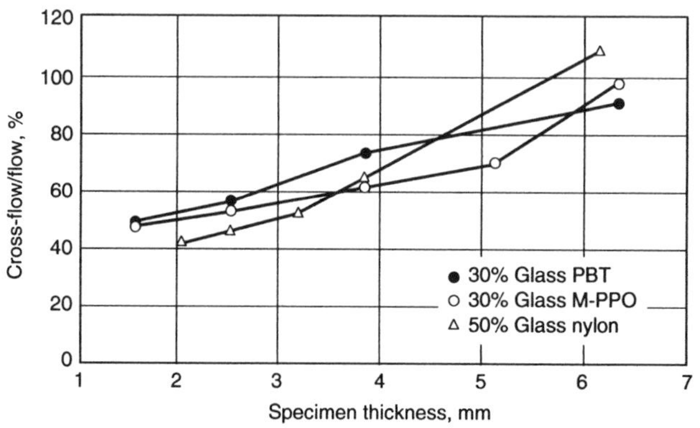

Fig. 5 Ratio of cross-flow/flow ultimate stress as a function of specimen thickness

various thicknesses. The ratio of the cross-flow/flow tensile modulus and strength of 30% glass-filled polybutylene terephthalate (PBT), 30% glass-filled modified polyphenylene oxide (M-PPO), and 50% glass-filled (long-glass fibers) nylon are plotted versus specimen thickness in Fig. 4 and 5. It is important to note the strong dependence of the cross-flow/flow ratio on specimen thickness and the small values of this ratio for small specimen thicknesses. These data clearly indicate that material selection and design for glass-filled materials that are based on injection-molded bars of a given thickness could be totally misleading—cross-flow properties could be only 50% of flow properties (small specimen thicknesses), and unless the thickness of the specimen is the same as the thickness of the part, the data could not be used for predicting part performance. However, for most parts (thickness less than 4 mm) with glass loadings of 30% or greater, a simple mold-filling analysis coupled with an anisotropic stress analysis with the cross-flow stiffness of 60% of the flow stiffness provides a reasonable prediction of part performance (Ref 3).

Part Strength and Impact Resistance. A number of test methods such as Izod (notched beam) and Gardner/Dynatup (disk) are available for measuring impact resistance (Ref 3). Such tests should not only measure the amount of energy absorbed, but also determine the effects of temperature on energy absorption. Additionally, they should be able to identify strain-rate-dependent transitions from ductile to brittle behavior. They should be applicable to a wide variety of geometric configurations. Unfortunately, these techniques provide only geometry-specific, single-point data for a specific temperature and strain rate. Also, each test provides a different ductile/brittle transition. Energy absorption, however measured, is made up of many complex processes involving elastic and plastic deformation, notch sensitivity, and fracture processes of crack initiation and propagation.

The prediction of strength and impact resistance of plastic parts is probably the most difficult challenge for the design engineer. Tensile stress-strain measurements as a function of temperature and strain rate provide one piece of useful information. Most unfilled engineering thermoplastics exhibit ductile behavior in these tensile tests, with increasing strength (maximum stress) as displacement rate increases and/or temperature decreases. However, stress-state effects must be added to the tensile behavior because the three-dimensional stress state created by notches, radii, holes, thick sections, and so forth increase the potential of brittle failure.

Ductile-to-brittle transitions in the fracture behavior of unfilled thermoplastics occur with increasing strain rates, decreasing temperatures, and increasingly constrained stress states. Figure 6 shows three common mechanical test techniques: uniaxial tension, biaxially stressed disks (usually clamped on the perimeter and loaded perpendicularly with a hemispherical tup), and notched beams loaded in bending. These three tests provide uniaxial, biaxial, and triaxial states of stress. Typical part geometries and loadings exhibit combinations of these states of stress. Thus, no one test is sufficient for part design and material selection. Furthermore, there are two competing failure modes: ductile and brittle (Fig. 6). With increasingly constrained stress states (uniaxial → biaxial → triaxial), the tendency for brittle failure tends to increase. Brittle failure occurs when the brittle failure mechanism occurs prior to ductile deformation (Fig. 6).

The calculation and measurement of the ductility ratio (Ref 6) is a method to characterize the ductility of a material for a relatively severe state of stress, for example, a beam with a notch radius of 0.25 mm (0.010 in.). The ductility ratio is defined as the ratio of the failure load in the notched-beam geometry ($P_{failure}$) to the maximum ductile, load-carrying capability in an unnotched-beam geometry where the height of the unnotched beam is equal to the net section height of the notched-beam geometry:

$$\text{Ductility ratio} = \frac{P_{failure}}{P_{ductile}} \quad \text{(Eq 1)}$$

where:

$$P_{ductile} = \frac{\sigma_f bh^2}{l} \quad \text{(Eq 2)}$$

and σ_f is the strength at appropriate rate and temperature, b is the beam thickness, h is the beam height, and l is the beam span.

This ductile load limit can be determined experimentally or with this plastic-hinge calculation assuming fully developed plasticity over the entire cross section and perfectly plastic material behavior. A ductility ratio of 1.0 corresponds to a ductile failure, while ductility numbers less than 1.0 correspond to varying levels of brittle behavior. Ductility ratios can be plotted as a function of strain rate at different temperatures to create fracture maps such as the one shown for polycarbonate (PC) in Fig. 7. This information is useful for material-selection and initial part design considerations.

Creep/Stress Relaxation—Time/Temperature Part Performance. Polymers exhibit time-dependent deformation (creep and stress relaxation) when subjected to loads. This deformation is significant in many polymers, even at room temperature, and is rapidly accelerated by small increases in temperature. Hence, the phenomenon is the source of many design problems. Development and application of methods are needed for predicting whether a component will sustain the required service life when subjected to loading, as the useful life of the part could be terminated by excessive deformation or even rupture. For most practical applications of polymers, pre-

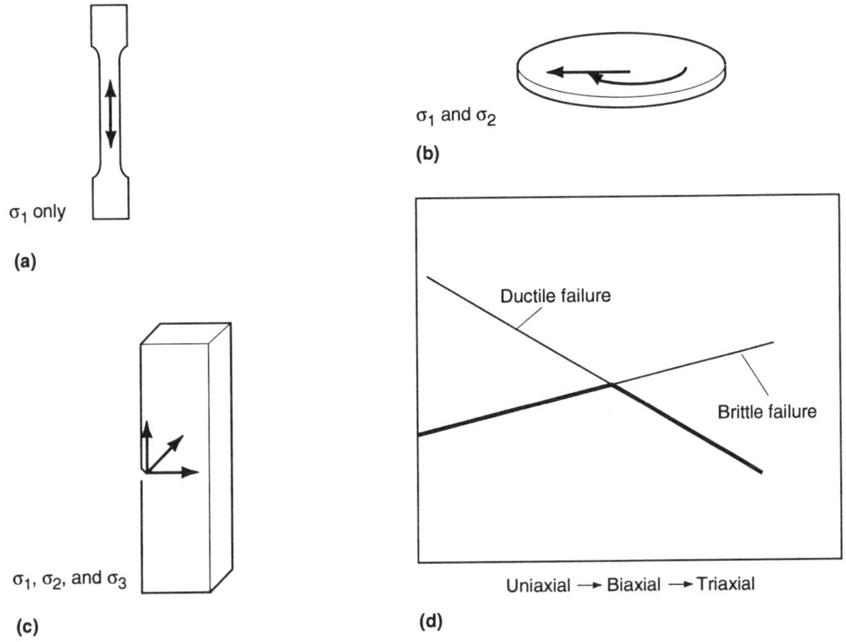

Fig. 6 Impact test methods exhibiting various states of stress. (a) Tensile test—uniaxial stress state. (b) Dynatup test—biaxial stress state. (c) Notch Izod test—triaxial stress state. (d) Competing failure modes

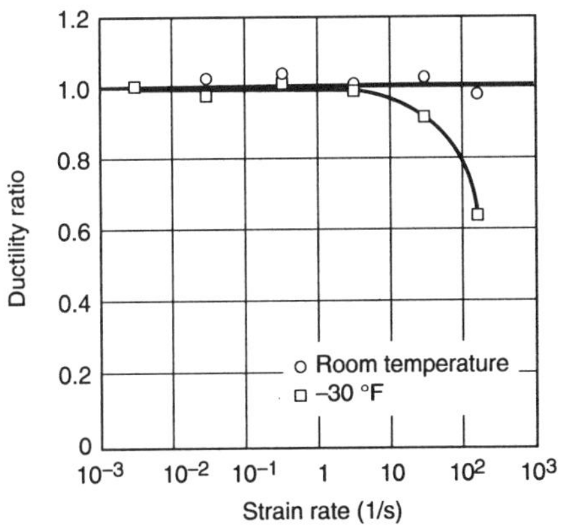

Fig. 7 Fracture map for polycarbonate

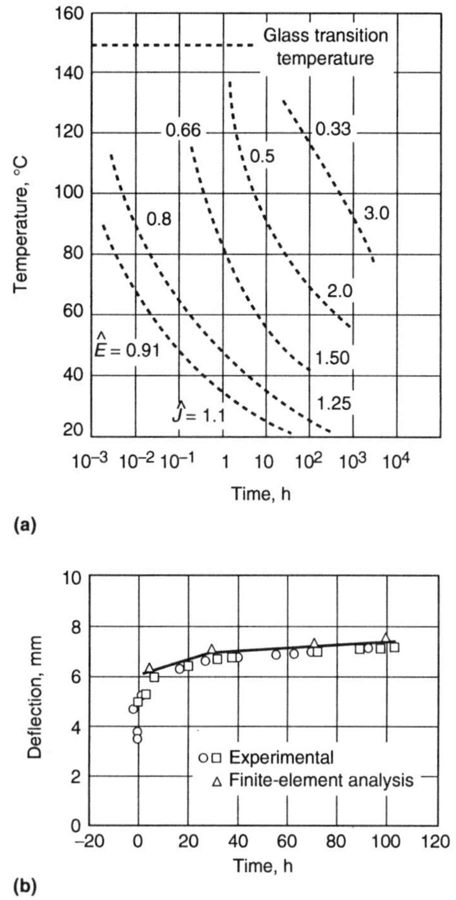

Fig. 8 Deformation map (a) used to predict PC part deformation at 82 °C (180 °F). $\hat{E} = E(T,t)/2350$ MPa. (b) Comparison of cathode-ray-tube housing creep prediction

dictive methods must account for part geometry, loading, and material behavior.

A common measure of heat resistance is the heat distortion temperature (HDT). For this test, bending specimen 127 by 12.7 mm (5 by 0.5 in.) with a thickness ranging from 3.2 to 12.7 mm (0.125 to 0.5 in.) is placed on supports 102 mm (4 in.) apart, and a load producing an outer fiber stress of 0.46 or 1.82 MPa (66 or 264 psi) is applied. The temperature in the chamber is increased at a rate of 2 °C/min (3.6 °F/min). The temperature at which the bar deflects an additional 0.25 mm (0.010 in.) is called the HDT or sometimes the deflection temperature under load (DTUL). Such a test, which involves variable temperature and arbitrary stress and deflection is of no use in predicting the structural performance of a thermoplastic at any temperature, stress, or time. In addition, it can be misleading when comparing materials. A material with a higher HDT than another material could exhibit more creep at a lower temperature. Also, some semicrystalline materials exhibit very different values of HDT at 0.46 and 1.82 MPa (66 and 264 psi). For example, with PBT, the HDT at 0.46 MPa (66 psi) is 154 °C (310 °F), and the HDT for 1.82 MPa (264 psi) is 54 °C (130 °F). The question of which HDT to use for comparison with another material that has the same HDT for both stress levels naturally arises. Another approach that is often used to account for the change in material modulus with temperature is the use of dynamic mechanical analysis (DMA) data (Ref 7). Although this approach may be a more useful indication of instantaneous modulus variation with temperature than HDT, it is unable to account for the time-dependent nature of most applications. For purposes of predicting part performance and for material selection, tensile creep data are the desired measurements.

To be useful for preliminary part design and material selection, creep data must be converted to simple information such as "deformation maps." A simple method is summarized where linear elastic part deformations are simply magnified by the use of deformation maps thus accounting for time and temperature effects.

A deformation map is produced directly from creep data (Ref 8). For a given temperature, T, the measured time-dependent strain, $\varepsilon(t)$ is divided by the applied stress, σ, to determine the creep compliance, J, as:

$$J(T,t) = \frac{\varepsilon(t)}{\sigma} \qquad \text{(Eq 3)}$$

The creep compliance is then normalized by dividing by the room temperature (T_0), instantaneous ($t \to 0$), elastic compliance, $J(T_0, 0)$:

$$\hat{J} = \frac{J(T,t)}{J(T_0, 0)} \qquad \text{(Eq 4)}$$

Because $J(T_0, 0)$ is the inverse of the room-temperature elastic modulus, E:

$$\hat{J} = J(T,t) \cdot E \qquad \text{(Eq 5)}$$

and

$$\hat{E} = \frac{1}{\hat{J}} \qquad \text{(Eq 6)}$$

Thus, a deformation map in time and temperature space can be produced from creep data with lines of constant compliance and modulus (Fig. 8). Thus, the design process involves calculating the linear elastic part deformation using E and then magnifying that deformation by $\hat{J}$ for the time of loading and ambient temperature. When a constant displacement is applied to a part, the calculated linear elastic stress using E is then reduced by $\hat{E}$ for the time of interest and ambient temperature. Thus, the deformation map provides a simple method to predict the time-dependent performance of plastic parts. As shown in Fig. 8, the deformation map provides the material response that can be combined with a linear elastic, time-independent analysis (in this case a finite-element stress analysis) to predict the time-dependent deformation. Validation of this approach is demonstrated by comparing it to experimentally measured part deformations (Fig. 8).

Fatigue-Cycle-Dependent Part Performance. An understanding of the deformation and fracture behavior of plastics subjected to cyclic loading is needed to predict the lifetime of structures fabricated from thermoplastics. This fatigue behavior is of concern because failure at fluctuating load levels can occur at much lower levels than failure under monotonic loading. A significant amount of information exists on the fatigue behavior of plastics. Unfortunately, very little has been documented about the application of this understanding to the prediction of the fatigue behavior of plastic parts.

There are two distinct approaches to treating and measuring the fatigue of polymers. The first approach is the traditional measurement of the number of cycles to failure (N) as a function of the fluctuating load or stress (S), that is, S-N. The "load" that is controlled is the minimum and maximum force or displacement in tension or bending. The fluctuations have a certain frequency and waveform. From a design viewpoint, it is difficult to predict part performance with these data because an enormous number of variables must be taken into consideration as well as various environmental conditions and a wide variety of materials.

The second approach to treating the fatigue of plastics is cyclic crack propagation. The use of fracture mechanics in cyclic fatigue involves the measurement of the amount of crack growth per cycle as a function of the stress-intensity factor. The fundamental addition here is the treatment of the crack length and thus an improved understanding of a fatigue mechanism. However, the same large number of variables that apply to the traditional fatigue (S-N) approach apply to the crack propagation approach. In addition, the design engineer is challenged with determining the initial or inherent flaw size.

Even though cycle-dependent part performance is not well understood, a general design-engineering approach can be applied to the fatigue of plastic parts. First, for material selection an awareness of the fatigue performance of numerous plastics is necessary. Materials should be compared under identical test conditions to determine their relative fatigue performance. This preliminary selection should be based on the general assessment of the relative fatigue performance, taking into account the overall severity of the part loading. Next, the part loading conditions should be determined and related to the appropriate laboratory data. This task is probably the most important, yet the most difficult due to the large number of variables involved. Establishing whether the part will experience load-controlled or displacement-controlled cyclic loadings is possibly the most significant factor. Next, the effects of frequency, waveform, and load level and type must be assessed to determine if part temperature will increase, leading to thermal fatigue, or if mechanical failure will occur with little or no temperature increase. Other conditions that should be considered or matched from the laboratory specimen to the component include environmental effects (e.g., temperature), stress state, stress concentrations, and mean stress. Finally, appropriate laboratory tests or full-scale component tests should be conducted. These laboratory tests must be carefully planned to achieve correspondence to the actual service conditions.

Fracture mechanics can be used to provide an approach to predicting the fatigue lifetime of components. The important additional feature is an understanding of crack growth through measurement of the amount of crack growth per cycle *(da/dN)* as a function of the cyclic range of stress-intensity factors (ΔK). Despite the fact that plastics are time-dependent materials, and that linear fracture mechanics only apply strictly to elastic materials, it appears that crack propagation rates in many polymers can be correlated with ΔK.

During the fatigue process, the stress amplitude ($\Delta\sigma$) usually remains constant and failure occurs as the result of crack growth from an initial, subcritical size to a critical size related to the fracture toughness (K_c) of the material. The lifetime of a component is thus dependent on the initial crack size, the rate of crack growth, and the critical crack size. The relation takes the power-law form:

$$\frac{da}{dN} = A\,\Delta K^n \qquad \text{(Eq 7)}$$

where A and n are material constants varying with temperature, environment, and frequency. The stress-intensity factor range is given as:

$$\Delta K = Y(\Delta\sigma)\sqrt{a} \qquad \text{(Eq 8)}$$

where Y is a crack and structural geometry factor and a is crack length. Typical crack propagation curves for a number of plastics (Ref 9) are shown in Fig. 9.

Fatigue lifetime of plastic parts can be calculated for design purposes by integrating the crack-growth rate expression (Eq 7) after substitution of Eq 8:

$$\frac{da}{dN} = AY^n\,\Delta\sigma^n\,a^{n/2} \qquad \text{(Eq 9)}$$

Assuming that the geometry factor Y does not change as the crack grows, this equation can be integrated to give the number of cycles to failure (N_f) that is necessary for the crack to grow from its initial size a_i to the critical size a_f. For $n \neq 2$:

$$N_f = \frac{2}{(n-2)AY^n\,\Delta\sigma^n}\left(\frac{1}{a_i^{(n-2)/2}} - \frac{1}{a_f^{(n-2)/2}}\right) \qquad \text{(Eq 10)}$$

This expression can be used to predict the fatigue lifetime of a component with an initial defect of known size.

The fatigue lifetime (number of cycles-to-failure) of a part is strongly dependent on the applied load. *S-N* curves have been generated for a number of thermoplastics (Ref 10) at room temperature with a standard tensile specimen with a net cross section of 12.7 by 3.2 mm (0.5 by 0.125 in.). The tensile load was varied from a very small load (nearly zero) to various maximum loads (stresses). A sinusoidal waveform with a frequency of 5 Hz was used. Very little or no specimen heating occurred. By choosing *S-N* curves for the same materials—polycarbonate (PC), modified polyphenylene ether (M-PPE), and acrylonitrile-butadiene-styrene (ABS)—whose fatigue crack propagation behavior is displayed in Fig. 9, the *S-N* data can be combined with the crack propagation data to compute the initial crack lengths (Eq 10). The final crack length a_f is computed from the fracture toughness of these materials. Thus, over the range of stresses for the *S-N* curves, the initial crack lengths can be computed. Ideally, these crack lengths would be independent of applied stress level. However, while there is some variation, the average crack length was computed and used in Eq 10 to "predict" the measured *S-N* data from the crack growth rate data. These results are shown in Fig. 10 for PC, M-PPE, and ABS. These data and this approach indicate the similarity of the *S-N* and crack growth rate methods of predicting part lifetime and suggest a method of utilizing both types of data.

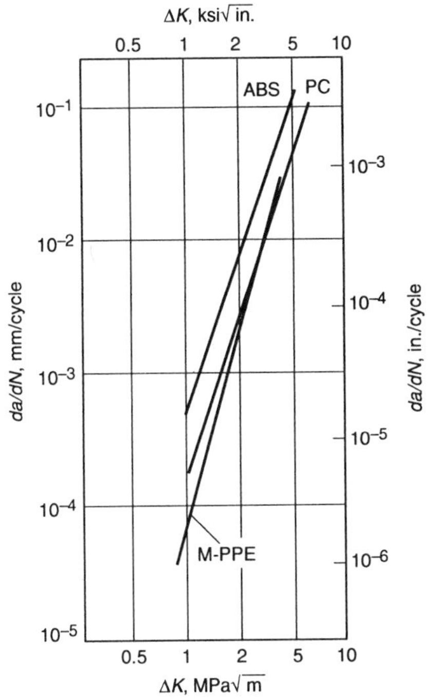

Fig. 9 Fatigue crack propagation behavior. ABS, acrylonitrile-butadiene-styrene; PC, polycarbonate; M-PPE, modified polyphenylene ether

Manufacturing Considerations

Flow Length Estimation. The ability to manufacture plastic parts using the injection-molding process is governed by the material behavior, part geometry, and processing conditions. Estimating the flow length of the resin into a mold of a given thickness is an important manufacturing consideration for the design engineer. An example is given here of a generic tool (Ref 11) capable of analyzing radial flow and quantifying effects of material, geometry, or process changes. This tool, Diskflow, is composed of a numerical flow analysis, automatic mesh generator, and menu-driven pre- and postprocessors. No knowledge of simulation techniques is required, though a knowledge of injection molding is needed when interpreting the results.

Diskflow utilizes modeling techniques common to most commercial analyses (see the article "Computational Fluid Dynamics" in this Volume), yet is much faster due to the radial flow assumption and subsequent numerical methods. Modeling, postprocessing, and computer analysis time have been minimized, reducing total analysis time. This analysis cannot replace three-dimensional filling analyses as it does not yield any information regarding knit-line and gas trap locations, cavity pressure and temperature distri-

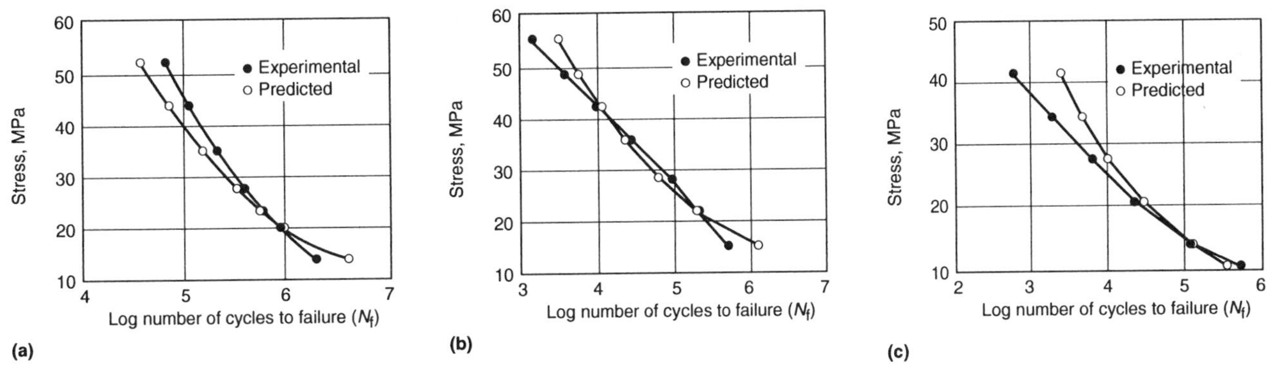

Fig. 10 *S-N* data compared to crack growth prediction. (a) PC; a_i = 0.013 mm. (b) M-PPE; a_i = 0.32 mm. (c) ABS; a_i = 0.23 mm

butions, or more complex mold geometries, but it does yield significant results regarding design feasibility, material performance, and process effects.

Diskflow uses complete viscosity versus shear rate and temperature data fit with a modified Cross model with Williams-Landel-Ferry (WLF) temperature dependence when available. If this advanced model is not available, the Arrhenius model is employed. Mold and melt temperatures can be chosen if default values are not adequate. The mold geometry is defined in terms of a nominal wall thickness and cavity radius that can be calculated by entering cavity volume or projected area. The sprue is defined by entering the sprue type (hot or cold) and then entering the length, upper diameter, and lower diameter.

For flow length estimation, an initial flow rate is assumed constant subject to some user-specified maximum pressure limit that mimics the capability of a molding machine. As the mold fills at a constant volumetric flow rate, the injection pressure rises due to the increasing flow resistance. When the injection pressure attains the user-specified maximum, the analysis switches over to a second phase in which the injection pressure is maintained at a constant value and the flow rate is allowed to vary; the flow rate eventually decays to zero at which point a final flow length is attained. The flow length may be defined as the farthest distance that a polymeric material travels in a mold of some nominal wall thickness given a set of processing conditions. The flow length capability examines the feasibility of manufacturing a desired design: if the distance from the gate to the corner of the part is greater than the predicted flow length, then the part may not be manufacturable. Figure 11 shows the dependence of flow length on wall thickness for a maximum injection pressure of 103.5 MPa (15 ksi) for PC. This information is useful for assessing manufacturability in the early stages of design and material selection.

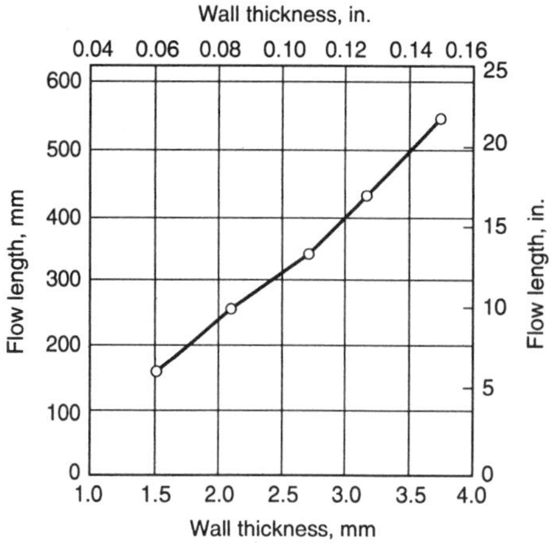

Fig. 11 Flow length versus wall thickness predicted by Diskflow mold-filling analysis. Material, unfilled PC; mold temperature, 82 °C; melt temperature, 335 °C; maximum injection pressure, 103.4 MPa

Cycle Time Estimation. The molding of thermoplastics consists of injecting a molten polymer into the cooled mold cavity. The injected resin is held in the cavity until the part solidifies (by heat transfer). The time for the melt to cool until it solidifies to the extent that the part can be removed from the mold and retain its dimensions is generally the majority of the total cycle time. The large impact of the cooling time on the total processing cost is obvious.

During the cooling phase, heat conduction is the prime mechanism of heat transfer. The development of a simplified mold-cooling program allows designers and molders to evaluate materials and process parameters in a rapid, convenient, and cost-efficient manner. Plastic parts are usually thin, and thus a one-dimensional, transient heat conduction analysis is adequate to approximate the cooling of the real part. The main assumption is that the mold surface is kept at a constant temperature throughout the cooling phase. Comparing calculated minimum cooling times for different material part geometries (i.e., thickness) and processing conditions help optimize the material-selection process.

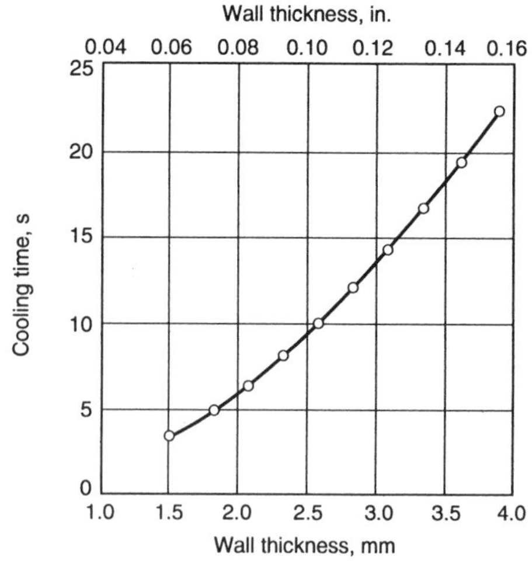

Material	Unfilled PC
Thickness, mm	1.62–3.81
Thermal conductivity, W/m · K	0.270
Specific heat, W · s/kg · K	1791
Melt temperature, °C	300
Mold temperature, °C	82
Ejection temperature, °C	112

Fig. 12 In-mold cooling time versus wall thickness predicted from one-dimensional, transient mold cooling analysis

Thermal material properties are strong functions of temperature. Because the thermoplastic material experiences a wide range of temperatures during the cooling phase, temperature-dependent material data such as specific heat and thermal conductivity are used for the computations. To perform the analysis the injection temperature, mold temperature, ejection temperature, material, and thickness must be chosen. The program uses a one-dimensional finite-difference

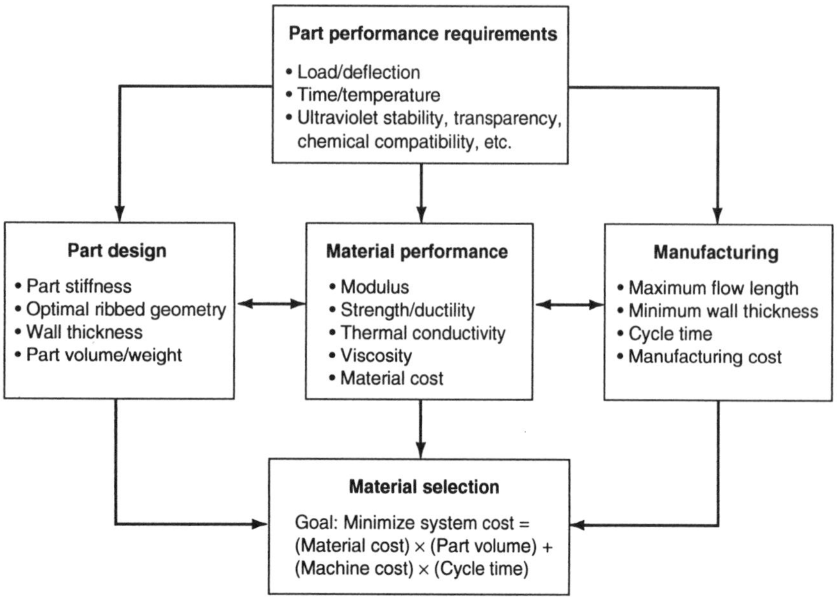

Fig. 13 Design-based material-selection process

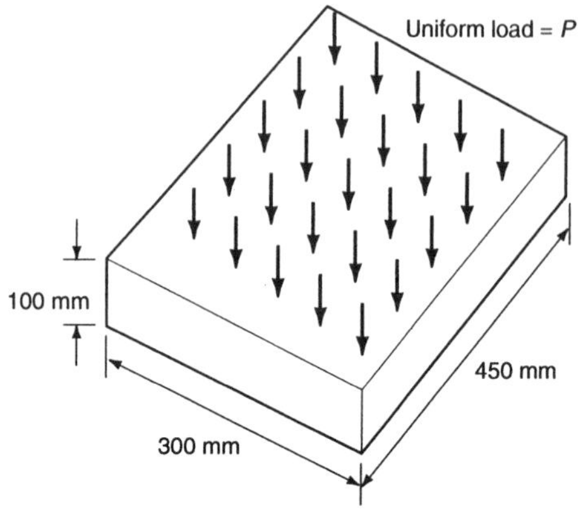

Fig. 14 Geometry of enclosure example

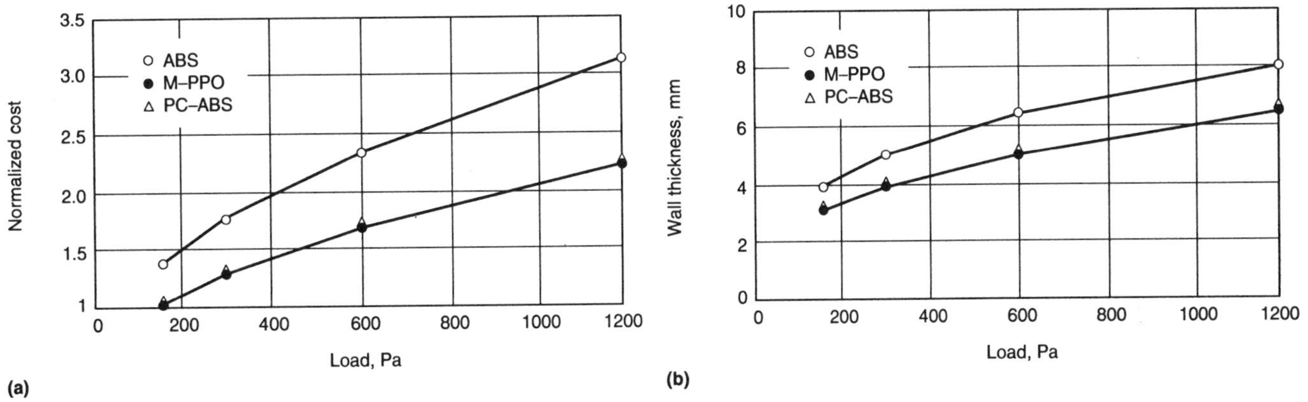

Fig. 15 Loading variation for 40 °C and 1000 h

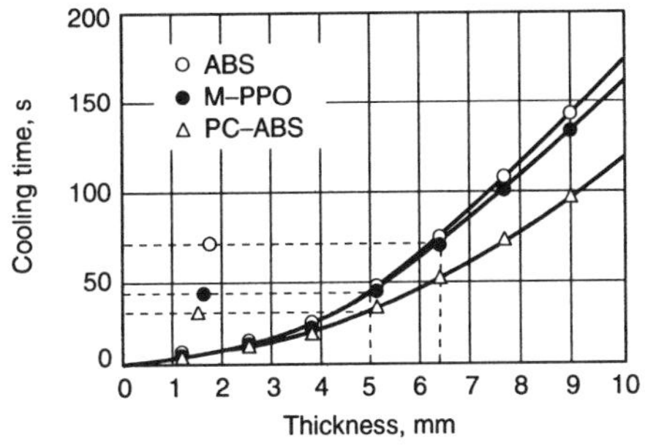

Fig. 16 Cooling time versus wall thickness

scheme to calculate temperature through the thickness as a function of time. When the center of the plate reaches the specified ejection temperature, the analysis is stopped and the results are displayed graphically. By performing the analysis for a range of part thicknesses, cooling time curves can be produced (Fig. 12). These curves can then be used to estimate cycle times in the early stages of material selection and design.

Design-Based Material Selection

Design-based material selection (Ref 12, 13) involves meeting the part performance requirements with a minimum system cost while considering preliminary part design, material performance, and manufacturing constraints (Fig. 13). Some performance requirements such as transparency, Food and Drug Administration (FDA) approval, or flammability rating are either met by the resin or not. Mechanical performance such as a deflection limit for a given load are more complicated requirements. Time- and temperature-reduced stiffness of the material is determined from the deformation map. Part design for stiffness involves meeting the deflection limit with optimal rib geometry and part thickness combined with the material stiffness. This part geometry can be used to compute the part volume that when multiplied by the material cost provides the first part of the system cost. The second half of the system cost is the injection-molding machine cost multiplied by the cycle time. This total system cost is a rough estimate used to rank materials/designs that meet the part performance requirements. In addition, the manufacturing constraint of flow length for the part thickness must be considered. The entire process is summarized in Fig. 13.

Example 1: Materials Selection for Plate Design. A simple example is presented to illustrate the design-based material-selection process. A 254 by 254 mm (10 by 10 in.) simply supported plate is loaded at room temperature with a uniform pressure of 760 Pa (0.11 psi). The maximum allowable deflection is 3.2 mm (0.125 in.). Using the program Ribstiff, described previously in the part stiffness section, the nonlinear load-displacement response of the plate can be computed. Through iteration, it is determined that a PC plate with a thickness of 2.5 mm (0.1 in.) satisfies the requirements (Fig. 3). From Fig. 11, the flow length is 320 mm (12.5 in.). Thus, the plate could be filled with a center gate or from the center of an edge. From Fig. 12, the in-mold cooling time is 10 s. The volume of the plate is 0.00016 m^3 (10 $in.^3$).

A second design can be produced by designing a rib-stiffened plate. Again, through iteration, a 1.5 mm (0.060 in.) thick plate with 10 ribs in each direction with a rib height of 4.5 mm (0.18 in.) and a rib thickness of 1.5 mm (0.060 in.) would meet the deflection requirement. From Fig. 11, the flow length is about 175 mm (7 in.). Thus, because a center-gated plate would have a flow length of 175 mm (7 in.), the part would probably fill if the ribs would serve as flow leaders to aid the flow. However, it is generally not recommended to push an injection-molding machine to its limits because this will exaggerate inconsistencies in the material and the process. A more thorough three-dimensional process simulation should be performed to determine the viability of this design before it is chosen. From Fig. 12, the in-mold cooling time is about 4 s, a considerable savings (6 s/part) in cycle time as compared to the plate with no ribs. In addition, the volume of the

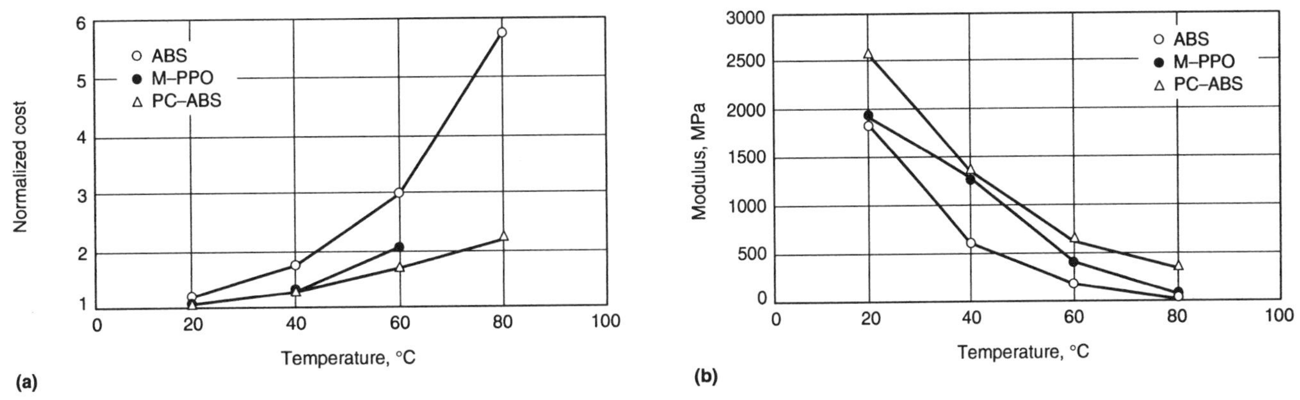

Fig. 17 Temperature variation

ribbed plate is 0.00013 m^3 (8 $in.^3$), a saving of 20% on material as compared to the plate with no ribs. The system cost of the ribbed plate is computed to be 73% of the plate with no ribs (Fig. 1). Because the ribs would produce a constrained, three-dimensional stress state, consideration of impact would be important for high rates of loading and low temperature (Fig. 7). The fracture map shows a tendency for brittle behavior with PC at low temperature and high loading rates for notched or constrained geometries.

If time/temperature performance were added to this example as a requirement, the optimal material may change or the initial design would need to be modified. If the same load were applied to the plate for 1000 h at a temperature of 79 °C (175 °F), the PC plate would exhibit a deformation as if its material stiffness were about 40% of the room-temperature modulus (Fig. 8). Simply increasing the thickness of the plate with no ribs to 3.5 mm (0.136 in.) would provide a design that would meet the deflection requirements. The penalty would be a 40% increase in material usage and an additional 8 s added to the cycle time. Choosing a material with more temperature resistance or initial stiffness is an option.

Example 2: Materials Selection for an Electrical Enclosure. The usefulness of this process can be demonstrated through another design example. In this case, a very simple five-sided box is chosen. The box is used as an electrical enclosure and must meet flammability requirements. This limits the number of candidate materials to examine more closely. Also, this enclosure is not painted and, therefore, the resin must be unfilled to maintain acceptable aesthetics. It is unribbed to minimize sink marks on the exposed surfaces. Finally, it must support a uniform load across its surface without deflecting more than 2.5 mm. The enclosure is a 300 mm wide by 450 mm long by 100 mm high box (Fig. 14).

A series of analyses is performed using three resins to see how they perform under different conditions. These resins are representative of what is currently used in electrical enclosures (computer housings, office equipment, etc.). They are an unfilled M-PPO resin, an unfilled ABS resin, and an unfilled PC-ABS resin blend.

To examine the relative performance of each resin, the application requirements are varied in loading, environment, and manufacturing. First, the uniform load is varied from 150 to 1200 Pa. Next, the ambient temperature the enclosure must withstand for 1000 h under load is varied from 20 to 80 °C. Finally, the gating scenario is changed from edge gated to center gated to multiple gates.

Using a center-gated box at 40 °C for 1000 h, the uniform load is varied from 150 to 1200 Pa. The Design-Based Material Selection Program determines for each resin the optimal wall thickness to support the load at the lowest variable system cost for each loading case. Figure 15(a) compares the normalized cost of the enclosure for each resin as the load is increased. As can be seen from this graph, the PC-ABS and M-PPO are virtually equivalent in cost, while the ABS is about 30% more expensive. While this may seem counterintuitive (ABS is less expensive per pound than PC-ABS or M-PPO), it is easily explained by examining Fig. 15(b), wall thickness versus loading. At this elevated temperature and long time (40 °C, 1000 h), the ABS requires significantly more material to support the required load within the specified 2.5 mm deflection than either the PC-ABS or the M-PPO. This added material far outweighs the price advantage of ABS.

The cooling time is another factor that will increase the variable system cost of the ABS resin enclosure. As the wall thickness increases, the time to cool the part to ejection temperature will increase. The cooling time is also influenced by the thermal properties of each resin. Figure 16 contains a graph of the cooling time versus wall thickness for the three example materials based on one-dimensional transient heat-transfer analyses. The wall thickness for each resin to support 600 Pa at a deflection of no more than 2.5 mm is indicated on the graph. From this graph, it can easily be seen that, in this case, the cooling time for each resin will be very different.

Using a center-gated box that must support a 300 Pa load within a 2.5 mm deflection of 1000 h, the temperature was varied from 20 to 80 °C. Figure 17(a) compares the normalized cost of these three resins as the temperature is increased. Initially, at 20 °C these resins have very similar variable system costs. As the temperature increases, the creep performance of each resin decreases. Figure 17(b) shows the creep modulus for each resin as the temperature changes. The creep modulus of the ABS resin decreases rapidly as temperature increases. The M-PPO maintains its stiffness longer, but eventually decreases rapidly while the PC-ABS performs better, because of the high creep resistance of the PC component of the blend.

The wall thickness to support the load must increase as temperature increases because the creep modulus decreases. This, in turn, increases the part volume and the cooling time, affecting the variable system cost. As the temperature increases, the cost rises to high levels (ABS at 80 °C, 1000 h). If the application must withstand these temperature extremes, a higher-performance thermoplastic may be a better choice.

The process to manufacture this enclosure can influence how the enclosure will be designed and what material will be used. Using a box that must support a 150 Pa load within a 2.5 mm deflection in a 40 °C environment for 1000 h, the gating

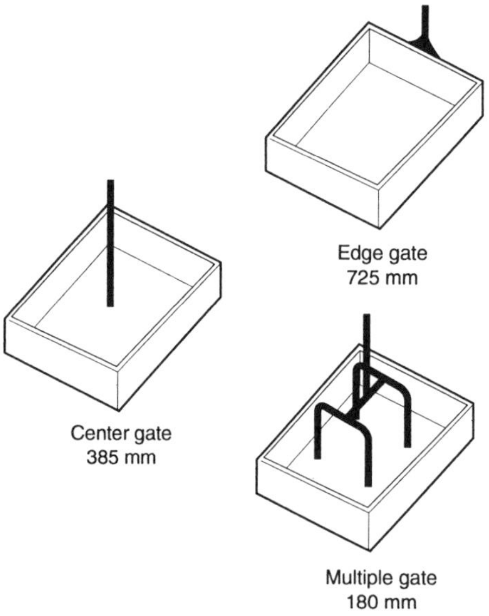

Fig. 18 Examples of gating scenarios

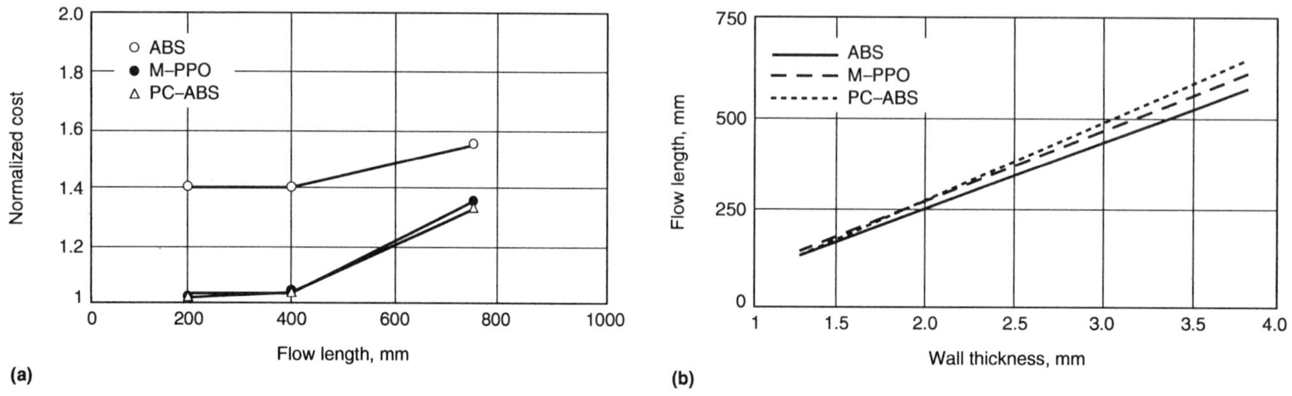

Fig. 19 Gating variation

scenario is varied choosing three common configurations (Fig. 18) edge gate, center gate, and four gates. The minimum flow length necessary to fill the part is determined for each case based on the geometry of the enclosure and the gate position. The minimum flow length necessary to fill the part is determined for each case based on the geometry of the enclosure and the gate position. The minimum wall thickness to allow each material to achieve this flow length, determined using the radial flow injection molding simulation, is then used as a lower bound on the thickness optimization and is shown in Fig. 19(b). Figure 19(a) details the normalized cost versus minimum flow length (i.e., gating scenario). Initially, as the flow length increases (from four gates to center gate) the normalized cost does not change. The wall thickness necessary to support the load within the specified deflection is greater than the minimum wall thickness dictated by the flow length constraint. As the flow length increases from the center-gated to the edge-gated case, the normalized cost increases because the wall thickness is now dictated by the manufacturing constraint rather than the loading condition. The gate placement now dictates the wall thickness that is necessary to fill the part.

There are other considerations that a design engineer can use to help determine the best material for an application. The strength of a resin over a range of temperatures may aid the engineer in determining if the part will fail under load. The impact performance of the resin, as indicated by the ductility ratio, can also be quite important. While it only indicates the impact performance for one specific geometry, and cannot be used in design, it does provide useful comparative information.

Conclusions

The goal of the design engineer is to achieve the lowest system (design/material/process) cost while meeting the part performance requirements. Material selection and engineering design of plastic parts is a difficult task because of a lack of effective and efficient design methods and the associated material data. However, there are now some new methods that will improve the design process by providing more accurate and effective predictive techniques. Fracture maps indicate the relative ductility of a material as a function of temperature and strain rate for a relatively severe stress state. A range of test data for different stress states from tensile tests, disk tests, and notched beams is used to predict part deformation and potential ductile-to-brittle behavior. For time-dependent deformation, such as creep or stress relaxation, deformation maps can be combined with linear-elastic calculations of part deformation to predict the time- and temperature-dependent deformation of the part. The cross-flow stiffness and strength of injection-molded glass-filled materials is sometimes only 50% of the stiffness and strength in the flow direction, especially for thin-walled parts. This must be accounted for in predicting part stiffness and strength. For predicting lifetime of parts subjected to cyclic loading, the combination of *S-N* data and crack growth rate data is useful because it provides two options: to use the *S-N* data directly or to use the initial defect size with the crack growth rate data. In either case, with the vast number of parameters that affect fatigue behavior, having more information is useful. The design methods and material data summarized here provide the design engineer with more effective and efficient techniques to select materials and design plastic parts.

REFERENCES

1. G.G. Trantina and D.A. Ysseldyke, An Engineering Design System for Thermoplastics, *1989 ANTEC Conf. Proc.*, Society of Plastics Engineers, p 635–639
2. E.H. Nielsen, J.R. Dixon, and M.K. Simmons, "GERES: A Knowledge Based Material Selection Program for Injection Molded Resins," ASME Computers in Engineering Conference (Chicago), American Society of Mechanical Engineers, July 1986
3. G.G. Trantina and R.P. Nimmer, *Structural Analysis of Thermoplastic Components*, McGraw-Hill, 1994
4. K.C. Sherman, R.J. Bankert, and R.P. Nimmer, "Engineering Performance Parameter Studies for Thermoplastic, Structural Panels," *1989 ANTEC Conf. Proc.*, Society of Plastics Engineers, p 640–644
5. G. Ambur and G.G. Trantina, Structural Failure Prediction with Short-Fiber Filled Injection Molded Thermoplastics, *1988 ANTEC Conf. Proc.*, Society of Plastics Engineers, p 1507
6. J.T. Woods and R.P. Nimmer, Design Aids for Preventing Brittle Failure in Polycarbonate and Polyetherimide, *1996 ANTEC Conf. Proc.*, Society of Plastics Engineers, p 3182–3186
7. M.P. Sepe, Material Selection for Elevated Temperature Applications: An Alternative to DTUL, *1991 ANTEC Conf. Proc.*, Society of Plastics Engineers, p 2257–2262
8. O.A. Hasan and G.G. Trantina, Use of Deformation Maps in Predicting the Time-Dependent Deformation of Thermoplastics, *1996 ANTEC Conf. Proc.*, Society of Plastics Engineers, p 3223–3228
9. R.W. Hertzberg and J.A. Manson, *Fatigue of Engineering Plastics*, Academic Press, 1990
10. G.G. Trantina, Material Properties for Part Design and Material Selection, *1996 ANTEC Conf. Proc.*, Society of Plastics Engineers, p 3170–3175
11. D.O. Kazmer, "Development and Application of an Axisymmetric Element for Injection Molding Analysis," *1990 RETEC Conf. Proc.*
12. G.G. Trantina, P.R. Oehler, M.D. Minnichelli, Selecting Materials for Optimum Performance, *Plast. Eng.*, Aug 1993, p 23–26
13. P.R. Oehler, C.M. Graichen, and G.G. Trantina, Design-Based Material Selection, *1994 ANTEC Conf. Proc.*, Society of Plastics Engineers, p 3092–3096

Design with Composites

R.J. Diefendorf, Clemson University

THE QUEST for improved performance has caused designers to search for materials with lower density, higher modulus, and higher theoretical strength than the common engineering metals. There are a number of elements and compounds, all centered around carbon in the periodic table, that have lower density and high modulus and strength. All are brittle ceramic materials with the exception of beryllium. The specific modulus (modulus/density) provides an indication of performance for a stiffness-limited application and for pure tensile applications in which weight is important (Fig. 1). The theoretical tensile strength of a brittle material is approximately 0.2 of the elastic modulus, so the same materials that have high modulus have the potential for high strength. Practically, many of these materials have been made in fiber form with high tensile strength, although with much lower strength than the theoretical values (Fig. 2). The ranking is more complicated for compressive loading as the failure mechanism may vary, but the general formula is $(\text{modulus})^n/\text{density}$, where $n = \frac{1}{2}$ or $\frac{1}{3}$. While the value of the modulus can be less important because of the fractional exponent, the density always is factored to the first power. The common engineering metals cannot match, on the basis of specific modulus and strength, the decrease in structural weight that can be achieved with these ceramic materials.

A high specific modulus and strength is insufficient to be a useful engineering material. Designers resist using brittle materials, because of poor impact, little damage tolerance, and difficulties with machining and assembly that may raise local stresses to high values. Structural integrity often depends on withstanding these high local stresses. Ductile metals yield and redistribute the stress, but composite materials exhibit a lower ductility. Stress analysis is often very good at the global scale, but is apt to be poorer in localized areas. The application of finite element analysis has decreased problems with stress concentrations, but providing the ceramic material with a pseudoductile behavior would be a breakthrough. Fiber-reinforced composite materials can provide better impact resistance and a more "graceful failure," or more yielding strain before failure. Additional information about performance characteristics of brittle materials is provided in the article "Design with Brittle Materials" in this Volume.

Fiber-Reinforced Composite Materials

Fibers can be made in small diameter with desirable microstructures, such that very high strength can be obtained. High loads can be carried by using a plurality of these high-strength fibers. Damage tolerance is also achieved, as the loss of a few fibers does not decrease the overall load-bearing capability much. However, if there is no coupling among a bundle of parallel fibers, fiber fractures accumulate with increasing load until the remaining fibers fail ("bundle strength") (Ref 2) (Fig. 3a). There are two problems: (1) no load is carried by a broken or discontinuous fiber, and (2) fibers buckle at very low loads in compression because of their small diameter. Man learned many centuries ago that twisting the continuous fiber bundle into a yarn provides a frictional force among the fibers that also allows discontinuous fibers to be used. Multiple breaks in a single fiber could even be generated in a yarn during loading. However, improving the compressive strength requires better fiber stabilization than coupling by friction. The problem is that in a strongly coupled system, failure of the whole structure occurs when the first fiber fails—a very low strength (Fig. 3b.). Noncatastrophic failure requires that the bond strength between the fiber and matrix be sufficiently low that when the fiber fractures, the fiber debonds from the matrix to arrest crack propagation by blunting (Fig. 3c). Damage tolerance is provided by the redundancy in load-carrying members. Localized impact may fracture a few fibers, but the load carried by the fractured fibers will be transferred through the matrix to the remaining fibers.

The whole basis for fiber-reinforced composites is that some of the potential strength and modulus are traded for a measure of damage tolerance and a more graceful failure. The penalty is that the strength transverse to the fiber bundle is low. The strength and modulus of a uniaxial fiber-reinforced composite is high in only one direction. Of course, fibers could be oriented in different directions to provide the required stiffness and strengths, but space is not necessarily filled very efficiently. Woven fabrics often are used to replace uniaxial plies or are added locally to improve transverse properties. For triaxially stressed parts, composites usually do not offer improved performance compared to the engineering metals. (There are important applications in which the engineering metals cannot be used, and composites are attractive.) Fortunately, structural elements are usually stressed predominantly

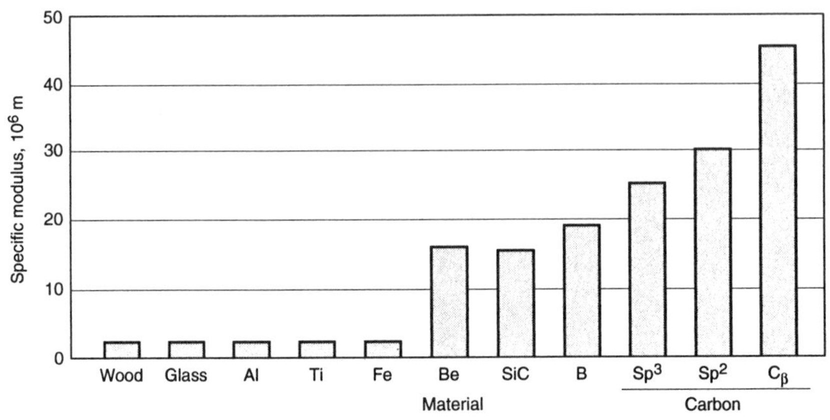

Fig. 1 Specific modulus (modulus/density) of selected materials for structural applications. Note the constancy of the common engineering metals and the high values for low-density ceramics. Source: Ref 1

in one (beams) or two directions (plates), and not three directions. For uniaxially and biaxially stressed elements, the performance levels possible with composites cannot be matched by other materials.

Structure/Property Relationships

A structure can be analyzed from the scale of atoms, unit cells, point and line defects, optical microstructure, beams and plates to the final level of the total structure. The dividing line between the interests of the materials engineer and the structural engineer has traditionally been between the microstructure, and beams and plates. Generally, the structural engineer was provided with a set of properties for the materials that were isotropic and homogeneous. The details of (micro)structure that determine the properties were invisible at the level of the structural engineer. Exceptions to assuming isotropy were made for metals and polymers with strong textures, and properties were provided for the different orientations. However, with composite materials, the structural engineer becomes much more the designer of the material. A composite is a structure in itself. In this case, the engineer must orient the fibers in the matrix to provide the stiffness and carry the loads. The fiber and polymer, metal, or ceramic matrix constituents are generally treated as homogeneous and isotropic, but sometimes as anisotropic.

Conceptual Design of a Composite Part. A composite part can be visualized as a group of ropes loosely coupled together by a matrix, or one step up the resolution scale as an anisotropic but homogeneous material in a beam or plate. The approach of the designer is quite different. For the rope analogy in tension, the designer attempts to have the ropes carry most of the load, while minimizing the loads carried by the matrix. By contrast, the stresses in the individual fiber and matrix constituents are not resolved in an anisotropic plate model, and the material can be treated much as a common engineering metal. (The term "black aluminum" is often used to describe carbon fiber/epoxy-matrix composites.) The design of composite materials is simplified by this approach, and much of the potential improvement in structural performance can often be achieved. Because composite laminates are frequently laid up against the surface of a mold, the fibers will follow the contour of the mold and provide high in-plane strength and stiffness. However, maximum performance is not attained, and high matrix or fiber stresses in localized regions may cause failure. The resolution of the individual constituents in a composite provides the designer with many more degrees of freedom, but more complexity.

Twist, Bend, and Splay Orientations. Fiber-reinforced laminates are usually produced by stacking a prescribed number of lamina or plies. Each ply is generally made up of uniaxially oriented fibers or woven fabric. The laminate is then built up by stacking plies at various rotations based on the predicted load state (Fig. 4). The result is much like plywood, although plies at angles other than 0/90° are also used. There are some advantages to using splay and bend fiber geometry to pick up off-axis loads (Ref 4). (Off-axis plies in Cartesian coordinates transform to logarithmic spirals for cylindrical coordinates.) Design and manufacture become more complicated, but machinery is now available to "steer" the fiber in these more complex patterns. Figure 5 shows an example of a pin-loaded joint. Not only is the stress concentration reduced at the hole, but also the maximum stress is moved from the free edge for the ply rotation case (Fig. 5b) toward the interior for bend/splay (Fig. 5c).

Generic Composite Behavior

The stress/strain and fracture behavior of uniaxial fiber-reinforced composites depends on the moduli and strains to failure of the constituents, the coupling between fiber and matrix, the residual stresses in the composite, and the loading directions. Generally, the modulus of the fiber is high, but the modulus of the matrix can vary from low to values even higher than that for the fiber.

In the case of resin-matrix composites, the fiber modulus can be 50 to even 100 times greater than that of the matrix. However, even though the strength of the fibers is usually very high, the strain to failure of the fibers (<2%) is frequently less than for the matrix. When loaded in uniaxial tension parallel to the fiber axis, the fibers carry most of the load because of their higher modulus. The weakest fibers start to fail as the load is increased, until a large crack forms that propagates through the composite.

If the fibers and matrix are well coupled, failure can even occur when the weakest fiber fails, and the fracture surface will be relatively flat and have the classic mirror/hackle appearance (Fig. 3b). For weaker coupling, fiber failures will accumulate until a few neighboring failed fibers generate a large enough stress concentration that catastrophic failure occurs through the remaining section of the composite (Ref 5). The composite strength will be much higher than in the first case, and damage tolerance will be improved but still poor. Still lower coupling will cause less overstress from broken fiber ends on neighboring

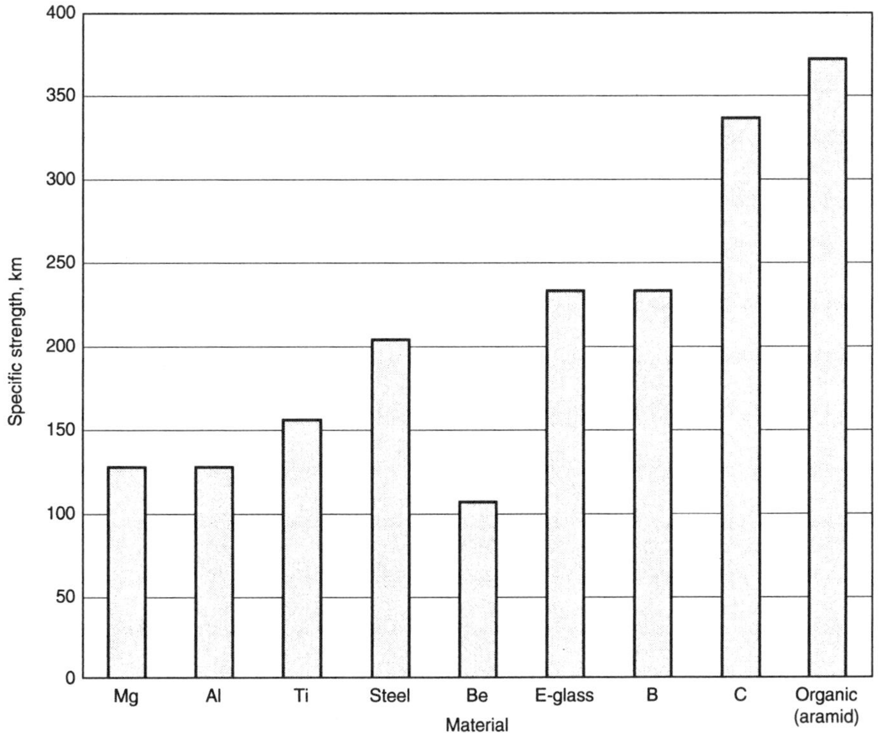

Fig. 2 Specific strength (strength/density) of selected materials for structural applications. Values are for high-strength wires. Materials with high specific moduli are often observed to have high specific strength also. Source: Ref 1

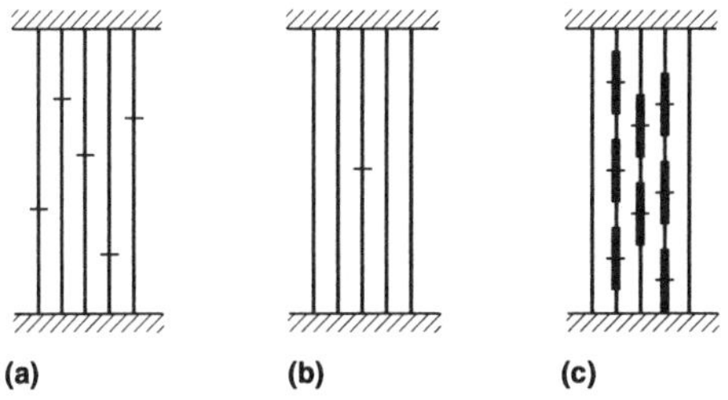

Fig. 3 Interfacial bonding. A bundle of fibers (a) usually will have about 70 to 80% of the average tensile strength determined on single fibers. With a very strong bond between fiber and matrix (b), the whole composite will fail in a brittle manner when the very weakest fiber in the composite fails. Tensile strength is low. An optimal interfacial strength (c) will produce multiple fractures in each filament, and the tensile strength will equal or even exceed the average fiber tensile strength. Source: Ref 3

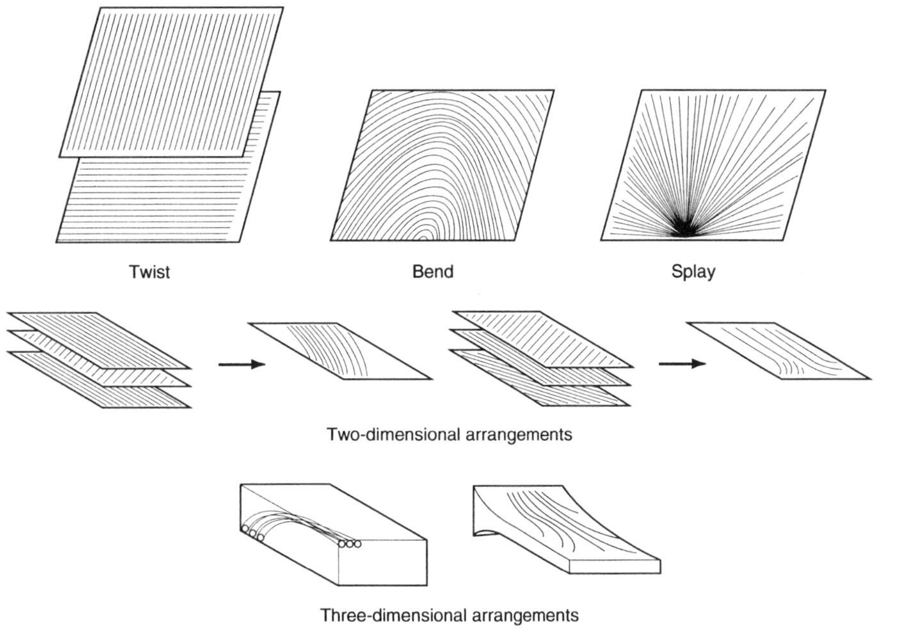

Fig. 4 Fiber arrangements. Composite laminates are generally fabricated by rotating lamina axes as shown by the twist example on the left. However, loads that are off-fiber axes cause large inter- and intralaminar matrix stresses. Safer structures can often be designed by using curved or splayed fiber paths to try to keep loads within the fibers as shown on the right. The design freedom that is added by using twist, bend, and splay is illustrated in the two- and three-dimensional design concepts. Fabric also can be used.

fibers, such that in the limit, failure might be conceived of as occurring when the total cross section, minus the broken fibers, can no longer support the increasing load (Fig. 3c.). Normally, this extreme is not used in composites as the tensile strength is lower, and more importantly, the off-fiber axis properties are necessarily much lower because of the poor coupling. The composite fracture surface will appear very fibrous with broken fibers projecting out from the surface to lengths of 20 diameters or more.

The strength goes through a maximum as the coupling between fibers and matrix is increased, and the fracture behavior becomes more brittle (Fig. 6). The length of the fibers projecting from the fracture surface at the maximum tensile strength is in the range of six to ten fiber diameters. The length of the projecting fibers indicates the length of fiber required to reach the breaking stress in the fiber. So, while a long fiber pullout provides toughness, a longer distance is also required to transfer stress into the fiber. In general, the fiber/matrix coupling has not been adjusted to tailor the properties of the composite toward toughness or transfer length. Polymer-matrix composites are usually optimized close to the maximum strength, which results in rather brittle systems. However, the development of ceramic-matrix composites has focused on toughness by manipulating fiber/matrix coupling.

Compressive strength of uniaxially oriented composites, loaded parallel to the fiber axis, is determined partly by the support provided by the matrix and neighboring fibers. Higher modulus matrices and tighter coupling between fiber and matrix can improve compressive strength (Ref 7). The overall fiber buckling load is proportional to the fiber modulus and diameter, but carbon and organic fibers buckle internally (Ref 8). In higher-modulus carbon or organic fibers, the compressive strength is a fraction of the tensile strength, which is design limiting for many applications.

The weak coupling between fiber and matrix, which is necessary to control fracture behavior when the composite is loaded parallel to the fiber axis, decreases off-axis and transverse tensile strength drastically. Any significant off-axis or transverse tensile loads require fibers to be oriented to pick up the loads. Fibers really are "reinforcing fibers" in low-modulus matrices.

Fibers and matrices can be combined in countless different ply orientations and sequences in composite materials. The cost to characterize the properties even at room temperature much less at high temperature would be prohibitive. Analytical procedures are required to calculate the properties, ideally from the properties of the constituent materials, and then validate the predictions by selected experiments (Ref 9). Unfortunately, the magnitude of the coupling between the fiber and matrix is often unknown, and the variable packing of the fibers is generally not accounted for. Calculations of composite properties from the constituent properties of the fiber and matrix have not always been successful. Progress is being made in the methodology and particularly in computational speed. The fracture behavior should be directly calculable for even composites with unusual fiber trajectories by the turn of the century. The short-term solution has been to determine the lamina properties experimentally and to calculate laminate properties from lamina. A number of computer programs for calculation of laminate properties for resin-matrix composites are available. Even the programs for personal computers provide a good first cut for most resin-matrix composite properties. Extensive experimental determination of the properties of many polymer- and carbon-matrix composite laminates has also been performed, and data are available in U.S. National Aeronautics and Space Administration (NASA) and Department of Defense (DoD) publications, including *MIL-HDBK-17.* Data for metal-matrix composites are less abundant, but still voluminous. Both metal- and ceramic-matrix composite data are much more scattered, and the values are frequently process and producer dependent. Additional information is provided in the article "Sources of Materials Properties Data and Information" in this Volume.

Lamina Properties

Fiber-reinforced composites have higher performance if the fibers are continuous, with the only fiber cuts occurring at the ends of the parts or subparts. While the geometrical lay-up of these fibers could be in any complex pattern, a local orientation of parallel fibers, such as a uniaxial tow or lamina, provides a high packing density. The lamina is the basic structural unit of many composites, as it is available in a form readily fabricated into sheet, tubes, or more complex shapes. Off-axis properties can be improved by including lamina with different orientations within the composite or by interleafing fabric into the lamina. Given the importance of lamina properties, but the large number of fibers and resin matrices, it is helpful to calculate lamina properties from the matrix and fiber constituent properties. Detailed information about the manufacture of composites is provided in the article "Design for Composite Manufacture" in this Volume.

Constituent Properties

The properties of carbon and other fibers are not changed when they are fabricated into a resin-matrix composite. Similarly, the resin properties do not appear to be much different in a composite than from a cured specimen of neat resin (resin with no fiber included), although there are notable exceptions. Hence, while the properties of thermosetting resin certainly change during cure, the final cured-state properties are often not affected by the presence of fibers. The system constituents are largely noninteracting, although fiber coupling agents can change properties. In general, though, constituent properties can be determined independently and used for any lamina calculation.

High-temperature composites are used and usually processed at high temperature, which may alter the constituent properties. For example, heating a carbon fiber to a temperature higher than the processing temperature of the fiber usually increases the fiber modulus and thermal conductivity, but decreases the strength, the strain to failure, and the coefficient of thermal expansion. If the constituents are noninteracting, the new properties can be determined after the appropriate heat treatment.

Constituents do interact chemically or mechanically in many high-temperature composite systems. The matrix may chemically attack the fiber surface to form a new compound, or may dissolve it. The attack is usually not very uniform, and the strength of the fiber is often degraded. Even in systems that are predicted to be thermodynamically stable, the matrix may dissolve and reprecipitate the fiber phase. There is always the same volume fraction of the fiber phase, but the microstructure is usually coarsening to minimize surface area. A number of studies in the past have measured the strength of the degraded fiber after etching away the matrix, and then these values were used for calculations to compare to experimental values. However, a poor composite strength value may be an indication of either fiber degradation, or too strong an interfacial coupling between fiber and matrix. The cause can be determined by scanning electron microscope observation of the fracture and fiber surfaces. Because of these difficulties, composite properties are often calculated from the original fiber strength to serve as an upper-limit benchmark. The observed strength can be compared to this benchmark to determine how effective the processing is.

Mechanical interactions during processing can often affect properties. For example, fibers inhibit sintering of a ceramic powder compact by restraining matrix shrinkage. An especially important example is carbon/carbon. The matrix often orients on the fiber surface and elongates during processing. The fiber is stretched when the matrix elongates, and the texture and modulus of the fiber increase. The increase in modulus is much greater than a simple heat treatment of the bare fibers. In contrast, some matrices do not wet the fibers and ball up. Because carbon fibers are not easily extracted from the carbon matrix, mechanical testing can often not be performed, and the fiber and matrix properties are determined by an inverse calculation from composite data. However, a single carbon fiber with a carbon-matrix coating can be tested.

Calculation of Lamina Properties

The modulus, strength, and coefficient of thermal expansion of matrices, and fibers parallel to the fiber axis can be measured relatively easily, but the fiber properties transverse to fiber axis are either difficult to measure or not measurable. Many fibers are isotropic, so that transverse properties are the same as the axial. However, carbon and organic fibers are extremely anisotropic. Some pitch precursor carbon fibers are even transversely anisotropic, for example, the transverse properties vary with direction. These transverse properties are not known very well for many carbon and organic fibers.

Parallel Properties. Estimates of modulus, strength, and coefficient of thermal expansion can be quickly made from a parallel model "rule-of-mixtures" type of calculation (Ref 10–12). Hence:

$$E_C = E_F V_F + E_M(1 - V_F) \qquad (Eq\ 1)$$

$$\sigma_C = \sigma_F V_F + \sigma_M(1 - V_F) \qquad (Eq\ 2)$$

$$\alpha_C = (\alpha_F E_F V_F + \alpha_M E_M V_M)/E_C \qquad (Eq\ 3)$$

where E is modulus, V is volume fraction, σ is strength, and α is coefficient of thermal expansion of the composite (C), fiber (F), and matrix (M).

The modulus estimate is exact for fibers and matrix that have the same Poisson's ratio, which for brittle fibers and matrices tends to be true. (The modulus of the composite may be about 5% higher than calculated by rule of mixtures, for ductile-matrix composites.) Although the derivation assumes isotropic constituents, the thermal expansion coefficient estimate is also in good agreement with experimental results, even for the case of anisotropic fibers.

The observed tensile strength is often close to the rule of mixtures estimate for a good "composite," although there is no fundamental argument that the estimate will always be close to the experimental value. However, if the interfacial coupling between the fiber and matrix is optimal, the stress concentration, produced by a broken fiber upon neighboring fibers, is offset by the higher strength of a fiber at short gage lengths. Strengths measured at 2 mm, which is still longer than the overstressed region next to a broken fiber in the composite, are usually significantly stronger than that at 25 mm gage length (a normal test length).

The compressive strength of composites loaded parallel to the fibers has been predicted by a number of models. Instability analysis of fibers supported in a medium with composite properties and a moderate-to-higher volume fraction of fibers (neighboring fibers buckle in-phase) results in (Ref 13):

$$\sigma_C = G_B/(1 - V_F) \qquad (Eq\ 4)$$

where σ_C is the longitudinal strength of the composite in compression, G_B is the shear modulus of the binder that is composed of matrix and fiber (a

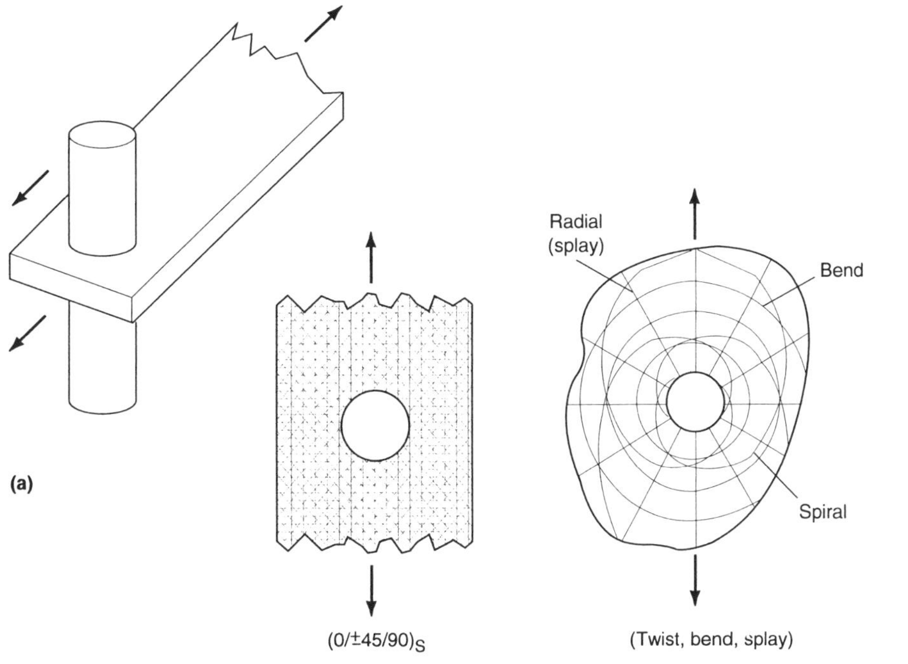

Fig. 5 Pin-loaded joint. (a) Configuration and load pattern. (b) Traditional laminate rotation fiber arrangement. (c) Twist, bend, and splay fiber arrangement

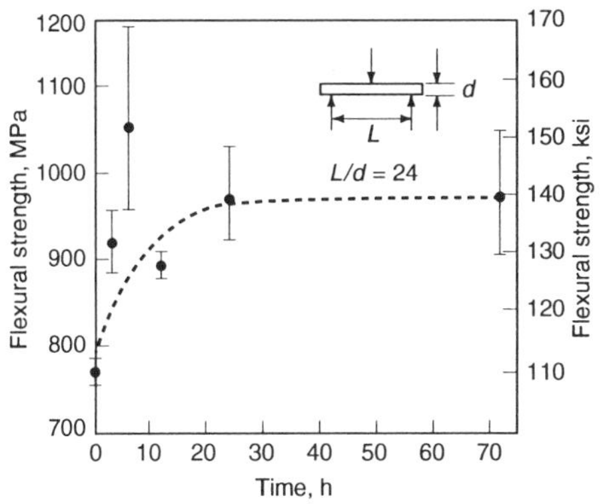

Fig. 6 Effect of interfacial bonding on strength. The coupling between fiber and matrix can be increased by the time that carbon fibers are etched in nitric acid. This first published data that bend or tensile strength has a maximum with respect to increasing interlaminar shear strength has been corroborated by other research, although the authors did not draw the curve to indicate a maximum. The bend strength is initially low, when the interlaminar shear strength is also low. The fracture surface, after the maximum, often has an identifiable brittle mirror/hackle pattern, because of too high an interfacial bond strength. The highest strengths in this brittle region can be higher than for poorer bonding, but the scatter also increases such that the average strength decreases. Source: Ref 6

lower-limit estimate would be the shear strength of the matrix), and V_F is the volume fraction of fibers.

The values predicted by this equation tend to be higher than the experimental strengths. However, for composites with very well aligned large-diameter fibers, homogeneously distributed, for example, a well-made boron filament composite, the observed strengths are only somewhat lower than predicted. The effects of binder yielding, and fiber misalignment have been used to formulate models that predict the experimental compressive strength better for some composites (Ref 14). The difficulty is determining which particular model should be applied. In general, the compressive strength for fine-diameter fibers is often observed to be similar to the tensile strength.

Two important classes of fibers produce very low compressive strengths because of internal fiber buckling. High-modulus organic fibers that have axially oriented linear molecules, and higher-modulus carbon fibers that have a molecular sheetlike structure oriented parallel to the fiber axis, fail by molecular buckling at low stresses. While the magnitude of stress for internal buckling can be predicted, coupling of the molecules by atomic bonding or microstructurally will alter the buckling stress. Unfortunately, experimental compressive strengths, reported in the literature, show much scatter. Testing is difficult and is very sensitive to specimen quality.

A number of other models have also been derived to more precisely describe composite properties. One that is useful for easy estimates of properties from isotropic constituents, because of its closed form, is the composite cylinder assemblage (CCA) model (Ref 15):

$$E_{CL} = E_{av} + 4V_F V_M(\nu_F - \nu_M)/(V_M/k_F + V_F/k_M + 1/G_M) \quad \text{(Eq 5)}$$

where E_{CL} is the longitudinal composite modulus; E_{av} is the volume-averaged modulus of the two phases (ROM average); $V_{F,M}$ is the volume fraction of the fiber and matrix, respectively; $\nu_{F,M}$ is the Poisson's ratio of the fiber and matrix, respectively; $k_{F,M}$ is the bulk modulus of the fiber and matrix, respectively; and G_M is the shear modulus of the matrix.

Similarly, the longitudinal composite Poisson's ratio (ν_{CL}), and shear modulus (G_{CL}) are:

$$\nu_{CL} = \nu_{av} + V_F V_M(\nu_F - \nu_M)(1/k_M - 1/k_F)/(V_M/k_F + V_F/k_M + 1/G_M) \quad \text{(Eq 6)}$$

$$G_{CL} = G_M[V_M G_M + (1 + V_F)G_F]/[(1 + V_F)G_M + V_M G_M] \quad \text{(Eq 7)}$$

where ν_{av} is the volume averaged Poisson's ratio for the fiber and matrix, G_F is the shear modulus of the fiber, and the other terms are as previously defined. These estimates of the longitudinal composite properties are all quite accurate.

Transverse properties depend on the constituent properties, the volume fractions, and the packing. The properties of carbon and organic

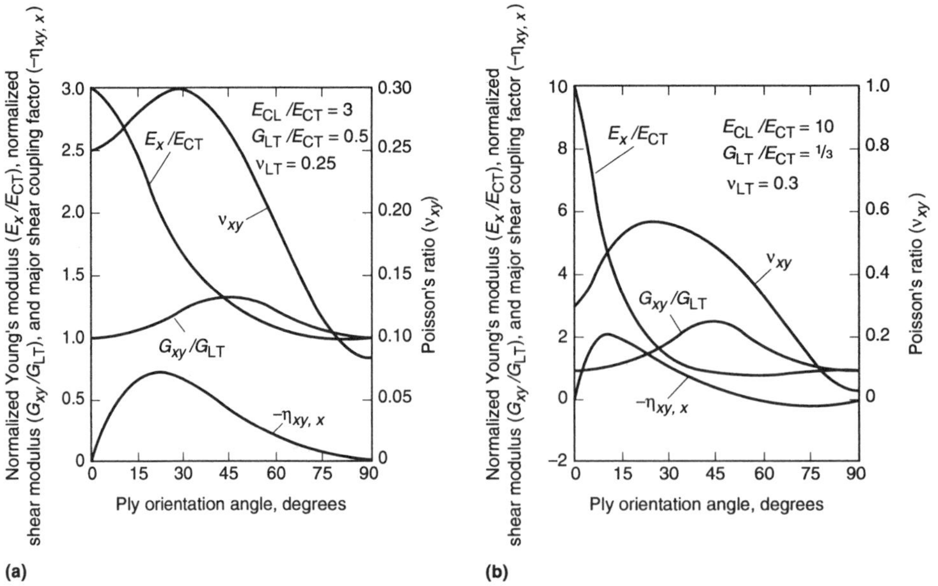

Fig. 7 Normalized moduli versus fiber orientation for a glass-fiber/epoxy-resin composite (a) and a boron-fiber/epoxy-resin composite (b). The normalized Young's modulus, shear modulus, Poisson's ratio, and major shear coupling factor are illustrated for loads applied at different angles with respect to the fiber orientation. 0°, load is applied parallel to the fiber axis; 90°, fiber oriented transverse to the load direction. Source: Ref 17

fibers are anisotropic, and often the transverse properties are not well known, if at all. In fact, experimental values of the transverse properties are used to perform the inverse calculation for obtaining fiber properties. The calculation can be performed assuming anisotropic constituents (Ref 9), and while small errors may be introduced from fiber-packing assumptions, these errors are canceled out when the same values are used for calculating composite properties. The transverse elastic properties can be estimated for isotropic constituents by the CCA model (Ref 15):

$$E_{CT} = 4k_{CT}G_{CT}/[k_{CT} + E_{CL}G_{CT}(1 + 4k_{CT}\nu_{CL}^2)] \quad \text{(Eq 8)}$$

$$\nu_{CT} = \tfrac{1}{2}(E_{CT}/G_{CT}) - 1 \quad \text{(Eq 9)}$$

where:

$$k_{CT} = (k_M k_F + k_{av}G_M)/(V_M k_F + V_F k_M + G_M)$$

$$G_{CT} = G_M[(\alpha + \beta_M V_F)(1 + \rho V_F^3) - V_F V_M^2 \beta_M^2]/[(\alpha - V_F)(1 + \rho V_F^3) - 3V_F V_M^2 \beta_M^2]$$

and:

$$\alpha = (\gamma + \beta_M)/(\gamma - 1) \qquad \beta = 1/(3 - 4\nu)$$

$$\rho = (\beta_M - \gamma\beta_F)/(1 + \gamma\beta_F) \qquad \gamma = G_F/G_M$$

The transverse thermal expansion coefficient can also be estimated:

$$\alpha_{CT} = \alpha_{av} + [(\alpha_M - \alpha_F)/(1/k_M - 1/k_F)]\,[3/(2k_{CT}) - (3\nu_{CL}(1 - 2\nu_{CL}))/E_{CL} - (1/k)_{av}] \quad \text{(Eq 10)}$$

where α_{av} is the volume average of the fiber and matrix thermal expansion coefficients and $(1/k)_{av}$ is the volume average of the reciprocal of the fiber and matrix bulk moduli.

Shear and Transverse Strength of a Uniaxial Resin-Matrix Composite. The two principal directions of applied shear stress are in a plane that contains the filaments and in a plane normal to the filaments. The shear strength for the first case is determined mainly by the shear strength of the matrix, the fiber/matrix interface, and the longitudinal shear strength of the fiber, while the second may have substantial reinforcement from the fibers. Generally, the matrix or interfacial strengths are the weak links, but in a few cases fiber shear failures have been observed. For resin-matrix composites, the shear strength for in-plane shear is typically equal to the shear strength of the resin. Similarly, the transverse tensile strength for resin-matrix composites is usually about one-half the tensile strength or equal to the shear strength of the resin. For off-fiber axis loading, the weak coupling between fiber and matrix—required to control axial fracture—causes the shear and transverse tensile strength to always be low, even with high-strength matrices and fibers.

Shear stresses are present within and between lamina, a distinction that is frequently not made. Care should be made in selecting stacking sequences of lamina, such that lower interlaminar shear stresses are generated at free edges and elsewhere.

Residual stresses usually result when lamina are used at temperatures that differ from the fabrication temperature because of the difference in the thermal expansion coefficients of the fiber and matrix. Warping occurs in curved laminates.

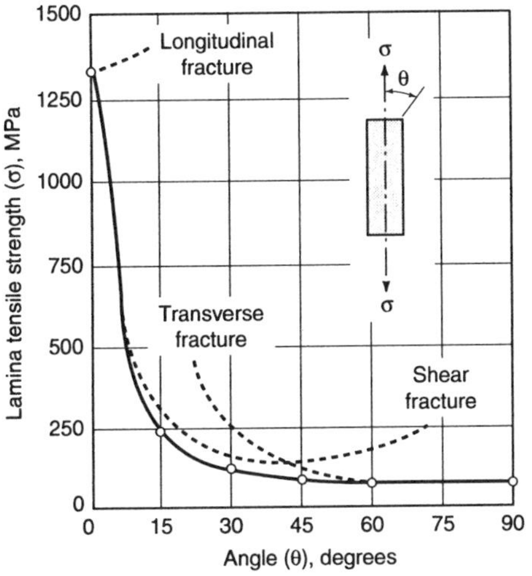

Fig. 8 Uniaxial lamina strength versus fiber orientation. The fracture stress curves, using the maximum stress criteria, are indicated by the dashed lines for a boron-fiber/epoxy-matrix composite. A fracture curve using a quadratic interaction criterion is shown with the solid line. Experimental points are indicated by open circles. Source: Ref 19

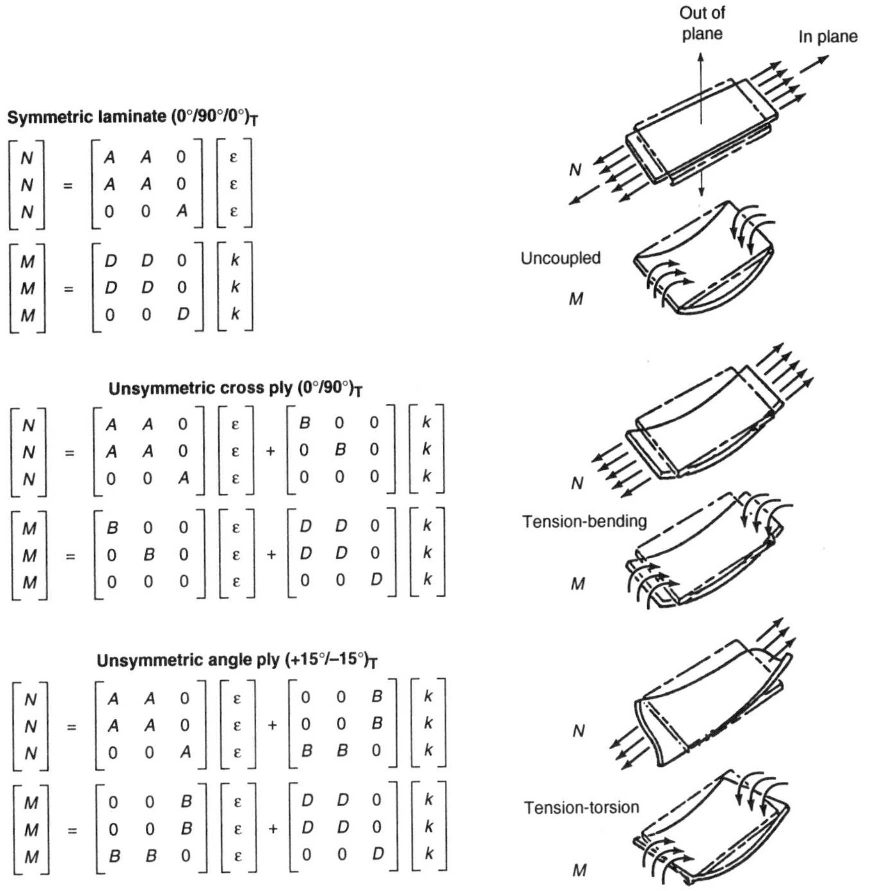

Fig. 9 Deformation behavior of composites. The deformations for composites with various symmetries are schematically illustrated for in-plane loading and bending. Generally, laminates that are symmetric through the thickness and symmetric in-plane with respect to the applied stress are used because no in-plane shear distortion or out-of-plane bending occurs. Less-symmetric laminates are useful for special applications. *N*, normal stress; *M*, bending moment. Source: Ref 21

However, residual stresses also exist in unwarped structures that have no net bending or twisting moments. Cracking can occur in the constituents, and thermal fatigue from thermal cycling has been observed in metal-matrix composites.

Processing residual stresses are due to the volume change that occurs upon curing of a resin, or within a chemically vapor deposited matrix. The volume change will cause stresses, which may not be annealed out at the processing temperature. The general conclusion is that a composite cannot be assumed to be stress-free before use.

General Lamina Properties

The elastic properties of a lamina in any in-plane direction can be calculated from the above mechanical properties measured in the principal materials directions by simple tensor transformations (Ref 16). The Young's modulus, shear modulus, Poisson's ratio, and the major shear coupling constant are shown in Fig. 7 for glass and boron/epoxy composites (Ref 12). The figure contrasts the behavior of a lower and higher modulus reinforcement. (E_X and G_{XY} are normalized by E_{CT} and G_{LT} to allow convenient plotting of the properties.) Both plots show the rapid drop-off in Young's modulus as the applied load is rotated away from the fiber axis of the composite because of the low matrix shear modulus. Interestingly, the lowest modulus for the boron fiber/epoxy resin composite is not at 0 or 90°, but at about 50°. Similarly, a maximum in Poisson's ratio occurs at about 25° for both composites. As might be expected from composite geometry, the shear modulus shows a maximum at 45°.

Lamina require an extra elastic constant, the major shear coupling factor, that relates the applied load to the in-plane shear deformation. The magnitude of the constant also provides an indication of the rotation of the fiber axis that occurs because of the shear deformation. There is no shear deformation when the load is applied parallel (0) or transverse (90) to the fiber axis. However, when the load is applied at any other angle in the glass-fiber composite, a shear deformation is induced, and the fibers rotate to become more parallel to the tensile load. Both composites show a maximum near 15°. For higher-modulus fibers, the sign of the factor reverses for high angles of loading, and the fiber rotation is toward a transverse orientation with respect to load. Hence, there are three angles for higher-modulus resin-matrix composites for which there is no shear deformation. These are 0, 90, and about 60° for boron-fiber/epoxy-matrix composites. Selection of lamina angles that have small major shear coupling factors minimizes shear stresses, especially at free edges.

For example, a $(+15/-15)_S$ composite is found to meet load and deflection requirements at a minimum weight; the subscript "S" indicates a symmetric (+15/–15/–15/+15) laminate. However, the large shear deformations that occur in each ply cause large interlaminar shear stresses at the free edge of the part, and may cause delamination. A $(0/+45/-45)_S$ composite might not be so structurally efficient, but the laminate has zero or small major shear coupling factors.

The strength of a lamina also varies with respect to the loading direction and the fiber axis. Failure, predicted by a maximum stress theory, occurs when any one of the stress components resolved to the material axes exceeds (Ref 18):

$$\sigma_{LT} > X_{LT} \text{ or } \sigma_{LC} > X_{LC}$$
$$\sigma_{TT} > X_{TT} \qquad \sigma_{TC} > X_{TC}$$
$$\sigma_S > X_S \qquad \text{(Eq 11)}$$

where X_{LT} is the tensile strength in the fiber direction, X_{TT} is the tensile strength transverse to the fibers, and X_S is the shear strength.

The strengths in compression, which are not generally the same as in tension, are indicated by the added subscript "C." A high tensile strength, fiber-dominated mode, is observed when a tensile load is applied nearly parallel to the fiber (Fig. 8). Failure shifts to a shear mode, usually in the matrix or at the fiber/matrix interface, as the load is applied more obliquely. Finally, failure occurs by transverse tension. For polymer-matrix composites, there is significant overprediction of tensile strength compared to the experimental val-

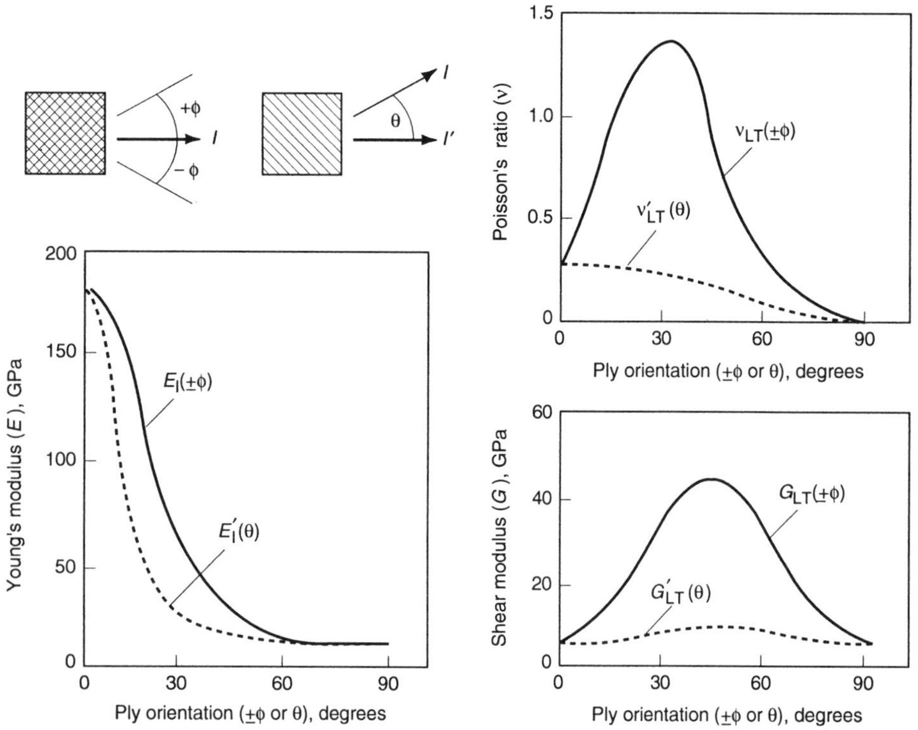

Fig. 10 Effect of balanced-angle ply construction on engineering mechanical constants. The Young's modulus and shear modulus of a laminate can be significantly higher if plies are oriented at both plus and minus an angle with respect to the stress axis rather than just at one of the directions. The very high Poisson's ratio for a balanced ply construction and small orientation angles provides a "Chinese finger puzzle" effect with a cylinder. Source: Ref 22

ues, especially near zero and 45°. A maximum strain criterion for failure has also been used, but with similar disagreement with experiment. Both sets of criteria neglect the interaction of failure modes.

Better failure models have been suggested, but they require more data points, particularly from biaxial tests (Ref 18). When the failure is complex, the value of the expression for the curve becomes less valuable, if a large number of data points are required to define it.

Discontinuous Fiber Lamina

Long discontinuous fibers offer some processing advantages with little penalty in axial properties. Variations on papermaking and other processes have been used to produce well-oriented fiber lamina, which can be impregnated with a metal, ceramic, or polymer. The initial work was performed on fine whiskers and later on carbon and aramid fibers. Mechanical tests show only a modest 10% loss in axial modulus, and a 15% loss in tensile strength in uniaxial composites (Ref 20). Shear and transverse properties were increased by as much as a factor of two, because of a substantial contribution from off-axis fibers.

The discontinuous fibers allow large deformations in a prepreg or a thermoplastic laminate, such that more complex parts can be directly molded to shape. Studies have been performed and computer programs developed that predict the proper movement of the fibers to occur during processing, such that a high-performance part can be achieved. There is much that is attractive, but chopping a continuous fiber and then reforming it to a discontinuous oriented composite has probably limited its application. However, there is significant use of chopped-fiber/resin composites for lower-cost, lower-performance applications.

Laminate Properties

Composite laminates are constructed from lamina with uniaxial fiber orientation and frequently from textile fabrics as well. The choice of fiber and matrix, ply orientation, and thickness is made to match the required loads and deflections. The extensional stiffness of the laminate is simply the sum of the individual laminate stiffnesses for the defined direction multiplied by the lamina thickness. The deformation of the composite can be quite complex with twisting, bending, or in-plane shearing occurring as well as simple extension under a tensile load. While twisting or bending for most applications would be undesirable, there are special cases where this behavior is beneficial. A conceptual approach to the types of deformation is discussed with respect to symmetries for in-plane and through-thickness loading directions (Fig. 9).

Symmetric In-Plane and Through-Thickness Laminates

A thin laminated composite can be fabricated that is symmetric to the applied load direction by balancing in-plane ply orientations and thickness and with the plies symmetrically distributed about the mid-plane through the thickness. Hence with a load applied at zero degrees, for every ply at +ϕ, there must be a ply at −ϕ, and the distribution of plies about the thickness midplane must be a mirror image both in angle and distance from the midplane. A symmetric in-plane and through-thickness laminate might be $(0_4/+45_2/-45_2/90)_S$, where the subscript "S" indicates that the laminate is symmetric about the mid-plane. In this 18-ply laminate, the 0° plies are on each outside, and the 90° plies are in the center. The effect of balanced construction of plies at ±ϕ instead of just +θ is quite strong (Fig. 10). The modulus for balanced ± ply construction decreases more slowly as the plies are rotated off the loading axis because of the inhibition of shear by alternate angle plies. Similarly, the shear modulus is much higher for the balanced ± composite, particularly at 45°. The Poisson's ratio effect can be positive and large (~1) for high-modulus fibers when θ is small (<30°). The large Poisson's ratio can be used in tubular structures to give a "Chinese finger puzzle" effect, because a composite tube tightens down on a rod when a tensile load is applied.

A tensile load applied at the 0 or 90° axis of a material results in a simple extension with no distortions in- or out-of-plane (as shown for the symmetric laminate in Fig. 9). No distortions make this "doubly" symmetrical composite desirable for many flat laminate applications. A special case is the $0_i/+45_j/-45_j/90_k$ composite laminate in a symmetric stacking sequence. The composite is isotropic in-the-plane when *i*, *j*, *k* are equal, but the numbers of 0, ±45, and 90° plies can be varied to provide optimal directional stiffness and strength. In addition to no distortions, the "scissoring" of the ±45 plies is small, which minimizes the likelihood of edge delamination.

The importance of the $0_i/+45_j/-45_j/90_k$ laminates in symmetric stacking sequences has lead to the development of a "carpet" plot that shows a property for variable *i*, *j*, *k*. The plots of composite properties for quite a few systems are available in government reports (Ref 23). While computer programs have eliminated some of the value of these plots, they provide a very easily comprehended visual input to the trends of composite properties with varying *i*, *j*, *k*. An example of a lower-modulus carbon fiber in an epoxy matrix is shown in Fig. 11. In Fig. 11(a), the tensile elastic modulus is plotted for various percentages of 0, ±45, 90° plies. The curve is read by selecting the percentage of 0° plies (which are next to the diagonal line), then picking the percentage of ±45° plies, and finally reading off the modulus from the appropriate curve. The intercepts on the modulus ordinate show how the modulus decreases as the 0° plies are replaced by 90° plies. Similarly, the intercepts on the diagonal line indicate the decrease in modulus as 0° plies are replaced by ±45° plies. Finally, the modulus would be read off one of the more horizontal curves when all three angles are present. The plot shows the example of a composite with 50% of the layers in the ±45° direction and with 25% each in the 0° and 90° directions. The general

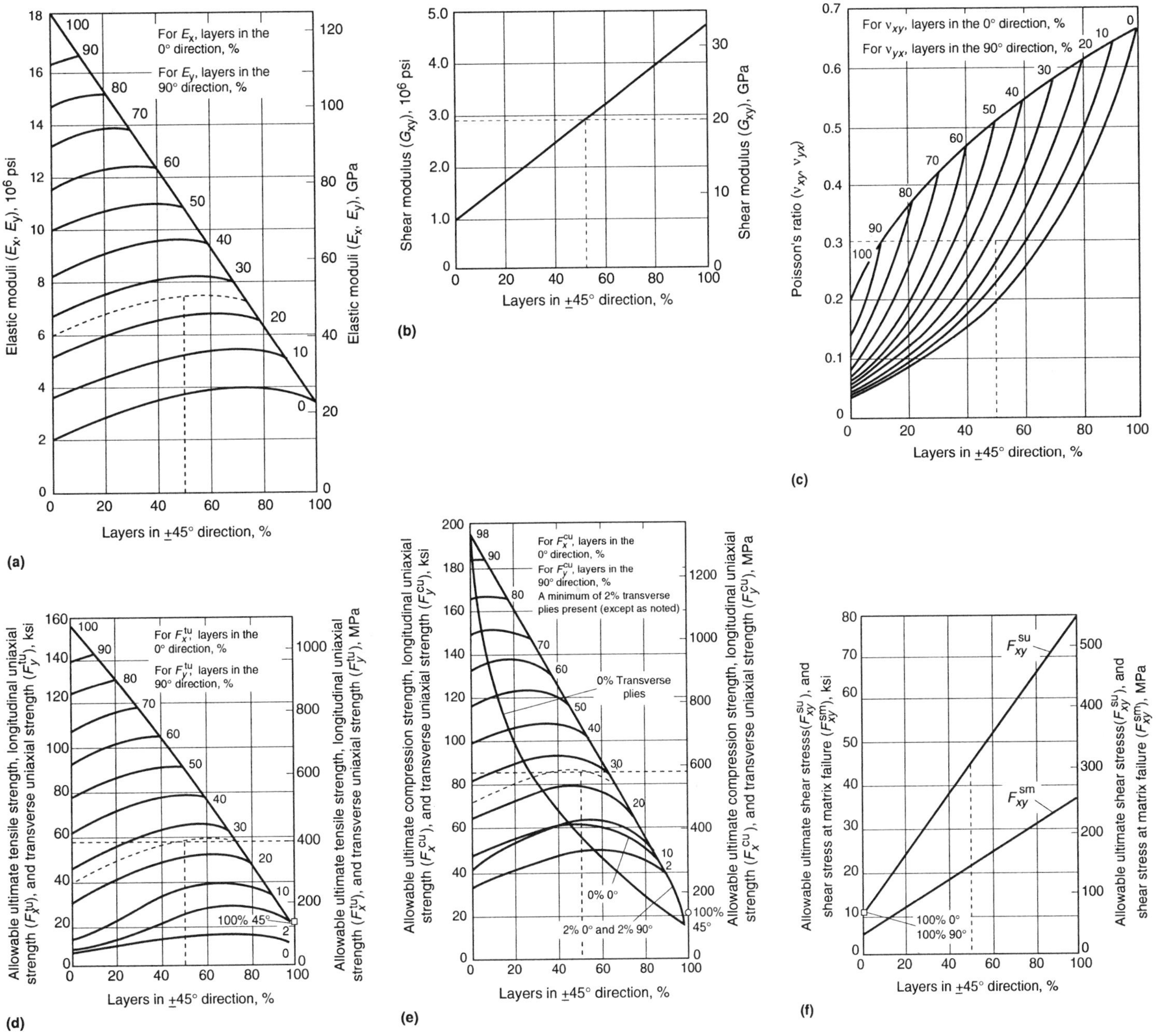

Fig. 11 Symmetric laminate properties for carbon-fiber (V_F = 65%)/epoxy-resin composites with variable percentages of plies at 0, ±45, and 90°. (a) Tensile modulus. (b) Shear modulus. (c) Poisson's ratio (tension). (d) A-basis tensile strength. (e) A-basis compression strength. (f) A-basis shear strength. A-basis values predict 99% chance of survival with 95% confidence level. The subscripts x and y indicate the measured property in the 0° and 90° fiber orientation, respectively.

trends are easily ascertained by looking at the plot. The other part of Fig. 11 show the results for shear modulus, Poisson's ratio, and tensile, compressive, and shear strength. Strength has to be considered carefully with the application. The accumulation of fiber, matrix, and even ply failures make the decision of what is considered failure difficult. Commonly, the design ultimate stress is defined as the maximum stress that does not result in rupture of any of the lamina in the laminate. Design limit stress is the maximum stress for which no laminate damage or degradation of stiffness occurs; it is normally used as the "yield" strength of the composite.

Asymmetric In-Plane and Symmetric Through-Thickness Laminates

An example of this class of composite is +15/0/+15° laminate. Tensile loading of this composite produces an in-plane shear deformation (distortion) with no out-of-plane deformation. Bending of these composites normally causes twisting as well as bending when the load is applied along the centerline of the laminate. However, there is an asymmetric loading point in bending that produces pure twist with no bending and another that produces pure bending with no twist.

Composites provide the ability to asymmetrically load a beam in bending, which results in no torsional deflection of the member. An application might be an automobile suspension for a wheel. The composite would allow the tire to remain fully in contact with the road over the full suspension travel. While this can be achieved with multilink suspensions, composites offer the possibility of very simple designs.

Asymmetric Laminates

Asymmetric laminates are a subclass of composite laminates that provide designers with a unique set of properties. Asymmetric laminates

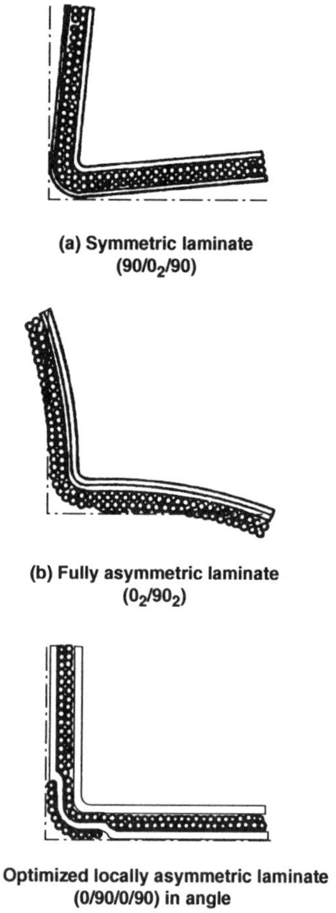

Fig. 12 Shape stability of right-angle laminates after cool-down from cure. All laminates are 90° before cure. (a) Symmetric laminate. The shrinkage in the matrix upon resin curing and the difference in coefficients of thermal expansion in-plane and through-plane cause a right angle to become acute in a symmetric laminate. (b) Fully asymmetric laminate. The use of an asymmetric laminate allows the thermal coefficients of expansion to be controlled to compensate for thermal distortions and preserve the right-angle. However, a fully asymmetric laminate would warp in the arms of the right angle. (c) Optimized locally asymmetric laminate. An asymmetric laminate is used in the right-angle region to prevent change of angle with temperature, while the arms are symmetric to eliminate warping. Making the mold at a slightly obtuse angle, which also helps laminate removal from the mold, accommodates the change in angle caused by resin shrinkage. Source: Ref 27

show a coupling between in-plane stresses and out-of-plane deflections that can be predicted using laminated plate theory (Fig. 9). Asymmetric laminates have an unsymmetrical stacking sequence with respect to the midplane of the laminate, and most react to mechanical stresses and to water and thermally induced (hygrothermal) stresses. Thus, laminates that are designed to twist or bend due to an in-plane load also have the potential to react to changes in temperature or humidity. This combined response to mechanical and hygrothermal stresses has been a barrier to their application (Ref 24). However, a family of asymmetric laminates has been found with mechanical tension/bending or tension/torsion coupling that is unaffected by hygrothermal changes (Ref 25). These laminates, when used as flat laminates or tubes with symmetric cross section, yield hygrothermal shape stability that was thought only possible through the use of symmetrically stacked laminates. While these laminates show no hygrothermal couple, they retain a strong mechanical coupling. Surprisingly, the solution for curvature stability is obtainable over a wide range of coupling coefficients and is independent of the thermal expansion and moisture swelling coefficients of the individual laminates.

Temperature changes warp most metal-, ceramic-, or polymer-matrix, asymmetric composites, while moisture absorption also warps many polymer-matrix asymmetric composites. Because the unsymmetrical laminates studied in the literature are prone to warping, the inductive conclusion that all unsymmetrical laminates warp is frequently stated in the literature. A solution to this problem is simply a $(0/90)_S$ symmetric laminate joined to a second laminate rotated by an angle θ. Each laminate is hygrothermally shape stable because it is symmetric and induces zero hygrothermal shear strain for all θ, and the combined laminate then is also hygrothermally shape stable. The system is not mechanically isotropic, and thus produces a tension/torsion or tension/bending coupling. The maximum value of the coupling coefficient is only reduced by about 15% from the classical $(\theta, -\theta)_T$ warping case. (The subscript "T" stands for a through-thickness stacking sequence.) Of course, temperature or moisture adsorption gradients, as may occur when only one side of the laminate is exposed, still causes warping just as with isotropic materials.

Some caution should be used when applying laminate theory to asymmetric laminates. For example, laminate theory predicts that a $(0/90)_T$ asymmetric laminate will assume a saddle shape as a result of an applied load or temperature change. In practice, the resulting shape usually is almost cylindrical. The reason for this unpredicted behavior is that the deflections are not small compared to the laminate thickness (as assumed in laminated plate theory). Better solutions have been derived (Ref 26), but are beyond the scope of this article.

The mechanical coupling observed in laminates that are not symmetric in-plane, with respect to an applied stress, or midplane would generally be a problem in most designs, because "warping" is usually undesirable. However, there are special cases where this coupling can be used. For example, an "infinite stiffness" beam can be designed for tension by having an out-of-plane curvature develop to compensate for the in-plane extension. An example of a hygrothermally stable laminate with this behavior would be a (0/90/90/0/90/0/0/90) laminate with the lamina materials selected properly to provide cancellation of the in-plane extension by bending. An application might be in machine tools to provide better dimensional tolerances with less weight. A second example is a tension/torsion member. The optimal pitch distribution along a blade of a tilt rotor aircraft is different for hover and forward flight. The difference in rotor revolutions per minute for hover and forward flight provides a spanwise tensile loading that could provide a torsional deflection in pitch by using, for example, a (+15/+105/+105/+15/–15/–105/–105/–15) hygrothermally stable laminate.

Curved Laminates

Often, large-scale warping or even delamination occurs during the manufacture of curved laminates, such as right-angle stiffeners or leading-edge components even when the laminate is of symmetric stacking sequence. This warping has often been considered a result of poor manufacturing such as resin accumulations or sticking of the laminate to the tooling. However, laminates are frequently made with fibers that have lower coefficients of thermal expansion than the matrix. Hence, the through-thickness thermal expansion coefficient is generally higher than the in-plane coefficient. While flat symmetric and special asymmetric laminates do not warp with temperature change, symmetric (even single-ply) out-of-plane curved laminates warp because of the difference of the inplane and through-thickness thermal expansion.

Hygrothermal stability requires that a curved laminate must be asymmetric through the thickness. Figure 12(a) illustrates a right-angle stiffener after cool-down using a conventional symmetric laminate sequence. The difference in inplane and out-of-plane thermal expansions causes the right angle to become acute. By contrast, a fully asymmetric laminate can be designed such that the right angle is maintained upon cool-down, but the nominally flat laminate away from the right angle warps because it is asymmetric (Fig. 12b). Only when an asymmetric laminate is used in the right-angle region and transitioned to a symmetrical laminate in the laminate arms does a truly shape stable laminate result (Fig. 12c). Resin shrinkage also causes a right angle to become acute during curing. Resin shrinkage can be accommodated by increasing the angle in a mold to greater than 90° for an interior lay-up. The increased angle also helps in part removal. Critical shape stable parts, such as optical benches, require very careful design. Many other factors also can influence the final shape of a complex composite component.

The warping of out-of-plane curved symmetric laminates could be capitalized upon by using them as temperature-sensitive elements. The sensitivity is maximized by having a low axial fiber thermal expansion coefficient and a high matrix and transverse fiber thermal expansion coefficient. A (0) laminate in a helix with a sharp radius of curvature would be desirable

Controlled Thermal Expansion Composites

The thermal expansion coefficient of higher modulus carbon and organic fibers near room temperature can be slightly negative parallel to the fiber axis, but positive transverse to the axis. Therefore, composites can be made with nearly zero thermal expansion in one direction by the proper combination of fiber/matrix properties, and fiber volume fraction and orientation (Ref 28). The axial modulus of the fiber must be sufficiently high compared to the transverse fiber and matrix modulus, such that the low negative axial expansion of the fiber compensates the higher positive expansion of the matrix and transverse expansion of the fiber. Commonly, a high volume fraction of high-modulus carbon fibers is used to achieve good mechanical properties, and the "excess" negative thermal expansion is eliminated by varying the fiber orientation with respect to the zero-expansion axis. The resulting laminates are good for stiffness-limited applications. However, the poor compression properties, which result from using high-modulus carbon fibers, are a design limitation.

As an example, a space structure for positioning dimensionally critical antennas is exposed to alternating heating/cooling cycles by the sun. This is mainly a stiffness-limited application that needs a zero thermal expansion material. A high-modulus carbon-fiber/resin-matrix composite fabricated into a tubular structure with the fibers oriented in a $\pm\theta$ (or more complicated) set of directions with respect to the tube axis is fabricated to produce a slightly negative coefficient of thermal expansion. The tubular structure is bonded together with positive expansion titanium fittings to form a stiff, lightweight structure with zero overall coefficient of thermal expansion.

As a second example, a semiconductor heat sink should have high thermal conductivity and match the thermal expansion of silicon. Very high thermal conductivity carbon fibers, which also have a negative coefficient of thermal expansion, can be placed into a copper or aluminum matrix to produce a $(0/90)_S$ laminate with a high thermal conductivity and a coefficient of thermal expansion that matches silicon with fiber-volume fractions of about 50%.

Fatigue

Fatigue of composites is often stated as being very good (Ref 29). A uniaxially oriented carbon-fiber composite, loaded parallel to the fibers in tension at low R values, usually exceeds a runout of 10^8 cycles at a tensile strength greater than 90%. If the brittle fibers are not broken on the first cycle, they tend to last many cycles. For higher R values, the slope of the S-N curve is increased. Damage accumulates from broken fibers and cracks in the matrix. Similarly, lower-modulus, but strong carbon-fiber composites are loaded to higher levels of strain for a given R value and also have higher S-N slopes. An associated decrease in modulus is observed in uniaxially oriented composites during the fatigue testing. In carbon-fiber/epoxy matrix composites, a gradual decrease in modulus occurs that amounts to about 10% before failure. Unfortunately, there may be no forewarning of impending failure, as indicated by a more rapid drop in modulus. Failure can occur in one additional cycle.

Fatigue testing that places high loads on the matrix or interface such as off-axis testing, shear, and/or compression testing generates much worse results. Many laminates show runouts at 10^8 cycles at 50 to 70% of the tensile strength, when loaded in tension. However, substantial microcracking in the matrix can accumulate in off-axis or transverse plies, such that strength in other directions is lowered. Fiber random mat/polymer composites have only 25% of the ultimate tensile strength at 10^8 cycles. Composite fatigue behavior in compression, loaded parallel to the fiber axis or with large matrix shear loading, also is poor. The worst results are noted in transverse tension in a uniaxially oriented specimen, because large stress concentrations occur on account of the moduli differences and poor interfacial bonding. This same response to fatigue loads is observed in composite bonded joints, limiting their long-term durability.

Composite fatigue resistance can be good or bad depending on the loading conditions and design. If the designer can keep the loads predominantly in the fibers and in tension and the loads in the matrix to a minimum, the composite is likely to have a high fatigue life. If not, the design may be more sensitive to fatigue failure than a corresponding metal part.

Energy Absorption

Composites have been considered for energy absorption applications. Tubular constructions with carefully designed crush-up initiators have the highest energy absorption per pound. The designs are very effective when the load is parallel to the cylindrical axis, but absorb very little energy when the tube is loaded off-axis. Generally, composite structures absorb less energy in a crash than a corresponding aluminum structure (Ref 30).

For example, the impact on the pilot of a helicopter that has lost power and suffers a vertical crash can be much reduced by using a tubular composite energy absorber under the pilot's seat. The helicopter, with its freewheeling rotor, hopefully can be controlled sufficiently to have a vertical impact such that the tubular composite will "crush up" and absorb most of the impact energy. An automobile bumper is not always subjected to impacts from the same direction. The highest-performance energy-absorbing composites are very sensitive to loading direction and may not be a good solution for this application.

Automated Design

The full realization of the potential of composite materials results in structures that may appear quite different than metal structures because of the difference in physical properties. However, the design optimization of a composite involves more variables than metals. While the design of a metal sheet may just consider thickness, a composite requires a minimum of six variables: three thicknesses and three orientations. Additionally, the proximity of yielding to the ultimate strength and the determination of load paths require that detailed analysis be performed early in the design process, often by the finite element method. The complexities of the design process have resulted in the development of computer programs to reduce both cost and time for designing a composite part. While large organizations may have internal capability for composite design, a number of companies have been formed specifically to provide composite design and analysis.

Metal-Matrix Composites

Metal-matrix composites have been developed to improve the specific strength and modulus and more particularly the maximum operating temperature of the lightweight alloys. A goal has been to extend the maximum operating temperature of a low-density metal, by using fiber reinforcement, to replace a denser, higher-operating-temperature metal, for example, an aluminum composite for titanium. Unfortunately, most of the potential fiber/matrix combinations are not thermodynamically stable and react or dissolve, unless coating protection is provided.

Eutectics provide stability, and several systems, such as tantalum-carbide/nickel, appeared attractive for high-temperature turbine applications (Ref 31). The phase diagram determines the volume fraction of reinforcement, and a filamentary microstructure only develops at low volume fractions. However, a low fiber volume fraction of 5 or 10% can substantially reduce creep. Directional solidification (DS) can be used to produce a uniaxially aligned composite. Two problems have limited the application of the DS eutectic composites. Firstly, the mismatch in coefficient of thermal expansion between phases causes thermal fatigue. Secondly, even though a low-energy interface usually exists between the fiber and matrix phases, the microstructure coarsens at high temperature and may even spheroidize, especially under stress. These difficulties have refocused attention upon reactive systems where success depends on development of fiber coatings or having slow kinetic interactions. Unfortunately, metal-matrix composites for replacement of the superalloys have been limited by the unavailability of creep-resistant fibers that are thermomechanically and thermochemically stable with the matrix alloy.

Magnesium, aluminum, and titanium are the most important metal matrices, although copper- and lead-matrix composites have niche applications. A problem with metal matrices is controlling the interaction with the fibers. On the one hand, the matrix may not wet the fiber; on the other, it may dissolve or react to form a compound with the fiber. Aluminum does not wet alumina or graphite, and it is hard to produce void-free composites. By contrast, aluminum reacts with boron and also graphite. Coatings can be used either to promote wetting or to prevent reaction. Silica can be coated on alumina to promote wetting, and silicon carbide can be deposited on boron or carbon fibers to inhibit reaction with aluminum. The matrix composition can also be adjusted to promote wetting by adding a reactive element or by lowering the surface energy. Lithium improves wetting of aluminum alloys on alumina (Ref 32). Similarly, the composition of the matrix can be changed to decrease the chemical potential of a species to decrease the rate of attack on a reinforcement. For example, adding silicon to an aluminum matrix can move the overall composition into a stable aluminum/silicon carbide two-phase field.

For example, the electronics in aircraft often require forced-air cooling to provide adequate lifetimes. A passive, lightweight system that fits into present bays and relies on high thermal conductivity is desirable. The aluminum racks could be replaced with high thermal conductivity carbon fibers in an aluminum/silicon alloy. Inspection for galvanic corrosion would be required, but fiber coatings can minimize degradation.

Large-diameter fibers are more commonly used in metal-matrix composites than in resin- or ceramic-matrix composites. Some of the desired fibers are produced by chemical vapor deposition on a substrate filament that necessitates relatively large diameters (~100 μm). A large fiber diameter also allows multimicron thick coatings or reaction layers to form without a large penalty in fiber properties. Finally, larger-diameter fibers provide greater thickness of matrix and enhanced plasticity.

The cost of metal-matrix composites compared to unreinforced metal has limited their application. Fiber, coating, and processing costs have led to selective placement of reinforcement, and melt techniques. The casting shrinkage is constrained by the fibers, and porosity can be avoided by squeeze casting (Ref 33). The microstructure of cast metal matrices can be quite different because of the presence of fibers. Nucleation can occur on the fiber surface, and the fiber geometry limits the microstructure that can form. In general, the properties of the matrix are poorer than the optimally processed unreinforced metal. For example, a diesel engine piston made from aluminum would reduce the reciprocating weight. However, the head of the piston is subjected to impulsive loading at a rather high temperature for aluminum. The head would also be less sensitive to thermal fatigue and provide a better cold fit with the cylinder if a lower coefficient of thermal expansion were possible. Finally, a greater wear resistance in the piston head at the cylinder interface would minimize piston slap and allow less expensive piston rings to be used. The selective use of fiber reinforcement just in the head allows the diesel pistons to be squeeze cast at low cost. However, thermal barrier coatings on the head of the piston may be an adequate and lower-cost solution.

Three other processing techniques have been extensively developed for metal-matrix composites. In the first, a layer of carefully spaced monofilament is wound on a metallic foil that is wrapped around a drum. The foil and fibers are slit to form a sheet, and these are stacked up with a cover sheet to make a preform, which is then vacuum hot pressed to produce the composite (Ref 34). The alternate process is plasma spraying the metallic alloy onto the drum to form the foil (Ref 35). In this case, no binding agent is required to hold the fiber to the foil. Either process can make excellent, but expensive, composites; however, care must be exercised during processing to make a fully consolidated specimen without too much interfacial reaction. An alternative low-cost process is the casting of aluminum with silicon carbide particulates for engine blocks and pistons. Strength, maximum service temperature, and wear resistance are increased, while the thermal expansion coefficient is decreased.

The high costs of continuous-fiber composites have driven the development of discontinuous-whisker and even particulate-reinforced composites. The potential performance of these discontinuous composites is much poorer, and the mechanism of strengthening may be only partially by fiber reinforcement with the dispersion strengthening being the major effect. Yet silicon-carbide-reinforced aluminum has a significantly higher modulus and strength, especially at higher temperatures. Although the normal compressive forming processes can shape the composite, the lower ductility and fracture toughness have limited its general application (Ref 36). It is competitive for applications such as snow tire studs, which should be wear and corrosion resistant, as well as low cost.

Properties of Metal-Matrix Composites

Metal-matrix composites offer significantly better mechanical properties than polymer-matrix composites for matrix-dominated properties, such as greater shear, compressive, and transverse tensile strengths. Reinforcement can increase the maximum-use temperature over that of the monolithic material. However, matrix behavior may not be as good as for the unreinforced

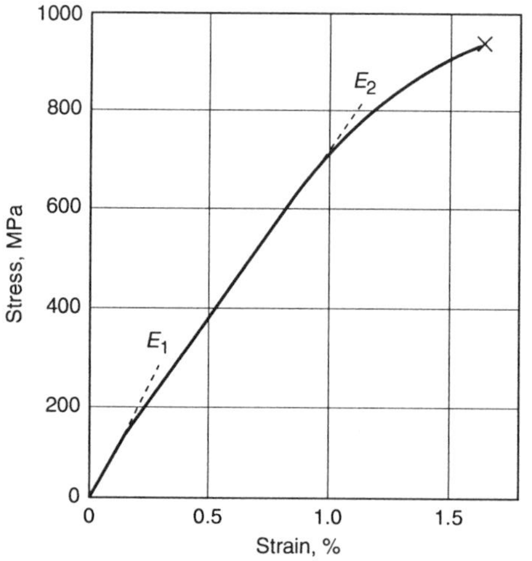

Fig. 13 Stress/strain curve for nicalon silicon carbide fiber in aluminum (1100) matrix. The material has an initial modulus (E_1) of 87 GPa, which is representative of both fiber and matrix elastically deforming. The secondary modulus (E_2) of 70 GPa is indicative of fiber elastic deformation and matrix micro-yielding. For a relatively low-ductility matrix, failure often occurs at the end of the secondary modulus straight line, when the first fibers begin to fail. For the higher-ductility 1100 aluminum matrix, fiber fractures accumulate during the curved part of the stress/strain plot until final failure occurs. Source: Ref 38

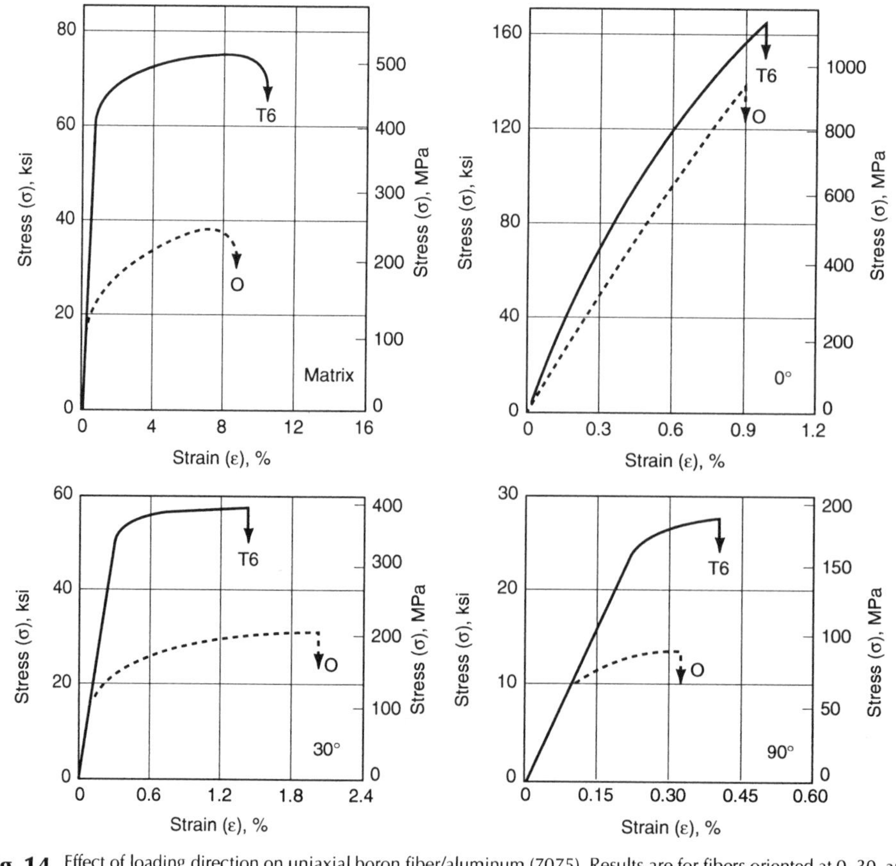

Fig. 14 Effect of loading direction on uniaxial boron fiber/aluminum (7075). Results are for fibers oriented at 0, 30, and 90° with respect to the load and pure matrix in the T6 and O conditions. Source: Ref 39

matrix. For example, the transverse tensile strength of a metal-matrix composite, modeled by a matrix with poorly bonded fibers, is only about one-third of the strength of the unreinforced matrix. Furthermore, the microstructure of the matrix may not be as desirable, and together with the physical constraints placed on the fibers by the matrix, ductility of the matrix is generally much reduced. Metal-matrix composites also can have good electrical and thermal conductivity.

The stress/strain behavior of metal-matrix composites is more complicated than resin-matrix composites because of work hardening and the change in the yield surface with different or multiple loading (Ref 37). (The same considerations make thermal expansion show hysteresis that changes with cycling.) If a residual stress-free metal-matrix composite is loaded parallel to the fiber axis, both the fiber and matrix will be elastically loaded initially (Fig. 13). Upon further loading, the matrix will finally be loaded to its yield stress and then plastically deform, but the fibers will still be loaded elastically. The effective modulus is decreased in this region. Then, substantial fiber failures occur and composites with high fiber volume fraction fail catastrophically. At low fiber volume fractions, fibers break up into critical lengths causing a substantial loss in modulus, and finally matrix failure occurs. These effects have been successfully included in describing the behavior of metal-matrix composites, but design of stable structures that are subjected to temperature cycles is more difficult. The details are beyond the scope of this article.

The plastic deformation that can occur in a metal-matrix composite is more apparent for off-axis and transverse loading to the fiber axis (Fig. 14). Loading at 30° produces a stress/strain curve that looks much the same as for the pure matrix except that the strain scale is reduced. The reason is that local matrix deformations around the fibers are much higher than the overall global strains, and failure finally occurs when local strains are similar to the failure strain in the unreinforced matrix. (Triaxial tensile strains are also present in the matrix, which also tends to reduce ductility.)

Fatigue properties in the axial direction of the fibers is excellent, but fatigue for off-axis loading, which relies on the fatigue behavior of the matrix, can be worse than for the unreinforced matrix. The accumulation of failed fibers and matrix cracks in fatigue testing results in loss of modulus. This may continue until composite failure occurs. However, in some laminate constructions or composites with low fiber volume fraction, damage may develop to a stable condition (shakedown), whereupon further fiber or matrix damage occurs slowly. The drop in modulus may be as great as a factor of two during shakedown, which may require part removal because of reduced stiffness.

Ceramic-Matrix Composites

The demand for composites that can operate at higher temperature has driven the development of ceramic-matrix composites. Tremendous barriers stand in the path for successful application of these composites in which the properties of the fiber and matrix may be very similar, and indeed the chemical composition of both may be identical. The lack of cost-effective and creep-resistant fibers, stable crack-stopping interfacial coatings, and high strain-to-failure matrices all lead to poor-extended-life, oxidation-prone, high-temperature composites. Yet, the successful application of car-bon-fiber/carbon-matrix and silicon-carbide/silicon-carbide composites to rocket nozzles and reentry vehicles illustrates the potential of these materials if environmental stability can be achieved. While ceramic composites might be used at lower temperatures because of their corrosion resistance, they are mainly considered for high and extreme temperatures often as the last resort.

The mechanical behavior of these composites, which have similar constituent moduli, is quite different from resin-matrix composites, especially because the strain to failure of the matrix is typically smaller than that for the fiber. The fibers are added not only to provide high-temperature creep resistance, but more importantly to provide damage tolerance and a more graceful failure at room temperature. When a ceramic-matrix composite with uniaxially oriented fibers is loaded in tension parallel to the fibers, the load is relatively evenly distributed between the fiber and matrix, the exact ratio being determined by the relative moduli and volume fractions. For relatively strong coupling between the fiber and the matrix, the composite fractures when the first crack occurs in the *matrix* (Fig. 15a). Often, there may already be surface cracks from specimen preparation. The composite strength may be lower than a monolithic specimen of pure matrix material, as it frequently is not possible to process the matrix to as high a quality with fibers present. The strength and toughness of commercially available composite products will generally be higher than for monolithic materials. Better processing eliminates many of the defects by filling in voids with more and better matrix material. Matrices with higher fracture toughness would help increase the strain at which first cracking occurs, as would lower modulus. However, the fracture toughness decreases as the refractoriness (bond strength) of the material increases. Low matrix fracture toughness is the likely result for the highest temperature applications. However, combining several techniques to improve fracture toughness, such as adding whiskers to the matrix and placing multiple crack-stopping interfaces within the matrix can increase the strain for first matrix failure.

Decreasing the coupling between the fiber and matrix sufficiently allows the first crack in the matrix to deflect around the fibers such that the fibers will bridge the cracked matrix and sustain the load if the fiber strength and volume fraction are high enough (Fig. 15b). As the load is increased, more matrix cracks form and finally yield a specimen with a relatively uniformly spaced set of cracks in the matrix, all bridged by the fibers. The spacing of the cracks is determined by the fiber/matrix coupling: the poorer the coupling, the wider the spacing. Finally, there is insufficient length between cracks to transfer the load, which only the fibers are carrying in the bridging regions, to the matrix to cause further cracking. A large drop (~50%) in modulus can occur with poor interfacial bonding when the matrix cracks. The design may be limited by this decrease in modulus or by resonant frequency changes. Further loading causes failure of fibers, and finally at one bridging section the remaining fibers can no longer support the load and the specimen fails. The probability for failure of the fibers is highest in the bridges, because the matrix carries no load. However, there is a distribution of strength-reducing flaws along the length of the fibers, and there will be fracture of fibers away from the bridged matrix crack, albeit with a decreasing frequency with distance from the crack (as the matrix picks up load) (Fig. 15c). The last step of the process is the pullout of the fractured fibers from their sockets in the matrix (Fig. 15d). Much energy can be absorbed, as the apparent

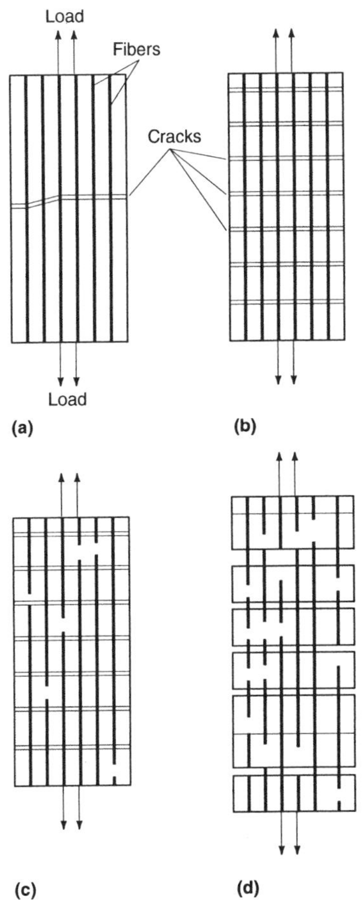

Fig. 15 Illustration of ceramic-matrix composite failure process. (a) The crack, which initiates in the matrix, propagates through both matrix and fibers in the composite when strong interfacial bonding exists. (b) For intermediate or low interfacial bonding, the matrix crack runs through the matrix but around the fibers. Multiple cracks accumulate if the fibers bridging the cracks can sustain the load. (c) After attaining multiple matrix cracks with an equilibrium spacing, fibers fail at flaws with increasing load, and not necessarily in the bridged regions of highest stress. (d) Fibers pull out of their matrix sockets with further extension.

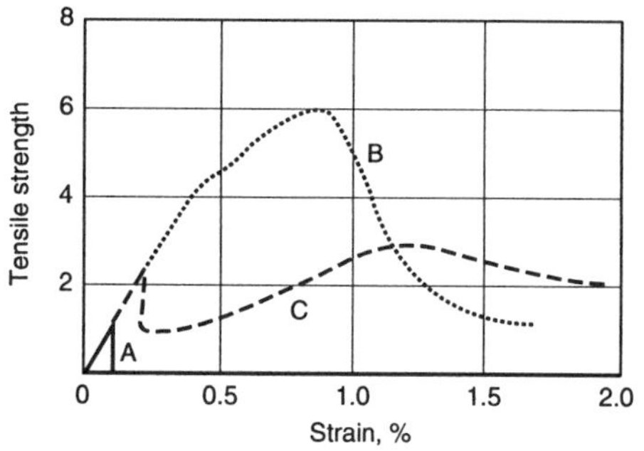

Fig. 16 Stress/strain curve for a typical uniaxial ceramic-matrix composite loaded parallel to the fibers. The solid line (A) shows the behavior for strong interfacial bonding and catastrophic failure with the first matrix crack. The dotted line (B) indicates intermediate bonding behavior such that the fibers bridge the matrix cracks, but with short fiber/matrix debonding near the cracks. The dashed line (C) illustrates the stress/strain curve for very weakly coupled fiber and matrix. In the limit, the matrix contribution to the modulus is completely lost after matrix fracture.

strain can be quite high. Studies have shown that this pseudometallic, stress-strain behavior can be achieved in ceramic composites.

Constituent Selection

Ceramic-matrix composites are likely to be used at high temperature and are usually fabricated at high temperature. In addition to the usual considerations of physical properties such as specific modulus and strength, creep and stress rupture become the limiting mechanical properties for high-temperature applications (Ref 40). The most creep-resistant ceramic is graphite, followed by silicon carbide, titanium diboride, and silicon nitride. Yttrium-aluminum-garnet (YAG), mullite, and sapphire appear to be the most attractive oxides (Ref 41). Both fibers *and* matrices must be creep resistant. For lower-temperature applications, glass-matrix or glass/ceramic-matrix provide a combination of thermal expansion coefficient, fracture toughness, and low modulus that allow very good composites to be produced. The selection of fiber and matrix material combinations is very much limited by thermochemical and thermomechanical incompatibilities. For example, carbon fibers in any oxide matrix are thermodynamically unstable above 1500 °C. Similarly, a mismatch of thermal expansion coefficients much above 2 μm/m · °C causes microcracking of the composite.

The problem of selection of a ceramic composite system is compounded because a weak interface is necessary to control the fracture. A third interfacial phase between the fiber and matrix is often added, especially when the matrix and fiber have the same chemical composition, such as with a silicon-carbide-fiber/silicon-carbide-matrix composite. The interfacial materials generally have a layer structure, although highly porous interfacial phases have also been used. However, no environmentally stable layers have been found for the higher-temperature composites (>1200 °C). Model systems usually have used graphite. Boron nitride, which has some oxidation resistance to 1100 °C, has been found to be an effective crack-stopping interfacial material also. Oxide-layer compounds, such as synthetic micas, are useful for temperatures as high as 1100 °C. However, the interface between fiber and matrix remains a problem, and present interfacial coatings add substantially to the composite cost.

Ceramic-matrix composites are not as well developed as polymer-matrix composites, with the exception of carbon-fiber/carbon-matrix composites. Carbon/carbon composites are used extensively for aircraft brakes, shuttle tiles, rocket motor nozzles, and reentry vehicles. For other ceramic-matrix composites, the designer must select one of the few systems that are available. Much stronger interaction with the materials supplier is required, because the processing and properties are very much producer dependent.

Properties of Ceramic-Matrix Composites

The elastic properties can be calculated using the equations described in the sections "Calculation of Lamina Properties" and "Symmetric In-Plane and Through-Thickness Laminates" in this article. The major difference with ceramic-matrix composites is that the moduli of the fiber and matrix are frequently similar, so that the elastic properties are more isotropic.

The fracture behavior of ceramic-matrix composites differs from that of polymer- or metal-matrix composites in that the failure strain of the matrix is less than that for the reinforcement. The stress/strain curve for a uniaxially aligned fiber specimen, loaded parallel to the fibers is shown in Fig. 16 (Ref 42). Both matrix and fibers are being loaded initially, until the failure strain of the matrix is reached, and the matrix starts fracturing. There are three different limiting behaviors depending on the interfacial bonding: (1) high interfacial bonding causes catastrophic failure, (2) intermediate bonding causes a change in slope, (3) poor bonding causes a load drop with a subsequent slope proportional to just the modulus and volume fraction of the fibers. For low or intermediate interfacial bonding and a volume fraction and fiber strength sufficiently high to support the applied load after matrix fracture, composite fracture does not occur, and multiple matrix cracks will accumulate. In the limit, as bond strength goes to zero, the ceramic matrix behaves as if it has holes. These "holes" act as stress raisers. The final crack spacing is determined by the strength of the matrix and inversely on the maximum shear stress that can be sustained at the fiber/matrix interface. The relationships for multiple matrix cracking, and the final matrix crack spacing are (Ref 43):

$$\sigma_{CMu} < \sigma_{Fu} V_F \quad \text{(Eq 12)}$$

$$L = V_M \sigma_{Mu} R / V_F \tau \quad \text{(Eq 13)}$$

where σ_{CMu} is the composite stress when first matrix cracking occurs; σ_{Mu} and σ_{Fu} are the matrix and fiber ultimate strengths, respectively; R is the radius of the fiber; τ is the shear stress at the fiber/matrix interface; and L is the spacing between matrix cracks.

In ceramic-matrix composite systems that retain high frictional or chemical bonding interaction between the fiber and the matrix after matrix fracture, a simple change in slope occurs in the stress/strain curve as the matrix begins fracturing (Fig. 16). In other systems, there is loss of the matrix stiffness if coupling between fiber and matrix is lost when matrix fracture occurs. The stress/strain curve under displacement control may show a load drop much like yielding, or become horizontal, until the load increases again with strain, but this time only with the modulus contribution from the fibers. (The actual curve may differ because of residual stresses between the fiber and matrix.)

The microcracking of the matrix not only results in loss of modulus, but also allows internal oxidation to occur. Parts could be designed to stress or strain levels below which matrix cracking occurs. While it might be expected that the matrix would crack at similar strains as for the unreinforced ceramic matrix (0.05 to 0.10%), the strain for matrix cracking was shown to be enhanced (Ref 43):

$$\varepsilon_{Mu} = [24\tau_O \gamma_M E_F^2 / E_C E_M^2 D_F (1 - V_F)]^{1/3} + E_F V_F \Delta\alpha \Delta T / E_C \quad \text{(Eq 14)}$$

where ε_{Mu} is the reinforced matrix strain at failure, τ_O is the interfacial shear strength, γ_M is the matrix fracture energy, D_F is the fiber diameter, $\Delta\alpha$ is the difference in thermal expansion coefficient between fiber and matrix, and ΔT is the difference between the stress-free temperature and the use temperature. The stress-free temperature is often assumed to be the processing temperature.

Although the fractional exponent on the first term minimizes the effect of the parameters, the wide range that some of the parameters can have produces significant changes in the strain. High matrix fracture energy, high fiber modulus and volume fraction, small fiber diameter, and low matrix modulus all increase the matrix failure strain. Increasing the interfacial shear strength also raises the matrix failure strain, but the value must not exceed that which causes brittle failure. One approach, which has doubled the matrix microcracking strain, is reinforcing the matrix with about 15% of fine whiskers.

The second term in the equation is the residual stress that arises from the mismatch in the coefficients of thermal expansion between the fiber and matrix. An axial compressive residual stress would be present in the matrix at room temperature if the fiber thermal expansion coefficient is larger than that for the matrix, because the composite is generally processed at high temperature. Unfortunately, the thermal expansion coefficient of the fiber is likely to be smaller than the matrix, placing the matrix in residual axial tension.

The average matrix strain-to-failure in a composite can frequently be increased to values of

0.4% or more, which combined with the 1% or better strain to failure of the fiber, can produce a stress/strain curve mimicking a ductile metal, albeit with very limited strain capability. A problem is that the first few matrix failures are observed at strains only slightly higher than the unreinforced matrix. Therefore, a prudent assumption is to assume that matrix microcracks will always be present that may allow internal oxidation and embrittlement.

The ultimate tensile strength of uniaxially aligned composite, loaded parallel to the fibers, is given by the bundle strength and volume fraction of the fibers bridging the matrix cracks:

$$\sigma_{Cu} \sim \sigma_{Fu} V_F \quad \text{(Eq 15)}$$

where σ_{Cu} and σ_{Fu} are the ultimate tensile strengths of the composite and fiber bundles, respectively.

The bundle strength of the fibers cannot be measured on a gage length that approximates the short gage length in the cracked matrix region of the composite, but might be approximated from resin- bonded strand tests. Because fiber volume fractions often are about 50% in a uniaxially aligned ceramic composite, the axial ultimate tensile strength will be about one-half of the fiber bundle strength, if the interfacial coupling has been properly adjusted to prevent brittle fracture. If a fabric is used, somewhat less than 25% of the tensile strength of a bundle can be obtained because of the over/under construction in a fabric and the effective fiber volume fraction in the load direction.

The toughness of a ceramic-matrix composite can be caused by at least five different mechanisms, which can act independently or in a combined manner. However, the work to fracture, the area under the stress/strain curve, is generally dominated by the "pull out" of broken fibers from their matrix sockets. Although it is generally desirable to have little scatter (high Weibull modulus) in the strengths of fibers in order to attain the highest composite, ultimate tensile strength, no pullout occurs when there is no variation in fiber strength. All fibers would fail simultaneously in the highest-loaded regions at the matrix cracks. Pullout requires that fibers fail at flaws away from the matrix cracks, and then the broken fibers drawn out from the matrix sockets. The work of pullout (W_P) for a multiple matrix cracked composite (in ft · lbf) is (Ref 44):

$$W_P = 0.083\,[(m-1)\,(m+2)/m^2(m+1)^{2/(m+1)}]$$

$$V_F(0.5D_F)^{3m/(m+1)}\,[\sigma_O^{2m/(m+1)} / \tau^{(m-1)/(m+1)}] \quad \text{(Eq 16)}$$

where m is the Weibull modulus, σ_O is the Weibull strength scale parameter, D_F is the fiber diameter, and τ is the interfacial shear strength.

For $m = 8$, typical for commercially produced fibers, the equation reduces to:

$$W_P = 0.008\,V_F\,D_F^{2.7}\,[\sigma_O^{1.8}/\tau^{0.8}] \quad \text{(Eq 17)}$$

The optimal Weibull modulus for maximizing work of fracture is about 4. (There will probably not be any development of fibers optimized for maximum work of pullout in the near future due to the limited market.) A high volume fraction of large-diameter fibers with high strength, but with a low interfacial strength produces a maximum work of pullout. A comparison with the matrix microcrack strain equation shows that high fiber volume fraction and strength are beneficial to both. However, the effects for fiber diameter and interfacial shear strength are opposite. A ceramic composite cannot be optimized for both the work of pullout and the matrix microcrack strain simultaneously.

The strength of a uniaxial ceramic-matrix composite that is loaded off-axis is low. This result is a consequence of the poor interfacial shear strength between the fiber and matrix that is required to control fracture. Multiple-ply laminates with off-axis oriented plies must be used to sustain off-axis loads.

The materials selected depend very much on the application. For inert atmosphere or very short time applications (tens of minutes), unprotected carbon fibers in a carbon matrix provide the best mechanical properties, especially at very high temperature. Creep is low until 2200 °C. Coated carbon/carbon composites with internal additives can have lifetimes in oxidizing environments up to 1700 °C for as long as 100 h, but not very reproducibly. Silicon carbide or nitride provide much better oxidation resistance, but would probably be creep limited at temperatures above 1500 °C for the carbide and somewhat lower for the nitride, even anticipating future improvements. At present, oxide systems appear limited to 1200 °C and up to 1300 °C in the future. The exact values all depend on stresses, time, and temperature, and the development of suitable interfacial materials.

Large Composite Structures: Joints, Connections, Cutting, and Repair

A major advantage of composites is that large integrated structures can often be designed with a major reduction in parts count and assembly time. The number of mechanical fasteners is often a small fraction of the number used in metal structures. While it is sometimes possible to design a large composite in which loads are predominantly carried by the fibers, usually there will be regions that rely only on the matrix for load transfer from one major volume to another. Several parts may be cocured together so that the matrix resin acts as the adhesive. This area behaves much as a bonded joint. Mechanical fasteners have been added to increase the reliability, but surprisingly performance has sometimes been decreased because of the stress concentrations and delamination damage at the free edge of holes. Extensive use of mechanical fasteners is costly with carbon-fiber/resin composites, because the fasteners are generally made of titanium to minimize galvanic corrosion; use of coated fasteners would avoid corrosion and lower cost.

Stitching has also been applied successfully using a tough fiber such as an aramid as the stitching yarn, as a method of improving interlaminar properties and minimizing edge delamination. Stitching pierces the laminate with a minimum of laminate fiber damage and displacement. Optimization of stitch spacing and yarn denier and tension has been performed for a number of different laminates and structures.

While composite materials are especially amenable for making large integrated structures, it should always be realized that damage is likely to occur during the lifetime of the structure. The structure must be designed for easy and efficient repair. Consideration of where cuts should be for removal of damaged material, and flanges for attachment or bonding of repaired sections, and so forth, is required.

Composite materials also offer the possibility for in situ structural-integrity monitoring by incorporating sensing fibers within the composite. A number of studies have shown that optical and strain-sensitive fibers can be used to measure loads and structural integrity. However, these techniques have yet to achieve wide application.

Variance and Scaling Considerations

Composites traditionally have used brittle fibers, because of their otherwise superior properties. For simplicity, consider a composite with the

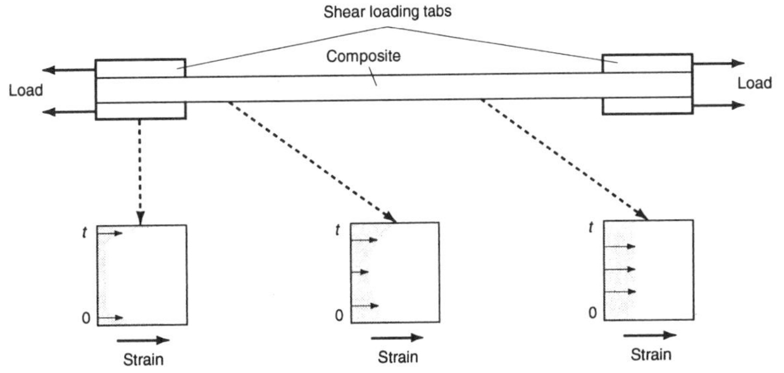

Fig. 17 Transfer of load into a uniaxial composite. The transfer of load through shear produces high stresses in the outer ply fibers at the grips. The distance along the composite that it takes for the load to diffuse throughout the cross section is much greater than in a metal. The stress profiles through the thickness are illustrated at several positions along the length of the composite.

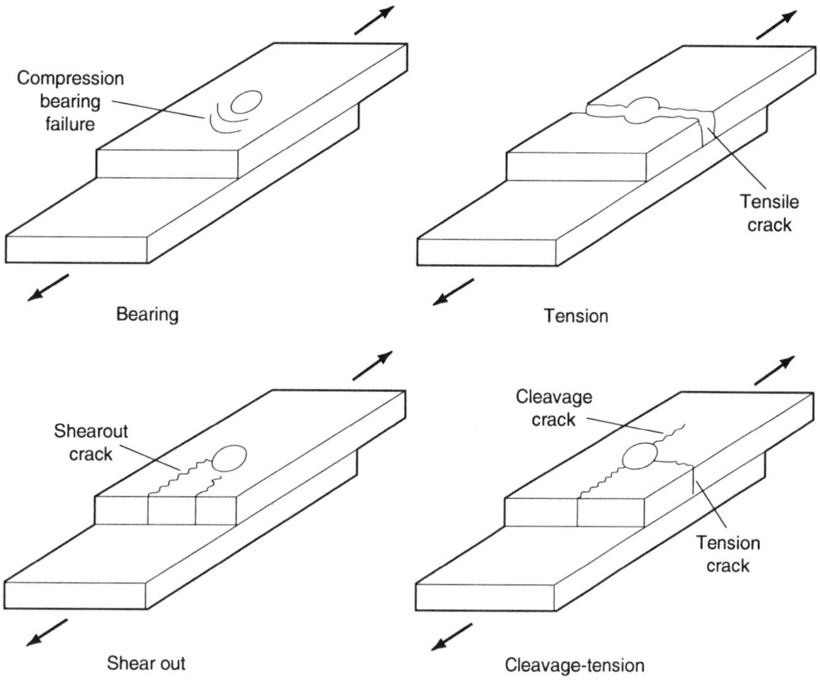

Fig. 18 Failure modes in bolted joints. Failure not only depends on the geometrical dimensions, but also on the laminate construction. Arrows show direction of tensile loading. Source: Ref 46

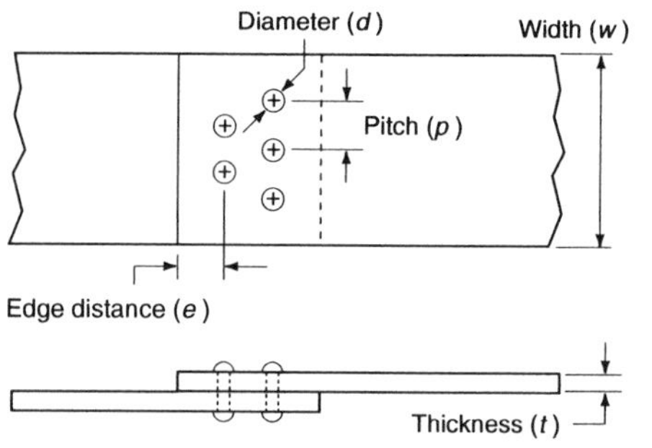

Fig. 19 Definition of bolted-joint geometry terms. Source: Ref 46

fibers all aligned parallel to the load direction. While the most probable state of the composite would be to have the worst flaws in each fiber randomly distributed along the fiber length, there is a small chance that all the worst flaws are located at the same distance along each fiber. (If there are periodic processing-induced defects, the odds become worse that the flaws correlate.) The question is how does the variability, as measured by the Weibull parameter, of composites compare to that of engineering metals? For well-developed composite systems, made in simple, flat laminates, the Weibull parameter for small specimens can be as high (20 to 24) as for aluminum sheet in the rolling direction, and even better than aluminum in the transverse direction. However, strengths of longer laminates have often been lower than predicted by simple Weibull scaling (Ref 45). There are two problems. Firstly, a large composite laminate may not be made in the same way as the smaller test laminates and more and worse flaws are introduced. Secondly, the small test specimens do not sample the worst flaws very well, which actually cause failure of the long laminate. By contrast, Weibull scaling of strength, when the width is increased, underpredicts the strength of the composite. (The strength does not drop off as fast as predicted.) The designer must be aware of the effect of scale on strength. Additional discussion of Weibull statistics is provided in the article "Design with Brittle Materials" in this Volume.

Load, Heat, and Electrical Current Transfer

The large difference in properties that often exist between fiber and matrix make transfer of load, heat, and electricity to a composite part very inefficient. The problem is easily illustrated in resin-matrix composites, because of the low modulus, and low thermal and electrical conductivity of most polymers. Most of the load, heat, or electrical current is usually transferred to a composite through the resin to the fibers from the sides and not directly to the fibers from the end (Fig. 17). In the case of mechanical loads, the load is transferred through shear stresses at the surface, and a tensile stress builds up in the composite from the end. A gradient in tensile stress also is present through the thickness, but disappears with sufficient length as the load is diffused throughout the composite. The distance along the composite that is required to reach approximately a constant stress is much longer in a composite than in a metal. While the stress can be considered constant in an isotropic metal within approximately two thicknesses for a sheet, the value for composites can vary from 6 to 40, the exact value depending on the composite architecture and materials. Hence, a designer must be more careful in transferring load to a composite to prevent overloading outer fibers. Load-transfer distances are much greater than in metals, and joints are less efficient.

Fiber-reinforced composites have also been considered for heat-transfer applications, because of the development of very high thermal conductivity carbon fibers (five times the thermal conductivity of copper at room temperature). The thermal conductivity is at its maximum value somewhat below room temperature and decreases to low values near 0 K. Above the maxima, the thermal conductivity decreases approximately inversely with temperature. Hence, the very high thermal conductivity can only be achieved in near-room-temperature applications.

These high thermal conductivity carbon fibers have been used in resin, metal (particularly aluminum), and carbon matrices. Flash laser thermal diffusivity experiments have shown that high thermal conductivity can be achieved in resin-matrix composites, when the laser beam directly impinges on the fiber ends in a uniaxial composite. However, the effective thermal conductivity in many real designs is much lower, because the heat must be transferred through the resin matrix, and the heat flow never becomes uniform through the cross section of the part. In the corresponding case for electrical conduction, uniform current flow is even more difficult to achieve. Adding an electrically or thermally conducting powder to the matrix resin can help alleviate the problem.

Metal-matrix composites allow better transfer of heat to the high thermal conductivity fibers, provided good fiber/matrix interfacial thermal contact can be maintained. Although aluminum can galvanically corrode with carbon fibers, the low density and high thermal conductivity makes this combination attractive especially for space and aeronautical applications. Protective coatings are available for carbon fibers to minimize corrosion.

Joint Design

Only one-half of the potential weight savings is often achieved when using composites because of the necessity of joints. The inherently inefficient load transfer into most composites, especially polymer-matrix composites, means that joint design is often a driving factor in composite design. Frequently, sections are thick and load paths complicated, which requires careful, and three-dimensional, analysis. Design of a joint is often more difficult than that of the rest of the composite part. A primary decision is whether to use bonded or mechanically fastened joints. Studies have shown that bonded joints can be more efficient than mechanically fastened joints. However, the questionable reliability and longevity of these bonded joints, especially if off-axis peeling stresses are present, have often resulted in the use of mechanical fasteners. Mechanical fasteners must be used if easy disassembly or access is required.

Mechanically Fastened Joints. The behavior of mechanically fastened joints is similar to but more complicated than the behavior of similar

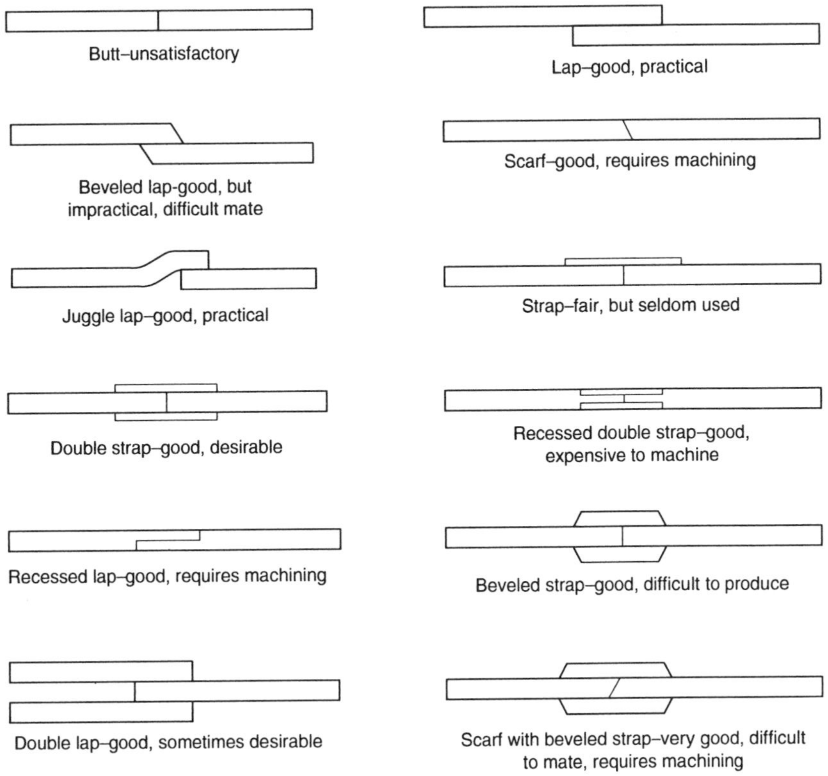

Fig. 20 Types of bonded joints. Many variations are possible, but difficult machining and assembly often eliminate more complex designs. The best bonded joints are those with the largest bond areas and minimum peeling stresses. Source: Ref 46

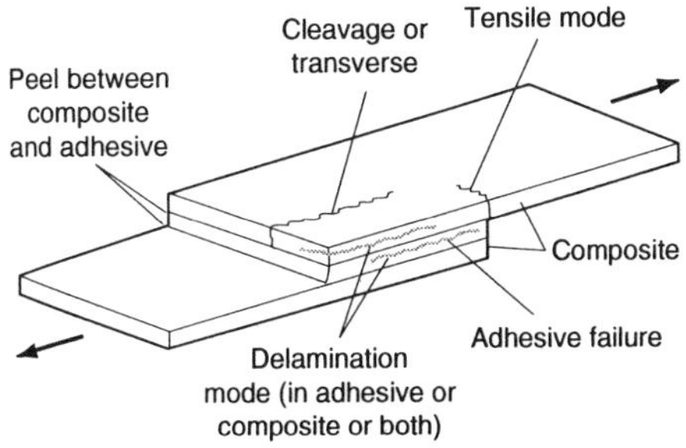

Fig. 21 Failure modes in bonded joints. Arrows show direction of tensile loading. Source: Ref 46

joints in metal. Not only is the strength dependent on geometry as with a metal, but also on the fiber orientations. Numerous composite joint configurations have been tested and evaluated. Only the major conclusions are presented. The important factors affecting joint strength are joint type, fastener, geometry, and failure mode.

Several types of mechanical joints have been used with composite materials. The choice of joint type depends on the application. Single- or double-lap designs with rivet (pinning) or bolting are frequently used. Riveted joints often provide adequate strength in carbon- or glass-fiber- reinforced resin-matrix composites, but bolts offer the greatest strength. Higher-bearing-strength bolted joints result if the clamping pressure is increased to an optimum by using a higher bolt tightening torque. Additionally, washers, and a carefully reamed hole that closely fits the bolt, produce higher-strength joints.

Mechanically fastened joints in composites display the same failure modes as observed in metals: net-section tension, edge-section shear, end-section bearing, and combined modes (Fig. 18). Geometrical factors as well as fiber orientations determine the failure mode. The geometrical factors of width (pitch), end distance, and diameter influence the behavior of composite joints, just as they do in metals (Ref 47, 48) (Fig. 19). End distance is defined as the distance from the end of the joint to the center of the closest hole, and width is the distance from the sides of the joint to the center of the nearest hole. End distance and width must be above a certain minimum if full bearing strength is to be obtained. Below these minima, tensile failure occurs if the width is too small, and shear-out or cleavage if the end distance is too small (Fig. 18). While all materials follow the same trend, the full strength occurs at differing w/d and e/d ratios for different materials and stacking sequences. However, values of w/d and e/d of 5 appear adequate to achieve a bearing failure in most resin-matrix composites. Adding fabric plies for reinforcement around holes is a very efficient way to reduce w/d and e/d values to below 5, although the thickness of the composite will be increased in the local region. The hole diameter compared to the laminate thickness, d/t, has a negligible effect if the hole is optimally clamped, but should be below 1 if not.

Bonded Joints. A bonded joint consists of three components with different mechanical properties: the adhesive or joining material, and the two parts that may be the same or different material. A large number of variables influence the load-carrying capabilities of an adhesive joint. Manufacturing variables include composition, preparation of the adhesive and adherends, and the joint geometry. Joints may also involve use of brazing with or without chemical reaction in metal or ceramic composites. The behavior of a bonded composite joint, as with mechanically fastened joints, is more complicated than its metal counterpart. The anisotropic nature of the material is significant, as are the low values of interlaminar and through-thickness modulus and strength for resin-matrix composites.

Various types of adhesively bonded joints are used in composites (Ref 49) (Fig. 20). The selection of joint shape is influenced by the type of assembly (thick/thin, etc.), bending moments, tooling requirements, and costs. While the scarf and beveled lap joint provide more uniform stress distributions and high strength, they present several fabrication problems, especially in thin sections. They are expensive to machine, the feather edges tear easily, the bonding jigs are expensive, and the pressure must be applied with care.

Compared to failure in metal joints, a larger number of failure modes can be identified in composites, due to their anisotropic nature. Failure can be tensile, interlaminar, or transverse in the adherends (Fig. 21). In the latter two cases, failure may either be in the resin or at the fiber/resin interface in a resin-matrix composite. A cohesive failure mode can also occur in the adhesive.

Most of the adhesives used for structural purposes are relatively strong in shear and weak in cleavage or peel. Therefore, the designer should attempt to place the adhesive in shear and minimize peel and cleavage stresses. The stress distribution depends on the joint geometry and the mechanical properties of the adhesive and the adherends. The strength of a simple, straight lap joint is directly proportional to the width of the bond, but increases to an asymptotic value with overlap length. The length required to reach the asymptotic strength depends on the particular material system, but values in excess of 40 are typical. Joint strengths are improved if identical adherends are used (if not, the in-plane and bending stiffness should be equalized), adherend stiffness is increased and adhesive moduli decreased, and as large an overlap as possible is used.

As a general rule, bonded joints are more suitable for lightly loaded joints where their high efficiency can be fully realized. Mechanically fastened joints are invariably used in high risk, highly loaded joints that are subjected to severe fatigue and environmental conditions. The major penalties involved in the use of mechanical joints are the increased weight, part count, and associated cost. However, there are many notable exceptions to this generality in which careful design has allowed the use of bonded structures in complex but critical parts. A combination of fasteners and bonding is the most structurally efficient and reliable arrangement in many cases.

Cut Edges

Resin-matrix composites are often sawed and then routed to final dimension, or cut by water jet. While the cut surface may appear smooth, interlaminar cracks may be introduced because of the poor interlaminar strength of resin-matrix composites. Even if no cracks are present, the free edge can act as an initiation site for cracking as high interlaminar stresses are often present at the free edge. Good design should minimize the number of free edges in the design if possible. Because free edges are often sites for failure in fatigue applications, a thin coating of a tough resin, even on the very smooth edges, has been found to increase fatigue life by a factor of ten.

A damaged surface often results when an abrasive wheel is used to cut metal or ceramic-matrix composites. The surface cracks may act as failure sites, especially in fatigue applications. A more important problem is that the cut edges expose reactive fibers and interfacial coatings to the environment.

Surface Coatings

Paints and wear-resistant coatings often have to be stripped and recoated. The similarity of paints to resins in resin-matrix composites often makes stripping difficult and usually more expensive than the application cost. However, the total life-cycle cost is often not minimized, because different groups manufacture and refurbish the composite structures. A goal of "paint for life" may be attractive for many applications.

Design for Manufacturing

Processing and fabrication costs could be reduced if structures were designed for manufacturing as well as performance. Composite materials are generally designed to maximize performance, and much touch labor is often used to manufacture these designs. Many composite laminates are fabricated by laying the laminate plies against a tooling surface, vacuum bagging, and then autoclave curing. While the surface molded against the tool can be good, the laminate surface nearest to the vacuum bag is often quite rough and the thickness is not well controlled because of variances from ply thickness and processing. The local roughness, often produced from "peel ply," can act as a beneficial spacer in bonding, but the variable thickness often requires shimming to maintain tolerances. Use of large closed metal molds can eliminate this problem, but they are too expensive to use for limited production runs. The use of high throughput, but limited geometry technology, such as pultrusion, filament winding, and braiding should be considered to reduce touch labor. Novel technology such as fast winding of a simple shape, with subsequent deformation to a more complex shape, should be considered for adding more versatility to these otherwise more limited processes.

High-temperature ceramic- and carbon-matrix composites often use chemical vapor infiltration to achieve very high performance. The process is lengthy as a precursor gas must be transported into and product gases out of the interior of a fibrous "preform" to form the matrix. A preform design that provides rapid transport of gases would drop costs and speed up design cycles without a large decrease in mechanical properties. Independent of the constituent materials, composites manufacture is complicated by the very tailorability that is the hallmark of these materials. While high performance and manufacturability are by no means mutually exclusive, it is critical that the designer include processing considerations as part of the design cycle.

REFERENCES

1. R.J. Diefendorf and E.W. Tokarsky, High Performance Carbon Fibers, *Polym. Sci. Eng.*, Vol 15 (No. 3), 1975, p 151
2. B.W. Rosen, Tensile Failure of Fibrous Composites, *AIAA J.*, Vol 2, 1964, p 1985
3. R.J. Diefendorf, The Chemical Nature of the Fiber/Resin Interface, *Tough Composite Materials,* NASA Langley, Noyes Publishing, 1985, p 192
4. C. Muser and N.J Hoff, Stress Concentrations in Cylindrically Orthotropic Plates with Radial Variation of the Compliances, *Progress in Science and Engineering of Composites,* T. Hayashi, K Kawata, and S. Umekawa, Ed., Japan Society for Composite Materials, 1982, p 389
5. C. Zweben and B.W. Rosen, A Statistical Theory of Materials Strength with Application to Fiber Composites, *J. Mech. Phys. Solids,* Vol 15, 1970, p 189
6. D.W. McKee and V.J. Mimeault, Surface Preparation of Carbon Fibers, *Chemistry and Physics of Carbon,* Vol 8, Marcel Dekker, 1973, p 201
7. H. Schuerch, Prediction of Compressive Strength in Uniaxial Boron Fiber-Metal Matrix Composite Materials, *AIAA J.*, Vol 4, 1966
8. S.J. DeTeresa, S.R. Allen, R.J. Farris, and R.S. Porter, Compressive and Torsional Behavior of Kevlar 49 Fibre, *J. Mater. Sci.*, Vol 19, 1984, p 57
9. Z. Hashin, Analysis of Properties of Fibrous Composites with Anisotropic Constituents, *J. Appl. Mech.*, Vol 46, Sept 1979
10. W. Voight, Lehrbuch der Kristallphysik, Teubner, Leipzig, 1910
11. D.L. McDanels, R.W. Jech, and J.W. Wheeton, Metals Reinforced with Fibers, *Met. Prog.*, Vol 78 (No. 6), Dec 1960, p 118–121
12. R.A. Schapery, *J. Compos. Mater.*, Vol 2, 1968, p 311
13. B.W. Rosen, Mechanics of Composite Strengthening, *Fiber Composite Materials,* American Society for Metals, 1965
14. H.T. Hahn and J.G. Williams, Compression Failure Mechanisms in Unidirectional Composites, *Composite Materials: Testing and Design (Seventh Conference),* STP 893, ASTM, 1986, p 115
15. Z. Hashin and B.W. Rosen, The Elastic Moduli of Fiber Reinforced Materials, *J. Appl. Mech.*, Vol 31, 1964, p 223
16. S.W. Tsai and H.T. Hahn, *Introduction to Composite Materials,* Technomic Press, 1980
17. R.M. Jones, *Mechanics of Composite Materials,* Scripta Book, 1975, p 55–56
18. R.E. Rowlands, Failure Mechanics of Composites, *Handbook of Composite Materials,* Vol 3, North-Holland, 1985, p 71
19. K.K. Chawla, *Composite Materials Science and Engineering,* Springer-Verlag, 1987, p 287
20. H. Edwards and N.P. Evans, A Method for the Production of High Quality Aligned Short Fibre Mats and Their Composites, *Advances in Composites (ICCM3),* Vol 2, Pergamon Press, 1980, p 1620
21. D.W. Radford, "Shape Stability in Composites," Ph.D. thesis, Rensselaer Polytechnic Institute, Troy, NY, May 1987, p 10
22. S.W. Tsai and H.T. Hahn, *Introduction to Composite Materials,* Vol 1, Technical Report AFML-TR-78-201, Air Force Materials Laboratory, 1978, p 138
23. "Structural Design Guide for Advanced Composite Applications," AFML Advanced Composites Division, Air Force Materials Laboratory, 1971
24. S.W. Tsai and H.T. Hahn, *Introduction to Composite Materials,* Technomic Press, 1980, p 221
25. R.J. Diefendorf, D.W. Radford, and S.J. Winckler, Asymmetric Composites-Hygrothermal Stability in Flat Plates, paper D4, *Verbundwerk '91—3rd International Conference on Reinforced Materials and Composite Technologies,* Demat, Frankfurt, Germany, 1991
26. M.W. Hyer, Calculations of the Room-Temperature Shapes of Unsymmetric Laminates, *J. Compos. Mater.*, Vol 15, July 1981
27. D.W. Radford and R.J. Diefendorf, Shape Instabilities in Composites Resulting From Laminate Anisotropy, *Reinf. Plast. Compos.*, Vol 12, Jan 1993, p 66
28. W.T. Freeman and G.C. Kuebeler, Mechanical and Physical Properties of Advanced Composites, *Composite Materials: Testing and Design (Third Conference),* STP 546, ASTM, 1974, p 435
29. H.T. Hahn and L. Lorenzo, *Advances in Fracture Research,* International Conference on Fracture No. 6, New Delhi, Vol 1, Pergamon, Oxford, 1984, p 549
30. A. Johnson, Modeling the Crash Response of a Composite Airframe, *Proceedings of the Tenth International Conference on Composite Materials,* Vol 6, Woodhouse Publishing, Cambridge, England, 1995, p 71
31. M. McLean, *Directionally Sloughed Materials for High Temperature Service,* The Metals Society, London, 1983
32. A.R. Champion, W.H. Kreuger, H.S. Hartman, and A.K. Dhringra, *Proceedings of the 1978 International Conference on Composite Materials (ICCM2),* TMS-AIME, 1978, p 883
33. T. Donomoto, N. Miura, K. Funitani, and N. Miyake, "Ceramic Fiber Reinforced Piston for High Performance Diesel Engine," Publication 83052, Society of Automotive Engineers, 1983

34. P.R. Smith and F.H. Froes, *J. Met.,* Vol 36, 1984, p 19
35. P.A. Selmers, M.R. Jackson, R.L. Mehan, and J.R. Rairden, Production of Composite Structures by Low-Pressure Plasma Deposition, *Ceram. Eng. Sci. Proc.,* Vol 6, 1985, p 896
36. W.A. Logsdon and P.K. Liaw, *Eng. Fract. Mech.,* Vol 24, 1986, p 737
37. Y.A. Bahei-El-Din and G.J. Dvorak, Plastic Deformation Behavior of Fibrous Composite Materials, *Proceedings of the 4th Japan U.S. Conference on Composite Materials,* Technomic Publishing, 1989, p 118
38. J. Tanaka, H. Ishikawa, T. Hayase, K. Okamura, and T. Matsuzawa, Mechanical Properties of SiC Fiber Reinforced Al Composites, *Progress in Science and Engineering of Composites,* ICCM-IV, Japan Society of Composite Materials, Tokyo, 1982, p 1410
39. G.D. Swanson and J.R. Hancock, Off-Axis and Transverse Tensile Properties of Boron Reinforced Aluminum Alloys, *Composite Materials: Testing and Design,* STP 497, ASTM, 1971, p 472
40. A. Kelly, Design of a Possible Microstructure for High Temperature Service, *Ceram. Trans.,* Vol 57, 1995, p 117
41. W.B. Hillig, A Methodology for Estimating the Mechanical Properties of Oxides at High Temperatures, *J. Am. Ceram. Soc.,* Vol 76 (No. 1), 1993, p 129
42. A.G. Evans and F.W. Zok, The Physics and Mechanics of Fibre Reinforced Brittle Matrix Composites, *J. Mater. Sci.,* Vol 29, 1994, p 3857
43. J. Aveston, G.A. Cooper, and A. Kelly, Single and Multiple Fracture, *Properties of Fiber Composites,* IPC Science and Technology, 1971, p 15
44. M. Sutco, Weibull Statistics Applied to Fiber Failure in Ceramic Composites and Work of Fracture, *Acta Metall.,* Vol 37 (No. 2), 1989, p 651
45. C. Zweben, Simple Design Oriented Composite Failure Criteria Incorporating Size Effects, *Tenth International Conference on Composite Materials,* Vol 1, Woodhouse Publishing, Cambridge, England, 1995, p 675
46. G. Ger, "The Joint Strength of Kevlar/Graphite Hybrid Composites," M.S. thesis, Rensselaer Polytechnic Institute, 1984
47. T.A. Collings, The Strength of Bolted Joints in Multidirectional CFRP Laminates, *Composites,* Vol 8 (No. 1), Jan 1977, p 43
48. E.W. Godwin and F.L. Matthews, A Review of the Strength of Joints in Fiber-Reinforced Plastics: Part 1, Mechanically Fastened Joints, *Composites,* Vol 11 (No. 3), July 1980, p 155
49. F.L. Matthews, P.F. Kilty, and E.W. Godwin, A Review of The Strength of Joints in Fiber-Reinforced Plastics: Part 2, Adhesively Bonded Joints, *Composites,* Vol 13 (No. 1), Jan 1982, p 29

Section 7: Manufacturing Aspects of Design

Introduction to Manufacturing and Design

Henry W. Stoll, Northwestern University

THIS ARTICLE introduces and describes general concepts and practices related to manufacturing and design. It is intended to:

- Place the activities of design and manufacturing in the context of the business system that they support
- Present an overview of the manufacturing technology field from a design and material selection perspective
- Provide insight into the complex relationship among design, material selection, and manufacturing
- Summarize modern design for manufacture practices being widely used in industry today

The main focus is on how design and manufacturing practices influence the properties and cost of engineered designs. Engineered designs are products, equipment, devices, and hardware that have been designed to meet specific end-user needs. In this context, nuts and bolts, computers, electrical transformers, portable phones, automobiles, machine tools, construction equipment, consumer products, and aircraft are all engineered designs. In engineered designs, materials and methods of manufacture are typically selected to meet functionality, performance, cost, and reliability objectives.

Manufacturing can be defined as the conversion of starting materials into finished parts, products, and goods that have value to end users. Starting materials for engineered designs are very often semifinished products such as coil steel or pelletized plastic as well as "off-the-shelf" components and hardware. The specific starting material used for a particular part or product depends on the manufacturing process and on the types, variety, and availability of semifinished products that are acceptable for use in the process.

Manufacturing is both a technical activity and an economic activity. As a technical activity, manufacturing involves the design, development, implementation, control, operation, and maintenance of a large variety of manufacturing processes that facilitate and perform the conversion of starting materials into finished products having greater value. Processes used in manufacture can be grouped into two basic types: (1) physical and chemical processes that transform the shape, properties, and/or appearance of starting materials into parts, and (2) assembly and joining processes that combine multiple parts into finished products. In addition, there are myriad ancillary processes and operations that support the basic manufacturing processes in a variety of ways such as material handling, quality control, testing, and so on.

In the economic sense, manufacturing is a commercial activity performed by companies that sell products to customers. Hence, economic considerations are generally the overriding constraint in manufacturing decision making. The costs of energy, material, purchased components, labor, tooling, and capital equipment must be minimized and properly controlled if the firm is to make a profit. Producing and selling products is generally a very complex activity involving a convoluted mix of people skills and disciplines, machines and equipment, tooling, computers, and automation working together to form a manufacturing system. A manufacturing system comprises a large number of distinct functions and activities (Fig. 1) that interact in a variety of ways to ultimately produce and sell the product. On the business level, therefore, the challenge is to organize, optimize, and operate the manufacturing system in a way that ensures both long-term customer satisfaction and economic viability and success of the manufacturing enterprise.

The Manufacturing Enterprise

Manufacturing usually is associated with a business activity or enterprise. To understand the relationship between manufacturing and business, consider the basic actions that occur when conducting a manufacturing-related business. Four groups of "main-line" work are generally necessary: decide what customers want, set up the factory to make it, produce it, and sell and

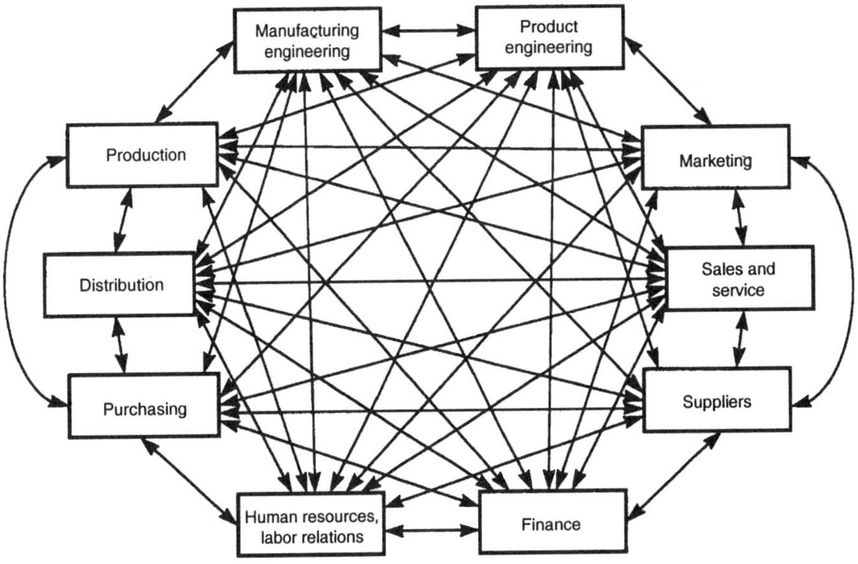

Fig. 1 The variety of ways the many different functions of a manufacturing system can interact

service it. Tracing the "main-line" generic work flows that connect these activities yields two major cycles on which most design and manufacturing practices focus: the order-to-delivery cycle that involves the receipt of orders and production of existing products, and the new product introduction cycle that involves planning and introducing new products and changes into production.

The order-to-delivery cycle is the operations side of the business. It generates the day-to-day cash flow for the company by producing and distributing products, ensures product quality, and provides the interface between the company and its customers in the form of sales and service. The new product introduction cycle, on the other hand, is the product development, engineering, and new technology side of the business. For many businesses, product development and engineering is the life blood of the company. Effective new product development increases the company's competitive edge in everchanging markets.

At first glance, it may appear that these two cycles are independent of each other, but in fact they are very interrelated and closely coupled. This is because the new products introduced into production by product development must map into the production environment with the least amount of disruption and expense and in a way that ensures high product quality. This requires close coordination and cooperation between the product development and operations activities.

This mapping is one of the principal motivations for design for manufacture (DFM). Most engineered designs are assemblages of parts. Direct and indirect cost as well as quality of assemblies are determined by the number of parts; the ease with which the parts and components are handled, assembled, procured, and inspected; and the functionality, serviceability, maintainability, reliability, and durability achieved by the finished device or product.

All of these factors are in turn dependent on the material, manufacturing processes, and detail design of the individual parts and components that make up the finished assembly. Therefore, each part must be designed so that it not only meets functional requirements, but also can be manufactured and assembled economically and with relative ease in the production environment of the company. Effective part design requires early consideration of the characteristics, capabilities, and limitations of materials, manufacturing processes, and related operations, machinery, and equipment (including consideration of part-to-part variability, dimensional accuracy, surface finish, processing time, and so forth). These issues are discussed in greater detail in the article "Design for Manufacture and Assembly" in this Volume.

The type of production environment the design must map into drives many product design, material selection, and manufacturing decisions. This often leads to different manufacturing strategies. For example, one possible strategy is to assemble products using parts and components purchased from suppliers and vendors. In this approach, the manufacturer might pay more for individual parts and components, but avoid the direct and long-term indirect cost of making them internally and gain the advantage that the design team is relatively free to select materials and processes that best meet the functional needs of the new product.

On the other hand, if a manufacturer chooses to make many of the parts used in its products, that is, if it is "vertically integrated," then new products may need to be designed to be compatible with available "in-house" tooling and equipment. This can impose constraints on material and manufacturing process selection. For example, the cost and difficulty of introducing a spot-welded structure may be overwhelming for a company that has historically manufactured arc-welded structures. Similarly, switching to an all-plastic housing may prove difficult for a company that has traditionally made sheet metal housings and, as a consequence, has invested heavily in sheet metal forming equipment and expertise. In many product developments, "make or buy" decisions can strongly influence design direction; for many businesses, they can form the basis for business and manufacturing strategy.

New Product Introduction Cycle

The new product introduction cycle involves two major activities: tactical planning and product development. Tactical planning determines the "what" and "when" for product design and development and quantifies the needs that guide design and manufacturing. Product development translates the "what" into "how" by defining the new product designs to be introduced into production and providing the physical hardware, systems, information, training, and guidance to make new processes, tooling, and equipment work in production. Tactical planning is typically performed by marketing, advanced product planning, and technology development groups, while product development is typically performed by product engineering and manufacturing engineering groups.

Designing new products and introducing product changes takes time and costs money. If a new product takes too long to design and launch, the manufacturing enterprise can lose financially because of lost sales and market share. Even if the product gets to market in a timely manner or if time-to-market is irrelevant, the firm can lose financially if the cost of developing the design is excessive. The need to develop new products efficiently has led to several major trends in design and manufacturing practice.

Design Process Reengineering. Organizational and procedural changes have been made to enhance communication between functions, encourage DFM, and simplify and optimize work flow. Many companies have adopted concurrent engineering practices and are using the team approach effectively. In some cases, design and manufacturing engineering have been located on the same premises, while in other situations technologies such as teleconferencing, E-mail, and the World Wide Web are used to overcome wide geographical separations. Many companies are also working to improve their product realization process so that the design team knows exactly what to do during each step of the process. Gate or design reviews have been instituted to ensure economic viability of design projects and to facilitate simultaneous achievement of product manufacturability and tight schedule commitments. See Ref 1 for a comprehensive treatment of management aspects of product design.

Integrated Design Systems. Computer-aided design (CAD), computer-aided manufacturing (CAM), and computer-aided engineering (CAE) systems have been integrated so that product and manufacturing engineers work on compatible systems that share information in a seamless environment. In addition, a wide variety of new CAD tools and methods have been introduced (see the Section "Design Tools" in this Volume), and their use has been made an integral part of the product realization process. Instead of computerizing manual practices such as drafting, design organizations are now using the computer to eliminate these practices where possible and to replace them with new computer-aided practices that reduce design time and effort. For example, parametric and feature-based CAD/CAM tools now enable design changes to be made and then propagated through the model with a few high-level commands. Using these tools, engineering analysis, manufacturing, and inspection can be automated and integrated with design to a higher degree than ever before.

Science Base. Traditionally, many engineered products have evolved and have been incrementally improved by using experimental or "test-and-fix" procedures involving costly and time-consuming construction and testing of many prototypes. This practice is being eliminated by developing a "science base" for the product technology involved. Once an appropriate science base is available, computer simulations and other analysis techniques can be used to more quickly select the best new design.

Product architectures are evolving that take advantage of commonalities present in different product models and variants. By modularizing and standardizing components and subsystems, creating building block parts, and rationalizing the variety of choices available for purchased components (such as threaded fasteners and ball bearings), many firms have greatly reduced design time for new products, especially those similar to existing products. These practices are also leading to significant savings of scope and scale. High-volume manufacturing processes and automation alternatives become economically feasible, and purchasing effort and the per part cost of supplied parts is reduced.

Taking time out of the product design cycle yields numerous benefits. Customer satisfaction is enhanced because improved product quality and short cycles give the customer what he wants when he wants it. Cost is reduced because waste and nonvalue activity is eliminated. Improved profitability results from increased market share combined with the ability to respond and adapt quickly to changing market conditions with minimum cost impact.

Order-to-Delivery Cycle

The order-to-delivery cycle typically involves three major activities: order fulfillment, production, and service. Order fulfillment is the sales arm of the business, responsible for bringing in orders for products and forecasting production requirements. Production converts sales orders into products. It encompasses all of the personnel, operations, processes, tooling, and equipment involved in the day-to-day manufacture of products including procurement, raw material and supplied component receiving and inspection, part manufacture, assembly, testing, and distribution of manufactured products. Service provides product-related maintenance, repair, disposal services, and support to the customer.

All of these activities influence design and manufacturing practices. For example, design for service and design for disassembly are important strategies in many companies. As another example, consider companies that sell specialized or customized products and services. These companies must often "quote" the job first, then manufacture and deliver the product or equipment after the order has been received. The time this process takes can contribute significantly to manufacturing lead time as well as strongly influencing customer satisfaction. Modular designs or designs that allow customization at the end of the production line help reduce order-to-delivery time for many of these types of customized products and equipment.

Marketing, manufacturing, and design strategy can also play an important role. Consider a company whose marketing strategy is to sell personal computers that are customized according to specifications provided by the customer via a phone order. Such a marketing strategy needs to be supported by a manufacturing strategy that facilitates easy manufacture and delivery of a customized product, together with a design strategy that makes the product easy to customize at remote distribution sites.

Another approach is to use computerization to take time out of the ordering process. One vision is that of a sales representative in the office of a customer with, for example, his or her portable computer having all product and pricing data available on disk, so that a complete purchase order or release can be negotiated on the spot and then inserted into the plant's production schedule by a modem and confirmed before the representative leaves the office.

An important measure of production efficiency and effectiveness is manufacturing lead time (MLT) or the time required to process the product through the plant. Manufacturing lead time is directly related to manufacturing cost and customer satisfaction. A short MLT implies less manufacturing time and labor and therefore lower manufacturing cost. Also, the shorter the MLT, the sooner the product can be sold and the company reimbursed for its investment in raw material and labor. Most importantly, a short MLT means the customer gets the product when needed.

Table 1 Manufacturing strategies to reduce manufacturing lead time and the order-to-delivery cycle

Strategy	Effect(a)
1. Specialization of operations	Reduce T_o
2. Combined operations	Reduce n_m, T_h, T_{no}
3. Simultaneous operations	Reduce n_m, T_o, T_h, T_{no}
4. Integration of operations	Reduce n_m, T_h, T_{no}
5. Increase flexibility	Reduce T_{su}, MLT, WIP; increase U
6. Improve material handling and storage	Reduce T_{no}, MLT, WIP
7. On-line inspection	Reduce T_{no}, q
8. Process control and optimization	Reduce T_o, q
9. Plant operation control	Reduce T_{no}, MLT; increase U
10. Computer-integrated manufacturing	Reduce MLT, design time, production planning time; increase U

(a) T_h, work-handling time; WIP, work in process; q, scrap rate or fraction defect; U, utilization

Reference 2 calculates MLT as,

$$\text{MLT} = \sum_{i=1}^{n_m} (T_{su} + QT_o + T_{no})_i$$

where n_m is the number of machines or operations and T_{su}, Q, T_o, and T_{no} are the setup time, batch quantity, operation time, and nonoperation time for each machine and/or operation, respectively. Nonoperation time includes handling, storage, inspections, and other nonvalue-added activities. Some manufacturing strategies given in Ref 2 that help reduce MLT and the order-to-delivery cycle are summarized in Table 1. These practices are often facilitated or made possible by design practices such as design for manufacture and assembly.

Manufacturing Processes

A general appreciation for the different manufacturing processes that are commonly used and for the design and manufacturing practices that are associated with them can be gained by considering the fundamental nature of manufacturing processes. A process can be defined as the change of properties of an object. Changes may concern geometry, hardness, strength, chemical composition, information content, and so forth. In general, three essential agents must be available to cause a change: energy, material, and information. Therefore, a process may be an energy process, a material process, or an information process. Most practical manufacturing processes are material processes.

A material process consists of a material flow on which shape information is impressed (information flow) and an energy flow that carries out the transformation of information through the tool/die and the pattern of movement for the tool/die and the material (Fig. 2). The energy flow includes both the energy necessary to carry out the process and the energy output (loss) that is produced by the process. The information flow, which includes both shape and property information, depends on the nature of the material, the type of process (e.g., mechanical, thermal, chemical), the characteristics of the tool/die, and the pattern of movement of the material and the tool/die. The final geometry or shape information of the part is the sum of the initial shape information of the starting material and the shape information impressed on it by the process. Similarly, the final material properties of the part or product is the result of the property information of the starting material and the property change caused by the process. In general, the nature of a material

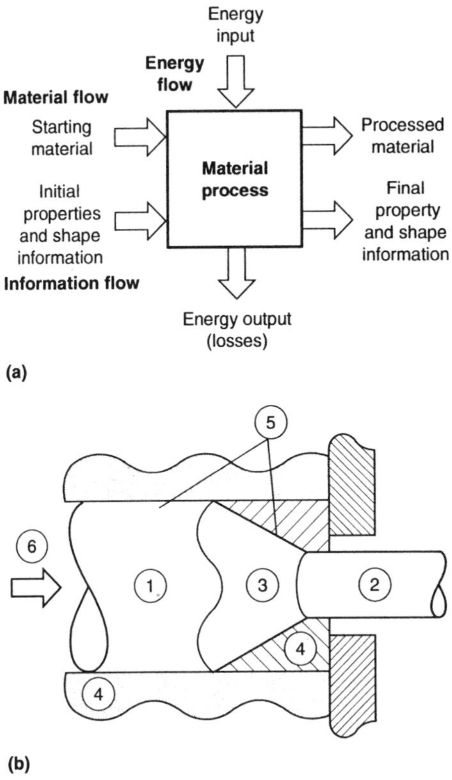

Fig. 2 Material process. (a) Model. (b) Schematic of one type of material process, bulk deformation by extrusion. 1, original material condition (shape and microstructure); 2, final material condition (shape and microstructure); 3, plastic deformation zone; 4, tooling effects; 5, friction effects; 6, equipment characteristics

process depends on whether the mass of the material is conserved, decreased, or increased.

Mass-Conserving Processes

In mass-conserving processes, the mass of the starting material is approximately equal to the mass of the final part or product. During the process, the starting material is "forced" to change its shape or properties in some way. Shape-replication processes are mass-conserving processes in which the part replicates the shape information stored in the die or mold by being forced to assume the shape of the surface of the tool cavity. Near-net-shape processes are shape-replication processes that produce parts requiring little or no subsequent processing (e.g., machining) to obtain the finished part. The nature and material requirements of shape-replication processes are generally determined by the state of the starting material.

Liquid or semifluid material is forced to flow into a mold cavity under pressure where it assumes the shape of the cavity. Hence, information is impressed primarily by the tool. However, most materials shrink as they solidify so the final geometry of the part is the sum of the tool shape plus distortions and dimensional changes due to material shrinkage. Representative processes include sand casting, die casting, and plastic injection molding. Process-related material considerations typically include melting temperature, flow behavior, ability of the material to assume the shape of the mold cavity, existing thermal and mechanical properties during solidification, resulting properties of the solidified material, and flaws such as porosity or voids that can occur as a result of the process.

Granular materials (metal powders, molding sand, etc.) are formed as a result of a flow process (filling, flow, placing) and a stabilization process (packing, plastic deformation, hardening, sintering, etc.). The geometry is primarily determined by the tool with some secondary geometry changes occurring due to shrinkage or forming occurring during stabilization. Representative processes include powder metallurgy and ceramics processing. Material considerations are process specific, but generally involve strength at various stages of the process, ability of the granular material to assume the desired shape, and density and final properties of the finished part.

Ductile Solid Materials. Forming takes place by plastic or elastic deformation of the starting material. The part shapes that are possible depend on the specific material, starting shape, and process to be used. In general, the material properties, the process parameters (state of stress, force, velocity, temperature, etc.) and the surface creation method must be analyzed together in order to establish the shape of the starting material, the tool design, and the process design. Representative processes include bulk deformation processes such as forging, extrusion, and rolling and sheet metal working processes such as stretching and drawing. Material considerations typically revolve around the ability of the material to flow plastically into the desired shape, property changes caused by plastic deformation, possible failure modes such as tears and splits, and dependency of mechanical properties on temperature and strain rate.

Near-net-shape processes are very desirable because little or no material is wasted and the final part can be produced quickly and consistently. However, large production volumes are often required to offset tooling cost. Design and manufacturing practice for mass-conserving processes generally focus on reducing tooling cost and development time and on setting process parameters to produce high-quality parts with low cycle times. Detail design of the part and material selection are of particular importance in these processes because the part features must be compatible with both the material properties and the manufacturing process.

Mass-Reducing Processes

A mass-reducing process is one in which the mass of the starting material is greater than the mass of the final part or product. Forming takes place by removal of material. Mass-reducing processes that alter the shape of the material are "shape-generation" processes because the part shape is determined by the pattern of relative motion between the cutting tool and the part. The starting material for mass-reducing processes is typically a ductile or brittle solid material. Methods of material removal include fracture, phase change such as melting or ablation, and chemical reaction. In general, material properties are not altered on a macroscopic level by mass-reducing processes. However, material properties are often a factor in determining processing speed and surface finish characteristics. The various processes that are commonly used can generally be classified according to the nature of the energy transfer medium or tooling used.

Conventional Machining Processes. In these processes, material is removed by fracture and the formation of chips created by relative movements between the workpiece material and the tool. In most conventional machining processes, the contour content of the tool is relatively small so the patterns of movement of the tool and workpiece material play a large role in the surface creation. The cutting tool can have a single cutting edge, such as the tool used in turning (lathe) or shaping operations, or it may have multiple cutting edges. Multiple-cutting-edge tools have either well-defined edge geometry such as milling cutters, drills, and saws, or randomly oriented cutting edges such as those in grinding wheels and abrasive paper.

Other Machining Processes. In these processes, material is removed by mechanical, thermal, electrical, and/or chemical action and often involve the use of granular, liquid, or gaseous media. In general, the surface creation is the result of the material removal mechanism, the geometry of the tool or cross section of the energy source, and the pattern of movement of the tool or energy source relative to the workpiece material. Examples include ultrasonic machining, electrical discharge machining (EDM), electrochemical machining (ECM), waterjet cutting, laser cutting, thermal cutting, and chemical etching.

Shearing processes involve separating adjacent parts of a sheet through controlled fracture. In some cases, such as punching, the material that is removed is scrap. In other cases, such as blanking, the material that remains is scrap. By varying the geometry of the tool and the pattern of movement, a large number of different processes are possible. Examples include punching, blanking, shearing, and slitting.

The complex patterns of movement required in machining processes typically result in excessive setup and cycle times. Hence, these processes are generally avoided if possible when large production quantities are involved. On the other hand, machining processes are capable of high precision and often they are the only option if tight tolerances are required. Also, because shape information is impressed through the pattern of movement rather than through the contour of the tool, machining is often the process of choice for low-volume manufacture where the cost of tooling must be kept low.

Design and manufacturing practices associated with mass-reducing processes generally focus on eliminating setup and processing time and on controlling process variation. For example, the use of numerical control (NC) makes it possible to change over from one part to another simply by downloading a different tool path program. CAD/CAM systems allow the rapid generation of the NC program directly from the solid model database. Also, by using modern sensor technology and feedback control techniques, NC systems can be designed to give extremely high accuracy and repeatability.

Assembly Processes

In mass-increasing processes, change is the result of the assembly or joining of components into a whole. Typically, the mass of the final part or product is approximately equal to the sum of the masses of the components. Assembly can be described as a series of joining processes in which parts are oriented and added to the build. The macrogeometry of the assembly is established by the positioning of the components. Joining processes are either permanent or nonpermanent and can be classified as shown in Fig. 3. Material properties and property changes produced by the process can be an important consideration in some joining processes such as solid-state and liquid-state welding.

Assembly imposes constraints on the design. Not only must the parts be designed so that they can be assembled and joined together to provide the needed function, they must also be designed so that they are easy to handle, insert, retain, and verify that they have been assembled correctly. Because assembly is an integrative process, problems with detail part designs often surface when they are assembled. Parts do not fit together properly, tools cannot reach in the space provided, parts can be incorrectly assembled, and so forth. These problems can often require extensive rework resulting in costly schedule slippage and undesirable design compromises.

The importance of assembly as a design constraint has resulted in a greatly increased emphasis on assembly in the design process. Design and manufacturing practice now focuses on ensuring that parts conform to specifications, on eliminating variability and randomness from the process, and on making nonvalue-added operations such as orienting and handling as simple and easy to perform as possible. Many product design departments now use design for assembly techniques to improve the ease with which products are assembled. Design for assembly, which is discussed in detail in the article "Design for Manufacture and Assembly," which follows in this Volume, seeks to ensure ease of assembly by developing designs that are easy to assemble.

Many excellent textbooks on manufacturing processes and methods are available. See, for example, Ref 3 to 5. Design issues related to specific processes are discussed in detail in the other articles in this Section of the handbook.

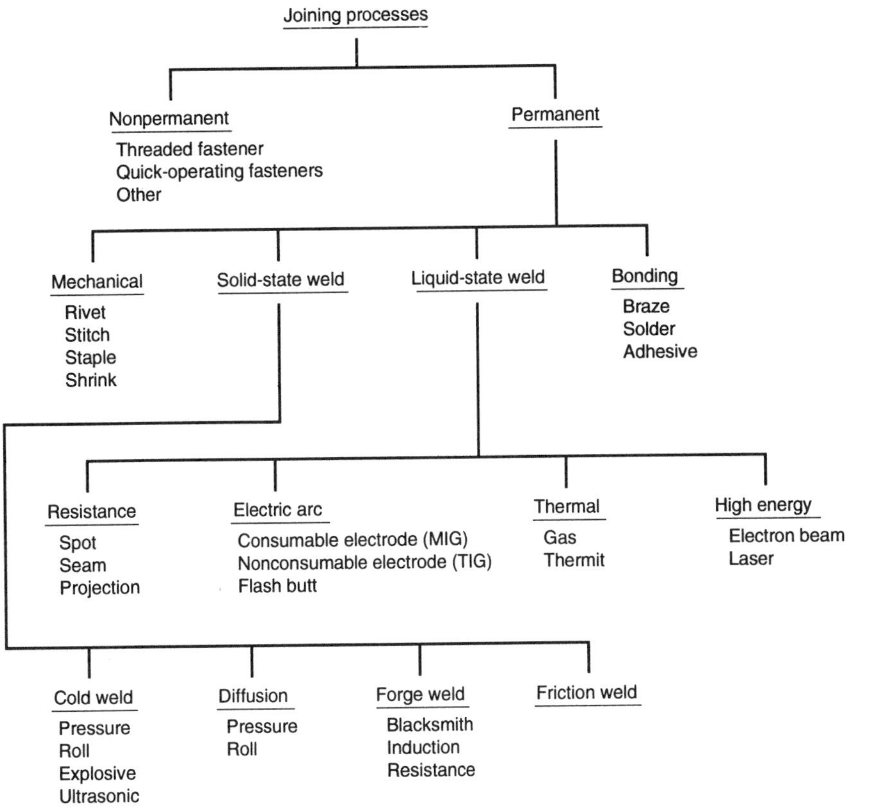

Fig. 3 Classification of joining processes (adapted from Fig. 9.1, in Ref 3). MIG, metal inert gas; TIG, tungsten inert gas

Production Systems

Manufacturing processes convert starting material into finished parts and products. Production systems link, coordinate, and integrate the various manufacturing processes used in production. Production systems are often designed and optimized to reduce manufacturing lead time, and as a result they may involve manual operations, mechanized operations, automated operations, or a mix of all three.

A distinction is generally made between mechanization and automation. According to Schey (Ref 3), mechanization means that something is done or operated by machinery and not by hand whereas automation means a system in which many or all of the processes involved in the production, movement, and inspection of parts and material are automatically performed or controlled by self-operating devices. Automation implies sensing, closed-loop control, and some degree of decision making in addition to mechanization. Flexible automation includes the added capability of being easily reprogrammed or adapted to meet varying or new production needs.

The kind of production systems used in modern factories depend greatly on the variety of products being produced and on production quantities. If production quantities are low and product variety is extensive, then the production systems are likely to be manual with an emphasis on accommodating variety. A glass blower who makes a wide variety of different ornamental decorations is an example. In this case, the glass blower probably does his or her work at one workbench with all needed tools readily at hand. He or she is ideally set up to make individual pieces, one at a time, and in any order necessary to meet customer demand.

At the other extreme, if production quantities are large and only one product is made, it is likely that dedicated mechanization or automation would be used. For example, a company that makes millions of a certain type of lightbulb each year is likely to manufacture the lightbulb using a production machine that is fully automated but capable of producing only that one particular type of lightbulb. Figures 4 and 5 depict the relationship among product complexity, production volume, and level of automation.

Although production systems are not of direct concern in material selection, they do impact part

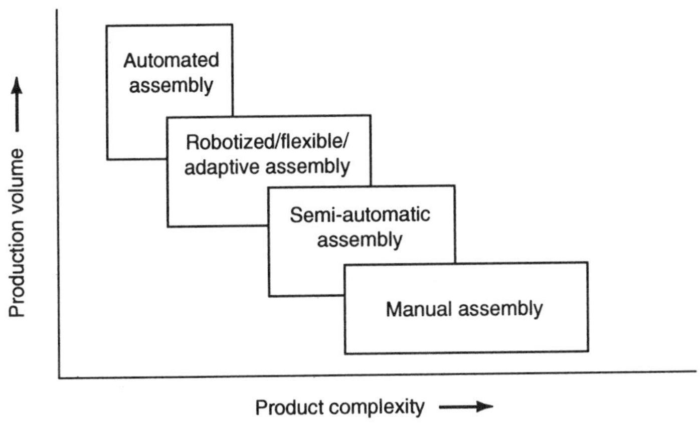

Fig. 4 Production volume versus product complexity. Source: Ref 6

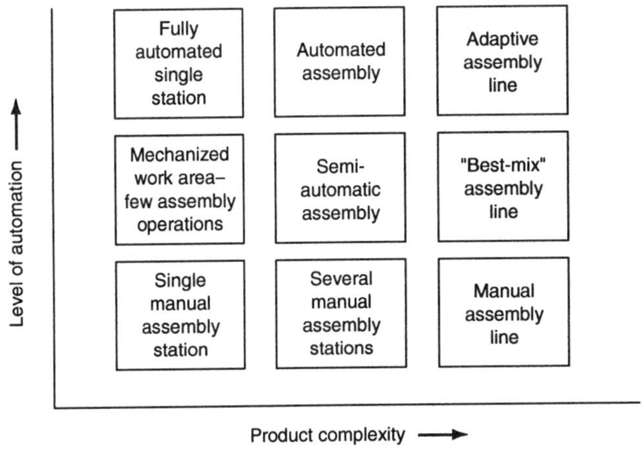

Fig. 5 Level of automation versus product complexity. Source: Ref 6

and product design in a variety of ways. For example, if a part is to be bowl fed, it must be properly designed for that environment. This implies that the material must hold up under bowl feeding conditions. Similarly, if a part is to be handled by a robot, it must have the appropriate design features that allow gripping and it must be capable of withstanding the gripping forces involved. Production systems therefore often superimpose additional constraints on design and material selection. Leading-edge companies that manufacture products in large quantities often utilize a "design for automation" approach in which the needs of the production system are considered very early in the product design and the systems integrator or supplier of the automation equipment is an important member of the concurrent engineering team.

Interaction between Design and Manufacturing

The design of a product and its method of manufacture are intimately connected. This interdependence cuts across all aspects of the manufacturing system and exists at the part, assembly, and system levels of product design and manufacturing. The interaction between design and manufacturing is easiest to see at the part level. Suppose, for example, that a part is to be designed as a casting. If the part is to be manufactured using a casting process, then a material suitable for the particular casting process being considered must be selected. This material must also have suitable functional properties such as strength, corrosion resistance, weight, and so forth. Given a specific material and casting process, the part must then be properly designed to satisfy both functional and processing requirements. This means that not only must the finished part have a certain desired shape and set of material properties, it must also have an acceptable parting line, sufficient draft, and a geometry that facilitates proper solidification of the material to achieve the desired material properties without porosity and other flaws.

The interaction between design and manufacturing at the assembly level should be one of ease and simplicity. How a product is assembled is determined by its design. If assembly is considered as part of design, then many potential difficulties and quality risks can usually be avoided. If assembly is not considered, manufacturing will still find a way to assemble the product, but the cost of assembly as well as indirect cost is likely to be higher and product quality may suffer.

Design and manufacturing interactions are perhaps hardest to delineate, quantify, and control at the systems level, and yet it is at this level that these interactions have the most long-term impact and far-reaching consequences. Consider, for example, the sheet metal used in making an automobile. From the design point of view a wide selection of different sheet metal thicknesses, roll widths, coatings, and material properties is desirable so that the most optimal design can be achieved with respect to structural rigidity, weight, and cost. From the manufacturing point of view, however, the ability to process a wide selection of different thicknesses and types of sheet metal means added cost and complexity in purchasing and supplier relations, material handling and storage, processing equipment and tooling, setups and operations, and so forth. What is needed is the right balance that gives design the flexibility it needs while still allowing manufacturing the ability to simplify and standardize its operations. How a company manages interactions such as this can greatly affect its long-term economic viability and competitiveness.

Design for Manufacture Practices

The interaction between design and manufacture has resulted in a variety of design and manufacturing practices. This section provides a brief overview of various practices that are of general importance to design and manufacturing. Some of these practices, such as standardization and group technology, span several organizations within a company and have their greatest effect at the systems level. Others, such as design for assembly, focus on the interaction between design and specific manufacturing processes. Still others, such as failure mode and effects analysis and value engineering, can be used at all levels and in all contexts to improve both product design and the method of manufacture. Related coverage is provided in the Sections "The Design Process" and "Criteria and Concepts in Design" in this Volume.

Design for life-cycle manufacturing is a philosophy of product design that includes the full range of policies, techniques, practices, and attitudes that cause a product to be designed for the optimal manufacturing cost, the optimal achievement of manufactured quality, and the optimal achievement of life-cycle support (serviceability, reliability, and maintainability). This philosophy is often implemented by companies that perceive product design to be a competitive advantage. The concepts of design for assembly, concurrent engineering (see the article "Concurrent Engineering" in this Volume), and other systematic design approaches that cause the design team, from the outset, to consider all elements of the product life cycle from conception through disposal are all considered to be part of this general, overarching philosophy. The philosophy is implemented both through the product realization process utilized within the company and through the general culture and attitudes embodied by the company employees.

The design for assembly (DFA) method was developed by Geoffrey Boothroyd while at the University of Massachusetts (Amherst). Details of the methodology are presented in Ref 7. Design for assembly is also discussed in detail in the article "Design for Manufacture and Assembly" in this Volume. Based largely on industrial engineering time-study methods, the DFA method seeks to minimize the cost of a product within constraints imposed by other design requirements. This is done by first reducing the number of parts and then ensuring that the remaining parts are easy to assemble.

Design for "X" (DFX) Methods. A variety of "design for" methods and approaches are being developed and used to improve the designs of parts and products with respect to specific manufacturing processes and activities. Examples of methods aimed at manufacturing activities include design for service, design for testing, design for disassembly, and so forth. Design for casting, design for plastic injection molding, and design for machining are examples of DFX methods directed toward improving the design of parts that are to be manufactured using a specific manufacturing process. Many companies have specialized manufacturing facilities that can only process parts or assemblies that have acceptable characteristics such as part spacing or size. Guidelines, developed by manufacturing engineers who are very familiar with the capability of the specific facility for use by design engineers, is another example of a DFX methodology. Design for "X" methods seek to provide guidance to the designer that helps ensure that parts and products are correctly designed to be compatible with a given process or activity. Benefits include lower tooling costs, reduced process cycle times, and improved process yields.

Design for Quality. Variability is the enemy of manufacturing. It is a major cause of poor quality resulting in unnecessary manufacturing cost, product unreliability, and ultimately, customer dissatisfaction and loss of market share. Variability reduction and robustness against variation of hard-to-control factors are therefore recognized as being of paramount importance in the quest for high-quality products. In a design for quality approach, the design team seeks to design the product and process in such a way that variation in hard-to-control manufacturing and operational parameters is minimal. The ideas behind this approach are largely attributable to the efforts of G. Taguchi and the cost-saving approaches to quality control pioneered in Japan. An important element of this approach is the extensive and innovative use of statistically designed experiments. The Taguchi method and other design for quality approaches are described in the articles "Robust Design" and "Design for Quality" in this Volume, and they are treated extensively in the literature (see, for example, Ref 8).

Process-Driven Design. In process-driven design, a method of assembly or other manufacturing process plan is developed prior to developing the product design. This plan is then used to guide the product design thereby ensuring a coordinated product and process that results in an optimization of the overall manufacturing system. Process-driven design is based on the recognition that product design decisions often inadvertently limit the manufacturing options available for use in production of the product. Process-driven design methods have been successfully applied in many different industries. For example, many of the modern innovative manufacturing and assembly methods now commonly used in the automotive, airplane, and farm machinery industries can be traced to process-driven design practices.

Failure mode and effects analysis (FMEA) is an important design and manufacturing engineering tool intended to help prevent failures and defects from occurring and reaching the customer. It provides the design team with a methodical way of studying the causes and effects of failures before the design is finalized. Similarly, it helps manufacturing engineers identify and correct potential manufacturing and/or process failures. In performing a FMEA, the product and/or production system is examined for all the ways in which failure can occur. For each failure, an estimate is made of its effect on the total system, its seriousness, and its frequency of occurrence. Corrective actions are then identified to prevent failures from occurring or reaching the customer. The great value of FMEA lies in its systematic approach in analyzing product and/or process performance in critical detail. Also, interaction between manufacturing and design engineering required while conducting a FMEA can be very effective in helping to ensure product and process conformance at an early stage of design. A more complete discussion of the FMEA approach is provided in the article "Failure Mode and Effects Analysis, Fault Tree Analysis" in this Volume and in Ref 9.

Value engineering provides a systematic approach to evaluating design alternatives that is often very useful and may even point the way to innovative new design approaches or ideas. Also called value analysis, value control, or value management, value engineering uses a multidiscipline team to analyze the functions provided by the product and the cost of each function. Based on results of the value analysis, where value is the ratio of function and cost, creative ways are sought to eliminate waste and undesired functions and to achieve required functions at the lowest possible cost while optimizing manufacturability, quality, and delivery. Value engineering is very broad in scope. A disadvantage of the value engineering approach is its reliance on cost data that often cannot be accurately determined until after design decisions have been made. This can detract from its usefulness as a tool for enhancing the quality of early design decisions. On the other hand, when used properly, value analysis can be a very useful tool for helping to make material and process selection decisions. Value engineering is widely discussed in the literature (see Ref 10, for example). See also the article "Use of Value Analysis in Materials Selection" in this Volume.

Standardization. This practice seeks to limit and/or reduce manufacturing information content of products by eliminating specials and standardizing wherever possible. Information content of the product and quality risks are reduced when part variations (e.g., in the types of screws used) are kept to a minimum. It is seldom justifiable, for example, to use several screw sizes or drive styles in one assembly. Minimizing part variations also simplifies manufacturing by reducing the information content of the production system required to produce the part. Substitution of standard (off-the-shelf) components in place of special-purpose designs helps reduce part variations as well as total information content of the manufacturing system. A stock item is always less expensive than a custom-made item. Standard components require little or no lead time and are more reliable because characteristics and weaknesses are well known. They can be ordered in any quantity at any time. They are usually easier to repair and replacements are easier to find. Use of standardized components puts the burden on the supplier and makes the supplier do more.

Standardization and rationalization (S&R) is a combined design and business approach or philosophy that specifically targets reduction of part proliferation companywide. In essence, S&R seeks to further leverage the benefits of standardization by minimizing the number of standard parts used. In the S&R approach, *standardization* is defined as the reduction in number of different parts used in current and former designs. *Rationalization* is the identification of the fewest number of parts required for use in future designs. Such an approach is ideal for material selection. Consider, for example, a company that buys a large variety of different thermoplastic materials. In the S&R approach, each of these different plastic materials would be evaluated with respect to volume purchased per year, required properties, and other relevant considerations. Based on this evaluation, a "rationalized" list is developed that is agreed upon by design and manufacturing personnel as being sufficient. Thermoplastic materials used in all new designs are then selected from this short list. Over time, the number of different materials used by the company decreases, saving time and money. Less time is needed to select materials, and product and process development time is reduced because the materials are well understood within the company. Money is saved because fewer materials are bought in higher volume from established suppliers.

Group Technology (GT) is an approach to design and manufacturing that seeks to reduce manufacturing system information content by identifying and exploiting the sameness or similarity of parts based on their geometrical shape and/or similarities in their production process. Group technology is implemented by using classification and coding systems to identify and understand part similarities and to establish parameters for action. Manufacturing engineers use GT to decide on more efficient ways to increase system flexibility by streamlining information flow, reducing setup time and floor space requirements, and standardizing procedures for batch-type production. Design engineers use GT to reduce design time and effort as well as part and tooling proliferation. With increasing emphasis on flexible and integrated manufacturing, GT is also an effective first step in structuring and building an integrated database. Standardized process planning, accurate cost estimation, efficient purchasing, and assessment of the impact of material costs are benefits that are often realized. A more complete discussion of GT, especially as it applies to manufacturing, is given in Ref 11.

REFERENCES

1. J.E. Ettlie and H.W. Stoll, *Managing the Design-Manufacturing Process*, McGraw-Hill, 1990
2. M.P. Groover, *Automation, Production Systems, and Computer Integrated Manufacturing*, Prentice-Hall, 1987, p 40
3. M.P. Groover, *Fundamentals of Modern Manufacturing*, Prentice-Hall, 1996
4. J.A. Schey, *Introduction to Manufacturing Processes*, McGraw-Hill, 2nd ed., 1987
5. S. Kalpakjian, *Manufacturing Processes for Engineering Materials*, Addison-Wesley, 2nd ed., 1991
6. M. Andreasen, S. Kahler, and T. Lund, *Design for Assembly*, IFS Publishing, 1983
7. G. Boothroyd and P. Dewhurst, *Product Design for Assembly*, Boothroyd Dewhurst, 1989
8. P.J. Ross, *Taguchi Techniques for Quality Engineering*, McGraw-Hill, 1988
9. J.J. Hollenback, *Failure Mode and Effects Analysis*, Society of Automotive Engineers, 1977
10. A.E. Mudge, *Value Engineering, A Systematic Approach*, McGraw-Hill, 1971
11. I. Ham, Group Technology, Chapter 7.8, *Handbook of Industrial Engineering*, G. Salvendy, Ed., John Wiley & Sons, 1982

Design for Manufacture and Assembly

Geoffrey Boothroyd, Boothroyd Dewhurst, Inc.

DURING THE 1980s AND 1990s, the United States has been losing millions of dollars per day to its foreign competitors. Competitiveness has been lost in many areas, but most notably in automobile manufacture, as highlighted by a worldwide study of this industry that was published in 1990 (Ref 1). The study, which showed that Japan had the most productive plants at that time, attempted to explain the wide variations in automotive assembly plant productivity throughout the world. It was found that automation could only account for one-third of the total difference in productivity among plants worldwide and that, at any level of automation, the difference between the most and least efficient plants was enormous.

The authors of the study concluded that no improvements in operation can make a plant fully competitive if the product design is defective. There is now overwhelming evidence to support the view that product design for manufacture and assembly can be the key to high productivity in all manufacturing industries.

Introduction to Design for Manufacture and Assembly

It has long been advocated that designers should give attention to possible manufacturing problems associated with a design. Traditionally, the idea was that a competent designer should be familiar with manufacturing processes to avoid adding unnecessarily to manufacturing costs. Until the 1960s, the supposed solution in education was to provide "shop" courses to familiarize engineering students with the ways products are manufactured. However, even this approach has now been abandoned by the colleges because the courses lacked academic content. Furthermore, with the increasingly complex technology incorporated within many products, the time pressures put on designers to get designs onto the shop floor, the "we design it, you manufacture it" attitude of designers, and the increasing sophistication of manufacturing techniques, designers have become less and less able to avoid unnecessary manufacturing costs.

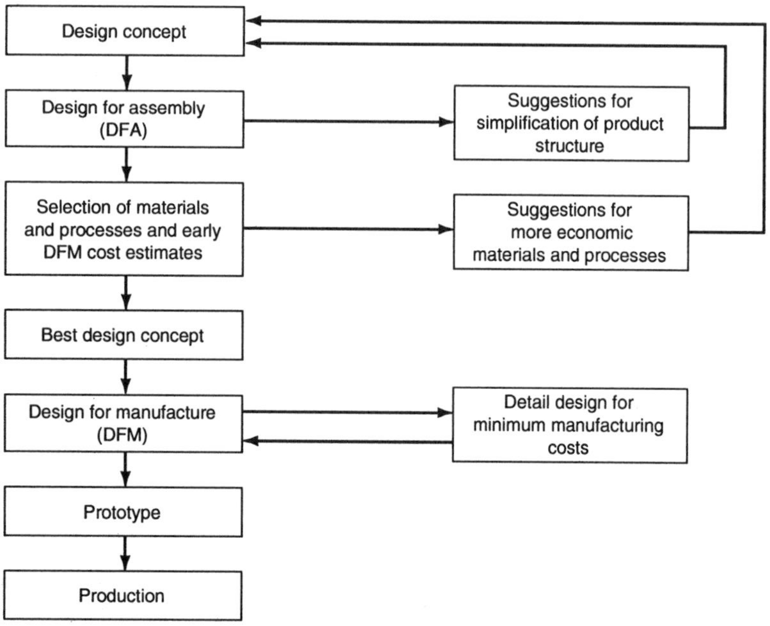

Fig. 1 Typical steps taken when using DFMA software in design

Since the early 1980s, it has become recognized that more effort is required to take manufacturing and assembly into account early in the product design cycle. One way of achieving this is for manufacturing engineers to be part of a simultaneous or concurrent engineering design team.

Within this teamworking, design for manufacture and assembly (DFMA) software analysis tools help in the evaluation of proposed designs. It is important that design teams have access to such tools in order to provide a focal point that helps identify problems from manufacturing and design perspectives.

Use of DFMA software allows a systematic procedure that aims to help companies keep the number of component parts in an assembly to a minimum and make the fullest use of the manufacturing processes that exist. It achieves this by enabling the analysis of design ideas. It is not a design system, and any innovation must come from the design team. However, it does provide quantification to help decision making at the early stages of design.

Figure 1 summarizes the steps taken when using DFMA software during design. The design-for-assembly (DFA) analysis is conducted first, leading to a simplification of the product structure. Then, early cost estimates for the parts are obtained for both the original design and the new design in order to make trade-off decisions. During this process, the best materials and processes to be used for the various parts are considered. For example, would it be better to manufacture a cover from plastic or sheet metal? Once the materials and processes have been finally selected, a more thorough analysis for design for manufacture (DFM) can be carried out for the detail design of the parts.

It should be remembered that DFMA is the integration of the separate but interrelated design issues of assembly and manufacturing processes. Therefore, there are two fundamental aspects to producing efficient designs: DFA to help simplify the product and quantify assembly costs and the early implementation of DFM to quantify parts cost and allow trade-off decisions to be made for design proposals and material and process selection.

Table 1 Examples of AEM symbols and penalty scores

Elemental operation	AEM symbol	Penalty score
Downward movement(a)	↓	0
Soldering	S	20

(a) As in a pressing operation. Source: Ref 3

		Easy to align and insert	Not easy to align or insert	Not easy to align and insert	Severe difficulties
		0	1	2	3
No access or vision difficulties	0	1.5	3	4.5	7.5
Obstructed access or restricted vision	1	3.7	5.2	6.7	9.7
Obstructed access and restricted vision	2	5.9	7.4	8.9	11.9

Fig. 2 Portion of database of manual insertion times (in seconds) without fastening for small parts where no holding down or regrasping is required. Copyright Boothroyd Dewhurst, Inc.; used with permission

Design for Assembly

The DFA analysis tool was developed by the author in the mid-1970s. The idea was to stress the economic implications of design decisions. This is crucial, because while design is usually a minor factor in the total cost of a product, the design process fixes between 70 and 95% of all costs.

The author and his colleague, Peter Dewhurst, developed a personal computer program for DFA that was first introduced in 1982. Since then, the DFA database of time standards has been expanded to include the assembly of large products, cable harnesses, and printed circuit boards.

Reducing the number of separate parts, thereby simplifying the product, is the greatest improvement provided by DFA. To give guidance in reducing the part count, the DFA software asks the following questions as each part is added to the product during assembly:

- Is the part or subassembly used only for fastening or securing other items?
- Is the part or subassembly used only for connecting other items?

If the answer is "yes" to either question then the part or subassembly is not considered theoretically necessary. If the answer is "no" to both questions the following criteria questions are then considered.

- During operation of the product, does the part move relative to all other parts already assembled?
- Must the part be of a different material than, or be isolated from, all other parts already assembled? Only fundamental reasons concerned with material properties are acceptable.
- Must the part be separate from all other parts already assembled because the necessary assembly or disassembly of other separate parts would otherwise be impossible?

If the answer to all three criteria questions is "no," the part cannot be considered theoretically necessary.

When these questions have been answered for all parts, a theoretical minimum part count for the product is obtained. It should be emphasized, however, that this theoretical minimum does not take into account practical considerations or cost considerations, but simply provides a basis for an independent measure of the quality of the design from an assembly viewpoint.

While answering these minimum part questions, the design team is challenged to justify the existence of each separate part and this is where brainstorming usually results in considerable product simplification. In fact, it is found from more than 70 published case studies that the average reduction in the number of parts is 50% as a result of using the DFA software.

Estimating Assembly Time. The next step is to estimate the assembly time for the product design and establish a DFA index in terms of difficulty of assembly.

To estimate assembly time, each part in the design is examined for two considerations: how the part is to be acquired (fetched if necessary), oriented, and made ready for insertion; and how it is inserted and/or fastened into the product.

The difficulty of these operations is rated, and from this rating standard times are determined for all the operations necessary to assemble each part. The DFA time standard is a classification of design features that affect part assembly. It is a system for use in product design similar to the standard time systems used by industrial engineers such as MTM, WorkFactor, or MOST and was developed from 12 years of industry and university experiments. Usage has proved the data to be quite accurate for the overall times that are generally within 6% of the actual times.

For estimation of the handling time the designer must specify the symmetry of the item, its major dimension, and its thickness. In addition, the designer must specify whether the item nests or tangles when in bulk, whether it is fragile, flexible, slippery, sticky, and whether it needs two hands, grasping tools, optical magnification, or mechanical assistance in the form of cranes.

A portion of the database for estimation of the time for insertion of small items is shown in Fig. 2. Here it is important to know whether the assembly worker's vision or access is restricted, and whether the item is difficult to align or insert. Other portions of the database deal with resistance to insertion, and whether the item requires holding down in order to maintain its position for subsequent assembly operations.

For fastening operations, further questions may be required. For example, the type of tool used for threaded fasteners and the number of revolutions required are important in a determination of the total fastening time. A further consideration relates to the location of the items that must be acquired. If turning, bending, or walking are needed to acquire the item, a different database for acquisition and handling is used.

Thus, for each classification of handling and insertion, an average time is given leading to an estimate of the total manual assembly time for an item.

Table 2 Examples of assemblability evaluation and improvement

Product structure	Assembly operations	Ability to assemble part evaluation score	Ability to assemblability evaluation score	Assembly cost ratio	Part to be improved
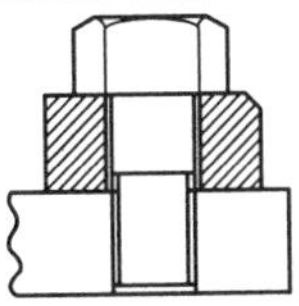	1. Set chassis 2. Bring down block and hold it to maintain its orientation 3. Fasten screw	100 50 65	73	1	Block
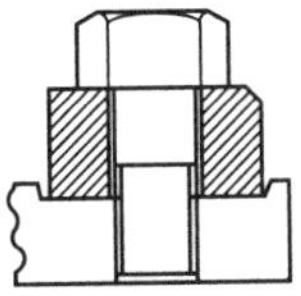	1. Set chassis 2. Bring down block (orientation is maintained by spot facing) 3. Fasten screw	100 100 65	88	~0.8	Screw
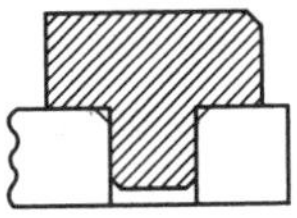	1. Set chassis 2. Bring down and press fit block	100 80	89	~0.5	Block

Source: Ref 3

Assembly procedure	FM	AM	KM
Insert receiver base in readiness	13.20	5.5	2.4
Insert holder for receiver cap in receiver base	1.48	16.5	0.09 ←
Insert holder for microphone in receiver base	1.48	16.5	0.09 ←
Insert microphone in receiver base	25.14	8.5	2.958
Insert receiver cap in receiver base	19.93	8.5	2.345
Clamp receiver cap and microphone with holder	1.82	31.0	0.059 ← Threshold level
Insert antikink device in receiver base, mount cable	6.30	202.5	0.031 ←
Lock receiver top in position	14.40	12.5	1.152
Screw receiver base to receiver top	2.13	43.0	0.049 ←

KM axis: 0 0.5 1.0 1.5 2.0 2.5 3.0 3.5

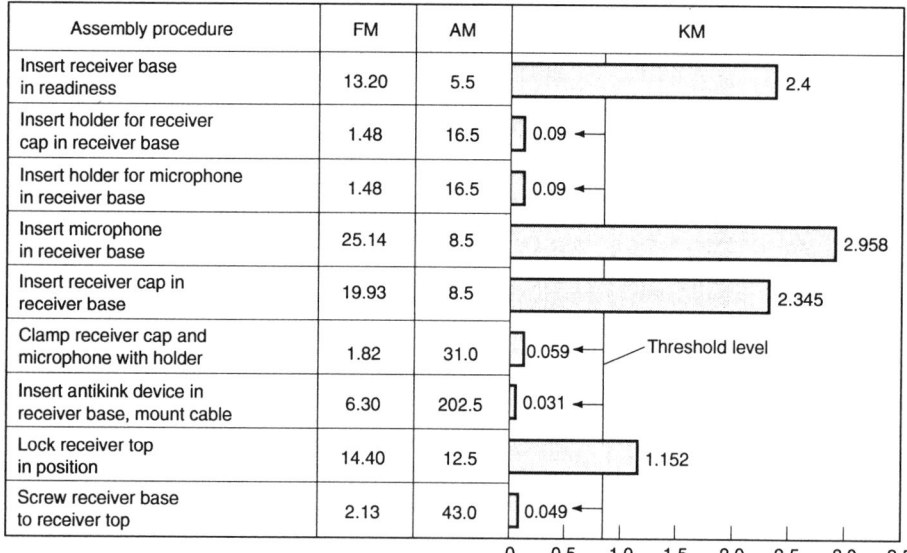

Fig. 3 Example of subassembly evaluation using assembly-oriented product design method. FM, functional content of procedure; AM, assembly expenditure; KM, characteristic value of procedure (KM = FM/AM). Characteristic value of subassembly (*K*) is 0.250. Arrows denote weak technical aspect of assembly. Source: Ref 4

Design Efficiency (Ease of Assembly). When all the items and operations for assembly of a product have been analyzed, the total assembly time for the product can be estimated, and using standard labor rates, so can assembly costs. Also, the efficiency of a design from an ease of assembly point of view can be determined; this is the DFA index.

Based on the assumption that all of the critical parts could be made easy to assemble—requiring only about 3 s each—the minimum assembly time is equal to the theoretical minimum number of parts times three. The assembly efficiency percentage or DFA index is equal to the minimum assembly time divided by the estimated total assembly time multiplied by 100.

The maximum value of the DFA index is 100%. However, excellent designs can have a DFA index much lower. For example, good designs of small electromechanical assemblies have values of around 20%. Thus, much depends on the type of assembly and, with experience, the values that are achievable in different circumstances will be known.

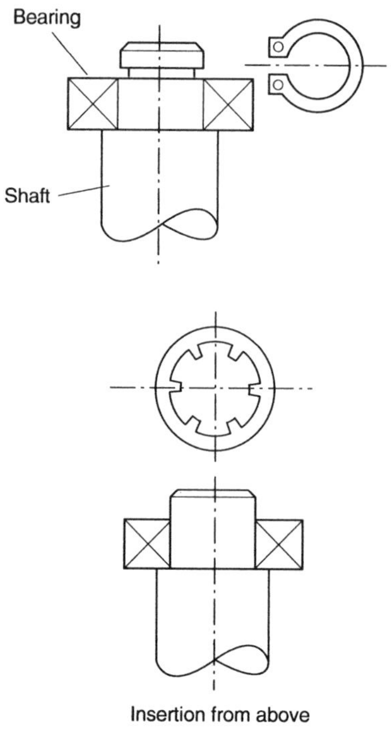

Before improvement

Core operation			DAC	50
Shaft press fit	P		80	
Bearing press fit	T		90	
C-ring			30	
Screw tightening			90	
			80	

After improvement

Core operation			DAC	50
Shaft press fit	P		80	
Bearing press fit	T		90	
O-ring			90	
Screw tightening			90	
			80	

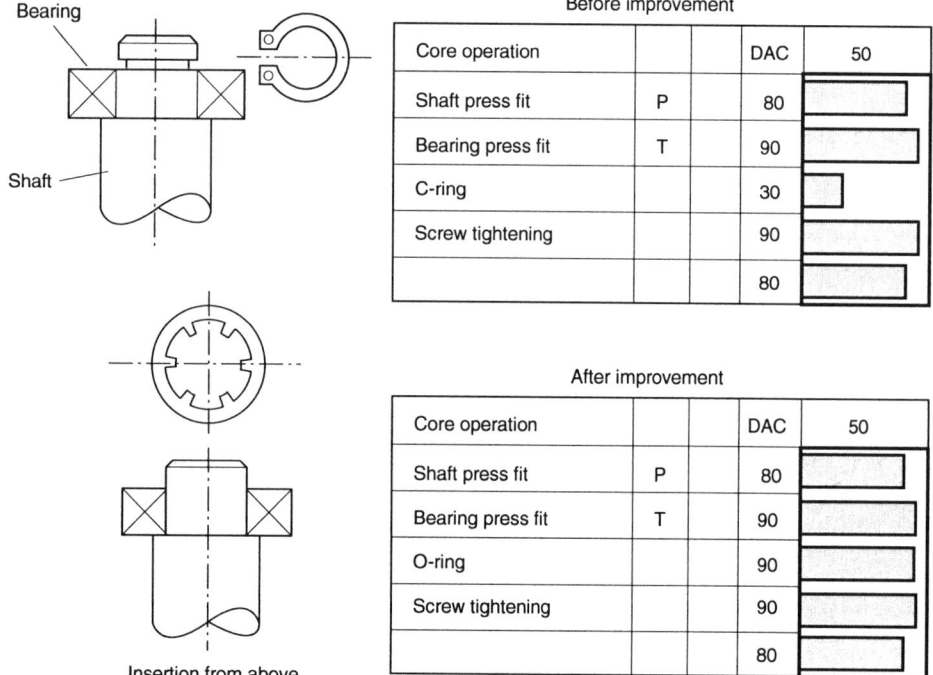

Fig. 4 Example of part improvement using design-for-assembly cost effectiveness (DAC) method. Source: Ref 6

At the DFA stage, part manufacturing costs are not brought into the analysis, but the DFA index and the estimated assembly times provide benchmarks against which further design iterations, previous estimates for an original product design, or a competitor's product can be compared.

Other Assembly Analysis Methods

Assemblability Evaluation Method (AEM). The AEM method, described in 1986 by Miyakawa and Ohashi (Ref 2), uses two indices at the earliest possible stage of design, namely the assembly-evaluation score, *E*, which is used to assess the design quality or the difficulty of assembly, and the assembly-cost ratio, *K*, which is used to project assembly costs relative to current assembly costs. The method does not distinguish between manual, robot, or automatic assembly, because according to Myakawa and Ohashi there is a strong correlation between the degrees of assembly difficulty using these three methods.

In the AEM, approximately 20 symbols represent the various assembly operations. Each symbol has an index to assess the ability to assemble the part under consideration. Examples of the symbols and penalty scores are given in Table 1, and examples of their application are given in Table 2.

Assembly-Oriented Product Design. Some ten years after the introduction of the Hitachi AEM and the Boothroyd Dewhurst DFA methods, variations started to appear. One of the first was that of Warnecke and Bassler at the University of Stuttgart, Germany. In their assembly-oriented product design method (Ref 4), they assess the usefulness or functional value of each part. Thus, both assembly difficulty and functional value are evaluated, and a combined rating is given. This means that parts that have little functional value (such as separate fasteners) and that are difficult to assemble are given the lowest ratings. These ratings are then used as guides to redesign (see Fig. 3).

Lucas Method. The Lucas method was developed at the University of Hull, United Kingdom, during the late 1980s (Ref 5). In the Lucas method, three steps are followed, as described below.

Step 1: Functional Analysis. Parts are categorized into A parts (demanded by the design specification) and B parts (required by that particular design solution). A target is set for design efficiency, which is A/(A + B) and is expressed as a percentage. The objective is to exceed an arbitrary 60% target value by the elimination of category B parts through redesign. The Lucas method authors emphasize assembly-cost reduction and parts-count reduction and include the use of the present author's minimum-parts criteria in a "truth" table to assist in part-count reduction.

Step 2: Handling and Feeding Analysis. The parts are scored on the basis of three factors: the size and weight of the part, handling difficulties, and the orientation of the part. The score is summed to give the total score for the part, and a handling/feeding ratio is calculated that is given by the total score divided by the number of A parts. A target of 2.5 is recommended.

Step 3: Fitting Analysis. Based on the proposed assembly sequence, each part is scored on the basis of whether it requires holding in a fixture, the assembly direction, alignment problems, restricted vision, and the required insertion force. The total score is divided by the number of A parts to give the fitting ratio. Again, it is recommended that this ratio approach 2.5 for an acceptable design.

DAC Method. Sony Corporation claims to have developed a unique set of rules for increased productivity involving design-for-assembly cost effectiveness (DAC). Yamigawa (Ref 6) reiterates the view that it is impossible to design for assembly ease unless one starts at the time of conception before the blueprint for the product is drawn up. The improvement of a design at its inception is referred to as the concept of feed-forward design, as opposed to making improvements later with feedback from the manufacturing process.

In the DAC method, factors for evaluation are classified into 30 key words. The evaluations are displayed on a diagram using a 100-point system for each operation, thus making judgment at a glance easy. A list of operations is presented on the DAC diagram, and a bar is drawn that represents the score for that particular operation (Fig. 4). Operations with low scores are easily identified. Since 1987, DAC has been introduced in various companies in Japan and overseas. Emphasis is given to the ease with which an operation can be carried out automatically, and the method is used to illustrate problems with the efficiency of the assembly system.

Design for Manufacture (DFM)

Design for assembly has generated a revolution in design practices, not principally because it usually reduces assembly costs, but because it has a far greater impact on the total manufacturing costs of a product. The reason is that DFA simplifies the product structure, reduces the number of parts, and thereby reduces the total cost of the parts. However, to judge the effects of DFA at the early design stage, companion methods for the early estimation of part costs must be made available, and accordingly, many of those who have developed DFA methods are now turning their attention to methods of assessment of part-manufacturing difficulties.

For example, the Hitachi researchers (Ref 7) have introduced a machining producibility evaluation method, which combined with their AEM, gives an overall producibility-evaluation method (PEM).

Similarly, Toshiba Corporation (Ref 8) has developed a processability evaluation method, which combined with other methods, including an assemblability-evaluation method, provides an overall producibility-evaluation method. Processability is defined as being proportional to the cost of the part. The cost of the part is determined by the selection of the part-processing method, and then by the design of the part shape. Various processing methods are considered for a particular part. The cost of the part is then determined for all combinations of the selected processing methods and suitable materials. Then the design of the part is evaluated to see whether it fits a particular processing method, and finally, a processability evaluation is carried out.

Since 1985, the author and his colleagues Dewhurst and Knight have developed methods for designers to obtain cost estimates for parts and tooling during the early phases of design. Studies have been completed for machined parts (Ref 9), injection-molded parts (Ref 10), sand-cast, investment-cast and die-cast parts (Ref 11–13), sheet-metal stampings (Ref 14), and powder-metal parts (Ref 15). The objective of these studies was to provide methods with which the designer or design team can quickly obtain information on costs before detailed design has taken place. For example, an analysis (Ref 16) of an injection-molded heater cover gave the results shown in Fig. 5. It was evident that certain wall thicknesses were too large, and that, through some fairly minor design changes, the processing cost could be reduced by 33%. If these studies had taken place at the early design stage, the designer could also have considered the cost for an equivalent sheet-metal part, for example. In fact, the use of these analysis techniques is now allowing designers and purchasing managers to challenge suppliers' estimates. In one example, it has been reported that Polaroid Corporation has saved $16,000 to $20,000 on the cost of tooling for an injection-molded part (Ref 17).

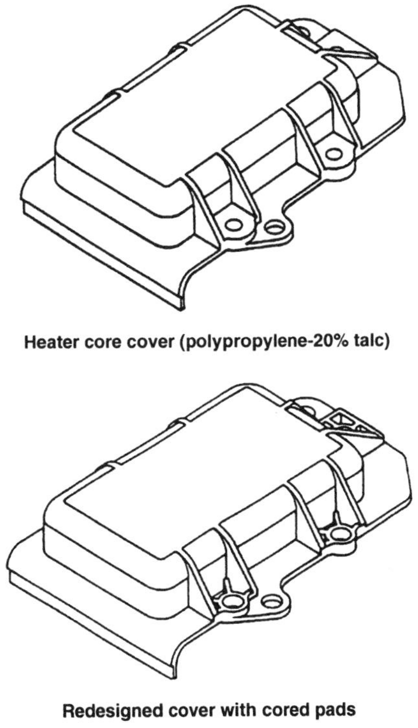

Manufacturing factor	Existing design	Redesign
Cost of one cavity and core	$8,032	$11,625
Cycle time, s	42.8	13.3
Number of cavities required	6	2
Cost of production mold	$36,383	$22,925
Cost per part(a)	$0.251	$0.168

(a) Includes $0.05 for material

Fig. 5 Design-for-manufacture analysis of injection-molded heater cover. Source: Ref 16

Early Cost Estimating. The problem of estimating part and tooling costs before the part has been fully detailed is discussed using machining as an example because this is one of the most common shape-forming processes. Several conventional cost estimating methods for machining are available both in handbook form, such as the *Machining Data Handbook* (Ref 18) and the *AM Cost Estimator* (Ref 19) and in software form. However, all of these methods are meant to be applied after the part has been detailed and its production has been planned, and they are not tailored for use by a designer. During the early stages of design, the designer will not wish to specify, for example, all the work-holding devices and tools that might be needed—a detailed design will not yet be available. Indeed, a final decision even on the work material might not have been made.

For early cost estimating an important assumption has to be made. The designer should be able to expect that, when the design is finalized, care will have been taken to avoid unnecessary manufacturing expense at the detail-design stage and that manufacturing will take place under efficient conditions.

To illustrate how such an assumption can help in providing reasonable estimates, consider the effect of the metal-removal rate on grinding costs, as shown in Fig. 6. These cost curves indicate that as the removal rate is increased the cost of grinding-wheel wear increases in proportion. At the same time, the cost of grinding decreases because the grinding cycle is shortened; in fact, the grinding costs are inversely proportional to the removal rate.

Under these circumstances, it is easy to show that the minimum cost condition occurs when the

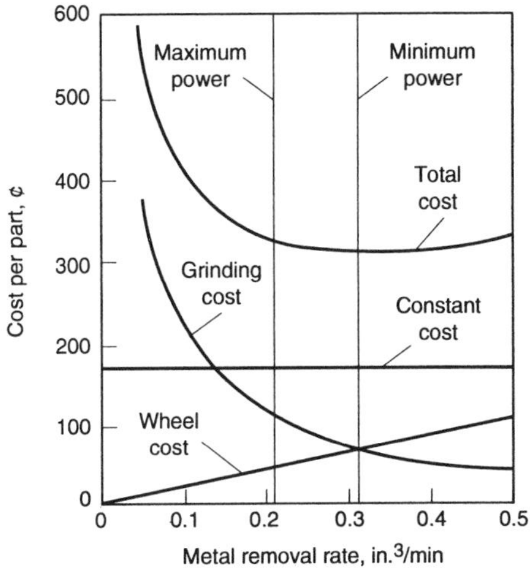

Fig. 6 Effect of metal removal rate on grinding costs

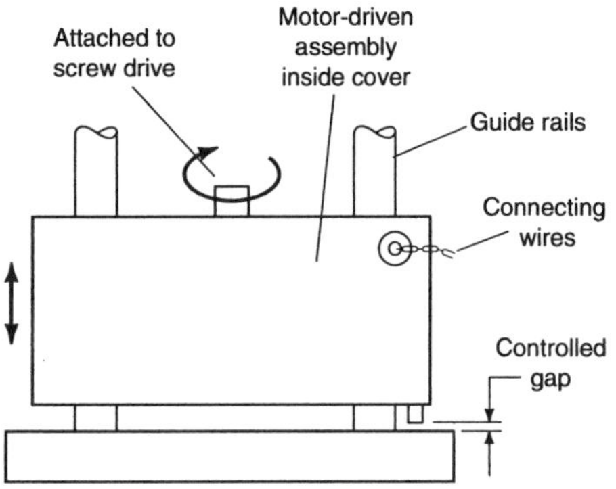

Fig. 7 Motor-drive assembly design requirements

grinding costs and wheel costs are equal. Hence, we can say that if economic grinding conditions are used in the manufacture of a part, wheel costs can be allowed for by doubling the estimated grinding costs. Even if economical conditions cannot be used—say, because of power limitations—it is still possible to adjust the grinding costs to allow for wheel costs. This simple example illustrates the general approach to early cost estimating.

In estimating machining costs, perhaps the simplest method would be to specify the shape and size of the original workpiece and the quantity of material to be removed by machining. An estimate could then be made of the material cost needed to manufacture the part, and if an approximate figure were available for the average cost of removal of each unit volume of material by machining, an estimate could also be made of the machining cost. Even the tool-replacement costs could be allowed for.

Unfortunately, this very simple approach does not account for the nonproductive costs associated with a series of machining operations. For example, if a small volume of material were to be removed in a single pass by a turning operation, the nonproductive costs would be relatively small. In this case, the part needs only to be loaded into the lathe and subsequently unloaded once. Compare this with the same volume of material removed by a sequence of operations, on different machines, for example: turning, threading, milling, and drilling. The nonproductive costs accumulate and become a significant factor in the ultimate cost of the machined part, especially for small parts.

Principal Cost Drivers. The principal factors in estimating machining costs for a particular part are:

- The amount of material removed (directly affects the material cost per unit volume of the finished part and the rough-machining time)
- The cost of using the machine tools and operators
- The power available for machining and the specific cutting energy (unit power) of the work material
- The nonproductive times—especially for smaller parts, for which they may become the most significant factors
- The surface area to be finish-machined
- The recommended finish-machining conditions, which are affected by the work material and tool material used and the surface roughnesses and tolerances required

Provided that values are available for the various material and machine-tool factors, the most time consuming aspects of applying the model are estimating the total quantity of material removed and the total surface area to be finished-machined. In fact, an estimate of the volume of material removed would probably be made by calculating the volumes removed by individual cutting operations. Similarly, the total surface area to be finish-machined is made up of the areas of the surfaces generated by individual cutting operations.

Clearly, if the dimensions of individual cuts must be specified, it is no additional effort for a computer to calculate the individual rough-machining and finish-machining times for each cut, making proper allowance for tool material, recommended conditions, machine-tool characteristics, and individual nonproductive times.

Thus, although experience shows that acceptable early cost estimates can be obtained using the model described above, in a computer implementation of the process it is more sensible to specify each volume or area element and to estimate the appropriate machining time and cost. This procedure has an added advantage: the results indicate the relative cost of each design feature, indicating to the designer where manufacturing costs might be reduced.

The importance of this is illustrated in the following example where a product design is analyzed for ease of assembly (DFA) as well as manufacture costs of the individual parts (DFM).

Example: Application of DFMA Software to the Design of a Motor-Drive Assembly. Figure 7 shows the requirements of a motor-drive assembly that must be designed to sense and control its position on two steel guide rails. The motor must be fully enclosed for aesthetic reasons and have a removable cover for access so that the position sensor can be adjusted. The principal requirement is a rigid base that is designed to slide up and

Table 3 Results of DFA analysis for initial design of motor-drive assembly shown in Fig. 8

Part	No.	Theoretical part count	Assembly time, s	Assembly cost, ¢
Base	1	1	3.5	2.9
Bushing	2	0	12.3	10.2
Motor subassembly	1	1	9.5	7.9
Motor screw	2	0	21.0	17.5
Sensor subassembly	1	1	8.5	7.1
Set screw	1	0	10.6	8.8
Standoff	2	0	16.0	13.3
End plate	1	1	8.4	7.0
End-plate screw	2	0	16.6	13.8
Plastic bushing	1	0	3.5	2.9
Thread leads	...	...	5.0	4.2
Reorient	...	...	4.5	3.8
Cover	1	0	9.4	7.9
Cover screw	4	0	31.2	26.0
Totals	**19**	**4**	**160.0**	**133.0**

DFA index = (4 × 3)/160 = 7.5%

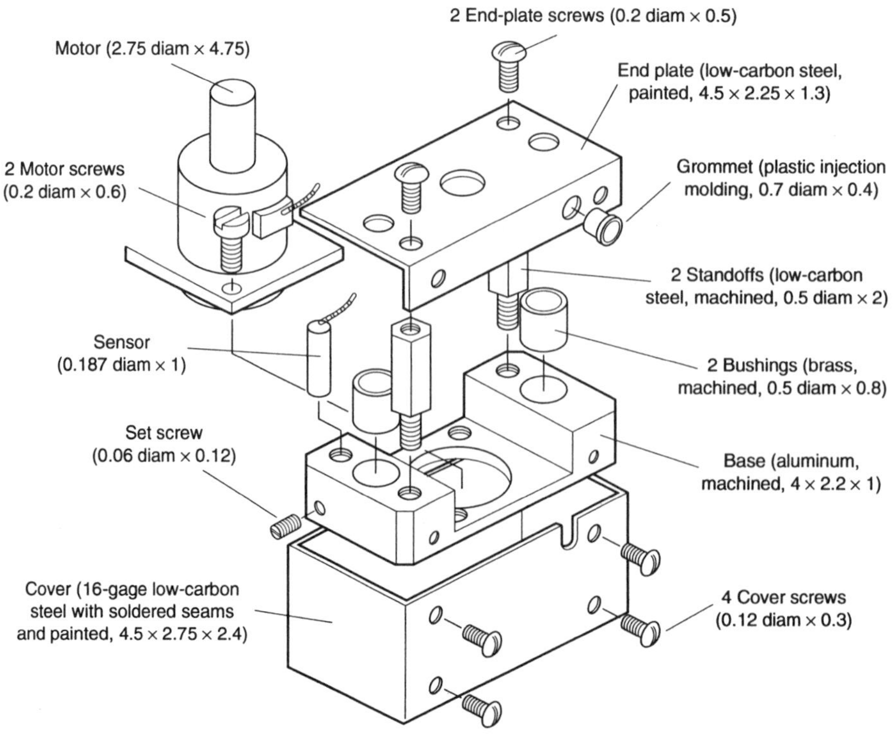

Fig. 8 Initial design of motor-drive assembly. Dimensions in inches

down the guide rails and that supports the motor and sensor. The motor and sensor have wires that connect them to a power supply and a control unit, respectively.

A proposed solution is shown in Fig. 8. The base is provided with two bushings to provide suitable friction and wear characteristics. The motor is secured to the base with two motor screws, and a hole in the base accepts the sensor, which is held in place with a set screw. To provide the required covers, an end plate is secured by two end-plate screws to two standoffs, which are, in turn, screwed into the base. This end plate is fitted with a plastic grommet through which the connecting wires pass. Finally, a box-shaped cover slides over the whole assembly from below the base and is held in place by four cover screws, two passing into the base and two into the end cover.

Two subassemblies are required, the motor and the sensor, and in this initial design there are eight additional main parts—and nine screws—making a total of 19 items to be assembled.

The application of the minimum part criteria to the proposed design proceeds as follows:

- The base is assembled into a fixture, and because there are no other parts with which to combine it, it is a theoretically necessary part.
- The two bushings do not satisfy the criteria and can theoretically be integral with the base.
- The motor is a standard subassembly of parts that is a purchased item. Thus, the criteria cannot be applied unless the assembly of the motor itself is considered as part of the analysis. In this example, it is assumed that the motor and sensor are not to be analyzed.
- Invariably, separate fasteners such as the two motor screws do not meet the criteria, because an integral fastening arrangement is always theoretically possible.

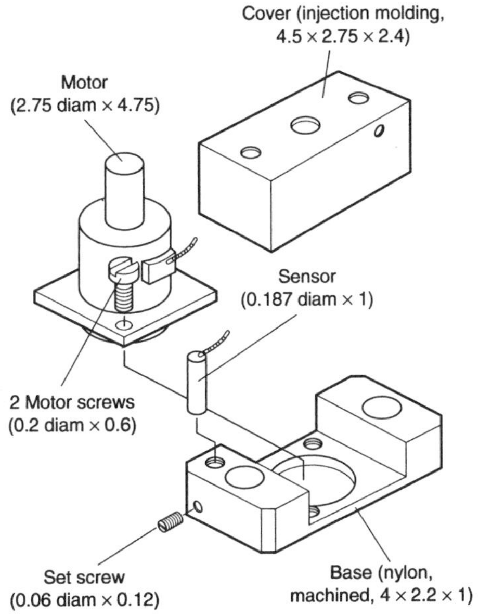

Fig. 9 Redesign of motor-drive assembly following DFA analysis. Dimensions in inches

- The sensor is a purchased item.
- The set screw is theoretically not necessary.
- The two standoffs do not meet the criteria; they could be incorporated into the base.
- The end plate must be separate for reasons of assembly.
- The two end-plate screws are theoretically not necessary.
- The plastic grommet can be of the same material as, and therefore combined with, the end plate.
- The cover can also be combined with the end plate.
- Finally, the four cover screws are theoretically not necessary.

From this analysis, it can be seen that, if the motor and sensor subassemblies can be arranged to snap or screw into the base, and a plastic cover can be designed to snap on, only four separate items will be needed, instead of 19. These four items represent the theoretical minimum number needed to satisfy the constraints of the product design without consideration of the practical limitations.

It is now necessary for the designer or design team to justify the existence of those parts that have not satisfied the criteria. Justification may arise from practical, technical, or economic considerations. In this example, it can be argued that the two motor screws are needed to secure the motor, and the set screw is needed to hold the sensor, because any alternatives would be impractical for a low-volume product such as this.

It can be argued that the two powder metal bushings are unnecessary, because the base could be machined from an alternative material with the necessary frictional characteristics. Finally, it is very difficult to justify the separate standoffs, end plate, cover, grommet, and associated six screws.

Before an alternative design can be considered, it is necessary to have estimates of the assembly times and costs, so that any possible savings can be taken into account when considering design alternatives. Using DFMA time standards and knowledge bases, it is possible to make estimates of assembly costs, and then to estimate the cost of the parts and associated tooling, without having final detail drawings of the parts.

First, Table 3 shows the results of the DFA analysis, where it is seen that the total assembly time is estimated to be 160 s. It is also possible to obtain a measure of the quality of the design for ease of assembly. The theoretical minimum number of parts is four and, if these parts were easy to assemble, they would take about 3 s each to assemble on average. Thus, the theoretical minimum (or ideal) assembly time is 12 s, a figure that can be compared with the estimated time of 160 s, giving a DFA index of 12/160, or 7.5%.

The elimination of parts not meeting the minimum part-count criteria, and that cannot be justified on practical grounds, results in the design concept shown in Fig. 9. Here, the bushings are combined with the base, and the standoffs, end plate, cover, plastic bushing, and six associated screws are replaced by one snap-on plastic cover. The eliminated items entailed an assembly time of 97.4 s. The new cover takes only 4 s to assemble, and it avoids the need for a reorientation. In addition, screws with pilot points are used and the base is redesigned so that the motor is self-aligning. Table 4 presents the results of a DFA analysis of the redesigned assembly. The new assembly time is only 46 s, and the DFA index has increased to 26%. Assuming an assembly labor rate of $30 per hour, the savings in assembly cost amount to $0.95.

Table 5 presents the results of DFM cost analyses of the parts for the two designs. There were five machined parts included in the original design. Their manufacturing costs were estimated using the machining cost estimating software. For small quantities, the standoffs cost $4.87 each and the bushings $1.53. Because these items are eliminated in the proposed redesign, a direct savings of $12.80 results.

The costs of the cover and end plate were estimated using the sheet metalworking software and assuming the use of a turret press. The cover cost was $3.73 and the end plate cost was $2.26, giving a total of $5.99. These items are replaced by a snap-on cover which, using the injection

Table 4 Results of DFA analysis for redesign of motor-drive assembly shown in Fig. 9

Part	No.	Theoretical part count	Assembly time, s	Assembly cost, ¢
Base	1	1	3.5	2.9
Motor subassembly	1	1	4.5	3.8
Motor screw	2	0	12.0	10.0
Sensor subassembly	1	1	8.5	7.1
Set screw	1	0	8.5	7.1
Thread leads	...	...	5.0	4.2
Plastic cover	1	1	4.0	3.3
Totals	**7**	**4**	**46.0**	**38.4**

DFA index = (4 × 3)/46 = 26%

Table 5 Comparison of part costs for motor-drive assembly initial design and redesign

Item	Cost, $
Initial design (Fig. 8)	
Base (aluminum)	15.29
Bushing (2)	3.06
Motor screw (2)	0.20(a)
Set screw	0.10(a)
Standoff (2)	9.74
End plate	2.26
End-plate screw (2)	0.20(a)
Plastic bushing	0.10(a)
Cover	3.73
Cover screw (4)	0.40
Total (initial design)	**35.08**
Redesign (Fig. 9)	
Base (nylon)	13.04
Motor screw (2)	0.20(a)
Set screw	0.10(a)
Plastic cover	8.66(b)
Total (redesign)	**22.00**

(a) Purchased in quantity. (b) Includes tooling cost for plastic cover ($8000)

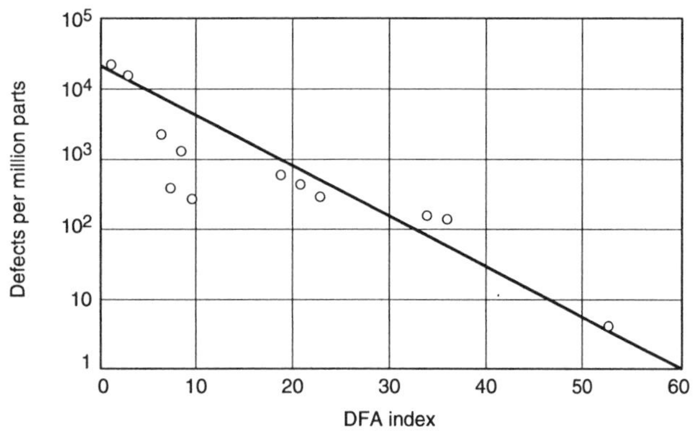

Fig. 10 Correlation between DFA index and part defects for an electronics manufacturer. Every 1 s of assembly penalty time (from DFA analysis) causes an average of 100 defects per million parts. Source: Ref 20

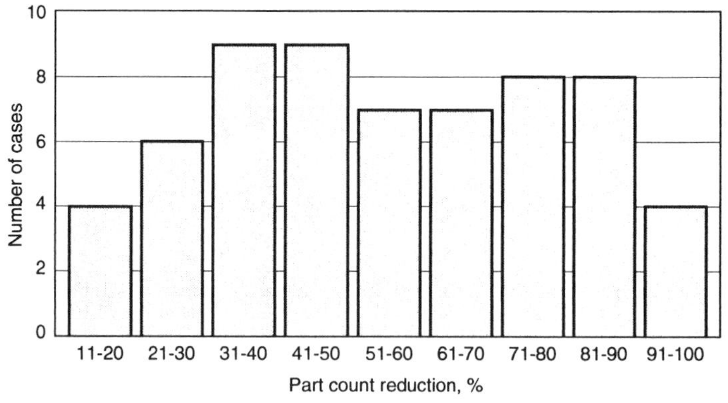

Fig. 11 Part count reductions when DFMA software was used (from 74 published case studies)

molding software was estimated to cost $8.66 for a total production of 1000. This figure includes tooling costs of $8000.

The shape of the base remains essentially unchanged for the redesigned motor assembly. However, it is now machined from nylon instead of aluminum, in order to eliminate the bushings. The effect of the change in material was to reduce the cost from $15.29 to $14.78 because the savings in material costs (nylon $0.49 compared to aluminum $2.34) more than offset increases in machining costs. However, in the conceptual redesign several drilled and tapped holes are eliminated reducing the cost of the base by a further $1.74 to $13.04. It might be noted that this saving alone is greater than the savings in assembly cost, emphasizing that it is important for the designer to estimate the cost of individual features on a machined component when considering various design proposals.

In conclusion, the example DFMA analysis of the motor assembly has shown a possible savings of $14.03, taking the cost of assembly and the cost of the eliminated screws and plastic bushing into account.

Results of DFMA Applications

DFMA software provides a systematic procedure for analyzing proposed designs from the point of view of assembly and manufacture. This procedure results in simpler and more reliable products that are less expensive to assemble and manufacture. In addition, any reduction in the number of parts in an assembly produces a snowball effect on cost reduction, because of the drawings and specifications that are no longer needed, the vendors that are no longer required, and the inventory that is eliminated. All of these factors have an important effect on overhead costs, which in many cases, form the largest proportion of the total product cost.

DFMA tools encourage dialogue between designers and the manufacturing engineers and any other individuals who play a part in determining final product costs during the early stages of design. This means that teamworking is encouraged, and the benefits of simultaneous or concurrent engineering can be achieved. The following selection of published case studies illustrates the results of DFMA applications.

Application of DFMA at an Electronics Manufacturer. The portable-products division of Motorola designs and manufactures portable radios for the landmobile-radio market. This includes such users as police and firemen, in addition to those in the construction and utility fields. DFMA methods have been used to simplify these products and reduce assembly costs (Ref 20).

The results of the first redesign efforts were so encouraging (Table 6) that Motorola surveyed several products that had been redesigned using the DFA methodology to see if there might be a general correlation of assembly efficiency with manufacturing quality. Figure 10 shows what they found. The defect levels are reported as defects per million parts assembled, which allows a quality evaluation to be made that is independent of the number of parts in the assembly. Each result in Fig. 10 represents a product with an analyzed assembly efficiency (DFA index) and a reported quality level.

Design-for-Assembly Implementation at an Automotive Manufacturer. Ford Motor Company has trained thousands of its engineers in the DFA methodology and requires its vendors to conduct DFA analysis prior to submitting bids on subcontracted products. This is not surprising, when Ford reported savings of more than $1 billion annually as a result of applying DFMA to the original Taurus line of cars.

The following example of the procedures for implementation of DFA comes from Ford Motor Company's Transmission and Chassis (T&C) Division (Ref 21).

- Provide DFA overview for senior management.
- Choose DFA champion/coordinator.
- Define objectives.
- Choose pilot program.
- Choose test case.
- Identify team structure.
- Identify team members.
- Coordinate training.
- Have first workshop.

During the workshop:

- Review the parts list and processes.
- Break up into teams.
- Analyze the existing design for manual assembly.
- Analyze the teams' redesigns for manual assembly.
- Teams present results of original design analysis versus redesign analysis.
- Prioritize redesign ideas A, B, C, and so forth.
- Incorporate all the A and B ideas into one analysis.
- Assign responsibilities and timing.

The combined results of the initial workshops held in the T&C Division of Ford indicated potential assembly labor savings of 29%, a reduction in part count of 20%, and a reduction in the number of operations of 23%.

Enormous cost benefits have been gained since the introduction of the DFA methodology in the T&C Division. Even more importantly, the changes resulting from DFA have brought substantial quality improvements. Moreover, the design lead time has been reduced by one-half and is expected to be halved again.

Manufacturability Benchmarking by an Automotive Manufacturer. A few years ago, General Motors (GM) made comparisons between its assembly plant for the Pontiac at Fairfax, Kansas, and Ford's subassembly plant for its Taurus and Mercury Sable models near Atlanta, Georgia. General Motors found that there was a

Table 6 Results from redesign of vehicular adaptor for a portable radio

Item	Old product	New product	Change, %
DFA index	4	36	800
Assembly time, s	2742	354	87
Assembly count	217	47	78
Fasteners	72	0	100

Source: Ref 20

large productivity gap between its plant and the Ford plant. General Motors concluded that 41% of the producibility gap could be traced to the manufacturability of the two designs. For example, the Ford car had many fewer parts (ten in its front bumper compared with 100 in the GM Pontiac), and the Ford parts fitted together more easily. The GM study found that the level of automation, which was actually much higher in the GM plant, was not a factor in explaining the productivity gap.

The subsequent application of DFMA can be seen in select areas of the 1992 Cadillac Seville and Eldorado (Ref 22). For example, the new bumper system reduced part count by half over the previous generation, and assembly time was about 19 min less than the pre-DFMA design. A further example was the Cadillac console. In this case a reduction of 40% in assembly time and a 33% reduction in part count was achieved by employing DFMA from the concept stage.

Application of DFMA at a Toy Manufacturer. Toy manufacturers must comply with some of the most demanding time-to-market schedules of any industry. With an average product life cycle of only one year, toys are serious business for the development teams in the promotional division of one of the largest toy companies in the world, Hasbro, Inc.

One case where DFMA software has cut redesign time and cost is for a toy fire truck (Ref 23). For example, the original ladder assembly was composed of 33 metal parts and subassemblies, with an assembly time of 198 s. The redesigned ladder brought the number of parts down to its theoretical minimum of only five parts—all plastic—with an assembly time of just 22 s.

Hasbro also has found that DFMA provides a nonthreatening way to get team members talking about a design without anyone feeling as though others are encroaching on his or her territory. It also allows Hasbro's tooling and manufacturing engineers to get involved at the concept stage, eliminating any surprises.

Application of DFMA at a Medical Devices Manufacturer. At Respironics Inc., Pennsylvania, DFMA software enables engineering teams to define goals at the earliest stages of product design and then manage priorities as the project progresses (Ref 24). This firm designs, manufactures, and markets medical devices used to treat respiratory and cardiopulmonary disorders.

A design team used DFMA in the early design stages of a manual resuscitator to make the product easier to manufacture and assemble and to improve the overall design process as well. Executives began the product-development project by establishing a team of engineering, product-development, and management personnel. The team evaluated customer requirements and created early product specifications. At this time, a multifunctional project team was also formed with representatives from key suppliers, engineering, purchasing, design-drafting, manufacturing, production and quality assurance, and accounting. According to the DFMA facilitator (Ref 24), the project benefited greatly from DFMA brainstorming sessions, and the most valuable success factor was how the process of DFMA focused the team on its objectives. The team created the vision of the product, all members had an early sense of ownership and what the product was going to be, and they challenged themselves to make it work.

The following table shows the improvements for the new design over the previous product:

Factor	Improvement
Assembly time	84%
Assembly cost	74%
Number of unique parts	65%
Total number of parts	66%
Number of operations	81%

The new product is greatly simplified. No separate fasteners or lubricants are used; the previous design used both. All parts snap or press fit together, with one heat-stake operation and one adhesive-bond joint. In addition, the adhesive joint is being redesigned to eliminate the adhesive.

The project produced innovative features and approaches that generated patent applications for six claims—all developed during the initial DFMA workshop.

Application of DFMA at an Aerospace Manufacturer. In 1991, the U.S. Navy presented McDonnell Douglas with a challenge: produce an aircraft to carry them into the 21st century. The company responded with a proposed new version of their FA-18 Hornet fighter jet (Ref 25).

One of the most direct ways DFMA helped reduce costs and improve quality was to force designers to reduce the number of parts and fasteners used in the aircraft. For each fastener there is the chance that a hole will either be drilled in the wrong place or be the wrong size. By reducing the number of parts and fasteners, the opportunity for defects is reduced.

Weight is also a critical aspect of aircraft design. There are 179,000 fasteners in the Hornet. A commercial transport like the MD-11 can have 1,300,000. Reference 26 explains how weight can be systematically reduced through part count reduction strategies. If two sheet metal parts are to be attached, they usually require a flange on one of them to allow space for a row of rivets or fasteners. This adds weight in the form of extra metal and the extra fasteners. By producing the two parts as one machined piece, the weight is reduced. When this principle is applied to an entire aircraft, overall weight reduction is very significant.

Other results of using DFMA were realized throughout the new aircraft design. In the previous model, several parts were made of multiple sheet metal components because machined parts were unable to reach minimum gage limits. With high-speed machining, however, the thinner gage limits became possible. One example: the nose barrel bulkhead—one of several large walls in the forward section of the aircraft between the cockpit and the nose radar dome—previously consisted of 90 separate parts. The redesigned one-piece wall in the new model cost $3746 less and cut 20 days from production time.

The Navy and McDonnell Douglas named the new model the Super Hornet. Rolled out on time and within budget, the Super Hornet was 25% larger than the previous version, yet it contained 42% fewer parts. It weighed in at 1000 lb under the Navy's weight specification and flew farther, with more payload, than its predecessor. Originally scheduled for first flight in December 1995, the first complete Super Hornet successfully took to the sky 32 days early, in November.

Roadblocks in Implementation of DFMA

The above examples have shown that the format for successful implementation of DFMA varies from company to company, but some major points stand out. First, DFMA software is a team tool and should be utilized as such.

Training is important. Implementation of DFMA efforts employ the software system, and for this reason some companies believe it is, for example, similar to using a spreadsheet program. This is not the case. It is important to train people in a workshop environment—a team using the system on an ongoing project with the company's "champion" or an outside system consultant providing help. In this way, one or two days provides useful training plus, often as not, real results. It is important to remember that it is often not the target, but the journey through the systematic procedure, that matters.

Experience has shown that there are many barriers to the implementation of DFMA. Reasons for resisting the implementation of DFMA are often put forward, but all can be answered.

Lack of Time. The most common complaint among designers is that they are not allowed sufficient time to carry out their work. Designers are usually constrained by the urgent need to minimize the design-to-manufacture time for a new product. However, more time spent in the initial stages of design will reap benefits later in terms of reduced engineering changes after the design has been released to manufacturing. Company executives and managers must be made to realize that the early stages of design are critical in determining not only manufacturing costs, but also the overall design-to-manufacturing cycle time.

"Not Invented Here" Syndrome. Enormous resistance can be encountered when new techniques are proposed to designers. Ideally, any proposal to implement DFMA should come from the designers themselves. However, more frequently it is the managers or executives who have heard of the successes resulting from DFMA and demand that their own designers implement the philosophy. Under these circumstances, great care must be taken to involve the designers in the decision to use these new techniques.

"Ugly Baby" Syndrome. Even greater difficulties exist when an outside group or a separate group within the company undertakes to analyze existing designs for ease of manufacture and assembly. Commonly, this group will find that significant improvements could be made to the original design, and when these improvements are brought to the attention of those who produced the design, this can result in extreme resistance. Telling a designer that his designs could be improved is much like telling a mother that her baby is ugly!

It is important, therefore, to involve the designers in the analysis and provide them with the incentive to produce better designs. If they perform the analysis of their own designs, they are less likely to feel criticized.

Perception that a Product with Low Assembly Costs Will Not Benefit from DFMA. The first step in the application of DFMA is a DFA analysis of the product or subassembly. Quite frequently, it will be suggested that because assembly costs for a particular product form only a small proportion of the total manufacturing cost, there is no point in performing a DFA analysis. However, like the sample analysis described earlier, the savings in part costs are invariably much greater than the savings in assembly cost.

Perception that a Low-Volume Product Will Not Benefit from DFMA. The view is often expressed that DFMA is only worthwhile when the product is manufactured in large quantities. It can be argued, though, that use of DFMA is even more important when the production quantities are small. This is often because reconsideration of an initial design is not usually carried out for low-volume production. Applying the philosophy "do it right the first time" becomes even more important, therefore, when the production quantities are small.

"The Database Doesn't Apply to Our Product." Everyone seems to think that their own company is unique and, therefore, in need of unique databases rather than the ones incorporated within the DFMA software. However, when one design is rated better than another using the software databases, it would almost certainly be rated in the same way using a customized database. Remembering that there is a need to apply DFMA at the early design stage before detailed design has taken place, there is a need for a generalized database for this purpose. Later, when more accurate estimates are desired, the user can employ a customized database if necessary.

"We've Been Doing It for Years." When this claim is made, it usually means that some procedure for "design for producibility" has been in use in the company. However, design for producibility usually means detailed design of the individual parts of an assembly for ease of manufacture. It was stated earlier that such a process should only occur at the end of the design phase. The important decisions affecting total manufacturing costs will already have been made. In fact, there is a great danger in implementing design for producibility in this way.

It has been found that the design of individual parts for ease of manufacture can mean, for example, limiting the number of bends in a sheet metal part. This invariably results in a more expensive assembly where several simple parts are fastened together, rather than a single, more complicated part. Again, experience has shown that it is important to combine as many features in one part as possible. In this way, full use is made of the abilities of the various manufacturing processes.

Confusion with Value Analysis. It is true that the objectives of DFMA and value analysis (VA) are the same (see the article "Value Analysis in Materials Selection and Design" in this Volume). However, it should be realized that DFMA is meant to be applied early in the design cycle, and that VA does not give proper attention to the structure of the product and its possible simplification. DFMA has the advantage that it is a systematic step-by-step procedure, which can be applied at all stages of design and challenges the designer or design team to justify the existence of all the parts and consider alternative designs. Value analysis, on the other hand, only looks at major parts; it is often the screws and washers—often not even shown on the drawings—that impose the difficulties during assembly.

Experience has shown that DFMA can still make significant improvements even after VA has been carried out.

Perception that DFMA is Only One among Many Techniques. Since the introduction of DFMA, many other acronyms have been proposed, for example, design for quality (DFQ), design for competitiveness (DFC), design for excellence (DFE), and now the all-inclusive design for "X" (DFX). Many have even suggested that design for performance is just as important as DFMA. One cannot argue with this. However, DFMA is the subject that has been neglected over the years while adequate consideration has always been given to the design of a product for performance, appearance, and so forth. The other factors, such as quality, reliability, and so forth, will follow when proper consideration is given to the manufacture and assembly of the product. The earlier example from Motorola (Fig. 10) illustrates how DFMA can lead to higher product quality.

Perception that DFMA Leads to Products that Are More Difficult to Service. It has been claimed that DFMA leads to products that are more difficult to service. However, experience shows that a product that is easy to assemble is usually easier to disassemble and reassemble. In fact, those products that need continuous servicing, involving the removal of inspection covers and the replacement of various items, should have DFMA applied even more rigorously during the design stage. How many times is an inspection cover originally fitted with numerous screws found with only two screws replaced after the first inspection?

"I Prefer Design Rules." There is a danger in using design rules, because they can guide the designer in the wrong direction. Generally, rules attempt to force the designer to think of simpler-shaped parts that are easier to manufacture. In an earlier example, it was pointed out that this can lead to more complicated product structures and a resulting increase in total product costs. In addition, in considering novel designs of parts that perform several functions, the designer needs to know the penalties when the rules are not followed. For these reasons, the systematic procedures embodied in DFMA software, which guide the designer to simpler product structures and provide quantitative data on the effect of any design changes or suggestions, are found to be the best approach.

General Resistance. Although a designer may not say that he refuses to use DFMA, if he does not have the incentive to adopt this philosophy and use the tools available, then no matter how useful the tools or how simple they are to apply, they will not work. Therefore, it is imperative that the designer or the design team is given the incentive and the necessary facilities to incorporate considerations of assembly and manufacture during design.

The main argument, however, against any reservations about adopting DFMA are the savings in manufacturing costs obtained by the hundreds of companies worldwide that have adopted the system, some examples of which have been described here.

Design for Automatic Assembly

For many years it has been felt that automatic assembly can significantly improve productivity, and during the early days of development of DFA, there was great interest in how to design products for automatic assembly. It is clearly unlikely that a product will be designed so that it will be impossible to assemble manually. However, if a product is to be assembled automatically, then it is essential that the product be designed with the automation in mind. Otherwise, there are likely to be many parts that would be impossibly expensive to feed and orient automatically and insert automatically. It can be said that, if a product is designed for ease of automatic assembly, then it will also be easy to assemble manually.

As DFA software for automatic and manual assembly became available in 1981, most companies were happy to use design for manual assembly because it attacked the most important aspects of design for assembly, namely, the reduction in part count. In addition, very few mechanical items were assembled automatically. Some years ago it was estimated that only about 6% of all mechanical assemblies were assembled on automatic machines.

Experience has shown that in product design for ease of automatic assembly, the most important aspect is the automatic feeding and orienting of the individual parts.

It has been seen that the most obvious way to facilitate the manual assembly process at the design stage is to reduce the number of different parts to a minimum. When considering product design for automatic assembly, it is even more important to consider reduction in the number of separate parts. For example, the elimination of a part can eliminate a complete station on an assembly machine, including the parts feeder, the special workhead, and the associated portion of the transfer device.

Apart from product simplification, automatic assembly can be facilitated by the introduction of guides and chamfers that directly facilitate assembly. Sharp corners should be removed so that the part to be assembled is guided into its correct position during assembly and requires less control by the placement device.

The types of screws used in automatic assembly are important. Screws such as those with a conical or oval point that tend to centralize themselves in the hole give the best results.

A major factor to be considered in design for automatic assembly is the difficulty of assembly from directions other than directly above. The aim of the designer should be to allow for assembly in sandwich or layer fashion, with each part placed or inserted from above. One advantage of this method is that gravity is useful in the feeding and placement of parts. It is also desirable to have workheads and feeding devices above the assembly station, where they will be accessible in the event of a fault due to the feeding of a defective part. Assembly from above may also assist in the problem of keeping parts in their correct positions during the machine index period, when dynamic forces in the horizontal plane might tend to displace them. In this case, with proper product design, where the parts are self-locating, gravity should be sufficient to retain the part until it is fastened or secured.

It is necessary in automatic assembly to have a base part on which the assembly can be built. This base part must have features that make it suitable for quick and accurate location on the work carrier.

Design of Parts for Feeding and Orienting. In automatic assembly, parts obtained in bulk are placed in feeders that will separate, orient, and deliver them to the placement device. If parts are to be automatically fed and oriented, it is essential that they be designed appropriately. For example:

- Parts should not tangle, nest, or shingle.
- Ideally, parts should be symmetrical.
- If parts cannot be made symmetrical, slight asymmetry or asymmetry resulting from small or nongeometrical features should be avoided.

While the asymmetrical feature of a part might be exaggerated to facilitate orientation, an alternative approach is to deliberately add asymmetrical features for the purpose of orienting.

Clearly, with some parts it will not be possible to make design changes that will enable them to be handled automatically; for example, very small parts or complicated shapes formed from thin strips are difficult to handle in an automatic environment. In these cases, it is sometimes possible to manufacture the parts on the assembly machine or to separate them from the strip at the moment of assembly. Operations such as spring winding or blanking out thin sections can be successfully introduced on assembly machines.

Product Design for Robot Assembly. Several important product design aspects will be affected if assembly robots are to be incorporated in the assembly system. There are two basic types of robot assembly systems:

- Single-station with one or two robot arms
- Multistation transfer machine with robots, special-purpose workheads, and manual assembly station as appropriate

For a single-station system, parts that require manual handling and assembly and that must be inserted during the assembly cycle present special problems. For reasons of safety, it is usually necessary to transfer the assembly to a location or fixture outside the working environment of the robot. This can be accomplished by having the robot place the assembly on a transfer device that carries the assembly to the manual station. After the manual operation has been completed, the assembly can be returned in a similar manner to within reach of the robot. Such an arrangement is costly and time consuming.

The use of special-purpose workheads for insertion or securing operations can present similar problems. Two different situations can be encountered. In the first, the robot inserts or places the part without securing it immediately. This operation is followed by transfer of the assembly to an external workstation where the securing operation is carried out. In the second situation, a special-purpose workhead is engineered to interact with the robot at the work fixture. This might take the form of equipment activated from the sides of, or underneath, the work fixture to carry out soldering, tab bending or twisting operations, spin riveting, and so forth, while the robot has to place and, if necessary, manipulate the part.

These major problems with single-station systems do not occur with the multistation system, where manual operations or special-purpose workheads can be assigned to individual stations as necessary. This illustrates why it is important to know the type of assembly system likely to be employed when the product is being designed.

It is also necessary to know which presentation method will be used for each part. In practice, there are usually only two choices: a special-purpose feeder or a manually or automatically loaded magazine pallet or part tray. If manually loaded magazines are used, the time to hand-load the parts into the magazines must be taken into account when the choice of presentation device is made.

Design Rules for Robot Assembly. Many of the rules for product design for manual assembly and special-purpose automatic assembly also apply to product design for robot assembly. However, when the suitability of a proposed design for robot assembly is considered, the need for any special-purpose equipment such as special grippers or special feeders should be noted. The cost of this special equipment must be amortized over the total life volume of the product and, for the midrange volumes to which robot assembly might be applied, this can add considerably to the cost of assembly.

Table 7 Various improvements due to application of DFMA software

Category	No. of cases	Average reduction, %
Part count	61	56
Assembly time	38	62
Product cost	21	50
Assembly cost	17	45
Assembly operations	14	57
Separate fasteners	12	72
Labor costs	8	42
Manufacturing cycle	6	58
Weight	6	31
Assembly tools	5	69
Part cost	3	56
Unique parts	3	57
Material cost	3	37
Manufacturing process steps	3	45
No. of suppliers	3	55
Assembly defects	3	68
Cost savings per year	6	$1,283,000

From 74 published case studies

The following are some specific rules to follow during product design:

- Reduce part count; this is a major strategy for reducing assembly, manufacture, and overhead costs, irrespective of the assembly system to be used.
- Include features such as leads, lips, and chamfers to make parts self-aligning in assembly. Because of the relatively poor repeatability of many robot manipulators compared to special-purpose workhead mechanisms, this is an important measure to ensure consistent fault-free part insertions.
- Ensure that parts that are not secured immediately on insertion are self-locating in the assembly. For multistation robot assembly systems or single-station systems with one robot arm, this is an essential design rule. Holding down of unsecured parts cannot be carried out by a single robot arm, and so special fixturing is required, which must be activated by the robot controller. This adds significantly to special-purpose tooling and, hence, assembly costs. With a two-arm single-station system, one arm can, in principle, hold down an unsecured part while the other continues the assembly and fastening processes. In practice, this requires one arm to change end-of-arm tooling to a hold-down device; the system then proceeds with 50% efficiency while one arm remains immobile.
- Design parts so that they can all be gripped and inserted using the same robot gripper. One major cause of inefficiency with robot assembly systems arises from the need for gripper or tool changes. Even with rapid gripper or tool-change systems, each change in a special gripper and then back to the standard gripper can be the equivalent of two assembly operations. Note that the use of screw fasteners results in the need for tool changes because robot wrists can seldom rotate more than one revolution.

- Design products so that they can be assembled in layer fashion from directly above (Z-axis assembly). This ensures that the simplest, least costly, and most reliable four-degree-of-freedom robot arms can accomplish the assembly tasks. It also simplifies the design of the special-purpose work fixture.
- Avoid the need for reorienting the partial assembly or manipulating previously assembled parts. These operations increase the robot assembly cycle time without adding value to the assembly. Moreover, if the partial assembly has to be rotated to a different resting aspect during the assembly process, this will usually result in increased work-fixture cost and the need to use a more expensive six-degree-of-freedom robot arm.
- If parts are to be presented using automatic feeders, ensure that they can be fed and oriented using simple tooling. Follow the rules for ease of automatic part feeding and orientation discussed earlier. Note, however, that feeding and orienting at high speed is seldom necessary in robot assembly and that the main concern is that the features defining part orientation can be easily detected.
- If parts are to be presented using automatic feeders, ensure that they can be delivered in an orientation from which they can be gripped and inserted without any manipulation. For example, avoid situations in which a part can be fed in only one orientation from which it must be turned over for insertion. This will require a six-degree-of-freedom robot and a special gripper or a special 180°-turn delivery track; both solutions lead to unnecessary cost increases.
- If parts are to be presented in magazines or part trays, ensure that they have a stable resting aspect from which they can be gripped and inserted without any manipulation by the robot. It should be noted that, if the production conditions are appropriate, the use of robots holds advantages over the use of special-purpose workheads and that some design rules can be relaxed. For example, a robot can be programmed to acquire parts presented in an array, such as a pallet or part tray that has been loaded manually, thus avoiding many of the problems arising with automatic feeding from bulk.

Conclusions

DFMA provides a systematic procedure for analyzing proposed designs from the point of view of assembly and manufacture. It encourages teamwork and a dialogue between designers and the manufacturing engineers, and any other individuals who play a part in determining final product costs during the early stages of design.

The use of DFMA software often produces a considerable reduction in part count, resulting in simpler and more reliable products that are less expensive to assemble and manufacture.

As discussed previously, there are many widely publicized DFMA case studies to illustrate these claims. Figure 11 shows the effect of DFA on part count reduction from published case studies, and Table 7 presents details of other improvements from the same case studies.

In spite of all the success stories, the major barrier to DFMA implementation continues to be human nature. People resist new ideas and unfamiliar tools or claim that they have always taken manufacturing into consideration during design. The DFMA methodology challenges the conventional product design hierarchy. It reorders the implementation sequence of other valuable manufacturing tools, such as statistical process control and Taguchi methods. Designers are traditionally under great pressure to produce results as quickly as possible and often perceive DFMA as yet another time delay. In fact, as numerous case studies have shown, the overall design development cycle is shortened through use of early manufacturing analysis tools, because designers can receive rapid feedback on the consequences of their design decisions where it counts—at the conceptual stage.

Overall, the facts are that DFMA is a subject that has been neglected over the years while adequate consideration has always been given to the design of a product for performance, appearance, and so forth. The other factors such as quality, reliability, and so forth will follow when proper consideration is given to the manufacture and assembly of the product. In order to remain competitive in the future, every manufacturing organization will have to adopt the DFMA philosophy and apply cost quantification tools at the early stages of product design.

For a more complete treatment of the subject of this article, see Ref 27.

ACKNOWLEDGMENT

DFMA is a registered trademark of Boothroyd Dewhurst, Inc.

REFERENCES

1. J.P. Womack, D.T. Jones, and D. Roos, *The Machine that Changed the World,* Macmillan, 1990
2. S. Miyakawa and T. Ohashi, "The Hitachi Assemblability Evaluation Method (AEM)," Proc. Int. Conf. Product Design for Assembly (Newport, RI), 15–17 April 1986
3. S. Miyakawa, T. Ohashi, and M. Iwata, The Hitachi New Assemblability Evaluation Method (AEM)," *Trans. North Am. Manuf. Res. Inst. SME,* 23–25 May 1990
4. H.J. Warnecke and R. Bassler, Design for Assembly—Part of the Design Process, *Ann. CIRP,* Vol 37 (No. l), 1988, p l
5. B.L. Miles and K.G. Swift, Working Together, *Manuf. Breakthrough,* March/April 1992, p 69
6. Y. Yamigawa, "An Assembly Ease Evaluation Method for Product Designers: DAC," *Technol. Jpn.,* Vol 21 (No. 12), 1988
7. S. Miyakawa, "Simultaneous Engineering and Producibility Evaluation Method," Int. Conf. Applications of Manufacture Technologies (Washington, D.C.), 17–19 April 1991, Society of Mechanical Engineers
8. K. Takahashi, I. Suzuki, and T. Suguro, "Producibility Evaluation Method," Tenth Int. Conf. Assembly Automation
9. G. Boothroyd and P. Radovanovic, Estimating the Cost of Machined Components During the Conceptual Design of a Product, *Ann. CIRP,* Vol 38 (No. 1), 1989, p 157
10. P. Dewhurst, "Computer-Aided Assessment of Injection Molding Cost—A Tool for DFA Analyses," Report 24, Department of Industrial & Manufacturing Engineering, University of Rhode Island, 1988
11. G.D. Kobrak and P. Dewhurst, "Design Rules and Early Cost Analysis for Sand Casting," Report 64, Department of Industrial & Manufacturing Engineering, University of Rhode Island, 1993
12. A.R. Mackay and G. Boothroyd, "Cost Estimating of Investment Castings," Report 89, Department of Industrial & Manufacturing Engineering, University of Rhode Island, 1995
13. P. Dewhurst and C. Blum, Supporting Analyses for the Economic Assessment of Die Casting in Product Design, *Ann. CIRP,* Vol 28 (No. 1), 1989, p 161
14. D. Zenger and P. Dewhurst, "Early Assessment of Tooling Costs in the Design of Sheet Metal Parts," Report 29, Department of Industrial & Manufacturing Engineering, University of Rhode Island, 1988
15. W.A. Knight, Design for Manufacture Analysis: Early Estimates of Tool Costs for Sintered Parts, *Ann. CIRP,* Vol 40 (No. 1), 1991
16. P. Dewhurst, Cutting Assembly Costs with Molded Parts, *Mach. Des.,* 21 July 1988
17. C. Kirkland, Design Watch, *Plast. World,* March 1992
18. *Machining Data Handbook,* Metcut Research Associates, Cincinnati, OH, 1980
19. P.F. Ostwald, *AM Cost Estimator,* Penton Education Division, Cleveland, OH, 1988
20. B. Branan, DFA Cuts Assembly Defects by 80%, *Appl. Manuf.,* Nov 1991
21. G.J. Burke and J.B. Carlson, DFA at Ford Motor Company, *DFMA Insight,* Vol 1 (No. 4), 1990
22. G. Kobe, DFMA at Cadillac, *Automotive Ind.,* May 1992, p 43
23. C. Kirkland, Hasbro Doesn't Toy with Time to Market, *Inject. Mold.,* Feb 1995, p 34
24. A. Benson, Software in Action, *Assembly,* Sept 1995
25. R. Mandel, Parts Reduction Receives High Machs, *Designfax,* July 1996
26. N.O. Weber, Flying High: Aircraft Design Takes Off with DFMA, *Assembly,* Sept 1994
27. G. Boothroyd, P. Dewhurst, and W.A. Knight, *Product Design for Manufacture and Assembly,* Marcel Dekker, 1994

Manufacturing Processes and Their Selection

John A. Schey, University of Waterloo

ENGINEERING DESIGN and manufacturing process selection are intricately interwoven; recognition of this interdependence is the basis of concurrent engineering. For this discussion, it is assumed that the functional requirements imposed on a discrete part have led to the definition of shape and properties and, together with this, to a preliminary choice of processes. This may have taken place formally, for example, with the aid of the material and process selection charts of Ashby (Ref 1) (see the articles "Overview of the Materials Selection Process," "Material Property Charts," and "Relationship between Materials Selection and Processing" in this Volume), or by the less formal approach practiced by many, in which acquaintance with production methods influences the choice. There are dangers with both approaches. The array of manufacturing processes is vast and is not easily fitted into a few charts; as pointed out by Ashby, his charts lead only to a preliminary selection. When process choice is based on the individual's knowledge, close familiarity with one process or one class of processes can lead to premature process choice and rejection of potentially valuable alternatives. Recently developed expert systems (see, for example, the article "Design for Manufacture and Assembly" in this Volume and Ref 2) facilitate process selection from a much broader perspective.

The treatment adopted in this article assumes some familiarity with processes normally described in introductory books on manufacturing processes (Ref 3–18). It emphasizes process characteristics that have an influence on process selection, and expands and updates textbook material (Ref 16). Once technical feasibility is established, process choice is further narrowed by cost and availability. In the case of a "buy" decision, cost is readily determined by comparing quotes from vendors. True costs of in-house production are often difficult to establish unless a well-designed accounting system is in place. In either case, costs associated with further processing of parts must form an integral part of informed decision making.

Detailed design rules for specific processes are given in later articles in this Section; this article explores the possibilities and limitations imposed by the process and material. Process groups are discussed in the sequence followed in the rest of this Section. For ease of discussion, main features of process groups are presented in tabular form. The tables are meant to give general guidelines only and are based on good standard practices. Inventive approaches often allow production to much higher quality, broader size range, lower cost, and so forth. Each table contains physical characteristics and ratings of relative cost and production factors. Examples of typical products

Table 1 General characteristics of casting processes

Characteristic	Casting process: Green sand	Resin-bonded sand	Plaster	Lost foam	Investment	Permanent mold	Die
Part							
Material (casting)	All	All	Zn to Cu	Al to cast iron	All	Zn to cast iron	Zn to Cu
Porosity and voids(a)	C–E	D–E	D–E	C–E	E	B–C	A–C
Shape(b)	All	All	All	All	All	Not T3, 5, F5 with solid core	Not T3, 5, F5
Size, kg	0.01–300,000	0.01–100	0.01–1000	0.01–100	0.01–100	0.1–100	<0.01 to 50
Minimum section, mm	3–6	2–4	1	2–4	1	2–4	0.5–1
Minimum core diameter, mm	4–6	3–6	10	4–6	0.5–1	4–6	3 (Zn: 0.8)
Surface detail(a)	C	B	A	C	A	B–C	A–B
Cost							
Equipment(a)	C–E	C	C–E	B–C	C–E	B	A
Die (or pattern)(a)	C–E	B–C	C–E	B–C	B–C	B	A
Labor(a)	A–C	C	A–B	C	A–B	C	E
Finishing(a)	A–C	B–D	C–D	C–D	C–D	B–D	C–E
Production							
Operator skill(a)	A–C	C	A–B	C	A–B	C	C–D
Lead time	Days	Weeks	Days	Weeks–months	Hours–weeks	Weeks	Weeks–months
Rates (piece/h · mold)	1–20	5–50	1–10	1–20	1–1000	5–50	20–200
Minimum quantity	~1–100	~100	~10	~500	~10–1000	~1000	~100,000

(a) Comparative ratings, with A indicating the highest value of the variable, E the lowest (e.g., investment casting gives very low porosity, produces excellent surface detail, involves moderate to low equipment cost, medium to high pattern cost, high labor cost, medium to low finishing cost, and high operator skill. It can be used for low or high production rates and requires a minimum quantity of 10 to 1000 to justify the cost of the pattern mold). (b) From Fig. 1. Source: Adapted from Ref 16 given in the text are meant to be only illustrative and in no way limiting.

Product Considerations

Some factors have an influence in all processes and need to be discussed further. The factors include size, shape complexity, tolerances and surface finish, and production volume.

Size. There are limits to size in both extremes. The minimum size is often dictated by physical constraints. For example, surface tension and freezing of a metal melt limit the smallest size of a mold cavity that can be filled. The minimum can be reduced but not eliminated by filling under pressure; the minimum is different for different alloys. Radically different techniques, mostly based on methods used for making semiconductor devices, push the limit much lower (nanotechnology). The maximum is more often set by the largest size of equipment available, although there may also be physical limitations. For example, the size of hot impression-die forgings or injection-molded plastic parts is limited by the largest press available, while the thickness of a plastic part is limited by loss of shape due to shrinkage.

Shape Complexity. Shape is an essential feature of all manufactured parts; complexity of this shape often determines what processes can be considered for making it. In the most general sense, increasing complexity narrows the range of processes and increases cost. A cardinal rule of design is, therefore, to keep the shape as simple as possible. This rule may, however, be broken if a more complex shape allows consolidation of several parts and/or elimination of one or more manufacturing steps.

There is no universally accepted shape classification system. The groupings given in Fig. 1 are used in the tables that list process attributes. Products of uniform cross section (spatial complexity = 0) are two-dimensional, all others are three-dimensional. With increasing spatial complexity, definition of the shape requires additional geometric parameters; in the terminology of Ashby (Ref 1), the shape has greater information content. A small increment in information content can, however, have significant manufacturing consequences. For example, moving from the solid shape R1 to the hollow shape T1 adds only one dimension (the diameter of the hole, with appropriate tolerances and specifications for concentricity), yet it immediately excludes some processes or necessitates extra operations in others. Adding a third diameter to the round product R1 would result in the same increase in information content yet does not impose limitations of process choice. One of the aims of group technology is classification of shapes suitable for processing by the same techniques (see the article "Introduction to Manufacturing and Design" in this Volume).

Limitations on shape are tightened by properties of the material and by interactions with the tooling. Wall or section thickness are affected, as shown in Fig. 2, and discussed further in conjunction with process groups. The aim is, generally, to produce a "net-shape" part ready for assembly; if this is not feasible, a "near-net-shape" part will need only minor finishing, usually by machining.

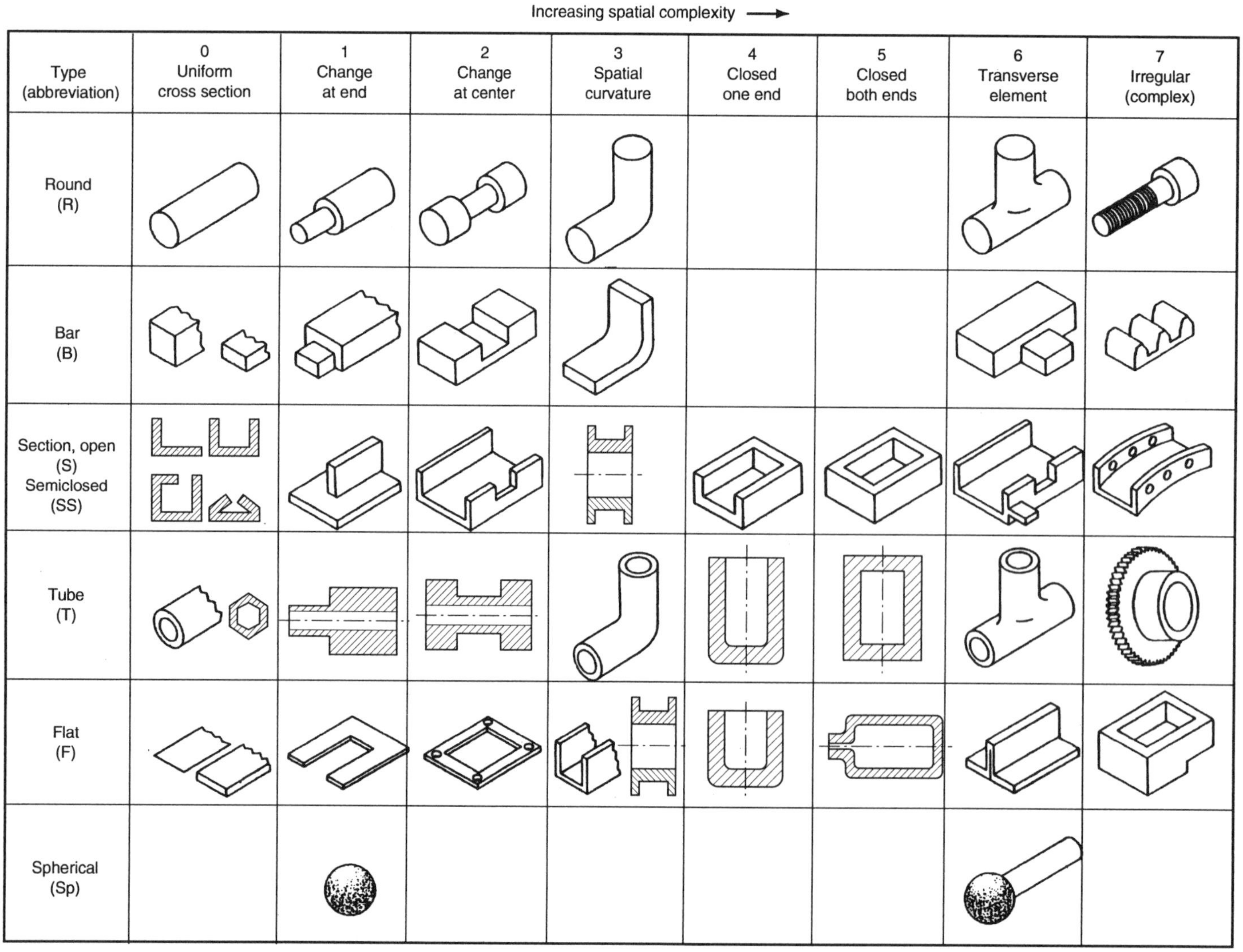

Increasing spatial complexity
Type (abbreviation)
0 Uniform cross section
1 Change at end
2 Change at center
3 Spatial curvature
4 Closed one end
5 Closed both ends
6 Transverse element
7 Irregular (complex)
Round (R)
Bar (B)
Section, open (S) Semiclosed (SS)
Tube (T)
Flat (F)
Spherical (Sp)

Fig. 1 The choice of possible manufacturing methods is aided by classifying shapes according to their geometric features. Source: Ref 16

Tolerances and Surface Finish. The article "Overview of the Materials Selection Process" in this Volume gives a general discussion of these aspects; the concern in this article is their manufacturing implication.

Tolerances. No manufacturing process can deliver a perfect geometrical shape. The permissible deviations are expressed by tolerances (see the article "Dimensional Management and Tolerance Analysis" in this Volume), and some processes are inherently more suitable for delivering parts to close tolerances (see Fig. 5 in the article "Overview of the Materials Selection Process" in this Volume).

Surface Topography. In addition to deviations from the ideal shape, the surface of manufactured parts is never perfectly smooth. Roughness is not necessarily undesirable. Sheet used for making an automobile body panel must have controlled, random roughness features to facilitate pressing and give desirable surface appearance after painting. A shaft of a journal bearing must have controlled circumferential roughness features to retain lubricant and prevent seizure. As shown by these examples, not only the degree of roughness but also the orientation (lay) of surface features relative to the direction of sliding motion are critical.

Measurement and characterization of surface topography is a rapidly developing field, and only the most commonly used approaches are discussed here. The profile of the surface (roughness and waviness) is measured by drawing a fine-pointed stylus across the surface (other, noncontacting measuring methods are available [Ref 19] and are gaining in popularity). For ease of visualization, recordings are made with greater magnification on the vertical (height) axis (Fig. 3). This makes peaks (asperities) appear much steeper than the 5 to 20° slope they really have. A centerline is computed so that the area filled with material equals the unfilled area. The average height deviation from this line is the centerline average (CLA) or arithmetic average (AA; R_a). The root-mean-square (RMS) value (R_q) is slightly smaller. When roughness-height distribution is not regular—for example, when local peaks would penetrate a lubricant film or when the surface is to be polished—the maximum roughness height (R_t) or average height difference between 5 highest peaks and 5 deepest valleys (R_{tm}) is more informative. Because R_t/R_a is typically between 7 and 14, roughness affects the closest tolerance that can be meaningfully specified.

Surfaces of the same R_a roughness may have very different asperity shapes (Fig. 4). Roughness height distribution is described by the Abbott curve (a plot of load-bearing areas produced by cuts taken at various levels from the top of the profile) and by the numerical values of skewness (R_{sq}) and kurtosis (R_k) (Ref 21). They affect the functional behavior of the surface. For example, automobile body sheet of positive skewness (peaky profile) is prone to form die pickup and suffer scoring.

Production Volume. Process choice is heavily influenced by the total number of parts to be made and by the required rate of production (i.e., the number of parts produced in a given time period). In general, a larger production quantity justifies greater investment in dies, equipment, and automation, whereas small quantities are often made with more labor input.

The total number of parts is frequently produced in lots. Lot size used to be determined to a large extent by total quantity, with lot size chosen to provide a supply for several days or weeks. Some standard components such as fasteners are still produced in large lots by hard automation. However, the spread of just-in-time delivery schedules and the increasing use of quick die-changing techniques and flexible automation have contributed to shrinking lot sizes even in mass production; the limit is reached when a single part constitutes a lot.

It is not feasible to state hard rules of what the economical production quantity and lot size is. Much depends on the equipment, degree of automation, and process control available in a given plant. Relative values shown in the tables are meant to serve only as initial guidelines and are heavily influenced by the cost of dies.

Casting Processes

In all casting processes (Table 1), a metal or alloy is melted (and possibly treated while molten), poured into a mold, allowed to solidify, and the part thus produced is released from the mold. Very broadly, casting processes fall into two groups (Fig. 5): expendable-mold and permanent-mold processes. Choice of process and mold material is greatly influenced by the melting point of the alloy. A vital issue is avoidance of voids (due to insufficient feeding of melt) and porosity (due to dendritic solidification and the presence

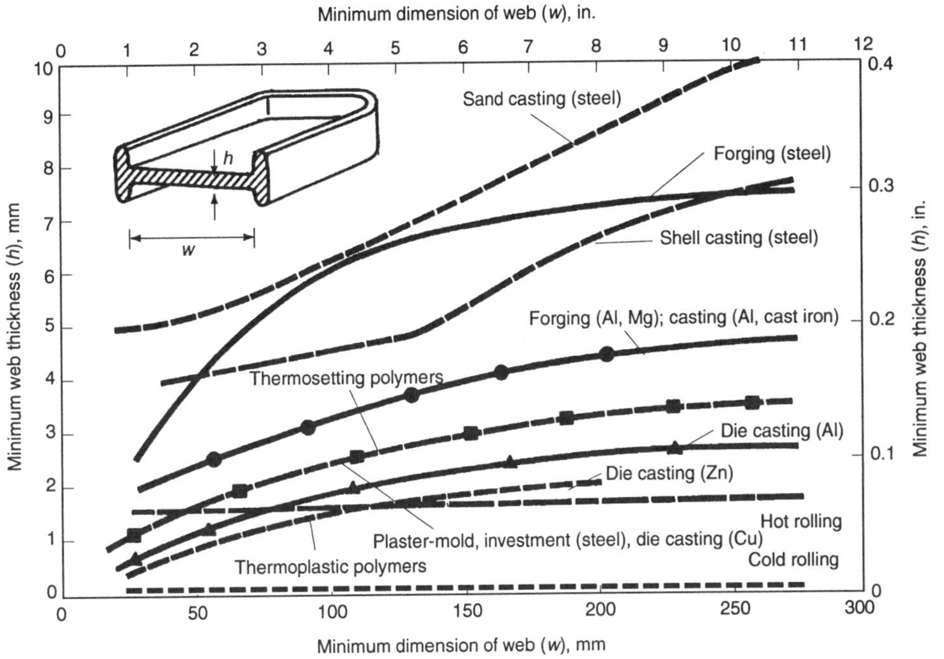

Fig. 2 For a given process, the minimum web thickness increases with the distance over which material must move. Source: Ref 16

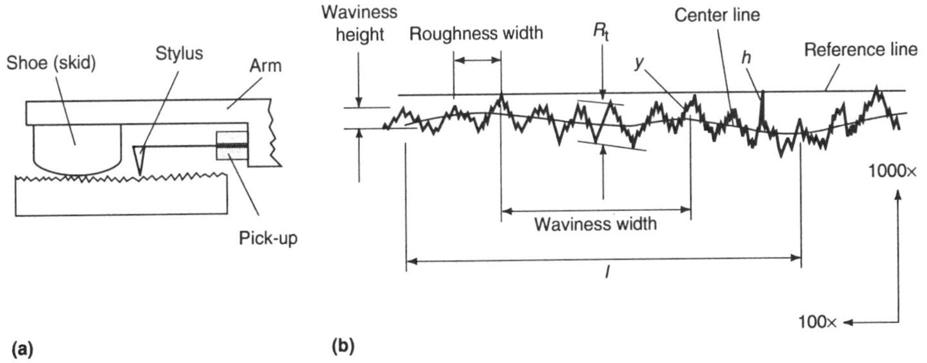

Fig. 3 Measurement (a) and characterization (b) of surface roughness. R_t, maximum roughness height; *l*, length; *y*, height from center line; *h*, distance of surface from reference surface. Source: Ref 20

of gases). Success depends, to a large extent, on the skill of the mold designer in choosing the position and size of sprue, runners, and risers, but there are very strong influences of the metal (alloy) and mold material. Expert systems and computer modeling of die filling and solidification are increasingly employed. Even with the best design, solidification on the mold walls closes passage of melt at a certain distance, and this leads to a minimum allowable section thickness. Because with proper design of runners a part can almost always be fed across the shorter dimension, minimum web thickness increases with the width of the web (Fig. 2). There are large differences in the fluidity of alloys; therefore lower minimum thickness can be allowed with zinc, aluminum, and cast iron than with steel. Streamlined flow is helpful; thus sharp corners are avoided. Thicker sections solidify last and must be fed adequately.

Expendable-Mold Processes

In expendable-mold casting processes, the mold is used only once and is destroyed to release the part; hence great complexity of shape is feasible. Expendable-mold processes are further divided into those using permanent patterns and those using expendable patterns.

Sand casting is used as a generic term for all processes in which an expendable mold is formed of a granular ceramic (ranging from natural sand to synthetic ceramics) by consolidation around a permanent pattern (wood, metal, or plastic), which is removed prior to filling the mold cavity with metal, necessitating the use of draft for the release of the pattern. The "sand" is bonded temporarily by clay (green clay or dry sand mold), sodium silicate (carbon dioxide, or CO_2, process), cement, or vacuum. In principle, all metals can be cast by these techniques provided the sand is refractory enough. Because cores are also destroyed upon completion of solidification, great shape complexity is possible. As a broad rule, a mold made with firmer bond allows the use of finer sand, the production of smaller parts with thinner walls, smaller cores, and greater surface detail. The low thermal conductivity of sands favors complete filling.

Costs can vary greatly. A great advantage of sand casting is that investment can be kept to a minimum with the use of wood patterns and hand molding, thus allowing even prototyping of one or a few parts, albeit at the expense of higher labor costs and greater skill. At the other end of scale, mass production with reduced labor cost and skill is possible with a substantial investment in sand preparation, distribution, mechanized molding, and sand reclamation. The ratings in Table 1 reflect these variables; the lowest-cost equipment and pattern will generally go with highest labor cost, skill requirement, and lowest productivity. The range of products is very broad, from small engine blocks to large machine-tool bases.

Resin-Bonded Sand Casting. The sand is bonded by a thermosetting polymer (resin). When the mold is a relatively thin shell, formed around a heated metal pattern, the process is referred to as shell casting. Pattern costs may be higher, but surface finish is better and tolerances are tighter.

Plaster Casting. In Table 1, this term stands for processes in which a fine-grained refractory is poured as a slurry over the pattern and is then allowed to set (plaster molding for nonferrous alloys) or is fired (ceramic-mold casting for higher-melting alloys). With proper control of refractory and shrinkages, complex parts can be cast with thin walls, close tolerances, and good surface finish, but at a higher cost.

Expendable-Pattern Casting. Patterns are melted out prior to casting in the ancient yet thoroughly modern lost-wax process for ceramic molds, or they burn up and evaporate during pouring, as does polystyrene foam in the lost-foam process for sand molds. Thus, the mold can be in one piece, eliminating parting lines. Draft is reduced to that required for releasing the pattern from its mold. Preheating of ceramic molds allows thin walls and small cores, as in superalloy turbine blades with internal cooling passages. By assembling several pattern pieces, shape complexity can reach the highest level of any manufacturing process, as exemplified by bronze castings of statues.

Permanent-Mold Processes

In permanent-mold casting processes, the mold (die) is made of a metal, and the shape is limited to geometries that can be released from two or more mold sections. Internal cavities and holes can be formed with retractable cores, as in casting aluminum-alloy pistons, but at the expense of higher cost and greater complexity of machinery. Even greater complexity is achievable when expendable cores are inserted. In all permanent-mold casting, porosity can be a problem with alloys of dendritic solidification pattern.

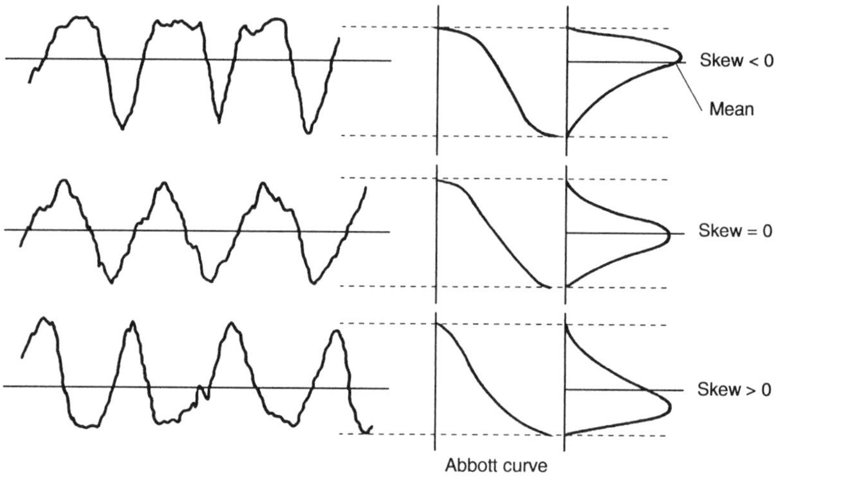

Fig. 4 Surface profiles, bearing area, and symmetry about mean for surfaces of similar average roughness. Source: after Ref 21

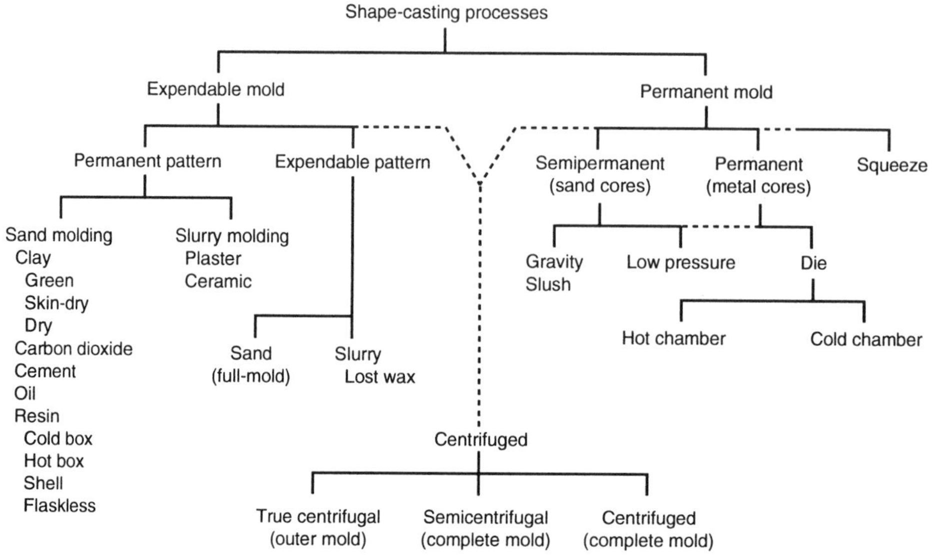

Fig. 5 General classification of shape-casting processes. Source: Ref 16

The metal mold can be fed by gravity (permanent-mold or gravity die casting) or under pressure (low- and high-pressure die casting). The latter ensures filling of thinner sections, especially with lower-melting alloys (Fig. 2). Highly complex shapes, such as office-machine housings and transmission cases are cast to close tolerances. The use of higher-melting alloys is limited by rapid wear of the die. Direct transfer of the melt (hot-chamber die casting) is also limited to lower-melting alloys (aluminum as the maximum) yet allows reducing or eliminating sprue and feeders. Porosity and voids are reduced or eliminated if high pressure is applied just prior to final solidification, or if the alloy is formed at a temperature between the liquidus and solidus. Such mushy-state forming (variously known as squeeze casting, melt forging, or semisolid metalworking) represents a transition between casting and hot forging processes.

It should be noted that when "All" is given for shape complexity in Table 1, the closed shape T5 can be cast only if the core is left in the casting, unless the part is made from a premeasured amount of melt that is swirled around the closed mold (slush casting, used also for complex shapes such as in casting small sculptures).

Design considerations are given in the article "Design for Casting" in this Volume.

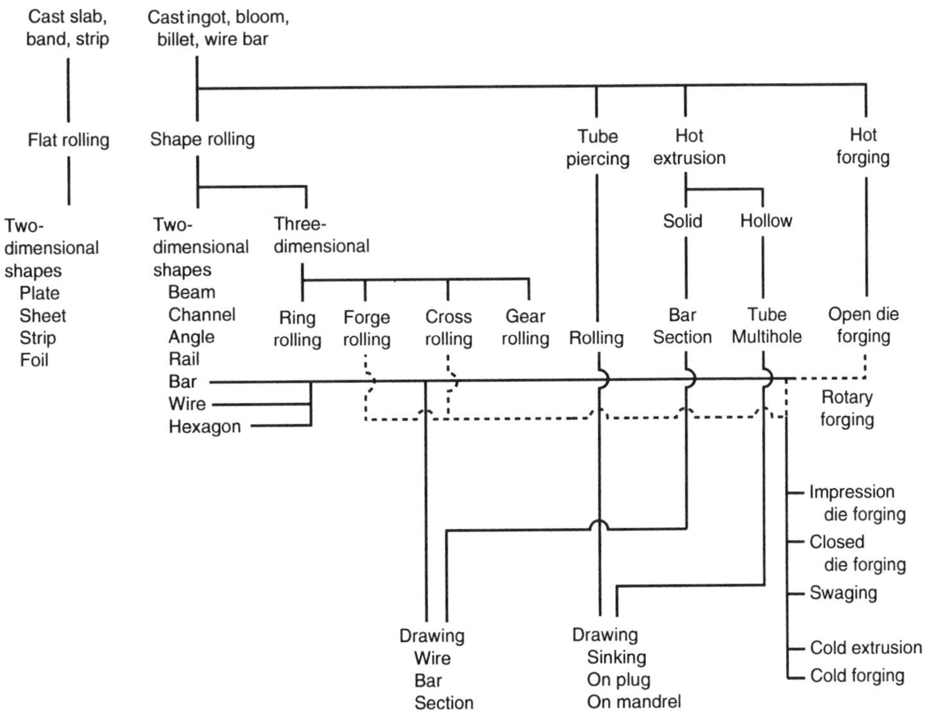

Fig. 6 General classification of bulk deformation processes

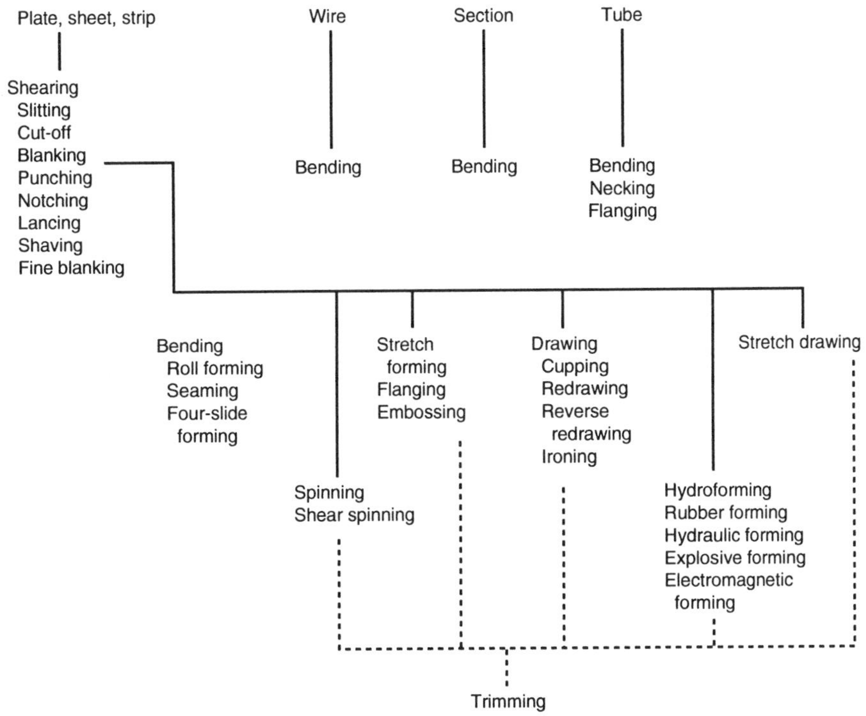

Fig. 7 General classification of sheet metalworking processes

Deformation Processes

All plastic deformation processes exploit the ability of metals and alloys to change shape by plastic flow. The ability to deform without fracture is highly variable. Bulk deformation processes (Table 2) impose material flow in all directions, and the ability to deform is expressed as workability; in sheet metalworking processes deformation is essentially two-dimensional, and limits of deformation are referred to as formability. Both are discussed in the article "Relationship between Materials Selection and Processing" in this Volume. As indicated there, it is important to recognize that the limits of plastic deformation are the outputs of a system and these are greatly affected by the stress state prevailing in the process. If stresses are highly compressive (in other words, if a hydrostatic pressure component prevails), even a material of low workability can be deformed to a significant extent. Thus, process selection is based on a knowledge of the workability or formability of the material and its variation with stress state, strain rate, and temperature. Because most processes involve sliding contact between the workpiece and a tool or die, friction affects material flow, die pressures, and force and energy requirements. In most instances lubricants are chosen to minimize friction, but in some cases control of (and not necessarily low) friction is the aim. Lubricants also fulfill the important functions of reducing die wear, cooling (or temperature control), and control of the surface topography of the product (Ref 19).

Bulk Deformation Processes

The starting material in bulk deformation processes is a slab, ingot, billet, and so forth, produced by casting into stationary molds or by continuous casting techniques (Fig. 6). Primary deformation processes such as hot rolling, tube piercing, extrusion, open-die forging are then used for converting the cast structure. The product may be suitable for immediate application, but in many cases it serves as the starting material for another deformation process, the so-called secondary deformation processes, such as drawing, hot and cold forging (Fig. 6), and sheet metalworking (Fig. 7).

Table 2 General characteristics of bulk deformation processes

Characteristic	Deformation process: Hot forging, Open die	Hot forging, Impression	Hot extrusion	Cold forging, extrusion	Shape drawing	Shape rolling	Transverse rolling
Part							
Material (wrought)	All	All	All	All	All	All	All
Shape(a)	R0–3; B; T1, 2; F0; Sp6	R; B; S; T1, 2, 4; (T6, 7); Sp	R; B; S; SS; T1, 4; Sp	Same as hot	R0; B0; S0; T0	R0; B0; S0	R1–2, 7; T1–2; Sp
Size, kg	0.1–200,000	0.01–100	1–500	0.001–50	10–1000	10–1000	0.001–10
Minimum section, mm	5	3	1	(0.005) 1	0.1	0.5	1
Minimum hole diameter, mm	(10) 20	10	20	(1) 5	0.1		
Surface detail(b)	E	C	B–C	A–B	A	A–B	A–C
Cost							
Equipment(b)	A–D	A–B	A–B	A–C	B–D	A–C	A–C
Die(b)	F	B–C	C–D	A–B	C–D	A–C	A–C
Labor(b)	A	B–D	B–C	C–E	C–E	C–E	C–E
Finishing(b)	A	B–C	C–D	D–E	E	E	D–E
Production							
Operator skill(b)	A	B–C	C–E	C–E	D–E	B	B–C
Lead time	Hours	Weeks	Days–weeks	Weeks	Days	Weeks	Weeks–months
Rates, pieces/machine	1–50 per h	10–300 per h	10–100 per h	100–10,000 per h	10–2000 m/min	20–500 m/min	100–1000 per h
Minimum quantity or length (m)	1	100–1000	1–10	1000–100,000	1000 m	50,000 m	1000–10,000

(a) From Fig. 1. (b) Comparative ratings with A indicating the highest value, E the lowest (see example in Table 1). Values in parentheses may be obtained by special techniques. Source: Ref 16

Much primary working is conducted at elevated temperature, in the hot-working temperature range (typically above one-half the melting point on the absolute temperature scale) where dynamic recovery and recrystallization ensure low flow stress and high workability. Because recovery processes take time, flow stress σ_f is a function of strain rate $\dot{\varepsilon}$:

$$\sigma_f = C\,\dot{\varepsilon}^m \qquad \text{(Eq 1)}$$

where C is the strength coefficient (decreasing with increasing temperature) and m is the strain-rate sensitivity exponent. A high m value means that an incipient neck becomes stronger and spreads to neighboring material, allowing more deformation in tension. This is important also for glasses and polymers, in which m can reach unity (Newtonian flow). In some very fine-grained metals the value of m may reach 0.4 or 0.5, but only at very low strain rates and in a limited temperature range, and then superplastic deformation is possible to large strains and with low stresses.

Cold working produces better surface finish, can impart higher strength, and allows thinner products, but at the expense of higher forces. Flow stress is now primarily a function of true strain ε:

$$\sigma_f = K\varepsilon^n \qquad \text{(Eq 2)}$$

where K is the strength coefficient and n is the strain-hardening exponent. A high n-value allows more tensile deformation before localization of a neck. The value of K increases and n decreases in the course of cold working; thus ductility is reduced and forces increase. Intermediate (process) anneals can restore ductility and reduce forces, but at an additional cost. Warm working in an intermediate temperature range can combine some of the benefits of hot and cold working. Many processes are capable of producing net-shape or near-net-shape products, and Table 2 refers to these. Limits of size and web thickness are often set by the pressure increase due to friction at high width-to-thickness ratios (Fig. 2). Design rules for components are given in the article "Design for Deformation Processes" in this Volume.

Rolling is the most important primary working process, practiced in highly specialized, high-productivity, capital-intensive plants. The products of flat hot rolling are relatively thick plate, sheet, and strip, typically down to 1.6 mm (0.060 in.). Cold rolling is capable of producing the thinnest of all metal products (strip and foil, to a few micrometers or a fraction of a thousandth of an inch) to very tight tolerances and controlled surface finish. Most of these products are further processed in sheet metalworking operations (Fig. 7). Hot-rolled bar and wire serve as starting materials for other bulk deformation processes or machining. Shapes are mostly hot rolled or hot extruded for structural applications, although cold-rolled precision shapes fill an important niche. All these processes yield products of two-dimensional configuration, in often very long lengths, and are capable of producing high-quality (close-tolerance) products at very high rates and low cost. Closer to final application are three-dimensional products of hot and cold ring rolling (rotating rings of jet engines, ball-bearing races), forge rolling for complex shapes (cutlery), gear rolling, and cross (transverse) rolling for shapes of axial symmetry (preforms for forging double-ended wrenches). Seamless tubes are made by hot piercing or hot extrusion and are further reduced by special rolling processes for immediate use (for example, in oil fields) and for further working, including cold reduction and drawing.

Drawing. A rolled preform is drawn through a converging gap, usually cold. Round bars and wires as well as two-dimensional sections of often complex cross-sectional shape are drawn through stationary dies or, less frequently, rollers. Tubes may be simply drawn through a die (tube sinking), but more often wall thickness reduction is achieved by drawing over a plug or a bar (also called mandrel, from which the tube is released by cross rolling). Die costs are relatively low and production rates can be high, making for low-cost, high-precision products of wide size and wall-thickness range.

Extrusion. A billet held in a container is pushed against a die (direct or forward extrusion), or the die penetrates the billet (indirect or reverse extrusion), either hot or cold. Friction over the container and die surfaces has a decisive influence on material flow and die pressures.

Only lower-melting alloys, including aluminum alloys, can be extruded hot without a lubricant, and then deformation occurs with shearing the alloy at the die-to-billet interface. Two-dimensional shapes of great complexity and thin walls (e.g., architectural extrusions) can be made at a relatively low die cost. In the absence of a lubricant, divided material strains can be reunited and welded in a bridge-type die, allowing extrusions with one or more closed cavities, in a very wide size range (multihole tubes are extruded with walls as thin as 0.25 mm, or 0.010 in.). Hot extrusion of higher-melting alloys requires a high-temperature lubricant, often a glass; shapes are more limited and wall thickness is greater. The purpose may be that of producing a two-dimensional semimanufactured product or, if part of the billet is retained at the starting diameter, that of making near-net-shape three-dimensional products such as large valves and fasteners. The skill of the die designer is called upon to ensure a sound product.

Cold extrusion is always conducted with an effective lubricant (often with the lubricant superimposed on or reacted with a conversion coating on the metal) and is normally used to create near-net-shape products such as fasteners, automotive components, and so forth. Impact extru-

Table 3 General characteristics of sheet metalworking processes including bar and tube bending

Characteristic	Forming process: Blanking	Bending	Spinning	Stretching	Deep drawing	Rubber forming
Part						
Material (wrought)	All	All	All	All	All	All
Shape(a)	F0–2; T7	R3; B3; S0, 3, 7; SS; T3; F3, 6	T1, 2, 4, 6; F4, 5	F4; S7	T4; F4, 7	Same as blanking, bending, deep drawing
Maximum thickness, mm	>10	>100	>25	>2	>10	2
Minimum hole diameter, mm	½–1 thickness				>3	50 (for h = 1 mm Al)
Cost						
Equipment(b)	B–D	C–E	B–D	B–C	A–C	A–C
Die(b)	B–D	B–E	B–D	A–C	A–B	C–D
Labor(b)	C–E	B–E	B–C	B–E	C–E	A–D
Finishing(b)	D–E	D–E	D–E	C–E	D–E	C–E
Production						
Operator skill(b)	D–E	B–E	B–C	B–E		C–E
Lead time	Days	Hours–days	Days	Days–months	Weeks–months	Days
Rates, pieces/h	10^2–10^5	10–10^4	10–10^2	10–10^4	10–10^4	10–10^2
Minimum quantity	10^2–10^4	1–10^4	1–10^2	10–10^5	10^3–10^5	10–10^2

(a) From Fig. 1. (b) Comparative ratings, with A indicating the highest value, E the lowest (see example in Table 1). Source: Ref 16

sion is a term employed for the cold extrusion of thin-walled products such as toothpaste tubes.

Forging. The aim in most forging operations is that of producing a part as close to the finished shape as possible.

Open-die forging uses simple tools in a programmed sequence of basic operations (upsetting, drawing out), mostly in the hot-working temperature range, and the products (ranging from the one-off products of the blacksmith to huge turbine rotors) usually require finishing by machining. Rotary forging and swaging on special-purpose machines produce parts of axial symmetry to much tighter tolerances (axles, gun barrels).

Hot impression-die forging (sometimes termed closed-die forging) shapes the part between two die halves; thus productivity is increased, albeit at the expense of higher die costs. Excess metal is allowed to escape in the flash; thus pressure is kept within safe limits while die filling is ensured. More complex shapes, thinner walls, and thinner webs may necessitate forging in a sequence of die cavities, as for connecting rods and crankshafts. Die design calls for a thorough knowledge of material flow and is greatly aided by computer models and expert systems. At high width-to-thickness ratios friction sets a limit to minimum web thickness (Fig. 2) that decreases with effective lubrication. In true closed-die forging the material is trapped in the die cavity.

With dies heated to or close to forging temperature (isothermal or hot-die forging), cooling is prevented and thin walls and webs can be produced, provided the die material is stronger than the workpiece material at the temperatures and strain rates prevailing in the process. This is relatively easy for aluminum alloys (airframe parts); low press speeds (low $\dot{\varepsilon}$) help to keep stresses and forces low. Titanium alloys and superalloys can be forged in the superplastic state (jet-engine fan blades and turbine disks).

The sequence of operations can be carried out by moving the heated end of a bar through the die cavities in an upsetter, achieving high production rates. Mechanized transfer between cavities in conventional presses is also possible. In all impression-die forging, die design calls for considerable knowledge and die cost can be high (Table 2), but the product often has superior properties because material flow can be directed to give the best orientation of the structure relative to loading direction in the service of the part.

Cold forging is related to cold extrusion and, when a complex shape is to be formed in a single step, requires special lubricants, often with a conversion coating, as in making spark-plug bodies. Alternatively, the shape is developed by moving the bar or slug through a sequence of cavities, using a liquid lubricant. Cold forging is often combined with cold extrusion. It is the preferred process for mass producing near-net-shape parts such as bolts, nuts, rivets, and many automotive and appliance components.

Sheet Metalworking Processes

The starting material is a flat-rolled product (hot-rolled plate, hot- or cold-rolled sheet or strip); hence there are no limitations on width-to-thickness ratios: sheet metalworking is the process of choice for relatively thin products. Deformation is mostly by tension or tension-compression, and limits are set by the formability of the material (see the article "Relationship between Materials Selection and Processing" in this Volume) and only rarely by force or die pressure. Most processes are conducted cold; only heavy-gage material is formed hot. Major process groups are shown in Fig. 7, and characteristics of major processes are given in Table 3. Forming of wires, sections, and tubes is governed by the same rules as the forming of sheet; therefore, it is included here.

Shearing. The purpose of shearing is controlled separation to: slit narrower strips, separate parts (cut-off, parting), cut out a part (blanking), create a hole (punching), produce notches, and so forth. The basic shearing process relies on cracks penetrating from both sides of the sheet. With proper clearance between punch and die, the cracks meet but the cut surface is slightly inclined and rough. Tool wear results in the formation of a ragged burr that causes stress concentrations and may lead to fracture during further forming or in service. In fine blanking, crack formation is suppressed by the hydrostatic pressure generated by impingement rings, and a smooth, perpendicular surface, suitable as the contact surface on a net-shape part, is produced, but at higher die and equipment cost.

Bending. In addition to sheets (or blanks), sections, tubes, and wires are bent to a great variety of shapes. Accuracy is improved with proper control of or compensation for springback. Limits are imposed by local necking, fracture, or—for aesthetic reasons—by orange-peel formation. Four-slide machines are capable of performing several operations on three-dimensional parts at high production rates. Roll forming allows the mass production of two-dimensional corrugated sheet, architectural sections, and lock-seam and welded tubes. With proper design, close tolerances can be achieved. Such products are in competition with extruded, shape-rolled, and cold-drawn products. Plates and sections are often bent in three-roll benders into complete circles. Tube bending presents the challenge of preventing collapse of the tube wall; complex shapes are now bent under computer control.

Bending a sheet along a curved line is termed flanging. Circular or other close-shaped flanges (collars) are mass produced in preparation for joining tubes and fasteners to sheet, as in heat exchangers. Limits are set by fracture in a stretch flange and by buckling in a shrink flange. Related processes are flanging and necking of tubes and cans, as on beverage cans.

Stretch Forming. The blank is firmly clamped at much or all of its circumference, and the shape is developed by the penetration of a punch, at the expense of thickness. Limits are set by necking or fracture, as given by the tension-tension side of the forming limit diagram (see Fig. 6 in the article "Relationship between Materials Selection and Processing" in this Volume). Small quantities can be produced with only a punch, although at higher material cost, as in the aircraft industry. Mating dies with carefully designed blankholders are more economical for the large quantities typical of the automotive and appliance industries. An important aspect is control of or compensation for elastic recovery (springback), which distorts the shape. Embossing is a highly localized form of stretch forming.

Drawing. In contrast to stretch forming, the shape is developed by drawing material into the die, and average thickness is approximately preserved. The flange is in circumferential compression and forms wrinkles, unless the blank is thick relative to the blank diameter or is adequately restrained with a blankholder (although in some applications, such as aluminum dinner trays, wrinkling has been exploited to increase stiffness of the part). Fracture occurs at the base or in the wall if stresses exceed the strength of the partly formed cup; therefore a limit is set to the attainable reduction in diameter or, as more frequently expressed, a limiting draw ratio (LDR = diameter of blank/diameter of cup) is reached. The LDR is a system property, affected by material (especially *r*-value; see the article "Relationship between Materials Selection and Processing" in this Volume), die design, lubricant, and so forth. Cups deeper than allowed by the LDR are produced by redrawing or reverse redrawing (food cans) or by thinning the wall in ironing (beverage cans, cartridge cases).

Stretch Drawing. Parts of nonaxial symmetry are mostly produced by a combination of stretching and drawing. The relative contribution of each is controlled by blank design, blankholder pressure, or the insertion of drawbeads into the blankholder surface. Strain distribution can be analyzed by imprinting a grid on the blank surface, and remedies for necking or fracture can be found with reference to the forming limit diagram of the material. Many deeper, complex-shape sheet metal parts of automobiles are made by these techniques. Great advances are being made in die design aided by computer modeling of material flow. Hydraulic forming allows stretching or drawing with only a punch or a die; the other die element is replaced by a fluid restrained by a diaphragm, reducing die costs. Rubber forming does the same, but with an elastomeric cushion.

Special processes are those that do not fit conveniently into the above categories, but are subject to similar limitations imposed by material properties.

Spinning can produce complex axially symmetrical shapes with programmed tool movement and a relatively simple inner die (form). Wall thickness can also be reduced (shear spinning) within the workability limits of the material (see Fig. 5 in the article "Relationship between Materials Selection and Processing" in this Volume).

Special Forming Processes. Explosive forming is used for bulging or reducing tubes and for forming shapes similar to those produced by stretching. Electromagnetic forming has similar applications, primarily to tubular parts. Hydroforming of tubes utilizes internal pressure to expand the tube; axial compression feeds material into the deformation zone and prevents excessive thinning. Originally used for making copper T-fittings, hydroforming is now growing into a mass production process for automotive components such as engine mounts.

Replacing one of the mating dies with a rubber cushion (rubber forming) or fluid contained by a rubber diaphragm (hydroforming) reduces die costs, but usually at the expense of greater material consumption. Greater depth and more complex shapes can often be obtained than with hard tooling.

Design guidelines are given in the article "Design for Deformation Processes" in this Volume.

Powder Processing

Discussion here is restricted to metal powders (powder metallurgy, P/M); ceramic powders are dealt with later in this article. Powders of metals and alloys are processed for several reasons. Most importantly, parts of complex shape, close tolerances, controlled density, and controlled (and often unusual) properties can be produced.

Powder processing involves a sequence of operations (Fig. 8). The powder is produced by various techniques: reduction of an oxide, thermal decomposition, electrolysis, hydrometallurgy, or breaking up (atomizing) a melt. Particle size is controlled during powder production or by comminution; size fractions are separated by sizing. After cleaning (and possibly annealing), alloying elements and lubricants are added by blending. There are several process routes to arrive at finished parts.

Most parts are produced by mixing the powder with a lubricant and pressing in a die (rigid tool). Friction between particles and on the die wall results in a gradual drop in density away from the punch face. Dies with several moving elements can become very complex, but allow manufacture of parts of greatly varying web thicknesses. The "green" compact has relatively low density and very low strength. Isostatic pressing in a deformable jacket gives better density distribution for complex shapes. Roll compaction is advantageous for strip and simple two-dimensional shapes. Sintering at hot-working temperatures increases density and strength by establishing a metallurgical bond, although some porosity remains and mechanical properties are below that of parts produced by deformation processes.

Shrinkage is considerable and must be taken into account in design. Properties are, nevertheless, adequate for many applications, such as lightly loaded gears. Hot restriking closes pores and improves properties. Hot forging with substantial deformation of green or sintered preforms produces near-net-shape parts of full density, free of segregation, and of often superior properties, such as steel connecting rods and superalloy jet engine disks. Hot rolling and extrusion are practiced for making P/M tool steels of very fine grain size and carbide distribution.

Adding a polymeric binder to very fine (around 10 μm) powder allows injection molding at low pressure. Upon heating, the binder is removed (debinding) and sintering establishes the metallurgical bond. Small parts of thin walls and com-

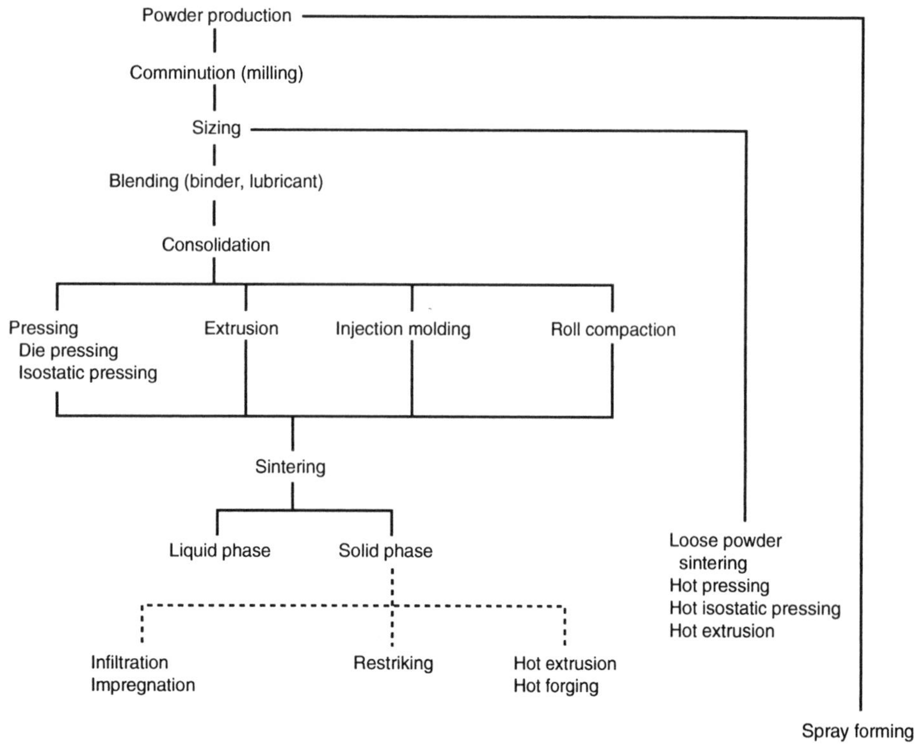

Fig. 8 General classification of powder metallurgy processes

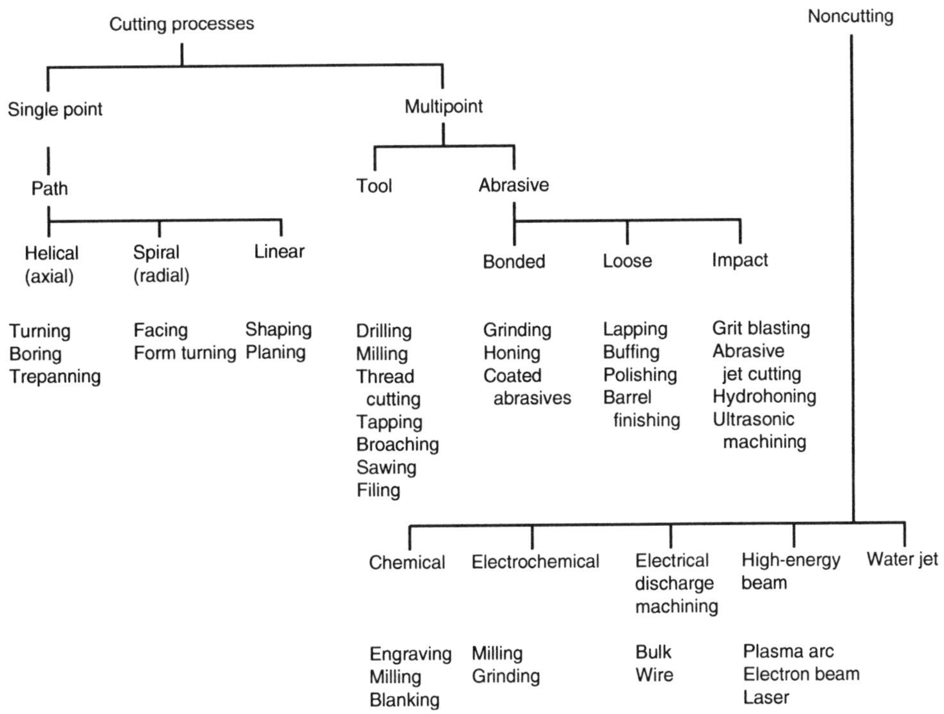

Fig. 9 General classification of metal removal processes

plex shape (even with cross-cored holes) can be produced at high density, but the high shrinkage must be taken into account in design.

Consolidation and sintering take place simultaneously during hot pressing in dies, hot isostatic pressing (HIP) in deformable jackets, and hot extrusion. Loose powder may be sintered into simple forms, as for filters. The most direct route is spray deposition of atomized powder, for sheet or finished part.

All P/M processes call for considerable knowledge and require substantial investment in equipment. Die costs increase with shape complexity and, when in competition with other processes, P/M usually becomes economical only at higher production volumes. Powder metallurgy is irreplaceable, however, when parts with unique properties are to be made for the following reasons:

- Porosity is intentionally kept high and interconnected for metal filters
- Composite structures are formed by infiltrating a porous compact with a different metal (e.g., the impact strength of steel parts is increased by infiltrating copper; self-lubricating bearings, welding electrodes, and electrical contacts are made by infiltrating a skeleton with a different metal)
- Impregnation with oil provides reservoirs of lubricant.

Carbide powders are cemented with 3 to 15% metal such as cobalt by liquid-phase sintering for hard, temperature-resistant tooling. Powder metallurgy is often the most favorable or feasible processing route for refractory metals and alloys.

Design guidelines are given in the article "Design for Powder Metallurgy" in this Volume.

Machining Processes

Parts produced by casting, deformation, and powder processing techniques frequently require some metal removal to create the final shape, meet tolerances, or obtain the requisite surface finish. Machining (Table 4) may also be an economical alternative to create the shape from a semifabricated product such as bar or plate. Metal may be removed by chip formation or by less conventional, noncutting techniques (Fig. 9).

Metal Cutting

In metal cutting, excess material is removed by a controlled chip-forming process. The relative ease or difficulty with which material is removed is termed machinability. An alloy of good machinability can be cut at higher metal removal rates with acceptable tool life and will give better surface finish at lower cost. Some processes are more suitable for producing tight tolerances and finer surface finish, but the general guidelines should be interpreted with caution: in any given process, a hard, leaded brass will be easier to cut and will give better finish than a soft copper or a hard superalloy. Because the shape is generated by removal of excess material, there is basically no limit on section thickness except as influenced by deflection of the part under cutting forces.

Table 4 General characteristics of machining processes

	Machining process						
Characteristic	Lathe turning	Automatic screw machine	Shaping, planing	Drilling	Milling	Grinding	Honing, lapping
Part							
Material	All(a)	All(a)	All(a)	All(a)	All(a)	All	All
Preferred material	...	Free-machining	...	Free-machining	Free-machining	Hard	Hard
Shape(b)	R0–2, 7; T0–2, 4, 5; Sp	Same as turning	B; S0–2; F0	T0	B; S; SS; F0–4, 7	Same as turning, shaping, milling	R0–2; T0–2, 4–7; F0–2; Sp
Minimum section, mm	<1 diameter	<1 diameter	<2	0.1 (hole diameter)	<1	<0.5	<0.5
Surface configuration	Axially symmetrical	As turning	Straight generatrix	Cylindrical	Three-dimensional	All (mostly flat, axially symmetrical)	Flat, cylindrical, three-dimensional
Cost							
Equipment(c)	B–D	A–C	B–D	D–E	A–C	A–C	B–D
Tooling(c)	D–E	A–D	D–E	D–E	A–D	B–D	A–E
Labor(c)	A–C	D–E	B–D	B–D	A–B	A–E	B–D
Production							
Operator skill(c)	A–C	D–E	B–D	B–E	A–B	A–D	C–E
Setup time(c)	C–D	A–C	C–D	C–E	A–C	B–D	C–E
Rates (pieces/h)	1–50	10–500	1–50	10–500	1–50	1–1000	10–1000
Minimum quantity	1	500	1	1	1	1	1

(a) Except most ceramics. (b) From Fig. 1. (c) Comparative ratings with A indicating the highest value, E the lowest (see example in Table 1). Source: Ref 16

Table 5 General characteristics of fusion welding processes

	Welding process						
	Arc welding						
Characteristic	Shielded metal	Flux cored	Submerged	Gas-metal	Gas-tungsten	Electron beam	Oxyacetylene
Part (assembly)							
Material	All but Zn	All steels	All steels	All but Zn	All but Zn	All but Zn	All but refractory metals
Preferred material	Steels	Low-C steels	Low-C steels	Steels; non-heat-treatable Al; Cu	All but Zn	All but Zn	Cast iron, steels
Thickness, minimum, mm	(1.5) 3	1.5	5	0.5	0.2	0.05	0.6
Single pass maximum	8–10	3–6	40	5	5	75	10
Multiple pass maximum	>25	>15	>200	>25	>6		>20
Unequal thickness	Difficult	Difficult	Very difficult	Difficult	Difficult	Easy	Difficult
Distortion(a)	A–B	A–C	A–B	B–C	B–C	C–E	B–D
Jigging needed	Minimum	Minimum	Full	Variable	Variable	Full	Minimum
Deslagging for multipass	Yes	Yes	Yes	No	No		No
Current							
Type	Alternating or direct	Direct (reverse polarity)	Alternating or direct	Direct (reverse polarity)	Alternating or direct (straight polarity)	...	...
Volts(b)	40 or 70 ↓	40 or 70 ↓	25–55	20–40 or 70 ↓	60–150	30–175 kV	...
Amperes	30–800	30–800	300–2500	70–700	100–500	0.05–1	...
Cost							
Equipment(a)	D	B–D	B–C	B–C	B–C	A	D–E
Labor(a)	A	A–D	B–D	A–C	A–C	A–D	A
Finishing(a)	A–B	A–C	A–C	B–D	B–E	C–E	A
Production							
Operator skill(a)	A	A–D	C–D	A–D	A–D	A–D	A
Welding rate, m/min	(1–6 kg/h)	0.02–1.5	0.1–5	0.2–15	0.2–1.5	0.2–2.5	(0.3–0.6 kg/h)
Operation	Manual	All	Automatic	All	All	All	Manual

(a) Comparative ratings, with A indicating the highest value, E the lowest (see example in Table 1). (b) Down arrow indicates dropping voltage. Source: Ref 16

Because heat generation increases with cutting speed and power consumed (see Table 11 in the article "Relationship between Materials Selection and Processing" in this Volume), the choice of tool material has decisive influence. Great advances in metal removal rates have been made by the introduction of heat-resistant tools, tool coatings, and improvements in machine tools. The number of processes is large but can be reasonably well categorized (Fig. 9).

Single-Point Machining. The tool has a well-defined geometry with only one cutting edge, and the part shape is developed by control of the tool path. In turning on a lathe, the tool path is helical or spiral; hence the machined surface always has axial symmetry. Under manual or computer numerical control (CNC), complex shapes can be produced singly or in batches. The size range is enormous, from tiny shafts to generator rotors weighing hundreds of tons. Mass production with hard-programmed (cam-driven) screw machines may still be economical, although CNC is also being applied. On a planer or shaper the straight tool path creates essentially two-dimensional shapes.

Multi-Point Machining. The tool has two or more cutting edges of well-defined geometry. The tool path is simple in drilling, broaching, thread cutting, tapping, and often also in sawing and filing. It can, however, be complex in milling, and CNC milling machines with five axes of control are capable of producing the widest variety of shapes, including undercut shapes (Table 4), in a wide range of sizes. Multihead milling machines remove more than 90% of the material for wing surfaces of jet aircraft. Metal removal rates can be very high on highly machinable materials such as heat-treated aluminum alloys, using spindle speeds in the tens of thousands of revolutions per minute (high-speed machining). Milling can be an economical choice even for rapid prototyping and for making parts directly from a billet (hogging out airframe components can be competitive with forging).

Abrasive Machining. Chips are formed by random encounters with abrasive grit. Shape control is achieved by binding the abrasive into a grinding wheel, a honing stick, or onto a backing paper or cloth (coated abrasives). Originally used only for finishing to close tolerances and good surface finish, creep-feed grinding rates have increased to where the process can be competitive with chip-forming processes, although disposal of sludge may be a problem. Loose abrasives give dimensional control and surface finish in lapping but only surface finish in buffing, polishing, and barrel finishing. Metal removal with control of surface finish is the aim of impact machining processes.

Noncutting Processes

Noncutting processes are also called nontraditional machining processes and include etching, electrical discharge machining (EDM), and high-energy-beam processes. Common to all these processes is insensitivity to the mechanical properties of the metal, making them excellent candidates for the manufacture of tools and parts from heat-treated steel, superalloys, and other low-machinability materials.

Etching. Metal removal by chemical action (etching) has a long history in art as engraving, but it is a relative newcomer for metal removal (chemical milling) or blanking, as in producing copper circuit boards or large aluminum-alloy aircraft components. Dissolution of metal is accelerated by electrochemical milling or grinding. Surface finish can be very good unless one phase of an alloy resists dissolution.

Electrical discharge machining relies on electrical discharge between the workpiece and a shaped electrode (bulk EDM, indispensable for sinking many forging dies and injection molds) or wire (wire-EDM for two-dimensional and, under multiaxis computer control, three-dimensional shapes). Material is removed by controlled melting and vaporization. Surface finish is less fine than with chemical machining, but production rates are much higher.

High-energy-beam processes melt and vaporize the metal by highly concentrated plasma arc, electron beam (EB), or laser beam. Electron beam machining requires vacuum, which increases cost but also protects the metal.

Design guidelines are given in the article "Design for Machining" in this Volume.

Joining Processes

Some joining processes can be regarded as assembly processes, others properly belong among manufacturing processes. Thus, welding of a tube is clearly manufacturing; brazing a heat exchanger from welded tube is assembly. The same physical principles govern both.

Nonpermanent joints allow easy disassembly. Screw joints are long established; snap fits are enjoying great growth (especially with plastics); and shrink fits can be nonpermanent, mainly with metals. The focus in this article is on methods for creating permanent load-bearing joints (Fig. 10 and Table 5).

Mechanical Joining

No metallurgical bond is established with mechanical joining processes such as riveting, stitching, stapling, and lap seams; hence their design presents special challenges. In some applications they may be used in conjunction with an adhesive bond (rivets in airframe construction) or solder (lap seams in tubes).

Solid-State Welding

When perfectly clean surfaces are brought into intimate contact, interatomic bonds form a joint. Bond strength is greatest when the mating metals are mutually soluble, but good welds can be obtained with dissimilar, not otherwise weldable metals and with highly differing thicknesses.

Cold Welding. Sufficient pressure must be exerted to establish conformance of surfaces. Sliding and deformation accompanied by surface expansion are needed to break up oxides and other adsorbed films. Cold welding processes differ only in the method of providing these conditions. Complex tubular parts such as refrigerator evaporator plates can be made by depositing a parting agent in a pattern to prevent welding; after roll bonding the passages may be inflated.

Diffusion Bonding. Elevated temperature accelerates diffusion and thus the establishment of bonds. Superplastic forming of titanium alloys, combined with diffusion bonding, permits the production of complex shapes.

Forge Welding. As a generic term, it applies to bonding by deformation at hot-working temperature. Large surface extension in hot-roll bonding creates strong bonds for cladding, as in bonding copper-nickel surface layers to a copper core for some U.S. coins (dimes and quarters). Heating is highly localized in high-frequency induction and high-frequency resistance welding. Some melting may take place, but the melt is squeezed out by compression, as in welding tubes from roll-formed strip.

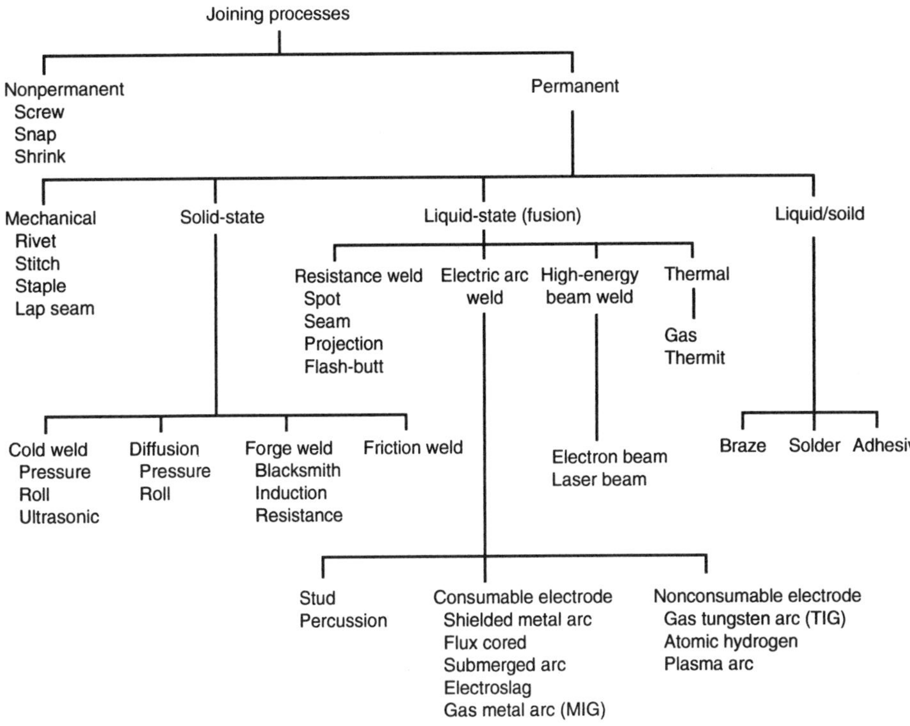

Fig. 10 General classification of joining processes (after Ref 16)

Friction Welding. Heat is produced by friction between a rotating and stationary part; again, some melt may form that is expelled together with oxidized metal. Localization of heat allows welding of dissimilar metals and of very different dimensions (e.g., a thin stem to a large head for an internal combustion engine valve).

Liquid-State (Fusion) Welding

A localized melt pool of the parent metal (autogenous weld) or of the filler and parent metal is formed and allowed to solidify. The weld metal has a cast structure of often complex composition, and if the temperature range of solidification is wide, it is prone to cracking in the presence of tensile stresses. Adjacent parent metal is affected by the rapid heating-cooling sequence, and properties in this heat-affected zone (HAZ) may deteriorate. Weldability is, therefore, highly specific to alloys (see the article "Relationship between Materials Selection and Processing" in this Volume). To maintain tolerances, distortion due to localized heating and cooling must be prevented or controlled during welding by the use of jigs and fixtures. Residual stresses lead to distortion after welding and are minimized by stress-relief anneal. Special techniques are used to mitigate problems arising from great differences in mate-

Table 6 General characteristics of ceramics manufacturing processes

	Manufacturing process							
	Pressing							
Characteristic	Hot	Isostatic	Dry	Plastic forming	Slip casting	Injection molding	Rolling	Extrusion
Part								
Shape(a)	R0, 1; B; S; T0, 1; F0–2, 6, 7; Sp	R; B; S; T0–2, 4, 7; F0–4, 6, 7; Sp	R; B; S; T0–2, 4, 7; F0–4, 6, 7; Sp	All but T3, 5, 6, F5	All	All but T5, F5	R0; B0; S0; F0	All 0
Size, kg	0.01–100	0.05–300	0.05–50	0.02–30	0.05–200	0.02–5	Not limited	Not limited
Minimum section, mm	1	1	1	2	1	1	0.25	1
Cost								
Equipment(b)	A–B	A–B	A–C	C–E	B–D	A–B	A–B	A–B
Tooling(b)	A–B	A–D	A–B	C–E	D–E	A–B	A–C	C–D
Labor(b)	C–E	C–E	C–E	B–D	A–C	C–E	D–E	C–E
Production								
Operator skill(b)	D–E	D–E	D–E	A–E	B–D	C–E	C–D	C–D
Lead time	Weeks	Weeks	Weeks	Days	Days	Weeks	Days	Weeks
Cycle time, min	5–10	1–10	1	1–10	5–60	1	Continuous	Continuous
Minimum quantity	10–100	1–10	100–1000	10–100	1	1000	...	...

(a) From Fig. 1. (b) Comparative ratings with A indicating the highest value, E the lowest (see example in Table 1) rial thickness. Several sources of heat may be drawn upon.

Resistance Welding. After the two parts have been pressed together, electric current passes through the joint to heat and melt the interface. Pressure is kept on until solidification of the melt is complete. Spot welding is widely employed in building car bodies using welding robots. Projection welding is commonly used to join small parts. Seam welding is used for making beams and box sections. Flash-butt welding differs in that the current is switched on during the approach of surfaces, drawing an arc; the technique is extensively employed in making rings and chain links from bent sections.

Electric-Arc Welding. In stud welding, an arc is formed between one part and a projection on the other; in capacitor-discharge welding the energy is stored in a capacitor. Parts of different materials and dimensions can be welded. Consumable electrode welding uses a filler rod (electrode). Protection is provided by a coating that forms a slag and gas cover and can also supply alloying elements. Flux-cored wire allows continuous and mechanized operation. The flux is supplied as a powder in submerged-arc welding for horizontal welds, and a resistive slag pool protects the weld zone in electroslag welding of thick plates. Protection is given by an inert gas in gas metal arc (previously referred to as metal inert gas, MIG) welding, suitable for welding in all positions. Autogeneous welds are produced by nonconsumable-electrode or gas tungsten-arc (tungsten inert gas, TIG) welding. Very high temperatures are created in plasma-arc welding. Seam-tracking robots allow automation even when the weld is to follow a complex path.

Thermal Welding. Heat is provided by an oxyfuel gas flame, mostly for manual welding, or by a thermit reaction for joining heavy sections such as rails. Portability is a great advantage.

High-Energy-Beam Welding. Highly concentrated beams of electrons impinge on the weld zone in electron-beam (EB) welding. When the workpiece is enclosed with the gun, the high vacuum required protects the surfaces, but increases cost and lowers production rates. Out-of-chamber welding is also possible. Gas (CO_2) or solid-state (Nd-YAG) lasers have seen increasing application for joining not only difficult metals and delicate parts, but also sheets of different thicknesses for "tailored" blanks used in autobody construction. All high-energy-beam processes have the advantage that the HAZ is small.

Liquid-Solid-State Bonding

Only the filler metal is melted, and joint strength relies on adhesion to the surfaces. Therefore, absence of interface films and wetting of surfaces are absolute requirements. Fluxes, protective atmospheres, or vacuum are applied. Control of gap is critical to ensure filling by capillary action.

Strong joints are formed by brazing with a relatively high-melting alloy, such as copper or silver alloys for steel and aluminum-silicon alloys for aluminum, often in the form of a cladding on one of the surfaces. Hundreds of parts may be simultaneously joined in products such as heat exchangers. Joining with low-melting lead-tin or, more recently, lead-free solders creates a weaker joint that is suitable for many applications, including radiators, water circuits, and electrical connections.

Adhesive Bonding. Structural adhesives are a special class of polymers, finding increasing application not only in the aircraft industry, but also for automobiles and consumer goods. Joint design must ensure the absence of out-of-line stresses that would cause peeling or cleavage. Extensive coverage is provided in *Adhesives and Sealants*, Volume 3 of the *Engineered Materials Handbook* published by ASM International.

Guidelines for joint design are given in the article "Design for Joining" in this Volume.

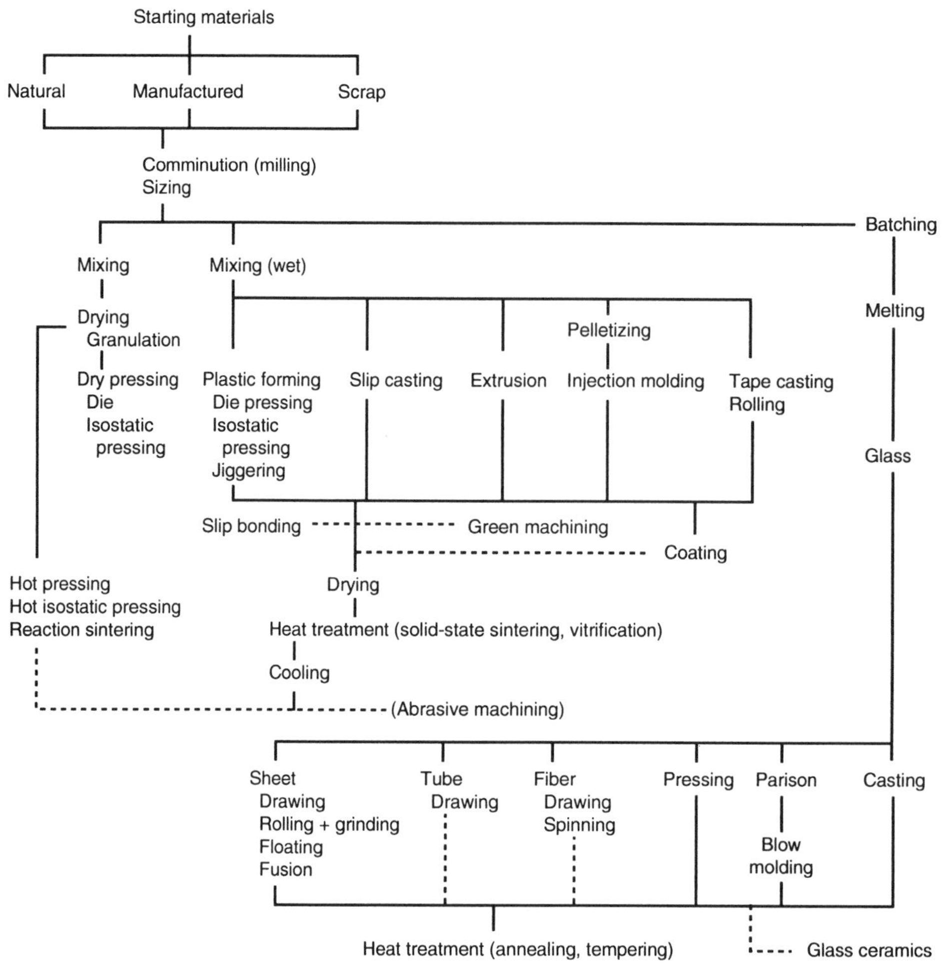

Fig. 11 General classification of ceramics processing

Processing of Ceramics and Glasses

The variety of ceramics is extremely large and the processing routes are highly varied. A simplified overview of the principal methods is given in Fig. 11, and Table 6 shows some characteristics. Starting materials are natural minerals or highly sophisticated manufactured products. After comminution by dry or wet milling, processing sequences diverge for crystalline ceramics and amorphous glasses.

Ceramics Processing

Crystalline ceramics are more or less free-flowing powders. When mixed with binders and lubricants (and possibly after granulation), they can be treated as metal powders are, by dry pressing in dies or by isostatic pressing. With the addition of some water, ceramic powders behave as plastic bodies and can be formed with techniques similar to those used for metals, such as pressing or isostatic pressing. Screw extrusion offers high productivity for two-dimensional shapes. Forming with a rotating template (jiggering) is unique to ceramics. More water makes a slurry (slip) that can be cast inside a plaster mold. As the porous mold absorbs water, a solid layer forms and the excess slurry is drained off to leave a body of reasonably well-controlled wall thickness and high shape complexity (e.g., toilet bowl). Several parts can be joined by slip bonding.

Mixed with resin, the ceramic can be injection molded for near-net shapes or cast into tapes that are rolled up and treated as metal strips (e.g., for ceramic substrates of microelectronic devices).

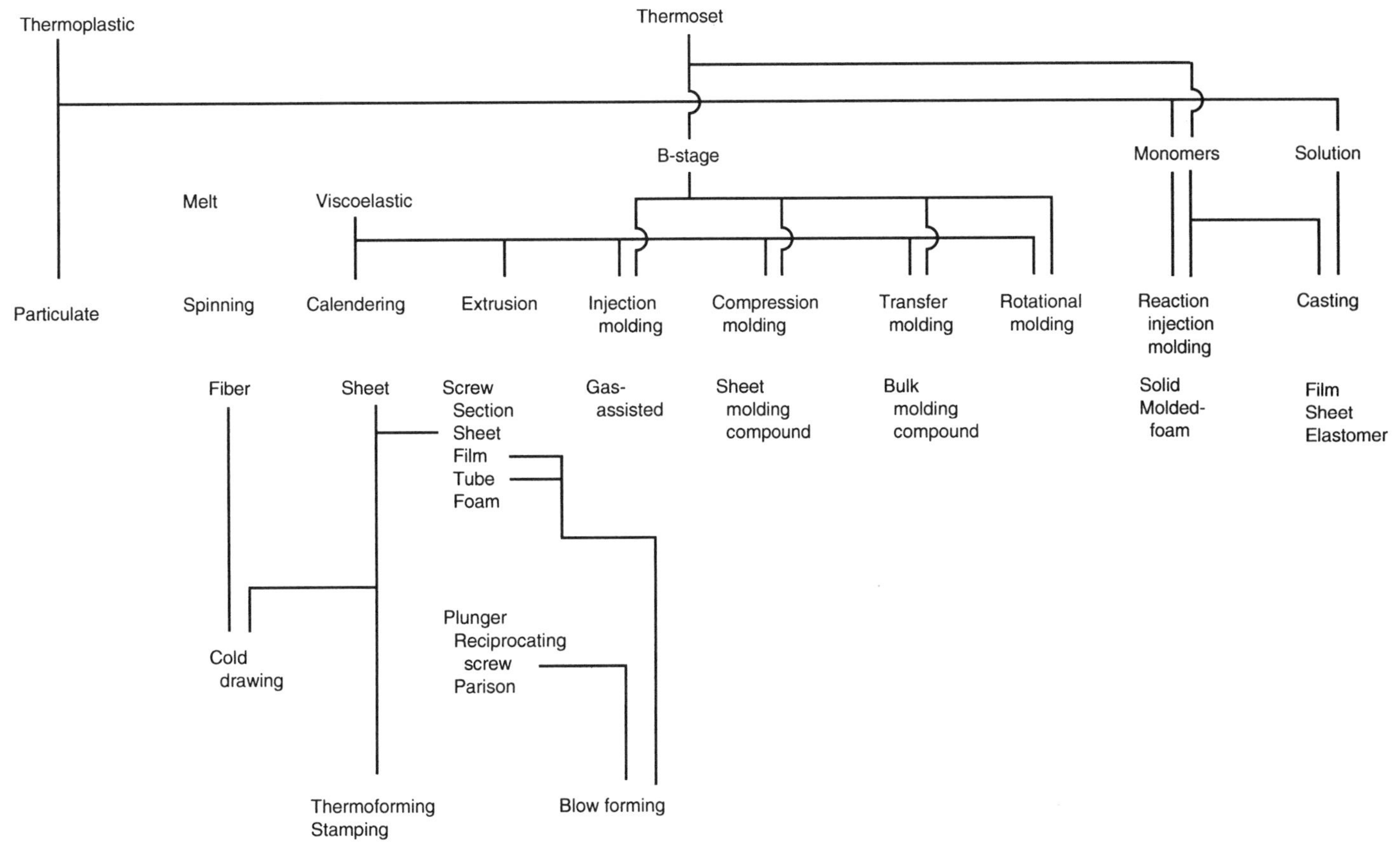

Fig. 12 General classification of polymer processing

Shape details can be developed and closer tolerances held by machining the green body. A crucial stage in all ceramic processing is removal of excess liquid (drying) or binder. During subsequent firing, bonds are established by solid-state sintering and possibly by melting (vitrification) at grain boundaries. Several processing steps are by-passed by HIP, by reaction sintering accompanied by chemical reactions, and by direct deposition of ceramics on metal or ceramic substrates by chemical vapor deposition (CVD) and physical vapor deposition (PVD).

Many ceramics are subject to cracking when exposed to thermal shock and must be cooled in a controlled manner. Removal of binder and buildup of stresses due to nonuniform heating and cooling set a practical limit to the size of ceramic parts. Surface finish and tolerances are controlled by the choice of particle size and by allowing for shrinkage. Tightest tolerances require final shaping by machining. Because ceramics are generally hard, only abrasive machining with a yet harder ceramic (most often, diamond) is feasible. High-energy beam machining, such as laser cutting, drilling, and scribing are widely practiced.

Glass Processing

In contrast to crystalline ceramics, a glass body is a single, giant three-dimensional network of an oxide (most frequently silicon dioxide, SiO_2) modified by other oxides. Glass can be regarded as a metastable, undercooled liquid of extremely high viscosity, exhibiting elastic-brittle behavior. Upon heating above the glass transition temperature (T_g), glass becomes a viscous or somewhat viscoelastic material which, because of its high strain-rate sensitivity (high m-value, Eq 1), can be formed in tension without localization of a neck. This allows processing with free surfaces, resulting in a very smooth finish.

Glass sheet is made by several techniques:

- Drawn for lower-quality requirements (window glass)
- Fusion drawn for higher quality
- Floated on a tin bath for plate and mirror glass

In earlier times, plate glass was rolled and then ground for flatness and surface finish. Continuous tubes and fibers (including fibers for glass-reinforced polymers and for fiber optics) are produced by stretching above T_g. Large quantities of glass wool are made by spinning at higher temperature.

Pressing in dies is practiced for mass production, but the die must be free of surface defects because all detail is faithfully reproduced. The same applies to blow molding of preforms (parisons) into bottles and lamp bulbs; die coatings that generate steam ensure a good surface. Casting is limited to special applications such as telescope mirrors. Special vapor-deposition processes are needed to produce bars of exceptional purity for drawing into optical fiber.

As with all ceramics, cooling rates are critical. Cooling is followed by annealing for the relief of internal stresses and if, required, by quenching to induce compressive surface stresses (physical tempering). Similar results are obtained by replacement of ions in the surface (chemical tempering). Extended heating converts the glass into a fully dense crystalline ceramic (glass ceramics). Glasses are applied also as glazes to ceramics and as enamels to metals. The relatively low hardness of many glasses permits extensive machining by abrasive processes. Chemical machining (etching with hydrofluoric acid, HF) has long been used.

Polymer Processing

Compared with metals, polymers exhibit a much greater variety of behavior that is further complicated by the extensive use of fillers and reinforcing particles in many plastics. There are, nevertheless, many similarities to processes practiced with metals and glasses. Differences arise from the molecular structure of polymers as opposed to the atomic structure of metals. A general overview is given in Fig. 12 and process characteristics are listed in Table 7.

Types of Polymers

Three basic groups may be distinguished: thermoplastics, thermosets, and elastomers.

Table 7 General characteristics of polymer manufacturing processes

	Manufacturing process						
	Molding					Thermoforming	
Characteristic	Compression	Transfer	Injection	Extrusion	Casting	Vacuum	Pressure
Part							
Material	Plastics, glass	Plastics	Plastics	Thermoplastics	Plastics, glass	Thermoplastics	Thermoplastics, glass
Preferred material	Thermosets	Thermosets	...	...	Thermosets	...	...
Shape(a)	All but T3, 5, 6 and F5	All but T3, 5, and F5	Same as transfer	Same as transfer	All	T4; F4, 7	T4, 5; F4, 5, 7
Minimum section, mm	(0.8) 1.5	(0.8) 1.5	0.4 thermoplastic, 1 thermoset	0.4	4	<1	<1
Cost							
Equipment(b)	B–C	B–C	A–C	A–B	D–E	B–D	B–D
Tooling(b)	A–C	A–C	A–C	A–C	B–E	B–C	B–C
Labor(b)	C–E	C–E	D–E	D–E	A–C	B–E	B–E
Production							
Operator skill(b)	D–E	D–E	D–E	D–E	B–E	B–E	A–E
Lead time	Weeks	Weeks	Weeks	Weeks	Days	Days–weeks	Days–weeks
Cycle time, s	20–600	10–300	10–60	10–60	...	10–60	(1)10–60
Minimum quantity	100–1000	100–1000	1000	10,000	1	10–1000	10–1000

(a) From Fig. 1. (b) Comparative ratings with A indicating the highest value, E the lowest (see example in Table 1). Values in parentheses may be obtained by special processing. Source: Ref 16

Thermoplastic Polymers. Most often, polymerization takes place in a primary process. The resulting polymer, usually in the form of particulates (pellets), is formed into final shape by consolidation at elevated temperature, and the shape is fixed by cooling below the T_g. The cycle may be repeated, allowing recycling of these plastics. Deformation is possible by moving polymer chains relative to each other. Above their melting point (T_m), low-molecular-weight polymers exhibit Newtonian or near-Newtonian flow; with increasing molecular weight and shape complexity, flow becomes non-Newtonian but strain rate is still highly dependent on the imposed stress. Strain-rate sensitivity (in terms of Eq 1, a high m-value) permits large tensile deformation. Most processing takes place in the solid state, just below T_m for partially crystalline polymers and above the T_g for amorphous polymers. Flow is viscoelastic and, upon releasing the forming stress, the part would regain its original shape. For the part to retain its shape, it must be cooled well below T_g. Thus, even though compression molding of thermoplastic polymers appears to be very similar to closed-die forming of a metal, cycle times are longer and the part has oriented molecules reflecting the direction of material flow. Properties in the transverse direction are lower, much more so than with metals. Shrinkage is higher too, especially with crystalline polymers, and contraction during cooling can be anisotropic. This, combined with slower cooling of thicker sections, induces residual stresses, and the part will warp.

Thermosetting Polymers. Polymerization takes place during manufacture of the part, creating a single, nondeformable, giant molecule. Polymerization may proceed in stages: the primary processor creates a linear, so-called A-stage resin and adds fillers, colorants, and catalysts to form a B-stage resin that is delivered to the manufacturer of finished products. This resin is thermoplastic for a limited time at elevated temperature, and it can be formed while cross-linking and polymerization take place. Thereafter, no shape change is possible; the plastic cannot be recycled. Shrinkage and differential cooling set up stresses that may lead to fracture in brittle plastics.

Elastomers. The part is again a single giant molecule, but it is formed by cross-linking the molecules of thermoplastic polymers. Once formed, the shape can be changed by loading, but is regained upon unloading.

Effects of Additives, Reinforcements, and Surface Topography

Additives. The properties of polymers are greatly changed by additives, such as antioxidants, flame retardants, plasticizers (for thermoplastic polymers), solvents, dyes, pigments, lubricants, and mold-release agents. From the manufacturing point of view, most significant are fillers. Particulate fillers are added for technical reasons (reduced shrinkage, less distortion, increased mechanical properties, color, etc.). Strictly speaking, reinforced polymers are composite structures, but—in keeping with general usage—plastics with chopped-fiber reinforcement, processed as though they were of a homogeneous mass, are included in this present discussion, while continuous-fiber-reinforced polymers are treated below.

Reinforcements. For reinforcements to have a beneficial effect, they must be wetted by the polymer so that load is transferred to them. Very short fibers and flakes increase the flexural modulus, but at the expense of impact resistance. Longer fibers increase impact strength, but properties become directional if the fiber is free to move during molding. Most importantly for crystalline and semicrystalline thermoplastic polymers, creep is greatly reduced to allow their application at temperatures close to the melting point, to the extent that many underhood engine components are now made of plastics.

Surface topography of plastics is highly variable. When the shape of a pure polymer is given by contact with a die, the die surface finish is faithfully reproduced. This is exploited in giving a controlled finish (pattern, textured surface) for aesthetic and/or technical purposes. The pattern must allow release of the part from the mold; thus typically 1° of draft is given for every 0.025 mm (0.001 in.) of texture depth. Shrinkage results in loss of surface quality. Fillers that reduce shrinkage are beneficial, but reinforcements can stand out of the surface of a high-shrinkage polymer. Surfaces not in contact with a die can be very poor, especially with fiber reinforcement.

Processes Restricted to Thermoplastic Polymers

In these processes, the shape is given while the polymer is heated, and shape is fixed by cooling below T_g.

Powder Processing. Some very long-chain polymers (polytetrafluoroethylene, ultrahigh molecular weight polyethylene) do not soften sufficiently to allow shaping and are compacted from powder by techniques similar to those used in P/M (cold pressing followed by sintering; hot pressing; hot extrusion).

Melt Processing. Techniques are based on viscous flow above T_m for crystalline and well above T_g for amorphous polymers. Highest temperatures are needed for melt spinning of fibers. Calendering is akin to rolling and produces sheet. Extrusion with continuously rotating screws is suitable for sheet, structural sections, tube, and thermoplastic foam products; a reciprocating screw or plunger is used for parisons. Because there is always some viscoelasticity, the material swells as it emerges from the die, and this must be taken into account to achieve the requisite tolerances.

Rubbery-State Processing. Around T_g (in the glass-transition temperature range), polymers are viscoelastic, and their high strain-rate sensitivity (high m-value, Eq 1) allows tensile deformation without localized necking. Thus, fibers are cold drawn for higher strength, sheet is stretched (thermoformed) by vacuum or pressure or it is stamped in matching dies, tubes are blown in diameter to give thin films, and parisons are blown into bottles and other closed-ended products. Unidirectional deformation results in the alignment of molecules, giving high strength in the longitudinal and often poor properties in the transverse direction. Biaxial stretching is needed for balanced properties, as is done for bottles blown from parisons.

Processes for Both Thermoplastics and Thermosets

For thermoplastics, the heated polymer is cooled in the mold. For thermosets, the mold is heated to promote polymerization (unless a catalyst can do it alone); the precursor can be cold or, if it is a B-stage resin, heated to make it flow. Fillers and, especially, fibrous reinforcement result in marked directionality.

Injection Molding. This is similar to die casting of metals, but the plastic is injected by a reciprocating screw or, less frequently, a plunger. The split die must be held closed during injection, and the press size sets a limit to the maximum allowable projected area. Gating and feeding are important, especially with polymers of high shrinkage. Filling of sections with various thicknesses (ribbed sections) can be promoted with the injection of gas (gas-assisted injection molding), which creates hollow ribs and forces plastic into thin sections. The need to retract solid cores limits shape complexity, although low-melting tin-bismuth alloy cores are used in the lost-core process to create shapes as complex as those made with lost-wax (investment) casting of metals, such as air intake manifolds.

Compression molding is similar to closed-die forging in that the plastic is deformed between mating dies. The preform can be shaped to ensure the best orientation of fillers. In sheet molding compounds (SMC) the chopped reinforcing fibers (random or oriented) are sandwiched between polymer paste layers. Large, exposed autobody panels, as well as load-bearing members, are formed with high surface quality. Bulk molding compounds (BMC) use shorter fibers in smaller concentrations.

Transfer Molding. A premeasured amount of polymer is placed in the transfer pot and pressed into the die; shear heating in the orifice helps to fill intricate details and to flow around delicate inserts, as in connectors for electronics.

Rotational Molding. A relative of slush casting. The mold is rotated around two axes to spread the polymer, for small squeeze bulbs to large containers. This is the only process capable of making totally enclosed hollow parts (shape T5).

Reaction Injection Molding (RIM). The reactants are injected into the mold at high velocity through an impingement head that ensures thorough mixing. The plastic polymerizes in the mold; because pressures are low, tooling and machine cost are lower than in injection molding.

Casting of polymers refers to filling a mold by gravity, using prepolymers or monomers that react in the mold, as for sheet casting or encapsulation in polymethyl methacrylate or potting in thermosets. Because pressure and temperature are low, even flexible rubber molds can be used, allowing a greatly increased shape complexity relative to metals. Slush casting of plastisols (i.e., a suspension of PVC particles in a plasticiser) is widespread, often by rotational molding, for industrial as well as consumer products (snow boots).

Machining of Plastics

Even though plastics are generally weaker than metals, this does not mean that they are easier to machine. Special tools are needed to prevent localized melting, smearing, or cleaving in coarse fragments, but machining at high rates with acceptable surface finish is then possible. Reinforced plastics can be hard on tools and carbide or diamond tools are extensively employed.

Manufacture of Composites

By definition, a composite is made up of two or more materials that retain their identity and exhibit an interface between one another. In this sense, all coated and clad metallic products, fiber-reinforced polymers, and many ceramics are composites. In the narrower sense used here, the composite is created by embedding long or continuous fibers in a matrix, which may be a plastic, a metal, or a ceramic. In all cases, the critical issue is bond strength at the interface, so that stresses are transferred to the fiber.

Polymer-Matrix Composites

Filled plastics are composites, as discussed in the preceding section. This section concerns plastics reinforced with long fibers, such as glass, carbon, boron, and other fibers. Manufacturing processes (Fig. 13) differ chiefly in the sequence of operations and in the size and shape complexity permitted. In many instances, metal assemblies formed of several parts can be replaced with a single-piece, integrated polymer-matrix composite part.

Open-Face Molding. In spray molding, the chopped fiber is mixed with the resin and sprayed on the form. In hand lay-up, the fiber is positioned first and then liquid resin is applied, often by hand. Optimal fiber orientation is achieved by filament winding (often under computer control) or by placing a cloth or braided preform. Curing takes place in air or, to reduce voids, under a vacuum bag (a cover). For higher properties, curing is performed in an autoclave at elevated temperature, under pressure or vacuum. While tolerances and surface finish are not the best, there is great freedom in size and shape complexity; for example, for complete boat hulls and filament-wound rocket motor cases. Almost totally closed shapes can be made if the mandrel is dissolved after curing the resin.

Prepreg Resins. As the name implies, the reinforcing fibers are preimpregnated with resin. Fibers may have unidirectional or multidirectional orientation or may be woven into a cloth. Prepregs used in aerospace applications contain carbon fibers in an epoxy or polyimide resin matrix. Sheet molding compounds of polyester or vinyl ester resin, reinforced with glass fiber, are extensively used in automotive and appliance manufacturing. Prepreg sheets can be rolled up into cylindrical or conical tubing. Resin-impregnated fiber bundles (prepreg tow) are useful for filament winding. Prepregs are used also for the construction of relatively low-cost tooling for other polymer processes.

Injection Molding. The process differs from injection molding of reinforced polymers in that the reinforcement is placed in the mold, the mold is closed, and only then is the resin injected. Such resin transfer molding (RTM) requires relatively low pressure, thus foam cores may be placed (and left) in totally enclosed sections of the part. There is greater freedom in shapes than with compression molding or metal stamping as, for example, in making deep spare tire wells. In structural reaction injection molding (SRIM), the components of the resin are mixed in a high-pressure mixing head from which the mixture flows at low

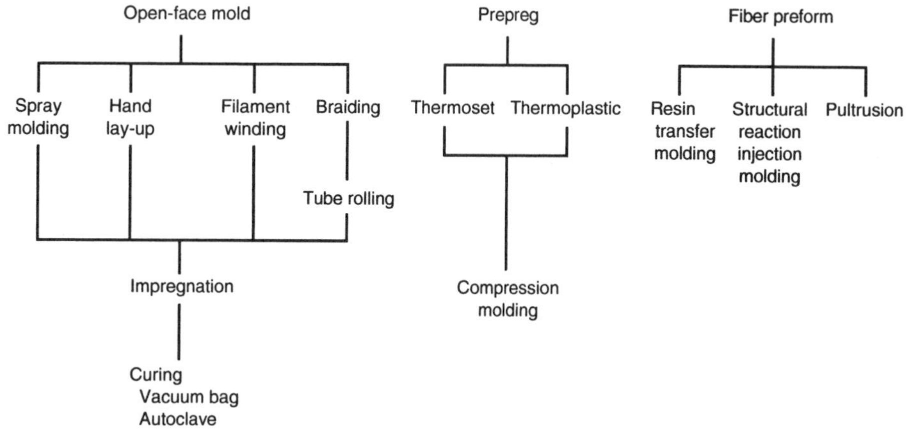

Fig. 13 General classification of processes for polymer-matrix composites

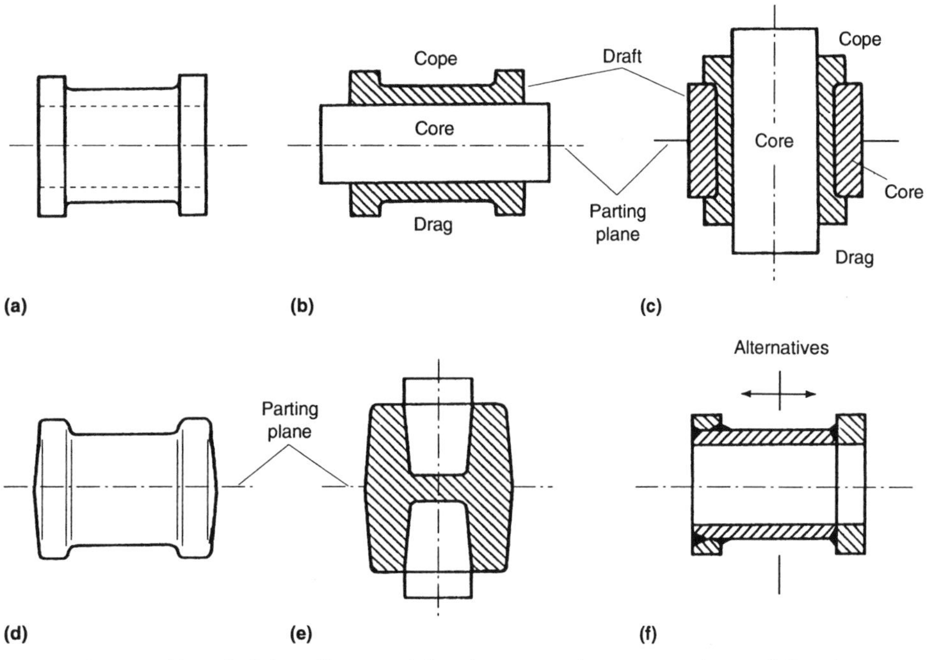

Fig. 14 Some possible methods for making a spool-shaped part. (a) Machining (turning) from a solid bar. (b) Casting in horizontal position. (c) Casting in vertical position. (d) Forging in horizontal position. (e) Forging in vertical position. (f) Welding a ring (left) or flange (right). Source: Ref 16

pressure into the mold. Viscosity is higher than for RTM resins; hence flow distances are limited to about 1 m (3 ft) from the inlet gate.

Pultrusion. A reinforcing fiber bundle (roving) is impregnated with a thermosetting resin and drawn through a heated die to make bars, sections, and tubes. Most often, glass fiber is reinforced with polyester or vinyl ester; for higher performance, aramid or carbon fibers are embedded in epoxy.

Metal- and Carbon-Matrix Composites

Metal-Matrix Composites (MMCs). Most frequently the matrix is a light metal (aluminum or magnesium), although titanium, copper, and superalloys are also used. The reinforcing fiber can be boron, tungsten, a ceramic such as silicon carbide (SiC) or alumina (Al_2O_3) or graphite (carbon). Processing of continuous-fiber-reinforced composites follows techniques used for polymer-matrix composite manufacture (infiltration, pultrusion) or metalworking (investment casting, hot isothermal rolling, diffusion bonding). Discontinuous-fiber reinforcement allows casting, squeeze casting, hot extrusion, hot forging, and P/M techniques.

Carbon-Matrix Composites. Carbon fiber is embedded in a carbon matrix, produced by pyrolysis of a resin or tar. Shapes are created from prepregs or by infiltration of fiber forms.

Ceramic-Matrix Composites

Glass, alumina, silica, or silicon nitride matrix is reinforced with refractory metal (molybdenum, tungsten, tantalum, niobium) or carbon fibers or ceramic (SiC) whiskers. Hot pressing, cold pressing and sintering, slip casting, plasma spraying, and other processes can be used.

Many processes for the manufacture of composites are under intense development, and their scope is continually being expanded; hence neither size nor shape limitations are set. Costs and productivities are also changing; some systems in the development stage are extremely expensive, while others have entered the production stage (e.g., aluminum-alloy diesel engine pistons reinforced with alumina or silicon carbide fibers or whiskers).

Machining of Composites

Machining of composites presents special challenges because many fibers are abrasive, reducing tool life, and delamination and fiber/matrix pull-out are always dangers. Ideally, the part would be made to net shape, but this is not often possible. Such finishing as is necessary is then done with special drills, diamond-coated circular saws, and special techniques such as water-jet cutting, water-jet abrasive cutting, and laser machining.

Manufacturing Process Selection: Example

The foregoing discussion has shown that, more often than not, several processes are potential candidates for making any one part. Indeed, many of these processes coexist for the manufacture of parts that meet the product specification (design) requirements. This is particularly true of parts of relatively little shape complexity, as demonstrated by the example of a spool-shaped part shown in Fig. 14(a) (shape T2, Fig. 1). Options will be a strong function of material, size, tolerances, and production volume.

For illustrative purposes, assume that the part is made from a ferrous material, with a minimum tensile strength of 180 MPa (40 ksi); ductility is of no major concern. Take first the following dimensions: bore diameter 50.00 mm (2.000 in.), wall thickness 2.5 mm (0.1 in.), flange diameter 75.0 mm (3.0 in.), flange width 10 mm (0.4 in.) and total length 100.00 mm (4.000 in.). Tolerances of ±0.025 mm (±0.001 in.) apply only to the bore diameter and total length. For a tolerance range of 0.05 mm (0.002 in.) over a 25 mm (1 in.) dimension, only a machining operation can provide required quality (see Fig. 5 in the article "Overview of the Materials Selection Process" in this Volume). This applies even more to the larger dimensions of the part. Because finish machining is needed irrespective of the method of manufacture, there is great freedom in choosing the initial process for creating the shape.

Principal Options. Seven options are described below.

Option 1. The shape is suitable for turning on a lathe (Table 4). Machining from a solid bar (Fig. 14a) results in a cutting-off loss of 6 mm (0.25 in.), and material utilization is only 18%. Starting with (a much more expensive) tube of 46 mm (1.85 in.) inside diameter, material utilization increases only slightly to 25%.

Option 2. Casting is feasible by all expendable-mold processes (Table 1). Casting in a horizontal position (Fig. 14b) requires an inexpensive cylindrical core. With a machining allowance of 2 mm (0.08 in.) on all surfaces, wall thickness becomes 6.5 mm (0.26 in.); this allows for possible eccentricity of the core. This thickness has to be fed over a distance of 76 mm (3.04 in.); according to Fig. 2, this should be possible even for a steel casting but, in view of the modest strength requirement, a nodular cast iron will do. Material utilization increases to 37%.

Option 3. Casting in a vertical position (Fig. 14c) complicates the mold but allows bottom feeding. In view of the small size of the part, the benefit is not worth the expense.

Option 4. If the strength requirement were raised to the point where cast iron is insufficient and, especially if the part is exposed to fatigue loading, a higher grade of nodular cast iron or cast steel could be considered. Forging also becomes an alternative. Forging in a horizontal position (Fig. 14d) saves little material and the flash forms at a position where the flow lines create a potential weakness under an axially rotating load.

Option 5. Forging in a vertical position (Fig. 14e) gives poor material utilization, and fiber orientation is still not optimum. Forging on a horizontal forging machine creates, however, a part with favorable material flow, very close to the finished shape, and machining allowances could be reduced. The tooling with sliding insert would, however, be fairly expensive.

Option 6. Powder metallurgy by die pressing could be used to produce a tube, with an outside diameter equal to flange diameter and an inside diameter equal to bore less machining allowance. Because there are flanges at both ends, the reduced outside diameter cannot be directly pressed, and the economy of the process is doubtful. Isostatic pressing in a deformable mold would come much closer to final shape.

Option 7. In principle, the part could be made by welding a ring (Fig. 14f, left) or flange (Fig. 14f, right) but finish machining would still be required. It is highly unlikely that this method would be competitive.

Choosing among Options. The final choice would most likely come down to machining from the solid for small-lot production and casting or horizontal upsetting for larger production volumes. Economies would change if the part were much longer, but with the same flange: upsetting of tube (pipe) ends or welding flanges onto a tube would become much more attractive.

Increasing the flange diameter to 100 mm (4.0 in.) would change the evaluation. Machining from a bar again gives only 18% material utilization, and a thick-walled tube is not worth considering. Casting becomes a clear favorite with 56% material utilization. Horizontal upsetting of the flange from 56 mm (2.2 in.) to 100 mm (4.0 in.) diameter is possible, but stretches the capabilities of the process (material has to be gathered in several die cavities).

Increasing the length of the part with the larger flange makes the machining option less favorable. Casting is still attractive. Horizontal upsetter work is feasible only to a certain length (typically, a length-to-hole diameter ratio under 4), although a solid spool could be forged. Assembly by welding may be the most economical, depending on the cost of tubing. Cold-drawn tube would eliminate the need for machining the bore, provided that welding were performed without causing distortion or loss of properties.

In general, the range of process options is narrowed with increasing shape complexity and, as shown by the above example, with increasing differences in dimensions among various portions of the part. It may then be necessary to return to the design stage and consider alternative geometries and material options.

The number of possible processes would increase greatly if the restriction on the material (ferrous alloy) were removed and the strength criteria were relaxed, because both nonferrous materials and reinforced plastics would become viable candidates.

A detailed examination of alternatives for cutting sheet materials (Ref 22) illustrates the very broad range of processes that would have to be considered even for a relatively simple task.

REFERENCES

1. M.F. Ashby, *Materials Selection in Mechanical Design*, Pergamon Press, 1992
2. Information Technology and the Global Materials Industry, *Adv. Mater. Process.*, Vol 151 (No. 1), 1997, p 20–23
3. L. Alting, *Manufacturing Engineering Processes*, 2nd ed., Dekker, 1994
4. E.P. DeGarmo, J.T. Black, and R.A. Kohser, *Materials and Processes in Manufacturing*, 7th ed., Macmillan, 1988
5. L.E. Doyle, *Manufacturing Processes and Materials for Engineers*, 3rd ed., Prentice Hall, 1985
6. L. Edwards and M. Endean, Ed., *Manufacturing with Materials*, Butterworths, 1990
7. M.M. Farag, *Selection of Materials and Manufacturing Processes for Engineering Design*, Prentice Hall, 1989
8. W.O. Fellers and W.W. Hunt, *Manufacturing Processes for Technology*, Prentice Hall, 1995
9. M.P. Groover, *Fundamentals of Modern Manufacturing*, Prentice Hall, 1996
10. S. Kalpakjian, *Manufacturing Processes for Engineering Materials*, 2nd ed., Addison-Wesley, 1991
11. S. Kalpakjian, *Manufacturing Engineering and Technology*, 3rd ed., Addison-Wesley, 1995
12. R.A. Lindberg, *Processes and Materials of Manufacture*, 4th ed., Allyn and Bacon, 1990
13. K.C. Ludema, R.M. Caddell, and A.G. Atkins, *Manufacturing Engineering: Economics and Processes*, Prentice Hall, 1987
14. B.W. Niebel, A. Draper, and R.Wysk, *Modern Manufacturing Process Engineering*, McGraw-Hill, 1989
15. H.D. Moore and D.R. Kibbey, *Manufacturing: Materials and Processes*, 3rd ed., Wiley, 1982
16. J.A. Schey, *Introduction to Manufacturing Processes*, 2nd ed., McGraw-Hill, 1987
17. R.H. Todd, D.K. Allen, and L. Alting, *Manufacturing Processes Reference Guide,* Industrial Press, 1994
18. T.F. Waters, *Fundamentals of Manufacturing for Engineers*, University College London, 1996
19. D.J. Whitehouse, Comparison between Stylus and Optical Methods for Measuring Surfaces, *Ann. CIRP*, Vol 37 (No. 2), 1988, p 649–653
20. J.A. Schey, *Tribology in Metalworking: Friction, Lubrication and Wear,* American Society for Metals, 1983
21. ANSI B 46.1-1985, *Surface Texture*, American National Standards Institute, 1985
22. C.A. van Luttervelt, On the Selection of Manufacturing Methods Illustrated by an Overview of Separation Techniques for Sheet Materials, *Ann. CIRP*, Vol 39 (No. 2), 1989, p 587–607

SELECTED REFERENCES

Casting Processes

- *Casting*, Vol 15, *ASM Handbook*, ASM International, 1988
- *Steel Castings Handbook*, 6th ed., ASM International, 1995

Deformation Processes

- T. Altan, S.-I. Oh, and H. Gegel, *Metal Forming: Fundamentals and Applications,* American Society for Metals, 1983
- T.Z. Blazynski, Ed., *Plasticity and Modern Metal Forming Technology*, Elsevier, 1989
- T.G. Byrer, Ed., *Forging Handbook*, Forging Industry Association, Cleveland, 1985
- *Forming*, Vol 2, *Tool and Manufacturing Engineers Handbook*, 4th ed., Society of Manufacturing Engineers, 1984
- *Forming and Forging*, Vol 14, *ASM Handbook*, ASM International, 1988
- S.K. Ghosh and M. Predeleanu, Ed., *Materials Processing Defects*, Elsevier, 1995
- W.F. Hosford and R.M. Caddell, *Metal Forming: Mechanics and Metallurgy*, 2nd ed., Prentice Hall, 1993
- K. Lange, Ed., *Handbook of Metal Forming*, McGraw-Hill, 1985 (now SME)
- O.D. Lascoe, *Handbook of Fabrication Processes*, ASM International, 1988
- Z. Marciniak and J.L. Duncan, *The Mechanics of Sheet Metal Forming*, Edward Arnold, 1992
- E. Mielnik, *Metalworking Science and Engineering*, McGraw-Hill, 1991
- R. Pearce, *Sheet Metal Forming*, Adam Hilger, 1991
- J.A. Schey, *Tribology in Metalworking: Friction, Lubrication and Wear,* American Society for Metals, 1983
- D.A. Smith, Ed., *Die Design Handbook*, 3rd ed., Society of Manufacturing Engineers, 1990
- R.H. Wagoner, K.S. Chan, and S.P. Keeler, Ed., *Forming Limit Diagrams*, TMS, Warrendale, PA, 1989
- R.A. Walsh, *Machining and Metalworking Handbook*, McGraw-Hill, 1994

Powder Processing

- H.V. Atkinson and B.A. Rickinson, *Hot Isostatic Pressing*, Adam Hilger, 1991
- G. Dowson, *Powder Metallurgy: The Process and its Products*, Adam Hilger, 1990
- R.M. German, *Powder Metallurgy Science*, Metal Powder Industries Federation, 1985
- C. Iliescu, *Cold-Pressing Technology*, Elsevier, 1990
- H.A. Kuhn and B.L. Ferguson, *Powder Forging*, Metal Powder Industries Federation, 1990
- M.H. Liebermann, *Rapidly Solidified Alloys*, Dekker, 1993
- *Powder Metallurgy*, Vol 7, *ASM Handbook*, American Society for Metals, 1984
- *Powder Metallurgy Design Manual*, 2nd ed., Metal Powder Industries Federation, 1995

Machining Processes

- G. Boothroyd and W.W. Knight, *Fundamentals of Machining and Machine Tools*, 2nd ed., Dekker, 1989
- *Machining*, Vol 1, *Tool and Manufacturing Engineers Handbook*, 4th ed., Society of Manufacturing Engineers, 1983
- *Machining*, Vol 16, *ASM Handbook*, ASM International, 1989
- S. Malkin, *Grinding Technology: Theory and Applications*, Ellis Horwood, 1989
- P.L.B. Oxley, *The Mechanics of Machining*, Ellis Horwood, 1989
- M.C. Shaw, *Metal Cutting Principles*, 4th ed., Oxford University Press, 1984
- D.A. Stephenson and J.S. Agapiov, *Metal Cutting Theory and Practice*, Dekker, 1996

- R.A. Walsh, *Machining and Metalworking Handbook*, McGraw-Hill, 1994

Joining Processes

- *Adhesives and Sealants*, Vol 3, *Engineered Materials Handbook*, ASM International, 1990
- *Brazing Handbook*, 4th ed., American Welding Society, 1991
- G. Humpston and D.M. Jacobson, *Principles of Soldering and Brazing*, ASM International, 1993
- D.L. Olson, R. Dixon, and A.L. Liby, Ed., *Welding Theory and Practice*, North Holland, 1990
- R.O. Parmley, Ed., *Standard Handbook of Fastening and Joining*, 3rd ed., McGraw-Hill, 1997
- A. Rahn, *The Basics of Soldering*, Wiley, 1993
- M. Schwartz, *Brazing*, ASM International, 1987
- *Welding, Brazing, and Soldering*, Vol 6, *ASM Handbook*, ASM International, 1993
- *Welding Handbook*, 8th ed., American Welding Society, 1996

Ceramics Processing

- *Ceramics and Glasses,* Vol 4, *Engineered Materials Handbook,* ASM International, 1991
- *Engineered Materials Handbook Desk Edition*, ASM International, 1995
- S. Musikant, *What Every Engineer Should Know about Ceramics*, Dekker, 1991
- G.C. Phillips, *A Concise Introduction to Ceramics*, Van Nostrand-Rheinhold, 1991
- J.S. Reed, *Principles of Ceramics Processing*, 2nd ed., Wiley, 1995
- M.M. Schwartz, *Ceramic Joining*, ASM International, 1993
- M.M. Schwartz, *Handbook of Structural Ceramics*, McGraw-Hill, 1992
- R.A. Terpstra, P.P.A.C. Pex, and A.H. DeVries, Ed., *Ceramic Processing*, Chapman & Hall, 1995

Polymer Processing

- R.J. Crawford, Ed., *Rotational Moulding of Plastics*, Wiley, 1992
- *Engineering Plastics*, Vol 2, *Engineered Materials Handbook*, ASM International, 1988
- R.G. Griskey, *Polymer Process Engineering*, Chapman & Hall, 1995
- *Handbook of Plastics Joining: A Practical Guide*, Plastics Design Library, 1996
- N.C. Lee, Ed., *Plastic Blow Molding Handbook*, Van Nostrand-Rheinhold, 1990
- N.G. McCrum, C.P. Buckley, and C.B. Bucknall, *Principles of Polymer Enginering*, Oxford University Press, 1989
- E.A. Muccio, *Plastics Processing Technology*, ASM International, 1994
- *Plastic Parts Manufacturing*, Vol 8, *Tool and Manufacturing Engineers Handbook*, Society of Manufacturing Engineers, 1995
- R.C. Progelhof and J.L. Throne, *Polymer Engineering Principles: Properties, Tests for Design*, Hanser, Munich, 1993
- G.W. Pye, *Injection Mold Design*, Longman/Wiley, 1989
- D.V. Rosato, D.P. DiMattia, and D.V. Rosato, *Designing with Plastics and Composites: A Handbook*, Van Nostrand-Rheinhold, 1991

Manufacture of Composites

- *Composites*, Vol 1, *Engineered Materials Handbook*, ASM International, 1987
- L. Hollaway, *Handbook of Polymer Composites for Engineers*, Woodhead, Cambridge, 1994
- B.Z. Jang, *Advanced Polymer Composites: Principles and Applications*, ASM International, 1994
- M.M. Schwartz, *Composite Materials Handbook*, McGraw-Hill, 1992
- M.M. Schwartz, *Handbook of Composite Ceramics*, McGraw-Hill, 1992
- M.M. Schwartz, *Joining of Composite Matrix Materials*, ASM International, 1994
- W.A. Woishnis, Ed., *Engineering Plastics and Composites*, 2nd ed., ASM International, 1993

Modeling of Manufacturing Processes

Anand J. Paul, Concurrent Technologies Corporation

MANUFACTURING PROCESSES typically involve the reshaping of materials from one form to another under a set of processing conditions. To minimize the production cost and shorten the time to market for the product, all iterations in terms of an appropriate set of operating conditions should not be done on the shop floor. Predictive models need to be used generously to perform numerical experiments to give an insight into the effect of the operating conditions on the properties of the final product.

Process models must be able to build the geometry of the product/process that is being modeled, accurately describe the physics of the process, and be able to analyze the results in a way that is comprehensible by manufacturing engineers. There are several types of models that are used by the industry. These include models used as a tool in the course of scientific research, models that are very generalized in nature and can be applied to a wide variety of processes, but may not be able to address the nuances of any one process, models that are very specific in nature and can address a narrow range of operating conditions, models that rely on gross phenomena and are about 90% accurate but take only 10% of the execution time or more accurate models. Irrespective of the complexity, most models are used to gain one or more of the following advantages:

- Reduce iterations on the shop floor
- Optimize an existing process
- Understand an existing process better
- Develop a new process
- Improve quality by reducing the variability in the product and process

Classification of Models

The following points need to be considered in order to classify modeling problems:

- The physical phenomena affecting the process under consideration
- Mathematical equations describing the physical process
- Data needed to solve the equations
- Numerical algorithm to solve the equations given the boundary conditions and the constitutive behavior
- Availability of the software to provide answers

One of the most common classification methods is by type of process or physics. This means that one must identify the major phenomena occurring in the process, for example, convection, radiation, chemical reaction, diffusion, deformation, and so forth. Once the phenomena has been identified, the process needs to be defined in terms of mathematical equations, typically partial differential equations. These equations are dependent on time, space, field variables, and internal states. Ordinary differential equations can be used if the problem can be simplified so that the shape is not important and a lumped-parameter model can be used. Several people have used lumped-parameter models for various materials processes (Ref 1, 2).

The requirements for particular data and the way in which it is gathered is an important step in the construction of a model. Researchers typically play down this step as an "industrial implementation detail." This means that the rest of the model needs to be robust and accurate before data are needed. Once accurate data are available, the result of the modeling effort will be good too. On the other hand, industrial practitioners place a greater emphasis on data gathering because they know the difficulties and time involved in gathering data on production-scale equipment.

Numerical algorithms to solve the differential equations consist of meshed-solution methods and lumped-parameter models. The major meshed-solution models consist of finite differences, finite elements, and boundary methods. Each of them is more appropriate for different types of equations and boundary conditions. Within these methods, one can use a structured or an unstructured mesh. Structured meshes are created by using rectilinear, bricklike elements. It is easy to use this type of mesh; however, fine geometry details may be missed. Unstructured meshes can be of any shape—tetrahedra, bricks, hexahedral, prisms, and so forth. Many of the disadvantages of using a structured mesh are eliminated through this type of a mesh.

Lumped-parameter models may help in understanding the effect of certain parameters on the process as along as the problem formulation does not change. These models do not model spatial variation directly, and the parameters may or may not be physically meaningful in themselves.

Choice of the appropriate software is an important aspect of the usefulness of the model. Almost without exception, research process models and all commercial software can be written directly in a third-generation language (Fortran, Lisp, Pascal, C, C++). User interfaces can be derived from various libraries. Because of the large number of calculations necessary to get the desired degree of detail, models may require parallel computing hardware for cost-effective solutions. Software developed for this has to be able to run and make use of the parallel-processing capabilities of the hardware.

Models for manufacturing processes can be classified in two primary ways, as shown in Fig. 1. One classification scheme considers whether the model is on-line or off-line; the other considers whether the model is empirical, mechanistic, or deterministic.

Fully on-line models are part of the bigger process control system in a plant. Sensors and feedback loops are characteristics of these models. They get their input directly from the system. These models implement changes in the plant on a continuous basis (Ref 3). Fully on-line models are extremely fast and reliable; therefore, these models need to be rather simple without the need to do any significant numerical calculations. For these models to be reliable, the physics of the process that they are addressing must be understood thoroughly. A good example of this type of model is the spray water system on a slab-casting machine, which is designed to deliver the same total amount of water to each portion of the strand surface. The flow rate changes to account for variations in the casting-speed history experi-

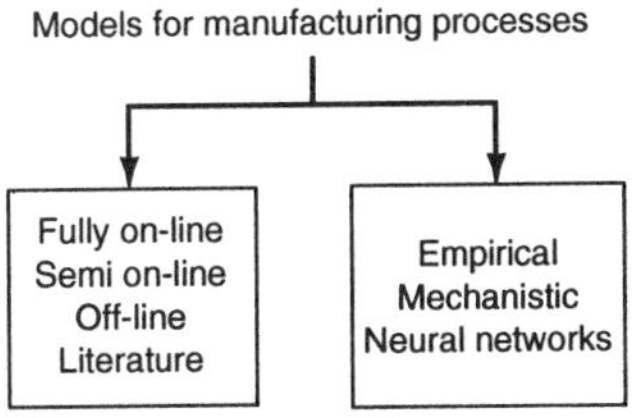

Fig. 1 Classification of models for manufacturing processes

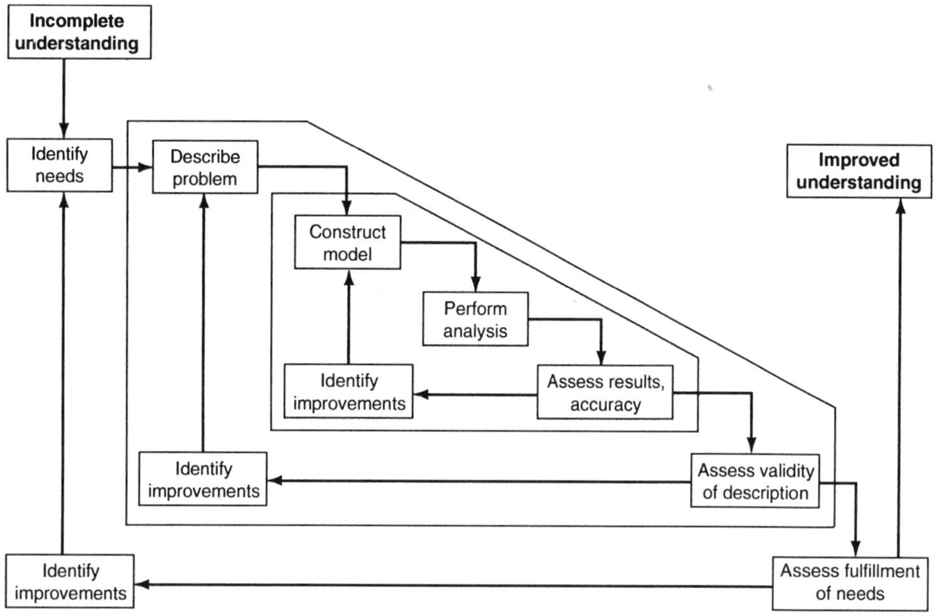

Fig. 2 Modeling cycles. Source: Ref 8

enced by each portion as it passes through the spray zones.

Semi-on-line models are similar to fully on-line models. The distinguishing factor is that rather than the model taking appropriate action, the process engineer analyzes the situation and performs any corrective action, if needed. These models are typically slightly more complex than the fully on-line models because they are essentially run by an operator. These models need to have an excellent user interface and should require minimum user intervention.

Off-line models are typically used in the premanufacturing stage, that is, during research, design, or process parameter determination. These models help to gain an insight into the process itself and thereby help optimize it. There are many general-purpose models as well as models designed to be used for very specific applications. These models are typically very complex and therefore need to be validated thoroughly before being used for any real predictions.

Literature models are those that exist primarily in the literature and are seldom used in conjunction with experiments. Typically these models are developed and run by the same individual. The advantage of literature models is that other developers can benefit from them instead of starting from scratch.

Empirical models are developed through statistical data gathering on a number of similar events. The model does not help one understand the process itself and may not be valid beyond the range of the available data.

Mechanistic models are based on the solution of the mathematical equations that represent the physics of the process that is being addressed. These models are very extensible and study the effect of a variety of external factors on the process.

Neural network models are based on artificial neural networks and provide a range of powerful techniques for solving problems in pattern recognition, data analysis, and control. Neural networks represent a complex, trainable, nonlinear transfer function between inputs and outputs. This allows an effective solution to be found to complex, nonlinear problems such as heat distribution.

Important Aspects of Modeling

There are several important issues that need to be addressed to understand what is important in modeling in general and what is important to the current problem in particular. Some of these are briefly discussed below.

Analytical versus Meshed Models. Developing an analytical or a closed-form solution model may be advantageous in many instances. However, it may be necessary to construct a discrete meshed model for finite element or finite difference calculations if the modeled volume:

- Has a complex shape (commonly found in many engineering applications)
- Contains different phases and grains, which are typically modeled by research groups (Ref 4)
- Contains discontinuous behavior such as a phase change, which can be handled easily with meshes using a volume-of-fluid (VOF) technique (Ref 5)
- Has nonlinear process physics such as when the heat transfer coefficient is a nonlinear function of the temperature

In many instances, meshed models are supplemented by some nonmeshed symbolic or analytical modeling. This is done in order to decide on appropriate boundary conditions for the meshed part of the problem because it is the boundary conditions that effectively model the physical problem and control the form of the final solution (Ref 6).

Analytic models are always useful for distinguishing between mechanisms that have to be modeled as a coupled set and mechanisms that can be modeled separately. These models no longer need closed-form solutions. Even simple computers can track evolving solutions and iterate to find solutions to implicit formulations (Ref 1).

Boundary Conditions and Multiphysics Models. Application of appropriate boundary conditions is a major part of the activity of process modeling. Boundary conditions are statements of symmetry, continuity and constancy of temperature, heat flux, strain stress, and so forth. Boundary conditions need to be set at a very early stage in analytical models. In meshed models, these are typically represented separate from the main equations and are decoupled to some extent from the model itself. Therefore, sensitivity analysis can be done much easier using meshed methods.

The type of boundary conditions used also determines what solving algorithm should be used for the partial differential equations. This determines the speed, accuracy, and robustness of the solution.

Material Properties. All process models require material properties to be simulated. Acquiring these properties can be difficult and expensive (Ref 7). A sensitivity analysis of the model with respect to these data provides information as to the importance of minor changes in them. In many instances, it may be possible to use models with doubtful material property information in order to predict trends, as opposed to determining actual values. A problem arises if the material properties are extrapolated beyond the range of their applicability where one does not know the behavior of the material at all. Related information is provided in the articles "Computer-Aided Materials Selection" and "Sources of Materials Properties Data and Information" in this Volume.

Modeling Process Cycles. The process of modeling is done in different cycles. Figure 2 attempts to distinguish those cycles (Ref 8). This figure shows three loops suggesting three levels of activities in any modeling effort. The outer loop is managed by someone close to the process who understands the business context of the problem and can concentrate on specifying the objective and providing the raw data. The innermost loop (shaded dark) requires mostly computational skills while the middle loop (shaded light) consists of activities balancing the other two. It may very well happen that all three of some combination of the activities can be done by the same person. However, very seldom is that the case. This highlights the need for forming modeling teams where all aspects of the problem can be addressed rigorously. It also emphasizes the importance of training and appropriate software tool development so that the input and output of the tools can be easily understood by all involved in the process.

Modeling of Deformation Processes

Finite element analysis (FEA) of deformation processes can provide an insight into the behavior of the product under various processing conditions and can help optimize the conditions to get the desired properties. It can also help understand the performance of the product before the part is put in actual use. Common problems solved by FEA include insufficient die filling, poor shape control, poor flow of material, cracks and voids that lead to fracture and poor final part properties.

The occurrence of typical processes in a forging operation are shown in Fig. 3. Figure 4 (Ref 9) shows a schematic representation of the interactions between the major process variables in metal forming. From Fig. 4 it can be seen that for a metal-forming analysis, one needs to satisfy the equilibrium conditions, compatibility equations/strain-displacement relations, constitutive equations, and, in some instances, the heat balance equation. In addition, one needs to apply appropriate boundary conditions. These may comprise displacement/velocity imposed on a part of the surface while stress is imposed on the remainder of the surface, heat transfer, or any other interface boundary condition.

Relevant Equations. The equilibrium equations describing the various forces acting on the body are given as:

$$\begin{aligned} \frac{\partial \sigma_x}{\sigma x} + \frac{\partial \tau_{xy}}{\partial y} + \frac{\partial \tau_{xz}}{\partial z} + F_x = 0 \\ \frac{\partial \tau_{xy}}{\partial x} + \frac{\partial \sigma_y}{\partial y} + \frac{\partial \tau_{yz}}{\partial z} + F_y = 0 \\ \frac{\partial \tau_{xz}}{\partial x} + \frac{\partial \tau_{yz}}{\partial y} + \frac{\partial \sigma_z}{\partial z} + F_z = 0 \end{aligned} \qquad \text{(Eq 1)}$$

where σ is the normal stress component, τ is the shear stress component, and F is the body force/unit volume component. Similarly, the strain-displacement relationships are given as:

$$\begin{aligned} \varepsilon_x = \frac{\partial u}{\partial x} \quad \gamma_{xy} = \frac{\partial u}{\partial y} + \frac{\partial v}{\partial x} \\ \varepsilon_y = \frac{\partial v}{\partial y} \quad \gamma_{yz} = \frac{\partial v}{\partial z} + \frac{\partial w}{\partial y} \\ \varepsilon_z = \frac{\partial w}{\partial z} \quad \gamma_{zx} = \frac{\partial w}{\partial x} + \frac{\partial u}{\partial z} \end{aligned} \qquad \text{(Eq 2)}$$

where ε is the normal strain, γ is the shear strain, and u, v, and w are the displacements in the x, y, and z directions, respectively.

Constitutive Theory. A constitutive equation relates the stress and strain behavior of a material. Schematic stress/strain curves for idealized materials are shown in Fig. 5. In addition to this information, a yield criterion and flow rules are also needed to adequately describe the material behavior. The stress/strain curve for a typical metal along with its major features is shown in Fig. 6.

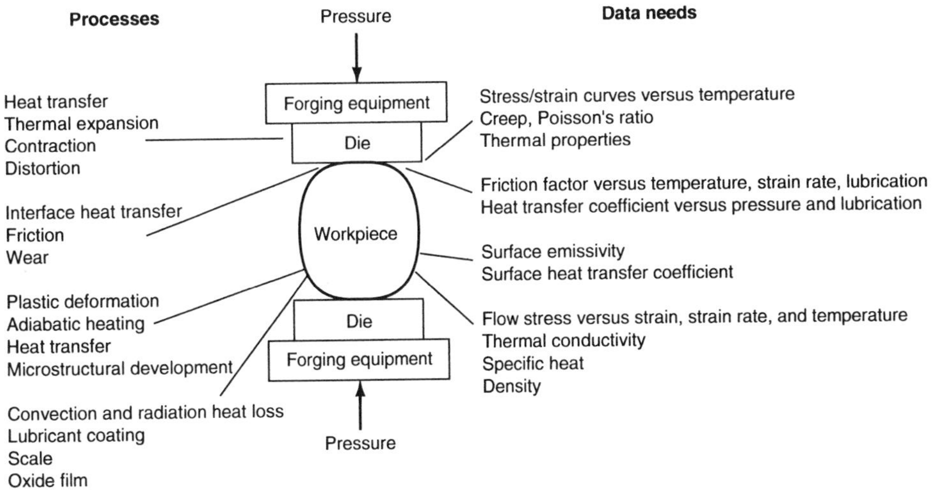

Fig. 3 Typical physical phenomena occurring during a forging operation

Finite Element Analyses. One of the most commonly used techniques to solve the various equations (as shown earlier) is finite element analysis (FEA). It is a numerical technique that approximates the mathematical equations that in turn approximate reality. It is a discrete representation of a continuous system. It breaks the bigger problem into a number of smaller ones and, therefore, is a piecewise representation of the problem.

Some of the common metal-forming problems that are solved by FEA are insufficient die fill, poor shape control, poor flow of materials, prediction of cracks and voids that lead to fracture, and poor final part properties. Finite element analysis has many advantages over the typical closed-form solutions. It provides greater insights into the behavior of the product and the process, gives a good understanding of the performance of the product before the actual usage, is useful in product and process optimization, is a powerful and mature design and analysis tool, and most importantly, gives solutions to irregular shapes, variable material properties, and irregular boundary conditions. However, FEA is not to be viewed as a solution to all problems or as a substitute for common sense and experience.

Typical costs of FEA can run into thousands of dollars. A rough estimate is that if brainstorming costs \$1, then refined hand calculations would cost \$2, "quick and dirty" process models would cost \$5, and detailed process models would cost \$30. However, the probability of success of detailed process models is much higher compared to that of brainstorming, and great savings in costs of experimentation and rework may be achieved.

There are several steps involved in conducting these analyses. A brief description of each is provided below. More detailed information is pro-

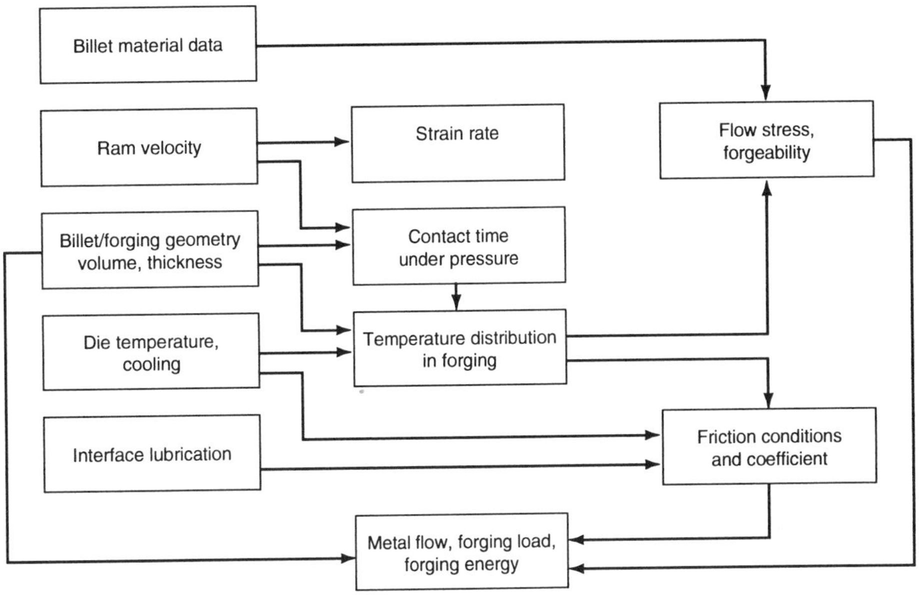

Fig. 4 Interaction among major process variables during forming. Source: Ref 9

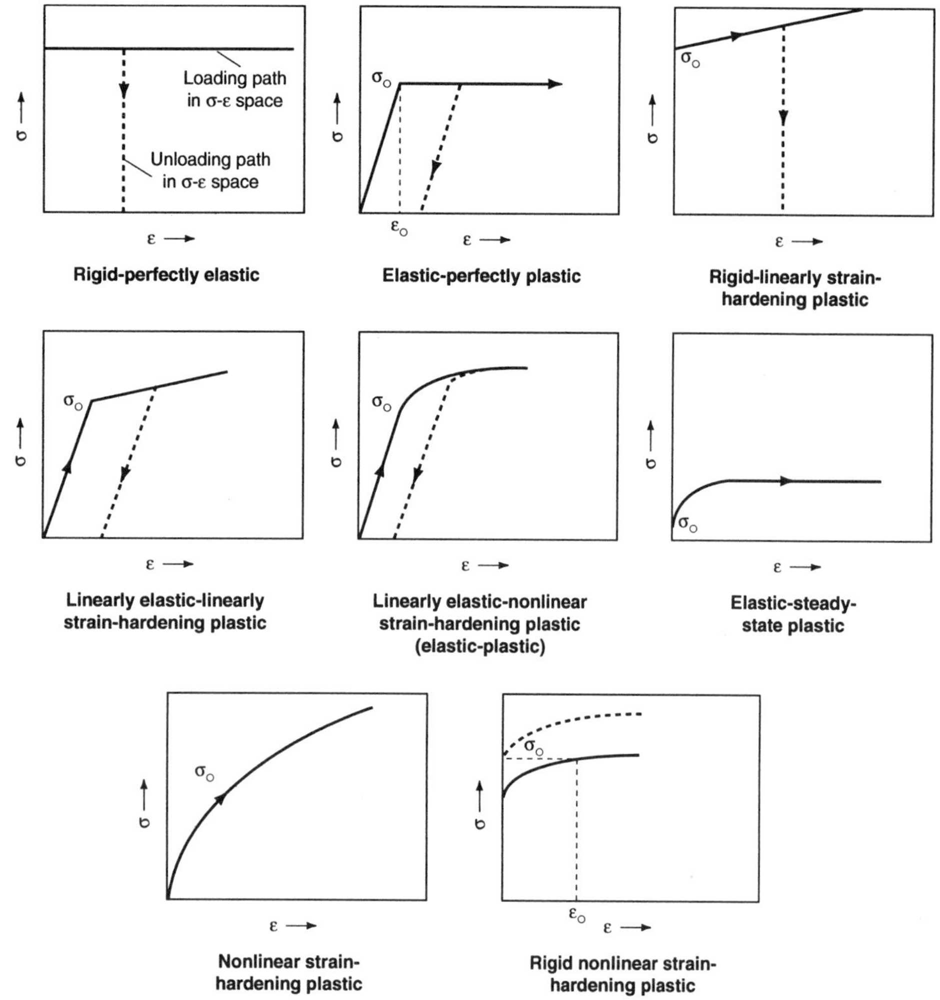

Fig. 5 Schematic stress/strain curves for various materials

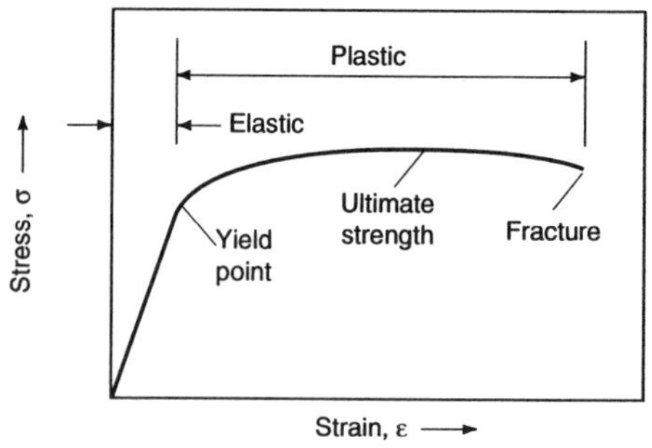

Fig. 6 Stress/strain curve for a typical metal

vided in the article "Finite Element Analysis" in this Volume.

Defining the Problem. Before a full-fledged FEA is undertaken, several questions need to be answered:

- Is FEA appropriate?
- What is desired from the FEA?
- When are results needed?
- What are the product and process limitations?

If it is determined that FEA would provide adequate answers, information is gathered to start the modeling process. The required information includes the geometry of interest, initial and boundary conditions, material properties and material behavior models, and an approximate solution to ensure that the finite element results are not physically absurd.

The preprocessing stage defines the physical problem and converts it into a form that the computer can solve. The definition of the physical problem involves fully defining the geometry as well as the material of the part to be simulated.

Next comes defining the physics of the process. This involves using the appropriate set of mathematical equations and the corresponding initial conditions and boundary conditions. Subsequently, the solution method needs to be determined. This involves choosing the appropriate algorithms to solve the numerical approximations of the mathematical equations. Finally, because FEA is a numerical approximation of reality, the problem needs to be discretized, (i.e., the continuum needs to be broken into many smaller pieces, the sum of which will represent the whole problem). In this stage, the shape of these smaller pieces, or elements, their size, and their complexity needs to be determined.

Discretization is the process of subdividing the whole geometry into discrete parts. The discrete parts are called elements. Many different types of elements exist depending on their shape, linearity, and order. The factors that impact the selection of the elements are the type of the problem, geometry, accuracy desired, availability within the algorithm, nature of the physical problem, and user familiarity. The size and number of elements are primarily determined by the various gradients (temperature, stress, etc.) in the system. For example, if the gradients are steep, a larger number of smaller-sized elements should be used. An example of discretization is shown in Fig. 7. For clarity, the node and element numbers are not shown in the mesh.

The mesh can be made coarser (fewer elements) or finer (more elements) depending on the needs of the problem. Coarser meshes use minimal computer resources in terms of storage and run time. However, because the representation is approximate, the results can be crude. Finer meshes provide a more accurate representation with improved results.

Typical elements are either linear or quadratic and can be one-, two- or three-dimensional. Figure 8 schematically shows some typically used elements.

Post Processing. Once the equations have been solved by the computer program, huge amounts of data are generated that are typically not in a user-friendly format. These data need to be organized and processed so as to make sense and give only the information that the user desires. The most frequently analyzed data are metal flow, time-history plots, ductile fracture tendency, stress/strain plots, and load-stroke curves. A good visualization scheme is crucial to understanding the outcome of the analysis.

Requirements of a Good FEA Code. For an FEA code to be useful, it must be easy to use and provide results that are easy to understand. A well-written user's manual or other documentation and good software support are also important. From a technical standpoint, the code must also be accurate, fast, and numerically stable under a wide range of conditions. It must be sufficient to capture the physics of the problem and should be capable of being interfaced with other codes (as appropriate).

Example 1: Use of FEA to Study Ductile Fracture during Forging. Brittle fracture is sometimes observed after forging and heat treating large nickel-copper alloy K-500 (K-Monel) shafts (Ref 10). Inspection of failed fracture surfaces revealed that the cracks were intergranular and occurred along a carbon film that had apparently developed during a slow cooling cycle (such as during casting of the ingot stock). Cracks were observed in both longitudinal and transverse directions and arrested just short of the outer bar diameter. Finite element analysis was used to determine the cause and remedy for this cracking. In addition, new processing parameters were determined to eliminate cracking.

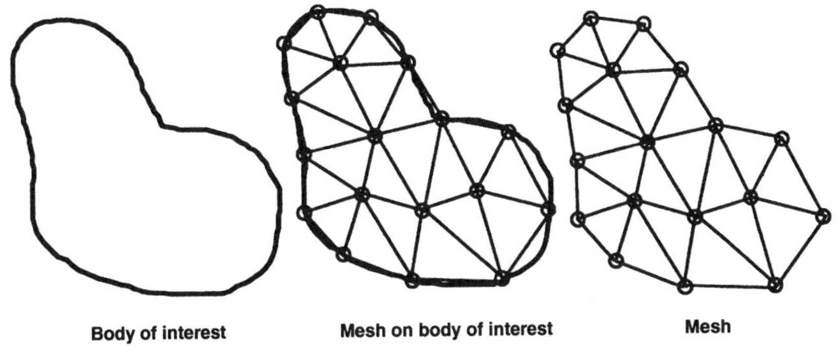

Fig. 7 The finite element analysis discretization process

A two-dimensional model of the bar was developed and discretized into a finite element mesh. NIKE2D, a public domain, nonlinear FEA code from Lawrence Livermore National Laboratory (Ref 11), was used to calculate the stress plots under the forging conditions. The stress/strain histories resulting from forging were required for ductile fracture analyses. In this particular case, the Oyane et al. ductile fracture criteria (Ref 12) were used. These criteria attempt to predict ductile fracture based on a porous plasticity model.

Figure 9 (Ref 10) shows in-process hydrostatic stress history at the center of a bar during forging with two rigid, parallel flat dies. (Hydrostatic stress is the mean of any three mutually perpendicular normal stresses at a given point. It is invariant with direction.) Various strain rate and temperature conditions are shown having similar trends. At very low reductions (1.25%), the center of the bar is still within the elastic range and, therefore, does not experience large stresses. At 2.5% reduction, the outward movement of the free sides of the bar produces a high tensile hydrostatic stress at the center of the bar. As reduction continues, the compressive stresses acting in the vertical direction begin to grow. This acts to reduce the hydrostatic stress at the center. When the bar has been reduced by 11%, the hydrostatic stress is compressive in all cases. Beyond this reduction, the hydrostatic stress at the center of the bar continues to grow in a compressive manner.

Figure 10(b) (Ref 10) shows a map of ductile fracture accumulation after one forging pass for air-melted material at 930 °C and a strain rate of 10 s^{-1}. Notice that the maximum damage is located at the center of the bar. After this pass, the bar would be rotated and further reductions would be taken. Damage would accumulate during these subsequent passes. Ductile fracture was predicted to occur at the bar center from the FEA results after six reducing passes under the specified conditions. One possible way of eliminating this ductile fracture is to use a V-shaped die assembly as shown in Fig. 10(a). In this case, the fracture does not occur at the center, but at the edges and gets an opportunity to heal as the bar is rotated during subsequent passes.

Example 2: Use of FEA for Modeling Superplastic Forming of Aluminum Assemblies. Superplastic forming (SPF) technology is being increasingly used for aerospace applications. The major benefits of using SPF include the ability to produce complex near-net shapes better than other forming operations and its good postformed mechanical properties. Even though titanium components have been formed by SPF for some time, SPF of aluminum components pose some challenges. These include the ability of the material to be superplastic only at elevated temperatures and within a small window of strain rates, cavitation, and amount and rate of application of pressure. Superplastic forming of aluminum is usually performed in a closed-die press capable of applying inert gas pressure to both sides of a superplastic sheet. Initially, an equilibrium pressure is applied to both sides of the sheet. Slowly, according to a predesigned forming schedule, pressure is released from one side of the sheet, with the pressure differential pushing the sheet into the die cavity. As the sheet freely forms into the die cavity, thinning occurs relatively uniformly. Once the sheet makes contact with the die, frictional effects begin to make the thinning less uniform.

The constitutive behavior of the superplastic material (aluminum alloy 7475) is expressed in the form of a simple power-law equation:

$$\dot{\varepsilon} = A\sigma^n \quad \text{(Eq 3)}$$

where $\dot{\varepsilon}$ is the strain rate, σ is the stress, and A and n (= $1/m$) are material constants. The above equation assumes that grain size remains constant during forging. Critical regions, where grain coarsening and an unacceptable amount of cavitation take place, can also be modeling by expressing the material behavior in a more rigorous form:

$$\dot{\varepsilon} = A' \frac{DGb}{kT} \left(\frac{\mathbf{b}}{d}\right)^p \left(\frac{\sigma}{G}\right)^n \exp\left(-\frac{Q}{RT}\right) \quad \text{(Eq 4)}$$

where A' is a material constant, D is the diffusion coefficient, G is the shear modulus, **b** is the Burgers vector, k is the Boltzmann's constant, T is the absolute temperature, d is the grain size, p is the grain size exponent, σ is the flow stress, n is the strain-hardening exponent, Q is the activation energy, and R is the universal gas constant. Cavitation may be expressed as a function of accumulated plastic strain as:

$$C = C_o \exp(K_3\bar{\varepsilon}) \quad \text{(Eq 5)}$$

where C_o is the initial void volume, K_3 is a constant, and $\bar{\varepsilon}$ is the accumulated plastic strain.

Several subroutines were developed and implemented in a commercially available FEA code (Ref 13). Several trial forming problems were

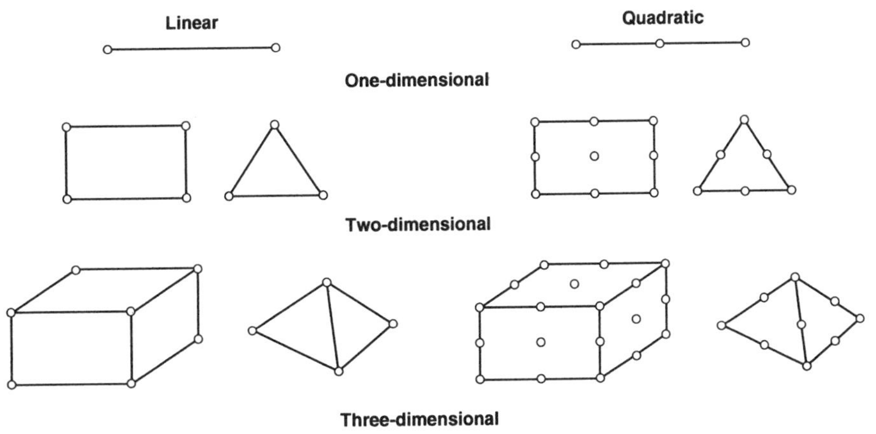

Fig. 8 Linear and quadratic elements used in typical finite element analyses

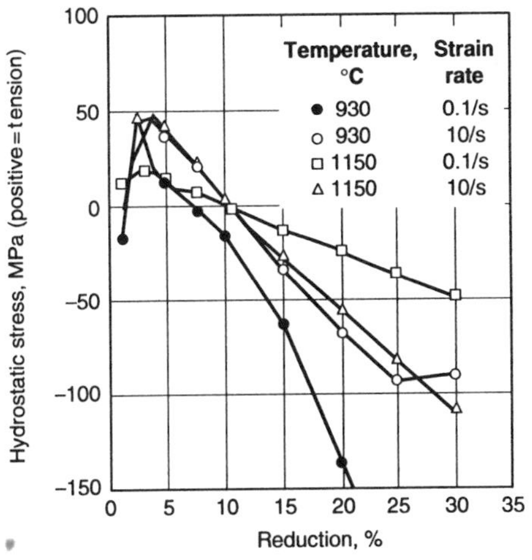

Fig. 9 Hydrostatic stress at the bar center during forging. Source: Ref 10

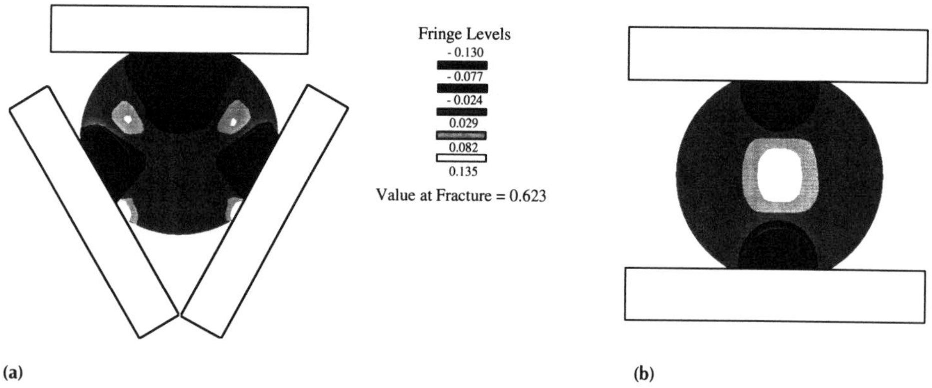

Fig. 10 Ductile fracture map of a nickel-copper alloy K-500 bar at 10% reduction, 930 °C, and a strain rate of 10.0 s^{-1} using the Oyane et al. ductile fracture criteria. (a) Three forging dies. (b) Two forging dies. Source: Ref 10

solved to show the benefits of various enhancements to the FEA code. Simple geometries were first taken to allow comparison of simple analytical solutions with the compared results. Subsequently, complex geometries were discussed to show where the simple analytical solutions broke down and where the FEA results give a significant insight into the forming process.

Modeling of Casting Operations

Of late, considerable developments have taken place in the field of solidification modeling of casting processes. In the current state-of-the-art solidification simulation, several software packages are available to analyze the solidification behavior in complex-shaped castings. These packages make use of several different approaches for solving the various problems associated with casting processes.

An overall architecture of a comprehensive solidification modeling system is shown in Fig. 11. This figure (Ref 14) depicts the various modules available in the current state-of-the-art solidification simulation of casting processes, the information available from each module, and the interconnection between the modules. It is evident from the figure that the initial casting design is linked to a module called the quick-analysis module. Here, one can make use of approximate analysis schemes, such as the modulus approach (Ref 15), which uses geometry-based considerations to provide valuable insights into the solidification times and, therefore, the propensity for defect formation during solidification.

The next stage is to design the rigging system for the casting, which includes the design of the gate, risers, downsprue, and so forth. This is currently based on the "rules of thumb" of foundry experts and empirical charts. Once the rigging design is established, the stage is set for solidification simulation. Here, the continuum mechanics problem of heat, mass, and momentum transfer are solved for the casting process simulation. Thus, one obtains the cooling history of the casting.

Subsequently, one can obtain information about the microstructure in the casting by coupling with the module for microstructure evolution. Further, the simulation data can be postprocessed using special-purpose models for defect prediction that enable one to visualize the defects under a given set of processing conditions. Apart from porosity-type defects, prediction of other defects such as macrosegregation is possible. Because macrosegregation primarily occurs due to the movement of solid phase by convection during solidification, solving the fluid-flow equations in the mushy zone provides a solution.

Modeling the development of stresses in the casting has been another area of great challenge. Of late, several researchers have addressed the issue of development of stresses during and after solidification, which is often the cause of distortion in castings. This is especially the case for highly nonequilibrium processes, such as die-casting. Special numerical algorithms and techniques are being developed for handling more complex casting processes. For example, in large structural thin-walled castings, the normal solution methods would require an extremely large number of elements or nodes in the mesh, which significantly increase the computation time.

Quick-Analysis Schemes. Traditionally, for sand-casting analysis, the use of geometric methods has been known as the section modulus approach. The fundamental basis of geometric modeling is the relationship between the solidification time (t_f) and a geometric parameter, called the section modulus (given by volume-to-surface area ratio, V/A), as given by Chvorinov's Rule (Ref 16):

$$t_f = C\left(\frac{V}{A}\right)^2 \quad \text{(Eq 6)}$$

where C is a constant for a given metal-mold material and mold temperature.

For simple shapes, the modulus in Eq 6 can be calculated from the ratio of volume and surface area involved in cooling. However, for complex shapes discretized in a three-dimensional grid, the continuous distribution of modulus can be determined using the concept of distance from the mold, as discussed in Ref 15. The modulus at each point in the casting is determined by the relation:

$$M = \frac{2}{\sum_{i=1}^{N} 1/d_i} \quad \text{(Eq 7)}$$

Recently, this technique has been extended to model the investment-casting process, taking radiation loss into account through the use of a novel approach for view-factor calculations (Ref 17).

Knowledge-Based Systems for Rigging Design. The starting step after the initial design of the casting is the design of gates and risers. This consists of proper orientation of the part and the determination of the parting plane and the size, number, and location of sprues, runners, gates, and risers. To this end, the use of knowledge-based design systems is growing for foundry applications. A feature-based design system has been developed for casting applications (Ref 18). Also, a system for design of gating and risering for light alloy castings has been developed (Ref 19). This system was extended to investment castings (Ref 20). In a similar vein, strategies have been developed for shape-feature abstrac-

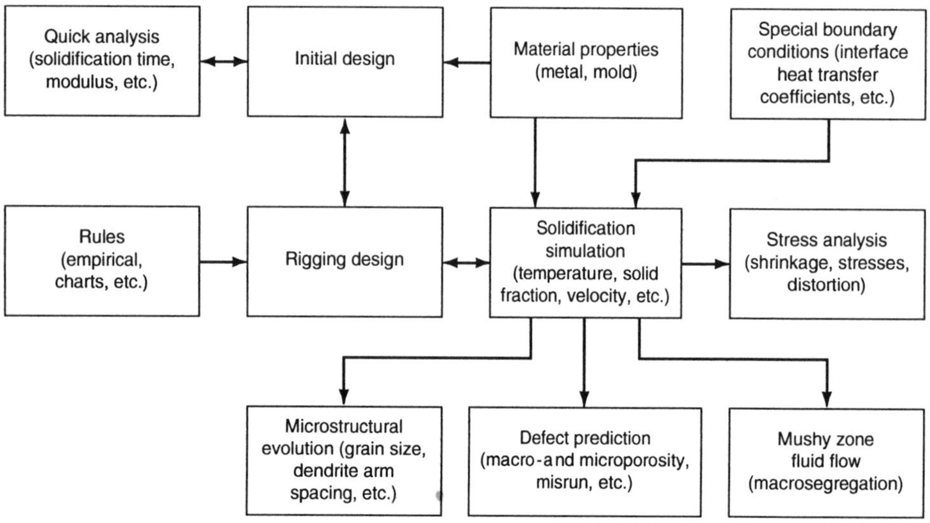

Fig. 11 Overall architecture of the modeling of casting processes. Source: Ref 14

tion in knowledge-based analysis of manufactured components (Ref 21).

However, the drawback of most of the currently available design tools is the lack of a fully integrated system for the design of gates and risers, followed by a comprehensive process simulation. More recently, an attempt has been made to close this gap by developing a system for automatic rigging design (Ref 22). The drawback of this system is that the final design (which includes the part and rigging) is in the form of a finite difference mesh model and not a solid model, which inhibits automatic pattern generation. Currently, efforts are underway to overcome this problem.

In these methods for rigging design, the first step is the generation of the solid model of the part. This solid model is used to generate a discretized description of the part, in the form of a finite difference or finite element mesh. The gating design is achieved by applying some empirical heuristics to the discretized solid model, as well as performing a geometric analysis to determine the natural flow paths for liquid metal. These empirical heuristics include rules for design of runners, sprues, and gates.

As a general rule, the casting should be gated and fed in a manner to ensure progressive solidification of the casting. There should be an adequate supply of molten metal to feed every section as it solidifies. The solidification should start at the location furthest from the ingate/casting junction and proceed toward the risers, which should solidify last. The design of risers involves a geometric analysis of the casting using empirical relations such as Chvorinov's Rule to obtain the solidification time profile, followed by determination of the size and location of the riser. The size of the riser is decided, based on the feed metal requirements as well as the solidification time.

The Comprehensive Problem: Fluid Flow and Heat Transfer. Knowledge about fluid flow during the filling of casting is important, for it affects heat transfer both during and after filling. The information obtained could help avoid problems of cold shuts, where the melt solidifies before filling a void, or where a molten front of liquid comes in contact with a solidified metal. A mold-filling simulation is, therefore, indispensable if a high-quality casting-solidification analysis is desired. There are several computational techniques to simulate fluid flow during mold filling, some of which follow.

Momentum Balance Technique: The Solution Algorithm-Volume of Fluid Approach. To obtain an accurate profile of the velocity distribution in the mold during the filling of castings, one has to solve the governing equations. The main governing equations that need to be solved to track the free surface and obtain the velocity distribution in the melt are the continuity equation, the Navier-Stokes equation, and the equation for free-surface tracking. The continuity equation is given by:

$$\nabla \cdot v = 0 \quad \text{(Eq 8)}$$

where v represents the velocity. The Navier-Stokes equation is:

$$\rho \frac{Dv}{Dt} = \nabla P - \nabla \cdot \tau + \rho g \quad \text{(Eq 9)}$$

where P is the fluid pressure, τ is the stress tensor, g is the acceleration due to gravity, and ρ is the density. The equation for free-surface tracking is:

$$\frac{\partial}{\partial t}\left(\int_{\Omega} \rho F d\Omega \right) = -\int_{s} \rho F v \cdot n dS \quad \text{(Eq 10)}$$

where F is a function that defines the fractional volume of the control element occupied by the fluid, Ω is the volume, S is the surface area, and n is the unit normal vector on the surface.

Several boundary conditions are applied to solve the equations, such as no-slip condition at the solid surfaces and atmospheric pressure at the free surface. The solutions to these equations, especially for three-dimensional cases, can be computationally very demanding. Over the past decade, effective algorithms have been developed for simulation of mold filling and solidification using both finite difference as well as finite element methods. Calculation of the location of the liquid and the orientation of its free surface is an integral part of this computational technique.

The thermal history inside the casting and mold is obtained by solving the energy equation:

$$C_P \rho \frac{dT}{dt} = -C_P \rho v \cdot \nabla T - \nabla \cdot q + \dot{Q} \quad \text{(Eq 11)}$$

where C_P is the specific heat, q is the heat flux, and $\dot{Q}$ is the rate of heat generation.

The boundary conditions can be of three types. In the first type, the temperature at the boundary is specified, and in the second, the heat flux is specified. More popularly, a third type of boundary condition is used, expressing the heat loss at the interface through a heat transfer coefficient:

$$q = -k \frac{dT}{dx}\Big|_{x = \text{bound}} = h(T - T_o) \quad \text{(Eq 12)}$$

where h is the effective heat transfer coefficient across the mold-metal interface, T is the casting surface temperature, and T_o is the mold surface temperature.

Recently, an alternative method of solving the volume-of-fluid equation through an analogy between the numerical treatment of filling and solidification has been developed (Ref 23). In this method, an alternative volume-of-fluid equation has been proposed, based on an enthalpy-type variable to determine the function F. Encouraging results have been obtained with this approach.

Modeling Microstructural Evolution. The solution to Eq 11 requires knowledge of the term $\dot{Q}$ (the rate of latent heat evolution during solidification). This term can be described in two ways: specific heat method or latent heat method.

Specific Heat Method. In this classical approach, the latent heat is released by assuming that the solid forms in a specified temperature range. The specified heat is modified as:

$$C_P^* = C_P - \frac{L}{\rho} \frac{df_s}{dT} \quad \text{(Eq 13)}$$

where L is the latent heat and df_s/dT is the rate of change of solid fraction with temperature, obtained from the phase diagram. Alternatively, one could use the Scheil equation to express solid fraction evolution as:

$$f_s = 1 - \left[\frac{T_f - T}{T_f - T_l} \right]^{1/k-1} \quad \text{(Eq 14)}$$

where T_f is the fusion temperature of the pure metal, T_l is the liquidus temperature, and k is the partition coefficient of the alloy. These parameters are obtained from the phase diagram for the alloy. Using the above equation, one can estimate the latent heat released due to the evolution of the primary phase.

The limitation of the above approach is that one cannot obtain information regarding the microstructure, such as grain size. New approaches were developed in the last decade to overcome this limitation (Ref 24, 25). These approaches incorporate the metallurgy of solidification into the simulations. In one such approach, known as the latent heat method, it has been shown that, in order to obtain microstructural information, the evolution of fraction of solid should take into account not only temperature but also the kinetics of nucleation and growth.

Latent Heat Method. In the latent heat method, the term $\dot{Q}$ in Eq 11 is evaluated based on the solidification kinetics of nucleation and growth as applicable to the transformations occurring in the system. The expression for $\dot{Q}$ is given by:

$$\dot{Q} = L \frac{df_s}{dt} \quad \text{(Eq 15)}$$

where L is the latent heat of the solidifying phase and is the rate of evolution of fraction of solid. Latent heat generation can be determined through the use of mathematical expressions to describe the evolution of the solid phase. The temperature history in the casting can then be obtained by solving Eq 11. Depending on the nature of the alloy, the expressions can vary. The following section describes the solidification kinetics for equiaxed eutectic structures, which are commonly found in cast iron systems and aluminum-silicon systems.

Probabilistic Models. Useful as they are, the deterministic models suffer from several shortcomings. They neglect any aspect related to crystallographic effects. So they are unable to account for the grain selection near the mold surface, which leads to the columnar region. Furthermore, they do not account for the random nature of equiaxed grains. One cannot visualize the actual evolution of the grains; one can only get an idea of the size of the grains.

To overcome these limitations, several researchers have used an altogether different track to model microstructural evolution (Ref 4, 26).

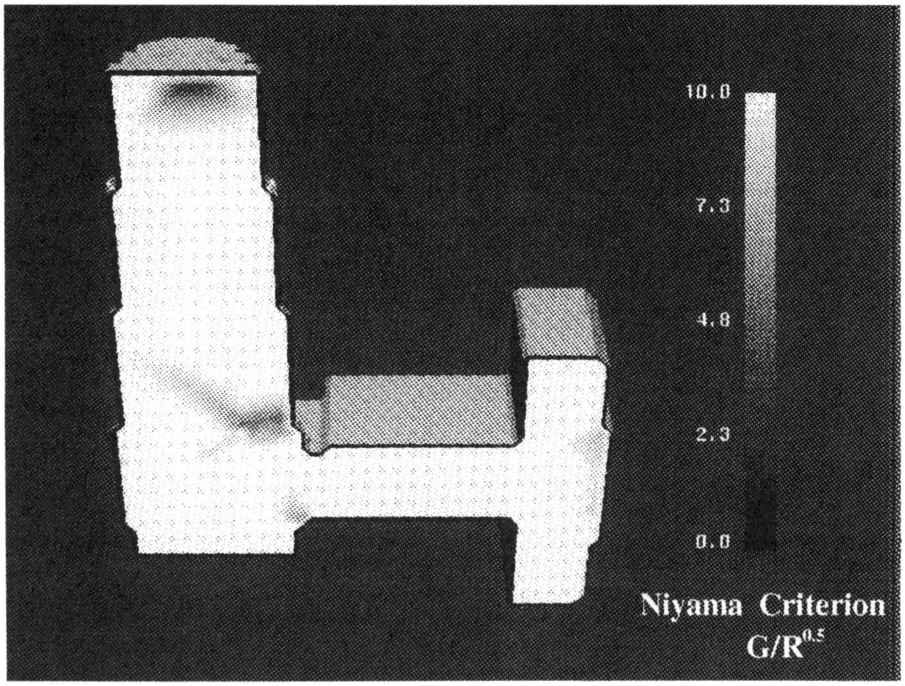

Fig. 12 Distribution of Niyama values in a ductile iron plate casting, showing propensity of defects only in the riser. Source: Ref 14

They make use of probabilistic models to simulate the evolution of grain structure. These simulations are more like numerical experiments. In one such work (Ref 26), a Monte Carlo procedure was used to simulate evolution of grain structure. This type of method is based on the principle of minimization of energy, where the energy of a given structural configuration is evaluated considering the present state of the various sites (whether solid or liquid). Transitions are allowed to take place according to randomly generated numbers. Using this technique, the researchers were able to compute two-dimensional microstructures that closely resembled those observed experimentally. The main drawback of such an approach was the lack of a physical basis.

In a more recent investigation (Ref 4), a new approach for modeling grain structure formation during solidification was proposed. Based on a two-dimensional cellular automata technique, the model includes the mechanisms of heterogeneous nucleation and of grain growth. Nucleation occurring at the mold wall, as well as in the liquid metal, is treated by using two distributions of nucleation sites. The location and the crystallographic orientation of the grains are chosen randomly among a large number of cells and a certain number of orientation classes. The model has been applied to small specimens of uniform temperature. The columnar-to-equiaxed transition, the selection and extension of columnar grains that occur in the columnar zone, and the impingement of equiaxed grains are clearly shown by this technique.

Prediction of Defects: The Porosity Problem. Analysis of the conditions leading to the occurrence of casting porosity has been the focus of a number of investigations in the past few decades. With the advances in computer modeling of the casting process in recent years, there has been considerable interest in the usage of numerical heat transfer and solidification models to predict casting porosity.

As far as the solidification parameters are concerned, the variables that control porosity may be narrowed down to the thermal gradient, the rate of solidification, the cooling rate, and the solidification time. Based on these, various approaches have been suggested to predict casting porosity, the oldest being the empirical criteria. Thermal parameters have also been formulated recently, from Ref 27 and 28. Many of these criteria are based on d'Arcy's Law, approximating the mushy region to a porous medium. The pressure drop in the mushy region is then expressed in terms of thermal criteria functions to predict the onset of porosity.

The current modeling practice is to calculate these criteria functions using the solidification model to predict porosity. Some of these functions are quite successful in predicting porosity in short-freezing-range alloys, though there are many difficulties in applying them for long-freezing-range-alloys. Figure 12 (Ref 14) shows the Niyama distribution for a ductile iron plate casting. The figure clearly shows the defects ending up in the riser, demonstrating adequacy of feeding.

A major limitation of the criteria functions (discussed previously) to predict porosity is that they ignore the effects of casting macrostructure and grain size on porosity. The resistance to liquid feeding in the mushy region depends on the available surface area of solid in the interdendritic region, which is dictated by the macrostructure and grain size. Recently, Suri et. al. have proposed a new number called the feeding resistance number (Ref 31), which takes into account the effect of casting macrostructure on final porosity. The validity of this proposed criterion to predict porosity is still under investigation.

Modeling Special Casting Processes: The Investment-Casting Process. Many critical and value-added components in automotive, aerospace, and other key industries are manufactured by special casting processes, such as the investment-casting process, lost-foam process, tilt-pour (Cosworth) process, and so forth. Simulation of such processes requires the application of suitable submodels to handle the phenomenological aspects specific to each process. In this section, some of the research efforts in modeling investment casting are reviewed.

A comprehensive solidification simulation of the investment-casting process involves a number of computationally intensive steps, particularly the calculation of view factors (to model the radiation loss), and the three-dimensional analysis of mold-filling and solidification.

The external heat loss is either purely radiative, or radiative as well as convective. For the general case, the heat transfer coefficient can be given by:

$$h = h_r + h_c \quad \text{(Eq 16)}$$

where h_r and h_c are the radiative and convective heat transfer coefficients, respectively. The radiative heat transfer coefficient is given by (Ref 32):

$$h_r = \sigma\varepsilon F_{m-a}(T_s^2 + T_o^2)(T_s + T_o) \quad \text{(Eq 17)}$$

where σ, ε, and F_{m-a} are the Stefan-Boltzmann constant, the mold emissivity, and the view factor of the mold with respect to air, respectively.

The convective heat transfer coefficient valid for natural convection at high temperature is given by:

$$h_c = c(T_s - T_o)^{1/3} \quad \text{(Eq 18)}$$

where c is a constant dependent on the surface geometry.

View factor is defined as the fraction of the radiation that leaves surface i in all directions and is intercepted by surface j. When two surfaces, dA_1 and dA_2 undergo radiation exchange, the view factor can be mathematically expressed as (Ref 32):

$$F_{1\text{-}2} = \int_{A_1}\int_{A_2} \frac{\cos\theta\cos\phi}{R_{1\text{-}2}^2}\, dA \quad \text{(Eq 19)}$$

were $R_{1\text{-}2}$ is the distance between the two surfaces, and θ and ϕ are the angles of the two surface normals with the line joining the two surfaces.

The calculation of view factors could become very complicated when multiple surfaces are involved in the radiation process. The presence of multiple surfaces can create a partial or full obstruction in the view path between any two surfaces. Thus, the view factors will now depend not only on the two surfaces exchanging heat, but also on the shadows cast by the other surfaces present in the model. Additional calculations are needed for determining the shadowing effects for any realistic three-dimensional geometries.

More recently, another technique has been proposed that enables quick calculation of the view factor distribution at the mold surface (Ref 17). This is a modified ray-racing technique, where a scheme is devised to send rays in various directions, and the number going into air without mold interception is estimated. The view factor is then computed by calculating the fraction of rays that go into air. This scheme has been successfully applied to three-dimensional finite difference geometries.

Modeling of Fusion Welding Processes

In fusion welding, parts are joined by melting and subsequent solidification of adjacent areas of two parts. Welding may be performed with or without the addition of a filler metal.

Figure 13 is a schematic diagram of the fusion welding process. Three distinct regions in the weldment are observed: the fusion zone, which undergoes melting and solidification; the heat-affected zone, which experiences significant thermal exposure and may undergo solid-state transformation, but no melting; and the base-metal zone, which is unaffected by the welding process.

The interaction of the material and the heat source leads to rapid heating, melting, and vigorous circulation of the molten metal driven by buoyancy, surface tension, impingement or friction, and when electric current is used, electromagnetic forces. The resulting heat transfer and fluid flow affect the size and shape of the weld pool, the cooling rate, and the kinetics and extent of various solid-state transformation reactions in the fusion zone and heat-affected zone. The weld geometry influences dendrite and grain-growth selection processes. Both the partitioning of nitrogen, oxygen, and hydrogen between the weld pool and its surroundings, and the vaporization of alloying elements from the weld-pool surface greatly influence the composition and the resulting microstructure and properties of the weld metal. In many processes, such as the arc welding and laser-beam welding, an electrically conducting, luminous gas plasma forms near the weld pool.

Energy Absorption. During welding, the workpiece absorbs only a portion of the total energy supplied by the heat source. The absorbed energy is responsible for the outcome of the welding process. The consequences of the absorbed energy include formation of the liquid pool, establishment of the time-dependent temperature field in the entire weldment, and the structure and properties of the weldment. Therefore, it is very important to understand the physical processes in the absorption of energy during the welding process. The physical phenomena that influence the energy absorption by the workpiece depends on the nature of the material, the type of heat source, and the parameters of the welding process.

For arc welding, the fraction of the arc energy transferred to the workpiece, η, commonly known as the arc efficiency, is given by (Ref 33):

$$\eta = \frac{q}{VI} = 1 - \frac{q_e + (1-n)q_p + mq_w}{VI} \qquad \text{(Eq 20)}$$

where q is the heat absorbed by the workpiece, I and V are the welding current and voltage, respectively, q_e is the heat transferred to the electrode from the heat source, q_p is the energy radiated and convected to the arc column per unit time (of which a proportion n is transferred to the workpiece), and q_w is the heat absorbed by the workpiece (of which a proportion m is radiated away). For a consumable electrode, the amount of energy transferred to the electrode is eventually absorbed by the workpiece. Thus, the above equation is simplified to:

$$\eta = 1 - \frac{(1-n)q_p + mq_w}{VI} \qquad \text{(Eq 21)}$$

Fluid Flow in the Weld Pool. The properties of the weld metal are strongly affected by the fluid flow and heat transfer in the weld pool. The flow is driven by surface tension, buoyancy, and, when electric current is used, electromagnetic forces (Ref 34–45). In some instances, aerodynamic drag forces of the plasma jet may also contribute to the convection in the weld pool (Ref 46). Buoyancy effects originate from the spatial variation of the liquid-metal density, mainly because of temperature variations and, to a lesser extent, from local composition variations. Electromagnetic effects are a consequence of the interaction between the divergent current path in the weld pool and the magnetic field that it generates. This effect is important in arc and electron-beam welding, especially when a large electric current passes through the weld pool. In arc welding, a high-velocity plasma stream impinges on the weld pool. The friction of the impinging jet on the weld-pool surface can cause significant fluid motion at high currents. Fluid flow and convective heat transfer are often very important in determining the size and shape of the weld pool, the weld macrostructures and microstructures, and the weldability of the material.

Marangoni Force. The spatial gradient of surface tension is a stress, known as the Marangoni stress. The spatial variation of the surface tension at the weld-pool surface can arise owing to variations of both temperature and composition. Frequently, the main driving force for convection is the spatial gradient of surface tension at the weld-pool surface. In most cases, the difference in surface tension is due to the temperature variation at the weld-pool surface. For such a situation, the Marangoni stress can be expressed as:

$$\tau = \frac{d\gamma}{dT}\frac{dT}{dy} \qquad \text{(Eq 22)}$$

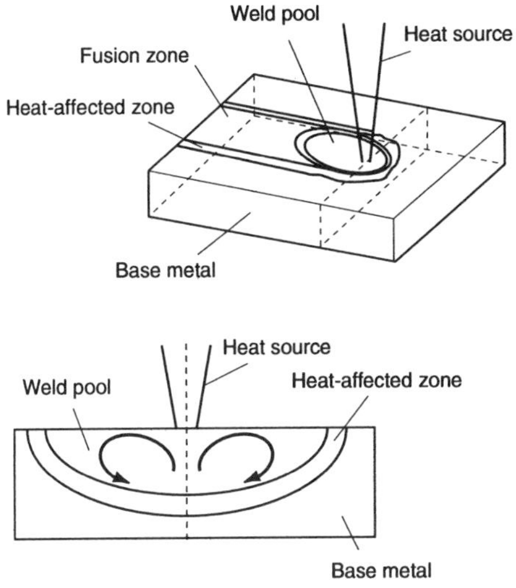

Fig. 13 Schematic representations of the fusion welding process

where τ is the shear stress due to temperature gradient, γ is the interfacial tension, T is the temperature, and y is the distance along the surface from the axis of the heat source. If a boundary layer develops, the shear stress τ can also be expressed as (Ref 47):

$$\tau = \frac{0.332\rho^{1/2}\mu^{1/2}u^{1/2}}{y^{1/2}} \qquad \text{(Eq 23)}$$

where ρ is the density, μ is the viscosity, and u is the local velocity.

Buoyancy and Electromagnetic Forces. When the surface-tension gradient is not the main driving force, the maximum velocities can be much smaller. For example, when the flow is buoyancy driven, the maximum velocity, u_m, can be approximated by the following relation (Ref 48):

$$u_m \approx \sqrt{g\beta\Delta T d} \qquad \text{(Eq 24)}$$

where g is the acceleration due to gravity, β is the coefficient of volume expansion, ΔT is the temperature difference, and d is the depth. For the values of $\Delta T = 600$ °C, $g = 981$ cm/s^2, $\beta = 3.5 \times 10^{-5}$/°C, and $d = 0.5$ cm, the value of u_m is 3.2 cm/s. The existence of electromagnetically driven flow was demonstrated by Woods and Milner (Ref 49), who observed flow of liquid metal when current was passed in the metal bath by means of a graphite electrode. In the case of electromagnetically driven flow in the weld pool, the velocity values reported in the literature are typically in the range of 2 to 20 cm/s (Ref 50). The magnitude of the velocities of both buoyancy and electromagnetically driven flows in the weld pool are commonly much smaller than those obtained for surface-tension-driven flows.

Convection Effects on Weld-Pool Shape and Size. Variable depth of penetration during the welding of different batches of a commercial material with composition within a prescribed range has received considerable attention. Often,

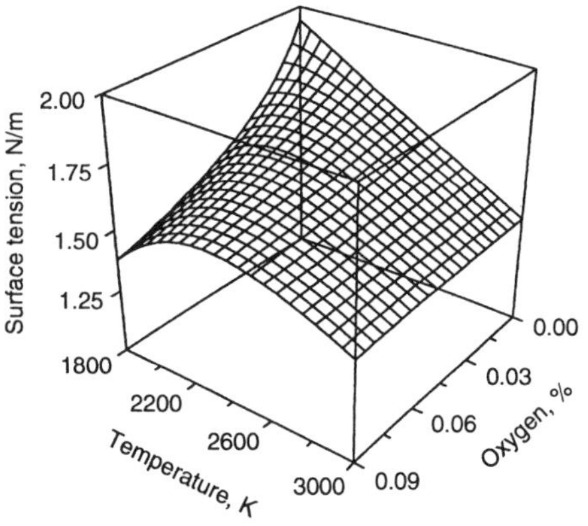

Fig. 14 Calculated values of surface tension for Fe-O alloys. Source: Ref 53

the penetration depth is strongly influenced by the concentration of surface-active elements such as oxygen or sulfur in steels. These surface-active impurity elements can affect the temperature coefficient of surface tension, $d\gamma/dT$, the resulting direction of convective flow in the weld pool (Ref 51), and the shape of the weld pool. The interfacial tension in these systems could be described by a formalism based on the combination of Gibbs and Langmuir absorption isotherms (Ref 52):

$$\gamma = \gamma_m^o - A(T - T_m) - RT\,\Gamma_s \ln (1 + k_1 a_i e^{-\Delta H^o/R'T}) \quad \text{(Eq 25)}$$

where γ is the interfacial tension as a function of composition and temperature, γ_m^o is the interfacial tension of the pure metal at the melting point T_m, A is the temperature coefficient of surface tension for the pure metal, R and R' are the gas constants in appropriate units, T is the absolute temperature, Γ_s is the surface excess of the solute at saturation solubility, k_1 is the entropy factor, a_i is the activity of the solute, and ΔH^o is the enthalpy of segregation. The calculated values (Ref 53) of surface tension for Fe-O alloys are shown in Fig. 14. It is seen that for certain concentrations of oxygen, $d\gamma/dT$ can change from a positive value at "low" temperature to a negative value at "high" temperature. This implies that in a weld pool containing fairly high oxygen contents, $d\gamma/dT$ can go through an inflection point on the surface of the pool. Under these conditions, the fluid flow in the weld pool is more complicated than a simple recirculation.

The calculated gas tungsten arc weld (GTAW) fusion-zone profiles (Ref 53) for pure iron, and an Fe-0.03O alloy are shown in Fig. 15. The results clearly show the significant effect of oxygen concentration on the weld-pool shape and the aspect ratio. Near the heat source, where the temperature is very high, the flow is radially outward. However, a short distance away from the heat source, where the temperature drops below the threshold value for the change in the sign of $d\gamma/dT$, the flow reverses in direction. The flow field is not a simple recirculation. Although the qualitative effects of the role of surface-active

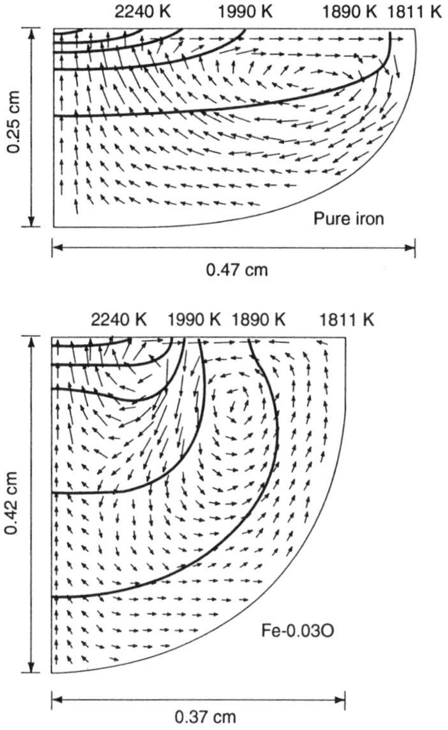

Fig. 15 Calculated GTAW fusion-zone profiles for pure iron and Fe-0.03O alloy. Source: Ref 53

elements are known, the numerical calculations provide a basis for quantitative assessment of their role in the development of weld-pool geometry.

Conclusions

Much work has been done in developing techniques to simulate various manufacturing processes. These efforts range from conducting basic materials research to advanced software engineering. Through an optimum combination of these, tools can be developed that reduce manufacturing costs and increase quality. In fact, many such commercial tools are available today and are being utilized in the manufacturing operations of many companies. However, for modeling to have a significant impact on the manufacturing industry, it should be geared toward the small- and medium-sized enterprises that make up the bulk of the industry. Issues such as personnel training, well-integrated software environments, and help in setting up problems are important. Today, software developers and vendors are working to address these issues in an effort to enable users to take full advantage of the benefits that modeling and simulation can provide to the manufacturing industry.

REFERENCES

1. M.F. Ashby, Physical Modeling of Materials Problems, *Mater. Sci. Technol.,* Vol 8 (No. 2), 1992, p 102–111
2. H.R. Shercliff and M.F. Ashby, Modeling Thermal Processing of Al Alloys, *Mater. Sci. Technol.,* Vol 7 (No. 1), 1991, p 85–88
3. B.G. Thomas, Comments on the Industrial Application of Process Models, *Materials Processing in the Computer Age—II,* V.R. Voller, S.P. Marsh, and N. El-Kaddah, Ed., The Minerals, Metals & Materials Society, 1995, p 3–19
4. M. Rappaz and Ch.-A. Gandin, Probabilistic Modeling of the Microstructure Formation in Solidification Processes, *Acta. Metall. Mater.,* Vol 41 (No. 2), 1993, p 345–360
5. J. Wang, S. Xun, R.W. Smith, and P.N. Hansen, Using SOLVA-VOF and Heat Convection and Conduction Technique to Improve the Casting Design of Cast Iron, *Modeling of Casting, Welding and Advanced Solidification Processes VI,* T. Piwonka, Ed., TMS, 1993, p 397–412
6. L. Edwards and M. Endean, Ed., *Manufacturing with Materials,* Materials in Action, Butterworth Scientific, 1990
7. S.C. Jain, Recognizing the Need for Materials Data: The Missing Link in Process Modeling, *J. Met.,* Oct 1991, p 6–7
8. K.-J. Bathe, Some Issues for Reliable Finite Element Analysis, *Reliability Methods for Engineering Analysis: Proc. First International Conference,* K.-J. Bathe and D.R.J. Owen, Ed., Pineridge Press, Swansea, U.K., 1986, p 139–157
9. S. Kobayashi, S. Oh, and T. Altan, *Metal Forming and the Finite Element Method,* Oxford University Press, 1989, p 27
10. M.L. Tims, J.D. Ryan, W.L. Otto, and M.E. Natishan, Crack Susceptibility of Nickel-Copper Alloy K-500 Bars During Forging and Quenching, *Proc. First International Conference on Quenching and Control of Distortion,* ASM International, 1992, p 243–250
11. J.O. Hallquist, "NIKE2D—A Vectorized Implicit, Finite Deformation Finite Element Code for Analyzing the Static and Dynamic Response of 2-D Solids with Interactive Rezoning and Graphics," User's Manual, UCID-19677, Rev. 1, Lawrence Livermore National Laboratory, 1986
12. S.E. Clift, P. Hartley, C.E.N. Sturgess, and G.W. Rowe, Fracture Prediction in Plastic Deformation Processes, *Int. J. Mech. Sci.,* Vol 32 (No. 1), p 1–17
13. D.K. Ebersole and M.G. Zelin, "Superplastic Forming of Aluminum Aircraft Assemblies—Simulation Software Enhancements," NCEMT Report No. TR 97-05, National Center for Excellence in Manufacturing Technologies, Johnstown, PA, 1997
14. G. Upadhya and A.J. Paul, Solidification Modeling: A Phenomenological Review, *AFS Trans.,* Vol 94, 1994, p 69–80
15. G. Upadhya, C.M. Wang, and A.J. Paul, Solidification Modeling: A Geometry Based Approached for Defect Prediction in Castings, *Light Metals 1992,* Proceedings of Light Metals Div. At 121st TMS Annual Meeting (San Diego), E.R. Cutshall, Ed., TMS, 1992, p 995–998
16. N. Chvorinov, Theory of Solidification of Castings, *Giesserei,* Vol 27, 1940, p 17–224

17. G. Upadhya, S. Das, U. Chandra, and A.J. Paul, Modeling the Investment Casting Process: A Novel Approach for View Factor Calculations and Defect Predictions, *Appl. Math. Model.*, Vol 19, 1995, p 354–362
18. S.C. Luby, J.R. Dixon, and M.K. Simmons, Designing with Features: Creating and Using Features Database for Evaluation of Manufacturability of Castings, *ASME Comput. Rev.*, 1988, p 285–292
19. J.L. Hill and J.T. Berry, Geometric Feature Extraction for Knowledge-Based Design of Rigging Systems for Light Alloys, *Modeling of Casting, Welding and Advanced Solidification Processes V*, TMS, 1990, p 321–328
20. J.L. Hill, J.T. Berry, and S. Guleyupoglu, Knowledge-Based Design of Rigging Systems for Light Alloy Castings, *AFS Trans.*, Vol 99, 1992, p 91–96
21. R. Gadh and F.B. Prinz, Shape Feature Abstraction in Knowledge-Based Analysis of Manufactured Products, *Proc. of the Seventh IEEE Conf. on AI Applications*, Institute of Electrical and Electronics Engineers, 1991, p 198–204
22. G. Upadhya, A.J. Paul, and J.L. Hill, Optimal Design of Gating and Risering in Castings: An Integrated Approach Using Empirical Heuristics and Geometric Analysis, *Modeling of Casting, Welding and Advanced Solidification Processes VI*, T.S. Piwonka, Ed., TMS, 1993, p 135–142
23. C.R. Swaminathan and V.R. Voller, An "Enthalpy Type" Formulation for the Numerical Modeling of Mold Filling, *Modeling of Casting, Welding and Advanced Solidification Processes VI*, T.S. Piwonka, Ed., TMS, 1993, p 365–372
24. D.M. Stefanescu and C.S. Kanetkar, Computer Modeling of the Solidification of Eutectic Alloys: The Case of Cast Iron, *Computer Simulation of Microstructural Evolution*, D.J. Srolovitz, Ed., TMS-AIME, 1985, p 171–188
25. D.M. Stefanescu, G. Upadhya, and D. Bandyopadhyay, Heat Transfer-Solidification Kinetics Modeling of Solidification of Castings, *Metall. Trans. A*, Vol 21A, 1990, p 997–1005
26. J.A. Spittle and S.G.R. Brown, *Acta Metall.*, Vol 37 (No. 7), 1989, p 1803–1810
27. E. Niyama, T. Uchida, M. Morikawa, and S. Saito, A Method of Shrinkage Prediction and Its Application to Steel Casting Practice, *AFS Cast Metal Res. J.*, Vol 7 (No. 3), 1982, p 52–63
28. Y.W. Lee, E. Chang, and C.F. Chieu, *Metall. Trans.*, Vol 21B, 1990, p 715–722
29. V.K. Suri, G. Huang, J.T. Berry, and J.L. Hill, Applicability of Thermal Parameter Based Porosity Criteria to Long Freezing Range Aluminum Alloys, *AFS Trans.*, 1992, p 399–407
30. P.N. Hansen and P.R. Sahm, How to Model and Simulate the Feeding Process in Casting to Predict Shrinkage and Porosity Formation, *Modeling of Casting and Welding Processes IV*, A. Giamei and G.J. Abbaschian, Ed., TMS, 1988, p 33–42
31. H. Huang, V.K. Suri, N. El-Kaddah, and J.T. Berry, *Modeling of Casting, Welding and Advanced Solidification Processes VI*, T.S. Piwonka, Ed., TMS Publications, 1993, p 219–226
32. G.H. Geiger and D.R. Poirier, *Transport Phenomena in Metallurgy*, Addison Wesley, 1975, p 332
33. J.F. Lancaster, *Metallurgy of Welding*, Allen and Unwin, London, 1980
34. S. Kou and Y. Le, *Metall. Trans. A.*, Vol 14, 1983, p 2245
35. C. Chan, J. Mazumdar, and M.M. Chen, *Metall. Trans. A*, Vol 15, 1984, p 2175
36. G.M. Oreper and J. Szekely, *J. Fluid Mech.*, Vol 147, 1984, p 53
37. J. Dowden, M. Davis, and P. Kapadia, *J. Appl. Phys.*, Vol 57 (No. 9), 1985, p 4474
38. A. Paul and T. DebRoy, *Advances in Welding Science and Technology*, S.A. David, Ed., American Society for Metals, 1986, p 29
39. M.C. Tsai and S. Kou, *J. Numer. Methods Phys.*, Vol 9, 1989, p 1503
40. T. Zacharia, S.A. David, J.M. Vitek, and T. DebRoy, *Weld. J. Res. Suppl.*, Vol 68, 1989, p 510s
41. T. Zacharia, S.A. David, J.M. Vitek, and T. DebRoy, *Metall. Trans. A*, Vol 20, 1989, p 957
42. J. Mazumdar and A. Kar, *Thermomechanical Aspects of Manufacturing and Materials Processing*, R.K. Shah, Ed., Hemisphere, 1992, p 283
43. K. Mundra and T. DebRoy, *Metall. Trans. B*, Vol 24, 1993, p 145
44. K. Mundra and T. DebRoy, *Weld. J. Res. Suppl.*, Vol 72, 1993, p 1s
45. R.T. Choo and Z. Szekely, *Weld. J. Res. Suppl.*, Vol 73, 1994, p 25s
46. A. Matsunawa, *International Trends in Welding Science and Technology*, S.A. David and J. Vitek, Ed., ASM International, 1993, p 3
47. C.J. Geankoplis, *Transport Processes and Unit Operations*, Allyn and Bacon, 1983, p 186
48. J. Szekely, *Advances in Welding Science and Technology*, S.A. David, Ed., ASM International, 1986, p 3
49. R.A. Woods and D.R. Milner, *Weld. J. Res. Suppl.*, Vol 50, 1971, p 163s
50. Y.H. Wang and S. Kou, *Advances in Welding Science and Technology*, S.A. David, Ed., ASM International, 1986, p 65
51. A.J. Paul and T. DebRoy, *Metall. Trans. B.*, Vol 19, 1988, p 851
52. M.J. McNallan and T. DebRoy, *Metall. Trans. B.*, Vol 22, 1991, p 557
53. T. DebRoy and S.A. David, Physical Processes in Fusion Welding, *Rev. Mod. Phys.*, Vol 67 (No. 1), 1995, p 85–112

Manufacturing Cost Estimating

David P. Hoult and C. Lawrence Meador, Massachusetts Institute of Technology

COST ESTIMATION is an essential part in the design, development, and use of products. In the development and design of a manufactured product, phases include concept assessment, demonstrations of key features, and detailed design and production. The next phase is the operation and maintenance of the product, and finally, its disposal. Cost estimation arises in each of these phases, but the cost impacts are greater in the development and design phases.

Anecdotes, experience, and some data (Ref 1, 2) support the common observation that by the time 20% of a product is specified, 70 to 80% of the costs are committed, even if those costs are unknown! Another common perception is that the cost of correcting a design error (cost overrun) rises very steeply as product development proceeds through its phases. What might cost one unit of effort to fix in the concept assessment phase might cost a thousand units in detailed engineering. These experiences of engineers and designers force the cost estimator to think carefully about the context, timing, and use of cost estimates.

General Concepts

In this article, the focus is on products defined by dimensions and tolerances, made from solid materials and, fabricated by some manufacturing process. Two issues should be apparent: first, accurate cost estimates of a product in its early stages of design is difficult; second, there are a very large number of manufacturing processes, so one must somehow restrict the discussion to achieve sensible results.

In dealing with the first issue a series of cost estimates corresponding to the phases in the product-development program should be considered. As more details of the product are specified, the cost estimates should become more accurate. Thinking this way, it is plausible that different tools for cost estimation may be employed in different phases of a program. In this review examples are given of methods of cost estimation that may be used in the contexts of the different phases of a development program.

Domain Limitation. The second issue gives rise to the important principle of *domain limitations* (Ref 3, 4), meaning data that form the basis for a cost estimate must be specific to the manufacturing process, the materials used, and so on. Cost estimates only apply within specific domains. This leads directly to another difficulty: most products, even if they are only moderately complex, like a dishwasher with 200 unique parts, have a least three or five domains in which cost estimates apply (i.e., sheet metal assembly, injection-molded plastic parts, formed sheet metal parts, etc.). More complex products, such as an aircraft jet engine with 25,000 unique parts, might have 200 different manufacturing processes, each of which defines one or more domains of cost estimation. Clearly, as one considers still more complex products, such as a modern military radar system with 10,000 to 20,000 unique parts, or a large commercial airliner with perhaps 5 million unique parts, the domains of cost estimation expand dramatically. So, although domain limitation is necessary for cost-estimates accuracy, it is not a panacea.

Database Commonality. Estimating the costs of a complex product through various phases of development and production requires organization of large amounts of data. If the data for design, manufacturing, and cost are linked, there is *database commonality.* It has been found (Ref 3) that having database commonality results in dramatic reductions in cost and schedule overruns in military programs. In the same study, domain limitation was found to be essential in achieving database commonality.

Having database commonality with domain limitation implies that the links between the design and specific manufacturing processes, with their associated costs, are understood and delineated. Focusing on specific manufacturing processes allows one to collect and organize data on where and how costs arise in specific processes. With this focus, the accuracy of cost estimates can be determined, provided that uniform methods of estimation are used, and provided that, over time, the cost estimates are compared with the actual costs as they arise in production. In this manner, the accuracy of complex cost estimates may be established and improved.

In present engineering and design practice, many organizations do not have adequate database commonality, and the accuracy of cost estimates is not well known. Database commonality requires an enterprise-wide description of cost-dominant manufacturing processes, a way of tracking actual costs for each part, and a way of giving this information—in an appropriate format—to designers and cost estimators. Most "empirical methods" of cost estimation, which are based on industrywide studies of statistical correlation of cost, may or may not apply to the experience of a specific firm (see the discussion in the sections that follow).

Costs are "rolled up" for a product when all elements of the cost of a product are accounted for. Criteria for cost estimation using database commonality is simple: speed (how long does it take to roll up a cost estimate on a new design), accuracy (what is the standard deviation of the estimate, based on comparison with actual costs) and risk (what is the probability distribution of the cost estimate; what fraction of the time will the estimate be more than 30% too low, for example). One excellent indicator of database commonality is the roll-up time criteria. World-class cost-estimation roll-up times are minutes to fractions of days. Organizations that have such rapid roll-up times have significantly less cost and schedule overruns on military projects (Ref 3).

Cost allocation is another general issue. Cost allocation refers to the process by which the components of a design are assigned target costs. The need for cost allocation is clear: how else would an engineer, working on a large project, know how much the part being designed should cost? And, if the cost is unknown and the target cost is not met, there will be time delays, and hence costs incurred due to unnecessary design iteration. It is generally recognized that having integrated product teams (IPTs) is a good industrial practice. Integrated product teams should allocate costs at the earliest stages of a development program. Cost estimates should be performed concurrently with the design effort throughout the development process. Clearly, estimating costs at early stages in a development program, for example, when the concept of the product is being assessed, requires quite different tools than when most or all the details of the design are specified. Various tools that can be used to estimate cost at different stages of the development process are described later in this section.

Elements of Cost. There are many elements of cost. The simplest to understand is the cost of material. For example, if a part is made of aluminum and is fabricated from 10 lb of the material, if the grade of aluminum costs \$2/lb, the material cost is \$20. The estimate gets only a bit more complex if, as in the case of some aerospace components, some 90% of the materials will be machined away; then the sale on scrap material is deducted from the material cost.

Tooling and fixtures are the next easiest items to understand. If tools are used for only one product, and the lifetime of the tool is known or can be estimated, then only the design and fabrication cost of the tool is needed. Estimates of the fabrication costs for tooling are of the same form as those for the fabricated parts. The design cost estimate raises a difficult and general problem: cost capture (Ref 4). For example, tooling design costs are often classified as overhead, even though the cost of tools relates to design features. In many accounting systems, manufacturing costs are assigned "standard values," and variances from the standard values are tabulated. This accounting methodology does not, in general, allow the cost engineer to determine the actual costs of various design features of a part. In the ledger entries of many accounting systems, there is no allocation of costs to specific activities or no activity-based accounting (ABC) (Ref 5). In such cases there are no data to support design cost estimates.

Direct labor for products or parts that have a high yield in manufacturing normally have straightforward cost estimates, based on statistical correlation to direct labor for past parts of a similar kind. However, for parts that have a large amount of rework the consideration is more complex, and the issues of cost capture and the lack of ABC arise again. Rework may be an indication of uncontrolled variation of the manufacturing process. The problem is that rework and its supervision may be classified all, or in part, as manufacturing overhead. For these reasons, the true cost of rework may not be well known, and so the data to support cost estimates for rework may be lacking.

The cost estimates of those parts of overheads that are associated with the design and production of a product are particularly difficult to estimate, due to the lack of ABC and the problem of cost capture. For products built in large volumes, of simple or moderate complexity, cost estimates of overheads are commonly done in the simplest possible way: the duration of the project and the level of effort are used to estimate the overhead. This practice does not lead to major errors because the overhead is a small fraction of the unit cost of the product.

For highly engineered, complex products built in low volume, cost estimation is very difficult. In such cases the problem of cost capture is also very serious (Ref 4).

Machining costs are normally related to the machine time required and a capital asset model for the machine, including depreciation, training, and maintenance. With a capital asset model, the focus of the cost estimate is the time to manufacture. A similar discussion holds for assembly costs: with a suitable capital asset model, the focus of the cost estimate is the time to assemble the product (Ref 1).

Methods of Cost Estimations. There are three methods of cost estimation discussed in the following sections of this article. The first is parametric cost estimation. Starting from the simplest description of the product, an estimate of its overall cost is developed. One might think that such estimates would be hopelessly inaccurate because so little is specified about the product, but this is not the case. The key to this method is a careful limitation of the *domain* of the estimate (see the previous section). This example deals with the estimate of the weight of an aircraft. The cost of the aircraft would then be calculated using dollars/pound typical of the aircraft type. Parametric cost estimation is the generally accepted method of cost estimation in the concept assessment phases of a development program. The accuracy is surprisingly good—about 30% (provided that recent product-design evolution has not been extensive).

The second method of cost estimation is empirically based: one identifies specific design features and then uses statistical correlation of costs of past designs to estimate the cost of the new design. This empirical method is by far the most common in use. For the empirical method to work well, the features of the product for which the estimate is made should be unambiguously related to features of prior designs, and the costs of prior designs unambiguously related to design features. Common practice is to account for only the major features of a design and to ignore details. Empirical methods are very useful in generating a rough ranking of the costs of different designs and are commonly used for that purpose (Ref 1, 6, 7). However, there are deficiencies inherent in the empirical methods commonly used.

The mapping of design features to manufacturing processes to costs is not one-to-one. Rather, the same design feature may be made in many different ways. This difficulty, the feature mapping problem, discussed in Ref 4, limits the accuracy of empirical methods and makes the assessment of risk very difficult. The problem is implicit in all empirical methods. The problem is that the data upon which the cost correlation is based may assume the use of manufacturing methods to generate the features of the design that do not apply to the new design. It is extraordinarily difficult to determine the implicit assumptions made about manufacturing processes used in a prior empirical correlation. A commonly stated accuracy goal of empirical cost estimates is 15 to 25%, but there is very little data published on the actual accuracy of the cost estimate when it is applied to new data.

The final method discussed in this article is based on the recent development called complexity theory. A mathematically rigorous definition of complexity in design has been formulated (Ref 8). In brief, complexity theory offers some improvement over traditional empirical methods: there is a rational way to assess the risk in a design, and there are ways of making the feature mapping explicit rather than implicit. Perhaps the most significant improvement is the capability to capture the cost impact of essentially all the design detail in a cost estimate. This allows designers and cost estimators to explore, in a new way, methods to achieve cost savings in complex parts and assemblies.

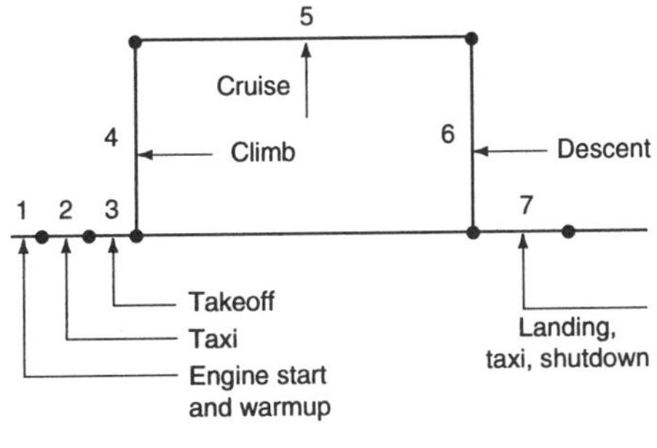

Fig. 1 Mission profile

Parametric Methods

An example for illustrating parametric cost estimation is that of aircraft. In Ref 9, Roskam—a widely recognized researcher in this field—describes a method to determine the size (weight) of an aircraft. Such a calculation is typical of parametric methods. To determine cost from weight, one would typically correlate costs (inflation adjusted) of past aircraft of similar complexity with their weight. Thus weight is surrogate for cost for a given level of complexity.

Most parametric methods are based on such surrogates. For another simple example, consider that large coal-fired power plants, based on a steam cycle, cost about \$1500/kW to be built. So, if the year the plant is to be built (for inflation adjustment) and its kW output is known, parametric cost estimate can be readily obtained.

Parametric cost estimates have the advantage that little needs to be known about the product to produce the estimate. Thus, parametric methods are often the only ones available in the initial (concept assessment) stages of product development.

The first step in a parametric cost estimation is to limit the domain of application. Roskam correlates statistical data for a dozen types of aircraft and fifteen sub types. The example he uses to explain the method is that of a twin-engine, propeller-driven airplane. The mission profile of this machine is given in Fig. 1 (Ref 9).

Inspection of the mission specifications and Fig. 1 shows that only a modest amount of information about the airplane is given. In particular, nothing is specified about the detailed design of the machine! The task is to estimate the total weight, W_{TO} or the empty weight, W_E, of the airplane. Roskam argues that the total weight is equal to the sum of the empty weight, fuel weight, W_F, payload and crew weight, $W_{PL} + W_{crew}$, and the trapped fuel and oil, which is

Table 1 Mission specification for a twin-engine, propeller-driven airplane

1. Payload	Six passengers at 175 lb each (including the pilot) and 200 lb total baggage
2. Range	1000 statute miles with maximum payload
3. Reserves	25% of mission fuel
4. Cruise speed	250 knots at 75% power at 10,000 ft and at takeoff weight
5. Climb	10 min to 10,000 ft at takeoff weight
6. Takeoff and landing	1500 ft ground fun at sea level, standard day. Landing at 0.95 of takeoff weight
7. Powerplants	Piston/propeller
8. Certification base	FAR23

modeled as a fraction, M_{tfo}, to the total weight. M_{tfo} is to be a small (constant) number, typically 0.001 to 0.005. Thus the fundamental equation for aircraft weight is:

$$W_{TO} = W_E + W_F + W_{PL} + W_{crew} + M_{tfo}W_{TO} \quad \text{(Eq 1)}$$

The basic idea of Roskam is that there is an empirical relationship between aircraft empty and total weights, which he finds to be:

$$\log_{10}W_E = \{ (\log_{10}W_{TO}) - A\} / B \quad \text{(Eq 2)}$$

The coefficients, A and B, depend on which of the dozen types and fifteen subtypes of aircraft fit the description in Table 1 and Fig. 1. It is at this point that the principle of domain limitation first enters. For the example used by Roskam, the correlation used to determine $A = 0.0966$ and $B = 1.0298$ for the twin-engine, propeller-driven aircraft spans a range of empty weights from 1000 to 7000 lb.

The method proceeds as follows to determine the weight of fuel required in the following way. The mission fuel, W_F, can be broken down into the weight of the fuel used and the reserve fuel:

$$W_F = W_{Fres} + W_{Fused} \quad \text{(Eq 3)}$$

Roskam models the reserve fuel as a fraction of the fuel used (see Table 1).The fuel used is modeled as a fraction of the total weight, and depends on the phase of the mission, as described in Fig. 1. For mission phases that are not fuel intensive, a fixed ratio of the weight at the end of the phase to that at the beginning of the phase is given. Again, these ratios are specific to the type of aircraft. For fuel-intensive phases, in this example the cruise phase, there is a relationship between the lift/drag ratio of the aircraft, the engine fuel efficiency, and the propeller efficiency. Again, these three parameters are specific to the type of aircraft.

When the fuel fraction of the total weight is determined by either a cruise calculation, or by the ratio of weight at the end of a mission phase to the beginning of a mission phase, the empty weight can be written in terms of the total weight. Then Eq 2 is used to find the total weight of the aircraft.

For the problem posed, Roskam obtains an estimated total weight of 7900 lb. The accuracy can be estimated from the scatter in the correlation used to determine the coefficients A and B, and is about 30%. For details of the method Roskam uses for obtaining the solution, refer to Ref 9.

Some limitations of the parametric estimating method are of general interest. For example, if the proposed aircraft does not fit any of the domains of the estimating model, the approach is of little use. Such an example might be the V-22, a tilt wing aircraft (Ref 10), which flies like a fixed-wing machine, but tilts its wings and propellers, allowing the craft to hover like a helicopter during take-off and landing. Such a machine might be considered outside the domain of Roskam's estimating model. The point is not that the model is inadequate (the V-22 is more recent than Roskam's 1986 article), but the limited product knowledge in the early stages of development makes it difficult to determine if a cost estimate for the V-22 fits in a well-established domain.

Conversely, even complex machines, such as aircraft, are amenable to parametric cost estimates with fairly good accuracy, provided they are within the domain of the cost model. In the same article, Roskam presents data for transport jets, such as those used by airlines. It should be emphasized that the weight (and hence cost) of such machines, with more than one million unique parts, can be roughly estimated by parametric methods.

Of course, cost is not the same as weight or, for that matter, any other engineering parameter. The details of the manufacturing process, inventory control, design change management, and so forth, all play a role in the relationship between weight and cost. The more complex the machine, the more difficult it is to understand if the domain of the parametric cost-estimating model is the same as that of the product being estimated.

Empirical Methods of Cost Estimation

Almost all the cost-estimating methods published in the literature are based on correlation of some feature or property of the part to be manufactured. Two examples are presented. The first is from the book by Boothroyd, Dewhurst, and Knight (Ref 1), hereafter referred to as BDK. Chapter 9 of this book is devoted to "Design for Sheet Metalworking." The first part of this chapter is devoted to estimates of the costs of the dies used for sheet metal fabrication. This example was chosen because the work of these authors is well recognized. (Boothroyd and Dewhurst Inc. sells widely used software for design for manufacture and design assembly.) In this chapter of the book, the concept of "complexity" of stamped sheet metal parts arises. The complexity of mechanical parts is discussed in the section "Complexity Theory" in this article.

Example 1: Cost Estimates for Sheet Metal Parts. Sheet metal comes in some 15 standard gages, ranging in thickness from 0.38 to 5.08 mm. It is commonly available in steel, aluminum, copper, and titanium. Typical prices for these materials are 0.80-0.90$/lb for low-carbon steel, $6.00-$7.00/lb for stainless steel, $3.00/lb for aluminum, $10.00/lb for copper, and $20.00/lb for titanium. It is typically shipped in large coils or large sheets.

Automobiles and appliances use large amounts of steel sheet metal. Aluminum sheet metal is used in commercial aircraft manufacture, but in lesser amounts due to the smaller number of units produced.

Sheet metal is fabricated by shearing and forming operations, carried out by dies mounted in presses. Presses have beds, which range in size from 50 by 30 cm to 210 by 140 cm (20 by 12 in. to 82 by 55 in.). The press force ranges from 200 to 4500 kN (45 to 1000 lbf). The speed ranges from 100 strokes/min to 15 strokes/min in larger sizes.

Dies typically have four components: a basic die set; a punch, held by the die set, which shears or forms the metal; a die plate through which or on which the punch acts; and a stripper plate, which removes the scrap at the end of the fabrication process.

BDK estimate the basic die set cost (C_{ds}, in U.S. dollars) as basically scaling with usable area (A_u, in cm^2):

$$C_{ds} = 120 + 0.36A_u \quad \text{(Eq 4)}$$

The coefficients (Eq 4) arise from correlating about 50 data points of die set cost with useable area. The tooling elements (the punch, die plate, and stripper plate) are estimated with a point system as follows: let the complexity of the part to be fabricated be X_p. Suppose that the profile has a perimeter P (cm), and that the part has an over width and length of W (cm) and L (cm) of the smallest dimensions which surround the punch. The complexity of the part is taken to be:

$$X_p = (P/L)(P/W) \quad \text{(Eq 5)}$$

The assessment of how part complexity affects cost arises repeatedly in cost estimating. The subject is discussed at length in the next section "Complexity Theory." From the data of BDK, the basic time to manufacture the die set (M, in hours) can be estimated by the following steps: Define the basic manufacturing points (M_{po}) as

$$M_{po} = 30 + 0.56X_p^{2/3} \quad \text{(Eq 6)}$$

Note that the manufacturing time increases a bit less than linearly with part complexity. This is consistent with the section "Complexity Theory." BDK goes on to add two correction factors to M_{po}. The first is a correction factor due to plate size and part complexity, f_{LW}. From BDK data it is found:

$$f_{LW} = 1 + 0.0276\,LW X_p^{0.093} \quad \text{(Eq 7)}$$

The second correction factor is to account for the die plate thickness. BDK cites Nordquist (Ref 11), who gives a recommended die thickness, h_d, as:

$$h_d = 9 + 2.5 \log_e(U/U_{ms}) V h^2 \quad \text{(Eq 8)}$$

where U is the ultimate tensile stress of the sheet metal, U_{ms} is the ultimate stress of mild steel, a reference value, V, is the required production volume, and h is the thickness (in mm) of the metal to be stamped. BDK recommends the second correction factor to be:

$$f_d = 0.5 + 0.02h_d \text{ or } f_d = 0.75 \quad \text{(Eq 9)}$$

whichever is greater.

The corrected labor hours, M_p, are then estimated as:

$$M_p = f_d f_{LW} M_{po} \quad \text{(Eq 10)}$$

The cost of the die is the sum of the corrected labor hours times the labor rate of the die fabricator plus the basic die set cost, from Eq 4.

As a typical example of the empirical cost estimating methods, the BDK method takes into account several factors such as the production volume, the strength of the material (relating to how durable the die needs to be), the die size, and complexity of the part. These factors clearly influence die cost. However, the specific form of the equations are chosen as convenient representations of the data at hand. (As, indeed, are Eq 6 and 7, derived by fitting BDK data.)

The die cost risk (i.e., uncertainty of the resulting estimate of die cost) is unknown, because it is not known how the model equations would change with different manufacturing processes or different die design methods.

It is worth noting carefully that only some features of the design of the part enter the cost estimate: the length and width of the punch area, the perimeter of the part to be made, the material, and the production volume. Thus, the product and die designers do not need to be complete in all details to make a cost estimate. Hence, the estimate can be made earlier in the product-development process. Cost trades between different designs can be made at an early stage in the product-development cycle with empirical methods.

Example 2: Assembly Estimate for Riveted Parts. The American Machinist Cost Estimator (Ref 7) is a very widely used tool for empirical cost estimation. It contains data on 126 different manufacturing processes. A spreadsheet format is used throughout for the cost analysis. One example is an assembly process. It is proposed to rivet the aluminum frame used on a powerboat. The members of the frame are made from 16-gage aluminum. The buttonhead rivets, which are sized according to recommendations in Ref 12, are 5/16 in. in diameter and conform to ANSI standards. Figure 2 shows the part.

There are 20 rivets in the assembly, five large members of the frame, and five small brackets.

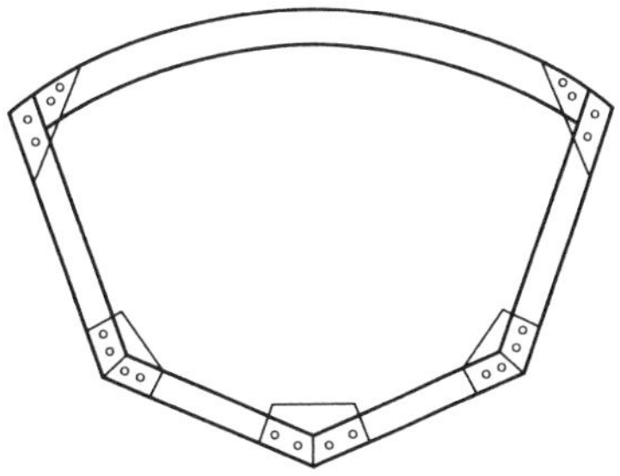

Fig. 2 Powerboat frame assembly

Chapter 21 in Ref 7 includes six tables for setup, handling, pressing in the rivets, and riveting. A simple spreadsheet (for the first unit) might look like Table 2. The pieces are placed in a frame, the rivets are inserted, and riveted. The total cycle time for the first unit is 18.6 min. There are several points to mention here. First, the thickness of the material and the size of the rivets play no direct part in this simple calculation. The methods of Ref 7 do not include such details.

Yet common sense suggests that some of the details must count. For example, if the rivet holes are sized to have a very small clearance, then the "press-in-hardware" task, where the rivets are placed in the rivet holes, would increase. In a like manner, if the rivets fit looser in the rivet holes, the cycle time for this task might decrease. The point of this elementary discussion is that there is some implied tolerance with each of the steps in the assembly process.

In fact, one can deduce the tolerance from the standard specification of the rivets. From Ref 12, in the tolerance on 5/16 in. diameter buttonhead rivets is 0.010 in. So the tolerance of the hole would be about the same size.

The second point is that there are 30 parts in this assembly. How the parts are stored and how they are placed in the riveting jig or fixture determines how fast the process is done. With experience, the process gets faster. There is a well-understood empirical model for process learning. The observation, often repeated in many different industries, is that inputs decrease by a fixed percentage each time the number of units produced doubles. So, for example, L_i is the labor in minutes of the ith unit produced, and L_0 is the labor of the first unit, then:

$$L_i = L_0 i^{\log\phi/\log 2} \quad \text{(Eq 11)}$$

Table 2 Spreadsheet example for assembly of frame (Fig. 2)

Source(a)	Process description	Table time, min	Setup, min
21.2-S	Setup		15
21.2-1	Get 5 frame members from skid	1.05	
21.2-1	Get 5 brackets from bench	0.21	
21.2-2	Press in hardware (20 rivets)	1.41	
21.2-3	Set 20 rivets	0.93	
Total cycle time (minutes)		**3.60**	**15**

(a) Tables in Ref 7, Chapter 21

The parameter ϕ measures the slope of the learning curve. The learning curve effects were first observed and documented in the aircraft industry, where a typical rate of improvement might be 20% between doubled quantities. This establishes an 80% learning function, that is, $\phi = 0.80$. Because this example is fabricated from aluminum, with rivets typical of aircraft construction, it is easy to work out that the 32nd unit will require 32.7% of the time (6.1 min) compared to the first unit (18.6 min).

Learning occurs in any well-managed manual assembly process. With automated assembly, "learning" occurs only when improvements are made to the robot used. In either case, there is evidence that, over substantial production runs and considerable periods of time, the improvement is a fixed percentage between doubled quantities. That is, if there is a 20% improvement between the tenth and twentieth unit, there will likewise be a 20% improvement between the hundredth and two hundredth unit.

The cost engineer should remember that, according to this rule, the percentage improvement from one unit to the next is a steeply falling function. After all, at the hundredth unit, it takes another hundred units to achieve the same improvement as arose between the 10th and 20th units (Ref 13).

Complexity Theory

Up to now this article has dealt with the cost-estimation tools that do not require a complete description of the part or assembly to make the desired estimates. What can be said if the design is fully detailed? Of course, one could build a prototype to get an idea of the costs, and this is often done, particularly if there is little experience

Table 3 Assembly times for piston example (Fig. 5)

Part ID No.	No. times operation carried out	Handling code	Time, s	Insertion code	Insertion time, s	Total time, s	Pneumatic piston
7	1	30	1.95	00	1.5	3.45	Block
6	1	15	2.25	22	6.5	8.75	Piston
5	1	10	1.50	00	1.5	3.0	Piston stop
4	1	80	4.10	00	1.5	5.6	Spring
3	1	28	3.18	09	7.5	10.7	Cover
2	2	68	8.00	39	8.0	29.0	Screw
						Total time: 60.5	

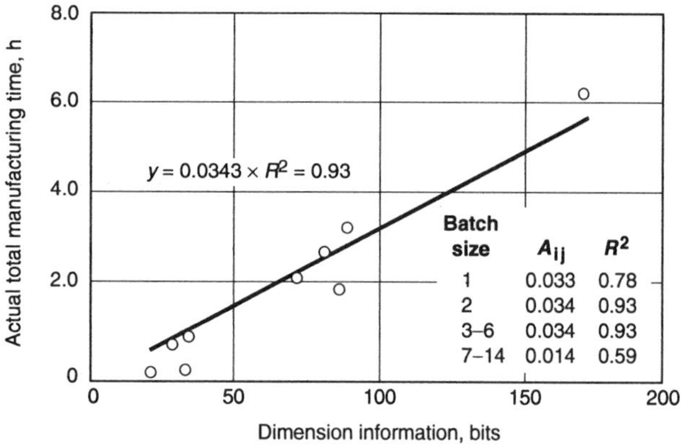

Fig. 3 Manufacturing time and dimension information for the lathe process (batch size 3 to 6 units)

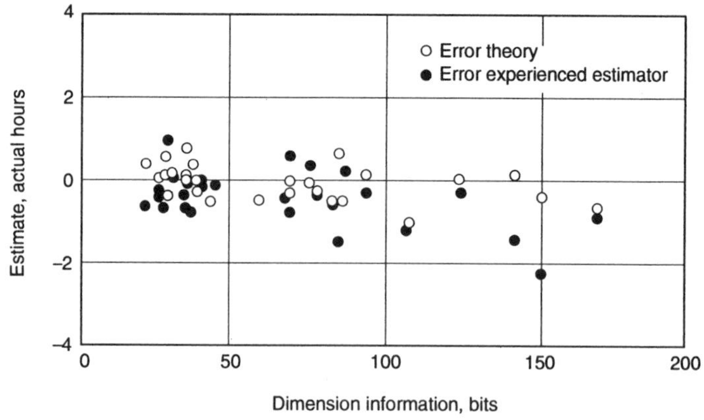

Fig. 4 Accuracy comparison for the lathe process

with the manufacturing methods to be used. For example, suppose there is a complex wave feed guide to be fabricated out of aluminum for a modern radar system. The part has some 600 dimensions. One could get a cost estimate by programming a numerically controlled milling machine to make the part, but is there a simpler way to get a statistically meaningful estimate of cost, while incorporating all of the design details? The method that fulfills this task is complexity theory.

There has been a long search for the "best" metric to measure how complex a given part or assembly is. The idea of using dimensions and tolerances as a metric comes from Wilson (Ref 14). The idea presented here is that the metric is a sum of log (d_i/t_i), where d_i is the ith dimension and t_i is its associated tolerance (i ranges over all the dimensions needed to describe the part). According to complexity theory, how complex a part is, I, is measured by:

$$I = \sum_i \log (d_i / t_i) \qquad \text{(Eq 12)}$$

Originally, the log function was chosen from an imperfect analogy with information theory. It is now understood that the log function arises from a limit process in which tolerance goes to zero while a given dimension remains fixed. In this limit, if good engineering practice is followed, that is, if the accuracy of the machine making the part is not greatly different than the accuracy required of the part, and if the "machine" can be modeled like a first-order damped system, then it can be shown that the log function is the correct metric. Because of historical reasons, the log is taken to the base 2, and I is measured in bits. Thus Eq 12a is written:

$$I = \sum_i \log_2(d_i / t_i) \qquad \text{(Eq 12a)}$$

There are two main attractions of the complexity theory. First, I will include all of the dimensions required to describe the part. Hence, the metric captures all of the information of the original design. For assemblies, the dimensions and tolerances refer to the placement of each part in the assembly, and second, the capability of making rigorous statements of how I effects costs. In Ref 8 it is *proven* that if the part is made by a single manufacturing process, the average time (T) to fabricate the part is:

$$T = A \cdot I, \qquad A = \text{const} \qquad \text{(Eq 13)}$$

Again, in many cases, the coefficient A must be determined empirically from past manufacturing data. The same formula applies to assemblies made with a single process, such as manual labor. The extension to multiple processes is given in Ref 8.

A final aspect of complexity theory worth mentioning is risk. Suppose a part with hundreds of dimensions is to be made on a milling machine. The exact sequence in which each feature of the part is cut out will determine the manufacturing time. But there are a large number of such sequences, each corresponding to some value of A. Hence there is a collection of As, which have a mean that corresponds to the average time to fabricate the part. That is the meaning of Eq 13.

It can be shown that the standard deviation of manufacturing time is:

$$s_T = \sigma_A I \qquad \text{(Eq 14)}$$

where σ_T is the standard deviation of the manufacturing time, and σ_A is the standard deviation of the coefficient A. σ_A can be determined from past data. These results have a simple interpretation: Parts or

Table 4 Assembly times for simplified piston design (Fig. 6)

Part ID No.	No. times operation carried out	Handling code	Time, s	Insertion code	Insertion time, s	Total time, s	Pneumatic piston
4	1	30	1.95	00	1.5	3.45	Block
3	1	15	2.25	00	1.5	3.75	Piston
2	1	80	4.10	00	1.5	5.6	Spring
1	1	10	1.50	30	2.0	3.5	Cover/stop
					Total time:	**16.3**	

Table 5 Original manual assembly design

Feature (No.)	Dimensions, mm	Tolerance	Bits	Notes
2	3	0.063	11.14693	Diameter dimension
2	30	0.1	16.45764	Horizontal location
2	30	0.1	16.45764	Horizontal location
2	2160	60	10.33985	6 turns to install screw
Subtotal			**54.40206**	
1	31	0.1	8.276124	Horizontal location of plate
1	31	0.1	8.276124	Horizontal location of plate
Subtotal			**70.95431**	
1	25	0.43	5.861448	Spring location
Subtotal			**76.81575**	
1	25	0.1	7.965784	Piston location
Subtotal			**84.78154**	
1	25	0.1	7.965784	Piston stop
Total:			**92.74732**	**1.5 bits/s**

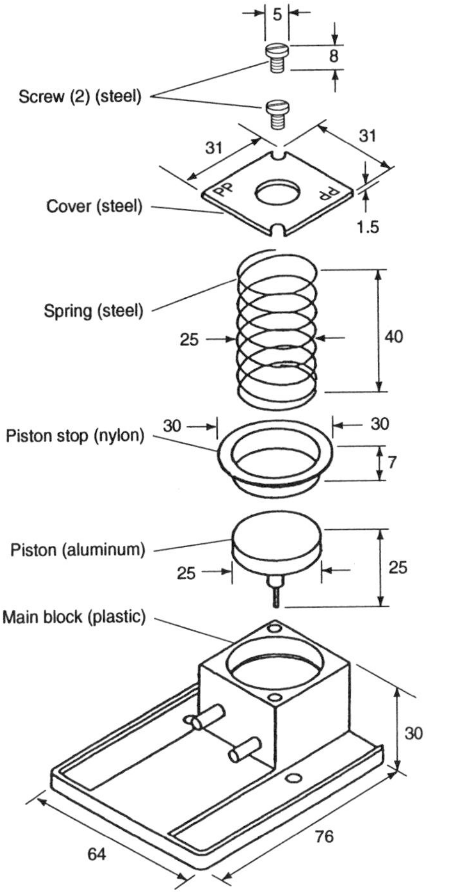

Fig. 5 Assembly of pneumatic piston. Dimensions in millimeters

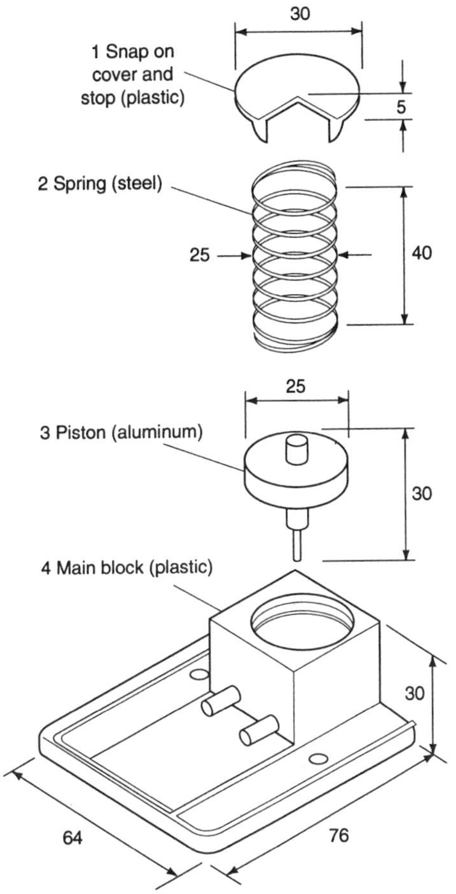

Fig. 6 Simplified assembly of pneumatic piston. Dimensions in millimeters

Table 6 Simplified manual assembly design

Feature (No.)	Dimensions, mm	Tolerance	Bits	Notes
1	25	0.1	7.965784	Piston location
1	25	0.43	5.861448	Spring location
1	25	0.1	7.965784	Piston cap
Total:			**21.79302**	**% reduction 76.50281**

assemblies with tighter (smaller) tolerances take longer to make or assemble because with dimensions fixed, the *log* functions increase as the tolerances decrease. More complex parts, larger I, take longer to make (Eq 13), and more complex parts have more cost risk (Eq 14). These trends are well known to experienced engineers.

In Eq 8, a large number of parts from three types of manufacturing processes were correlated according to Eq 13. The results of the following manual lathe process are typical of all the processes studied in Eq 8. Figure 3 shows the correlation of time with I, the dimension information, measured in bits. An interesting fact, shown in Fig. 4 is that the accuracy of the estimate is no different than that of an experienced estimator.

In Eq 13, the coefficient, A, is shown to depend on the machine properties such as speed, operation range, and time to reach steady-state speed. Can one estimate their value from first principals? It turns out that for manual processes one can make rough estimates of the coefficient.

The idea is based on the basic properties of human performance, known as Fitts' law. Fitts and Posner reported the maximum human information capacity for discrete, one-dimensional positioning tasks at about 12 bits/s (Ref 15). Other experiments have reported from 8 to 15 bits/s for assembly tasks (Ref 16).

The rivet insertion process discussed previously in this article is an example. The tolerance of the holes for the rivets is estimated to be 0.010 in., that is, the same as the handbook value of the tolerance of the barrel of the rivet (Ref 12). Then it is found that $d/t \approx 0.312/0.010 = 31.2$ and $\log_2 = 4.815$ bits for each insertion. The initial rate of insertion (Ref 7) was 20 units in 1.41 min. That corresponds to $A = 1.14$ bits/s. Clearly, there is some considerable improvement available if the maximum values quoted (Ref 15, 16) can be achieved for rivet insertion.

Example 3: Manual Assembly of a Pneumatic Piston. In Ref 1 there is an extensive and helpful section on manual assembly. The method BDK used categorizes the difficulty of assembling parts by a number of parameters, such as the need to use one or two hands, the need to use mechanical tools, part symmetry, and so on. Figure 5 (reproduced from an example in Ref 17) shows the assembly of a small pneumatic piston. Table 3 lists assembly times.

Consider the entries for the two screws. The handling code, 68, describes a part with 360° symmetry that can be handled with standard tools. The insertion code, 39, describes a part not easy to align or position. The time for assembly of the screws is nominally 32 s, less an allowance of 31 s for repetitive operations.

Now consider a simplified design (Fig. 6 and Table 4). The same tables from Chapter 21 in Ref 7 are used as in Table 2. Software is available to automate the table look-up process. For the same problem using complexity theory, there is only one coefficient, $A = 1.5$ bits/s for the small manual assembly. This value is found by calculating the bits of information in the initial design and using the time found in Ref 17 to determine A.

The tolerances were obtained in the following way. For the screws, the size was chosen to be M3X0.5—a coarse thread metric size consistent with insertion into molded nylon. As reported in Ref 12, the tolerance is ANSI B1.13M-1979.

The spring tolerance is derived using a standard wire size, 0.076 in. (0.193 mm), which gives a spring index $D/d = 12.95$, (D = spring diameter, d = wire diameter) well within the Spring Manufacturers Institute recommended range. The tolerance quoted is the standard tolerance for this wire diameter and spring index.

The plastic parts are presumed to be injection molded, with a typical tolerance of 0.1 mm. The screws are assumed to tighten to a tolerance of $^{60}/_{360} = {}^{1}/_{6}$th of a turn.

These data, which can be easily verified in practice, give essentially the same results as the other empirical methods. Calculations for the original design (Fig. 5) and the simplified design (Fig. 6) are compared in Tables 5 and 6, respectively. This compares well with the results of Ref 17 (73% reduction) versus 76% reduction here.

There are two comments to make:

- This method requires only one coefficient for hand assembly of small parts. $A \approx 1.6$ bits/s and no look-up tables.
- One can make small changes in design, for example, change the screw size, and get an indication of the change in assembly time.

If one started with a preliminary design, the assembly time estimate would grow more accurate as more details of the design itself and the manufacturing process to build the design become known. However, the methods of Boothroyd, and others (Ref 1, 6, 17), require less data than does the complexity theory.

In the simplified manufacturing process, the coefficient $A = 1.6$ bits/s is substantially less than the Fitts' law (discussed earlier in this section on complexity theory) value of 8 bits/s. The discrepancy may lie in the time it takes an assembler to pick up and orient each part before the part is assembled. Jigs and trays, and so forth, that reduce this pick-and-place orientation effort would save assembly time. As before, there is some considerable improvement available if the maximum values quoted (Ref 15, 16) can be achieved for manual assembly. The value obtained here ($A = 1.16$) is close to that deduced from Ref 7 for the hand insertion of rivets.

Using complexity theory and a single assembly process, the ratio of the assembly times can be calculated without any knowledge of the coefficient, A. Thus complexity theory offers advantages when a single process is used, even if little or nothing is known about the performance of the process.

Cost Estimation Recommendations

Which type of cost estimate one uses depends on how much is know about the design. In the early stages of concept assessment of a new part or product, parametric methods, based upon past experience, are preferred. Risk is hard to quantify for these methods, because it can be very difficult to determine whether the new product really is similar to those used to establish the parametric cost model.

If there is some detailed information about the part or product, and the method of manufacturing is well known, then the empirical methods should be used. They can indicate the relative cost between different designs and give estimates of actual costs.

If detailed designs are specified, and a single manufacturing process is to be used, complexity theory should be used to compare the relative costs and cost risks of the different designs, even if the manufacturing process is poorly understood.

If there are detailed designs available, and well-known manufacturing methods are used, either complexity theory or empirical methods can be used to generate cost estimates. If a rigorous risk assessment is needed, complexity theory should be used.

REFERENCES

1. G. Boothroyd, P. Dewhurst, and W. Knight, *Product Design for Manufacture and Assembly,* Marcel Dekker, 1994, Chapt. 1
2. K.T. Ulrich and S.A. Pearson, "Does Product Design Really Determine 80% of Manufacturing Cost?," working paper 3601-93-MSA, MIT Sloan School, Nov 1994
3. D.P. Hoult and C.L. Meador, "Cost Awareness in Design: the Role of Database Commonality," SAE 96008, Society of Automotive Engineers, 1996
4. D.P. Hoult and C.L. Meador, "Methods of Integrating Design and Cost Information to Achieve Enhanced Manufacturing Cost/Performance Trade-Offs," Save International Conference Proceedings, Society for American Value Engineers, 1996, p 95–99
5. H.T. Johnson and R.S. Kaplan, *Relevance Lost, the Rise and Fall of Management Accounting,* Harvard Business School Press, 1991
6. G. Boothroyd, *Assembly Automation and Product Design,* Marcel Dekker, 1992
7. P.F. Ostwald, "American Machinist Cost Estimator," Penton Educational Division, Penton Publishing, 1988
8. D.P. Hoult and C.L. Meador, "Predicting Product Manufacturing Costs from Design Attributes: A Complexity Theory Approach," No. 960003, Society of Automotive Engineers, 1996
9. J. Roskam, Rapid Sizing Method for Airplanes, *J. Aircraft,* Vol 23 (No.7), July 1986, p 554–560
10. The Bell-Boeing V-22 Osprey entered Low Rate Initial Production with the MV-22 contract signed June 7, 1996, *Tiltrotor Times,* Vol 1 (No. 5), Aug 1996
11. W.N. Nordquist, *Die Designing and Estimating,* 4th ed., Huebner Publishing, 1955
12. E. Oberg, F.D. Jones, and H.L. Horton, *Machinery's Handbook,* 22nd ed., Industrial Press, 1987, p 1188–1205
13. G.J. Thuesen, and W.J. Fabrycky, *Engineering Economy,* Prentice Hall, 1989, p 472–474
14. D.R. Wilson, "An Exploratory Study of Complexity in Axiomatic Design," Doctoral Thesis, Massachusetts Institute of Technology, 1980
15. P.M. Fitts, and M.I. Posner, Human Performance, Brooks/Cole Publishing, Basic Concepts in Psychology Series, 1967
16. J. Annett, C.W. Golby, and H. Kay, The Measurement of Elements in an Assembly Task — The Information Output of the Human Motor System, *Quart. J. Experimental Psychology,* Vol 10, 1958
17. G. Boothroyd and P. Dewhurst, *Product Design for Assembly,* Boothroyd Dewhurst, 1989

Design for Casting

Thomas S. Piwonka, The University of Alabama

CASTING offers the designer cost advantages over other manufacturing methods for most components, especially those having complex geometries. Casting properties are usually isotropic, and castings may be designed for function rather than for ease of assembly, like built-up structures. Fillet radii are usually generous, decreasing stress concentration factors. Converting a built-up assembly to a casting is usually accompanied by a decrease in part count, assembly time, and inventory, and a savings in weight. With the development of rapid prototyping, expensive tooling is not necessary, and parts can be delivered within days after the order is placed. The casting process is well understood, and high integrity castings are as reliable as forgings. Indeed, castings prove their quality everyday in applications as demanding as prostheses, automotive chassis components, primary aircraft structures, and rotating hardware in gas turbine engines.

Designers face a number of challenges in the design of castings. To begin with, there are a wide variety of casting alloys. Complete mechanical property data, especially for dynamic properties, are sometimes lacking for these alloys. Static property data, though often found in handbooks, may consist of "typical values," which are of little help in creating an efficient design. These typical values obscure a crucial fact: casting properties are determined during solidification and subsequent heat treatment. This means that properties will vary depending on how quickly the casting solidifies and how the casting is heat treated.

Thus, to create effective designs, designers should be acquainted with fundamental information about how casting design influences casting solidification and how casting solidification influences casting properties. This article approaches design by reviewing the aspects of castings with which designers should be familiar. It also reviews methods used by foundries to produce high-integrity castings. Specification of casting quality levels should be based on solid knowledge of the effect of casting discontinuities and on component testing. Procurement of high-quality castings requires the involvement of the designer, the purchasing agent, and the foundry in a cooperative effort.

Design Considerations

Casting design begins with the determination of the stresses that must be supported by the cast component and the geometrical constraints on that component. The designer then arranges the cast material to support the stresses in the most efficient way. However, in designing a part that will be cast, the designer will optimize the design by taking into account characteristics of the casting process. Because the properties of the casting determine its performance, those features of the casting process that affect the casting properties are reviewed here. This discussion emphasizes those concepts that designers and foundries can use to obtain maximum performance from cast parts.

Designers must begin the design process with a thorough understanding of what properties the component must have in addition to strength and ductility. Fundamental properties such as Young's modulus, Poisson's ratio, density, thermal conductivity, and coefficient of thermal expansion vary between alloy families and within alloy families. Cast alloy selection should begin by taking these differences into account.

How the Casting Process Affects Casting Properties. The properties of a component depend on the way it is made because the microstructure of the component depends on the manufacturing method, and the properties depend in turn on the microstructure. Thus the choice of making a part by casting, by forging, or by machining will affect the performance of the part. Once the decision is made on how the part is to be manufactured, the details of the processing parameters employed also affect component properties.

Component design is usually approached with the assumption that the material is uniform and isotropic and that there is an inherent small and random scatter in properties. Exceptions are made for fiber-reinforced composite materials and for components where directionality of properties is beneficial (directionally solidified gas turbine blades, for example). In castings, properties will be uniform in specific casting sections from casting to casting provided that the casting variables are constant for each casting. However, they may vary in a predictable manner from point to point within the casting.

Casting microstructure is determined by cooling rate, that is, how fast each part of the casting freezes. The cooling rate is roughly proportional to the ratio of the square of the surface area of the casting to the square of its volume (a consequence of what metal casters know as Chvorinov's law). In other words, bulky castings freeze much slower than thin castings—a sphere of a given volume will freeze more slowly than a thin plate of the same volume because the plate has much more surface area to transfer the same quantity of heat into the mold. Because the sphere solidifies more slowly, its microstructure will be coarser than that of the plate even if both are poured from the same melt at the same temperature. Because microstructure determines casting properties, the properties of the sphere and the plate will be different.

Casting is the solidification of liquid confined in a mold, which shapes the final component. A cavity is made in a mold, which may be a reusable aggregate, such as sand, or a metal mold, used in permanent mold and die casting. The metal is delivered to the mold cavity through channels in the mold or die, which are called *runners*. The passages between the runner and the mold cavity, where the metal enters the mold, are called *gates*.

Castings are frequently complex shapes made up of some bulky sections and some thin sections. Obviously, the thin sections will solidify faster than the thick sections; therefore, their properties will differ from those of the thick sections. Certain other geometric features will also influence solidification rate. For instance, concave sections, or reentrant angles, solidify more slowly than fins or protrusions, again affecting the resultant local structure and the properties. In other words, property variation within a casting, which is caused by local differences in cooling rate, is natural, expected, and entirely reproducible—it is not "random" scatter. It can be predicted and should be taken into account during component design to enhance the component performance.

One way to take account of this during the design phase is to apply a "section size" effect, a factor by which the local casting properties are adjusted based on local dimensions. While this technique is often effective, the designer should remember that casting properties depend on cooling rate, not section size. Cooling rate depends not only on section size but also on pouring temperature, mold material, gating system, the presence or absence of chills in the mold, mold coatings, and insulation. Thus, within limits (some very broad), the properties can be controlled by the metal caster to produce those which the designer desires.

Cast iron provides the most dramatic example of the differences in properties caused by differences in structure resulting from cooling rate differences. When cast iron (which is essentially a solution of carbon and silicon in iron) solidifies, the carbon can take different forms, depending on its composition and solidification rate and the way the metal has been treated during melting. Chill cast iron (white iron) has properties completely different from gray iron of the same composition, caused solely by the accelerated cooling rate in the "chilled" iron. White iron freezes so quickly that the carbon combines with the iron to solidify as the compound Fe_3C, known as *cementite* because it is hard and brittle. In gray iron, which solidifies more slowly, the carbon appears as graphite flakes; this iron is easy to machine.

Ductile iron also has properties significantly different from gray iron; in this case the carbon solidifies as tiny spheres in a steel matrix. Since the spheres of graphite are less effective as stress raisers, compared to the flakes of carbon in gray iron, ductile iron has significant ductility, whereas gray iron does not, even though they both may have nearly identical compositions. In this case, the difference in structure, which produces the property difference, is caused not by cooling rate, but by chemical treatment of the melt, which alters the undercooling at the beginning of solidification, which in turn affects the way the carbon solidifies. It is important here to realize that in castings, similar compositions, in similar geometries, can have very different properties, depending on the way the castings are made.

Designers frequently use handbook data in designing components. Very often, these data are developed using Gaussian statistics. As noted above, such data can be misleading if not corrected to reflect differences in cooling rate from section to section within a casting. Gaussian statistics are appropriate for those materials that are truly ductile, meaning those that have a tensile elongation over 8 to 10%. Many casting alloys, however, when completely free of discontinuities, have elongations of only 5 to 10%. Indeed, this is often the reason that these alloys are commonly cast: their low ductilities make them hard to form by forging or machining. Use of Gaussian statistics to determine design allowables in these alloys is incorrect, leads to "casting factors," and causes significant overdesign, waste of material, and weight penalties. Casting factors are arbitrary increases in casting section thickness applied to compensate for perceived lack of reproducibility of casting properties. Such factors are unnecessary and wasteful (Ref 1). For this reason, it is important that designers recognize that designing with low ductility (not brittle) materials requires a different approach than designing with ductile materials. Also, casting versions of wrought alloys may have different compositions than the wrought alloys to aid in solidification.

Reproducibility of properties produced by a process is most important for the designer because the components that are designed must behave according to the design. Reliability is often evaluated using Weibull statistics (Ref 2), which consider the probability of the existence of a discontinuity that would cause failure. This is particularly appropriate for the design of low ductility materials. In analyzing data using Weibull statistics, one evaluates the value of the Weibull modulus. The higher the Weibull modulus is, the more reliable the material will be (that is, the higher the Weibull modulus, the less the variation in property within a given section of the component).

For tensile properties, ceramics typically have Weibull moduli of approximately 10. Aluminum alloy forgings typically have Weibull moduli of approximately 50. Conventionally cast aluminum castings have Weibull moduli of approximately 30; using techniques for "premium quality castings," castings with Weibull moduli of 50 are routinely produced for use as primary structures for commercial aircraft (Ref 3, 4). For carefully made thin sections, the Weibull modulus can be above 80 (Ref 5), well above that expected for forgings.

Ductile iron, most cast steel alloys, and copper-base alloys commonly have high ductilities, and the use of Gaussian statistics to determine design allowables is appropriate. However, gray and compacted graphite iron, superalloys, some tool steels, and many aluminum alloys are low ductility alloys and should be approached using Weibull statistics.

Basic Features of Solidification. Solid and liquid alloys usually have different densities, which means when solidification is complete, the solid will occupy less space than the liquid (in most alloys). In other words, the solid "shrinks." Because solid metal is more dense than liquid metal, the metal shrinks when it solidifies, and the final casting will not fill the mold unless this shrinkage is compensated. To do this, the foundry adds a *riser* to the casting. This riser is a reservoir of molten metal, which supplies liquid metal to the solidifying casting and compensates for the shrinkage that occurs. If the riser is to function properly, the liquid metal that it contains must be able to flow through the solidifying casting to reach the areas in the casting where solidification and hence shrinkage is occurring.

In alloy solidification, the picture is complicated by the fact that alloys freeze over a range of temperatures. This means that small nuclei of solid grains form at the beginning of solidification and grow (forming *dendrites*, the term for solidifying grains) as the temperature in the casting falls and solidification progresses. As these grains grow, they form a "mush" of liquid and solid, which becomes progressively more solid until solidification is complete. Because each small dendrite that forms is a site where shrinkage occurs during solidification, feeding the shrinkage means finding a way to cause metal to flow between the dendrites to the location where solidification (and shrinkage) is taking place. As the dendrites grow, the paths that deliver liquid metal from the riser to the areas where solidification is taking place become smaller until they are at last too narrow to allow metal to pass. (There is one exception to this rule. In cast iron, both gray and ductile, the graphite that solidifies expands on solidification. Because graphite forms at the end of solidification, its expansion often (but not always) compensates for the shrinkage of the iron. For this reason, cast iron often needs very little in the way of risers.)

Liquid metal can dissolve much more gas in solution than solid metal. Therefore, when metal solidifies, gas that is present in the liquid is rejected and forms bubbles. A commonly encountered example is that of hydrogen in aluminum. If these bubbles are trapped in the casting when it freezes, the result is a pore. Pores that result from gas may be spherical, indicating that they formed early in solidification when the metal was mostly liquid, or they may be interdendritic in shape, showing that they formed late in solidification, when the liquid that remained was present between dendrites.

As the solidification rate increases, the microstructure of the casting is refined. That is, the grains are smaller, and the spacing between the arms of the dendrites that make up grains is finer. Mechanical properties usually improve as the microstructure becomes finer. Because the properties depend on the microstructure, which depends on the solidification rate, which in turn depends on the processing variables used by the foundry and the casting design, designers have a major influence on the final properties of the casting.

Using solidification simulation programs, foundries predict how fast each section of each casting will solidify and, therefore, what will be the properties of each section. The use of these programs combined with the application of statistical process control techniques has transformed casting into a high technology manufacturing process capable of reliably producing critical components for the most demanding applications.

General Design Considerations for Castings. Casting design influences the way the casting solidifies because the geometry of the casting influences how fast each section solidifies; therefore, it is a major factor in determining how castings will perform. The overarching principle of good casting design is that casting sections should freeze progressively, allowing the risers to supply liquid metal to feed shrinkage that occurs during solidification. There are a number of excellent summaries of principles of good casting design (Ref 6–8), and a comprehensive booklet was published on design of premium quality aluminum alloy castings (Ref 9). This latter publication also includes general information on the specification and process of approving foundry

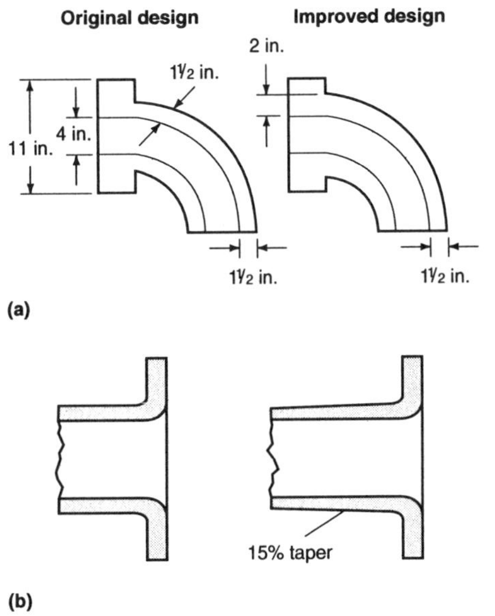

Fig. 1 Redesign of castings to provide progressive solidification through the use of tapered walls. (a) Elbow design. (b) Valve fitting design. Source: Ref 10

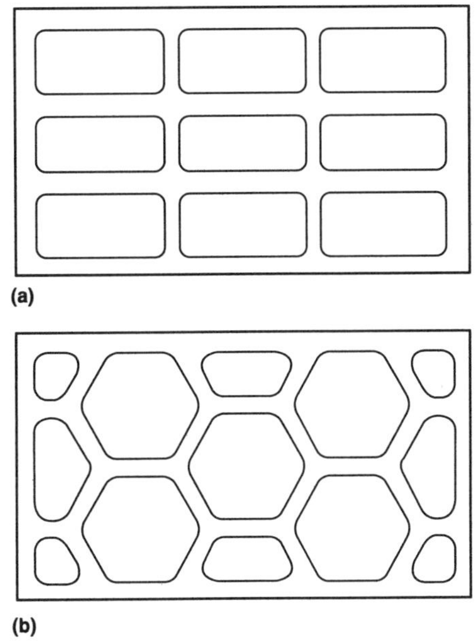

Fig. 2 Redesign of a casting to minimize heat concentration. (a) Design has numerous hot spots (X junctions) that will cause the casting to distort. (b) Improved design using Y junctions. Source: Ref 6

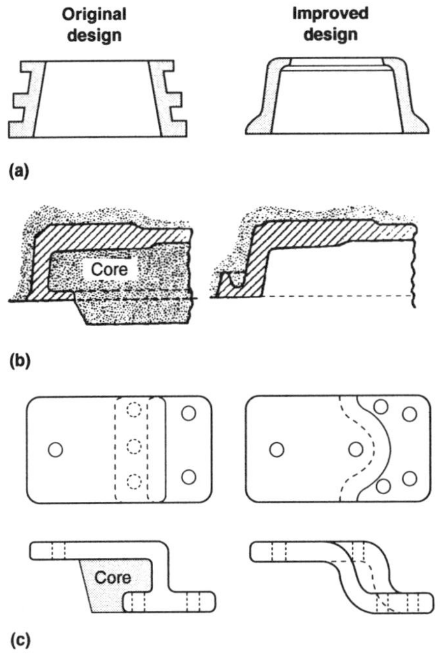

Fig. 3 Redesign of castings to eliminate cores. (a) Casting redesigned to eliminate outside cores. (b) Simplification of a base plate design to eliminate a core. (c) Redesign of a bracket to eliminate a core and to decrease stress problems. Source: Ref 10

sources, which is applicable to castings made from any metal.

Designing for progressive solidification requires tapering walls so that they freeze from one end to the other, as shown in Fig. 1, and avoiding situations where two heavy sections are separated by a thin section. (This is a poor design because metal must feed one heavy section through the thin section; when the thin section freezes before the heavy section, the flow path will be cut off and shrinkage may form in the heavy section.)

Junctions also concentrate heat, leading to areas in the casting where heat is retained. These areas solidify more slowly than others, thus having a coarser structure and different properties from other sections, and solidify after the rest of the casting has solidified, so that shrinkage cannot be fed. Minimizing the concentration of heat in junctions, therefore, aids in improving casting properties. Examples of this are shown in Fig. 2.

Concave corners concentrate heat, so they freeze later and more slowly than straight sections, while convex corners lose heat faster, and freeze sooner and more quickly than straight sections. In designing a casting, the designer should use properties from test bars that have solidified at the actual cooling rate in that section of the casting. This cooling rate can be determined by instrumenting a casting and measuring the cooling rate in various sections or by simulating its solidification using a commercial solidification simulation program.

The hollow spaces in castings are formed by *cores*, refractory shapes placed in the molds and around which the casting freezes. These cores are later removed from the casting, usually by thermal or mechanical means. However, each core requires tooling to form it and time to place it in the mold. Casting designs that minimize cores are preferable to minimize costs. Some examples are given in Fig. 3.

The designer must provide surfaces for the attachment of gates and risers. The casting must solidify toward the riser in order to be sound, and gates must be located so that the mold fills from the bottom to the top so that oxide films that form are swept to the top surface of the casting or into risers where they will not affect casting properties. Gate and riser locations must be accessible for easy removal to minimize processing costs. The position of gate and riser contacts may also add costs if they are placed where subsequent machining will be required to remove gate stubs or riser pads in the finished component.

The Effect of Casting Discontinuities on Properties

Poor casting design can interfere with the ability of the foundry to use the best techniques to produce reliable castings. The designer also specifies the quality requirements that ensure that the cast component will perform as desired. Over specification causes needless expense and can be avoided by understanding the effect of discontinuities on casting performance and the effect of casting design on the tendency for discontinuities to form during the casting process. Important types of casting discontinuities include porosity, inclusions, oxide films, second phases, hot tears, metal penetration, and surface defects.

Porosity is a common defect in castings and takes many forms. Pores may be connected to the surface, where they can be detected by dye penetrant techniques, or they may be wholly internal, where they require radiographic techniques to discover. *Macroporosity* refers to pores that are large enough to see with the unaided eye on radiographic inspection, while *microporosity* refers to pores that are not visible without magnification.

Both macroporosity and microporosity are caused by the combined action of metal shrinkage and gas evolution during solidification. It has been shown (Ref 11, 12) that nucleation of pores is difficult in the absence of some sort of substrate, such as a nonmetallic inclusion, a grain refiner, or a second phase particle. This is why numerous investigations have shown that clean castings, those castings that are free from inclusions, have fewer pores than castings that contain inclusions. Microporosity is found not only in castings, but also in heavy section forgings that have not been worked sufficiently to close it up.

When the shrinkage and the gas combine to form macroporosity, properties are deleteriously affected. Static properties are reduced at least by the portion of the cross-sectional area that is taken up with the pores (since there is no metal in the pores, there is no metal to support the load there, and the section acts as though its area was reduced). Because the pores may also cause a stress concentration in the remaining material (Ref 13, 14), static properties may be reduced by more than the percentage of cross-sectional area that is caused by the macroporosity.

Dynamic properties are also affected. A study of aluminum alloys showed that fatigue properties in some were reduced 11% when specimens having x-ray quality equivalent to ASTM E 155 level 4 were tested, and that they were reduced 17% when specimens having quality of ASTM E 155 level 8 were tested (Ref 15).

Static properties are mostly unaffected by microporosity. Microporosity is found between den-

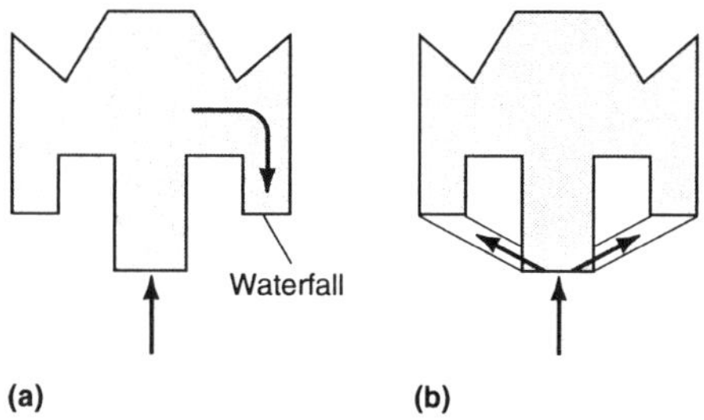

Fig. 4 Redesign of a casting to avoid waterfalling. (a) In this design, waterfalling results when casting is filled from the bottom. (b) Improved design provides a path for the metal to follow as it fills the mold.

drites and, like macroporosity, is caused by the inability of feed metal to reach the interdendritic areas of the casting where shrinkage is occurring and where gas is being evolved. However, because this type of porosity occurs late in solidification, particularly in long-range freezing (mushy-freezing) alloys, it is particularly difficult to eliminate. The most effective method is to increase the thermal gradient (often accomplished by increasing the solidification rate), which decreases the length of the mushy zone. This technique may be limited by alloy and mold thermal properties, and by casting geometry, that is, the design of the casting.

As long as the micropores are less than 0.2 mm in length, there is no effect on dynamic properties; fatigue properties of castings with pores that size or smaller are in the same range as those of castings where no micropores were found (Ref 16-18). The shape of the micropore is as important as its size, with elongated pores having a greater effect than round pores (Ref 19). Areas where microporosity is expected can be predicted by solidification modeling, similar to the prediction of macroporosity (see below). Microporosity can be healed by hot isostatic pressing (HIP). In one study comparing HIP and non-HIP samples, no difference was found in fatigue lives of HIP and non-HIP samples (Ref 20). However, the HIP samples showed a lower crack growth rate than non-HIP samples. In another study (Ref 21), HIP improved fatigue crack growth resistance only close to threshold levels. (Additional information about HIP is provided in the section "Hot Isostatic Pressing" in this article.) As noted above, the design of the casting directly affects its tendency to solidify in a progressive manner, thereby affecting both the quality and the price of the cast component.

Porosity and casting costs are minimized in casting designs that emphasize progressive solidification toward a gate or riser, tapered walls, and the avoidance of hot spots.

Inclusions are nonmetallic particles that are found in the casting. They may form during solidification as some elements (notably manganese and sulfur in steel) precipitate from solution in the liquid. More frequently, they are formed before solidification begins. The former are sometimes called *indigenous* inclusions, and the latter are called *exogenous* inclusions. Inclusions are ceramic phases; they have little ductility. A crack may form in the inclusion and propagate from the inclusion into the metal, or a crack may form at the interface between the metal and the inclusion. In addition, because the inclusion and the metal have different coefficients of thermal expansion, thermally induced stresses may appear in the metal surrounding the inclusion during solidification (Ref 22). As a result, the inclusion acts as a stress concentration point and reduces dynamic properties. As in the case of microporosity, the size of the inclusion and its location determine its effect (Ref 23, 24). Small inclusions that are located well within the center of the cross section of the casting have little effect, whereas larger inclusions and those located near the surface of the casting may be particularly detrimental to properties. Inclusions may also be a problem when machining surfaces, causing excessive tool wear and tool breakage.

Exogenous inclusions are mostly oxides or mixtures of oxides and are primarily slag or dross particles, which are the oxides that result when the metal reacts with oxygen in the air during melting. These are removed from the melt before pouring by filtration. Most inclusions found in steel castings arise from the oxidation of metal during the pouring operation (Ref 25). This is known as *reoxidation*, and takes place when the turbulent flow of the metal in the gating system causes the metal to break up into small droplets, which then react with the oxygen in the air in the gating system or casting cavity to form oxides. Metal casters use computer analysis of gating systems to indicate when reoxidation can be expected in a gating system and to eliminate them. However, casting designs that require molten metal to "jet" through a section of the casting to fill other sections will recreate these inclusions and should be avoided.

Oxide films are similar to inclusions and have been found to reduce casting properties (Ref 3-5). These form on the surface of the molten metal as it fills the mold. If this surface film is trapped within the casting instead of being carried into a riser, it is a linear discontinuity and an obvious site for crack initiation. It has been shown (Ref 26, 27) that elimination of oxide films, in addition to substantially improving static properties, results in a five-fold improvement of fatigue life in axial tension-tension tests.

Oxide films are of particular concern in nonferrous castings, although they also must be controlled in steel and stainless steel castings (because of the high carbon content of cast iron, oxide films do not form on that metal). If the film folds over on itself as a result of turbulent flow or "waterfalling" (when molten metal falls to a lower level in the casting during mold filling), the effects are particularly damaging. Casting design influences how the metal fills the mold, and features of the design that require the metal to fall from one level to another while the mold is filling should be avoided so that waterfalls are eliminated. Oxide films are avoided by filling the casting from the bottom, in a controlled manner, by pumping the metal into the mold using pneumatic or electromagnetic pumps. If the casting is poorly designed, waterfalling will result. An example is given in Fig. 4.

Second phases, which form during solidification, may also nucleate cracks if they have the proper size and morphology (Ref 28). An example is aluminum silicon alloys, where the silicon eutectic is present as large platelets, which nucleate cracks, and along which cracks propagate (Ref 29, 30). The size of these platelets may be significantly reduced by modifying the alloy with additions of sodium or strontium. However, such additions increase the size of micropores (Ref 31), and for this reason, many foundrymen rely on accelerated solidification of the casting to refine the silicon. As noted above, solidification rates normally increase, and the structure is thus refined, in thin sections. Heavy sections are to be avoided if a fine structure is desired. Generally speaking, however, secondary phases in the structure of castings become important in limiting mechanical behavior of castings only in the absence of nonmetallic inclusions and microporosity (Ref 32).

Hot tears form when casting sections are constrained by the mold from shrinking as they cool near the end of solidification. These discontinuities are fairly large and are most often weld repaired. If not repaired, their effect is not readily predictable (Ref 17). While generally they are detrimental to casting properties, under some circumstances they do not affect them. Hot tears are caused by a combination of factors, including alloy type, metal cleanliness, and mold and core hardness. However, poor casting design is the primary cause. Castings should be designed so that solidifying sections are not subjected to tensile forces caused by shrinkage during solidification, as the solidifying alloy has little strength before it solidifies. An example is given in Fig. 5, and an extensive discussion on how to prevent hot tears through casting design is provided in Ref 34.

Metal Penetration. Molten metal may penetrate the surface of the mold, forming a rough surface or, in extreme cases, actually becoming intimately mixed with the sand in the mold. In iron castings, this is normally the result of the combination of metallostatic head (the pressure exerted on the molten iron at the bottom of the mold by the weight of the metal on top of it) and the surface tension relationships between the liquid iron and molding materials (Ref 35). In cast iron, it is frequently also the result of the expansion of graphite at the end of solidification, forcing liquid metal into the mold if the casting is not properly designed with a tapered wall to promote directional solidification and avoid hot spots.

Surface Defects. Surface finish is also an important specification. Surface discontinuities affect fatigue life (Ref 36, 37), and obviously smoother surfaces are superior to rough surfaces. Designers should be certain that fatigue data used in design calculations has been taken from as-cast surfaces rather than machined surfaces, as most surfaces on castings where stress concentrations might be expected are not machined. Surface finish in castings is controlled by the application

Table 1 Factors affecting selection of casting process for aluminum alloys

Factor	Sand casting	Permanent mold casting	Die casting
Cost of equipment	Lowest cost if only a few items required	Less than die casting	Highest
Casting rate	Lowest rate	11 kg/h (25 lb/h) common; higher rates possible	4.5 kg/h (10lb/h) common; 45 kg/h (100 lb/h) possible
Size of casting	Largest of any casting method	Limited by size of machine	Limited by size of machine
External and internal shape	Best suited for complex shapes where coring required	Simple sand cores can be used, but more difficult to insert than in sand castings	Cores must be able to be pulled because they are metal; undercuts can be formed only by collapsing cores or loose pieces
Minimum wall thickness	3.0-5.0 mm (0.125-0.200 in.) required; 4.0 mm (0.150 in.) normal	3.0-5.0 mm (0.125-0.200 in.) required; 3.5 mm (0.140 in.) normal	1.0-2.5 mm (0.100-0.040 in.); depends on casting size
Type of cores	Complex baked sand cores can be used	Reuseable cores can be made of steel, or nonreuseable baked cores can be used	Steel cores; must be simple and straight so they can be pulled
Tolerance obtainable	Poorest; best linear tolerance is 300 mm/m (300 mils/in.)	Best linear tolerance is 10 mm/m (10 mils/in.)	Best lienar tolerance is 4 mm/m (4 mils/in.)
Surface finish	6.5-12.5 μm (250-500 μin.)	4.0-10 μm (150-400 μin.)	1.5 μm (50 μin.); best finish of the three casting processes
Gas porosity	Lowest porosity possible with good technique	Best pressure tightness; low porosity possible with good technique	Porosity may be present
Cooling rate	0.1-0.5 °C/s (0.2-0.9 °F/s)	0.3-1.0 °C/s (0.5-1.8 °F/s)	50-500 °C/s (90-900 °F/s)
Grain size	Coarse	Fine	Very fine on surface
Strength	Lowest	Excellent	Highest, usually used in the as-cast condition
Fatigue properties	Good	Good	Excellent
Wear resistance	Good	Good	Excellent
Overall quality	Depends on foundry technique	Highest quality	Tolerance and repeatability very good
Remarks	Very versatile as to size, shape, internal configurations	...	Excellent for fast production rates

Source: Ref 42

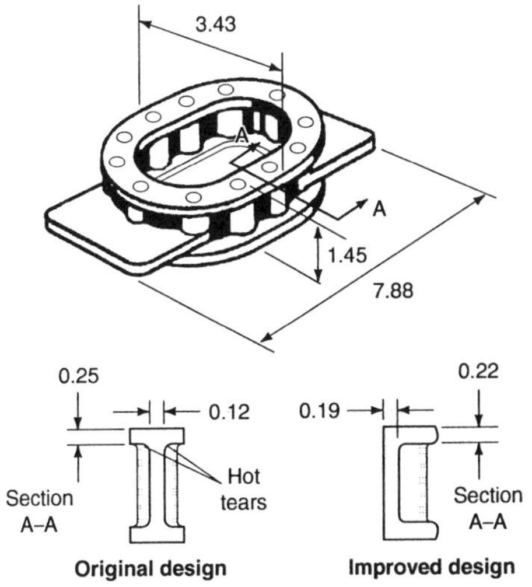

Fig. 5 Redesign of a casting to eliminate hot tears. Mold restraint coupled with nonuniform freezing of the various sections of this aluminum alloy 356 casting resulted in hot tears. Moving the wall and increasing its thickness corrected the problem. Part dimensions in inches. Source: Ref 33

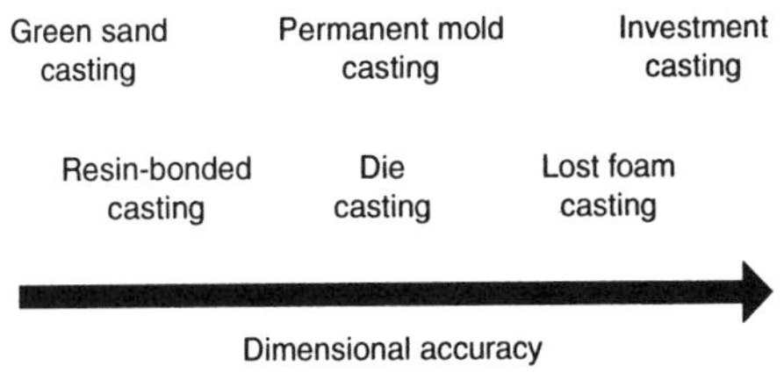

Fig. 6 General relationship between dimensional accuracy and casting process

of coatings to the mold as well as proper selection of mold materials. Metal mold casting processes generally produce better surfaces than sand casting processes.

Design and Service Considerations. The existence of casting discontinuities does not, in and of itself, indicate that casting performance in service will be affected. Equally important are the size, location, and distribution of these discontinuities (Ref 18, 38–40). Those discontinuities that are small and located near the center of the casting have little effect, while those located at or near the surface of the casting are usually damaging. Clustered discontinuities and those that occur in a regular array have a greater effect on properties than those that are isolated and randomly distributed.

In specifying acceptable levels of discontinuities, such as microporosity and inclusion sizes and distribution, the designer should determine the critical flaw size that will deleteriously affect performance in service. This permits the foundry to design a casting practice that will eliminate such discontinuities at minimum cost.

Casting Tolerances

Casting tolerances depend on the alloy being poured, the size of the casting, and the molding method used. Generally speaking, casting dimensional accuracy increases as one moves to the right in Fig. 6 and as the casting dimensions decrease. However, there is significant overlap between processes, and new developments are continually being made so that even this general trend may not apply for a specific cast component. In addition, evidence is mounting that most foundries produce castings that are more accurate dimensionally than called for in casting specifications. For this reason, designers should consider more than one casting process during the design stage and select a process that offers the best combination of dimensions, properties, and cost. Each casting process has its particular strengths, and designers should acquaint themselves with each (Ref 41). Table 1 lists factors that affect the selection of an appropriate casting process for aluminum alloy parts. Additional information on the selection of casting processes is provided in the article "Selection of Manufacturing Processes" in this Volume.

Casting design also affects tolerances. Critical dimensions should not cross parting lines in molds or injection or core dies. The choice of pattern materials also can affect dimensional tolerances.

Stresses that arise from unequal contraction of parts of the casting as it cools from solidification temperatures (Ref 43), as well as those that result from heat treating (Ref 44) can distort castings so that they no longer conform to design dimensions. This also affects tolerances and may lead to extra costs to straighten castings. Solidification stresses are caused by shrinkage during casting solidification when different sections of the casting solidify at different times and rates. Because casting solidification rates depend on the casting design, the designer should consider the effect of the component design on casting distortion. Some examples are given in Fig. 7.

Hot Isostatic Pressing

Occasionally, there are applications where designers require that casting soundness be 100% of theoretical density. For those castings, HIP is recommended. In HIP treatments, the castings are placed in a pressure vessel, and an inert gas is introduced, partially pressurized, and heated while it is confined to the pressure vessel. The gas, unable to expand, increases in pressure and exerts this pressure on the surface of the casting. As the casting is softened by the heat, internal voids collapse under the pressure.

Hot isostatic pressing treatment is relatively inexpensive. If the porosity to be healed is relatively large (macroporosity), HIP will form small dimples on the surface of the casting that may require weld repair. Because HIP is a thermal as well as a pressure treatment, it can alter the microstructure (and therefore the properties) of components if not carefully designed. Hot isostatic pressing will not affect inclusions or

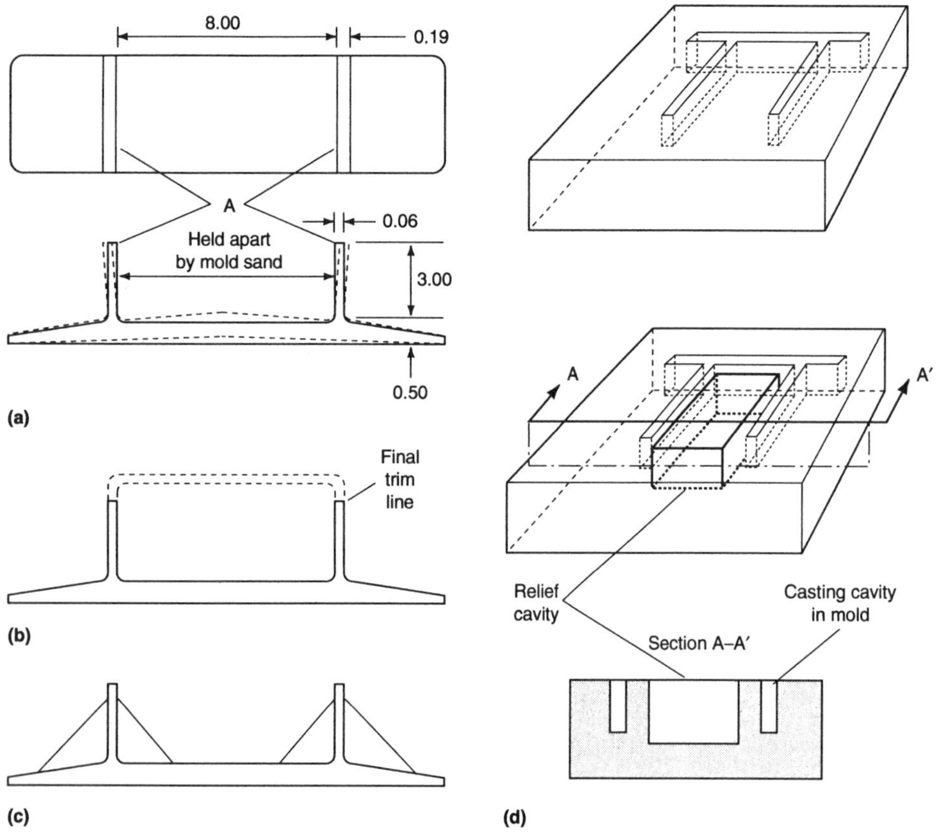

Fig. 7 Designing castings to prevent distortion caused by mold restraint. The original design, shown in top and front views in (a), was altered to three possible preventative designs, as follows. (b) Preferred method incorporating a tie bar. (c) Less effective method than that shown in (b). (d) Open cavity created in the molding media to relieve restraint upon flanges during casting solidification by allowing solid sand mass to collapse during shrinkage and minimize restraint. Dimensions given in inches. Source: Ref 45

oxide folds and, therefore, will not repair castings that have those types of discontinuities. It also has no effect on pores that are connected to the surface. Hot isostatic pressing has made possible the use of alloys that, because of their composition, do not solidify pore free and, without HIP, could not be used for high-integrity applications.

Solidification Simulation and Its Use in Designing Castings

This article has emphasized the importance of the casting design in influencing the way the casting solidifies and, thus, the cast structure and the properties that are produced. Since the cast structure controls the heat treated structure and the final structure controls the properties of the component, it is clear that successful design of castings means that the designer must be able to predict the effect of the design on the structure of the final part. This is done today by using solidification simulation models. Most progressive foundries today use simulation models in engineering the casting process. Designers are encouraged to work with foundries and have them run solidification models on their designs in a concurrent design process to engineer castings that perform satisfactorily and are reasonably priced.

Solidification simulation models have been extensively developed over the last twenty years, and today there are a large variety of them (Ref 46). The more elementary models are based on Chvorinov's law and operate by comparing surface area to volume ratios of adjoining casting sections to predict the order in which the sections solidify (that is, how progressive the solidification will be). Those models are satisfactory for predicting whether or not macroporosity will form on solidification; some models are also effective at predicting the location of microporosity.

Advanced models use finite element or finite difference methods to predict how the casting fills and, from those results, are able to establish the temperature of each point in the mold and each point in the liquid metal at the start of solidification. From this, a more accurate picture of the way the casting solidifies is possible. These models are also capable of predicting the grain size and structure and are now being used to predict casting properties that result from a specific casting design and manufacturing process. Solidification models today are capable of predicting the distortion that will occur during casting and heat treatment and can be used to minimize residual stresses. Some of the commercial packages include expert systems to help orient the casting in the mold to minimize casting defects and to suggest gating and risering schemes.

These models are a powerful tool for concurrent engineering of cast components. The designer, working with the foundry, can see how the casting will solidify before the casting is poured. If problem areas are found with the design, it can be altered, and the new design can be simulated. This iteration can be repeated until a satisfactory design is developed. When these simulations are combined with rapid prototyping methods, expedited delivery of cast components can be achieved.

The designer must begin the design of a casting with an understanding of what properties are desired and required throughout the cast component. Flaw sizes must be kept below the critical level, and the designer should know what that level is. Critical flaw sizes will vary according to location in the casting; that information should be reflected in specifications that also vary by location in order to minimize manufacturing costs.

Casting design influences how the castings solidify and, hence, the properties that are obtained. Many collections of casting data in handbooks overlook the effect of cooling rate on casting performance and, as a result, do not provide adequate information for designers. Designers are therefore encouraged to work with foundries to develop data bases of reliable mechanical property data that reflect the effect of cooling rates of the alloy and the sections that the designer intends to use. This is particularly important for areas of the casting where stress concentrations are likely; use of a solidification simulation program should ensure that these areas are not also areas of slow solidification.

Designers should be aware that different casting alloys have different levels of "castability," meaning that development of reliable casting properties may be more difficult in some alloys than others. Specification of a casting alloy solely on the results of test bar data (test bars are easy to cast and often do not reflect casting problems encountered in actual castings) without consulting the metal caster can lead to high casting costs as the foundry compensates for castability difficulties with the alloy.

Conclusions

The casting process offers the designer exceptional design flexibility at reasonable cost. The designer is the key to successful application of the process. Because casting design affects solidification rate, the design also affects the properties that can be obtained and the discontinuities that can arise from the process. Close cooperation between designer and foundry is essential to realize the benefits of the process.

Today's technology, both in component design and manufacturing process design, makes extensive use of computer simulation for optimization. In the future, the combination of these technologies will enable designers and metal casters to produce highly engineered cast components.

REFERENCES

1. J. Gruner, "Structural Aluminum Aircraft Casting with No Casting Factor," presented at Aeromat '96 (Dayton, OH), ASM International, June 1996
2. W. Weibull, A Statistical Distribution Function of Wide Applicability, *J. Appl. Mech.*, Vol 18, 1951, p 293
3. J. Campbell, J. Runyoro, and S.M.A. Boutorabi, Critical Gate Velocities for Film-Forming Alloys, *AFS Trans.*, Vol 100, 1992, p 225
4. N.R. Green and J. Campbell, *Proc. Spring 1993 Meeting* (Strasbourg, France), European Division Materials Research Society, 4–7 May, 1993
5. N.R. Green and J. Campbell, The Influence of Oxide Film Filling Defects on the Strength of Al-7Si-Mg Alloy Castings, *AFS Trans.*, Vol 102, 1994, p 341
6. Manufacturing Considerations in Design, *Steel Castings Handbook*, 5th ed., P.F. Wieser, Ed., Steel Founders' Society of America, 1980, p 5-6
7. *Materials Handbook*, Vol 15, *Casting*, ASM International, 1988, p 598
8. *Investment Casting*, P.R. Beeley and R.F. Smart, Ed., The Institute of Materials, 1995, p 334
9. *Design and Procurement of High-Strength Structural Aluminum Castings*, S.P. Thomas, Ed., American Foundrymen's Society, 1995
10. Manufacturing Design Considerations, Chap. 7, *Steel Castings Handbook*, 6th ed., M. Blair and T.L. Stevens, Ed., Steel Founders' Society of America and ASM International, 1995
11. E.L. Rooy, Hydrogen: The One-Third Solution, *AFS Trans.*, Vol 101, 1993, p 961
12. N. Roy, A.M. Samuel, and F.H. Samuel, Porosity Formation in Al-9 Wt Pct Mg - 3 Wt Pct Cu Alloy Systems: Metallographic Observations, *Met. Mater. Trans.*, Vol 27A, Feb 1996, p 415
13. M.K. Surappa, E. Blank, and J.C. Jaquet, Effect of Macro-porosity on the Strength and Ductility of Cast Al-7Si-0.3Mg Alloy, *Scr. Metall.*, Vol 20, 1986, p 1281
14. C.H. Cáceres, On the Effect of Macroporosity on the Tensile Properties of the Al-7%Si-0.4%Mg Casting Alloy, submitted to *Scr. Metall.*, 1994
15. C.M. Sonsino and K. Dietrich, Einfluβ der Porosität auf das Schwingfestigkeitverhalten von Aluminium-Guβwerkstoffen, *Giessereiforschung*, Vol 43 (No. 3 and 4), 1992, p 119–140
16. B. Skallerud, T. Iveland, and G. Härkegård, Fatigue Life Assessment of Aluminum Alloys with Casting Defects, *Eng. Fract. Mech.*, Vol 44 (No. 6), 1993, p 857
17. M.J. Couper, A.E. Neeson, and J.R. Griffiths, Casting Defects and the Fatigue Behaviour of an Aluminium Casting Alloy, *Fatigue Fract. Eng. Mater. Struct.*, Vol 13 (No. 3), 1990, p 213
18. J.C. Ting and F.V. Lawrence, Jr., Modeling the Long-Life Fatigue Behavior of a Cast Aluminum Alloy, *Fatigue Fract. Eng. Mater. Struct.*, Vol 16 (No. 6), 1993, p 631
19. J.T. Berry, Linking Solidification Conditions and Mechanical Behavior in Al Castings—A Quarter Century of Evolving Evidence, *AFS Trans.*, Vol 103, 1995, p 837
20. S. Kennerknecht, "Design and Specification of Aluminum Airframe Structural Castings," presented at Aeromat '95 (Anaheim, CA), ASM International, May 1995
21. G.O. Rading, J. Li, and J.T. Berry, Fatigue Crack Growth in Cast Al-Cu Alloy A206 with Different Levels of Porosity, *AFS Trans.*, Vol 102, 1994, p 57
22. I.P. Volchok, Non-Metallic Inclusions and the Failure of Ferritic-Pearlitic Cast Steel, *Cast Metals*, Vol 6 (No. 3), 1993, p 162
23. P. Heuler, C. Berger, and J. Motz, Fatigue Behaviour of Steel Casting Containing Near-Surface Defects, *Fatigue Fract. Eng. Mater. Struct.*, Vol 16 (No. 1), 1992, p 115
24. J. Motz et al., Einfluss oberflächener Fehlstellen im Stahlguβ auf die Riβeinleitung bei Schwingungsbeanspruchung, *Geissereiforschung*, Vol 43, 1991, p 37
25. C.E. Bates and C. Wanstall, Clean Steel Castings, in *Metalcasting Competitiveness Research*, Final Report, DOE/ID/13163-1 (DE95016652), Department of Energy, Aug 1994, p 51
26. C. Nyahumwa, N.R. Green, and J. Campbell, "The Effect of Oxide Film Filling Defects on the Fatigue Life Distributions of Al-7Si-Mg Alloy Castings," presented at the International Symposium on Solidification Science and Processing (Honolulu, HI), Japan Institute of Metals and TMS, Dec 1995
27. J. Campbell, The Mechanical Strength of Non-Ferrous Castings, *Proc. 61st World Foundry Cong.* (Beijing), 1995, p 104; available from the American Foundrymen's Society, Des Plaines, IL
28. K.E. Höner and J. Gross, Bruch verhalten und mechanische Eigenschafter von Aluminium-Silicium-Guβlegierungen in unterschiedlichen Behandlungszuständen, *Giessereiforschung*, Vol 44 (No. 4), 1992, p 146
29. F.T. Lee, J.F. Major, and F.H. Samuel, Effect of Silicon Particles on the Fatigue Crack Growth Characteristics of Al-12 Wt Pct - 0.35 Wt Pct Mg - (0 to 0.02) Wt Pct Sr Casting Alloys, *Met. Mater. Trans.*, Vol 26A (No. 6), June 1995, p 1553
30. J.F. Major, F.T. Lee, and F.H. Samuel, Fatigue Crack Growth and Fracture Behavior of Al-12 wt% Si-0.35 wt% Mg (0-0.02) % Sr Casting Alloys, Paper 96-027, *AFS Trans.*, Vol 104, 1996
31. D. Argo and J.E. Gruzleski, Porosity in Modified Aluminum Alloy Castings, *AFS Trans.*, Vol 96, 1988, p 65
32. T.L. Reinhart, "The Influence of Microstructure on the Fatigue and Fracture Properties of Aluminum Alloy Castings," presented at Aeromat '96 (Dayton, OH), *ASM International*, June 1996
33. Permanent Mold Casting, *Forging and Casting*, Vol 5, *Metals Handbook*, 8th ed., American Society for Metals, 1970, p 279
34. A.L. Kearney and J. Raffin, *Heat Tear Control Handbook for Aluminum Foundrymen and Casting Designers*, American Foundrymen's Society, 1987
35. D.M. Stefanescu et al., Cast Iron Penetration in Sand Molds: Part I: Physics of Penetration Defects and Penetration Model, Paper 96-206, *AFS Trans.*, Vol 104, 1996
36. R.L. Naro and J.F. Wallace, Effect of Mold-Steel Interface Reactions on Casting Surface and Properties, *AFS Trans.*, Vol 75, 1967, p 741
37. R.L. Naro and J.F. Wallace, Effect of Mold-Steel Interface Reactions on Casting Surfaces, *AFS Trans.*, Vol 100, 1992, p 797
38. E.M. Dubensky and D.A. Koss, Void/Pore Distributions and Ductile Fracture, *Metall. Trans. A*, Vol 18, 1987, p 1887
39. L. Sidanin, S. Milicev, and N. Matovic, Fatigue Failure of Ductile Iron Crankshafts, *Cast Metals*, Vol 4 (No. 1), 1991, p 50
40. A. Needleman and V. Tvergaard, A Numerical Study of Void Distribution Effects on Dynamic, Ductile Crack Growth, *Eng. Fracture Mech.*, Vol 38 (No. 2/3), 1991, p 157
41. *ASM Handbook*, Vol 15, *Casting*, D.M. Stefanescu, Ed., ASM International, 1988
42. A. Kearny and E.L. Rooy, Aluminum Foundry Products, *ASM Handbook*, Vol 2, *Properties and Selection: Nonferrous Alloys and Special-Purpose Materials*, ASM International, 1990, p 139
43. J. Cech et al., Rationalizing Foundry Production and Assuring Quality of Castings with the Aid of Computer Science, Paper 10, *Proc. 62nd World Foundry Congress* (Philadelphia), 1996; available from the American Foundrymen's Society, Des Plaines, IL
44. J. Campbell, Review of Reliable Processes for Aluminum Aerospace Castings, Paper 96-158, *AFS Trans.*, Vol 104, 1996
45. D.E. Groteke, Dimensional Tolerances and Allowances, *ASM Handbook*, Vol 15, *Casting*, ASM International, 1988, p 617
46. *Modeling of Casting, Welding and Advanced Solidification Processes VII*, M. Cross and J. Campbell, Ed., TMS/AIME, 1995

Design for Deformation Processes

B. Lynn Ferguson, Deformation Control Technology, Inc.

DEFORMATION PROCESSING involves changing the shape of a workpiece by plastic deformation through application of compressive forces. In practice, deformation processes are used to transform an initial material form (for example, cast ingot, continuously cast slab or billet, or powder) into an intermediate form (for example, plate, strip, bar, or porous preform) and then into a part. In addition to the workpiece, two tools are therefore required: a machine to generate the force and an anvil or die to support the workpiece as the force is applied. In its most primitive form, the blacksmith applied the force through a hammer blow and the workpiece rested on the anvil as the force was applied. In its modern implementation, a machine has replaced the blacksmith, and the anvil has been incorporated into the machine as a replaceable tool component to control the shape change of the workpiece.

The goal of a deformation process is the same as other manufacturing processes in that a desired geometrical form composed of a particular material that possesses certain mechanical and/or physical characteristics must be achieved at a minimum cost. Because a product form may be produced by a wide variety of methods (as described in the article "Manufacturing Processes and Their Selection" in this Volume), what characteristics of deformation processing provide the bases for the designer to select a deformation process as the desired production method? In other words, what are the advantages that controlled plastic deformation impart to a workpiece as opposed to achieving the shape by solidification, molding of powder, or by machining? This article introduces reasons behind the selection of a deformation process as the method of choice for producing a part or product form. Some fundamental aspects of plastic flow are presented since it is the ability of a material to flow plastically that allows these processes to be used. Because machinery and tooling must be involved to impart shape change of the workpiece, it is necessary to consider the effect of friction at the tool/workpiece interface, as well as the ability of the tool to withstand the loads and temperatures required to deform the workpiece into the desired geometry. For a more complete discussion of equipment used for deformation processes, see Ref 1 to 4.

Why Use a Deformation Process?

Cost, dimensions and tolerances, surface finish, throughput, available equipment, and part performance requirements dictate the material and process selection for production. The simplest practice should be used that achieves the desired product form. Deformation offers many advantages, especially in terms of microstructural benefits, but these processes also have disadvantages, the main ones being the cost of equipment and tooling.

Advantages and Disadvantages of Deformation Processes. Discussions of process advantages must be approached carefully because most final parts are generally subjected to more than one type of manufacturing process. For example, a screw-machined part has been first cast, hot rolled to bar stock, and possibly cold rolled or drawn prior to screw machining. A forging may have been cast, hot rolled to bar form, cropped into a billet, forged through multiple stations, and then finish machined. There are steps needed to produce the starting material (ingot or cast shape), intermediate steps needed to shape the material into a manageable interim form (bar, plate, tube, sheet, wire), and then steps needed to make the final part. Both processes to make interim product forms or stock and processes to make parts are included here.

The objective of hot forging or hot rolling of cast materials is to refine the structure that results from solidification. To alter the inhomogeneous structure due to solidification and to produce a more workable microstructure, cast ingots and continuously cast slabs and blooms are typically hot worked into interim product forms, that is, plate, bars, tubes, or sheet. Large deformation in combination with heat is very effective for refining the microstructure of a metal, breaking up macrosegregation patterns, collapsing and sealing porosity, and refining the grain size.

Many design decisions are required in order to take advantage of the benefits of deformation processing while avoiding potential problems of flow-related defects: fracture or poor microstructure. Some of the advantages of a part produced by deformation are listed in Table 1. In addition, the process can be tailored to achieve tight control of dimensions for mass production, and typically some net surfaces can be achieved. While the goal is to achieve a net shape, it is rare that a totally net shape is produced by bulk-deformation processes, and some machining is typical to produce a usable part. Sheet-forming processes, however, often result in net functional surfaces. The production rate for many deformation processes can be high, so that high-volume production requirements can be met with efficient machinery utilization.

Disadvantages of deformation processes are listed in Table 2. It is interesting to note that many of the advantages of deformation also show up on the list of disadvantages. If the deformation process is poorly designed and/or poorly executed, the sought-after advantages will not be realized, and instead an inferior part will be produced. The

Table 1 Advantages of deformation processing

Improved internal quality due to compressive deformation
Uniform grain structure
Elimination of casting porosity
Breakup of macrosegregation patterns
Beneficial grain-flow pattern for improved part performance
Improved toughness due to grain flow and fibering
Improved fatigue resistance due to grain-flow pattern
Controlled surface quality
Burnished surface can have improved fatigue resistance due to quality of as-forged surface
High throughput due to potentially high rates of forming
Ability to produce a net-shape or near-net-shape part

Table 2 Potential disadvantages of deformation processing

Fracture-related problems
Internal bursts or chevron cracks
Cracks on free surfaces
Cracks on die contacted surfaces
Metal-flow-related problems
End grain and poor surface performance
Inhomogeneous grain size
Shear bands and locally weakened structures
Cold shuts, folds, and laps
Flow-through defect
Control, material selection, and utilization problems
Underfill, part distortion, and poor dimensional control
Tool overload and breakage
Excessive tool wear
High initial investment due to equipment cost
Poor material utilization and high scrap loss

categorization in Table 2 is somewhat arbitrary because metal flow, fracture, die wear, and tool stresses are so interlinked. The decisions that the designer must make concerning the preform or initial workpiece geometry, the deformation temperature, amount of force and forging speed, the friction conditions, and the metallurgical condition of the workpiece are all interrelated. Decisions about the deformation process must be made to accentuate the advantages listed in Table 1 and to overcome or avoid disadvantages listed in Table 2. Some of these design decisions are listed in Table 3.

Categories of Deformation Processes

Dieter (Ref 5) has categorized deformation processes into five broad classes:

- *Direct Compression Processes:* Force is applied directly to the surface of the workpiece and material flow is normal to the application of the compressive force; examples are open-die forging and rolling.
- *Indirect Compression Processes:* Deformation is imposed by compressive loads generated as the workpiece is pushed or pulled through a converging die. The direction of the external load applied to the workpiece is in the direction of workpiece motion; examples include extrusion, wire drawing, and deep drawing.
- *Tension-Based Processes:* Tensile loading is developed in the workpiece to cause thinning, with stretch forming being a primary example.
- *Bending Processes:* A bending moment is applied to cause a geometry change, the deformation being limited to the local region of the bend. Sheet bending, rod bending and coiling, and plate bending are example processes.
- *Shearing Processes:* Metal deformation is highly localized in a workpiece as offset blades moving in opposite directions generate a plane of intense shear to intentionally cause a shear failure. Hole punching, plate shearing, blanking, and slitting are examples of shearing processes.

Other terminology is also recognized in the industry. *Bulk-forming processes* are processes that have large volumes of material participating in the deformation and may be termed three-dimensional processes. A typical goal is to alter a cast grain structure to a more uniform, sound structure, with hot rolling if ingots, slabs, or billets being a primary example. Table 2 in the article "Manufacturing Processes and Their Selection" in this Volume is a list of some general characteristics of bulk-deformation processes such as hot forging, hot extrusion, cold forging, cold extrusion, wire and strip drawing, and rolling. The remaining deformation processes are defined by an initially large surface-area-to-volume ratio such that the volume of material in the deformation zone at any given time during the process is small. *Sheet-forming processes*—blanking, shearing, bending, spinning, and stretching—fall into this category (see Table 3 in the article "Manufacturing Processes and Their Selection").

Table 3 Design decisions associated with deformation processes

Part-related decisions
Part or product material selection
Geometry and dimensions to be produced
Required properties (mechanical, physical, and metallurgical)
Process-related decisions
Equipment selection (type, rate, and load requirements)
Starting material geometry (plate, bar, sheet, etc.)
Workpiece temperature and tooling temperature
Orientation of part during deformation step(s)
Location of flash or scrap loss
Number of deformation steps
Lubrication and method of application
Starting microstructure and control of microstructure during forging sequence (preheat practice and intermediate heating steps, if any)

Another general way of referring to deformation processes are as hot-, warm-, or cold-working processes. Cold working typically refers to processes that are conducted at or near room temperature. Hot-working processes are conducted at temperatures above the recrystallization temperature, which is roughly a homologous temperature of 0.5. Homologous temperature is the ratio of workpiece temperature to its absolute melting point. Warm working processes are conducted at intermediate temperatures. These designations really relate to the deformation mechanisms involved in plastic flow and the effect that the working operation has on the grain structure. Hot working produces a recrystallized grain structure, while the grain structure due to cold working is unrecrystallized and retains the effects of the working operation.

Fundamentals of Deformation Processing

There are four major design considerations in applying a deformation process. The first consideration is the workpiece material and its flow stress behavior. The second consideration is the fracture behavior of the material and the effects of temperature, stress state, and strain rate on fracture; this combined view of ductility and stress state is termed workability for bulk-forming processes and formability for sheet-forming processes. The third major consideration is a determination of the desired final microstructure needed to produce an acceptable product and a determination of which process should be used to produce this microstructure. A fourth consideration involves added constraints of available equipment and economics in addition to flow stress, forming, and part performance considerations. The fourth consideration usually dominates the other considerations, sometimes to the detriment of the material being worked.

Historically, deformation processing has fallen in the gap between the traditional disciplines of metallurgy and mechanics, and, as a result, this area has often been neglected in an academic sense. A main reason for this is the difference in length scales by which mechanics and materials science view a material. The length scale of deformation is at the atomic level, and it is at this level that materials science addresses deformation. In manufacturing, deformation effects are related or measured at a macroscopic level, and continuum mechanics is applied to analyze and explain plasticity quantitatively. However, to understand the macroscopic response of materials to temperatures and rates of deformation, consideration must be at a lower length scale level, at least to qualitatively explain material behavior. The empirical equations used to address plasticity do not generally capture the microscopic aspects of deformation. Fortunately, the metals that are commonly processed by bulk-deformation methods have many grains per unit volume, and microscopic events are suitably averaged at the macroscopic level. The macroscopic or continuum mechanics approach begins to break down when the grain size approaches the physical size of the workpiece or when a dominant crystallographic texture is present in a workpiece. Examples of the former include fine wire drawing, bending of fine wire, and sheet-forming processes. Primary examples of the latter are sheet-metal-forming processes or bending of heavily drawn wire where crystallographic texture plays a dominant role. An additional complication is the fact that most metals have more than one phase present in their microstructure. The second phase may be present due to alloying, that is, cementite in iron, or it may be an unwanted

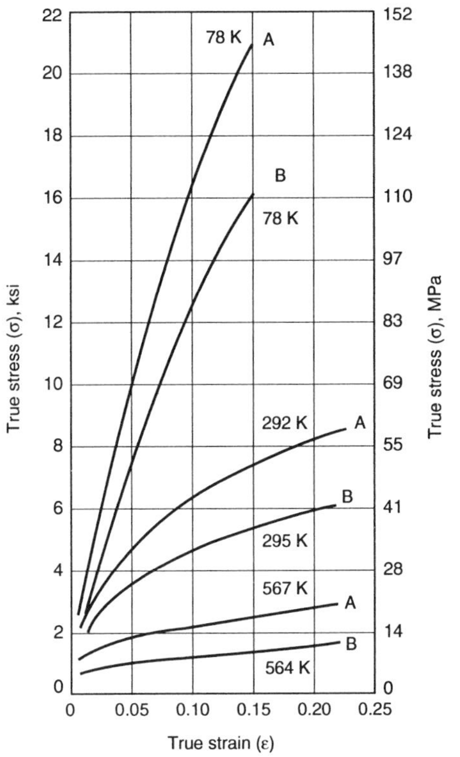

Fig. 1 True stress-true strain curves for commercial-purity aluminum as a function of temperature and strain rate. Strain rates: A, 0.167/s, B, 0.00194/s. Source: Ref 6

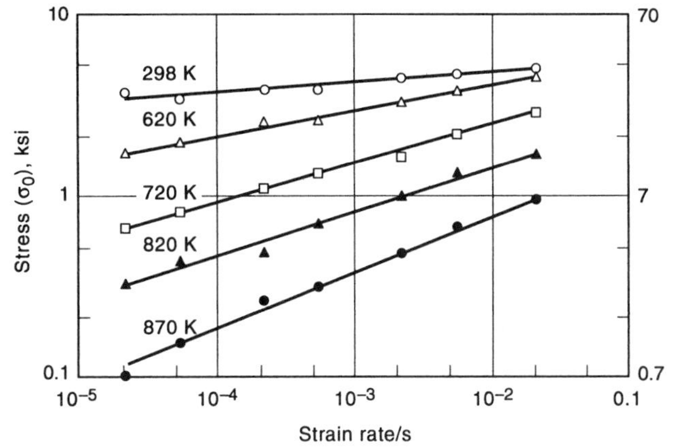

Fig. 2 Flow stress for 6063-O temper aluminum at a strain of 0.2%. Source: Ref 7

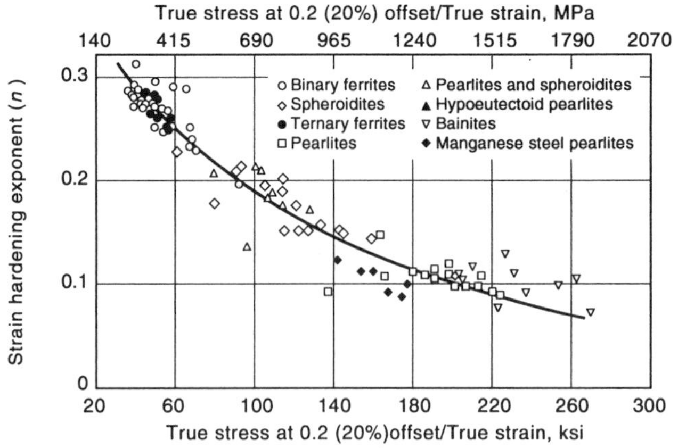

Fig. 3 Relationship between yield strength and the strain-hardening exponent (n) for a variety of steel microstructures. Source: Ref 10

phase, that is, sulfide or silicate inclusions in steel. The effects of these second phases are again averaged at the macroscopic level, and the material has not been adequatelly described as having separate phases from a mathematical sense. The materials science community and the mechanics community are trying very hard to bridge this length scale problem, but for now the most useful analysis tools are combinations of continuum mechanics and empirical results.

Flow Stress

Flow stress behavior refers to the effects of temperature, deformation rate, and work- hardening behavior on the stress needed to cause further plastic deformation. Flow stress characterizes the dynamic stress-strain behavior of a material, where

$$\sigma_f = f(\varepsilon, \dot{\varepsilon}, T, M, G)$$

where σ_f is flow stress, ε is the local strain level, $\dot{\varepsilon}$ is the local strain rate, T is the local temperature, M is the microstructure in terms of phases and distributions of phases, and G is grain size.

The stress-strain curves for commercially pure aluminum over a range of temperatures and at two strain rates are given in Fig. 1. These curves show that at low temperatures work hardening is pronounced as flow stress increases rapidly with strain. At high temperatures, there is negligible change in flow stress with strain. To better examine strain rate effects, Fig. 2 shows a plot of stress at 0.2% strain versus strain rate data for 6063-O aluminum plotted on a log-log scale. Strain rate is the time dependence of strain, $\dot{\varepsilon} = d\varepsilon/dt$. As temperature increases, the flow stress decreases, and furthermore, the increase in slope of these lines with temperature shows that strain rate becomes more influential at higher temperatures. Laboratory tests such as the compression test and tensile test have been developed to measure flow stress, and atlases of flow stress data are available (Ref 1, 3–5, 8, 9). Typically, the data are reported for a constant temperature and constant strain rate, and these data are vital for numerical calculations regarding deformation processes. However, care must be exercised when applying these data to a specific problem because the grain size, second-phase volume fraction, and the distribution of second phases may be different for a specific case. For this reason, the pedigree of any data that are used for any numerical calculations is necessary information.

Dislocations, or defects in the crystal structure, enable ductile behavior of metals as these defects participate actively in deformation at the microscopic level. For polycrystalline metals, there is a difference in ease of deformation between grain interiors and grain boundaries. The equicohesive temperature is the temperature at which the grain interiors and the grain boundaries have the same strength or resistance to deformation. This temperature is not a fixed temperature, but is sensitive to the rate of stress application. A second temperature of significance is the homologous temperature, which is the ratio of temperature to the absolute melting point. The sections that follow address deformation process temperature effects.

Cold Working

Cold working results in a deformed, unrecrystallized grain structure with the grains being elongated in the direction of metal flow. Deformation begins in the grain interiors first when the critical resolved shear stress of the material is exceeded. At cold-working temperatures, the grain boundaries are more resistant to deformation, so workpieces with fine grains and a large amount of grain boundary area are stronger than coarse-grained material of the same alloy. This dependence of yield strength on grain size and the amount of grain-boundary area in a material is captured by the well-known Hall-Petch equation:

$$\sigma_y = \sigma_i + k/\sqrt{D} \quad \text{(Eq 1)}$$

where σ_y is the yield stress of polycrystalline metal, σ_i is the stress related to resistance of dislocation motion within a grain, k is the parameter relating grain-boundary hardening effect, and D is the average grain diameter. Equation 1 indicates that the yield strength of an alloy increases as the grain size becomes finer, where yield strength is the initial flow stress.

Once deformation is initiated, the moving dislocations interact with each other and with the grain boundaries to make continued yielding more difficult. This is work hardening, and a further feature of cold forming is that work hardening effects continue to build with continued deformation. An empirical relationship for cold working between flow stress and plastic strain is:

$$\sigma_f = K \cdot \varepsilon^n \quad \text{(Eq 2)}$$

where σ_f is the flow stress, K is the strength coefficient (stress when $\varepsilon = 1.0$), ε is the plastic strain, and n is the work-hardening exponent. From Eq 2, a high-strength coefficient indicates a high initial resistance to plastic flow. Metals with a high K require large machines for deformation. Work hardening is a measure of how the resistance to plastic flow increases as the metal is deformed. Typically, n has values of 0.1 to 0.5 for cold working, with 0 being a perfectly plastic metal (no work hardening). A metal with a high work-hardening exponent but a low strength coefficient will achieve a high strength level after a large amount of deformation. Copper, brasses, and low-carbon steels are typical examples

Table 4 Values for the work-hardening exponent and strength coefficient for selected metals

Metal	Condition	Work-hardening exponent (n)	Strength coefficient (K) MPa	ksi
0.05% C steel	Annealed	0.26	531	77
4340 steel	Annealed	0.15	641	93
0.6% C steel	Quenched and tempered 540 °C (1000 °F)	0.10	1572	228
0.6% C steel	Quenched and tempered 700 °C (1300 °F)	0.19	1227	178
Copper	Annealed	0.54	317	46
70/30 brass	Annealed	0.49	896	130

Source: Ref 5

of metals that are cold worked to produce improved hardness and strength in the formed part. Table 4 contains some values of K and n for these metals (Ref 5). For steels, K increases with carbon content, while n generally decreases. Both copper and brass have a much higher work hardening exponent than steel. Both K and n are affected not only by chemistry, but also by prior history and the microstructure. This is shown in Fig. 3 for the work-hardening exponent for a variety of steels and microstructures.

The work-hardening exponent indicates further information about the metal. Figure 4 shows an ideal tensile stress versus strain curve for cold deformation. The work-hardening exponent corresponds to the strain at which strain becomes localized and results in necking. The higher the value of n, the greater the strain before necking begins. This is especially important for metals undergoing deformation processes that are tensile in nature, such as many sheet-forming processes and wire drawing.

Over the range of strain rates at which cold-deformation processes are conducted (0.1 to 100/s), the sensitivity to strain rate for most metals is low, as supported by the low slopes at low temperatures in Fig. 2. Strain level rather than strain rate controls the flow stress, in addition to the initial strength coefficient K.

Grain growth is not a factor in cold working. However, grain flow and a change in grain aspect ratio is very much a factor. As the grains distort, a well-defined grain-flow pattern is developed due to grain-boundary alignment. Nonmetallic inclusions may also participate and further define a definite directionality in the microstructure and mechanical properties due to this mechanical fibering. Extremely deformed microstructures, as are present in cold-rolled sheet products, may also show alignment of crystallographic planes or texture, as well as grain-boundary alignment. The result is anisotropic behavior of the deformed material, either in service or in subsequent deformation steps. The designer must be aware of the effects of microstructural features such as fibering or preferred orientation on mechanical properties, and the relationship of microstructural alignment and performance stresses. This is especially critical in cases where fatigue and fracture toughness are design issues.

Hot Working

Hot working takes place roughly above a homologous temperature of 0.5, with typical hot-working temperatures being 70 to 80% of the absolute melting temperature. At these temperatures, there is a high amount of internal energy available and a number of deformation mechanisms, in addition to slip, are available. These additional mechanisms include power law creep mechanisms such as dislocation glide and climb, and diffusional flow such as diffusion of vacancies and boundary motions. Referring again to Fig. 1, the room-temperature curves for aluminum show typical work-hardening behavior associated with cold deformation. However, at high temperature work hardening is low and the flow stress curve becomes very different from that of cold deformation. Hot working involves recovery and recrystallization of the microstructure. These may be dynamic or static in nature, with dynamic recovery and dynamic recrystallization occurring during the deformation step, and with static recovery and recrystallization occurring after the deformation step while the workpiece is still hot.

Figure 5 shows schematically grain structure changes during hot rolling to a moderate level and deformation and extrusion with a high level of deformation (Ref 3). Stacking-fault energy, as mentioned in Fig. 5 as being high or low, relates to the dislocation structure of the crystal. Low stacking-fault energy results in wide stacking faults that have a relatively high resistance to thermally activated mechanisms, and these metals strain harden rapidly. Metals with high stacking-fault energy have narrower stacking faults, the dislocations are more mobile, and as a result the rate of work hardening is low. In comparing the effect of hot deformation in metals having low or high stacking-fault energies, Fig. 5 shows that dynamic recovery occurs in all cases. However, a high deformation level is required to produce recrystallization in metals with high stacking-fault energy, while metals with low stacking-fault energy can recrystallize at a lower level of deformation. Metals with low stacking-fault energies include brass and austenitic stainless steels, and metals having high stacking-fault energy include aluminum and nickel alloys. In practice, most conventional hot-working processes are too fast for dynamic recrystallization to occur.

Dynamic recovery occurs when there is sufficient atomic mobility to balance or nearly balance work hardening. That is, dislocations are sufficiently active to move in response to local stresses associated with dislocation tangles and forests, the presence of second phases, and other local stress concentrations. A metal that is undergoing dynamic recovery during hot working will exhibit negligible work hardening, with most low-carbon and low-alloy steels being primary examples. Figure 6(a) shows a schematic flow stress-strain curve for a metal that dynamically recovers during hot working (Ref 3).

Recrystallization occurs if a critical level of strain energy is achieved so that a new set of grains forms. If recrystallization occurs during hot deformation, the result is flow softening as shown in Fig. 6(b) (Ref 3). Examples of metals that may flow soften during hot working include

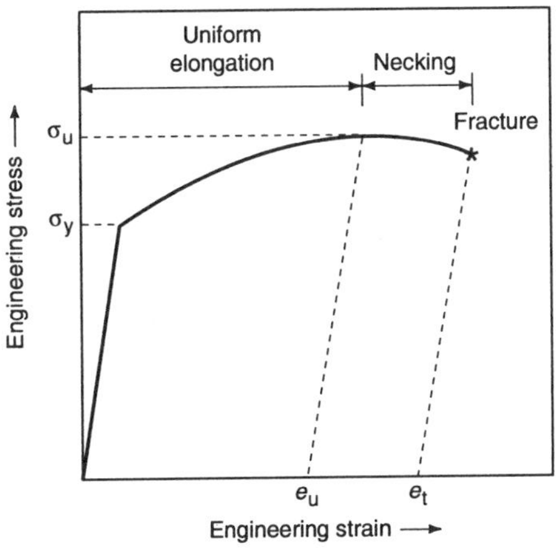

Fig. 4 Schematic engineering tensile stress-strain curve to show uniform elongation, ultimate strength, and necking. e_t, tensile elongation; e_u, uniform elongation; σ_u, ultimate tensile strength; σ_y, yield strength

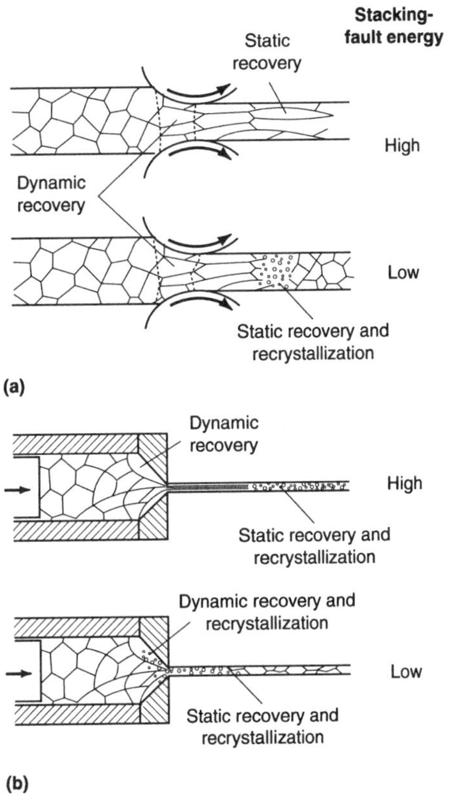

Fig. 5 Hot-working effects on microstructure. (a) Rolling with a thickness strain of 50%. (b) Extrusion with a strain of 99%. Source: after Ref 3

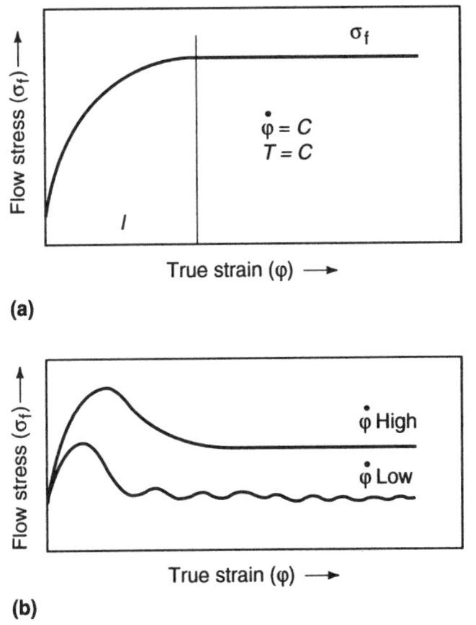

Fig. 6 Schematic flow stress curves representative of (a) dynamic recovery during hot working and (b) dynamic recovery and dynamic recrystallization. $\dot{\varphi}$, shear strain rate; T, temperature; C, constant. Source: Ref 3

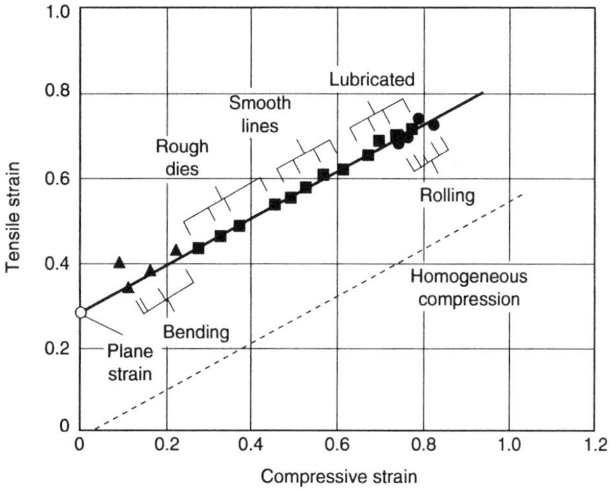

Fig. 7 Workability limit for upsetting, bending, and rolling with varying aspect ratios and friction conditions. Source: Ref 11

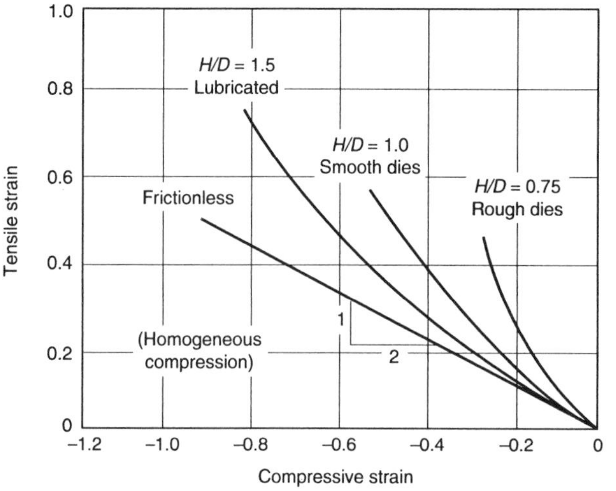

Fig. 8 Surface strain paths observed in upset tests for different friction conditions and aspect (height/depth, or *H/D*) ratios. Source: Ref 11

nickel-aluminum bronze, commercial purity titanium, super α_2 titanium aluminide, Ti-6Al-4V, and 7075 aluminum. Whether or not recrystallization and flow softening occurs is a function of the working temperature and strain rate and the local deformation level. Flow softening can lead to strain localization and highly nonuniform microstructures, and in extreme cases, shear cracking. This is mainly true in cases where this phenomenon occurs over a narrow range of temperatures or strain rates.

Recovery and recrystallization are thermally activated, and therefore these mechanisms are very dependent on temperature. They are also very dependent on the level of deformation because strain is indicative of a growing volume fraction of defects in the crystal structure and greater internal energy or stress within the grains. The benefit of recovery is that work-hardening effects can be minimized by allowing atomic rearrangement to reduce the internal stress within the grain. This is why hot-working processes can accomplish large deformation levels while maintaining relatively low working loads. For processes such as hot rolling, extrusion, and forging, the time in the deformation zone is short. Grain refinement is accomplished by static recrystallization after hot working. A high level of hot deformation followed by a hold time at an elevated temperature causes static recovery and recrystallization to result in a fine grain size. This may occur in hot rolling, where there is time between roll passes, or after hot forging where the workpiece slowly cools in a bin. Metallurgical specifications are more frequently including grain size limits, and therefore it is becoming more critical to control workpiece temperature, deformation rate, the amount of deformation per working step, and the time between steps in order to control the microstructure of the deformed workpiece.

The other microstructural phenomenon that can occur during hot-working processes is grain growth. The natural drive for a polycrystalline material is to minimize internal energy, and because grain boundaries are regions of higher internal energy, grain growth is a way for nature to minimize energy by minimizing the grain-boundary content. Grain growth is also thermally driven, and because hot-working processes may hold a workpiece at a high temperature for a long time, grain growth can occur. In fact, in an extended hot working process such as ingot breakdown rolling, a cyclic history of grain deformation, recrystallization and growth is established for each deformation step. The ability to put work into the grains at a level sufficient to cause recrystallization is the reason that fine grains can be developed from a coarse-grained structure by hot working. Hot-working processes must balance recovery and recrystallization against grain growth in order to be effective in refining large-grained microstructures or in homogenizing microstructures of mixed grain sizes. For the designer, this is important because grain size has such a pronounced effect on mechanical performance of the part.

At temperatures above the equicohesive temperature, the grain interiors are more resistant to deformation than the grain boundaries, and the grain boundaries can sustain deformation. If a very fine grain size can be achieved and main-

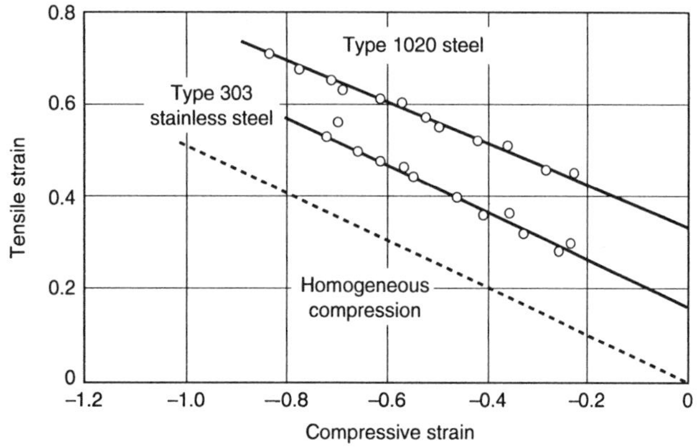

Fig. 9 Workability limits for free-surface fracture for 1020 steel bar and 303 stainless steel bar at room temperature. Source: Ref 11

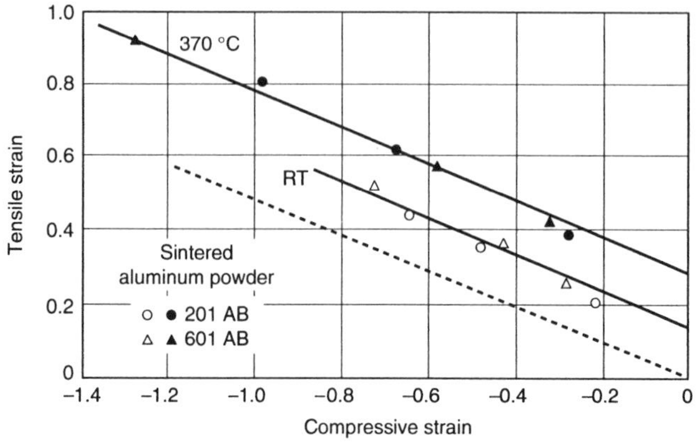

Fig. 10 Workability limits for sintered aluminum alloys at room temperature and at 370 °C (700 °F). Source: Ref 12

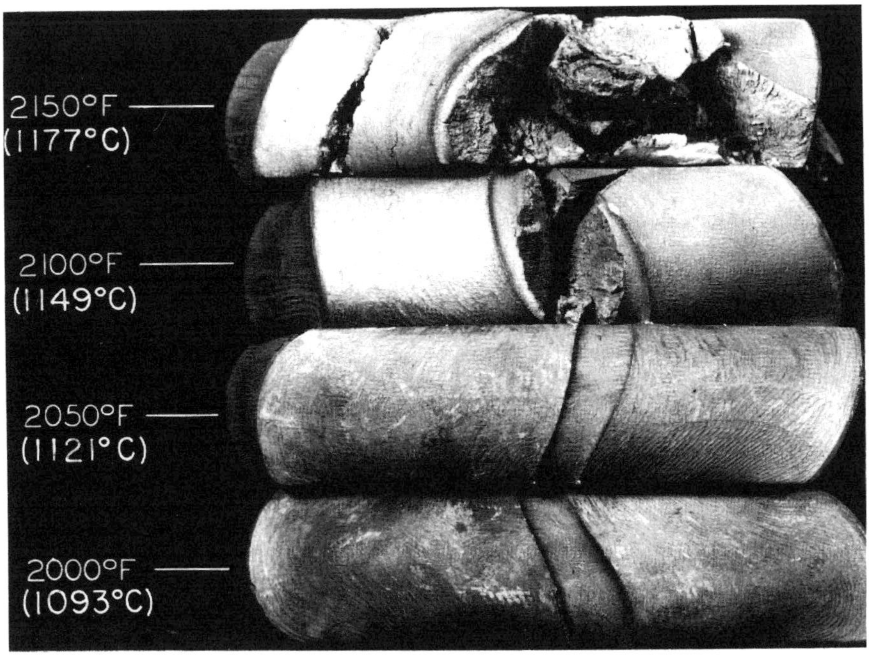

Fig. 11 Notched bar upset testpieces for IN718 showing the workability sensitivity to deformation temperature. Source: Ref 13

tained during deformation at a low strain rate, grain-boundary sliding can occur. This mechanism, in combination with other thermally activated deformation mechanisms, is used successfully to deform very-fine-grained metals to large deformation levels, and this is termed superplastic behavior. Superplastic behavior is not a primary feature in most hot-working processes because of the required low deformation rates and extremely fine grain size. Creep forming, hot-die forging, isothermal forging and sizing, and isothermal rolling are processes that rely in part on

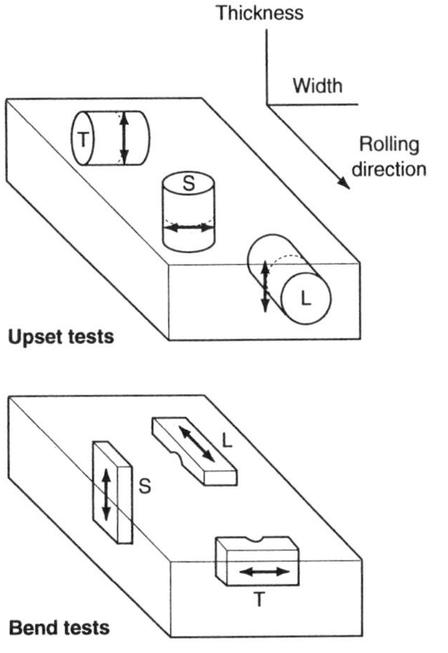

Fig. 13 Orientation of bend samples and compression samples in a rolled plate showing the relationship between tensile stress developed during testing and the original orientation in the plate. Source: Ref 16

grain-boundary sliding and other thermally activated deformation mechanisms.

Some general observations concerning recovery, recrystallization, and grain growth include:

- If the metal has been previously deformed, grain growth is preceded by recovery and by recrystallization.
- The driving force for recrystallization exceeds that of grain growth, thus recrystallization can occur at lower temperatures than grain growth.
- If the workpiece is stress free, grain growth begins without recovery or recrystallization occurring, but the temperature must be relatively high.
- The rate of grain growth is a function of the starting grain size, grain shape, and, most importantly, temperature. The driving force for grain growth is the minimization of internal energy, with the amount of grain-boundary area representing internal energy. Therefore, a fine-grained workpiece will experience a higher rate of grain growth than a coarse-grained workpiece at the same temperature.
- As the workpiece is held at temperature, a saturation grain size will be reached, and holding the workpiece for a longer time will not result in further appreciable grain growth.
- Grain growth is retarded by the presence of stable second-phase particles, as these tend to hinder grain-boundary movement. Aluminum-killed steels have finer grain size than nonkilled steels because of the presence of AlN precipitates that lock up grain boundaries. Other common second phases that can be used to control grain size in iron and other metals include thermally and chemically stable carbides, nitrides, and oxides.

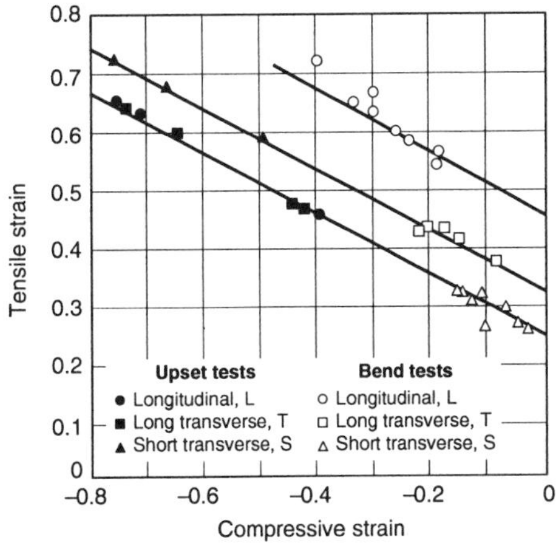

Fig. 14 Workability limits for the sample configurations shown in Fig. 13 indicating the directionality in structure developed during plate rolling. The material is 1045 hot-rolled plate. Source: Ref 16

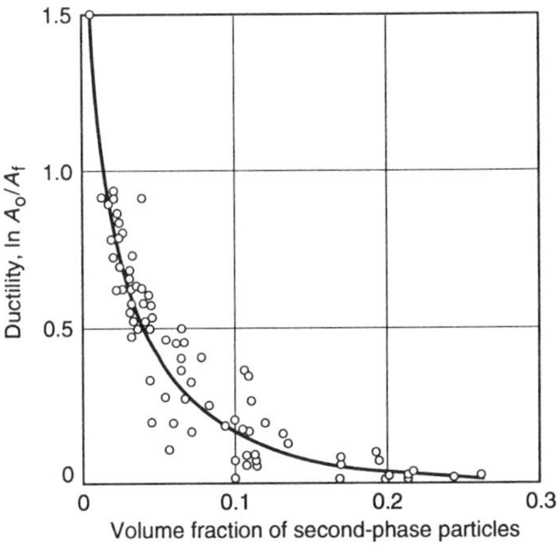

Fig. 12 Ductility of copper alloys as a function of the volume fraction of second-phase particles. Source: Ref 15

Workability

Workability is a measure of the ability of a metal to endure deformation without cracking. This term is applied to bulk-forming processes such as forging, rolling, extrusion, and bending of thick sections. Ductility as measured by the ten-

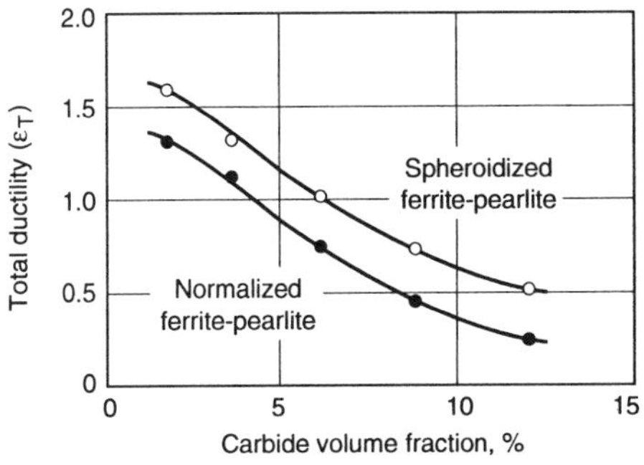

Fig. 15 Effect of carbon content and spheroidization on ductility. Source: Ref 17

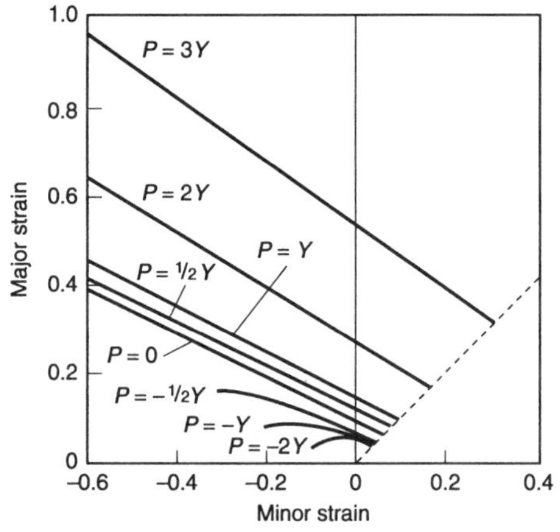

Fig. 16 Effect of hydrostatic pressure (*P*) and tension on the workability limit. *Y*, yield strength. Source: Ref 12

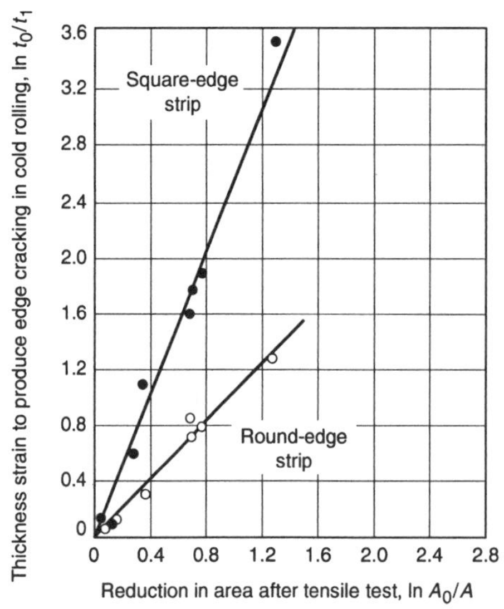

Fig. 17 Thickness strains to produce edge cracking in flat rolling versus tensile ductility. Notice the increased rolling strains possible with the square-edged strip. Source: Ref 18

sile test is a good way to compare the inherent resistance of different materials or different microstructures for the same material. However, ductility alone is insufficient when judging fracture susceptibility of a deformation process because of fracture dependence on both the local strain and stress states. Workability is not a well-defined, specific property like tensile reduction of area at fracture, but rather it must be defined in combination with the stress state that a particular deformation process imposes on the material. Some common cracking modes are discussed below. (Workability is also discussed in the articles "Relationship between Materials Selection and Processing" and "Manufacturing Processes and Their Selection" in this Volume.)

Free-Surface Cracking

Cracking can occur on a free surface as it expands due to the deformation process. For forging processes, this would include cracking on the exposed surfaces during upsetting and cracking on surfaces at the leading edge of localized extrusion during forging. For rolling processes, this may be edge cracking of rolled slabs, plates, or rings. Figure 7 shows that free-surface cracking occurs in bending, upset compression of cylinders, and rolling. Uniaxial compression tests are used to construct workability limits that may then be used for evaluating other deformation processes that may suffer from cracking on free surfaces (Ref 11).

Experiments have shown that the strain state developed on the free surface of a workpiece during deformation dictates the overall level of deformation that may be achieved before a surface crack occurs (Ref 8). The strain path that is experienced at a free surface is a function of friction and deformation-zone geometry, which for upsetting is the workpiece aspect ratio (height/diameter). This effect is shown in Fig. 8, where a steeper strain path is observed for either higher friction or lower workpiece aspect ratio. Figures 7 and 8 in combination indicate that for a given material and set of process conditions (temperature and strain rate), a forming limit can be defined by the set of fracture points for various strain paths. The slope of this forming limit is approximately −0.5 in major-minor strain space. Different metals have different workability limits, as shown by comparing 1020 steel and 303 stainless steel in Fig. 9. These limits change with chemistry, grain size, temperature, second-phase content, and possibly with strain rate.

The concept of a forming limit for free-surface fracture is important because the workability of a metal may be characterized for a particular set of process conditions. In general, as the working temperature is increased, the location of the fracture line will move upward, indicating that higher deformation can be accommodated before fracture. Figure 10 shows the improvement in workability with temperature for sintered preforms for powder forging (Ref 12). Figure 11 shows that higher temperatures are not always beneficial and that, for IN718, a nickel-base superalloy, a temperature limit of about 1120 °C (2050 °F) for hot working exists (Ref 13). Experimental observations have shown that the slope of the line increases with strain rate for some metals, most notably some brasses and austenitic stainless steels (Ref 14).

Experiments have also shown that the position of the line drops (lower formability) as the second-phase content increases, much like the tensile ductility decreases with second phase as shown in Fig. 12 (Ref 15). For example, a rolled steel bar may have a higher fraction of inclusions near its centerline (especially for conventional

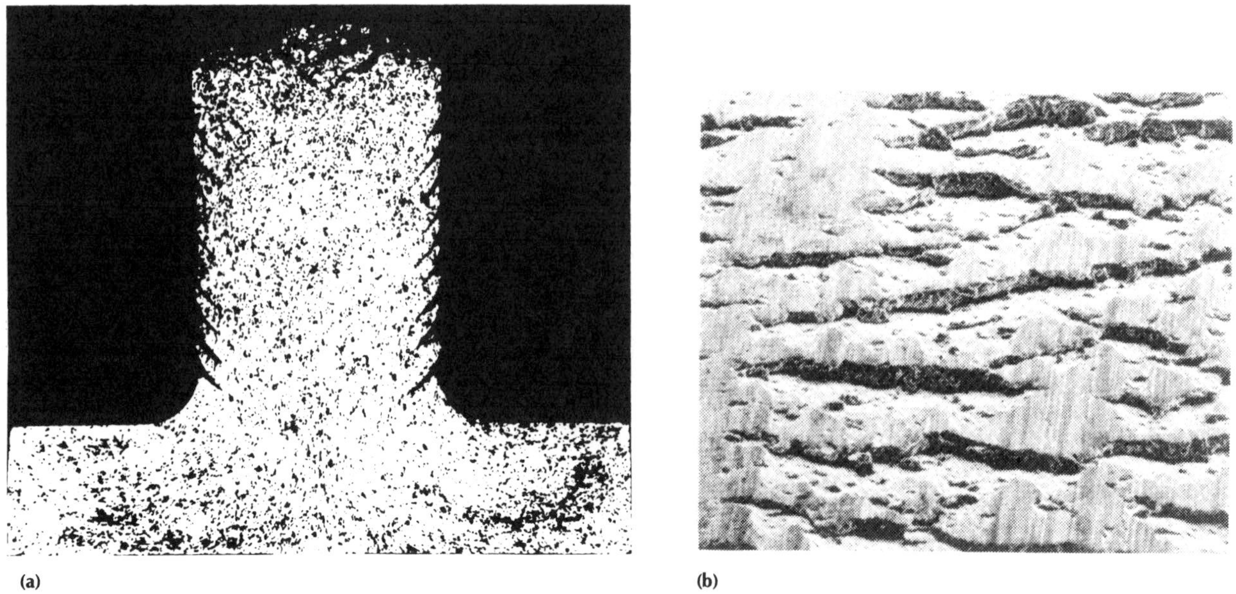

(a) (b)

Fig. 18 Die contact fractures during rib-web forging. (a) Cross section showing the crack angle. (b) Magnified view of the cracks on the rib surface. Source: Ref 12 ingot metallurgy material). Thus, workpieces machined from the center of the bar may have reduced workability in comparison to the outer bar material. Free-machining steels with high manganese sulfide content have reduced workability in comparison to equivalent steel of normal sulfide content. The orientation of the second phases with regard to the stress state is also critical, especially for nonspherical second phases. Figure 13 shows the orientation of some bend and compression test samples sectioned from hot-rolled steel bar. Figure 14 shows the fracture lines that were experimentally determined for these samples. Tensile stresses perpendicular to the plate thickness resulted in fracture at the lowest strain values. Highest strains could be achieved in the rolling direction before free-surface cracking was observed. Process design should include an assessment of the stress state developed during deformation and the orientation of initial mechanical fibering due to inclusions and the initial grain structure. Similarly, the performance of a part in service will be dependent on the orientation relationship between the stress state and the microstructure, including the resultant grain-flow pattern. Sintered preforms for powder forging have reduced workability in comparison to bar stock workpieces, making preform design a critical design issue for successful forging.

The fracture line location is sensitive to the microstructure. For example, a spheroidized structure for a high-carbon steel has a higher fracture line position than a pearlitic microstructure, as shown by the tensile ductility improvement by spheroidization in Fig. 15 (Ref 17). This figure also shows the decrease in ductility with increasing content of second phase, carbides in this case, as was shown in Fig. 12 for a different material system.

The fracture line for free-surface cracking can also be used to show the benefit of hydrostatic pressure on fracture resistance, see Fig. 16 (Ref 12). This figure also shows the detrimental effect of hydrostatic tension on fracture resistance. As an example, continuously cast slabs may be rectangular in cross section with perpendicular sides or they may have rounded corners in the thickness direction. The susceptibility to edge cracking during rolling is dependent on the geometry of the free edge. A rounded side has less vertical free pressure on it from the rolls than a straight side. Consequently, there is more compression on the straight side and edge cracking is less of a problem. Figure 17 shows thickness strains needed to produce edge cracking for cold strip rolling, and as shown, square-edged strip can endure higher reductions than round-edged strip without cracking (Ref 19). This relates directly back to the effect of hydrostatic pressure on the workability line shown in Fig. 16.

The fracture line location is not exact—its location is statistical in nature. Metals with exceptional workability may never experience free-surface fracture during conventional deformation processes. Indeed, producers of cold forgings often report that free-surface cracking is low on their list of defect problems. This is due partly to the fact that past history has already limited cold-forging processes to readily cold-forged materials. For metals with lower workability such as sintered metals, nickel-base superalloys, and oxide dispersion-strengthened alloys, the use of these diagrams can save much time and development cost in designing a process to produce parts successfully from these materials. Fracture limits are also useful in identifying forging conditions to avoid, such as inappropriate temperature ranges, strain-rate ranges, or workpiece geometries. Workability lines are also helpful in designing process sequences that may require annealing or other in-process heating steps to prevent cracking problems.

Cracking on Die Contact Surface

Cracking on surfaces in contact with a die is a common problem. These cracks usually do not propagate deeply into the workpiece, but instead result in unacceptable surface quality or unacceptable machining depths if that surface was to be finish machined. As the forging community has moved closer to net-shape forming, this type of defect has become an increasing problem.

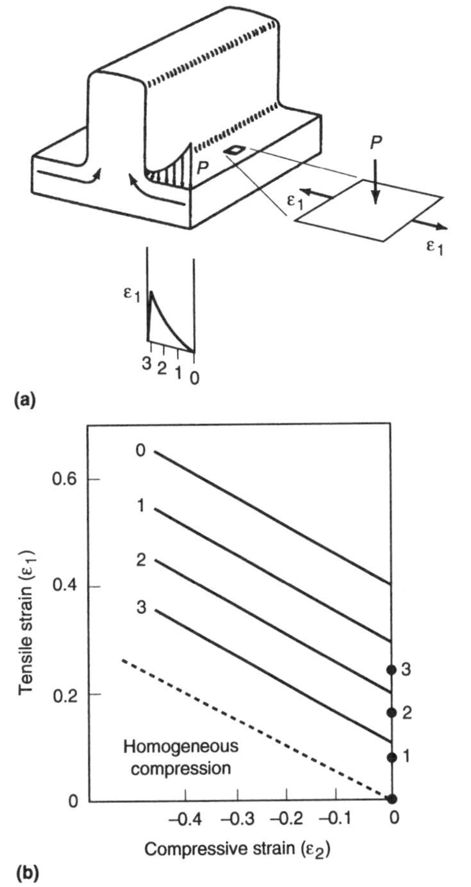

Fig. 19 Surface strains leading to die contact fractures during rib-web forging. (a) Surface strains and pressure at the die corner where fractures occurred. (b) Superposition of the surface strain path and the workability limit with progressive deformation from under the die (0) to near the die corner (3). Source: Ref 12

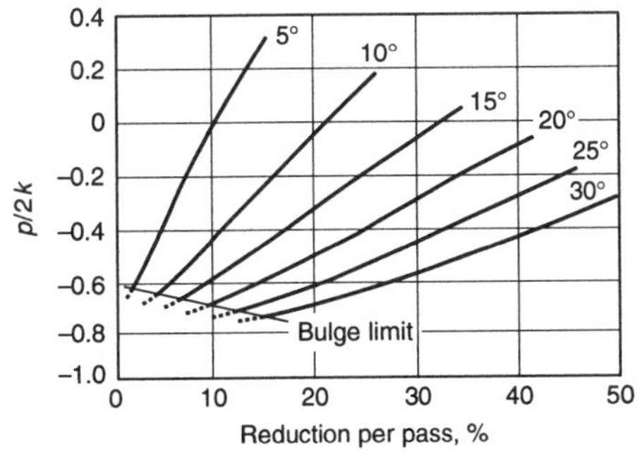

Fig. 20 Ratio of hydrostatic pressure (*p*) to the tensile yield stress (2*k*) at the centerline of cold-drawn strip. The angles are semi-die angles; a positive *p*/2*k* value indicates pressure. Source: Ref 20

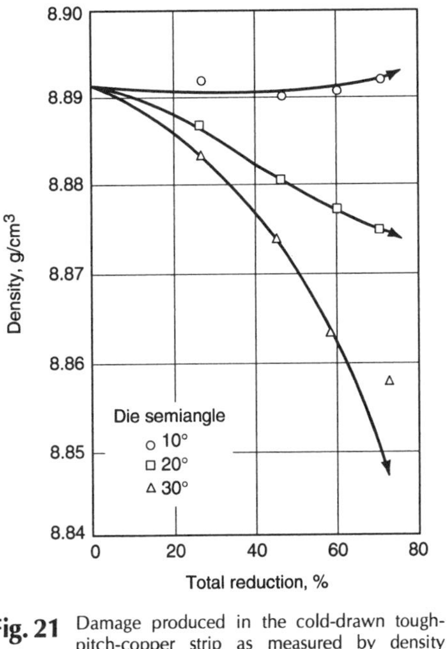

Fig. 21 Damage produced in the cold-drawn tough-pitch-copper strip as measured by density change with drawing reduction. Source: Ref 20

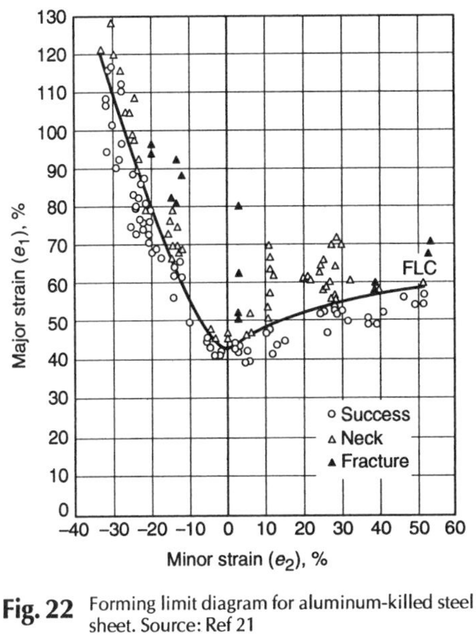

Fig. 22 Forming limit diagram for aluminum-killed steel sheet. Source: Ref 21

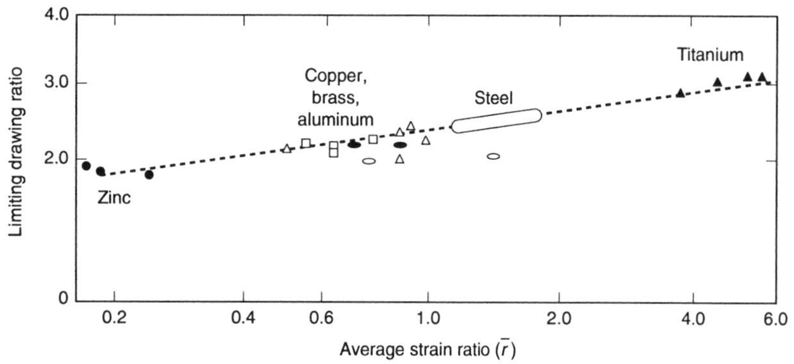

Fig. 23 Correlation between the limiting draw ratio (LDR) and $\bar{r}$ for selected sheet metals. Source: Ref 5

Causes of this problem include nonuniform lubrication, flow around die corners, and an improper starting workpiece surface.

Lubrication variations causes local velocity gradients and secondary tensile stresses in the plane of the flowing surface. Even though there is compression normal to the workpiece surface, the local in-plane tension may be high enough to cause cracking. Once a small crack forms, the tension is relieved so the crack does not propagate.

A typical location for this type of cracking is near a punch or die corner. Here, there is a drop in internal pressure as metal flows around a corner, and this pressure drop often results in surface cracking. Again, experimental studies have shown that combinations of shear deformation combined with tension or low values of hydrostatic pressure in the vicinity of a die corner are responsible for surface cracking (Ref 12). Figure 18 shows cracks along a die contact surface on a rib/web forging. Figure 19(a) shows schematically the drop in hydrostatic pressure experienced by the workpiece as it flows around a die corner into a rib. The beneficial effect of hydrostatic pressure on the fracture line is also shown schematically, with position "0" being a point of high pressure in Fig. 19(b). As metal flows around the die corner (moves from point 0 to 1 to 2 to 3), it experiences a drop in hydrostatic pressure while the surface is being progressively stretched. The fracture line drops as the level of hydrostatic pressure drops, with fracture occurring when the tensile strain exceeds the fracture intercept value for this plane-strain deformation. The severity of the drop in hydrostatic pressure and the magnitude of shear strain is a function of friction, rib width, and die corner radius. Here, wide ribs are more susceptible to cracking than narrow ribs, and friction on the die wall actually helps provide back pressure to retard the die contact fractures. A draft angle on the rib also helps promote back pressure. However, both increased surface friction and draft angle generate higher tensile strains on the top of the rib and may thus promote free-surface cracking.

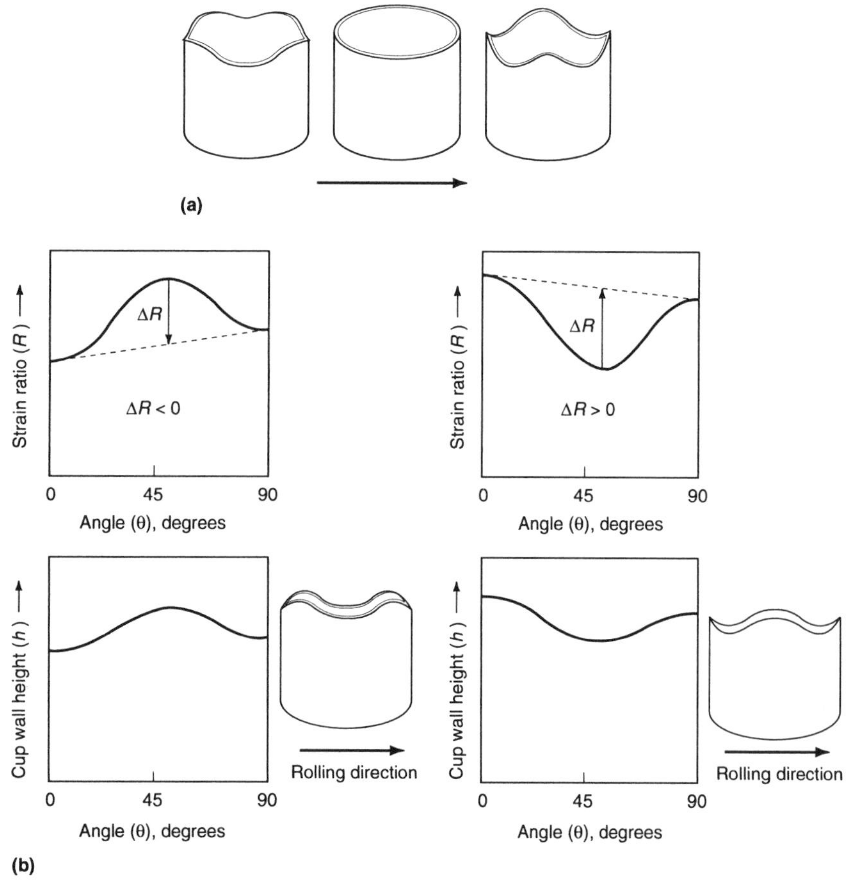

Fig. 24 Ear formation. (a) Drawn copper cups showing earing. The rolling direction is indicated by the arrow. Source: Ref 22. (b) The relation of ear formation to the direction of rolling. Source: Ref 19

Central Burst (Chevron Cracking)

A central burst, or chevron crack, is a dangerous type of defect because it cannot be seen visually. On the outside, the part looks like an acceptable part, but the interior rupture will cause unacceptable part performance. A central burst is detected by x-ray or ultrasonic inspection methods.

Central bursting is most commonly associated with extrusion and drawing operations, although it can be generated by forging and rolling processes as well. Rotary piercing is possible because of centerline fracture. The cause of a burst is hydrostatic tension generated along the workpiece centerline. A change in deformation zone geometry is usually sufficient to eliminate the problem.

Figure 20 shows a series of die angles and reductions for cold drawing that produced either sound strip or strip with bursts (Ref 18). In this figure, the negative ratios indicate hydrostatic tension at the strip centerline. Figure 21 from this same study shows that die angles of 20 and 30° actually caused a density loss upon strip drawing due to porosity development along the strip centerline. At a 10° die angle there was no centerline porosity developed in the strip. From Fig. 20, it is clear that both the 20 and 30° die angles produced hydrostatic tension along the strip centerline over the range of reductions while the 10° die angle produced hydrostatic pressure at reductions of 20% or more per pass. In this example, light reductions and higher die angles are shown to promote central bursts due to the resulting hydrostatic tension along the centerline.

The conservative design approach is to ensure that no hydrostatic tension develops; this ensures that bursting will not occur. Often, however, the part or tooling design cannot be changed sufficiently to eliminate hydrostatic tension. If the level of hydrostatic tension can be kept below a critical level, bursting can likely be avoided. This may be accomplished by a change in lubricant, die profile, temperature, deformation level, or process rate.

Formability

Formability is a term that is similar to workability, only it is generally associated with sheet metal behavior. Forming sheet metal involves locally deforming a region or regions of the sheet, and the stress state that develops in the sheet and the strain path that various locations experience dictate whether or not the process is successful. Keeler and Goodwin developed the concept of a forming limit diagram (FLD); it divides major-minor surface strain space into safe and failure regions. An FLD for aluminum-killed steel sheet is shown in Fig. 22, with failure indicating that either a neck or a tear has occurred. Several features of the FLD are important. First, the lowest point on the curve is at the plane strain position where the minor strain is 0. Second, while a sharp line is drawn, the FLD position is statistical in nature, just as with the workability limit discussed above. Third, the limit moves up as the sheet thickness is increased. In addition to thickness, the vertical position of the FLD and to some extent its shape are functions of chemistry, second-phase content, grain size and shape, and the presence of a crystallographic texture. The strain path followed during the forming operation is a function of friction, die and punch geometry, hold-down pressure, and the work-hardening behavior of the metal.

Two metallurgical parameters that are used to rank sheet forming characteristics are the work hardening exponent, n, and the resistance to thinning in the thickness direction, r (normal anisotropy). The average of r in the plane of the rolled sheet is $\bar{r}$. It seems reasonable that a metal with high work hardening has a high formability because work hardening prevents strain localization under tension and distributes deformation throughout the material. However, the contact area between the sheet and the punch may be so large that friction minimizes the natural ability of the metal to distribute strain. The result may be strain localization and tearing unless the sheet metal has a high resistance to strain in its through thickness direction. A high $\bar{r}$ indicates a high yield strength in the through thickness and good deep drawability as measured by the limiting draw ratio (LDR), that is, the ratio of the largest blank diameter to punch diameter that can be drawn through a die without tearing. Figure 23 shows typical LDR values versus $\bar{r}$ values for some common sheet metals. This normal anisotropic behavior is beneficial in this case.

Rather than average r over a plane, another way to examine the effect of planar anisotropy is by Δr, which indicates the variation in r with direction in the plane of the sheet. For values of Δr less than or greater than 1, nonuniform straining during deep drawing will occur in directions parallel and normal to the rolling direction, resulting in the formation of ears. Ears are shown in Fig. 24(a), while 24(b) shows the orientation dependence on the ears with rolling direction for positive and negative Δr values. It is important for the process designer to know whether or not the sheet metal is subject to earing because an extra step is needed to trim away the ears.

For more information on sheet metal forming processes, see *Forming and Forging*, Volume 14 of *ASM Handbook* (Ref 1).

Microstructural Effects on Metal Flow

Single-Phase versus Multiphase Microstructures. There is a considerable difference in mechanical behavior between pure metals and their alloys. A pure metal is by definition a single-phase, homogeneous material. An alloy may be a single phase that is strengthened by interstitial and/or substitutional methods. An alloy may also be a multiphase metal, with the alloy properties being a function of the properties of the individual phases. The actual alloy properties will be a function of the volume fraction of phases, the distribution of the phases, and the size/shape of the minor phase(s). Properties that are especially influenced include flow behavior, ductility, and toughness.

The four figures that are contained in Fig. 25 show schematically how workability is affected by temperature for single-phase and two-phase metals (Ref 13).

In Fig. 25(a), the workability for pure metals and single-phase alloys of aluminum, tantalum, and niobium are represented by curve 1. As temperature increases, workability increases for these metals. Curve 2 represents a single-phase metal that experiences rapid grain growth at elevated temperatures. For metals such as beryllium, magnesium, tungsten, and β-titanium alloys, workability increases with temperature until grain growth becomes excessive, and then it decreases.

Figure 25(b) contains two schematic curves for a single-phase metal matrix that contains a dispersion of second-phase particles. Curve 3 is representative for metals that contain insoluble particles; examples are iron with MnS inclusions and Se additions in stainless steel. Here, workability is generally low due to the deleterious effect of the second-phase compound on ductility and workability, and temperature has only a minor effect. If the second-phase compound is soluble at higher temperatures, a behavior like curve 4 results. At low temperatures, workability is poor due to the presence of the second phase, and once it has dissolved at higher temperatures, the workability is improved; examples include molybdenum oxides in molybdenum metal and stainless steel strengthened by vanadium carbides and nitrides.

Figure 25(c) contains schematic curves for a metal that is single phase at lower temperatures and two phases at higher temperatures. Curve 5 represents a metal that forms a ductile second phase. Curve 6 represents a curve for a metal that forms either a brittle second phase or a second phase that melts and causes hot shortness; sulfur in iron is an example where FeS may form and melt. Leaded steels can have a similar problem, although lead is always present as a second phase.

Figure 25(d) contains two curves for metals that are two phases at lower temperatures and a single phase at higher temperatures. The second phase at low temperature will be formed upon cooling from the high temperature, single-phase state. For curve 7, the second phase has some ductility, so a measure of workability is retained

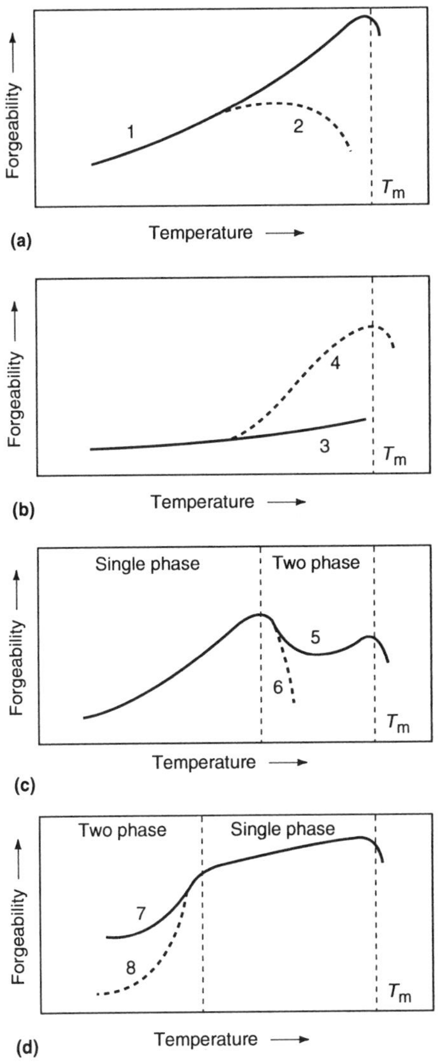

Fig. 25 Schematic curves depicting the forgeability of single-phase and two-phase alloys as a function of temperature. (a) Curve 1: pure metals and single-phase alloys (aluminum alloys, tantalum alloys, niobium alloys). Curve 2: pure metals and single-phase alloys exhibiting rapid grain growth (beryllium, magnesium alloys, tungsten alloys, all-β titanium alloys). (b) Curve 3: alloys containing elements that form insoluble compounds (resulfurized steel, stainless steel containing selenium). Curve 4: alloys containing elements that form soluble compounds (molybdenum alloys containing oxides, stainless steel containing soluble carbides or nitrides). (c) Curve 5: alloys forming ductile second phase on heating (high-chromium stainless steels). Curve 6: alloys forming low-melting second phase on heating (iron containing sulfur, magnesium alloys containing zinc). (d) Curve 7: alloys forming ductile second phase on cooling (carbon and low-alloy steels, α-β and α titanium alloys). Curve 8: alloys forming brittle second phase on cooling (superalloys, precipitation-hardenable stainless steels). T_m, melting temperature. Source: Ref 13

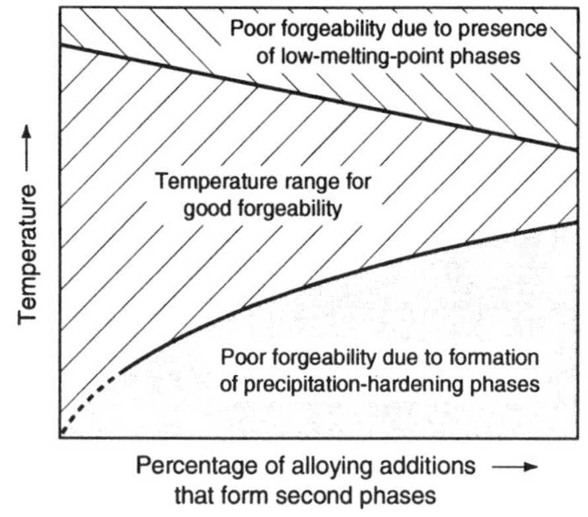

Fig. 26 Schematic effect of second phases on the working temperature range of alloys. Source: Ref 5

after it has formed. Low-carbon steels and low-alloy steels with low-carbon content are examples as these steels can be cold forged successfully. For curve 8, a brittle second phase is formed and cold workability is nil. Superalloys, which precipitation-harden, tend to behave in this manner, as do precipitation hardened stainless steels (the PH series).

While the curves in Fig. 25 are only of descriptive use, it is clear that the microstructural nature of the workpiece is important to its workability. Temperature and strain rate conditions must be accurately selected and controlled for acceptable forging.

As the curves in Fig. 25 indicate, the presence of a second phase is usually detrimental to ductility. Figure 26 shows schematically that as the percentage of alloying additions that contribute to the formation of a second phase increases, the acceptable forging window decreases. On the high-temperature side, workability is reduced by the increasing likelihood of low-melting phases. On the low-temperature side, a higher temperature is required due to the increasing volume fraction of second-phase particles. Figure 27 conveys a similar message for a variety of metals. As the inherent strength is increased (left to right), a higher forging load is required. As alloy additions are made that result in increasing amounts of second phases (top to bottom), the workability is decreased.

Hot Shortness and Burning. Time at temperature allows second phases to either dissolve or grow. If they dissolve, their effect disappears and workability improves. However, many of the second phases grow instead, and this can lead to decreased workability. In severe cases, local chemistry changes can interact with second phases to result in low-melting phases. These low-melting phases promote hot shortness. For example, Fig. 28 shows the severity of surface cracking in a copper bearing steel that has been oxidized. Alloying elements such as copper in the case of Fig. 28 or nickel and tin become enriched on the surface as other elements are preferentially oxidized. Copper has a limited solubility in austenite, and when its solubility limit is exceeded, a copper-rich phase is precipitated at austenite grain boundaries. This phase has a melting point lower than typical preheating temperatures for hot forging, and the liquid penetrates along grain boundaries and weakens them. During forging, the grain boundaries open and surface cracking occurs.

Burning is excessive overheating that results in very large austenite grains and excessive surface enrichment due to oxidation. As mentioned above, the surface enrichment can lead to the precipitation of low-melting phases. These cause local melting along grain boundaries. This effect cannot be removed except by physically removing the surface layer to the depth that burning was

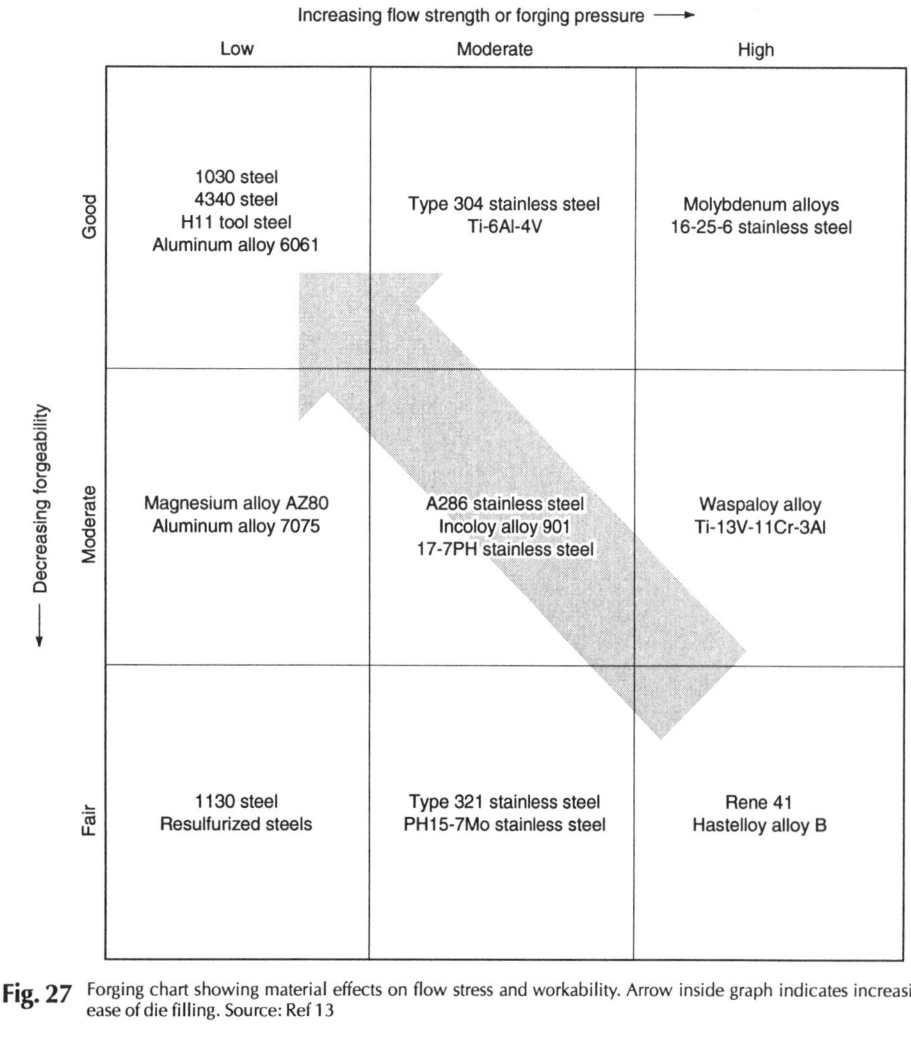

Fig. 27 Forging chart showing material effects on flow stress and workability. Arrow inside graph indicates increasing ease of die filling. Source: Ref 13

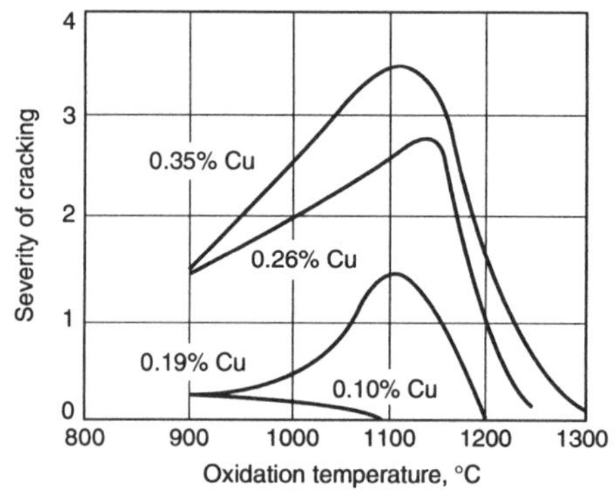

Fig. 28 Hot bend test results for low-carbon steels that have been oxidized prior to bending. Source: Ref 23

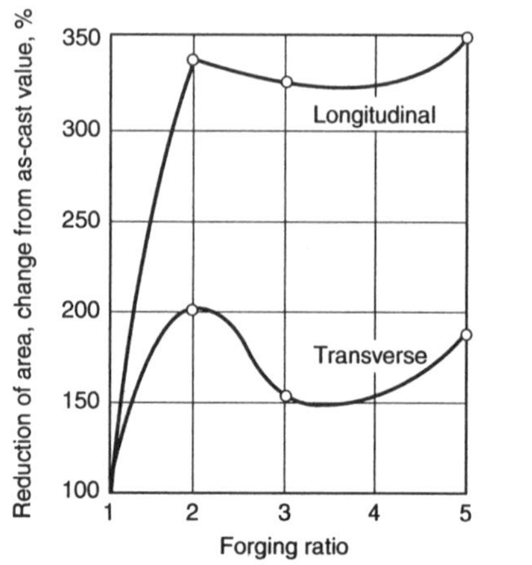

Fig. 29 Effect of forging reduction ratio on the percentage increase in tensile ductility from that of the initial cast material for a manganese steel. Source: Ref 24

experienced. It is best to scrap burned workpieces rather than attempt to salvage them.

The process designer and the process engineer should anticipate the effects of environment on workpiece performance, especially high-temperature environments as in these cases. The interaction between the workpiece and the heating equipment is critical to the success of the deformation process because unanticipated reactions or conditions can lead to defects.

Grain Flow Pattern

Grain flow pattern refers to the combination of grain shape and stringering of nonmetallic inclusions that develop with metal flow. Together, these can produce a "fiber" in the workpiece that results in anisotropic behavior in the workpiece. The anisotropy is most easily identified by differences in ductility, toughness, and fatigue resistance for longitudinal, transverse, and short-transverse directions. Both hot and cold deformation result in directionality of mechanical properties in the forging, respectively.

Grain flow can be used effectively to produce a better forging. If the grain flow can be aligned in the direction of maximum tension in the part during service, superior performance will be achieved. The resistance to crack initiation and propagation is highest in the direction normal to the grain flow because the grain boundaries and elongated inclusions act to blunt advancing cracks. If the tension is normal to the fiber direction, the fiber will enhance crack propagation, hence transverse and especially through-thickness directions have lower dynamic properties such as toughness and fatigue than the longitudinal direction. This is shown in Fig. 29 for a manganese steel where ductility is increased from the level of the initial casting by forging deformation, with the longitudinal direction increasing the most. Figure 30 shows grain-flow patterns for a channel section forged with different parting-line locations. Because the grain-flow pattern breaks out to the part surface at the parting line, any stress during service that is normal to the parting line will act perpendicular to the weakest planes. These planes are the easiest to fracture along, for example, the short-transverse equivalent position. Because a channel will usually be stressed in a direction normal to it (flexing the channel legs to open or close), the grain-flow pattern in Fig. 30(b) is least detrimental.

Flow-Related Defects in Bulk Forming

Underfill may not seem like a flow-related defect, but aside from simple insufficient starting mass, the reasons for underfill are flow related.

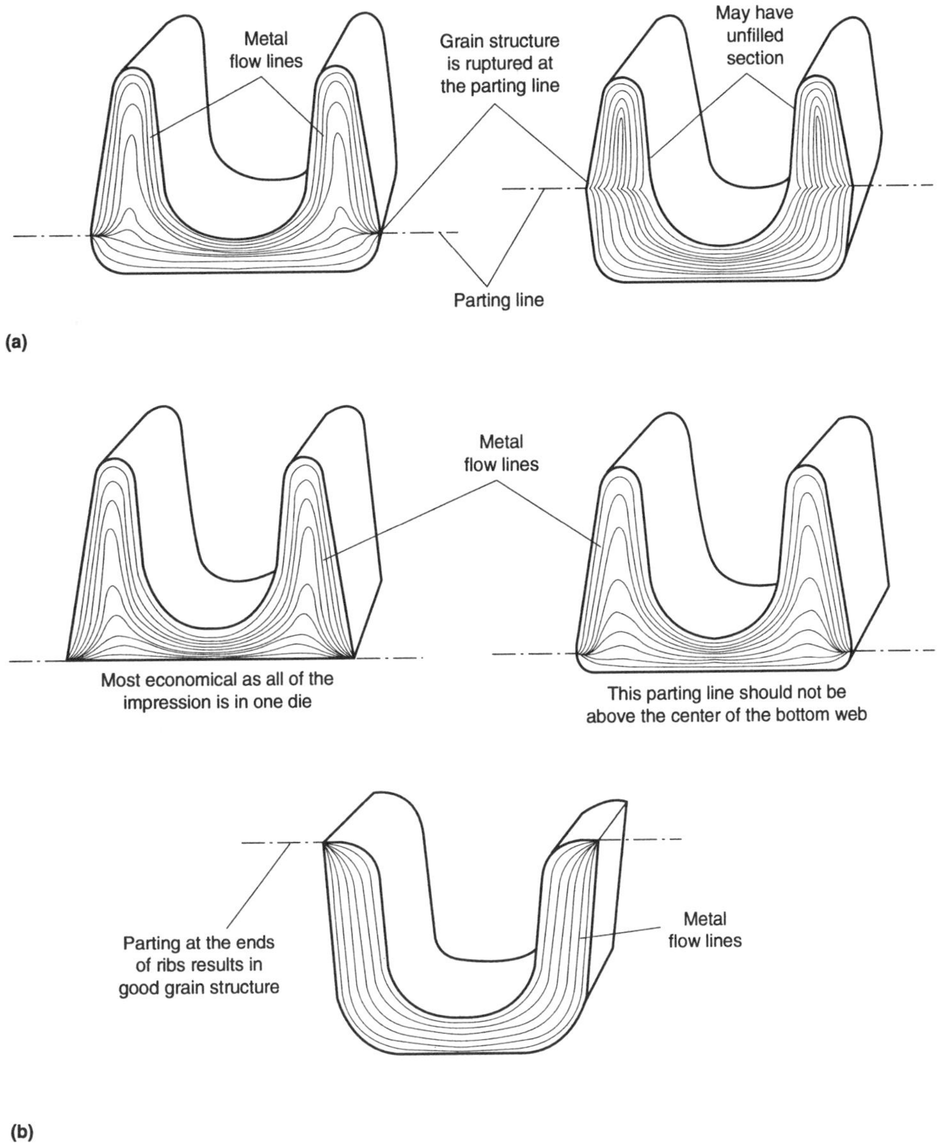

Fig. 30 Parting-line location and its influence on grain-flow pattern in a channel section forging. (a) Parting lines resulting in metal-flow patterns that cause forging defects. (b) Parting lines resulting in smooth flow lines at stressed sections. Source: Ref 24

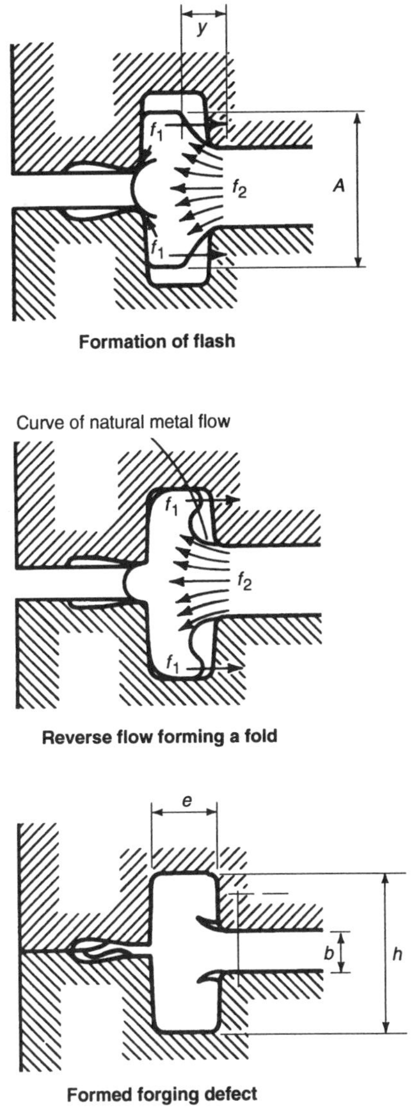

Fig. 31 Lap formation in a rib-web forging caused by improper radius in the preform die. Source: Ref 25

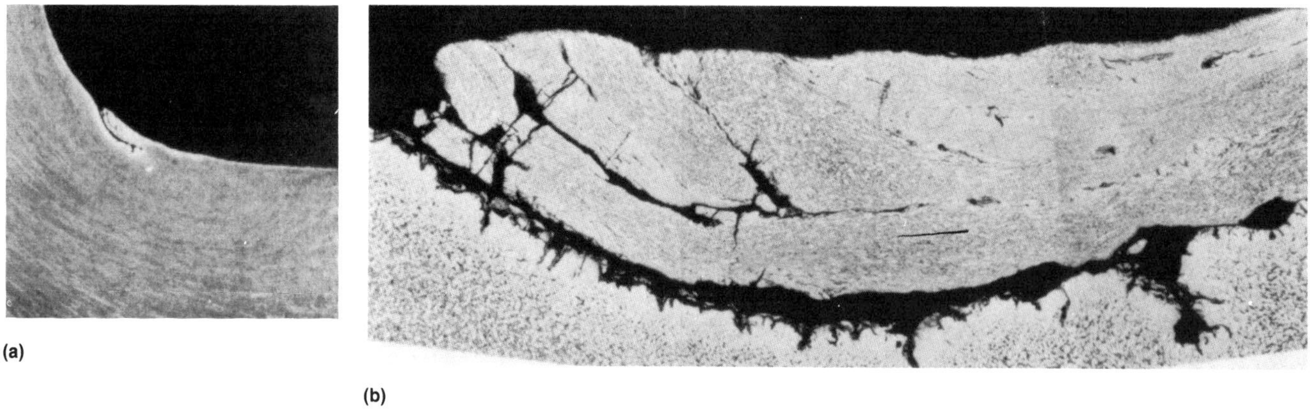

Fig. 32 Lap in a Ti-6Al-4V bulkhead forging. (a) 3.5×. (b) 50×. Source: Ref 26

These include improper fill sequence, insufficient forging pressure, insufficient preheat temperature, lubricant build up in die corners, poor or uneven lubrication, and excessive die chill. An improper fill sequence may result in excessive flash loss, or it may be the result of extraordinary pressure requirements to fill a particular section. Sometimes, venting may eliminate the problem; more often than not, a change in the incoming workpiece shape or a change in the deformation sequence is required.

Important points to keep in mind are that the forging load increases very rapidly as the sharpness of details is increased, that flash geometry directly influences the amount of back pressure experienced by the metal in the die cavity, and that die wear increases directly with die pressure.

A lap or fold results from an improper progression in fill sequence. Normally, a lap or fold is associated with flow around a die corner, as in the case of an upper rib or lower rib, or with a reversal in metal-flow direction. A general rule of thumb is to keep metal moving in the same direction. The die corner radius is a critical tool dimension, and it should be as generous as possible. In progressing through a forging sequence, the die corners should become tighter, so that the workpiece fillets are initially large and progressively become smaller as the forging steps are completed. Figure 31 shows schematically a lap forming as metal flows around a die corner. An actual forging lap is shown in Fig. 32.

Extrusion Defect, Pipe, or Suck-in. The tail of an extrusion is unusable because of nonuniform flow through the extrusion die. This results in a center-to-surface-velocity gradient, with metal from the workpiece interior moving through the die at a slightly higher velocity than the outer material. The result shows up at the tail of the extrusion as a suck-in or pipe, and, for extrusions, the tail is simply cut off and discarded. Alternatively, a follower block of cheaper material may be added so that most of the defect falls in the cheaper material, and less length of the extruded workpiece is lost.

For forgings that involve forward or backward extrusion to fill a part section, the same situation can develop. Metal flow into a rib or hub can result in a suck-in defect, which in a worst-case scenario would show up as a fold on the face opposite to the rib, and a best case would be a depression on what otherwise should be a flat surface. As shown schematically in Fig. 33, metal flowing up into the hub does not generate sufficient back pressure to maintain underside contact. One method of eliminating this type of defect is to position more material on the back face initially. Another method is to change the rib geometry (aspect ratio and/or angles). If neither of these changes can be accomplished, an extra forging step may be needed to limit the amount of extrusion that is done in any one step.

Shear-Related or Flow-Through Defects. Shearing defects are also known as flow-through defects because they result from excessive metal flow past a filled detail of the part. An example of this is shown in Fig. 34 for a trapped-die forging that has a rib on the top surface. The rib denoted by "2" is filled early in the forging sequence, and significant mass must flow past the rib in order to fill the inner hub, zone "4." The result can be a complete shearing-off of the rib in the worst case, with a lesser case being the formation of a shear-type crack.

Defects in Sheet Metal Parts

Defects in sheet-metal formed parts are often appearance related rather than mechanical in nature. A primary example of a cosmetic defect is stretcher strains, or Lüder's bands, in steel parts.

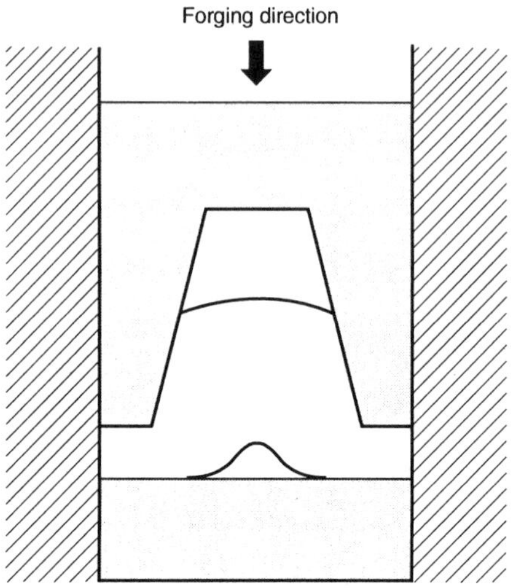

Fig. 33 Schematic showing the formation of a "suck-in" defect

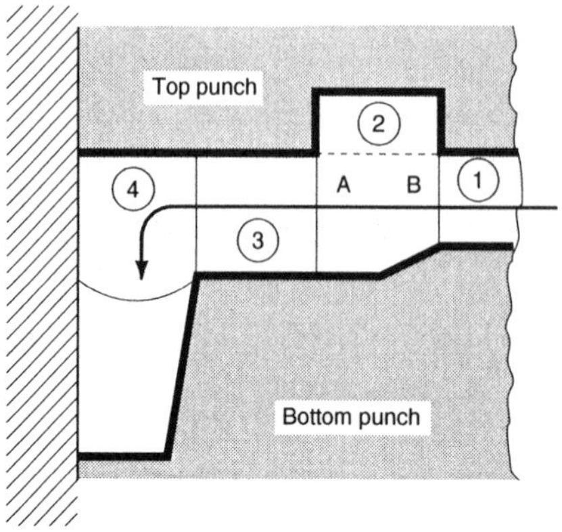

Fig. 34 Schematic of a flow through crack at the base of a rib in a trapped-die forging. Excessive metal flow past region 2 causes a shear crack to form at A and propagate toward B. Source: Ref 12

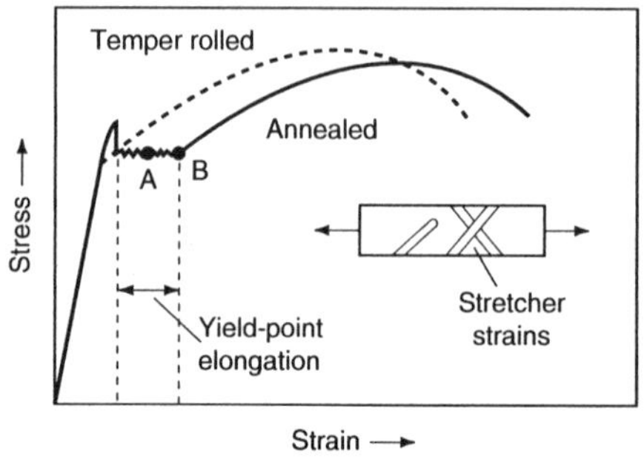

Fig. 35 Tensile stress-strain curve for a low-carbon steel showing yield drop and stretches strains (Lüder's strains). Temper rolling removes the discontinuous yield behavior. Source: Ref 5

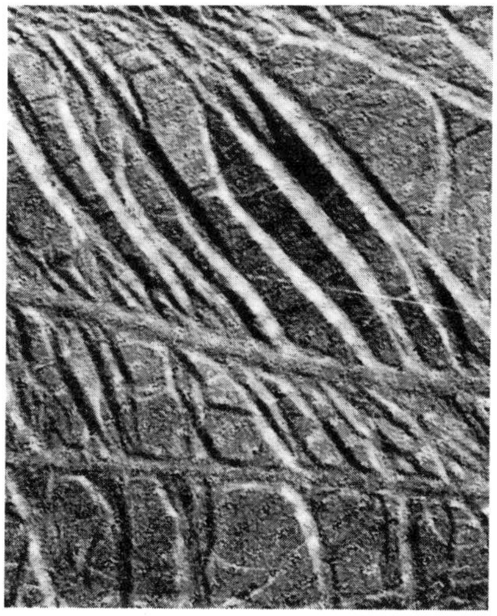

Fig. 36 Stretcher strains (Lüder's bands) in rimmed 1008 steel

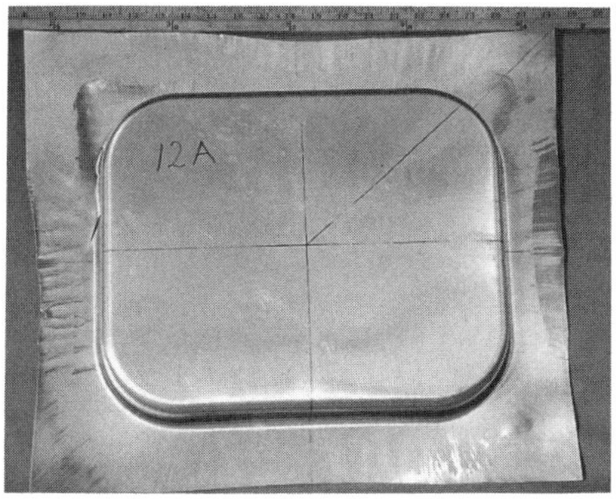

Fig. 37 Flat-bottomed pan stamped from 2008-T4 aluminum sheet showing wrinkling on the flange. Source: Ref 27

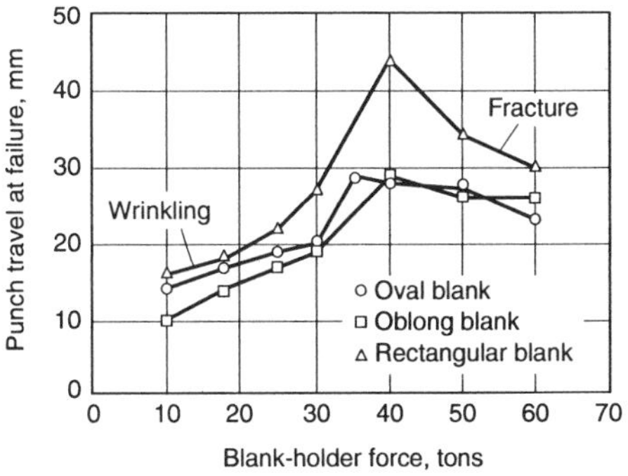

Fig. 38 Blank-holder force versus punch travel, with wrinkling and fracture limits for three blank geometries. Source: Ref 27

Mold steel displays discontinuous yield behavior as shown in Fig. 35. A sharp yield drop, followed by yield extension, known as Lüder's strain, at a lower level of stress, followed by normal work hardening in a tension test is the indicator of this behavior. If the part-forming operation imposes a low level of strain, Lüder's strain is often visible; Fig. 36 shows Lüder's strain as concentrated bands of deformation. Painting will not hide these bands, so for a smooth, mirrorlike finish, they must be avoided.

A mechanism known as strain aging is responsible for the discontinuous yield behavior. It can be temporarily eliminated by temper rolling the sheet or by stretcher leveling it such that it is deformed into the work-hardening region of the stress-strain curve. As Fig. 35 depicts, temper-rolled sheet will not exhibit discontinuous yielding or Lüder's strains if it is formed within a certain time period, for example, a few days. However, if the sheet is stored for several weeks before it is processed into parts, strain aging will have occurred and the Lüder's strain will be present in the yield behavior of the sheet. Again, the process engineer and the process designer should be aware of this phenomenon so that it can be avoided, or at least anticipated.

Necks, both diffuse and localized, are another surface-related defect, and these indicate mechanical instability of the sheet. Necks are thinned regions of the sheet, so they result in a mechanically weakened zone. In the extreme case, a neck leads to cracking. The corner of flat-bottomed cans is a typical area that experiences problems of localized thinning and tearing. A material with a high r lessens this problem.

Another defect is called wrinkling, although it is actually a buckle. Wrinkles tend to be problems along the flanges of deep-drawn shapes. As a blank is drawn into a die cavity, sheet thickening occurs in the flange because the flow is converging in nature. If the hold-down pressure is not sufficient, the sheet in the flange area will wrinkle, as shown in Fig. 37. Using an initially equal hold-down pressure around the flange, the corners thicken more than the main lengths of the flange, so the resultant hold-down becomes nonuniform. The main lengths of the flange do not adequately clamp the blank and it wrinkles. The sheet must experience a minimum hold-down pressure to avoid wrinkling. However, if the hold-down pressure is too high, it will cause tearing by restricting metal thickening that must occur as the blank edges are drawn toward the die. Figure 38 shows a balance between the wrinkling hold-down pressure limit and the excessive hold-down limit for drawing the aluminum pan shown in Fig. 37. The initial blank shape affects the volume of metal being drawn toward the die cavity and, thus the pressure needed to counter wrinkling. As shown, a rectangular blank requires a higher hold-down pressure than an oblong or oval blank because it has slightly greater metal volume, especially in the corners where the metal flow is more axisymmetric in nature.

Conclusions

Deformation processing has its origins in ancient manufacturing techniques, and although these processes may be referred to as "low-tech" or "rust-belt" technology, this field is complex. These processes have the capability of producing a precision part or product form with a specified microstructure at high rates of production *if* the process is properly matched with the workpiece material and it is well controlled. There are many process variables, both mechanical and metallurgical in nature, and it is the interaction of the process conditions of rate, temperature, and shape with the metallurgical conditions of flow stress, workability, and microstructure that dictate the success or failure of the process. Both the process designer and the product designer have many analytical tools available, and these tools (as discussed in the article "Modeling of Manufacturing Processes" in this Volume) should be used to assist in making critical geometry, material, and deformation process decisions.

REFERENCES

1. *Forming and Forging,* Vol 14, *ASM Handbook* (formerly 9th ed. *Metals Handbook*), ASM International, 1988
2. T.G. Byrer, Ed., *Forging Handbook,* Forging Industry Association and American Society for Metals, 1985
3. K. Lange, *Handbook of Metal Forming,* McGraw-Hill, 1985
4. S. Kalpakjian, *Manufacturing Processes for Engineering Materials,* 3rd ed., Addison-Wesley, 1996
5. G.E. Dieter, *Mechanical Metallurgy,* 3rd ed., McGraw-Hill, 1986
6. W.T. Becker, Special Applications of Tension and Compression Testing, Lesson 5, MEI Course 12, *Mechanical Testing of Metals,* American Society for Metals, 1983
7. R. Mignogna et al., *Metall. Trans.,* Vol 1, 1970, p 1771
8. Atlas of Formability (available for variety of materials), Concurrent Technology Corp., Johnstown, PA
9. G.E. Dieter, Ed., *Workability Testing Techniques,* American Society for Metals, 1984
10. M. Gensamer, Strength and Ductility, *Trans. ASM,* Vol 36, 1945, p 30–60
11. P.W. Lee and H.A. Kuhn, Fracture in Cold Upset Forging—A Criterion and Model, *Metall. Trans.,* Vol 4, 1973, p 969–974
12. H.A. Kuhn and B.L. Ferguson, *Powder Forging,* Metal Powder Industries Federation, 1990
13. S.L. Semiatin, Workability in Forging, *Workability Testing Techniques,* G.E. Dieter, Ed., American Society for Metals, 1984, p 197–247
14. H.A. Kuhn and G.E. Dieter, Workability in Bulk Forming Processes, *Fracture 1977,* Vol 1, ICF4, Waterloo, Canada, 1977, p 307–324
15. B.I. Edelson and W.M. Baldwin, The Effect of Second Phases on the Mechanical Properties of Alloys, *Trans. ASM,* Vol 55, 1962, p 230–250
16. S.K. Suh and H.A. Kuhn, Anisotropy of Ductile Fracture in Hot-Rolled Steel Plates—Complementary Results from Bend Tests, *Metall. Trans.,* Vol 6A, 1975, p 2157
17. F.B. Pickering, The Effect of Composition and Microstructure on Ductility and Toughness, *To-*

ward Improved Ductility and Toughness, Climax Molybdenum Development Co. Ltd., 1971, p 9–31

18. M.G. Cockcroft and D.J. Latham, *J. Inst. Met.*, Vol 96, 1968, p 33–39
19. W.F. Hosford and R.M. Caddell, *Metal Forming: Mechanics and Metallurgy*, 2nd ed., Prentice-Hall, 1993, p 268
20. L.F. Coffin and H.C. Rogers, Influence of Pressure on the Structural Damage in Metal Forming Processes, *Trans. ASM*, Vol 60, 1967, p 672–686
21. S. Hecker, *Proc. 7th Biennial Congress of Int. Deep Drawing Research Group*, Amsterdam, 9–10 Oct 1972
22. D.V. Wilson and R.D. Butler, *J. Inst. Metals,* 1961–1962, p 473–483
23. K. Mayland et al., Influence of Microstructural and Compositional Variables on Hot Working of Steel, *Met. Technol.,* Aug 1976, p 350–357
24. S.L. Semiatin, Material Characteristics, Chapter 3, *Forging Handbook*, T.G. Byrer, Ed., Forging Industry Association and American Society for Metals, 1985
25. A. Chamouard, *Closed Die Forging*, Part 1, Dunod, Paris, 1964
26. F.N. Lake and D.J. Moracz, "Comparison of Major Forging Systems," Air Force Report AFML-TR-71-112, Air Force Materials Laboratory, May 1971
27. M. Ahmetoglu, T. Broek, G. Kinzel, T. Altan, and K. Chandorkar, "Deep Drawing of Rectangular Pans from Aluminum Alloy 2008-T4," paper 950694, Society of Automotive Engineers, 1995

SELECTED REFERENCES

- W.A. Backofen, *Deformation Processing*, Addison-Wesley, 1972
- A. Chaudhary and S. Doraivelu, "Processing Science Research to Develop Fundamental Analytical, Physical, and Material Modeling Techniques for Difficult-to-Process Materials," final report, WL-TR-91-4018, Wright-Patterson Air Force Base, 1990
- H.J. Frost and M.F. Ashby, *Deformation-Mechanism Maps: The Plasticity and Creep of Metals and Ceramics*, Pergamon Press, 1982
- G.D. Lahoti and T. Altan, "Research to Develop Process Models for Producing a Dual Property Titanium Alloy Compressor Disk," interim report for 1 Aug 1979 to 31 July 1980, AFWAL-TR-80-4162, Air Force Wright Aeronautical Laboratory, 1980
- S.W. Lui and M.K. Das, Interactive Design of Axisymmetric Forging Dies Using a Desk-Top Computer, *J. Mech. Work. Technol.*, Vol 5, June 1981
- R.P. McDermott and A.N. Bramley, Forging Analysis—A New Approach, *Metall. Met. Form.*, May 1974
- C.M. Sellars and G.J. Davies, Ed., *Hot Working and Forming Processes*, The Metals Society, London, 1979
- J.P. Tang, S.I. Oh, D.W. Birch, and K. Hoang-Vu, "Manufacturing Science Program to Develop a Computer-Aided Engineering (CAE) System for Die (Mold) Design and Manufacturing," AFWAL-TR-86-4058, interim report for Jan 1985 to Jan 1986, Air Force Wright Aeronautical Laboratory
- J.R.L. Trasorras, "Computer-Based Design and Modeling of P/M Forging Processes," Ph.D. Thesis, University of Pittsburgh, 1991

Design for Powder Metallurgy*

Howard I. Sanderow, Management & Engineering Technologies

THE POWDER METALLURGY (P/M) process is a near-net or net-shape manufacturing process that combines the features of shape-making technology for powder compaction with the development of final material and design properties (physical and mechanical) during subsequent densification or consolidation processes (e.g., sintering). It is critical to recognize this interrelationship at the outset of the design process because a subtle change in the manufacturing process can cause a significant change in material properties.

General P/M Design Considerations

To begin a design using powder processing, six key design considerations must be recognized. With the variety of powder processing schemes available, the selection of the appropriate method depends to a great extent on these design constraints.

Size. Due to the physical nature of the processes and the physical limits of commercial manufacturing equipment, product size has certain critical boundaries. For some powder processes, the product size is quite limited (such as metal injection molding, MIM), while for hot-isostatic pressing (HIP) size is not considered a serious constraint.

Shape Complexity. Powder metallurgy is a flexible process capable of producing complex shapes. The ability to develop complex shapes in powder processing is determined by the method used to consolidate the powders. Because a die or mold provides the container for the consolidation step, the ease of manufacture of the container and the ability to remove a green compact (unsintered) from the container, in most cases, determines the allowable shape complexity of a given part.

Tolerances. Control of dimensional tolerances, a demanding feature of all near-net or net-shape manufacturing processes, is a complex issue in powder processing. Tolerances are determined by such process parameters as powder characteristics, compaction parameters, and the sinter cycle. The amount of densification during sintering and the uniformity of that shrinkage controls dimensional tolerance in most P/M products. Due to the very small amount of size change during sintering conventional press-and-sinter P/M parts, these products typically have the closest dimensional tolerances, as compared to HIP parts, which require the largest spread in tolerances.

Material Systems. Powder shape, size, and purity are important factors in the application of a powder processing technique. For some consolidation processes or steps, powders must be smooth, spherical particles, but for other processes a much more irregular powder shape is required. Nearly every material and alloy system are available in powder form. For some materials such as cemented carbides, copper-tungsten composites and the refractory metals (tungsten, molybdenum, tantalum, etc.) powder processing is the only commercially viable manufacturing process.

As an example, for "press and sinter" processing an irregular powder shape and distribution of particle sizes are desired for adequate green strength and sinter response. Hot isostatic pressing requires spherical powders (gas atomized) for lowest impurities and good particle packing. The MIM process also prefers spherical particles, but very small particle size (10–20 μm) is needed to ensure proper rheology, homogeneous distribution in the plastic binder, and excellent sinter response.

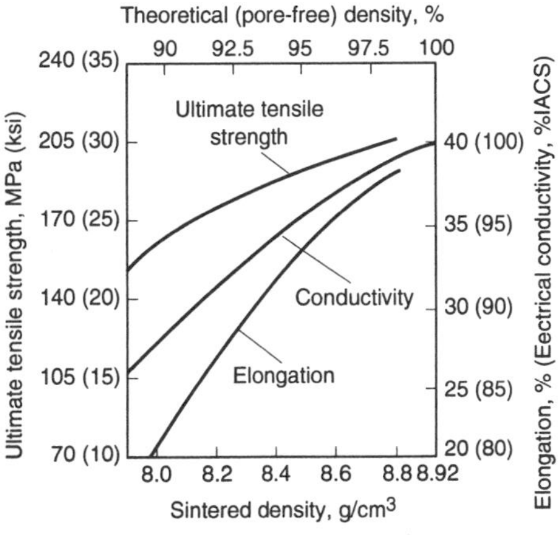

Fig. 1 Properties of pure copper. Source: Ref 1

Properties. The functional response of any product is determined by its physical or mechanical properties. In powder processing these properties are influenced directly by the product density, the raw material (powder), and the processing conditions (most often the sintering cycle). As P/M materials deviate from full density, the properties decrease (as shown for the tensile properties—and electrical conductivity—of pure copper in Fig. 1). The mechanical response for 4% Ni steels is found in Fig. 2.

Quantity and Cost. The economic feasibility of P/M processing is typically a function of the number of pieces being produced. For conventional press-and-sinter processing, production quantities of at least 1,000 to 10,000 pieces are desired in order to amortize the tooling investment. In contrast, isostatic processing can be feasible for much lower quantities, in some cases as small as 1 to 10 pieces. On a per pound basis, the approximate costs for steel P/M parts produced by various methods are roughly as follows:

Condition	Density range, g/cm^3	1997 selling price(a), $/lb
Pressed and sintered	6.0–7.1	2.45–2.70
Pressed, sintered, sized	6.0–7.1	2.90–3.20
Copper infiltrated	7.3–7.5	3.50–3.55
Warm formed	7.2–7.4	3.10–3.30
Double pressed and sintered	7.2–7.4	4.00–4.10
Metal injection molded	7.5–7.6	45.0–70.0
Hot forged	7.8	5.00–5.50
Double press and sinter + HIP	7.87	6.00–7.00

(a) These numbers are only averages; smaller parts are more expensive and larger parts less expensive per pound.

Powder Processing Techniques

In order to understand the design restrictions of each powder processing method, it is best to review these processes individually. The P/M manufacturing methods can be divided into two main categories: (1) conventional press-and-sinter methods and (2) full-density processes.

Conventional (Press-and-Sinter) Processes. The conventional press-and-sinter process technologies follow the steps outlined in Fig. 3. The various powder ingredients are selected to satisfy

*Portions of this article have been adapted from *Powder Metallurgy Design Manual,* 2nd ed., Metal Powder Industries Federation, 1995. Used with permission.

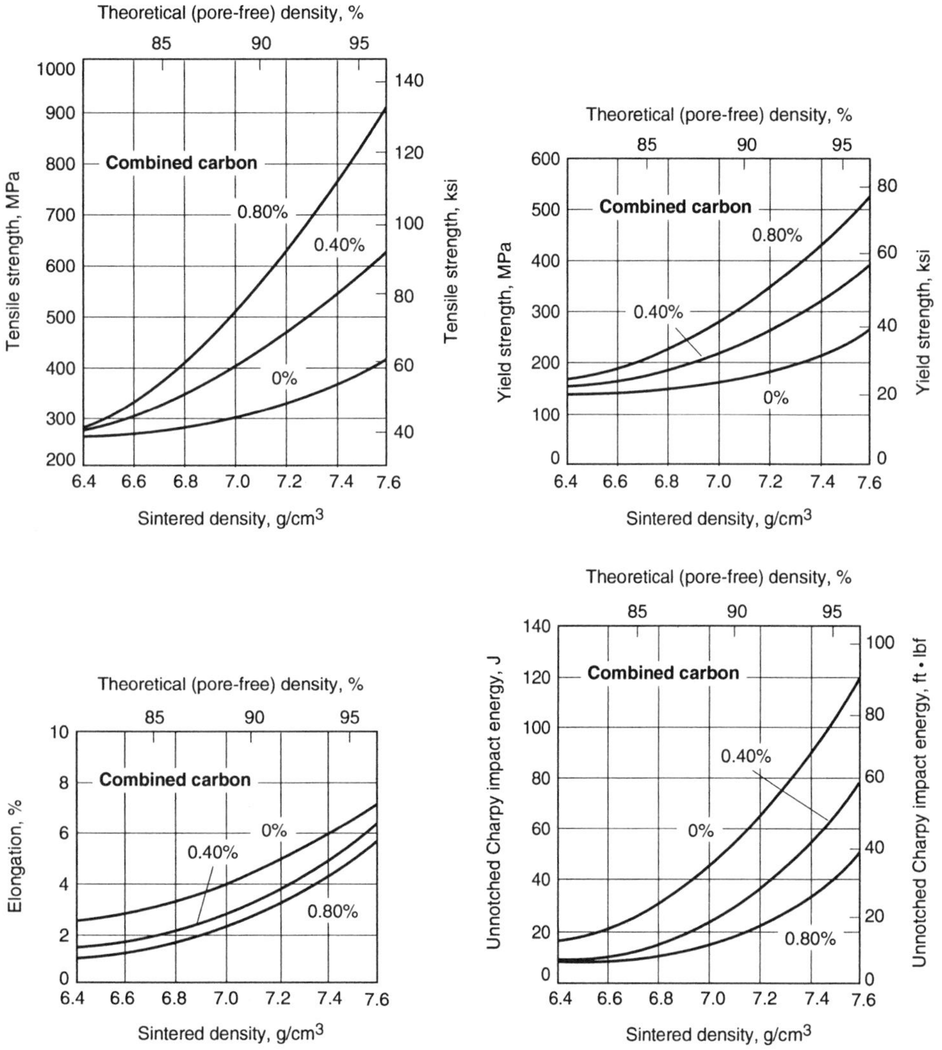

Fig. 2 Effects of density on mechanical properties of as-sintered 4% Ni steel. Source: Ref 2

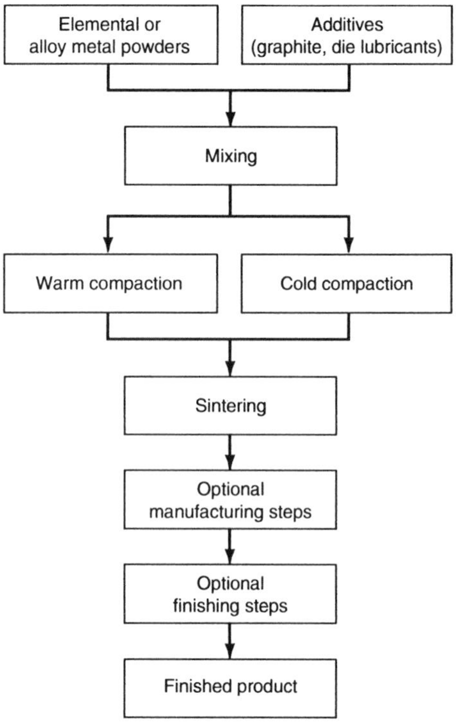

Fig. 3 General steps in the P/M process

the process constraints and still meet the requirements of the end product. For example, in cold compaction irregularly shaped powders are used to ensure adequate green strength and structural integrity of the as-pressed product. Special solid lubricants are added to the powder blend to reduce friction between the powder particles and the tooling. If these lubricants might contaminate the metal powder particles, then an alternate consolidation method would be needed.

Because powder is compacted in hard tooling using a vertical compaction motion, the product size and shape are limited by the constraints of available press capacity, powder compressibility, and the density level required in the product. For most conventional P/M products these limitations have a maximum size of about 160 cm^2 (25 $in.^2$) compaction area, part thickness of about 75 mm (3 in.), and a weight of 2.2 kg (5 lb). However, parts as large as 200 mm (8 in.) diameter by 100 mm (4 in.) thick, weighing 14.5 kg (32 lb) have been produced on conventional equipment. Even parts 380 mm (15 in.) in diameter by 6 mm (1/4 in.) thick have been produced by conventional P/M methods.

After compaction the green compact is sintered in a controlled-atmosphere furnace. Dimensional tolerance control is determined by the maximum temperature of the sintering cycle and the metallurgical changes that occur during sintering. If solid-state diffusion is the primary sintering mechanism, very little densification occurs, dimensional change is minimal and tolerance control is very good. This practice is followed for most P/M steels where size change during sintering is held to less than 0.3%. In contrast, other alloy systems utilize liquid-phase formation as the primary sintering mechanism, causing a significant increase in density, large dimensional changes, and much lower tolerance control. Examples of these material systems include cemented carbides where dimensional changes of 6 to 8% are typical and tolerance control is in the range of ±0.25 mm (±0.010 in.). In addition to its effects on dimensional tolerance levels, the sintering step also plays a significant role in determining the final physical and mechanical properties of the product. Higher sintering temperatures and longer sintering times promote pore rounding and increase densification, thereby improving critical mechanical properties such as tensile strength, ductility, impact resistance, and fatigue limit (see Table 1). The sintering process is extremely important in determining the magnetic response of soft magnetic P/M alloys. As shown in Table 2 for the Fe-0.45 wt% P alloy, increasing

Table 1 Effect of sintering conditions on the mechanical properties of two P/M nickel steels

	MPIF FN0205(a)		MPIF FN0208(b)	
	Belt(c)	Vacuum(d)	Belt(c)	Vacuum(d)
Tensile strength, MPa (ksi)	380 (55)	552 (80)	448 (65)	758 (110)
Yield strength, MPa (ksi)	193 (28)	414 (60)	331 (48)	586 (85)
Elongation, %	4	7	2	4
Impact energy, J (ft · lbf)	19 (14)	38 (28)	11 (8)	33 (24)
Hardness, HRB	64	80	80	90
Sintered density, g/cm^3	(e)	7.32	(e)	7.30

All samples pressed to a green density of 7.2 g/cm^3. (a) 1–3% Ni, 0.3–0.6% C. (b) 1–3% Ni, 0.6–0.9% C. (c) Belt, 30 min at 1125 °C (2060 °F) in nitrogen/endo atmosphere. (d) Vacuum, 2 h at 1260 °C (2300 °F) with nitrogen backfill. (e) Density not reported but estimable by MPIF Standard 35. Source: Ref 3

Table 2 Effect of sintering conditions on the properties of magnetic P/M iron (0.45 wt% P)

Sintering conditions		Maximum magnetic	Coercive	Maximum	Tensile strength		Elongation,
Atmosphere(a)	Temperature, °C (°F)	induction (B_{max}), kG	force (H_c), Oe	permeability	MPa	ksi	%
10% H_2	1120 (2050)	13.2	2.3	2620	345	50	3
75% H_2	1120 (2050)	13.3	2.0	3220	355	52	7
100% H_2	1120 (2050)	13.4	1.7	3680	372	54	5
100% H_2	1200 (2200)	13.7	1.3	5710	400	58	14

(a) Balance N_2. Source: Ref 4

Table 3 Properties of powder forged steels

Alloy	Hardness, HRC	Tensile strength MPa	Tensile strength ksi	Yield strength MPa	Yield strength ksi	Elongation, %	Impact toughness J	Impact toughness ft · lbf
10C60	23	793	115	690	100	11	2.7	2
11C60	28	895	130	620	90	11	4	3
4620	28	965	140	895	130	24	81	60
	38	1310	190	1070	155	20	47	35
4640	38	1310	190	1070	155	17	34	25
	48	1585	230	1310	190	11	16	12
4660	38	1310	190	1070	155	15	27	20
	48	1585	230	1310	190	10	13.5	10

Source: Ref 6

Table 4 Comparison of powder processing methods

Characteristic	Conventional	MIM	HIP	P/F
Size	Good	Fair	Excellent	Good
Shape complexity	Good	Excellent	Very good	Good
Density	Fair	Very good	Excellent	Excellent
Dimensional tolerance	Excellent	Good	Poor	Very good
Production rate	Excellent	Good	Poor	Excellent
Cost	Excellent	Good	Poor	Very good

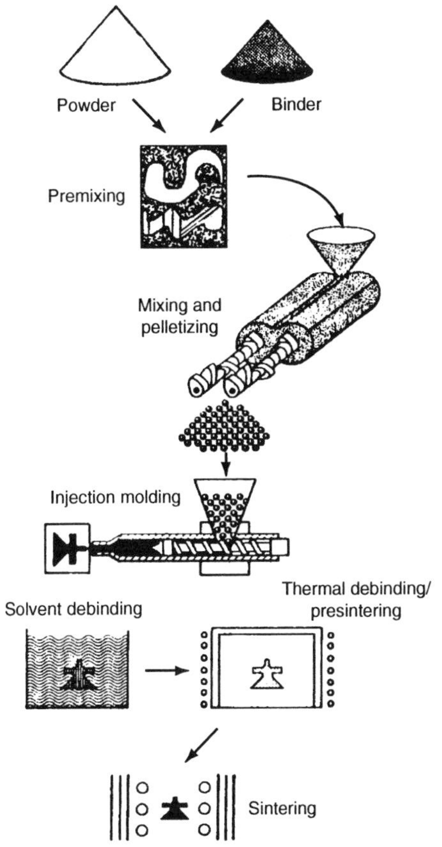

Fig. 4 Metal injection molding process. Source: Ref 7

the hydrogen content in the sintering atmosphere and raising the sintering temperature improved the maximum permeability more than 100%, the tensile strength more than 15%, and the ductility more than 300%. In a similar manner, the mechanical properties and corrosion resistance of P/M stainless steels are strongly dependent on the sintering process parameters (Ref 5).

Warm compaction is used to increase the green density and green strength of P/M steel parts. When combined with high-temperature sintering, this process can provide mechanical properties equivalent to double press-double sinter processing at a lower cost. Due to the much higher green strength, warm compacted parts can be machined in the green condition. This technique can also be used to produce insulated magnetic cores, a composite material suitable for high-frequency electromagnetic systems.

Full-Density Processes. The second group of powder process technologies are formulated specifically to yield a product as close to full density as possible. This contrasts significantly with the previous conventionally processed products where attainment of full density was not the primary goal. The full-density processes include powder forging (P/F), metal injection molding (MIM), hot isostatic pressing (HIP), roll compaction, hot pressing and extrusion.

Powder Forging. In P/F a preform is manufactured using conventional P/M process techniques and then hot formed in confined dies to cause sufficient material deformation that nearly all the porosity is eliminated. Due to the high costs in developing the preform design and maintaining forging tools and automated production systems of the P/F process, it has been limited, in most commercial practices, to high-volume products such as automotive connecting rods and transmission components. The P/F process has been successful in developing mechanical properties in P/F steel comparable to wrought steels (see Table 3). This process successfully overcomes the mechanical property limitations imposed by the residual porosity in conventional P/M products.

Metal Injection Molding. The MIM process combines the structural benefits of metallic materials with the shape complexity of plastic injection molding technology. A uniform mixture of powder and binders is prepared and injected into a mold (see Fig. 4). The MIM powders are typically spherical in shape and much finer in particle size than those used for conventional cold-die compaction (MIM powder, 10–20 μm; conventional die-compaction powders, 50–150 μm). The binders are formulated specially to provide the proper rheological properties during injection molding as well as ease of binder removal after the molding step. Once the part is ejected from the mold, the binder material is removed using either solvent extraction or thermal processes (or both). After the debinding step the part is then sintered to complete the process. Due to the large amount of binder in the MIM starting material (up to 40% by volume), the MIM part undergoes a large reduction in size (as much as 20% linear shrinkage) during sintering. Dimensional tolerances, therefore, are not as good as in conven-

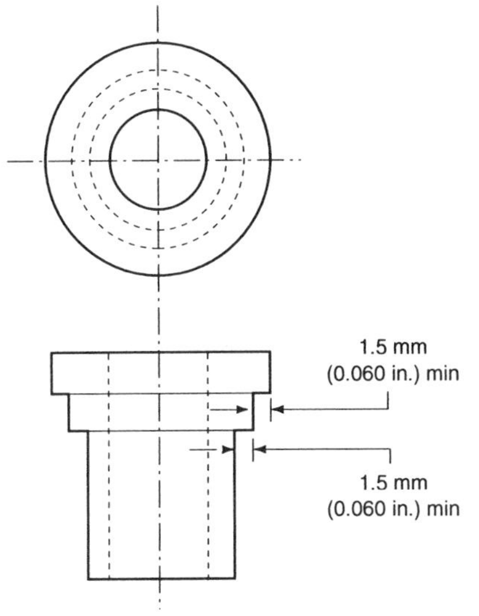

Fig. 5 Minimum wall thickness for conventional die compaction. Source: Ref 8

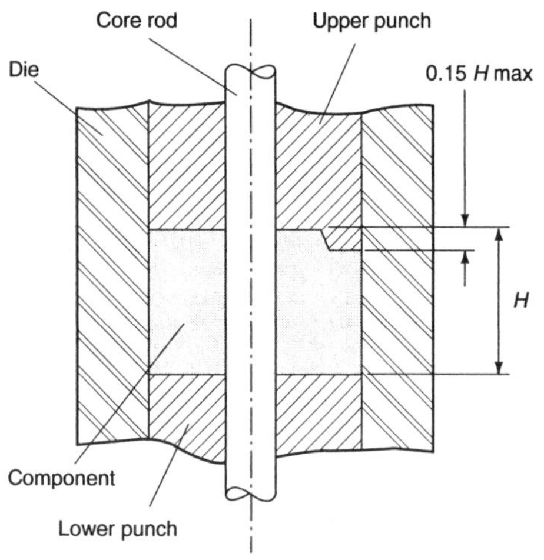

Fig. 6 Level change limit for P/M parts. Source: Ref 8

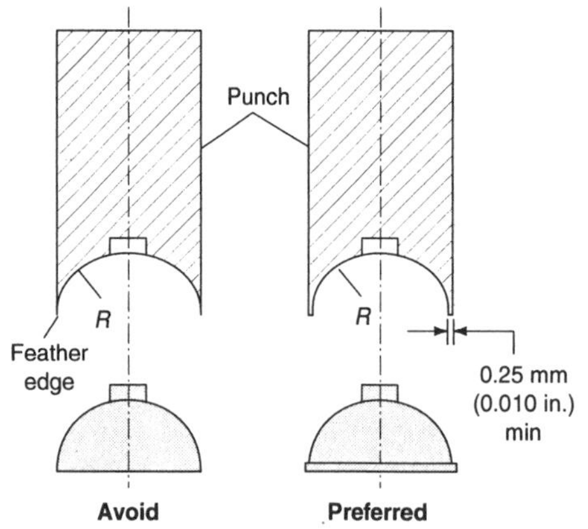

Fig. 7 Proper design of spherical shapes in P/M parts. Source: Ref 8

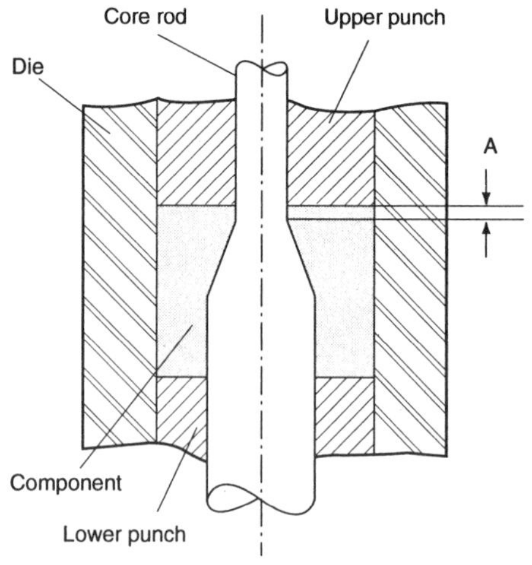

Fig. 8 Tapered hole design for P/M parts. Source: Ref 8

tional die compaction and a straightening or coining step is sometimes needed.

Hot Isostatic Pressing. This fully dense process method is the least constrained technique. However, due to its very low production rate, costly equipment, and unique tool requirements, the HIP process is normally relegated to expensive materials such as tool steels, superalloys, titanium, and so forth. The process also requires high-purity powders (generally spherical in shape), and it is considered only a near-net-shape process. The powders are vibrated in place in a container, which is then evacuated and sealed. These metal or ceramic containers are placed in the HIP vessel, which applies an isostatic pressure (using a gaseous medium) and temperature to the container and the powder mass. This combination of heat and pressure on the container consolidates the powder to its final shape, as defined by the initial container configuration. The container must be removed from the HIP part after the process cycle, typically by machining or chemical etching.

Other Full-Density Methods. The remaining full-density consolidation methods are used infrequently in commercial practice and are limited to specialty materials. For example, roll compaction is used to form certain soft magnetic alloys, composite materials, and compositions unique to powder metallurgy. Hot pressing is used when the deformation characteristics of the base powder require high temperatures to achieve plastic flow and adequate consolidation. The powder extrusion process requires a container, and it is similar to roll compaction and limited to specialty materials not suitable for conventional extrusion methods, such as composites, titanium, and nuclear materials.

Comparison of Powder Processing Methods

Effective application of powder processing methods requires a general comparison of the major design features, focusing on the similarities, differences, advantages, and disadvantages of each method. Table 4 provides a qualitative comparison, while Table 5 offers more specific design information. Characteristics for each processing method are summarized below.

Conventional die compaction:

- Widest range of most frequently used engineering materials, including iron, steel, stainless steel, brass, bronze, copper, and aluminum
- Most applicable to medium-to-high production volumes; small- to medium-size parts such as gears, sprockets, pulleys, cams, levers, and pressure plates (automotive, appliances, power tools, sporting equipment, office machines, and garden tractors are typical markets)
- Greatest density range, including high-porosity filters, self-lubricating bearings, and high-performance structural parts

Table 5 Application of powder processing methods

	Conventional die compaction	MIM	HIP	P/F
Material	Steel, stainless steel, brass, copper	Steel, stainless steel	Superalloys, titanium, stainless steel, tool steel	Steel
Production quantity	>5000	>5000	1–1000	>10,000
Size, lb	<5	<¼	5–5000	<5
Dimensional tolerance	±0.001 in./in.	±0.003 in./in.	±0.020 in./in.	±0.0015 in./in.
Mechanical properties	~80–90% wrought	~90–95% wrought	Greater than wrought	Equal to wrought
Price per pound	$0.50–5.00	$1–10	>$100	$1–5

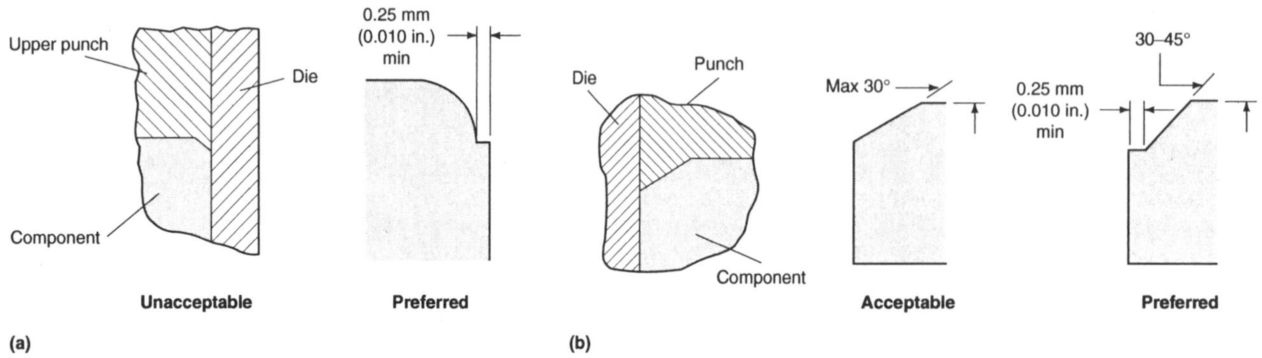

Fig. 9 Chamfer (a) and edge radii (b) design features. Source: Ref 8

- Limited physical and mechanical properties caused by residual porosity
- Most cost-competitive of the powder processes
- Wide range of applications from low- to high-stress applications

Powder forging:

- Potentially applicable to all engineering materials now hot forged, but actual applications currently limited to low-alloy steels
- Product applications limited to high-volume products such as automotive connecting rods and transmission components as well as power tool parts
- Mechanical properties equivalent to wrought steel
- Most cost-competitive of the full-density processes for medium-to-large parts

Metal injection molding:

- Limited range of materials, though most standard engineering alloys available as well as several specialty alloys
- Limited to relatively small, highly complex shaped products for medium-to-high production volumes
- Greatest range in shape complexity including high aspect ratios
- More costly than conventional die-compaction processes
- Superior physical and mechanical properties as compared to conventional process, due to higher density

Hot isostatic pressing:

- Materials limited only by the inherent cost of the process, therefore typically applied only to expensive materials
- Most suited for low-to-medium production volumes
- Competitive against large casting or forging products where substantial machining is needed to obtain the final product
- Much shape detail is machined after HIP processing; not normally a "net-shape" manufacturing process
- Physical and mechanical properties meet or exceed those of cast or wrought materials

Design Issues

The specific design issues for the powder process technologies would require an extensive discussion well beyond the limits of this article. An excellent source for additional details is the manual on P/M product design prepared by the Metal Powder Industries Federation (MPIF) trade association (Ref 8). Much of the sections that follow have been taken from that manual.

Conventional Die Compaction

In order to use die-compaction technology effectively, the designer must be familiar with the limitations and design constraints related to product shape and special features. The requirements for secondary operations focus on the effect of residual porosity on these process steps. Material selection and the design approach to be used conclude this discussion. Throughout this section structural parts are used to describe these design issues. Specific information on bearings and porous products is given in a separate section.

Shapes and Features. A shape or feature can be die compacted provided that (1) it can be ejected from the tooling and (2) the tools that form the feature have sufficient strength to withstand the repeated compaction loads. Due to the vertical closure of the tooling and the lack of tool motions perpendicular to the pressing direction, part removal from the tools controls many features. Examples of features that cannot be accommodated in die compaction, and therefore require secondary machining operations, include: undercuts, reverse taper (larger on bottom than on top), annular grooves and threads. The following guidelines provide assistance with many possible features in die compaction.

Wall Thickness. Minimum wall thickness is governed by overall part size and shape. For parts of any appreciable length, walls should be not less than 1.5 mm (0.060 in.) thick (Fig. 5). A maximum length-to-wall thickness ratio of 8 to 1 should be followed to ensure reasonable density uniformity and adequate tool life. Separate tool members (punches) should be used to provide density uniformity and proper ejection.

Steps. Simple steps or level changes not exceeding 15% of the overall part height (H) (Fig. 6) can be formed by face contours in the punches. A draft of 5° or more is needed to release this contour from the punch face during ejection. Features such as countersinks and counterbores can be similarly formed. This tooling method, as compared to multiple punches, will result in slight density variations from level to level. However, this approach offers the simplest tooling, lower-cost tooling, and closer axial tolerances than multiple punches.

Spherical Shapes. Complete spheres cannot normally be made because the punches would have to feather to zero width (Fig. 7). Spherical parts require a flat area around a major diameter to allow the punch to terminate in a flat section (Fig. 7). Parts that must fit into ball sockets are repressed after sintering to remove the flats.

Taper and Draft. Draft is not generally required or desired on straight-through parts. While tapered side walls can be produced where required, the tools may demand a short straight surface (A in Fig. 8) at the end of the taper to prevent the punch from running into the taper in the die wall or on the core rod.

Holes. Through holes in the pressing direction are produced with core rods extending through the punches. Round holes require the least expensive tooling but many other shapes such as splines, keys, keyways, D-shapes, squares, and so forth can readily be produced. Blind holes, blind steps in holes, and tapered holes are also readily pressed. For very large parts, lightening holes are

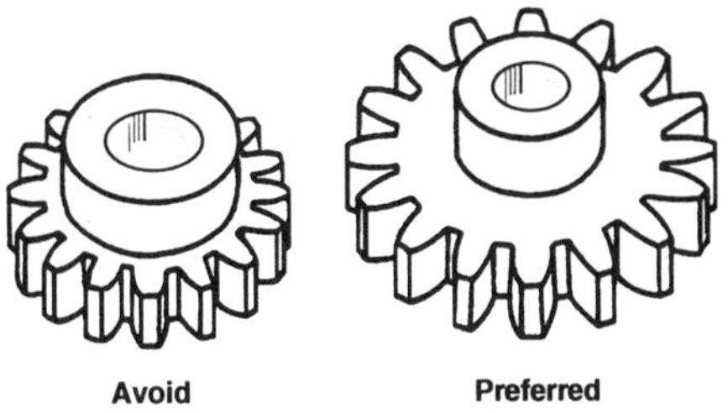
Avoid
Preferred

Fig. 10 Hub feature. Source: Ref 8

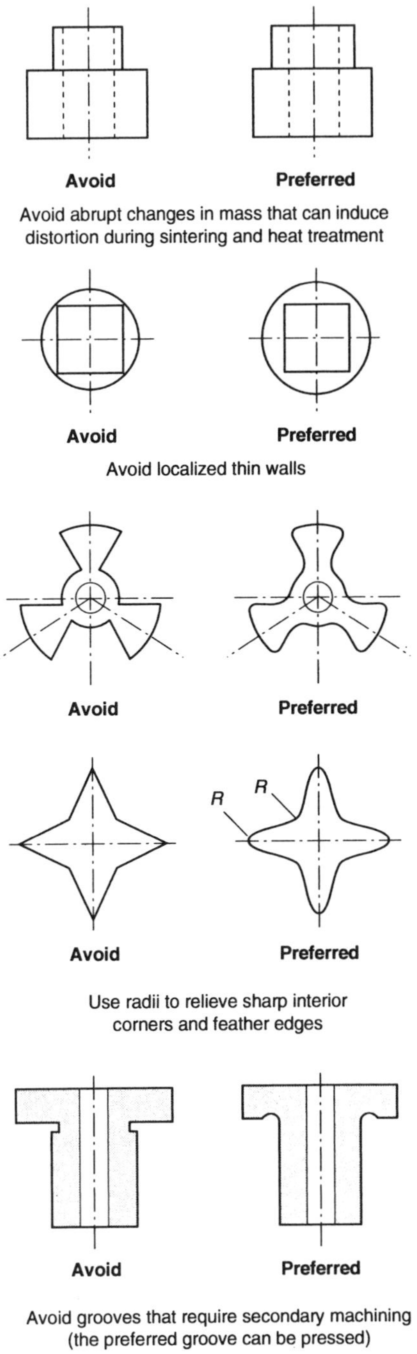
Avoid
Preferred
Avoid abrupt changes in mass that can induce distortion during sintering and heat treatment
Avoid
Preferred
Avoid localized thin walls
Avoid
Preferred
R
R
Avoid
Preferred
Use radii to relieve sharp interior corners and feather edges
Avoid
Preferred
Avoid grooves that require secondary machining (the preferred groove can be pressed)

Fig. 11 Example of undesirable design features and preferred alternatives. Source: Ref 8

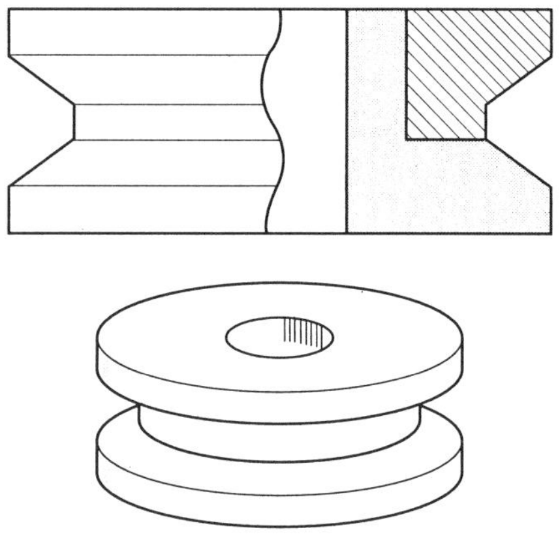

Fig. 12 Multipiece pulley design composed of bonded P/M parts. Source: Ref 8

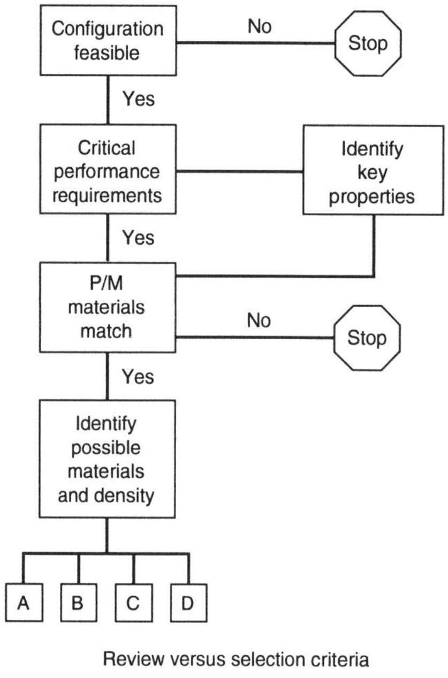

Fig. 13 Material selection method

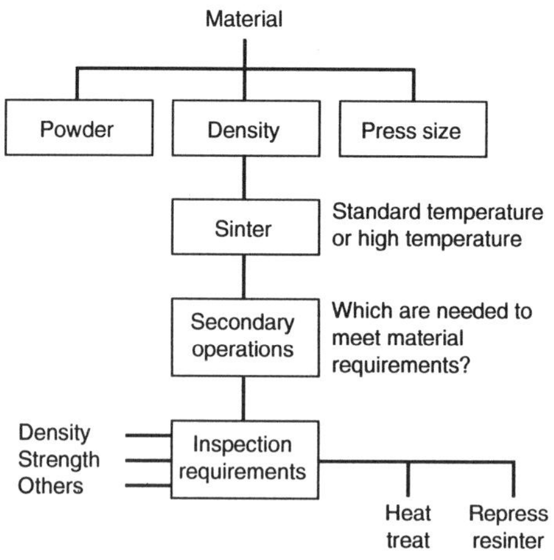

Fig. 14 How material selection affects process

added to reduce weight and the area of compacted surface.

Flanges. A small flange, step, or overhang can be produced by a shelf or step in the die. Separate lower punches are used if the amount of overhang becomes too great to permit ejection without breaking the flange.

Alphanumeric Characters. Numbers, lettering, logos, and other characters can be pressed into surfaces oriented perpendicular to the pressing direction. Recessed lettering is preferred because raised letters are fragile, easily damaged in the green compact, and prevent stacking of parts for sintering.

Chamfers, Radii, and Bevels. Chamfers are preferred rather than radii on part edges to prevent burring. It is common practice to add a 0.25 mm (0.010 in.) flat at a 45° chamfer (Fig. 9); lower chamfer angles may not require the tooling flat.

Hubs and Bosses. Hubs or bosses that provide for drive or alignment rigidity in gears, sprockets, and cams can be readily produced. However, the design should ensure the maximum permissible material between the outside diameter of the hub and the root diameter of gear or sprocket features (Fig. 10).

Other Features. Additional information on these and other features (slots, grooves, knurls, studs, fillets, countersinks, etc.) can be found in Ref 8. Examples of undesirable features that can cause tooling or powder compaction problems are shown in Fig. 11. Because shape complexity is a recognized limitation of die compaction, multipiece assembly is a useful alternative, especially where extensive machining would be required. Pulleys, spools, and sprockets have been produced using sinter bonding, brazing and welding techniques (Fig. 12).

Secondary Operations. A variety of secondary manufacturing and finishing operations may be required to complete the part, to improve properties, or to calibrate dimensional tolerances. Because die-compacted parts have residual porosity that may affect the response to these secondary operations, several guidelines are provided in this section.

Repressing. Die-compacted parts can be repressed in a second tool set in order to reduce the amount of porosity or reduce dimensional variation. By increasing the part density, critical physical and mechanical properties will be improved. This secondary densification step can be performed on the entire part or just specific features, for example, gear teeth. The sizing of P/M parts for improved dimensional tolerances is widely performed, especially for the manufacture of bearings.

Impregnation and Infiltration. Several procedures are used to fill the residual porosity in conventional P/M parts. Oil impregnation adds an internal lubricant to the product, which is useful in bearing and wear applications. The part can be resin impregnated to seal the pores for pressure tightness and improved machinability or to prevent the intrusion of undesired plating chemicals. Infiltration with a lower melting metal than the base material (such as copper-infiltrated steel or copper-infiltrated tungsten) is also used to seal porosity, but most often to increase mechanical properties or create unique composite structures.

Steam treating, also known as steam oxidizing, is a low-temperature (540 °C, or 1000 °F, 1 to 2 h) heat treatment process in which P/M steel parts are exposed to superheated steam. The process can be conducted in batch, pressurized furnaces or continuous belt furnaces. The steam reacts with the iron surface converting it to an adherent, protective blue-grey iron oxide (Fe_3O_4). Because the steam can penetrate the porosity, the oxide layer can be 0.020 to 0.050 in. deep depending on the processing conditions. Steam treating enhances the product by:

- Increasing wear resistance
- Increasing surface hardness
- Improving corrosion resistance
- Increasing compressive yield strength
- Providing low-pressure leak tightness

Formation of this oxide layer unfortunately reduces tensile strength and ductility, about 10 to 20%, depending on the material system and processing conditions.

Heat Treatment. Due to the presence of residual porosity in die-compacted P/M parts, heat-treat practices should utilize gaseous or noncorrosive liquid media (such as quench oil rather than water or fused salts). In addition, because the porosity allows for the penetration of gaseous media (i.e., carburizing gas), special steps must be considered when trying to develop a carburized case in P/M steels. The surface (or entire part) must be high enough in density to prevent the carburizing gas from penetrating the pore network. A density of 7.2 g/cm^3 or greater is desired. Copper infiltration can also be used to seal the part for a case-carburizing treatment. Heat treatment of P/M steels is most effective in improving mechanical properties when parts have a density greater than 7.0 g/cm^3.

Finishing. The variety of finishing operations—machining, plating, deburring, joining—must also consider the effects of residual porosity. Improvements in machinability have been achieved through resin impregnation (seals the porosity) and by adding machinability aids to the original powder blend. Surface finishing may also require resin impregnation if the part density is low enough to allow entrapment of plating or finishing chemicals. These liquids can cause internal corrosion of the P/M part if allowed to penetrate the open porosity network. For welding and brazing, precautions are necessary to ensure a sound joint (Ref 8).

Material Selection. Once the die-compaction processing method has been selected the designer must consider the product configuration and material requirements. Previous sections have discussed the design constraints regarding shape and configuration. The next step in the procedure (see Fig. 13) is comparing the critical part performance requirements with the available P/M materials. Reference 5 serves as a very useful source of

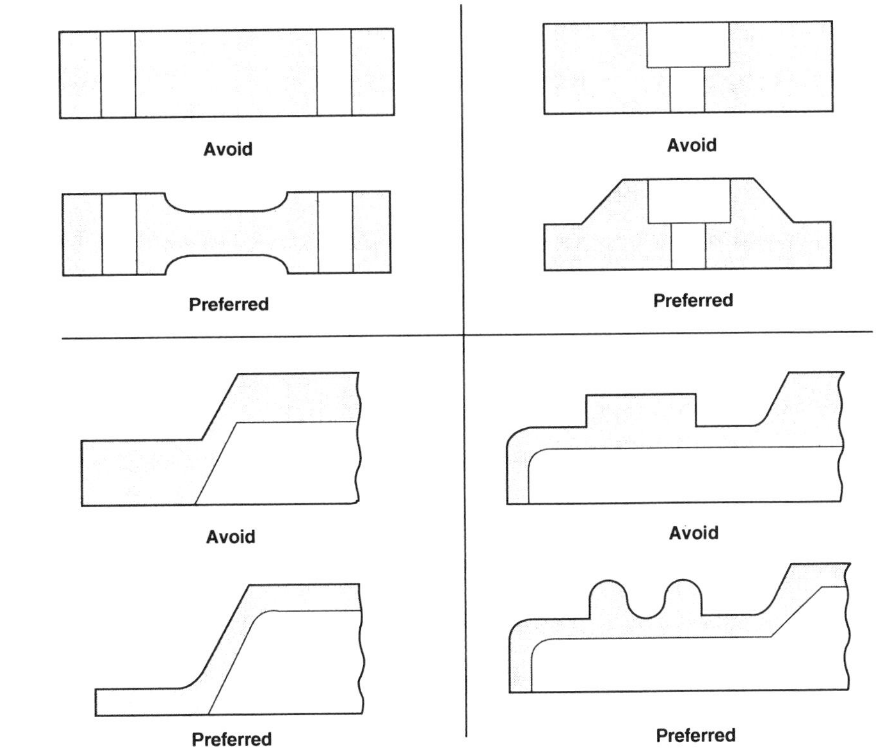

Fig. 15 Designs for maintaining uniform wall thickness in MIM parts. Source: Ref 8

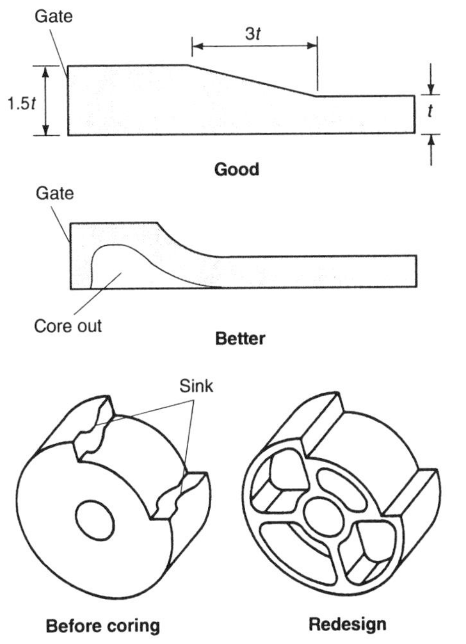

Fig. 16 Redesign of a MIM part to establish uniform wall thickness. Source: Ref 8

material property data. In many cases, more than one material can satisfy the design performance requirements. Therefore, the various production methods must be considered because these affect the ultimate cost of the product.

Figure 14 provides the second evaluation scheme. Starting from the material selected, this leads directly to the density needed in the critical section of the part. Knowing the desired density leads to a preliminary selection of the powder and press size necessary to reach the required density. From this point the remainder of the processing steps are considered. For example, is a conventional sintering process or a "high-temperature" process needed? Will a repressing-resintering process be necessary to achieve the density selected? What secondary processing treatments are needed, and how will the density affect these processes? Using a standard press-and-sinter process as a baseline, the cost of high-temperature sintering may add a 10 to 20% premium, while a double-press/double-sinter process could result in as much as a 40% premium. Therefore, the designer must recognize that selecting a material and density level leads directly to a processing cycle and the cost associated with the process. A judicious review and selection of materials can ensure the most cost-effective product design.

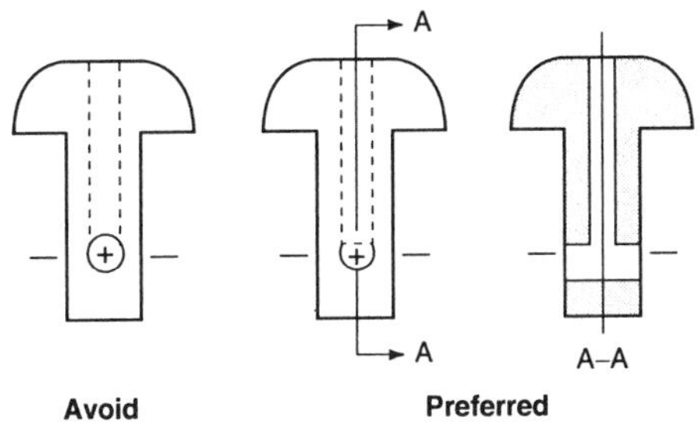

Fig. 17 Intersection holes in MIM parts. Source: Ref 8

Bearings

Due to the unique self-lubricating characteristics of P/M bearings, these products have additional design guidelines. After conventional die compaction and sintering, bearings are vacuum-oil impregnated to fill the interconnected porosity with oil. The design of P/M bearings and selection of the optimal alloy depend primarily on shaft velocity and bearing load. Other design factors include type of operation (stop-start versus continuous), shaft surface finish, type of lubricant, and conditions of uneven loading. The MPIF bearing standard (Ref 9) provides a systematic guide for these design issues:

- Recommended load as a function of shaft velocity
- Environmental factors that can reduce permissible loads
- Recommended press fits
- Running clearances
- Dimensional tolerances

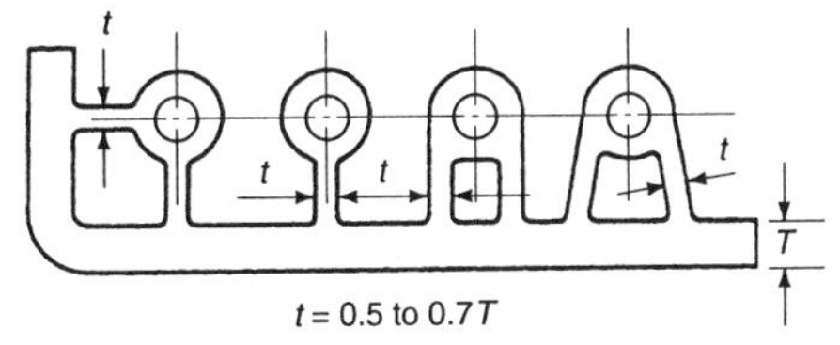

Fig. 18 Design of ribs and webs for MIM parts. Source: Ref 8

The selection of a bearing alloy normally considers the following factors:

- Chemical compatibility with the lubricant
- Thermal conductivity to dissipate heat
- Mechanical properties to maintain structural integrity and withstand press-fit installation forces
- Antigalling with the shaft material
- Wear characteristics with the shaft material

Six commercial alloys dominate the self-lubricating bearing designs: bronze, diluted bronze, plain carbon steel, iron-copper alloys, copper steels, and iron-graphite. Graphite can be used as a solid lubricant and is added to the powder mix prior to compacting. More information on these materials can be found in Ref 9.

In some products the designer can incorporate bearing surfaces into the part, thereby eliminating the need for separately fabricated and installed bearings. The following design restrictions must be followed:

- Bearing areas should have a minimum of 10% interconnected porosity to provide sufficient oil capacity.
- Infiltration is not allowed in the bearing areas because the infiltration will fill the open pores.
- Machining on the bearing surfaces must be performed so as to avoid sealing or closing off the open porosity.

Metal Injection Molding

The MIM process is similar to plastic injection molding: (1) material is fed into the mold cavity through gates, which may leave a visible mark on the part; (2) the mold must open, leaving a parting line on the part; and (3) the part must be ejected from the mold cavity using ejector pins, which also may leave visible marks on the part. Location of the parting line is critical to the orientation of part features. To the extent possible, all features should be oriented perpendicular to the plane of the parting line to facilitate removing the part from the mold. Features not so oriented require additional tooling, such as side pulls.

The optimal location of gates is a design balance between product and processing requirements. In general, gates are located on the parting line, positioned to direct flow onto a core rod or cavity wall. Where wall thicknesses vary, gates are located so the material flows from thicker sections to thinner. Ejector pins push the molded part off of the mold cores. There must be enough pins to free the part off the core without damaging or distorting the as-molded part. Typically, these pins are located near features requiring the highest ejection forces, such as cored holes. Because the pins leave permanent marks, they should be positioned on recessed features, surfaces that will be machined, or unused areas such as runners.

Injection-molded parts are normally placed on flat plates or shelves for thermal debinding and sintering. A geometric feature that allows the part to rest securely with no additional support is preferred. Long spans, cantilevers, or delicate resting points may require special fixtures or setters to minimize distortion during sintering. A postsintering straightening operation may also be required.

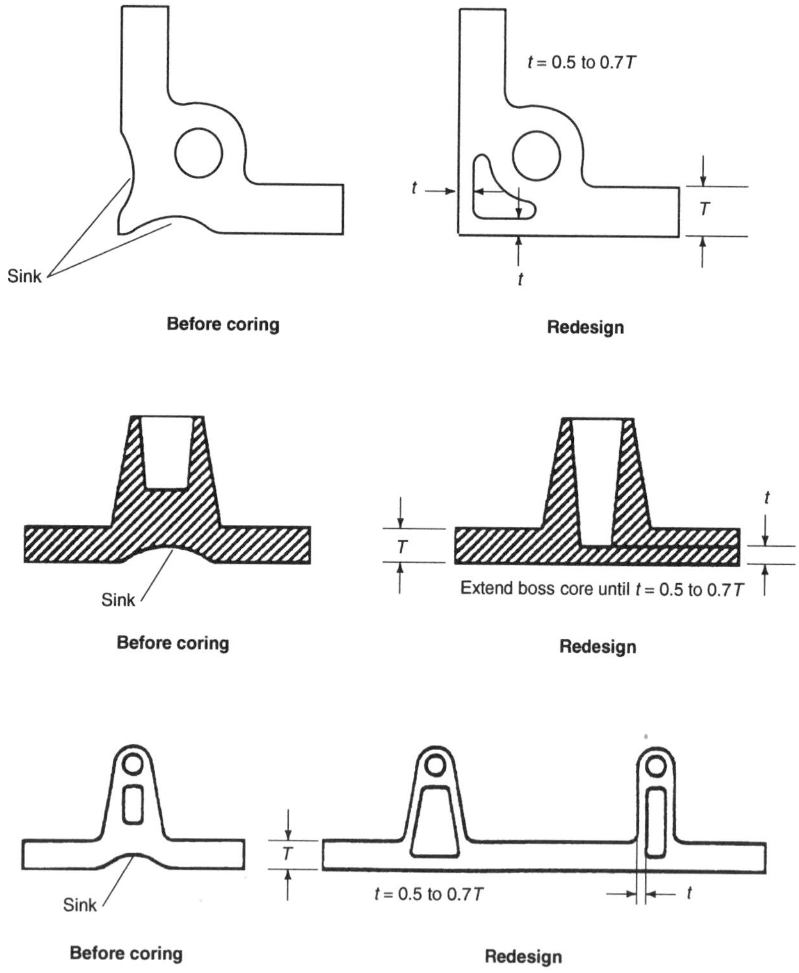

Fig. 19 Design of bosses and studs for MIM parts. Source: Ref 8

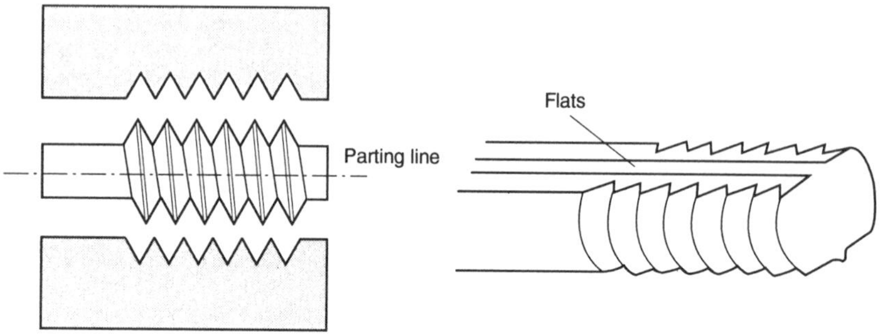

Fig. 20 Design of threads on MIM parts. Source: Ref 8

Shape and Features. The design features that can be produced by MIM are similar to those made by conventional plastic injection molding or die casting. Dimensions and proportions are different due to the small size of MIM parts and molding characteristics of the MIM materials.

Draft. To facilitate release and ejection of the molded part, draft is normally in the range of ½ to 2°. Draft should be uniform on all features of the part to facilitate mold fabrication and near the upper range with more complex part features or multiple cores.

Wall Thickness. Where feasible, walls should be of uniform thickness throughout the part (Fig. 15). Variations in thickness can lead to distortion, internal stresses, voids, cracking, and sink marks. In addition, thickness variations cause nonuniform shrinkage during sintering, leading to greater size variation and looser tolerances. A thickness in the range of 1.25 to 6.35 mm (0.050 to 0.250 in.) is preferred. Common ways of modifying a part to make wall thickness more uniform or ease the transition from thick to thin are shown in Fig. 16.

Holes should be cored to reduce material usage, promote uniform wall thickness, and reduce or eliminate machining operations. Preferred direction is parallel to the direction of mold opening. Through holes are desired over blind holes. Internally connected holes are possible, preferably perpendicular to each other and, where possible, should be D-shaped, to facilitate core pin sealing and minimize flash (see Fig. 17).

Ribs and webs are useful for reinforcing thin walls and avoiding thin sections (Fig. 18). Rib thickness should not exceed that of the adjoining wall. If a thicker rib is required, multiple ribs should be used.

Fillets and radii are advantageous to the molding process because they eliminate sharp corners that can cause cracking, distortion, or erosion of mold features. A fillet or radius of 0.40 to 0.75 mm (0.015 to 0.030 in.) is preferred.

Bosses and Studs. The maximum length of a stud or hole should not exceed five times the adjoining wall thickness. A stud diameter should not exceed the thickness of the adjoining wall. A boss with a blind or through hole should have twice the diameter of the hole and a wall thickness not exceeding that of the adjoining surface. As a result, hole diameters are limited to two times the adjoining wall thickness (see Fig. 19).

Threads. Both internal and external threads can be formed by MIM; however, tapping internal threads is usually more precise and cost effective than forming with core pins that must be unscrewed. Optimal location for external threads is on a parting line (Fig. 20). Adding flats to the threads on the parting surface (Fig. 20) allows for 0.05 mm (0.002 in.) flash, improves thread quality, and reduces mold maintenance.

Undercuts. External undercuts can be formed readily on the parting line. Internal undercuts, though feasible with collapsible cores, are not considered economically practical and should be avoided.

Other features such as knurls, logos, and part numbers can easily be molded in place. The very fine powder size and low porosity produce typical surfaces of 0.8 μm (32 μin.) rms (root mean square) roughness.

Material Selection. The material systems currently in commercial use include ferrous alloys (low-alloy steel, stainless steels, soft magnetic alloys), nonferrous alloys (brass, bronze), tungsten carbide, pure nickel, electronic alloys (Invar, Kovar) and tungsten-copper composites. The physical and mechanical properties of several MIM engineering alloys (low-alloy steel and stainless steels) have been standardized by MPIF (Ref 10). After sintering, the residual porosity in MIM parts is very low and not interconnected. With densities typically in excess of 96% of theoretical density, the resultant mechanical properties are superior to conventional die-compacted P/M materials and more closely match the properties of investment castings in similar alloys. Therefore, the limitations discussed previously regarding the residual porosity in conventional die-compacted products do not apply to MIM parts. Secondary operations can be performed with no restrictions.

Powder Forging

The design issues in powder forging (P/F) are similar to the requirement of any precision, closed-die forging. The difference is the starting preform; in the case of P/F, the preform is a sintered powder metal part, typically 80 to 85% of theoretical density, with a shape similar to the final part configuration. By contrast, in a precision closed-die forging the preform is a wrought steel blank with very little shape detail. Preform design for P/F fabrication determines the extent of product shape detail required to meet the performance requirements of the finished P/F part. Preform design is a complex, iterative process currently modeled by computer simulation software programs to help reduce design time and development costs.

In the forging step, the P/M preform is removed from the reheat (or sintering) furnace, coated with a die lube, and forged in a heated, closed die operation. The forging process reduces the preform height and forces metal into the recesses of the closed die. This step also brings all features to their final tolerances and densities.

Configuration guidelines, typical of precision closed-die forged parts also apply to P/F parts as follows:

- Radii on inside corners of the forging as large as possible to promote metal flow around corners in the tool and promote complete fill of all details.
- Radii of at least 1 mm (0.040 in.) on all outside corners of the forging to aid in material flow to define features.
- Shape of the forging should be such that, when placed in the die, the lateral forces will be balanced. Shapes that are symmetrical along a vertical plane, such as connecting rods and shapes that are axisymmetric (or nearly so), are preferred.
- Zero draft is possible on surfaces formed by the die and core rod, but not by the upper punch.
- Re-entrant angles (undercuts) cannot be forged.
- Axial tolerances—in the direction of forging—are driven by variations in the mass of metal in the preform. Lateral tolerances are driven by metal flow as the cavity fills. Typical axial tolerance of 0.25 to 0.5 mm (0.010 to 0.020 in.) are encountered, with diametric tolerances of 0.003 to 0.005 mm/mm (in./in.) of diameter.
- Concentricity of a P/F part is determined by the quality and density distribution in the preform. Concentricity is normally double that of the preform.

REFERENCES

1. F.V. Lenel, *Powder Metallurgy—Principles and Applications,* Metal Powder Industries Federation, 1980, p 426
2. L.F. Pease III and V.C. Potter, Mechanical Properties of P/M Materials, *Powder Metallurgy,* Vol 7, *ASM Handbook* (formerly *Metals Handbook,* 9th ed.), American Society for Metals, 1984, p 467
3. H.I. Sanderow, H. Rodrigues, and J.D. Ruhkamp, New High Strength 4100 Alloy P/M Steels, *Prog. Powder Metall.,* Vol 41, Metal Powder Industries Federation, 1985, p 283
4. D. Gay and H. Sanderow, The Effect of Sintering Conditions on the Magnetic and Mechanical Properties of Warm Compacted Fe-P P/M Steels, *Advances in Powder Metallurgy and Particulate Materials—1996,* Vol 6, Metal Powder Industries Federation, 1996, p 20–127
5. "Material Standards for P/M Structural Parts," Standard 35 1994 edition, Metal Powder Industries Federation, 1994
6. "Standard Specification for Powder Forged (P/F) Ferrous Structural Parts," B 848-94, *Annual Book of ASTM Standards,* American Society for Testing and Materials
7. R.M. German, *Powder Metallurgy Science,* 2nd ed., Metal Powder Industries Federation, 1994, p 193
8. *Powder Metallurgy Design Manual,* 2nd ed., Metal Powder Industries Federation, 1995
9. "Material Standards for P/M Self-Lubricating Bearings," Standard 35 1991–1992 edition, Metal Powder Industries Federation, 1991
10. "Material Standards for Metal Injection Molded Parts," Standard 35 1993–1994 edition, Metal Powder Industries Federation, 1993

Design for Machining

D.A. Stephenson, General Motors Powertrain Group

MACHINING or material removal processes are secondary manufacturing operations used to achieve precise tolerances or to impart controlled surface finishes to a part. There are three basic classes of machining operations: conventional, abrasive, and nontraditional. *Conventional machining operations* include turning, drilling, milling, boring, planing, shaping, reaming, and tapping. In these operations material is removed from the workpiece in the form of a chip, which for metal parts is produced through plastic deformation. In *abrasive processes*, such as grinding and honing, small particles are removed from the workpiece by abrasion with a hard tool. Abrasive processes have lower material removal rates and higher energy requirements than conventional processes, but can produce closer tolerances and finer surface finishes. *Nontraditional machining processes* include electrodischarge machining (EDM), electrochemical machining (ECM), and ultrasonic machining (USM). In these operations, material is removed by a variety of physical mechanisms, often on an atomic scale with substantially less or no mechanical contact. They tend to remove material much more slowly than conventional or abrasive processes and are generally used when the part geometry or material does not permit the use of any other operation.

Machining operations are costly compared to most other manufacturing processes because they are relatively slow to perform and often require substantial investment in overhead and skilled labor. It has long been known that machining costs can be reduced through proper part design (Ref 1). Many recent books on mechanical design list simple rules that can be used to avoid designing parts with unnecessary machined content or features that are difficult to machine accurately (Ref 2, 3). Although in some cases these lists are overly simplified because they do not take into account part volume and equipment concerns, they indicate a growing awareness of the importance of considering machining issues in the design stage.

This article summarizes rules for designing parts to improve machined part quality and reduce machining costs in mass and batch production environments. Only conventional and abrasive processes such as turning, boring, milling, drilling, and grinding are considered. These processes are described in detail in *Machining,* Volume 16 of *ASM Handbook.* Rules for designing for nontraditional machining operations are discussed in Ref 4.

Machining Costs

As discussed previously, design rules for machining are intended to improve machined part quality and reduce machining costs. Quality issues are discussed below as the various rules are described. The total cost of a machining operation includes contributions from some or all of the following components (Ref 5):

- *Raw material costs:* The cost of unmachined stock, which may be in the form of a standard bar or slab, casting, or forged blank
- *Labor costs:* The wages for the machine operator, usually measured in units of standard hours
- *Setup costs:* The cost of special fixtures or tool setups and the wages paid to setup personnel
- *Tooling costs:* The cost of perishable tooling, including inventory, and any special tooling required for the operation
- *Equipment costs:* The cost of the machine tools, including required capital expenditures, facilities costs, and machine depreciation
- *Scrap and rework costs:* The cost of repairing or disposing of finished or partially finished parts of unacceptable quality
- *Programming costs:* The cost of writing numerical control (NC) programs to generate the required toolpaths
- *Engineering costs:* salaries paid to engineers for process design, validation, and other overhead functions

Design-for-machining rules may address any of the first seven cost categories. The most significant cost component in a given application depends on several factors. Material costs are most significant for parts made of expensive materials or machined from complex castings or forgings. In high-volume production, material, labor, tooling, and scrap costs are generally the most significant. In small- or medium-lot production, setup and programming costs are proportionally more significant. Equipment and tooling costs are of greatest significance when producing complex or precision parts that require special tooling or investment in precision machines. Specific design rules generally have a strong impact on one or a few cost components; therefore, the rules that should be emphasized in a given application vary depending on the complexity of the part, the cost and form of the raw material, and the required production volume.

Engineering costs are generally not reduced through use of design rules for machining; in fact, the application of these rules usually increases overhead as engineering effort is invested up front to realize benefits in later stages of manufacture.

Types of Machining Equipment

Individual machining operations are performed using machine tools. When multiple operations are required to produce a part, they may be performed on a collection of machines arranged as a machining system. Numerous types of machine tools and machining systems are used for different types of parts and production schedules (Ref 5). In applying design-for-machining rules, it is useful to distinguish between three types of machining systems: general-purpose machine tools, production machining systems, and computer numerically controlled (CNC) machining systems.

General-purpose machine tools include lathes, milling machines, drill presses, and similar machines normally found in tool rooms and small job shops. They can perform only a limited range of operations and are used for small batch production of relatively simple parts. The general design-for-machining rules discussed in the next section are applicable to general-purpose machine tools.

Production machining systems (or transfer machines) are complex systems consisting of simpler machine tools connected by an automated material-handling system. They are dedicated systems used for high-volume production of one or a few parts. They operate with a fixed cycle time, during which all operations at a particular station must be completed. The machining time is typically two-thirds to three-quarters of the cycle time for bottleneck operations. Because individual stations in the system resemble conventional machine tools, the general design-for-machining rules discussed in the section "General Design for Machining Rules" in this article are broadly valid for these systems. However, due to cycle time restrictions and tool replacement considerations, a number of special design rules also apply, as discussed in the section "Special Considerations" in this article.

Computer numerically controlled machining systems are composed of CNC machining and turning centers operating alone or arranged in cellular or flexible systems. Computer numerically controlled machine tools are capable of a broad range of operations. They can follow tool paths requiring simultaneous motions along several axes and are usually equipped with automatic tool changing systems, so they can often manufacture complex parts with minimal part transfer and refixturing. They are used for medium to large batch production of a variety of parts. As with production machining systems, the general design-for-machining rules discussed in the next section of this article are broadly applicable to these systems. In CNC systems, however, the actual machining time consumes a smaller fraction of the total processing time. Reducing the time required for noncutting functions such as tool changes, axis motions, and pallet rotations thus becomes a priority in operating these systems efficiently, resulting in a number of special design rules as described in the section "Special Considerations" in this article.

Design Considerations. To design a part properly for machining, it is essential to know what kind of equipment will be used to manufacture it. As a first step, the designer should know the type of equipment currently available in the intended manufacturing facility, because design-for-machining rules do not apply equally to all machining systems, and special rules may be critical for certain classes of equipment. The designer should also determine the capabilities (i.e., the range of possible operations and the achievable tolerances) of the available equipment, so that features which will require investment in new or precision equipment can be avoided to the extent possible.

General Design-for-Machining Rules

The design-for-machining rules discussed in this section are broadly applicable to all parts, regardless of the type of equipment used to produce them.

Choose Materials for Optimal Machinability. Nothing has a greater impact on machining costs and quality than the nature of the work material itself. The nature of the work material determines machining system characteristics such as motor power and bearing sizes, the tool materials and geometries that can be used, the range of cutting speeds and other cutting conditions, the perishable tooling costs, and the tolerances and surface finishes that can be achieved. The material choice is determined largely by material property and other functional requirements independent of machining, but in many cases at least a narrow spectrum of materials is available that satisfy these requirements. When this is the case, the relative machinability of the candidate materials and the standard forms in which each material can be obtained should be considered.

"Machinability" is a loosely defined term reflecting the ease with which materials can be machined (Ref 5, 6). In most cases tool wear rates or tool life under typical conditions are the most significant practical constraints on machining operations, and in practice most tables of machinability ratings are based on tool life test results. For example, a reference material such as 10L14 steel may be assigned a rating of 100, and ratings for other materials are computed from ratios of tool lives obtained with the material in question and the reference material in a standardized test (Ref 7). Other criteria used to rank material machinability include machining forces and power consumption, chip form, and achievable surface finish.

Machinability varies most significantly between different material classes or base chemistries. The common classes of metallic work ma-

Size range, mm	Achievable tolerance, ± mm							
0-15	0.004	0.005	0.008	0.013	0.02	0.031	0.051	0.08
15-25	0.004	0.0065	0.01	0.015	0.025	0.038	0.064	0.10
25-38	0.005	0.008	0.013	0.02	0.031	0.051	0.08	0.13
38-70	0.0065	0.01	0.015	0.025	0.038	0.064	0.10	0.15
70-115	0.008	0.013	0.02	0.031	0.051	0.08	0.13	0.20
115-200	0.01	0.015	0.025	0.038	0.064	0.10	0.15	0.25
200-350	0.013	0.02	0.031	0.051	0.08	0.13	0.20	0.30
350-500	0.015	0.025	0.038	0.064	0.10	0.15	0.25	0.38

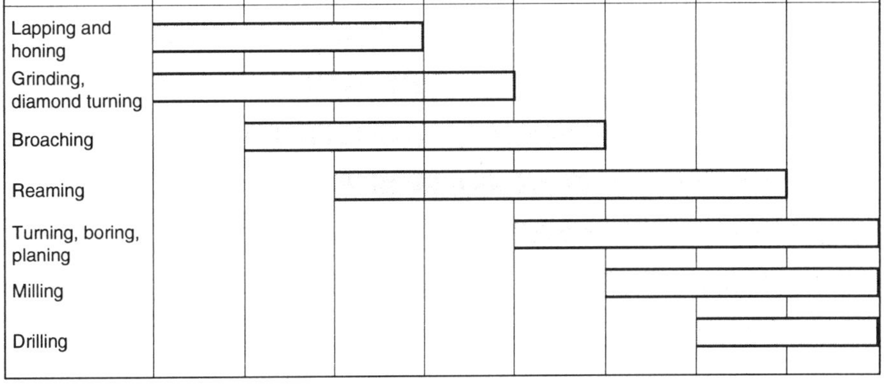

Fig. 1 Dimensional tolerance achievable through various machining operations under general machining conditions as a function of feature size

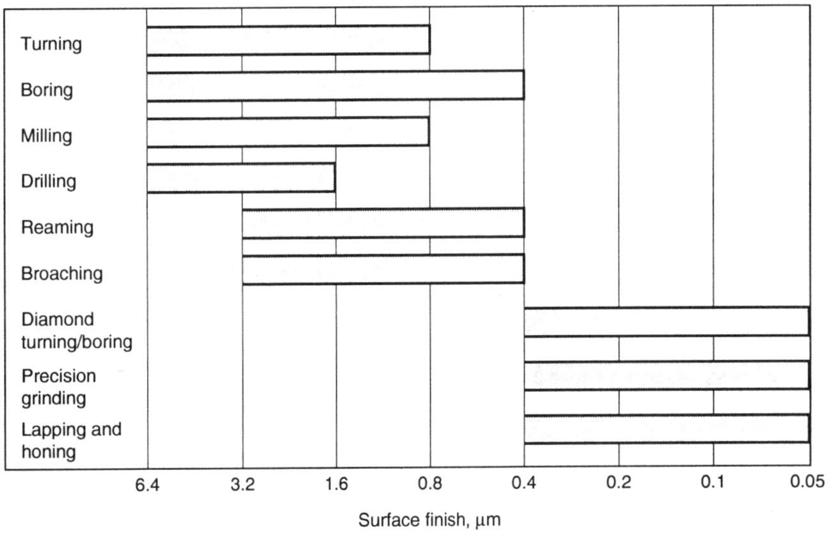

Fig. 2 Surface finish achievable through various machining operations

terials, in order of decreasing machinability, include:

- Magnesium alloys
- Aluminum alloys
- Copper alloys
- Cast irons
- Nodular irons
- Carbon steels
- Low-alloy steels
- Stainless steels
- Hardened and high-alloy steels
- Nickel-base superalloys
- Titanium alloys

When possible, materials from classes higher on this list should be substituted for those lower on the list, if machinability is a primary consideration.

Machinability generally decreases with increasing penetration hardness and yield strength and with increasing ductility. Homogeneous materials with fine grain structures are easier to machine than nonuniform materials or materials with coarse grain structures. Annealing and tempering also usually increase machinability, although annealed materials are more prone to chip-control problems. Hard alloying elements such as chromium, nickel, and silicon typically reduce machinability. Metal-matrix composites are generally less machinable than the base-matrix metal; similarly, powder metals are less machinable than cast or wrought alloys of equivalent chemistry.

Most material classes include "free-machining" alloys, which contain alloy additives that increase tool life or chip formation characteristics. Free-machining additives are typically substances that form inclusions in the matrix that serve to break chips and provide internal lubrication; examples include lead in brasses and steels, sulfur compounds in steels and powder metals, and insoluble metals such as beryllium, selenium, and tellurium in steels. The addition of free-machining additives often compromises material properties such as hardness or strength; nonetheless, free-machining alloys should be chosen over standard alloys whenever possible to simplify required machining operations.

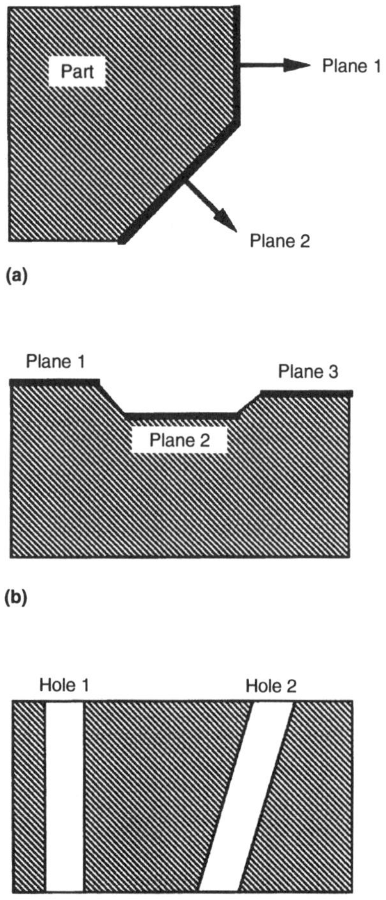

Fig. 3 Examples of features with different machined orientations. (a) Planes with normals in different directions, which may require refixturing or pallet rotations to complete. (b) Parallel planes at different depths, which may require multiple passes, spindle extensions, or tool changes to complete. (c) Holes with axes in different directions, which may require refixturing or pallet rotations to complete

Minimize the Number of Machined Features. Features should be machined only when they require dimensional or surface finish tolerances that cannot be produced by a primary operation. This is most often the case when the feature in question is used for bearing, locking, locating, press fitting, or dynamic balance, or when subsequent assembly considerations dictate a close dimensional tolerance. Alternative methods of producing features such as casting in holes, chamfers, and undercuts should be considered to minimize the number of machining operations required.

Minimizing the number of machined features reduces machining cycle times, tool costs, and the total number of production operations required to finish a part. To date, elimination of unnecessary machining operations has had a greater practical impact than any of the other design-for-machining rules (Ref 2).

Minimize the Machined Stock Allowance. The amount of material that must be machined away to produce the final part should be minimized. Excessive stock to be removed increases material costs, equipment costs, and fixed tooling costs (since tool wear per part increases with the stock allowance), and increases machining cycle times.

For parts produced from cast or forged blanks, the stock allowance can be controlled by controlling the blank dimensions. Most machined features require roughly 1.5 mm (0.06 in.) of stock for roughing and finishing in two passes. A greater amount of stock should be provided only when primary process tolerances are of this magnitude or when the primary process leaves a thick surface layer of material with unacceptable mechanical properties. An excessively thick surface layer is more likely to occur with cast than with forged blanks. For surfaces with relatively open tolerances requiring only a single machining pass, the stock allowance should be reduced to 1 mm (0.04 in.) when primary process tolerances permit.

For parts made from bar stock or other standard cold-worked shapes, machined part dimensions should be chosen when possible to be slightly smaller than the dimension of a standard available size. Rotational parts made from cold-rolled steel bars, for example, should have finished diameters roughly 1.25 mm (0.05 in.) smaller than the diameter of a standard bar size. The stock allowance should be increased for larger-diameter bars to account for increased out-of-roundness tolerances in the raw bar. Similarly, if hot-rolled raw stock is used, the machined stock allowance should be increased (to roughly 2.5 mm, or 0.1 in.) to ensure that surface scale is removed.

Optimize Dimensional and Surface Finish Tolerances. The most open dimensional and surface finish tolerances compatible with the part function should be specified for all machined features. Excessively stringent dimensional or surface finish tolerances increase machining costs by requiring the use of additional finishing passes and/or reduced feed rates, which increase machining time, and by dictating more frequent tool changes to avoid the degradation of surface finish that accompanies tool wear.

Figures 1 and 2 (Ref 4) show the range of dimensional and surface finish tolerances that can be achieved using various machining processes under general machining conditions. It is particularly desirable to avoid the use of final grinding and honing operations, as these operations are relatively slow and capital intensive and may produce environmentally hazardous byproducts. In general, dimensional tolerances less than 0.025 mm (0.001 in.) and surface finish tolerances less than 0.4 μm (15 μin.) often require the use of grinding or honing operations. (As discussed in the section "Special Considerations" in this article, dimensional tolerances less than 0.05 mm (0.002 in.) can be difficult to achieve consistently without grinding in mass production applications.) Aluminum and magnesium workpieces are generally less likely to require grinding and polishing operations because very fine finishes can be produced through turning and milling with polycrystalline diamond (PCD) tooling.

Standardize Features. Machined features should be standardized to the extent possible. For example, hole diameters should be selected from a limited range of sizes, and the minimum number of different diameters (ideally, one diameter) should be used on a given part. Thread forms should also be standardized based on the hole diameter and work material, and consideration should also be given to standardizing the depth of blind holes based on their intended function. Standardization simplifies and reduces the cost of maintaining a tool inventory. It also promotes interchangeability of tools between operations, reducing the likelihood of tool shortages. As discussed in the section "Special Considerations" in this article, it also reduces the number of tool changes required when CNC equipment is used, reducing machining cycle times.

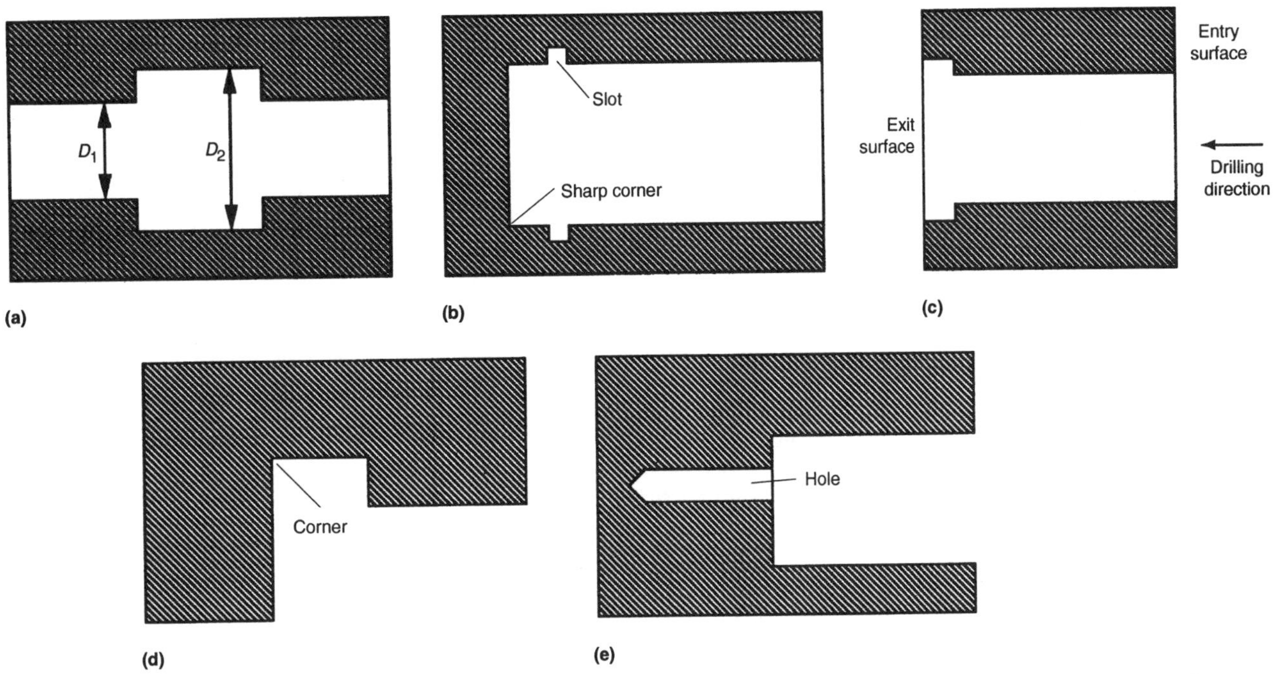

Fig. 4 Machined features that are difficult to access and should be avoided. (a) Increased diameter on an internal bore. (b) A slot and a sharp corner at the bottom of an internal bore. (c) A counterbore on the exit surface of a through hole. (d) A corner blocked by an overhanging surface. (e) A hole on an internal surface

In standardizing features, dimensions that can be produced from standard rather than special tools should be chosen. This is easily done for hole diameters and tap styles. Internal corner radii used on rotational parts should be chosen to match the nose radii available on standard inserts; this simplifies the CNC programming required to generate radii and reduces the need to stock special inserts.

Minimize the Number of Machined Orientations. Parts should be designed so that the number of machined orientations is minimized. A machined orientation is defined as spatial or directional orientation of a feature with respect to a reference plane or orientation on the part. For example, planes at distinct angles to a reference plane, parallel planes at different depths from a reference position, or holes with axes at different angles to a reference direction (often a normal to a reference plane) can all define distinct machined orientations (Fig. 3).

Features with different machined orientations generally cannot be machined with a single tool in a single fixturing or pallet orientation. Therefore, minimizing the number of machined orientations minimizes the number of fixturings, pallet rotations, tool changes, and machining passes required to finish a part. This in turn minimizes processing time and increases machined accuracy.

Provide Adequate Accessibility. In general, a feature becomes more difficult to machine when it becomes more difficult to access with standard tools. Accessibility is most often a problem for features located in internal cavities or on remote faces of the part. Examples of features that should be avoided are shown in Fig. 4 and include increased diameters in an internal bore, slots and sharp corners at the bottom of a bore, countersinks or counterbores on the exit surfaces of holes (especially when they must be highly concentric), holes on internal surfaces, and surfaces or corners blocked by overhanging features.

Inaccessible features often require the use of special tooling, tools with long overhangs (length/diameter ratios >5) that may be prone to deflection or unstable vibration, specialized machine attachments such as right-angle drives, or excessive fixturings or pallet rotations. The machining cost is increased, and the achievable tolerance is often limited by these requirements.

Provide Adequate Strength and Stiffness. The cutting forces generated during machining act between the tool and the part and can cause breakage, deflection, or unstable vibrations if the strength and stiffness of the system is inadequate. The part may be the weakest or most compliant element of the system, particularly if it is made of a material such as aluminum—which is relatively weak, a material with a high yield strength and comparatively low elastic modulus such as titanium, or if its geometry is structurally weak.

Care should be taken to design the part so that it has adequate strength and stiffness in the expected directions of loading. This can be done by thickening sections over which heavy loading is expected or by adding ribs or other structurally stiffening features to support thin sections. Because cutting forces increase with the metal removal rate, particular attention should be paid to surfaces that will be subjected to roughing cuts. Thin wall sections and areas where large diameter holes are to be drilled should also be examined and stiffened with ribs or other structural features when possible.

If it is not possible to stiffen the part significantly, cutting forces can be reduced by proper design of the tool and by removing large amounts of stock in multiple passes, although this increases cutting time. Inadequate part stiffness can also be compensated for by designing supporting elements into the fixture, although this significantly increases fixture costs and setup times and reduces the robustness of the process. In extreme cases, stress-sensitive parts may be impossible to machine using conventional operations and may have to be processed using a nontraditional operation such as EDM.

Provide Surfaces for Clamping and Fixturing. Parts must be clamped to a chuck or fixtured securely before they can be machined. In designing a part it is important to consider the possible ways it may be held and to determine if the most likely workholding methods present access problems or part deflection concerns.

Rotational parts are held in lathes between centers or in chucks or collets. Parts finished on both ends and held in chucks or collets must be reversed at some point to complete all required operations. In this case, a clear section of the part with a constant diameter and without a tight surface finish tolerance should be provided for clamping when possible. If this is not possible, an additional grinding operation may be required to produce the part.

Prismatic parts with irregular or curved surfaces may present fixturing difficulties if clamps must be applied to curved surfaces. This can result in point loading and surface deformation or damage, particularly when fixturing for roughing cuts. When possible, clamping pads with flat surfaces should be designed into such parts.

Clamping and fixturing concerns are particularly critical for structurally weak parts, especially when clamping stresses are transmitted through the part (e.g., when the part is held in a vise). In many applications clamping forces ex-

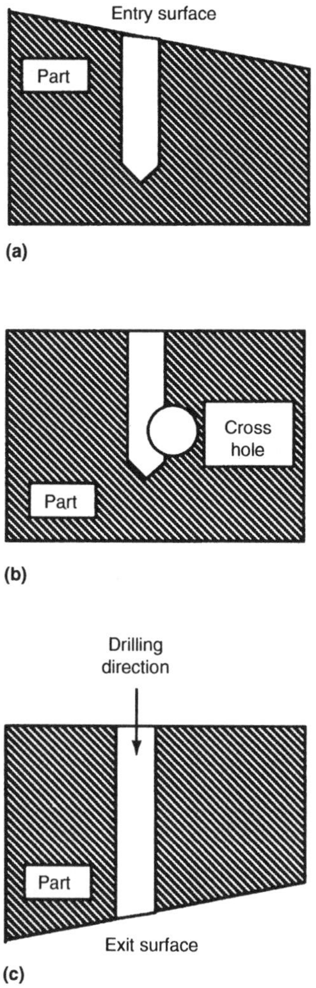

Fig. 5 Examples of hole geometries that should be avoided when possible. (a) A hole drilled into an inclined entry surface. (b) Intersecting holes. (c) A through hole with an inclined exit surface

ceed machining forces and can contribute significantly to deflections and form errors. Once the principal clamping force directions are determined, the structural stiffness of the part in these directions should be examined, and—if necessary—ribs or other stiffening elements should be added when possible.

Special Considerations

Special Considerations for Production Machining Systems (Transfer Machines). The fixed cycle times and specialized nature of individual mechanisms characteristic of production machining systems result in a number of special design rules for parts manufactured using such systems. In addition, some of the general design-for-machining rules, such as using open tolerances when possible, are more critical for parts made using transfer machines, while other rules, such as maintaining adequate accessibility, are less critical.

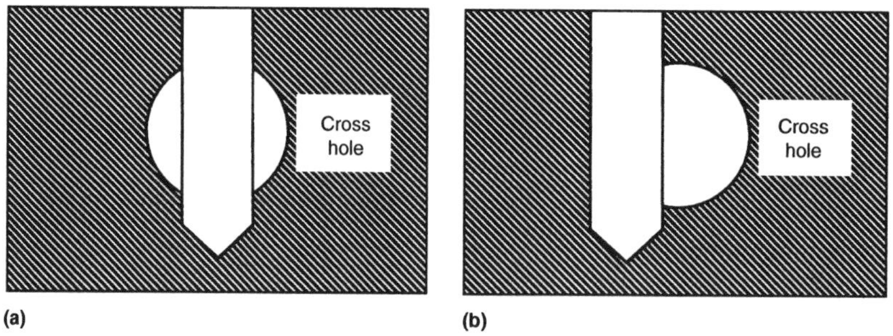

Fig. 6 On-center (a) and scalloped (b) intersecting holes. When intersecting holes are unavoidable, the on-center configuration should be used to minimize drill chipping and burr formation.

Stations on transfer machines can be equipped with multiple spindles, so that patterns of holes can often be drilled simultaneously at a single station. To facilitate this, a minimum separation between holes should be maintained to leave room for distinct spindles. The magnitude of the separation required varies with the hole diameter and the particular type of equipment; for 12 mm (0.5 in.) holes, a minimum separation of 50 mm (2 in.) is typically adequate. Because a single feed slide is normally used on multispindle heads, it is also desirable to make all holes in a given pattern of roughly equal depth; a great variation in depth complicates tool setup and may require an excessive feed stroke length, which can increase cycle time.

Features on parts designed for transfer machine manufacture should be grouped so that multiple features can be machined simultaneously at a single station. Features that require close dimensional tolerances relative to each other should be machined at a single station. When possible, flat surfaces on a given part face should be at equal depths so that all can be machined with a single milling cutter. The amount of stock removed at each station should also be balanced to the extent possible so that all tools wear out at roughly the same time. Light cuts at odd depths that would require additional unbalanced stations should be avoided.

It is more critical to avoid tight tolerances on transfer machine parts than on parts to be manufactured on CNC equipment. The use of tight tolerances that increase the frequency of required tool changes has an exaggerated impact on transfer machine operations, because all or part of a line may need to be shut down to perform a tool change on a single station. Ideally, tools on all stations should be changed on schedule between production shifts to maximize machine utilization. Tolerancing becomes a more critical issue as the machinability of the part material decreases; fairly tight tolerances can be consistently achieved on aluminum and magnesium parts because tool wear rates are low, but it is difficult to achieve dimensional tolerances less than 0.05 mm (0.002 in.) consistently when machining iron or steel parts in high volumes.

Maintaining adequate access to features such as interior holes is not as critical for parts manufactured on transfer machines because special fixtures and jigs (fixtures incorporating guiding components such as bushings) can be more easily incorporated; the additional cost of such jigs and fixtures is more easily justified for the high production runs typical of transfer machines.

Special Considerations for CNC Machining Systems. In operations on CNC machining and turning centers, tools are typically cutting for a smaller fraction of the processing time than in production machining systems; more of the processing time is taken up by noncutting functions such as tool changes, pallet rotations, and axis moves. Most actions that consume noncutting time also have a negative impact on part quality because they introduce additional tolerance components due to the finite accuracy and repeatability of the system. In designing parts to be manufactured on CNC machinery, therefore, the primary concern from a machining viewpoint should be to minimize noncutting functions by appropriate standardization and grouping of features.

Standardizing feature dimensions can reduce the number of required tool changes and thus the processing time. A minimum number of hole diameters should be used on a given part, and holes used for functions such as clearance or mounting should have the same diameter. Care should also be taken to avoid designing in pockets of widely varying dimensions, so that tool changes between end mills of different diameters can be avoided. Standardizing features to eliminate tool changes also simplifies tool magazine management as discussed below.

Pallet rotations can be minimized by reducing the number of machining orientations as discussed in the section "General Design-for-Machining Rules" in this article. Standardizing features reduces axis motion time by eliminating excess tool changes; tool changes consume axis motion time because the spindle or table must generally be moved to a set position for a tool change. In addition to standardizing features, axis motion time can be reduced by reducing the distance between features that must be machined using the same tool. For example, when drilling a pattern of holes with equal diameter, the holes should be placed as close together as possible to minimize transit time between holes. (Note that in this respect hole patterns should be designed differently for parts to be manufactured on NC machines than for those manufactured on transfer machines.)

To further minimize tool changes for medium to long production runs, multidiameter or stepped holes should be designed so that they can be machined using stepped or combination holemaking tools rather than discrete tools as discussed below. The added cost for these special tools is generally not justified for small batch production.

Finally, tool magazine management can be of concern when producing complex parts on CNC machines. Generally, the tool magazine on a CNC machine has a finite capacity (i.e., can hold only a specified number of tools). Some magazine types, such as belt or chain magazines, have a minimum access time to position tools for tool change, and this time often depends on the number of slots between tools used in succeeding operations. Operations should be grouped so that each tool retrieved from the magazine is used for some minimum period greater than the tool access time, so that the machine does not have to wait during the tool access cycle. This becomes a more difficult constraint to satisfy as the access time increases. Parts should be designed so that the number of tools required for the operations scheduled for a given machine does not exceed the number of slots in the tool magazine; if this cannot be arranged, additional time will be required to reload the tool magazine. If the number of slots available exceeds the number of tools required, the surplus slots can be loaded with redundant tools of the type that wear most rapidly, and the NC program can be written to access these additional tools after a specified number of parts have been produced, reducing the number of magazine reloads required. Tool magazine management can be simplified through standardization of features to reduce the number of tools required and through the use of stepped or combination tools to produce multidiameter holes.

Special Considerations for Holemaking Operations. Holemaking operations such as drilling, reaming, and tapping are time-consuming operations often used to produce critical features such as locating holes. As a result, quality issues and machining time constraints are often particularly critical, so that special care should be taken in the design phase to simplify required holemaking operations.

When possible, holes with increased internal diameters, interrupted holes, and holes intersecting inclined entry and exit surfaces should be avoided (Fig. 4a and 5). Holes with increased internal diameters require additional boring operations to produce and cannot be produced using step or compound tools. Holes drilled into inclined entry surfaces or at compound angles often represent unique machined orientations that may require additional fixturings or pallet rotations. It is also more difficult to maintain location accuracy and adequate tool life for these features; the drill has a tendency to "walk" at the entry of such holes, and unbalanced loading that may result in cutting-edge chipping or margin rubbing is also present. Interrupted or intersecting holes also generate unbalanced loads on the drill and may result in straightness errors, excessive vibration, or drill chipping, especially when drilling iron or steel with solid carbide drills. Burrs can also form at intersections, requiring an additional deburring operation. Intersecting holes are often used to produce lubrication passages; when they are unavoidable, an on-center rather than a scalloped design should be chosen (Fig. 6) (Ref 8) to minimize load unbalance, burr formation, and the likelihood that straightness errors will cause the drill to miss the target hole in deep-drilling applications. Drilling through inclined exit surfaces also results in unbalanced loading, which can cause excessive vibration, drill chipping, burr formation, and straightness errors. If such holes are unavoidable, it is advisable to drill them through relatively thick sections of the part (i.e., at depths greater than two drill diameters), so that the initial hole can act as a bushing and support the drill during exit.

For large batch or mass production, multidiameter holes should be designed to be manufactured with step or combination drills rather than discrete drills and counterbores. Specifically, the diameter of such holes should decrease in a stepwise fashion with hole depth, all steps in the hole should have a minimum axial length generally greater than the step diameter, and the difference in the diameters of adjacent steps should not exceed 50% of the larger diameter (Ref 5). As noted above, interior steps with increased diameter (Fig. 3a) require use of a boring bar after drilling and should be avoided, especially when they are greater than three times the drill diameter below the surface. Stepped and combination tools are particularly attractive for CNC equipment because they eliminate tool changes. Cost analyses can be used to determine the conditions under which the additional cost of a stepped or combination tool over discrete standard tools is justified (Ref 5).

Lists of simple design rules often state that blind holes should be avoided (Ref 3). It is often difficult to remove chips from blind holes in parts manufactured on vertical spindle machines, and deep blind holes should be avoided when using such equipment. Blind holes should also be avoided when drilling magnesium parts, because fines that can result in a fire hazard may be generated during spindle reversal at the bottom of such holes. When drilling materials other than magnesium on horizontal spindle equipment, however, the preference for through versus blind holes is not as easily justified. In these applications, chip removal does not present as serious a problem, and the burr formation and additional tool wear generated by vibration and feed surging at exit make through hole drilling less attractive.

Application of Design-for-Machining Rules

Application of the design-for-machining rules requires cooperation between part designers and manufacturing engineers. Ideally this cooperation should begin early in the design process, with the designer consulting with a manufacturing engineer from the intended manufacturing site periodically to develop an understanding of the type and capabilities of the equipment and tooling available at the site, similar information on proposed new equipment purchases, facilities issues, and operating policies that can influence production decisions.

For simple parts, only one designer and a shop foreman or process engineer may be responsible for the part design and manufacturing plan, and communication issues generally do not arise. In large organizations or for complex parts, additional personnel may be involved on both sides. Complicated parts intended for high-volume production may be designed by a team of designers and engineers, and the manufacturing process, tooling, and materials-handling plans may be developed by different manufacturing engineers. In these cases, periodic DFM workshops may be useful in ensuring that input is received from all parties and that decisions are rapidly communicated. These workshops seldom focus on machining issues alone, but also cover manufacturability concerns in casting, assembly, and other operations.

The rules themselves can be applied through various mechanisms (Ref 9). Commonly, they are applied as rules of thumb or incorporated into best practice or design-rule books tailored to the business needs and capabilities of a particular organization. When formal DFM workshops involving a number of participants are held, computer programs such as those described in the section "Computer Aids" in this article can also be used as a supplement to group discussion to help apply the rules in a structured fashion.

Computer Aids

The design-for-machining rules discussed above can be incorporated, at least in simplified forms, into computer programs to aid in automating the design process. These programs may also include tool path generation, functional requirement evaluation, and economic analysis modules to help quantify some of the trade-offs often required to apply the rules, although some subjective input is required to adapt algorithms to the specific application, production volume, and type of equipment under consideration. Two basic types of programs can be used for this purpose: computer-aided process planning (CAPP) and design-for-manufacturing or -manufacturability (DFM) programs. A detailed review of the literature on such programs is beyond the scope of this article; this section briefly describes the structure and typical uses of such programs.

Computer-aided process planning programs (Ref 10–13) work with CAD systems to extract the relevant geometric features of a part and produce a process sequence or set of tool paths that is optimal in some sense. Two steps are involved in this approach: feature extraction and optimization. In the feature-extraction stage, the CAD data for the part are analyzed to identify and classify the features that require machining. The extraction algorithm is specific to the CAD system used and particularly to how data are stored and whether the data are parameterized or not. Many systems can be used to classify features. Based on the extraction results, order and precedence constraints are established, and feasible tool paths are generated. The initial tool paths are then refined using an optimization algorithm until a solution that is optimal according to some criterion is achieved. The simplest criterion is minimum processing or cutting time, although if tool life and economic analyses are performed, a minimum cost criterion can also be used.

Design-for-machining rules and concepts can be applied in principle with CAPP programs as heuristics based on data-extraction results or as constraints based on identified features. Also, if a breakdown of processing time or cost is output as a function of individual features, those features that result in the maximum time or cost can be identified and considered for redesign. The second approach is more straightforward and easily applied.

Computer-aided process planning programs are most easily applied to production with CNC equipment and medium to large production volumes. It is an optimization approach that is normally used at a relatively late stage of the design process when a complete initial design is available.

Design-for-Manufacturability programs (Ref 9, 14) are artificial intelligence programs written specifically to apply DFM rules. General-purpose DFM programs include modules for assembly, stamping, and other processes as well as machining. Because desirable machining practices vary depending on the volume of production and the machine tools available, it is difficult to write a widely applicable general-purpose design-for-machining module. Some large companies have proprietary in-house codes used to apply design-for-machining rules in a manner tailored to their business operations.

A DFM program typically has an input module and an analysis module. Data input is not as automated as in CAPP programs; rather than reading required geometric information from a CAD file, part features and dimensions must generally be input manually according to some format and classification scheme. This is partly because DFM programs are intended to be applied at an earlier stage of the design process (when no complete CAD model of the part may be available), and partly because additional subjective information, such as the perceived relative machinability of various materials or the relative penalty associated with given undesirable features, is often required.

Once data are input, the analysis module is used to compute a relative machinability score for the design as entered. The algorithm used to compute the score varies from program to program, but in general the score depends on the complexity of the design and the penalties associated with difficult-to-machine materials or features. In DFM workshops, some rough estimate of the machining cost can also be computed (e.g., using a spreadsheet) for the given design. The output of the program is a detailed breakdown of components of the score due to individual features, which often clearly identifies the feature(s) most responsible for complexity or excessive cost.

Unlike CAPP programs, DFM programs are used for comparison rather than formal optimization. Usually several design alternatives are compared to a benchmark design, and based on the DFM score the best design is chosen and refined. For complex parts, the process may be repeated at various stages of the design (e.g., at an early stage and before fabrication of the first prototype). Design for manufacturability programs can be used for parts manufactured on either CNC or dedicated production equipment. They are well suited for designing complex parts for mass production and are currently more widely used than CAPP programs in these applications.

Related information is provided in the article "Design for Manufacture and Assembly" in this Volume.

Design-for-Machining Examples

Example 1: Redesign of a Shaft Support Bracket. Figure 7 shows an initial design of a shaft support bracket. This part was designed to be bolted to a mating housing wall to provide support and lubrication for a long shaft. The features to be machined include the shaft bore, an oil hole for lubrication, and mounting holes for dowel pins and bolts. To prevent binding of the shaft, the diameter of the bore must be machined accurately, and the location of the bore center with respect to the dowel pin hole centers must be held to a close tolerance.

The part, made of nodular cast iron, was to be machined in high volumes on a horizontal spindle CNC machining center. The critical tolerances dictated that the dowel holes be machined first using short drills, and that the bore be produced by a single-point boring bar without refixturing the part.

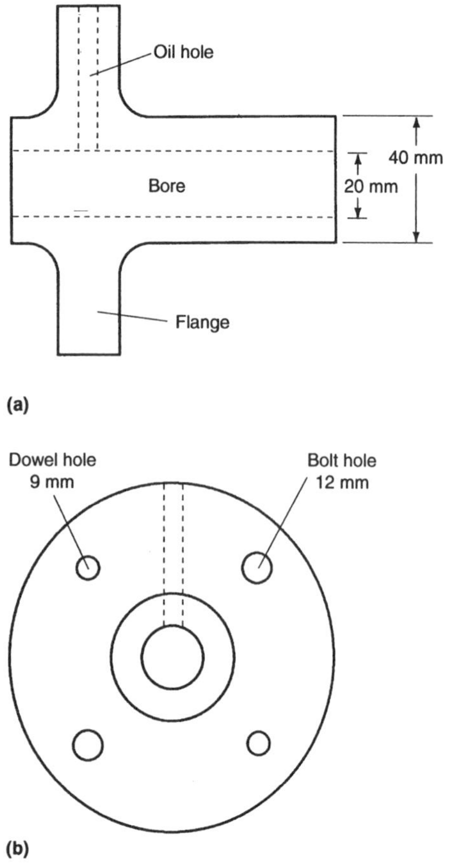

Fig. 7 Side (a) and end (b) views of the initial design for a shaft support bracket

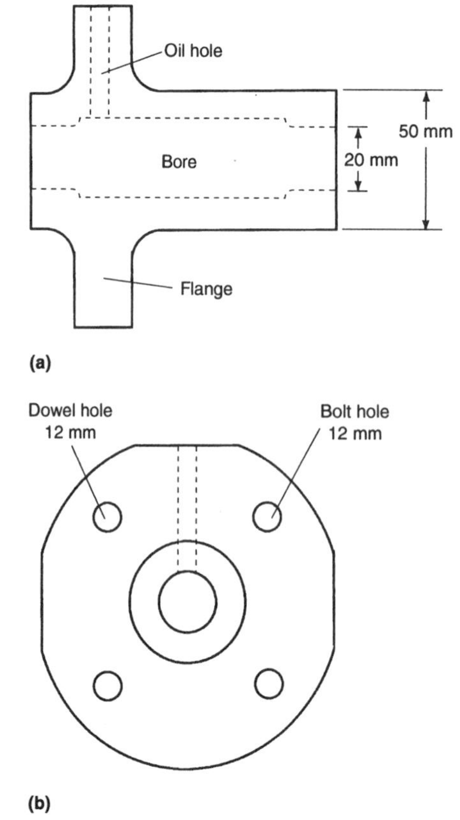

Fig. 8 Side (a) and end (b) views of a shaft support bracket redesigned to simplify machining

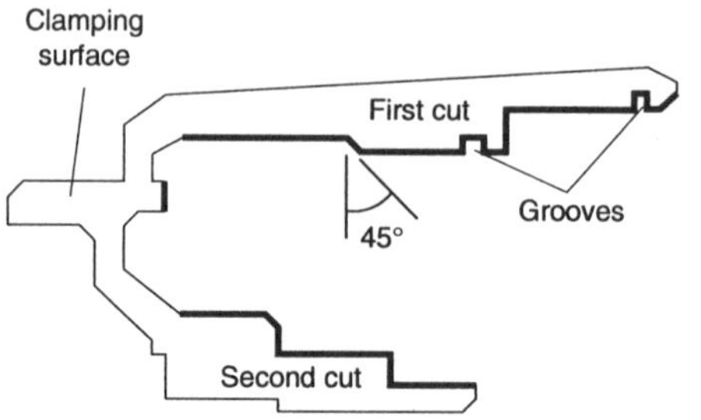

Fig. 9 Cross section of the initial design for a rotor housing

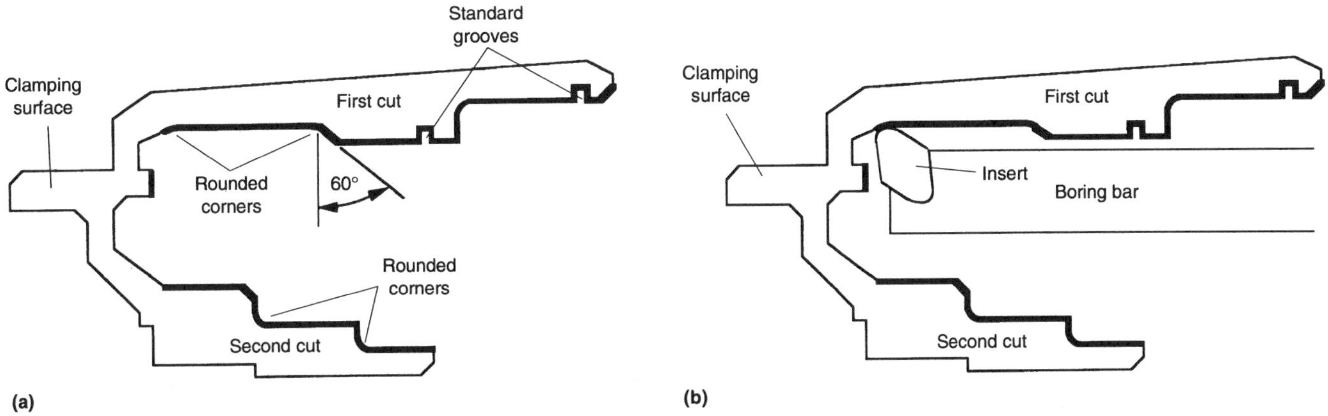

Fig. 10 (a) Cross section of a rotor housing redesigned to simplify machining. (b) Access to internal features using a boring bar with a standard 55° boring insert

The initial design presented a number of difficulties from a machining viewpoint. Different diameters were used for the dowel and bolt holes, so that a tool change would be required to produce these features. This would increase processing time and reduce the location accuracy of the holes relative to one another. The bore was long (roughly 100 mm, or 4 in.) and would require a boring step of significant length. The oil hole was also long (roughly 50 mm, or 2 in.) relative to its diameter and would require a long processing step. Finally, there is no obvious way to fixture the part automatically; presumably, it would be held in a vise on the outer surfaces of the flange, which are contoured and would not present flat clamping surfaces. To prevent rotation of the part due to the torques involved in the boring operations, a high clamping pressure would be required; because the clamping stresses are transmitted through the part, this could result in distortion of the bore and an out-of-roundness error upon unclamping.

The part was redesigned to address these difficulties as shown in Fig. 8. Three significant changes were made. First, the diameters of the dowel and bolt holes were standardized to eliminate the tool change. Second, the casting was changed so that the center of the bore had a larger diameter than the ends. This increased the mass of the part, but reduced the length of the bore, which would have to be machined to roughly 40 mm (1.6 in.). In addition, the depth of the oil hole was reduced by 5 mm (0.2 in.), and the hole no longer exited on a machined portion of the bore, eliminating a possible exit burr condition. Finally, flat surfaces were cast on three faces of the flange to provide clamping pads and to eliminate a contoured entry surface when drilling the oil hole. The revised part could easily be fixtured by clamping in a vise on the flat surfaces parallel to the oil hole; by mounting the vise on an angle plate on the machine table, the hole could be drilled horizontally from the side of the vise. A lower clamping pressure could be used because the flats provide a positive stop that would resist rotation. The flat surface provided for the oil hole entry also reduced the hole depth by an additional 5 mm (0.2 in.).

With these changes, the processing time required to machine these features was reduced from 173 to 119 s, a reduction of more than 33%; in addition, the design changes should improve part quality by simplifying achievement of the critical tolerances and by permitting the use of reduced clamping pressures.

Example 2: Redesign of a Rotor Housing. Figure 9 shows a cross section of the initial design of a die-cast aluminum rotor housing to be machined in several steps on CNC turning centers. In the operation shown, internal surfaces are to be cut using boring and grooving tools. To minimize axial tolerances between normal surfaces, two long boring passes (labeled "first cut" and "second cut" in Fig. 9) are planned. In addition, two grooves for retaining rings are also required.

The boring passes are difficult to plan for the initial design. There are several internal sharp corners; in addition, there is an internal angled cut at 45° to the part axis which cannot be accessed by a standard tool that will clear other internal surfaces. The grooves also have different axial widths. For the initial design, therefore, two grooving tools, two standard boring tools, and additional special tools to reach the internal sharp corners would be required to make the desired cuts.

Figure 10(a) shows a revised design that simplifies the machining. Upon consultation, it was determined that the dimensions of the grooves could be standardized, eliminating the need for one of the grooving tools. The internal sharp corners were replaced by radiused corners to permit machining with a standard boring insert. Finally, the initial 45° angled cut was replaced with a 60° angled cut that could be produced with a standard 55° boring insert mounted in a standard –5° lead boring bar as shown in Fig. 10(b). In the revised design, the required cuts can be made with two standard boring tools and a single grooving tool, saving at least three tool changes. Because the special tools required to machine the sharp internal corners would wear rapidly, the revised design also results in increased tool life and improved part quality.

REFERENCES

1. O.W. Boston, *Metal Processing*, 2nd ed., Wiley, 1951, p 1–8
2. J.G. Bralla, *Design for Excellence*, McGraw Hill, 1996, p 46–47
3. C.V. Starkey, *Engineering Design Decisions*, Edward Arnold, London, 1992, p 178–179
4. R. Bakerjian, ed., Chapter 11, *Tool and Manufacturing Engineer's Handbook*, Vol VI, *Design for Manufacturability*, 4th ed., Society of Manufacturing Engineers, 1992
5. D.A. Stephenson and J.S. Agapiou, Chapters 2, 11, and 13, *Metal Cutting Theory and Practice*, Marcel Dekker, 1996
6. *Machining*, Vol 16, *ASM Handbook*, ASM International, 1989
7. Bar Products Group, American Iron and Steel Institute, *Steel Bar Product Guidelines*, Iron and Steel Society, Warrendale, PA, 1994, p 164–166
8. J.S. Agapiou, An Evaluation of Advanced Drill Body and Point Geometries in Drilling Cast Iron, *Trans. NAMRI/SME*, Vol 19, 1991, p 79–89
9. H.W. Stoll, Tech Report: Design for Manufacture, *Manuf. Eng.*, Vol 100 (No. 1), 1988, p 67–73
10. H.A. ElMaraghy, Evolution and Perspectives of CAPP, *CIRP Ann.*, Vol 42 (No.2), 1993, p 1–13
11. L. Alting and H. Zhang, Computer Aided Process Planning: The State of the Art Survey, *Int. J. Prod. Res.*, Vol 26, 1989, p 999–1014
12. S.K. Gupta and D.S. Nau, Systematic Approach to Analysing the Manufacturability of Machined Parts, *Comput.-Aided Des.*, Vol 27 (No. 5), 1995, p 323–342
13. F.G. Mill, J.C. Naish, and C.J. Salmon, Design for Machining with a Simultaneous-Engineering Workstation, *Comput.-Aided Des.*, Vol 26 (No. 7), 1994, p 521–527
14. G. Boothroyd, Product Design for Manufacture and Assembly, *Comput.-Aided Des.*, Vol 26 (No. 7), 1994, p 505–520

Design for Joining

K. Sampath, Concurrent Technologies Corporation

JOINING is an important manufacturing activity employed in assembling parts to make components. The individual parts of a component meet at *joints*. Joints primarily transmit or distribute forces generated during service from one part to the other parts of an assembly. A joint can be either temporary or permanent. Commonly, five joint types are used in the joining of parts: butt, tee, corner, lap, and edge (Fig. 1).

The selection of an appropriate design to join parts is based on a concurrent understanding of several considerations related to product and joining process. Product-related considerations include codes and standards, fitness for service, aesthetics, manufacturability, repairability, reliability, inspectability, safety, and unit cost of fabrication. Considerations related to joining process include material types and thicknesses, joint (part) geometry, joint location and accessibility, handling, jigging and fixturing, distortion control, productivity, and initial investment. Additional considerations include whether the joint is fabricated in a shop or at a remote site, possibilities for premature failure, and containment in case of a catastrophic failure (this is applicable, for example, to components subjected to nuclear radiation).

The term *joint design* emphasizes designing of a joint based on product-related considerations for meeting structural design requirements. The design or selection of appropriate joint type is determined primarily from the type of service loading. For example, butt joints are preferred over tee, corner, lap, or edge joints in components subjected to fatigue loading. The specific joint design aspects, such as the size, length, and relative orientation of the joint, are based on stress calculations that are derived from an evaluation of service loads, properties of materials, properties of sections, and appropriate structural design requirements. An ideal joint is one that effectively transmits forces among the joint members and throughout the assembly, meets all structural design requirements, and can still be produced at minimal cost (Ref 1). Individual articles in various Sections of this Volume specifically address design of parts or components based on an understanding of several product-related considerations vis-à-vis appropriate structural design requirements.

The term *design for joining* refers to creating a mechanism that allows the fabrication of a joint using a suitable joining process, at minimal cost. In this context, design for joining emphasizes how to design a joint or conduct a joining process so that components can be produced most efficiently and without defects. This involves selection and application of good design practices based on an understanding of process-related manufacturing aspects such as accessibility, quality, productivity, and overall manufacturing cost. This article provides a brief description of various joining processes, a summary of good design practices from a joining process standpoint, and several examples of selected parts and joining processes to illustrate or highlight the advantages of a specific design practice in improving manufacturability.

Joining Processes

Joining processes include mechanical fastening, adhesive bonding, welding, brazing, and soldering. Mechanical fastening and adhesive bonding are often (but not always) used to produce temporary or semi-permanent joints, while welding, brazing, and soldering processes are used to provide permanent joints. Mechanical fastening and adhesive bonding usually do not cause metallurgical reactions. Consequently, these methods are preferred when joining dissimilar combinations of materials, and for joining metal-matrix, ceramic-matrix, and polymer-matrix composites that are sensitive to metallurgical phase changes or polymerization reactions.

Mechanical Fastening (Ref 2). The selection and satisfactory use of a particular fastener are dictated by the design requirements and conditions under which the fastener will be used. Consideration must be given to the purpose of the fastener, the type and thickness of materials to be joined, the configuration and total thickness of the joint to be fastened, the operating environment of the installed fastener, and the type of loading to which the fastener will be subjected in service.

Threaded fasteners are considered to be any threaded part that, after assembly of the joint, may be removed without damage to the fastener or to the members being joined.

Rivets are permanent one-piece fasteners that are installed by mechanically upsetting one end.

Blind fasteners are usually multiple-piece devices that can be installed in a joint that is accessible from only one side. When a blind fastener is being installed, a self-contained mechanism, an explosive, or other device forms an upset on the inaccessible side.

Pin fasteners are one-piece fasteners, either solid or tubular, that are used in assemblies in which the load is primarily shear. A malleable collar is sometimes swaged or formed on the pin to secure the joint.

Special-purpose fasteners, many of which are proprietary, such as retaining rings, latches, slotted springs, and studs, are designed to allow easy, quick removal and replacement and show little or no deterioration with repeated use.

Adhesive Bonding (Ref 3). An adhesive is a substance (usually an organic or silicone polymer) capable of holding materials together in a functional manner by surface attachment. The

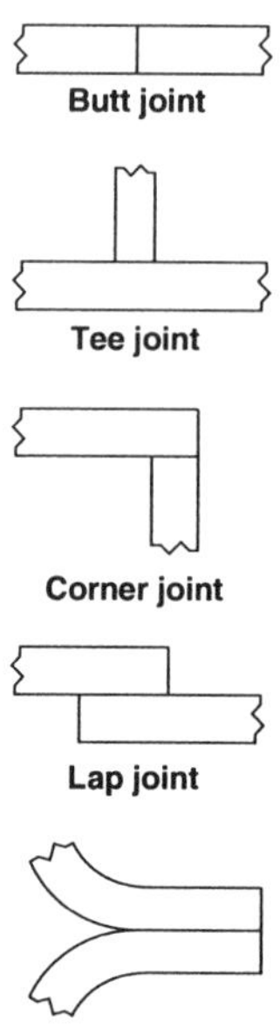

Fig. 1 Types of joints. Source: Ref 1

capability of holding materials together is not an intrinsic property of a substance but, rather, depends on the context in which that substance is used. Two important, basic facts about adhesive materials are that a substance called an adhesive does not perform its function independent of a context of use and that an adhesive does not exist that will bond "anything to anything" with (implied) equal utility.

The major function of adhesives is for mechanical fastening. Because an adhesive can transmit loads from one member of a joint to another, it allows a more uniform stress distribution than is obtained using a mechanical fastener. Thus, adhesives often permit the fabrication of structures that are mechanically equivalent or superior to conventional assemblies and, furthermore, have cost and weight benefits.

Although the major function of adhesives is to fasten, sometimes they are also required to seal and insulate. Formulations that are good electrical and/or thermal conductors are also available. Further, adhesives prevent electrochemical corrosion in joints between dissimilar metals and resist vibration and fatigue. In addition, unlike mechanical fasteners, adhesives do not generally change the contours of the parts that they join.

Detailed information on adhesives and adhesive bonding is available in *Adhesives and Sealants,* Volume 3 of the *Engineered Materials Handbook* published by ASM International.

Welding includes both fusion welding and solid-state welding processes.

Fusion welding processes involve localized melting and solidification and are normally used when joining similar material combinations or materials belonging to the same family (e.g., joining one type of stainless steel with another type). Figure 2 illustrates the type of welds commonly used with fusion welding processes such as arc welding (Ref 1).

Fusion welding processes also include electron beam welding and laser welding. These two welding processes require precise joint gap and positioning. Joint designs and clearances that overwhelmingly trap the beam energy within the joint cavity are preferred for increasing process efficiency. Figure 3 shows preferred and non-recommended joint designs for electron beam welding (Ref 4). When joining thick sections, the preferred joint designs allow the weld metal to freely shrink without causing cracking.

Solid-state welding processes preclude melting and solidification and therefore are suitable for joining dissimilar materials. However, the process conditions may allow solid-state metallurgical reactions to occur in the weld zone. When metallurgical reactions occur, they can either benefit or adversely affect the properties of the joint. From a metallurgical perspective, the application of both fusion welding and solid-state welding processes must be evaluated using appropriate weldability testing methods for their ability to either *recreate* or *retain* base metal characteristics across the joint. These weldability evaluations combine material, process, and procedure aspects to identify combinations that would provide a weld joint with an acceptable set of properties.

Solid-state welding processes also have special joint design or part cross-section requirements. For example, continuous-drive and inertia friction welding processes require that one of the parts exhibit a circular or near-circular cross section. Diffusion bonding is another solid-state welding process that allows joining of a variety of structural materials, both metals and nonmetals. However, diffusion bonding requires an extremely smooth surface finish (8 μm) to provide intimate contact of parts, a high temperature, and a high pressure, first to allow intimate contact of the parts along the bond interface, followed by plastic deformation of the surface asperities (on a microscopic scale), and second to promote diffusion across the bond interface. The need to apply pressure while maintaining part alignment imposes a severe limitation on joint design.

Alternatively, when exceptional surface finish is difficult to achieve, a metallurgically compatible, low-melting interlayer can be inserted between the parts to produce a transient liquid phase on heating. On subsequent cooling this liquid phase undergoes progressive solidification, aided by diffusion across the solid/liquid interfaces, and thereby joins the parts. This process has characteristics similar to those of the brazing process.

Brazing (Ref 5, 6) is a process for joining solid metals in close proximity by introducing a liquid metal that melts above 450 °C (840 °F). A sound brazed joint generally results when an appropriate filler alloy is selected, the parent metal surfaces are clean and remain clean during heating to the flow temperature of the brazing alloy, and a suitable joint design that allows capillary action is used.

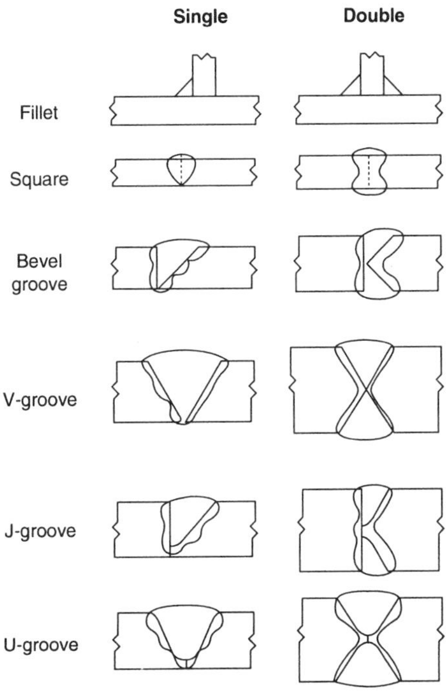

Fig. 2 Types of welds. Source: Ref 1

Strong, uniform, leakproof joints can be made rapidly, inexpensively, and even simultaneously. Joints that are inaccessible and parts that may not be joinable at all by other methods often can be joined by brazing. Complicated assemblies comprising thick and thin sections, odd shapes, and differing wrought and cast alloys can be turned into integral components by a single trip through a brazing furnace or a dip pot. Metal as thin as

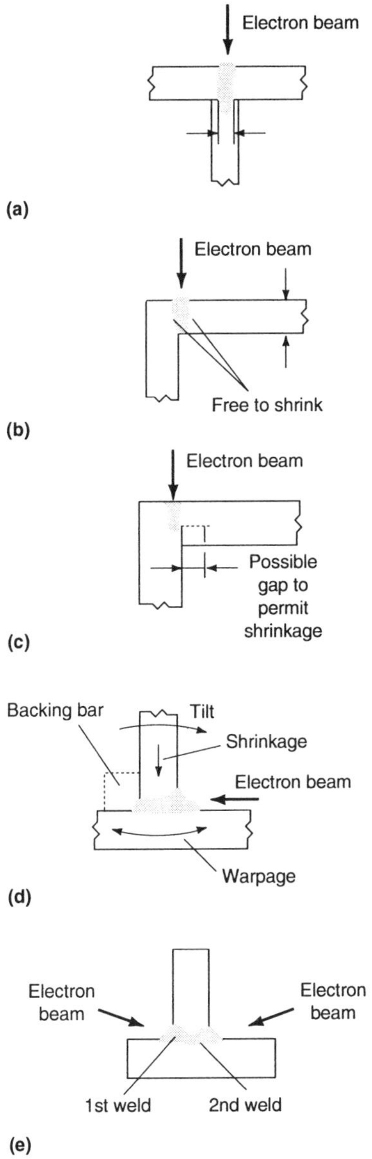

Fig. 3 Optimum versus least desirable weld configurations. (a) Not recommended—maximum confinement of molten metal, minimum joining cross section (arrows); wastes beam energy for melting, nonfunctional metal. (b) Most favorable—volume of melt not confined; maximum joining cross section (arrows). (c) Not recommended—maximum confinement of melt (unless gap is provided); joining cross section less than plate cross section. (d) Most favorable—minimum constraint and confinement of melt; minimum internal stresses; warpage can be offset by bending prior to welding; tilt can be offset by location of T-arm at less than 90° to base prior to welding. Fillet obtained by placing wire in right corner and melting it with the beam. (e) Not recommended—two successive welds; second weld is fully constrained by the first weld and shows strong tendency to crack. Source: Ref 4

0.01 mm (0.0004 in.) and as thick as 150 mm (6 in.) can be brazed.

Brazed joint strength is high. The nature of the interatomic (metallic) bond is such that even a simple joint, when properly designed and made, will have strength equal to or greater than that of the as-brazed parent metal.

The mere fact that brazing does not involve any substantial melting of the base metals offers several advantages over other welding processes. It is generally possible to maintain closer assembly tolerances and to produce a cosmetically neater joint without costly secondary operations. Even more important, however, is that brazing makes it possible to join dissimilar metals (or metals to ceramics) that, because of metallurgical incompatibilities, cannot be joined by traditional fusion welding processes. (If the base metals do not have to be melted to be joined, it does not matter that they have widely different melting points. Therefore, steel can be brazed to copper as easily as to another steel.)

Brazing also generally produces less thermally induced distortion, or warping, than fusion welding. An entire part can be brought up to the same brazing temperature, thereby preventing the kind of localized heating that causes distortion in welding.

Finally, and perhaps most important to the manufacturing engineer, brazing readily lends itself to mass production techniques. It is relatively easy to automate, because the application of heat does not have to be localized, as in fusion welding, and the application of filler metal is less critical. In fact, given the proper clearance conditions and heat, a brazed joint tends to "make itself" and is not dependent on operator skill, as are most fusion welding processes.

Automation is also simplified by the fact that there are many means of applying heat to the joint, including torches, furnaces, induction coils, electrical resistance, and dipping. Several joints in one assembly often can be produced in one multiple-braze operation during one heating cycle, further enhancing production automation.

Soldering (Ref 7) is a joining process by which two substrates are bonded together using a filler metal (solder) with a liquidus temperature that does not exceed 450 °C (840 °F). The substrate materials remain solid during the bonding process. The solder is usually distributed between the properly fitted surfaces of the joint by capillary action.

The bond between solder and base metal is more than adhesion or mechanical attachment, although these do contribute to bond strength.

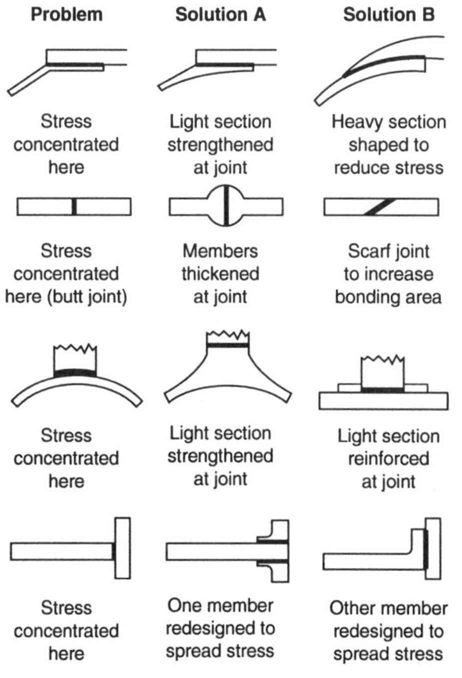

Problem
Solution A
Solution B
Stress concentrated here
Light section strengthened at joint
Heavy section shaped to reduce stress
Stress concentrated here (butt joint)
Members thickened at joint
Scarf joint to increase bonding area
Stress concentrated here
Light section strengthened at joint
Light section reinforced at joint
Stress concentrated here
One member redesigned to spread stress
Other member redesigned to spread stress

Fig. 4 Design of a brazed joint to redistribute stress. Source: Ref 8

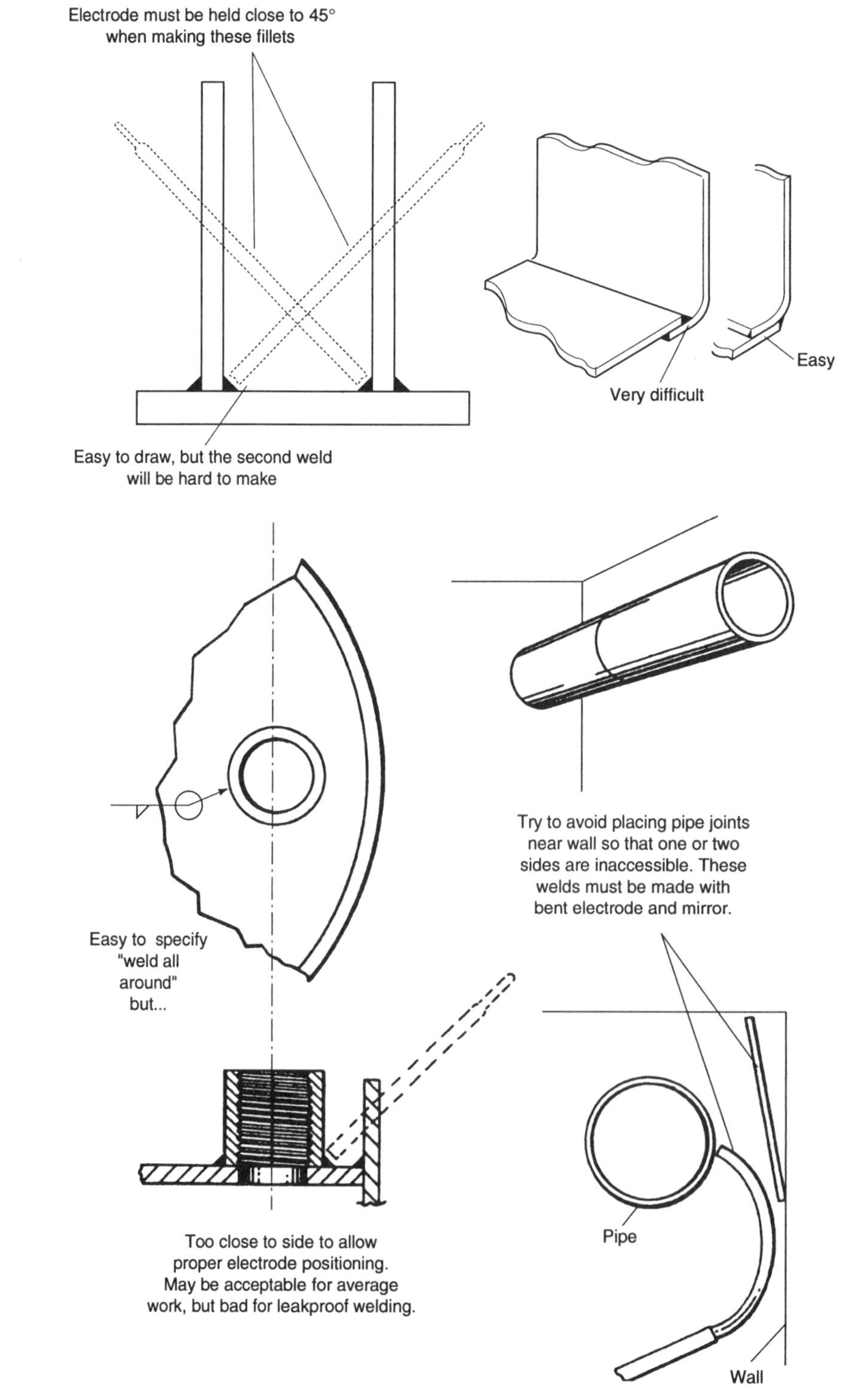

Electrode must be held close to 45° when making these fillets
Easy to draw, but the second weld will be hard to make
Easy
Very difficult
Try to avoid placing pipe joints near wall so that one or two sides are inaccessible. These welds must be made with bent electrode and mirror.
Easy to specify "weld all around" but...
Too close to side to allow proper electrode positioning. May be acceptable for average work, but bad for leakproof welding.
Pipe
Wall

Fig. 5 Effect of joint location on accessibility. Source: Ref 1

Rather, the essential feature of the soldered joint is that a metallurgical bond is produced at the filler-metal/base-metal interface. The solder reacts with the base metal surface and wets the metal by intermetallic compound formation. Upon solidification, the joint is held together by the same attraction, between adjacent atoms, that holds a piece of solid metal together. When the joint is completely solidified, diffusion between the base metal and soldered joint continues until the completed part is cooled to room temperature. Mechanical properties of soldered joints, therefore, are generally related to, but not equivalent to, the mechanical properties of the soldering alloy.

Mass soldering by wave, drag, or dip machines has been a preferred method for making high-quality, reliable connections for many decades. Correctly controlled, soldering is one of the least expensive methods for fabricating electrical connections.

Advantages of brazing and soldering include the following:

- The joint forms itself by the nature of the flow, wetting, and subsequent crystallization process, even when the heat and the braze or solder are not directed precisely to the places to be joined.
- The process temperature is relatively low, so there is no need for the heat to be applied locally, as in welding.
- Brazing and soldering allow considerable freedom in the dimensioning of joints, so that it is possible to obtain good results even if a variety of components are used on the same product.
- The brazed or soldered connections can be disconnected if necessary, thus facilitating repair.
- The equipment for both manual and machine brazing/soldering is relatively simple.
- The processes can be easily automated, offering the possibility of in-line arrangements of brazing/soldering machines with other equipment.

Basic Design Considerations

When designing a joint, one should initially consider manufacturability of the joint, whether at a shop or at a remote site. For example, consider the need for a high integrity, high-performance joint between two dissimilar materials such as a low-carbon steel and an aluminum alloy. If this joint has to be produced at a remote site, the available choice of joining processes is extremely limited. A viable alternative would be to produce at a shop a transition piece involving the two dissimilar materials. Using controlled process conditions at a shop, one could produce a high-integrity transition piece using one of the solid-state welding processes. The selection of the appropriate solid-state welding process would depend on joint (part) geometry. A transition joint between a plate and a pipe is best produced using a friction welding process, while a joint between two large plate surfaces is best produced using explosive bonding. Because these joining processes preclude melting and solidification, they provide high-integrity joints free from porosity or solidification-related defects. Transition pieces so produced could be used at a remote site to make similar metal joints between component parts with no undue quality assurance or quality control concerns.

Good Design Practices

A joint must be designed to benefit from the inherent advantages of the selected method of joining. For example, braze joints perform very well when subjected to shear loading, but not when subjected to pure tensile loading. When using a brazing process to join parts, it would be beneficial to employ innovative design features that would convert a joint subjected to tensile loading to shear loading. For example, use of butt-lap joints instead of butt joints can provide a beneficial effect in flat parts and tubular sections.

Joints must be designed to reduce stress concentration. Sharp changes in part geometry near the joint tend to increase stress concentration or notch effects. Smooth contours and radiused corners tend to reduce stress concentration effects. Figure 4 shows a number of ways to redistribute stresses in a brazed joint (Ref 8).

When determining appropriate joint designs, one should initially consider standard or recommended joint designs. In practice, several standard joint designs may be suitable for producing a joint. Subtle or innovative features could be added to the recommended joint designs to improve productivity through mechanization or automation, to enhance joint performance, and to ensure safety.

Orientation and Alignment. Design features that promote self-location and maintain the relative orientation and alignment of component parts save valuable time during fit-up and enhance the ability to produce a high-quality joint. For example, operations involving furnace brazing or diffusion bonding with interlayers benefit from such a type of joint design, because they also require pre-placement of the brazing filler or the interlayer in the joint.

The pin-socket type of temporary joints in modern electrical, telephone, and computer connectors allow temporary joining of cables in only one way. These joint designs strongly discourage any inadvertent misalignment or wrong orientation of the connectors and thereby eliminate a variety of hazards. The snap-on interlocking features in twisted, threaded, or non-threaded adapter joint designs, commonly used in children's toys, often allow the snapping sound of a latch to indicate the satisfactory completion of the joint and its safety for the intended use.

Jigging and fixturing can also be used to maintain relative orientation of parts. When necessary, the fixturing devices should be designed for the least possible thermal mass and pin-point or knife-line contact with the parts. Fixtures of low thermal mass and minimal contact with the parts reduce the overall thermal load during joining. Further, arc welding processes generally allow higher deposition rates when joining is performed in the downhand position, where gravity effects tend to support a large volume of molten weld metal at the joint region. When joining parts that exhibit a nonplanar joint contour, positioning

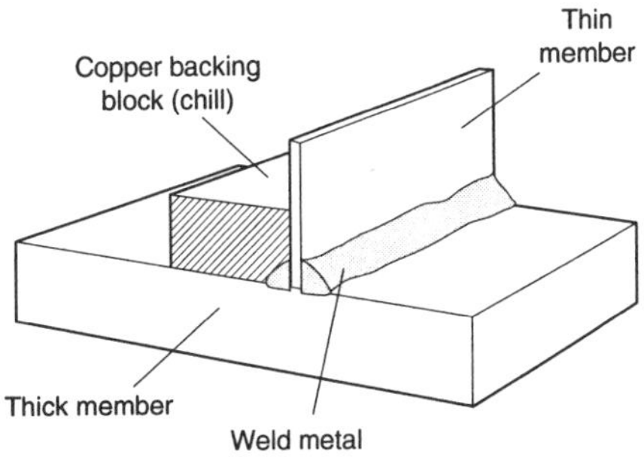

Fig. 6 Use of copper backing bar as a chill to minimize heat sink differential. Source: Ref 9

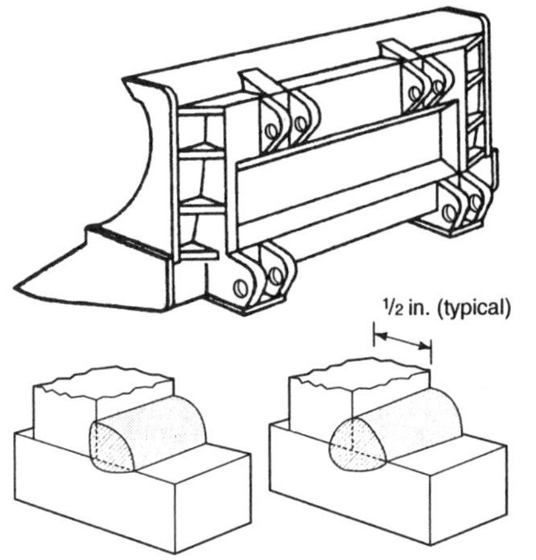

Auxiliary gas shielded FCAW	
Joint type	Corner
Weld type	Fillet; some groove
Weld size	6.35 to 12.7 mm (1/4 to 1/2 in.)
Welding position	Flat; horizontal
Number of passes for fillet welds:	
For 6.35 to 7.94 mm (1/4 and 5/16 in.), flat position	1
For 2.03 mm (3/8 in.), horizontal position	2
For 0.44 and 12.7 mm (7/16 and 1/2 in.), flat position	1
For 0.44 and 12.7 mm (7/16 and 1/2 in.), horizontal position	3
Shielding gas	Carbon dioxide at 1 m^3/h (35 ft^3/h)
Electrode	2.4 mm (3/32 in.) diam flux cored wire
Current	450 A
Voltage	30–32 V
Electrode feed	494 cm (206 in.) per min

Fig. 7 Bulldozer blade and comparison of joint penetration (and actual throat depth) of fillet welds made by shielded metal arc welding and by auxiliary gas shielded flux cored arc welding (FCAW). Low-carbon base metal; low-carbon steel filler metal. Source: Ref 11

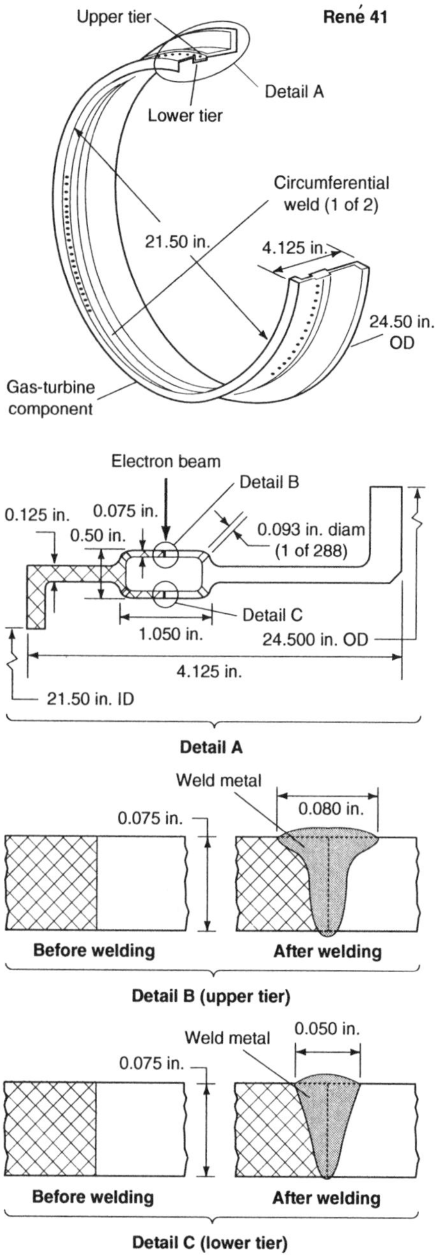

Joint type	Circumferential, two-tier butt
Weld type	Square groove
Machine capacity	150 kV at 40 mA
Gun type	Fixed
Maximum vacuum	1.3×10^{-3} Pa (10^{-5} torr)
Fixtures	Bolted end plates; rotating positioner
Preheat	None
Welding power	125 kV at 9.3 mA
Welding vacuum	1.3×10^{-2} Pa (10^{-4} torr) minimum
Working distance	31 cm (12 in.)
Beam focal point	Midway between tiers(a)
Welding speed (at 0.42 rpm):	
Upper tier	76 cm/min (30 in./min)
Lower tier	73 cm/min (28.7 in./min)
Beam oscillation	None
Number of passes	One, plus 30° downslope
Postweld heat treatment	Aged 16 h at 760 °C (1400 °F)

(a) Indicated machine setting. See text of example for explanation of beam focal point.

Fig. 8 Tiered welds made simultaneously using the electron beam welding process. Source: Ref 12

equipment can be used to continuously manipulate the parts so that the welding is performed in the downhand position. In such cases, the design of the joint and fixtures should be complementary to the positioning equipment used, and it should not interfere with the functioning of the positioning equipment.

Joint Location and Accessibility. From a structural integrity standpoint, joint locations should be chosen such that they are not in regions subjected to maximum stress. Concurrently, joints must be placed in locations that will allow operators to readily make the joints using the selected method of joining. Figure 5 illustrates the effect of location of a joint on accessibility (Ref 1). Limited accessibility can reduce the overall quality of the joint, decrease productivity, or both. Invariably, limited accessibility to produce joints also limits accessibility to perform nondestructive evaluation of joint quality, either during the time the joint is made or afterwards.

Weld joint designs employ bevel angles and root openings to enhance accessibility to the welding torch (or electrode) and provide adequate weld penetration. The best bevel angles provide adequate accessibility while reducing the amount of weld metal required to complete the joint. Currently, computer-based software tools are available to facilitate the selection of a weld joint for minimizing the amount of weld metal. Use of such computer-based selection of joint designs increases welding productivity (joint completion rate), improves quality, and reduces overall fabrication cost, but such designs must be used only when they are consistent with structural design requirements. For this reason, codes such as the ASME Boiler and Pressure Vessel Code, Section IX: Welding Qualifications, and ANSI/AWS D1.1 Structural Welding Code provide flexibility to a welding manufacturer (fabricator) to select or change weld joint design for fabrication, but they require the manufacturer to qualify the welding procedure to meet design performance requirements whenever changes are made to a previously qualified, nonstandard weld joint design. In recent years, the use of narrow-gap gas-metal arc welding and submerged arc welding techniques in the place of conventional welding techniques for welding thick-section pressure vessel steels has contributed significantly to increased weld joint completion rates.

Unequal Section Thickness. When constituent members of an assembly exhibit unequal section thicknesses, modifications to the recommended joint designs will be necessary for a variety of technical reasons (Ref 9), but mainly to provide a smooth flow of stress patterns through the unequal sections. When making a fillet weld using an arc welding process, if thicknesses of the members are not greatly different, directing the arc toward the thicker member may produce acceptable penetration. However, special designs for joining will be required when the components to be welded exhibit a large heat sink differential (difference in heat-dissipating capacities). When a thick member is joined to a thin member, the welding heat input (mainly current) needed to obtain a good penetration into the thick member is sometimes too much for the thin member and results in undercutting of the thin member and a poor weld. Similarly, if the proper amount of current for the thin member is used, the heat is insufficient to provide adequate fusion in the thick member, and again a poor weld results. Too little heat input can also cause underbead cracking in certain structural materials.

A widely applicable method of minimizing heat sink differential is to place a copper backing block (Fig. 6) against the thin member during fusion welding (Ref 9). The block serves as a chill, or heat sink, for the thin member. The block can be beveled along one edge so that it can be used when horizontal fillet welds are deposited on both sides of a thin member. Copper backing bars or strips are made in a variety of shapes and sizes to dissipate heat as needed. Often some experimentation and proof testing are required to obtain the optimum backing location and design. Another way to obtain equalized heating and smooth transfer of stress where unequal section thicknesses are being welded is to taper one or both members to obtain an equal width or thickness at the joint. Commonly, when two pipes of dissimilar internal diameter and wall thickness are to be joined, a convenient way is to introduce a "reducer" between the two pipes. One end of the reducer will have the same size and wall thickness as the larger pipe, while the other end of the reducer will have the same size and wall thickness as the smaller pipe.

Distortion Control. Design of an appropriate weld joint can also help reduce welding-related distortion. Fusion welding processes employ localized melting and solidification to join component parts, which can result in excessive thermal strains. These thermal strains are dependent on the type of material, the welding process, and the welding procedure. Thermal strains produced by fusion welding processes can cause residual stresses and distortion, leading to transverse and angular shrinkage. Reducing the overall length of the weld or the amount of weld metal that needs to be deposited to complete a joint reduces both residual stresses and distortion. For example, intermittent welding instead of continuous welding reduces the overall length of a weld. Similarly, the use of a double-V groove instead of a single-V groove results in the reduction of the amount of weld metal and minimizes transverse shrinkage (Ref 10). Further, the amount of angular shrinkage is strongly influenced by the ratio of the weld metal in the top and the bottom sides of the plate. To minimize the out-of-plane distortion in fillet welded joints, efforts should be directed to using the minimum size of the welds that is consistent with strength considerations.

Case Histories

The case histories in this section illustrate the selection of design for joining, based on one or more of the good design practices discussed above. The case histories were compiled from Volume 6 of the 9th Edition of *Metals Handbook.*

The examples highlight the application of industrial engineering principles to design practices pertaining to fusion welding, diffusion welding, and brazing as they are widely used in the manufacture of components. The basic principles of these design practices are also applicable to other methods of joining such as mechanical fastening, adhesive bonding, soldering, and so on.

Example 1: Process Selection Obviates the Need for Joint Preparation. Bulldozer blades (Ref 11) were assembled from several low-carbon steel components that had relatively thick sections, generally 13 mm (½ in.) or more (Fig. 7). Most welds were 6.4 to 13 mm (¼ to ½ in.) fillet welds. A few were groove welds.

To produce fillet welds involving thicker plates, flux cored arc welding (FCAW) was selected in preference to shielded metal arc welding (SMAW), for three reasons: deeper joint penetration, permitting the use of smaller fillets without decreasing the strength of the joint; higher deposition rate; and greater visibility of the arc to the welder, resulting in a better weld. The difference in joint penetration for the two processes is shown in Fig. 7. Use of SMAW to achieve the same level of penetration as FCAW would have required beveling the edge of the vertical member, or an additional number of weld passes. Details for FCAW are given in the table accompanying Fig. 7.

Example 2: Two-Tier Welding. Electron beam welding provided a unique solution to the problem of designing and fabricating a component of an aircraft gas turbine engine (Ref 12). As indicated in Fig. 8, the component consisted of a cylinder with an external flange on one end, an internal flange on the other, and a tubular annulus between. The components were assembled by welding the trough shape ends of the two subcomponent cylinders by a single two-tier circumferential weld.

The components were made of René 41, for service at elevated temperature. The chief welding objectives were to obtain sound welds and to avoid distortion of the part, especially the alignment of the 288 holes located in the annulus less than 13 mm (½ in.) from the joints. The holes had to be drilled before welding because they could not be deburred if drilled after welding.

Arc welding was rejected as a joining method because it would have been necessary to use internal chills with gas backing in the annulus to minimize distortion and avoid atmospheric contamination. Electron beam welding not only met the basic requirements but also made both joints simultaneously, even though the two joints were separated by approximately 13 mm (½ in.), as shown in detail A of Fig. 8.

Fixturing was relatively simple. The joints were accurately machined square and the components were assembled between two aluminum plates fitted over the flanges. The plates were connected and forced together by bolts located inside the inner flange. This fixture was then mounted on the faceplate of a welding positioner in the vacuum chamber so that the part would rotate with its axis horizontal. The electron beam gun was in a fixed, overhead position.

Success of the two-tier welding operation depended on careful control of part alignment, beam alignment, beam focal point, power adjustment, and travel speed. The joints for the upper-tier and the lower-tier welds had to rotate in the same vertical plane, although by direct viewing they could not be observed simultaneously. In addition, beam impingement on the joint of the lower-tier weld could be verified only by emergence of the weld bead from the underside, or by sectioning of the test pieces.

Alignment of the part for true horizontal-axis rotation was done with the aid of a precision level (sensitivity of 0.0004 mm per centimeter, or 0.0005 in. per foot) and precision spacer blocks placed on the face of the flange (62 cm, or 24.5 in. OD). Beam alignment was done by centering the beam spot on the reticle of the scope and moving the joint to this position. Beam focal point, beam power, and travel speed were adjusted by trial and error on test components until a satisfactory welding procedure was established. The final settings are shown in the table with Fig. 8. By adjusting the beam focus for an indicated setting midway between the two tiers, the weld shapes of Fig. 8, details B and C, were obtained.

The mushroom-head shape of the upper-tier weld was caused by the defocused condition of the beam at that point, while the somewhat oversize root reinforcement resulted from the excess of power needed to penetrate to and through the lower tier. The relatively narrow face of the lower-tier weld, as well as the narrowing of the weld in its progress through the joints, was explained as an effect of a charged plasma that surrounded and refocused the beam on its passage through the material. The plasma, having a net negative charge, repelled the beam electrons, causing the beam to constrict and to change its focal point. The net result on the lower tier was to produce a weld closely approaching the contour of a normal single-thickness weld made with a tight, surface-focused beam. Thus, the indicated focal-point setting was more virtual than real.

Both welds were satisfactory as to soundness and shape. Because of lack of access to the interior of the joint, weld spatter and undercutting were of concern, and the spatter associated with penetration of the upper tier was a problem. Most of the particles were loosened with pipe cleaners and were flushed out with solvent at high pressure. The few small particles that remained were judged to be acceptable after radiographic examination. Undercutting was not a problem. The René 41 material was capable of withstanding considerable excess beam power, which was especially important in making the upper-tier weld.

Components produced using the two-tier welding procedure met all test requirements.

Example 3: Design for Diffusion Bonding. Design for diffusion bonding must facilitate intimate contact of parts and local (microscopic) plastic deformation to promote the formation of a joint. The following example illustrates the use of an innovative joint design that exploits the differences in thermal expansion between parts and tooling (the tooling material exhibits a higher strength than the part material at the bonding temperature) to promote intimate contact, incrementally increase pressure at the joint interface, cause localized plastic deformation, and thereby produce a diffusion bond.

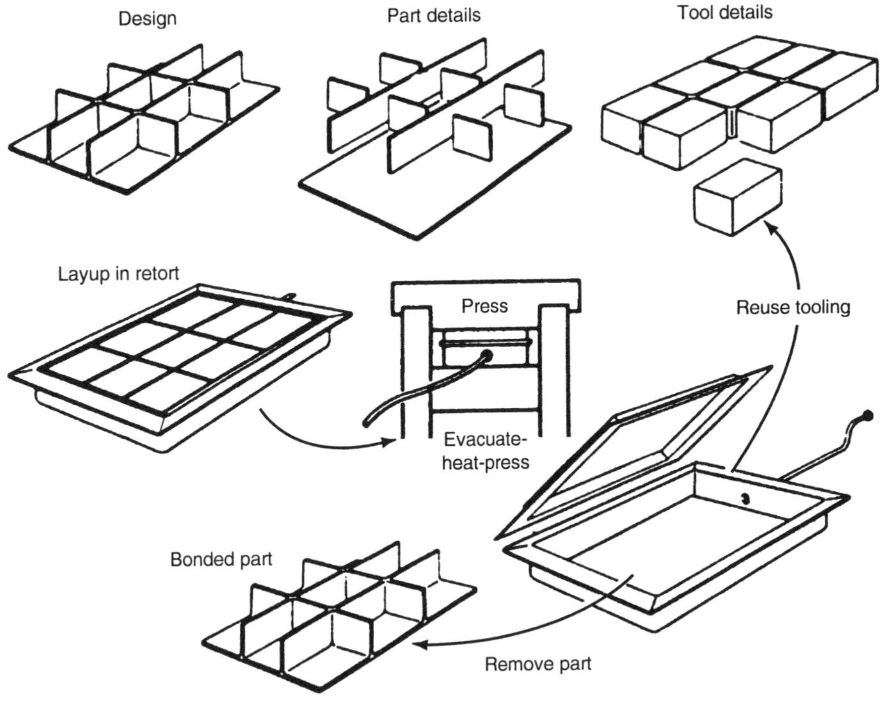

Fig. 9 Processing sequence during diffusion bonding of a titanium part using stainless steel tooling. Source: Ref 13

Figure 9 shows diffusion bonding of a titanium part using tooling blocks and spacers of 22-4-9 stainless steel (Ref 13). Initially, the parts and the tooling are fitted into a welded retort made of 1.6 mm (0.063 gage) muffler steel and conforming to the shape of the part. The retort contains an end rail of 22-4-9 spanning the entire width. This end rail contains machined grooves that allow the air to escape when a vacuum pump is turned on. Similar 7.6 cm (3 in.) thick, 22-4-9 plates line the bottom, walls, and opposite end of the retort, and one covers the filled retort before a lid is welded to the retort to seal the container and make it leakproof. To prevent sticking to titanium, all 22-4-9 tooling is surface oxidized by oven baking at 760 °C (1400 °F) for 4 h. During the loading of the retort, titanium slip shims are inserted to separate the tooling blocks slightly from the titanium parts. Later, these shims are removed to create a vacuum path for the air to escape during evacuation of the retort. A steel tube is used to connect the retort with a vacuum pump.

Initially, the container is evacuated to 0.13 Pa (10^{-3} torr) vacuum and checked for leaks. A higher vacuum, below 1.3×10^{-4} Pa (10^{-6} torr), is then obtained. The leakproof container is placed within reusable ceramic blocks, which in turn are covered by steel plates containing thermocouples. The entire assembly is placed in a bonding press. The ceramic blocks that press against the top and bottom of the retort contain heating coils that bring the entire assembly to about 927 °C (1700 °F). The ceramic blocks on the sides and ends of the retort transmit heat and pressure to the assembly during bonding. A press is used to apply about 13.8 MPa (2000 psi) pressure on the retort in all directions, and the retort is held at the bonding temperature and pressure for about 2 to 12 h. Thermal expansion of the titanium part against the relatively rigid stainless steel tooling allows intimate contact of the titanium parts across the joint line, and it facilitates localized plastic deformation and the formation of a diffusion bond. Following the bonding cycle, the entire assembly is cooled slowly and dismantled, and then the retort is cut open to retrieve the diffusion-bonded titanium part. Generally, the tooling is reused, while the retort (made from cheap muffler steel) is scrapped.

Example 4: Revision of Joint Design to Reduce Cost. The longitudinal butt joints in 6.1 m (20 ft) long sections of SA-106, grade B carbon steel pipe (Ref 14) used for power-boiler headers were originally designed as shown at lower left in Fig. 10. With this design, the root pass and the second pass were made by SMAW using a backing bar, and then the weld was completed by submerged arc welding (SAW).

To reduce the cost, the joint design was revised to that shown at lower right in Fig. 10. This permitted making the root pass by gas-tungsten arc welding (GTAW), using a consumable insert, instead of by SMAW with a backing bar. Then, as with the original joint design, the second pass was made by SMAW, and the weld was completed by SAW. The SMAW process was used for the second pass to provide a deposit thick enough to ensure against melt-through by SAW. The improved joint design and change in welding procedure resulted in a 25% saving in cost (material, labor, and overhead) per foot of seam welded.

Example 5: Submerged Arc Welding of a Large Piston. The large hydraulic-jack piston shown in Fig. 11 was assembled by welding three low-carbon steel castings (head, piston body, and seat) at girth joints (Ref 15). When similar smaller pistons with wall thicknesses of 7.6 to 12.7 cm (3 to 5 in.) had been assembled by SMAW, about one welded joint in eight was found to be defective and had to be reworked.

Because of the experience with the pistons with 7.6 to 12.7 cm (3 to 5 in.) wall, it was decided to use SAW to assemble four large pistons, in which

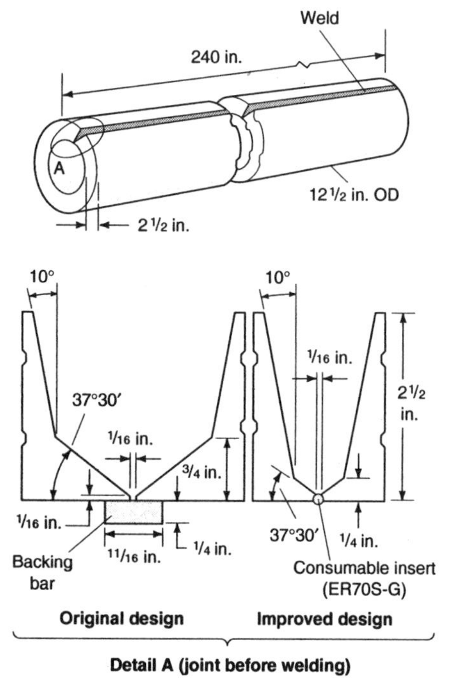

Welding conditions for improved design

Welding process:	
Root pass	GTAW
Second pass	SMAW
Remainder	SAW
Power supply	
GTAW, SMAW	200 A transformer-rectifier(a)
SAW	600 A motor-generator(b)
Edge preparation	Machined
Preheat	121 °C (250 °F) min
Filler metal (low-carbon steel)	
GTAW (argon shielding)	ER70S-G consumable insert
SMAW	E7018
SAW	EL12
Power setting:	
GTAW	90 A (DCEN); 12 V
SMAW	121 A (DCEP); 23 V
SAW	450 A (DCEP); 30 V
Interpass temperature	260 °C (500 °F) max
Postheat	Stress relieve at 621 ± 25 °C (1150 ± 25 °F)(c)

With the original joint design, the root and second passes were made by SMAW using a backing bar. Weld was then completed by SAW. With the improved joint design, the root pass was made by GTAW using a consumable insert (no backing), the second pass by SMAW, and the remainder by SAW. (a) With high-frequency start, and slope control. (b) Weld head on boom-type manipulator, workpiece supported on power and idler rolls for turning. (c) In furnace. 1 h per 25 mm (1 in.) of section

Fig. 10 Revision of joint design. Use of a consumable insert permitted change to a lower-cost method of welding boiler-header pipes. Carbon steel (SA-106, grade B; 0.30 max C) base metal; low-carbon steel filler metal. GTAW, gas-tungsten arc welding; SMAW, shielded metal arc welding; SAW, submerged arc welding. Source: Ref 14

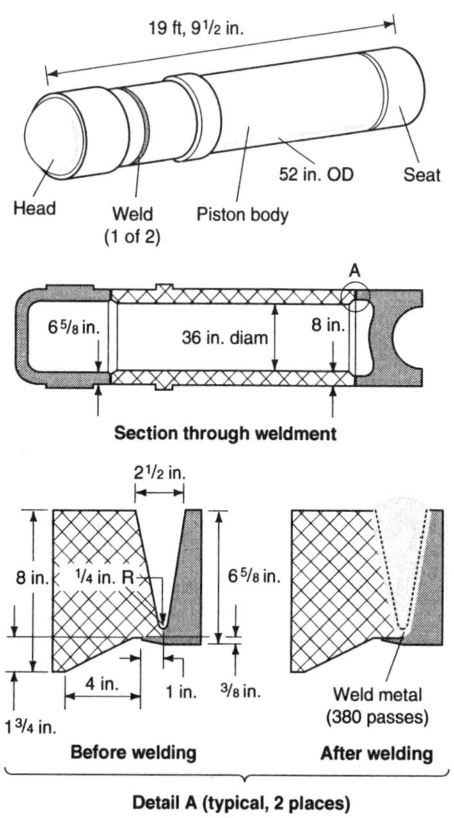

Conditions for SAW

Joint type	Circumferential butt
Weld type	Single-U-groove, integral backing
Joint preparation	Machining
Power supply	1000 A transformer
Wire feed	Fully automatic, constant speed
Welding head	Machine held, air cooled
Fixture	50 ton variable-speed roll
Auxiliary equipment	Exhaust fan, vacuum flux remover, positioning arm
Electrode wire	2.4 mm (3/32 in.) diam EL12(a)
Flux	F71(a)
Welding position	Flat (horizontal-rolled pipe position)
Number of passes	380
Current and voltage:	
Passes 1 through 3	700 A, ac; 38 V
Remaining passes	750 A, ac; 40 V
Preheat	204 °C (400 °F) (by torch)
Postheat (stress relief)	7 h at 600 °C (1115 °F), furnace cool to 315 °C (600 °F)
Welding speed	24 cm (9 1/2 in.) per min

(a) Electrode and flux yielded a weld deposit containing 0.12 C, 0.84 Mn, 0.72 Si, 0.018 S

Fig. 11 Large piston assembled by submerged arc welding (SAW). Low-carbon steel base metal; low-carbon steel filler metal (EL12). Source: Ref 15

a 19.2 cm (8 in.) wall was to be joined to a 16.8 cm (6⅝ in.) wall, using the joint design shown in detail A. The outside surfaces of the three castings to be welded were rough machined and the joints were prepared by machining. The joints were of the interlocking type (see Fig. 11, detail A) and provided support for the unwelded components during positioning on variable-speed welding rolls. Joint areas were preheated to 204 °C (400 °F) with the gas torches as the piston was rotated. The welds were made in 380 passes and were produced oversize and machined to size after magnetic-particle inspection and stress-relief. The welded pistons were stress relieved at 600 °C (1115 °F) for 7 h and furnace cooled to 315 °C (600 °F).

Each welded joint was ultrasonically inspected for a distance of 7.6 cm (3 in.) on each side of the weld. After inspection, the pistons were hydrostatically tested at 2 MPa (300 psi). There were no rejections. Production time for welding the large piston was 101 h, which was a considerable improvement over the production time of 212 h for the smaller pistons assembled by SMAW.

Example 6: Use of an Offset to Eliminate Backing Rings. A component of a heat-exchanger shell assembly (Ref 15) was initially made by SAW: a medium-carbon steel pipe cap with a wall thickness of 6.4 mm (¼ in.) was attached to a low-carbon steel pipe of the same wall thickness by means of a circumferential butt joint, supported and aligned by a backing ring, as shown in the "original design" in Fig. 12.

When it became apparent that the wall thickness of the pipe cap could be less than that of the pipe without adversely affecting service performance, the joint was rede-signed as a joggled lap joint (see "improved design," Fig. 12). The offset incorporated in the pipe for the redesigned joint took the place of the backing ring previously used and furnished a locating surface for the cap. The redesigned joint was made by SAW under the

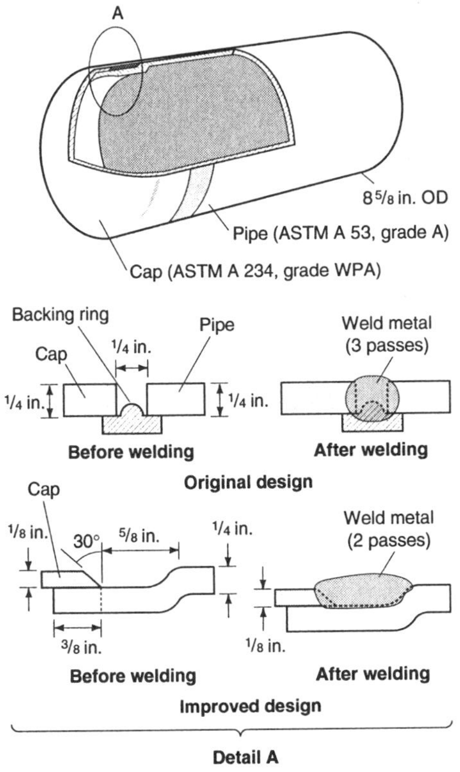

Joint type	Joggled lap
Weld type, original design	Square-groove, with backing ring
Weld type, improved design	Modified single-V-groove, with integral backing
Joint preparation:	
Original design	Backing ring machined
Improved design	Cap end machined, pipe end reduced
Electrode wire	3.2 mm (⅛ in.) diam EL12
Flux	F62
Welding position	Flat (horizontal-rolled pipe)
Welding voltage	25 to 26 V
Welding current	350 to 410 A (DCEN)
Welding speed	46 to 51 mm (18 to 20 in.) per min
Number of passes, original design	Three
Number of passes, improved design	Two
Power supply	40 V, 600 A transformer-rectifier (constant-voltage)
Fixturing	Chuck-type turning rolls; alignment clamps for tack welding

Fig. 12 Cap-to-pipe weldment. Low-carbon steel welded to medium-carbon steel; low-carbon steel filler metal (EL12). Source: Ref 15

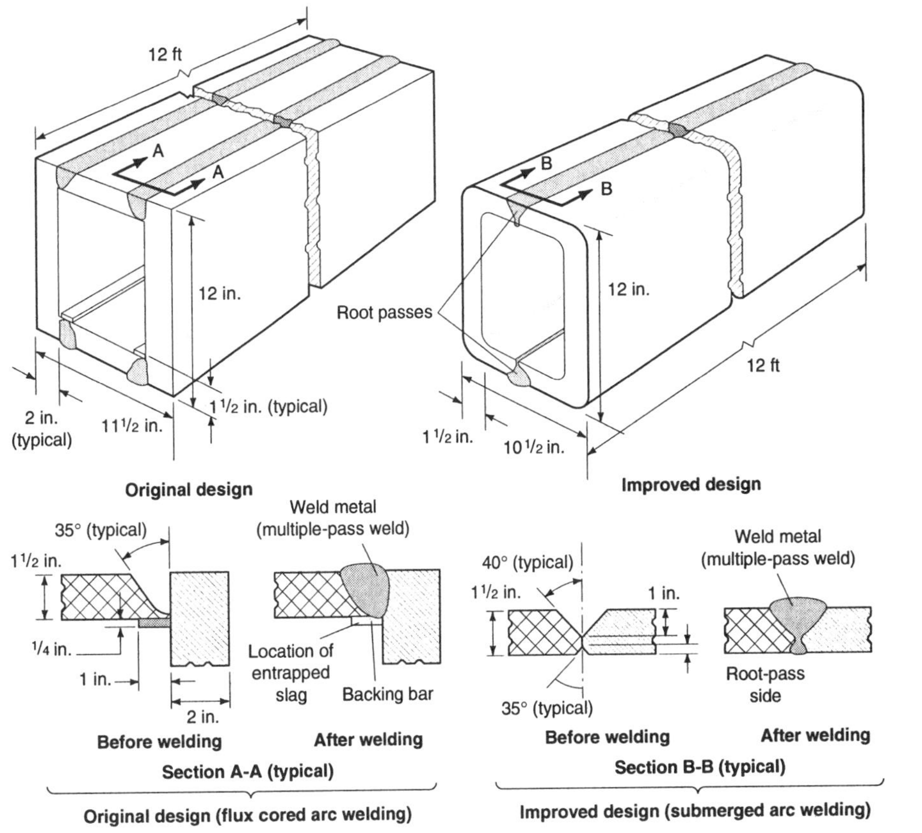

	Original design	Improved design
Welding process	Manual FCAW	Automatic SAW
Electrode	2.4 mm (³⁄₃₂ in.) diam flux cored wire	2.4 mm (³⁄₃₂ in.) diam solid wire
Flux	...	F72
Welding position	Flat	Flat
Root-pass welding conditions		
Welding current(a), A	375–425	460–480
Number of passes per joint	1	1
Welding speed, cm/min (in./min)	15 (6)	20 (8)
Filler-pass welding conditions		
Welding current(a), A	375–425	400–600
Deposition rate, kg/h (lb/h)	2.7 (6)	8.2 (18)
Number of passes per joint	7–8	5
Welding speed, cm/min (in./min)	25 (10)	56 (22)(b)

(a) Power supply for welding of both designs was an 80 V (open-circuit) transformer-rectifier. (b) Welding speed for the first filler pass was 71 cm/min (28 in./min).

Fig. 13 Submerged arc welding (SAW) setup for heat-exchanger header. Carbon steel, 0.35% max C (ASTM A 515, grade 70) base metal; carbon steel filler metals. FCAW, flux cored arc welding. Source: Ref 15 same conditions as those for the original joint, except that only two passes were required, rather than three.

Cost reduction was realized from eliminating the backing ring, from the savings in material resulting from the use of thinner pipe caps, and from eliminating one circumferential welding pass. The change in joint design led to a savings of approximately 35% in total factory cost. All joints were inspected visually and radiographically to check for full penetration and absence of slag inclusions. The rejection rate was less than 1%.

Example 7: Elimination of Backing Bars. A 3.7 m (12 ft) long header assembly for a large high-pressure heat exchanger (Ref 15) was manufactured to Section VIII, Division I, of the ASME Boiler and Pressure Vessel Code (Fig. 13).

As originally designed (see upper left in Fig. 13), for manual FCAW from the outside, the assembly consisted of four steel components and was welded at corner joints that incorporated backing bars, as shown in Fig. 13, section A-A. It was difficult to ensure a uniformly tight fit of the backing in the joint. Under radiographic inspection, slag was revealed that ran between the backing bars and the adjacent 38 mm (1½ in.) thick components.

The problem was eliminated by redesigning the header assembly for automatic SAW without backing bars. The redesigned assembly, shown in the upper right of Fig. 13, consisted of two 38 mm (1½ in.) thick channels formed in a press brake. The two components were welded at two longitudinal butt joints of the double-V-groove design (Fig. 13, section B-B).

For this improved design, the welding was done by the use of a boom-mounted automatic welding head. The formed channels were held stationary while the welding head was advanced along the joints. First, root passes were made along the inside grooves of the two joints, then filler passes were made along the outside grooves. After the root passes, the outside grooves were machined-out to sound metal before the filler passes were begun.

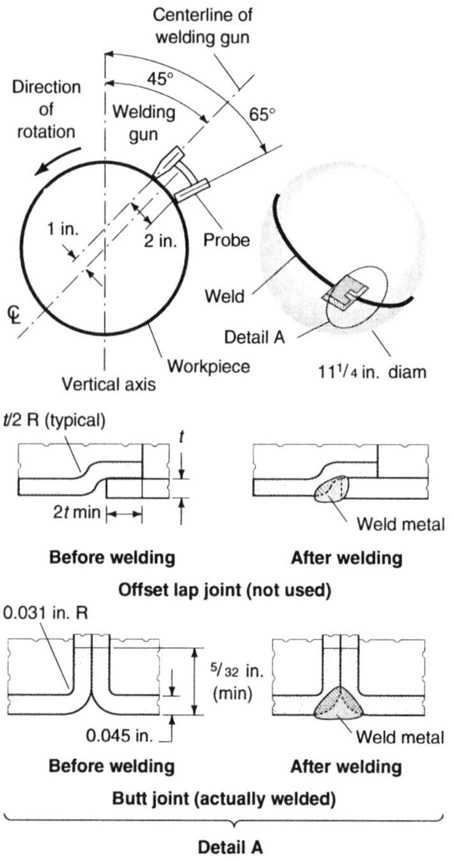

Joint type	Circumferential modified butt
Weld type	Single-flare V-groove
Power supply	300-A transformer-rectifier
Electrode wire(a)	0.162 mm (0.030 in.) diam ER70S-3
Welding gun	Mechanized, fixed, water cooled
Wire feed	Push-type motor, on welding gun
Current	170–190 A (DCEP)
Voltage	22–23 V
Shielding gas(b)	98% argon–2% oxygen, 1 m^3/h (35 ft^3/h)
Number of passes	1
Wire-feed rate	863 to 965 cm (340–380 in.) per min
Electrode extension	6.35 to 9.5 mm (¼–⅜ in.)
Welding speed	118 cm (46.6 in.) per min
Weld time per container	42 s

(a) Selection of wire wound to a large diameter eliminated need for wire straighteners and reduced leakage rate. (b) Argon of 99.999% purity from bulk-liquid holder

Fig. 14 Girth welded refrigerant container. Labor and tooling costs were reduced by use of a modified butt joint instead of an offset lap joint. Low-carbon steel (ASTM A 620) 0.045 in. base metal; low-carbon steel filler metal (ER70S-3). Source: Ref 16

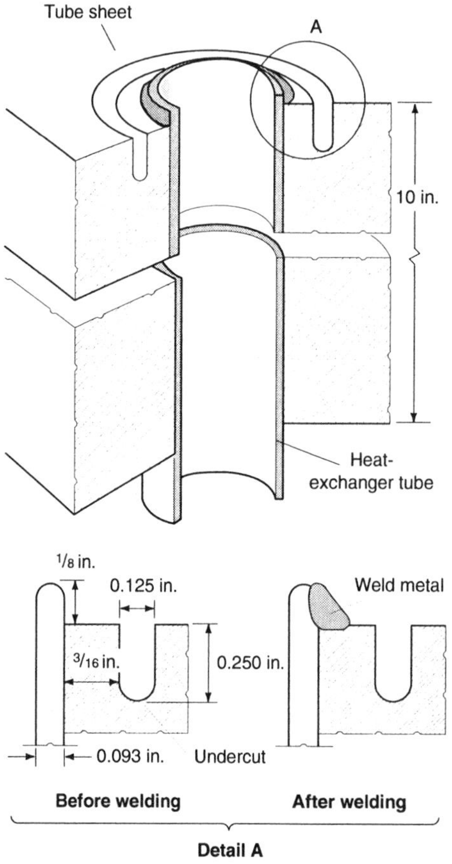

Fig. 15 Thick tube sheet with machined groove. Minimizes heat sink differential during welding of thin-walled heat-exchanger tube to the tube sheet. Low-carbon steel base metal; low-carbon steel filler metal. Source: Ref 9

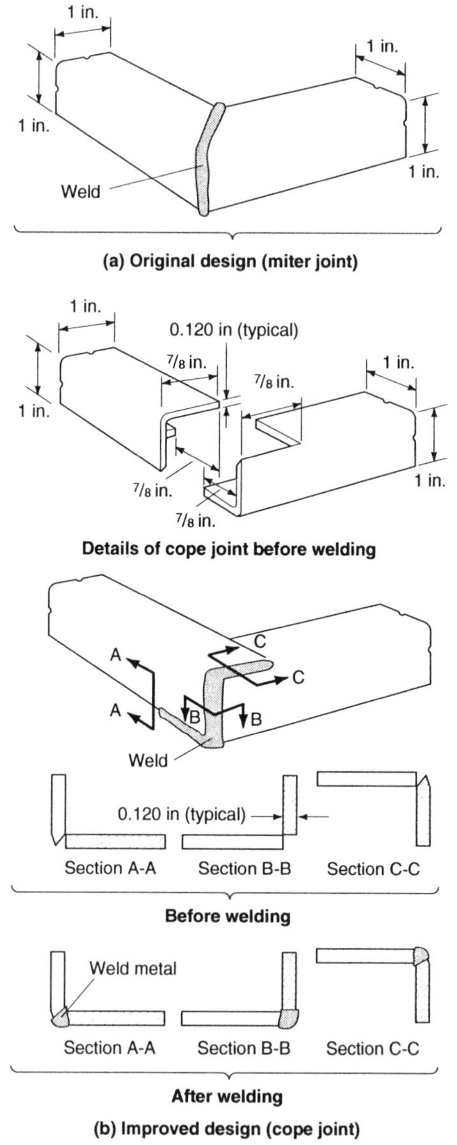

Joint type	Corner
Weld type	Fillet and V
Power supply	200 A, constant-voltage rectifier
Electrode wire(a)	0.76 mm (0.030 in.) diam ER70S-2
Wire feed	Constant feed
Current	80–85 A (DCEN)
Voltage	26 V
Shielding gas	75% argon–25% carbon dioxide, at 1.1 m^3/h (40 ft^3/h)
Wire-feed rate	76–254 cm (30–100 in.) per min
Welding speed	25 cm (10 in.) per min

(a) 0.04% C; triple deoxidized, with flash copper plating

Fig. 16 Corner section of a rectangular frame. A cope joint was substituted for a miter joint to improve dimensional control. Low-carbon steel base metal; low-carbon steel filler metal (ER70S-2). Source: Ref 17

A major benefit of the change to two-piece design was that only about one-third as many filler passes were required for the entire weldment (10 passes, as compared with 28 to 32 passes for the original four-piece design).

Example 8: Use of Modified Butt Joint to Save Tooling and Labor Costs. An offset lap joint (middle view in Fig. 14) is frequently used in welding the components of a variety of pressure cylinders and spheres. However, the use of this lap joint for welding the spherical refrigerant container shown in Fig. 14 would have resulted in high labor and tooling costs. The lip of the offset hemisphere would have caused interference in assembly, and additional tooling would have been needed to offset the lip.

The modified butt joint shown at the bottom in Fig. 14 allowed use of identical forming tools for both halves of the sphere and was the best compromise among weldability, tooling costs, and labor costs (Ref 16). To reduce labor costs still further, each welder operated two girth welding machines and thus was not able to observe the welding operation; therefore, automatic seam tracking was necessary.

The automatic seam tracking system consisted of two recirculating ball-screw cross-slides mounted at right angles and driven by reversible alternating current motors. A probe, which was mounted to move with the welding gun (see view at upper left in Fig. 14), sensed the location of the joint in relation to the gun. A movement of the probe tip caused the appropriate slide to bring the probe and gun back to the neutral position. The probe was mounted on a small screw-adjusted slide to provide quick and accurate adjustment of both the horizontal position of the welding gun and the distance between it and the work. At the end of the welding cycle, the probe and the welding gun were raised by the vertical slide. After the next assembly was in place, the probe and gun were lowered by the same means and welding was started automatically.

The hemispheres were held in a special welding machine (lathe) consisting of one fixture rotated by a continuously variable drive and a second fixture mounted on the tailstock. Both fixtures were mounted on air-operated slides. Thus, the parts were held together and rotated under the welding gun. When the gun was retracted after completing the weld, the air cylinders separated, releasing the welded workpiece.

The operator loaded the hemispheres in the machine and pushed a button to close the fixtures. The operator then had an option of using automatic start, whereby the weld started as soon as the welding gun was in position, or manual start, whereby the position of the gun could be observed, corrections could be made (if required), and the weld could be started by pushing a button. While one container was being welded, the operator loaded and started a second machine. After each container had been welded, the operator checked it for visible defects. The containers requiring repair welding were set aside, and those with no visible defects were transferred on a conveyor to a testing area.

The workpiece was a disposable refrigerant container with a water capacity of 11.6 kg (25.5 lb), produced under a special permit that specified two types of pressure tests. Each welded container was tested by subjecting it to 2.1 MPa (300 psi) internal air pressure while within a heavy steel safety chamber. The pressure in the container was then reduced to 0.7 MPa (100 psi), the chamber was opened, and the sphere was forced under water to check for leaks. If repairs were required, the spheres were resettled after repair. A destructive test was required on one

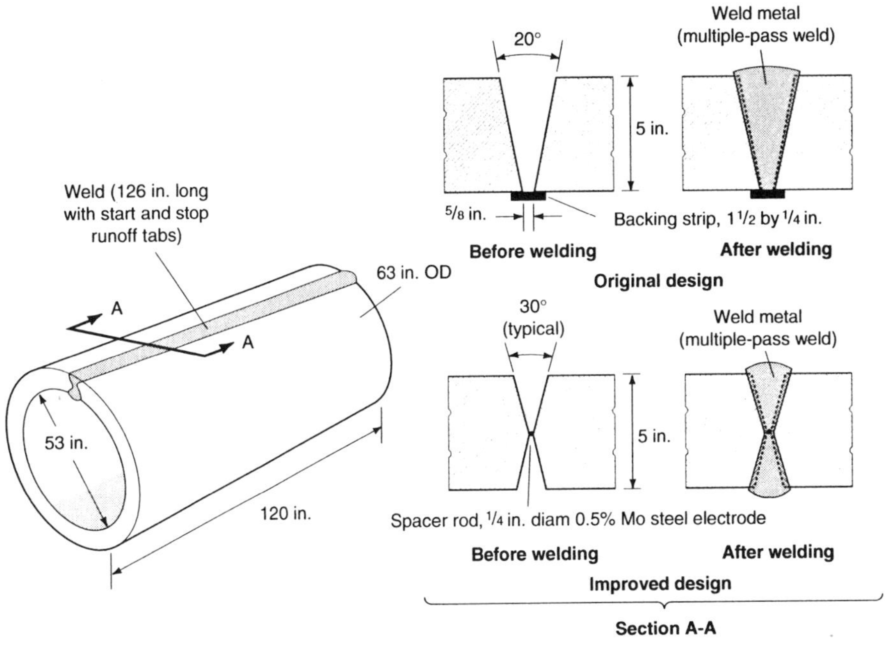

Welding conditions for both joint designs	
Weight of electrode and flux deposited per hour	7.1 kg (15.6 lb)
Weight of electrode and flux deposited per foot of weld:	
Original design (single-V-groove)	11.8 kg (26 lb)
Improved design (double-V-groove)	6.4 kg (14 lb)
Deposition efficiency	98%
Length of weld (including runoff tabs at ends)	3.2 m ($10\frac{1}{2}$ ft)
Time for installing and removing backing strip (original design)	12 h
Joint type	Butt
Weld types	Single-V-groove (original); double-V-groove (improved)
Welding position	Flat(a)
Arc starting	Touch and retract
Preheat	79 °C (175 °F), then 121 °C (250 °F) (propane torch)
Interpass temperature	260 °C (500 °F)
Postheat	621 ± 25 °C (1150 ± 25 °F) (furnace), 1 h/25 mm (1 in.) of section
Root passes (SMAW):	
Power supply	300 A motor-generator
Electrode	4.8 mm ($\frac{3}{16}$ in.) E7018
Current and voltage	250 A (DCEP); 24 V
Intermediate passes (single-electrode SAW):	
Power supply	900 A motor-generator
Electrode wire	5.6 mm ($\frac{7}{32}$ in.) diam 0.5% Mo steel(b)
Current and voltage	700 A (DCEP); 30 V
Travel speed	48 cm (20 in.) per min
Final passes (tandem SAW)(c)	
Leading head:	
Power supply	900 A motor-generator
Electrode wire	5.6 mm ($\frac{7}{32}$ in.) diam 0.5% Mo steel(b)
Current and voltage	800 A; 30 V
Trailing head:	
Power supply	1000 A transformer
Electrode wire	2.4 mm ($\frac{3}{32}$ in.) diam 0.5% Mo steel(b)
Current and voltage	700 A, ac; 35 V
Travel speed	76 cm (30 in.) per min

(a) Workpiece supported on one power roll and one idler roll. (b) Electrode wire contained 0.11% C, 0.50% Mo, 0.85% Mn, and was used at a 1-to-1 ratio of wire to flux. (c) Tandem welding head was mounted on a boom-type manipulator.

Fig. 17 Submerged arc welding setup for steam-drum shell course. Base metal: carbon steel, 0.35% max C (ASTM A 515, grade 70), normalized. Filler metals: low-carbon steel (E7018) for root passes (shielded metal arc welding); 0.5% Mo steel for remaining passes (submerged arc welding). Source: Ref 15 container out of each lot of 1000, with a minimum of one per day (although, in practice, at least one container was tested from each machine during each shift). This test consisted of filling the container with water, connecting it to a high-pressure pump, and increasing the pressure until the container burst. The minimum bursting strength was 5.5 MPa (800 psi). Fewer than 5% of the spheres required weld repairs.

The guidance system caused some problems, primarily because of the maintenance required. Repair and adjustment of the probe switch were difficult, and improperly adjusted probes sometimes caused misplaced welds. Spare systems were available for replacement of defective probe units.

Although the guidance system added to the machine cost and caused maintenance problems, these disadvantages were soon canceled out by decreased welding costs. Satisfactory welds were difficult to produce manually, because the horizontal variance of the welding gun position had to be held to 0.8 mm (1/32 in.) to prevent melt-through. In addition, it would have been necessary for the manual operator to correct for differences in the heights of the weld seams, limiting the welder to operating one machine and thus doubling the labor cost.

The hemispheres were press formed and vapor degreased. No edge preparation or postweld finishing was done. Welding conditions are given in the table with Fig. 14.

Example 9: Joining Sections of Unequal Thickness. An application involving components of unequal section thicknesses (Ref 9) is the welding of heat-exchanger tubes having 2.4 mm (0.093 in.) wall thickness to a tube sheet as thick as 25.4 cm (10 in.). The usual method of avoiding difficulty is to cut a circular groove, 6.4 mm (1/4 in.) deep, in the upper surface of the tube (Fig. 15). By restricting heat transfer, this groove minimizes heat sink differential between the thin tube wall and thick tube sheet.

Example 10: Redesign of a Joint to Improve Dimensional Control. The corner-welded channel sections shown in Fig. 16 were parts of rectangular frames for data-processing machines (Ref 17). Originally, 45° miter joints were used (Fig. 16a), but dimensions after welding were unsatisfactory because of joint location and weld restraint.

To provide a more positive joint location with less weld restraint, cope joints (Fig. 16b) were substituted for the miter joints. Tolerances of ±0.25 mm (±0.010 in.) on length and width, and ±0.813 (±0.032 in.) on squareness, were met on channel sections welded with the improved joint design.

The channel sections were contour roll formed from 3.05 mm (0.120 in.) thick low-carbon steel strip. Pieces were cut to length by a cutoff die in a press. The cut lengths were also coped by a die in a press, and all parts were inspected. Tolerances on individual pieces were held to ±0.127 mm (±0.005 in.).

Example 11: Change in Joint Design to Reduce Distortion and Cost. Figure 17 shows a 305 cm (120 in.) long steam-drum shell course, roll formed with a welded longitudinal seam (Ref 15). Originally, the butt joint for this seam was of single-V-groove design and was welded with the use of a backing strip (see "original design" in section A-A in Fig. 17). Fit-up and removal of the backing strips were time-consuming operations, and welding from one side distorted the weldment.

The joint was changed to a double-V-groove design (shown as "improved design" in section A-A of Fig. 17). This change resulted in the need for much less weld metal; the need for a backing strip was eliminated; and distortion was reduced by sequential deposition of weld beads on the inside and outside of the joint. The amount of back gouging needed was less than that required to remove the backing strip from the single-V-groove weld. As a result of these improvements, electrode, flux, and labor costs were reduced by 46%, and the total cost of welding was reduced by 62%. Welding procedures and post-weld operations for the two designs are described below. For both designs, the shell courses were hot roll formed into a cylinder and descaled, and the joint grooves were flame cut.

Originally, the single-V-groove joint was preheated to 79 °C (175 °F) with a propane torch, the backing strip was installed, and the temperature of the joint was raised to 121 °C (250 °F). At least two root passes were made, using SMAW. This operation was followed by depositing six single-pass layers, each 3.2 mm (1/8 in.) thick, by single-electrode SAW. Tandem SAW was used to complete the weld, single-pass layers 3.2 mm (1/8 in.) thick being deposited to a weld level of 38 mm (1 1/2 in.), followed by two-pass (split) layers 3.2 mm (1/8 in.) thick. Then the backing strip was removed by air carbon arc gouging and grinding, and back welding was done, as required, to provide a flush joint.

In the improved design, the double-V-groove joint was also preheated in two stages (79 and 121 °C, or 175 and 250 °F) with a propane torch, except that instead of a backing strip being installed between stages, a spacer rod of 6.4 mm (1/4 in.) diameter 0.5% Mo steel electrode material was tacked in place and seal welded by SMAW. Shielded metal arc welding was used also for root passes. The first increment of single-electrode submerged arc welds consisted of eight 3.2 mm (1/8 in.) thick single-pass welds on the outside of the weldment. The workpiece was rotated 180°, and the joint was back gouged and ground to a radius of 6.4 to 9.5 mm (1/4 to 3/8 in.). The first increment of welding on the inside of the joint consisted of 3.2 mm (1/8 in.) thick single-pass welds to a 38 mm (1 1/2 in.) level, using single-electrode SAW. Then the workpiece was again rotated 180°, and the remainder of the outside welding was completed using tandem SAW to deposit two-pass (split) layers of 3.2 mm (1/8 in.) thickness. After a final 180° rotation of the workpiece, the inside welding was completed using the same sequence of two-pass tandem SAW of 3.2 mm (1/8 in.) thickness.

Future Directions

The foregoing examples illustrate that value engineering, methods study, and time study principles can be applied to select the best design for joining of parts. Future efforts could be directed toward developing computer-based simulations with graphic user interfaces that would integrate appropriate part design and manufacturing databases. Such efforts would allow one to effectively consolidate existing knowledge on basic design practices, design criteria for joining, and appropriate case examples involving parts and processes. These computer-based simulations can serve as powerful learning tools, and their effective use can be expected to eliminate or minimize trial-and-error methods of design for joining, and thereby facilitate agile manufacturing at minimal cost.

ACKNOWLEDGMENTS

The following sections in this article were adapted from handbooks published by ASM International (as cited in the list of References): "Mechanical Fastening" (Ref 2), "Adhesive Bonding" (Ref 3), "Brazing" (Ref 5, 6), and "Soldering" (Ref 7). The numbered examples were compiled from Volume 6 of the 9th Edition *Metals Handbook*.

REFERENCES

1. O.W. Blodgett, Joint Design and Preparation, *Welding, Brazing, and Soldering,* Vol 6, 9th ed., *Metals Handbook,* American Society for Metals, 1983, p 60–72
2. W.J. Jensen, Failures of Mechanical Fasteners, *Failure Analysis and Prevention,* Vol 11, *ASM Handbook* (formerly 9th ed. *Metals Handbook*), American Society for Metals, 1986, p 529–549
3. Adhesives, *Engineered Materials Handbook Desk Edition,* M. Gauthier, Ed., ASM International, 1995, p 633–671
4. Procedure Development and Practice Considerations for Electron-Beam Welding, *Welding, Brazing, and Soldering,* Vol 6, *ASM Handbook,* ASM International, 1993, p 851–873
5. M.M. Schwartz, Fundamentals of Brazing, *Welding, Brazing, and Soldering,* Vol 6, *ASM Handbook,* ASM International, 1993, p 114–125
6. M.M. Schwartz, Introduction to Brazing and Soldering, *Welding, Brazing, and Soldering,* Vol 6, *ASM Handbook,* ASM International, 1993, p 109–113
7. M.M. Schwartz, Fundamentals of Soldering, *Welding, Brazing, and Soldering,* Vol 6, *ASM Handbook,* ASM International, 1993, p 126–137
8. *The Brazing Book,* Handy & Harmon, 1983, p 10

9. G.L. Serangeli, et al., Shielded Metal Arc Welding, *Welding, Brazing, and Soldering,* Vol 6, 9th ed., *Metals Handbook,* American Society for Metals, 1983, p 91
10. K. Masubuchi, Residual Stresses and Distortion, *Welding, Brazing, and Soldering,* Vol 6, 9th ed., *Metals Handbook,* American Society for Metals, 1983, p 887
11. G.L. Serangeli, et al., Flux Cored Arc Welding, *Welding, Brazing, and Soldering,* Vol 6, 9th ed., *Metals Handbook,* American Society for Metals, 1983, p 108
12. E.A. Metzbower, et al., Electron Beam Welding, *Welding, Brazing, and Soldering,* Vol 6, 9th ed., *Metals Handbook,* American Society for Metals, 1983, p 609–646
13. S. Bangs, Diffusion Bonding: No Longer a Mysterious Process, *Source Book on Innovative Welding Processes*, American Society for Metals, 1981, p 259–262
14. D. Hauser, et al., Gas Tungsten Arc Welding, *Welding, Brazing, and Soldering,* Vol 6, 9th ed., *Metals Handbook,* American Society for Metals, 1983, p 202–203
15. D.L. Olson, et al., Submerged Arc Welding, *Welding, Brazing, and Soldering,* Vol 6, 9th ed., *Metals Handbook,* American Society for Metals, 1983, p 114–152
16. D. Hauser, et al., Gas Metal Arc Welding (MIG Welding), *Welding, Brazing, and Soldering,* Vol 6, 9th ed., *Metals Handbook,* American Society for Metals, 1983, p 165
17. D. Hauser, et al., Gas Metal Arc Welding (MIG Welding), *Welding, Brazing, and Soldering,* Vol 6, 9th ed., *Metals Handbook,* American Society for Metals, 1983, p 166

SELECTED REFERENCES

- *Brazing Handbook,* 4th ed., American Welding Society, 1991
- P.W. Marshall, *Design of Welded Tubular Connections: Basis and Use of AWS Code Provisions,* Elsevier, 1992
- R.C. Juvinall and K.M. Marshek, Rivets, Welding, and Bonding, Chapter 11, *Fundamentals of Machine Component Design,* 2nd ed., John Wiley & Sons, 1991
- R.O. Parmley, Ed., *Standard Handbook of Fastening & Joining,* 3rd ed., McGraw-Hill, 1997
- D. Radaj, *Design and Analysis of Fatigue Resistant Welded Structures,* Halsted Press/Woodhead Publishing, 1990
- M.M. Schwartz, *Brazing,* ASM International, 1987
- M.M. Schwartz, *Ceramic Joining,* ASM International, 1990
- M.M. Schwartz, *Joining of Composite Matrix Materials,* ASM International, 1994
- J.E. Shigley and C.R. Mischke, Welded, Brazed, and Bonded Joints, Chapter 9, *Mechanical Engineering Design,* 5th ed., McGraw-Hill, 1989
- *Weld Integrity and Performance,* ASM International, 1997

Design for Heat Treatment

William E. Dowling, Jr. and Nagendra Palle, Ford Motor Company

THE SELECTION OF MATERIALS and manufacturing processes for a component design is a complex process and often involves iterative decision making. The component is designed to provide a specific mechanical function, and its design is often limited by space and cost considerations. The component must be able to survive extreme external loading conditions from thermal and/or applied mechanical forces. Therefore, a high level of performance needs to be achieved at a minimum cost. Based on the design and loading conditions, a material and manufacturing process are selected to cost effectively provide adequate properties for the operating environment. Very often a low-cost material and processing combination requires heat treatment after component shaping to enable the part to meet its design criteria. The relationship between performance at minimum cost, design, material, and the manufacturing process is analogous to a three-legged stool with all legs having equal importance. In order for a component to fulfill its cost and performance criteria, its design should accommodate all the loading conditions of the component, its material properties should meet the expectations of the design, and its manufacturing processes should produce the component at a minimum cost.

The mechanical design process has evolved from an experience-based process using design factors, such as stress-concentration factors, and design rules (based on experimental data) to the current reliance on analytical processes. The ability to analytically design components combined with the constant desire for system cost reduction while achieving greater performance has led to decreased design times and increased component complexity. The designer, when provided with accurate component loading information, can accurately assess part life with reliable input of experimentally determined material properties. These material property databases are continually growing, both in the open and in proprietary databases (see the article "Sources of Materials Properties Data and Information" in this Volume). In addition to these databases, the mechanical properties resulting from a broad range of heat treatment processes, for many classes of materials, are well documented. The ability to select materials and process parameters to achieve the desired property goals is becoming more automated as evidenced by the success of Jominy hardness and carburizing prediction programs (Ref 1). The foundation of the currently used design practices requires a close relationship among analytical design procedures, material property databases, and the ability of the heat treat process to achieve desired mechanical properties.

In addition to providing appropriate physical and mechanical properties to meet design requirements, heat treatment also produces dimensional changes and residual stress patterns that in some cases can lead to component cracking. The dimensional changes and residual stresses produced by heat treatment are very sensitive to geometric and processing specifics. Currently, the relationship between design, dimensional changes, residual stresses, and cracking is determined by experience and results in the development of general rules for design. Prototype component designs must be experimentally evaluated and iterated to bring the component within acceptable tolerances for dimensions and residual stresses. A clear need for reducing development costs has motivated significant corporate and academic research in the area of analytical prediction of the response of a component to heat treatment.

This article presents an overview of techniques that are currently in use to design for heat treatment. The primary design criteria addressed in this article are the minimization of distortion and undesirable residual stresses. The article presents both theoretical and empirical guidelines to understand sources of common heat treat defects and how they can be controlled. A simple example is presented to demonstrate how thermal and phase-transformation-induced strains cause dimensional changes and residual stresses. This example also serves as a representation of a typical "process model." The final sections of the article describe the state-of-the-art in heat treatment process modeling technology.

Overview of Component Heat Treatment

Component heat treatment is often the most cost-effective method for a manufacturing process to produce the desired material properties. (Detailed information about heat treating processes is provided in *Heat Treating,* Volume 4 of *ASM Handbook,* Ref 2.) However, in addition to material strength, heat treatment can result in the development of residual stresses (both compressive and tensile), dimensional changes (with respect to size and shape), and, in an extreme situation, component cracking, often referred to as quench cracking. These factors (residual stresses and dimensional changes) have the greatest influence on the design process of a component. Often, the inability to produce components with acceptable dimensions and residual stress patterns will cause changes in design, materials, and process selection leading to additional cost and lower material strength.

A typical component manufacturing process includes the following five steps, all of which can influence dimensional changes and residual stress patterns in heat-treated components:

1. Metalworking, machining, or other forming operations
2. Component heat-up
3. Hold at temperature for through-heating, solutionizing, or thermal chemical treatments such as carburizing or nitriding
4. Quenching from elevated temperature
5. Postquench tempering or aging treatment

Within these five process steps there are seven major factors that lead to size and shape changes and the development of residual stresses in heat treated components (Ref 3, 4):

- Variation in structure and material composition throughout the component, leading to anisotropy in properties and transformation behavior
- Movement due to relief of residual stresses from prior machining and forming operations
- Creep of the part at elevated temperature under its own weight or as a result of fixturing
- Large differences in section size and asymmetric distribution of material causing differential heating and cooling during quenching
- Volume changes caused by phase transformation
- Nonuniform heat extraction from the part during quenching
- Thermal expansion

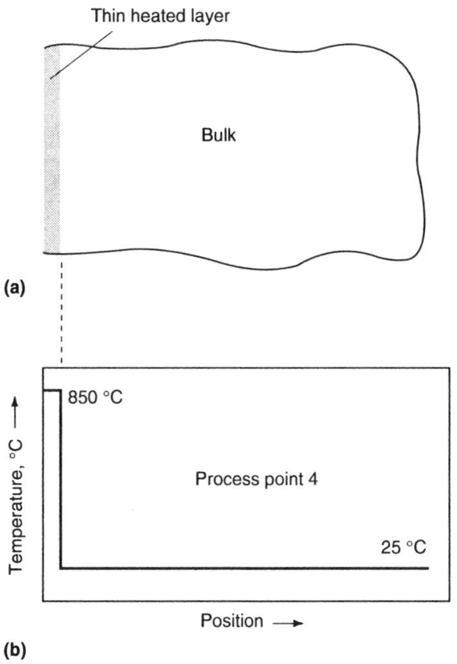

Fig. 1 Schematic (a) of a large, thick plate and assumed temperature distribution (b) at process point 4 (Fig. 2)

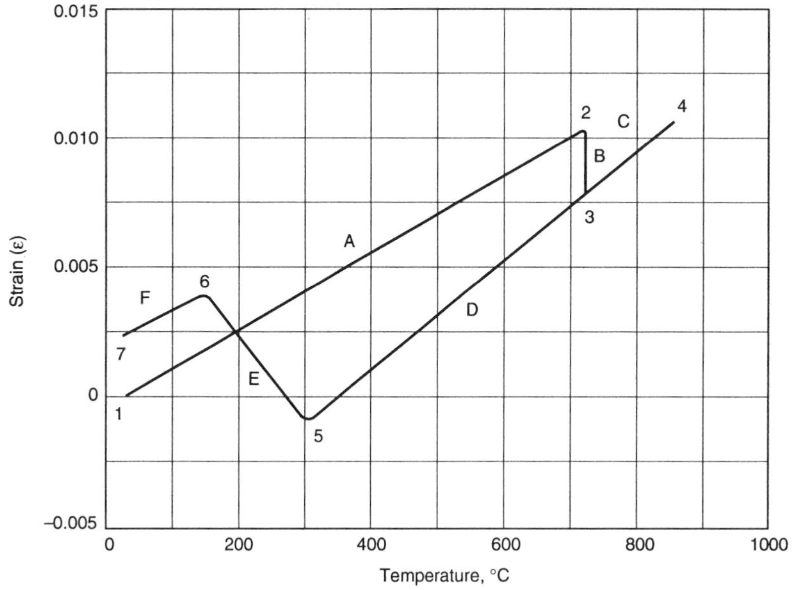

Fig. 2 Strain history during heating and cooling for a large, thick plate. See text for discussion of the labeled process segments.

All of these factors, except relief of prior residual stresses (second item) and creep at elevated temperature (third item), can be directly related to thermal and transformation-induced strains in the component. Residual stresses from forming operations can be reduced by stress relief prior to final shaping operations. Creep at elevated temperature can be addressed by appropriate component loading in the furnace. Neither of these two factors are directly associated with the design of the component. The other five factors are directly related to component design, and more precisely, strains introduced by transformations and nonuniform temperature distributions in the component. The relationship among design, material properties (yield strength at temperature, coefficient of expansion, thermal conductivity, etc.), and heating and cooling processes determine the distortion and residual stress patterns in heat treated components.

Thermal and Transformation-Induced Strains in Heat Treated Components

Thermal strain is developed in a component when differential thermal expansion (or contraction) occurs. The magnitude of strain is directly proportional to the thermal expansion coefficient of the material (α) and the temperature difference between two points (ΔT). This strain translates directly to a thermal stress σ_{th} (Ref 5):

$$\sigma_{th} = \alpha \Delta T E \qquad \text{(Eq 1)}$$

where E is the elastic modulus of the material. However, if the thermally induced stress is greater than the flow strength of either the cooler or hotter material, permanent (plastic) deformation occurs. This plastic flow causes permanent shape change (distortion) and impacts the magnitude and distribution of residual stresses. Without plastic deformation, the component would return to its original dimensions once the part has thermally equilibrated. In addition to thermal strains, many materials systems undergo phase transformations as a function of temperature. Often, the new phase(s) that form have different volumes and different coefficients of expansion as well as different mechanical behavior(s) than the parent phase(s). These differences increase the complexity of understanding the effect of thermal gradients on strains produced and the resulting plastic deformation.

Example: Thermal and Transformation-Induced Strains in a Large, Thick Plate. This simplified example demonstrates how thermal and transformation-induced strains can result in substantial plastic deformation and residual stresses. A large, thick plate (Fig. 1a) has a very thin layer uniformly heated from room temperature to 850 °C and then cooled. It is assumed that the bulk of the material is unheated and does not strain because of its much greater thickness. The heating and cooling processes can be broken down into 6 segments (shown schematically as A to F in Fig. 2; key process points are labeled 1 to 7). During the heating and cooling cycle, the thin layer of heated material will undergo a phase transformation from phase 1 to phase 2 upon heating and from phase 2 to phase 3 upon cooling. The thermal profiles through the sample for process point 4 is shown in Fig. 1(b) to illustrate that the heated layer is considered to be at the same temperature all the way through, with a thermal step change from the heated layer to the unheated bulk.

The strain produced by thermal ($\alpha \Delta T = \varepsilon_{th}$) and transformation ($\varepsilon_{tr}$) strain is calculated for each segment identified in Fig. 2. The total induced strain must be accommodated in the thin layer through either elastic (ε_{el}) or plastic (ε_{pl}) strain, which sums to the total strain:

$$\varepsilon_t = \varepsilon_{th} + \varepsilon_{tr} = -(\varepsilon_{el} + \varepsilon_{pl}) \qquad \text{(Eq 2)}$$

The induced strain and the accommodation strain must equalize in the thin layer because the assumption has been made that the bulk material will not accommodate any strain. The values for α (thermal expansion coefficient) for phases 1, 2, and 3 are chosen to be 15, 21, and 13 × 10^{-6}/°C, respectively, and are assumed to be independent of temperature for this example. The transformation strains chosen are $\varepsilon_{tr\ 1\text{–}2}$ = 0.0025 for the transformation from phase 1 to phase 2 at 725 °C

Table 1 Material condition at each stage in the thermal cycle shown in Fig. 2

Process segment (Fig. 2)	Temperature range, °C	Phase transformation	Total strain (ε_t)	Accommodation strain ε_{el}	Accommodation strain ε_{pl}	Phases	Remaining stress, MPa	Yield strength, MPa
A	25–725	No	0.0105	−0.0007	−0.0098	Phase 1	−100	100
B	725	Yes	−0.0025	0.0017	0.0008	Phase 2	100	100
C	725–850	No	0.0026	−0.0017	−0.0009	Phase 2	−50	50
D	850–300	No	−0.0116	0.0017	0.0099	Phase 2	250	250
E	300–150	Yes	0.0050	−0.0050	None	Phase 2 + phase 3	−775	1700
F	150–25	No	−0.0016	0.0016	None	Phase 3	−440	1800

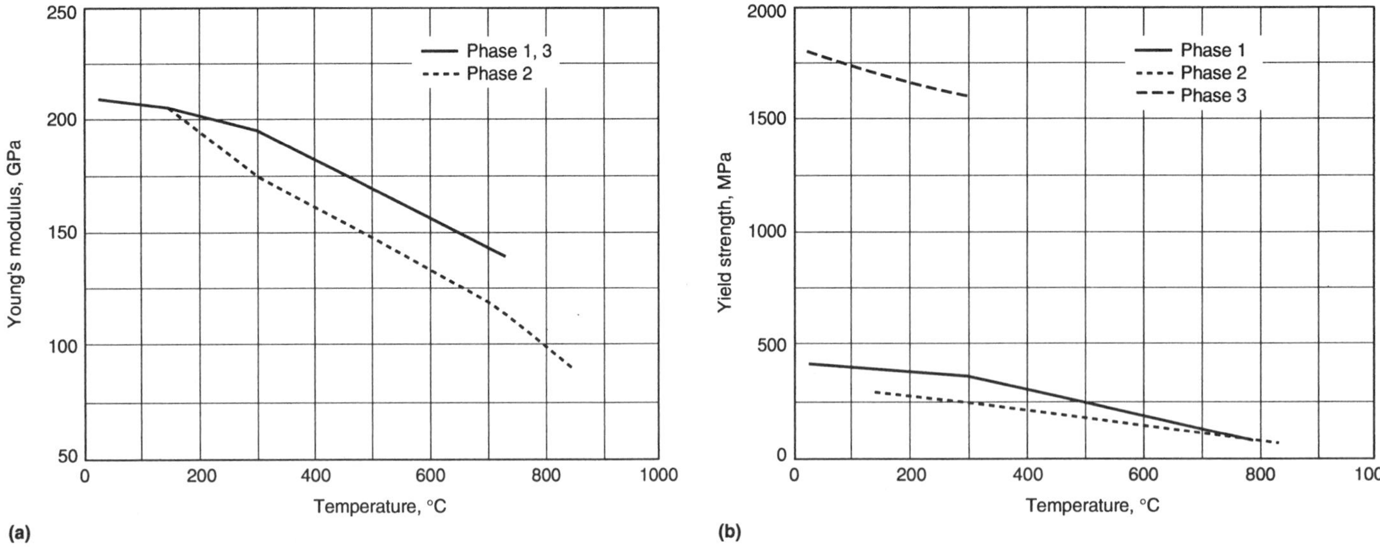

Fig. 3 Young's modulus (a) and yield strength (b) as a function of phase and temperature for the large, thick plate example

and $\varepsilon_{tr\ 2\text{–}3}$ = 0.0075 for the transformation from phase 2 to phase 3 over the temperature range 300 to 150 °C.

In order to determine the accommodation strain values, Young's modulus (E), and the yield strength (σ_{ys}) are required as a function of phase and temperature; the data for this example are shown in Fig. 3. For simplicity, the flow behavior is assumed to be elastic-plastic in nature with no work hardening. Thus, the yield strength is the flow stress at any plastic strain level.

The total strain and how it is accommodated is described below for each segment of the thermal cycle shown in Fig. 2.

Segment A. The thin layer is heated from 25 to 725 °C without phase transformation. This heat-up induces a linear thermal strain of $\alpha\Delta T$ (ε_{th}) that equals 0.0105. However, the thin layer is constrained by the bulk and is not allowed to expand. This constraint produces a compressive stress in the heated layer equal to the flow stress of the material. This compressive stress accommodates the elastic strain, with the remaining applied strain accommodated through plasticity in the heated layer. For cases in which the yield strength of the material is exceeded, the strain distributions follow the relationship:

$$\varepsilon_{el} = (\sigma_{ys} + \sigma_{old})/E \quad \text{or} \quad \varepsilon_{el} = (\sigma_{ys} - \sigma_{old})/E \qquad \text{(Eq 3)}$$

The stresses are additive if the stress from the prior history, σ_{old}, is in the opposite direction of the new stress. The difference in the stresses must be taken if the prior history stress and the new stress are the same direction. The results are shown in Table 1.

Segment B. The thin layer undergoes a complete phase transformation from phase 1 to phase 2 with no temperature change, resulting in a transformation strain of –0.0025 and no thermal strain (Fig. 2). This results in a stress reversal, which causes the heated layer to be in tension. The magnitude of the transformation strain is sufficient to cause the yield strength of phase 2 to be exceeded and induce a plastic strain of 0.0008.

Segment C. The layer is heated from 725 to 850 °C, causing a thermal strain of 0.0026. This again reverses the stress back to compression in the thin layer and causes a compressive plastic strain of –0.0009.

Segment D. Heating is stopped and the layer uniformly cools from 850 to 300 °C. This causes contraction inducing a thermal strain of –0.0116. This strain accommodation is broken down into 0.0017 elastic and 0.0099 plastic. The resulting deformation is predominantly plastic because of the low yield strength (250 MPa) of phase 2 at 300 °C and its high modulus (175 GPa).

Segment E. The layer is cooled from 300 to 150 °C. During this period, a phase transformation occurs from phase 2 to phase 3 that causes an expansion strain of 0.0075. In addition, a thermal

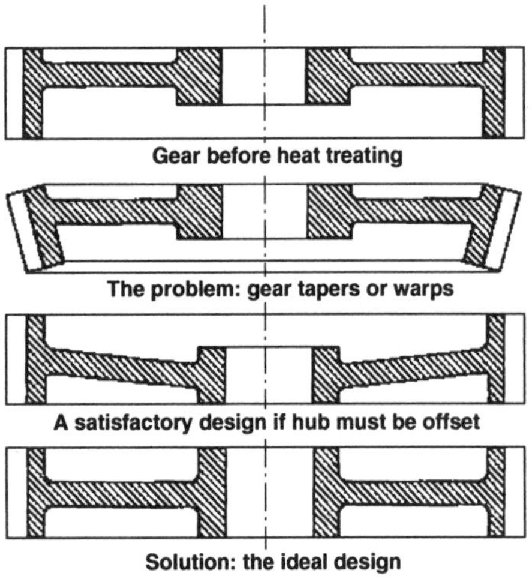

Fig. 4 A typical problem caused by lack of symmetry in design, illustrated by a gear that warped during heat treating. Design modifications can solve the problem. Source: Ref 6

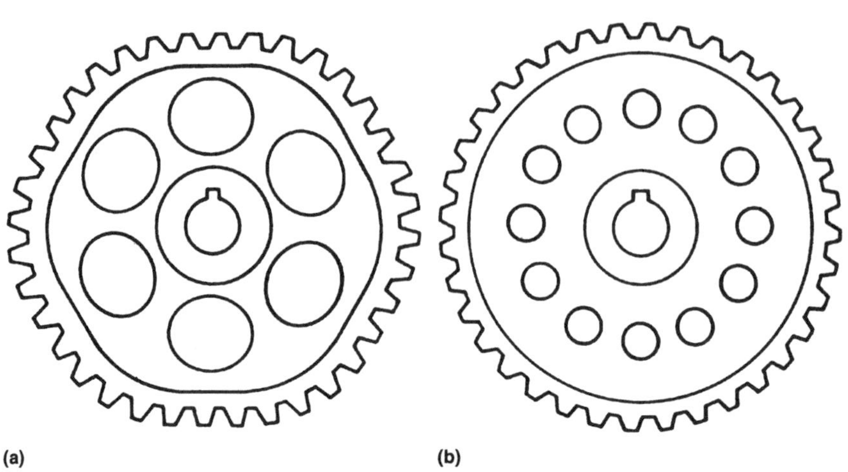

Fig. 5 Problem caused by the use of holes to reduce weight of a gear. (a) If the designer specifies large holes in the web, heat treatment may produce a flat spot for each hole. (b) Keeping the hole diameter to one-third of the web width eliminates the problem. Source: Ref 6

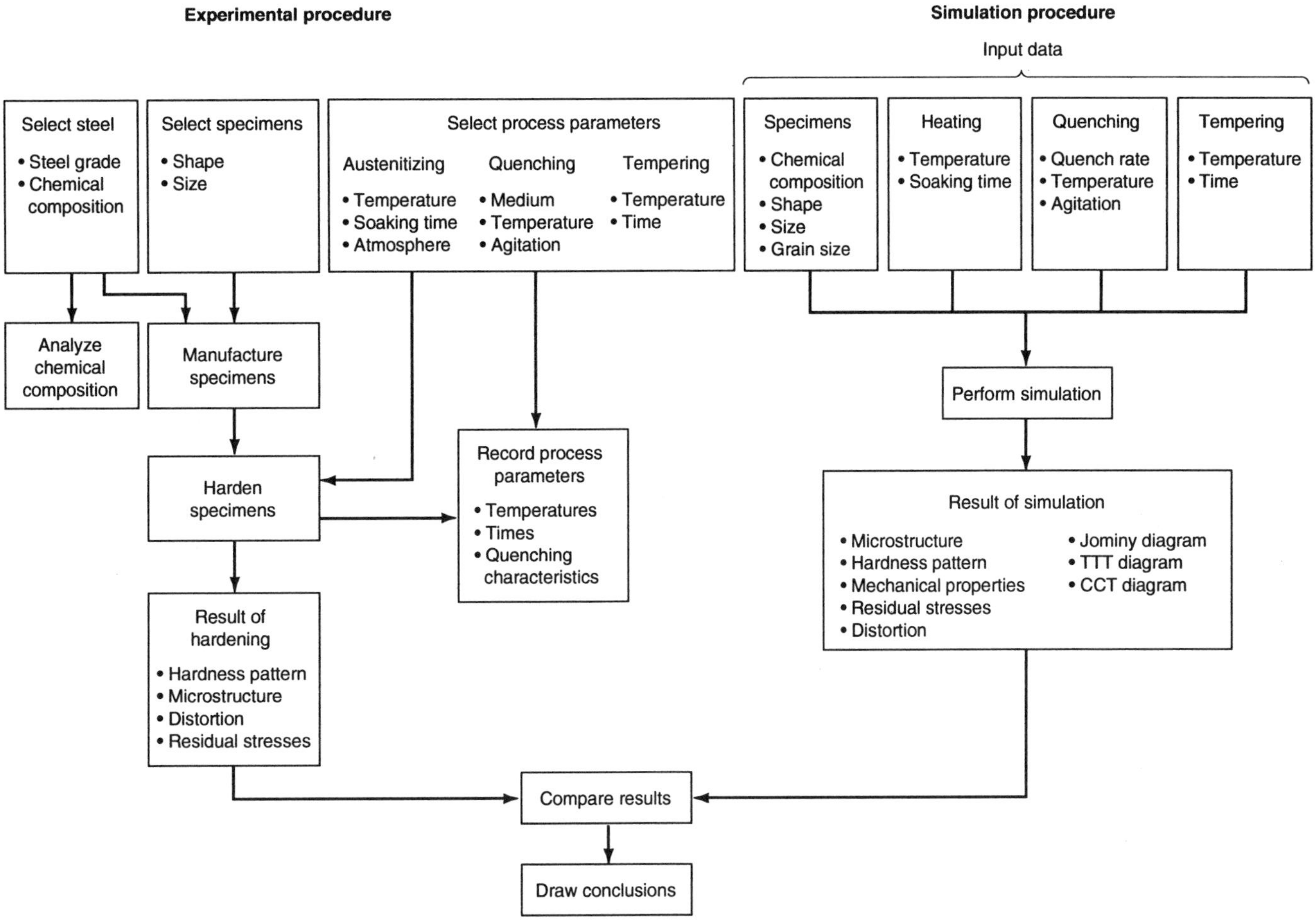

Fig. 6 Comparison of experimental and simulation procedures for heat treat design. Source: Ref 10

contraction of –0.0025 occurs. The thermal strain was determined by averaging the expansion coefficients. The net total strain was 0.0050. However, phase 3 has a much higher yield strength than phase 2, and this strain is all accommodated elastically. The resultant residual stress at 150 °C is –775 MPa.

Segment F. The final stage of the process is a cooling from 150 to 25 °C. This is all thermal contraction strain, which reduces the compressive residual stress to –440 MPa.

Conclusions. A summary of the results from this example is shown in Table 1. Most of the plastic deformation occurred during the heat-up and cool-down stages of the process. This was largely the result of the low strength of the material at high temperature and the low strength of phase 2 at lower temperature (300 °C). A substantial residual stress was produced mainly from the expansion during the final phase transformation and the simultaneous change to a high-strength material. This example problem is a simplistic representation of surface hardening of steel, which usually produces a large compressive stress at the surface.

From this example, it is readily seen that even for a simple problem, a substantial plastic deformation and a substantial residual stress can result from heat treat processes. During the actual quenching of a complex part, not only can similar transitions in phases and phase strengths happen, but the thermal distribution can vary greatly depending on the component geometry and the heat treating process.

Component Design for Heat Treating: Experience-Based Design Rules

Several general experience-based design guides have been developed (Ref 5–9). They are all based on the concept of minimizing nonuniform heating and cooling of components. Large thermal gradients and thermal asymmetry produce distortion and have the potential of causing tensile stresses and thus cracking of the component. The concept of reducing the opportunity for large thermal gradients to minimize distortion is counter to the need for rapid quenching in order to form martensite in steel or to prevent precipitation after solution treatment of age-hardenable materials. The preferred method is to quench the material only as rapidly as absolutely necessary to form the desired phase(s). Optimization of this manufacturing system is a combination of material selection, heat treatment process, and component design for heat treatment. This optimization necessitates that the heat treater work in conjunction with the designer to avoid geometric features that can be difficult to heat treat and aid in the selection of appropriate heat treat processes.

Over the years, some basic rules have been developed for component design for heat treatment; these are summarized below.

Design the Component To Be As Symmetrical As Possible. Part asymmetry causes asymmetrical thermal and transformation gradients, which often result in nonuniform plastic deformation and distortion. A classic example of this is shown in Fig. 4 for the carburizing and quenching of a gear (Ref 6). The thinner section in the web cools more rapidly than the gear section. As the web section contracts the gear section is in circumferential compression. Then, as the gear section cools and contracts, the web restricts uniform contraction because it is not centered. This results in a tapered gear. The problem can be resolved by offsetting the hub, or ideally, by making the gear symmetrical.

Maintain Uniform Section Thicknesses. Large changes in section thickness cause large thermal gradients during both heating and cooling. If these section changes are abrupt they will act as strain concentrators causing distortion. Figure 5 shows the result of placing large holes in the web of a gear to reduce weight. If the holes are too large they cause flat spots on the gears. A general rule that has been developed is to keep the diameter of such holes to no more than one-third the web width.

Minimize Holes, Deep Splines, and Keyways. These features often destroy the symmetry of the component and act as strain and stress concentrators during heat treatment. In addition, these features can entrap vapor resulting in slower local cooling rates, making it difficult to effectively quench the component. If they must be used, small radii at the corners of keyways should be avoided and spline transition should be gradual.

Avoid Sharp Corners. Corners should be designed with as large a radius as possible. If sharp corners cannot be avoided, it is good practice to provide relief notches in place of sharp edges (Ref 5).

Avoid Long, Thin Sections. The definition of "long, thin section" varies and depends greatly on quenching media; however, any section length greater than 15 times the diameter is almost always characterized as such and the slightest nonuniformity in quench will cause it to distort. As the quenching medium becomes more severe (i.e., water, caustic quench), this criterion is reduced to as low as 5 times the diameter (Ref 7). For larger length-to-diameter ratios, consideration should be given to fixture quenching or induction hardening.

Beware of Part Complexity. The current trend leans toward placing more individual components together to make a single part or toward adding many geometric features to a part in order for the component to serve several functions. This can often lead to improved package efficiency or reduced cost in other stages of the manufacturing process (see the article "Design for Manufacture and Assembly" in this Volume). However, the component may become more difficult to heat treat, especially if the design rules mentioned above concerning uniform section thickness, symmetry, and the avoidance of holes and keyways are violated.

All these simple rules are intended to encourage designers to produce component designs that are readily heat treatable. However, components will continue to be designed that are asymmetrical, have nonuniform sections, and contain many features. When faced with such components, heat treaters must use their experience to find solutions to the problem. Analytical solutions to improve designing for heat treatment are on the horizon with the development of finite element approaches to simulation and with the vast improvements in computer power that are occurring every day.

Computer Modeling as a Design Tool for Heat Treated Components

A computer process simulation model allows a particular design to be tested under a specific set of process conditions. The computer software can graphically display not only the resulting residual stresses and distortions in the component, but also the associated transient evolution of temperature, metallurgical phases, volume changes, and stresses. These analytical models can be used for point-by-point comparison (with experiments) and to view distributions of quantities such as residual stresses or phases throughout the part. Based on the transient response of the component design to the specified process, design parameters and processing conditions can be adjusted so as to minimize the undesirable outcomes that are predicted. Of course, like any analytical tool, accuracy and confidence in the modeling technique is of utmost importance. There are inherent inaccuracies in any computational technique, and the user of the process model must be aware of the limitations as well as the strengths of the model. Given the state-of-the art of these modeling techniques, qualitative results are probably the most valuable product of process modeling.

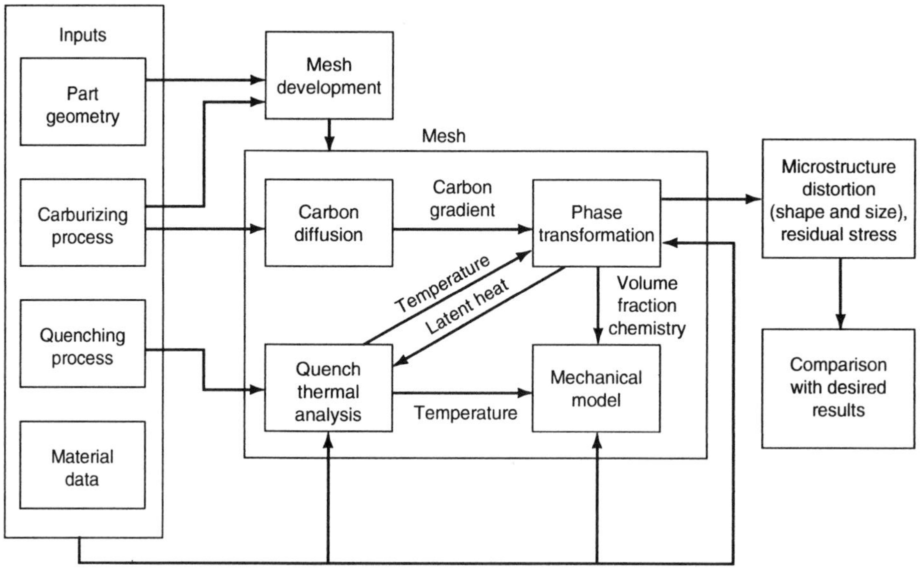

Fig. 7 Process model for a carburizing problem. Source: Ref 11

The thermal and transformation strain example described in a previous section is in fact a process model that includes the various transformations that a point in the material undergoes during heat treatment and the associated changes in dimensions and stress patterns. The example illustrates how the various stages of transformation that a material experiences can be analytically described (or modeled!). It is important to realize that all the changes in the thin section of the plate are caused by the imposed thermal profile, which in turn results entirely from the "process" conditions that the plate experiences. The process conditions usually include the furnace atmosphere (for example, temperature and carbon potential), heating rates, and quench conditions. Modeling of heat treat processes, like other materials processes such as casting and welding, is quite complex due to the tight coupling of various metallurgical transformations and the associated changes in thermal and mechanical states.

A process simulation model is usually exercised for a specific material and a specific process. Both the material and process characteristics are inputs to the model. The process specifics manifest themselves in "initial" conditions (e.g., temperature of the component at the start of heat treatment) and boundary conditions (e.g., the quench temperature and level of agitation in the bath) of the model. The material properties needed are the modulus, strength, coefficient of thermal expansion, yield strength, conductivity, specific heat, and density. Because the material undergoes drastic changes in temperature, these properties should be temperature and phase dependent. It is extremely important for the user to have accurate information relating to these data, thus generating this information often requires extensive characterization of the material and process under consideration.

The thin section in the example experienced several different stress states primarily because the yield strength of the material changed depending on the temperature and phase. Also, for materials such as the one illustrated, accurate transformation kinetics are required to identify the phases present anywhere in the component at any time in the process. The solvers (using finite or boundary element or finite difference techniques; see the article "Finite Element Analysis" in this Volume) are used to solve thermal and structural governing equations subject to the specified boundary conditions and material characteristics. A comparison of typical experimental and simulation procedures is shown in Fig. 6. Typically, each of the five steps of the heat treat process (described in the section "Overview of Component Heat Treatment" in this article) represents a computer analysis. For example, carbon diffusion during a carburizing process requires a mass diffusion analysis, the heat-up in the furnace requires a thermal analysis, and the quenching requires a thermal and structural analysis.

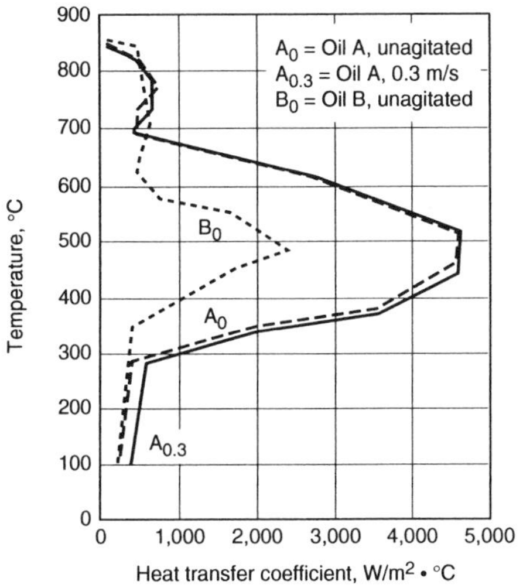

(a)

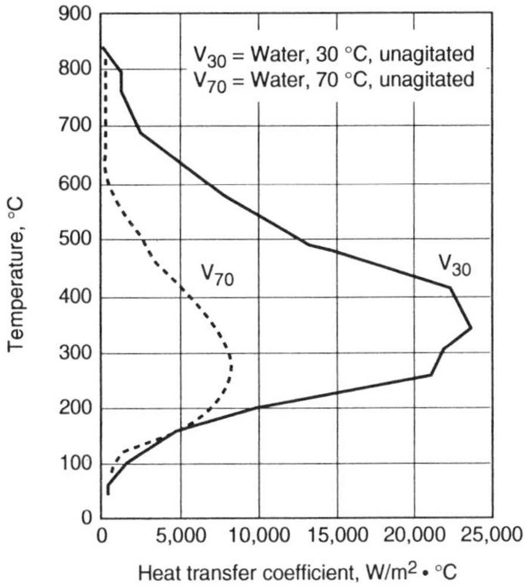

(b)

Fig. 8 Example of heat transfer coefficient boundary conditions for oil and water quenches. Source: Ref 10

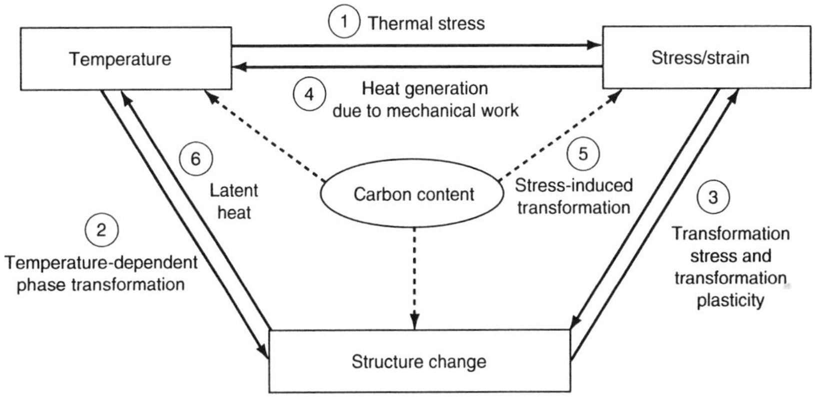

Fig. 9 Phenomenological coupling in heat treat simulation. Source: Ref 14

Some of these analyses depend on each other very closely, leading to "coupled" analyses.

All seven of the factors that lead to size and shape changes and the development of residual stresses (listed in the section "Overview of Component Heat Treatment" in this article) can be input into (or output from) the model. The variation in structure and composition, the differential heating and cooling, and the quench agitation can be included in the model using "initial" and "boundary" conditions. The movement due to stress relief, the creep of the part under its own weight, the volume changes due to phase transformation, and thermal expansion can be output from the model based on the boundary conditions and material description.

A typical process model, such as one for heat treatment, has several major aspects to it. A process model for a carburizing process is shown in Fig. 7. The flow chart shown in Fig. 7 represents the different pieces of a simulation tool that are used to analyze the carburizing process. The part geometry represents the initial design. If the solver is finite element based, the geometry has to be discretized into a finite element mesh (see the article "Finite Element Analysis" in this Volume). The various constituents in the model are described below.

Boundary conditions represent the process conditions under which the part is being heat treated. To simulate a carburizing process, the carburizing cycle parameters constitute the boundary conditions. These are expressed using a surface carbon potential as a function of time and temperature. To simulate a quenching process, convection boundary conditions are used to describe the part-quench interface. Heat transfer coefficients for a water quench and oil quench are shown schematically in Fig. 8. Some examples for salt quenches are given in Ref 12. These values of heat transfer coefficients are typical for small parts in a large quench tank. The boundary conditions are extremely important because they represent the uniqueness of a particular part (and quench). Extensive experiments are sometimes required to estimate these boundary conditions. Occasionally, the supplier of the quenchant can also provide information about the rate of heat removal from a part surface for a particular quench.

The deformation model (stress-strain relationships) used for computing distortion and residual stresses requires an accurate description of the phase transformation kinetics and the associated microstructure and volume changes. The deformation model uses the temperature- and phase-dependent mechanical properties to determine the deformation changes, given the temperature changes. These volume changes (e.g., from austenite to martensite) typically lead to transformation-induced plasticity (TRIP). This is caused by a soft phase (such as austenite) being deformed by a harder phase (such as martensite) during the transformation. Several deformation models have been developed with reasonable success (Ref 13). Implementation of these models requires a good understanding of mechanics of the materials. The close interaction between various parameters in the deformation model is referred to as "metallo-thermo-mechanical" coupling. Figure 9 shows the various couplings considered in the modeling of phase transformations during the heat treatment of steels.

Analysis Software. An analysis package capable of performing thermal and structural analyses is required to execute the process model. This analysis package should integrate not only the above-mentioned pieces, but also provide tools to graphically display the thermal and phase evolution, the final size and shape of the part, and the resulting residual stresses. This information can be used by the process engineer to make either a design or process change recommendation. From a practical standpoint, even if the computer model is not completely accurate, the analysis tool can be used to perform sensitivity studies to understand which process or material parameter has the greatest (or least) impact on the resulting part distortion or residual stress. Examples of commercial software systems capable of performing heat treatment analyses are HEARTS (Ref 14), TRAST, and MetalCore (Ref 15). TRAST is a deformation model written in the form of a FORTRAN user subroutine that has to be used with ABAQUS (Ref 18). TRAST is based on some original work carried out by Sjostrom at Linköping University in 1982 (Ref 17). Another software system will result from a collaborative effort under the auspices of the National Center of Manufacturing Sciences (Ref 18).

The examples presented in these references represent the state-of-the art in the area of modeling of heat treatment processes. Several open issues need to be resolved before computer modeling of heat treat processes becomes routine. Much needs to be done in the areas of numerical analysis, material characterization and modeling, quench tank characterization, and prediction of phase transformation behavior. However, successes demonstrated in recent literature show that heat treatment modeling is very much a viable technology.

REFERENCES

1. M.A.H. Howes, Factors Affecting Distortion in Hardened Steel Components, *Quenching and Distortion Control: Proceedings of the First*

International Conference on Quenching and Control of Distortion, ASM International, 1992, p 251–258
2. *Heat Treating,* Vol 4, *ASM Handbook,* ASM International, 1991
3. J.S. Kirkaldi, Quantitative Prediction of Transformation Hardening in Steels, *Heat Treating,* Vol 4, *ASM Handbook,* ASM International, 1991, p 20–34
4. H. Walton, Dimensional Changes During Hardening and Tempering of Through-Hardened Bearing Steels, *Quenching and Distortion Control: Proceedings of the First International Conference on Quenching and Control of Distortion,* ASM International, 1992, p 265–273
5. A. Kumar Sinha, Defects and Distortion in Heat-Treated Parts, *Heat Treating,* Vol 4, *ASM Handbook,* ASM International, 1991, p 601–619
6. *Materials and Processes,* Vol I, *Source Book on Heat Treating,* American Society for Metals, 1975
7. R.F. Kern and M.E. Seuss, *Steel Selection: A Guide for Improving Performance and Profits,* John Wiley & Sons, 1979
8. T. Bell, *Survey of the Heat Treatment of Engineering Components,* The Iron & Steel Institute, London, 1973
9. K.E. Thelning, *Steel and Its Heat Treatment,* English ed., Butterworths, 1975
10. J. Bodin and S. Segerberg, Benchmark Testing of Computer Programs for Determination of Hardening Performance, *Quenching and Distortion Control: Proceedings of the First International Conference on Quenching and Control of Distortion,* ASM International, 1992, p 133–139
11. W. Dowling, T. Pattok, B.L. Ferguson, D. Shick, Y.-H. Gu, and M. Howes, Development of a Carburizing and Quenching Simulation Tool: Program Overview, *Quenching and the Control of Distortion: Proceedings of the Second International Conference,* ASM International, 1996, p 349–355
12. D. Shick, D. Chenoweth, N. Palle, C. Mack, W. Copple, W.-T. Lee, W. Elliot, J. Park, G.M. Ludtka, R. Lenarduzzi, H. Walton, and M. Howes, Development of a Carburizing and Quenching Simulation Tool: Determination of Heat Transfer Boundary Conditions in Salt, *Quenching and the Control of Distortion: Proceedings of the Second International Conference,* ASM International, 1996, p 357–366
13. D. Bamman, V. Prantil, A. Kumar, J. Lathrop, D. Mosher, M. Callabresi, H.-J. Lou, M. Lusk, G. Krauss, B. Elliot, Jr., G. Ludtka, T. Lowe, W. Dowling, D. Shick, and D. Nikkel, Development of a Carburizing and Quenching Tool: A Material Model for Carburizing Steels Undergoing Phase Transformations, *Quenching and the Control of Distortion: Proceedings of the Second International Conference,* ASM International, 1996, p 367–375
14. T. Inoue and D.-Y. Ju, Metallo-Thermo-Mechanical Simulation of Quenching Process: Theory and Implementation of Computer Code "HEARTS," *Quenching and Distortion Control: Proceedings of the First International Conference,* ASM International, 1992
15. N.J. Marchand and E. Malenfant, Modeling of Microstructural Transformations Using MetalCore, *Heat Treating: Proceedings of the 16th Conference,* ASM International, 1996, p 411–418
16. ABAQUS Users Manual, HKS Inc.
17. S. Sjostrom, "The Calculation of Quench Stresses in Steel," Ph.D. Thesis, Linköping University, Sweden, 1982
18. C. Anderson, P. Goldman, P. Rangaswamy, G. Petrus, B.L. Ferguson, J. Lathrop, and D. Nikkel, Jr., Development of a Carburizing and Quenching Simulation Tool: Numerical Simulations of Rings and Gears, *Quenching and the Control of Distortion: Proceedings of the Second International Conference,* ASM International, 1996, p 377–383

Design for Ceramic Processing

Victor A. Greenhut, Rutgers—The State University of New Jersey

THE CERAMICS AND GLASSES INDUSTRY has estimated annual sales of between $100 and $200 billion worldwide. Ceramic production accounts for about 45% of the entire world market for manufactured ceramics and glasses, not including cements and ceramic powders. One estimate of ceramic sales in various categories and subcategories is provided in Fig. 1; Fig. 2 provides an alternative estimate. It may be noted that dollar values differ in several categories. This is for two reasons. First, the market is rather fragmented with production ranging from large corporations to individual artisans. Much production activity is conducted within the industry for internal consumption. Low estimates tend to count only larger-scale industrial activity, while higher values attempt to include the small-shop and internal-company production. Second, there are significant overlaps between various categories—for example, electrical porcelains might be included with technical ceramics or whitewares. From the data of Fig. 1 and 2 it can be seen that the major fraction of ceramics produced include clay (as well as beneficiated and/or chemically derived materials) among their raw materials. For this reason, the discussions of materials and process design first concentrates on these materials.

Many technical ceramics contain clay as part of their formulation, but tend to employ a higher proportion of high-purity, processed, or derived raw materials. Most refractory brick and parts are processed similarly to the clay-based systems. Technical ceramics include oxide and nonoxide materials used for a number of industrial applications such as electrical porcelains, electronic substrates, wear parts, and chemically resistant parts. It also includes advanced ceramics providing particularly demanding performance in engine parts, biomedical prostheses, high-temperature bearings, industrial cutting tools, and so forth. Such materials are estimated to account for as much as 5% of ceramic sales. Consideration in this article is given to some of the special issues in process design and materials selection for those technical ceramics that differ from clay-based systems. The article "Effects of Composition, Processing, and Structure on Properties of Ceramics and Glasses" in this Volume provides a brief discussion of many of the plastic cements (monolithics, gunning mixes, ramming compounds) that are applied and fired in situ. These materials are not discussed here. Neither are porcelain enamels, which are chiefly composed of glass frit (fine particulate) and combine glass technology with metal-coating methods. Detailed information about all major types of these materials can be found in *Ceramics and Glasses,* Volume 4 of the *Engineered Materials Handbook.*

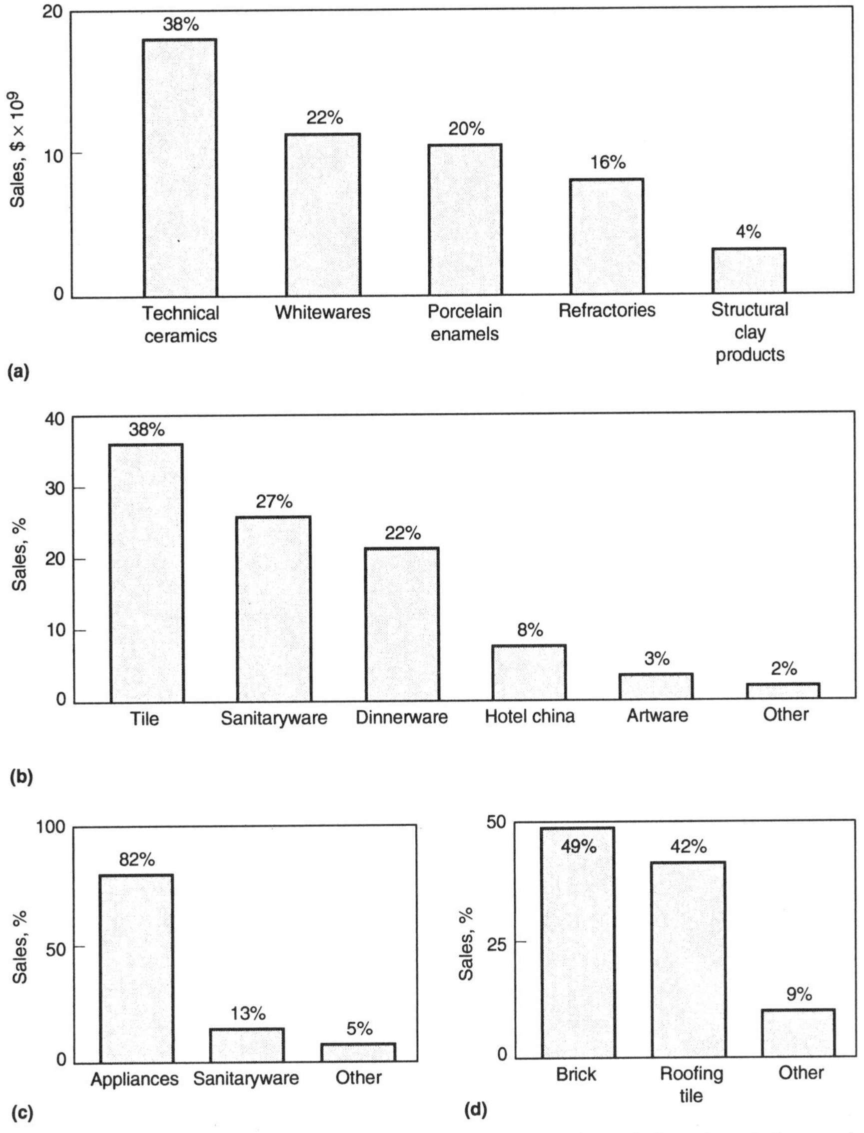

Fig. 1 Estimated worldwide sales of ceramics. Does not include sales of glass products, which is estimated to be approximately $58 billion (1996). (a) Sales of traditional/industrial ceramics (1996). (b) Distribution of whiteware sales (1992). (c) Distribution of porcelain enamel sales (1992). (d) Distribution of structural clay products sales (1992). Source: Ref 1–3

Table 1 Relative productivity and product value for ceramic product categories

Product category	Relative productivity(a)	Relative value(b)
Bricks	1.05	0.15
Sewer pipe	1.00	1.00
Roofing tile	0.91	1.05
Tile	0.79	1.40
Refractories	0.80	1.50
Sanitaryware	0.81	8.90
Tableware	0.41	20.00
Technical ceramics	0.31	10–10,000

(a) Relative pounds per employee-year, indexed to sewer pipe. (b) Relative price per pound

This article provides an overview of the processing steps used in ceramics processing and related mechanical design considerations. The breadth of the topics presented prevents a detailed discussion of each. A wide variety of products and ceramics compositions are represented. The references listed at the end of this article provide more detailed information about ceramics processing issues.

Costs

The production of ceramic products covers a broad spectrum of materials and applications manufactured by facilities that vary over a broad scale. Contrast the production of brick and tile at hundreds of tons per employee each year with the exclusive art of an artisan potter or the special-purpose advanced ceramic valves for a racing car. All ceramic production is capital intensive. The required equipment for materials handling, grinding, mixing, forming, drying, firing, and suitable packaging necessitates a relatively large investment in capital equipment and physical plant at any scale. Long-term investment strategies are necessary. Large-scale production is undergoing major change in terms of automation, controls, and continuous processing, which require flexible plant and equipment selection and design to anticipate future developments.

In order to examine the various factors in production, the manufacture of tile will be used as a model. Tile is chosen both because it has a substantial level of production (see Fig. 1 and 2) and because the economics of tile manufacture has been examined over the past two decades as automated tile manufacture became a major industrial trend. This approach to continuous tile production initiated chiefly in Italy. Figure 3 shows the typical distribution of costs in 1971, before significant automation, and in 1989, when most of the Italian tile manufacturing was highly automated. The graph is for an indexed Lira (It). Figure 4 shows the relative change in costs per square meter of tile. The net result of automation was a 30% reduction in cost per square meter, a tripling of productivity per employee, and a 150% increase in production above the 1971 level. The sales price of tile decreased by about 15%, due in large part to an increase in the total market (competition with other surfacing materials) and capture of a greater international market share. It is clear that automation of tile manufacture has led to improved manufacturing practice at a reduced cost. The capital investment in an automated facility is similar to conventional operations. The technology has now been implemented worldwide in turnkey tile plants "exported" from Italy.

It may be noted that the cost of labor is a major factor across the two-decade span, but with automation the productivity per employee nearly tripled and employee cost per square meter of tile decreased by about 35%. There was a shift to a more highly trained labor force providing greater turnover and product value. Table 1 shows that tile manufacture is intermediate among major ceramic products in terms of employee productivity and sales value. In general, products lower in value tend to be manufactured in highly automated plants. The use of high-volume, low-cost production is necessary for profitability in such markets except where specialty shapes or properties provide higher-than-usual value added. Significant advances are being made in the volume and yield of sanitaryware and tableware. Many process steps are becoming more automated; where batch manufacture is still employed concepts such as "just-in-time" manufacturing, inventory control, and plant flow of product are improving productivity and lowering rejects. Technical ceramics vary widely in labor (and other) costs depending on the level of value added in manufacture, applicability of mass production methods, need for high-cost raw materials, and special property or shape requirements. It is therefore virtually impossible to make generalizations about technical and advanced ceramics.

Costs of body (Fig. 3 and 4) and glaze materials were influenced by a short-term increase in manufacturing costs and the cost of shipping raw materials in the late 1980s. Assuming constant material costs, the body costs would decrease slightly (about 10%) with lowered plant losses offset partially by a small increase to provide the more consistent body formulation required for automation. The glazing cost decreased more substantially because automation replaced the considerable hand labor that is required for traditional decorating. Loss of glazing material and defect losses decreased substantially. Packaging costs (Fig. 4) were also lower because of the elimination of hand labor. The greater reliance on mechanical systems caused increases in both machine-power consumption and equipment maintenance costs, while automated packaging machines lowered that cost segment.

The change in energy consumption is largely attributable to the incorporation of "fast-firing" technology. Historically, large volumes of material were loaded and heated simultaneously for both drying and firing steps. The newer approach employs a belt or roller mechanism in which a very thin support is used. This significantly reduces the thermal load—in particular the nonproductive heating of kiln furniture. The throughput speed is maximized, and residual heat from the firing cycle is used to dry the tile. A 40 to 60% productive usage of heat energy results as compared to an energy efficiency about half of this in conventional firing. This fast-firing paradigm is being applied increasingly to ceramic production. Even when a batch process cannot be converted to a continuous, automated one, considerable savings can be achieved by reducing the thermal load of support materials and firing as rapidly as possible in an energy-efficient furnace. Many plants established firing profiles designed for the largest ware to be fired. Separate firing of large pieces, which require a slower firing profile, can improve efficiency by permitting optimal firing profiles for each shape and size. It has been found

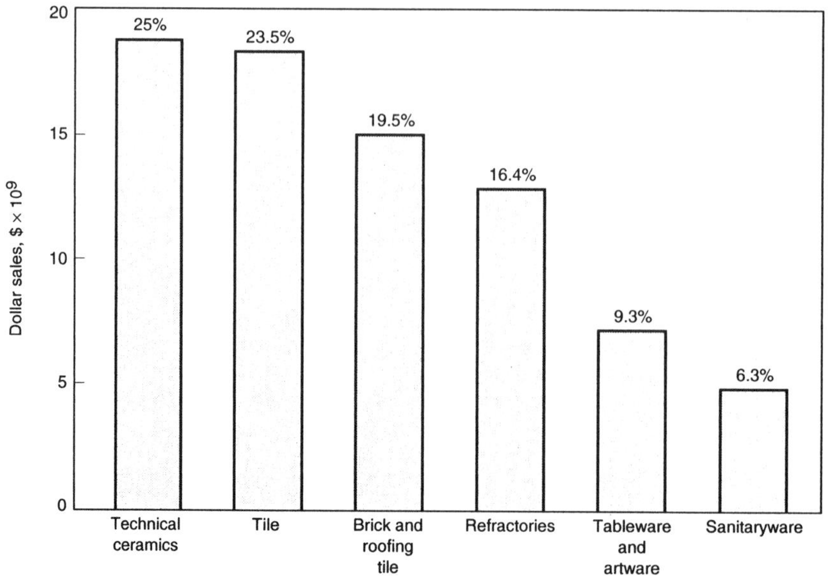

Fig. 2 Estimated worldwide sales of traditional/industrial ceramics (1992). Source: Ref 1–3

to both decrease fuel costs and increase productivity. The latter factor is quite important in a capital-intensive industry in which the investment return of major production equipment is an important consideration.

Shipping is also an important cost factor, although not included in Fig. 3 and 4. For lower-cost, lower-profit margin-products such as construction materials, shipping costs of both finished product and raw materials can be a major factor. Shipping needs often dictate that the plant be placed close to sources or customers, or where lower-cost shipping methods are available.

Design Approaches

Ceramics are frequently specified by design engineers because of their mechanical strength at high temperature, thermal stability, electrical and thermal insulating properties, hardness, and resistance to harsh chemical environments (durability). Special electrical, magnetic, and optical properties also dictate ceramics for various applications. Structural applications of ceramics that are particularly important are construction materials (brick, pipe, tile, etc.), applications where chemical inertness is required (labware, plant piping, tile, etc.), and structures for use at temperatures above about 500 °C (900 °F). Ceramics might be effective design choices for structures for which metals or plastics may be effective at lower temperatures or in less-aggressive environments.

The specific material to be selected depends on its basic properties. Thus a material may be selected based on the chemistry required to achieve the overall behavior desired. Tables 2 and 3 provide key mechanical and other properties for many important whiteware and technical/advanced ceramics, respectively, employed in structural applications. These tables may form an initial guide to materials selection. The properties described refer to materials fired to near-full density. As fired density decreases (due to shorter firing time or lower soak temperature) a given ceramic will generally show poorer mechanical properties, although well-designed high-porosity materials may show adequate strength and are important for thermal insulation applications. As the amount of lower-softening-point glass increases for a whiteware or sintering additive increases for a particular ceramic, full firing can be accomplished more easily. However, high-temperature strength is usually proportionately lower because of a less thermally resistant glass for whitewares or an increase in the amount of grain-boundary phase (usually glassy material) for technical ceramics. Strength and chemical durability is also usually lowered at ambient temperatures as the additive content and consequent firing temperature are reduced.

Oxide materials are usually somewhat less expensive than nonoxide ceramics, as their raw material and production costs are lower. Earthenwares and lower-quality whitewares, made from naturally occurring raw materials, generally have the lowest properties and cost with higher porosity (water absorption). The lowest strength is exhibited by many construction materials for which increased load-bearing mass in a structure compensates for lower relative strength. As the density, continuity, and softening point of the vitreous (glassy) phase increases in such materials, the strength, hot strength, and chemical durability usually increase—as does the cost. These property improvements often involve the use of chemically derived raw materials such as alumina. These materials—used as fine china, porcelain, and electrical porcelain—are at the upper end of whiteware properties. Oxide ceramics made chiefly from chemically derived components occupy the next level of resistance to mechanical, chemical, and thermal stresses. Generally, as the amount of second-phase additive and porosity decreases the properties improve. Nonoxide ceramics are usually even more costly and difficult to process, but provide superior mechanical properties and thermal stability, making them particularly applicable for elevated-temperature mechanical applications. However, caution must be used as many nonoxide ceramics oxidize at sufficiently elevated temperatures in air unless a protective oxide "skin" forms.

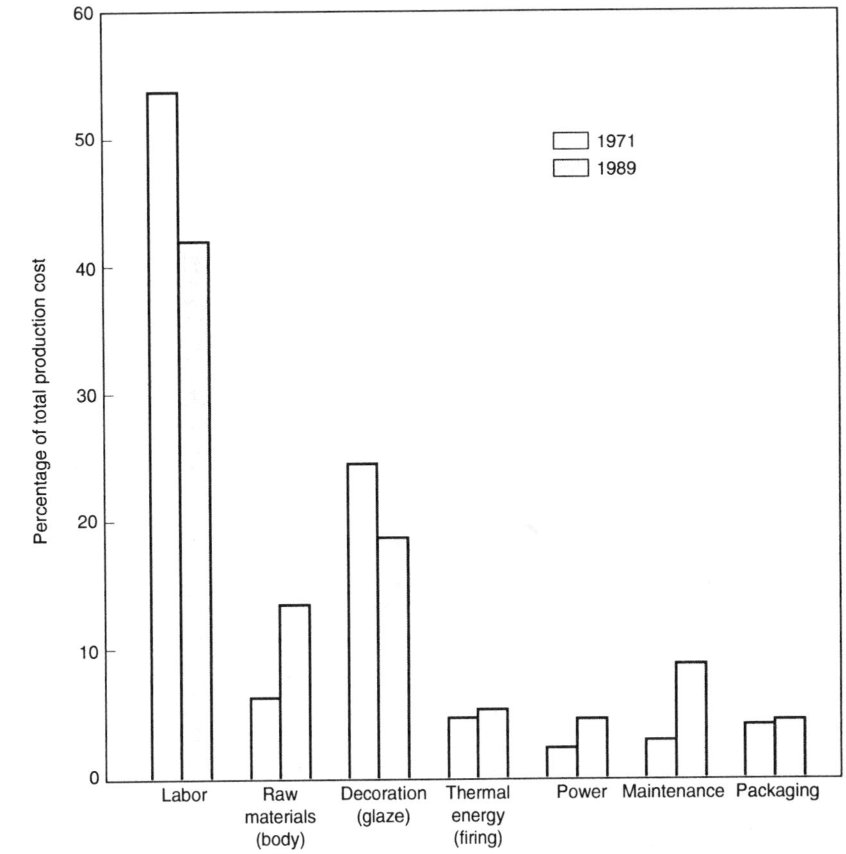

Fig. 3 Change in costs of Italian tile production 1989 versus 1971 (square meter basis). Source: Ref 1–3

A useful approach to material selection in ceramic design is to consider a high-performance technical alumina (aluminum oxide), such as 96% alumina, as a starting material. This type of material usually consists of a high-quality alumina raw material into which is milled a 4% (by weight) additive usually containing oxide combinations of silicon, magnesium, and calcium; water and organics that "burn out" on firing are not considered in this formulation. (Caution must be taken in evaluating "96% alumina" or any other ceramic "designation." This is not a specific designation, and significant differences can exist in composition, microstructure, porosity, and consequent properties from manufacturer to manufacturer and with the particular production process selected. This situation holds for most ceramics, and understanding of or advisement in the basics of ceramic engineering can be vital for critical evaluation of materials selection and design.) This is a technical ceramic formulation with moderate cost used for applications such as electronic substrates, mechanical parts, and high-performance chemical ware. Such a material offers very good mechanical performance in terms of strength, toughness, hardness, and wear resistance as well as good chemical durability, high-temperature stability, and thermal shock resistance. If the properties are nearly satisfactory a higher grade of alumina can be used with lower additive content and a somewhat higher material cost.

For still higher performance at elevated temperature, the material selected would move toward silicon carbide or silicon nitride (hot pressed) at increased cost. Ceramic composite materials reinforced with particulate, fiber/whisker, or continuous fiber offer increased

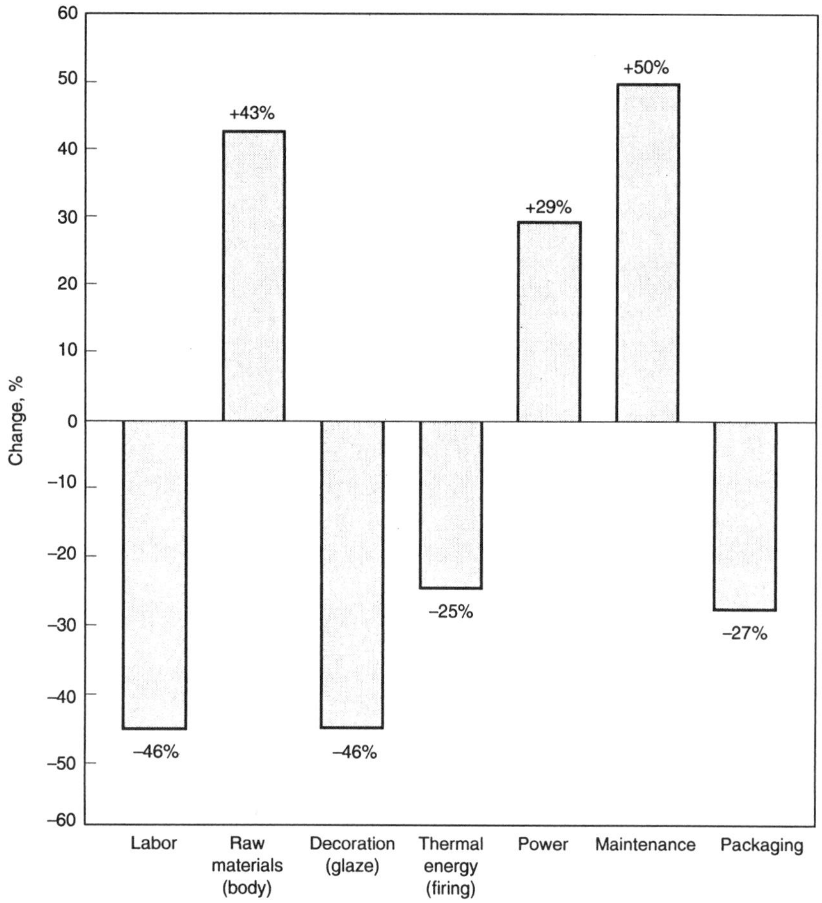

Fig. 4 Relative change in tile productions cost factors (1989 relative to 1971). Source: Ref 1–3

protection against catastrophic failure at elevated temperature, but with increased cost. If greater ambient temperature toughness or wear resistance is required, a transformation (zirconia) toughened alumina or zirconia might be employed, each with greater performance and greater accompanying cost. On the other hand, if 96% alumina exceeds requirements a higher glass content alumina can be used with some material economy. If a significantly lower set of mechanical (or other) properties will satisfy requirements, a clay-based whiteware formulation can be chosen. Technical whitewares such as electrical porcelains offer slightly lower performance, with porcelains, stoneware, earthenware, and structural clay ceramics yielding progressively lower mechanical properties at lower cost. A similar logic can be followed for thermal and durability properties with some consideration of specific behaviors of particular materials.

Empirical Design. Most ceramic design is performed empirically, that is by a trial-and-error approach based on past experience. Some consideration of the loads to be experienced and the properties of the ceramic may be included. If the material fails, a modification is made in material, part design, or the applied forces. This is done iteratively until satisfactory performance is achieved. Mechanics models of the stresses in the object and average of material strength or strength distribution may provide a guideline to material selection and design modification. If a part performs properly, it is often used although it may exceed design requirements. However, consideration should be given to whether a ceramic with reduced properties or a modified design might be more cost effective. This approach has been used quite successfully, particularly when failure of the ceramic device or component does not lead to a catastrophic failure hazard. Many specifications for construction materials rely on this empirical approach; experience shows that brick or tile will perform correctly if conventional formulations are used and a minimum test standard is met. A manufacturer of plates or crucibles will similarly use a new material or modify product shape using empirical testing methods.

Deterministic design employs average properties of a material in a particular condition. This method is often applied to metals because an alloy will show, for a particular heat and working treatment, a rather well-defined yield strength and/or ultimate strength with a standard deviation within a few percent of the mean value. The yield or ultimate strength is divided by a safety factor and is used to provide a satisfactory design in calculated continuum mechanics models such as those provided by finite element analysis (FEA) methods. An additional safety factor preventing catastrophic failure is provided by the ductility of the metal. Deterministic methods can be used for ceramics, but the typically brittle nature of ceramic materials can result in a wide variability in failure strength. The strength of a part is dictated by very small flaws manufactured in and induced during use on the object. A deterministic approach requires a very high safety factor because of the flaw sensitivity of ceramics. Much effort toward improving mechanical performance of ceramic materials centers on the elimination of intrinsic flaws produced in forming and firing and extrinsic (surface) flaws formed by postmachining, mechanical abrasion, or environmental attack.

Probabilistic Design (Weibull Analysis). The most rapidly growing approach to ceramic design is the probabilistic method (see the article "Design with Brittle Materials" in this Volume). It permits a fully quantitative approach, as does deterministic design, but deals with the wide variability of strengths dictated by the distribution of flaws in a brittle ceramic. The method accounts for the wide distribution of strengths in conducting the design protocol. It is often integrated with computer FEA models to allow the effect(s) of materials modification, change in flaw population, and part design to be iterated to a successful design in terms of the stresses calculated by finite element analysis. The finite element net is combined with the known statistical failure behavior (flaw distribution) of the ceramic. This protocol is most frequently based on Weibull statistics (Ref 4), which consider the cumulative probability of failure versus failure stress for each element (usually each sample or part).

A typical Weibull plot constructed on a log-probability scale is shown in Fig. 5(a). This plot considers 20 reaction-bonded silicon nitride samples. (While 20 elements appear adequate for this plot, it would usually be considered a small base for design. A common rule of convenience is to test more than 50 samples.) The plot is constructed by ordering the n strength values and assigning each element a probability equal to its rank (from 1 to n) divided by $[n + 1]$. This approximation to true probability usually leads to less than 5% relative error and allows for simple treatment of a moderate number of failure samples. The slope of the line in the plot (two lines in Fig. 5a) is termed the "Weibull modulus," m. (This "modulus" should not be confused with other mechanical moduli such as the modulus of elasticity.) A higher slope indicates a narrower distribution of failure stresses, that is, a narrower distribution of flaws or a more fracture-tough (damage-tolerant) material. A mean stress for failure can also be determined from the plot. The desired failure probability, for example, one part per million, is used as a design criterion, and the Weibull probability with strength is used to design to this performance requirement. Changes in selected material (fracture toughness), fabrication (intrinsic flaw distribution), finishing (extrinsic flaw distribution), and part configuration or applied forces (stress distribution), can be

Table 2 Properties of ceramics

Material	Crystal structure	Theoretical density, g/cm^3	Knoop or Vickers hardness, GPa	Knoop or Vickers hardness, 10^6 psi	Transverse rupture strength, MPa	Transverse rupture strength, ksi	Fracture toughness, $MPa\sqrt{m}$	Fracture toughness, $ksi\sqrt{in.}$	Young's modulus, GPa	Young's modulus, 10^6 psi	Poisson's ratio	Thermal expansion, $10^{-6}/K$	Thermal conductivity, $W/m \cdot K$
Glass-ceramics	Variable	2.4–5.9	6–7	0.9–1.0	70–350	10–51	2.4	2.2	83–138	12–20	0.24	5–17	2.0–5.4(a), 2.7–3.0(b)
Pyrex glass	Amorphous	2.52	5	0.7	69	10	0.75	0.7	70	10	0.2	4.6	1.3(a) 1.7(c)
TiO_2	Rutile tetragonal	4.25	7–11	1.0–1.6	69–103	10–15	2.5	2.3	283	41	0.28	9.4	8.8(a)
	Anatase tetragonal	3.84	…	…	…	…	…	…	…	…	…	…	3.3(d)
	Brookite orthorhombic	4.17	…	…	…	…	…	…	…	…	…	…	…
Al_2O_3	Hexagonal	3.97	18–23	2.6–3.3	276–1034	40–150	2.7–4.2	2.5–3.8	380	55	0.26	7.2–8.6	27.2(a), 5.8(d)
Cr_2O_3	Hexagonal	5.21	29	4.2	>262	>38	3.9	3.5	>103	>15	…	7.5	10–33(e)
Mullite	Orthorhombic	2.8	…	…	185	27	2.2	2.0	145	21	0.25	5.7	5.2(a), 3.3(d)
Partially stabilized ZrO_2	Cubic monoclinic, tetragonal	5.70–5.75	10–11	1.5–1.6	600–700	87–102	(f)	(f)	205	30	0.23	8.9–10.6	1.8–2.2
Fully stabilized ZrO_2	Cubic	5.56–6.1	10–15	1.5–2.2	245	36	2.8	2.5	97–207	14–30	0.23–0.32	13.5	1.7(a), 1.9(g)
Plasma-sprayed ZrO_2	Cubic, monoclinic, tetragonal	5.6–5.7	…	…	6–80	0.9–12	1.3–3.2	1.2–2.9	48(h)	7	0.25	7.6–10.5	0.69–2.4
CeO_2	Cubic	7.28	…	…	…	…	…	…	172	25	0.27–0.31	13	9.6(a), 1.2(d)
TiB_2	Hexagonal	4.5–4.54	15–45	1.5–6.5	700–1000	102–145	6–8	5.5–7.3	514–574	75–83	0.09–0.13	8.1	65–120(i), 33–80(j), 54–122(k)
TiC	Cubic	4.92	28–35	4.0–5.1	241–276	35–40	…	…	430	62	0.19	7.4–8.6	33(a), 43(d)
TaC	Cubic	14.4–14.5	16–24	2.3–3.5	97–290	14–42	…	…	285	41	0.24	6.7	32(a), 40(d)
Cr_3C_2	Orthorhombic	6.70	10–18	1.5–2.6	49	7.1	…	…	373	54	…	9.8	19
Cemented carbides	Variable	5.8–15.2	8–20	1.2–2.9	758–3275	110–475	5–18	4.6–16.4	396–654	57–95	0.2–0.29	4.0–8.3	16.3–119
SiC	α, hexagonal	3.21	20–30	2.9–4.4	(l)	(l)	(m)	(m)	207–438	30–70	0.19	4.3–5.6	63–155(a), 21–33(d)
	β, cubic	3.21	…	…	…	…	…	…	…	…	…	…	…
SiC (CVD)	β, cubic	3.21	28–44	4.1–6.4	(n)	(n)	5–7	4.6–6.4	415–441	60–64	…	5.5	121(a), 34.6(g)
Si_3N_4	α, hexagonal	3.18	8–19	1.2–2.8	(o)	(o)	(p)	(p)	304	44	0.24	3.0	9–30(a)
	β, hexagonal	3.19	…	…	…	…	…	…	…	…	…	…	…
TiN	Cubic	5.43–5.44	16–20	2.3–2.9	…	…	…	…	251	36	…	8.0	24(a), 67.8(q), 56.9(r)

(a) At 400 K. (b) At 1200 K. (c) At 800 K. (d) At 1400 K. (e) At 350 K. (f) 8–9 (7.3–8.2 at 293 K, 6–6.5 (5.5–5.9) at 723 K, and 5 (4.6 at 1073 K, in units of $MPa\sqrt{m}$ ($ksi\sqrt{in.}$). (g) At 1600 K. (h) 21 (3) at 1373 K, GPa (10^6 psi). (I) At 300 K. (j) At 1100 K. (k) At 2300 K. (l) Sintered: 96–520 (14–75) at 300 K, and 250 (36) at 1273 K. Hot pressed: 230–825 (33–120) at 300 K, and 398–743 (58–108) at 1273 K, MPa (ksi). (m) Sintered: 4.8 (4.4) at 300 K, and 2.6–5.0 (2.4–4.6) at 1273 K. Hot pressed: 4.8–6.1 (4.4–5.6) at 300 K, and 4.1–5.0 (3.7–4.6) at 1273 K, $MPa\sqrt{m}$ ($ksi\sqrt{in.}$). (n) 1034–1380 (150–200) at 300 K, and 2060–2400 (300–350) at 1473 K, MPa (ksi). (o) Sintered: 414–650 (60–94). Hot pressed: 700–1000 (100–145). Reaction bonded: 250–345 (36–50), MPa (ksi). (p) Sintered: 5.3 (4.8). Hot pressed: 4.1–6.0 (3.7–5.5). Reaction bonded: 3.6 (3.3), $MPa\sqrt{m}$ ($ksi\sqrt{in.}$). (q) At 1773 K. (r) At 2573 K

treated iteratively to yield a final design in terms of material, production requirements, and part configuration for a particular set of applied forces and environment. Weibull statistics are discussed in greater detail in the article "Design with Brittle Materials" in this Volume.

The Weibull plot can also be used in several other approaches to improved material or part performance. In Fig. 5(a) two line segments are shown (solid and dashed). The high-strength failures (dashed line) have a higher Weibull modulus (slope) and mean strength than those elements on the solid line, which exhibit greater variability in strength and a lower mean strength. Usually the lower slope and mean indicates a different, more severe flaw for corresponding samples. The overall performance can be improved by examining these severe failure-inducing flaws and/or the specifics of production for the low m samples and correcting matters. There are two ways to improve performance by material improvements and/or flaw reduction, as shown in Fig. 5(b). The mean failure value can be increased to yield a higher overall strength, or the Weibull modulus can be increased, improving the reliability by decreasing the distribution of failure values. The improvement in modulus decreases the low-strength portion of the population and allows a higher-use stress at the same failure probability. Some caution should be taken as a high modulus is not necessarily good—severely abraded material may have very low strength but a high modulus because similar large flaws have been introduced throughout all parts. Naturally, a change in material or production method that increases both the slope and mean would be most desirable.

It should be noted that the Weibull modulus approach provides useful tools for improved design of materials and processes. Nondestructive evaluation methods can be used to eliminate samples with large flaws and consequent low strength. If the flaws belong to a separable group of low strength values such that the solid line in Fig. 5(a) can be drawn, the Weibull model can be used unmodified. Another application of this concept is to mechanical proof testing. As diagrammed in Fig. 5(c), all produced parts are subjected to stress at a predetermined proof test level.

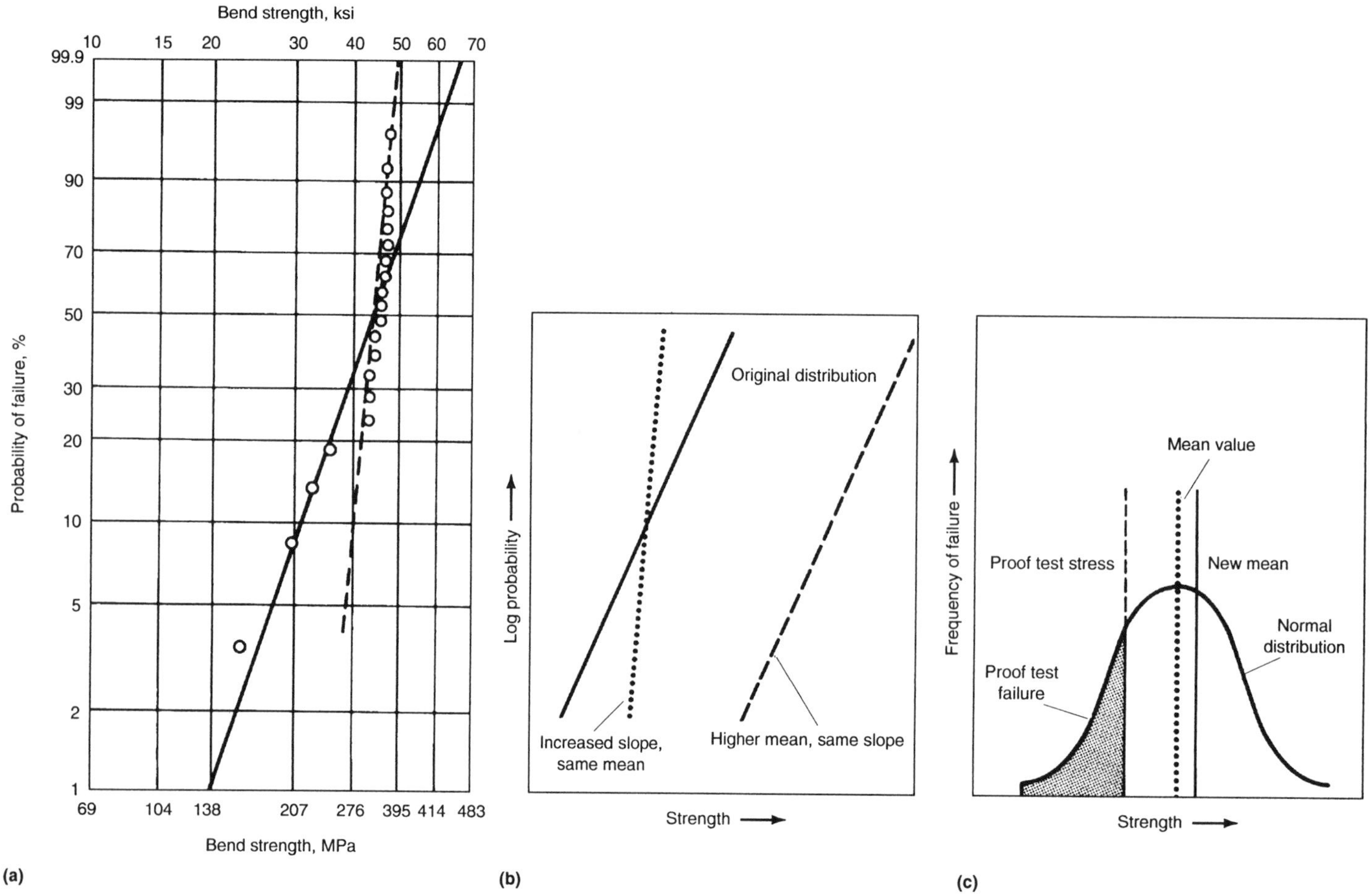

Fig. 5 Statistical aspects of material design. (a) Weibull plot for reaction-bonded silicon nitride. (b) Improved Weibull distributions. (c) Modification of normal distribution by proof testing

Parts with large, effective flaws fail (shaded portion) and are eliminated from the population. A new, higher mean value of strength results. There are two cautions:

- If the initial distribution is normal, as shown (one flaw type), normal statistics can no longer be applied.
- If flaws, particularly cracks, propagate during proof testing, the strength and/or Weibull modulus will deteriorate relative to application.

Fortunately, there is significant evidence that little or no deterioration need occur during proof tests if nonimpact types of proof testing in dry environments (prevents chemical attack of flaws) are employed.

General Process Design

Figure 6 shows the overall scheme usually used in ceramic processing. It is a simplified rendering of the steps that are frequently developed by an iterative process involving design, materials, testing (often nondestructive evaluation), and process engineering. Each step and its interaction with others in the process flow stream need to be compatible and optimized to produce a sound, high-quality product. For example, the raw materials selected and their milled particle size strongly affects whether the rheology is appropriate for a plastic-forming process. This in turn dictates drying and firing conditions and the level of defects, density, homogeneity, and anisotropy of the final part. After the overall scheme is introduced, each process area is discussed in overview. An overview of relationships between structure, processing, and properties can be found in the article "Effects of Composition, Processing, and Structure on Properties of Ceramics and Glasses" in this Volume. Detailed presentations of design aspects of ceramic processing can be found in *Ceramics and Glasses,* Volume 4 of the *Engineered Materials Handbook,* and other selected references provided at the end of this article.

Figure 6 is divided into various stages. First, the starting materials need to be selected. Subtle, even trace, differences in starting chemistry or structure can alter the mechanical strength, chemical durability, thermal resistance, conductivity, and aesthetics of a ceramic part. For example, a fraction of a percent of additive can substantially alter the high-temperature strength of a ceramic, decreasing the practical temperature limit of use by many hundreds of degrees. Starting materials are usually powders of natural or derived materials.

The raw materials and appropriate additives are next blended, and particle size distribution adjusted in a milling and mixing step. In blenders, the particles are agitated by either motion of stirrers or of the container. Mills provide particle reduction and break-up of agglomerates by the inclusion of "media" or by other means of applying kinetic energy to the ceramic particles. Various types of mills are available, including ball mills, attrition mills, and jet mills. The powder can then be further prepared for forming by classifying particle sizes and eliminating large particles and agglomerates. The powder can be aggregated by methods such as granulation or spray drying for ease of handling or good forming properties.

The material is then formed by one of three general approaches with decreasing liquid phase (water) content: wet (casting, tape casting), plastic (extrusion, plastic pressing, jiggering, injection molding), or dry (dry pressing, isostatic pressing). The shaped object is then dried. After forming and drying, the part must have sufficient green (unfired) strength for handling and to resist

Table 3 Physical properties of whitewares

Property	Earthenware	Hard porcelain	Bone china	Hotel china	Normal electrical porcelain	High-strength electrical porcelain
Water absorption, %	6–8	0.0–0.5	0.0–1.0	0.1–0.3	0.0	0.0
Specific gravity	2.6	...	2.75	2.6	2.4	2.8
Bulk density, kg/m^3	2200	2400	2700	2600	2400	2770
Compressive strength, MPa (ksi)	...	400 (58)	...	...	700 (102)	700 (102)
Tensile strength, MPa (ksi)	...	23–34 (3–5)	...	...	35 (5)	56 (8)
Modulus of rupture, MPa (ksi)	55–72 (8–10)	39–69 (5.5–10)	97–111 (14–16)	82–96 (12–14)	105 (15)	175 (25)
Modulus of elasticity, GPa (10^6 psi)	55 (8)	69–79 (5.5–11.5)	96 (13.9)	82 (12)	...	...
Linear thermal expansion, μm/m · K						
20–500 °C	7.3–8.3	...	8.4	7.3–8.3	5.7	6.7
20–1000 °C	...	3.5–4.5	...	...	...	...
Thermal conductivity, W/m · K at 20–100 °C	1.26	...	1.26	...	...	...
Dielectric constant at 1 megacycle	...	...	...	...	5.6	6.9
Power factor at 1 megacycle	...	...	...	...	0.8	0.7

the formation of defects during further processing. The part can then be surface finished by green machining to near final shape (accounting for firing shrinkage), and surface finishes can be applied such as glazes. The part is then fired to the desired density and structure. Final machining to tolerance or postfiring coatings can then be applied. Often, the ceramic may go through a partial process and return to prior process steps. For example, powder can be formed to pellets, fired, and then be reground to become part of a ceramic formulation. Fired parts or partially (bisque) fired parts can be coated with raw glaze and refired. The chart given in Fig. 6 should be viewed as an introductory guide to design processing. It should be recognized that many variations are made in practice on the steps shown.

Ceramic Raw Materials and Formulation

Traditional ceramics are usually made from beneficiated raw materials extracted from the earth and partially purified. As the density, strength, chemical durability, and other properties improve, the raw materials are required to have more controlled composition, greater freedom from foreign matter, and consistent particle size. Traditional ceramics are usually formulated to yield a triaxial body based on the combination of silica, clay, and flux. For some applications, refractory minerals may also be included. A similar concept can be employed for technical oxide ceramics made from chemically derived materials.

The term *triaxial* refers to triaxial phase diagrams based on common mineral raw materials such as those shown in Fig. 7. The corners of the diagram indicate either pure, single metal/oxide materials (Fig. 7a) or minerals (Fig. 7b). In Fig. 7(a) the various phases and stability fields are indicated. A particular composition can be determined by combining raw materials to yield a final result determined by the proportional distances from the "pure" phase vertices. The equilibrium phase diagram shown in Fig. 7(a) is usually used for technical ceramics and pure, chemically processed raw materials would be used. For example, alumina-based ceramics would be produced if the major constituent Al_2O_3, chemically derived from bauxite ore, were combined with minor additions of finely divided quartz (silica, SiO_2) and magnesia (MgO). The complex composition plane of Fig. 7(b) is more useful for whiteware ceramics, which are derived from ceramic raw materials involving more than three oxides. Raw materials and pure minerals are indicated along the sides and vertices, for example, silica, metakaolin (kaolin with its structural water removed), mullite, and feldspar (as leucite or potash feldspar). Regions for various whiteware types are shown inside, for example, hard porcelain, sanitaryware, and electrical porcelain. As with Fig. 7(a), the regions of thermal phase stability are shown. Such diagrams are used to design final ceramic formulations.

The starting materials may combine several raw materials; for example, alumina may be part of a clay, flux, and refractory oxide, which together supply the total amount of alumina supplied. The actual reaction behavior must be understood because some of the raw material may not react fully during firing. Techniques of batch formulation are used to arrive at the final composition required and involve considerations of organic and water loss as well as cost factors. Sophisticated computer programs have been developed, many company proprietary, to optimize formulation in terms of green forming, firing, final properties, and costs.

Traditional Ceramics. For traditional ceramics, the silica fraction of the triaxial body is usually obtained from rather pure deposits of quartz in the form of beach sand or crushed quartz rock, which is washed, ground, and classified (usually 200 to 400 mesh). A wide variety of clay and mineral materials and hydrated minerals can be used, with kaolin ($Al_2O_3{\cdot}2SiO_2{\cdot}2H_2O$) and talc ($3MgO{\cdot}3SiO_2{\cdot}2H_2O$) being the most common mineral forms. Clays usually provide the plastic fraction of a ceramic formulation when mixed with a small amount of water. China clays, such as kaolinite, are chiefly composed of kaolin and are a principle ingredient of most whiteware ceramics. Ball clays are composed of very fine clay particulate with significant quantities of quartz and organic material. They are particularly useful in traditional ceramic formulations because they provide excellent plasticity and green strength to the body, which are useful for working (shaping) the green ceramic. Fluxes contain alkali, which promotes the fusion of silica, alumina, and aluminosilicates thereby allowing firing at relatively low temperatures. Feldspars and nepheline syenite are common fluxing ingredients. Refractory minerals are usually added for elevated-temperature use when high-temperature strength and chemical resistance are required. A variety of oxides, carbides, and other materials are used.

Advanced ceramics tend to use processed, high-purity, chemically derived powders such as alumina, zirconia, silicon nitride, aluminum nitride, and silicon carbide. Technical ceramics use raw materials both chemically derived and of traditional formulation. As the performance requirements become more critical, chemical purity, controlled particle size (distribution), reactivity, and freedom from agglomeration become more important. The optimal particle size distribution is still a matter of disagreement. An extremely fine, uniform, submicrometer particle size can yield ordered packing, provide freedom from pore defects, provide high sintering or reaction rates, lower sintering time, and prevent exaggerated grain-growth defects. Achieving full density may be difficult because uniform spheres pack with about a third void space. Forming times are long, it is difficult to obtain good forming rheology, and prevention of agglomeration is critical. A broad distribution of particle sizes with many orders of magnitude of particle size down to extremely fine (nanometer) scale will provide denser green packing, and desirable forming rheology is far easier to achieve. Forming time is shortened, although defects (large pores and exaggerated grain growth) require careful control to prevent.

Preparation of Materials

The various raw materials are typically ground together to provide intimate mixing and particle size reduction. The configuration, speed, and time duration of milling can significantly alter the intimacy of mixing and ultimate particle size distribution. As milling time increases, the relative improvement in mixedness or particle size reduction decreases in a quasi-logarithmic fashion.

Milling is usually done wet with the water or other liquid often being part of the final formulation. In some cases dry milling is done, although many materials are dried after milling.

For slip casting, powders are usually dispersed in water or nonaqueous media resulting in a "slip," a fluid slurry with thixotropic rheology. It is desired to have a material that flows when stirred yet maintains its shape after discharge into a mold. A clear understanding of colloid chemistry and rheological control is vital to proper slip preparation (a similar understanding is important to plastic forming). Factors that affect slip (or plastic working) rheology include: solids fraction, liquid fraction, dispersant(s) type and quantity, particle size distribution, particle surface chemistry, pH, and order (timing) of additions. Both traditional clay ceramics and nonclay technical ceramics can be prepared as slips and plastic masses; however, nonclay materials require greater additive and time control because they have no "natural" plasticity.

Ceramic powders used in plastic forming employ similar additives and liquids to those used in slip casting, but the fluid content is reduced and additive formulation adjusted. A stiff plastic body must be prepared suitable to the particular plastic process, applied stress, and forming rate. Preparation is usually performed in pug mills or sigma blade mixers that provide high shear. The body must show a suitable yield point (stress at which flow begins) and maintain shape after forming. Bingham viscoelastic behavior (linear relation between shear stress and shear rate with a yield stress) is usually desired. Clay bodies usually can provide the required rheology with minor additives such as dispersants (deflocculants) and through control of clay fractions and particle size distribution. For technical and advanced ceramics that contain no clay fraction organic binders, plasticizers and lubricants are added to obtain appropriate plastic rheology.

Dry pressed powders must usually be granulated or spray dried to provide free-flowing powders that will readily fill a mold. Spray drying usually provides spherical aggregates of powder particles ideal for flow and die filling. Spray drying also permits the uniform incorporation of multiphase materials, binders, lubricants, and other additives. Service firms will spray dry material for those who cannot justify the capital investment of a spray dryer.

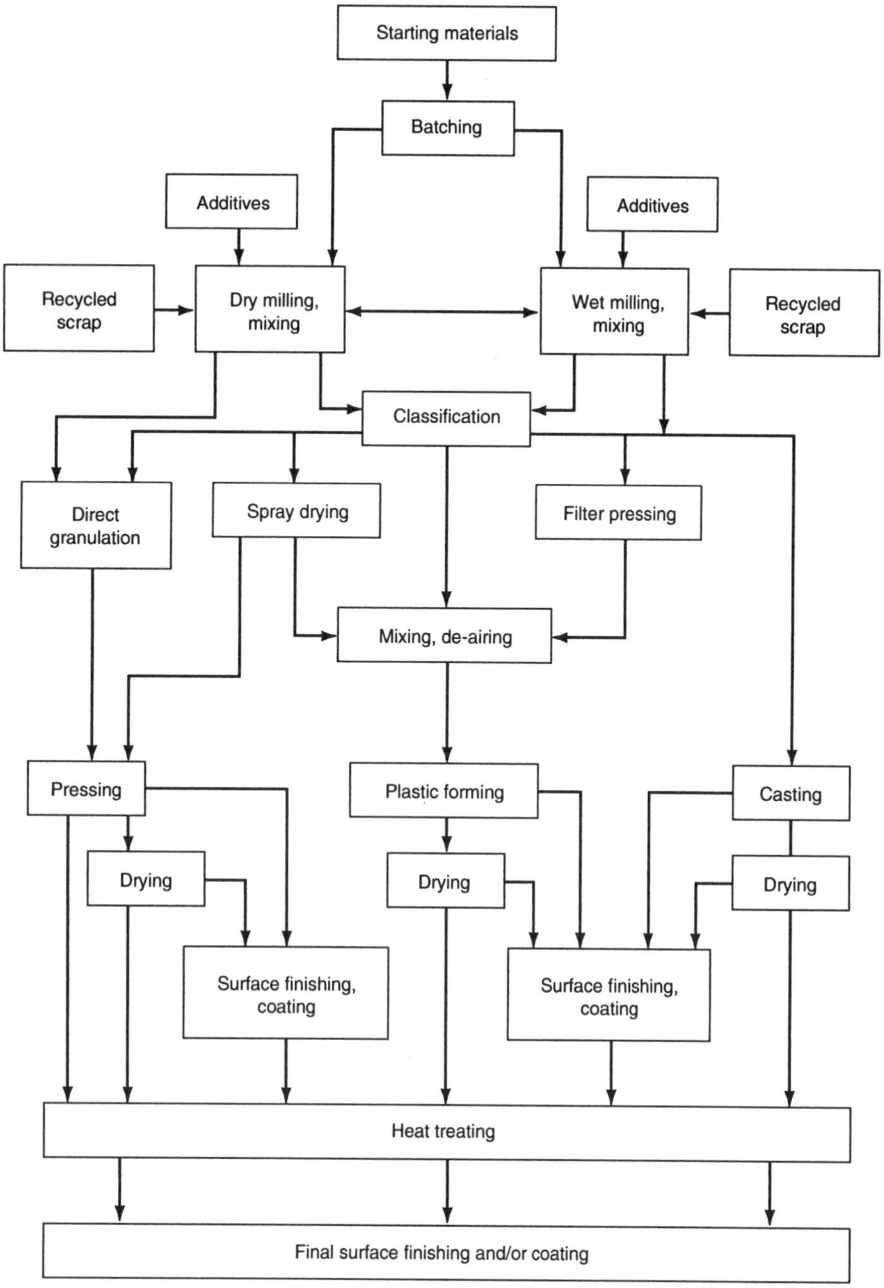

Fig. 6 General ceramic processing flow chart

Forming Processes

Many forming processes used in ceramics such as dry pressing, extrusion, plastic forming, and injection molding are similar in principle to processes used in the metals, plastics, and food industries. Experience in these areas can be used insofar as rheological behavior is similar. The abrasiveness of ceramics requires that production equipment otherwise similar to other industries be hardfaced to avoid wear and contamination. Slip casting, tape casting, and jiggering are forming methods more specifically developed for ceramic processing. Table 4 summarizes the suitability of various forming processes. The great variability of practices and methods for those described and the number of specialty processes used means that there is significant variation from Table 4, and the descriptions in both Table 4 and this section should be used as a preliminary guide only.

Wet processing methods include slip casting, pressure casting, and tape casting.

Slip casting provides great flexibility of both part shape and size. The major incremental cost for a new shape is development of a master mold. Multiple-use molds made of plaster and/or porous plastics (more recently) are produced from the master with controlled mold porosity—fine pores provide capillarity, coarse pores a larger liquid reservoir. Slip casting most commonly uses a low viscosity, clay-based ceramic slurry, but can also be used for nonclay ceramics with suitable additives for rheological control. Water is the common liquid vehicle. A layer of the slurry deposits on the mold surface as water is drawn off by the porous nature of the mold wall. Capillary suction continues to draw off water and deposit material from the slurry, building up the thickness of the cast structure. Fine fractions may migrate through the drying cake so that the surface microstructure may differ from the bulk. This difference may be desirable for surface finish or fired densification. However, inappropriate chemical segregation or drying/firing stresses can result if segregation is not controlled. Molds must be conditioned with slip before use and dried between casts to maintain casting behavior. After a number of casts either mold definition is lost or capillaries clog, preventing indefinite mold use. Vacuum suction can be applied to the outside of the mold to speed the capillary effect somewhat. Usually, many similar molds are used in sequence to maintain productivity. As a result, space usage is high, considerable labor is required and cycle times are long, although the molds are low-cost forming equipment with great shape and composition flexibility.

Mold time depends on the liquid content of the slurry and thickness of the cross section cast. The ideal slip maximizes solids content so that there is as little water to draw off as possible. Thick cross sections can take a very long time to produce because a parabolic decrease in casting rate occurs with time. For solid casting, a liquid reservoir must be provided in the mold design to provide additional slurry as water is drawn off and shrinkage occurs. Drain casting is used to provide a hollow internal cavity for shapes such as a vase or bowl. After the cake has built up to a controlled thickness the residual slurry is poured off and reused. For all shapes, edges and mold seams are trimmed green (unfired) with a knife and smoothed with a wet sponge. Shapes such as handles can be attached by applying a small amount of slurry to cast pieces and pressing pieces together until attached.

Pressure casting can be used to accelerate the casting speed of thick-walled or solid pieces. Gas pressure injects the slurry into the mold and helps to force the liquid through the mold capillaries. The pressure also allows a slightly lower fluid content in the slip. The pressure also causes an intimate conformance with the mold wall to provide superior definition and surface quality. If yet greater precision and definition is desired, fine particulate can be sprayed on a solid mold made of wax or plastic. This is done in layers with alternate furnace bakeouts and eventual buildup with coarser particulate slurry. Finally the mold is dissolved or burned out. The final product is used as a refractory mold for precision metal casting (investment casting, lost-wax casting).

Tape casting is used to produce thin sheets about 0.1 to 2 mm (4 to 100 mils) thick of ceramic, particularly for electronic substrates and multilayer capacitors. A volatile organic liquid is usually used with appropriate binders and plasticizers that provide rheological control, green strength, and flexibility to the tape produced. The material is drawn across a thin sheet of flexible, nonporous plastic using a doctor blade to control thickness. The continuous ribbon of slurry on plastic passes through a drying oven to evaporate the vehicle. The resulting green ceramic can be cut into pieces and fired after removal from the tape. It can also be metallized, stacked, and punched to produce multilayer capacitors and electronic substrates. Very fine particle sizes, usually of nonclay materials such as high alumina, titanates, and aluminum nitride are usual materials produced in this way. Fine particle size increases strength and permits a thinner tape to be produced.

Plastic Forming. The plastic nature of a clay-based composition makes possible the mechanical shaping of the mass by plastic forming. This has developed from the artisan's hand and potter's wheel shaping of moist clay into processes that allow the mass production of many identical objects. Nonclay ceramics are brought to a sufficiently plastic state by the addition of organic plasticizers. Jiggering represents an automated development of the potter's wheel. The plastic mass is pressed into a die (usually plaster to increase yield of the mass after forming) and

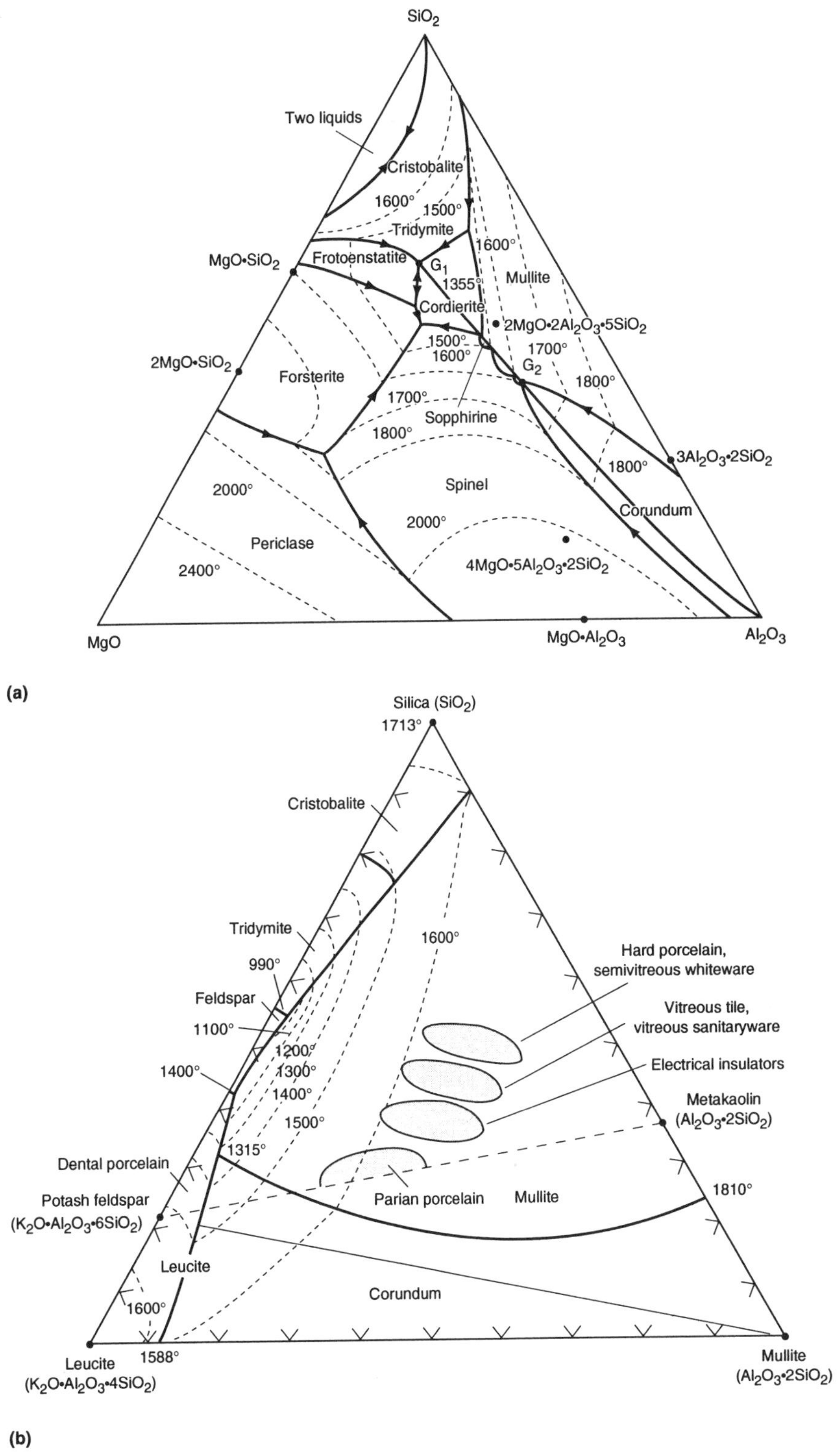

Fig. 7 Ternary phase diagrams for ceramic materials. (a) Diagram of the system MgO-Al_2O_3-SiO_2. Dashed lines show isotherms in degrees centigrade for the presence of liquid phase. (b) Areas of triaxial whiteware compositions shown on the silica-leucite-mullite phase-equilibrium diagram

formed on the reverse side with a jiggering tool that is a template of the shape desired. Only axially symmetric cross sections can be made, such as dinner plates or oval platters. In plastic pressing the material is pressed by dies in order to form the object. This can be automated to quite high speeds, especially for small, simple-shaped objects. The die can be made of metal or plaster. Plaster is used in order to partially dewater the plastic mass, thereby increasing its yield point (shape retention) at the expense of more rapid tool wear.

Extrusion rapidly forms a uniform cross section, such as for circular or square pipe or thermocouple tubes, by forcing the material out of a die aperture of appropriate shape using a ram or screw auger. The feed angles of the die must be appropriate to provide sound and efficient flow of the plastic mass, but this depends on the rheology of the ceramic at operating temperature and the extrusion rate. If the die angle is improper either inefficient production or defects such as central burst and surface cracks can occur. Aside from the use of abrasion-resistant linings, equipment is virtually identical to that used in plastic and food extrusion. A large cross section such as a large, high-tension insulator blank may be postmachined green to provide contours and varied external or internal diameters.

Injection Molding. Complex objects can be made to precise, near-net shape (accounting for firing shrinkage) using injection molding. A ram or screw feed forces the plastic mass into a mold. The process is similar to that used to form high-volume parts of engineering plastics. This method is not much used for clay-based ceramics because of equipment cost, the need for very high volume production, and the existence of many cost-effective alternate methods. Injection molding has been used particularly for making precise, complex, advanced ceramic shapes, such as turbocharger rotors, by incorporating materials such as silicon nitride into an organic plastic and solvent. Subsequent to forming the binder, material must be burnt out before sintering. This limits the object size to relatively thin cross sections. Work has been done to minimize the amount of binder and to design binders that will burn out rapidly. New water-based systems are being developed that may lower production costs and decrease binder content, burnout cost, and environmental impact of nonaqueous solvents.

Dry Processing. A nearly dry, free-flowing powder (usually spray dried) fills a metal mold and is compacted uniaxially in the process of dry pressing. This method is most suitable for high-volume production of small, simple, low-aspect-ratio shapes with fairly uniform cross section. Die and interparticle friction preclude forming large, long, and complex objects. Internal (in the ceramic) and external (on the die) lubricants are used to decrease friction, thereby facilitating forming and reducing density variations. Multisegment dies, multidirectional pressing, floating die cavities, and high time/pressure cycles can reduce pressure gradients and increase the flexibility of size and shape with high capital costs and part cycle times.

For minimization of density variations and production of more complex (or larger) shapes, isostatic pressing is used. This is often termed cold isostatic pressing (or CIPing) to distinguish it from hot isostatic pressing, a firing process. The nominally dry powder with internal lubricants is filled into a rubbery mold with the desired shape features. The mold is sealed and placed into a hydraulic fluid, which is then pressurized. A part is produced with very uniform density because of the multidirectional pressure and because significantly higher pressure can be achieved compared to that obtained by uniaxial pressing.

Machining. Whenever possible, it is desirable to machine a ceramic while it is still green (unfired). Conventional machining methods can be used and the piece may be machined either before or after drying depending on the strength, shape, and integrity of the part. In some cases, parts can be bisque (partially) fired to provide enough strength for the machining operation while avoiding firing a fully matured piece. Green and bisque firing may not provide critical tolerances, although careful control of both the machining and firing shrinkage may prove adequate. Machining after firing is quite costly (as much as 80% of total cost) and can introduce critical strength-limiting flaws in the material. Postfire machining usually requires diamond tools and slow material removal. New approaches are being developed to increase the machining speed and reduce the damage, thereby improving machining costs and productivity. One successful approach involves very stiff, very-high-speed, multiple-head, multiple-axis machines that are quite capital intensive.

Table 4 Evaluation factors for different ceramic forming methods

Forming methods	Large component sizes	Complex component shape	Independence of plasticity	Tolerance	Ceramic problems (defects)	Production volume	Production speed	Required plant space	Equipment cost
Wet forming									
Slip casting	1	1	3	5	5	5	5	5	1
Pressure casting	2	2	3	4	4	5	4	5	2
Tape casting	3	5(a)	4	1	2	1	1	2	4
Plastic forming									
Extrusion	2	3(b)	5	3	3	2	2	4	3
Plastic pressing	3	3	4	2	3	2	3	3	2
Jiggering	2	4(c)	5	4	4	3	4	3	2
Injection molding	5	5(d)	3	1	2	2	2	3	5
Dry forming									
Dry pressing	4	5(e)	1	1	1	1	1	3	3
Cold isostatic pressing	3(f)	4	1	1	2	4	3	4	4

Note: Under usual circumstances 1 is best, and 5 is worst. Variability in processing methods may deviate from common factors indicated. (a) Thin sheets only. (b) Uniform cross section. (c) Axially symmetric shapes. (d) Size limited by binder burnout. (e) Small to medium size, short aspect ratio. (f) Some very large shapes made

Drying

Drying involves the removal of free, nonstructural water or other solvents from a formed ceramic. It is a two-step process controlled both by internal flow of liquid to the surface and surface evaporation of the liquid. The internal flow can be increased by increasing permeability with a coarser structure, decreasing the viscosity of the liquid or increasing the concentration gradient to the surface (more rapid removal of surface liquid). Evaporation can be hastened by air motion (removal of saturated vapor), increased temperature, lower vapor pressure liquid, increased drying surface area, decreased air humidity, or reduced pressure.

Drying occurs in three stages, as shown in Fig. 8. At high moisture content, liquid is the continuous phase both in bulk and on the surface and evaporates almost as if it were free, continuous liquid. This is the constant rate period, region A of Fig. 8. In this region, evaporation dominates the drying rate. Most drying shrinkage occurs during this stage. An excessive drying rate can cause differential shrinkage in thick cross sections. Warping or cracking may result, dependent on whether the body can accommodate the drying stresses or not. When the surface film of liquid becomes discontinuous, the first falling-rate period with a linear drying rate begins, region B. Liquid is still continuous in the ceramic as large pores are being drained. There is a small amount of shrinkage in this region, so that the transition from the constant rate to the first falling-rate period is often considered the critical liquid content for shrinkage damage. However, some care must also be taken not to dry too rapidly in the

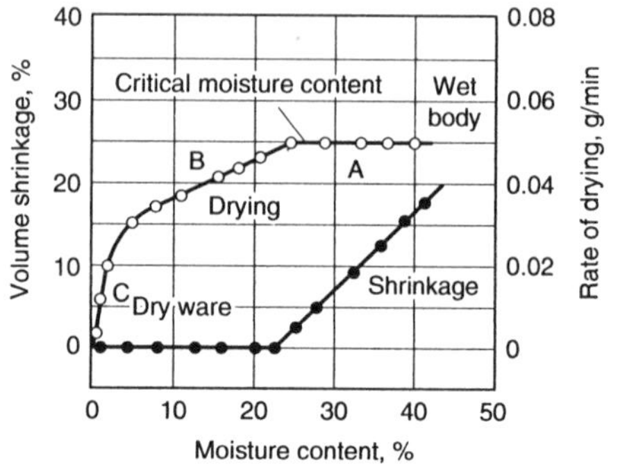

Fig. 8 Change in bulk volume on drying a ceramic body

first falling-rate period. In the second falling-rate period, region C, the drying rate is nonlinear. Air is the continuous phase, and liquid pockets are removed by evaporation. If liquid is distributed uniformly in the body, there is no shrinkage in the second falling-rate period. Additional causes of damage due to differential shrinkage are: mechanical restraint of the body, preferred particle orientation, uneven drying, and varying cross sections of the body.

Drying can be performed as a separate or integrated operation with firing. A drying oven or heat lamps can be used to remove liquid. There has been some success with industrial microwave drying, because of the ability to heat the object more uniformly in depth and thereby both speed up liquid removal and avoid drying stresses. Another trend is to reduce the mass or load and fast-dry parts, thereby increasing throughput and lowering fuel costs. The use of firing-waste heat for drying can also lead to cost reduction.

Firing

When a dry, green ceramic is exposed to elevated temperature, chemical and physical changes occur that consolidate the material, cause an irreversible shrinkage and increase strength and other properties. Vitrification and sintering are the two processes that generally pertain to ceramic firing.

Vitrification. Traditional clay-containing ceramics and other systems that are based on triaxial body formulations most commonly vitrify. Constituent materials react at elevated temperature to form a large amount of liquid phase. This liquid consolidates the body through capillary forces. Most of the liquid forms a continuous glassy phase upon cooling. Figure 9 shows a typical firing curve for a whiteware ceramic. The heating rates and temperature holds are dictated by the transformations that occur during heating and cooling. The correct amount and composition of viscous glass is needed to mature the part at an optimal rate and energy cost while preventing it from slumping and flowing. Usually, a support structure is used to prevent excessive deformation, and some slumping may be precalculated to bring the object to its desired final shape during firing.

Some of the glassy material may crystallize on cooling, forming microstructural constituents that depend on liquid chemistry and cooling rate. During cooling, crystal phase changes that will create internal cracks must be avoided. In addition to the dictates of material change that determine a complex firing profile, the shape of thick or large parts must be considered. The thermal profile of the object can introduce stresses on heating or cooling that will warp or fracture the part. Large, thick-walled, and complex shapes must be fired more slowly to prevent damage. Economies can be obtained by separating by size to optimize firing speed and minimize damage. Several manufacturers have developed computer algorithms to predict optimal firing curves for both traditional and advanced ceramics.

Sintering. Ceramics that do not produce a great deal of glassy phase during firing mature by the process of sintering. This is the densification by solid-state diffusion at elevated temperature driven by reduction of surface area. Several processes or stages can occur including evaporation-condensation and surface and bulk diffusion. Usually a small amount of sintering additive (sintering aid) is introduced to accelerate the sintering process by introducing liquid-phase sintering and/or accelerating diffusion rates. Other additives function as grain-growth inhibitors, preventing the competitive process of crystal growth (large crystals may deteriorate strength and other properties). The liquid phase usually forms a thin glassy layer on the grain (crystal) boundaries of the fired ceramic and has a substantial effect on mechanical, electrical, thermal, and magnetic properties. This grain-boundary film has been identified in virtually all manufactured ceramics— even those with trace amounts of additive or impurities. A careful balance must be struck between pore reduction by densification and grain growth. The sintering additives make ceramics prone to exaggerated grain growth, which may deteriorate properties. It is even possible for large, agglomeration pores to grow during firing. As with traditional ceramics, firing stresses must be controlled. The firing process must be understood and carefully optimized if a proper microstructure and properties are to result.

Firing Process Factors. Peak temperature during the soak period is usually considered the most important factor in either vitrification or sintering

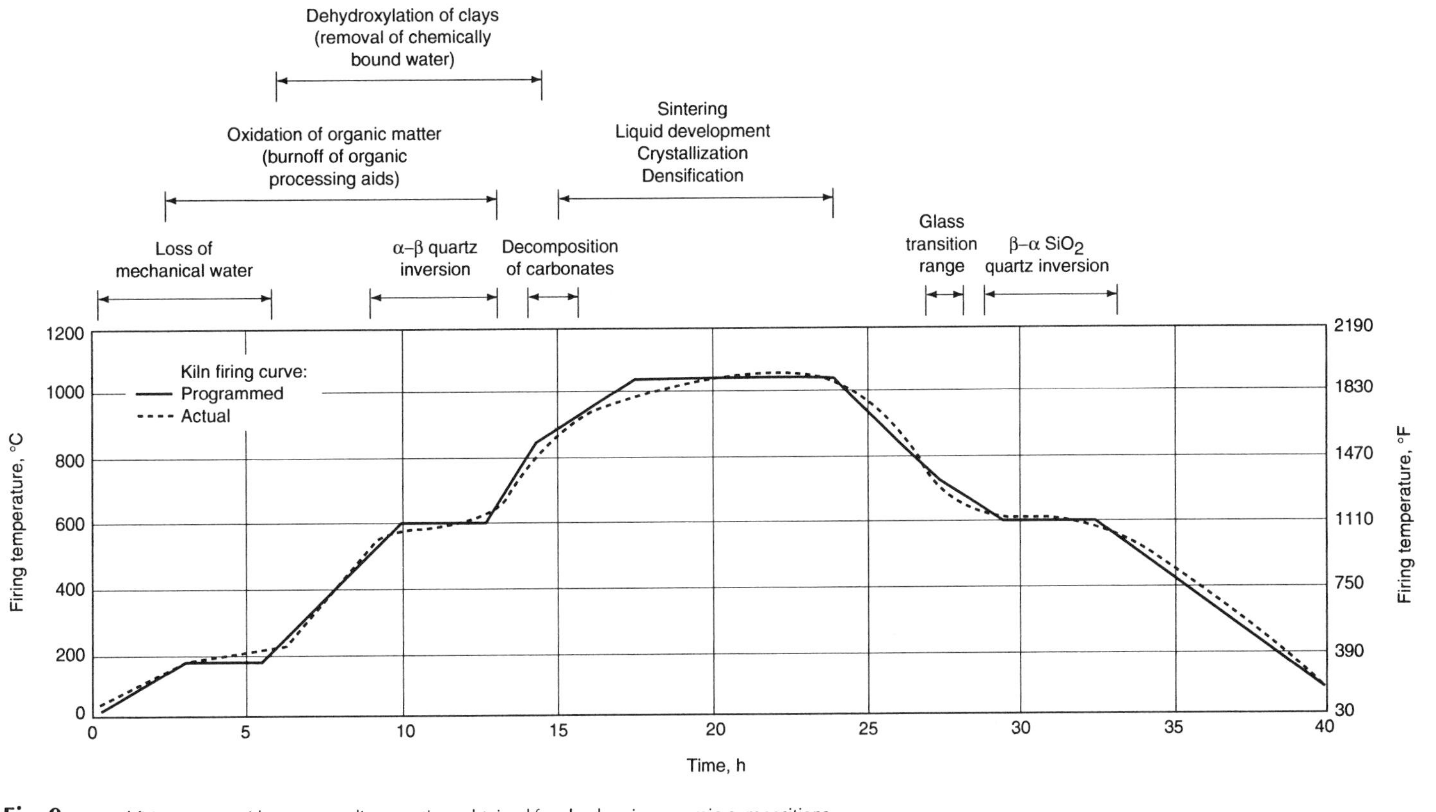

Fig. 9 Typical firing curves with corresponding reactions obtained for clay-bearing ceramic compositions

because a critical temperature is needed to promote reactions between phases, production of glassy liquid, and activation of transport mechanisms. Time, the precise firing profile, atmosphere, and pressure are also important. All of these factors must be balanced in order to maximize productivity, minimize production costs, and optimize properties. Sufficient time must be provided for appropriate transport, reaction, and densification. Heating and cooling rates must not cause excessive body stresses. Oxidizing or reducing atmospheres may promote the formation of specific phases, chemistries, or defect structures. For example, a fraction of a percent of oxygen deficiency will change titania and zirconia from white to black, and atmosphere has a critical effect on high-temperature superconductivity. While most ceramics are fired in conventional furnaces at near-atmospheric pressure, significant mechanical and other property improvement can be accomplished by firing under high pressure. This may be done with a unidirectional hot press in which rams apply pressure in a die or with a hot isostatic press (HIP) in which gas applies pressure to a sealed ceramic at elevated temperature.

Most ceramics are fired in electric or gas kilns using programmable controllers or computers connected to calibrated thermocouples or pyrometers. Pyrometric cones are a convenient means of monitoring the thermal work input to the fired parts. They are tall, triangular pyramids of ceramic of varied composition that soften and slump at controlled combinations of temperature and time. A series of such cones can be used to verify that the correct firing conditions have been achieved. In addition to computer control of firing, efficient use of fuel, and use of waste heat for drying, several new trends in firing have improved economy and productivity. Fast firing of a small thermal mass in a continuous process is growing in application. Microwave firing and use of chemical reaction energy to augment other energy input are being used on a small scale and are actively being researched.

Conclusions

Both traditional and technical ceramics are based on powder raw materials that are formed into green (unfired) ceramics, dried, and fired. All process steps are critical for producing a material with superior and reliable strength properties. Generally, the more highly fired material will have lower porosity and improved mechanical performance. The transition from traditional to advanced technical ceramics is a continuum; in general, technical ceramics are made from higher quality, more chemically derived raw materials processed with similar techniques as traditional ceramics, but with stricter control.

ACKNOWLEDGMENT

The statistical information used in this article is taken primarily from Ref 1 to 3.

REFERENCES

1. *Bulletin of the American Ceramic Society* (monthly), American Ceramic Society
2. *Ceramic Industry* (monthly), Corcoran Publications
3. *Ceramic Forum International,* German Ceramic Society
4. W. Weibull, A Statistical Theory of the Strength of Materials, *Ing. Venterst. Akad.,* No. 151, 1939, p 1–45

SELECTED REFERENCES

- *Ceramic Data Book* (annual), Corcoran Publications
- *Ceramic Source* (annual), American Ceramic Society
- *Engineered Materials Handbook,* Vol 4, *Ceramics and Glasses,* S.J. Schneider, Ed., ASM International, 1991
- *Engineered Materials Handbook Desk Edition,* ASM International, 1995
- V.E. Henkes, G.Y. Onoda, and W.M. Carty, *Science of Whitewares,* American Ceramic Society, 1996
- J.T. Jones and M.F. Berard, *Ceramics—Industrial Processing and Testing,* Iowa State University Press, 1972
- W.D. Kingery, H.K. Bowen, and D.R. Uhlmann, *Introduction to Ceramics,* 2nd ed., John Wiley & Sons, 1976
- J.S. Reed, *Principles of Ceramic Processing,* 2nd ed., John Wiley & Sons, 1995
- D.W. Richerson, *Modern Ceramic Engineering,* 2nd ed., Marcel Dekker, 1992
- *Uhlmann's Encyclopedia of Industrial Chemistry, Vol A6, Verlag Chemie*

Design for Plastics Processing

Edward A. Muccio, Ferris State University

DESIGNING A PLASTIC PART that will meet customer application demands and be durable enough to survive years of use requires the product designer to consider several factors, such as:

- Plastic material(s) to be used
- Product shape and features
- Production process
- End-use applications

The relationship among these and other factors that influence the production of quality plastic parts is represented in Fig. 1.

The product designer must also consider that the plastic molding or forming process influences the plastic part performance. The physical, mechanical, and chemical properties of the material can be affected by the molding/forming process. The part designer needs to understand the rudiments of plastic processing methods in order to select a plastic material, define the specific shape of the part, and define the process used to manufacture the plastic product.

This article describes key processing methods and related design, manufacturing, and application considerations for plastic parts; it concludes with a discussion of a materials selection methodology for plastics. More detailed coverage of the processes discussed in this article, including process considerations for different types of plastics (that is, thermoplastics and thermosets) is provided in *Engineering Plastics*, Volume 2 of the *Engineered Materials Handbook*, and the *Engineered Materials Handbook Desk Edition*, both published by ASM International. Related coverage in this Volume is contained in the articles "Effects of Composition, Processing, and Structure on Properties of Engineering Plastics," "Design with Plastics," and "Design for Composite Manufacture."

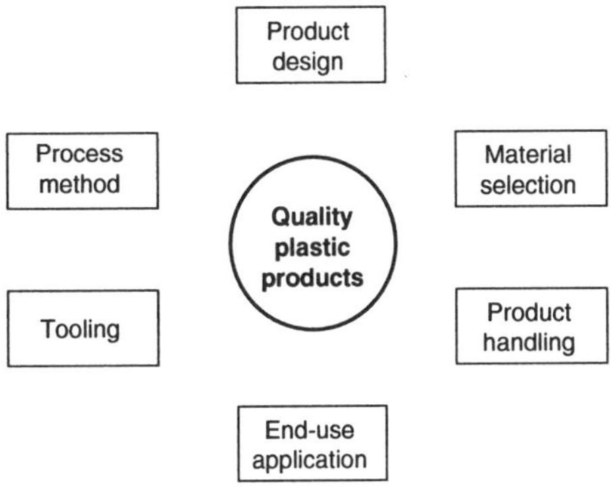

Fig. 1 Key factors in the development and production of quality plastic parts

Plastics Processing Methods

The primary plastics processing methods are:

- Injection molding
- Extrusion
- Thermoforming
- Blow molding
- Rotational molding
- Compression molding/transfer molding
- Composites processing
- Casting

Other plastics processing methods exist, but most are variants of these processes. Table 1 lists characteristics and capacities of processing methods used for thermoplastic and thermoset parts.

Plastics processing is a form conversion process. The material that enters the process as plastic pellets or powder is basically the same material that exits the process as a plastic part. The plastic process converts the shape of the plastic material. However, this simple explanation of plastic processing needs to be slightly modified. Although the plastic entering the process is the same plastic exiting the process, the properties of the plastic material may be affected by the rigorous activities that occur during the process. The resulting properties of the plastic part may be different from the properties of the plastic material as defined by the plastic material manufacturer.

Each processing method can have a different effect on the final properties. Following is a brief description of the primary plastic processing methods and a summary of how each process influences part design and the properties of the plastic part.

Injection Molding

Injection molding, and all its variants, is the most popular process for producing plastic products. Designers prefer the injection molding process because, in addition to being fast and cost effective, it allows the designer the opportunity to create true three-dimensional part shapes. (Many plastic processes, such as extrusion, blow molding, thermoforming, and rotational molding, do not allow the designer to control all surfaces of the plastic part being manufactured. One surface is a function of the process, not the product design; some examples include the inside of a hollow container produced by blow or rotational molding, the length of an extruded profile, and the outer surface of a thermoformed part produced on a female mold.)

Product designers desire control over all aspects of the design of a product, and injection molding allows this to occur. Additionally, injection molding allows the designer to incorporate product design features such as holes, snaps, color, texture, and symbolization that might demand secondary operations if the design were manufactured using materials such as metal, wood, or ceramic.

The injection molding process involves several steps:

- Feed and melting of the plastic pellets
- Metering of the plastic melt
- Injection of the plastic melt into the mold
- Cooling and solidifying of the plastic in the mold
- Ejection or removal of the molded part from the mold

The following description of these steps is based on the processing required to mold a simple part such as the polystyrene poker chip shown in Fig. 2.

Feed and Melting of the Plastic Pellets. The polystyrene, in the form of pellets, is fed into the throat of the injection molding machine (Fig. 3). Initially, the plastic pellets are heated by the electric heater bands; however, the shear and friction created by turning the injection molding machine screw will provide the majority of the energy required to melt the plastic.

As the screw turns, the plastic pellets melt, and the melted material is conveyed toward the discharge end of the injection unit.

Metering of the Plastic Melt. As the plastic melt is conveyed forward through the barrel of the molding machine, it is allowed to pass through a nonreturn valve that prevents the plas-

Table 1 Thermoplastics and thermoset processing comparison

Process	Process pressure MPa	Process pressure ksi	Maximum equipment pressure MN	Maximum equipment pressure tonf	Maximum size m^2	Maximum size ft^2	Pressure limited	Ribs	Bosses	Vertical walls	Spherical shape	Box sections	Slides/ cores	Weldable	Good finish, both sides	Varying cross section
Thermoplastics																
Injection	15–45	2–7	30	3370	0.75	8.0	y	y	y	y	n	n	y	y	y	y
Injection compression	20	2.9	30	3370	1.5	16	y	y	y	n	n	n	y	y	y	y
Hollow injection	15	2.2	30	3370	2.0	20	y	y	y	y	n	y	y	y	y	y
Foam injection	5	0.7	15	1690	3.0	30	y	y	y	...	n	y	y	y	y	y
Sandwich molding	20	2.9	30	3370	1.5	16	y	y	y	y	n	n	y	y	y	y
Compression	20	2.9	30	3370	1.5	16	y	y	y	y	n	n	y	y	y	y
Stamping	20	2.9	30	3370	1.5	16	y	n	n	n	n	n	n	y	y	n
Extrusion	n/a	n/a	n/a	n/a	n/a	n/a	n/a	y	n	n/a	n	y	n	y	y	y
Blow molding	1	0.15	10	1120	2.0	20	n	n	n	y	y	y	y	y	n	n
Twin-sheet forming	1	0.15	10	1120	6.0	65	n	n	n	y	y	y	n	y	n	n
Twin-sheet stamping	1	0.15	30	3370	6.0	65	n	n	n	n	n	y	n	y	y	n
Thermoforming	0.1	0.015	n/a	n/a	...	...	n	n	n	y	n	n	y	y	n	n
Filament winding	0	0	n/a	n/a	...	...	n/a	y	n	y	y	y	n	y	n	y
Rotational casting	0.1	0.015	n/a	n/a	...	...	n	n	n	y	y	n	n	y	n	n
Thermoset plastics																
Compression																
Powder	60	8.7	30	3370	0.5	5	y	y	y	y	n	n	y	n	y	y
Sheet molding compound	6–20	0.85–3	30	3370	4–5	45–55	y	y	y	y	n	n	y	n	y	y
Cold-press molding	1	0.15	30	3370	...	...	n	n	y	y	n	n	n	n	y	y
Hot-press molding	5	0.75	30	3370	6.0	65	y	n	y	y	n	n	n	n	y	y
High-strength sheet molding compound	4–10	0.60–1.5	30	3370	3.0	30	y	y	y	y	n	n	n	n	y	y
Prepreg	0.5–5	0.07–0.75	30	3370	6.0	65	y	n	n	y	n	n	n	n	y	y
Vacuum bag	0.1	0.015	n/a	n/a	...	...	n	n	y	y	n	y	n	n	n	y
Hand lay-up	0	0	n/a	n/a	...	...	n	n	y	y	n	y	n	n	n	y
Injection																
Powder	100	14.5	10	1120	0.1	1.1	y	y	y	y	n	n	y	n	y	y
Bulk molding compound	30	4.5	30	3370	1.0	11	y	y	y	y	n	n	y	n	y	y
ZMC	30	4.5	30	3370	1.0	11	y	y	y	y	n	n	y	n	y	y
Stamping	3	0.45	30	3370	6.0	65	y	n	n	y	n	n	n	n	y	n
Reaction injection molding	1	0.15	10	1120	...	...	y	y	y	n	n	y	n	y	y	
Resin transfer molding	0.1	0.015	10	1120	...	...	n	y	n	y	n	y	n	n	y	y
High-speed resin transfer molding or fast resinject	2	0.3	30	3370	...	...	n	y	n	y	n	y	n	n	y	y
Foam polyurethane	0.5	0.07	n/a	...	...	...	n	y	y	y	y	y	n	n	y	y
Reinforced foam	1	0.15	30	3370	3.0	30	y	y	y	y	n	y	n	n	y	y
Filament winding	n/a	n/a	n/a	n/a	...	...	n/a	y	n	y	y	y	n	n	(a)	y
Pultrusion	n/a	n/a	n/a	n/a	n/a	n/a	n/a	y	n	n/a	n	y	n	n	y	y

Note: y, yes; n, no; n/a, not applicable. (a) One side of filament-wound article will exhibit a strong fiber pattern.

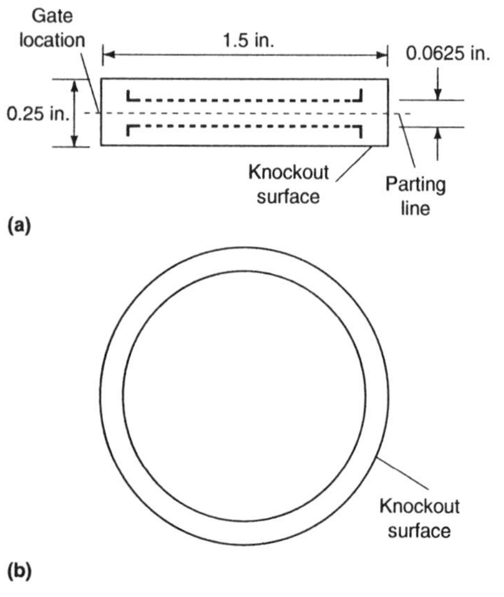

Fig. 2 Polystyrene poker chip. (a) Side view. (b) Bottom view

tic melt from traveling rearward or back through the valve. The plastic melt that moves through the valve and in front of the screw will push the screw rearward. This rearward motion of the screw, while the screw is turning, creates more shear and facilitates the melting of the plastic pellets. The amount of plastic melt that is allowed to move through the valve and reside in front of the screw is defined by a limit switch or stopping point assigned by the molding technician. The plastic melt in front of the screw will be the material that is injected into the mold to produce the plastic parts.

Injection of the Plastic Melt into the Mold. Injecting the plastic melt into the closed mold requires high pressures (between 35 and 205 MPa, or 5 and 30 ksi, on the plastic material) and often high speeds. The specific values for injecting the plastic melt are a function of the melt viscosity of the plastic material, the mold design, and the plastic product design. To allow the injection of plastic into the mold, the part designer must consider design features such as the wall thickness and gate type and location.

Wall thickness, the thickness of the major portion of the wall of the plastic part, depends on the melt characteristics (melt viscosity) of the plastic. A plastic part with thin walls (<1 mm, or 0.04 in., thick) will usually require higher molding pressures than a plastic part with a wall thickness of about 4 mm (0.16 in.). Thicker wall sections (>6 mm, or 0.25 in., thick) may result in poor part quality and molding defects such as underfill or sink marks.

Gate Type and Location. The gate (Fig. 4) is the point where the plastic melt is allowed to enter the cavity to form the part. The gate is designed to cool or freeze after the cavity has been filled and packed with plastic. This cooling prevents any plastic melt from exiting the filled cavity.

Cooling and Solidifying of the Plastic in the Mold. Plastic materials are thermal insulators; that is, they tend *not* to absorb or release thermal energy at a rapid rate. The plastic part designer must avoid thick wall sections to avoid cooling problems in the mold. Specifically, parts with thicker wall sections require a longer cooling time within the mold; additionally, the thick sections may distort, have sink marks, or contain voids (Fig. 5).

To avoid these problems, the designer must strive for a nearly constant thickness of every section of the part. This *nominal* thickness must meet the application requirements of the part, ensure nearly uniform cooling, and be fillable by

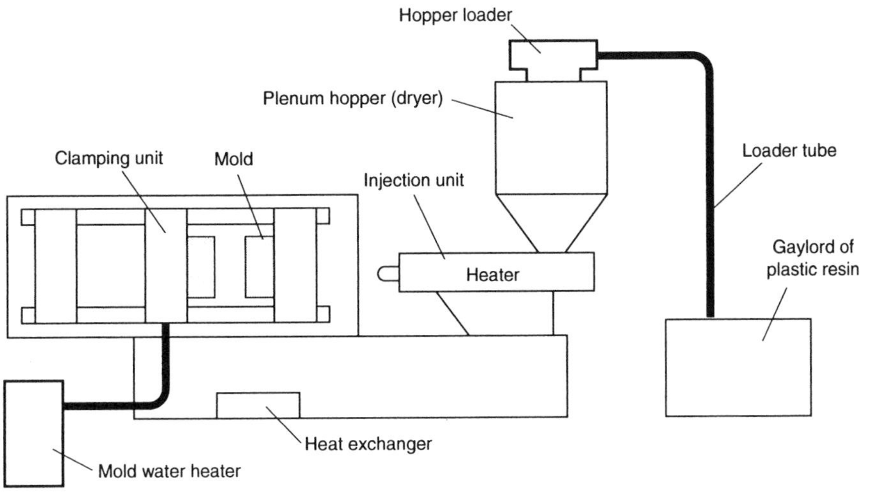

Fig. 3 Injection molding machine

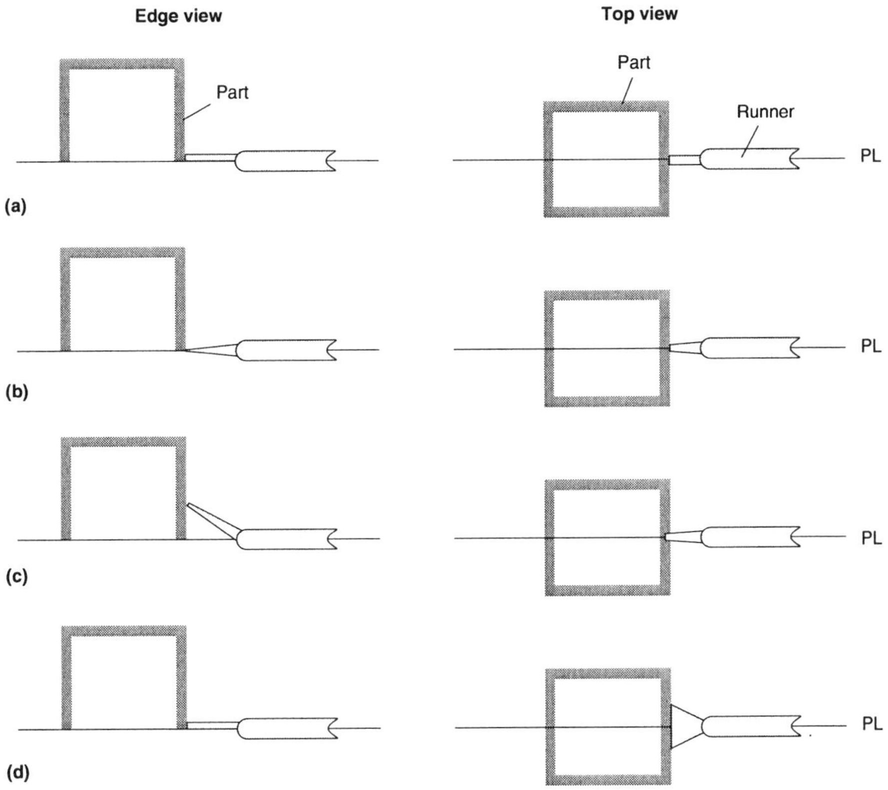

Fig. 4 Types of injection molding gates. (a) Tab gate. (b) Pinpoint gate. (c) Sub gate. (d) Fan gate. PL, parting line

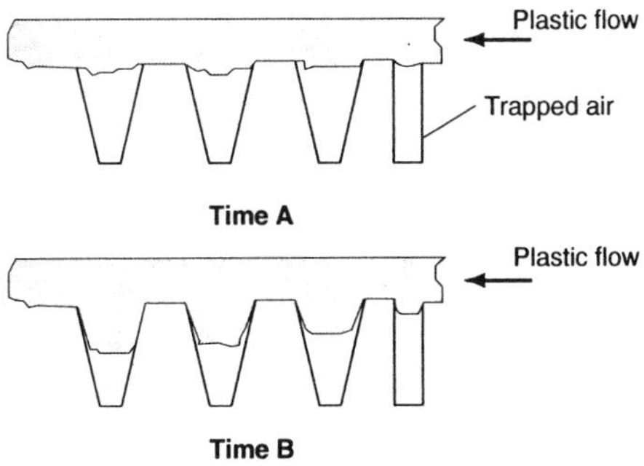

Fig. 5 Problems in cooling and solidification caused by the rib fill rate for an injection-molded part

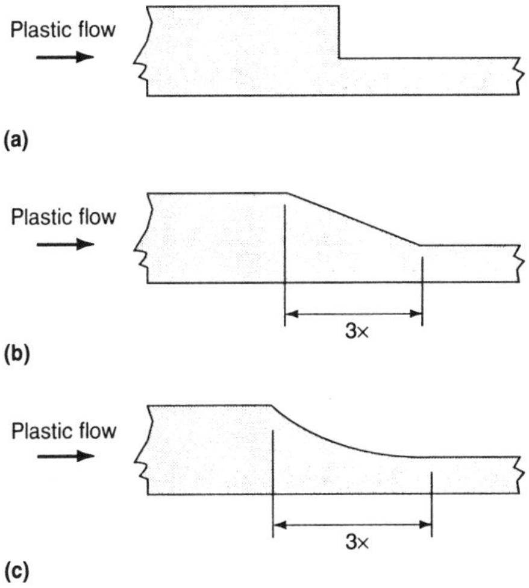

Fig. 6 Wall transitions in a plastic part. (a) Poor (sharp) transition. (b) Better (gradual) transition. (c) Best (smooth) transition

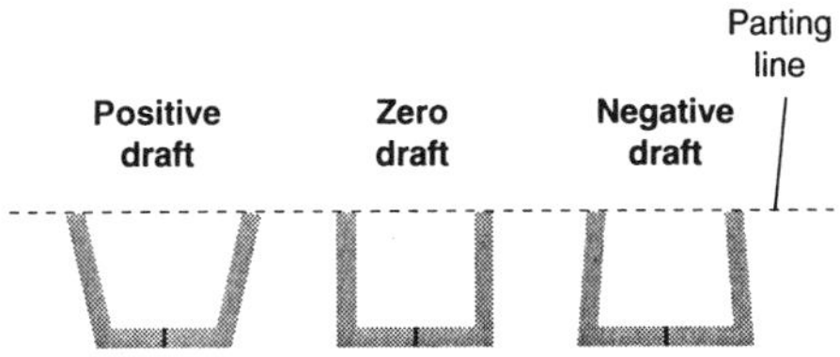

Fig. 7 Types of draft in plastic injection-molded parts

the plastic material selected. As an example, a plastic material manufacturer may suggest a nominal wall thickness of 4.5 mm (0.18 in.) for a specific plastic. The plastic part designer is not bound to make all the walls this thickness, but should design the wall to average this dimension. The wall thickness may vary, but only at a reasonable rate of change (Fig. 6).

Ejection or Removal of the Molded Part from the Mold. To allow an injection-molded part to be removed from the mold requires that the part designer consider ejection surfaces and draft.

Ejection surfaces on the part provide an allowance for ejector pins to push the part out of the mold (Fig. 2). The ejector pins or other mold components such as inserts and slides will leave a witness mark on the plastic part, which the plastic part designer needs to respect. Often the part specification will include a note that states "knockout witness to be flush to or 0.125 mm (0.005 in.) below the molding surface."

Draft is the angle in the wall design that facilitates ejection from the mold (Fig. 7).

Details and design considerations for injection molding include shrinkage, postmold shrinkage, and size and location of holes and other features.

Shrinkage occurs because the plastic melt volume is greater than the solid volume, and the plastic melt is packed into the mold under high pressures. Shrinkage needs to be understood in order to produce plastic parts with a high degree of dimensional stability. Different plastics experience different amounts of shrinkage (Table 2). Additives affect the shrinkage rate. Rate and direction of flow of the melt into the mold can influence shrinkage and may cause the same material to exhibit two different types of shrinkage depending on the part geometry.

Postmold Shrinkage. It is best to have any shrinkage occur while the plastic part is constrained by the mold. Shrinkage that occurs out-

Table 2 Shrinkage of selected plastic materials

Material	Shrinkage, %
Amorphous plastics	
Acrylic	0.6
Polycarbonate	0.6
Acrylonitrile-butadiene-styrene (ABS)	0.6
Polycarbonate (40% glass filled)	0.3(a) 0.5(b)
Semicrystalline plastics	
Polyethylene	2.0
Polypropylene	2.0
Nylon 6/6	1.5
Nylon (40% glass filled)	0.8(a) 0.3(b)

(a) Flow direction. (b) Transverse direction

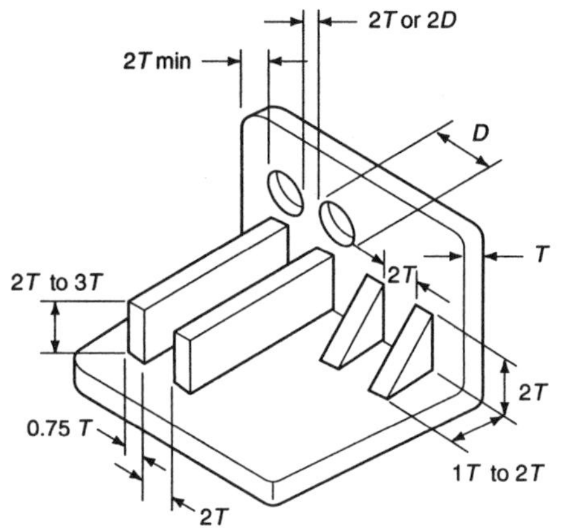

Fig. 8 Good design practice for holes and projections in injection-molded parts

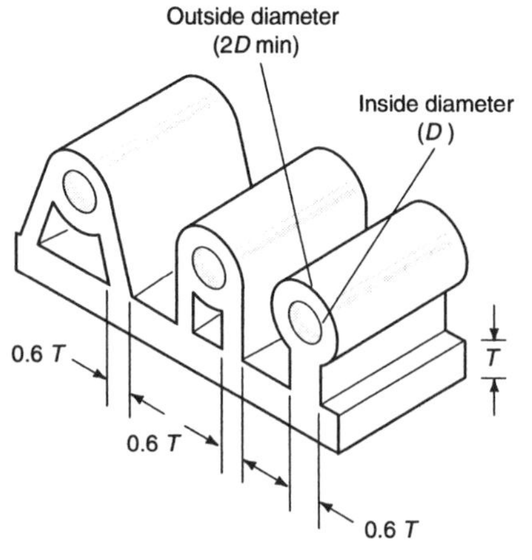

Fig. 9 Boss configurations for injection-molded plastic parts

side the confines of the mold after the part is ejected, known as postmold shrinkage, may be uncontrolled and/or unpredictable. The result could be a major dimensional problem for an injection-molded part. Postmold shrinkage is a function of both the plastic material and the process. Several semicrystalline plastics tend to exhibit a higher potential for postmold shrinkage.

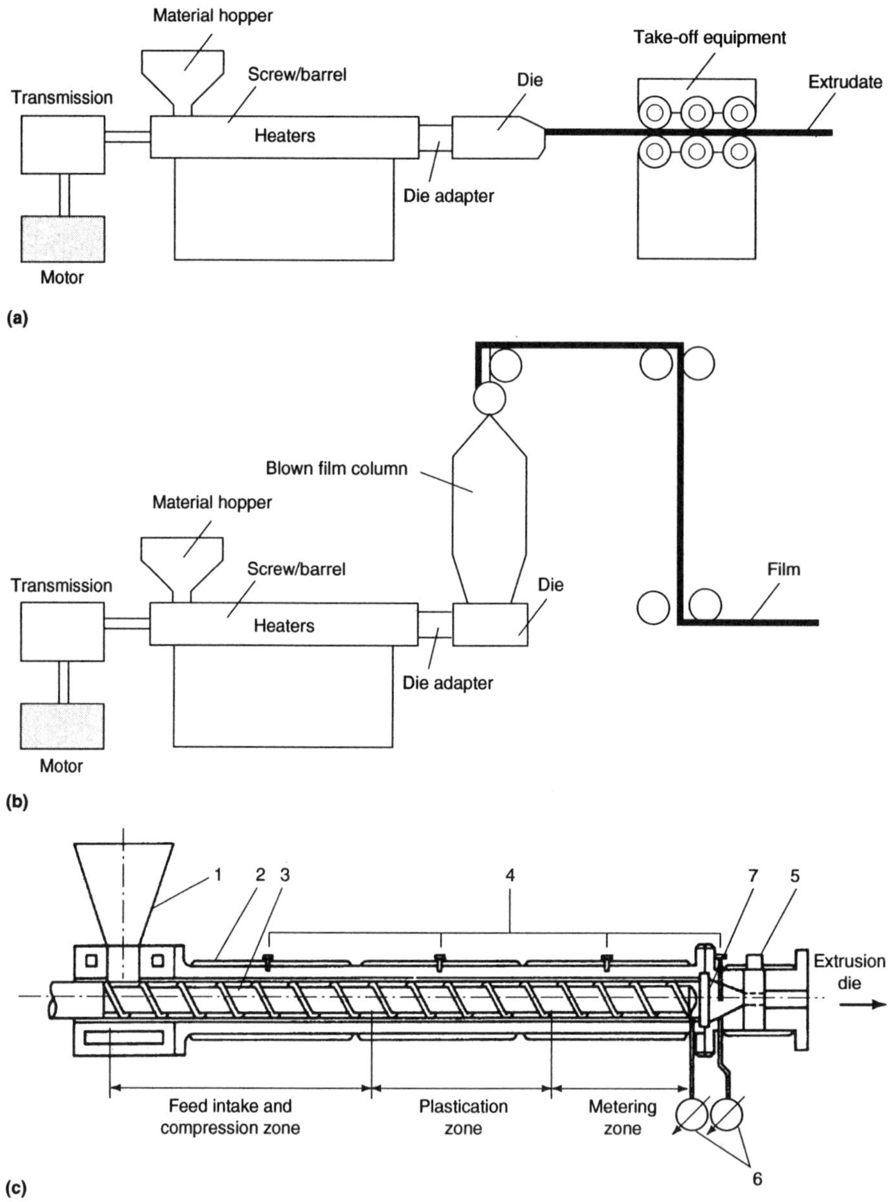

Fig. 10 Extrusion processes. (a) Profile/sheet extrusion. (b) Blown film extrusion. (c) Construction arrangement of the plastication barrel of an extruder. 1, feed hopper; 2, barrel heating; 3, screw; 4, thermocouples; 5, back pressure regulating valve; 6, pressure-measuring instruments; 7, breaker plate and screen pack

If the injection molding process is not optimized, it can contribute to postmold shrinkage. For example, consider an injection molding process that has the plastic melt in the barrel at 260 °C (500 °F) and a mold temperature of 82 °C (180 °F). The desire for productivity gains, that is, output of more parts per hour, leads to cooling the mold to 38 °C (100 °F) and speeding up the cycle. The result of this process change may not be immediately visible. While output gains may be achieved, the lower mold temperature may cause a higher degree of molded-in (residual) stress. This increased stress may be relieved after the part is removed from the mold. Over the next hours, days, or weeks, the relieving of the stress may manifest itself as postmold shrinkage.

Holes and Other Features. Injection-molded part features can be expressed as a function of the nominal wall thickness (*T*) as shown in Fig. 8 and 9.

Extrusion

The extrusion of plastic material is, surprisingly, the process that utilizes the most plastic material, even more than injection molding. One reason for this great material consumption is that extrusion is one of the few continuous plastics processes. Other plastics processes are batch processes, relying upon repetition. Extrusion of plastic material is continuous, and the plastic product is cut and formed in a secondary process. Another reason is that extrusion is used to compound and produce the plastic pellets used in most other thermoplastic processing operations. For example, most plastic pellets used in the

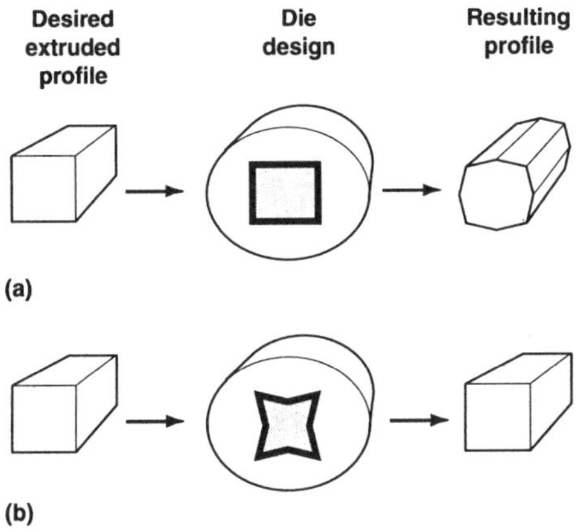

Fig. 11 Die swell in extrusion. (a) Incorrect die design for intended profile. (b) Correct die design

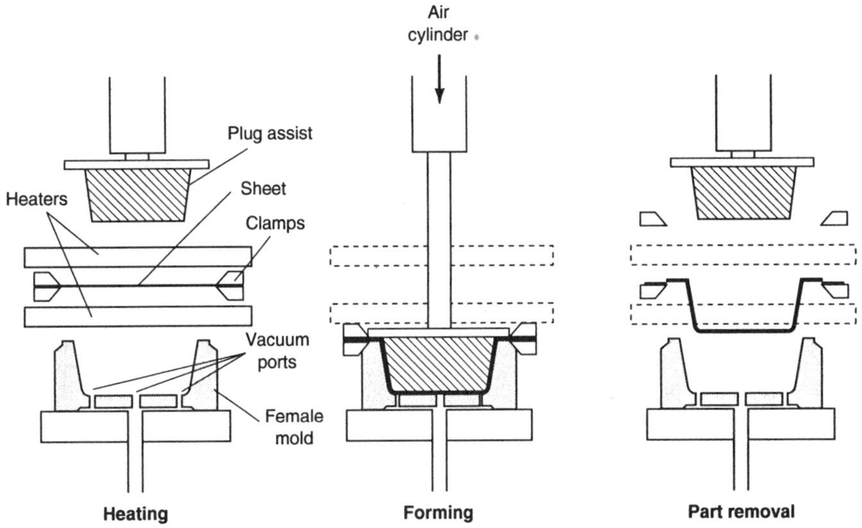

Fig. 12 Thermoforming (vacuum forming)

injection molding process are produced in an extruder at the plant of the material manufacturer.

The extruded product is designed as a two-dimensional cross-section shape, which is extruded in the third dimension. The third dimension is usually controlled by a cut-off operation. As an example, polyvinyl chloride (PVC) pipe is designed as two simple concentric circles. A die is fabricated, and the plastic melt is extruded through the die on a continuous basis (Fig. 10a). The length of the pipe is defined and created by cutting the continuous extrudate to the desired length.

Types of extruded parts can be categorized as follows:

- *Sheet* is a flat extruded profile greater than 0.0004 mm (0.010 in.) thick.
- *Film* is a flat extruded profile less than 0.0004 mm (0.010 in.) thick. Blown film (Fig. 10b) is a volume product used for trash bags, packaging, and wrappings. Cast film is a high-volume, high-tolerance product used for carrier material in the printing and audio/video recording industry.
- *Profile* is a shaped extruded profile.
- *Fiber* is a cylindrical or tubular profile less than 0.0004 mm (0.010 in.) thick.

Details and design considerations for extruded parts include die swell and orientation.

Die swell (Fig. 11) is the phenomenon where an extrudate swells to a size greater than the die from which it came. As the plastic exits the die, it tends to swell. This is associated with the reduction in pressure as well as the nature of the polymer itself. Die swell has to be considered by the product designer as well as the die designer in order to produce extrusions that meet the customer requirements.

Orientation is the phenomenon where the polymer molecules are aligned as a result of the high degree of laminar flow as well as the pulling of the extrusion take-off apparatus. Orientation is often desirable, if controlled, because it can improve the properties of the extruded product. Biaxial orientation is orientation in two directions and improves strength in film materials.

Orientation also allows an extruded product to shrink when exposed to heat. Shrink-wrap materials for packaging and dunnage have become very important products that incorporate this phenomenon of shrinking due to controlled orientation and heating.

Thermoforming

Thermoforming, also referred to as vacuum forming, forms plastic sheet into shapes. The plastic sheet is placed into a clamp frame to hold it securely on all edges. The sheet material is placed into the clamp frame manually, robotically for high-volume processing, or continuously if the sheet material is produced by an in-line extruder. Thermal energy, usually in the form of convection and radiant heat from electrical heating elements, is applied for a sufficient amount of time to soften (not melt) the plastic sheet. Once the sheet is sufficiently softened, a mold is brought in contact with the sheet, and a vacuum is applied that draws the softened sheet onto the mold. After the sheet cools, it will retain the shape of the mold when the mold is removed. The thermoforming process sequence is shown in Fig. 12.

Historically, thermoforming has been considered a one-sided process, that is, the softened sheet will either conform to a male mold with the inside becoming the critical surface and the outside the noncritical surface, or conform to a female mold with the outside becoming the critical surface and the inside the noncritical surface. This one-sided approach to thermoforming was satisfactory for decades when the process was used primarily for simple packaging parts. Today, process advancements have enabled the production of thermoformed parts that have two critical sides and sufficient dimensional accuracy to allow them to be used in key automotive, building, and construction applications. This dimensional control is accomplished by having two dies or molds, one forming either side of the sheet.

Typical Thermoformed Parts. The majority of thermoformed products are produced for the packaging market; however, broader applications include:

- Blister packages
- Foam food containers
- Refrigerator and dishwasher door liners
- Auto interior panels
- Tub/shower shells, which are later fiber reinforced
- Pickup truck bed liners
- Internally lighted acrylic and cellulose acetate butyrate (CAB) signs

The thermoforming process offers some unique tooling advantages over other conventional plastic processes, primarily because the thermoform molds are relatively simple in design and construction as well as lower in cost.

Prototypes produced using the thermoforming process can be made quickly by using simple molds made from inexpensive materials, such as wood, plaster, and epoxy. Many designers will insist upon a product design review that includes the development of one or more thermoformed prototype parts.

Blow Molding

Blow molding has historically been associated with simple geometries such as bottles and containers. However, there have been significant developments in the blow molding process and its variants in the past two decades. These developments allow the blow molding of more complex shapes such as air ducts and automobile fuel tanks.

Basic blow molding equipment (Fig. 13a) is essentially a profile extruder attached to a blowing station. The extruder produces a tube referred to as a parison. The parison can be controlled in both its size and shape. At the blowing station (Fig. 13b), the mold captures the parison and seals it by pinching either end. A blow pin is then

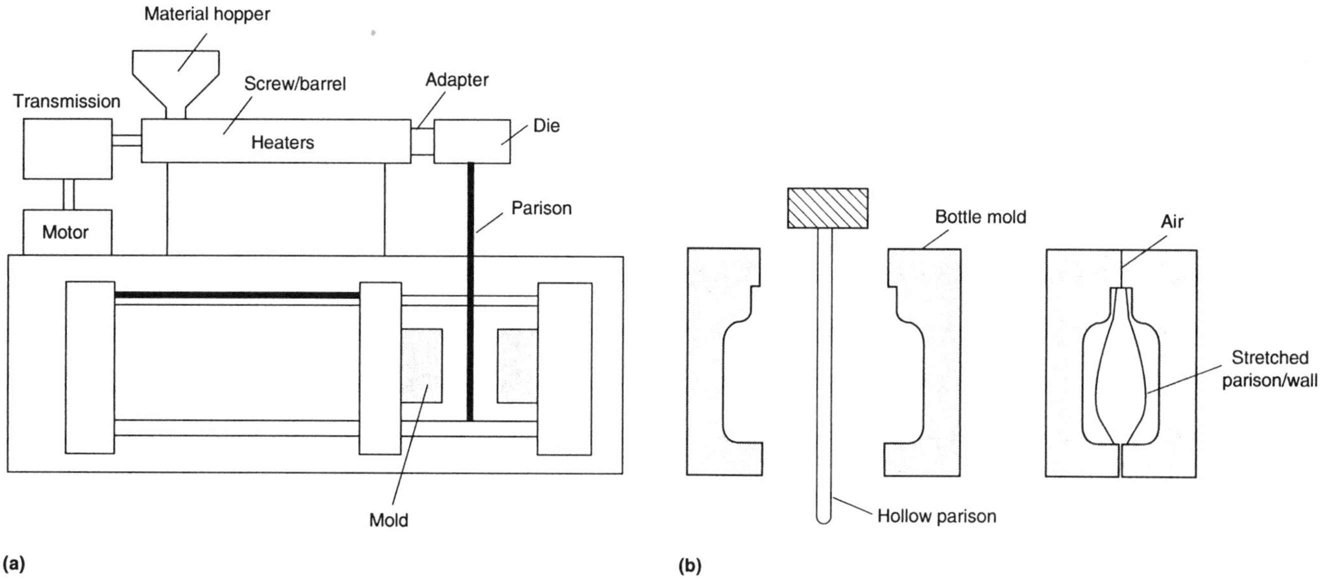

Fig. 13 Blow molding. (a) Equipment configuration. (b) Sequence at blowing station for a bottle mold

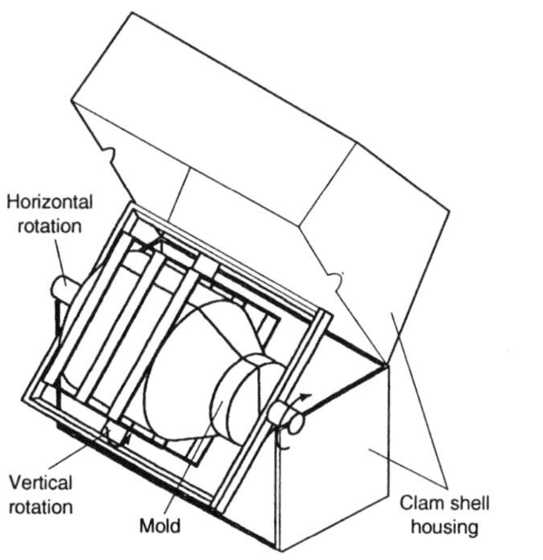

Fig. 14 Rotational molding equipment (clam shell type)

inserted into the parison, and air is introduced at about 700 kPa (100 psi). The air causes the pinched parison to expand and take the shape of the mold.

This basic process results in a product that is dimensionally defined on the exterior surfaces. The interior surfaces are not controlled as they do not contact a mold surface. As a result, the wall thickness of a conventionally blow molded part may vary. The nature of the conventional blow molding process also does not lend itself to incorporating design features such as holes, sharp corners, and narrow ribs.

Rotational Molding

Like blow molding, rotational molding produces a hollow product. Unlike blow molding, however, rotational molding is a relatively slow process that begins with plastic in the form of a powder, not a parison. The advantage of rotational molding is that it can produce large objects, with capacities from 1 to more than 500 gal. Additionally, the wall thickness is a function of how much plastic powder is placed into the mold, and thick (2.5 to 12 mm, or 0.10 to 0.4 in.) wall sections can be formed.

The rotational molding process uses a mold made of sheet metal or cast aluminum. Because rotational molding is a low-pressure process, tooling can be lower in strength than that used for the other molding processes.

Another advantage of rotational molding over other plastic processes is that it results in a very low stressed product. Since the rotational process is low in pressure, and the plastic is not forced through narrow channels, it does not induce a significant amount of internal stress. The result is a high degree of dimensional stability in the final product.

Processing Sequence. The plastic powder is placed directly into the mold by the operator. The mold is attached to the rotational process equipment where it passes through three distinct process stages: loading, heating, and cooling.

Loading is the stage of the process where the plastic powder is loaded into the mold and the mold is attached to the process equipment. After loading is completed, the mold begins to rotate along three axes. Although the rotation speed is relatively slow (<20 rpm), it is sufficient to force the plastic powder to the mold walls.

Heating is the stage of the process where heat, usually generated by a natural gas fired heater, brings the rotating mold to a temperature high enough to fuse the plastic powder.

Cooling is the stage of the process where the mold, which is still rotating, is allowed to cool, and the fused plastic in the mold solidifies. Once the contents of the mold are sufficiently cooled, the mold is separated and the product removed.

Processing Systems. Advances in the rotational molding equipment have resulted in two distinct styles of processing systems:

- *A conventional rotational system* has the mold located on the end of one of several arms. Several molds are at various stages of the process simultaneously.
- *A clam shell system* (Fig. 14) has the mold located on a rotating apparatus that is housed within a chestlike chamber. The clam shell system has a fixed and compact footprint. The chamber environment changes to heat as well as cool the mold. Especially good for small rotational molded parts, the clam shell system provides superior control and safety.

Compression Molding and Transfer Molding

Compression molding and transfer molding (a variant of compression molding) are two of the oldest plastic processing methods.

Compression molding is a simple process that offers the manufacturer an excellent method for producing low stress plastic parts. As shown in Fig. 15, the plastic material, usually in the form of a powder or preformed pill, is placed in the cavity or female section of the mold. Since most compression molding is associated with thermosetting plastics, the mold temperature will be relatively high (149 to 177 °C, or 300 to 350 °F). The male or core portion of the mold is located on the upper portion of the mold. After the plastic material is loaded, the male portion of the mold is lowered and compresses the heated plastic in the cavity. The heat and pressure cause the material to flow and fill the cavity details. In many compression molding processes, excess plastic forms a flash that will be removed in a secondary operation. The plastic material will then be allowed to cure or set, and the plastic part will be ejected from the mold.

Because the thermosetting plastic flows only a short distance and the flow rate is relatively slow, little shear stress is developed in the process. This low stress level will result in low internal or molded-in stress in the part, which thus will have a high degree of dimensional stability.

Transfer molding is a variant of the compression molding process and is a precursor to the modern injection molding process. The main difference between compression and transfer molding is in the method by which the plastic material enters the mold. As stated above, compression molding requires the material to be preloaded into the mold in the form of either a powder or a preform. Transfer molding utilizes a transfer ram or plunger where powder or preform is loaded into a chamber above the mold. The material in this chamber is forced through a sprue and runner/gate system to fill the mold. The transfer process allows the plastic to enter the mold in a molten or fluid state; thus, the process is adaptable to insert or overmolding. The downside of the transfer process is that there is often a need for secondary operations to separate the product from the gates/runners. There is also a certain amount of waste with the sprue and runners.

The ability to overmold or insert mold product allows many unique products to be transfer molded. These include electrical outlets, semiconductors such as integrated circuits, plastic handles for products such as knives, and plastic exteriors for metal parts.

Composites Processing

The term *composite* applies to plastic materials that are reinforced with glass, mica, metal, carbon fibers, or other materials. Composite materials provide the plastics part designer and plastics processor with the opportunity to customize a plastic compound by adjusting the type and volume of reinforcements added to a specific plastic matrix. Such composites usually exhibit synergistic behavior. Synergism, in the context of plastic materials, is when the strength of the composite is greater than the sum of the strength of the individual matrix and reinforcements. Additionally, composite plastics also distribute loads better than conventional plastics and result in superior strength-to-weight ratios. Other advantages of composite plastics are that they can be used to fabricate large parts and they offer excellent product durability and superior chemical resistance.

Although the development of composite materials and fabrication methods have made great strides over the past 20 years, there are very few high-volume processes available to manufacturers. Additionally, there are few design rules available to composite product designers. Products designed for conventional processes, such as injection molding, blow molding, extrusion, and thermoforming, can utilize a series of design guidelines based on an understanding of the materials and processes involved. Since composite product fabrication was limited, these design guidelines are still being developed. Some other disadvantages of composite processing include high product cost, high equipment cost (and limited availability), and low production rates.

Composite production processes include casting, reaction injection molding (RIM), structural reaction injection molding (SRIM), resin transfer molding (RTM), matched metal molding, filament winding, and pultrusion. Some of these processes are also used for noncomposite materials; casting is described separately below in the section "Casting" in this article.

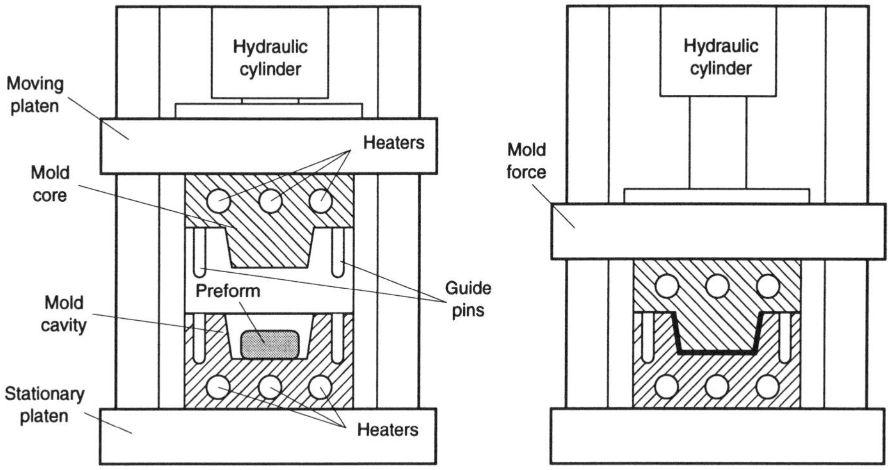

Fig. 15 Compression molding

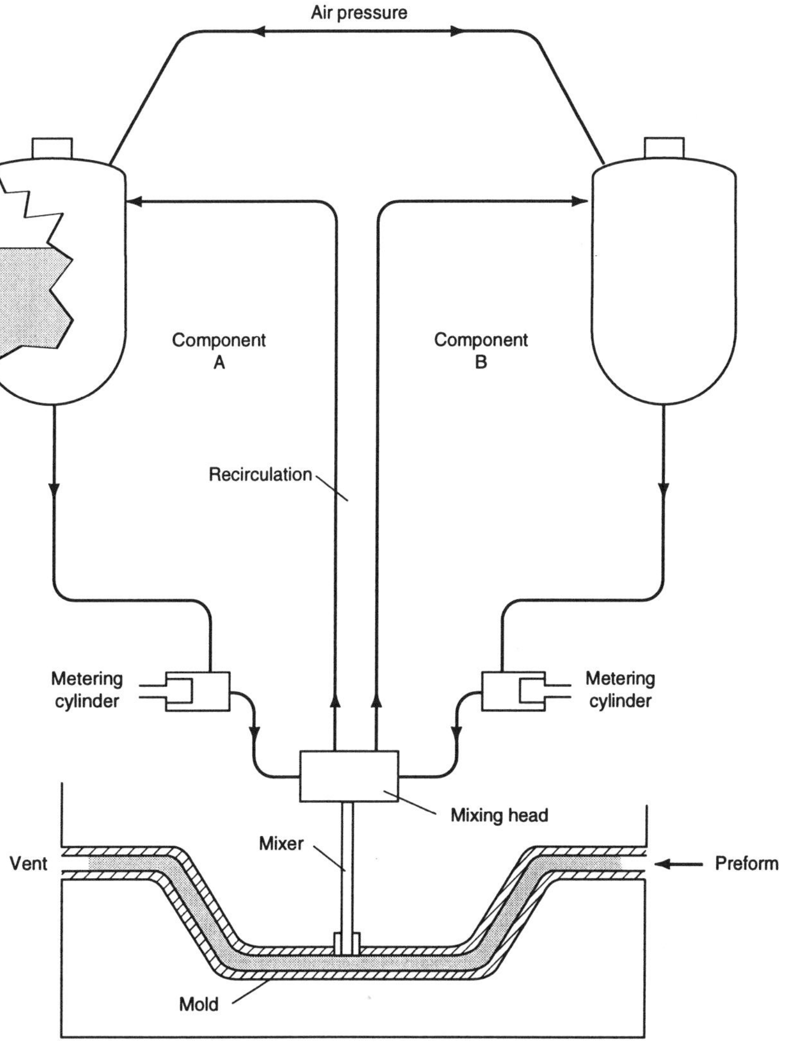

Fig. 16 Schematic of the structural reaction injection molding process

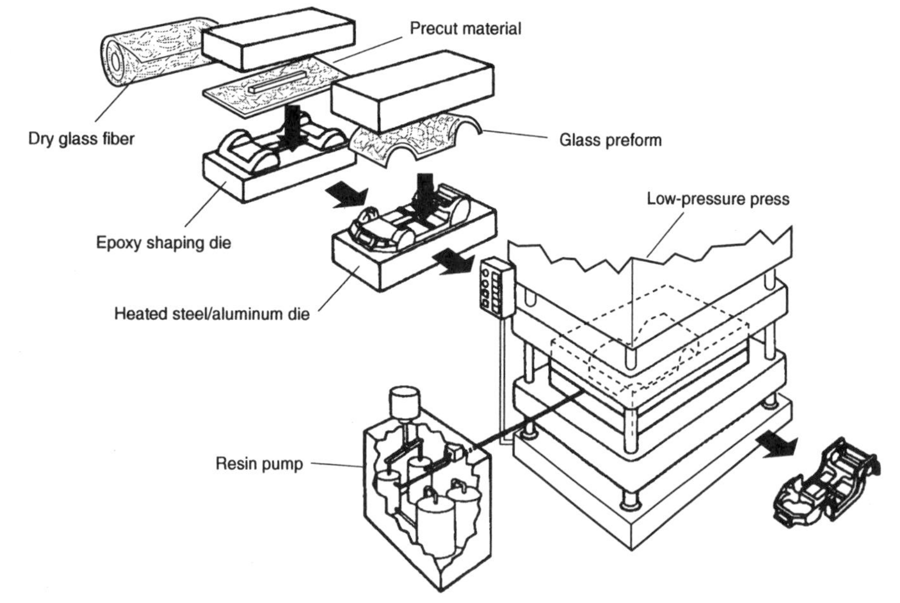

Fig. 17 Schematic of the high-speed resin transfer molding process

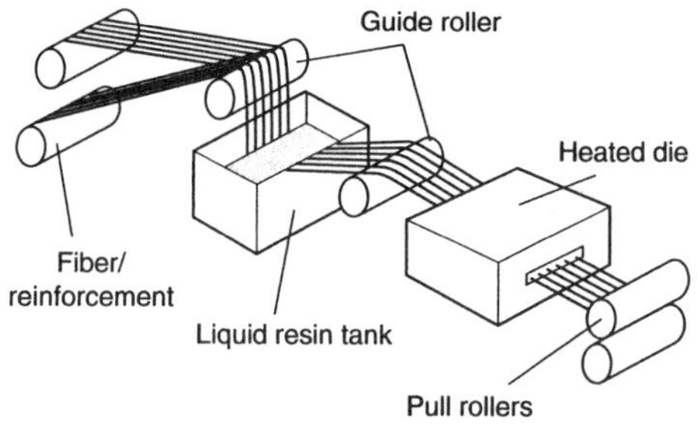

Fig. 18 Pultrusion

Reaction injection molding is a variant of injection molding (described previously in the section "Injection Molding" in this article). It is used for molding polyurethane, epoxy, and other liquid chemical systems. Mixing two to four components in the proper ratio is accomplished by a high-pressure impingement-type mixing head, from which the mixed material is delivered into the mold at low pressure, where it reacts (cures).

Structural reaction injection molding is in its technological infancy. As a consequence, the process is undergoing dramatic material and process improvements and rapid industrial growth. In many ways, SRIM is the natural evolution of two more established molding processes: RIM (described above) and RTM (described below). Like RIM, SRIM uses the fast polymerization reactions of RIM-type polymers, its intensive resin mixing procedures, and its rapid resin reaction rates. Like RTM, SRIM also employs preforms preplaced in the cavity of a compression mold to obtain optimum composite mechanical properties. A schematic of the SRIM process is shown in Fig. 16.

The ability of SRIM to fabricate large, lightweight composite parts consisting of all types of precisely located inserts and selected reinforcements is an advantage that other competitive manufacturing processes find difficult to match. In addition, large SRIM parts can often be molded in 2 to 3 min with clamping pressures as low as 700 kPa (100 psi). Thus the capital requirements of SRIM are relatively low, allowing the economical manufacture of parts with annual production volumes below 10,000 units. These advantages, when coupled with the concurrent development of a large family of commercially available SRIM resins, have led to predictions of a high SRIM annual growth rate.

Applications of SRIM. The first SRIM part commercially produced was the cover of the spare tire well in several automobiles produced by General Motors. Other SRIM automotive structural parts include foam door panels, sunshades, instrument panel inserts, and rear window decks. Nonautomotive SRIM applications include seat shells for the furniture market and satellite dishes for home entertainment centers.

Structural reaction injection molding is an attractive process for the economical production of large, complex structural parts. The main drawback of this process is that it is relatively new so that some aspects of this technology are poorly understood and some crucial equipment remains to be fully developed.

Resin transfer molding is a process by which catalyzed resin is transferred or injected into an enclosed mold in which reinforcement has been placed (Fig. 17). The fiberglass reinforcement is usually woven, nonwoven, or knitted fabric. Resin transfer molding is primarily used for prototyping; low-volume production of large, relatively complex parts; or low-to-medium-volume production of small, simple parts.

Prototyping. Resin transfer molding is an excellent process choice for making prototype components. Unlike processes such as compression molding and injection molding, which require tools and equipment approaching production level to accurately simulate the physical properties achievable in the production level component, RTM allows representative prototypes to be molded at low cost. It should be noted that in some cases, RTM can be used to prototype components designed for other processes; the RTM component will typically have properties that exceed those of the production-level product.

When prototyping with RTM, less reactive resins are generally used allowing long fill times and easier control of the vents. Tooling is usually low-cost epoxy, but could be made with an impervious material that would contain the resin. Prototype preforms are made by cut-and-sew methods, and any foam cores used are machined to shape. Sizes can range from small components to very large, complex three-dimensional structures. RTM provides two finished surfaces and controlled thickness, and it requires no auxiliary vacuum or autoclave equipment. Other processes used for prototyping such as hand layup and wet molding give only a single finished surface, and dimensions in the thickness direction are controlled.

Matched metal molding, also known as matched-die press forming, is the most common and probably the most widely used composites forming system. This is because forming presses are readily available from very simple, hand-operated small presses to fairly sophisticated, computer-controlled hydraulic systems. For simple forming operations, standard heated platen presses that have generally been used for flat panel molding have proven to be adequate. However, in operations where the control of deformation rate and pressure history are important, high quality stamping presses are used.

The dies used in this forming method are generally made of metal, which can be internally heated and/or cooled. When metals are used, the dies are generally designed to fixed gap (thickness) of close tolerance. High pressures can easily be applied to the workpiece. A disadvantage of this forming method is that when there is a thickness mismatch between the formed piece and the premachined cavity, nonuniform pressure is produced on the part resulting in nonuniform consolidation.

When heating or cooling is desired, the dies usually have such a high heat content that heat transfer times are long. Finally, matched-die fabrication costs are high because of the requirement that the two close-tolerance die halves have to match. Substituting an elastomeric material for one of the die halves will usually reduce the tooling cost and enable the application of a more uniform consolidation pressure than in an all-metal die set.

Pultrusion (Fig. 18) is a composite process that has many similarities to extrusion. Pultrusion begins with strands of reinforcement, usually glass or carbon fibers, that have been wetted in a resin tank. The resin used is most often an epoxy or polyester. The next step in the process is to pull

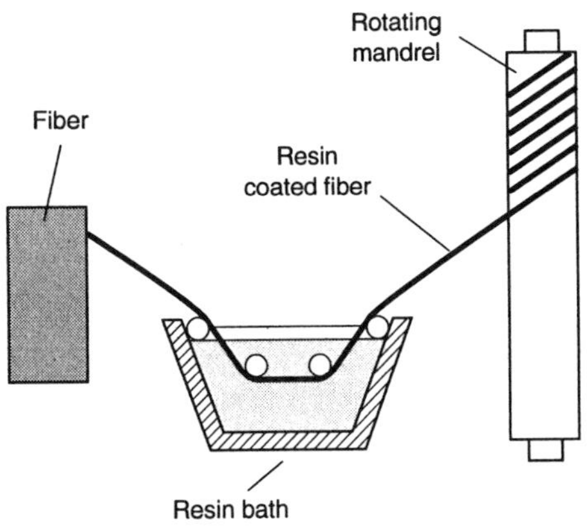

Fig. 19 Filament winding

the resin-soaked strands through a heated shaping die. The die may be in the shape of a rod, tube, I-beam, or other geometric shapes. After the resin is cured and pulled through the die, the resulting profile has a high strength-to-weight ratio and is very durable, especially in a chemical environment.

Applications for pultruded products include structural beams for electrical and chemical environments (for example, ladders for use near electrical wires), poles, and shafts. The future of the pultrusion process may be in space applications where long structural components for space stations could be manufactured in space, as required, eliminating the need to transport long beams from earth.

Filament winding is primarily used to manufacture large structural containers or tanks. The process involves several spools of reinforcing materials such as glass or carbon strands. The strands are directed into a resin bath of either polyester or epoxy. The wetted strands are wound over a turning mandrel (Fig. 19) in different patterns to provide different strengths. After the resin has cured, the filament wound part is removed from the mandrel and machined or assembled as required.

Applications for filament wound composites include gasoline storage tanks, septic tanks, large-diameter drainage pipes, chemical storage systems, and sporting equipment such as golf club shafts and bike frames.

Additional information about composites processing is provided in the article "Design for Composite Manufacture" in this Volume.

Casting

Casting is the process of pouring liquid plastic into a mold. The recent development of a wide variety of casting resins and rapid tooling fabrication has allowed the casting process to be considered as a viable process for both prototyping and low-volume production. Typical casting resins include casting acrylic, casting polycarbonate, epoxy, polyurethane, and polyester. The casting process produces plastic parts with the lowest level of internal stress and a high degree of dimensional stability.

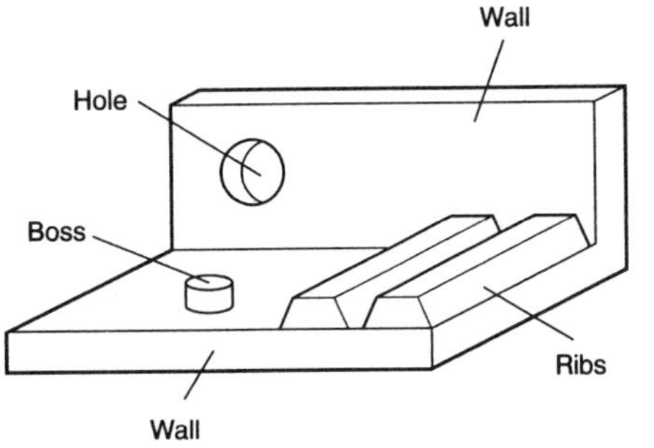

Fig. 20 Plastic part wall components

Design Features and Process Considerations

A design feature is an aspect of the shape of a product. Principal features incorporated into the design of a plastic part are:

- *Walls* are the predominant features of the shape of a product. The nominal wall for a plastic part is the platform on which all other design features reside. Most other design features are configured and sized as a function of the size of the nominal wall (Fig. 20).
- *Projections* are design features that rise from the nominal wall. They include ribs, gussets, threads, snap-fits, and bosses (Fig. 21).
- *Depressions* are design features that enter into or reside within the nominal wall. They include through holes, blind holes, and slots (Fig. 22).

Design for assembly (DFA), or design for manufacturing and assembly (DFMA), is a design methodology that embraces the concept that well-engineered assemblies will take advantage of the high functionality of materials, such as plastics, and integrate several design features, such as snaps, alignment features, and locks, to facilitate assembly. These integral features help to eliminate conventional assembly components such as screws, washers, and nuts. Additionally, the product designers that utilize DFA techniques go beyond the design of the individual part to consider the optimization of the design of several parts in an assembly (Fig. 23). General DFA concepts are described in the articles "Introduction to Manufacturing and Design" and "Design for Manufacture and Assembly" in this Volume.

Design for Optimum Properties and Performance. While many product designers understand that it is possible to degrade or lower material properties through processing operations, opportunities to optimize properties of the plastic and the plastic product are often overlooked. The most important step in the optimizing process is to understand how the properties of a plastic will vary when processed and to alter the product design to avoid or manage this variation.

The best resource for determining these variations is direct experience. Certainly, material suppliers can provide generalizations as to the effects of processing; however, there are simply too many variables (design, process equipment,

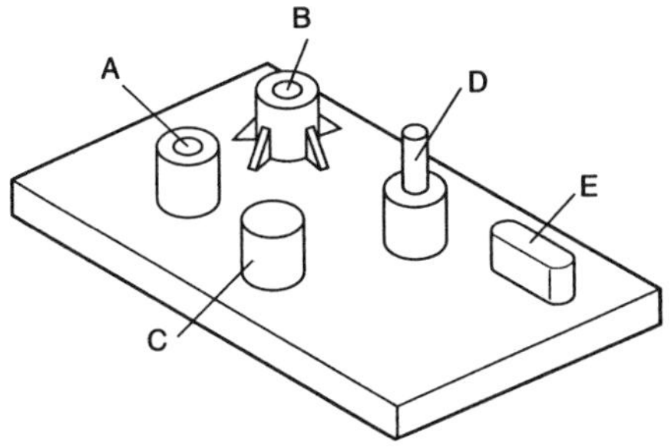

Fig. 21 Boss designs for plastic parts. A, hollow boss; B, gussetted boss; C, solid boss; D, stepped boss; E, elongated boss

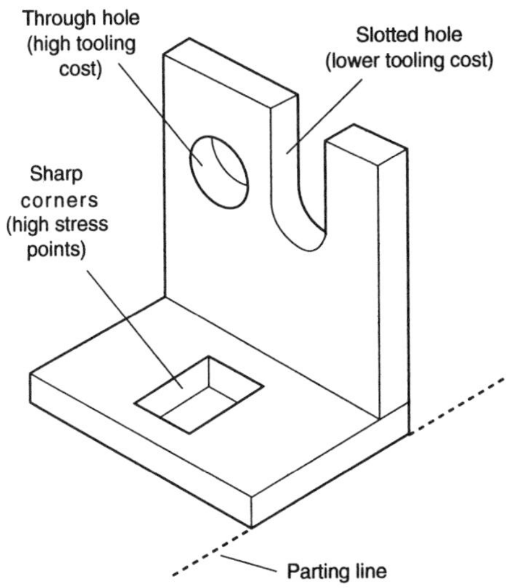

Fig. 22 Depression designs for plastic parts

additives, etc.) for detailed and reliable property variations to be published.

An example of such a property variation is a polycarbonate part designed for high-impact strength that has a wall section that is too thick. The polycarbonate shows impact degradation at a critical thickness value; thus the impact properties of the resulting plastic product may be much lower than those published by the material manufacturer.

A second example is a polyacetal part molded with a low mold temperature, to improve the cycle time, resulting in an induced stress level and postmold shrinkage. The product may be dimensionally correct when measured directly after molding but too small when inspected 1 week later after delivery to the customer.

A third example is a polyphenylene sulfide part processed with a mold temperature of 93 °C (200 °F), the highest temperature the molder's water controller can safely attain. As a result, the polymer does not achieve optimum crystallinity, and the part performance suffers in its end-use application. While water-heated molds may be acceptable for many plastics, the molder did not recognize that an electric or oil-heated mold would have allowed a more appropriate mold temperature of 121 °C (250 °F) to maximize the crystallinity of the polymer.

Other Plastics Design and Processing Considerations

Stress in plastic parts is most often related to shear stress, especially for molded or extruded parts. During molding, a plastic part is subjected to a rigorous process environment. When melted plastic is forced into a closed mold, three major stress contributors are present: high pressure, high speed, and constricted flow areas.

The stress level in a part depends on the processing method used. The three major stress-producing components are present in injection molding; thus injection-molded parts usually have a high degree of molded-in stress. Compression molding usually has only the high-pressure component and thus produces parts with much lower molded-in stress.

Casting involves none of the major stress-contributing factors and, therefore, results in parts with the lowest levels of molded-in stress.

Stress levels, in a plastic part, affect the properties and performance of the product. In general, as the molded-in stress level increases, dimensional stability decreases, chemical properties lower, mechanical properties decrease, and optical properties diminish. A good example of this relationship is the compact disc (CD). A CD needs to be optically pure and mechanically strong. For efficient production, the best processing method for CDs is high-volume injection molding. However, by its nature, the injection molding process will lower the required properties, especially because the discs are extremely thin and will promote high shear stress. This problem was solved, in large part, by the development of a low-viscosity plastic (polycarbonate) that required lower molding pressures to fill the mold and thus lowered the resulting molded-in stress in the CDs.

Published versus Actual Product Properties. It is technically naive to believe that the material manufacturer's published properties will be duplicated in the final product. Published properties are derived from parts that are molded and tested under highly controlled conditions. The environment is controlled to either "dry as molded" or 50% relative humidity conditions. The samples tested are usually ASTM/ISO test specimens that do not have the design features that often can serve to lower properties in molded products.

The property effects of additives such as colorants and regrind are often not published simply because there are too many variables involved. Also, time-related properties such as weatherability and creep often are interpretive, and published data should serve only as guidelines for the designer.

It is important that the product design team along with the manufacturing team develop and test the plastic products prior to production. Using the manufacturer's published data as a reference, the product tests should always emulate the end-use application environment as closely as is practical.

End-Use Concerns. Even well-designed and manufactured plastic products can fail because one or more aspects of the end-use application was not considered. The longer the list of end-use environmental conditions that are assessed and respected, the more successful the plastic product will be.

Many new plastic product designers tend to focus on details, but overlook some obvious end-use considerations. Examples include:

- Effects of household chemicals (for example, milk, water, cleaners, makeup, and makeup removers) on plastics used for consumer applications
- Shipping and handling factors (for example, high compartment temperatures and damage from dropped boxes)
- Time-related issues such as creep, stress relaxation, and ultraviolet weatherability

Materials Selection Methodology

Selecting a plastic material for a specific end-use application is a challenge. It requires a thorough understanding of the end-use application, a knowledge of available plastic materials and their properties, and a methodology to sort and select all the data to make a prudent decision.

Understanding the End-Use Application. As noted above in the section "Other Plastics Design and Processing Considerations," it is important for the designer to understand the intended application of a part including physical loads that will be applied, chemical resistance and exposure factors, and temperature. However, a designer must investigate the end-use environment even further if a good material selection is to be made.

For example, a design specification from a furniture maker noted that their products typically experience temperatures between 16 and 27 °C (60 and 80 °F). This is a narrow range, but the furniture maker understood that their customer base would always properly control the environment of the office where the furniture was to be located. However, the shipping and storage of the furniture was overlooked. Temperatures in the back of a semitrailer truck or a warehouse can be significantly different from those in the ultimate end-use environment. Temperatures below freezing and over 43 °C (110 °F) are certainly possible. The plastic materials selected for this application must be able to withstand the shipping and storage as well as the end-use application.

In another example, a manufacturer carefully studied the end-use requirements for a new taillight lens and designed and processed the part to meet the requirements. The plastic lenses were molded in Texas and shipped (by truck) to the automaker in Detroit. The automaker was dismayed to discover upon delivery that more than 20% of the snap tabs used to hold the lens in its assembly were cracked or broken. The cause of the problem was the cyclical loading that occurred during shipping, which was different from

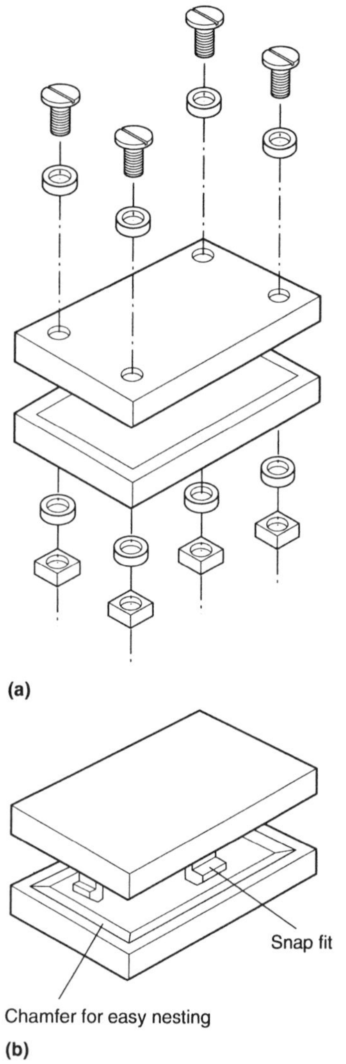

Fig. 23 Designing for product assembly. (a) Original design. (b) Improved design for ease of assembly

any loading tested for the end-use application of the part.

A third and final example involves plastic material selection for a simple coat hook. The design engineer assessed the end-use application and determined that a 10 lb coat hanging for 8 to 10 h on the hook would be no problem. However, it was later discovered that the same 10 lb coat left on the hook for 30 days resulted in unacceptable deformation of the hook. The designer failed to consider the time-related phenomenon of creep.

Understanding the Properties of the Plastic Material. Thousands of plastic materials are available today, and the list is growing rapidly. It is challenging for designers to select the right plastic material, that is, one that meets all the property, processing, and cost requirements and goals. The material selection process is further complicated by the fact that much of the information published about plastic materials focuses on their positive attributes. Very few plastics manufacturers overtly comment on the weaknesses or drawbacks of their materials.

Selecting a plastic material requires an understanding of the balance of properties; that is, how well a material meets the overall property requirements for a particular application and any effects that different properties may have on one another. For example, for a plastic material considered for a roof gutter system on a residential building, the properties that might be assessed include water resistance, impact resistance, temperature resistance (resistance to heat and cold), and weatherability (ability to withstand the rigors of outdoors).

Unfortunately, however, a simple list of attributes, although an excellent starting point, is not adequate to differentiate among the thousands of plastic materials available in order to make a proper selection. The selection process must have more depth. Consideration must go well beyond attributes and consider the specific variable or number value of each property. Once that is determined, there must be an assessment of interaction.

A more detailed property assessment might appear as follows:

Property	Requirement
Water resistance	Must not absorb >0.05% water when exposed to 100% humidity
Physical properties	Must be measured at 50% relative humidity
Impact resistance	Izod notched impact strength must be >133 J/m (>2.5 ft · lbf/in.) from –17 to 66 °C (–20 to 150 °F)
Temperature resistance	Temperature range is –40 to 71 °C (–40 to 160 °F)
Weatherability	<5% degradation in physical properties after 15 yr ultraviolet exposure (UVA and UVB)

It is clear that the process of identifying the end-use application requirements is more challenging than it may initially appear.

Other end-use properties that should not be overlooked include:

- Cost (material, labor, overhead, and yield)
- Processability (easy or difficult, inexpensive or costly)
- Maintenance (low or high, durability)
- Ease of assembly

While some of these attributes are a direct function of the product shape and design, the material selection aspect needs to be considered. As an example, when considering ease of assembly, it may be important to have a plastic material that can be bonded using solvents. This requirement would eliminate many plastics from the list that might otherwise have been candidates.

Desired properties	Rating	PE	PP	PVC	ABS	PS	
Impact strength	9	3	4	5	2	1	Rank
		27	**36**	**45**	**18**	**9**	Rating × rank
Low cost	6	5	4	3	2	4	Rank
		30	**24**	**18**	**12**	**24**	Rating × rank
Flexural modulus	3	1	2	3	4	5	Rank
		3	**6**	**9**	**12**	**15**	Rating × rank
Low flammability	9	2	3	5	4	1	Rank
		18	**27**	**45**	**36**	**9**	Rating × rank
Tensile strength	6	2	3	4	5	1	Rank
		12	**18**	**24**	**30**	**6**	Rating × rank
Total (sum of products)		**90**	**111**	**141**	**108**	**63**	

Fig. 24 Example of a material selection matrix. PE, polyethylene; PP, polypropylene; PVC, polyvinyl chloride; ABS, acrylonitrile-butadiene-styrene; PS, polystyrene

Material Selection Matrix. In order to simplify the material selection process, many product designers utilize a material selection matrix (Fig. 24). The matrix allows a direct comparison between the end-use properties desired and the actual properties available from the candidate materials. Additionally, the matrix helps prioritize and sort these properties to make the final selection.

Developing a material selection matrix involves a number of steps:

1. Identify as many material properties and attributes as possible that are required to meet the demands of the application.
2. Assign a rating to each property and attribute (where 9 is critical, 6 is desirable, and 3 is optional).
3. List the candidate materials.
4. Rank the materials, relative to one another, as to how well they meet each property/attribute requirement. (If there are four materials being compared, the material that best meets the property/attribute requirement receives a 4. The material the least meets the property/attribute requirement receives a 1.)
5. Multiply the property rating by the material ranking.
6. Add the products of Step 5.

The material with the highest sum is the top candidate. In Fig. 24, for example, PVC is the top candidate for the application represented by this matrix. This matrix analysis tool also can be readily adapted to compare processes as a means for optimizing process selection.

SELECTED REFERENCES

- J.F. Agassant, P. Avenas, J. Sergent, and P.J. Carreau, *Polymer Processing: Principles and Modeling,* Hanser Gardner Publications, 1991
- R.D. Beck, *Plastic Product Design*, 2nd ed., Van Nostrand Reinhold, 1980
- M.L. Berins, Ed., *Plastics Engineering Handbook of the Society of the Plastics Industry, Inc.*, 5th ed., Van Nostrand Reinhold, 1991
- J.-M. Charrier, *Polymeric Materials and Processing: Plastics, Elastomers, and Composites*, Hanser Gardner Publications, 1990
- *Engineered Materials Handbook Desk Edition*, ASM International, 1995
- *Engineering Plastics*, Vol 2, *Engineered Materials Handbook,* ASM International, 1988
- S. Levy and J.H. DuBois, *Plastic Product Design Engineering Handbook,* 2nd ed., Chapman and Hall, 1985
- E. Miller, *Plastic Product Design Handbook,* Part A, Marcel Dekker, 1981
- P. Mitchell, Ed., *Plastic Part Manufacturing*, Vol 8, *Tool and Manufacturing Engineers Handbook*, Society of Manufacturing Engineers, 1996
- E.A. Muccio, *Plastic Part Technology*, ASM International, 1991
- E.A. Muccio, *Plastics Processing Technology*, ASM International, 1994
- *Polymer Engineering Principles: Properties, Process, and Tests for Design,* Hanser Gardner Publications,1993
- P.A. Tres, *Designing Plastic Parts for Assembly*, 2nd ed., Hanser Gardner Publications, 1995

Design for Composite Manufacture

Anthony J. Vizzini, University of Maryland

THE USE OF COMPOSITE MATERIALS provides the designer with great flexibility with which to meet part performance requirements. Even after a composite material has been selected, the design must still address the microstructure of the component. The type of reinforcement—continuous fiber, short fiber, whiskers, chopped roving, woven or braided fabrics, or preforms—and its orientation greatly affect the performance of the component. Indeed, the ability to tailor a component increases its range of functionality as well as the complexity of the design. For example, the skin of an aircraft wing near its root may involve a hundred plies. Although the plies may all be the same material in the same format, it is the collective orientations of the plies that yield the desired extensional, flexural, and torsional stiffnesses (Ref 1). The design at the microscale is more important than the simple orientation of the reinforcement. Failure mechanisms such as delamination, fiber microbuckling, fiber kinking, and transverse cracking all occur on the microscale and are sensitive to manufacturing variations (Ref 2).

Just as the design must address the microstructure, it must also address the macroscale. Composite materials can be formed into near-net shapes. Construction of components that often require many individual parts can be made in one step. Stiffeners can be staged, cocured, precured and integrated, or bonded. Sandwich construction with honeycomb or foam results in lightweight alternatives to stiffened panels. As the part count is reduced, the complexity of the components is increased. The structural design may be as extensive as the aft fuselage of the V-22 Osprey (built by Boeing Defense and Space Group, Helicopters Division) or as simple as composite dimensional lumber made from recycled plastics.

Designing composites for manufacturing is more extensive than being able to fabricate a part within specified dimensional tolerances. Indeed, the desired orientations of the reinforcement must be met with special attention paid to the presence and effects of anomalies or defects. Moreover, the prospect of large, integrated structures necessitates designing the manufacturing process concurrently with the component.

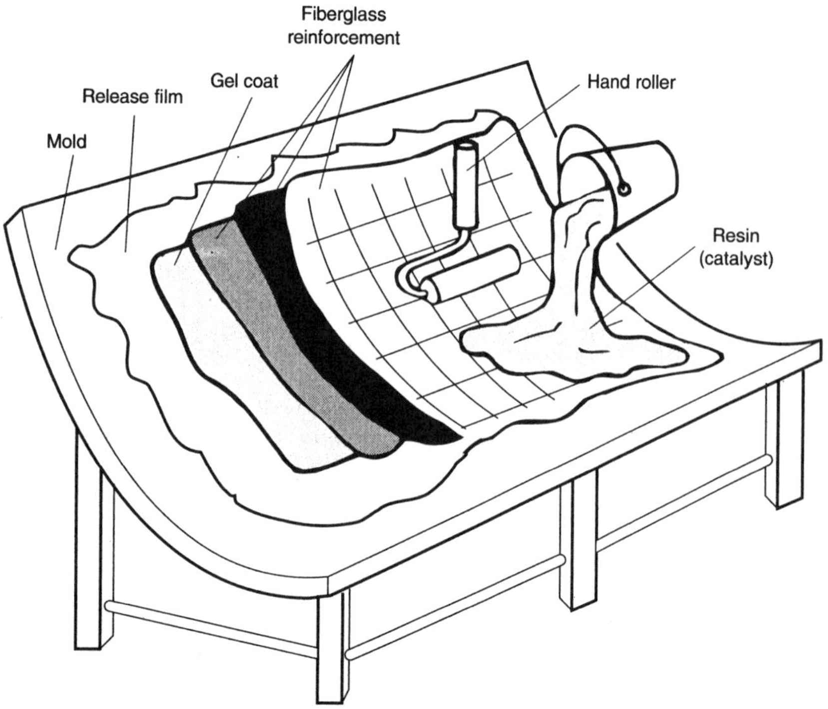

Fig. 1 Hand lay-up process for fabricating a boat hull

Composites Manufacturing Processes

Often the design of a composite component is done in tandem with an equivalent metallic component to assess the cost/benefit of alternate materials. This could result in an unfair comparison as the designer employs such constraints as *black aluminum,* the use of quasi-isotropic layups to degenerate the behavior of an orthotropic material to that of an isotropic material. Or the designer develops the manufacturing process of the composite component concurrently with that for the metallic component. This design process does not take into account the tailorability of the composite material nor the potential advantages of the manufacturing process to increase performance, lower part count and weight, and decrease overall costs. The differences between composites and the materials they would potentially be replacing are significant enough to warrant independent development of the component and the manufacturing process.

Prior to the design of the composite component, many constraints may be imposed. Selection of the manufacturing process may be automatic based on available equipment and prior experience. The number of parts to be made is known, thus limiting the time to manufacture and the expected costs per part. Overall dimensions (such as those for a boat hull) or dimensional constraints are specified. Surface finish and dimensional stability may be specified. The operating environment and general loadings should be known. Thus, many selections are made prior to the actual design of the component. The material system is most likely chosen based on the operating temperature, desired mechanical properties, material availability, and cost considerations. The manufacturing process is selected based on surface finish, cycle time, and part cost.

The choice of material greatly influences the ability to design for manufacturing. The greatest latitude is afforded by polymeric resins. Structures that require ceramic, metallic, or carbon matrices involve forming processes that introduce matrix into the reinforcement. In metal-matrix composites, for example, reinforcements increase strength, decrease the coefficient of thermal expansion, and improve wear resistance at a cost of a reduction in ductility and in fracture toughness. Thus, the design of a component often begins with the manufacturing process itself and extends the available technology to incorporate the reinforcement. The goal is to improve the overall performance of the metal or ceramic matrix rather than to create a material with different response than the base matrix. Thus, this article concentrates on design for manufacturing of polymeric composites.

Thermoplastics that are reinforced with short-chopped, randomly oriented fibers are easily fabricated using conventional techniques. Injection molding and extrusion are widely employed. Composites based on thermosetting plastics are processed using specially developed methods, including:

- Contact molding, which employs single-surface molds, as in hand lay-up, spray-up, and filament winding
- Compression-type molding, as in sheet molding, bulk molding, preform molding, and cold molding
- Resin-injection molding, which is similar to the process used for nonreinforced materials and reinforced thermoplastics
- Pultrusion, which is a modification of the extrusion process

Each of these methods is briefly described below. More detailed information can be found in *Composites*, Volume 1 of the *Engineered Materials Handbook* published by ASM International.

Contact molding processes are used for liquid resin systems and various forms of fibrous reinforcement. As shown in Fig. 1, a mixture of the resin (liquid monomer and catalyst) is applied in liquid form to the mold. The reinforcing phase is usually placed in the mold in the form of a mat, a woven-roving cloth, or a preformed fibrous shape. The reinforcement is saturated with the liquid resin, which is then cured by heating. A male or female single-surface mold is used, thus providing only one smooth surface. The mold is first coated with a release film to prevent the finished part from sticking permanently. A thin (about 0.4 to 0.5 mm, or 0.015 to 0.020 in.) pigmented gel coat is then sprayed onto the mold surface to provide a smooth finish.

Hand lay-up is used for manufacturing low-to-medium production volumes of large, relatively thin-walled parts, such as boat hulls (Fig. 1). Spray-up uses a continuous-strand glass roving that is fed through a combination fiber chopper and spray gun. The chopped roving and liquid resin are sprayed on the mold surface to build up the required wall thickness. Curing under pressure can be achieved by placing pressure bags over the surface of the molding.

Filament winding is used for manufacturing highly reinforced (50 to 60% fiber volume fraction), high-strength parts, which are usually axisymmetric in shape. Continuous strands, roving, or woven tapes are drawn through a resin bath and wound onto a rotating mandrel (Fig. 2). The wound part is then cured by heating. Prepreg, in which the fibers have been coated with resin and stored at low temperature, can also be used for filament winding. Examples of filament-wound products are tubes and storage tanks.

High-pressure laminates are made by stacking layers of resin-impregnated sheets (prepregs), then applying heat and pressure to bond the layers together. The pressures applied range from 8 to 14 MPa (1200 to 2000 psi).

In most laminated designs, the plies (layers) are arranged at a variety of angles. In high-performance applications, plies used in the laminated stack are oriented at fixed angles to the direction of major load. The most common fixed load angles are 0, 45, –45, and 90° to the major load axis (Fig. 3). By selecting the proper number of plies at each of these angles, composite properties and their anisotropies can be designed for strength, modulus, or even the degree of thermal expansion along each principal direction.

In addition to making flat and curved laminated sheets, high-pressure laminating can be used to produce laminated tubing by winding the impregnated sheet under pressure on a mandrel, then curing the resin. Because of the high pressures used, the size of parts made by this method is limited by the sizes of available presses.

Compression molding, sometimes referred to as matched-metal molding, employs either bulk molding compound (BMC) or sheet molding compound (SMC). Bulk molding compound is a prepared mixture of thermosetting resin, chopper fibers (glass, carbon, or aramid, with lengths ranging from 3 to 50 mm, or ⅛ to 2 in.), filler, catalyst, and other additives. It is supplied in bulk form or as extruded "rope." The required amount of the mixture is placed in a heated mold, which is then closed, subjected to pressure, and cured. Sheet molding is similar to bulk molding, but the reinforcement is longer glass fibers, 25 to 75 mm (1 to 3 in.), and the compound is supplied in sheet form. The sheet is first cut to a shape similar to that of the mold and is then pressed and heated in the mold for curing.

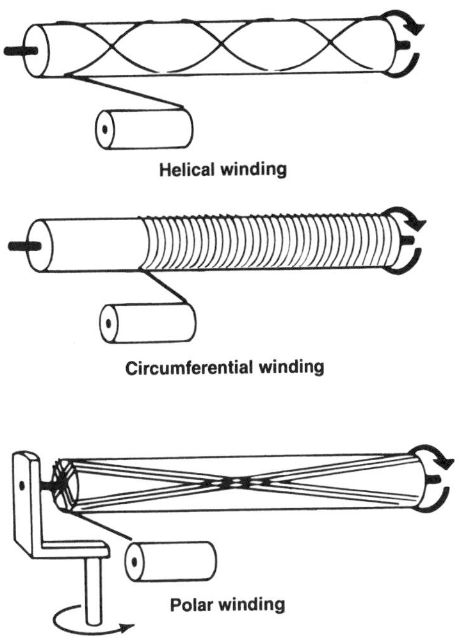

Fig. 2 Schematic representations of winding techniques

Preform molding involves spraying a mixture of chopped glass roving and weak binder on a pattern resembling the product. The preform is then placed in a heated metal mold and saturated by a mixture of resin, catalyst, fillers, and pigment. Curing then takes place under pressure. Cold-press molding is similar to preform molding, except that curing takes place at room temperature and under low mold pressure. This allows the use of inexpensive mold materials such as plastics and plaster. Preform molding gives more consistent and more attractive products than those produced by lay-up or spray-up processes.

Resin transfer molding (RTM) is a matched-mold process that uses low-cost molds. The reinforcement is laid up dry in the mold, and liquid-catalyzed resin is pumped in until the mold is filled and the pressure reaches about 170 kPa (25 psi). The process can be vacuum assisted (VARTM) to prevent trapped air pockets and to assist with the flow of resin. The curing time is a function of the resin and can vary from about 15 min to on the order of 1 h. This process is considered intermediate between spray-up and the faster pressure-molding methods. It is suitable for medium-volume production and can be used for manufacturing complex parts for automobile and truck bodies. Tailoring composites by placing extra reinforcement in highly stressed areas makes it possible to consolidate several smaller parts into one large structure, thus reducing assembly operations and cost.

Pultrusion is used for reinforced thermosetting plastics to make shapes similar to extruded shapes. The reinforcing fibers are immersed in a liquid-resin bath and are then pulled through a long heated die to cure the resin (Fig. 4). Rubber-faced rolls, which are shaped to the required form

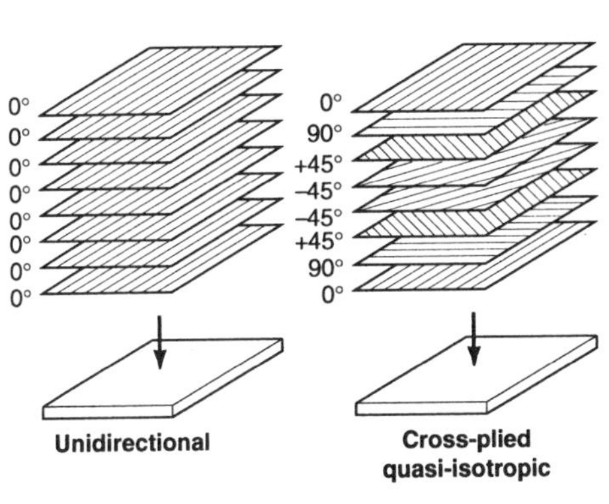

Fig. 3 Typical composite laminate configurations

of the pultrusion, pull the cured and finished shape from the die. The pulling speed ranges from 50 to 5000 mm/min (2 to 200 in./min), depending on the resin compound, shape and thickness of the section, mold temperature, and mold length. The most commonly pultruded materials are glass-reinforced unsaturated polyester resins. Standard pultruded shapes are available in the form of:

- Solid bar of round, rectangular, and dogbone cross section
- Hollow shapes of round and rectangular cross section
- Structural shapes, such as channels and I-beams
- Sheet of flat and corrugated cross section

Epoxy resins can also be used in pultrusions, but they require longer curing time and do not release as cleanly from the dies.

Process Considerations

Independent of selections made prior to the design, many aspects of the manufacturing process are still unknown and are a function of the microscale and macroscale design of the component. The general composite manufacturing process can be subdivided into three separate stages. The first is preparation and involves handling the raw materials as well as mold fabrication and preparation. The second stage is the actual forming process. It is in this process that chemical reactions occur or the constituents are consolidated into the desired component. Finally, additional fabrication may be required such as trimming, machining, bonding, finishing, or coating.

Preparation

Two aspects of manufacturing that interact with the design are the handling qualities of the raw materials and characteristics of the mold. Prior to the manufacturing step, materials are available as individual constituents or in mixed form. As individual constituents, they are either combined prior to the manufacturing process or combined as a result of the process. Materials premixed such as preimpregnated fabrics and tapes, SMCs, and BMCs may be assembled prior to the manufacturing process.

Handling Qualities. *Tack* is the ability of the material to stick to itself and to other materials. A tacky material will stay in place once it is situated on a mold or on top of another layer. Thermoplastics with little tack may need to be held in place in the mold through the use of a local heat source such as a heat gun. Materials with too much tack increase the effort required by the operator to place the material in the correct position because errors in placement cannot be easily remedied. The degree of tack is influenced by room temperature, relative humidity, and dryness of the material. Tack affects the design by limiting the size of the individual ply that can be handled and the ability to place it in the part. Although templates can be used for positioning individual plies, tacky materials must be placed accurately the first time, whereas low-tack materials may shift prior to consolidation resulting in a disturbance in the location of a ply.

Drape is the ability of the material (fiber or fabric, with or without resin) to conform to the shape of the mold. The greater the drape, the more contour that can be designed into the part. Therefore, drape limits the radius of curvature in any given direction. This is separate from the concern of laying a planar material on a doubly curved mold or the reverse problem of representing in two dimensions a three-dimensional surface without distortions. Typically, thermoplastic materials have little drape and thus are made to conform to the shape of the mold during the manufacturing process as is done in thermoforming. However, thermoplastic materials that have been cut into long, narrow strips and then loosely laid as a stretchable fabric do exhibit some degree of drape.

Because thermosets cure at any temperature, preimpregnated staged materials have a finite workable time out of storage, and resin wet-out systems, once mixed, have a finite pot life. Beyond these time limits, the mechanical properties of the composite structure decrease and handling of the constituents becomes difficult. Thus, the time to complete the assembly must be addressed in the design of the component. If intricate plies are required, then automation must be employed in cutting the individual plies. The size of the structure may be limited by how much material can be applied without significant degradation of mechanical properties. However, there are manufacturing processes and materials well suited for long assembly times. Many of the closed-mold processes, VARTM, or structural reaction injection molding (SRIM), when used with preforms can reduce the handling time of the resin. The dry reinforcement is placed in the desired orientation, and then the resin is introduced.

The ability to properly wet out the fibers, to consolidate individual layers, and to decrease void content is related to the viscosity of the resin. There is a trade-off between increased wet-out and handling characteristics. Low-viscosity resins tend to pool in resin-rich areas for large parts, resulting in a lack of uniformity in fiber volume. Resin systems must be tailored to provide low viscosity during winding processes or in flow periods of autoclave processes, but must be able to remain where placed within the structure during the process.

Tooling Considerations. The role of the tool or mold is to impart the shape to the composite part during the manufacturing process. Tool design is based on the requirements for the tool to maintain its shape during the manufacturing process and is a function of the processing temperature, applied loads, and curing time. In addition the number of replicates per tool affects its material of construction.

Here, consideration of the manufacturing process is very important. The geometry of the part is constrained by the ability to form the necessary tooling. Highly contoured surfaces will require a highly contoured tool. Making use of available geometries such as flats and tubes results in low-cost tooling. Complex tooling requires manufacturing the desired master model shape using a rigid medium such as wood. The mold is made from this machined master model shape or plug, often using composite materials.

One key aspect that must be incorporated into the design is the phenomenon referred to as *springback*. Because of differences in thermal expansion, residual stresses arise during the manufacturing process. When the part is removed from the tool, the part deforms to the state governed by the residual stress state. For example, springback results in a one or two degree change

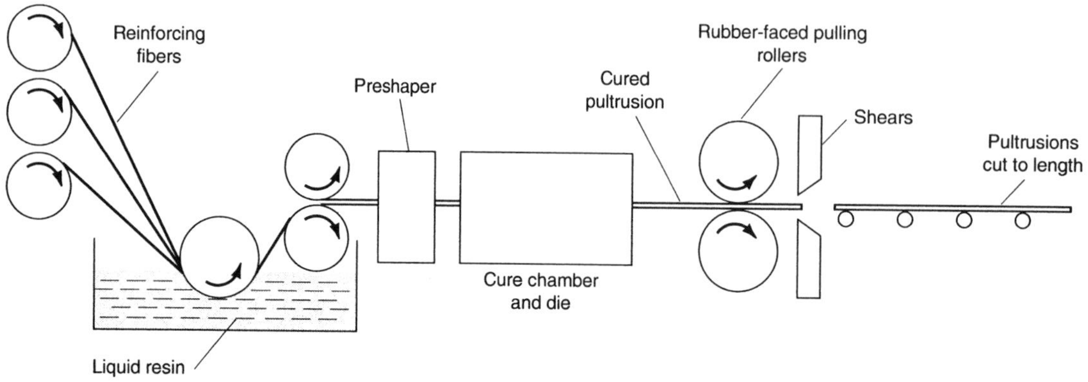

Fig. 4 Schematic of the pultrusion process for fabricating composite products

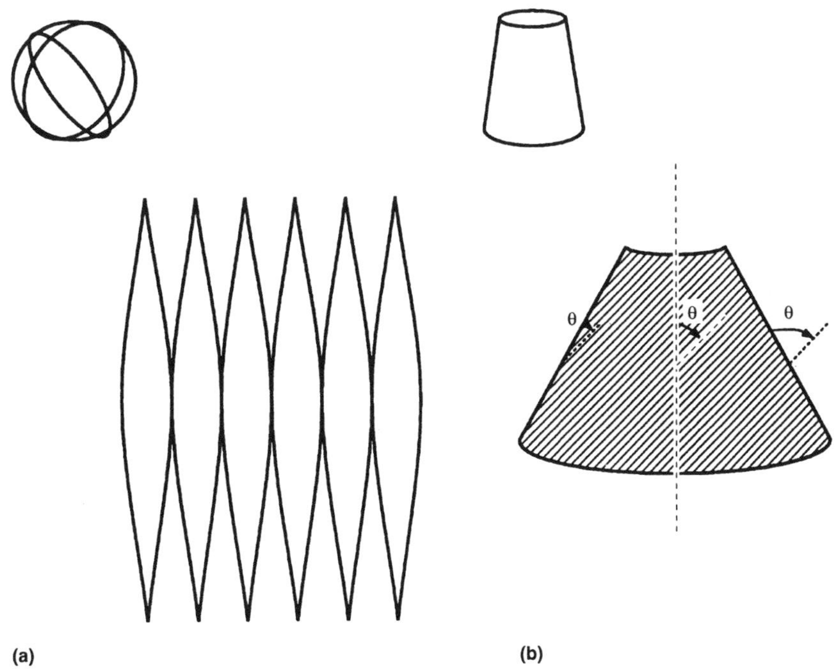

Fig. 5 Flattened contours (a) sphere and (b) truncated cone

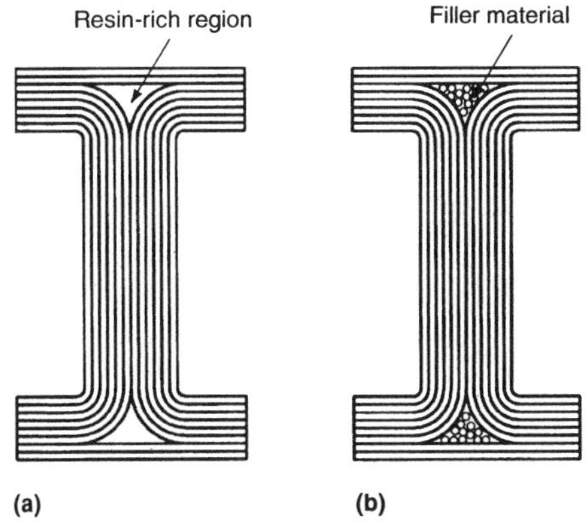

Fig. 6 Schematic of a C-channel with (a) a resin-rich region and (b) filler material to maintain fiber volume fraction

in an "L"-section and can be addressed by considering the residual stress state in the design of the tool. Also, dimensional stability (shrinkage) must be calculated and accommodated in the mold design.

Forming Processes

Because the manufacturing process is to be designed concurrently with the component, design guidelines are flexible. Manufacturing of a designed component is often only limited by the creativity of the engineer and the budget of the project. Although all conceivable parts currently cannot be made with composites, the constant stream of new products using composite materials attests to the ingenuity of manufacturing engineers.

In general, the manufacturing process can impose limitations on the microscale, such as the orientation of the reinforcement, and on the macroscale, such as the ability to form an integrated structure. Although an individual forming process has specific requirements on types of resins and reinforcements that can be used, the design considerations can be discussed among three areas. Laminated hand laid-up structures require extensive labor and tooling and thus are used only in those applications where the added performance can justify the relatively high costs. A second group of forming processes are those that make use of dry reinforcements in a semicontinuous or continuous fashion such as filament winding, braiding, or pultrusion. The third group includes the closed-mold processes as discussed in the article "Effects of Composition, Processing, and Structure on Properties of Composites" in this Volume.

Preimpregnated Tapes and Fabrics. Laminated preimpregnated materials often offer the greatest mechanical properties. High specific strength and stiffness are important in weight-constrained structures within the aerospace industry. Preimpregnation provides a consistent product with a uniform distribution of reinforcement and resin with high fiber-volume fractions possible. Such laminated structures typically require an excessive amount of hand labor and autoclave processing, resulting in a relatively high cost of manufacturing. Nevertheless, proper design of a composite component can decrease the overall manufacturing costs.

In laminated structures, the orientation of each ply is a design variable. This increases the complexity of the design process, and thus ply angles are often limited. Several common approaches include allowing only orientations that are multiples of 15° or, more stringently, laminates made up entirely of 0, +45, –45, and 90° plies. These design strategies limit the orientations that are involved if hand lay-up is to be used. However, limitation of angles may not be necessary. Templates can be used to cut material with the desired orientations and to place the plies on the tool. A trained worker can just as easily lay a 15° ply as a 17° ply. However, if the design requires many plies (shape and orientation), then additional care must be taken to sort the individual plies and to ensure that each ply is laid down in the correct sequence. It is also important that the tolerance for a given angle is not greater than ±3°. However, variations between specified and as-manufactured orientation will always occur.

The cutting of plies can be automated to increase efficiency and quality. Techniques include water jets, lasers, reciprocating knife, and die cutting. Using these methods, intricate plies can be cut with repeatability with no limitation on orientation of the reinforcement. A water jet introduces moisture to the uncured composite and extended exposure needs to be avoided. Laser cutting results in a heat-affected zone and may alter the local mechanical properties.

The contour of the tool is of great concern when using preimpregnated tapes and fabrics. Wherever there is a double radius of curvature, flat materials cannot be used without stretching or overlap. The inverse problem is to take any shape with a compound curvature such as a hemisphere and flatten it without excessive straining of the surface (Fig. 5a). One can see that the initial planar material requires many discontinuities. Even if cutting is reduced to a minimum as in the cone in Fig. 5(b), the orientation about the structure cannot be prescribed. Thus, a requirement for all of the fibers in a layer of a cone to be in one direction is impossible. The orientation of fibers in structures with realistic surface contours is a continuum problem that the designer must be aware of.

Much of the above effort can be reduced by using tow placement. Individual tows or strips of preimpregnated material can be laid down using a multiaxis, multi-degree-of-freedom robot arm. The material is tacked to the tool by applying pressure through the head of the tow placing arm. If this head is heated or an external heat source such as a laser is used, then thermoplastics can be used. Individual strips can be placed on the tool in the desired orientation. This process removes the variations inherent in hand processing, although it requires extensive programming. Errors in orientations caused by the surface contour are limited by the width of the tow or strip being used.

Because most laminated structures are placed in a vacuum bag and additional autoclave pressure is applied, a major concern is maintaining the integrity of the bag during the forming process and proper transfer of the applied pressure to the part. In any concave portion of the part, the pressure bag can form a bridge if locally it is of

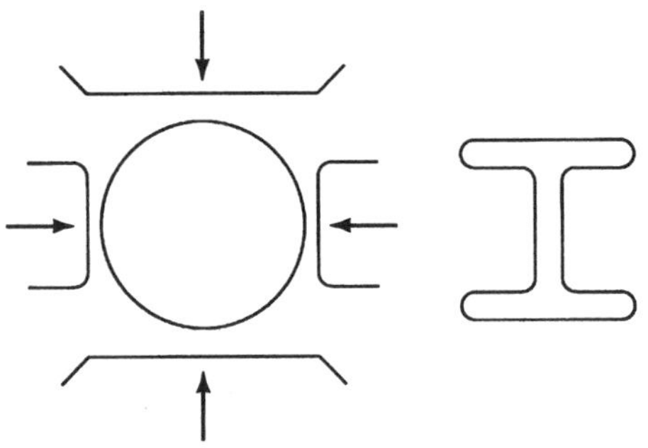

Fig. 7 Post-winding forming

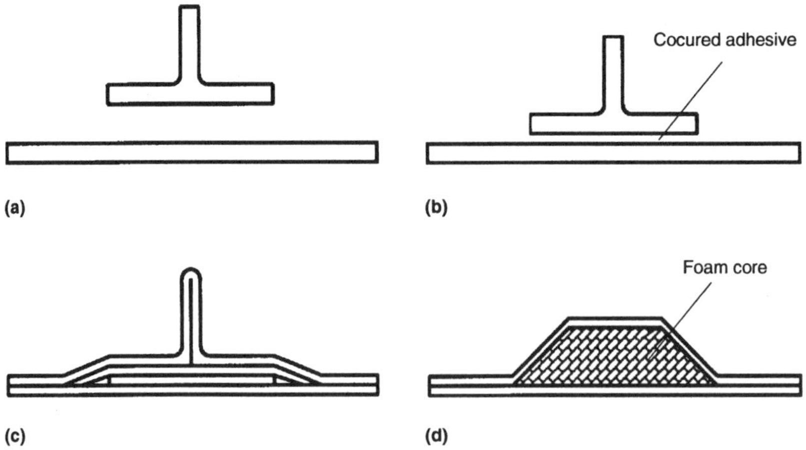

Fig. 8 Integration of stiffener and skin. (a) Mechanical/adhesive attachment. (b) Cocured. (c) Integrated lay-up. (d) Redesign

inadequate size. Thus, insufficient consolidation will occur in that section. One approach that reduces this effect is a reusable silicone rubber bag. The bag is originally formed on the tool. Even if the bag conforms as it should, a fillet should still be allowed in a corner; however, the fiber volume in the fillet will be lower than the surrounding material. For example, if an I-beam is constructed by laying up two separate C-channels, placing them together, and applying additional plies on the flanges, a small resin rich pocket will form (Fig. 6a). To alleviate this resin-rich zone, additional material can be placed in that area to maintain the same fiber volume fraction (Fig. 6b). In some cases, this remedy cannot be applied. Because plies have discrete thicknesses, the thickness of a part is changed using terminating plies, resulting in a resin-rich area that cannot be filled. A possible remedy is the use of a preform of BMC partially staged that can be cured with overlaid flange plies.

Dry Fibers and Broad Goods. One major disadvantage of the preimpregnated tapes and fabrics processes is that they are batch processes that involve significant hand labor or, in the case of tape layers, significant programming and capital investment. Mixing the reinforcement with the resin system in situ during the fabrication process allows for three-dimensional reinforcements or continuous processing while increasing material utilization.

In filament winding, fibers are drawn through a resin bath and wound on a rotating mandrel. This is the primary manufacturing method for cylindrical or spherical pressure vessels because the fibers are easily placed in the hoop direction. Low-cost constant cross-section components can be made using a filament winder with two degrees of freedom, the rotation of the mandrel and the translation of the payout eye. Increasing the number of degrees of freedom in the process increases its complexity as well as the complexity of parts that can be created. Additional payout eyes increase the speed of the process as well as introduce braiding. Variations to the winding process can include varying cross-sectional geometries, post-winding forming that can alter a convex outer surface to an I-beam shape as in Fig. 7, and inclusion of fixtures.

One concern in filament winding is the angle of the fibers. Hoop (circumferential radial) winding involves laying fibers at nearly 90° with respect to the longitudinal axis of the mandrel. This results in high hoop strength. Helical (off-axis) or polar (axial, longitudinal) windings provide reinforcement along the length of the structure. As these fibers are wound, they have a tendency to slip. One exception is the geodesic path, which is a straight line on the flattened equivalent surface. As the part is formed, its thickness increases, thus changing the geometry of the process.

Pultrusion is used to produce indefinite-length constant cross-section components. Fibers, fabrics, and mats are pulled through a die with resin introduced at one end. The die is tapered to apply hydrostatic pressure to the resin resulting in fiber wet-out. An exothermic reaction can occur or heat is added to result in curing of the component. Design limitations include the need for sufficient reinforcement in the longitudinal (axial) direction to enable the process to be run. Annular cross sections or holes that increase the complexity and cost of the die, however, are preferred to post forming machining.

Molding Processes. The last set of forming processes involve forming with a closed mold. Such processes include matched-die molding, resin-transfer molding, and extrusion. These processes are used to fabricate parts in reasonable quantities within reasonable cycle times.

The constraints that are posed by the forming process are the methods by which the resin is introduced into the mold and how the chemical reaction is to occur. The flow characteristics are often non-Newtonian. Mold design is based on sophisticated software to locate entry and exit ports and to account for flow patterns, velocity shrinkage, and thermal expansion. Often the resin is curing as it is being introduced into the mold; thus its rheology is a nonlinear function of time. Clearly, there are limitations in the shape of the component. A high degree of compartmentalization of the mold increases the need for many entry ports.

Often dry reinforcement is placed in the mold before the resin is introduced. Fabrics, mats, and preforms of various cross sections and geometries are possible by employing the expertise of textile engineers.

Fabrication Processes

Once the consolidation process is completed, the component usually requires additional fabrication steps to complete. The part is trimmed, holes are drilled, the surface may be prepped and painted. In addition, several parts may be bonded to form a larger component or secondary bonding is performed to yield a sandwich structure. The design of the component must also specify these steps and must anticipate possible adverse effects.

Machining. The shape and form of a metallic structure is realized in the machining process. Composite structures are formed prior to being machined. Machining is necessary to facilitate mechanical joints, cut small parts from stock items, and remove excess material. To avoid damage, composites should not be machined using metal techniques that use cutting, tearing, or shaving tools. Thus, if a cavity is required, it should be formed in the consolidation process and not by machining unless it is a thick-walled shape or has a "strong-back" composite or metal liner bonded in the hole cavity.

Cured composite materials can be cut using a number of techniques including mechanical methods, water jet, or laser. Mechanical methods require an increase in tool speed, a decrease in the feed rate, and a cooling lubricant. Often carbide tip tooling is used; however, to prevent excessive damage or delamination to the composite structure, diamond-grit tooling should be used. Thus, the composite structure is machined using abrasives. Care must be taken to allow for particle removal and to prevent heating the part.

Water jets and lasers provide for a greater amount of complexity in cutting. Water jets can cut through a 25 mm thick structure at relatively fast feed rates (1.5 m/min) and can be used for intricate details. The starting point of the water jet must be away from the part edge because of the possibility of delamination. Lasers require no start-up location, but damage occurs because of local heating of the structure. This is especially true in low thermal conducting materials such as composites with glass reinforcement.

Because of the potential threat of damage during machining, designs should limit the amount of required postfabrication forming. Near-net-shape fabrication can result in a minimum amount of machining. Whereas holes can be drilled effectively, it is often better to incorporate a fixture into the component during its forming rather than to attach it through mechanical means later.

Bonding. The ability to reduce the number of parts in an assembly is a clear advantage of using composite materials. Yet if the manufacturing and design processes are developed independently of each other, this advantage will be severely diminished. An illustrative example of the choices available is a stiffened plate. In Fig. 8(a), the plate is made in three separate steps. The skin is fabricated, the stiffener is fabricated, and a secondary bond is performed. Although the stiffener fabrication may make use of a high-yield low-cost method such as pultrusion, some cost benefit will be lost by performing the secondary bond. One alternative is to cocure the stiffener with the skin. Film adhesive that has a similar cure cycle as the resin can be used to increase the integrity of the bond line. In Fig. 8(b), the stiffener may or may not be precured. A second variation has the stiffener now integrated with the skin. In Fig. 8(c), the macroscale of the design has increased resulting in additional specifications on the microscale. Finally, the whole concept of skin/stiffener construction can be abandoned and either a foam-filled hat section (Fig. 8d) or a sandwich construction can be used.

Again, if one approaches a composite structure with the same limitations imposed by metal construction the potential of the composite material is not achieved. For example, an aircraft fuselage is constructed using a thin skin stiffened by longerons along its length and hoop stiffeners about its circumference. The intersections of these stiffeners are at right angles out of fabrication necessity. This limitation is imposed primarily because the metal structure must be made in many pieces and assembled. The true design objective of flexural and torsional stiffness can be met more efficiently if the artificial constraint of right-angle stiffeners is removed.

Sandwich structures provide high bending stiffness at low weight. Various core materials are used such as balsa wood, aluminum or Nomex honeycomb, or syntactic closed-cell foams. The face sheets can be formed separately and bonded to the core, cocured with the core, or cured in separate steps layer by layer. Forming core to different contours requires additional fabrication. Foam and honeycomb cores can readily conform to flat and circular sections. Compound curvatures are more difficult to attempt. The minimum radius of curvature is determined by the core material. Lower density core materials tend to conform easily but may offer lower through-the-thickness stiffness and strength. The use of in-place foaming or scored foam boards allows complex curvature or shapes while maintaining structural integrity.

The manufacturing process determines the consolidation pressures seen by the face sheets during their forming process. If cured separately, then high pressures can be used. Secondary bonding then occurs at temperatures lower than the cure temperature of the face sheets and at significantly lower pressures so as not to compromise the core material. Curing face sheets on the core directly results in a slight dimpling if honeycomb cores are used. If a resin infusion method is being used, honeycomb cores must first be sealed.

When using sandwich structures, the edges must be closed out to allow for joining and to prevent water intrusion. Several techniques are used, and some require machining of the core. Solid cores such as foam and balsa wood can be readily cut or sanded into the desired shape. Honeycomb cores require special cutting tools. The machined-out core is then backfilled with a resin/filler edging compound and cured.

Repair

An often overlooked aspect of manufacturing with composites is repair. Repair is necessary if the component is damaged during its manufacture or if a quality assurance check indicates that certain specifications are not met. The choice between repair or rejection depends on the ability to repair the structure effectively. This requires both access and proper procedures. Clearly, a proper design would anticipate likely damage events such as delicate areas being overstrained prior to final assembly or a tool drop on a composite surface. Poorly designed manufacturing processes yield a systematic flaw that may go undetected until the component is in the field.

General guidelines apply for composite repair as they do for repair of metallic structures. Convenient access should be provided. Thus there is a trade-off between fully integrated structures and structures that are assembled with mechanical fasteners. Composite repair requires significant surface preparation. The damaged material is first removed, and the surrounding good material is scarfed, chemically cleaned, and primed. A filler or partially staged or cured bonded preform and a surface patch of several plies of fabric/resin prepreg is applied to return the strength and stiffness, to restore the contour of the surface, or to prevent water intrusion. Application is accomplished using a portable heating and vacuum bag unit. The core of sandwich structures is replaced if it is damaged. This necessitates an additional bonding cycle.

Conclusions

The scale of design of a composite structure is substantially broader than an equivalent metal structure. This in turn increases the difficulty in the design process and is partly attributable to some of the advantages of composites, notably the ability to tailor the microstructure of the component and to integrate individual components into a single structure. This additional complexity is compounded by the relative scarcity of individuals who have knowledge of or experience with such materials. Design of metallic structures is aided by an older and larger pool of readily available experience. Clearly, the manufacturing practices of today involving milling machines, lathes, and drill presses have been honed in the preceding decades. Advanced composite materials are relatively new materials that pose new challenges to engineers. Moreover, it is unlikely that the necessary expertise in materials, structures, and manufacturing will reside in a single individual. Even simple components may require a design team and consultants to cover adequately the many facets involved in the process.

A key point is the need to codesign the manufacturing process with the component. Failure to coordinate both activities will result in higher costs and poorer performance. Indeed, manufacturing may offer alternatives to the design that an engineer biased by his or her experience with metals may not contemplate. The emergence of composite tooling for composite components clearly indicates the inability of conventional materials to be formed in other than conventional shapes.

One drawback of using composite materials is the relative lack of guidelines to assist in the design for manufacturability. New applications result in new challenges with little or no previous experience on which to build. An exposition in composite materials (International Composites EXPO '97 in Nashville, TN) displayed many new products in major and emerging end-use markets. Examples included a front-end apron for passenger cars that replaces 44 steel parts at a 25% weight savings and a sheet molding compound cross-vehicle beam that consolidates 20 steel and plastic pieces with a 10% weight savings. Such examples indicate the advantages of integrating large structures through the use of composites.

Almost all processes can be improved once implemented by controlling key parameters that affect the occurrence and development of defects and damage. Temperatures and pressures can be monitored and controlled to reduce process variations, yet mechanical performance may not be as sensitive to the processing variables as it is to the process itself. Care must be taken not to simply make incremental progress along a given path for an old process, but to explore alternative manufacturing methods. New raw materials, new processing materials, new methods are continuously being developed. The increased role of textile science due to the emergence of braided preforms is an example of how advantages can be seen by those who are willing to exploit both old and new technology in different areas.

REFERENCES

1. R.M. Jones, *Mechanics of Composite Materials,* Hemisphere Publishing, 1987

2. *Effects of Defects in Composite Materials,* ASTM STP 893, D.J. Wilkins, Ed., American Society for Testing and Materials, 1984

SELECTED REFERENCES

Books

- *Delaware Composites Design Encyclopedia,* Vol 3, *Processing and Fabrication Technology,* Technomic Publishing, 1990
- *Engineered Materials Handbook,* Vol 1, *Composites,* ASM International, 1987
- *Engineered Materials Handbook Desk Edition,* ASM International, 1995
- G. Lubin, Ed., *Handbook of Composites,* Van Nostrand Reinhold, 1982
- D. Hull, *An Introduction to Composite Materials,* Cambridge University Press, 1981
- B.Z. Jang, *Advanced Polymer Composites: Principles and Applications,* ASM International, 1994
- M.M. Schwartz, *Composite Materials Handbook,* McGraw-Hill, 1984
- M.M. Schwartz, *Joining of Composite Matrix Materials,* ASM International, 1994
- A.B. Strong, *Fundamentals of Composite Manufacturing: Materials, Methods, and Applications,* Society of Manufacturing Engineers, 1989

Periodicals

- *Composites Design & Application,* Composites Institute of the Society of Plastics Industry, Fort Collins, CO
- *High-Performance Composites,* Ray Publishing, Wheat Ridge, CO
- *SAMPE Journal,* SAMPE International Business Office, Covina, CA

Control of Residual Stresses

U. Chandra, Concurrent Technologies Corporation

ALL THERMO-MECHANICAL manufacturing processes—such as forging, extrusion, casting, heat treatment, welding, coating, and machining—create residual stresses in industrial products. There are situations when such stresses can be beneficial and are intentionally created, for example, compressive stresses on the outer surface of a component subjected to fatigue loads, autofrettage in gun barrels, and prestressed pressure vessels; even bolted connections and prestressed concrete can be included in this category. In many other situations, however, the presence of residual stress is detrimental to the integrity of the product under service conditions. Examples in this category include: tensile stresses on the outer surface of a component subjected to fatigue loads, tensile stresses on the inner surface of an austenitic stainless steel pipe caused by welding leading to intergranular stress-corrosion cracking in boiling water reactors, interlaminar stresses in coatings leading to their spallation, premature yielding or fracture (especially in brittle materials), and part distortion or dimensional instability. The presence of residual stresses in a part is also known to affect its machinability. For these reasons, mechanical and manufacturing engineers have long been interested in understanding the source of such stresses, their control, and relief (Ref 1–4).

In the case of metallic products, the selection of material is generally dictated by functional requirements such as the ability of the product to withstand service loads, resistance to wear or corrosion, and so forth. It is rare that the magnitude and distribution of residual stresses is a matter of primary consideration while selecting the material to manufacture a metallic part. The issue facing the manufacturing engineer is to control the residual stresses in the product once the material (and often the manufacturing process) has already been selected. In the case of a composite or coated product, however, minimization of residual stresses is a prime consideration while selecting the constituent materials. This article primarily deals with metallic products.

A logical way to control residual stresses in a product should consist of the following steps:

1. Understanding the fundamental sources of stress generation
2. Identifying the parameters that can cause residual stresses in a particular manufacturing process
3. Understanding the relative significance of each one of these parameters
4. Experimenting with the most significant process parameters until a suitable combination is obtained that results in the desired magnitude and distribution of residual stresses

If the residual stresses in the product are still higher than acceptable, the only recourse left is the use of one of the various techniques of stress relief or the inducement of a stress pattern more favorable than the original.

The fundamental sources of residual stresses in a product can best be understood with the help of the disciplines of solid mechanics, heat transfer, and metallurgy, as discussed in the section "Fundamental Sources of Residual Stresses" in this article. The identification of process parameters comes from an engineer's knowledge of and experience with a particular manufacturing process. Finally, to understand the relative significance of each process parameter and to arrive at their optimal combination one can use either of two approaches: trial and error on the shop floor combined with a suitable method of stress measurements, or computer simulation, especially one based on the finite element method. Both of these approaches have some advantages and limitations. In general, computer simulation offers a more economical and efficient means of performing a parametric study. It also helps in understanding the behavior of the product at various intermediate stages of its manufacture. For example, in case of forging, it is possible to monitor stresses during the entire duration of the die motion, as well as after the die removal. Computer simulation also helps in monitoring other phenomena of interest, such as the flow of material during forging. However, the trial-and-error approach offers only one opportunity to see the effect of changing a process parameter, for example, after the die removal. On the other hand, trial and error requires less investment in personnel training and development of a material properties database. Also, the use of physical modeling in a laboratory coupled with residual-stress measurements is still needed to verify the results of computer simulation until full confidence in the accuracy of simulation code and personnel skill is attained.

Until recently, the trial-and-error approach has played a more prominent role in the control of residual stresses in industrial products, as compared to its simulation counterpart. However, in recent years, considerable progress has been made in computer simulation of several manufacturing processes, for example, casting, forging, quenching, and postweld heat treatment. In such processes, it is expected that computer simulation will soon be used routinely on the shop floor. Until then, a judicious combination of computer simulation and trial and error appears to be the most prudent approach to control the residual stresses. On the other hand, in the case of shot peening and some other processes, computer simulation is either not mature enough or is too uneconomical to be used as an alternative to trial and error, at least in the foreseeable future.

It is important to note that, before attaining its final shape, a product often undergoes a series of primary and secondary operations; for example, it may involve forging as the primary operation and quenching, aging, and machining as the secondary operations. Each of these operations affects the state of residual stresses in the part. Therefore, in order to control the residual stresses in a finished product, it appears logical to apply the aforementioned techniques (either trial and error or computer simulation) for each operation. In some cases (e.g., forging followed by quenching), it may be argued that the final manufacturing operation is mainly responsible for the state of residual stresses in the part and, hence, the effort to control them should focus on the final operation. However, this argument does not apply if the sequence of quenching and machining is considered. Hence, it should not be used as a general rule.

This article is an introduction to the subject of control of residual stresses. Its objectives are to:

- Introduce the various fundamental sources of residual stresses common to most manufacturing processes
- Explain the effect of material removal on residual stresses and distortions in a part

- Provide a summary of commonly used techniques of measuring residual stresses
- Provide a summary of the finite element method used for predicting residual stresses
- Demonstrate the application of the above to a few selected manufacturing processes

Because the number of manufacturing processes is too large, it is not possible to cover them all in a short article. Also, a listing of rules for control of residual stresses in various manufacturing processes is not attempted in this article, because an indiscriminate application of such rules without proper appreciation of the basic concepts could lead to adverse results.

It is hoped that the article provides some insight to a manufacturing engineer into the cause of residual stresses in a product and assists in identifying key process parameters responsible for such stresses. The article is also expected to assist in making a choice between the trial-and-error and computer-simulation approaches (or a combination of the two) for the control of residual stresses.

Fundamental Sources of Residual Stresses

In manufacturing processes, residual stresses are caused by a combination of some or all of the following fundamental sources:

- Inhomogeneous plastic deformation in different portions of the product due to mechanical loads or constraints
- Inhomogeneous plastic deformation due to thermal loads
- Volumetric changes and transformation plasticity during solid-state phase transformation
- A mismatch in the coefficients of thermal expansion

The mechanics related to these sources is explained in the remainder of this section. Also, an important concept related to the effect of material removal (e.g., a casting mold, a forging die, or the material removed during machining) on the magnitude and distribution of residual stresses in the workpiece is introduced.

Mechanical Loads. Generation of residual stresses due to mechanical loads can be understood by consideration of the example shown in Fig. 1 (Ref 5). It consists of an assembly of three bars, each 254 mm (10 in.) long with a cross section of 25.4 mm (1 in.) square. The bars are spaced 25.4 mm (1 in.) apart, center to center. The two outer bars are made of the same material with an elastic modulus of 207 GPa (30×10^6 psi), yield strength of 207 MPa (30 ksi), and plastic modulus of 41 GPa (6×10^6 psi). The middle bar has a higher yield strength of 414 MPa (60 ksi), but the same elastic and plastic moduli as the two outer bars. The upper ends of the three bars are fixed and not allowed to move in any degree of freedom; their lower ends are tied to a rigid but a weightless block. Also, the assembly carried a load *P* in the center, as shown in Fig. 1.

Now consider two different loading histories. In the first case, the load *P* is gradually increased from 0 to 400 kN (90 kips), and then brought back to zero. When *P* is 400 kN (90 kips), each bar shares 133 kN (30 kips) and the stress in none of them exceeds the yield strength. On unloading, the original zero stress state in each bar is restored, and no residual stresses are introduced.

In the second case, *P* is increased from 0 to 534 kN (120 kips) and then brought back to zero. The entire history of stresses in the three bars is shown in Fig. 1(b). When *P* exceeds 400 kN (90 kips), the two outer bars deform plastically and, because of the reduced modulus, begin to share less load. The stress in the two outer bars follows the path ABCD, whereas that in the middle bar follows the path ABEF. It can be seen that when *P* is again zero (unloading), the stresses in the three bars do not go back to zero. Instead, the middle bar has a residual tensile stress of 78.8 MPa (11.4 ksi), and each of the two outer bars has a residual compressive stress of 39.4 MPa (5.7 ksi). Because there is no external load on the assembly, the residual stresses in the three bars are in self-equilibrium. A comparison of the two loading histories indicates that the presence of inhomogeneous plastic deformation in the three bars is responsible for the generation of residual stresses. Similarly, mechanical residual stresses occur in any component when the distribution of plastic deformation in the material is inhomogeneous, such as the surface deformation in shot-peening operation.

Thermal Loads. A similar three-bar model explaining the generation of residual stresses due to inhomogeneous plastic deformation caused by thermal loads is discussed by Masubuchi (Ref 6, presumably adopted from Ref 7). In this model, three carbon-steel bars of equal length and cross-sectional area are connected to two rigid blocks at their ends. The middle bar is heated to 593 °C (1100 °F) and then cooled to room temperature, while the two outer bars are kept at room temperature. Some of the details are not clearly explained in Ref 6, but the problem is very similar to the previous example. When the temperature in the middle bar is raised, the requirements of compatibility and equilibrium imply that a compressive stress be generated in the middle bar and tensile stresses in the two outer bars; the stress in each of the two outer bars being half of that in the middle bar. If the temperature in the middle bar is so high that its stress exceeds yield but in the two outer bars the stresses are still below yield, residual stresses will occur in the three bars when the temperature of the middle bar is brought back to room temperature (i.e., on unloading). Similarly, if the stresses in all three bars exceed yield but by different amounts, residual stresses will still occur when the temperature of the middle bar is

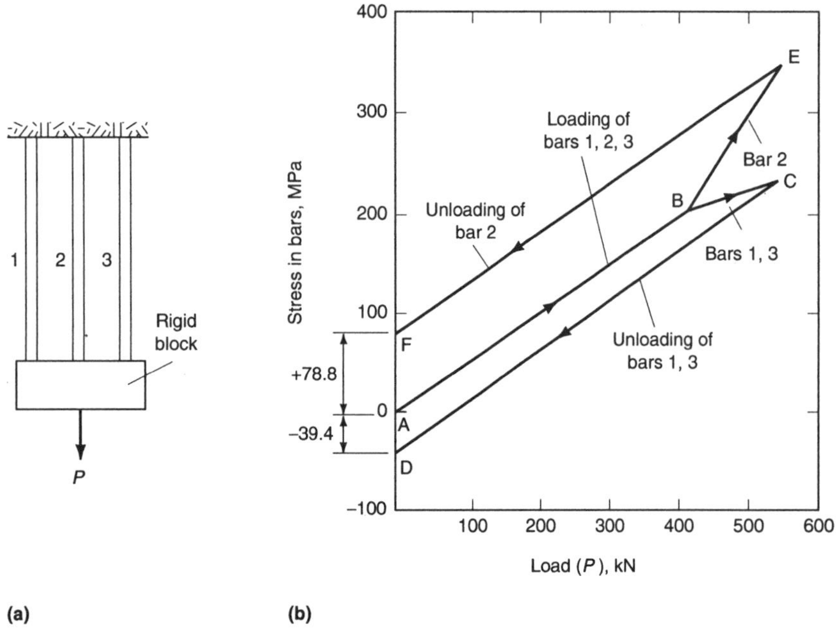

Fig. 1 Residual stresses caused by mechanical loads in an assembly of three bars. Source: Ref 5

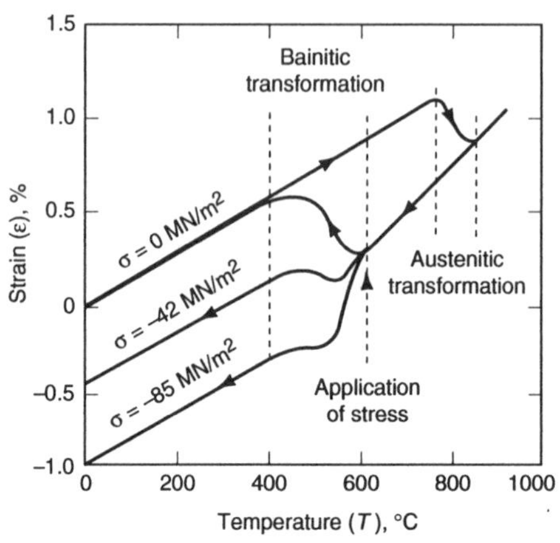

Fig. 2 Transformation plasticity. Source: Ref 9

brought back to room temperature. Indeed, this case is very similar to that of a cylinder immersed vertically in a quenchant where, during the initial stages of quenching, the temperature in the outer layer is much lower than that in the inner core.

The three-bar model can be further utilized to explain the generation of residual stresses due to the mismatch in coefficients of thermal expansion. For example, suppose the two outer bars represent the layers of matrix in a composite lamina and the inner bar represents a layer of fibers. The coefficient of thermal expansion of the two outer bars is equal but, in general, different from that of the middle bar. It is assumed that the initial temperature of all the three bars is equal, which corresponds to a certain processing temperature much higher than room temperature. When the assembly is brought to room temperature, the requirements of compatibility and equilibrium will be satisfied if a system of forces (residual stresses) is established such that the sum of the forces in the two outer bars is equal and opposite to that in the middle bar. In this case, the presence of unequal plastic deformation is not a prerequisite for the generation of residual stresses. This explains why, while selecting the constituent materials for a composite or for a coating, the designers try to minimize the mismatch between their coefficients of thermal expansion.

Solid-State Transformation. In quenching, welding, and casting processes, many metals such as steels undergo one or more solid-state transformations. These transformations are accompanied by a release of latent heat, a change in volume, and a pseudoplasticity effect (transformation plasticity). All of these affect the state of residual stresses in the part. The release of latent heat during solid-state transformation is similar to that during the liquid-to-solid transformation, albeit of a smaller amount. The change (increase) in volume occurs due to the difference in mass densities of the parent phase (e.g., austenite) and the decomposed phases (pearlite, ferrite, bainite, and martensite). In steels, the volumetric change due to phase transformation is in contrast to the normal contraction or shrinkage during cooling (Ref 8).

A simple example of transformation plasticity is shown in Fig. 2, which is based on the results of a constrained dilatometry experiment (Ref 9). The figure shows that during cooling in the phase transformation regime, the presence of even a very low stress may result in residual plastic strains. Two widely accepted mechanisms for transformation plasticity were developed by Greenwood and Johnson (Ref 10) and Magee (Ref 11). According to the former, the difference in volume between two coexisting phases in the presence of an external load generates microscopic plasticity in the weaker phase. This leads to macroscopic plastic flow, even if the external load is insufficient to cause plasticity on its own. According to the Magee mechanism, if martensite transformation occurs under an external load, martensitic plates are formed with a preferred orientation affecting the overall shape of the body.

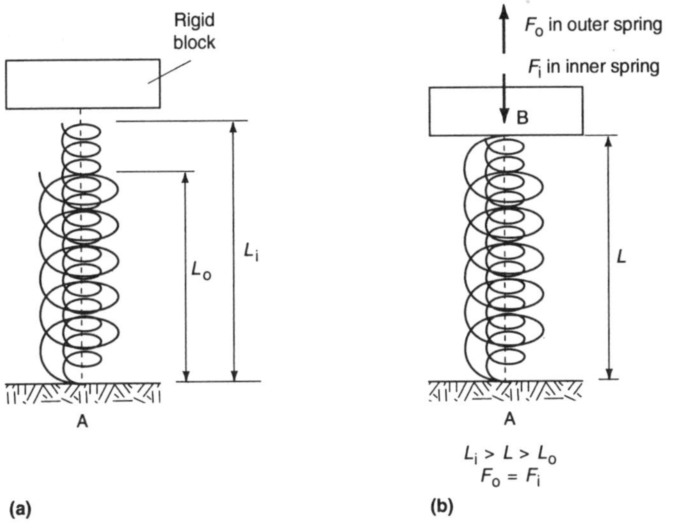

Fig. 3 Residual stresses in an assembly of two springs with unequal initial lengths. Source: Ref 5

Material Removal. A fact that is often overlooked in discussing residual stresses caused by various manufacturing processes is the effect of material removal on the state of stresses in the product. Consider, for example, that a casting mold must be finally broken and removed, or a forging die must be retracted. Likewise, in making a machined part some of the material has to be removed. All of these operations change the state of stress in the part. In order to fully understand this concept, three examples discussed in Ref 5 should be considered.

The first example entails an assembly of two concentric springs of slightly different lengths, L_i and L_o, as shown in Fig. 3(a); the subscripts i and o refer to inner and outer springs, respectively. The bottom ends of the two springs are fixed. Then, the upper ends are tied to a rigid block that is free to move only in the vertical direction. The two springs adopt a compromise length, L, which is in between L_i and L_o, as shown in Fig. 3(b). As a result, the two springs develop equal and opposite forces: compressive in the longer inner spring and tensile in the outer shorter spring. The assembly of the two springs may be viewed as analogous to the assembly of a cast part and its mold or to the assembly of the forged part and the die, or to a machined part before some portion of it is removed. Then, the removal of the outer spring becomes analogous to removal of material during machining (Ref 5, 12), of the casting mold (Ref 13–16), or of the forging die (Ref 17). Two cases are considered. In the first case, the stresses in both springs are assumed to be within their elastic limits. When the outer spring is removed, the force acting on it is transferred to the inner spring in order to satisfy equilibrium and the inner spring returns to its original length. In the second case, it is assumed that the inner spring has undergone a certain amount of plastic deformation. When the outer spring is removed, the inner spring does not return to its original length, L_i. In either case, because the two springs and, therefore, the forces, are concentric, the residual stress in the inner spring becomes zero when the outer spring is removed.

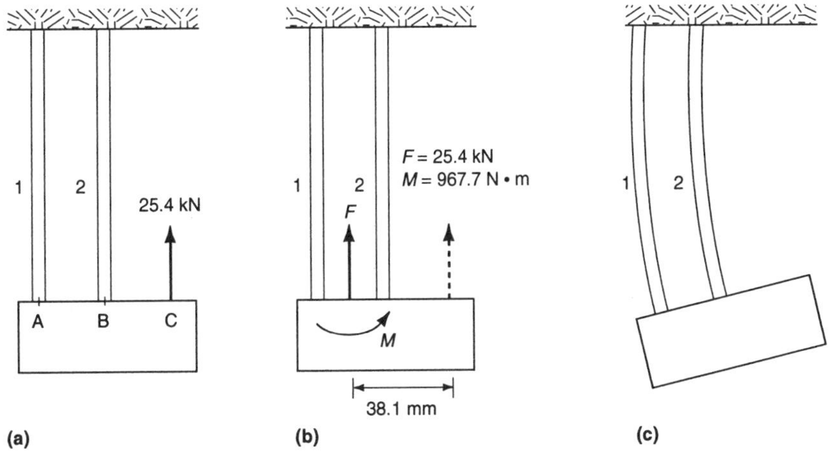

Fig. 4 Effect of asymmetric material removal in the three-bar model of Fig. 1. Source: Ref 5

For the second example, reconsider the three-bar model from the section "Mechanical Loads" in this article. After creating residual stresses in the three bars by loading and unloading the assembly, bar 3 is removed, by (for example) machining. As shown in Fig. 4(a) and (b), a redistribution of stresses in the remaining two bars takes place. The resultant stresses at the centroids of the two bars become −14.8 MPa (−2.14 ksi) in bar 1, and 14.8 MPa (2.14 ksi) in bar 2. Also, the assembly rotates (distorts) by an angle of 4.3×10^{-3} radians.

The third example in Ref 5 is of a thick-walled cylinder with an internal diameter of 101.6 mm (4 in.) and an outer diameter of 203.2 mm (8 in.) as shown in Fig. 5(a). Both ends of the cylinder are restrained axially, and the cylinder is subjected to an internal pressure. A 25.4 mm (1 in.) thick (along the axis) slice of the cylinder is analyzed by subdividing it into 10 equal finite elements (5.08 mm, or 0.2 in., thick each) in the radial direction (Fig. 5b). The residual stresses are created by increasing the pressure from zero to 345 MPa (50 ksi), and then back to zero. The elements 1 and 2 are removed successively. The variation of the three stress components along the radius is shown in Fig. 6, before material removal (i.e., the residual stresses) and after removing the two layers. It may be noted that in an overall sense, the level of residual stresses goes down as the material is removed. However, this is not necessarily true in a local sense. Consider, for example, the circumferential stress at the centroids of elements 3 and 4 in Fig. 6(b); it increases as the material is removed.

Important conclusions from the three examples discussed above can be summarized as follows:

- When the material removal is symmetric with respect to the stress distribution (Fig. 3), the residual stresses in the remainder of the assembly or part are very small or even zero.
- When the material removal is not symmetric with respect to the stress distribution (Fig. 4, 6), the residual stresses in the remainder of the assembly or part are not necessarily small.
- Material removal may result in an increase in stresses at some locations of the assembly or the part (Fig. 6).

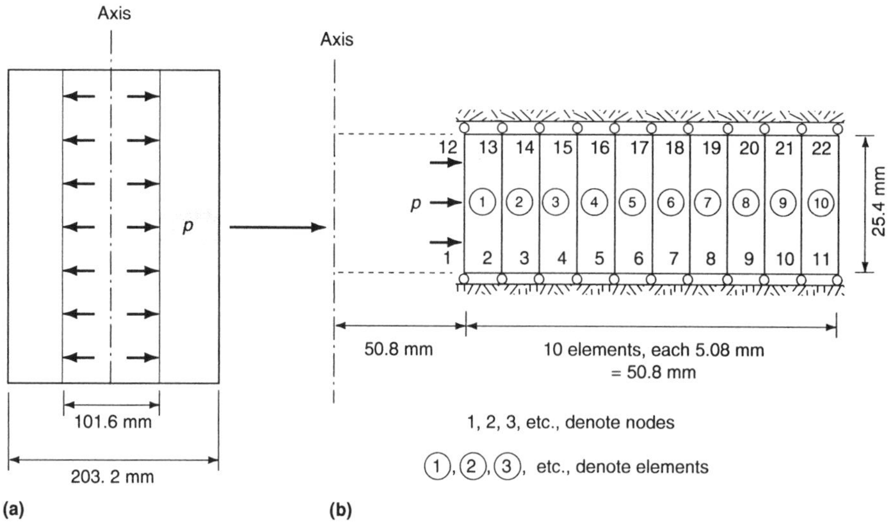

Fig. 5 Cylinder with internal pressure and its finite element mesh. Source: Ref 5

Computer Prediction of Residual Stresses

In recent years, the finite element method has become the preeminent method for computer prediction of residual stresses caused by various manufacturing processes. A transient, nonlinear, thermomechanical analysis software is generally employed for that purpose. Some of the mathematics that form the basis of such software is common for all manufacturing processes. Such common mathematics is summarized by this section. However, because every process is unique, some mathematical requirements are, in turn, dependent on the process. Also, for the simulation of certain processes a sequential thermomechanical analysis is adequate, whereas for others a coupled analysis may be preferred or even essential. Such subtleties are pointed out later when individual processes are discussed.

Ignoring convection, the following conduction heat-transfer equation is solved with appropriate initial and boundary conditions:

$$\nabla \cdot (k\nabla T) + \dot{Q}_c = \rho C_p \frac{\partial T}{\partial t} \quad \text{(Eq 1)}$$

where T is the temperature at an arbitrary location in the workpiece at time t, k is the thermal conductivity of the material, $\dot{Q}_c$ is the rate of heat generated per unit volume, ρ is the density, C_p is the specific heat, and ∇ is the differential operator; all material properties are assumed to vary with temperature. The term $\dot{Q}_c$ accounts for the release of latent heat during liquid-to-solid transformation in casting and welding processes or during solid-state phase transformation in quenching, welding, or casting processes. It also accounts for the heat of plastic deformation in forging and other bulk deformation processes. The initial and boundary conditions are process dependent. Details of converting Eq 1 into its finite element form and of numerical solution are available in a number of technical papers and textbooks and are not repeated here. For a general treatment of the subject, the reader is referred to Ref 18 to 21.

The transient temperatures computed above are used as loading for the subsequent transient stress/displacement analysis. Using the incremental theory, the total strain increment $\{\Delta\varepsilon\}$ at time t can be divided into various components (Ref 22–26):

$$\{\Delta\varepsilon\} = \{\Delta\varepsilon^e\} + \{\Delta\varepsilon^t\} + \{\Delta\varepsilon^p\} + \{\Delta\varepsilon^{cr}\} + \{\Delta\varepsilon^v\} + \{\Delta\varepsilon^{tr}\} \quad \text{(Eq 2)}$$

where superscripts e, t, p, cr, v, and tr refer to elastic, thermal, plastic, creep, volumetric change, and transformation plasticity components, respectively. The first three strain terms are needed in the simulation of every manufacturing process discussed here, whereas the use of the other three terms is dependent on the process and are pointed out as appropriate. Also, mathematical details for the first four strain terms are discussed in most standard references (Ref 22, 23), whereas the details for the last two terms are discussed often in the context of the simulation of quenching and welding processes (Ref 24–26).

In forging and other large deformation processes, the term $\dot{Q}_c$ in Eq 1 represents the heat of plastic deformation and leads to a coupling between Eq 1 and 2.

At present, no single computer code is capable of predicting residual stresses caused by all manufacturing processes. However, several general-purpose finite element codes are capable of predicting these stresses to a reasonable degree of accuracy for at least some of the manufacturing processes (Ref 27–29). In addition, some of these codes permit customized enhancements leading to more reliable results for a specific process. Before attempting to predict residual stresses due to a manufacturing process, it is advisable to compare the capabilities of two or three leading codes and use the one most suited for the simulation of the process in consideration. Examples of such comparisons are given in Ref 12 and 13 for forging, quenching, and casting processes. It must be noted that, due to continuous enhancement in these codes, it is always advisable to compare the capabilities of their latest versions.

Measurement of Residual Stresses

It is generally not possible to measure residual stresses in a product during its manufacture; instead, they are measured after the manufacturing process is complete. Smith et al. (Ref 30) have divided the residual stress measurement methods into two broad categories: mechanical and physical. The mechanical category includes the stress-relaxation methods of layer removal, cutting,

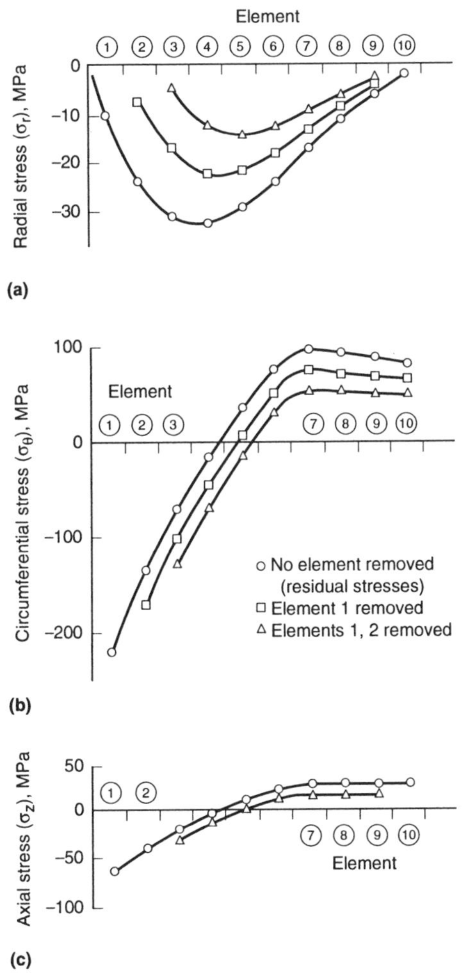

Fig. 6 Effect of removing layers or elements of material from the inside of the cylinder. Source: Ref 5

hole drilling, and trepanning, whereas the physical category includes x-ray diffraction (XRD), neutron diffraction, acoustic, and magnetic. The layer-removal technique as originally proposed by Mesnager and Sachs (Ref 4) is only applicable to simple geometries such as a cylinder with no stress variation along its axis or circumference, or to a plate with no variation along its length or width. Thus, whereas it could be used to measure quench-induced residual stresses in a cylinder or a plate, it is not suitable for measuring complex stress patterns such as those caused by welding. The layer removal and cutting techniques, however, have been applied to pipe welds in combination with conventional strain gages and XRD measurements. The layer-removal technique is also used to measure residual stresses in coatings.

Hole-drilling and trepanning techniques can be used in situations where the stress variation is nonuniform, but they are generally restricted to stress levels of less than one-third of the material yield strength. Also, these two techniques can be unreliable in areas of steep stress gradients and require extreme care while drilling a hole or ring in terms of its alignment as well as the heat and stress generation during drilling (Ref 31). For such reasons, and others, these two techniques have found little application in the measurement of weld-induced residual stresses.

Of the methods in the physical category, XRD is probably the most widely used method, the neutron diffraction method being relatively new. These two methods measure changes in the dimensions of the lattice of the crystals, and from these measurements the components of strains and stress are computed. The XRD technique has undergone many improvements in recent years. With the development of small portable x-ray diffractometers, the technique can be used for on-site measurement of residual stresses. It should be noted, however, that this technique is capable of measuring strains in only a shallow layer (approximately 0.0127 mm, or 0.0005 in., thick) at the specimen surface. To measure subsurface residual stresses in a workpiece, thin layers of materials are successively removed and XRD measurements are made at each exposed layer. For reasons discussed in the section "Material Removal" in this article, the measurements at an inner layer should be corrected to account for the material removed in all the previous layers. Reference 32 gives analytical expressions for such corrections in cases of simple geometries and stress distributions. For more complex cases, it still remains difficult to determine subsurface residual stresses accurately.

In contrast to the x-rays, neutrons can penetrate deeper into the metals. For example, in iron the relative depth of penetration at the 50% absorption thickness is about 2000 times greater for neutrons than for x-rays. Only a few materials, such as cadmium and boron, absorb neutrons strongly. However, to gain the advantage of greater penetration of neutrons requires the component to be transported to a high flux neutron source (Ref 30), which limits the use of the technique.

Residual Stresses Caused by Various Manufacturing Processes

Casting. In the past, little attention has been paid to the control of residual stresses in casting; much of the interest was focused on the prediction and control of porosity, misruns, and segregation. A review of the transactions of the American Foundrymen's Society or of earlier textbooks on casting (e.g., Ref 33) reveals practically no information on the subject; even the *ASM Handbook* on casting (Ref 34) provides little insight. In a recent book, Campbell (Ref 35) has included a brief discussion of residual stresses summarizing the work done by Dodd (Ref 36) with simple sand-mold castings. Dodd studied the effect of two process parameters: mold strength, by changing water content of sand or by ramming to different levels, and casting temperature. The conclusions of these costly experiments could have been more economically and easily arrived at by using the basic concepts discussed in the section "Fundamental Sources of Residual Stresses" and further amplified in the following paragraphs.

When a casting is still in its mold, the stresses are caused by a combination of the mechanical constraints imposed by the mold, thermal gradients, and solid-state phase transformation. Also, creep at elevated temperature affects these stresses. Finally, when the casting is taken out of its mold, it experiences springback that modifies the residual stresses.

As discussed in Ref 13 to 16, the computer prediction of residual stresses in castings requires a software that is capable of performing coupled transient nonlinear thermomechanical analysis (see the section "Computer Prediction of Residual Stresses" in this article). In addition, it should be able to account for the following:

- Release of latent heat during liquid-to-solid transformation, that is, in the mushy region
- Mechanical behavior of the cast metal in the mushy region
- Transfer of heat and forces at the mold-metal interface
- Creep at elevated temperatures under condition of varying stress
- Enclosure radiation at the mold surface to model the investment-casting process
- Mold withdrawal to model directional solidification
- Mold (material) removal

The author and his coworkers have recently modified a commercial finite element code and have analyzed simple sand-mold castings (Ref 15, 16). Computer simulation of these castings indicates that: (1) for an accurate prediction of transient and residual stresses, consideration of creep is important; creep is also found to make the stress distribution more uniform; and (2) just prior to mold removal the stresses in the casting can be extremely high, but after the mold removal they become very small (owing to the springback discussed in the section "Material Removal") except in the areas of stress concentration. The residual stresses after the mold removal will not necessarily be small if the casting is complex and the mold removal is asymmetric with respect to the stress distribution. Also, small variations in mold rigidity are not found to have any noticeable effect on residual stresses, which confirms the observations based on trial and error using greensand molds with various water contents (Ref 35).

Although very little work is published thus far on the subject of control of residual stresses in castings, finite element simulation methodology is now sufficiently advanced to enable the study of the effect of various process and design parameters on the residual stresses in castings, for example, superheat, stiffness and design of the mold, design of the feeding system and risers, and the design of the part itself. Also, residual stresses caused by different casting practices such as sand-mold, permanent-mold, investment casting, and so forth, can be determined. As the manufacturers and end-users of cast products become more aware of the status and benefits of the computer-simulation methodology, it can be expected

Table 1 Dimensional check of a machined part

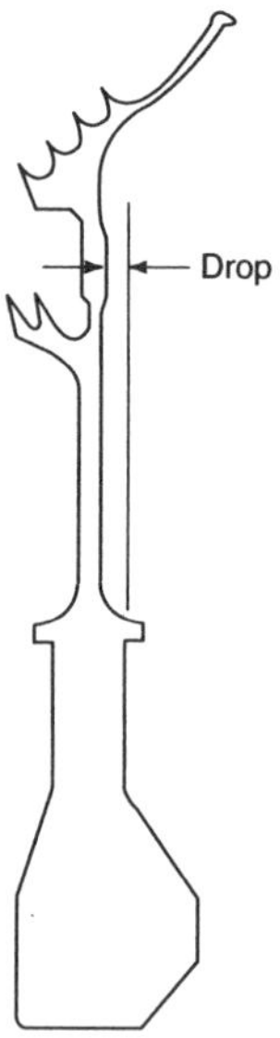

No.	Drop, mm	Drop, in.
Vendor A		
1	2.692	0.1060
2	2.769	0.1090
3	2.725	0.1073
4	2.756	0.1085
5	2.738	0.1078
6	2.667	0.1050
7	2.769	0.1090
8	2.743	0.1080
9	2.680	0.1055
10	2.743	0.1080
Vendor B		
1	2.921	0.1150
2	2.906	0.1144
3	2.997	0.1180
4	2.941	0.1158
5	2.936	0.1156
6	2.870	0.1130
7	2.972	0.1170
8	2.954	0.1163
9	2.926	0.1152
10	2.954	0.1163
Vendor C		
1	3.294	0.1297
2	3.124	0.1230
3	3.200	0.1260
4	3.195	0.1258
5	3.213	0.1265
6	3.251	0.1280
7	3.294	0.1297
8	3.127	0.1231
9	3.251	0.1280
10	3.277	0.1290

Source: Ref 12

to play a very important role in controlling residual stresses in complex industrial castings. At present, the biggest limiting factor in the use of simulation is the lack of thermophysical and mechanical properties data for the cast metal and the mold materials.

Forging. As with the casting process, little attention has been paid in the past to the control of residual stresses caused by forging; most of the interest was in predicting the filling and the direction of material flow. Now, due to recent advances in computer-simulation techniques, it is possible to predict and control the residual stresses in forged parts.

Large plastic flow of the workpiece material is inherent in the forging process. The material flow is influenced by a number of factors including the die shape and material, forging temperature, die speed, and lubrication at the die/workpiece interface. Therefore, finite element simulation software used to predict and control residual stresses in the part should be capable of accounting for these factors. Because a significant amount of energy is dissipated during forging in the form of heat due to plastic deformation, a coupled thermomechanical analysis becomes necessary especially for nonisothermal forging. Other factors contributing to the complexity of the finite element simulation of this class of problems are: temperature-dependent thermal and mechanical properties of the materials (especially for a nonisothermal forging); the choice of solution algorithm and remeshing due to large plastic deformation in the workpiece; and mathematical treatment of the die/workpiece interface that includes heat transfer, lubrication, and contact. The last two terms in Eq 2 need not be considered in the simulation of the forging process.

Finite element simulation of the forging process with simple geometries and of a two-dimensional idealization of the thread-rolling process (Ref 17) showed that, although the stresses in the workpiece are high during the deformation stage, the stresses after retraction of the die (residual stresses) are no longer high except in the regions of stress concentration. Again, similar to the simulation of the casting processes, it is premature to generalize this conclusion but it is clear that the technique of computer simulation of forging and many other bulk deformation processes has advanced to a stage where it can assist in controlling the residual stresses in the part by performing a detailed parametric study with much less investment of time and capital than trial and error on the shop floor.

Quenching involves heating of the workpiece to the heat treatment temperature followed by rapid cooling in a quenchant (e.g., air, water, oil, or salt bath) in order to impart the desired metallurgical and mechanical properties. The choice of a quench medium is the key element; it should be such that it removes the heat fast enough to produce the desired microstructure, but not too fast to cause transient and residual stresses of excessive magnitude or of an adverse nature (e.g., tensile instead of compressive). The heat removal characteristic of a quenchant is known to be affected by a number of factors including the size, shape, orientation of the workpiece (even for simple shapes such as plates and cylinders, the heat removal is different at the bottom, top, and side surfaces); the use of trays and fixtures to hold the workpiece in the quenchant; composition of the quenchant; size of the pool and its stirring, and so forth (Ref 37–39). Additional difficulties arise when, due to economic reasons, quenching is performed in a batch process.

In the past, using trial and error, shop-floor personnel have come up with some interesting strategies to control the residual stresses (and warpage), for example, air delay or an intentional delay while transporting the workpiece from the heating furnace to the quenchant, and time quenching or performing the quenching operation in two steps. In the first step, the part is quenched in a medium such as a salt bath until the part has cooled below the nose of time-temperature transformation curve, followed by quenching in second medium such as air to slow the cooling rate. Obviously, perfecting the quenching operation by trial and error can be an extremely time-consuming task.

At first glance, computer simulation of the quenching process may appear to be simple. It involves an uncoupled transient nonlinear small deformation thermomechanical analysis (as outlined in the section "Computer Prediction of Residual Stresses" in this article), with due consideration to solid-state transformation effects (Ref 9, 24, 25); creep is generally ignored. However, the major difficulty lies (for reasons discussed in the preceding paragraph) in a lack of knowledge of the heat removal characteristic of various quenchants, which is mathematically represented as the convective heat transfer coefficient at the outer boundary of the workpiece. Other difficulties arise due to the lack of thermophysical and mechanical properties of the workpiece material at elevated temperatures. Still, at least in the United States, major aircraft engine manufacturers and their forging vendors have been using computer simulation to control quench-related cracking and residual stresses for some time. One such example involving a turbine disk is discussed in Ref 40. The reported work was performed without the benefit of sophisticated simulation software that could account for solid-state transformation effects. For proprietary reasons, few such cases are published in the open literature.

Machining. Many complex parts in aerospace and other key industries are made by machining forgings, castings, bars, or plates to their net shapes. The presence of residual stresses in the workpiece affects its machinability and, on the other hand, the machining process also creates residual stresses and undesired distortions in the part and alters the already existing stress state. In order to minimize or eliminate these adverse effects, machine-shop personnel often experiment with a number of process parameters, for example, depth of cut, speed of the cutting tool, and coolant. For single-point turning, they frequently flip the workpiece in order to balance the distortions and stresses evenly on the two sides. This trial and error is frequently combined with statistical process control.

A serious problem associated with machining and residual stresses is often manifested in the form of part distortion. For example, consider the example in Table 1 (Ref 12). The table shows the results of a dimensional check on 30 samples of an aircraft engine part that was made by machining heat treated forgings procured from three different vendors (10 samples each). The location at which the dimensional check was performed is identified on the figure included in the table. It was found that: (1) for all forgings from any one vendor, the drop was almost identical; (2) the drop in forgings from vendor B was within the specifications, but not so in the case of the other two vendors; and (3) the drop in forgings from vendors A and C was on the two opposite sides of that from vendor B. It was recognized that all heat treated forgings contained residual stresses. When these forgings were machined to net shapes, distortions occurred for two reasons: the release of residual stresses from the removed portion of the workpiece and the machining process itself. The former can be easily modeled using the finite element method if the magnitude of residual stresses in the forgings prior to machining is known (Ref 5). However, to this author's knowledge, no serious attempt has so far been made to predict residual stresses in a workpiece due to the machining process itself. If an attempt of this type is to provide reliable results, it must take into account such factors as: depth of cut, speed of the cutting tool, interaction between the tool and the workpiece (heat and force), coolant, and clamping/unclamping of the workpiece. It must also recognize the fact that the location of contact between the tool and the workpiece moves as the machining process progresses. If such technology could be developed, it would become possible to predict and control the overall distortions and residual stresses in a part after machining, thereby reducing scrap.

Welding. The residual stress distribution in welded joints depends on a number of process and design parameters such as the heat input, speed of the welding arc, preheat, thickness of the welded part, groove geometry, and weld schedule. Welding engineers have long used trial and error to obtain a suitable combination of these parameters in order to control the residual stresses.

The role of computer simulation in the prediction of residual stresses in weldments is the subject of a recent review (Ref 14). Major elements of computer simulation of the process are:

- Mathematical representation of the heat input from the welding source
- A transient thermal analysis
- A transient stress/displacement analysis; the flow of molten metal and thermal convection in the weld pool are generally ignored

Following Rosenthal (Ref 41), a semisteady state approach is often used, although some attempts at full three-dimensional analysis have also been made. As mentioned earlier, it is now possible to account for volumetric change and transformation plasticity effects. Because of the short time periods involved, creep is ignored. In the case of a single-pass weld or a weld with few passes (e.g., four or five), it is now possible to predict residual stresses with reasonable accuracy. But, as the number of passes increases (e.g., 20 or 30), it becomes computationally intractable to model each pass. The scheme of lumping several passes into one layer has been employed with less than satisfactory results. In addition to excessive computation time, other major difficulties with the simulation of a multipass weld are: the numerical errors tend to accumulate with each pass, and the changes in metallurgical and mechanical properties of material in previously deposited layers during deposition of a subsequent layer are difficult to quantify and to account for in the finite element analysis. The technique of lumping several layers together aggravates these problems.

Between the mid 1970s and early 1980s, the Electric Power Research Institute (EPRI) in the United States sponsored a program to systematically study the effects of various process and geometric parameters such as the heat input, welding method (gas tungsten arc welding, submerged arc welding, laser, and plasma), speed of the welding arc, diameter and thickness of the pipe, and groove geometry, on residual stresses in pipe welds (Ref 42–51). In addition, various thermal processes such as heat-sink welding, backlay welding, and induction heat treatment were investigated to verify if the residual stresses on the inner surface of the pipe could be changed from tensile to compressive to avoid intergranular stress-corrosion cracking. Both experimental and finite element methods were used in the study. The results of this effort are summarized in Ref 52.

A very interesting effort related to in-process control and reduction of residual stresses and distortions in weldments is being pursued (Ref 53, 54). The effort aims at moving beyond mere analysis of residual stresses and distortions to aggressively controlling and reducing them. To accomplish this objective, the effort is subdivided into the development of the following three primary capabilities: prediction, sensing, and control. For prediction purposes, a series of computer programs have been developed, include simple but fast one-dimensional programs that analyze only the most important stress component, that is, the one parallel to the weld line. Sensing capability refers to a set of devices including a laser interferometer to measure minute amounts of distortions, a laser vision system to measure large amounts of distortions, and a mechanical system to measure radi of curvature. Finally, to control the residual stresses, various techniques including changes in heating pattern and application of additional forces have been attempted. References 53 and 54 provide further examples of the application of this methodology in reduction of residual stresses in weldments in high-strength steels and girth-welded pipes.

Coating. Coatings are being used extensively in aerospace, marine, automobile, biomedical, electronics, and other industries. For example, in modern jet aircraft engines, approximately 75% of all components are coated. Some of the reasons for the application of coatings are: thermal barrier, wear resistance, corrosion resistance, oxidation protection, electrical resistance, and repair or dimensional restoration of worn parts. A variety of methods are used for the deposition of coatings on a substrate; the following discussion is limited primarily to the thermal spray process.

The prediction of residual stresses in a coating/substrate system is in its infancy. These stresses result from the difference in the coefficient of thermal expansion of the coating and substrate materials and from plastic deformation of the substrate material. The limited number of numerical studies conducted thus far have been related to small button-type specimens where the coating material was assumed to be fully molten and deposited instantaneously. These efforts have ignored several important factors, for example:

- The presence of partially molten particles in the spray
- A nonuniform deposition of coating material normal to the axis of the plasma jet
- A liquid-to-solid and solid-state transformation
- Imperfect bond between the coating and the substrate
- The relative motion between the spray and the substrate

Also, because a layer of coating consists of several successive passes, the effect of any new pass on its adjacent previously deposited pass in terms of partial remelting, additional material buildup, solute diffusion, and redistribution of residual stresses could be important and should be accounted for. If more than one layer is involved, for example, in functionally graded coatings, modeling the effect of a whole new layer of material on the previously deposited layer would be computationally prohibitive. Also, due to the morphology of the coating material on deposition, its thermal and mechanical properties are extremely difficult to measure and, thus, are generally unavailable for simulation purposes. Due to such reasons, end-users of coated products still rely on the methods of trial and error and statistical process control for the selection of an optimal combination of process parameters in order to control the residual stresses in coated parts (Ref 55).

Stress-Relief Methods

The basic premise of a stress-relief method is to produce rearrangement of atoms or molecules from their momentary equilibrium position (higher residual stress state) to more stable positions associated with lower potential energy or stress state. These methods can be classified into three broad categories: thermal, mechanical, and chemical (Ref 4, p 134). The following concern the methods in the first two categories.

Thermal stress-relief methods include annealing, aging, reheat treatment (e.g., postweld heat treatment), and others. In general, a stress-relief operation involves heating the part to a certain temperature, holding at the elevated temperature for a specified length of time, followed by cooling to room temperature. Primary reduction in residual stresses takes place during the holding period due to creep and relaxation. Thus, computer simulation of a thermal stress-relief method generally entails a thermal-elastic-plastic-creep analysis of the part. A simple, one-dimensional computer analysis of residual stresses in thin plates along with experimental verification is discussed by Agapakis and Masubuchi (Ref 56). More sophisticated thermal-elastic-plastic-creep simulations of the annealing of single pass and multipass girth-butt welds in pipes are presented in Ref 57 and 58.

A number of subcategories of mechanical stress-relief methods are listed in Ref 4. Of these, the methods in the static-stressing subcategory such as stretching, upsetting, bending and straightening, and autofrettage are common, and these should not pose much difficulty in simulation by the finite element method. Similarly, in the mechanical surface treatment subcategory, it should be possible to model the surface-rolling method. However, within the same subcategory, shot peening (a frequently used stress-relief method) is likely to be difficult to simulate; and to this author's knowledge, no realistic attempt has yet been made to do so. The obvious reason is that, whereas it should be possible to model a single impact, modeling multiple impacts will be difficult, just as it is for modeling multipass welding.

In recent years, the method of vibratory stress relief (especially in the subresonant region) has received considerable attention (Ref 59, 60). The basic premise of this method is that the presence of residual stresses in a part changes (increases) its natural resonant frequency. When the part is subjected to vibrations below its new frequency, the metal absorbs energy. During this process, the stresses redistribute gradually and the resonant frequency shifts back to the point corresponding to a residual stress-free (or almost free) state. The process does not change the metallurgical or mechanical properties of the material. The technique has been found successful in relieving residual stresses induced by thermal processes such as welding and casting, but not those induced by cold working. It has also been applied to reduce residual stresses in parts prior to machining in order to minimize distortions. It has been found particularly beneficial in low- and medium-carbon steels, stainless steels, and aluminum alloys, but not in copper alloys. In view of the fact that the technique is much simpler, quicker, and more inexpensive than the thermal-relief methods, it merits further study.

ACKNOWLEDGMENT

The author would like to thank H.A. Kuhn for providing many valuable technical suggestions.

REFERENCES

1. E. Heyn, Internal Strains in Cold Wrought Metals, and Some Troubles Caused Thereby, *J. Inst. Met.*, Vol 12, 1914, p 1–37
2. E. Orowan, Classification and Nomenclature of Internal Stresses, *Proc. Symposium on Internal Stresses*, Institute of Metals, 1948, p 47–59
3. W.M. Baldwin, "Residual Stresses in Metals," Edgar Marburg Lecture, American Society for Testing Materials, 1949
4. R.G. Treuting, J.J. Lynch, H.B. Wishart, and D.G. Richards, *Residual Stress Measurements,* American Society of Metals, 1952
5. U. Chandra, Validation of Finite Element Codes for Prediction of Machining Distortions in Forgings, *Commun. Numer. Meth. Eng.*, Vol 9, 1993, p 463–473
6. K. Masubuchi, *Analysis of Welded Structures,* Pergamon Press, 1980, p 94–96
7. W.M. Wilson and C.C. Hao, Residual Stresses in Welded Structures, *Weld. J.,* Vol 26 (No. 5), Research Supplement, 1974, p 295s–320s
8. W.K.C. Jones and P.J. Alberry, "The Role of Phase Transformation in the Development of Residual Welding Stresses," Central Electricity Generating Board, London, 1977
9. J.-B. Leblond, G. Mottet, J. Devaux, and J.-C. Devaux, Mathematical Models of Anisothermal Phase Transformations in Steels, and Predicted Plastic Behavior," *Mater. Sci. Technol.,* Vol 1, 1985, p 815–822
10. G.W. Greenwood and R.H. Johnson, The Deformation of Metals under Small Stresses During Phase Transformation, *Proc. Royal Soc.,* Vol 283, 1965, p 403–422
11. C.L. Magee, "Transformation Kinetics, Microplasticity and Aging of Martensite in FE31 Ni," Ph.D. Thesis, Carnegie Institute of Technology, 1966
12. U. Chandra, S. Rachakonda, and S. Chandrasekharan, Total Quality Management of Forged Products through Finite Element Simulation, *Proc. Third International SAMPE Metals and Metals Processing Conf.,* Vol 3, F.H. Froes, W. Wallace, R.A. Cull, and E. Struckholt, Ed., SAMPE International, 1992, p M379–M393
13. U. Chandra, Computer Prediction of Hot Tears, Hot Cracks, Residual Stresses and Distortions in Precision Castings: Basic Concepts and Approach, *Proc. Light Metals,* J. Evans, Ed., TMS, 1995, p 1107–1117
14. U. Chandra, Computer Simulation of Manufacturing Processes—Casting and Welding, *Comput. Model. Simul. Eng.,* Vol 1, 1996, p 127–174
15. U. Chandra, R. Thomas, and S. Cheng, Shrinkage, Residual Stresses, and Distortions in Castings, *Comput. Model. Simul. Eng.,* Vol 1, 1996, p 369–383
16. A. Ahmed and U. Chandra, Prediction of Hot Tears, Residual Stresses and Distortions in Castings Including the Effect of Creep, *Comput. Model. Simul. Eng.,* to be published
17. U. Chandra, S. Chandrasekharan, and R. Thomas, "Finite Element Analysis of the Thread Rolling Process," Concurrent Technologies Corporation, submitted to Knolls Atomic Power Lab, Schenectady, NY, 1995
18. G. Comini, S. Del Guidice, R.W. Lewis, and O.C. Zienkiewicz, Finite Element Solution of Non-linear Heat Conduction Problems with Special Reference to Phase Change, *Int. J. Numer. Methods Eng.,* Vol 8, 1974, p 613–624
19. R.W. Lewis, K. Morgan, and R.H. Gallagher, Finite Element Analysis of Solidification and Welding Processes, *ASME Numerical Modeling of Manufacturing Processes,* PVP-PB-025, R.F. Jones, Jr., H. Armen, and J.T. Fong, Ed., American Society of Mechanical Engineers, 1977, p 67–80
20. A.B. Shapiro, "TOPAZ—A Finite Element Heat Conduction Code for Analyzing 2-D Solids," Lawrence Livermore National Laboratory, Livermore, CA, 1984
21. B.G. Thomas, I. Samarasekera, and J.K. Brimacombe, Comparison of Numerical Modeling Techniques for Complex, Two-Dimensional, Transient Heat-Conduction Problems, *Metall. Trans.*, Vol 15B, 1984, p 307–318
22. A. Levy and A.B. Pifko, On Computational Strategies for Problems Involving Plasticity and Creep, *Int. J. Numer. Methods Eng.,* Vol 17, 1981, p 747–771
23. M.D. Snyder and K.-J. Bathe, A Solution Procedure for Thermo-Elastic-Plastic and Creep Problems, *Nuc. Eng. Des.,* Vol 64, 1981, p 49–80
24. S. Sjöström, Interactions and Constitutive Models for Calculating Quench Stresses in Steel, *Mater. Sci. Technol,* Vol 1, 1985, p 823–829
25. S. Das, G. Upadhya, and U. Chandra, Prediction of Macro- and Micro-Residual Stress in Quenching Using Phase Transformation Kinetics, *Proc. First International Conf. Quenching and Control of Distortion,* G.E. Totten, Ed., ASM International, 1992, p 229–234
26. S. Das, G. Upadhya, U. Chandra, M.J. Kleinosky, and M.L. Tims, Finite Element Modeling of a Single-Pass GMA Weldment, Proc. Engineering Foundation Conference on *Modeling of Casting, Welding and Advanced Solidification Processes VI,* T.S. Piwonka, V. Voller, and L. Katgerman, Ed., TMS, 1993, p 593–600
27. "ABAQUS," Version 5.5, Hibbitt, Karlsson and Sorenson, Pawtucket, RI, 1995
28. "ANSYS," Release 5.3, ANSYS, Inc., Houston, PA, 1996
29. "MARC," Version K 6.2, MARC Analysis Research Corporation, Palo Alto, CA, 1996
30. D.J. Smith, G.A. Webster, and P.J. Webster, *Measurement of Residual Stress and the Effects of Prior Deformation Using the Neutron Diffraction Technique,* The Welding Institute, Cambridge, UK, 1987
31. C.O. Ruud, A Review of Nondestructive Methods for Residual Stress Measurement, *J. Met.,* Vol 33 (No. 7), 1981, p 35–40
32. M.E. Hilley, J.A. Larson, C.F. Jatczak, and R.E. Ricklefs, Ed., "Residual Stress Measurement by X-Ray Diffraction," SAE J784a, Society of Automotive Engineers, 1971
33. R.W. Heine, C.R. Loper, and P.C. Rosenthal, *Principles of Metal Casting,* Tata McGraw-Hill, New Delhi, India, 1976
34. *Casting,* Vol 15, *ASM Handbook,* (formerly *Metals Handbook,* 9th ed.), ASM International, 1988
35. J. Campbell, *Casting,* Butterworth Heinmann, Oxford, U.K., 1991
36. R.A. Dodd, Ph.D. Thesis, Department of Industrial Metallurgy, University of Birmingham, U.K., 1950
37. H.E. Boyer and P.R. Cary, *Quenching and Control of Distortion,* ASM International, 1988, p 11
38. T.V. Rajan, C.P. Sharma, and A. Sharma, *Heat Treatment—Principles and Techniques,* Prentice-Hall of India Private Ltd., 1988
39. S. Segerberg and J. Bodin, Variation in the Heat Transfer Coefficient around Components of Different Shapes During Quenching, *Proc. First International Conf. Quenching and Control of Distortion,* G.E. Totten, Ed., ASM International, 1992, p 165–170
40. R.A. Wallis, N.M. Bhathena, P.R. Bhowal, and E.L. Raymond, Application of Process Modelling to Heat Treatment of Superalloys, *Ind. Heat.,* 1988, p 30–33

41. D. Rosenthal, The Theory of Moving Sources of Heat and Its Application to Metal Treatments, *Trans. ASME,* Nov 1946, p 849–866
42. R.M. Chrenko, "Residual Stress Studies of Austenitic and Ferritic Steels," Conference on Residual Stresses in Welded Construction and Their Effects, London, Nov 1977
43. R.M. Chrenko, "Weld Residual Stress Measurements on Austenitic Stainless Steel Pipes," Lake George Conference, General Electric, 1978, p 195–205
44. W.A. Ellingson and W.J. Shack, Residual Stress Measurements on Multi-Pass Weldments of Stainless Steel Piping, *Exp. Mech.,* Vol 19 (No. 9), 1979, p 317–323
45. W.J. Shack, W.A. Ellingson, and L.E. Pahis, "Measurement of Residual Stresses in Type-303 Stainless Steel Piping Butt Weldments," Report NP-1413, Electric Power Research Institute, June 1980
46. E.F. Rybicki and P.A. McGuire, "A Computational Model for Improving Weld Residual Stresses in Small Diameter Pipes by Induction Heating," 80-C2-PVP-152, Century 2 Pressure Vessels and Piping Conference, San Francisco, CA, Aug 1980, American Society of Mechanical Engineers
47. A.F. Bush and F.J. Kromer, Residual Stresses in a Shaft after Weld Repair and Subsequent Stress Relief, *Exp. Tech.,* Vol 5 (No. 2), 1981, p 6–12
48. F.W. Brust and R.W. Stonesifer, "Effect of Weld Parameters on Residual Stresses in BWR Piping Systems," Report NP-1743, Electric Power Research Institute, March 1981
49. F.W. Brust and E.F. Rybicki, A Computational Model of Backlay Welding for Controlling Residual Stresses in Welded Pipes, *J. Pressure Vessel Technol. (Trans. ASME),* Vol 103, 1981, p 226–232
50. R.M. Chrenko, Thermal Modification of Welding Residual Stresses, *Residual Stress and Stress Relaxation,* E. Kula and V. Weiss, Ed., Plenum Publishing, 1982, p 61–70
51. E.F. Rybicki, P.A. McGuire, E. Merrick, and J. Wert, The Effect of Pipe Thickness on Residual Stresses due to Girth Welds, *J. Pressure Vessel Technol. (Trans. ASME),* Vol 104, 1982, p 204–209
52. U. Chandra, Determination of Residual Stresses due to Girth-Butt Welds in Pipes, *J. Pressure Vessel Technol. (Trans. ASME),* Vol 107, 1985, p 178–184
53. K. Masubuchi, In-Process Control and Reduction of Residual Stresses and Distortion in Weldments, *Proc. Practical Applications of Residual Stress Technology,* C.O. Ruud, Ed., ASM International, 1991, p 95–101
54. K. Masubuchi, Research Activities Examine Residual Stresses and Distortion in Welded Structures, *Weld. J.,* Dec 1991, p 41–47
55. R.V. Hillery, Coatings Producibility, *The Leading Edge,* GE Aircraft Engines, Cincinnati, Ohio, Fall 1989, p 4–9
56. J.E. Agapakis and K. Masubuchi, Analytical Modeling of Thermal Stress Relieving in Stainless and High Strength Steel Weldments, *Weld. J. Res. Suppl.,* 1984, p 187s–196s
57. B.L. Josefson, Residual Stresses and Their Redistribution During Annealing of a Girth-Butt Welded Thin-Walled Pipe, *J. Pressure Vessel Technol. (Trans. ASME),* Vol 104, 1982, p 245–250
58. B.L. Josefson, Stress Redistribution During Annealing of a Multi-Pass Butt-Welded Pipe, *J. Pressure Vessel Technol. (Trans. ASME),* Vol 105, 1983, p 165–170
59. R.A. Claxton, Vibratory Stress Relieving—Its Advantages and Limitations as an Alternative to Thermal Treatments, *Heat. Treat. Met.,* 1974, p 131–137
60. A.G. Hebel, Jr., Subresonant Vibrations Relieve Residual Stress, *Met. Prog.,* Nov 1985, p 51–55

Design for Surface Finishing

Eric W. Brooman, Concurrent Technologies Corporation

AN AXIOM among surface-finishing industry professionals is that the quality of a finish is only as good as the quality of the substrate and its pretreatment. That is, to obtain a finish of high quality, meeting all performance specifications and customer expectations, great care has to be taken in identifying and using the appropriate pretreatment. The latter could be grinding, heating, buffing, cleaning, or a number of other processes to prepare the surface properly. Although not articulated in quite the same way, the surface-finishing industry also recognizes that the design of the part (component or assembly) can have a significant influence on the ability to use satisfactory pretreatments and obtain quality finishes. The overall part design, and the design of surface features and their size, can have an impact not only on the choice of pretreatments and finishes, but also on the efficacy of these processes and the results obtained. This article provides some guidelines about general design principles for different types of surface-finishing processes, which include cleaning, organic coatings, and inorganic coatings applied by a variety of techniques.

Many of the guidelines discussed here apply equally as well to other articles in this section and vice versa. Therefore, although what is presented here is a fairly comprehensive summary of the topic of design for surface finishing, useful information can be found in other articles such as "Design for Machining" and "Design for Heat Treatment." As is stressed elsewhere and discussed in this article, design must be considered an integral part of the overall manufacturing process and cannot be considered in isolation.

Definitions and more detailed descriptions of the processes discussed in this article can be found in *Surface Engineering,* Volume 5 of *ASM Handbook.*

Design as an Integral Part of Manufacturing

In recent years, manufacturing processes have been evaluated in terms of life-cycle costs and their impact on the environment, and even ecology in general. Several scenarios have been proposed for the life cycle of materials, which of necessity incorporates manufacturing processes. In the article "Introduction to Manufacturing and Design" in this Volume, an example of one such scheme is given where a material flows through the extraction, refining (preparation), shapemaking and structural treatments (manufacturing), and surfacing (surface-finishing) stages before being assembled and placed in use. Manufacturing also can be viewed as part of an "integrated product- and process-development" process, also described in that article. This approach places emphasis on "product design" and "process design" in a concurrent-engineering environment and forces product developers to consider both simultaneously, rather than sequentially and separately (Ref 1). If these concepts are extrapolated, with a focus on design in relation to surface finishing, the iterative process shown in Fig. 1 results. The scheme presented in this figure is entirely consistent with another modern manufacturing concept, namely that of continuous improvement.

Having stressed the importance of considering design precepts as an integral part of the manufacturing process and product improvement, the following guidelines should be considered as just

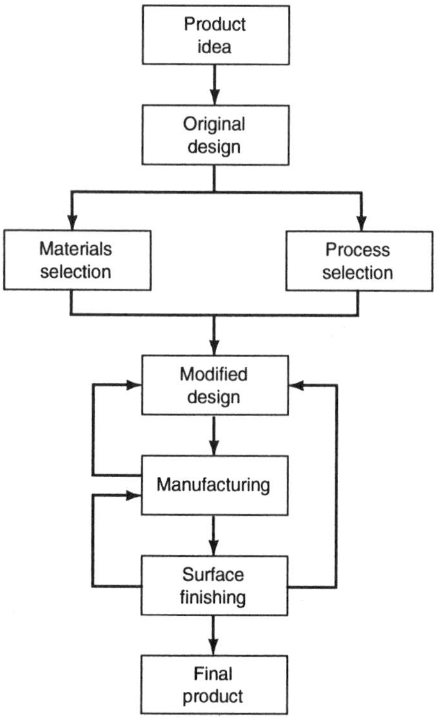

Fig. 1 Flow diagram for incorporating design principles into surface-finishing operations

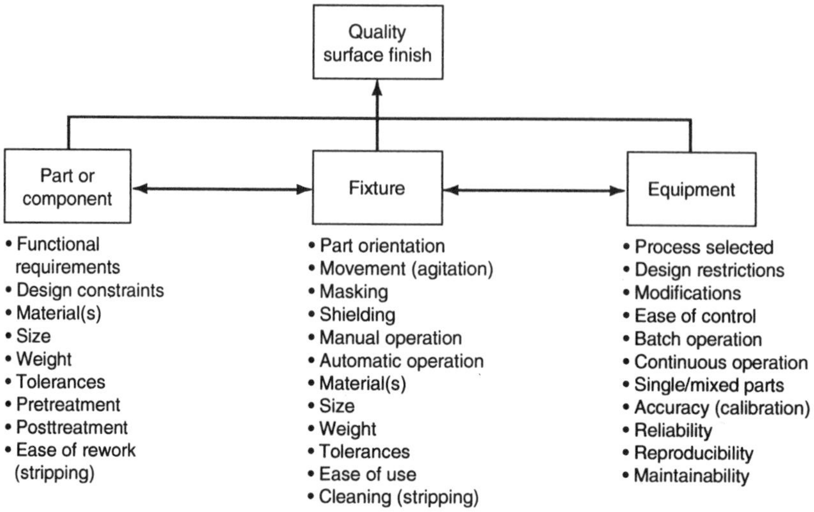

Fig. 2 Interrelation between part design, equipment limitations, and fixturing

that—guidelines. Each application should be treated on an individual basis, and surface-finishing processes should be flexible enough to permit design improvements, and vice versa. In addition, it should be obvious from Fig. 1 that the earlier in the product-development cycle manufacturing-related design considerations are addressed, the more efficient will be the manufacturing process and the better the quality of the surface finish will be. Communication with and input from the manufacturing and surface-finishing staff are very important in establishing a satisfactory product design.

Design of the part or component and pretreatment selections are important, but in keeping with the concept of design being an integral part of manufacturing and surface finishing, process equipment design restrictions and fixturing design also are very important. Both can influence the quality of the resulting finish. Figure 2 shows their interrelated roles schematically. In this overview, emphasis is placed on issues relating to part design and process equipment. Fixturing must be designed and tailored for each individual application and is beyond the scope of this article.

General Design Principles Related to Surface Finishing

There are a number of general design principles that apply to a variety of finishing processes, while others are specific to individual finishing techniques. These general principles are discussed in this section, and the following three sections discuss design aspects relating to: (1) surface-preparation techniques, including cleaning; (2) organic finishing techniques; and (3) inorganic finishing techniques. References 2 and 3 provide some background material on the various finishing techniques discussed, while Tables 1 through 3 summarize the important design limitations for these three categories. The subject of materials selection is covered elsewhere in this Volume. The choice of materials can limit the choices for surface finishing. Where appropriate, significant limitations are described.

Fabrication Processes. Some methods of fabrication such as the forging, extrusion, molding, and casting of metals and ceramics can lead to surface defects that must be removed by sub-

Table 1 Summary of design limitations for selected surface-preparation processes

Process	Design limitations
Blasting/deburring	Avoid recesses, holes, channels, and similar features (such as closely spaced ribs) that could trap blasting media Avoid thin cross sections (such as fins, louvers, walls) that could be distorted by the blasting media Avoid intricate designs and surface features
Broaching/honing	Typically used for inside diameters of tubes and other cylindrical parts, or for grooves, large holes, and other cavities Surfaces must be accessible to tools and withstand the local pressure and heat buildup Avoid very thin cross sections/wall thickness
Brushing/burnishing	Surfaces must be accessible to tools and withstand the local pressure and heat buildup Avoid very thin cross sections/wall thickness that could deflect Avoid sharp corners and edges Avoid intricate designs and surface features
Chemical milling	Avoid features (e.g., small recesses, blind holes, cavities) that would trap process chemicals or prevent satisfactory rinsing Provide good natural drainage or use drainage holes on nonsignificant surfaces to minimize carryover Avoid features that could trap air or evolved gases and prevent chemical action from occurring or cause uneven attack Mask areas not to be attacked
Conversion coating	Avoid features (e.g., small recesses, blind holes, cavities) that would trap process chemicals or prevent satisfactory rinsing Provide good natural drainage or use drainage holes on nonsignificant surfaces to minimize carryover Avoid features that could trap air and prevent surface chemical reactions from occurring or cause staining Mask areas not to be attacked
Electrocleaning	Allow for electrical contact to be made on nonsignificant surfaces Avoid features that would trap process chemicals or prevent satisfactory rinsing Provide good natural drainage or use drainage holes on nonsignificant surfaces to minimize carryover Avoid features that could trap air or evolved gases and would prevent cleaning from occurring or cause staining
Electropolishing	Allow for electrical contact to be made on nonsignificant surfaces Avoid features (e.g., small recesses, blind holes, cavities) that would trap process chemicals or prevent satisfactory rinsing Provide good natural drainage or use drainage holes on nonsignificant surfaces to minimize carryover Avoid features that could trap air or evolved gases and prevent polishing action from occurring or cause staining Mask areas not to be attacked
Etching	Avoid features (e.g., small recesses, blind holes, cavities) that would trap process chemicals or prevent satisfactory rinsing Provide good natural drainage or use drainage holes on nonsignificant surfaces to minimize carryover Avoid features that could trap air and prevent etching action from occurring Avoid sharp corners and edges Avoid shallow intricate designs and surface features Mask areas not to be attacked
Grinding	Surfaces must be accessible to tools and withstand the local pressure and heat buildup
Grinding (continued)	Avoid very thin cross sections/wall thickness Avoid sharp corners, edges, and protuberances Avoid intricate designs and surface features
Lapping/buffing	Surfaces must be accessible to tools (preferably flat or simple, curved contours Avoid very thin cross sections/wall thickness that cannot withstand the local pressure and heat buildup Avoid sharp corners and edges Avoid intricate designs and surface features that would trap the lapping/buffing compounds
Pickling	Avoid features (e.g., small recesses, blind holes, cavities) that would trap process chemicals or prevent satisfactory rinsing Provide good natural drainage or use drainage holes on nonsignificant surfaces to minimize carryover Avoid features that could trap air and prevent pickling action Avoid flat surfaces on small parts that could stick together, exclude the acid, and prevent the pickling action
Polishing	Surfaces must be accessible to tools and withstand the local pressure and heat buildup Avoid very thin cross sections/wall thickness Avoid sharp corners, edges, and protuberances Avoid intricate designs and surface features that could trap the polishing compound
Solvent cleaning, immersion	Avoid features (e.g., small recesses, blind holes, cavities) that would trap process chemicals or prevent satisfactory rinsing Provide good natural drainage or use drainage holes on nonsignificant surfaces to minimize carryover Avoid features that could trap air and prevent cleaning from occurring Avoid flat or curved surfaces on small parts that could stick together during immersion and prevent cleaning of those surfaces
Solvent cleaning, ultrasonic	Avoid features (e.g., small recesses, blind holes, cavities) that would trap process chemicals or prevent satisfactory rinsing Provide good natural drainage or use drainage holes on nonsignificant surfaces to minimize carryover Avoid features that could trap air and prevent cleaning from occurring Avoid thin cross sections that could be damaged by the energy released during cavitation
Stripping, chemical	Avoid features (e.g., small recesses, blind holes, cavities) that would trap smut and process chemicals or prevent satisfactory rinsing Provide good natural drainage or use drainage holes on nonsignificant surfaces to minimize carryover Avoid features that could trap air and prevent coating removal from occurring Mask areas not to be attacked
Stripping, mechanical	Avoid recesses, holes, channels, and similar features that could trap blasting media Avoid thin cross sections or intricate designs that could be damaged by the stripping media Mask areas not to be attacked
Stripping, thermal	Avoid thin cross sections or intricate designs that could be distorted by the thermal cycling Try to provide uniform cross-sectional mass throughout the part to help provide a uniform temperature distribution during heating

sequent surface-finishing techniques, such as grinding, lapping, and polishing or electropolishing, or hidden by techniques such as applying a leveling copper deposit before a decorative plated finish. Defects include laps, tears, cracks, pores, shrinkage cavities, gating and venting residues, ejection marks, and parting lines. Careful design of the casting or molding operation—including the dies, gates, vents, and overflows—will minimize finishing problems by ensuring such defects are avoided, occur on nonsignificant surfaces, or are hidden by specially incorporated design features, such as steps or ridges at parting lines.

When polymeric materials are being cast, molded, extruded, or formed, it is especially important that the design and tooling lend themselves to producing an acceptable surface finish because any finishing operation that removes surface layers could expose porosity and other defects and remove aesthetic qualities such as smoothness and luster. Also, the selection of finishing tools and conditions is much more critical because of the physical properties (e.g., softness, plasticity) of the polymeric materials and the potential for damage (e.g., heat distortion) caused by heat generated during the finishing operation. This topic is discussed more fully in Ref 4. As with metals and ceramics, some undesirable attributes of the fabrication process, such as parting lines, can be hidden by added design features.

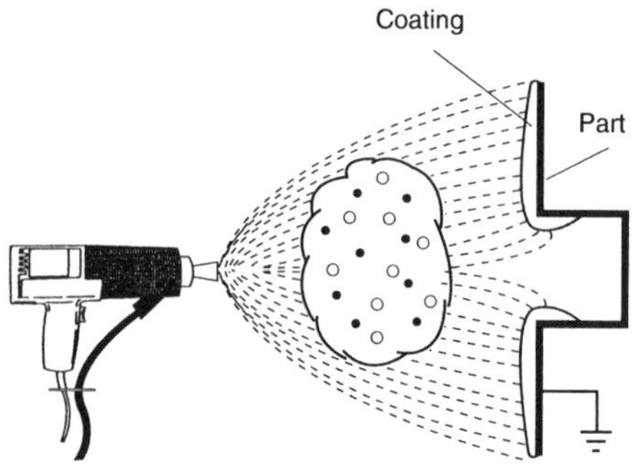

Fig. 3 Faraday cage effect in powder coating. Adapted from Ref 7

Whatever the type of material being cast or molded, dimensional and warpage allowances must be made in the design of the tooling (i.e., dies) to accommodate shrinkage and distortion during solidification and cooling. Otherwise, parts may be undersized or require excessive machining to obtain the specified dimensional tolerances.

Control of fastening or joining processes also can influence surface finishing. For example, two flat surfaces riveted together produce cavities that can entrap processing solutions, impair coating, and lead to corrosion (Ref 5). Spot or tack welding is no better in this regard. However, a continuous weld—with a smooth bead and no weld spatter—will prevent this problem and make surface finishing easier. Also, the elimination of sharp edges and corners will prolong the life of grinding, polishing, and buffing belts and wheels.

Size and Weight. The design, dimensions, and weight of a part to be surface finished have a direct influence on fixturing and the size and type of equipment that is used. As Ref 6 points out, size has little to do with production rate. Large parts do not necessarily mean low production rates, but they will dictate the size of the cleaning tanks, spray booths, plating tanks, vacuum chambers, and the like. Nevertheless, the end-use application often dictates the size of a part or component and materials selection will dictate its weight. Consequently, as a variable, only design is left to influence ease of finishing, efficacy of finishing, and finishing cost. Whatever can be done in terms of optimizing a design to facilitate surface finishing will have a positive impact. In this context, design optimization includes facilitating the handling of parts between the various surface-finishing steps, as well as during the processes themselves. These operations will be most efficient when the parts are designed to prevent "nesting" or flat surfaces from sticking together (because of surface tension effects) and preventing those surfaces from being treated or coated. Similarly, small parts may be light in weight and unstable in sprays during pretreatment or painting. The design of these parts should be such that clips or magnets can be used to secure and stabilize them during processing.

Aesthetics and Function. Another general consideration is that not all surfaces may require the same high standard of surface finish. While surfaces exposed to view must be aesthetically pleasing, and surfaces subjected to more aggressive conditions of exposure or use require durable coatings, hidden (internal) surfaces or less-exposed surfaces may not need such a high-quality finish. Specifications for surface finishes for a part depend not only on the design and end-use application, but also must take into account that the requirements may differ for different areas or surfaces on that part. A design should take this into consideration, as well as the fact that different types of equipment or equipment operation settings may be necessary for those areas and surfaces.

Functional requirements of a part also influence the selection of surface preparation processes. For example, grinding processes can introduce stresses that could have a negative impact on fatigue properties. Choosing an alternative process, such as chemical milling, or mitigating the stresses by shot peening can alleviate the problem.

Table 2 Summary of design limitations for selected organic finishing processes

Process	Design limitations
Electrocoating	Allow for electrical contact to be made on nonsignificant surfaces
	Avoid features that could trap air and prevent wetting by process solutions
	Provide good natural drainage or use drainage holes on nonsignificant surfaces to minimize carryover
	Avoid thin cross sections or intricate designs that could become distorted during drying/curing cycle
Electropolymerization	Allow for electrical contact to be made on nonsignificant surfaces
	Avoid features that could trap air and prevent wetting by process solutions
	Provide good natural drainage or use drainage holes on nonsignificant surfaces to minimize carryover
	Avoid thin cross sections or intricate designs that could become distorted during drying/curing cycle
Painting, brushing or dipping	Surfaces must be accessible to application tools (preferably flat or simple, curved contours)
	Avoid features that would trap excess paint
	Provide good natural drainage or use drainage holes on nonsignificant surfaces to minimize carryover
	Avoid features that could trap air and prevent coating from occurring
	Avoid thin cross sections or intricate designs that could become distorted during drying/curing cycle
Painting, solvent spraying	Surfaces must be accessible to application tools (preferably flat or simple, curved contours)
	Allow for fixturing/racking on nonsignificant surfaces
	Avoid features that would trap excess paint
Painting, solvent spraying (cont'd)	Provide good natural drainage or use drainage holes on nonsignificant surfaces to minimize carryover
	Avoid features that could trap air and prevent coating from occurring
	Avoid thin cross sections or intricate designs that could become distorted during drying/curing cycle
Powder coating	Allow for fixturing/racking on nonsignificant surfaces
	Allow for electrical contact to be made on nonsignificant surfaces
	Avoid deep recesses and blind holes that cause the "Faraday cage" effect
	Avoid thin cross sections or intricate designs that could become distorted during drying/curing cycle
Sol-gel coating	Allow for fixturing/racking on nonsignificant surfaces
	Avoid features (e.g., small recesses, blind holes, cavities) that would trap process chemicals
	Avoid thin cross sections or intricate designs that could be distorted by the thermal cycling
	Try to provide uniform cross-sectional mass throughout the part to help provide a uniform temperature distribution during heating cycle
Solution coating	Allow for fixturing/racking on nonsignificant surfaces
	Avoid features (e.g., small recesses, blind holes, cavities) that would trap process chemicals
	Provide good natural drainage or use drainage holes on nonsignificant surfaces to minimize carryover
	Avoid features that could trap air and prevent coating from occurring
	Avoid thin cross sections or intricate designs that could become distorted during drying/curing cycle

Design Features. Shape and features such as recesses, holes, threads, keyways, slots, fins, and louvers can present problems to the finisher, and the severity of the problem can depend on the finishing technique. For example, when holes are included in thin sections that require a finishing operation such as grinding, if too much pressure is applied edges and corners might be chipped. If only a light pressure is used to avoid this possibility, then the desired finish might not be obtained. Another example is when paint is applied by conventional solvent spraying or when a part is electroplated, bowl-shaped recesses, blind holes, and similar features can trap the paint or plating solution, leading to areas that sag or do not cure properly (in the case of paint), or carry over trapped chemicals to subsequent processing steps (in electroplating). The latter can cause problems such as rinse-water contamination and increased waste-treatment costs. Also, solutions that are trapped can lead to blistering or delamination of the plated coating, especially if there is a posttreatment step that requires the part to be heated (such as for electroless nickel, cadmium, and hard chromium deposition).

For parts that will be sprayed, especially with paint, another problem with deep recesses, closely spaced, large fins or partitions, and the like is the entrapment of air. The back pressure of entrapped air causes incomplete coverage at the bottom of the recesses. One way to avoid this problem, if a change of design is not possible, is to use an "airless" spraying technique (Ref 6). During electrostatic powder coating there is the problem associated with "Faraday cage" effect, in which the charged components of the powder-coating system are attracted by the high fields at the edges and corners of parts, causing excessive coverage there and incomplete coverage in other areas (Ref 7), as shown in Fig. 3. Rounding corners and edges, tapering the sides and decreasing the depth of recesses, minimizing the use of louvers or fins, or changing their dimensions are ways to avoid the Faraday cage effect. In electroplating, a similar phenomenon exists whereby the depositing metal or alloy ions are attracted to the high-current-density areas at edges and corners, and thicker coatings are obtained in those locations. Rounding such edges, changing dimensions to allow for the excessive buildup, or using shields and current "robbers" or "thieves" will help the finisher to obtain the desired coating thickness distribution. Reference 8 provides some examples of the use of such devices.

In conventional paint spraying and many vacuum-deposition techniques, such as ion plating, ion implantation, physical vapor deposition, and sputtering, attention has to be paid to the limitations imposed by the "line-of-sight" deposition process. Certain features, such as ridges, flanges, and fins, can shadow or mask areas behind them leading to incomplete or nonuniform coverage, as shown in Fig. 4 and 5. Similarly, if the aspect ratio of holes and recesses is too high (i.e., the depth is much greater than the diameter of the opening), it is not possible with line-of-sight limited techniques to penetrate to the bottom surfaces and coat them (Fig. 5). Decreasing the aspect ratio, providing rounded edges, and tapering the sides of ridges and fins or holes will help to facilitate finishing, as will lowering the height of features such as fins. Of course, rotating or translating a part in the spray plume also will help to obtain complete and more uniform coverage, but this approach usually requires longer times and more sophisticated finishing equipment and fixturing; hence, it often leads to higher costs. The same can be said for using multiple line-of-sight sources to obtain better coverage.

Finally, as a general rule of thumb, parts of the same size, weight, design, and material should always be finished at the same time so that the finishing process(es) can be optimized for those parts. Batches of mixed parts should be avoided unless they share some common features, such as shape and substrate material.

Surface-Preparation Processes

To facilitate surface preparation—including cleaning—prior to subsequent coating operations, design precepts should have been considered during the product-design and manufacturing stages, as indicated in Fig. 1. Abrupt changes in surface contours should be avoided, and features such as fine grooves, recesses, surface patterning, blind holes, and reentrant areas should be avoided because they will be inaccessible to polishing media or would trap polishing media, making subsequent cleaning more difficult. Such features also would entrap cleaning chemicals, making rinsing more difficult, or could possibly entrap air, preventing cleaning of these areas.

Sharp corners and edges or protrusions can cause excessive wear of polishing wheels and belts and lead to uneven polishing because the high areas are polished at the expense of the surrounding lower areas. As mentioned earlier, rounding edges and corners is a good design precept, while minimizing the height of protuberances is beneficial, as is decreasing the aspect ratio of holes, grooves, and recesses.

Large expanses of flat surfaces may be a problem if these are significant surfaces, especially if these surfaces must be polished to a reflective, mirrorlike finish. Imperfections are exaggerated. Minimizing the area of such surfaces and providing a slightly rounded contour will help to attain the desired finish and help with visual appearance.

Simpler designs lend themselves to automatic finishing processes, while more complex designs may require manual surface-preparation techniques. If parts are to be mass finished (e.g., by tumbling or vibratory finishing) significant flat

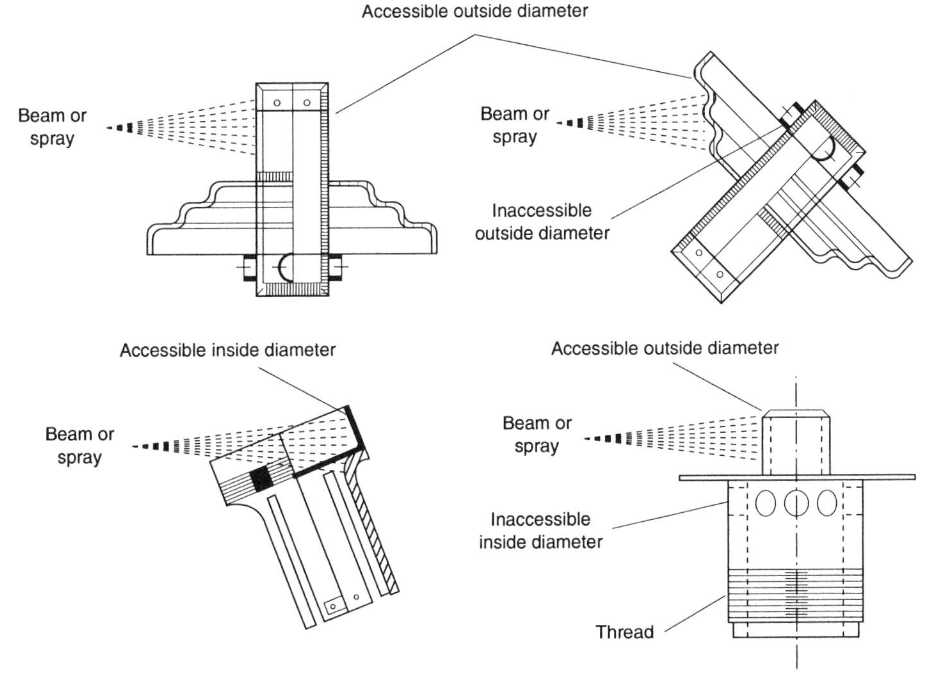

Fig. 4 Some examples of line-of-sight limitations in spraying or ion-beam coating processes

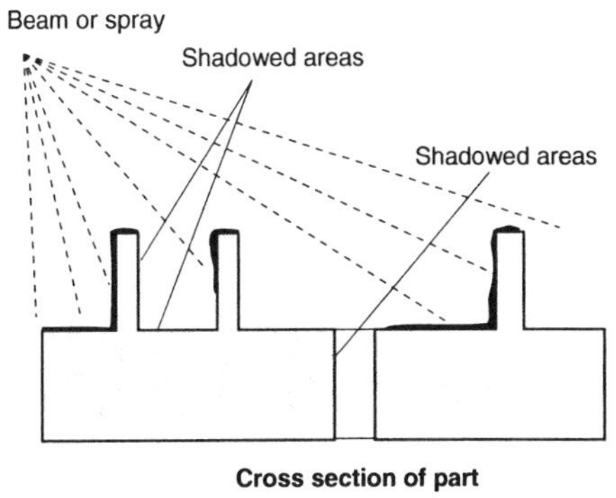

Fig. 5 Design features that cause shadowing or poor coverage because of line-of-sight limitations

Table 3 Summary of design limitations for selected inorganic finishing processes

Process	Design limitations
Anodizing	Allow for electrical contact to be made on nonsignificant surfaces Avoid, if possible, sharp edges and corners, ridges, blind holes, etc., that would prevent uniform density distribution Avoid features (e.g., small recesses, blind holes, cavities) that would trap process chemicals Provide good natural drainage or use drainage holes on nonsignificant surfaces to minimize carryover Avoid features that could trap air and prevent electrochemical reactions from occurring Avoid features that could trap evolved gases and cause staining Mask areas not to be anodized
Cementation/diffusion	Surfaces must be thoroughly deburred and cleaned before cladding, so design principles for these processes also apply Avoid thin cross sections or intricate designs that could become distorted during thermal cycling Mask areas not to be coated
Cladding	Only for relatively simple shapes, especially with flat surfaces Surfaces must be thoroughly cleaned before cladding, so design principles for cleaning also apply
Electroless plating	Allow for fixturing/racking on nonsignificant surfaces Avoid features (e.g., small recesses, blind holes, cavities) that would trap process chemicals or prevent satisfactory rinsing Provide good natural drainage or use drainage holes on nonsignificant surfaces to minimize carryover Avoid features that could trap air and prevent chemical reactions from occurring or cause staining Mask areas not to be coated
Electrophoretic plating	Allow for electrical contact to be made on nonsignificant surfaces Avoid features (e.g., small recesses, blind holes, cavities) that would trap process chemicals or prevent satisfactory rinsing Provide good natural drainage or use drainage holes on nonsignificant surfaces to minimize carryover Avoid features that could trap air and prevent surface chemical reactions from occurring or cause staining Mask areas not to be coated
Electroplating (plating, electrodeposition)	Allow for electrical contact to be made on nonsignificant surfaces Avoid, if possible, sharp edges and corners, ridges, blind holes, etc., that would prevent uniform current density distribution; or use current robbers and/or shields Avoid features (e.g., small recesses, blind holes, cavities) that would trap process chemicals or prevent satisfactory rinsing Provide good natural drainage or use drainage holes on nonsignificant surfaces to minimize carryover Avoid features that could trap air and prevent deposition from occurring Avoid features that could trap evolved gases and cause staining Avoid thin cross sections (such as fins, louvers, walls) that could be distorted by internal stress in the coating Mask areas not to be coated
Hot dipping, galvanizing	Allow for fixturing/racking on nonsignificant surfaces for discrete, small parts Best for relatively simple shapes (e.g., tubing) and flat surfaces Allow for excess coating material to drain quickly Allow for doctor blades or air knives to be used to obtain uniform coating thickness Avoid thin cross sections that could become distorted during thermal cycling
Inorganic painting, slurry coating	Surfaces must be accessible (preferably flat or simple, curved contours) Allow for fixturing/racking on nonsignificant surfaces Avoid features that would trap excess paint Provide good natural drainage or use drainage holes on nonsignificant surfaces to minimize carryover Avoid features that could trap air and prevent coating from occurring Avoid thin cross sections or intricate designs that could become distorted during drying/fusing cycle Mask areas not to be coated
Ion implantation	Allow for electrical contact to be made on nonsignificant surfaces or use a conductive screen Avoid features that would shield the surface from the beam (line-of-sight limited) unless multiple beams are used or part is rotated/translated in beam Avoid high aspect ratio holes and recesses, grooves, etc., that would not allow the beam to reach the bottom surfaces Mask areas not to be coated
Ion plating	Allow for electrical contact to be made on nonsignificant surfaces or use a conductive screen Avoid features that would shield the surface from the beam (line-of-sight limited) unless multiple beams are used or part is rotated/translated in beam Avoid high aspect ratio holes and recesses, grooves, etc., that would not allow the beam to reach the bottom surfaces Mask areas not to be coated
Laser glazing	Allow for fixturing/racking on nonsignificant surfaces Avoid features that would shield the surface from the laser beam (line-of-sight limited) unless multiple beams are used or part is rotated/translated in beam Avoid high aspect ratio holes and recesses, grooves, etc., that would not allow the beam to reach the bottom surfaces Avoid thin cross sections or intricate designs that could be damaged by local heating during glazing Mask areas not to be treated
Mechanical (peen) plating	Allow for fixturing/racking on nonsignificant surfaces on large parts Avoid features that could trap air and prevent activation by the process chemicals from occurring Provide good natural drainage or use drainage holes on nonsignificant surfaces to minimize carryover of activating solutions Avoid recesses, holes, channels, and similar features that could trap peening media Avoid thin cross sections (such as fins, louvers, walls) that could be distorted by the peening action Avoid sharp edges and corners that could be damaged by the peening media Avoid intricate designs and small surface features that cannot be reached by the peening media Provide good natural drainage or use drainage holes on nonsignificant surfaces to minimize carryover Mask areas not to be coated
Passivation	Allow for fixturing/racking on nonsignificant surfaces Avoid features (e.g., small recesses, blind holes, cavities) that would trap process chemicals or prevent satisfactory rinsing Avoid features that could trap air and prevent surface chemical reactions from occurring or cause staining Mask areas not to be attacked
Thermal spraying	Allow for fixturing/racking on nonsignificant surfaces Design should allow for surface roughening to promote adhesion, so blasting design precepts also apply Avoid features that would shield the surface from the spray (line-of-sight limited) unless multiple sprays are used or part is rotated/translated in spray plume Avoid high aspect ratio holes and recesses, grooves, etc., that would not allow the spray to reach the bottom surfaces Avoid thin cross sections (such as fins, louvers, walls) that could be distorted by the local heating and kinetic energy Mask areas not to be coated
Vapor deposition (CVD, PVD, RVD)	Allow for fixturing/racking on nonsignificant surfaces Avoid thin cross sections (such as fins, louvers, walls) that could be distorted by heating, if needed prior to coating deposition Vacuum processes are line-of-sight limited, so similar design precepts to those for ion plating will apply Mask areas not to be coated

CVD, chemical vapor deposition; PVD, physical vapor deposition; RVD, reactive vapor deposition

Table 4 Design features that influence electroplating

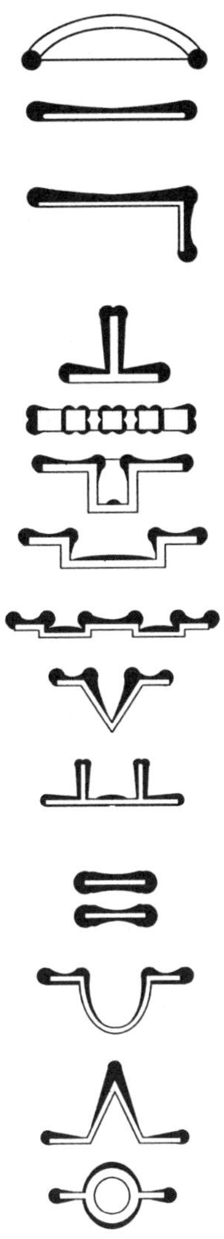

Design feature	Influence on platability	Better design
	Convex surface. Ideal shape. Easy to plate to uniform thickness, especially where edges are rounded.	
	Flat surface. Not as desirable as crowned surface. Use 0.015 mm/mm (0.015 in./in.) crown to hide undulations caused by uneven buffing.	
	Sharply angled edge. Undesirable. Reduced thickness of plate at center areas. Requires increased plating time for depositing minimum thickness of durable plate. All edges should be rounded. Edges that contact painted surfaces should have a 0.8 mm ($\frac{1}{32}$ in.) min radius.	
	Flange. Large flange with sharp inside angles should be avoided to minimize plating costs. Use generous radius on inside angles and taper abutment.	
	Slots. Narrow, closely spaced slots and holes cannot be plated properly with some metals (e.g., nickel and chromium) unless corners are rounded.	
	Blind hole. Must usually be exempted from minimum thickness requirements.	
	Sharply angled indentation. Increases plating time and cost for attaining a specified minimum thickness and reduces the durability of the plated part.	
	Flat-bottom groove. Inside and outside angles should be rounded generously to minimize plating costs.	
	V-shaped groove. Deep grooves cannot be plated satisfactorily; should be avoided. Shallow, rounded grooves are better.	
	Fins. Increase plating time and costs for attaining a specified minimum thickness and reduce the durability of the plated part.	
	Ribs. Narrow ribs with sharp angles usually reduce platability; wide ribs with rounded edges impose no problem. Taper each rib from its center to both sides and round off edges. Increase spacing, if possible.	
	Deep scoop. Increases time and cost for plating specified minimum thickness.	
	Spearlike jut. Buildup on jut robs corners of electroplate. Crown base and round all corners.	
	Ring. Platability depends on dimensions. Round corners; crown from center line, sloping toward both sides.	

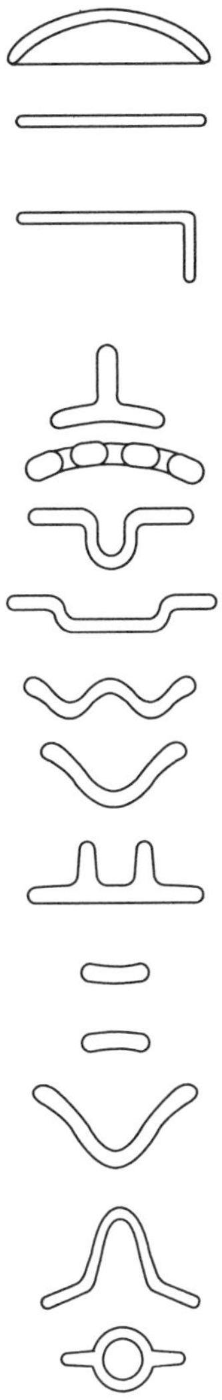

Note: Distribution of electroplate on design shapes is intentionally exaggerated by solid black outline. Cross-hatched areas indicate part before plating.

areas should be avoided. Otherwise, parts may stick together, and these occluded surfaces will not be finished. Designs that prevent access by the deburring or polishing media (such as small recesses and holes) or that entrap the media (such as narrowly spaced ribs) should be avoided as mentioned above.

When it is impractical or impossible to use mechanical polishing, chemical etching, chemical milling, or electropolishing can be used. The design principles for the latter are similar to those for electroplating, which is discussed later. In electropolishing, the workpiece is the anode, which is the opposite of electroplating. Current density distribution is extremely important, as is the original surface of the part being electropolished. In high-current-density areas on susceptible materials, the surface layers may be removed and etching of the substrate can occur. Polishing occurs on a microscopic scale, so macro features such as large grooves or scratch marks will not be removed but will receive a luster and become more noticeable. Similarly, parting lines can be smoothed, but not removed; therefore, parting lines must be minimized by good die design and careful molding operations.

Solvent cleaning is a fairly forgiving surface-finishing process, but part design can influence its efficacy, as already alluded to. If agitation or other cleaning aids are used, such as ultrasonic energy, care must be taken to prevent soft materials or thin and fragile features or cross sections from being damaged. The energy released during cavitation, for example, in ultrasonic cleaning is very large. If techniques such as plastic media blasting are used, the blasting parameters should be tailored to the part material and design, and the part should be designed to allow easy access by the media and easy removal of the media once the desired finish (cleanliness) is obtained.

If a power spray washing technique is used, the part design should allow for proper drainage to conserve chemicals and minimize carryover to the next process step. Providing drainage holes may be necessary. These should be either a natural feature of the design or located on nonsignificant surfaces. As the design of a part becomes more complex, rinsing requirements become more stringent, and several rinsing stages may be necessary. If an air knife is used afterward to remove excess water, the part must be capable of withstanding the pressure or must be fixtured such that the air pressure does not distort any delicate design features while holding the part steady.

Table 1 provides a summary of the design limitations of some surface-preparation and cleaning processes and indicates which design features to avoid.

Organic Finishing Processes

Organic finishes are applied by a variety of techniques, such as dipping, brushing, spraying, airless spraying, or electrostatic spraying. In addition, some primers are deposited using electrophoretic techniques, while electropolymerization is being looked at for certain types of organic coatings. Table 2 summarizes these techniques and the design limitations associated with each.

Most of the techniques are line-of-sight limited, and the guidelines provided in the previous section, "Surface-Preparation Techniques," will apply. Allowance for drainage is important for processes that involve dripping or spraying. Avoiding sags and runs on large, flat, vertical surfaces can be accomplished by applying good coating practices and by minimizing such surfaces in the design of the part.

A few organic coating techniques use electric or electrostatic fields. Designing the fixtures and electrical grounding, such that points of contact are on nonsignificant surfaces, will improve the appearance of the coated part and give the impression of a better quality product. With spraying techniques, proper fixturing and racking of parts can improve the use of coating material because less empty space exists during a run. However, the parts should not be racked so closely together that they shield some surfaces and prevent some areas from being coated.

Avoiding thin cross sections and good fixturing will help prevent distortion during the curing and baking steps used after paint or powder is applied.

Optimizing a design for surface finishing, such as painting, becomes very important as coating thickness is reduced to 30 μm or less. Access to all surfaces must be possible and any features that would prevent this should be avoided (see the section "Design Features" in this article). This is because the dimensions of the solid components in the coating formulation (e.g., powder particle) are similar to the dimensions of the desired dry film thickness (Ref 9). For example, during the first part of curing, when the particles liquefy, the surface tension of the film formed will tend to pull it away from sharp corners or edges, resulting in poor coverage. If a design modification is not possible, the powder formulation should be changed to include higher-viscosity resins, and no, or only small amounts, of surfactants (Ref 9). Thin-film coatings are best applied to parts with simple geometries, with flat or curved surfaces, and few sharp edges.

Earlier, the problem with the Faraday cage effect was mentioned. This phenomenon is further complicated by back-ionization with traditional corona-charging systems (Ref 7). Not only does the design of a recess, hole, or channel control the distribution of coating thickness, but the buildup of back-ionization at the areas of high field intensity lowers the effective charge of the powder particles, further reducing their ability to reach the bottom surfaces. Some possible design modifications were mentioned earlier but, if these are not possible, changing to a turbocharging system will help. Back-ionization is greatly reduced, and the absence of free ions between the gun and the part promotes better coverage of all surfaces (Ref 7).

Inorganic Finishing Processes

Inorganic finishes—including metal- and ceramic-based coatings—are applied by a variety of techniques, such as electroplating, electroless plating, thermal spraying, hot dipping, ion plating, and various vapor-deposition techniques. Other techniques, such as ion implantation and laser glazing, modify surface properties. Table 3 summarizes design limitations for these and other types of inorganic coating processes.

Electroplating is widely used in industry to apply inorganic coatings, especially metals and alloys. Like some organic finishing processes, satisfactory coatings are only obtained when a uniform current density can be established on all surfaces to be finished. Phenomena like the Faraday cage effect occur when design features prevent the establishment of a uniform current density distribution. As mentioned earlier, techniques relating to fixturing and racking can alleviate some of the problems. General design approaches are discussed in Ref 8 and 10 and summarized in Table 4.

With the plating of fasteners, some special considerations apply, particularly in respect to threads (Ref 10). As might be expected, electroplated metals build up faster on apexes of the threads, and coverage can be minimal at the bottom of the grooves. ANSI Specification B 1.1 states that, compared to flat surfaces, plating thickness builds up six times faster on the major

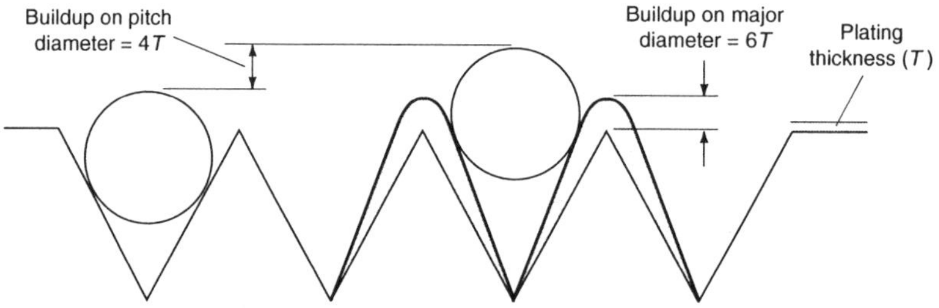

Fig. 6 Rule of Four and Six as applied to coating external threads. Source: Ref 10

Table 5 Surface-finishing technologies grouped by similar design limitations

Design limitation	Surface-finishing process
Allow for fixturing/racking on nonsignificant surfaces	Painting, powder coating, sol-gel coating, solution coating, electroplating, electroless plating, anodizing, electrocoating, electropolymerization, electrophoretic plating, hot dipping, ion implantation, laser glazing, passivation, thermal spraying, electric arc spraying
Allow for electrical contact to be made on nonsignificant surfaces or use a conductive screen	Elecroplating, electropolishing, anodizing, electrocleaning, electrocoating, electropolymerization, electrophoretic plating, ion implantation, ion plating, powder coating
Surfaces must be accessible (preferably flat or simple, curved contours)	Grinding, polishing, lapping, burnishing, buffing, painting, electroplating, electroless plating, ion vapor deposition, cladding, ion implantation, hot dipping, slurry coating, laser glazing, mechanical plating, thermal spraying, electric arc spraying
Avoid intricate designs and surface features	Blasting, peening, brushing, burnishing, polishing, lapping, buffing, grinding, mechanical plating, hot dipping, cladding, cementation, slurry coating, mechanical plating, thermal spraying, electric arc spraying
Surfaces must be accessible to tools and withstand the local pressure and heat build up	Broaching, honing, grinding, brushing, burnishing, polishing, lapping
Avoid sharp corners, edges, and protuberances to prolong tool life	Blasting, lapping, brushing, burnishing, polishing, buffing, grinding
Allow for surface roughening to promote adhesion	Thermal spraying, electric arc spraying, sputtering, ion vapor deposition, ion plating, mechanical plating
Avoid thin cross sections (such as fins, louvers, walls) that could be distorted by heating or kinetic energy	Brushing, burnishing, grinding, lapping, solution coating, polishing, thermal stripping, hot dipping, cementation, laser glazing, mechanical plating, thermal spraying, electric arc spraying, diffusion coating
Avoid features that would shield the surface from beams or sprays (line-of-sight limited) unless multiple sources are used, or part is rotated/translated in beam or spray	Chemical cleaning, chemical stripping, ion implantation, sputtering, ion plating, laser glazing, thermal spraying, electric arc spraying
Avoid high aspect ratio holes and recesses, grooves, etc., that would not allow beams or sprays to reach bottom surfaces	Spray painting, ion plating, ion implantation, chemical cleaning, sputtering, laser glazing, thermal spraying, electric arc spraying
Avoid features that could trap air and prevent surface chemical reactions or coating from occurring or cause staining	Chemical milling, chemical cleaning, chemical stripping, etching, pickling, electroplating, spray painting, hot dipping, diffusion coating, electrocoating, sol-gel coating, solution coating, anodizing, electroless plating, electrophoretic plating, chemical vapor deposition, passivation
Avoid features (e.g., small recesses, blind holes, cavities) that would trap process chemicals or prevent satisfactory rinsing	Chemical milling, chemical cleaning, pickling, electrocleaning, chemical stripping, etching, electroplating, electroless plating, anodizing, mechanical plating, dipping (painting), conversion coating, solution coating, passivation, electrophoretic plating, electropolishing, sol-gel coating, electrocoating
Avoid recesses, holes, channels, and similar features that could trap materials	Blasting, peening, grinding, polishing, lapping, mechanical plating, thermal spraying, electric arc spraying
Provide good drainage or use drainage holes on nonsignificant surfaces to minimize carryover of process solutions	Chemical cleaning, chemical stripping, chemical milling, etching, electocleaning, pickling, electropolishing, conversion coating, electroplating, anodizing, electrophoretic plating, electroless plating, spray painting, sol-gel coating, solution coating, electrocoating, mechanical plating, passivation
Avoid, if possible, sharp edges and corners, ridges, blind holes, etc., that would prevent uniform coating thickness or metal removal	Chemical milling, etching, electropolishing, pickling, chemical stripping, sol-gel coating, solution coating, hot dipping, ion implantation, ion plating, thermal spraying, electric arc spraying
Avoid features that could trap evolved gases and cause staining or uneven attack	Chemical milling, etching, chemical cleaning, chemical stripping, electropolishing, pickling, electroplating, electrocleaning, electrocoating, anodizing, electroless plating, electrophoretic plating, passivation

diameter than the minor diameter and that this results in a fourfold buildup on the pitch diameter, as illustrated in Fig. 6. This is known as the "Rule of Four and Six."

Similarly, plating inside holes can be difficult. The general rule of thumb is that if the hole diameter is x, the plating will occur to a depth of $2x$. However, for blind holes, plating will only occur to a depth of x. Agitation, solution flow, maximizing the throwing power of the plating bath, and other aids can improve the situation somewhat, but the best approach is to eliminate or minimize holes with high aspect ratios during the product-design stage.

In plasma-coating processes, the part design will have considerable influence over the operating parameters of the coating-deposition equipment. Complex shapes, blind holes, fins, slots, and similar features will dictate that a high vacuum pressure, low part temperature, and light plasma density be used (Ref 11). The converse will be true for simple geometries. In plasma processing, consideration also must be given to heating of the part by the plasma itself. Some design features with thin cross sections and low mass, such as fins, louvers, and bosses will heat up faster than the bulk material in the part. For parts that have been heat treated, or otherwise finished to provide desirable mechanical properties, overheating could destroy those properties or at least change the values detrimentally. Reference 11 provides some examples of process and equipment modifications to avoid such problems during plasma nitriding. Reference 12 discusses the effect of part geometry on the growth of the nitride layer during ion nitriding and how coating uniformity can be improved for grooved surfaces. As in electroplating, decreasing the aspect ratio (depth of groove to width of groove) has a positive effect.

Conclusions

From the above discussion, it may seem that design for surface finishing is an art rather than a science. Recently there has been a trend to modeling some of the coating processes in order to predict coating composition or distribution, and these models can be useful in avoiding bad design features. However, most of the principles in use and described in the article are derived from good common sense and have resulted from the cumulative experience gained by the finishing industry over many years.

Some finishing processes may impose their own special design limitations, but for many the general precepts outlined above should provide some guidance. Table 5 groups together the surface-treatment techniques discussed and a few others—not included here because of their similarity to these processes—that have similar design limitations associated with them. It cannot be stressed enough that product design is not a stand-alone activity, but must be integrated with other manufacturing operations to provide the requisite surface finish quality.

REFERENCES

1. H.A. Kuhn, Concurrent Technologies Corp., Johnstown, PA, personal communication, 1996
2. M.F. Browning et al., "Applying Inorganic Coatings: A Vital Technology for Industry," Report 38, *Battelle Technical Inputs to Planning,* Battelle, Columbus, OH, 1983
3. E.W. Brooman, "Plating, Finishing and Coating: State-of-the-Art Assessment," Report EM-4569, Electric Power Research Institute, 1969
4. C.A. Harper, Ed., *Handbook of Plastics and Elastomers,* McGraw-Hill, 1975, Chap. 12
5. M. Henthorne, *Corrosion Causes and Control,* Chemical Engineering Series, McGraw-Hill, 1972, Part 7
6. D.L. Stauffer, Ed., *Finishing Systems Design and Implementation,* Society of Manufacturing Engineers, 1993, Chap. 1
7. S. Guskov, Faraday Cage, Finish Quality, and Recoating: New Technology for More Effective Powder Coating, *Powder Coat.,* 1996, p 82–91
8. W.H. Safranek and E.W. Brooman, *Finishing and Electroplating Die Cast and Wrought Zinc,* Zinc Institute, 1973, Chap. 7
9. B. Fawer, Thin-Film Powder Coatings: Design and Application Issues, *Powder Coat.,* Vol 7 (No. 7), 1996, p 56–63
10. L.W. Flott, Quality Control: Becoming a Better Customer, *Met. Finish.,* Vol 94 (No. 2), 1996, p 79–82
11. R. Gunn, Industrial Advances for Plasma Nitriding, *Ion Nitriding and Ion Carburizing,* T. Spalvins and W.L. Kovacs, Ed., ASM International, 1990, p 157–163
12. M.J. Park et al., Effect of Geometry on Growth of Nitride Layer in Ion Nitriding, *Ion Nitriding and Ion Carburizing,* T. Spalvins and W.L. Kovacs, Ed., ASM International, 1990, p 203–209

Glossary of Terms

A

accuracy. In measurement, the lack of deviation of a reading from a known input. A voltmeter that reads 10 V when subjected to 10 V is accurate. Accuracy is related to and governed by statistical *bias*.

activity-based costing (ABC). A cost-accounting approach that assumes that products incur costs by the activities they require for design, manufacture, and marketing. To implement ABC one must identify the major activities and their cost drivers, for example, hours of engineering services, number of production setups, etc. This is in contrast to conventional cost accounting where overhead costs are allocated solely through hours of direct labor or machine time.

activity diagram. A diagram that lays out the tasks of a project in sequential order and shows when each must take place. Examples are Gantt charts and milestones charts.

adhesive. A substance capable of holding materials together by surface attachment. Adhesive is a general term and includes, among others, cement, glue, mucilage, and paste. These terms are loosely used interchangeably.

advanced ceramics. Ceramic materials that exhibit superior mechanical properties, corrosion/oxidation resistance, or electrical, optical, and/or magnetic properties. This term includes many monolithic ceramics as well as particulate-, whisker-, and fiber-reinforced glass, glass-ceramics, and ceramic-matrix composites. Also known as engineering, fine, or technical ceramics.

affinity diagram. A diagram that organizes large number of ideas into their natural relationships. Often used as a follow up to group creativity activities such as *brainstorming*.

allotropic transformation. The ability of a material to transform from one crystal structure to another. Closely synonymous with polymorphism.

amorphous material. A material that lacks the long-range atomic periodicity that is characteristic of a *crystalline* solid.

anion. A negatively charged ion that migrates toward the anode (positive electrode) under the influence of a potential gradient.

anisotropy. The characteristic of exhibiting different values of a property in different directions with respect to a fixed reference system in the material.

assembly. A collection of two or more parts. The assembly process is a series of joining processes (either permanent or nonpermanent) in which parts are oriented and added to the build. It is a mass-increasing process in which the macrogeometry is established by the positioning of the components.

assembly design. The process of creating groups of parts that operate together.

assembly drawing. An engineering drawing that shows how the individual elements of a product are to be put together, i.e., it shows the structure of the finished product. An assembly drawing usually includes a parts list to aid in the identification of the individual components.

austempering. A heat treatment for ferrous alloys in which a part is quenched from the austenitizing temperature at a rate fast enough to avoid formation of ferrite or pearlite and then held at a temperature just above M_s until transformation to bainite is complete. Although designated as bainite in both austempered steel and austempered ductile iron, austempered steel consists of two-phase mixtures containing ferrite and carbide, while austempered ductile iron consists of two-phase mixtures containing ferrite and austenite.

austenite. A high-temperature form of iron. In steel heat treating, the steel is heated into the austenite region before rapidly cooling (quenching).

austenite stabilizer. An alloying element that when added to iron increases the region of the phase diagram in which austenite (face-centered cubic iron) is stable. Strong austenite stabilizers are carbon, nickel, and manganese.

availability. In reliability theory, a measure of the degree to which an item is in operable and commitable state at the start of a mission, when the mission is called for at an unknown (random) time.

B

bake-hardening steels. Very-low carbon, fully ferritic steels (0.001% C) that are very formable, but harden during the paint-curing cycle. The strengthening results from the precipitation of titanium/niobium carbonitrides at 175 °C (350 °F).

benchmarking. An improvement process in which a product design, process, or service is measured against best-of-class operations in other companies.

bending moment. The moments (force × distance) that tend to bend a beam in the plane of the loads.

benefit-cost analysis. A form of economic analysis in which the economic benefits and costs of a project are determined with due allowance for the time value of money.

bias. In statistics, the difference in a set of measurements between the mean value of a sample and the parameter from the large population that is used.

bill of materials. A document used to track the parts in an assembly. It consists of the part number, part name, and quantity needed for the complete assembly. It may also contain information on procurement tracking or special assembly instructions. Also called the parts list.

blow molding. A method of fabricating plastics in which a warm plastic parison (hollow tube) is placed between the two halves of a mold (cavity) and forced to assume the shape of that mold cavity by use of air pressure. The air pressure is introduced through the inside of the parison and forces the plastic against the surface of the mold, which defines the shape of the product.

blueprint. The common shop term for a detailed engineering drawing giving dimensions and manufacturing details for a part. The term comes from an early printing process that gave a drawing with white lines on a blue background.

Boolean operation. In solid modeling, three operators for combining sets of objects. The operators are union (+), difference (–), and intersection (*).

brainstorming. A technique used by teams to generate ideas on a subject. Each person on the team is asked to think creatively and suggest as many ideas as possible. No criticism or discussion of the ideas is allowed until after the brainstorming session.

brazing. A group of welding processes that join solid materials together by heating them to a suitable temperature and using a filler metal having a liquidus above 450 °C (840 °F) and below the solidus of the base materials. The filler metal is distributed between the closely fitted surfaces of the joint by capillary attraction.

break-even analysis. An economic analysis in which costs are determined as a function of units of output or volume of production. The break-even point (minimum batch size) occurs at the number of units for which the revenues equal the total costs.

break-in. The early period of operating an engineering system, during which early equipment failures may be detected. With electronic systems this is usually known as burn-in. The break-in process is useful for weeding out design and manufacturing defects.

brittle material. A material that exhibits brittleness, i.e., the tendency to fracture without first undergoing significant plastic deformation. Brittle materials exhibit low ductility.

buckling. A form of mechanical instability in which failure occurs by unstable lateral deflection due to compressive forces applied to the structural element.

bulk forming. Forming processes, such as extrusion, forging, rolling, and drawing, in which the input material is in billet, rod, or slab form and a considerable increase in surface-to-volume ratio in the formed part occurs under the action of largely compressive loading. Compare with *sheet forming.*

burden. See *overhead costs.*

burn in. See *break-in.*

C

CAD/CAM. An abbreviation for computer-aided design/computer-aided manufacturing.

Cartesian coordinate system. A coordinate system in which the position of a point in a plane is determined by its distance and direction from the x and y axis, which are perpendicular to each other. In three dimensions a Cartesian coordinate system is defined by the x, y, and z axes.

casting. (1) Metal object cast to the required shape by pouring of injecting liquid metal into a mold, as distinct from one shaped by a mechanical process. (2) Pouring molten metal into a mold to produce an object of desired shape. (3) Ceramic forming process in which a body slip is introduced into a porous mold, which absorbs sufficient water from the slip to produce a semirigid circle. See also *polymer casting.*

casting modulus. A simplified approach to determining solidification time. The time is proportional to the square of the section modulus (the ratio of volume to surface area), known as Chvorinov's rule.

cation. A positively charged ion that migrates through the electrolyte toward a cathode (negative electrode) under the influence of a potential gradient.

censored distribution. A frequency distribution of observations that have either a lower or upper cutoff. For example, fatigue tests that are stopped unbroken after 10^6 cycles represent a censored frequency distribution.

central tendency of data. In statistics this is described by the central limit theorem. This theorem states the fact that if one draws samples of size n from the population, and calculates the mean of these samples, the means will form a distribution which tends toward normality regardless of the form of the original sample distribution.

ceramic(s). Any of a class of inorganic nonmetallic products that are subjected to a high temperature during manufacture or use (high temperature usually means a temperature above a barely visible red, approximately 540 °C, or 1000 °F). Typically, but not exclusively, a ceramic is a metallic oxide, boride, carbide, or nitride, or a mixture of compound of such materials; that is, it includes anions that play important roles in atomic structures and properties. See also *advanced ceramics.*

Charpy test. An impact test in which a V-notched, keyhole-notched, or U-notched specimen, supported at both ends, is struck behind the notch by a strike mounted at the lower end of a bar that can swing as a pendulum. The energy that is absorbed in fracture is calculated from the height to which the striker would have risen had there been no specimen and the height to which it actually rises after fracture of the specimen. Contrast with *Izod test.*

checklist. A tool used to ensure that all important steps or actions in an operation have been taken. Checklists contain those items that are important to the situation.

Chvorinov's rule. See *casting modulus.*

coefficient of thermal expansion. (1) Change in unit of length (or volume) accompanying a unit change of temperature, at a specified temperature. (2) The linear or volume expansion of a given material per degree rise of temperature, expressed at an arbitrary base temperature or as a more complicated equation applicable to a wide range of temperature.

coherent precipitate. A crystalline precipitate that forms from solid solution with an orientation that maintains continuity between the crystal lattice of the precipitate and the lattice of the matrix, usually accompanied by some strain in both lattices. Because the lattices fit at the interface between precipitate and matrix, there is no discernible phase boundary.

colligative properties. Properties of plastics based on the number of molecules present. Most important are certain solution properties extensively used in molecular weight characterization.

column. A structural element that is long compared to its cross section. A column is often loaded axially in compression.

comminution. The act of reducing a material to a powder, or reducing its particle size, by mechanical means (usually by grinding or milling).

commodity plastics. Polymers such as polyethylene, polypropylene, and polyvinyl chloride that are used in high volume.

component. A generic term for special-purpose parts, standard parts, and standard assemblies such as electric motors, transmissions, switch boxes, pulleys, etc.

composite material. A combination of two or more materials (reinforcing elements, fillers, and composite matrix binder), differing in form or composition on a macroscale. The constituents retain their identities, i.e., they do not dissolve or merge completely into one another although they act in concert. Normally, the components can be physically identified and exhibit an interface between one another.

composite structure. A structural member (such as a panel, plate, pipe, or other shape) that is built up by bonding together two or more distinct components, each of which may be made of a metal, alloy, nonmetal, or *composite material.* Examples of composite structures include: honeycomb panels, clad plate, electrical contacts, sleeve bearings, carbide-tipped drills or lathe tools, and weldments constructed of two or more different alloys.

compression molding. A technique of thermoset molding in which the plastic molding compound (generally preheated) is placed in the heated open mold cavity, the mold is closed under pressure (usually in a hydraulic press), causing the material to flow and completely fill the cavity, and then pressure is held until the material has cured.

computational fluid dynamics (CFD). An area of computer-aided engineering devoted to the numerical solution and visualization of fluid-flow problems.

computer-aided design (CAD). Any design activity that involves the effective use of the computer to create or modify an engineering design. Often used synonymously with the more general term computer-aided engineering (CAE).

computer-aided materials selection system. A computerized database of materials properties operated on by an appropriate knowledge base of decision rules through an *expert system* to select the most appropriate materials for an application. See also *knowledge base.*

computer-aided process planning (CAPP) program. A computer program that extracts the relevant geometric features from a CAD drawing and produces a process sequence or set of tool paths (for machining) that is optimal with respect to some criterion such as minimum processing time.

computer modeling. The use of computers to simulate a physical system. Computers perform the numerical analysis and often graphically display the results.

concept selection method. A group process of selecting a design concept or material in which the selection criteria and concepts are arranged in matrix form, and for each criterion the concepts are compared one at a time to a datum (reference) concept. Often called the Pugh concept selection method, after its originator, Stuart Pugh.

conceptual design. The initial stage of the engineering design process in which a physical concept of the product is developed. The physical concept includes the principles by which the product will work and an abstract physical embodiment that will employ the principles to achieve the desired functions.

concurrent engineering. A style of product design and development that is done by concurrently utilizing all of the relevant information in making each decision. It replaces a sequential approach to product development in which one type of information was predominant in making each sequential decision. Concurrent engineering is carried out by a multifunctional team that integrates the specialties or functions needed to solve the problem. Sometimes called simultaneous engineering.

configuration design. The stage after conceptual design in the engineering design process in which the features of a part and their arrangement and connectivity are determined. Qualitative reasoning based on fundamental principles is used to make decisions between alternatives. A sketch of the part and preliminary decisions on material selection and manufacturing methods are made in this stage. Final dimensions and tolerances are not determined in this stage. Also known as embodiment design.

constitutive equation. A mathematical relationship that describes the flow stress of a material in terms of the plastic strain, the strain rate, and temperature.

constraint modeling. A form of computer modeling in which constraints are used to create a set of rules that control how changes are made to a group of geometric elements (lines, arcs, form features, etc.). These rules are typically embodied in a set of equations. Constraint models are defined as either parametric or variational.

continuum mechanics. The science of mathematically describing the behavior of continuous media. The same basic approach can apply to descriptions of stress, heat, mass, and momentum transfer.

controlled rolling. A hot-rolling process in which the temperature of the steel is closely controlled, particularly during the final rolling passes, to produce a fine-grain microstructure.

convergent thinking. See *vertical thinking.*

conversion coating. A coating consisting of a compound of the surface metal, produced by chemical or electrochemical treatments of the metal. Examples include chromate coatings on zinc, cadmium, magnesium, and aluminum, and oxide and phosphate coatings on steel.

copolymer. A long-chain molecule formed by the reaction of two or more dissimilar monomers. See also *polymer.*

corrosion. The chemical or electrochemical reaction between a material, usually a metal, and its environment that produces a deterioration of the material and its properties.

cost allocation. In design, the process by which the components of a design are assigned a target cost.

covalent bond. A bond in which two atoms share a pair of electrons. Contrast with *ionic bond.*

crack opening displacement (COD). On a K_{Ic} specimen, the opening displacement of the notched surfaces at the notch and in the direction perpendicular to the plane of the notch and the crack. The displacement at the tip is called the crack tip opening displacement (CTOD); at the mouth, it is called the crack mouth opening displacement (CMOD). See also *stress-intensity factor* for definition of K_{Ic}.

creative process. A total problem resolution process that yields a truly novel solution to the problem. A creative solution requires both *vertical thinking* (convergent thinking) and *lateral thinking* (divergent thinking).

creep. Time-dependent strain occurring under stress. The creep strain occurring at a diminishing rate is called primary creep; that occurring at a minimum and almost constant rate, secondary creep; and that occurring at an accelerating rate, tertiary creep.

creep rupture. Failure at elevated temperature under load (stress) conditions greater than found in creep tests so that rupture occurs, generally at shorter times and larger deformations than for a creep test.

criterion function. See *objective function.*

cross-functional team. A team whose members are from several units of the organization that interface with, and need input from, one another.

cross linking. With thermosetting and certain thermoplastic polymers, the setting up of chemical links between the molecular chains. When extensive, as in most thermosetting resins, cross linking makes an infusible supermolecule of all the chains. In rubbers, the cross linking is just enough to join all molecules into a network.

crystalline. That form of a substance comprised predominantly of (one or more) crystals, as opposed to glassy or amphorous.

crystalline defects. The deviations from a perfect three-dimensional atomic packing that are responsible for much of the structure-sensitive properties of materials. Crystal defects can be point defects (vacancies), line defects (dislocations), or surface defects (grain boundaries).

cumulative distribution function (CDF). A frequency distribution arranged to give the number of observations that are less than given values. 100% of the observations will be less than the largest class interval of the observations.

customer. The recipient or beneficiary of the output of an organization's work effort or the purchaser of a product or service. A customer may be either external or internal to the organization. A chief objective of total quality management is to exceed customer expectations, each and every time.

customer delight. The result of delivering a product or service that exceeds customer expectations.

customer importance rating. An indication of the priority of importance the customer places on a certain want. Determined by dividing the number of times customers mention this need by the number of subjects interviewed or surveyed.

cycle time. The time to produce a part. Cycle time is the reciprocal of production rate.

D

damage. Impairment of the useful life of a part.

damage tolerance. The ability of a component to resist failure due to the presence of flaws, cracks, or other structural damage for a specified period of usage.

data associativity. The ability of a computer-aided engineering system to share design information among a variety of computer-based applications (such as design, drafting, and numerical-controlled machining) without each application having to translate or transfer the data. Associativity also requires that applications can recognize when geometry or other information has been changed so that the application can adjust its own data to these changes.

database. A collection of specific information, such as alloy compositions or mechanical properties. Many databases are accessible through on-line connections or with a CD-ROM.

data integration. See *data associativity.*

decision matrix. A formalized process by which design concepts or materials are ranked prior to a selection decision.

decomposition. A general term used in design to describe the process of breaking the product into subassemblies, or smaller units; for example, the automobile can be decomposed into the engine, drive train, chassis, etc. Decomposition can be done in terms of physical embodiments, as above, or in terms of functions that the design must perform.

defect. The nonfulfillment of a product or service with respect to an intended requirement or reasonable expectation of use, including safety considerations. There are four classes of defects: (1) very serious, leading directly to severe injury or catastrophic economic loss; (2) serious, leading directly to significant injury or significant economic loss; (3) major, related to major problems with respect to intended normal or reasonable foreseeable use; (4) minor, related to minor problems with respect to intended normal or reasonably foreseeable use.

deflection temperature under load (DTUL). See *heat distortion temperature (HDT).*

deformation curve. See *stress-strain curve.*

deformation processing. A class of manufacturing operation that involves changing the shape of a workpiece by plastic deformation through the application of a compressive force. Often carried out at elevated temperature.

delamination. The separation of layers in a laminate or composite material because of failure of the matrix, either in the matrix itself or at the interface between the matrix and the fiber.

design changes. Changes in a design that come about because of (1) mistakes on the drawings, (2) design- or manufacturing-related problems, or (3) product improvements.

design code. A set of rules or regulations that tell the designer what to do and when and under what circumstances. While codes are often developed by technical or professional groups, they often are adopted by governmental jurisdictions and become legal requirements. Examples are the building code for the structural safety of buildings and the fire code.

design constraint. A design parameter (variable) that is required to fall within a fixed range of values.

design defect. Any aspect of a design or product that causes it to fail to perform as safely or reliably as an ordinary consumer would expect. See also *defect*.

design features. The elements that comprise a part, such as walls, holes, and grooves, and the solid elements, such as rods, cubes, and tubes.

design for assembly (DFA). A methodology for examining the assembled parts of a product in terms of the time and cost for assembly of each part. The overriding strategy is to produce an acceptable design with the minimum number of parts.

design for disassembly (DFD). A methodology for designing a product so that it can be economically disassembled and the material from which the individual parts are made can be recovered and reused.

design for the environment (DFE). A design methodology in which environmental factors are considered of equal importance to performance factors. Sometimes called green design.

design for function and fit. An expression that exemplifies the two central rules of design: (1) first and foremost, the design must perform the function for which it is intended; (2) the design must be designed to tolerances appropriate to the application.

design for manufacture, design for manufacturability (DFM). An approach toward integrating the product and process concepts so that the best match is made between product and process requirements. DFM includes detailed examination of the design so that it is most easily and inexpensively manufactured by the intended process.

design for "X" (DFX). A catch phrase that describes the understanding that engineering designs increasingly must be consciously designed for factors other than function and tolerances. These issues include design for assembly (DFA), design for manufacture (DFM), and design for installation, maintenance, safety, disposal, etc.

design intent. A general term used with respect to describing and documenting the aspects of some part of the design that are crucial to the success of the design.

design layout. An engineering drawing, either two-dimensional or three-dimensional, that shows the individual parts of a system in relationship to one another. An *assembly drawing* often serves the same purpose.

design optimization. See *optimization*.

design of experiments (DOE). A methodology involving statistically designed experiments in which the character and sequence of individual experiments are planned in advance so that data are taken in a way that will provide the most unbiased and precise results commensurate with the available time and money. The chief classes of statistically designed experiments are: (1) blocking designs to remove the effect of background variables from the experimental error, (2) factorial designs, in which all levels of each factor in an experiment are combined with all levels of every other factor, and (3) response surface designs, which determine the empirical relation between the factors (independent variables) and the response (performance variable).

design parameters. A term used to describe the chief variables of a design problem. Sometimes called the design attributes or design characteristics. Design parameters define the object or process that is being designed and thus provide a means for changing the design to improve performance.

design performance measures. Performance measures that are functions of the design parameters and are used to quantify the effectiveness of a given design. Performance measures often are grouped into an *objective function* and the problem constraints.

design review. A group examination and discussion intended to ensure that the design, and all of its components, will meet customer requirements.

design rules. A set of guidelines developed from experience that tell the designer how to make decisions in typical situations. Often these rules suggest how to design to minimize cost of manufacture. For example: "Try to design the part so it can be machined on one machine tool only."

design sensitivity. See *sensitivity analysis*.

design standards. Agreed-upon descriptions of best solutions to repetitive design problems, arrived at by general consent of an appropriate broad-based technical group. While standards do not have the force of law, they often are incorporated into a *design code*.

design team. A multi-functional group that contains not only the technical expertise to design the product but also expertise in manufacturing and marketing.

detail design. The final phase of design in which all information on materials, manufacturing processes, dimensions, and tolerances are supplied. The output of this design phase is a complete set of drawings and product specifications. Also known as parametric design.

detail drawing. An engineering drawing that provides sufficient information to make the part. This information usually includes plan, front, and side views; auxiliary views and sections through the part; dimensions; and tolerances.

die swell. The expansion of the polymer melt that occurs as the extruded melt exits the die. This is due to the aligned polymer chains being released from the confines of the die and returning to their random coil configuration.

dimensional management. An engineering design methodology that aims to absorb as much variation as possible without affecting the function of the product. It does this through the optimal selection of datums, feature controls, assembly methods, and assembly sequence. It differs from the conventional practice of assigning tolerances to drawings prior to release in that it is more systematic and more global in outlook and employs three-dimensional tolerance analysis CAE tools.

directionally solidified (DS) eutectic composite. A metal-matrix composite material produced by directional solidification of the eutectic composition of an alloy.

directional solidification. Controlled solidification of molten metal in a casting so as to provide feed metal to the solidifying front of the casting. Usually this results in the metal solidifying in a preferred direction. In the limit, the solidification can be controlled to grow as a single grain (single crystal casting).

dislocation. A linear imperfection in a crystalline array of atoms. Two basic types are recognized: (1) an edge dislocation corresponds to the row of mismatched atoms along the edge formed by an extra, partial plane of atoms within the body of a crystal; (2) a screw dislocation corresponds to the axis of a spiral structure in a crystal, characterized by a distortion that joins normally parallel planes together to form a continuous helical ramp (with a pitch of one interplanar distance) winding about the dislocation. Most prevalent is the so-called mixed dislocation, which is any combination of an edge dislocation and a screw dislocation.

dislocation density. The total length of dislocation lines per unit volume, or the number of dislocation lines that cut through a unit cross-sectional area.

dispersion strengthening. The strengthening of a metal or alloy by incorporating chemically stable submicron size particles of a nonmetallic phase that impede dislocation movement at elevated temperature.

divergent thinking. See *lateral thinking*.

Dominic method. A qualitative method for evaluating and choosing between alternative design concepts or materials. The method uses an evaluation table in which the evaluation criteria are grouped according to priority (high, moderate, low), and the alternative solutions are ranked as excellent, good, fair, poor, or unacceptable.

draft. An angle or taper on the surface of a pattern, core box, punch, or die (or of the parts made with them) that facilitates removal of the parts from a mold or die cavity, or a core from a casting.

drape. The ability of a polymer sheet or composite sheet to conform to the shape of the mold.

drawing. A term used for a variety of forming operations, such as deep drawing a sheet metal blank; redrawing a tubular part; and drawing rod, wire, and tube. The usual drawing process with regard to sheet metal working in a press is a method for producing a cuplike form from a sheet metal disk by holding it firmly between blankholding surfaces to prevent the formation of wrinkles while the punch travel produces the required shape.

ductility. The ability of a material to deform plastically without fracturing.

ductile-to-brittle transition temperature. See *transition temperature.*

dynamic recovery. A process occurring in hot working of metals in which a fine subgrain structure forms within the elongated grains due to annihilation of dislocations due to easy cross slip and climb. It results in a lowering of the flow stress. Dynamic recovery, as opposed to dynamic recrystallization that occurs in hot working, occurs in metals of high stacking fault energy such as aluminum, α iron, and most bcc metals.

dynamic strain aging. A behavior in metals in which solute atoms are sufficiently mobile to move toward and interact with dislocations. This results in strengthening over a specific range of elevated temperature and strain rate.

E

economy of scale. Variation of cost with production volume. Usually, unit cost decreases with increasing volume.

economy of scope. The concept that value is created more by the ability to create products with very many functions or variations from a common theme than by reducing the unit cost from increased volume of production.

elasticity. The property of a material by virtue of which deformation caused by stress disappears upon removal of the stress. A perfectly elastic body completely recovers it original shape and dimensions after release of stress.

elastomer. Any elastic, rubberlike substance, such as natural or synthetic rubber.

electrical permittivity. The material property that directly relates the voltage and geometry of the system to the electrical field energy. Magnetic permeability is the magnetic field energy counterpart.

electrical resistivity. The electrical resistance offered by a material to the flow of current, times the cross-sectional area of current flow and per unit length of current path; the reciprocal of the conductivity. Also called resistivity or specific resistance.

electronegativity. The relative tendency of an atom to attract electrons. Carbon, the basic building block of polymers, is neutral in this respect with a Pauling electronegativity of 2.5. Metallic atoms on the left side of the periodic table have a propensity to lose electrons when forming covalent bonds, and their electronegativities are less than 2.5. The atoms on the right side of the periodic table tend to gain electrons, and their electronegativities tend to be greater than 2.5.

electrophoretic paint (e-coat). A paint that is applied by electrophoresis, in which colloidal charged particles are transported within an electric field.

elongation. (1) A term used in mechanical testing to describe the amount of extension of a test piece when stressed. (2) In tensile testing, the increase in the gage length, measured after fracture of the specimen within the gage length, usually expressed as a percentage of the original gage length.

embodiment design. See *configuration design.*

end-grain attack. Preferential corrosive attack of grains exposed by cutting through the cross section of a rod or plate.

endurance limit. See *fatigue limit.*

engineering design specification. A detailed description of the intended and unintended uses that a product will be put to, a list of any special features required or desired, and a detailed list of the functional requirements with qualitative or quantitative goals and limits for each. Also known as product design specification.

engineering plastics. A general term covering all plastics, with or without fillers or reinforcements, that have mechanical, chemical, and thermal properties suitable for use as construction materials, machine components, and chemical processing equipment components. Included are acrylonitrile-butadiene-styrene, acetal, acrylic, fluorocarbon, nylon, phenoxy, polybutylene, polyaryl ether, polycarbonate, polyether (chlorinated), polyether sulfone, polyphenylene oxide, polysulfone, polyimide, rigid polyvinyl chloride, polyphenylene sulfide, thermoplastic urethane elastomers, and many other reinforced plastics.

environmentally assisted cracking. Brittle fracture of a normally ductile material in which the corrosive effect of the environment is a causative factor. Environmental cracking is a general term that includes corrosion fatigue, high-temperature hydrogen attack, hydrogen blistering, hydrogen embrittlement, liquid metal embrittlement, solid metal embrittlement, stress-corrosion cracking, and sulfide stress cracking.

environmental impact assessment. A system-wide determination of the multiple impacts that an engineering product has on the environment.

erosion. (1) Loss of material from a solid surface due to relative motion in contact with a fluid that contains solid particles. Erosion in which the relative motion of particles is nearly parallel to the solid surface is called abrasive erosion. Erosion in which the relative motion of the solid particles is nearly normal to the solid surface is called impingement erosion or impact erosion. (2) Progressive loss of original material from a solid surface due to mechanical interaction between that surface and a fluid, a multicomponent fluid, and impinging liquid, or solid particles. (3) Loss of material from the surface of an electrical contact due to an electrical discharge (arcing).

evaluation of property data. The process of establishing the accuracy and integrity of property data by appraisal of the data presented, assessment of the experimental technique and its associated errors, and checking for consistency of values and comparison with other experimental or theoretical values.

event tree analysis (ETA). A binary logic tree for tracing all the possible consequences of an event. Unlike the fault tree, the event tree starts with an initiating event and traces all possible consequences. See also *fault tree analysis.*

expert system. A computer-based system that captures the knowledge of experts through the integration of databases and knowledge bases using search and logic deduction algorithms.

extrusion (ceramics). The process of forcing a mixture of plastic binder and ceramic powder through the opening(s) of a die at relatively high pressure. The material may thus be compacted and emerges in elongated cylindrical or ribbon (or wire, etc.) form having the cross section of the die opening. Ordinarily followed by drying, curing, activating, or firing.

extrusion (metals). The conversion of an ingot or billet into long lengths by forcing metal to flow through a die orifice. The cross section of the extrusion may be circular or highly irregular and specific in shape.

extrusion (plastics). Compacting a plastic material into a powder or granules in a uniform melt and forcing it through an orifice in a more or less continuous fashion to yield a desired shape. While held in the desired shape, the melt must be cooled to a solid state.

F

factor of safety. The ratio of the material strength to the computed stress. The factor of safety can vary from about 1.3, where component performance is well known and material properties show little variability, to 3 or 4 for an untried material or where stresses or environment are uncertain.

fail-safe design. A design that seeks to ensure that a failure will not affect the product or change it to a state in which injury or damage will occur.

failure mode and effect analysis (FMEA). A systematic, detailed method of analysis of the malfunctions or defects that can be produced in the components of an engineering system. The analysis examines each potential failure mode to assess the reliability of the system elements and to ascertain the consequences of the failure upon the entire system.

failure mode, effect, and criticality analysis (FMECA). An extension of FMEA in which the criticality of each assembly is examined and the components and assemblies to which special attention should be given are identified.

fatigue. The phenomenon leading to fracture under repeated or fluctuating stresses having a maximum value less than the ultimate tensile strength of the material. Fatigue failure generally occurs at loads that applied statically would produce little perceptible effect. Fatigue fractures are progressive, beginning as minute cracks that grow under the action of the fluctuating stress.

fatigue limit. The maximum cyclic stress that a material can withstand for an infinitely large number of stress cycles. Also called endurance limit.

fatigue-strength reduction factor. The ratio of the fatigue strength of a member or specimen with no stress concentration to the fatigue strength with stress concentration. This factor has no meaning unless the stress range and the shape, size, and material of the member or specimen are stated.

fault hazard analysis (FHA). A qualitative analysis of design hazards other than malfunctions. FMEA, FMECA, and FTA consider only malfunctions.

fault tree analysis (FTA). An analysis method that provides a systematic description of the combinations of possible occurrences in a system that can result in failure. It is a graphical representation of the Boolean logic that relates to the output (top) event.

features. In computer modeling, higher level constructs (than points, curves, and surfaces) that allow designers to work more naturally with geometries such as slots, through holes, bosses, etc. Form features such as these allow designers to add relatively complex yet common shapes without having to apply Boolean operations on blocks and other shapes.

ferrite stabilizer. An alloying element that, when added to iron, increases the region of the phase diagram in which ferrite is the stable phase. The strongest ferrite stabilizers are silicon, chromium, and molybdenum.

filament winding. A process for fabricating a reinforced plastic or composite structure in which continuous reinforcements (filament, wire, yarn, tape, and the like), either previously impregnated with a matrix material or impregnated during the winding, are placed over a rotating and removable form or mandrel in a prescribed way to meet certain stress conditions. Generally the shape is a surface of revolution and may or may not include end closures. When the required number of layers is applied, the wound form is cured and the mandrel is removed.

finite element analysis. See *finite element modeling (FEM).*

finite element modeling (FEM). A numerical technique in which the analysis of a complex part is represented by a mesh of elements interconnected at node points. The coordinates of the nodes are combined with the elastic properties of the material to produce a stiffness matrix, and this matrix is combined with the applied loads to determine the deflections at the nodes, and hence, the stresses. All of the above is done with special FEM software. The FEM approach also may be used to solve other field problems in heat transfer, fluid flow, acoustics, etc. Also known as finite element analysis (FEA).

firing. The controlled heat treatment of ceramic ware in a kiln or furnace, during the process of manufacture, to develop the desired properties.

fixed costs. Those costs that do not increase with the volume of production, e.g., the cost of capital equipment, overhead labor costs, etc. See also *variable costs.*

fixture. A device for holding a part to be joined, assembled, or machined. Often, two or more parts are held by the fixture in proper relation to each other.

flanging. Bending a sheet along a curved line.

flexural modulus. The ratio, within the elastic limit, of the applied stress on a reinforced plastic test specimen in flexure to the corresponding strain in the outermost fibers of the specimen.

flexural strength. A property of solid material that indicates its ability to withstand a flexural or transverse load.

flow stress. The stress required to produce plastic deformation in a solid material.

forging. The process of working metal to a desired shape by impact or pressure in hammers, forging machines (upsetters), presses, rolls, and related forming equipment. Forging hammers, counterblow equipment, and high-energy-rate forging machines apply impact to the workpiece, while most other types of forging equipment apply squeeze pressure in shaping the stock. Some metals can be forged at room temperature, but most are made more plastic for forging by heating.

formability. The ease with which a material can be shaped through plastic deformation. Closely related to the term workability. In general practice, formability refers to the shaping of sheet materials while workability refers to shaping by bulk forming processes such as forging or extrusion.

forming. (1) Making a change, with the exception of shearing or blanking, in the shape or contour of a metal part without intentionally altering its thickness. (2) The plastic deformation of a billet or a blanked sheet between tools (dies) to obtain the final configuration. Metal forming processes are typically classified as *bulk forming* and *sheet forming.* Also referred to as metalworking. (3) A process in which the shape of plastic pieces such as sheets, rods, or tubes is changed to a desired configuration. The use of the term forming in plastics technology does not include such operations as molding, casting, or extrusion, in which shapes or pieces are made from molding materials or liquids.

forming limit diagram (FLD). A diagram in which the major strains at the onset of necking in sheet metal are plotted vertically and the corresponding minor strains are plotted horizontally. The onset-of-failure line divides all possible strain combinations into two zones: the safe zone (in which failure during forming is not expected) and the failure zone (in which failure during forming is expected).

fracture. (1) The irregular surface produced when a piece of metal is broken. (2) In composites, the separation of a body. Defined both as rupture of the surface without complete separation of the laminate and as complete separation of a body because of external or internal forces. Fractures in continuous-fiber-reinforced composites can be divided into three basic fracture types: intralaminar, interlaminar, and translaminar. Translaminar fractures are those oriented transverse to the laminated plane in which conditions of fiber fracture are generated. Interlaminar fracture, on the other hand, describes failures oriented between plies, whereas intralaminar fractures are those located internally within a ply.

fracture toughness. A generic term for measures of resistance to extension of a crack. The term is sometimes restricted to results of fracture mechanics tests, which are directly applicable in fracture control. However, the term commonly includes results from simple tests of notched or precracked specimens not based on fracture mechanics analysis. Results from tests of the latter type are often useful for fracture control, based on either service experience or empirical correlations with fracture mechanics tests. See also *stress-intensity factor.*

function analysis. A technique used in the discipline of value analysis that focuses on the reason-for-being of each element of the product.

functional decomposition. A hierarchical breakdown of a design or product into the basic functions that must be achieved. This is done by asking the question how? For example, the basic function of a copier is to make copies. Asking how leads to the next level of functions: feed paper, make marks, handle documents, etc. Each function is stated by a verb and a noun.

functional requirements. Those elements of the design that describe its performance behavior, including its human interface and the environment in which it must function.

fuzzy logic. The use of fuzzy sets in the representation and manipulation of vague information for the purpose of making decisions or taking actions. Fuzzy logic enables computers to make decisions based on information that is not clearly defined.

G

galvanic corrosion. Corrosion associated with the current of a galvanic cell consisting of two dissimilar conductors in an electrolyte or two similar conductors in dissimilar electrolytes. Where the two dissimilar metals are in contact, the resulting reaction is referred to as couple action.

gate. In a metal casting, the portion of the runner in a mold through which molten metal enters the cavity. In injection molding and transfer molding of plastics, the gate is the orifice through which the viscous melt enters the mold cavity.

geometric dimensioning and tolerancing (GD&T). A comprehensive system of standards for the location of features on an engineering drawing and describing their tolerances with respect to form, profile, orientation, and runout.

glass. An amorphous solid made by fusing silica (SiO_2) with a basic oxide. Its characteristic room-temperature properties are transparency to light, hardness and rigidity, and resistance to weathering and most chemicals.

glass ceramics. A family of fine-grained crystalline materials made by a process of controlled crystallization from special glass compositions containing nucleating agents.

glass transition temperature (T_g). The temperature at which an amorphous polymer (or the amorphous regions in a partially crystalline polymer) changes from a hard and relatively brittle condition to a viscous or rubbery condition. In this temperature region, many physical properties, such as hardness, brittleness, thermal expansion, and specific heat, undergo significant, rapid changes.

glaze. A continuous adherent layer of glass applied to the surface of a ceramic body as a suspension of ingredients and then fired to yield a hard, nonabsorbent surface layer.

go/no-go decision. A type of decision where an absolute lower (or upper) limit has been established as necessary for the functioning of the design. All materials or components that do not exceed the lower limit specification are rejected (no-go), while those that do are accepted and subjected to further screening. See *screening property*.

grain boundary. A narrow zone in a metal or ceramic corresponding to the transition from one crystallographic orientation to another, thus separating one grain from another; the atoms in each grain are arranged in an orderly pattern.

granulation. The intentional agglomeration of fine ceramic particles into larger clusters in order to improve such properties as bulk density and powder flowability and to cause a reduction of dusting losses.

granulated metal. Small pellets produced by pouring liquid metal through a screen or by dropping it onto a revolving disk and, in both instances, chilling with water.

green design. See *design for the environment*.

green strength. The strength of a powdered material in the compacted state before sintering or firing. Green strength must be sufficient to allow the part to be removed from the compaction die and handled before sintering.

group technology. An approach to design and manufacturing that seeks to reduce manufacturing system information content by identifying and exploiting the sameness or similarity of parts based on their geometrical shape and/or similarities in their production process.

guided iteration. A general problem-solving methodology applicable to engineering design. The steps in this process are: formulation of the problem, generation of alternative solutions, and evaluation of the alternatives. If none of the alternatives is acceptable, the process continues with redesign, guided by the results of the evaluation.

Guinier-Preston (G-P) zone. A small precipitation domain in a supersaturated metallic solid solution. A G-P zone has no well-defined crystalline structure of its own and contains an abnormally high concentration of solute atoms. The formation of G-P zones constitutes the first stage of precipitation and is usually accompanied by a change in properties of the solid solution in which they occur.

H

Hall-Petch relationship. A general relationship for metals that shows that the yield strength is linearly related to the reciprocal of the square root of the grain diameter.

hard constraint. A design requirement that cannot be relaxed, i.e., a must requirement.

hardfacing alloys. Wear-resistant materials available as bare welding rod, flux-coated rod, long-length solid wires, long-length tubular wires, or powders that are deposited by hardfacing. Hardfacing materials include a wide variety of alloys, ceramics, and combinations of these materials. Conventional hardfacing alloys are normally classified as steels or low-alloy ferrous materials, chromium white irons, high-alloy ferrous materials, carbides, nickel-base alloys, or cobalt-base alloys.

hardness. A measure of the resistance of a material to surface indentation or abrasion; may be thought of as a function of the stress required to produce some specified type of surface deformation. There is no absolute scale for hardness; therefore, to express hardness quantitatively, each type of test has its own scale of arbitrarily defined hardness. Indentation hardness can be measured by Brinell, Rockwell, Vickers, Knoop, and Scleroscope hardness tests.

hazard. Any aspect of technology or human activity that produces risk, i.e., the potential for harm or damage to people, property, or the environment.

heat-affected zone (HAZ). That portion of the base metal that was not melted during brazing, cutting, or welding, but whose microstructure and mechanical properties were altered by the heat.

heat distortion temperature (HDT). The temperature at which a bend specimen made from a plastic material deflects a specified amount under a fixed static load. This temperature is sometimes called the deflection temperature under load (DTUL).

high-cycle fatigue. Fatigue that occurs at relatively large numbers of cycles. The arbitrary, but commonly accepted, dividing line between high-cycle fatigue and *low-cycle fatigue* is considered to be about 10^4 to 10^5 cycles. In practice, this distinction is made by determining whether the dominant component of the *strain* imposed during cyclic loading is elastic (high cycle) or plastic (low cycle), which in turn depends on the properties of the metal and on the magnitude of the nominal stress.

Hooke's law. A generalization applicable to all solid material, which states that stress is directly proportional to strain and is expressed as:

$$\frac{\text{Stress}}{\text{Strain}} = \frac{\sigma}{\varepsilon} = \text{constant} = E$$

where E is the modulus of elasticity (Young's modulus). The constant relationship between stress and strain applies only below the proportional limit. See also *modulus of elasticity.*

hot isostatic pressing (HIP). (1) A process for simultaneously heating and forming a compact in which the powder is contained in a sealed flexible sheet metal or glass enclosure and the so-contained powder is subjected to equal pressure from all directions at a temperature high enough to permit plastic deformation and sintering to take place. (2) A process that subjects a component (casting, powder forgings, etc.) to both elevated temperature and isostatic gas pressure in an autoclave. The most widely used pressurizing gas is argon. When castings are hot isostatically pressed, the simultaneous application of heat and pressure virtually eliminates internal voids and microporosity through a combination of plastic deformation, creep, and diffusion.

house of quality. See *quality function deployment*.

hygrothermal effect. Change in properties and shape change of a material (particularly plastics and polymer matrix composites) due to moisture absorption and temperature change.

hypereutectic alloy. In an alloy system exhibiting a eutectic, any alloy whose composition has an excess of alloying element compared with the eutectic composition and whose equilibrium microstructure contains some eutectic structure.

hysteresis loop. In dynamic mechanical measurement, the closed curve representing successive stress-strain status of a material during a deformation cycle.

I

impact strength. A measure of the resiliency or toughness of a solid. The maximum force or energy of a blow (given by a fixed procedure) that can be withstood without fracture, as opposed to fracture strength under a steady applied force.

inclusion. (1) A physical and mechanical discontinuity occurring within a material or part, usually consisting of solid, encapsulated foreign material. Inclusions are often capable of transmitting some structural stresses and energy fields, but to a noticeably different degree than from the parent material.

indirect costs. See *overhead costs.*

industrial design. The first broadly functional description of a product with emphasis on visual and aesthetic features (shape, color, texture).

Initial Graphics Exchange Standard (IGES). A protocol for transferring graphics data between CAD systems.

injection molding (ceramics). A process for forming ceramic articles in which a granular ceramic-binder mix is heated until softened and then forced into a mold cavity, where it cools and resolidifies to produce a part of the desired shape.

injection molding (metals). A process similar to plastic injection molding using a plastic-coated metal powder of fine particle size (~10 μm).

injection molding (plastics). Method of forming a plastic to the desired shape by forcing the heat-softened plastic into a relatively cool cavity under pressure.

inoculant. Materials that, when added to molten metal, modify the structure and thus change the physical and mechanical properties to a degree not explained on the basis of the change in composition resulting from their use. Ferrosilicon-base alloys are commonly used to inoculate gray irons and ductile irons.

integrated product development team (IPDT). A flexible, collaborative, multidisciplinary team assigned with the responsibility of developing a product and the process to manufacture it. In addition to design and manufacturing engineers with the appropriate technical backgrounds, the team should include members with financial, marketing, field service, and purchasing experience.

intellectual property. Knowledge-based property, usually represented by patents, copyrights, trademarks, or trade secrets.

interlock. A device provided to ensure that an event does not occur inadvertently or that a specific sequence of events is followed where the sequence is important or necessary and a wrong sequence of events could cause a mishap.

interstitial-free steels. Steels where carbon and nitrogen are removed in the steelmaking process to very low levels, and any remaining interstitial carbon and nitrogen is tied up with small amounts of alloying elements that form carbides and nitrides, i.e., titanium and niobium. Although these steels have low strength they exhibit exceptional formability.

ionic bond. (1) A type of chemical bonding in which one or more electrons are transferred completely from one atom to another, thus converting the neutral atoms into electrically charged ions. These ions are approximately spherical and attract each other because of their opposite charges. (2) A primary bond arising from the electrostatic attraction between two oppositely charged ions. Contrast with *covalent bond.*

ISO 9000. A set of five individual but related international standards on quality management and *quality assurance*. The standards were developed by the International Organization for Standardization (ISO).

ISO 14000. A generic international standard for environmental management systems.

isothermal transformation (IT) diagram. A diagram that shows the isothermal time required for transformation of austenite to begin and to finish as a function of temperature.

isotropy. The condition in which the properties are independent of the direction in which they are measured.

Izod test. A type of impact test in which a V-notched specimen, mounted vertically, is subjected to a sudden blow delivered by the weight at the end of a pendulum arm. The energy required to break off the free end is a measure of the impact strength or toughness of the material. Contrast with *Charpy test.*

J

jig. A device for both holding a part and guiding the tool during a machining or assembly operation.

jiggering. A plastic forming process used with clay-based ceramics in which a plastic mass is pressed under relatively low pressure into a plaster die, and the final shape is achieved by rotating the die and shaping the plastic mass with a fixed tool. Jiggering is a mechanical version of the potter's wheel and produces shapes with circular cross sections such as dinner plates and coffee mugs.

***J*-integral.** A mathematical expression involving a line or surface integral that encloses the crack front from one crack surface to the other, used to characterize the *fracture toughness* of a material having appreciable plasticity before fracture. The *J*-integral eliminates the need to describe the behavior of the material near the crack tip by considering the local stress-strain field around the crack front; J_{Ic} is the critical value of the *J*-integral required to initiate growth of a preexisting crack.

K

kinematic analysis. An analysis to determine the position, velocity, and acceleration of all bodies in a mechanical system at a sequence of user-specified time steps. This type of analysis is often used to check the range of motion and to determine whether the design will allow motion/velocity/acceleration to be in the expected range.

knowledge base. A collection of formulas, rules, and lessons learned that, when connected to a computerized database and manipulated with artificial intelligence methods, allows decisions to be made that approximate those made by humans, but much faster.

L

lamina. A single ply or layer in a laminate, which is made up of a series of layers.

laminate. A sheet comprised of two or more layers of material (lamina) bonded together. A reinforced laminate usually refers to superimposed layers of resin-impregnated or resin-coated fabrics or fibrous reinforcements that have been bonded, especially by heat and/or pressure.

lateral thinking. A thought process in which the mind moves in many different directions, combining different bits of information into new patterns until several different solution concepts evolve. Also called divergent thinking.

learning curve. A curve that shows the way that production time or manufacturing cost decreases with cumulative production because of greater worker experience.

least commitment policy of design. The concept that it is best to allow as much freedom as possible for downstream decisions in the design process so that engineers are free to develop the best possible solutions unconstrained by previous unnecessary commitments (decisions).

life-cycle assessment (LCA). A "cradle-to-grave" assessment of the energy requirements and environmental impacts of a given product design.

life-cycle cost. A cost analysis that considers not only the cost of producing a product, but the cost of using, servicing, and disposing or recycling the product.

linear elastic fracture mechanics. A method of fracture analysis that can determine the stress (or load) required to induce fracture instability in a structure containing a crack-like flaw of known size and shape. See also *stress-intensity factor.*

linear programming. A type of optimization algorithm that is very effective for problems where the objective and constraint functions are linear.

lockin. A procedure in maintenance that maintains an event or prohibits an individual, an object, force, or other factor from leaving a safe zone.

lockout. A procedure in maintenance that prevents an individual, an object, force, or other factor from entering a dangerous zone.

low-cycle fatigue. *Fatigue* that occurs at relatively small numbers of cycles ($<10^4$ cycles). Low-cycle fatigue may be accompanied by some plastic, or permanent, deformation. Compare with *high-cycle fatigue.*

lumped-parameter model. A mathematical model in which the distributed properties of physical quantities are replaced with their lumped equivalents. When a problem can be analyzed in terms of a finite number of discrete elements, it can be expressed by ordinary differential equations. To describe the more realistic case of distributed parameters having many values spread over a field in space requires the use of partial differential equations.

M

machinability. The relative ease with which material is removed from a solid by controlled chip-forming in a machining process.

magnetically hard alloy. A ferromagnetic alloy capable of being magnetized permanently because of its ability to retain induced magnetization and magnetic poles after removal of externally applied fields; an alloy with high coercive force. The name is based on the fact that the quality of the early permanent magnets was related to their hardness.

magnetically soft alloy. A ferromagnetic alloy that becomes magnetized readily upon application of a field and that returns to practically a nonmagnetic condition when the field is removed; an alloy with the properties of high magnetic permeability, low coercive force, and low magnetic hysteresis loss.

maintainability. In reliability theory, the measure of an item to be retained in, or restored to, a specified condition when maintenance is performed by personnel having specified skill levels and using prescribed procedures and resources, at each prescribed level of maintenance and repair.

malleability. The characteristic of metals that permits *plastic deformation* in compression without fracture. See also *ductility.*

manufacturing. The conversion of materials into finished parts, products, and goods that have value to end-users. Manufacturing involves the design, development, implementation, control, operation, and maintenance of processes that facilitate and perform the conversion of starting materials into finished products having greater value.

manufacturing flexibility. The ease with which a process can be adapted to produce different products or variations of the same product. Process flexibility is influenced greatly by the time to set up or change tooling.

manufacturing lead time. The time required to process the product through the manufacturing plant. Related to process cycle time.

martensite. A generic term for microstructures formed by diffusionless phase transformation in which the parent and product phases have a specific crystallographic relationship. Martensite is characterized by an acicular pattern in the microstructure in both ferrous and nonferrous alloys. In alloys where the solute atoms occupy interstitial positions in the martensitic lattice (such as carbon in iron), the structure is hard and highly strained; but where the solute atoms occupy substitutional positions (such as nickel in iron), the martensite is soft and ductile. The amount of high-temperature phase that transforms to martensite on cooling depends to a large extent on the lower temperature attained, there being a rather distinct beginning temperature (M_s) and a temperature at which the transformation is essentially complete (M_f).

mass-conserving process. A manufacturing process in which the mass of the starting material is approximately equal to the mass of the final product or part. Examples are casting, precision forming, and powder processes.

mass-reducing process. A manufacturing process in which the mass of the starting material is greater than the mass of the final product or part, and forming takes place by the removal of material, e.g., machining.

master product model. An approach toward design in which the product design is centered on an information-rich computer model that supports not only the physical design, but analysis, testing, design, optimization, simulation, prototyping, manufacturing, and maintenance.

materials performance index. A combination of materials properties formulated in such a way that the largest value of the index designates the best material for the application. See also *performance index* .

material substitution. A class of material selection problem where a design already exists and the task is to replace the existing material with a more suitable one.

material utilization. The percentage of the material processed that ends up in the product. A high material utilization means high product yield and near-net shape processing.

mechanism dynamics. The procedure for determining the detailed motions, velocities, and accelerations of the parts in a complex mechanical system. Computer-aided methods, based on the engineering sciences of dynamics and kinematics, are used.

melt index. The amount, in grams, of a thermoplastic resin that can be forced through a 2.0955 mm (0.0825 in.) orifice when subjected to 20.7 N (2160 gf) in 10 min at 190 °C (375 °F).

melt viscosity. The bulk property of a fluid or semifluid (melted) polymer that causes it to resist flow. It is given by the shear stress acting on the fluid divided by the rate of shear. Melt viscosity in polymers is usually measured by the *melt index.*

metadata. Descriptive data about the material for which data are reported. Metadata include a complete description of the material (producer, heat number, grade, temper, etc.), a complete description of the test method, and information about the test plan.

metal. An opaque lustrous elemental chemical substance that is a good conductor of heat and electricity and, when polished, a good reflector of light. Most elemental metals are malleable and ductile and are, in general, denser than other elemental substances. As to structure, metals may be distinguished from nonmetals by their atomic binding and electron availability. Metallic atoms tend to lose electrons from the outer shells, the positive ions thus formed being held together by the electron gas produced by the separation. The ability of these "free electrons" to carry an electric current, and the fact that this ability decreases as temperature increases, establish the prime distinctions of a metallic solid.

metal injection molding (MIM). A powder process in which a metal powder is mixed with up to 40% polymer binder and shaped by injection molding. The part is removed from the mold, the binder is removed using either solvent extraction or thermal processes, and the part is sintered to decrease porosity.

metal working. See *forming.*

microalloying. A steelmaking technology in which small amounts of alloying elements (vanadium, niobium, or titanium) are used to retard recrystallization of austenite, thereby allowing a wider range of rolling temperatures for controlled rolling. Without retarding recrystallization, as in normal hot rolling, the pancake-type grains do not form and a fine grain size cannot be developed in these ferrite-pearlite steels.

microstructure. The structure of an object, organism, or material as revealed by a microscope at magnifications greater than 25.

minimum batch size. See *break-even point.*

mistake minimization. A process utilized in design to minimize mistakes on system elements for which satisfactory design experience is available. The process involves rigorous application of standards and computerized design methods, reuse of successful designs where possible, and root-cause analysis to eliminate problems from prototypes.

model. See *process model.*

modeling. See *computer modeling.*

modulus of elasticity (*E*). (1) The measure of rigidity or stiffness of a material; the ratio of stress, below the proportional limit, to the corresponding strain. If a tensile stress of 13.8 MPa (2.0 ksi) results in an elongation of 1.0%, the modulus of elasticity is 13.8 MPa (2.0 ksi) divided by 0.01, or 1380 MPa (200 ksi). (2) In terms of the stress-strain curve, the modulus of elasticity is the slope of the stress-strain curve in the range of linear proportionally of stress to strain. Also known as Young's modulus. For materials that do not conform to *Hooke's law* throughout the elastic range, the slope of either the tangent to the stress-strain curve at the origin or at low stress, the secant drawn from the origin to any specified point on the stress-strain curve, or the chord connecting any two specific points on the stress-strain curve is usually taken to be the modulus of elasticity. In these cases, the modulus is referred to as the tangent modulus, secant modulus, or chord modulus, respectively.

modulus of rigidity. See *shear modulus.*

modulus of rupture. Nominal stress at fracture in a bend test or torsion test. In bending, modulus of rupture is the bending moment at fracture divided by the section modulus. In torsion, modulus of rupture is the torque at fracture divided by the polar section modulus.

moment of inertia. A mathematical quantity used to calculate the stress in bending of beams. The moment of inertia of a plane surface is the sum of the products obtained by multiplying the area of each element of the surface by the square of its distance from the axis.

monotonic loading. A testing procedure in which the load on the specimen progressively increases or decreases, but does not oscillate in value.

Monte Carlo simulation. A method of generating information for a simulation when events occur in a random way. The Monte Carlo method uses unrestricted random sampling in a computer simulation in which the results are run off repeatedly to develop statistically reliable answers.

morphological matrix. In a formalized process for developing design concepts, a list of all functions needed by the design, including all of the conceivable ways of accomplishing each function.

morphology. The characteristic shape, form, or surface texture or contours of the crystals, grains, or particles of (or in) a material, generally on a microscopic scale.

***m*-value.** See *strain-rate sensitivity (m-value).*

N

natural strain. See *true strain.*

near-net-shape processes. Shape replication processes that produce parts requiring little or no subsequent processing to obtain the finished part.

necking. The process by which a material in tension reduces locally its cross-sectional area. Important in creating imperfections and voids in metal deformation processes. A high *strain-rate sensitivity* in the material means that an incipient neck becomes stronger and spreads the deformation to neighboring material, thereby resisting neck formation.

nesting. (1) A condition found when surface finishing or machine assembling small parts where they cluster together, making them difficult to treat as individual parts. (2) Orientation of sheet metal parts so as to minimize waste when cut from a large sheet.

net surface. A surface in a manufactured part that requires no finishing in a subsequent operation.

Newtonian fluid. A fluid exhibiting Newtonian viscosity wherein the shear stress is proportional to the rate of shear.

noise factors. Uncontrollable factors in the design environment that cause quality characteristics to deviate from target values. Noise factors cause variability and quality loss. *Robust design* corrects for noise factors.

normal distribution. The *probability density function* used to describe the mechanical properties of materials and the distribution of most random variables encountered in engineering design.

numerical control programming. The process of developing a set of instructions that controls a machine tool or other manufacturing device to produce a part.

O

objective function. In design optimization, the grouping of design parameters that is attempted to be maximized or minimized, subject to the problem constraints. Also known as criterion function.

OEM vendors. Suppliers of components to original equipment manufacturers (OEM), i.e., those who manufacture and sell the completely assembled product. For example, a typical OEM vendor might provide small electrical motors to the producer of home appliances.

operating hazard analysis (OHA). A hazard analysis that focuses on operating the product, including testing, maintaining, repairing, transporting, and handling. Emphasis is on people performing tasks. Note that FMEA, FTA, and FHA deal primarily with problems of hardware.

optimization. The process of searching for the best combination of design parameters. Design optimization suggests that for a given set of possible designs and design criteria there exists a single design that is best or optimal.

orange peel. (1) In metals, a surface roughening in the form of a pebble-grained pattern that occurs when a metal of unusually coarse grain size is stressed beyond its elastic limits. (2) In injection molding of plastics, a part with an undesirable uneven surface somewhat resembling the skin of an orange. (3) In ceramics, a pitted texture of a fired glaze resembling the surface of a rough orange peel.

orthogonal array. A balanced fractional factorial type of design of experiments. Used as a key part of *robust design* to determine the effect each factor has on the mean and the variability in an experiment where many factors are involved.

overhead costs. Any cost not directly associated with the production of identifiable goods or services. Sometimes called burden or indirect costs.

oxidation reaction. In electrochemistry, a chemical reaction that occurs at an anodic site by the liberation of electrons.

oxidation resistance. Ability of a material to resist degradation by reacting with air or another gas to form a surface oxide.

oxide-dispersion-strengthened (ODS) alloys. A class of materials in which fine oxide particles are incorporated in metal powders, compacted, and then fabricated into finished forms by deformation processing. The resulting material has improved thermal softening resistance with excellent thermal and electrical conductivity. Examples are ODS copper alloys and sintered aluminum powder (SAP).

P

parameter design. The second stage of the *robust design* process, as defined by Genichi Taguchi, in which a robust design is achieved by optimizing performance of the system through experimentation that minimizes variation in the presence of uncontrolled user and environmental factors (noise).

parametric design. The third stage of the engineering design process, usually called *detail design.* In this stage of design, detailed analysis determines the final shape, dimensions, and properties of parts. Optimization methods, especially those employed in *robust design* are used to enhance quality. Final decisions are made on materials and manufacturing processes, and the part cost is determined.

parison. The hollow plastic tube from which a plastic component is blow molded.

part. Designed object that has no assembly operations in its manufacture. Parts may be either standard or special purpose. A standard part, like a screw or spring, has a generic function and is manufactured in quantity without the intention of being used in a particular product. A special-purpose part is designed and manufactured for a specific purpose in a specific product.

particle strengthening. A common form of strengthening in metals, in which small second-phase particles are distributed in a ductile matrix. Examples are precipitation-hardening (age-hardening) systems and dispersion-hardened alloys.

parts list. See *bill of materials.*

patent. A grant made by a government to an inventor that permits him or her the sole right to make, sell, and use the invention for a period of time.

PDES/STEP. A standard data protocol for the transfer of digital data between different software and machines. PDES is more comprehensive than the *Initial Graphics Exchange Standard (IGES)* in that it includes product attributes such as material type and form features in addition to solid model topology.

performance index. Groupings of material properties that when maximized optimizes some aspect of performance; for example, $\sigma_f / E\alpha$ leads to improved thermal shock resistance. See also *materials performance index.*

performance standards. Standards that state design requirements but not the methods that are to be used to achieve the objective.

permeability. A general term used to express various relationships between magnetic induction and magnetizing force. These relationships are either "absolute permeability," which is a change in magnetic induction divided by the corresponding change in magnetizing force, or "specific (relative) permeability," the ratio of the absolute permeability to the permeability of free space.

pH. The negative logarithm of the hydrogen-ion activity; it denotes the degree of acidity or basicity of a solution. At 25 °C (77 °F), 7.0 is the neutral value. Decreasing values below 7.0 indicates increasing acidity; increasing values above 7.0, increasing basicity. The pH values range from 0 to 14.

phase. A physically homogeneous and distinct portion of a material system.

phase diagram. A graphical representation of the temperature and composition limits of phase fields in an alloy or ceramic system as they actually exist under the specific conditions of heating or cooling. A phase diagram may be an equilibrium diagram, or a representation of metastable conditions or phases. Synonymous with constitution diagram.

plastic. A material that contains as an essential ingredient an organic polymer of large molecular weight; is solid in its finished state; and, at some stage in its manufacture or its processing into finished articles, can be shaped by flow. Although materials such as rubber, textiles, adhesives, and paint may in some cases meet this definition, they are not considered plastics. The terms plastic, resin, and polymer are somewhat synonymous, but the terms resins and polymers most often denote the basic material as polymerized, while the term plastic encompasses compounds containing plasticizers, stabilizers, fillers, and other additives.

plastic deformation. The permanent (inelastic) distortion of materials under applied stresses that strain the material beyond its elastic limit.

plasticity. The property of a material that allows it to be repeatedly deformed without rupture when acted upon by a force sufficient to cause deformation and that allows it to retain its shape after the applied force has been removed.

plastic strain ratio (*r*-value). In formability testing of metals, the ratio of the true width strain to the true thickness strain in a sheet tensile test, $r = \varepsilon_w/\varepsilon_t$. A formability parameter that relates to drawing, it is also known as the anisotropy factor. A high *r*-value indicates a material with good drawing properties.

Poisson's ratio (ν). The absolute value of the ratio of transverse (lateral) strain to the corresponding axial strain resulting from uniformly distributed axial stress below the proportional limit of the material.

polar moment of inertia. The moment of inertia where the distance is measured from an axis perpendicular to the plane of the area.

polymer. A high-molecular-weight organic compound, natural or synthetic, with a structure that can be represented by a repeated small unit, the mer. Examples include polyethylene, rubber, and cellulose. Synthetic polymers are formed by addition or condensation polymerization of monomers. Some polymers are elastomers, some are plastics, and some are fibers. When two or more dissimilar monomers are involved, the product is called a copolymer. The chain lengths of commercial thermoplastics vary from ~1000 to >100,000 repeating units. Thermosetting polymers approach infinity after curing, their resin precursors, often called prepolymers, may be relatively short—6 to 100 repeating units—before curing. The lengths of polymer chains, usually measured by molecular weight, have very significant effects on the performance properties of plastics and profound effects on processibility.

polymer blend. A mixture of different types of polymer chains. In miscible blends the polymers mix on a molecular level to produce a single phase, as in the blend of polyphenylene oxide and polystyrene. In immiscible blends the polymers cannot mix on a molecular level and separate into two phases, as in blends of ABS and polystyrene.

polymer casting. To form a "plastic" object by pouring a fluid monomer-polymer solution into an open mold where it finishes polymerizing. Forming plastic film and sheet by pouring the liquid resin onto a moving belt or by precipitation in a chemical bath.

population. A statistical concept describing the total set of objects or observations under consideration.

porosity. Fine holes or pores within a solid; the amount of these pores is expressed as a percentage of the total volume of the solid.

powder forging (P/F). The plastic deformation of a powder metallurgy compact or preform into a fully dense finished shape by using compressive force; usually done hot and within closed dies.

powder metallurgy (P/M). The technology and art of producing metal powders and utilizing metal powders for production of mass materials and shaped objects.

precipitation hardening. Hardening in metals caused by the precipitation of a constituent from a supersaturated solid solution.

precision. In testing, a measure of the variability that can be expected among test results. The precision of an instrument indicates its ability to reproduce a certain reading. Precision is the inverse of standard deviation. A decrease in the scatter of test results is represented by a smaller standard deviation, leading directly to an increase in precision.

preimpregnation (reinforced plastics). The practice of mixing resin and reinforcement and effecting partial cure before use or shipment to the user. See also *prepreg.*

prepreg. In composites fabrication, either ready-to-mold material in sheet form or ready-to-wind material in roving form, which may be cloth, mat, unidirectional fiber, or paper, impregnated with resin and stored for use. The resin is partially cured to a B-stage and supplied to the fabricator, who lays up the finished shape and completes the cure with heat and pressure. The two distinct types of prepreg available are commercial prepregs, in which the roving is coated with a hot melt or solvent system to produce a specific product to meet specific customer requirements; and wet prepreg, in which the basic resin is installed without solvents or preservatives but has limited room-temperature shelf life.

prescriptive standard. A standard that states the requirements in terms of specific details and leaves no discretion to the designer. Sometimes referred to as a design standard.

probability density function (PDF). A mathematical function that, when integrated between two limits, gives the probability that a random variable assumes a value between these limits.

process capability. A statistical measure of the inherent process variability with respect to some characteristic.

process-driven design. A design approach in which the manufacturing process plan or method of assembly is developed prior to developing the product design.

process model. A mathematical description of the physical behavior underlying a manufacturing process that is used to predict performance of the process in terms of operating parameters. Most often process models are reduced to software and are manipulated with computers.

product design specification (PDS). See *engineering design specification.*

product documentation. All of the documentation, whether on paper or in digital form, that defines the steps in designing the product and the specifies the steps required for its manufacture. These include: product design specification, QFD matrices, concept selection results, geometric dimensioning and tolerancing schemes, detail drawings, *bill of materials*, and manufacturing process plans.

product integrity. In the manufacturing sense, it is the absence of voids, pores, cracks, and harmful inclusions.

products liability. A legal term that describes the action whereby an injured party (plaintiff) seeks to recover damages for personal injuries or property loss from a producer and/or seller (defendant) when the plaintiff alleges that a defective product or design caused the injury or loss.

property of unique interest. A material property that is of such importance or uniqueness to the application that this single property predominates all other considerations in materials selection.

prototype. A full-scale working model of the design that is technically and visually complete. See also *rapid prototyping.*

Pugh concept selection method. See *concept selection method.*

Q

quality. The characteristics of a product or service that bear on its ability to satisfy stated or implied needs. A quality product or service is free from deficiencies.

quality assurance. All the planned and systematic activities implemented within a quality system that provide confidence that a product or service will deliver the required quality.

quality control. The operational techniques and activities used to fulfill the requirements for quality. Often quality assurance and quality control are used interchangeably.

quality function deployment (QFD). A product planning tool that uses a matrix to map the customer wants to the quantified specifications of the design (product). Often called the house of quality (because of the house-shaped QFD matrices). The QFD approach that is used at the first stage of product planning can be deployed all the way down the stage of production planning on the shop floor.

quality loss function. An approach to evaluating quality, introduced by Genichi Taguchi, in which quality is measured by the dollar loss due to some quality characteristic deviating from the target value. Mathematically, the loss function is a quadratic function of the deviation from the target value.

quench cracking. Fracture of a metal during quenching from elevated temperature. Most frequently observed in hardened carbon steel, alloy steel, or tool steel parts of high hardness and low toughness. Cracks often emanate from fillets, holes, corners, or other stress raisers and result from high stresses due to the volume changes accompanying transformation to martensite.

R

random variable. A parameter or variable that deviates from some measure of average performance or central tendency in a predictable way.

rapid prototyping. A process by which a design created by computer solid modeling can be created quickly as a physical object or assembly that can be used for visualization and functional testing.

rationalization. See *standardization and rationalization.*

***R*-curve.** In fracture mechanics, a plot of crack-extension resistance as a function of stable crack extension, which is the difference between either the physical crack size or the effective crack size and the original crack size. *R*-curves normally depend on specimen thickness and, for some materials, on temperature and strain rate.

readiness review. A final review of an engineered system from the viewpoint of readiness for safe operation. All critical personnel are brought together and all safety hazards are reexamined to see whether the hazards have been adequately addressed. Operating procedures are assessed to ensure that personnel are properly trained, not only for normal operation, but also for emergency situations.

recycling. The reassimilation of a material into the raw materials stream in a matter such that it may be used for a purpose identical or similar to that of its first use. A material must be recovered or reclaimed from a waste stream before it is available for recycling. Recycling is different from reuse or *remanufacture.*

redesign. The process of redoing an existing design to improve some deficiency in the existing design.

reduction in area (RA). The difference between the original cross-sectional area of a tensile specimen and the smallest area at or after fracture as specified for the material undergoing testing. Also known as reduction of area.

reduction reaction. In electrochemistry, a chemical reaction taking place at cathodic sites by the consumption of electrons.

reentrant angle. An interior angle of a polygon greater than 180°.

refractory. A material (usually an inorganic, nonmetallic, ceramic material) of very high melting point with properties that make it suitable for such uses as furnace linings and kiln construction.

reliability. A measure of the capacity of equipment or a system to operate without failure in the service environment. Reliability is always expressed as a probability of performing for a given time without failure.

remanufacture. A form of green design in which parts and components are rebuilt or refurbished after a portion of their life so that they may be reissued and used as essentially new parts.

repair. Corrective action made on a damaged product so that the product will fulfill the original specifications.

requirements list. See *engineering design specification.*

residual stress. A stress that exists in a solid body without an imposed external force. Residual stresses often result from forming or thermal processing and are caused by such factors as cold working, phase changes, temperature gradients, or rapid cooling.

resin. A solid, semisolid, or pseudosolid organic material that has an indefinite and often high molecular weight, exhibits a tendency to flow when subjected to stress, usually has a softening or melting range, and usually fracture conchoidally. (2) In reinforced plastics, the material used to bind together the reinforcement material; the matrix.

resistivity. See *electrical resistivity.*

response surface modeling. A statistical, mathematical, or graphical model that describes the variation of the response variable in terms of the parameters of the problem.

retained austenite. An amount of the high-temperature face-centered cubic phase of iron (austenite) that does not transform to martensite (is retained) when quenched to room temperature.

reverse engineering. The process of determining design concepts or details by disassembling a similar or competitive product.

rework. An action taken on nonconforming products or services to allow them to meet the original specifications.

rheology. The science of deformation and the flow of matter.

risk. The probability of damage or injury. The potential for harm to people, property, or the environment.

risk/benefit analysis. A decision-making technique that uses a common value scale, usually dollars, to balance the risk against benefit.

robust design. An integrated system of tools and techniques that are aimed at reducing product or process performance variability while simultaneously guiding that performance toward an optimal setting. Robustness optimization is chiefly done for design concepts that are new so that the best values of the critical functional parameters are unknown. Robust design follows the methods first proposed by Genichi Taguchi.

robustness. Performance of a system that is always acceptably close to the ideal function of the system. It refers to how consistently a component or product performs under variable conditions in its environment and as it wears during its lifetime.

rotational casting. A method used to make hollow articles from thermoplastic materials. The material is charged into a hollow mold capable of being rotated in one or two planes. The hot mold fuses the material into a gel after the rotation has caused it to cover all surfaces. The mold is then chilled, and the product is stripped out.

rotational molding. The preferred term for a variation of the *rotational casting* process that uses dry, finely divided (35 mesh, or 500 μ) plastic powders, such as polyethylene, rather than fluid materials. After the powders are heated, they are fused against the mold walls forming a hollow item with uniform wall thickness.

rules of thumb. Empirical relationships developed from practical experience. In design, often used to improve ease of manufacture or to relate performance to costs. Rules of thumb can be very useful when employed by an experienced person, but they may easily be misinterpreted or misused by less experienced personnel.

***r*-value.** See *plastic strain ratio (r-value).*

***R*-value.** In fatigue testing, the stress ratio determined by the minimum stress in the stress cycle divided by the maximum stress. For a completely reversed stress cycle, $R = -1$.

S

safety factor. See *factor of safety.*

sales point. A feature of a product that gives it a competitive advantage. This term is used in the *quality function deployment* methodology.

sample. In statistics, part of a population that is withdrawn for study or observation. The most important type of sample is a random sample, which is defined as a collection of objects from a population selected in such a way that each item in the population has the same likelihood of being a member of the sample. Other types of samples are stratified or biased samples.

sandwich construction. A panel composed of a lightweight core material, such as honeycomb, cellular or foamed plastic, and so forth, to which two relatively thin, dense, high-strength or high-stiffness faces, or skins, are adhered.

scanning electron microscope (SEM). A high-power magnifying and imaging instrument using an accelerated electron beam as an optical device and containing circuitry which causes the beam to traverse or scan an area of sample in the same manner as does an oscilloscope or TV tube. May utilize reflected or transmitted electron optics. The scanning electron microscope provides two outstanding improvements over the optical microscope; it extends the resolution limits so that picture magnifications can be increased from 1000 to 2000× up to 30,000 to 60,000×, and it improves the depth-of-field resolution more dramatically, by a factor of approximately 300, thus facilitating its use in fracture studies.

screening property. Any material property for which an absolute lower (or upper) limit can be established in the material selection process. These are often referred to as go/no-go limits.

secondary manufacturing operation. A manufacturing operation used to impart close tolerances or controlled surface finish to a part. Machining operations are a good example.

sensitivity analysis. An analytical method for determining the importance of each design parameter to the performance of the design (output measure) and the critical ranges of those parameters.

sequential engineering. An older style of product development in which design decisions are made in a rigid linear pattern, e.g., all decisions on functionality are made before any decisions of production method, etc. This is being supplanted by *concurrent engineering*.

shape design parameter. A design parameter that describes the boundary position in a numerical (FEM) model, and thus defines nodal location. Optimizing a shape design parameter may require that the entire finite element model be remeshed.

shape generation process. A mass-reducing process in which the shape of the part is determined by the pattern of relative motion between the cutting tool and the part. Shape information is impressed by the process.

shape memory alloy (SMA). A type of metallic material that demonstrates the ability to return to some previously defined shape or size when subjected to the appropriate thermal treatment. These materials can be plastically deformed at some relatively low temperature and, upon exposure to some higher temperature, will return to their shape prior to the deformation. Commercially promising systems are alloys of Ni-Ti, Cu-Zn-Al, and Cu-Al-Ni.

shape replication process. A *mass-conserving process* in which the part replicates the shape information stored in the die or mold by being forced to assume the shape of the surface of the tool cavity.

shear modulus (*G*). The ratio of shear stress to the corresponding shear strain for shear stresses below the proportional limit of the material. Values of shear modulus are usually determined by torsion testing. Also known as modulus of rigidity.

sheet forming. The plastic deformation of a piece of sheet metal by tensile loads into a three-dimensional shape, often without significant changes in sheet thickness or surface characteristics. Compare with *bulk forming*.

shelf energy. In the Charpy impact test, shelf energy is the upper (plateau) value of energy absorbed at a temperature well above the ductile-to-brittle transition temperature.

shot peening. A method of cold working metals in which compressive stresses are induced in the exposed surface layers of parts by impingement of a stream of shot (small spherical particles), directed at the metal surface at high velocity under controlled conditions. It differs from blast cleaning in primary purpose and in the extent to which it is controlled to yield accurate and reproducible results. Although shot peening cleans the surface being peened, this function is incidental. The major purpose of shot peening is to increase fatigue strength. Shot for peening is made of iron, steel, or glass.

shrinkage. A common manufacturing term that denotes changes in a dimension on cooling from temperature, e.g., in forgings, castings, or molded plastics. Shrinkage in castings can result in internal voids.

signal-to-noise ratio. A transformation of the data of an experimental design that indicates how much variation around the mean is shown by the data. Large values represent more robust behavior.

simulation. The representation of a real system by a set of models, usually mathematical or computer based, and subjecting these models to various inputs or environmental conditions to assess how the real system might behave.

simultaneous engineering. See *concurrent engineering*.

sintering. The bonding of adjacent surfaces of particles in a mass of powder or a compact by heating. Sintering strengthens a powder mass and normally produces densification and, in powdered metals, recrystallization.

size effect. The behavior in which the dimensions of the test specimen affect the value of the mechanical property measured. Most prominent for fatigue properties and strength of brittle materials, where strength is lower for large section size.

sizing design parameter. A design parameter that does not alter the location of nodes in a numerical (FEM) model of the design. Examples are material properties and boundary conditions.

slip casting. The ceramic forming process consisting of filling or coating a porous mold with a slip, allowing to dry, and removing for subsequent firing.

slip system. The combination of slip plane with the close-packed slip directions that lie in the plane.

social costs. The costs of a product that are borne by society at large. An example is the cost of emissions resulting from extraction, processing, using, and disposing of a material or product.

solid modeling. A form of computer modeling in which the three-dimensional features of the part or object are represented. With solid modeling, a cut through the model reveals interior details. The method also permits accurate calculation of mass properties (e.g., mass and moment of inertia), and, with full color, shading, and shadowing, it creates realistic displays. Solid models may be integrated with motion analysis software to create realistic simulations. Solid models may also be linked with finite element models.

soldering. A group of processes that join metals by heating them to a suitable temperature below the solidus of the base metals and applying a filler metal having a liquidus not exceeding 450 °C (840 °F). Molten filler metal is distributed between the closely fitted surfaces of the joint by capillary action.

solvation. The process of swelling, gelling, or dissolving a resin by a solvent or plasticizer.

spallation. The cracking or rupturing of a solid body, which usually results in the detachment of a portion of the solid.

special-purpose part. Parts that are designed and manufactured especially for use in a particular product.

specification. A detailed statement describing the materials, dimensions, manufacturing process. See *engineering design specification*.

specific modulus. The material elastic modulus divided by the material density.

specific strength. The material strength divided by the material density.

spinoidal hardening. Strengthening caused by the formation of a periodic array of coherent face-centered cubic solid-solution phases on a submicrostructural size level.

spray drying. A powder-producing process in which a slurry of liquids and solids or a solution is atomized into droplets in a chamber through which heated gases, usually air, are passed. The liquids are evaporated from the droplets, and the solids are collected continuously from the chamber. The resulting powder consists of free-flowing, spherical agglomerates.

springback. (1) The elastic recovery of metal after stressing. (2) The extent to which metal tends to return to its original shape or contour after undergoing a forming operation. This is compensated for by overbending or by a secondary operation of restriking.

standard deviation. A measure of the dispersion of observed values or results from the average expressed as the positive square root of the variance.

standardization. (1) The process of establishing, by common agreement, engineering criteria, terms, principles, practices, materials, items, processes, and equipment parts and components. (2) The adoption of generally accepted uniform procedures, dimensions, materials, or parts that directly affect the design of a product or a facility.

standardization and rationalization. A combined design and business approach that specifically targets reduction in the number of parts company-wide. Standardization refers to the reduction in number of parts used in current and former designs. Rationalization is the identification of the fewest number of parts required for use in future designs.

standard module. A product manufactured in quantity for use in a number of other products. Examples are motors, clutches, switches, pumps, etc. Standard modules are usually offered for sale through the catalogs of their manufacturers.

standard part. See *part.*

standard time. Time required to perform a step in a manufacturing or assembly operation of a product. These times, usually in seconds, are obtained from empirical observations.

stochastic search methods. A large group of optimization techniques that utilizes probabilistic methods. Two common methods are genetic algorithms and simulated annealing.

stoichiometric. Having the precise weight relation of the elements in a chemical compound; or (quantities of reacting elements or compounds) being in the same weight relation as the theoretical combining weight of the elements involved.

strain. The unit of change in the size or shape of a body due to force. Also known as normal strain. The term is also used in a broader sense to denote a dimensionless number that characterizes the change in dimensions of an object during a deformation or flow process.

strain hardening. An increase in hardness and strength of metals caused by plastic deformation at temperatures below the recrystallization range. Also known as work hardening.

strain-rate sensitivity (*m*-value). The increase in stress (σ) needed to cause a certain increase in plastic strain rate ($\dot{\varepsilon}$) at a given level of plastic strain (ε) and a given temperature (T):

$$\text{Strain-rate sensitivity} = m = \left(\frac{\Delta \log \sigma}{\Delta \log \dot{\varepsilon}}\right)_{\varepsilon T}$$

stress. The intensity of the internally distributed forces or components of forces that resist a change in the volume or shape of a material that is or has been subjected to external forces. Stress is expressed in force per unit area. Stress can be normal (tension or compression) or shear.

stress concentration. On a macromechanical level, the magnification of the level of an applied stress in the region of a notch, void, hole, or inclusion.

stress concentration factor (K_t). A multiplying factor for applied stress that allows for the presence of a structural discontinuity such as a notch or hole, K_t equals the ratio of the greatest stress in the region of the discontinuity to the nominal stress for the entire section. Also called theoretical stress concentration factor.

stress-intensity factor. A scaling factor, usually denoted by the symbol K, used in linear-elastic fracture mechanics to describe the intensification of applied stress at the tip of a crack of known size and shape. At the onset of rapid crack propagation in any structure containing a crack, the factor is called the critical stress-intensity factor, or the fracture toughness. Various subscripts are used to denote different loading conditions or fracture toughnesses:

K_c	Plane-stress fracture toughness. The value of stress intensity at which crack propagation becomes rapid in sections thinner than those in which plane-strain conditions prevail.
K_I	Stress-intensity factor for a loading condition that displaces the crack faces in a direction normal to the crack plane (also known as the opening mode of deformation).
K_{Ic}	Plane-strain fracture toughness. The minimum value of K_c for any given material and condition, which is attained when rapid crack propagation in the opening mode is governed by plane-strain conditions.
K_{Id}	Dynamic fracture toughness. The fracture toughness determined under dynamic loading conditions; it is used as an approximation of K_{Ic} for very tough materials.
K_{ISCC}	Threshold stress intensity factor for stress-corrosion cracking. The critical plane-strain stress intensity at the onset of stress-corrosion cracking under specified conditions.
K_Q	Provisional value for plane-strain fracture toughness.
ΔK	Threshold stress intensity for stress-corrosion cracking. The critical stress intensity at the onset of stress-corrosion cracking under specified conditions.

stress relaxation. The time-dependent decrease in stress in a solid under constant constraint at constant temperature.

stress relief. The removal or reduction of residual stress by thermal treatment, mechanical treatment (shot peening, surface rolling, stretching, bending, and straightening), or vibratory stress relief.

stress-strain curve. A graph in which corresponding values of stress and strain from a tension, compression, or torsion test are plotted against each other. Values of stress are usually plotted vertically (ordinates or y-axis) and values of strain horizontally (abscissas or x-axis). Also known as deformation curve and stress-strain diagram.

stretch forming. The forming of a sheet material by the penetration of a sheet into a blank clamped firmly at its circumference. Deformation occurs chiefly by reducing thickness and increasing surface area.

structure-insensitive properties. Material properties that are determined more by the composition of the material than by its structure. Examples are density, elastic modulus, and thermal conductivity.

structure-sensitive properties. Material properties that are strongly dependent on microstructural features and defect structure. Examples are yield strength and fracture toughness.

subassembly. An assembly that is included within another assembly or subassembly. See also *assembly.*

subcritical crack growth (SCG). A failure process in which a crack initiates at a preexisting flaw and grows until it attains a critical length. At that point the crack grows in an unstable fashion leading to catastrophic failure. Typical examples of SCG processes are fatigue failure and stress corrosion.

suboptimization. The optimization of a part of a large, complex engineering system with many parameters. Suboptimization of all subparts of a large system does not in general lead to optimization of the whole system. Suboptimization is successful in improving quality only when the subsystems are loosely coupled.

superplasticity. The ability of certain metals (most notably aluminum- and titanium-base alloys) to develop extremely high tensile elongations under controlled rates of deformation at elevated temperatures.

surface engineering. Treatment of the surface and near-surface regions of a material to allow the surface to perform functions that are distinct from those functions demanded from the bulk material. Examples are case hardening or galvanizing of a steel surface.

surface finish. (1) The geometric irregularities in the surface of a solid material. Measurement of surface finish shall not include inherent structural irregularities unless these are the characteristics being measured. (2) Condition of a surface as a result of a final treatment.

surface finishing. The process of applying a surface treatment to clean the surface, to improve its appearance, or to give it a special property such as corrosion or wear resistance.

surface hardening. A localized heat treating process that produces a hard-quenched surface in steel without introducing additional alloying elements. Surface hardening can be produced by flame, induction, or laser or electron beam thermal treatments.

surface integrity. A technology that involves the specification and manufacture of unimpaired or enhanced surfaces through the control of the many possible alterations produced in a surface layer during manufacture. Surface integrity is achieved by the proper selection and control of manufacturing processes and the ability to estimate their effects on the significant engineering properties of work materials.

surface topography. The fine-scale features of a surface as defined by the size and distribution of asperities. Surface topography is measured by surface roughness and the direction of surface features (lay).

synectics. A creative problem-solving process in which the problem statement or product design specification is transformed through the use of analogies.

system architecture. The total plan for the design or product that lays out the type of subsystems and ensures that they are properly related. A complex system would contain the following levels: total system, module, subsystem, subassembly, piece part, and piece-part feature.

system design. The primary product-development stage in which the basic architecture of the product or process is determined. The first stage of *robust design,* as defined by Genichi Taguchi.

system verification test. A final verification of robustness of the system that is usually performed on the first total-system prototypes, which are made after the detail design has been completed.

T

tack. The ability of a material (usually a polymer) to stick to itself and to other materials. A tacky material will stay in place when positioned in a mold or on top of another layer of material.

Taguchi methods. See *robust design.*

taxonomy. The principles and process of classification.

temper. In nonferrous alloys and in some ferrous alloys (steels that cannot be hardened by heat treatment), the hardness and strength produced by mechanical or thermal treatment, or both, and characterized by a certain structure, mechanical properties, or reduction in area during cold working.

tempering. (1) In heat treatment, reheating hardened steel to some temperature below the eutectoid temperature to decrease hardness and/or increase toughness. (2) The process of rapidly cooling glass from near its softening point to induce compressive stresses on the surface balanced by interior tension, thereby imparting increased strength.

tensile strength. In tensile testing, the ratio of maximum load to original cross-sectional area. Also called ultimate strength. Compare with *yield strength.*

texture. In a polycrystalline aggregate, the state of distribution of crystal orientations. In the usual sense, it is synonymous with preferred orientation, in which the distribution is not random.

thermal barrier coating (TBC). A thick (125–250 μm) insulating layer, designed to reduce the temperature at the metal interface. A plasma sprayed layer of yttria-stabilized zirconia is frequently used.

thermal conductivity. (1) Ability of a material to conduct heat. (2) The rate of heat flow, under steady conditions, through unit area, per unit temperature gradient in the direction perpendicular to the area. Usually expressed in English units as Btu per square feet per degrees Fahrenheit ($Btu/ft^2 \cdot °F$). It is given in SI units as watts per meter Kelvin ($W/m \cdot K$).

thermal fatigue. Fracture resulting from the presence of temperature gradients that vary with time in such a manner as to produce cyclic stresses in a structure.

thermoforming. The process of forming a thermoplastic sheet into a three-dimensional shape after heating it to the point at which it is soft and flowable, and then applying differential pressure to make the sheet conform to the shape of the mold or die.

thermomechanical working. A general term covering a variety of metal forming processes combining controlled thermal and deformation treatments to obtain synergistic effects, such as improvement in strength without loss of toughness. Same as thermal-mechanical treatment.

thermoplastic. Capable of being repeatedly softened by an increase in temperature and hardened by a decrease in temperature. Those polymeric materials that, when heated, undergo a substantially physical rather than chemical change and that in the softened stage can be shaped into articles by molding or extrusion.

thermoset. A resin that is cured, set, or hardened, usually by heating, into a permanent shape. The polymerization reaction is an irreversible reaction known as cross linking. Once set, a thermosetting plastic cannot be remelted, although most soften with the application of heat.

thixotropy. A property of certain gels or slurry mixtures to thin upon stirring agitation (shearing) and to thicken upon subsequent rest.

throw-away product. Small, inexpensive products that are designed without any consideration of repair.

time-temperature-transformation (TTT) diagram. See *isothermal transformation (IT) diagram.*

tolerance. The total amount by which a specific dimension may be allowed to vary without impairing performance. The tolerance is the difference between the allowable limits.

tolerance analysis. See *dimensional management.*

tolerance design. The third stage of *robust design,* as defined by Genichi Taguchi, in which it is decided how much performance levels of certain parameters will have to be increased to meet the quality characteristic. The *quality loss function* is used to find the most cost-effective way of determining which tolerances will have to be tightened and which can be left as is or opened up, in order to reduce cost.

tolerance stackup. The situation with a subassembly of several parts by which all parts are within the tolerance specification, but the cumulative effect of the tolerance of each part results in a critical dimension of the assembly being out-of-tolerance.

topology optimization. Design optimization carried out early in *configuration design* that computes the best geometric configuration or layout of a structure.

torsion. The state of stress produced by applying torque to a bar or shaft.

total quality management (TQM). A management approach to long-term success that emphasizes customer satisfaction. The elements of TQM are continuous improvement, involving all members of the organization, and decision making based on data, using a set of simple TQM analysis tools.

toughness. Ability of a material to absorb energy and deform plastically before fracturing. Toughness is proportional to the area under the *stress-strain curve* from the origin to the breaking point. In metals, toughness is usually measured by the energy absorbed in a notch impact test.

tow. An untwisted bundle of continuous filaments, usually referring to man-made fibers, particularly carbon and graphite, but also fiberglass and aramid. A tow designated as 140 K has 140,000 filaments.

trade-off. A situation commonly found in design where more than one functional characteristic of the design is of value to the user. Often it is found that these characteristics oppose one another, and a compromise must be achieved. For example, with many materials, increasing the strength reduces fracture toughness, and a trade-off must be reached. Trade-off studies are best done after gathering data from a *sensitivity analysis.*

transformation-induced plasticity (TRIP). A phenomenon, occurring chiefly in certain highly alloyed steels that have been heat treated to produce metastable austenite or metastable austenite plus martensite, whereby, on subsequent deformation, part of the austenite undergoes strain-induced transformation to martensite. Steels capable of transforming in this manner, commonly referred to as TRIP steels, are highly plastic after heat treatment, but exhibit a very high rate of strain hardening and thus have high tensile and yield strengths after plastic deformation at temperatures between 20 and 500 °C (70 and 930 °F). Cooling to –195 °C (–320 °F) may or may not be required to complete the transformation to martensite. Tempering usually is done following transformation.

transformation temperature. The temperature at which a change in phase occurs. This term is sometimes used to denote the limiting temperature of a transformation range.

transition temperature. (1) An arbitrarily defined temperature that lies within the temperature range in which metal fracture characteristics (as usually determined by tests of notched specimens) change rapidly, such as the ductile-to-brittle transition temperature (DBTT). The DBTT can be assessed in several ways, the most common being the temperature for 50% ductile and 50% brittle fracture (50% fracture appearance transition temperature, or FATT), or the lowest temperature at which the fracture is 100% ductile (100% fibrous criterion). The DBTT is commonly associated with temper embrittlement and radiation damage (neutron irradiation) of low-alloy steels. (2) Sometimes used to denote an arbitrarily defined temperature within a range in which the ductility changes rapidly with temperature.

true strain. (1) The ratio of the change in dimension, resulting from a given load increment, to the magnitude of the dimension immediately prior to applying the load increment. (2) In a body subjected to axial force, the natural logarithm of the ratio of the gage length at the moment of observation to the original gage length. Also known as natural strain.

true stress. The value obtained by dividing the load applied to a member at a given instant by the cross-sectional area over which it acts.

U

ultimate strength. See *tensile strength.*

uncertainty. The condition of being in doubt. In statistical terms it refers to the bounds within which the "true" engineering result, such as the average fatigue limit of a component, can be expected to fall. Generally described in terms of confidence limits, tolerance limits, or prediction limits.

undercut. In welding, a groove melted into the base metal adjacent to the root of the weld and left unfilled by weld metal.

undercut shape. A shape in which material has been removed to create an overhang.

V

validation. The process of substantiating that material property test data have been generated according to standard methods and practices, or other indices of quality, reliability, and precision. The validation process is the first step toward ratification or confirmation of the data, making them legally effective and binding in some specified application.

value. The worth of a product or service, equal to function divided by cost or quality divided by cost.

value analysis. A problem-solving methodology that focuses on identifying the key function(s) of a design so that unnecessary costs can be removed without compromising the quality of the design. Also known as value engineering.

value engineering. See *value analysis.*

variability of data. The degree to which random variables deviate from a central value or mean. In statistical terms, this is measured by the sample standard deviation or sample variance.

variable costs. Those costs that can be directly associated with the production of a unit of output, and whose total increases roughly linearly with the total number of units produced. A good example of variable costs is material costs. See also *fixed costs.*

vertical thinking. A thought process that moves forward in sequential steps after a positive decision has been made about the idea.

virtual manufacturer. A company that focuses on the design and marketing of a product and outsources the manufacturing and assembly of the product to other companies.

viscoelasticity. A property involving a combination of elastic and viscous behavior that makes deformation dependent upon both temperature and strain rate. A material having this property is considered to combine the features of a perfectly elastic solid and a perfect fluid.

vitrification. The formation of a glassy phase in a ceramic body that serves to bind the particulate material together and close the pores.

W

weakest-link failure theory. A concept of failure in which failure is caused by the weakest element in a body, as in the failure of the weakest link in a chain.

wear. Damage to a solid surface, generally involving progressive loss of material, due to a relative motion between that surface and a contacting surface or substance.

wear rate. The rate of material removal or dimensional change due to *wear* per unit of exposure parameter—for example, quantity of material removed (mass, volume, thickness) in unit distance of sliding or unit time.

Weibull distribution. The *probability density function* commonly used to describe the strength of brittle materials such as ceramics and glasses. Also used to describe the fatigue performance of complex assemblies.

weighted property index. An index that uses weighted values of the required properties to arrive at a relative value for selection of a material for a specific application.

weldability. A specific or relative measure of the ability of a material to be welded under a given set of conditions. Implicit in this definition is the ability of the completed weldment to fulfill all functions for which the part was designed.

weld-overlay coatings. Surface coatings applied for corrosion, wear, or erosion resistance by melting and solidifying a molten alloy produced by a welding process. Sometimes referred to as hardfacing.

welding. Joining two or more pieces of material by applying heat or pressure, or both, with or without filler material, to produce a localized union through fusion or recrystallization across the interface. The thickness of the filler material is much greater than the capillary dimensions encountered in *brazing.*

whiteware. A group of ceramic products characterized by a white or light-colored body with a fine-grained structure that consist primarily of clay materials, feldspars, and quartz. Most whiteware products are glazed—in whole or in part—and whiteware glazes may range from clear to completely opaque, white or colored. Examples of whiteware products are sanitaryware, tableware, electrical porcelain, artware, stoneware, and tile.

workability. See *formability.*

work hardening. See *strain hardening.*

Y

yield strength. The stress at which a material exhibits a specified deviation from proportionality of stress and strain. An offset of 0.2% is used for many materials, particularly metals. Compare with *tensile strength.*

Young's modulus. See *modulus of elasticity* (*E*).

Metric Conversion Guide

This Section is intended as a guide for expressing weights and measures in the Système International d'Unités (SI). The purpose of SI units, developed and maintained by the General Conference of Weights and Measures, is to provide a basis for worldwide standardization of units and measure. For more information on metric conversions, the reader should consult the following references:

- "Standard for Metric Practice," E 380, *Annual Book of ASTM Standards,* American Society for Testing and Materials, 100 Barr Harbor Drive, West Conshohocken, PA 19428-2959
- "Metric Practice," ANSI/IEEE 268-1982, American National Standards Institute, 1430 Broadway, New York, NY 10018
- *The International System of Units,* SP 330, 1986, National Institute of Standards and Technology. Order from Superintendent of Documents, U.S. Government Printing Office, Washington, DC 20402-9325
- *Metric Editorial Guide,* 4th ed. (revised), 1985, American National Metric Council, 1010 Vermount Avenue NW, Suite 1000, Washington, DC 20005-4960
- *ASME Orientation and Guide for Use of SI (Metric) Units,* ASME Guide SI 1, 9th ed., 1982, The American Society of Mechanical Engineers, 345 East 47th Street, New York, NY 10017

Base, supplementary, and derived SI units

Measure	Unit	Symbol
Base units		
Amount of substance	mole	mol
Electric current	ampere	A
Length	meter	m
Luminous intensity	candela	cd
Mass	kilogram	kg
Thermodynamic temperature	kelvin	K
Time	second	s
Supplementary units		
Plane angle	radian	rad
Solid angle	steradian	sr
Derived units		
Absorbed dose	gray	Gy
Acceleration	meter per second squared	m/s^2
Activity (of radionuclides)	becquerel	Bq
Angular acceleration	radian per second squared	rad/s^2
Angular velocity	radian per second	rad/s
Area	square meter	m^2
Capacitance	farad	F
Concentration (of amount of substance)	mole per cubic meter	mol/m^3
Conductance	siemens	S
Current density	ampere per square meter	A/m^2
Density, mass	kilogram per cubic meter	kg/m^3
Electric charge density	coulomb per cubic meter	C/m^3
Electric field strength	volt per meter	V/m
Electric flux density	coulomb per square meter	C/m^2
Electric potential, potential difference, electromotive force	volt	V
Electric resistance	ohm	Ω
Energy, work, quantity of heat	joule	J
Energy density	joule per cubic meter	J/m^3
Entropy	joule per kelvin	J/K
Force	newton	N
Frequency	hertz	Hz
Heat capacity	joule per kelvin	J/K
Heat flux density	watt per square meter	W/m^2
Illuminance	lux	lx
Inductance	henry	H
Irradiance	watt per square meter	W/m^2
Luminance	candela per square meter	cd/m^2
Luminous flux	lumen	lm
Magnetic field strength	ampere per meter	A/m
Magnetic flux	weber	Wb
Magnetic flux density	tesla	T
Molar energy	joule per mole	J/mol
Molar entropy	joule per mole kelvin	J/mol · K
Molar heat capacity	joule per mole kelvin	J/mol · K
Moment of force	newton meter	N · m
Permeability	henry per meter	H/m
Permittivity	farad per meter	F/m
Power, radiant flux	watt	W
Pressure, stress	pascal	Pa
Quantity of electricity, electric charge	coulomb	C
Radiance	watt per square meter steradian	$W/m^2 \cdot sr$
Radiant intensity	watt per steradian	W/sr
Specific heat capacity	joule per kilogram kelvin	J/kg · K
Specific energy	joule per kilogram	J/kg
Specific entropy	joule per kilogram kelvin	J/kg · K
Specific volume	cubic meter per kilogram	m^3/kg
Surface tension	newton per meter	N/m
Thermal conductivity	watt per meter kelvin	W/m · K
Velocity	meter per second	m/s
Viscosity, dynamic	pascal second	Pa · s
Viscosity, kinematic	square meter per second	m^2/s
Volume	cubic meter	m^3
Wavenumber	1 per meter	1/m

Conversion factors

To convert from	to	multiply by
Angle		
degree	rad	1.745 329 E−02
Area		
in.^{2}	mm^2	6.451 600 E+02
in.^{2}	cm^2	6.451 600 E+00
in.^{2}	m^2	6.451 600 E−04
ft^2	m^2	9.290 304 E−02
Bending moment or torque		
lbf · in.	N · m	1.129 848 E−01
lbf · ft	N · m	1.355 818 E+00
kgf · m	N · m	9.806 650 E+00
ozf · in.	N · m	7.061 552 E−03
Bending moment or torque per unit length		
lbf · in./in.	N · m/m	4.448 222 E+00
lbf · ft/in.	N · m/m	5.337 866 E+01
Current density		
A/in.^{2}	A/cm^2	1.550 003 E−01
A/in.^{2}	A/mm^2	1.550 003 E−03
A/ft^2	A/m^2	1.076 400 E+01
Electricity and magnetism		
gauss	T	1.000 000 E−04
maxwell	μWb	1.000 000 E−02
mho	S	1.000 000 E+00
Oersted	A/m	7.957 700 E+01
Ω · cm	Ω · m	1.000 000 E−02
Ω circular-mil/ft	μΩ · m	1.662 426 E−03
Energy (impact, other)		
ft · lbf	J	1.355 818 E+00
Btu (thermochemical)	J	1.054 350 E+03
cal (thermochemical)	J	4.184 000 E+00
kW · h	J	3.600 000 E+06
W · h	J	3.600 000 E+03
Flow rate		
ft^3/h	L/min	4.719 475 E−01
ft^3/min	L/min	2.831 000 E+01
gal/h	L/min	6.309 020 E−02
gal/min	L/min	3.785 412 E+00
Force		
lbf	N	4.448 222 E+00
kip (1000 lbf)	N	4.448 222 E+03
tonf	kN	8.896 443 E+00
kgf	N	9.806 650 E+00
Force per unit length		
lbf/ft	N/m	1.459 390 E+01
lbf/in.	N/m	1.751 268 E+02
Fracture toughness		
ksi $\sqrt{\text{in.}}$	MPa $\sqrt{\text{m}}$	1.098 800 E+00
Heat content		
Btu/lb	kJ/kg	2.326 000 E+00
cal/g	kJ/kg	4.186 800 E+00
Heat input		
J/in.	J/m	3.937 008 E+01
kJ/in.	kJ/m	3.937 008 E+01
Length		
Å	nm	1.000 000 E−01
μin.	μm	2.540 000 E−02
mil	μm	2.540 000 E+01
in.	mm	2.540 000 E+01
in.	cm	2.540 000 E+00
ft	m	3.048 000 E−01
yd	m	9.144 000 E−01
mile	km	1.609 300 E+00
Mass		
oz	kg	2.834 952 E−02
lb	kg	4.535 924 E−01
ton (short, 2000 lb)	kg	9.071 847 E+02
ton (short, 2000 lb)	kg × 10^3(a)	9.071 847 E−01
ton (long, 2240 lb)	kg	1.016 047 E+03
Mass per unit area		
oz/in.^{2}	kg/m^2	4.395 000 E+01
oz/ft^2	kg/m^2	3.051 517 E−01
oz/yd^2	kg/m^2	3.390 575 E−02
lb/ft^2	kg/m^2	4.882 428 E+00
Mass per unit length		
lb/ft	kg/m	1.488 164 E+00
lb/in.	kg/m	1.785 797 E+01
Mass per unit time		
lb/h	kg/s	1.259 979 E−04
lb/min	kg/s	7.559 873 E−03
lb/s	kg/s	4.535 924 E−01
Mass per unit volume (includes density)		
g/cm^3	kg/m^3	1.000 000 E+03
lb/ft^3	g/cm^3	1.601 846 E−02
lb/ft^3	kg/m^3	1.601 846 E+01
lb/in.^{3}	g/cm^3	2.767 990 E+01
lb/in.^{3}	kg/m^3	2.767 990 E+04
Power		
Btu/s	kW	1.055 056 E+00
Btu/min	kW	1.758 426 E−02
Btu/h	W	2.928 751 E−01
erg/s	W	1.000 000 E−07
ft · lbf/s	W	1.355 818 E+00
ft · lbf/min	W	2.259 697 E−02
ft · lbf/h	W	3.766 161 E−04
hp (550 ft · lbf/s)	kW	7.456 999 E−01
hp (electric)	kW	7.460 000 E−01
Power density		
W/in.^{2}	W/m^2	1.550 003 E+03
Press capacity		
See Force		
Pressure (fluid)		
atm (standard)	Pa	1.013 250 E+05
bar	Pa	1.000 000 E+05
in. Hg (32 °F)	Pa	3.386 380 E+03
in. Hg (60 °F)	Pa	3.376 850 E+03
lbf/in.^{2} (psi)	Pa	6.894 757 E+03
torr (mm Hg, 0 °C)	Pa	1.333 220 E+02
Specific heat		
Btu/lb · °F	J/kg · K	4.186 800 E+03
cal/g · °C	J/kg · K	4.186 800 E+03
Stress (force per unit area)		
tonf/in.^{2} (tsi)	MPa	1.378 951 E+01
kgf/mm^2	MPa	9.806 650 E+00
ksi	MPa	6.894 757 E+00
lbf/in.^{2} (psi)	MPa	6.894 757 E−03
MN/m^2	MPa	1.000 000 E+00
Temperature		
°F	°C	5/9 · (°F − 32)
°R	°K	5/9
Temperature interval		
°F	°C	5/9
Thermal conductivity		
Btu · in./s · ft^2 · °F	W/m · K	5.192 204 E+02
Btu/ft · h · °F	W/m · K	1.730 735 E+00
Btu · in./h · ft^2 · °F	W/m · K	1.442 279 E−01
cal/cm · s · °C	W/m · K	4.184 000 E+02
Thermal expansion		
in./in. · °C	m/m · K	1.000 000 E+00
in./in. · °F	m/m · K	1.800 000 E+00
Velocity		
ft/h	m/s	8.466 667 E−05
ft/min	m/s	5.080 000 E−03
ft/s	m/s	3.048 000 E−01
in./s	m/s	2.540 000 E−02
km/h	m/s	2.777 778 E−01
mph	km/h	1.609 344 E+00
Velocity of rotation		
rev/min (rpm)	rad/s	1.047 164 E−01
rev/s	rad/s	6.283 185 E+00
Viscosity		
poise	Pa · s	1.000 000 E−01
stokes	m^2/s	1.000 000 E−04
ft^2/s	m^2/s	9.290 304 E−02
in.^{2}/s	mm^2/s	6.451 600 E+02
Volume		
in.^{3}	m^3	1.638 706 E−05
ft^3	m^3	2.831 685 E−02
fluid oz	m^3	2.957 353 E−05
gal (U.S. liquid)	m^3	3.785 412 E−03
Volume per unit time		
ft^3/min	m^3/s	4.719 474 E−04
ft^3/s	m^3/s	2.831 685 E−02
in.^{3}/min	m^3/s	2.731 177 E−07
Wavelength		
Å	nm	1.000 000 E−01

(a) kg × 10^3 = 1 metric ton or 1 megagram (Mg)

Abbreviations and Symbols

a crack length; thermal diffusivity of a product
A ampere
A area; ratio of the alternating stress amplitude to the mean stress
Å angstrom
ABC activity-based costing
ac alternating current
AI artificial intelligence
AMS Aerospace Material Specification
ANSI American National Standards Institute
ASTM American Society for Testing and Materials
at.% atomic percent
atm atmosphere (pressure)
b Burgers vector
bal balance
bcc body-centered cubic
bct body-centered tetragonal
Btu British thermal unit
C Coulomb; heat capacity
C-C carbon-carbon
CAD computer-aided design
CAE computer-aided engineering
CAM computer-aided manufacturing
CAPP computer-aided process planning
CARES Ceramic Analysis and Reliability Evaluation of Structures
cd candela
CDF cumulative distribution function
CFC chlorofluorocarbon
CFD computational fluid dynamics
CFRP carbon-fiber-reinforced plastic
CIP cold isostatic pressing
cm centimeter
CMC ceramic-matrix composite
CNC computer numerical control
cpm cycles per minute
cps cycles per second
CPS creative problem solving
CSA Canadian Standards Association
cSt centiStokes
CTE coefficient of thermal expansion
CTOD crack tip opening displacement
CVD chemical vapor deposition
CVI chemical vapor infiltration (impregnation)
CVN Charpy V-notch (impact test or specimen)
d an operator used in mathematical expressions involving a derivative (denotes rate of change)
d depth; diameter
DAE differential algebraic equation
dB decibel
DBTT ductile-brittle transition temperature
dc direct current
DFA design for assembly
DFD design for disassembly
DFE design for the environment
DFM design for manufacture
DFMA design for manufacture and assembly
dhcp double hexagonal close-packed
diam diameter
DIN Deutsche Institut für Normung
DOE design of experiments
DOF degrees of freedom
DTA differential thermal analysis
DTUL deflection temperature under load
DWTT drop weight transition temperature
e charge of an electron; natural log base, 2.71828
E modulus of elasticity; Young's modulus; potential
E_s secant modulus
EDM electrical discharge machining
emf electromotive force
EMI electromagnetic interference
Eq equation
ESC environmental stress cracking
esu electrostatic units
et al. and others
ETA event tree analysis
eV electron volt
f fiber
f focal length; frequency
f_n normal load
F Faraday constant; force
fcc face-centered cubic
fct face-centered tetragonal
FEA finite-element analysis
FEM finite-element modeling
FHA fault hazard analysis
Fig. figure
FLD forming limit diagram
FM figure of merit
FMEA failure modes and effects analysis
FMECA failure modes, effects, and criticality analysis
ft foot
FTA fault tree analysis
g gram
G gauss
G mean grain size; shear modulus
G_{Ic} interlaminar fracture toughness (mode I, peel; mode II, shear; mode III, scissor shear)
gal gallon
GD&T geometric dimensioning and tolerancing
GPa gigapascal
gpd grams per denier
gr grain
h hour
h thickness
H henry
H change in height; degree of homogenization; enthalpy; hardness; height; magnetic field
HAZ heat-affected zone
HB Brinell hardness
hcp hexagonal close-packed
HDT heat-deflection temperature
HERF high-energy-rate forging
HIP hot isostatic pressing
HK Knoop hardness
hp horsepower
HR Rockwell hardness (requires a scale designation, such as HRC for Rockwell C hardness)
HTML hypertext markup language
HV Vickers hardness
Hz hertz
i current (measure of number of electrons)
I current; emergent intensity
IC integrated circuit
ID inside diameter
in. inch
IPD integrated product development
IPTS International Practical Temperature Scale
IR infrared (radiation)
ISO International Organization for Standardization
ITS International Temperature Scale
IV intrinsic viscosity
J joule
JIS Japanese Industrial Standard
k karat
k Boltzmann constant; notch sensitivity factor; thermal conductivity; wave number
K Kelvin
K bulk modulus of elasticity; coefficient of thermal conductivity; empirical constant; interface reaction; mean integrated thermal conductivity

K_I stress-intensity factor
K_{Ic} mode I critical stress-intensity factor; plane-strain fracture toughness
K_{Id} dynamic fracture toughness
K_{Iscc} threshold stress intensity for stress-corrosion cracking
K_c plane-stress fracture toughness
K_f stress-concentration factor
K_K^∞ stress-concentration factor for infinite plate
K_t stress-concentration factor
K_{th} threshold crack tip stress-intensity factor
kg kilogram
km kilometer
kN kilonewton
kPa kilopascal
ksi kips (1000 lb) per square inch
kV kilovolt
kW kilowatt
l mean free path; length
L liter; longitudinal direction; liquid
L length
lb pound
lbf pound-force
LCA life cycle analysis (or assessment)
L/D length-to-diameter ratio
LED light-emitting diode
LEFM linear-elastic fracture mechanics
ln natural logarithm (base *e*)
log common logarithm (base 10)
m matrix; meter
m ion mass; Weibull modulus
M metal atom
M molecular weight
mA milliampere
MeV megaelectronvolt
mg milligram
Mg megagram
MIM metal injection molding
min minimum; minute
MJ megajoule
mL milliliter
MLE maximum likelihood estimator
MLT marketing lead time
mm millimeter
MMC metal-matrix composite
mod modified
MOD metallographic deposition
MOE modulus of elasticity
mol% mole percent
MOR modulus of resilience; modulus of rupture
mPa millipascal
MPa megapascal
mpg miles per gallon
mph miles per hour
ms millisecond
MS megasiemens
MSDS material safety data sheet
mT millitesla
MTBF mean time between failures
MTTF mean time to failure
mV millivolt
MV megavolt
MW molecular weight
n growth exponent
n integral number
N Newton
N fatigue life (number of cycles)
NASA National Aeronautics and Space Administration
NBS National Bureau of Standards
NC numerical control
NDE nondestructive evaluation
NDI nondestructive inspection
NDT nondestructive testing
NIST National Institute of Standards and Technology
nm nanometer
No. number
NTC negative temperature coefficient
OD outside diameter
Oe oersted
OEM original equipment manufacturer
OHA operating hazards analysis
ORNL Oak Ridge National Laboratory
OSHA Occupational Safety and Health Administration
oz ounce
p page
p pressure
P applied load; power; pressure
Pa pascal
PCE pyrometric cone equivalent
PDS product design specification
PDT product design team
P/F powder forging
pH negative logarithm of hydrogen-ion activity
P/M powder metallurgy
ppb parts per billion
ppba parts per billion atomic
ppm parts per million
ppt parts per trillion
psi pounds per square inch
psia pounds per square inch absolute
psid pounds per square inch differential
psig pounds per square inch gage
PTC positive temperature coefficient
PVD physical vapor deposition
QC quality control
QFD quality function deployment
QLF quality loss function
r radius vector in a plane normal to the axis
r particle radius; radius of curvature; rate of reaction; reflectivity; spherical particle radius
R roentgen
R average particle radius; gas constant; radius; ratio of the minimum stress to the maximum stress; resistance; reliability
R_a surface roughness in terms of arithmetic average
RA reduction of area
rad absorbed radiation dose; radian
RCF rolling contact fatigue
RE rare earth
Ref reference
rem remainder
rf, RF radio frequency
RH relative humidity
rms, RMS root mean square
ROM rough order of magnitude; rule of mixtures
rpm revolutions per minute
RT room temperature
RTI relative thermal index
s second
S siemens
S distance traveled
SAE Society of Automotive Engineers
scfm standard cubic foot per minute
SEM scanning electron microscopy
sfm surface feet per minute
SG standard grade
SI Système International d'Unités
sineh sine hyperbolic
S/N signal-to-noise (ratio)
S-N stress-number of cycles
SPC statistical process control
SPF superplastic forming
sp gr specific gravity
SRIM structural reaction injection molding
std standard
STEM scanning transmission electron microscopy
STM scanning tunneling microscopy
Sv sievert
t thickness; time
T Tesla; transverse direction
T absolute temperature; temperature; tenacity; total dispersion; transmittance
T_g glass transition temperature
T_m melt temperature
tan equal to ratio of the loss modulus to the storage modulus
TBC thermal barrier coating
TEM transmission electron microscopy
TGA thermogravimetric analysis
TMA thermomechanical analysis
TP thermoplastic
TQM total quality management
TS thermoset
tsi tons per square inch
TTT time-temperature-transformation
UL Underwriters' Laboratories
UNS Unified Numbering System
UTS ultimate tensile strength
UV ultraviolet
v' workpiece velocity
V volt
V_f volume fraction of fiber
V_m volume fraction of matrix
V_v volume fraction of void content
VA value analysis
VE value engineering

VI viscosity index
vol volume
vol% volume percent
w whisker
W watt
W total radiation; width
wt% weight percent
x axial distance
X neck diameter
Y scale of microstructural segregation
YAG yttrium aluminum garnet
yr year
z ion charge
Z atomic number; impedance
° angular measure; degree
°C degree Celsius (centigrade)
°F degree Fahrenheit
0° fiber direction
90° perpendicular to fiber direction
α coefficient of thermal expansion
γ shear strain; surface energy; surface tension
γ_b interfacial energy
γ_s surface energy
δ an increment; a range; change in quantity; grain boundary width; thickness of liquid boundary
Δ an increment; a range; change in quantity
ΔT temperature difference or change
ε strain
$\dot{\varepsilon}$ strain rate
η loss coefficient; viscosity
θ angle; geometrical constant
θ_i angle of incidence
θ_r angle of refraction
κ dielectric constant
λ thermal conductivity
μ friction coefficient; linear attenuation coefficient; magnetic permeability; the mean (or average) of a distribution
μin. microinch
μm micrometer (micron)
μs microsecond
ν Poisson's ratio; velocity
π pi (3.141592)
ρ density; resistivity
σ Stefan-Boltzmann constant; strength; stress; tensile stress; tensile strength
σ_a applied stress
σ_f applied axial load
Σ summation of
τ shear stress
ϕ angle of internal friction; dihedral angle; porosity; power
Φ reaction rate (kinetic function); volume concentration of impurity ions
ψ damping; geometrical constant
ω frequency
Ω atomic volume; electrical conductivity; molecular volume of solute; ohm
∂ partial derivative
$\partial\eta/\partial t$ kinetic function
$\leftrightarrows$ direction of reaction
$\div$ divided by
$=$ equals
$\approx$ approximately equals
$\neq$ not equal to
$\equiv$ identical with
$>$ greater than
$>>$ much greater than
$\geq$ greater than or equal to
∞ infinity
$\propto$ is proportional to; varies as
$\int$ integral of
$<$ less than
$<<$ much less than
$\leq$ less than or equal to
$\pm$ maximum deviation
$-$ minus; negative ion charge
$\times$ diameters (magnification); multiplied by
$\cdot$ multiplied by
/ per
% percent
+ plus; positive ion charge
$\sqrt{}$ square root of
~ approximately; similar to

Greek Alphabet

Α, α alpha
Β, β beta
Γ, γ gamma
Δ, δ delta
Ε, ε epsilon
Ζ, ζ zeta
Η, η eta
Θ, θ theta
Ι, ι iota
Κ, κ kappa
Λ, λ lambda
Μ, μ mu
Ν, ν nu
Ξ, ξ xi
Ο, ο omicron
Π, π pi
Ρ, ρ rho
Σ, σ sigma
Τ, τ tau
Υ, υ upsilon
Φ, ϕ phi
Χ, χ chi
Ψ, ψ psi
Ω, ω omega

Index

A

B

C

E

F

G

H

N

O

P

Q

R

S

T

U

V

W

X

Y

Z